BUSINESS STATISTICS OF THE UNITED STATES

BUSINESS STATISTICS OF THE UNITED STATES

Patterns of Economic Change

Twenty-Fifth Edition
2020

Edited by Susan Ockert

Bernan Press

Lanham • Boulder • New York • London

Published by Bernan Press
An imprint of The Rowman & Littlefield Publishing Group, Inc.
4501 Forbes Boulevard, Suite 200, Lanham, Maryland 20706
www.rowman.com
800-462-6420

6 Tinworth Street, London SE11 5AL, United Kingdom

ISBN: 978-1-64143-446-1
E-ISBN: 978-1-64143-447-8

CONTENTS

PREFACE

Business Statistics of the United States: Patterns of Economic Change, 25th Edition, 2020 is a basic desk reference for anyone requiring statistics on the U.S. economy. It contains over 3,000 economic time series portraying the period from World War II to December 2019 in industry, product, and demographic detail. In the case of about 200 key series, the period from 1929 through 1948 is depicted as well.

Additionally, for three important statistical series that have been compiled monthly on a continuous basis, monthly and annual data are included going back to 1919 for industrial production and 1913 for consumer and producer prices. Annual data for money and credit have also been taken back before 1929, including money supply figures back to 1892. Annual estimates of the unemployment rate from 1929 to 1948 are also included.

The National Income and Product Accounts (NIPAs) were comprehensively revised in July 2013, changing some measurement concepts, changing base periods for price indexes and real output, and as a result altering all the data back to 1929. *Business Statistics* incorporates fully revised data for over 700 series from this data set and productivity and costs series that are based on the revised NIPAs, all extended and updated through 2019. The important new concepts introduced in the NIPA revision are explained in the Notes and Definitions to Chapters 1, 4, 5, and 6. In all, this edition provides, in one volume, unparalleled, up-to-date background information on the course of the U.S. economy.

The data are predominantly from federal government sources. Of equal importance are the extensive background notes for each chapter, which help users to understand the data, use them appropriately, and, if desired, seek additional information from the source agencies.

Chapters are organized according to subject matter. For most time series, users will find an initial page displaying the latest two years of monthly data, or three years of quarterly data, for the most recent completed years, along with annual averages for the last completed 40, 50, or even 60 years.

For the most important series for which we have long-term historical data, this initial page will be labeled with "A" after the table number, and identified in the title by the additional words "Recent Data". It will be immediately followed by pages labeled with "B" after the table number and identified by the words "Historical Data", which will show all available historical data annually back as far as possible, and monthly or quarterly back to the earliest available postwar year. In some cases, additional tables of historical data will be included.

The leading article "Cycle and Growth Perspectives" provides other information and techniques to assist in using and interpreting the data.

THE PLAN OF THE BOOK

The history of the U.S. economy is told in major U.S. government sets of statistical data: the national income and product accounts compiled by the Bureau of Economic Analysis (BEA); the data on labor force, employment, hours, earnings, and productivity compiled by the Bureau of Labor Statistics (BLS); the price indexes collected by BLS; and the financial market and industrial production data compiled primarily by the Board of Governors of the Federal Reserve System (FRB). All of these sets exist in annual and either monthly or quarterly form beginning in 1946, 1947, or 1948; many are available, at least annually, as far back as 1929, and a few even farther back; and three are available monthly back to the teens of the 20th century.

In Part A, *Business Statistics* presents the aggregate United States economy in a number of important dimensions. The presentations begin with the national income and product accounts, or NIPAs. The NIPAs comprise a comprehensive, thorough, and internally consistent body of data. They measure the value of the total output of the U.S. economy (the gross domestic product, or GDP) and they allocate that value between its quantity, or "real," and price components. They show how the value of aggregate demand is generated by consumers, business investors, government, and foreign customers; how much of aggregate demand is supplied by imports and how much by domestic production; and how the income generated in domestic production is distributed between labor and capital.

Production estimates covering only the sectors of the economy traditionally labeled "industrial"—manufacturing, mining, and utilities—are shown in Chapter 2, after the presentation of the overall NIPAs in Chapter 1.

Median income—the best measure of the economic well-being of the typical American—income distribution, and poverty statistics from a Census household survey are presented in Chapter 3.

Then, more detail from the NIPAs is presented for the demand components of economic activity. GDP, by its product-side definition, consists of the sum of consumption expenditures, business investment in fixed capital and inventories, government purchases of goods and services, and exports minus imports—as in the elementary economics blackboard identity "GDP = C + I + G + X – M." Chapters on each of these components—consumption, investment, government, and foreign trade--are presented in Part A.

Following these chapters, there are a chapter on prices, two chapters on the compensation of labor and capital inputs and the amount and productivity of labor input, one chapter on energy inputs into production and consumption, and one chapter on money, interest, assets, and debt.

At the end of Part A, comparisons of output, prices, and labor markets among major industrial countries are presented, along with statistics on the value of the dollar against other currencies.

While GDP is initially defined and measured by adding up its demand categories and subtracting imports, this output is produced in industries—some in the old-line heavy industries such as manufacturing, mining, and utilities, but an increasing share in the huge and heterogeneous group known as "service-providing" industries. Part A gives a number of measures of activity classified by industry or industrial sector: industrial production, profits, and employment-related data. Further industry information is provided in Part B, including new quarterly indicators of GDP by industry.

Industry data collection is important because demands for goods and services are channeled into demands for labor and capital through the industries responsible for producing the requested goods and services. These data are reported using the North American Industry Classification System (NAICS). This system, introduced in 1997 to replace the older Standard Industrial Classification System (SIC), delineates industries that are better defined in relation to today's demands and more closely related to each other by technology. Notable examples include more detailed data available on service industries, a more rational grouping of the Computer and electronic product manufacturing subsector, and the creation of the Information sector. See the References at the end of this Preface.

NAICS industries are groupings of producing units—not of products as such—and are grouped according to similarity of production processes. This is done in order to collect consistent data on inputs and outputs, which are then used to measure important concepts, such as productivity and input-output parameters. Emphasis on the production process helps to explain a number of ways in which the NAICS differs from the SIC.

• Manufacturing activities at retail locations, such as bakeries, have been classified separately from retail activity and put into the Food manufacturing industry.

• Central administrative offices of companies have a new sector of their own, Management of companies and enterprises (sector 55). For example, the headquarters office of a food-producing corporation is considered part of the new sector instead of part of the Food manufacturing industry.

• Reproduction of packaged software, which was classified as a business service in the SIC, is now classified in sector 334, Computer and electronic product manufacturing, as a manufacturing process.

• Electronic markets and agents and brokers, formerly undifferentiated components of wholesale trade industries, have a sector of their own (425).

• Retail trade in NAICS (sectors 44 and 45) now includes establishments such as office supply stores, computer and software stores, building materials dealers, plumbing supply stores, and electrical supply stores, that display merchandise and use mass-media advertising to sell to individuals as well as to businesses, which were formerly classified in wholesale trade.

In Part B, *Business Statistics* shows GDP, income, employment, hours, and earnings by industry, followed by statistics for key sectors such as petroleum, housing, manufacturing, retail trade, and services.

Notes and definitions. Productive use of economic data requires accurate knowledge about the sources and meaning of the data. The notes and definitions for each chapter, shown immediately after that chapter's tables, contain definitions, descriptions of recent data revisions, and references to sources of additional technical information. They also include information about data availability and revision and release schedules, which helps users to readily access the latest current values if they need to keep up with the data month by month or quarter by quarter.

A NOTE ON THE IMPORTANCE OF ECONOMIC STATISTICS

To retrieve money and credit data for earlier years, the editor of *Business Statistics* had to consult old printed volumes, where she encountered some inspiring prefatory words in the Federal Reserve Board volume *Banking and Monetary Statistics* (1943). These words were written by the Fed's longtime statistics chief E. A. Goldenweiser, in the stately cadences of an earlier era, about the financial statistics collected in that volume. But they well express the hope and expectation of statisticians and economists that their work can lead to better economic decisions:

"These serried ranks of organized statistics on banking and finance, even though they may inspire awe, should also inspire confidence. They are an augury that credit policy can be based in the future, as in the past, on fact rather than on fancy."

THE HISTORY OF *BUSINESS STATISTICS*

The history of *Business Statistics* began with the publication, many years ago, of the first edition of a volume with the same name by the U.S. Department of Commerce's Bureau of Economic Analysis (BEA). After 27 periodic editions, the last of which appeared in 1992, BEA found it necessary, for budgetary and other reasons, to discontinue both the publication and the maintenance of the database from which the publication was derived.

The individual statistical series gathered together here are all publicly available. However, the task of gathering them from the numerous different sources within the government and assembling them into one coherent database is impractical for most data users. Even when current data are readily available, obtaining the full historical time series is often time-consuming and difficult. Definitions and other documentation can also be inconvenient to find. Believing that a *Business Statistics* compilation was too valuable to be lost to the public, Bernan Press published the first edition of the present publication, edited by Dr. Courtenay M. Slater, in 1995. The first edition received a warm welcome from users of economic data. Dr. Slater, formerly chief economist of the Department of Commerce, continued to develop *Business Statistics* through four subsequent annual editions. Cornelia Strawser, a previous editor worked with Dr. Slater on the fourth and fifth editions. In subsequent editions, she has continued in the tradition established by Dr. Slater of ensuring high-quality data, while revising and expanding the book's scope to include significant new aspects of the U.S. economy and longer historical background.

Nearly all of the statistical data in this book are from federal government sources and all are available in the public domain. Sources are given in the applicable notes and definitions.

The data in this volume meet the publication standards of the federal statistical agencies from which they were obtained. Every effort has been made to select data that are accurate, meaningful, and useful. All statistical data are subject to error arising from sampling variability, reporting errors, incomplete coverage, imputation, and other causes. The responsibility of the editor and publisher of this volume is limited to reasonable care in the reproduction and presentation of data obtained from established sources.

The 2020 edition has been edited by Susan Ockert.

Susan Ockert has worked as an economist in the military, for the federal government, and at regional and state levels. She earned her Masters of Economics at George Mason University in Fairfax, Virginia as well as her Masters of International Management at the Thunderbird University in Glendale, Arizona. She currently teaches economics at the collegiate level.

References

The most recent edition of NAICS is explained and laid out in *North American Industry Classification System: United States, 2017,* from the Executive Office of the President, Office of Management and Budget. This presents the latest five-year updating of the system, which was first introduced in 1997. The latest changes introduced in these updatings have been relatively minor and have not affected the definitions of the industry divisions presented in *Business Statistics.*

Other editions of *The North American Industry Classification System: United States* are available in 2002, 2007, and 2012.

These volumes are available from Bernan Press and fully describe the development and application of the new classification system and are the sources for the material presented in this volume. Information is also available on the NAICS Web site at http://www.census.gov/naics.

CYCLE AND GROWTH PERSPECTIVES

This twenty-fifth edition of *Business Statistics of the United States* presents comprehensive and detailed data on U.S. economic performance through December 2019, going back to the early 20th century. These numbers show changes in many dimensions—economic growth, employment, inflation, income, to name a few. There are patterns of fluctuation, and, simultaneously, patterns of enduring change. This article will present some ways of looking at these movements and trying to make sense of them. In addition, policy instruments are also described.

The Federal Reserve conducts the nation's monetary policy by managing the level of short-term interest rates and influencing the overall availability and cost of credit in the economy. Monetary policy directly affects short-term interest rates; it indirectly affects longer-term interest rates, currency exchange rates, and prices of equities and other assets and thus wealth. Through these channels, monetary policy influences household spending, business investment, production, employment, and inflation in the United States.

Governments use spending and taxing powers to promote stable and sustainable growth. Fiscal policy is the use of government spending and taxation to influence the economy. Governments typically use fiscal policy to promote strong and sustainable growth and reduce poverty. Fiscal policy aims to stabilize economic growth, avoiding a boom and bust economic cycle. Fiscal policy is often used in conjunction with monetary policy. In fact, governments often prefer monetary policy for stabilizing the economy.

The U.S. economy, even in its infancy, experienced boom or bust periods. The first recorded downturn was the Panic of 1785, which followed the American Revolution when the economy changed from military to industrial. Attempts have been made to date recessions in America beginning in the late 18th century. Economists and historians, because of the lack of economic statistics, didn't always agree on the start and stop of recessions, but they agree that volatility in economic activity and unemployment did exist.

BUSINESS CYCLES IN THE U.S. ECONOMY: RECESSIONS AND EXPANSIONS

The study of economic fluctuations in the United States was pioneered by Wesley C. Mitchell and Arthur F. Burns early in the twentieth century, and was carried on subsequently by other researchers affiliated with the National Bureau of Economic Research (NBER). NBER is an independent, nonpartisan research organization founded in 1920 and is committed to providing unbiased economic research. These analysts observed that indicators of the general state of business activity tended to move up and down over periods that were longer than a year and were therefore not accounted for by seasonal variation. Although these periods of expansion and contraction were not uniform in length, and thus not "cycles" in any strict mathematical sense, their recurrent nature and certain generic similarities caused them to be called "business cycles." NBER has identified 32 complete peak-to-peak business cycles over the period beginning with December 1854.

The first NBER-established business cycle dates, identifying the months in which peaks and troughs in general economic activity occurred, were published in 1929. The dates of current cycles are established, typically within a year or so of their occurrence, by the NBER Business Cycle Dating Committee. This group, first formed in 1978, consists of eight economists who are university professors, associated with research organizations, or both. The peak of the cycle signals the start of a downturn of an economy while the trough signals the beginning of a upturn.

Business cycle dates are based on monthly data, and have been identified for periods long before the availability of quarterly data on real gross national product (GDP). It is important to understand that even in the period since 1947 for which quarterly real GDP exists, the NBER identification of a recession does not always coincide with the frequently cited definition, "two consecutive quarters of decline in real GDP." Because a recession is a broad contraction of the economy, NBER emphasizes economy-wide measures of economic activity such as production, employment, real income, real manufacturing, wholesale-retail sales, and many more indicators. Federal agencies collect the economic data used by NBER. Employment data is provided by the Department of Labor, Bureau of Labor Statistics while the Department of Commerce provides data from the Census Bureau and Bureau of Economic Analysis along with the Federal Reserve Board.

The NBER monthly and quarterly dates of the cycles from 1854 to the latest announced turning point—the June 2009 trough ending the recession that began in December 2007—are shown in Table A-1 below. The quarterly turning points are identified by Roman numerals. NBER considers that the trough month is both the end of the decline and the beginning of the expansion, based on the concept that the actual turning point was some particular day within that month. The longest running contraction, during the Great Depression, lasted from August 1929 to November 1933, for a total of 43 months. The latest recession ended in June 2009, after 19 months.

Meanwhile, the current expansion began in June 2009 and ended in February 2020, the longest expansion in U.S. history, lasting for 10 years and 8 months. The NBER Business Cycle

xvi BUSINESS STATISTICS OF THE UNITED STATES

Dating Committee announced that the U.S. economy had officially entered another recession. See the Great Recession article for more information on the data used by the NBER.

TABLE A-1 BUSINESS CYCLE REFERENCE DATES 1854–2009

Peak (Recession)	Trough (Recovery)
June 1857(II)	December 1854 (IV)
October 1860(III)	December 1858 (IV)
April 1865(I)	June 1861 (III)
June 1869(II)	December 1867 (I)
October 1873(III)	December 1870 (IV)
March 1882(I)	March 1879 (I)
March 1887(II)	May 1885 (II)
July 1890(III)	April 1888 (I)
January 1893(I)	May 1891 (II)
December 1895(IV)	June 1894 (II)
June 1899(III)	June 1897 (II)
September 1902(IV)	December 1900 (IV)
May 1907(II)	August 1904 (III)
January 1910(I)	June 1908 (II)
January 1913(I)	January 1912 (IV)
August 1918(III)	December 1914 (IV)
January 1920(I)	March 1919 (I)
May 1923(II)	July 1921 (III)
October 1926(III)	July 1924 (III)
August 1929(III)	November 1927 (IV)
May 1937(II)	March 1933 (I)
February 1945(I)	June 1938 (II)
November 1948(IV)	October 1945 (IV)
July 1953(II)	October 1949 (IV)
August 1957(III)	May 1954 (II)
April 1960(II)	April 1958 (II)
December 1969(IV)	February 1961 (I)
November 1973(IV)	November 1970 (IV)
January 1980(I)	March 1975 (I)
July 1981(III)	July 1980 (III)
July 1990(III)	November 1982 (IV)
March 2001(I)	March 1991(I)
December 2007 (IV)	November 2001 (IV)
February 2020 (I)	June 2009 (II)

Between 1854 to 2009, the American economy experienced 33 separate recessions (contractions) and 32 expansions (recoveries). Before the Great Depression, recessions occurred more frequently and in conjunction with financial panics. With such financial instability, the Federal Reserve System was established in 1913 to assert central control of the monetary system. Financial crises during the 1800s occurred due to crop failures, banking failures, drops in cotton prices, rapid speculation in land, sudden plunges in the stock market and currency and credit crises, The United States of America during this period was a very young nation and thus these panics devastated its economy.

THE U.S. ECONOMY 1929–1948

While there has been much discussion in recent years of the "Great Depression," there is surprisingly little familiarity with the basic statistical record of the period between 1929 and the end of World War II. Reproducing material from Chapter 18 of the 15th edition of *Business Statistics,* we present here an explanation of that period from the NBER business cycle perspective, along with a graphic overview and narration of its economic developments as recorded in the statistics now presented in Part A of this volume.

The NBER chronology, presented in Table A-1 in the preceding article "Cycle and Growth Perspectives," may surprise readers who are looking for "The Great Depression" and are not familiar with the NBER approach to business cycles. As NBER perceives it, a downtrend in economic activity began in August 1929 (before the stock market crash) and lasted until March 1933. This 43-month period has been called the "Great Contraction": it was the longest period of economic decline since the 1870s, and more than double the length of the longest recession identified and completed since that time (the one from December 2007 to June 2009, which lasted 18 months). For the rest of the 1930s—excepting a 13-month recession in 1937–1938—the economy is viewed by the NBER as having been in an expansion phase. The NBER chronology does not use the term "depression."

However, the term "Great Depression" is often colloquially used for the entire 1929–1939 period, even though the economy was expanding for most of the period following March 1933. This is because economic activity during that time, though increasing, remained below the likely capacity of the economy, as is indicated in Figures A-1 and A-2 below.

It should also be noted that NBER construes the entire period from June 1938 through February 1945 as a business cycle expansion. The recovery from the 1937–1938 recession merged into a further, continued rise in activity that reflected the outbreak of war in Europe in September 1939, a consequent preparedness effort in the United States, and an increase in demand from abroad for U.S. output. The United States entered the war after being attacked by Japan in December 1941, launching an all-out war production effort at that time.

The February 1945 end of the "wartime expansion" (as NBER terms the period June 1938–February 1945) preceded the end of the war (the European war ended in May 1945 and the Pacific war concluded in August 1945). A brief recession associated with demobilization occurred from February to October 1945, followed by the first postwar expansion, which lasted from October 1945 to November 1948.

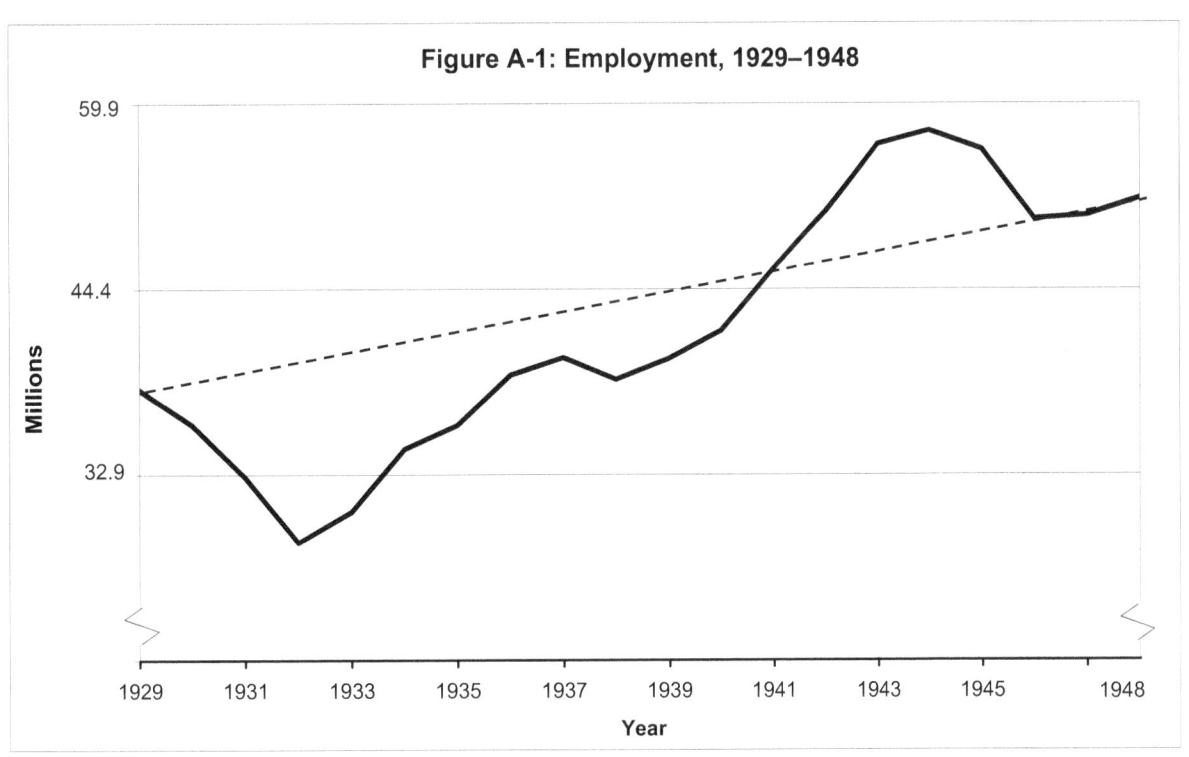

Figure A-1: Employment, 1929–1948

In the graphs that follow, summary annual statistics for the years 1929 through 1948 are presented, depicting the economy during two of its most tumultuous decades. This period encompasses the "great contraction," the economic collapse that began in August 1929 and lasted until March 1933; a subsequent period of recovery, which failed to return many important economic indicators to their expected trend levels and was interrupted by a new, though less severe, recession; apprehension of coming war, then the outbreak of war in Europe in September 1939, leading to increased war-related production; United States participation in all-out war, beginning with the Japanese attack on Pearl Harbor on December 7, 1941, and ending in 1945, with supercharged production and employment rates; then, rapid demobilization, and return to a high peacetime rate of economic activity by 1948.

For a first overall view of the economy during these decades, Figure A-1 shows total U.S. employment as calculated by the Bureau of Economic Analysis, including all private and government jobs, both civilian and military. In this graph a trend line is shown connecting the two peacetime high employment levels of 1929 and 1948.

- More than one-fifth of all the jobs held in the U.S. economy in 1929 were gone by 1932. In the subsequent recovery, total employment was back at the 1929 level by 1936, but only because of government employment, including over 3 ½ million work relief jobs; private industry employment would not recover to its 1929 level until 1941. (Table 10-21)

- Because the recovery was incomplete and economic activity did not recover to a trend level until after the decade's end, the entire decade of the 1930s is often described as "The Great Depression."

- During the war, men were drafted into the armed forces, practically all of the unemployed were put back to work, and women were drawn into the labor force, resulting in a period of what might be called "super-employment." After the war, employment fell back to a more normal trend level. (Tables 10-1C and 10-21)

- The unemployment rates for the prewar period calculated by the Bureau of Labor Statistics, unlike the employment data used in Figure A-1, count people on government work relief programs as unemployed. By this reckoning, unemployment rose from 3.2 percent of the civilian labor force in 1929 to a peak of 24.9 percent in 1933, and got no lower than 14.3 percent for the rest of the 1930s, as shown in Figure A-2. (Table 10-1C)

- State and local governments started hiring people for work relief in 1930. The Federal government's programs began in 1933, and in 1936 and 1938 agencies such as the WPA and the CCC employed over 3 ½ million people. When workers in these programs are counted as employed rather than unemployed, the high point for unemployment was 22.5 percent in 1932, and it was reduced to 9.1 percent in 1937, as shown by the dashed line in Figure A-2. (Tables 10-1C and

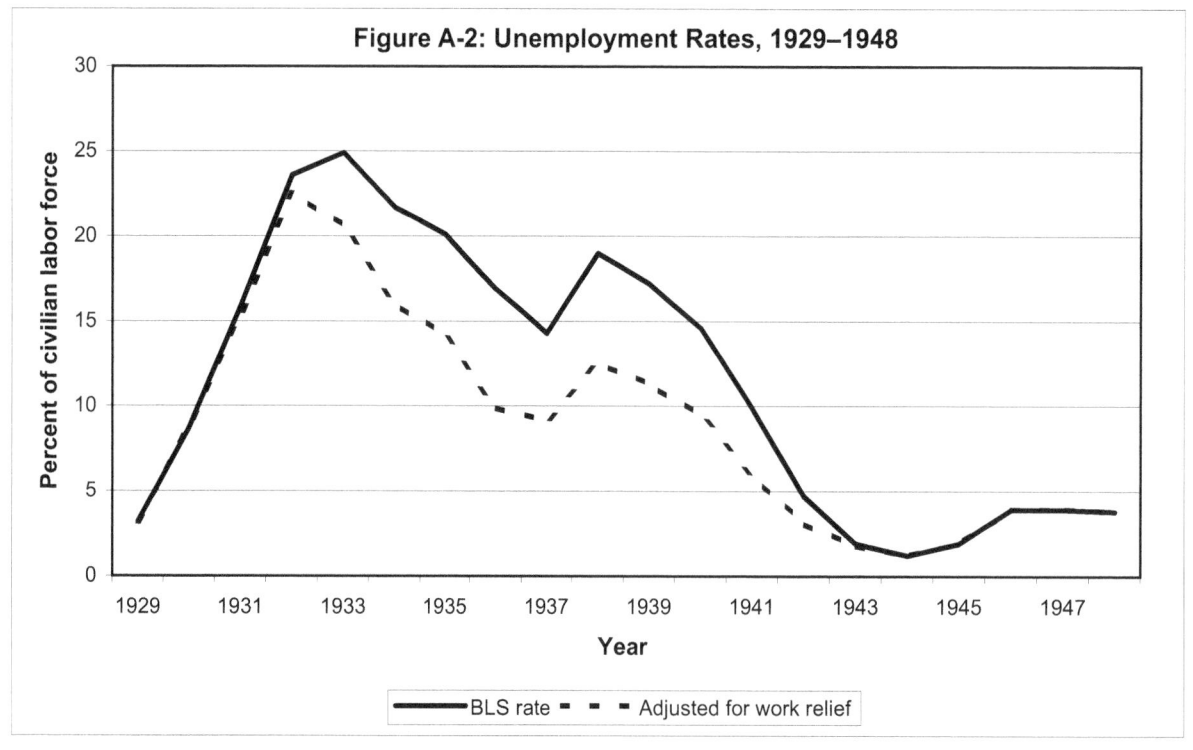

Figure A-2: Unemployment Rates, 1929–1948

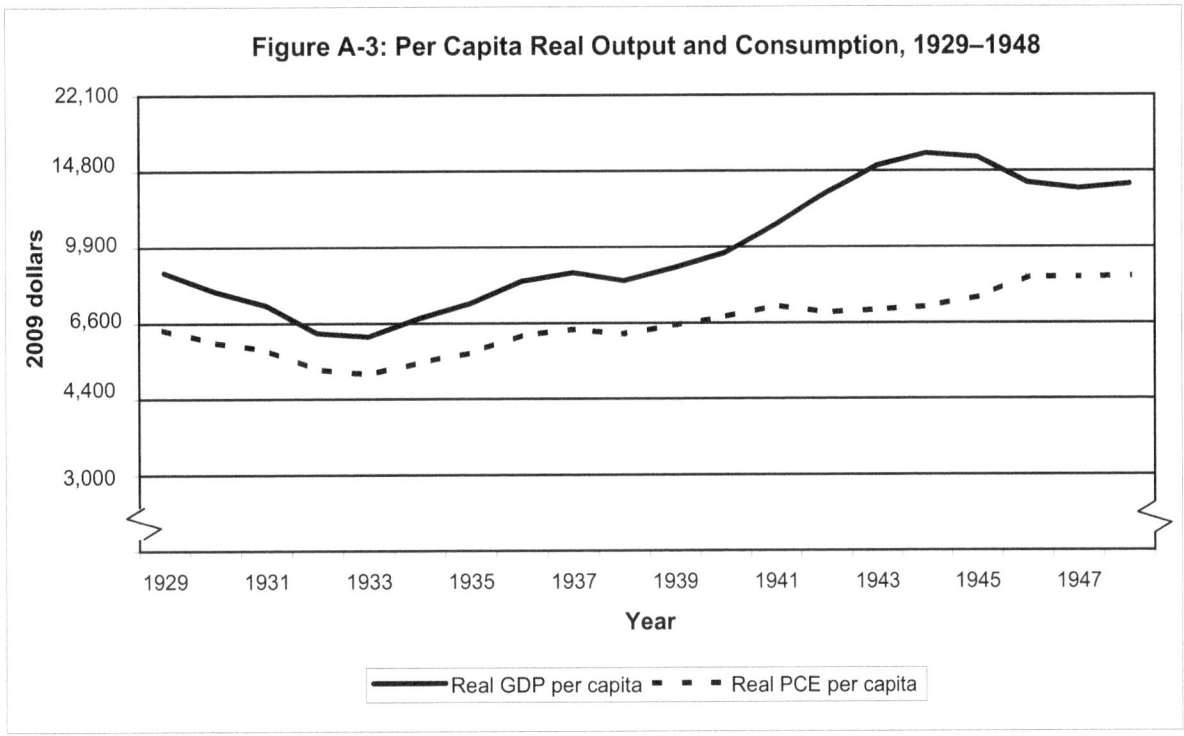

Figure A-3: Per Capita Real Output and Consumption, 1929–1948

10-21; see Notes and Definitions to Chapter 10 for further explanation.)

- By 1943, the economy reached a state of over-full employment, which would have been associated with runaway inflation if not for comprehensive price, wage, and production controls; the years of full wartime production, 1943 through 1945, all saw unemployment rates below 2 percent. After demobilization, the unemployment rate returned to just under 4 percent. (Table 10-1C)

- Real GDP declined 26.7 percent from 1929 to 1933—28.9 percent on a per capita basis. Per capita output recovered almost to the 1929 level in 1937 but fell back 4.1 percent in the 1938 recession. Output then nearly doubled from 1939 to the peak war production year, 1944; the per capita annual growth rate was 12.4 percent. Output fell back during the demobilization, but in 1948 was still at a per capita level representing a 5.0 percent per year growth rate since 1939. (Tables 1-3B and 4-1B)

- Per capita personal consumption expenditures fell 20.9 percent from 1929 to 1933. The decline would have been greater if consumers had not dipped into their assets to keep their living standards from declining as steeply as their incomes; the personal saving rate was negative in 1932 and

1933. Real per capita consumption recovered to the 1929 level by 1937. Despite rationing and shortages, real per capita consumption spending declined little during the war years. In 1948, it was 29.3 percent above the 1939 level. (Table 1-3B)

- Why was the 1929-1933 contraction so deep and long-lasting? By some, blame is placed on U.S. government tax increases and imposition of new trade barriers (the Smoot-Hawley Tariff), which undoubtedly made their contribution. A substantial number of well-regarded economists, however, point primarily to deflation and its interaction with debt. The price index for personal consumption expenditures declined 27.2 percent from 1929 to 1933, for an annual average deflation rate of 7.6 percent. Current and prospective price declines make debt more burdensome and debtors more likely to default, as interest payments remain fixed while incomes and asset values decline. (Table 1-6B)

- Current and prospective price declines also make borrowing prohibitively expensive; a low nominal interest rate becomes high in real terms (since the real rate is the nominal rate minus the inflation rate, and subtraction means changing the sign and adding). A dollar in the hands of a prospective lender will be worth more in terms of purchasing power if he simply holds on to it than if he invests it in some real

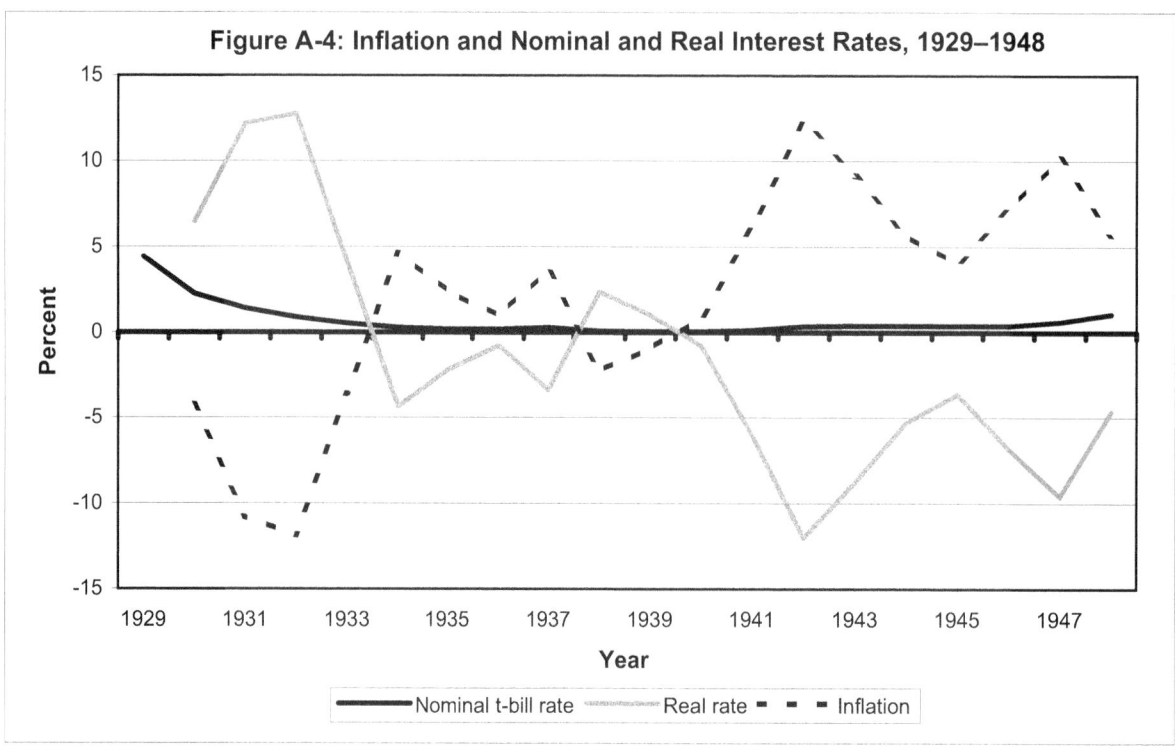

Figure A-4: Inflation and Nominal and Real Interest Rates, 1929–1948

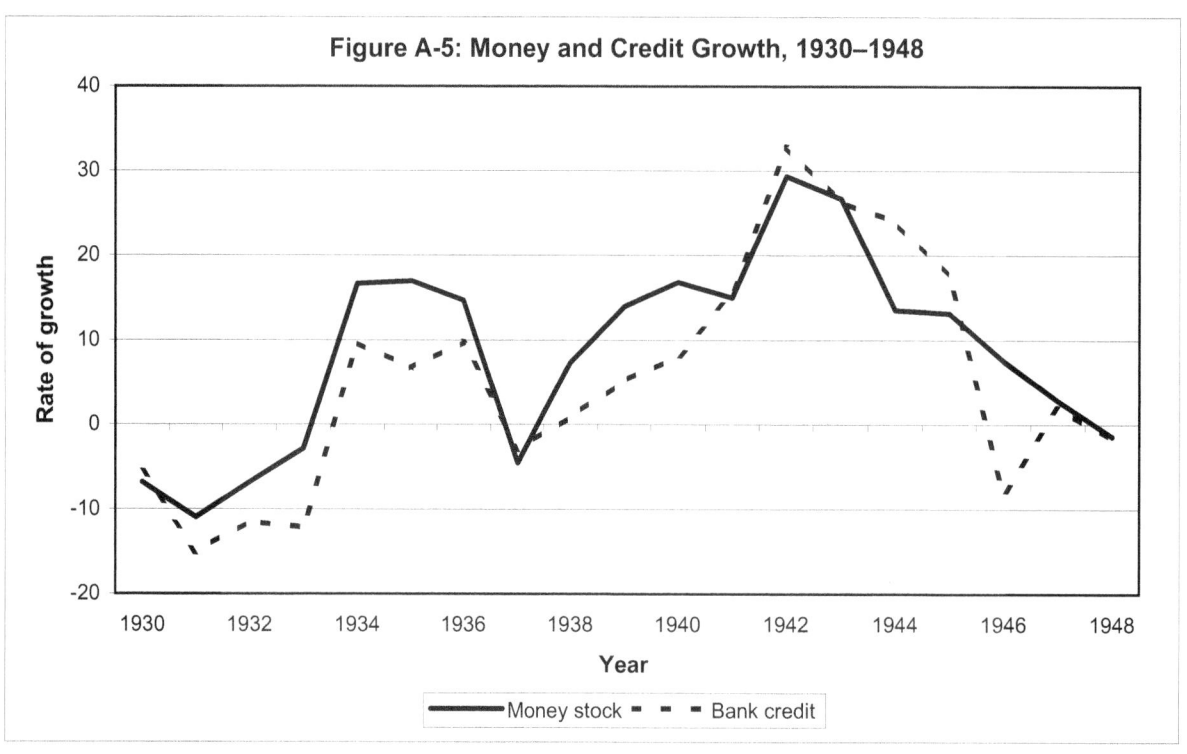

Figure A-5: Money and Credit Growth, 1930–1948

economic asset or activity whose price will be lower at the end of the year. There is no feasible way for a central bank to lower interest rates below zero in order to reduce real rates in the presence of deflation. As Figure A-4 shows, real rates were high from 1930 through 1933 despite near-zero market nominal rates on Treasury bills. (Tables 12-9B and 1-6B)

- Deflation was rooted in deep and sustained declines in the dollar volumes of money and credit, as shown in Tables 12-1B and 12-4B and illustrated in Figure A-5. The money stock (currency and demand deposits) declined 27 percent from June 30, 1929, to June 30, 1933. Loans and investments at all commercial banks declined 29 percent over the same period. Commercial paper outstanding plunged 76 percent between December 1929 and December 1932, and bankers' acceptances were down 59 percent.

- Positive money and credit growth after 1933 was reflected in moderate rates of price increase and moderately

negative—which means stimulative--real interest rates. As can be seen in Figures A-5, there was a monetary component to the 1937-38 recession. There was also a premature federal budget retrenchment at that time.

- During the war, monetary policy was pre-empted by wartime needs. Interest rates were held low to facilitate the financing of huge federal deficits.

- Worker incomes increased during the wartime boom as employment, average hours, and hourly wages all rose. But inflation and consumer spending were held down by price and production controls and rationing. Personal saving rates soared when there was little to buy and the government emphasized the sale of "war bonds"; the saving rate reached 26 percent in 1943 and 1944, then fell back to 4.2 percent in 1947, almost equal to its 1929 level.

THE GREAT RECESSION AND RECOVERY

Determining the beginning and ending of economic recessions and expansions depends on data such as GDP growth, unemployment and household income. The National Bureau of Economic Research (NBER) committee also examines other statistics such as industrial production, wholesale-retail sales and real disposal income. The former statistics are provided by the U.S. Department of Commerce, Bureau of Census and Bureau of Economics Analysis (BEA) and Department of Labor, Bureau of Labor Statistics (BLS). Economists at the NBER analyzes this data before announcing any inflection point in the U.S. economy.

The U.S. economy experienced an eight-month decline from March 2001 to November 2001. The dot.com bubble burst was mainly fueled by the excessive speculation in dot.com companies. The NASDAQ sunk significantly between March 2000 and December 2002. One lesson learned was that excessive speculation in internet-related companies with no profits is not sound financial strategy.

The economy recovered after 2001 but the Great Recession began in December 2007 and lasted until June 2009. Numerous consequential actions led to this downturn. Housing prices soared while mortgage lending standards deteriorated. For example, as house prices rose, banks started offering subprime mortgages without down payment or income requirements. Housing prices collapsed between 2007 and 2010 as demand for residential construction dropped between 2006 to 2011.

During the Great Recession, demand for construction materials such as wood, carpeting, appliances, etc. fell, which led to layoffs which led to increased unemployment. Without jobs, more and more people ended up losing their homes through foreclosures. The domino effect continued as household income, the stock market, and wealth all continued to fall. Millions of people fell below the poverty level in between 2007 to 2009 and household income did not return to its 2007 level until 2016.

Both President Bush and President Obama enacted fiscal and monetary policies to counteract the recession. An initial fiscal policy introduced by President Bush in 2008 was the Economic Stimulus Package Act which cost approximately $830 million. As part of the Act, individuals received checks from the government with the amount varying on family size. Next, the Troubled Assets Relief Program (TARP) was a program to purchase assets and equity from financial institutions to strengthen the financial sector. With mortgage foreclosures, high unemployment and declining income created a credit crunch where banks did not have enough reserves to provide loans to businesses and consumers. TARP provided funding to banks to the tune of $431 billion.

The Federal Reserve created the Quantitative Easing (QE) program as an expansionary monetary policy designed to lower interest rates and spur economic growth. QE works by the Federal Reserve buying bank's bonds added to reserves to ease the credit crunch caused by toxic assets.

President Obama continued the QE program and the automobile buyout under TARP, then introduced the American Recovery and Reinvestment Act (ARRA) with a price tag of $787 billion. This program cut taxes, extended unemployment benefits and funded public works projects.

The Financial Crisis Inquiry Commission (FCIC) was a ten-member commission appointed by Congressional leaders with a mission of determining the cause of the financial crisis of 2007 – 2010. Its final report, published in January 2011, concluded that the financial crisis was avoidable. Failures of financial regulation and corporate governance plus risky management actions including excessive borrowing, risky investments and lack of transparency put the financial system on a collision course with the crisis.

Recovery from the Great Recession was slow, sluggish, and at times weak. Between 2010 and 2019, both GDP and median household income grew at a much slower rate than it had historically. Unemployment rose to 9.6 percent in 2010 then receded to 3.7 percent in 2019.

February 2020 became another inflection point. The longest recovery ended after 10 years and 2 months due to the global pandemic. Coronavirus (COVID-19) and its resulting "stay-at-home" orders that were enacted throughout the country caused consumer and business demand to plunge, except for the essential activities, such as health care, groceries, deliveries, truck drivers, etc. Businesses rearranged their work environment such that employees could work at home and many of those companies only retained a small workforce. The unemployment rate reached 14.7 percent in April, 13.3 percent in May and 11.1 percent in June 2020. In the second quarter of 2020, GDP dropped to levels never experienced in the U.S. Personal income fared better with the CARES Act through July 2020 but the economic fallout from COVID-19 remains to be seen.

The upcoming sixteen chapters explore topics such as saving and investment, employment, household income, and Gross Domestic Product (GDP). Specifically, Chapters 1, 4, 5, 6, 10 and 12 provide extensive analytics of GDP, Chapter 3 includes detailed information on household income, poverty and health insurance coverage. Meanwhile, Chapter 8 includes information on inflation while Chapter 12 discusses interest rates and housing prices and Chapter 16 addresses key sector statistics. Historical data extending back to 1892 are available for certain series.

GENERAL NOTES

These notes provide general information about the data in Tables 1-1 through 16-14. Specific notes with information about data sources, definitions, methodology, revisions, and sources of additional information follow the tables in each chapter.

MAIN DIVISIONS OF THE BOOK

The tables are presented in two parts:

Part A (Tables 1-1 through 13-2) pertains principally to the U.S. economy as a whole. Generally, each table presents, on its initial page, annual averages as far back as data availability and space permit, and quarterly or monthly values for the most recent year or years. For many important series, this initial page is followed by full annual and quarterly or monthly histories as far back as they are available on a continuous, consistent basis. Some chapters present data for the United States only in aggregate, while others—such as the chapters concerning industrial production and capacity utilization (chapter 2), capital expenditures (chapter 5), profits (chapter 9), and employment, hours, and earnings (chapter 10)—also have detail for industry groups.

Data by industry are classified using the North American Industry Classification System (NAICS), as far back as such data are made available by the source agencies.

Part B focuses on the individual industries that together produce the gross domestic product (GDP).

- Chapter 14 contains data on the value of GDP, quantity production trends, and factor income by NAICS industry group.

- Chapter 15 provides further detail on payroll employment, hours, and earnings classified according to NAICS.

- Chapter 16 presents various data sets for key economic sectors. Some of the tables are based on definitions of products, rather than of producing establishments, and are valid for either classification system. This is the case for Tables 16-1, Petroleum and Petroleum Products; 16-2, New Construction; 16-3, Housing Starts and Building Permits, New House Sales, and Prices; and 16-8, Motor Vehicle Sales and Inventories. Tables 16-4 through 16-7 and 16-9, 16-11, and 16-12, which cover manufacturing and retail and wholesale trade, show data classified according to NAICS. Table 16-15 presents data for services industries classified according to NAICS.

Characteristics of the Tables and the Data

The subtitles or column headings for the data tables normally indicate whether the data are *seasonally adjusted, not seasonally adjusted,* or *at a seasonally adjusted annual rate.* These descriptions refer to the monthly or quarterly data, rather than the annual data; annual data by definition require no seasonal adjustment. Annual values are normally calculated as totals or averages, as appropriate, of unadjusted data. Such annual values are shown in either or both adjusted or unadjusted data columns.

Seasonal adjustment removes from a monthly or quarterly time series the average impact of variations that normally occur at about the same time each year, due to occurrences such as weather, holidays, and tax payment dates.

A simplified example of the process of seasonal adjustment, or deseasonalizing, can indicate its importance in the interpretation of economic time series. Statisticians compare actual monthly data for a number of years with "moving average" trends of the monthly data for the 12 months centered on each month's data. For example, they may find that in November, sales values are usually about 95 percent of the moving average, while in December, usual sales values are 110 percent of the average. Suppose that actual November sales in the current year are $100 and December sales are $105. The seasonally adjusted value for November will be $105 ($100/0.95) while the value for December will be $95 ($105/1.10). Thus, an apparent increase in the unadjusted data turns out to be a decrease when adjusted for the usual seasonal pattern.

The statistical method used to achieve the seasonal adjustment may vary from one data set to another. Many of the data are adjusted by a computer method known as X-12-ARIMA, developed by the Census Bureau. A description of the method is found in "New Capabilities and Methods of the X-12-ARIMA Seasonal Adjustment Program," by David F. Findley, Brian C. Monsell, William R. Bell, Mark C. Otto and Bor-Chung Chen (*Journal of Business and Economic Statistics*, April 1998). This article can be downloaded from the Bureau of the Census Web site at <http://www.census.gov>.

Production and sales data presented at *annual rates*—such as NIPA data in dollars, or motor vehicle data in number of units—show values at their annual equivalents: the values that would be registered if the seasonally adjusted rate of activity measured during a particular month or quarter were maintained for a full year. Specifically, seasonally adjusted monthly values are multiplied by 12 and quarterly values by 4 to yield seasonally adjusted annual rates.

Percent changes at seasonally adjusted annual rates for quarterly time periods are calculated using a compound interest formula, by raising the quarter-to-quarter percent change in a seasonally adjusted series to the fourth power. See the article

"Cycle and Growth Perspectives" for an explanation of compound annual growth rates.

Indexes. In many of the most important data sets presented in this volume, aggregate measures of prices and quantities are expressed in the form of indexes. The most basic and familiar form of index, the original Consumer Price Index, begins with a "market basket" of goods and services purchased in a base period, with each product category valued at its dollar prices—the amount spent on that category by the average consumer. The value weight ascribed to each component of the market basket is moved forward by the observed change in the price of the item selected to represent that component. These weighted component prices—which constitute the quantities in the base period repriced in the prices of subsequent periods—are aggregated, divided by the base period aggregate, and multiplied by 100 to provide an index number. An index calculated in this way is known as a *Laspeyres index.* In general, economists believe that Laspeyres price indexes have an upward bias, showing more price increase than they would if account were taken of consumers' ability to change spending patterns and maintain the same level of satisfaction in response to changing relative prices.

A *Paasche index* is one that uses the weights of the current period. Since the weights in the Paasche index change in each period, Paasche indexes only provide acceptable indications of change relative to the base period. Paasche indexes for two periods neither of which is the base period cannot be correctly compared: for example, a Paasche price index for a recent period might increase from the period just preceding even if no prices changed between those two periods, if there was a change in the composition of output toward prices that had previously increased more from the base period. When the national income and product account (NIPA) measures of real output were Laspeyres measures, using the weights of a single base year, the implicit deflators (current-dollar values divided by constant-dollar values) were Paasche indexes. Just as Laspeyres price indexes are upward-biased, Paasche price indexes are downward-biased because they overestimate consumers' ability to maintain the same level of satisfaction by changing spending patterns.

In recent years, government statisticians—with the aid of complex computer programs—have developed measures of real output and prices that minimize bias by using the weights of both periods and updating the weights for each period-to-period comparison. Such measures are described as chained indexes and are used in the NIPAs, the index of industrial production, and an experimental consumer price index. Chained measures are discussed more fully in the notes and definitions for Chapter 1, Chapter 2, and Chapter 8. The "Fisher Ideal" index, the "superlative" index, and the "Tornqvist formula" are all types of chained indexes that use weights for both periods under comparison.

Detail may not sum to totals due to rounding. Since annual data are typically calculated by source agencies as the annual totals or averages of not-seasonally-adjusted data, they therefore will not be precisely equal to the annual totals or averages of monthly seasonally-adjusted data. Also, seasonal adjustment procedures are typically multiplicative rather than additive, which may also prevent seasonally adjusted data from adding or averaging to the annual figure. Percent changes and growth rates may have been calculated using unrounded data and therefore differ from those using the published figures.

The data in this volume are from federal government sources and may be reproduced freely. A list of data sources is shown below.

The tables in this volume incorporate data revisions and corrections released by the source agencies through mid-2014, including the July annual revision of the NIPAs and the resulting August revision of productivity and costs.

DATA SOURCES

The source agencies for the data in this volume are listed below. The specific source or sources for each particular data set are identified at the beginning of the notes and definitions for the relevant data pages.

BOARD OF GOVERNORS OF THE FEDERAL RESERVE SYSTEM

20th Street & Constitution Avenue NW
Washington, DC 20551

Data Inquiries and Publication Sales:
Publications Services
Mail Stop 127
Board of Governors of the Federal Reserve System
Washington, DC 20551
Phone: (202) 452-3245

Quarterly Publication:
As of 2006, the *Federal Reserve Bulletin* is available free of charge and only on the Federal Reserve Web site.

URL:
http://www.federalreserve.gov

Census Bureau
U.S. Department of Commerce
4700 Silver Hill Road
Washington, DC 20233

URL:
http://www.census.gov

Ordering Data Products:
Call Center: (301) 763-INFO (4636)

E-mail Questions:
webmaster@census.gov

Bureau of Economic Analysis
U.S. Department of Commerce
Washington, DC 20230

Data Inquiries:
Public Information Office
Phone: (202) 606-9900

Monthly Publication:
Survey of Current Business
Available online.

URL:
http://www.bea.gov

Bureau of Labor Statistics
U.S. Department of Labor
2 Massachusetts Avenue NE
Washington, DC 20212-0001
(202) 691-5200

URL:
http://www.bls.gov

Data Inquiries:
Blsdata_staff@bls.gov

Monthly Publications available online:
Monthly Labor Review
Employment and Earnings
Compensation and Working Conditions
Producer Price Indexes
CPI Detailed Report

Employment and Training Administration
U.S. Department of Labor
200 Constitution Avenue NW
Washington, DC 20210
(877) US2-JOBS

URL:
http://www.doi.gov/agencies/eta
http://www.itsc.state.md.us

Energy Information Administration
U.S. Department of Energy
1000 Independence Avenue SW
Washington, DC 20585

Data Inquiries and Publications:
National Energy Information Center
Phone: (202) 586-8800
E-mail: infoctr@eia.doe.gov

Monthly Publication:
Monthly Energy Review

URL:
http://www.eia.doe.gov

Federal Housing Finance Agency
FHFAinfo@FHFA.gov
(202) 414-6921,6922
(202) 414-6376

URL:
http://www.fhfa.gov/hpi

U.S. Department of the Treasury
Office of International Affairs
Treasury International Capital System

URL:
http://www.treas.gov/tic

To order government publications
Superintendent of Documents
Government Printing Office
Washington, DC 20402
(202) 512-1800

URL:
http://bookstore.gpo.gov

PART A: THE U.S. ECONOMY

CHAPTER 1: NATIONAL INCOME AND PRODUCT

SECTION 1A: GROSS DOMESTIC PRODUCT: VALUES, QUANTITIES, AND PRICES

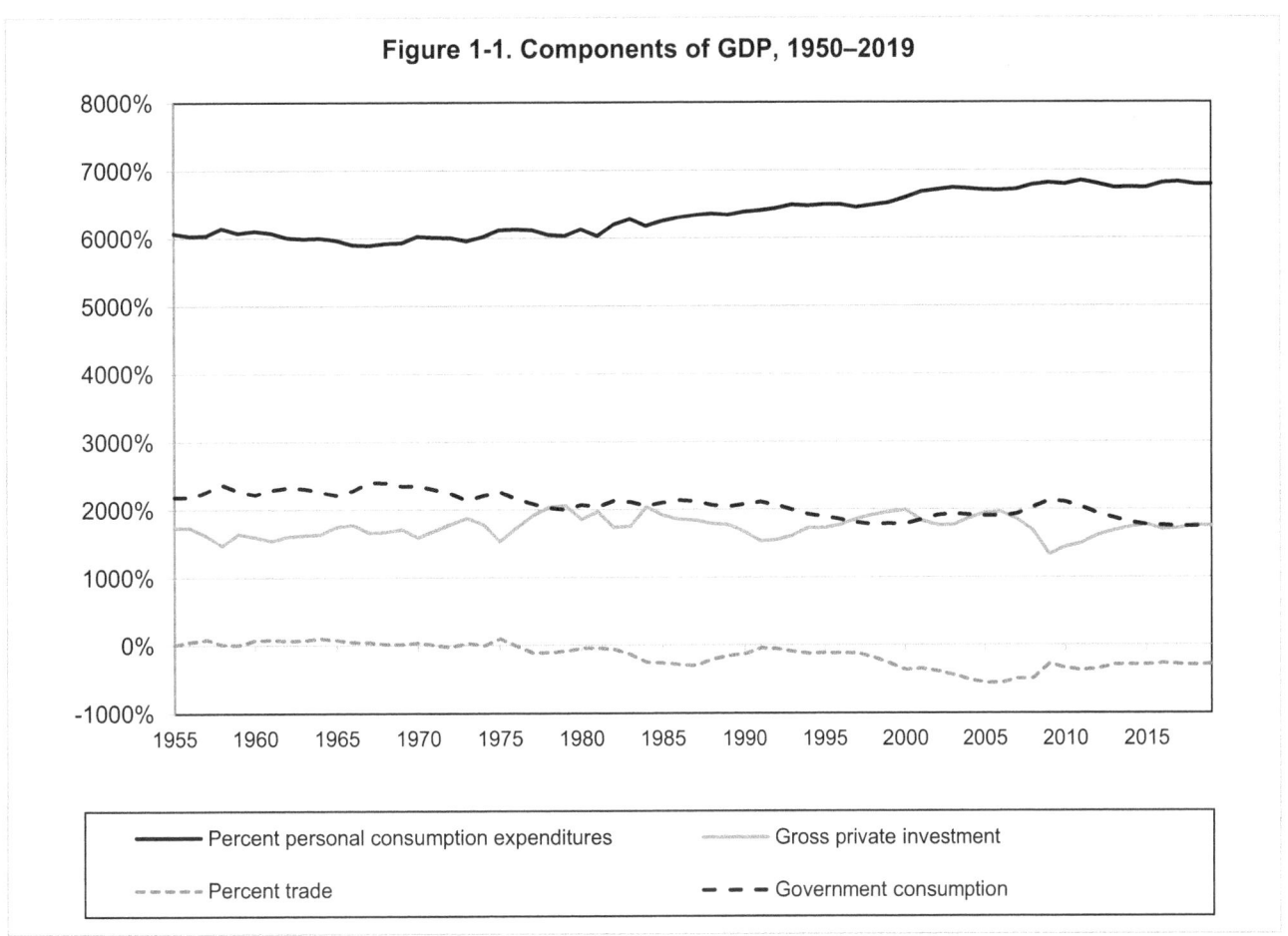

Figure 1-1. Components of GDP, 1950–2019

- As presented in the preface, the output approach to GDP consists of four components: consumption, business investment, government taxes and spending and the foreign trade balance (exports – imports). GDP is one of the most comprehensive and closely watched economic statistics in not only the United States but also globally. Figure 1-1 illustrates the trend of these four components. For example, consumption's share of GDP rose from 61 percent in 1955 to nearly 68 percent in 2019. Government's percent share of GDP fell in the same time period. For more details of GDP, see the Notes and Definitions at the end of this chapter. (Table 1-1A)

- Real GDP per capita—the constant-dollar average value of production for each man, woman, and child in the population—rose from $17,372 (2012 dollars) in 1955 to $51,794 in 2007. Reflecting the recession that began in December 2007, it declined 4.4 percent from 2007 to 2009. Over the ten years since then, per capita GDP recovered with a 1.4 annual rate and surpassed its 2007 peak in 2013. (Table 1-3A)

- With the impositions of tariffs in January 2018, imports became more expensive. Tariffs have been imposed on solar panels, washing machines, steel and aluminum as well as products from Canada, Mexico, China and the European Union. China then imposed their own tariffs. Americans enjoy a vast selection of goods, many of which are not available in the United States. Tariffs raise prices for those consumers and businesses. The impact of these tariffs on the U.S. trade balance has been negative.

Table 1-1A. Gross Domestic Product: Recent Data

(Billions of dollars, quarterly data are at seasonally adjusted annual rates.)   **NIPA Tables 1.1.5, 1.2.5**

Year and quarter	Gross domestic product	Personal consumption expenditures	Gross private domestic investment Total	Fixed investment Nonresidential	Fixed investment Residential	Change in private inventories	Net exports	Exports	Imports	Government Total	Federal	State and local	Addendum: final sales of domestic product
1955	425.5	258.3	73.8	43.4	25.4	5.0	0.5	17.7	17.2	93.0	61.0	31.9	420.5
1956	449.4	271.1	77.7	49.7	24.0	4.0	2.4	21.3	18.9	98.2	63.1	35.1	445.4
1957	474.0	286.3	76.5	53.1	22.6	0.8	4.1	24.0	19.9	107.2	68.5	38.7	473.3
1958	481.2	295.6	70.9	48.5	22.8	-0.4	0.5	20.6	20.0	114.1	71.5	42.6	481.6
1959	521.7	317.1	85.7	53.1	28.6	3.9	0.4	22.7	22.3	118.5	73.6	44.9	517.7
1960	542.4	331.2	86.5	56.4	26.9	3.2	4.2	27.0	22.8	120.5	72.9	47.6	539.1
1961	562.2	341.5	86.6	56.6	27.0	3.0	4.9	27.6	22.7	129.2	77.4	51.9	559.2
1962	603.9	362.6	97.0	61.2	29.6	6.1	4.1	29.1	25.0	140.3	85.5	54.8	597.8
1963	637.5	382.0	103.3	64.8	32.9	5.6	4.9	31.1	26.1	147.2	87.9	59.3	631.8
1964	684.5	410.6	112.2	72.2	35.1	4.8	6.9	35.0	28.1	154.8	90.3	64.5	679.6
1965	742.3	443.0	129.6	85.2	35.2	9.2	5.6	37.1	31.5	164.1	93.2	70.9	733.0
1966	813.4	479.9	144.2	97.2	33.4	13.6	3.9	40.9	37.1	185.4	106.6	78.9	799.8
1967	860.0	506.7	142.7	99.2	33.6	9.9	3.6	43.5	39.9	207.0	120.0	87.0	850.1
1968	940.7	556.9	156.9	107.7	40.2	9.1	1.4	47.9	46.6	225.5	128.0	97.6	931.6
1969	1 017.6	603.6	173.6	120.0	44.4	9.2	1.4	51.9	50.5	239.0	131.2	107.8	1 008.4
1970	1 073.3	646.7	170.0	124.6	43.4	2.0	3.9	59.7	55.8	252.6	132.8	119.8	1 071.3
1971	1 164.9	699.9	196.8	130.4	58.2	8.3	0.6	63.0	62.3	267.5	134.5	133.0	1 156.6
1972	1 279.1	768.2	228.1	146.6	72.4	9.1	-3.4	70.8	74.2	286.2	141.6	144.5	1 270.0
1973	1 425.4	849.6	266.9	172.7	78.3	15.9	4.1	95.3	91.2	304.8	146.2	158.6	1 409.5
1974	1 545.2	930.2	274.5	191.1	69.5	14.0	-0.8	126.7	127.5	341.4	158.8	182.5	1 531.3
1975	1 684.9	1 030.5	257.3	196.8	66.7	-6.3	16.0	138.7	122.7	381.1	173.7	207.4	1 691.2
1976	1 873.4	1 147.7	323.2	219.3	86.8	17.1	-1.6	149.5	151.1	404.2	184.8	219.4	1 856.3
1977	2 081.8	1 274.0	396.6	259.1	115.2	22.3	-23.1	159.3	182.4	434.3	200.3	234.0	2 059.5
1978	2 351.6	1 422.3	478.4	314.6	138.0	25.8	-25.4	186.9	212.3	476.3	218.9	257.4	2 325.8
1979	2 627.3	1 585.4	539.7	373.8	147.8	18.0	-22.5	230.1	252.7	524.8	240.6	284.2	2 609.3
1980	2 857.3	1 750.7	530.1	406.9	129.5	-6.3	-13.1	280.8	293.8	589.6	274.9	314.7	2 863.6
1981	3 207.0	1 934.0	631.2	472.9	128.5	29.8	-12.5	305.2	317.8	654.4	314.0	340.4	3 177.2
1982	3 343.8	2 071.3	581.0	485.1	110.8	-14.9	-20.0	283.2	303.2	711.5	348.3	363.1	3 358.7
1983	3 634.0	2 281.6	637.5	482.2	161.1	-5.8	-51.6	277.0	328.6	766.6	382.4	384.2	3 639.8
1984	4 037.6	2 492.3	820.1	564.3	190.4	65.4	-102.7	302.4	405.1	827.9	411.8	416.1	3 972.2
1985	4 339.0	2 712.8	829.7	607.8	200.1	21.8	-114.0	303.2	417.2	910.5	452.9	457.6	4 317.2
1986	4 579.6	2 886.3	849.1	607.8	234.8	6.6	-131.9	321.0	452.9	976.1	481.7	494.4	4 573.1
1987	4 855.2	3 076.3	892.2	615.2	249.8	27.1	-144.8	363.9	508.7	1 031.5	502.8	528.7	4 828.1
1988	5 236.4	3 330.0	937.0	662.3	256.2	18.5	-109.4	444.6	554.0	1 078.9	511.4	567.4	5 218.0
1989	5 641.6	3 576.8	999.7	716.0	256.0	27.7	-86.7	504.3	591.0	1 151.9	534.1	617.8	5 613.9
1990	5 963.1	3 809.0	993.4	739.2	239.7	14.5	-77.9	551.9	629.7	1 238.6	562.4	676.2	5 948.6
1991	6 158.1	3 943.4	944.3	723.6	221.2	-0.4	-28.6	594.9	623.5	1 299.0	582.9	716.0	6 158.5
1992	6 520.3	4 197.6	1 013.0	741.9	254.7	16.3	-34.7	633.1	667.8	1 344.5	588.5	756.0	6 504.0
1993	6 858.6	4 452.0	1 106.8	799.2	286.8	20.8	-65.2	654.8	720.0	1 364.9	580.2	784.8	6 837.7
1994	7 287.2	4 721.0	1 256.5	868.9	323.8	63.8	-92.5	720.9	813.4	1 402.3	574.7	827.6	7 223.5
1995	7 639.7	4 962.6	1 317.5	962.2	324.1	31.2	-89.8	812.8	902.6	1 449.4	576.7	872.7	7 608.6
1996	8 073.1	5 244.6	1 432.1	1 043.2	358.1	30.8	-96.4	867.6	964.0	1 492.8	579.2	913.7	8 042.3
1997	8 577.6	5 536.8	1 595.6	1 149.1	375.6	70.9	-102.0	953.8	1 055.8	1 547.1	583.3	963.8	8 506.6
1998	9 062.8	5 877.2	1 736.7	1 254.1	418.8	63.7	-162.7	953.0	1 115.7	1 611.6	585.5	1 026.1	8 999.1
1999	9 630.7	6 279.1	1 887.1	1 364.5	461.8	60.8	-255.8	992.8	1 248.6	1 720.4	611.3	1 109.0	9 569.8
2000	10 252.3	6 762.1	2 038.4	1 498.4	485.4	54.5	-375.1	1 096.3	1 471.3	1 826.8	633.7	1 193.1	10 197.8
2001	10 581.8	7 065.6	1 934.8	1 460.1	513.1	-38.3	-367.9	1 024.6	1 392.6	1 949.3	670.1	1 279.2	10 620.1
2002	10 936.4	7 342.7	1 930.4	1 352.8	557.6	20.0	-425.4	998.7	1 424.1	2 088.7	743.0	1 345.7	10 916.4
2003	11 458.2	7 723.1	2 027.1	1 375.9	637.1	14.1	-503.1	1 036.2	1 539.3	2 211.2	826.3	1 384.9	11 444.2
2004	12 213.7	8 212.7	2 281.3	1 467.4	749.8	64.1	-619.1	1 177.6	1 796.7	2 338.9	891.7	1 447.1	12 149.7
2005	13 036.6	8 747.1	2 534.7	1 621.0	856.2	57.5	-721.2	1 305.2	2 026.4	2 476.0	947.5	1 528.5	12 979.1
2006	13 814.6	9 260.3	2 701.0	1 793.8	838.2	69.0	-770.9	1 472.6	2 243.5	2 624.2	1 000.7	1 623.5	13 745.6
2007	14 451.9	9 706.4	2 673.0	1 948.6	690.5	34.0	-718.4	1 660.9	2 379.3	2 790.8	1 050.5	1 740.3	14 417.9
2008	14 712.8	9 976.3	2 477.6	1 990.9	516.0	-29.2	-723.1	1 837.1	2 560.1	2 982.0	1 150.6	1 831.4	14 742.1
2009	14 448.9	9 842.2	1 929.7	1 690.4	390.0	-150.8	-396.5	1 582.0	1 978.4	3 073.5	1 218.2	1 855.3	14 599.7
2010	14 992.1	10 185.8	2 165.5	1 735.0	376.6	53.9	-513.9	1 846.3	2 360.2	3 154.6	1 297.9	1 856.7	14 938.1
2011	15 542.6	10 641.1	2 332.6	1 907.5	378.8	46.3	-579.5	2 103.0	2 682.5	3 148.4	1 298.9	1 849.4	15 496.3
2012	16 197.0	11 006.8	2 621.8	2 118.5	432.0	71.2	-568.6	2 191.3	2 759.9	3 137.0	1 286.5	1 850.5	16 125.8
2013	16 784.9	11 317.2	2 826.0	2 211.5	510.0	104.5	-490.8	2 273.4	2 764.2	3 132.4	1 226.6	1 905.8	16 680.3
2014	17 527.3	11 822.8	3 044.2	2 400.1	560.2	84.0	-507.7	2 371.7	2 879.4	3 168.0	1 215.0	1 953.0	17 443.3
2015	18 238.3	12 297.5	3 237.2	2 466.6	633.8	136.8	-526.6	2 265.9	2 792.4	3 230.2	1 220.8	2 009.4	18 101.5
2016	18 745.1	12 770.0	3 188.3	2 460.5	699.5	28.4	-512.5	2 227.2	2 739.7	3 299.3	1 234.7	2 064.6	18 716.7
2017	19 543.0	13 340.4	3 351.1	2 574.5	760.3	16.3	-555.5	2 374.6	2 930.1	3 407.0	1 263.9	2 143.2	19 526.7
2018	20 611.9	13 993.3	3 632.9	2 776.7	798.5	57.7	-609.5	2 528.7	3 138.2	3 595.2	1 339.4	2 255.7	20 554.1
2019	21 433.2	14 544.6	3 751.2	2 895.0	807.1	49.1	-610.5	2 514.8	3 125.2	3 747.9	1 419.2	2 328.7	21 384.1
2017													
1st quarter	19 237.4	13 153.2	3 266.2	2 532.5	746.0	-12.3	-543.6	2 326.4	2 869.9	3 361.6	1 246.5	2 115.1	19 249.7
2nd quarter	19 379.2	13 241.3	3 313.3	2 555.9	753.3	4.0	-559.5	2 333.1	2 892.6	3 384.2	1 257.9	2 126.3	19 375.2
3rd quarter	19 617.3	13 370.9	3 378.8	2 575.2	758.5	45.1	-543.6	2 370.1	2 913.7	3 411.1	1 262.7	2 148.4	19 572.2
4th quarter	19 938.0	13 596.0	3 446.3	2 634.2	783.6	28.5	-575.5	2 468.7	3 044.1	3 471.1	1 288.3	2 182.9	19 909.5
2018													
1st quarter	20 242.2	13 755.5	3 555.0	2 716.2	794.3	44.5	-589.8	2 507.2	3 097.0	3 521.5	1 308.1	2 213.4	20 197.7
2nd quarter	20 552.7	13 939.9	3 580.9	2 765.9	804.3	10.7	-548.1	2 550.3	3 098.4	3 580.0	1 329.3	2 250.7	20 541.9
3rd quarter	20 742.7	14 086.3	3 671.7	2 792.6	800.7	78.4	-646.4	2 523.9	3 170.3	3 631.2	1 352.0	2 279.1	20 664.4
4th quarter	20 909.9	14 191.4	3 723.9	2 831.9	794.7	97.3	-653.4	2 533.4	3 186.9	3 648.0	1 368.4	2 279.6	20 812.5
2019													
1st quarter	21 115.3	14 276.6	3 772.8	2 878.4	795.8	98.6	-615.5	2 523.5	3 139.0	3 681.5	1 388.8	2 292.7	21 016.7
2nd quarter	21 329.9	14 497.3	3 739.7	2 891.3	795.3	53.1	-644.7	2 514.6	3 159.4	3 737.6	1 410.6	2 327.0	21 276.8
3rd quarter	21 540.3	14 645.3	3 759.8	2 908.0	810.5	41.3	-631.8	2 505.2	3 137.1	3 767.1	1 429.3	2 337.8	21 499.0
4th quarter	21 747.4	14 759.2	3 732.6	2 902.3	827.0	3.4	-549.8	2 515.7	3 065.4	3 805.3	1 447.9	2 357.4	21 744.0

Table 1-1B. Gross Domestic Product: Historical Data

(Billions of dollars, quarterly data are at seasonally adjusted annual rates.)   **NIPA Tables 1.1.5, 1.2.5**

Year and quarter	Gross domestic product	Personal consumption expenditures	Gross private domestic investment				Exports and imports of goods and services			Government consumption expenditures and gross investment			Addendum: Final sales of domestic product
			Total	Fixed investment		Change in private inventories	Net exports	Exports	Imports	Total	Federal	State and local	
				Nonresidential	Residential								
1929	104.6	77.4	17.2	11.6	4.1	1.5	0.4	5.9	5.6	9.6	1.9	7.7	103.0
1930	92.2	70.1	11.4	9.2	2.5	-0.2	0.3	4.4	4.1	10.3	2.0	8.3	92.4
1931	77.4	60.7	6.5	5.8	1.9	-1.1	0.0	2.9	2.9	10.2	2.0	8.1	78.5
1932	59.5	48.7	1.8	3.3	0.9	-2.4	0.0	2.0	1.9	9.0	2.0	7.0	61.9
1933	57.2	45.9	2.3	3.0	0.7	-1.4	0.1	2.0	1.9	8.9	2.4	6.5	58.6
1934	66.8	51.5	4.3	3.8	1.0	-0.6	0.3	2.6	2.2	10.7	3.4	7.3	67.4
1935	74.3	55.9	7.4	4.8	1.4	1.1	-0.2	2.8	3.0	11.2	3.6	7.6	73.1
1936	84.9	62.2	9.4	6.4	1.8	1.2	-0.1	3.0	3.2	13.4	5.8	7.7	83.7
1937	93.0	66.8	13.0	8.2	2.2	2.6	0.1	4.0	4.0	13.1	5.3	7.9	90.4
1938	87.4	64.3	7.9	6.3	2.2	-0.6	1.0	3.8	2.8	14.2	5.9	8.3	88.0
1939	93.5	67.2	10.2	6.9	3.2	0.2	0.8	4.0	3.1	15.2	6.2	9.0	93.3
1940	102.9	71.3	14.6	8.5	3.6	2.4	1.5	4.9	3.4	15.6	6.8	8.8	100.6
1941	129.4	81.1	19.4	10.8	4.2	4.3	1.0	5.5	4.4	27.9	19.1	8.8	125.0
1942	166.0	89.0	11.8	7.5	2.4	1.9	-0.3	4.4	4.6	65.5	56.7	8.8	164.1
1943	203.1	99.9	7.4	6.6	1.6	-0.7	-2.2	4.0	6.3	98.1	89.5	8.6	203.9
1944	224.6	108.6	9.2	8.6	1.5	-0.9	-2.0	4.9	6.9	108.7	100.1	8.6	225.5
1945	228.2	120.0	12.4	12.1	1.8	-1.5	-0.8	6.8	7.5	96.6	87.3	9.3	229.6
1946	227.8	144.3	33.1	19.2	8.0	6.0	7.2	14.2	7.0	43.2	32.1	11.1	221.8
1947	249.9	162.0	37.1	25.5	12.2	-0.6	10.8	18.7	7.9	40.0	25.8	14.2	250.5
1948	274.8	175.0	50.3	28.9	15.8	5.7	5.5	15.5	10.1	44.0	27.1	16.9	269.1
1949	272.8	178.5	39.1	26.9	14.8	-2.7	5.2	14.5	9.2	50.0	30.5	19.6	275.5
1947													
1st quarter	243.2	156.2	35.9	24.8	10.5	0.5	10.9	18.4	7.5	40.3	27.0	13.3	242.7
2nd quarter	246.0	160.0	34.5	25.2	10.6	-1.2	11.3	19.5	8.2	40.1	26.4	13.7	247.2
3rd quarter	249.6	163.5	34.9	25.4	12.5	-2.9	11.8	19.4	7.7	39.4	25.0	14.3	252.5
4th quarter	259.7	167.7	43.3	26.5	15.3	1.5	9.3	17.6	8.3	39.5	24.5	15.0	258.3
1948													
1st quarter	265.7	170.4	47.2	28.2	15.3	3.6	7.3	16.9	9.6	40.9	25.4	15.5	262.1
2nd quarter	272.6	174.1	50.3	28.1	16.4	5.9	5.2	15.2	10.0	42.9	26.6	16.3	266.7
3rd quarter	279.2	177.1	52.5	29.1	16.3	7.2	4.9	15.4	10.5	44.7	27.5	17.2	272.0
4th quarter	280.4	177.9	51.3	30.2	15.2	6.0	4.5	14.6	10.1	46.6	28.7	17.9	274.4
1949													
1st quarter	275.0	176.8	43.1	28.6	14.2	0.4	6.5	16.1	9.6	48.6	30.2	18.4	274.6
2nd quarter	271.4	178.4	36.2	27.5	13.9	-5.1	6.3	15.6	9.4	50.4	31.2	19.1	276.5
3rd quarter	272.9	177.8	39.5	26.1	14.7	-1.3	5.2	14.1	8.9	50.4	30.5	19.9	274.2
4th quarter	270.6	180.2	37.5	25.6	16.5	-4.7	3.0	12.1	9.1	49.9	29.8	20.1	275.3
1950													
1st quarter	280.8	182.9	46.7	26.4	18.3	2.0	2.2	11.7	9.5	49.0	28.5	20.4	278.8
2nd quarter	290.4	186.8	52.3	28.9	20.6	2.8	1.6	11.9	10.2	49.6	28.9	20.7	287.6
3rd quarter	308.2	200.5	58.6	31.9	22.6	4.2	-0.7	12.3	13.0	49.7	28.4	21.3	304.0
4th quarter	319.9	197.9	68.4	32.9	21.5	14.0	-0.2	13.5	13.7	53.7	31.8	21.9	305.9
1951													
1st quarter	336.0	209.2	64.6	33.1	21.1	10.4	0.2	15.0	14.9	62.0	39.6	22.5	325.6
2nd quarter	344.1	204.9	67.4	34.1	18.5	14.8	1.9	17.1	15.2	69.8	46.6	23.2	329.3
3rd quarter	351.4	207.6	62.0	34.8	17.4	9.7	3.7	18.1	14.3	78.0	54.3	23.7	341.6
4th quarter	356.2	211.6	57.1	34.6	17.7	4.7	4.2	18.2	14.0	83.3	59.3	24.0	351.5
1952													
1st quarter	359.8	213.0	58.1	35.1	18.3	4.7	3.7	18.7	15.0	85.0	60.9	24.1	355.1
2nd quarter	361.0	217.1	53.0	35.8	18.8	-1.5	2.0	16.6	14.6	89.0	64.1	24.9	362.6
3rd quarter	367.7	219.6	57.2	32.9	18.8	5.6	0.0	15.2	15.3	90.9	66.2	24.8	362.1
4th quarter	380.8	227.7	60.7	35.8	19.6	5.3	-1.0	15.3	16.3	93.4	68.1	25.3	375.5
1953													
1st quarter	388.0	231.2	61.7	37.8	20.0	3.9	-0.7	15.1	15.8	95.8	69.7	26.1	384.1
2nd quarter	391.7	233.0	62.1	38.5	20.1	3.6	-1.3	15.2	16.4	97.9	72.0	26.0	388.2
3rd quarter	391.2	233.7	61.4	39.6	19.5	2.3	-0.6	15.8	16.3	96.6	70.0	26.7	388.8
4th quarter	386.0	233.1	56.4	39.2	19.3	-2.0	-0.3	15.2	15.5	96.7	69.5	27.2	388.0
1954													
1st quarter	385.3	235.2	55.7	38.3	19.4	-2.0	-0.4	14.4	14.8	94.8	66.7	28.1	387.3
2nd quarter	386.1	237.9	55.4	38.2	20.7	-3.4	0.3	16.4	16.2	92.5	63.7	28.8	389.5
3rd quarter	391.0	240.3	59.0	38.9	22.2	-2.1	0.6	15.9	15.3	91.1	61.4	29.8	393.1
4th quarter	399.7	245.1	62.1	38.9	23.6	-0.3	1.1	16.6	15.5	91.4	61.2	30.1	400.1
1955													
1st quarter	413.1	251.4	68.7	39.5	25.4	3.8	1.1	17.3	16.2	91.9	60.7	31.1	409.3
2nd quarter	421.5	256.5	72.7	42.1	26.0	4.6	-0.2	16.9	17.1	92.5	60.8	31.7	416.9
3rd quarter	430.2	260.7	74.7	44.8	25.6	4.3	0.7	18.1	17.4	94.1	61.9	32.2	425.9
4th quarter	437.1	264.6	78.9	47.1	24.6	7.2	0.2	18.3	18.1	93.3	60.6	32.8	429.9
1956													
1st quarter	439.7	266.2	78.3	47.8	24.1	6.4	0.4	19.4	18.9	94.8	61.0	33.9	433.3
2nd quarter	446.0	268.8	77.0	49.1	24.4	3.6	1.9	20.9	19.0	98.2	63.5	34.7	442.4
3rd quarter	451.2	272.1	78.3	50.7	24.0	3.6	2.6	21.8	19.3	98.3	62.8	35.5	447.6
4th quarter	460.5	277.4	77.1	51.4	23.5	2.2	4.5	23.1	18.5	101.3	65.2	36.2	458.2
1957													
1st quarter	469.8	281.9	77.7	52.5	23.1	2.2	4.8	24.9	20.1	105.3	67.9	37.5	467.6
2nd quarter	472.0	284.2	77.9	52.6	22.6	2.7	4.1	24.4	20.3	105.8	67.5	38.3	469.4
3rd quarter	479.5	288.8	79.3	54.1	22.5	2.8	4.0	23.8	19.8	107.4	68.4	39.1	476.7
4th quarter	474.9	290.4	71.0	53.2	22.3	-4.5	3.4	23.0	19.6	110.1	70.0	40.0	479.4
1958													
1st quarter	467.5	289.9	66.7	49.4	21.4	-4.0	1.1	20.5	19.5	109.8	68.7	41.1	471.5
2nd quarter	472.0	292.8	65.1	47.8	21.5	-4.2	0.5	20.5	20.1	113.6	71.5	42.1	476.2
3rd quarter	485.8	297.9	72.0	47.4	23.0	1.5	0.9	20.6	19.7	115.1	71.9	43.2	484.3
4th quarter	499.6	301.8	80.0	49.3	25.4	5.2	-0.3	20.6	20.8	118.0	73.9	44.1	494.3

Table 1-1B. Gross Domestic Product: Historical Data—*Continued*

(Billions of dollars, quarterly data are at seasonally adjusted annual rates.)

NIPA Tables 1.1.5, 1.2.5

| Year and quarter | Gross domestic product | Personal consumption expenditures | Gross private domestic investment | | | | Exports and imports of goods and services | | | Government consumption expenditures and gross investment | | | Addendum: Final sales of domestic product |
| | | | Total | Fixed investment | | Change in private inventories | Net exports | Exports | Imports | Total | Federal | State and local | |
				Nonresidential	Residential								
1959													
1st quarter	510.3	309.4	83.2	50.9	28.3	3.9	0.5	21.9	21.4	117.2	72.4	44.8	506.5
2nd quarter	522.7	315.5	89.4	52.7	29.4	7.3	-0.8	21.8	22.5	118.5	73.6	45.0	515.4
3rd quarter	525.0	320.7	83.6	54.4	28.8	0.4	1.2	24.1	22.9	119.5	74.5	45.0	524.7
4th quarter	528.6	322.8	86.5	54.4	28.0	4.1	0.6	23.1	22.5	118.6	73.8	44.8	524.5
1960													
1st quarter	542.6	326.4	96.5	56.3	29.0	11.2	2.9	26.1	23.3	117.0	71.0	46.0	531.4
2nd quarter	541.1	332.2	87.1	57.2	26.7	3.2	3.4	26.9	23.5	118.4	71.1	47.3	537.9
3rd quarter	545.6	332.1	86.4	56.2	25.9	4.3	4.7	27.6	22.9	122.4	74.2	48.3	541.3
4th quarter	540.2	334.0	76.0	55.9	25.9	-5.8	5.9	27.6	21.7	124.3	75.3	49.0	546.0
1961													
1st quarter	545.0	334.5	78.4	55.0	25.9	-2.5	5.9	27.6	21.7	126.2	75.3	50.9	547.6
2nd quarter	555.5	339.5	84.1	56.2	26.1	1.8	4.7	26.6	21.9	127.3	76.2	51.1	553.8
3rd quarter	567.7	342.3	90.9	56.7	27.6	6.7	4.5	27.8	23.3	130.0	78.1	51.8	561.0
4th quarter	580.6	349.6	92.9	58.5	28.5	6.0	4.6	28.4	23.9	133.5	79.9	53.6	574.6
1962													
1st quarter	594.0	354.8	98.1	59.7	29.0	9.4	4.0	28.3	24.3	137.1	83.3	53.8	584.6
2nd quarter	600.4	360.5	96.7	61.4	29.9	5.4	4.8	29.7	24.9	138.4	84.1	54.3	594.9
3rd quarter	609.0	364.3	98.2	62.1	29.9	6.2	4.5	29.6	25.1	142.1	87.0	55.0	602.8
4th quarter	612.3	370.6	95.0	61.8	29.8	3.4	3.1	28.7	25.6	143.6	87.6	56.0	608.9
1963													
1st quarter	621.7	374.3	99.7	62.0	30.8	6.9	4.0	29.2	25.2	143.7	86.2	57.5	614.8
2nd quarter	629.8	378.4	101.7	63.9	32.9	4.8	5.6	31.4	25.9	144.1	85.8	58.4	624.9
3rd quarter	644.4	385.4	104.6	65.7	33.2	5.7	4.5	31.2	26.7	149.9	89.9	60.1	638.7
4th quarter	653.9	390.0	107.2	67.7	34.5	5.1	5.7	32.5	26.8	151.0	89.7	61.3	648.9
1964													
1st quarter	669.8	399.6	110.5	69.1	36.2	5.1	7.2	34.2	27.0	152.6	90.1	62.5	664.7
2nd quarter	678.7	407.5	110.5	71.1	35.0	4.5	6.3	34.0	27.7	154.3	90.1	64.2	674.2
3rd quarter	692.0	416.4	112.6	73.4	34.5	4.7	6.9	35.4	28.4	156.0	90.8	65.2	687.3
4th quarter	697.3	419.0	115.0	75.3	34.6	5.0	7.2	36.5	29.3	156.2	90.0	66.2	692.3
1965													
1st quarter	717.8	429.7	126.5	80.2	34.8	11.5	4.5	33.0	28.5	157.0	89.5	67.5	706.2
2nd quarter	730.2	436.6	127.1	83.4	35.1	8.6	6.7	38.4	31.7	159.8	90.1	69.7	721.6
3rd quarter	749.3	445.8	131.2	86.7	35.3	9.3	5.5	37.6	32.0	166.8	94.5	72.3	740.0
4th quarter	771.9	459.7	133.8	90.6	35.5	7.6	5.8	39.7	33.9	172.6	98.7	73.9	764.3
1966													
1st quarter	795.7	470.1	144.2	94.4	35.9	13.9	4.4	39.4	35.0	177.0	101.2	75.8	781.9
2nd quarter	805.0	475.2	143.5	96.8	34.4	12.3	4.6	40.9	36.2	181.7	104.0	77.6	792.7
3rd quarter	819.6	484.3	143.2	98.3	33.0	11.9	2.7	40.9	38.2	189.5	109.9	79.5	807.8
4th quarter	833.3	490.1	145.9	99.2	30.2	16.5	3.8	42.6	38.8	193.6	111.2	82.5	816.8
1967													
1st quarter	844.2	494.3	142.8	98.0	29.4	15.4	4.5	43.9	39.4	202.6	117.9	84.7	828.7
2nd quarter	849.0	503.5	137.5	98.3	32.9	6.3	4.2	43.2	39.0	203.8	117.9	85.9	842.6
3rd quarter	865.2	510.7	142.8	98.8	34.8	9.3	3.3	42.8	39.5	208.4	121.0	87.4	856.0
4th quarter	881.4	518.2	147.7	101.7	37.5	8.4	2.2	43.9	41.7	213.3	123.2	90.1	873.0
1968													
1st quarter	909.4	536.3	152.3	105.6	38.3	8.4	1.1	45.5	44.4	219.7	126.4	93.3	900.9
2nd quarter	934.3	550.0	158.9	105.3	39.6	14.1	1.7	47.2	45.4	223.6	127.2	96.5	920.3
3rd quarter	950.8	566.1	155.7	107.6	40.4	7.7	1.7	49.9	48.2	227.4	128.5	98.8	943.1
4th quarter	968.0	575.0	160.8	112.2	42.6	6.0	0.9	49.1	48.2	231.3	129.7	101.6	962.0
1969													
1st quarter	993.3	587.0	172.4	116.0	44.9	11.5	0.2	44.0	43.8	233.7	129.6	104.1	981.8
2nd quarter	1 009.0	598.3	172.7	118.4	45.1	9.2	1.1	53.8	52.7	236.9	129.8	107.1	999.8
3rd quarter	1 030.0	608.6	177.6	122.4	45.0	10.2	1.2	53.6	52.4	242.5	133.4	109.1	1 019.8
4th quarter	1 038.1	620.6	171.6	123.3	42.5	5.8	3.1	56.3	53.1	242.8	132.0	110.8	1 032.4
1970													
1st quarter	1 051.2	631.7	168.1	123.8	42.5	1.8	3.5	57.0	53.5	247.9	133.6	114.3	1 049.4
2nd quarter	1 067.4	641.6	171.5	125.0	41.4	5.1	5.2	60.4	55.2	249.1	131.8	117.4	1 062.3
3rd quarter	1 086.1	653.5	173.9	126.3	42.6	5.1	4.1	60.5	56.4	254.6	132.4	122.2	1 081.0
4th quarter	1 088.6	660.2	166.8	123.5	47.2	-4.0	3.0	60.9	57.9	258.7	133.5	125.2	1 092.6
1971													
1st quarter	1 135.2	679.2	189.5	126.3	51.0	12.3	4.6	63.2	58.7	261.9	133.3	128.6	1 122.9
2nd quarter	1 156.3	693.2	197.3	129.5	57.0	10.9	-0.4	62.9	63.3	266.1	134.3	131.9	1 145.4
3rd quarter	1 177.7	705.6	202.1	131.2	60.7	10.2	0.2	65.7	65.5	269.8	135.6	134.2	1 167.5
4th quarter	1 190.3	721.7	198.4	134.7	64.0	-0.3	-1.9	60.0	61.9	272.1	134.7	137.4	1 190.6
1972													
1st quarter	1 230.6	738.9	213.0	140.6	69.2	3.2	-3.5	68.6	72.2	282.2	141.4	140.8	1 227.4
2nd quarter	1 266.4	757.4	226.8	144.0	70.8	12.0	-4.3	67.2	71.4	286.5	144.2	142.2	1 254.4
3rd quarter	1 290.6	775.8	233.1	147.0	72.4	13.7	-2.6	71.5	74.1	284.3	138.8	145.6	1 276.9
4th quarter	1 328.9	800.5	239.7	155.0	77.3	7.5	-3.1	76.1	79.2	291.7	142.2	149.6	1 321.4
1973													
1st quarter	1 377.5	825.0	254.3	162.8	80.9	10.6	-1.4	84.0	85.4	299.6	146.4	153.2	1 366.9
2nd quarter	1 413.9	840.5	268.2	171.3	78.7	18.2	2.5	91.9	89.5	302.7	146.5	156.2	1 395.7
3rd quarter	1 433.8	858.9	264.3	176.6	77.9	9.8	6.4	97.6	91.1	304.2	144.2	159.9	1 424.1
4th quarter	1 476.3	873.9	280.9	180.1	75.7	25.0	9.0	107.6	98.7	312.6	147.6	165.0	1 451.3
1974													
1st quarter	1 491.2	891.9	268.4	183.4	72.4	12.5	6.4	116.7	110.3	324.6	152.7	171.9	1 478.7
2nd quarter	1 530.1	920.4	277.4	188.8	71.2	17.4	-2.7	126.7	129.4	335.0	154.9	180.1	1 512.6
3rd quarter	1 560.0	949.3	271.0	194.5	70.9	5.6	-7.0	126.6	133.6	346.7	160.4	186.3	1 554.4
4th quarter	1 599.7	959.1	281.3	197.6	63.3	20.4	0.0	136.6	136.6	359.2	167.4	191.9	1 579.2

Table 1-1B. Gross Domestic Product: Historical Data—*Continued*

(Billions of dollars, quarterly data are at seasonally adjusted annual rates.) NIPA Tables 1.1.5, 1.2.5

| Year and quarter | Gross domestic product | Personal consump-tion expen-ditures | Gross private domestic investment | | | | Exports and imports of goods and services | | | Government consumption expenditures and gross investment | | | Addendum: Final sales of domestic product |
| | | | Total | Fixed investment | | Change in private inventories | Net exports | Exports | Imports | Total | Federal | State and local | |
				Nonresi-dential	Residential								
1975													
1st quarter	1 616.1	985.2	244.3	193.1	61.2	-10.0	16.5	141.4	124.9	370.1	168.6	201.5	1 626.1
2nd quarter	1 651.9	1 013.6	243.3	193.3	64.0	-14.0	21.6	136.8	115.2	373.4	169.4	204.0	1 665.8
3rd quarter	1 709.8	1 047.2	265.2	197.8	68.8	-1.4	12.0	134.1	122.1	385.4	176.1	209.3	1 711.2
4th quarter	1 761.8	1 076.2	276.2	202.9	73.0	0.3	13.8	142.5	128.7	395.6	180.8	214.8	1 761.5
1976													
1st quarter	1 820.5	1 109.9	304.6	209.5	80.4	14.7	4.7	143.6	138.9	401.3	181.6	219.7	1 805.8
2nd quarter	1 852.3	1 129.5	322.3	215.0	84.8	22.4	-0.5	146.6	147.1	401.0	182.5	218.5	1 829.9
3rd quarter	1 886.6	1 158.8	328.3	222.6	84.9	20.8	-4.1	151.8	155.8	403.5	184.9	218.6	1 865.8
4th quarter	1 934.3	1 192.4	337.7	230.2	96.9	10.5	-6.6	156.1	162.7	410.8	190.2	220.6	1 923.8
1977													
1st quarter	1 988.6	1 228.2	360.3	243.3	102.2	14.8	-21.1	155.4	176.4	421.2	194.2	227.0	1 973.8
2nd quarter	2 055.9	1 256.0	389.7	253.7	116.5	19.5	-21.1	161.9	183.0	431.4	198.9	232.4	2 036.4
3rd quarter	2 118.5	1 286.9	414.1	263.3	120.0	30.9	-20.6	162.3	182.9	438.0	201.9	236.1	2 087.6
4th quarter	2 164.3	1 324.8	422.3	275.9	122.2	24.1	-29.6	157.8	187.4	446.7	206.3	240.5	2 140.1
1978													
1st quarter	2 202.8	1 354.1	434.8	282.4	126.9	25.5	-38.7	164.6	203.3	452.6	208.8	243.8	2 177.3
2nd quarter	2 331.6	1 411.4	470.6	309.4	137.0	24.3	-22.6	186.2	208.8	472.3	217.0	255.3	2 307.4
3rd quarter	2 395.1	1 442.2	492.4	325.1	142.3	25.0	-23.8	191.3	215.1	484.2	222.1	262.2	2 370.1
4th quarter	2 476.9	1 481.4	515.8	341.4	145.8	28.5	-16.4	205.4	221.8	496.2	227.8	268.4	2 448.5
1979													
1st quarter	2 526.6	1 517.1	525.8	356.7	145.3	23.9	-18.2	211.7	229.8	501.8	231.7	270.1	2 502.8
2nd quarter	2 591.2	1 557.6	539.3	364.3	147.6	27.4	-22.2	220.9	243.1	516.5	237.6	278.9	2 563.8
3rd quarter	2 667.6	1 611.9	545.6	383.0	150.5	12.1	-23.0	234.3	257.3	533.1	243.7	289.4	2 655.4
4th quarter	2 723.9	1 655.0	547.9	391.3	148.0	8.6	-26.8	253.7	280.5	547.8	249.3	298.4	2 715.3
1980													
1st quarter	2 789.8	1 702.3	554.6	404.5	140.2	9.9	-35.8	268.5	304.3	568.8	261.1	307.7	2 779.9
2nd quarter	2 797.4	1 704.7	519.3	394.7	116.9	7.8	-15.2	277.4	292.6	588.5	276.5	312.0	2 789.6
3rd quarter	2 856.5	1 763.8	495.1	405.7	123.2	-33.9	5.5	284.7	279.2	592.2	276.1	316.1	2 890.3
4th quarter	2 985.6	1 831.9	551.5	422.8	137.8	-9.1	-6.7	292.5	299.2	608.9	285.8	323.1	2 994.7
1981													
1st quarter	3 124.2	1 885.7	619.4	443.0	137.6	38.8	-14.3	305.5	319.7	633.4	297.2	336.1	3 085.5
2nd quarter	3 162.5	1 917.5	609.8	462.9	135.3	11.7	-13.5	308.5	322.0	648.7	311.9	336.8	3 150.8
3rd quarter	3 260.6	1 958.1	652.3	482.1	126.2	44.0	-7.6	302.3	309.9	657.8	317.4	340.3	3 216.7
4th quarter	3 280.8	1 974.4	643.4	503.8	114.8	24.8	-14.8	304.7	319.4	677.7	329.3	348.4	3 256.0
1982													
1st quarter	3 274.3	2 014.2	588.3	500.1	109.7	-21.5	-16.3	293.2	309.5	688.1	334.9	353.2	3 295.8
2nd quarter	3 332.0	2 039.6	593.6	490.1	107.7	-4.2	-4.4	294.7	299.1	703.1	342.9	360.2	3 336.1
3rd quarter	3 366.3	2 085.7	593.0	478.7	108.4	5.8	-29.6	279.6	309.3	717.3	351.5	365.8	3 360.5
4th quarter	3 402.6	2 145.6	549.2	471.5	117.5	-39.8	-29.6	265.3	294.9	737.4	364.1	373.3	3 442.4
1983													
1st quarter	3 473.4	2 184.6	565.5	462.1	138.5	-35.1	-24.5	270.7	295.3	747.9	370.5	377.4	3 508.5
2nd quarter	3 578.8	2 249.4	613.8	466.5	155.0	-7.7	-45.4	272.5	317.9	761.1	380.3	380.7	3 586.5
3rd quarter	3 689.2	2 319.9	652.3	485.4	171.1	-4.2	-65.2	278.2	343.4	782.2	394.4	387.8	3 693.4
4th quarter	3 794.7	2 372.5	718.5	514.7	179.9	23.9	-71.4	286.6	358.0	775.1	384.2	390.9	3 770.8
1984													
1st quarter	3 908.1	2 418.2	790.9	531.5	186.4	73.0	-95.0	293.0	388.0	794.0	392.4	401.6	3 835.1
2nd quarter	4 009.6	2 475.9	818.9	558.3	191.3	69.3	-104.3	302.2	406.5	819.1	408.3	410.8	3 940.3
3rd quarter	4 084.3	2 513.5	838.9	576.6	190.9	71.3	-103.8	305.7	409.6	835.7	414.0	421.7	4 012.9
4th quarter	4 148.6	2 561.8	831.7	590.9	192.9	48.0	-107.8	308.6	416.4	862.8	432.5	430.2	4 100.6
1985													
1st quarter	4 230.2	2 636.0	809.9	599.4	194.2	16.2	-91.3	306.0	397.3	875.6	434.8	440.8	4 213.9
2nd quarter	4 294.9	2 681.8	827.0	609.1	196.3	21.6	-114.4	304.1	418.6	900.5	447.3	453.2	4 273.2
3rd quarter	4 386.8	2 754.1	822.2	604.4	201.4	16.3	-116.9	297.3	414.2	927.4	463.1	464.3	4 370.5
4th quarter	4 444.1	2 779.4	859.5	618.1	208.4	33.1	-133.4	305.4	438.9	938.6	466.4	472.1	4 411.0
1986													
1st quarter	4 507.9	2 823.6	863.5	613.5	219.5	30.4	-126.0	313.4	439.4	946.8	464.0	482.8	4 477.5
2nd quarter	4 545.3	2 851.5	855.2	605.0	234.6	15.7	-128.9	315.1	444.0	967.5	477.8	489.7	4 529.7
3rd quarter	4 607.7	2 917.2	835.8	602.0	240.9	-7.0	-139.0	320.5	459.4	993.6	495.1	498.5	4 614.7
4th quarter	4 657.6	2 952.8	842.1	610.6	244.3	-12.8	-133.6	335.0	468.6	996.4	489.8	506.6	4 670.4
1987													
1st quarter	4 722.2	2 983.5	871.2	596.6	246.7	28.0	-141.2	336.5	477.7	1 008.7	492.1	516.5	4 694.2
2nd quarter	4 806.2	3 053.3	874.6	608.4	249.7	16.5	-147.0	355.4	502.3	1 025.2	501.2	524.0	4 789.6
3rd quarter	4 884.6	3 117.4	876.5	625.5	250.0	1.0	-145.5	371.9	517.3	1 036.2	504.1	532.1	4 883.5
4th quarter	5 008.0	3 150.9	946.5	630.6	252.8	63.1	-145.4	392.1	537.5	1 056.0	513.7	542.3	4 944.9
1988													
1st quarter	5 073.4	3 231.9	908.6	641.5	250.1	17.0	-124.0	418.7	542.7	1 056.9	505.8	551.1	5 056.4
2nd quarter	5 190.0	3 291.7	934.5	659.4	255.5	19.6	-106.6	439.5	546.1	1 070.4	506.9	563.5	5 170.4
3rd quarter	5 282.8	3 361.9	942.0	666.3	257.5	18.2	-99.3	453.6	552.8	1 078.2	507.4	570.8	5 264.7
4th quarter	5 399.5	3 434.5	962.7	681.9	261.7	19.1	-107.7	466.6	574.3	1 109.9	525.6	584.3	5 380.4
1989													
1st quarter	5 511.3	3 490.2	1 005.5	696.5	260.9	48.1	-101.0	485.2	586.2	1 116.6	519.9	596.7	5 463.1
2nd quarter	5 612.5	3 553.8	1 001.0	709.0	255.8	36.3	-88.2	507.2	595.4	1 145.8	534.3	611.5	5 576.2
3rd quarter	5 695.4	3 609.4	996.5	731.1	255.5	9.8	-75.1	509.4	584.4	1 164.6	541.4	623.2	5 685.5
4th quarter	5 747.2	3 653.7	995.8	727.4	251.9	16.6	-82.8	515.4	598.2	1 180.5	540.8	639.7	5 730.7
1990													
1st quarter	5 872.7	3 737.9	1 010.8	740.9	256.0	14.0	-88.5	538.2	626.8	1 212.5	553.7	658.8	5 858.7
2nd quarter	5 960.0	3 783.4	1 014.7	734.1	246.9	33.7	-68.8	545.9	614.8	1 230.7	563.9	666.8	5 926.3
3rd quarter	6 015.1	3 846.6	1 000.8	744.4	234.5	21.9	-75.0	555.1	630.1	1 242.6	562.2	680.3	5 993.3
4th quarter	6 004.7	3 867.9	947.5	737.5	221.3	-11.3	-79.1	568.2	647.3	1 268.5	569.7	698.8	6 016.1

Table 1-1B. Gross Domestic Product: Historical Data—*Continued*

(Billions of dollars, quarterly data are at seasonally adjusted annual rates.)   **NIPA Tables 1.1.5, 1.2.5**

| Year and quarter | Gross domestic product | Personal consumption expenditures | Gross private domestic investment | | | | Exports and imports of goods and services | | | Government consumption expenditures and gross investment | | | Addendum: Final sales of domestic product |
| | | | Total | Fixed investment | | Change in private inventories | Net exports | Exports | Imports | Total | Federal | State and local | |
				Nonresidential	Residential								
1991													
1st quarter	6 035.2	3 873.6	924.6	729.8	210.3	-15.5	-47.1	573.2	620.3	1 284.2	581.4	702.8	6 050.7
2nd quarter	6 126.9	3 926.9	926.5	726.8	217.8	-18.0	-23.2	590.7	613.9	1 296.6	586.6	709.9	6 144.9
3rd quarter	6 205.9	3 973.3	947.5	720.2	226.5	0.8	-21.1	600.6	621.7	1 306.3	586.3	719.9	6 205.1
4th quarter	6 264.5	4 000.0	978.8	717.6	230.1	31.1	-23.1	615.2	638.3	1 308.8	577.4	731.4	6 233.4
1992													
1st quarter	6 363.1	4 100.4	956.8	714.2	242.4	0.2	-20.5	625.3	645.8	1 326.4	580.3	746.1	6 362.9
2nd quarter	6 470.8	4 155.7	1 013.1	736.7	253.2	23.2	-32.8	626.2	659.0	1 334.8	580.9	753.9	6 447.6
3rd quarter	6 566.6	4 227.0	1 024.2	748.5	255.1	20.5	-38.5	639.4	677.9	1 354.0	594.2	759.8	6 546.1
4th quarter	6 680.8	4 307.2	1 058.0	768.3	268.3	21.3	-47.1	641.4	688.5	1 362.8	598.4	764.4	6 659.5
1993													
1st quarter	6 729.5	4 349.5	1 083.8	776.5	271.4	35.9	-55.7	643.6	699.3	1 351.8	580.3	771.5	6 693.5
2nd quarter	6 808.9	4 418.6	1 094.5	792.4	278.0	24.1	-63.2	653.1	716.3	1 359.1	576.7	782.3	6 784.8
3rd quarter	6 882.1	4 487.2	1 095.9	798.3	290.9	6.6	-68.4	650.9	719.3	1 367.4	578.7	788.7	6 875.5
4th quarter	7 013.7	4 552.7	1 153.1	829.6	306.9	16.6	-73.4	671.6	745.0	1 381.4	584.9	796.5	6 997.1
1994													
1st quarter	7 115.7	4 621.2	1 201.7	840.7	315.6	45.4	-80.6	681.2	761.8	1 373.4	567.0	806.3	7 070.3
2nd quarter	7 246.9	4 683.2	1 264.9	855.6	327.9	81.4	-90.6	707.0	797.6	1 389.4	569.4	820.0	7 165.5
3rd quarter	7 331.1	4 752.8	1 251.7	872.1	326.4	53.2	-96.9	736.9	833.8	1 423.4	586.5	836.9	7 277.9
4th quarter	7 455.3	4 826.7	1 307.6	907.0	325.4	75.1	-101.9	758.6	860.6	1 422.9	575.8	847.1	7 380.1
1995													
1st quarter	7 522.3	4 862.4	1 327.6	944.6	321.8	61.2	-105.3	781.6	886.9	1 437.6	579.1	858.5	7 461.1
2nd quarter	7 581.0	4 933.6	1 304.0	956.8	313.5	33.8	-109.5	798.9	908.3	1 452.9	581.0	871.9	7 547.2
3rd quarter	7 683.1	4 998.7	1 303.2	965.5	326.4	11.3	-74.4	831.4	905.8	1 455.7	579.3	876.3	7 671.8
4th quarter	7 772.6	5 055.7	1 335.1	982.1	334.6	18.4	-69.8	839.4	909.2	1 451.6	567.3	884.3	7 754.2
1996													
1st quarter	7 868.5	5 130.6	1 355.4	1 003.7	344.7	6.9	-88.8	847.9	936.7	1 471.3	579.8	891.5	7 861.6
2nd quarter	8 032.8	5 220.5	1 418.4	1 026.5	361.4	30.5	-93.7	859.0	952.8	1 487.7	582.1	905.5	8 002.3
3rd quarter	8 131.4	5 274.5	1 474.4	1 059.0	364.3	51.1	-114.2	859.6	973.8	1 496.7	577.8	919.0	8 080.3
4th quarter	8 259.8	5 352.8	1 480.1	1 083.6	361.8	34.7	-88.8	903.8	992.6	1 515.7	576.9	938.8	8 225.0
1997													
1st quarter	8 362.7	5 433.1	1 522.4	1 107.3	365.4	49.7	-108.8	918.4	1 027.2	1 516.0	570.7	945.3	8 312.9
2nd quarter	8 518.8	5 471.3	1 590.2	1 129.6	372.3	88.4	-85.2	954.5	1 039.7	1 542.5	587.2	955.4	8 430.5
3rd quarter	8 662.8	5 579.2	1 625.3	1 178.4	379.0	67.9	-96.8	974.1	1 070.9	1 555.2	586.0	969.2	8 594.9
4th quarter	8 765.9	5 663.6	1 644.5	1 181.1	385.8	77.7	-117.0	968.3	1 085.3	1 574.8	589.2	985.6	8 688.3
1998													
1st quarter	8 866.5	5 721.3	1 712.3	1 212.4	394.8	105.1	-135.2	963.0	1 098.2	1 568.0	572.2	995.9	8 761.4
2nd quarter	8 969.7	5 832.6	1 695.8	1 247.1	411.3	37.3	-162.3	947.3	1 109.6	1 603.7	587.1	1 016.6	8 932.4
3rd quarter	9 121.1	5 926.8	1 741.6	1 261.7	427.6	52.4	-174.6	935.3	1 109.9	1 627.3	588.6	1 038.6	9 068.7
4th quarter	9 294.0	6 028.2	1 797.0	1 295.4	441.5	60.0	-178.7	966.3	1 145.0	1 647.5	594.2	1 053.2	9 234.0
1999													
1st quarter	9 417.3	6 102.5	1 853.1	1 322.3	447.4	83.4	-207.7	962.4	1 170.1	1 669.4	595.5	1 073.9	9 333.9
2nd quarter	9 524.2	6 225.3	1 848.3	1 354.0	459.3	35.1	-244.7	973.5	1 218.2	1 695.2	599.8	1 095.4	9 489.0
3rd quarter	9 681.9	6 328.9	1 893.7	1 386.6	466.6	40.5	-275.3	1 003.2	1 278.5	1 734.5	614.9	1 119.6	9 641.3
4th quarter	9 899.4	6 459.6	1 953.1	1 395.0	473.8	84.2	-295.6	1 032.0	1 327.7	1 782.3	635.2	1 147.1	9 815.1
2000													
1st quarter	10 002.9	6 613.6	1 950.7	1 450.3	484.2	16.2	-352.1	1 054.2	1 406.3	1 790.7	620.4	1 170.4	9 986.7
2nd quarter	10 247.7	6 707.5	2 075.8	1 498.7	486.6	90.4	-358.7	1 092.5	1 451.2	1 823.1	642.0	1 181.1	10 157.2
3rd quarter	10 319.8	6 815.4	2 060.0	1 519.7	483.1	57.2	-387.8	1 124.3	1 512.1	1 832.3	634.1	1 198.3	10 262.6
4th quarter	10 439.0	6 912.1	2 067.2	1 525.1	487.8	54.3	-401.5	1 114.1	1 515.6	1 861.2	638.4	1 222.9	10 384.7
2001													
1st quarter	10 472.9	6 986.9	1 971.3	1 505.2	496.7	-30.6	-390.8	1 095.1	1 485.8	1 905.4	653.1	1 252.3	10 503.5
2nd quarter	10 597.8	7 036.3	1 973.0	1 473.6	511.0	-11.6	-358.5	1 052.6	1 411.1	1 947.0	666.1	1 280.9	10 609.4
3rd quarter	10 596.3	7 064.7	1 944.9	1 452.6	522.4	-30.1	-366.0	996.4	1 362.4	1 952.7	674.3	1 278.4	10 626.4
4th quarter	10 660.3	7 174.7	1 850.1	1 408.9	522.1	-80.8	-356.4	954.5	1 310.9	1 992.0	686.8	1 305.2	10 741.1
2002													
1st quarter	10 789.0	7 209.9	1 912.7	1 374.0	538.3	0.3	-372.6	970.7	1 343.3	2 038.9	713.9	1 325.0	10 788.7
2nd quarter	10 893.2	7 302.1	1 933.3	1 357.3	554.8	21.2	-415.7	1 003.9	1 419.6	2 073.5	734.7	1 338.8	10 872.0
3rd quarter	10 992.1	7 390.9	1 933.2	1 348.9	558.9	25.4	-432.5	1 017.2	1 449.7	2 100.4	748.2	1 352.2	10 966.6
4th quarter	11 071.5	7 467.7	1 942.5	1 331.2	578.3	33.0	-480.8	1 003.2	1 484.0	2 142.0	775.1	1 366.9	11 038.5
2003													
1st quarter	11 183.5	7 555.8	1 960.2	1 332.7	601.4	26.1	-504.9	1 008.2	1 513.0	2 172.4	792.3	1 380.0	11 157.4
2nd quarter	11 312.9	7 642.6	1 972.4	1 366.8	612.0	-6.5	-501.5	1 007.9	1 509.5	2 199.4	825.5	1 374.0	11 319.4
3rd quarter	11 567.3	7 802.6	2 044.3	1 392.1	651.7	0.4	-500.8	1 038.1	1 538.9	2 221.2	832.7	1 388.5	11 566.9
4th quarter	11 769.3	7 891.5	2 131.3	1 411.9	683.1	36.3	-505.4	1 090.5	1 595.8	2 251.8	854.6	1 397.3	11 732.9
2004													
1st quarter	11 920.2	8 027.7	2 154.1	1 401.8	706.2	46.0	-549.0	1 137.7	1 686.7	2 287.3	871.3	1 416.0	11 874.1
2nd quarter	12 109.0	8 133.0	2 262.6	1 445.5	743.2	74.0	-608.0	1 167.8	1 775.8	2 321.4	884.2	1 437.2	12 035.0
3rd quarter	12 303.3	8 264.3	2 318.3	1 490.5	763.6	64.2	-636.4	1 182.7	1 819.1	2 357.2	902.2	1 455.0	12 239.2
4th quarter	12 522.4	8 425.6	2 390.1	1 531.7	786.3	72.1	-682.9	1 222.3	1 905.2	2 389.7	909.3	1 480.3	12 450.3
2005													
1st quarter	12 761.3	8 523.0	2 486.1	1 568.3	815.3	102.5	-674.6	1 264.8	1 939.4	2 426.9	931.5	1 495.4	12 658.9
2nd quarter	12 910.0	8 671.4	2 476.5	1 603.8	843.8	28.8	-690.8	1 296.0	1 986.8	2 452.9	939.0	1 513.9	12 881.2
3rd quarter	13 142.9	8 849.2	2 531.1	1 643.1	875.6	12.4	-732.5	1 307.1	2 039.6	2 495.1	956.1	1 539.0	13 130.5
4th quarter	13 332.3	8 944.9	2 645.3	1 668.7	890.2	86.4	-786.9	1 353.0	2 139.9	2 529.1	963.3	1 565.8	13 245.9
2006													
1st quarter	13 603.9	9 090.7	2 709.7	1 735.3	896.2	78.2	-777.2	1 413.0	2 190.2	2 580.7	996.6	1 584.1	13 525.7
2nd quarter	13 749.8	9 210.2	2 709.3	1 774.4	859.3	75.5	-780.6	1 460.0	2 240.6	2 610.9	996.6	1 614.3	13 674.3
3rd quarter	13 867.5	9 333.0	2 709.4	1 815.9	814.7	78.8	-805.6	1 477.8	2 283.5	2 630.7	994.9	1 635.7	13 788.7
4th quarter	14 037.2	9 407.5	2 675.4	1 849.5	782.4	43.5	-720.3	1 539.6	2 259.9	2 674.7	1 014.6	1 660.1	13 993.8

Table 1-1B. Gross Domestic Product: Historical Data—*Continued*

(Billions of dollars, quarterly data are at seasonally adjusted annual rates.)

NIPA Tables 1.1.5, 1.2.5

Year and quarter	Gross domestic product	Personal consumption expenditures	Gross private domestic investment				Exports and imports of goods and services			Government consumption expenditures and gross investment			Addendum: Final sales of domestic product
			Total	Fixed investment		Change in private inventories	Net exports	Exports	Imports	Total	Federal	State and local	
				Nonresidential	Residential								
2007													
1st quarter	14 208.6	9 549.4	2 664.3	1 892.0	750.9	21.4	-724.3	1 577.1	2 301.4	2 719.2	1 017.2	1 702.0	14 187.2
2nd quarter	14 382.4	9 644.7	2 699.2	1 937.7	719.3	42.2	-731.9	1 622.1	2 354.0	2 770.3	1 042.0	1 728.3	14 340.1
3rd quarter	14 535.0	9 753.8	2 686.0	1 967.4	673.7	44.9	-713.8	1 683.7	2 397.5	2 809.0	1 058.3	1 750.7	14 490.1
4th quarter	14 681.5	9 877.8	2 642.6	1 997.1	618.2	27.3	-703.7	1 760.5	2 464.2	2 864.9	1 084.6	1 780.3	14 654.2
2008													
1st quarter	14 651.0	9 934.3	2 563.7	2 013.7	566.7	-16.7	-756.2	1 811.0	2 567.2	2 909.3	1 110.3	1 799.0	14 667.7
2nd quarter	14 805.6	10 052.8	2 540.6	2 024.0	538.8	-22.3	-758.9	1 909.7	2 668.7	2 971.1	1 145.5	1 825.6	14 827.9
3rd quarter	14 835.2	10 081.0	2 498.2	2 007.0	507.1	-15.8	-771.6	1 921.8	2 693.3	3 027.5	1 168.7	1 858.9	14 851.0
4th quarter	14 559.5	9 837.3	2 307.9	1 918.7	451.4	-62.2	-605.7	1 705.7	2 311.4	3 020.0	1 177.9	1 842.2	14 621.8
2009													
1st quarter	14 394.5	9 756.1	2 014.9	1 761.4	404.5	-151.0	-396.1	1 514.3	1 910.4	3 019.7	1 183.0	1 836.7	14 545.5
2nd quarter	14 352.9	9 760.2	1 863.7	1 685.0	374.7	-196.0	-338.6	1 518.3	1 856.9	3 067.6	1 210.8	1 856.7	14 548.9
3rd quarter	14 420.3	9 895.4	1 841.4	1 656.0	389.4	-204.0	-405.5	1 591.1	1 996.6	3 089.0	1 225.5	1 863.5	14 624.3
4th quarter	14 628.0	9 957.1	1 998.7	1 659.3	391.5	-52.1	-445.6	1 704.3	2 149.8	3 117.8	1 253.4	1 864.4	14 680.1
2010													
1st quarter	14 721.4	10 040.5	2 038.2	1 660.0	379.4	-1.2	-489.2	1 746.9	2 236.1	3 131.9	1 275.7	1 856.2	14 722.6
2nd quarter	14 926.1	10 131.8	2 148.8	1 715.7	396.3	36.7	-519.1	1 810.0	2 329.1	3 164.7	1 302.6	1 862.1	14 889.4
3rd quarter	15 079.9	10 220.6	2 236.5	1 762.4	361.2	112.9	-535.1	1 865.6	2 400.7	3 157.9	1 302.3	1 855.6	14 967.0
4th quarter	15 240.8	10 350.5	2 238.4	1 801.9	369.3	67.3	-512.2	1 962.6	2 474.8	3 164.1	1 311.1	1 853.0	15 173.5
2011													
1st quarter	15 285.8	10 485.4	2 206.0	1 805.1	368.8	32.1	-561.5	2 030.3	2 591.7	3 156.0	1 304.7	1 851.2	15 253.8
2nd quarter	15 496.2	10 612.1	2 297.4	1 862.0	374.3	61.0	-581.9	2 105.1	2 686.9	3 168.6	1 311.8	1 856.7	15 435.1
3rd quarter	15 591.9	10 705.4	2 322.8	1 953.8	381.1	-12.0	-573.9	2 138.8	2 712.6	3 137.5	1 288.0	1 849.5	15 603.9
4th quarter	15 796.5	10 761.6	2 504.1	2 009.0	391.2	104.0	-600.7	2 137.8	2 738.5	3 131.4	1 291.2	1 840.3	15 692.5
2012													
1st quarter	16 019.8	10 922.4	2 567.8	2 073.4	414.1	80.3	-615.1	2 164.6	2 779.7	3 144.7	1 295.6	1 849.0	15 939.5
2nd quarter	16 152.3	10 964.9	2 636.9	2 126.2	419.3	91.4	-580.5	2 192.1	2 772.6	3 131.0	1 288.2	1 842.9	16 060.9
3rd quarter	16 257.2	11 014.2	2 644.1	2 126.0	433.7	84.4	-540.8	2 201.8	2 742.6	3 139.6	1 293.3	1 846.3	16 172.7
4th quarter	16 358.9	11 125.7	2 638.3	2 148.6	460.9	28.7	-537.8	2 206.6	2 744.5	3 132.7	1 269.1	1 863.7	16 330.1
2013													
1st quarter	16 569.6	11 223.2	2 738.2	2 170.9	485.1	82.3	-516.9	2 239.2	2 756.0	3 125.0	1 240.0	1 885.0	16 487.3
2nd quarter	16 637.9	11 239.6	2 775.3	2 180.3	507.2	87.8	-508.9	2 250.7	2 759.6	3 132.0	1 232.3	1 899.6	16 550.2
3rd quarter	16 848.7	11 330.9	2 880.0	2 220.7	523.1	136.2	-496.3	2 269.4	2 765.7	3 134.1	1 218.4	1 915.7	16 712.5
4th quarter	17 083.1	11 475.1	2 910.5	2 274.0	524.6	111.9	-441.1	2 334.4	2 775.5	3 138.5	1 215.6	1 923.0	16 971.3
2014													
1st quarter	17 104.6	11 574.2	2 899.2	2 314.9	532.1	52.3	-506.3	2 335.9	2 842.2	3 137.4	1 211.0	1 926.4	17 052.3
2nd quarter	17 432.9	11 756.9	3 030.4	2 383.6	550.4	96.4	-507.6	2 387.1	2 894.8	3 153.3	1 209.0	1 944.2	17 336.5
3rd quarter	17 721.7	11 915.4	3 107.6	2 439.8	568.4	99.4	-492.3	2 391.9	2 884.2	3 190.9	1 228.2	1 962.7	17 622.2
4th quarter	17 849.9	12 044.5	3 139.5	2 462.0	589.7	87.8	-524.4	2 371.9	2 896.3	3 190.3	1 211.7	1 978.6	17 762.1
2015													
1st quarter	18 003.4	12 099.1	3 245.1	2 464.3	605.1	175.6	-528.8	2 295.0	2 823.8	3 188.1	1 214.9	1 973.2	17 827.8
2nd quarter	18 223.6	12 255.5	3 245.8	2 472.5	623.3	150.0	-508.3	2 296.2	2 804.5	3 230.6	1 219.6	2 011.0	18 073.6
3rd quarter	18 347.4	12 389.3	3 251.7	2 479.2	645.7	126.8	-542.7	2 255.5	2 798.2	3 249.1	1 220.2	2 028.9	18 220.6
4th quarter	18 378.8	12 446.0	3 206.1	2 450.3	661.1	94.7	-526.5	2 216.7	2 743.2	3 253.2	1 228.5	2 024.7	18 284.1
2016													
1st quarter	18 470.2	12 551.6	3 166.0	2 430.2	686.4	49.4	-513.9	2 177.3	2 691.2	3 266.4	1 229.4	2 036.9	18 420.7
2nd quarter	18 656.2	12 707.5	3 157.9	2 448.6	692.4	16.8	-492.3	2 207.8	2 700.2	3 283.1	1 227.1	2 056.0	18 639.4
3rd quarter	18 821.4	12 841.2	3 167.0	2 473.0	698.9	-4.9	-497.7	2 258.2	2 755.9	3 310.9	1 237.8	2 073.1	18 826.2
4th quarter	19 032.6	12 979.5	3 262.4	2 490.2	720.2	52.0	-546.1	2 265.4	2 811.5	3 336.7	1 244.5	2 092.3	18 980.6
2017													
1st quarter	19 237.4	13 153.2	3 266.2	2 532.5	746.0	-12.3	-543.6	2 326.4	2 869.9	3 361.6	1 246.5	2 115.1	19 249.7
2nd quarter	19 379.2	13 241.3	3 313.3	2 555.9	753.3	4.0	-559.5	2 333.1	2 892.6	3 384.2	1 257.9	2 126.3	19 375.2
3rd quarter	19 617.3	13 370.9	3 378.8	2 575.2	758.5	45.1	-543.6	2 370.1	2 913.7	3 411.1	1 262.7	2 148.4	19 572.2
4th quarter	19 938.0	13 596.0	3 446.3	2 634.2	783.6	28.5	-575.5	2 468.7	3 044.1	3 471.1	1 288.3	2 182.9	19 909.5
2018													
1st quarter	20 242.2	13 755.5	3 555.0	2 716.2	794.3	44.5	-589.8	2 507.2	3 097.0	3 521.5	1 308.1	2 213.4	20 197.7
2nd quarter	20 552.7	13 939.9	3 580.9	2 765.9	804.3	10.7	-548.1	2 550.3	3 098.4	3 580.0	1 329.3	2 250.7	20 541.9
3rd quarter	20 742.7	14 086.3	3 671.7	2 792.6	800.7	78.4	-646.4	2 523.9	3 170.3	3 631.2	1 352.0	2 279.1	20 664.4
4th quarter	20 909.9	14 191.4	3 723.9	2 831.9	794.7	97.3	-653.4	2 533.4	3 186.9	3 648.0	1 368.4	2 279.6	20 812.5
2019													
1st quarter	21 115.3	14 276.6	3 772.8	2 878.4	795.8	98.6	-615.5	2 523.5	3 139.0	3 681.5	1 388.8	2 292.7	21 016.7
2nd quarter	21 329.9	14 497.3	3 739.7	2 891.3	795.3	53.1	-644.7	2 514.6	3 159.4	3 737.6	1 410.6	2 327.0	21 276.8
3rd quarter	21 540.3	14 645.3	3 759.8	2 908.0	810.5	41.3	-631.8	2 505.2	3 137.1	3 767.1	1 429.3	2 337.8	21 499.0
4th quarter	21 747.4	14 759.2	3 732.6	2 902.3	827.0	3.4	-549.8	2 515.7	3 065.4	3 805.3	1 447.9	2 357.4	21 744.0

Table 1-2A. Real Gross Domestic Product: Recent Data

(Billions of chained [2012] dollars, quarterly data are at seasonally adjusted annual rates.)   **NIPA Tables 1.1.6, 1.2.6**

Year and quarter	Gross domestic product	Personal consumption expenditures	Gross private domestic investment Total	Fixed investment Nonresidential	Fixed investment Residential	Change in private inventories	Net exports	Exports	Imports	Gov't Total	Federal	State and local	Residual	Addendum: final sales of domestic product
1955	2 871.2	1 740.0	331.1	. . .	. . .	. . .	. . .	79.6	102.4	971.6	. . .	. . .	-169.3	2 866.6
1956	2 932.4	1 790.8	330.2	. . .	. . .	. . .	. . .	92.7	110.7	973.9	. . .	. . .	-159.9	2 936.4
1957	2 994.1	1 835.2	317.8	. . .	. . .	. . .	. . .	100.8	115.3	1 018.7	. . .	. . .	-166.9	3 014.2
1958	2 972.0	1 851.1	294.8	. . .	. . .	. . .	. . .	87.2	120.8	1 052.9	. . .	. . .	-192.2	2 997.1
1959	3 178.2	1 956.7	351.4	. . .	. . .	. . .	. . .	96.2	133.5	1 079.6	. . .	. . .	-190.2	3 180.0
1960	3 260.0	2 010.4	352.7	. . .	. . .	. . .	. . .	112.9	135.3	1 086.4	. . .	. . .	-182.4	3 266.5
1961	3 343.5	2 051.6	353.8	. . .	. . .	. . .	. . .	113.5	134.4	1 145.2	. . .	. . .	-199.9	3 352.4
1962	3 548.4	2 153.1	395.9	. . .	. . .	. . .	. . .	119.2	149.7	1 220.1	. . .	. . .	-218.6	3 540.2
1963	3 702.9	2 241.9	422.8	. . .	. . .	. . .	. . .	127.7	153.7	1 249.7	. . .	. . .	-211.6	3 698.6
1964	3 916.3	2 375.3	456.1	. . .	. . .	. . .	. . .	142.8	161.9	1 280.3	. . .	. . .	-198.7	3 918.1
1965	4 170.8	2 526.0	519.0	. . .	. . .	. . .	. . .	146.8	179.1	1 321.3	. . .	. . .	-206.0	4 149.2
1966	4 445.9	2 669.3	565.9	. . .	. . .	. . .	. . .	156.9	205.7	1 436.4	. . .	. . .	-237.8	4 402.8
1967	4 567.8	2 749.1	546.1	. . .	. . .	. . .	. . .	160.5	220.7	1 550.4	. . .	. . .	-261.4	4 546.7
1968	4 792.3	2 907.5	578.7	. . .	. . .	. . .	. . .	173.2	253.6	1 603.2	. . .	. . .	-256.4	4 777.1
1969	4 942.1	3 015.9	610.9	. . .	. . .	. . .	. . .	181.6	268.0	1 604.8	. . .	. . .	-242.3	4 927.8
1970	4 951.3	3 086.9	573.8	. . .	. . .	. . .	. . .	201.0	279.4	1 571.0	. . .	. . .	-209.2	4 971.9
1971	5 114.3	3 204.8	632.9	. . .	. . .	. . .	. . .	204.5	294.4	1 541.5	. . .	. . .	-207.2	5 108.2
1972	5 383.3	3 401.0	704.2	. . .	. . .	. . .	. . .	220.4	327.5	1 533.6	. . .	. . .	-181.7	5 375.9
1973	5 687.2	3 569.4	781.3	. . .	. . .	. . .	. . .	261.9	342.7	1 528.7	. . .	. . .	-161.7	5 655.4
1974	5 656.5	3 539.5	729.5	. . .	. . .	. . .	. . .	282.7	334.9	1 562.3	. . .	. . .	-160.2	5 639.3
1975	5 644.8	3 619.7	611.4	. . .	. . .	. . .	. . .	280.8	297.7	1 596.7	. . .	. . .	-149.5	5 698.0
1976	5 949.0	3 821.5	728.0	. . .	. . .	. . .	. . .	293.1	355.9	1 605.0	. . .	. . .	-187.0	5 925.8
1977	6 224.1	3 983.0	831.9	. . .	. . .	. . .	. . .	300.2	394.8	1 624.3	. . .	. . .	-176.2	6 187.7
1978	6 568.6	4 157.3	928.1	. . .	. . .	. . .	. . .	331.8	429.0	1 670.9	. . .	. . .	-149.4	6 526.5
1979	6 776.6	4 256.1	960.7	. . .	. . .	. . .	. . .	364.7	436.2	1 701.0	. . .	. . .	-105.9	6 761.6
1980	6 759.2	4 242.8	864.0	. . .	. . .	. . .	. . .	404.0	407.1	1 731.8	. . .	. . .	-64.1	6 804.7
1981	6 930.7	4 301.6	940.1	. . .	. . .	. . .	. . .	408.9	417.8	1 748.5	. . .	. . .	-101.0	6 900.1
1982	6 805.8	4 364.6	822.0	. . .	. . .	. . .	. . .	377.6	412.5	1 780.0	. . .	. . .	-100.4	6 865.5
1983	7 117.7	4 611.7	898.7	. . .	. . .	. . .	. . .	367.8	464.5	1 846.4	. . .	. . .	-132.3	7 159.9
1984	7 632.8	4 854.3	1 144.0	. . .	. . .	. . .	. . .	397.8	577.6	1 911.3	. . .	. . .	-197.5	7 541.5
1985	7 951.1	5 105.6	1 143.2	. . .	. . .	. . .	. . .	411.1	615.1	2 038.8	. . .	. . .	-165.0	7 940.2
1986	8 226.4	5 316.4	1 145.0	. . .	. . .	. . .	. . .	442.6	667.5	2 149.5	. . .	. . .	-170.3	8 241.0
1987	8 511.0	5 496.9	1 177.6	. . .	. . .	. . .	. . .	491.0	707.2	2 211.9	. . .	. . .	-201.8	8 492.5
1988	8 866.5	5 726.5	1 206.6	. . .	. . .	. . .	. . .	570.6	735.0	2 239.4	. . .	. . .	-169.6	8 860.6
1989	9 192.1	5 893.5	1 255.4	. . .	. . .	. . .	. . .	636.6	767.3	2 303.0	. . .	. . .	-168.1	9 172.3
1990	9 365.5	6 012.2	1 223.0	. . .	. . .	. . .	. . .	692.8	794.8	2 376.7	. . .	. . .	-165.5	9 365.1
1991	9 355.4	6 023.0	1 142.1	. . .	. . .	. . .	. . .	738.6	793.6	2 405.2	. . .	. . .	-157.8	9 379.0
1992	9 684.9	6 244.7	1 225.3	. . .	. . .	. . .	. . .	789.8	849.2	2 416.4	. . .	. . .	-164.9	9 683.1
1993	9 951.5	6 462.2	1 323.3	. . .	. . .	. . .	. . .	815.7	922.6	2 396.6	. . .	. . .	-151.9	9 943.6
1994	10 352.4	6 712.6	1 479.8	. . .	. . .	. . .	. . .	887.7	1 032.7	2 398.6	. . .	. . .	-182.1	10 284.1
1995	10 630.3	6 910.7	1 527.3	. . .	. . .	. . .	. . .	979.0	1 115.3	2 410.8	. . .	. . .	-122.7	10 608.5
1996	11 031.4	7 150.5	1 660.8	. . .	. . .	. . .	. . .	1 059.0	1 212.3	2 433.1	. . .	. . .	-99.5	11 008.6
1997	11 521.9	7 419.7	1 850.2	. . .	. . .	. . .	. . .	1 185.2	1 375.6	2 473.0	. . .	. . .	-126.0	11 442.2
1998	12 038.3	7 813.8	2 026.3	. . .	. . .	. . .	. . .	1 212.8	1 536.4	2 533.0	. . .	. . .	-103.1	11 966.9
1999	12 610.5	8 225.4	2 198.7	. . .	. . .	. . .	. . .	1 273.3	1 710.0	2 615.4	. . .	. . .	-79.7	12 543.1
2000	13 131.0	8 643.4	2 346.7	. . .	. . .	. . .	. . .	1 379.5	1 930.3	2 663.0	. . .	. . .	-50.0	13 068.0
2001	13 262.1	8 861.1	2 214.6	. . .	. . .	. . .	. . .	1 299.7	1 876.2	2 762.3	. . .	. . .	53.8	13 309.5
2002	13 493.1	9 088.7	2 195.5	1 472.7	692.6	24.3	-667.3	1 277.1	1 944.4	2 885.2	969.6	1 927.6	-86.1	13 476.4
2003	13 879.1	9 377.5	2 290.4	1 509.4	755.5	19.9	-735.0	1 305.0	2 040.1	2 947.2	1 032.7	1 922.2	-47.5	13 864.7
2004	14 406.4	9 729.3	2 502.6	1 594.0	830.9	82.6	-841.4	1 431.2	2 272.6	2 992.7	1 077.5	1 920.1	-5.9	14 335.7
2005	14 912.5	10 075.9	2 670.6	1 716.4	885.4	63.7	-887.8	1 533.2	2 421.0	3 015.5	1 099.1	1 920.1	35.1	14 852.3
2006	15 338.3	10 384.5	2 752.4	1 854.2	818.9	87.1	-905.0	1 676.4	2 581.5	3 063.5	1 125.0	1 941.6	31.8	15 263.0
2007	15 626.0	10 615.3	2 684.1	1 982.1	665.8	40.6	-823.6	1 822.3	2 646.0	3 118.6	1 147.0	1 974.7	25.7	15 588.7
2008	15 604.7	10 592.8	2 462.9	1 994.2	504.6	-32.7	-661.6	1 925.4	2 587.1	3 195.6	1 218.8	1 978.7	-2.4	15 639.7
2009	15 208.8	10 460.0	1 942.0	1 704.3	395.3	-177.3	-484.8	1 763.8	2 248.6	3 307.3	1 293.0	2 015.6	-15.5	15 373.0
2010	15 598.8	10 643.0	2 216.5	1 781.0	383.0	57.3	-565.9	1 977.9	2 543.8	3 307.2	1 346.1	1 961.3	-11.3	15 546.6
2011	15 840.7	10 843.8	2 362.1	1 935.4	382.5	46.7	-568.1	2 119.0	2 687.1	3 203.3	1 311.1	1 892.2	-3.7	15 796.5
2012	16 197.0	11 006.8	2 621.8	2 118.5	432.0	71.2	-568.6	2 191.3	2 759.9	3 137.0	1 286.5	1 850.5	0.0	16 125.8
2013	16 495.4	11 166.9	2 801.5	2 206.0	485.5	108.7	-532.8	2 269.6	2 802.4	3 061.0	1 215.3	1 845.3	-0.4	16 386.2
2014	16 912.0	11 497.4	2 959.2	2 365.3	504.1	86.3	-577.2	2 365.3	2 942.5	3 033.4	1 183.8	1 848.6	0.7	16 822.3
2015	17 432.2	11 934.3	3 121.8	2 420.3	555.4	137.6	-719.5	2 375.2	3 094.8	3 088.2	1 183.8	1 902.9	11.3	17 290.1
2016	17 730.5	12 264.6	3 074.8	2 433.0	592.1	24.5	-763.6	2 382.3	3 145.9	3 144.4	1 190.5	1 952.0	22.3	17 686.9
2017	18 144.1	12 587.2	3 183.4	2 524.2	615.7	15.8	-816.8	2 475.5	3 292.4	3 172.3	1 194.1	1 976.2	20.8	18 107.2
2018	18 687.8	12 928.1	3 384.9	2 698.9	612.0	53.4	-877.7	2 549.5	3 427.2	3 229.8	1 227.8	2 000.2	1.1	18 613.8
2019	19 091.7	13 240.2	3 442.6	2 776.8	601.5	48.5	-917.6	2 546.6	3 464.2	3 303.9	1 277.2	2 025.5	-25.0	19 021.1
2017														
1st quarter	17 977.3	12 477.3	3 120.4	2 492.6	614.4	-18.9	-792.3	2 446.0	3 238.3	3 156.9	1 186.4	1 968.4	32.5	17 970.8
2nd quarter	18 054.1	12 533.1	3 149.1	2 507.3	612.7	0.6	-815.0	2 451.9	3 266.9	3 169.0	1 192.7	1 974.2	26.0	18 031.4
3rd quarter	18 185.6	12 604.5	3 207.5	2 520.3	610.1	56.1	-813.0	2 468.0	3 281.0	3 170.6	1 191.3	1 977.2	9.1	18 115.0
4th quarter	18 359.4	12 733.7	3 256.7	2 576.4	625.5	25.3	-847.0	2 536.2	3 383.2	3 192.8	1 206.0	1 984.9	15.7	18 311.4
2018														
1st quarter	18 530.5	12 798.1	3 342.5	2 651.5	620.3	47.3	-833.0	2 553.2	3 386.1	3 204.3	1 211.7	1 990.7	7.7	18 463.4
2nd quarter	18 654.4	12 898.1	3 333.3	2 691.9	617.6	-4.9	-820.2	2 565.2	3 385.4	3 227.3	1 222.3	2 003.0	2.1	18 630.7
3rd quarter	18 752.4	12 983.0	3 415.4	2 709.5	609.1	79.1	-920.3	2 531.0	3 451.3	3 247.4	1 235.8	2 009.9	3.2	18 655.1
4th quarter	18 813.9	13 033.4	3 448.3	2 742.6	601.0	92.3	-937.3	2 548.8	3 486.0	3 240.2	1 241.6	1 997.1	-8.7	18 705.9
2019														
1st quarter	18 950.3	13 093.2	3 481.3	2 770.8	598.4	101.7	-907.4	2 560.4	3 467.8	3 260.0	1 245.8	2 012.7	-19.6	18 833.1
2nd quarter	19 020.6	13 212.8	3 429.9	2 771.0	595.2	49.4	-951.4	2 531.4	3 482.9	3 300.3	1 273.6	2 025.5	-18.5	18 949.6
3rd quarter	19 141.7	13 301.3	3 445.7	2 783.9	601.9	44.0	-950.2	2 536.6	3 486.8	3 317.7	1 288.5	2 028.3	-24.6	19 075.2
4th quarter	19 254.0	13 353.7	3 413.3	2 781.5	610.5	-1.1	-861.5	2 557.8	3 419.3	3 337.5	1 301.1	2 035.6	-37.3	19 226.6

Note: Chained (2012) dollar series are calculated as the product of the chain-type quantity index and the 2012 current-dollar value of the corresponding series, divided by 100. Because the formula for the chain-type quantity indexes uses weights from more than one period, the corresponding chained-dollar estimates are usually not additive. The residual column is the difference between the total and the sum of the most detailed components shown in the Bureau of Economic Analysis (BEA) published data.

. . . = Not available.

Table 1-2B. Real Gross Domestic Product: Historical Data

(Billions of chained [2012] dollars, quarterly data are at seasonally adjusted annual rates.)　　　　　**NIPA Tables 1.1.6, 1.2.6**

| Year and quarter | Gross domestic product | Personal consumption expenditures | Gross private domestic investment | | | | Exports and imports of goods and services | | | Government consumption expenditures and gross investment | | | Residual | Addendum: Final sales of domestic product |
| | | | Total | Fixed investment | | Change in private inventories | Net exports | Exports | Imports | Total | Federal | State and local | | |
				Nonresidential	Residential									
1929	1 109.4	830.8	120.4	...	...	...	...	43.5	58.0	180.4	...	...	-20.9	1 121.3
1930	1 015.1	786.3	82.0	...	...	...	...	36.0	50.5	198.7	...	...	-31.5	1 045.6
1931	950.0	761.8	53.3	...	...	...	...	29.9	44.0	206.8	...	...	-44.0	984.8
1932	827.5	693.6	19.7	...	...	...	...	23.4	36.5	200.2	...	...	-42.0	874.0
1933	817.3	678.3	26.6	...	...	...	...	23.6	38.1	193.8	...	...	-51.1	848.1
1934	905.6	726.7	44.0	...	...	...	...	26.2	38.9	217.8	...	...	-62.6	928.4
1935	986.2	770.8	76.5	...	...	...	...	27.6	50.9	224.6	...	...	-72.1	986.5
1936	1 113.3	849.2	96.8	...	...	...	...	29.0	50.4	259.9	...	...	-81.3	1 115.3
1937	1 170.3	880.6	119.2	...	...	...	...	36.5	56.7	249.7	...	...	-75.8	1 162.0
1938	1 131.6	866.6	82.0	...	...	...	...	36.1	44.0	268.7	...	...	-73.8	1 146.6
1939	1 222.4	915.0	103.1	...	...	...	...	38.1	46.3	292.4	...	...	-82.0	1 230.9
1940	1 330.2	962.4	140.4	...	...	...	...	43.4	47.4	302.9	...	...	-92.5	1 315.3
1941	1 565.8	1 030.5	171.9	...	...	...	...	44.5	58.3	509.8	...	...	-166.7	1 535.2
1942	1 861.5	1 006.0	95.7	...	...	...	...	29.4	52.9	1 184.0	...	...	-414.0	1 861.3
1943	2 178.4	1 034.1	59.7	...	...	...	...	24.8	66.7	1 775.0	...	...	-643.7	2 207.2
1944	2 351.6	1 063.5	71.3	...	...	...	...	26.6	69.8	1 992.3	...	...	-726.7	2 383.2
1945	2 328.6	1 129.2	91.7	...	...	...	...	37.3	74.2	1 749.6	...	...	-594.5	2 367.0
1946	2 058.4	1 268.8	220.3	...	...	...	...	80.5	61.5	616.0	...	...	-98.6	2 029.2
1947	2 034.8	1 293.1	212.1	...	...	...	...	91.8	58.4	522.6	...	...	-21.7	2 065.1
1948	2 118.5	1 322.3	267.3	...	...	...	...	72.3	68.1	551.8	...	...	-52.0	2 099.3
1949	2 106.6	1 359.1	206.7	...	...	...	...	71.6	65.7	611.1	...	...	-63.0	2 146.9
1947														
1st quarter	2 033.1	1 275.0	218.1	...	...	...	...	97.8	60.3	524.1	...	...	-28.0	2 049.7
2nd quarter	2 027.6	1 296.2	201.4	...	...	...	...	96.5	61.4	523.6	...	...	-24.2	2 060.3
3rd quarter	2 023.5	1 300.4	195.5	...	...	...	...	91.6	54.5	523.5	...	...	-15.6	2 074.0
4th quarter	2 055.1	1 300.8	233.3	...	...	...	...	81.2	57.4	519.3	...	...	-19.2	2 076.3
1948														
1st quarter	2 086.0	1 307.3	257.2	...	...	...	...	77.5	64.7	527.0	...	...	-32.9	2 084.0
2nd quarter	2 120.5	1 322.5	273.1	...	...	...	...	70.3	67.3	545.6	...	...	-51.5	2 095.8
3rd quarter	2 132.6	1 324.4	275.9	...	...	...	...	71.9	70.8	557.0	...	...	-58.4	2 100.3
4th quarter	2 135.0	1 335.0	262.7	...	...	...	...	69.4	69.7	577.5	...	...	-65.1	2 117.1
1949														
1st quarter	2 105.6	1 337.2	223.5	...	...	...	...	77.7	67.6	592.8	...	...	-55.5	2 125.3
2nd quarter	2 098.4	1 357.7	193.8	...	...	...	...	77.0	66.6	618.6	...	...	-59.5	2 152.6
3rd quarter	2 120.0	1 360.8	209.9	...	...	...	...	70.6	63.9	621.8	...	...	-74.2	2 150.2
4th quarter	2 102.3	1 380.7	199.5	...	...	...	...	61.1	64.9	611.0	...	...	-62.6	2 159.7
1950														
1st quarter	2 184.9	1 403.7	246.9	...	...	...	...	60.1	66.7	599.6	...	...	-68.3	2 193.0
2nd quarter	2 251.5	1 426.8	273.3	...	...	...	...	60.8	70.3	610.6	...	...	-63.6	2 252.2
3rd quarter	2 338.5	1 500.1	296.8	...	...	...	...	62.4	86.8	600.7	...	...	-52.9	2 332.4
4th quarter	2 383.3	1 454.9	334.5	...	...	...	...	67.5	87.0	643.2	...	...	-88.0	2 311.9
1951														
1st quarter	2 415.7	1 490.0	300.4	...	...	...	...	70.9	87.0	711.6	...	...	-106.4	2 381.7
2nd quarter	2 457.5	1 447.9	307.3	...	...	...	...	77.8	84.1	806.5	...	...	-150.5	2 395.4
3rd quarter	2 508.2	1 464.8	284.7	...	...	...	...	79.7	77.0	895.1	...	...	-173.9	2 475.6
4th quarter	2 513.7	1 473.3	262.0	...	...	...	...	79.0	75.0	942.4	...	...	-183.0	2 512.8
1952														
1st quarter	2 540.6	1 476.7	268.9	...	...	...	...	83.2	83.7	970.5	...	...	-193.8	2 535.2
2nd quarter	2 546.0	1 505.7	248.6	...	...	...	...	74.1	83.2	1 003.6	...	...	-198.1	2 578.0
3rd quarter	2 564.4	1 512.9	261.9	...	...	...	...	68.2	88.3	1 013.7	...	...	-224.7	2 555.3
4th quarter	2 648.6	1 566.3	280.8	...	...	...	...	68.7	96.2	1 030.6	...	...	-222.2	2 639.5
1953														
1st quarter	2 697.9	1 584.7	286.8	...	...	...	...	67.5	93.9	1 063.3	...	...	-227.3	2 696.0
2nd quarter	2 718.7	1 594.2	288.4	...	...	...	...	67.9	98.7	1 085.1	...	...	-234.7	2 717.7
3rd quarter	2 703.4	1 590.4	282.4	...	...	...	...	70.7	98.4	1 073.6	...	...	-225.2	2 712.1
4th quarter	2 662.5	1 579.7	262.0	...	...	...	...	68.3	93.5	1 070.9	...	...	-216.4	2 697.9
1954														
1st quarter	2 649.8	1 585.6	260.1	...	...	...	...	65.3	88.1	1 043.0	...	...	-208.5	2 682.2
2nd quarter	2 652.6	1 606.1	259.5	...	...	...	...	74.6	96.0	1 007.7	...	...	-187.1	2 692.4
3rd quarter	2 682.6	1 627.9	272.0	...	...	...	...	72.2	90.2	985.9	...	...	-175.5	2 719.3
4th quarter	2 735.1	1 662.0	284.0	...	...	...	...	75.6	91.1	980.0	...	...	-171.8	2 763.6
1955														
1st quarter	2 813.2	1 699.3	312.3	...	...	...	...	78.4	96.5	980.6	...	...	-175.4	2 816.4
2nd quarter	2 859.0	1 731.9	331.3	...	...	...	...	76.5	102.0	970.9	...	...	-171.2	2 852.8
3rd quarter	2 897.6	1 753.3	336.4	...	...	...	...	81.6	103.7	977.2	...	...	-166.7	2 894.7
4th quarter	2 915.0	1 775.3	344.2	...	...	...	...	81.9	107.4	957.8	...	...	-163.8	2 902.4
1956														
1st quarter	2 903.7	1 778.2	334.4	...	...	...	...	85.8	111.9	957.5	...	...	-163.0	2 897.8
2nd quarter	2 927.7	1 784.1	331.8	...	...	...	...	91.6	111.4	976.9	...	...	-161.1	2 930.9
3rd quarter	2 925.0	1 788.1	328.8	...	...	...	...	94.7	112.3	968.3	...	...	-154.7	2 933.0
4th quarter	2 973.2	1 812.8	325.7	...	...	...	...	98.9	107.2	992.9	...	...	-160.8	2 983.9
1957														
1st quarter	2 992.2	1 825.4	320.7	...	...	...	...	105.3	115.7	1 014.2	...	...	-163.6	3 010.6
2nd quarter	2 985.7	1 828.6	320.4	...	...	...	...	102.2	116.6	1 008.9	...	...	-166.3	3 000.3
3rd quarter	3 014.9	1 843.0	327.7	...	...	...	...	99.4	114.3	1 017.5	...	...	-170.7	3 023.5
4th quarter	2 983.7	1 843.8	302.5	...	...	...	...	96.4	114.7	1 034.3	...	...	-167.0	3 022.5

Note: Chained (2012) dollar series are calculated as the product of the chain-type quantity index and the 2012 current-dollar value of the corresponding series, divided by 100. Because the formula for the chain-type quantity indexes uses weights from more than one period, the corresponding chained-dollar estimates are usually not additive. The residual column is the difference between the total and the sum of the most detailed components shown in the Bureau of Economic Analysis (BEA) published data.

. . . = Not available.

Table 1-2B. Real Gross Domestic Product: Historical Data—*Continued*

(Billions of chained [2012] dollars, quarterly data are at seasonally adjusted annual rates.)   **NIPA Tables 1.1.6, 1.2.6**

Year and quarter	Gross domestic product	Personal consumption expenditures	Gross private domestic investment Total	Fixed investment Nonresidential	Fixed investment Residential	Change in private inventories	Net exports	Exports	Imports	Government Total	Federal	State and local	Residual	Addendum: Final sales of domestic product
1958														
1st quarter	2 906.3	1 818.3	282.1	. . .	. . .	. . .	. . .	86.8	116.5	1 025.3	. . .	. . .	-175.5	2 948.4
2nd quarter ...	2 925.4	1 833.3	276.4	. . .	. . .	. . .	. . .	87.3	121.4	1 050.9	. . .	. . .	-188.6	2 965.0
3rd quarter ...	2 993.1	1 863.7	297.6	. . .	. . .	. . .	. . .	87.5	119.6	1 057.0	. . .	. . .	-199.5	3 008.1
4th quarter ...	3 063.1	1 889.1	323.0	. . .	. . .	. . .	. . .	87.2	125.8	1 078.3	. . .	. . .	-205.3	3 066.8
1959														
1st quarter ...	3 121.9	1 923.7	340.6	. . .	. . .	. . .	. . .	93.5	129.1	1 067.3	. . .	. . .	-187.6	3 127.0
2nd quarter ...	3 192.4	1 953.4	367.1	. . .	. . .	. . .	. . .	92.8	135.3	1 080.5	. . .	. . .	-200.4	3 173.1
3rd quarter ...	3 194.7	1 973.8	343.4	. . .	. . .	. . .	. . .	101.7	136.6	1 089.8	. . .	. . .	-181.5	3 216.6
4th quarter ...	3 203.8	1 976.0	354.6	. . .	. . .	. . .	. . .	96.7	133.1	1 080.8	. . .	. . .	-191.3	3 203.2
1960														
1st quarter ...	3 275.8	1 994.9	390.2	. . .	. . .	. . .	. . .	109.1	138.2	1 062.4	. . .	. . .	-192.9	3 239.4
2nd quarter ...	3 258.1	2 020.1	353.8	. . .	. . .	. . .	. . .	112.3	139.2	1 074.3	. . .	. . .	-177.6	3 266.5
3rd quarter ...	3 274.0	2 012.0	352.9	. . .	. . .	. . .	. . .	114.7	135.1	1 100.6	. . .	. . .	-191.4	3 272.9
4th quarter ...	3 232.0	2 014.6	313.8	. . .	. . .	. . .	. . .	115.3	128.7	1 108.2	. . .	. . .	-167.7	3 287.1
1961														
1st quarter ...	3 253.8	2 013.9	322.0	. . .	. . .	. . .	. . .	114.5	128.1	1 124.7	. . .	. . .	-182.4	3 292.8
2nd quarter ...	3 309.1	2 043.8	344.5	. . .	. . .	. . .	. . .	109.2	129.7	1 127.5	. . .	. . .	-195.7	3 323.3
3rd quarter ...	3 372.6	2 053.8	371.3	. . .	. . .	. . .	. . .	114.1	138.3	1 151.6	. . .	. . .	-210.8	3 359.9
4th quarter ...	3 438.7	2 095.1	377.5	. . .	. . .	. . .	. . .	116.0	141.6	1 176.8	. . .	. . .	-210.7	3 433.5
1962														
1st quarter ...	3 500.1	2 117.3	397.8	. . .	. . .	. . .	. . .	115.3	145.9	1 199.6	. . .	. . .	-224.5	3 475.6
2nd quarter ...	3 531.7	2 143.3	394.7	. . .	. . .	. . .	. . .	121.9	149.0	1 205.6	. . .	. . .	-210.2	3 527.1
3rd quarter ...	3 575.1	2 160.6	401.4	. . .	. . .	. . .	. . .	121.7	150.9	1 233.4	. . .	. . .	-220.9	3 565.5
4th quarter ...	3 586.8	2 191.2	389.7	. . .	. . .	. . .	. . .	117.8	152.8	1 242.0	. . .	. . .	-218.6	3 592.7
1963														
1st quarter ...	3 626.0	2 206.5	410.2	. . .	. . .	. . .	. . .	119.6	149.2	1 225.1	. . .	. . .	-220.9	3 609.6
2nd quarter ...	3 666.7	2 227.3	416.7	. . .	. . .	. . .	. . .	129.1	152.8	1 226.0	. . .	. . .	-202.7	3 665.5
3rd quarter ...	3 747.3	2 257.5	429.5	. . .	. . .	. . .	. . .	128.5	156.6	1 280.4	. . .	. . .	-219.3	3 742.3
4th quarter ...	3 771.8	2 276.3	434.7	. . .	. . .	. . .	. . .	133.5	156.1	1 267.2	. . .	. . .	-203.7	3 776.9
1964														
1st quarter ...	3 851.4	2 321.0	451.0	. . .	. . .	. . .	. . .	140.3	156.0	1 271.1	. . .	. . .	-197.3	3 854.4
2nd quarter ...	3 893.3	2 362.1	449.4	. . .	. . .	. . .	. . .	139.7	159.5	1 282.5	. . .	. . .	-201.3	3 897.9
3rd quarter ...	3 954.1	2 405.7	459.9	. . .	. . .	. . .	. . .	144.1	163.8	1 284.1	. . .	. . .	-200.2	3 954.2
4th quarter ...	3 966.3	2 412.6	464.2	. . .	. . .	. . .	. . .	147.0	168.2	1 283.5	. . .	. . .	-196.1	3 965.8
1965														
1st quarter ...	4 062.3	2 466.4	508.5	. . .	. . .	. . .	. . .	129.9	162.1	1 282.0	. . .	. . .	-215.8	4 028.1
2nd quarter ...	4 113.6	2 493.6	509.8	. . .	. . .	. . .	. . .	151.6	181.3	1 297.2	. . .	. . .	-195.8	4 096.5
3rd quarter ...	4 205.1	2 536.4	527.6	. . .	. . .	. . .	. . .	148.2	182.0	1 340.2	. . .	. . .	-208.7	4 181.8
4th quarter ...	4 302.0	2 607.7	531.1	. . .	. . .	. . .	. . .	157.3	190.9	1 365.6	. . .	. . .	-203.7	4 290.4
1966														
1st quarter ...	4 406.7	2 646.2	573.5	. . .	. . .	. . .	. . .	153.4	195.8	1 391.1	. . .	. . .	-222.8	4 362.4
2nd quarter ...	4 421.7	2 653.0	565.0	. . .	. . .	. . .	. . .	158.0	200.7	1 417.4	. . .	. . .	-229.7	4 382.3
3rd quarter ...	4 459.2	2 683.3	560.8	. . .	. . .	. . .	. . .	156.1	211.9	1 456.0	. . .	. . .	-237.3	4 426.7
4th quarter ...	4 495.8	2 694.5	564.1	. . .	. . .	. . .	. . .	160.0	214.3	1 481.1	. . .	. . .	-261.5	4 439.8
1967														
1st quarter ...	4 535.6	2 710.0	550.3	. . .	. . .	. . .	. . .	162.0	217.7	1 543.7	. . .	. . .	-279.9	4 486.0
2nd quarter ...	4 538.4	2 747.1	530.4	. . .	. . .	. . .	. . .	159.8	215.9	1 537.3	. . .	. . .	-249.8	4 533.7
3rd quarter ...	4 581.3	2 761.1	545.9	. . .	. . .	. . .	. . .	158.6	218.8	1 554.4	. . .	. . .	-259.7	4 563.7
4th quarter ...	4 615.9	2 778.0	557.7	. . .	. . .	. . .	. . .	161.5	230.3	1 566.2	. . .	. . .	-256.0	4 603.4
1968														
1st quarter ...	4 710.0	2 844.6	568.9	. . .	. . .	. . .	. . .	166.1	243.9	1 594.2	. . .	. . .	-255.3	4 699.0
2nd quarter ...	4 788.7	2 887.9	590.6	. . .	. . .	. . .	. . .	168.9	247.6	1 602.5	. . .	. . .	-275.5	4 746.1
3rd quarter ...	4 825.8	2 942.0	574.5	. . .	. . .	. . .	. . .	180.9	262.5	1 607.3	. . .	. . .	-250.2	4 817.4
4th quarter ...	4 844.8	2 955.3	580.8	. . .	. . .	. . .	. . .	176.9	260.3	1 608.7	. . .	. . .	-244.7	4 845.8
1969														
1st quarter ...	4 920.6	2 988.1	615.0	. . .	. . .	. . .	. . .	156.3	235.7	1 612.4	. . .	. . .	-262.5	4 896.1
2nd quarter ...	4 935.6	3 007.4	611.4	. . .	. . .	. . .	. . .	190.5	282.2	1 607.3	. . .	. . .	-238.2	4 920.5
3rd quarter ...	4 968.2	3 022.0	623.8	. . .	. . .	. . .	. . .	187.4	278.6	1 610.6	. . .	. . .	-242.5	4 947.6
4th quarter ...	4 943.9	3 046.2	593.7	. . .	. . .	. . .	. . .	192.3	275.5	1 588.9	. . .	. . .	-226.2	4 946.8
1970														
1st quarter ...	4 936.6	3 065.1	576.0	. . .	. . .	. . .	. . .	194.8	274.6	1 581.3	. . .	. . .	-210.4	4 960.0
2nd quarter ...	4 943.6	3 079.0	577.2	. . .	. . .	. . .	. . .	202.3	280.0	1 562.9	. . .	. . .	-218.7	4 949.1
3rd quarter ...	4 989.2	3 106.0	586.6	. . .	. . .	. . .	. . .	203.2	279.4	1 569.5	. . .	. . .	-217.6	4 995.0
4th quarter ...	4 935.7	3 097.5	555.5	. . .	. . .	. . .	. . .	203.6	283.7	1 570.3	. . .	. . .	-189.9	4 983.8
1971														
1st quarter ...	5 069.7	3 157.0	620.2	. . .	. . .	. . .	. . .	205.4	280.4	1 547.3	. . .	. . .	-229.2	5 045.8
2nd quarter ...	5 097.2	3 186.0	637.8	. . .	. . .	. . .	. . .	203.9	301.5	1 543.7	. . .	. . .	-214.8	5 079.2
3rd quarter ...	5 139.1	3 211.4	645.5	. . .	. . .	. . .	. . .	214.1	308.2	1 542.9	. . .	. . .	-206.6	5 125.0
4th quarter ...	5 151.2	3 264.7	628.2	. . .	. . .	. . .	. . .	194.3	287.3	1 532.1	. . .	. . .	-178.4	5 183.0
1972														
1st quarter ...	5 246.0	3 307.8	669.6	. . .	. . .	. . .	. . .	216.4	329.3	1 540.5	. . .	. . .	-173.5	5 258.3
2nd quarter ...	5 365.0	3 370.7	707.6	. . .	. . .	. . .	. . .	210.1	317.6	1 546.7	. . .	. . .	-199.8	5 341.4
3rd quarter ...	5 415.7	3 422.7	717.6	. . .	. . .	. . .	. . .	222.8	324.2	1 517.3	. . .	. . .	-191.1	5 389.0
4th quarter ...	5 506.4	3 503.0	722.1	. . .	. . .	. . .	. . .	232.2	338.8	1 529.8	. . .	. . .	-162.3	5 514.7

Note: Chained (2012) dollar series are calculated as the product of the chain-type quantity index and the 2012 current-dollar value of the corresponding series, divided by 100. Because the formula for the chain-type quantity indexes uses weights from more than one period, the corresponding chained-dollar estimates are usually not additive. The residual column is the difference between the total and the sum of the most detailed components shown in the Bureau of Economic Analysis (BEA) published data.

. . . = Not available.

Table 1-2B. Real Gross Domestic Product: Historical Data—*Continued*

(Billions of chained [2012] dollars, quarterly data are at seasonally adjusted annual rates.)   **NIPA Tables 1.1.6, 1.2.6**

Year and quarter	Gross domestic product	Personal consump-tion expen-ditures	Gross private domestic investment				Exports and imports of goods and services			Government consumption expenditures and gross investment			Residual	Adden-dum: Final sales of domestic product
			Total	Fixed investment		Change in private inventories	Net exports	Exports	Imports	Total	Federal	State and local		
				Nonresi-dential	Residential									
1973														
1st quarter	5 642.7	3 567.0	764.5	...	...	...	...	249.2	354.5	1 543.0	...	...	-159.8	5 631.7
2nd quarter ...	5 704.1	3 565.3	797.0	...	...	...	...	261.2	344.3	1 532.1	...	...	-172.5	5 656.4
3rd quarter ...	5 674.1	3 577.9	768.1	...	...	...	...	262.1	334.8	1 514.1	...	...	-144.4	5 662.9
4th quarter	5 728.0	3 567.2	795.5	...	...	...	...	275.0	337.3	1 525.5	...	...	-170.0	5 670.5
1974														
1st quarter	5 678.7	3 535.3	750.1	...	...	...	...	278.5	325.8	1 553.0	...	...	-147.6	5 663.3
2nd quarter ...	5 692.2	3 548.0	747.1	...	...	...	...	292.1	343.1	1 560.8	...	...	-158.9	5 668.7
3rd quarter ...	5 638.4	3 563.3	708.4	...	...	...	...	276.4	337.1	1 563.9	...	...	-151.2	5 652.6
4th quarter	5 616.5	3 511.2	712.7	...	...	...	...	283.6	333.7	1 571.4	...	...	-183.0	5 572.7
1975														
1st quarter	5 548.2	3 540.6	597.9	...	...	...	...	285.6	300.2	1 588.5	...	...	-142.3	5 612.8
2nd quarter ...	5 587.8	3 598.9	580.1	...	...	...	...	277.3	276.2	1 575.1	...	...	-128.9	5 667.2
3rd quarter ...	5 683.4	3 650.0	625.0	...	...	...	...	272.6	299.2	1 604.5	...	...	-163.4	5 720.9
4th quarter	5 760.0	3 689.3	642.5	...	...	...	...	288.0	315.2	1 618.6	...	...	-163.6	5 791.0
1976														
1st quarter	5 889.5	3 763.0	704.5	...	...	...	...	286.0	334.9	1 621.1	...	...	-190.9	5 871.0
2nd quarter ...	5 932.7	3 797.7	732.4	...	...	...	...	288.9	349.4	1 603.1	...	...	-201.0	5 890.3
3rd quarter ...	5 965.3	3 837.7	734.9	...	...	...	...	297.3	363.7	1 598.4	...	...	-194.5	5 930.0
4th quarter	6 008.5	3 887.4	740.4	...	...	...	...	300.3	375.7	1 597.4	...	...	-161.5	6 011.8
1977														
1st quarter	6 079.5	3 933.3	774.7	...	...	...	...	295.5	393.5	1 611.0	...	...	-175.8	6 069.8
2nd quarter ...	6 197.7	3 954.6	830.3	...	...	...	...	303.4	397.1	1 625.7	...	...	-170.5	6 165.3
3rd quarter ...	6 309.5	3 992.0	872.3	...	...	...	...	305.8	391.6	1 632.3	...	...	-191.2	6 232.3
4th quarter	6 309.7	4 052.0	850.4	...	...	...	...	296.0	397.1	1 628.4	...	...	-167.3	6 283.4
1978														
1st quarter	6 329.8	4 074.8	867.0	...	...	...	...	303.0	423.8	1 628.8	...	...	-176.3	6 290.2
2nd quarter ...	6 574.4	4 161.9	923.3	...	...	...	...	334.5	425.1	1 669.6	...	...	-145.3	6 535.3
3rd quarter ...	6 640.5	4 179.4	950.2	...	...	...	...	338.2	430.7	1 685.1	...	...	-138.1	6 597.0
4th quarter	6 729.8	4 213.1	972.1	...	...	...	...	351.6	436.6	1 700.2	...	...	-137.9	6 683.3
1979														
1st quarter	6 741.9	4 234.9	973.8	...	...	...	...	351.9	435.4	1 683.4	...	...	-117.6	6 705.0
2nd quarter ...	6 749.1	4 232.2	972.9	...	...	...	...	352.8	437.8	1 699.8	...	...	-128.4	6 705.7
3rd quarter ...	6 799.2	4 273.3	956.5	...	...	...	...	365.5	431.1	1 704.8	...	...	-91.1	6 803.3
4th quarter	6 816.2	4 284.0	939.5	...	...	...	...	388.5	440.5	1 716.0	...	...	-86.6	6 832.4
1980														
1st quarter	6 837.6	4 277.9	933.1	...	...	...	...	399.4	440.8	1 741.2	...	...	-90.0	6 845.3
2nd quarter ...	6 696.8	4 181.5	853.8	...	...	...	...	406.9	408.8	1 744.6	...	...	-95.7	6 710.3
3rd quarter ...	6 688.8	4 227.4	797.4	...	...	...	...	406.0	379.3	1 720.9	...	...	-21.5	6 799.9
4th quarter	6 813.5	4 284.5	871.7	...	...	...	...	403.6	399.7	1 720.5	...	...	-49.5	6 863.2
1981														
1st quarter	6 947.0	4 298.8	952.4	...	...	...	...	411.2	416.7	1 743.5	...	...	-113.9	6 891.4
2nd quarter ...	6 895.6	4 299.2	912.9	...	...	...	...	413.4	417.4	1 746.9	...	...	-77.4	6 907.5
3rd quarter ...	6 978.1	4 319.0	964.4	...	...	...	...	404.6	412.6	1 740.8	...	...	-112.2	6 916.2
4th quarter	6 902.1	4 289.5	930.4	...	...	...	...	406.4	424.4	1 763.0	...	...	-100.5	6 885.3
1982														
1st quarter	6 794.9	4 321.1	842.4	...	...	...	...	388.8	412.3	1 761.0	...	...	-70.8	6 869.5
2nd quarter ...	6 825.9	4 334.3	841.8	...	...	...	...	391.1	405.5	1 767.7	...	...	-96.5	6 861.7
3rd quarter ...	6 799.8	4 363.2	834.1	...	...	...	...	373.8	424.3	1 781.9	...	...	-136.6	6 817.5
4th quarter	6 802.5	4 439.7	769.6	...	...	...	...	356.6	407.7	1 809.3	...	...	-97.7	6 913.3
1983														
1st quarter	6 892.1	4 483.6	796.3	...	...	...	...	362.4	417.0	1 826.7	...	...	-97.0	6 991.7
2nd quarter ...	7 049.0	4 574.9	866.8	...	...	...	...	363.5	449.1	1 841.8	...	...	-139.0	7 094.1
3rd quarter ...	7 189.9	4 657.0	921.1	...	...	...	...	368.9	484.1	1 873.6	...	...	-143.2	7 228.0
4th quarter	7 339.9	4 731.2	1 010.7	...	...	...	...	376.4	507.8	1 843.6	...	...	-149.8	7 325.7
1984														
1st quarter	7 483.4	4 770.5	1 108.3	...	...	...	...	384.2	548.6	1 863.1	...	...	-207.8	7 378.3
2nd quarter ...	7 612.7	4 837.3	1 144.4	...	...	...	...	393.8	571.6	1 903.6	...	...	-201.6	7 513.5
3rd quarter ...	7 686.1	4 873.2	1 169.1	...	...	...	...	402.5	586.4	1 919.0	...	...	-202.9	7 582.3
4th quarter	7 749.2	4 936.3	1 154.0	...	...	...	...	410.7	603.6	1 959.3	...	...	-177.7	7 692.0
1985														
1st quarter	7 824.2	5 020.2	1 122.3	...	...	...	...	411.8	590.3	1 980.6	...	...	-142.1	7 822.8
2nd quarter ...	7 893.1	5 066.3	1 141.4	...	...	...	...	410.9	619.3	2 024.6	...	...	-162.6	7 885.2
3rd quarter ...	8 013.7	5 162.5	1 133.7	...	...	...	...	404.9	613.3	2 070.7	...	...	-171.5	8 010.3
4th quarter	8 073.2	5 173.6	1 175.5	...	...	...	...	417.0	637.3	2 079.3	...	...	-184.0	8 042.6
1986														
1st quarter	8 148.6	5 218.9	1 175.1	...	...	...	...	430.0	636.7	2 095.7	...	...	-178.1	8 122.9
2nd quarter ...	8 185.3	5 275.7	1 154.6	...	...	...	...	434.6	663.8	2 139.4	...	...	-176.5	8 187.5
3rd quarter ...	8 263.6	5 369.0	1 123.5	...	...	...	...	444.6	682.2	2 186.0	...	...	-165.2	8 300.9
4th quarter	8 308.0	5 402.0	1 126.9	...	...	...	...	461.2	687.5	2 177.0	...	...	-161.2	8 352.7
1987														
1st quarter	8 369.9	5 407.4	1 157.3	...	...	...	...	461.6	683.4	2 189.7	...	...	-209.7	8 347.8
2nd quarter ...	8 460.2	5 481.2	1 157.9	...	...	...	...	480.6	700.5	2 206.7	...	...	-192.1	8 460.5
3rd quarter ...	8 533.6	5 543.7	1 158.1	...	...	...	...	501.5	714.2	2 209.0	...	...	-162.2	8 560.1
4th quarter	8 680.2	5 555.5	1 237.0	...	...	...	...	520.2	730.6	2 242.1	...	...	-243.2	8 601.6

Note: Chained (2012) dollar series are calculated as the product of the chain-type quantity index and the 2012 current-dollar value of the corresponding series, divided by 100. Because the formula for the chain-type quantity indexes uses weights from more than one period, the corresponding chained-dollar estimates are usually not additive. The residual column is the difference between the total and the sum of the most detailed components shown in the Bureau of Economic Analysis (BEA) published data.

. . . = Not available.

Table 1-2B. Real Gross Domestic Product: Historical Data—*Continued*

(Billions of chained [2012] dollars, quarterly data are at seasonally adjusted annual rates.)   **NIPA Tables 1.1.6, 1.2.6**

Year and quarter	Gross domestic product	Personal consumption expenditures	Gross private domestic investment				Exports and imports of goods and services			Government consumption expenditures and gross investment			Residual	Adden-dum: Final sales of domestic product
			Total	Fixed investment		Change in private inventories	Net exports	Exports	Imports	Total	Federal	State and local		
				Nonresidential	Residential									
1988														
1st quarter	8 725.0	5 653.6	1 177.6	. . .	. . .	. . .	. . .	549.1	727.2	2 223.4	. . .	. . .	-173.4	8 726.3
2nd quarter ...	8 839.6	5 695.3	1 205.7	. . .	. . .	. . .	. . .	564.5	718.8	2 230.1	. . .	. . .	-167.4	8 834.1
3rd quarter ...	8 891.4	5 745.9	1 212.4	. . .	. . .	. . .	. . .	575.6	735.7	2 230.3	. . .	. . .	-162.2	8 884.5
4th quarter ...	9 009.9	5 811.3	1 230.9	. . .	. . .	. . .	. . .	593.2	758.2	2 273.6	. . .	. . .	-175.4	8 997.5
1989														
1st quarter ...	9 101.5	5 838.2	1 272.6	. . .	. . .	. . .	. . .	610.9	761.7	2 263.6	. . .	. . .	-190.5	9 048.8
2nd quarter ...	9 171.0	5 865.5	1 260.3	. . .	. . .	. . .	. . .	637.6	765.3	2 298.4	. . .	. . .	-177.9	9 136.6
3rd quarter ...	9 238.9	5 922.3	1 249.0	. . .	. . .	. . .	. . .	644.2	764.3	2 319.1	. . .	. . .	-145.8	9 247.1
4th quarter ...	9 257.1	5 948.0	1 239.7	. . .	. . .	. . .	. . .	653.9	778.1	2 330.8	. . .	. . .	-158.2	9 256.6
1990														
1st quarter ...	9 358.3	5 998.1	1 252.1	. . .	. . .	. . .	. . .	682.3	802.8	2 368.0	. . .	. . .	-158.6	9 360.8
2nd quarter ...	9 392.3	6 016.3	1 252.4	. . .	. . .	. . .	. . .	690.8	801.4	2 371.8	. . .	. . .	-185.9	9 362.9
3rd quarter ...	9 398.5	6 040.2	1 228.2	. . .	. . .	. . .	. . .	696.0	798.2	2 375.5	. . .	. . .	-174.8	9 387.3
4th quarter ...	9 312.9	5 994.2	1 159.5	. . .	. . .	. . .	. . .	702.1	776.8	2 391.4	. . .	. . .	-142.7	9 349.5
1991														
1st quarter ...	9 269.4	5 971.7	1 120.9	. . .	. . .	. . .	. . .	706.4	767.9	2 405.7	. . .	. . .	-143.9	9 313.8
2nd quarter ...	9 341.6	6 021.2	1 120.9	. . .	. . .	. . .	. . .	732.2	781.6	2 415.5	. . .	. . .	-138.7	9 390.6
3rd quarter ...	9 388.8	6 051.2	1 143.1	. . .	. . .	. . .	. . .	749.5	803.6	2 408.7	. . .	. . .	-161.2	9 411.6
4th quarter ...	9 421.6	6 048.2	1 183.6	. . .	. . .	. . .	. . .	766.4	821.3	2 391.1	. . .	. . .	-187.2	9 399.9
1992														
1st quarter ...	9 534.3	6 161.4	1 161.7	. . .	. . .	. . .	. . .	780.2	827.3	2 412.3	. . .	. . .	-157.1	9 555.9
2nd quarter ...	9 637.7	6 203.2	1 227.4	. . .	. . .	. . .	. . .	780.6	840.9	2 407.6	. . .	. . .	-170.0	9 625.5
3rd quarter ...	9 733.0	6 269.7	1 237.2	. . .	. . .	. . .	. . .	797.2	854.1	2 422.9	. . .	. . .	-167.2	9 726.0
4th quarter ...	9 834.5	6 344.4	1 274.9	. . .	. . .	. . .	. . .	801.2	874.6	2 422.7	. . .	. . .	-165.4	9 825.0
1993														
1st quarter ...	9 851.0	6 368.8	1 305.0	. . .	. . .	. . .	. . .	803.0	893.5	2 392.6	. . .	. . .	-178.9	9 816.2
2nd quarter ...	9 908.3	6 426.7	1 312.5	. . .	. . .	. . .	. . .	812.6	912.0	2 392.8	. . .	. . .	-160.7	9 892.2
3rd quarter ...	9 955.6	6 498.2	1 303.6	. . .	. . .	. . .	. . .	810.5	924.0	2 396.2	. . .	. . .	-131.3	9 974.9
4th quarter ...	10 091.0	6 555.3	1 372.1	. . .	. . .	. . .	. . .	836.6	961.1	2 404.7	. . .	. . .	-136.7	10 091.0
1994														
1st quarter ...	10 189.0	6 630.3	1 424.4	. . .	. . .	. . .	. . .	844.8	983.5	2 374.9	. . .	. . .	-166.0	10 145.1
2nd quarter ...	10 327.0	6 681.8	1 494.4	. . .	. . .	. . .	. . .	873.3	1 020.1	2 386.7	. . .	. . .	-204.9	10 232.0
3rd quarter ...	10 387.4	6 732.8	1 470.0	. . .	. . .	. . .	. . .	906.0	1 048.8	2 426.9	. . .	. . .	-171.8	10 336.0
4th quarter ...	10 506.4	6 805.6	1 530.5	. . .	. . .	. . .	. . .	926.9	1 078.3	2 405.8	. . .	. . .	-185.7	10 423.3
1995														
1st quarter ...	10 543.6	6 822.5	1 546.5	. . .	. . .	. . .	. . .	944.8	1 102.0	2 414.2	. . .	. . .	-167.4	10 478.2
2nd quarter ...	10 575.1	6 882.3	1 514.4	. . .	. . .	. . .	. . .	958.0	1 112.5	2 422.3	. . .	. . .	-138.0	10 547.3
3rd quarter ...	10 665.1	6 944.7	1 505.4	. . .	. . .	. . .	. . .	999.5	1 116.3	2 416.0	. . .	. . .	-95.2	10 673.0
4th quarter ...	10 737.5	6 993.1	1 542.8	. . .	. . .	. . .	. . .	1 013.7	1 130.6	2 390.7	. . .	. . .	-90.1	10 735.4
1996														
1st quarter ...	10 817.9	7 057.6	1 567.6	. . .	. . .	. . .	. . .	1 025.9	1 167.3	2 406.5	. . .	. . .	-74.8	10 832.8
2nd quarter ...	10 998.3	7 133.6	1 642.9	. . .	. . .	. . .	. . .	1 042.6	1 193.4	2 436.4	. . .	. . .	-98.5	10 981.6
3rd quarter ...	11 097.0	7 176.8	1 717.5	. . .	. . .	. . .	. . .	1 051.0	1 232.1	2 436.4	. . .	. . .	-129.5	11 035.5
4th quarter ...	11 212.2	7 233.9	1 715.2	. . .	. . .	. . .	. . .	1 116.6	1 256.4	2 453.2	. . .	. . .	-95.1	11 184.4
1997														
1st quarter ...	11 284.6	7 310.2	1 749.9	. . .	. . .	. . .	. . .	1 137.8	1 310.1	2 440.5	. . .	. . .	-80.9	11 253.0
2nd quarter ...	11 472.1	7 343.1	1 857.0	. . .	. . .	. . .	. . .	1 183.3	1 354.1	2 472.1	. . .	. . .	-173.8	11 354.1
3rd quarter ...	11 615.6	7 468.2	1 883.0	. . .	. . .	. . .	. . .	1 210.7	1 404.0	2 483.5	. . .	. . .	-121.8	11 539.4
4th quarter ...	11 715.4	7 557.4	1 911.0	. . .	. . .	. . .	. . .	1 209.0	1 434.2	2 496.0	. . .	. . .	-127.6	11 622.4
1998														
1st quarter ...	11 832.5	7 633.9	1 994.4	. . .	. . .	. . .	. . .	1 214.6	1 488.0	2 487.1	. . .	. . .	-165.7	11 705.7
2nd quarter ...	11 942.0	7 768.3	1 982.1	. . .	. . .	. . .	. . .	1 201.2	1 522.1	2 530.6	. . .	. . .	-74.5	11 902.4
3rd quarter ...	12 091.6	7 869.6	2 033.0	. . .	. . .	. . .	. . .	1 195.6	1 542.7	2 549.7	. . .	. . .	-84.9	12 036.4
4th quarter ...	12 287.0	7 983.3	2 095.8	. . .	. . .	. . .	. . .	1 239.9	1 592.7	2 564.7	. . .	. . .	-87.5	12 222.9
1999														
1st quarter ...	12 403.3	8 060.8	2 153.3	1 427.0	632.0	118.0	. . .	1 238.3	1 633.9	2 582.2	. . .	. . .	-115.4	12 314.3
2nd quarter ...	12 498.7	8 178.3	2 154.7	1 464.9	641.9	50.0	. . .	1 250.1	1 680.0	2 591.9	. . .	. . .	-46.2	12 463.7
3rd quarter ...	12 662.4	8 270.6	2 213.8	1 506.9	646.7	62.4	. . .	1 286.5	1 742.0	2 623.3	. . .	. . .	-52.2	12 616.4
4th quarter ...	12 877.6	8 391.8	2 273.1	1 513.6	651.4	119.2	. . .	1 318.2	1 784.0	2 664.2	. . .	. . .	-105.0	12 777.9
2000														
1st quarter ...	12 924.2	8 520.7	2 257.1	1 569.0	654.6	21.4	. . .	1 337.2	1 853.9	2 645.4	. . .	. . .	-3.7	12 912.0
2nd quarter ...	13 160.8	8 603.0	2 390.7	1 618.0	651.3	123.6	. . .	1 375.2	1 911.4	2 671.3	. . .	. . .	-91.5	13 055.1
3rd quarter ...	13 178.4	8 687.5	2 367.3	1 635.6	641.1	87.3	. . .	1 408.8	1 978.1	2 659.7	. . .	. . .	-54.0	13 109.9
4th quarter ...	13 260.5	8 762.2	2 371.8	1 641.8	641.8	82.3	. . .	1 396.8	1 977.8	2 675.7	. . .	. . .	-50.6	13 194.8
2001														
1st quarter ...	13 222.7	8 797.3	2 264.2	1 627.1	644.6	-45.6	. . .	1 376.2	1 946.1	2 716.8	. . .	. . .	60.0	13 261.1
2nd quarter ...	13 300.0	8 818.1	2 267.4	1 593.4	654.6	-2.7	. . .	1 330.8	1 886.6	2 764.1	. . .	. . .	8.9	13 305.3
3rd quarter ...	13 244.8	8 848.3	2 218.4	1 574.6	658.6	-46.7	. . .	1 266.0	1 849.0	2 761.4	. . .	. . .	46.5	13 285.9
4th quarter ...	13 280.9	8 980.6	2 108.5	1 529.3	654.4	-117.9	. . .	1 225.9	1 823.1	2 807.0	. . .	. . .	99.8	13 386.0
2002														
1st quarter ...	13 397.0	9 008.1	2 170.7	1 492.6	675.9	-7.8	-623.9	1 252.8	1 876.7	2 855.5	946.9	1 921.6	-102.7	13 413.4
2nd quarter ...	13 478.2	9 054.3	2 206.4	1 475.8	692.6	33.1	-652.1	1 287.8	1 939.9	2 877.5	965.3	1 924.2	-91.6	13 453.2
3rd quarter ...	13 538.1	9 119.9	2 203.1	1 471.0	694.7	28.7	-669.8	1 294.5	1 964.3	2 892.8	974.8	1 929.8	-67.9	13 513.1
4th quarter ...	13 559.0	9 172.4	2 201.7	1 451.2	707.2	43.4	-723.4	1 273.4	1 996.8	2 915.1	991.3	1 934.7	-82.1	13 525.7

Note: Chained (2012) dollar series are calculated as the product of the chain-type quantity index and the 2012 current-dollar value of the corresponding series, divided by 100. Because the formula for the chain-type quantity indexes uses weights from more than one period, the corresponding chained-dollar estimates are usually not additive. The residual column is the difference between the total and the sum of the most detailed components shown in the Bureau of Economic Analysis (BEA) published data.

. . . = Not available.

Table 1-2B. Real Gross Domestic Product: Historical Data—*Continued*

(Billions of chained [2012] dollars, quarterly data are at seasonally adjusted annual rates.)   **NIPA Tables 1.1.6, 1.2.6**

Year and quarter	Gross domestic product	Personal consump-tion expen-ditures	Gross private domestic investment				Exports and imports of goods and services			Government consumption expenditures and gross investment			Residual	Adden-dum: Final sales of domestic product
			Total	Fixed investment		Change in private inventories	Net exports	Exports	Imports	Total	Federal	State and local		
				Nonresi-dential	Residential									
2003														
1st quarter	13 634.3	9 215.5	2 218.7	1 458.4	720.2	39.9	-712.0	1 274.7	1 986.6	2 918.5	1 002.2	1 926.2	-79.9	13 604.0
2nd quarter ...	13 751.5	9 319.0	2 234.0	1 498.7	731.7	-9.2	-741.9	1 272.8	2 014.7	2 946.1	1 036.7	1 916.7	-49.1	13 761.4
3rd quarter ...	13 985.1	9 455.7	2 315.4	1 530.5	773.6	2.8	-739.9	1 308.2	2 048.0	2 953.1	1 036.4	1 924.3	-33.2	13 985.7
4th quarter ...	14 145.6	9 519.8	2 393.4	1 549.8	796.6	46.2	-746.4	1 364.4	2 110.8	2 971.0	1 055.7	1 921.6	-27.7	14 107.8
2004														
1st quarter	14 221.1	9 604.5	2 398.2	1 534.4	805.9	58.7	-776.9	1 401.6	2 178.5	2 984.4	1 067.2	1 922.7	-21.3	14 171.3
2nd quarter ...	14 329.5	9 664.3	2 493.9	1 572.8	831.4	97.7	-844.0	1 423.2	2 267.1	2 992.4	1 073.6	1 924.0	-14.1	14 245.7
3rd quarter ...	14 465.0	9 771.1	2 532.6	1 617.8	837.4	83.1	-863.2	1 434.0	2 297.2	2 997.9	1 085.5	1 916.6	0.1	14 394.7
4th quarter ...	14 609.9	9 877.4	2 585.5	1 651.0	848.9	91.1	-881.6	1 465.9	2 347.5	2 996.2	1 083.6	1 917.0	11.9	14 531.1
2005														
1st quarter	14 771.6	9 935.0	2 658.2	1 675.0	868.9	124.9	-869.8	1 500.2	2 370.1	3 011.2	1 095.7	1 919.3	11.2	14 661.9
2nd quarter ...	14 839.8	10 047.8	2 622.8	1 702.0	885.0	24.7	-874.7	1 527.0	2 401.7	3 009.5	1 094.5	1 918.8	38.2	14 815.2
3rd quarter ...	14 972.1	10 145.3	2 657.5	1 737.5	894.7	7.9	-886.8	1 530.7	2 417.4	3 019.4	1 102.9	1 920.0	54.0	14 959.2
4th quarter ...	15 066.6	10 175.4	2 743.8	1 750.9	892.9	97.4	-919.8	1 574.9	2 494.7	3 021.8	1 103.3	1 922.1	36.8	14 972.8
2006														
1st quarter	15 267.0	10 288.9	2 784.6	1 809.7	884.6	102.7	-913.0	1 634.7	2 547.7	3 060.1	1 131.9	1 930.8	32.1	15 183.2
2nd quarter ...	15 302.7	10 341.0	2 766.6	1 841.3	840.3	92.7	-908.8	1 666.7	2 575.5	3 059.2	1 124.1	1 938.2	32.0	15 222.6
3rd quarter ...	15 326.4	10 403.8	2 756.0	1 872.9	793.8	97.3	-932.5	1 667.2	2 599.7	3 054.8	1 113.9	1 944.5	32.6	15 239.1
4th quarter ...	15 456.9	10 504.5	2 702.5	1 892.7	756.9	55.7	-866.0	1 737.1	2 603.1	3 079.9	1 130.2	1 952.9	30.5	15 407.3
2007														
1st quarter	15 493.3	10 563.3	2 683.9	1 926.8	722.7	27.5	-872.9	1 761.4	2 634.2	3 084.5	1 123.5	1 964.6	39.8	15 461.4
2nd quarter ...	15 582.1	10 582.8	2 713.5	1 969.6	694.3	56.4	-862.3	1 788.4	2 650.8	3 112.5	1 141.9	1 973.8	27.6	15 531.5
3rd quarter ...	15 666.7	10 642.5	2 686.1	2 000.5	650.1	43.4	-819.4	1 842.1	2 661.5	3 126.5	1 151.7	1 977.8	21.2	15 629.2
4th quarter ...	15 762.0	10 672.8	2 653.1	2 031.3	596.1	35.0	-740.0	1 897.4	2 637.4	3 150.8	1 170.8	1 982.5	14.2	15 732.7
2008														
1st quarter	15 671.4	10 644.4	2 583.3	2 039.4	548.7	1.6	-732.1	1 913.5	2 645.6	3 157.7	1 188.4	1 971.4	3.9	15 673.6
2nd quarter ...	15 752.3	10 661.7	2 536.4	2 043.5	523.8	-25.4	-647.1	1 974.7	2 621.8	3 184.5	1 213.6	1 972.8	1.0	15 782.2
3rd quarter ...	15 667.0	10 581.9	2 485.5	2 005.4	497.2	-19.3	-626.8	1 961.2	2 588.0	3 209.5	1 228.8	1 982.5	5.0	15 682.0
4th quarter ...	15 328.0	10 483.4	2 246.4	1 888.6	448.8	-87.6	-640.4	1 852.4	2 492.8	3 230.5	1 244.3	1 987.8	-19.3	15 421.2
2009														
1st quarter	15 155.9	10 459.7	1 986.8	1 746.3	405.3	-179.0	-543.8	1 702.7	2 246.5	3 266.2	1 260.1	2 007.7	-23.7	15 328.3
2nd quarter ...	15 134.1	10 417.3	1 872.3	1 693.2	380.4	-228.3	-445.2	1 707.9	2 153.1	3 313.2	1 289.7	2 024.9	-20.9	15 343.9
3rd quarter ...	15 189.2	10 489.2	1 868.0	1 683.3	398.1	-245.1	-473.0	1 770.0	2 242.9	3 321.9	1 301.3	2 021.8	-1.0	15 410.3
4th quarter ...	15 356.1	10 473.6	2 040.7	1 694.5	397.2	-56.8	-477.3	1 874.5	2 351.8	3 328.0	1 321.0	2 007.9	-16.3	15 409.4
2010														
1st quarter	15 415.1	10 525.4	2 087.2	1 706.4	384.4	-5.0	-505.9	1 902.6	2 408.6	3 315.2	1 336.1	1 979.5	-13.8	15 420.0
2nd quarter ...	15 557.3	10 609.1	2 196.7	1 762.3	404.4	33.3	-572.6	1 947.6	2 520.2	3 326.5	1 353.9	1 972.8	-11.1	15 528.1
3rd quarter ...	15 672.0	10 683.3	2 294.7	1 810.1	368.6	128.8	-611.0	2 001.6	2 612.6	3 303.9	1 348.1	1 955.8	-14.4	15 555.1
4th quarter ...	15 750.6	10 754.0	2 287.4	1 845.2	374.7	72.1	-574.1	2 059.6	2 633.7	3 283.3	1 346.2	1 937.0	-6.1	15 683.0
2011														
1st quarter	15 712.8	10 799.7	2 244.2	1 842.7	373.4	32.8	-573.7	2 077.1	2 650.8	3 243.2	1 327.7	1 915.5	-5.0	15 685.0
2nd quarter ...	15 825.1	10 823.7	2 336.1	1 890.4	377.6	71.3	-555.8	2 110.8	2 666.6	3 221.4	1 322.9	1 898.4	-4.6	15 757.0
3rd quarter ...	15 820.7	10 866.0	2 343.8	1 978.9	384.4	-17.7	-563.8	2 133.2	2 697.0	3 175.5	1 294.4	1 881.1	-4.2	15 840.6
4th quarter ...	16 004.1	10 885.9	2 524.4	2 029.4	394.4	100.4	-579.0	2 155.2	2 734.1	3 173.2	1 299.4	1 873.8	-0.9	15 903.4
2012														
1st quarter	16 129.4	10 973.3	2 577.2	2 081.3	418.3	72.8	-580.5	2 168.5	2 749.0	3 159.6	1 299.4	1 860.1	4.6	16 051.7
2nd quarter ...	16 198.8	10 989.6	2 636.5	2 128.0	421.9	87.1	-570.4	2 192.2	2 762.5	3 143.0	1 289.1	1 854.0	-0.3	16 112.2
3rd quarter ...	16 220.7	11 007.5	2 648.5	2 120.9	432.7	98.0	-573.7	2 203.6	2 777.3	3 138.2	1 291.7	1 846.5	-2.8	16 125.8
4th quarter ...	16 239.1	11 056.9	2 624.9	2 144.0	455.2	27.0	-549.8	2 200.8	2 750.6	3 107.2	1 265.9	1 841.4	-1.5	16 213.4
2013														
1st quarter	16 383.0	11 114.2	2 722.8	2 171.6	471.8	81.2	-532.7	2 225.9	2 758.6	3 079.4	1 236.9	1 842.3	-3.2	16 304.1
2nd quarter ...	16 403.2	11 122.2	2 753.2	2 177.5	486.8	85.8	-545.8	2 252.7	2 798.5	3 074.0	1 226.8	1 846.8	2.1	16 315.2
3rd quarter ...	16 531.7	11 167.4	2 859.8	2 214.7	495.5	147.6	-551.9	2 266.8	2 818.6	3 057.4	1 209.1	1 847.8	0.6	16 383.6
4th quarter ...	16 663.6	11 263.6	2 870.1	2 260.0	487.7	120.0	-500.9	2 333.2	2 834.1	3 033.4	1 188.2	1 844.4	-1.1	16 541.7
2014														
1st quarter	16 616.5	11 308.0	2 838.3	2 291.7	484.3	58.3	-549.5	2 316.7	2 866.2	3 020.8	1 187.0	1 833.1	2.8	16 553.6
2nd quarter ...	16 841.5	11 431.8	2 958.1	2 353.3	499.8	98.6	-572.2	2 366.8	2 939.0	3 024.6	1 179.9	1 843.8	3.8	16 736.5
3rd quarter ...	17 047.1	11 554.8	3 018.6	2 400.8	507.1	106.3	-569.2	2 377.3	2 946.6	3 044.9	1 193.0	1 851.0	-1.3	16 936.3
4th quarter ...	17 143.0	11 695.0	3 021.9	2 415.5	525.2	82.0	-617.9	2 400.3	3 018.2	3 043.4	1 175.5	1 866.6	-2.7	17 062.8
2015														
1st quarter	17 305.8	11 798.3	3 127.3	2 412.6	535.5	177.3	-683.5	2 382.0	3 065.6	3 059.0	1 180.4	1 877.2	5.7	17 128.9
2nd quarter ...	17 422.8	11 892.3	3 136.7	2 423.6	548.9	156.5	-696.3	2 388.0	3 084.4	3 085.0	1 182.7	1 900.8	9.9	17 262.0
3rd quarter ...	17 486.0	11 991.2	3 130.4	2 432.4	563.7	125.2	-743.8	2 367.4	3 111.1	3 098.9	1 181.5	1 915.6	12.4	17 356.1
4th quarter ...	17 514.1	12 055.4	3 092.7	2 412.8	573.6	91.4	-754.5	2 363.5	3 118.0	3 110.0	1 190.5	1 918.0	17.1	17 413.3
2016														
1st quarter	17 613.3	12 148.1	3 073.9	2 406.0	593.0	54.0	-757.0	2 357.9	3 114.9	3 138.8	1 194.6	1 942.4	17.6	17 545.7
2nd quarter ...	17 668.2	12 225.8	3 047.0	2 420.2	590.5	8.5	-746.0	2 366.6	3 112.6	3 133.8	1 185.6	1 946.3	23.7	17 637.9
3rd quarter ...	17 764.4	12 304.5	3 048.1	2 448.4	587.4	-16.5	-745.2	2 406.6	3 151.8	3 148.4	1 190.8	1 955.7	23.6	17 756.3
4th quarter ...	17 876.2	12 380.0	3 130.1	2 457.4	597.6	52.3	-806.3	2 397.9	3 204.2	3 156.6	1 191.2	1 963.4	24.3	17 807.9
2017														
1st quarter	17 977.3	12 477.3	3 120.4	2 492.6	614.4	-18.9	-792.3	2 446.0	3 238.3	3 156.9	1 186.4	1 968.4	32.5	17 970.8
2nd quarter ...	18 054.1	12 533.1	3 149.1	2 507.3	612.7	0.6	-815.0	2 451.9	3 266.9	3 169.0	1 192.7	1 974.2	26.0	18 031.4
3rd quarter ...	18 185.6	12 604.5	3 207.5	2 520.3	610.1	56.1	-813.0	2 468.0	3 281.0	3 170.6	1 191.3	1 977.2	9.1	18 115.0
4th quarter ...	18 359.4	12 733.7	3 256.7	2 576.4	625.5	25.3	-847.0	2 536.2	3 383.2	3 192.8	1 206.0	1 984.9	15.7	18 311.4

Note: Chained (2012) dollar series are calculated as the product of the chain-type quantity index and the 2012 current-dollar value of the corresponding series, divided by 100. Because the formula for the chain-type quantity indexes uses weights from more than one period, the corresponding chained-dollar estimates are usually not additive. The residual column is the difference between the total and the sum of the most detailed components shown in the Bureau of Economic Analysis (BEA) published data.

Table 1-3A. U.S. Population and Per Capita Product and Income: Recent Data

(Dollars, except as noted; quarterly data are at seasonally adjusted annual rates.)   **NIPA Table 7.1**

Year and quarter	Population (mid-period, thousands)	Current dollars							Chained (2012) dollars					
		Gross domestic product	Personal income	Dispos-able personal income	Personal consumption expenditures				Gross domestic product	Dispos-able personal income	Personal consumption expenditures			
					Total	Durable goods	Nondur-able goods	Services			Total	Durable goods	Nondur-able goods	Services
1955	165 275	2 574	1 961	1 763	1 563	247	645	671	17 372	11 873	10 528	474	3 872	6 457
1956	168 221	2 671	2 068	1 851	1 612	239	666	707	17 432	12 223	10 646	448	3 938	6 649
1957	171 274	2 768	2 149	1 921	1 672	245	686	740	17 482	12 317	10 715	444	3 945	6 756
1958	174 141	2 763	2 176	1 955	1 698	227	701	770	17 066	12 241	10 630	404	3 932	6 860
1959	177 130	2 945	2 274	2 036	1 790	253	721	816	17 943	12 560	11 047	444	4 039	7 082
1960	180 760	3 001	2 335	2 080	1 832	252	727	853	18 035	12 629	11 122	444	4 024	7 212
1961	183 742	3 060	2 398	2 141	1 858	241	733	885	18 197	12 861	11 166	422	4 039	7 355
1962	186 590	3 237	2 513	2 236	1 943	265	748	930	19 017	13 281	11 539	463	4 099	7 585
1963	189 300	3 367	2 603	2 315	2 018	286	760	971	19 561	13 585	11 843	498	4 128	7 795
1964	191 927	3 566	2 752	2 481	2 140	310	796	1 034	20 405	14 350	12 376	537	4 262	8 146
1965	194 347	3 819	2 936	2 640	2 279	342	840	1 098	21 460	15 052	12 998	596	4 421	8 491
1966	196 599	4 137	3 155	2 818	2 441	365	905	1 171	22 614	15 672	13 577	638	4 614	8 809
1967	198 752	4 327	3 350	2 982	2 549	372	931	1 246	22 982	16 181	13 832	641	4 662	9 071
1968	200 745	4 686	3 641	3 208	2 774	423	995	1 356	23 873	16 748	14 483	705	4 811	9 458
1969	202 736	5 019	3 948	3 432	2 977	446	1 057	1 475	24 377	17 148	14 876	724	4 895	9 780
1970	205 089	5 233	4 218	3 715	3 153	439	1 116	1 599	24 142	17 734	15 051	697	4 947	10 048
1971	207 692	5 609	4 491	4 002	3 370	493	1 154	1 723	24 625	18 321	15 430	757	4 976	10 267
1972	209 924	6 093	4 880	4 291	3 659	555	1 226	1 878	25 644	18 999	16 201	842	5 123	10 740
1973	211 939	6 725	5 383	4 758	4 009	616	1 350	2 043	26 834	19 989	16 841	921	5 220	11 140
1974	213 898	7 224	5 852	5 146	4 349	609	1 502	2 238	26 445	19 583	16 547	854	5 046	11 250
1975	215 981	7 801	6 340	5 657	4 771	658	1 617	2 497	26 136	19 869	16 759	848	5 040	11 561
1976	218 086	8 590	6 890	6 098	5 262	773	1 732	2 757	27 278	20 306	17 523	945	5 229	11 941
1977	220 289	9 450	7 532	6 634	5 783	871	1 854	3 058	28 254	20 740	18 081	1 019	5 295	12 310
1978	222 629	10 563	8 371	7 340	6 388	958	2 022	3 408	29 505	21 455	18 674	1 061	5 428	12 739
1979	225 106	11 672	9 252	8 057	7 043	1 005	2 273	3 765	30 104	21 630	18 907	1 044	5 507	12 985
1980	227 726	12 547	10 204	8 888	7 688	994	2 518	4 176	29 681	21 542	18 631	950	5 434	13 045
1981	230 008	13 943	11 326	9 823	8 408	1 061	2 719	4 628	30 132	21 849	18 702	950	5 448	13 116
1982	232 218	14 399	12 021	10 494	8 919	1 090	2 783	5 047	29 308	22 113	18 795	939	5 452	13 261
1983	234 333	15 508	12 721	11 216	9 737	1 259	2 897	5 581	30 374	22 669	19 680	1 064	5 579	13 816
1984	236 394	17 080	13 929	12 330	10 543	1 447	3 052	6 043	32 289	24 016	20 535	1 205	5 759	14 221
1985	238 506	18 192	14 779	13 027	11 374	1 595	3 175	6 605	33 337	24 518	21 407	1 314	5 880	14 816
1986	240 683	19 028	15 510	13 691	11 992	1 751	3 217	7 024	34 179	25 219	22 089	1 427	6 035	15 136
1987	242 843	19 993	16 313	14 297	12 668	1 820	3 353	7 494	35 047	25 548	22 636	1 442	6 084	15 680
1988	245 061	21 368	17 479	15 414	13 589	1 939	3 519	8 131	36 181	26 508	23 368	1 511	6 187	16 236
1989	247 387	22 805	18 698	16 403	14 458	1 998	3 757	8 703	37 157	27 027	23 823	1 530	6 294	16 595
1990	250 181	23 835	19 641	17 264	15 225	1 987	3 974	9 264	37 435	27 250	24 031	1 506	6 297	16 891
1991	253 530	24 290	20 056	17 734	15 554	1 882	4 024	9 648	36 900	27 086	23 757	1 406	6 195	16 928
1992	256 922	25 379	21 099	18 714	16 338	1 978	4 107	10 253	37 696	27 841	24 306	1 467	6 232	17 373
1993	260 282	26 350	21 738	19 245	17 104	2 119	4 191	10 795	38 234	27 935	24 828	1 557	6 308	17 679
1994	263 455	27 660	22 574	19 943	17 919	2 305	4 325	11 290	39 295	28 356	25 479	1 661	6 475	18 001
1995	266 588	28 658	23 600	20 792	18 615	2 385	4 426	11 805	39 875	28 954	25 923	1 706	6 558	18 310
1996	269 714	29 932	24 762	21 658	19 445	2 507	4 603	12 335	40 900	29 528	26 511	1 812	6 669	18 622
1997	272 958	31 424	25 984	22 550	20 284	2 621	4 730	12 933	42 211	30 246	27 183	1 937	6 782	18 990
1998	276 154	32 818	27 545	23 806	21 283	2 822	4 813	13 647	43 593	31 651	28 295	2 145	6 953	19 623
1999	279 328	34 478	28 647	24 666	22 479	3 063	5 125	14 292	45 146	32 312	29 447	2 393	7 224	20 137
2000	282 398	36 305	30 640	26 262	23 945	3 232	5 455	15 259	46 498	33 568	30 607	2 570	7 373	20 920
2001	285 225	37 100	31 574	27 230	24 772	3 301	5 554	15 917	46 497	34 149	31 067	2 678	7 427	21 175
2002	287 955	37 980	31 807	28 153	25 499	3 422	5 603	16 474	46 858	34 847	31 563	2 848	7 493	21 361
2003	290 626	39 426	32 645	29 192	26 574	3 502	5 866	17 206	47 756	35 446	32 267	3 026	7 685	21 641
2004	293 262	41 648	34 219	30 643	28 004	3 685	6 211	18 109	49 125	36 302	33 176	3 247	7 865	22 094
2005	295 993	44 044	35 806	31 710	29 552	3 813	6 603	19 136	50 381	36 526	34 041	3 395	8 052	22 600
2006	298 818	46 231	38 089	33 548	30 990	3 876	6 965	20 148	51 330	37 621	34 752	3 512	8 238	22 996
2007	301 696	47 902	39 801	34 854	32 173	3 938	7 222	21 013	51 794	38 118	35 186	3 645	8 298	23 214
2008	304 543	48 311	40 855	35 905	32 758	3 608	7 436	21 715	51 240	38 124	34 783	3 403	8 091	23 291
2009	307 240	47 028	39 250	35 499	32 034	3 294	7 056	21 684	49 501	37 727	34 045	3 167	7 887	23 012
2010	309 774	48 397	40 518	36 523	32 881	3 386	7 324	22 171	50 355	38 162	34 357	3 316	7 945	23 105
2011	312 010	49 814	42 713	38 054	34 105	3 505	7 771	22 829	50 770	38 778	34 755	3 460	7 958	23 339
2012	314 212	51 548	44 588	39 784	35 030	3 642	7 936	23 453	51 548	39 784	35 030	3 642	7 936	23 453
2013	316 357	53 057	44 826	39 527	35 774	3 760	8 031	23 983	52 142	39 002	35 298	3 838	8 024	23 440
2014	318 631	55 008	47 050	41 450	37 105	3 898	8 225	24 981	53 077	40 309	36 084	4 085	8 176	23 836
2015	320 918	56 832	48 998	42 953	38 320	4 075	8 150	26 095	54 320	41 684	37 188	4 364	8 394	24 458
2016	323 186	58 001	50 004	43 946	39 513	4 178	8 194	27 141	54 862	42 207	37 949	4 585	8 546	24 864
2017	325 220	60 091	52 114	45 821	41 019	4 338	8 491	28 190	55 790	43 234	38 703	4 872	8 714	25 200
2018	326 949	63 043	54 601	48 223	42 800	4 531	8 840	29 428	57 158	44 553	39 542	5 177	8 901	25 592
2019	328 527	65 240	56 469	49 763	44 272	4 670	9 064	30 537	58 113	45 301	40 302	5 402	9 136	25 935
2017														
1st quarter	324 496	59 284	51 260	45 093	40 534	4 263	8 397	27 874	55 401	42 776	38 451	4 736	8 622	25 150
2nd quarter	324 948	59 638	51 788	45 616	40 749	4 290	8 421	28 038	55 560	43 176	38 570	4 805	8 698	25 140
3rd quarter	325 475	60 273	52 344	46 038	41 081	4 341	8 493	28 247	55 874	43 399	38 727	4 899	8 722	25 193
4th quarter	325 963	61 166	53 060	46 532	41 710	4 456	8 654	28 600	56 324	43 581	39 065	5 049	8 815	25 315
2018														
1st quarter	326 325	62 031	53 776	47 385	42 153	4 479	8 745	28 928	56 785	44 087	39 219	5 093	8 829	25 414
2nd quarter	326 703	62 909	54 331	48 013	42 668	4 538	8 834	29 296	57 099	44 424	39 480	5 175	8 879	25 554
3rd quarter	327 167	63 401	54 946	48 526	43 055	4 550	8 883	29 622	57 317	44 725	39 683	5 208	8 926	25 680
4th quarter	327 602	63 827	55 348	48 966	43 319	4 558	8 898	29 863	57 429	44 970	39 784	5 233	8 971	25 717
2019														
1st quarter	327 923	64 391	56 009	49 390	43 536	4 557	8 907	30 072	57 789	45 296	39 927	5 240	9 035	25 791
2nd quarter	328 270	64 977	56 298	49 528	44 163	4 679	9 083	30 401	57 942	45 139	40 250	5 393	9 142	25 888
3rd quarter	328 730	65 526	56 574	49 890	44 551	4 724	9 128	30 699	58 229	45 312	40 463	5 469	9 199	25 983
4th quarter	329 186	66 064	56 991	50 244	44 835	4 721	9 138	30 976	58 490	45 459	40 566	5 503	9 169	26 079

Table 1-3B. U.S. Population and Per Capita Product and Income: Historical Data

(Dollars, except as noted; quarterly data are at seasonally adjusted annual rates.)   NIPA Table 7.1

Year and quarter	Population (mid-period, thousands)	Current dollars							Chained (2012) dollars					
		Gross domestic product	Personal income	Disposable personal income	Personal consumption expenditures				Gross domestic product	Disposable personal income	Personal consumption expenditures			
					Total	Durable goods	Nondurable goods	Services			Total	Durable goods	Nondurable goods	Services
1929	121 878	858	700	686	635	81	278	276	9 103	7 361	6 817	251	2 769	4 063
1930	123 188	748	621	609	569	62	248	260	8 240	6 822	6 383	206	2 597	3 936
1931	124 149	623	529	521	489	48	208	233	7 652	6 540	6 136	177	2 549	3 806
1932	124 949	476	403	397	390	32	161	197	6 623	5 648	5 551	133	2 310	3 553
1933	125 690	455	376	369	366	30	159	177	6 502	5 452	5 396	129	2 288	3 401
1934	126 485	528	428	421	407	36	189	182	7 160	5 938	5 745	147	2 459	3 546
1935	127 362	583	478	469	439	43	205	191	7 744	6 468	6 052	177	2 582	3 644
1936	128 181	662	540	530	485	53	228	205	8 685	7 237	6 625	214	2 858	3 844
1937	128 961	721	579	564	518	57	240	221	9 075	7 437	6 829	224	2 911	4 000
1938	129 969	672	532	517	495	47	230	217	8 706	6 974	6 668	184	2 930	3 937
1939	131 028	713	562	551	513	55	235	224	9 329	7 496	6 984	217	3 039	4 057
1940	132 122	779	601	588	540	63	245	232	10 068	7 942	7 284	246	3 143	4 179
1941	133 402	969	734	716	608	77	279	252	11 737	9 106	7 724	281	3 299	4 384
1942	134 860	1 231	940	903	660	57	322	282	13 803	10 212	7 460	176	3 272	4 575
1943	136 739	1 485	1 142	1 020	731	55	357	319	15 931	10 556	7 562	156	3 248	4 873
1944	138 397	1 622	1 226	1 099	785	56	381	348	16 992	10 754	7 684	142	3 276	5 083
1945	139 928	1 629	1 256	1 117	857	65	418	374	16 642	10 516	8 070	156	3 441	5 297
1946	141 389	1 609	1 291	1 169	1 020	121	489	410	14 558	10 289	8 974	279	3 713	5 471
1947	144 126	1 732	1 349	1 212	1 123	151	538	433	14 118	9 684	8 972	322	3 621	5 389
1948	146 631	1 872	1 456	1 325	1 193	167	566	460	14 448	10 019	9 018	337	3 586	5 437
1949	149 188	1 826	1 415	1 303	1 195	178	546	471	14 120	9 927	9 110	358	3 577	5 472
1947														
1st quarter	143 143	1 699	1 325	1 191	1 091	145	523	423	14 203	9 727	8 907	311	3 603	5 380
2nd quarter	143 790	1 711	1 319	1 184	1 113	148	535	430	14 101	9 590	9 014	317	3 648	5 434
3rd quarter	144 449	1 728	1 369	1 233	1 132	151	544	437	14 008	9 801	9 002	319	3 648	5 397
4th quarter	145 122	1 790	1 383	1 240	1 155	162	551	443	14 161	9 618	8 964	339	3 584	5 344
1948														
1st quarter	145 709	1 824	1 417	1 272	1 169	162	560	448	14 316	9 758	8 972	336	3 581	5 380
2nd quarter	146 289	1 863	1 449	1 319	1 190	164	569	458	14 495	10 021	9 040	337	3 602	5 442
3rd quarter	146 921	1 900	1 480	1 356	1 205	172	568	465	14 515	10 140	9 015	341	3 565	5 448
4th quarter	147 607	1 899	1 478	1 353	1 205	169	567	469	14 464	10 150	9 044	334	3 595	5 477
1949														
1st quarter	148 254	1 855	1 430	1 310	1 193	165	558	470	14 202	9 906	9 020	327	3 593	5 488
2nd quarter	148 847	1 823	1 417	1 303	1 199	177	550	471	14 098	9 916	9 121	354	3 588	5 493
3rd quarter	149 485	1 826	1 408	1 299	1 189	183	537	470	14 182	9 938	9 103	369	3 542	5 458
4th quarter	150 167	1 802	1 404	1 299	1 200	189	539	472	13 999	9 949	9 194	382	3 583	5 450
1950														
1st quarter	150 786	1 862	1 501	1 391	1 213	195	540	478	14 490	10 672	9 309	395	3 614	5 495
2nd quarter	151 319	1 919	1 505	1 388	1 235	197	547	490	14 879	10 603	9 429	397	3 644	5 614
3rd quarter	151 973	2 028	1 553	1 429	1 319	246	570	503	15 388	10 690	9 871	489	3 691	5 677
4th quarter	152 658	2 096	1 604	1 457	1 297	218	568	511	15 612	10 707	9 530	426	3 592	5 687
1951														
1st quarter	153 291	2 192	1 658	1 499	1 365	232	600	532	15 759	10 673	9 720	438	3 643	5 819
2nd quarter	153 902	2 236	1 710	1 538	1 332	200	594	538	15 968	10 868	9 408	374	3 576	5 835
3rd quarter	154 610	2 273	1 730	1 551	1 343	195	603	545	16 223	10 941	9 474	365	3 634	5 881
4th quarter	155 344	2 293	1 752	1 561	1 362	196	614	552	16 181	10 870	9 484	362	3 659	5 868
1952														
1st quarter	155 976	2 307	1 758	1 561	1 365	197	608	561	16 288	10 821	9 468	363	3 614	5 918
2nd quarter	156 587	2 306	1 777	1 574	1 386	198	617	571	16 259	10 916	9 616	369	3 684	5 986
3rd quarter	157 267	2 338	1 814	1 609	1 396	187	626	583	16 306	11 088	9 620	345	3 733	6 038
4th quarter	157 986	2 410	1 848	1 639	1 441	213	635	593	16 765	11 274	9 914	403	3 770	6 096
1953														
1st quarter	158 574	2 447	1 869	1 658	1 458	222	634	602	17 013	11 365	9 993	415	3 779	6 133
2nd quarter	159 160	2 461	1 886	1 676	1 464	220	632	612	17 082	11 471	10 016	411	3 783	6 183
3rd quarter	159 889	2 447	1 877	1 670	1 461	216	624	621	16 908	11 364	9 947	407	3 731	6 191
4th quarter	160 639	2 403	1 868	1 663	1 451	210	622	619	16 574	11 273	9 834	393	3 719	6 116
1954														
1st quarter	161 328	2 389	1 860	1 673	1 458	204	628	626	16 425	11 282	9 828	379	3 739	6 147
2nd quarter	161 983	2 384	1 851	1 666	1 469	208	626	635	16 376	11 247	9 916	396	3 714	6 227
3rd quarter	162 731	2 403	1 854	1 670	1 477	204	628	644	16 485	11 314	10 004	396	3 744	6 306
4th quarter	163 524	2 445	1 878	1 692	1 499	214	633	652	16 726	11 475	10 164	418	3 788	6 351
1955														
1st quarter	164 204	2 516	1 909	1 718	1 531	234	636	661	17 132	11 613	10 349	454	3 812	6 399
2nd quarter	164 864	2 557	1 947	1 751	1 556	247	644	665	17 341	11 822	10 505	477	3 866	6 419
3rd quarter	165 612	2 598	1 983	1 782	1 574	256	646	672	17 496	11 986	10 587	490	3 875	6 456
4th quarter	166 419	2 626	2 004	1 799	1 590	249	655	686	17 516	12 067	10 668	475	3 934	6 553
1956														
1st quarter	167 109	2 632	2 029	1 817	1 593	237	661	694	17 376	12 139	10 641	451	3 962	6 588
2nd quarter	167 788	2 658	2 055	1 839	1 602	237	663	702	17 449	12 204	10 633	450	3 936	6 625
3rd quarter	168 573	2 677	2 077	1 858	1 614	235	667	712	17 352	12 210	10 608	440	3 921	6 663
4th quarter	169 416	2 718	2 112	1 888	1 638	245	671	722	17 550	12 337	10 700	452	3 932	6 717
1957														
1st quarter	170 172	2 761	2 129	1 903	1 656	250	678	728	17 583	12 321	10 727	458	3 939	6 713
2nd quarter	170 869	2 762	2 148	1 920	1 663	247	682	735	17 473	12 357	10 702	446	3 937	6 743
3rd quarter	171 638	2 794	2 165	1 937	1 682	244	695	743	17 566	12 364	10 738	441	3 974	6 762
4th quarter	172 417	2 754	2 150	1 925	1 684	241	690	753	17 305	12 226	10 694	433	3 930	6 804

Table 1-3B. U.S. Population and Per Capita Product and Income: Historical Data—*Continued*

(Dollars, except as noted; quarterly data are at seasonally adjusted annual rates.)   NIPA Table 7.1

| Year and quarter | Population (mid-period, thousands) | Current dollars | | | | | | | Chained (2012) dollars | | | | | |
| | | Gross domestic product | Personal income | Disposable personal income | Personal consumption expenditures | | | | Gross domestic product | Disposable personal income | Personal consumption expenditures | | | |
					Total	Durable goods	Nondurable goods	Services			Total	Durable goods	Nondurable goods	Services
1958														
1st quarter	173 064	2 702	2 145	1 925	1 675	228	693	755	16 793	12 073	10 506	403	3 880	6 769
2nd quarter	173 729	2 717	2 151	1 934	1 685	222	698	766	16 839	12 110	10 553	395	3 901	6 842
3rd quarter	174 483	2 784	2 193	1 970	1 707	226	705	776	17 154	12 323	10 681	402	3 955	6 907
4th quarter	175 288	2 850	2 215	1 990	1 722	232	708	782	17 475	12 455	10 777	415	3 991	6 922
1959														
1st quarter	176 045	2 899	2 238	2 006	1 758	248	716	794	17 734	12 473	10 927	435	4 025	6 978
2nd quarter	176 727	2 957	2 277	2 039	1 785	257	719	809	18 064	12 625	11 053	450	4 043	7 055
3rd quarter	177 481	2 958	2 280	2 040	1 807	261	722	824	18 000	12 554	11 121	457	4 041	7 120
4th quarter	178 268	2 965	2 302	2 057	1 811	247	726	837	17 972	12 588	11 085	433	4 048	7 174
1960														
1st quarter	179 319	3 026	2 324	2 071	1 820	253	723	844	18 268	12 659	11 125	444	4 035	7 197
2nd quarter	180 401	2 999	2 338	2 083	1 842	257	731	853	18 060	12 664	11 198	452	4 054	7 233
3rd quarter	181 301	3 009	2 341	2 084	1 832	253	726	853	18 059	12 625	11 098	446	4 010	7 188
4th quarter	182 019	2 968	2 339	2 084	1 835	246	727	862	17 756	12 570	11 068	433	3 997	7 229
1961														
1st quarter	182 634	2 984	2 355	2 101	1 832	231	730	870	17 816	12 646	11 027	408	4 008	7 276
2nd quarter	183 337	3 030	2 380	2 124	1 852	236	732	883	18 049	12 789	11 148	415	4 046	7 353
3rd quarter	184 103	3 083	2 409	2 151	1 859	242	731	886	18 319	12 905	11 156	423	4 032	7 345
4th quarter	184 894	3 140	2 446	2 186	1 891	252	736	902	18 598	13 101	11 331	442	4 069	7 444
1962														
1st quarter	185 553	3 201	2 475	2 209	1 912	257	743	912	18 863	13 183	11 411	449	4 088	7 487
2nd quarter	186 203	3 224	2 507	2 233	1 936	263	745	927	18 967	13 279	11 511	460	4 090	7 572
3rd quarter	186 926	3 258	2 522	2 242	1 949	265	749	935	19 126	13 298	11 558	461	4 106	7 607
4th quarter	187 680	3 262	2 546	2 260	1 975	275	754	946	19 111	13 363	11 675	480	4 113	7 672
1963														
1st quarter	188 299	3 302	2 566	2 279	1 988	279	757	952	19 257	13 434	11 718	488	4 122	7 681
2nd quarter	188 906	3 334	2 584	2 296	2 003	285	756	962	19 410	13 514	11 791	497	4 122	7 738
3rd quarter	189 631	3 398	2 611	2 323	2 032	288	766	978	19 761	13 610	11 905	501	4 144	7 843
4th quarter	190 362	3 435	2 651	2 361	2 049	293	763	993	19 814	13 779	11 958	507	4 123	7 919
1964														
1st quarter	190 954	3 508	2 693	2 411	2 093	304	778	1 010	20 169	14 005	12 155	525	4 181	8 020
2nd quarter	191 560	3 543	2 733	2 474	2 127	310	791	1 026	20 324	14 337	12 331	537	4 247	8 107
3rd quarter	192 256	3 600	2 773	2 505	2 166	319	806	1 041	20 567	14 471	12 513	553	4 314	8 185
4th quarter	192 938	3 614	2 809	2 533	2 172	308	808	1 056	20 558	14 584	12 504	534	4 307	8 269
1965														
1st quarter	193 467	3 710	2 858	2 564	2 221	336	815	1 070	20 997	14 716	12 748	582	4 336	8 341
2nd quarter	193 994	3 764	2 900	2 599	2 251	334	829	1 088	21 205	14 841	12 854	581	4 366	8 444
3rd quarter	194 647	3 850	2 964	2 671	2 290	343	843	1 105	21 604	15 196	13 031	599	4 417	8 531
4th quarter	195 279	3 953	3 022	2 724	2 354	354	873	1 127	22 030	15 452	13 354	622	4 566	8 647
1966														
1st quarter	195 763	4 065	3 080	2 765	2 402	369	889	1 143	22 510	15 566	13 518	650	4 587	8 715
2nd quarter	196 277	4 101	3 122	2 788	2 421	355	903	1 163	22 528	15 566	13 517	623	4 619	8 787
3rd quarter	196 877	4 163	3 180	2 835	2 460	367	913	1 179	22 650	15 708	13 629	641	4 638	8 832
4th quarter	197 481	4 220	3 239	2 882	2 482	368	913	1 200	22 766	15 844	13 644	639	4 611	8 901
1967														
1st quarter	197 967	4 264	3 286	2 926	2 497	359	921	1 217	22 911	16 043	13 689	625	4 651	8 961
2nd quarter	198 455	4 278	3 314	2 956	2 537	375	927	1 234	22 869	16 131	13 842	651	4 674	9 025
3rd quarter	199 012	4 348	3 375	3 004	2 566	375	933	1 257	23 020	16 239	13 874	644	4 657	9 123
4th quarter	199 572	4 417	3 423	3 043	2 597	379	941	1 276	23 129	16 311	13 920	643	4 665	9 174
1968														
1st quarter	199 995	4 547	3 513	3 120	2 682	404	968	1 310	23 551	16 548	14 223	682	4 746	9 299
2nd quarter	200 452	4 661	3 604	3 196	2 744	414	987	1 343	23 889	16 782	14 407	695	4 794	9 422
3rd quarter	200 997	4 731	3 688	3 231	2 817	436	1 009	1 372	24 009	16 790	14 637	724	4 857	9 516
4th quarter	201 538	4 803	3 759	3 283	2 853	436	1 017	1 400	24 039	16 875	14 664	718	4 845	9 595
1969														
1st quarter	201 955	4 919	3 824	3 316	2 907	446	1 034	1 427	24 365	16 882	14 796	730	4 889	9 658
2nd quarter	202 419	4 985	3 906	3 384	2 956	447	1 048	1 461	24 383	17 008	14 857	726	4 888	9 756
3rd quarter	202 986	5 074	3 998	3 485	2 998	446	1 064	1 488	24 475	17 306	14 888	722	4 893	9 807
4th quarter	203 584	5 099	4 061	3 542	3 048	446	1 079	1 523	24 284	17 387	14 963	717	4 911	9 900
1970														
1st quarter	204 086	5 151	4 124	3 612	3 095	439	1 100	1 556	24 189	17 526	15 019	705	4 941	9 982
2nd quarter	204 721	5 214	4 201	3 686	3 134	445	1 107	1 583	24 148	17 688	15 040	711	4 923	10 012
3rd quarter	205 419	5 287	4 258	3 768	3 181	448	1 118	1 615	24 288	17 909	15 120	711	4 941	10 091
4th quarter	206 130	5 281	4 287	3 795	3 203	424	1 138	1 641	23 945	17 805	15 027	661	4 982	10 108
1971														
1st quarter	206 763	5 490	4 371	3 895	3 285	474	1 141	1 670	24 520	18 104	15 269	730	4 987	10 147
2nd quarter	207 362	5 576	4 471	3 985	3 343	487	1 152	1 704	24 581	18 315	15 364	744	4 984	10 220
3rd quarter	208 000	5 662	4 525	4 033	3 392	497	1 156	1 739	24 707	18 354	15 439	762	4 958	10 285
4th quarter	208 642	5 705	4 598	4 092	3 459	514	1 168	1 777	24 689	18 511	15 647	792	4 977	10 413
1972														
1st quarter	209 142	5 884	4 717	4 144	3 533	529	1 181	1 823	25 083	18 551	15 816	808	4 981	10 563
2nd quarter	209 637	6 041	4 790	4 202	3 613	544	1 213	1 856	25 592	18 700	16 079	827	5 104	10 666
3rd quarter	210 181	6 140	4 901	4 310	3 691	559	1 237	1 894	25 767	19 014	16 284	846	5 161	10 782
4th quarter	210 737	6 306	5 111	4 507	3 799	586	1 272	1 941	26 129	19 724	16 623	888	5 244	10 946

Table 1-3B. U.S. Population and Per Capita Product and Income: Historical Data—*Continued*

(Dollars, except as noted; quarterly data are at seasonally adjusted annual rates.)   NIPA Table 7.1

Year and quarter	Population (mid-period, thousands)	Current dollars							Chained (2012) dollars					
		Gross domestic product	Personal income	Disposable personal income	Personal consumption expenditures				Gross domestic product	Disposable personal income	Personal consumption expenditures			
					Total	Durable goods	Nondurable goods	Services			Total	Durable goods	Nondurable goods	Services
1973														
1st quarter	211 192	6 522	5 185	4 587	3 906	624	1 304	1 978	26 718	19 831	16 890	943	5 264	11 062
2nd quarter	211 663	6 680	5 314	4 704	3 971	620	1 329	2 021	26 949	19 954	16 844	929	5 208	11 131
3rd quarter	212 191	6 757	5 432	4 800	4 048	616	1 366	2 065	26 741	19 999	16 862	919	5 220	11 177
4th quarter	212 708	6 940	5 598	4 940	4 108	603	1 399	2 106	26 929	20 164	16 771	895	5 186	11 191
1974														
1st quarter	213 144	6 996	5 662	4 992	4 184	594	1 449	2 141	26 643	19 788	16 586	872	5 106	11 147
2nd quarter	213 602	7 163	5 772	5 075	4 309	611	1 489	2 209	26 649	19 562	16 611	875	5 062	11 235
3rd quarter	214 147	7 285	5 926	5 203	4 433	637	1 530	2 266	26 330	19 530	16 639	878	5 060	11 266
4th quarter	214 700	7 451	6 048	5 314	4 467	592	1 541	2 334	26 160	19 454	16 354	792	4 957	11 352
1975														
1st quarter	215 135	7 512	6 128	5 394	4 579	613	1 563	2 404	25 789	19 385	16 458	808	4 956	11 435
2nd quarter	215 652	7 660	6 260	5 698	4 700	634	1 599	2 467	25 911	20 233	16 688	821	5 054	11 544
3rd quarter	216 289	7 905	6 410	5 704	4 842	679	1 646	2 517	26 277	19 880	16 876	870	5 088	11 564
4th quarter	216 848	8 125	6 561	5 830	4 963	707	1 659	2 597	26 562	19 984	17 013	893	5 063	11 700
1976														
1st quarter	217 314	8 377	6 701	5 953	5 107	751	1 691	2 665	27 101	20 182	17 316	936	5 154	11 811
2nd quarter	217 776	8 506	6 802	6 024	5 187	762	1 713	2 712	27 242	20 254	17 438	938	5 222	11 864
3rd quarter	218 338	8 641	6 954	6 148	5 307	778	1 743	2 786	27 321	20 359	17 577	946	5 249	11 977
4th quarter	218 917	8 836	7 102	6 268	5 447	801	1 780	2 866	27 446	20 434	17 758	959	5 290	12 109
1977														
1st quarter	219 427	9 063	7 241	6 380	5 597	838	1 807	2 952	27 706	20 433	17 925	993	5 274	12 229
2nd quarter	219 956	9 347	7 424	6 535	5 710	860	1 835	3 014	28 177	20 575	17 979	1 014	5 261	12 245
3rd quarter	220 573	9 604	7 606	6 706	5 834	879	1 858	3 096	28 605	20 803	18 098	1 026	5 272	12 341
4th quarter	221 201	9 784	7 854	6 912	5 989	907	1 915	3 167	28 525	21 140	18 318	1 044	5 370	12 423
1978														
1st quarter	221 719	9 935	8 020	7 063	6 107	894	1 944	3 269	28 549	21 256	18 378	1 014	5 381	12 587
2nd quarter	222 281	10 490	8 270	7 266	6 350	973	2 001	3 376	29 577	21 427	18 724	1 087	5 408	12 744
3rd quarter	222 933	10 743	8 490	7 430	6 469	972	2 044	3 453	29 787	21 530	18 747	1 069	5 434	12 798
4th quarter	223 583	11 078	8 704	7 598	6 626	994	2 099	3 532	30 100	21 609	18 844	1 075	5 488	12 828
1979														
1st quarter	224 152	11 272	8 947	7 815	6 768	996	2 162	3 610	30 077	21 815	18 893	1 059	5 499	12 929
2nd quarter	224 737	11 530	9 101	7 936	6 931	988	2 225	3 718	30 031	21 561	18 832	1 033	5 474	12 975
3rd quarter	225 418	11 834	9 364	8 145	7 151	1 024	2 317	3 809	30 163	21 594	18 957	1 058	5 521	12 984
4th quarter	226 117	12 046	9 593	8 331	7 319	1 013	2 385	3 921	30 145	21 565	18 946	1 028	5 533	13 052
1980														
1st quarter	226 754	12 303	9 843	8 587	7 507	1 023	2 469	4 014	30 154	21 579	18 866	1 009	5 515	13 037
2nd quarter	227 389	12 302	9 980	8 696	7 497	932	2 489	4 076	29 451	21 329	18 389	897	5 425	12 919
3rd quarter	228 070	12 525	10 297	8 972	7 733	988	2 528	4 218	29 328	21 505	18 535	934	5 396	13 029
4th quarter	228 689	13 055	10 690	9 296	8 010	1 033	2 585	4 392	29 794	21 742	18 735	960	5 402	13 192
1981														
1st quarter	229 155	13 634	10 948	9 504	8 229	1 074	2 680	4 476	30 316	21 667	18 760	986	5 439	13 090
2nd quarter	229 674	13 770	11 145	9 653	8 349	1 049	2 712	4 588	30 023	21 643	18 719	945	5 450	13 150
3rd quarter	230 301	14 158	11 546	9 997	8 502	1 094	2 732	4 677	30 300	22 050	18 754	971	5 448	13 116
4th quarter	230 903	14 209	11 661	10 134	8 551	1 026	2 753	4 772	29 892	22 016	18 577	899	5 454	13 109
1982														
1st quarter	231 395	14 150	11 789	10 265	8 704	1 065	2 765	4 874	29 365	22 023	18 674	926	5 453	13 150
2nd quarter	231 906	14 368	11 954	10 403	8 795	1 075	2 754	4 966	29 434	22 107	18 690	927	5 431	13 195
3rd quarter	232 498	14 479	12 097	10 591	8 971	1 084	2 795	5 092	29 247	22 156	18 767	932	5 441	13 263
4th quarter	233 074	14 599	12 244	10 714	9 205	1 134	2 818	5 254	29 186	22 171	19 049	972	5 483	13 435
1983														
1st quarter	233 546	14 872	12 385	10 882	9 354	1 151	2 813	5 389	29 511	22 335	19 198	980	5 490	13 586
2nd quarter	234 028	15 292	12 571	11 035	9 612	1 237	2 876	5 499	30 120	22 443	19 549	1 050	5 543	13 743
3rd quarter	234 603	15 725	12 793	11 320	9 889	1 289	2 934	5 665	30 647	22 724	19 850	1 087	5 613	13 911
4th quarter	235 153	16 137	13 134	11 622	10 089	1 358	2 962	5 769	31 213	23 176	20 120	1 138	5 668	14 022
1984														
1st quarter	235 605	16 587	13 493	11 960	10 264	1 407	2 998	5 858	31 762	23 594	20 248	1 179	5 672	14 057
2nd quarter	236 082	16 984	13 821	12 252	10 487	1 446	3 060	5 982	32 246	23 937	20 490	1 204	5 784	14 139
3rd quarter	236 657	17 258	14 102	12 479	10 621	1 445	3 062	6 114	32 478	24 194	20 592	1 201	5 774	14 282
4th quarter	237 232	17 487	14 297	12 628	10 799	1 492	3 089	6 218	32 665	24 333	20 808	1 237	5 805	14 404
1985														
1st quarter	237 673	17 798	14 538	12 720	11 091	1 548	3 124	6 419	32 920	24 224	21 122	1 278	5 829	14 638
2nd quarter	238 176	18 032	14 676	13 045	11 260	1 567	3 161	6 532	33 140	24 644	21 271	1 290	5 862	14 741
3rd quarter	238 789	18 371	14 838	13 072	11 534	1 660	3 185	6 689	33 560	24 503	21 619	1 369	5 895	14 901
4th quarter	239 387	18 564	15 062	13 270	11 610	1 603	3 231	6 777	33 725	24 701	21 612	1 319	5 933	14 983
1986														
1st quarter	239 861	18 794	15 291	13 513	11 772	1 633	3 249	6 891	33 972	24 976	21 758	1 343	5 997	15 015
2nd quarter	240 368	18 910	15 430	13 643	11 863	1 694	3 193	6 976	34 053	25 242	21 949	1 388	6 037	15 074
3rd quarter	240 962	19 122	15 592	13 768	12 106	1 850	3 200	7 057	34 294	25 341	22 282	1 502	6 035	15 162
4th quarter	241 539	19 283	15 728	13 839	12 225	1 826	3 225	7 173	34 396	25 318	22 365	1 475	6 069	15 292
1987														
1st quarter	242 009	19 512	15 933	14 071	12 328	1 729	3 295	7 304	34 585	25 502	22 344	1 385	6 065	15 511
2nd quarter	242 520	19 818	16 153	14 044	12 590	1 811	3 349	7 430	34 885	25 210	22 601	1 439	6 099	15 631
3rd quarter	243 120	20 091	16 406	14 394	12 822	1 894	3 376	7 552	35 101	25 598	22 802	1 494	6 089	15 732
4th quarter	243 721	20 548	16 758	14 678	12 928	1 846	3 392	7 690	35 615	25 879	22 794	1 450	6 081	15 847

Table 1-3B. U.S. Population and Per Capita Product and Income: Historical Data—*Continued*

(Dollars, except as noted; quarterly data are at seasonally adjusted annual rates.)   NIPA Table 7.1

Year and quarter	Population (mid-period, thousands)	Current dollars							Chained (2012) dollars					
		Gross domestic product	Personal income	Disposable personal income	Personal consumption expenditures				Gross domestic product	Disposable personal income	Personal consumption expenditures			
					Total	Durable goods	Nondurable goods	Services			Total	Durable goods	Nondurable goods	Services
1988														
1st quarter	244 208	20 775	17 037	14 981	13 234	1 926	3 433	7 875	35 728	26 207	23 151	1 516	6 135	16 031
2nd quarter	244 716	21 208	17 305	15 271	13 451	1 934	3 488	8 030	36 122	26 421	23 273	1 513	6 170	16 139
3rd quarter	245 354	21 531	17 633	15 568	13 702	1 917	3 549	8 236	36 239	26 607	23 419	1 490	6 197	16 339
4th quarter	245 966	21 952	17 937	15 834	13 963	1 977	3 603	8 384	36 631	26 792	23 626	1 525	6 246	16 435
1989														
1st quarter	246 460	22 362	18 410	16 166	14 161	1 974	3 662	8 526	36 929	27 042	23 688	1 515	6 262	16 514
2nd quarter	247 017	22 721	18 593	16 299	14 387	1 997	3 756	8 634	37 127	26 902	23 745	1 531	6 257	16 543
3rd quarter	247 698	22 993	18 767	16 459	14 572	2 041	3 780	8 750	37 299	27 006	23 909	1 562	6 302	16 605
4th quarter	248 374	23 139	19 018	16 684	14 710	1 980	3 830	8 900	37 271	27 161	23 948	1 510	6 356	16 717
1990														
1st quarter	248 936	23 591	19 358	17 018	15 016	2 070	3 913	9 032	37 593	27 307	24 095	1 570	6 334	16 765
2nd quarter	249 711	23 868	19 620	17 238	15 151	1 996	3 928	9 227	37 612	27 412	24 093	1 514	6 321	16 916
3rd quarter	250 595	24 003	19 787	17 390	15 350	1 970	4 003	9 378	37 505	27 306	24 103	1 493	6 309	16 991
4th quarter	251 482	23 877	19 797	17 408	15 380	1 912	4 051	9 417	37 032	26 977	23 836	1 447	6 224	16 891
1991														
1st quarter	252 258	23 925	19 791	17 489	15 356	1 870	4 020	9 465	36 746	26 962	23 673	1 403	6 207	16 826
2nd quarter	253 063	24 211	19 972	17 656	15 518	1 878	4 038	9 602	36 914	27 072	23 793	1 405	6 224	16 939
3rd quarter	253 965	24 436	20 109	17 785	15 645	1 907	4 034	9 704	36 969	27 086	23 827	1 422	6 212	16 952
4th quarter	254 835	24 583	20 351	18 001	15 697	1 874	4 005	9 818	36 971	27 218	23 734	1 394	6 138	16 995
1992														
1st quarter	255 585	24 896	20 728	18 424	16 043	1 941	4 060	10 042	37 304	27 685	24 107	1 444	6 211	17 223
2nd quarter	256 439	25 233	21 036	18 668	16 205	1 954	4 084	10 168	37 583	27 866	24 190	1 450	6 216	17 299
3rd quarter	257 386	25 513	21 199	18 805	16 423	1 990	4 122	10 311	37 815	27 892	24 359	1 475	6 230	17 418
4th quarter	258 277	25 867	21 430	18 957	16 677	2 025	4 161	10 490	38 077	27 923	24 565	1 499	6 271	17 551
1993														
1st quarter	259 039	25 979	21 467	19 085	16 791	2 038	4 166	10 587	38 029	27 945	24 586	1 509	6 259	17 566
2nd quarter	259 826	26 206	21 691	19 214	17 006	2 108	4 181	10 716	38 135	27 946	24 735	1 554	6 290	17 601
3rd quarter	260 714	26 397	21 777	19 249	17 211	2 135	4 191	10 886	38 186	27 875	24 925	1 566	6 330	17 745
4th quarter	261 547	26 816	22 014	19 432	17 407	2 194	4 226	10 987	38 582	27 980	25 063	1 598	6 354	17 803
1994														
1st quarter	262 250	27 133	22 147	19 578	17 621	2 245	4 259	11 117	38 852	28 089	25 282	1 632	6 422	17 901
2nd quarter	263 020	27 553	22 480	19 827	17 805	2 276	4 289	11 240	39 263	28 289	25 404	1 644	6 454	17 976
3rd quarter	263 870	27 783	22 657	20 021	18 012	2 309	4 357	11 346	39 366	28 362	25 516	1 656	6 488	18 037
4th quarter	264 678	28 167	23 007	20 342	18 236	2 387	4 394	11 455	39 695	28 681	25 713	1 709	6 537	18 089
1995														
1st quarter	265 388	28 344	23 291	20 561	18 322	2 341	4 397	11 584	39 729	28 849	25 708	1 670	6 543	18 165
2nd quarter	266 142	28 485	23 486	20 680	18 538	2 356	4 422	11 760	39 735	28 849	25 860	1 682	6 557	18 294
3rd quarter	267 000	28 776	23 705	20 888	18 722	2 407	4 433	11 882	39 944	29 020	26 010	1 724	6 559	18 368
4th quarter	267 820	29 022	23 913	21 038	18 877	2 435	4 449	11 992	40 092	29 100	26 111	1 748	6 572	18 414
1996														
1st quarter	268 487	29 307	24 288	21 302	19 109	2 458	4 511	12 141	40 292	29 302	26 287	1 763	6 599	18 550
2nd quarter	269 251	29 834	24 698	21 579	19 389	2 512	4 604	12 273	40 848	29 487	26 494	1 814	6 664	18 604
3rd quarter	270 128	30 102	24 899	21 776	19 526	2 515	4 614	12 396	41 080	29 630	26 568	1 821	6 690	18 644
4th quarter	270 991	30 480	25 158	21 971	19 753	2 545	4 680	12 527	41 375	29 692	26 694	1 850	6 723	18 689
1997														
1st quarter	271 709	30 778	25 540	22 220	19 996	2 597	4 715	12 685	41 532	29 896	26 904	1 895	6 745	18 805
2nd quarter	272 487	31 263	25 771	22 409	20 079	2 556	4 689	12 833	42 102	30 075	26 948	1 884	6 729	18 893
3rd quarter	273 391	31 687	26 099	22 657	20 407	2 644	4 745	13 018	42 487	30 328	27 317	1 961	6 814	19 055
4th quarter	274 246	31 964	26 520	22 991	20 652	2 688	4 769	13 195	42 719	30 679	27 557	2 005	6 840	19 204
1998														
1st quarter	274 950	32 248	27 040	23 417	20 809	2 683	4 752	13 373	43 035	31 245	27 765	2 016	6 872	19 376
2nd quarter	275 703	32 534	27 425	23 716	21 155	2 790	4 787	13 578	43 315	31 588	28 176	2 112	6 936	19 573
3rd quarter	276 564	32 980	27 726	23 954	21 430	2 839	4 826	13 765	43 721	31 807	28 455	2 165	6 965	19 745
4th quarter	277 400	33 504	27 984	24 134	21 731	2 975	4 885	13 871	44 293	31 962	28 779	2 286	7 039	19 795
1999														
1st quarter	278 103	33 863	28 230	24 354	21 943	2 948	4 980	14 015	44 600	32 169	28 985	2 286	7 151	19 907
2nd quarter	278 864	34 153	28 402	24 475	22 324	3 065	5 086	14 173	44 820	32 154	29 327	2 388	7 199	20 041
3rd quarter	279 751	34 609	28 700	24 694	22 623	3 114	5 148	14 362	45 263	32 270	29 564	2 438	7 209	20 197
4th quarter	280 592	35 280	29 250	25 138	23 021	3 123	5 284	14 614	45 894	32 657	29 907	2 460	7 334	20 403
2000														
1st quarter	281 304	35 559	30 064	25 767	23 510	3 274	5 305	14 932	45 944	33 198	30 290	2 590	7 252	20 668
2nd quarter	282 002	36 339	30 467	26 105	23 785	3 198	5 444	15 143	46 669	33 482	30 507	2 536	7 377	20 869
3rd quarter	282 769	36 496	30 912	26 500	24 102	3 224	5 508	15 370	46 605	33 779	30 723	2 572	7 406	21 009
4th quarter	283 518	36 820	31 111	26 671	24 380	3 230	5 564	15 585	46 771	33 810	30 905	2 583	7 457	21 132
2001														
1st quarter	284 169	36 854	31 615	27 034	24 587	3 261	5 533	15 792	46 531	34 038	30 958	2 620	7 396	21 186
2nd quarter	284 838	37 207	31 645	27 050	24 703	3 228	5 584	15 891	46 693	33 899	30 958	2 611	7 413	21 182
3rd quarter	285 584	37 104	31 517	27 618	24 738	3 235	5 572	15 930	46 378	34 591	30 983	2 634	7 428	21 156
4th quarter	286 311	37 233	31 518	27 216	25 059	3 478	5 525	16 056	46 386	34 067	31 367	2 847	7 472	21 175
2002														
1st quarter	286 935	37 601	31 567	27 820	25 127	3 402	5 526	16 199	46 690	34 759	31 394	2 809	7 487	21 251
2nd quarter	287 574	37 880	31 814	28 159	25 392	3 399	5 597	16 396	46 868	34 916	31 485	2 821	7 475	21 340
3rd quarter	288 303	38 127	31 822	28 201	25 636	3 475	5 608	16 554	46 958	34 798	31 633	2 898	7 469	21 381
4th quarter	289 007	38 309	32 023	28 430	25 839	3 412	5 681	16 746	46 916	34 919	31 738	2 865	7 540	21 472

Table 1-3B. U.S. Population and Per Capita Product and Income: Historical Data—*Continued*

(Dollars, except as noted; quarterly data are at seasonally adjusted annual rates.)   NIPA Table 7.1

Year and quarter	Population (mid-period, thousands)	Current dollars							Chained (2012) dollars					
		Gross domestic product	Personal income	Disposable personal income	Personal consumption expenditures				Gross domestic product	Disposable personal income	Personal consumption expenditures			
					Total	Durable goods	Nondurable goods	Services			Total	Durable goods	Nondurable goods	Services
2003														
1st quarter	289 609	38 616	32 164	28 638	26 090	3 365	5 811	16 914	47 078	34 929	31 821	2 861	7 585	21 517
2nd quarter	290 253	38 976	32 447	28 930	26 331	3 472	5 754	17 105	47 378	35 276	32 106	2 985	7 629	21 588
3rd quarter	290 974	39 754	32 794	29 527	26 815	3 580	5 927	17 309	48 063	35 783	32 497	3 109	7 750	21 694
4th quarter	291 669	40 351	33 171	29 670	27 056	3 590	5 972	17 494	48 499	35 792	32 639	3 146	7 776	21 763
2004														
1st quarter	292 237	40 789	33 459	29 995	27 470	3 642	6 088	17 741	48 663	35 887	32 866	3 195	7 821	21 886
2nd quarter	292 875	41 345	33 986	30 480	27 770	3 640	6 156	17 974	48 927	36 219	32 998	3 196	7 835	22 008
3rd quarter	293 603	41 905	34 382	30 757	28 148	3 696	6 220	18 232	49 267	36 365	33 280	3 271	7 872	22 161
4th quarter	294 334	42 545	35 042	31 333	28 626	3 761	6 379	18 486	49 637	36 733	33 559	3 324	7 932	22 316
2005														
1st quarter	294 957	43 265	35 051	31 077	28 896	3 763	6 414	18 718	50 081	36 225	33 683	3 333	7 990	22 377
2nd quarter	295 588	43 676	35 540	31 493	29 336	3 850	6 494	18 993	50 204	36 491	33 992	3 414	8 036	22 541
3rd quarter	296 340	44 351	35 978	31 843	29 862	3 887	6 702	19 273	50 523	36 506	34 235	3 472	8 045	22 704
4th quarter	297 086	44 877	36 648	32 421	30 109	3 752	6 799	19 558	50 715	36 881	34 251	3 360	8 137	22 777
2006														
1st quarter	297 736	45 691	37 582	33 147	30 533	3 876	6 861	19 795	51 277	37 517	34 557	3 486	8 203	22 864
2nd quarter	298 408	46 077	37 948	33 421	30 865	3 851	6 960	20 054	51 281	37 524	34 654	3 479	8 201	22 973
3rd quarter	299 180	46 352	38 202	33 662	31 195	3 879	7 062	20 254	51 228	37 524	34 774	3 520	8 227	23 019
4th quarter	299 946	46 799	38 620	33 961	31 364	3 899	6 977	20 488	51 532	37 922	35 021	3 561	8 320	23 129
2007														
1st quarter	300 609	47 266	39 376	34 498	31 767	3 922	7 083	20 762	51 540	38 161	35 140	3 601	8 330	23 189
2nd quarter	301 284	47 737	39 753	34 789	32 012	3 935	7 190	20 887	51 719	38 173	35 126	3 630	8 293	23 174
3rd quarter	302 062	48 119	39 887	34 926	32 291	3 946	7 244	21 101	51 866	38 108	35 233	3 665	8 298	23 237
4th quarter	302 829	48 481	40 185	35 204	32 618	3 948	7 372	21 298	52 049	38 037	35 244	3 683	8 270	23 255
2008														
1st quarter	303 494	48 275	40 577	35 520	32 733	3 801	7 423	21 509	51 637	38 060	35 073	3 556	8 188	23 313
2nd quarter	304 160	48 677	41 566	36 463	33 051	3 741	7 581	21 730	51 790	38 671	35 053	3 523	8 183	23 337
3rd quarter	304 902	48 656	40 889	35 979	33 063	3 594	7 649	21 820	51 384	37 766	34 706	3 396	8 044	23 269
4th quarter	305 616	47 640	40 391	35 664	32 188	3 298	7 091	21 799	50 154	38 006	34 302	3 139	7 948	23 244
2009														
1st quarter	306 237	47 005	39 151	35 226	31 858	3 280	6 908	21 670	49 491	37 767	34 156	3 138	7 929	23 115
2nd quarter	306 866	46 772	39 376	35 691	31 806	3 241	6 973	21 592	49 318	38 094	33 948	3 109	7 866	22 999
3rd quarter	307 573	46 884	39 140	35 450	32 173	3 365	7 131	21 676	49 384	37 577	34 103	3 252	7 873	22 989
4th quarter	308 285	47 450	39 332	35 632	32 298	3 290	7 211	21 798	49 811	37 481	33 974	3 168	7 880	22 944
2010														
1st quarter	308 901	47 657	39 740	35 883	32 504	3 306	7 268	21 930	49 903	37 616	34 074	3 201	7 906	22 984
2nd quarter	309 468	48 231	40 368	36 449	32 739	3 373	7 262	22 104	50 271	38 166	34 282	3 295	7 934	23 062
3rd quarter	310 088	48 631	40 750	36 699	32 960	3 394	7 297	22 269	50 540	38 361	34 453	3 341	7 948	23 171
4th quarter	310 718	49 050	41 201	37 053	33 312	3 471	7 468	22 373	50 691	38 498	34 610	3 427	7 991	23 196
2011														
1st quarter	311 237	49 113	42 277	37 695	33 689	3 495	7 650	22 543	50 485	38 825	34 699	3 451	8 004	23 247
2nd quarter	311 763	49 705	42 556	37 920	34 039	3 473	7 799	22 766	50 760	38 675	34 718	3 421	7 974	23 326
3rd quarter	312 389	49 912	42 902	38 193	34 269	3 492	7 801	22 976	50 644	38 766	34 784	3 445	7 931	23 412
4th quarter	313 002	50 468	43 065	38 368	34 382	3 554	7 824	23 004	51 131	38 811	34 779	3 520	7 913	23 346
2012														
1st quarter	313 520	51 097	43 971	39 289	34 838	3 630	7 944	23 264	51 446	39 472	35 000	3 600	7 958	23 442
2nd quarter	314 041	51 434	44 411	39 675	34 915	3 610	7 906	23 400	51 582	39 765	34 994	3 601	7 936	23 457
3rd quarter	314 660	51 666	44 224	39 427	35 004	3 628	7 914	23 462	51 550	39 402	34 982	3 641	7 924	23 417
4th quarter	315 277	51 887	45 651	40 667	35 289	3 691	7 963	23 635	51 508	40 415	35 070	3 716	7 909	23 446
2013														
1st quarter	315 750	52 477	44 343	39 119	35 544	3 765	8 053	23 726	51 886	38 739	35 199	3 806	8 002	23 394
2nd quarter	316 251	52 610	44 696	39 377	35 540	3 748	7 949	23 843	51 868	38 966	35 169	3 817	7 981	23 374
3rd quarter	316 877	53 171	44 913	39 629	35 758	3 752	8 020	23 986	52 171	39 057	35 242	3 842	8 017	23 388
4th quarter	317 515	53 803	45 214	39 867	36 140	3 762	8 076	24 303	52 481	39 133	35 474	3 875	8 072	23 533
2014														
1st quarter	317 838	53 815	46 059	40 571	36 416	3 787	8 136	24 493	52 280	39 638	35 578	3 932	8 082	23 570
2nd quarter	318 370	54 757	46 772	41 251	36 928	3 893	8 240	24 795	52 899	40 110	35 907	4 067	8 146	23 708
3rd quarter	319 009	55 552	47 397	41 761	37 351	3 936	8 283	25 132	53 438	40 497	36 221	4 133	8 198	23 906
4th quarter	319 651	55 842	47 917	42 174	37 680	3 972	8 233	25 475	53 631	40 950	36 587	4 202	8 270	24 132
2015														
1st quarter	320 058	56 250	48 512	42 561	37 803	4 011	8 074	25 718	54 071	41 503	36 863	4 273	8 334	24 279
2nd quarter	320 578	56 846	48 885	42 824	38 230	4 085	8 166	25 979	54 348	41 554	37 096	4 356	8 365	24 403
3rd quarter	321 205	57 121	49 210	43 147	38 571	4 105	8 225	26 241	54 439	41 761	37 332	4 406	8 433	24 525
4th quarter	321 831	57 107	49 382	43 276	38 672	4 097	8 134	26 442	54 420	41 918	37 459	4 422	8 442	24 625
2016														
1st quarter	322 354	57 298	49 544	43 573	38 937	4 129	8 091	26 718	54 640	42 172	37 686	4 476	8 508	24 738
2nd quarter	322 871	57 782	49 759	43 734	39 358	4 150	8 208	27 000	54 722	42 076	37 866	4 528	8 564	24 816
3rd quarter	323 473	58 185	50 132	44 038	39 698	4 207	8 213	27 278	54 918	42 198	38 039	4 638	8 562	24 890
4th quarter	324 048	58 734	50 580	44 438	40 054	4 225	8 263	27 566	55 165	42 385	38 204	4 698	8 550	25 009
2017														
1st quarter	324 496	59 284	51 260	45 093	40 534	4 263	8 397	27 874	55 401	42 776	38 451	4 736	8 622	25 150
2nd quarter	324 948	59 638	51 788	45 616	40 749	4 290	8 421	28 038	55 560	43 176	38 570	4 805	8 698	25 140
3rd quarter	325 475	60 273	52 344	46 038	41 081	4 341	8 493	28 247	55 874	43 399	38 727	4 899	8 722	25 193
4th quarter	325 963	61 166	53 060	46 532	41 710	4 456	8 654	28 600	56 324	43 581	39 065	5 049	8 815	25 315

Table 1-4. Contributions to Percent Change in Real Gross Domestic Product: Recent Data

(Percent, percentage points.)   NIPA Table 1.1.2

Year and quarter	Percent change at seasonally adjusted annual rate, real GDP	Personal consump-tion expen-ditures	Gross private domestic investment				Exports and imports of goods and services			Government consumption expenditures and gross investment		
			Total	Fixed investment		Change in private inventories	Net exports	Exports	Imports	Total	Federal	State and local
				Nonresi-dential	Residential							
1955	7.1	4.50	3.45	1.07	0.88	1.50	-0.04	0.43	-0.47	-0.78	-1.31	0.54
1956	2.1	1.77	-0.05	0.70	-0.47	-0.28	0.36	0.69	-0.33	0.05	-0.19	0.25
1957	2.1	1.50	-0.64	0.20	-0.31	-0.53	0.24	0.41	-0.18	1.01	0.55	0.46
1958	-0.7	0.52	-1.16	-1.04	0.05	-0.17	-0.87	-0.67	-0.19	0.76	0.08	0.68
1959	6.9	3.51	2.83	0.81	1.18	0.83	0.00	0.44	-0.44	0.60	0.30	0.30
1960	2.6	1.67	0.06	0.56	-0.37	-0.13	0.70	0.76	-0.06	0.14	-0.23	0.37
1961	2.6	1.25	0.05	0.08	0.02	-0.05	0.05	0.03	0.03	1.21	0.66	0.54
1962	6.1	3.00	1.82	0.81	0.46	0.55	-0.21	0.25	-0.45	1.51	1.24	0.27
1963	4.4	2.48	1.08	0.59	0.57	-0.07	0.23	0.34	-0.11	0.57	0.03	0.53
1964	5.8	3.57	1.27	1.09	0.31	-0.12	0.35	0.57	-0.22	0.57	-0.05	0.62
1965	6.5	3.80	2.26	1.76	-0.13	0.64	-0.29	0.14	-0.44	0.73	0.10	0.63
1966	6.6	3.38	1.56	1.40	-0.40	0.56	-0.28	0.35	-0.63	1.94	1.34	0.60
1967	2.7	1.76	-0.62	-0.04	-0.11	-0.47	-0.21	0.12	-0.33	1.81	1.32	0.49
1968	4.9	3.39	0.99	0.55	0.53	-0.09	-0.29	0.40	-0.68	0.82	0.21	0.61
1969	3.1	2.20	0.93	0.79	0.14	0.00	-0.03	0.25	-0.28	0.02	-0.34	0.36
1970	0.2	1.39	-1.03	-0.10	-0.23	-0.70	0.33	0.54	-0.21	-0.50	-0.80	0.30
1971	3.3	2.29	1.63	-0.01	1.08	0.56	-0.18	0.10	-0.28	-0.45	-0.80	0.35
1972	5.3	3.66	1.90	0.97	0.87	0.06	-0.19	0.42	-0.61	-0.12	-0.37	0.25
1973	5.6	2.97	1.95	1.51	-0.04	0.48	0.80	1.08	-0.28	-0.07	-0.39	0.32
1974	-0.5	-0.50	-1.24	0.10	-1.08	-0.26	0.73	0.56	0.17	0.47	0.06	0.41
1975	-0.2	1.36	-2.91	-1.13	-0.54	-1.24	0.86	-0.05	0.91	0.49	0.05	0.43
1976	5.4	3.41	2.91	0.66	0.88	1.37	-1.05	0.36	-1.41	0.12	0.01	0.10
1977	4.6	2.59	2.47	1.26	0.97	0.24	-0.70	0.19	-0.89	0.26	0.21	0.05
1978	5.5	2.68	2.22	1.72	0.38	0.12	0.05	0.80	-0.76	0.60	0.23	0.37
1979	3.2	1.44	0.72	1.34	-0.22	-0.40	0.64	0.80	-0.16	0.36	0.20	0.16
1980	-0.3	-0.19	-2.07	0.00	-1.19	-0.89	1.64	0.95	0.69	0.36	0.38	-0.02
1981	2.5	0.85	1.64	0.87	-0.37	1.13	-0.15	0.12	-0.26	0.20	0.43	-0.23
1982	-1.8	0.88	-2.46	-0.43	-0.72	-1.31	-0.59	-0.71	0.12	0.37	0.35	0.01
1983	4.6	3.51	1.60	-0.06	1.38	0.28	-1.32	-0.22	-1.10	0.79	0.65	0.14
1984	7.2	3.30	4.73	2.18	0.65	1.90	-1.54	0.61	-2.16	0.74	0.33	0.41
1985	4.2	3.20	-0.01	0.91	0.11	-1.03	-0.39	0.24	-0.63	1.37	0.78	0.59
1986	3.5	2.58	0.03	-0.24	0.58	-0.31	-0.29	0.53	-0.82	1.14	0.61	0.53
1987	3.5	2.15	0.53	0.01	0.10	0.41	0.17	0.77	-0.60	0.62	0.38	0.24
1988	4.2	2.65	0.45	0.63	-0.05	-0.13	0.81	1.23	-0.41	0.26	-0.15	0.42
1989	3.7	1.86	0.72	0.71	-0.16	0.17	0.51	0.97	-0.46	0.58	0.15	0.43
1990	1.9	1.28	-0.45	0.14	-0.38	-0.21	0.40	0.78	-0.37	0.65	0.20	0.45
1991	-0.1	0.12	-1.09	-0.48	-0.35	-0.26	0.62	0.61	0.01	0.25	0.01	0.24
1992	3.5	2.36	1.11	0.33	0.49	0.28	-0.04	0.66	-0.70	0.10	-0.15	0.25
1993	2.8	2.24	1.24	0.84	0.32	0.07	-0.56	0.31	-0.87	-0.17	-0.32	0.15
1994	4.0	2.51	1.90	0.91	0.38	0.61	-0.41	0.84	-1.25	0.02	-0.31	0.32
1995	2.7	1.91	0.55	1.15	-0.15	-0.44	0.12	1.02	-0.90	0.10	-0.21	0.31
1996	3.8	2.26	1.49	1.13	0.35	0.02	-0.15	0.86	-1.01	0.18	-0.09	0.27
1997	4.4	2.45	2.01	1.38	0.11	0.52	-0.31	1.26	-1.57	0.30	-0.06	0.36
1998	4.5	3.42	1.76	1.44	0.38	-0.07	-1.14	0.26	-1.39	0.44	-0.06	0.50
1999	4.8	3.42	1.62	1.36	0.29	-0.03	-0.87	0.52	-1.39	0.58	0.13	0.46
2000	4.1	3.32	1.31	1.31	0.03	-0.03	-0.83	0.86	-1.69	0.33	0.02	0.31
2001	1.0	1.66	-1.11	-0.31	0.04	-0.84	-0.22	-0.61	0.39	0.67	0.24	0.43
2002	1.7	1.71	-0.16	-0.94	0.29	0.48	-0.64	-0.17	-0.47	0.82	0.47	0.35
2003	2.9	2.13	0.76	0.30	0.47	-0.02	-0.45	0.20	-0.64	0.41	0.45	-0.03
2004	3.8	2.53	1.64	0.67	0.57	0.41	-0.67	0.88	-1.55	0.30	0.31	-0.01
2005	3.5	2.39	1.26	0.92	0.41	-0.07	-0.29	0.69	-0.97	0.15	0.15	0.00
2006	2.9	2.05	0.60	1.00	-0.50	0.10	-0.10	0.94	-1.04	0.30	0.17	0.13
2007	1.9	1.49	-0.48	0.89	-1.13	-0.25	0.53	0.93	-0.41	0.34	0.14	0.20
2008	-0.1	-0.14	-1.52	0.08	-1.14	-0.46	1.04	0.66	0.38	0.48	0.46	0.02
2009	-2.5	-0.85	-3.52	-1.95	-0.74	-0.83	1.13	-1.01	2.14	0.70	0.47	0.23
2010	2.6	1.20	1.86	0.52	-0.08	1.42	-0.49	1.35	-1.84	0.00	0.35	-0.35
2011	1.6	1.29	0.94	1.00	0.00	-0.05	-0.01	0.90	-0.91	-0.66	-0.23	-0.44
2012	2.2	1.03	1.64	1.16	0.31	0.17	0.00	0.46	-0.46	-0.42	-0.16	-0.26
2013	1.8	0.99	1.11	0.54	0.34	0.23	0.22	0.48	-0.26	-0.47	-0.44	-0.03
2014	2.5	1.99	0.95	0.95	0.12	-0.12	-0.25	0.57	-0.81	-0.17	-0.19	0.02
2015	3.1	2.55	0.95	0.32	0.33	0.31	-0.76	0.06	-0.81	0.33	0.00	0.33
2016	1.7	1.87	-0.27	0.07	0.23	-0.57	-0.21	0.04	-0.25	0.32	0.04	0.28
2017	2.3	1.79	0.60	0.49	0.15	-0.04	-0.22	0.47	-0.68	0.16	0.02	0.14
2018	3.0	1.85	1.08	0.91	-0.02	0.20	-0.25	0.36	-0.62	0.32	0.18	0.13
2019	2.2	1.64	0.30	0.39	-0.07	-0.02	-0.18	-0.01	-0.16	0.40	0.26	0.14
2017												
1st quarter	2.3	2.15	-0.23	0.75	0.43	-1.41	0.36	0.98	-0.62	0.01	-0.10	0.11
2nd quarter	1.7	1.23	0.61	0.31	-0.04	0.34	-0.39	0.13	-0.52	0.27	0.14	0.13
3rd quarter	2.9	1.57	1.26	0.28	-0.07	1.05	0.08	0.33	-0.25	0.04	-0.03	0.07
4th quarter	3.9	2.82	1.07	1.18	0.39	-0.50	-0.49	1.36	-1.85	0.49	0.32	0.17
2018												
1st quarter	3.8	1.40	1.83	1.55	-0.13	0.41	0.29	0.34	-0.05	0.26	0.12	0.13
2nd quarter	2.7	2.13	-0.19	0.82	-0.07	-0.94	0.25	0.24	0.01	0.50	0.23	0.27
3rd quarter	2.1	1.79	1.72	0.36	-0.22	1.58	-1.83	-0.66	-1.17	0.44	0.29	0.15
4th quarter	1.3	1.05	0.69	0.66	-0.21	0.23	-0.27	0.34	-0.61	-0.16	0.12	-0.28
2019												
1st quarter	2.9	1.25	0.71	0.56	-0.06	0.21	0.55	0.22	0.33	0.43	0.09	0.34
2nd quarter	1.5	2.47	-1.04	0.01	-0.08	-0.97	-0.79	-0.54	-0.25	0.86	0.58	0.28
3rd quarter	2.6	1.83	0.34	0.25	0.17	-0.09	0.04	0.10	-0.06	0.37	0.31	0.06
4th quarter	2.4	1.07	-0.64	-0.04	0.22	-0.82	1.52	0.39	1.13	0.42	0.26	0.16

Table 1-5A. Chain-Type Quantity Indexes for Gross Domestic Product and Domestic Purchases: Recent Data

(Index numbers, 2012 = 100.)   NIPA Tables 1.1.3, 1.4.3, 2.3.3

| Year and quarter | Gross domestic product | | | | | | | | | | | Gross domestic purchases |
| | Gross domestic product, total | Personal consumption expenditures | | Private fixed investment | | | Exports and imports of goods and services | | Government consumption expenditures and gross investment | | | |
		Total	Excluding food and energy	Total	Nonresidential	Residential	Exports	Imports	Total	Federal	State and local	
1955	17.7	15.8	12.9	12.8	7.8	48.9	3.6	3.7	31.0	40.1	21.7	17.5
1956	18.1	16.3	13.2	13.0	8.3	45.0	4.2	4.0	31.0	39.6	22.4	17.8
1957	18.5	16.7	13.5	12.9	8.5	42.4	4.6	4.2	32.5	41.1	23.7	18.1
1958	18.3	16.8	13.6	12.1	7.7	42.8	4.0	4.4	33.6	41.3	25.7	18.2
1959	19.6	17.8	14.4	13.7	8.3	53.6	4.4	4.8	34.4	42.2	26.6	19.4
1960	20.1	18.3	14.9	13.9	8.8	49.9	5.2	4.9	34.6	41.5	27.7	19.8
1961	20.6	18.6	15.3	13.9	8.8	50.1	5.2	4.9	36.5	43.5	29.4	20.3
1962	21.9	19.6	16.2	15.1	9.5	54.9	5.4	5.4	38.9	47.4	30.3	21.6
1963	22.9	20.4	17.0	16.3	10.1	61.3	5.8	5.6	39.8	47.6	32.1	22.5
1964	24.2	21.6	18.1	17.8	11.2	65.0	6.5	5.9	40.8	47.4	34.2	23.7
1965	25.8	23.0	19.3	19.7	13.0	63.4	6.7	6.5	42.1	47.7	36.5	25.3
1966	27.4	24.3	20.5	20.9	14.6	58.0	7.2	7.5	45.8	52.8	38.7	27.1
1967	28.2	25.0	21.1	20.7	14.6	56.5	7.3	8.0	49.4	58.1	40.7	27.9
1968	29.6	26.4	22.4	22.1	15.3	64.2	7.9	9.2	51.1	59.0	43.1	29.4
1969	30.5	27.4	23.3	23.4	16.4	66.2	8.3	9.7	51.2	57.6	44.6	30.3
1970	30.6	28.0	23.8	23.0	16.2	62.7	9.2	10.1	50.1	54.0	45.8	30.2
1971	31.6	29.1	24.8	24.5	16.2	79.4	9.3	10.7	49.1	50.6	47.3	31.3
1972	33.2	30.9	26.6	27.3	17.6	93.2	10.1	11.9	48.9	49.0	48.3	33.0
1973	35.1	32.4	28.3	29.7	19.9	92.6	12.0	12.4	48.7	47.2	49.7	34.6
1974	34.9	32.2	28.2	28.0	20.1	74.5	12.9	12.1	49.8	47.5	51.5	34.2
1975	34.9	32.9	28.8	25.3	18.3	65.5	12.8	10.8	50.9	47.8	53.4	33.8
1976	36.7	34.7	30.5	27.8	19.3	80.0	13.4	12.9	51.2	47.9	53.8	36.0
1977	38.4	36.2	32.0	31.5	21.4	96.4	13.7	14.3	51.8	48.9	54.1	37.9
1978	40.6	37.8	33.7	35.2	24.4	102.8	15.1	15.5	53.3	50.0	55.8	40.0
1979	41.8	38.7	34.7	37.2	26.8	98.9	16.6	15.8	54.2	51.1	56.6	41.0
1980	41.7	38.5	34.7	35.0	26.8	78.2	18.4	14.8	55.2	53.3	56.5	40.2
1981	42.8	39.1	35.4	35.9	28.5	71.8	18.7	15.1	55.7	55.7	55.4	41.3
1982	42.0	39.7	35.9	33.7	27.6	58.8	17.2	14.9	56.7	57.7	55.4	40.8
1983	43.9	41.9	38.3	36.3	27.5	83.5	16.8	16.8	58.9	61.3	56.2	43.2
1984	47.1	44.1	40.6	42.1	32.1	95.8	18.2	20.9	60.9	63.2	58.3	46.9
1985	49.1	46.4	42.9	44.4	34.2	98.0	18.8	22.3	65.0	68.1	61.6	49.0
1986	50.8	48.3	44.9	45.2	33.6	110.1	20.2	24.2	68.5	72.1	64.7	50.8
1987	52.5	49.9	46.7	45.5	33.6	112.3	22.4	25.6	70.5	74.7	66.1	52.4
1988	54.7	52.0	48.7	47.0	35.3	111.3	26.0	26.6	71.4	73.6	68.7	54.1
1989	56.8	53.5	50.3	48.5	37.3	107.7	29.1	27.8	73.4	74.8	71.4	55.8
1990	57.8	54.6	51.4	47.8	37.7	98.6	31.6	28.8	75.8	76.4	74.3	56.6
1991	57.8	54.7	51.5	45.3	36.2	89.8	33.7	28.8	76.7	76.4	75.9	56.2
1992	59.8	56.7	53.7	47.8	37.3	102.2	36.0	30.8	77.0	75.2	77.5	58.2
1993	61.4	58.7	55.8	51.5	40.1	110.5	37.2	33.4	76.4	72.6	78.5	60.1
1994	63.9	61.0	58.1	55.7	43.2	120.4	40.5	37.4	76.5	69.9	80.7	62.7
1995	65.6	62.8	60.0	59.1	47.4	116.3	44.7	40.4	76.9	68.1	82.9	64.3
1996	68.1	65.0	62.3	64.3	51.7	125.9	48.3	43.9	77.6	67.3	84.9	66.8
1997	71.1	67.4	64.9	69.8	57.3	128.9	54.1	49.8	78.8	66.7	87.6	70.0
1998	74.3	71.0	68.7	77.0	63.5	140.0	55.3	55.7	80.7	66.1	91.5	73.9
1999	77.9	74.7	72.5	84.0	69.8	148.8	58.1	62.0	83.4	67.4	95.1	77.9
2000	81.1	78.5	76.6	89.9	76.3	149.8	63.0	69.9	84.9	67.6	97.6	81.7
2001	81.9	80.5	78.8	88.7	74.6	151.2	59.3	68.0	88.1	70.2	101.2	82.7
2002	83.3	82.6	81.0	85.6	69.5	160.3	58.3	70.5	92.0	75.4	104.2	84.6
2003	85.7	85.2	83.7	89.4	71.2	174.9	59.6	73.9	93.9	80.3	103.9	87.3
2004	88.9	88.4	87.1	95.7	75.2	192.3	65.3	82.3	95.4	83.8	103.8	91.0
2005	92.1	91.5	90.3	102.7	81.0	204.9	70.0	87.7	96.1	85.4	103.8	94.3
2006	94.7	94.3	93.3	105.3	87.5	189.5	76.5	93.5	97.7	87.4	104.9	96.9
2007	96.5	96.4	95.6	104.0	93.6	154.1	83.2	95.9	99.4	89.2	106.7	98.1
2008	96.3	96.2	95.7	98.0	94.1	116.8	87.9	93.7	101.9	94.7	106.9	97.0
2009	93.9	95.0	94.3	82.3	80.5	91.5	80.5	81.5	105.4	100.5	108.9	93.6
2010	96.3	96.7	96.0	84.9	84.1	88.7	90.3	92.2	105.4	104.6	106.0	96.4
2011	97.8	98.5	98.2	90.9	91.4	88.5	96.7	97.4	102.1	101.9	102.3	97.9
2012	100.0	100.0	100.0	100.0	100.0	100.0	100.0	100.0	100.0	100.0	100.0	100.0
2013	101.8	101.5	101.4	105.6	104.1	112.4	103.6	101.5	97.6	94.5	99.7	101.6
2014	104.4	104.5	104.7	112.5	111.6	116.7	107.9	106.6	96.7	92.0	99.9	104.3
2015	107.6	108.4	109.0	116.8	114.2	128.6	108.4	112.1	98.4	92.0	102.8	108.2
2016	109.5	111.4	112.1	118.9	114.8	137.1	108.7	114.0	100.2	92.5	105.5	110.2
2017	112.0	114.4	115.1	123.4	119.1	142.5	113.0	119.3	101.1	92.8	106.8	113.0
2018	115.4	117.5	118.3	129.8	127.4	141.7	116.3	124.2	103.0	95.4	108.1	116.5
2019	117.9	120.3	121.4	132.2	131.1	139.2	116.2	125.5	105.3	99.3	109.5	119.2
2017												
1st quarter	111.0	113.4	114.2	122.2	117.7	142.2	111.6	117.3	100.6	92.2	106.4	111.8
2nd quarter	111.5	113.9	114.6	122.6	118.4	141.8	111.9	118.4	101.0	92.7	106.7	112.4
3rd quarter	112.3	114.5	115.3	123.0	119.0	141.2	112.6	118.9	101.1	92.6	106.8	113.2
4th quarter	113.4	115.7	116.4	125.8	121.6	144.8	115.7	122.6	101.8	93.7	107.3	114.4
2018												
1st quarter	114.4	116.3	117.0	128.4	125.2	143.6	116.5	122.7	102.1	94.2	107.6	115.4
2nd quarter	115.2	117.2	118.0	129.8	127.1	143.0	117.1	122.7	102.9	95.0	108.2	116.0
3rd quarter	115.8	118.0	118.9	130.1	127.9	141.0	115.5	125.1	103.5	96.1	108.6	117.1
4th quarter	116.2	118.4	119.3	130.9	129.5	139.1	116.3	126.3	103.3	96.5	107.9	117.6
2019												
1st quarter	117.0	119.0	120.0	131.8	130.8	138.5	116.8	125.7	103.9	96.8	108.8	118.3
2nd quarter	117.4	120.0	121.1	131.7	130.8	137.8	115.5	126.2	105.2	99.0	109.5	118.9
3rd quarter	118.2	120.8	121.9	132.5	131.4	139.3	115.8	126.3	105.8	100.2	109.6	119.6
4th quarter	118.9	121.3	122.5	132.8	131.3	141.3	116.7	123.9	106.4	101.1	110.0	119.9

Table 1-5B. Chain-Type Quantity Indexes for Gross Domestic Product and Domestic Purchases: Historical Data

(Index numbers, 2012 = 100.)   **NIPA Tables 1.1.3, 1.4.3, 2.3.3**

Year and quarter	Gross domestic product, total	Personal consumption expenditures		Private fixed investment			Exports and imports of goods and services		Government consumption expenditures and gross investment			Gross domestic purchases
		Total	Excluding food and energy	Total	Nonresi-dential	Residential	Exports	Imports	Total	Federal	State and local	
1929	6.9	7.5	6.2	5.9	4.1	18.6	2.0	2.1	5.8	2.3	11.3	6.8
1930	6.3	7.1	5.8	4.6	3.4	11.5	1.6	1.8	6.3	2.5	12.4	6.2
1931	5.9	6.9	5.5	3.3	2.3	9.8	1.4	1.6	6.6	2.6	13.0	5.8
1932	5.1	6.3	5.0	2.0	1.4	5.5	1.1	1.3	6.4	2.6	12.4	5.1
1933	5.0	6.2	4.8	1.8	1.3	4.6	1.1	1.4	6.2	3.2	11.1	5.0
1934	5.6	6.6	5.1	2.3	1.7	6.0	1.2	1.4	6.9	4.2	11.6	5.6
1935	6.1	7.0	5.4	2.9	2.1	8.2	1.3	1.8	7.2	4.3	12.0	6.1
1936	6.9	7.7	5.9	3.8	2.7	10.2	1.3	1.8	8.3	6.4	12.0	6.9
1937	7.2	8.0	6.2	4.4	3.2	11.0	1.7	2.1	8.0	5.8	12.0	7.2
1938	7.0	7.9	6.0	3.6	2.5	11.0	1.6	1.6	8.6	6.4	12.7	6.9
1939	7.5	8.3	6.4	4.2	2.7	15.4	1.7	1.7	9.3	6.9	14.0	7.4
1940	8.2	8.7	6.7	5.0	3.3	17.3	2.0	1.7	9.7	7.8	13.5	8.1
1941	9.7	9.4	7.3	5.8	4.0	18.4	2.0	2.1	16.3	20.6	12.7	9.6
1942	11.5	9.1	7.1	3.4	2.5	9.5	1.3	1.9	37.7	61.6	11.6	11.5
1943	13.4	9.4	7.5	2.7	2.1	5.9	1.1	2.4	56.6	98.1	10.6	13.6
1944	14.5	9.7	7.7	3.2	2.6	5.1	1.2	2.5	63.5	111.7	10.2	14.7
1945	14.4	10.3	8.2	4.3	3.6	5.8	1.7	2.7	55.8	96.5	10.6	14.4
1946	12.7	11.5	9.1	7.6	5.2	23.5	3.7	2.2	19.6	27.0	11.6	12.2
1947	12.6	11.7	9.3	9.1	6.0	30.1	4.2	2.1	16.7	20.2	13.2	12.0
1948	13.1	12.0	9.6	9.9	6.3	35.9	3.3	2.5	17.6	21.3	14.0	12.7
1949	13.0	12.3	9.9	9.1	5.7	33.3	3.3	2.4	19.5	23.0	16.0	12.6
1947												
1st quarter	12.6	11.6	. . .	8.9	6.1	27.6	4.5	2.2	16.7	20.6	12.8	11.9
2nd quarter	12.5	11.8	. . .	8.7	6.0	26.2	4.4	2.2	16.7	20.4	13.0	11.9
3rd quarter	12.5	11.8	. . .	9.0	5.9	30.3	4.2	2.0	16.7	20.2	13.3	11.9
4th quarter	12.7	11.8	. . .	9.7	6.0	36.1	3.7	2.1	16.6	19.7	13.5	12.2
1948												
1st quarter	12.9	11.9	. . .	10.0	6.4	35.6	3.5	2.3	16.8	20.2	13.5	12.5
2nd quarter	13.1	12.0	. . .	10.0	6.2	37.7	3.2	2.4	17.4	21.0	13.9	12.8
3rd quarter	13.2	12.0	. . .	9.9	6.2	36.5	3.3	2.6	17.8	21.5	14.1	12.8
4th quarter	13.2	12.1	. . .	9.8	6.3	33.8	3.2	2.5	18.4	22.5	14.4	12.9
1949												
1st quarter	13.0	12.1	. . .	9.3	6.0	31.3	3.5	2.4	18.9	22.9	15.0	12.6
2nd quarter	13.0	12.3	. . .	9.0	5.8	30.8	3.5	2.4	19.7	23.7	15.9	12.6
3rd quarter	13.1	12.4	. . .	9.0	5.5	33.4	3.2	2.3	19.8	23.3	16.5	12.7
4th quarter	13.0	12.5	. . .	9.3	5.5	37.5	2.8	2.4	19.5	22.3	16.8	12.7
1950												
1st quarter	13.5	12.8	. . .	9.9	5.6	41.7	2.7	2.4	19.1	21.3	17.2	13.2
2nd quarter	13.9	13.0	. . .	10.7	6.1	45.7	2.8	2.5	19.5	21.8	17.4	13.7
3rd quarter	14.4	13.6	. . .	11.5	6.6	48.2	2.8	3.1	19.1	21.2	17.4	14.3
4th quarter	14.7	13.2	. . .	11.3	6.6	45.9	3.1	3.2	20.5	23.8	17.4	14.5
1951												
1st quarter	14.9	13.5	. . .	10.8	6.4	43.6	3.2	3.2	22.7	28.2	17.2	14.7
2nd quarter	15.2	13.2	. . .	10.4	6.5	37.7	3.6	3.0	25.7	33.8	17.5	14.9
3rd quarter	15.5	13.3	. . .	10.2	6.6	35.4	3.6	2.8	28.5	39.3	17.6	15.1
4th quarter	15.5	13.4	. . .	10.1	6.5	35.7	3.6	2.7	30.0	42.3	17.6	15.1
1952												
1st quarter	15.7	13.4	. . .	10.3	6.5	36.6	3.8	3.0	30.9	44.1	17.6	15.3
2nd quarter	15.7	13.7	. . .	10.5	6.6	37.3	3.4	3.0	32.0	45.8	18.0	15.4
3rd quarter	15.8	13.7	. . .	9.9	6.1	37.0	3.1	3.2	32.3	46.9	17.5	15.6
4th quarter	16.4	14.2	. . .	10.6	6.7	38.9	3.1	3.5	32.9	47.6	17.9	16.2
1953												
1st quarter	16.7	14.4	. . .	11.1	7.0	39.6	3.1	3.4	33.9	49.3	18.3	16.5
2nd quarter	16.8	14.5	. . .	11.2	7.1	39.7	3.1	3.6	34.6	50.7	18.3	16.7
3rd quarter	16.7	14.4	. . .	11.2	7.2	38.2	3.2	3.6	34.2	49.5	18.8	16.5
4th quarter	16.4	14.4	. . .	11.1	7.2	37.9	3.1	3.4	34.1	48.9	19.2	16.3
1954												
1st quarter	16.4	14.4	. . .	10.9	7.0	38.3	3.0	3.2	33.2	46.4	19.9	16.2
2nd quarter	16.4	14.6	. . .	11.1	6.9	40.8	3.4	3.5	32.1	44.1	20.0	16.2
3rd quarter	16.6	14.8	. . .	11.5	7.1	43.3	3.3	3.3	31.4	42.2	20.5	16.3
4th quarter	16.9	15.1	. . .	11.8	7.1	46.0	3.4	3.3	31.2	41.7	20.6	16.6
1955												
1st quarter	17.4	15.4	. . .	12.2	7.2	49.4	3.6	3.5	31.3	40.9	21.5	17.1
2nd quarter	17.7	15.7	. . .	12.7	7.7	50.1	3.5	3.7	31.0	40.0	21.7	17.5
3rd quarter	17.9	15.9	. . .	13.0	8.0	49.0	3.7	3.8	31.1	40.4	21.7	17.7
4th quarter	18.0	16.1	. . .	13.1	8.3	47.0	3.7	3.9	30.5	39.1	21.8	17.8

. . . = Not available.

Table 1-5B. Chain-Type Quantity Indexes for Gross Domestic Product and Domestic Purchases: Historical Data—*Continued*

(Index numbers, 2012 = 100.)   NIPA Tables 1.1.3, 1.4.3, 2.3.3

Year and quarter	Gross domestic product, total	Personal consumption expenditures		Private fixed investment			Exports and imports of goods and services		Government consumption expenditures and gross investment			Gross domestic purchases
		Total	Excluding food and energy	Total	Nonresi- dential	Residential	Exports	Imports	Total	Federal	State and local	
1956												
1st quarter	17.9	16.2	. . .	12.9	8.2	45.8	3.9	4.1	30.5	38.9	22.1	17.7
2nd quarter	18.1	16.2	. . .	13.0	8.3	45.6	4.2	4.0	31.1	39.8	22.4	17.8
3rd quarter	18.1	16.2	. . .	13.0	8.4	44.7	4.3	4.1	30.9	39.2	22.5	17.8
4th quarter	18.4	16.5	. . .	12.9	8.4	44.0	4.5	3.9	31.7	40.5	22.7	18.0
1957												
1st quarter	18.5	16.6	. . .	12.9	8.4	43.4	4.8	4.2	32.3	41.3	23.2	18.1
2nd quarter	18.4	16.6	. . .	12.8	8.4	42.4	4.7	4.2	32.2	40.8	23.4	18.1
3rd quarter	18.6	16.7	. . .	13.0	8.6	41.8	4.5	4.1	32.4	41.0	23.8	18.3
4th quarter	18.4	16.8	. . .	12.7	8.4	41.8	4.4	4.2	33.0	41.5	24.4	18.1
1958												
1st quarter	17.9	16.5	. . .	12.0	7.9	40.1	4.0	4.2	32.7	40.2	25.0	17.7
2nd quarter	18.1	16.7	. . .	11.7	7.6	40.3	4.0	4.4	33.5	41.5	25.4	17.9
3rd quarter	18.5	16.9	. . .	11.9	7.5	43.2	4.0	4.3	33.7	41.4	25.9	18.3
4th quarter	18.9	17.2	. . .	12.6	7.8	47.7	4.0	4.6	34.4	42.3	26.4	18.8
1959												
1st quarter	19.3	17.5	14.1	13.4	8.0	53.1	4.3	4.7	34.0	41.4	26.5	19.1
2nd quarter	19.7	17.7	14.4	13.8	8.2	54.9	4.2	4.9	34.4	42.2	26.6	19.6
3rd quarter	19.7	17.9	14.6	13.9	8.5	53.9	4.6	5.0	34.7	42.8	26.6	19.5
4th quarter	19.8	18.0	14.6	13.7	8.4	52.4	4.4	4.8	34.5	42.3	26.5	19.6
1960												
1st quarter	20.2	18.1	14.8	14.2	8.7	54.0	5.0	5.0	33.9	40.7	26.9	19.9
2nd quarter	20.1	18.4	15.0	13.9	8.9	49.5	5.1	5.0	34.2	40.8	27.6	19.8
3rd quarter	20.2	18.3	14.9	13.6	8.7	48.1	5.2	4.9	35.1	42.1	28.0	19.9
4th quarter	20.0	18.3	15.0	13.6	8.7	48.0	5.3	4.7	35.3	42.3	28.3	19.6
1961												
1st quarter	20.1	18.3	14.9	13.5	8.6	48.2	5.2	4.6	35.9	42.4	29.3	19.7
2nd quarter	20.4	18.6	15.2	13.7	8.8	48.4	5.0	4.7	35.9	42.8	29.1	20.1
3rd quarter	20.8	18.7	15.3	14.1	8.8	51.0	5.2	5.0	36.7	44.1	29.3	20.5
4th quarter	21.2	19.0	15.7	14.5	9.1	52.8	5.3	5.1	37.5	44.9	30.1	20.9
1962												
1st quarter	21.6	19.2	15.8	14.8	9.3	53.7	5.3	5.3	38.2	46.5	29.9	21.3
2nd quarter	21.8	19.5	16.1	15.2	9.6	55.4	5.6	5.4	38.4	46.7	30.1	21.5
3rd quarter	22.1	19.6	16.2	15.3	9.7	55.3	5.6	5.5	39.3	48.1	30.4	21.7
4th quarter	22.1	19.9	16.5	15.3	9.6	55.2	5.4	5.5	39.6	48.4	30.8	21.9
1963												
1st quarter	22.4	20.0	16.6	15.5	9.7	57.2	5.5	5.4	39.1	46.7	31.3	22.1
2nd quarter	22.6	20.2	16.9	16.2	10.0	61.3	5.9	5.5	39.1	46.5	31.6	22.2
3rd quarter	23.1	20.5	17.1	16.6	10.2	62.5	5.9	5.7	40.8	49.1	32.4	22.8
4th quarter	23.3	20.7	17.3	17.1	10.5	64.4	6.1	5.7	40.4	47.9	32.9	22.9
1964												
1st quarter	23.8	21.1	17.7	17.6	10.7	68.4	6.4	5.7	40.5	47.6	33.4	23.3
2nd quarter	24.0	21.5	18.0	17.6	11.0	65.0	6.4	5.8	40.9	47.6	34.1	23.6
3rd quarter	24.4	21.9	18.3	17.9	11.4	63.9	6.6	5.9	40.9	47.3	34.5	23.9
4th quarter	24.5	21.9	18.4	18.1	11.6	62.8	6.7	6.1	40.9	46.9	34.8	24.0
1965												
1st quarter	25.1	22.4	18.9	18.9	12.3	63.1	5.9	5.9	40.9	46.5	35.2	24.7
2nd quarter	25.4	22.7	19.1	19.4	12.8	63.7	6.9	6.6	41.4	46.6	36.0	25.0
3rd quarter	26.0	23.0	19.4	19.9	13.3	64.0	6.8	6.6	42.7	48.3	37.1	25.6
4th quarter	26.6	23.7	19.9	20.4	13.8	62.7	7.2	6.9	43.5	49.4	37.6	26.1
1966												
1st quarter	27.2	24.0	20.3	21.2	14.4	64.3	7.0	7.1	44.3	50.6	38.0	26.8
2nd quarter	27.3	24.1	20.3	20.9	14.6	59.3	7.2	7.3	45.2	52.0	38.3	26.9
3rd quarter	27.5	24.4	20.6	21.0	14.8	57.3	7.1	7.7	46.4	54.0	38.8	27.2
4th quarter	27.8	24.5	20.7	20.4	14.8	51.3	7.3	7.8	47.2	54.6	39.8	27.4
1967												
1st quarter	28.0	24.6	20.8	20.0	14.6	49.9	7.4	7.9	49.2	58.1	40.3	27.7
2nd quarter	28.0	25.0	21.1	20.5	14.5	55.6	7.3	7.8	49.0	57.5	40.5	27.7
3rd quarter	28.3	25.1	21.3	20.8	14.5	58.6	7.2	7.9	49.6	58.5	40.6	28.0
4th quarter	28.5	25.2	21.4	21.4	14.8	62.0	7.4	8.3	49.9	58.5	41.3	28.2
1968												
1st quarter	29.1	25.8	21.9	21.9	15.2	62.2	7.6	8.8	50.8	59.5	42.1	28.9
2nd quarter	29.6	26.2	22.3	21.8	15.0	63.7	7.7	9.0	51.1	59.2	42.9	29.3
3rd quarter	29.8	26.7	22.7	22.2	15.2	65.0	8.3	9.5	51.2	58.9	43.5	29.6
4th quarter	29.9	26.9	22.8	22.7	15.7	65.9	8.1	9.4	51.3	58.6	43.9	29.7

. . . = Not available.

Table 1-5B. Chain-Type Quantity Indexes for Gross Domestic Product and Domestic Purchases: Historical Data—*Continued*

(Index numbers, 2012 = 100.)   **NIPA Tables 1.1.3, 1.4.3, 2.3.3**

Year and quarter	Gross domestic product, total	Personal consumption expenditures		Private fixed investment			Exports and imports of goods and services		Government consumption expenditures and gross investment			Gross domestic purchases
		Total	Excluding food and energy	Total	Nonresidential	Residential	Exports	Imports	Total	Federal	State and local	
1969												
1st quarter	30.4	27.1	23.0	23.3	16.1	68.1	7.1	8.5	51.4	58.4	44.3	30.2
2nd quarter	30.5	27.3	23.2	23.4	16.2	67.5	8.7	10.2	51.2	57.7	44.7	30.3
3rd quarter	30.7	27.5	23.4	23.8	16.6	66.9	8.6	10.1	51.3	57.8	44.8	30.5
4th quarter	30.5	27.7	23.5	23.2	16.5	62.3	8.8	10.0	50.7	56.5	44.7	30.3
1970												
1st quarter	30.5	27.8	23.6	23.1	16.4	62.4	8.9	10.0	50.4	55.5	45.1	30.2
2nd quarter	30.5	28.0	23.8	22.6	16.3	58.5	9.2	10.1	49.8	54.0	45.3	30.2
3rd quarter	30.8	28.2	24.0	23.0	16.4	61.9	9.3	10.1	50.0	53.4	46.4	30.5
4th quarter	30.5	28.1	23.8	23.0	15.8	68.2	9.3	10.3	50.1	53.2	46.6	30.2
1971												
1st quarter	31.3	28.7	24.4	23.5	15.9	71.7	9.4	10.2	49.3	51.5	46.8	30.9
2nd quarter	31.5	28.9	24.6	24.4	16.1	78.6	9.3	10.9	49.2	50.9	47.1	31.2
3rd quarter	31.7	29.2	24.9	24.8	16.2	82.2	9.8	11.2	49.2	50.6	47.3	31.5
4th quarter	31.8	29.7	25.4	25.5	16.5	85.3	8.9	10.4	48.8	49.3	47.9	31.6
1972												
1st quarter	32.4	30.1	25.9	26.5	17.1	90.9	9.9	11.9	49.1	49.7	48.0	32.2
2nd quarter	33.1	30.6	26.3	27.0	17.3	92.6	9.6	11.5	49.3	50.2	47.9	32.9
3rd quarter	33.4	31.1	26.8	27.3	17.6	93.0	10.2	11.7	48.4	47.9	48.3	33.2
4th quarter	34.0	31.8	27.4	28.5	18.4	96.3	10.6	12.3	48.8	48.0	49.0	33.7
1973												
1st quarter	34.8	32.4	28.1	29.6	19.2	99.6	11.4	12.8	49.2	48.7	49.2	34.5
2nd quarter	35.2	32.4	28.2	29.8	19.9	94.4	11.9	12.5	48.8	47.9	49.3	34.7
3rd quarter	35.0	32.5	28.3	29.8	20.2	90.3	12.0	12.1	48.3	46.2	49.8	34.5
4th quarter	35.4	32.4	28.3	29.5	20.4	86.2	12.5	12.2	48.6	46.3	50.4	34.7
1974												
1st quarter	35.1	32.1	28.3	28.9	20.4	80.6	12.7	11.8	49.5	47.4	51.1	34.3
2nd quarter	35.1	32.2	28.3	28.5	20.3	77.4	13.3	12.4	49.8	47.2	51.7	34.3
3rd quarter	34.8	32.4	28.3	28.1	20.1	74.8	12.6	12.2	49.9	47.5	51.6	34.1
4th quarter	34.7	31.9	27.8	26.6	19.6	65.1	12.9	12.1	50.1	48.0	51.6	33.9
1975												
1st quarter	34.3	32.2	28.2	25.1	18.5	61.4	13.0	10.9	50.6	47.5	53.1	33.2
2nd quarter	34.5	32.7	28.5	24.8	18.0	63.1	12.7	10.0	50.2	47.0	52.8	33.3
3rd quarter	35.1	33.2	29.0	25.4	18.2	67.3	12.4	10.8	51.1	48.2	53.4	34.1
4th quarter	35.6	33.5	29.6	25.9	18.4	70.2	13.1	11.4	51.6	48.4	54.2	34.6
1976												
1st quarter	36.4	34.2	30.1	26.9	18.8	76.8	13.1	12.1	51.7	47.9	54.8	35.5
2nd quarter	36.6	34.5	30.3	27.4	19.1	78.7	13.2	12.7	51.1	47.8	53.8	35.9
3rd quarter	36.8	34.9	30.6	27.7	19.5	77.5	13.6	13.2	51.0	47.8	53.4	36.1
4th quarter	37.1	35.3	30.9	29.0	19.9	86.8	13.7	13.6	50.9	47.9	53.3	36.4
1977												
1st quarter	37.5	35.7	31.4	30.0	20.6	89.4	13.5	14.3	51.4	48.3	53.8	37.1
2nd quarter	38.3	35.9	31.8	31.6	21.2	99.2	13.8	14.4	51.8	48.9	54.1	37.7
3rd quarter	39.0	36.3	32.1	32.0	21.6	98.9	14.0	14.2	52.0	49.4	54.1	38.3
4th quarter	39.0	36.8	32.6	32.5	22.3	97.8	13.5	14.4	51.9	49.0	54.2	38.5
1978												
1st quarter	39.1	37.0	32.8	32.8	22.5	98.7	13.8	15.4	51.9	49.1	54.2	38.8
2nd quarter	40.6	37.8	33.8	35.1	24.2	103.4	15.3	15.4	53.2	50.0	55.8	39.9
3rd quarter	41.0	38.0	34.0	36.0	25.0	104.6	15.4	15.6	53.7	50.3	56.4	40.3
4th quarter	41.5	38.3	34.3	36.8	25.8	104.4	16.0	15.8	54.2	50.8	57.0	40.8
1979												
1st quarter	41.6	38.5	34.4	37.1	26.4	101.9	16.1	15.8	53.7	50.7	56.0	40.9
2nd quarter	41.7	38.5	34.5	36.9	26.4	100.1	16.1	15.9	54.2	51.2	56.5	40.9
3rd quarter	42.0	38.8	34.9	37.6	27.2	98.9	16.7	15.6	54.3	51.3	56.7	41.0
4th quarter	42.1	38.9	35.0	37.2	27.2	94.9	17.7	16.0	54.7	51.4	57.4	41.0
1980												
1st quarter	42.2	38.9	34.9	36.7	27.5	87.6	18.2	16.0	55.5	52.7	57.6	41.0
2nd quarter	41.3	38.0	34.0	33.7	26.3	71.4	18.6	14.8	55.6	53.8	56.9	39.8
3rd quarter	41.3	38.4	34.6	34.1	26.4	73.5	18.5	13.7	54.9	53.2	56.0	39.5
4th quarter	42.1	38.9	35.2	35.4	27.0	80.4	18.4	14.5	54.8	53.4	55.7	40.4
1981												
1st quarter	42.9	39.1	35.5	35.7	27.5	78.4	18.8	15.1	55.6	54.5	56.1	41.3
2nd quarter	42.6	39.1	35.3	36.0	28.1	75.9	18.9	15.1	55.7	55.9	55.1	41.0
3rd quarter	43.1	39.2	35.6	36.0	28.8	70.0	18.5	14.9	55.5	55.7	54.9	41.5
4th quarter	42.6	39.0	35.2	35.9	29.4	62.7	18.5	15.4	56.2	56.6	55.3	41.2

Table 1-5B. Chain-Type Quantity Indexes for Gross Domestic Product and Domestic Purchases: Historical Data—*Continued*

(Index numbers, 2012 = 100.)   **NIPA Tables 1.1.3, 1.4.3, 2.3.3**

Year and quarter	Gross domestic product												Gross domestic purchases
	Gross domestic product, total	Personal consumption expenditures		Private fixed investment			Exports and imports of goods and services		Government consumption expenditures and gross investment			
		Total	Excluding food and energy	Total	Nonresidential	Residential	Exports	Imports	Total	Federal	State and local	
1982												
1st quarter	42.0	39.3	35.4	34.9	28.8	59.1	17.7	14.9	56.1	56.7	55.2	40.6
2nd quarter	42.1	39.4	35.6	33.8	27.9	57.2	17.8	14.7	56.4	56.9	55.4	40.7
3rd quarter	42.0	39.6	35.9	33.0	27.1	57.2	17.1	15.4	56.8	57.9	55.4	40.9
4th quarter	42.0	40.3	36.7	33.1	26.6	61.7	16.3	14.8	57.7	59.2	55.8	40.9
1983												
1st quarter	42.6	40.7	37.1	33.8	26.2	72.3	16.5	15.1	58.2	60.2	56.0	41.5
2nd quarter	43.5	41.6	37.9	35.1	26.6	80.7	16.6	16.3	58.7	61.3	55.9	42.7
3rd quarter	44.4	42.3	38.6	37.0	27.7	88.6	16.8	17.5	59.7	62.9	56.4	43.7
4th quarter	45.3	43.0	39.3	39.1	29.4	92.3	17.2	18.4	58.8	60.8	56.4	44.8
1984												
1st quarter	46.2	43.3	39.9	40.3	30.3	95.1	17.5	19.9	59.4	61.3	57.1	45.9
2nd quarter	47.0	43.9	40.4	41.9	31.8	96.7	18.0	20.7	60.7	63.2	57.9	46.8
3rd quarter	47.5	44.3	40.8	42.7	32.7	95.6	18.4	21.2	61.2	63.1	58.9	47.3
4th quarter	47.8	44.8	41.4	43.5	33.5	95.7	18.7	21.9	62.5	65.2	59.5	47.7
1985												
1st quarter	48.3	45.6	42.1	43.9	33.9	95.9	18.8	21.4	63.1	65.8	60.2	48.0
2nd quarter	48.7	46.0	42.6	44.5	34.4	96.7	18.8	22.4	64.5	67.5	61.3	48.7
3rd quarter	49.5	46.9	43.5	44.3	34.0	98.5	18.5	22.2	66.0	69.5	62.3	49.4
4th quarter	49.8	47.0	43.5	45.1	34.6	100.7	19.0	23.1	66.3	69.5	62.8	49.8
1986												
1st quarter	50.3	47.4	43.9	45.2	34.2	104.8	19.6	23.1	66.8	69.3	63.9	50.2
2nd quarter	50.5	47.9	44.5	45.3	33.5	110.8	19.8	24.1	68.2	71.6	64.5	50.6
3rd quarter	51.0	48.8	45.5	45.1	33.2	112.3	20.3	24.7	69.7	74.1	65.1	51.1
4th quarter	51.3	49.1	45.7	45.4	33.5	112.6	21.0	24.9	69.4	73.2	65.4	51.3
1987												
1st quarter	51.7	49.1	45.8	44.6	32.7	112.4	21.1	24.8	69.8	73.7	65.7	51.6
2nd quarter	52.2	49.8	46.5	45.3	33.3	112.8	21.9	25.4	70.3	74.7	65.9	52.1
3rd quarter	52.7	50.4	47.2	46.1	34.3	111.9	22.9	25.9	70.4	74.6	66.1	52.5
4th quarter	53.6	50.5	47.3	46.1	34.2	112.1	23.7	26.5	71.5	75.8	67.0	53.4
1988												
1st quarter	53.9	51.4	48.1	46.1	34.5	110.1	25.1	26.3	70.9	73.7	67.7	53.4
2nd quarter	54.6	51.7	48.4	47.0	35.3	111.4	25.8	26.0	71.1	73.1	68.5	53.9
3rd quarter	54.9	52.2	48.9	47.2	35.5	111.5	26.3	26.7	71.1	72.8	68.9	54.2
4th quarter	55.6	52.8	49.5	47.8	36.0	112.2	27.1	27.5	72.5	74.8	69.7	55.0
1989												
1st quarter	56.2	53.0	49.7	48.2	36.6	111.1	27.9	27.6	72.2	73.4	70.2	55.4
2nd quarter	56.6	53.3	50.1	48.3	37.1	107.7	29.1	27.7	73.3	74.9	71.0	55.6
3rd quarter	57.0	53.8	50.6	49.1	38.0	107.2	29.4	27.7	73.9	75.5	71.7	56.0
4th quarter	57.2	54.0	50.7	48.5	37.6	104.9	29.8	28.2	74.3	75.2	72.7	56.1
1990												
1st quarter	57.8	54.5	51.4	49.0	38.1	105.8	31.1	29.1	75.5	76.4	73.8	56.7
2nd quarter	58.0	54.7	51.4	48.1	37.6	101.6	31.5	29.0	75.6	76.5	73.9	56.8
3rd quarter	58.0	54.9	51.6	47.7	37.9	96.2	31.8	28.9	75.7	76.2	74.4	56.8
4th quarter	57.5	54.5	51.3	46.4	37.3	90.6	32.0	28.1	76.2	76.3	75.2	56.1
1991												
1st quarter	57.2	54.3	51.1	45.2	36.5	85.8	32.2	27.8	76.7	77.2	75.4	55.7
2nd quarter	57.7	54.7	51.4	45.3	36.3	88.5	33.4	28.3	77.0	77.5	75.7	56.1
3rd quarter	58.0	55.0	51.7	45.4	36.0	91.5	34.2	29.1	76.8	76.4	76.1	56.4
4th quarter	58.2	54.9	51.8	45.6	36.0	93.3	35.0	29.8	76.2	74.5	76.6	56.6
1992												
1st quarter	58.9	56.0	53.0	46.1	35.9	98.6	35.6	30.0	76.9	74.8	77.6	57.2
2nd quarter	59.5	56.4	53.3	47.6	37.0	102.0	35.6	30.5	76.7	74.6	77.5	57.9
3rd quarter	60.1	57.0	54.0	48.1	37.6	102.0	36.4	30.9	77.2	75.7	77.5	58.5
4th quarter	60.7	57.6	54.6	49.5	38.6	105.9	36.6	31.7	77.2	75.7	77.5	59.2
1993												
1st quarter	60.8	57.9	54.9	49.9	39.0	106.0	36.6	32.4	76.3	73.2	77.8	59.4
2nd quarter	61.2	58.4	55.5	50.8	39.7	107.4	37.1	33.0	76.3	72.5	78.4	59.8
3rd quarter	61.5	59.0	56.1	51.6	40.1	111.4	37.0	33.5	76.4	72.2	78.8	60.2
4th quarter	62.3	59.6	56.6	53.7	41.5	117.2	38.2	34.8	76.7	72.3	79.2	61.1
1994												
1st quarter	62.9	60.2	57.3	54.3	41.9	118.9	38.6	35.6	75.7	69.8	79.5	61.7
2nd quarter	63.8	60.7	57.7	55.4	42.6	122.8	39.9	37.0	76.1	69.5	80.4	62.6
3rd quarter	64.1	61.2	58.3	55.8	43.3	121.0	41.3	38.0	77.4	71.2	81.4	62.9
4th quarter	64.9	61.8	59.0	57.2	45.0	119.1	42.3	39.1	76.7	69.3	81.7	63.7
1995												
1st quarter	65.1	62.0	59.2	58.4	46.7	116.4	43.1	39.9	77.0	69.1	82.3	64.0
2nd quarter	65.3	62.5	59.7	58.4	47.1	112.8	43.7	40.3	77.2	68.9	82.9	64.1
3rd quarter	65.8	63.1	60.3	59.2	47.4	117.0	45.6	40.4	77.0	68.3	83.0	64.4
4th quarter	66.3	63.5	60.8	60.3	48.3	119.2	46.3	41.0	76.2	66.0	83.5	64.8

Table 1-5B. Chain-Type Quantity Indexes for Gross Domestic Product and Domestic Purchases: Historical Data—*Continued*

(Index numbers, 2012 = 100.)   **NIPA Tables 1.1.3, 1.4.3, 2.3.3**

Year and quarter	Gross domestic product											Gross domestic purchases
	Gross domestic product, total	Personal consumption expenditures		Private fixed investment			Exports and imports of goods and services		Government consumption expenditures and gross investment			
		Total	Excluding food and energy	Total	Nonresidential	Residential	Exports	Imports	Total	Federal	State and local	
1996												
1st quarter	66.8	64.1	61.3	61.9	49.6	122.3	46.8	42.3	76.7	67.3	83.3	65.5
2nd quarter	67.9	64.8	62.1	63.8	50.9	127.7	47.6	43.2	77.7	67.9	84.5	66.6
3rd quarter	68.5	65.2	62.6	65.2	52.5	127.4	48.0	44.6	77.7	67.0	85.2	67.4
4th quarter	69.2	65.7	63.1	66.3	53.8	126.0	51.0	45.5	78.2	66.7	86.4	67.8
1997												
1st quarter	69.7	66.4	63.9	67.5	55.1	126.6	51.9	47.5	77.8	65.8	86.5	68.5
2nd quarter	70.8	66.7	64.2	68.8	56.2	128.4	54.0	49.1	78.8	67.2	87.1	69.6
3rd quarter	71.7	67.9	65.4	71.2	58.7	129.6	55.2	50.9	79.2	66.9	88.0	70.5
4th quarter	72.3	68.7	66.2	71.8	59.1	131.0	55.2	52.0	79.6	66.8	88.8	71.3
1998												
1st quarter	73.1	69.4	67.0	73.9	61.0	133.7	55.4	53.9	79.3	65.2	89.6	72.3
2nd quarter	73.7	70.6	68.2	76.4	63.0	138.3	54.8	55.2	80.7	66.5	91.1	73.3
3rd quarter	74.7	71.5	69.2	77.9	64.0	142.4	54.6	55.9	81.3	66.2	92.3	74.3
4th quarter	75.9	72.5	70.4	80.0	65.9	145.7	56.6	57.7	81.8	66.6	92.9	75.5
1999												
1st quarter	76.6	73.2	71.0	81.4	67.4	146.3	56.5	59.2	82.3	66.6	93.9	76.5
2nd quarter	77.2	74.3	72.1	83.4	69.1	148.6	57.0	60.9	82.6	66.5	94.5	77.2
3rd quarter	78.2	75.1	72.9	85.3	71.1	149.7	58.7	63.1	83.6	67.6	95.4	78.4
4th quarter	79.5	76.2	74.1	85.8	71.4	150.8	60.2	64.6	84.9	68.9	96.7	79.7
2000												
1st quarter	79.8	77.4	75.6	88.2	74.1	151.5	61.0	67.2	84.3	66.7	97.4	80.3
2nd quarter	81.3	78.2	76.2	90.2	76.4	150.8	62.8	69.3	85.2	68.7	97.2	81.8
3rd quarter	81.4	78.9	77.0	90.6	77.2	148.4	64.3	71.7	84.8	67.4	97.6	82.1
4th quarter	81.9	79.6	77.6	90.8	77.5	148.6	63.7	71.7	85.3	67.5	98.4	82.7
2001												
1st quarter	81.6	79.9	78.0	90.3	76.8	149.2	62.8	70.5	86.6	69.0	99.6	82.4
2nd quarter	82.1	80.1	78.4	89.2	75.2	151.5	60.7	68.4	88.1	70.0	101.5	82.8
3rd quarter	81.8	80.4	78.6	88.6	74.3	152.5	57.8	67.0	88.0	70.4	101.0	82.6
4th quarter	82.0	81.6	80.0	86.6	72.2	151.5	55.9	66.1	89.5	71.3	102.9	82.9
2002												
1st quarter	82.7	81.8	80.3	85.8	70.5	156.4	57.2	68.0	91.0	73.6	103.8	83.8
2nd quarter	83.2	82.3	80.6	85.7	69.7	160.3	58.8	70.3	91.7	75.0	104.0	84.4
3rd quarter	83.6	82.9	81.2	85.6	69.4	160.8	59.1	71.2	92.2	75.8	104.3	84.8
4th quarter	83.7	83.3	81.7	85.2	68.5	163.7	58.1	72.4	92.9	77.1	104.6	85.3
2003												
1st quarter	84.2	83.7	82.1	86.0	68.8	166.7	58.2	72.0	93.0	77.9	104.1	85.7
2nd quarter	84.9	84.7	83.2	88.1	70.7	169.4	58.1	73.0	93.9	80.6	103.6	86.5
3rd quarter	86.3	85.9	84.5	91.0	72.2	179.1	59.7	74.2	94.1	80.6	104.0	87.9
4th quarter	87.3	86.5	85.0	92.6	73.2	184.4	62.3	76.5	94.7	82.1	103.8	88.9
2004												
1st quarter	87.8	87.3	85.8	92.4	72.4	186.5	64.0	78.9	95.1	83.0	103.9	89.5
2nd quarter	88.5	87.8	86.4	94.9	74.2	192.5	64.9	82.1	95.4	83.5	104.0	90.5
3rd quarter	89.3	88.8	87.5	96.9	76.4	193.8	65.4	83.2	95.6	84.4	103.6	91.5
4th quarter	90.2	89.7	88.5	98.7	77.9	196.5	66.9	85.1	95.5	84.2	103.6	92.4
2005												
1st quarter	91.2	90.3	88.9	100.4	79.1	201.1	68.5	85.9	96.0	85.2	103.7	93.3
2nd quarter	91.6	91.3	90.0	102.1	80.3	204.9	69.7	87.0	95.9	85.1	103.7	93.8
3rd quarter	92.4	92.2	91.0	103.9	82.0	207.1	69.9	87.6	96.3	85.7	103.8	94.6
4th quarter	93.0	92.4	91.3	104.3	82.6	206.7	71.9	90.4	96.3	85.8	103.9	95.4
2006												
1st quarter	94.3	93.5	92.6	106.3	85.4	204.8	74.6	92.3	97.5	88.0	104.3	96.5
2nd quarter	94.5	94.0	92.9	105.7	86.9	194.5	76.1	93.3	97.5	87.4	104.7	96.7
3rd quarter	94.6	94.5	93.4	105.0	88.4	183.7	76.1	94.2	97.4	86.6	105.1	97.0
4th quarter	95.4	95.4	94.4	104.3	89.3	175.2	79.3	94.3	98.2	87.8	105.5	97.4
2007												
1st quarter	95.7	96.0	94.9	104.2	91.0	167.3	80.4	95.4	98.3	87.3	106.2	97.6
2nd quarter	96.2	96.1	95.2	104.7	93.0	160.7	81.6	96.0	99.2	88.8	106.7	98.1
3rd quarter	96.7	96.7	95.8	104.1	94.4	150.5	84.1	96.4	99.7	89.5	106.9	98.4
4th quarter	97.3	97.0	96.2	103.1	95.9	138.0	86.6	95.6	100.4	91.0	107.1	98.5
2008												
1st quarter	96.8	96.7	96.0	101.5	96.3	127.0	87.3	95.9	100.7	92.4	106.5	97.9
2nd quarter	97.3	96.9	96.3	100.7	96.5	121.2	90.1	95.0	101.5	94.3	106.6	97.9
3rd quarter	96.7	96.1	95.8	98.1	94.7	115.1	89.5	93.8	102.3	95.5	107.1	97.2
4th quarter	94.6	95.2	94.6	91.7	89.1	103.9	84.5	90.3	103.0	96.7	107.4	95.2
2009												
1st quarter	93.6	95.0	94.4	84.4	82.4	93.8	77.7	81.4	104.1	97.9	108.5	93.7
2nd quarter	93.4	94.6	93.9	81.3	79.9	88.1	77.9	78.0	105.6	100.2	109.4	93.0
3rd quarter	93.8	95.3	94.6	81.6	79.5	92.1	80.8	81.3	105.9	101.1	109.3	93.4
4th quarter	94.8	95.2	94.3	82.0	80.0	92.0	85.5	85.2	106.1	102.7	108.5	94.5

Table 1-5B. Chain-Type Quantity Indexes for Gross Domestic Product and Domestic Purchases:
   Historical Data—*Continued*

(Index numbers, 2012 = 100.)   NIPA Tables 1.1.3, 1.4.3, 2.3.3

Year and quarter	Gross domestic product, total	Personal consumption expenditures		Private fixed investment			Exports and imports of goods and services		Government consumption expenditures and gross investment			Gross domestic purchases
		Total	Excluding food and energy	Total	Nonresi-dential	Residential	Exports	Imports	Total	Federal	State and local	
2010												
1st quarter	95.2	95.6	94.8	82.0	80.5	89.0	86.8	87.3	105.7	103.9	107.0	95.0
2nd quarter	96.1	96.4	95.7	85.0	83.2	93.6	88.9	91.3	106.0	105.2	106.6	96.2
3rd quarter	96.8	97.1	96.4	85.4	85.4	85.3	91.3	94.7	105.3	104.8	105.7	97.1
4th quarter	97.2	97.7	97.1	87.0	87.1	86.7	94.0	95.4	104.7	104.6	104.7	97.4
2011												
1st quarter	97.0	98.1	97.7	86.9	87.0	86.4	94.8	96.1	103.4	103.2	103.5	97.1
2nd quarter	97.7	98.3	98.1	88.9	89.2	87.4	96.3	96.6	102.7	102.8	102.6	97.7
3rd quarter	97.7	98.7	98.4	92.7	93.4	89.0	97.3	97.7	101.2	100.6	101.7	97.7
4th quarter	98.8	98.9	98.8	95.0	95.8	91.3	98.4	99.1	101.2	101.0	101.3	98.9
2012												
1st quarter	99.6	99.7	99.8	98.0	98.2	96.8	99.0	99.6	100.7	101.0	100.5	99.7
2nd quarter	100.0	99.8	99.7	100.0	100.4	97.6	100.0	100.1	100.2	100.2	100.2	100.0
3rd quarter	100.1	100.0	99.9	100.1	100.1	100.2	100.6	100.6	100.0	100.4	99.8	100.2
4th quarter	100.3	100.5	100.5	101.9	101.2	105.4	100.4	99.7	99.1	98.4	99.5	100.1
2013												
1st quarter	101.1	101.0	100.9	103.7	102.5	109.2	101.6	100.0	98.2	96.1	99.6	100.9
2nd quarter	101.3	101.0	101.0	104.5	102.8	112.7	102.8	101.4	98.0	95.4	99.8	101.1
3rd quarter	102.1	101.5	101.5	106.3	104.5	114.7	103.4	102.1	97.5	94.0	99.9	101.9
4th quarter	102.9	102.3	102.3	107.7	106.7	112.9	106.5	102.7	96.7	92.4	99.7	102.4
2014												
1st quarter	102.6	102.7	102.6	108.8	108.2	112.1	105.7	103.9	96.3	92.3	99.1	102.4
2nd quarter	104.0	103.9	104.1	111.8	111.1	115.7	108.0	106.5	96.4	91.7	99.6	103.9
3rd quarter	105.2	105.0	105.4	114.0	113.3	117.4	108.5	106.8	97.1	92.7	100.0	105.1
4th quarter	105.8	106.3	106.6	115.3	114.0	121.6	109.5	109.4	97.0	91.4	100.9	105.9
2015												
1st quarter	106.8	107.2	107.5	115.6	113.9	124.0	108.7	111.1	97.5	91.8	101.4	107.3
2nd quarter	107.6	108.0	108.6	116.6	114.4	127.1	109.0	111.8	98.3	91.9	102.7	108.0
3rd quarter	108.0	108.9	109.5	117.6	114.8	130.5	108.0	112.7	98.8	91.8	103.5	108.7
4th quarter	108.1	109.5	110.3	117.3	113.9	132.8	107.9	113.0	99.1	92.5	103.6	108.9
2016												
1st quarter	108.7	110.4	111.1	117.9	113.6	137.3	107.6	112.9	100.1	92.9	105.0	109.5
2nd quarter	109.1	111.1	111.7	118.3	114.2	136.7	108.0	112.8	99.9	92.2	105.2	109.8
3rd quarter	109.7	111.8	112.4	119.3	115.6	136.0	109.8	114.2	100.4	92.6	105.7	110.3
4th quarter	110.4	112.5	113.2	120.1	116.0	138.3	109.4	116.1	100.6	92.6	106.1	111.3
2017												
1st quarter	111.0	113.4	114.2	122.2	117.7	142.2	111.6	117.3	100.6	92.2	106.4	111.8
2nd quarter	111.5	113.9	114.6	122.6	118.4	141.8	111.9	118.4	101.0	92.7	106.7	112.4
3rd quarter	112.3	114.5	115.3	123.0	119.0	141.2	112.6	118.9	101.1	92.6	106.8	113.2
4th quarter	113.4	115.7	116.4	125.8	121.6	144.8	115.7	122.6	101.8	93.7	107.3	114.4
2018												
1st quarter	114.4	116.3	117.0	128.4	125.2	143.6	116.5	122.7	102.1	94.2	107.6	115.4
2nd quarter	115.2	117.2	118.0	129.8	127.1	143.0	117.1	122.7	102.9	95.0	108.2	116.0
3rd quarter	115.8	118.0	118.9	130.1	127.9	141.0	115.5	125.1	103.5	96.1	108.6	117.1
4th quarter	116.2	118.4	119.3	130.9	129.5	139.1	116.3	126.3	103.3	96.5	107.9	117.6
2019												
1st quarter	117.0	119.0	120.0	131.8	130.8	138.5	116.8	125.7	103.9	96.8	108.8	118.3
2nd quarter	117.4	120.0	121.1	131.7	130.8	137.8	115.5	126.2	105.2	99.0	109.5	118.9
3rd quarter	118.2	120.8	121.9	132.5	131.4	139.3	115.8	126.3	105.8	100.2	109.6	119.6
4th quarter	118.9	121.3	122.5	132.8	131.3	141.3	116.7	123.9	106.4	101.1	110.0	119.9

Table 1-6A. Chain-Type Price Indexes for Gross Domestic Product and Domestic Purchases: Recent Data

(Index numbers, 2012 = 100.)   **NIPA Tables 1.1.4, 1.6.4, 2.3.4**

Year and quarter	Gross domestic product											Gross domestic purchases
	Gross domestic product, total	Personal consumption expenditures		Private fixed investment			Exports and imports of goods and services		Government consumption expenditures and gross investment			
		Total	Excluding food and energy	Total	Nonresidential	Residential	Exports	Imports	Total	Federal	State and local	
1955	14.8	14.8	15.3	22.0	26.3	12.0	22.2	16.8	9.6	11.8	8.0	14.5
1956	15.3	15.1	15.7	23.3	28.2	12.3	22.9	17.1	10.1	12.4	8.5	14.9
1957	15.8	15.6	16.2	24.1	29.6	12.4	23.8	17.3	10.5	12.9	8.8	15.4
1958	16.2	16.0	16.5	24.1	29.8	12.3	23.6	16.6	10.8	13.4	9.0	15.8
1959	16.4	16.2	16.9	24.4	30.2	12.4	23.6	16.7	11.0	13.6	9.1	16.0
1960	16.6	16.5	17.2	24.5	30.4	12.5	24.0	16.9	11.1	13.7	9.3	16.2
1961	16.8	16.6	17.4	24.5	30.3	12.5	24.3	16.9	11.3	13.8	9.5	16.4
1962	17.0	16.8	17.6	24.5	30.3	12.5	24.4	16.7	11.5	14.0	9.8	16.6
1963	17.2	17.0	17.9	24.4	30.3	12.4	24.3	17.0	11.8	14.4	10.0	16.8
1964	17.5	17.3	18.1	24.6	30.5	12.5	24.5	17.4	12.1	14.8	10.2	17.0
1965	17.8	17.5	18.4	25.0	30.9	12.8	25.3	17.6	12.4	15.2	10.5	17.3
1966	18.3	18.0	18.8	25.5	31.3	13.3	26.1	18.0	12.9	15.7	11.0	17.8
1967	18.8	18.4	19.4	26.1	32.1	13.8	27.1	18.1	13.4	16.0	11.6	18.3
1968	19.6	19.2	20.2	27.1	33.2	14.5	27.7	18.4	14.1	16.8	12.2	19.1
1969	20.6	20.0	21.1	28.4	34.6	15.5	28.6	18.8	14.9	17.7	13.1	20.0
1970	21.7	21.0	22.1	29.6	36.3	16.0	29.7	20.0	16.1	19.1	14.1	21.1
1971	22.8	21.8	23.2	31.1	38.0	16.9	30.8	21.2	17.4	20.7	15.2	22.2
1972	23.8	22.6	23.9	32.4	39.3	18.0	32.1	22.7	18.7	22.5	16.2	23.2
1973	25.1	23.8	24.8	34.2	40.9	19.6	36.4	26.6	19.9	24.1	17.2	24.5
1974	27.3	26.3	26.8	37.6	44.9	21.6	44.8	38.1	21.9	26.0	19.2	27.0
1975	29.8	28.5	29.0	42.1	50.8	23.6	49.4	41.2	23.9	28.3	21.0	29.5
1976	31.5	30.0	30.8	44.4	53.6	25.1	51.0	42.5	25.2	30.0	22.0	31.1
1977	33.4	32.0	32.8	47.7	57.1	27.7	53.1	46.2	26.7	31.9	23.4	33.1
1978	35.8	34.2	34.9	51.5	60.9	31.1	56.3	49.5	28.5	34.0	24.9	35.5
1979	38.8	37.3	37.5	56.1	65.8	34.6	63.1	57.9	30.9	36.6	27.1	38.6
1980	42.3	41.3	40.9	61.4	71.6	38.3	69.5	72.2	34.0	40.1	30.1	42.6
1981	46.3	45.0	44.5	67.1	78.5	41.4	74.7	76.1	37.4	43.8	33.2	46.5
1982	49.1	47.5	47.4	70.7	82.9	43.6	75.0	73.5	40.0	46.9	35.4	49.2
1983	51.1	49.5	49.8	70.9	82.8	44.7	75.3	70.8	41.5	48.5	37.0	50.9
1984	52.9	51.3	51.9	71.7	83.0	46.0	76.0	70.1	43.3	50.6	38.5	52.6
1985	54.6	53.1	54.0	72.5	83.9	47.3	73.8	67.8	44.7	51.7	40.1	54.2
1986	55.7	54.3	55.9	74.2	85.4	49.4	72.5	67.8	45.4	52.0	41.3	55.3
1987	57.0	56.0	57.7	75.7	86.3	51.5	74.1	71.9	46.6	52.3	43.2	56.9
1988	59.1	58.2	60.1	77.6	88.5	53.3	77.9	75.4	48.2	54.0	44.6	58.9
1989	61.4	60.7	62.6	79.6	90.6	55.0	79.2	77.0	50.0	55.5	46.8	61.2
1990	63.7	63.4	65.2	81.3	92.5	56.3	79.7	79.2	52.1	57.3	49.2	63.7
1991	65.8	65.5	67.5	82.6	94.3	57.0	80.5	78.6	54.0	59.3	51.0	65.7
1992	67.3	67.2	69.5	82.6	94.0	57.7	80.2	78.6	55.6	60.8	52.7	67.2
1993	68.9	68.9	71.4	83.6	94.2	60.1	80.3	78.0	57.0	62.2	54.0	68.7
1994	70.4	70.3	73.0	84.9	94.9	62.2	81.2	78.8	58.5	63.9	55.4	70.1
1995	71.9	71.8	74.6	86.2	95.8	64.5	83.0	80.9	60.1	65.8	56.9	71.7
1996	73.2	73.3	76.0	86.2	95.3	65.9	81.9	79.5	61.4	66.9	58.2	72.9
1997	74.4	74.6	77.4	86.2	94.7	67.4	80.5	76.8	62.6	68.0	59.5	74.0
1998	75.3	75.2	78.4	85.6	93.2	69.2	78.6	72.6	63.6	68.8	60.6	74.5
1999	76.3	76.3	79.4	85.7	92.3	71.8	78.0	73.0	65.8	70.5	63.0	75.6
2000	78.1	78.2	80.8	86.8	92.7	75.0	79.5	76.2	68.6	72.9	66.0	77.6
2001	79.8	79.7	82.3	87.6	92.3	78.6	78.8	74.2	70.6	74.2	68.3	79.0
2002	81.0	80.8	83.6	87.8	91.9	80.5	78.2	73.2	72.4	76.6	69.8	80.1
2003	82.6	82.4	84.8	88.6	91.2	84.3	79.4	75.5	75.0	80.0	72.1	81.8
2004	84.8	84.4	86.5	91.1	92.1	90.2	82.3	79.1	78.2	82.8	75.4	84.1
2005	87.4	86.8	88.4	94.8	94.4	96.7	85.1	83.7	82.1	86.2	79.6	87.0
2006	90.1	89.2	90.4	98.2	96.7	102.4	87.8	86.9	85.7	88.9	83.6	89.8
2007	92.5	91.4	92.4	99.7	98.3	103.7	91.1	89.9	89.5	91.6	88.1	92.2
2008	94.3	94.2	94.2	100.5	99.8	102.2	95.4	99.0	93.3	94.4	92.6	94.8
2009	95.0	94.1	95.3	99.3	99.2	98.7	89.7	88.0	92.9	94.2	92.0	94.6
2010	96.1	95.7	96.6	97.7	97.4	98.3	93.3	92.8	95.4	96.4	94.7	95.9
2011	98.1	98.1	98.1	98.7	98.6	99.0	99.2	99.8	98.3	99.1	97.7	98.2
2012	100.0	100.0	100.0	100.0	100.0	100.0	100.0	100.0	100.0	100.0	100.0	100.0
2013	101.8	101.3	101.5	101.0	100.3	105.1	100.2	98.6	102.3	100.9	103.3	101.5
2014	103.6	102.8	103.1	102.9	101.5	111.1	100.3	97.9	104.4	102.6	105.6	103.1
2015	104.6	103.0	104.4	103.5	101.9	114.1	95.4	90.0	104.6	103.1	105.6	103.4
2016	105.7	104.1	106.1	103.5	101.1	118.1	93.5	86.9	104.9	103.7	105.8	104.2
2017	107.8	106.0	107.9	105.2	102.0	123.5	95.9	88.8	107.4	105.8	108.5	106.1
2018	110.3	108.2	110.0	107.2	102.9	130.5	99.2	91.3	111.3	109.1	112.8	108.6
2019	112.3	109.9	111.9	109.0	104.3	134.2	98.8	90.0	113.4	111.1	115.0	110.3
2017												
1st quarter	107.0	105.4	107.3	104.5	101.6	121.4	95.1	88.4	106.5	106.5	107.5	105.5
2nd quarter	107.4	105.7	107.6	105.1	101.9	123.0	95.2	88.3	106.8	106.8	107.7	105.8
3rd quarter	108.0	106.1	108.0	105.6	102.2	124.3	96.0	88.6	107.6	107.6	108.7	106.3
4th quarter	108.6	106.8	108.6	105.8	102.2	125.3	97.3	89.8	108.7	108.7	110.0	107.0
2018												
1st quarter	109.3	107.5	109.2	106.4	102.4	128.0	98.2	91.2	109.9	109.9	111.2	107.7
2nd quarter	110.2	108.1	109.8	107.1	102.8	130.2	99.4	91.3	110.9	110.9	112.4	108.4
3rd quarter	110.7	108.5	110.2	107.6	103.1	131.4	99.7	91.6	111.8	111.8	113.4	108.9
4th quarter	111.2	108.9	110.7	107.8	103.3	132.2	99.4	91.2	112.6	112.6	114.1	109.3
2019												
1st quarter	111.5	109.0	111.1	108.5	103.9	133.0	98.6	90.3	112.9	112.9	113.9	109.6
2nd quarter	112.2	109.7	111.7	109.0	104.3	133.6	99.3	90.5	113.3	113.3	114.9	110.2
3rd quarter	112.6	110.1	112.2	109.2	104.5	134.7	98.8	89.7	113.5	113.5	115.3	110.6
4th quarter	113.0	110.5	112.6	109.3	104.3	135.5	98.4	89.4	114.0	114.0	115.8	110.9

Table 1-6B. Chain-Type Price Indexes for Gross Domestic Product and Domestic Purchases: Historical Data

(Index numbers, 2012 = 100.)   NIPA Tables 1.1.4, 1.6.4, 2.3.4

Year and quarter	Gross domestic product												Gross domestic purchases
	Gross domestic product, total	Personal consumption expenditures		Private fixed investment			Exports and imports of goods and services		Government consumption expenditures and gross investment			
		Total	Excluding food and energy	Total	Nonresidential	Residential	Exports	Imports	Total	Federal	State and local	
1929	9.4	9.3	9.6	14.0	13.5	5.1	13.7	9.6	5.3	6.6	3.7	9.1
1930	9.0	8.9	9.2	13.4	12.8	5.0	12.4	8.2	5.2	6.3	3.6	8.8
1931	8.1	8.0	8.5	12.3	11.9	4.5	9.7	6.6	4.9	6.2	3.4	7.9
1932	7.2	7.0	7.5	10.4	10.8	3.7	8.4	5.3	4.5	5.9	3.0	7.0
1933	7.0	6.8	7.2	9.5	10.6	3.6	8.4	5.1	4.6	6.0	3.1	6.8
1934	7.4	7.1	7.3	9.5	10.9	4.0	9.8	5.8	4.9	6.4	3.4	7.1
1935	7.5	7.3	7.4	9.4	11.0	4.0	10.0	5.9	5.0	6.4	3.4	7.3
1936	7.6	7.3	7.5	9.5	11.0	4.1	10.4	6.3	5.2	7.0	3.4	7.4
1937	7.9	7.6	7.8	10.2	11.9	4.5	11.1	7.0	5.3	7.1	3.5	7.6
1938	7.7	7.4	7.8	9.9	12.0	4.7	10.5	6.5	5.3	7.1	3.5	7.5
1939	7.6	7.3	7.7	9.9	11.9	4.7	10.4	6.8	5.2	7.0	3.5	7.4
1940	7.7	7.4	7.8	10.1	12.1	4.9	11.3	7.2	5.1	6.8	3.5	7.5
1941	8.2	7.9	8.2	10.8	12.9	5.3	12.3	7.6	5.5	7.2	3.7	7.9
1942	8.9	8.8	9.1	12.0	14.3	5.7	14.9	8.7	5.5	7.2	4.1	8.6
1943	9.3	9.7	9.8	12.6	14.9	6.2	16.3	9.4	5.5	7.1	4.4	9.0
1944	9.5	10.2	10.5	13.1	15.4	6.8	18.4	9.9	5.5	7.0	4.5	9.2
1945	9.8	10.6	11.0	13.4	15.8	7.2	18.2	10.2	5.5	7.0	4.7	9.4
1946	11.0	11.4	11.7	14.7	17.4	7.8	17.6	11.3	7.0	9.2	5.1	10.7
1947	12.3	12.5	12.7	17.4	20.1	9.4	20.4	13.6	7.6	9.9	5.8	11.9
1948	12.9	13.2	13.4	18.8	21.8	10.2	21.5	14.8	7.9	9.9	6.5	12.6
1949	12.9	13.1	13.4	19.0	22.3	10.3	20.2	14.1	8.2	10.3	6.5	12.6
1947												
1st quarter	12.0	12.3	. . .	16.5	19.3	8.9	18.7	12.5	7.7	10.2	5.6	11.7
2nd quarter	12.1	12.3	. . .	17.2	19.9	9.4	20.1	13.3	7.7	10.1	5.7	11.8
3rd quarter	12.3	12.6	. . .	17.7	20.4	9.6	21.1	14.0	7.5	9.7	5.8	12.0
4th quarter	12.6	12.9	. . .	18.0	20.8	9.8	21.7	14.5	7.6	9.7	6.0	12.2
1948												
1st quarter	12.7	13.0	. . .	18.2	20.9	10.0	21.8	14.9	7.8	9.8	6.2	12.4
2nd quarter	12.9	13.2	. . .	18.6	21.5	10.1	21.7	14.9	7.9	9.8	6.4	12.5
3rd quarter	13.1	13.4	. . .	19.1	22.2	10.3	21.4	14.8	8.0	9.9	6.6	12.7
4th quarter	13.1	13.3	. . .	19.2	22.5	10.4	21.1	14.5	8.1	10.0	6.7	12.7
1949												
1st quarter	13.1	13.2	. . .	19.1	22.4	10.5	20.7	14.2	8.2	10.3	6.6	12.7
2nd quarter	13.0	13.1	. . .	19.1	22.3	10.4	20.3	14.1	8.1	10.3	6.5	12.6
3rd quarter	12.9	13.1	. . .	18.9	22.3	10.2	20.0	14.0	8.1	10.2	6.5	12.5
4th quarter	12.9	13.1	. . .	18.8	22.1	10.2	19.8	14.0	8.2	10.4	6.5	12.6
1950												
1st quarter	12.8	13.0	. . .	18.8	22.2	10.2	19.5	14.3	8.2	10.4	6.4	12.5
2nd quarter	12.9	13.1	. . .	19.1	22.4	10.5	19.5	14.6	8.1	10.3	6.4	12.6
3rd quarter	13.2	13.4	. . .	19.6	22.8	10.9	19.7	15.1	8.3	10.4	6.6	12.9
4th quarter	13.4	13.6	. . .	20.0	23.6	10.8	20.1	15.8	8.4	10.4	6.8	13.1
1951												
1st quarter	13.8	14.0	. . .	20.8	24.4	11.2	21.3	17.1	8.7	10.9	7.0	13.5
2nd quarter	13.9	14.2	. . .	21.1	24.8	11.3	22.0	18.0	8.7	10.7	7.2	13.6
3rd quarter	13.9	14.2	. . .	21.2	25.0	11.4	22.7	18.5	8.7	10.7	7.3	13.7
4th quarter	14.1	14.4	. . .	21.4	25.3	11.5	23.0	18.6	8.8	10.9	7.4	13.8
1952												
1st quarter	14.1	14.4	. . .	21.5	25.4	11.5	22.5	17.9	8.8	10.7	7.4	13.9
2nd quarter	14.2	14.4	. . .	21.5	25.4	11.6	22.4	17.6	8.9	10.9	7.5	13.9
3rd quarter	14.3	14.5	. . .	21.5	25.4	11.8	22.4	17.3	9.0	11.0	7.6	14.0
4th quarter	14.4	14.5	. . .	21.4	25.4	11.7	22.3	17.0	9.1	11.1	7.6	14.0
1953												
1st quarter	14.4	14.6	. . .	21.4	25.4	11.7	22.4	16.8	9.0	11.0	7.7	14.0
2nd quarter	14.4	14.6	. . .	21.5	25.6	11.7	22.4	16.7	9.0	11.0	7.7	14.1
3rd quarter	14.5	14.7	. . .	21.7	25.8	11.8	22.3	16.6	9.0	11.0	7.7	14.1
4th quarter	14.5	14.8	. . .	21.6	25.8	11.8	22.2	16.6	9.0	11.1	7.7	14.2
1954												
1st quarter	14.6	14.8	. . .	21.7	25.9	11.7	22.1	16.8	9.1	11.2	7.6	14.2
2nd quarter	14.6	14.8	. . .	21.7	26.0	11.7	22.0	16.9	9.2	11.2	7.8	14.3
3rd quarter	14.6	14.8	. . .	21.7	25.8	11.9	22.0	17.0	9.2	11.3	7.9	14.3
4th quarter	14.6	14.7	. . .	21.7	25.9	11.9	22.0	17.0	9.3	11.4	7.9	14.3
1955												
1st quarter	14.7	14.8	. . .	21.7	25.8	11.9	22.0	16.8	9.4	11.5	7.8	14.3
2nd quarter	14.7	14.8	. . .	21.9	26.0	12.0	22.1	16.8	9.5	11.8	7.9	14.4
3rd quarter	14.8	14.9	. . .	22.1	26.4	12.1	22.2	16.8	9.6	11.9	8.0	14.5
4th quarter	14.9	14.9	. . .	22.4	26.9	12.1	22.4	16.9	9.7	12.0	8.1	14.6
1956												
1st quarter	15.1	15.0	. . .	22.9	27.6	12.2	22.6	16.9	9.9	12.2	8.3	14.7
2nd quarter	15.2	15.1	. . .	23.1	27.9	12.4	22.8	17.0	10.1	12.4	8.4	14.9
3rd quarter	15.4	15.2	. . .	23.5	28.5	12.4	23.1	17.1	10.2	12.5	8.5	15.0
4th quarter	15.5	15.3	. . .	23.7	28.9	12.4	23.3	17.3	10.2	12.5	8.6	15.1
1957												
1st quarter	15.7	15.4	. . .	23.9	29.3	12.3	23.7	17.4	10.4	12.8	8.7	15.3
2nd quarter	15.8	15.5	. . .	24.0	29.5	12.3	23.8	17.4	10.5	12.9	8.8	15.4
3rd quarter	15.9	15.7	. . .	24.1	29.7	12.4	23.9	17.3	10.6	13.0	8.9	15.5
4th quarter	16.0	15.7	. . .	24.2	29.9	12.4	23.9	17.1	10.6	13.1	8.9	15.6

. . . = Not available.

Table 1-6B. Chain-Type Price Indexes for Gross Domestic Product and Domestic Purchases: Historical Data—*Continued*

(Index numbers, 2012 = 100.)   NIPA Tables 1.1.4, 1.6.4, 2.3.4

Year and quarter	Gross domestic product, total	Personal consumption expenditures		Private fixed investment			Exports and imports of goods and services		Government consumption expenditures and gross investment			Gross domestic purchases
		Total	Excluding food and energy	Total	Nonresidential	Residential	Exports	Imports	Total	Federal	State and local	
1958												
1st quarter	16.1	15.9	. . .	24.0	29.6	12.3	23.6	16.7	10.7	13.3	8.9	15.7
2nd quarter	16.2	16.0	. . .	24.1	29.8	12.3	23.5	16.5	10.8	13.4	9.0	15.8
3rd quarter	16.2	16.0	. . .	24.2	29.9	12.3	23.5	16.5	10.9	13.5	9.0	15.8
4th quarter	16.3	16.0	. . .	24.2	30.0	12.3	23.6	16.6	10.9	13.6	9.0	15.8
1959												
1st quarter	16.3	16.1	16.7	24.2	30.0	12.4	23.4	16.6	11.0	13.6	9.1	15.9
2nd quarter	16.4	16.2	16.8	24.3	30.2	12.4	23.5	16.7	11.0	13.6	9.1	16.0
3rd quarter	16.4	16.3	16.9	24.4	30.3	12.4	23.7	16.7	11.0	13.6	9.1	16.0
4th quarter	16.5	16.3	17.0	24.5	30.4	12.4	23.9	16.9	11.0	13.6	9.1	16.1
1960												
1st quarter	16.5	16.4	17.1	24.5	30.4	12.4	24.0	16.9	11.0	13.5	9.2	16.1
2nd quarter	16.6	16.4	17.2	24.5	30.4	12.5	23.9	16.8	11.0	13.5	9.3	16.2
3rd quarter	16.7	16.5	17.2	24.5	30.4	12.5	24.0	16.9	11.1	13.7	9.3	16.2
4th quarter	16.7	16.6	17.3	24.5	30.3	12.5	24.0	16.9	11.2	13.9	9.4	16.3
1961												
1st quarter	16.8	16.6	17.3	24.5	30.3	12.4	24.1	16.9	11.2	13.8	9.4	16.3
2nd quarter	16.8	16.6	17.4	24.5	30.3	12.5	24.4	16.9	11.3	13.8	9.5	16.4
3rd quarter	16.8	16.7	17.4	24.5	30.3	12.5	24.3	16.9	11.3	13.8	9.6	16.4
4th quarter	16.9	16.7	17.5	24.5	30.3	12.5	24.5	16.9	11.3	13.8	9.6	16.4
1962												
1st quarter	17.0	16.8	17.6	24.5	30.3	12.5	24.6	16.7	11.4	13.9	9.7	16.5
2nd quarter	17.0	16.8	17.6	24.5	30.3	12.5	24.3	16.7	11.5	14.0	9.8	16.6
3rd quarter	17.0	16.9	17.7	24.5	30.3	12.5	24.3	16.6	11.5	14.1	9.8	16.6
4th quarter	17.1	16.9	17.7	24.5	30.3	12.5	24.3	16.7	11.6	14.1	9.8	16.6
1963												
1st quarter	17.2	17.0	17.8	24.5	30.3	12.5	24.4	16.9	11.7	14.3	9.9	16.7
2nd quarter	17.2	17.0	17.8	24.4	30.3	12.4	24.3	17.0	11.8	14.3	10.0	16.7
3rd quarter	17.2	17.1	17.9	24.4	30.3	12.3	24.3	17.1	11.7	14.2	10.0	16.8
4th quarter	17.3	17.1	18.0	24.4	30.3	12.4	24.3	17.1	11.9	14.6	10.1	16.9
1964												
1st quarter	17.4	17.2	18.1	24.4	30.4	12.3	24.4	17.3	12.0	14.7	10.1	16.9
2nd quarter	17.4	17.3	18.1	24.6	30.5	12.5	24.4	17.4	12.0	14.7	10.2	17.0
3rd quarter	17.5	17.3	18.2	24.6	30.5	12.5	24.6	17.4	12.2	14.9	10.2	17.1
4th quarter	17.6	17.4	18.2	24.8	30.7	12.8	24.8	17.4	12.2	14.9	10.3	17.1
1965												
1st quarter	17.7	17.4	18.3	24.8	30.7	12.8	25.4	17.6	12.3	15.0	10.4	17.2
2nd quarter	17.7	17.5	18.3	24.9	30.8	12.8	25.3	17.5	12.3	15.0	10.5	17.3
3rd quarter	17.8	17.6	18.4	24.9	30.9	12.8	25.3	17.6	12.4	15.2	10.5	17.4
4th quarter	17.9	17.6	18.5	25.2	31.0	13.1	25.2	17.8	12.6	15.5	10.6	17.5
1966												
1st quarter	18.1	17.8	18.6	25.1	31.0	12.9	25.7	17.9	12.7	15.6	10.8	17.6
2nd quarter	18.2	17.9	18.7	25.5	31.3	13.4	25.9	18.1	12.8	15.6	11.0	17.8
3rd quarter	18.4	18.0	18.9	25.5	31.4	13.3	26.2	18.0	13.0	15.8	11.1	17.9
4th quarter	18.5	18.2	19.0	25.8	31.6	13.6	26.6	18.1	13.1	15.8	11.2	18.0
1967												
1st quarter	18.6	18.2	19.1	25.9	31.8	13.6	27.1	18.1	13.1	15.8	11.4	18.1
2nd quarter	18.7	18.3	19.3	26.0	32.0	13.7	27.1	18.1	13.3	15.9	11.5	18.2
3rd quarter	18.9	18.5	19.4	26.1	32.2	13.8	27.0	18.1	13.4	16.1	11.6	18.4
4th quarter	19.1	18.7	19.6	26.4	32.5	14.0	27.2	18.1	13.6	16.4	11.8	18.6
1968												
1st quarter	19.3	18.9	19.9	26.7	32.7	14.3	27.4	18.2	13.8	16.5	12.0	18.8
2nd quarter	19.5	19.0	20.1	26.9	33.1	14.4	27.9	18.3	14.0	16.7	12.2	19.0
3rd quarter	19.7	19.2	20.3	27.1	33.3	14.4	27.6	18.4	14.1	17.0	12.3	19.2
4th quarter	20.0	19.5	20.5	27.7	33.8	15.0	27.8	18.5	14.4	17.2	12.5	19.4
1969												
1st quarter	20.2	19.6	20.8	28.0	34.1	15.3	28.2	18.6	14.5	17.3	12.7	19.6
2nd quarter	20.4	19.9	21.0	28.3	34.4	15.5	28.2	18.7	14.7	17.5	13.0	19.9
3rd quarter	20.7	20.1	21.3	28.5	34.8	15.5	28.6	18.8	15.1	17.9	13.2	20.2
4th quarter	21.0	20.4	21.5	28.9	35.2	15.8	29.3	19.3	15.3	18.2	13.4	20.4
1970												
1st quarter	21.3	20.6	21.7	29.1	35.6	15.8	29.3	19.5	15.7	18.7	13.7	20.7
2nd quarter	21.6	20.8	22.0	29.7	36.2	16.3	29.9	19.7	15.9	18.9	14.0	21.0
3rd quarter	21.8	21.0	22.2	29.7	36.4	15.9	29.8	20.2	16.2	19.3	14.2	21.2
4th quarter	22.1	21.3	22.5	30.0	36.9	16.0	29.9	20.4	16.5	19.5	14.5	21.5
1971												
1st quarter	22.4	21.5	22.8	30.5	37.4	16.5	30.8	20.9	16.9	20.1	14.8	21.8
2nd quarter	22.7	21.8	23.1	31.0	37.9	16.8	30.8	21.0	17.2	20.5	15.1	22.1
3rd quarter	22.9	22.0	23.3	31.3	38.2	17.1	30.7	21.2	17.5	20.8	15.3	22.3
4th quarter	23.1	22.1	23.4	31.6	38.5	17.4	30.9	21.6	17.8	21.2	15.5	22.5
1972												
1st quarter	23.5	22.3	23.7	31.9	38.9	17.6	31.7	21.9	18.3	22.1	15.8	22.9
2nd quarter	23.6	22.5	23.8	32.2	39.2	17.7	32.0	22.5	18.5	22.3	16.0	23.0
3rd quarter	23.8	22.7	24.0	32.5	39.4	18.0	32.1	22.9	18.7	22.5	16.3	23.3
4th quarter	24.1	22.9	24.2	33.0	39.7	18.6	32.8	23.4	19.1	23.0	16.5	23.5

. . . = Not available.

Table 1-6B. Chain-Type Price Indexes for Gross Domestic Product and Domestic Purchases: Historical Data—*Continued*

(Index numbers, 2012 = 100.)   **NIPA Tables 1.1.4, 1.6.4, 2.3.4**

Year and quarter	Gross domestic product, total	Personal consumption expenditures Total	Personal consumption expenditures Excluding food and energy	Private fixed investment Total	Private fixed investment Nonresi-dential	Private fixed investment Residential	Exports and imports of goods and services Exports	Exports and imports of goods and services Imports	Government consumption expenditures and gross investment Total	Government consumption expenditures and gross investment Federal	Government consumption expenditures and gross investment State and local	Gross domestic purchases
1973												
1st quarter	24.4	23.1	24.3	33.3	40.0	18.8	33.8	24.1	19.4	23.4	16.8	23.8
2nd quarter	24.8	23.6	24.7	33.8	40.6	19.3	35.3	25.9	19.8	23.8	17.1	24.3
3rd quarter	25.3	24.0	25.0	34.6	41.2	19.9	37.3	27.2	20.1	24.3	17.4	24.7
4th quarter	25.7	24.5	25.3	35.0	41.7	20.3	39.2	29.2	20.5	24.8	17.7	25.2
1974												
1st quarter	26.2	25.2	25.8	35.7	42.5	20.7	41.9	33.9	20.9	25.0	18.2	25.8
2nd quarter	26.8	25.9	26.4	36.8	43.8	21.2	43.4	37.7	21.5	25.5	18.8	26.6
3rd quarter	27.7	26.6	27.2	38.2	45.6	21.9	45.8	39.7	22.2	26.2	19.5	27.4
4th quarter	28.5	27.3	27.8	39.6	47.6	22.5	48.1	40.9	22.9	27.1	20.1	28.2
1975												
1st quarter	29.1	27.8	28.4	41.0	49.3	23.1	49.5	41.6	23.3	27.6	20.5	28.8
2nd quarter	29.6	28.2	28.8	41.9	50.6	23.5	49.4	41.7	23.7	28.0	20.9	29.2
3rd quarter	30.1	28.7	29.2	42.4	51.2	23.7	49.2	40.8	24.0	28.4	21.2	29.7
4th quarter	30.6	29.2	29.7	43.0	51.9	24.1	49.5	40.8	24.4	29.0	21.4	30.2
1976												
1st quarter	30.9	29.5	30.2	43.4	52.5	24.3	50.2	41.5	24.8	29.5	21.7	30.5
2nd quarter	31.2	29.7	30.5	44.1	53.2	25.0	50.8	42.1	25.0	29.7	21.9	30.8
3rd quarter	31.6	30.2	31.0	44.7	53.9	25.4	51.1	42.9	25.2	30.0	22.1	31.2
4th quarter	32.2	30.7	31.5	45.4	54.7	25.8	52.0	43.3	25.7	30.8	22.4	31.7
1977												
1st quarter	32.7	31.2	32.0	46.3	55.7	26.5	52.6	44.9	26.1	31.3	22.8	32.3
2nd quarter	33.2	31.8	32.5	47.2	56.6	27.2	53.4	46.1	26.5	31.6	23.2	32.9
3rd quarter	33.7	32.2	33.0	48.1	57.6	28.1	53.1	46.7	26.8	31.8	23.6	33.4
4th quarter	34.2	32.7	33.5	49.1	58.6	29.0	53.3	47.2	27.4	32.7	24.0	33.9
1978												
1st quarter	34.8	33.2	34.0	50.0	59.4	29.8	54.4	48.0	27.8	33.1	24.3	34.5
2nd quarter	35.5	33.9	34.6	51.0	60.4	30.7	55.7	49.1	28.3	33.7	24.7	35.2
3rd quarter	36.1	34.5	35.2	52.0	61.4	31.5	56.6	49.9	28.7	34.3	25.1	35.8
4th quarter	36.8	35.2	35.9	53.1	62.5	32.3	58.5	50.8	29.2	34.9	25.5	36.5
1979												
1st quarter	37.5	35.8	36.4	54.2	63.8	33.0	60.2	52.8	29.8	35.5	26.1	37.2
2nd quarter	38.4	36.8	37.2	55.5	65.2	34.1	62.7	55.6	30.4	36.1	26.7	38.1
3rd quarter	39.2	37.7	37.8	56.9	66.6	35.2	64.2	59.7	31.3	36.9	27.6	39.1
4th quarter	39.9	38.6	38.6	58.0	67.8	36.1	65.3	63.7	31.9	37.7	28.1	40.0
1980												
1st quarter	40.8	39.8	39.6	59.4	69.3	37.0	67.3	69.0	32.7	38.5	28.9	41.1
2nd quarter	41.8	40.8	40.5	60.8	70.9	37.9	68.2	71.5	33.7	40.0	29.6	42.1
3rd quarter	42.7	41.7	41.4	62.0	72.4	38.8	70.1	73.5	34.4	40.4	30.5	43.1
4th quarter	43.8	42.8	42.4	63.4	73.9	39.6	72.5	74.8	35.4	41.6	31.3	44.1
1981												
1st quarter	45.0	43.9	43.3	65.2	76.0	40.6	74.3	76.7	36.3	42.4	32.3	45.3
2nd quarter	45.8	44.6	44.1	66.6	77.8	41.2	74.6	77.1	37.1	43.4	33.0	46.2
3rd quarter	46.7	45.3	44.9	67.7	79.2	41.7	74.7	75.1	37.8	44.4	33.5	46.9
4th quarter	47.5	46.0	45.7	69.0	80.9	42.3	75.0	75.3	38.4	45.2	34.0	47.7
1982												
1st quarter	48.2	46.6	46.4	69.8	81.9	42.9	75.4	75.1	39.1	45.9	34.6	48.3
2nd quarter	48.8	47.1	47.0	70.6	82.9	43.6	75.4	73.8	39.8	46.8	35.2	48.9
3rd quarter	49.5	47.8	47.8	71.1	83.3	43.9	74.8	72.9	40.3	47.2	35.7	49.6
4th quarter	50.0	48.3	48.5	71.2	83.5	44.1	74.4	72.3	40.8	47.8	36.2	50.0
1983												
1st quarter	50.4	48.7	49.1	71.0	83.1	44.4	74.7	70.8	40.9	47.9	36.4	50.3
2nd quarter	50.8	49.2	49.5	70.8	82.7	44.5	75.0	70.8	41.3	48.2	36.8	50.7
3rd quarter	51.3	49.8	50.2	70.8	82.6	44.7	75.4	70.9	41.8	48.8	37.2	51.2
4th quarter	51.7	50.2	50.6	71.0	82.6	45.1	76.1	70.5	42.0	49.1	37.5	51.5
1984												
1st quarter	52.2	50.7	51.1	71.2	82.7	45.4	76.2	70.7	42.6	49.7	38.0	52.0
2nd quarter	52.7	51.2	51.7	71.6	83.0	45.8	76.7	71.1	43.0	50.2	38.4	52.5
3rd quarter	53.2	51.6	52.2	71.9	83.2	46.2	76.0	69.8	43.6	51.0	38.7	52.9
4th quarter	53.5	51.9	52.6	72.0	83.3	46.6	75.1	69.0	44.0	51.6	39.1	53.2
1985												
1st quarter	54.1	52.5	53.3	72.2	83.5	46.9	74.3	67.3	44.2	51.4	39.6	53.7
2nd quarter	54.4	52.9	53.8	72.3	83.7	47.0	74.0	67.6	44.5	51.5	40.0	54.0
3rd quarter	54.8	53.4	54.3	72.6	84.0	47.3	73.4	67.5	44.8	51.8	40.3	54.4
4th quarter	55.0	53.7	54.7	73.1	84.4	47.9	73.3	68.9	45.1	52.1	40.7	54.8
1986												
1st quarter	55.3	54.1	55.3	73.4	84.6	48.5	72.9	69.0	45.2	52.0	40.8	55.1
2nd quarter	55.5	54.1	55.7	73.9	85.1	49.0	72.5	66.9	45.2	51.9	41.0	55.1
3rd quarter	55.8	54.3	56.1	74.5	85.7	49.7	72.1	67.3	45.5	51.9	41.4	55.4
4th quarter	56.1	54.7	56.5	75.0	86.1	50.2	72.6	68.2	45.8	52.0	41.9	55.8
1987												
1st quarter	56.4	55.2	56.9	75.3	86.1	50.8	72.9	69.9	46.1	51.9	42.5	56.2
2nd quarter	56.8	55.7	57.4	75.5	86.1	51.2	74.0	71.7	46.5	52.2	43.0	56.7
3rd quarter	57.2	56.2	57.9	75.7	86.1	51.7	74.2	72.5	46.9	52.5	43.5	57.1
4th quarter	57.7	56.7	58.5	76.4	87.0	52.2	75.4	73.6	47.1	52.7	43.8	57.6

Table 1-6B. Chain-Type Price Indexes for Gross Domestic Product and Domestic Purchases: Historical Data—*Continued*

(Index numbers, 2012 = 100.)   **NIPA Tables 1.1.4, 1.6.4, 2.3.4**

Year and quarter	Gross domestic product											Gross domestic purchases
	Gross domestic product, total	Personal consumption expenditures		Private fixed investment			Exports and imports of goods and services		Government consumption expenditures and gross investment			
		Total	Excluding food and energy	Total	Nonresi-dential	Residential	Exports	Imports	Total	Federal	State and local	
1988												
1st quarter	58.1	57.2	59.1	76.9	87.8	52.6	76.3	74.6	47.5	53.4	44.0	58.1
2nd quarter	58.7	57.8	59.8	77.4	88.2	53.1	77.9	76.0	48.0	53.9	44.4	58.6
3rd quarter	59.4	58.5	60.5	77.7	88.6	53.5	78.8	75.1	48.3	54.2	44.8	59.2
4th quarter	60.0	59.1	61.2	78.5	89.4	54.0	78.7	75.7	48.8	54.6	45.3	59.8
1989												
1st quarter	60.5	59.8	61.8	78.9	89.9	54.3	79.4	77.0	49.3	55.0	45.9	60.4
2nd quarter	61.2	60.6	62.4	79.4	90.3	55.0	79.5	77.8	49.9	55.4	46.5	61.1
3rd quarter	61.7	61.0	62.9	79.8	90.8	55.2	79.1	76.5	50.2	55.7	47.0	61.5
4th quarter	62.1	61.4	63.4	80.2	91.3	55.6	78.8	76.9	50.7	55.9	47.6	62.0
1990												
1st quarter	62.8	62.3	64.2	80.7	91.8	56.0	78.9	78.1	51.2	56.3	48.3	62.7
2nd quarter	63.5	62.9	64.9	81.0	92.1	56.2	79.0	76.7	51.9	57.3	48.8	63.3
3rd quarter	64.0	63.7	65.5	81.5	92.8	56.4	79.8	78.9	52.3	57.4	49.4	64.0
4th quarter	64.5	64.5	66.1	81.8	93.4	56.5	80.9	83.3	53.0	58.0	50.2	64.7
1991												
1st quarter	65.1	64.9	66.7	82.6	94.3	56.7	81.1	80.7	53.4	58.6	50.4	65.1
2nd quarter	65.6	65.2	67.2	82.7	94.4	57.0	80.7	78.5	53.7	58.9	50.7	65.4
3rd quarter	66.1	65.7	67.8	82.8	94.3	57.3	80.1	77.3	54.2	59.6	51.1	65.9
4th quarter	66.5	66.1	68.3	82.5	94.1	57.1	80.3	77.7	54.7	60.2	51.6	66.3
1992												
1st quarter	66.7	66.6	68.8	82.4	94.0	56.9	80.1	78.1	55.0	60.3	51.9	66.6
2nd quarter	67.1	67.0	69.3	82.5	93.9	57.5	80.2	78.4	55.4	60.5	52.6	67.0
3rd quarter	67.5	67.4	69.7	82.7	94.0	57.9	80.2	79.4	55.9	61.0	52.9	67.4
4th quarter	67.9	67.9	70.3	83.0	94.0	58.6	80.1	78.7	56.3	61.4	53.3	67.8
1993												
1st quarter	68.3	68.3	70.7	83.2	94.1	59.3	80.1	78.3	56.5	61.6	53.6	68.2
2nd quarter	68.7	68.8	71.3	83.6	94.1	59.9	80.4	78.5	56.8	61.9	53.9	68.6
3rd quarter	69.1	69.1	71.7	83.8	94.1	60.4	80.3	77.8	57.1	62.3	54.1	68.9
4th quarter	69.5	69.5	72.0	84.0	94.4	60.6	80.3	77.5	57.4	62.9	54.4	69.2
1994												
1st quarter	69.9	69.7	72.4	84.5	94.6	61.4	80.7	77.5	57.8	63.2	54.8	69.5
2nd quarter	70.2	70.1	72.9	84.7	94.8	61.8	81.0	78.2	58.2	63.7	55.1	69.9
3rd quarter	70.6	70.6	73.3	85.0	95.0	62.5	81.4	79.5	58.7	64.0	55.6	70.4
4th quarter	71.0	70.9	73.6	85.4	95.1	63.3	81.9	79.8	59.1	64.6	56.1	70.8
1995												
1st quarter	71.4	71.3	74.0	85.9	95.5	64.0	82.7	80.5	59.5	65.1	56.4	71.1
2nd quarter	71.7	71.7	74.5	86.2	95.8	64.3	83.4	81.7	60.0	65.5	56.8	71.5
3rd quarter	72.0	72.0	74.8	86.5	96.1	64.6	83.2	81.1	60.2	65.9	57.0	71.8
4th quarter	72.4	72.3	75.2	86.4	95.9	65.0	82.8	80.4	60.7	66.8	57.3	72.1
1996												
1st quarter	72.7	72.7	75.5	86.2	95.6	65.2	82.6	80.2	61.1	67.0	57.8	72.5
2nd quarter	73.0	73.2	75.8	86.0	95.2	65.5	82.4	79.8	61.1	66.6	57.9	72.7
3rd quarter	73.4	73.5	76.2	86.3	95.2	66.2	81.8	79.0	61.4	67.0	58.3	73.0
4th quarter	73.7	74.0	76.6	86.2	95.0	66.5	80.9	79.0	61.8	67.2	58.7	73.4
1997												
1st quarter	74.0	74.3	76.9	86.2	94.9	66.8	80.7	78.4	62.1	67.4	59.1	73.7
2nd quarter	74.4	74.5	77.3	86.3	94.9	67.1	80.7	76.7	62.4	67.9	59.3	73.9
3rd quarter	74.6	74.7	77.5	86.3	94.8	67.7	80.5	76.2	62.6	68.0	59.5	74.1
4th quarter	74.8	74.9	77.8	86.2	94.4	68.2	80.1	75.6	63.1	68.5	60.0	74.3
1998												
1st quarter	74.9	74.9	78.0	85.8	93.8	68.4	79.3	73.8	63.1	68.2	60.1	74.2
2nd quarter	75.1	75.1	78.2	85.6	93.4	68.8	78.9	72.9	63.4	68.7	60.3	74.3
3rd quarter	75.4	75.3	78.5	85.5	93.0	69.5	78.2	71.9	63.8	69.1	60.8	74.6
4th quarter	75.6	75.5	78.8	85.6	92.8	70.1	77.9	71.9	64.2	69.3	61.3	74.8
1999												
1st quarter	75.9	75.7	79.0	85.6	92.7	70.8	77.7	71.6	64.7	69.5	61.8	75.0
2nd quarter	76.2	76.1	79.3	85.7	92.4	71.6	77.9	72.5	65.4	70.1	62.7	75.4
3rd quarter	76.5	76.5	79.5	85.6	92.0	72.2	78.0	73.4	66.1	70.8	63.4	75.8
4th quarter	76.9	77.0	79.9	85.8	92.2	72.7	78.3	74.5	66.9	71.6	64.1	76.3
2000												
1st quarter	77.4	77.6	80.3	86.4	92.4	74.0	78.8	75.9	67.7	72.3	65.0	76.9
2nd quarter	77.8	78.0	80.6	86.7	92.6	74.7	79.4	75.9	68.3	72.6	65.6	77.3
3rd quarter	78.3	78.5	81.0	87.0	92.9	75.3	79.8	76.4	68.9	73.1	66.3	77.8
4th quarter	78.7	78.9	81.3	87.2	92.9	76.0	79.8	76.6	69.6	73.5	67.2	78.2
2001												
1st quarter	79.2	79.4	81.8	87.2	92.5	77.1	79.6	76.4	70.1	73.6	68.0	78.7
2nd quarter	79.8	79.8	82.1	87.5	92.5	78.1	79.1	74.8	70.4	74.0	68.2	79.0
3rd quarter	80.0	79.8	82.3	87.7	92.3	79.3	78.7	73.7	70.7	74.4	68.4	79.2
4th quarter	80.3	79.9	82.8	87.8	92.1	79.8	77.9	72.0	71.0	74.9	68.5	79.3
2002												
1st quarter	80.5	80.0	83.0	87.9	92.1	79.7	77.5	71.6	71.4	75.4	69.0	79.5
2nd quarter	80.8	80.7	83.5	87.7	92.0	80.1	78.0	73.2	72.1	76.1	69.6	79.9
3rd quarter	81.2	81.0	83.9	87.6	91.7	80.5	78.6	73.8	72.6	76.8	70.1	80.3
4th quarter	81.6	81.4	84.2	88.1	91.7	81.8	78.8	74.3	73.5	78.2	70.7	80.8

Table 1-6B. Chain-Type Price Indexes for Gross Domestic Product and Domestic Purchases: Historical Data—*Continued*

(Index numbers, 2012 = 100.)   **NIPA Tables 1.1.4, 1.6.4, 2.3.4**

Year and quarter	Gross domestic product, total	Personal consumption expenditures		Private fixed investment			Exports and imports of goods and services		Government consumption expenditures and gross investment			Gross domestic purchases
		Total	Excluding food and energy	Total	Nonresidential	Residential	Exports	Imports	Total	Federal	State and local	
2003												
1st quarter	82.0	82.0	84.4	88.5	91.4	83.5	79.1	76.2	74.4	79.1	71.6	81.4
2nd quarter	82.3	82.0	84.6	88.4	91.2	83.7	79.2	74.9	74.7	79.6	71.7	81.5
3rd quarter	82.7	82.5	85.0	88.4	91.0	84.3	79.4	75.1	75.2	80.4	72.2	81.9
4th quarter	83.2	82.9	85.3	89.0	91.1	85.8	79.9	75.6	75.8	81.0	72.7	82.3
2004												
1st quarter	83.8	83.6	85.9	89.8	91.4	87.7	81.2	77.4	76.6	81.7	73.6	83.1
2nd quarter	84.5	84.2	86.4	90.8	91.9	89.4	82.1	78.4	77.6	82.4	74.7	83.8
3rd quarter	85.1	84.6	86.7	91.6	92.1	91.2	82.5	79.2	78.6	83.1	75.9	84.4
4th quarter	85.7	85.3	87.2	92.4	92.8	92.7	83.4	81.2	79.8	83.9	77.2	85.2
2005												
1st quarter	86.4	85.8	87.8	93.4	93.6	93.9	84.3	81.9	80.6	85.0	77.9	85.9
2nd quarter	87.0	86.3	88.2	94.3	94.2	95.4	84.9	82.8	81.5	85.8	78.9	86.5
3rd quarter	87.8	87.2	88.5	95.3	94.6	97.9	85.4	84.4	82.6	86.7	80.2	87.5
4th quarter	88.5	87.9	89.1	96.4	95.3	99.7	85.9	85.8	83.7	87.3	81.5	88.3
2006												
1st quarter	89.1	88.4	89.6	97.3	95.9	101.3	86.4	86.0	84.3	88.0	82.0	88.9
2nd quarter	89.8	89.1	90.2	97.9	96.4	102.2	87.6	87.0	85.3	88.7	83.3	89.6
3rd quarter	90.5	89.7	90.7	98.4	97.0	102.6	88.7	87.8	86.1	89.3	84.1	90.3
4th quarter	90.8	89.6	91.1	99.2	97.7	103.3	88.7	86.8	86.8	89.8	85.0	90.4
2007												
1st quarter	91.8	90.4	91.7	99.7	98.2	103.9	89.6	87.4	88.2	90.5	86.6	91.3
2nd quarter	92.3	91.1	92.1	99.7	98.4	103.6	90.7	88.8	89.0	91.3	87.6	91.9
3rd quarter	92.7	91.7	92.5	99.7	98.3	103.6	91.4	90.1	89.9	91.9	88.5	92.4
4th quarter	93.2	92.6	93.1	99.6	98.3	103.7	92.8	93.4	90.9	92.6	89.8	93.2
2008												
1st quarter	93.6	93.3	93.7	99.7	98.7	103.4	94.7	97.1	92.1	93.4	91.3	94.0
2nd quarter	93.9	94.3	94.1	99.8	99.0	102.9	96.8	101.8	93.3	94.4	92.5	94.8
3rd quarter	94.7	95.3	94.5	100.3	100.0	102.1	98.0	104.1	94.3	95.1	93.8	95.7
4th quarter	94.9	93.8	94.6	102.1	101.6	100.6	92.1	92.8	93.5	94.6	92.7	94.9
2009												
1st quarter	95.0	93.3	94.7	101.3	100.9	99.8	89.0	85.1	92.5	93.9	91.5	94.2
2nd quarter	94.9	93.7	95.1	99.7	99.5	98.5	88.9	86.3	92.6	93.9	91.7	94.3
3rd quarter	94.9	94.3	95.4	98.3	98.4	97.8	89.9	89.1	93.0	94.2	92.2	94.6
4th quarter	95.3	95.1	96.0	98.0	97.9	98.5	91.0	91.5	93.7	94.9	92.9	95.2
2010												
1st quarter	95.5	95.4	96.3	97.6	97.3	98.7	91.9	92.9	94.5	95.5	93.8	95.5
2nd quarter	95.9	95.5	96.5	97.6	97.4	98.0	93.0	92.4	95.1	96.2	94.4	95.7
3rd quarter	96.3	95.7	96.7	97.6	97.4	98.0	93.2	91.9	95.6	96.6	94.9	95.9
4th quarter	96.8	96.3	96.9	98.0	97.7	98.6	95.3	94.0	96.4	97.4	95.7	96.5
2011												
1st quarter	97.3	97.1	97.4	98.2	98.0	98.8	97.8	97.8	97.3	98.3	96.6	97.3
2nd quarter	98.0	98.0	98.0	98.7	98.5	99.1	99.7	100.8	98.4	99.2	97.8	98.2
3rd quarter	98.5	98.5	98.4	98.9	98.7	99.1	100.3	100.6	98.8	99.5	98.3	98.6
4th quarter	98.7	98.9	98.8	99.0	99.0	99.2	99.2	100.2	98.7	99.4	98.2	98.9
2012												
1st quarter	99.3	99.5	99.4	99.4	99.6	99.0	99.8	101.1	99.5	99.7	99.4	99.5
2nd quarter	99.7	99.8	99.9	99.9	99.9	99.4	100.0	100.4	99.6	99.9	99.4	99.8
3rd quarter	100.3	100.1	100.1	100.3	100.2	100.3	99.9	98.7	100.0	100.1	100.0	100.1
4th quarter	100.7	100.6	100.6	100.4	100.2	101.3	100.3	99.8	100.8	100.2	101.2	100.6
2013												
1st quarter	101.1	101.0	101.0	100.5	100.0	102.8	100.6	99.9	101.5	100.2	102.3	101.0
2nd quarter	101.4	101.1	101.3	100.8	100.1	104.2	99.9	98.6	101.9	100.4	102.9	101.2
3rd quarter	102.0	101.5	101.7	101.1	100.3	105.6	100.1	98.1	102.5	100.8	103.7	101.6
4th quarter	102.6	101.9	102.2	101.6	100.6	107.6	100.1	97.9	103.5	102.3	104.3	102.1
2014												
1st quarter	103.0	102.4	102.5	102.3	101.0	109.9	100.8	99.1	103.9	102.0	105.1	102.6
2nd quarter	103.5	102.8	103.0	102.6	101.3	110.2	100.9	98.5	104.3	102.5	105.4	103.1
3rd quarter	104.0	103.1	103.4	103.2	101.6	112.1	100.6	97.9	104.8	102.9	106.0	103.4
4th quarter	104.1	103.0	103.6	103.6	101.9	112.3	98.8	95.9	104.8	103.1	106.0	103.4
2015												
1st quarter	104.1	102.6	103.8	103.6	102.1	113.0	96.3	91.9	104.2	102.9	105.1	103.0
2nd quarter	104.6	103.1	104.3	103.5	102.0	113.6	96.2	90.7	104.7	103.1	105.8	103.4
3rd quarter	104.9	103.3	104.6	103.6	101.9	114.6	95.3	89.7	104.8	103.3	105.9	103.6
4th quarter	104.9	103.2	104.9	103.4	101.6	115.3	93.8	87.8	104.6	103.2	105.6	103.5
2016												
1st quarter	104.9	103.3	105.3	102.9	101.0	115.8	92.3	86.2	104.1	102.9	104.9	103.4
2nd quarter	105.6	103.9	105.9	103.4	101.2	117.3	93.3	86.5	104.8	103.5	105.6	104.0
3rd quarter	105.9	104.4	106.4	103.6	101.0	119.0	93.8	87.2	105.2	104.0	106.0	104.4
4th quarter	106.5	104.8	106.8	104.2	101.3	120.5	94.5	87.5	105.7	104.5	106.6	104.9
2017												
1st quarter	107.0	105.4	107.3	104.5	101.6	121.4	95.1	88.4	106.5	105.1	107.5	105.5
2nd quarter	107.4	105.7	107.6	105.1	101.9	123.0	95.2	88.3	106.8	105.5	107.7	105.8
3rd quarter	108.0	106.1	108.0	105.6	102.2	124.3	96.0	88.6	107.6	106.0	108.7	106.3
4th quarter	108.6	106.8	108.6	105.8	102.2	125.3	97.3	89.8	108.7	106.8	110.0	107.0

Table 1-7. Final Sales

(Quarterly dollar data are at seasonally adjusted annual rates.)   **NIPA Tables 1.4.4, 1.4.5, 1.4.6**

Year and quarter	Final sales of domestic product			Final sales to domestic purchasers		
	Billions of dollars	Billions of chained (2012) dollars	Chain-type price index, 2012 = 100	Billions of dollars	Billions of chained (2012) dollars	Chain-type price index, 2012 = 100
1955	420.5	2 866.6	14.7	420.0	2 929.8	14.3
1956	445.4	2 936.4	15.2	443.0	2 990.5	14.8
1957	473.3	3 014.2	15.7	469.2	3 063.0	15.3
1958	481.6	2 997.1	16.1	481.1	3 072.3	15.7
1959	517.7	3 180.0	16.3	517.3	3 259.9	15.9
1960	539.1	3 266.5	16.5	534.9	3 325.7	16.1
1961	559.2	3 352.4	16.7	554.3	3 412.0	16.2
1962	597.8	3 540.2	16.9	593.7	3 612.0	16.4
1963	631.8	3 698.6	17.1	626.9	3 766.2	16.6
1964	679.6	3 918.1	17.3	672.7	3 977.9	16.9
1965	733.0	4 149.2	17.7	727.4	4 226.9	17.2
1966	799.8	4 402.8	18.2	795.9	4 499.5	17.7
1967	850.1	4 546.7	18.7	846.5	4 657.1	18.2
1968	931.6	4 777.1	19.5	930.2	4 907.7	19.0
1969	1 008.4	4 927.8	20.5	1 007.0	5 064.4	19.9
1970	1 071.3	4 971.9	21.5	1 067.4	5 093.0	21.0
1971	1 156.6	5 108.2	22.6	1 156.0	5 242.4	22.1
1972	1 270.0	5 375.9	23.6	1 273.4	5 527.0	23.0
1973	1 409.5	5 655.4	24.9	1 405.4	5 769.1	24.4
1974	1 531.3	5 639.3	27.2	1 532.1	5 710.1	26.8
1975	1 691.2	5 698.0	29.7	1 675.2	5 719.9	29.3
1976	1 856.3	5 925.8	31.3	1 857.9	6 011.3	30.9
1977	2 059.5	6 187.7	33.3	2 082.6	6 318.8	33.0
1978	2 325.8	6 526.5	35.6	2 351.2	6 657.9	35.3
1979	2 609.3	6 761.6	38.6	2 631.9	6 851.8	38.4
1980	2 863.6	6 804.7	42.1	2 876.7	6 783.2	42.4
1981	3 177.2	6 900.1	46.0	3 189.8	6 887.9	46.3
1982	3 358.7	6 865.5	48.9	3 378.7	6 894.2	49.0
1983	3 639.8	7 159.9	50.8	3 691.4	7 278.9	50.7
1984	3 972.2	7 541.5	52.7	4 074.9	7 772.3	52.4
1985	4 317.2	7 940.2	54.4	4 431.2	8 203.0	54.0
1986	4 573.1	8 241.0	55.5	4 704.9	8 528.9	55.2
1987	4 828.1	8 492.5	56.9	4 972.8	8 767.5	56.7
1988	5 218.0	8 860.6	58.9	5 327.3	9 067.0	58.8
1989	5 613.9	9 172.3	61.2	5 700.6	9 333.9	61.1
1990	5 948.6	9 365.1	63.5	6 026.5	9 489.7	63.5
1991	6 158.5	9 379.0	65.7	6 187.1	9 445.2	65.5
1992	6 504.0	9 683.1	67.2	6 538.7	9 753.9	67.0
1993	6 837.7	9 943.6	68.8	6 902.9	10 069.3	68.6
1994	7 223.5	10 284.1	70.2	7 315.9	10 451.4	70.0
1995	7 608.6	10 608.5	71.7	7 698.3	10 764.1	71.5
1996	8 042.3	11 008.6	73.1	8 138.7	11 182.0	72.8
1997	8 506.6	11 442.2	74.3	8 608.6	11 652.3	73.9
1998	8 999.1	11 966.9	75.2	9 161.8	12 313.5	74.4
1999	9 569.8	12 543.1	76.3	9 825.7	13 001.0	75.6
2000	10 197.8	13 068.0	78.0	10 572.9	13 635.7	77.5
2001	10 620.1	13 309.5	79.8	10 988.0	13 908.0	79.0
2002	10 916.4	13 476.4	81.0	11 341.8	14 162.1	80.1
2003	11 444.2	13 864.7	82.5	11 947.3	14 615.3	81.7
2004	12 149.7	14 335.7	84.8	12 768.7	15 184.1	84.1
2005	12 979.1	14 852.3	87.4	13 700.3	15 745.1	87.0
2006	13 745.6	15 263.0	90.1	14 516.5	16 172.2	89.8
2007	14 417.9	15 588.7	92.5	15 136.3	16 418.1	92.2
2008	14 742.1	15 639.7	94.3	15 465.2	16 306.5	94.8
2009	14 599.7	15 373.0	95.0	14 996.2	15 863.7	94.5
2010	14 938.1	15 546.6	96.1	15 452.0	16 112.5	95.9
2011	15 496.3	15 796.5	98.1	16 075.8	16 364.6	98.2
2012	16 125.8	16 125.8	100.0	16 694.4	16 694.4	100.0
2013	16 680.3	16 386.2	101.8	17 171.1	16 919.3	101.5
2014	17 443.3	16 822.3	103.7	17 951.0	17 398.0	103.2
2015	18 101.5	17 290.1	104.7	18 628.1	17 998.6	103.5
2016	18 716.7	17 686.9	105.9	19 229.2	18 437.9	104.3
2017	19 526.7	18 107.2	107.9	20 082.2	18 903.0	106.2
2018	20 554.1	18 613.8	110.5	21 163.6	19 463.5	108.7
2019	21 384.1	19 021.1	112.5	21 994.6	19 910.9	110.5
2017						
1st quarter	19 249.7	17 970.8	107.2	19 793.2	18 746.0	105.6
2nd quarter	19 375.2	18 031.4	107.5	19 934.7	18 826.1	105.9
3rd quarter	19 572.2	18 115.0	108.1	20 115.8	18 908.4	106.4
4th quarter	19 909.5	18 311.4	108.8	20 485.0	19 131.6	107.1
2018						
1st quarter	20 197.7	18 463.4	109.4	20 787.5	19 273.1	107.9
2nd quarter	20 541.9	18 630.7	110.3	21 090.1	19 431.2	108.5
3rd quarter	20 664.4	18 655.1	110.8	21 310.8	19 542.8	109.1
4th quarter	20 812.5	18 705.9	111.3	21 466.0	19 607.0	109.5
2019						
1st quarter	21 016.7	18 833.1	111.6	21 632.2	19 710.2	109.8
2nd quarter	21 276.8	18 949.6	112.3	21 921.6	19 866.8	110.3
3rd quarter	21 499.0	19 075.2	112.8	22 130.8	19 993.1	110.7
4th quarter	21 744.0	19 226.6	113.1	22 293.7	20 073.6	111.1

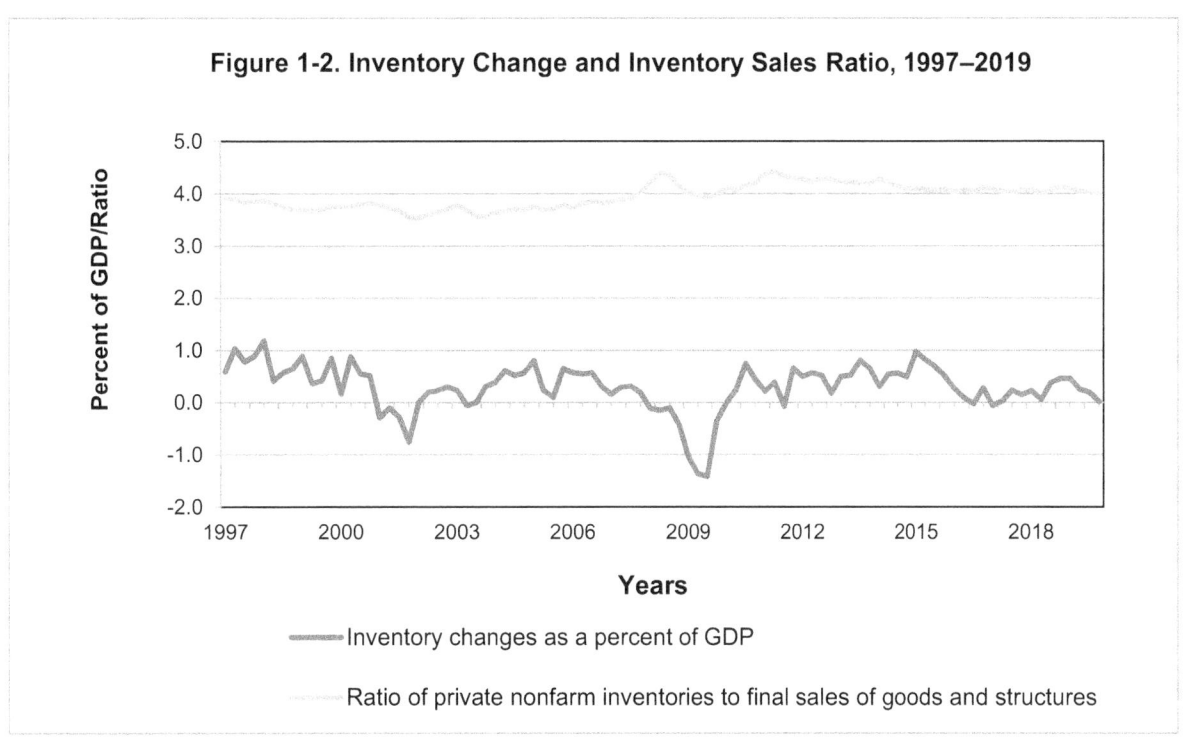

Figure 1-2. Inventory Change and Inventory Sales Ratio, 1997–2019

— Inventory changes as a percent of GDP

— Ratio of private nonfarm inventories to final sales of goods and structures

- The change in inventories is part of business investment component that, although small, can be volatile and even negative, reflecting the fact that production is frequently more variable than sales. When inventory change is positive, the economy is producing more than is sold. This is the normal condition, as inventories grow to accommodate a growing level of sales. When inventory change is negative, production is below the rate of sales—which could be because sales rose faster than anticipated, causing involuntary inventory reduction, or on the other hand because production was deliberately cut back in order to get rid of unwanted stocks. Furthermore, it is the <u>change</u> in inventory —the change in an already volatile number—that drives the GDP's <u>growth rate</u>.

- In 2009, following a rapid and unwanted rise in inventories, stocks were reduced at a record rate of $144 billion (2012 dollars), contributing to the 2.5 percent decline in GDP. In 2010, the resumption of inventory-building contributed significantly to the 2.6 percent GDP increase. (Tables 1-2A, 1-8A and 1-4).

- Inventory investment has been moderate and variable during the recovery. (Tables 1-8A and B)

Table 1-8A. Inventory Change and Inventory-Sales Ratios: Recent Data

(Quarterly dollar data are at seasonally adjusted annual rates.)   **NIPA Tables 5.7.5B, 5.7.6B, 5.8.5B, 5.8.6B**

Year and quarter	Change in private inventories				Ratio, inventories at end of quarter to monthly rate of sales during the quarter					
	Billions of current dollars		Billions of chained dollars [1]		Total private inventories to final sales of domestic business		Nonfarm inventories to final sales of domestic business		Nonfarm inventories to final sales of goods and structures	
	Farm	Nonfarm	Farm	Nonfarm	Current dollars	Chained dollars[1]	Current dollars	Chained dollars[1]	Current dollars	Chained dollars[1]
OLD BASIS										
1960	0.6	2.7	1.7	9.9	4.22	3.37	2.89	2.34	3.93	4.10
1961	0.9	2.1	2.5	7.8	4.12	3.30	2.81	2.29	3.84	4.03
1962	0.6	5.5	1.7	20.4	4.14	3.32	2.82	2.34	3.87	4.11
1963	0.5	5.1	1.4	19.0	3.95	3.25	2.78	2.32	3.81	4.07
1964	-1.2	6.0	-3.8	21.8	3.79	3.17	2.75	2.31	3.79	4.07
1965	0.8	8.4	2.5	30.5	3.77	3.09	2.72	2.29	3.71	3.99
1966	-0.5	14.1	-1.4	50.2	3.92	3.26	2.92	2.50	4.01	4.38
1967	0.9	9.0	2.5	31.5	3.90	3.35	2.99	2.59	4.14	4.57
1968	1.4	7.7	4.1	26.5	3.79	3.33	2.90	2.58	4.03	4.56
1969	0.0	9.2	0.0	30.9	3.88	3.40	2.98	2.68	4.18	4.76
1970	-0.8	2.8	-2.4	8.6	3.81	3.39	2.96	2.68	4.23	4.83
1971	1.7	6.6	4.1	20.6	3.76	3.33	2.88	2.64	4.12	4.75
1972	0.3	8.8	0.4	26.7	3.74	3.18	2.77	2.55	3.94	4.53
1973	1.5	14.4	1.8	40.5	4.20	3.22	2.98	2.62	4.22	4.67
1974	-2.8	16.8	-4.4	39.0	4.52	3.46	3.54	2.88	5.11	5.26
1975	3.4	-9.6	5.8	-21.4	4.04	3.26	3.16	2.66	4.59	4.88
1976	-0.8	18.0	-1.6	37.6	3.95	3.23	3.19	2.68	4.68	4.92
1977	4.5	17.8	7.2	35.7	3.89	3.22	3.16	2.67	4.63	4.87
1978	1.4	24.4	2.1	44.9	3.98	3.14	3.15	2.63	4.54	4.73
1979	3.6	14.4	4.2	23.8	4.19	3.16	3.34	2.64	4.85	4.75
1980	-6.1	-0.2	-7.2	-0.4	4.24	3.13	3.44	2.65	5.08	4.82
1981	8.8	21.0	10.1	28.2	4.17	3.28	3.49	2.75	5.20	5.03
1982	5.8	-20.7	7.5	-27.4	3.97	3.23	3.30	2.68	5.10	4.99
1983	-15.4	9.6	-17.4	10.7	3.69	2.98	3.08	2.52	4.77	4.64
1984	5.7	59.7	6.3	72.3	3.72	3.05	3.16	2.60	4.89	4.72
1985	5.8	16.1	7.0	18.4	3.51	2.98	3.01	2.53	4.74	4.65
1986	-1.5	8.0	-2.3	10.4	3.25	2.90	2.82	2.47	4.49	4.53
1987	-6.4	33.6	-9.1	41.1	3.33	2.90	2.90	2.50	4.65	4.62
1988	-11.9	30.4	-13.1	35.4	3.29	2.81	2.86	2.47	4.63	4.55
1989	0.0	27.7	0.0	30.8	3.23	2.80	2.83	2.47	4.63	4.56
1990	2.4	12.2	2.5	14.0	3.22	2.83	2.82	2.49	4.71	4.66
1991	-1.3	0.9	-1.8	0.4	3.06	2.83	2.70	2.50	4.62	4.74
1992	6.3	10.1	7.1	11.0	2.92	2.73	2.56	2.39	4.41	4.53
1993	-6.2	27.0	-7.5	29.8	2.85	2.68	2.51	2.38	4.33	4.47
1994	12.0	51.8	13.7	55.9	2.89	2.72	2.56	2.40	4.39	4.47
1995	-11.1	42.2	-13.0	44.3	2.88	2.69	2.58	2.40	4.45	4.46
1996	8.6	22.1	8.3	23.2	2.76	2.62	2.47	2.33	4.24	4.28
1997	3.3	67.7	3.7	73.6	2.70	2.65	2.42	2.37	4.17	4.32
NEW BASIS										
1997	3.3	67.7	6.2	88.0	2.60	2.97	2.34	2.57	3.86	4.64
1998	1.3	62.5	2.8	86.5	2.47	2.93	2.25	2.55	3.69	4.56
1999	-2.7	63.5	-6.3	87.7	2.49	2.91	2.27	2.55	3.75	4.55
2000	-1.4	55.9	-3.0	77.6	2.51	2.90	2.29	2.56	3.82	4.60
2001	0.0	-38.3	0.2	-51.4	2.33	2.81	2.13	2.47	3.55	4.43
2002	-2.5	22.5	-5.2	27.1	2.38	2.83	2.16	2.50	3.71	4.55
2003	0.1	14.0	0.2	19.2	2.33	2.71	2.10	2.39	3.57	4.28
2004	8.8	55.3	13.5	70.3	2.39	2.73	2.17	2.40	3.70	4.30
2005	0.2	57.3	0.4	61.5	2.42	2.71	2.21	2.39	3.79	4.27
2006	-3.6	72.6	-6.3	89.1	2.42	2.71	2.21	2.41	3.82	4.31
2007	0.6	33.3	-1.4	40.6	2.51	2.69	2.28	2.40	4.01	4.31
2008	-0.5	-28.7	-0.6	-31.5	2.50	2.76	2.28	2.45	4.14	4.56
2009	-6.7	-144.1	-10.0	-165.2	2.37	2.58	2.17	2.29	3.99	4.28
2010	-12.2	66.1	-17.8	70.2	2.49	2.58	2.27	2.31	4.21	4.31
2011	-2.1	48.3	-2.2	48.4	2.58	2.57	2.32	2.30	4.30	4.27
2012	-18.7	89.9	-18.7	89.9	2.56	2.57	2.33	2.33	4.27	4.28
2013	11.7	92.8	10.4	98.2	2.52	2.60	2.30	2.36	4.19	4.27
2014	-3.2	87.2	-3.5	90.1	2.46	2.58	2.23	2.36	4.10	4.26
2015	1.5	135.3	1.1	136.5	2.39	2.67	2.20	2.45	4.09	4.45
2016	-5.9	34.3	-6.7	30.2	2.35	2.63	2.19	2.41	4.11	4.38
2017	-5.1	21.5	-6.4	21.0	2.34	2.55	2.18	2.35	4.07	4.21
2018	-6.8	64.5	-7.8	59.3	2.36	2.53	2.21	2.35	4.13	4.19
2019	-14.5	63.6	-16.9	62.2	2.29	2.50	2.15	2.33	4.00	4.12
2017										
1st quarter	-5.0	-7.3	-6.0	-13.6	2.35	2.59	2.18	2.38	4.09	4.31
2nd quarter	-4.8	8.8	-5.3	5.3	2.34	2.58	2.17	2.37	4.05	4.28
3rd quarter	-5.6	50.7	-7.1	61.4	2.32	2.58	2.17	2.37	4.05	4.28
4th quarter	-5.2	33.6	-7.3	30.9	2.34	2.55	2.18	2.35	4.07	4.21
2018										
1st quarter	-4.1	48.7	-4.2	50.7	2.34	2.53	2.19	2.34	4.08	4.18
2nd quarter	-5.0	15.7	-5.6	0.1	2.31	2.50	2.16	2.31	4.01	4.11
3rd quarter	-7.8	86.2	-9.3	85.5	2.32	2.52	2.19	2.33	4.08	4.17
4th quarter	-10.2	107.5	-12.2	100.8	2.36	2.53	2.21	2.35	4.13	4.19
2019										
1st quarter	-17.8	116.4	-15.3	114.4	2.36	2.54	2.21	2.35	4.11	4.19
2nd quarter	-15.6	68.6	-16.3	63.2	2.31	2.53	2.18	2.35	4.05	4.18
3rd quarter	-14.5	55.8	-20.4	59.8	2.28	2.52	2.16	2.35	4.02	4.17
4th quarter	-10.4	13.8	-15.6	11.5	2.29	2.50	2.15	2.33	4.00	4.12

[1]Before 1997, 2005 dollars; 1997 forward, 2012 dollars.

Table 1-8B. Inventory Change and Inventory-Sales Ratios: Historical Data

(Quarterly dollar data are at seasonally adjusted annual rates.)   **NIPA Tables 5.7.5B, 5.7.6B, 5.8.5B, 5.8.6B**

Year and quarter	Change in private inventories				Ratio, inventories at end of quarter to monthly rate of sales during the quarter					
	Billions of current dollars		Billions of chained dollars¹		Total private inventories to final sales of domestic business		Nonfarm inventories to final sales of domestic business		Nonfarm inventories to final sales of goods and structures	
	Farm	Nonfarm	Farm	Nonfarm	Current dollars	Chained dollars¹	Current dollars	Chained dollars¹	Current dollars	Chained dollars¹
OLD BASIS										
1929	-0.1	1.7	-0.6	10.6	. . .	. . .	. . .	. . .	. . .	. . .
1930	-0.3	0.0	-1.7	-2.4	. . .	. . .	. . .	. . .	. . .	. . .
1931	0.5	-1.6	3.1	-13.4	. . .	. . .	. . .	. . .	. . .	. . .
1932	0.1	-2.5	0.8	-21.2	. . .	. . .	. . .	. . .	. . .	. . .
1933	-0.1	-1.3	-1.1	-9.4	. . .	. . .	. . .	. . .	. . .	. . .
1934	-0.8	0.2	-5.7	1.0	. . .	. . .	. . .	. . .	. . .	. . .
1935	0.7	0.4	4.0	2.5	. . .	. . .	. . .	. . .	. . .	. . .
1936	-0.9	2.0	-5.4	15.5	. . .	. . .	. . .	. . .	. . .	. . .
1937	0.9	1.7	4.1	7.9	. . .	. . .	. . .	. . .	. . .	. . .
1938	0.4	-1.0	2.8	-6.0	. . .	. . .	. . .	. . .	. . .	. . .
1939	-0.1	0.3	-1.1	2.7	. . .	. . .	. . .	. . .	. . .	. . .
1940	0.5	1.9	2.8	12.4	. . .	. . .	. . .	. . .	. . .	. . .
1941	0.4	3.9	1.9	24.3	. . .	. . .	. . .	. . .	. . .	. . .
1942	1.3	0.6	5.6	3.0	. . .	. . .	. . .	. . .	. . .	. . .
1943	-0.2	-0.5	-1.1	-2.3	. . .	. . .	. . .	. . .	. . .	. . .
1944	-0.3	-0.6	-1.5	-2.5	. . .	. . .	. . .	. . .	. . .	. . .
1945	-0.9	-0.6	-3.3	-3.8	. . .	. . .	. . .	. . .	. . .	. . .
1946	-0.2	6.2	-1.0	31.8	. . .	. . .	. . .	. . .	. . .	. . .
1947	-1.8	1.2	-4.5	4.4	. . .	. . .	. . .	. . .	. . .	. . .
1948	2.7	3.0	6.1	12.4	. . .	. . .	. . .	. . .	. . .	. . .
1949	-0.6	-2.1	-1.9	-8.6	. . .	. . .	. . .	. . .	. . .	. . .
1950	-0.1	5.9	-0.3	22.8	5.81	3.67	3.04	2.30	3.94	3.97
1951	1.0	8.9	2.1	28.4	5.85	3.76	3.15	2.46	4.07	4.24
1952	1.4	2.1	3.1	7.4	5.24	3.74	3.06	2.45	3.98	4.23
1953	0.7	1.2	2.0	4.5	5.04	3.74	3.08	2.45	4.04	4.22
1954	0.2	-2.1	0.5	-7.5	4.74	3.52	2.88	2.28	3.81	3.94
1947										
1st quarter	-1.1	1.6	-0.4	6.6	5.93	3.88	2.82	2.27	3.71	3.99
2nd quarter	-2.7	1.5	-4.5	4.2	5.89	3.84	2.82	2.27	3.69	4.00
3rd quarter	-2.5	-0.4	-6.8	-3.7	6.02	3.76	2.76	2.23	3.61	3.93
4th quarter	-0.8	2.3	-6.3	10.5	6.30	3.74	2.81	2.25	3.65	3.93
1948										
1st quarter	1.3	2.4	1.6	11.7	6.00	3.76	2.87	2.27	3.71	3.96
2nd quarter	2.8	3.0	7.7	11.9	6.02	3.81	2.89	2.30	3.76	4.03
3rd quarter	3.4	3.7	8.3	15.0	5.88	3.88	2.97	2.34	3.85	4.10
4th quarter	3.1	2.9	6.6	11.1	5.74	3.91	3.00	2.36	3.90	4.13
1949										
1st quarter	-0.2	0.6	-1.2	-0.4	5.66	3.90	2.97	2.36	3.88	4.14
2nd quarter	-1.2	-4.0	-3.0	-15.1	5.38	3.82	2.83	2.30	3.70	4.02
3rd quarter	-0.9	-0.4	-2.3	-0.6	5.39	3.81	2.82	2.29	3.69	4.00
4th quarter	-0.2	-4.5	-0.9	-18.4	5.23	3.73	2.77	2.23	3.61	3.87
1950										
1st quarter	-0.1	2.2	-1.0	9.7	5.26	3.69	2.76	2.22	3.61	3.84
2nd quarter	-1.3	4.2	-2.5	16.2	5.27	3.62	2.75	2.20	3.57	3.79
3rd quarter	0.5	3.7	0.6	15.2	5.21	3.47	2.72	2.12	3.50	3.63
4th quarter	0.7	13.4	1.5	50.2	5.81	3.67	3.04	2.30	3.94	3.97
1951										
1st quarter	1.2	9.2	2.6	29.0	5.94	3.68	3.12	2.33	4.01	4.01
2nd quarter	0.9	13.8	2.5	44.3	5.99	3.81	3.25	2.45	4.23	4.28
3rd quarter	0.8	9.0	1.6	29.3	5.89	3.79	3.20	2.47	4.16	4.29
4th quarter	1.1	3.6	1.7	11.0	5.85	3.76	3.15	2.46	4.07	4.24
1952										
1st quarter	1.0	3.8	2.7	12.8	5.77	3.79	3.16	2.49	4.11	4.31
2nd quarter	1.9	-3.4	3.6	-10.5	5.63	3.75	3.08	2.44	4.01	4.24
3rd quarter	2.2	3.4	4.5	11.2	5.63	3.85	3.15	2.51	4.13	4.38
4th quarter	0.5	4.8	1.5	16.2	5.24	3.74	3.06	2.45	3.98	4.23
1953										
1st quarter	0.8	3.1	2.5	10.6	5.06	3.69	3.02	2.42	3.93	4.17
2nd quarter	-0.7	4.3	-0.2	14.1	5.00	3.69	3.06	2.43	3.99	4.20
3rd quarter	0.7	1.6	1.8	5.8	4.99	3.72	3.09	2.45	4.05	4.23
4th quarter	2.1	-4.1	3.8	-12.3	5.04	3.74	3.08	2.45	4.04	4.22
1954										
1st quarter	0.8	-2.8	2.1	-9.2	5.02	3.72	3.04	2.42	4.01	4.20
2nd quarter	-0.2	-3.2	0.5	-10.8	4.91	3.66	2.98	2.38	3.93	4.11
3rd quarter	0.7	-2.8	0.8	-9.1	4.85	3.59	2.93	2.32	3.89	4.04
4th quarter	-0.5	0.2	-1.5	-0.9	4.74	3.52	2.88	2.28	3.81	3.94
1955										
1st quarter	-0.1	3.9	-1.7	14.0	4.68	3.46	2.86	2.25	3.77	3.88
2nd quarter	-1.1	5.7	-2.1	20.0	4.57	3.44	2.87	2.26	3.73	3.83
3rd quarter	-1.3	5.6	-2.0	17.9	4.47	3.42	2.89	2.26	3.77	3.85
4th quarter	0.3	6.9	-1.6	22.7	4.43	3.44	2.95	2.29	3.87	3.94
1956										
1st quarter	0.0	6.4	-2.9	20.8	4.50	3.48	3.00	2.33	3.96	4.02
2nd quarter	-1.5	5.0	-3.4	16.2	4.57	3.48	3.03	2.35	3.99	4.04
3rd quarter	-0.8	4.3	-3.2	13.2	4.50	3.48	3.02	2.36	3.98	4.07
4th quarter	-1.6	3.9	-3.3	12.1	4.47	3.45	3.03	2.35	4.00	4.07

¹Before 1997, 2005 dollars; 1997 forward, 2009 dollars.
. . . = Not available.

Table 1-8B. Inventory Change and Inventory-Sales Ratios: Historical Data—*Continued*

(Quarterly dollar data are at seasonally adjusted annual rates.)   **NIPA Tables 5.7.5B, 5.7.6B, 5.8.5B, 5.8.6B**

Year and quarter	Change in private inventories				Ratio, inventories at end of quarter to monthly rate of sales during the quarter					
	Billions of current dollars		Billions of chained dollars [1]		Total private inventories to final sales of domestic business		Nonfarm inventories to final sales of domestic business		Nonfarm inventories to final sales of goods and structures	
	Farm	Nonfarm	Farm	Nonfarm	Current dollars	Chained dollars [1]	Current dollars	Chained dollars [1]	Current dollars	Chained dollars [1]
1957										
1st quarter	0.3	1.9	-0.9	5.7	4.44	3.44	3.02	2.35	3.97	4.05
2nd quarter	0.7	2.0	-0.2	6.8	4.48	3.48	3.04	2.38	4.00	4.11
3rd quarter	0.5	2.3	1.8	7.5	4.46	3.47	3.03	2.38	3.99	4.10
4th quarter	-1.0	-3.5	1.1	-10.2	4.46	3.48	3.01	2.37	4.01	4.14
1958										
1st quarter	2.2	-6.3	7.6	-20.4	4.67	3.56	3.04	2.41	4.06	4.20
2nd quarter	1.6	-5.8	6.9	-18.6	4.65	3.56	3.00	2.38	4.00	4.17
3rd quarter	2.2	-0.7	6.2	-2.5	4.58	3.48	2.92	2.32	3.92	4.07
4th quarter	1.8	3.4	2.5	10.6	4.50	3.44	2.90	2.30	3.88	4.00
1959										
1st quarter	-0.3	4.2	-3.5	14.9	4.37	3.35	2.83	2.25	3.79	3.93
2nd quarter	-1.9	9.2	-5.3	33.8	4.33	3.34	2.86	2.27	3.85	3.98
3rd quarter	-2.3	2.6	-5.4	9.6	4.25	3.31	2.84	2.27	3.82	3.96
4th quarter	-1.8	5.9	-4.6	21.1	4.26	3.37	2.90	2.32	3.94	4.08
1960										
1st quarter	0.6	10.6	0.4	39.0	4.29	3.37	2.93	2.35	3.96	4.10
2nd quarter	0.9	2.4	1.9	9.0	4.21	3.36	2.91	2.34	3.95	4.11
3rd quarter	1.0	3.3	3.1	12.1	4.27	3.41	2.95	2.38	3.99	4.15
4th quarter	-0.2	-5.6	1.5	-20.2	4.22	3.37	2.89	2.34	3.93	4.10
1961										
1st quarter	0.6	-3.1	2.5	-11.3	4.19	3.35	2.86	2.32	3.90	4.07
2nd quarter	0.6	1.2	2.2	4.7	4.12	3.33	2.83	2.30	3.89	4.07
3rd quarter	0.9	5.7	2.6	21.0	4.17	3.35	2.85	2.32	3.89	4.08
4th quarter	1.4	4.6	2.8	16.8	4.12	3.30	2.81	2.29	3.84	4.03
1962										
1st quarter	1.5	7.9	2.5	29.0	4.15	3.32	2.82	2.32	3.85	4.07
2nd quarter	0.2	5.2	0.8	19.2	4.09	3.29	2.80	2.31	3.84	4.06
3rd quarter	0.2	5.9	1.4	22.0	4.17	3.32	2.83	2.33	3.87	4.09
4th quarter	0.4	3.0	2.2	11.3	4.14	3.32	2.82	2.34	3.87	4.11
1963										
1st quarter	1.9	4.9	7.5	18.2	4.12	3.34	2.83	2.35	3.87	4.12
2nd quarter	0.1	4.8	0.7	17.4	4.04	3.29	2.79	2.32	3.83	4.08
3rd quarter	-0.6	6.4	-1.7	23.6	4.00	3.27	2.79	2.33	3.81	4.07
4th quarter	0.6	4.5	-0.8	16.7	3.95	3.25	2.78	2.32	3.81	4.07
1964										
1st quarter	-0.6	5.8	-3.9	21.1	3.85	3.19	2.74	2.29	3.75	4.01
2nd quarter	-1.2	5.7	-4.3	20.7	3.79	3.17	2.73	2.29	3.75	4.02
3rd quarter	-1.8	6.5	-4.2	23.8	3.78	3.15	2.73	2.29	3.74	4.00
4th quarter	-1.1	6.1	-3.1	21.7	3.79	3.17	2.75	2.31	3.79	4.07
1965										
1st quarter	0.3	11.2	0.9	40.8	3.81	3.17	2.76	2.33	3.79	4.08
2nd quarter	1.1	7.4	2.9	26.9	3.82	3.15	2.75	2.32	3.79	4.08
3rd quarter	0.8	8.5	3.1	30.5	3.78	3.14	2.76	2.32	3.78	4.06
4th quarter	1.0	6.6	3.2	23.7	3.77	3.09	2.72	2.29	3.71	3.99
1966										
1st quarter	0.6	13.2	1.1	46.8	3.80	3.09	2.73	2.31	3.71	4.00
2nd quarter	-1.7	14.0	-2.7	50.2	3.86	3.16	2.80	2.38	3.83	4.16
3rd quarter	-0.6	12.4	-1.7	43.7	3.91	3.19	2.84	2.42	3.88	4.22
4th quarter	-0.4	16.9	-2.3	59.9	3.92	3.26	2.92	2.50	4.01	4.38
1967										
1st quarter	1.7	13.7	3.4	48.6	3.94	3.32	2.98	2.55	4.11	4.50
2nd quarter	2.1	4.2	7.6	14.7	3.92	3.31	2.95	2.54	4.07	4.46
3rd quarter	0.5	8.7	1.4	29.7	3.92	3.34	2.97	2.57	4.10	4.52
4th quarter	-0.9	9.3	-2.2	33.1	3.90	3.35	2.99	2.59	4.14	4.57
1968										
1st quarter	3.0	5.4	7.3	19.1	3.87	3.32	2.94	2.57	4.08	4.51
2nd quarter	4.3	9.8	13.4	33.9	3.87	3.34	2.93	2.58	4.07	4.55
3rd quarter	0.2	7.5	1.2	25.1	3.81	3.32	2.90	2.56	4.03	4.51
4th quarter	-2.0	8.0	-5.5	27.8	3.79	3.33	2.90	2.58	4.03	4.56
1969										
1st quarter	1.8	9.7	3.9	33.0	3.79	3.32	2.90	2.59	4.02	4.54
2nd quarter	1.4	7.8	4.4	26.6	3.83	3.35	2.91	2.61	4.06	4.61
3rd quarter	-1.2	11.4	-2.5	38.6	3.83	3.38	2.94	2.64	4.10	4.67
4th quarter	-1.9	7.7	-5.8	25.5	3.88	3.40	2.98	2.68	4.18	4.76
1970										
1st quarter	1.6	0.2	3.4	0.0	3.86	3.38	2.96	2.66	4.18	4.74
2nd quarter	0.4	4.7	1.9	14.9	3.85	3.40	2.96	2.68	4.20	4.80
3rd quarter	-1.8	6.9	-4.4	21.9	3.84	3.39	2.97	2.67	4.22	4.79
4th quarter	-3.5	-0.5	-10.3	-2.2	3.81	3.39	2.96	2.68	4.23	4.83
1971										
1st quarter	2.5	9.7	6.6	31.3	3.83	3.38	2.95	2.68	4.21	4.81
2nd quarter	4.2	6.7	10.7	20.6	3.81	3.39	2.94	2.68	4.20	4.81
3rd quarter	2.3	7.9	5.8	25.2	3.80	3.38	2.93	2.68	4.18	4.80
4th quarter	-2.3	2.0	-6.8	5.4	3.76	3.33	2.88	2.64	4.12	4.75

[1]Before 1997, 2005 dollars; 1997 forward, 2009 dollars.

Table 1-8B. Inventory Change and Inventory-Sales Ratios: Historical Data—*Continued*

(Quarterly dollar data are at seasonally adjusted annual rates.)   **NIPA Tables 5.7.5B, 5.7.6B, 5.8.5B, 5.8.6B**

Year and quarter	Change in private inventories				Ratio, inventories at end of quarter to monthly rate of sales during the quarter					
	Billions of current dollars		Billions of chained dollars [1]		Total private inventories to final sales of domestic business		Nonfarm inventories to final sales of domestic business		Nonfarm inventories to final sales of goods and structures	
	Farm	Nonfarm	Farm	Nonfarm	Current dollars	Chained dollars[1]	Current dollars	Chained dollars[1]	Current dollars	Chained dollars[1]
1972										
1st quarter	-0.5	3.7	0.1	11.1	3.72	3.30	2.84	2.62	4.06	4.69
2nd quarter	2.0	10.0	6.8	30.6	3.75	3.27	2.83	2.60	4.03	4.64
3rd quarter	1.0	12.7	1.9	38.5	3.77	3.26	2.83	2.60	4.05	4.66
4th quarter	-1.4	8.9	-7.1	26.5	3.74	3.18	2.77	2.55	3.94	4.53
1973										
1st quarter	-4.2	14.8	-10.7	42.9	3.86	3.11	2.79	2.52	3.92	4.45
2nd quarter	5.0	13.2	12.0	38.3	4.03	3.15	2.86	2.54	4.02	4.51
3rd quarter	2.6	7.2	3.5	20.0	4.11	3.16	2.87	2.56	4.05	4.55
4th quarter	2.8	22.2	2.3	60.9	4.20	3.22	2.98	2.62	4.22	4.67
1974										
1st quarter	-3.3	15.9	-5.2	39.7	4.26	3.27	3.13	2.68	4.42	4.76
2nd quarter	1.0	16.4	0.9	39.6	4.29	3.31	3.28	2.72	4.67	4.86
3rd quarter	-0.1	5.7	1.7	11.7	4.45	3.34	3.40	2.75	4.84	4.93
4th quarter	-8.7	29.2	-15.0	64.9	4.52	3.46	3.54	2.88	5.11	5.26
1975										
1st quarter	7.1	-17.1	16.4	-38.3	4.32	3.42	3.41	2.82	4.94	5.16
2nd quarter	3.1	-17.1	5.4	-38.6	4.26	3.35	3.30	2.75	4.81	5.06
3rd quarter	0.8	-2.2	-1.0	-5.0	4.18	3.31	3.24	2.71	4.69	4.95
4th quarter	2.5	-2.1	2.3	-3.7	4.04	3.26	3.16	2.66	4.59	4.88
1976										
1st quarter	-0.5	15.2	0.1	33.2	4.00	3.22	3.14	2.64	4.56	4.83
2nd quarter	-1.8	24.3	-1.9	51.8	4.07	3.26	3.21	2.69	4.67	4.91
3rd quarter	2.0	18.8	5.4	38.9	4.04	3.28	3.23	2.70	4.74	4.96
4th quarter	-3.0	13.6	-9.8	26.5	3.95	3.23	3.19	2.68	4.68	4.92
1977										
1st quarter	-1.3	16.1	-7.0	33.0	3.96	3.22	3.21	2.68	4.72	4.92
2nd quarter	5.3	14.2	11.3	27.8	3.88	3.19	3.16	2.65	4.62	4.84
3rd quarter	4.4	26.5	16.3	53.8	3.86	3.22	3.16	2.67	4.63	4.88
4th quarter	9.8	14.4	8.4	28.1	3.89	3.22	3.16	2.67	4.63	4.87
1978										
1st quarter	-0.6	26.1	-3.6	48.9	4.02	3.26	3.21	2.72	4.78	5.00
2nd quarter	0.4	23.9	0.2	44.2	3.92	3.14	3.12	2.62	4.55	4.75
3rd quarter	7.2	17.8	11.1	32.3	3.95	3.15	3.13	2.62	4.54	4.74
4th quarter	-1.5	30.0	0.5	54.1	3.98	3.14	3.15	2.63	4.54	4.73
1979										
1st quarter	4.2	19.6	6.6	33.0	4.15	3.17	3.22	2.65	4.66	4.78
2nd quarter	3.1	24.3	5.3	40.4	4.19	3.21	3.29	2.69	4.78	4.87
3rd quarter	6.6	5.5	7.0	8.2	4.16	3.16	3.28	2.64	4.74	4.73
4th quarter	0.3	8.3	-1.9	13.4	4.19	3.16	3.34	2.64	4.85	4.75
1980										
1st quarter	-0.6	10.5	-2.6	17.4	4.25	3.17	3.45	2.66	5.02	4.78
2nd quarter	-5.2	13.0	-5.9	19.3	4.40	3.28	3.57	2.76	5.26	5.02
3rd quarter	-12.5	-21.4	-14.2	-32.6	4.34	3.18	3.50	2.68	5.15	4.88
4th quarter	-6.1	-3.0	-6.0	-5.8	4.24	3.13	3.44	2.65	5.08	4.82
1981										
1st quarter	6.8	31.9	10.3	44.6	4.26	3.17	3.48	2.67	5.13	4.84
2nd quarter	9.9	1.8	10.9	1.7	4.24	3.18	3.47	2.67	5.15	4.87
3rd quarter	11.8	32.1	14.5	42.6	4.20	3.22	3.48	2.71	5.16	4.93
4th quarter	6.5	18.3	4.5	23.8	4.17	3.28	3.49	2.75	5.20	5.03
1982										
1st quarter	5.1	-26.5	6.7	-34.3	4.19	3.28	3.46	2.74	5.20	5.03
2nd quarter	4.3	-8.5	6.4	-12.2	4.15	3.28	3.42	2.74	5.17	5.04
3rd quarter	9.0	-3.2	12.0	-5.5	4.12	3.33	3.43	2.77	5.26	5.15
4th quarter	4.6	-44.4	4.9	-57.5	3.97	3.23	3.30	2.68	5.10	4.99
1983										
1st quarter	-7.3	-27.8	-10.4	-37.3	3.89	3.15	3.19	2.62	4.98	4.89
2nd quarter	-13.0	5.3	-12.6	6.5	3.81	3.09	3.14	2.58	4.89	4.80
3rd quarter	-32.4	28.2	-37.7	34.3	3.73	3.03	3.12	2.55	4.86	4.73
4th quarter	-8.8	32.7	-9.0	39.4	3.69	2.98	3.08	2.52	4.77	4.64
1984										
1st quarter	5.5	67.5	7.4	81.9	3.78	3.02	3.15	2.56	4.87	4.71
2nd quarter	5.6	63.7	6.4	77.2	3.76	3.03	3.16	2.57	4.88	4.69
3rd quarter	6.8	64.5	8.1	78.9	3.75	3.06	3.18	2.60	4.93	4.75
4th quarter	5.0	43.0	3.3	51.0	3.72	3.05	3.16	2.60	4.89	4.72
1985										
1st quarter	7.9	8.3	8.4	8.8	3.60	3.00	3.06	2.55	4.77	4.65
2nd quarter	4.2	17.4	4.8	20.3	3.55	2.99	3.04	2.55	4.76	4.66
3rd quarter	6.6	9.7	9.1	12.0	3.47	2.96	2.98	2.51	4.67	4.60
4th quarter	4.3	28.8	5.9	32.7	3.51	2.98	3.01	2.53	4.74	4.65
1986										
1st quarter	2.5	27.8	1.7	33.3	3.42	2.97	2.94	2.52	4.65	4.63
2nd quarter	-3.5	19.1	-5.9	22.3	3.37	2.97	2.92	2.53	4.63	4.65
3rd quarter	-3.1	-3.9	-4.3	-5.5	3.29	2.92	2.85	2.49	4.52	4.55
4th quarter	-1.8	-11.0	-0.6	-8.5	3.25	2.90	2.82	2.47	4.49	4.53

[1] Before 1997, 2005 dollars; 1997 forward, 2009 dollars.

Table 1-8B. Inventory Change and Inventory-Sales Ratios: Historical Data—*Continued*

(Quarterly dollar data are at seasonally adjusted annual rates.) **NIPA Tables 5.7.5B, 5.7.6B, 5.8.5B, 5.8.6B**

Year and quarter	Change in private inventories				Ratio, inventories at end of quarter to monthly rate of sales during the quarter					
	Billions of current dollars		Billions of chained dollars [1]		Total private inventories to final sales of domestic business		Nonfarm inventories to final sales of domestic business		Nonfarm inventories to final sales of goods and structures	
	Farm	Nonfarm	Farm	Nonfarm	Current dollars	Chained dollars[1]	Current dollars	Chained dollars[1]	Current dollars	Chained dollars[1]
1987										
1st quarter	-7.7	35.6	-9.7	44.5	3.31	2.94	2.86	2.51	4.62	4.66
2nd quarter	-10.7	27.2	-14.5	33.1	3.30	2.90	2.86	2.49	4.60	4.60
3rd quarter	-4.0	5.0	-5.2	4.1	3.25	2.85	2.82	2.45	4.52	4.51
4th quarter	-3.3	66.4	-7.0	82.5	3.33	2.90	2.90	2.50	4.65	4.62
1988										
1st quarter	-4.4	21.4	-7.6	25.1	3.30	2.86	2.86	2.47	4.61	4.56
2nd quarter	-12.2	31.9	-15.2	38.0	3.29	2.83	2.85	2.46	4.60	4.52
3rd quarter	-11.2	29.4	-11.4	32.6	3.30	2.83	2.85	2.47	4.63	4.57
4th quarter	-19.7	38.8	-18.3	45.9	3.29	2.81	2.86	2.47	4.63	4.55
1989										
1st quarter	7.2	41.0	7.8	46.6	3.32	2.83	2.89	2.48	4.70	4.58
2nd quarter	2.9	33.2	3.6	37.6	3.28	2.83	2.87	2.48	4.65	4.56
3rd quarter	-5.5	15.5	-5.8	16.9	3.21	2.79	2.82	2.45	4.55	4.49
4th quarter	-4.6	21.2	-5.6	22.3	3.23	2.80	2.83	2.47	4.63	4.56
1990										
1st quarter	1.8	12.1	1.7	13.3	3.18	2.78	2.78	2.45	4.53	4.50
2nd quarter	-1.5	35.3	-2.6	39.9	3.17	2.81	2.78	2.47	4.58	4.60
3rd quarter	4.0	17.9	4.4	20.3	3.20	2.82	2.82	2.48	4.67	4.63
4th quarter	5.2	-16.6	6.7	-17.7	3.22	2.83	2.82	2.49	4.71	4.66
1991										
1st quarter	0.0	-15.6	0.7	-18.1	3.18	2.85	2.78	2.51	4.65	4.71
2nd quarter	-0.7	-17.3	-0.5	-20.6	3.08	2.81	2.70	2.47	4.55	4.66
3rd quarter	-10.9	11.7	-13.8	13.1	3.04	2.81	2.69	2.48	4.54	4.67
4th quarter	6.4	24.7	6.1	27.1	3.06	2.83	2.70	2.50	4.62	4.74
1992										
1st quarter	8.0	-7.8	10.5	-7.5	3.00	2.78	2.63	2.44	4.52	4.64
2nd quarter	9.4	13.8	10.9	13.2	2.99	2.77	2.62	2.43	4.50	4.61
3rd quarter	5.1	15.4	4.7	16.6	2.96	2.75	2.60	2.41	4.47	4.57
4th quarter	2.5	18.9	2.4	21.6	2.92	2.73	2.56	2.39	4.41	4.53
1993										
1st quarter	-5.6	41.6	-6.2	46.5	2.95	2.75	2.58	2.42	4.47	4.60
2nd quarter	-4.9	29.0	-5.3	32.3	2.92	2.74	2.57	2.41	4.44	4.56
3rd quarter	-12.5	19.1	-14.2	20.6	2.89	2.72	2.54	2.40	4.42	4.55
4th quarter	-1.7	18.3	-4.1	19.9	2.85	2.68	2.51	2.38	4.33	4.47
1994										
1st quarter	16.5	28.8	19.4	31.1	2.87	2.69	2.51	2.38	4.33	4.48
2nd quarter	17.1	64.3	21.0	70.0	2.86	2.71	2.53	2.39	4.37	4.49
3rd quarter	10.7	42.5	10.6	46.1	2.86	2.71	2.53	2.39	4.36	4.48
4th quarter	3.7	71.5	3.7	76.3	2.89	2.72	2.56	2.40	4.39	4.47
1995										
1st quarter	-5.8	67.0	-5.9	71.2	2.93	2.75	2.62	2.43	4.49	4.52
2nd quarter	-13.7	47.4	-13.6	50.1	2.94	2.75	2.64	2.44	4.55	4.55
3rd quarter	-20.4	31.7	-26.4	32.8	2.90	2.71	2.60	2.41	4.49	4.50
4th quarter	-4.4	22.8	-6.2	22.9	2.88	2.69	2.58	2.40	4.45	4.46
1996										
1st quarter	1.1	5.8	-2.9	6.2	2.84	2.66	2.55	2.38	4.39	4.42
2nd quarter	11.8	18.6	7.4	19.8	2.81	2.63	2.51	2.35	4.30	4.35
3rd quarter	17.1	34.0	27.4	35.0	2.81	2.65	2.50	2.35	4.29	4.34
4th quarter	4.5	30.2	1.3	31.7	2.76	2.62	2.47	2.33	4.24	4.28
1997										
1st quarter	-0.5	50.3	-10.8	53.3	2.74	2.62	2.45	2.33	4.21	4.28
2nd quarter	0.8	87.6	8.2	95.7	2.73	2.65	2.44	2.36	4.21	4.33
3rd quarter	8.3	59.6	10.1	65.5	2.69	2.63	2.41	2.34	4.14	4.27
4th quarter	4.7	73.1	7.2	79.8	2.70	2.65	2.42	2.37	4.17	4.32
NEW BASIS										
1997										
1st quarter	-0.6	50.3	-13.5	45.4	2.66	2.93	2.37	2.54	3.91	4.61
2nd quarter	0.8	87.6	13.1	130.9	2.64	2.97	2.36	2.57	3.90	4.67
3rd quarter	8.3	59.6	15.1	82.4	2.60	2.94	2.33	2.55	3.83	4.60
4th quarter	4.6	73.0	10.0	93.4	2.60	2.97	2.34	2.57	3.86	4.64
1998										
1st quarter	5.6	99.5	9.4	144.4	2.60	3.00	2.34	2.61	3.87	4.71
2nd quarter	-4.9	42.2	-4.0	57.0	2.55	2.97	2.30	2.58	3.81	4.65
3rd quarter	0.5	51.9	4.5	65.5	2.51	2.96	2.28	2.57	3.75	4.62
4th quarter	3.8	56.2	1.3	79.1	2.47	2.93	2.25	2.55	3.69	4.56
1999										
1st quarter	4.1	79.3	-2.9	114.9	2.48	2.95	2.25	2.57	3.71	4.61
2nd quarter	-0.2	35.3	1.6	46.9	2.46	2.92	2.24	2.55	3.68	4.56
3rd quarter	-9.1	49.6	-18.1	71.1	2.47	2.91	2.25	2.54	3.71	4.54
4th quarter	-5.7	90.0	-5.9	118.0	2.49	2.91	2.27	2.55	3.75	4.55
2000										
1st quarter	-17.7	33.9	-30.9	40.7	2.48	2.87	2.26	2.52	3.75	4.51
2nd quarter	5.9	84.5	10.4	112.1	2.48	2.88	2.27	2.53	3.76	4.54
3rd quarter	-1.1	58.4	-3.5	86.1	2.48	2.89	2.28	2.55	3.79	4.58
4th quarter	7.3	47.0	11.8	71.5	2.51	2.90	2.29	2.56	3.82	4.60

[1] Before 1997, 2005 dollars; 1997 forward, 2009 dollars.

Table 1-8B. Inventory Change and Inventory-Sales Ratios: Historical Data—*Continued*

(Quarterly dollar data are at seasonally adjusted annual rates.) **NIPA Tables 5.7.5B, 5.7.6B, 5.8.5B, 5.8.6B**

| Year and quarter | Change in private inventories | | | | Ratio, inventories at end of quarter to monthly rate of sales during the quarter | | | | | |
| | Billions of current dollars | | Billions of chained dollars [1] | | Total private inventories to final sales of domestic business | | Nonfarm inventories to final sales of domestic business | | Nonfarm inventories to final sales of goods and structures | |
	Farm	Nonfarm	Farm	Nonfarm	Current dollars	Chained dollars[1]	Current dollars	Chained dollars[1]	Current dollars	Chained dollars[1]
2001										
1st quarter	5.6	-36.2	10.0	-50.7	2.48	2.88	2.26	2.53	3.79	4.56
2nd quarter	-3.0	-8.6	-5.0	0.8	2.44	2.88	2.23	2.53	3.72	4.55
3rd quarter	1.8	-31.9	2.6	-46.8	2.41	2.88	2.20	2.53	3.68	4.55
4th quarter	-4.3	-76.6	-6.9	-108.8	2.33	2.81	2.13	2.47	3.55	4.43
2002										
1st quarter	4.1	-3.8	4.1	-10.2	2.31	2.81	2.10	2.47	3.54	4.45
2nd quarter	-10.3	31.5	-16.4	43.0	2.32	2.82	2.12	2.48	3.60	4.50
3rd quarter	-2.9	28.3	-5.6	31.5	2.34	2.82	2.14	2.48	3.64	4.50
4th quarter	-0.8	33.8	-3.0	44.1	2.38	2.83	2.16	2.50	3.71	4.55
2003										
1st quarter	2.8	23.2	4.9	35.1	2.42	2.83	2.20	2.49	3.78	4.53
2nd quarter	-1.4	-5.1	-1.5	-7.9	2.37	2.79	2.16	2.45	3.69	4.45
3rd quarter	0.9	-0.4	0.3	2.5	2.33	2.72	2.10	2.40	3.57	4.30
4th quarter	-2.1	38.4	-2.9	47.3	2.33	2.71	2.10	2.39	3.57	4.28
2004										
1st quarter	4.4	41.6	8.6	50.7	2.36	2.73	2.13	2.40	3.64	4.31
2nd quarter	18.4	55.5	23.7	77.1	2.39	2.74	2.16	2.41	3.68	4.33
3rd quarter	7.7	56.5	12.6	71.5	2.40	2.73	2.17	2.40	3.71	4.31
4th quarter	4.7	67.4	9.0	82.0	2.39	2.73	2.17	2.40	3.70	4.30
2005										
1st quarter	-6.7	109.2	-7.6	126.7	2.42	2.74	2.20	2.42	3.75	4.32
2nd quarter	0.9	27.9	1.2	23.0	2.38	2.71	2.17	2.39	3.69	4.25
3rd quarter	4.3	8.0	5.0	4.1	2.39	2.68	2.17	2.37	3.70	4.20
4th quarter	2.3	84.1	2.9	92.4	2.42	2.71	2.21	2.39	3.79	4.27
2006										
1st quarter	3.7	74.5	2.6	98.0	2.39	2.70	2.18	2.38	3.72	4.22
2nd quarter	-8.1	83.6	-11.0	97.5	2.42	2.71	2.22	2.40	3.81	4.28
3rd quarter	-6.3	85.1	-9.4	101.2	2.45	2.74	2.24	2.43	3.85	4.34
4th quarter	-3.6	47.1	-7.1	59.7	2.42	2.71	2.21	2.41	3.82	4.31
2007										
1st quarter	7.9	13.4	5.3	22.6	2.44	2.71	2.22	2.40	3.85	4.32
2nd quarter	-4.1	46.4	5.9	50.4	2.46	2.71	2.24	2.41	3.89	4.32
3rd quarter	-1.7	46.6	-15.5	54.3	2.46	2.70	2.24	2.40	3.91	4.32
4th quarter	0.4	26.9	-1.4	35.2	2.51	2.69	2.28	2.40	4.01	4.31
2008										
1st quarter	-8.6	-8.1	-6.1	6.5	2.61	2.71	2.37	2.42	4.21	4.37
2nd quarter	3.8	-26.1	2.1	-26.6	2.71	2.69	2.46	2.39	4.39	4.32
3rd quarter	4.1	-20.0	2.3	-20.7	2.68	2.71	2.44	2.41	4.36	4.38
4th quarter	-1.5	-60.8	-0.8	-85.0	2.50	2.76	2.28	2.45	4.14	4.56
2009										
1st quarter	-4.8	-146.2	-8.6	-168.1	2.43	2.75	2.21	2.44	4.02	4.55
2nd quarter	-6.3	-189.7	-9.5	-215.3	2.40	2.68	2.19	2.38	3.98	4.44
3rd quarter	-10.9	-193.1	-13.5	-228.3	2.35	2.60	2.14	2.30	3.91	4.28
4th quarter	-4.6	-47.4	-8.4	-49.0	2.37	2.58	2.17	2.29	3.99	4.28
2010										
1st quarter	-6.6	5.3	-14.0	6.1	2.42	2.58	2.20	2.29	4.10	4.28
2nd quarter	-11.7	48.4	-16.5	45.2	2.40	2.57	2.19	2.28	4.07	4.26
3rd quarter	-16.5	129.4	-20.6	141.9	2.45	2.59	2.23	2.31	4.17	4.34
4th quarter	-13.9	81.2	-20.0	87.7	2.49	2.58	2.27	2.31	4.21	4.31
2011										
1st quarter	-2.1	34.2	-2.1	34.3	2.60	2.58	2.34	2.31	4.38	4.34
2nd quarter	-5.7	66.8	-4.1	74.4	2.62	2.59	2.37	2.32	4.42	4.34
3rd quarter	-1.2	-10.8	-2.2	-15.4	2.58	2.55	2.33	2.29	4.34	4.27
4th quarter	0.7	103.2	-0.4	100.2	2.58	2.57	2.32	2.30	4.30	4.27
2012										
1st quarter	-6.4	86.7	-10.8	83.7	2.57	2.55	2.32	2.29	4.28	4.23
2nd quarter	-15.4	106.8	-16.4	103.3	2.54	2.56	2.30	2.31	4.23	4.25
3rd quarter	-29.9	114.3	-25.1	122.8	2.58	2.59	2.34	2.34	4.29	4.30
4th quarter	-23.1	51.8	-22.5	49.8	2.56	2.57	2.33	2.33	4.27	4.28
2013										
1st quarter	2.4	79.9	0.1	80.7	2.52	2.57	2.31	2.33	4.22	4.25
2nd quarter	15.0	72.7	12.4	73.7	2.52	2.59	2.30	2.35	4.22	4.28
3rd quarter	17.4	118.8	15.0	132.5	2.52	2.61	2.30	2.37	4.21	4.30
4th quarter	11.9	100.0	14.1	105.9	2.52	2.60	2.30	2.36	4.19	4.27
2014										
1st quarter	-3.5	55.7	-5.1	63.3	2.57	2.62	2.33	2.38	4.28	4.32
2nd quarter	0.0	96.3	-3.1	102.0	2.54	2.60	2.30	2.37	4.21	4.27
3rd quarter	-2.0	101.5	-1.5	108.1	2.51	2.59	2.27	2.36	4.15	4.25
4th quarter	-7.4	95.2	-4.4	87.0	2.46	2.58	2.23	2.36	4.10	4.26
2015										
1st quarter	5.9	169.7	4.8	173.1	2.44	2.62	2.22	2.40	4.10	4.36
2nd quarter	2.2	147.8	2.2	154.7	2.43	2.64	2.22	2.41	4.09	4.38
3rd quarter	-1.9	128.7	-1.5	126.4	2.40	2.65	2.21	2.43	4.07	4.41
4th quarter	-0.1	94.8	-1.2	91.8	2.39	2.67	2.20	2.45	4.09	4.45
2016										
1st quarter	-13.9	63.4	-13.5	66.2	2.37	2.66	2.18	2.44	4.05	4.43
2nd quarter	-3.7	20.5	-4.8	12.7	2.36	2.65	2.19	2.43	4.07	4.41
3rd quarter	-0.1	-4.8	-1.2	-15.5	2.33	2.62	2.17	2.41	4.06	4.36
4th quarter	-6.0	58.0	-7.3	57.3	2.35	2.63	2.19	2.41	4.11	4.38

[1]Before 1997, 2005 dollars; 1997 forward, 2009 dollars.

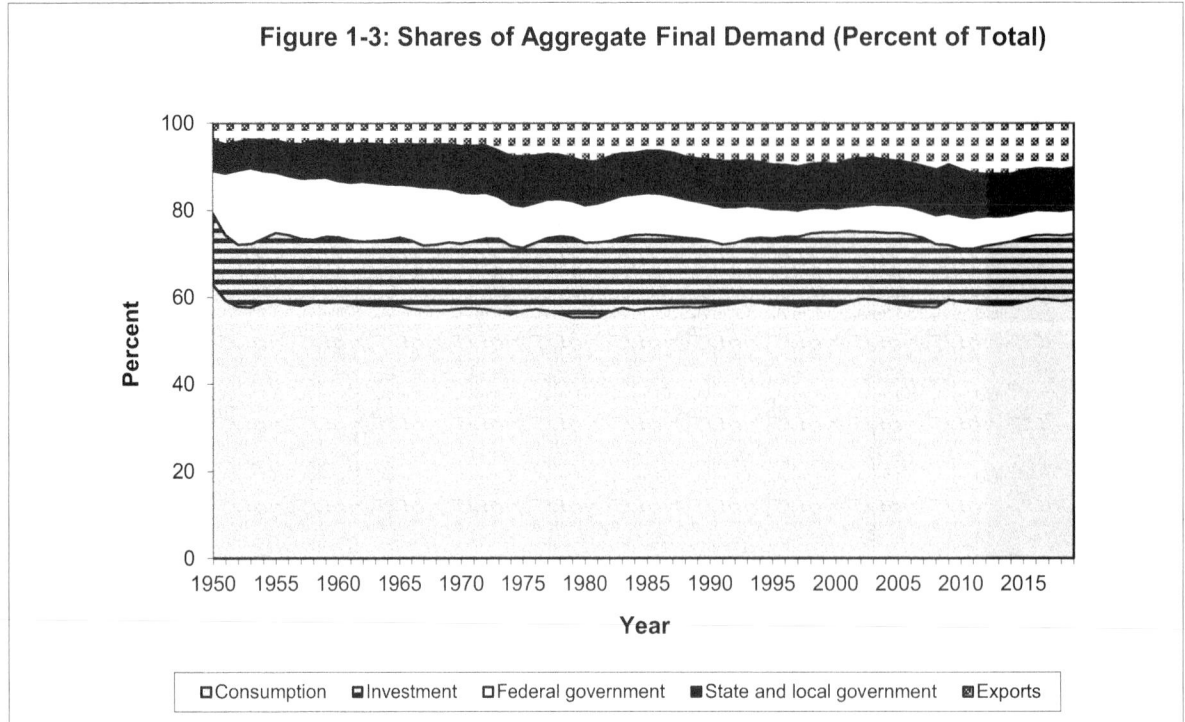

Figure 1-3: Shares of Aggregate Final Demand (Percent of Total)

□ Consumption ▣ Investment □ Federal government ■ State and local government ▨ Exports

- People often want to know how various sources of demand, especially consumption spending, affect GDP. Often, it is said that consumption "accounts for" about 70 percent, or sometimes two-thirds, of GDP. That is based on a NIPA table showing the components of GDP as a percent of total GDP. In such a calculation, personal consumption expenditures (PCE) amounted to nearly 68 percent of total GDP in 2019, up from a range of 60 to 65 percent earlier in the postwar period. (Table 1-1A)

- Is "amounted to" the same as "accounted for"? No, because the value of PCE includes imported goods and services—coffee, cocoa, and bananas; crude oil from the Middle East, transformed into gasoline; clothing, cars, and toys manufactured overseas— even the money that U.S. consumers spend when they travel. These imports included in PCE don't account for demand for U.S. GDP, which excludes imports; instead, they account for demand for GDP in China and other countries that export to us.

- It is not possible with NIPA data to precisely estimate and subtract out the import content of PCE, and so there is no way to precisely estimate PCE's contribution to GDP. What we can do is calculate PCE and the other sources of aggregate final demand for U.S. output as shares of <u>total final demand</u>; this is shown in Table 1-10 and Figure 1-3. These shares also approximate their contributions to GDP (excluding inventory change) if each final demand source has, in aggregate, about the same percentage import content as total GDP.

- The share of imports in the total supply of goods and services in the U.S. market rose from around 4 percent in the 1950s to a high of 14.8 percent in 2008. By 2019, the import share decreased to 12.7 percent. Only in 2009, the year of maximum inventory liquidation, did the import share temporarily dip back to 12.0 percent. (Table 1-9)

Table 1-9. Shares of Aggregate Supply

(Billions of dollars, quarterly dollar data are at seasonally adjusted rates, percents.)   NIPA Table 1.1.5

Year and quarter	Aggregate supply (billions of dollars)			Shares of aggregate supply (percent)	
	Total	GDP	Imports	Domestic production (GDP)	Imports
1950	311.4	299.8	11.6	96.3	3.7
1951	361.5	346.9	14.6	96.0	4.0
1952	382.6	367.3	15.3	96.0	4.0
1953	405.2	389.2	16.0	96.1	3.9
1954	405.9	390.5	15.4	96.2	3.8
1955	442.7	425.5	17.2	96.1	3.9
1956	468.3	449.4	18.9	96.0	4.0
1957	493.9	474.0	19.9	96.0	4.0
1958	501.2	481.2	20.0	96.0	4.0
1959	544.0	521.7	22.3	95.9	4.1
1960	565.2	542.4	22.8	96.0	4.0
1961	584.9	562.2	22.7	96.1	3.9
1962	628.9	603.9	25.0	96.0	4.0
1963	663.6	637.5	26.1	96.1	3.9
1964	712.6	684.5	28.1	96.1	3.9
1965	773.8	742.3	31.5	95.9	4.1
1966	850.5	813.4	37.1	95.6	4.4
1967	899.9	860.0	39.9	95.6	4.4
1968	987.3	940.7	46.6	95.3	4.7
1969	1 068.1	1 017.6	50.5	95.3	4.7
1970	1 129.1	1 073.3	55.8	95.1	4.9
1971	1 227.2	1 164.9	62.3	94.9	5.1
1972	1 353.3	1 279.1	74.2	94.5	5.5
1973	1 516.6	1 425.4	91.2	94.0	6.0
1974	1 672.7	1 545.2	127.5	92.4	7.6
1975	1 807.6	1 684.9	122.7	93.2	6.8
1976	2 024.5	1 873.4	151.1	92.5	7.5
1977	2 264.2	2 081.8	182.4	91.9	8.1
1978	2 563.9	2 351.6	212.3	91.7	8.3
1979	2 880.0	2 627.3	252.7	91.2	8.8
1980	3 151.1	2 857.3	293.8	90.7	9.3
1981	3 524.8	3 207.0	317.8	91.0	9.0
1982	3 647.0	3 343.8	303.2	91.7	8.3
1983	3 962.6	3 634.0	328.6	91.7	8.3
1984	4 442.7	4 037.6	405.1	90.9	9.1
1985	4 756.2	4 339.0	417.2	91.2	8.8
1986	5 032.5	4 579.6	452.9	91.0	9.0
1987	5 363.9	4 855.2	508.7	90.5	9.5
1988	5 790.4	5 236.4	554.0	90.4	9.6
1989	6 232.6	5 641.6	591.0	90.5	9.5
1990	6 592.8	5 963.1	629.7	90.4	9.6
1991	6 781.6	6 158.1	623.5	90.8	9.2
1992	7 188.1	6 520.3	667.8	90.7	9.3
1993	7 578.6	6 858.6	720.0	90.5	9.5
1994	8 100.6	7 287.2	813.4	90.0	10.0
1995	8 542.3	7 639.7	902.6	89.4	10.6
1996	9 037.1	8 073.1	964.0	89.3	10.7
1997	9 633.4	8 577.6	1 055.8	89.0	11.0
1998	10 178.5	9 062.8	1 115.7	89.0	11.0
1999	10 879.3	9 630.7	1 248.6	88.5	11.5
2000	11 723.6	10 252.3	1 471.3	87.5	12.5
2001	11 974.4	10 581.8	1 392.6	88.4	11.6
2002	12 360.5	10 936.4	1 424.1	88.5	11.5
2003	12 997.5	11 458.2	1 539.3	88.2	11.8
2004	14 010.4	12 213.7	1 796.7	87.2	12.8
2005	15 063.0	13 036.6	2 026.4	86.5	13.5
2006	16 058.1	13 814.6	2 243.5	86.0	14.0
2007	16 831.2	14 451.9	2 379.3	85.9	14.1
2008	17 272.9	14 712.8	2 560.1	85.2	14.8
2009	16 427.3	14 448.9	1 978.4	88.0	12.0
2010	17 352.3	14 992.1	2 360.2	86.4	13.6
2011	18 225.1	15 542.6	2 682.5	85.3	14.7
2012	18 956.9	16 197.0	2 759.9	85.4	14.6
2013	19 549.1	16 784.9	2 764.2	85.9	14.1
2014	20 406.7	17 527.3	2 879.4	85.9	14.1
2015	21 030.7	18 238.3	2 792.4	86.7	13.3
2016	21 484.8	18 745.1	2 739.7	87.2	12.8
2017	22 473.1	19 543.0	2 930.1	87.0	13.0
2018	23 750.1	20 611.9	3 138.2	86.8	13.2
2019	24 558.4	21 433.2	3 125.2	87.3	12.7
2017					
1st quarter	22 107.3	19 237.4	2 869.9	87.0	13.0
2nd quarter	22 271.8	19 379.2	2 892.6	87.0	13.0
3rd quarter	22 531.0	19 617.3	2 913.7	87.1	12.9
4th quarter	22 982.1	19 938.0	3 044.1	86.8	13.2
2018					
1st quarter	23 339.2	20 242.2	3 097.0	86.7	13.3
2nd quarter	23 651.1	20 552.7	3 098.4	86.9	13.1
3rd quarter	23 913.0	20 742.7	3 170.3	86.7	13.3
4th quarter	24 096.8	20 909.9	3 186.9	86.8	13.2
2019					
1st quarter	24 254.3	21 115.3	3 139.0	87.1	12.9
2nd quarter	24 489.3	21 329.9	3 159.4	87.1	12.9
3rd quarter	24 677.4	21 540.3	3 137.1	87.3	12.7
4th quarter	24 812.8	21 747.4	3 065.4	87.6	12.4

Table 1-10. Shares of Aggregate Final Demand

(Billions of dollars, quarterly dollar data are at seasonally adjusted rates, percents.)   NIPA Table 1.1.5

Year and quarter	Aggregate final demand (billions of dollars)							Shares of aggregate final demand (percent)					
	Total	Consumption (PCE)	Nonresidential fixed investment	Residential investment	Exports	Government consumption and gross investment Federal	State and local	PCE	Nonresidential investment	Residential investment	Exports	Federal government	State and local government
1955	437.7	258.3	43.4	25.4	17.7	61.0	31.9	59.0	9.9	5.8	4.0	13.9	7.3
1956	464.3	271.1	49.7	24.0	21.3	63.1	35.1	58.4	10.7	5.2	4.6	13.6	7.6
1957	493.2	286.3	53.1	22.6	24.0	68.5	38.7	58.0	10.8	4.6	4.9	13.9	7.8
1958	501.6	295.6	48.5	22.8	20.6	71.5	42.6	58.9	9.7	4.5	4.1	14.3	8.5
1959	540.0	317.1	53.1	28.6	22.7	73.6	44.9	58.7	9.8	5.3	4.2	13.6	8.3
1960	562.0	331.2	56.4	26.9	27.0	72.9	47.6	58.9	10.0	4.8	4.8	13.0	8.5
1961	582.0	341.5	56.6	27.0	27.6	77.4	51.9	58.7	9.7	4.6	4.7	13.3	8.9
1962	622.8	362.6	61.2	29.6	29.1	85.5	54.8	58.2	9.8	4.8	4.7	13.7	8.8
1963	658.0	382.0	64.8	32.9	31.1	87.9	59.3	58.1	9.8	5.0	4.7	13.4	9.0
1964	707.7	410.6	72.2	35.1	35.0	90.3	64.5	58.0	10.2	5.0	4.9	12.8	9.1
1965	764.6	443.0	85.2	35.2	37.1	93.2	70.9	57.9	11.1	4.6	4.9	12.2	9.3
1966	836.9	479.9	97.2	33.4	40.9	106.6	78.9	57.3	11.6	4.0	4.9	12.7	9.4
1967	890.0	506.7	99.2	33.6	43.5	120.0	87.0	56.9	11.1	3.8	4.9	13.5	9.8
1968	978.3	556.9	107.7	40.2	47.9	128.0	97.6	56.9	11.0	4.1	4.9	13.1	10.0
1969	1 058.9	603.6	120.0	44.4	51.9	131.2	107.8	57.0	11.3	4.2	4.9	12.4	10.2
1970	1 127.0	646.7	124.6	43.4	59.7	132.8	119.8	57.4	11.1	3.9	5.3	11.8	10.6
1971	1 219.0	699.9	130.4	58.2	63.0	134.5	133.0	57.4	10.7	4.8	5.2	11.0	10.9
1972	1 344.1	768.2	146.6	72.4	70.8	141.6	144.5	57.2	10.9	5.4	5.3	10.5	10.8
1973	1 500.7	849.6	172.7	78.3	95.3	146.2	158.6	56.6	11.5	5.2	6.4	9.7	10.6
1974	1 658.8	930.2	191.1	69.5	126.7	158.8	182.5	56.1	11.5	4.2	7.6	9.6	11.0
1975	1 813.8	1 030.5	196.8	66.7	138.7	173.7	207.4	56.8	10.9	3.7	7.6	9.6	11.4
1976	2 007.5	1 147.7	219.3	86.8	149.5	184.8	219.4	57.2	10.9	4.3	7.4	9.2	10.9
1977	2 241.9	1 274.0	259.1	115.2	159.3	200.3	234.0	56.8	11.6	5.1	7.1	8.9	10.4
1978	2 538.1	1 422.3	314.6	138.0	186.9	218.9	257.4	56.0	12.4	5.4	7.4	8.6	10.1
1979	2 861.9	1 585.4	373.8	147.8	230.1	240.6	284.2	55.4	13.1	5.2	8.0	8.4	9.9
1980	3 157.5	1 750.7	406.9	129.5	280.8	274.9	314.7	55.4	12.9	4.1	8.9	8.7	10.0
1981	3 495.0	1 934.0	472.9	128.5	305.2	314.0	340.4	55.3	13.5	3.7	8.7	9.0	9.7
1982	3 661.8	2 071.3	485.1	110.8	283.2	348.3	363.1	56.6	13.2	3.0	7.7	9.5	9.9
1983	3 968.5	2 281.6	482.2	161.1	277.0	382.4	384.2	57.5	12.2	4.1	7.0	9.6	9.7
1984	4 377.3	2 492.3	564.3	190.4	302.4	411.8	416.1	56.9	12.9	4.3	6.9	9.4	9.5
1985	4 734.4	2 712.8	607.8	200.1	303.2	452.9	457.6	57.3	12.8	4.2	6.4	9.6	9.7
1986	5 026.0	2 886.3	607.8	234.8	321.0	481.7	494.4	57.4	12.1	4.7	6.4	9.6	9.8
1987	5 336.7	3 076.3	615.2	249.8	363.9	502.8	528.7	57.6	11.5	4.7	6.8	9.4	9.9
1988	5 771.9	3 330.0	662.3	256.2	444.6	511.4	567.4	57.7	11.5	4.4	7.7	8.9	9.8
1989	6 205.0	3 576.8	716.0	256.0	504.3	534.1	617.8	57.6	11.5	4.1	8.1	8.6	10.0
1990	6 578.4	3 809.0	739.2	239.7	551.9	562.4	676.2	57.9	11.2	3.6	8.4	8.5	10.3
1991	6 782.0	3 943.4	723.6	221.2	594.9	582.9	716.0	58.1	10.7	3.3	8.8	8.6	10.6
1992	7 171.8	4 197.6	741.9	254.7	633.1	588.5	756.0	58.5	10.3	3.6	8.8	8.2	10.5
1993	7 557.8	4 452.0	799.2	286.8	654.8	580.2	784.8	58.9	10.6	3.8	8.7	7.7	10.4
1994	8 036.9	4 721.0	868.9	323.8	720.9	574.7	827.6	58.7	10.8	4.0	9.0	7.2	10.3
1995	8 511.1	4 962.6	962.2	324.1	812.8	576.7	872.7	58.3	11.3	3.8	9.5	6.8	10.3
1996	9 006.4	5 244.6	1 043.2	358.1	867.6	579.2	913.7	58.2	11.6	4.0	9.6	6.4	10.1
1997	9 562.4	5 536.8	1 149.1	375.6	953.8	583.3	963.8	57.9	12.0	3.9	10.0	6.1	10.1
1998	10 114.7	5 877.2	1 254.1	418.8	953.0	585.5	1 026.1	58.1	12.4	4.1	9.4	5.8	10.1
1999	10 818.5	6 279.1	1 364.5	461.8	992.8	611.3	1 109.0	58.0	12.6	4.3	9.2	5.7	10.3
2000	11 669.0	6 762.1	1 498.4	485.4	1 096.3	633.7	1 193.1	57.9	12.8	4.2	9.4	5.4	10.2
2001	12 012.7	7 065.6	1 460.1	513.1	1 024.6	670.1	1 279.2	58.8	12.2	4.3	8.5	5.6	10.6
2002	12 340.5	7 342.7	1 352.8	557.6	998.7	743.0	1 345.7	59.5	11.0	4.5	8.1	6.0	10.9
2003	12 983.5	7 723.1	1 375.9	637.1	1 036.2	826.3	1 384.9	59.5	10.6	4.9	8.0	6.4	10.7
2004	13 946.3	8 212.7	1 467.4	749.8	1 177.6	891.7	1 447.1	58.9	10.5	5.4	8.4	6.4	10.4
2005	15 005.5	8 747.1	1 621.0	856.2	1 305.2	947.5	1 528.5	58.3	10.8	5.7	8.7	6.3	10.2
2006	15 989.1	9 260.3	1 793.8	838.2	1 472.6	1 000.7	1 623.5	57.9	11.2	5.2	9.2	6.3	10.2
2007	16 797.2	9 706.4	1 948.6	690.5	1 660.9	1 050.5	1 740.3	57.8	11.6	4.1	9.9	6.3	10.4
2008	17 302.3	9 976.3	1 990.9	516.0	1 837.1	1 150.6	1 831.4	57.7	11.5	3.0	10.6	6.6	10.6
2009	16 578.1	9 842.2	1 690.4	390.0	1 582.0	1 218.2	1 855.3	59.4	10.2	2.4	9.5	7.3	11.2
2010	17 298.3	10 185.8	1 735.0	376.6	1 846.3	1 297.9	1 856.7	58.9	10.0	2.2	10.7	7.5	10.7
2011	18 178.7	10 641.1	1 907.5	378.8	2 103.0	1 298.9	1 849.4	58.5	10.5	2.1	11.6	7.1	10.2
2012	18 885.6	11 006.8	2 118.5	432.0	2 191.3	1 286.5	1 850.5	58.3	11.2	2.3	11.6	6.8	9.8
2013	19 444.5	11 317.2	2 211.5	510.0	2 273.4	1 226.6	1 905.8	58.2	11.4	2.6	11.7	6.3	9.8
2014	20 322.8	11 822.8	2 400.1	560.2	2 371.7	1 215.0	1 953.0	58.2	11.8	2.8	11.7	6.0	9.6
2015	20 894.0	12 297.5	2 466.6	633.8	2 265.9	1 220.8	2 009.4	58.9	11.8	3.0	10.8	5.8	9.6
2016	21 456.5	12 770.0	2 460.5	699.5	2 227.2	1 234.7	2 064.6	59.5	11.5	3.3	10.4	5.8	9.6
2017	22 456.9	13 340.4	2 574.5	760.3	2 374.6	1 263.9	2 143.2	59.4	11.5	3.4	10.6	5.6	9.5
2018	23 692.3	13 993.3	2 776.7	798.5	2 528.7	1 339.4	2 255.7	59.1	11.7	3.4	10.7	5.7	9.5
2019	24 509.4	14 544.6	2 895.0	807.1	2 514.8	1 419.2	2 328.7	59.3	11.8	3.3	10.3	5.8	9.5
2017													
1st quarter	22 119.7	13 153.2	2 532.5	746.0	2 326.4	1 246.5	2 115.1	59.5	11.4	3.4	10.5	5.6	9.6
2nd quarter	22 267.8	13 241.3	2 555.9	753.3	2 333.1	1 257.9	2 126.3	59.5	11.5	3.4	10.5	5.6	9.5
3rd quarter	22 485.8	13 370.9	2 575.2	758.5	2 370.1	1 262.7	2 148.4	59.5	11.5	3.4	10.5	5.6	9.6
4th quarter	22 953.7	13 596.0	2 634.2	783.6	2 468.7	1 288.3	2 182.9	59.2	11.5	3.4	10.8	5.6	9.5
2018													
1st quarter	23 294.7	13 755.5	2 716.2	794.3	2 507.2	1 308.1	2 213.4	59.0	11.7	3.4	10.8	5.6	9.5
2nd quarter	23 640.4	13 939.9	2 765.9	804.3	2 550.3	1 329.3	2 250.7	59.0	11.7	3.4	10.8	5.6	9.5
3rd quarter	23 834.6	14 086.3	2 792.6	800.7	2 523.9	1 352.0	2 279.1	59.1	11.7	3.4	10.6	5.7	9.6
4th quarter	23 999.4	14 191.4	2 831.9	794.7	2 533.4	1 368.4	2 279.6	59.1	11.8	3.3	10.6	5.7	9.5
2019													
1st quarter	24 155.8	14 276.6	2 878.4	795.8	2 523.5	1 388.8	2 292.7	59.1	11.9	3.3	10.4	5.7	9.5
2nd quarter	24 436.1	14 497.3	2 891.3	795.3	2 514.6	1 410.6	2 327.0	59.3	11.8	3.3	10.3	5.8	9.5
3rd quarter	24 636.1	14 645.3	2 908.0	810.5	2 505.2	1 429.3	2 337.8	59.4	11.8	3.3	10.2	5.8	9.5
4th quarter	24 809.5	14 759.2	2 902.3	827.0	2 515.7	1 447.9	2 357.4	59.5	11.7	3.3	10.1	5.8	9.5

SECTION 1B: INCOME AND VALUE ADDED

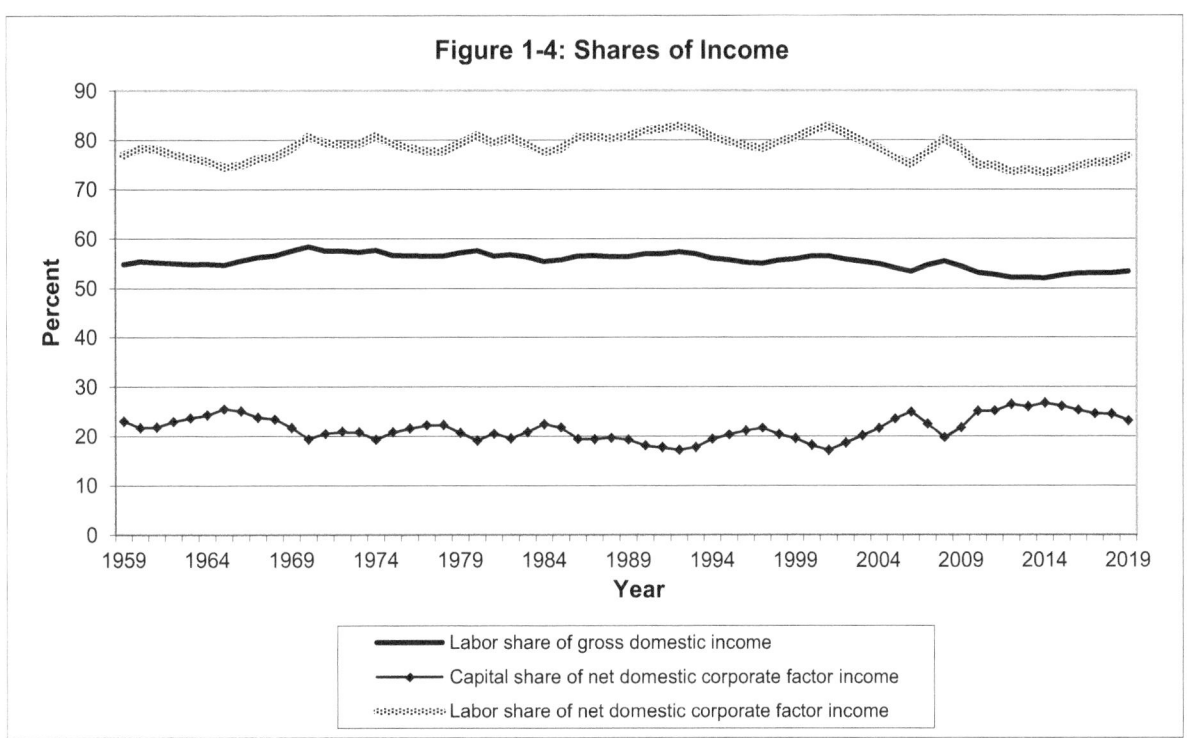

Figure 1-4: Shares of Income

- Recent years have seen a decline in the share of national income paid to workers—not just the share paid out as worker wages, but in total compensation, including all health and other benefits. One indicator of this can be seen in a later chapter, in Tables 9-3A and B, where Bureau of Labor Statistics calculations (based on BEA data) document a declining labor share in the value of output of private nonfarm business. In the tables in Section 1B immediately following this page, BEA compilations show the composition of domestic income, national income, and corporate value added. Table 1-16 shows income share percentages calculated by the editors based on these data. Labor shares rise temporarily in recession years (when the "pie" is smaller) but aside from that effect they have declined over much of the period.

- Employee compensation has fallen since the early 1970s to a record low of 52.0 percent in 2014, as shown in the middle line in Figure 1-4. In 2019, the share grew to 53.4 percent. One factor in this decline has been a rising share of output required to replace used-up capital ("consumption of fixed capital"), reflecting increased use of capital per unit of production. In addition, the share of "net operating surplus" which is largely profits, interest, and proprietor's income, reached a peak in 2014 but has declined each year since then. (Tables 1-12 and 1-16)

- Narrowing the focus to the corporate business sector, net income shares become clearer. (Proprietor's, i.e. unincorporated business; income includes a labor compensation component of uncertain size.) Total net factor income (that is, net of capital consumption) can be completely attributed to either labor or capital. These shares are also charted in Figure 1-4. Between 1959 and 2004, the shares vary cyclically but seem stable over time, averaging around 78 percent for labor and 22 percent for capital. Since then, they have broken out of that range. (Tables 1-14 and 1-16)

Table 1-11. Relation of Gross Domestic Product, Gross and Net National Product, National Income, and Personal Income

(Billions of dollars, quarterly data are at seasonally adjusted annual rates.)   **NIPA Table 1.7.5**

Year and quarter	Gross domestic product	Plus: Income receipts from the rest of the world	Less: Income payments to the rest of the world	Equals: Gross national product	Less: Consumption of fixed capital — Total	Private — Total	Domestic business — Total	Capital consumption allowances	Less: Capital consumption adjustment	Households and institutions	Government — Total	General government	Government enterprises	Equals: Net national product
1960	542.4	4.9	1.8	545.5	67.9	48.2	40.0	39.7	-0.3	8.2	19.7	18.6	1.1	477.6
1961	562.2	5.3	1.8	565.7	70.6	49.8	41.3	41.7	0.4	8.5	20.8	19.7	1.1	495.1
1962	603.9	5.9	1.8	608.0	74.1	51.8	42.9	46.6	3.6	8.8	22.3	21.1	1.2	533.9
1963	637.5	6.5	2.1	641.9	78.0	54.2	44.9	49.9	5.0	9.3	23.9	22.6	1.3	563.9
1964	684.5	7.2	2.3	689.4	82.4	57.3	47.4	53.2	5.8	9.9	25.1	23.7	1.4	607.0
1965	742.3	7.9	2.6	747.6	88.0	61.6	50.9	57.0	6.1	10.7	26.4	25.0	1.5	659.6
1966	813.4	8.1	3.0	818.5	95.3	67.2	55.6	61.9	6.2	11.5	28.1	26.5	1.6	723.2
1967	860.0	8.7	3.3	865.4	103.6	73.3	60.9	66.9	6.0	12.4	30.2	28.5	1.8	761.8
1968	940.7	10.1	4.0	946.7	113.4	80.6	67.0	72.5	5.6	13.6	32.8	30.8	2.0	833.4
1969	1 017.6	11.8	5.7	1 023.7	124.9	89.4	74.2	79.7	5.5	15.2	35.5	33.3	2.2	898.8
1970	1 073.3	12.8	6.4	1 079.7	136.8	98.3	81.8	85.9	4.1	16.5	38.6	36.1	2.5	942.9
1971	1 164.9	14.0	6.4	1 172.4	148.9	107.6	89.4	92.3	2.9	18.2	41.3	38.5	2.8	1 023.5
1972	1 279.1	16.3	7.7	1 287.7	161.0	117.5	97.2	102.3	5.2	20.3	43.5	40.4	3.1	1 126.7
1973	1 425.4	23.5	10.9	1 438.0	178.7	131.5	108.2	111.9	3.7	23.3	47.2	43.7	3.5	1 259.3
1974	1 545.2	29.8	14.3	1 560.8	206.9	153.2	126.1	123.9	-2.2	27.0	53.7	49.4	4.3	1 353.9
1975	1 684.9	28.0	15.0	1 697.9	238.5	178.8	147.9	135.9	-12.0	30.9	59.7	54.6	5.1	1 459.4
1976	1 873.4	32.4	15.5	1 890.3	260.2	196.5	162.2	147.4	-14.8	34.3	63.7	58.2	5.5	1 630.0
1977	2 081.8	37.2	16.9	2 102.1	289.8	221.1	181.5	167.0	-14.5	39.6	68.7	62.7	6.0	1 812.3
1978	2 351.6	46.3	24.7	2 373.2	327.2	252.1	205.5	189.4	-16.2	46.6	75.1	68.5	6.6	2 046.0
1979	2 627.3	68.3	36.4	2 659.3	373.9	290.7	236.4	216.1	-20.2	54.4	83.1	75.6	7.5	2 285.4
1980	2 857.3	79.1	44.9	2 891.5	428.4	335.0	272.8	245.8	-26.9	62.2	93.5	84.7	8.7	2 463.1
1981	3 207.0	92.0	59.1	3 240.0	487.2	381.9	313.5	301.2	-12.3	68.5	105.3	95.4	9.9	2 752.7
1982	3 343.8	101.0	64.5	3 380.3	537.0	420.4	347.5	343.1	-4.4	72.9	116.6	105.7	10.8	2 843.3
1983	3 634.0	101.9	64.8	3 671.1	562.6	438.8	362.8	385.6	22.8	76.0	123.8	112.4	11.4	3 108.5
1984	4 037.6	121.9	85.6	4 073.9	598.4	463.5	382.7	435.6	52.9	80.9	134.9	122.8	12.1	3 475.5
1985	4 339.0	112.7	87.3	4 364.3	640.1	496.4	410.3	493.5	83.2	86.2	143.7	130.9	12.8	3 724.2
1986	4 579.6	111.3	94.4	4 596.6	685.3	531.6	438.1	512.0	73.9	93.4	153.7	140.1	13.6	3 911.3
1987	4 855.2	123.3	105.8	4 872.7	730.4	566.3	464.5	533.1	68.6	101.8	164.1	149.6	14.5	4 142.3
1988	5 236.4	152.1	129.5	5 259.1	784.5	607.9	499.2	563.7	64.5	108.8	176.6	161.1	15.5	4 474.6
1989	5 641.6	177.7	152.9	5 666.4	838.3	649.6	532.6	589.5	56.9	117.0	188.6	172.0	16.6	4 828.1
1990	5 963.1	188.8	154.2	5 997.8	888.5	688.4	564.6	596.0	31.4	123.8	200.1	182.3	17.8	5 109.3
1991	6 158.1	168.4	136.8	6 189.7	932.4	721.5	592.3	611.6	19.3	129.2	210.9	192.1	18.8	5 257.3
1992	6 520.3	152.1	121.0	6 551.4	960.2	742.9	608.2	628.4	20.2	134.7	217.4	197.6	19.8	5 591.2
1993	6 858.6	155.6	124.4	6 889.7	1 003.5	778.2	634.2	658.9	24.7	144.0	225.3	204.2	21.1	5 886.2
1994	7 287.2	184.5	161.6	7 310.2	1 055.6	822.5	669.0	701.6	32.6	153.5	233.1	210.9	22.2	6 254.6
1995	7 639.7	229.8	201.9	7 667.7	1 122.4	880.7	717.8	753.9	36.1	162.9	241.7	218.2	23.5	6 545.3
1996	8 073.1	246.4	215.5	8 104.0	1 175.3	929.1	758.5	808.3	49.8	170.6	246.2	221.6	24.6	6 928.7
1997	8 577.6	280.1	256.8	8 600.9	1 239.3	987.8	807.9	877.0	69.2	179.9	251.6	225.9	25.7	7 361.5
1998	9 062.8	286.8	269.4	9 080.2	1 309.7	1 052.2	860.1	941.4	81.3	192.1	257.6	230.8	26.8	7 770.5
1999	9 630.7	320.2	294.7	9 656.2	1 398.9	1 132.2	924.0	1 023.8	99.8	208.2	266.7	238.3	28.4	8 257.3
2000	10 252.3	380.6	345.6	10 287.4	1 511.2	1 231.5	1 004.6	1 096.1	91.5	227.0	279.7	249.4	30.3	8 776.1
2001	10 581.8	324.1	275.3	10 630.6	1 599.5	1 311.7	1 064.9	1 180.1	115.2	246.8	287.8	256.0	31.8	9 031.1
2002	10 936.4	314.8	269.6	10 981.7	1 658.0	1 361.8	1 098.8	1 283.9	185.1	263.0	296.2	262.9	33.2	9 323.7
2003	11 458.2	353.8	295.4	11 516.6	1 719.1	1 411.9	1 125.4	1 303.5	178.1	286.6	307.1	272.2	34.9	9 797.5
2004	12 213.7	446.9	368.8	12 291.9	1 821.8	1 497.1	1 178.8	1 332.0	153.2	318.3	324.7	286.9	37.9	10 470.1
2005	13 036.6	566.0	488.1	13 114.6	1 971.0	1 622.6	1 266.4	1 169.5	-97.0	356.2	348.4	307.0	41.4	11 143.6
2006	13 814.6	712.0	661.5	13 865.1	2 124.1	1 751.8	1 362.6	1 245.1	-117.4	389.2	372.3	327.6	44.7	11 740.9
2007	14 451.9	866.6	757.6	14 560.9	2 252.8	1 852.5	1 446.7	1 326.8	-119.8	405.8	400.3	352.1	48.2	12 308.1
2008	14 712.8	848.8	694.2	14 867.5	2 358.8	1 931.8	1 524.9	1 584.1	59.3	407.0	427.0	375.7	51.3	12 508.6
2009	14 448.9	647.8	505.8	14 590.9	2 371.5	1 928.7	1 531.2	1 562.1	30.9	397.5	442.8	389.8	53.0	12 219.4
2010	14 992.1	715.2	519.5	15 187.8	2 390.9	1 933.8	1 537.1	1 605.9	68.8	396.7	457.2	402.2	55.0	12 796.8
2011	15 542.6	789.2	552.8	15 779.0	2 474.5	1 997.3	1 595.1	1 831.3	236.2	402.2	477.2	419.3	57.8	13 304.5
2012	16 197.0	799.7	567.4	16 429.3	2 576.0	2 082.4	1 670.4	1 665.9	-4.5	412.0	493.6	433.0	60.7	13 853.3
2013	16 784.9	823.4	592.7	17 015.6	2 681.2	2 176.6	1 738.8	1 753.6	14.8	437.8	504.6	441.8	62.9	14 334.4
2014	17 527.3	853.5	612.5	17 768.3	2 815.0	2 298.5	1 831.2	1 866.4	35.2	467.3	516.6	451.4	65.2	14 953.3
2015	18 238.3	860.8	640.4	18 458.7	2 911.4	2 388.5	1 903.1	1 981.9	78.8	485.4	522.9	455.8	67.1	15 547.3
2016	18 745.1	893.5	661.5	18 977.1	2 986.6	2 458.3	1 949.5	2 074.4	124.9	508.9	528.3	459.2	69.0	15 990.5
2017	19 543.0	1 032.7	740.4	19 835.3	3 112.9	2 569.7	2 031.8	2 302.1	270.3	537.8	543.2	471.2	72.0	16 722.4
2018	20 611.9	1 142.9	858.2	20 896.6	3 265.0	2 699.0	2 125.0	2 578.0	453.0	574.0	566.1	490.0	76.1	17 631.6
2019	21 433.2	1 169.8	900.2	21 702.9	3 420.9	2 832.7	2 233.6	2 586.3	352.6	599.0	588.3	508.1	80.1	18 281.9
2017														
1st quarter	19 237.4	964.9	688.2	19 514.1	3 062.8	2 526.0	2 000.0	2 168.1	168.3	526.0	536.8	465.9	70.9	16 451.3
2nd quarter	19 379.2	987.7	720.1	19 646.8	3 097.2	2 556.5	2 022.2	2 197.6	175.3	534.3	540.6	469.0	71.6	16 549.7
3rd quarter	19 617.3	1 058.2	756.8	19 918.6	3 131.2	2 585.7	2 043.2	2 224.3	181.0	542.5	545.5	473.1	72.4	16 787.5
4th quarter	19 938.0	1 120.2	796.6	20 261.5	3 160.3	2 610.4	2 061.8	2 618.4	556.5	548.6	549.9	476.7	73.2	17 101.2
2018														
1st quarter	20 242.2	1 116.2	801.7	20 556.8	3 204.8	2 648.6	2 086.9	2 543.5	456.7	561.7	556.2	482.0	74.3	17 351.9
2nd quarter	20 552.7	1 153.4	861.9	20 844.2	3 248.5	2 684.6	2 113.1	2 567.1	454.1	571.5	563.9	488.3	75.6	17 595.7
3rd quarter	20 742.7	1 129.2	869.2	21 002.7	3 286.8	2 717.4	2 138.9	2 590.0	451.2	578.5	569.4	492.7	76.7	17 715.9
4th quarter	20 909.9	1 172.6	899.9	21 182.6	3 319.9	2 745.2	2 161.1	2 611.4	450.2	584.2	574.7	496.9	77.8	17 862.7
2019														
1st quarter	21 115.3	1 148.0	901.5	21 361.8	3 361.6	2 781.3	2 191.9	2 549.8	357.6	589.4	580.3	501.6	78.7	18 000.2
2nd quarter	21 329.9	1 184.3	913.2	21 601.0	3 404.4	2 818.6	2 223.7	2 573.4	349.4	594.9	585.8	506.0	79.8	18 196.6
3rd quarter	21 540.3	1 181.2	901.4	21 820.1	3 443.3	2 851.9	2 248.9	2 598.2	349.3	603.0	591.4	510.7	80.7	18 376.8
4th quarter	21 747.4	1 165.9	884.8	22 028.5	3 474.4	2 878.9	2 269.9	2 623.6	354.2	608.9	595.6	514.2	81.4	18 554.0

Table 1-12. Gross Domestic Income by Type of Income

(Billions of dollars, quarterly data are at seasonally adjusted annual rates.)

NIPA Tables 1.10

Year and quarter	Gross domestic income	Compensation of employees			Taxes on production and imports	Less: Subsidies	Net operating surplus					
		Total	Wages and salaries	Supple-ments to wages and salaries				Private enterprises				
							Total	Total	Net interest and miscel-laneous payments, domestic industries	Business current transfer payments, net	Proprietors' income with IVA and CCAdj	Rental income of persons with CCAdj
1960	543.7	301.4	273.0	28.4	44.5	1.1	130.9	130.5	10.1	1.7	50.6	16.5
1961	563.1	310.6	280.7	29.9	47.0	2.0	137.0	136.7	11.7	1.9	53.2	17.2
1962	603.9	332.3	299.5	32.8	50.4	2.3	149.4	149.0	13.4	2.1	55.2	18.0
1963	638.9	350.5	314.9	35.6	53.4	2.2	159.3	158.5	14.4	2.5	56.4	18.7
1964	684.5	376.0	337.8	38.2	57.3	2.7	171.6	170.7	16.5	3.0	59.1	18.8
1965	741.5	405.4	363.8	41.7	60.7	3.0	190.4	189.6	18.6	3.5	63.7	19.3
1966	808.3	449.2	400.3	48.9	63.2	3.9	204.5	204.2	21.3	3.4	67.9	19.9
1967	856.6	481.8	429.0	52.8	67.9	3.8	207.2	206.9	24.3	3.6	69.5	20.3
1968	937.5	530.7	472.0	58.8	76.4	4.2	221.2	220.7	26.5	4.2	73.8	20.1
1969	1 016.0	584.4	518.3	66.1	83.9	4.5	227.4	227.2	33.2	4.8	77.0	20.3
1970	1 068.0	623.3	551.5	71.8	91.4	4.8	221.2	222.2	40.2	4.4	77.8	20.7
1971	1 155.3	665.0	584.5	80.4	100.5	4.7	245.6	247.1	44.4	4.2	83.9	21.8
1972	1 271.9	731.3	638.8	92.5	107.9	6.6	278.3	279.2	49.0	4.8	95.1	22.7
1973	1 419.2	812.7	708.8	103.9	117.2	5.2	315.9	317.8	58.0	5.7	112.5	23.1
1974	1 537.8	887.7	772.3	115.4	124.9	3.3	321.6	324.0	73.6	6.8	112.2	23.2
1975	1 671.6	947.3	814.9	132.4	135.3	4.5	355.0	359.0	85.3	9.0	118.2	22.3
1976	1 852.8	1 048.4	899.8	148.6	146.4	5.1	402.9	405.3	87.1	9.1	131.0	20.3
1977	2 062.4	1 165.9	994.2	171.7	159.7	7.1	454.1	457.1	101.9	8.1	144.5	15.9
1978	2 328.3	1 316.9	1 120.7	196.2	170.9	8.9	522.3	524.6	116.0	10.4	166.0	16.5
1979	2 582.3	1 477.3	1 253.4	223.9	180.1	8.5	559.5	562.2	139.6	12.8	179.4	16.1
1980	2 812.9	1 622.3	1 373.5	248.8	200.3	9.8	571.6	576.0	183.4	14.0	171.6	19.0
1981	3 169.0	1 792.7	1 511.5	281.2	235.6	11.5	664.9	669.6	231.4	16.9	179.7	23.8
1982	3 335.0	1 893.2	1 587.7	305.5	240.9	15.0	678.9	682.3	270.7	19.3	171.2	23.8
1983	3 577.1	2 012.7	1 677.7	335.0	263.3	21.3	759.8	761.7	284.6	21.7	186.3	24.4
1984	3 996.0	2 216.1	1 845.1	371.0	289.8	21.1	912.8	913.7	330.3	29.2	228.2	24.7
1985	4 284.7	2 387.5	1 982.8	404.8	308.1	21.4	970.3	968.4	350.7	34.1	241.1	26.2
1986	4 499.6	2 543.8	2 104.1	439.7	323.4	24.9	972.0	969.4	373.8	36.0	256.5	18.3
1987	4 811.4	2 723.8	2 257.6	466.1	347.5	30.3	1 040.0	1 037.2	382.9	33.3	286.5	16.6
1988	5 233.4	2 948.9	2 440.6	508.2	374.5	29.5	1 155.1	1 149.6	411.3	32.8	325.5	22.5
1989	5 573.6	3 140.9	2 584.3	556.6	398.9	27.4	1 223.0	1 215.8	467.7	38.3	341.1	21.5
1990	5 867.6	3 342.7	2 743.5	599.2	425.0	27.0	1 238.4	1 234.7	472.5	39.2	353.2	28.2
1991	6 065.2	3 453.3	2 817.2	636.0	457.1	27.5	1 249.9	1 241.7	433.9	38.9	354.2	38.6
1992	6 404.4	3 671.2	2 968.5	702.7	483.4	30.1	1 319.6	1 309.2	404.7	39.7	400.2	60.6
1993	6 702.6	3 820.6	3 082.7	737.9	503.1	36.7	1 412.1	1 400.6	395.5	39.4	428.0	90.1
1994	7 147.3	4 010.2	3 240.6	769.6	545.2	32.5	1 568.7	1 556.1	398.3	40.7	456.6	113.7
1995	7 546.7	4 202.2	3 422.1	780.1	557.9	34.8	1 699.1	1 682.6	418.3	45.0	481.2	124.9
1996	8 015.0	4 421.1	3 620.6	800.5	580.8	35.2	1 873.1	1 855.4	428.9	52.6	543.8	142.5
1997	8 566.0	4 713.2	3 881.2	832.0	611.6	33.8	2 035.6	2 017.5	474.3	50.1	584.0	147.1
1998	9 118.1	5 075.7	4 186.2	889.5	639.5	36.4	2 129.5	2 112.6	537.5	64.1	640.2	165.2
1999	9 663.7	5 409.9	4 465.2	944.8	673.6	45.2	2 226.4	2 209.6	553.8	67.7	696.4	178.5
2000	10 348.6	5 854.6	4 832.4	1 022.2	708.6	45.8	2 320.1	2 308.6	645.2	85.2	753.9	183.5
2001	10 695.0	6 046.3	4 961.6	1 084.7	727.7	58.7	2 380.1	2 374.5	651.9	99.2	831.0	202.4
2002	11 009.1	6 143.4	5 004.1	1 139.3	760.0	41.4	2 489.2	2 481.6	565.1	80.8	869.8	211.1
2003	11 471.7	6 362.3	5 147.0	1 215.3	805.6	49.1	2 633.8	2 628.6	526.9	76.0	896.9	231.5
2004	12 236.1	6 729.3	5 430.8	1 298.5	868.1	46.4	2 863.2	2 863.2	475.8	82.2	962.0	248.9
2005	13 089.3	7 077.7	5 703.0	1 374.7	942.4	60.9	3 159.1	3 163.6	598.9	91.7	978.0	232.0
2006	14 021.8	7 491.3	6 068.3	1 422.9	997.0	51.5	3 460.9	3 467.9	728.5	80.9	1 049.6	202.3
2007	14 434.5	7 889.4	6 407.3	1 482.1	1 036.8	54.6	3 310.0	3 324.2	851.9	98.5	994.0	184.4
2008	14 530.3	8 067.7	6 545.9	1 522.7	1 049.7	52.6	3 105.6	3 123.8	896.3	114.3	960.9	256.7
2009	14 259.0	7 767.2	6 257.3	1 509.9	1 026.8	58.3	3 151.9	3 167.9	737.4	126.6	938.5	327.3
2010	14 932.1	7 933.0	6 380.1	1 552.9	1 063.1	55.8	3 600.9	3 621.0	647.2	127.9	1 108.7	394.2
2011	15 599.4	8 234.0	6 634.0	1 600.0	1 103.7	60.0	3 847.2	3 866.6	629.8	131.7	1 229.3	478.6
2012	16 436.8	8 575.4	6 936.1	1 639.2	1 136.1	58.0	4 207.4	4 222.8	668.1	97.3	1 347.3	518.0
2013	16 941.8	8 843.6	7 122.6	1 721.0	1 188.7	59.7	4 288.0	4 303.9	624.6	106.9	1 403.6	557.0
2014	17 813.9	9 259.7	7 485.8	1 773.9	1 240.8	58.1	4 556.5	4 567.4	669.3	130.4	1 447.7	604.6
2015	18 475.6	9 709.5	7 869.6	1 839.9	1 275.2	57.2	4 636.7	4 642.0	750.5	154.2	1 423.0	649.0
2016	18 837.3	9 974.8	8 100.1	1 874.7	1 311.8	61.7	4 625.8	4 630.1	754.2	164.6	1 424.8	682.7
2017	19 674.4	10 433.1	8 482.0	1 951.1	1 364.0	59.9	4 824.4	4 829.6	830.8	150.7	1 509.0	721.9
2018	20 669.9	10 960.6	8 904.7	2 055.9	1 444.8	63.3	5 062.8	5 068.7	836.4	156.6	1 585.9	759.3
2019	21 420.4	11 444.5	9 321.4	2 123.1	1 491.4	73.9	5 137.4	5 145.0	797.2	158.0	1 657.7	787.1
2017												
1st quarter	19 352.2	10 233.3	8 319.0	1 914.2	1 340.0	59.5	4 775.6	4 780.7	821.0	165.8	1 493.4	709.0
2nd quarter	19 572.2	10 345.9	8 410.4	1 935.4	1 357.4	58.1	4 829.9	4 835.2	827.2	156.0	1 502.7	713.2
3rd quarter	19 750.4	10 487.6	8 526.1	1 961.6	1 368.7	61.9	4 824.8	4 830.2	834.1	137.2	1 507.8	725.9
4th quarter	20 022.8	10 665.6	8 672.5	1 993.0	1 389.8	60.0	4 867.1	4 872.4	840.7	143.7	1 532.1	739.6
2018												
1st quarter	20 319.8	10 786.4	8 766.7	2 019.7	1 418.3	58.2	4 968.4	4 973.3	851.2	149.2	1 557.7	745.3
2nd quarter	20 533.2	10 892.6	8 846.6	2 046.0	1 433.1	57.8	5 016.8	5 022.1	838.9	147.5	1 570.7	752.4
3rd quarter	20 847.3	11 044.9	8 973.9	2 070.9	1 448.7	57.9	5 124.8	5 130.9	834.8	173.9	1 588.4	768.2
4th quarter	20 979.1	11 118.5	9 031.6	2 086.9	1 479.0	79.5	5 141.2	5 148.2	820.9	155.6	1 627.0	771.2
2019												
1st quarter	21 147.8	11 346.9	9 240.3	2 106.6	1 473.8	71.3	5 036.8	5 043.7	790.6	152.1	1 627.5	776.6
2nd quarter	21 347.2	11 403.6	9 286.8	2 116.8	1 480.7	61.1	5 119.6	5 127.3	795.4	159.7	1 628.5	786.7
3rd quarter	21 465.3	11 450.1	9 323.4	2 126.6	1 501.6	82.0	5 152.4	5 160.4	793.7	168.1	1 677.0	789.7
4th quarter	21 721.2	11 577.4	9 435.1	2 142.4	1 509.6	81.1	5 240.8	5 248.7	808.9	152.0	1 697.7	795.5

Table 1-13A. National Income by Type of Income: Recent Data

(Billions of dollars, quarterly data are at seasonally adjusted annual rates.)   **NIPA Tables 1.7.5, 1.12**

Year and quarter	National income, total	Compensation of employees							Proprietors' income with IVA and CCAdj			Rental income of persons with CCAdj
		Total	Wages and salaries			Supplements to wages and salaries			Total	Farm	Nonfarm	
			Total	Government	Other	Total	Employer contributions for:					
							Employee pension and insurance funds	Government social insurance				
1960	478.9	301.3	272.9	49.2	223.7	28.4	19.2	9.3	50.6	10.6	39.9	16.5
1961	496.0	310.4	280.5	52.5	228.0	29.9	20.3	9.6	53.2	11.2	42.0	17.2
1962	533.9	332.2	299.4	56.3	243.0	32.8	21.7	11.2	55.2	11.2	44.0	18.0
1963	565.4	350.4	314.9	60.0	254.8	35.6	23.2	12.4	56.4	11.0	45.4	18.7
1964	607.0	376.0	337.8	64.9	272.9	38.2	25.6	12.6	59.1	9.8	49.4	18.8
1965	658.8	405.4	363.8	69.9	293.8	41.7	28.6	13.1	63.7	12.0	51.6	19.3
1966	718.1	449.2	400.3	78.4	321.9	48.9	32.1	16.8	67.9	13.0	54.9	19.9
1967	758.4	481.8	429.0	86.5	342.5	52.8	34.8	18.0	69.5	11.6	57.8	20.3
1968	830.2	530.8	472.0	96.7	375.3	58.8	38.8	20.0	73.8	11.7	62.2	20.1
1969	897.2	584.5	518.3	105.6	412.7	66.1	43.4	22.8	77.0	12.8	64.2	20.3
1970	937.5	623.3	551.6	117.2	434.3	71.8	47.9	23.8	77.8	12.9	64.9	20.7
1971	1 014.0	665.0	584.5	126.8	457.8	80.4	54.0	26.4	83.9	13.4	70.5	21.8
1972	1 119.5	731.3	638.8	137.9	500.9	92.5	61.4	31.2	95.1	17.0	78.1	22.7
1973	1 253.2	812.7	708.8	148.8	560.0	103.9	64.1	39.8	112.5	29.1	83.4	23.1
1974	1 346.4	887.7	772.3	160.5	611.8	115.4	70.7	44.7	112.2	23.5	88.7	23.2
1975	1 446.0	947.2	814.8	176.2	638.6	132.4	85.7	46.7	118.2	22.0	96.2	22.3
1976	1 609.4	1 048.3	899.7	188.9	710.8	148.6	94.2	54.4	131.0	17.2	113.8	20.3
1977	1 792.8	1 165.8	994.2	202.6	791.6	171.7	110.6	61.1	144.5	16.0	128.5	15.9
1978	2 022.7	1 316.8	1 120.6	220.0	900.6	196.2	124.7	71.5	166.0	19.9	146.1	16.5
1979	2 240.3	1 477.2	1 253.3	237.1	1 016.2	223.9	141.3	82.6	179.4	22.2	157.3	16.1
1980	2 418.6	1 622.2	1 373.4	261.5	1 112.0	248.8	159.9	88.9	171.6	11.7	159.9	19.0
1981	2 714.7	1 792.5	1 511.4	285.8	1 225.5	281.2	177.5	103.6	179.7	19.0	160.7	23.8
1982	2 834.5	1 893.0	1 587.5	307.5	1 280.0	305.5	195.7	109.8	171.2	13.3	157.9	23.8
1983	3 051.5	2 012.5	1 677.5	324.8	1 352.7	335.0	215.1	119.9	186.3	6.2	180.1	24.4
1984	3 433.9	2 215.9	1 844.9	348.1	1 496.8	371.0	231.9	139.0	228.2	20.9	207.3	24.7
1985	3 669.9	2 387.3	1 982.6	373.9	1 608.7	404.8	257.0	147.7	241.1	21.0	220.1	26.2
1986	3 831.2	2 542.1	2 102.3	397.2	1 705.1	439.7	281.9	157.9	256.5	22.8	233.7	18.3
1987	4 098.5	2 722.4	2 256.3	423.1	1 833.2	466.1	299.9	166.3	286.5	28.9	257.6	16.6
1988	4 471.6	2 948.0	2 439.8	452.0	1 987.7	508.2	323.6	184.6	325.5	26.8	298.7	22.5
1989	4 760.1	3 139.6	2 583.1	481.1	2 101.9	556.6	362.9	193.7	341.1	33.0	308.1	21.5
1990	5 013.8	3 340.4	2 741.2	519.0	2 222.2	599.2	392.7	206.5	353.2	32.2	321.0	28.2
1991	5 164.4	3 450.5	2 814.5	548.8	2 265.7	636.0	420.9	215.1	354.2	26.8	327.4	38.6
1992	5 475.2	3 668.2	2 965.5	572.0	2 393.5	702.7	474.3	228.4	400.2	34.8	365.4	60.6
1993	5 730.3	3 817.3	3 079.3	589.0	2 490.3	737.9	498.3	239.7	428.0	31.4	396.6	90.1
1994	6 114.6	4 006.2	3 236.6	609.5	2 627.1	769.6	515.5	254.1	456.6	34.7	422.0	113.7
1995	6 452.3	4 198.1	3 418.0	629.0	2 789.0	780.1	515.9	264.1	481.2	22.0	459.2	124.9
1996	6 870.6	4 416.9	3 616.5	648.1	2 968.4	800.5	525.7	274.8	543.8	37.3	506.4	142.5
1997	7 349.9	4 708.8	3 876.8	671.9	3 205.0	832.0	542.4	289.6	584.0	32.4	551.6	147.1
1998	7 825.7	5 071.1	4 181.6	701.3	3 480.3	889.5	582.3	307.2	640.2	28.5	611.7	165.2
1999	8 290.3	5 402.8	4 458.0	733.8	3 724.2	944.8	621.4	323.3	696.4	28.1	668.3	178.5
2000	8 872.4	5 848.1	4 825.9	779.8	4 046.1	1 022.2	677.0	345.2	753.9	31.5	722.4	183.5
2001	9 144.2	6 039.1	4 954.4	822.0	4 132.4	1 084.7	726.7	358.0	831.0	32.1	798.9	202.4
2002	9 396.4	6 135.6	4 996.3	872.9	4 123.4	1 139.3	773.2	366.0	869.8	19.9	849.8	211.1
2003	9 811.0	6 354.1	5 138.7	914.0	4 224.8	1 215.3	832.8	382.5	896.9	36.5	860.4	231.5
2004	10 492.4	6 720.1	5 421.6	952.3	4 469.2	1 298.5	889.7	408.8	962.0	51.5	910.5	248.9
2005	11 196.3	7 066.6	5 691.9	991.3	4 700.6	1 374.7	946.7	428.1	978.0	46.8	931.2	232.0
2006	11 948.1	7 479.9	6 057.0	1 034.5	5 022.4	1 422.9	975.6	447.3	1 049.6	33.1	1 016.6	202.3
2007	12 290.7	7 878.9	6 396.8	1 088.5	5 308.2	1 482.1	1 020.4	461.7	994.0	40.3	953.8	184.4
2008	12 326.1	8 057.0	6 534.2	1 143.9	5 390.4	1 522.7	1 051.3	471.4	960.9	40.2	920.7	256.7
2009	12 029.5	7 758.5	6 248.6	1 175.2	5 073.4	1 509.9	1 051.8	458.1	938.5	28.1	910.5	327.3
2010	12 736.8	7 924.9	6 372.1	1 191.2	5 180.9	1 552.9	1 083.9	469.0	1 108.7	39.0	1 069.7	394.2
2011	13 361.3	8 225.9	6 625.9	1 194.9	5 431.1	1 600.0	1 107.3	492.7	1 229.3	64.9	1 164.4	478.6
2012	14 093.1	8 566.7	6 927.5	1 198.3	5 729.2	1 639.2	1 125.9	513.3	1 347.3	60.9	1 286.4	518.0
2013	14 491.3	8 834.2	7 113.2	1 208.0	5 905.2	1 721.0	1 194.7	526.3	1 403.6	88.3	1 315.3	557.0
2014	15 239.9	9 249.1	7 475.2	1 236.9	6 238.3	1 773.9	1 227.5	546.4	1 447.7	69.8	1 377.9	604.6
2015	15 784.6	9 699.4	7 859.5	1 275.8	6 583.7	1 839.9	1 270.6	569.4	1 423.0	56.2	1 366.7	649.0
2016	16 082.7	9 963.9	8 089.1	1 308.2	6 780.9	1 874.7	1 293.5	581.2	1 424.8	36.0	1 388.7	682.7
2017	16 853.9	10 422.0	8 471.5	1 347.7	7 123.7	1 951.1	1 346.0	605.1	1 509.0	41.5	1 467.4	721.9
2018	17 689.6	10 950.0	8 894.2	1 402.5	7 491.7	2 055.9	1 430.7	625.2	1 585.9	43.0	1 542.9	759.3
2019	18 269.1	11 432.0	9 309.3	1 450.8	7 858.5	2 123.1	1 474.0	649.1	1 657.7	49.7	1 608.0	787.1
2017												
1st quarter	16 566.1	10 223.0	8 308.8	1 331.5	6 977.3	1 914.2	1 319.2	595.1	1 493.4	47.3	1 446.1	709.0
2nd quarter	16 742.6	10 335.0	8 399.9	1 340.1	7 059.8	1 935.4	1 334.6	600.8	1 502.7	44.7	1 457.9	713.2
3rd quarter	16 920.6	10 476.0	8 515.3	1 352.7	7 162.7	1 961.6	1 353.7	607.9	1 507.8	37.1	1 470.7	725.9
4th quarter	17 186.1	10 654.0	8 661.8	1 366.7	7 295.1	1 993.0	1 376.4	616.6	1 532.1	37.0	1 495.1	739.6
2018												
1st quarter	17 429.5	10 776.0	8 756.4	1 380.2	7 376.1	2 019.7	1 402.3	617.4	1 557.7	40.8	1 516.9	745.3
2nd quarter	17 576.2	10 882.0	8 836.3	1 393.9	7 442.4	2 046.0	1 424.4	621.6	1 570.7	42.3	1 528.4	752.4
3rd quarter	17 820.5	11 034.0	8 963.2	1 412.5	7 550.7	2 070.9	1 441.7	629.2	1 588.4	34.0	1 554.4	768.2
4th quarter	17 932.0	11 107.0	9 021.0	1 423.3	7 597.6	2 086.9	1 454.3	632.6	1 627.0	55.0	1 572.0	771.2
2019												
1st quarter	18 032.7	11 335.0	9 228.7	1 432.3	7 796.4	2 106.6	1 462.5	644.1	1 627.5	44.2	1 583.4	776.6
2nd quarter	18 214.0	11 391.0	9 274.9	1 442.3	7 832.5	2 116.8	1 469.8	646.9	1 628.5	36.9	1 591.6	786.7
3rd quarter	18 301.8	11 438.0	9 311.3	1 459.0	7 852.3	2 126.6	1 477.6	649.0	1 677.0	58.9	1 618.1	789.7
4th quarter	18 527.8	11 564.0	9 422.5	1 469.5	7 953.0	2 142.4	1 486.1	656.3	1 697.7	58.7	1 639.0	795.5

Table 1-13A. National Income by Type of Income: Recent Data—*Continued*

(Billions of dollars, quarterly data are at seasonally adjusted annual rates.) **NIPA Tables 1.7.5, 1.12**

| Year and quarter | Corporate profits with IVA and CCAdj | | Profits after tax | | | Net interest and miscellaneous payments | Taxes on production and imports | Less: Subsidies | Business current transfer payments, net | | | Current surplus of government enterprises | Addendum: Net national factor income |
	Total	Taxes on corporate income	Total	Net dividends	Undistributed corporate profits				Total [1]	To persons	To government		
1960	54.7	21.9	32.8	14.3	18.5	10.3	44.5	1.1	1.7	1.3	0.4	. . .	433.3
1961	55.9	22.2	33.7	14.6	19.2	12.1	47.0	2.0	1.9	1.4	0.5	. . .	448.8
1962	64.0	23.3	40.7	15.8	24.9	13.8	50.4	2.3	2.1	1.5	0.6	. . .	483.3
1963	70.5	25.5	45.0	17.1	27.9	14.8	53.4	2.2	2.5	1.9	0.7	. . .	510.9
1964	77.7	26.6	51.1	19.8	31.4	17.0	57.3	2.7	3.0	2.2	0.8	. . .	548.6
1965	89.3	29.8	59.5	21.5	38.0	19.1	60.7	3.0	3.5	2.3	1.2	. . .	596.8
1966	96.1	32.2	63.9	22.3	41.6	21.8	63.2	3.9	3.4	2.1	1.3	. . .	655.1
1967	93.9	31.0	62.9	23.4	39.5	24.9	67.9	3.8	3.6	2.3	1.4	. . .	690.4
1968	101.7	37.2	64.6	26.0	38.6	27.0	76.4	4.2	4.2	2.8	1.4	. . .	753.4
1969	98.4	37.0	61.5	27.3	34.2	32.7	83.9	4.5	4.8	3.3	1.5	. . .	813.0
1970	86.2	31.3	55.0	27.8	27.2	39.5	91.4	4.8	4.4	2.9	1.4	. . .	847.6
1971	100.6	34.8	65.8	28.4	37.5	44.2	100.5	4.7	4.2	2.7	1.5	. . .	915.5
1972	117.2	39.1	78.1	30.1	48.0	48.0	107.9	6.6	4.8	3.1	1.7	. . .	1 014.4
1973	133.4	45.6	87.8	34.2	53.5	55.7	117.2	5.2	5.7	3.9	1.8	. . .	1 137.4
1974	125.7	47.2	78.5	38.8	39.7	71.7	124.9	3.3	6.8	4.7	2.1	. . .	1 220.5
1975	138.9	46.3	92.6	38.3	54.3	83.7	135.3	4.5	9.0	6.8	2.2	. . .	1 310.3
1976	174.3	59.4	114.9	44.9	70.0	87.4	146.4	5.1	9.1	6.7	2.4	. . .	1 461.4
1977	205.8	68.5	137.3	50.7	86.6	103.2	159.7	7.1	8.1	5.1	3.0	. . .	1 635.2
1978	238.6	77.9	160.7	57.8	102.9	114.8	170.9	8.9	10.4	6.5	3.9	. . .	1 852.7
1979	249.0	80.7	168.2	66.8	101.4	137.0	180.1	8.5	12.8	8.2	4.5	. . .	2 058.7
1980	223.6	75.5	148.1	75.8	72.3	182.2	200.3	9.8	14.0	8.6	5.4	. . .	2 218.6
1981	247.5	70.3	177.2	87.8	89.4	234.8	235.6	11.5	16.9	11.2	5.7	. . .	2 478.3
1982	229.9	51.3	178.6	92.9	85.6	274.8	240.9	15.0	19.3	12.4	6.9	. . .	2 592.7
1983	279.8	66.4	213.3	97.7	115.7	286.8	263.3	21.3	21.7	13.8	7.9	. . .	2 789.8
1984	337.9	81.5	256.4	106.9	149.5	330.2	289.8	21.1	29.2	19.7	9.4	. . .	3 136.9
1985	354.5	81.6	272.9	115.3	157.5	338.2	308.1	21.4	34.1	22.3	11.8	. . .	3 347.2
1986	324.4	91.9	232.5	124.0	108.5	353.1	323.4	24.9	36.0	22.9	13.0	. . .	3 494.3
1987	366.0	112.7	253.3	130.1	123.2	353.7	347.5	30.3	33.3	20.2	13.1	. . .	3 745.2
1988	414.5	124.3	290.2	147.3	142.9	377.9	374.5	29.5	32.8	20.6	12.2	. . .	4 088.3
1989	414.3	124.4	289.9	179.6	110.3	426.6	398.9	27.4	38.3	23.2	15.1	. . .	4 343.2
1990	417.7	121.8	295.9	192.7	103.2	433.4	425.0	27.0	39.2	22.2	17.0	. . .	4 572.9
1991	452.6	117.8	334.8	201.3	133.5	391.8	457.1	27.5	38.9	17.6	21.3	. . .	4 687.6
1992	477.2	131.9	345.3	206.3	139.0	365.6	483.4	30.1	39.7	16.3	23.6	-0.1	4 971.8
1993	524.6	155.0	369.5	221.3	148.2	353.1	503.1	36.7	39.4	14.1	25.6	-0.3	5 213.0
1994	624.8	172.7	452.1	256.4	195.7	347.3	545.2	32.5	40.7	13.3	27.9	-0.4	5 548.6
1995	706.2	194.4	511.8	282.3	229.4	357.4	557.9	34.8	45.0	18.7	24.9	1.4	5 867.7
1996	789.5	211.4	578.1	323.6	254.5	361.9	580.8	35.2	52.6	22.9	30.7	-1.0	6 254.7
1997	869.7	224.8	645.0	360.1	284.9	394.4	611.6	33.8	50.1	19.4	29.8	0.8	6 704.0
1998	808.5	221.8	586.6	383.6	203.0	456.7	639.5	36.4	64.1	26.0	34.3	3.8	7 141.7
1999	834.9	227.4	607.5	373.5	234.0	464.8	673.6	45.2	67.7	34.0	36.5	-2.7	7 577.4
2000	786.6	233.4	553.1	410.2	142.9	541.1	708.6	45.8	85.2	42.4	41.9	0.8	8 113.1
2001	758.7	170.1	588.6	397.9	190.7	539.1	727.7	58.7	99.2	46.8	43.9	8.6	8 370.3
2002	911.7	160.7	751.0	424.9	326.2	461.4	760.0	41.4	80.8	34.2	46.2	0.3	8 589.4
2003	1 056.3	213.8	842.5	456.0	386.5	434.6	805.6	49.1	76.0	26.3	48.3	1.5	8 973.2
2004	1 289.3	278.5	1 010.8	582.2	428.6	368.1	868.1	46.4	82.2	16.8	52.4	13.1	9 588.4
2005	1 488.6	379.7	1 108.9	602.0	506.9	462.3	942.4	60.9	91.7	25.8	53.4	12.5	10 227.0
2006	1 646.3	430.1	1 216.1	755.1	461.1	550.6	997.0	51.5	80.9	20.8	57.5	2.6	10 928.0
2007	1 533.2	391.8	1 141.4	853.5	287.9	633.6	1 036.8	54.6	98.5	30.8	60.9	6.9	11 224.0
2008	1 285.8	255.9	1 029.9	840.3	189.6	672.4	1 049.7	52.6	114.3	35.8	69.5	9.0	11 232.0
2009	1 386.8	203.9	1 182.9	622.1	560.8	539.3	1 026.8	58.3	126.6	39.0	86.1	1.5	10 950.0
2010	1 728.7	272.3	1 456.5	643.2	813.3	465.2	1 063.1	55.8	127.9	43.7	84.1	0.0	11 621.8
2011	1 809.8	280.8	1 529.0	779.1	749.9	461.7	1 103.7	60.0	131.7	48.5	86.8	-3.7	12 205.3
2012	1 997.4	334.6	1 662.8	948.7	714.1	503.7	1 136.1	58.0	97.3	40.4	69.0	-12.1	12 933.2
2013	2 010.7	362.6	1 648.1	1 009.0	639.1	465.9	1 188.7	59.7	106.9	38.4	84.7	-16.2	13 271.4
2014	2 120.2	407.1	1 713.1	1 096.1	617.1	516.1	1 240.8	58.1	130.4	42.9	98.1	-10.6	13 937.8
2015	2 060.5	396.3	1 664.2	1 164.9	499.3	585.8	1 275.2	57.2	154.2	50.3	110.5	-6.6	14 417.7
2016	2 023.7	376.2	1 647.6	1 189.4	458.2	577.3	1 311.8	61.7	164.6	59.7	105.5	-0.5	14 672.4
2017	2 114.5	311.3	1 803.2	1 270.4	532.8	636.4	1 364.0	59.9	150.7	48.8	99.6	2.2	15 404.3
2018	2 243.0	282.9	1 960.1	1 390.1	570.0	619.1	1 444.8	63.3	156.6	47.4	101.0	8.2	16 157.4
2019	2 250.5	298.7	1 951.8	1 360.8	591.0	573.4	1 491.4	73.9	158.0	47.2	89.1	21.7	16 701.2
2017													
1st quarter	2 064.1	313.3	1 750.8	1 235.7	515.1	635.2	1 340.0	59.5	165.8	52.9	116.3	-3.3	15 124.8
2nd quarter	2 103.0	319.6	1 783.3	1 268.4	514.9	638.5	1 357.4	58.1	156.0	49.2	93.5	13.2	15 292.7
3rd quarter	2 136.0	322.1	1 813.9	1 276.2	537.7	635.5	1 368.7	61.9	137.2	47.0	93.9	-3.8	15 482.0
4th quarter	2 155.0	290.2	1 864.7	1 301.2	563.5	636.3	1 389.8	60.0	143.7	46.2	94.8	2.8	15 717.8
2018													
1st quarter	2 206.0	255.8	1 950.2	1 339.9	610.3	640.0	1 418.3	58.2	149.2	46.9	104.6	-2.2	15 925.1
2nd quarter	2 225.3	277.4	1 947.9	1 377.0	570.9	628.0	1 433.1	57.8	147.5	47.3	96.6	3.5	16 058.8
3rd quarter	2 258.1	288.2	1 969.8	1 413.0	556.8	613.0	1 448.7	57.9	173.9	47.6	114.2	12.2	16 261.8
4th quarter	2 282.5	310.1	1 972.4	1 430.4	542.0	595.3	1 479.0	79.5	155.6	47.8	88.5	19.3	16 383.8
2019													
1st quarter	2 181.2	294.6	1 886.6	1 369.3	517.3	564.4	1 473.8	71.3	152.1	47.7	84.4	19.9	16 485.0
2nd quarter	2 263.2	304.9	1 958.2	1 369.3	588.9	572.3	1 480.7	61.1	159.7	47.5	86.2	26.0	16 642.3
3rd quarter	2 246.5	283.0	1 963.4	1 348.5	615.0	571.0	1 501.6	82.0	168.1	47.1	102.1	18.9	16 722.1
4th quarter	2 311.3	312.3	1 998.9	1 356.3	642.7	586.0	1 509.6	81.1	152.0	46.5	83.5	22.1	16 955.3

[1] Includes net transfer payments to the rest of the world, not shown separately.
. . . = Not available.

Table 1-13B. National Income by Type of Income: Historical Data

(Billions of dollars, quarterly data are at seasonally adjusted annual rates.)   **NIPA Tables 1.7.5, 1.12**

Year and quarter	National income, total	Compensation of employees							Proprietors' income with IVA and CCAdj			Rental income of persons with CCAdj
		Total	Wages and salaries			Supplements to wages and salaries			Total	Farm	Nonfarm	
			Total	Government	Other	Total	Employer contributions for:					
							Employee pension and insurance funds	Government social insurance				
1929	94.2	51.4	50.5	5.0	45.5	0.9	0.9	0.0	14.0	5.7	8.4	6.1
1930	83.1	47.2	46.2	5.2	41.0	1.0	0.9	0.0	10.9	3.9	7.0	5.4
1931	67.7	40.1	39.2	5.3	33.9	0.9	0.9	0.0	8.3	3.0	5.3	4.4
1932	51.3	31.3	30.5	5.0	25.5	0.8	0.8	0.0	5.0	1.8	3.3	3.6
1933	48.9	29.8	29.0	5.2	23.9	0.8	0.8	0.0	5.3	2.2	3.0	2.9
1934	58.3	34.6	33.7	6.1	27.6	0.8	0.8	0.0	7.0	2.6	4.4	2.5
1935	66.3	37.7	36.7	6.5	30.2	0.9	0.9	0.0	10.1	4.9	5.2	2.6
1936	75.1	43.3	42.0	7.9	34.1	1.3	1.1	0.2	10.4	3.9	6.4	2.7
1937	83.7	48.3	46.1	7.5	38.6	2.2	1.1	1.0	12.5	5.6	6.9	3.0
1938	77.0	45.4	43.0	8.3	34.8	2.4	1.2	1.2	10.6	4.0	6.6	3.5
1939	82.4	48.6	46.0	8.2	37.7	2.6	1.3	1.3	11.1	4.0	7.1	3.7
1940	91.5	52.7	49.9	8.5	41.4	2.9	1.5	1.4	12.2	4.1	8.2	3.8
1941	117.3	66.2	62.1	10.2	51.9	4.1	2.4	1.7	16.7	6.0	10.6	4.4
1942	152.4	88.0	82.1	16.0	66.1	5.9	3.9	2.0	23.3	9.7	13.7	5.5
1943	187.2	112.7	105.8	26.6	79.2	6.9	4.6	2.3	28.2	11.5	16.7	6.0
1944	200.9	124.3	116.7	33.0	83.8	7.6	5.0	2.5	29.3	11.4	17.9	6.3
1945	201.3	126.3	117.5	34.9	82.6	8.8	5.3	3.5	30.8	11.8	19.0	6.6
1946	201.3	122.5	112.0	20.7	91.3	10.5	5.4	5.1	35.7	14.2	21.5	6.9
1947	218.7	132.4	123.1	17.5	105.6	9.3	5.3	3.9	34.6	14.3	20.2	6.9
1948	244.8	144.3	135.6	19.0	116.5	8.8	5.8	3.0	39.3	16.7	22.6	7.5
1949	239.7	144.3	134.7	20.8	113.9	9.6	6.3	3.3	34.7	12.0	22.7	7.8
1947												
1st quarter	213.0	129.0	118.9	17.1	101.7	10.1	5.5	4.7	36.2	16.0	20.2	6.7
2nd quarter	215.6	130.8	121.3	17.3	103.9	9.5	5.2	4.3	33.3	12.4	20.9	6.8
3rd quarter	219.2	132.8	124.0	17.6	106.4	8.8	5.2	3.5	33.9	14.1	19.8	7.0
4th quarter	227.2	136.9	128.3	18.0	110.3	8.6	5.4	3.2	35.0	14.9	20.0	7.2
1948												
1st quarter	235.8	140.4	131.7	18.3	113.4	8.7	5.5	3.2	36.4	14.5	21.9	7.3
2nd quarter	244.0	142.1	133.4	18.6	114.7	8.7	5.7	3.0	40.2	18.0	22.3	7.5
3rd quarter	248.8	146.8	138.1	19.3	118.8	8.8	5.8	2.9	40.6	17.9	22.7	7.5
4th quarter	250.7	147.9	139.1	19.9	119.2	8.8	6.0	2.9	39.9	16.4	23.5	7.6
1949												
1st quarter	243.3	146.3	136.7	20.4	116.4	9.5	6.0	3.5	35.4	12.7	22.7	7.5
2nd quarter	239.7	144.7	135.0	20.7	114.3	9.7	6.1	3.5	34.9	12.1	22.8	7.7
3rd quarter	239.9	143.6	134.0	21.1	112.9	9.6	6.3	3.3	34.2	11.6	22.6	7.9
4th quarter	236.0	142.8	133.2	21.2	112.0	9.6	6.5	3.0	34.2	11.5	22.7	8.1
1950												
1st quarter	246.3	147.0	136.9	21.2	115.6	10.1	6.9	3.3	36.2	12.2	24.0	8.5
2nd quarter	258.5	153.6	143.0	21.8	121.2	10.6	7.3	3.3	36.8	12.2	24.6	8.7
3rd quarter	275.6	162.3	151.0	23.0	128.1	11.3	7.9	3.4	38.5	13.1	25.4	8.8
4th quarter	286.1	170.2	158.1	24.5	133.5	12.1	8.5	3.6	38.5	13.9	24.6	9.1
1951												
1st quarter	297.8	178.6	165.3	26.7	138.6	13.2	9.2	4.0	40.9	15.0	25.9	9.3
2nd quarter	305.2	184.9	171.0	28.5	142.5	13.9	9.8	4.1	42.5	15.4	27.1	9.6
3rd quarter	310.6	187.9	173.5	30.2	143.3	14.4	10.3	4.1	43.2	15.1	28.1	9.9
4th quarter	317.0	191.5	176.6	31.5	145.1	14.9	10.7	4.1	43.7	15.7	28.0	10.1
1952												
1st quarter	318.5	196.1	181.0	32.4	148.6	15.1	11.0	4.1	41.7	13.7	28.0	10.4
2nd quarter	320.4	197.4	182.1	33.1	149.0	15.3	11.2	4.1	43.0	14.5	28.4	10.7
3rd quarter	327.0	201.5	186.0	33.8	152.2	15.6	11.4	4.1	44.7	16.0	28.8	10.9
4th quarter	338.5	209.3	193.5	34.2	159.3	15.9	11.6	4.2	42.8	13.0	29.7	11.3
1953												
1st quarter	344.5	213.1	197.1	34.3	162.7	16.1	11.9	4.2	42.9	13.0	29.9	11.6
2nd quarter	347.3	216.2	200.0	34.4	165.7	16.2	12.0	4.2	42.3	12.4	29.9	11.9
3rd quarter	345.6	216.5	200.1	34.3	165.9	16.4	12.1	4.3	41.2	11.7	29.5	12.2
4th quarter	338.0	215.1	198.7	34.4	164.3	16.4	12.2	4.2	41.6	11.5	30.0	12.5
1954												
1st quarter	339.6	213.1	196.4	34.4	162.0	16.7	12.1	4.6	42.7	12.9	29.8	12.8
2nd quarter	339.9	212.7	196.0	34.7	161.2	16.7	12.1	4.6	42.1	11.7	30.4	13.0
3rd quarter	343.3	213.2	196.3	35.1	161.2	16.9	12.3	4.6	42.2	11.8	30.5	13.1
4th quarter	352.2	217.5	200.4	35.4	165.0	17.2	12.5	4.6	42.0	10.7	31.3	13.3
1955												
1st quarter	364.6	221.9	204.1	35.5	168.6	17.8	12.8	5.0	43.7	10.9	32.7	13.3
2nd quarter	374.0	228.1	209.9	36.2	173.6	18.2	13.1	5.1	44.3	11.3	33.0	13.4
3rd quarter	381.4	233.9	215.2	37.4	177.8	18.7	13.4	5.2	44.6	10.7	33.9	13.4
4th quarter	387.7	238.4	219.5	37.3	182.2	18.9	13.7	5.3	44.5	10.0	34.5	13.5
1956												
1st quarter	391.2	242.8	223.3	37.8	185.4	19.5	14.0	5.6	44.9	10.2	34.7	13.6
2nd quarter	397.0	247.5	227.5	38.5	189.0	20.0	14.3	5.7	45.3	10.4	34.9	13.6
3rd quarter	401.9	250.4	230.0	39.2	190.8	20.5	14.8	5.7	46.4	10.9	35.5	13.7
4th quarter	410.1	256.4	235.4	39.7	195.7	21.0	15.2	5.8	46.8	10.9	35.8	13.8
1957												
1st quarter	416.8	260.2	238.3	40.2	198.0	22.0	15.6	6.4	46.9	9.9	37.0	13.9
2nd quarter	418.9	262.1	239.6	40.7	198.9	22.4	16.0	6.4	47.7	10.5	37.2	14.0
3rd quarter	422.6	264.7	241.8	41.5	200.3	22.9	16.5	6.4	48.8	11.0	37.8	14.1
4th quarter	415.8	263.3	240.1	41.6	198.5	23.2	16.9	6.3	47.7	10.7	37.0	14.3

Table 1-13B. National Income by Type of Income: Historical Data—*Continued*

(Billions of dollars, quarterly data are at seasonally adjusted annual rates.) NIPA Tables 1.7.5, 1.12

Year and quarter	Corporate profits with IVA and CCAdj					Net interest and miscellaneous payments	Taxes on production and imports	Less: Subsidies	Business current transfer payments, net			Current surplus of government enterprises	Addendum: Net national factor income
			Profits after tax										
	Total	Taxes on corporate income	Total	Net dividends	Undistributed corporate profits				Total ¹	To persons	To government		
1929	10.8	1.4	9.5	5.8	3.7	4.6	6.8	0.0	0.5	0.4	0.1	. . .	86.9
1930	7.5	0.8	6.7	5.5	1.2	4.8	7.0	0.1	0.5	0.4	0.1	. . .	75.7
1931	3.0	0.5	2.5	4.1	-1.6	4.8	6.7	0.1	0.5	0.4	0.1	. . .	60.6
1932	-0.2	0.4	-0.6	2.5	-3.1	4.5	6.6	0.1	0.6	0.5	0.1	. . .	44.2
1933	-0.2	0.5	-0.7	2.0	-2.7	4.0	6.9	0.2	0.5	0.4	0.1	. . .	41.7
1934	2.5	0.7	1.8	2.6	-0.8	4.0	7.6	0.5	0.5	0.4	0.1	. . .	50.6
1935	4.0	1.0	3.1	2.8	0.2	4.1	8.0	0.6	0.5	0.4	0.1	. . .	58.4
1936	6.2	1.4	4.8	4.5	0.3	3.8	8.5	0.3	0.5	0.4	0.1	. . .	66.4
1937	7.1	1.5	5.6	4.7	0.9	3.7	8.9	0.3	0.5	0.4	0.1	. . .	74.5
1938	5.0	1.0	4.0	3.2	0.8	3.6	8.9	0.5	0.4	0.3	0.2	. . .	68.1
1939	6.6	1.4	5.2	3.8	1.4	3.6	9.1	0.8	0.4	0.3	0.2	. . .	73.6
1940	9.9	2.8	7.0	4.0	3.0	3.3	9.8	0.7	0.5	0.3	0.2	. . .	82.0
1941	15.7	7.6	8.1	4.4	3.7	3.3	11.1	0.5	0.5	0.4	0.2	. . .	106.2
1942	20.8	11.4	9.4	4.3	5.1	3.2	11.5	0.5	0.5	0.3	0.2	. . .	140.8
1943	24.9	14.1	10.8	4.4	6.4	2.9	12.4	0.6	0.6	0.4	0.3	. . .	174.7
1944	25.0	12.9	12.0	4.6	7.4	2.4	13.7	1.0	0.8	0.4	0.4	. . .	187.4
1945	20.5	10.7	9.8	4.6	5.2	2.3	15.1	1.1	0.9	0.5	0.4	. . .	186.5
1946	18.2	9.1	9.1	5.6	3.6	1.8	16.8	1.4	0.7	0.4	0.3	. . .	185.1
1947	24.2	11.2	13.0	6.4	6.6	2.3	18.1	0.4	0.7	0.4	0.3	. . .	200.3
1948	31.4	12.3	19.1	7.2	11.9	2.5	19.7	0.5	0.7	0.4	0.3	. . .	225.0
1949	29.1	10.0	19.0	7.4	11.6	2.8	20.9	0.5	0.7	0.4	0.3	. . .	218.6
1947													
1st quarter	21.5	11.5	10.0	6.1	3.9	2.0	17.7	0.5	0.5	0.2	0.3	. . .	195.3
2nd quarter	24.5	10.9	13.6	6.4	7.2	2.3	17.7	0.4	0.7	0.4	0.3	. . .	197.6
3rd quarter	24.6	10.7	13.9	6.6	7.3	2.5	18.0	0.4	0.8	0.5	0.3	. . .	200.7
4th quarter	26.1	11.7	14.5	6.6	7.9	2.5	19.0	0.3	0.9	0.5	0.3	. . .	207.6
1948													
1st quarter	30.0	12.0	18.0	7.2	10.8	2.4	19.0	0.4	0.7	0.4	0.4	. . .	216.5
2nd quarter	31.8	12.6	19.1	6.8	12.3	2.4	19.7	0.3	0.7	0.4	0.3	. . .	224.0
3rd quarter	31.2	12.4	18.8	7.3	11.5	2.5	20.0	0.6	0.7	0.4	0.3	. . .	228.7
4th quarter	32.7	12.0	20.7	7.6	13.1	2.5	20.3	0.9	0.7	0.4	0.3	. . .	230.6
1949													
1st quarter	31.0	10.8	20.2	7.4	12.7	2.7	20.4	0.6	0.7	0.3	0.3	. . .	222.9
2nd quarter	28.7	9.5	19.1	7.4	11.7	2.7	20.8	0.4	0.7	0.3	0.3	. . .	218.7
3rd quarter	29.9	9.9	20.0	7.3	12.7	2.8	21.3	0.5	0.7	0.4	0.3	. . .	218.5
4th quarter	26.6	9.7	16.9	7.6	9.3	2.8	21.2	0.5	0.7	0.4	0.3	. . .	214.5
1950													
1st quarter	30.1	13.4	16.6	8.5	8.1	3.0	21.6	0.7	0.7	0.4	0.3	. . .	224.8
2nd quarter	33.9	16.1	17.8	8.6	9.2	3.1	22.5	0.8	0.8	0.5	0.3	. . .	236.0
3rd quarter	38.4	19.7	18.7	9.4	9.4	3.1	24.4	0.8	0.9	0.6	0.3	. . .	251.1
4th quarter	41.8	21.6	20.2	9.7	10.5	3.1	23.4	1.1	0.9	0.7	0.3	. . .	262.8
1951													
1st quarter	40.7	26.0	14.7	8.6	6.1	3.3	25.1	1.2	1.1	0.8	0.3	. . .	272.8
2nd quarter	40.6	22.0	18.6	8.9	9.7	3.5	24.1	1.1	1.2	0.9	0.3	. . .	281.1
3rd quarter	41.1	19.9	21.2	8.8	12.4	3.7	24.5	0.9	1.2	0.9	0.3	. . .	285.8
4th quarter	42.4	21.3	21.1	9.0	12.2	3.8	25.3	1.0	1.2	0.9	0.3	. . .	291.5
1952													
1st quarter	40.0	19.5	20.5	8.4	12.1	3.8	26.1	0.8	1.2	0.9	0.3	. . .	292.0
2nd quarter	38.1	18.4	19.8	9.0	10.8	3.9	26.9	0.8	1.2	0.9	0.3	. . .	293.1
3rd quarter	38.1	18.3	19.8	8.9	10.9	4.0	27.3	0.8	1.2	0.9	0.3	. . .	299.3
4th quarter	42.4	20.2	22.2	9.1	13.1	4.1	28.1	0.7	1.2	0.9	0.3	. . .	309.9
1953													
1st quarter	43.3	21.3	22.1	8.8	13.3	4.3	28.7	0.7	1.2	0.9	0.3	. . .	315.3
2nd quarter	42.5	21.4	21.1	9.5	11.6	4.4	29.2	0.5	1.2	0.9	0.3	. . .	317.4
3rd quarter	41.3	20.9	20.4	9.4	11.0	4.6	29.3	0.7	1.2	0.8	0.3	. . .	315.8
4th quarter	34.0	16.2	17.8	9.2	8.5	4.8	29.2	0.2	1.1	0.8	0.3	. . .	307.9
1954													
1st quarter	36.5	16.1	20.4	9.7	10.7	5.0	28.7	0.1	1.0	0.6	0.3	. . .	310.1
2nd quarter	37.8	16.6	21.2	9.2	12.1	5.3	28.8	0.7	0.9	0.5	0.3	. . .	310.9
3rd quarter	39.9	17.7	22.2	9.6	12.7	5.5	28.7	0.3	0.9	0.5	0.4	. . .	314.0
4th quarter	43.6	19.0	24.6	9.7	14.8	5.7	29.3	0.2	0.9	0.6	0.4	. . .	322.1
1955													
1st quarter	48.8	21.1	27.7	10.3	17.4	5.8	30.2	0.3	1.1	0.7	0.4	. . .	333.5
2nd quarter	50.1	21.4	28.7	10.4	18.3	6.0	31.2	0.3	1.2	0.9	0.4	. . .	341.8
3rd quarter	50.2	21.9	28.3	11.0	17.2	5.9	31.9	0.0	1.4	1.0	0.4	. . .	348.1
4th quarter	51.6	22.8	28.8	11.2	17.6	5.9	32.5	0.2	1.5	1.1	0.4	. . .	354.0
1956													
1st quarter	49.4	21.7	27.7	11.5	16.2	6.2	33.0	0.3	1.6	1.2	0.4	. . .	356.9
2nd quarter	49.6	22.1	27.5	11.6	16.0	6.4	33.6	0.6	1.6	1.2	0.4	. . .	362.3
3rd quarter	49.3	20.7	28.6	11.6	17.0	6.7	34.6	0.9	1.7	1.3	0.4	. . .	366.5
4th quarter	50.2	21.9	28.4	12.0	16.3	6.7	35.7	1.2	1.7	1.3	0.4	. . .	373.8
1957													
1st quarter	51.7	22.3	29.3	12.1	17.2	7.5	36.1	1.3	1.8	1.4	0.4	. . .	380.2
2nd quarter	50.3	21.4	28.9	12.3	16.6	7.7	36.6	1.2	1.8	1.4	0.4	. . .	381.7
3rd quarter	49.3	20.9	28.5	12.5	16.0	7.8	37.0	1.1	1.8	1.4	0.4	. . .	384.8
4th quarter	45.3	19.0	26.3	12.3	14.0	7.6	36.7	1.0	1.8	1.4	0.4	. . .	378.2

¹Includes net transfer payments to the rest of the world, not shown separately.
. . . = Not available.

Table 1-13B. National Income by Type of Income: Historical Data—*Continued*

(Billions of dollars, quarterly data are at seasonally adjusted annual rates.) NIPA Tables 1.7.5, 1.12

Year and quarter	National income, total	Compensation of employees								Proprietors' income with IVA and CCAdj			Rental income of persons with CCAdj
		Total	Wages and salaries			Supplements to wages and salaries				Total	Farm	Nonfarm	
			Total	Government	Other	Total	Employer contributions for:						
							Employee pension and insurance funds	Government social insurance					
1958													
1st quarter	409.7	259.8	236.7	42.0	194.7	23.0	16.7	6.3		50.3	13.3	36.9	14.6
2nd quarter	411.6	259.4	236.3	43.1	193.2	23.2	16.9	6.3		50.3	13.0	37.3	14.8
3rd quarter	424.5	267.2	243.8	46.1	197.7	23.4	17.0	6.4		50.1	12.1	38.0	14.9
4th quarter	437.2	272.3	248.4	45.3	203.2	23.8	17.4	6.4		49.9	11.2	38.7	15.0
1959													
1st quarter	448.3	279.7	254.0	45.5	208.6	25.7	18.0	7.8		50.1	10.9	39.2	15.0
2nd quarter	462.4	286.5	260.6	45.8	214.7	25.9	18.0	7.9		50.3	9.9	40.5	15.4
3rd quarter	459.6	286.9	260.9	46.2	214.7	26.0	18.2	7.9		50.4	9.4	41.0	15.8
4th quarter	464.8	290.2	263.9	46.7	217.2	26.3	18.4	7.9		50.6	10.1	40.5	16.0
1960													
1st quarter	478.1	298.7	270.7	47.7	223.0	28.0	18.7	9.3		49.7	9.6	40.1	16.2
2nd quarter	478.8	301.7	273.4	48.6	224.8	28.3	19.1	9.3		50.7	10.6	40.1	16.4
3rd quarter	480.2	302.6	274.0	49.9	224.1	28.6	19.3	9.3		50.9	11.0	39.9	16.5
4th quarter	478.5	302.1	273.3	50.5	222.8	28.8	19.6	9.2		51.0	11.3	39.7	16.7
1961													
1st quarter	480.3	303.1	273.8	51.1	222.7	29.3	19.9	9.5		52.3	11.5	40.8	16.9
2nd quarter	489.9	307.3	277.6	51.8	225.8	29.7	20.1	9.6		52.6	10.9	41.8	17.1
3rd quarter	499.7	312.3	282.3	52.8	229.5	30.1	20.4	9.6		53.3	11.0	42.3	17.3
4th quarter	514.2	318.9	288.4	54.2	234.2	30.5	20.7	9.8		54.5	11.5	43.0	17.5
1962													
1st quarter	523.4	325.5	293.3	55.3	238.0	32.2	21.2	11.0		55.3	11.9	43.5	17.7
2nd quarter	530.5	331.4	298.7	56.0	242.8	32.7	21.5	11.2		55.0	11.1	43.9	18.0
3rd quarter	536.9	334.2	301.2	56.5	244.7	33.1	21.8	11.2		54.9	10.7	44.2	18.2
4th quarter	544.9	337.7	304.2	57.6	246.6	33.4	22.1	11.3		55.8	11.3	44.5	18.3
1963													
1st quarter	551.8	342.7	308.0	58.6	249.3	34.7	22.5	12.2		56.0	11.4	44.6	18.5
2nd quarter	560.9	347.6	312.4	59.4	253.0	35.2	22.9	12.3		55.8	10.8	45.0	18.7
3rd quarter	569.4	352.6	316.8	60.2	256.7	35.8	23.3	12.5		56.2	10.7	45.6	18.8
4th quarter	579.4	358.8	322.2	61.9	260.3	36.5	23.9	12.6		57.6	11.1	46.5	18.8
1964													
1st quarter	591.7	365.2	328.2	63.1	265.1	37.0	24.5	12.4		57.9	9.9	48.0	18.8
2nd quarter	601.8	372.6	334.8	64.2	270.7	37.8	25.2	12.6		58.8	9.5	49.3	18.8
3rd quarter	613.2	380.0	341.4	65.7	275.7	38.6	25.9	12.7		59.3	9.3	50.0	18.9
4th quarter	621.4	386.1	346.7	66.8	279.9	39.4	26.6	12.8		60.5	10.4	50.1	18.8
1965													
1st quarter	640.0	392.9	352.8	67.5	285.4	40.1	27.3	12.8		61.9	11.4	50.5	19.1
2nd quarter	651.2	399.9	358.9	68.5	290.3	41.0	28.1	13.0		63.2	12.0	51.2	19.3
3rd quarter	662.4	408.3	366.2	70.5	295.7	42.1	29.0	13.2		64.0	12.2	51.8	19.5
4th quarter	681.8	420.5	377.1	73.2	304.0	43.3	29.9	13.4		65.6	12.5	53.0	19.5
1966													
1st quarter	702.0	433.1	385.8	74.9	310.8	47.3	30.9	16.5		69.2	14.9	54.4	19.8
2nd quarter	711.8	444.4	395.9	76.9	319.0	48.5	31.8	16.7		67.2	12.7	54.5	19.8
3rd quarter	723.1	455.7	406.1	79.8	326.3	49.5	32.5	17.0		67.2	12.3	54.9	20.0
4th quarter	735.5	463.8	413.5	82.0	331.5	50.3	33.1	17.2		68.1	12.2	55.9	20.0
1967													
1st quarter	741.8	470.0	418.8	83.5	335.4	51.2	33.6	17.6		68.9	11.9	57.0	20.3
2nd quarter	748.4	475.6	423.6	85.0	338.6	52.0	34.3	17.8		68.5	11.1	57.4	20.4
3rd quarter	763.8	485.3	432.0	87.1	344.9	53.3	35.1	18.2		70.5	12.0	58.6	20.4
4th quarter	779.7	496.2	441.6	90.4	351.2	54.7	36.1	18.6		70.0	11.5	58.4	20.3
1968													
1st quarter	800.5	510.9	454.2	92.8	361.5	56.7	37.2	19.4		71.7	11.6	60.1	20.1
2nd quarter	821.5	524.0	465.9	95.2	370.7	58.1	38.3	19.8		73.2	11.2	61.9	20.1
3rd quarter	841.0	537.8	478.3	98.7	379.6	59.5	39.4	20.1		75.0	11.8	63.2	20.2
4th quarter	858.0	550.3	489.4	100.1	389.4	60.9	40.4	20.5		75.5	12.0	63.5	20.0
1969													
1st quarter	874.6	562.8	499.1	101.3	397.8	63.7	41.6	22.1		75.6	11.6	64.1	20.2
2nd quarter	890.1	576.7	511.4	103.1	408.3	65.3	42.8	22.5		77.1	12.5	64.6	20.3
3rd quarter	908.1	593.4	526.4	108.2	418.3	67.0	44.0	23.0		78.0	13.1	64.8	20.4
4th quarter	916.3	604.9	536.5	109.8	426.6	68.5	45.1	23.4		77.3	14.1	63.2	20.4
1970													
1st quarter	921.1	615.0	545.1	114.4	430.7	69.9	46.2	23.7		77.1	13.5	63.5	20.4
2nd quarter	933.2	620.2	549.1	116.5	432.6	71.2	47.3	23.9		76.7	12.4	64.3	20.2
3rd quarter	947.7	628.2	555.7	118.3	437.4	72.5	48.5	23.9		78.5	13.2	65.3	20.9
4th quarter	948.0	630.0	556.4	119.7	436.6	73.7	49.8	23.9		78.9	12.5	66.4	21.2
1971													
1st quarter	984.6	647.9	570.5	123.9	446.6	77.4	51.3	26.1		80.3	13.1	67.2	21.1
2nd quarter	1 004.4	659.8	580.4	125.6	454.8	79.3	52.9	26.4		82.8	13.3	69.6	21.7
3rd quarter	1 022.4	670.1	588.8	128.0	460.8	81.3	54.8	26.5		84.6	13.0	71.6	22.0
4th quarter	1 044.7	682.2	598.4	129.6	468.8	83.7	56.9	26.8		87.9	14.2	73.6	22.5
1972													
1st quarter	1 077.9	708.4	618.5	134.2	484.3	89.9	59.3	30.6		87.9	13.1	74.8	23.1
2nd quarter	1 098.2	722.5	630.4	135.7	494.7	92.1	61.1	31.0		91.3	15.4	75.9	20.2
3rd quarter	1 127.9	735.7	642.3	138.4	503.9	93.4	62.3	31.2		95.5	17.2	78.4	23.9
4th quarter	1 174.0	758.8	664.0	143.3	520.7	94.8	63.0	31.8		105.6	22.4	83.3	23.8

Table 1-13B. National Income by Type of Income: Historical Data—*Continued*

(Billions of dollars, quarterly data are at seasonally adjusted annual rates.)   NIPA Tables 1.7.5, 1.12

| Year and quarter | Corporate profits with IVA and CCAdj | | | | | Net interest and miscellaneous payments | Taxes on production and imports | Less: Subsidies | Business current transfer payments, net | | | Current surplus of government enterprises | Addendum: Net national factor income |
| | Total | Taxes on corporate income | Profits after tax | | | | | | | | | | |
			Total	Net dividends	Undistributed corporate profits				Total 1	To persons	To government		
1958													
1st quarter	39.5	16.3	23.1	12.1	11.0	8.1	36.8	1.1	1.8	1.3	0.5	. . .	372.2
2nd quarter	40.6	16.7	23.9	12.2	11.7	8.7	37.3	1.3	1.7	1.3	0.5	. . .	373.8
3rd quarter	44.8	18.9	25.9	12.1	13.7	9.5	37.8	1.5	1.7	1.2	0.5	. . .	386.5
4th quarter	50.6	21.8	28.7	11.8	16.9	10.6	38.9	1.6	1.6	1.2	0.4	. . .	398.3
1959													
1st quarter	54.3	22.8	31.5	12.8	18.7	8.5	39.8	1.1	1.6	1.3	0.3	. . .	407.5
2nd quarter	59.1	25.0	34.1	13.2	20.9	9.5	40.3	0.9	1.7	1.3	0.3	. . .	420.8
3rd quarter	54.2	22.1	32.1	13.8	18.3	9.5	41.7	1.1	1.7	1.3	0.3	. . .	416.8
4th quarter	54.5	21.4	33.1	14.1	19.0	9.8	42.4	1.1	1.7	1.4	0.3	. . .	421.2
1960													
1st quarter	58.6	24.2	34.4	14.1	20.3	9.8	43.7	1.0	1.7	1.3	0.4	. . .	433.1
2nd quarter	54.8	22.1	32.7	14.2	18.5	9.9	44.3	1.3	1.7	1.3	0.4	. . .	433.5
3rd quarter	53.8	21.1	32.7	14.4	18.2	10.5	44.9	1.0	1.7	1.3	0.4	. . .	434.3
4th quarter	51.5	20.1	31.4	14.3	17.1	10.9	45.3	1.2	1.8	1.3	0.5	. . .	432.3
1961													
1st quarter	50.2	20.0	30.3	14.3	16.0	11.3	45.8	1.6	1.9	1.3	0.5	. . .	433.9
2nd quarter	54.5	21.4	33.1	14.3	18.7	11.9	46.5	2.0	1.9	1.3	0.5	. . .	443.4
3rd quarter	57.2	22.7	34.5	14.6	19.9	12.2	47.3	2.2	1.9	1.4	0.5	. . .	452.4
4th quarter	61.8	24.7	37.1	15.2	21.9	12.9	48.4	2.3	2.0	1.4	0.6	. . .	465.7
1962													
1st quarter	62.6	23.1	39.4	15.3	24.1	12.7	49.4	2.3	2.0	1.5	0.6	. . .	473.8
2nd quarter	62.5	22.9	39.7	15.8	23.9	13.7	49.9	2.4	2.1	1.5	0.6	. . .	480.6
3rd quarter	64.2	23.6	40.7	16.0	24.7	14.2	50.9	2.2	2.1	1.6	0.6	. . .	485.7
4th quarter	66.9	23.6	43.2	16.2	27.0	14.6	51.3	2.2	2.2	1.6	0.6	. . .	493.2
1963													
1st quarter	67.2	23.7	43.5	16.6	26.9	14.3	52.0	2.0	2.4	1.8	0.6	. . .	498.7
2nd quarter	70.2	25.4	44.8	16.9	27.9	14.5	52.9	2.2	2.5	1.8	0.7	. . .	506.8
3rd quarter	71.7	26.1	45.5	17.2	28.3	15.0	53.9	2.3	2.5	1.9	0.7	. . .	514.4
4th quarter	73.0	26.7	46.2	17.7	28.5	15.5	54.7	2.4	2.6	2.0	0.7	. . .	523.6
1964													
1st quarter	77.1	26.4	50.7	19.0	31.7	16.2	55.7	2.7	2.7	2.1	0.6	. . .	535.1
2nd quarter	77.3	26.4	50.9	19.6	31.3	16.7	56.7	2.9	2.8	2.2	0.6	. . .	544.3
3rd quarter	78.4	27.0	51.5	20.0	31.4	17.5	57.9	2.6	3.2	2.2	1.0	. . .	554.1
4th quarter	78.0	26.5	51.5	20.5	31.0	17.6	58.8	2.7	3.3	2.3	1.0	. . .	561.1
1965													
1st quarter	85.8	28.3	57.5	20.4	37.1	18.6	60.2	2.9	3.5	2.3	1.2	. . .	578.4
2nd quarter	88.0	29.2	58.8	21.2	37.7	19.0	60.3	3.0	3.5	2.3	1.2	. . .	589.4
3rd quarter	89.3	29.8	59.5	21.9	37.6	19.5	60.5	3.0	3.5	2.2	1.2	. . .	600.6
4th quarter	94.0	31.7	62.3	22.6	39.7	19.5	61.8	3.1	3.4	2.2	1.3	. . .	619.0
1966													
1st quarter	97.4	32.7	64.7	22.7	41.9	20.7	61.4	3.6	3.4	2.2	1.3	. . .	640.2
2nd quarter	96.4	32.6	63.7	22.5	41.3	21.3	62.8	3.9	3.4	2.1	1.3	. . .	649.1
3rd quarter	94.8	32.2	62.7	22.2	40.4	22.1	63.7	4.1	3.4	2.1	1.3	. . .	659.8
4th quarter	95.9	31.4	64.5	21.8	42.6	23.2	64.9	4.2	3.4	2.1	1.3	. . .	671.1
1967													
1st quarter	93.3	30.6	62.7	22.9	39.8	24.0	65.6	4.0	3.5	2.2	1.3	. . .	676.4
2nd quarter	92.2	30.3	61.8	23.5	38.3	24.8	66.8	3.9	3.6	2.2	1.4	. . .	681.5
3rd quarter	93.2	30.6	62.6	23.9	38.7	25.2	68.7	3.7	3.7	2.3	1.4	. . .	694.7
4th quarter	96.8	32.4	64.4	23.2	41.1	25.7	70.6	3.7	3.8	2.4	1.4	. . .	709.0
1968													
1st quarter	98.1	36.5	61.7	24.9	36.8	26.0	73.4	4.0	3.9	2.5	1.4	. . .	726.7
2nd quarter	102.1	37.1	65.1	25.8	39.2	26.3	75.4	4.2	4.1	2.7	1.4	. . .	745.7
3rd quarter	102.6	37.1	65.5	26.5	39.0	27.1	77.7	4.2	4.3	2.9	1.4	. . .	762.8
4th quarter	104.1	38.0	66.1	26.8	39.3	28.4	79.1	4.2	4.4	3.0	1.4	. . .	778.3
1969													
1st quarter	103.5	38.7	64.8	26.7	38.1	31.4	80.6	4.3	4.7	3.3	1.5	. . .	793.5
2nd quarter	100.3	37.5	62.9	27.1	35.8	32.2	82.9	4.5	4.8	3.4	1.5	. . .	806.6
3rd quarter	97.4	36.1	61.4	27.3	34.0	33.3	85.2	4.7	4.8	3.4	1.5	. . .	822.6
4th quarter	92.4	35.6	56.8	27.9	29.0	34.1	86.7	4.7	4.8	3.3	1.5	. . .	829.2
1970													
1st quarter	84.7	31.1	53.6	27.9	25.7	36.2	88.5	4.7	4.6	3.1	1.5	. . .	833.3
2nd quarter	88.5	31.2	57.3	27.8	29.6	38.5	90.5	4.8	4.5	3.0	1.5	. . .	844.2
3rd quarter	88.4	32.1	56.2	27.8	28.4	40.9	92.5	4.7	4.2	2.8	1.4	. . .	856.8
4th quarter	83.5	30.7	52.8	27.7	25.1	42.4	94.1	4.8	4.1	2.8	1.4	. . .	855.9
1971													
1st quarter	96.6	34.4	62.2	28.5	33.8	43.4	97.7	4.8	4.1	2.7	1.5	. . .	889.4
2nd quarter	98.7	35.3	63.4	28.2	35.2	44.3	98.9	4.8	4.1	2.7	1.4	. . .	907.4
3rd quarter	101.4	34.6	66.9	28.5	38.4	44.4	101.7	4.5	4.2	2.7	1.5	. . .	922.6
4th quarter	105.8	35.0	70.8	28.3	42.5	44.5	103.7	4.6	4.2	2.8	1.5	. . .	942.8
1972													
1st quarter	111.5	37.1	74.4	29.3	45.0	45.4	104.6	6.1	4.5	2.9	1.6	. . .	976.3
2nd quarter	113.4	37.5	76.0	29.6	46.3	46.6	106.8	6.2	4.7	3.0	1.7	. . .	994.0
3rd quarter	118.2	38.8	79.4	30.3	49.1	48.9	108.9	7.2	4.9	3.2	1.7	. . .	1 022.2
4th quarter	125.7	43.1	82.7	31.0	51.7	51.1	111.5	7.1	5.0	3.3	1.7	. . .	1 065.0

1Includes net transfer payments to the rest of the world, not shown separately.
. . . = Not available.

Table 1-13B. National Income by Type of Income: Historical Data—*Continued*

(Billions of dollars, quarterly data are at seasonally adjusted annual rates.)   **NIPA Tables 1.7.5, 1.12**

Year and quarter	National income, total	Compensation of employees							Proprietors' income with IVA and CCAdj			Rental income of persons with CCAdj
		Total	Wages and salaries			Supplements to wages and salaries			Total	Farm	Nonfarm	
			Total	Government	Other	Total	Employer contributions for:					
							Employee pension and insurance funds	Government social insurance				
1973												
1st quarter	1 211.0	785.4	683.4	145.0	538.3	102.1	63.3	38.7	104.1	21.8	82.4	23.5
2nd quarter	1 235.7	803.2	700.2	147.2	553.0	103.0	63.7	39.3	109.9	27.6	82.4	23.3
3rd quarter	1 264.0	820.4	716.2	149.6	566.6	104.2	64.3	39.9	113.8	29.5	84.3	22.3
4th quarter	1 301.9	841.7	735.3	153.4	582.0	106.4	65.3	41.1	122.2	37.5	84.7	23.3
1974												
1st quarter	1 313.9	858.3	748.2	155.6	592.5	110.2	66.6	43.6	115.7	28.6	87.1	23.5
2nd quarter	1 334.6	878.5	765.3	158.2	607.1	113.2	68.7	44.5	108.6	20.1	88.5	22.9
3rd quarter	1 360.6	900.2	783.1	161.3	621.8	117.1	71.8	45.3	111.0	21.5	89.5	23.3
4th quarter	1 376.5	913.8	792.5	166.9	625.6	121.3	75.9	45.4	113.5	23.7	89.8	23.0
1975												
1st quarter	1 383.4	918.4	791.9	170.5	621.4	126.5	80.7	45.7	112.7	19.7	93.0	22.7
2nd quarter	1 413.8	931.2	800.4	174.5	625.9	130.8	84.8	46.0	114.3	20.1	94.2	22.4
3rd quarter	1 471.7	956.0	821.3	177.8	643.5	134.7	87.7	47.0	120.7	23.8	96.9	22.2
4th quarter	1 515.3	983.4	845.8	182.2	663.6	137.6	89.4	48.2	125.2	24.3	100.8	21.9
1976												
1st quarter	1 565.6	1 014.7	871.2	184.8	686.4	143.5	90.4	53.1	126.4	19.4	107.0	21.6
2nd quarter	1 591.3	1 035.5	889.4	187.3	702.1	146.1	92.2	53.9	128.6	17.0	111.6	20.5
3rd quarter	1 623.6	1 058.4	908.5	189.5	719.0	149.9	95.1	54.9	132.7	16.1	116.6	20.0
4th quarter	1 656.9	1 084.8	929.9	194.1	735.8	154.9	99.2	55.7	136.2	16.3	119.9	19.1
1977												
1st quarter	1 704.1	1 113.3	950.0	196.7	753.3	163.2	104.2	59.1	138.6	16.0	122.6	16.4
2nd quarter	1 770.3	1 150.3	980.9	199.7	781.2	169.4	108.8	60.6	140.5	14.1	126.4	16.3
3rd quarter	1 825.0	1 182.1	1 007.5	203.7	803.8	174.6	112.9	61.7	142.4	11.8	130.6	15.6
4th quarter	1 871.9	1 217.6	1 038.2	210.3	827.9	179.4	116.4	63.0	156.5	22.0	134.5	15.1
1978												
1st quarter	1 910.5	1 250.9	1 063.8	213.6	850.3	187.1	119.4	67.6	158.5	18.5	140.0	15.9
2nd quarter	2 006.3	1 299.1	1 106.0	217.3	888.8	193.1	122.6	70.4	166.1	20.9	145.2	15.6
3rd quarter	2 054.8	1 336.9	1 138.0	222.4	915.6	198.9	126.3	72.6	168.3	20.8	147.6	17.1
4th quarter	2 119.1	1 380.2	1 174.4	226.8	947.6	205.8	130.3	75.4	170.9	19.5	151.4	17.5
1979												
1st quarter	2 172.6	1 422.5	1 208.6	231.3	977.2	214.0	134.7	79.2	180.1	23.8	156.3	17.3
2nd quarter	2 212.5	1 454.7	1 234.4	233.5	1 000.9	220.4	139.0	81.4	179.2	21.8	157.3	15.6
3rd quarter	2 264.9	1 496.7	1 269.4	239.4	1 030.0	227.3	143.5	83.9	180.7	22.3	158.5	15.4
4th quarter	2 311.3	1 534.9	1 300.9	244.3	1 056.6	234.0	148.1	85.9	177.8	20.8	157.0	16.1
1980												
1st quarter	2 353.0	1 573.6	1 333.8	250.2	1 083.6	239.8	152.9	86.9	167.6	13.3	154.3	15.8
2nd quarter	2 356.2	1 599.2	1 353.9	260.6	1 093.3	245.3	157.7	87.6	160.3	3.1	157.2	17.1
3rd quarter	2 422.8	1 628.6	1 377.3	264.2	1 113.1	251.3	162.2	89.1	173.4	11.7	161.7	19.9
4th quarter	2 542.6	1 687.6	1 428.7	270.8	1 157.9	258.9	166.7	92.2	185.0	18.7	166.3	23.2
1981												
1st quarter	2 630.5	1 739.6	1 467.7	277.3	1 190.3	271.9	171.0	100.9	188.2	17.6	170.6	24.0
2nd quarter	2 676.3	1 774.8	1 496.7	282.6	1 214.1	278.1	175.3	102.7	176.4	18.2	158.2	23.3
3rd quarter	2 769.4	1 815.1	1 530.6	288.6	1 241.9	284.5	179.7	104.8	182.3	23.3	158.9	23.1
4th quarter	2 782.5	1 840.7	1 550.6	294.8	1 255.8	290.1	184.1	106.0	171.8	17.0	154.9	24.7
1982												
1st quarter	2 786.1	1 864.7	1 567.5	300.8	1 266.8	297.1	188.5	108.6	165.2	14.4	150.8	23.5
2nd quarter	2 833.6	1 884.4	1 581.7	305.4	1 276.3	302.6	193.2	109.5	169.4	12.8	156.6	21.1
3rd quarter	2 852.9	1 904.8	1 596.4	309.5	1 286.9	308.3	198.0	110.3	170.7	11.9	158.8	24.4
4th quarter	2 865.4	1 918.1	1 604.4	314.5	1 289.9	313.8	203.1	110.7	179.5	14.2	165.3	26.3
1983												
1st quarter	2 924.4	1 947.2	1 622.4	318.7	1 303.7	324.8	208.4	116.4	183.3	12.9	170.4	25.8
2nd quarter	3 004.2	1 986.3	1 654.7	322.6	1 332.1	331.6	213.2	118.4	182.6	7.9	174.8	24.7
3rd quarter	3 085.1	2 029.6	1 691.4	326.7	1 364.7	338.2	217.5	120.7	183.8	0.2	183.6	23.7
4th quarter	3 192.4	2 086.8	1 741.5	331.2	1 410.4	345.3	221.3	124.0	195.3	3.8	191.5	23.5
1984												
1st quarter	3 320.8	2 143.8	1 784.0	337.8	1 446.2	359.8	224.7	135.1	221.9	19.5	202.4	23.2
2nd quarter	3 408.8	2 195.2	1 828.1	344.8	1 483.4	367.0	228.9	138.1	231.8	21.3	210.5	21.7
3rd quarter	3 474.3	2 241.6	1 867.1	351.9	1 515.2	374.6	234.0	140.5	232.4	20.4	211.9	24.4
4th quarter	3 531.6	2 282.9	1 900.5	358.1	1 542.4	382.4	240.0	142.4	226.9	22.5	204.4	29.3
1985												
1st quarter	3 593.1	2 323.5	1 932.1	363.9	1 568.2	391.4	247.0	144.4	240.7	22.8	217.8	26.9
2nd quarter	3 641.2	2 363.6	1 963.6	370.4	1 593.2	400.0	253.8	146.2	238.5	20.4	218.0	25.3
3rd quarter	3 702.9	2 406.4	1 997.4	378.1	1 619.3	409.0	260.4	148.6	240.7	19.3	221.5	27.1
4th quarter	3 742.5	2 455.7	2 037.0	383.0	1 654.0	418.7	266.9	151.8	244.3	21.4	223.0	25.6
1986												
1st quarter	3 793.2	2 491.1	2 063.5	387.9	1 675.6	427.6	273.1	154.5	245.6	18.3	227.4	22.0
2nd quarter	3 804.1	2 517.2	2 081.8	393.6	1 688.3	435.4	279.1	156.3	251.6	19.6	232.0	21.1
3rd quarter	3 842.3	2 555.2	2 111.5	399.4	1 712.0	443.8	284.9	158.9	264.4	26.6	237.8	16.6
4th quarter	3 885.3	2 604.7	2 152.5	407.9	1 744.6	452.2	290.4	161.8	264.2	26.6	237.6	13.4
1987												
1st quarter	3 953.9	2 647.8	2 192.7	413.5	1 779.1	455.1	292.4	162.7	275.4	27.3	248.2	14.5
2nd quarter	4 053.5	2 692.0	2 229.9	419.9	1 810.0	462.1	297.5	164.6	282.3	29.0	253.3	14.1
3rd quarter	4 146.6	2 740.3	2 270.8	425.1	1 845.7	469.5	302.4	167.1	289.5	28.8	260.7	17.5
4th quarter	4 239.9	2 809.6	2 331.7	433.9	1 897.8	477.8	307.2	170.6	298.6	30.4	268.2	20.3

Table 1-13B. National Income by Type of Income: Historical Data—*Continued*

(Billions of dollars, quarterly data are at seasonally adjusted annual rates.)   **NIPA Tables 1.7.5, 1.12**

Year and quarter	Corporate profits with IVA and CCAdj					Net interest and miscellaneous payments	Taxes on production and imports	Less: Subsidies	Business current transfer payments, net			Current surplus of government enterprises	Addendum: Net national factor income
	Total	Taxes on corporate income	Profits after tax						Total [1]	To persons	To government		
			Total	Net dividends	Undistributed corporate profits								
1973													
1st quarter	133.5	46.0	87.5	32.1	55.4	51.6	114.6	5.9	5.6	3.6	1.9	. . .	1 098.1
2nd quarter	131.2	46.0	85.2	33.5	51.7	53.4	116.2	5.7	5.9	3.9	2.0	. . .	1 121.1
3rd quarter	133.0	44.0	89.0	35.0	54.0	57.3	118.4	4.7	5.6	4.1	1.6	. . .	1 146.8
4th quarter	135.8	46.5	89.3	36.4	52.9	60.5	119.7	4.6	5.8	4.2	1.6	. . .	1 183.5
1974													
1st quarter	130.3	44.6	85.7	37.7	48.0	64.9	120.8	3.6	6.3	4.3	2.0	. . .	1 192.8
2nd quarter	128.9	46.7	82.3	38.8	43.5	69.8	124.1	2.9	6.6	4.5	2.1	. . .	1 208.8
3rd quarter	124.3	51.5	72.8	39.4	33.4	73.4	127.1	3.2	7.0	4.9	2.1	. . .	1 232.2
4th quarter	119.3	46.2	73.2	39.2	33.9	78.5	127.7	3.6	7.3	5.2	2.1	. . .	1 248.1
1975													
1st quarter	117.8	38.3	79.4	38.4	41.0	82.6	128.8	4.2	8.4	6.1	2.2	. . .	1 254.1
2nd quarter	128.9	41.4	87.5	38.0	49.5	83.1	133.0	4.3	9.1	6.9	2.2	. . .	1 279.9
3rd quarter	149.8	52.0	97.7	38.0	59.7	84.3	138.2	4.6	9.3	7.1	2.2	. . .	1 333.0
4th quarter	159.0	53.3	105.7	38.9	66.8	84.7	141.1	4.9	9.4	7.2	2.2	. . .	1 374.1
1976													
1st quarter	175.1	60.8	114.3	41.9	72.3	84.1	141.7	5.1	9.5	7.0	2.5	. . .	1 422.0
2nd quarter	173.2	59.4	113.8	43.9	69.9	86.6	144.9	4.8	9.4	7.0	2.4	. . .	1 444.4
3rd quarter	174.7	59.0	115.8	45.8	69.9	88.4	147.7	5.1	9.1	6.7	2.4	. . .	1 474.3
4th quarter	174.3	58.5	115.9	47.9	68.0	90.4	151.3	5.5	8.6	6.1	2.5	. . .	1 504.9
1977													
1st quarter	184.1	62.9	121.2	48.6	72.6	97.1	154.8	5.8	8.2	5.5	2.7	. . .	1 549.6
2nd quarter	204.9	68.4	136.5	49.9	86.6	101.0	158.0	5.9	7.8	5.0	2.8	. . .	1 613.0
3rd quarter	219.6	70.8	148.8	51.5	97.3	105.3	161.5	6.4	8.0	4.9	3.2	. . .	1 665.1
4th quarter	214.5	71.8	142.6	52.8	89.8	109.3	164.3	10.3	8.3	5.0	3.3	. . .	1 713.0
1978													
1st quarter	209.9	66.2	143.7	54.6	89.2	110.8	166.9	8.7	9.4	5.6	3.7	. . .	1 746.1
2nd quarter	239.7	80.0	159.7	56.1	103.7	113.7	173.1	8.4	10.0	6.2	3.9	. . .	1 834.3
3rd quarter	246.2	80.2	166.0	59.0	107.0	115.4	169.7	8.3	10.7	6.7	4.0	. . .	1 884.0
4th quarter	258.4	85.0	173.4	61.6	111.8	119.4	173.9	10.4	11.4	7.3	4.1	. . .	1 946.4
1979													
1st quarter	249.7	81.9	167.8	63.6	104.2	125.0	176.4	8.4	12.3	7.9	4.4	. . .	1 994.5
2nd quarter	252.1	82.1	170.0	65.8	104.3	130.8	178.5	8.8	12.7	8.2	4.5	. . .	2 032.5
3rd quarter	250.1	81.0	169.0	67.6	101.4	138.9	180.9	8.1	13.0	8.4	4.6	. . .	2 081.8
4th quarter	244.0	77.9	166.1	70.2	95.9	153.1	184.6	8.9	13.1	8.5	4.6	. . .	2 126.0
1980													
1st quarter	237.5	85.4	152.1	73.4	78.7	168.6	189.5	9.2	13.2	8.2	5.0	. . .	2 163.1
2nd quarter	207.0	64.9	142.1	76.6	65.5	176.1	196.9	9.6	13.4	8.3	5.1	. . .	2 159.8
3rd quarter	214.6	72.1	142.5	75.7	66.8	182.9	204.3	10.1	13.7	8.7	5.0	. . .	2 219.4
4th quarter	235.1	79.5	155.6	77.4	78.2	201.3	210.6	10.3	15.7	9.3	6.4	. . .	2 332.2
1981													
1st quarter	244.4	78.1	166.2	81.7	84.5	203.8	230.8	10.6	16.4	10.4	6.1	. . .	2 399.9
2nd quarter	240.8	69.1	171.8	86.3	85.5	223.7	235.5	10.7	16.6	11.0	5.6	. . .	2 439.0
3rd quarter	257.8	71.6	186.2	90.4	95.8	252.3	237.5	11.1	17.1	11.5	5.6	. . .	2 530.6
4th quarter	246.9	62.4	184.5	92.8	91.8	259.5	238.8	13.5	17.5	11.9	5.7	. . .	2 543.7
1982													
1st quarter	221.8	50.6	171.2	92.8	78.4	272.2	237.4	14.0	18.6	12.0	6.6	. . .	2 547.5
2nd quarter	236.2	52.7	183.5	92.2	91.2	282.1	238.3	13.6	19.2	12.3	6.9	. . .	2 593.1
3rd quarter	233.7	53.2	180.5	92.5	87.9	274.6	241.8	13.0	19.5	12.5	7.0	. . .	2 608.1
4th quarter	227.7	48.6	179.1	94.1	85.0	270.3	246.3	19.4	20.0	12.7	7.2	. . .	2 621.9
1983													
1st quarter	244.6	50.2	194.3	95.0	99.4	275.1	250.7	19.9	20.3	12.8	7.5	. . .	2 676.0
2nd quarter	274.7	65.3	209.4	96.0	113.4	277.4	261.2	21.6	20.9	13.2	7.7	. . .	2 745.7
3rd quarter	291.2	74.0	217.3	98.6	118.6	291.2	267.5	22.2	21.9	14.0	7.9	. . .	2 819.7
4th quarter	308.6	76.1	232.4	101.2	131.3	303.6	273.7	21.5	23.6	15.3	8.4	. . .	2 917.8
1984													
1st quarter	336.6	88.4	248.2	104.0	144.1	309.8	281.6	21.2	26.7	17.6	9.1	. . .	3 035.4
2nd quarter	338.2	87.1	251.1	106.6	144.5	327.6	287.7	21.0	28.6	19.2	9.4	. . .	3 114.5
3rd quarter	333.6	74.7	258.8	107.4	151.5	341.7	292.2	20.9	30.1	20.6	9.6	. . .	3 173.7
4th quarter	343.1	75.6	267.5	109.4	158.1	341.9	297.5	21.2	31.2	21.5	9.7	. . .	3 224.1
1985													
1st quarter	347.9	80.5	267.4	113.6	153.8	341.3	301.0	21.1	32.2	21.9	10.3	. . .	3 280.2
2nd quarter	351.5	78.8	272.7	115.2	157.6	337.3	305.7	21.0	38.1	22.1	16.0	. . .	3 316.3
3rd quarter	368.9	84.7	284.1	115.8	168.3	334.1	311.9	21.3	32.8	22.3	10.5	. . .	3 377.2
4th quarter	349.7	82.4	267.3	116.8	150.5	340.0	313.9	22.0	33.1	22.6	10.5	. . .	3 415.3
1986													
1st quarter	340.7	87.8	252.9	122.0	130.9	354.9	317.5	23.1	42.1	23.3	18.7	. . .	3 454.3
2nd quarter	326.6	88.4	238.2	124.3	113.9	355.3	319.5	24.2	34.5	23.5	11.1	. . .	3 471.8
3rd quarter	315.6	90.0	225.6	124.8	100.8	354.0	326.2	25.5	33.2	23.0	10.2	. . .	3 505.8
4th quarter	314.7	101.2	213.4	124.9	88.6	348.1	330.4	26.8	34.0	22.0	12.0	. . .	3 545.1
1987													
1st quarter	325.9	101.5	224.3	126.0	98.3	345.6	336.0	28.3	34.0	21.3	12.7	. . .	3 609.1
2nd quarter	362.9	115.1	247.8	127.5	120.2	351.0	344.4	30.4	34.4	20.6	13.8	. . .	3 702.4
3rd quarter	386.4	119.9	266.5	131.3	135.2	356.9	352.4	31.3	32.5	19.7	12.8	. . .	3 790.5
4th quarter	388.8	114.3	274.5	135.3	139.2	361.4	357.4	31.1	32.3	19.3	13.0	. . .	3 878.7

[1]Includes net transfer payments to the rest of the world, not shown separately.
. . . = Not available.

Table 1-13B. National Income by Type of Income: Historical Data—*Continued*

(Billions of dollars, quarterly data are at seasonally adjusted annual rates.)   **NIPA Tables 1.7.5, 1.12**

Year and quarter	National income, total	Compensation of employees							Proprietors' income with IVA and CCAdj			Rental income of persons with CCAdj
		Total	Wages and salaries			Supplements to wages and salaries			Total	Farm	Nonfarm	
			Total	Government	Other	Total	Employer contributions for:					
							Employee pension and insurance funds	Government social insurance				
1988												
1st quarter	4 329.2	2 857.6	2 366.2	441.7	1 924.4	491.4	312.1	179.2	319.2	33.3	285.9	21.1
2nd quarter	4 421.4	2 923.1	2 421.5	448.5	1 973.0	501.6	318.6	183.0	322.5	27.4	295.1	20.3
3rd quarter	4 514.3	2 975.8	2 462.9	454.5	2 008.3	513.0	326.8	186.2	333.6	28.8	304.8	21.2
4th quarter	4 621.4	3 035.4	2 508.6	463.4	2 045.2	526.9	336.8	190.0	326.8	17.7	309.2	27.4
1989												
1st quarter	4 697.6	3 079.1	2 540.8	469.5	2 071.3	538.3	348.1	190.1	348.1	36.7	311.4	23.1
2nd quarter	4 732.4	3 114.6	2 563.5	476.9	2 086.6	551.1	358.8	192.3	339.5	32.6	306.9	21.2
3rd quarter	4 785.4	3 155.0	2 592.3	485.1	2 107.1	562.8	368.2	194.6	337.2	30.3	306.9	20.0
4th quarter	4 825.0	3 209.7	2 635.7	493.0	2 142.7	574.1	376.4	197.7	339.7	32.5	307.2	21.8
1990												
1st quarter	4 927.3	3 272.8	2 684.4	503.5	2 180.9	588.3	386.3	202.0	345.6	34.5	311.1	23.1
2nd quarter	5 014.2	3 334.7	2 738.8	518.7	2 220.1	595.9	390.3	205.6	351.1	32.7	318.4	26.7
3rd quarter	5 042.9	3 372.5	2 769.3	523.9	2 245.5	603.2	394.7	208.5	358.9	31.8	327.2	31.6
4th quarter	5 070.6	3 381.5	2 772.2	530.1	2 242.1	609.3	399.6	209.6	357.0	29.7	327.3	31.3
1991												
1st quarter	5 093.1	3 387.5	2 771.7	539.6	2 232.1	615.8	405.2	210.6	347.5	26.2	321.3	33.0
2nd quarter	5 133.7	3 427.6	2 800.1	546.0	2 254.1	627.5	413.6	213.9	353.0	27.9	325.2	35.5
3rd quarter	5 185.5	3 469.9	2 828.1	551.6	2 276.5	641.8	425.1	216.8	353.6	24.5	329.1	39.4
4th quarter	5 245.1	3 517.0	2 858.0	557.8	2 300.2	659.0	439.7	219.3	362.6	28.6	333.9	46.4
1992												
1st quarter	5 369.1	3 595.7	2 912.9	564.2	2 348.7	682.8	456.8	226.0	378.7	32.9	345.8	50.0
2nd quarter	5 451.9	3 651.9	2 952.0	569.8	2 382.2	699.9	471.0	229.0	395.2	35.4	359.8	57.3
3rd quarter	5 492.3	3 688.5	2 976.6	575.2	2 401.4	712.0	481.4	230.6	407.9	36.7	371.1	64.3
4th quarter	5 587.6	3 736.9	3 020.8	578.8	2 442.0	716.1	488.1	228.0	418.8	34.1	384.8	70.7
1993												
1st quarter	5 596.5	3 745.6	3 015.9	582.8	2 433.1	729.6	491.7	238.0	418.9	28.6	390.3	80.7
2nd quarter	5 696.8	3 800.2	3 066.5	586.6	2 479.8	733.8	495.6	238.2	429.6	34.9	394.6	86.9
3rd quarter	5 744.2	3 834.1	3 093.6	591.8	2 501.9	740.5	500.2	240.3	423.4	26.1	397.4	92.5
4th quarter	5 883.6	3 889.3	3 141.4	594.8	2 546.5	747.9	505.7	242.2	440.0	35.8	404.2	100.2
1994												
1st quarter	5 949.8	3 913.4	3 152.7	601.1	2 551.6	760.7	511.4	249.3	450.2	41.8	408.4	109.0
2nd quarter	6 073.8	3 991.5	3 223.3	607.2	2 616.2	768.2	515.4	252.8	454.1	36.7	417.5	113.5
3rd quarter	6 163.9	4 031.8	3 258.7	612.5	2 646.2	773.0	517.6	255.5	456.8	32.1	424.6	116.3
4th quarter	6 271.1	4 088.1	3 311.6	617.2	2 694.4	776.5	517.4	259.0	465.5	28.1	437.4	115.9
1995												
1st quarter	6 331.9	4 134.4	3 358.1	622.4	2 735.7	776.3	515.6	260.8	467.4	20.3	447.1	118.3
2nd quarter	6 397.0	4 172.9	3 395.2	627.5	2 767.7	777.7	514.9	262.8	472.0	18.4	453.6	121.6
3rd quarter	6 497.6	4 219.7	3 438.9	630.9	2 808.0	780.8	515.6	265.3	484.5	21.4	463.1	125.7
4th quarter	6 582.7	4 265.3	3 479.9	635.3	2 844.6	785.4	517.8	267.7	501.0	28.1	472.9	134.0
1996												
1st quarter	6 701.0	4 314.5	3 524.4	640.0	2 884.4	790.1	521.0	269.1	523.3	36.9	486.4	139.5
2nd quarter	6 827.4	4 387.5	3 590.0	645.6	2 944.3	797.6	524.4	273.2	547.5	44.2	503.3	141.6
3rd quarter	6 909.7	4 451.2	3 647.2	650.7	2 996.5	804.0	527.3	276.7	547.5	33.5	514.0	143.7
4th quarter	7 044.3	4 514.5	3 704.3	656.0	3 048.2	810.2	530.0	280.3	556.8	34.7	522.1	145.3
1997												
1st quarter	7 159.1	4 592.4	3 776.4	662.3	3 114.1	816.0	532.7	283.3	575.6	37.8	537.8	144.2
2nd quarter	7 273.8	4 660.0	3 835.6	668.0	3 167.7	824.3	537.4	286.9	575.1	28.1	547.0	144.6
3rd quarter	7 422.5	4 738.9	3 903.0	674.1	3 228.9	835.9	544.7	291.2	588.5	32.4	556.1	147.4
4th quarter	7 544.4	4 844.0	3 992.4	683.2	3 309.2	851.6	554.7	297.0	596.6	31.2	565.4	152.1
1998												
1st quarter	7 641.9	4 942.0	4 074.9	689.8	3 385.1	867.1	566.7	300.4	617.4	27.9	589.5	156.8
2nd quarter	7 767.1	5 028.3	4 145.5	697.4	3 448.1	882.8	578.0	304.9	629.6	26.4	603.2	162.6
3rd quarter	7 898.7	5 114.9	4 217.4	705.6	3 511.8	897.4	587.9	309.5	645.0	26.6	618.4	168.0
4th quarter	7 995.3	5 199.4	4 288.7	712.4	3 576.2	910.7	596.6	314.1	668.7	33.0	635.7	173.4
1999												
1st quarter	8 142.8	5 288.0	4 365.7	718.1	3 647.6	922.3	604.1	318.2	681.9	33.6	648.3	176.1
2nd quarter	8 215.0	5 344.3	4 409.7	727.0	3 682.7	934.6	613.9	320.7	690.1	28.1	661.9	179.6
3rd quarter	8 309.7	5 423.3	4 473.0	738.6	3 734.5	950.3	626.3	324.0	699.8	26.2	673.6	179.4
4th quarter	8 494.2	5 555.5	4 583.6	751.4	3 832.3	971.8	641.4	330.5	713.9	24.5	689.4	178.9
2000												
1st quarter	8 730.1	5 750.0	4 752.3	765.2	3 987.1	997.7	657.1	340.6	720.9	27.1	693.9	179.8
2nd quarter	8 824.0	5 792.7	4 780.0	778.6	4 001.4	1 012.7	671.3	341.4	751.3	33.4	717.9	180.2
3rd quarter	8 942.1	5 904.9	4 872.3	783.8	4 088.4	1 032.7	684.6	348.1	760.8	31.9	728.9	182.9
4th quarter	8 994.3	5 944.7	4 898.9	791.5	4 107.4	1 045.8	694.9	350.8	782.4	33.7	748.7	191.0
2001												
1st quarter	9 137.5	6 050.8	4 980.1	801.8	4 178.4	1 070.7	712.9	357.8	809.8	34.0	775.7	197.1
2nd quarter	9 168.7	6 041.7	4 962.2	814.0	4 148.2	1 079.5	721.6	358.0	827.7	32.5	795.2	202.6
3rd quarter	9 121.3	6 025.4	4 936.8	828.9	4 107.9	1 088.6	731.2	357.4	845.6	32.6	813.0	204.8
4th quarter	9 149.4	6 038.6	4 938.6	843.3	4 095.3	1 100.0	741.3	358.7	840.9	29.2	811.7	205.1
2002												
1st quarter	9 257.0	6 061.1	4 946.2	855.1	4 091.1	1 114.9	753.2	361.6	859.7	17.9	841.8	209.1
2nd quarter	9 356.4	6 129.3	4 998.5	868.2	4 130.3	1 130.8	764.8	366.0	863.7	12.8	850.9	211.0
3rd quarter	9 417.0	6 157.8	5 013.0	879.8	4 133.2	1 144.8	777.3	367.5	872.4	21.3	851.2	210.4
4th quarter	9 554.9	6 194.2	5 027.5	888.4	4 139.2	1 166.6	797.6	369.0	883.2	27.7	855.5	213.6

Table 1-13B. National Income by Type of Income: Historical Data—*Continued*

(Billions of dollars, quarterly data are at seasonally adjusted annual rates.) NIPA Tables 1.7.5, 1.12

Year and quarter	Corporate profits with IVA and CCAdj					Net interest and miscellaneous payments	Taxes on production and imports	Less: Subsidies	Business current transfer payments, net			Current surplus of government enterprises	Addendum: Net national factor income
	Total	Taxes on corporate income	Profits after tax						Total ¹	To persons	To government		
			Total	Net dividends	Undistributed corporate profits								
1988													
1st quarter	394.2	112.6	281.6	138.5	143.1	369.1	365.2	30.4	31.1	19.6	11.5	. . .	3 961.1
2nd quarter	408.6	120.0	288.6	143.1	145.5	366.4	372.5	29.8	31.7	20.1	11.6	. . .	4 040.9
3rd quarter	415.2	129.1	286.1	150.3	135.8	380.4	377.5	29.2	33.2	20.8	12.4	. . .	4 126.3
4th quarter	439.8	135.4	304.5	157.3	147.2	395.5	382.6	28.6	35.1	21.9	13.2	. . .	4 225.0
1989													
1st quarter	423.7	137.2	286.5	169.2	117.3	416.3	391.0	28.0	37.2	23.3	14.0	. . .	4 290.2
2nd quarter	415.8	123.0	292.9	177.8	115.1	426.5	397.5	27.4	37.7	23.1	14.5	. . .	4 317.6
3rd quarter	416.6	118.9	297.7	182.5	115.2	433.4	403.9	27.1	39.1	23.2	15.9	. . .	4 362.2
4th quarter	401.0	118.6	282.4	189.0	93.4	430.3	403.0	27.3	39.0	23.1	16.0	. . .	4 402.6
1990													
1st quarter	413.9	116.8	297.1	192.7	104.4	435.0	419.5	27.1	39.4	23.3	16.1	. . .	4 490.3
2nd quarter	433.9	121.7	312.2	193.1	119.1	432.8	419.5	27.0	39.0	22.9	16.0	. . .	4 579.2
3rd quarter	407.7	125.1	282.6	194.7	87.9	430.0	426.8	26.9	38.9	21.9	17.0	. . .	4 600.8
4th quarter	415.4	123.7	291.7	190.1	101.6	435.9	434.2	27.0	39.5	20.7	18.8	. . .	4 621.1
1991													
1st quarter	455.0	120.2	334.8	196.7	138.1	407.7	444.0	27.1	39.8	19.2	20.6	. . .	4 630.6
2nd quarter	452.7	116.0	336.7	200.9	135.7	393.8	451.6	27.2	38.7	17.9	20.7	. . .	4 662.6
3rd quarter	450.7	116.9	333.8	203.6	130.2	390.9	461.3	27.5	38.5	17.0	21.5	. . .	4 704.6
4th quarter	452.0	118.2	333.8	203.9	129.9	374.8	471.5	28.1	38.7	16.4	22.2	. . .	4 752.7
1992													
1st quarter	478.2	130.4	347.8	202.9	144.8	370.3	476.4	28.6	38.6	16.5	22.4	-0.2	4 872.9
2nd quarter	479.6	132.4	347.3	205.5	141.7	367.0	481.2	29.2	38.4	16.5	22.0	-0.1	4 951.0
3rd quarter	465.2	127.8	337.4	207.3	130.1	360.6	486.0	30.4	38.9	16.3	22.7	-0.1	4 986.5
4th quarter	485.7	137.1	348.6	209.4	139.2	364.6	489.9	32.2	43.0	15.9	27.1	0.0	5 076.7
1993													
1st quarter	483.6	141.7	341.9	213.0	128.9	363.5	489.7	35.5	40.2	15.1	25.1	-0.1	5 092.4
2nd quarter	511.9	154.1	357.8	217.1	140.7	357.3	497.6	37.6	39.3	14.4	25.1	-0.2	5 185.8
3rd quarter	526.3	146.4	379.9	223.3	156.5	350.1	504.9	37.7	38.7	13.7	25.4	-0.4	5 226.5
4th quarter	576.5	178.0	398.5	231.8	166.7	341.5	520.3	36.0	39.3	13.1	26.9	-0.6	5 347.4
1994													
1st quarter	584.3	155.9	428.4	244.2	184.2	341.5	531.5	33.6	40.7	12.8	28.7	-0.8	5 398.4
2nd quarter	607.3	164.1	443.2	253.3	189.9	342.9	544.4	32.4	39.7	12.8	27.7	-0.8	5 509.3
3rd quarter	641.4	180.2	461.2	260.9	200.3	345.1	550.5	31.9	41.3	13.2	28.4	-0.4	5 591.3
4th quarter	666.0	190.4	475.6	267.4	208.3	359.6	554.6	32.2	41.2	14.2	26.8	0.2	5 695.2
1995													
1st quarter	667.8	194.7	473.0	270.6	202.4	362.6	555.3	34.0	43.6	16.0	25.9	1.7	5 750.5
2nd quarter	692.4	191.0	501.4	277.0	224.4	358.1	553.6	34.6	44.9	17.8	25.1	1.9	5 816.9
3rd quarter	726.4	198.0	528.4	285.4	243.0	355.1	558.9	35.1	45.8	19.6	24.6	1.6	5 911.4
4th quarter	738.3	194.0	544.3	296.4	247.9	353.7	563.8	35.5	45.5	21.4	23.9	0.2	5 992.2
1996													
1st quarter	772.3	201.7	570.6	306.2	264.4	352.5	570.4	35.5	46.8	22.8	24.7	-0.7	6 102.1
2nd quarter	784.8	213.4	571.4	318.7	252.6	358.2	577.7	35.4	48.2	23.5	25.8	-1.1	6 219.6
3rd quarter	790.0	213.7	576.3	329.9	246.4	363.6	581.8	35.2	48.9	23.2	26.8	-1.2	6 296.0
4th quarter	810.9	216.8	594.1	339.6	254.5	373.4	593.2	34.8	66.7	21.9	45.5	-0.8	6 401.0
1997													
1st quarter	839.0	218.2	620.8	347.7	273.2	381.8	595.7	34.4	46.9	19.7	27.8	-0.6	6 532.9
2nd quarter	861.6	222.5	639.2	356.2	282.9	389.8	610.4	33.6	47.4	18.7	28.5	0.1	6 631.1
3rd quarter	896.6	234.2	662.4	364.4	298.0	396.4	616.6	33.4	53.5	19.0	33.4	1.2	6 767.8
4th quarter	881.6	224.2	657.4	372.2	285.3	409.8	623.8	33.8	52.5	20.4	29.6	2.5	6 884.2
1998													
1st quarter	812.7	222.1	590.5	383.5	207.0	442.0	629.1	33.8	58.9	22.8	31.1	5.0	6 970.9
2nd quarter	809.4	218.9	590.5	386.4	204.1	459.0	635.5	35.0	60.4	25.1	30.4	4.9	7 088.9
3rd quarter	817.6	225.5	592.2	385.2	207.0	467.0	643.0	36.8	62.6	27.1	31.8	3.7	7 212.6
4th quarter	794.1	220.7	573.4	379.4	194.0	458.8	650.3	39.9	74.3	28.9	43.9	1.6	7 294.3
1999													
1st quarter	847.1	226.4	620.6	368.7	251.9	451.4	657.5	42.4	65.4	31.2	35.6	-1.4	7 444.5
2nd quarter	840.2	223.5	616.7	366.6	250.1	454.5	667.1	45.0	67.0	32.8	36.2	-2.0	7 508.6
3rd quarter	823.7	227.6	596.1	372.1	224.1	466.3	679.0	46.4	67.8	34.8	36.2	-3.2	7 592.5
4th quarter	828.7	231.9	596.8	386.6	210.2	487.0	690.7	46.9	71.1	37.1	37.8	-3.7	7 763.9
2000													
1st quarter	808.5	243.3	565.2	400.6	164.7	521.3	698.6	45.1	81.2	39.8	41.0	0.4	7 980.6
2nd quarter	798.1	241.7	556.5	412.0	144.5	542.0	707.3	45.5	84.2	41.5	41.8	0.9	8 064.3
3rd quarter	781.2	222.8	558.5	416.0	142.5	550.4	711.3	45.8	86.5	43.3	42.0	1.1	8 180.1
4th quarter	758.4	225.9	532.5	412.3	120.2	550.9	717.1	47.0	89.6	45.2	42.8	1.6	8 227.4
2001													
1st quarter	748.0	188.6	559.4	404.2	155.2	558.1	724.2	55.2	98.5	47.9	42.8	7.8	8 363.8
2nd quarter	778.0	182.6	595.3	396.5	198.9	551.5	724.1	62.0	100.5	48.3	43.2	8.9	8 401.5
3rd quarter	746.7	162.8	583.9	393.7	190.3	535.1	725.3	71.2	100.3	47.0	44.2	9.1	8 357.6
4th quarter	762.2	146.4	615.8	397.1	218.6	511.6	737.1	46.4	97.6	43.9	45.2	8.5	8 358.4
2002													
1st quarter	852.5	147.3	705.2	412.3	292.9	480.9	744.0	42.6	87.7	39.1	46.5	2.1	8 463.3
2nd quarter	883.2	153.6	729.6	424.3	305.3	470.8	751.3	39.8	81.4	35.2	45.9	0.4	8 558.0
3rd quarter	914.6	161.8	752.8	428.2	324.7	446.8	768.5	41.3	77.4	32.3	45.8	-0.6	8 602.0
4th quarter	996.5	179.8	816.7	434.7	382.1	446.9	776.3	41.9	76.2	30.2	46.8	-0.8	8 734.4

¹Includes net transfer payments to the rest of the world, not shown separately.
. . . = Not available.

Table 1-13B. National Income by Type of Income: Historical Data—*Continued*

(Billions of dollars, quarterly data are at seasonally adjusted annual rates.) NIPA Tables 1.7.5, 1.12

Year and quarter	National income, total	Compensation of employees							Proprietors' income with IVA and CCAdj			Rental income of persons with CCAdj
		Total	Wages and salaries			Supplements to wages and salaries			Total	Farm	Nonfarm	
			Total	Government	Other	Total	Employer contributions for:					
							Employee pension and insurance funds	Government social insurance				
2003												
1st quarter	9 593.5	6 213.6	5 027.4	901.2	4 126.3	1 186.2	812.2	374.0	874.9	28.4	846.4	223.0
2nd quarter	9 723.3	6 307.7	5 101.8	911.9	4 189.8	1 205.9	826.2	379.8	891.5	36.4	855.1	227.7
3rd quarter	9 881.9	6 394.1	5 169.1	917.8	4 251.4	1 225.0	839.9	385.0	905.0	39.4	865.5	232.0
4th quarter	10 046.0	6 500.8	5 256.5	924.9	4 331.6	1 244.3	852.8	391.4	916.2	41.6	874.5	243.3
2004												
1st quarter	10 225.4	6 542.0	5 275.2	936.6	4 338.6	1 266.8	868.0	398.8	945.7	55.4	890.3	247.5
2nd quarter	10 416.9	6 669.1	5 381.4	948.3	4 433.1	1 287.7	881.6	406.1	962.1	54.3	907.8	252.1
3rd quarter	10 614.6	6 804.3	5 493.4	956.6	4 536.7	1 310.9	897.1	413.9	963.2	46.1	917.1	248.5
4th quarter	10 711.8	6 864.8	5 536.4	967.9	4 568.5	1 328.4	912.2	416.2	977.1	50.2	926.8	247.4
2005												
1st quarter	10 937.4	6 926.9	5 575.3	977.8	4 597.5	1 351.6	930.6	421.1	956.6	47.1	909.5	240.1
2nd quarter	11 083.5	7 006.0	5 638.3	986.1	4 652.2	1 367.7	943.1	424.6	963.4	50.4	913.0	237.3
3rd quarter	11 265.5	7 122.9	5 738.6	995.5	4 743.0	1 384.4	953.4	430.9	986.8	48.2	938.5	225.6
4th quarter	11 508.4	7 210.6	5 815.4	1 005.8	4 809.6	1 395.2	959.5	435.6	1 005.4	41.4	963.9	225.0
2006												
1st quarter	11 799.7	7 380.2	5 971.9	1 014.6	4 957.3	1 408.3	964.7	443.6	1 048.1	33.8	1 014.2	216.7
2nd quarter	11 898.5	7 428.1	6 012.7	1 025.8	4 986.9	1 415.4	970.1	445.3	1 050.9	32.7	1 018.2	206.3
3rd quarter	12 011.3	7 487.2	6 061.8	1 041.9	5 019.8	1 425.4	978.5	447.0	1 047.7	32.4	1 015.3	199.5
4th quarter	12 085.8	7 624.0	6 181.4	1 055.8	5 125.7	1 442.6	989.2	453.3	1 051.9	33.3	1 018.5	186.8
2007												
1st quarter	12 183.1	7 806.8	6 344.5	1 071.6	5 272.9	1 462.2	1 002.5	459.7	1 016.2	41.8	974.4	174.2
2nd quarter	12 309.2	7 845.4	6 370.5	1 080.5	5 290.0	1 474.8	1 015.0	459.8	995.9	36.6	959.3	183.8
3rd quarter	12 304.0	7 885.1	6 397.4	1 093.6	5 303.9	1 487.6	1 026.8	460.9	985.1	37.2	947.9	187.5
4th quarter	12 365.3	7 978.2	6 474.6	1 108.5	5 366.1	1 503.6	1 037.3	466.3	978.9	45.4	933.5	191.9
2008												
1st quarter	12 377.2	8 055.3	6 539.8	1 124.8	5 415.0	1 515.5	1 045.8	469.7	964.7	48.6	916.1	220.5
2nd quarter	12 401.1	8 054.8	6 532.6	1 136.0	5 396.6	1 522.2	1 051.9	470.4	968.6	45.5	923.1	245.0
3rd quarter	12 440.9	8 074.0	6 546.7	1 151.2	5 395.6	1 527.2	1 054.4	472.8	958.2	35.5	922.8	268.1
4th quarter	12 083.9	8 043.8	6 517.9	1 163.4	5 354.4	1 526.0	1 053.3	472.7	952.2	31.2	921.0	293.0
2009												
1st quarter	11 863.4	7 729.8	6 225.0	1 162.8	5 062.2	1 504.8	1 048.8	456.0	914.3	24.0	890.3	303.4
2nd quarter	11 918.8	7 761.4	6 253.9	1 176.0	5 077.8	1 507.5	1 048.1	459.4	902.3	24.4	878.0	318.4
3rd quarter	12 056.5	7 746.2	6 236.9	1 179.0	5 057.9	1 509.3	1 051.5	457.8	935.3	27.4	907.9	338.5
4th quarter	12 270.2	7 796.7	6 278.6	1 183.1	5 095.6	1 518.1	1 058.9	459.2	1 002.2	36.5	965.7	349.0
2010												
1st quarter	12 402.8	7 767.2	6 232.8	1 185.8	5 047.0	1 534.4	1 070.3	464.1	1 056.9	31.5	1 025.4	371.8
2nd quarter	12 633.0	7 910.1	6 360.2	1 196.6	5 163.6	1 549.9	1 080.4	469.5	1 104.0	35.3	1 068.7	387.9
3rd quarter	12 894.2	7 981.8	6 422.1	1 191.3	5 230.8	1 559.7	1 088.9	470.8	1 129.4	41.3	1 088.2	399.5
4th quarter	13 013.3	8 040.7	6 473.3	1 191.1	5 282.2	1 567.4	1 095.9	471.5	1 144.4	47.8	1 096.7	417.6
2011												
1st quarter	13 092.4	8 170.0	6 577.2	1 193.1	5 384.2	1 592.8	1 102.0	490.8	1 181.0	66.8	1 114.2	453.4
2nd quarter	13 286.7	8 206.1	6 607.9	1 199.2	5 408.7	1 598.2	1 106.4	491.8	1 212.2	61.2	1 151.0	471.6
3rd quarter	13 446.7	8 287.1	6 681.5	1 197.6	5 483.8	1 605.7	1 109.5	496.2	1 253.5	66.4	1 187.1	485.0
4th quarter	13 605.2	8 240.4	6 637.1	1 189.5	5 447.6	1 603.3	1 111.1	492.2	1 270.4	65.1	1 205.3	504.6
2012												
1st quarter	13 999.6	8 455.9	6 835.8	1 198.7	5 637.2	1 620.1	1 112.1	508.0	1 322.8	60.9	1 261.9	509.9
2nd quarter	14 076.7	8 503.0	6 875.2	1 195.8	5 679.5	1 627.8	1 117.8	510.0	1 356.5	59.0	1 297.5	517.5
3rd quarter	14 043.4	8 545.3	6 905.3	1 196.8	5 708.5	1 640.0	1 128.8	511.2	1 345.8	60.3	1 285.5	520.5
4th quarter	14 258.9	8 762.6	7 093.6	1 202.0	5 891.6	1 669.0	1 145.1	523.9	1 364.2	63.4	1 300.8	524.0
2013												
1st quarter	14 310.7	8 731.7	7 043.0	1 206.4	5 836.6	1 688.7	1 168.4	520.4	1 388.3	94.1	1 294.2	537.9
2nd quarter	14 444.4	8 818.7	7 102.1	1 206.8	5 895.3	1 716.6	1 190.7	525.9	1 411.8	95.3	1 316.5	550.2
3rd quarter	14 521.8	8 844.7	7 112.0	1 204.8	5 907.2	1 732.7	1 205.9	526.8	1 412.5	94.3	1 318.2	564.6
4th quarter	14 701.9	8 941.7	7 195.6	1 214.0	5 981.6	1 746.2	1 214.0	532.1	1 401.7	69.6	1 332.0	575.2
2014												
1st quarter	14 826.7	9 113.0	7 357.4	1 221.9	6 135.6	1 755.6	1 216.1	539.5	1 413.4	68.3	1 345.1	588.4
2nd quarter	15 148.7	9 171.3	7 408.2	1 230.6	6 177.6	1 763.1	1 221.3	541.8	1 452.7	79.0	1 373.6	599.2
3rd quarter	15 436.0	9 279.6	7 502.0	1 241.5	6 260.5	1 777.6	1 229.9	547.7	1 468.3	66.1	1 402.2	609.9
4th quarter	15 558.5	9 432.5	7 633.3	1 253.7	6 379.6	1 799.2	1 242.5	556.7	1 456.6	65.9	1 390.7	621.0
2015												
1st quarter	15 641.7	9 563.3	7 745.0	1 263.2	6 481.8	1 818.4	1 257.0	561.4	1 432.9	54.0	1 378.9	623.1
2nd quarter	15 770.0	9 667.2	7 830.2	1 272.8	6 557.4	1 837.0	1 269.5	567.5	1 406.3	55.7	1 350.5	647.7
3rd quarter	15 851.5	9 748.7	7 898.2	1 280.1	6 618.1	1 850.5	1 278.3	572.2	1 425.9	58.9	1 367.0	658.1
4th quarter	15 888.3	9 813.4	7 953.3	1 286.4	6 666.9	1 860.0	1 284.2	575.8	1 423.7	55.5	1 368.2	663.5
2016												
1st quarter	15 925.2	9 843.5	7 982.8	1 294.2	6 688.5	1 860.7	1 286.5	574.2	1 415.2	36.5	1 378.7	669.9
2nd quarter	15 952.3	9 900.1	8 032.1	1 302.5	6 729.6	1 868.0	1 290.5	577.5	1 410.2	38.3	1 371.9	680.2
3rd quarter	16 083.1	9 993.2	8 112.2	1 314.1	6 798.1	1 881.1	1 298.0	583.1	1 429.5	36.5	1 393.0	683.6
4th quarter	16 253.9	10 104.5	8 206.9	1 321.2	6 885.7	1 897.5	1 307.5	590.0	1 440.0	31.2	1 408.9	692.1
2017												
1st quarter	16 475.3	10 227.6	8 310.6	1 331.4	6 979.2	1 917.0	1 320.4	596.6	1 494.8	44.5	1 450.3	707.4
2nd quarter	16 611.7	10 334.2	8 397.7	1 340.3	7 057.4	1 936.5	1 334.3	602.2	1 512.2	42.1	1 470.1	709.9
3rd quarter	16 753.1	10 456.7	8 497.9	1 353.0	7 144.9	1 958.8	1 350.8	607.9	1 523.1	34.1	1 489.0	722.0
4th quarter	16 995.2	10 628.0	8 642.0	1 367.2	7 274.9	1 985.9	1 370.0	615.9	1 542.9	31.8	1 511.1	736.0

Table 1-13B. National Income by Type of Income: Historical Data—*Continued*

(Billions of dollars, quarterly data are at seasonally adjusted annual rates.)   **NIPA Tables 1.7.5, 1.12**

Year and quarter	Corporate profits with IVA and CCAdj					Net interest and miscellaneous payments	Taxes on production and imports	Less: Subsidies	Business current transfer payments, net			Current surplus of government enterprises	Addendum: Net national factor income
	Total	Taxes on corporate income	Profits after tax						Total 1	To persons	To government		
			Total	Net dividends	Undistributed corporate profits								
2003													
1st quarter	988.6	199.9	788.7	435.3	353.4	467.2	788.6	47.1	77.7	29.6	47.1	0.9	8 767.2
2nd quarter	1 024.8	196.4	828.4	435.6	392.8	446.7	800.0	57.1	76.4	28.0	47.7	0.7	8 898.4
3rd quarter	1 075.7	217.6	858.1	457.8	400.3	428.0	813.0	45.9	75.9	25.5	48.6	1.8	9 034.8
4th quarter	1 135.9	241.0	894.9	495.4	399.5	396.5	820.9	46.0	75.1	22.0	49.8	3.3	9 192.6
2004													
1st quarter	1 242.4	251.4	991.0	532.7	458.3	362.7	847.3	44.2	80.1	17.5	51.3	11.3	9 340.2
2nd quarter	1 271.7	271.6	1 000.1	560.3	439.7	365.2	859.9	43.7	79.6	15.4	52.5	11.6	9 520.1
3rd quarter	1 330.4	292.7	1 037.7	567.2	470.5	364.5	871.3	45.4	78.5	15.7	53.0	9.8	9 711.0
4th quarter	1 312.9	298.4	1 014.5	668.5	346.0	380.2	893.8	52.3	89.7	18.4	52.6	18.6	9 782.3
2005													
1st quarter	1 442.1	375.8	1 066.2	575.2	491.1	421.7	915.1	56.7	94.6	23.7	52.0	18.9	9 987.4
2nd quarter	1 457.0	364.0	1 092.9	592.1	500.9	449.4	937.3	60.7	97.8	26.7	53.2	17.9	10 113.1
3rd quarter	1 483.0	370.3	1 112.7	598.4	514.3	469.5	952.1	62.0	93.2	27.3	53.5	12.4	10 287.8
4th quarter	1 572.5	409.1	1 163.3	642.3	521.0	508.5	965.3	64.2	91.1	25.6	55.0	10.5	10 521.9
2006													
1st quarter	1 629.6	421.7	1 208.0	714.1	493.9	523.0	981.8	55.7	79.5	21.6	56.0	2.0	10 797.6
2nd quarter	1 643.1	432.9	1 210.2	747.2	463.0	558.0	991.7	51.5	77.8	19.7	57.1	1.0	10 886.4
3rd quarter	1 688.7	451.5	1 237.1	765.6	471.5	558.4	1 004.1	49.9	83.9	19.9	58.0	6.0	10 981.5
4th quarter	1 623.6	415.6	1 208.1	793.4	414.7	562.9	1 010.5	48.7	85.2	22.1	58.9	4.2	11 049.2
2007													
1st quarter	1 528.8	418.9	1 109.9	820.3	289.6	592.6	1 025.9	49.5	101.1	26.5	59.2	15.3	11 118.7
2nd quarter	1 593.0	413.6	1 179.4	858.8	320.6	631.7	1 033.1	58.2	98.1	30.0	59.9	8.2	11 249.8
3rd quarter	1 525.9	376.8	1 149.1	868.1	281.0	655.0	1 035.8	55.9	99.1	32.5	60.8	5.8	11 238.6
4th quarter	1 485.0	359.0	1 126.0	866.8	259.2	655.2	1 052.6	54.7	94.8	34.1	63.7	-2.9	11 289.2
2008													
1st quarter	1 388.4	298.2	1 090.2	903.0	187.2	657.0	1 045.7	51.9	115.0	34.7	65.2	15.1	11 285.9
2nd quarter	1 364.6	285.5	1 079.2	866.5	212.7	675.7	1 054.7	51.7	107.9	35.4	66.1	6.3	11 308.8
3rd quarter	1 372.6	270.9	1 101.7	820.8	280.9	673.6	1 058.5	52.0	106.5	36.1	65.3	5.0	11 346.5
4th quarter	1 017.8	170.0	847.8	770.9	76.9	683.2	1 040.0	54.6	126.5	36.9	81.2	8.4	10 990.1
2009													
1st quarter	1 255.1	172.2	1 082.8	702.8	380.0	595.7	1 015.9	55.4	121.8	37.8	82.3	1.7	10 798.3
2nd quarter	1 283.8	195.6	1 088.2	620.7	467.4	566.7	1 017.3	55.5	140.1	38.6	99.1	2.4	10 832.6
3rd quarter	1 456.5	206.6	1 249.9	584.1	665.8	516.7	1 028.8	67.1	116.7	39.4	81.2	-4.0	10 993.2
4th quarter	1 551.9	242.3	1 309.6	580.5	729.1	477.9	1 045.3	55.5	119.0	40.2	81.8	-3.1	11 177.7
2010													
1st quarter	1 642.6	256.6	1 386.0	593.9	792.1	464.4	1 044.6	54.8	127.8	41.0	82.4	4.3	11 302.9
2nd quarter	1 640.8	262.5	1 378.3	613.0	765.3	477.8	1 062.1	55.5	125.6	42.4	81.4	1.8	11 520.7
3rd quarter	1 802.8	279.4	1 523.4	661.7	861.7	459.8	1 069.1	56.0	128.9	44.1	86.1	-1.3	11 773.4
4th quarter	1 828.8	291.6	1 537.2	704.3	832.9	458.8	1 076.4	56.9	125.0	47.4	86.5	-8.9	11 890.3
2011													
1st quarter	1 670.4	285.4	1 384.9	748.0	636.9	466.9	1 091.5	58.9	139.1	48.8	88.9	1.5	11 941.7
2nd quarter	1 791.8	285.4	1 506.4	761.0	745.3	454.0	1 105.5	59.9	125.2	49.5	87.2	-11.5	12 135.7
3rd quarter	1 818.5	256.7	1 561.7	787.5	774.2	451.3	1 103.9	60.2	126.7	48.9	87.9	-10.1	12 295.3
4th quarter	1 958.5	296.8	1 661.7	819.8	841.9	474.6	1 114.0	61.1	121.4	47.0	83.3	-8.9	12 448.5
2012													
1st quarter	2 025.6	320.1	1 705.5	879.7	825.8	516.8	1 130.9	58.4	112.2	43.5	77.0	-8.4	12 831.1
2nd quarter	2 006.9	334.5	1 672.4	902.1	770.3	530.3	1 133.9	58.1	102.1	41.0	69.9	-8.8	12 914.2
3rd quarter	1 985.4	342.0	1 643.4	906.9	736.6	495.0	1 131.3	56.3	91.3	39.1	63.0	-10.8	12 892.1
4th quarter	1 971.7	342.8	1 628.9	1 106.1	522.8	472.8	1 148.4	59.4	89.9	38.1	66.0	-14.2	13 095.3
2013													
1st quarter	1 983.5	360.8	1 622.7	870.8	751.9	467.2	1 174.6	59.4	102.1	37.8	78.8	-14.5	13 108.6
2nd quarter	2 000.2	357.3	1 642.9	1 127.8	515.2	453.8	1 180.8	60.1	105.3	37.9	80.6	-13.2	13 234.7
3rd quarter	2 011.1	364.9	1 646.2	958.4	687.9	465.7	1 195.0	60.0	104.4	38.5	77.3	-11.4	13 298.6
4th quarter	2 047.9	368.1	1 679.8	1 079.0	600.8	477.0	1 204.1	59.4	129.3	39.5	102.0	-12.2	13 443.6
2014													
1st quarter	1 967.4	403.7	1 563.8	1 052.4	511.3	485.4	1 220.5	58.7	109.8	40.9	81.8	-12.9	13 567.7
2nd quarter	2 138.3	425.8	1 712.4	1 093.2	619.3	504.4	1 237.5	58.5	115.3	42.2	82.3	-9.3	13 865.9
3rd quarter	2 191.0	398.3	1 792.7	1 105.6	687.2	532.8	1 248.4	58.2	174.6	43.6	137.9	-6.9	14 081.6
4th quarter	2 184.2	401.5	1 782.7	1 133.1	649.6	541.7	1 257.0	57.0	132.0	45.0	90.2	-3.2	14 236.0
2015													
1st quarter	2 132.0	418.8	1 713.1	1 150.9	562.2	564.0	1 263.8	56.0	126.0	46.3	85.3	-5.6	14 315.2
2nd quarter	2 103.7	420.1	1 683.7	1 131.4	552.3	596.5	1 275.0	56.4	136.5	48.4	92.4	-4.4	14 421.5
3rd quarter	2 062.3	389.1	1 673.2	1 157.2	516.1	602.9	1 277.8	57.8	138.5	51.3	87.7	-0.6	14 497.9
4th quarter	1 948.0	358.4	1 589.7	1 220.1	369.6	583.9	1 291.9	58.8	226.1	55.0	172.5	-1.5	14 432.6
2016													
1st quarter	2 022.2	373.3	1 649.0	1 168.9	480.1	573.9	1 297.1	60.7	166.2	59.5	103.9	2.8	14 524.7
2nd quarter	1 998.1	373.8	1 624.3	1 166.7	457.6	556.0	1 301.2	62.4	172.0	61.3	108.7	2.0	14 544.7
3rd quarter	2 013.0	391.7	1 621.3	1 183.3	438.0	552.6	1 322.3	63.1	154.7	60.6	90.7	3.4	14 671.9
4th quarter	2 012.6	371.5	1 641.0	1 184.8	456.2	557.6	1 330.5	61.0	179.9	57.2	111.8	10.9	14 806.7
2017													
1st quarter	1 995.4	322.8	1 672.5	1 219.5	453.1	607.7	1 340.8	59.9	163.6	51.2	113.5	-1.1	15 032.8
2nd quarter	2 008.0	314.1	1 693.9	1 246.8	447.1	603.6	1 355.1	58.7	149.5	47.6	90.7	11.2	15 168.0
3rd quarter	2 019.0	335.3	1 683.7	1 242.7	441.0	596.3	1 371.4	63.2	130.5	46.2	91.4	-7.1	15 317.0
4th quarter	2 001.4	305.4	1 696.0	1 249.5	446.5	624.5	1 390.6	62.8	138.0	47.3	92.8	-2.1	15 532.6

1Includes net transfer payments to the rest of the world, not shown separately.

Table 1-14. Gross and Net Value Added of Domestic Corporate Business

(Billions of dollars, quarterly data are at seasonally adjusted annual rates.)   **NIPA Table 1.14**

Year and quarter	Gross value added of corporate business, total	Consumption of fixed capital	Net value added											Gross value added of financial corporate business
			Total	Compensation of employees	Taxes on production and imports less subsidies	Net operating surplus								
						Total	Net interest and miscellaneous payments	Business current transfer payments	Corporate profits with IVA and CCAdj					
									Total	Taxes on corporate income	Profits after tax			
											Total	Net dividends	Undistributed	
1960	299.9	28.5	271.4	190.7	27.8	52.9	-0.2	1.5	51.5	21.9	29.7	12.3	17.4	17.4
1961	308.7	29.5	279.2	195.6	28.9	54.7	0.4	1.7	52.6	22.2	30.4	12.2	18.2	18.3
1962	336.0	30.8	305.1	211.0	31.2	62.9	0.8	1.9	60.3	23.3	37.0	13.2	23.8	19.1
1963	357.4	32.4	324.9	222.7	33.2	69.1	0.4	2.3	66.4	25.5	40.9	14.4	26.5	19.5
1964	386.0	34.4	351.6	239.2	35.6	76.8	0.8	2.7	73.2	26.6	46.7	16.6	30.1	21.5
1965	423.8	37.1	386.6	259.9	37.8	89.0	1.2	3.1	84.6	29.8	54.8	18.2	36.6	23.0
1966	465.1	40.9	424.2	288.5	38.9	96.8	2.3	2.9	91.6	32.2	59.4	19.5	39.9	25.0
1967	491.2	45.1	446.1	308.4	41.4	96.3	4.0	3.1	89.1	31.0	58.1	20.2	37.9	28.0
1968	542.1	50.0	492.1	340.2	47.8	104.1	4.3	3.7	96.1	37.2	58.9	22.6	36.3	31.2
1969	590.7	55.7	534.9	377.5	52.9	104.6	8.5	4.3	91.8	37.0	54.9	23.4	31.4	36.0
1970	612.1	61.8	550.3	398.0	57.0	95.3	12.5	3.7	79.1	31.3	47.9	23.9	24.0	39.2
1971	660.9	67.6	593.3	421.7	62.8	108.8	12.6	3.4	92.8	34.8	58.0	23.7	34.3	42.8
1972	732.9	73.4	659.5	468.2	67.3	124.0	12.4	3.9	107.7	39.1	68.6	25.2	43.4	47.0
1973	820.2	82.0	738.2	526.1	74.1	138.0	14.7	4.8	118.5	45.6	72.9	27.4	45.4	51.4
1974	889.6	96.0	793.5	577.3	78.6	137.6	23.1	6.3	108.2	47.2	61.0	29.0	32.0	59.5
1975	965.5	113.5	852.0	607.8	84.5	159.7	27.2	8.2	124.2	46.3	78.0	31.8	46.2	67.3
1976	1 087.3	125.0	962.3	682.8	91.4	188.1	22.7	7.6	157.8	59.4	98.4	35.9	62.4	72.6
1977	1 232.5	140.5	1 092.1	771.7	100.0	220.3	27.7	6.0	186.7	68.5	118.2	39.6	78.5	85.1
1978	1 406.9	159.3	1 247.6	884.7	108.7	254.2	30.6	7.9	215.7	77.9	137.8	46.6	91.2	103.0
1979	1 564.3	183.4	1 380.9	1 004.4	115.0	261.5	36.3	10.8	214.4	80.7	133.6	50.8	82.9	114.0
1980	1 702.2	212.1	1 490.1	1 102.0	128.6	259.6	59.4	12.0	188.1	75.5	112.6	59.0	53.6	127.9
1981	1 934.8	244.9	1 689.9	1 220.6	154.4	314.9	82.8	14.3	217.8	70.3	147.5	72.3	75.2	146.1
1982	2 018.4	272.5	1 745.9	1 275.1	161.3	309.4	95.5	16.6	197.3	51.3	146.0	76.5	69.4	163.7
1983	2 172.4	285.9	1 886.5	1 353.0	177.4	356.1	91.9	19.5	244.7	66.4	178.3	85.5	92.7	186.6
1984	2 434.6	303.2	2 131.5	1 501.1	195.6	434.8	106.6	26.9	301.3	81.5	219.8	94.6	125.3	208.8
1985	2 601.3	326.9	2 274.4	1 615.9	209.0	449.5	102.2	30.9	316.4	81.6	234.8	103.5	131.3	232.3
1986	2 706.6	349.9	2 356.7	1 723.4	219.6	413.7	99.5	29.3	284.9	91.9	193.0	106.1	86.9	241.9
1987	2 894.9	371.3	2 523.6	1 847.6	233.4	442.6	99.7	24.9	318.0	112.7	205.3	113.3	91.9	253.2
1988	3 142.4	399.9	2 742.5	2 002.3	252.0	488.2	104.4	26.4	357.5	124.3	233.2	115.4	117.8	268.6
1989	3 320.0	427.1	2 892.8	2 119.3	267.5	506.0	125.2	33.6	347.2	124.4	222.7	148.0	74.7	299.9
1990	3 467.1	454.7	3 012.5	2 234.9	284.5	493.1	117.4	34.0	341.6	121.8	219.8	167.7	52.1	306.6
1991	3 554.4	479.6	3 074.8	2 277.8	307.9	489.1	79.3	33.7	376.1	117.8	258.3	177.0	81.3	323.5
1992	3 741.9	494.3	3 247.6	2 420.5	325.9	501.2	63.2	34.0	404.1	131.9	272.2	178.3	93.9	365.9
1993	3 918.3	517.2	3 401.2	2 515.7	343.8	541.7	62.3	31.8	447.6	155.0	292.6	200.7	91.9	383.9
1994	4 209.4	547.6	3 661.8	2 649.0	376.1	636.7	57.7	32.2	546.8	172.7	374.1	218.3	155.9	393.5
1995	4 473.4	590.4	3 882.9	2 787.9	384.8	710.3	60.2	36.9	613.3	194.4	418.8	249.6	169.2	432.7
1996	4 769.1	626.3	4 142.8	2 953.7	398.4	790.7	59.4	43.7	687.5	211.4	476.1	283.2	192.9	473.8
1997	5 141.8	670.1	4 471.8	3 176.6	416.9	878.3	81.7	34.4	762.2	224.8	537.4	312.9	224.5	533.8
1998	5 473.4	715.0	4 758.4	3 447.5	430.8	880.2	121.6	52.8	705.7	221.8	483.9	341.4	142.5	600.1
1999	5 802.1	769.6	5 032.5	3 684.3	454.0	894.2	127.2	53.8	713.2	227.4	485.8	331.8	154.0	632.3
2000	6 214.6	838.6	5 376.0	4 008.9	479.8	887.2	174.0	72.4	640.9	233.4	407.4	380.8	26.7	702.0
2001	6 202.0	888.6	5 313.4	4 013.8	469.4	830.2	155.4	84.8	589.9	170.1	419.8	357.0	62.8	735.9
2002	6 293.7	914.6	5 379.1	3 972.7	497.5	908.8	92.3	61.6	754.9	160.7	594.3	376.8	217.5	754.3
2003	6 515.3	932.5	5 582.8	4 040.9	524.3	1 017.6	70.0	50.3	897.3	213.8	683.6	423.9	259.7	791.2
2004	6 948.5	973.0	5 975.5	4 240.2	569.4	1 165.9	16.9	54.8	1 094.2	278.5	815.7	519.8	295.8	825.5
2005	7 471.0	1 042.0	6 429.0	4 443.0	616.3	1 369.7	50.3	56.5	1 262.9	379.7	883.2	341.1	542.0	911.0
2006	8 008.0	1 119.3	6 888.8	4 681.2	654.5	1 553.0	98.9	47.6	1 406.5	430.1	976.4	677.2	299.2	1 005.6
2007	8 180.9	1 188.9	6 992.0	4 894.2	676.3	1 421.4	156.5	69.6	1 195.4	391.8	803.6	713.1	90.5	960.9
2008	8 094.5	1 256.6	6 837.9	4 940.3	685.0	1 212.6	220.8	96.2	895.7	255.9	639.7	659.8	-20.1	807.5
2009	7 804.2	1 263.8	6 540.4	4 607.5	655.5	1 277.4	141.4	98.0	1 038.0	203.9	834.1	503.9	330.2	939.3
2010	8 231.9	1 271.2	6 960.8	4 700.8	684.9	1 575.1	132.0	100.1	1 343.0	272.3	1 070.7	521.8	548.9	988.4
2011	8 627.8	1 324.6	7 303.2	4 928.0	718.7	1 656.5	151.0	108.3	1 397.2	280.8	1 116.5	624.2	492.3	1 012.1
2012	9 182.7	1 392.5	7 790.3	5 182.7	743.2	1 864.4	210.2	62.2	1 592.1	334.6	1 257.5	768.8	488.7	1 123.2
2013	9 478.0	1 450.2	8 027.8	5 352.4	793.1	1 882.3	202.4	68.0	1 611.9	362.6	1 249.2	870.6	378.6	1 103.5
2014	10 056.3	1 527.5	8 528.7	5 645.2	828.3	2 055.2	245.1	94.8	1 715.3	407.1	1 308.3	934.4	373.8	1 269.0
2015	10 470.1	1 588.2	8 881.9	5 944.4	838.1	2 099.4	309.9	124.3	1 665.3	396.3	1 269.0	1 002.1	266.9	1 340.9
2016	10 642.5	1 626.1	9 016.4	6 098.5	855.7	2 062.2	315.9	142.5	1 603.8	376.2	1 227.6	1 003.7	223.9	1 419.8
2017	11 094.6	1 695.8	9 398.8	6 418.7	886.6	2 093.4	355.2	121.0	1 617.3	311.3	1 306.0	1 053.7	252.2	1 458.8
2018	11 683.6	1 773.6	9 909.9	6 762.4	949.9	2 197.6	332.7	134.5	1 730.4	282.9	1 447.5	474.1	973.5	1 582.8
2019	12 056.5	1 867.9	10 188.6	7 075.9	978.2	2 134.4	258.1	131.2	1 745.1	298.7	1 446.4	895.8	550.6	1 598.3
2017														
1st quarter	10 899.1	1 669.2	9 229.9	6 286.6	870.7	2 072.6	344.0	137.1	1 591.5	313.3	1 278.1	1 036.1	242.0	1 393.8
2nd quarter	11 048.0	1 687.7	9 360.3	6 361.2	882.2	2 116.9	355.4	125.4	1 636.1	319.6	1 316.5	1 075.3	241.2	1 419.2
3rd quarter	11 141.2	1 705.2	9 436.0	6 453.9	889.8	2 092.3	360.9	106.2	1 625.3	322.1	1 303.1	999.5	303.6	1 485.4
4th quarter	11 290.1	1 721.2	9 568.9	6 573.0	903.9	2 092.0	360.5	115.3	1 616.3	290.2	1 326.1	1 104.1	221.9	1 536.9
2018														
1st quarter	11 482.4	1 741.6	9 740.8	6 658.0	933.7	2 149.1	354.2	124.9	1 670.0	255.8	1 414.2	30.9	1 383.3	1 523.9
2nd quarter	11 606.1	1 763.3	9 842.7	6 717.8	943.5	2 181.4	343.1	125.7	1 712.6	277.4	1 435.2	446.8	988.4	1 563.9
3rd quarter	11 800.6	1 785.3	10 015.3	6 815.8	953.9	2 245.6	327.1	152.9	1 765.6	288.2	1 477.4	882.5	594.8	1 625.7
4th quarter	11 845.3	1 804.3	10 041.0	6 858.0	968.7	2 214.3	306.3	134.5	1 773.5	310.1	1 463.4	536.1	927.3	1 617.8
2019														
1st quarter	11 923.3	1 831.6	10 091.7	7 019.6	967.2	2 104.8	280.7	127.4	1 696.8	294.6	1 402.2	840.2	562.0	1 577.8
2nd quarter	12 038.1	1 859.7	10 178.4	7 052.4	974.4	2 151.6	261.4	133.2	1 756.9	304.9	1 452.0	878.5	573.5	1 603.3
3rd quarter	12 055.1	1 881.2	10 174.0	7 070.3	982.7	2 120.9	248.5	140.5	1 731.9	283.0	1 448.9	895.7	553.2	1 581.8
4th quarter	12 209.5	1 899.3	10 310.2	7 161.3	988.4	2 160.5	242.0	123.9	1 794.6	312.3	1 482.3	968.7	513.6	1 630.2

Table 1-15. Gross Value Added of Nonfinancial Domestic Corporate Business in Current and Chained Dollars

(Billions of dollars, quarterly data are at seasonally adjusted annual rates.)   **NIPA Table 1.14**

Year and quarter	Current-dollar gross value added														Gross value added in billions of chained (2009) dollars
	Total	Consumption of fixed capital	Net value added												
			Total	Compensation of employees	Taxes on production and imports less subsidies	Net operating surplus									
						Total	Net interest and miscellaneous payments	Business current transfer payments	Corporate profits with IVA and CCAdj						
									Total	Taxes on corporate income	Profits after tax				
											Total	Net dividends	Undistributed		
1960	282.5	27.6	254.9	180.4	26.6	47.9	3.3	1.3	43.3	19.1	24.2	10.5	13.7		1 288.2
1961	290.4	28.6	261.9	184.5	27.6	49.7	3.8	1.4	44.5	19.4	25.1	10.6	14.5		1 319.1
1962	316.8	29.8	287.0	199.3	29.9	57.9	4.5	1.6	51.8	20.6	31.2	11.6	19.6		1 430.8
1963	337.8	31.3	306.5	210.1	31.7	64.7	4.8	1.6	58.3	22.8	35.5	12.4	23.1		1 519.0
1964	364.5	33.3	331.3	225.7	33.9	71.7	5.3	1.9	64.6	23.9	40.7	14.0	26.7		1 625.5
1965	400.7	35.8	364.9	245.4	36.0	83.5	6.0	2.1	75.4	27.1	48.3	16.2	32.1		1 762.0
1966	440.1	39.4	400.7	272.9	37.0	90.9	7.2	2.6	81.1	29.5	51.6	16.8	34.8		1 892.2
1967	463.2	43.5	419.7	291.1	39.3	89.4	8.6	2.7	78.1	27.8	50.2	17.3	33.0		1 946.4
1968	510.9	48.1	462.9	320.9	45.5	96.4	10.1	2.9	83.4	33.5	49.8	19.0	30.8		2 072.7
1969	554.6	53.5	501.1	356.1	50.2	94.9	13.5	3.0	78.3	33.3	45.0	19.0	25.9		2 155.6
1970	572.9	59.3	513.6	374.5	54.2	84.9	17.9	3.2	63.9	27.3	36.6	18.3	18.3		2 136.8
1971	618.1	64.7	553.4	396.2	59.5	97.7	18.9	3.6	75.2	30.0	45.2	18.1	27.0		2 221.0
1972	685.9	70.2	615.8	439.9	63.7	112.2	20.0	3.9	88.3	33.8	54.5	19.7	34.8		2 389.1
1973	768.8	78.2	690.7	495.1	70.1	125.5	23.7	4.5	97.2	40.4	56.8	20.8	36.1		2 534.8
1974	830.0	91.4	738.7	542.9	74.4	121.3	30.1	3.7	87.4	42.8	44.6	21.5	23.0		2 496.5
1975	898.3	107.7	790.6	569.0	80.2	141.3	32.4	4.7	104.3	41.9	62.4	24.6	37.8		2 461.9
1976	1 014.7	118.3	896.4	640.0	86.7	169.7	30.0	6.7	133.0	53.5	79.6	27.8	51.8		2 663.6
1977	1 147.4	132.6	1 014.8	723.3	94.6	196.9	33.0	8.7	155.1	60.6	94.6	30.9	63.6		2 860.6
1978	1 303.9	150.2	1 153.8	829.5	102.7	221.6	36.6	9.2	175.8	67.6	108.2	35.9	72.3		3 046.3
1979	1 450.3	172.4	1 277.8	942.4	108.8	226.7	43.6	9.0	174.0	70.6	103.4	37.6	65.8		3 144.1
1980	1 574.4	198.9	1 375.5	1 030.7	121.5	223.3	58.0	9.6	155.7	68.2	87.5	44.7	42.8		3 113.9
1981	1 788.7	229.1	1 559.6	1 139.9	146.7	273.1	72.5	10.7	189.9	66.0	123.9	52.5	71.4		3 240.6
1982	1 854.7	253.9	1 600.8	1 183.3	152.9	264.6	83.2	8.1	173.3	48.8	124.5	54.1	70.5		3 170.0
1983	1 985.8	264.6	1 721.2	1 250.1	168.0	303.1	82.6	10.0	210.5	61.7	148.8	63.2	85.6		3 324.9
1984	2 225.9	279.0	1 946.9	1 388.2	185.0	373.7	93.8	10.9	268.9	75.9	193.0	67.2	125.8		3 621.0
1985	2 369.0	299.0	2 070.0	1 490.1	196.6	383.3	96.9	15.4	271.1	71.1	200.0	72.0	128.0		3 787.4
1986	2 464.7	317.9	2 146.8	1 578.2	204.6	364.1	107.0	26.9	230.2	76.2	154.0	72.9	81.1		3 886.1
1987	2 641.7	334.7	2 307.0	1 685.5	216.8	404.7	115.7	29.7	259.4	94.2	165.1	76.3	88.9		4 092.1
1988	2 873.7	358.2	2 515.5	1 825.3	233.8	456.4	135.4	27.0	294.0	104.0	190.0	82.2	107.8		4 343.6
1989	3 020.0	380.2	2 639.9	1 934.8	248.2	456.9	159.4	23.5	274.0	101.2	172.8	105.4	67.5		4 426.2
1990	3 160.5	402.6	2 757.9	2 037.5	263.5	456.8	169.5	24.9	262.5	98.5	164.0	118.3	45.7		4 490.3
1991	3 231.0	423.6	2 807.4	2 071.1	285.7	450.6	156.4	26.1	268.1	88.6	179.5	125.5	54.0		4 467.0
1992	3 376.1	436.3	2 939.8	2 188.7	302.5	448.6	130.9	30.7	287.1	94.4	192.6	134.3	58.4		4 602.9
1993	3 534.4	455.8	3 078.6	2 271.0	319.3	488.2	117.0	29.5	341.7	108.0	233.7	149.2	84.5		4 716.7
1994	3 815.9	482.3	3 333.5	2 398.7	350.7	584.1	116.3	34.7	433.2	132.4	300.8	158.0	142.9		5 007.3
1995	4 040.6	520.8	3 519.8	2 524.6	358.7	636.6	125.6	30.2	480.8	140.3	340.5	178.0	162.5		5 249.1
1996	4 295.3	554.2	3 741.1	2 667.7	371.7	701.7	118.9	37.3	545.5	152.9	392.6	197.6	195.1		5 556.6
1997	4 608.0	594.6	4 013.4	2 862.6	388.9	762.0	126.6	38.5	596.9	161.4	435.5	215.9	219.6		5 928.1
1998	4 873.3	634.1	4 239.2	3 093.8	402.9	742.5	146.4	34.1	562.0	158.7	403.3	241.0	162.3		6 250.7
1999	5 169.8	679.1	4 490.7	3 310.0	424.6	756.1	159.8	45.6	550.7	166.5	384.3	224.7	159.6		6 570.1
2000	5 512.6	737.5	4 775.2	3 597.3	449.9	728.0	198.0	45.4	484.5	165.1	319.4	251.3	68.1		6 886.2
2001	5 466.1	781.5	4 684.6	3 582.3	441.5	660.8	219.8	55.2	385.8	106.2	279.6	245.4	34.2		6 713.9
2002	5 539.3	802.9	4 736.4	3 540.5	467.0	728.9	199.8	52.0	477.1	91.1	386.0	254.8	131.2		6 782.8
2003	5 724.1	816.1	4 907.9	3 594.3	491.1	822.5	171.2	61.6	589.6	126.5	463.2	293.4	169.8		6 930.6
2004	6 123.0	849.7	5 273.3	3 761.3	531.7	980.2	162.5	62.4	755.3	179.2	576.1	364.5	211.6		7 273.1
2005	6 560.0	911.0	5 649.0	3 928.7	575.7	1 144.6	176.9	79.1	888.7	262.5	626.2	170.8	455.4		7 537.3
2006	7 002.5	981.5	6 020.9	4 127.5	611.8	1 281.7	186.9	68.3	1 026.4	296.1	730.4	471.1	259.3		7 825.7
2007	7 220.0	1 044.0	6 176.0	4 307.1	631.3	1 237.6	250.3	58.8	928.5	277.4	651.1	484.6	166.5		7 896.4
2008	7 287.0	1 105.0	6 182.0	4 364.3	637.4	1 180.3	310.2	46.1	824.0	207.8	616.2	474.2	142.0		7 809.5
2009	6 864.9	1 110.7	5 754.1	4 094.9	608.5	1 050.7	284.2	60.3	706.2	162.1	544.1	351.4	192.7		7 255.4
2010	7 243.5	1 119.8	6 123.7	4 166.6	638.2	1 318.9	283.3	71.6	964.0	203.8	760.2	375.5	384.7		7 568.4
2011	7 615.7	1 169.3	6 446.3	4 372.7	670.8	1 402.8	283.2	79.7	1 039.9	209.0	830.9	441.0	389.9		7 774.4
2012	8 059.5	1 230.6	6 828.9	4 608.3	695.1	1 525.5	291.2	82.1	1 152.1	245.4	906.7	517.9	388.7		8 059.5
2013	8 374.5	1 279.7	7 094.8	4 768.1	742.0	1 584.7	280.1	84.3	1 220.2	263.5	956.7	531.9	424.8		8 261.9
2014	8 787.3	1 349.2	7 438.0	5 026.2	767.6	1 644.2	294.0	80.4	1 269.8	290.7	979.1	597.5	381.5		8 524.0
2015	9 129.2	1 401.7	7 727.5	5 292.6	772.8	1 662.2	309.0	95.5	1 257.7	283.2	974.5	641.1	333.4		8 776.5
2016	9 222.7	1 429.3	7 793.4	5 429.6	787.3	1 576.6	327.9	68.6	1 180.1	261.9	918.2	690.7	227.5		8 796.6
2017	9 635.8	1 489.3	8 146.5	5 703.9	825.3	1 617.3	331.8	86.1	1 199.3	224.2	975.2	720.7	254.4		9 026.1
2018	10 100.8	1 557.1	8 543.6	6 018.3	873.6	1 651.8	262.0	83.4	1 306.4	196.4	1 109.9	259.8	850.1		9 260.6
2019	10 458.2	1 639.2	8 819.0	6 300.9	914.7	1 603.5	243.3	77.1	1 283.0	212.4	1 070.7	636.6	434.1		9 405.1
2017															
1st quarter	9 505.3	1 465.8	8 039.5	5 618.6	810.5	1 610.4	342.8	77.5	1 190.1	232.3	957.8	716.6	241.1		8 959.0
2nd quarter	9 628.8	1 481.9	8 146.9	5 666.7	821.2	1 658.9	341.0	93.8	1 224.1	234.8	989.3	746.7	242.6		9 048.5
3rd quarter	9 655.9	1 497.9	8 157.9	5 726.6	828.2	1 603.1	330.9	87.1	1 185.1	229.1	956.0	666.4	289.6		9 029.1
4th quarter	9 753.2	1 511.6	8 241.5	5 803.5	841.4	1 596.6	312.5	86.0	1 198.1	200.5	997.5	753.2	244.3		9 067.3
2018															
1st quarter	9 958.5	1 529.6	8 428.9	5 951.0	858.7	1 619.1	285.8	92.1	1 241.3	182.0	1 059.3	-176.9	1 236.2		9 214.6
2nd quarter	10 042.2	1 548.2	8 494.0	5 989.6	867.8	1 636.7	265.6	84.4	1 286.7	197.6	1 089.1	213.6	875.5		9 219.0
3rd quarter	10 174.9	1 567.1	8 607.8	6 052.7	877.3	1 677.8	251.9	81.1	1 344.8	205.6	1 139.1	641.8	497.3		9 302.2
4th quarter	10 227.5	1 583.7	8 643.8	6 079.9	890.4	1 673.4	244.7	76.0	1 352.7	200.6	1 152.1	360.8	791.3		9 305.8
2019															
1st quarter	10 345.5	1 607.8	8 737.7	6 264.7	904.5	1 568.5	244.0	81.8	1 242.6	212.5	1 030.1	606.1	424.1		9 373.4
2nd quarter	10 434.8	1 632.1	8 802.7	6 285.8	911.4	1 605.6	243.5	68.9	1 293.2	219.7	1 073.5	605.6	467.9		9 397.8
3rd quarter	10 473.3	1 650.4	8 822.9	6 297.3	918.7	1 606.9	243.1	88.6	1 275.2	195.6	1 079.5	672.8	406.8		9 394.4
4th quarter	10 579.3	1 666.6	8 912.7	6 355.7	924.1	1 632.9	242.8	68.9	1 321.2	221.7	1 099.5	661.8	437.8		9 454.3

Table 1-16. Shares of Income

(Billions of dollars, percent, quarterly data are at seasonally adjusted annual rates.)   **NIPA Tables 1.10, 1.14**

Year and quarter	Gross domestic income, total economy									Net factor income, domestic corporate business				
	Billions of dollars					Percent of total				Billions of dollars			Percent of total	
	Total	Compensation of employees	Taxes on production and imports less subsidies	Net operating surplus	Consumption of fixed capital	Compensation of employees	Taxes on production and imports less subsidies	Net operating surplus	Consumption of fixed capital	Total	Compensation of employees	Net operating surplus	Compensation	Net operating surplus
1959	521.5	285.9	40.0	130.1	65.4	54.8	7.7	24.9	12.5	234.3	180.3	54.0	77.0	23.0
1960	543.7	301.4	43.4	130.9	67.9	55.4	8.0	24.1	12.5	243.6	190.7	52.9	78.3	21.7
1961	563.1	310.6	45.0	137.0	70.6	55.2	8.0	24.3	12.5	250.3	195.6	54.7	78.1	21.9
1962	603.9	332.3	48.1	149.4	74.1	55.0	8.0	24.7	12.3	273.9	211.0	62.9	77.0	23.0
1963	638.9	350.5	51.2	159.3	78.0	54.9	8.0	24.9	12.2	291.8	222.7	69.1	76.3	23.7
1964	684.5	376.0	54.6	171.6	82.4	54.9	8.0	25.1	12.0	316.0	239.2	76.8	75.7	24.3
1965	741.5	405.4	57.7	190.4	88.0	54.7	7.8	25.7	11.9	348.9	259.9	89.0	74.5	25.5
1966	808.3	449.2	59.3	204.5	95.3	55.6	7.3	25.3	11.8	385.3	288.5	96.8	74.9	25.1
1967	856.6	481.8	64.1	207.2	103.6	56.2	7.5	24.2	12.1	404.7	308.4	96.3	76.2	23.8
1968	937.5	530.7	72.2	221.2	113.4	56.6	7.7	23.6	12.1	444.3	340.2	104.1	76.6	23.4
1969	1 016.0	584.4	79.4	227.4	124.9	57.5	7.8	22.4	12.3	482.1	377.5	104.6	78.3	21.7
1970	1 068.0	623.3	86.6	221.2	136.8	58.4	8.1	20.7	12.8	493.3	398.0	95.3	80.7	19.3
1971	1 155.3	665.0	95.8	245.6	148.9	57.6	8.3	21.3	12.9	530.5	421.7	108.8	79.5	20.5
1972	1 271.9	731.3	101.3	278.3	161.0	57.5	8.0	21.9	12.7	592.2	468.2	124.0	79.1	20.9
1973	1 419.2	812.7	112.0	315.9	178.7	57.3	7.9	22.3	12.6	664.1	526.1	138.0	79.2	20.8
1974	1 537.8	887.7	121.6	321.6	206.9	57.7	7.9	20.9	13.5	714.9	577.3	137.6	80.8	19.2
1975	1 671.6	947.3	130.8	355.0	238.5	56.7	7.8	21.2	14.3	767.5	607.8	159.7	79.2	20.8
1976	1 852.8	1 048.4	141.3	402.9	260.2	56.6	7.6	21.7	14.0	870.9	682.8	188.1	78.4	21.6
1977	2 062.4	1 165.9	152.6	454.1	289.8	56.5	7.4	22.0	14.1	992.0	771.7	220.3	77.8	22.2
1978	2 328.3	1 316.9	162.0	523.3	327.2	56.6	7.0	22.4	14.1	1 138.9	884.7	254.2	77.7	22.3
1979	2 582.3	1 477.3	171.6	559.5	373.9	57.2	6.6	21.7	14.5	1 265.9	1 004.4	261.5	79.3	20.7
1980	2 812.9	1 622.3	190.5	571.6	428.4	57.7	6.8	20.3	15.2	1 361.6	1 102.0	259.6	80.9	19.1
1981	3 169.0	1 792.7	224.1	664.9	487.2	56.6	7.1	21.0	15.4	1 535.5	1 220.6	314.9	79.5	20.5
1982	3 335.0	1 893.2	225.9	678.9	537.0	56.8	6.8	20.4	16.1	1 584.5	1 275.1	309.4	80.5	19.5
1983	3 577.1	2 012.7	242.0	759.8	562.6	56.3	6.8	21.2	15.7	1 709.1	1 353.0	356.1	79.2	20.8
1984	3 996.0	2 216.1	268.7	912.8	598.4	55.5	6.7	22.8	15.0	1 935.9	1 501.1	434.8	77.5	22.5
1985	4 284.7	2 387.5	286.7	970.3	640.1	55.7	6.7	22.6	14.9	2 065.4	1 615.9	449.5	78.2	21.8
1986	4 499.6	2 543.8	298.5	972.0	685.3	56.5	6.6	21.6	15.2	2 137.1	1 723.4	413.7	80.6	19.4
1987	4 811.4	2 723.8	317.2	1 040.0	730.4	56.6	6.6	21.6	15.2	2 290.2	1 847.6	442.6	80.7	19.3
1988	5 233.4	2 948.9	345.0	1 155.1	784.5	56.3	6.6	22.1	15.0	2 490.5	2 002.3	488.2	80.4	19.6
1989	5 573.6	3 140.9	371.5	1 223.0	838.3	56.4	6.7	21.9	15.0	2 625.3	2 119.3	506.0	80.7	19.3
1990	5 867.6	3 342.7	398.0	1 238.4	888.5	57.0	6.8	21.1	15.1	2 728.0	2 234.9	493.1	81.9	18.1
1991	6 065.2	3 453.3	429.6	1 249.9	932.4	56.9	7.1	20.6	15.4	2 766.9	2 277.8	489.1	82.3	17.7
1992	6 404.4	3 671.2	453.3	1 319.6	960.2	57.3	7.1	20.6	15.0	2 921.7	2 420.5	501.2	82.8	17.2
1993	6 702.6	3 820.6	466.4	1 412.1	1 003.5	57.0	7.0	21.1	15.0	3 057.4	2 515.7	541.7	82.3	17.7
1994	7 147.3	4 010.2	512.7	1 568.7	1 055.6	56.1	7.2	21.9	14.8	3 285.7	2 649.0	636.7	80.6	19.4
1995	7 546.7	4 202.2	523.1	1 699.1	1 122.4	55.7	6.9	22.5	14.9	3 498.2	2 787.9	710.3	79.7	20.3
1996	8 015.0	4 421.1	545.6	1 873.1	1 175.3	55.2	6.8	23.4	14.7	3 744.4	2 953.7	790.7	78.9	21.1
1997	8 566.0	4 713.2	577.8	2 035.6	1 239.3	55.0	6.7	23.8	14.5	4 054.9	3 176.6	878.3	78.3	21.7
1998	9 118.1	5 075.7	603.1	2 129.5	1 309.7	55.7	6.6	23.4	14.4	4 327.7	3 447.5	880.2	79.7	20.3
1999	9 663.7	5 409.9	628.4	2 226.4	1 398.9	56.0	6.5	23.0	14.5	4 578.5	3 684.3	894.2	80.5	19.5
2000	10 348.6	5 854.6	662.8	2 320.1	1 511.2	56.6	6.4	22.4	14.6	4 896.1	4 008.9	887.2	81.9	18.1
2001	10 695.0	6 046.3	669.0	2 380.1	1 599.5	56.5	6.3	22.3	15.0	4 844.0	4 013.8	830.2	82.9	17.1
2002	11 009.1	6 143.4	718.6	2 489.2	1 658.0	55.8	6.5	22.6	15.1	4 881.5	3 972.7	908.8	81.4	18.6
2003	11 471.7	6 362.3	756.5	2 633.8	1 719.1	55.5	6.6	23.0	15.0	5 058.5	4 040.9	1 017.6	79.9	20.1
2004	12 236.1	6 729.3	821.7	2 863.2	1 821.8	55.0	6.7	23.4	14.9	5 406.1	4 240.2	1 165.9	78.4	21.6
2005	13 089.3	7 077.7	881.5	3 159.1	1 971.0	54.1	6.7	24.1	15.1	5 812.7	4 443.0	1 369.7	76.4	23.6
2006	14 021.8	7 491.3	945.5	3 460.9	2 124.1	53.4	6.7	24.7	15.1	6 234.2	4 681.2	1 553.0	75.1	24.9
2007	14 434.5	7 889.4	982.2	3 310.0	2 252.8	54.7	6.8	22.9	15.6	6 315.6	4 894.2	1 421.4	77.5	22.5
2008	14 530.3	8 068.7	997.1	3 105.6	2 358.8	55.5	6.9	21.4	16.2	6 152.9	4 940.3	1 212.6	80.3	19.7
2009	14 259.0	7 767.2	968.5	3 151.9	2 371.5	54.5	6.8	22.1	16.6	5 884.9	4 607.5	1 277.4	78.3	21.7
2010	14 932.1	7 933.0	1 007.3	3 600.9	2 390.9	53.1	6.7	24.1	16.0	6 275.9	4 700.8	1 575.1	74.9	25.1
2011	15 599.4	8 234.0	1 043.7	3 847.2	2 474.5	52.8	6.7	24.7	15.9	6 584.5	4 928.0	1 656.5	74.8	25.2
2012	16 436.8	8 575.4	1 078.1	4 207.4	2 576.0	52.2	6.6	25.6	15.7	7 047.1	5 182.7	1 864.4	73.5	26.5
2013	16 941.8	8 843.6	1 129.0	4 288.0	2 681.2	52.2	6.7	25.3	15.8	7 234.7	5 352.4	1 882.3	74.0	26.0
2014	17 813.9	9 259.7	1 182.7	4 556.5	2 815.0	52.0	6.6	25.6	15.8	7 700.4	5 645.2	2 055.2	73.3	26.7
2015	18 475.6	9 709.5	1 218.0	4 636.7	2 911.4	52.6	6.6	25.1	15.8	8 043.8	5 944.4	2 099.4	73.9	26.1
2016	18 837.3	9 974.8	1 250.1	4 625.8	2 986.6	53.0	6.6	24.6	15.9	8 160.7	6 098.5	2 062.2	74.7	25.3
2017	19 674.4	10 433.1	1 304.1	4 824.4	3 112.9	53.0	6.6	24.5	15.8	8 512.1	6 418.7	2 093.4	75.4	24.6
2018	20 669.9	10 960.6	1 381.5	5 062.8	3 265.0	53.0	6.7	24.5	15.8	8 960.0	6 762.4	2 197.6	75.5	24.5
2019	21 420.4	11 444.5	1 417.5	5 137.4	3 420.9	53.4	6.6	24.0	16.0	9 210.3	7 075.9	2 134.4	76.8	23.2
2017														
1st quarter	19 352.2	10 233.3	1 280.5	4 775.6	3 062.8	52.9	6.6	24.7	15.8	8 359.2	6 286.6	2 072.6	75.2	24.8
2nd quarter	19 572.2	10 345.9	1 299.3	4 829.9	3 097.2	52.9	6.6	24.7	15.8	8 478.1	6 361.2	2 116.9	75.0	25.0
3rd quarter	19 750.4	10 487.6	1 306.8	4 824.8	3 131.2	53.1	6.6	24.4	15.9	8 546.2	6 453.9	2 092.3	75.5	24.5
4th quarter	20 022.8	10 665.6	1 329.8	4 867.1	3 160.3	53.3	6.6	24.3	15.8	8 665.0	6 573.0	2 092.0	75.9	24.1
2018														
1st quarter	20 319.8	10 786.4	1 360.1	4 968.4	3 204.8	53.1	6.7	24.5	15.8	8 807.1	6 658.0	2 149.1	75.6	24.4
2nd quarter	20 533.2	10 892.6	1 375.3	5 016.8	3 248.5	53.0	6.7	24.4	15.8	8 899.2	6 717.8	2 181.4	75.5	24.5
3rd quarter	20 847.3	11 044.9	1 390.8	5 124.8	3 286.8	53.0	6.7	24.6	15.8	9 061.4	6 815.8	2 245.6	75.2	24.8
4th quarter	20 979.1	11 118.5	1 399.5	5 141.2	3 319.9	53.0	6.7	24.5	15.8	9 072.3	6 858.0	2 214.3	75.6	24.4
2019														
1st quarter	21 147.8	11 346.9	1 402.5	5 036.8	3 361.6	53.7	6.6	23.8	15.9	9 124.4	7 019.6	2 104.8	76.9	23.1
2nd quarter	21 347.2	11 403.6	1 419.6	5 119.6	3 404.4	53.4	6.7	24.0	15.9	9 204.0	7 052.4	2 151.6	76.6	23.4
3rd quarter	21 465.3	11 450.1	1 419.6	5 152.4	3 443.3	53.3	6.6	24.0	16.0	9 191.2	7 070.3	2 120.9	76.9	23.1
4th quarter	21 721.2	11 577.4	1 428.5	5 240.8	3 474.4	53.3	6.6	24.1	16.0	9 321.8	7 161.3	2 160.5	76.8	23.2

NOTES AND DEFINITIONS, CHAPTER 1

TABLES 1-1 THROUGH 1-16

National Income and Product

SOURCE: U.S. DEPARTMENT OF COMMERCE, BUREAU OF ECONOMIC ANALYSIS (BEA)

The mission of the Bureau of Economic Analysis (BEA) of the Department of Commerce is to promote a better understanding of the U.S. economy by providing timely, relevant, and accurate economic accounts. The BEA quarterly and annually collects and produces data on how the U.S. economy is functioning. Featured in the NIPAs is gross domestic product (GDP), which measures the value of the goods and services produced by the U.S. economy in a given time period. GDP is one of the most comprehensive and closely watched economic statistics.

The National Income and Product Accounts (NIPA) are a set of economic accounts that provide information on the value and composition of output produced in the United States during a given period and on the types and uses of the income generated by that production. NIPA was created during the Great Depression to try to understand why the U.S. economy was in freefall and how to stop it.

U.S. input – output (I-O) accounts are a primary component of the U.S. economic accounts. I-O analysis is an economic tool that measures the relationships between various industries in the economy. These tables provides a detailed snapshot of the economy. More specifically, I-O tables show the commodity inputs that are used by each industry to produce its outputs; the commodities produced by each industry and the use of commodities by consumers.

NIPA is a snapshot of the economy at a certain time period. Over the past eight decades, new products have been created that use different inputs. To incorporate these changes, BEA conducts comprehensive revisions, called benchmarks, every five years. These comprehensive updates seek to introduce major improvements to maintain and improve the NIPA data.

The 2020 annual update of the National Income and Product Accounts was released on July 30, 2020. Data for 2015 through 2019 were revised. The reference year remained 2012. The updated statistics largely reflect the incorporation of newly available and revised source data and improvements to existing methodologies.

The 2018 comprehensive benchmark revision in July 2018 incorporated three major types of revisions: 1) statistical changes to introduce new and improved technologies and newly available source data; 2) changes in definitions to more accurately portray the evolving U.S. economy to more accurately portray the evolving U.S. economy and to provide consistent comparisons with data for other national economies; and 3) changes in presentations to reflect the definitional and statistical changes, where necessary, or to provide additional data or perspective for users. Also output and price measures will be expressed with 2012 equal to 100.

Major statistical changes to the NIPA data included the 2012 benchmark input-output (I-O) accounts which incorporated the quinquennial economic census. The I-O accounts tracks the flows of detailed inputs and outputs throughout the economy. Another change was the incorporation of the improved methodology for seasonal adjustments over historical time spans and extending back improvements in previous annual updates. Other changes incorporated the handling of issues such as

- fixed investment in software, medical equipment, and communications equipment

- savings institutions and credit unions

- state and local defined benefit pension plans

Major changes in definitions in this comprehensive update are as follows:

- reclassifications of research and development (R & D) for software originals

- recognition of capital services

- reclassification of payments by Federal Reserve banks to the U.S. government

- reclassification of 'other' state and local personal taxes

The above revisions were incorporated into the presentations produced by BEA.

Currently, BEA has adopted an expansion of the basic concept of production embodied in the accounts and changed the method of accounting for an important category of pension plans. Also, the constant-dollar or "real" estimates were restated to 2012 dollars, so that the levels shown are different even if the indicated rates of change are the same.

Fixed investment, one of the basic building blocks of the estimation of gross domestic product, was expanded in this revision by recognizing expenditures for research and development by business, government, and nonprofit institutions serving households (NPISHs) as investment.

The workers and capital that produce this R&D have always been reported in the labor and capital tabulations, but were treated as if they were costs of producing this year's output only. In the national income accounting as in conventional business accounting, the full costs of the R&D were subtracted from current-year business receipts in calculating business profit.

Now, in the NIPAs, the estimated value of the research over the whole life of the product—for example, from the estimated future sales of a new drug—is added to the value produced this year. However, this R&D will lose value over time and, as with a piece of machinery, its use and obsolescence, called consumption of fixed capital (CFC), is estimated and charged against GDP in each year that this capital is in use. For that reason, this change adds more to gross national product than to net national product and income.

The basic concept of the NIPAs requires equivalence (except for measurement error) between the value of output and the income and other charges against that value. The higher values of production are balanced on the income side of the accounts by the higher capital consumption allowances and by addition, to profits and other operating surplus (nonlabor income) components, of the net R&D investment—the difference between gross investment value and current year's CFC. (This does not change the tax accounting treatment. The firm still gets to charge off the entire expenditure in the year it is made, which is the most favorable treatment possible.)

Similarly, BEA is now capitalizing private spending for certain entertainment, literary, and other artistic originals as fixed investment. This includes movies, books, music, and long-lived TV programs such as situation comedies and drama (but not news, sports, games, or reality shows).

In a third expansion of the GDP investment definition, all ownership transfer costs for residential and nonresidential structures are capitalized (instead of only broker's commissions as in the previous treatment).

The other important conceptual change introduced in this revision was the adoption of accrual instead of cash accounting for defined benefit pension plans. This does not increase income and saving (or dis-saving) overall, but relocates it. Pension benefits are credited to the labor income and saving of employees when earned instead of when paid—matched by an equal and opposite change in the surplus or deficit of the employing government or business.

All the quarterly NIPA data published here, including indexes of quantity and price, are seasonally adjusted; this is not specifically noted on each table. All of the quarterly level values in current and constant dollars are also expressed at annual rates, which means that seasonally adjusted quarterly levels have been multiplied by 4 so that their scale will be comparable with the annual values shown, and this is noted on the pertinent *Business Statistics* tables. Where quarterly percent changes are shown, they are expressed as seasonally adjusted annual rates, which means that the quarter-to-quarter change in the level of GDP is calculated and then raised to the 4th power in order to express what that rate of change would be if continued over an entire year; this is noted in the column headings. For further information on these concepts see "General Notes" at the front of this volume.

Each table is notated, just above the table at the right-hand side, with the numbers of the NIPA tables from which the data are drawn.

All data and references are available on the BEA website, <http://www.bea.gov>.

DEFINITIONS AND NOTES ON THE DATA:

Basic concepts of total output and income (Tables 1-1, 1-11, 1-12, and 1-13)

The NIPAs depict the U.S. economy in several different dimensions. The basic concept, and the measure that is now most frequently cited, is gross domestic product (GDP), which is the market value of all goods and services produced by labor and property located in the United States.

In principle, GDP can be measured by summing the values created by each industry in the economy. However, it can also, and more readily, be measured by summing all the final demands for the economy's output. This final-demand approach also has the advantage of depicting the origins of demand for economic production, whether from consumers, businesses, or government.

Since production for the market necessarily generates incomes equal to its value, there is also an income total that corresponds to the production value total. This income can be measured and its distribution among labor, capital, and other income recipients can be depicted.

The relationships of several of these major concepts are illustrated in Table 1-11. The definitions of these concepts are as follows:

Gross domestic product (GDP), the featured measure of the value of U.S. output, is the market value of the goods and services produced by labor and property located in the United States. Market values represent output valued at the prices paid by the final customer, and therefore include taxes on production and imports, such as sales taxes, customs duties, and taxes on property.

The term "gross" in gross domestic product and gross national product is used to indicate that capital consumption allowances (economic depreciation) have not been deducted.

GDP is primarily measured by summing the values of all of the final demands in the economy, net of the demands met by imports; this is shown in Table 1-1. Specifically, GDP is the sum of personal consumption expenditures (PCE), gross private domestic investment (including change in private inventories and before deduction of charges for consumption of fixed capital), net exports of goods and services, and government consumption expenditures and gross investment. GDP measured in this way excludes duplication involving "intermediate" purchases of goods and services (goods and services purchased by industries and used in production), the value of which is already included in the value of the final products. Production of any intermediate goods that are not used in further production in the current period is captured in the measurement of inventory change.

In concept, GDP is equal to the sum of the economic value added by (also referred to as "gross product originating in") all industries in the United States. This, in turn, also makes it the conceptual equivalent of *gross domestic income (GDI)*, a new concept introduced in the 2003 revision. GDI is the sum of the incomes earned in each domestic industry, plus the taxes on production and imports and less the subsidies that account for the difference between output value and factor input value. This derivation is shown in Table 1-12. Since the incomes and taxes can be measured directly, they can be summed to a total that is equivalent to GDP in concept but differs due to imperfections in measurement. The difference between the two is known as the *statistical discrepancy*. It is defined as GDP minus GDI, and is shown in Tables 1-11 and 1-12.

Gross national product (GNP) refers to all goods and services produced by labor and property supplied by U.S. residents— whether located in the United States or abroad—expressed at market prices. It is equal to GDP, plus *income receipts from the rest of the world*, less *income payments to the rest of the world*. *Domestic* production and income refer to the location of the factors of production, with only factors located in the United States included; *national* production and income refer to the ownership of the factors of production, with only factors owned by United States residents included.

Before the comprehensive NIPA revisions that were made in 1991, GNP was the commonly used measure of U.S. production. However, GDP is clearly preferable to GNP when used in conjunction with indicators such as employment, hours worked, and capital utilized—for example, in the calculation of labor and capital productivity—because it is confined to production taking place within the borders of the United States.

The income-side aggregate corresponding to GNP is *gross national income (GNI)*, shown as an addendum to Table 1-11. It consists of gross domestic income plus income receipts from the rest of the world, less income payments to the rest of the world. It is used as the denominator for a national saving-income ratio, presented in Chapter 5. National income is the preferred measure for calculating and comparing saving, since it is the income aggregate from which that saving arises. As with GDP and gross domestic income, the statistical discrepancy indicates the difference between the product-side and income-side measurement of the same concept.

Net national product is the market value, net of depreciation, of goods and services attributable to the labor and property supplied by U.S. residents. It is equal to GNP minus the *consumption of fixed capital (CFC)*. CFC relates only to fixed capital located in the United States. (Investment in that capital is measured by private fixed investment and government gross investment.) As of the 2009 comprehensive revision, CFC represents only the normal using-up of capital in the process of production, and no longer includes extraordinary disaster losses such as those caused by Hurricane Katrina and the 9/11 attacks. These losses are still estimated and used to write down the estimates of the capital stock, but they no longer have negative effects on our calculation of current income from production.

National income has been redefined and now includes all net incomes (net of the consumption of fixed capital) earned by U.S. residents, and also includes not only "factor incomes"—net incomes received by labor and capital as a result of their participation in the production process, but also "nonfactor charges"— taxes on production and imports, business transfer payments, and the current surplus of government enterprises, less subsidies. This change has been made to conform with the international guidelines for national accounts, *System of National Accounts (SNA) 1993*. According to *SNA 1993,* these charges cannot be eliminated from the input and output prices.

Since national income now includes the nonfactor charges, it is conceptually equivalent to *net national product* and differs only by the amount of the statistical discrepancy.

The concept formerly known as "national income," which excludes the nonfactor charges, is still included in the accounts as an addendum item, now called "net national factor income." It is shown in Table 1-13. *Net national factor income* consists of compensation of employees, proprietors' income with inventory valuation and capital consumption adjustments (IVA and CCadj, respectively), rental income of persons with capital consumption adjustment, corporate profits with inventory valuation and capital consumption adjustments, and net interest.

By definition, national income and its components exclude all income from capital gains (increases in the value of owned assets). Such increases have no counterpart on the production side of the accounts. This exclusion is partly accomplished by means of the inventory valuation and capital consumption adjustments, which will be described in the definitions of the components of product and income.

DEFINITIONS AND NOTES ON THE DATA:

Imputation

The term *imputation* will appear from time to time in the definitions of product and income components. Imputed values are values estimated by BEA statisticians for certain important product and income components that are not explicitly valued in the source data, usually because a market transaction in money terms is not involved. Imputed values appear on both the product and income side of the accounts; they add equal amounts to income and spending, so that no imputed saving is created.

One important example is the imputed rent on owner-occupied housing. The building of such housing is counted as investment, yet in the monetary accounts of the household sector, there is no income from that investment nor any rental paid for it. In the NIPAs, the rent that each such dwelling would earn if rented is estimated and added to both national and personal income (as part of rental income receipts) and to personal consumption expenditures (as part of expenditures on housing services).

Another important example is imputed interest. For example, many individuals keep monetary balances in a bank or other financial institution, receiving either no interest or below-market interest, but receiving the institution's services, such as clearing checks and otherwise facilitating payments, with little or no charge. In this case, where is the product generated by the institution's workers and capital? In the NIPAs, the depositor is imputed a market-rate-based interest return on his or her balance, which is then imputed as a service charge received by the institution, and therefore included in the value of the institution's output.

DEFINITIONS AND NOTES ON THE DATA:

Components of product (Tables 1-1 through 1-6)

Personal consumption expenditures (PCE) is goods and services purchased by persons residing in the United States. PCE consists mainly of purchases of new goods and services by individuals from businesses. It includes purchases that are financed by insurance, such as government-provided and private medical insurance. In addition, PCE includes purchases of new goods and services by nonprofit institutions, net purchases of used goods ("net" here indicates purchases of used goods from business less sales of used goods to business) by individuals and nonprofit institutions, and purchases abroad of goods and services by U.S. residents traveling or working in foreign countries. PCE also includes purchases for certain goods and services provided by government agencies. (See the notes and definitions for Chapter

4 for additional information.) In the 2009 revision, new detail was provided on the allocation of PCE between the household and nonprofit sectors.

Gross private domestic investment consists of gross private fixed investment and change in private inventories.

Private fixed investment consists of both nonresidential and residential fixed investment. The term "residential" refers to the construction and equipping of living quarters for permanent occupancy. Hotels and motels are included in *nonresidential fixed investment*, as described subsequently in this section.

Private fixed investment consists of purchases of fixed assets, which are commodities that will be used in a production process for more than one year, including replacements and additions to the capital stock, and now also including research and development and other intellectual property. It is measured "gross," before a deduction for consumption of existing fixed capital. It covers all investment by private businesses and nonprofit institutions in the United States, regardless of whether the investment is owned by U.S. residents. The residential component includes investment in owner-occupied housing; the homeowner is treated equivalently to a business in these investment accounts. (However, when GDP by sector is calculated, owner-occupied housing is no longer included in the business sector. It is allocated to the households and institutions sector.) Private fixed investment does not include purchases of the same types of equipment and structures by government agencies, which are included in government gross investment, nor does it include investment by U.S. residents in other countries.

Nonresidential fixed investment is the total of nonresidential structures, nonresidential equipment, and intellectual property.

Nonresidential structures consists of new construction, brokers' commissions and other ownership transfer costs on sales of structures, and net purchases of used structures by private business and by nonprofit institutions from government agencies (that is, purchases of used structures from government minus sales of used structures to government). New construction also includes hotels and motels and mining exploration, shafts, and wells.

Nonresidential equipment consists of private business purchases on capital account of new machinery, equipment, and vehicles; dealers' margins on sales of used equipment; and net purchases of used equipment from government agencies, persons, and the rest of the world (that is, purchases of such equipment minus sales of such equipment). It does not include the estimated personal-use portion of equipment purchased for both business and personal use, which is allocated to PCE.

Intellectual property consists of purchases and in-house production of software; research and development; and private expenditures for specified entertainment, literary, and artistic originals. (See above for further description.)

Residential private fixed investment consists of both residential structures and residential producers' durable equipment (including such equipment as appliances owned by landlords and rented to tenants). Investment in structures consists of new units, improvements to existing units, purchases of manufactured homes, brokers' commissions and other ownership transfer costs on the sale of residential property, and net purchases of used residential structures from government agencies (that is, purchases of such structures from government minus sales of such structures to government). As noted above, it includes investment in owner-occupied housing.

Change in private inventories is the change in the physical volume of inventories held by businesses, with that <u>change</u> being valued at the average price of the period. It differs from the change in the book value of inventories reported by most businesses; an *inventory valuation adjustment (IVA)* converts book value change using historical cost valuations to the change in physical volume, valued at average replacement cost.

Net exports of goods and services is *exports of goods and services* less *imports of goods and services*. It does not include income payments or receipts or transfer payments to and from the rest of the world.

Government consumption expenditures is the estimated value of the services produced by governments (federal, state, and local) for current consumption. Since these are generally not sold, there is no market valuation and they are priced at the cost of inputs. The input costs consist of the compensation of general government employees; the estimated consumption of general government fixed capital including software and R&D (CFC, or economic depreciation); and the cost of goods and services purchased by government less the value of sales to other sectors. The value of investment in equipment and structures produced by government workers and capital is also subtracted, and is instead included in government investment. Government sales to other sectors consist primarily of receipts of tuition payments for higher education and receipts of charges for medical care.

This definition of government consumption expenditures differs in concept—but not in the amount contributed to GDP—from the treatment in existence before the 2003 revision of the NIPAs. In the current definition, goods and services purchased by government are considered to be intermediate output. In the previous definition, they were considered to be final sales. Since their value is added to the other components to yield total government consumption expenditures, the dollar total contributed to GDP is the same. The only practical difference is that the goods purchased disappear from the goods account and appear in the services account instead. In the industry sector accounts, the value added by government is also unchanged. It continues to be measured as the sum of compensation and CFC, or equivalently as gross government output less the value of goods and services purchased.

Gross government investment consists of general government and government enterprise expenditures for fixed assets (structures, equipment, software, and R&D). Government inventory investment is included in government consumption expenditures.

DEFINITIONS AND NOTES ON THE DATA:

Real values, quantity and price indexes (Tables 1-2 through 1-7)

Real, or chained (2012) dollar, estimates are estimates from which the effect of price change has been removed. Prior to the 1996 comprehensive revision, constant-dollar measures were obtained by combining real output measures for different goods and services using the relative prices of a single year as weights for the entire time span of the series. In the recent environment of rapid technological change, which has caused the prices of computers and electronic components to decline rapidly relative to other prices, this method distorts the measurement of economic growth and causes excessive revisions of growth rates at each benchmark revision. The current, chained-dollar measure changes the relative price weights each year, as relative prices shift over time. As a result, recent changes in relative prices do not change historical growth rates.

Chained-dollar estimates, although expressed for continuity's sake as if they had occurred according to the prices of a single year (currently 2012), are usually not additive. This means that because of the changes in price weights each year, the chained (2012) dollar components in any given table for any year other than 2012 usually do not add to the chained (2012) dollar total. The amount of the difference for the major components of GDP is called the *residual* and is shown in Table 1-2. It is specific to each individual BEA tabulation, corresponding to the sum of the lowest level of aggregation <u>on that tabulation.</u> In time periods close to the base year, residuals are usually quite small; over longer periods, the differences become much larger. For this reason, BEA no longer publishes chained-dollar estimates prior to 1999, except for selected aggregate series. For the more detailed components of GDP, historical trends and fluctuations in real volumes are represented by *chain-type quantity indexes,* which are presented in Tables 1-5, 4-3, 5-4, 5-6, 6-6, 6-7, 6-11, 6-16, 7-2, 7-5, and 14-2.

Chain-weighting leads to complexity in estimating the contribution of economic sectors to an overall change in output: it becomes difficult, for someone without access to the complicated statistical methods that BEA uses, to find the correct answers to questions such as "How much are government spending cuts contributing to the drag on GDP growth?" Because of this, BEA is now calculating and publishing estimates of the arithmetic contribution of each major component to the total change in real GDP. *Business Statistics* reproduces these calculations in Table 1-4. (As will be explained later, users of these calculations should, however, bear in mind that imports are treated as a negative contribution to GDP instead of being subtracted from the demand components that give rise to them.) For further information, see J. Steven Landefeld, Brent R. Moulton, and Cindy M. Vojtech, "Chained-Dollar Indexes: Issues, Tips on Their Use, and Upcoming Changes," *Survey of Current Business* (November 2003); and J. Steven Landefeld and Robert P. Parker, "BEA's Chain Indexes, Time Series, and Measures of Long-Term Economic Growth," *Survey of Current Business* (May 1997).

GDP price indexes measure price changes between any two adjacent years (or quarters) for a fixed "market basket" of goods and services consisting of the average quantities purchased in those two years (or quarters). The annual measures are chained together to form an index with prices in 2012 set to equal 100. Using average quantities as weights while changing weights each period eliminates the substitution bias that arises in more conventional indexes, in which weights are taken from a single base period that usually takes place early in the period under measurement. Generally, using a single, early base period leads to an overstatement of price increase. The CPI-U and the CPI-W are examples of such conventional indexes, technically known as "Laspeyres" indexes. See the "General Notes" at the beginning of this volume and the notes and definitions for Chapter 8 for further explanation.

The chain-type formula guarantees that a GDP price index change will differ only trivially from the change in the implicit deflator (ratio of current-dollar to real value, expressed as a percent). Therefore, *Business Statistics* is no longer publishing a separate table of implicit deflators.

DEFINITIONS AND NOTES ON THE DATA:

BEA Aggregates of sales and purchases (Tables 1-1 and 1-5 through 1-7)

Final sales of domestic product is GDP minus change in private inventories. It is the sum of personal consumption expenditures, gross private domestic fixed investment, government consumption expenditures and gross investment, and net exports of goods and services.

Gross domestic purchases is the market value of goods and services purchased by U.S. residents, regardless of where those goods and services were produced. It is GDP minus net exports (that is, minus exports plus imports) of goods and services; equivalently, it is the sum of personal consumption expenditures, gross private domestic investment, and government consumption expenditures and gross investment. The price index for gross domestic purchases is therefore a measure of price change for goods and services purchased by (rather than produced by) U.S. residents.

Final sales to domestic purchasers is gross domestic purchases minus change in private inventories.

DEFINITIONS AND NOTES ON THE DATA:

U.S. Population and per capita product and income estimates (Table 1-3)

In Table 1-3, annual and quarterly measures of product, income, and consumption spending are expressed in per capita terms—the aggregate dollar amount divided by the U.S. population. Population data from 1991 forward reflect the results of the 2000 and 2010 Censuses.

Definitions and notes on the data:

Inventory-sales ratios (Table 1-8)

Inventories to sales ratios. The ratios shown in Table 1-8 are based on the inventory estimates underlying the measurement of inventory change in the NIPAs. They include data and estimates for not only the inventories held in manufacturing and trade (which are shown in Chapter 16), but also stocks held by all other businesses in the U.S. economy.

For the current-dollar ratios, inventories at the end of each quarter are valued in the prices that prevailed at the end of that quarter. For the constant-dollar ratios, they are valued in chained (2012) dollars. In both cases, the inventory-sales ratio is the value of the inventories at the end of the quarter divided by quarterly total sales at <u>monthly</u> rates (quarterly totals divided by 3). In other words, they represent how many months' supply businesses had on hand at the end of the period. This makes them comparable in concept and order of magnitude to the ratios shown in Chapter 16. Annual ratios are those for the fourth quarter.

Inventory data consistent with the 2013 revision are only available from 1997 to date. Tables 1-8A and B show the old estimates for earlier years with an overlap period.

DEFINITIONS AND NOTES ON THE DATA:

Shares of aggregate supply and demand (Tables 1-9 and 1-10)

These tables, developed by the editor of *Business Statistics*, are not official NIPA calculations. They are components of current-dollar GDP rearranged in order to highlight relationships that are not always apparent in the official presentation of the NIPAs.

Aggregate supply combines domestic production (GDP) and imports, the two sources that, between them, supply the goods and services demanded by consumers, businesses, and governments in the U.S. economy. In this table the user can observe the growing share of imports that satisfy demands in the U.S. marketplace.

Aggregate final demand is the sum of all final (that is, excluding inventory change) demands for goods and services in the U.S. market—consumption spending, fixed investment, exports, and government consumption and investment. It is different from *final sales of domestic product* (Tables 1-1 and 1-7) because imports are not subtracted; in this table, imports are considered a source of supply, not a negative element of demand. It is different from *gross domestic purchases* (Tables 1-5 and 1-6) because it includes exports, since they are a source of demand for U.S. output, but excludes inventory change. It is like *final sales to domestic purchasers* (Table 1-7) in excluding inventory change, but different because it also includes exports.

Table 1-10 provides alternative data on the question of the relative importance of consumption spending and other final demands to the U.S. economy. NIPA statistics on PCE as a percent of GDP are frequently cited, but there is a problem with this, since PCE includes the value of imports while GDP does not.

DEFINITIONS AND NOTES ON THE DATA:

Components of income (Tables 1-11, 1-12 and 1-13)

There are now two different presentations of aggregate income for the United States: *gross domestic income* (Table 1-12) and *national income* (Table 1-13). As noted above, domestic income refers to income generated from production within the United States, while national income refers to income received by residents of the United States. This means that some of the income components differ between the two tables. Domestic income payments include payments to the rest of the world from domestic industries. National income payments exclude payments to the rest of the world but include payments received by U.S. residents from the rest of the world. These differences are seen in employee compensation, interest, and corporate profits. Taxes on production and imports, taxes on corporate income, business transfer payments, subsidies, proprietors' income, rental income, and the current surplus of government enterprises are the same in both accounts.

All income entries are now calculated on an accrual basis, associating the income with the period in which it was earned rather than when it was paid.

A third important income aggregate is *personal income*. The derivation of this well-known statistic from national income is shown in Table 1-11. See Chapter 4 and its notes and definitions for data and more information.

Compensation of employees is the income accruing to employees as remuneration for their work. It is the sum of wage and salary accruals and supplements to wages and salaries. In the domestic income account, it refers to all payments generated by domestic production, including those to workers residing in the "rest of the

world." In the national and personal income accounts, there is a slightly different "compensation of employees," including that received by U.S. residents from the rest of the world but excluding that paid from the domestic production account to workers residing elsewhere.

Wages and salaries consists of the monetary remuneration of employees, including the compensation of corporate officers; corporate directors' fees paid to directors who are also employees of the corporation; commissions, tips, and bonuses; voluntary employee contributions to certain deferred compensation plans, such as 401(k) plans; and receipts-in-kind that represent income. As of the 2003 revision, it also includes judicial fees to jurors and witnesses, compensation of prison inmates, and marriage fees to justices of the peace, all of which were formerly included in "other labor income."

In concept, wages and salaries include the value of the exercise by employees of "nonqualified stock options," in which an employee is allowed to buy stock for less than its current market price. (Actual measurement of these values involves a number of problems, particularly in the short run. Such stock options are not included in the monthly wage data from the Bureau of Labor Statistics, which are the main source for current extrapolations of wages and salaries, and are not consistently reported in corporate financial statements. They are, however, generally included in the unemployment insurance wage data that are used to correct the preliminary wage and salary estimates.) Another form of stock option, the "incentive stock option," leads to a capital gain only and is thus not included in the definition of wages and salaries.

Supplements to wages and salaries consists of *employer contributions for employee pension and insurance funds* and *employer contributions for government social insurance*.

Employer contributions for employee pension and insurance funds consists of employer payments (including payments-in-kind) to private pension and profit-sharing plans, private group health and life insurance plans, privately administered workers' compensation plans, government employee retirement plans, and supplemental unemployment benefit plans. They are now measured in the period in which the employee earns the obligation rather than when the employer makes the cash payment into the fund. This includes the major part of what was once called "other labor income." The remainder of "other labor income" has been reclassified as wages and salaries, as noted above.

Employer contributions for government social insurance consists of employer payments under the following federal, state, and local government programs: old-age, survivors, and disability insurance (Social Security); hospital insurance (Medicare); unemployment insurance; railroad retirement; pension benefit guaranty; veterans' life insurance; publicly administered workers' compensation; military medical insurance; and temporary disability insurance.

Taxes on production and imports is included in the gross domestic income account to make it comparable in concept to gross domestic product. It consists of federal excise taxes and customs duties and of state and local sales taxes, property taxes (including residential real estate taxes), motor vehicle license taxes, severance taxes, special assessments, and other taxes. It is equal to the former "indirect business taxes and nontax liabilities" less most of the nontax liabilities, which have now been reclassified as "business transfer payments."

Subsidies (payments by government to business other than purchases of goods and services) are now presented separately from the current surplus of government enterprises, which is presented as a component of net operating surplus. However, for data representing the years before 1959, subsidies continue to be presented as net of the current surplus of government enterprises, since detailed data to separate the series for this period are not available.

Net operating surplus is a new aggregate introduced in the 2003 NIPA revision—a grouping of the business income components of the gross domestic income account. It represents the net income accruing to business capital. It is equal to gross domestic income minus compensation of employees, taxes on production and imports less subsidies (that is, the taxes are taken out of income and the subsidies are put in), and consumption of fixed capital (CFC). Net operating surplus consists of the surplus for private enterprises and the current surplus of government enterprises. The net operating surplus of private enterprises comprises net interest and miscellaneous payments, business current transfer payments, proprietors' income, rental income of persons, and corporate profits.

Net interest and miscellaneous payments, domestic industries consists of interest paid by domestic private enterprises and of rents and royalties paid by private enterprises to government, less interest received by domestic private enterprises. Interest received does not include interest received by noninsured pension plans, which are recorded as being directly received by persons in personal income. Both interest categories include monetary and imputed interest. In the *national* account, interest paid to the rest of the world is subtracted from the interest paid by domestic industries and interest received from the rest of the world is added. Interest payments on mortgage and home improvement loans and on home equity loans are included as net interest in the private enterprises account.

It should be noted that net interest does not include interest paid by federal, state, or local governments. In fact, government interest does not enter into the national and domestic income accounts, though it does appear as a component of personal income. The NIPAs draw a distinction between interest paid by government and that paid by business.

The reasoning behind this distinction is that interest paid by business is one of the income counterparts of the production side of the account. The value of business production (as measured by its output of goods and services) includes the value added by business capital, and interest paid by business to its lenders is part of the total return to business capital.

However, there is no product flow in the accounts that is a counterpart to the payment of interest by government. The output of government does not have a market value. For purposes of GDP measurement, BEA estimates the government contribution to GDP as the sum of government's compensation of employees, purchases of goods and services, and consumption of government fixed capital. (See above, and also the notes and definitions to Chapter 6.) This implies an estimate (described as "conservative" by BEA) that the net return to government capital is zero—that is, that the gross return is just sufficient to pay down the depreciation. Consequently, this assumption generates no income, imputed or actual, that might correspond to the government's interest payment.

Supporting the distinction between business and government interest payments, it may be noted that most federal government debt was not incurred to finance investment, but rather to finance wars, to avoid tax increases and spending cuts during recessions, or to stimulate the economy. Furthermore, some of the largest and most productive government investments—highways—are typically financed by taxes on a pay-as-you-go basis and not by borrowing.

Business current transfer payments, net consists of payments to persons, government, and the rest of the world by private business for which no current services are performed. Net insurance settlements—actual insured losses (or claims payable) less a normal level of losses—are treated as transfer payments. Payments to government consist of federal deposit insurance premiums, fines, regulatory and inspection fees, tobacco settlements, and other miscellaneous payments previously classified as "nontaxes." Taxes paid by domestic corporations to foreign governments, formerly classified as transfer payments, are now counted as taxes on corporate income.

In the NIPAs, capital income other than interest—corporate profits, proprietors' income, and rental income—is converted from the basis usually shown in the books of business, and reported to the Internal Revenue Service, to a basis that more closely represents income from current production. In the business accounts that provide the source data, depreciation of structures and equipment typically reflects a historical cost basis and a possibly arbitrary service life allowed by law to be used for tax purposes. BEA adjusts these values to reflect the average actual life of the capital goods and the cost of replacing them in the current period's prices. This conversion is done for all three forms of capital income. In addition, corporate and proprietors' incomes also require an adjustment for inventory valuation to exclude any profits or losses that might appear in the books, should the cost of inventory acquisition not be valued in the current period's prices. These two adjustments are called the *capital consumption adjustment (CCAdj)* and the *inventory valuation adjustment (IVA)*. They are described in more detail below.

Proprietors' income with inventory valuation and capital consumption adjustments is the current-production income (including income-in-kind) of sole proprietorships and partnerships and of tax-exempt cooperatives. The imputed net rental income of owner-occupants of farm dwellings is included, but the imputed net rental income of owner-occupants of nonfarm dwellings is included in rental income of persons. Fees paid to outside directors of corporations are included. Proprietors' income excludes dividends and monetary interest received by nonfinancial business and rental incomes received by persons not primarily engaged in the real estate business; these incomes are included in dividends, net interest, and rental income of persons, respectively.

Rental income of persons with capital consumption adjustment is the net current-production income of persons from the rental of real property (except for the income of persons primarily engaged in the real estate business), the imputed net rental income of owner-occupants of nonfarm dwellings, and the royalties received by persons from patents, copyrights, and rights to natural resources. Consistent with the classification of investment in owner-occupied housing as business investment, the homeowner is considered to be paying himself or herself the rental value of the house (classified as PCE for services) and receiving as net income the amount of the rental that remains after paying interest and other costs.

Corporate profits with inventory valuation and capital consumption adjustments, often referred to as "economic profits," is the current-production income, net of economic depreciation, of organizations treated as corporations in the NIPAs. These organizations consist of all entities required to file federal corporate tax returns, including mutual financial institutions and cooperatives subject to federal income tax; private noninsured pension funds; nonprofit institutions that primarily serve business; Federal Reserve Banks, which accrue income stemming from the conduct of monetary policy; and federally sponsored credit agencies. This income is measured as receipts less expenses as defined in federal tax law, except for the following differences: receipts exclude capital gains and dividends received; expenses exclude depletion and capital losses and losses resulting from bad debts; inventory withdrawals are valued at replacement cost; and depreciation is on a consistent accounting basis and is valued at replacement cost.

Since *national* income is defined as the income of U.S. residents, its profits component includes income earned abroad by U.S. corporations and excludes income earned by the rest of the world within the United States.

Taxes on corporate income consists of taxes on corporate income paid to government and to the rest of the world.

Taxes on corporate income paid to government is the sum of federal, state, and local income taxes on all income subject to taxes. This income includes capital gains and other income excluded from profits before tax. These taxes are measured on an accrual basis, net of applicable tax credits.

Taxes on corporate income paid to the rest of the world consists of nonresident taxes, which are those paid by domestic corporations to foreign governments. These taxes were formerly classified as "business transfer payments to the rest of the world."

Profits after tax is total corporate profits with IVA and CCAdj less taxes on corporate income. It consists of dividends and undistributed corporate profits.

Dividends is payments in cash or other assets, excluding those made using corporations' own stock, that are made by corporations to stockholders. In the domestic account, these are payments by domestic industries to stockholders in the United States and abroad; in the national account, these are dividends received by U.S. residents from domestic and foreign industries. The payments are measured net of dividends received by U.S. corporations. Dividends paid to state and local government social insurance funds and general government are included.

Undistributed profits is corporate profits after tax with IVA and CCAdj less dividends.

The *inventory valuation adjustment (IVA)* is the difference between the cost of inventory withdrawals valued at replacement cost and the cost as valued in the source data used to determine profits before tax, which in many cases charge inventories at acquisition cost. It is calculated separately for corporate profits and for nonfarm proprietors' income. Its behavior is determined by price changes, especially for materials. When prices are rising, which has been typical of much of the postwar period, the business-reported value of inventory change will include a capital gains component, which needs to be removed from reported inventory change on the product side in order to correctly measure the change in the volume of inventories, and from reported profits on the income side of the accounts in order to remove the capital gains element. At such times, the IVA will be a negative figure, which is added to reported profits to yield economic profits. Occasionally, falling prices—especially for petroleum and products—will result in a positive IVA. No adjustment is needed for farm proprietors' income, as farm inventories are measured on a current-market-cost basis.

Consumption of fixed capital (CFC) is a charge for the using-up of private and government fixed capital located in the United States. It is not based on the depreciation schedules allowed in tax law, but instead on studies of prices of used equipment and structures in resale markets and other service life information.

For general government and for nonprofit institutions that primarily serve individuals, CFC on their capital assets is recorded in government consumption expenditures and in personal consumption expenditures, respectively. It is considered to be the value of the current services of the fixed capital assets owned and used by these entities.

Private capital consumption allowances consists of tax-return-based depreciation charges for corporations and nonfarm proprietorships; BEA estimates for R&D and other intellectual property; and historical cost depreciation (calculated by BEA using a geometric pattern of price declines) for farm proprietorships, rental income of persons, and nonprofit institutions.

The *private capital consumption adjustment (CCAdj)* is the difference between private capital consumption allowances and private consumption of fixed capital. The CCAdj has two parts:

- The first component of CCAdj converts tax-return-based depreciation to consistent historical cost accounting based on actual service lives of capital. In the postwar period, this has usually been a large positive number, that is, a net addition to profits and subtraction from reported depreciation. This is the case because U.S. tax law typically allows depreciation periods shorter than actual service lives. Tax depreciation was accelerated even further for 2001 through 2004. This component is a reallocation of gross business saving from depreciation to profits; gross saving is unchanged, with exactly offsetting changes in capital consumption and net saving.

- The second component is analogous to the IVA: it converts reported business capital consumption allowances from the historical cost basis to a replacement cost basis. It is determined by the price behavior of capital goods. These prices have had an upward drift in the postwar period, which has been much more stable than the changes in materials prices. Hence, this component is consistently negative (serving to reduce economic profits relative to the reported data) but less volatile than the IVA.

In 1982 through 2004, positive values for the first component outweighed negative values for the second, resulting in a net positive CCAdj, an addition to profits. However, when the accelerated depreciation expired in 2005, there was a sharp decline in the consistent-accounting adjustment and it was outweighed by the price adjustment. This led to negative CCAdjs in 2005 through 2007, reducing profits from the reported numbers.

DEFINITIONS AND NOTES ON THE DATA:

Gross value added of domestic corporate business (Tables 1-14 and 1-15)

Gross value added is the term now used for what was formerly called "gross domestic product originating." It represents the share of the GDP that is produced in the specified sector or industry. Tables 1-14 and 1-15 show the current-dollar value of gross value added for all domestic corporate business and its financial and nonfinancial components. For the total and for nonfinancial corporations, consumption of fixed capital and net value added are shown, as is the allocation of net value added among employee compensation, taxes and transfer payments, and capital income. Constant-dollar values are also shown for nonfinancial corporations.

The data for nonfinancial corporations are often considered to be somewhat sturdier than data for the other sectors of the economy, since they exclude sectors whose outputs are difficult to evaluate—households, institutions, general government, and financial business—as well as excluding all noncorporate business, in which the separate contributions of labor and capital are not readily measured.

DATA AVAILABILITY AND REVISIONS

Annual data are available beginning with 1929. Quarterly data begin with 1946 for current-dollar values and 1947 for quantity and price measures such as real GDP and the GDP price index. Not all data are available for all time periods.

New data are normally released toward the end of each month. The "advance" estimate of GDP for each calendar quarter is released at the end of the month after the quarter's end. The "second" estimate, including more complete product data and the first estimates of corporate profits, is released at the end of the second month after the quarter's end, and a "third" estimate including still more complete data at the end of the third month. Wage and salary and related income-side components may be revised for previous quarters as well.

At the end of each July, there is an "annual" revision, incorporating more complete data and other improvements, affecting at least the previous 3 years. Every five years, there is a "comprehensive" revision, such as the 2013 revision incorporated in this volume, corresponding with updated statistics from the quinquennial benchmark input-output accounts and incorporating a revision of the base year for constant-price and index numbers.

The most recent data are published each month in the *Survey of Current Business*. Current and historical data may be obtained from the BEA Web site at <http://www.bea.gov> and the STAT-USA subscription Web site at <http://www.stat-usa.gov>.

REFERENCES

The 2020 annual revision of the NIPAs is presented and described in a July 30, 2020 release on the BEA Web site.

Other documentation available on the BEA Web site at <http://www.bea.gov> includes the following: "NIPA Handbook: Concepts and Methods of the U.S. National Income and Product

Accounts, October 2009"; separate chapters from the Handbook on Personal Consumption Expenditures, Private Fixed Investment, and Change in Private Inventories; "Measuring the Economy: A Primer on GDP and the National Income and Product Accounts"; "An Introduction to the National Income and Product Accounts"; and "Taking the Pulse of the Economy: Measuring GDP," *Journal of Economic Perspectives*, Spring 2008.

The treatment of employee stock options is discussed in Carol Moylan, "Treatment of Employee Stock Options in the U.S.

National Economic Accounts," available on the BEA Web site at <http://www.bea.gov>.

The data for 1929 through 1946 published here have been calculated after the fact and differ from the national income data that were currently available during the 1930s and 1940s. For an article on what was available at that time and the history of the NIPAs during that period, see Rosemary D. Marcuss and Richard E. Kane, "U.S. National Income and Product Statistics: Born of the Great Depression and World War II," *Survey of Current Business*, February 2007, pp. 32-46.

CHAPTER 2: INDUSTRIAL PRODUCTION AND CAPACITY UTILIZATION

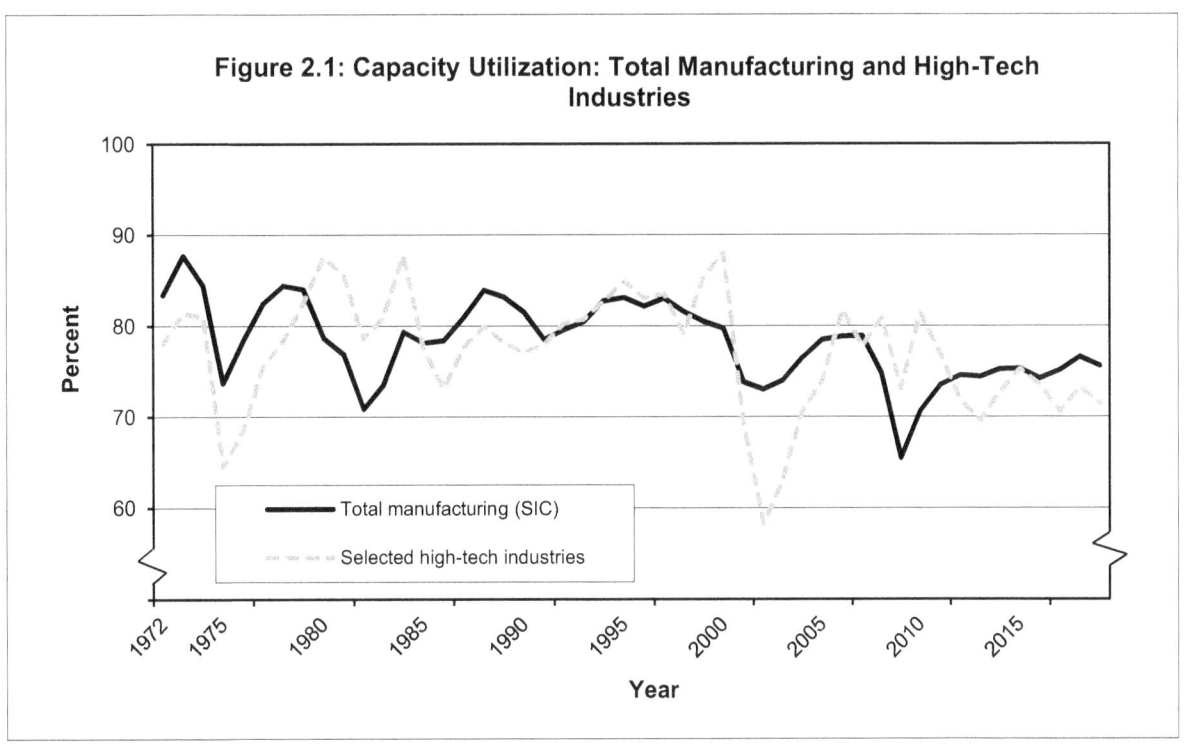

Figure 2.1: Capacity Utilization: Total Manufacturing and High-Tech Industries

- Manufacturing capacity utilization is a key statistic for the U.S. economy, despite being limited to a sector that by some measures has diminished in importance. The Federal Reserve also provides measures of capacity utilization for "total industry"—manufacturing, mining, and utilities. However, mining and utilities are less significant in the context of business cycle analysis, and much of the variation in capacity use by utilities is a result of transitory weather variations, not economic factors. Manufacturing utilization is an important indicator of inflationary pressure and also measures an important element in the demand for new capital goods. (Tables 2-3 and 2-4A)

- In 2019, industrial production reached an all-time high of 109.4 while increasing for most categories listed on Table 2-2. However, it decreased for printing and support, petroleum and coal products, plastics and rubber products, and other manufacturing. (Table 2-2)

- Manufacturing capacity utilization plunged in 2008 and 2009 following the Great Recession. In November 2007, capacity utilization reached 81.1 but by June 2009, it collapsed to 66.7. Manufacturing capacity has fluctuated monthly in the years since and it reached a high of 79.6 in November 2018 but it never recovered to its high in 2007. (Table 2-4A)

- Capacity utilization in the high-tech industries (computers and office equipment, communications equipment, and semiconductors and related electronic components) peaked in 2000, before the Tech Bubble, at 88.0 but fell nearly 34 percent by 2002. Capacity utilization has recovered but not to its 2000 level as seen in Figure 2-1. (Tables 2-3 and 2-4A)

Table 2-1. Industrial Production Indexes by Market Groups

(Seasonally adjusted, 2012 = 100.)

Year and month	Total industrial production	Final products and nonindustrial supplies							
		Total	Consumer goods						Nondurable consumer goods
			Total	Durable consumer goods					Total
				Total	Automotive products	Home electronics	Appliances, furniture, and carpeting	Miscellaneous durable goods	
1970	38.6	39.7	52.8	39.6	35.3	1.0	76.1	61.0	59.4
1971	39.1	40.2	55.9	44.9	45.0	1.1	80.3	64.3	61.1
1972	42.9	44.0	60.4	50.3	48.5	1.2	94.3	72.2	65.0
1973	46.4	47.3	63.1	54.1	52.7	1.5	101.9	75.5	66.9
1974	46.3	47.1	61.3	49.2	45.6	1.3	92.9	72.8	67.0
1975	42.2	43.8	58.8	44.7	44.0	1.2	79.9	64.2	65.7
1976	45.5	46.9	63.6	50.5	50.0	1.4	90.1	71.6	69.9
1977	48.9	50.7	67.5	56.7	56.6	1.6	101.2	79.4	72.4
1978	51.6	53.8	69.7	58.1	56.2	1.8	106.6	82.2	74.9
1979	53.2	55.6	68.6	56.0	50.6	1.8	107.2	82.9	74.4
1980	51.8	54.8	65.9	48.7	39.0	1.8	99.8	76.1	74.5
1981	52.5	55.9	66.4	49.4	40.2	1.9	98.5	77.1	74.8
1982	49.8	54.4	66.2	46.6	39.0	1.7	89.0	72.7	76.1
1983	51.1	56.0	68.6	51.7	45.3	2.5	98.3	74.8	77.0
1984	55.7	60.7	71.8	57.9	50.6	3.1	109.3	82.9	78.5
1985	56.4	62.2	72.5	57.9	50.5	3.2	108.8	83.1	79.5
1986	56.9	63.4	74.9	61.6	54.4	4.0	114.1	85.8	81.3
1987	59.9	66.6	78.1	65.3	58.2	4.0	120.3	91.1	84.2
1988	63.0	69.8	81.1	68.5	61.0	5.1	122.3	95.2	87.0
1989	63.6	70.6	81.4	70.1	63.0	5.7	123.7	95.7	86.7
1990	64.2	71.4	81.8	68.0	59.3	6.2	120.7	94.7	88.2
1991	63.2	70.3	81.7	65.0	55.7	7.2	112.4	91.3	89.5
1992	65.1	72.2	84.2	71.3	64.6	8.3	118.8	94.4	90.3
1993	67.2	74.5	86.8	77.1	71.0	10.6	125.6	98.5	91.5
1994	70.8	77.8	90.4	84.0	78.0	13.9	134.5	104.2	93.7
1995	74.0	81.0	92.9	87.2	79.5	20.3	133.7	107.4	96.0
1996	77.4	84.2	94.6	89.9	80.8	24.8	135.0	111.1	97.2
1997	83.0	89.8	98.1	96.1	87.2	35.0	141.3	113.2	99.6
1998	87.8	95.0	101.7	103.1	93.6	45.9	150.5	117.1	101.9
1999	91.7	97.9	103.9	111.3	104.1	58.6	155.4	120.1	101.8
2000	95.2	100.9	105.9	114.5	105.7	67.7	159.7	123.7	103.3
2001	92.3	98.4	104.8	109.2	101.6	69.5	152.8	114.7	103.8
2002	92.6	98.1	107.0	116.2	112.4	71.3	156.2	118.2	104.2
2003	93.8	99.4	108.4	119.8	117.9	79.4	156.4	118.3	104.9
2004	96.4	101.5	109.6	121.5	117.1	92.4	161.3	120.6	105.9
2005	99.6	105.7	112.5	122.0	113.6	105.4	164.4	124.2	109.5
2006	101.8	108.3	113.1	121.4	109.5	127.8	159.4	126.3	110.5
2007	104.4	110.3	113.1	120.9	112.6	137.1	149.6	122.2	110.6
2008	100.8	105.6	107.2	106.6	93.3	153.0	130.6	113.1	107.2
2009	89.2	93.2	99.2	86.0	74.9	132.6	98.9	93.1	102.9
2010	94.1	96.2	100.3	94.2	90.2	87.1	97.7	99.6	101.9
2011	97.1	98.2	101.4	97.7	96.4	91.4	97.1	100.4	102.3
2012	100.0	100.0	100.0	100.0	100.0	100.0	100.0	100.0	100.0
2013	102.0	100.8	100.7	105.5	108.8	104.9	102.3	102.5	99.5
2014	105.2	102.0	101.5	110.7	118.5	101.8	103.9	103.8	99.2
2015	104.1	101.4	102.9	115.0	126.2	108.3	108.5	103.5	99.9
2016	102.1	100.3	103.5	117.5	130.1	107.4	110.0	104.8	100.0
2017	104.4	102.3	104.1	119.3	131.8	107.6	110.2	107.5	100.3
2018	108.6	104.7	106.3	122.0	135.6	109.9	108.9	110.5	102.3
2019	109.4	104.5	105.3	120.1	132.7	125.2	105.0	109.6	101.5
2018									
January	106.3	103.7	105.9	120.0	133.5	106.0	109.2	107.7	102.3
February	106.6	103.9	105.8	122.2	136.0	109.1	110.5	109.9	101.6
March	107.3	104.1	106.1	122.7	137.7	106.2	108.6	109.8	102.0
April	108.2	105.1	107.5	121.9	135.9	105.5	109.4	109.8	103.8
May	107.4	103.5	105.0	117.9	127.9	105.4	108.0	109.8	101.7
June	108.2	104.3	105.8	121.1	134.5	106.8	108.4	109.8	101.9
July	108.7	104.4	106.0	120.9	132.6	109.4	110.0	110.8	102.2
August	109.5	105.0	106.4	122.8	136.7	110.3	108.7	111.2	102.3
September	109.7	105.0	106.5	123.9	138.4	110.5	109.0	111.8	102.2
October	109.9	105.4	106.7	122.7	136.0	113.3	108.5	111.9	102.6
November	110.5	105.8	107.4	123.0	136.8	115.3	108.5	111.5	103.4
December	110.6	105.7	106.8	125.0	140.6	121.5	107.8	111.8	102.3
2019									
January	110.1	105.2	105.4	119.5	129.4	121.4	106.9	111.6	101.9
February	109.6	104.8	105.8	121.0	133.5	123.4	105.7	111.1	102.0
March	109.7	105.3	106.5	119.2	131.5	125.2	104.1	109.1	103.2
April	109.0	103.8	104.5	118.2	129.3	120.7	104.3	109.2	101.0
May	109.2	104.3	105.0	120.2	132.3	126.4	106.2	110.0	101.2
June	109.3	104.7	105.5	121.7	136.1	127.1	104.1	109.8	101.5
July	109.1	104.5	105.6	122.5	138.1	127.2	105.5	109.0	101.3
August	109.9	104.7	105.5	122.2	137.0	124.6	105.3	109.7	101.2
September	109.5	104.1	104.7	118.7	129.5	127.0	105.2	109.7	101.2
October	109.0	103.8	104.7	115.4	123.4	127.6	103.1	109.3	101.9
November	110.0	105.3	106.7	122.5	138.5	126.8	104.5	108.8	102.7
December	109.7	104.7	105.5	119.3	130.9	125.0	106.6	109.1	102.0

Table 2-1. Industrial Production Indexes by Market Groups—*Continued*

(Seasonally adjusted, 2012 = 100.)

Year and month	Consumer goods—*Continued*						Business equipment				Defense and space equipment	Construction supplies	Business supplies
	Nondurable consumer goods—*Continued*												
	Nondurable non-energy consumer goods					Consumer energy products	Total	Transit	Information processing	Industrial and other			
	Total	Foods and tobacco	Clothing	Chemical products	Paper products								
1970	62.1	60.8	498.0	31.8	78.9	48.9	15.8	50.2	1.0	47.9	47.4	62.7	38.7
1971	63.7	62.6	495.1	33.6	80.5	51.1	15.0	48.6	1.0	45.8	42.6	64.7	39.9
1972	67.9	66.1	538.8	36.7	81.2	54.0	17.1	53.0	1.1	52.1	41.4	73.5	43.9
1973	70.1	67.8	549.1	39.5	83.9	54.6	19.8	63.3	1.3	59.2	45.4	79.8	46.6
1974	69.8	68.3	514.9	41.7	82.5	56.2	20.9	61.7	1.6	61.7	46.9	77.9	46.4
1975	68.0	67.0	501.2	40.1	77.8	57.5	18.7	54.3	1.5	54.0	47.3	65.9	42.6
1976	72.4	71.4	527.7	43.7	80.9	60.6	19.9	56.2	1.8	56.2	45.9	70.9	45.4
1977	75.0	72.7	553.5	45.5	88.4	62.8	23.0	66.7	2.4	61.2	41.1	77.1	49.3
1978	77.9	75.4	566.8	48.3	92.7	63.8	25.9	77.3	3.0	65.5	41.9	81.5	52.0
1979	76.8	74.8	535.0	48.4	93.8	65.5	29.2	90.2	3.8	69.6	44.8	83.6	53.8
1980	77.7	76.0	544.3	48.3	94.9	62.7	29.9	85.1	4.8	67.3	53.2	77.4	52.5
1981	78.2	76.2	544.4	49.0	97.2	62.2	30.8	79.7	5.7	67.1	57.6	76.1	53.7
1982	79.8	78.8	542.8	49.1	99.8	62.6	28.1	61.1	6.4	58.1	68.9	69.1	53.0
1983	80.8	78.9	558.4	49.7	104.2	62.9	28.3	60.1	7.4	53.8	69.3	73.9	55.6
1984	82.2	79.8	560.4	50.8	109.5	64.9	32.5	62.0	9.1	61.3	79.4	80.3	60.5
1985	83.5	82.2	537.0	51.3	115.1	64.7	33.7	64.1	9.8	62.1	88.9	82.2	61.9
1986	85.6	83.5	537.0	55.3	116.6	65.8	33.2	58.6	9.9	61.8	94.4	84.9	63.9
1987	88.6	85.4	542.4	59.6	123.3	68.2	35.5	59.5	11.3	64.0	96.4	90.4	67.8
1988	91.2	87.9	534.5	63.2	126.6	71.8	39.2	66.4	12.7	69.5	97.2	92.4	70.3
1989	90.7	87.3	510.3	64.5	127.1	72.2	40.5	69.8	13.1	71.8	97.3	92.1	71.3
1990	92.8	89.8	501.4	66.9	129.6	71.5	42.0	74.7	14.2	71.3	93.9	91.4	72.9
1991	93.9	90.3	499.7	69.3	130.2	73.5	41.4	77.6	14.4	67.5	87.0	86.4	71.8
1992	95.2	91.7	511.5	69.1	131.8	72.7	43.2	75.8	16.3	68.7	80.8	90.1	73.6
1993	95.9	91.0	521.0	71.0	134.9	75.8	45.3	70.9	17.9	73.3	76.4	94.1	75.9
1994	98.4	94.5	530.5	72.5	133.4	76.7	48.5	68.6	20.4	78.2	71.8	101.0	78.8
1995	100.7	96.9	527.9	75.6	134.3	78.7	53.1	68.0	24.6	83.1	69.3	103.3	82.0
1996	101.4	96.6	514.0	79.3	133.7	82.1	58.3	71.9	30.1	86.1	67.6	108.0	85.2
1997	104.6	98.2	512.7	83.8	145.5	81.5	67.1	84.4	38.4	91.4	66.7	113.2	91.2
1998	107.5	100.9	487.1	88.0	152.4	81.2	74.8	99.1	46.0	94.3	69.7	119.2	96.4
1999	106.7	98.9	466.9	89.5	155.4	84.2	78.9	97.0	55.8	92.1	67.6	122.2	100.5
2000	108.2	100.3	451.6	92.9	154.8	85.8	85.0	86.1	68.0	97.6	60.1	124.8	104.4
2001	108.5	100.4	403.1	98.0	150.4	86.9	80.1	83.0	66.9	88.3	66.0	119.1	100.6
2002	108.2	98.5	308.9	106.2	148.2	90.6	75.0	75.1	59.2	86.8	66.7	119.0	100.7
2003	109.0	99.8	291.0	108.7	142.8	91.1	75.2	70.8	62.7	86.4	71.0	118.7	102.6
2004	109.7	100.9	246.2	111.5	143.6	92.7	79.4	74.8	69.3	88.7	69.7	121.5	105.1
2005	113.4	104.3	236.4	117.3	142.1	96.1	85.4	81.9	76.0	93.6	76.4	127.4	109.0
2006	114.7	104.6	229.3	121.5	141.6	96.5	93.4	95.1	87.0	97.3	76.1	130.6	110.8
2007	113.5	104.4	184.2	122.4	135.6	100.5	99.0	98.3	97.3	101.2	90.7	129.4	112.1
2008	109.1	100.9	148.8	118.1	126.9	100.0	96.7	88.3	100.6	99.1	98.0	117.5	107.2
2009	103.5	99.5	108.8	108.9	112.3	100.4	80.0	71.1	90.4	79.1	93.8	90.4	95.0
2010	101.8	99.8	107.6	103.9	106.8	102.3	86.1	80.2	92.6	85.6	100.9	93.6	98.0
2011	102.0	99.7	105.2	105.7	103.9	103.1	91.1	82.9	92.9	93.8	98.0	95.9	99.0
2012	100.0	100.0	100.0	100.0	100.0	100.0	100.0	100.0	100.0	100.0	100.0	100.0	100.0
2013	97.6	101.5	92.2	90.9	95.6	105.3	99.9	105.9	100.2	97.3	97.2	103.1	101.8
2014	97.3	101.2	88.2	91.5	94.2	104.9	101.7	115.5	97.8	97.6	93.9	106.4	102.8
2015	98.8	103.1	83.8	93.6	91.7	102.8	99.6	120.7	98.9	91.3	91.7	107.1	101.4
2016	98.4	103.9	80.0	91.8	89.2	105.1	94.4	110.5	98.7	86.1	89.1	108.1	102.0
2017	98.9	106.0	73.4	91.2	86.1	104.2	97.8	109.7	104.2	90.4	90.9	111.8	104.3
2018	100.0	107.4	71.4	93.6	80.5	109.7	101.0	111.6	108.2	93.8	93.1	114.9	105.2
2019	99.3	107.0	63.6	93.4	76.5	108.3	101.4	107.8	113.5	94.1	101.7	116.5	104.1
2018													
January	99.2	106.0	73.2	93.2	82.3	112.3	99.4	109.6	106.9	92.4	90.2	111.9	105.1
February	100.9	108.4	73.2	93.8	84.5	103.4	99.4	110.5	106.7	92.0	90.2	115.1	105.1
March	99.9	107.4	73.0	92.7	83.7	108.3	99.6	111.3	107.3	91.7	90.6	114.4	105.3
April	100.4	108.3	73.6	93.1	81.6	114.6	100.2	111.3	107.9	92.6	92.1	114.7	106.0
May	99.8	107.6	73.1	93.1	79.7	107.3	98.5	105.6	107.5	92.0	91.5	115.1	105.2
June	99.9	108.0	71.1	93.2	77.6	108.1	100.3	112.0	108.1	92.5	92.3	114.7	105.2
July	100.6	108.6	73.7	94.0	78.3	107.0	100.7	111.5	109.3	92.9	92.9	114.5	105.1
August	100.3	108.1	71.5	93.8	79.8	108.4	102.0	112.2	108.7	95.1	93.9	115.3	105.1
September	100.1	107.9	69.8	93.5	80.3	108.5	102.5	114.1	108.5	95.3	94.5	114.8	104.8
October	99.4	106.2	69.1	93.8	80.9	113.0	102.8	112.3	108.5	96.6	95.3	115.3	105.6
November	99.4	105.9	68.3	94.6	80.4	116.3	103.0	112.2	108.4	97.1	95.4	115.5	105.4
December	99.7	107.0	67.0	94.2	77.7	110.1	103.2	116.0	110.2	95.3	98.1	117.5	104.8
2019													
January	99.6	107.2	65.2	93.7	77.9	108.9	103.2	111.8	111.3	96.5	98.8	117.8	105.3
February	99.8	106.9	63.1	94.3	79.6	108.7	101.8	111.5	111.6	94.0	98.8	115.9	104.7
March	100.4	107.6	63.9	94.9	79.4	112.1	102.7	112.9	112.3	94.7	99.9	116.1	104.3
April	99.1	106.3	65.0	93.1	78.9	106.7	100.7	108.1	112.6	93.1	100.2	115.2	103.8
May	98.5	106.3	64.5	91.6	77.8	109.8	101.0	106.1	112.4	94.5	100.6	115.9	104.4
June	100.1	109.2	64.7	92.2	77.0	105.2	101.6	107.9	113.8	94.3	101.9	117.0	103.6
July	99.1	106.0	63.9	94.4	76.1	108.1	100.9	108.3	113.5	93.0	102.0	116.0	103.9
August	99.1	105.8	63.4	94.9	75.6	107.6	101.7	108.3	113.5	94.5	102.4	117.0	104.2
September	98.9	106.4	63.6	93.3	75.9	108.1	100.5	104.2	115.4	93.3	103.3	117.1	104.2
October	99.2	107.2	62.9	93.2	75.6	110.5	99.7	100.5	114.4	93.8	103.6	116.4	104.1
November	99.0	107.2	62.0	93.0	73.2	115.0	101.8	107.1	115.3	94.3	104.1	116.3	104.5
December	99.9	109.8	61.9	92.3	71.3	108.2	101.3	104.4	115.9	94.4	105.4	117.5	104.1

Table 2-1. Industrial Production Indexes by Market Groups—*Continued*

(Seasonally adjusted, 2012 = 100.)

Year and month	Total	Materials										Energy materials
		Non-energy materials										
		Total	Durable				Nondurable					
			Total	Consumer parts	Equipment parts	Other	Total	Textile	Paper	Chemicals		
1970	36.6	30.8	20.5	45.7	3.6	64.2	61.6	151.6	66.7	66.7		65.5
1971	37.2	31.3	20.6	50.6	3.6	61.8	64.2	158.8	69.7	69.7		66.1
1972	40.9	35.1	23.3	56.3	4.1	69.8	70.8	167.2	74.3	74.3		68.6
1973	44.6	38.9	26.6	65.3	4.9	77.5	74.3	163.0	80.1	80.1		70.5
1974	44.5	38.8	26.3	57.8	5.1	77.2	75.4	152.1	84.3	84.3		70.3
1975	39.7	33.3	22.0	46.6	4.4	64.5	67.5	148.9	73.3	73.3		69.9
1976	43.1	37.0	24.5	59.5	4.8	69.2	74.8	166.3	80.6	80.6		71.6
1977	46.2	40.1	26.7	65.2	5.4	73.4	80.2	177.2	84.2	84.2		74.0
1978	48.5	42.7	28.9	69.0	6.0	78.3	83.2	175.2	88.3	88.3		74.9
1979	49.8	43.9	29.8	65.0	6.6	80.2	84.7	173.8	91.9	91.9		76.9
1980	47.9	41.3	27.6	50.1	6.8	74.1	81.9	169.7	92.6	92.6		77.6
1981	48.2	41.4	27.6	47.5	7.0	74.3	82.6	165.9	94.4	94.4		78.4
1982	44.5	37.2	24.0	40.4	6.4	62.5	78.3	151.6	95.1	95.1		75.1
1983	45.7	39.8	25.7	49.1	6.5	65.5	83.8	169.9	101.1	101.1		72.8
1984	50.0	44.4	29.8	58.4	7.8	72.6	87.4	169.4	107.3	107.3		77.4
1985	49.9	44.4	30.0	60.8	7.8	72.0	86.7	160.3	106.6	106.6		77.2
1986	49.9	45.2	30.3	60.2	8.0	73.4	89.4	166.7	111.1	111.1		74.4
1987	52.6	48.2	32.5	61.9	8.7	78.6	94.7	186.0	116.4	116.4		76.3
1988	55.5	51.3	35.1	66.9	9.5	84.2	98.2	184.2	120.3	120.3		79.2
1989	55.9	51.6	35.2	63.5	9.9	84.5	99.0	188.0	120.2	120.2		80.1
1990	56.3	51.7	35.3	59.1	10.2	85.0	99.3	178.7	120.6	120.6		81.7
1991	55.5	50.6	34.3	55.8	10.2	81.3	98.2	178.7	118.1	118.1		82.0
1992	57.3	53.1	36.6	62.8	10.7	85.7	100.7	189.3	120.9	120.9		81.2
1993	59.3	55.5	39.0	71.0	11.4	89.1	101.8	197.2	120.9	120.9		81.5
1994	63.0	59.9	43.3	82.0	13.0	95.6	104.9	209.0	125.5	125.5		83.0
1995	66.4	63.8	47.6	85.2	15.7	99.2	105.9	205.3	128.7	128.7		84.4
1996	69.9	67.9	52.8	87.5	19.4	102.2	104.9	199.4	124.6	124.6		85.7
1997	75.4	74.9	60.0	94.3	24.5	107.3	109.9	208.6	126.0	126.0		85.5
1998	79.8	80.4	66.4	97.7	30.0	110.2	111.3	206.9	127.2	127.2		86.1
1999	84.8	86.9	74.3	107.3	37.2	111.9	112.7	203.3	128.9	128.9		86.0
2000	89.0	91.9	81.1	107.7	45.7	112.9	112.0	195.2	126.9	126.9		87.5
2001	85.5	87.3	77.7	95.9	46.6	105.0	104.8	172.6	120.3	120.3		86.6
2002	86.6	89.1	79.3	100.9	47.8	105.1	106.7	175.3	120.5	120.5		86.4
2003	87.7	90.7	82.0	100.5	53.1	104.3	105.8	166.3	116.6	116.6		86.4
2004	90.7	95.0	87.5	101.6	60.3	108.3	107.8	158.8	117.9	117.9		86.5
2005	92.8	98.9	92.5	104.1	67.9	110.7	109.1	162.5	117.4	117.4		85.8
2006	94.7	101.3	96.1	103.9	73.8	113.0	109.3	147.3	116.4	116.4		86.8
2007	98.0	105.6	100.2	104.3	83.4	112.9	114.1	134.8	116.4	116.4		88.6
2008	95.6	100.7	97.3	90.7	87.3	107.7	105.7	116.8	111.3	111.3		89.3
2009	84.9	83.8	76.5	64.9	71.9	84.5	95.4	95.8	97.8	97.8		86.7
2010	91.9	93.0	87.9	82.2	82.4	94.0	101.1	106.4	99.7	99.7		90.6
2011	95.9	97.3	95.4	89.5	93.7	98.6	100.3	102.2	99.3	99.3		94.1
2012	100.0	100.0	100.0	100.0	100.0	100.0	100.0	100.0	100.0	100.0		100.0
2013	103.3	101.9	102.4	103.8	101.6	102.5	101.1	104.5	100.1	100.1		105.1
2014	108.5	103.9	106.6	109.1	107.7	105.1	99.7	103.3	98.5	98.5		114.0
2015	106.9	102.1	104.4	110.9	103.9	102.5	98.5	96.8	97.2	97.2		112.6
2016	103.7	100.9	102.1	108.5	101.1	100.5	99.0	94.2	96.0	96.0		105.9
2017	106.5	102.9	104.0	109.1	103.2	102.6	101.1	94.5	94.7	94.7		110.1
2018	112.8	106.1	107.6	112.2	108.2	105.7	103.7	97.7	93.0	93.0		121.3
2019	115.0	105.7	107.4	106.6	110.9	105.7	103.1	99.5	89.0	89.0		127.8
2018												
January	108.9	103.5	105.1	109.3	105.1	103.7	101.1	96.3	93.9	99.7		115.4
February	109.5	104.5	106.6	113.1	105.5	105.0	101.4	98.1	92.2	99.8		115.2
March	110.5	105.1	106.8	114.1	105.7	105.0	102.5	95.9	93.6	102.4		117.1
April	111.5	105.8	107.4	112.9	107.3	105.5	103.5	96.5	93.0	103.6		118.3
May	111.6	105.4	106.5	108.8	107.0	105.3	103.7	96.2	93.0	104.5		119.3
June	112.4	106.0	107.2	110.9	107.9	105.5	104.1	96.9	92.4	105.7		120.4
July	113.3	106.5	107.5	110.7	108.8	105.7	104.8	96.6	93.9	106.1		122.0
August	114.5	106.9	108.0	112.0	110.0	105.6	105.1	99.6	92.6	107.1		124.4
September	114.8	106.9	108.5	112.6	110.2	106.1	104.4	98.2	92.6	105.8		125.1
October	114.9	106.9	108.6	112.2	110.3	106.4	104.3	101.4	92.8	105.7		125.3
November	115.7	107.3	109.1	113.6	110.0	107.0	104.6	100.0	92.8	106.6		126.9
December	115.9	108.1	110.0	115.6	110.3	107.9	105.3	96.8	93.1	107.8		125.9
2019												
January	115.6	107.2	108.7	109.9	110.8	107.0	101.1	99.0	92.7	107.0		126.9
February	114.8	106.6	108.5	109.6	111.0	106.7	101.4	100.2	90.3	105.6		125.5
March	114.5	105.9	107.9	108.1	111.5	105.9	102.5	101.4	88.5	104.8		125.9
April	114.8	105.6	107.3	106.3	110.2	106.0	103.5	100.0	89.2	104.6		127.5
May	114.8	105.4	107.2	107.1	110.6	105.4	103.7	100.0	88.1	104.5		127.6
June	114.4	105.3	107.5	107.5	111.1	105.4	104.1	100.7	86.2	103.8		126.7
July	114.2	105.0	107.2	109.5	110.6	104.6	104.8	97.3	87.1	103.3		126.6
August	115.6	106.0	107.8	107.8	111.3	105.8	105.1	100.2	89.2	105.4		128.9
September	115.5	105.6	106.9	103.5	110.7	105.8	104.4	99.3	88.5	105.6		129.6
October	114.9	104.8	105.8	100.1	110.6	105.0	104.3	98.7	89.7	105.2		129.1
November	115.3	105.3	107.0	106.2	111.1	104.9	104.6	99.5	89.1	104.3		129.5
December	115.2	105.9	107.2	105.6	111.2	105.6	105.3	99.2	89.6	105.5		128.1

Table 2-1. Industrial Production Indexes by Market Groups—*Continued*

(Seasonally adjusted, 2012 = 100.)

Year and month	Special aggregates											
	Energy						Non-energy					Total non-energy, excluding high-tech
									Selected high-tech			
	Total	Consumer energy products	Commercial energy products	Oil and gas well drilling	Converted fuels	Primary energy	Total	Total	Computers and peripheral equipment	Communications equipment	Semiconductors and related components	
1970	57.5	48.9	34.1	. . .	61.1	71.6	36.0	0.1	. . .	. . .	. . .	57.4
1971	58.6	51.1	35.8	. . .	62.7	71.0	36.5	0.1	. . .	. . .	. . .	58.5
1972	61.3	54.0	37.8	62.0	66.6	72.2	40.3	0.1	0.0	2.1	0.1	64.3
1973	62.9	54.6	39.7	58.4	69.3	73.5	44.0	0.1	0.0	2.3	0.1	69.8
1974	63.3	56.2	39.8	67.8	68.1	74.1	43.8	0.1	0.1	2.5	0.1	69.0
1975	63.6	57.5	41.1	76.5	65.8	75.3	39.1	0.1	0.1	2.4	0.1	61.5
1976	65.7	60.6	43.2	86.2	69.5	75.4	42.6	0.2	0.1	2.6	0.1	66.7
1977	68.3	62.8	44.6	109.4	72.2	77.8	46.2	0.2	0.2	3.1	0.1	71.7
1978	69.4	63.8	45.9	121.9	71.6	79.7	49.2	0.2	0.2	3.4	0.1	75.7
1979	71.5	65.5	48.2	130.0	74.2	81.4	50.6	0.3	0.3	4.1	0.1	77.1
1980	71.7	62.7	47.1	153.7	73.0	83.1	48.9	0.4	0.5	4.9	0.1	73.4
1981	72.8	62.2	48.0	184.7	72.1	84.7	49.5	0.5	0.7	5.3	0.2	73.6
1982	70.2	62.6	48.6	162.5	66.5	82.4	46.7	0.5	0.8	5.5	0.2	68.5
1983	68.1	62.9	49.5	125.9	66.3	78.9	48.9	0.6	1.1	5.9	0.2	71.1
1984	72.1	64.9	52.2	137.1	70.4	84.0	53.8	0.8	1.6	6.3	0.3	77.2
1985	71.9	64.7	53.9	125.6	70.2	83.7	54.7	0.9	1.9	6.1	0.3	78.3
1986	69.4	65.8	55.3	61.2	67.9	80.6	56.0	0.9	2.0	5.8	0.3	80.1
1987	71.5	68.2	58.2	59.0	71.4	81.4	59.2	1.1	2.6	6.4	0.4	83.8
1988	74.5	71.8	60.3	70.8	74.6	84.1	62.4	1.3	3.2	7.5	0.4	87.6
1989	75.3	72.2	62.4	61.7	77.4	83.8	62.9	1.4	3.4	7.9	0.4	88.0
1990	76.5	71.5	64.1	65.3	77.8	86.2	63.4	1.5	3.6	9.4	0.5	88.1
1991	76.9	73.5	65.0	50.8	77.9	86.6	62.2	1.7	3.8	10.1	0.6	85.8
1992	76.1	72.7	64.6	40.5	79.2	84.5	64.6	2.0	4.7	12.0	0.7	88.1
1993	77.3	75.8	66.4	51.0	81.0	83.9	66.9	2.4	5.8	14.2	0.8	90.4
1994	78.7	76.7	69.1	51.2	82.3	85.5	70.9	3.0	7.3	17.8	1.0	94.4
1995	80.3	78.7	71.5	49.8	83.4	87.1	74.5	4.2	10.2	22.0	1.4	96.8
1996	82.3	82.1	73.7	56.7	84.9	88.3	78.2	6.0	14.5	28.0	2.2	98.3
1997	82.7	81.5	76.7	66.9	86.4	87.2	84.8	8.8	20.8	39.1	3.4	103.2
1998	83.0	81.2	77.8	61.5	87.8	87.3	90.6	12.2	29.0	47.4	4.9	107.0
1999	83.7	84.2	80.9	50.0	89.4	86.1	95.1	17.7	38.4	60.0	8.0	108.3
2000	85.6	85.8	84.1	60.2	91.6	87.2	99.1	24.5	46.1	81.1	11.8	109.1
2001	85.6	86.9	85.1	71.6	89.5	87.0	95.4	25.9	47.7	76.7	13.3	103.9
2002	85.9	90.6	86.6	52.8	91.3	85.7	95.8	26.4	48.8	58.4	15.8	104.2
2003	86.7	91.1	91.5	61.7	91.6	85.5	97.1	31.9	54.0	63.3	20.9	104.2
2004	87.5	92.7	94.4	68.3	93.8	84.7	100.1	38.4	56.3	74.1	27.0	106.2
2005	88.1	96.1	97.1	75.8	95.4	83.1	104.2	45.6	68.6	76.9	33.4	109.5
2006	89.2	96.5	97.7	87.1	95.1	84.6	106.9	54.8	87.9	96.1	38.3	111.2
2007	91.6	100.5	100.2	91.1	99.6	85.5	109.6	65.3	104.5	99.5	48.8	112.8
2008	92.2	100.0	100.1	98.2	97.5	87.1	104.2	75.4	127.2	100.1	58.4	106.0
2009	89.2	100.4	98.2	63.3	92.0	85.3	89.4	66.6	113.7	87.5	51.5	90.7
2010	92.7	102.3	99.4	71.5	97.5	88.7	94.7	80.9	93.6	92.4	75.4	95.5
2011	95.9	103.1	100.4	88.4	96.3	93.5	97.5	91.1	80.8	100.7	90.6	97.9
2012	100.0	100.0	100.0	100.0	100.0	100.0	100.0	100.0	100.0	100.0	100.0	100.0
2013	104.8	105.3	102.8	101.9	99.3	106.5	100.8	110.7	100.3	116.8	111.0	100.4
2014	111.8	104.9	104.0	110.4	100.5	117.4	102.3	122.7	102.6	109.4	134.3	101.5
2015	109.3	102.8	104.5	77.3	101.0	115.4	101.8	126.7	111.8	117.5	134.7	100.9
2016	104.1	105.1	106.7	45.2	99.5	106.4	100.7	133.9	111.7	121.4	145.7	99.6
2017	107.5	104.2	107.3	65.2	97.9	113.1	102.7	137.4	127.8	117.1	149.2	101.6
2018	117.0	109.7	111.3	75.8	102.6	126.8	105.1	146.1	134.1	120.4	161.4	103.7
2019	120.7	108.3	110.6	71.0	101.9	136.0	104.9	153.6	133.0	134.5	168.2	103.5
2018												
January	113.3	112.3	112.0	68.0	103.7	118.3	103.3	142.4	135.7	116.9	155.9	102.0
February	111.1	103.4	108.5	70.5	96.7	120.7	104.5	142.3	134.3	115.9	156.6	103.3
March	113.7	108.3	111.1	72.8	100.4	121.9	104.4	142.7	134.9	114.8	157.6	103.2
April	116.0	114.6	113.1	74.9	102.1	122.8	104.9	143.3	135.8	114.4	158.8	103.7
May	115.1	107.3	110.5	77.4	103.2	123.8	104.1	143.4	133.3	115.2	159.4	102.8
June	116.1	108.1	111.2	79.2	100.6	126.4	104.8	146.5	135.9	117.7	162.9	103.5
July	116.9	107.0	110.7	77.0	103.3	127.4	105.2	148.3	135.1	121.0	164.7	103.9
August	119.0	108.4	112.1	77.3	104.3	130.4	105.7	150.5	134.5	124.0	167.5	104.3
September	119.3	108.5	110.7	77.0	105.3	131.0	105.7	150.1	131.1	125.6	167.1	104.3
October	120.6	113.0	113.4	77.7	105.6	131.2	105.6	149.2	131.9	126.0	165.0	104.2
November	122.4	116.3	113.2	78.9	107.2	132.7	105.8	147.5	133.7	125.7	161.4	104.5
December	120.1	110.1	109.9	79.0	98.9	134.7	106.6	147.4	132.8	127.7	160.5	105.3
2019												
January	120.8	108.9	112.1	79.0	103.0	118.3	105.8	149.7	134.6	131.0	162.3	104.5
February	119.6	108.7	110.4	77.9	102.9	120.7	105.5	150.9	135.9	132.6	163.3	104.1
March	120.6	112.1	110.4	78.2	103.8	121.9	105.3	152.1	136.9	134.8	163.9	103.9
April	120.5	106.7	110.3	77.2	99.5	122.8	104.5	150.3	130.6	133.6	163.5	103.1
May	121.1	109.8	110.9	74.1	100.4	123.8	104.6	149.7	131.0	131.8	163.2	103.2
June	119.2	105.2	107.4	74.4	97.8	126.4	105.2	151.6	131.7	134.2	165.2	103.8
July	120.2	108.1	112.0	72.0	101.9	127.4	104.7	152.9	131.3	135.9	166.9	103.2
August	121.3	107.6	110.6	70.2	102.4	130.4	105.3	155.4	134.4	136.0	170.5	103.8
September	121.8	108.1	111.1	66.3	104.9	131.0	104.7	155.5	130.5	136.7	171.7	103.2
October	122.0	110.5	111.8	63.0	102.7	131.2	104.1	156.1	128.6	137.0	173.3	102.5
November	123.2	115.0	112.5	60.5	105.2	132.7	105.0	159.1	135.1	137.2	176.5	103.4
December	120.7	108.2	111.9	60.1	98.9	134.7	105.2	159.5	134.1	136.1	178.1	103.7

. . . = Not available.

Table 2-2. Industrial Production Indexes by NAICS Industry Groups

(Seasonally adjusted, 2012 = 100.)

Year and month	Total industrial production	Manu-facturing (SIC)	Manufacturing (NAICS)										
			Total	Durable goods manufacturing									
				Total	Wood products	Nonmetallic mineral products	Primary metals	Fabricated metal products	Machinery	Computer and electronic products	Electrical equipment, appliances, and components	Motor vehicles and parts	Aerospace and miscellaneous transport equipment
1970	38.6	35.9	. . .	. . .	. . .	. . .	. . .	. . .	. . .	. . .	. . .	. . .	. . .
1971	39.1	36.5	. . .	. . .	. . .	. . .	. . .	. . .	. . .	. . .	. . .	. . .	. . .
1972	42.9	40.3	38.1	25.5	100.0	94.8	115.4	69.1	54.9	0.5	80.7	45.9	53.5
1973	46.4	43.9	41.7	28.7	96.8	101.8	134.3	76.3	63.4	0.6	90.9	52.5	60.9
1974	46.3	43.8	41.6	28.5	87.9	100.7	137.6	75.0	66.5	0.7	88.7	45.1	62.0
1975	42.2	39.2	37.1	24.8	81.0	90.1	106.7	64.8	57.9	0.6	71.2	39.3	58.8
1976	45.5	42.7	40.5	27.1	91.3	95.1	113.2	69.4	60.4	0.7	80.4	50.2	55.1
1977	48.9	46.4	44.0	29.8	98.4	101.3	114.6	75.3	65.9	0.9	88.6	57.1	55.5
1978	51.6	49.2	46.7	32.1	99.7	108.0	122.0	79.0	70.9	1.2	94.0	59.5	61.2
1979	53.2	50.7	48.2	33.7	96.1	107.8	124.7	82.5	74.9	1.4	97.9	54.4	71.3
1980	51.8	48.9	46.3	32.2	89.0	97.2	109.3	77.8	71.2	1.7	92.1	40.1	76.7
1981	52.5	49.4	46.8	32.5	87.0	93.1	109.4	77.3	70.5	2.0	90.9	39.0	73.0
1982	49.8	46.7	44.1	29.7	78.2	82.4	76.8	69.2	58.9	2.2	81.9	35.2	68.3
1983	51.1	49.0	46.2	31.2	90.5	88.6	79.0	69.7	53.2	2.6	84.7	44.9	65.3
1984	55.7	53.7	50.9	35.6	97.0	95.6	86.9	75.9	62.0	3.2	95.3	53.8	69.7
1985	56.4	54.6	51.6	36.4	98.0	97.3	80.0	76.9	62.2	3.4	93.8	55.9	74.4
1986	56.9	55.8	52.8	37.0	106.5	101.3	78.3	76.4	61.2	3.5	95.6	55.9	78.3
1987	59.9	59.0	55.8	39.2	115.9	106.9	84.5	77.9	62.5	4.0	96.8	57.9	81.2
1988	63.0	62.1	58.9	42.1	115.7	109.1	94.5	81.8	68.8	4.4	101.5	61.9	85.4
1989	63.6	62.6	59.5	42.6	113.8	108.2	92.3	81.2	71.4	4.6	100.0	61.2	90.6
1990	64.2	63.1	60.0	42.7	112.5	106.6	91.3	80.2	69.6	5.0	97.4	57.6	90.9
1991	63.2	61.9	58.9	41.4	105.2	98.1	85.7	76.6	65.3	5.2	92.4	55.1	87.7
1992	65.1	64.2	61.3	43.6	111.1	102.4	88.2	79.0	65.2	5.9	97.9	62.7	81.1
1993	67.2	66.5	63.6	46.1	112.4	104.6	92.4	82.0	70.0	6.5	104.0	69.3	75.6
1994	70.8	70.4	67.6	50.0	119.1	110.5	99.4	89.1	76.7	7.7	111.6	79.6	67.9
1995	74.0	74.0	71.2	54.1	122.1	113.7	100.5	94.6	82.1	9.8	114.2	81.9	64.5
1996	77.4	77.6	74.9	59.1	126.0	121.1	102.8	98.0	84.9	12.8	117.6	82.6	67.0
1997	83.0	84.2	81.2	66.1	129.4	125.2	107.1	102.4	89.6	17.1	122.0	89.1	74.8
1998	87.8	89.8	86.6	73.0	135.1	131.5	108.9	105.7	92.0	21.9	126.4	93.7	86.7
1999	91.7	94.3	91.1	79.3	140.1	132.7	108.8	106.5	90.1	29.0	128.4	103.8	83.5
2000	95.2	98.2	95.1	85.0	138.3	132.8	105.0	110.7	94.8	37.6	134.9	103.3	73.2
2001	92.3	94.6	91.8	81.6	129.6	127.8	95.0	102.7	83.9	39.1	121.5	94.4	77.9
2002	92.6	95.1	92.4	82.0	135.2	127.9	95.5	100.5	80.8	38.8	111.9	103.9	73.7
2003	93.8	96.4	93.9	84.2	134.9	129.1	93.8	99.4	80.4	44.6	109.1	107.6	70.4
2004	96.4	99.4	97.0	88.2	138.4	133.0	101.8	99.8	83.7	52.5	111.0	108.1	69.4
2005	99.6	103.4	101.2	93.4	147.7	138.6	99.1	104.4	88.9	59.7	112.9	108.8	77.3
2006	101.8	106.1	104.0	97.8	148.9	141.2	101.9	110.2	93.1	68.0	113.6	107.1	82.0
2007	104.4	109.0	107.3	102.7	139.2	139.7	104.0	114.9	97.1	78.4	118.4	106.3	96.3
2008	100.8	103.8	102.4	99.2	119.0	123.4	104.2	110.7	94.5	84.8	113.9	85.1	98.2
2009	89.2	89.5	88.4	80.6	90.9	93.0	77.5	85.2	73.5	75.6	89.5	62.3	89.4
2010	94.1	94.7	94.2	89.2	94.1	95.9	95.1	90.7	82.1	85.6	93.1	82.7	90.6
2011	97.1	97.5	97.2	94.7	94.3	97.4	102.0	97.1	92.5	92.6	97.6	90.4	90.5
2012	100.0	100.0	100.0	100.0	100.0	100.0	100.0	100.0	100.0	100.0	100.0	100.0	100.0
2013	102.0	100.9	101.1	102.1	105.8	105.2	100.3	101.8	95.4	103.2	100.0	107.2	102.9
2014	105.2	102.0	102.3	105.1	108.4	109.1	104.0	103.6	96.7	107.4	101.8	117.1	106.8
2015	104.1	101.5	101.9	103.9	112.7	109.8	96.8	100.2	89.0	108.1	101.3	123.2	106.8
2016	102.1	100.7	101.1	101.7	116.9	111.3	92.5	96.5	82.2	110.4	101.0	124.8	99.5
2017	104.4	102.7	103.2	104.0	124.1	115.3	93.7	97.9	87.9	115.2	101.8	124.7	99.9
2018	108.6	105.0	106.0	107.5	127.1	119.6	97.6	102.5	92.6	120.9	103.7	129.9	100.1
2019	109.4	104.8	105.9	108.2	126.7	119.4	96.9	103.5	92.2	127.2	103.3	127.0	101.9
2018													
January	106.2	100.8	101.7	102.4	121.7	105.3	93.7	98.4	87.7	116.5	101.1	120.0	98.2
February	106.3	103.4	104.4	106.3	126.1	111.3	98.1	100.1	91.5	114.7	102.2	138.2	97.7
March	107.3	104.7	105.8	108.0	128.9	114.5	97.6	100.8	92.8	119.3	101.8	141.1	99.1
April	106.8	105.3	106.3	108.6	130.5	120.6	97.3	101.8	96.1	118.1	103.1	137.4	99.7
May	106.3	104.3	105.3	106.3	128.5	123.5	97.4	101.8	93.3	118.4	102.5	122.0	98.6
June	109.7	107.0	107.9	109.9	132.5	126.0	98.5	103.7	96.0	122.3	104.3	135.4	100.7
July	108.5	103.8	104.4	103.7	126.4	124.8	95.9	102.6	89.2	120.7	102.4	104.6	98.6
August	111.7	107.3	108.0	109.5	131.7	126.0	98.5	103.6	94.0	122.0	104.8	135.3	100.3
September	110.3	106.9	107.8	109.6	129.2	122.8	98.5	103.3	93.9	124.4	105.1	136.8	100.9
October	109.8	107.0	108.0	110.7	129.5	125.2	98.5	103.9	94.9	123.5	106.1	140.8	101.5
November	109.7	105.5	106.5	108.5	122.6	119.3	99.2	104.1	93.5	123.3	106.1	128.6	101.8
December	109.9	104.5	105.5	107.0	117.9	116.0	97.4	105.6	88.4	127.2	105.1	118.7	104.1
2019													
January	109.7	103.2	104.3	106.0	122.0	111.1	97.1	103.7	91.6	122.2	101.9	119.7	102.5
February	109.3	104.3	105.5	108.3	122.4	108.2	100.0	103.3	94.1	120.9	102.4	136.4	102.2
March	109.8	105.5	106.9	110.1	123.6	112.8	98.9	103.6	96.5	125.4	103.1	137.9	104.6
April	107.4	104.4	105.7	108.7	127.2	118.8	99.7	103.8	94.9	123.7	101.9	128.3	102.1
May	107.9	104.6	105.7	108.5	127.2	121.4	97.8	103.1	95.2	124.4	103.9	127.9	99.8
June	110.8	107.1	108.2	111.3	130.9	125.4	96.9	104.6	97.5	128.6	104.8	137.8	100.9
July	108.8	103.0	103.8	104.8	126.8	123.6	95.2	102.8	87.6	125.4	102.6	109.0	99.6
August	112.1	106.8	107.7	110.1	130.5	126.6	98.2	104.2	92.4	127.0	104.7	135.7	100.7
September	110.1	105.7	106.7	108.7	128.6	125.8	96.5	103.3	90.3	131.3	103.7	127.0	101.3
October	108.8	105.2	106.3	108.1	133.3	124.9	93.9	103.0	90.8	129.8	102.8	125.3	101.9
November	109.2	104.6	105.8	108.3	126.2	118.9	94.2	102.8	88.7	132.0	104.2	129.4	102.7
December	109.2	103.3	104.5	105.9	122.3	114.8	94.9	103.7	86.2	135.2	103.1	109.3	104.7

. . . = Not available.

Table 2-2. Industrial Production Indexes by NAICS Industry Groups—*Continued*

(Seasonally adjusted, 2012 = 100.)

Year and month	Durable goods manufacturing—*Continued*		Manufacturing (NAICS)—*Continued* Nondurable goods manufacturing									Other manufacturing (non-NAICS)
	Furniture and related products	Miscellaneous manufacturing	Total	Food, beverage, and tobacco products	Textile and product mills	Apparel and leather	Paper	Printing and support	Petroleum and coal products	Chemicals	Plastics and rubber products	
1970	...	...	...	...	...	...	...	...	...	...	...	...
1971	...	...	...	...	...	...	...	...	...	...	...	...
1972	77.6	39.6	64.2	64.0	158.4	499.6	76.4	66.3	68.7	48.3	39.6	119.0
1973	81.3	40.8	67.2	64.6	157.1	507.7	82.6	69.7	67.5	52.9	44.7	122.8
1974	75.1	39.9	67.5	65.8	144.4	477.7	86.2	67.6	72.8	54.9	43.1	123.4
1975	64.5	37.1	62.6	64.4	139.6	466.3	74.7	63.1	71.8	48.3	36.4	117.4
1976	71.9	40.5	68.3	68.8	155.7	490.3	82.4	67.7	79.5	54.0	40.2	121.1
1977	82.4	44.1	73.0	70.3	170.0	514.0	86.0	73.3	85.1	58.8	47.5	132.7
1978	88.8	45.1	75.6	72.8	169.6	525.3	89.9	77.6	86.0	61.8	49.6	137.3
1979	88.6	45.3	76.1	72.3	169.2	492.9	91.2	80.0	91.8	63.2	48.9	140.2
1980	85.5	43.0	73.7	73.5	162.3	502.3	91.0	80.6	81.4	59.7	43.8	145.0
1981	84.8	44.6	74.4	74.2	158.8	502.2	92.2	82.6	77.6	60.7	46.3	148.4
1982	79.9	45.1	73.3	76.4	147.5	499.2	90.7	88.9	74.2	56.8	45.6	150.2
1983	87.6	45.1	76.7	76.5	165.8	513.3	96.5	95.5	75.4	60.7	49.6	154.5
1984	98.1	48.8	80.2	77.6	169.8	512.9	101.4	104.0	77.0	64.3	57.3	161.6
1985	99.2	49.5	80.7	80.1	163.9	490.9	99.2	108.2	75.9	63.8	59.5	168.0
1986	103.6	50.7	83.0	81.1	170.1	489.4	103.1	113.6	75.5	66.7	62.0	171.4
1987	111.1	54.5	87.4	82.7	186.7	495.3	106.4	122.0	79.1	71.8	68.7	181.2
1988	109.9	59.6	90.4	85.0	185.0	487.0	110.6	125.9	81.5	76.0	71.7	180.4
1989	109.5	60.4	90.9	84.6	187.8	465.9	111.8	126.4	80.7	77.4	74.2	177.9
1990	107.3	63.3	92.4	87.0	180.2	456.4	111.7	131.2	80.8	79.2	76.3	175.8
1991	99.2	64.5	92.1	87.7	177.8	453.5	111.9	127.1	79.6	79.0	75.4	168.6
1992	107.0	67.3	94.5	89.0	187.6	466.6	114.7	134.0	79.2	80.1	81.2	165.1
1993	111.5	71.2	95.9	88.8	194.8	475.9	116.0	134.4	79.8	81.1	86.9	166.3
1994	115.2	71.7	99.2	91.6	205.2	481.7	121.0	135.9	82.0	83.2	94.2	164.9
1995	117.1	74.4	100.9	94.2	203.0	478.2	122.7	137.9	83.5	84.6	96.6	164.8
1996	118.1	78.0	101.2	93.5	198.1	467.2	118.9	138.9	85.5	86.3	99.8	163.3
1997	130.8	80.2	105.0	95.6	209.1	466.7	121.4	141.7	88.3	91.4	105.9	177.1
1998	140.1	84.9	106.7	98.6	207.5	444.3	122.4	143.5	86.6	92.9	109.7	187.6
1999	144.7	86.8	107.3	96.8	208.3	423.2	123.3	142.9	90.0	94.8	115.5	193.0
2000	147.2	91.6	107.8	98.1	204.1	408.0	120.1	144.2	89.2	96.2	116.7	192.5
2001	138.0	90.7	104.7	98.0	182.7	358.6	113.3	139.4	88.7	94.5	110.0	180.0
2002	142.7	95.9	106.0	97.0	183.6	282.4	114.6	136.2	93.5	99.9	113.9	173.9
2003	140.5	98.8	106.2	98.9	176.3	262.9	111.9	130.8	92.1	101.6	114.0	169.0
2004	144.5	98.8	107.8	99.1	174.1	230.8	112.9	131.2	97.5	105.6	115.4	169.7
2005	149.8	105.3	110.5	102.7	176.6	225.7	112.4	131.0	100.1	109.3	116.6	169.2
2006	147.7	108.7	111.2	102.9	160.2	220.6	111.2	129.5	101.9	111.9	117.4	167.2
2007	142.2	105.9	112.5	103.1	141.7	179.8	111.7	132.2	104.7	117.5	114.1	157.7
2008	128.3	107.7	105.8	100.2	124.5	144.8	106.9	123.7	100.1	108.6	103.4	143.9
2009	93.1	99.8	97.7	99.5	98.7	106.8	95.4	103.6	98.8	98.1	86.4	120.4
2010	91.9	103.4	99.8	99.8	105.4	107.8	97.4	103.5	97.9	101.3	94.2	111.3
2011	95.0	102.9	99.9	99.6	103.1	106.0	97.3	101.8	100.2	101.4	95.5	106.1
2012	100.0	100.0	100.0	100.0	100.0	100.0	100.0	100.0	100.0	100.0	100.0	100.0
2013	100.8	103.2	100.0	101.7	103.7	92.7	100.2	100.3	104.0	96.6	101.2	95.0
2014	101.3	100.0	99.3	101.6	105.8	88.8	99.3	98.5	100.2	95.6	103.9	93.8
2015	106.3	100.4	99.6	103.8	100.9	84.7	98.6	97.5	98.1	95.2	105.8	90.4
2016	106.6	100.7	100.4	105.0	99.0	80.9	97.7	99.2	104.4	94.7	107.1	88.0
2017	106.3	99.5	102.3	107.6	99.7	74.6	97.0	99.8	107.1	96.6	109.3	87.5
2018	106.3	99.4	104.3	109.7	100.9	72.5	96.0	97.6	106.9	100.4	110.2	78.9
2019	105.7	101.7	103.5	109.5	98.3	64.7	93.0	93.7	105.8	100.5	108.3	73.2
2018												
January	104.1	95.8	100.8	104.7	97.5	72.7	95.4	97.8	100.0	97.7	108.9	77.3
February	104.0	97.8	102.3	108.8	99.2	74.9	95.1	97.9	99.4	98.0	110.6	74.8
March	105.0	98.6	103.4	109.0	100.0	74.3	96.4	96.9	102.9	99.5	110.9	74.1
April	104.7	98.9	103.9	109.2	102.3	75.6	96.1	97.8	106.3	99.7	110.5	76.1
May	105.7	100.0	104.1	108.4	99.3	74.4	95.5	97.6	109.3	100.8	109.5	78.0
June	107.3	100.0	105.8	112.1	103.4	72.9	96.5	98.3	112.4	101.0	110.1	81.2
July	108.6	98.3	105.1	109.7	101.6	73.2	96.5	98.0	112.4	101.3	109.0	84.9
August	109.0	99.5	106.4	113.5	103.9	71.9	97.3	97.5	113.4	101.1	110.2	86.6
September	106.9	98.7	105.8	113.8	101.4	70.8	96.1	97.6	109.4	100.5	110.6	81.8
October	107.0	101.1	105.2	111.9	104.3	69.7	96.6	97.5	106.4	100.9	110.2	79.1
November	106.4	100.9	104.3	108.4	101.2	70.1	95.9	96.4	105.8	102.0	110.6	77.9
December	107.4	102.9	103.9	106.5	97.3	69.8	94.3	98.2	105.2	102.7	111.2	75.2
2019												
January	104.8	99.6	102.6	106.1	96.2	64.6	94.8	95.2	101.5	101.2	109.7	69.9
February	103.1	100.8	102.5	107.8	96.6	64.5	94.1	93.3	97.3	101.1	109.7	68.1
March	103.5	101.5	103.6	109.5	100.3	64.8	92.7	93.3	101.5	101.8	108.8	66.4
April	104.2	102.1	102.5	107.6	101.3	66.6	92.5	93.1	103.9	100.3	107.1	69.3
May	105.9	102.4	102.7	107.4	98.1	65.6	91.6	93.2	107.4	100.1	108.4	72.3
June	106.8	102.7	104.8	113.0	101.5	66.2	91.8	93.1	110.9	99.7	110.7	78.3
July	107.8	101.3	102.7	107.6	97.1	63.7	91.5	92.6	112.4	99.7	105.3	81.5
August	107.9	102.5	105.1	111.5	99.2	63.8	94.7	94.0	113.1	101.0	109.0	82.1
September	105.8	101.6	104.5	112.3	98.2	64.7	93.3	94.2	108.8	100.5	107.8	77.1
October	105.1	100.8	104.3	112.7	98.8	63.5	94.2	93.9	104.9	100.4	107.5	73.6
November	105.8	102.1	103.1	109.6	97.7	63.5	93.1	94.0	104.1	100.1	108.0	71.1
December	108.1	102.8	103.0	109.3	94.9	64.8	92.0	94.7	103.4	100.4	107.9	69.2

. . . = Not available.

Table 2-3. Capacity Utilization by NAICS Industry Groups

(Output as a percent of capacity, seasonally adjusted.)

Year and month	Total industry	Total manufac-turing (SIC)	Manufacturing (NAICS) Total	Durable goods manufacturing Total	Wood products	Nonmetallic mineral products	Primary metals	Fabricated metal products	Machinery	Computer and electronic products	Electrical equipment, appliances, and components	Motor vehicles and parts	Aerospace and miscel-laneous transpor-tation equipment
1970	81.2	79.4	. . .	77.7	. . .	73.5	79.4	78.4	79.7	. . .	. . .	66.3	75.4
1971	79.6	77.9	. . .	75.4	. . .	74.9	73.0	78.4	73.8	. . .	. . .	79.1	67.0
1972	84.7	83.4	83.3	82.0	92.2	79.0	82.9	85.0	83.1	80.7	89.8	84.3	65.6
1973	88.3	87.7	87.8	88.6	87.9	83.9	94.7	90.9	92.4	85.2	97.7	91.8	73.5
1974	85.1	84.4	84.5	84.7	77.8	81.4	96.5	86.2	91.9	83.1	91.5	76.7	74.0
1975	75.8	73.7	73.5	71.8	70.0	72.6	75.2	72.2	77.7	68.7	70.7	66.1	70.7
1976	79.8	78.4	78.4	76.5	79.5	77.4	78.6	75.9	79.9	72.1	78.8	81.8	66.2
1977	83.4	82.5	82.5	81.1	86.0	82.4	79.3	79.9	85.4	77.8	85.5	89.9	66.2
1978	85.1	84.4	84.4	83.8	84.8	86.4	84.2	80.7	88.6	80.4	87.7	90.7	72.5
1979	85.0	84.0	84.0	84.0	79.8	84.4	86.1	81.7	90.1	83.5	88.8	80.9	81.4
1980	80.8	78.7	78.4	77.5	72.4	75.1	76.0	75.2	83.5	86.8	81.7	59.7	84.1
1981	79.5	76.9	76.5	75.1	70.6	72.4	77.0	72.8	80.5	84.7	78.8	57.3	76.7
1982	73.6	70.9	70.3	66.4	63.8	64.8	56.5	64.7	66.6	81.1	69.6	50.7	69.8
1983	74.9	73.5	72.9	68.8	75.2	70.2	59.8	66.7	60.5	82.0	72.8	67.3	66.5
1984	80.4	79.4	79.0	76.9	81.1	75.4	69.5	73.6	70.9	87.6	82.3	81.7	69.8
1985	79.2	78.1	77.6	75.8	79.7	75.7	67.6	74.1	70.5	80.4	79.0	83.5	72.4
1986	78.6	78.4	78.0	75.4	83.7	78.1	69.6	73.8	70.0	77.5	80.2	79.1	74.0
1987	81.1	80.9	80.5	77.6	85.7	80.6	78.3	75.4	71.8	79.9	82.4	78.0	75.4
1988	84.2	83.9	83.7	81.9	84.3	81.5	87.9	80.1	80.1	81.3	87.2	82.6	79.7
1989	83.7	83.2	83.1	81.7	82.1	80.6	85.1	79.7	83.8	78.8	86.3	81.0	85.3
1990	82.4	81.5	81.4	79.3	80.2	79.2	85.6	77.6	81.4	78.6	84.0	71.9	84.7
1991	79.9	78.6	78.4	75.4	76.0	73.1	80.8	73.9	76.3	78.3	79.2	64.0	83.8
1992	80.6	79.6	79.6	77.1	78.7	76.8	81.9	76.0	75.1	78.9	82.3	72.4	78.9
1993	81.5	80.5	80.4	78.6	78.8	78.5	85.0	76.9	78.6	78.3	86.7	78.7	74.3
1994	83.5	82.8	82.8	81.5	82.8	81.4	90.6	81.2	83.2	79.9	91.9	86.1	68.1
1995	83.9	83.1	83.2	82.1	80.8	81.7	88.9	83.1	85.3	82.9	91.3	83.4	65.5
1996	83.4	82.1	82.2	81.6	81.2	85.0	88.7	82.6	84.4	81.4	89.9	80.5	68.1
1997	84.1	83.0	82.9	82.3	80.9	84.0	90.0	81.7	84.8	81.5	88.4	81.8	74.2
1998	82.8	81.6	81.3	80.7	81.0	84.4	86.4	80.0	81.8	77.6	85.7	79.2	82.6
1999	81.8	80.5	80.2	80.2	81.4	82.2	84.4	78.1	76.3	81.2	83.3	83.5	76.9
2000	81.5	79.7	79.3	79.7	78.2	79.3	82.4	79.2	77.7	83.7	86.0	81.7	66.6
2001	76.2	73.8	73.3	71.6	72.4	74.4	73.8	73.2	68.4	69.5	77.0	72.8	70.8
2002	74.9	73.0	72.6	70.1	75.1	73.8	75.8	72.2	67.1	60.3	73.4	78.7	67.1
2003	76.0	74.0	73.6	71.1	76.1	74.2	74.1	73.6	68.8	63.9	74.4	78.5	64.2
2004	78.2	76.5	76.2	74.2	79.2	75.1	81.9	75.8	72.9	71.1	77.0	77.0	63.6
2005	80.1	78.5	78.3	76.7	80.8	76.3	79.5	79.5	77.0	73.1	80.7	77.3	70.3
2006	80.6	78.8	78.8	77.9	78.2	75.2	78.8	84.4	80.3	77.6	82.5	72.3	73.6
2007	80.8	78.9	79.1	78.8	73.0	70.9	79.8	86.8	83.4	76.8	87.5	71.9	85.3
2008	77.8	74.7	74.6	74.9	63.1	61.9	77.6	81.9	81.0	77.7	85.8	58.0	84.4
2009	68.5	65.5	65.3	61.4	50.3	47.9	54.1	64.6	63.1	70.8	69.1	43.8	74.0
2010	73.5	70.7	70.9	68.8	55.8	51.8	68.7	71.5	70.4	77.0	74.4	59.1	73.4
2011	76.1	73.5	73.8	72.6	60.6	54.9	76.9	77.3	77.1	76.3	79.5	64.1	71.1
2012	76.9	74.5	75.0	75.1	67.9	57.9	76.1	77.7	81.8	75.8	80.9	69.6	76.7
2013	77.2	74.4	74.9	74.9	72.8	61.8	75.8	78.2	77.1	73.9	79.9	71.2	77.0
2014	78.6	75.2	75.7	76.2	72.4	64.8	73.0	80.0	77.0	73.9	80.5	75.8	81.4
2015	76.9	75.3	75.8	75.3	74.4	66.4	66.8	78.2	71.1	72.3	77.4	79.4	82.0
2016	75.0	74.2	74.6	73.1	76.4	66.2	63.9	75.9	66.5	69.9	75.7	79.5	75.7
2017	76.5	75.1	75.4	74.2	79.3	66.8	67.1	77.2	72.3	69.9	75.9	77.4	75.9
2018	78.7	76.6	77.0	76.1	79.5	68.2	72.0	80.5	77.3	71.7	76.1	78.5	75.6
2019	77.8	75.6	76.1	75.6	76.3	67.1	71.6	80.9	77.4	72.4	74.5	76.2	75.7
2018													
January	77.6	75.5	75.9	74.7	80.1	66.7	69.5	78.4	75.1	71.1	75.3	77.2	74.7
February	77.8	76.3	76.7	75.6	81.5	69.7	70.2	79.8	74.9	71.0	76.0	79.2	74.5
March	78.2	76.3	76.7	75.7	81.2	68.8	71.0	79.9	75.0	71.3	75.0	80.5	74.3
April	78.8	76.6	77.0	76.0	80.5	69.1	71.1	80.0	76.3	71.6	75.9	79.2	75.3
May	78.1	76.0	76.4	75.0	80.0	69.0	71.0	79.7	75.5	71.4	75.8	73.5	75.0
June	78.6	76.5	77.0	75.9	79.9	68.3	71.0	80.6	76.0	71.9	76.4	78.0	75.5
July	78.8	76.7	77.2	75.9	79.2	68.4	71.1	80.5	77.0	72.6	77.1	76.5	75.7
August	79.3	77.0	77.5	76.7	79.9	68.1	72.3	81.0	78.7	72.5	76.8	78.6	76.3
September	79.3	76.9	77.4	76.8	79.7	66.9	73.2	81.0	79.0	72.2	76.6	80.1	76.2
October	79.3	76.8	77.3	76.9	78.3	67.9	73.7	81.4	79.7	71.8	76.2	78.5	76.4
November	79.6	76.9	77.4	77.0	77.3	67.0	75.2	81.7	80.9	71.3	76.1	78.9	76.1
December	79.5	77.3	77.8	77.5	77.0	68.9	75.0	81.8	79.3	72.0	76.2	82.1	76.9
2019													
January	79.0	76.7	77.2	76.6	77.6	69.3	73.0	82.1	79.7	72.3	74.5	76.1	77.0
February	78.5	76.3	76.7	76.2	76.4	66.6	72.2	81.7	78.0	72.4	74.8	77.4	76.7
March	78.4	76.2	76.6	76.1	75.2	66.8	72.5	81.6	78.8	72.5	74.7	76.3	77.2
April	77.8	75.4	75.9	75.4	75.6	67.0	73.2	80.9	76.1	72.2	73.7	75.0	75.8
May	77.8	75.4	75.9	75.5	76.1	66.8	71.5	80.4	77.8	71.9	75.5	76.7	74.5
June	77.7	75.7	76.2	75.7	76.1	67.1	69.9	80.9	77.5	72.5	75.3	78.8	74.5
July	77.4	75.3	75.8	75.6	76.4	66.5	70.5	80.2	76.2	72.2	75.9	79.5	75.1
August	77.8	75.7	76.2	75.9	76.1	67.2	71.7	80.8	77.8	72.1	75.4	78.5	75.3
September	77.4	75.1	75.6	75.0	76.3	67.6	71.5	80.8	76.3	72.8	74.4	73.8	75.2
October	77.0	74.6	75.1	74.1	77.3	66.9	70.1	80.4	76.4	72.2	72.8	69.6	75.5
November	77.6	75.2	75.8	75.5	76.3	65.9	71.1	80.5	76.9	72.6	73.7	78.1	75.6
December	77.2	75.3	75.8	75.2	76.6	67.2	72.6	80.2	77.7	72.7	73.8	74.2	76.0

. . . = Not available.

Table 2-3. Capacity Utilization by NAICS Industry Groups—*Continued*

(Output as a percent of capacity, seasonally adjusted.)

Year and month	Durable goods manufacturing —*Continued*		Manufacturing (NAICS)—*Continued* Nondurable goods manufacturing									Other manufac- turing (non- NAICS)
	Furniture and related products	Miscel- laneous manufac- turing	Total	Food, beverage, and tobacco products	Textile and product mills	Apparel and leather	Paper	Printing and support	Petroleum and coal products	Chemicals	Plastics and rubber products	
1970	83.6	. . .	82.2	84.1	. . .	. . .	86.2	. . .	96.0	76.3	79.3	. . .
1971	84.3	. . .	81.8	84.1	. . .	. . .	86.9	. . .	94.6	75.6	79.8	. . .
1972	93.6	80.2	85.3	85.2	88.8	81.6	91.4	92.3	93.1	80.3	88.5	85.6
1973	94.6	79.5	86.6	84.7	86.3	82.2	94.9	94.0	90.3	83.8	92.5	84.7
1974	83.5	74.6	84.2	83.9	76.4	76.3	95.0	88.0	92.5	83.9	84.2	82.7
1975	69.7	67.6	76.1	80.1	72.7	74.6	80.7	79.6	83.7	71.7	70.1	77.3
1976	76.3	72.4	81.2	83.2	81.6	78.2	87.7	82.3	86.1	77.9	77.3	77.6
1977	84.6	77.6	84.4	82.9	88.9	81.7	90.2	86.0	87.7	81.4	88.4	83.2
1978	86.1	78.8	85.3	83.6	88.1	84.2	92.2	87.4	86.4	82.6	88.7	85.1
1979	81.0	78.4	83.9	81.2	87.6	79.2	90.9	86.0	88.5	83.0	83.7	85.6
1980	75.0	73.4	79.7	80.9	83.6	80.0	88.3	83.7	76.2	76.7	74.1	86.8
1981	71.8	75.7	78.8	80.3	80.7	78.8	86.8	81.2	72.8	76.1	77.5	87.5
1982	66.3	73.7	76.4	81.3	74.4	77.9	83.9	82.4	71.2	69.3	74.5	87.4
1983	72.2	70.8	79.4	80.8	84.1	81.1	88.9	84.4	74.4	72.9	81.5	88.0
1984	79.0	75.9	82.1	81.3	85.8	81.2	91.1	87.0	78.4	76.0	91.2	89.5
1985	77.2	73.9	80.5	82.6	81.6	77.9	87.8	84.8	80.0	73.4	87.0	90.4
1986	78.9	73.9	81.8	82.8	84.2	79.7	90.1	85.5	81.8	76.0	85.3	88.8
1987	82.7	77.1	84.7	83.7	91.3	82.1	90.2	89.1	82.3	81.0	89.5	90.5
1988	80.4	81.7	86.2	85.3	88.7	82.1	91.3	90.3	83.2	84.4	89.0	88.6
1989	79.0	79.7	84.9	83.7	88.3	80.2	90.4	88.9	84.2	83.3	87.1	85.4
1990	76.1	79.5	84.2	83.9	83.4	79.0	89.2	88.9	84.7	82.9	83.8	83.7
1991	70.6	78.2	82.3	83.0	81.6	80.3	87.5	84.2	82.7	81.1	78.8	80.8
1992	76.5	77.1	82.7	82.6	85.6	82.8	88.3	86.4	85.4	79.7	82.1	80.1
1993	79.1	77.4	82.7	81.1	88.0	83.8	88.9	84.9	89.1	79.1	86.6	81.4
1994	80.5	77.2	84.6	83.3	90.2	84.9	90.7	84.1	89.0	80.5	91.3	81.5
1995	79.9	79.6	84.5	84.3	86.2	84.5	89.6	83.3	89.6	80.8	89.6	82.2
1996	78.7	81.1	83.1	82.4	82.3	83.2	85.5	83.1	91.8	80.4	88.3	80.6
1997	83.3	79.4	83.8	82.4	84.5	83.0	87.4	81.4	95.3	81.3	89.0	85.6
1998	83.0	79.8	82.2	82.8	81.8	77.4	87.3	79.8	92.9	78.6	87.6	86.8
1999	80.6	77.5	80.1	78.4	81.7	76.0	86.5	77.7	90.7	77.2	86.3	87.2
2000	77.5	77.1	78.9	77.3	79.4	78.8	84.3	77.0	89.5	76.4	82.3	87.5
2001	70.8	73.8	75.7	76.2	71.5	75.1	80.0	74.8	88.5	72.1	76.6	82.9
2002	72.6	74.6	75.9	75.4	73.8	66.8	82.0	74.9	88.9	73.8	78.7	81.6
2003	71.5	75.4	76.8	77.1	73.0	69.5	81.8	75.4	89.8	73.9	79.2	81.5
2004	76.9	76.0	78.7	77.2	75.0	70.2	83.2	77.5	92.8	76.5	83.1	82.4
2005	80.2	78.7	80.3	79.2	78.4	75.4	83.7	77.5	92.0	77.1	83.9	81.9
2006	79.1	77.9	79.8	78.9	75.4	75.3	83.6	77.5	89.4	77.4	82.0	79.8
2007	75.3	73.8	79.3	78.4	72.4	75.9	83.7	77.0	87.6	78.0	78.6	76.3
2008	71.1	74.0	74.1	76.3	67.9	75.2	82.0	69.6	81.7	71.3	70.0	77.3
2009	57.6	69.7	69.8	76.2	56.6	59.9	77.1	60.0	78.5	66.3	60.2	69.6
2010	61.2	75.3	73.3	77.1	63.8	66.8	81.5	63.0	80.5	70.3	68.7	66.2
2011	66.9	76.8	75.2	78.7	65.8	70.4	81.2	63.4	84.5	72.1	69.9	65.4
2012	71.7	74.8	75.0	78.5	66.4	69.2	83.1	63.6	82.9	70.5	75.3	63.1
2013	73.4	77.0	74.9	78.4	70.5	65.6	83.0	66.0	84.1	68.4	79.3	62.2
2014	75.9	75.6	75.1	76.6	72.7	65.8	83.8	67.5	81.8	70.5	81.8	63.7
2015	80.3	78.4	76.3	76.8	69.6	66.6	85.7	69.5	81.4	72.3	84.4	63.8
2016	79.5	80.9	76.2	76.5	69.0	66.8	87.0	73.2	77.8	72.6	84.2	64.2
2017	78.6	80.5	76.8	76.8	70.7	65.0	87.4	75.6	79.0	73.6	83.2	66.3
2018	77.7	79.2	78.0	77.0	72.3	66.2	86.9	73.9	80.6	76.7	81.7	62.3
2019	76.3	78.3	76.6	75.5	70.1	61.5	84.7	70.7	79.5	76.6	77.1	59.1
2018												
January	77.3	77.9	77.2	76.5	72.3	66.6	87.3	75.0	80.7	74.5	82.7	64.5
February	77.7	79.3	77.9	78.1	73.3	66.9	85.9	75.6	80.1	75.0	83.7	65.4
March	77.4	79.2	77.8	77.3	72.2	66.9	86.6	73.9	80.4	75.5	83.2	64.8
April	77.5	79.6	78.2	77.9	72.3	67.7	87.3	74.5	80.4	76.1	82.4	63.8
May	77.7	80.0	78.0	77.3	71.7	67.4	86.7	74.6	80.5	76.6	81.2	62.3
June	77.7	79.3	78.2	77.5	72.2	65.8	86.2	74.5	80.9	76.9	81.1	60.5
July	78.0	79.2	78.7	77.8	72.2	68.3	87.5	74.6	80.7	77.6	81.8	60.5
August	77.6	79.2	78.4	77.3	73.3	66.6	86.8	73.2	81.3	77.6	81.0	60.8
September	78.2	78.7	78.1	77.1	72.2	65.3	86.8	73.1	81.0	77.1	80.8	61.2
October	78.3	79.7	77.8	75.8	73.1	64.9	87.1	73.3	80.5	77.4	80.5	61.8
November	78.0	79.2	77.8	75.4	72.1	64.3	87.0	72.6	79.7	78.1	80.8	61.6
December	77.2	79.4	78.1	76.0	70.7	63.4	87.6	72.3	80.6	78.1	81.1	60.6
2019												
January	77.3	79.0	77.9	76.1	71.2	61.9	87.4	72.5	82.1	77.5	80.2	60.8
February	76.4	79.5	77.4	75.9	71.1	60.1	85.7	71.4	78.5	77.6	79.7	61.4
March	75.5	79.2	77.2	76.2	72.2	60.9	83.9	70.6	79.2	77.4	78.4	60.5
April	76.2	79.6	76.4	75.3	71.3	62.2	84.7	70.5	78.4	76.6	76.7	60.2
May	76.7	79.2	76.2	75.1	70.6	61.9	83.8	70.8	78.9	76.1	77.4	59.3
June	76.0	78.5	76.7	76.8	70.9	62.4	82.4	70.3	79.3	75.8	78.1	59.6
July	76.3	78.6	76.0	74.8	68.7	61.9	83.5	70.2	80.3	76.0	75.7	59.0
August	75.9	78.3	76.5	74.6	69.5	61.6	84.9	70.3	80.8	77.0	76.6	58.7
September	76.2	77.6	76.2	74.8	69.3	62.0	84.8	70.4	79.9	76.5	75.8	58.5
October	75.9	76.6	76.1	75.1	69.0	61.5	85.3	70.5	78.9	76.4	75.4	58.4
November	76.6	76.9	75.9	74.9	68.9	60.9	84.7	70.7	78.5	76.1	75.8	56.9
December	76.6	76.2	76.4	76.4	68.5	61.1	85.7	70.0	79.7	75.8	75.4	56.0

. . . = Not available.

Table 2-3. Capacity Utilization by NAICS Industry Groups—*Continued*

(Output as a percent of capacity, seasonally adjusted.)

Year and month	Mining	Utilities	Selected high-tech industries				Measures excluding selected high-tech industries		Stage-of-process groups		
			Total	Computers and peripheral equipment	Communications equipment	Semiconductors and related electronic components	Total industry	Manufacturing	Crude	Primary and semi-finished	Finished
1970	89.5	96.4	82.9	. . .	. . .	. . .	81.0	79.3	84.6	81.4	78.1
1971	88.1	95.0	73.6	. . .	. . .	. . .	80.0	78.3	83.6	81.6	75.7
1972	90.8	95.4	78.0	81.4	73.3	81.4	84.9	83.6	88.4	88.1	79.6
1973	91.6	93.1	81.4	81.9	75.1	88.5	88.6	87.9	90.0	92.1	83.2
1974	91.1	86.7	81.0	88.3	73.3	83.5	85.2	84.6	91.0	87.3	80.3
1975	89.5	85.2	64.5	69.0	62.1	63.3	76.2	74.1	84.0	75.2	73.7
1976	89.6	85.7	68.5	76.9	62.0	68.6	80.3	78.8	87.0	80.2	76.9
1977	89.5	86.9	75.6	77.8	72.5	76.7	83.8	82.8	89.1	84.6	79.9
1978	89.7	87.2	78.4	77.7	77.2	80.3	85.4	84.7	88.7	86.3	82.3
1979	91.2	87.2	82.6	78.1	85.3	85.1	85.1	84.1	90.0	85.9	81.7
1980	91.3	85.5	87.5	87.0	91.6	84.6	80.5	78.2	89.4	78.8	79.4
1981	90.9	84.4	85.5	85.3	89.3	83.1	79.3	76.4	89.3	77.1	77.5
1982	84.1	80.0	78.6	70.4	87.6	82.0	73.3	70.4	82.3	70.4	73.1
1983	79.8	79.3	81.0	75.9	86.0	82.7	74.6	73.0	79.9	74.5	73.0
1984	85.8	81.9	87.8	85.4	85.9	91.3	80.0	78.8	85.8	81.2	77.2
1985	84.4	81.7	77.5	76.2	79.6	77.1	79.3	78.2	83.8	79.8	76.6
1986	77.6	80.9	73.1	72.6	76.5	71.3	78.9	78.8	79.2	79.7	77.1
1987	80.3	83.5	77.6	73.7	80.3	80.0	81.4	81.2	82.8	82.8	78.7
1988	84.1	86.8	79.9	76.2	83.6	81.5	84.5	84.2	86.3	85.8	81.6
1989	85.1	86.8	78.2	74.8	80.3	79.8	84.0	83.6	86.8	84.6	81.6
1990	86.9	86.6	77.1	71.4	81.8	79.5	82.7	81.9	87.9	82.6	80.5
1991	85.4	87.8	78.0	73.7	78.3	80.9	80.0	78.6	85.5	80.0	78.2
1992	85.2	86.4	80.3	78.2	79.4	82.0	80.6	79.6	85.9	81.5	78.2
1993	85.8	88.2	80.8	79.9	81.9	80.5	81.6	80.5	85.8	83.3	78.4
1994	86.8	88.3	82.7	74.3	84.8	85.8	83.6	82.8	87.8	86.3	79.2
1995	87.6	89.3	84.9	78.2	79.7	90.4	83.8	83.0	89.0	86.4	79.7
1996	90.5	90.7	83.1	87.0	76.4	84.5	83.4	82.1	89.1	85.6	79.3
1997	91.8	90.1	83.5	83.4	79.3	85.6	84.1	83.0	90.4	86.0	80.3
1998	89.3	92.6	79.1	79.3	84.6	76.3	83.2	81.9	87.1	84.2	80.3
1999	86.2	94.2	85.3	84.5	87.5	84.5	81.5	80.0	86.1	84.3	78.0
2000	90.5	94.3	88.0	81.3	90.2	89.5	80.9	78.9	88.5	84.0	76.9
2001	89.8	90.1	69.3	70.8	70.3	68.1	76.7	74.3	85.5	77.4	72.6
2002	86.0	87.6	58.2	68.7	43.1	63.7	76.3	74.4	83.2	77.4	70.5
2003	87.8	85.7	63.1	72.5	44.0	72.2	76.9	74.9	85.0	78.2	71.3
2004	88.2	84.5	70.5	77.3	53.6	77.9	78.6	76.9	86.5	80.2	73.4
2005	88.5	85.1	73.9	75.8	58.9	80.8	80.5	78.8	86.7	81.9	75.7
2006	90.1	83.7	81.8	78.2	77.8	85.6	80.5	78.7	88.1	81.5	76.4
2007	89.4	85.9	77.8	76.4	76.5	78.0	80.9	79.0	88.7	81.2	77.1
2008	90.0	84.2	80.9	79.1	81.6	81.6	77.6	74.3	87.5	77.0	73.9
2009	80.3	80.6	73.1	89.9	80.3	64.7	68.3	65.1	77.9	65.8	68.1
2010	83.9	83.0	81.4	99.4	81.5	77.8	73.3	70.2	83.2	71.8	71.2
2011	85.9	81.4	76.9	75.6	81.9	75.9	76.1	73.3	84.5	74.4	73.7
2012	87.3	78.4	71.8	68.2	77.3	71.2	77.1	74.7	85.5	74.7	74.8
2013	87.2	79.9	69.6	63.4	82.6	66.8	77.4	74.6	86.0	75.5	73.8
2014	90.5	80.8	72.7	70.5	70.3	74.2	78.7	75.3	88.4	76.7	74.6
2015	84.2	79.9	75.2	80.3	72.2	75.3	77.0	75.3	82.7	76.3	75.1
2016	77.6	78.8	73.5	78.1	71.7	73.1	75.1	74.2	78.4	75.2	73.6
2017	84.3	77.0	70.6	76.7	65.2	71.4	76.6	75.3	83.7	75.7	74.2
2018	90.2	79.3	73.2	73.9	62.1	78.3	78.8	76.7	88.8	77.5	75.4
2019	90.4	76.9	72.8	73.4	62.5	78.1	77.9	75.7	88.6	75.9	74.7
2018											
January	86.6	81.7	72.3	75.4	62.7	75.8	77.7	75.6	85.2	77.2	74.6
February	88.1	75.8	72.1	74.2	61.8	76.2	78.0	76.5	86.3	76.6	75.3
March	88.6	78.6	72.2	74.3	60.8	76.8	78.3	76.5	87.2	77.3	75.0
April	89.0	81.7	72.4	74.7	60.2	77.4	79.0	76.8	87.5	78.2	75.4
May	89.0	79.2	72.3	73.2	60.2	77.7	78.2	76.1	87.9	77.2	74.5
June	90.2	78.5	73.7	74.6	61.1	79.3	78.7	76.5	89.1	77.1	75.2
July	90.4	78.3	74.4	74.3	62.4	80.1	78.9	76.8	89.5	77.2	75.5
August	91.8	79.2	75.2	74.0	63.4	81.3	79.4	77.0	90.6	77.6	75.8
September	92.1	78.7	74.8	72.2	63.7	80.9	79.3	77.0	90.7	77.4	75.8
October	91.6	80.6	74.1	72.6	63.4	79.6	79.4	76.9	90.2	77.9	75.6
November	91.8	82.6	72.9	73.7	62.7	77.7	79.7	77.0	90.4	78.5	75.5
December	93.3	76.8	72.5	73.2	63.1	76.9	79.6	77.4	91.5	77.4	76.0
2019											
January	92.4	77.3	73.3	74.3	64.2	77.4	79.1	76.8	90.6	77.2	75.4
February	90.7	77.6	73.5	75.2	64.4	77.6	78.6	76.4	89.1	76.7	75.3
March	90.1	78.7	73.7	75.7	64.8	77.5	78.5	76.2	88.2	76.7	75.5
April	91.9	75.9	72.4	72.3	63.6	76.9	77.9	75.5	89.6	75.7	74.5
May	91.3	77.2	71.7	72.5	62.1	76.4	77.9	75.5	89.1	76.0	74.4
June	91.3	73.9	72.1	72.8	62.5	76.9	77.8	75.8	88.7	75.2	75.2
July	88.9	76.9	72.3	72.6	62.7	77.3	77.5	75.4	86.8	75.6	74.8
August	90.6	76.3	73.1	74.2	62.1	78.6	77.9	75.8	88.6	75.8	74.9
September	90.2	77.2	72.7	72.0	61.7	78.7	77.5	75.2	88.4	75.7	74.1
October	89.7	77.3	72.5	71.0	61.2	79.1	77.1	74.7	88.3	75.1	73.6
November	88.8	79.5	73.4	74.5	60.7	80.1	77.7	75.3	87.7	75.8	74.7
December	89.2	74.8	73.2	73.9	59.6	80.4	77.3	75.3	87.9	74.9	74.6

. . . = Not available.

Table 2-4A. Industrial Production and Capacity Utilization, Historical Data, 1955–2019

(Seasonally adjusted.)

Year and month	Production indexes, 2007 = 100										Capacity utilization (output as percent of capacity)	
	Total industry	Manufac-turing (SIC)	Market groups								Total industry	Manufac-turing (SIC)
			Consumer goods			Business equipment	Defense and space equipment	Construction supplies	Business supplies	Materials		
			Total	Durable	Nondurable							
1955	21.2	19.9	28.8	22.2	32.0	7.5	29.0	41.0	18.9	20.6	. . .	87.0
1956	22.1	20.7	29.8	21.5	34.0	8.7	28.4	42.2	20.0	21.2	. . .	86.1
1957	22.4	20.9	30.5	21.5	35.1	9.0	29.6	41.6	20.3	21.2	. . .	83.6
1958	21.0	19.5	30.3	19.1	36.1	7.6	29.7	40.2	20.1	19.1	. . .	75.0
1959	23.5	21.9	33.2	22.6	38.6	8.6	31.3	45.0	21.9	22.0	. . .	81.6
1960	24.0	22.4	34.5	23.9	39.9	8.8	32.2	43.9	22.6	22.3	. . .	80.1
1961	24.1	22.4	35.2	23.5	41.2	8.5	32.7	44.3	23.3	22.3	. . .	77.3
1962	26.1	24.4	37.5	26.6	43.1	9.3	37.9	47.0	24.8	24.3	. . .	81.4
1963	27.7	25.9	39.6	28.9	45.1	9.7	40.9	49.2	26.4	25.8	. . .	83.5
1964	29.6	27.7	41.8	31.0	47.3	10.9	39.6	52.2	28.3	27.9	. . .	85.6
1965	32.5	30.7	45.1	36.4	49.3	12.5	43.8	55.4	30.1	31.1	. . .	89.5
1966	35.4	33.5	47.4	38.5	51.7	14.4	51.4	57.7	32.4	33.9	. . .	91.1
1967	36.1	34.1	48.6	37.0	54.3	14.7	58.7	59.3	34.2	33.6	87.0	87.2
1968	38.1	36.0	51.5	41.2	56.5	15.4	58.8	62.3	36.2	35.8	87.3	87.1
1969	39.9	37.6	53.4	43.0	58.4	16.4	56.0	65.0	38.5	38.0	87.4	86.6
1970	38.6	35.9	52.8	39.6	59.4	15.8	47.4	62.7	38.7	36.6	81.2	79.4
1971	39.1	36.5	55.9	44.9	61.1	15.0	42.6	64.7	39.9	37.2	79.6	77.9
1972	42.9	40.3	60.4	50.3	65.0	17.1	41.4	73.5	43.9	40.9	84.7	83.4
1973	46.4	43.9	63.1	54.1	66.9	19.8	45.4	79.8	46.6	44.6	88.3	87.7
1974	46.3	43.8	61.3	49.2	67.0	20.9	46.9	77.9	46.4	44.5	85.1	84.4
1975	42.2	39.2	58.8	44.7	65.7	18.7	47.3	65.9	42.6	39.7	75.8	73.7
1976	45.5	42.7	63.6	50.5	69.9	19.9	45.9	70.9	45.4	43.1	79.8	78.4
1977	48.9	46.4	67.5	56.7	72.4	23.0	41.1	77.1	49.3	46.2	83.4	82.5
1978	51.6	49.2	69.7	58.1	74.9	25.9	41.9	81.5	52.0	48.5	85.1	84.4
1979	53.2	50.7	68.6	56.0	74.4	29.2	44.8	83.6	53.8	49.8	85.0	84.0
1980	51.8	48.9	65.9	48.7	74.5	29.9	53.2	77.4	52.5	47.9	80.8	78.7
1981	52.5	49.4	66.4	49.4	74.8	30.8	57.6	76.1	53.7	48.2	79.5	76.9
1982	49.8	46.7	66.2	46.6	76.1	28.1	68.9	69.1	53.0	44.5	73.6	70.9
1983	51.1	49.0	68.6	51.7	77.0	28.3	69.3	73.9	55.6	45.7	74.9	73.5
1984	55.7	53.7	71.8	57.9	78.5	32.5	79.4	80.3	60.5	50.0	80.4	79.4
1985	56.4	54.6	72.5	57.9	79.5	33.7	88.9	82.2	61.9	49.9	79.2	78.1
1986	56.9	55.8	74.9	61.6	81.3	33.2	94.4	84.9	63.9	49.9	78.6	78.4
1987	59.9	59.0	78.1	65.3	84.2	35.5	96.4	90.4	67.8	52.6	81.1	80.9
1988	63.0	62.1	81.1	68.5	87.0	39.2	97.2	92.4	70.3	55.5	84.2	83.9
1989	63.6	62.6	81.4	70.1	86.7	40.5	97.3	92.1	71.3	55.9	83.7	83.2
1990	64.2	63.1	81.8	68.0	88.2	42.0	93.9	91.4	72.9	56.3	82.4	81.5
1991	63.2	61.9	81.7	65.0	89.5	41.4	87.0	86.4	71.8	55.5	79.9	78.6
1992	65.1	64.2	84.2	71.3	90.3	43.2	80.8	90.1	73.6	57.3	80.6	79.6
1993	67.2	66.5	86.8	77.1	91.5	45.3	76.4	94.1	75.9	59.3	81.5	80.5
1994	70.8	70.4	90.4	84.0	93.7	48.5	71.8	101.0	78.8	63.0	83.5	82.8
1995	74.0	74.0	92.9	87.2	96.0	53.1	69.3	103.3	82.0	66.4	83.9	83.1
1996	77.4	77.6	94.6	89.9	97.2	58.3	67.6	108.0	85.2	69.9	83.4	82.1
1997	83.0	84.2	98.1	96.1	99.6	67.1	66.7	113.2	91.2	75.4	84.1	83.0
1998	87.8	89.8	101.7	103.1	101.9	74.8	69.7	119.2	96.4	79.8	82.8	81.6
1999	91.7	94.3	103.9	111.3	101.8	78.9	67.6	122.2	100.5	84.8	81.8	80.5
2000	95.2	98.2	105.9	114.5	103.3	85.0	60.1	124.8	104.4	89.0	81.5	79.7
2001	92.3	94.6	104.8	109.2	103.8	80.1	66.0	119.1	100.6	85.5	76.2	73.8
2002	92.6	95.1	107.0	116.2	104.2	75.0	66.7	119.0	100.7	86.6	74.9	73.0
2003	93.8	96.4	108.4	119.8	104.9	75.2	71.0	118.7	102.6	87.7	76.0	74.0
2004	96.4	99.4	109.6	121.5	105.9	79.4	69.7	121.5	105.1	90.7	78.2	76.5
2005	99.6	103.4	112.5	122.0	109.5	85.4	76.4	127.4	109.0	92.8	80.1	78.5
2006	101.8	106.1	113.1	121.4	110.5	93.4	76.1	130.6	110.8	94.7	80.6	78.8
2007	104.4	109.0	113.1	120.9	110.6	99.0	90.7	129.4	112.1	98.0	80.8	78.9
2008	100.8	103.8	107.2	106.6	107.2	96.7	98.0	117.5	107.2	95.6	77.8	74.7
2009	89.2	89.5	99.2	86.0	102.9	80.0	93.8	90.4	95.0	84.9	68.5	65.5
2010	94.1	94.7	100.3	94.2	101.9	86.1	100.9	93.6	98.0	91.9	73.5	70.7
2011	97.1	97.5	101.4	97.7	102.3	91.1	98.0	95.9	99.0	95.9	76.1	73.5
2012	100.0	100.0	100.0	100.0	100.0	100.0	100.0	100.0	100.0	100.0	76.9	74.5
2013	102.0	100.9	100.7	105.5	99.5	99.9	97.2	103.1	101.8	103.3	77.2	74.4
2014	105.2	102.0	101.5	110.7	99.2	101.7	93.9	106.4	102.8	108.5	78.6	75.2
2015	104.1	101.5	102.9	115.0	99.9	99.6	91.7	107.1	101.4	106.9	76.9	75.3
2016	102.1	100.7	103.5	117.5	100.0	94.4	89.1	108.1	102.0	103.7	75.0	74.2
2017	104.4	102.7	104.1	119.3	100.3	97.8	90.9	111.8	104.3	106.5	76.5	75.1
2018	108.6	105.0	106.3	122.0	102.3	101.0	93.1	114.9	105.2	112.8	78.7	76.6
2019	109.4	104.8	105.3	120.1	101.5	101.4	101.7	116.5	104.1	115.0	77.8	75.6
1955												
January	19.9	18.7	27.6	21.0	30.9	6.9	29.4	38.5	18.0	19.0	. . .	83.5
February	20.1	18.9	27.8	21.2	30.9	7.0	29.4	39.1	18.2	19.4	. . .	84.1
March	20.6	19.3	28.3	21.8	31.4	7.1	29.3	40.4	18.7	20.0	. . .	85.8
April	20.8	19.6	28.5	22.2	31.6	7.3	29.2	40.7	18.6	20.3	. . .	86.7
May	21.2	20.0	28.9	22.7	31.9	7.4	29.2	40.8	18.9	20.7	. . .	87.9
June	21.2	20.0	28.6	22.3	31.8	7.5	28.9	41.6	19.1	20.8	. . .	87.6
July	21.4	20.1	28.7	22.6	31.7	7.5	28.9	41.6	19.0	21.0	. . .	87.7
August	21.3	20.1	28.8	22.6	31.8	7.6	28.7	41.6	18.9	21.0	. . .	87.3
September	21.5	20.2	29.0	22.7	32.1	7.6	28.8	41.9	19.3	21.3	. . .	87.5
October	21.8	20.4	29.5	22.8	32.9	8.0	28.6	41.9	19.5	21.5	. . .	88.4
November	21.9	20.5	29.6	22.6	33.1	8.0	28.6	42.3	19.8	21.4	. . .	88.3
December	22.0	20.8	29.7	22.5	33.4	8.1	28.7	42.6	19.6	21.5	. . .	89.0

. . . = Not available.

Table 2-4A. Industrial Production and Capacity Utilization, Historical Data, 1955–2019—*Continued*

(Seasonally adjusted.)

Year and month	Production indexes, 2007 = 100										Capacity utilization (output as percent of capacity)	
	Total industry	Manufac-turing (SIC)	Market groups								Total industry	Manufac-turing (SIC)
			Consumer goods			Business equipment	Defense and space equipment	Construction supplies	Business supplies	Materials		
			Total	Durable	Nondurable							
1956												
January	22.1	20.7	29.8	22.3	33.6	8.2	28.4	43.3	19.8	21.6	. . .	88.2
February	21.9	20.6	29.7	21.9	33.7	8.3	28.1	43.1	19.8	21.2	. . .	87.4
March	21.9	20.5	29.7	21.9	33.7	8.4	27.5	43.0	20.0	21.1	. . .	87.0
April	22.1	20.8	29.8	22.2	33.7	8.7	27.7	42.8	20.2	21.3	. . .	87.8
May	21.9	20.6	29.6	21.7	33.8	8.6	27.7	42.2	20.0	20.9	. . .	86.3
June	21.7	20.4	29.5	21.2	33.9	8.7	27.7	41.8	20.0	20.6	. . .	85.3
July	21.0	19.6	29.6	21.2	34.0	8.7	27.7	39.6	20.1	19.0	. . .	81.5
August	21.9	20.5	29.8	21.2	34.2	8.8	28.1	41.6	20.1	20.7	. . .	84.9
September	22.4	20.8	29.7	20.9	34.3	8.8	28.4	42.7	20.1	21.8	. . .	86.0
October	22.6	21.0	30.0	21.2	34.5	8.9	29.1	42.4	20.3	22.2	. . .	86.5
November	22.4	20.9	29.8	20.9	34.4	9.0	29.5	42.1	20.3	21.7	. . .	85.8
December	22.7	21.3	30.1	21.7	34.3	9.1	30.2	42.8	20.4	22.1	. . .	86.8
1957												
January	22.6	21.2	30.3	21.8	34.5	9.3	30.3	42.2	20.4	21.6	. . .	86.2
February	22.8	21.5	30.6	22.2	34.9	9.5	30.5	43.5	20.5	21.8	. . .	87.0
March	22.8	21.4	30.7	22.1	35.1	9.4	30.4	42.8	20.4	21.7	. . .	86.4
April	22.5	21.1	30.4	21.4	34.9	9.2	30.5	41.9	20.4	21.4	. . .	85.0
May	22.4	21.0	30.5	21.3	35.0	9.1	30.1	41.7	20.6	21.3	. . .	84.2
June	22.5	21.1	30.6	21.7	35.1	9.1	30.2	42.0	20.4	21.4	. . .	84.6
July	22.6	21.1	30.7	21.5	35.4	9.1	29.9	42.1	20.5	21.5	. . .	84.3
August	22.6	21.1	30.9	22.1	35.4	9.1	29.9	41.7	20.5	21.6	. . .	84.2
September	22.4	21.0	30.9	22.0	35.4	9.0	29.3	41.5	20.5	21.3	. . .	83.2
October	22.1	20.6	30.4	21.2	35.1	8.8	28.6	41.0	20.2	21.0	. . .	81.4
November	21.6	20.1	30.3	21.1	34.9	8.5	27.7	40.5	20.0	20.2	. . .	79.4
December	21.2	19.7	30.1	20.2	35.3	8.3	27.5	39.8	20.0	19.5	. . .	77.5
1958												
January	20.8	19.3	29.7	19.4	35.2	8.1	27.8	39.3	19.9	19.0	. . .	75.7
February	20.3	18.9	29.5	18.7	35.2	7.8	28.0	38.1	19.8	18.4	. . .	73.8
March	20.1	18.7	29.3	18.1	35.2	7.6	28.6	37.9	19.9	17.9	. . .	72.7
April	19.7	18.4	29.0	17.4	35.2	7.5	29.0	37.4	19.7	17.4	. . .	71.3
May	19.9	18.6	29.4	18.0	35.5	7.3	29.3	38.6	19.7	17.7	. . .	71.9
June	20.5	19.1	30.0	18.5	36.1	7.3	30.2	40.2	19.9	18.4	. . .	73.9
July	20.8	19.3	30.4	18.8	36.5	7.4	30.2	39.9	19.9	18.9	. . .	74.3
August	21.2	19.7	30.5	19.1	36.6	7.5	30.5	41.7	20.2	19.5	. . .	75.7
September	21.4	19.9	30.2	17.8	36.8	7.5	30.7	41.8	20.4	19.9	. . .	76.2
October	21.6	20.0	30.5	18.7	36.8	7.6	30.6	41.8	20.8	20.3	. . .	76.4
November	22.3	20.7	31.9	21.9	37.2	7.8	30.9	43.4	20.9	20.8	. . .	79.1
December	22.3	20.8	32.1	22.0	37.2	7.8	30.9	42.8	20.8	20.9	. . .	79.0
1959												
January	22.6	21.1	32.4	22.1	37.7	8.0	31.0	43.7	21.3	21.3	. . .	80.2
February	23.1	21.5	32.7	22.2	38.1	8.1	30.8	44.9	21.5	21.9	. . .	81.4
March	23.4	21.9	32.7	22.6	37.9	8.2	30.9	46.0	21.7	22.5	. . .	82.5
April	23.9	22.3	33.2	22.7	38.6	8.4	31.1	47.3	21.7	23.1	. . .	84.0
May	24.3	22.7	33.4	23.2	38.6	8.7	31.3	47.7	21.7	23.7	. . .	84.9
June	24.3	22.7	33.2	23.4	38.3	8.9	31.4	47.5	22.0	23.6	. . .	84.8
July	23.7	22.3	33.6	23.8	38.6	8.9	31.5	45.7	22.2	22.2	. . .	83.0
August	22.9	21.4	33.7	23.2	39.0	8.8	31.4	42.8	22.1	20.5	. . .	79.5
September	22.9	21.3	33.6	22.5	39.3	8.8	31.5	42.3	22.3	20.4	. . .	79.0
October	22.7	21.1	33.4	23.0	38.8	8.7	31.5	42.4	22.2	20.2	. . .	78.2
November	22.8	21.3	32.7	20.2	39.3	8.5	31.5	43.9	22.2	21.0	. . .	78.5
December	24.3	22.8	33.7	22.7	39.5	8.7	31.7	47.3	22.4	23.3	. . .	83.6
1960												
January	24.9	23.4	34.8	25.2	39.6	9.0	31.9	47.0	22.7	24.0	. . .	85.6
February	24.7	23.2	34.4	24.9	39.2	9.1	32.1	46.5	22.7	23.7	. . .	84.6
March	24.5	22.9	34.5	24.4	39.6	9.1	32.3	45.1	22.6	23.3	. . .	83.2
April	24.3	22.7	34.7	24.4	40.0	9.0	32.1	45.1	23.0	22.7	. . .	82.3
May	24.2	22.6	34.9	24.6	40.1	9.0	32.4	44.6	23.1	22.5	. . .	81.5
June	23.9	22.3	34.7	24.3	39.9	8.9	31.6	43.8	22.9	22.2	. . .	80.2
July	23.8	22.3	34.3	23.4	40.0	8.8	32.4	44.2	22.9	22.2	. . .	79.7
August	23.8	22.2	34.4	23.7	40.0	8.7	32.6	43.1	22.7	22.1	. . .	79.1
September	23.6	22.0	34.3	23.4	39.9	8.6	32.5	42.7	22.6	21.7	. . .	77.9
October	23.5	21.9	34.6	23.6	40.3	8.5	32.2	42.8	22.7	21.6	. . .	77.5
November	23.2	21.5	34.0	22.9	39.9	8.5	32.3	42.3	22.7	21.1	. . .	75.8
December	22.8	21.1	33.7	22.1	39.8	8.3	31.8	41.8	22.3	20.5	. . .	74.3
1961												
January	22.8	21.2	33.5	21.3	40.0	8.4	32.1	41.4	22.6	20.7	. . .	74.1
February	22.8	21.1	33.7	21.2	40.2	8.3	31.9	41.4	22.7	20.5	. . .	73.5
March	22.9	21.2	33.7	21.2	40.2	8.3	31.8	42.3	22.9	20.7	. . .	73.9
April	23.4	21.7	34.5	22.7	40.6	8.4	31.9	43.4	23.0	21.3	. . .	75.4
May	23.7	22.1	34.9	23.4	40.9	8.4	31.9	43.5	23.1	22.0	. . .	76.4
June	24.1	22.4	35.3	24.2	41.0	8.4	31.9	44.5	23.3	22.3	. . .	77.3
July	24.3	22.7	35.6	24.6	41.2	8.5	32.1	45.2	23.5	22.6	. . .	78.1
August	24.6	23.0	35.8	24.7	41.6	8.5	32.3	45.7	23.7	23.1	. . .	79.0
September	24.5	22.8	35.2	23.3	41.3	8.7	33.0	46.1	23.6	23.1	. . .	78.2
October	25.0	23.3	36.2	24.6	42.1	8.6	33.7	46.4	23.9	23.5	. . .	79.6
November	25.4	23.7	36.8	25.7	42.5	8.9	34.5	46.1	24.1	23.8	. . .	80.8
December	25.6	24.0	36.9	26.2	42.5	8.9	35.1	46.4	24.4	24.2	. . .	81.6

. . . = Not available.

Table 2-4A. Industrial Production and Capacity Utilization, Historical Data, 1955–2019—*Continued*

(Seasonally adjusted.)

Year and month	Production indexes, 2007 = 100										Capacity utilization (output as percent of capacity)	
	Total industry	Manufac-turing (SIC)	Market groups								Total industry	Manufac-turing (SIC)
			Consumer goods			Business equipment	Defense and space equipment	Construction supplies	Business supplies	Materials		
			Total	Durable	Nondurable							
1962												
January	25.4	23.7	36.6	25.6	42.3	8.9	35.5	44.3	24.4	24.0	. . .	80.2
February	25.8	24.1	36.8	25.7	42.6	9.0	36.2	46.8	24.6	24.4	. . .	81.4
March	25.9	24.3	37.1	26.1	42.8	9.1	36.6	47.4	24.5	24.4	. . .	81.9
April	26.0	24.3	37.4	26.7	42.9	9.2	37.0	46.8	24.5	24.4	. . .	81.7
May	26.0	24.3	37.6	26.9	43.2	9.2	37.2	46.7	24.9	24.0	. . .	81.3
June	25.9	24.2	37.4	26.5	43.0	9.3	37.6	47.1	24.9	24.0	. . .	80.9
July	26.2	24.4	38.0	27.0	43.6	9.3	38.3	46.9	24.8	24.2	. . .	81.5
August	26.2	24.5	37.6	26.7	43.2	9.4	38.9	47.7	25.0	24.2	. . .	81.4
September	26.4	24.7	37.8	27.0	43.4	9.4	38.9	48.3	25.2	24.4	. . .	81.8
October	26.4	24.6	37.8	27.1	43.2	9.5	39.1	47.5	25.2	24.4	. . .	81.4
November	26.5	24.8	38.0	27.1	43.4	9.5	39.5	47.7	25.3	24.6	. . .	81.8
December	26.5	24.9	38.2	27.4	43.6	9.4	39.6	48.1	25.2	24.5	. . .	81.7
1963												
January	26.7	25.0	38.6	27.6	44.2	9.4	41.5	46.9	25.4	24.6	. . .	81.9
February	27.0	25.2	39.1	27.9	44.7	9.5	41.2	47.0	25.6	25.0	. . .	82.4
March	27.2	25.4	39.2	28.0	45.0	9.5	40.9	47.5	25.4	25.3	. . .	82.6
April	27.4	25.7	39.4	28.2	45.1	9.5	40.9	49.2	26.2	25.6	. . .	83.5
May	27.7	25.9	39.5	28.7	45.0	9.5	40.9	50.2	26.5	26.2	. . .	84.0
June	27.8	26.0	39.7	29.2	45.0	9.6	40.9	50.0	26.4	26.3	. . .	83.9
July	27.7	25.9	39.6	29.1	44.8	9.7	40.4	49.8	26.5	26.0	. . .	83.3
August	27.8	26.0	39.9	29.2	45.3	9.9	40.6	50.0	26.7	25.8	. . .	83.5
September	28.1	26.2	40.0	29.7	45.2	9.9	40.8	49.7	27.0	26.3	. . .	83.8
October	28.2	26.5	40.3	29.8	45.5	10.1	40.8	50.4	27.2	26.5	. . .	84.3
November	28.4	26.5	40.3	30.0	45.5	10.1	40.6	51.0	27.5	26.6	. . .	84.3
December	28.3	26.5	40.6	30.1	45.9	10.1	40.7	50.2	27.3	26.5	. . .	84.0
1964												
January	28.6	26.8	41.0	30.2	46.4	10.4	40.3	50.5	27.6	26.7	. . .	84.5
February	28.8	26.9	40.9	30.4	46.2	10.3	40.0	51.9	27.7	27.1	. . .	84.7
March	28.8	26.9	40.7	30.1	46.0	10.5	39.9	52.0	27.9	27.1	. . .	84.4
April	29.2	27.4	41.7	30.9	47.1	10.7	39.8	52.2	28.3	27.4	. . .	85.6
May	29.4	27.5	42.0	31.1	47.6	10.8	39.1	52.5	28.5	27.6	. . .	85.6
June	29.5	27.6	42.0	31.4	47.3	10.9	38.9	52.1	28.6	27.8	. . .	85.4
July	29.7	27.8	42.6	32.0	47.9	11.0	38.8	53.1	28.6	27.9	. . .	85.9
August	29.9	28.0	42.5	32.2	47.6	11.0	39.0	52.6	28.4	28.4	. . .	86.1
September	30.0	28.1	42.0	31.4	47.3	11.1	39.2	52.2	28.5	28.9	. . .	86.2
October	29.6	27.7	41.2	28.2	47.9	11.0	39.5	52.4	28.5	28.3	. . .	84.6
November	30.5	28.5	42.7	32.0	48.1	11.4	39.9	53.6	28.7	29.2	. . .	86.8
December	30.8	29.1	43.6	33.9	48.3	11.6	40.2	52.9	29.0	29.6	. . .	88.0
1965												
January	31.2	29.4	44.2	34.5	49.0	11.6	40.7	53.3	29.3	29.9	. . .	88.6
February	31.4	29.6	44.3	34.9	48.9	11.8	41.1	54.7	29.4	30.0	. . .	88.7
March	31.8	29.9	44.7	35.9	49.0	11.9	41.8	55.1	29.7	30.5	. . .	89.3
April	31.9	30.1	44.6	35.8	48.8	12.0	42.4	54.3	29.7	30.8	. . .	89.3
May	32.2	30.4	44.9	36.1	49.1	12.2	43.3	54.9	30.0	30.9	. . .	89.4
June	32.4	30.5	45.0	36.4	49.2	12.4	43.8	54.9	30.2	31.3	. . .	89.5
July	32.7	31.0	45.0	36.7	49.0	12.6	44.5	56.5	30.2	31.6	. . .	90.3
August	32.9	31.0	45.0	36.3	49.2	12.6	44.9	55.9	30.4	31.9	. . .	89.9
September	33.0	31.1	45.6	37.0	49.8	12.8	44.9	55.4	30.5	31.6	. . .	89.6
October	33.3	31.4	45.9	37.3	49.9	13.0	45.5	56.3	30.8	31.9	. . .	89.8
November	33.4	31.5	46.1	37.6	50.2	13.3	45.9	57.1	31.0	31.8	. . .	89.6
December	33.8	32.0	46.4	38.3	50.2	13.5	46.4	58.3	31.5	32.2	. . .	90.5
1966												
January	34.2	32.3	46.6	38.5	50.5	13.8	47.4	58.2	31.4	32.6	. . .	90.9
February	34.4	32.5	46.8	38.4	50.7	13.8	48.1	57.6	31.8	33.0	. . .	90.9
March	34.9	32.9	47.1	38.7	51.1	14.0	48.5	58.8	32.1	33.6	. . .	91.6
April	34.9	33.1	47.3	39.4	51.0	14.2	49.5	58.8	31.8	33.5	. . .	91.5
May	35.3	33.4	47.3	38.8	51.3	14.3	50.4	59.1	32.3	33.9	. . .	91.6
June	35.4	33.5	47.5	38.8	51.6	14.5	51.2	58.2	32.7	34.1	. . .	91.5
July	35.6	33.7	47.4	38.1	51.9	14.7	51.8	58.8	33.1	34.2	. . .	91.4
August	35.6	33.7	47.2	37.5	52.0	14.8	52.5	57.1	32.9	34.5	. . .	91.1
September	36.0	34.0	47.4	37.7	52.1	14.9	53.2	57.1	33.1	34.8	. . .	91.2
October	36.2	34.3	48.4	39.6	52.5	14.9	54.1	57.1	33.1	34.9	. . .	91.6
November	36.0	34.0	48.1	38.4	52.8	14.7	55.0	57.2	33.2	34.4	. . .	90.1
December	36.1	34.1	48.0	38.0	52.9	14.9	55.6	57.1	33.3	34.4	. . .	90.0
1967												
January	36.2	34.2	48.5	37.1	54.1	14.8	56.6	58.6	34.0	34.3	89.4	89.8
February	35.8	33.9	47.8	36.1	53.7	14.8	57.2	58.1	33.7	33.6	88.0	88.4
March	35.6	33.7	48.0	36.4	53.7	14.8	57.8	58.2	33.8	33.0	87.1	87.5
April	35.9	33.9	48.8	36.7	54.8	14.7	58.4	58.0	34.1	33.3	87.5	87.7
May	35.6	33.7	47.8	36.2	53.6	14.8	58.8	58.9	33.5	33.0	86.4	86.6
June	35.6	33.6	48.0	35.8	54.1	14.7	58.7	59.2	33.7	32.9	86.0	86.1
July	35.6	33.5	47.9	36.2	53.7	14.4	59.0	59.5	33.8	32.9	85.4	85.3
August	36.2	34.1	48.4	36.7	54.2	14.7	59.1	60.1	34.7	33.9	86.6	86.5
September	36.2	34.1	48.5	36.5	54.4	14.6	59.3	60.6	34.8	33.6	86.1	86.1
October	36.5	34.4	49.1	37.2	55.1	14.5	59.7	60.3	35.1	34.1	86.4	86.4
November	37.0	35.0	50.2	39.3	55.5	14.9	59.8	60.7	35.1	34.4	87.3	87.5
December	37.4	35.4	51.0	40.9	55.9	15.0	59.8	60.7	35.1	34.9	87.8	88.0

. . . = Not available.

Table 2-4A. Industrial Production and Capacity Utilization, Historical Data, 1955–2019—*Continued*

(Seasonally adjusted.)

Year and month	Production indexes, 2007 = 100											Capacity utilization (output as percent of capacity)	
	Total industry	Manufac-turing (SIC)	Market groups									Total industry	Manufac-turing (SIC)
			Consumer goods			Business equipment	Defense and space equipment	Construction supplies	Business supplies	Materials		
			Total	Durable	Nondurable							
1968													
January	37.3	35.3	50.3	39.7	55.5	15.1	59.6	61.1	35.2	35.0		87.4	87.4
February	37.5	35.4	50.6	40.3	55.6	15.1	60.3	61.7	35.5	35.0		87.3	87.4
March	37.6	35.5	50.9	40.2	56.1	15.2	59.1	61.8	35.5	35.2		87.3	87.2
April	37.7	35.5	50.8	40.2	56.0	15.2	57.9	62.2	35.8	35.4		87.1	86.9
May	38.1	36.0	51.1	40.8	56.1	15.4	58.7	62.4	36.2	36.0		87.7	87.6
June	38.2	36.1	51.4	41.1	56.4	15.4	59.0	62.4	36.3	36.1		87.7	87.4
July	38.2	35.9	51.2	40.8	56.4	15.2	59.0	62.5	36.3	36.2		87.2	86.7
August	38.3	36.1	51.9	41.3	57.0	15.3	59.2	62.8	36.7	35.9		87.1	86.8
September	38.4	36.1	52.1	41.8	57.1	15.5	59.1	62.3	36.8	36.0		87.1	86.5
October	38.5	36.4	52.4	42.3	57.3	15.6	57.6	62.2	37.0	36.1		86.9	86.6
November	39.0	36.9	53.1	43.3	57.7	15.7	58.2	63.7	37.4	36.7		87.7	87.5
December	39.1	36.9	52.8	43.7	57.0	15.8	57.9	64.9	37.7	36.9		87.6	87.1
1969													
January	39.3	37.1	53.1	43.6	57.5	16.0	57.9	65.3	37.8	37.0		87.8	87.3
February	39.6	37.4	53.5	43.5	58.2	16.0	57.5	65.9	37.7	37.4		88.1	87.7
March	39.9	37.7	53.9	43.8	58.7	16.2	57.8	66.2	38.9	37.6		88.5	88.0
April	39.8	37.5	53.1	42.6	58.1	16.4	57.3	65.5	38.4	37.7		87.8	87.3
May	39.6	37.4	52.7	42.1	57.8	16.2	57.2	65.1	38.7	37.6		87.2	86.6
June	40.0	37.6	53.3	43.5	57.9	16.4	56.4	65.5	39.0	38.1		87.7	86.8
July	40.2	37.9	54.2	43.5	59.3	16.6	56.2	64.9	38.8	38.2		87.9	87.1
August	40.3	37.9	54.1	43.9	58.9	16.5	55.4	64.8	39.0	38.5		87.8	86.9
September	40.3	37.9	53.7	43.5	58.5	16.7	55.0	64.8	38.9	38.6		87.4	86.4
October	40.3	37.9	53.7	43.7	58.4	16.7	54.5	65.1	39.0	38.6		87.1	86.2
November	39.9	37.5	53.3	41.9	58.8	16.3	53.5	64.7	38.8	38.4		86.0	85.0
December	39.8	37.3	53.3	41.7	59.0	16.3	52.9	64.3	39.3	38.2		85.5	84.2
1970													
January	39.1	36.5	52.4	39.5	58.9	16.1	52.1	62.3	39.2	37.3		83.6	82.1
February	39.0	36.5	53.0	40.1	59.4	16.1	51.3	62.1	38.9	37.1		83.3	81.9
March	39.0	36.4	52.9	40.5	59.0	16.2	50.1	62.6	39.1	37.0		82.9	81.4
April	38.9	36.3	53.1	40.4	59.4	16.1	49.0	63.3	38.9	36.8		82.5	80.9
May	38.9	36.2	53.4	40.5	59.8	16.1	47.9	63.3	38.8	36.7		82.1	80.4
June	38.7	36.1	53.5	41.3	59.5	16.0	47.1	63.0	38.8	36.5		81.6	79.9
July	38.8	36.2	53.7	41.3	59.7	16.0	46.3	63.8	38.9	36.7		81.5	79.9
August	38.8	36.0	52.8	40.1	59.0	15.9	46.0	63.5	38.6	37.1		81.1	79.1
September	38.5	35.7	52.6	39.0	59.3	15.6	45.5	63.4	38.9	36.9		80.3	78.2
October	37.7	34.9	51.9	36.7	59.7	15.1	44.9	62.7	38.7	35.9		78.5	76.2
November	37.5	34.7	51.6	36.8	59.1	15.0	44.5	61.9	38.8	35.7		77.8	75.5
December	38.3	35.6	53.8	41.1	60.1	15.2	44.0	62.6	38.8	36.5		79.3	77.3
1971													
January	38.6	35.9	54.6	42.7	60.4	14.9	44.4	62.7	39.0	37.0		79.7	77.7
February	38.6	35.9	54.6	43.7	59.8	15.0	43.3	63.1	39.4	36.8		79.4	77.6
March	38.5	35.8	54.7	43.7	60.0	14.8	42.9	62.8	39.1	36.8		79.1	77.2
April	38.7	36.0	55.2	44.0	60.5	14.7	43.0	63.4	39.5	37.1		79.3	77.4
May	38.9	36.3	55.2	44.7	60.2	14.6	43.5	63.7	39.5	37.5		79.5	77.7
June	39.1	36.4	55.6	45.1	60.6	14.7	42.8	64.4	39.4	37.7		79.6	77.7
July	39.0	36.4	56.5	45.9	61.5	14.7	42.5	65.0	40.4	36.8		79.2	77.7
August	38.8	36.0	55.8	45.4	60.7	14.9	42.5	63.9	39.9	36.6		78.5	76.5
September	39.4	36.7	56.3	45.4	61.6	15.3	42.1	66.3	40.6	37.3		79.6	77.9
October	39.7	37.3	57.1	46.2	62.3	15.4	41.8	67.2	40.8	37.4		80.0	78.9
November	39.9	37.4	57.6	46.7	62.7	15.5	41.5	67.4	41.3	37.4		80.2	79.0
December	40.3	37.7	57.9	46.9	63.2	15.6	40.8	68.5	41.4	38.2		80.9	79.5
1972													
January	41.3	38.7	58.8	48.3	63.8	16.1	40.9	70.4	42.2	39.4		82.6	81.3
February	41.7	39.0	59.1	48.5	64.2	16.3	41.0	70.7	42.9	39.8		83.2	81.7
March	42.0	39.3	59.1	48.1	64.4	16.5	41.3	71.2	43.3	40.2		83.6	82.1
April	42.4	39.7	59.9	49.5	64.7	16.8	41.5	72.0	43.3	40.6		84.3	82.8
May	42.4	39.8	59.6	49.1	64.5	16.8	41.1	72.4	43.5	40.7		84.0	82.7
June	42.5	39.9	59.6	49.1	64.5	16.9	41.1	73.0	44.0	40.7		84.1	82.8
July	42.5	39.9	60.1	50.2	64.6	16.9	41.1	74.1	44.0	40.5		83.8	82.6
August	43.1	40.5	60.7	50.6	65.3	17.2	41.0	74.6	44.6	41.1		84.7	83.4
September	43.4	40.8	61.0	51.1	65.5	17.4	41.3	75.2	44.5	41.5		85.1	83.9
October	44.0	41.4	61.9	52.3	66.2	17.7	41.3	76.4	45.2	41.9		86.0	84.9
November	44.5	41.9	62.4	53.5	66.2	18.1	42.5	76.9	45.4	42.5		86.7	85.7
December	45.0	42.5	62.9	54.6	66.3	18.3	43.2	76.7	45.5	43.2		87.5	86.6
1973													
January	45.3	42.7	62.7	54.4	66.2	18.6	43.8	77.5	45.8	43.6		87.8	86.9
February	46.0	43.4	63.5	55.4	66.9	19.0	44.7	79.0	46.3	44.2		88.8	88.0
March	46.0	43.5	63.7	55.5	67.0	19.1	44.5	79.5	46.3	44.1		88.5	87.9
April	45.9	43.5	63.1	54.7	66.6	19.3	44.3	79.3	46.3	44.2		88.1	87.5
May	46.2	43.7	63.4	54.5	67.2	19.5	44.6	79.7	46.6	44.4		88.3	87.7
June	46.3	43.7	62.9	54.4	66.6	19.6	45.0	79.8	46.7	44.5		88.1	87.4
July	46.4	43.9	62.8	54.4	66.4	19.9	46.0	80.5	46.9	44.7		88.2	87.5
August	46.4	43.8	62.2	52.7	66.4	19.9	46.0	80.6	46.9	44.8		87.7	87.0
September	46.8	44.2	63.2	54.5	67.0	20.3	46.0	80.5	47.0	45.0		88.2	87.5
October	47.1	44.5	63.4	54.1	67.5	20.6	47.0	80.4	47.4	45.3		88.5	87.9
November	47.3	44.9	63.7	54.1	67.9	20.7	46.6	80.9	47.4	45.6		88.7	88.3
December	47.2	44.9	62.6	52.9	66.9	20.8	46.3	81.6	47.1	45.8		88.3	88.1

Table 2-4A. Industrial Production and Capacity Utilization, Historical Data, 1955–2019—*Continued*

(Seasonally adjusted.)

Year and month	Production indexes, 2007 = 100										Capacity utilization (output as percent of capacity)	
	Total industry	Manufac-turing (SIC)	Market groups								Total industry	Manufac-turing (SIC)
			Consumer goods			Business equipment	Defense and space equipment	Construction supplies	Business supplies	Materials		
			Total	Durable	Nondurable							
1974												
January	46.9	44.5	61.7	50.2	67.2	20.8	46.1	81.8	47.2	45.4	87.4	87.1
February	46.8	44.4	61.5	50.1	66.9	20.7	46.5	80.8	46.9	45.4	86.9	86.5
March	46.8	44.3	61.6	50.1	67.0	20.9	46.5	81.0	47.0	45.2	86.7	86.2
April	46.6	44.1	61.4	49.8	66.9	20.7	46.4	80.3	47.0	45.1	86.2	85.6
May	47.0	44.5	62.0	50.0	67.7	21.0	46.6	80.6	47.3	45.4	86.6	86.0
June	46.9	44.5	62.2	50.6	67.7	21.0	45.9	80.2	47.4	45.2	86.4	85.8
July	46.9	44.4	62.1	50.5	67.6	21.0	46.5	78.6	47.1	45.4	86.2	85.5
August	46.5	44.1	62.1	50.5	67.6	21.0	47.4	77.7	46.9	44.6	85.2	84.6
September	46.5	44.1	61.6	50.4	66.9	21.4	47.5	77.1	46.6	44.8	85.1	84.4
October	46.3	43.8	61.6	49.7	67.2	21.4	48.2	75.7	46.4	44.5	84.6	83.6
November	44.8	42.5	59.8	47.3	65.9	21.1	47.9	73.3	45.3	42.6	81.7	80.9
December	43.2	40.6	57.9	43.4	65.2	20.2	47.4	69.9	44.3	41.0	78.6	77.2
1975												
January	42.6	39.7	56.6	41.6	64.1	19.8	47.8	69.6	43.6	40.6	77.4	75.5
February	41.7	38.6	55.8	40.4	63.6	19.1	45.3	67.4	42.7	39.7	75.5	73.2
March	41.2	38.1	56.0	40.9	63.6	18.8	45.7	64.9	42.1	39.1	74.6	72.1
April	41.2	38.1	57.1	42.5	64.5	18.6	45.5	64.3	42.2	38.9	74.5	71.9
May	41.2	38.0	57.4	43.6	64.3	18.4	47.7	64.4	42.0	38.6	74.2	71.8
June	41.4	38.4	58.2	44.0	65.3	18.3	48.4	64.0	42.1	38.9	74.6	72.2
July	41.8	38.9	59.7	46.2	66.3	18.4	47.5	64.9	42.5	39.0	75.2	73.1
August	42.3	39.3	60.1	46.9	66.5	18.3	47.4	65.7	42.9	39.7	75.8	73.7
September	42.8	39.9	60.9	47.8	67.3	18.5	48.8	66.4	43.0	40.2	76.6	74.8
October	43.0	40.1	61.0	47.6	67.6	18.6	48.7	66.8	43.3	40.4	76.8	75.0
November	43.1	40.2	61.2	47.7	67.9	18.6	46.8	67.1	43.4	40.6	76.8	75.1
December	43.6	40.8	61.8	48.5	68.2	18.9	48.1	67.1	43.8	41.2	77.6	75.9
1976												
January	44.2	41.3	62.6	49.3	69.0	19.1	48.1	68.8	44.3	41.8	78.6	76.7
February	44.7	41.9	62.8	50.0	69.1	19.2	48.1	69.9	44.4	42.4	79.2	77.7
March	44.7	42.0	62.7	49.9	68.9	19.3	48.1	68.8	44.6	42.6	79.1	77.6
April	45.0	42.3	62.9	49.9	69.1	19.5	47.2	69.7	44.8	42.9	79.4	78.0
May	45.2	42.5	63.4	50.0	69.9	19.7	46.7	70.5	45.1	42.9	79.6	78.2
June	45.2	42.5	63.1	49.7	69.7	19.7	46.1	71.0	44.8	43.1	79.4	78.0
July	45.5	42.8	63.6	50.0	70.2	19.9	44.9	72.6	45.5	43.2	79.7	78.5
August	45.8	43.1	63.6	50.7	69.9	20.2	44.9	71.8	45.6	43.7	80.1	78.8
September	45.9	43.2	63.7	50.4	70.1	20.2	44.5	72.3	46.6	43.7	80.1	78.8
October	45.9	43.2	64.2	51.0	70.5	20.2	44.5	72.2	46.9	43.5	80.0	78.6
November	46.6	43.7	65.3	52.7	71.2	20.8	44.2	72.5	47.1	44.1	80.9	79.3
December	47.1	44.2	65.9	54.0	71.6	21.2	43.4	73.0	47.5	44.6	81.6	80.0
1977												
January	46.8	44.1	65.8	53.8	71.4	21.3	42.8	71.8	47.3	44.2	81.0	79.6
February	47.5	44.9	66.6	54.4	72.3	21.8	42.7	73.4	47.9	44.8	82.0	80.8
March	48.1	45.5	66.7	55.9	71.6	22.1	41.8	75.0	48.2	45.7	82.8	81.7
April	48.6	46.0	67.1	56.3	72.0	22.4	42.0	76.7	48.8	46.1	83.4	82.4
May	49.0	46.4	67.2	56.6	71.9	22.9	41.9	77.8	49.3	46.5	83.9	82.9
June	49.3	46.8	67.7	57.7	72.0	23.4	41.9	78.4	49.8	46.6	84.3	83.3
July	49.4	46.8	67.8	57.7	72.2	23.5	41.8	78.5	49.9	46.7	84.2	83.1
August	49.4	47.0	67.9	57.8	72.4	23.6	41.5	79.1	50.2	46.6	84.0	83.3
September	49.7	47.1	68.0	58.0	72.4	23.7	41.6	78.8	50.4	46.9	84.1	83.1
October	49.8	47.2	68.6	58.0	73.3	23.6	37.9	78.8	50.4	47.1	84.1	83.1
November	49.8	47.3	68.6	58.0	73.4	23.6	37.5	79.2	50.5	47.1	83.9	83.0
December	49.9	47.8	69.3	58.4	74.1	23.9	40.1	80.0	50.9	46.7	83.8	83.6
1978												
January	49.2	47.1	67.5	55.3	73.2	23.5	40.6	78.4	50.7	46.3	82.4	82.2
February	49.5	47.3	68.6	56.7	74.1	24.0	38.5	78.2	50.9	46.2	82.6	82.2
March	50.4	48.1	69.9	58.4	75.2	24.7	42.0	79.6	51.6	46.9	83.9	83.4
April	51.4	48.9	70.4	59.5	75.3	25.3	41.8	81.5	51.8	48.3	85.4	84.5
May	51.6	49.0	69.9	58.6	75.1	25.4	42.0	81.3	52.1	48.7	85.4	84.5
June	52.0	49.4	70.4	58.9	75.5	25.8	42.4	82.1	52.5	49.0	85.8	84.9
July	52.0	49.4	70.0	59.0	74.9	26.0	42.4	82.0	52.4	49.0	85.5	84.6
August	52.2	49.6	69.9	58.7	74.9	26.5	42.8	82.1	52.5	49.1	85.6	84.7
September	52.3	49.8	70.0	58.3	75.3	26.8	42.8	82.4	52.6	49.2	85.6	84.8
October	52.7	50.2	69.9	58.6	75.1	27.3	42.4	83.2	52.9	49.7	86.1	85.2
November	53.1	50.7	70.1	58.7	75.3	27.8	42.1	83.7	53.2	50.1	86.5	85.7
December	53.4	51.1	70.3	58.9	75.5	28.1	42.6	85.1	53.5	50.3	86.7	86.2
1979												
January	53.0	50.6	70.1	59.3	74.9	28.4	42.9	83.0	53.5	49.7	85.9	85.2
February	53.3	50.8	69.6	58.7	74.6	28.7	43.8	83.5	54.0	50.1	86.2	85.3
March	53.5	51.1	69.9	58.5	75.1	28.9	43.4	84.6	54.3	50.1	86.2	85.5
April	52.9	50.3	68.6	55.6	74.8	28.3	42.4	83.3	54.0	49.8	85.1	83.9
May	53.3	50.9	69.1	57.3	74.5	29.1	43.1	83.8	54.2	50.1	85.5	84.7
June	53.3	51.0	68.8	56.6	74.5	29.2	43.5	84.1	53.9	50.1	85.3	84.6
July	53.2	51.0	68.3	55.9	74.1	29.5	44.3	84.1	54.0	50.0	85.0	84.5
August	52.9	50.4	67.5	53.5	74.2	29.2	45.0	83.4	54.2	49.7	84.3	83.2
September	52.9	50.5	67.9	55.4	73.8	30.0	45.4	83.5	53.5	49.4	84.2	83.2
October	53.2	50.7	68.1	55.1	74.2	29.6	46.9	84.1	54.0	49.8	84.5	83.2
November	53.2	50.6	67.9	54.2	74.4	29.6	48.1	83.9	54.3	49.7	84.2	82.9
December	53.3	50.7	67.9	53.9	74.6	29.8	49.1	84.3	54.3	49.6	84.1	82.9

Table 2-4A. Industrial Production and Capacity Utilization, Historical Data, 1955–2019—*Continued*

(Seasonally adjusted.)

Year and month	Production indexes, 2007 = 100										Capacity utilization (output as percent of capacity)	
	Total industry	Manufac-turing (SIC)	Market groups								Total industry	Manufac-turing (SIC)
			Consumer goods			Business equipment	Defense and space equipment	Construction supplies	Business supplies	Materials		
			Total	Durable	Nondurable							
1980												
January	53.5	51.0	67.7	52.9	74.8	30.3	49.7	84.0	53.9	50.0	84.3	83.1
February	53.5	50.9	67.9	52.6	75.3	30.4	51.7	82.7	54.0	49.9	84.2	82.8
March	53.3	50.5	67.4	51.7	75.1	30.2	52.2	81.7	53.8	49.9	83.8	82.0
April	52.2	49.4	66.3	49.5	74.6	29.9	52.7	78.0	52.9	48.7	81.9	80.1
May	51.0	48.0	64.9	46.4	74.2	29.4	52.9	74.8	51.9	47.3	79.8	77.6
June	50.3	47.3	64.6	45.6	74.2	29.1	53.4	73.3	51.2	46.5	78.6	76.2
July	49.9	46.8	64.6	45.5	74.1	29.1	53.6	72.7	51.4	45.7	77.8	75.2
August	50.1	47.1	64.8	45.9	74.3	29.1	53.8	73.7	51.6	45.9	78.0	75.6
September	50.9	47.9	65.4	48.0	74.0	29.6	53.8	75.6	52.3	46.8	79.1	76.6
October	51.6	48.7	65.9	48.9	74.4	30.2	54.5	77.2	52.5	47.4	79.9	77.7
November	52.5	49.6	66.2	50.1	74.2	30.7	55.1	78.8	53.1	48.5	81.1	79.0
December	52.8	49.8	66.2	49.4	74.5	30.7	55.3	78.7	53.7	49.0	81.3	79.0
1981												
January	52.5	49.6	66.2	49.2	74.7	30.8	55.0	78.5	53.7	48.4	80.7	78.5
February	52.2	49.3	66.1	49.0	74.5	30.4	54.9	77.7	53.2	48.3	80.1	77.8
March	52.5	49.5	66.1	49.7	74.1	30.7	55.0	78.0	53.2	48.6	80.3	77.8
April	52.3	49.7	66.2	50.3	74.0	31.1	55.0	78.0	53.5	47.9	79.8	78.0
May	52.6	50.0	66.9	51.1	74.6	31.1	55.6	78.0	54.1	48.2	80.0	78.2
June	52.8	49.7	66.4	50.6	74.1	30.9	56.0	76.8	54.3	48.8	80.2	77.5
July	53.2	49.8	66.9	50.9	74.7	31.0	56.9	77.0	54.6	49.2	80.5	77.5
August	53.2	49.9	66.9	50.6	74.9	31.0	57.8	76.8	54.3	49.1	80.3	77.3
September	52.9	49.7	66.3	49.4	74.7	31.0	59.1	76.3	54.4	48.7	79.6	76.8
October	52.5	49.2	66.7	49.2	75.4	30.9	60.5	74.0	54.0	48.1	78.9	75.9
November	51.9	48.6	66.6	48.4	75.8	30.4	62.1	72.8	53.7	47.2	77.8	74.8
December	51.3	47.8	65.9	46.3	75.8	30.0	63.7	71.5	53.6	46.6	76.7	73.3
1982												
January	50.3	46.6	64.7	45.0	74.7	28.8	63.2	69.0	52.8	45.8	75.0	71.4
February	51.3	47.9	66.5	46.7	76.6	29.8	66.6	71.5	53.9	46.3	76.4	73.2
March	50.9	47.5	66.1	46.5	76.0	29.4	67.5	70.0	53.6	46.0	75.7	72.5
April	50.5	47.2	66.1	47.4	75.5	29.0	68.2	69.5	53.4	45.4	74.8	71.8
May	50.1	47.1	66.2	47.4	75.6	28.8	69.3	69.9	53.1	44.9	74.2	71.6
June	50.0	47.0	66.6	47.7	76.1	28.3	69.4	69.3	53.1	44.8	73.9	71.4
July	49.8	47.0	66.8	48.0	76.2	28.2	70.3	69.2	53.1	44.5	73.6	71.2
August	49.4	46.6	66.6	47.3	76.4	27.5	70.0	69.2	53.1	44.0	72.8	70.5
September	49.2	46.5	66.6	46.6	76.7	27.4	70.8	69.4	53.3	43.8	72.5	70.3
October	48.8	45.9	66.6	46.0	77.1	26.8	70.4	68.4	53.0	43.3	71.8	69.4
November	48.6	45.6	66.5	45.9	76.9	26.7	70.7	68.0	53.0	43.1	71.5	68.9
December	48.2	45.5	65.6	45.9	75.6	26.9	70.2	67.4	52.7	42.7	70.9	68.6
1983												
January	49.2	46.6	67.1	47.8	76.7	27.0	69.8	69.9	53.5	43.7	72.2	70.2
February	48.9	46.6	66.2	47.7	75.4	26.9	68.7	69.5	53.4	43.5	71.7	70.1
March	49.3	47.0	66.5	48.4	75.5	27.1	68.8	70.3	54.2	43.9	72.3	70.7
April	49.9	47.5	67.7	49.4	76.8	27.2	68.2	71.2	54.8	44.4	73.1	71.4
May	50.2	48.1	68.0	50.4	76.8	27.5	68.1	72.6	54.9	44.8	73.6	72.3
June	50.5	48.5	68.3	51.3	76.7	27.8	67.7	73.9	55.2	45.1	74.0	72.8
July	51.3	49.2	69.2	52.4	77.4	28.3	68.5	75.4	55.9	45.8	75.1	73.8
August	51.8	49.6	69.7	53.4	77.7	28.6	69.0	75.7	56.5	46.5	75.9	74.4
September	52.6	50.5	70.6	54.4	78.5	29.4	69.9	76.7	57.4	47.1	77.0	75.7
October	53.1	51.1	70.3	55.1	77.6	29.8	70.6	77.9	57.7	47.7	77.6	76.5
November	53.3	51.3	70.3	55.1	77.6	30.0	71.0	77.7	57.9	48.0	77.8	76.7
December	53.5	51.4	70.4	56.4	77.1	30.3	71.8	77.9	58.0	48.2	78.1	76.9
1984												
January	54.6	52.4	71.9	57.8	78.6	31.0	73.8	78.4	59.1	49.2	79.6	78.2
February	54.8	52.9	71.7	57.9	78.2	31.2	75.3	80.2	59.3	49.4	79.8	78.9
March	55.1	53.2	72.0	58.1	78.6	31.5	75.8	79.7	59.8	49.7	80.1	79.1
April	55.5	53.5	72.2	58.0	79.0	31.8	77.6	80.2	59.8	50.0	80.5	79.4
May	55.7	53.6	71.9	57.5	78.8	32.0	78.3	80.4	60.6	50.4	80.7	79.4
June	55.9	53.8	71.8	57.7	78.6	32.3	79.3	81.0	61.0	50.5	80.9	79.6
July	56.1	54.1	71.9	58.3	78.3	32.7	78.5	80.7	61.2	50.6	81.0	79.8
August	56.1	54.2	71.5	58.7	77.6	33.1	81.0	81.0	61.3	50.6	80.9	79.7
September	56.0	54.0	71.3	57.9	77.7	33.2	82.8	81.3	61.2	50.3	80.5	79.3
October	55.9	54.2	71.9	57.4	78.8	33.4	83.4	81.0	61.5	49.9	80.2	79.4
November	56.2	54.4	72.1	58.3	78.7	33.7	82.9	80.8	61.8	50.1	80.3	79.5
December	56.2	54.6	72.5	59.0	78.9	34.0	84.4	81.6	61.4	49.9	80.2	79.5
1985												
January	56.1	54.4	72.0	58.2	78.6	33.8	84.7	80.2	61.4	50.0	79.9	79.0
February	56.3	54.2	72.5	57.7	79.5	33.6	85.5	80.3	62.0	50.2	80.0	78.5
March	56.4	54.7	72.4	58.4	79.1	34.0	87.0	82.5	61.9	50.1	79.9	78.9
April	56.3	54.4	72.0	57.5	78.9	33.6	87.2	82.4	62.2	50.1	79.5	78.3
May	56.3	54.6	72.1	57.6	79.1	33.7	87.7	82.7	62.4	50.1	79.4	78.2
June	56.4	54.7	72.4	57.5	79.6	33.8	89.0	83.2	62.0	50.0	79.3	78.2
July	56.0	54.3	72.1	57.5	79.1	33.6	88.3	82.8	61.6	49.7	78.6	77.5
August	56.3	54.6	72.5	58.1	79.4	33.7	89.7	83.0	62.2	49.8	78.7	77.8
September	56.5	54.7	72.9	58.1	79.9	33.6	90.3	82.9	62.6	50.0	78.9	77.7
October	56.3	54.6	72.7	57.8	79.9	33.5	91.3	83.0	62.1	49.7	78.5	77.4
November	56.5	54.9	73.1	59.2	79.8	33.8	92.4	83.0	62.3	49.8	78.6	77.8
December	57.0	55.1	73.9	59.3	80.9	33.8	93.3	82.7	63.2	50.5	79.3	78.0

Table 2-4A. Industrial Production and Capacity Utilization, Historical Data, 1955–2019—*Continued*

(Seasonally adjusted.)

Year and month	Production indexes, 2007 = 100										Capacity utilization (output as percent of capacity)	
	Total industry	Manufac-turing (SIC)	Market groups								Total industry	Manufac-turing (SIC)
			Consumer goods			Business equipment	Defense and space equipment	Construction supplies	Business supplies	Materials		
			Total	Durable	Nondurable							
1986												
January	57.3	55.7	74.7	60.7	81.4	33.8	93.9	84.5	63.6	50.6	79.6	78.8
February	56.9	55.4	74.1	60.3	80.6	33.5	92.6	83.7	63.1	50.4	78.9	78.2
March	56.5	55.3	73.8	60.3	80.2	33.5	93.4	83.9	62.8	49.8	78.3	77.9
April	56.6	55.5	74.3	60.2	81.0	33.2	93.5	84.7	63.4	49.7	78.3	78.2
May	56.7	55.6	74.7	60.4	81.5	33.1	93.9	85.2	63.9	49.7	78.4	78.2
June	56.5	55.4	74.7	61.1	81.2	32.8	94.3	84.4	64.3	49.5	78.1	77.9
July	56.8	55.6	75.1	61.8	81.5	33.0	94.8	84.8	64.3	49.8	78.4	78.2
August	56.7	55.8	75.1	62.1	81.3	33.1	95.0	85.7	64.4	49.6	78.2	78.3
September	56.9	55.9	75.2	62.7	81.1	33.1	94.8	85.8	64.5	49.8	78.3	78.4
October	57.1	56.2	75.5	62.7	81.6	33.1	95.1	85.7	64.9	50.1	78.5	78.6
November	57.4	56.4	76.1	63.6	82.0	33.1	95.5	86.0	65.0	50.3	78.8	78.8
December	57.9	56.9	76.8	64.7	82.4	33.4	95.7	86.5	65.9	50.7	79.3	79.3
1987												
January	57.7	56.7	76.0	64.3	81.5	33.5	96.0	87.5	65.4	50.6	79.0	78.9
February	58.4	57.6	77.0	65.4	82.4	34.2	96.6	89.1	65.9	51.2	79.8	79.9
March	58.5	57.6	77.2	65.1	82.9	34.2	96.3	88.7	66.3	51.3	79.8	79.7
April	58.9	57.9	77.1	64.7	83.0	34.5	96.4	89.2	67.0	51.8	80.2	80.0
May	59.3	58.3	77.7	65.0	83.6	34.8	96.1	89.9	67.8	52.0	80.5	80.4
June	59.5	58.5	77.8	64.4	84.2	34.9	95.8	90.1	68.1	52.4	80.7	80.4
July	60.0	59.0	78.4	64.3	85.0	35.2	95.5	90.5	68.6	52.7	81.1	80.9
August	60.5	59.4	78.8	64.9	85.4	35.8	96.4	91.2	68.9	53.2	81.7	81.2
September	60.6	59.7	78.5	65.4	84.6	36.3	96.8	91.6	69.1	53.4	82.1	81.5
October	61.5	60.6	79.9	67.6	85.8	37.1	96.5	92.9	69.6	54.1	82.8	82.6
November	61.8	61.0	80.0	67.5	85.9	37.4	96.9	92.9	69.6	54.6	83.1	82.9
December	62.1	61.4	80.1	66.9	86.3	37.8	97.7	93.7	69.8	54.9	83.4	83.3
1988												
January	62.1	61.3	80.4	66.6	87.0	37.8	99.8	92.1	70.1	54.7	83.4	83.1
February	62.4	61.4	80.8	66.6	87.5	38.1	98.3	92.9	70.6	54.9	83.7	83.2
March	62.5	61.5	80.7	67.2	87.1	38.5	97.6	93.3	70.5	55.1	83.8	83.4
April	62.9	62.1	81.3	68.7	87.2	38.9	96.7	92.9	70.5	55.5	84.3	84.1
May	62.8	62.0	80.9	68.7	86.7	39.1	96.6	93.1	70.0	55.5	84.1	83.9
June	63.0	62.1	80.9	69.0	86.6	39.4	95.8	92.5	70.2	55.7	84.3	84.0
July	63.0	62.0	80.9	67.4	87.2	39.1	96.9	92.5	70.5	55.8	84.3	84.0
August	63.3	62.1	81.4	68.0	87.7	39.2	96.6	91.8	71.0	56.1	84.7	84.0
September	63.1	62.3	80.9	69.3	86.4	39.6	97.0	92.2	70.6	55.8	84.4	84.2
October	63.4	62.6	81.6	70.0	87.1	40.0	97.0	92.5	70.9	55.9	84.7	84.6
November	63.5	62.8	81.7	70.8	86.8	40.0	96.8	92.9	70.9	56.1	84.8	84.7
December	63.8	63.1	82.1	71.6	87.0	40.1	97.4	93.2	71.1	56.4	85.1	85.0
1989												
January	64.0	63.6	82.2	73.3	86.4	40.5	97.4	94.7	71.1	56.5	85.2	85.6
February	63.7	63.0	82.2	72.5	86.7	40.3	97.3	92.4	71.4	56.1	84.7	84.6
March	63.9	62.9	82.3	71.4	87.4	40.1	96.7	92.6	72.0	56.3	84.8	84.4
April	63.9	63.1	82.2	72.0	87.0	40.8	98.0	92.7	71.5	56.3	84.7	84.4
May	63.5	62.5	81.3	70.4	86.5	40.2	98.5	91.8	71.2	56.1	84.0	83.5
June	63.5	62.6	81.4	69.6	87.0	40.8	98.4	92.1	71.5	55.8	83.8	83.5
July	62.9	61.9	79.7	67.4	85.5	40.3	98.8	92.1	70.9	55.6	82.9	82.3
August	63.5	62.5	81.0	69.9	86.2	40.8	99.3	92.1	71.3	56.0	83.5	82.8
September	63.3	62.3	80.7	69.8	85.8	40.6	98.8	91.8	71.5	55.7	83.0	82.4
October	63.3	62.2	80.9	68.5	86.8	40.1	95.1	92.2	71.5	55.8	82.7	82.1
November	63.5	62.3	81.3	69.0	87.0	40.4	93.8	92.2	71.9	56.0	82.8	82.0
December	63.8	62.4	82.5	69.7	88.6	41.0	95.9	91.2	72.2	56.0	83.1	81.9
1990												
January	63.4	62.2	80.7	65.8	87.8	40.5	96.3	93.0	72.8	55.8	82.4	81.5
February	64.0	63.1	81.8	69.9	87.4	41.3	96.0	93.5	72.7	56.3	83.0	82.5
March	64.4	63.5	82.3	71.2	87.6	42.0	95.5	93.3	73.1	56.5	83.2	82.7
April	64.3	63.3	82.1	69.7	87.9	41.8	94.8	92.5	73.1	56.5	83.0	82.3
May	64.4	63.4	81.9	70.2	87.5	42.3	94.0	91.9	73.4	56.6	82.9	82.2
June	64.6	63.6	82.8	71.2	88.2	42.4	93.7	92.2	73.3	56.7	83.0	82.3
July	64.5	63.5	82.1	69.2	88.2	42.6	94.3	91.4	73.6	56.7	82.8	82.0
August	64.7	63.7	82.3	69.0	88.6	42.8	92.8	91.4	73.5	57.0	82.9	82.0
September	64.8	63.6	83.0	69.4	89.4	42.9	92.5	91.1	73.5	56.9	82.9	81.9
October	64.3	63.1	81.7	67.0	88.6	42.6	92.8	90.0	73.4	56.6	82.1	81.1
November	63.6	62.4	80.9	63.5	89.1	41.7	91.3	90.0	73.1	56.0	81.1	80.0
December	63.2	62.0	80.3	62.4	88.7	41.4	92.4	89.4	72.6	55.6	80.4	79.3
1991												
January	62.9	61.5	80.7	62.5	89.2	41.1	91.3	86.1	72.4	55.2	80.0	78.6
February	62.4	61.1	79.9	61.1	88.8	40.9	90.3	85.8	71.8	55.0	79.3	78.0
March	62.1	60.7	80.0	61.4	88.7	40.9	89.9	84.8	70.8	54.5	78.8	77.3
April	62.2	60.9	80.0	62.8	88.1	40.8	87.4	85.4	71.3	54.8	78.9	77.5
May	62.9	61.3	81.4	63.9	89.6	41.1	85.8	85.5	71.9	55.3	79.6	78.0
June	63.4	61.9	82.5	65.6	90.4	41.6	86.2	87.1	72.3	55.6	80.2	78.7
July	63.5	62.1	82.1	66.6	89.4	41.7	85.5	86.7	71.9	56.0	80.2	78.8
August	63.6	62.3	82.2	65.7	89.9	41.5	86.1	87.7	72.3	56.0	80.2	78.9
September	64.1	63.0	83.4	68.5	90.4	42.2	85.7	88.0	72.7	56.3	80.8	79.6
October	64.0	62.8	83.2	68.1	90.2	41.8	85.9	87.1	72.5	56.4	80.5	79.3
November	63.9	62.7	83.3	68.3	90.3	41.8	85.4	87.9	72.6	56.2	80.3	79.1
December	63.7	62.6	82.3	67.7	89.1	41.8	84.7	87.8	72.5	56.2	79.8	78.8

Table 2-4A. Industrial Production and Capacity Utilization, Historical Data, 1955–2019—*Continued*

(Seasonally adjusted.)

Year and month	Production indexes, 2007 = 100										Capacity utilization (output as percent of capacity)	
	Total industry	Manufac-turing (SIC)	Market groups								Total industry	Manufac-turing (SIC)
			Consumer goods			Business equipment	Defense and space equipment	Construction supplies	Business supplies	Materials		
			Total	Durable	Nondurable							
1992												
January	63.3	62.2	81.4	64.8	89.2	41.1	83.6	88.0	72.4	56.1	79.2	78.1
February	63.8	62.8	82.3	67.6	89.2	41.9	83.1	88.7	72.5	56.3	79.6	78.6
March	64.3	63.4	83.1	69.0	89.7	42.4	82.6	89.3	73.0	56.8	80.2	79.3
April	64.8	63.8	83.8	70.3	90.2	42.8	81.2	90.0	73.5	57.2	80.6	79.5
May	65.0	64.1	84.4	72.4	90.1	43.2	80.9	90.8	73.7	57.3	80.7	79.7
June	65.0	64.3	83.9	71.4	89.9	43.2	80.7	90.4	73.7	57.5	80.5	79.8
July	65.6	64.9	85.0	73.3	90.6	43.7	79.9	91.1	74.2	57.9	81.0	80.3
August	65.3	64.6	85.1	72.7	91.0	43.5	79.7	91.3	74.1	57.3	80.5	79.7
September	65.4	64.7	84.6	72.4	90.4	43.7	79.6	91.2	74.3	57.8	80.5	79.6
October	65.9	65.1	85.9	74.0	91.5	43.8	79.4	91.3	74.5	58.1	80.9	79.9
November	66.2	65.4	86.1	74.7	91.5	44.2	79.4	91.0	74.8	58.4	81.1	80.1
December	66.3	65.3	86.0	75.5	91.1	44.6	79.3	91.3	75.1	58.3	81.1	79.8
1993												
January	66.6	65.9	86.5	76.6	91.3	44.8	78.7	91.6	75.2	58.6	81.3	80.4
February	66.9	66.1	86.6	76.3	91.6	44.8	78.2	93.2	75.6	59.0	81.5	80.5
March	66.8	65.9	86.6	76.7	91.4	44.8	77.3	92.4	76.1	58.8	81.3	80.2
April	67.0	66.3	86.7	77.1	91.4	45.2	77.4	92.7	76.1	59.1	81.4	80.5
May	66.8	66.2	86.1	77.2	90.5	45.2	76.5	93.8	75.8	58.9	81.1	80.3
June	66.9	66.1	86.2	76.4	91.0	44.9	76.0	93.7	75.9	59.2	81.1	80.0
July	67.1	66.3	86.9	76.3	92.1	44.9	76.9	94.3	76.0	59.2	81.2	80.1
August	67.0	66.2	86.8	75.5	92.2	44.5	75.4	94.5	76.1	59.2	81.0	79.8
September	67.3	66.6	87.1	77.0	92.0	45.3	75.7	95.0	76.4	59.4	81.3	80.2
October	67.9	67.2	87.6	79.0	91.8	46.3	75.1	95.9	76.5	59.9	81.8	80.8
November	68.1	67.4	87.6	79.7	91.6	46.5	75.0	96.7	76.6	60.3	82.0	80.9
December	68.5	67.8	87.8	80.0	91.7	46.8	74.4	97.8	77.1	60.7	82.3	81.2
1994												
January	68.8	67.9	88.5	81.3	92.1	47.1	73.6	97.5	77.5	60.8	82.4	81.2
February	68.8	68.0	88.6	81.2	92.3	46.6	72.4	96.8	77.7	61.0	82.2	81.1
March	69.5	68.9	89.4	81.9	93.1	47.1	73.2	98.6	78.2	61.7	82.9	81.9
April	69.9	69.5	89.5	82.9	92.9	47.5	73.4	100.1	78.5	62.1	83.1	82.4
May	70.2	69.9	90.1	83.2	93.7	47.6	72.3	100.9	78.6	62.5	83.3	82.6
June	70.7	70.1	90.8	83.9	94.3	47.9	71.4	101.0	79.3	63.0	83.6	82.6
July	70.8	70.4	90.3	84.0	93.6	48.5	71.0	102.0	79.0	63.2	83.4	82.7
August	71.2	71.0	91.4	85.5	94.5	48.6	70.0	101.9	79.2	63.6	83.7	83.0
September	71.5	71.3	90.9	85.8	93.7	49.1	70.7	102.9	79.6	64.0	83.7	83.1
October	72.1	72.0	91.8	86.9	94.6	50.0	70.5	103.3	80.2	64.4	84.2	83.7
November	72.5	72.6	91.9	86.7	94.7	50.6	71.4	103.5	80.4	65.0	84.4	84.0
December	73.3	73.4	92.6	87.6	95.3	51.1	71.6	104.6	81.0	65.9	85.0	84.6
1995												
January	73.4	73.5	92.5	88.0	95.0	51.4	71.5	104.6	81.2	66.0	84.9	84.4
February	73.3	73.3	92.7	87.5	95.5	51.4	70.3	103.3	81.3	65.9	84.5	83.9
March	73.4	73.5	92.6	87.3	95.4	51.9	70.4	103.3	81.5	65.9	84.3	83.8
April	73.4	73.3	92.3	87.1	95.1	51.9	69.9	102.7	81.4	66.1	84.0	83.3
May	73.6	73.5	92.4	86.3	95.6	52.4	69.9	102.0	81.7	66.2	83.9	83.1
June	73.9	73.8	92.9	86.9	96.0	53.1	70.3	102.2	82.1	66.2	83.9	83.2
July	73.6	73.4	92.4	85.0	96.1	52.9	69.6	102.0	82.1	66.0	83.3	82.3
August	74.5	74.2	93.8	87.8	96.9	53.8	69.3	102.9	83.0	66.7	84.0	82.8
September	74.8	74.8	93.9	88.7	96.7	54.3	68.9	104.5	83.0	67.0	84.0	83.2
October	74.7	74.8	93.1	87.7	96.1	54.3	67.9	104.4	83.2	67.2	83.6	82.7
November	74.9	74.8	93.5	87.8	96.5	54.5	66.6	104.5	83.6	67.3	83.4	82.4
December	75.2	75.1	93.8	88.4	96.7	54.9	66.5	105.0	83.5	67.6	83.4	82.3
1996												
January	74.7	74.5	92.6	86.0	96.0	54.2	65.7	103.4	83.2	67.5	82.5	81.2
February	75.8	75.7	94.2	88.4	97.2	55.6	67.5	104.6	84.1	68.4	83.4	82.1
March	75.8	75.6	93.6	85.6	97.5	55.5	67.8	105.8	84.3	68.5	82.9	81.5
April	76.5	76.4	94.4	89.6	97.1	56.8	67.7	106.4	84.2	69.0	83.3	82.0
May	77.0	77.0	94.6	90.1	97.1	57.6	67.8	107.5	85.0	69.6	83.5	82.2
June	77.7	77.8	95.3	91.9	97.4	58.5	67.4	109.3	85.3	70.2	83.8	82.6
July	77.6	78.0	94.5	92.1	96.2	59.2	68.0	108.6	85.3	70.1	83.3	82.3
August	78.0	78.5	94.4	91.2	96.4	59.5	68.0	109.7	86.2	70.7	83.4	82.3
September	78.6	79.1	95.4	92.2	97.5	60.2	68.1	110.0	86.6	71.0	83.6	82.5
October	78.5	79.0	94.8	90.1	97.5	60.0	68.0	110.2	86.8	71.2	83.1	82.0
November	79.2	79.7	95.9	91.4	98.5	60.9	67.6	111.0	87.6	71.7	83.5	82.2
December	79.7	80.4	96.2	92.8	98.3	62.0	67.6	110.4	88.1	72.2	83.6	82.5
1997												
January	79.8	80.5	95.9	92.2	98.1	62.3	66.7	109.4	88.7	72.4	83.3	82.1
February	80.8	81.6	96.4	93.7	98.2	63.4	67.1	111.6	89.6	73.4	83.9	82.8
March	81.3	82.4	97.2	94.9	98.8	64.4	66.7	112.6	89.8	73.8	84.0	83.2
April	81.4	82.3	96.2	92.5	98.4	64.8	66.7	112.3	90.1	74.1	83.6	82.5
May	81.8	82.9	96.9	93.5	99.0	65.6	66.5	113.0	90.7	74.4	83.7	82.7
June	82.2	83.5	96.9	95.3	98.3	66.7	66.3	112.8	91.0	74.8	83.6	82.7
July	82.9	84.0	97.7	95.3	99.4	67.1	66.4	112.9	91.6	75.5	83.8	82.7
August	83.7	85.1	98.6	97.4	99.8	68.6	66.7	113.7	91.9	76.3	84.2	83.2
September	84.5	85.8	99.3	98.2	100.5	69.1	66.5	114.2	93.0	77.1	84.4	83.4
October	85.2	86.6	100.9	99.4	102.2	70.1	66.8	114.9	94.0	77.3	84.6	83.5
November	85.9	87.5	101.2	101.9	101.7	71.5	66.4	115.4	94.4	78.2	84.8	83.8
December	86.2	87.9	100.7	101.6	101.1	71.8	67.4	116.7	94.6	78.7	84.5	83.5

Table 2-4A. Industrial Production and Capacity Utilization, Historical Data, 1955–2019—*Continued*

(Seasonally adjusted.)

Year and month	Total industry	Manufac- turing (SIC)	Consumer goods Total	Durable	Nondurable	Business equipment	Defense and space equipment	Construction supplies	Business supplies	Materials	Capacity utilization Total industry	Manufac- turing (SIC)
1998												
January	86.6	88.6	101.2	102.1	101.6	72.8	68.0	117.5	94.6	78.9	84.4	83.5
February	86.8	88.7	101.1	101.9	101.5	73.1	68.4	118.1	94.9	79.0	83.9	83.0
March	86.8	88.6	101.2	102.2	101.6	73.5	68.1	117.3	95.4	78.9	83.5	82.3
April	87.1	89.0	101.9	102.6	102.3	73.7	68.2	117.8	95.7	79.1	83.2	82.1
May	87.7	89.5	102.3	103.3	102.6	74.1	69.0	119.2	96.4	79.7	83.2	81.9
June	87.1	88.8	101.1	98.4	102.8	74.1	69.3	118.7	96.5	79.1	82.2	80.7
July	86.8	88.5	100.2	94.4	103.1	73.7	70.4	119.4	97.1	78.8	81.4	79.9
August	88.6	90.6	103.1	105.6	102.9	76.0	70.7	119.8	97.8	80.3	82.7	81.4
September	88.5	90.4	102.2	105.6	101.6	76.0	70.2	119.4	97.8	80.4	82.1	80.7
October	89.2	91.3	102.7	107.6	101.6	76.9	71.6	120.7	98.1	81.1	82.3	81.0
November	89.1	91.4	102.1	107.3	100.8	76.9	71.4	120.9	98.4	81.3	81.9	80.7
December	89.4	91.9	102.1	108.2	100.4	77.1	70.8	122.3	98.4	81.9	81.8	80.8
1999												
January	89.9	92.2	103.1	108.2	101.9	77.1	70.5	121.6	99.2	82.1	81.8	80.6
February	90.3	93.0	103.5	109.1	102.1	77.6	71.1	121.7	99.4	82.7	81.9	80.9
March	90.5	92.9	103.3	108.8	101.8	77.5	70.7	120.4	99.8	83.3	81.7	80.4
April	90.7	93.3	103.2	110.1	101.3	78.0	69.9	120.8	100.0	83.6	81.6	80.4
May	91.4	94.1	104.2	111.1	102.3	78.9	69.1	121.2	100.6	84.1	81.8	80.7
June	91.2	93.8	103.1	110.5	101.0	78.5	68.1	121.3	100.5	84.5	81.4	80.1
July	91.8	94.2	102.7	110.8	100.3	79.1	67.8	122.2	101.2	85.6	81.6	80.1
August	92.2	94.8	104.1	113.4	101.3	79.7	67.5	122.0	101.2	85.6	81.7	80.3
September	91.8	94.4	103.3	111.7	100.9	79.3	65.5	122.1	101.2	85.4	81.0	79.6
October	93.0	95.8	105.4	115.3	102.3	80.0	65.1	123.6	102.1	86.4	81.8	80.5
November	93.4	96.4	105.2	114.5	102.5	80.0	63.7	124.3	102.6	87.3	81.9	80.7
December	94.2	97.1	106.4	114.5	104.0	80.6	62.8	125.6	103.3	87.9	82.2	80.9
2000												
January	94.2	97.2	105.0	116.4	101.5	82.0	62.8	126.0	103.2	88.0	82.1	80.7
February	94.5	97.4	105.7	116.1	102.5	82.5	61.3	126.2	103.2	88.2	82.0	80.5
March	94.8	98.0	105.3	115.6	102.2	83.6	60.7	126.3	103.9	88.7	82.0	80.7
April	95.5	98.7	106.3	117.2	102.9	84.7	59.5	126.9	105.1	89.1	82.3	80.9
May	95.6	98.6	106.3	116.6	103.2	85.3	58.8	124.9	105.1	89.4	82.2	80.5
June	95.7	98.8	106.4	116.2	103.5	85.4	59.4	124.5	104.9	89.5	82.0	80.3
July	95.6	98.9	106.0	114.4	103.5	86.0	60.5	125.0	105.0	89.2	81.6	80.1
August	95.3	98.2	105.5	114.2	102.9	85.6	59.0	124.2	104.7	89.2	81.1	79.3
September	95.7	98.7	106.4	114.5	104.1	86.4	56.9	124.3	104.6	89.4	81.2	79.3
October	95.4	98.4	105.4	113.2	103.2	86.5	59.1	123.8	104.3	89.3	80.7	78.8
November	95.4	98.1	105.8	110.9	104.5	86.3	61.0	123.5	104.5	89.1	80.4	78.3
December	95.2	97.5	106.4	109.2	105.8	85.7	61.6	121.9	104.1	88.6	80.0	77.6
2001												
January	94.5	97.0	105.4	107.2	105.2	85.6	63.3	122.2	104.0	87.7	79.2	76.8
February	93.9	96.3	104.7	106.9	104.4	85.2	63.1	120.9	102.5	87.3	78.4	76.1
March	93.7	96.1	104.7	109.4	103.5	84.7	64.9	121.1	101.9	86.9	78.0	75.7
April	93.4	95.8	105.1	109.6	104.0	82.8	65.5	120.8	101.2	86.7	77.6	75.2
May	92.9	95.2	105.1	110.5	103.6	81.4	65.8	120.1	100.7	86.1	76.9	74.5
June	92.3	94.6	105.0	109.8	103.8	80.5	66.9	119.2	100.2	85.3	76.2	73.8
July	91.8	94.2	104.5	110.1	103.0	79.7	67.8	119.2	100.1	84.6	75.6	73.3
August	91.7	93.7	104.7	109.4	103.5	78.2	66.8	117.9	99.7	84.9	75.3	72.8
September	91.3	93.5	104.3	108.5	103.2	77.2	67.3	117.8	99.8	84.7	74.8	72.5
October	90.9	93.0	104.7	107.4	104.1	76.0	67.2	116.4	99.2	84.2	74.3	71.9
November	90.5	92.7	104.7	109.3	103.5	75.4	66.5	116.1	98.5	83.6	73.8	71.6
December	90.5	93.0	105.3	111.9	103.5	74.8	66.5	117.1	98.7	83.4	73.7	71.7
2002												
January	91.1	93.5	106.4	112.2	104.8	74.5	65.9	116.7	98.4	84.3	74.0	72.1
February	91.1	93.5	105.6	112.5	103.6	74.4	65.5	117.1	98.5	84.8	73.9	71.9
March	91.8	94.2	106.5	113.6	104.4	74.8	65.4	118.9	99.5	85.4	74.4	72.4
April	92.2	94.4	106.3	115.3	103.6	74.4	65.3	118.8	100.1	86.4	74.6	72.5
May	92.6	94.9	106.6	115.8	103.9	74.8	65.4	119.4	100.6	86.8	74.9	72.9
June	93.4	96.0	107.9	117.2	105.1	75.4	66.1	120.3	101.2	87.6	75.5	73.6
July	93.2	95.6	107.7	118.5	104.4	75.1	66.3	118.6	101.2	87.4	75.3	73.4
August	93.2	95.8	107.2	117.6	104.0	75.5	66.5	119.2	101.2	87.6	75.3	73.5
September	93.4	96.0	107.4	117.6	104.3	75.5	67.7	120.2	101.8	87.5	75.4	73.6
October	93.1	95.6	107.2	116.9	104.2	75.2	68.2	119.8	102.3	86.9	75.2	73.3
November	93.6	96.0	108.0	119.7	104.3	75.5	67.8	119.7	102.0	87.6	75.6	73.6
December	93.1	95.6	107.1	117.9	103.7	74.6	70.5	118.8	101.9	87.3	75.2	73.3
2003												
January	93.8	96.2	108.4	119.7	104.9	74.8	70.5	119.3	102.9	87.8	75.8	73.8
February	94.0	96.2	108.8	117.9	106.1	74.8	71.4	118.3	103.1	87.8	75.9	73.7
March	93.7	96.3	108.8	118.3	106.0	75.0	71.4	117.9	103.0	87.3	75.8	73.8
April	93.1	95.5	107.9	117.1	105.1	74.2	70.8	116.6	101.9	87.0	75.3	73.2
May	93.1	95.6	107.6	117.2	104.7	74.2	71.5	118.2	102.3	86.9	75.3	73.3
June	93.2	96.0	107.9	118.3	104.6	74.5	71.7	118.6	101.8	87.1	75.5	73.7
July	93.7	96.3	108.8	121.3	104.9	74.8	71.0	118.0	102.4	87.3	75.8	73.9
August	93.5	95.9	108.0	119.0	104.6	75.0	71.4	118.8	102.2	87.4	75.8	73.6
September	94.1	96.6	108.9	122.0	104.7	75.8	71.5	118.6	102.2	87.9	76.2	74.2
October	94.2	96.8	108.2	121.2	104.1	75.7	71.4	119.2	102.7	88.5	76.3	74.3
November	94.9	97.7	109.0	122.7	104.7	77.1	70.8	120.7	103.3	89.0	76.9	75.1
December	94.9	97.5	109.0	122.6	104.7	76.8	69.1	120.5	103.0	89.1	76.9	75.0

Table 2-4A. Industrial Production and Capacity Utilization, Historical Data, 1955–2019—*Continued*

(Seasonally adjusted.)

Year and month	Production indexes, 2007 = 100										Capacity utilization (output as percent of capacity)	
	Total industry	Manufac-turing (SIC)	Market groups								Total industry	Manufac-turing (SIC)
			Consumer goods			Business equipment	Defense and space equipment	Construction supplies	Business supplies	Materials		
			Total	Durable	Nondurable							
2004												
January	95.1	97.5	109.5	124.0	104.9	77.1	67.2	120.5	103.5	89.2	77.1	75.0
February	95.7	98.3	110.0	123.7	105.7	78.2	68.5	120.4	104.4	89.7	77.6	75.6
March	95.2	98.1	108.8	122.6	104.5	78.0	68.6	120.4	103.5	89.5	77.2	75.5
April	95.6	98.5	109.6	122.7	105.5	78.1	68.6	120.2	104.2	89.8	77.6	75.8
May	96.4	99.3	110.0	121.7	106.4	79.0	69.2	121.6	105.0	90.7	78.2	76.4
June	95.6	98.5	108.4	118.9	105.2	78.8	68.3	120.7	104.9	90.0	77.6	75.9
July	96.3	99.4	108.8	119.5	105.4	80.5	69.5	122.2	105.5	90.7	78.2	76.6
August	96.4	99.9	109.3	120.4	105.8	80.2	70.1	122.3	105.5	90.6	78.2	76.9
September	96.5	99.9	109.2	119.7	106.0	80.4	71.0	121.3	105.3	90.8	78.3	76.9
October	97.4	100.9	110.3	122.3	106.6	81.0	71.2	123.3	105.9	91.7	79.0	77.6
November	97.6	100.8	110.1	120.6	106.9	80.6	71.9	122.9	106.3	92.3	79.1	77.5
December	98.3	101.5	110.9	121.7	107.5	81.4	72.7	122.9	107.1	93.0	79.7	77.9
2005												
January	98.8	102.3	111.3	120.8	108.4	82.9	72.6	124.4	108.1	93.0	80.0	78.4
February	99.5	103.1	111.9	123.6	108.3	83.7	75.2	125.2	108.1	93.7	80.5	79.0
March	99.3	102.6	111.4	121.2	108.3	83.7	76.6	123.9	108.2	93.8	80.3	78.4
April	99.5	103.0	111.3	120.1	108.6	84.6	77.7	126.1	108.8	93.6	80.3	78.6
May	99.6	103.4	112.0	120.2	109.3	85.4	77.4	126.3	108.8	93.3	80.3	78.7
June	100.0	103.5	112.9	120.8	110.4	85.2	77.9	125.3	109.2	93.6	80.5	78.6
July	99.7	103.1	112.5	118.6	110.5	84.7	77.4	126.2	108.7	93.3	80.2	78.2
August	99.9	103.6	113.0	121.9	110.1	85.8	78.2	126.7	109.0	93.3	80.3	78.4
September	98.1	102.6	113.2	124.5	109.7	83.3	75.3	128.5	109.2	89.7	78.7	77.5
October	99.3	104.1	113.2	125.8	109.3	87.8	75.9	131.1	109.6	90.8	79.6	78.5
November	100.3	104.9	113.0	124.0	109.6	88.9	76.4	131.8	110.0	92.6	80.3	78.9
December	100.9	105.1	113.8	122.8	111.0	88.2	76.7	132.9	110.5	93.5	80.7	78.9
2006												
January	101.1	105.9	113.0	124.3	109.5	89.9	75.1	133.7	110.6	93.8	80.7	79.4
February	101.1	105.6	112.4	122.1	109.3	90.2	75.8	132.6	110.6	94.1	80.6	79.1
March	101.3	105.6	112.9	123.1	109.8	91.0	74.0	132.4	111.0	94.1	80.6	78.9
April	101.7	106.1	113.3	123.0	110.3	92.6	74.5	131.8	111.0	94.3	80.8	79.2
May	101.6	105.6	113.0	122.1	110.2	92.5	73.9	130.6	110.8	94.5	80.6	78.7
June	102.0	106.0	113.6	122.9	110.7	93.3	74.6	130.1	110.9	94.8	80.8	78.9
July	101.9	105.7	112.6	118.7	110.7	94.0	75.9	130.4	110.9	95.0	80.6	78.5
August	102.3	106.4	113.6	121.9	111.0	94.8	75.7	129.6	110.7	95.2	80.8	78.9
September	102.1	106.4	113.1	119.8	110.9	94.8	76.4	129.2	110.6	95.1	80.5	78.8
October	102.1	106.0	113.0	118.2	111.3	95.1	77.6	127.8	110.7	95.0	80.2	78.3
November	102.0	106.1	113.1	118.8	111.3	95.5	78.9	127.5	110.3	94.7	80.0	78.2
December	103.0	107.6	113.7	121.5	111.2	97.0	80.4	131.2	111.1	95.7	80.6	79.1
2007												
January	102.5	107.0	112.9	118.4	111.1	95.1	82.1	128.6	111.0	95.6	80.0	78.5
February	103.5	107.5	114.2	119.9	112.4	96.0	82.8	129.1	111.9	96.6	80.6	78.6
March	103.8	108.4	113.3	120.5	111.0	97.8	82.1	130.7	112.3	96.9	80.6	79.1
April	104.5	109.1	113.9	122.6	111.2	98.9	84.6	130.1	113.3	97.6	81.0	79.4
May	104.5	109.0	113.5	121.9	110.8	99.0	86.7	129.9	112.8	98.0	80.9	79.1
June	104.6	109.4	113.6	123.3	110.6	99.0	90.2	130.5	112.1	97.9	80.8	79.2
July	104.5	109.5	113.4	122.1	110.7	99.0	92.0	130.0	111.4	98.0	80.6	79.1
August	104.8	109.2	113.2	121.6	110.5	99.1	94.1	129.4	111.6	98.5	80.7	78.7
September	105.2	109.7	113.3	120.1	111.0	100.7	96.8	129.6	112.3	98.7	81.0	78.9
October	104.7	109.3	112.1	119.7	109.7	100.1	97.2	128.4	112.0	98.8	80.6	78.6
November	105.3	110.0	112.1	120.2	109.5	101.0	99.5	128.3	112.5	99.6	81.1	78.9
December	105.3	110.1	111.8	120.2	109.1	102.1	100.2	128.5	112.0	99.6	81.1	78.9
2008												
January	105.1	109.6	111.4	117.3	109.5	102.4	100.8	128.4	111.7	99.2	80.9	78.6
February	104.7	108.9	111.2	116.5	109.4	102.4	100.5	125.9	111.5	98.9	80.7	78.1
March	104.5	108.6	109.7	113.5	108.3	103.2	100.9	124.4	111.1	99.1	80.6	77.8
April	103.7	107.4	108.9	111.0	108.1	101.1	100.4	122.1	110.0	98.6	80.0	77.0
May	103.1	106.8	108.3	109.8	107.6	101.5	99.6	121.0	109.3	97.9	79.6	76.7
June	102.8	106.1	108.1	110.4	107.2	100.9	100.3	119.6	108.5	97.9	79.5	76.2
July	102.3	104.8	107.3	107.9	107.0	99.0	98.1	119.3	107.6	97.8	79.0	75.4
August	100.7	103.6	105.5	103.3	105.9	97.0	97.4	116.9	106.9	96.2	77.8	74.6
September	96.4	100.0	104.8	102.4	105.2	89.3	94.8	113.7	105.4	90.0	74.4	72.1
October	97.3	99.3	105.1	99.9	106.3	86.7	94.8	111.9	104.4	92.5	75.0	71.8
November	96.1	97.0	104.2	96.1	106.4	88.0	94.5	106.4	102.0	91.2	74.0	70.2
December	93.3	93.6	102.0	91.7	104.8	88.6	94.1	100.2	98.4	87.7	71.7	67.8
2009												
January	91.0	90.8	99.8	81.9	104.8	83.4	94.1	96.0	96.9	86.2	70.0	65.9
February	90.5	90.6	99.8	83.6	104.4	83.1	93.8	94.3	95.7	85.6	69.4	65.9
March	89.0	89.0	99.3	83.3	103.8	81.5	92.3	91.1	94.8	84.0	68.3	64.7
April	88.3	88.3	98.9	83.4	103.3	79.9	91.7	89.4	94.4	83.4	67.7	64.4
May	87.4	87.4	97.7	80.0	102.7	77.6	91.8	89.0	93.9	82.9	67.0	63.8
June	87.1	87.1	97.2	79.3	102.3	76.9	92.1	89.0	93.9	82.6	66.7	63.7
July	88.0	88.4	98.4	87.0	101.5	78.0	93.4	89.4	93.8	83.6	67.4	64.7
August	89.0	89.4	99.3	88.0	102.4	79.0	94.3	90.1	94.3	84.8	68.2	65.6
September	89.7	90.2	100.0	91.6	102.2	79.5	95.4	89.7	94.6	85.7	68.8	66.2
October	90.0	90.4	100.4	90.3	103.1	79.9	95.0	88.5	95.4	85.9	69.1	66.5
November	90.3	91.2	99.8	91.8	101.9	79.8	95.7	90.1	95.6	86.8	69.5	67.2
December	90.6	91.1	99.6	91.1	101.9	81.2	95.8	88.0	96.7	87.1	69.9	67.2

Table 2-4A. Industrial Production and Capacity Utilization, Historical Data, 1955–2019—*Continued*

(Seasonally adjusted.)

Year and month	Production indexes, 2007 = 100										Capacity utilization (output as percent of capacity)	
	Total industry	Manufac-turing (SIC)	Market groups								Total industry	Manufac-turing (SIC)
			Consumer goods			Business equipment	Defense and space equipment	Construction supplies	Business supplies	Materials		
			Total	Durable	Nondurable							
2010												
January	91.7	92.1	100.4	92.3	102.5	82.4	98.2	89.8	97.0	88.3	70.8	68.0
February	92.0	92.0	99.5	90.7	101.9	82.0	98.8	89.9	96.9	89.5	71.2	68.1
March	92.6	93.1	99.7	91.8	101.9	83.4	101.2	91.4	96.8	90.0	71.8	69.0
April	92.9	93.9	99.0	92.3	100.8	84.5	102.2	94.2	97.5	90.4	72.3	69.7
May	94.3	95.2	100.9	95.5	102.2	86.2	102.1	94.6	98.4	91.7	73.5	70.8
June	94.4	95.1	100.5	94.5	102.1	86.6	101.8	94.6	98.4	92.2	73.8	70.9
July	94.9	95.7	101.2	98.3	101.8	87.6	103.2	94.3	98.4	92.4	74.2	71.4
August	95.1	95.7	100.9	95.4	102.3	87.5	103.4	94.8	98.4	93.2	74.6	71.6
September	95.4	95.7	100.5	95.5	101.8	87.8	101.3	94.8	98.4	93.8	74.9	71.7
October	95.1	95.8	100.5	96.0	101.7	88.1	100.1	95.0	97.7	93.4	74.8	71.9
November	95.1	95.9	100.0	94.6	101.3	87.9	99.7	95.7	98.8	93.5	74.9	72.0
December	96.1	96.3	101.0	94.2	102.8	88.6	99.1	94.5	99.3	94.8	75.6	72.5
2011												
January	95.9	96.5	101.1	95.3	102.7	89.4	98.7	94.1	98.9	94.4	75.6	72.7
February	95.5	96.6	100.9	97.5	101.8	89.7	98.0	93.7	98.9	93.7	75.3	72.8
March	96.5	97.1	101.1	99.1	101.6	89.4	97.1	94.5	99.1	95.4	76.0	73.3
April	96.1	96.6	101.0	95.4	102.4	88.7	96.7	94.4	98.8	95.0	75.7	72.9
May	96.3	96.7	101.3	96.7	102.4	89.5	96.8	95.6	98.9	95.0	75.8	73.0
June	96.6	96.8	101.1	96.1	102.4	89.5	96.2	95.9	98.8	95.6	75.9	73.1
July	97.1	97.4	101.9	97.6	102.9	90.7	97.1	96.8	99.1	95.8	76.2	73.5
August	97.7	97.8	102.3	98.5	103.3	91.6	98.0	96.6	99.3	96.4	76.6	73.8
September	97.6	98.1	101.9	98.7	102.7	92.1	98.0	96.8	99.6	96.3	76.4	74.0
October	98.3	98.7	102.1	100.3	102.6	93.4	99.1	96.9	99.2	97.3	76.8	74.3
November	98.2	98.4	101.0	98.6	101.6	94.1	100.6	96.9	98.3	97.7	76.6	74.0
December	98.8	99.1	100.9	99.2	101.4	95.6	99.9	98.4	98.9	98.3	76.9	74.5
2012												
January	99.4	99.9	100.9	101.8	100.7	97.2	100.3	99.1	99.2	99.1	77.2	75.0
February	99.6	100.2	100.7	101.1	100.7	97.8	101.0	100.2	99.7	99.3	77.2	75.1
March	99.2	99.8	99.2	100.0	98.9	98.6	101.0	100.1	99.1	99.1	76.7	74.7
April	99.9	100.3	100.0	100.5	99.8	99.6	100.0	101.3	100.2	99.7	77.2	75.0
May	100.1	99.9	100.4	99.8	100.6	100.1	98.9	100.3	100.5	99.8	77.2	74.6
June	100.1	100.1	100.1	99.8	100.2	101.5	98.1	99.6	100.5	99.8	77.0	74.7
July	100.3	100.0	100.3	99.9	100.3	100.4	100.4	99.2	100.4	100.4	77.1	74.5
August	99.9	99.8	99.7	98.8	99.9	101.0	100.4	99.7	100.0	99.7	76.6	74.2
September	99.9	99.7	99.5	97.7	100.0	100.9	99.9	99.6	100.0	99.9	76.5	74.1
October	100.1	99.4	99.4	98.6	99.6	99.8	99.8	99.2	99.9	100.7	76.6	73.8
November	100.6	100.1	99.9	100.0	99.9	101.1	100.1	100.6	100.1	101.0	76.8	74.2
December	101.0	100.9	100.0	102.1	99.4	102.1	99.9	101.2	100.4	101.4	77.0	74.7
2013												
January	100.8	100.6	100.0	101.4	99.6	99.9	99.2	101.5	100.6	101.6	76.7	74.4
February	101.4	101.0	100.5	103.6	99.7	100.8	99.1	103.2	100.6	102.1	77.1	74.7
March	101.8	100.9	101.0	104.5	100.1	101.3	98.8	102.8	101.0	102.6	77.3	74.5
April	101.6	100.5	100.7	104.2	99.8	100.5	98.4	102.2	101.2	102.5	77.1	74.2
May	101.7	100.8	100.5	104.7	99.5	100.0	97.5	102.0	101.7	102.9	77.0	74.4
June	102.0	101.1	100.8	106.1	99.5	100.0	97.5	102.9	101.9	103.0	77.1	74.5
July	101.5	100.1	99.8	103.8	98.8	97.9	96.2	102.6	101.7	103.2	76.7	73.8
August	102.2	101.0	100.3	106.5	98.8	99.6	96.6	103.2	102.2	103.8	77.2	74.4
September	102.7	101.2	100.7	107.2	99.1	100.4	96.5	104.3	102.4	104.4	77.5	74.5
October	102.5	101.2	101.0	107.3	99.4	99.9	96.5	104.2	102.6	104.0	77.3	74.5
November	102.8	101.2	101.0	108.5	99.1	99.7	95.1	104.9	102.5	104.8	77.4	74.5
December	103.2	101.2	101.9	108.5	100.2	98.9	94.5	103.4	103.1	105.1	77.6	74.5
2014												
January	102.7	100.0	100.1	104.9	98.9	98.9	93.7	103.1	102.5	105.3	77.2	73.6
February	103.6	101.1	101.4	108.6	99.5	100.4	93.8	103.8	103.1	106.0	77.8	74.4
March	104.6	101.9	101.9	109.4	100.0	101.6	94.7	105.2	103.3	107.3	78.5	75.0
April	104.6	101.8	101.4	108.6	99.6	101.2	94.1	104.4	102.9	107.8	78.4	74.9
May	105.0	102.0	101.3	110.7	99.0	101.9	94.1	106.1	103.0	108.3	78.6	75.2
June	105.4	102.4	101.4	111.3	98.9	101.5	94.1	106.8	103.0	109.1	78.8	75.5
July	105.6	102.9	101.5	114.3	98.2	102.8	94.5	107.7	102.8	109.1	78.9	75.8
August	105.5	102.3	100.9	111.1	98.3	101.9	93.4	107.7	102.5	109.4	78.7	75.5
September	105.8	102.3	101.2	111.0	98.8	101.3	93.9	108.1	102.6	109.8	78.8	75.5
October	105.8	102.2	101.0	110.7	98.5	102.4	93.9	107.7	102.4	109.8	78.7	75.5
November	106.7	103.0	103.0	114.1	100.2	103.9	93.4	107.7	103.1	110.0	79.2	76.1
December	106.5	102.7	102.7	113.4	99.9	102.6	93.1	108.5	102.1	110.3	79.0	75.9
2015												
January	106.0	102.2	102.8	113.4	100.1	102.2	91.9	107.8	102.2	109.4	78.6	75.7
February	105.4	101.5	102.5	110.7	100.4	101.5	93.4	106.5	102.1	109.1	78.1	75.2
March	105.1	101.9	103.5	113.2	101.1	101.6	93.1	105.9	101.4	108.3	77.8	75.5
April	104.5	101.7	102.6	114.0	99.7	100.7	92.5	106.4	101.8	107.9	77.3	75.5
May	104.1	101.7	102.4	116.5	98.9	101.0	92.1	106.8	101.4	107.0	76.9	75.5
June	103.7	101.3	102.3	113.8	99.4	100.3	92.0	107.2	101.1	106.5	76.6	75.2
July	104.3	102.0	104.0	119.8	100.0	100.3	91.7	107.0	100.8	106.9	77.0	75.8
August	104.2	101.7	103.8	116.6	100.6	99.9	91.6	107.6	101.1	106.6	76.9	75.5
September	103.8	101.2	103.3	116.0	100.1	98.6	90.7	106.4	101.2	106.5	76.6	75.2
October	103.4	101.2	103.0	116.0	99.7	97.8	90.4	107.8	101.5	105.9	76.3	75.1
November	102.7	100.9	102.8	115.3	99.6	96.5	90.1	107.6	101.6	104.8	75.7	74.9
December	102.1	100.6	102.2	115.1	98.9	95.3	90.4	108.2	101.1	104.3	75.3	74.6

Table 2-4A. Industrial Production and Capacity Utilization, Historical Data, 1955–2019—*Continued*

(Seasonally adjusted.)

Year and month	Production indexes, 2007 = 100										Capacity utilization (output as percent of capacity)	
			Market groups									
	Total industry	Manufac-turing (SIC)	Consumer goods			Business equipment	Defense and space equipment	Construction supplies	Business supplies	Materials	Total industry	Manufac-turing (SIC)
			Total	Durable	Nondurable							
2016												
January	103.0	101.3	103.7	116.8	100.3	95.9	89.6	108.9	101.9	105.0	75.9	75.1
February	102.2	100.7	103.4	116.3	100.1	95.0	88.3	108.2	101.6	104.1	75.3	74.6
March	101.4	100.5	102.4	115.0	99.2	94.4	87.7	108.0	100.7	103.3	74.7	74.3
April	101.5	100.1	103.2	116.1	100.0	94.0	87.1	108.0	101.1	103.1	74.7	74.0
May	101.4	100.1	103.1	115.1	100.1	94.0	88.3	107.3	101.3	103.0	74.6	73.9
June	101.9	100.4	103.9	117.5	100.5	94.1	87.8	107.1	101.6	103.5	74.9	74.0
July	102.1	100.7	104.0	119.2	100.1	94.3	88.3	108.0	102.2	103.7	75.1	74.1
August	102.0	100.3	103.9	118.1	100.4	93.8	89.1	107.3	102.3	103.6	75.0	73.8
September	102.0	100.7	103.9	118.7	100.1	93.7	89.6	107.8	102.8	103.5	74.9	74.0
October	102.2	101.0	103.6	119.8	99.5	94.2	90.5	108.4	102.5	103.9	75.0	74.1
November	102.1	101.1	102.9	118.7	98.8	94.2	91.3	109.0	102.6	103.7	74.9	74.1
December	102.9	101.4	104.6	118.9	100.9	95.2	91.2	108.9	103.4	104.1	75.5	74.4
2017												
January	103.0	102.0	103.4	120.1	99.2	95.7	91.8	110.9	103.0	104.8	75.5	74.8
February	102.6	102.0	102.3	120.0	97.8	95.4	90.5	111.4	103.0	104.6	75.2	74.7
March	103.3	101.7	103.0	117.7	99.3	96.2	90.9	110.8	103.8	105.4	75.7	74.5
April	104.3	102.8	103.7	120.0	99.6	98.7	91.5	111.5	104.3	106.2	76.4	75.3
May	104.4	102.6	104.6	119.5	100.8	98.3	91.5	111.2	104.5	106.0	76.5	75.1
June	104.6	102.8	104.1	119.1	100.3	98.1	91.2	111.5	104.4	106.7	76.6	75.2
July	104.5	102.6	104.1	118.0	100.6	97.8	91.1	111.3	104.8	106.6	76.5	75.1
August	104.0	102.3	103.9	118.1	100.3	97.7	91.0	110.7	104.3	105.9	76.2	74.8
September	104.1	102.1	104.0	119.3	100.2	98.5	90.5	112.5	104.0	105.5	76.1	74.7
October	105.6	103.5	105.2	120.3	101.4	99.2	90.4	112.9	104.8	108.0	77.3	75.7
November	106.2	103.8	105.2	119.8	101.5	99.4	90.5	113.1	105.4	109.0	77.6	75.9
December	106.5	103.7	105.8	119.8	102.2	98.8	89.9	113.8	105.6	109.5	77.9	75.8
2018												
January	106.3	103.3	105.9	120.0	102.3	99.4	90.2	111.9	105.1	108.9	77.6	75.5
February	106.6	104.4	105.8	122.2	101.6	99.4	90.2	115.1	105.1	109.5	77.8	76.3
March	107.3	104.5	106.1	122.7	102.0	99.6	90.6	114.4	105.3	110.5	78.2	76.3
April	108.2	104.9	107.5	121.9	103.8	100.2	92.1	114.7	106.0	111.5	78.8	76.6
May	107.4	104.1	105.0	117.9	101.7	98.5	91.5	115.1	105.2	111.6	78.1	76.0
June	108.2	104.8	105.8	121.1	101.9	100.3	92.3	114.7	105.2	112.4	78.6	76.5
July	108.7	105.2	106.0	120.9	102.2	100.7	92.9	114.5	105.1	113.3	78.8	76.7
August	109.5	105.7	106.4	122.8	102.3	102.0	93.9	115.3	105.1	114.5	79.3	77.0
September	109.7	105.7	106.5	123.9	102.2	102.5	94.5	114.8	104.8	114.8	79.3	76.9
October	109.9	105.6	106.7	122.7	102.6	102.8	95.3	115.3	105.6	114.9	79.3	76.8
November	110.5	105.8	107.4	123.0	103.4	103.0	95.4	115.5	105.4	115.7	79.6	76.9
December	110.6	106.4	106.8	125.0	102.3	103.2	98.1	117.5	104.8	115.9	79.5	77.3
2019												
January	110.1	105.8	105.4	119.5	101.9	103.2	98.8	117.8	105.3	115.6	79.0	76.7
February	109.6	105.3	105.8	121.0	102.0	101.8	98.8	115.9	104.7	114.8	78.5	76.3
March	109.7	105.2	106.5	119.2	103.2	102.7	99.9	116.1	104.3	114.5	78.4	76.2
April	109.0	104.3	104.5	118.2	101.0	100.7	100.2	115.2	103.8	114.8	77.8	75.4
May	109.2	104.4	105.0	120.2	101.2	101.0	100.6	115.9	104.4	114.8	77.8	75.4
June	109.3	105.0	105.5	121.7	101.5	101.6	101.9	117.0	103.6	114.4	77.7	75.7
July	109.1	104.6	105.6	122.5	101.3	100.9	102.0	116.0	103.9	114.2	77.4	75.3
August	109.9	105.2	105.5	122.2	101.2	101.7	102.4	117.0	104.2	115.6	77.8	75.7
September	109.5	104.5	104.7	118.7	101.2	100.5	103.3	117.1	104.2	115.5	77.4	75.1
October	109.0	103.9	104.7	115.4	101.9	99.7	103.6	116.4	104.1	114.9	77.0	74.6
November	110.0	104.9	106.7	122.5	102.7	101.8	104.1	116.3	104.5	115.3	77.6	75.2
December	109.7	105.1	105.5	119.3	102.0	101.3	105.4	117.5	104.1	115.2	77.2	75.3

Table 2-4B. Industrial Production: Historical Data, 1919–1947

(Seasonally adjusted, 2012 = 100.)

Year and month	January	February	March	April	May	June	July	August	September	October	November	December	Annual averages
1919													
Industrial production, total ...	5.01	4.79	4.65	4.74	4.76	5.07	5.37	5.46	5.34	5.29	5.21	5.29	5.08
Manufacturing	4.95	4.85	4.67	4.77	4.74	5.08	5.37	5.49	5.29	5.18	5.37	5.26	5.08
1920													
Industrial production, total ..	5.79	5.79	5.68	5.37	5.51	5.57	5.43	5.46	5.26	5.04	4.62	4.35	5.32
Manufacturing	5.81	5.81	5.68	5.34	5.52	5.52	5.31	5.34	5.21	4.92	4.41	4.10	5.25
1921													
Industrial production, total ..	4.10	4.02	3.90	3.90	4.02	3.99	3.96	4.10	4.13	4.38	4.32	4.29	4.09
Manufacturing	3.89	3.89	3.78	3.78	3.91	3.89	3.89	4.04	4.10	4.28	4.33	4.28	4.00
1922													
Industrial production, total ..	4.46	4.65	4.90	4.74	4.98	5.23	5.23	5.12	5.40	5.70	5.95	6.12	5.21
Manufacturing	4.38	4.48	4.69	4.90	5.16	5.39	5.44	5.24	5.39	5.62	5.83	6.01	5.21
1923													
Industrial production, total ..	5.98	6.06	6.26	6.40	6.48	6.42	6.37	6.26	6.12	6.09	6.09	5.95	6.21
Manufacturing	5.91	5.96	6.22	6.30	6.43	6.35	6.25	6.09	6.06	5.96	5.96	5.88	6.11
1924													
Industrial production, total ...	6.09	6.20	6.09	5.90	5.65	5.40	5.32	5.51	5.70	5.84	5.95	6.12	5.82
Manufacturing	5.99	6.12	6.01	5.86	5.57	5.29	5.21	5.39	5.60	5.75	5.88	6.09	5.73
1925													
Industrial production, total ..	6.31	6.31	6.31	6.37	6.34	6.29	6.45	6.34	6.26	6.51	6.65	6.73	6.41
Manufacturing	6.27	6.30	6.32	6.32	6.27	6.25	6.38	6.27	6.30	6.56	6.74	6.89	6.41
1926													
Industrial production, total ..	6.62	6.62	6.70	6.70	6.65	6.73	6.76	6.84	6.95	6.95	6.92	6.90	6.78
Manufacturing	6.74	6.69	6.66	6.66	6.61	6.69	6.69	6.76	6.89	6.84	6.79	6.76	6.73
1927													
Industrial production, total ..	6.87	6.92	7.01	6.84	6.90	6.87	6.78	6.78	6.67	6.54	6.54	6.56	6.77
Manufacturing	6.71	6.76	6.82	6.79	6.82	6.82	6.76	6.71	6.61	6.51	6.48	6.56	6.70
1928													
Industrial production, total ..	6.70	6.76	6.81	6.78	6.87	6.92	7.01	7.14	7.20	7.34	7.48	7.62	7.05
Manufacturing	6.69	6.76	6.79	6.76	6.87	6.95	7.05	7.21	7.23	7.36	7.52	7.65	7.07
1929													
Industrial production, total ..	7.73	7.70	7.73	7.86	8.00	8.06	8.17	8.09	8.03	7.89	7.50	7.17	7.83
Manufacturing	7.70	7.67	7.75	7.85	8.01	8.11	8.19	8.14	7.98	7.91	7.46	7.02	7.82
1930													
Industrial production, total ..	7.17	7.14	7.03	6.98	6.87	6.67	6.37	6.23	6.12	5.95	5.82	5.68	6.50
Manufacturing	7.10	7.05	6.82	6.92	6.79	6.61	6.30	6.06	6.01	5.81	5.65	5.52	6.39
1931													
Industrial production, total ..	5.65	5.68	5.79	5.82	5.73	5.59	5.51	5.32	5.07	4.87	4.82	4.79	5.39
Manufacturing	5.52	5.57	5.68	5.68	5.62	5.42	5.34	5.16	4.92	4.69	4.61	4.59	5.23
1932													
Industrial production, total ..	4.65	4.54	4.49	4.18	4.04	3.90	3.79	3.90	4.15	4.29	4.29	4.21	4.20
Manufacturing	4.54	4.41	4.25	3.99	3.89	3.76	3.63	3.73	3.99	4.10	4.12	4.04	4.04
1933													
Industrial production, total ..	4.13	4.15	3.90	4.18	4.87	5.62	6.15	5.90	5.57	5.29	4.98	5.01	4.98
Manufacturing	3.99	3.89	3.65	4.04	4.74	5.52	6.09	5.73	5.44	5.13	4.79	4.85	4.82
1934													
Industrial production, total ..	5.18	5.43	5.68	5.68	5.79	5.68	5.29	5.23	4.93	5.15	5.21	5.54	5.40
Manufacturing	5.00	5.29	5.49	5.60	5.68	5.55	5.11	5.08	4.79	4.98	5.05	5.42	5.25
1935													
Industrial production, total ..	5.98	6.09	6.06	5.95	5.95	6.04	6.04	6.26	6.42	6.62	6.76	6.84	6.25
Manufacturing	5.91	6.01	5.94	5.88	5.86	5.83	6.01	6.25	6.40	6.58	6.74	6.82	6.19
1936													
Industrial production, total ..	6.73	6.56	6.65	7.06	7.20	7.34	7.48	7.59	7.73	7.84	8.06	8.31	7.38
Manufacturing	6.69	6.48	6.61	7.02	7.21	7.36	7.49	7.62	7.75	7.88	8.09	8.35	7.38
1937													
Industrial production, total ..	8.28	8.39	8.58	8.58	8.61	8.50	8.56	8.50	8.22	7.62	6.87	6.26	8.08
Manufacturing	8.35	8.48	8.53	8.66	8.71	8.55	8.60	8.50	8.19	7.52	6.64	5.94	8.05
1938													
Industrial production, total ..	6.12	6.06	6.06	5.95	5.82	5.87	6.20	6.54	6.73	6.90	7.17	7.26	6.39
Manufacturing	5.83	5.81	5.83	5.65	5.62	5.62	5.96	6.35	6.56	6.76	7.08	7.13	6.18

Table 2-4B. Industrial Production: Historical Data, 1919–1947—*Continued*

(Seasonally adjusted, 2012 = 100.)

Year and month	January	February	March	April	May	June	July	August	September	October	November	December	Annual averages
1939													
Industrial production, total	7.26	7.31	7.34	7.31	7.28	7.45	7.67	7.78	8.25	8.67	8.89	8.89	7.84
Products	7.71	7.74	7.83	7.83	7.89	7.97	8.12	8.20	8.46	8.69	8.87	8.92	8.19
Consumer goods	12.08	12.08	12.28	12.28	12.32	12.44	12.60	12.64	12.84	12.96	13.07	13.11	12.56
Materials	6.51	6.62	6.62	6.41	6.38	6.69	6.98	7.05	7.86	8.60	8.76	8.76	7.27
Manufacturing	6.79	6.84	6.84	6.76	6.76	6.92	7.13	7.33	7.75	8.19	8.35	8.45	7.34
1940													
Industrial production, total	8.78	8.50	8.31	8.47	8.72	9.00	9.11	9.17	9.36	9.50	9.72	10.05	9.06
Products	8.84	8.75	8.66	8.75	8.87	9.07	9.07	9.18	9.44	9.61	9.79	10.13	9.18
Consumer goods	13.07	13.03	12.88	12.96	13.03	13.15	13.11	13.11	13.47	13.71	13.99	14.42	13.33
Materials	8.53	8.06	7.75	7.98	8.35	8.79	9.02	9.10	9.20	9.30	9.48	9.82	8.78
Manufacturing	8.32	8.06	7.88	7.96	8.29	8.66	8.76	8.89	9.07	9.20	9.36	9.69	8.68
1941													
Industrial production, total	10.30	10.61	10.94	10.97	11.46	11.55	11.69	11.82	11.82	11.94	11.99	12.18	11.44
Products	10.45	10.76	11.05	11.28	11.66	11.77	11.86	12.00	12.03	12.15	12.20	12.38	11.63
Consumer goods	14.70	15.13	15.53	15.81	16.28	16.32	16.32	16.36	16.28	16.28	16.32	16.32	15.97
Materials	9.95	10.23	10.59	10.36	11.01	11.16	11.32	11.40	11.40	11.45	11.52	11.76	11.01
Manufacturing	9.87	10.24	10.50	10.78	11.14	11.27	11.40	11.48	11.46	11.56	11.59	11.74	11.09
1942													
Industrial production, total	12.43	12.66	12.79	12.43	12.46	12.49	12.79	13.18	13.49	13.93	14.26	14.59	13.13
Products	12.75	13.07	13.21	12.38	12.41	12.46	12.81	13.18	13.64	14.13	14.59	15.05	13.31
Consumer goods	16.64	16.44	16.44	14.18	14.02	13.79	13.99	14.10	14.14	14.38	14.46	14.78	14.78
Materials	11.78	11.96	12.07	12.38	12.35	12.38	12.61	12.97	13.13	13.49	13.70	13.85	12.72
Manufacturing	12.00	12.29	12.47	12.10	12.16	12.23	12.62	13.04	13.43	13.92	14.31	14.70	12.94
1943													
Industrial production, total	14.73	15.12	15.23	15.42	15.54	15.45	15.92	16.26	16.67	16.92	17.14	16.92	15.94
Products	15.11	15.46	15.63	15.83	15.97	16.15	16.58	16.87	17.33	17.59	17.85	17.47	16.49
Consumer goods	14.34	14.46	14.50	14.66	14.90	15.13	15.37	15.45	15.57	15.45	15.41	15.09	15.03
Materials	14.06	14.42	14.57	14.70	14.68	14.32	14.86	15.14	15.53	15.79	15.92	15.92	14.99
Manufacturing	14.90	15.21	15.34	15.52	15.63	15.68	16.04	16.38	16.82	17.18	17.42	17.13	16.11
1944													
Industrial production, total	17.11	17.25	17.22	17.22	17.11	17.06	17.03	17.25	17.14	17.20	17.06	17.00	17.14
Products	17.70	17.87	17.87	17.93	17.90	17.96	18.10	18.31	18.16	18.19	17.96	17.82	17.98
Consumer goods	15.25	15.33	15.53	15.61	15.73	15.81	15.89	16.24	15.89	15.93	15.89	15.89	15.75
Materials	16.07	16.20	16.12	16.07	15.84	15.61	15.37	15.56	15.53	15.58	15.61	15.68	15.77
Manufacturing	17.39	17.52	17.49	17.47	17.34	17.31	17.31	17.55	17.44	17.49	17.34	17.34	17.42
1945													
Industrial production, total	16.84	16.78	16.67	16.37	15.92	15.56	15.20	13.62	12.41	11.91	12.35	12.41	14.67
Products	17.67	17.47	17.21	16.90	16.44	16.12	15.86	14.13	12.66	12.35	12.52	12.55	15.16
Consumer goods	16.12	16.01	15.97	16.09	16.12	16.24	16.20	15.49	16.24	16.48	16.96	17.15	16.26
Materials	15.50	15.66	15.68	15.45	15.04	14.65	14.11	12.79	11.83	11.19	11.96	12.09	13.83
Manufacturing	17.11	17.03	16.85	16.48	15.97	15.52	15.16	13.32	11.84	11.38	11.69	11.82	14.51
1946													
Industrial production, total	11.71	11.13	12.30	12.07	11.63	12.35	12.77	13.24	13.49	13.74	13.82	13.90	12.68
Products	12.41	12.29	12.58	12.78	12.64	12.78	13.07	13.50	13.87	14.10	14.25	14.39	13.22
Consumer goods	17.91	18.62	18.50	18.70	18.78	18.82	19.25	19.89	20.28	20.52	20.80	20.92	19.42
Materials	10.59	9.33	11.73	10.96	10.13	11.55	12.22	12.69	12.76	13.05	13.02	13.10	11.76
Manufacturing	11.02	10.32	11.61	11.64	11.09	11.72	12.10	12.65	12.91	13.19	13.35	13.40	12.08
1947													
Industrial production, total	14.07	14.15	14.23	14.12	14.18	14.18	14.10	14.18	14.29	14.43	14.62	14.68	14.27
Products	14.54	14.59	14.65	14.65	14.68	14.65	14.62	14.71	14.85	15.05	15.26	15.34	14.80
Consumer goods	20.48	20.40	20.48	20.40	20.25	20.28	20.36	20.52	20.72	21.04	21.35	21.39	20.64
Materials	13.10	13.20	13.64	13.31	13.36	13.23	13.10	13.08	13.26	13.36	13.67	13.51	13.32
Manufacturing	13.35	13.40	13.45	13.48	13.37	13.37	13.30	13.37	13.43	13.61	13.81	13.84	13.48

NOTES AND DEFINITIONS, CHAPTER 2

TABLES 2-1 THROUGH 2-4

SOURCE: BOARD OF GOVERNORS OF THE FEDERAL RESERVE SYSTEM

Ever since the Federal Reserve was founded in 1913, understanding current business conditions has been a central focus in pursuing its mission of a stable and secure financial system. The index of industrial production is one of the oldest continuous statistical series maintained by the federal government, and one of the few economic indicators for which monthly data are available before the post-World-War-II period.

The Federal Reserve's monthly index of industrial production and the related capacity indexes and capacity utilization rates cover manufacturing, mining, and electric and gas utilities. The industrial sector, together with construction, accounts for the bulk of the variation in national output over the course of the business cycle. The industrial detail provided by these measures helps illuminate structural developments in the economy.

Around the 15th day of each month, the Federal Reserve issues estimates of industrial production and capacity utilization for the previous month. The production estimates are in the form of index numbers (2012 = 100) that reflect the monthly levels of total output of the nation's factories, mines, and gas and electric utilities expressed as a percent of the monthly average in the 2012 base year. Capacity estimates are expressed as index numbers, 2012 output (not 2012 capacity) = 100, and capacity utilization is measured by the production index as a percent of the capacity index. Since, for each component industry, the bases of those two indexes are the same, this procedure yields production as a percent of capacity. Monthly estimates are subject to revision in subsequent months, as well as to annual and comprehensive revisions in subsequent years.

Definitions and notes on the data

The *industrial production index* measures changes in the physical volume or quantity of output of manufacturing, mining, and electric and gas utilities. *Capacity utilization* is calculated by dividing a seasonally adjusted industrial production index for an industry or group of industries by a related index of productive capacity.

The index of industrial production measures a large portion of the goods output of the national economy on a monthly basis. This portion, together with construction, has also accounted for the bulk of the variation in output over the course of many historical business cycles. The substantial industrial detail included in the index illuminates structural developments in the economy.

The total industrial production index and the indexes for its major components are constructed from individual industry series (312 series for data from 1997 forward) based on the 2012 North American Industry Classification System (NAICS). See the Preface to this volume for information on NAICS.

The Federal Reserve has been able to provide a longer continuous historical series on the NAICS basis than some other government agencies. In a major research effort, the Fed and the Census Bureau's Center for Economic Studies re-coded data from seven Censuses of Manufactures, beginning in 1963, to establish benchmark NAICS data for output, value added, and capacity utilization. The resulting indexes are shown annually for the last 48 years (52 years for aggregate levels) in Tables 2-1 through 2-4.

The Fed's featured indexes for total industry and total manufacturing are on the Standard Industrial Classification (SIC) basis and do not observe the reclassifications under NAICS of the logging industry to the Agriculture sector and the publishing industry to the Information sector. (The reason cited by the Fed was to avoid "changing the scope or historical continuity of these statistics.") One advantage of the SIC index for capacity utilization is that it is a continuous series back to 1949 (shown in Table 2-4A). On the new NAICS basis, production and capacity utilization are shown back to 1970 in Tables 2-2 and 2-3.

The individual series components of the indexes are grouped in two ways: market groups and industry groups.

Market groups. For analyzing market trends and product flows, the individual series are grouped into two major divisions: *final products and nonindustrial supplies* and *materials. Final products* consists of products purchased by consumers, businesses, or government for final use. *Nonindustrial supplies* are expected to become inputs in nonindustrial sectors: the two major subgroups are *construction supplies* and *business supplies. Materials* comprises industrial output that requires further processing within the industrial sector. This twofold division distinguishes between products that are ready to ship outside the industrial sector and those that will stay within the sector for further processing.

Final products are divided into *consumer goods* and *equipment*, and *equipment* is divided into *business equipment* and *defense and space equipment*. Further subdivisions of each market group are based on type of product and the market destination for the product.

Industry groups are typically groupings by 3-digit NAICS industries and major aggregates of these industries—for example, *durable goods* and *nondurable goods manufacturing, mining*, and *utilities*. Indexes are also calculated for *stage-of-process* industry groups—*crude*, *primary and semifinished*, and *finished* processing. The stage-of-process grouping was a new feature in the 2002 revision, replacing the two narrower and less well-defined "primary processing manufacturing" and "advanced processing manufacturing" groups that were previously published. *Crude processing* consists of logging, much of mining, and certain basic manufacturing activities in the chemical, paper, and metals industries. *Primary and semifinished processing* represents industries that produce materials and parts used as inputs by other industries. *Finished processing* includes industries that produce goods in their finished form for use by consumers, business investment, or government.

The indexes of industrial production are constructed using data from a variety of sources. Current monthly estimates of production are based on measures of physical output where possible and appropriate. For a few high-tech industries, the estimated value of nominal output is deflated by a corresponding price index. For industries in which such direct measurement is not possible on a monthly basis, output is inferred from production-worker hours, adjusted for trends in worker productivity derived from annual and benchmark revisions. (Between the 1960s and 1997, electric power consumption was used as a monthly output indicator for some industries instead of hours. However, the coverage of the electric power consumption survey deteriorated, and in the 2005 revision, the decision was made to resume the use of hours in those industries, beginning with the data for 1997.)

In annual and benchmark revisions, the individual indexes are revised using data from the quinquennial Economic Census which includes Manufactures and Mineral Industries, the Annual Survey of Manufactures, and the quarterly Survey of Plant Capacity, prepared by the Census Bureau; deflators from the Producer Price Indexes and other sources; the *Minerals Yearbook*, prepared by the Department of the Interior; publications from the Department of Energy; and other sources.

The weights used in compiling the indexes are based on Census value added—the difference between the value of production and the cost of materials and supplies consumed. Census value added differs in some respects from the economic concept of industry value added used in the national income and product accounts (NIPAs). Industry value added as defined in the NIPAs is not available in sufficient detail for the industrial production indexes. See Chapter 14 for data and a description of NIPA value added (equivalently, gross domestic product) by major industry group.

Before 1972, a linked-Laspeyres formula (base period prices) is used to compute the weighted individual indexes. Beginning with 1972, the index uses a version of the Fisher-ideal index formula—a chain-weighting (continually updated average price) system similar to that in the NIPAs. See the "General Notes" article at the front of this book and the notes and definitions for Chapter 1 for more information. Chain-weighting keeps the index from being distorted by the use of obsolete relative prices.

For the purpose of these value-added weights, value added per unit of output is based on data from the Censuses of Manufacturing and Mineral Industries, the Census Bureau's Annual Survey of Manufactures, and revenue and expense data reported by the Department of Energy and the American Gas Association, which are projected into recent years by using changes in relevant Producer Price Indexes.

To separate seasonal movements from cyclical patterns and underlying trends, each component of the index is seasonally adjusted by the Census X-12-ARIMA method.

The index does not cover production on farms, in the construction industry, in transportation, or in various trade and service industries. A number of groups and subgroups include data for individual series not published separately.

Capacity utilization is calculated for the manufacturing, mining, and electric and gas utilities industries. Output is measured by seasonally adjusted indexes of industrial production. The capacity indexes attempt to capture the concept of sustainable maximum output, which is defined as the greatest level of output that a plant can maintain within the framework of a realistic work schedule, taking account of normal downtime and assuming sufficient availability of inputs to operate the machinery and equipment in place. The 89 individual industry capacity indexes are based on a variety of data, including capacity data measured in physical units compiled by government agencies and trade associations, Census Bureau surveys of utilization rates and investment, and estimates of growth of the capital stock.

In the "Explanatory Note" to its monthly release, the statistics cover output, capacity, and capacity utilization in the U.S. industrial sector, which is defined by the Federal Reserve to comprise manufacturing, mining, and electric and gas utilities. Mining is defined as all industries in sector 21 of the North American Industry Classification System (NAICS); electric and gas utilities are those in NAICS sectors 2211 and 2212. Manufacturing comprises NAICS manufacturing industries (sector 31-33) plus the logging industry and the newspaper, periodical, book, and directory publishing industries. Logging and publishing are classified elsewhere in NAICS (under agriculture and information respectively), but historically they were considered to be manufacturing and were included in the industrial sector under the Standard Industrial Classification (SIC) system. In December 2002 the Federal Reserve reclassified all its industrial output data from the SIC system to NAICS.

Revisions

Revisions normally occur annual with the newest release on March 31, 2019. New annual data have been incorporated in addition to the data from the Annual Survey of Manufacturing. The IP indexes for publishing reflect new data for 2017 and revised data for 2013 from the Census Bureau's Service Annual Survey. The Census Bureau recently benchmarked the Service Annual Survey to the 2012 Economic Census, which resulted in updated estimates for 2008 through 2012. For logging, the IP indexes were updated with 2015 data from the U.S. Forest Service. In addition, the indexes for metallic and nonmetallic minerals were updated with revised annual data for 2015 from the Department of the Interior's U.S. Geological Survey (USGS). Data on prices from the Bureau of Labor Statistics (BLS) were also incorporated into most of the manufacturing indexes.

Data availability

Data are available monthly in Federal Reserve release G.17. Current and historical data and background information are available on the Federal Reserve Web site at <http://www.federalreserve.gov>.

Chain-weighting makes it difficult for the user to analyze in detail the sources of aggregate output change. An "Explanatory Note," included in each month's index release, provides some assistance for the user, including a reference to a an Internet location with the exact contribution of a monthly change in a component index to the monthly change in the total index.

References

The G.17 release each month contains extensive explanatory material, as well as references for further detail.

An earlier detailed description of the industrial production index, together with a history of the index, a glossary of terms, and a bibliography is presented in *Industrial Production—1986 Edition*, available from Publication Services, Mail Stop 127, Board of Governors of the Federal Reserve System, Washington, DC 20551.

CHAPTER 3: INCOME DISTRIBUTION AND POVERTY

SECTION 3A: HOUSEHOLD AND FAMILY INCOME

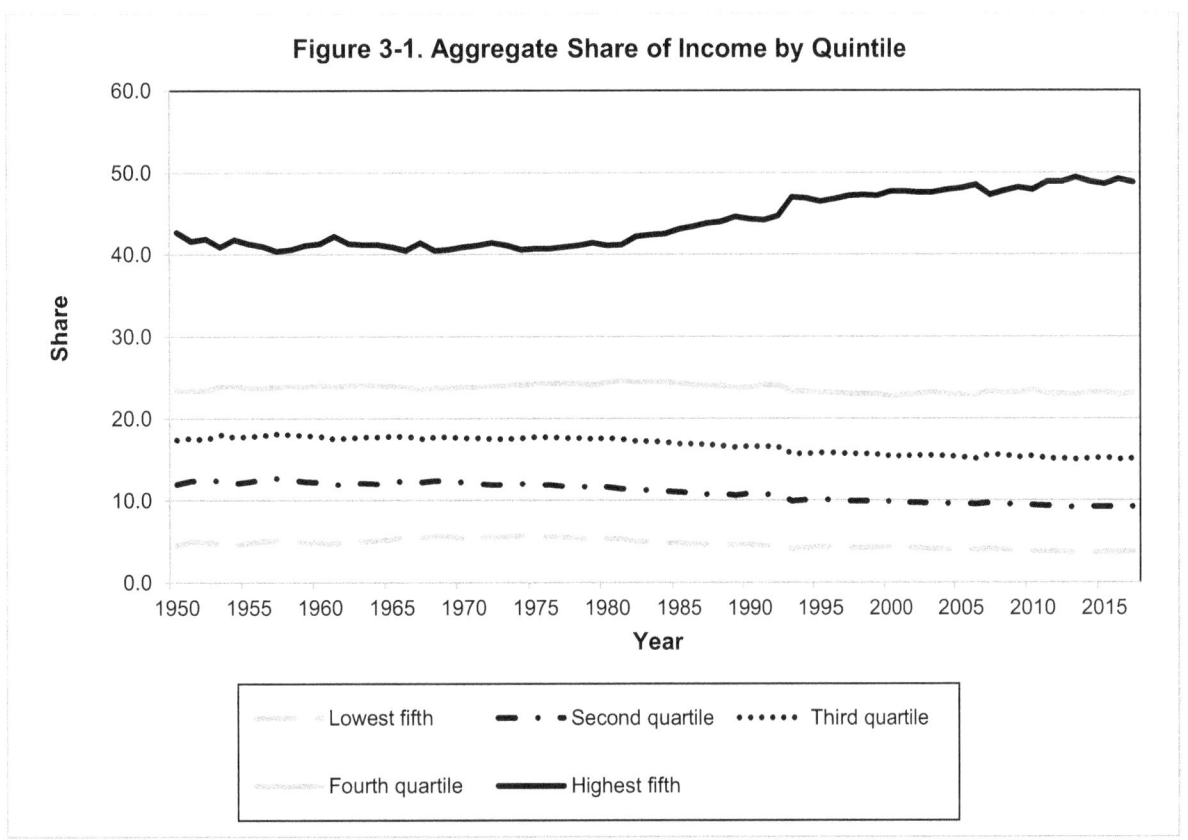

Figure 3-1. Aggregate Share of Income by Quintile

- Census Bureau statistics can shed some light on income inequality in the United States. Income inequality is the disparity of income distributions with concentration of income usually in the hands of a small percentage of the population. These statistics include the Gini Index and the shares of aggregate income of households income received by each quintiles (five equal groups).

- The Gini index is a statistical measure of income inequality ranging from 0 to 1, with a measure of 1 indicating perfect inequality (one household having all the income and the rest having none) and a measure of 0 indicating perfect equality (all households having an equal share of income). Household income inequality, as measured by the "Gini coefficient," increased slightly from .463 in 2017 to .464 in 2018. (Table 3-3)

- Recovering from the Great Recession has been sluggish, slow and uneven. Despite the longest recovery (more than 10 years) in U.S. history, income dispersion has become even more stark. Table 3-3 depicts how the gap in U.S. income has widen for the bottom four-fifths of population. (Table 3-3)

- Median household income varied by race. Asians had the highest median income at $87,194 followed by Non-Hispanic Whites at $70,642 and Blacks at $41,361. (Table 3-1)

Table 3-1. Median and Mean Household Income and Median Earnings

(2018 dollars.)

Year	Number of households (thousands)	Average size of household (number of people)	Median household income — All races	White Total	White Not Hispanic	Black	Asian¹	Hispanic (any race)	Mean household income, all races	Median earnings — Male workers	Female workers	Ratio, female to male
1975	72 867	2.89	50 214	52 512	52 908	31 524	...	37 725	58 636	54 291	31 933	0.588
1976	74 142	2.86	51 048	53 474	54 565	31 797	...	38 505	60 045	54 142	32 590	0.602
1977	76 030	2.81	51 371	54 020	55 091	31 878	...	40 299	60 939	55 360	32 620	0.589
1978	77 330	2.78	53 359	55 470	56 515	33 335	...	41 808	62 802	55 718	33 119	0.594
1979	80 776	2.76	53 257	55 839	56 625	32 784	...	42 195	63 264	55 046	32 842	0.597
1980	82 368	2.73	51 528	54 361	55 325	31 318	...	39 718	61 283	54 152	32 578	0.602
1981	83 527	2.72	50 709	53 577	54 351	30 065	...	40 675	60 580	53 862	31 905	0.592
1982	83 918	2.73	50 571	52 943	53 830	30 005	...	38 053	60 946	52 843	32 628	0.617
1983	85 407	2.71	50 216	52 662	...	29 885	...	38 245	61 075	52 611	33 458	0.636
1984	86 789	2.69	51 742	54 586	55 719	31 096	...	39 224	63 397	53 596	34 118	0.637
1985	88 458	2.67	52 709	55 588	56 838	33 072	...	38 977	64 868	53 997	34 869	0.646
1986	89 479	2.66	54 608	57 411	58 716	33 076	...	40 252	67 465	55 395	35 603	0.643
1987	91 124	2.64	55 260	58 222	59 823	33 231	68 332	41 000	68 723	55 016	35 858	0.652
1988	92 830	2.62	55 716	58 900	60 523	33 577	66 034	41 664	69 615	54 551	36 030	0.660
1989	93 347	2.63	56 678	59 619	60 901	35 456	70 787	42 982	71 607	53 590	36 802	0.687
1990	94 312	2.63	55 952	58 359	59 693	34 898	71 848	41 726	69 892	51 720	37 040	0.716
1991	95 669	2.62	54 318	56 920	58 279	33 909	65 718	40 912	68 374	53 047	37 058	0.699
1992	96 426	2.66	53 897	56 665	58 566	32 995	66 502	39 754	68 330	53 125	37 605	0.708
1993	97 107	2.67	53 610	56 560	58 641	33 519	65 804	39 273	71 091	52 179	37 318	0.715
1994	98 990	2.65	54 233	57 198	59 044	35 344	68 047	39 369	72 503	51 863	37 325	0.720
1995	99 627	2.65	55 931	58 705	61 023	36 755	66 662	37 522	73 760	51 696	36 926	0.714
1996	101 018	2.64	56 744	59 413	62 012	37 543	69 189	39 819	75 340	51 391	37 907	0.738
1997	102 528	2.62	57 911	60 990	63 501	39 202	70 813	41 672	77 766	52 698	39 082	0.742
1998	103 874	2.61	60 040	63 170	65 528	39 143	72 010	43 743	80 067	54 574	39 932	0.732
1999	106 434	2.60	61 526	63 989	66 759	42 196	77 044	46 484	82 754	55 018	39 786	0.723
2000	108 209	2.58	61 399	64 216	66 712	43 380	81 530	48 500	83 545	54 471	40 156	0.737
2001	109 297	2.58	60 038	63 293	65 835	41 899	76 256	47 721	82 758	54 418	41 537	0.763
2002	111 278	2.57	59 360	...	...	...	...	46 334	80 975	55 189	42 275	0.766
2003	112 000	2.57	59 286	...	...	...	...	45 160	80 840	55 659	42 049	0.755
2004	113 343	2.57	59 080	...	...	...	...	45 670	80 578	54 365	41 631	0.766
2005	114 384	2.57	59 712	...	...	...	...	46 360	81 647	53 345	41 063	0.770
2006	116 011	2.56	60 178	...	...	...	...	47 169	83 111	52 762	40 594	0.769
2007	116 783	2.56	60 985	...	...	...	...	46 958	82 081	54 769	42 616	0.778
2008	117 181	2.57	58 811	...	...	...	...	44 326	79 997	54 210	41 791	0.771
2009	117 538	2.59	58 400	...	...	...	...	44 628	79 751	55 290	42 562	0.770
2010	119 927	2.56	56 873	...	...	...	...	43 433	77 783	55 344	42 575	0.769
2011	121 084	2.55	56 006	...	...	...	...	43 217	77 962	53 934	41 532	0.770
2012	122 459	2.54	55 900	...	...	...	...	42 738	78 095	54 126	41 408	0.765
2013	122 952	2.55	56 079	...	...	...	...	44 228	78 431	54 021	42 278	0.783
2014	124 587	2.54	56 969	...	...	...	...	45 114	80 413	53 493	42 067	0.786
2015	125 819	2.53	59 901	...	...	...	...	47 852	84 011	54 280	43 183	0.796
2016	126 224	2.54	61 779	...	...	...	...	49 887	87 001	54 036	43 482	0.805
2017	127 586	2.53	62 868	...	...	...	...	51 717	88 322	53 417	43 000	0.805
2018	128 579	2.52	63 179	...	...	...	...	51 450	90 021	55 291	45 097	0.816
By race												
Race alone												
2003	...	...	...	62 451	65 388	40 573	76 231	...	...	...	...	...
2004	...	...	...	62 177	65 178	40 105	76 631	...	...	...	...	...
2005	...	...	...	62 584	65 458	39 774	78 747	...	...	...	...	...
2006	...	...	...	63 264	65 449	39 913	80 200	...	...	...	...	...
2007	...	...	...	63 270	66 676	41 176	80 252	...	...	...	...	...
2008	...	...	...	61 160	64 923	40 006	76 739	...	...	...	...	...
2009	...	...	...	60 845	63 895	38 228	76 810	...	...	...	...	...
2010	...	...	...	59 682	62 857	37 077	74 167	...	...	...	...	...
2011	...	...	...	58 423	62 001	36 061	72 874	...	...	...	...	...
2012	...	...	...	58 846	62 465	36 510	75 205	...	...	...	...	...
2013	...	...	...	59 661	62 915	37 356	72 411	...	...	...	...	...
2014	...	...	...	60 376	63 976	37 583	78 883	...	...	...	...	...
2015	...	...	...	63 710	66 721	39 108	81 788	...	...	...	...	...
2016	...	...	...	64 729	68 059	41 323	85 210	...	...	...	...	...
2017	...	...	...	66 864	69 806	41 239	83 314	...	...	...	...	...
2018	...	...	...	66 943	70 642	41 361	87 194	...	...	...	...	...
Race alone or in combination												
2003	...	...	...	...	...	40 633	75 632	...	...	...	...	...
2004	...	...	...	...	...	40 292	76 557	...	...	...	...	...
2005	...	...	...	...	...	39 898	78 688	...	...	...	...	...
2006	...	...	...	...	...	40 116	79 778	...	...	...	...	...
2007	...	...	...	...	...	41 388	79 977	...	...	...	...	...
2008	...	...	...	...	...	40 154	76 657	...	...	...	...	...
2009	...	...	...	...	...	38 423	76 345	...	...	...	...	...
2010	...	...	...	...	...	37 114	73 322	...	...	...	...	...
2011	...	...	...	...	...	36 215	72 724	...	...	...	...	...
2012	...	...	...	...	...	36 945	74 707	...	...	...	...	...
2013	...	...	...	...	...	37 547	72 736	...	...	...	...	...
2014	...	...	...	...	...	37 854	79 448	...	...	...	...	...
2015	...	...	...	...	...	39 440	81 359	...	...	...	...	...
2016	...	...	...	...	...	41 924	84 573	...	...	...	...	...
2017	...	...	...	...	...	41 584	82 935	...	...	...	...	...
2018	...	...	...	...	...	41 692	86 815	...	...	...	...	...

¹For 1987 through 2001, Asian and Pacific Islander.
. . . = Not available.

Table 3-2. Median Family Income by Type of Family

(2018 dollars, except as noted.)

Year	All families	Married couples			Male householder [1]	Female householder [1]	4-person families	Average size of family (number of people)
		Total	Wife in paid labor force	Wife not in paid labor force				
1950	39 929	31 465	35 681	29 549	28 443	17 550	33 556	3.54
1951	40 578	32 470	38 256	30 020	29 212	18 786	34 881	3.54
1952	40 832	33 747	39 751	30 924	30 041	18 573	36 340	3.53
1953	41 202	36 080	43 554	33 175	33 951	20 265	. . .	3.59
1954	41 951	35 450	42 617	32 354	32 840	18 768	. . .	3.59
1955	42 889	37 794	45 101	34 704	34 432	20 306	40 423	3.58
1956	43 497	40 241	47 056	36 692	33 719	22 285	43 041	3.60
1957	43 696	40 404	46 968	36 964	35 891	21 647	42 997	3.64
1958	44 232	40 526	46 253	37 090	32 481	20 899	43 347	3.65
1959	45 111	42 818	49 499	39 252	34 885	20 902	45 904	3.67
1960	45 539	43 699	50 119	40 095	36 162	22 084	46 839	3.70
1961	46 418	44 472	51 691	40 213	37 341	22 048	47 418	3.67
1962	47 059	45 682	53 125	41 041	41 655	22 837	49 277	3.68
1963	47 540	47 434	54 705	42 414	41 081	23 102	51 355	3.70
1964	47 956	49 203	56 610	43 916	41 111	24 544	53 149	3.70
1965	48 509	50 786	58 668	44 985	42 978	24 691	54 526	3.69
1966	49 214	53 281	61 357	47 302	43 723	27 259	56 700	3.67
1967	50 111	55 641	64 066	48 976	44 917	28 305	59 287	3.63
1968	50 823	58 001	66 169	50 868	46 437	28 755	62 378	3.60
1969	51 586	60 729	68 934	52 632	50 643	29 280	64 506	3.58
1970	52 227	60 858	69 353	52 563	52 154	29 474	64 625	3.57
1971	53 296	60 931	69 564	52 737	48 357	28 353	64 457	3.53
1972	54 373	64 072	73 025	55 469	55 470	28 755	68 943	3.48
1973	55 053	65 997	75 350	56 464	54 416	29 366	69 451	3.44
1974	55 698	64 119	72 924	54 986	53 688	29 879	68 936	3.42
1975	56 245	63 266	71 606	52 974	55 300	29 124	67 441	3.39
1976	56 710	65 200	73 579	54 724	51 748	29 017	69 675	3.37
1977	57 215	66 678	74 890	55 658	54 951	29 391	70 868	3.33
1978	57 804	68 505	76 450	55 865	56 554	30 239	72 359	3.31
1979	59 550	69 330	78 520	55 922	54 380	31 965	72 834	3.29
1980	60 309	67 329	76 344	53 886	50 972	30 282	70 794	3.27
1981	61 019	66 636	75 903	52 749	52 875	29 137	69 850	3.25
1982	61 393	65 233	74 261	52 128	50 493	28 792	69 244	3.26
1983	61 997	65 607	75 362	51 380	52 525	28 346	70 171	3.24
1984	62 706	68 355	78 122	53 140	53 843	29 554	71 783	3.23
1985	63 558	69 407	79 370	53 499	50 487	30 486	73 150	3.21
1986	64 491	71 953	82 105	55 248	54 751	29 933	76 145	3.19
1987	65 204	73 958	84 353	55 144	53 451	31 134	78 638	3.17
1988	65 837	74 470	85 323	54 380	54 901	31 405	79 917	3.16
1989	66 090	75 582	86 644	55 025	54 601	32 239	79 927	3.17
1990	66 322	74 549	85 328	55 208	54 276	31 639	77 456	3.18
1991	67 173	73 915	84 783	52 936	51 118	30 096	77 631	3.17
1992	68 216	73 696	85 484	51 821	48 514	29 952	77 850	3.19
1993	68 506	73 797	85 776	50 620	45 418	29 932	77 497	3.20
1994	69 313	75 572	87 475	51 157	46 647	30 653	79 023	3.19
1995	69 597	77 246	89 446	51 875	49 829	32 320	81 555	3.20
1996	70 241	79 471	91 118	52 672	50 522	31 833	82 366	3.19
1997	70 884	80 738	92 685	55 039	51 581	32 900	83 491	3.18
1998	71 551	83 657	96 092	56 013	55 093	34 221	86 561	3.18
1999	73 206	85 421	98 113	56 792	56 451	35 925	90 341	3.15
2000	73 778	86 417	98 829	57 072	55 166	37 603	91 638	3.14
2001	74 340	85 782	98 312	56 602	52 022	36 603	89 966	3.15
2002	75 616	85 563	99 481	54 795	52 823	36 984	87 806	3.13
2003	76 232	85 239	100 431	54 941	52 051	36 337	89 087	3.13
2004	76 866	84 789	99 979	54 918	53 786	35 939	87 983	3.13
2005	77 418	84 950	99 096	55 939	52 990	35 116	90 629	3.13
2006	78 454	86 650	100 900	55 767	52 241	35 992	91 657	3.13
2007	77 908	88 127	102 439	56 092	53 853	36 781	91 873	3.15
2008	78 874	85 047	98 862	55 356	50 941	35 225	89 404	3.15
2009	78 867	84 034	98 437	54 573	48 690	34 927	87 295	3.16
2010	79 559	83 379	98 472	54 908	49 868	33 650	86 275	3.14
2011	80 529	82 564	97 232	55 066	48 190	33 857	84 556	3.13
2012	80 944	82 764	98 170	54 424	46 412	33 623	87 325	3.12
2013	81 217	82 424	99 393	54 639	48 020	33 911	86 761	3.14
2014	81 730	85 803	103 629	56 776	50 535	33 731	88 562	3.14
2015	82 199	89 375	107 295	57 952	52 753	36 170	93 422	3.14
2016	82 854	90 840	108 363	59 956	53 961	38 359	94 957	3.14
2017	83 103	92 346	110 893	61 901	54 241	38 002	97 110	3.14
2018	83 508	93 329	. . .	. . .	54 336	40 233	97 631	3.14

[1]No spouse present.
. . . = Not available.

Table 3-3. Shares of Aggregate Income Received by Each Fifth and Top 5 Percent of Households

Year	Money income							Equivalence-adjusted income						
	Share of aggregate income (percent)						Gini coefficient	Share of aggregate income (percent)						Gini coefficient
	Lowest fifth	Second fifth	Third fifth	Fourth fifth	Highest fifth	Top 5 percent		Lowest fifth	Second fifth	Third fifth	Fourth fifth	Highest fifth	Top 5 percent	
1967	4.0	10.8	17.3	24.2	43.6	17.2	0.397	5.6	12.0	17.1	23.2	42.1	. . .	0.362
1968	4.2	11.1	17.6	24.5	42.6	16.3	0.386	5.8	12.3	17.4	23.4	41.1	. . .	0.351
1969	4.1	10.9	17.5	24.5	43.0	16.6	0.391	5.8	12.2	17.3	23.4	41.3	. . .	0.353
1970	4.1	10.8	17.4	24.5	43.3	16.6	0.394	5.7	12.1	17.3	23.4	41.5	. . .	0.357
1971	4.1	10.6	17.3	24.5	43.5	16.7	0.396	5.7	12.0	17.2	23.4	41.7	. . .	0.359
1972	4.1	10.4	17.0	24.5	43.9	17.0	0.401	5.6	11.9	17.2	23.4	41.9	. . .	0.362
1973	4.2	10.4	17.0	24.5	43.9	16.9	0.400	5.6	12.0	17.2	23.5	41.7	. . .	0.360
1974	4.3	10.6	17.0	24.6	43.5	16.5	0.395	5.8	12.1	17.3	23.6	41.2	. . .	0.354
1975	4.3	10.4	17.0	24.7	43.6	16.5	0.397	5.6	11.9	17.3	23.6	41.6	. . .	0.359
1976	4.3	10.3	17.0	24.7	43.7	16.6	0.398	5.6	11.8	17.4	23.8	41.5	. . .	0.359
1977	4.2	10.2	16.9	24.7	44.0	16.8	0.402	5.5	11.7	17.3	23.7	41.7	. . .	0.362
1978	4.2	10.2	16.8	24.7	44.1	16.8	0.402	5.4	11.8	17.3	23.7	41.8	. . .	0.363
1979	4.1	10.2	16.8	24.6	44.2	16.9	0.404	5.3	11.7	17.2	23.8	41.9	. . .	0.366
1980	4.2	10.2	16.8	24.7	44.1	16.5	0.403	5.2	11.6	17.3	24.0	41.9	. . .	0.367
1981	4.1	10.1	16.7	24.8	44.3	16.5	0.406	5.0	11.4	17.2	24.0	42.4	. . .	0.373
1982	4.0	10.0	16.5	24.5	45.0	17.0	0.412	4.7	11.1	17.0	23.9	43.2	. . .	0.384
1983	4.0	9.9	16.4	24.6	45.1	17.0	0.414	4.6	11.0	16.9	24.0	43.5	. . .	0.389
1984	4.0	9.9	16.3	24.6	45.2	17.1	0.415	4.6	11.0	16.8	24.0	43.6	. . .	0.389
1985	3.9	9.8	16.2	24.4	45.6	17.6	0.419	4.6	10.9	16.7	23.7	44.1	. . .	0.394
1986	3.8	9.7	16.2	24.3	46.1	18.0	0.425	4.5	10.8	16.6	23.8	44.3	. . .	0.397
1987	3.8	9.6	16.1	24.3	46.2	18.2	0.426	4.4	10.8	16.7	23.8	44.4	. . .	0.399
1988	3.8	9.6	16.0	24.2	46.3	18.3	0.426	4.4	10.7	16.5	23.7	44.7	. . .	0.402
1989	3.8	9.5	15.8	24.0	46.8	18.9	0.431	4.4	10.5	16.3	23.4	45.4	. . .	0.408
1990	3.8	9.6	15.9	24.0	46.6	18.5	0.428	4.4	10.6	16.3	23.5	45.1	. . .	0.406
1991	3.8	9.6	15.9	24.2	46.5	18.1	0.428	4.3	10.6	16.5	23.7	45.0	. . .	0.406
1992	3.8	9.4	15.8	24.2	46.9	18.6	0.433	4.1	10.3	16.3	23.7	45.5	. . .	0.413
1993	3.6	9.0	15.1	23.5	48.9	21.0	0.454	3.9	9.8	15.6	23.0	47.7	. . .	0.436
1994	3.6	8.9	15.0	23.4	49.1	21.2	0.456	4.0	9.8	15.6	22.8	47.8	. . .	0.436
1995	3.7	9.1	15.2	23.3	48.7	21.0	0.450	4.1	9.9	15.6	22.8	47.6	. . .	0.433
1996	3.6	9.0	15.1	23.3	49.0	21.4	0.455	4.0	9.8	15.5	22.7	47.9	. . .	0.437
1997	3.6	8.9	15.0	23.2	49.4	21.7	0.459	4.0	9.8	15.4	22.6	48.3	. . .	0.440
1998	3.6	9.0	15.0	23.2	49.2	21.4	0.456	4.0	9.8	15.4	22.7	48.1	. . .	0.439
1999	3.6	8.9	14.9	23.2	49.4	21.5	0.458	4.0	9.7	15.3	22.6	48.4	. . .	0.441
2000	3.6	8.9	14.8	23.0	49.8	22.1	0.462	4.1	9.8	15.2	22.3	48.6	. . .	0.442
2001	3.5	8.7	14.6	23.0	50.1	22.4	0.466	4.0	9.6	15.2	22.4	48.8	. . .	0.446
2002	3.5	8.8	14.8	23.3	49.7	21.7	0.462	4.0	9.6	15.2	22.7	48.4	. . .	0.443
2003	3.4	8.7	14.8	23.4	49.8	21.4	0.464	3.9	9.5	15.2	22.8	48.6	. . .	0.445
2004	3.4	8.7	14.7	23.2	50.1	21.8	0.466	3.8	9.6	15.2	22.7	48.7	. . .	0.447
2005	3.4	8.6	14.6	23.0	50.4	22.2	0.469	3.8	9.5	15.1	22.6	49.1	. . .	0.450
2006	3.4	8.6	14.5	22.9	50.5	22.3	0.470	3.8	9.4	14.9	22.5	49.3	. . .	0.452
2007	3.4	8.7	14.8	23.4	49.7	21.2	0.463	3.8	9.5	15.3	22.9	48.5	. . .	0.444
2008	3.4	8.6	14.7	23.3	50.0	21.5	0.466	3.7	9.4	15.1	22.8	48.9	21.4	0.450
2009	3.4	8.6	14.6	23.2	50.3	21.7	0.468	3.6	9.3	15.0	22.9	49.4	21.7	0.456
2010	3.3	8.5	14.6	23.4	50.3	21.3	0.470	3.4	9.2	15.0	23.1	49.2	21.0	0.456
2011	3.2	8.4	14.3	23.0	51.1	22.3	0.477	3.4	9.0	14.8	22.8	50.0	22.1	0.463
2012	3.2	8.3	14.4	23.0	51.0	22.3	0.477	3.4	9.0	14.8	22.9	49.9	22.1	0.463
2013	3.1	8.2	14.3	23.0	51.4	22.2	0.482	3.5	8.8	14.7	22.8	50.3	22.1	0.467
2014	3.1	8.2	14.3	23.2	51.2	21.9	0.480	3.3	9.0	14.8	22.9	50.0	21.8	0.464
2015	3.1	8.2	14.3	23.2	51.1	22.1	0.479	3.4	9.0	14.8	22.9	49.8	21.8	0.462
2016	3.1	8.3	14.2	22.9	51.5	22.6	0.481	3.5	9.1	14.7	22.5	50.2	22.4	0.464
2017	3.1	8.2	14.3	23.0	51.5	22.3	0.482	3.5	9.0	14.7	22.7	50.1	21.8	0.463
2018	3.1	8.3	14.1	22.6	52.0	23.1	0.486	3.5	9.1	14.7	22.4	50.3	22.5	0.464

. . . = Not available.

Table 3-4. Shares of Aggregate Income Received by Each Fifth and Top 5 Percent of Families

Year	Number of families (thousands)	Share of aggregate income (percent)						Mean family income (2017 dollars)						Gini coefficient
		Lowest fifth	Second fifth	Third fifth	Fourth fifth	Highest fifth	Top 5 percent	Lowest fifth	Second fifth	Third fifth	Fourth fifth	Highest fifth	Top 5 percent	
1950	39 929	4.5	12.0	17.4	23.4	42.7	17.3	. . .	. . .	. . .	. . .	. . .	. . .	0.379
1951	40 578	5.0	12.4	17.6	23.4	41.6	16.8	. . .	. . .	. . .	. . .	. . .	. . .	0.363
1952	40 832	4.9	12.3	17.4	23.4	41.9	17.4	. . .	. . .	. . .	. . .	. . .	. . .	0.368
1953	41 202	4.7	12.5	18.0	23.9	40.9	15.7	. . .	. . .	. . .	. . .	. . .	. . .	0.359
1954	41 951	4.5	12.1	17.7	23.9	41.8	16.3	. . .	. . .	. . .	. . .	. . .	. . .	0.371
1955	42 889	4.8	12.3	17.8	23.7	41.3	16.4	. . .	. . .	. . .	. . .	. . .	. . .	0.363
1956	43 497	5.0	12.5	17.9	23.7	41.0	16.1	. . .	. . .	. . .	. . .	. . .	. . .	0.358
1957	43 696	5.1	12.7	18.1	23.8	40.4	15.6	. . .	. . .	. . .	. . .	. . .	. . .	0.351
1958	44 232	5.0	12.5	18.0	23.9	40.6	15.4	. . .	. . .	. . .	. . .	. . .	. . .	0.354
1959	45 111	4.9	12.3	17.9	23.8	41.1	15.9	. . .	. . .	. . .	. . .	. . .	. . .	0.361
1960	45 539	4.8	12.2	17.8	24.0	41.3	15.9	. . .	. . .	. . .	. . .	. . .	. . .	0.364
1961	46 418	4.7	11.9	17.5	23.8	42.2	16.6	. . .	. . .	. . .	. . .	. . .	. . .	0.374
1962	47 059	5.0	12.1	17.6	24.0	41.3	15.7	. . .	. . .	. . .	. . .	. . .	. . .	0.362
1963	47 540	5.0	12.1	17.7	24.0	41.2	15.8	. . .	. . .	. . .	. . .	. . .	. . .	0.362
1964	47 956	5.1	12.0	17.7	24.0	41.2	15.9	. . .	. . .	. . .	. . .	. . .	. . .	0.361
1965	48 509	5.2	12.2	17.8	23.9	40.9	15.5	. . .	. . .	. . .	. . .	. . .	. . .	0.356
1966	49 214	5.6	12.4	17.8	23.8	40.5	15.6	15 562	34 547	49 432	66 175	112 939	173 419	0.349
1967	50 111	5.4	12.2	17.5	23.5	41.4	16.4	15 804	35 341	50 797	68 127	120 295	190 326	0.358
1968	50 823	5.6	12.4	17.7	23.7	40.5	15.6	17 035	37 153	53 079	70 986	121 297	186 983	0.348
1969	51 586	5.6	12.4	17.7	23.7	40.6	15.6	17 576	38 815	55 543	74 405	127 429	195 723	0.349
1970	52 227	5.4	12.2	17.6	23.8	40.9	15.6	17 310	38 275	55 342	74 652	128 373	195 477	0.353
1971	53 296	5.5	12.0	17.6	23.8	41.1	15.7	17 325	37 713	55 178	74 728	128 710	196 250	0.355
1972	54 373	5.5	11.9	17.5	23.9	41.4	15.9	18 082	39 453	57 949	79 136	137 248	210 410	0.359
1973	55 053	5.5	11.9	17.5	24.0	41.1	15.5	18 554	40 190	59 051	80 701	138 461	208 960	0.356
1974	55 698	5.7	12.0	17.6	24.1	40.6	14.8	18 882	39 858	58 192	79 726	134 339	195 947	0.355
1975	56 245	5.6	11.9	17.7	24.2	40.7	14.9	18 141	38 352	57 025	78 078	131 576	193 013	0.357
1976	56 710	5.6	11.9	17.7	24.2	40.7	14.9	18 580	39 321	58 656	80 143	134 886	197 419	0.358
1977	57 215	5.5	11.7	17.6	24.3	40.9	14.9	18 501	39 547	59 478	82 047	138 222	201 506	0.363
1978	57 804	5.4	11.7	17.6	24.2	41.1	15.1	18 762	40 689	61 097	84 178	142 841	209 318	0.363
1979	59 550	5.4	11.6	17.5	24.1	41.4	15.3	18 944	40 958	61 768	85 039	145 859	215 905	0.365
1980	60 309	5.3	11.6	17.6	24.4	41.1	14.6	18 209	39 639	59 882	83 078	139 878	198 848	0.365
1981	61 019	5.3	11.4	17.5	24.6	41.2	14.4	17 656	38 394	58 723	82 446	138 335	193 300	0.369
1982	61 393	5.0	11.3	17.2	24.4	42.2	15.3	16 655	37 789	57 701	81 855	141 441	202 366	0.380
1983	62 015	4.9	11.2	17.2	24.5	42.4	15.3	16 400	37 734	58 056	82 741	143 466	206 850	0.382
1984	62 706	4.8	11.1	17.1	24.5	42.5	15.4	16 948	38 804	59 869	85 612	148 792	215 126	0.383
1985	63 558	4.8	11.0	16.9	24.3	43.1	16.1	17 144	39 340	60 738	87 063	154 725	231 221	0.389
1986	64 491	4.7	10.9	16.9	24.1	43.4	16.5	17 639	40 727	63 096	90 154	162 382	246 587	0.392
1987	65 204	4.6	10.7	16.8	24.0	43.8	17.2	17 609	41 215	63 999	91 596	167 393	262 044	0.393
1988	65 837	4.6	10.7	16.7	24.0	44.0	17.2	17 770	41 382	64 397	92 484	169 692	265 138	0.395
1989	66 090	4.6	10.6	16.5	23.7	44.6	17.9	18 115	42 145	65 474	94 199	177 367	284 126	0.401
1990	66 322	4.6	10.8	16.6	23.8	44.3	17.4	17 937	41 837	64 433	92 662	172 207	270 201	0.396
1991	67 173	4.5	10.7	16.6	24.1	44.2	17.1	17 133	40 668	63 102	91 521	168 144	260 175	0.397
1992	68 216	4.3	10.5	16.5	24.0	44.7	17.6	16 463	39 708	62 732	91 184	169 684	267 156	0.404
1993	68 506	4.1	9.9	15.7	23.3	47.0	20.3	16 315	39 182	62 092	92 044	185 973	320 983	0.429
1994	69 313	4.2	10.0	15.7	23.3	46.9	20.1	17 044	40 325	63 680	94 132	189 702	325 451	0.426
1995	69 597	4.4	10.1	15.8	23.2	46.5	20.0	18 050	41 588	65 113	95 268	191 401	328 254	0.421
1996	70 241	4.2	10.0	15.8	23.1	46.8	20.3	17 774	41 901	66 280	96 847	196 072	339 235	0.425
1997	70 884	4.2	9.9	15.7	23.0	47.2	20.7	18 420	43 161	68 098	99 856	205 150	359 046	0.429
1998	71 551	4.2	9.9	15.7	23.0	47.3	20.7	18 881	44 438	70 334	103 145	212 298	371 581	0.430
1999	73 206	4.3	9.9	15.6	23.0	47.2	20.3	19 641	45 655	72 063	106 362	217 990	375 855	0.429
2000	73 778	4.3	9.8	15.4	22.7	47.7	21.1	20 158	46 091	72 438	106 760	223 993	396 919	0.433
2001	74 340	4.2	9.7	15.4	22.9	47.7	21.0	19 460	45 061	71 531	106 379	221 574	389 053	0.435
2002	75 616	4.2	9.7	15.5	23.0	47.6	20.8	19 153	44 436	70 873	105 410	217 663	380 936	0.434
2003	76 232	4.1	9.6	15.5	23.2	47.6	20.5	18 532	43 951	70 832	106 500	218 206	376 053	0.436
2004	76 866	4.0	9.6	15.4	23.0	47.9	20.9	18 472	43 967	70 565	105 426	219 414	382 253	0.438
2005	77 418	4.0	9.6	15.3	22.9	48.1	21.1	18 581	44 212	70 749	105 815	221 824	388 350	0.440
2006	78 454	4.0	9.5	15.1	22.9	48.5	21.5	18 939	44 813	71 280	107 732	228 384	404 269	0.444
2007	77 908	4.1	9.7	15.6	23.3	47.3	20.1	19 043	45 396	72 821	108 894	221 067	375 243	0.432
2008	78 874	4.0	9.6	15.5	23.1	47.8	20.5	18 154	43 513	70 285	105 184	217 308	373 130	0.438
2009	78 867	3.9	9.4	15.3	23.2	48.2	20.7	17 511	42 428	68 612	104 179	217 019	372 250	0.443
2010	79 559	3.8	9.4	15.4	23.5	47.9	20.0	16 814	41 588	67 784	103 417	210 831	352 523	0.440
2011	80 529	3.8	9.3	15.1	23.0	48.9	21.3	16 692	41 022	66 664	101 836	216 198	376 326	0.450
2012	80 944	3.8	9.2	15.1	23.0	48.9	21.3	16 616	40 843	66 813	102 122	216 663	376 871	0.451
2013	82 316	3.6	9.1	15.0	22.9	49.4	21.2	16 979	41 648	67 368	102 458	217 851	378 098	0.455
2014	81 730	3.6	9.2	15.1	23.2	48.9	20.8	16 697	42 164	69 338	106 875	224 934	383 579	0.452
2015	82 199	3.7	9.2	15.2	23.2	48.6	20.9	17 969	44 181	72 945	111 246	233 092	400 244	0.448
2016	82 854	3.7	9.2	15.0	22.9	49.2	21.4	18 593	45 910	74 472	113 642	244 636	426 060	0.452
2017	83 103	3.8	9.2	15.1	23.1	48.8	20.7	18 944	46 346	75 840	115 834	245 039	416 639	0.449
2018	83 508	3.8	9.3	14.9	22.6	49.4	21.6	20 378	49 214	78 966	119 904	261 762	457 189	0.452

. . . = Not available.

SECTION 3B: POVERTY

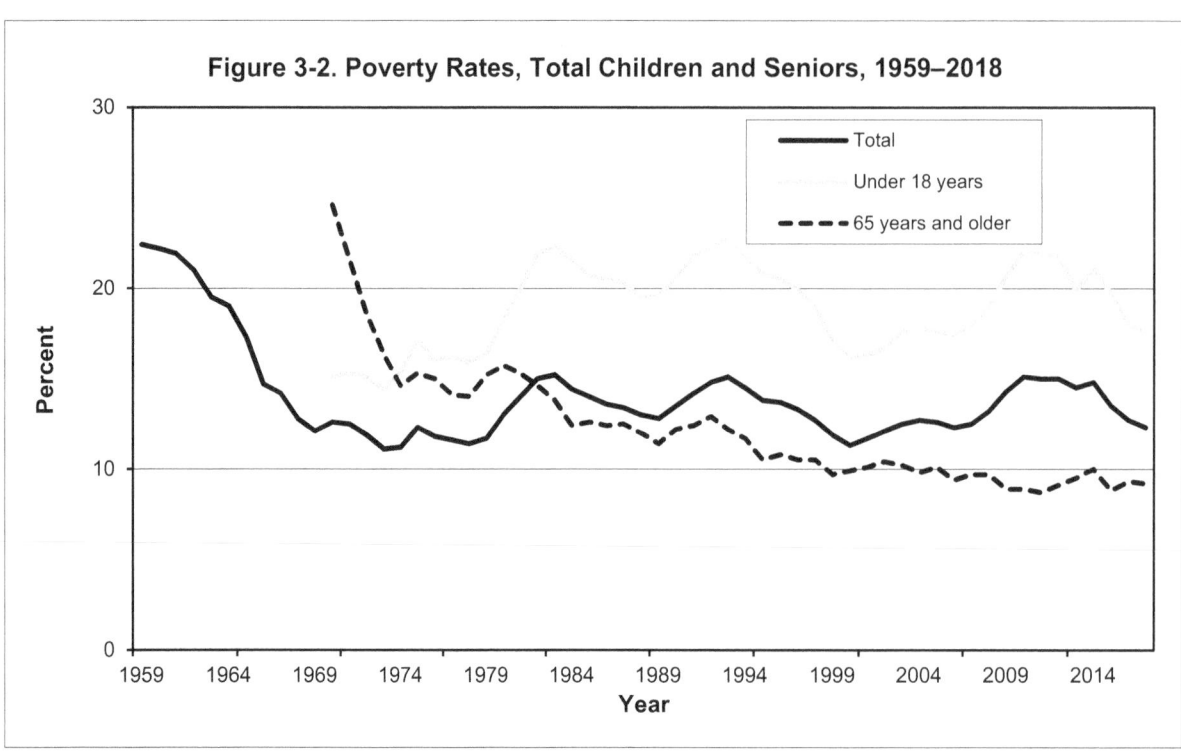

Figure 3-2. Poverty Rates, Total Children and Seniors, 1959–2018

- Poverty measures the ability of families and individuals who do not earn an income to cover the cost of a minimum standard of living. There is not one poverty amount but rather a range of income depending on age and number of people in families. In 2018, an individual aged 65 and over lived in poverty if their income was less than $12,043. Meanwhile the poverty threshold for a family of four was $25,701.

- The official poverty rate in 2018 was 11.8 percent, down 0.5 percentage points from 2017. This is the fourth consecutive year that the poverty rate has declined. Since 2014, the number of people in poverty has declined from over 46 million to over 38 million in 2018. (Table 3-6)

- According to the official calculation, 5.8 percent of children under 18 in married-couple families lived below the poverty level in 2018, continuing a downward trend since 2015. The percent of persons aged 65 and over in poverty actually increased in 2018 to its higher level since 2014. However, the poverty rate for seniors has been consistently lower than the poverty rate for the general population. (Table 3-7 and Table 3-8)

- The official poverty rate among Hispanics (who may be of any race) decreased for the fourth consecutive year in 2018 to 17.6 percent. However, this was still significantly higher than the poverty rate for the entire population (11.8 percent). (Table 3-6)

Table 3-5. Weighted Average Poverty Thresholds by Family Size

(Dollars.)

Year	Unrelated individuals			Families of 2 people			Families, all ages							CPI-U, all items (1982–1984 = 100)
	All ages	Under 65 years	65 years and older	All ages	House-holder under 65 years	House-holder 65 years and older	3 people	4 people	5 people	6 people	7 people	8 people	9 people or more	
1959	1 467	1 503	1 397	1 894	1 952	1 761	2 324	2 973	3 506	3 944	. . .	. . .	. . .	29.2
1960	1 490	1 526	1 418	1 924	1 982	1 788	2 359	3 022	3 560	4 002	. . .	. . .	. . .	29.6
1961	1 506	1 545	1 433	1 942	2 005	1 808	2 383	3 054	3 597	4 041	. . .	. . .	. . .	29.9
1962	1 519	1 562	1 451	1 962	2 027	1 828	2 412	3 089	3 639	4 088	. . .	. . .	. . .	30.3
1963	1 539	1 581	1 470	1 988	2 052	1 850	2 442	3 128	3 685	4 135	. . .	. . .	. . .	30.6
1964	1 558	1 601	1 488	2 015	2 079	1 875	2 473	3 169	3 732	4 193	. . .	. . .	. . .	31.0
1965	1 582	1 626	1 512	2 048	2 114	1 906	2 514	3 223	3 797	4 264	. . .	. . .	. . .	31.5
1966	1 628	1 674	1 556	2 107	2 175	1 961	2 588	3 317	3 908	4 388	. . .	. . .	. . .	32.5
1967	1 675	1 722	1 600	2 168	2 238	2 017	2 661	3 410	4 019	4 516	. . .	. . .	. . .	33.4
1968	1 748	1 797	1 667	2 262	2 333	2 102	2 774	3 553	4 188	4 706	. . .	. . .	. . .	34.8
1969	1 840	1 893	1 757	2 383	2 458	2 215	2 924	3 743	4 415	4 958	. . .	. . .	. . .	36.7
1970	1 954	2 010	1 861	2 525	2 604	2 348	3 099	3 968	4 680	5 260	. . .	. . .	. . .	38.8
1971	2 040	2 098	1 940	2 633	2 716	2 448	3 229	4 137	4 880	5 489	. . .	. . .	. . .	40.5
1972	2 109	2 168	2 005	2 724	2 808	2 530	3 339	4 275	5 044	5 673	. . .	. . .	. . .	41.8
1973	2 247	2 307	2 130	2 895	2 984	2 688	3 548	4 540	5 358	6 028	. . .	. . .	. . .	44.4
1974	2 495	2 562	2 364	3 211	3 312	2 982	3 936	5 038	5 950	6 699	. . .	. . .	. . .	49.3
1975	2 724	2 797	2 581	3 506	3 617	3 257	4 293	5 500	6 499	7 316	. . .	. . .	. . .	53.8
1976	2 884	2 959	2 730	3 711	3 826	3 445	4 540	5 815	6 876	7 760	. . .	. . .	. . .	56.9
1977	3 075	3 152	2 906	3 951	4 072	3 666	4 833	6 191	7 320	8 261	. . .	. . .	. . .	60.6
1978	3 311	3 392	3 127	4 249	4 383	3 944	5 201	6 662	7 880	8 891	. . .	. . .	. . .	65.2
1979	3 689	3 778	3 479	4 725	4 878	4 390	5 784	7 412	8 775	9 914	. . .	. . .	. . .	72.6
1980	4 190	4 290	3 949	5 363	5 537	4 983	6 565	8 414	9 966	11 269	12 761	14 199	16 896	82.4
1981	4 620	4 729	4 359	5 917	6 111	5 498	7 250	9 287	11 007	12 449	14 110	15 655	18 572	90.9
1982	4 901	5 019	4 626	6 281	6 487	5 836	7 693	9 862	11 684	13 207	15 036	16 719	19 698	96.5
1983	5 061	5 180	4 775	6 483	6 697	6 023	7 938	10 178	12 049	13 630	15 500	17 170	20 310	99.6
1984	5 278	5 400	4 979	6 762	6 983	6 282	8 277	10 609	12 566	14 207	16 096	17 961	21 247	103.9
1985	5 469	5 593	5 156	6 998	7 231	6 503	8 573	10 989	13 007	14 696	16 656	18 512	22 083	107.6
1986	5 572	5 701	5 255	7 138	7 372	6 630	8 737	11 203	13 259	14 986	17 049	18 791	22 497	109.6
1987	5 778	5 909	5 447	7 397	7 641	6 872	9 056	11 611	13 737	15 509	17 649	19 515	23 105	113.6
1988	6 022	6 155	5 674	7 704	7 958	7 157	9 435	12 092	14 304	16 146	18 232	20 253	24 129	118.3
1989	6 310	6 451	5 947	8 076	8 343	7 501	9 885	12 674	14 990	16 921	19 162	21 328	25 480	124.0
1990	6 652	6 800	6 268	8 509	8 794	7 905	10 419	13 359	15 792	17 839	20 241	22 582	26 848	130.7
1991	6 932	7 086	6 532	8 865	9 165	8 241	10 860	13 924	16 456	18 587	21 058	23 582	27 942	136.2
1992	7 143	7 299	6 729	9 137	9 443	8 487	11 186	14 335	16 952	19 137	21 594	24 053	28 745	140.3
1993	7 363	7 518	6 930	9 414	9 728	8 740	11 522	14 763	17 449	19 718	22 383	24 838	29 529	144.5
1994	7 547	7 710	7 108	9 661	9 976	8 967	11 821	15 141	17 900	20 235	22 923	25 427	30 300	148.2
1995	7 763	7 929	7 309	9 933	10 259	9 219	12 158	15 569	18 408	20 804	23 552	26 237	31 280	152.4
1996	7 995	8 163	7 525	10 233	10 564	9 491	12 516	16 036	18 952	21 389	24 268	27 091	31 971	156.9
1997	8 183	8 350	7 698	10 473	10 805	9 712	12 802	16 400	19 380	21 886	24 802	27 593	32 566	160.5
1998	8 316	8 480	7 818	10 634	10 972	9 862	13 003	16 660	19 680	22 228	25 257	28 166	33 339	163.0
1999	8 499	8 667	7 990	10 864	11 213	10 075	13 289	17 030	20 128	22 730	25 918	28 970	34 436	166.6
2000	8 791	8 959	8 259	11 235	11 589	10 418	13 740	17 604	20 815	23 533	26 750	29 701	35 150	172.2
2001	9 039	9 214	8 494	11 569	11 920	10 715	14 128	18 104	21 405	24 195	27 517	30 627	36 286	177.1
2002	9 183	9 359	8 628	11 756	12 110	10 885	14 348	18 392	21 744	24 576	28 001	30 907	37 062	179.9
2003	9 393	9 573	8 825	12 015	12 384	11 133	14 680	18 810	22 245	25 122	28 544	31 589	37 656	184.0
2004	9 646	9 827	9 060	12 335	12 714	11 430	15 066	19 307	22 830	25 787	29 233	32 641	39 062	188.9
2005	9 973	10 160	9 367	12 755	13 145	11 815	15 577	19 971	23 613	26 683	30 249	33 610	40 288	195.3
2006	10 294	10 488	9 669	13 167	13 569	12 201	16 079	20 614	24 382	27 560	31 205	34 774	41 499	201.6
2007	10 590	10 787	9 944	13 540	13 954	12 550	16 530	21 203	25 080	28 323	32 233	35 816	42 739	207.3
2008	10 991	11 201	10 326	14 051	14 489	13 030	17 163	22 025	26 049	29 456	33 529	37 220	44 346	215.3
2009	10 956	11 161	10 289	13 991	14 439	12 982	17 098	21 954	25 991	29 405	33 372	37 252	44 366	214.5
2010	11 137	11 344	10 458	14 216	14 676	13 194	17 373	22 315	26 442	29 904	34 019	37 953	45 224	218.1
2011	11 484	11 702	10 788	14 657	15 139	13 609	17 916	23 021	27 251	30 847	35 085	39 064	46 572	224.9
2012	11 720	11 945	11 011	14 937	15 450	13 892	18 284	23 492	27 827	31 471	35 743	39 688	47 297	229.6
2013	11 888	12 119	11 173	15 142	15 679	14 095	18 552	23 834	28 265	31 925	36 384	40 484	48 065	233.0
2014	12 071	12 316	11 354	15 379	15 934	14 326	18 850	24 230	28 695	32 473	36 927	40 968	49 021	236.7
2015	12 082	12 331	11 367	15 391	15 952	14 342	18 871	24 257	28 741	32 542	36 998	41 029	49 177	237.0
2016	12 228	12 486	11 511	15 569	16 151	14 522	19 105	24 563	29 111	32 928	37 458	41 781	49 721	240.0
2017	12 488	12 752	11 756	15 877	16 493	14 828	19 515	25 094	29 714	33 618	38 173	42 684	50 681	245.1
2018	12 784	13 064	12 043	16 247	16 889	15 193	19 985	25 701	30 459	34 533	39 194	43 602	51 393	251.1

. . . = Not available.

Table 3-6. Poverty Status of People by Race and Hispanic Origin

(Thousands of people, percent of population.)

Year	Number of people, all races	Below poverty level												
		All races		White		White, not Hispanic		Black		Asian [1]		Hispanic (any race)	
		Number	Poverty rate (percent)	Number	Poverty rate (percent)	Number	Poverty rate (percent)	Number	Poverty rate (percent)	Number	Poverty rate (percent)	Number	Poverty rate (percent)
1970	202 183	25 420	12.6	17 484	9.9	. . .	. . .	7 548	33.5	. . .	. . .	. . .	. . .
1971	204 554	25 559	12.5	17 780	9.9	. . .	. . .	7 396	32.5	. . .	. . .	. . .	. . .
1972	206 004	24 460	11.9	16 203	9.0	. . .	. . .	7 710	33.3	. . .	. . .	2 414	22.8
1973	207 621	22 973	11.1	15 142	8.4	12 864	7.5	7 388	31.4	. . .	. . .	2 366	21.9
1974	209 362	23 370	11.2	15 736	8.6	13 217	7.7	7 182	30.3	. . .	. . .	2 575	23.0
1975	210 864	25 877	12.3	17 770	9.7	14 883	8.6	7 545	31.3	. . .	. . .	2 991	26.9
1976	212 303	24 975	11.8	16 713	9.1	14 025	8.1	7 595	31.1	. . .	. . .	2 783	24.7
1977	213 867	24 720	11.6	16 416	8.9	13 802	8.0	7 726	31.3	. . .	. . .	2 700	22.4
1978	215 656	24 497	11.4	16 259	8.7	13 755	7.9	7 625	30.6	. . .	. . .	2 607	21.6
1979	222 903	26 072	11.7	17 214	9.0	14 419	8.1	8 050	31.0	. . .	. . .	2 921	21.8
1980	225 027	29 272	13.0	19 699	10.2	16 365	9.1	8 579	32.5	. . .	. . .	3 491	25.7
1981	227 157	31 822	14.0	21 553	11.1	17 987	9.9	9 173	34.2	. . .	. . .	3 713	26.5
1982	229 412	34 398	15.0	23 517	12.0	19 362	10.6	9 697	35.6	. . .	. . .	4 301	29.9
1983	231 700	35 303	15.2	23 984	12.1	19 538	10.8	9 882	35.7	. . .	. . .	4 633	28.0
1984	233 816	33 700	14.4	22 955	11.5	18 300	10.0	9 490	33.8	. . .	. . .	4 806	28.4
1985	236 594	33 064	14.0	22 860	11.4	17 839	9.7	8 926	31.3	. . .	. . .	5 236	29.0
1986	238 554	32 370	13.6	22 183	11.0	17 244	9.4	8 983	31.1	. . .	. . .	5 117	27.3
1987	240 982	32 221	13.4	21 195	10.4	16 029	8.7	9 520	32.4	1 021	16.1	5 422	28.0
1988	243 530	31 745	13.0	20 715	10.1	15 565	8.4	9 356	31.3	1 117	17.3	5 357	26.7
1989	245 992	31 528	12.8	20 785	10.0	15 599	8.3	9 302	30.7	939	14.1	5 430	26.2
1990	248 644	33 585	13.5	22 326	10.7	16 622	8.8	9 837	31.9	858	12.2	6 006	28.1
1991	251 192	35 708	14.2	23 747	11.3	17 741	9.4	10 242	32.7	996	13.8	6 339	28.7
1992	256 549	38 014	14.8	25 259	11.9	18 202	9.6	10 827	33.4	985	12.7	7 592	29.6
1993	259 278	39 265	15.1	26 226	12.2	18 882	9.9	10 877	33.1	1 134	15.3	8 126	30.6
1994	261 616	38 059	14.5	25 379	11.7	18 110	9.4	10 196	30.6	974	14.6	8 416	30.7
1995	263 733	36 425	13.8	24 423	11.2	16 267	8.5	9 872	29.3	1 411	14.6	8 574	30.3
1996	266 218	36 529	13.7	24 650	11.2	16 462	8.6	9 694	28.4	1 454	14.5	8 697	29.4
1997	268 480	35 574	13.3	24 396	11.0	16 491	8.6	9 116	26.5	1 468	14.0	8 308	27.1
1998	271 059	34 476	12.7	23 454	10.5	15 799	8.2	9 091	26.1	1 360	12.5	8 070	25.6
1999	276 208	32 791	11.9	22 169	9.8	14 735	7.7	8 441	23.6	1 285	10.7	7 876	22.7
2000	278 944	31 581	11.3	21 645	9.5	14 366	7.4	7 982	22.5	1 258	9.9	7 747	21.5
2001	281 475	32 907	11.7	22 739	9.9	15 271	7.8	8 136	22.7	1 275	10.2	7 997	21.4
2002	285 317	34 570	12.1	. . .	. . .	. . .	. . .	. . .	. . .	. . .	. . .	8 555	21.8
2003	287 699	35 861	12.5	. . .	. . .	. . .	. . .	. . .	. . .	. . .	. . .	9 051	22.5
2004	290 617	37 040	12.7	. . .	. . .	. . .	. . .	. . .	. . .	. . .	. . .	9 122	21.9
2005	293 135	36 950	12.6	. . .	. . .	. . .	. . .	. . .	. . .	. . .	. . .	9 368	21.8
2006	296 450	36 460	12.3	. . .	. . .	. . .	. . .	. . .	. . .	. . .	. . .	9 243	20.6
2007	298 699	37 276	12.5	. . .	. . .	. . .	. . .	. . .	. . .	. . .	. . .	9 890	21.5
2008	301 041	39 829	13.2	. . .	. . .	. . .	. . .	. . .	. . .	. . .	. . .	10 987	23.2
2009	303 820	43 569	14.3	. . .	. . .	. . .	. . .	. . .	. . .	. . .	. . .	12 350	25.3
2010	306 130	46 343	15.1	. . .	. . .	. . .	. . .	. . .	. . .	. . .	. . .	13 522	26.5
2011	308 456	46 247	15.0	. . .	. . .	. . .	. . .	. . .	. . .	. . .	. . .	13 244	25.3
2012	310 648	46 496	15.0	. . .	. . .	. . .	. . .	. . .	. . .	. . .	. . .	13 616	25.6
2013	312 965	45 318	14.5	. . .	. . .	. . .	. . .	. . .	. . .	. . .	. . .	12 744	23.5
2014	315 804	46 657	14.8	. . .	. . .	. . .	. . .	. . .	. . .	. . .	. . .	13 104	23.6
2015	318 454	43 123	13.5	. . .	. . .	. . .	. . .	. . .	. . .	. . .	. . .	12 133	21.4
2016	319 911	40 616	12.7	. . .	. . .	. . .	. . .	. . .	. . .	. . .	. . .	11 137	19.4
2017	322 549	39 698	12.3	. . .	. . .	. . .	. . .	. . .	. . .	. . .	. . .	10 790	18.3
2018	323 847	38 146	11.8	. . .	. . .	. . .	. . .	. . .	. . .	. . .	. . .	10 526	17.6
By race													
Race alone													
2003	. . .	. . .	. . .	24 272	10.5	15 902	8.2	8 781	24.4	1 401	11.8	. . .	. . .
2004	. . .	. . .	. . .	25 327	10.8	16 908	8.7	9 014	24.7	1 201	9.8	. . .	. . .
2005	. . .	. . .	. . .	24 872	10.6	16 227	8.3	9 168	24.9	1 402	11.1	. . .	. . .
2006	. . .	. . .	. . .	24 416	10.3	16 013	8.2	9 048	24.3	1 353	10.3	. . .	. . .
2007	. . .	. . .	. . .	25 120	10.5	16 032	8.2	9 237	24.5	1 349	10.2	. . .	. . .
2008	. . .	. . .	. . .	26 990	11.2	17 024	8.6	9 379	24.7	1 576	11.8	. . .	. . .
2009	. . .	. . .	. . .	29 830	12.3	18 530	9.4	9 944	25.8	1 746	12.5	. . .	. . .
2010	. . .	. . .	. . .	31 083	13.0	19 251	9.9	10 746	27.4	1 899	12.2	. . .	. . .
2011	. . .	. . .	. . .	30 849	12.8	19 171	9.8	10 929	27.6	1 973	12.3	. . .	. . .
2012	. . .	. . .	. . .	30 816	12.7	18 940	9.7	10 911	27.2	1 921	11.7	. . .	. . .
2013	. . .	. . .	. . .	29 936	0.1	18 796	0.1	11 041	27.2	1 785	10.5	. . .	. . .
2014	. . .	. . .	. . .	31 088	0.1	19 653	0.1	10 755	26.2	2 137	12.0	. . .	. . .
2015	. . .	. . .	. . .	28 566	11.6	17 786	9.1	10 020	24.1	2 078	11.4	. . .	. . .
2016	. . .	. . .	. . .	27 113	11.0	17 263	8.8	9 234	22.0	1 908	10.1	. . .	. . .
2017	. . .	. . .	. . .	26 436	10.7	16 993	8.7	8 993	21.2	1 953	10.0	. . .	. . .
2018	. . .	. . .	. . .	24 945	10.1	15 725	8.1	8 884	20.8	1 996	10.1	. . .	. . .
Race alone or in combination													
2010	. . .	. . .	. . .	. . .	. . .	. . .	. . .	11 597	27.4	2 064	12.0	. . .	. . .
2011	. . .	. . .	. . .	. . .	. . .	. . .	. . .	11 730	27.5	2 189	12.3	. . .	. . .
2012	. . .	. . .	. . .	. . .	. . .	. . .	. . .	11 809	27.1	2 072	11.4	. . .	. . .
2013	. . .	. . .	. . .	. . .	. . .	. . .	. . .	11 162	27.1	1 974	10.4	. . .	. . .
2014	. . .	. . .	. . .	. . .	. . .	. . .	. . .	11 581	26.0	2 268	11.5	. . .	. . .
2015	. . .	. . .	. . .	. . .	. . .	. . .	. . .	10 797	23.9	2 234	11.1	. . .	. . .
2016	. . .	. . .	. . .	. . .	. . .	. . .	. . .	9 965	21.8	2 062	9.9	. . .	. . .
2017	. . .	. . .	. . .	. . .	. . .	. . .	. . .	9 820	21.2	2 104	9.8	. . .	. . .
2018	. . .	. . .	. . .	. . .	. . .	. . .	. . .	9 695	20.7	2 166	9.8	. . .	. . .

[1] For 1987 through 2001, Asian and Pacific Islander.
. . . = Not available.

Table 3-7. Poverty Status of Families by Type of Family

(Thousands of families, percent.)

| | Married couple families | | | | Families with no spouse present | | | | | | | Unrelated individuals | |
| | Number of families | | Poverty rate (percent) | | Male householder | | | Female householder | | | | | |
Year	Total	Total below poverty level	Total	With children under 18 years	Familes below poverty level	Poverty rate (percent) Total	Poverty rate (percent) With children under 18 years	Familes below poverty level	Poverty rate (percent) Total	Poverty rate (percent) With children under 18 years		Below poverty level	Poverty rate
1959	39 335	...	...	...	...	...	...	1 916	42.6	59.9		4 928	46.1
1960	39 624	...	...	...	...	...	...	1 955	42.4	56.3		4 926	45.2
1961	40 405	...	...	...	...	...	...	1 954	42.1	56.0		5 119	45.9
1962	40 923	...	...	...	...	...	...	2 034	42.9	59.7		5 002	45.4
1963	41 311	...	...	...	...	...	...	1 972	40.4	55.7		4 938	44.2
1964	41 648	...	...	...	...	...	...	1 822	36.4	49.7		5 143	42.7
1965	42 107	...	...	...	...	...	...	1 916	38.4	52.2		4 827	39.8
1966	42 553	...	...	...	...	...	...	1 721	33.1	47.1		4 701	38.3
1967	43 292	...	...	...	...	...	...	1 774	33.3	44.5		4 998	38.1
1968	43 842	...	...	...	...	...	...	1 755	32.3	44.6		4 694	34.0
1969	44 436	...	...	...	...	...	...	1 827	32.7	44.9		4 972	34.0
1970	44 739	...	...	...	...	...	...	1 952	32.5	43.8		5 090	32.9
1971	45 752	...	...	...	...	...	...	2 100	33.9	44.9		5 154	31.6
1972	46 314	...	...	...	...	...	...	2 158	32.7	44.5		4 883	29.0
1973	46 812	2 482	5.3	...	154	10.7	...	2 193	32.2	43.2		4 674	25.6
1974	47 069	2 474	5.3	6.0	125	8.9	15.4	2 324	32.1	43.7		4 553	24.1
1975	47 318	2 904	6.1	7.2	116	8.0	11.7	2 430	32.5	44.0		5 088	25.1
1976	47 497	2 606	5.5	6.4	162	10.8	15.4	2 543	33.0	44.1		5 344	24.9
1977	47 385	2 524	5.3	6.3	177	11.1	14.8	2 610	31.7	41.8		5 216	22.6
1978	47 692	2 474	5.2	5.9	152	9.2	14.7	2 654	31.4	42.2		5 435	22.1
1979	49 112	2 640	5.4	6.1	176	10.2	15.5	2 645	30.4	39.6		5 743	21.9
1980	49 294	3 032	6.2	7.7	213	11.0	18.0	2 972	32.7	42.9		6 227	22.9
1981	49 630	3 394	6.8	8.7	205	10.3	14.0	3 252	34.6	44.3		6 490	23.4
1982	49 908	3 789	7.6	9.8	290	14.4	20.6	3 434	36.3	47.8		6 458	23.1
1983	50 081	3 815	7.6	10.1	268	13.2	20.2	3 564	36.0	47.1		6 740	23.1
1984	50 350	3 488	6.9	9.4	292	13.1	18.1	3 498	34.5	45.7		6 609	21.8
1985	50 933	3 438	6.7	8.9	311	12.9	17.1	3 474	34.0	45.4		6 725	21.5
1986	51 537	3 123	6.1	8.0	287	11.4	17.8	3 613	34.6	46.0		6 846	21.6
1987	51 675	3 011	5.8	7.7	340	12.0	16.8	3 654	34.2	45.5		6 857	20.8
1988	52 100	2 897	5.6	7.2	336	11.8	18.0	3 642	33.4	44.7		7 070	20.6
1989	52 317	2 931	5.6	7.3	348	12.1	18.1	3 504	32.2	42.8		6 760	19.2
1990	52 147	2 981	5.7	7.8	349	12.0	18.8	3 768	33.4	44.5		7 446	20.7
1991	52 457	3 158	6.0	8.3	392	13.0	19.6	4 161	35.6	47.1		7 773	21.1
1992	53 090	3 385	6.4	8.6	484	15.8	22.5	4 275	35.4	46.2		8 075	21.9
1993	53 181	3 481	6.5	9.0	488	16.8	22.5	4 424	35.6	46.1		8 388	22.1
1994	53 865	3 272	6.1	8.3	549	17.0	22.6	4 232	34.6	44.0		8 287	21.5
1995	53 570	2 982	5.6	7.5	493	14.0	19.7	4 057	32.4	41.5		8 247	20.9
1996	53 604	3 010	5.6	7.5	531	13.8	20.0	4 167	32.6	41.9		8 452	20.8
1997	54 321	2 821	5.2	7.1	507	13.0	18.7	3 995	31.6	41.0		8 687	20.8
1998	54 778	2 879	5.3	6.9	476	12.0	16.6	3 831	29.9	38.7		8 478	19.9
1999	56 290	2 748	4.9	6.4	485	11.8	16.3	3 559	27.8	35.7		8 400	19.1
2000	56 598	2 637	4.7	6.0	485	11.3	15.3	3 278	25.4	33.0		8 653	19.0
2001	56 755	2 760	4.9	6.1	583	13.1	17.7	3 470	26.4	33.6		9 226	19.9
2002	57 327	3 052	5.3	6.8	564	12.1	16.6	3 613	26.5	33.7		9 618	20.4
2003	57 725	3 115	5.4	7.0	636	13.5	19.1	3 856	28.0	35.5		9 713	20.4
2004	57 983	3 216	5.5	7.0	657	13.4	17.1	3 962	28.3	35.9		9 926	20.4
2005	58 189	2 944	5.1	6.5	669	13.0	17.6	4 044	28.7	36.2		10 425	21.1
2006	58 964	2 910	4.9	6.4	671	13.2	17.9	4 087	28.3	36.5		9 977	20.0
2007	58 395	2 849	4.9	6.7	696	13.6	17.5	4 078	28.3	37.0		10 189	19.7
2008	59 137	3 261	5.5	7.5	723	13.8	17.6	4 163	28.7	37.2		10 710	20.8
2009	58 428	3 409	5.8	8.3	942	16.9	23.7	4 441	29.9	38.5		11 678	22.0
2010	58 667	3 681	6.3	9.0	892	15.8	24.1	4 827	31.7	40.9		12 449	22.9
2011	58 963	3 652	6.2	8.8	950	16.1	21.9	4 894	31.2	40.9		12 416	22.8
2012	59 224	3 705	6.3	8.9	1 023	16.4	22.6	4 793	30.9	40.9		12 558	22.4
2013	59 692	3 476	5.8	7.6	1 008	15.9	19.7	4 646	30.6	39.6		13 181	23.3
2014	60 015	3 735	6.2	8.2	969	15.7	22.0	4 646	30.6	39.8		13 374	23.1
2015	60 258	3 245	5.4	7.5	939	14.9	22.1	4 404	28.2	36.5		12 671	21.5
2016	60 821	3 096	5.1	6.6	847	13.1	17.3	4 136	26.6	35.6		12 336	21.0
2017	61 254	3 005	4.9	6.4	793	12.4	16.2	3 959	25.7	34.4		12 593	20.7
2018	61 971	2 938	4.7	5.8	824	12.7	16.6	3 742	24.9	33.8		12 287	20.2

. . . = Not available.

Table 3-8. Poverty Status of People by Sex and Age

(Thousands of people, percent of population.)

Year	Poverty status of people by sex				Poverty status of people by age					
	Males below poverty level		Females below poverty level		Children under 18 years below poverty level		People 18 to 64 years below poverty level		People 65 years and older below poverty level	
	Number (thousands)	Poverty rate (percent)	Number (thousands)	Poverty rate (percent)	Number (thousands)	Poverty rate (percent)	Number (thousands)	Poverty rate (percent)	Number (thousands)	Poverty rate (percent)
1959	. . .	. . .	. . .	. . .	17 552	27.3	16 457	17.0	5 481	35.2
1966	12 225	13.0	16 265	16.3	12 389	17.6	11 007	10.5	5 114	28.5
1967	11 813	12.5	15 951	15.8	11 656	16.6	10 725	10.0	5 388	29.5
1968	10 793	11.3	14 578	14.3	10 954	15.6	9 803	9.0	4 632	25.0
1969	10 292	10.6	13 978	13.6	9 691	14.0	9 669	8.7	4 787	25.3
1970	10 879	11.1	14 632	14.0	10 440	15.1	10 187	9.0	4 793	24.6
1971	10 708	10.8	14 841	14.1	10 551	15.3	10 735	9.3	4 273	21.6
1972	10 190	10.2	14 258	13.4	10 284	15.1	10 438	8.8	3 738	18.6
1973	9 642	9.6	13 316	12.5	9 642	14.4	9 977	8.3	3 354	16.3
1974	9 945	9.8	13 429	12.5	10 156	15.4	10 132	8.3	3 085	14.6
1975	10 908	10.7	14 970	13.8	11 104	17.1	11 456	9.2	3 317	15.3
1976	10 373	10.1	14 603	13.4	10 273	16.0	11 389	9.0	3 313	15.0
1977	10 340	10.0	14 381	13.0	10 288	16.2	11 316	8.8	3 177	14.1
1978	10 017	9.6	14 480	13.0	9 931	15.9	11 332	8.7	3 233	14.0
1979	10 861	10.1	15 211	13.2	10 377	16.4	12 014	8.9	3 682	15.2
1980	12 207	11.2	17 065	14.7	11 543	18.3	13 858	10.1	3 871	15.7
1981	13 360	12.1	18 462	15.8	12 505	20.0	15 464	11.1	3 853	15.3
1982	14 842	13.4	19 556	16.5	13 647	21.9	17 000	12.0	3 751	14.6
1983	15 296	13.6	20 006	16.8	13 911	22.3	17 767	12.4	3 625	13.8
1984	14 537	12.8	19 163	15.9	13 420	21.5	16 952	11.7	3 330	12.4
1985	14 140	12.3	18 923	15.6	13 010	20.7	16 598	11.3	3 456	12.6
1986	13 721	11.8	18 649	15.2	12 876	20.5	16 017	10.8	3 477	12.4
1987	13 781	11.8	18 439	14.9	12 843	20.3	15 815	10.6	3 563	12.5
1988	13 599	11.5	18 146	14.5	12 455	19.5	15 809	10.5	3 481	12.0
1989	13 366	11.2	18 162	14.4	12 590	19.6	15 575	10.2	3 363	11.4
1990	14 211	11.7	19 373	15.2	13 431	20.6	16 496	10.7	3 658	12.2
1991	15 082	12.3	20 626	16.0	14 341	21.8	17 586	11.4	3 781	12.4
1992	16 222	12.9	21 792	16.6	15 294	22.3	18 793	11.9	3 928	12.9
1993	16 900	13.3	22 365	16.9	15 727	22.7	19 781	12.4	3 755	12.2
1994	16 316	12.8	21 744	16.3	15 289	21.8	19 107	11.9	3 663	11.7
1995	15 683	12.2	20 742	15.4	14 665	20.8	18 442	11.4	3 318	10.5
1996	15 611	12.0	20 918	15.4	14 463	20.5	18 638	11.4	3 428	10.8
1997	15 187	11.6	20 387	14.9	14 113	19.9	18 085	10.9	3 376	10.5
1998	14 712	11.1	19 764	14.3	13 467	18.9	17 623	10.5	3 386	10.5
1999	14 079	10.4	18 712	13.2	12 280	17.1	17 289	10.1	3 222	9.7
2000	13 536	9.9	18 045	12.6	11 587	16.2	16 671	9.6	3 323	9.9
2001	14 327	10.4	18 580	12.9	11 733	16.3	17 760	10.1	3 414	10.1
2002	15 162	10.9	19 408	13.3	12 133	16.7	18 861	10.6	3 576	10.4
2003	15 783	11.2	20 078	13.7	12 866	17.6	19 443	10.8	3 552	10.2
2004	16 399	11.5	20 641	13.9	13 041	17.8	20 545	11.3	3 453	9.8
2005	15 950	11.1	21 000	14.1	12 896	17.6	20 450	11.1	3 603	10.1
2006	16 000	11.0	20 460	13.6	12 827	17.4	20 239	10.8	3 394	9.4
2007	16 302	11.1	20 973	13.8	13 324	18.0	20 396	10.9	3 556	9.7
2008	17 698	12.0	22 131	14.4	14 068	19.0	22 105	11.7	3 656	9.7
2009	19 475	13.0	24 094	15.6	15 451	20.7	24 684	12.9	3 433	8.9
2010	20 893	14.0	25 451	16.3	16 286	22.0	26 499	13.8	3 558	8.9
2011	20 501	13.6	25 746	16.3	16 134	21.9	26 492	13.7	3 620	8.7
2012	20 656	13.6	25 840	16.3	16 073	21.8	26 497	13.7	3 926	9.1
2013	20 119	13.1	25 199	15.8	14 659	19.9	26 429	13.6	4 231	9.5
2014	20 708	13.4	25 949	16.1	15 540	21.1	26 527	13.5	4 590	10.0
2015	19 037	12.2	24 086	14.8	14 509	19.7	24 414	12.4	4 201	8.8
2016	17 685	11.3	22 931	14.0	13 253	18.0	22 795	11.6	4 568	9.3
2017	17 365	11.0	22 333	13.6	12 808	17.5	22 209	11.2	4 681	9.2
2018	16 782	10.6	21 363	12.9	11 869	16.2	21 130	10.7	5 146	9.7

. . . = Not available.

Table 3-9. Working-Age Poor People by Work Experience

(Thousands of people, percent of population [poverty rate], percent of total poor people.)

Year	Total number of working-age poor people	Worked		Worked year-round, full-time			Worked less than year-round or full-time			Did not work		
		Number	Percent of total poor	Number	Poverty rate (percent)	Percent of total poor	Number	Poverty rate (percent)	Percent of total poor	Number	Poverty rate (percent)	Percent of total poor
16 Years and Over												
1978	16 914	6 599	39.0	1 309	. . .	7.7	5 290	. . .	31.3	10 315	. . .	61.0
1979	16 803	6 601	39.3	1 394	. . .	8.3	5 207	. . .	31.0	10 202	. . .	60.7
1980	18 892	7 674	40.6	1 644	. . .	8.7	6 030	. . .	31.9	11 218	. . .	59.4
1981	20 571	8 524	41.4	1 881	. . .	9.1	6 643	. . .	32.3	12 047	. . .	58.6
1982	22 100	9 013	40.8	1 999	. . .	9.0	7 014	. . .	31.7	13 087	. . .	59.2
1983	22 741	9 329	41.0	2 064	. . .	9.1	7 265	. . .	31.9	13 412	. . .	59.0
1984	21 541	8 999	41.8	2 076	. . .	9.6	6 923	. . .	32.1	12 542	. . .	58.2
1985	21 243	9 008	42.4	1 972	. . .	9.3	7 036	. . .	33.1	12 235	. . .	57.6
1986	20 688	8 743	42.3	2 007	. . .	9.7	6 736	. . .	32.6	11 945	. . .	57.7
1987	20 546	8 258	40.2	1 821	2.4	8.9	6 436	12.5	31.3	12 288	21.6	59.8
1988	20 323	8 363	41.2	1 929	2.4	9.5	6 434	12.7	31.7	11 960	21.2	58.8
1989	19 952	8 376	42.0	1 908	2.4	9.6	6 468	12.5	32.4	11 576	20.8	58.0
1990	21 242	8 716	41.0	2 076	2.6	9.8	6 639	12.6	31.3	12 526	22.1	59.0
1991	22 530	9 208	40.9	2 103	2.6	9.3	7 105	13.4	31.5	13 322	22.8	59.1
1992	23 951	9 739	40.6	2 211	2.7	9.2	7 529	14.1	31.4	14 212	23.7	59.3
1993	24 832	10 144	40.8	2 408	2.9	9.7	7 737	14.6	31.2	14 688	24.2	59.1
1994	24 108	9 829	40.8	2 520	2.9	10.5	7 309	13.9	30.3	14 279	23.6	59.2
1995	23 077	9 484	41.1	2 418	2.7	10.5	7 066	13.7	30.6	13 593	22.3	58.9
1996	23 472	9 586	40.8	2 263	2.5	9.6	7 322	14.1	31.2	13 886	22.7	59.2
1997	22 753	9 444	41.5	2 345	2.5	10.3	7 098	13.8	31.2	13 309	21.7	58.5
1998	22 256	9 133	41.0	2 804	2.9	12.6	6 330	12.7	28.4	13 123	21.1	59.0
1999	21 762	9 251	42.5	2 559	2.6	11.8	6 692	13.2	30.8	12 511	19.9	57.5
2000	21 080	8 511	40.4	2 439	2.4	11.6	6 072	12.1	28.8	12 569	19.8	59.6
2001	22 245	8 530	38.3	2 567	2.6	11.5	5 964	11.8	26.8	13 715	20.6	61.7
2002	23 601	8 954	37.9	2 635	2.6	11.2	6 318	12.4	26.8	14 647	21.0	62.1
2003	24 266	8 820	36.3	2 636	2.6	10.9	6 183	12.2	25.5	15 446	21.5	63.7
2004	25 256	9 384	37.2	2 891	2.8	11.4	6 493	12.8	25.7	15 872	21.7	62.8
2005	25 381	9 340	36.8	2 894	2.8	11.4	6 446	12.8	25.4	16 041	21.8	63.2
2006	24 896	9 181	36.9	2 906	2.7	11.7	6 275	12.6	25.2	15 715	21.1	63.1
2007	25 297	9 089	35.9	2 768	2.5	10.9	6 320	12.7	25.0	16 208	21.5	64.1
2008	27 216	10 085	37.1	2 754	2.6	10.1	7 331	13.5	26.9	17 131	22.0	62.9
2009	29 625	10 680	36.1	2 641	2.7	8.9	8 039	14.5	27.1	18 945	22.7	63.9
2010	31 731	10 742	33.9	2 640	2.7	8.3	8 102	15.0	25.5	20 989	23.9	66.1
2011	31 630	10 588	33.5	2 770	2.7	8.8	7 818	14.9	24.7	21 042	23.6	66.5
2012	31 933	10 977	34.4	2 904	2.8	9.1	8 073	15.0	25.3	20 956	23.6	65.6
2013	32 011	11 025	34.4	2 812	2.7	8.8	8 213	15.8	25.7	20 986	23.2	65.6
2014	32 508	10 469	32.2	3 142	2.9	9.7	7 327	14.3	22.5	22 039	24.2	67.8
2015	30 123	9 795	32.5	2 576	2.3	8.8	7 220	13.9	24.0	20 328	22.5	67.5
2016	28 696	9 109	31.7	2 448	2.2	8.5	6 661	13.0	21.6	19 587	21.6	68.3
2017	28 167	8 503	32.2	2 468	2.1	8.8	6 035	12.0	21.4	19 644	21.4	69.7
2018	27 496	8 213	29.9	2 591	2.2	9.4	5 621	11.4	20.4	19 283	20.9	70.1

. . . = Not available.

Table 3-10. Number and Percent of People in Poverty Using the Supplemental Poverty Measure, 2016–2018, and Comparison With Official Measures in 2018

(Numbers in thousands; percent of population.)

Characteristic	SPM 2016		SPM 2017		SPM 2018		Official 2018[1]	
	Number	Percent	Number	Percent	Number	Percent	Number	Percent
All People	44 752	14.0	42 075	13.0	41 420	12.8	38 200	11.8
Sex								
Male	20 693	13.2	19 505	12.3	19 269	12.1	16 820	10.6
Female	24 059	14.7	22 570	13.7	22 151	13.4	21 380	12.9
Age								
Under 18 years	11 281	15.2	10 532	14.2	10 096	13.7	11 924	16.2
18 to 64 years	26 303	13.3	24 582	12.4	24 151	12.2	21 130	10.7
65 years and older	7 168	14.5	6 960	13.6	7 174	13.6	5 146	9.7
Type of Unit								
In married-couple unit	16 516	8.6	14 899	7.6	15 043	7.7	10 530	5.4
In female householder unit	11 655	27.3	10 621	25.3	3 659	13.9	6 374	24.2
In male householder unit	2 635	17.5	2 488	17.3	10 390	25.0	10 506	25.3
Cohabiting	3 261	13.0	3 877	14.9	2 197	15.1	1 684	11.6
Unrelated individuals	10 685	23.6	10 191	22.3	10 132	21.9	9 105	19.7
Race and Hispanic Origin								
White	30 717	12.5	28 380	11.5	27 820	11.2	24 984	10.1
White, not Hispanic	19 446	9.9	17 689	9.0	16 932	8.7	15 742	8.1
Black	9 086	21.6	8 775	20.6	8 727	20.4	8 891	20.8
Asian	2 774	14.7	2 743	14.0	2 749	13.9	2 004	10.1
Hispanic, any race	12 670	22.0	12 146	20.5	12 216	20.3	10 548	17.6
Nativity								
Native born	35 515	12.8	33 314	12.0	32 540	11.7	31 878	11.4
Foreign born	9 237	21.1	8 761	19.3	8 880	19.4	6 322	13.8
Naturalized citizen	3 205	15.7	3 238	14.8	3 297	14.8	2 215	9.9
Not a citizen	6 032	25.7	5 522	23.4	5 584	23.7	4 107	17.5
Educational Attainment								
Total, age 25 and over	27 929	12.9	25 990	11.8	26 158	11.8	21 916	9.9
No high school diploma	6 356	28.2	6 137	27.4	6 320	28.8	5 693	25.9
High school, no college	10 139	16.2	9 500	15.2	9 272	14.9	7 925	12.7
Some college	6 615	11.5	5 879	10.2	5 599	9.7	4 812	8.4
Bachelor's degree or higher	4 819	6.5	4 474	5.8	4 967	6.2	3 486	4.4
Tenure								
Owner/ mortgage	10 122	7.4	8 588	6.4	7 831	5.9	5 249	3.9
Owner/no-mortgage/rent-free	9 825	12.7	9 967	11.7	10 146	11.8	9 773	11.3
Renter	24 806	23.3	23 521	22.4	23 443	22.4	23 179	22.1
Residence								
Inside metropolitan statistical areas	39 120	14.1	36 790	13.1	36 249	12.9	31 978	11.3
Inside principal cities	17 971	17.4	17 413	16.7	16 818	16.0	15 309	14.6
Outside principal cities	21 148	12.2	19 377	11.0	19 431	11.0	16 669	9.4
Outside metropolitan statistical areas	5 633	12.9	5 285	12.3	5 171	12.2	6 222	14.7
Region								
Northeast	6 874	12.4	7 218	12.9	6 768	12.2	5 689	10.3
Midwest	7 424	11.1	6 874	10.2	6 223	9.2	7 008	10.4
South	17 966	14.8	16 846	13.8	17 219	13.9	16 786	13.6
West	12 489	16.3	11 137	14.4	11 211	14.4	8 716	11.2
Health Insurance Coverage								
With private insurance	17 898	8.3	13 552	6.2	12 747	5.9	8 376	3.8
With public, no private insurance	19 646	25.8	21 707	27.6	21 805	27.8	23 520	30.0
Not insured	7 208	25.7	6 816	26.0	6 868	24.4	6 305	22.4
Work Experience								
Total, 18 to 64 years	26 303	13.3	24 582	12.4	24 151	12.2	21 130	10.7
All workers	12 111	8.0	11 319	7.4	10 959	7.2	7 781	5.1
Full-time, year-round	5 099	4.7	4 925	4.5	4 847	4.3	2 544	2.3
Less than full-time, year-round	7 012	16.3	6 394	15.0	6 112	14.9	5 237	12.7
Did not work at least 1 week	14 193	30.8	13 263	29.0	13 191	29.4	13 349	29.7
Disability Status								
Total, 18 to 64 years	26 303	13.3	24 582	12.4	24 151	12.2	21 130	10.7
With a disability	3 905	25.4	3 429	22.7	3 609	24.3	3 818	25.7
With no disability	22 350	12.4	21 116	11.6	20 500	11.3	17 279	9.5

[1]Differs from published official rates because these figures include unrelated individuals under 15 years of age as does the SPM.

SECTION 3C: HEALTH INSURANCE COVERAGE

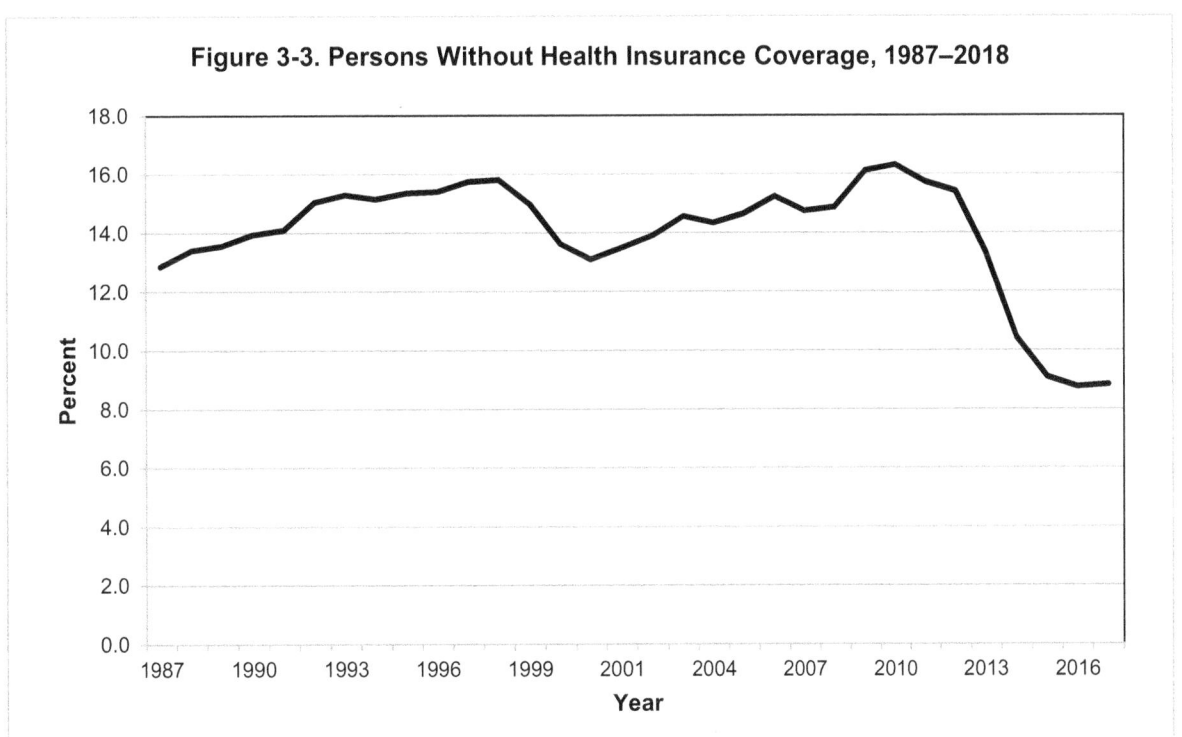

Figure 3-3. Persons Without Health Insurance Coverage, 1987–2018

- Health insurance is a means for financing a person's health care expenses. Over time, changes in the rate of health insurance coverage and the distribution of coverage types may reflect economic trends, shifts in the demographic composition of the population, and policy changes that impact access to care. Such a policy change was the Patient Protection and Affordable Care Act (ACA) in 2010. A new trend began during the COVID-19 pandemic earlier this year, when millions of people lost their jobs and their employer-paid health insurance.

- Employer based private health care insurance continues to be, by far, the most common method for people to obtain health insurance. The emergence of COVID-19 has revitalized the debate on health care coverage in this country.

- The percentage of people without health insurance has fallen steadily from 16.3 percent in 2010 to 8.5 percent in 2018. The number of people without health insurance declined again in 2018 after increasing slightly in 2017. In 2018, more than 27 million people did not have health insurance compared with nearly 50 million people in 2010. (Table 3-11)

- In 2018, 17.8 percent of Hispanics (who may be of any race) lacked insurance compared with 9.7 percent of Blacks, 6.8 percent of Asians, and 5.4 percent of Non-Hispanic Whites. (Table 3-12)

Table 3-11. Health Insurance Coverage

(Number in thousands. People as of March of the following year.)

Year	Total	Total covered	Private health insurance			Government health insurance				Not covered
			Total	Employment	Direct purchase	Total	Medicaid	Medicare	Military	
NUMBER										
1987	241 187	210 161	182 160	149 739	. . .	56 282	20 211	30 458	10 542	31 026
1988	243 685	211 005	182 019	150 940	. . .	56 850	20 728	30 925	10 105	32 680
1989	246 191	212 807	183 610	151 644	. . .	57 382	21 185	31 495	9 870	33 384
1990	248 886	214 167	182 135	150 215	. . .	60 965	24 261	32 260	9 922	34 719
1991	251 447	216 003	181 375	150 077	. . .	63 882	26 880	32 907	9 820	35 444
1992	256 830	218 189	181 466	148 796	. . .	66 244	29 416	33 230	9 510	38 641
1993	259 753	220 040	182 351	148 318	. . .	68 554	31 749	33 097	9 560	39 713
1994	262 105	222 387	184 318	159 634	31 349	70 163	31 645	33 901	11 165	39 718
1995	264 314	223 733	185 881	161 453	30 188	69 776	31 877	34 655	9 375	40 581
1996	266 792	225 699	188 224	164 096	28 419	69 000	31 451	35 227	8 712	41 093
1997	269 094	226 735	189 955	166 419	27 431	66 685	28 956	35 590	8 527	42 359
1998	271 743	228 800	192 507	170 105	26 165	66 087	27 854	35 887	8 747	42 943
1999	550 891	472 175	398 557	349 227	56 608	133 279	55 243	73 056	17 056	78 716
2000	279 517	242 932	205 575	181 862	28 432	68 183	28 062	37 787	8 937	36 585
2001	282 082	244 059	204 142	179 984	28 398	70 330	30 166	37 870	9 580	38 023
2002	285 933	246 157	204 163	179 563	29 287	72 825	31 934	38 359	9 892	39 776
2003	288 280	246 332	201 989	177 362	28 826	76 119	34 326	39 284	10 124	41 948
2004	291 166	249 414	203 014	177 924	29 161	79 480	38 055	39 757	10 584	41 752
2005	293 834	250 799	203 205	178 391	28 980	80 283	38 191	40 167	11 164	43 035
2006	296 824	251 610	203 942	178 880	29 033	80 343	38 370	40 336	10 543	45 214
2007	299 106	255 018	203 903	178 971	28 500	83 147	39 685	41 387	10 955	44 088
2008	301 483	256 702	202 626	177 543	28 513	87 586	42 831	43 031	11 562	44 781
2009	304 280	255 295	196 245	170 762	29 098	93 245	47 847	43 434	12 414	48 985
2010	306 553	256 603	196 147	169 372	30 347	95 525	48 533	44 906	12 927	49 950
2011	308 827	260 214	197 323	170 102	30 244	99 497	50 835	46 922	13 712	48 613
2012	311 116	263 165	198 812	170 877	30 622	101 493	50 903	48 884	13 702	47 951
2013	313 401	271 606	201 038	174 418	35 755	108 287	54 919	49 020	14 016	41 795
2014	316 168	283 200	208 600	175 027	46 165	115 470	61 650	50 546	14 143	32 968
2015	318 868	289 903	214 238	177 540	52 057	118 395	62 384	51 865	14 849	28 966
2016	320 372	292 320	216 203	178 455	51 961	119 361	62 303	53 372	14 638	28 052
2017	323 156	294 613	217 007	181 036	51 821	121 965	62 492	55 623	15 532	28 543
2018	323 668	296 206	217 780	178 350	34 846	111 330	57 819	57 720	11 754	27 462

. . . = Not available.

Table 3-12. Percentage of People by Type of Health Insurance by Selected Demographics, 2017–2018

(Number in thousands. Percent. People as of March of the following year.)

Characteristic	Total		Any health insurance		Private health insurance		Government health insurance		Uninsured	
	2017	2018	2017	2018	2017	2018	2017	2018	2017	2018
Total	323 156	323 668	91.2	91.5	67.2	67.3	37.7	34.4	8.8	8.5
Age										
Under age 19	78 106	77 333	94.6	94.5	63.3	61.8	41.9	35.7	5.4	5.5
Aged 19 to 64	193 971	193 548	87.8	88.5	72.9	73.5	21.3	17.6	12.2	11.7
Aged 65 and older	51 080	52 788	98.7	99.1	51.1	52.4	93.7	94.1	1.3	0.9
Marital Status										
Married	101 580	101 805	90.9	91.7	79.7	82.3	18.3	12.6	9.1	8.3
Widowed	3 586	3 385	86.6	86.3	57.2	55.6	36.0	34.9	13.4	13.7
Divorced	19 460	18 683	86.4	87.0	65.4	64.7	26.3	25.3	13.6	13.0
Separated	4 495	4 200	79.7	80.1	55.4	52.4	26.9	29.7	20.3	19.9
Never married	63 537	65 475	84.0	84.0	66.6	64.7	23.1	21.6	16.0	16.0
Disability Status										
With a disability	14 957	15 438	91.2	90.4	44.8	44.7	57.8	53.9	8.8	9.6
With no disability	178 063	195 434	87.5	88.5	75.5	74.9	17.8	16.0	12.5	11.5
Work Experience										
All workers	150 487	155 221	88.7	89.3	90.2	805.0	14.0	11.1	11.3	10.7
Worked full-time, year-round	109 511	111 950	90.2	90.5	84.4	85.1	10.9	7.2	9.8	9.5
Worked less than full-time, year-round	40 976	43 271	84.6	86.2	68.9	68.5	22.4	21.3	15.4	13.8
Did not work at least one week	43 484	55 573	84.9	86.9	47.9	51.3	46.5	40.2	15.1	13.1
Educational Attainment										
Total, 26 to 64 years old	164 049	164 250	88.1	88.7	73.4	74.2	20.9	17.5	11.9	11.3
No high school diploma	15 150	15 197	73.7	71.0	42.4	37.0	37.5	36.9	26.3	29.0
High school graduate (includes equivalency)	44 772	44 573	84.5	85.1	65.4	73.8	26.3	24.4	15.5	14.9
Some college, no degree	26 109	24 977	88.0	89.3	70.6	73.8	24.7	19.3	12.0	10.7
Associate degree	17 659	17 735	90.5	91.0	77.2	78.7	19.5	15.8	9.5	9.0
Bachelor's degree	38 465	39 255	2.8	93.8	85.5	87.6	12.4	8.5	7.2	6.2
Family Status										
In families	260 709	261 336	91.7	92.1	68.3	68.7	36.9	32.9	8.3	7.9
Householder	83 103	83 508	91.2	92.4	70.0	72.0	37.0	34.4	8.8	7.6
Related children under 18	72 532	71 750	94.7	94.7	63.4	62.0	41.8	35.6	5.3	5.3
Related children under 6	23 574	22 720	94.0	94.8	59.2	59.7	44.9	38.0	6.0	5.2
In unrelated subfamilies	1 054	1 069	87.7	86.9	52.5	50.0	45.3	42.4	12.3	13.1
Unrelated individuals	61 393	61 264	88.8	88.0	62.5	61.5	41.2	40.6	11.2	11.1
Residence										
Inside metropolitan statistical areas	280 049	281 369	91.2	91.6	68.0	68.1	36.6	33.2	8.8	8.4
Inside principal cities	104 068	104 716	89.6	90.4	63.1	63.4	38.2	35.3	10.4	9.6
Outside principal cities	175 980	176 653	92.2	92.3	70.8	70.9	35.6	32.0	7.8	7.7
Outside metropolitan statistical areas	43 108	42 300	90.8	90.9	61.9	62.0	45.5	42.4	9.2	9.1
Race and Hispanic Origin										
White	247 695	247 472	91.5	91.8	69.0	69.3	37.1	33.8	8.5	8.2
White, not Hispanic	195 530	194 679	93.7	94.6	73.2	74.8	36.6	33.2	6.3	5.4
Black	42 564	42 758	89.4	90.3	56.5	55.4	44.1	41.2	10.6	9.7
Asian	19 484	19 770	92.7	93.2	72.2	73.1	29.6	26.1	7.3	6.8
Hispanic (any race)	59 227	59 925	83.9	82.2	53.5	49.6	39.5	36.5	16.1	17.8
Race and Hispanic Origin										
Native born	277 748	277 848	92.5	93.2	68.2	69.1	38.7	34.9	7.5	6.8
Foreign born	45 408	45 820	83.2	81.1	60.6	56.0	32.0	31.2	16.8	18.9
Naturalized citizen	21 854	22 296	91.1	91.2	65.6	64.0	37.5	36.4	8.9	8.8
Not a citizen	23 554	23 524	75.9	71.4	55.9	48.4	27.0	26.2	24.1	28.6

. . . = Not available.

NOTES AND DEFINITIONS, CHAPTER 3

TABLES 3-1 THROUGH 3-12

Income Distribution and Poverty

SOURCE: U.S. DEPARTMENT OF COMMERCE, BUREAU OF THE CENSUS

All data in this chapter are derived from the Current Population Survey (CPS), which is also the source of the data on labor force, employment, and unemployment used in Chapter 10. (See the notes and definitions for Tables 10-1 through 10-5.) In March of each year (with some data also collected in February and April), the households in this monthly survey are asked additional questions concerning earnings and other income in the previous year. This additional information, informally known as the "March Supplement," is now formally known as the Current Population Survey Annual Social and Economic Supplement (CPS-ASEC). It was previously called the Annual Demographic Supplement.

Collected annually, CPS-ASEC is one of the most widely used socioeconomic surveys, and provides detailed information on income, work experience, poverty, health insurance coverage, and living arrangements in the U.S. For the past several years, the Census Bureau has engaged in implementing improvements to the CPS-ASEC, especially to improve data quality, using redesigned income questions. More information on these changes in the section *Notes on the Data.*

Researchers should use caution when comparing results over time. Due to the changes, comparison of variables between 2013 and 2018 are not comparable. The 2014 CPS ASEC included redesigned questions for income and health insurance coverage. All of the approximately 98,000 addresses were eligible to receive the redesigned set of health insurance coverage questions. The redesigned income questions were implemented to a subsample of these 98,000 addresses using a probability split panel design.

The population represented by the income and poverty survey is the civilian noninstitutional population of the United States and members of the armed forces in the United States living off post or with their families on post, but excluding all other members of the armed forces. This is slightly different from the population base for the civilian employment and unemployment data, which excludes those armed forces households. As it is a survey of households, homeless persons are not included.

TABLES 3-1 THROUGH 3-10

Definitions: Racial classification and Hispanic origin

In 2002 and all earlier years, the CPS required respondents to report identification with only one race group. Since 2003, the CPS has allowed respondents to choose more than one race group. Income data for 2002 were collected in early 2003; thus, in the data for 2002 and all subsequent years, an individual could report identification with more than one race group. In the 2010 census, about 2.9 percent of people reported identification with more than one race.

Therefore, data from 2002 onward that are classified by race are not strictly comparable with race-classified data for 2001 and earlier years. As alternative approaches to dealing with this problem, the Census Bureau has tabulated two different race concepts for each racial category in a number of cases. In the case of Blacks, for example, this means there is one income measure for "Black alone," consisting of persons who report Black and no other race, and one for "Black alone or in combination," which includes all the "Black alone" reporters plus those who report Black in combination with any other race. The tables in this volume show both the "alone" and the "alone or in combination" values where available.

The race classifications now used in the CPS are *White, Black, Asian, American Indian and Alaska Native,* and *Native Hawaiian and Other Pacific Islander.* (Native Hawaiians and other Pacific Islanders were included in the "Asian" category in the data for 1987 through 2001.) The last two of these five racial groups are too small to provide reliable data for a single year, but in new Census Bureau tables available on the website, household income and poverty data for all five groups are presented in 2- and 3-year averages. Table 3-1 displays some of these data.

Hispanic origin is a separate question in the survey—not a racial classification—and Hispanics may be of any race. A subgroup of *White non-Hispanic* is shown in some tables. According to the Census 2010 population results, "Being Hispanic was reported by 11.6 percent of White householders who reported only one race, 4.5 percent of Black householders who reported only one race, and 3.5 percent of Asian householders who reported only one race. Data users should exercise caution when interpreting aggregate results for the Hispanic population or for race groups because these populations consist of many distinct groups that differ in socioeconomic characteristics, culture, and recent immigration status." ("Income, and Poverty in the United States: 2017", footnote 2, p. 3 and "Health Insurance Coverage in the United States: 2017" footnote 2, p. 3.)

Definitions: General

A *household* consists of all persons who occupy a housing unit. A household includes the related family members and all the unrelated persons, if any (such as lodgers, foster children, wards, or employees), who share the housing unit. A person living alone in a housing unit or a group of unrelated persons sharing a housing unit as partners is also counted as a household. The count of households excludes group quarters.

Earnings includes all income from work, including wages, salaries, armed forces pay, commissions, tips, piece-rate payments, and cash bonuses, before deductions such as taxes, bonds, pensions, and union dues. This category also includes net income from nonfarm self-employment and farm self-employment. Wage and salary supplements that are paid directly by the employer, such as the employer share of Social Security taxes and the cost of employer-provided health insurance, are not included.

Income, in the official definition used in the survey, is money income, including *earnings* from work as defined above; unemployment compensation; workers' compensation; Social Security; Supplemental Security Income; cash public assistance (welfare payments); veterans' payments; survivor benefits; disability benefits; pension or retirement income; interest income; dividends (but not capital gains); rents, royalties, and payments from estates or trusts; educational assistance, such as scholarships or grants; child support; alimony; financial assistance from outside of the household; and other cash income regularly received, such as foster child payments, military family allotments, and foreign government pensions. Receipts not counted as income include capital gains or losses, withdrawals of bank deposits, money borrowed, tax refunds, gifts, and lump-sum inheritances or insurance payments.

A *year-round, full-time worker* is a person who worked 35 or more hours per week and 50 or more weeks during the previous calendar year.

A *family* is a group of two or more persons related by birth, marriage, or adoption who reside together.

Unrelated individuals are persons 15 years old and over who are not living with any relatives. In the official poverty measure, the poverty status of unrelated individuals is determined independently of and is not affected by the incomes of other persons with whom they may share a household.

Median income is the amount of income that divides a ranked income distribution into two equal groups, with half having incomes above the median, and half having incomes below the median. The median income for persons is based on persons 15 years old and over with income. Since median income is updated annually to inflation adjusted current dollars, comparison between editions is not possible.

Mean income is the amount obtained by dividing the total aggregate income of a group by the number of units in that group. In this survey, as in most surveys of incomes, means are higher than medians because of the skewed nature of the income distribution;

see the section "Whose standard of living?" in the article at the beginning of this book.

Historical income figures are shown in constant *2018 dollars*. All constant-dollar figures are converted from current-dollar values using the *CPI-U-RS* (the Consumer Price Index, All Urban, Research Series), which measures changes in prices for past periods using the methodologies of the current CPI, and is similar in concept and behavior to the deflators used in the NIPAs for consumer income and spending. See Chapter 8 for CPI-U-RS data and the corresponding notes and definitions.

Definitions: Income distribution

Income distribution is portrayed by dividing the total ranked distribution of families or households into *fifths*, also known as *quintiles,* and also by separately tabulating the top 5 percent (which is included in the highest fifth). The households or families are arrayed from those with the lowest income to those with the highest income, then divided into five groups, with each group containing one-fifth of the total number of households. Within each quintile, incomes are summed and calculated as a share of total income for all quintiles, and are averaged to show the average (mean) income within that quintile.

A statistical measure that summarizes the dispersion of income across the entire income distribution is the *Gini coefficient* (also known as Gini ratio, Gini index, or index of income concentration), which can take values ranging from 0 to 1. A Gini value of 1 indicates "perfect" inequality; that is, one household has all the income and the rest have none. A value of 0 indicates "perfect" equality, a situation in which each household has the same income.

A new "equivalence-adjusted" measure of household income inequality was introduced recently and is displayed in Table 3-3. For a Census-defined household, a given level of money income can have different implications for that household's well-being, depending on the size of the household and how many children, if any, are in the household. Since there have been substantial changes over past decades in the average size and composition of households, some have questioned the pertinence of standard income distribution tables. In response, the Census Bureau now also reports measures of income inequality for households using "equivalence-adjusted" income.

The equivalence adjustment is based on a three-parameter scale reflecting the size of the household and the facts that children consume less than adults; that as family size increases, expenses do not increase at the same rate; and that the increase in expenses is larger for the first child of a single-parent family than the first child of a two-adult family.

As can be seen in Table 3-3, the equivalence-adjusted measures generally show somewhat less inequality than the raw money income measures, but they show a greater rise in inequality over the period 1967-2018.

Definitions: Poverty

The *number of people below poverty level,* or the number of poor people, is the number of people with family or individual incomes below a specified level that is intended to measure the cost of a minimum standard of living. These minimum levels vary by size and composition of family and are known as *poverty thresholds.*

The official poverty thresholds are based on a definition developed in 1964 by Mollie Orshansky of the Social Security Administration. She calculated food budgets for families of various sizes and compositions, using an "economy food plan" developed by the U.S. Department of Agriculture (the cheapest of four plans developed). Reflecting a 1955 Department of Agriculture survey that found that families of three or more persons spent about one-third of their after-tax incomes on food, Orshansky multiplied the costs of the food plan by 3 to arrive at a set of thresholds for poverty income for families of three or larger. For 2-person families, the multiplier was 3.7; for 1-person families, the threshold was 80 percent of the two-person threshold.

These poverty thresholds have been adjusted each year for price increases, using the percent change in the Consumer Price Index for All Urban Consumers (CPI-U). See Chapter 8 for additional information on the Consumer Price Index.

For more information on the Orshansky thresholds (the description of which has been simplified here), see Gordon Fisher, "The Development of the Orshansky Thresholds and Their Subsequent History as the Official U.S. Poverty Measure" (May 1992), available on the Census Bureau Web site at https://www.census.gov/library/working-papers/1997/demo/fisher-02.html.

The *poverty rate* for a demographic group is the number of poor people or poor families in that group expressed as a percentage of the total number of people or families in the group.

Average poverty thresholds. The thresholds used to calculate the official poverty rates vary not only with the size of the family but with the number of children in the family. For example, the threshold for a three-person family in 2018 was $19,642 if there were no children in the family but $20,231 if the family consisted of 1 adults and 2 children. There are 48 different threshold values depending on size of household, number of children, and whether the householder is 65 years old or over (with lower thresholds for the older householders). To give a general sense of the "poverty line," the Census Bureau also publishes the average threshold for each size family, based on the actual mix of family types in that year. These are the values shown in Table 3-5 to represent the history of poverty thresholds.

A person with *work experience* (Table 3-9) is one who, during the preceding calendar year and on a part-time or full-time

basis, did any work for pay or profit or worked without pay on a family-operated farm or business at any time during the year). A *year-round* worker is one who worked for 50 weeks or more during the preceding calendar year. A person is classified as having worked *full time* if he or she worked 35 hours or more per week during a majority of the weeks worked. A *year-round, full-time worker* is a person who worked 35 or more hours per week and 50 or more weeks during the previous calendar year.

Working poor: Those workers employed full- or -part time whose income still falls below the poverty thresholds of their family size.

Toward better measures of income and poverty

The definition of the official poverty rate is established by the Office of Management of Budget in the Executive Office of the President and has not been substantially changed since 1969. Criticisms of the current definition are legion. In response to these criticisms, the Census Bureau has published extensive research work illustrating the effects of various ways of modifying income definitions and poverty thresholds. Some of the results of this work are published here in Tables 3-10 and 3-11 and explained in the notes and definitions below.

Improving the income concept

One type of criticism accepted the general concept of the Orshansky threshold but recommended making the income definition more realistic by including some or all of the following: capital gains; taxes and tax credits; noncash food, housing, and health benefits provided by government and employers; and the value of homeownership. There is debate about whether it is appropriate to use income data augmented in this way in conjunction with the official thresholds. The original 1964 thresholds made no allowance for health insurance or other health expenses—in effect, they assumed that the poor would get free medical care, or at least that the poverty calculation was not required to allow for medical needs—and because of the imprecision of Orshansky's multiplier it is not clear to what extent they include housing expenses in a way that is comparable with the inclusion of a homeownership component in income. Nevertheless, the Census Bureau has calculated and published income and poverty figures based on broadened income definitions and either the official thresholds or thresholds that are closely related to the official ones. Some of these calculations are presented in Table 3-10 and described below.

Still accepting the validity of the basic Orshansky threshold concept, some critics have also argued that use of the CPI-U in the official measure to update the thresholds each year has overstated the price increase, and that an inflator such as the CPI-U-RS should be used instead. (See the notes and definitions for Chapter 8.) Use of the CPI-U-RS leads to lower poverty thresholds beginning in the late 1970s, when the CPI began to be distorted by housing and other biases subsequently corrected by new methods; these newer methods were not carried backward to revise the official CPI-U. Use of the CPI-U-RS, which does carry current methods back and thereby revises the CPI time series, eliminates a presumed upward bias in the poverty rate relative to the poverty rates estimated before the bias emerged. This is a bias in the behavior of the time series given the concept of the Orshansky threshold, not necessarily a bias in the current level of poverty, since all the other criticisms of the Orshansky thresholds need to be considered when assessing the general adequacy of today's poverty measurements.

Improving the concepts of income and poverty together

Another type of criticism argues that the official thresholds are also no longer relevant to today's needs, and that the concepts of income (or "resources") and of the threshold level that depicts a minimum adequate standard of living need to be rethought together. These critics cite the availability of more up-to-date and detailed information about consumer spending at various income levels. The Consumer Expenditure Survey (CEX), originally designed to provide the weights for the Consumer Price Index, is now conducted quarterly and provides extensive data on consumer spending patterns. To give just one example of the information available now that was not available to Mollie Orshansky, the CEX indicates that food now accounts for one-sixth, not one-third, of total family expenditures, even among low-income families. (For data from, and notes on, the Consumer Expenditure Survey, see Bernan Press, *Handbook of U.S. Labor Statistics.*)

A special panel of the National Academy of Sciences (NAS) undertook a study that reconsidered both resources and thresholds. The Census Bureau has calculated and published experimental poverty rates for 1999 through 2015, developed following NAS recommendations, which are presented in Table 3-10. Beginning in November 2011 the Bureau has issued "The Research Supplemental Poverty Measure" or SPM for the most recent two years, a further refinement of the NAS recommendations; SPM measures are presented in Tables 3-10.

Supplemental Poverty Measure—Tables 3-10

In March of 2010, an Interagency Technical Working Group produced suggestions for a Supplemental Poverty Measure (SPM), which were first implemented in the "Research Supplemental Poverty Measure: 2010" issued by the Census Bureau in November 2011. The SPM goes beyond the experimental measures just described, and takes advantage of new data from questions introduced in the CPS-ASEC in recent years. In the new SPM:

- The measurement unit now covers not just the family but "all related individuals who live at the same address, including any co-resident unrelated children who are cared for by the family (such as foster children) and any co-habitors and their

children." This redefinition adds unrelated individuals under the age of 15 to the universe measured by the official rate. The "official" measures shown in Table 3-10 have been adjusted to include these individuals, and are therefore slightly higher than the regular published official rates shown in Tables 3-6 through 3-9.

- The SPM thresholds are calculated separately for three housing status groups: owners with mortgages, owners without mortgages, and renters. For each of these three groups the basic threshold represents the 33rd percentile of expenditures on food, clothing, shelter, and utilities (FCSU), averaged over a five-year period, by consumer units with two children, multiplied by 1.2 to allow for other needs. (Before averaging, the expenditures are converted to their dollar values in the prices of the threshold year using the CPI-U.) The consumer units selected for calculating the threshold include not only two-adult two-child families but also other types with two children, such as single-parent families. But they are adjusted to a four-person basis, using the equivalence scales, before averaging them to determine the basic threshold appropriate to a two-adult two-child household. Then that threshold is used as the basis for calculating thresholds for other size measurement units, again using the three-parameter equivalence scales for family size and composition, and for geographic differences.

- The thresholds are updated each year, using an updated five-year moving average of FCSU at the 33rd percentile, expressed in the prices of the new threshold year. Thus the thresholds are adjusted for inflation, but in addition, the "real" (constant-dollar) purchasing power of the poverty threshold will change (gradually) over time as the real standard of FCSU spending in the 33rd percentile changes. The real value of the threshold will likely change more slowly and gradually than in the "experimental" measures described above, because of using five instead of three years in the average.

- Family resources include cash income, plus in-kind benefits that families can use to meet their FCSU needs (for example, the Supplemental Nutrition Assistance Program [SNAP] formerly known as food stamps), minus income and payroll taxes, plus tax credits, minus childcare and other work-related expenses, minus child support payments to another household, and minus out-of-pocket medical expenses.

The SPM is not intended to replace the official poverty measure and is not to be used in calculations affecting program eligibility and funding distribution. According to the report referenced above, it is "designed to provide information on aggregate levels of economic need at a national level or within large subpopulations or areas...providing further understanding of economic conditions and trends." Census Bureau presentation of the SPM has focused on the different distribution of the poverty population indicated by the new measures, as shown in Table 3-10.

Notes on the data

The following are the principal changes that may affect year-to-year comparability of all income and poverty data from the CPS.

- Beginning in 1952, the estimates are based on 1950 census population controls. Earlier figures were based on 1940 census population controls.

- Beginning in 1962, 1960 census–based sample design and population controls are fully implemented.

- With 1971 and 1972 data, 1970 census–based sample design and population controls were introduced.

- With 1983–1985 data, 1980 census–based sample design was introduced; 1980 population controls were introduced; and these were extended back to 1979 data.

- With 1993 data, there was a major redesign of the CPS, including the introduction of computer-assisted interviewing. The limits used to "code" reported income amounts were changed, resulting in reporting of higher income values for the highest-income families and, consequently, an exaggerated year-to-year increase in income inequality. (It is possible that this jump actually reflects in one year an increase that had emerged more gradually, so that the distribution measures for 1993 and later years may be properly comparable with data for decades earlier even if they should not be directly compared with 1992.) In addition, 1990 census–based population controls were introduced, and these were extended back to the 1992 data.

- With 1995 data, the 1990 census–based sample design was implemented and the sample was reduced by 7,000 households.

- Data for 2001 implemented population controls based on the 2000 census, which were carried back to 2000 and 1999 data as well. Data from 2000 forward also incorporate results from a 28,000-household sample expansion.

- Beginning with the data for 2010 presented here, population controls based on the 2010 census were implemented. (This resulted in some revision of CPS data initially published for 2010.)

- In 2014, the Census Bureau introduced redesigned income and health insurance questions in the CPS-ASEC. The second phase of the implementation in 2019, updated the processing system that imputes missing data, determines family relationships and constructs key health insurance measures.

For more information on these and other changes that could affect comparability, see "Current Population Survey Technical Paper 63RV: Design and Methodology" (March 2002) and footnotes to CPS historical income tables, both available on the Census Bureau Web site at www.census.gov/hhes/income.

Data availability

Data embodying the official definitions of income and poverty are published annually in late summer or early fall by the Census Bureau, as part of a series with the general title *Current Population Reports: Consumer Income, P60*. Most of the data up to 2012 in this chapter were derived from report P60-245, "Income, Poverty, and Health Insurance Coverage in the United States: 2012" (September 2013), and from the "Historical Income Tables" and "Historical Poverty Tables" on the Census Web site (see below for the address). In 2013, the Census Bureau split the previous report in two. Health insurance coverage is now published separately from income and poverty. The historical tables can be found in the "CPS-ASEC" section under the "Data" category.

The SPM data in Tables 3-10 are from reports entitled "The Research Supplemental Poverty Measure." Reports for 2012, 2011, and 2010 respectively are numbered P60-247, P60-244, and P60-241, and are available on the Census Web site.

All these reports and related data, including historical tabulations, used in *Business Statistics* are available on the Census Bureau Web site at <http://www.census.gov>, under the general headings of "Income" and "Poverty."

References

Definitions and descriptions of the concepts and data of all series are provided in the source documents listed above and in the references contained therein.

TABLES 3-11 AND 3-12

Health Insurance Coverage

SOURCE: U.S. DEPARTMENT COMMERCE, BUREAU OF THE CENSUS

The Current Population Survey Annual Social and Economic Supplement (CPS ASEC) and the American Community Survey (ACS) is used to produce official estimates of income and poverty, and it serves as the most widely-cited source of estimates on health insurance coverage.

Due to questions of the validity of the health insurance data in previous reports, the Census Bureau implemented changes in 2014 to CPS ACS, including a complete redesign of the health insurance questions that replaced the existing questions in the CPS ASEC. Due the differences in measurement, health insurance estimates from calendar year 2013 are different from estimates in previous years.

Health insurance coverage in the CPS ASEC refers to comprehensive coverage during the calendar year. The American Community Survey (ACS) health insurance coverage status is at the time the individual is interviewed. Therefore, two uninsured measures are reported.

Since the passage of the Patient Protection and Affordable Care Act (ACA) in 2010, several provisions of the ACA have gone into effect at different times. For example, in 2010, the Young Adult Provision enabled adults under age 26 to remain as dependents on their parents' health insurance plans. In 2014 policy changes associated with the ACA provide the option for states to expand Medicaid to people whose income-to-poverty ratio full under a particular threshold. The decreases in the uninsured rates in 2013 and 2014 are consistent with what some of the provisions of the ACA intended.

Definitions

Health insurance coverage refers to comprehensive coverage during the calendar year for the civilian, noninstitutionalized population.

Private coverage is a plan provided through market forces.

Employment-based provided by an employer or union.

Direct-purchase: Coverage purchased directly from an insurance company or a federal or state marketplace or exchange.

TRICARE: Coverage through TRICARE, formerly known as Civilian Health and Medical Program of the Uniformed Services (e.g. healthcare.gov).

Government health insurance: Coverage provided by the federal government.

Medicare: Federal program that helps to pay health care costs for people aged 65 years and older and for certain people under age 65 with long-term disabilities.

Medicaid: Medicaid, the Children's Health Insurance Program (CHIP), and individual state health plans.

CHAMPVA or *VA*: Civilian Health and Medical Program of the Department of Veterans Affairs, as well as care provided by the Department of Veterans Affairs and the military.

Uninsured are people if, for an entire year, they were not covered by any type of health insurance. Additionally, people were considered uninsured if they only had coverage through the Indian Health Service (IHS).

Data availability

Detailed health insurance questions have been asked in the CPS since 1988 as part of a mandate to collect data on non-cash basis. However, as noted, comparing older results with newer ones is uncertain due to the changes in questions in 2013, 2014 and 2018.

CHAPTER 4: CONSUMER INCOME AND SPENDING

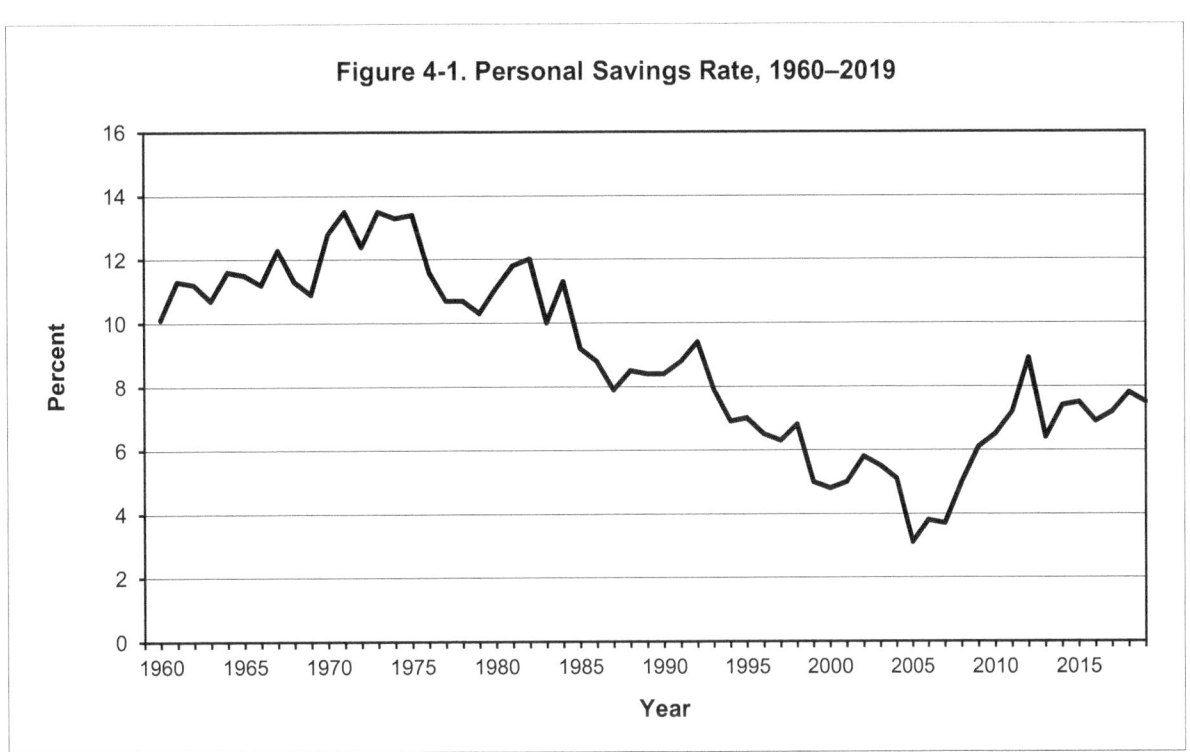

Figure 4-1. Personal Savings Rate, 1960–2019

- The personal saving rate—now defined to include saving in individual pension plans as the earned benefits rather than the employer contributions—got as high as 13.5 percent in the early 1970s. (Benefits from social insurance systems such as Social Security, unlike those in private and government employee retirement plans, do not figure in personal saving; any excess of receipts over payments is counted as government saving.) Since then the personal saving rate entered a downtrend, and hit the lowest rates since the Great Depression in the real estate bubble years 2005 through 2007, saving as a percent of disposable personal income has increased somewhat since then and reached a post-recession high of 8.9 percent in 2012. It has declined steadily since then. In 2019, personal savings as a percent of disposable personal income decreased to 7.5 percent after increasing in 2018. (Tables 4-1A and B)

- It should be noted that personal income does not, by definition, include any capital gains. Despite that, taxes on realized capital gains are deducted (along with all other income taxes) from personal income to get after-tax income. Capital gains, both realized and unrealized, can be a source of spending power in addition to current disposable income; during the housing price bubble they were converted into cash by asset sales, refinancing, and home equity loans, all of which were reflected in low saving rates. (Table 12-7)

- Government social benefits were 5.8 percent of personal income in 1960, rising to 14.1 percent in the high-employment year 2007. In the recession they rose further, to 18.2 percent in 2010, falling back somewhat to 16.6 percent by 2019. It should be understood that this category counts as income all spending financed by government health insurance programs such as Medicare and Medicaid, simultaneously counting that spending as personal consumption expenditures rather than direct government spending in the NIPAs. (Table 4-1)

Table 4-1A. Personal Income and Its Disposition: Recent Data

(Billions of current dollars, except as noted; quarterly data are at seasonally adjusted annual rates.)   NIPA Table 2.1

Year and quarter	Personal income				Personal income receipts on assets		Personal current transfer receipts					
							Government social benefits to persons					
	Total	Compen-sation of employees	Propri-etors' income with IVA and CCAdj	Rental income of persons with CCAdj	Personal interest income	Personal dividend income	Total	Social Security and Medicare	Govern-ment unem-ployment insurance	Veterans	Medicaid	Other government benefits to persons
1960	422.1	301.3	50.6	16.5	31.1	13.4	24.4	11.1	3.1	4.5	. . .	5.7
1961	440.6	310.4	53.2	17.2	33.4	13.9	28.1	12.6	4.4	4.9	. . .	6.2
1962	468.8	332.2	55.2	18.0	37.1	15.0	28.8	14.3	3.2	4.6	. . .	6.8
1963	492.8	350.4	56.4	18.7	40.5	16.2	30.3	15.2	3.1	4.8	. . .	7.3
1964	528.2	376.0	59.1	18.8	44.9	18.2	31.3	16.0	2.8	4.6	. . .	7.9
1965	570.7	405.4	63.7	19.3	49.3	20.2	33.9	18.1	2.4	4.8	. . .	8.6
1966	620.3	449.2	67.9	19.9	54.3	20.7	37.5	20.8	1.9	4.8	1.9	8.1
1967	665.7	481.8	69.5	20.3	59.5	21.5	45.8	25.8	2.2	5.5	2.7	9.4
1968	730.9	530.8	73.8	20.1	65.3	23.5	53.3	30.5	2.2	5.9	4.0	10.8
1969	800.3	584.5	77.0	20.3	76.1	24.2	59.0	33.1	2.3	6.6	4.6	12.4
1970	865.0	623.3	77.8	20.7	90.6	24.3	71.7	38.7	4.2	7.5	5.5	16.0
1971	932.8	665.0	83.9	21.8	100.1	25.0	85.4	44.6	6.2	8.4	6.7	19.4
1972	1 024.5	731.3	95.1	22.7	109.8	26.8	94.8	49.7	6.0	9.4	8.2	21.4
1973	1 140.8	812.7	112.5	23.1	125.5	29.9	108.6	60.9	4.6	10.2	9.6	23.3
1974	1 251.8	887.7	112.2	23.2	147.4	33.2	128.6	70.3	7.0	11.6	11.2	28.4
1975	1 369.4	947.2	118.2	22.3	168.0	32.9	163.1	81.5	18.1	14.0	13.9	35.7
1976	1 502.6	1 048.3	131.0	20.3	181.0	39.0	177.6	93.3	16.4	13.8	15.5	38.7
1977	1 659.2	1 165.8	144.5	15.9	206.9	44.7	189.5	105.3	13.1	13.3	16.7	40.9
1978	1 863.7	1 316.8	166.0	16.5	235.1	50.7	203.4	116.9	9.4	13.6	18.6	44.9
1979	2 082.7	1 477.2	179.4	16.1	269.7	57.4	227.3	132.5	9.7	14.2	21.1	49.9
1980	2 323.6	1 622.2	171.6	19.0	332.9	64.0	271.5	154.8	16.1	14.7	23.9	62.1
1981	2 605.1	1 792.5	179.7	23.8	412.2	73.6	307.8	182.1	15.9	15.8	27.7	66.3
1982	2 791.6	1 893.0	171.2	23.8	479.5	77.6	343.1	204.6	25.2	16.3	30.2	66.8
1983	2 981.1	2 012.5	186.3	24.4	516.3	83.3	370.5	222.2	26.4	16.4	33.9	71.5
1984	3 292.7	2 215.9	228.2	24.7	590.1	90.6	380.9	237.7	16.0	16.3	36.6	74.3
1985	3 524.9	2 387.3	241.1	26.2	628.9	97.4	403.1	253.0	15.9	16.6	39.7	78.0
1986	3 733.1	2 542.1	256.5	18.3	662.1	106.0	428.6	268.9	16.5	16.6	43.6	83.0
1987	3 961.6	2 722.4	286.5	16.6	679.0	112.2	447.9	282.6	14.6	16.5	47.8	86.4
1988	4 283.4	2 948.0	325.5	22.5	721.7	129.7	476.9	300.2	13.3	16.7	53.0	93.6
1989	4 625.6	3 139.6	341.1	21.5	806.5	157.8	521.1	325.6	14.4	17.2	60.8	103.1
1990	4 913.8	3 340.4	353.2	28.2	836.5	168.8	574.7	351.7	18.2	17.7	73.1	113.9
1991	5 084.9	3 450.5	354.2	38.6	823.5	180.2	650.5	381.7	26.8	18.1	96.9	127.0
1992	5 420.9	3 668.2	400.2	60.6	809.8	189.1	731.8	414.4	39.6	18.6	116.2	142.9
1993	5 657.9	3 817.3	428.0	90.1	802.3	204.7	778.9	444.7	34.8	19.3	130.1	150.0
1994	5 947.1	4 006.2	456.6	113.7	814.6	235.2	815.7	476.6	23.9	19.7	139.4	156.1
1995	6 291.4	4 198.1	481.2	124.9	878.6	258.0	864.7	508.9	21.7	20.5	149.6	164.0
1996	6 678.5	4 416.9	543.8	142.5	899.0	302.2	906.3	536.9	22.3	21.4	158.2	167.6
1997	7 092.5	4 708.8	584.0	147.1	947.1	337.9	935.4	563.5	20.1	22.3	163.1	166.4
1998	7 606.7	5 071.1	640.2	165.2	1 015.6	355.4	957.9	574.8	19.7	23.3	170.2	170.0
1999	8 001.9	5 402.8	696.4	178.5	1 012.7	346.6	992.2	588.6	20.5	24.1	184.6	174.4
2000	8 652.6	5 848.1	753.9	183.5	1 102.2	383.5	1 044.9	620.5	20.7	25.0	199.5	179.1
2001	9 005.6	6 039.1	831.0	202.4	1 104.3	369.3	1 145.8	667.7	31.9	26.6	227.3	192.4
2002	9 159.0	6 135.6	869.8	211.1	1 010.1	398.8	1 251.0	706.6	53.5	29.5	250.0	211.3
2003	9 487.5	6 354.1	896.9	231.5	1 005.0	432.1	1 321.0	740.2	53.2	31.8	264.5	231.2
2004	10 035.1	6 720.1	962.0	248.9	950.4	561.7	1 404.5	789.9	36.4	34.1	289.8	254.3
2005	10 598.2	7 066.6	978.0	232.0	1 100.4	577.8	1 490.9	844.8	31.8	36.4	304.4	273.5
2006	11 381.7	7 479.9	1 049.6	202.3	1 235.8	722.8	1 593.0	943.2	30.4	38.9	299.1	281.5
2007	12 007.8	7 878.9	994.0	184.4	1 368.6	815.3	1 697.3	1 003.9	32.7	41.7	324.2	294.9
2008	12 442.2	8 057.0	960.9	256.7	1 396.3	804.6	1 919.3	1 067.1	51.1	45.0	338.3	417.7
2009	12 059.1	7 758.5	938.5	327.3	1 299.3	553.0	2 107.7	1 157.5	131.2	51.5	369.6	398.0
2010	12 551.6	7 924.9	1 108.7	394.2	1 238.5	543.9	2 281.4	1 203.6	138.9	58.0	396.9	484.2
2011	13 326.8	8 225.9	1 229.3	478.6	1 269.4	681.5	2 310.1	1 248.9	107.2	63.3	406.0	484.8
2012	14 010.1	8 566.7	1 347.3	518.0	1 330.5	835.1	2 322.6	1 316.8	83.6	70.1	417.5	434.4
2013	14 181.1	8 834.2	1 403.6	557.0	1 273.0	793.3	2 385.9	1 371.8	62.5	79.1	440.0	432.5
2014	14 991.7	9 249.1	1 447.7	604.6	1 349.0	953.2	2 498.6	1 434.6	35.5	84.2	490.9	453.5
2015	15 724.2	9 699.4	1 423.0	649.0	1 439.1	1 033.1	2 635.1	1 506.7	32.5	92.6	535.9	467.4
2016	16 160.7	9 963.9	1 424.8	682.7	1 474.3	1 077.4	2 717.1	1 558.6	32.3	96.8	562.7	466.8
2017	16 948.6	10 422.5	1 509.0	721.9	1 577.6	1 160.8	2 806.2	1 618.4	30.3	111.5	573.7	472.4
2018	17 851.8	10 950.1	1 585.9	759.3	1 641.6	1 305.1	2 922.9	1 706.6	27.9	119.7	589.8	478.9
2019	18 551.5	11 432.4	1 657.7	787.1	1 677.4	1 290.4	3 078.0	1 814.4	27.7	130.9	614.0	490.9
2017												
1st quarter	16 633.7	10 223.1	1 493.4	709.0	1 551.7	1 101.6	2 778.8	1 596.0	30.9	105.9	572.4	473.7
2nd quarter	16 828.4	10 335.3	1 502.7	713.2	1 572.8	1 155.2	2 789.3	1 609.5	30.2	110.7	567.9	471.0
3rd quarter	17 036.6	10 476.9	1 507.8	725.9	1 585.4	1 175.9	2 822.2	1 625.3	30.3	113.9	578.8	473.9
4th quarter	17 295.6	10 654.9	1 532.1	739.6	1 600.7	1 210.7	2 834.7	1 642.6	29.8	115.4	575.8	471.0
2018												
1st quarter	17 548.6	10 776.1	1 557.7	745.3	1 619.2	1 255.6	2 890.9	1 676.8	29.2	118.0	581.8	485.0
2nd quarter	17 750.3	10 882.3	1 570.7	752.4	1 634.5	1 299.9	2 916.0	1 695.8	27.9	118.6	592.6	481.1
3rd quarter	17 976.5	11 034.1	1 588.4	768.2	1 650.1	1 321.8	2 935.4	1 716.4	27.4	119.9	595.1	476.6
4th quarter	18 132.0	11 107.8	1 627.0	771.2	1 662.5	1 343.0	2 949.4	1 737.4	27.0	122.3	589.8	472.8
2019												
1st quarter	18 366.7	11 335.3	1 627.5	776.6	1 652.0	1 298.7	3 037.5	1 786.3	28.0	127.3	599.4	496.5
2nd quarter	18 480.9	11 391.7	1 628.5	786.7	1 682.6	1 287.6	3 071.0	1 806.4	27.5	129.8	615.0	492.4
3rd quarter	18 597.6	11 438.0	1 677.0	789.7	1 681.7	1 288.3	3 094.8	1 824.2	27.6	131.9	622.3	488.9
4th quarter	18 760.8	11 564.8	1 697.7	795.5	1 693.4	1 287.0	3 108.7	1 840.9	27.9	134.5	619.4	486.0

. . . = Not available.

Table 4-1A. Personal Income and Its Disposition: Recent Data—*Continued*

(Billions of current dollars, except as noted; quarterly data are at seasonally adjusted annual rates.)   **NIPA Table 2.1**

Year and quarter	Personal income—Continued				Less: Personal outlays						Equals: Personal saving		Billions of chained (2009) dollars	
	Personal current transfer receipts—Continued: From business, net	Less: Contributions for government social insurance, domestic	Less: Personal current taxes	Equals: Disposable personal income	Total	Personal consumption expenditures	Personal interest payments	Personal current transfer payments			Billions of dollars	Percent of disposable personal income	Percent income excluding current transfers	Disposable personal income
								Total	To government	To the rest of the world, net				
1960	1.3	16.4	46.1	376.1	338.2	331.2	6.2	0.8	0.3	0.5	37.9	10.1	2 406.3	2 282.9
1961	1.4	17.0	47.3	393.3	348.9	341.5	6.5	1.0	0.5	0.5	44.4	11.3	2 470.1	2 363.2
1962	1.5	19.1	51.6	417.3	370.6	362.6	6.9	1.1	0.5	0.6	46.7	11.2	2 603.7	2 478.1
1963	1.9	21.7	54.6	438.2	391.1	382.0	7.9	1.2	0.5	0.7	47.1	10.7	2 702.8	2 571.7
1964	2.2	22.4	52.1	476.1	420.8	410.6	8.9	1.3	0.6	0.7	55.3	11.6	2 861.6	2 754.2
1965	2.3	23.4	57.7	513.0	454.2	443.0	9.9	1.4	0.6	0.8	58.8	11.5	3 047.8	2 925.3
1966	2.1	31.3	66.4	554.0	492.1	479.9	10.6	1.6	0.8	0.8	61.9	11.2	3 229.9	3 081.1
1967	2.3	34.9	73.0	592.8	519.8	506.7	11.1	2.0	1.0	1.0	73.0	12.3	3 351.4	3 216.1
1968	2.8	38.7	87.0	643.9	571.0	556.9	12.1	2.0	1.0	1.0	72.9	11.3	3 523.4	3 362.1
1969	3.3	44.1	104.5	695.8	619.8	603.6	13.9	2.2	1.1	1.1	76.1	10.9	3 687.4	3 476.5
1970	2.9	46.4	103.1	762.0	664.4	646.7	15.1	2.6	1.3	1.3	97.6	12.8	3 772.6	3 637.0
1971	2.7	51.2	101.7	831.1	719.2	699.9	16.4	2.8	1.5	1.4	111.9	13.5	3 867.5	3 805.2
1972	3.1	59.2	123.6	900.8	789.3	768.2	18.0	3.2	1.8	1.4	111.5	12.4	4 102.3	3 988.4
1973	3.9	75.5	132.4	1 008.4	872.6	849.6	19.6	3.4	1.8	1.6	135.8	13.5	4 319.9	4 236.5
1974	4.7	85.2	151.0	1 100.8	954.5	930.2	20.9	3.4	2.1	1.4	146.3	13.3	4 256.2	4 188.7
1975	6.8	89.3	147.6	1 221.8	1 057.8	1 030.5	23.4	3.8	2.5	1.3	164.0	13.4	4 212.8	4 291.4
1976	6.7	101.3	172.7	1 330.0	1 175.6	1 147.7	23.5	4.4	3.0	1.4	154.4	11.6	4 389.7	4 428.5
1977	5.1	113.1	197.9	1 461.4	1 305.4	1 274.0	26.6	4.8	3.5	1.4	155.9	10.7	4 579.1	4 568.8
1978	6.5	131.3	229.6	1 634.1	1 459.0	1 422.3	31.3	5.4	3.9	1.6	175.1	10.7	4 834.2	4 776.4
1979	8.2	152.7	268.9	1 813.8	1 627.0	1 585.4	35.5	6.0	4.3	1.7	186.8	10.3	4 958.6	4 869.1
1980	8.6	166.2	299.5	2 024.1	1 800.1	1 750.7	42.5	6.9	5.0	2.0	224.1	11.1	4 952.5	4 905.6
1981	11.2	195.7	345.8	2 259.3	1 993.9	1 934.0	48.4	11.5	6.0	5.6	265.5	11.8	5 085.0	5 025.4
1982	12.4	208.9	354.7	2 436.9	2 143.5	2 071.3	58.5	13.8	7.1	6.7	293.3	12.0	5 133.5	5 135.0
1983	13.8	226.0	352.9	2 628.2	2 364.2	2 281.6	67.4	15.1	8.1	7.0	264.0	10.0	5 248.7	5 312.2
1984	19.7	257.5	377.9	2 914.8	2 584.5	2 492.3	75.0	17.1	9.2	7.9	330.3	11.3	5 632.9	5 677.1
1985	22.3	281.4	417.8	3 107.1	2 822.1	2 712.8	90.6	18.8	10.4	8.3	284.9	9.2	5 833.4	5 847.6
1986	22.9	303.4	437.8	3 295.3	3 004.7	2 886.3	97.3	21.1	12.0	9.1	290.6	8.8	6 044.4	6 069.8
1987	20.2	323.1	489.6	3 472.0	3 196.6	3 076.3	97.1	23.2	13.2	10.0	275.4	7.9	6 242.4	6 204.1
1988	20.6	361.5	505.9	3 777.5	3 457.0	3 330.0	101.3	25.6	14.8	10.8	320.5	8.5	6 510.6	6 496.0
1989	23.2	385.2	567.7	4 057.8	3 717.9	3 576.8	113.1	28.0	16.5	11.6	340.0	8.4	6 724.9	6 686.2
1990	22.2	410.1	594.7	4 319.1	3 958.0	3 809.0	118.4	30.6	18.4	12.2	361.1	8.4	6 813.9	6 817.4
1991	17.6	430.2	588.9	4 496.0	4 100.0	3 943.4	119.9	36.7	22.6	14.1	396.0	8.8	6 746.0	6 867.0
1992	16.3	455.0	612.8	4 808.1	4 354.2	4 197.6	116.1	40.5	26.0	14.5	453.9	9.4	6 951.8	7 152.9
1993	14.1	477.4	648.8	5 009.2	4 611.5	4 452.0	113.9	45.6	28.6	17.1	397.7	7.9	7 061.7	7 271.1
1994	13.3	508.2	693.1	5 254.0	4 890.6	4 721.0	119.9	49.8	30.9	18.9	363.4	6.9	7 277.4	7 470.6
1995	18.7	532.8	748.4	5 543.0	5 155.9	4 962.6	140.4	52.9	32.6	20.3	387.1	7.0	7 530.8	7 718.9
1996	22.9	555.1	837.1	5 841.4	5 459.2	5 244.6	157.0	57.6	34.9	22.6	382.3	6.5	7 838.6	7 964.2
1997	19.4	587.2	931.8	6 160.7	5 770.4	5 536.8	169.7	63.9	38.2	25.7	390.3	6.3	8 224.9	8 255.8
1998	26.0	624.7	1 032.4	6 574.2	6 127.7	5 877.2	180.9	69.5	39.9	29.7	446.5	6.8	8 804.9	8 740.4
1999	34.0	661.3	1 111.9	6 890.0	6 542.9	6 279.1	187.5	76.3	44.1	32.3	347.1	5.0	9 137.9	9 025.6
2000	42.4	705.8	1 236.3	7 416.3	7 060.2	6 762.1	214.8	83.2	48.8	34.5	356.1	4.8	9 669.9	9 479.5
2001	46.8	733.2	1 239.0	7 766.5	7 377.2	7 065.6	220.0	91.5	52.9	38.6	389.4	5.0	9 798.3	9 740.1
2002	34.2	751.5	1 052.2	8 106.8	7 635.1	7 342.7	195.7	96.7	56.4	40.4	471.7	5.8	9 746.0	10 034.5
2003	26.3	779.3	1 003.5	8 484.0	8 015.1	7 723.1	190.9	101.1	60.5	40.6	468.9	5.5	9 884.0	10 301.4
2004	16.8	829.2	1 048.7	8 986.3	8 525.8	8 212.7	202.2	110.9	66.9	44.0	460.6	5.1	10 204.6	10 645.9
2005	25.8	873.3	1 212.5	9 385.8	9 095.7	8 747.1	230.5	118.1	71.6	46.5	290.1	3.1	10 461.1	10 811.6
2006	20.8	922.5	1 357.0	10 024.7	9 643.7	9 260.3	258.4	124.9	75.3	49.6	381.0	3.8	10 953.7	11 241.7
2007	30.8	961.4	1 492.5	10 515.3	10 129.4	9 706.4	284.6	138.4	79.2	59.2	385.9	3.7	11 242.3	11 499.9
2008	35.8	988.4	1 507.5	10 934.7	10 389.8	9 976.3	268.8	144.6	81.0	63.7	544.9	5.0	11 135.2	11 610.4
2009	39.0	964.3	1 152.4	10 906.7	10 240.5	9 842.2	254.0	144.3	80.1	64.3	666.2	6.1	10 534.5	11 591.3
2010	43.7	983.7	1 237.6	11 314.0	10 573.6	10 185.8	242.8	145.0	79.4	65.6	740.3	6.5	10 685.4	11 821.8
2011	48.5	916.7	1 453.7	11 873.1	11 024.0	10 641.1	232.1	150.8	82.2	68.6	849.1	7.2	11 177.0	12 099.3
2012	40.4	950.5	1 509.6	12 500.6	11 394.0	11 006.8	232.4	154.8	86.3	68.5	1 106.6	8.9	11 647.1	12 500.6
2013	38.4	1 104.3	1 676.4	12 504.7	11 705.0	11 317.2	229.5	158.3	87.5	70.8	799.7	6.4	11 600.6	12 338.6
2014	42.9	1 153.6	1 784.6	13 207.1	12 236.2	11 822.8	243.8	169.6	93.4	76.3	970.9	7.4	12 107.6	12 843.7
2015	50.3	1 204.7	1 939.9	13 784.3	12 745.6	12 297.5	264.7	183.5	100.9	82.5	1 038.7	7.5	12 653.7	13 377.2
2016	59.7	1 239.1	1 957.6	14 202.8	13 227.8	12 770.0	273.0	184.8	103.4	81.5	975.0	6.9	12 854.3	13 640.8
2017	48.8	1 298.4	2 046.7	14 931.9	13 830.9	13 340.4	297.3	193.3	107.1	86.1	1 071.0	7.2	13 297.8	14 060.5
2018	47.4	1 360.4	2 085.3	15 766.5	14 529.2	13 993.3	332.9	203.0	111.4	91.6	1 237.3	7.8	13 748.8	14 566.4
2019	47.2	1 418.8	2 202.9	16 348.6	15 117.4	14 544.6	362.3	210.5	115.2	95.3	1 231.2	7.5	14 042.9	14 882.5
2017														
1st quarter	52.9	1 276.8	2 001.1	14 632.7	13 625.1	13 153.2	284.0	187.9	105.7	82.2	1 007.6	6.9	13 092.8	13 880.7
2nd quarter	49.2	1 289.3	2 005.6	14 822.8	13 728.1	13 241.3	293.1	193.8	106.6	87.2	1 094.7	7.4	13 241.7	14 030.1
3rd quarter	47.0	1 304.5	2 052.3	14 984.2	13 867.0	13 370.9	302.0	194.1	107.6	86.5	1 117.3	7.5	13 355.3	14 125.4
4th quarter	46.2	1 323.2	2 127.9	15 167.8	14 103.4	13 596.0	310.1	197.3	108.7	88.6	1 064.3	7.0	13 500.5	14 205.8
2018														
1st quarter	46.9	1 343.1	2 085.6	15 463.0	14 274.1	13 755.5	318.3	200.2	109.8	90.4	1 188.8	7.7	13 593.8	14 386.7
2nd quarter	47.3	1 352.8	2 064.4	15 685.9	14 467.9	13 939.9	325.9	202.0	110.9	91.1	1 218.0	7.8	13 681.9	14 513.6
3rd quarter	47.6	1 369.1	2 100.5	15 876.1	14 628.2	14 086.3	338.6	203.3	111.9	91.4	1 247.9	7.9	13 819.2	14 632.6
4th quarter	47.8	1 376.8	2 090.7	16 041.3	14 746.8	14 191.4	348.9	206.5	112.9	93.6	1 294.5	8.1	13 899.8	14 732.3
2019														
1st quarter	47.7	1 408.7	2 170.7	16 196.0	14 841.5	14 276.6	355.1	209.8	114.2	95.6	1 354.5	8.4	14 014.8	14 853.5
2nd quarter	47.5	1 414.7	2 222.5	16 258.4	15 072.3	14 497.3	364.7	210.3	114.9	95.4	1 186.1	7.3	14 001.1	14 817.8
3rd quarter	47.1	1 419.0	2 197.1	16 400.5	15 219.9	14 645.3	364.9	209.7	115.5	94.2	1 180.6	7.2	14 037.4	14 895.4
4th quarter	46.5	1 432.9	2 221.2	16 539.6	15 335.8	14 759.2	364.6	212.0	116.0	96.0	1 203.8	7.3	14 119.5	14 964.5

Table 4-1B. Personal Income and Its Disposition: Historical Data

(Billions of current dollars, except as noted; quarterly data are at seasonally adjusted annual rates.)   **NIPA Table 2.1**

Year and quarter	Personal income							Less: Personal current taxes	Equals: Disposable personal income	Less: Personal outlays	Equals: Personal saving		Billions of chained (2009) dollars	
	Total	Compensation of employees	Proprietors' income with IVA and CCAdj	Rental income of persons with CCAdj	Personal income receipts on assets	Personal current transfer receipts	Less: Contributions for government social insurance, domestic				Billions of dollars	Percent of disposable personal income	Personal income excluding current transfers	Disposable personal income
1929	85.3	51.4	14.0	6.1	12.7	1.2	0.1	1.7	83.6	79.6	4.0	4.7	902.9	897.1
1930	76.5	47.2	10.9	5.4	12.0	1.2	0.1	1.6	75.0	71.6	3.3	4.5	844.3	840.4
1931	65.7	40.1	8.3	4.4	10.6	2.3	0.1	1.0	64.7	61.8	2.8	4.4	795.0	811.9
1932	50.3	31.3	5.0	3.6	8.7	1.7	0.1	0.7	49.6	49.7	-0.1	-0.2	691.2	705.7
1933	47.2	29.8	5.3	2.9	7.7	1.7	0.1	0.8	46.4	46.8	-0.3	-0.7	672.3	685.3
1934	54.1	34.6	7.0	2.5	8.3	1.8	0.1	0.9	53.2	52.3	0.9	1.7	739.1	751.1
1935	60.8	37.7	10.1	2.6	8.6	2.0	0.1	1.1	59.8	56.8	3.0	5.1	811.3	823.8
1936	69.2	43.3	10.4	2.7	10.1	3.1	0.3	1.3	67.9	63.1	4.8	7.1	903.2	927.6
1937	74.7	48.3	12.5	3.0	10.4	2.0	1.5	1.9	72.8	67.9	4.9	6.7	958.3	959.0
1938	69.1	45.4	10.6	3.5	8.8	2.4	1.6	1.9	67.2	65.3	2.0	2.9	899.1	906.4
1939	73.6	48.6	11.1	3.7	9.5	2.5	1.8	1.5	72.1	68.2	3.9	5.4	968.2	982.2
1940	79.4	52.7	12.2	3.8	9.8	2.7	1.9	1.7	77.7	72.4	5.3	6.8	1 035.9	1 049.3
1941	97.9	66.2	16.7	4.4	10.3	2.7	2.3	2.3	95.6	82.3	13.3	13.9	1 210.3	1 214.7
1942	126.7	88.0	23.3	5.5	10.1	2.7	2.9	4.9	121.8	90.0	31.9	26.2	1 402.3	1 377.1
1943	156.2	112.7	28.2	6.0	10.5	2.5	3.8	16.7	139.5	100.8	38.6	27.7	1 590.9	1 443.4
1944	169.7	124.3	29.3	6.3	11.0	3.1	4.3	17.7	152.1	109.7	42.4	27.9	1 630.8	1 488.3
1945	175.8	126.3	30.8	6.6	11.8	5.6	5.3	19.4	156.3	121.2	35.2	22.5	1 601.4	1 471.5
1946	182.5	122.5	35.7	6.9	13.5	10.5	6.6	17.2	165.3	145.8	19.6	11.8	1 513.4	1 454.7
1947	194.5	132.4	34.6	6.9	15.4	10.8	5.6	19.8	174.7	163.7	11.0	6.3	1 467.3	1 395.7
1948	213.5	144.3	39.3	7.5	16.8	10.3	4.6	19.2	194.3	177.1	17.2	8.8	1 536.8	1 469.1
1949	211.1	144.3	34.7	7.8	17.9	11.2	4.9	16.7	194.3	180.7	13.6	7.0	1 523.1	1 481.0
1950	233.7	158.3	37.5	8.8	20.7	14.0	5.5	18.9	214.8	194.8	20.0	9.3	1 655.1	1 617.9
1951	264.2	185.7	42.6	9.7	21.4	11.4	6.6	27.1	237.2	211.3	25.9	10.9	1 782.7	1 672.4
1952	282.5	201.1	43.0	10.8	22.6	11.9	6.9	32.0	250.5	222.6	27.8	11.1	1 869.6	1 730.5
1953	299.2	215.2	42.0	12.0	24.5	12.5	7.1	33.2	266.0	236.8	29.2	11.0	1 955.3	1 814.0
1954	302.2	214.1	42.3	13.1	26.5	14.3	8.1	30.2	272.1	243.9	28.1	10.3	1 947.0	1 839.9
1955														
1st quarter	313.5	221.9	43.7	13.3	28.2	15.3	8.9	31.4	282.1	255.9	26.2	9.3	2 015.9	1 906.9
2nd quarter	321.1	228.1	44.3	13.4	28.7	15.6	9.0	32.4	288.6	261.2	27.5	9.5	2 062.7	1 949.1
3rd quarter	328.5	233.9	44.6	13.4	29.8	15.9	9.2	33.4	295.1	265.6	29.5	10.0	2 102.7	1 985.1
4th quarter	333.6	238.4	44.5	13.5	30.4	16.0	9.3	34.2	299.4	269.8	29.6	9.9	2 130.7	2 008.2
1956														
1st quarter	339.0	242.8	44.9	13.6	31.2	16.3	9.9	35.4	303.6	271.6	32.1	10.6	2 155.6	2 028.5
2nd quarter	344.8	247.5	45.3	13.6	31.8	16.6	10.0	36.2	308.6	274.3	34.2	11.1	2 178.2	2 047.7
3rd quarter	350.1	250.4	46.4	13.7	32.5	17.1	10.0	36.9	313.2	277.6	35.5	11.3	2 188.6	2 058.3
4th quarter	357.8	256.4	46.8	13.8	33.6	17.3	10.1	37.9	319.9	283.1	36.8	11.5	2 224.2	2 090.1
1957														
1st quarter	362.4	260.2	46.9	13.9	34.5	18.2	11.4	38.6	323.8	287.7	36.1	11.1	2 228.6	2 096.7
2nd quarter	367.1	262.1	47.7	14.0	35.4	19.4	11.4	39.0	328.1	290.1	38.1	11.6	2 237.2	2 111.5
3rd quarter	371.7	264.7	48.8	14.1	35.9	19.6	11.5	39.2	332.5	294.7	37.8	11.4	2 247.0	2 122.2
4th quarter	370.8	263.3	47.7	14.3	36.0	20.8	11.3	38.8	332.0	296.4	35.6	10.7	2 222.2	2 108.0
1958														
1st quarter	371.3	259.8	50.3	14.6	35.9	22.1	11.3	38.2	333.1	295.9	37.2	11.2	2 190.1	2 089.4
2nd quarter	373.7	259.4	50.3	14.8	36.6	23.9	11.3	37.7	336.0	298.8	37.2	11.1	2 190.2	2 103.9
3rd quarter	382.6	267.2	50.1	14.9	37.6	24.2	11.4	38.9	343.7	303.8	39.9	11.6	2 242.4	2 150.1
4th quarter	388.2	272.3	49.9	15.0	38.9	23.7	11.5	39.4	348.8	307.8	41.0	11.8	2 281.6	2 183.1
1959														
1st quarter	394.0	279.7	50.1	15.0	38.8	24.0	13.7	40.8	353.2	315.4	37.9	10.7	2 299.9	2 195.8
2nd quarter	402.4	286.5	50.3	15.4	40.1	23.9	13.9	42.0	360.4	321.6	38.8	10.8	2 343.0	2 231.2
3rd quarter	404.7	286.9	50.4	15.8	41.4	24.2	13.9	42.7	362.1	327.1	35.0	9.7	2 342.0	2 228.2
4th quarter	410.4	290.2	50.6	16.0	42.6	24.8	13.9	43.7	366.6	329.4	37.2	10.2	2 359.7	2 244.1
1960														
1st quarter	416.7	298.7	49.7	16.2	43.7	24.6	16.4	45.3	371.4	333.1	38.3	10.3	2 396.2	2 270.0
2nd quarter	421.7	301.7	50.7	16.4	44.1	25.3	16.5	46.0	375.7	339.2	36.5	9.7	2 410.8	2 284.6
3rd quarter	424.4	302.6	50.9	16.5	44.9	26.0	16.5	46.5	377.8	339.3	38.6	10.2	2 413.1	2 289.0
4th quarter	425.8	302.1	51.0	16.7	45.3	27.0	16.4	46.4	379.4	341.3	38.0	10.0	2 405.1	2 288.0
1961														
1st quarter	430.2	303.1	52.3	16.9	45.7	28.8	16.7	46.5	383.7	341.9	41.7	10.9	2 416.1	2 309.7
2nd quarter	436.4	307.3	52.6	17.1	46.6	29.7	16.9	46.9	389.4	346.8	42.6	10.9	2 448.6	2 344.7
3rd quarter	443.4	312.3	53.3	17.3	47.7	29.9	17.1	47.4	396.0	349.8	46.2	11.7	2 481.1	2 375.9
4th quarter	452.3	318.9	54.5	17.5	49.3	29.4	17.3	48.1	404.2	357.2	47.0	11.6	2 534.5	2 422.3
1962														
1st quarter	459.3	325.5	55.3	17.7	49.7	30.0	18.9	49.4	409.9	362.5	47.4	11.6	2 561.9	2 446.1
2nd quarter	466.8	331.4	55.0	18.0	51.5	30.0	19.1	50.9	415.8	368.4	47.5	11.4	2 597.0	2 472.5
3rd quarter	471.5	334.2	54.9	18.2	53.0	30.4	19.2	52.3	419.2	372.5	46.7	11.1	2 615.7	2 485.8
4th quarter	477.8	337.7	55.8	18.3	54.1	31.2	19.3	53.6	424.2	379.0	45.2	10.7	2 640.5	2 507.9
1963														
1st quarter	483.2	342.7	56.0	18.5	54.7	32.6	21.3	54.1	429.1	382.9	46.2	10.8	2 656.2	2 529.7
2nd quarter	488.1	347.6	55.8	18.7	55.9	31.7	21.5	54.3	433.7	387.3	46.4	10.7	2 686.1	2 553.0
3rd quarter	495.2	352.6	56.2	18.8	57.3	32.0	21.8	54.6	440.6	394.7	45.9	10.4	2 713.5	2 580.9
4th quarter	504.6	358.8	57.6	18.8	58.9	32.5	22.0	55.2	449.4	399.6	49.8	11.1	2 755.2	2 622.9

Table 4-1B. Personal Income and Its Disposition: Historical Data—*Continued*

(Billions of current dollars, except as noted; quarterly data are at seasonally adjusted annual rates.) NIPA Table 2.1

Year and quarter	Personal income							Less: Personal current taxes	Equals: Disposable personal income	Less: Personal outlays	Equals: Personal saving		Billions of chained (2009) dollars	
	Total	Compensation of employees	Proprietors' income with IVA and CCAdj	Rental income of persons with CCAdj	Personal income receipts on assets	Personal current transfer receipts	Less: Contributions for government social insurance, domestic				Billions of dollars	Percent of disposable personal income	Personal income excluding current transfers	Disposable personal income
1964														
1st quarter	514.1	365.2	57.9	18.8	60.7	33.6	22.0	53.8	460.4	409.3	51.1	11.1	2 791.5	2 674.3
2nd quarter ...	523.6	372.6	58.8	18.8	62.4	33.2	22.3	49.7	473.8	417.6	56.3	11.9	2 842.1	2 746.4
3rd quarter	533.2	380.0	59.3	18.9	64.0	33.5	22.5	51.6	481.6	426.8	54.7	11.4	2 886.6	2 782.2
4th quarter	541.9	386.1	60.5	18.8	65.4	33.7	22.7	53.2	488.7	429.6	59.0	12.1	2 926.1	2 813.8
1965														
1st quarter	553.0	392.9	61.9	19.1	66.9	35.0	22.9	57.0	496.0	440.5	55.5	11.2	2 972.9	2 847.0
2nd quarter	562.6	399.9	63.2	19.3	68.7	34.6	23.2	58.4	504.1	447.9	56.3	11.2	3 014.9	2 879.1
3rd quarter	576.9	408.3	64.0	19.5	70.5	38.1	23.6	57.0	519.9	457.3	62.6	12.0	3 065.1	2 957.8
4th quarter	590.2	420.5	65.6	19.5	71.8	36.9	24.0	58.3	532.0	471.4	60.6	11.4	3 138.6	3 017.4
1966														
1st quarter	602.9	433.1	69.2	19.8	73.3	37.8	30.4	61.5	541.4	481.9	59.5	11.0	3 180.5	3 047.2
2nd quarter ...	612.8	444.4	67.2	19.8	74.4	37.9	30.8	65.6	547.2	487.3	59.9	11.0	3 209.9	3 055.3
3rd quarter	626.0	455.7	67.2	20.0	75.5	39.6	31.9	67.9	558.2	496.6	61.6	11.0	3 249.2	3 092.6
4th quarter	639.6	463.8	68.1	20.0	76.7	43.1	32.2	70.6	569.1	502.6	66.5	11.7	3 279.6	3 128.9
1967														
1st quarter	650.5	470.0	68.9	20.3	78.9	46.1	33.6	71.2	579.3	507.1	72.2	12.5	3 313.6	3 176.0
2nd quarter	657.6	475.6	68.5	20.4	80.5	47.2	34.6	70.9	586.7	516.8	69.9	11.9	3 330.7	3 201.2
3rd quarter	671.6	485.3	70.5	20.4	81.9	48.7	35.2	73.8	597.8	523.8	74.0	12.4	3 367.5	3 231.7
4th quarter	683.2	496.2	70.0	20.3	82.6	50.0	36.0	75.9	607.3	531.4	75.9	12.5	3 393.8	3 255.3
1968														
1st quarter	702.5	510.9	71.7	20.1	85.1	52.4	37.6	78.6	624.0	549.8	74.2	11.9	3 448.2	3 309.6
2nd quarter	722.4	524.0	73.2	20.1	87.6	55.9	38.4	81.7	640.7	563.9	76.8	12.0	3 499.2	3 364.0
3rd quarter	741.3	537.8	75.0	20.2	90.0	57.3	39.1	91.9	649.4	580.6	68.8	10.6	3 554.2	3 374.8
4th quarter	757.5	550.3	75.5	20.0	92.7	58.7	39.7	95.9	661.7	589.8	71.9	10.9	3 591.9	3 400.9
1969														
1st quarter	772.4	562.8	75.6	20.2	96.2	60.4	42.9	102.6	669.8	602.3	67.4	10.1	3 623.9	3 409.3
2nd quarter	790.6	576.7	77.1	20.3	98.7	61.5	43.7	105.7	685.0	614.3	70.6	10.3	3 664.5	3 442.7
3rd quarter	811.6	593.4	78.0	20.4	101.5	62.9	44.6	104.1	707.5	625.0	82.5	11.7	3 717.6	3 512.9
4th quarter	826.8	604.9	77.3	20.4	105.0	64.3	45.3	105.6	721.1	637.4	83.7	11.6	3 742.4	3 539.7
1970														
1st quarter	841.7	615.0	77.1	20.4	109.2	66.1	46.0	104.6	737.2	648.9	88.3	12.0	3 763.6	3 576.9
2nd quarter	860.0	620.2	76.7	20.2	113.1	76.1	46.3	105.5	754.5	659.1	95.4	12.6	3 762.3	3 621.0
3rd quarter	874.7	628.2	78.5	20.9	117.5	76.3	46.7	100.7	774.0	671.3	102.7	13.3	3 794.4	3 678.9
4th quarter	883.7	630.0	78.9	21.2	120.0	80.1	46.5	101.5	782.2	678.2	103.9	13.3	3 770.6	3 670.1
1971														
1st quarter	903.7	647.9	80.3	21.1	122.8	82.0	50.5	98.3	805.3	697.6	107.7	13.4	3 819.1	3 743.3
2nd quarter ...	927.1	659.8	82.8	21.7	124.2	89.6	51.0	100.7	826.4	712.1	114.2	13.8	3 849.0	3 797.8
3rd quarter	941.1	670.1	84.6	22.0	126.1	89.6	51.3	102.3	838.8	725.1	113.7	13.6	3 875.6	3 817.7
4th quarter	959.3	682.2	87.9	22.5	127.4	91.3	51.9	105.5	853.8	741.8	112.1	13.1	3 926.5	3 862.2
1972														
1st quarter	986.5	708.4	87.9	23.1	131.0	94.3	58.1	119.8	866.8	759.4	107.4	12.4	3 993.7	3 879.9
2nd quarter	1 004.2	722.5	91.3	20.2	134.1	94.9	58.8	123.4	880.8	778.4	102.4	11.6	4 046.9	3 920.2
3rd quarter	1 030.1	735.7	95.5	23.9	138.4	96.0	59.5	124.3	905.8	797.2	108.7	12.0	4 121.0	3 996.4
4th quarter	1 077.0	758.8	105.6	23.8	142.8	106.4	60.4	127.1	949.9	822.2	127.6	13.4	4 247.3	4 156.6
1973														
1st quarter	1 095.1	785.4	104.1	23.5	146.6	109.1	73.6	126.4	968.7	847.0	121.7	12.6	4 263.1	4 188.1
2nd quarter	1 124.9	803.2	109.9	23.3	151.6	111.5	74.7	129.2	995.7	863.0	132.7	13.3	4 298.7	4 223.5
3rd quarter	1 152.5	820.4	113.8	22.3	158.8	113.3	76.1	134.1	1 018.4	881.8	136.6	13.4	4 329.2	4 242.6
4th quarter	1 190.7	841.7	122.2	23.3	164.6	116.5	77.6	140.0	1 050.7	898.4	152.3	14.5	4 384.9	4 289.0
1974														
1st quarter	1 206.8	858.3	115.7	23.5	170.5	121.9	83.1	142.8	1 064.0	915.3	148.7	14.0	4 300.7	4 217.8
2nd quarter	1 232.9	878.5	108.6	22.9	177.6	129.9	84.7	148.9	1 084.0	944.4	139.6	12.9	4 251.8	4 178.5
3rd quarter	1 269.1	900.2	111.0	23.3	183.9	137.1	86.4	154.9	1 114.2	973.9	140.2	12.6	4 249.1	4 182.3
4th quarter	1 298.5	913.8	113.5	23.0	190.4	144.3	86.6	157.6	1 140.9	984.3	156.5	13.7	4 225.4	4 176.7
1975														
1st quarter	1 318.4	918.4	112.7	22.7	196.4	155.9	87.6	158.0	1 160.4	1 011.8	148.6	12.8	4 177.9	4 170.5
2nd quarter	1 350.0	931.2	114.3	22.4	198.5	171.5	88.0	121.1	1 228.9	1 040.4	188.5	15.3	4 184.1	4 363.4
3rd quarter	1 386.5	956.0	120.7	22.2	202.5	174.8	89.8	152.8	1 233.6	1 074.8	158.9	12.9	4 223.3	4 299.9
4th quarter	1 422.7	983.4	125.2	21.9	206.5	177.6	91.8	158.5	1 264.2	1 104.0	160.1	12.7	4 268.1	4 333.5
1976														
1st quarter	1 456.1	1 014.7	126.4	21.6	210.5	181.8	98.9	162.5	1 293.6	1 137.3	156.3	12.1	4 320.6	4 385.9
2nd quarter ...	1 481.2	1 035.5	128.6	20.5	217.1	180.0	100.4	169.3	1 311.9	1 157.1	154.8	11.8	4 374.8	4 410.8
3rd quarter	1 518.4	1 058.4	132.7	20.0	222.8	186.7	102.2	176.1	1 342.3	1 186.9	155.4	11.6	4 410.1	4 445.2
4th quarter	1 554.9	1 084.8	136.2	19.1	229.7	188.8	103.8	182.7	1 372.1	1 221.1	151.0	11.0	4 453.5	4 473.4
1977														
1st quarter	1 588.8	1 113.3	138.6	16.4	238.8	191.0	109.3	188.8	1 400.0	1 258.2	141.9	10.1	4 476.5	4 483.5
2nd quarter	1 633.0	1 150.3	140.5	16.3	246.6	191.4	112.1	195.7	1 437.3	1 287.0	150.4	10.5	4 539.0	4 525.6
3rd quarter	1 677.7	1 182.1	142.4	15.6	255.4	196.6	114.3	198.6	1 479.2	1 318.8	160.4	10.8	4 594.6	4 588.5
4th quarter	1 737.4	1 217.6	156.5	15.1	265.5	199.3	116.7	208.5	1 528.9	1 357.8	171.1	11.2	4 704.1	4 676.2

Table 4-1B. Personal Income and Its Disposition: Historical Data—*Continued*

(Billions of current dollars, except as noted; quarterly data are at seasonally adjusted annual rates.)　　　　　　　NIPA Table 2.1

| Year and quarter | Personal income | | | | | | | Less: Personal current taxes | Equals: Disposable personal income | Less: Personal outlays | Equals: Personal saving | | Billions of chained (2009) dollars | |
	Total	Compensation of employees	Proprietors' income with IVA and CCAdj	Rental income of persons with CCAdj	Personal income receipts on assets	Personal current transfer receipts	Less: Contributions for government social insurance, domestic				Billions of dollars	Percent of disposable personal income	Personal income excluding current transfers	Disposable personal income
1978														
1st quarter	1 778.1	1 250.9	158.5	15.9	273.3	203.3	123.9	212.0	1 566.1	1 388.5	177.6	11.3	4 738.9	4 712.9
2nd quarter ...	1 838.2	1 299.1	166.1	15.6	281.3	205.1	129.0	223.1	1 615.1	1 447.5	167.6	10.4	4 815.7	4 762.7
3rd quarter	1 892.6	1 336.9	168.3	17.1	289.8	213.9	133.4	236.3	1 656.3	1 479.8	176.5	10.7	4 864.7	4 799.8
4th quarter	1 946.0	1 380.2	170.9	17.5	298.9	217.2	138.8	247.2	1 698.7	1 520.2	178.5	10.5	4 916.7	4 831.4
1979														
1st quarter	2 005.4	1 422.5	180.1	17.3	308.8	222.7	146.0	253.6	1 751.8	1 556.7	195.0	11.1	4 976.1	4 889.8
2nd quarter ...	2 045.4	1 454.7	179.2	15.6	318.8	227.4	150.3	262.0	1 783.4	1 598.3	185.1	10.4	4 939.6	4 845.6
3rd quarter	2 110.9	1 496.7	180.7	15.4	330.4	243.0	155.4	274.8	1 836.0	1 653.9	182.1	9.9	4 952.0	4 867.6
4th quarter	2 169.0	1 534.9	177.8	16.1	350.4	249.1	159.4	285.2	1 883.8	1 698.9	184.9	9.8	4 969.6	4 876.2
1980														
1st quarter	2 231.9	1 573.6	167.6	15.8	377.5	259.4	161.9	284.8	1 947.1	1 751.3	195.8	10.1	4 956.9	4 893.1
2nd quarter ...	2 269.4	1 599.2	160.3	17.1	391.2	264.6	162.9	292.2	1 977.3	1 753.6	223.6	11.3	4 917.8	4 850.1
3rd quarter	2 348.5	1 628.6	173.4	19.9	397.0	296.6	167.0	302.2	2 046.3	1 812.2	234.1	11.4	4 918.0	4 904.5
4th quarter	2 444.7	1 687.6	185.0	23.2	421.9	300.0	173.0	318.9	2 125.8	1 883.0	242.9	11.4	5 016.2	4 972.1
1981														
1st quarter	2 508.9	1 739.6	188.2	24.0	440.5	306.5	189.9	330.9	2 178.0	1 941.4	236.6	10.9	5 020.8	4 965.1
2nd quarter ...	2 559.8	1 774.8	176.4	23.3	469.0	309.9	193.6	342.7	2 217.1	1 975.7	241.4	10.9	5 044.3	4 970.9
3rd quarter	2 659.1	1 815.1	182.3	23.1	508.5	328.5	198.4	356.9	2 302.2	2 019.6	282.6	12.3	5 140.8	5 078.1
4th quarter	2 692.7	1 840.7	171.8	24.7	525.4	331.1	201.0	352.7	2 340.0	2 038.8	301.2	12.9	5 130.6	5 083.7
1982														
1st quarter	2 727.8	1 864.7	165.2	23.5	544.1	336.3	206.0	352.5	2 375.3	2 082.1	293.2	12.3	5 130.7	5 095.9
2nd quarter ...	2 772.3	1 884.4	169.4	21.1	559.9	345.5	208.0	359.7	2 412.5	2 110.9	301.6	12.5	5 156.9	5 126.7
3rd quarter	2 812.5	1 904.8	170.7	24.4	560.8	362.1	210.3	350.1	2 462.3	2 159.3	303.0	12.3	5 126.2	5 151.3
4th quarter	2 853.8	1 918.1	179.5	26.3	563.2	377.9	211.2	356.6	2 497.2	2 221.7	275.5	11.0	5 123.4	5 167.5
1983														
1st quarter	2 892.4	1 947.2	183.3	25.8	574.3	380.7	218.9	350.9	2 541.5	2 263.2	278.4	11.0	5 154.9	5 216.2
2nd quarter ...	2 942.1	1 986.3	182.6	24.7	584.4	386.9	222.9	359.6	2 582.5	2 330.6	251.9	9.8	5 196.7	5 252.4
3rd quarter	3 001.2	2 029.6	183.8	23.7	609.2	382.5	227.7	345.4	2 655.8	2 403.9	251.8	9.5	5 256.8	5 331.2
4th quarter	3 088.6	2 086.8	195.3	23.5	630.2	387.0	234.3	355.7	2 732.9	2 458.9	274.0	10.0	5 387.3	5 449.9
1984														
1st quarter	3 178.9	2 143.8	221.9	23.2	645.6	393.9	249.5	361.2	2 817.8	2 506.0	311.8	11.1	5 494.2	5 558.8
2nd quarter ...	3 262.9	2 195.2	231.8	21.7	671.4	398.3	255.5	370.4	2 892.4	2 566.5	325.9	11.3	5 596.8	5 651.1
3rd quarter	3 337.4	2 241.6	232.4	24.4	698.4	401.0	260.5	384.1	2 953.3	2 607.2	346.1	11.7	5 693.0	5 725.7
4th quarter	3 391.7	2 282.9	226.9	29.3	707.7	409.4	264.5	395.9	2 995.7	2 658.1	337.6	11.3	5 746.5	5 772.5
1985														
1st quarter	3 455.4	2 323.5	240.7	26.9	718.6	420.0	274.3	432.3	3 023.1	2 740.5	282.7	9.3	5 780.7	5 757.4
2nd quarter ...	3 495.5	2 363.6	238.5	25.3	723.8	422.6	278.3	388.5	3 107.0	2 789.6	317.3	10.2	5 805.1	5 869.5
3rd quarter	3 543.1	2 406.4	240.7	27.1	724.1	427.9	283.2	421.5	3 121.5	2 865.2	256.4	8.2	5 839.1	5 851.1
4th quarter	3 605.5	2 455.7	244.3	25.6	738.7	430.9	289.6	428.9	3 176.6	2 893.3	283.3	8.9	5 909.4	5 913.0
1986														
1st quarter	3 667.6	2 491.1	245.6	22.0	762.3	443.2	296.7	426.3	3 241.3	2 940.1	301.2	9.3	5 959.5	5 990.8
2nd quarter ...	3 708.8	2 517.2	251.6	21.1	770.2	449.1	300.4	429.4	3 279.4	2 969.3	310.0	9.5	6 031.1	6 067.4
3rd quarter	3 757.1	2 555.2	264.4	16.6	770.3	456.1	305.5	439.5	3 317.7	3 036.0	281.7	8.5	6 075.5	6 106.1
4th quarter	3 798.8	2 604.7	264.2	13.4	769.8	457.8	311.1	456.0	3 342.8	3 073.2	269.5	8.1	6 112.1	6 115.4
1987														
1st quarter	3 856.0	2 647.8	275.4	14.5	771.1	463.1	315.9	450.7	3 405.3	3 102.2	303.0	8.9	6 149.3	6 171.7
2nd quarter ...	3 917.5	2 692.0	282.3	14.1	781.0	468.1	320.0	511.7	3 405.8	3 172.7	233.2	6.8	6 192.2	6 114.0
3rd quarter	3 988.5	2 740.3	289.5	17.5	796.9	469.2	324.8	489.0	3 499.6	3 238.6	260.9	7.5	6 258.6	6 223.4
4th quarter	4 084.3	2 809.6	298.6	20.3	815.4	472.2	331.7	507.0	3 577.4	3 273.0	304.4	8.5	6 368.7	6 307.3
1988														
1st quarter	4 160.6	2 857.6	319.2	21.1	824.0	489.9	351.1	502.1	3 658.5	3 355.5	303.0	8.3	6 421.1	6 399.9
2nd quarter ...	4 234.8	2 923.1	322.5	20.3	833.7	493.5	358.3	497.8	3 737.0	3 417.4	319.6	8.6	6 473.2	6 465.8
3rd quarter	4 326.3	2 975.8	333.6	21.2	860.5	499.6	364.5	506.7	3 819.6	3 489.6	330.0	8.6	6 540.3	6 528.3
4th quarter	4 411.9	3 035.4	326.8	27.4	887.5	506.8	372.0	517.2	3 894.7	3 565.3	329.4	8.5	6 607.5	6 589.8
1989														
1st quarter	4 537.2	3 079.1	348.1	23.1	933.6	531.4	378.0	552.9	3 984.4	3 626.3	358.0	9.0	6 700.7	6 664.9
2nd quarter ...	4 592.9	3 114.6	339.5	21.2	961.5	538.8	382.6	566.7	4 026.2	3 693.9	332.3	8.3	6 691.3	6 645.2
3rd quarter	4 648.5	3 155.0	337.2	20.0	975.6	547.9	387.2	571.6	4 076.9	3 751.7	325.2	8.0	6 728.3	6 689.4
4th quarter	4 723.7	3 209.7	339.7	21.8	986.6	558.8	393.1	579.8	4 143.9	3 799.5	344.4	8.3	6 780.1	6 746.0
1990														
1st quarter	4 818.9	3 272.8	345.6	23.1	999.5	579.5	401.6	582.5	4 236.3	3 884.5	351.9	8.3	6 802.7	6 797.8
2nd quarter ...	4 899.2	3 334.7	351.1	26.7	1 003.1	590.5	406.9	594.6	4 304.6	3 930.6	374.0	8.7	6 851.7	6 845.1
3rd quarter	4 958.5	3 372.5	358.9	31.6	1 010.0	600.0	414.6	600.7	4 357.8	3 997.1	360.7	8.3	6 843.8	6 842.7
4th quarter	4 978.6	3 381.5	357.0	31.3	1 008.7	617.5	417.4	600.8	4 377.8	4 019.8	357.9	8.2	6 758.5	6 784.3
1991														
1st quarter	4 992.5	3 387.5	347.5	33.0	1 003.9	641.6	421.0	580.8	4 411.7	4 028.3	383.4	8.7	6 707.5	6 801.3
2nd quarter ...	5 054.1	3 427.6	353.0	35.5	1 004.2	661.4	427.7	585.9	4 468.2	4 083.2	384.9	8.6	6 735.3	6 851.0
3rd quarter	5 107.1	3 469.9	353.6	39.4	1 006.8	670.8	433.5	590.2	4 516.8	4 130.5	386.3	8.6	6 756.4	6 879.0
4th quarter	5 186.0	3 517.0	362.6	46.4	999.9	698.8	438.6	598.7	4 587.4	4 158.2	429.1	9.4	6 784.9	6 936.2

Table 4-1B. Personal Income and Its Disposition: Historical Data—*Continued*

(Billions of current dollars, except as noted; quarterly data are at seasonally adjusted annual rates.) NIPA Table 2.1

Year and quarter	Personal income							Less: Personal current taxes	Equals: Disposable personal income	Less: Personal outlays	Equals: Personal saving		Billions of chained (2009) dollars	
	Total	Compensation of employees	Proprietors' income with IVA and CCAdj	Rental income of persons with CCAdj	Personal income receipts on assets	Personal current transfer receipts	Less: Contributions for government social insurance, domestic				Billions of dollars	Percent of disposable personal income	Personal income excluding current transfers	Disposable personal income
1992														
1st quarter	5 297.8	3 595.7	378.7	50.0	996.8	727.0	450.4	588.9	4 708.9	4 255.1	453.8	9.6	6 868.3	7 075.8
2nd quarter ...	5 394.4	3 651.9	395.2	57.3	1 000.4	745.6	456.0	607.2	4 787.2	4 311.2	476.0	9.9	6 939.3	7 146.0
3rd quarter	5 456.3	3 688.5	407.9	64.3	997.1	757.6	459.1	616.2	4 840.1	4 388.0	452.1	9.3	6 969.4	7 179.1
4th quarter	5 535.0	3 736.9	418.8	70.7	1 000.9	762.0	454.4	638.9	4 896.1	4 462.5	433.5	8.9	7 030.6	7 211.8
1993														
1st quarter	5 560.7	3 745.6	418.9	80.7	1 007.7	781.6	473.8	617.0	4 943.7	4 508.3	435.4	8.8	6 997.7	7 238.8
2nd quarter ...	5 635.8	3 800.2	429.6	86.9	1 007.4	786.0	474.2	643.5	4 992.3	4 578.0	414.3	8.3	7 053.9	7 261.2
3rd quarter	5 677.6	3 834.1	423.4	92.5	1 006.4	799.9	478.8	659.2	5 018.4	4 646.1	372.3	7.4	7 063.7	7 267.5
4th quarter	5 757.7	3 889.3	440.0	100.2	1 006.7	804.3	482.9	675.3	5 082.4	4 713.7	368.7	7.3	7 132.2	7 318.0
1994														
1st quarter	5 807.9	3 913.4	450.2	109.0	1 015.2	818.1	498.0	673.7	5 134.3	4 783.9	350.4	6.8	7 159.1	7 366.3
2nd quarter ...	5 912.7	3 991.5	454.1	113.5	1 035.9	822.9	505.1	697.8	5 214.9	4 850.1	364.8	7.0	7 262.2	7 440.5
3rd quarter	5 978.4	4 031.8	456.8	116.3	1 057.7	826.9	511.0	695.4	5 283.0	4 925.8	357.2	6.8	7 297.6	7 483.9
4th quarter	6 089.4	4 088.1	465.5	115.9	1 090.4	848.0	518.5	705.4	5 384.0	5 002.8	381.2	7.1	7 390.3	7 591.3
1995														
1st quarter	6 181.2	4 134.4	467.4	118.3	1 112.7	874.0	525.5	724.6	5 456.6	5 042.6	414.0	7.6	7 446.7	7 656.2
2nd quarter ...	6 250.7	4 172.9	472.0	121.6	1 130.9	883.5	530.0	746.8	5 503.9	5 124.1	379.9	6.9	7 487.3	7 677.9
3rd quarter	6 329.3	4 219.7	484.5	125.7	1 144.3	890.4	535.4	752.2	5 577.0	5 195.4	381.6	6.8	7 556.3	7 748.3
4th quarter	6 404.3	4 265.3	501.0	134.0	1 158.4	886.0	540.3	770.0	5 634.3	5 261.6	372.7	6.6	7 633.1	7 793.5
1996														
1st quarter	6 521.0	4 314.5	523.3	139.5	1 170.8	916.0	543.2	801.7	5 719.2	5 337.7	381.5	6.7	7 710.1	7 867.3
2nd quarter ...	6 649.8	4 387.5	547.5	141.6	1 188.5	936.4	551.6	839.6	5 810.3	5 432.5	377.8	6.5	7 807.2	7 939.5
3rd quarter	6 725.9	4 451.2	547.5	143.7	1 210.7	931.8	559.0	843.5	5 882.3	5 492.0	390.4	6.6	7 883.7	8 003.8
4th quarter	6 817.5	4 514.5	556.8	145.3	1 234.7	932.6	566.5	863.5	5 954.0	5 574.5	379.4	6.4	7 953.0	8 046.4
1997														
1st quarter	6 939.4	4 592.4	575.6	144.2	1 251.4	950.2	574.4	902.1	6 037.2	5 658.1	379.1	6.3	8 058.3	8 123.0
2nd quarter ...	7 022.3	4 660.0	575.1	144.6	1 274.5	950.0	581.9	916.2	6 106.0	5 702.5	403.6	6.6	8 149.7	8 195.0
3rd quarter	7 135.3	4 738.9	588.5	147.4	1 294.9	956.2	590.5	941.1	6 194.2	5 815.8	378.4	6.1	8 271.3	8 291.5
4th quarter	7 273.0	4 844.0	596.6	152.1	1 319.4	963.1	602.2	967.8	6 305.2	5 905.2	400.0	6.3	8 419.9	8 413.6
1998														
1st quarter	7 434.6	4 942.0	617.4	156.8	1 353.9	974.7	610.3	996.1	6 438.4	5 961.5	476.9	7.4	8 619.3	8 590.7
2nd quarter ...	7 561.1	5 028.3	629.6	162.6	1 379.1	981.1	619.7	1 022.4	6 538.7	6 082.6	456.1	7.0	8 763.8	8 708.8
3rd quarter	7 668.1	5 114.9	645.0	168.0	1 385.2	984.5	629.5	1 043.2	6 624.9	6 181.1	443.8	6.7	8 874.5	8 796.5
4th quarter	7 762.9	5 199.4	668.7	173.4	1 365.3	995.4	639.2	1 068.0	6 694.9	6 285.6	409.3	6.1	8 962.3	8 866.2
1999														
1st quarter	7 850.9	5 288.0	681.9	176.1	1 340.5	1 014.6	650.2	1 077.9	6 773.0	6 354.8	418.1	6.2	9 030.1	8 946.3
2nd quarter ...	7 920.4	5 344.3	690.1	179.6	1 344.4	1 017.8	655.7	1 095.2	6 825.3	6 484.9	340.4	5.0	9 068.1	8 966.5
3rd quarter	8 028.8	5 423.3	699.8	179.4	1 358.4	1 030.9	663.0	1 120.6	6 908.2	6 593.6	314.6	4.6	9 144.8	9 027.7
4th quarter	8 207.3	5 555.5	713.9	178.9	1 394.0	1 041.4	676.2	1 154.0	7 053.4	6 729.1	324.3	4.6	9 309.5	9 163.2
2000														
1st quarter	8 457.2	5 750.0	720.9	179.8	1 445.8	1 056.7	696.0	1 208.8	7 248.5	6 892.1	356.3	4.9	9 534.5	9 338.7
2nd quarter ...	8 591.8	5 792.7	751.3	180.2	1 482.3	1 083.7	698.4	1 230.2	7 361.6	6 995.9	365.7	5.0	9 629.8	9 442.0
3rd quarter	8 741.0	5 904.9	760.8	182.9	1 506.0	1 098.0	711.6	1 247.7	7 493.3	7 119.9	373.3	5.0	9 742.4	9 551.6
4th quarter	8 820.4	5 944.7	782.4	191.0	1 508.9	1 110.8	717.3	1 258.7	7 561.7	7 224.0	337.7	4.5	9 773.2	9 585.7
2001														
1st quarter	8 984.0	6 050.8	809.8	197.1	1 503.0	1 155.6	732.3	1 301.9	7 682.1	7 298.0	384.1	5.0	9 856.7	9 672.6
2nd quarter ...	9 013.6	6 041.7	827.7	202.6	1 487.3	1 187.4	733.1	1 308.9	7 704.7	7 347.0	357.7	4.6	9 807.9	9 655.7
3rd quarter	9 000.8	6 025.4	845.6	204.8	1 465.2	1 192.2	732.4	1 113.6	7 887.2	7 376.2	511.0	6.5	9 780.1	9 878.5
4th quarter	9 024.0	6 038.6	840.9	205.1	1 439.3	1 235.2	735.0	1 231.8	7 792.3	7 478.6	313.7	4.0	9 749.4	9 753.7
2002														
1st quarter	9 057.8	6 061.1	859.7	209.1	1 411.0	1 260.0	743.1	1 075.1	7 982.7	7 507.2	475.5	6.0	9 742.5	9 973.5
2nd quarter ...	9 148.8	6 129.3	863.7	211.0	1 414.1	1 282.3	751.5	1 051.0	8 097.9	7 593.0	504.8	6.2	9 754.3	10 041.1
3rd quarter	9 174.4	6 157.8	872.4	210.4	1 396.0	1 292.1	754.3	1 044.1	8 130.3	7 678.3	452.0	5.6	9 726.2	10 032.3
4th quarter	9 254.8	6 194.2	883.2	213.6	1 414.4	1 306.4	757.0	1 038.4	8 216.4	7 753.9	462.5	5.6	9 762.7	10 091.9
2003														
1st quarter	9 315.1	6 213.6	874.9	223.0	1 442.7	1 324.2	763.3	1 021.3	8 293.8	7 841.5	452.2	5.5	9 746.1	10 115.6
2nd quarter ...	9 417.8	6 307.7	891.5	227.7	1 424.2	1 340.6	773.9	1 020.8	8 397.0	7 929.4	467.6	5.6	9 848.9	10 238.9
3rd quarter	9 542.3	6 394.1	905.0	232.0	1 434.0	1 361.0	783.8	950.6	8 591.6	8 090.9	500.8	5.8	9 914.6	10 411.9
4th quarter	9 675.1	6 500.8	916.2	243.3	1 447.7	1 363.2	796.1	1 021.3	8 653.7	8 188.3	465.4	5.4	10 026.9	10 439.3
2004														
1st quarter	9 777.9	6 542.0	945.7	247.5	1 454.8	1 397.1	809.2	1 012.2	8 765.7	8 328.7	437.0	5.0	10 026.9	10 487.4
2nd quarter ...	9 953.6	6 669.1	962.1	252.1	1 478.0	1 416.0	823.6	1 026.7	8 926.9	8 439.3	487.6	5.5	10 145.0	10 607.6
3rd quarter	10 094.7	6 804.3	963.2	248.5	1 492.3	1 425.5	839.2	1 064.3	9 030.4	8 576.7	453.7	5.0	10 249.8	10 676.9
4th quarter	10 314.1	6 864.8	977.1	247.4	1 623.4	1 446.3	844.9	1 091.5	9 222.5	8 745.7	476.8	5.2	10 395.8	10 811.7
2005														
1st quarter	10 338.5	6 926.9	956.6	240.1	1 584.8	1 488.1	858.1	1 172.2	9 166.3	8 839.9	326.3	3.6	10 316.7	10 684.9
2nd quarter ...	10 505.2	7 006.0	963.4	237.3	1 652.1	1 512.6	866.3	1 196.3	9 308.9	9 015.1	293.8	3.2	10 419.8	10 786.5
3rd quarter	10 661.7	7 122.9	986.8	225.6	1 677.6	1 528.3	879.5	1 225.4	9 436.3	9 199.8	236.4	2.5	10 471.1	10 818.3
4th quarter	10 887.6	7 210.6	1 005.4	225.0	1 798.2	1 537.9	889.5	1 255.7	9 631.9	9 301.6	330.2	3.4	10 635.9	10 956.9

Table 4-1B. Personal Income and Its Disposition: Historical Data—*Continued*

(Billions of current dollars, except as noted; quarterly data are at seasonally adjusted annual rates.)   **NIPA Table 2.1**

Year and quarter	Personal income — Total	Compensation of employees	Proprietors' income with IVA and CCAdj	Rental income of persons with CCAdj	Personal income receipts on assets	Personal current transfer receipts	Less: Contributions for government social insurance, domestic	Less: Personal current taxes	Equals: Disposable personal income	Less: Personal outlays	Equals: Personal saving — Billions of dollars	Percent of disposable personal income	Billions of chained (2009) dollars — Personal income excluding current transfers	Disposable personal income
2006														
1st quarter	11 189.5	7 380.2	1 048.1	216.7	1 869.5	1 588.3	913.2	1 320.3	9 869.2	9 451.7	417.5	4.2	10 866.8	11 170.1
2nd quarter	11 324.0	7 428.1	1 050.9	206.3	1 953.9	1 602.8	918.1	1 351.0	9 973.0	9 582.5	390.4	3.9	10 914.7	11 197.4
3rd quarter	11 429.4	7 487.2	1 047.7	199.5	1 989.3	1 628.3	922.6	1 358.5	10 070.9	9 718.5	352.4	3.5	10 925.5	11 226.4
4th quarter	11 583.9	7 624.0	1 051.9	186.8	2 021.6	1 635.9	936.2	1 397.3	10 186.6	9 804.5	382.1	3.8	11 108.1	11 374.5
2007														
1st quarter	11 836.7	7 806.8	1 016.2	174.2	2 088.5	1 706.7	955.7	1 466.3	10 370.4	9 951.9	418.6	4.0	11 205.5	11 471.4
2nd quarter	11 977.0	7 845.4	995.9	183.8	2 198.8	1 710.4	957.3	1 495.6	10 481.3	10 065.1	416.2	4.0	11 265.2	11 500.8
3rd quarter	12 048.4	7 885.1	985.1	187.5	2 218.6	1 732.7	960.6	1 498.6	10 549.8	10 177.4	372.4	3.5	11 255.6	11 511.0
4th quarter	12 169.1	7 978.2	978.9	191.9	2 229.5	1 762.7	972.1	1 508.3	10 660.8	10 301.3	359.4	3.4	11 244.0	11 518.8
2008														
1st quarter	12 315.0	8 055.3	964.7	220.5	2 255.6	1 802.9	984.0	1 534.8	10 780.2	10 357.7	422.4	3.9	11 263.6	11 550.8
2nd quarter	12 642.6	8 054.8	968.6	245.0	2 212.0	2 148.4	986.2	1 552.1	11 090.5	10 469.9	620.6	5.6	11 129.8	11 762.2
3rd quarter	12 467.2	8 074.0	958.2	268.1	2 217.0	1 941.4	991.5	1 497.2	10 970.0	10 497.9	472.0	4.3	11 048.8	11 515.0
4th quarter	12 344.0	8 043.8	952.2	293.0	2 118.9	1 927.8	991.7	1 444.6	10 899.4	10 234.8	664.7	6.1	11 100.4	11 615.3
2009														
1st quarter	11 989.6	7 729.8	914.3	303.4	1 962.2	2 039.7	959.8	1 202.1	10 787.5	10 155.4	632.1	5.9	10 667.5	11 565.5
2nd quarter	12 083.2	7 761.4	902.3	318.4	1 888.7	2 178.6	966.3	1 130.8	10 952.4	10 158.6	793.8	7.2	10 571.4	11 689.8
3rd quarter	12 038.3	7 746.2	935.3	338.5	1 805.8	2 176.3	963.8	1 135.0	10 903.4	10 293.5	609.8	5.6	10 453.8	11 557.6
4th quarter	12 125.3	7 796.7	1 002.2	349.0	1 752.3	2 192.3	967.2	1 140.4	10 984.9	10 354.6	630.3	5.7	10 448.3	11 554.8
2010														
1st quarter	12 275.9	7 767.2	1 056.9	371.8	1 750.3	2 303.2	973.6	1 191.5	11 084.4	10 434.4	650.0	5.9	10 454.3	11 619.8
2nd quarter	12 492.6	7 910.1	1 104.0	387.9	1 764.0	2 311.1	984.5	1 212.9	11 279.7	10 518.3	761.4	6.8	10 661.3	11 811.2
3rd quarter	12 636.0	7 981.8	1 129.4	399.5	1 776.5	2 336.1	987.4	1 255.9	11 380.1	10 605.8	774.2	6.8	10 766.2	11 895.3
4th quarter	12 801.9	8 040.7	1 144.4	417.6	1 838.5	2 350.1	989.5	1 288.8	11 513.1	10 735.3	777.8	6.8	10 859.2	11 962.0
2011														
1st quarter	13 158.2	8 170.0	1 181.0	453.4	1 903.8	2 361.8	911.8	1 426.1	11 732.1	10 866.5	865.7	7.4	11 120.1	12 083.9
2nd quarter	13 267.3	8 206.1	1 212.2	471.6	1 930.2	2 361.6	914.5	1 445.4	11 821.9	10 994.3	827.7	7.0	11 123.1	12 057.6
3rd quarter	13 402.0	8 287.1	1 253.5	485.0	1 947.3	2 352.1	922.9	1 470.9	11 931.1	11 087.8	843.3	7.1	11 215.8	12 110.2
4th quarter	13 479.5	8 240.4	1 270.4	504.6	2 022.3	2 359.3	917.4	1 470.4	12 009.2	11 146.4	862.7	7.2	11 248.7	12 147.9
2012														
1st quarter	13 785.7	8 455.9	1 322.8	509.9	2 097.0	2 340.4	940.3	1 467.8	12 317.9	11 309.1	1 008.9	8.2	11 498.6	12 375.3
2nd quarter	13 946.8	8 503.0	1 356.5	517.5	2 151.7	2 362.8	944.7	1 487.1	12 459.7	11 353.5	1 106.1	8.9	11 610.1	12 487.8
3rd quarter	13 915.4	8 545.3	1 345.8	520.5	2 086.7	2 364.7	947.6	1 509.5	12 406.0	11 398.0	1 008.0	8.1	11 543.7	12 398.4
4th quarter	14 392.6	8 762.6	1 364.2	524.0	2 327.1	2 384.2	969.4	1 571.4	12 821.2	11 513.9	1 307.3	10.2	11 934.1	12 741.9
2013														
1st quarter	14 001.2	8 731.7	1 388.3	537.9	2 030.5	2 403.5	1 090.6	1 649.3	12 351.9	11 608.6	743.2	6.0	11 485.1	12 231.9
2nd quarter	14 135.0	8 818.7	1 411.8	550.2	2 041.1	2 416.2	1 103.1	1 681.9	12 453.1	11 625.5	827.7	6.6	11 596.4	12 323.0
3rd quarter	14 232.0	8 844.7	1 412.5	564.6	2 082.0	2 434.5	1 106.3	1 674.5	12 557.5	11 716.7	840.9	6.7	11 627.3	12 376.3
4th quarter	14 356.2	8 941.7	1 401.7	575.2	2 111.6	2 443.2	1 117.2	1 697.7	12 658.5	11 864.9	793.6	6.3	11 693.5	12 425.2
2014														
1st quarter	14 639.3	9 113.0	1 413.4	588.4	2 190.1	2 474.1	1 139.8	1 744.4	12 894.9	11 976.2	918.6	7.1	11 885.4	12 598.3
2nd quarter	14 890.8	9 171.3	1 452.7	599.2	2 285.3	2 526.8	1 144.5	1 757.8	13 133.0	12 164.8	968.2	7.4	12 022.2	12 769.9
3rd quarter	15 120.0	9 279.6	1 468.3	609.9	2 350.1	2 568.2	1 156.1	1 797.9	13 322.0	12 332.5	989.5	7.4	12 172.0	12 919.0
4th quarter	15 316.8	9 432.5	1 456.6	621.0	2 383.4	2 597.1	1 173.8	1 835.8	13 481.0	12 474.3	1 006.7	7.5	12 350.6	13 089.7
2015														
1st quarter	15 526.6	9 563.8	1 437.1	622.1	2 448.5	2 644.0	1 188.9	1 904.6	13 622.1	12 535.5	1 086.6	8.0	12 562.3	13 283.4
2nd quarter	15 671.3	9 666.2	1 413.5	648.9	2 461.4	2 682.3	1 200.9	1 943.0	13 728.3	12 701.9	1 026.5	7.5	12 604.0	13 321.4
3rd quarter	15 806.4	9 751.7	1 425.3	660.1	2 480.9	2 699.1	1 210.9	1 947.3	13 859.1	12 842.5	1 016.6	7.3	12 686.1	13 413.7
4th quarter	15 892.6	9 816.0	1 416.0	665.1	2 497.8	2 716.1	1 218.3	1 964.9	13 927.7	12 902.6	1 025.1	7.4	12 762.9	13 490.5
2016														
1st quarter	15 970.8	9 850.4	1 412.7	671.3	2 516.7	2 745.2	1 225.5	1 925.0	14 045.8	13 004.3	1 041.5	7.4	12 800.4	13 594.2
2nd quarter	16 065.6	9 903.1	1 411.3	681.5	2 533.8	2 767.6	1 231.8	1 945.2	14 120.4	13 158.8	961.6	6.8	12 793.9	13 585.2
3rd quarter	16 216.2	9 995.7	1 430.7	685.0	2 560.8	2 786.8	1 242.9	1 971.0	14 245.1	13 300.6	944.6	6.6	12 868.1	13 649.8
4th quarter	16 390.3	10 106.2	1 444.5	692.8	2 595.6	2 807.5	1 256.3	1 990.4	14 399.9	13 447.4	952.5	6.6	12 955.4	13 734.8
2017														
1st quarter	16 633.7	10 223.1	1 493.4	709.0	2 653.3	2 831.7	1 276.8	2 001.1	14 632.7	13 625.1	1 007.6	6.9	13 092.8	13 880.7
2nd quarter	16 828.4	10 335.3	1 502.7	713.2	2 728.0	2 838.5	1 289.3	2 005.6	14 822.8	13 728.1	1 094.7	7.4	13 241.7	14 030.1
3rd quarter	17 036.6	10 476.9	1 507.8	725.9	2 761.3	2 869.2	1 304.5	2 052.3	14 984.2	13 867.0	1 117.3	7.5	13 355.3	14 125.4
4th quarter	17 295.6	10 654.9	1 532.1	739.6	2 811.4	2 880.9	1 323.2	2 127.9	15 167.8	14 103.4	1 064.3	7.0	13 500.5	14 205.8
2018														
1st quarter	17 548.6	10 776.1	1 557.7	745.3	2 874.8	2 937.8	1 343.1	2 085.6	15 463.0	14 274.1	1 188.8	7.7	13 593.8	14 386.7
2nd quarter	17 750.3	10 882.3	1 570.7	752.4	2 934.4	2 963.3	1 352.8	2 064.4	15 685.9	14 467.9	1 218.0	7.8	13 681.9	14 513.6
3rd quarter	17 976.5	11 034.1	1 588.4	768.2	2 971.8	2 983.0	1 369.1	2 100.5	15 876.1	14 628.2	1 247.9	7.9	13 819.2	14 632.6
4th quarter	18 132.0	11 107.8	1 627.0	771.2	3 005.6	2 997.1	1 376.8	2 090.7	16 041.3	14 746.8	1 294.5	8.1	13 899.8	14 732.3
2019														
1st quarter	18 366.7	11 335.3	1 627.5	776.6	2 950.7	3 085.2	1 408.7	2 170.7	16 196.0	14 841.5	1 354.5	8.4	14 014.8	14 853.5
2nd quarter	18 480.9	11 391.7	1 628.5	786.7	2 970.2	3 118.6	1 414.7	2 222.5	16 258.4	15 072.3	1 186.1	7.3	14 001.1	14 817.8
3rd quarter	18 597.6	11 438.0	1 677.0	789.7	2 970.1	3 141.9	1 419.0	2 197.1	16 400.5	15 219.9	1 180.6	7.2	14 037.4	14 895.4
4th quarter	18 760.8	11 564.8	1 697.7	795.5	2 980.4	3 155.2	1 432.9	2 221.2	16 539.6	15 335.8	1 203.8	7.3	14 119.5	14 964.5

Table 4-2. Personal Consumption Expenditures by Major Type of Product

(Billions of dollars, quarterly data are at seasonally adjusted rates.)

NIPA Table 2.3.5

Year and quarter	Personal consumption expenditures, total	Goods										
		Total goods	Durable goods					Nondurable goods				
			Durable goods, total	Motor vehicles and parts	Furnishings and household equipment	Recreational goods and vehicles	Other durable goods	Nondurable goods, total	Food and beverages off-premises	Clothing and footwear	Gasoline and other energy goods	Other nondurable goods
1960	331.2	177.0	45.6	19.6	15.4	6.4	4.3	131.4	62.6	25.9	15.8	27.1
1961	341.5	178.8	44.2	17.7	15.6	6.6	4.3	134.6	63.7	26.6	15.7	28.6
1962	362.6	189.0	49.5	21.4	16.4	6.9	4.7	139.5	64.7	27.9	16.3	30.7
1963	382.0	198.2	54.2	24.2	17.5	7.6	4.9	143.9	65.9	28.6	16.9	32.5
1964	410.6	212.3	59.6	25.8	19.5	8.8	5.5	152.7	69.5	31.1	17.7	34.5
1965	443.0	229.7	66.4	29.6	20.7	10.1	6.0	163.3	74.4	32.7	19.1	37.1
1966	479.9	249.6	71.7	29.9	22.6	12.3	6.9	177.9	80.6	35.8	20.7	40.8
1967	506.7	259.0	74.0	29.6	23.7	13.6	7.1	185.0	82.6	37.5	21.9	43.0
1968	556.9	284.6	84.8	35.4	26.1	15.3	8.0	199.8	88.8	41.3	23.2	46.5
1969	603.6	304.7	90.5	37.4	27.6	16.7	8.7	214.2	95.4	44.3	25.0	49.5
1970	646.7	318.8	90.0	34.5	28.2	17.9	9.4	228.8	103.5	45.5	26.3	53.6
1971	699.9	342.1	102.4	43.2	29.9	19.2	10.1	239.7	107.1	49.0	27.6	55.9
1972	768.2	373.8	116.4	49.4	33.5	22.6	11.0	257.4	114.5	53.5	29.4	60.0
1973	849.6	416.6	130.5	54.4	38.0	25.3	12.9	286.1	126.7	59.2	34.3	65.8
1974	930.2	451.5	130.2	48.2	40.9	26.6	14.5	321.4	143.0	62.4	43.8	72.1
1975	1 030.5	491.3	142.2	52.6	42.6	30.4	16.5	349.2	156.6	66.9	48.0	77.7
1976	1 147.7	546.3	168.6	68.2	47.2	34.2	19.1	377.7	167.3	72.2	53.0	85.2
1977	1 274.0	600.4	192.0	79.8	53.4	37.7	21.1	408.4	179.8	79.3	57.8	91.5
1978	1 422.3	663.6	213.3	89.2	59.0	41.7	23.5	450.2	196.1	89.3	61.5	103.3
1979	1 585.4	737.9	226.3	90.2	65.3	45.8	25.1	511.6	218.4	96.4	80.4	116.5
1980	1 750.7	799.8	226.4	84.4	67.8	46.5	27.6	573.4	239.2	103.0	101.9	129.3
1981	1 934.0	869.4	243.9	93.0	71.5	49.9	29.6	625.4	255.3	113.2	113.4	143.5
1982	2 071.3	899.3	253.0	100.0	71.8	51.3	29.9	646.3	267.1	116.7	108.4	154.0
1983	2 281.6	973.8	295.0	122.9	79.8	58.7	33.7	678.8	277.0	126.4	106.5	168.8
1984	2 492.3	1 063.7	342.2	147.2	88.8	67.8	38.3	721.5	291.1	137.6	108.2	184.6
1985	2 712.8	1 137.6	380.4	170.1	94.6	74.1	41.6	757.2	303.0	146.8	110.5	196.9
1986	2 886.3	1 195.6	421.4	187.5	103.5	83.0	47.5	774.2	316.4	157.2	91.2	209.4
1987	3 076.3	1 256.3	442.0	188.2	109.5	91.8	52.5	814.3	324.3	167.7	96.4	225.9
1988	3 330.0	1 337.3	475.1	202.2	115.2	99.9	57.8	862.3	342.8	178.2	99.9	241.4
1989	3 576.8	1 423.8	494.3	207.8	121.4	103.9	61.3	929.5	365.4	190.4	110.4	263.3
1990	3 809.0	1 491.3	497.1	205.1	120.9	105.6	65.5	994.2	391.2	195.2	124.2	283.6
1991	3 943.4	1 497.4	477.2	185.7	118.8	107.7	64.9	1 020.3	403.0	199.1	121.1	297.1
1992	4 197.6	1 563.3	508.1	204.8	124.3	111.0	68.0	1 055.2	404.5	211.2	125.0	314.5
1993	4 452.0	1 642.3	551.5	224.7	131.4	123.5	72.0	1 090.8	413.5	219.1	126.9	331.4
1994	4 721.0	1 746.6	607.2	249.8	140.5	140.3	76.5	1 139.4	432.1	227.4	129.2	350.6
1995	4 962.6	1 815.5	635.7	255.7	146.7	153.7	79.6	1 179.8	443.7	231.2	133.4	371.4
1996	5 244.6	1 917.7	676.3	273.5	153.5	164.9	84.3	1 241.4	461.9	239.5	144.7	395.2
1997	5 536.8	2 006.5	715.5	293.1	160.5	174.6	87.3	1 291.0	474.8	247.5	147.7	421.0
1998	5 877.2	2 108.4	779.3	320.2	173.6	191.4	94.2	1 329.1	487.4	257.8	132.4	451.5
1999	6 279.1	2 287.1	855.6	350.7	191.2	210.9	102.7	1 431.5	515.5	271.1	146.5	498.3
2000	6 762.1	2 453.2	912.6	363.2	208.1	230.9	110.4	1 540.6	540.6	280.8	184.5	534.7
2001	7 065.6	2 525.6	941.5	383.3	214.9	234.9	108.4	1 584.1	564.0	277.9	178.0	564.2
2002	7 342.7	2 598.8	985.4	401.3	225.9	244.8	113.4	1 613.4	575.1	278.8	167.9	591.7
2003	7 723.1	2 722.6	1 017.8	401.5	235.2	259.5	121.7	1 704.8	599.6	285.3	196.4	623.5
2004	8 212.7	2 902.0	1 080.6	409.3	254.3	284.8	132.1	1 821.4	632.6	297.4	232.7	658.7
2005	8 747.1	3 082.9	1 128.6	410.0	271.3	306.4	141.0	1 954.3	668.2	310.5	283.8	691.8
2006	9 260.3	3 239.7	1 158.3	394.9	283.6	326.3	153.5	2 081.3	700.3	320.0	319.7	741.4
2007	9 706.4	3 367.0	1 188.0	400.6	283.5	339.2	164.8	2 179.0	737.3	323.5	345.5	772.6
2008	9 976.3	3 363.2	1 098.8	343.3	264.3	328.1	163.0	2 264.5	769.1	317.4	391.1	786.9
2009	9 842.2	3 180.0	1 012.1	318.6	238.3	297.5	157.7	2 167.9	772.9	304.0	287.0	803.9
2010	10 185.8	3 317.8	1 049.0	344.5	240.9	298.6	165.0	2 268.9	786.9	316.6	336.7	828.7
2011	10 641.1	3 518.1	1 093.5	365.2	246.9	305.4	176.1	2 424.6	819.5	332.6	413.8	858.7
2012	11 006.8	3 637.7	1 144.2	396.6	253.9	311.8	181.9	2 493.5	846.2	345.2	421.9	880.2
2013	11 317.2	3 730.0	1 189.4	417.5	263.6	321.6	186.7	2 540.6	864.0	350.5	418.2	907.8
2014	11 822.8	3 863.0	1 242.1	442.0	276.2	329.9	194.0	2 620.9	896.9	360.8	403.3	959.9
2015	12 297.5	3 923.0	1 307.6	475.3	294.2	336.5	201.6	2 615.4	921.0	368.7	309.4	1 016.3
2016	12 770.0	3 998.4	1 350.2	485.6	309.4	351.4	203.9	2 648.1	939.9	376.4	275.0	1 056.9
2017	13 340.4	4 172.3	1 410.7	503.6	324.7	374.2	208.1	2 761.6	970.2	380.0	309.0	1 102.5
2018	13 993.3	4 371.9	1 481.6	523.2	343.3	399.0	216.0	2 890.3	998.8	394.2	349.2	1 148.0
2019	14 544.6	4 512.2	1 534.4	521.8	357.4	433.4	221.7	2 977.9	1 025.7	403.5	335.4	1 213.3
2017												
1st quarter	13 153.2	4 108.1	1 383.2	494.9	319.4	363.9	205.0	2 724.9	956.5	378.4	307.8	1 082.1
2nd quarter	13 241.3	4 130.3	1 394.0	493.9	321.3	372.6	206.2	2 736.3	964.8	379.0	295.3	1 097.2
3rd quarter	13 370.9	4 177.3	1 413.0	503.4	324.7	375.3	209.6	2 764.3	972.8	379.0	305.1	1 107.5
4th quarter	13 596.0	4 273.4	1 452.6	522.4	333.5	385.0	211.7	2 820.8	986.4	383.5	327.8	1 123.0
2018												
1st quarter	13 755.5	4 315.6	1 461.8	517.7	337.8	392.0	214.3	2 853.8	991.7	387.9	343.7	1 130.6
2nd quarter	13 939.9	4 368.8	1 482.6	523.2	344.3	397.1	218.0	2 886.2	996.7	395.2	349.6	1 144.7
3rd quarter	14 086.3	4 394.8	1 488.7	525.6	346.0	401.4	215.8	2 906.1	1 001.2	395.7	356.4	1 152.8
4th quarter	14 191.4	4 408.3	1 493.2	526.4	345.3	405.5	215.9	2 915.1	1 005.6	398.2	347.3	1 164.0
2019												
1st quarter	14 276.6	4 415.2	1 494.5	508.5	349.7	418.6	217.6	2 920.7	1 011.7	399.0	321.9	1 188.1
2nd quarter	14 497.3	4 517.7	1 536.0	524.9	357.4	432.2	221.5	2 981.7	1 023.4	404.4	344.9	1 209.0
3rd quarter	14 645.3	4 553.6	1 552.8	525.7	361.8	441.0	224.3	3 000.8	1 035.2	405.6	334.5	1 225.5
4th quarter	14 759.2	4 562.4	1 554.1	528.2	360.9	441.6	223.4	3 008.2	1 032.4	404.9	340.4	1 230.5

Table 4-2. Personal Consumption Expenditures by Major Type of Product—*Continued*

(Billions of dollars, quarterly data are at seasonally adjusted rates.) **NIPA Table 2.3.5**

Year and quarter	Services	Household services, total	Housing and utilities	Health care	Transportation services	Recreation services	Food services and accommodations	Financial services and insurance	Other services	Final consumption expenditures	Gross output	Less: receipts from sales of goods and services
										Nonprofit institutions serving households (NPISHs)		
1960	154.2	149.1	56.7	16.0	9.2	6.5	20.5	13.2	27.1	5.1	14.8	9.8
1961	162.7	157.4	60.3	17.1	9.6	6.9	21.0	14.3	28.3	5.2	15.8	10.5
1962	173.6	168.1	64.5	19.1	10.1	7.5	22.3	14.8	29.9	5.5	16.9	11.5
1963	183.9	178.1	68.2	21.0	10.6	7.9	23.3	15.5	31.6	5.8	18.3	12.6
1964	198.4	191.9	72.1	24.1	11.4	8.5	24.9	17.2	33.8	6.4	20.3	13.8
1965	213.3	206.3	76.6	26.0	12.1	9.0	27.2	18.8	36.6	7.0	22.1	15.1
1966	230.3	222.8	81.2	28.7	13.1	9.8	29.6	20.6	39.9	7.5	24.3	16.9
1967	247.7	239.6	86.3	31.9	14.3	10.5	31.0	22.1	43.6	8.1	27.0	18.9
1968	272.2	263.4	92.7	36.6	15.9	11.7	34.6	25.2	46.8	8.8	30.3	21.6
1969	299.0	289.5	101.0	42.1	18.0	12.9	37.5	27.7	50.5	9.4	34.0	24.6
1970	327.9	317.5	109.4	47.7	20.0	14.0	41.6	30.1	54.6	10.5	38.1	27.6
1971	357.8	346.1	120.0	53.7	22.4	15.1	43.8	33.1	58.1	11.7	42.8	31.1
1972	394.3	381.5	131.2	59.8	24.5	16.3	48.9	37.1	63.7	12.8	47.4	34.5
1973	432.9	419.2	143.5	67.2	26.1	18.3	54.8	39.9	69.3	13.8	51.8	38.1
1974	478.6	463.1	158.6	76.1	28.5	20.9	60.6	44.1	74.4	15.5	58.8	43.3
1975	539.2	522.2	176.5	89.0	32.0	23.7	68.8	51.8	80.3	17.0	67.2	50.2
1976	601.4	582.4	194.7	101.8	36.2	26.5	77.8	56.8	88.6	19.0	76.3	57.3
1977	673.6	653.0	217.8	115.7	41.4	29.6	85.7	65.1	97.7	20.6	85.2	64.5
1978	758.7	735.7	244.3	131.2	45.2	32.9	97.1	76.7	108.3	23.0	96.7	73.6
1979	847.5	821.4	273.4	148.8	50.7	36.7	110.9	83.6	117.2	26.1	109.4	83.2
1980	950.9	920.8	312.5	171.7	55.4	40.8	121.7	91.7	127.1	30.0	126.1	96.1
1981	1 064.6	1 030.4	352.1	201.9	59.8	47.1	133.9	98.5	137.2	34.1	145.7	111.5
1982	1 172.0	1 134.0	387.5	225.2	61.7	52.5	142.5	113.7	150.9	38.0	163.8	125.8
1983	1 307.8	1 267.1	421.2	253.1	68.9	59.4	153.6	141.0	169.9	40.8	179.2	138.5
1984	1 428.6	1 383.3	457.5	276.5	80.0	66.1	164.9	150.8	187.5	45.4	194.4	149.0
1985	1 575.2	1 527.3	500.6	302.2	90.1	74.0	174.3	178.2	207.9	47.9	209.3	161.4
1986	1 690.7	1 638.0	537.0	330.2	95.0	80.4	186.7	187.7	221.2	52.6	227.3	174.7
1987	1 820.0	1 764.3	571.6	366.0	103.1	87.3	204.4	189.5	242.4	55.7	246.9	191.2
1988	1 992.7	1 929.4	614.4	410.1	114.2	99.2	225.8	202.9	262.8	63.3	275.5	212.2
1989	2 153.0	2 084.9	655.2	451.2	121.8	110.7	242.6	222.3	281.1	68.0	301.2	233.2
1990	2 317.7	2 241.8	696.5	506.2	126.4	121.8	262.7	230.8	297.5	75.9	334.4	258.6
1991	2 446.0	2 365.9	735.2	555.8	123.7	127.2	273.4	250.1	300.5	80.1	362.9	282.8
1992	2 634.3	2 546.4	771.1	612.8	133.6	139.8	286.3	277.0	325.7	87.9	396.4	308.5
1993	2 809.6	2 719.6	814.9	648.8	145.7	153.4	298.4	314.0	344.3	90.1	418.7	328.7
1994	2 974.4	2 876.6	863.3	680.5	161.0	164.7	308.3	327.9	371.0	97.8	440.1	342.3
1995	3 147.1	3 044.7	913.7	719.9	176.4	181.1	316.1	347.0	390.6	102.3	458.5	356.2
1996	3 326.9	3 216.9	962.4	752.1	192.7	195.6	326.6	372.1	415.5	110.0	483.1	373.1
1997	3 530.3	3 424.7	1 009.8	790.9	211.8	208.3	343.4	408.9	451.5	105.6	502.6	397.0
1998	3 768.8	3 645.0	1 065.5	832.0	225.2	220.2	361.8	446.1	494.2	123.8	542.5	418.7
1999	3 992.0	3 853.8	1 123.1	863.6	241.3	238.1	380.3	486.4	520.9	138.2	576.3	438.1
2000	4 309.0	4 150.9	1 198.6	918.4	261.3	254.4	408.8	543.0	566.5	158.0	621.6	463.6
2001	4 540.0	4 361.0	1 287.5	996.6	259.8	262.3	419.7	525.7	609.5	179.1	676.4	497.4
2002	4 743.9	4 545.5	1 333.6	1 082.9	251.9	271.4	436.3	534.7	634.8	198.3	737.1	538.7
2003	5 000.5	4 795.0	1 394.1	1 154.0	259.6	288.9	462.7	560.3	675.4	205.5	774.6	569.1
2004	5 310.6	5 104.3	1 469.1	1 238.9	271.2	311.5	498.2	605.5	709.9	206.4	819.0	612.6
2005	5 664.2	5 453.9	1 583.6	1 320.5	283.9	328.1	533.6	659.0	745.1	210.3	868.5	658.2
2006	6 020.7	5 781.5	1 682.4	1 391.9	297.1	351.3	570.6	695.0	793.3	239.2	932.2	693.0
2007	6 339.4	6 090.6	1 758.2	1 478.2	307.6	375.6	601.5	737.2	832.4	248.8	983.1	734.4
2008	6 613.1	6 325.8	1 835.4	1 555.3	312.7	389.1	620.2	756.6	856.6	287.3	1 047.4	760.1
2009	6 662.2	6 373.0	1 877.7	1 632.7	297.4	388.4	612.7	711.3	852.9	289.2	1 088.3	799.1
2010	6 868.0	6 573.6	1 903.9	1 699.6	305.2	403.7	635.7	754.4	871.1	294.4	1 128.6	834.2
2011	7 123.0	6 811.1	1 955.9	1 757.1	328.4	409.0	669.5	797.9	893.3	311.9	1 173.5	861.7
2012	7 369.1	7 027.5	1 996.3	1 821.3	341.1	430.8	704.9	820.1	913.0	341.5	1 236.5	895.0
2013	7 587.2	7 234.6	2 055.3	1 858.2	359.9	447.1	732.3	858.4	923.5	352.6	1 271.7	919.1
2014	7 959.8	7 594.2	2 149.9	1 940.5	383.0	466.6	776.9	908.1	969.1	365.6	1 322.4	956.8
2015	8 374.5	8 002.9	2 257.9	2 057.3	398.7	491.7	832.9	957.3	1 007.2	371.6	1 383.3	1 011.7
2016	8 771.6	8 370.8	2 358.5	2 165.1	419.4	518.3	873.2	984.0	1 052.5	400.8	1 466.3	1 065.5
2017	9 168.1	8 751.4	2 459.5	2 248.3	440.3	538.5	913.7	1 052.4	1 098.6	416.7	1 523.4	1 106.7
2018	9 621.4	9 182.7	2 570.2	2 345.0	466.7	561.8	961.2	1 119.5	1 158.2	438.8	1 596.9	1 158.2
2019	10 032.0	9 593.2	2 681.2	2 450.8	483.4	580.4	999.5	1 176.1	1 221.8	439.2	1 658.1	1 218.9
2017												
1st quarter	9 045.1	8 622.4	2 412.8	2 222.6	429.1	537.0	906.9	1 023.0	1 091.0	422.7	1 516.3	1 093.6
2nd quarter	9 111.0	8 692.0	2 450.5	2 224.3	434.8	538.8	906.9	1 044.2	1 092.4	419.0	1 508.6	1 089.6
3rd quarter	9 193.6	8 781.1	2 469.5	2 258.7	441.9	539.5	913.0	1 061.3	1 097.1	412.6	1 525.5	1 112.9
4th quarter	9 322.7	8 910.1	2 505.2	2 287.6	455.5	538.7	928.0	1 081.1	1 114.0	412.5	1 543.0	1 130.5
2018												
1st quarter	9 440.0	9 016.1	2 528.5	2 306.7	465.8	555.2	941.9	1 097.3	1 120.7	423.9	1 555.4	1 131.5
2nd quarter	9 571.1	9 137.2	2 559.5	2 332.6	465.7	557.1	956.5	1 109.9	1 156.0	434.0	1 589.7	1 155.7
3rd quarter	9 691.4	9 255.5	2 580.3	2 370.3	465.4	566.5	974.5	1 127.7	1 170.9	436.0	1 615.1	1 179.2
4th quarter	9 783.1	9 321.9	2 612.5	2 370.5	469.9	568.5	972.0	1 143.3	1 185.2	461.2	1 627.5	1 166.3
2019												
1st quarter	9 861.4	9 424.9	2 639.2	2 405.9	471.1	571.6	979.0	1 151.6	1 206.6	436.6	1 641.3	1 204.7
2nd quarter	9 979.6	9 542.2	2 668.9	2 440.2	480.1	579.8	997.9	1 168.7	1 206.6	437.4	1 651.2	1 213.8
3rd quarter	10 091.0	9 647.4	2 698.8	2 457.0	489.1	579.7	1 009.5	1 184.4	1 228.8	444.3	1 661.8	1 217.5
4th quarter	10 196.0	9 758.5	2 717.8	2 500.3	493.5	590.7	1 011.5	1 199.5	1 245.2	438.3	1 678.0	1 239.7

Table 4-3. Chain-Type Quantity Indexes for Personal Consumption Expenditures by Major Type of Product

(Index numbers, 2012 = 100.)   NIPA Table 2.3.3

Year and quarter	Personal consumption expenditures, total	Goods										
		Total goods	Durable goods					Nondurable goods				
			Durable goods, total	Motor vehicles and parts	Furnishings and household equipment	Recreational goods and vehicles	Other durable goods	Nondurable goods, total	Food and beverages off-premises	Clothing and footwear	Gasoline and other energy goods	Other nondurable goods
1960	18.3	18.3	7.0	16.7	12.2	0.8	7.9	29.2	44.7	14.5	53.2	18.5
1961	18.6	18.4	6.8	15.0	12.2	0.8	8.0	29.8	45.4	14.8	52.9	19.5
1962	19.6	19.3	7.5	17.8	12.9	0.9	8.8	30.7	45.7	15.5	54.6	20.9
1963	20.4	20.1	8.2	20.1	13.8	0.9	9.1	31.3	46.0	15.7	56.5	21.8
1964	21.6	21.3	9.0	21.2	15.3	1.1	9.9	32.8	47.6	16.9	59.5	22.8
1965	23.0	22.8	10.1	24.5	16.3	1.3	10.9	34.5	50.0	17.6	62.3	24.2
1966	24.3	24.3	11.0	24.9	17.7	1.6	12.6	36.4	51.9	18.8	65.7	26.1
1967	25.0	24.7	11.1	24.3	18.1	1.7	12.7	37.2	53.1	19.0	67.4	26.8
1968	26.4	26.3	12.4	28.1	19.2	1.9	13.9	38.7	55.5	19.7	70.2	27.9
1969	27.4	27.1	12.8	29.2	19.6	2.0	14.4	39.8	57.0	20.1	73.3	28.7
1970	28.0	27.3	12.5	26.1	19.6	2.2	15.1	40.7	58.6	19.9	76.1	29.7
1971	29.1	28.4	13.7	31.2	20.5	2.2	15.8	41.4	59.3	20.7	78.7	29.8
1972	30.9	30.3	15.5	35.5	22.6	2.6	16.7	43.1	60.5	22.2	82.7	31.2
1973	32.4	31.9	17.1	38.9	25.0	2.9	19.0	44.4	59.4	23.8	87.5	33.3
1974	32.2	30.7	16.0	32.5	24.9	2.9	19.9	43.3	58.2	23.3	80.4	33.2
1975	32.9	30.9	16.0	32.3	23.6	3.1	21.1	43.7	59.3	23.9	82.2	32.1
1976	34.7	33.1	18.0	38.9	25.1	3.4	23.2	45.7	62.2	25.0	86.6	33.5
1977	36.2	34.5	19.6	43.0	27.4	3.6	24.5	46.8	63.1	26.4	88.2	34.0
1978	37.8	35.9	20.7	44.9	28.8	3.9	25.8	48.5	62.8	29.1	89.6	36.4
1979	38.7	36.5	20.5	42.2	30.0	4.1	25.5	49.7	63.7	30.8	87.0	38.5
1980	38.5	35.6	18.9	36.8	28.8	3.9	23.3	49.6	64.4	31.8	79.3	39.2
1981	39.1	36.0	19.1	37.7	28.3	4.0	23.7	50.3	64.1	33.8	78.3	40.1
1982	39.7	36.2	19.1	38.7	27.0	4.0	23.7	50.8	65.4	34.4	78.4	40.0
1983	41.9	38.6	21.8	46.1	29.3	4.7	25.9	52.4	67.0	36.7	80.2	41.0
1984	44.1	41.3	24.9	53.6	32.3	5.5	28.8	54.6	68.4	39.8	82.3	43.2
1985	46.4	43.5	27.4	60.6	34.0	6.1	31.0	56.2	70.4	41.5	83.9	44.4
1986	48.3	45.9	30.0	65.0	36.8	7.0	34.5	58.2	71.8	44.6	88.2	45.4
1987	49.9	46.8	30.6	62.4	38.3	7.9	35.7	59.2	71.4	46.3	89.9	47.1
1988	52.0	48.5	32.4	66.0	39.7	8.6	36.9	60.8	73.3	47.7	92.4	48.1
1989	53.5	49.7	33.1	65.9	41.8	9.0	37.4	62.4	74.2	50.2	93.9	49.7
1990	54.6	50.0	32.9	64.8	41.2	9.2	37.6	63.2	75.7	49.8	92.2	51.0
1991	54.7	49.0	31.2	57.1	40.2	9.5	36.0	63.0	75.6	49.9	91.5	50.7
1992	56.7	50.6	32.9	61.8	41.6	10.1	36.7	64.2	75.3	52.5	95.1	51.8
1993	58.7	52.7	35.4	65.2	43.7	11.7	38.5	65.8	75.9	54.8	97.5	53.6
1994	61.0	55.5	38.2	69.3	46.1	13.6	40.3	68.4	78.1	57.8	99.1	56.3
1995	62.8	57.1	39.7	68.2	48.2	15.5	41.2	70.1	78.5	60.2	100.9	58.7
1996	65.0	59.7	42.7	71.5	50.3	17.9	44.0	72.1	79.2	63.1	102.5	61.3
1997	67.4	62.5	46.2	76.3	52.6	20.7	46.2	74.2	79.9	65.2	104.6	64.6
1998	71.0	66.7	51.8	84.3	57.0	24.8	50.7	77.0	81.0	69.2	107.5	68.0
1999	74.7	72.0	58.4	92.1	63.6	29.9	56.8	80.9	84.3	73.7	110.0	72.2
2000	78.5	75.7	63.4	95.0	69.7	34.9	61.8	83.5	86.4	77.3	107.1	75.8
2001	80.5	78.0	66.8	99.9	73.3	38.2	60.5	85.0	87.6	78.0	107.0	78.2
2002	82.6	81.0	71.7	105.1	78.6	42.5	64.4	86.5	88.0	80.4	107.9	80.5
2003	85.2	85.0	76.8	108.2	84.2	48.1	70.7	89.6	90.0	84.4	108.0	84.6
2004	88.4	89.3	83.2	111.2	92.2	55.5	77.1	92.5	92.1	88.2	108.9	88.3
2005	91.5	93.0	87.8	109.7	98.3	63.2	83.1	95.6	95.6	92.9	108.4	91.4
2006	94.3	96.5	91.7	105.6	103.3	71.6	89.7	98.7	98.6	96.2	108.1	95.9
2007	96.4	99.2	96.1	107.7	104.1	79.9	94.5	100.4	99.9	98.1	107.9	98.5
2008	96.2	96.2	90.6	94.1	97.8	81.3	91.2	98.8	98.2	97.0	103.7	98.0
2009	95.0	93.2	85.0	87.4	88.5	78.3	88.0	97.2	97.5	92.2	104.3	95.8
2010	96.7	95.8	89.8	90.8	93.4	84.5	92.4	98.7	99.0	96.6	103.8	97.0
2011	98.5	97.9	94.4	93.3	97.2	92.4	96.1	99.6	99.1	99.8	101.4	99.0
2012	100.0	100.0	100.0	100.0	100.0	100.0	100.0	100.0	100.0	100.0	100.0	100.0
2013	101.5	103.1	106.1	104.7	105.8	108.9	104.9	101.8	101.1	100.5	101.9	103.0
2014	104.5	107.4	113.8	110.8	114.8	116.9	113.5	104.5	103.0	103.2	101.9	107.6
2015	108.4	112.5	122.4	119.2	125.4	124.2	122.3	108.0	104.6	106.7	106.6	112.6
2016	111.4	116.5	129.5	129.5	135.4	136.3	124.7	110.8	107.8	109.3	107.1	115.6
2017	114.4	121.2	138.5	129.3	146.1	151.0	129.0	113.7	111.5	111.0	106.5	119.3
2018	117.5	126.2	147.9	134.9	156.2	167.4	136.4	116.7	114.1	115.0	105.9	123.6
2019	120.3	130.9	155.1	134.2	161.5	189.4	143.3	120.4	116.1	119.3	105.5	130.1
2017												
1st quarter	113.4	119.0	134.3	126.1	141.2	145.3	126.1	112.2	110.3	109.6	105.5	117.4
2nd quarter	113.9	120.4	136.4	126.6	143.7	150.1	127.5	113.3	110.8	110.8	107.4	118.9
3rd quarter	114.5	121.6	139.4	129.9	147.3	151.5	130.8	113.8	111.6	110.8	106.9	119.6
4th quarter	115.7	123.9	143.8	134.7	152.3	157.3	131.7	115.2	113.1	112.7	106.3	121.3
2018												
1st quarter	116.3	124.6	145.2	133.9	154.0	162.0	133.1	115.5	113.6	112.6	106.1	121.8
2nd quarter	117.2	125.9	147.8	135.2	155.9	166.1	136.8	116.3	114.0	114.2	106.3	122.9
3rd quarter	118.0	126.7	148.9	135.1	158.1	169.2	136.7	117.1	114.3	116.4	105.5	124.1
4th quarter	118.4	127.5	149.8	135.4	156.8	172.2	139.0	117.9	114.7	117.0	105.9	125.4
2019												
1st quarter	119.0	128.3	150.2	131.3	157.3	178.8	140.9	118.8	114.6	116.8	105.9	128.1
2nd quarter	120.0	130.7	154.7	134.8	161.1	187.9	142.3	120.4	115.8	120.0	105.9	129.9
3rd quarter	120.8	132.1	157.1	134.9	163.2	194.3	144.9	121.3	117.2	119.4	105.4	131.4
4th quarter	121.3	132.3	158.3	136.0	164.3	196.6	145.0	121.0	116.6	121.1	104.7	130.9

Table 4-3. Chain-Type Quantity Indexes for Personal Consumption Expenditures by Major Type of Product—*Continued*

(Index numbers, 2012 = 100.)   NIPA Table 2.3.3

Year and quarter	Services	Household consumption expenditures for services								Nonprofit institutions serving households (NPISHs)		
		Household services, total	Housing and utilities	Health care	Transportation services	Recreation services	Food services and accommodations	Financial services and insurance	Other services	Final consumption expenditures	Gross output	Less: receipts from sales of goods and services
1960	17.7	18.9	20.9	14.0	20.6	10.3	26.4	12.1	25.2	2.9	12.6	19.3
1961	18.3	19.6	22.0	14.7	20.9	10.7	26.5	12.7	26.1	2.9	13.1	20.4
1962	19.2	20.5	23.2	16.0	21.7	11.3	27.5	12.7	27.1	3.0	13.8	21.7
1963	20.0	21.4	24.3	17.0	22.6	11.7	28.2	13.2	28.1	3.1	14.6	23.1
1964	21.2	22.6	25.5	19.0	23.8	12.2	29.6	14.1	29.3	3.5	15.7	24.6
1965	22.4	23.9	26.9	19.8	25.1	12.7	31.7	15.0	31.0	3.7	16.7	26.1
1966	23.5	25.1	28.1	20.9	26.5	13.4	33.3	15.6	32.8	3.8	17.8	28.2
1967	24.5	26.1	29.4	21.9	27.9	13.9	33.0	16.3	34.8	4.0	18.8	30.1
1968	25.8	27.5	30.8	23.6	29.7	14.7	35.0	17.3	35.7	4.2	20.1	32.4
1969	26.9	28.8	32.5	25.3	31.7	15.6	35.8	17.5	36.8	4.2	21.2	34.6
1970	28.0	29.9	33.8	26.7	32.8	16.2	37.0	18.1	38.0	4.5	22.2	36.2
1971	28.9	30.9	35.2	28.6	34.2	16.7	37.0	18.8	38.1	4.8	23.8	39.0
1972	30.6	32.7	37.1	30.6	36.1	17.6	39.8	19.8	39.8	5.0	25.2	41.3
1973	32.0	34.2	38.8	32.8	36.8	19.1	41.8	20.9	40.9	5.0	25.9	42.8
1974	32.7	34.9	40.7	33.9	37.0	20.4	41.4	21.4	39.6	5.0	26.2	43.8
1975	33.9	36.3	42.0	35.7	37.5	21.5	43.1	23.5	39.9	5.0	27.2	45.9
1976	35.3	37.8	43.2	37.2	39.3	22.9	45.7	24.8	41.4	5.3	28.7	48.3
1977	36.8	39.4	44.4	39.0	42.1	24.4	46.9	26.3	43.5	5.4	29.8	50.5
1978	38.5	41.2	46.5	40.8	43.1	25.6	48.9	27.6	45.8	5.6	31.4	53.4
1979	39.7	42.5	48.1	42.2	44.4	26.8	50.4	28.4	46.2	5.8	32.3	54.8
1980	40.3	43.1	49.8	43.4	41.8	27.7	50.5	28.8	45.9	6.1	33.3	56.1
1981	40.9	43.7	50.5	45.5	40.3	30.1	51.0	28.7	44.9	6.6	34.7	57.7
1982	41.8	44.4	50.9	45.6	39.4	31.8	51.3	31.1	45.9	7.8	36.2	58.1
1983	43.9	46.5	52.0	47.0	42.3	34.5	53.0	35.4	48.8	9.0	37.6	58.4
1984	45.6	48.1	53.7	47.7	47.3	36.8	54.4	35.6	51.3	10.5	38.9	58.3
1985	48.0	50.6	55.9	49.2	52.3	39.4	55.2	39.1	54.9	11.4	40.2	59.6
1986	49.4	52.0	57.2	50.7	54.5	41.2	56.8	40.5	55.5	12.9	42.2	61.1
1987	51.7	54.3	58.8	52.8	56.8	43.1	59.7	42.9	59.4	13.9	43.9	62.8
1988	54.0	56.5	60.8	54.9	60.1	47.2	63.3	43.7	61.8	16.0	46.3	64.6
1989	55.7	58.1	62.3	55.5	61.4	50.2	65.0	46.2	63.7	18.4	47.9	64.9
1990	57.3	59.5	63.5	57.4	61.3	52.3	67.4	46.9	64.8	22.4	50.8	66.1
1991	58.2	60.2	64.8	58.5	58.4	52.0	67.4	50.8	62.4	25.6	52.7	66.6
1992	60.6	62.3	66.1	60.3	61.4	55.4	69.1	54.2	64.9	30.9	55.5	67.6
1993	62.4	64.1	67.9	60.6	64.7	58.9	70.6	58.3	66.3	33.4	56.9	68.2
1994	64.4	66.0	70.2	61.0	70.7	61.9	71.7	59.7	68.9	36.4	58.1	68.5
1995	66.2	67.8	72.3	62.3	75.8	66.6	71.9	61.6	70.2	38.1	58.9	68.8
1996	68.2	69.7	74.0	63.6	82.0	69.6	72.3	63.6	72.6	40.3	60.4	69.9
1997	70.3	72.2	75.6	65.6	88.0	71.8	73.8	67.1	76.7	37.1	61.3	72.9
1998	73.5	75.4	77.9	67.7	92.0	74.1	75.7	72.9	82.2	40.7	64.2	75.4
1999	76.3	78.2	80.2	68.8	97.3	77.6	77.6	79.9	84.6	43.9	66.4	76.9
2000	80.2	82.0	82.8	71.1	102.2	79.8	81.1	88.3	90.4	47.5	68.9	78.7
2001	82.0	83.6	85.0	74.7	101.1	79.7	81.0	85.9	94.2	51.5	72.2	81.6
2002	83.5	84.9	85.5	79.1	98.1	80.2	82.2	85.4	94.3	56.7	76.5	85.2
2003	85.3	86.6	86.7	81.2	99.0	82.8	85.4	85.9	96.5	60.9	78.3	85.6
2004	87.9	89.3	88.9	84.1	102.1	87.0	89.0	88.8	98.0	61.9	80.3	88.1
2005	90.8	92.3	92.5	86.9	103.1	89.1	92.3	93.6	99.4	61.7	82.3	91.0
2006	93.3	94.5	94.3	88.9	103.5	92.3	95.5	95.8	102.2	69.3	85.4	92.0
2007	95.0	96.3	95.2	91.0	104.9	96.0	96.9	98.6	103.8	71.3	87.4	94.0
2008	96.3	97.0	96.2	93.2	101.0	96.5	96.2	100.6	102.5	81.7	90.7	94.3
2009	95.9	96.5	97.3	95.3	93.8	95.2	92.9	98.7	99.5	84.7	93.0	96.2
2010	97.1	97.6	98.5	96.7	94.1	98.0	95.1	98.8	99.3	87.5	94.9	97.7
2011	98.8	99.2	99.8	98.2	98.2	97.6	97.7	101.4	100.0	91.7	96.7	98.6
2012	100.0	100.0	100.0	100.0	100.0	100.0	100.0	100.0	100.0	100.0	100.0	100.0
2013	100.6	100.6	100.5	100.6	104.5	102.0	101.7	99.4	98.9	101.2	100.7	100.5
2014	103.1	103.2	102.2	103.9	109.8	104.6	105.2	99.7	102.3	101.1	102.4	102.9
2015	106.5	106.9	104.7	109.5	113.8	108.5	109.7	102.1	105.6	99.1	105.0	107.3
2016	109.0	109.3	106.3	113.9	118.7	111.7	112.1	100.1	109.3	103.5	109.1	111.3
2017	111.2	111.6	107.1	116.6	123.1	112.9	114.9	102.2	113.5	103.7	110.6	113.3
2018	113.5	113.9	108.6	119.4	127.8	115.3	118.1	102.5	117.3	106.9	113.2	115.7
2019	115.6	116.2	109.9	122.6	129.7	116.8	119.5	104.6	122.1	103.6	114.7	119.1
2017												
1st quarter	110.7	111.0	106.4	115.9	120.6	113.5	114.8	101.7	112.9	106.1	111.0	112.9
2nd quarter	110.9	111.2	107.2	115.6	121.6	113.4	114.4	101.9	113.2	104.2	109.7	111.9
3rd quarter	111.3	111.7	107.2	117.0	123.8	112.6	114.8	102.6	113.2	102.3	110.5	113.8
4th quarter	112.0	112.5	107.9	117.8	126.6	112.0	115.6	102.5	114.7	102.1	111.0	114.6
2018												
1st quarter	112.5	113.0	108.0	118.3	129.3	114.8	116.8	102.3	114.4	104.1	111.2	114.0
2nd quarter	113.3	113.6	108.5	118.9	128.1	114.7	117.7	102.1	117.4	106.1	113.0	115.7
3rd quarter	114.0	114.4	108.6	120.4	126.9	115.9	119.4	102.5	118.2	106.2	113.4	117.5
4th quarter	114.3	114.5	109.1	119.9	126.7	115.9	118.6	103.0	119.2	110.9	114.3	115.6
2019												
1st quarter	114.8	115.3	109.3	121.2	127.5	115.9	118.1	104.1	121.0	104.9	114.8	118.7
2nd quarter	115.3	115.9	109.7	122.4	129.0	117.0	119.4	104.1	120.9	103.6	114.6	119.0
3rd quarter	115.9	116.5	110.2	122.7	131.2	116.7	120.2	104.6	122.5	103.7	114.5	118.7
4th quarter	116.5	117.2	110.2	124.1	131.3	117.6	120.1	105.6	123.8	102.3	115.0	120.1

Table 4-4. Chain-Type Price Indexes for Personal Consumption Expenditures by Major Type of Product

(Index numbers, 2012 =1000.)   NIPA Table 2.3.4

Year and quarter	Personal consumption expend-itures, total	Total goods	Goods									
			Durable goods					Nondurable goods				
			Durable goods, total	Motor vehicles and parts	Furnishings and household equipment	Recreational goods and vehicles	Other durable goods	Nondurable goods, total	Food and beverages off-premises	Clothing and footwear	Gasoline and other energy goods	Other nondurable goods
1960	16.5	26.6	56.9	29.6	50.0	266.2	29.7	18.1	16.5	51.7	7.0	16.6
1961	16.6	26.7	57.0	29.8	50.2	264.8	29.6	18.1	16.6	52.0	7.1	16.6
1962	16.8	26.9	57.3	30.2	50.1	261.9	29.5	18.2	16.7	52.2	7.1	16.7
1963	17.0	27.1	57.5	30.5	50.1	261.3	29.6	18.4	16.9	52.6	7.1	16.9
1964	17.3	27.4	57.8	30.7	50.2	258.2	30.4	18.7	17.3	53.1	7.0	17.2
1965	17.5	27.7	57.3	30.5	49.8	254.6	30.2	19.0	17.6	53.6	7.3	17.4
1966	18.0	28.3	57.2	30.3	50.3	251.8	30.2	19.6	18.3	55.1	7.5	17.8
1967	18.4	28.8	58.1	30.7	51.4	253.3	30.9	20.0	18.4	57.4	7.7	18.2
1968	19.2	29.8	59.9	31.7	53.4	258.3	31.6	20.7	18.9	60.6	7.8	18.9
1969	20.0	30.9	61.6	32.4	55.3	264.2	33.2	21.6	19.8	63.9	8.1	19.6
1970	21.0	32.1	63.0	33.4	56.5	264.6	34.0	22.6	20.9	66.3	8.2	20.5
1971	21.8	33.1	65.1	35.0	57.5	274.0	35.0	23.2	21.4	68.6	8.3	21.3
1972	22.6	33.9	65.9	35.1	58.4	278.0	36.1	23.9	22.4	69.9	8.4	21.8
1973	23.8	35.9	66.8	35.2	59.7	282.2	37.3	25.9	25.2	72.2	9.3	22.4
1974	26.3	40.4	71.2	37.4	64.6	294.7	40.2	29.8	29.0	77.5	12.9	24.6
1975	28.5	43.7	77.6	41.1	71.0	314.2	43.2	32.1	31.2	81.0	13.8	27.5
1976	30.0	45.4	81.8	44.2	74.1	325.1	45.3	33.1	31.8	83.6	14.5	28.9
1977	32.0	47.8	85.5	46.8	76.8	334.4	47.3	35.0	33.7	87.2	15.5	30.6
1978	34.2	50.8	90.3	50.1	80.7	346.4	49.9	37.3	36.9	88.8	16.3	32.2
1979	37.3	55.6	96.3	53.8	85.7	361.8	54.1	41.3	40.5	90.6	21.9	34.3
1980	41.3	61.8	104.6	57.8	92.7	381.9	65.1	46.3	43.9	93.9	30.4	37.5
1981	45.0	66.4	111.6	62.3	99.6	399.0	68.7	49.9	47.1	97.0	34.3	40.6
1982	47.5	68.2	116.0	65.2	104.7	408.2	69.4	51.0	48.3	98.4	32.7	43.8
1983	49.5	69.4	118.3	67.3	107.1	403.3	71.6	51.9	48.8	99.7	31.5	46.8
1984	51.3	70.7	120.1	69.2	108.4	397.6	73.1	53.0	50.3	100.3	31.1	48.5
1985	53.1	71.9	121.4	70.8	109.6	390.6	73.9	54.0	50.9	102.4	31.2	50.4
1986	54.3	71.5	122.7	72.7	110.8	378.3	75.7	53.3	52.1	102.0	24.5	52.4
1987	56.0	73.8	126.2	76.1	112.5	373.5	80.8	55.1	53.7	104.9	25.4	54.5
1988	58.2	75.8	128.3	77.3	114.1	371.1	86.1	56.9	55.3	108.3	25.6	57.0
1989	60.7	78.7	130.6	79.6	114.5	370.1	90.1	59.7	58.2	109.9	27.9	60.1
1990	63.4	81.9	131.9	79.8	115.6	367.1	95.8	63.1	61.1	113.5	31.9	63.2
1991	65.5	83.9	133.9	82.1	116.3	362.4	99.3	65.0	63.0	115.6	31.4	66.5
1992	67.2	84.9	134.8	83.6	117.7	351.3	101.9	65.9	63.5	116.5	31.1	69.0
1993	68.9	85.7	136.1	86.9	118.3	338.7	102.7	66.4	64.3	115.9	30.8	70.2
1994	70.3	86.6	138.8	90.9	120.1	331.6	104.4	66.8	65.4	113.9	30.9	70.7
1995	71.8	87.4	139.8	94.5	119.9	317.7	106.1	67.5	66.8	111.3	31.3	71.9
1996	73.3	88.3	138.4	96.4	120.3	295.2	105.5	69.0	68.9	109.9	33.5	73.3
1997	74.6	88.2	135.4	96.8	120.1	270.5	104.0	69.7	70.2	110.0	33.5	74.1
1998	75.2	86.9	131.5	95.7	120.0	247.6	102.1	69.2	71.1	108.0	29.2	75.5
1999	76.3	87.3	128.0	96.0	118.4	226.1	99.5	70.9	72.2	106.5	31.6	78.4
2000	78.2	89.1	125.7	96.4	117.6	211.9	98.2	74.0	73.9	105.3	40.8	80.1
2001	79.7	89.0	123.3	96.8	115.5	197.5	98.5	74.8	76.1	103.2	39.4	82.0
2002	80.8	88.2	120.1	96.3	113.2	184.7	96.9	74.8	77.2	100.5	36.9	83.5
2003	82.4	88.1	115.8	93.5	110.0	173.0	94.7	76.3	78.7	98.0	43.1	83.7
2004	84.4	89.3	113.5	92.8	108.6	164.5	94.2	79.0	81.2	97.6	50.7	84.7
2005	86.8	91.1	112.3	94.2	108.6	155.6	93.3	82.0	82.6	96.8	62.0	86.0
2006	89.2	92.3	110.4	94.3	108.1	146.2	94.1	84.6	84.0	96.4	70.1	87.8
2007	91.4	93.3	108.0	93.8	107.2	136.2	95.9	87.0	87.2	95.5	75.9	89.1
2008	94.2	96.1	106.0	92.0	106.5	129.4	98.3	91.9	92.6	94.7	89.4	91.2
2009	94.1	93.8	104.0	91.9	106.1	121.8	98.6	89.5	93.7	95.6	65.2	95.3
2010	95.7	95.2	102.1	95.7	101.5	113.4	98.2	92.2	93.9	94.9	76.9	97.0
2011	98.1	98.8	101.3	98.7	100.0	106.0	100.8	97.7	97.7	96.5	96.7	98.5
2012	100.0	100.0	100.0	100.0	100.0	100.0	100.0	100.0	100.0	100.0	100.0	100.0
2013	101.3	99.4	98.0	100.5	98.1	94.7	97.9	100.1	101.0	101.0	97.3	100.2
2014	102.8	98.9	95.4	100.6	94.7	90.5	94.0	100.6	102.9	101.3	93.8	101.3
2015	103.0	95.9	93.4	100.5	92.4	86.9	90.6	97.1	104.1	100.1	68.8	102.5
2016	104.1	94.3	91.1	99.4	90.0	82.7	89.9	95.9	103.0	99.8	60.8	103.8
2017	106.0	94.6	89.0	98.2	87.5	79.4	88.7	97.4	102.9	99.2	68.7	105.0
2018	108.2	95.2	87.5	97.8	86.6	76.4	87.1	99.3	103.4	99.3	78.1	105.6
2019	109.9	94.8	86.5	98.0	87.2	73.4	85.1	99.2	104.4	97.9	75.4	106.0
2017												
1st quarter	105.4	94.9	90.0	98.9	89.1	80.3	89.4	97.4	102.5	100.0	69.1	104.7
2nd quarter	105.7	94.3	89.3	98.3	88.0	79.6	88.9	96.8	102.9	99.1	65.1	104.9
3rd quarter	106.1	94.4	88.6	97.7	86.8	79.4	88.1	97.4	103.0	99.0	67.7	105.2
4th quarter	106.8	94.8	88.3	97.8	86.2	78.5	88.4	98.2	103.1	98.6	73.1	105.2
2018												
1st quarter	107.5	95.2	88.0	97.5	86.4	77.6	88.5	99.0	103.1	99.8	76.8	105.5
2nd quarter	108.1	95.4	87.7	97.5	87.0	76.7	87.6	99.5	103.4	100.2	78.0	105.8
3rd quarter	108.5	95.3	87.4	98.1	86.1	76.0	86.8	99.5	103.5	98.5	80.1	105.5
4th quarter	108.9	95.0	87.1	98.0	86.7	75.5	85.4	99.2	103.6	98.6	77.7	105.4
2019												
1st quarter	109.0	94.6	87.0	97.7	87.5	75.0	84.9	98.6	104.3	98.9	72.1	105.3
2nd quarter	109.7	95.0	86.7	98.2	87.3	73.7	85.6	99.4	104.4	97.6	77.2	105.8
3rd quarter	110.1	94.8	86.4	98.3	87.3	72.7	85.1	99.2	104.4	98.4	75.2	106.0
4th quarter	110.5	94.8	85.8	98.0	86.5	72.0	84.7	99.7	104.6	96.8	77.1	106.8

Table 4-4.   Chain-Type Price Indexes for Personal Consumption Expenditures by Major Type of Product—*Continued*

(Index numbers, 2012 =1000.)

NIPA Table 2.3.4

Year and quarter	Services	Household services, total	Housing and utilities	Health care	Transporta- tion services	Recreation services	Food services and accommo- dations	Financial services and insurance	Other services	Final consumption expenditures	Gross output	Less: receipts from sales of goods and services
1960	11.8	11.2	13.6	6.3	13.0	14.6	11.0	13.3	11.7	51.5	9.6	5.7
1961	12.0	11.4	13.7	6.4	13.4	15.0	11.2	13.7	11.9	52.1	9.7	5.8
1962	12.3	11.7	13.9	6.6	13.7	15.4	11.5	14.2	12.1	52.8	9.9	5.9
1963	12.5	11.8	14.0	6.8	13.8	15.7	11.7	14.3	12.3	53.7	10.2	6.1
1964	12.7	12.1	14.2	7.0	14.0	16.1	11.9	14.9	12.6	53.9	10.4	6.3
1965	12.9	12.3	14.3	7.2	14.2	16.5	12.2	15.3	12.9	55.4	10.7	6.5
1966	13.3	12.6	14.5	7.5	14.5	16.9	12.6	16.1	13.3	57.4	11.1	6.7
1967	13.7	13.1	14.7	8.0	15.0	17.5	13.3	16.5	13.7	59.3	11.6	7.0
1968	14.3	13.6	15.1	8.5	15.7	18.4	14.0	17.8	14.3	61.5	12.2	7.4
1969	15.1	14.3	15.6	9.1	16.6	19.2	14.9	19.3	15.0	65.4	13.0	7.9
1970	15.9	15.1	16.2	9.8	17.9	20.1	16.0	20.3	15.8	68.3	13.9	8.5
1971	16.8	16.0	17.1	10.3	19.2	21.0	16.8	21.5	16.7	72.1	14.5	8.9
1972	17.5	16.6	17.7	10.7	19.9	21.5	17.4	22.8	17.5	75.1	15.2	9.4
1973	18.3	17.4	18.5	11.3	20.8	22.3	18.6	23.3	18.6	79.9	16.2	10.0
1974	19.9	18.9	19.5	12.3	22.6	23.8	20.8	25.1	20.6	91.5	18.1	11.1
1975	21.6	20.5	21.1	13.7	25.1	25.6	22.6	26.9	22.1	99.9	19.9	12.2
1976	23.1	21.9	22.6	15.0	26.9	26.9	24.1	27.9	23.4	105.1	21.5	13.3
1977	24.8	23.6	24.6	16.3	28.8	28.1	25.9	30.2	24.6	112.2	23.1	14.3
1978	26.8	25.4	26.3	17.7	30.7	29.8	28.2	33.9	25.9	119.6	24.9	15.4
1979	29.0	27.5	28.5	19.4	33.5	31.8	31.2	35.9	27.8	131.3	27.4	17.0
1980	32.0	30.4	31.4	21.7	38.8	34.2	34.2	38.8	30.3	143.5	30.6	19.1
1981	35.3	33.5	34.9	24.4	43.5	36.4	37.3	41.9	33.5	150.6	34.0	21.6
1982	38.1	36.3	38.1	27.1	46.0	38.4	39.4	44.5	36.1	142.8	36.6	24.2
1983	40.4	38.7	40.6	29.6	47.7	40.0	41.1	48.5	38.1	132.1	38.6	26.5
1984	42.5	40.9	42.7	31.8	49.6	41.8	43.0	51.6	40.0	126.3	40.5	28.6
1985	44.6	43.0	44.9	33.7	50.6	43.6	44.8	55.6	41.5	123.5	42.1	30.3
1986	46.4	44.8	47.0	35.7	51.1	45.2	46.7	56.5	43.7	119.5	43.6	32.0
1987	47.8	46.3	48.7	38.0	53.2	47.0	48.5	53.8	44.7	117.2	45.5	34.0
1988	50.1	48.6	50.7	41.0	55.7	48.8	50.6	56.6	46.6	115.8	48.2	36.7
1989	52.4	51.0	52.7	44.6	58.2	51.2	52.9	58.7	48.3	108.3	50.8	40.2
1990	54.8	53.6	55.0	48.4	60.4	54.0	55.3	60.0	50.3	99.2	53.2	43.7
1991	57.0	55.9	56.8	52.2	62.1	56.8	57.5	60.1	52.7	91.7	55.7	47.4
1992	59.0	58.2	58.4	55.8	63.8	58.6	58.8	62.3	55.0	83.3	57.7	51.0
1993	61.1	60.4	60.1	58.8	66.1	60.5	59.9	65.7	56.9	78.9	59.5	53.8
1994	62.7	62.1	61.6	61.2	66.8	61.8	61.0	67.0	59.0	78.7	61.2	55.8
1995	64.5	63.9	63.3	63.4	68.2	63.1	62.4	68.7	60.9	78.7	62.9	57.9
1996	66.2	65.6	65.2	64.9	68.9	65.2	64.1	71.3	62.7	80.0	64.7	59.6
1997	68.1	67.5	66.9	66.2	70.5	67.3	66.0	74.3	64.5	83.4	66.3	60.9
1998	69.5	68.8	68.5	67.5	71.8	69.0	67.8	74.6	65.8	89.1	68.3	62.1
1999	71.0	70.2	70.2	69.0	72.7	71.2	69.5	74.2	67.4	92.3	70.2	63.6
2000	72.9	72.0	72.5	70.9	75.0	74.0	71.5	75.0	68.6	97.4	72.9	65.8
2001	75.2	74.2	75.9	73.3	75.4	76.4	73.5	74.6	70.9	101.9	75.7	68.1
2002	77.1	76.2	78.1	75.2	75.3	78.6	75.3	76.4	73.7	102.3	77.9	70.7
2003	79.5	78.8	80.6	78.0	76.9	81.0	76.9	79.6	76.6	98.7	80.0	74.3
2004	82.0	81.3	82.8	80.9	77.8	83.1	79.4	83.1	79.3	97.6	82.5	77.7
2005	84.7	84.1	85.8	83.5	80.7	85.5	82.0	85.8	82.1	99.7	85.4	80.8
2006	87.6	87.1	89.4	86.0	84.1	88.4	84.8	88.4	85.0	101.0	88.3	84.1
2007	90.5	90.0	92.5	89.2	86.0	90.8	88.0	91.2	87.8	102.2	91.0	87.2
2008	93.2	92.8	95.5	91.6	90.8	93.6	91.5	91.7	91.6	102.9	93.3	90.1
2009	94.2	94.0	96.6	94.1	93.0	94.7	93.5	87.9	93.9	99.9	94.6	92.8
2010	96.0	95.8	96.8	96.5	95.1	95.7	94.8	93.1	96.1	98.5	96.2	95.4
2011	97.8	97.7	98.1	98.2	98.1	97.3	97.2	96.0	97.9	99.6	98.1	97.6
2012	100.0	100.0	100.0	100.0	100.0	100.0	100.0	100.0	100.0	100.0	100.0	100.0
2013	102.3	102.3	102.4	101.4	101.0	101.7	102.1	105.3	102.3	102.0	102.1	102.2
2014	104.8	104.8	105.4	102.5	102.3	103.6	104.8	111.0	103.7	105.9	104.4	103.9
2015	106.7	106.5	108.1	103.1	102.7	105.2	107.7	114.3	104.5	109.8	106.5	105.3
2016	109.2	109.0	111.2	104.3	103.6	107.7	110.5	119.9	105.5	113.4	108.7	107.0
2017	111.9	111.6	115.0	105.9	104.9	110.8	112.8	125.6	106.0	117.7	111.4	109.2
2018	115.0	114.7	118.6	107.9	107.1	113.1	115.4	133.2	108.1	120.2	114.1	111.9
2019	117.7	117.5	122.3	109.8	109.3	115.3	118.7	137.1	109.6	124.1	116.9	114.3
2017												
1st quarter	110.8	110.6	113.6	105.3	104.4	109.8	112.1	122.7	105.9	116.7	110.5	108.3
2nd quarter	111.5	111.2	114.5	105.7	104.9	110.3	112.5	124.9	105.7	117.7	111.2	108.8
3rd quarter	112.1	111.9	115.4	106.0	104.7	111.2	112.8	126.2	106.1	118.0	111.6	109.3
4th quarter	113.0	112.7	116.3	106.6	105.5	111.7	113.9	128.6	106.4	118.3	112.4	110.3
2018												
1st quarter	113.8	113.6	117.3	107.1	105.7	112.3	114.4	130.9	107.3	119.2	113.1	110.9
2nd quarter	114.7	114.4	118.1	107.7	106.6	112.7	115.3	132.5	107.8	119.7	113.7	111.6
3rd quarter	115.4	115.1	119.0	108.1	107.5	113.5	115.7	134.1	108.5	120.2	114.3	112.2
4th quarter	116.1	115.9	120.0	108.6	108.7	113.9	116.3	135.4	108.9	121.7	115.2	112.8
2019												
1st quarter	116.6	116.4	120.9	109.0	108.3	114.4	117.6	134.9	109.2	121.9	115.7	113.4
2nd quarter	117.4	117.2	121.8	109.5	109.2	115.0	118.5	136.9	109.3	123.6	116.6	114.0
3rd quarter	118.2	117.8	122.7	110.0	109.3	115.4	119.2	138.0	109.9	125.4	117.4	114.6
4th quarter	118.8	118.5	123.6	110.6	110.2	116.6	119.4	138.6	110.1	125.4	118.0	115.3

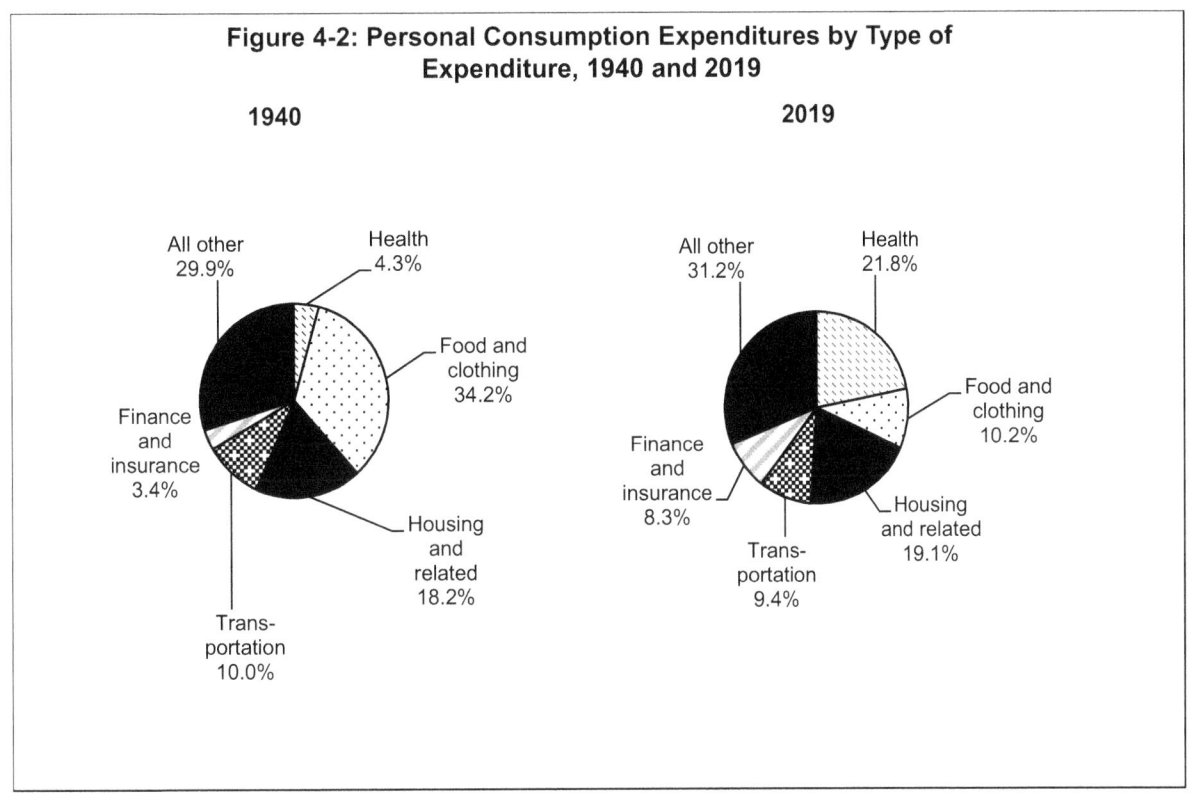

Figure 4-2: Personal Consumption Expenditures by Type of Expenditure, 1940 and 2019

- Figure 4-2 compares the composition of household consumer spending (excluding the net spending of nonprofit institutions) in 1940 and 2019, the latest available data at the time this edition of *Business Statistics* was compiled. (Table 4-5)

- Household spending on health in 2019 made up 21.8 percent of all household consumption spending—more than five times the percentage in 1940. It is important to recognize that household consumption spending, as presented in these data, includes (in health spending and in the total) medical care payments made by government and private insurance on behalf of individuals as well as out-of-pocket consumer payments. (Table 4-5)

- In 2019, a much smaller share was required for food and clothing than in 1940, (10.2 percent vs 34.2 percent respectively) as seen in Figure 4-2. The shares of housing-related spending and transportation were similar in the two years. The share of finance and insurance spending nearly tripled, and households also spent a larger share on the remaining category, here labeled "all other," which includes communication, recreation, education, food services, accommodations, and other miscellaneous goods and services. (Table 4-5)

Table 4-5. Household Consumption Expenditures by Function

(Billions of dollars.)   NIPA Table 2.5.5

Year	Household consumption expenditures, total	Food and beverages off-premises	Clothing, footwear, and related services	Housing, utilities, and fuels	Furnishings, household equipment, and maintenance	Health	Transportation	Communication	Recreation	Education	Food services and accommodations	Financial services and insurance	Other goods and services	Foreign travel and expenditures, net
1940	70.2	15.9	8.1	12.8	6.0	3.0	7.0	0.7	3.9	0.6	4.4	2.4	5.3	0.1
1941	80.0	18.2	9.5	13.7	7.1	3.2	8.3	0.8	4.5	0.6	5.3	2.5	6.0	0.1
1942	87.7	21.6	11.8	14.8	7.4	3.6	5.4	1.0	4.9	0.7	6.9	2.6	6.8	0.2
1943	98.4	24.3	14.4	15.5	7.4	4.0	5.4	1.2	5.3	0.8	9.1	2.8	8.0	0.3
1944	107.0	26.0	15.7	16.2	8.0	4.5	5.7	1.3	5.8	0.9	10.9	3.0	8.5	0.6
1945	118.2	28.2	17.6	16.9	8.9	4.8	6.6	1.4	6.5	0.9	12.7	3.2	9.3	1.2
1946	142.2	34.9	19.7	18.8	12.7	5.9	12.1	1.6	9.0	1.0	12.7	3.7	10.3	-0.1
1947	159.7	40.2	20.4	21.2	15.5	6.6	15.4	1.7	9.8	1.2	12.5	4.1	11.1	0.0
1948	172.6	41.8	21.8	23.9	16.8	7.6	18.0	1.9	10.3	1.3	12.6	4.6	11.7	0.3
1949	176.0	40.4	21.1	25.6	16.3	7.9	21.2	2.1	10.6	1.4	12.4	4.8	11.8	0.6
1950	189.6	41.4	21.4	28.4	18.7	8.5	24.6	2.3	11.7	1.5	12.7	5.4	12.2	0.7
1951	205.7	46.0	23.1	31.6	19.7	9.3	24.2	2.6	12.3	1.6	15.0	6.2	13.3	0.9
1952	216.4	48.4	23.9	34.6	19.3	10.1	24.4	2.9	13.0	1.7	16.0	6.6	14.3	1.1
1953	229.7	49.3	24.2	37.9	20.0	11.0	28.0	3.2	13.8	1.8	16.5	7.5	15.0	1.5
1954	236.4	50.6	24.3	40.9	19.9	11.9	27.4	3.3	14.3	1.9	16.6	8.3	15.4	1.5
1955	254.9	52.0	25.4	43.8	22.3	12.8	33.4	3.6	15.4	2.1	17.0	9.2	16.3	1.6
1956	267.3	54.2	26.5	46.8	23.6	14.1	32.7	3.9	16.4	2.4	17.7	10.0	17.5	1.7
1957	282.3	57.1	26.8	50.0	23.9	15.5	35.7	4.3	16.8	2.6	18.5	10.6	18.8	1.7
1958	291.3	59.8	27.1	53.3	24.1	17.1	33.7	4.5	17.3	2.9	18.7	11.2	19.9	1.9
1959	312.3	61.6	28.6	56.9	25.6	18.8	38.8	4.8	18.8	3.1	19.7	12.1	21.4	2.0
1960	326.1	62.6	29.3	60.5	26.2	20.4	40.8	5.2	19.7	3.4	20.5	13.2	22.4	2.1
1961	336.2	63.7	29.9	64.0	26.8	21.8	39.3	5.6	20.5	3.6	21.0	14.3	23.7	2.0
1962	357.1	64.7	31.3	68.2	28.3	24.3	44.1	5.9	22.1	4.0	22.3	14.8	24.9	2.3
1963	376.3	65.9	32.1	72.2	29.8	26.4	47.8	6.4	23.9	4.3	23.3	15.5	26.2	2.5
1964	404.2	69.5	34.7	76.2	32.7	29.8	50.7	6.9	26.3	4.8	24.9	17.2	27.8	2.6
1965	436.0	74.4	36.5	81.0	34.7	32.0	56.5	7.5	28.9	5.5	27.2	18.8	30.2	2.9
1966	472.5	80.6	39.8	85.8	37.8	35.0	59.0	8.1	33.4	6.2	29.6	20.6	33.3	3.1
1967	498.6	82.6	41.7	91.1	39.7	38.4	61.0	8.9	35.8	6.9	31.0	22.1	35.8	3.8
1968	548.1	88.8	45.6	97.4	43.2	43.8	69.8	9.7	39.9	7.7	34.6	25.2	38.6	3.7
1969	594.2	95.4	48.8	105.5	45.4	50.0	75.9	10.7	43.7	8.7	37.5	27.7	41.0	4.0
1970	636.3	103.5	49.9	113.8	46.6	56.8	76.5	11.6	47.0	9.9	41.6	30.1	44.3	4.5
1971	688.2	107.1	53.4	124.5	48.8	63.4	88.7	12.8	50.1	10.9	43.8	33.1	47.0	4.8
1972	755.3	114.5	58.0	136.2	53.5	70.4	98.3	14.4	56.2	11.7	48.9	37.1	51.0	5.2
1973	835.8	126.7	63.7	149.7	59.9	78.7	108.6	16.1	63.0	13.0	54.8	39.9	56.9	4.7
1974	914.7	143.0	66.9	166.3	64.6	88.7	112.8	17.7	68.9	14.2	60.6	44.1	62.2	4.7
1975	1 013.5	156.6	71.4	184.8	67.8	102.7	124.4	20.0	77.1	15.9	68.8	51.8	67.8	4.4
1976	1 128.7	167.3	76.9	204.6	75.0	116.8	147.4	22.4	86.2	17.4	77.8	56.8	76.3	3.8
1977	1 253.3	179.8	84.3	228.7	84.0	131.6	168.1	24.3	94.9	18.8	85.7	65.1	83.8	4.3
1978	1 399.2	196.1	94.6	255.6	93.6	149.4	184.5	27.0	106.3	20.9	97.1	76.7	93.2	4.3
1979	1 559.3	218.4	101.9	287.5	104.3	169.9	207.1	29.4	118.9	22.9	110.9	83.6	100.4	4.1
1980	1 720.6	239.2	108.8	327.7	110.7	195.5	226.5	31.8	127.4	25.4	121.7	91.7	110.7	3.5
1981	1 899.8	255.3	119.1	367.6	118.3	228.9	250.7	36.0	141.6	28.3	133.9	98.5	121.2	0.4
1982	2 033.3	267.1	122.7	401.7	121.0	254.9	255.9	41.1	151.4	31.0	142.5	113.7	127.8	2.5
1983	2 240.8	277.0	132.9	434.6	132.3	287.1	284.9	44.8	169.3	34.3	153.6	141.0	143.4	5.4
1984	2 447.0	291.1	144.7	471.1	147.1	315.2	321.8	48.4	190.0	37.7	164.9	150.8	157.7	6.6
1985	2 664.9	303.0	154.3	513.9	156.0	345.3	357.4	53.1	207.2	41.2	174.3	178.2	173.3	7.7
1986	2 833.7	316.4	165.1	548.1	168.7	378.4	362.6	56.9	225.4	44.5	186.7	187.7	190.0	3.3
1987	3 020.6	324.3	176.4	582.6	176.9	419.9	376.6	60.0	247.0	48.8	204.4	189.5	208.2	6.1
1988	3 266.7	342.8	188.1	625.9	186.6	470.7	404.8	63.1	273.4	54.4	225.8	202.9	225.4	2.8
1989	3 508.1	365.4	201.2	666.9	197.5	519.0	428.3	67.3	295.7	60.6	242.6	222.3	245.7	-3.7
1990	3 733.1	391.2	206.5	709.3	200.6	583.7	442.9	70.1	314.7	66.0	262.7	230.8	262.3	-7.7
1991	3 863.3	403.0	210.1	747.5	199.1	638.4	418.3	73.9	326.3	70.6	273.4	250.1	268.0	-15.2
1992	4 109.6	404.5	223.0	783.3	209.4	700.4	451.3	81.1	346.8	76.4	286.3	277.0	290.1	-20.0
1993	4 361.9	413.5	231.1	827.3	221.9	741.7	485.0	85.8	378.4	81.1	298.4	314.0	304.5	-20.7
1994	4 623.2	432.1	240.1	876.1	238.6	779.9	527.3	93.3	414.0	86.4	308.3	327.9	316.9	-17.6
1995	4 860.3	443.7	244.7	926.7	251.7	826.0	552.5	98.9	449.8	92.3	316.1	347.0	332.7	-21.9
1996	5 134.6	461.9	253.5	976.7	263.7	868.3	596.7	108.3	481.5	99.6	326.6	372.1	350.8	-25.2
1997	5 431.2	474.8	262.0	1 023.1	277.3	919.9	639.3	120.1	509.5	107.1	343.4	408.9	368.4	-22.5
1998	5 753.4	487.4	273.1	1 077.0	297.3	979.7	666.2	128.8	544.3	115.2	361.8	446.1	393.7	-17.3
1999	6 140.8	515.5	287.2	1 135.5	320.6	1 033.3	726.3	140.2	589.9	123.9	380.3	486.4	429.4	-27.7
2000	6 604.1	540.6	297.5	1 214.5	344.0	1 109.6	793.1	153.8	633.7	134.3	408.8	543.0	458.5	-27.2
2001	6 886.6	564.0	294.6	1 303.0	353.2	1 209.4	805.6	159.1	647.0	143.6	419.7	525.7	477.9	-16.2
2002	7 144.3	575.1	295.2	1 347.9	366.3	1 317.1	806.8	161.1	669.3	149.5	436.3	534.7	496.8	-11.8
2003	7 517.6	599.6	301.4	1 411.6	381.1	1 410.7	840.1	165.1	704.3	159.5	462.7	560.3	526.5	-5.3
2004	8 006.3	632.6	313.4	1 488.4	406.6	1 514.9	893.8	171.2	758.8	169.0	498.2	605.5	559.1	-5.3
2005	8 536.9	668.2	326.4	1 606.0	430.1	1 612.3	955.3	177.3	805.2	180.5	533.6	659.0	589.5	-6.6
2006	9 021.2	700.3	336.2	1 706.1	448.7	1 715.0	987.9	190.5	855.7	193.1	570.6	695.0	624.5	-2.3
2007	9 457.7	737.3	339.5	1 783.8	455.0	1 823.2	1 028.0	203.0	897.0	206.0	601.5	737.2	657.6	-11.6
2008	9 689.0	769.1	333.1	1 865.5	438.7	1 909.3	1 017.0	214.0	900.0	216.4	620.2	756.6	667.9	-18.8
2009	9 553.0	772.9	319.1	1 900.9	405.9	1 999.3	879.7	212.8	863.4	226.1	612.7	711.3	666.4	-17.4
2010	9 891.4	786.9	331.7	1 928.5	410.7	2 078.1	961.8	220.7	884.4	240.3	635.7	754.4	687.2	-28.9
2011	10 329.3	819.5	347.1	1 982.9	423.7	2 153.2	1 080.3	228.1	901.6	248.2	669.5	797.9	712.1	-34.9
2012	10 665.3	846.2	359.5	2 020.5	438.6	2 230.7	1 135.5	232.5	934.4	250.6	704.9	820.1	725.4	-33.6
2013	10 964.6	864.0	365.3	2 080.5	454.3	2 288.3	1 170.5	237.5	967.4	256.7	732.3	858.4	740.1	-50.5
2014	11 457.2	896.9	376.0	2 176.2	476.6	2 409.1	1 202.0	254.0	1 004.5	263.3	776.9	908.1	766.2	-52.8
2015	11 925.9	921.0	384.1	2 277.6	500.1	2 564.8	1 163.7	259.8	1 045.6	271.9	832.9	957.3	797.9	-50.8
2016	12 369.2	939.9	391.8	2 374.3	520.7	2 696.8	1 164.1	268.5	1 096.9	281.9	873.2	984.0	822.9	-45.7
2017	12 923.7	970.2	395.8	2 477.2	542.9	2 806.3	1 235.3	266.1	1 151.1	290.3	913.7	1 052.4	857.6	-35.3
2018	13 554.5	998.8	410.6	2 590.8	575.3	2 925.6	1 318.6	273.0	1 209.7	299.4	961.2	1 119.5	893.4	-21.3
2019	14 105.4	1 025.7	420.0	2 700.2	599.6	3 071.8	1 321.7	279.9	1 277.9	311.3	999.5	1 176.1	924.5	-2.7

NOTES AND DEFINITIONS, CHAPTER 4

TABLES 4-1 THROUGH Table 4-5

SOURCE: U.S. DEPARTMENT OF COMMERCE, BUREAU OF ECONOMIC ANALYSIS (BEA)

All personal income and personal consumption expenditure series are from the national income and product accounts (NIPAs). All quarterly series are shown at seasonally adjusted annual rates. Current and constant dollar values are in billions of dollars. Indexes of price and quantity are based on the average for the year 2012, which equals 100.

Tables 4-1 through 4-4 cover all income and spending by the personal sector, which includes nonprofit institutions serving households (NPISHs). In a new feature introduced in the 2009 comprehensive revision of the NIPAs, Tables 4-2, 4-3, and 4-4 show the services component of personal consumption expenditures broken down into separate aggregates for "household consumption expenditures for services" and "final consumption expenditures" by NPISHs.

The last table gives further details of the separate accounts for households and NPISHs that are only available annually. Table 4-5 shows a more detailed functional breakdown of household consumption expenditures, not including NPISHs.

In several cases, the notes and definitions below will refer to *imputations* or *imputed values*. See the notes and definitions to Chapter 1 for an explanation of imputation and the role it plays in national and personal income measurement.

TABLES 4-1 THROUGH 4-4

Sources and Disposition of Personal Income; Personal Consumption Expenditures by Major Type of Product

Definitions

Personal income is the income received by persons residing in the United States from participation in production, from government and business transfer payments, and from government interest, which is treated similarly to a transfer payment rather than as income from participation in production. *Persons* denotes the total for individuals, *nonprofit institutions that primarily serve households (NPISHs),* private noninsured welfare funds, and private trust funds. Personal income, outlays, and saving excluding NPISHs are referred to as *household* income, outlays, and saving. All proprietors' income is treated as received by individuals. Life insurance carriers and private noninsured pension funds are not counted as persons, but their saving is credited to persons.

Income from the sale of illegal goods and services is excluded by definition from national and personal income, and the value of purchases of illegal goods and services is not included in personal consumption expenditures.

Personal income is the sum of compensation received by employees, proprietors' income with inventory valuation and capital consumption adjustments (IVA and CCAdj), rental income of persons with capital consumption adjustment, personal receipts on assets, and personal current transfer receipts, less contributions for social insurance.

Personal income differs from national income in that it includes current transfer payments and interest received by persons, regardless of source, while it excludes the following national income components: employee and employer contributions for social insurance; business transfer payments, interest payments, and other payments on assets other than to persons; taxes on production and imports less subsidies; the current surplus of government enterprises; and undistributed corporate profits with IVA and CCAdj. The relationships of GDP, gross and net national product, national income, and personal income are displayed in Table 1-11.

Compensation of employees is the sum of wages and salaries and supplements to wages and salaries, as defined in the *national income account* (see Table 1-13 and the notes and definitions to Chapter 1).

As in *national* income, the *compensation of employees* component of personal income refers to compensation received by residents of the United States, including compensation from the rest of the world, but excludes compensation from domestic industries to workers residing in the rest of the world.

Wages and salaries consists of the monetary remuneration of employees, including wages and salaries as conventionally defined; the compensation of corporate officers; corporate directors' fees paid to directors who are also employees of the corporation; the value of employee exercise of "nonqualified stock options"; commissions, tips, and bonuses; voluntary employee contributions to certain deferred-compensation plans, such as 401(k) plans; and receipts in kind that represent income. This category also now includes judicial fees to jurors and witnesses, compensation of prison inmates, and marriage fees to justices of the peace, which earlier were classified as "other labor income". As of the 2013 revision, wages and salaries are now measured on an accrual basis, that is to say when earned rather than when paid, consistent with the treatment in the gross domestic income and national income tables.

Supplements to wages and salaries consists of employer contributions to employee pension and insurance funds and to government social insurance funds. In a substantial change introduced in the 2013 revision, defined benefit pension plan transactions are now recorded on an accrual basis instead of a cash transactions basis: employees are now credited with defined pension benefits, based on actuarial estimates of pension costs, at the time they earn them. (This was already the case with defined contribution plans.)

The following two categories, *proprietors' income* and *rental income,* are both measured net of depreciation of the capital (structures, equipment, and intellectual property products) involved. BEA calculates normal depreciation, based on the estimated life of the capital, and subtracts it from the estimated value of receipts to yield net income.

Proprietors' income with inventory valuation and capital consumption adjustments is the currentproduction income (including income-in-kind) of sole proprietors and partnerships and of taxexempt cooperatives. The imputed net rental income of owneroccupants of farm dwellings is included. Dividends and monetary interest received by proprietors of nonfinancial business and rental incomes received by persons not primarily engaged in the real estate business are excluded. These incomes are included in personal income receipts on assets and rental income of persons, respectively. Fees paid to outside directors of corporations are included. The two valuation adjustments are designed to obtain income measures that exclude any element of capital gains: inventory withdrawals are valued at replacement cost, rather than historical cost, and charges for depreciation are on an economically consistent accounting basis and are valued at replacement cost.

Rental income of persons with capital consumption adjustment consists of the net currentproduction income of persons from the rental of real property (other than the incomes of persons primarily engaged in the real estate business), the imputed net rental income of owner occupants of nonfarm dwellings, and the royalties received by persons from patents, copyrights, and rights to natural resources. The capital consumption adjustment converts charges for depreciation to an economically consistent accounting basis valued at replacement cost. Rental income is net of interest and other expenses, and hence is affected by changing indebtedness and interest payments on owner-occupied and other housing.

Personal income receipts on assets consists of personal interest income and personal dividend income.

Personal interest income is the interest income (monetary and imputed) of persons from all sources, including interest paid by government to government employee retirement plans as well as government interest paid directly to persons.

Personal dividend income is the dividend income of persons from all sources, excluding capital gains distributions. It equals net dividends paid by corporations (dividends paid by corporations minus dividends received by corporations) less a small amount of corporate dividends received by general government. Dividends received by government employee retirement systems are included in personal dividend income.

Personal current transfer receipts is income payments to persons for which no current services are performed. It consists of government social benefits to persons and net receipts from business.

Government social benefits to persons consists of benefits from the following categories of programs:

- *Social Security and Medicare,* consisting of federal oldage, survivors, disability, and health insurance benefits distributed from the Social Security and Medicare trust funds;

- *Medicaid,* the federal-state means-tested program covering medical expenses for lower-income children and adults as well as nursing care expenses;

- *Unemployment insurance;*

- *Veterans' benefits;*

- *Other government benefits to persons*, which includes pension benefit guaranty; workers' compensation; military medical insurance; temporary disability insurance; food stamps; Black Lung benefits; supplemental security income; family assistance, which consists of aid to families with dependent children and (beginning in 1996) assistance programs operating under the Personal Responsibility and Work Opportunity Reconciliation Act of 1996; educational assistance; and the earned income credit. Government payments to nonprofit institutions, other than for work under research and development contracts, also are included. Payments from government employee retirement plans are not included.

Note that the value of Medicare and Medicaid spending, though in practice it is usually paid directly from the government to the health care provider, is treated as if it were cash income to the consumer which is then expended in personal consumption expenditures; this value is not treated in the national accounts as a government purchase of medical services but as a government benefit paid to persons, which then finances personal consumption spending.

Contributions for government social insurance, domestic, which is subtracted to arrive at personal income, includes payments by U.S. employers, employees, selfemployed, and other individuals who participate in the following programs: oldage, survivors, and disability insurance (Social Security); hospital insurance and supplementary medical insurance (Medicare); unemployment insurance; railroad retirement; veterans' life insurance; and temporary disability insurance. Contributions to government employee retirement plans are not included in this item.

In the 2009 comprehensive revision, most transactions between the U.S. government and economic agents in Guam, the U.S. Virgin Islands, American Samoa, Puerto Rico, and the Northern Mariana Islands are treated as government transactions with the rest of the world. Since the NIPAs only cover the 50 states and the District of Columbia, the *domestic* contributions to government social insurance funds are the only ones that need to be subtracted from NIPA payroll data to calculate personal income. The social insurance receipts of governments, shown in Chapter 6, will be somewhat larger than this personal income entry because they will include contributions from residents of those territories and commonwealths.

Personal current taxes is tax payments (net of refunds) by persons residing in the United States that are not chargeable to business expenses, including taxes on income, on realized net capital gains, and on personal property. As of the 1999 revisions, estate and gift taxes are classified as capital transfers and are not included in personal current taxes.

Disposable personal income is personal income minus personal current taxes. It is the income from current production that is available to persons for spending or saving. However, it is not the cash flow available, since it excludes realized capital gains. Disposable personal income in chained (2009) dollars represents the inflation-adjusted value of disposable personal income, using the implicit price deflator for personal consumption expenditures.

Personal income excluding current transfer receipts, also shown in chained (2009) dollars using the implicit price deflator for personal consumption expenditures, is an important business cycle indicator, to which particular attention is paid because it is calculated monthly as well as quarterly and annually, and thus can help establish monthly cycle turning point dates. As a pre-income-tax measure which excludes transfer payments, it is a better measure of income generated by the economy than alternative monthly income indicators such as total or disposable personal income, which are more oriented toward purchasing power.

Personal outlays is the sum of *personal consumption expenditures* (defined below), *personal interest payments,* and *personal current transfer payments.*

Personal interest payments is nonmortgage interest paid by households. As noted above in the definition of rental income, mortgage interest has been subtracted from gross rental or imputed rental receipts of persons to yield a net rental income estimate; hence, it is not included as an interest outlay in this category.

Personal current transfer payments to government includes donations, fees, and fines paid to federal, state, and local governments.

Personal current transfer payments to the rest of the world (net) is personal remittances in cash and in kind to the rest of the world less such remittances from the rest of the world.

Personal saving is derived by subtracting personal outlays from disposable personal income. It is the current net saving of individuals (including proprietors), nonprofit institutions that primarily serve individuals, life insurance carriers, retirement funds (including those of government employees), private noninsured welfare funds, and private trust funds. Conceptually, personal saving may also be viewed as the sum of the net acquisition of financial assets and the change in physical assets less the sum of net borrowing and consumption of fixed capital. In either case, it is defined to exclude both realized and unrealized capital gains.

Note that in the context of national income accounting, the term just defined is *saving,* not "savings." *Saving* refers to a flow of income during a particular time span (such as a year or a quarter) that is not consumed. It is therefore available to finance a commensurate flow of investment during that time span. Strictly defined, "savings" denotes an accumulated stock of monetary funds—possibly the cumulative effects of successive periods of *saving*—available to the owner in asset form, such as in a bank savings account.

Personal consumption expenditures (PCE) is goods and services purchased by persons residing in the United States. Persons are defined as individuals and nonprofit institutions that primarily serve individuals. PCE mostly consists of purchases of new goods and services by individuals from business, including purchases financed by insurance (such as both private and government medical insurance). In addition, PCE includes purchases of new goods and services by nonprofit institutions, net purchases of used goods by individuals and nonprofit institutions, and purchases abroad of goods and services by U.S. residents traveling or working in foreign countries. PCE also includes purchases for certain goods and services provided by the government, primarily tuition payments for higher education, charges for medical care, and charges for water and sanitary services. Finally, PCE includes imputed purchases that keep PCE invariant to changes in the way that certain activities are carried out. For example, to take account of the value of the services provided by owner-occupied housing, PCE includes an imputation equal to what (estimated) rent homeowners would pay if they rented their houses from themselves. (See the discussion of imputation in the notes and definitions to Chapter 1.) Actual purchases of residential structures by individuals are classified as gross private domestic investment.

In the 2009 comprehensive revision, the classification system used for breakdowns of PCE was revised, and new calculations were introduced separating the consumption spending of the household sector proper from that of nonprofit institutions serving households.

Goods is the sum of *Durable* and *Nondurable goods.*

The PCE category *Durable goods* is subdivided into *Motor vehicles and parts, Furnishings and household equipment* (which includes appliances), *Recreational goods and vehicles* (which includes video, audio, photographic, and information processing equipment and media), and *Other durable goods.*

Nondurable goods encompasses *Food and beverages off-premises, Clothing and footwear, Gasoline and other energy goods,* and *Other nondurable goods.* This food and beverages category no longer includes meals and beverages purchased for consumption on the premises, which are now included in services, since they have a high service component and since their prices are much more stable than those of off-premises food and beverages.

Services, total is subdivided into *Household services* and *Final consumption expenditures of nonprofit institutions serving households (NPISHs).* The latter is the difference between the *gross output* of NPISHs and the amounts that they receive from sales of goods and services to households—for example, payments for services of a nonprofit hospital. Such sales of goods and services appear in the appropriate category of household expenditures—for example, *Health* in the case of the hospital services.

The components of *Household services* are *Housing and utilities, Health care, Transportation services, Recreation services, Food services and accommodations* (which includes meals and beverages purchased for consumption on premises, as well as payments for hotels and similar accommodations), *Financial services and insurance,* and *Other services.*

These are the categories used in the quarterly data presented in Tables 4-2, 4-3, and 4-4. This classification system is not particularly helpful with respect to the objective of spending. For example, the *Health care* component of services does not include drugs and medicines, which are included instead in nondurable goods. For a more precise classification of consumption spending by objective, see Table 4-5, Household Consumption Expenditures by Function, described in more detail below. This classification by type of expenditure is only available on an annual basis, and later than the principal quarterly NIPA data.

Data availability

Monthly data on personal income and spending are made available in a BEA press release, usually distributed the first business day following the monthly release of the latest quarterly national income and product account (NIPA) estimates. Monthly and quarterly data are subsequently published each month in the BEA's *Survey of Current Business.* Current and historical data are available on the BEA Web site at <http://www.bea.gov>, and may also be obtained from the STAT-USA subscription Web site at <http://www.stat-usa.gov>.

References

References can be found in the notes and definitions to Chapter 1. A discussion of monthly estimates of personal income and its disposition appears in the November 1979 edition of the *Survey of Current Business.* Additional and more recent information can be found in the articles listed in the notes and definitions for Chapter 1.

TABLE 4-5

Household Consumption Expenditures by Function

SOURCE: BUREAU OF ECONOMIC ANALYSIS (BEA)

In this table, also derived from the NIPAs, annual estimates of the current-dollar value of PCE by households—excluding the "final" consumption expenditures of nonprofit institutions serving households (NPISHs); see definition above--are presented by function.

Definitions

Food and beverages includes food and beverages (including alcoholic beverages) purchased for home consumption and food produced and consumed on farms.

Clothing, footwear, and related services includes purchases, rental, cleaning, and repair of clothing and footwear.

Housing, utilities, and fuels includes rents paid for rental housing, imputed rent of owner-occupied dwellings, and purchase of fuels and utility services.

Furnishings, household equipment, and routine household maintenance includes furniture, floor coverings, household textiles, appliances, tableware etc., tools, and other supplies and services.

Health includes drugs, other medical products and equipment, and outpatient, hospital, and nursing home services.

Transportation includes the purchase and operation of motor vehicles and public transportation services.

Communication includes telephone equipment, postal and delivery services, telecommunication services, and Internet access.

Recreation includes video and audio equipment, computers, and related services; sports vehicles and other sports goods and services; memberships and admissions; magazines, newspapers, books, and stationery; pets and related goods and services; photo goods and services; tour services; and legal gambling. (As noted earlier, purchases of goods and services that are illegal and the incomes from such purchases are outside the scope of the national income and product accounts.)

Education includes educational services and books.

Food services and accommodations includes meals and beverages purchased for consumption on the premises, food furnished to employees, and hotel and other accommodations including housing at schools.

Financial services and insurance consists of financial services and life, household, medical care, motor vehicle, and other insurance.

Other goods and services includes personal goods (such as cosmetics, jewelry, and luggage) and services, social services and religious activities, professional and other services, and tobacco.

Foreign travel and expenditures, net consists of foreign travel spending and other expenditures abroad by U.S. residents minus expenditures in the United States by nonresidents. A negative figure indicates that foreigners spent more here than U.S. residents spent abroad. Positive values for this foreign travel category, indicating that U.S. residents spent more abroad than foreigners spent here, appear in the 1980s when the dollar was strong against other major currencies. (The international value of the dollar is shown in Table 13-2.) Negative values in subsequent years have resulted in part from the weakening of the dollar, which discouraged U.S. residents' travel abroad and encouraged tourism by foreigners in the United States. The negative sign does not indicate a drain on GDP—these effects of a weaker dollar are in fact positive for real GDP—but rather reflects the fact that the goods and services purchased by foreigners in the United States must be subtracted from the total consumer purchases recorded in the other columns of this table in order to be added to other exports and classified in the category of exports rather than in the consumption spending of U.S. residents.

Data availability and revisions

Data are updated once a year, after the general midyear revision of the NIPAs, and are available on the BEA Web site at <http://www.bea.gov>.

CHAPTER 5: SAVING AND INVESTMENT

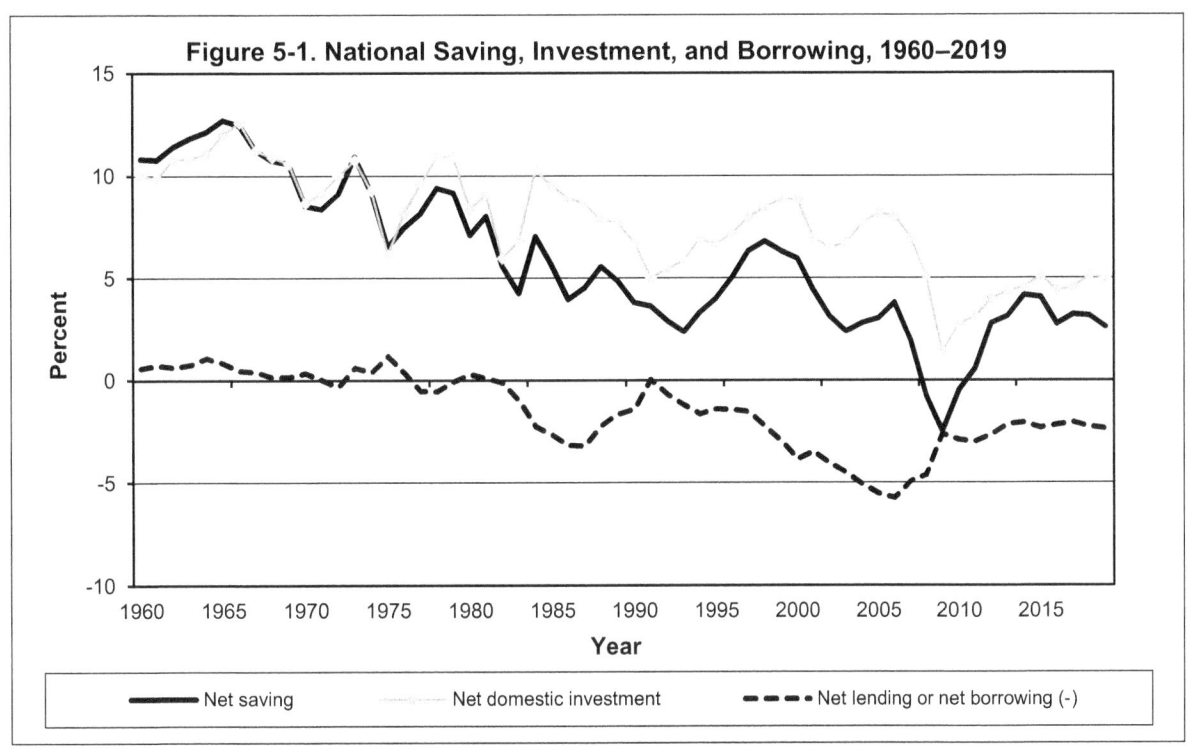

Figure 5-1. National Saving, Investment, and Borrowing, 1960–2019

Legend: Net saving — Net domestic investment — Net lending or net borrowing (-)

- Net national saving—the middle line in Figure 5-1—averaged a much lower rate of gross national income (GNI) in the years 1984 through 2000, than in the rates seen in the 1950s and 1960s. Net saving is gross saving by U.S. persons, businesses, and governments, less the consumption of fixed capital. Thus, it represents saving available for investment over and above the replacement of the existing capital stock. (Tables 5-1A and B)

- In the recession that began at the end of 2007, net national saving fell into negative territory, for the first time since the depression years 1931 through 1934. Government deficits soared as the Great Recession cut deeply into tax revenues and called forth both automatic and discretionary anti-recession spending such as unemployment compensation, food stamps (Supplemental Nutrition Assistance Program – SNAP, and temporary assistance for needy families - TANF. In one year, 2008 to 2009, the budget deficit doubled. (Tables 5-1A and B and Table 6-14B) The global pandemic that began in 2020 is having a grave impact on the economy as millions of people lost their jobs and states and local areas are seeing their tax revenues decline.

- Net saving is currently much lower than it was in the 1960s and 1970s. In 1965, net saving got as high as 12.7 percent. It gradually declined through the 1980 and 1990s. In 2019, the national saving rate declined to its lowest level since 2011. (Table 5-1A)

Table 5-1A. Saving and Investment: Recent Data

(Billions of dollars, except as noted; quarterly data are at seasonally adjusted annual rates.)   **NIPA Tables 1.7.5, 5.1**

Year and quarter	Total	Net saving						Consumption of fixed capital					
		Total	Private			Government		Total	Private			Government	
			Total	Personal saving	Domestic corporate business	Federal	State and local		Total	Domestic business	House-holds and institutions	Federal	State and local
1960	127.0	59.1	56.4	37.9	18.5	0.2	2.5	67.9	48.2	40.0	8.2	15.2	4.5
1961	131.6	61.0	63.5	44.4	19.2	-4.7	2.2	70.6	49.8	41.3	8.5	16.0	4.8
1962	143.4	69.3	71.6	46.7	24.9	-5.4	3.1	74.1	51.8	42.9	8.8	17.2	5.1
1963	154.1	76.1	75.0	47.1	27.9	-2.2	3.3	78.0	54.2	44.9	9.3	18.4	5.5
1964	166.1	83.7	86.6	55.3	31.4	-6.9	3.9	82.4	57.3	47.4	9.9	19.2	5.9
1965	182.9	94.9	96.8	58.8	38.0	-5.5	3.7	88.0	61.6	50.9	10.7	20.0	6.4
1966	196.5	101.2	103.4	61.9	41.6	-7.0	4.7	95.3	67.2	55.6	11.5	21.0	7.1
1967	200.5	96.9	112.5	73.0	39.5	-19.5	4.0	103.6	73.3	60.9	12.4	22.5	7.8
1968	214.6	101.2	111.5	72.9	38.6	-13.8	3.5	113.4	80.6	67.0	13.6	24.2	8.6
1969	233.1	108.2	110.3	76.1	34.2	-5.1	3.1	124.9	89.4	74.2	15.2	25.8	9.6
1970	228.2	91.4	124.8	97.6	27.2	-34.8	1.4	136.8	98.3	81.8	16.5	27.6	10.9
1971	246.1	97.2	149.4	111.9	37.5	-50.9	-1.3	148.9	107.6	89.4	18.2	29.1	12.2
1972	277.6	116.6	159.6	111.5	48.0	-49.0	6.1	161.0	117.5	97.2	20.3	30.2	13.3
1973	335.3	156.6	189.3	135.8	53.5	-38.3	5.6	178.7	131.5	108.2	23.3	32.4	14.8
1974	349.2	142.3	186.0	146.3	39.7	-41.3	-2.3	206.9	153.2	126.1	27.0	35.4	18.4
1975	348.1	109.6	218.3	164.0	54.3	-97.9	-10.7	238.5	178.8	147.9	30.9	38.7	21.0
1976	399.3	139.1	224.4	154.4	70.0	-80.9	-4.4	260.2	196.5	162.2	34.3	41.7	22.0
1977	459.4	169.6	242.5	155.9	86.6	-73.4	0.5	289.8	221.1	181.5	39.6	45.3	23.4
1978	548.0	220.8	278.0	175.1	102.9	-62.0	4.9	327.2	252.1	205.5	46.6	49.7	25.4
1979	613.5	239.6	288.2	186.8	101.4	-47.4	-1.2	373.9	290.7	236.4	54.4	54.6	28.6
1980	630.1	201.7	296.4	224.1	72.3	-88.8	-5.9	428.4	335.0	272.8	62.2	60.4	33.1
1981	743.9	256.6	354.9	265.5	89.4	-88.1	-10.2	487.2	381.9	313.5	68.5	67.5	37.8
1982	725.8	188.9	379.0	293.3	85.6	-167.4	-22.8	537.0	420.4	347.5	72.9	75.4	41.2
1983	716.7	154.1	379.7	264.0	115.7	-207.2	-18.4	562.6	438.8	362.8	76.0	81.1	42.7
1984	881.6	283.2	479.9	330.3	149.5	-196.5	-0.2	598.4	463.5	382.7	80.9	90.6	44.3
1985	881.0	240.8	442.5	284.9	157.5	-199.2	-2.4	640.1	496.4	410.3	86.2	97.0	46.8
1986	864.5	179.2	399.1	290.6	108.5	-215.9	-4.0	685.3	531.6	438.1	93.4	103.6	50.1
1987	948.9	218.5	398.6	275.4	123.2	-165.7	-14.4	730.4	566.3	464.5	101.8	110.1	54.0
1988	1 076.6	292.1	463.4	320.5	142.9	-160.0	-11.3	784.5	607.9	499.2	108.8	118.9	57.7
1989	1 109.8	271.5	450.2	340.0	110.3	-159.4	-19.3	838.3	649.6	532.6	117.0	126.9	61.8
1990	1 113.4	224.8	464.4	361.1	103.2	-203.3	-36.2	888.5	688.4	564.6	123.8	133.5	66.6
1991	1 153.4	221.0	529.5	396.0	133.5	-248.4	-60.1	932.4	721.5	592.3	129.2	140.1	70.8
1992	1 147.6	187.4	592.8	453.9	139.0	-334.5	-71.0	960.2	742.9	608.2	134.7	143.3	74.1
1993	1 163.4	159.9	545.9	397.7	148.2	-313.5	-72.5	1 003.5	778.2	634.2	144.0	147.0	78.3
1994	1 295.1	239.5	559.0	363.4	195.7	-255.6	-63.9	1 055.6	822.5	669.0	153.5	150.3	82.8
1995	1 426.3	303.9	616.5	387.1	229.4	-242.1	-70.4	1 122.4	880.7	717.8	162.9	153.5	88.1
1996	1 578.9	403.6	636.8	382.3	254.5	-179.4	-53.8	1 175.3	929.1	758.5	170.6	153.7	92.5
1997	1 780.5	541.2	675.1	390.3	284.9	-92.0	-42.0	1 239.3	987.8	807.9	179.9	154.3	97.3
1998	1 930.6	620.8	649.5	446.5	203.0	1.4	-30.1	1 309.7	1 052.2	860.1	192.1	155.4	102.2
1999	2 010.3	611.3	581.1	347.1	234.0	69.1	-38.9	1 398.9	1 132.2	924.0	208.2	158.2	108.6
2000	2 129.2	618.0	499.0	356.1	142.9	159.7	-40.6	1 511.2	1 231.5	1 004.6	227.0	163.1	116.6
2001	2 075.6	476.1	580.1	389.4	190.7	15.0	-119.0	1 599.5	1 311.7	1 064.9	246.8	164.4	123.4
2002	2 005.2	347.2	797.9	471.7	326.2	-267.8	-182.9	1 658.0	1 361.8	1 098.8	263.0	166.8	129.3
2003	1 995.8	276.7	855.4	468.9	386.5	-397.4	-181.3	1 719.1	1 411.9	1 125.4	286.6	172.2	134.9
2004	2 168.6	346.7	889.2	460.6	428.6	-393.5	-149.0	1 821.8	1 497.1	1 178.8	318.3	180.5	144.2
2005	2 371.4	400.4	796.9	290.1	506.9	-293.8	-102.8	1 971.0	1 622.6	1 266.4	356.2	190.5	157.9
2006	2 659.3	535.1	842.1	381.0	461.1	-221.9	-85.0	2 124.1	1 751.8	1 362.6	389.2	201.3	171.0
2007	2 537.6	284.8	673.8	385.9	287.9	-259.7	-129.3	2 252.8	1 852.5	1 446.7	405.8	212.8	187.5
2008	2 247.6	-111.2	734.6	544.9	189.6	-624.9	-220.9	2 358.8	1 931.8	1 524.9	407.0	225.9	201.2
2009	2 013.9	-357.5	1 227.0	666.2	560.8	-1 243.2	-341.3	2 371.5	1 928.7	1 531.2	397.5	233.6	209.1
2010	2 318.7	-72.2	1 553.6	740.3	813.3	-1 318.4	-307.5	2 390.9	1 933.8	1 537.1	396.7	243.7	213.4
2011	2 564.3	89.8	1 599.0	849.1	749.9	-1 234.1	-275.1	2 474.5	1 997.3	1 595.1	402.2	254.9	222.2
2012	3 041.3	465.3	1 820.8	1 106.6	714.1	-1 072.7	-282.8	2 576.0	2 082.4	1 670.4	412.0	261.6	232.0
2013	3 222.9	541.7	1 438.8	799.7	639.1	-631.8	-265.3	2 681.2	2 176.6	1 738.8	437.8	265.9	238.8
2014	3 567.8	752.8	1 588.0	970.9	617.1	-597.4	-237.9	2 815.0	2 298.5	1 831.2	467.3	270.2	246.3
2015	3 673.4	762.0	1 538.0	1 038.7	499.3	-560.2	-215.8	2 911.4	2 388.5	1 903.1	485.4	271.5	251.4
2016	3 511.5	524.9	1 433.2	975.0	458.2	-669.1	-239.2	2 986.6	2 458.3	1 949.5	508.9	271.8	256.5
2017	3 755.4	642.5	1 603.8	1 071.0	532.8	-722.4	-238.8	3 112.9	2 569.7	2 031.8	537.8	277.0	266.2
2018	3 927.0	661.9	1 807.3	1 237.3	570.0	-931.7	-213.7	3 265.0	2 699.0	2 125.0	574.0	285.8	280.3
2019	3 988.4	567.5	1 822.2	1 231.2	591.0	-1 047.0	-207.7	3 420.9	2 832.7	2 233.6	599.0	294.8	293.5
2017													
1st quarter	3 665.2	602.4	1 522.6	1 007.6	515.1	-671.5	-248.7	3 062.8	2 526.0	2 000.0	526.0	274.8	262.0
2nd quarter	3 735.0	637.8	1 609.6	1 094.7	514.9	-698.1	-273.7	3 097.2	2 556.5	2 022.2	534.3	276.1	264.6
3rd quarter	3 807.4	676.2	1 655.0	1 117.3	537.7	-731.0	-247.8	3 131.2	2 585.7	2 043.2	542.5	277.6	267.8
4th quarter	3 813.9	653.6	1 627.8	1 064.3	563.5	-789.1	-185.1	3 160.3	2 610.4	2 061.8	548.6	279.6	270.3
2018													
1st quarter	3 910.2	705.3	1 799.1	1 188.8	610.3	-917.2	-176.5	3 204.8	2 648.6	2 086.9	561.7	282.3	273.9
2nd quarter	3 868.6	620.1	1 788.9	1 218.0	570.9	-942.3	-226.5	3 248.5	2 684.6	2 113.1	571.5	284.9	279.0
3rd quarter	3 964.0	677.2	1 804.7	1 247.9	556.8	-907.2	-220.3	3 286.8	2 717.4	2 138.9	578.5	287.1	282.3
4th quarter	3 965.0	645.1	1 836.5	1 294.5	542.0	-960.0	-231.4	3 319.9	2 745.2	2 161.1	584.2	288.8	285.9
2019													
1st quarter	4 003.9	642.3	1 871.8	1 354.5	517.3	-1 016.0	-213.5	3 361.6	2 781.3	2 191.9	589.4	292.1	288.2
2nd quarter	3 971.2	566.8	1 775.0	1 186.1	588.9	-1 033.0	-175.2	3 404.4	2 818.6	2 223.7	594.9	293.1	292.7
3rd quarter	3 934.9	491.6	1 795.5	1 180.6	615.0	-1 084.1	-219.8	3 443.3	2 851.9	2 248.9	603.0	295.8	295.6
4th quarter	4 043.6	569.1	1 846.5	1 203.8	642.7	-1 054.9	-222.4	3 474.4	2 878.9	2 269.9	608.9	298.3	297.3

Table 5-1A. Saving and Investment: Recent Data—*Continued*

(Billions of dollars, except as noted; quarterly data are at seasonally adjusted annual rates.) **NIPA Tables 1.7.5, 5.1**

Year and quarter	Gross domestic investment, capital account transactions, and net lending, NIPAs						Statistical discrepancy	Net domestic investment	Gross national income	Gross saving as a percent of gross national income	Net saving as a percent of gross national income
	Total	Gross domestic investment			Capital account transactions, net	Net lending or net borrowing (-), NIPAs					
		Total	Private	Government							
1960	125.7	122.5	86.5	36.0	. . .	3.2	-1.3	54.6	546.8	23.2	10.8
1961	130.7	126.5	86.6	39.9	. . .	4.2	-0.9	55.9	566.6	23.2	10.8
1962	143.4	139.6	97.0	42.6	. . .	3.8	0.0	65.5	608.0	23.6	11.4
1963	152.7	147.7	103.3	44.4	. . .	4.9	-1.5	69.7	643.4	24.0	11.8
1964	166.0	158.5	112.2	46.4	. . .	7.5	0.0	76.2	689.4	24.1	12.1
1965	183.7	177.5	129.6	47.9	. . .	6.2	0.7	89.5	746.8	24.5	12.7
1966	201.6	197.8	144.2	53.6	. . .	3.8	5.1	102.4	813.4	24.2	12.4
1967	203.9	200.4	142.7	57.7	. . .	3.5	3.4	96.8	862.0	23.3	11.2
1968	217.7	216.2	156.9	59.2	. . .	1.5	3.1	102.8	943.6	22.7	10.7
1969	234.7	233.1	173.6	59.5	0.0	1.6	1.6	108.2	1 022.1	22.8	10.6
1970	233.6	229.8	170.0	59.8	0.0	3.7	5.3	93.0	1 074.4	21.2	8.5
1971	255.6	255.3	196.8	58.5	0.0	0.3	9.5	106.4	1 162.9	21.2	8.4
1972	284.8	288.8	228.1	60.7	0.0	-4.1	7.2	127.8	1 280.5	21.7	9.1
1973	341.4	332.6	266.9	65.6	0.0	8.8	6.1	153.9	1 431.8	23.4	10.9
1974	356.6	350.7	274.5	76.2	0.0	5.9	7.4	143.8	1 553.3	22.5	9.2
1975	361.5	341.7	257.3	84.4	0.1	19.8	13.3	103.1	1 684.6	20.7	6.5
1976	420.0	412.9	323.2	89.6	0.1	7.0	20.7	152.6	1 869.6	21.4	7.4
1977	478.9	489.8	396.6	93.2	0.1	-11.0	19.4	199.9	2 082.7	22.1	8.1
1978	571.3	583.9	478.4	105.6	0.1	-12.7	23.3	256.7	2 349.9	23.3	9.4
1979	658.6	659.8	539.7	120.1	0.1	-1.3	45.1	285.9	2 614.2	23.5	9.2
1980	674.6	666.0	530.1	135.9	0.1	8.4	44.4	237.6	2 847.1	22.1	7.1
1981	781.9	778.6	631.2	147.3	0.1	3.3	38.1	291.3	3 201.9	23.2	8.0
1982	734.7	738.0	581.0	156.9	0.1	-3.4	8.8	201.0	3 371.4	21.5	5.6
1983	773.6	808.7	637.5	171.2	0.1	-35.2	57.0	246.1	3 614.2	19.8	4.3
1984	923.2	1 013.3	820.1	193.2	0.1	-90.2	41.6	414.9	4 032.3	21.9	7.0
1985	935.2	1 049.5	829.7	219.9	0.1	-114.5	54.3	409.4	4 310.1	20.4	5.6
1986	944.6	1 087.2	849.1	238.1	0.1	-142.8	80.1	401.9	4 516.5	19.1	4.0
1987	992.7	1 146.8	892.2	254.6	0.1	-154.2	43.8	416.4	4 828.9	19.7	4.5
1988	1 079.6	1 195.4	937.0	258.4	0.1	-115.9	3.0	410.9	5 256.1	20.5	5.6
1989	1 177.8	1 270.1	999.7	270.4	0.3	-92.7	68.0	431.9	5 598.4	19.8	4.9
1990	1 208.9	1 283.8	993.4	290.4	7.4	-82.3	95.5	395.3	5 902.3	18.9	3.8
1991	1 246.3	1 238.4	944.3	294.1	5.3	2.6	93.0	306.0	6 096.8	18.9	3.6
1992	1 263.6	1 309.1	1 013.0	296.1	-1.3	-44.3	115.9	348.9	6 435.5	17.8	2.9
1993	1 319.3	1 398.7	1 106.8	291.9	0.9	-80.2	156.0	395.2	6 733.8	17.3	2.4
1994	1 435.1	1 550.7	1 256.5	294.2	1.3	-116.9	140.0	495.0	7 170.3	18.1	3.3
1995	1 519.3	1 625.2	1 317.5	307.7	0.4	-106.3	93.0	502.8	7 574.7	18.8	4.0
1996	1 637.0	1 752.0	1 432.1	320.0	0.2	-115.2	58.1	576.7	8 045.9	19.6	5.0
1997	1 792.1	1 922.2	1 595.6	326.6	0.5	-130.6	11.6	682.9	8 589.3	20.7	6.3
1998	1 875.3	2 080.7	1 736.7	344.0	0.2	-205.6	-55.2	770.9	9 135.5	21.1	6.8
1999	1 977.3	2 255.5	1 887.1	368.5	6.7	-285.0	-33.0	856.6	9 689.2	20.7	6.3
2000	2 033.0	2 427.3	2 038.4	388.9	4.6	-398.9	-96.3	916.0	10 383.7	20.5	6.0
2001	1 962.5	2 346.7	1 934.8	411.9	-11.9	-372.3	-113.1	747.2	10 743.7	19.3	4.4
2002	1 932.4	2 374.1	1 930.4	443.7	4.2	-445.8	-72.7	716.1	11 054.4	18.1	3.1
2003	1 982.3	2 491.3	2 027.1	464.2	8.8	-517.7	-13.4	772.2	11 530.1	17.3	2.4
2004	2 146.2	2 767.5	2 281.3	486.2	4.6	-625.8	-22.3	945.6	12 314.3	17.6	2.8
2005	2 318.7	3 048.0	2 534.7	513.3	-0.7	-728.6	-52.7	1 077.0	13 167.3	18.0	3.0
2006	2 452.1	3 251.8	2 701.0	550.9	7.7	-807.5	-207.2	1 127.7	14 072.3	18.9	3.8
2007	2 555.0	3 265.0	2 673.0	592.0	6.4	-716.3	17.4	1 012.2	14 543.5	17.4	2.0
2008	2 430.2	3 107.2	2 477.6	629.6	0.8	-677.8	182.6	748.4	14 684.9	15.3	-0.8
2009	2 203.9	2 572.6	1 929.7	642.9	6.3	-375.0	189.9	201.1	14 401.0	14.0	-2.5
2010	2 378.7	2 810.0	2 165.5	644.5	7.4	-438.7	60.0	419.1	15 127.8	15.3	-0.5
2011	2 507.4	2 969.2	2 332.6	636.6	9.5	-471.2	-56.8	494.7	15 835.8	16.2	0.6
2012	2 801.4	3 242.8	2 621.8	621.0	-0.5	-440.8	-239.8	666.8	16 669.1	18.2	2.8
2013	3 066.0	3 426.4	2 826.0	600.4	7.0	-367.4	-156.9	745.2	17 172.5	18.8	3.2
2014	3 281.2	3 646.7	3 044.2	602.6	6.9	-372.5	-286.6	831.7	18 055.0	19.8	4.2
2015	3 436.1	3 859.8	3 237.2	622.6	8.3	-432.0	-237.3	948.4	18 696.0	19.6	4.1
2016	3 419.3	3 826.8	3 188.3	638.4	7.0	-414.5	-92.2	840.2	19 069.3	18.4	2.8
2017	3 623.9	4 015.5	3 351.1	664.3	16.0	-407.5	-131.4	902.6	19 966.7	18.8	3.2
2018	3 869.0	4 336.8	3 632.9	703.9	4.6	-472.4	-58.0	1 071.7	20 954.6	18.7	3.2
2019	4 001.2	4 504.0	3 751.2	752.8	6.7	-509.5	12.8	1 083.1	21 690.0	18.4	2.6
2017											
1st quarter	3 550.5	3 918.4	3 266.2	652.2	8.9	-376.8	-114.8	855.5	19 628.9	18.7	3.1
2nd quarter	3 542.0	3 975.0	3 313.3	661.7	8.4	-441.4	-193.0	877.8	19 839.8	18.8	3.2
3rd quarter	3 674.2	4 043.5	3 378.8	664.7	39.5	-408.8	-133.2	912.4	20 051.8	19.0	3.4
4th quarter	3 729.0	4 124.9	3 446.3	678.7	7.2	-403.1	-84.8	964.6	20 346.3	18.7	3.2
2018											
1st quarter	3 832.6	4 240.5	3 555.0	685.4	6.0	-413.9	-77.6	1 035.6	20 634.3	18.9	3.4
2nd quarter	3 888.1	4 284.5	3 580.9	703.6	12.1	-408.5	19.5	1 036.0	20 824.8	18.6	3.0
3rd quarter	3 859.4	4 386.7	3 671.7	715.0	2.2	-529.5	-104.6	1 099.9	21 107.3	18.8	3.2
4th quarter	3 895.7	4 435.3	3 723.9	711.5	-1.9	-537.7	-69.3	1 115.4	21 251.9	18.7	3.0
2019											
1st quarter	3 971.4	4 503.5	3 772.8	730.7	10.6	-542.7	-32.5	1 141.9	21 394.3	18.7	3.0
2nd quarter	3 953.8	4 489.6	3 739.7	749.9	3.8	-539.6	-17.4	1 085.2	21 618.4	18.4	2.6
3rd quarter	4 009.9	4 517.7	3 759.8	757.9	3.8	-511.5	75.0	1 074.4	21 745.1	18.1	2.3
4th quarter	4 069.8	4 505.4	3 732.6	772.8	8.5	-444.2	26.2	1 031.0	22 002.3	18.4	2.6

. . . = Not available.

Table 5-1B. Saving and Investment: Historical Data

(Billions of dollars, except as noted; quarterly data are at seasonally adjusted annual rates.)   **NIPA Tables 1.7.5, 5.1**

Year and quarter	Gross saving								Gross domestic investment and net lending, NIPAs					Net domestic invest-ment	Gross national income	Net saving as a percent-age of gross national income
	Total	Net saving			Consumption of fixed capital				Gross domestic investment			Net lending or net borrow-ing (-), NIPAs			
		Private	Government		Total	Private	Government		Total	Private	Govern-ment				
			Federal	State and local			Federal	State and local							
1929	20.1	7.6	0.8	1.3	10.4	9.4	0.2	0.8	20.1	17.2	2.9	0.8	9.7	104.6	9.3
1930	15.8	4.5	0.0	1.1	10.2	9.2	0.2	0.8	14.7	11.4	3.3	0.7	4.5	93.3	6.0
1931	9.1	1.2	-2.4	0.7	9.5	8.6	0.2	0.7	9.6	6.5	3.1	0.2	0.1	77.2	-0.5
1932	4.0	-3.2	-1.6	0.4	8.3	7.5	0.2	0.7	4.1	1.8	2.3	0.2	-4.3	59.6	-7.3
1933	3.9	-3.1	-1.2	0.2	8.0	7.0	0.2	0.7	4.3	2.3	2.0	0.2	-3.7	56.9	-7.2
1934	7.0	0.1	-2.6	1.0	8.4	7.3	0.3	0.9	7.0	4.3	2.7	0.4	-1.5	66.7	-2.1
1935	10.4	3.2	-2.3	1.0	8.5	7.3	0.3	0.9	10.2	7.4	2.9	-0.1	1.7	74.8	2.6
1936	12.3	5.1	-3.6	2.1	8.8	7.5	0.4	0.9	13.6	9.4	4.2	-0.1	4.8	83.9	4.2
1937	17.1	5.8	-0.2	1.8	9.8	8.4	0.4	1.0	16.9	13.0	3.9	0.2	7.1	93.4	7.9
1938	12.7	2.8	-1.8	1.7	10.0	8.6	0.4	1.0	12.2	7.9	4.3	1.2	2.2	87.1	3.1
1939	14.6	5.3	-2.6	1.7	10.1	8.6	0.5	1.1	14.8	10.2	4.6	1.0	4.7	92.5	4.8
1940	19.5	8.3	-1.0	1.6	10.6	9.0	0.5	1.1	19.0	14.6	4.4	1.5	8.4	102.1	8.7
1941	31.1	16.9	0.6	1.5	12.1	10.0	0.9	1.2	30.2	19.4	10.8	1.3	18.1	129.4	14.7
1942	41.6	37.0	-11.4	1.1	14.9	11.2	2.2	1.4	40.7	11.8	29.0	-0.1	25.8	167.3	16.0
1943	47.0	45.0	-17.1	1.0	18.0	11.5	5.0	1.5	47.3	7.4	39.9	-2.1	29.3	205.2	14.1
1944	42.8	49.8	-29.4	1.1	21.3	12.0	7.8	1.5	47.3	9.2	38.2	-2.0	26.0	222.2	9.7
1945	32.6	40.4	-32.0	1.1	23.1	12.5	9.2	1.4	37.7	12.4	25.3	-1.3	14.6	224.4	4.2
1946	41.5	23.1	-8.1	0.8	25.7	14.2	9.9	1.6	37.8	33.1	4.7	4.9	12.1	227.0	7.0
1947	49.2	17.6	2.5	0.1	29.1	17.7	9.5	1.9	42.8	37.1	5.7	9.3	13.7	247.9	8.1
1948	61.5	29.1	1.0	0.1	31.3	20.8	8.4	2.1	58.8	50.3	8.5	2.4	27.5	276.2	10.9
1949	49.6	25.2	-8.5	0.5	32.3	22.6	7.5	2.2	50.4	39.1	11.3	0.9	18.1	272.0	6.3
1950	64.9	29.3	1.9	0.3	33.4	24.3	6.9	2.2	68.0	56.5	11.5	-1.8	34.6	300.0	10.5
1951	79.8	35.9	4.5	1.6	37.7	27.7	7.4	2.6	82.3	62.8	19.6	0.9	44.6	345.4	12.2
1952	79.9	39.5	-2.1	1.9	40.6	29.5	8.3	2.8	81.9	57.3	24.6	0.6	41.3	366.7	10.7
1953	82.0	40.3	-4.2	2.3	43.5	31.3	9.3	2.9	87.1	60.4	26.7	-1.3	43.6	387.3	9.9
1954	80.7	40.7	-8.0	2.0	46.0	33.0	10.1	3.0	83.5	58.1	25.4	0.2	37.5	389.7	8.9
1947															
1st quarter	47.1	16.5	2.5	0.1	28.1	16.3	10.0	1.8	40.8	35.9	4.9	9.4	12.7	241.1	7.9
2nd quarter	46.7	15.7	2.1	0.2	28.7	17.3	9.6	1.8	39.6	34.5	5.1	9.9	10.9	244.3	7.4
3rd quarter	49.1	19.9	-0.4	0.0	29.5	18.2	9.4	1.9	41.0	34.9	6.1	10.1	11.5	248.7	7.9
4th quarter	54.0	18.3	5.6	-0.1	30.2	19.1	9.1	2.0	49.8	43.3	6.6	7.8	19.6	257.4	9.3
1948															
1st quarter	59.0	23.6	5.1	-0.3	30.6	19.9	8.7	2.0	54.7	47.2	7.5	4.9	24.1	266.5	10.6
2nd quarter	62.7	29.0	2.6	0.1	31.1	20.6	8.4	2.1	58.6	50.3	8.2	3.0	27.5	275.1	11.5
3rd quarter	62.0	31.4	-1.1	0.1	31.7	21.2	8.3	2.2	61.3	52.5	8.8	0.9	29.6	280.4	10.8
4th quarter	62.4	32.6	-2.7	0.5	32.0	21.7	8.1	2.2	60.7	51.3	9.4	0.8	28.7	282.7	10.8
1949															
1st quarter	54.7	27.9	-6.0	0.6	32.2	22.1	8.0	2.2	53.4	43.1	10.3	2.2	21.1	275.6	8.2
2nd quarter	48.6	24.9	-9.1	0.4	32.3	22.4	7.6	2.2	47.5	36.2	11.3	1.7	15.3	272.0	6.0
3rd quarter	50.1	26.6	-9.4	0.6	32.3	22.8	7.3	2.2	51.5	39.5	11.9	0.6	19.2	272.1	6.6
4th quarter	44.8	21.5	-9.4	0.3	32.4	23.1	7.1	2.2	49.3	37.5	11.8	-1.0	16.9	268.4	4.6
1950															
1st quarter	53.2	32.3	-11.5	0.0	32.4	23.4	6.9	2.1	57.4	46.7	10.7	-1.0	25.0	278.7	7.4
2nd quarter	61.7	29.8	-0.6	-0.3	32.8	23.9	6.8	2.1	63.4	52.3	11.1	-1.2	30.6	291.3	9.9
3rd quarter	67.1	23.1	9.6	0.8	33.6	24.6	6.8	2.2	70.4	58.6	11.7	-2.7	36.8	309.1	10.8
4th quarter	77.6	32.0	10.0	0.8	34.8	25.5	7.0	2.3	80.8	68.4	12.3	-2.5	45.9	320.9	13.3
1951															
1st quarter	74.3	23.6	12.7	1.7	36.3	26.6	7.2	2.5	79.5	64.6	14.9	-1.7	43.2	334.1	11.4
2nd quarter	82.5	38.5	5.1	1.5	37.3	27.5	7.2	2.6	85.6	67.4	18.2	0.4	48.3	342.5	13.2
3rd quarter	80.7	41.6	-0.5	1.4	38.3	28.2	7.4	2.7	83.1	62.0	21.1	2.2	44.9	348.8	12.2
4th quarter	81.5	40.1	0.7	1.7	39.0	28.7	7.6	2.7	81.2	57.1	24.1	2.7	42.1	356.0	11.9
1952															
1st quarter	82.0	39.4	1.2	1.8	39.5	29.0	7.9	2.7	82.2	58.1	24.1	3.5	42.7	358.1	11.9
2nd quarter	76.0	36.9	-2.4	1.3	40.2	29.3	8.1	2.8	77.3	53.0	24.3	1.1	37.1	360.6	9.9
3rd quarter	79.3	40.9	-4.8	2.1	41.0	29.7	8.4	2.9	82.1	57.2	24.8	-1.1	41.1	368.0	10.4
4th quarter	82.3	40.8	-2.5	2.3	41.7	30.2	8.7	2.8	86.0	60.7	25.3	-1.2	44.3	380.2	10.7
1953															
1st quarter	84.0	41.2	-1.5	1.7	42.5	30.7	8.9	2.9	88.2	61.7	26.5	-1.3	45.7	387.0	10.7
2nd quarter	84.2	41.4	-3.3	2.8	43.3	31.1	9.2	2.9	89.3	62.1	27.2	-1.8	46.0	390.6	10.5
3rd quarter	83.7	40.2	-2.6	2.3	43.8	31.6	9.3	2.9	88.5	61.4	27.0	-1.2	44.6	389.4	10.2
4th quarter	76.0	38.4	-9.2	2.4	44.4	32.0	9.5	2.9	82.4	56.4	25.9	-0.9	38.0	382.4	8.3
1954															
1st quarter	78.9	41.2	-9.7	2.4	44.9	32.4	9.7	2.8	82.1	55.7	26.4	-0.5	37.2	384.6	8.8
2nd quarter	79.2	39.8	-8.4	2.1	45.7	32.8	10.0	3.0	81.4	55.4	25.9	0.3	35.7	385.6	8.7
3rd quarter	80.2	39.7	-7.6	1.8	46.3	33.2	10.2	3.0	83.7	59.0	24.7	-0.1	37.4	389.6	8.7
4th quarter	84.7	42.1	-6.3	1.9	47.0	33.5	10.4	3.0	86.6	62.1	24.5	0.9	39.6	399.2	9.4
1955															
1st quarter	89.9	43.7	-2.9	1.7	47.5	34.0	10.5	3.0	92.8	68.7	24.1	0.5	45.4	412.1	10.3
2nd quarter	95.4	45.8	-0.5	2.0	48.2	34.5	10.6	3.1	97.4	72.7	24.8	-0.2	49.3	422.1	11.2
3rd quarter	97.8	46.8	-0.7	2.5	49.3	35.3	10.8	3.2	99.0	74.7	24.3	0.8	49.7	430.7	11.3
4th quarter	102.3	47.2	1.9	2.6	50.6	36.2	11.1	3.3	103.2	78.9	24.3	0.4	52.5	438.3	11.8
1956															
1st quarter	105.6	48.3	2.3	2.9	52.1	37.3	11.4	3.5	103.9	78.3	25.6	0.9	51.7	443.4	12.1
2nd quarter	107.5	50.2	0.7	3.1	53.5	38.4	11.6	3.6	103.6	77.0	26.6	2.2	50.1	450.5	12.0
3rd quarter	111.6	52.5	1.2	3.1	54.8	39.3	11.8	3.7	106.0	78.3	27.8	3.0	51.2	456.7	12.4
4th quarter	113.4	53.1	1.4	2.9	56.0	40.2	12.0	3.8	105.8	77.1	28.7	4.5	49.8	466.1	12.3

Table 5-1B. Saving and Investment: Historical Data—*Continued*

(Billions of dollars, except as noted; quarterly data are at seasonally adjusted annual rates.)   **NIPA Tables 1.7.5, 5.1**

Year and quarter	Gross saving									Gross domestic investment and net lending, NIPAs				Net domestic invest-ment	Gross national income	Net saving as a percent-age of gross national income
	Total	Net saving			Consumption of fixed capital				Gross domestic investment			Net lending or net borrow-ing (-), NIPAs			
		Private	Government		Total	Private	Government		Total	Private	Govern-ment				
			Federal	State and local			Federal	State and local							
1957															
1st quarter	114.3	53.3	0.4	3.4	57.3	41.1	12.4	3.8	107.5	77.7	29.8	5.6	50.2	474.2	12.0
2nd quarter	114.4	54.7	-1.6	2.8	58.4	41.9	12.6	4.0	107.6	77.9	29.7	4.8	49.2	477.3	11.7
3rd quarter	114.2	53.8	-1.6	2.5	59.5	42.7	12.8	4.0	109.9	79.3	30.6	4.9	50.5	482.1	11.3
4th quarter	104.3	49.6	-7.7	1.9	60.4	43.4	13.1	4.0	101.9	71.0	30.8	3.6	41.4	476.2	9.2
1958															
1st quarter	100.1	48.2	-10.2	1.0	61.1	44.0	13.1	4.0	98.0	66.7	31.3	1.5	36.9	470.8	8.3
2nd quarter	96.8	48.9	-15.0	0.9	62.0	44.7	13.3	4.1	97.0	65.1	32.0	0.8	35.0	473.6	7.3
3rd quarter	105.0	53.6	-12.5	0.9	62.9	45.2	13.5	4.2	104.9	72.0	32.9	1.0	42.0	487.4	8.6
4th quarter	113.1	57.9	-10.6	1.9	63.8	45.7	13.8	4.2	114.3	80.0	34.3	-0.1	50.5	501.0	9.8
1959															
1st quarter	117.8	56.6	-4.6	1.3	64.5	46.2	14.0	4.3	119.4	83.2	36.2	-1.4	54.9	512.7	10.4
2nd quarter	124.9	59.7	-1.8	2.0	65.1	46.6	14.2	4.3	125.5	89.4	36.2	-2.9	60.4	527.5	11.3
3rd quarter	117.3	53.3	-4.3	2.5	65.8	47.0	14.4	4.3	119.7	83.6	36.1	0.0	54.0	525.4	9.8
4th quarter	120.3	56.2	-5.0	2.6	66.4	47.4	14.6	4.4	121.5	86.5	34.9	-0.8	55.1	531.2	10.1
1960															
1st quarter	132.9	58.6	4.7	2.6	67.0	47.7	14.9	4.4	131.6	96.5	35.1	1.9	64.6	545.1	12.1
2nd quarter	126.3	55.0	1.2	2.5	67.6	48.0	15.1	4.5	122.3	87.1	35.2	1.7	54.7	546.4	10.8
3rd quarter	127.1	56.8	-0.4	2.5	68.2	48.4	15.3	4.6	123.2	86.4	36.8	4.2	55.0	548.5	10.7
4th quarter	121.5	55.1	-4.9	2.5	68.8	48.7	15.5	4.6	112.8	76.0	36.9	4.9	44.0	547.3	9.6
1961															
1st quarter	124.6	57.8	-4.7	2.1	69.5	49.1	15.7	4.7	118.0	78.4	39.6	5.4	48.5	549.8	10.0
2nd quarter	127.0	61.3	-6.4	1.9	70.2	49.5	15.9	4.8	122.6	84.1	38.5	3.2	52.4	560.1	10.1
3rd quarter	134.6	66.1	-4.9	2.4	71.0	50.0	16.2	4.8	130.8	90.9	39.9	4.3	59.8	570.7	11.1
4th quarter	140.2	68.9	-2.9	2.5	71.8	50.5	16.4	4.9	134.5	92.9	41.6	3.9	62.8	586.0	11.7
1962															
1st quarter	141.6	71.5	-5.4	2.8	72.7	51.0	16.7	5.0	140.0	98.1	42.0	3.1	67.3	596.1	11.6
2nd quarter	142.0	71.3	-5.7	2.8	73.5	51.5	16.9	5.1	138.6	96.7	41.9	3.7	65.0	604.1	11.3
3rd quarter	144.4	71.4	-4.9	3.4	74.5	52.1	17.3	5.2	141.3	98.2	43.1	4.8	66.7	611.5	11.4
4th quarter	145.5	72.3	-5.5	3.2	75.6	52.6	17.7	5.3	138.4	95.0	43.5	3.6	62.8	620.5	11.3
1963															
1st quarter	149.1	73.1	-3.5	3.0	76.5	53.2	18.0	5.4	142.9	99.7	43.3	4.0	66.4	628.4	11.5
2nd quarter	153.8	74.3	-1.1	3.1	77.5	53.8	18.3	5.5	145.2	101.7	43.6	4.3	67.7	638.4	12.0
3rd quarter	154.7	74.3	-1.7	3.6	78.5	54.5	18.5	5.5	150.4	104.6	45.8	5.2	71.9	647.9	11.8
4th quarter	158.9	78.4	-2.4	3.4	79.5	55.2	18.7	5.7	152.3	107.2	45.1	6.3	72.8	658.9	12.1
1964															
1st quarter	161.9	82.8	-5.5	4.1	80.6	55.9	18.9	5.8	156.3	110.5	45.8	8.2	75.6	672.4	12.1
2nd quarter	163.0	87.6	-10.2	3.9	81.8	56.8	19.1	5.9	157.0	110.5	46.5	6.0	75.3	683.6	11.9
3rd quarter	166.4	86.2	-6.7	4.0	82.9	57.7	19.3	6.0	159.2	112.6	46.6	8.1	76.3	696.1	12.0
4th quarter	172.9	90.0	-5.1	3.7	84.2	58.7	19.5	6.1	161.7	115.0	46.7	7.6	77.5	705.6	12.6
1965															
1st quarter	180.8	92.6	-1.0	3.5	85.6	59.7	19.7	6.2	172.7	126.5	46.2	5.7	87.1	725.6	13.1
2nd quarter	182.7	93.9	-2.0	3.7	87.1	60.9	19.9	6.4	174.0	127.1	46.9	6.3	86.8	738.3	12.9
3rd quarter	183.3	100.2	-9.4	3.7	88.8	62.1	20.1	6.5	180.1	131.2	48.9	6.6	91.3	751.2	12.6
4th quarter	185.0	100.3	-9.6	3.8	90.5	63.5	20.4	6.6	183.2	133.8	49.5	6.1	92.7	772.2	12.2
1966															
1st quarter	194.3	101.4	-4.1	4.7	92.3	65.0	20.6	6.8	196.1	144.2	51.9	4.6	103.8	794.3	12.8
2nd quarter	194.9	101.2	-5.7	5.0	94.3	66.4	20.9	7.0	195.4	143.5	51.9	3.3	101.1	806.1	12.5
3rd quarter	195.3	102.0	-7.9	4.9	96.3	67.9	21.2	7.2	197.4	143.2	54.2	3.2	101.1	819.4	12.1
4th quarter	201.4	109.1	-10.3	4.3	98.3	69.4	21.5	7.4	202.1	145.9	56.2	4.1	103.8	833.8	12.4
1967															
1st quarter	197.6	112.0	-19.3	4.6	100.3	70.9	21.8	7.5	200.7	142.8	57.9	4.3	100.4	842.0	11.6
2nd quarter	194.0	108.2	-20.5	4.0	102.4	72.5	22.2	7.6	194.1	137.5	56.6	3.3	91.7	850.8	10.8
3rd quarter	201.9	112.6	-18.8	3.5	104.6	74.1	22.7	7.9	200.8	142.8	57.9	3.6	96.1	868.5	11.2
4th quarter	208.5	117.0	-19.5	4.0	107.0	75.8	23.1	8.1	206.0	147.7	58.3	2.8	99.0	886.7	11.4
1968															
1st quarter	207.0	111.0	-17.0	3.5	109.5	77.6	23.6	8.3	210.6	152.3	58.4	1.6	101.1	909.9	10.7
2nd quarter	213.6	116.0	-18.7	4.3	111.9	79.5	23.9	8.5	218.5	158.9	59.5	2.0	106.5	933.5	10.9
3rd quarter	215.5	107.8	-10.1	3.2	114.5	81.6	24.3	8.6	215.2	155.7	59.5	1.8	100.7	955.5	10.6
4th quarter	222.1	111.2	-9.3	2.8	117.5	83.7	24.8	8.9	220.3	160.8	59.6	0.7	102.9	975.5	10.7
1969															
1st quarter	231.0	105.5	2.2	2.8	120.5	86.1	25.2	9.2	233.9	172.4	61.5	1.7	113.4	995.1	11.1
2nd quarter	231.3	106.4	-1.6	3.0	123.5	88.3	25.6	9.5	232.4	172.7	59.7	0.5	108.9	1 013.6	10.6
3rd quarter	237.7	116.5	-8.7	3.6	126.3	90.6	26.0	9.8	237.5	177.6	59.9	1.7	111.2	1 034.4	10.8
4th quarter	232.5	112.7	-12.6	3.2	129.3	92.8	26.4	10.1	228.6	171.6	57.0	2.5	99.3	1 045.6	9.9
1970															
1st quarter	226.9	113.9	-22.3	3.0	132.2	94.9	27.0	10.4	227.1	168.1	59.0	3.9	94.9	1 053.4	9.0
2nd quarter	229.8	125.0	-32.7	2.3	135.3	97.1	27.4	10.8	230.3	171.5	58.8	5.0	95.0	1 068.5	8.8
3rd quarter	231.3	131.2	-39.6	1.4	138.3	99.4	27.9	11.1	234.1	173.9	60.2	3.7	95.8	1 086.1	8.6
4th quarter	224.9	129.0	-44.7	-0.9	141.5	101.7	28.3	11.5	227.9	166.8	61.1	2.2	86.3	1 089.5	7.7
1971															
1st quarter	238.7	141.4	-45.6	-1.8	144.6	104.1	28.7	11.8	247.5	189.5	58.1	4.5	102.9	1 129.3	8.3
2nd quarter	243.2	149.4	-52.5	-1.3	147.6	106.5	29.0	12.1	255.7	197.3	58.3	-0.2	108.1	1 152.0	8.3
3rd quarter	249.0	152.1	-51.8	-1.7	150.4	108.8	29.2	12.3	261.2	202.1	59.2	0.0	110.9	1 172.7	8.4
4th quarter	253.6	154.6	-53.7	-0.4	153.1	111.1	29.3	12.7	256.9	198.4	58.5	-3.1	103.8	1 197.8	8.4

Table 5-1B. Saving and Investment: Historical Data—*Continued*

(Billions of dollars, except as noted; quarterly data are at seasonally adjusted annual rates.)   **NIPA Tables 1.7.5, 5.1**

Year and quarter	Gross saving — Total	Net saving — Private	Net saving — Government — Federal	Net saving — Government — State and local	Consumption of fixed capital — Total	Consumption of fixed capital — Private	Consumption of fixed capital — Government — Federal	Consumption of fixed capital — Government — State and local	Gross domestic investment — Total	Gross domestic investment — Private	Gross domestic investment — Government	Net lending or net borrowing (-), NIPAs	Net domestic investment	Gross national income	Net saving as a percentage of gross national income
1972															
1st quarter	263.0	152.4	-45.8	0.4	155.9	113.4	29.6	12.9	273.0	213.0	60.0	-5.1	117.0	1 233.8	8.7
2nd quarter	265.7	148.8	-51.1	9.1	159.0	116.0	29.9	13.1	287.7	226.8	60.9	-4.8	128.8	1 257.2	8.5
3rd quarter	279.8	157.8	-40.9	0.2	162.7	118.8	30.5	13.4	292.3	233.1	59.2	-3.6	129.6	1 290.6	9.1
4th quarter	302.1	179.3	-58.3	14.7	166.4	121.8	30.8	13.8	302.3	239.7	62.6	-2.8	135.9	1 340.5	10.1
1973															
1st quarter	316.0	177.1	-40.5	9.0	170.5	125.2	31.2	14.2	320.5	254.3	66.2	2.2	150.0	1 381.5	10.5
2nd quarter	324.8	184.4	-40.8	5.8	175.4	129.0	31.8	14.6	333.7	268.2	65.5	5.4	158.3	1 411.1	10.6
3rd quarter	339.0	190.7	-37.3	4.4	181.2	133.4	32.8	15.0	329.0	264.3	64.6	12.4	147.7	1 445.2	10.9
4th quarter	361.4	205.2	-34.5	3.1	187.6	138.3	33.7	15.6	347.1	280.9	66.2	15.3	159.5	1 489.5	11.7
1974															
1st quarter	355.8	196.7	-36.1	0.9	194.3	143.8	34.0	16.5	339.3	268.4	71.0	16.5	145.0	1 508.2	10.7
2nd quarter	346.4	183.1	-38.1	-0.7	202.1	149.7	34.6	17.8	353.4	277.4	76.0	2.6	151.3	1 536.8	9.4
3rd quarter	345.9	173.6	-37.0	-1.9	211.2	156.1	36.0	19.1	349.2	271.0	78.2	0.3	138.0	1 571.7	8.6
4th quarter	348.7	190.5	-54.2	-7.6	220.0	163.0	36.9	20.0	360.8	281.3	79.5	4.4	140.9	1 596.5	8.1
1975															
1st quarter	329.8	189.7	-76.2	-12.3	228.7	170.3	37.8	20.6	327.3	244.3	83.0	17.8	98.6	1 612.1	6.3
2nd quarter	330.0	237.9	-132.9	-10.8	235.8	176.6	38.3	20.9	322.4	243.3	79.2	21.4	86.7	1 649.6	5.7
3rd quarter	360.1	218.6	-90.9	-9.7	242.1	182.0	39.1	21.1	350.7	265.2	85.5	18.3	108.6	1 713.8	6.9
4th quarter	372.6	226.9	-91.7	-10.0	247.4	186.3	39.8	21.4	366.1	276.2	89.9	21.6	118.7	1 762.7	7.1
1976															
1st quarter	392.0	228.6	-81.9	-6.6	251.9	189.6	40.7	21.7	397.5	304.6	92.9	13.1	145.6	1 817.5	7.7
2nd quarter	399.3	224.7	-76.8	-5.3	256.7	193.7	41.1	22.0	411.3	322.3	89.0	8.9	154.6	1 848.0	7.7
3rd quarter	402.1	225.3	-80.8	-5.2	262.7	198.6	42.0	22.1	417.0	328.3	88.7	2.4	154.3	1 886.3	7.4
4th quarter	403.8	219.1	-84.2	-0.7	269.6	204.2	42.9	22.4	425.6	337.7	88.0	3.8	156.1	1 926.5	7.0
1977															
1st quarter	416.0	214.5	-73.5	-2.5	277.5	210.7	44.0	22.8	451.7	360.3	91.4	-8.1	174.2	1 981.7	7.0
2nd quarter	454.4	237.0	-67.6	-0.7	285.7	217.5	45.0	23.2	483.9	389.7	94.2	-8.9	198.3	2 056.0	8.2
3rd quarter	479.3	257.7	-75.1	2.9	293.8	224.5	45.6	23.6	507.8	414.1	93.6	-7.9	214.0	2 118.8	8.8
4th quarter	488.0	260.9	-77.7	2.4	302.4	231.8	46.5	24.0	515.7	422.3	93.4	-19.0	213.3	2 174.3	8.5
1978															
1st quarter	500.5	266.8	-81.6	3.7	311.6	239.4	47.7	24.5	528.3	434.8	93.5	-25.1	216.7	2 222.1	8.5
2nd quarter	540.6	271.3	-62.3	9.6	322.0	247.5	49.4	25.1	575.5	470.6	105.0	-12.9	253.5	2 328.3	9.4
3rd quarter	561.5	283.5	-56.0	2.0	332.0	256.2	50.1	25.7	602.0	492.4	109.7	-11.5	270.1	2 386.7	9.6
4th quarter	589.4	290.3	-48.3	4.1	343.2	265.4	51.4	26.3	629.9	515.8	114.1	-1.4	286.7	2 462.3	10.0
1979															
1st quarter	608.8	299.2	-45.1	-0.1	354.8	275.2	52.5	27.1	635.5	525.8	109.7	-1.9	280.8	2 527.4	10.1
2nd quarter	612.5	289.4	-42.6	-1.6	367.2	285.3	53.9	28.0	656.2	539.3	116.9	-2.8	289.0	2 579.7	9.5
3rd quarter	615.0	283.5	-48.5	-0.3	380.3	295.8	55.5	29.0	671.0	545.6	125.4	2.3	290.7	2 645.2	8.9
4th quarter	617.8	280.8	-53.3	-2.9	393.2	306.6	56.6	30.1	676.2	547.9	128.4	-2.7	283.0	2 704.5	8.3
1980															
1st quarter	611.5	274.4	-66.2	-3.5	406.8	317.7	57.9	31.2	690.3	554.6	135.7	-10.8	283.5	2 759.7	7.4
2nd quarter	607.9	289.2	-93.3	-9.0	421.0	329.0	59.6	32.4	654.0	519.3	134.7	9.7	233.0	2 777.2	6.7
3rd quarter	624.8	300.8	-104.1	-7.3	435.4	340.7	61.0	33.7	629.4	495.1	134.4	28.0	194.1	2 858.1	6.6
4th quarter	676.3	321.1	-91.5	-3.9	450.6	352.5	63.1	35.0	690.5	551.5	139.0	6.6	239.9	2 993.2	7.5
1981															
1st quarter	707.9	321.1	-74.3	-4.5	465.5	364.7	64.7	36.2	765.8	619.4	146.4	1.5	300.2	3 096.0	7.8
2nd quarter	719.4	326.9	-78.7	-9.2	480.4	376.4	66.7	37.3	755.1	609.8	145.3	0.1	274.7	3 156.7	7.6
3rd quarter	775.7	378.4	-86.8	-10.5	494.5	387.8	68.4	38.3	798.0	652.3	145.7	7.0	303.5	3 263.9	8.6
4th quarter	772.5	393.0	-112.4	-16.5	508.4	398.8	70.4	39.3	795.4	643.4	152.0	4.5	287.0	3 291.0	8.0
1982															
1st quarter	736.9	371.6	-136.7	-20.0	521.9	409.4	72.5	40.1	738.6	588.3	150.2	0.5	216.6	3 308.0	6.5
2nd quarter	762.0	392.9	-143.3	-21.5	533.9	418.0	74.9	41.0	750.0	593.6	156.4	17.3	216.1	3 367.5	6.8
3rd quarter	730.6	390.9	-179.3	-23.7	542.7	424.7	76.4	41.7	750.4	593.0	157.4	-14.1	207.6	3 395.7	5.5
4th quarter	673.7	360.5	-210.2	-25.9	549.3	429.5	77.8	42.0	713.0	549.2	163.8	-17.4	163.7	3 414.7	3.6
1983															
1st quarter	691.6	377.7	-208.6	-29.8	552.3	432.3	77.7	42.3	730.2	565.5	164.7	-8.0	177.9	3 476.7	4.0
2nd quarter	700.9	365.3	-203.2	-20.2	559.1	436.0	80.5	42.5	780.8	613.8	167.1	-27.7	221.8	3 563.3	4.0
3rd quarter	703.1	370.5	-219.8	-13.7	566.2	440.6	82.7	42.8	826.8	652.3	174.5	-48.0	260.6	3 651.3	3.8
4th quarter	771.1	405.3	-197.1	-10.1	572.9	446.2	83.6	43.1	896.9	718.5	178.4	-56.9	324.0	3 765.3	5.3
1984															
1st quarter	854.4	455.9	-182.3	-1.5	582.2	452.6	86.1	43.5	975.0	790.9	184.1	-78.6	392.7	3 903.1	7.0
2nd quarter	873.6	470.4	-191.0	2.7	591.5	459.5	87.9	44.0	1 008.4	818.9	189.5	-88.0	416.9	4 000.3	7.1
3rd quarter	898.1	497.5	-199.6	-3.0	603.2	467.0	91.6	44.6	1 032.6	838.9	193.8	-90.5	429.4	4 077.5	7.2
4th quarter	900.3	495.7	-213.1	1.0	616.7	474.9	96.7	45.0	1 037.1	831.7	205.4	-103.5	420.5	4 148.3	6.8
1985															
1st quarter	889.6	436.4	-171.0	0.6	623.6	483.4	94.5	45.8	1 018.1	809.9	208.3	-89.5	394.5	4 216.8	6.3
2nd quarter	887.3	474.9	-221.4	-0.7	634.6	492.0	96.2	46.4	1 045.4	827.0	218.4	-111.5	410.9	4 275.8	5.9
3rd quarter	865.9	424.7	-200.6	-3.3	645.2	500.7	97.4	47.1	1 048.4	822.2	226.3	-121.4	403.2	4 348.0	5.1
4th quarter	880.9	433.9	-203.9	-6.2	657.2	509.6	99.7	47.8	1 086.1	859.5	226.6	-135.3	428.9	4 399.7	5.1
1986															
1st quarter	895.5	432.0	-206.5	1.8	668.2	518.6	100.9	48.6	1 092.5	863.5	229.0	-128.4	424.3	4 461.4	5.1
2nd quarter	871.0	423.9	-228.0	-4.8	679.9	527.4	102.9	49.6	1 089.7	855.2	234.4	-142.0	409.8	4 484.0	4.3
3rd quarter	838.6	382.5	-232.3	-2.4	691.0	536.0	104.4	50.6	1 081.5	835.8	245.7	-150.3	390.6	4 533.2	3.3
4th quarter	852.8	358.1	-196.9	-10.4	702.1	544.3	106.1	51.7	1 085.3	842.1	243.2	-150.4	383.2	4 587.5	3.3

Table 5-1B. Saving and Investment: Historical Data—*Continued*

(Billions of dollars, except as noted; quarterly data are at seasonally adjusted annual rates.)   **NIPA Tables 1.7.5, 5.1**

Year and quarter	Gross saving Total	Net saving Private	Net saving Government Federal	Net saving Government State and local	Consumption of fixed capital Total	Consumption of fixed capital Private	Consumption of fixed capital Government Federal	Consumption of fixed capital Government State and local	Gross domestic investment Total	Gross domestic investment Private	Gross domestic investment Government	Net lending or net borrowing (-), NIPAs	Net domestic investment	Gross national income	Net saving as a percentage of gross national income
1987															
1st quarter	897.9	401.3	-197.6	-18.4	712.6	552.3	107.7	52.6	1 118.8	871.2	247.7	-151.1	406.2	4 666.5	4.0
2nd quarter	926.6	353.4	-144.0	-6.6	723.8	561.2	109.1	53.5	1 128.0	874.6	253.5	-154.5	404.3	4 777.3	4.2
3rd quarter	961.2	396.1	-155.6	-15.4	736.1	570.7	110.9	54.5	1 134.8	876.5	258.3	-154.2	398.7	4 882.7	4.6
4th quarter	1 009.8	443.6	-165.8	-17.1	749.1	581.0	112.8	55.3	1 205.6	946.5	259.1	-157.2	456.5	4 989.0	5.2
1988															
1st quarter	1 029.9	446.1	-166.6	-13.4	763.8	591.9	115.6	56.3	1 162.1	908.6	253.5	-127.0	398.3	5 093.1	5.2
2nd quarter	1 069.0	465.1	-158.7	-15.1	777.7	602.6	117.9	57.2	1 192.7	934.5	258.2	-110.1	415.0	5 199.1	5.6
3rd quarter	1 096.4	465.8	-153.3	-7.5	791.4	613.3	119.9	58.1	1 200.1	942.0	258.1	-106.6	408.7	5 305.7	5.7
4th quarter	1 111.3	476.6	-161.3	-9.1	805.1	623.9	122.2	59.1	1 226.5	962.7	263.8	-119.7	421.4	5 426.5	5.6
1989															
1st quarter	1 145.0	475.4	-140.1	-8.7	818.4	634.4	124.0	60.1	1 269.3	1 005.5	263.8	-106.8	450.9	5 516.1	5.9
2nd quarter	1 104.7	447.4	-161.5	-12.9	831.7	644.7	125.8	61.2	1 268.3	1 001.0	267.2	-93.7	436.6	5 564.1	4.9
3rd quarter	1 100.5	440.3	-165.6	-19.1	844.9	654.8	127.8	62.3	1 271.5	996.5	275.0	-81.8	426.6	5 630.3	4.5
4th quarter	1 089.0	437.8	-170.3	-36.6	858.0	664.6	129.9	63.5	1 271.5	995.8	275.6	-88.5	413.4	5 683.0	4.1
1990															
1st quarter	1 102.7	456.3	-198.6	-24.9	870.0	674.3	131.0	64.7	1 297.7	1 010.8	286.8	-90.9	427.7	5 797.3	4.0
2nd quarter	1 134.9	493.1	-208.0	-32.4	882.2	683.8	132.4	66.0	1 303.0	1 014.7	288.3	-73.0	420.8	5 896.4	4.3
3rd quarter	1 105.6	448.6	-199.6	-38.4	895.0	693.1	134.4	67.4	1 292.4	1 000.8	291.6	-82.5	397.3	5 937.9	3.5
4th quarter	1 110.2	459.5	-206.9	-49.3	906.9	702.3	136.1	68.4	1 242.2	947.5	294.7	-82.7	335.3	5 977.5	3.4
1991															
1st quarter	1 199.7	521.6	-185.8	-55.0	919.0	711.4	138.2	69.4	1 215.7	924.6	291.1	42.1	296.7	6 012.1	4.7
2nd quarter	1 144.5	520.7	-244.6	-60.4	928.9	719.0	139.5	70.5	1 222.6	926.5	296.0	14.2	293.6	6 062.6	3.6
3rd quarter	1 122.7	516.5	-273.3	-58.3	937.8	725.3	141.1	71.4	1 243.5	947.5	296.0	-29.0	305.7	6 123.3	3.0
4th quarter	1 146.4	559.1	-290.0	-66.6	943.9	730.2	141.6	72.1	1 272.0	978.8	293.2	-17.0	328.1	6 189.0	3.3
1992															
1st quarter	1 156.0	598.7	-325.6	-65.2	948.2	733.7	141.8	72.7	1 256.3	956.8	299.5	-22.0	308.2	6 317.3	3.3
2nd quarter	1 175.0	617.7	-328.7	-69.2	955.2	738.8	142.7	73.7	1 310.2	1 013.1	297.1	-39.8	355.0	6 407.0	3.4
3rd quarter	1 125.4	582.3	-344.2	-76.0	963.4	745.4	143.4	74.5	1 318.3	1 024.2	294.1	-46.8	354.9	6 455.7	2.5
4th quarter	1 134.0	572.7	-339.5	-73.4	974.3	753.6	145.1	75.5	1 351.7	1 058.0	293.7	-68.6	377.4	6 561.9	2.4
1993															
1st quarter	1 131.0	564.3	-337.5	-82.0	986.2	763.4	146.1	76.8	1 372.3	1 083.8	288.5	-60.0	386.2	6 582.6	2.2
2nd quarter	1 168.0	555.1	-308.8	-75.8	997.5	773.2	146.5	77.9	1 387.7	1 094.5	293.2	-74.7	390.1	6 694.4	2.5
3rd quarter	1 143.8	528.8	-318.9	-74.8	1 008.7	783.1	146.9	78.7	1 386.5	1 095.9	290.7	-77.9	377.8	6 752.9	2.0
4th quarter	1 210.9	535.4	-289.0	-57.1	1 021.6	793.2	148.7	79.7	1 448.3	1 153.1	295.2	-108.5	426.7	6 905.2	2.7
1994															
1st quarter	1 232.5	534.6	-269.1	-66.3	1 033.3	803.3	149.0	81.0	1 485.4	1 201.7	283.7	-93.1	452.1	6 983.1	2.9
2nd quarter	1 297.0	554.7	-238.6	-66.1	1 047.0	815.0	149.9	82.0	1 555.2	1 264.9	290.3	-112.8	508.3	7 120.7	3.5
3rd quarter	1 305.9	557.5	-256.8	-57.1	1 062.3	828.4	150.5	83.4	1 553.8	1 251.7	302.1	-122.8	491.5	7 226.2	3.4
4th quarter	1 345.0	589.4	-258.0	-66.3	1 079.9	843.3	151.9	84.7	1 608.2	1 307.6	300.6	-138.8	528.3	7 351.0	3.6
1995															
1st quarter	1 394.2	616.4	-254.4	-66.9	1 099.2	859.8	153.1	86.2	1 634.6	1 327.6	307.0	-119.5	535.4	7 431.0	4.0
2nd quarter	1 394.1	604.2	-243.0	-82.9	1 115.8	874.8	153.4	87.6	1 614.4	1 304.0	310.4	-119.1	498.6	7 512.7	3.7
3rd quarter	1 437.0	624.6	-247.0	-71.3	1 130.8	888.2	153.8	88.8	1 610.4	1 303.2	307.1	-101.8	479.6	7 628.4	4.0
4th quarter	1 479.9	620.6	-224.0	-60.6	1 143.9	900.1	153.8	90.0	1 641.4	1 335.1	306.3	-84.8	497.5	7 726.6	4.3
1996															
1st quarter	1 520.1	645.9	-224.8	-56.6	1 155.6	910.4	154.0	91.1	1 672.2	1 355.4	316.8	-103.1	516.6	7 856.6	4.6
2nd quarter	1 559.9	630.4	-180.2	-57.9	1 167.6	922.0	153.6	91.9	1 736.1	1 418.4	317.8	-109.5	568.6	7 995.0	4.9
3rd quarter	1 596.2	636.8	-170.6	-51.9	1 181.9	934.9	153.9	93.0	1 795.4	1 474.4	321.0	-133.7	613.5	8 091.5	5.1
4th quarter	1 639.4	633.9	-141.9	-48.8	1 196.2	949.0	153.1	94.1	1 804.4	1 480.1	324.2	-114.5	608.2	8 240.5	5.4
1997															
1st quarter	1 696.4	652.3	-121.5	-47.8	1 213.3	964.4	153.6	95.3	1 841.7	1 522.4	319.3	-132.2	628.4	8 372.4	5.8
2nd quarter	1 767.8	686.5	-104.0	-45.9	1 231.2	979.9	154.6	96.7	1 914.2	1 590.2	324.0	-102.5	683.0	8 505.0	6.3
3rd quarter	1 818.3	676.4	-68.5	-37.4	1 247.8	995.5	154.5	97.8	1 956.7	1 625.3	331.5	-123.5	708.9	8 670.3	6.6
4th quarter	1 839.6	685.3	-73.8	-36.8	1 265.0	1 011.2	154.6	99.2	1 976.2	1 644.5	331.7	-164.1	711.2	8 809.4	6.5
1998															
1st quarter	1 910.1	683.9	-24.5	-30.7	1 281.3	1 027.0	154.2	100.2	2 043.6	1 712.3	331.2	-165.1	762.2	8 923.2	7.0
2nd quarter	1 915.7	660.1	-10.4	-34.1	1 300.0	1 043.3	155.5	101.2	2 036.8	1 695.8	341.1	-196.3	736.8	9 067.1	6.8
3rd quarter	1 958.0	650.8	19.9	-31.5	1 318.8	1 060.4	155.6	102.8	2 095.8	1 741.6	354.2	-224.9	777.0	9 217.4	6.9
4th quarter	1 938.5	603.3	20.6	-24.2	1 338.8	1 078.0	156.4	104.4	2 146.5	1 797.0	349.5	-236.0	807.7	9 334.1	6.4
1999															
1st quarter	2 039.7	670.0	46.3	-35.3	1 358.7	1 096.3	156.6	105.9	2 210.1	1 853.1	357.1	-234.0	851.4	9 501.5	7.2
2nd quarter	1 995.7	590.5	66.2	-44.2	1 383.2	1 117.9	157.5	107.7	2 210.9	1 848.3	362.5	-263.6	827.6	9 598.2	6.4
3rd quarter	1 981.2	538.7	71.5	-39.6	1 410.7	1 143.1	158.3	109.3	2 263.8	1 893.7	370.0	-298.6	853.1	9 720.3	5.9
4th quarter	2 024.6	534.5	83.5	-36.6	1 443.2	1 171.6	160.2	111.4	2 337.4	1 953.1	384.3	-335.2	894.2	9 937.3	5.9
2000															
1st quarter	2 146.1	521.0	173.0	-26.9	1 479.0	1 203.5	162.0	113.4	2 339.8	1 950.7	389.2	-371.3	860.8	10 209.1	6.5
2nd quarter	2 133.1	510.2	144.1	-27.4	1 506.1	1 227.5	162.8	115.8	2 461.6	2 075.8	385.8	-379.5	955.5	10 330.1	6.1
3rd quarter	2 150.5	515.8	154.5	-44.7	1 525.0	1 243.5	163.7	117.7	2 446.2	2 060.0	386.2	-416.1	921.2	10 467.1	6.0
4th quarter	2 079.5	457.9	150.3	-63.6	1 534.8	1 251.5	163.9	119.5	2 461.5	2 067.2	394.2	-420.3	926.6	10 529.1	5.2
2001															
1st quarter	2 153.4	539.3	127.4	-79.8	1 566.4	1 281.2	164.0	121.2	2 372.2	1 971.3	400.9	-412.4	805.8	10 703.9	5.5
2nd quarter	2 137.0	556.5	93.1	-106.7	1 594.0	1 307.1	164.3	122.7	2 392.0	1 973.0	418.9	-373.4	797.9	10 762.7	5.0
3rd quarter	2 053.6	701.3	-130.1	-129.2	1 611.6	1 322.8	164.8	124.0	2 348.4	1 944.9	403.5	-355.2	736.8	10 732.8	4.1
4th quarter	1 963.6	532.3	-34.4	-160.3	1 626.0	1 335.7	164.8	125.5	2 274.3	1 850.1	424.3	-339.1	648.4	10 775.4	3.1

Table 5-1B. Saving and Investment: Historical Data—*Continued*

(Billions of dollars, except as noted; quarterly data are at seasonally adjusted annual rates.)   **NIPA Tables 1.7.5, 5.1**

Year and quarter	Gross saving								Gross domestic investment and net lending, NIPAs				Net domestic invest-ment	Gross national income	Net saving as a percent-age of gross national income
	Total	Net saving			Consumption of fixed capital				Gross domestic investment			Net lending or net borrow-ing (-), NIPAs			
		Private	Government		Total	Private	Government		Total	Private	Govern-ment				
			Federal	State and local			Federal	State and local							
2002															
1st quarter	2 005.8	768.4	-233.4	-165.6	1 636.3	1 343.8	165.5	127.1	2 347.4	1 912.7	434.7	-397.6	711.1	10 893.4	3.4
2nd quarter	2 009.9	810.1	-264.1	-185.1	1 648.9	1 354.0	166.2	128.7	2 371.7	1 933.3	438.4	-444.4	722.9	11 005.3	3.3
3rd quarter	1 980.5	776.7	-278.4	-183.0	1 665.2	1 367.8	167.2	130.1	2 380.3	1 933.2	447.1	-447.1	715.0	11 082.2	2.8
4th quarter	2 018.1	844.5	-310.1	-197.8	1 681.5	1 381.7	168.4	131.4	2 397.0	1 942.5	454.5	-485.7	715.5	11 236.4	3.0
2003															
1st quarter	1 947.2	805.7	-341.1	-212.8	1 695.5	1 392.7	169.7	133.0	2 417.3	1 960.2	457.1	-533.0	721.9	11 288.9	2.2
2nd quarter	1 982.0	860.4	-390.8	-196.1	1 708.5	1 403.1	171.2	134.2	2 430.7	1 972.4	458.3	-517.6	722.2	11 431.8	2.4
3rd quarter	1 985.9	901.0	-474.0	-167.2	1 726.1	1 417.5	173.0	135.5	2 513.4	2 044.3	469.1	-516.0	787.3	11 608.0	2.2
4th quarter	2 051.8	864.9	-410.3	-149.1	1 746.3	1 434.4	175.1	136.8	2 603.7	2 131.3	472.4	-494.8	857.4	11 792.3	2.6
2004															
1st quarter	2 068.0	895.3	-443.6	-157.8	1 774.1	1 458.2	177.3	138.5	2 628.4	2 154.1	474.3	-541.4	854.3	11 999.5	2.4
2nd quarter	2 168.2	927.3	-399.2	-163.4	1 803.5	1 482.3	179.4	141.8	2 745.9	2 262.6	483.3	-616.0	942.4	12 220.4	3.0
3rd quarter	2 239.4	924.2	-369.8	-150.7	1 835.8	1 508.4	181.2	146.2	2 808.4	2 318.3	490.1	-620.1	972.6	12 450.4	3.2
4th quarter	2 181.7	822.8	-391.0	-124.0	1 873.9	1 539.5	184.0	150.4	2 887.2	2 390.1	497.1	-712.3	1 013.3	12 585.6	2.4
2005															
1st quarter	2 310.5	817.4	-315.1	-103.3	1 911.5	1 572.4	186.7	152.5	2 987.6	2 486.1	501.6	-683.8	1 076.1	12 849.0	3.1
2nd quarter	2 324.5	794.7	-312.0	-106.7	1 948.5	1 603.4	189.1	156.0	2 986.7	2 476.5	510.2	-707.1	1 038.2	13 031.9	2.9
3rd quarter	2 335.7	750.7	-297.8	-107.1	1 989.0	1 638.1	191.7	160.1	3 048.4	2 531.1	517.3	-676.7	1 058.6	13 255.3	2.6
4th quarter	2 492.4	851.2	-298.9	-94.1	2 034.2	1 676.6	194.7	163.0	3 169.3	2 645.3	524.0	-830.2	1 135.1	13 542.6	3.4
2006															
1st quarter	2 668.9	911.4	-247.2	-66.0	2 070.7	1 708.4	197.4	164.9	3 249.4	2 709.7	539.7	-793.3	1 178.7	13 870.4	4.3
2nd quarter	2 651.8	853.4	-243.0	-65.9	2 107.2	1 737.9	200.0	169.3	3 261.1	2 709.3	551.8	-814.3	1 153.8	14 005.8	3.9
3rd quarter	2 646.3	824.0	-224.8	-93.8	2 140.9	1 765.7	202.7	172.5	3 259.5	2 709.4	550.1	-858.6	1 118.6	14 152.2	3.6
4th quarter	2 664.7	796.8	-195.4	-114.4	2 177.7	1 795.2	205.1	177.4	3 237.4	2 675.4	562.0	-749.4	1 059.7	14 263.5	3.4
2007															
1st quarter	2 585.0	708.1	-226.8	-111.2	2 214.9	1 823.9	208.4	182.6	3 238.6	2 664.3	574.3	-791.0	1 023.7	14 398.0	2.6
2nd quarter	2 632.0	736.7	-242.4	-103.8	2 241.3	1 844.2	211.2	185.9	3 288.9	2 699.2	589.6	-749.2	1 047.5	14 550.6	2.7
3rd quarter	2 507.4	653.4	-271.7	-139.4	2 265.0	1 861.7	214.3	189.0	3 281.9	2 686.0	596.0	-677.5	1 016.9	14 569.0	1.7
4th quarter	2 422.2	618.6	-323.6	-162.8	2 290.0	1 880.2	217.3	192.6	3 250.8	2 642.6	608.2	-624.9	960.8	14 655.3	0.9
2008															
1st quarter	2 336.4	609.6	-424.9	-168.2	2 319.9	1 903.0	221.0	196.0	3 175.3	2 563.7	611.6	-723.7	855.4	14 697.1	0.1
2nd quarter	2 236.1	833.3	-777.9	-166.5	2 347.3	1 923.9	224.5	198.9	3 168.2	2 540.6	627.6	-709.4	820.9	14 748.4	-0.8
3rd quarter	2 234.6	753.0	-646.9	-247.4	2 376.0	1 945.7	227.7	202.6	3 136.0	2 498.2	637.7	-674.4	759.9	14 816.9	-1.0
4th quarter	2 157.5	741.5	-674.7	-301.4	2 392.1	1 954.7	230.3	207.1	2 949.3	2 307.9	641.4	-603.6	557.2	14 476.0	-1.6
2009															
1st quarter	2 008.1	1 012.1	-1 059.6	-331.1	2 386.7	1 946.0	231.3	209.3	2 654.9	2 014.9	640.0	-385.8	268.3	14 250.1	-2.7
2nd quarter	1 982.2	1 261.2	-1 313.5	-335.0	2 369.5	1 927.9	232.5	209.0	2 510.2	1 863.7	646.5	-348.4	140.7	14 288.2	-2.7
3rd quarter	1 949.0	1 275.6	-1 329.8	-355.8	2 359.1	1 916.4	234.0	208.7	2 486.4	1 841.4	644.9	-361.7	127.3	14 415.6	-2.8
4th quarter	2 093.9	1 359.4	-1 292.8	-343.5	2 370.7	1 924.5	236.7	209.5	2 638.8	1 998.7	640.1	-394.9	268.1	14 640.9	-1.9
2010															
1st quarter	2 109.2	1 442.1	-1 374.8	-333.7	2 375.6	1 925.3	239.3	211.0	2 671.3	2 038.2	633.2	-425.5	295.8	14 778.3	-1.8
2nd quarter	2 255.9	1 526.7	-1 320.1	-333.8	2 383.1	1 927.8	242.5	212.7	2 796.8	2 148.8	648.0	-436.2	413.7	15 016.1	-0.8
3rd quarter	2 438.9	1 635.9	-1 299.4	-290.1	2 392.6	1 933.5	244.9	214.2	2 885.9	2 236.5	649.4	-465.9	493.3	15 286.8	0.3
4th quarter	2 444.9	1 610.7	-1 306.0	-272.3	2 412.5	1 948.5	248.2	215.8	2 885.8	2 238.4	647.4	-422.0	473.4	15 425.8	0.2
2011															
1st quarter	2 397.1	1 502.6	-1 262.7	-279.1	2 436.4	1 966.7	251.8	217.9	2 845.3	2 206.0	639.4	-474.7	408.9	15 528.8	-0.3
2nd quarter	2 480.2	1 573.0	-1 298.1	-256.8	2 462.2	1 987.2	254.2	220.7	2 933.6	2 297.4	636.3	-489.6	471.4	15 748.9	0.1
3rd quarter	2 610.0	1 617.6	-1 207.9	-287.6	2 487.9	2 007.7	256.4	223.8	2 958.0	2 322.8	635.1	-446.9	470.0	15 934.6	0.8
4th quarter	2 740.1	1 704.6	-1 199.1	-276.8	2 511.4	2 027.6	257.3	226.5	3 139.8	2 504.1	635.7	-457.6	628.5	16 116.6	1.4
2012															
1st quarter	2 974.4	1 834.6	-1 117.2	-279.6	2 536.5	2 047.9	259.7	228.9	3 194.6	2 567.8	626.8	-488.9	658.1	16 536.1	2.6
2nd quarter	3 054.7	1 876.4	-1 104.7	-281.5	2 564.4	2 071.8	261.2	231.4	3 264.0	2 636.9	627.1	-471.4	699.5	16 641.1	2.9
3rd quarter	2 990.6	1 744.6	-1 057.5	-287.5	2 591.0	2 095.2	262.6	233.2	3 263.5	2 644.1	619.4	-423.9	672.5	16 634.4	2.4
4th quarter	3 124.4	1 830.2	-1 035.1	-282.7	2 612.0	2 114.6	263.0	234.4	3 249.0	2 638.3	610.8	-382.3	637.0	16 871.0	3.0
2013															
1st quarter	3 091.7	1 495.1	-776.8	-261.7	2 635.0	2 135.5	263.6	235.9	3 336.9	2 738.2	598.7	-407.8	701.9	16 945.7	2.7
2nd quarter	3 222.8	1 342.9	-537.1	-247.2	2 664.2	2 161.3	265.1	237.7	3 373.7	2 775.3	598.5	-394.6	709.5	17 108.5	3.3
3rd quarter	3 239.2	1 528.8	-706.7	-277.0	2 694.2	2 188.4	266.2	239.6	3 483.2	2 880.0	603.2	-370.6	789.0	17 216.0	3.2
4th quarter	3 319.1	1 394.4	-531.0	-275.8	2 731.4	2 221.0	268.5	241.9	3 511.8	2 910.5	601.3	-304.5	780.4	17 433.3	3.4
2014															
1st quarter	3 403.1	1 461.3	-575.3	-251.3	2 768.4	2 254.8	269.8	243.9	3 489.5	2 901.6	587.8	-373.7	721.1	17 634.9	3.6
2nd quarter	3 572.6	1 585.9	-591.0	-220.5	2 798.2	2 283.0	269.7	245.6	3 630.3	3 028.8	601.5	-345.9	832.1	17 950.7	4.3
3rd quarter	3 637.0	1 667.7	-632.3	-233.1	2 834.6	2 317.2	270.1	247.4	3 705.0	3 099.0	606.0	-355.1	870.3	18 266.3	4.4
4th quarter	3 643.9	1 627.8	-608.5	-242.1	2 866.6	2 347.7	270.4	248.5	3 738.3	3 126.3	612.0	-429.3	871.8	18 397.8	4.2
2015															
1st quarter	3 715.1	1 610.2	-551.6	-229.8	2 886.4	2 366.0	271.3	249.1	3 822.9	3 216.8	606.1	-427.4	936.5	18 521.9	4.5
2nd quarter	3 691.7	1 593.4	-571.6	-237.7	2 907.6	2 385.9	270.8	251.0	3 852.8	3 225.9	626.9	-394.7	945.2	18 673.8	4.2
3rd quarter	3 639.7	1 576.2	-611.0	-256.4	2 930.9	2 407.4	271.2	252.3	3 860.6	3 229.6	631.0	-458.3	929.7	18 786.7	3.8
4th quarter	3 610.9	1 377.7	-541.3	-170.5	2 945.0	2 421.8	270.3	252.9	3 797.5	3 175.5	622.0	-417.4	852.5	18 819.5	3.5
2016															
1st quarter	3 543.8	1 478.2	-638.0	-250.0	2 953.5	2 431.4	269.1	253.1	3 778.5	3 142.1	636.4	-463.3	825.0	18 841.2	3.1
2nd quarter	3 448.4	1 382.9	-668.8	-245.2	2 979.6	2 453.5	270.0	256.1	3 782.4	3 152.2	630.2	-414.6	802.8	18 933.5	2.5
3rd quarter	3 421.0	1 335.0	-674.9	-240.5	3 001.5	2 473.5	270.8	257.1	3 784.8	3 157.7	627.1	-449.3	783.3	19 082.7	2.2
4th quarter	3 516.6	1 415.4	-678.6	-247.7	3 027.5	2 495.7	272.7	259.1	3 859.8	3 227.6	632.2	-457.9	832.4	19 340.4	2.5
2017															
1st quarter	3 648.2	1 505.0	-655.9	-265.8	3 064.9	2 529.3	273.9	261.7	3 915.7	3 278.6	637.1	-462.8	850.7	19 590.1	3.0
2nd quarter	3 659.3	1 490.6	-661.5	-270.9	3 101.1	2 561.9	275.0	264.2	3 980.2	3 337.9	642.3	-508.0	879.1	19 762.1	2.8
3rd quarter	3 733.3	1 513.8	-660.5	-254.7	3 134.8	2 590.9	276.8	267.1	4 054.1	3 413.9	640.2	-448.4	919.3	19 957.4	3.0
4th quarter	3 686.3	1 570.9	-803.6	-244.9	3 163.9	2 616.4	278.3	269.2	4 094.6	3 441.4	653.2	-485.5	930.7	20 179.2	2.6

Table 5-2. Gross Private Fixed Investment by Type

(Billions of dollars, quarterly data are at seasonally adjusted annual rates.)   NIPA Table 5.3.5

Year and quarter	Total gross private fixed investment	Nonresidential											
			Structures						Equipment				
											Information processing equipment		
		Total	Total	Comm-ercial and health care	Manufac-turing	Power and communi-cation	Mining explor-ation, shafts, and wells	Other non-residential structures	Total	Total	Computers and peripheral equipment	Other information processing	Industrial equipment
1960	83.2	56.4	19.6	4.8	2.9	4.4	2.3	5.2	29.7	4.7	0.2	4.6	9.4
1961	83.6	56.6	19.7	5.5	2.8	4.1	2.3	5.0	28.9	5.1	0.3	4.8	8.8
1962	90.9	61.2	20.8	6.2	2.8	4.1	2.5	5.2	32.1	5.5	0.3	5.1	9.3
1963	97.7	64.8	21.2	6.1	2.9	4.4	2.3	5.6	34.4	6.1	0.7	5.4	10.0
1964	107.3	72.2	23.7	6.8	3.6	4.8	2.4	6.2	38.7	6.8	0.9	5.9	11.4
1965	120.4	85.2	28.3	8.2	5.1	5.4	2.4	7.2	45.8	7.8	1.2	6.7	13.7
1966	130.6	97.2	31.3	8.3	6.6	6.3	2.5	7.8	53.0	9.7	1.7	8.0	16.2
1967	132.8	99.2	31.5	8.2	6.0	7.1	2.4	7.8	53.7	10.1	1.9	8.2	16.9
1968	147.9	107.7	33.6	9.4	6.0	8.3	2.6	7.3	58.5	10.6	1.9	8.7	17.3
1969	164.4	120.0	37.7	11.7	6.8	8.7	2.8	7.8	65.2	12.8	2.4	10.4	19.1
1970	168.0	124.6	40.3	12.5	7.0	10.2	2.8	7.8	66.4	14.3	2.7	11.6	20.3
1971	188.6	130.4	42.7	14.9	6.3	11.0	2.7	7.9	69.1	14.9	2.8	12.2	19.5
1972	219.0	146.6	47.2	17.6	5.9	12.1	3.1	8.6	78.9	16.7	3.5	13.2	21.4
1973	251.0	172.7	55.0	19.8	7.9	13.8	3.5	9.9	95.1	19.9	3.5	16.3	26.0
1974	260.5	191.1	61.2	20.6	10.0	15.1	5.2	10.3	104.3	23.1	3.9	19.2	30.7
1975	263.5	196.8	61.4	17.7	10.6	15.7	7.4	10.1	107.6	23.8	3.6	20.2	31.3
1976	306.1	219.3	65.9	18.1	10.1	18.2	8.6	11.0	121.2	27.5	4.4	23.1	34.1
1977	374.3	259.1	74.6	20.3	11.1	19.3	11.5	12.5	148.7	33.7	5.7	28.0	39.4
1978	452.6	314.6	93.6	25.3	16.2	21.4	15.4	15.2	180.6	42.3	7.6	34.8	47.7
1979	521.7	373.8	117.7	33.5	22.0	24.6	19.0	18.5	208.1	50.3	10.2	40.2	56.2
1980	536.4	406.9	136.2	41.0	20.5	27.3	27.4	20.0	216.4	58.9	12.5	46.4	60.7
1981	601.4	472.9	167.3	48.3	25.4	30.0	42.5	21.2	240.9	69.6	17.1	52.5	65.5
1982	595.9	485.1	177.6	55.8	26.1	29.6	44.8	21.3	234.9	74.2	18.9	55.3	62.7
1983	643.3	482.2	154.3	55.8	19.5	25.8	30.0	23.3	246.5	83.7	23.9	59.8	58.9
1984	754.7	564.3	177.4	70.6	20.9	26.5	31.3	28.1	291.9	101.2	31.6	69.6	68.1
1985	807.8	607.8	194.5	84.1	24.1	26.5	27.9	31.8	307.9	106.6	33.7	72.9	72.5
1986	842.6	607.8	176.5	80.9	21.0	28.3	15.7	30.7	317.7	111.1	33.4	77.7	75.4
1987	865.0	615.2	174.2	80.8	21.2	25.4	13.1	33.7	320.9	112.2	35.8	76.4	76.7
1988	918.5	662.3	182.8	86.3	23.2	25.0	15.7	32.5	346.8	120.8	38.0	82.8	84.2
1989	972.0	716.0	193.7	88.3	28.8	27.5	14.9	34.3	372.2	130.7	43.1	87.6	93.3
1990	978.9	739.2	202.9	87.5	33.6	26.3	17.9	37.6	371.9	129.6	38.6	90.9	92.1
1991	944.7	723.6	183.6	68.9	31.4	31.6	18.5	33.2	360.8	129.2	37.7	91.5	89.3
1992	996.7	741.9	172.6	64.5	29.0	33.9	14.2	31.0	381.7	142.1	44.0	98.1	93.0
1993	1 086.0	799.2	177.2	69.4	23.6	33.2	16.6	34.5	425.1	153.3	47.9	105.4	102.2
1994	1 192.7	868.9	186.8	75.4	28.9	31.2	16.4	34.9	476.4	167.0	52.4	114.6	113.6
1995	1 286.3	962.2	207.3	83.1	35.5	33.1	15.0	40.6	528.1	188.4	66.1	122.3	129.0
1996	1 401.3	1 043.2	224.6	91.5	38.2	29.2	16.8	48.9	565.3	204.7	72.8	131.9	136.5
1997	1 524.7	1 149.1	250.3	104.3	37.6	28.8	22.4	57.2	610.9	222.8	81.4	141.4	140.4
1998	1 673.0	1 254.1	276.0	116.0	40.5	34.2	22.3	63.1	660.0	240.1	87.9	152.2	147.4
1999	1 826.2	1 364.5	285.7	125.4	35.1	40.4	18.3	66.4	713.6	259.8	97.2	162.5	149.1
2000	1 983.9	1 498.4	321.0	139.3	37.6	48.1	23.7	72.2	766.1	293.8	103.2	190.6	162.9
2001	1 973.1	1 460.1	333.5	137.3	37.8	51.1	34.6	72.7	711.5	265.9	87.6	178.4	151.9
2002	1 910.4	1 352.8	287.0	119.4	22.7	51.0	30.2	63.6	659.6	236.7	79.7	157.0	141.7
2003	2 013.0	1 375.9	286.6	115.0	21.4	48.1	38.5	63.6	670.6	242.7	79.9	162.8	143.4
2004	2 217.2	1 467.4	307.7	125.3	23.2	43.1	47.3	68.8	721.9	255.8	84.2	171.6	144.2
2005	2 477.2	1 621.0	353.0	135.9	28.4	48.1	69.4	71.2	794.9	267.0	84.2	182.8	162.4
2006	2 632.0	1 793.8	425.2	156.3	32.3	55.8	96.0	84.7	862.3	288.5	92.6	195.9	181.6
2007	2 639.1	1 948.6	510.3	181.8	40.2	81.6	102.2	104.5	893.4	310.9	95.4	215.5	194.1
2008	2 506.9	1 990.9	571.1	181.9	53.0	95.6	120.3	120.3	845.4	306.3	93.9	212.4	194.3
2009	2 080.4	1 690.4	455.8	126.8	56.8	95.8	79.2	97.3	670.3	275.6	88.9	186.7	153.7
2010	2 111.6	1 735.0	379.8	92.1	40.3	83.8	93.5	70.1	777.0	307.5	99.6	207.9	155.2
2011	2 286.3	1 907.5	404.5	93.4	39.6	81.8	124.7	65.0	881.3	313.3	95.6	217.7	191.5
2012	2 550.5	2 118.5	479.4	103.8	46.8	102.4	152.9	73.6	983.4	331.2	103.5	227.7	211.2
2013	2 721.5	2 211.5	492.5	109.8	49.9	98.9	155.6	78.3	1 027.0	341.7	102.1	239.6	209.3
2014	2 960.2	2 400.1	577.6	127.4	58.1	115.3	188.3	88.4	1 091.9	346.0	101.9	244.1	218.8
2015	3 100.4	2 466.6	584.4	145.4	79.6	121.1	137.1	101.1	1 119.5	352.8	101.3	251.5	218.2
2016	3 160.0	2 460.5	560.3	172.1	76.3	124.3	75.8	111.9	1 088.6	353.3	99.0	254.3	214.3
2017	3 334.8	2 574.5	599.1	181.6	67.6	121.3	108.4	120.1	1 122.2	371.3	106.1	265.2	227.9
2018	3 575.1	2 776.7	631.4	191.2	69.6	118.6	124.9	127.1	1 213.4	395.9	119.3	276.6	251.5
2019	3 702.1	2 895.0	650.2	195.6	77.1	129.2	120.6	127.7	1 241.0	397.2	121.6	275.7	260.9
2017													
1st quarter	3 278.5	2 532.5	600.1	181.7	69.5	132.1	99.2	117.7	1 094.3	359.1	100.6	258.5	219.4
2nd quarter	3 309.2	2 555.9	604.5	182.6	68.8	122.3	110.4	120.5	1 107.5	368.9	105.2	263.7	225.5
3rd quarter	3 333.8	2 575.2	592.3	180.0	66.2	116.4	109.0	120.7	1 124.7	372.3	110.0	262.3	229.2
4th quarter	3 417.8	2 634.2	599.3	182.2	66.1	114.6	114.9	121.5	1 162.4	384.9	108.6	276.3	237.7
2018													
1st quarter	3 510.5	2 716.2	629.2	192.2	68.5	121.6	120.1	126.8	1 189.6	395.1	117.1	278.0	245.0
2nd quarter	3 570.2	2 765.9	640.7	193.1	68.3	123.7	127.2	128.3	1 197.0	392.6	119.9	272.7	247.0
3rd quarter	3 593.3	2 792.6	634.2	191.8	70.8	116.0	128.3	127.2	1 219.6	400.0	121.4	278.6	252.5
4th quarter	3 626.5	2 831.9	621.5	187.6	70.8	113.1	124.1	125.9	1 247.6	396.0	118.8	277.2	261.3
2019													
1st quarter	3 674.2	2 878.4	640.1	192.1	77.6	116.0	123.7	130.7	1 256.5	401.6	120.4	281.2	260.5
2nd quarter	3 686.6	2 891.3	649.7	193.2	76.9	125.6	126.8	127.1	1 243.1	399.2	124.0	275.2	261.7
3rd quarter	3 718.5	2 908.0	658.8	197.6	77.4	133.9	122.9	127.0	1 234.9	396.1	119.6	276.5	263.7
4th quarter	3 729.2	2 902.3	652.3	199.3	76.5	141.4	109.1	125.9	1 229.3	392.1	122.3	269.7	257.8

Table 5-2. Gross Private Fixed Investment by Type—*Continued*

(Billions of dollars, quarterly data are at seasonally adjusted annual rates.)　　　　　　　　　　**NIPA Table 5.3.5**

Year and quarter	Equipment—Continued Transportation equipment	Other nonresidential equipment	Intellectual property Total	Software [1]	Research and development [2]	Entertainment, literary, and artistic originals	Residential Total	Residential structures Total	Permanent site Total	Single family	Multifamily	Other residential structures	Residential equipment
1960	8.5	7.1	7.1	0.1	4.9	2.1	26.9	26.3	17.5	14.9	2.6	8.8	0.5
1961	8.0	7.0	8.0	0.2	5.2	2.7	27.0	26.5	17.4	14.1	3.3	9.1	0.5
1962	9.8	7.5	8.4	0.2	5.6	2.6	29.6	29.1	19.9	15.1	4.8	9.2	0.5
1963	9.4	8.8	9.2	0.4	6.0	2.8	32.9	32.3	22.4	16.0	6.4	9.8	0.6
1964	10.6	9.9	9.8	0.5	6.5	2.8	35.1	34.5	24.1	17.6	6.4	10.4	0.6
1965	13.2	11.0	11.1	0.7	7.2	3.2	35.2	34.5	23.8	17.8	6.0	10.6	0.7
1966	14.5	12.7	12.8	1.0	8.1	3.7	33.4	32.7	21.8	16.6	5.2	10.9	0.7
1967	14.3	12.4	14.0	1.2	9.0	3.8	33.6	32.9	21.5	16.8	4.7	11.4	0.7
1968	17.6	13.0	15.6	1.3	9.9	4.3	40.2	39.3	26.7	19.5	7.2	12.6	0.9
1969	18.9	14.4	17.2	1.8	11.0	4.4	44.4	43.4	29.2	19.7	9.5	14.1	1.0
1970	16.2	15.6	17.9	2.3	11.5	4.1	43.4	42.3	27.1	17.5	9.5	15.2	1.1
1971	18.4	16.3	18.7	2.4	11.9	4.4	58.2	56.9	38.7	25.8	12.9	18.2	1.3
1972	21.8	19.0	20.6	2.8	12.9	4.9	72.4	70.9	50.1	32.8	17.2	20.8	1.5
1973	26.6	22.6	22.7	3.2	14.6	4.9	78.3	76.6	54.6	35.2	19.4	22.0	1.7
1974	26.3	24.3	25.5	3.9	16.4	5.2	69.5	67.6	43.4	29.7	13.7	24.2	1.9
1975	25.2	27.4	27.8	4.8	17.5	5.5	66.7	64.8	36.3	29.6	6.7	28.5	1.9
1976	30.0	29.6	32.2	5.2	19.6	7.4	86.8	84.6	50.8	43.9	6.9	33.9	2.1
1977	39.3	36.3	35.8	5.5	21.8	8.6	115.2	112.8	72.2	62.2	10.0	40.6	2.4
1978	47.3	43.2	40.4	6.3	24.9	9.1	138.0	135.3	85.6	72.8	12.8	49.7	2.7
1979	53.6	47.9	48.1	8.1	29.1	10.9	147.8	144.7	89.3	72.3	17.0	55.4	3.2
1980	48.4	48.3	54.4	9.8	34.2	10.3	129.5	126.1	69.6	52.9	16.7	56.5	3.4
1981	50.6	55.2	64.8	11.8	39.7	13.2	128.5	124.9	69.4	52.0	17.5	55.4	3.6
1982	46.8	51.2	72.7	14.0	44.8	13.9	110.8	107.2	57.0	41.5	15.5	50.2	3.7
1983	53.5	50.4	81.3	16.4	49.6	15.3	161.1	156.9	95.0	72.5	22.4	61.9	4.2
1984	64.4	58.2	95.0	20.4	56.9	17.8	190.4	185.6	114.6	86.4	28.2	71.0	4.7
1985	69.0	59.9	105.3	23.8	63.0	18.6	200.1	195.0	115.9	87.4	28.5	79.1	5.1
1986	70.5	60.7	113.5	25.6	66.5	21.4	234.8	229.3	135.2	104.1	31.0	94.1	5.5
1987	68.1	64.0	120.1	29.0	69.2	21.9	249.8	244.0	142.7	117.2	25.5	101.3	5.8
1988	72.9	69.0	132.7	33.3	76.4	23.0	256.2	250.1	142.4	120.1	22.3	107.7	6.1
1989	67.9	80.2	150.1	40.6	84.1	25.4	256.0	249.9	143.2	120.9	22.3	106.6	6.1
1990	70.0	80.2	164.4	45.4	91.5	27.5	239.7	233.7	132.1	112.9	19.3	101.5	6.0
1991	71.5	70.8	179.1	48.7	101.0	29.4	221.2	215.4	114.6	99.4	15.1	100.8	5.7
1992	74.7	72.0	187.7	51.1	105.4	31.2	254.7	248.8	135.1	122.0	13.1	113.8	5.9
1993	89.4	80.2	196.9	57.2	106.3	33.4	286.8	280.7	150.9	140.1	10.8	129.8	6.1
1994	107.7	88.1	205.7	60.4	109.2	36.1	323.8	317.6	176.4	162.3	14.1	141.2	6.2
1995	116.1	94.7	226.8	65.5	121.2	40.2	324.1	317.7	171.4	153.5	17.9	146.3	6.3
1996	123.2	101.0	253.3	74.5	134.5	44.3	358.1	351.7	191.1	170.8	20.3	160.6	6.3
1997	135.5	112.1	288.0	93.8	148.1	46.1	375.6	369.3	198.1	175.2	22.9	171.3	6.3
1998	147.1	125.4	318.1	109.2	160.6	48.3	418.8	412.1	224.0	199.4	24.6	188.2	6.7
1999	174.4	130.4	365.1	136.6	177.5	51.0	461.8	454.5	251.3	223.8	27.4	203.2	7.3
2000	170.8	138.6	411.3	156.8	199.0	55.6	485.4	477.7	265.0	236.8	28.3	212.7	7.7
2001	154.2	139.5	415.0	157.7	202.7	54.7	513.1	505.2	279.4	249.1	30.3	225.8	7.8
2002	141.6	139.6	406.2	152.5	196.1	57.6	557.6	549.6	298.8	265.9	33.0	250.7	8.0
2003	134.1	150.5	418.7	155.0	201.0	62.7	637.1	628.8	345.7	310.6	35.1	283.1	8.3
2004	159.2	162.7	437.8	166.3	207.4	64.1	749.8	740.8	417.5	377.6	39.9	323.3	9.0
2005	179.6	186.0	473.1	178.6	224.7	69.8	856.2	846.6	480.8	433.5	47.3	365.8	9.6
2006	194.3	198.0	506.3	189.5	245.6	71.2	838.2	828.1	468.8	416.0	52.8	359.3	10.0
2007	188.8	199.6	544.8	206.4	268.0	70.4	690.5	680.6	354.1	305.2	49.0	326.5	9.9
2008	148.7	196.1	574.4	223.8	284.2	66.4	516.0	506.4	230.1	185.8	44.3	276.3	9.6
2009	74.9	166.1	564.4	226.0	274.6	63.7	390.0	381.2	133.9	105.3	28.5	247.3	8.8
2010	135.8	178.5	578.2	226.4	282.4	69.4	376.6	367.4	127.3	112.6	14.7	240.2	9.2
2011	177.8	198.7	621.7	249.8	303.4	68.6	378.8	369.1	123.2	108.2	15.0	245.9	9.8
2012	215.3	225.7	655.7	272.1	313.4	70.2	432.0	421.5	154.5	132.0	22.5	267.0	10.5
2013	242.5	233.6	691.9	283.7	337.9	70.3	510.0	499.0	202.3	170.8	31.5	296.7	11.0
2014	272.8	254.4	730.5	297.5	359.5	73.4	560.2	548.8	235.2	193.6	41.6	313.7	11.3
2015	306.3	242.2	762.7	307.1	378.3	77.3	633.8	622.1	273.6	221.1	52.5	348.5	11.8
2016	292.0	229.0	811.7	327.3	403.4	81.0	699.5	687.3	304.4	242.5	61.9	382.9	12.2
2017	292.0	231.0	853.2	349.2	420.0	84.0	760.3	747.9	332.6	270.2	62.5	415.2	12.5
2018	309.5	256.5	931.8	382.7	461.3	87.8	798.5	785.5	355.0	289.6	65.5	430.5	13.0
2019	310.3	272.5	1 003.8	411.2	501.9	90.7	807.1	793.9	347.6	280.0	67.6	446.3	13.2
2017													
1st quarter	290.3	225.5	838.0	340.1	414.8	83.1	746.0	733.6	321.6	259.9	61.7	412.0	12.4
2nd quarter	284.8	228.3	843.9	346.7	413.6	83.6	753.3	741.0	329.5	267.3	62.2	411.5	12.3
3rd quarter	292.4	230.8	858.2	352.4	421.6	84.2	758.5	746.2	335.3	273.2	62.1	410.9	12.4
4th quarter	300.4	239.5	872.5	357.5	430.0	85.0	783.6	770.8	344.2	280.2	64.0	426.6	12.7
2018													
1st quarter	305.4	244.0	897.4	367.8	443.5	86.1	794.3	781.5	354.5	290.6	63.9	427.0	12.8
2nd quarter	303.0	254.3	928.3	380.9	460.0	87.4	804.3	791.3	359.3	295.1	64.1	432.0	13.0
3rd quarter	307.4	259.6	938.9	386.9	463.6	88.4	800.7	787.7	356.1	291.0	65.1	431.5	13.1
4th quarter	322.2	268.1	962.8	395.3	478.2	89.3	794.7	781.7	350.3	281.6	68.7	431.5	13.0
2019													
1st quarter	324.3	270.0	981.9	401.3	490.9	89.7	795.8	782.7	345.4	275.5	69.9	437.3	13.1
2nd quarter	309.0	273.3	998.5	407.6	500.5	90.5	795.3	782.1	343.2	274.3	68.9	438.9	13.2
3rd quarter	300.0	275.1	1 014.2	416.3	506.8	91.1	810.5	797.2	346.5	279.8	66.7	450.7	13.3
4th quarter	307.8	271.7	1 020.7	419.6	509.4	91.7	827.0	813.6	355.3	290.3	65.0	458.4	13.3

[1]Excludes software "embedded," or bundled, in computers and other equipment. Includes software development expenditures.
[2]Excludes software development.

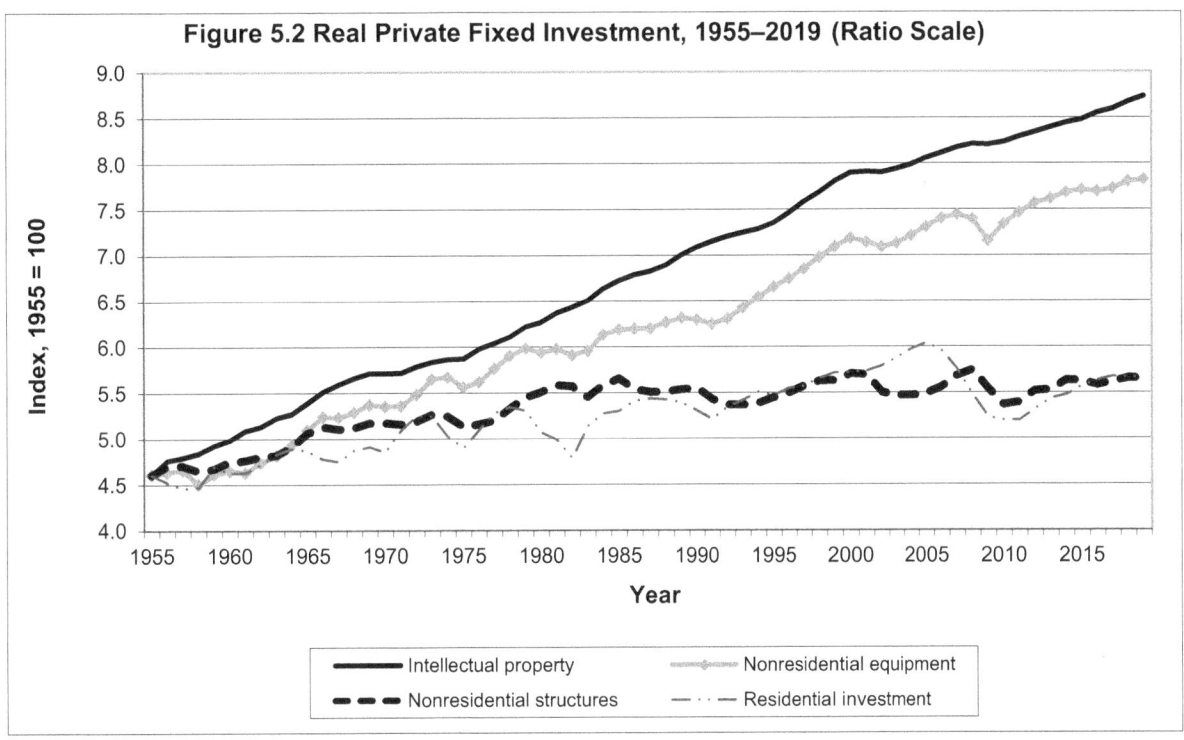

Figure 5.2 Real Private Fixed Investment, 1955–2019 (Ratio Scale)

- Between 1955 and 2006, which was the recent high point for total real (constant-dollar) investment, the quantity of real gross private fixed investment increased eightfold, with an average annual growth rate of 4.2 percent. Every major type of investment grew in real terms between those two years, but as Figure 5-2 indicates, by far the fastest-growing major category has been "intellectual property products." Computer software is the main driver of growth in this category, but it also includes research and development and "entertainment, literary, and artistic originals." (Table 5-3)

- Between 2006 and 2009, overall gross investment fell 21.8 percent in real terms—a collapse unprecedented in the postwar period. (Table 5-3) By 2019, gross investment surpassed the 2006 level by 25.5 percent, achieving a new all-time high. (Table 5-4)

- Intellectual property investment barely hesitated in 2009 and went on to new record peaks in each year through 2019. Equipment purchases continued to increase and went on to a new record in 2019. On the other hand, residential investment dropped 40.6 percent between 2007 and 2009. Residential investment increased steadily from 2012 to 2017 before declining slightly in 2018 and 2019. (Table 5-4)

Table 5-3. Real Gross Private Fixed Investment by Type

(Billions of chained [2012] dollars, quarterly data are at seasonally adjusted annual rates.)   NIPA Table 5.3.6

Year and quarter	Total gross private fixed investment	Nonresidential											
		Total	Structures						Equipment				
			Total	Commercial and health care	Manufacturing	Power and communication	Mining exploration, shafts, and wells	Other nonresidential structures	Total	Information processing equipment			Industrial equipment
										Total	Computers and peripheral equipment[1]	Other information processing	
2003	2 280.6	1 509.4	456.6	161.7	29.9	76.7	97.6	79.7	634.3	150.4	40.2	111.1	182.2
2004	2 440.7	1 594.0	456.3	165.7	30.7	64.0	103.6	82.0	688.6	169.4	45.7	124.7	178.8
2005	2 618.7	1 716.4	466.1	164.3	34.9	66.4	113.4	79.7	760.0	187.6	51.8	136.5	194.2
2006	2 686.8	1 854.2	501.7	174.2	37.2	70.4	122.5	89.1	832.6	217.0	64.7	152.4	210.6
2007	2 653.5	1 982.1	568.6	191.6	44.0	98.1	120.5	105.0	865.8	247.2	73.9	173.3	217.3
2008	2 499.4	1 994.2	605.4	184.6	55.1	107.8	129.4	118.9	824.4	260.6	79.7	180.9	208.3
2009	2 099.8	1 704.3	492.2	128.0	57.9	109.3	92.4	98.1	649.7	247.5	81.1	166.5	162.7
2010	2 164.2	1 781.0	412.8	96.2	42.2	91.7	108.3	72.6	781.2	289.1	94.1	195.1	162.5
2011	2 317.8	1 935.4	424.1	95.8	40.6	84.6	136.7	66.5	886.2	303.2	93.9	209.3	194.9
2012	2 550.5	2 118.5	479.4	103.8	46.8	102.4	152.9	73.6	983.4	331.2	103.5	227.7	211.2
2013	2 692.1	2 206.0	485.5	107.5	48.7	97.8	155.4	76.1	1 029.2	351.8	103.0	248.8	208.4
2014	2 869.2	2 365.3	538.8	121.3	55.0	112.5	167.8	82.4	1 101.1	370.2	102.9	267.7	216.5
2015	2 979.0	2 420.3	534.1	136.0	74.0	116.4	119.9	91.0	1 134.6	393.3	103.4	291.0	216.7
2016	3 032.2	2 433.0	510.5	158.7	70.4	118.9	69.4	97.5	1 115.1	410.8	102.6	310.1	213.7
2017	3 147.4	2 524.2	531.7	163.8	60.9	113.5	96.3	101.3	1 150.3	441.2	110.5	332.7	225.4
2018	3 310.4	2 698.9	551.1	166.0	59.9	107.0	120.6	102.6	1 242.2	479.3	124.0	356.6	243.9
2019	3 371.7	2 776.8	547.7	162.2	62.6	110.7	118.0	99.2	1 267.7	493.9	130.7	363.8	249.1
2010													
1st quarter	2 091.0	1 706.4	405.1	100.7	47.7	83.0	94.4	76.1	721.1	276.6	93.1	183.6	151.7
2nd quarter	2 167.1	1 762.3	416.8	96.6	45.6	91.7	107.1	73.8	768.1	285.8	96.3	189.6	161.3
3rd quarter	2 178.7	1 810.1	410.7	94.1	40.2	89.4	114.8	71.1	809.5	295.6	95.4	200.2	164.7
4th quarter	2 220.0	1 845.2	418.6	93.5	35.3	102.7	116.8	69.4	825.9	298.5	91.4	207.1	172.2
2011													
1st quarter	2 216.2	1 842.7	387.8	90.5	34.3	77.5	122.8	62.7	846.3	292.9	87.8	205.1	182.8
2nd quarter	2 268.0	1 890.4	414.8	95.0	39.0	83.7	130.5	66.4	858.0	302.1	93.6	208.5	185.4
3rd quarter	2 363.3	1 978.9	439.4	98.2	43.6	88.7	141.1	67.7	909.7	308.2	97.5	210.7	199.1
4th quarter	2 423.7	2 029.4	454.5	99.2	45.5	88.3	152.4	69.3	930.7	309.6	96.8	212.8	212.1
2012													
1st quarter	2 499.4	2 081.3	476.0	101.7	44.5	102.0	156.4	71.4	959.6	327.3	105.3	222.0	210.3
2nd quarter	2 549.8	2 128.0	487.4	104.0	46.6	104.8	158.1	73.9	986.7	333.5	104.6	228.9	215.0
3rd quarter	2 553.6	2 120.9	481.8	105.2	47.4	103.2	151.1	74.8	983.8	327.3	100.6	226.7	207.5
4th quarter	2 599.4	2 144.0	472.5	104.1	48.6	99.4	145.9	74.4	1 003.4	336.8	103.6	233.2	212.1
2013													
1st quarter	2 643.9	2 171.6	462.1	106.3	47.0	81.9	153.3	73.4	1 021.1	349.9	105.9	244.0	210.1
2nd quarter	2 665.3	2 177.5	475.8	103.4	46.2	91.3	159.3	75.4	1 019.0	348.6	100.2	248.5	207.7
3rd quarter	2 711.3	2 214.7	499.8	106.9	51.0	104.2	159.9	77.7	1 019.0	352.1	101.2	251.0	209.2
4th quarter	2 748.0	2 260.0	504.2	113.2	50.7	113.9	148.9	77.7	1 057.9	356.5	104.8	251.8	206.4
2014													
1st quarter	2 775.6	2 291.7	521.9	113.1	49.6	126.1	155.5	78.5	1 066.9	358.9	101.3	257.9	211.4
2nd quarter	2 852.8	2 353.3	540.3	119.2	50.4	119.0	170.9	81.2	1 093.6	372.1	103.5	269.0	217.1
3rd quarter	2 907.3	2 400.8	542.2	124.2	54.6	109.4	171.7	82.4	1 127.0	370.6	102.3	268.7	222.0
4th quarter	2 941.2	2 415.5	551.0	128.8	65.6	95.5	173.1	87.4	1 116.9	379.3	104.7	275.1	215.3
2015													
1st quarter	2 949.5	2 412.6	545.1	129.4	72.4	106.6	152.1	85.8	1 126.0	381.8	103.0	279.5	215.8
2nd quarter	2 974.9	2 423.6	549.1	138.1	76.8	122.9	122.7	92.1	1 130.3	388.4	103.5	285.7	217.6
3rd quarter	2 999.8	2 432.4	531.8	137.1	74.7	121.3	108.1	94.5	1 148.3	399.2	105.6	294.5	216.0
4th quarter	2 991.8	2 412.8	510.6	139.6	72.1	114.6	96.6	91.5	1 133.6	403.8	101.3	304.3	217.4
2016													
1st quarter	3 006.8	2 406.0	491.8	149.0	70.9	109.8	74.0	92.1	1 126.6	406.5	103.9	304.1	212.9
2nd quarter	3 018.0	2 420.2	500.9	154.6	72.6	114.2	64.6	99.1	1 114.5	406.0	102.5	305.2	214.7
3rd quarter	3 041.8	2 448.4	520.6	164.3	71.2	121.8	68.5	99.3	1 110.5	412.7	101.0	314.0	212.4
4th quarter	3 062.2	2 457.4	528.5	167.0	66.9	129.7	70.3	99.6	1 108.7	418.1	103.1	317.2	214.9
2017													
1st quarter	3 115.5	2 492.6	538.5	166.0	63.7	124.4	88.7	100.5	1 121.5	425.1	105.0	322.4	218.0
2nd quarter	3 127.7	2 507.3	537.6	165.6	62.6	114.5	96.8	102.2	1 135.5	438.1	109.7	330.4	223.3
3rd quarter	3 137.1	2 520.3	522.3	161.1	58.9	108.5	96.2	101.4	1 152.6	442.8	114.6	329.4	226.4
4th quarter	3 209.2	2 576.4	528.4	162.5	58.7	106.4	103.4	101.1	1 191.4	458.9	112.6	348.8	233.9
2018													
1st quarter	3 275.2	2 651.5	554.8	169.7	60.2	112.2	112.9	104.1	1 220.3	474.2	121.5	354.3	239.6
2nd quarter	3 310.6	2 691.9	561.6	168.4	59.2	112.5	122.4	104.0	1 227.7	474.1	124.6	350.3	240.0
3rd quarter	3 317.0	2 709.5	553.2	166.1	60.8	103.7	125.7	102.2	1 245.9	484.8	126.1	359.9	244.2
4th quarter	3 338.7	2 742.6	534.9	159.7	59.3	99.6	121.2	100.1	1 274.8	483.9	123.8	361.8	251.7
2019													
1st quarter	3 362.3	2 770.8	545.5	162.1	64.3	100.7	120.2	102.9	1 281.1	493.5	126.4	368.6	249.7
2nd quarter	3 358.6	2 771.0	547.8	160.7	62.4	107.9	122.9	99.4	1 268.6	494.8	132.6	362.4	250.2
3rd quarter	3 378.9	2 783.9	552.6	162.6	62.3	114.2	120.9	98.0	1 263.3	494.3	129.5	365.7	251.4
4th quarter	3 387.2	2 781.5	545.1	163.4	61.3	120.1	107.9	96.4	1 258.0	492.9	134.4	358.3	245.3

[1]See notes and definitions.

Table 5-3. Real Gross Private Fixed Investment by Type—*Continued*

(Billions of chained [2012] dollars, quarterly data are at seasonally adjusted annual rates.)   **NIPA Table 5.3.6**

Year and quarter	Nonresidential—*Continued*						Residential						
	Equipment—*Continued*		Intellectual property						Residential structures				
										Permanent site			
	Transportation equipment	Other nonresidential equipment	Total	Software ²	Research and development ³	Entertainment, literary, and artistic originals	Total	Total	Total	Single family	Multifamily	Other residential structures	Residential equipment
2003	150.3	180.2	437.7	133.5	246.1	64.6	755.5	747.7	416.1	362.0	49.6	329.2	8.1
2004	171.2	193.3	459.2	149.3	248.1	65.6	830.9	822.1	464.3	405.4	53.4	355.4	9.1
2005	192.1	211.9	493.1	163.4	261.6	71.0	885.4	876.3	497.0	432.8	58.7	376.7	9.5
2006	206.4	220.7	521.5	173.5	279.6	71.8	818.9	809.5	453.6	390.4	59.1	353.5	9.7
2007	197.7	218.0	554.3	191.1	296.1	70.0	665.8	656.6	337.7	283.5	52.3	317.7	9.4
2008	155.0	207.4	575.3	206.7	304.8	66.0	504.6	495.7	224.2	178.1	46.0	270.9	9.1
2009	72.5	165.3	572.4	212.9	297.4	63.4	395.3	386.9	134.5	105.3	29.3	252.3	8.4
2010	141.5	186.2	588.1	220.9	298.5	69.5	383.0	373.8	129.1	114.3	14.7	244.7	9.3
2011	181.8	206.2	624.8	245.2	311.0	68.9	382.5	372.4	124.5	109.1	15.4	247.9	10.0
2012	215.3	225.7	655.7	272.1	313.4	70.2	432.0	421.5	154.5	132.0	22.5	267.0	10.5
2013	238.5	230.6	691.4	287.2	333.8	70.3	485.5	474.1	192.4	161.8	30.6	281.7	11.4
2014	265.0	248.9	724.8	305.3	346.9	72.7	504.1	491.8	210.4	171.8	38.8	281.4	12.4
2015	292.8	232.4	752.4	320.2	357.1	75.5	555.4	542.0	235.2	191.5	44.0	306.8	13.4
2016	275.7	219.4	809.8	345.9	386.7	77.6	592.1	577.6	250.5	201.3	49.3	327.1	14.6
2017	271.4	220.6	844.2	374.6	392.3	79.3	615.7	600.3	261.4	214.8	47.0	338.9	15.6
2018	287.0	241.7	910.2	416.4	416.0	81.8	612.0	596.6	266.2	220.7	46.0	330.1	15.8
2019	285.7	250.1	968.2	449.3	440.5	83.7	601.5	586.0	253.3	206.8	46.8	332.6	16.0
2010													
1st quarter	112.8	177.2	581.9	220.5	295.8	66.4	384.4	375.5	131.9	115.6	16.2	243.6	8.9
2nd quarter	131.2	187.9	579.0	217.0	293.4	69.3	404.4	395.1	136.1	121.6	14.5	258.9	9.4
3rd quarter	159.9	187.8	590.3	220.9	299.6	70.6	368.6	359.3	126.6	112.6	14.0	232.7	9.3
4th quarter	162.2	191.7	601.3	225.3	305.1	71.7	374.7	365.1	121.7	107.6	14.1	243.4	9.5
2011													
1st quarter	172.0	198.5	607.4	232.1	306.5	69.3	373.4	363.6	122.1	108.2	14.0	241.5	9.8
2nd quarter	170.2	199.8	617.4	240.2	309.2	68.3	377.6	367.7	121.7	107.2	14.5	246.0	10.0
3rd quarter	188.0	214.3	629.9	249.8	311.7	68.5	384.4	374.4	125.5	109.4	16.0	248.9	10.1
4th quarter	197.1	212.0	644.3	258.5	316.5	69.5	394.4	384.1	128.7	111.7	17.0	255.4	10.3
2012													
1st quarter	205.6	216.5	645.7	263.1	312.3	70.3	418.3	407.8	139.3	120.5	18.8	268.5	10.4
2nd quarter	215.6	222.7	653.9	272.1	311.5	70.2	421.9	411.5	147.6	126.2	21.4	263.9	10.4
3rd quarter	216.1	232.9	655.2	272.8	312.4	70.1	432.7	422.1	158.1	134.5	23.6	264.1	10.6
4th quarter	223.9	230.6	668.0	280.2	317.6	70.2	455.2	444.5	173.1	146.9	26.2	271.4	10.7
2013													
1st quarter	232.2	229.0	688.6	288.3	330.1	70.2	471.8	460.7	184.3	156.0	28.2	276.4	11.1
2nd quarter	236.6	226.2	682.9	280.2	331.9	70.8	486.8	475.5	192.6	163.0	29.5	282.9	11.3
3rd quarter	237.2	220.7	695.8	288.3	337.1	70.3	495.5	483.9	196.5	165.7	30.8	287.4	11.6
4th quarter	248.2	246.4	698.1	292.1	336.1	70.0	487.7	476.2	196.3	162.6	33.8	279.9	11.5
2014													
1st quarter	252.4	243.8	702.5	296.2	335.3	71.2	484.3	472.6	199.9	165.0	35.0	272.8	11.7
2nd quarter	260.5	243.8	718.5	303.3	343.3	72.1	499.8	487.5	207.6	169.5	38.3	279.9	12.4
3rd quarter	275.6	257.5	731.4	309.8	348.7	73.2	507.1	494.7	209.9	169.9	40.3	284.8	12.6
4th quarter	271.6	250.5	746.6	311.9	360.2	74.4	525.2	512.4	224.1	182.8	41.6	288.3	12.8
2015													
1st quarter	286.6	241.4	741.1	315.2	351.5	74.7	535.5	522.5	227.4	185.4	42.3	295.0	13.0
2nd quarter	292.4	232.1	743.6	318.2	350.9	75.0	548.9	535.6	229.6	187.3	42.5	306.0	13.4
3rd quarter	302.8	230.9	753.6	320.3	357.8	75.8	563.7	550.1	239.0	194.4	44.8	311.1	13.6
4th quarter	289.4	225.2	771.3	327.1	368.1	76.3	573.6	559.9	244.8	198.7	46.3	315.0	13.7
2016													
1st quarter	285.6	224.4	791.7	336.2	379.1	76.7	593.0	578.7	254.0	205.0	49.2	324.5	14.3
2nd quarter	276.3	221.2	808.0	342.3	389.4	76.5	590.5	576.2	251.5	202.1	49.6	324.6	14.4
3rd quarter	272.8	217.8	818.7	349.5	391.5	78.1	587.4	572.8	245.3	195.7	49.7	327.6	14.8
4th quarter	268.0	214.1	820.8	355.7	387.0	79.1	597.6	582.8	251.2	202.6	48.8	331.7	14.9
2017													
1st quarter	269.3	216.1	832.6	363.0	391.7	79.2	614.4	599.2	256.3	209.0	47.7	343.0	15.3
2nd quarter	264.6	218.3	834.8	369.8	388.0	79.0	612.7	597.5	259.7	212.9	47.2	337.7	15.4
3rd quarter	271.8	220.1	848.4	379.3	392.4	79.1	610.1	594.8	262.0	216.2	46.3	332.5	15.6
4th quarter	279.8	227.9	860.9	386.5	397.1	79.9	625.5	609.7	267.4	221.1	46.9	342.2	16.1
2018													
1st quarter	284.0	232.7	879.3	399.0	403.1	80.5	620.3	604.6	270.6	225.4	45.9	333.7	16.1
2nd quarter	281.8	241.5	905.2	413.2	414.4	81.4	617.6	602.0	269.8	225.3	45.2	331.9	15.9
3rd quarter	284.1	243.6	914.9	420.8	416.5	82.0	609.1	593.8	265.4	220.6	45.4	328.1	15.7
4th quarter	298.0	249.2	941.5	432.6	430.0	83.4	601.0	585.9	258.9	211.6	47.6	326.8	15.4
2019													
1st quarter	299.1	248.6	951.9	438.8	434.4	83.4	598.4	583.2	252.9	204.7	48.4	330.3	15.5
2nd quarter	283.7	251.2	961.5	443.6	439.4	83.3	595.2	580.0	251.4	203.9	47.7	328.4	15.7
3rd quarter	277.1	252.2	974.0	452.9	442.7	83.9	601.9	586.4	252.2	206.3	46.3	334.1	16.1
4th quarter	283.0	248.5	985.2	461.8	445.3	84.3	610.5	594.6	256.8	212.3	45.0	337.7	16.5

²Excludes software "embedded," or bundled, in computers and other equipment.
³Excludes software development.

Table 5-4. Chain-Type Quantity Indexes for Private Fixed Investment by Type

(Index numbers, 2012 = 100, quarterly data are seasonally adjusted.)   NIPA Table 5.3.3

Year and quarter	Total gross private fixed investment	Nonresidential											
		Total	Structures						Equipment				
			Total	Commercial and health care	Manufacturing	Power and communication	Mining exploration, shafts, and wells	Other nonresidential structures	Total	Information processing equipment			Industrial equipment
										Total	Computers and peripheral equipment1	Other information processing	
1960	13.9	8.8	46.3	47.3	60.9	36.1	42.0	62.1	5.4	0.2	0.0	2.3	28.7
1961	13.9	8.8	46.9	54.0	59.6	33.8	42.8	60.6	5.3	0.2	0.0	2.5	27.2
1962	15.1	9.5	49.1	60.2	60.2	34.0	44.9	61.9	5.9	0.3	0.0	2.7	28.6
1963	16.3	10.1	49.7	58.3	60.4	36.3	42.1	65.5	6.4	0.3	0.0	2.8	30.9
1964	17.8	11.2	54.8	63.4	72.6	39.5	45.1	71.8	7.2	0.3	0.0	3.0	34.9
1965	19.7	13.0	63.6	74.3	100.4	43.7	44.6	80.7	8.5	0.4	0.0	3.4	41.1
1966	20.9	14.6	67.9	72.5	124.8	49.8	42.2	83.8	9.8	0.5	0.0	4.0	47.3
1967	20.7	14.6	66.2	69.7	109.8	54.5	40.4	82.0	9.7	0.6	0.0	4.0	47.5
1968	22.1	15.3	67.1	76.3	105.7	60.6	40.6	72.4	10.3	0.6	0.0	4.1	46.5
1969	23.4	16.4	70.7	88.0	111.2	60.6	42.3	71.8	11.1	0.7	0.0	4.8	49.6
1970	23.0	16.2	70.9	89.2	107.3	66.3	39.8	67.5	10.9	0.8	0.0	5.2	50.3
1971	24.5	16.2	69.8	97.5	88.6	66.9	36.9	63.3	11.0	0.8	0.0	5.2	46.2
1972	27.3	17.6	72.0	106.5	77.4	69.7	39.5	64.0	12.4	1.0	0.0	5.5	49.9
1973	29.7	19.9	77.8	111.2	95.9	74.2	42.2	69.0	14.7	1.1	0.0	6.7	58.9
1974	28.0	20.1	76.2	104.3	109.3	69.5	50.2	63.1	15.0	1.3	0.0	7.4	63.5
1975	25.3	18.3	68.2	81.3	104.8	64.0	58.6	55.6	13.4	1.3	0.0	7.1	53.8
1976	27.8	19.3	69.8	80.2	96.8	69.8	63.2	58.4	14.3	1.4	0.0	7.8	54.1
1977	31.5	21.4	72.7	84.4	99.4	68.7	71.4	61.6	16.5	1.8	0.0	9.4	57.3
1978	35.2	24.4	83.2	96.2	132.6	72.0	81.4	70.0	19.0	2.3	0.1	11.2	63.8
1979	37.2	26.8	93.7	114.7	162.9	74.7	86.7	76.9	20.5	2.8	0.1	12.6	68.3
1980	35.0	26.8	99.2	126.1	136.2	75.4	121.9	74.7	19.6	3.3	0.1	13.5	65.4
1981	35.9	28.5	107.2	136.5	154.7	77.4	141.7	72.3	20.3	3.8	0.2	14.3	64.4
1982	33.7	27.6	105.4	148.7	149.3	73.1	131.0	68.2	19.0	4.0	0.3	14.4	58.8
1983	36.3	27.5	94.0	144.3	108.0	62.8	108.4	72.0	19.9	4.6	0.4	15.1	54.1
1984	42.1	32.1	107.1	175.6	111.8	64.2	123.3	84.4	23.8	5.9	0.7	17.2	61.9
1985	44.4	34.2	114.7	202.8	125.0	63.5	109.8	92.8	25.1	6.5	0.9	17.8	64.8
1986	45.2	33.6	102.2	188.4	105.0	67.8	64.4	87.2	25.4	7.0	1.0	18.5	64.5
1987	45.5	33.6	99.2	181.5	102.4	60.3	62.9	92.4	25.5	7.3	1.3	17.7	62.9
1988	47.0	35.3	99.9	187.0	108.0	56.8	69.0	86.2	27.2	7.9	1.4	19.0	66.1
1989	48.5	37.3	101.9	184.6	129.4	59.0	62.4	87.9	28.6	8.7	1.7	19.8	70.7
1990	47.8	37.7	103.4	177.3	146.6	54.9	72.1	93.3	28.0	8.8	1.7	20.3	66.6
1991	45.3	36.2	91.9	137.7	135.0	65.2	70.5	81.0	26.7	9.0	1.9	20.2	62.3
1992	47.8	37.3	86.3	128.0	124.0	69.2	58.4	75.0	28.3	10.3	2.6	21.4	63.4
1993	51.5	40.1	86.1	133.2	97.4	66.0	67.8	81.0	31.9	11.7	3.3	23.1	68.8
1994	55.7	43.2	87.7	139.4	115.3	59.8	65.9	79.2	35.8	13.3	4.1	25.2	75.0
1995	59.1	47.4	93.2	147.6	135.9	61.0	56.0	88.7	40.1	16.1	6.1	27.1	82.3
1996	64.3	51.7	98.5	159.0	142.9	52.5	59.6	104.3	44.0	19.4	8.9	29.7	85.3
1997	69.8	57.3	105.8	175.7	136.5	50.6	70.2	118.3	48.9	23.4	12.9	32.3	87.2
1998	77.0	63.5	111.5	186.7	141.1	59.4	61.4	126.5	55.3	28.9	18.7	36.3	91.0
1999	84.0	69.8	112.0	193.5	117.7	70.1	51.9	128.4	62.1	35.3	26.5	40.5	91.5
2000	89.9	76.3	121.1	206.4	121.4	80.8	64.8	134.1	68.2	42.6	32.2	48.8	99.4
2001	88.7	74.6	119.6	196.4	118.0	83.4	76.4	129.5	65.3	42.1	33.0	47.3	92.1
2002	85.6	69.5	98.8	166.4	68.9	81.3	55.9	110.9	61.8	40.3	34.6	43.2	85.9
2003	89.4	71.2	95.2	155.8	63.8	74.9	63.9	108.2	64.5	45.4	38.9	48.8	86.3
2004	95.7	75.2	95.2	159.7	65.7	62.6	67.8	111.4	70.0	51.2	44.1	54.8	84.7
2005	102.7	81.0	97.2	158.3	74.6	64.8	74.2	108.3	77.3	56.6	50.0	60.0	91.9
2006	105.3	87.5	104.6	167.9	79.6	68.8	80.1	121.0	84.7	65.5	62.5	66.9	99.7
2007	104.0	93.6	118.6	184.6	94.1	95.8	78.8	142.6	88.0	74.6	71.4	76.1	102.9
2008	98.0	94.1	126.3	177.9	117.9	105.4	84.6	161.4	83.8	78.7	77.0	79.4	98.6
2009	82.3	80.5	102.7	123.3	123.9	106.8	60.4	133.2	66.1	74.7	78.3	73.1	77.0
2010	84.9	84.1	86.1	92.7	90.2	89.6	70.8	98.6	79.4	87.3	90.9	85.7	76.9
2011	90.9	91.4	88.5	92.3	86.8	82.6	89.4	90.4	90.1	91.5	90.7	91.9	92.3
2012	100.0	100.0	100.0	100.0	100.0	100.0	100.0	100.0	100.0	100.0	100.0	100.0	100.0
2013	105.6	104.1	101.3	103.6	104.2	95.6	101.6	103.3	104.7	106.2	99.5	109.3	98.6
2014	112.5	111.6	112.4	116.9	117.7	109.9	109.7	111.9	112.0	111.8	99.4	117.6	102.5
2015	116.8	114.2	111.4	131.1	158.2	113.7	78.4	123.6	115.4	118.7	99.8	127.8	102.6
2016	118.9	114.8	106.5	153.0	150.5	116.1	45.4	132.5	113.4	124.0	99.1	136.2	101.2
2017	123.4	119.1	110.9	157.8	130.3	110.8	62.9	137.6	117.0	133.2	106.7	146.1	106.7
2018	129.8	127.4	115.0	159.9	128.0	104.5	78.8	139.4	126.3	144.7	119.7	156.6	115.5
2019	132.2	131.1	114.2	156.3	133.8	108.2	77.2	134.7	128.9	149.1	126.3	159.7	117.9
2017													
1st quarter	122.2	117.7	112.3	160.0	136.2	121.6	58.0	136.5	114.0	128.3	101.4	141.6	103.2
2nd quarter	122.6	118.4	112.1	159.6	133.8	111.9	63.3	138.8	115.5	132.3	105.9	145.1	105.7
3rd quarter	123.0	119.0	108.9	155.3	125.8	106.0	62.9	137.7	117.2	133.7	110.7	144.7	107.2
4th quarter	125.8	121.6	110.2	156.6	125.4	104.0	67.6	137.3	121.1	138.5	108.8	153.2	110.7
2018													
1st quarter	128.4	125.2	115.7	163.5	128.6	109.6	73.8	141.4	124.1	143.2	117.3	155.6	113.4
2nd quarter	129.8	127.1	117.1	162.3	126.6	109.9	80.1	141.3	124.8	143.1	120.3	153.8	113.6
3rd quarter	130.1	127.9	115.4	160.0	129.9	101.3	82.2	138.8	126.7	146.4	121.8	158.1	115.6
4th quarter	130.9	129.5	111.6	153.9	126.8	97.3	79.3	135.9	129.6	146.1	119.6	158.9	119.2
2019													
1st quarter	131.8	130.8	113.8	156.2	137.4	98.4	78.6	139.7	130.3	149.0	122.1	161.9	118.2
2nd quarter	131.7	130.8	114.2	154.8	133.4	105.4	80.4	134.9	129.0	149.4	128.1	159.2	118.4
3rd quarter	132.5	131.4	115.3	156.7	133.3	111.6	79.0	133.2	128.5	149.2	125.1	160.6	119.0
4th quarter	132.8	131.3	113.7	157.4	131.1	117.3	70.5	130.9	127.9	148.8	129.8	157.3	116.1

1See notes and definitions.

Table 5-4. Chain-Type Quantity Indexes for Private Fixed Investment by Type—*Continued*

(Index numbers, 2012 = 100, quarterly data are seasonally adjusted.)   NIPA Table 5.3.3

Year and quarter	Nonresidential—*Continued*						Residential						
	Equipment—*Continued*		Intellectual property				Total	Residential structures				Other residential structures	Residential equipment
	Transportation equipment	Other nonresidential equipment	Total	Software [2]	Research and development [3]	Entertainment, literary, and artistic originals		Total	Permanent site				
									Total	Single family	Multifamily		
1960	16.5	18.9	3.5	0.0	8.0	17.5	49.9	51.7	97.3	92.6	114.5	25.8	8.6
1961	15.6	18.7	3.9	0.0	8.5	21.1	50.1	51.9	96.8	87.7	144.3	26.4	8.6
1962	19.2	20.0	4.0	0.0	9.2	20.2	54.9	56.9	110.4	93.5	209.8	26.7	9.1
1963	18.6	23.2	4.4	0.1	9.9	21.7	61.3	63.6	125.2	99.9	281.6	28.8	10.2
1964	21.0	25.8	4.6	0.1	10.6	20.6	65.0	67.4	133.7	109.6	279.7	30.0	10.9
1965	26.3	28.6	5.2	0.1	11.9	22.9	63.4	65.5	127.8	107.1	251.4	30.3	12.2
1966	29.0	32.2	5.9	0.2	13.3	25.0	58.0	59.9	112.1	95.4	210.8	30.2	12.5
1967	28.0	30.4	6.4	0.3	14.7	24.8	56.5	58.3	106.9	93.5	183.1	30.6	12.8
1968	33.7	30.8	6.9	0.3	15.9	25.9	64.2	66.1	125.6	102.7	264.5	32.3	15.3
1969	35.0	32.8	7.2	0.4	16.9	24.5	66.2	68.0	129.2	97.6	329.1	33.3	17.4
1970	28.7	34.3	7.2	0.5	17.0	21.4	62.7	64.2	116.9	84.8	322.1	34.1	18.9
1971	31.0	34.2	7.2	0.6	16.9	21.1	79.4	81.6	157.5	117.7	409.9	38.6	21.3
1972	36.1	38.7	7.8	0.7	17.7	22.8	93.2	95.7	190.6	139.9	513.2	42.3	25.5
1973	43.4	45.2	8.1	0.7	18.9	21.8	92.6	94.8	189.7	136.8	527.6	41.4	28.6
1974	39.2	44.7	8.4	0.9	19.0	22.1	74.5	75.8	137.3	105.0	339.7	40.8	29.2
1975	33.8	41.3	8.5	1.0	18.7	21.4	65.5	66.5	105.1	95.9	151.5	44.1	27.2
1976	37.6	41.7	9.4	1.1	19.9	27.2	80.0	81.6	138.1	133.5	146.8	49.1	28.6
1977	45.7	46.9	10.0	1.1	20.9	30.3	96.4	98.5	178.2	171.5	194.6	53.2	31.3
1978	50.5	51.7	10.7	1.4	22.6	29.8	102.8	105.0	186.4	177.1	219.2	58.7	34.0
1979	52.7	52.7	11.9	1.7	24.2	34.2	98.9	100.8	174.4	157.0	267.8	58.7	37.3
1980	43.0	47.3	12.5	2.0	25.9	31.0	78.2	79.3	122.5	103.2	239.7	54.1	37.3
1981	41.8	48.6	13.9	2.3	27.5	37.4	71.8	72.6	113.0	93.9	230.6	49.1	37.3
1982	37.1	42.1	14.8	2.7	29.0	37.7	58.8	59.1	88.1	72.0	187.9	42.2	35.8
1983	41.8	40.5	15.9	3.2	30.6	39.7	83.5	84.6	143.9	125.0	257.3	50.6	40.0
1984	49.7	45.9	18.1	4.0	33.7	44.3	95.8	97.1	169.1	145.1	314.5	56.1	44.3
1985	51.7	46.3	19.7	4.7	36.6	44.1	98.0	99.2	166.9	143.8	306.4	60.4	47.8
1986	49.7	45.5	21.1	5.3	38.1	48.7	110.1	111.6	186.3	164.6	315.4	68.7	51.7
1987	47.3	46.8	21.9	6.1	38.8	47.9	112.3	113.8	188.8	177.9	248.9	70.6	53.8
1988	50.0	49.0	23.5	7.0	40.9	48.4	111.3	112.7	182.1	175.7	212.4	72.5	55.7
1989	45.1	54.7	26.2	9.0	43.7	50.6	107.7	108.9	177.6	170.7	211.8	69.3	56.0
1990	44.5	52.7	28.4	10.5	46.5	52.4	98.6	99.5	160.4	155.1	184.8	64.3	54.3
1991	43.3	45.0	30.3	11.3	49.9	52.9	89.8	90.5	138.1	135.9	142.4	62.5	52.2
1992	44.3	44.8	32.0	13.0	50.8	55.5	102.2	103.3	161.0	165.1	121.0	69.5	53.8
1993	52.8	49.0	33.4	14.7	50.6	58.5	110.5	111.8	171.9	180.8	97.6	76.6	54.8
1994	61.8	53.0	34.7	16.1	50.9	61.6	120.4	122.1	192.4	200.1	126.1	81.1	55.2
1995	66.5	55.8	37.1	17.6	53.7	66.7	116.3	117.8	179.7	181.6	156.3	81.5	55.9
1996	68.7	57.9	41.3	20.6	58.8	71.5	125.9	127.7	196.6	198.4	172.8	87.2	55.6
1997	73.7	63.5	46.7	26.9	62.8	73.6	128.9	130.8	197.7	197.8	186.1	91.4	55.2
1998	79.7	70.0	51.8	32.6	67.1	77.1	140.0	142.2	216.5	219.4	184.1	98.5	59.1
1999	94.0	71.9	58.6	40.6	72.4	79.4	148.8	151.0	232.5	235.6	198.7	103.2	65.3
2000	92.1	75.9	64.1	45.8	78.3	84.1	149.8	151.9	234.4	238.2	195.3	103.4	69.4
2001	83.4	75.3	64.8	46.3	79.9	81.5	151.2	153.2	235.8	238.5	203.8	104.7	70.7
2002	75.4	74.7	64.3	46.1	77.9	85.6	160.3	162.6	245.3	247.8	214.5	113.8	72.8
2003	69.8	79.8	66.8	49.1	78.5	92.0	174.9	177.4	269.3	274.2	220.2	123.3	77.3
2004	79.5	85.6	70.0	54.9	79.2	93.4	192.3	195.1	300.5	307.1	237.3	133.1	86.2
2005	89.2	93.9	75.2	60.1	83.5	101.2	204.9	207.9	321.7	327.8	260.9	141.1	90.0
2006	95.9	97.8	79.5	63.8	89.2	102.3	189.5	192.1	293.6	295.8	262.6	132.4	91.7
2007	91.8	96.6	84.5	70.2	94.5	99.8	154.1	155.8	218.5	214.7	232.6	119.0	89.1
2008	72.0	91.9	87.7	76.0	97.2	94.1	116.8	117.6	145.1	134.9	204.5	101.5	86.3
2009	33.7	73.3	87.3	78.2	94.9	90.4	91.5	91.8	87.0	79.7	130.1	94.5	80.1
2010	65.7	82.5	89.7	81.2	95.2	99.0	88.7	88.7	83.5	86.6	65.3	91.7	87.9
2011	84.5	91.4	95.3	90.1	99.2	98.2	88.5	88.4	80.6	82.6	68.2	92.9	95.4
2012	100.0	100.0	100.0	100.0	100.0	100.0	100.0	100.0	100.0	100.0	100.0	100.0	100.0
2013	110.8	102.2	105.4	105.6	106.5	100.2	112.4	112.5	124.5	122.6	136.0	105.5	107.9
2014	123.1	110.3	110.5	112.2	110.7	103.6	116.7	116.7	136.1	130.1	172.3	105.4	117.3
2015	136.0	103.0	114.7	117.7	113.9	107.5	128.6	128.6	152.2	145.0	195.4	114.9	127.5
2016	128.1	97.2	123.5	127.2	123.4	110.6	137.1	137.0	162.1	152.5	219.2	122.5	138.5
2017	126.0	97.8	128.7	137.7	125.2	113.0	142.5	142.4	169.2	162.7	208.8	126.9	148.2
2018	133.3	107.1	138.8	153.1	132.7	116.6	141.7	141.5	172.2	167.2	204.4	123.7	150.0
2019	132.7	110.8	147.7	165.1	140.5	119.3	139.2	139.0	163.9	156.6	208.1	124.6	151.7
2017													
1st quarter	125.1	95.8	127.0	133.4	125.0	112.8	142.2	142.2	165.9	158.3	211.8	128.5	145.3
2nd quarter	122.9	96.7	127.3	135.9	123.8	112.6	141.8	141.8	168.1	161.3	209.8	126.5	146.4
3rd quarter	126.3	97.5	129.4	139.4	125.2	112.7	141.2	141.1	169.6	163.8	205.5	124.6	148.4
4th quarter	129.9	101.0	131.3	142.1	126.7	113.9	144.8	144.7	173.1	167.5	208.2	128.2	152.6
2018													
1st quarter	131.9	103.1	134.1	146.7	128.6	114.8	143.6	143.4	175.1	170.7	203.7	125.0	153.2
2nd quarter	130.9	107.0	138.1	151.9	132.2	116.0	143.0	142.8	174.6	170.6	200.9	124.3	151.2
3rd quarter	132.0	107.9	139.5	154.7	132.9	116.9	141.0	140.9	171.7	167.1	201.6	122.9	149.0
4th quarter	138.4	110.4	143.6	159.0	137.2	118.8	139.1	139.0	167.5	160.3	211.4	122.4	146.5
2019													
1st quarter	138.9	110.2	145.2	161.3	138.6	118.8	138.5	138.4	163.6	155.0	214.9	123.7	147.5
2nd quarter	131.8	111.3	146.6	163.0	140.2	118.7	137.8	137.6	162.7	154.5	211.9	123.0	149.4
3rd quarter	128.7	111.8	148.6	166.5	141.2	119.6	139.3	139.1	163.2	156.2	205.6	125.2	152.9
4th quarter	131.4	110.1	150.2	169.7	142.1	120.1	141.3	141.1	166.2	160.8	199.9	126.5	157.0

[2]Excludes software "embedded," or bundled, in computers and other equipment.
[3]Excludes software development.

Table 5-5A. Current-Cost Net Stock of Fixed Assets: Recent Data

(Billions of dollars, year-end estimates.)

Year	Total	Private						Government							
		Total	Nonresidential				Residential	Total	Nonresidential				Residential	Federal	State and local
			Total	Equipment	Structures	Intellectual property products			Total	Equipment	Structures	Intellectual property products			
1955	1 197.2	888.8	457.0	141.2	296.3	19.6	431.7	308.4	301.1	68.3	219.5	13.3	7.3	158.3	150.0
1956	1 301.6	958.5	505.5	158.6	324.9	22.1	452.9	343.1	334.5	72.4	246.2	15.9	8.6	173.1	170.0
1957	1 369.5	1 008.9	540.6	172.9	343.3	24.4	468.3	360.6	351.6	74.7	257.9	19.0	9.0	182.4	178.2
1958	1 415.1	1 034.0	551.3	178.0	346.9	26.4	482.6	381.1	371.3	76.2	273.0	22.1	9.8	191.4	189.7
1959	1 471.2	1 078.1	573.9	186.6	358.4	29.0	504.1	393.1	382.4	80.5	277.0	25.0	10.7	198.0	195.1
1960	1 520.5	1 111.0	586.9	192.7	363.0	31.1	524.1	409.6	398.2	83.5	286.7	28.1	11.3	205.2	204.4
1961	1 580.2	1 147.6	604.3	196.1	374.6	33.7	543.3	432.6	420.5	87.5	301.4	31.6	12.1	215.7	216.8
1962	1 652.8	1 190.4	626.8	203.7	387.2	35.9	563.6	462.4	449.5	94.6	319.6	35.3	12.9	230.3	232.2
1963	1 717.5	1 228.7	650.6	212.3	399.5	38.9	578.1	488.8	475.6	96.7	338.3	40.6	13.2	240.7	248.1
1964	1 829.7	1 314.0	689.3	224.9	422.5	41.9	624.6	515.7	501.8	98.9	356.6	46.3	14.0	251.0	264.8
1965	1 954.3	1 402.9	738.8	242.5	450.6	45.8	664.0	551.5	536.8	100.6	384.0	52.2	14.7	263.0	288.5
1966	2 119.9	1 522.1	804.3	270.0	483.7	50.5	717.8	597.8	582.0	104.3	418.6	59.1	15.7	278.9	318.9
1967	2 285.4	1 636.6	871.2	296.2	518.7	56.3	765.3	648.8	632.3	110.2	455.6	66.6	16.5	299.8	349.0
1968	2 511.4	1 804.7	957.9	326.5	568.5	62.8	846.9	706.6	688.1	114.1	499.0	75.1	18.5	319.6	387.1
1969	2 744.0	1 962.9	1 055.8	360.1	625.2	70.5	907.1	781.1	760.5	117.7	558.0	84.8	20.6	343.0	438.0
1970	2 990.8	2 121.0	1 161.7	394.8	689.4	77.5	959.3	869.8	847.8	123.2	629.9	94.7	21.9	370.2	499.6
1971	3 294.5	2 352.7	1 275.1	422.5	770.1	82.5	1 077.6	941.9	917.3	123.3	691.3	102.6	24.6	391.2	550.7
1972	3 621.8	2 594.0	1 385.7	455.1	841.9	88.7	1 208.2	1 027.9	1 000.3	127.7	761.8	110.8	27.5	424.7	603.1
1973	4 110.6	2 946.9	1 560.5	507.1	955.0	98.4	1 386.4	1 163.7	1 132.6	135.9	872.9	123.8	31.1	470.8	692.9
1974	4 887.7	3 468.3	1 893.8	625.5	1 156.4	111.9	1 574.5	1 419.4	1 384.7	149.1	1 096.7	138.9	34.7	539.7	879.7
1975	5 277.3	3 786.4	2 083.8	716.2	1 246.2	121.5	1 702.6	1 490.9	1 452.8	165.0	1 138.2	149.6	38.0	572.1	918.8
1976	5 747.3	4 168.9	2 280.1	792.5	1 354.7	132.9	1 888.7	1 578.5	1 536.0	180.7	1 194.5	160.8	42.5	619.5	958.9
1977	6 415.5	4 735.7	2 532.5	891.6	1 495.1	145.8	2 203.2	1 679.7	1 630.9	198.3	1 258.6	174.1	48.9	656.0	1 023.7
1978	7 261.1	5 412.9	2 870.3	1 018.0	1 689.2	163.2	2 542.5	1 848.2	1 791.7	222.5	1 378.7	190.6	56.5	722.1	1 126.1
1979	8 360.7	6 264.5	3 313.9	1 181.9	1 945.2	186.8	2 950.5	2 096.2	2 028.9	236.7	1 578.9	213.3	67.3	800.2	1 296.0
1980	9 512.5	7 118.0	3 800.4	1 369.9	2 216.6	213.8	3 317.6	2 394.5	2 320.8	260.3	1 819.6	240.9	73.7	888.6	1 505.9
1981	10 490.4	7 860.3	4 300.0	1 528.5	2 526.7	244.9	3 560.2	2 630.1	2 549.3	286.7	1 992.5	270.1	80.8	960.0	1 670.1
1982	11 083.9	8 297.0	4 588.9	1 619.9	2 694.9	274.1	3 708.0	2 787.0	2 701.3	315.1	2 092.4	293.8	85.7	1 024.5	1 762.5
1983	11 468.9	8 599.6	4 744.7	1 670.5	2 770.2	304.1	3 854.8	2 869.3	2 773.3	345.2	2 110.8	317.3	96.0	1 084.6	1 784.7
1984	12 140.1	9 112.5	5 041.0	1 758.8	2 943.8	338.5	4 071.5	3 027.5	2 928.0	404.3	2 182.7	341.0	99.5	1 182.2	1 845.4
1985	12 759.0	9 619.0	5 332.4	1 858.0	3 101.8	372.6	4 286.6	3 140.0	3 039.6	400.2	2 274.6	364.7	100.4	1 210.4	1 929.6
1986	13 548.2	10 230.6	5 603.6	1 964.7	3 232.5	406.4	4 627.0	3 317.6	3 213.7	413.1	2 415.0	385.6	103.9	1 260.1	2 057.5
1987	14 342.5	10 843.5	5 919.1	2 055.3	3 415.1	448.7	4 924.4	3 499.0	3 385.6	431.0	2 538.2	416.3	113.4	1 318.2	2 180.8
1988	15 253.0	11 560.9	6 335.9	2 178.1	3 641.8	516.1	5 224.9	3 692.1	3 563.5	466.3	2 648.8	448.4	128.6	1 405.8	2 286.3
1989	16 120.6	12 230.0	6 726.7	2 300.6	3 858.7	567.4	5 503.2	3 890.7	3 753.2	498.4	2 780.9	474.0	137.4	1 477.7	2 413.0
1990	16 885.5	12 802.8	7 100.5	2 423.7	4 055.4	621.4	5 702.3	4 082.8	3 940.9	534.7	2 907.9	498.2	141.9	1 541.0	2 541.8
1991	17 305.0	13 090.9	7 274.6	2 482.2	4 116.7	675.6	5 816.3	4 214.1	4 071.3	559.1	2 994.6	517.6	142.8	1 592.5	2 621.5
1992	18 034.2	13 643.6	7 521.7	2 544.3	4 251.0	726.4	6 121.9	4 390.6	4 239.5	585.5	3 120.0	534.0	151.1	1 652.1	2 738.4
1993	18 937.5	14 358.0	7 874.3	2 642.4	4 458.5	773.4	6 483.7	4 579.6	4 417.8	608.9	3 262.5	546.4	161.8	1 704.8	2 874.8
1994	20 066.8	15 242.3	8 312.5	2 784.9	4 696.8	830.7	6 929.8	4 824.5	4 652.0	634.4	3 452.8	564.8	172.6	1 770.4	3 054.1
1995	21 042.0	15 993.5	8 765.6	2 959.2	4 913.2	893.2	7 228.0	5 048.5	4 869.7	641.7	3 648.7	579.2	178.8	1 811.0	3 237.5
1996	22 046.6	16 812.9	9 197.5	3 104.1	5 136.2	957.2	7 615.4	5 233.7	5 047.1	633.9	3 826.7	586.4	186.6	1 833.8	3 399.9
1997	23 219.0	17 752.7	9 723.7	3 235.2	5 445.7	1 042.8	8 029.0	5 466.3	5 271.2	629.5	4 043.3	598.5	195.1	1 864.3	3 602.0
1998	24 525.4	18 828.3	10 279.7	3 384.7	5 759.2	1 135.8	8 548.6	5 697.1	5 492.1	639.3	4 240.2	612.6	205.0	1 902.5	3 794.6
1999	26 101.0	20 085.2	10 907.6	3 578.7	6 063.8	1 265.0	9 177.6	6 015.7	5 798.0	660.8	4 503.0	634.2	217.8	1 967.3	4 048.4
2000	27 823.6	21 482.6	11 672.4	3 805.2	6 468.4	1 398.7	9 810.2	6 341.0	6 111.4	655.2	4 799.9	656.3	229.6	2 003.8	4 337.2
2001	29 376.5	22 772.5	12 255.5	3 913.4	6 880.7	1 461.4	10 517.0	6 604.0	6 359.8	650.9	5 040.4	668.5	244.2	2 026.4	4 577.6
2002	30 805.4	23 906.8	12 684.4	3 959.3	7 212.5	1 512.5	11 222.4	6 898.7	6 641.1	666.8	5 285.2	689.1	257.6	2 077.6	4 821.1
2003	32 468.3	25 270.5	13 109.5	4 000.4	7 514.3	1 594.9	12 161.0	7 197.7	6 922.0	685.7	5 510.1	726.2	275.7	2 149.8	5 047.9
2004	35 787.7	27 811.3	14 227.1	4 225.9	8 334.0	1 667.2	13 584.3	7 976.3	7 673.3	711.2	6 199.5	762.6	303.0	2 289.1	5 687.2
2005	39 374.9	30 662.1	15 511.7	4 423.0	9 311.4	1 777.3	15 150.4	8 712.8	8 380.8	735.3	6 842.1	803.4	332.0	2 432.9	6 279.9
2006	42 596.0	32 986.8	16 793.6	4 729.2	10 180.9	1 883.6	16 193.2	9 609.2	9 262.1	768.0	7 654.8	839.3	347.0	2 579.5	7 029.7
2007	44 509.5	34 154.6	17 747.4	4 952.1	10 772.1	2 023.2	16 407.2	10 354.9	10 010.1	803.6	8 316.4	890.1	344.8	2 709.8	7 645.1
2008	46 019.3	34 981.3	18 885.6	5 200.2	11 557.2	2 128.2	16 095.7	11 038.0	10 702.1	855.2	8 919.1	927.8	335.9	2 828.7	8 209.3
2009	45 191.4	34 101.1	18 343.1	5 120.2	11 029.7	2 193.2	15 758.0	11 090.2	10 760.4	885.7	8 918.4	956.4	329.8	2 838.4	8 251.8
2010	46 099.4	34 582.2	18 799.2	5 224.0	11 286.3	2 288.9	15 783.0	11 517.2	11 184.7	923.4	9 249.1	1 012.2	332.5	2 955.8	8 561.4
2011	47 690.3	35 557.7	19 665.1	5 459.7	11 799.6	2 405.8	15 892.7	12 132.5	11 795.5	954.8	9 800.5	1 040.2	337.0	3 061.3	9 071.2
2012	49 215.9	36 693.1	20 340.8	5 699.6	12 126.7	2 514.4	16 352.4	12 522.8	12 175.1	969.0	10 142.0	1 064.1	347.6	3 126.7	9 396.0
2013	51 642.8	38 699.6	21 205.3	5 897.2	12 641.7	2 666.5	17 494.3	12 943.2	12 572.0	979.9	10 492.1	1 100.0	371.1	3 204.8	9 738.4
2014	53 723.9	40 485.0	22 176.0	6 138.8	13 245.0	2 792.1	18 309.0	13 238.9	12 854.9	992.4	10 747.2	1 115.3	384.0	3 248.7	9 990.2
2015	55 038.6	41 605.7	22 645.7	6 304.7	13 448.3	2 892.7	18 960.0	13 432.9	13 041.1	993.4	10 928.0	1 119.6	391.9	3 252.7	10 180.3
2016	57 296.1	43 475.0	23 255.4	6 452.1	13 739.8	3 063.4	20 219.6	13 821.1	13 407.2	1 005.6	11 254.2	1 147.3	413.9	3 315.4	10 505.7
2017	59 806.5	45 443.4	24 283.5	6 690.7	14 322.5	3 270.3	21 160.0	14 363.0	13 936.5	1 022.6	11 725.7	1 188.3	426.5	3 405.6	10 957.4
2018	62 889.4	47 831.7	25 474.7	7 007.7	15 023.6	3 443.4	22 357.0	15 057.7	14 614.2	1 059.6	12 335.0	1 219.6	443.5	3 503.8	11 553.9

Table 5-5B. Current-Cost Net Stock of Fixed Assets: Historical Data

(Billions of dollars, year-end estimates.)

Year	Total	Private						Government							
		Total	Nonresidential				Resi-dential	Total	Nonresidential				Resi-dential	Federal	State and local
			Total	Equip-ment	Struc-tures	Intellect-ual property products			Total	Equip-ment	Struc-tures	Intellect-ual property products			
1925	261.7	223.4	127.0	31.1	94.0	1.9	96.4	38.3	38.3	2.6	35.6	0.1	0.0	8.9	29.4
1926	270.5	231.6	131.5	32.8	96.7	2.0	100.0	38.9	38.9	2.6	36.2	0.1	0.0	8.8	30.1
1927	277.2	236.9	134.2	33.5	98.5	2.1	102.7	40.3	40.3	2.6	37.7	0.1	0.0	8.6	31.7
1928	289.2	248.6	137.8	34.2	101.3	2.3	110.8	40.7	40.7	2.5	38.0	0.1	0.0	8.3	32.4
1929	294.4	251.7	137.1	34.3	100.3	2.5	114.6	42.7	42.7	2.5	40.0	0.2	0.0	8.9	33.8
1930	281.7	239.4	130.2	32.7	95.0	2.5	109.2	42.3	42.3	2.4	39.7	0.2	0.0	8.4	33.9
1931	245.5	205.4	116.2	29.7	84.0	2.5	89.2	40.0	40.0	2.4	37.4	0.3	0.0	7.8	32.3
1932	222.3	187.6	107.6	26.7	78.6	2.4	79.9	34.7	34.7	2.3	32.1	0.3	0.0	7.5	27.2
1933	236.3	196.5	109.3	26.4	80.5	2.4	87.2	39.8	39.8	2.3	37.1	0.4	0.0	8.5	31.3
1934	246.0	198.6	110.6	26.5	81.6	2.6	87.9	47.4	47.4	2.5	44.5	0.4	0.0	9.4	38.0
1935	247.3	199.5	110.3	25.9	81.6	2.8	89.2	47.7	47.7	2.6	44.6	0.4	0.0	10.5	37.2
1936	275.1	220.4	121.4	27.8	90.4	3.1	99.1	54.7	54.5	2.7	51.3	0.5	0.1	12.1	42.5
1937	289.9	231.7	126.5	29.9	93.1	3.4	105.2	58.1	57.9	3.0	54.4	0.5	0.3	13.4	44.7
1938	292.7	232.0	125.2	29.9	91.6	3.7	106.8	60.7	60.4	3.1	56.7	0.6	0.3	14.2	46.5
1939	298.2	235.6	125.5	30.5	91.0	4.0	110.1	62.6	62.2	3.3	58.3	0.6	0.4	14.8	47.7
1940	320.5	254.4	133.6	32.8	96.5	4.3	120.8	66.1	65.6	3.7	61.3	0.6	0.6	16.8	49.3
1941	365.5	282.8	151.0	37.6	108.2	5.1	131.9	82.7	81.6	7.1	73.6	0.9	1.1	27.1	55.6
1942	424.9	302.3	160.5	38.5	115.9	6.0	141.8	122.5	120.6	21.0	98.3	1.3	1.9	57.5	65.0
1943	478.7	317.1	163.3	38.9	117.6	6.8	153.8	161.6	158.6	46.5	109.9	2.2	3.0	92.5	69.1
1944	513.4	329.3	165.2	38.8	119.0	7.5	164.1	184.0	180.7	67.4	109.8	3.5	3.3	117.0	67.1
1945	554.9	353.0	180.2	44.2	127.9	8.1	172.8	201.9	198.3	78.5	115.5	4.3	3.6	135.7	66.2
1946	639.7	434.0	221.0	53.3	158.6	9.1	213.0	205.7	201.4	70.7	125.7	5.0	4.3	132.7	73.0
1947	734.2	516.6	265.7	65.5	189.6	10.6	250.9	217.6	212.0	60.0	146.4	5.5	5.6	124.2	93.3
1948	780.0	562.2	291.6	80.3	199.9	11.3	270.6	217.9	212.9	50.2	156.6	6.2	5.0	116.2	101.7
1949	785.1	581.1	296.7	84.6	200.4	11.7	284.4	204.0	198.9	42.4	149.7	6.8	5.1	106.7	97.2

Table 5-6A. Chain-Type Quantity Indexes for Net Stock of Fixed Assets: Recent Data

(Index numbers, 2012 = 100.)

Year	Total	Private						Government							
		Total	Nonresidential				Residential	Total	Nonresidential				Residential	Federal	State and local
			Total	Equipment	Structures	Intellectual property products			Total	Equipment	Structures	Intellectual property products			
1950	17.3	16.4	14.5	8.3	24.0	3.1	18.9	20.5	20.6	28.2	22.3	4.7	17.1	34.7	14.5
1951	18.0	17.0	15.0	8.8	24.6	3.2	19.6	21.4	21.4	30.3	23.0	5.0	18.3	36.2	15.1
1952	18.7	17.5	15.5	9.2	25.1	3.4	20.4	22.5	22.6	34.2	23.8	5.4	19.7	38.9	15.6
1953	19.4	18.2	16.0	9.7	25.8	3.7	21.1	23.8	23.9	38.4	24.7	5.9	20.9	41.7	16.2
1954	20.2	18.8	16.5	10.0	26.5	4.0	21.9	24.9	25.0	40.5	25.7	6.4	21.5	43.5	17.0
1955	21.0	19.6	17.1	10.5	27.2	4.3	22.9	25.8	25.9	41.3	26.8	7.1	21.9	44.5	17.9
1956	21.8	20.3	17.8	11.0	28.1	4.6	23.8	26.8	26.9	41.8	27.8	8.3	22.5	45.7	18.8
1957	22.5	21.0	18.4	11.4	28.9	5.0	24.5	27.8	27.9	42.5	28.8	9.6	23.5	46.9	19.7
1958	23.2	21.6	18.8	11.6	29.6	5.3	25.3	29.0	29.1	43.2	30.0	10.8	25.4	48.4	20.8
1959	24.1	22.3	19.3	11.9	30.4	5.6	26.3	30.3	30.4	45.0	31.2	12.1	27.5	50.3	21.9
1960	24.9	23.0	19.9	12.2	31.2	6.0	27.3	31.5	31.6	46.3	32.4	13.5	29.1	52.0	23.0
1961	25.8	23.7	20.5	12.5	32.0	6.4	28.2	33.0	33.1	48.3	33.7	15.0	31.0	54.1	24.1
1962	26.7	24.5	21.1	12.9	32.9	6.8	29.2	34.5	34.6	50.4	35.0	16.8	33.1	56.5	25.3
1963	27.8	25.4	21.8	13.4	33.8	7.2	30.3	36.1	36.1	51.6	36.4	19.2	34.2	58.5	26.7
1964	28.9	26.4	22.7	14.1	34.9	7.6	31.6	37.6	37.7	52.4	37.9	21.8	35.3	60.4	28.1
1965	30.1	27.5	23.8	15.1	36.2	8.2	32.7	39.2	39.2	52.5	39.4	24.4	36.6	62.0	29.6
1966	31.5	28.7	25.2	16.4	37.6	8.8	33.7	40.9	41.0	53.2	41.1	27.5	37.9	64.1	31.3
1967	32.7	29.8	26.4	17.4	39.0	9.5	34.6	42.8	42.8	54.4	42.8	30.5	39.4	66.1	33.0
1968	34.0	31.0	27.6	18.6	40.3	10.1	35.7	44.4	44.5	54.4	44.5	33.4	40.9	67.4	34.8
1969	35.3	32.3	28.9	19.8	41.7	10.8	36.8	45.8	45.9	53.9	46.0	36.0	40.9	68.3	36.5
1970	36.4	33.4	30.0	20.8	43.1	11.3	37.8	47.0	47.0	53.2	47.3	37.8	44.6	68.7	38.0
1971	37.5	34.5	31.1	21.6	44.4	11.6	39.3	47.9	47.9	51.0	48.6	39.3	46.4	68.3	39.3
1972	38.8	36.0	32.3	22.8	45.7	12.0	41.0	48.7	48.7	49.2	49.7	40.8	48.0	68.1	40.6
1973	40.2	37.5	33.7	24.4	47.2	12.5	42.7	49.5	49.5	47.7	50.8	42.1	49.5	68.0	41.8
1974	41.3	38.8	35.1	25.9	48.6	12.9	43.8	50.4	50.3	47.2	51.9	42.9	50.9	68.0	43.0
1975	42.2	39.7	36.0	26.8	49.6	13.2	44.7	51.2	51.1	46.8	52.8	43.7	52.6	68.0	44.1
1976	43.2	40.8	36.9	27.7	50.7	13.8	45.9	52.0	52.0	46.7	53.8	44.6	53.9	68.3	45.2
1977	44.5	42.2	38.2	29.1	51.8	14.4	47.6	52.8	52.7	46.7	54.6	45.7	55.1	68.6	46.2
1978	46.0	43.8	39.8	31.0	53.2	15.1	49.3	53.7	53.6	46.5	55.6	47.0	56.2	69.0	47.3
1979	47.6	45.6	41.6	33.0	54.9	16.1	50.9	54.7	54.6	46.9	56.7	48.4	57.2	69.6	48.5
1980	48.9	46.9	43.2	34.4	56.8	17.0	51.9	55.7	55.6	47.4	57.7	49.9	58.4	70.3	49.6
1981	50.1	48.3	44.9	35.8	59.0	18.2	52.7	56.6	56.5	48.2	58.5	51.5	59.9	71.2	50.5
1982	51.1	49.3	46.3	36.5	61.0	19.4	53.2	57.5	57.3	49.3	59.2	53.0	61.3	72.3	51.3
1983	52.2	50.4	47.5	37.2	62.5	20.6	54.3	58.4	58.3	51.1	59.8	54.7	63.0	73.7	52.0
1984	53.7	52.1	49.3	38.7	64.5	22.3	55.8	59.6	59.4	53.2	60.6	56.8	64.4	75.3	53.0
1985	55.4	53.8	51.2	40.2	66.8	24.1	57.2	61.0	60.9	56.0	61.6	59.5	66.2	77.4	54.2
1986	57.0	55.5	52.8	41.3	68.6	25.8	58.9	62.6	62.5	59.6	62.6	62.1	68.0	79.9	55.4
1987	58.6	57.0	54.2	42.2	70.3	27.5	60.7	64.3	64.1	63.3	63.8	64.9	70.0	82.5	56.7
1988	60.2	58.6	55.7	43.3	71.9	29.2	62.4	65.8	65.6	65.9	64.9	67.1	71.7	84.2	58.1
1989	61.7	60.1	57.2	44.5	73.5	31.2	63.8	67.2	67.0	68.7	66.0	68.8	73.3	85.6	59.5
1990	63.1	61.5	58.7	45.4	75.3	33.3	65.1	68.7	68.5	71.6	67.2	70.4	74.9	87.0	61.1
1991	64.2	62.5	59.8	45.9	76.5	35.5	66.0	70.1	69.9	73.7	68.5	71.3	76.4	87.9	62.7
1992	65.3	63.6	60.8	46.5	77.3	37.6	67.2	71.4	71.1	75.4	69.8	72.0	77.9	88.5	64.2
1993	66.6	64.9	62.1	47.9	78.2	39.5	68.6	72.4	72.2	76.1	71.1	72.3	79.3	88.7	65.7
1994	68.0	66.5	63.6	49.8	79.1	41.3	70.3	73.3	73.1	76.0	72.3	72.4	80.4	88.4	67.1
1995	69.5	68.2	65.4	52.1	80.2	43.3	71.8	74.4	74.1	76.0	73.6	72.4	81.8	88.2	68.7
1996	71.4	70.2	67.6	54.8	81.6	45.9	73.6	75.5	75.3	76.0	75.1	72.5	83.2	88.4	70.3
1997	73.3	72.4	70.1	57.8	83.2	49.3	75.4	76.6	76.3	75.5	76.6	72.8	84.6	87.9	72.0
1998	75.5	74.9	72.9	61.4	84.9	53.0	77.5	77.8	77.5	75.4	78.1	73.3	85.9	87.5	73.9
1999	77.9	77.6	75.9	65.5	86.6	57.3	79.7	79.2	78.9	75.8	79.7	73.9	87.1	87.3	75.9
2000	80.3	80.4	79.2	69.8	88.6	61.9	81.9	80.6	80.3	76.1	81.3	74.7	88.2	87.1	78.0
2001	82.5	82.7	81.7	72.6	90.3	65.5	84.0	82.1	81.8	76.4	83.1	76.0	89.3	87.0	80.2
2002	84.3	84.5	83.2	74.4	91.3	68.1	86.3	83.9	83.7	77.6	85.0	77.9	90.6	87.6	82.5
2003	86.3	86.6	84.7	76.3	92.2	70.7	89.0	85.8	85.6	78.7	86.9	80.2	91.7	88.5	84.8
2004	88.5	88.8	86.3	78.8	92.9	73.3	91.8	87.6	87.4	80.3	88.7	82.7	92.9	89.6	86.9
2005	90.6	91.0	88.1	82.0	93.6	76.5	94.6	89.2	89.0	82.3	90.1	85.4	93.8	90.7	88.6
2006	92.9	93.5	90.4	86.0	94.7	80.0	97.3	90.9	90.8	84.8	91.7	88.1	94.7	92.0	90.6
2007	95.0	95.7	92.9	90.0	96.3	83.9	99.1	92.7	92.6	87.8	93.3	90.6	95.5	93.3	92.5
2008	96.5	97.2	95.2	92.6	98.0	87.7	99.7	94.5	94.4	91.4	94.9	92.9	96.6	94.8	94.4
2009	97.2	97.6	95.8	91.9	98.8	90.5	99.8	96.2	96.1	94.5	96.4	95.0	97.6	96.4	96.1
2010	98.0	98.1	96.7	93.3	99.0	93.1	99.8	97.8	97.7	97.1	97.9	97.0	98.8	98.1	97.7
2011	98.9	98.9	98.1	96.1	99.4	96.5	99.8	99.1	99.1	98.8	99.1	98.7	99.7	99.3	99.0
2012	100.0	100.0	100.0	100.0	100.0	100.0	100.0	100.0	100.0	100.0	100.0	100.0	100.0	100.0	100.0
2013	101.2	101.4	102.0	104.0	100.8	103.8	100.7	100.7	100.8	100.2	100.8	101.0	100.0	100.0	101.0
2014	102.7	103.1	104.5	108.5	101.9	107.8	101.4	101.4	101.4	100.2	101.5	101.5	100.0	99.8	101.9
2015	104.1	104.8	106.7	112.8	102.9	111.6	102.5	102.1	102.0	100.3	102.4	102.1	100.7	99.6	103.0
2016	105.5	106.4	108.6	116.2	103.7	116.5	103.7	102.9	103.0	100.8	103.3	102.7	100.3	99.4	104.1
2017	106.9	107.9	110.6	120.0	104.5	120.9	104.7	103.8	103.9	101.9	104.1	103.5	100.3	99.5	105.2
2018	108.4	109.7	113.1	124.5	105.6	126.2	105.7	104.6	104.7	103.5	104.9	104.2	100.1	99.5	106.3

Table 5-6B. Chain-Type Quantity Indexes for Net Stock of Fixed Assets: Historical Data

(Index numbers, 2012 = 100.)

Year	Total	Private						Government							
		Total	Nonresidential				Resi-dential	Total	Nonresidential				Resi-dential	Federal	State and local
			Total	Equip-ment	Struc-tures	Intellect-ual property products			Total	Equip-ment	Struc-tures	Intellect-ual property products			
1925	10.6	11.5	10.6	4.8	19.8	0.7	13.1	5.5	5.7	2.5	8.0	0.10	0.00	4.5	6.7
1926	11.0	12.0	10.9	5.0	20.4	0.7	13.7	5.8	6.0	2.4	8.3	0.10	0.00	4.4	7.0
1927	11.4	12.4	11.2	5.1	21.0	0.8	14.2	6.0	6.2	2.4	8.7	0.10	0.00	4.3	7.5
1928	11.8	12.8	11.5	5.2	21.6	0.8	14.7	6.3	6.5	2.4	9.2	0.10	0.00	4.3	7.9
1929	12.1	13.2	11.9	5.4	22.2	0.9	15.0	6.6	6.8	2.5	9.6	0.10	0.00	4.3	8.4
1930	12.4	13.3	12.1	5.4	22.7	1.0	15.0	7.0	7.2	2.4	10.2	0.20	0.00	4.3	9.0
1931	12.5	13.3	12.0	5.2	22.8	1.1	15.1	7.3	7.6	2.5	10.7	0.30	0.10	4.4	9.6
1932	12.4	13.2	11.8	4.9	22.6	1.1	15.0	7.6	7.9	2.4	11.2	0.30	0.10	4.6	10.0
1933	12.4	13.0	11.6	4.7	22.4	1.2	14.9	7.9	8.1	2.4	11.5	0.40	0.10	4.8	10.2
1934	12.4	12.9	11.4	4.5	22.2	1.2	14.8	8.1	8.4	2.4	11.9	0.40	0.10	5.1	10.4
1935	12.4	12.9	11.3	4.5	22.0	1.3	14.8	8.4	8.7	2.6	12.3	0.40	0.20	5.6	10.7
1936	12.6	12.9	11.4	4.6	22.0	1.4	14.8	8.9	9.2	2.6	13.1	0.50	0.70	6.1	11.2
1937	12.8	13.0	11.6	4.8	22.1	1.5	14.9	9.3	9.6	2.7	13.6	0.50	1.30	6.5	11.6
1938	12.9	13.1	11.6	4.7	22.0	1.7	15.0	9.8	10.0	2.9	14.3	0.50	1.50	6.9	12.1
1939	13.2	13.2	11.6	4.8	22.0	1.8	15.2	10.3	10.5	3.1	15.0	0.60	1.90	7.3	12.8
1940	13.4	13.3	11.8	5.0	22.1	1.9	15.4	10.8	11.0	3.2	15.7	0.60	2.80	7.9	13.2
1941	13.9	13.6	12.0	5.2	22.2	2.1	15.7	12.3	12.5	6.1	17.0	0.90	5.20	11.9	13.5
1942	14.6	13.5	11.9	5.2	22.0	2.2	15.7	16.3	16.5	19.2	19.8	1.20	8.10	24.0	13.5
1943	15.2	13.4	11.8	5.1	21.8	2.3	15.6	20.8	21.0	45.0	20.9	1.70	12.00	38.7	13.4
1944	15.8	13.4	11.8	5.1	21.6	2.4	15.6	24.4	24.8	70.1	21.2	2.40	13.00	51.0	13.2
1945	16.1	13.5	12.0	5.4	21.6	2.5	15.5	25.8	26.2	79.2	21.4	3.00	13.40	55.8	13.1
1946	16.1	13.9	12.4	5.9	22.1	2.6	15.9	23.9	24.2	64.6	21.2	3.40	14.90	49.7	13.1
1947	16.2	14.5	13.0	6.7	22.5	2.8	16.5	22.3	22.5	50.9	21.1	3.70	15.30	43.8	13.3
1948	16.5	15.1	13.6	7.4	23.0	2.9	17.3	21.1	21.3	39.7	21.3	4.00	15.50	39.2	13.5
1949	16.8	15.7	14.0	7.8	23.4	3.0	17.9	20.8	20.9	33.9	21.8	4.30	16.40	36.9	14.0

Table 5-7. Capital Expenditures

(Millions of dollars, except totals are shown in billions.)

Capital expenditures	All companies												
	2006	2007	2008	2009	2010	2011	2012	2013	2014	2015	2016	2017	2018
TOTAL (billions)	1 309.9	1 354.7	1 374.2	1 090.7	1 105.7	1 243.0	1 424.2	1 491.3	1 597.9	1 642.0	1 574.8	1 678.8	1 697.9
Structures	488 701	525 273	562 381	449 545	428 713	471 779	570 489	581 836	643 590	642 859	599 707	665 642	644 387
New	448 861	480 839	522 999	422 780	394 517	442 745	534 696	549 359	606 980	589 689	543 758	616 850	608 121
Used	39 840	44 434	39 382	26 765	34 196	29 034	35 793	32 477	36 610	53 171	55 948	48 792	36 266
Equipment	821 238	829 455	811 779	641 149	676 989	771 177	853 661	909 477	954 261	999 122	975 138	101 318	105 349
New	777 059	790 407	765 279	606 576	639 214	730 033	800 519	856 012	898 581	941 297	924 954	956 582	100 039
Used	44 179	39 048	46 501	34 572	37 775	41 144	53 142	53 466	55 680	57 825	50 184	56 607	53 100
Not distributed as structures or equipment	0	0	0	0	0	. . .	0	0	0	0	0	0	0
CAPITALIZED COMPUTER SOFTWARE [1]	. . .	. . .	. . .	. . .	. . .	. . .	. . .	. . .	. . .	. . .	. . .	. . .	. . .
Prepackaged	. . .	. . .	. . .	. . .	. . .	. . .	. . .	. . .	. . .	. . .	. . .	. . .	. . .
Vendor-customized	. . .	. . .	. . .	. . .	. . .	. . .	. . .	. . .	. . .	. . .	. . .	. . .	. . .
Internally-developed	. . .	. . .	. . .	. . .	. . .	. . .	. . .	. . .	. . .	. . .	. . .	. . .	. . .
CAPITAL LEASE AND CAPITALIZED INTEREST EXPENSES [1]													
Capital leases	24 442	20 210	20 169	17 410	15 780	20 509	26 087	26 252	31 114	35 350	36 219	36 745	35 362

Capital expenditures	Companies with employees												
	2006	2007	2008	2009	2010	2011	2012	2013	2014	2015	2016	2017	2018
TOTAL (billions)	1 217.1	1 270.5	1 294.5	1 015.3	1 036.2	1 169.6	1 334.9	1 400.9	1 506.6	1 548.1	1 479.4	1 577.8	1 697.9
Structures	453 893	490 779	529 393	414 051	395 531	440 893	533 115	546 056	608 698	600 766	552 637	612 025	644 387
New	420 090	457 233	500 474	395 022	366 853	414 958	501 420	518 430	577 743	557 641	508 523	576 460	608 121
Used	33 802	33 546	28 919	19 030	28 678	25 935	31 695	27 625	30 955	43 126	44 114	35 565	36 266
Equipment	763 215	779 744	765 098	601 270	640 631	728 710	801 820	854 827	897 885	947 291	926 802	965 760	105 349
New	734 160	750 353	728 322	577 051	612 441	697 766	759 632	813 972	853 888	900 519	887 884	920 780	100 039
Used	29 055	29 391	36 776	24 219	28 190	30 944	42 189	40 855	43 996	46 772	38 918	44 981	53 100
Not distributed as structures or equipment	0	0	0	. . .	0	0	0	0	0	0	0	0	. . .
CAPITALIZED COMPUTER SOFTWARE [1]	58 522	63 116	72 241	. . .	63 780	71 068	87 474	89 893	90 023	93 985	100 110	101 872	107 436
Prepackaged	21 181	21 777	26 260	. . .	21 749	23 242	27 868	28 986	27 901	26 759	27 296	27 294	28 748
Vendor-customized	16 912	17 990	19 259	. . .	17 264	20 346	23 507	24 207	23 638	25 547	26 003	25 340	26 652
Internally-developed	20 433	23 350	26 723	. . .	24 768	27 480	36 099	36 701	38 484	41 679	46 811	49 238	52 036
CAPITAL LEASE AND CAPITALIZED INTEREST EXPENSES [1]													
Capital leases	23 923	19 432	19 422	. . .	15 212	20 145	25 301	25 550	30 400	34 570	35 408	35 973	35 362

Capital expenditures	Companies without employees												
	2006	2007	2008	2009	2010	2011	2012	2013	2014	2015	2016	2017	2018
TOTAL (billions)	92.8	84.2	79.7	75.4	69.5	73.4	89.2	90.4	91.3	93.9	95.4	101.0	. . .
Structures	34 809	34 494	32 988	35 493	33 182	30 886	37 374	35 780	34 892	42 093	47 070	53 617	. . .
New	28 771	23 606	22 525	27 758	27 664	27 787	33 276	30 929	29 237	32 048	35 236	40 390	. . .
Used	6 038	10 888	10 463	7 735	5 518	3 099	4 098	4 851	5 655	10 045	11 834	13 227	. . .
Equipment	58 023	49 711	46 681	39 878	36 357	42 467	51 840	54 650	56 376	51 831	48 336	47 429	. . .
New	42 899	40 054	36 957	29 525	26 773	32 267	40 887	42 040	44 693	40 778	37 070	35 802	. . .
Used	15 124	9 657	9 724	10 353	9 585	10 200	10 953	12 610	11 684	11 053	11 266	11 627	. . .
Not distributed as structures or equipment	0	0	0	0	0	0	0	0	0	0	0	0	. . .
CAPITALIZED COMPUTER SOFTWARE [1]	. . .	. . .	. . .	. . .	. . .	. . .	. . .	. . .	. . .	. . .	. . .	. . .	. . .
Prepackaged	. . .	. . .	. . .	. . .	. . .	. . .	. . .	. . .	. . .	. . .	. . .	. . .	. . .
Vendor-customized	. . .	. . .	. . .	. . .	. . .	. . .	. . .	. . .	. . .	. . .	. . .	. . .	. . .
Internally-developed	. . .	. . .	. . .	. . .	. . .	. . .	. . .	. . .	. . .	. . .	. . .	. . .	. . .
CAPITAL LEASE AND CAPITALIZED INTEREST EXPENSES [1]													
Capital leases	519	778	747	577	568	365	786	702	714	781	812	771	. . .

[1]Included in structures and equipment data shown above.
. . . = Not available.

Table 5-8. Capital Expenditures for Structures and Equipment for Companies with Employees by Major Industry Sector

(Millions of dollars.)

Year and type of expenditure	Total	Forestry, fishing, and agricultural services (113–115)	Mining (21)	Utilities (22)	Construction (23)	Manufacturing (31–33) Total	Durable goods industries (321, 327, 33)	Nondurable goods industries (31, 322–326)	Wholesale trade (42)	Retail trade (44–45)	Transportation and warehousing (48–49)	Information (51)
2001												
Total expenditures	1 052 344	1 532	51 278	82 823	24 802	192 835	118 875	73 959	29 981	66 917	57 756	144 793
Structures, total	346 221	226	32 678	38 093	3 859	39 815	22 032	17 784	6 932	30 010	16 594	41 742
New	323 871	149	31 825	36 504	3 389	38 001	20 701	17 301	5 357	29 118	14 479	41 384
Used	22 349	77	853	1 588	470	1 814	1 331	483	1 575	892	2 116	358
Equipment, total	706 123	1 306	18 600	44 731	20 943	153 019	96 844	56 176	23 049	36 906	41 161	103 051
New	679 090	1 091	17 567	42 939	17 432	148 397	94 251	54 145	20 757	35 074	38 521	102 410
Used	27 033	215	1 033	1 792	3 511	4 623	2 592	2 030	2 292	1 833	2 640	641
2002												
Total expenditures	917 490	1 910	42 467	65 502	24 773	157 243	84 062	73 181	26 789	59 316	47 124	88 156
Structures, total	325 168	184	30 685	29 893	1 890	32 643	15 133	17 510	5 885	26 286	14 498	33 607
New	299 941	118	29 775	29 008	1 254	31 022	14 396	16 626	5 447	25 051	13 870	33 472
Used	25 227	66	910	886	456	1 622	737	885	438	1 234	628	135
Equipment, total	592 321	1 726	11 783	35 609	23 063	124 600	68 929	55 671	20 904	33 030	32 626	54 550
New	564 218	1 319	10 262	34 816	19 257	118 621	66 112	52 510	18 562	31 157	29 178	54 247
Used	28 103	407	1 520	793	3 806	5 978	2 817	3 161	2 342	1 873	3 447	303
2003												
Total expenditures	886 846	1 894	50 548	54 569	23 159	149 065	80 226	68 839	26 014	65 868	44 460	80 524
Structures, total	314 021	202	36 617	24 841	1 676	31 108	13 330	17 778	5 615	29 675	13 005	30 765
New	281 892	177	35 897	24 580	1 424	29 315	12 631	16 685	4 921	27 393	11 779	30 406
Used	32 128	25	720	261	251	1 793	700	1 093	694	2 282	1 226	358
Equipment, total	572 825	1 692	13 931	29 729	21 484	117 956	66 895	51 061	20 399	36 193	31 454	49 759
New	540 611	1 267	12 135	29 044	16 170	112 102	62 810	49 292	19 457	32 162	26 786	47 857
Used	32 214	425	1 796	685	5 313	5 855	4 086	1 769	942	4 031	4 668	1 902
2004												
Total expenditures	953 171	2 081	51 253	50 409	28 627	156 651	85 119	71 532	32 314	72 170	46 054	83 488
Structures, total	335 405	324	34 564	24 398	4 511	31 823	13 606	18 217	7 133	33 308	13 992	28 636
New	300 371	309	33 583	23 626	4 167	30 016	12 818	17 198	6 555	31 486	13 018	26 253
Used	35 034	15	982	772	345	1 807	788	1 019	578	1 822	975	2 384
Equipment, total	617 766	1 757	16 689	26 011	24 115	124 828	71 513	53 315	25 181	38 862	32 062	54 852
New	588 110	1 507	15 415	25 724	18 939	120 481	68 904	51 576	21 888	36 965	28 472	53 120
Used	29 656	250	1 274	286	5 176	4 347	2 609	1 738	3 293	1 897	3 590	1 732
2005												
Total expenditures	1 062 536	2 702	66 746	58 032	30 072	165 634	92 180	73 455	40 578	73 531	56 926	91 373
Structures, total	368 791	344	46 433	24 186	2 544	34 132	14 735	19 397	9 184	34 119	17 855	31 977
New	341 223	283	45 655	23 485	2 247	32 564	14 033	18 531	8 830	33 360	16 954	31 716
Used	27 568	61	777	701	297	1 569	703	866	355	759	901	262
Equipment, total	693 745	2 358	20 313	33 847	27 528	131 502	77 444	54 058	31 394	39 412	39 072	59 396
New	664 648	2 016	18 495	33 083	22 082	126 387	73 889	52 498	28 224	38 301	34 953	59 071
Used	29 096	341	1 818	764	5 446	5 115	3 555	1 560	3 169	1 111	4 119	325
2006												
Total expenditures	1 217 107	2 672	99 309	69 757	30 257	192 364	106 843	85 521	36 600	86 735	68 021	104 373
Structures, total	453 893	391	68 662	30 587	2 556	41 617	17 515	24 103	10 375	43 188	20 852	31 947
New	420 090	316	67 322	29 294	2 217	39 419	16 243	23 176	9 956	41 985	19 765	31 621
Used	33 802	75	1 340	1 293	338	2 198	1 272	926	419	1 203	1 087	326
Equipment, total	763 215	2 281	30 647	39 170	27 701	150 747	89 328	61 419	26 226	43 547	47 168	72 425
New	734 160	1 846	28 813	37 617	23 276	146 551	86 637	59 914	24 366	41 944	41 258	71 830
Used	29 055	435	1 833	1 553	4 425	4 196	2 692	1 504	1 860	1 604	5 911	595
2007												
Total expenditures	1 270 522	2 149	120 681	85 354	36 692	197 298	107 664	89 633	30 776	82 511	67 351	106 084
Structures, total	490 779	469	85 242	40 178	3 529	42 458	17 879	24 579	7 526	41 527	23 712	29 081
New	457 233	320	83 206	37 647	2 704	41 247	17 431	23 816	7 091	40 471	22 808	28 304
Used	33 546	149	2 036	2 531	824	1 211	447	763	436	1 056	904	777
Equipment, total	779 744	1 681	35 440	45 176	33 164	154 840	89 786	65 054	23 250	40 983	43 639	77 003
New	750 353	1 368	33 095	44 107	27 325	150 333	87 020	63 313	21 690	39 668	39 238	76 143
Used	29 391	313	2 344	1 069	5 838	4 507	2 766	1 741	1 560	1 316	4 400	861
2008												
Total expenditures	1 294 491	2 337	149 272	98 668	40 838	213 117	103 022	110 095	32 370	73 234	79 617	103 327
Structures, total	529 393	421	109 683	43 515	11 129	49 346	18 896	30 450	8 387	36 147	30 078	27 376
New	500 474	417	105 064	41 746	10 525	48 184	18 168	30 015	7 958	35 342	29 214	27 080
Used	28 919	4	4 618	1 769	603	1 162	728	434	430	805	863	296
Equipment, total	765 098	1 917	39 590	55 154	29 709	163 771	84 125	79 645	23 982	37 087	49 539	75 951
New	728 322	1 610	35 928	53 486	22 404	158 104	80 690	77 414	22 554	36 328	41 877	75 342
Used	36 776	307	3 662	1 668	7 305	5 667	3 435	2 232	1 429	759	7 662	609
2009												
Total expenditures	1 015 322	2 168	100 564	103 024	19 751	155 153	76 039	79 114	25 252	58 428	55 702	88 373
Structures, total	414 051	460	72 255	45 973	4 556	35 735	13 054	22 680	5 485	28 205	22 088	21 764
New	395 022	453	69 942	45 168	4 215	34 537	12 273	22 264	5 131	27 220	21 128	21 410
Used	19 030	7	2 313	805	341	1 198	782	416	354	985	961	354
Equipment, total	601 270	1 708	28 308	57 051	15 195	119 418	62 985	56 434	19 767	30 223	33 614	66 609
New	577 051	1 428	26 790	54 848	11 815	115 470	60 699	54 772	18 512	29 063	30 182	65 879
Used	24 219	279	1 518	2 204	3 380	3 948	2 286	1 662	1 255	1 160	3 432	731

Table 5-8. Capital Expenditures for Structures and Equipment for Companies with Employees by Major Industry Sector—*Continued*

(Millions of dollars.)

Year and type of expenditure	Finance and insurance (52)	Real estate and rental and leasing (53)	Professional, scientific, and technical services (54)	Management of companies and enterprises (55)	Administrative and support and waste management (56)	Educational services (61)	Health care and social assistance (62)	Arts, entertainment, and recreation (71)	Accommodation and food services (72)	Other services, except public administration (81)	Structure and equipment expenditures serving multiple industries
2001											
Total expenditures	131 105	82 674	30 464	3 035	15 785	17 377	52 932	14 974	21 365	29 006	911
Structures, total	22 744	20 489	7 258	933	3 527	12 852	27 030	8 998	12 248	20 031	163
New	19 571	17 325	6 793	869	3 367	11 860	25 241	8 157	11 402	18 918	162
Used	3 173	3 164	465	64	160	991	1 789	841	846	1 112	0
Equipment, total	108 361	62 185	23 206	2 102	12 258	4 525	25 902	5 976	9 117	8 976	749
New	107 268	60 295	22 330	2 019	11 644	4 238	24 573	5 590	7 921	8 300	725
Used	1 093	1 891	876	83	613	287	1 329	386	1 196	676	24
2002											
Total expenditures	128 444	94 529	25 864	3 430	14 719	19 532	59 311	13 169	22 409	21 269	1 532
Structures, total	24 308	35 579	7 129	933	3 276	14 655	30 291	7 758	12 157	13 261	250
New	19 748	30 227	6 424	913	2 948	13 601	27 273	7 332	10 848	11 363	248
Used	4 739	5 352	706	21	328	1 055	3 018	425	1 309	1 899	2
Equipment, total	103 956	58 949	18 735	2 497	11 443	4 876	29 021	5 412	10 252	8 007	1 282
New	103 421	56 847	18 021	2 481	10 585	4 690	28 196	5 132	9 290	6 858	1 276
Used	535	2 102	714	16	857	186	825	280	962	1 149	6
2003											
Total expenditures	120 787	87 952	24 703	3 298	16 612	16 667	61 151	11 029	21 036	26 035	1 476
Structures, total	26 200	25 028	5 314	925	3 976	11 984	30 996	6 800	10 568	18 518	209
New	17 908	16 446	4 671	869	3 213	11 569	28 885	6 532	9 417	16 288	202
Used	8 292	8 583	643	56	763	415	2 111	268	1 151	2 230	7
Equipment, total	94 587	62 923	19 389	2 373	12 636	4 683	30 155	4 229	10 468	7 517	1 267
New	94 205	61 253	18 675	2 368	11 374	4 569	29 497	4 038	9 684	6 706	1 263
Used	383	1 671	714	5	1 262	114	658	192	783	811	4
2004											
Total expenditures	153 629	91 606	26 688	2 825	17 455	18 919	64 561	12 165	20 641	19 701	1 572
Structures, total	43 919	27 277	6 007	860	2 567	13 728	32 608	7 360	9 860	12 278	321
New	30 216	21 610	5 714	798	2 309	12 781	30 668	7 196	9 126	10 867	307
Used	13 703	5 667	293	62	259	947	1 939	164	734	1 411	13
Equipment, total	109 710	64 329	20 681	1 965	14 888	5 190	31 953	4 804	10 781	7 423	1 252
New	109 244	61 947	20 081	1 931	12 692	4 965	31 280	4 677	10 373	6 788	1 248
Used	466	2 382	600	34	2 196	225	673	128	408	635	3
2005											
Total expenditures	161 389	103 022	33 066	2 809	18 194	17 484	73 825	14 165	30 718	20 105	2 163
Structures, total	39 383	24 791	8 717	857	3 051	12 711	39 089	9 242	17 679	12 036	460
New	31 023	17 341	7 633	795	2 759	11 913	37 493	8 805	16 567	11 350	452
Used	8 360	7 450	1 084	62	292	798	1 597	436	1 112	686	8
Equipment, total	122 005	78 231	24 350	1 951	15 143	4 773	34 736	4 924	13 039	8 069	1 703
New	121 511	76 894	23 887	1 917	13 523	4 597	34 110	4 757	11 950	7 209	1 681
Used	494	1 337	463	34	1 620	176	626	166	1 089	861	22
2006											
Total expenditures	163 069	132 073	30 284	3 306	19 231	22 615	75 296	17 156	36 217	25 959	1 813
Structures, total	41 326	40 794	6 971	875	3 613	17 537	41 197	11 733	22 585	16 621	467
New	34 028	30 240	6 375	799	3 485	16 203	37 765	11 326	21 774	15 754	446
Used	7 298	10 554	596	76	128	1 334	3 433	406	811	866	21
Equipment, total	121 743	91 280	23 313	2 432	15 618	5 078	34 099	5 424	13 632	9 339	1 346
New	121 157	88 957	22 867	2 188	14 932	4 983	33 508	5 121	13 161	8 463	1 322
Used	586	2 323	446	244	686	95	591	303	472	875	24
2007											
Total expenditures	172 894	117 969	31 804	4 542	18 167	23 238	84 160	18 769	38 021	29 680	2 380
Structures, total	45 159	41 222	7 453	1 472	3 859	17 910	45 361	12 589	22 211	19 165	656
New	36 507	34 026	6 944	1 404	3 677	17 322	43 522	11 894	21 226	16 264	650
Used	8 652	7 196	509	69	182	588	1 839	695	984	2 901	6
Equipment, total	127 735	76 747	24 351	3 070	14 308	5 328	38 799	6 180	15 810	10 515	1 725
New	126 763	74 735	23 931	2 943	13 599	5 253	38 258	5 916	14 671	9 593	1 724
Used	972	2 012	420	127	709	75	541	264	1 139	922	1
2008											
Total expenditures	132 913	106 910	32 980	4 567	16 552	27 426	90 248	17 109	40 519	28 312	4 175
Structures, total	25 058	45 929	8 836	1 304	4 133	21 698	50 105	11 594	24 947	18 673	1 035
New	22 640	36 532	8 481	1 241	3 789	20 632	47 205	11 224	24 271	17 899	1 030
Used	2 418	9 397	355	63	344	1 065	2 900	370	675	775	6
Equipment, total	107 855	60 981	24 144	3 264	12 420	5 728	40 143	5 515	15 572	9 638	3 139
New	107 108	59 117	23 349	3 146	11 687	5 672	39 161	5 340	14 012	8 974	3 124
Used	747	1 864	794	117	733	56	981	175	1 560	664	15
2009											
Total expenditures	99 466	72 902	28 163	4 719	19 234	28 018	79 370	16 265	26 439	29 296	3 034
Structures, total	21 825	26 564	6 669	1 371	5 876	22 388	44 375	11 035	14 735	22 054	641
New	20 419	22 350	5 975	1 334	5 361	21 577	42 188	10 804	13 607	21 564	639
Used	1 406	4 214	694	36	515	810	2 187	231	1 128	490	2
Equipment, total	77 641	46 339	21 494	3 349	13 359	5 630	34 994	5 230	11 704	7 243	2 393
New	77 169	44 594	20 980	3 311	12 471	5 403	34 271	5 073	10 858	6 545	2 389
Used	473	1 745	514	38	888	227	723	157	846	698	4

Table 5-8. Capital Expenditures for Structures and Equipment for Companies with Employees by Major Industry Sector—*Continued*

(Millions of dollars.)

Year and type of expenditure	Total	Forestry, fishing, and agricultural services (113–115)	Mining (21)	Utilities (22)	Construction (23)	Manufacturing (31–33)			Wholesale trade (42)	Retail trade (44–45)	Transportation and warehousing (48–49)	Information (51)
						Total	Durable goods industries (321, 327, 33)	Nondurable goods industries (31, 322–326)				
2010												
Total expenditures	1 036 153	3 255	115 749	94 462	17 856	160 798	86 570	74 227	31 075	65 252	58 952	97 150
Structures, total	396 398	679	85 221	43 896	2 633	31 166	14 099	17 068	6 879	29 169	23 401	21 771
New	367 759	671	82 147	42 853	2 361	30 069	13 492	16 576	5 709	27 608	22 622	20 696
Used	28 639	8	3 074	1 044	272	1 098	606	491	1 170	1 561	778	1 074
Equipment, total	639 755	2 576	30 528	50 566	15 223	129 631	72 472	57 160	24 196	36 084	35 551	75 379
New	611 573	1 979	28 820	48 675	11 377	125 911	70 692	55 219	21 756	34 457	29 912	74 657
Used	28 182	597	1 708	1 891	3 846	3 720	1 780	1 940	2 440	1 626	5 639	723
2011												
Total expenditures	1 169 604	3 063	165 693	98 047	21 778	192 441	110 151	82 290	35 745	68 131	72 722	100 057
Structures, total	440 893	529	124 508	46 729	2 867	36 339	17 394	18 946	7 988	27 332	29 026	20 152
New	414 958	518	120 421	45 882	2 590	34 737	16 655	18 081	7 512	26 419	28 148	19 734
Used	25 935	11	4 086	847	277	1 602	738	864	476	912	878	418
Equipment, total	728 710	2 534	41 185	51 319	18 910	156 102	92 758	63 344	27 757	40 799	43 696	79 905
New	697 766	2 050	39 373	48 843	13 895	151 094	89 389	61 705	26 699	39 575	37 612	79 686
Used	30 944	484	1 813	2 476	5 015	5 009	3 369	1 639	1 058	1 224	6 084	219
2012												
Total expenditures	1 334 421	3 149	196 653	124 958	23 555	203 119	113 299	89 820	40 852	77 567	81 792	106 537
Structures, total	533 042	496	150 653	70 940	1 975	43 080	20 181	22 900	9 367	33 443	32 397	23 263
New	501 764	452	146 020	69 578	1 821	41 765	19 563	22 202	8 324	30 978	31 393	22 844
Used	31 279	44	4 633	1 362	154	1 316	618	697	1 043	2 464	1 003	419
Equipment, total	801 378	2 653	46 000	54 017	21 580	160 039	93 118	66 921	31 485	44 124	49 395	83 274
New	759 355	2 266	43 985	50 358	16 446	153 572	89 830	63 742	29 297	42 466	42 663	82 762
Used	42 023	387	2 015	3 660	5 134	6 467	3 288	3 179	2 187	1 658	6 732	512
2013												
Total expenditures	1 400 883	2 965	202 213	111 310	27 562	221 345	121 494	99 850	37 505	77 520	92 616	123 881
Structures, total	546 056	492	152 441	58 051	2 545	46 594	20 303	26 291	9 659	33 865	37 645	33 297
New	518 430	474	148 917	56 353	2 102	44 100	19 536	24 564	9 223	33 160	36 454	32 663
Used	27 625	18	3 523	1 698	442	2 494	767	1 727	435	705	1 191	634
Equipment, total	854 827	2 473	49 773	53 260	25 018	174 751	101 191	73 560	27 846	43 655	54 971	90 584
New	813 972	2 124	47 372	49 935	20 641	169 310	98 239	71 071	26 704	42 561	47 951	90 180
Used	40 855	349	2 401	3 324	4 377	5 441	2 952	2 489	1 142	1 094	7 020	404
2014												
Total expenditures	1 506 582	3 985	230 776	118 895	30 277	231 089	124 797	106 292	44 758	82 402	111 010	132 049
Structures, total	608 698	952	187 348	63 844	3 815	50 481	19 520	30 962	12 288	35 564	40 557	33 315
New	577 743	792	183 822	61 741	2 873	49 175	18 898	30 276	10 679	34 099	39 136	32 785
Used	30 955	160	3 526	2 103	942	1 307	621	685	1 609	1 466	1 421	530
Equipment, total	897 885	3 033	43 428	55 052	26 462	180 608	105 277	75 331	32 470	46 838	70 453	98 734
New	853 888	2 488	39 552	51 676	19 642	174 845	101 761	73 084	31 021	45 233	62 902	98 307
Used	43 996	545	3 876	3 375	6 820	5 763	3 516	2 247	1 449	1 605	7 551	426
2015												
Total expenditures	1 548 057	3 322	174 069	130 514	33 276	245 123	127 714	117 409	42 378	85 963	116 633	132 671
Structures, total	600 766	598	135 552	71 277	2 600	54 716	22 845	31 871	12 624	37 005	42 490	30 161
New	557 641	571	133 903	68 610	2 232	53 804	22 515	31 288	12 514	34 581	41 325	29 929
Used	43 126	26	1 649	2 667	368	913	330	583	110	2 423	1 164	233
Equipment, total	947 291	2 724	38 517	59 237	30 676	190 407	104 869	85 537	29 754	48 959	74 144	102 510
New	900 519	2 392	35 869	55 598	24 385	184 904	101 952	82 951	27 848	47 578	65 280	101 595
Used	46 772	333	2 649	3 639	6 291	5 503	2 917	2 586	1 906	1 381	8 864	915
2016												
Total expenditures	1 479 439	4 561	92 641	133 455	35 979	243 595	122 898	120 698	43 768	86 931	109 650	142 911
Structures, total	552 637	559	75 865	73 533	4 302	57 171	22 759	34 412	13 290	39 098	35 717	32 303
New	508 523	. . .	74 049	69 539	3 284	55 628	22 361	33 267	12 957	38 103	35 120	31 541
Used	44 114	. . .	1 816	3 993	1 018	1 543	398	1 145	333	995	597	763
Equipment, total	926 802	4 002	16 776	59 922	31 676	186 425	100 139	86 286	30 478	47 833	73 933	110 608
New	887 884	. . .	14 698	57 256	23 993	181 333	97 266	84 067	28 730	46 090	66 228	110 145
Used	38 918	. . .	2 078	2 666	7 683	5 092	2 873	2 219	1 748	1 743	7 705	462
2017												
Total expenditures	1 577 785	4 497	134 380	133 888	34 989	247 013	129 121	117 891	43 977	90 493	108 538	158 888
Structures, total	612 025	750	108 370	66 586	3 251	57 732	24 206	33 526	14 000	38 739	40 439	34 432
New	576 460	736	106 014	66 179	2 891	56 476	23 435	33 041	13 093	37 750	39 501	34 418
Used	35 565	14	2 356	407	360	1 255	771	484	907	989	938	14
Equipment, total	965 760	3 747	26 011	67 302	31 738	189 281	104 915	84 366	29 977	51 754	68 098	124 456
New	920 780	3 264	23 437	59 866	24 325	184 325	101 899	82 426	28 284	50 496	60 370	124 106
Used	44 981	483	2 573	7 435	7 413	4 956	3 016	1 940	1 693	1 258	7 728	350
2018												
Total expenditures	1 697 882	4 653	153 390	146 453	38 979	259 157	138 398	120 759	42 460	89 279	122 404	175 692
Structures, total	644 387	716	113 623	73 915	4 045	59 362	26 679	32 682	12 720	38 885	44 645	43 475
New	608 121	694	110 771	73 226	3 539	57 653	26 068	31 585	12 180	38 297	43 291	43 342
Used	36 266	22	2 852	690	505	1 709	611	1 097	540	589	1 353	133
Equipment, total	1 053 495	3 937	39 768	72 538	34 935	199 796	111 719	88 077	29 740	50 394	77 759	132 217
New	1 000 395	3 582	37 467	63 187	27 961	192 893	107 786	85 107	28 589	49 385	68 840	131 455
Used	53 100	355	2 301	9 351	6 974	6 902	3 933	2 970	1 151	1 008	8 920	762

. . . = Not available.

Table 5-8. Capital Expenditures for Structures and Equipment for Companies with Employees by Major Industry Sector—Continued

(Millions of dollars.)

Year and type of expenditure	Finance and insurance (52)	Real estate and rental and leasing (53)	Professional, scientific, and technical services (54)	Management of companies and enterprises (55)	Administrative and support and waste management (56)	Educational services (61)	Health care and social assistance (62)	Arts, entertainment, and recreation (71)	Accommodation and food services (72)	Other services, except public administration (81)	Structure and equipment expenditures serving multiple industries
2010											
Total expenditures	103 093	81 282	28 203	4 946	16 873	23 368	78 381	12 121	19 915	20 951	2 471
Structures, total	16 209	32 375	5 706	1 567	3 552	17 673	42 861	7 709	9 765	13 605	562
New	13 653	23 085	5 365	1 517	3 466	16 678	40 113	7 551	9 051	11 987	558
Used	2 556	9 290	341	50	86	995	2 748	158	714	1 618	4
Equipment, total	86 884	48 908	22 497	3 379	13 321	5 695	35 520	4 412	10 151	7 346	1 909
New	86 401	47 593	21 889	3 359	12 626	5 553	34 808	4 164	8 856	6 879	1 902
Used	483	1 315	608	20	695	142	713	248	1 294	467	7
2011											
Total expenditures	109 229	91 124	28 109	5 449	18 423	21 019	83 114	11 430	24 833	16 064	3 133
Structures, total	14 405	33 540	6 733	1 368	3 536	15 037	42 910	6 589	12 021	8 675	608
New	12 238	24 764	6 340	1 308	3 422	14 444	40 534	. . .	11 352	7 867	. . .
Used	2 167	8 776	393	60	114	593	2 376	. . .	669	807	. . .
Equipment, total	94 823	57 583	21 376	4 081	14 887	5 982	40 204	4 841	12 811	7 389	2 524
New	94 551	55 163	20 822	4 020	14 340	5 910	39 342	. . .	10 892	7 031	. . .
Used	272	2 420	554	61	546	72	862	. . .	1 920	358	. . .
2012											
Total expenditures	130 168	115 652	31 605	6 916	18 811	21 629	88 860	13 296	28 802	16 789	3 711
Structures, total	17 749	42 034	7 536	2 555	3 578	15 721	45 981	8 055	14 154	9 078	988
New	14 141	31 344	7 098	2 429	3 452	15 449	43 773	8 031	13 359	8 537	977
Used	3 608	10 690	438	126	126	272	2 208	24	794	541	12
Equipment, total	112 419	73 618	24 069	4 361	15 233	5 908	42 879	5 241	14 649	7 711	2 722
New	107 539	70 375	23 183	4 290	14 059	5 806	41 973	5 052	13 471	7 110	2 683
Used	4 880	3 243	886	71	1 174	102	905	189	1 178	601	40
2013											
Total expenditures	137 824	114 181	35 655	6 147	21 760	22 619	94 181	14 981	34 532	18 615	3 472
Structures, total	15 458	39 359	8 571	1 577	4 123	16 583	49 364	8 263	16 304	10 999	868
New	13 167	31 813	8 346	1 523	3 995	15 417	47 088	7 969	14 303	10 505	860
Used	2 291	7 547	225	54	129	1 166	2 276	294	2 000	494	8
Equipment, total	122 366	74 821	27 084	4 570	17 637	6 036	44 817	6 718	18 228	7 616	2 604
New	117 827	69 780	26 419	4 526	16 709	5 891	43 564	6 221	16 509	7 166	2 582
Used	4 539	5 041	665	44	927	145	1 253	497	1 719	450	22
2014											
Total expenditures	153 260	121 919	30 383	5 366	22 568	25 823	89 011	19 550	29 836	20 304	3 321
Structures, total	17 843	44 677	7 228	1 237	4 112	19 710	47 481	12 153	14 181	11 075	837
New	16 127	33 225	6 603	1 214	3 984	18 871	45 661	11 987	13 568	10 574	828
Used	1 716	11 452	625	24	128	839	1 819	167	614	500	8
Equipment, total	135 417	77 241	23 156	4 128	18 456	6 114	41 531	7 397	15 654	9 230	2 484
New	130 479	74 946	22 303	4 093	17 451	5 863	40 180	7 007	14 760	8 681	2 461
Used	4 938	2 296	853	36	1 005	251	1 351	390	895	549	23
2015											
Total expenditures	164 594	151 875	33 327	5 180	26 183	33 176	93 755	16 142	33 066	23 195	3 613
Structures, total	19 787	62 188	8 393	1 047	6 547	25 906	49 234	8 822	17 331	13 092	1 395
New	18 137	37 404	8 179	1 045	5 159	25 213	46 504	8 760	16 490	11 886	1 392
Used	1 651	24 783	215	2	1 388	693	2 730	62	841	1 206	3
Equipment, total	144 807	89 688	24 933	4 133	19 636	7 270	44 521	7 321	15 734	10 103	2 218
New	139 004	86 649	23 755	4 117	17 758	7 155	43 423	7 155	14 407	9 447	2 199
Used	5 803	3 038	1 179	16	1 878	115	1 098	165	1 327	656	19
2016											
Total expenditures	161 653	150 686	31 728	6 161	27 672	30 424	93 588	22 544	29 855	27 249	4 389
Structures, total	19 485	62 906	8 736	1 618	6 867	23 999	48 238	13 534	16 795	16 794	1 827
New	17 927	38 222	8 450	1 610	6 251	23 329	46 199	13 358	15 837	14 881	. . .
Used	1 558	24 684	285	8	616	670	2 039	177	958	1 913	. . .
Equipment, total	142 168	87 779	22 992	4 543	20 805	6 425	45 350	9 010	13 059	10 456	2 562
New	141 604	84 182	22 367	4 532	19 753	6 324	44 441	8 670	12 164	9 380	. . .
Used	564	3 597	625	11	1 052	101	908	340	895	1 075	. . .
2017											
Total expenditures	163 038	161 396	37 213	6 918	26 505	36 951	104 628	22 448	35 828	21 667	4 530
Structures, total	22 742	73 028	11 569	1 674	4 948	30 171	56 711	13 972	18 295	13 178	1 439
New	20 773	54 325	11 349	1 645	4 778	28 144	54 188	13 825	16 722	12 247	1 406
Used	1 969	18 703	221	29	170	2 026	2 523	147	1 573	931	33
Equipment, total	140 297	88 368	25 643	5 244	21 557	6 781	47 917	8 476	17 533	8 489	3 091
New	140 088	84 208	24 987	5 097	19 352	6 648	46 824	7 918	16 430	7 671	3 081
Used	208	4 160	656	146	2 204	133	1 093	558	1 103	818	10
2018											
Total expenditures	181 582	173 962	42 486	6 750	29 171	36 339	108 443	24 740	36 013	22 245	3 683
Structures, total	25 588	70 474	13 583	1 782	6 937	29 486	59 827	14 760	17 385	12 392	790
New	24 899	50 760	13 296	1 762	6 779	28 598	56 784	14 512	15 211	11 748	780
Used	689	19 713	287	20	158	888	3 043	248	2 174	644	10
Equipment, total	155 995	103 488	28 903	4 968	22 235	6 853	48 616	9 979	18 628	9 853	2 894
New	155 654	96 996	27 404	4 917	20 402	6 705	47 058	9 618	17 011	8 403	2 867
Used	340	6 492	1 500	51	1 833	148	1 558	361	1 617	1 451	27

. . . = Not available.

NOTES AND DEFINITIONS, CHAPTER 5

TABLES 5-1 THROUGH 5-4

Gross Saving and Investment Accounts

SOURCE: U.S. DEPARTMENT OF COMMERCE, BUREAU OF ECONOMIC ANALYSIS (BEA)

All of the data in these tables are from the July 30, 2020 National Income and Product Accounts NIPA), publication. Explanation of NIPA data are described in the Notes and Definitions to Chapter 1. All quarterly series are shown at seasonally adjusted annual rates. Current and constant dollar values are in billions of dollars. Constant dollar values are in 2012 dollars. Indexes of quantity are based on the average for the year 2012, set to equal 100.

The 2013 revision included a major expansion of the investment accounts. Expenditures for research and development are now recognized as fixed investment. Before 2013, they were treated as if they were costs of producing this year's output. R&D spending by business, government, and nonprofit institutions serving households (NPISHs) is now counted as fixed investment. It is depreciated (with the estimated depreciation added to capital consumption allowances), so that there can be either positive or negative net investment in R&D. A similar treatment is now given to expenditures by private enterprises for the creation of entertainment, literary, and artistic originals. Finally, an expanded set of ownership transfer costs for residential fixed assets is recognized as fixed investment.

Results from the 2017 Annual Update of NIPA slightly revised the average growth rates of GDP between 2013 and 2016. Other revisions were incorporated include the Census Bureau's annual retail sales, construction, manufacturing plus others.

Definitions: Table 5-1

Gross saving is saving before the deduction of allowances for the consumption of fixed capital. It represents the amount of saving available to finance gross investment. *Net saving* is gross saving less allowances for fixed capital consumption. It represents the amount of saving available for financing expansion of the capital stock, and comprises net private saving (the sum of personal saving, undistributed corporate profits, and wage accruals less disbursements) and the net saving of federal, state, and local governments.

Personal saving is derived by subtracting personal outlays from disposable personal income. (See Chapter 4 for more information.) It is the current net saving of individuals (including proprietors of unincorporated businesses), nonprofit institutions that primarily serve individuals, life insurance carriers, retirement funds, private noninsured welfare funds, and private trust funds.

Conceptually, personal saving may also be viewed as the sum for all persons (including institutions as previously defined) of the net acquisition of financial assets and the change in physical assets, less the sum of net borrowing and consumption of fixed capital. In either case, it is defined to exclude capital gains. That is, it excludes profits on the increase in the value of homes, securities, and other property—whether realized or unrealized—and therefore includes the noncorporate inventory valuation adjustment and the capital consumption adjustment (IVA and CCAdj, respectively). (See notes and definitions to Chapter 1.)

The net saving of *Domestic corporate business* is corporate profits after tax less dividends plus the corporate IVA and corporate CCAdj. (See notes and definitions for Chapter 1.)

Government net saving was formerly called "current surplus or deficit (-) of general government." (See Chapter 6 for further detail from the government accounts.) Where current receipts of government exceed current expenditures, government has a current surplus (indicated by a positive value) and saving is made available to finance investment by government or other sectors—for example, by the repayment of debt, which can free up funds for private investment. Where current expenditures exceed current receipts, there is a government deficit (indicated by a negative value) and government must borrow, drawing on funds that would otherwise be available for private investment. In these accounts, current expenditures are defined to include a charge for the consumption of fixed capital.

Consumption of fixed capital is an accounting charge for the using-up of private and government fixed capital, including software, located in the United States. It is based on studies of prices of used equipment and structures in resale markets. As of the 2013 revision, it also includes estimated charges for the using-up of the research and development and other intellectual property capital now defined as investment spending.

For general government and nonprofit institutions that primarily serve individuals, consumption of fixed capital is recorded in government consumption expenditures and in personal consumption expenditures (PCE), respectively, and taken to be the value of the current services of the fixed capital assets owned and used by these entities and the estimated using-up of R&D and other intellectual property.

Private consumption of fixed capital consists of tax-return-based depreciation charges for corporations and nonfarm proprietorships and historical-cost depreciation (calculated by the Bureau of Economic Analysis [BEA] using a geometric pattern of price declines) for farm proprietorships, rental income of persons, and nonprofit institutions, minus the capital consumption adjustments. (In other words, in the NIPA treatment of saving, the amount of the CCAdj is taken out of book depreciation and added to income and profits—a reallocation from one form of gross saving to another.) It also includes the charges for the using-up of private R&D and other intellectual property, as described above.

Gross private domestic investment consists of gross private fixed investment and change in private inventories. (See the notes and definitions for Chapter 1.)

Gross government investment consists of federal, state, and local general government and government enterprise expenditures for fixed assets (structures, equipment, and intellectual property). Government inventory investment is included in government consumption expenditures. For further detail, see Chapter 6.

Capital account transactions, net are the net cash or in-kind transfers between the United States and the rest of the world that are linked to the acquisition or disposition of assets rather than the purchase or sale of currently-produced goods and services. When positive, it represents a net transfer from the United States to the rest of the world; when negative, it represents a net transfer to the United States from the rest of the world. This is a definitional category that was introduced in the 1999 revision of the NIPAs. Estimates are available only from 1982 forward. With the new treatment of disaster losses and disaster insurance introduced in the 2009 revision, this line will include disaster-related insurance payouts to the rest of the world less what is received from the rest of the world.

Net lending or net borrowing (-), NIPAs is equal to the international balance on current account as measured in the NIPAs (see Chapter 7) less capital account transactions, net. When positive, this represents net investment by the United States in the rest of the world; when negative, it represents net borrowing by the United States from the rest of the world. For data before 1982, net lending or net borrowing equals the NIPA balance on current account, because estimates of capital account transactions are not available.

By definition, gross national saving must equal the sum of gross domestic investment, capital account transactions, and net international lending (where net international borrowing appears as negative lending). In practice, due to differences in measurement, these two aggregates differ by the *statistical discrepancy* calculated in the product and income accounts. (See Chapter 1.) Gross saving is therefore equal to the sum of gross domestic investment, capital account transactions, and net international lending minus the statistical discrepancy. Where the statistical discrepancy is negative, it means that the sum of measured investment, capital transactions, and net international lending has fallen short of measured saving.

Net domestic investment is gross domestic investment minus consumption of fixed capital, calculated by the editors from the data shown in the table.

Gross national income is national income plus the consumption of fixed capital. (See Chapter 1 for further information.) This is a new concept introduced in the 2003 revision. It is conceptually equal to gross national product, but differs by the statistical discrepancy. Gross national income is an appropriate denominator for the national saving ratios. Saving was previously shown as a percentage of gross national product; in the revision, it is instead shown as a percentage of the income-side equivalent of gross national product. Since saving is measured as a residual from income, it is appropriate to involve consistent measurements— and consistent imperfections in those measurements—in both the numerator and the denominator of the fraction.

Definitions: Tables 5-2 through 5-4

Gross private fixed investment comprises both nonresidential and residential fixed investment. It consists of purchases of fixed assets, which are commodities that will be used in a production process for more than one year, including replacements and additions to the capital stock, and intellectual property, including software, research and development, and entertainment, literary, and artistic originals. It is "gross" in the sense that it is measured before a deduction for consumption of fixed capital. It covers investment by private businesses and nonprofit institutions in the United States, regardless of whether the investment is owned by U.S. residents. It does not include purchases of the same types of equipment, structures, or intellectual property by government agencies, which are included in government gross investment. It also does not include investment by U.S. residents in other countries.

Gross nonresidential fixed investment consists of structures, equipment, and intellectual property that are not related to personal residences.

Nonresidential structures consists of new construction, brokers' commissions on sales of structures, and net purchases (purchases less sales) of used structures by private business and by nonprofit institutions from government agencies. New construction includes hotels, motels, and mining exploration, shafts, and wells.

Other nonresidential structures consists primarily of religious, educational, vocational, lodging, railroads, farm, and amusement and recreational structures, net purchases of used structures, and brokers' commissions on the sale of structures.

Nonresidential equipment consists of private business purchases—on capital account—of new machinery, equipment, and vehicles; dealers' margins on sales of used equipment; and net purchases (purchases less sales) of used equipment from government agencies, persons, and the rest of the world. (However, it does not include the personal-use portion of equipment purchased for both business and personal use. This is included in PCE.)

Computers have displayed phenomenal growth in numbers and power not easily represented by index numbers or constant-dollar estimates at the scale shown in these tables. Zero entries shown in the quantity indexes for early years actually represent very small quantities. Because of rapid growth in computing power and declines in its price, constant-dollar measures are deemed not meaningful and are not calculated by BEA.

Other information processing includes communication equipment, nonmedical instruments, medical equipment and instruments, photocopy and related equipment, and office and accounting equipment.

Other nonresidential equipment consists primarily of furniture and fixtures, agricultural machinery, construction machinery, mining and oilfield machinery, service industry machinery, and electrical equipment not elsewhere classified.

Intellectual property comprises information processing software, research and development, and entertainment, literary, and artistic originals.

Software excludes the value of software "embedded," or bundled, in computers and other equipment, which is instead included in the value of that equipment.

Research and development consists of expenditures for both purchased and own-account R&D by businesses, NPISHs, and general governments. Government R&D expenditures are treated as investment regardless of whether the R&D is protected or made freely available to the public. Investment is measured as the sum of production costs.

Entertainment, literary, and other artistic originals include theatrical movies, long-lived television programs, books, music, and other miscellaneous entertainment.

Residential private fixed investment consists of both *structures* and residential producers' durable *equipment*—that is, equipment owned by landlords and rented to tenants. Investment in *structures* consists of new units, improvements to existing units, manufactured homes, brokers' commissions and other ownership transfer costs on the sale of residential property, and net purchases (purchases less sales) of used structures from government agencies.

Other residential structures consists primarily of manufactured homes, improvements, dormitories, net purchases of used structures, and brokers' commissions on the sale of residential structures.

Real gross private investment (Table 5-3) and *chain-type quantity indexes for private fixed investment* (Table 5-4) are defined and explained in the notes and definitions to Chapter 1. The chained-dollar (2012) estimates in Table 5-3 are constructed by applying the changes in the chain-type quantity indexes, as shown in Table 5-4, to the 2012 current-dollar values. Thus, they do not contain any information about time trends that is not already present in the quantity indexes.

In Table 5-4, the user may wish to distinguish between the use of the "not available" symbol (...) and the publication of zero values (0.0). The "not available" values shown for computers and software mean that BEA has no separate estimates of their values; they are included in total "information processing equipment and software." The zeroes indicate quantities so small relative to the 2009 base that they round to zero, but they do exist and are included in the higher-level aggregates.

As the quantity indexes are chain-weighted at the basic level of aggregation, chained constant-dollar components generally do not add to the chained constant-dollar totals. For this reason, BEA only makes available year-2009-dollar estimates back to 1999 (except for the very highest levels of aggregation of gross domestic product [GDP]), since the addition problem is less severe for years close to the base year. However, the addition problem is so severe for computers that BEA does not even publish recent year-2012-dollar values for this component. BEA notes that "The quantity index for computers can be used to accurately measure the real growth rate of this component. However, because computers exhibit rapid changes in prices relative to other prices in the economy, the chained-dollar estimates should not be used to measure the component's relative importance or its contribution to the growth rate of more aggregate series." (Footnote to BEA Table 5.3.6, *Survey of Current Business*, available on the BEA Web site at <http://www.bea.gov>.) Accurate estimates of these contributions are shown in BEA Table 5.3.2, which is published in the *Survey of Current Business* and can be found on the BEA Web site.

Data availability, revisions, and references

See the information on the NIPAs at the end of the notes and definitions to Chapter 1. All current and historical data are available on the BEA Web site at <http://www.bea.gov> or the STAT-USA subscription Web site at <http://www.stat-usa.gov>.

Tables 5-5 and 5-6

Current-Cost Net Stock of Fixed Assets; Chain-Type Quantity Indexes for Net Stock of Fixed Assets

SOURCE: U.S. DEPARTMENT OF COMMERCE, BUREAU OF ECONOMIC ANALYSIS (BEA)

The Bureau of Economic Analysis (BEA) calculates annual, end-of-year measurements, integrated with the national income and product accounts (NIPAs), of the level of the stock of fixed assets in the U.S. economy, or what is commonly called the "capital stock." (The fixed investment component of the GDP is a flow, or the increment of new capital goods into the capital stock.) Data on consumer stocks of durable goods are also included in the accounts, but are not shown here. Historical data are available back to 1901, with detailed estimates of net stocks, depreciation, and investment by type and by NAICS (North American Industry Classification System) industry. From this data system, *Business Statistics* presents time series data on the net stock of fixed assets valued in current dollars and also as constant-dollar quantity indexes.

The expanded definition of investment introduced in the 2013 comprehensive revision of the NIPAs, explained at the beginning of the Notes and Definitions to this chapter, was not incorporated in the fixed assets accounts until October 2013, so this edition of *Business Statistics* now publishes for the first time fixed assets data including the new NIPA categories.

Definitions and methods

The definitions of capital stock categories are now the same as the fixed investment categories listed earlier in these Notes and Definitions.

The values of fixed capital and depreciation typically reported by businesses are inadequate for economic analysis and are not typically used in these measures. In business reports, capital is generally valued at historical costs—each year's capital acquisition in the prices of the year acquired—and the totals thus represent a mixture of pricing bases. Reported depreciation is generally based on historical cost and on depreciation rates allowable by federal income tax law, rather than on a realistic rate of economic depreciation.

In these data, the *net stock of fixed assets* is measured by a perpetual inventory method. In other words, net stock at any given time is the cumulative value of past gross investment less the cumulative value of past depreciation (measured by "consumption of fixed capital," the component of the NIPAs that is subtracted from GDP in order to yield net domestic product) and also less damages from disasters and war losses that exceed normal depreciation (such as Hurricane Katrina and the terrorist attacks of September 11, 2001).

The initial calculations using this perpetual inventory method are performed in real terms for each type of asset. They are then aggregated to higher levels using an annual-weighted Fisher-type index. (See the definition of *real or chained-dollar estimates* in the notes and definitions for Chapter 1.) This provides the *chain-type quantity indexes* shown in Table 5-6. Growth rates in these indexes measure real growth in the capital stock.

The real values are then converted to a *current-cost* basis to yield the values shown in Table 5-5. They are converted by multiplying the real values by the appropriate price index for the period under consideration. A major use of the current-cost net stock figures is comparison with the value of output in that year; for example, the current-cost net stock of fixed assets for the total economy divided by the current-dollar value of GDP yields a capital-output ratio for the entire economy. Growth rates in current-cost values will reflect both the real growth measured by the quantity indexes and the increase in the value at current prices of the existing stock.

Data availability and references

The 2018 Annual Capital Expenditure Survey was released on January 16, 2020. Full historical data are available on the BEA website at www.bea.gov (select Surveys and Programs.

TABLES 5-7 AND 5-8

Annual Capital Expenditures

SOURCE: U.S. DEPARTMENT OF COMMERCE, CENSUS BUREAU

These data are from the Census Bureau's Annual Capital Expenditures Survey (ACES). The survey provides detailed information on capital investment in new and used structures and equipment by nonfarm businesses.

The survey is based on a stratified random sample of approximately 45,000 companies with employees and 30,000 non-employer businesses (businesses with an owner but no employees). For companies with employees, the Census Bureau reports data for 132 separate industry categories from the North American Industry Classification System (NAICS). Major exclusions are foreign operations of U.S. businesses, businesses in U.S. territories, government operations (including the U.S. Postal Service, agricultural production companies and private households.

Table 5-8 shows these data for the major NAICS sectors. Total capital expenditures, with no industry detail, are reported for the nonemployer businesses and are shown in Table 5-7, where they can be compared with the totals for companies with employees. The 1999 ACES was the first to use NAICS, providing data for the years 1998 forward on that basis.

Definitions

Capital expenditures include all capitalized costs during the year for both new and used structures and equipment, including software, that were chargeable to fixed asset accounts for which depreciation or amortization accounts are ordinarily maintained. For projects lasting longer than one year, this definition includes gross additions to construction-in-progress accounts, even if the asset was not in use and not yet depreciated. For *capital leases*, the company using the asset (lessee) is asked to include the cost or present value of the leased assets in the year in which the lease was entered. Also included in capital expenditures are capitalized leasehold improvements and capitalized interest charges on loans used to finance capital projects.

Structures consist of the capitalized costs of buildings and other structures and all necessary expenditures to acquire, construct, and prepare the structures. The costs of any machinery and equipment that is integral to or built-in features of the structures are classified as structures. Also included are major additions and alterations to existing structures and capitalized repairs and improvements to buildings.

New structures include new buildings and other structures not previously owned, as well as buildings and other structures that have been previously owned but not used or occupied.

Used structures are buildings and other structures that have been previously owned and occupied.

Equipment includes machinery, furniture and fixtures, computers, and vehicles used in the production and distribution of goods and services. Expenditures for machinery and equipment that is housed in structures and can be removed or replaced without significantly altering the structure are classified as equipment.

New equipment consists of machinery and equipment purchased new, as well as equipment produced in the company for the company's own use.

Used equipment is secondhand machinery and equipment.

Capital leases consist of new assets acquired under capital lease arrangements entered into during the year. Capital leases are defined by the criteria in the Financial Accounting Standards (FASB) Number 13.

Capitalized computer software consists of costs of materials and services directly related to the development or acquisition of software; payroll and payroll-related costs for employees directly associated with software development; and interest cost incurred while developing the software. Capitalized computer software is defined by the criteria in Statement of Position 98-1, Accounting for the Costs of Computer Software Developed or Obtained for Internal Use.

Prepackaged software is purchased off-the-shelf through retailers or other mass-market outlets for internal use by the company and includes the cost of licensing fees and service/maintenance agreements.

Vendor-customized software is externally developed by vendors and customized for the company's use.

Internally-developed software is developed by the company's employees for internal use and includes loaded payroll (salaries, wages, benefits, and bonuses related to all software development activities).

Data availability and references

Current and historical data and references are available on the Census Bureau Web site at https://www.census.gov/programs-surveys/aces.html.

CHAPTER 6: GOVERNMENT

SECTION 6A: FEDERAL GOVERNMENT IN THE NATIONAL INCOME AND PRODUCT ACCOUNTS

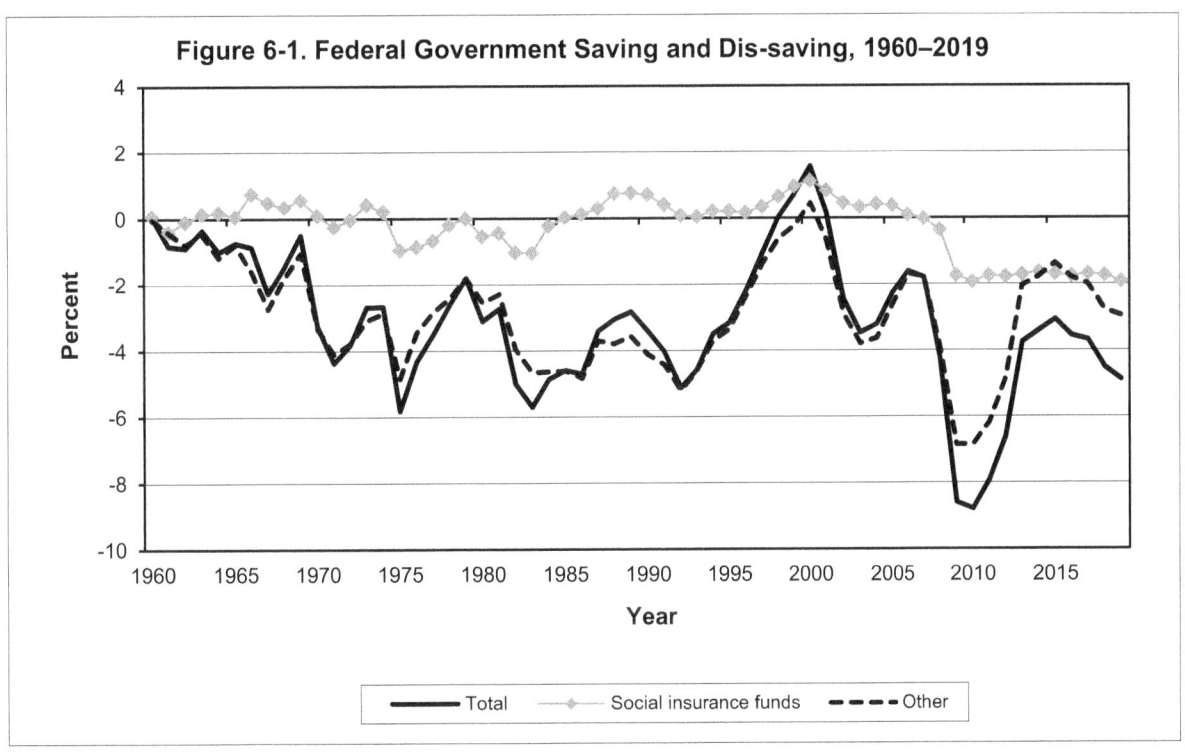

Figure 6-1. Federal Government Saving and Dis-saving, 1960–2019

- In the recovery and economic expansion of the 2000s, the federal budget never got back to a surplus such as those achieved in 1999 and 2000. With the onset of the most severe recession of the postwar period beginning in December 2007, deficits grew rapidly, due both to the "automatic stabilizers"; such as unemployment insurance, built into the budget structure and to anti-recession programs, such as the American Recovery and Reinvestment Act of 2009. (Table 6-1)

- In calendar 2009, the NIPA budget deficit peaked at 10.7 percent of GDP, including over a $250 billion deficit in social insurance funds. This was not as great as World War II deficits but exceeded in scale the deficits of the 1930s. By 2015, the deficit shrank to less than a third from its peak and was 2.6 percent of GDP, the smallest since 2007. However, since 2015 the deficit has increased each year. (Tables 6-1, 1-1, and 6-14B)

- The main contributors to the rise in the deficit from 2007 to 2010 were increasing social benefits to persons, such as Social Security and unemployment insurance, and a drop in non-social-insurance tax receipts. Federal consumption spending and grants to state and local governments also rose. The deficit shrinkage from 2010 to 2015 was powered mainly by a resurgence in non-social-insurance tax receipts while the increase in the deficit in recent years is due to tax cuts. (Table 6-1)

- Total federal government current expenditures have increased from 17.2 percent of gross domestic product (GDP) in 1960 to 24.1 percent in 2009, decreasing again to 21.7 percent in 2017 and then increasing to 22.2 percent in 2019. The composition of current expenditures changed significantly over the last several decades. "Consumption" spending on defense and nondefense programs fell significantly while benefits to persons, such as Social Security and Medicare, increased significantly. (Tables 6-1, 1-1 and 6-3)

Table 6-1. Federal Government Current Receipts and Expenditures

(National income and product accounts, calendar years, billions of dollars, quarterly data are at seasonally adjusted annual rates.)

NIPA Table 3.2

Year and quarter	Total current receipts	Current tax receipts							Contributions for government social insurance	Income receipts on assets			Current transfer receipts	Current surplus of government enterprises
		Total tax receipts [1]	Personal current taxes	Taxes on production and imports		Taxes on corporate income		Taxes from the rest of the world		Interest receipts	Non-federal reserve bank dividends	Rents and royalties		
				Excise taxes	Customs duties	Federal Reserve banks	Other corporate taxes							
1960	93.6	75.6	41.8	12.0	1.1	0.9	20.6	0.1	16.0	1.3	. . .	0.2	0.3	-0.7
1961	95.1	76.8	42.7	12.1	1.0	0.7	20.8	0.1	16.6	1.4	. . .	0.2	0.4	-0.9
1962	103.2	82.5	46.5	12.9	1.2	0.8	21.7	0.1	18.6	1.6	. . .	0.2	0.4	-0.9
1963	111.3	87.7	49.1	13.5	1.2	0.9	23.7	0.2	21.1	1.7	. . .	0.2	0.5	-0.7
1964	111.3	86.2	46.0	14.1	1.3	1.6	24.6	0.2	21.8	1.7	. . .	0.2	0.6	-0.8
1965	120.4	94.3	51.1	13.8	1.6	1.3	27.6	0.2	22.7	1.8	. . .	0.2	0.9	-0.9
1966	137.4	103.1	58.6	12.6	1.9	1.6	29.8	0.2	30.6	2.0	. . .	0.2	1.0	-1.2
1967	146.3	107.9	64.4	13.3	1.9	1.9	28.1	0.2	34.1	2.3	. . .	0.3	1.0	-1.3
1968	170.6	127.2	76.4	14.6	2.3	2.5	33.6	0.3	37.9	2.7	. . .	0.3	1.0	-1.0
1969	191.8	143.0	91.7	15.4	2.4	3.0	33.0	0.4	43.3	2.5	. . .	0.3	1.0	-1.3
1970	185.1	134.4	88.9	15.6	2.5	3.5	27.1	0.4	45.5	2.8	. . .	0.4	1.0	-2.5
1971	190.7	135.3	85.8	15.9	3.1	3.4	30.1	0.4	50.3	3.1	. . .	0.5	0.9	-2.8
1972	219.0	155.0	102.8	15.5	3.0	3.2	33.4	0.4	58.3	3.3	. . .	0.5	1.1	-2.4
1973	249.2	168.7	109.6	16.5	3.3	4.3	38.9	0.4	74.5	3.4	. . .	0.7	1.0	-3.4
1974	278.5	186.5	126.5	16.4	3.7	5.6	39.6	0.4	84.1	3.6	. . .	0.9	1.1	-3.2
1975	276.8	181.5	120.7	16.2	5.9	5.4	38.2	0.5	88.1	4.3	. . .	1.0	1.1	-4.4
1976	322.6	212.4	141.6	16.8	4.6	5.9	48.7	0.7	99.8	5.2	. . .	1.1	1.2	-2.9
1977	363.9	241.6	162.5	17.3	5.4	5.9	55.7	0.7	111.1	5.8	. . .	1.3	1.6	-3.4
1978	423.8	279.8	189.2	18.2	7.1	7.0	64.4	1.0	128.7	7.4	. . .	1.5	2.2	-2.9
1979	487.0	316.9	224.9	18.2	7.5	9.3	65.1	1.1	149.8	9.2	. . .	2.1	2.5	-2.7
1980	533.7	344.5	250.6	26.5	7.2	11.7	58.6	1.6	163.6	11.3	. . .	3.0	3.1	-3.6
1981	621.1	394.3	291.2	41.3	8.6	14.0	51.7	1.5	193.0	14.8	. . .	4.4	3.1	-2.6
1982	618.7	371.8	295.6	32.3	8.6	15.2	33.8	1.4	206.0	18.3	. . .	5.1	4.4	-2.1
1983	644.8	379.5	286.8	35.3	9.1	14.2	47.1	1.2	223.1	20.4	. . .	4.3	5.2	-1.9
1984	711.2	409.7	301.9	35.4	11.9	16.1	59.2	1.3	254.1	22.8	. . .	4.9	6.3	-2.7
1985	775.7	442.9	336.5	33.9	12.2	17.8	58.5	1.9	277.9	25.6	. . .	4.6	8.6	-1.6
1986	817.9	462.0	350.6	30.0	13.7	17.8	66.0	1.7	298.9	28.8	. . .	3.3	7.6	-0.6
1987	899.5	526.4	393.0	30.4	15.5	17.7	85.4	2.0	317.4	25.1	. . .	3.1	10.2	-0.5
1988	962.4	549.8	403.8	33.4	16.4	17.4	93.8	2.4	354.8	27.4	. . .	2.8	9.6	0.7
1989	1 042.5	601.1	453.1	32.3	17.5	21.6	95.6	2.7	378.0	25.9	. . .	3.0	11.8	1.1
1990	1 087.6	620.6	472.1	33.4	17.5	23.6	94.5	3.0	402.0	26.9	. . .	3.5	13.3	-2.3
1991	1 107.8	617.1	463.6	44.9	16.8	20.8	89.2	2.6	420.6	26.2	. . .	3.8	17.2	2.1
1992	1 154.4	645.4	477.5	45.0	18.3	16.8	102.0	2.6	444.0	22.2	. . .	3.5	18.4	4.1
1993	1 231.0	699.3	507.7	46.5	19.8	16.0	122.5	2.7	465.5	22.9	. . .	3.8	20.2	3.5
1994	1 329.3	763.5	545.1	57.5	21.4	20.5	136.3	3.1	496.2	20.1	. . .	3.6	21.6	4.0
1995	1 417.4	825.7	590.3	55.7	19.8	23.4	155.9	3.9	521.9	20.9	. . .	3.0	17.5	5.0
1996	1 536.3	917.0	668.4	53.6	19.2	20.1	170.5	5.2	545.4	22.6	. . .	4.1	22.7	4.5
1997	1 667.4	1 015.1	749.8	58.2	19.6	20.7	182.3	5.1	579.4	21.0	. . .	4.5	20.1	6.6
1998	1 789.8	1 095.3	831.2	60.2	19.6	26.6	177.7	5.7	617.4	17.4	. . .	4.2	21.5	7.6
1999	1 906.6	1 174.5	897.4	63.3	19.2	25.4	187.6	6.1	654.8	17.2	. . .	4.0	23.4	7.2
2000	2 068.4	1 288.5	999.6	65.1	21.1	25.3	194.1	7.6	698.6	19.4	. . .	6.3	24.9	5.3
2001	2 032.2	1 226.8	996.0	63.7	20.6	27.1	137.6	7.9	723.3	18.1	. . .	8.4	25.7	2.8
2002	1 870.8	1 053.2	832.3	65.9	19.9	24.5	126.0	8.1	739.4	15.4	. . .	6.5	25.0	6.8
2003	1 895.6	1 053.9	778.6	67.7	21.5	22.0	175.8	9.2	763.3	16.6	. . .	7.0	26.2	6.6
2004	2 027.7	1 140.6	803.0	70.9	23.3	18.1	232.2	10.1	809.0	16.6	0.1	8.7	29.8	4.8
2005	2 304.4	1 367.8	937.2	72.7	25.3	21.5	319.5	11.7	853.4	17.1	0.2	9.7	33.7	0.9
2006	2 538.3	1 534.8	1 055.8	71.5	26.7	29.1	366.0	13.7	905.7	18.6	0.3	9.9	38.1	1.9
2007	2 667.8	1 607.7	1 170.9	64.8	28.8	34.6	328.2	13.9	947.3	22.1	0.2	11.1	42.7	2.1
2008	2 580.7	1 489.5	1 176.6	63.7	29.2	31.7	202.0	16.9	974.5	19.8	0.6	13.8	49.7	1.1
2009	2 239.5	1 123.7	866.6	67.3	23.1	47.4	153.0	12.8	950.7	23.8	18.7	7.0	67.2	0.9
2010	2 444.0	1 273.6	943.6	67.2	28.6	79.3	219.4	13.9	970.9	29.5	17.0	8.0	68.1	-2.4
2011	2 572.8	1 478.4	1 130.8	75.7	31.9	75.4	224.0	15.1	903.2	26.3	18.8	9.9	67.1	-6.3
2012	2 700.3	1 573.0	1 166.4	80.6	33.5	88.4	274.7	16.7	938.0	21.4	21.1	10.2	56.1	-7.8
2013	3 139.0	1 744.9	1 302.9	89.0	35.5	79.6	298.4	18.1	1 091.8	22.9	131.3	9.3	69.3	-10.1
2014	3 292.0	1 900.1	1 403.7	97.8	37.4	96.9	339.6	20.4	1 140.1	23.8	40.7	10.3	87.3	-7.1
2015	3 448.0	2 024.2	1 532.6	101.0	38.1	110.4	329.1	22.2	1 190.8	27.1	16.0	6.7	76.2	-3.2
2016	3 463.3	2 020.3	1 548.0	97.7	37.5	91.5	311.9	24.1	1 224.6	29.1	14.8	4.8	79.7	-1.4
2017	3 524.3	2 015.5	1 614.6	91.2	38.5	80.6	245.4	24.5	1 283.7	29.3	23.2	6.1	85.2	0.9
2018	3 567.6	2 017.1	1 617.5	108.2	53.3	65.3	210.6	26.3	1 344.6	35.8	13.7	7.8	83.9	-0.6
2019	3 711.2	2 131.7	1 713.0	94.7	77.8	54.9	217.3	27.7	1 402.2	39.0	8.9	8.5	67.8	-1.9
2017														
1st quarter	3 523.7	1 979.5	1 581.1	88.4	37.5	88.4	247.8	23.6	1 262.2	28.5	40.0	5.3	119.1	0.6
2nd quarter	3 504.1	2 013.0	1 601.7	90.9	39.0	87.2	255.8	24.3	1 274.6	28.6	20.2	5.8	73.5	1.1
3rd quarter	3 531.0	2 035.4	1 625.0	92.0	38.0	73.7	254.5	24.6	1 289.7	29.2	20.6	6.3	75.0	1.1
4th quarter	3 538.5	2 034.1	1 650.8	93.5	39.5	72.8	223.5	25.6	1 308.2	30.9	11.8	6.8	73.0	0.8
2018														
1st quarter	3 492.4	1 962.9	1 598.6	107.1	41.8	78.3	188.7	25.4	1 327.8	33.5	0.2	7.3	81.9	0.5
2nd quarter	3 527.5	1 995.0	1 608.9	107.1	46.9	67.1	204.7	26.2	1 337.2	35.1	3.9	7.7	81.7	-0.1
3rd quarter	3 617.6	2 030.5	1 628.3	108.5	52.3	60.7	214.2	26.0	1 353.1	36.8	24.4	8.0	105.2	-1.0
4th quarter	3 633.0	2 080.0	1 634.2	110.2	72.1	55.2	234.7	27.5	1 360.3	37.9	26.3	8.3	66.9	-1.9
2019														
1st quarter	3 674.1	2 108.5	1 695.5	95.4	75.5	44.9	213.8	27.1	1 391.9	37.9	19.1	8.4	64.9	-1.6
2nd quarter	3 704.5	2 123.4	1 703.1	95.0	72.1	58.8	224.2	27.8	1 397.8	38.6	16.3	8.5	63.0	-1.9
3rd quarter	3 702.4	2 117.7	1 713.2	94.4	79.4	53.4	201.6	27.9	1 402.3	39.5	0.2	8.5	82.9	-2.1
4th quarter	3 763.7	2 177.1	1 740.2	94.0	84.0	62.4	229.7	27.9	1 416.9	40.1	0.2	8.4	60.5	-2.0

[1]Does not include "contributions for government social insurance" (the taxes that fund Social Security, Medicare, and unemployment insurance), which are shown separately in column 9.
. . . = Not available.

Table 6-1. Federal Government Current Receipts and Expenditures—*Continued*

(National income and product accounts, calendar years, billions of dollars, quarterly data are at seasonally adjusted annual rates.)

NIPA Table 3.2

Year and quarter	Total current expenditures	Consumption expenditures	Government social benefits		Other current transfer payments		Interest payments			Subsidies	Net federal government saving, NIPA (surplus + / deficit -)		
			Benefits to persons	Benefits to the rest of the world	Grants-in-aid to state and local governments	Payments to the rest of the world (net)	Total	To persons and business	To the rest of the world		Total	Social insurance funds	Other
1960	93.4	51.0	19.9	0.2	3.8	3.5	13.9	13.5	0.3	1.1	0.2	0.4	-0.2
1961	99.8	52.7	23.1	0.3	4.3	3.5	13.9	13.6	0.3	2.0	-4.7	-2.3	-2.4
1962	108.6	59.0	23.5	0.3	4.7	3.6	15.1	14.8	0.3	2.3	-5.4	-0.6	-4.8
1963	113.5	61.2	24.6	0.3	5.2	3.6	16.3	15.9	0.4	2.2	-2.2	0.8	-3.0
1964	118.2	63.1	25.2	0.3	6.0	3.4	17.5	17.0	0.5	2.7	-6.9	1.2	-8.1
1965	125.9	66.5	27.3	0.5	6.6	3.6	18.6	18.1	0.5	3.0	-5.5	0.4	-5.9
1966	144.4	76.5	29.9	0.5	9.4	4.0	20.3	19.7	0.5	3.9	-7.0	6.0	-13.0
1967	165.8	88.1	36.5	0.6	10.9	4.0	21.9	21.3	0.6	3.8	-19.5	4.1	-23.6
1968	184.3	96.8	41.9	0.6	11.8	4.5	24.6	23.9	0.7	4.1	-13.8	3.2	-16.9
1969	197.0	100.5	45.8	0.6	13.7	4.5	27.5	26.7	0.8	4.5	-5.1	5.7	-10.8
1970	219.9	102.3	55.6	0.7	18.3	4.8	33.3	32.4	0.9	4.8	-34.8	1.0	-35.8
1971	241.6	106.7	66.1	0.8	22.1	6.0	35.2	33.5	1.8	4.6	-50.9	-3.0	-47.9
1972	268.0	112.3	72.9	1.0	30.5	7.2	37.6	35.1	2.6	6.6	-49.0	-0.6	-48.4
1973	287.6	114.5	84.5	1.2	33.5	5.4	43.3	39.6	3.7	5.1	-38.3	5.9	-44.2
1974	319.8	123.2	103.3	1.3	34.9	6.0	48.0	43.9	4.0	3.2	-41.3	3.2	-44.5
1975	374.8	133.9	132.3	2.0	43.6	6.1	52.5	48.1	4.3	4.3	-97.9	-16.3	-81.6
1976	403.5	140.0	143.5	2.5	49.1	4.5	58.9	54.6	4.3	4.9	-80.9	-16.0	-64.9
1977	437.3	151.1	152.4	2.6	54.8	4.2	65.3	60.0	5.2	6.9	-73.4	-14.1	-59.4
1978	485.9	164.2	162.7	2.7	63.5	5.0	79.1	70.9	8.2	8.7	-62.0	-4.7	-57.3
1979	534.4	178.8	183.0	3.0	64.0	5.8	91.5	80.9	10.6	8.2	-47.4	-0.1	-47.3
1980	622.5	204.6	220.3	3.5	69.7	7.3	107.5	95.7	11.9	9.4	-88.8	-15.8	-73.0
1981	709.1	233.6	250.7	4.3	69.4	6.7	133.4	117.2	16.3	11.1	-88.1	-14.3	-73.8
1982	786.0	258.4	281.9	4.1	66.3	8.1	152.7	134.7	18.0	14.6	-167.4	-34.4	-132.9
1983	851.9	280.0	303.6	3.6	67.9	8.9	167.0	148.9	18.1	20.9	-207.2	-37.9	-169.3
1984	907.7	296.2	309.7	3.7	72.3	11.2	193.8	173.7	20.2	20.7	-196.5	-9.4	-187.1
1985	975.0	320.4	325.9	4.0	76.2	13.8	213.7	191.6	22.2	21.0	-199.2	0.8	-200.1
1986	1 033.8	339.8	344.3	4.4	82.4	14.2	224.2	200.9	23.3	24.6	-215.9	5.6	-221.5
1987	1 065.2	350.3	357.2	4.3	78.4	12.7	232.3	208.2	24.1	30.0	-165.7	14.4	-180.2
1988	1 122.4	363.5	378.4	4.6	85.7	13.2	247.9	219.2	28.8	29.2	-160.0	39.0	-198.9
1989	1 201.8	382.9	411.7	5.1	91.8	13.5	269.7	235.1	34.5	27.1	-159.4	42.6	-201.9
1990	1 290.9	404.2	447.0	6.2	104.4	13.5	289.0	253.0	36.1	26.6	-203.3	42.3	-245.6
1991	1 356.2	426.1	494.0	6.3	124.0	-25.6	304.4	268.3	36.1	27.1	-248.4	24.6	-273.0
1992	1 488.9	432.1	551.8	6.2	141.7	20.6	306.9	272.6	34.3	29.7	-334.5	5.3	-339.8
1993	1 544.6	429.5	583.7	6.2	155.7	21.7	311.4	277.3	34.2	36.3	-313.5	2.8	-316.3
1994	1 585.0	429.3	609.0	6.8	166.8	19.9	320.8	284.5	36.4	32.2	-255.6	15.3	-270.9
1995	1 659.5	429.0	647.1	6.8	174.5	15.4	352.3	305.2	47.1	34.5	-242.1	15.5	-257.6
1996	1 715.7	429.4	682.1	7.6	181.5	20.4	360.0	303.2	56.8	34.9	-179.4	12.8	-192.2
1997	1 759.4	439.6	707.9	7.9	188.1	17.0	365.6	295.9	69.7	33.4	-92.0	28.8	-120.8
1998	1 788.4	437.8	722.1	8.2	200.8	18.0	365.6	293.5	72.1	35.9	1.4	58.2	-56.8
1999	1 837.5	456.5	739.9	8.7	219.2	16.3	352.2	285.4	66.7	44.8	69.1	92.2	-23.1
2000	1 908.7	476.3	773.4	8.9	233.1	17.9	353.7	289.3	64.4	45.3	159.7	114.0	45.7
2001	2 017.2	506.2	840.8	9.8	261.3	16.9	331.2	274.1	57.1	51.1	15.0	87.7	-72.6
2002	2 138.6	560.8	918.0	9.9	288.7	19.4	301.3	252.0	49.3	40.5	-267.8	48.8	-316.6
2003	2 293.0	631.1	967.5	10.4	321.7	21.7	291.7	244.3	47.4	49.0	-397.4	37.5	-434.9
2004	2 421.2	683.3	1 019.5	11.0	332.3	23.0	306.0	251.2	54.8	46.0	-393.5	50.8	-444.3
2005	2 598.1	725.4	1 084.4	11.6	343.5	28.3	344.4	276.4	68.0	60.5	-293.8	48.0	-341.7
2006	2 760.2	765.1	1 189.1	12.4	341.0	29.3	372.2	286.3	85.9	51.1	-221.9	11.8	-233.7
2007	2 927.5	800.5	1 263.3	13.3	359.1	35.9	408.0	308.9	99.0	47.5	-259.7	-3.2	-256.5
2008	3 205.6	878.9	1 463.2	15.4	371.2	38.9	388.4	291.3	97.0	49.6	-624.9	-53.6	-571.3
2009	3 482.6	935.6	1 614.5	16.0	458.1	46.9	354.5	265.7	88.8	56.9	-1 243.2	-251.7	-991.5
2010	3 762.4	1 000.7	1 757.5	16.5	505.2	46.8	381.5	288.9	92.6	54.2	-1 318.4	-290.6	-1 027.0
2011	3 806.9	1 003.3	1 779.5	17.0	472.5	49.8	425.4	328.0	97.5	59.5	-1 234.1	-271.5	-962.7
2012	3 773.0	999.3	1 781.8	18.0	444.4	49.3	422.6	326.7	95.9	57.6	-1 072.7	-287.7	-785.0
2013	3 770.8	956.9	1 821.5	18.9	450.1	47.7	416.3	319.8	96.5	59.2	-631.8	-288.6	-343.1
2014	3 889.4	951.2	1 881.1	19.5	495.0	45.8	439.1	345.5	93.6	57.6	-597.4	-287.2	-310.1
2015	4 008.3	954.2	1 969.9	20.4	533.1	44.8	429.3	334.3	95.0	56.7	-560.2	-310.9	-249.4
2016	4 132.5	966.6	2 024.5	20.9	556.8	48.2	454.1	348.6	105.5	61.2	-669.1	-330.4	-338.7
2017	4 246.8	985.1	2 098.8	21.8	559.8	46.0	475.9	359.4	116.6	59.3	-722.4	-330.6	-391.8
2018	4 499.3	1 043.5	2 195.7	22.8	582.6	51.5	540.5	403.1	137.4	62.7	-931.7	-358.2	-573.5
2019	4 758.1	1 097.3	2 323.5	24.0	608.1	50.4	581.6	434.6	147.0	73.3	-1 047.0	-411.8	-635.2
2017													
1st quarter	4 195.3	973.9	2 075.1	21.5	561.0	45.3	459.6	350.8	108.9	58.9	-671.5	-329.4	-342.2
2nd quarter	4 202.2	980.0	2 089.0	21.7	542.3	41.7	470.0	357.2	112.8	57.6	-698.1	-330.5	-367.6
3rd quarter	4 261.9	984.4	2 106.6	21.9	562.9	44.1	480.6	361.4	119.2	61.3	-731.0	-331.9	-399.0
4th quarter	4 327.7	1 002.3	2 124.3	22.2	572.9	53.0	493.5	368.2	125.3	59.4	-789.1	-330.7	-458.4
2018													
1st quarter	4 409.6	1 021.4	2 173.4	22.4	581.5	44.5	508.8	377.2	131.5	57.6	-917.2	-345.5	-571.7
2nd quarter	4 469.8	1 038.3	2 186.7	22.8	578.0	53.7	533.0	397.1	135.9	57.2	-942.3	-354.5	-587.8
3rd quarter	4 524.8	1 054.1	2 202.1	23.0	584.3	52.1	551.8	412.0	139.8	57.3	-907.2	-359.4	-547.8
4th quarter	4 593.0	1 060.1	2 220.5	23.0	586.5	55.5	568.6	426.3	142.3	78.9	-960.0	-373.3	-586.8
2019													
1st quarter	4 690.1	1 076.0	2 298.8	23.6	594.2	52.3	574.5	429.1	145.4	70.7	-1 016.0	-392.8	-623.2
2nd quarter	4 737.5	1 094.9	2 315.8	23.8	612.5	46.3	583.6	436.7	146.9	60.5	-1 033.0	-407.3	-625.7
3rd quarter	4 786.4	1 104.6	2 331.4	24.3	610.3	50.5	583.9	436.2	147.7	81.4	-1 084.1	-421.8	-662.2
4th quarter	4 818.6	1 113.7	2 347.7	24.4	615.4	52.3	584.5	436.5	148.0	80.5	-1 054.9	-425.3	-629.6

Table 6-2. Federal Government Consumption Expenditures and Gross Investment

(National income and product accounts, calendar years, billions of dollars, quarterly data are at seasonally adjusted annual rates.)

NIPA Tables 3.9.5, 3.10.5

Year and quarter	Total	Federal government consumption expenditures and gross investment											
		Consumption expenditures					Gross investment						
		Total	Compensation of general government employees	Consumption of general government fixed capital	Intermediate goods and services purchased [1]	Less: Own-account investment and sales to other sectors	Total	National defense			Nondefense		
								Structures	Equipment	Intellectual property	Structures	Equipment	Intellectual property
1960	72.9	51.0	26.6	15.0	12.0	2.7	21.9	2.2	10.1	5.9	1.7	0.3	1.7
1961	77.4	52.7	27.7	15.9	11.9	2.8	24.6	2.4	11.5	6.5	1.9	0.3	2.1
1962	85.5	59.0	29.5	17.0	15.6	3.1	26.5	2.0	12.5	6.7	2.1	0.3	2.9
1963	87.9	61.2	30.8	18.2	15.6	3.4	26.7	1.6	10.9	6.9	2.3	0.4	4.5
1964	90.3	63.1	33.0	19.0	14.9	3.8	27.1	1.3	10.1	6.9	2.5	0.6	5.7
1965	93.2	66.5	34.8	19.8	16.0	4.2	26.7	1.1	8.9	6.7	2.8	0.5	6.8
1966	106.6	76.5	39.8	20.8	20.6	4.7	30.1	1.3	10.4	7.0	2.8	0.6	7.9
1967	120.0	88.1	43.8	22.2	26.5	4.5	31.9	1.2	12.2	7.9	2.2	0.6	7.8
1968	128.0	96.8	48.3	23.9	29.3	4.6	31.1	1.2	10.8	7.9	2.1	0.5	8.6
1969	131.2	100.5	51.4	25.5	28.7	5.1	30.7	1.5	9.7	8.2	1.9	0.6	8.8
1970	132.8	102.3	55.2	27.3	25.3	5.4	30.5	1.3	9.6	8.1	2.1	0.7	8.8
1971	134.5	106.7	58.8	28.7	25.0	5.8	27.8	1.8	5.5	8.4	2.5	0.8	8.9
1972	141.6	112.3	62.6	29.7	26.9	6.9	29.3	1.8	5.3	9.2	2.7	1.1	9.2
1973	146.2	114.5	64.6	31.8	25.8	7.7	31.7	2.1	6.0	9.7	3.1	1.1	9.6
1974	158.8	123.2	67.8	34.7	29.2	8.6	35.7	2.2	8.2	10.0	3.4	1.5	10.4
1975	173.7	133.9	72.5	38.0	31.8	8.4	39.9	2.3	9.9	10.5	4.1	1.7	11.5
1976	184.8	140.0	76.5	40.8	31.3	8.5	44.8	2.1	12.1	11.3	4.6	1.8	12.7
1977	200.3	151.1	80.9	44.3	35.6	9.7	49.2	2.4	13.3	12.3	5.0	2.3	14.0
1978	218.9	164.2	87.1	48.6	39.3	10.8	54.7	2.5	14.3	13.3	6.1	2.7	15.9
1979	240.6	178.8	92.0	53.4	45.8	12.4	61.7	2.5	17.5	14.9	6.3	2.7	17.9
1980	274.9	204.6	100.8	59.1	58.4	13.6	70.2	3.2	19.8	17.2	7.1	3.2	19.8
1981	314.0	233.6	112.2	66.0	69.6	14.3	80.4	3.2	24.1	20.9	7.7	3.3	21.2
1982	348.3	258.4	122.0	73.7	77.6	14.9	90.0	4.0	29.1	25.2	6.8	3.9	21.1
1983	382.4	280.0	127.7	79.4	88.8	15.9	102.3	4.8	35.2	29.8	6.7	4.3	21.6
1984	411.8	296.2	136.5	88.7	88.3	17.4	115.6	4.9	41.1	35.0	7.0	4.7	22.8
1985	452.9	320.4	145.0	95.0	99.1	18.7	132.5	6.2	47.9	41.3	7.3	5.1	24.6
1986	481.7	339.8	149.1	101.5	108.7	19.6	141.9	6.8	52.7	43.9	8.0	4.8	25.7
1987	502.8	350.3	153.1	108.0	110.2	20.9	152.4	7.7	55.0	48.1	9.0	5.2	27.4
1988	511.4	363.5	161.9	116.6	107.4	22.5	148.0	7.4	49.6	48.5	6.8	6.1	29.6
1989	534.1	382.9	168.5	124.3	114.0	23.9	151.2	6.4	51.5	47.3	6.9	7.2	31.9
1990	562.4	404.2	176.0	130.7	123.1	25.6	158.2	6.1	54.6	46.6	8.0	8.1	34.8
1991	582.9	426.1	186.5	137.0	129.4	26.8	156.8	4.6	53.4	43.3	9.2	8.9	37.4
1992	588.5	432.1	189.6	140.0	129.4	26.9	156.4	5.2	50.7	41.3	10.3	10.0	38.9
1993	580.2	429.5	188.0	143.4	125.1	27.2	150.7	5.3	44.7	39.2	11.2	9.9	40.4
1994	574.7	429.3	186.1	146.6	126.1	29.3	145.4	5.8	41.9	38.1	10.2	7.5	41.9
1995	576.7	429.0	183.3	149.6	123.2	27.3	147.7	6.7	39.8	38.4	10.8	8.6	43.4
1996	579.2	429.4	182.2	149.6	125.2	27.6	149.8	6.3	39.9	38.9	11.3	9.7	43.8
1997	583.3	439.6	183.5	150.1	131.6	25.6	143.7	6.1	33.6	39.2	9.9	10.0	44.9
1998	585.5	437.8	185.8	150.9	127.7	26.6	147.7	5.8	34.1	39.4	10.8	10.2	47.5
1999	611.3	456.5	191.7	153.3	137.1	25.6	154.9	5.4	36.8	39.3	10.7	11.9	50.7
2000	633.7	476.3	202.4	157.9	144.2	28.2	157.4	5.4	37.8	40.2	8.4	10.8	54.8
2001	670.1	506.2	211.2	159.1	165.5	29.6	163.8	5.3	39.9	42.5	8.1	10.0	58.1
2002	743.0	560.8	236.3	161.5	194.4	31.5	182.2	6.1	46.4	47.9	10.0	11.9	59.8
2003	826.3	631.1	267.7	167.0	229.9	33.5	195.2	7.1	50.0	53.0	10.3	12.3	62.5
2004	891.7	683.3	284.3	175.1	258.6	34.6	208.4	6.9	55.7	58.3	9.6	13.3	64.5
2005	947.5	725.4	303.5	184.9	276.3	39.2	222.0	7.2	61.4	62.9	8.1	15.2	67.3
2006	1 000.7	765.1	314.5	195.4	295.2	40.0	235.6	7.8	66.0	65.7	9.4	17.4	69.2
2007	1 050.5	800.5	329.1	206.7	304.8	40.1	250.0	10.0	71.8	69.4	11.4	16.1	71.4
2008	1 150.6	878.9	349.6	219.5	352.2	42.3	271.7	13.7	82.6	70.8	11.4	17.5	75.7
2009	1 218.2	935.6	375.8	227.2	376.2	43.6	282.6	17.1	86.6	69.6	12.0	17.8	79.5
2010	1 297.9	1 000.7	402.8	237.2	407.1	46.5	297.2	16.7	89.3	70.2	16.0	19.2	85.9
2011	1 298.9	1 003.3	411.6	248.1	394.4	50.9	295.7	13.3	87.3	71.3	16.7	18.4	88.5
2012	1 286.5	999.3	409.1	254.6	382.5	47.0	287.3	8.1	85.2	70.5	14.5	18.7	90.2
2013	1 226.6	956.9	399.8	258.6	345.5	46.9	269.6	6.5	78.8	67.7	11.6	16.2	88.8
2014	* 1 215.0	951.2	404.2	262.7	333.4	49.0	263.8	5.3	74.7	64.8	12.0	17.4	89.6
2015	1 220.8	954.2	409.9	263.7	330.0	49.4	266.6	5.0	72.6	65.3	12.7	18.8	92.2
2016	1 234.7	966.6	420.7	263.8	333.6	51.5	268.1	4.8	72.3	61.6	12.3	18.9	98.1
2017	1 263.9	985.1	431.5	268.8	337.7	52.8	278.7	5.6	77.1	62.4	10.1	19.1	104.5
2018	1 339.4	1 043.5	449.7	277.0	369.7	52.9	295.9	6.0	84.1	67.9	12.1	20.1	105.7
2019	1 419.2	1 097.3	465.5	285.6	402.3	56.2	321.9	7.4	93.4	75.1	14.7	20.2	111.1
2017													
1st quarter	1 246.5	973.9	427.4	266.7	330.9	51.2	272.6	5.4	73.2	61.2	10.8	19.0	103.0
2nd quarter	1 257.9	980.0	428.8	267.8	336.4	53.0	277.9	6.2	76.6	61.8	9.9	18.8	104.6
3rd quarter	1 262.7	984.4	432.5	269.3	335.6	53.0	278.3	5.1	77.1	62.6	9.3	19.1	105.1
4th quarter	1 288.3	1 002.3	437.2	271.2	347.8	53.9	286.0	5.6	81.4	63.8	10.6	19.3	105.2
2018													
1st quarter	1 308.1	1 021.4	442.5	273.8	356.5	51.3	286.6	5.4	80.7	65.4	10.3	20.0	104.8
2nd quarter	1 329.3	1 038.3	447.3	276.3	367.7	52.9	291.0	5.7	81.0	67.1	11.8	20.2	105.2
3rd quarter	1 352.0	1 054.1	452.6	278.3	375.4	52.1	297.9	6.4	83.6	68.7	13.0	20.4	105.7
4th quarter	1 368.4	1 060.1	456.5	279.9	379.4	55.6	308.3	6.5	90.9	70.5	13.3	20.0	107.1
2019													
1st quarter	1 388.8	1 076.0	460.0	283.0	386.5	53.6	312.8	7.4	91.4	72.2	14.1	19.6	108.2
2nd quarter	1 410.6	1 094.9	463.8	283.9	403.1	55.8	315.7	7.3	89.3	74.0	14.6	20.4	110.1
3rd quarter	1 429.3	1 104.6	467.8	286.5	406.3	56.1	324.7	7.2	94.0	76.0	15.1	20.3	112.1
4th quarter	1 447.9	1 113.7	470.5	289.0	413.5	59.2	334.3	7.9	98.9	78.0	15.1	20.5	113.8

[1]Includes general government intermediate inputs for goods and services sold to other sectors and for own-account investment.

Table 6-3. Federal Government Defense and Nondefense Consumption Expenditures by Type

(National income and product accounts, calendar years, billions of dollars, quarterly data are at seasonally adjusted annual rates.)

NIPA Table 3.10.5

Year and quarter	Defense consumption expenditures [1]						Nondefense consumption expenditures [1]					
	Total	Compensation of general government employees	Consumption of general government fixed capital	Intermediate goods and services purchased [2]			Total	Compensation of general government employees	Consumption of general government fixed capital	Intermediate goods and services purchased [2]		
				Durable goods	Nondurable goods	Services				Durable goods	Nondurable goods	Services
1960	42.6	21.1	13.8	4.4	1.7	3.5	8.4	5.6	1.3	0.0	0.5	1.9
1961	44.2	21.7	14.5	3.6	2.2	4.0	8.6	6.1	1.4	0.1	-0.3	2.3
1962	48.6	22.9	15.4	4.6	2.8	4.7	10.5	6.5	1.6	0.1	1.1	2.4
1963	50.5	23.6	16.3	4.7	2.5	5.2	10.7	7.2	1.9	0.1	0.6	2.6
1964	51.4	25.2	16.7	4.0	2.8	4.7	11.7	7.8	2.3	0.1	0.3	3.0
1965	53.9	26.4	17.1	4.2	3.2	5.1	12.6	8.3	2.8	0.1	0.4	3.1
1966	63.8	30.8	17.5	6.1	4.7	7.0	12.7	9.0	3.3	0.1	-0.7	3.3
1967	73.7	34.2	18.4	6.2	7.3	9.7	14.3	9.7	3.9	0.1	0.1	3.1
1968	81.4	37.6	19.4	7.4	8.4	10.5	15.4	10.7	4.5	0.1	0.8	2.0
1969	82.6	40.0	20.3	6.4	7.6	10.5	17.9	11.5	5.2	0.1	1.4	2.6
1970	81.8	41.9	21.3	6.1	5.4	9.6	20.6	13.3	6.0	0.1	0.5	3.6
1971	82.3	43.4	21.9	4.6	4.4	10.6	24.3	15.5	6.7	0.1	0.9	4.4
1972	84.4	45.6	22.4	5.5	4.7	9.5	27.9	17.0	7.4	0.2	1.2	5.8
1973	84.9	46.1	23.6	5.3	4.3	9.3	29.7	18.4	8.2	0.2	1.0	5.8
1974	89.5	47.7	25.3	5.0	5.2	10.7	33.6	20.1	9.4	0.3	1.4	6.7
1975	95.3	50.2	27.3	5.8	5.1	10.9	38.5	22.3	10.6	0.3	2.1	7.6
1976	99.3	51.8	29.2	5.7	4.4	11.7	40.7	24.7	11.6	0.3	2.1	7.1
1977	106.5	53.9	31.5	7.6	4.5	12.7	44.7	27.0	12.8	0.4	2.5	7.8
1978	115.1	57.6	34.4	9.0	4.9	13.3	49.0	29.5	14.3	0.5	3.2	8.4
1979	125.7	60.9	37.3	10.8	6.2	14.9	53.1	31.1	16.1	0.6	3.3	10.0
1980	143.2	66.7	40.6	12.3	10.0	18.7	61.4	34.0	18.5	0.8	5.5	11.2
1981	165.1	76.3	45.0	15.7	11.8	22.0	68.4	35.9	21.1	0.8	8.7	10.5
1982	187.2	84.8	50.4	19.1	11.5	28.1	71.2	37.2	23.3	0.8	7.5	10.7
1983	202.3	89.3	55.2	24.6	11.3	28.9	77.7	38.4	24.2	0.8	10.1	13.1
1984	217.7	95.9	62.3	26.6	10.4	30.7	78.5	40.7	26.4	0.9	5.8	13.9
1985	234.1	102.4	67.2	29.0	9.9	34.7	86.3	42.6	27.8	0.9	8.9	15.7
1986	249.0	106.2	72.6	31.7	10.2	38.3	90.8	42.9	28.9	0.9	11.7	16.1
1987	261.6	109.5	77.7	33.6	10.2	40.9	88.7	43.6	30.2	1.0	6.2	18.3
1988	276.6	113.8	84.4	33.4	10.6	45.4	86.8	48.1	32.2	1.1	-0.8	17.7
1989	286.1	117.9	90.1	32.0	10.8	46.5	96.8	50.6	34.2	1.2	5.1	18.2
1990	297.7	121.2	94.5	31.6	11.0	51.8	106.5	54.8	36.2	1.4	5.0	22.2
1991	312.7	126.8	98.5	31.0	10.7	58.1	113.4	59.7	38.6	1.5	5.9	22.3
1992	309.3	126.8	99.8	28.4	9.4	57.6	122.8	62.8	40.2	1.6	6.5	25.9
1993	302.4	121.0	101.3	26.4	8.5	58.1	127.1	67.1	42.1	1.6	7.6	23.0
1994	296.3	117.4	102.7	22.9	7.6	60.2	133.0	68.7	43.9	1.5	6.1	27.7
1995	292.2	114.1	103.3	20.9	6.3	59.8	136.8	69.2	46.4	1.4	7.5	27.3
1996	292.3	112.0	101.7	20.8	7.6	62.5	137.1	70.2	47.9	1.5	6.6	26.0
1997	293.0	111.4	100.4	20.9	7.6	64.6	146.6	72.0	49.7	1.6	8.5	28.5
1998	289.5	110.8	99.7	20.9	7.0	62.6	148.3	75.1	51.3	1.5	9.4	26.3
1999	301.8	112.7	99.4	22.2	8.2	70.9	154.7	79.0	53.9	1.7	7.2	27.0
2000	309.2	116.9	100.7	22.1	10.4	71.4	167.2	85.5	57.2	1.8	9.3	29.1
2001	325.5	123.2	99.8	22.3	10.3	84.2	180.7	88.0	59.3	1.9	12.2	34.7
2002	358.5	139.3	100.4	23.4	11.4	100.0	202.3	97.0	61.1	2.2	12.6	44.7
2003	411.2	160.8	103.5	26.0	13.5	124.6	219.9	106.9	63.5	2.2	15.4	48.2
2004	448.9	172.2	108.8	28.5	17.1	139.8	234.4	112.1	66.3	2.5	17.1	53.7
2005	478.0	186.0	115.2	29.8	21.0	145.3	247.5	117.5	69.7	2.7	19.0	58.6
2006	501.3	193.0	122.1	32.8	22.3	152.5	263.8	121.5	73.3	3.0	20.3	64.4
2007	528.1	201.5	129.6	37.0	24.0	157.9	272.4	127.6	77.1	3.0	20.2	62.8
2008	583.2	214.3	138.3	43.2	30.3	180.0	295.7	135.2	81.2	3.3	23.6	71.9
2009	614.3	229.2	143.5	47.1	24.5	193.0	321.3	146.6	83.7	3.5	26.6	81.4
2010	651.8	244.0	149.5	47.4	26.8	207.8	348.9	158.8	87.7	3.9	28.4	92.7
2011	662.0	250.8	156.3	46.2	33.0	201.8	341.3	160.8	91.8	3.7	24.8	84.9
2012	650.3	248.0	159.6	45.2	31.9	191.6	348.9	161.1	95.0	3.7	24.0	86.2
2013	611.2	239.6	160.5	39.9	28.2	169.4	345.7	160.2	98.1	3.5	22.6	81.9
2014	598.7	239.1	160.9	36.5	26.0	162.8	352.5	165.0	101.8	3.6	22.0	82.5
2015	587.4	239.0	159.6	35.6	21.0	158.8	368.9	172.2	104.0	3.8	23.4	88.2
2016	588.3	242.5	157.6	35.3	20.6	159.8	380.3	180.0	106.2	4.0	23.7	90.6
2017	600.0	246.6	158.5	37.3	22.8	162.9	392.6	186.3	110.3	4.0	24.5	92.3
2018	637.4	258.7	162.0	39.5	26.4	178.4	419.5	193.0	115.3	4.5	28.2	103.1
2019	584.6	241.2	157.2	35.5	19.3	158.7	376.2	177.0	104.9	4.0	23.7	90.2
2017												
1st quarter	582.4	242.2	157.5	34.2	20.5	155.6	378.9	179.2	105.7	4.0	23.7	90.6
2nd quarter	592.3	243.2	157.7	35.4	21.1	162.6	380.5	181.2	106.5	4.0	23.8	90.7
3rd quarter	593.8	243.5	158.2	36.2	21.6	162.4	385.5	182.6	107.7	4.0	23.7	90.9
4th quarter	590.8	244.4	158.0	36.7	22.9	156.8	387.4	184.3	108.7	4.0	23.8	90.0
2018												
1st quarter	599.8	244.9	158.1	38.2	22.1	164.6	387.4	185.3	109.7	3.9	23.6	89.7
2nd quarter	599.6	247.2	158.5	35.8	22.2	164.0	393.5	186.7	110.8	4.0	24.4	92.4
3rd quarter	609.7	250.0	159.3	38.3	24.0	166.1	402.2	188.7	112.0	4.2	26.0	97.0
4th quarter	618.1	253.5	160.5	37.6	25.5	168.8	415.8	190.5	113.6	4.4	27.7	102.7
2019												
1st quarter	634.9	257.1	161.6	39.9	25.7	178.2	417.7	192.2	114.9	4.4	28.1	102.9
2nd quarter	644.9	260.8	162.8	41.6	26.7	180.5	423.2	193.8	116.0	4.5	28.6	103.9
3rd quarter	651.8	263.3	163.2	39.1	27.7	185.9	421.3	195.3	116.8	4.5	28.4	103.0
4th quarter	...	...	...	...	...	...	...	...	...	...	...	...

[1]Excludes government sales to other sectors and government own-account investment (construction and software).
[2]Includes general government intermediate inputs for goods and services sold to other sectors and for own-account investment.
. . . = Not available.

Table 6-4. National Defense Consumption Expenditures and Gross Investment: Selected Detail

(National income and product accounts, calendar years, billions of dollars, quarterly data are at seasonally adjusted annual rates.)

NIPA Table 3.11.5

Year and quarter	Defense consumption expenditures								Defense gross investment				
	Compensation of general government employees		Intermediate goods and services purchased [1]						Aircraft	Missiles	Ships	Software	Research and development
			Durable goods	Nondurable goods		Services							
	Military	Civilian	Aircraft	Petroleum products	Ammunition	Installation support	Weapons support	Personnel support					
1975	32.0	18.2	2.1	2.9	1.1	4.8	1.7	1.6	3.6	1.3	2.2	0.5	10.0
1976	32.6	19.2	2.0	2.5	0.6	5.3	1.9	1.7	3.4	1.4	2.4	0.5	10.8
1977	33.7	20.2	3.3	2.4	0.8	5.8	1.9	1.9	3.7	1.2	3.1	0.6	11.7
1978	35.5	22.2	3.5	2.5	1.0	5.9	2.1	2.1	3.7	1.1	3.9	0.7	12.6
1979	37.4	23.5	4.8	3.6	1.2	6.6	2.7	2.1	4.8	1.8	4.3	0.8	14.0
1980	41.5	25.2	5.6	6.8	1.4	8.3	3.9	2.3	6.1	2.3	4.1	1.0	16.2
1981	48.8	27.6	7.7	7.7	1.6	9.6	4.8	2.9	7.5	2.8	5.1	1.3	19.6
1982	54.9	29.9	10.2	6.8	2.1	12.6	6.3	3.6	8.4	3.4	6.2	1.4	23.7
1983	57.8	31.5	13.6	6.4	2.5	12.4	7.0	3.9	10.1	4.6	7.1	1.7	28.1
1984	62.0	33.9	14.0	5.9	2.2	13.8	7.2	3.7	10.9	5.7	8.0	2.2	32.8
1985	66.2	36.1	15.4	5.8	1.3	15.1	8.4	5.0	13.4	6.6	9.0	2.7	38.6
1986	69.4	36.8	17.2	3.6	3.6	16.0	9.4	6.2	17.9	7.9	8.9	3.2	40.7
1987	72.2	37.3	18.2	3.9	2.8	17.7	9.2	6.8	17.6	8.7	8.8	3.6	44.5
1988	74.8	39.0	17.9	3.5	3.5	18.3	10.3	9.5	13.5	7.8	8.6	4.3	44.3
1989	76.5	41.4	16.4	4.2	3.1	18.0	10.5	10.2	12.2	8.8	10.0	4.8	42.5
1990	78.5	42.7	14.8	5.3	2.8	21.4	11.6	10.1	12.0	11.2	10.8	5.2	41.4
1991	82.4	44.4	13.6	4.7	2.7	22.9	10.0	9.5	9.2	10.8	10.2	5.4	37.9
1992	80.9	45.9	12.2	3.5	2.6	23.1	9.4	13.7	8.3	10.6	10.1	5.6	35.6
1993	75.1	45.9	10.7	3.2	2.5	25.2	8.8	14.7	9.3	7.9	8.7	5.4	33.7
1994	71.7	45.7	9.2	3.0	1.8	26.2	9.4	16.3	10.5	5.7	8.1	5.3	32.8
1995	69.5	44.6	8.9	2.8	1.2	24.9	9.5	17.0	9.0	4.7	8.0	5.3	33.1
1996	67.7	44.3	8.8	3.4	1.4	25.6	9.0	19.3	9.2	4.1	6.8	5.5	33.4
1997	67.8	43.6	9.4	2.9	1.7	24.7	10.5	21.3	5.8	2.9	6.1	5.6	33.6
1998	68.0	42.8	9.9	2.1	1.9	23.5	10.0	20.6	5.8	3.3	6.4	5.9	33.5
1999	69.8	42.9	10.5	2.6	1.9	24.5	11.3	26.0	7.0	2.9	6.7	5.9	33.3
2000	72.9	44.1	9.8	4.1	1.8	24.7	11.7	26.1	7.8	2.7	6.6	6.0	34.2
2001	78.4	44.8	9.8	4.2	2.1	27.1	14.8	32.4	8.5	3.3	7.2	5.8	36.7
2002	89.7	49.6	9.8	4.6	2.4	30.6	18.0	41.3	9.6	3.3	8.4	5.6	42.3
2003	107.4	53.4	11.4	5.3	2.7	35.8	22.6	49.5	9.2	3.5	9.3	5.7	47.3
2004	113.9	58.3	12.0	7.0	3.6	36.8	23.8	62.4	11.1	4.0	9.8	5.8	52.5
2005	123.3	62.7	10.8	10.1	4.0	36.0	26.5	66.9	13.5	4.0	9.6	6.1	56.8
2006	128.1	64.9	11.1	11.4	4.1	38.3	27.8	71.4	13.6	4.5	10.5	6.4	59.3
2007	133.4	68.1	11.5	12.2	4.2	38.4	29.0	70.7	12.9	4.3	10.3	6.8	62.5
2008	143.0	71.3	12.9	17.6	4.3	41.6	31.3	86.5	13.5	4.3	11.0	7.4	63.4
2009	152.5	76.8	14.9	10.5	4.2	44.2	32.8	93.4	13.8	5.1	11.1	7.5	62.1
2010	159.1	84.9	16.2	13.7	4.2	46.5	37.4	99.3	16.7	5.6	11.8	8.1	62.1
2011	160.9	89.9	18.0	19.2	4.2	44.5	35.6	97.1	20.2	5.1	11.7	8.7	62.6
2012	159.1	88.9	19.3	18.3	4.3	41.9	33.6	97.3	20.2	6.9	12.0	9.0	61.5
2013	153.2	86.3	17.9	15.0	3.6	35.5	28.9	89.9	21.7	6.4	12.5	9.1	58.5
2014	150.0	89.1	16.3	13.3	3.1	39.2	31.1	78.0	19.2	6.6	13.3	9.4	55.4
2015	147.5	91.2	15.8	8.4	2.8	36.5	32.3	75.8	17.6	6.6	13.5	9.4	55.9
2016	148.2	93.8	15.8	6.7	3.1	38.0	33.6	73.8	16.4	4.8	14.4	10.1	51.6
2017	149.1	97.4	16.8	7.9	3.7	39.4	35.6	73.4	18.7	4.7	14.1	10.7	51.7
2018	155.9	102.0	17.1	10.2	4.7	40.2	39.0	82.5	19.7	4.5	15.5	11.7	56.2
2019	163.1	105.5	17.9	9.1	5.6	40.1	43.3	102.0	23.3	5.1	17.2	12.6	62.5
2013													
1st quarter	154.3	88.0	17.8	16.3	3.8	36.4	28.9	94.7	21.9	5.2	11.6	9.4	59.9
2nd quarter	153.6	87.4	18.0	15.1	3.7	35.6	29.3	94.7	19.4	6.5	12.6	9.0	59.1
3rd quarter	152.9	81.8	18.4	14.3	3.5	34.3	28.2	89.5	21.1	7.9	12.8	9.0	58.1
4th quarter	152.2	88.1	17.5	14.1	3.4	35.9	29.2	80.7	24.3	6.2	13.2	9.1	57.0
2014													
1st quarter	151.3	87.6	17.2	14.7	3.3	37.8	30.3	79.1	17.3	6.1	12.4	9.2	55.9
2nd quarter	150.7	88.7	16.7	13.5	3.2	39.2	30.4	74.1	20.6	6.4	14.0	9.3	55.2
3rd quarter	149.6	89.6	15.2	13.2	3.0	40.8	33.0	87.2	19.3	6.6	13.6	9.5	55.1
4th quarter	148.5	90.6	16.1	11.7	2.9	39.0	30.9	71.4	19.5	7.4	13.0	9.4	55.4
2015													
1st quarter	147.5	90.6	16.0	9.1	2.8	37.3	31.7	77.4	15.6	6.5	13.8	9.4	56.3
2nd quarter	147.2	90.8	15.3	8.9	2.8	37.0	32.8	76.8	19.2	6.3	13.1	9.4	56.5
3rd quarter	147.4	91.4	15.7	8.1	2.8	35.5	32.1	74.5	18.1	5.8	13.6	9.4	56.0
4th quarter	147.9	92.0	16.2	7.4	2.9	36.2	32.8	74.3	17.6	7.7	13.6	9.5	54.9
2016													
1st quarter	148.2	92.4	16.3	6.1	3.0	37.0	33.5	74.2	18.7	5.1	14.4	9.8	53.0
2nd quarter	148.3	93.3	14.7	6.5	3.0	37.6	32.6	70.4	16.6	5.6	13.9	10.0	51.7
3rd quarter	148.3	94.3	15.8	7.0	3.1	38.3	34.2	75.1	15.2	5.0	14.7	10.1	50.9
4th quarter	147.9	95.2	16.3	7.2	3.2	39.0	34.0	75.3	15.3	3.7	14.5	10.3	50.6
2017													
1st quarter	148.1	96.2	16.8	8.0	3.4	38.5	33.5	70.8	17.6	5.1	12.4	10.4	50.7
2nd quarter	148.2	96.7	18.3	7.3	3.6	39.3	35.1	75.1	18.4	4.8	14.5	10.6	51.2
3rd quarter	149.2	97.8	15.6	7.6	3.8	39.9	36.5	72.2	18.2	4.7	14.3	10.7	51.9
4th quarter	150.7	99.0	16.5	8.8	4.0	39.9	37.2	75.3	20.5	4.2	15.2	10.9	52.9
2018													
1st quarter	152.9	100.2	15.5	10.3	4.3	39.7	37.7	76.0	17.8	4.7	15.5	11.3	54.1
2nd quarter	154.7	101.6	17.6	10.2	4.5	40.6	38.9	80.6	17.4	4.1	15.9	11.7	55.5
3rd quarter	157.2	102.7	19.4	10.4	4.8	40.2	39.2	83.8	19.7	4.1	15.0	11.9	56.9
4th quarter	159.0	103.6	16.0	9.9	5.0	40.4	40.3	89.5	24.1	5.3	15.7	12.1	58.3
2019													
1st quarter	160.9	104.1	17.0	8.8	5.3	40.0	41.5	96.3	23.8	4.8	16.4	12.3	59.9
2nd quarter	162.5	105.1	17.4	9.3	5.5	40.2	42.8	102.6	20.9	4.0	16.7	12.4	61.6
3rd quarter	164.1	105.6	17.7	8.9	5.7	40.3	43.9	104.3	22.9	4.9	17.7	12.7	63.3
4th quarter	164.7	107.4	19.7	9.3	5.9	40.1	45.2	104.9	25.8	6.7	17.8	12.8	65.2

[1] Includes general government intermediate inputs for goods and services sold to other sectors and for own-account investment.

Table 6-5. Federal Government Output, Lending and Borrowing, and Net Investment

(National income and product accounts, calendar years, billions of dollars, quarterly data are at seasonally adjusted annual rates.)

NIPA Tables 3.2, 3.10.5

Year and quarter	Output						Net lending (net borrowing -)							Net investment
	Gross		Value added		Intermediate goods and services purchased [1]		Net saving, current (surplus +, deficit -)	Plus: capital transfer receipts	Minus			Plus: Consumption of fixed capital	Equals: Net lending (borrowing -)	
	Defense	Non-defense	Defense	Non-defense	Defense	Non-defense			Gross investment	Capital transfer payments	Net purchases of non-produced assets			
1960	44.4	9.3	34.8	6.8	9.6	2.4	0.2	1.8	21.9	2.6	0.5	15.2	-7.9	6.7
1961	46.0	9.6	36.2	7.5	9.8	2.1	-4.7	2.0	24.6	2.9	0.5	16.0	-14.7	8.6
1962	50.4	11.7	38.3	8.2	12.1	3.5	-5.4	2.1	26.5	3.1	0.6	17.2	-16.3	9.3
1963	52.3	12.3	39.9	9.1	12.4	3.2	-2.2	2.2	26.7	3.6	0.5	18.4	-12.3	8.3
1964	53.3	13.6	41.9	10.1	11.4	3.5	-6.9	2.6	27.1	4.1	0.6	19.2	-16.9	7.9
1965	55.9	14.7	43.5	11.1	12.4	3.6	-5.5	2.8	26.7	4.0	0.5	20.0	-13.9	6.7
1966	66.1	15.1	48.3	12.4	17.9	2.7	-7.0	3.0	30.1	4.4	0.6	21.0	-18.0	9.1
1967	75.7	16.8	52.5	13.5	23.2	3.3	-19.5	3.1	31.9	4.3	-0.2	22.5	-30.0	9.4
1968	83.4	18.0	57.0	15.1	26.4	2.9	-13.8	3.1	31.1	6.0	-0.9	24.2	-22.7	6.9
1969	84.9	20.7	60.3	16.6	24.6	4.1	-5.1	3.6	30.7	5.9	0.1	25.8	-12.5	4.9
1970	84.2	23.5	63.2	19.3	21.0	4.3	-34.8	3.7	30.5	5.3	-0.3	27.6	-39.0	2.9
1971	84.9	27.6	65.3	22.2	19.6	5.4	-50.9	4.6	27.8	5.9	-0.4	29.1	-50.6	-1.3
1972	87.6	31.6	68.0	24.4	19.7	7.2	-49.0	5.4	29.3	6.1	-0.7	30.2	-48.1	-0.9
1973	88.6	33.6	69.7	26.6	18.9	6.9	-38.3	5.1	31.7	6.0	-3.2	32.4	-35.3	-0.7
1974	93.8	37.8	73.0	29.5	20.8	8.3	-41.3	4.8	35.7	7.9	-5.7	35.4	-39.1	0.3
1975	99.4	42.9	77.6	32.9	21.8	10.0	-97.9	4.9	39.9	9.7	-0.4	38.7	-103.4	1.2
1976	102.8	45.8	81.0	36.2	21.8	9.5	-80.9	5.6	44.8	10.6	-2.4	41.7	-86.6	3.1
1977	110.3	50.5	85.4	39.8	24.8	10.8	-73.4	7.2	49.2	11.2	-1.4	45.3	-80.0	3.9
1978	119.2	55.8	92.0	43.8	27.2	12.1	-62.0	5.2	54.7	12.0	-0.6	49.7	-73.2	5.0
1979	130.1	61.1	98.2	47.2	31.9	13.9	-47.4	5.5	61.7	14.5	-2.8	54.6	-60.7	7.1
1980	148.2	70.0	107.3	52.5	40.9	17.5	-88.8	6.5	70.2	16.7	-4.1	60.4	-104.8	9.8
1981	170.9	77.0	121.3	57.0	49.6	20.1	-88.1	6.9	80.4	15.7	-5.5	67.5	-104.3	12.9
1982	193.9	79.5	135.2	60.5	58.7	19.0	-167.4	7.5	90.0	14.7	-3.7	75.4	-185.4	14.6
1983	209.3	86.6	144.4	62.6	64.8	24.0	-207.2	5.8	102.3	15.6	-4.9	81.1	-233.3	21.2
1984	225.9	87.6	158.2	67.1	67.7	20.5	-196.5	6.0	115.6	17.8	-4.0	90.6	-229.4	25.0
1985	243.2	95.8	169.6	70.4	73.6	25.4	-199.2	6.4	132.5	19.6	-1.2	97.0	-246.8	35.5
1986	258.9	100.4	178.8	71.8	80.1	28.6	-215.9	7.0	141.9	20.1	-3.0	103.6	-264.3	38.3
1987	272.0	99.3	187.2	73.8	84.8	25.5	-165.7	7.2	152.4	19.1	-0.4	110.1	-219.5	42.3
1988	287.6	98.3	198.2	80.3	89.4	18.0	-160.0	7.6	148.0	19.8	-0.2	118.9	-201.0	29.1
1989	297.4	109.3	208.0	84.8	89.4	24.5	-159.4	8.9	151.2	20.2	-0.7	126.9	-194.3	24.3
1990	310.1	119.6	215.7	91.0	94.5	28.6	-203.3	11.6	158.2	28.2	-0.8	133.5	-243.8	24.7
1991	324.9	128.0	225.2	98.3	99.7	29.7	-248.4	11.0	156.8	26.5	0.1	140.1	-280.8	16.7
1992	322.0	137.0	226.6	103.0	95.4	34.0	-334.5	11.3	156.4	22.6	0.2	143.3	-359.2	13.1
1993	315.3	141.3	222.3	109.2	92.9	32.2	-313.5	12.9	150.7	24.3	0.2	147.0	-328.7	3.7
1994	310.8	147.9	220.0	112.6	90.7	35.3	-255.6	15.1	145.4	26.0	0.1	150.3	-261.7	-4.9
1995	304.4	151.8	217.4	115.6	87.0	36.2	-242.1	14.9	147.7	27.9	-7.9	153.5	-241.3	-5.8
1996	304.7	152.2	213.7	118.1	91.0	34.2	-179.4	17.5	149.8	28.4	-4.8	153.7	-181.7	-3.9
1997	304.9	160.3	211.8	121.7	93.0	38.6	-92.0	20.6	143.7	29.2	-8.8	154.3	-81.2	-10.6
1998	301.0	163.5	210.4	126.4	90.5	37.2	1.4	25.2	147.7	28.9	-6.0	155.4	11.4	-7.7
1999	313.3	168.8	212.0	133.0	101.3	35.8	69.1	28.8	154.9	38.8	-1.2	158.2	63.6	-3.3
2000	321.6	183.0	217.6	142.7	104.0	40.3	159.7	28.1	157.4	41.2	-0.6	163.1	152.9	-5.7
2001	339.8	196.0	223.0	147.3	116.8	48.8	15.0	28.0	163.8	43.6	-1.5	164.4	1.5	-0.6
2002	374.6	217.7	239.7	158.1	134.9	59.6	-267.8	25.3	182.2	53.4	-0.3	166.8	-310.9	15.4
2003	428.3	236.3	264.3	170.4	164.1	65.9	-397.4	22.0	195.2	69.8	-0.8	172.2	-467.2	23.0
2004	466.3	251.6	281.0	178.4	185.4	73.3	-393.5	24.6	208.4	71.4	-0.6	180.5	-467.5	27.9
2005	497.2	267.5	301.2	187.2	196.0	80.3	-293.8	25.0	222.0	97.5	-1.8	190.5	-396.0	31.5
2006	522.6	282.5	315.1	194.8	207.5	87.7	-221.9	27.8	235.6	76.7	-14.2	201.3	-290.9	34.3
2007	550.0	290.6	331.1	204.6	218.9	86.0	-259.7	26.5	250.0	85.9	-3.3	212.8	-353.0	37.2
2008	606.1	315.2	352.6	216.4	253.5	98.7	-624.9	28.3	271.7	152.1	-20.4	225.9	-774.2	45.8
2009	637.3	341.8	372.7	230.3	264.6	111.5	-1 243.2	20.6	282.6	212.6	-8.9	233.6	-1 475.3	49.0
2010	675.6	371.5	393.5	246.5	282.1	125.0	-1 318.4	15.1	297.2	148.1	-1.0	243.7	-1 504.0	53.5
2011	688.1	366.0	407.1	252.6	281.0	113.4	-1 234.1	9.6	295.7	131.2	-0.9	254.9	-1 395.6	40.8
2012	676.3	369.9	407.6	256.1	268.7	113.9	-1 072.7	14.1	287.3	105.0	-2.0	261.6	-1 187.3	25.7
2013	637.6	366.3	400.1	258.3	237.5	108.0	-631.8	20.9	269.6	85.1	-2.5	265.9	-697.3	3.7
2014	625.3	374.9	400.0	266.8	225.3	108.1	-597.4	18.8	263.8	83.7	-2.6	270.2	-653.2	-6.4
2015	613.3	390.3	398.4	275.3	214.9	115.0	-560.2	20.2	266.6	80.1	-30.9	271.5	-584.4	-4.9
2016	615.8	402.4	399.5	285.0	216.2	117.4	-669.1	20.1	268.1	80.6	-8.9	271.8	-717.0	-3.7
2017	628.7	409.2	404.5	295.8	224.3	113.4	-722.4	273.2	278.7	91.2	-2.2	277.0	-540.0	1.7
2018	663.4	433.1	419.0	307.8	244.4	125.3	-931.7	22.7	295.9	82.2	-0.8	285.8	-1 000.6	10.1
2019	705.6	447.9	433.4	317.7	272.2	130.2	-1 047.0	16.2	321.9	82.3	-2.7	294.8	-1 137.4	27.1
2017														
1st quarter	619.7	405.4	402.1	292.0	217.6	113.3	-671.5	22.1	272.6	84.3	-0.6	274.8	-730.9	-2.2
2nd quarter	628.2	404.9	402.6	294.1	225.6	110.8	-698.1	22.8	277.9	84.8	-0.6	276.1	-761.3	1.8
3rd quarter	628.2	409.2	405.0	296.9	223.2	112.4	-731.0	23.2	278.3	115.8	-7.4	277.6	-816.9	0.7
4th quarter	638.9	417.3	408.3	300.2	230.6	117.2	-789.1	1 024.4	286.0	80.0	-0.4	279.6	149.3	6.4
2018														
1st quarter	645.4	427.3	412.6	303.7	232.8	123.6	-917.2	24.0	286.6	79.2	-0.3	282.3	-976.5	4.3
2nd quarter	659.7	431.5	417.0	306.6	242.8	124.9	-942.3	23.7	291.0	85.2	-0.4	284.9	-1 009.4	6.1
3rd quarter	670.3	436.0	421.6	309.3	248.7	126.6	-907.2	22.5	297.9	83.6	-0.4	287.1	-978.7	10.8
4th quarter	678.1	437.6	424.8	311.5	253.3	126.1	-960.0	20.5	308.3	80.8	-2.1	288.8	-1 037.7	19.5
2019														
1st quarter	690.2	439.4	428.6	314.5	261.6	124.9	-1 016.0	17.9	312.8	85.3	-6.4	292.1	-1 097.6	20.7
2nd quarter	702.4	448.3	431.4	316.3	271.0	132.1	-1 033.0	15.7	315.7	77.6	-1.9	293.1	-1 115.6	22.6
3rd quarter	709.8	450.9	434.9	319.4	274.9	131.4	-1 084.1	15.4	324.7	80.0	-1.8	295.8	-1 175.8	28.9
4th quarter	719.9	453.1	438.7	320.7	281.1	132.3	-1 054.9	15.6	334.3	86.1	-0.9	298.3	-1 160.5	36.0

[1]Includes general government intermediate inputs for goods and services sold to other sectors and for own-account investment.

Table 6-6. Chain-Type Quantity Indexes for Federal Government Defense and Nondefense Consumption Expenditures and Gross Investment

(Seasonally adjusted, 2012 = 100.)   NIPA Tables 3.9.3, 3.10.3

Year and quarter	Defense consumption expenditures [1]						Defense gross investment	Nondefense consumption expenditures [1]						Non-defense gross investment
	Total	Compensation of general government employees	Consumption of general government fixed capital	Intermediate goods and services purchased [2]				Total	Compensation of general government employees	Consumption of general government fixed capital	Intermediate goods and services purchased [2]			
				Durable goods	Non-durable goods	Services					Durable goods	Non-durable goods excluding CCC inventory change	Services	
1960	60.2	118.5	49.5	39.1	64.2	18.7	45.7	20.9	48.3	6.6	1.9	6.8	15.8	12.7
1961	61.9	121.1	51.9	32.1	81.9	20.8	50.5	20.5	49.5	7.3	2.6	10.5	18.9	15.0
1962	67.2	126.3	54.7	39.6	107.2	23.7	51.9	24.4	52.1	8.2	3.1	15.7	18.7	18.5
1963	67.5	124.0	56.9	39.9	95.8	24.7	47.5	24.3	55.2	9.5	3.5	15.8	19.7	24.7
1964	66.1	124.1	58.3	33.7	109.1	20.4	44.6	25.3	56.5	11.3	4.5	15.9	22.2	29.6
1965	67.1	124.2	59.0	35.1	123.2	20.9	40.4	26.1	57.5	13.4	4.7	16.8	21.8	33.8
1966	76.2	136.9	59.8	51.0	175.7	26.8	44.9	25.3	59.8	15.9	5.0	17.0	22.6	37.8
1967	85.7	149.0	61.4	50.4	269.3	34.2	50.4	28.0	62.8	18.2	4.9	19.0	20.6	34.8
1968	89.6	151.0	63.0	58.3	309.3	35.4	46.0	28.3	64.2	20.5	3.2	12.6	12.6	35.5
1969	86.5	151.2	63.3	48.6	274.6	33.1	42.9	30.9	64.8	22.6	3.2	13.1	15.3	34.2
1970	79.4	140.2	62.9	43.5	191.2	30.7	39.2	32.2	65.7	24.4	3.9	16.5	21.0	32.8
1971	73.5	129.5	61.2	31.4	156.4	31.2	30.7	35.0	67.8	26.0	4.8	20.3	24.5	32.7
1972	68.8	119.7	59.0	38.3	159.1	26.8	27.6	38.1	69.8	27.3	7.9	24.8	32.2	33.8
1973	64.1	113.6	57.2	35.0	118.3	24.7	28.8	38.2	70.0	28.7	7.2	23.8	30.6	34.0
1974	62.3	111.9	55.9	30.4	102.4	26.0	31.0	40.7	73.5	30.0	7.7	25.1	31.7	34.1
1975	60.9	110.4	54.9	32.2	84.8	23.9	32.3	42.5	74.8	31.2	8.1	22.8	32.0	35.1
1976	59.5	108.2	54.3	29.1	67.5	23.9	34.4	42.3	78.5	32.6	8.5	22.6	26.8	37.4
1977	59.8	107.2	54.1	36.5	64.5	23.8	35.4	44.0	80.2	34.1	11.1	22.3	29.2	39.5
1978	60.2	108.0	53.9	40.1	65.4	23.0	35.7	45.7	82.3	35.9	12.8	32.9	30.0	43.6
1979	60.6	107.1	53.9	43.7	67.4	23.5	39.1	46.5	82.2	37.8	15.0	32.3	32.8	44.1
1980	62.1	107.7	54.4	46.0	73.7	26.0	41.9	49.7	84.4	39.7	16.9	31.4	33.4	45.2
1981	65.1	111.1	55.5	54.1	78.4	28.1	46.5	50.9	81.9	41.4	15.8	58.7	28.4	44.5
1982	68.6	113.6	57.3	59.4	78.1	33.7	52.5	49.9	80.4	42.8	14.6	43.8	26.9	41.3
1983	71.5	115.4	60.1	70.9	81.4	34.0	60.5	53.4	81.6	44.1	15.6	54.8	32.5	41.2
1984	73.2	117.0	63.7	72.0	77.0	35.5	68.1	51.7	81.6	45.4	16.9	57.0	33.4	42.4
1985	76.8	119.2	68.5	77.7	75.4	39.1	80.4	54.7	81.8	46.6	17.4	49.6	36.2	44.7
1986	80.8	119.6	73.9	83.7	95.8	41.7	89.5	56.5	80.5	47.9	17.2	42.3	36.0	45.8
1987	83.6	120.7	79.3	88.9	93.9	42.6	97.0	55.1	82.0	49.3	19.6	49.0	40.5	48.8
1988	85.3	119.2	83.8	91.2	91.2	44.8	90.6	51.9	84.1	50.7	20.7	48.8	38.2	48.2
1989	85.8	119.2	86.9	87.7	88.2	44.8	88.3	56.2	84.5	52.2	23.4	43.6	37.8	50.6
1990	86.1	118.5	89.1	84.5	78.3	47.5	88.3	60.2	88.7	53.8	26.7	46.4	44.9	54.5
1991	87.4	118.0	90.1	81.1	79.4	51.2	81.7	60.6	88.5	55.7	28.4	40.6	44.5	58.1
1992	83.6	111.3	90.0	72.8	73.6	49.5	77.0	64.1	90.1	57.3	30.9	50.5	50.1	61.6
1993	80.8	106.5	88.9	65.9	67.6	48.9	68.9	63.6	90.0	58.9	30.1	55.0	42.9	62.7
1994	77.2	101.0	87.1	56.5	61.5	49.4	64.4	64.2	87.5	60.3	28.3	46.2	50.4	59.7
1995	74.1	95.5	85.0	51.5	50.3	47.6	61.8	63.6	84.0	61.7	27.6	52.1	48.6	61.0
1996	72.7	91.4	83.0	51.1	55.5	48.7	61.3	62.3	81.8	63.3	31.4	45.0	45.8	62.9
1997	71.6	88.1	81.2	51.3	56.0	49.4	57.0	65.1	81.3	65.0	33.8	54.9	48.8	62.4
1998	69.7	85.3	79.4	51.7	57.8	47.0	57.1	64.6	82.2	66.8	34.0	59.7	44.7	65.9
1999	70.7	83.2	77.9	55.0	64.2	52.1	57.7	65.5	82.4	69.1	38.1	44.4	45.3	69.9
2000	69.7	82.6	76.7	54.4	67.3	51.1	58.3	68.1	84.9	71.5	41.9	50.5	47.5	69.0
2001	71.6	83.1	75.7	54.7	69.6	58.0	61.6	72.0	84.9	73.5	45.1	67.1	54.4	70.6
2002	75.5	85.3	75.7	57.3	80.0	67.4	70.8	77.2	86.4	75.6	54.9	73.3	68.2	75.8
2003	81.7	88.2	76.9	62.9	87.3	81.0	76.6	80.4	88.6	77.5	56.5	88.0	71.8	78.3
2004	85.7	89.9	78.9	68.3	97.9	88.0	83.0	82.3	87.8	79.5	62.6	98.5	77.3	79.2
2005	86.7	90.2	81.5	70.0	99.0	88.8	88.5	83.5	87.7	81.7	68.9	99.7	81.8	80.4
2006	87.4	88.9	84.5	75.4	96.3	90.3	92.6	86.4	87.8	84.4	77.2	98.4	87.0	84.2
2007	89.0	88.6	87.6	84.8	98.3	90.8	98.6	86.3	88.4	87.1	79.0	95.6	82.0	85.4
2008	94.8	91.9	91.2	97.9	100.1	100.2	106.6	91.2	91.7	89.6	88.6	106.3	90.7	88.5
2009	100.8	97.8	94.5	106.1	105.9	107.1	110.4	98.0	96.5	92.3	96.7	123.2	102.1	92.6
2010	104.4	101.0	97.3	106.7	102.3	113.4	111.1	103.3	100.7	94.8	106.5	129.8	113.1	100.6
2011	102.9	102.0	99.2	103.0	104.3	106.7	105.9	98.5	99.5	97.5	99.8	106.9	100.0	101.0
2012	100.0	100.0	100.0	100.0	100.0	100.0	100.0	100.0	100.0	100.0	100.0	100.0	100.0	100.0
2013	93.4	96.5	99.7	87.6	88.9	87.2	93.0	97.4	97.1	102.3	94.7	92.3	93.9	93.7
2014	90.1	94.5	98.5	79.7	83.3	82.4	86.9	97.2	97.0	104.3	97.9	90.2	93.0	94.1
2015	88.0	92.2	97.0	77.5	84.6	80.0	85.8	100.1	98.1	106.2	102.7	95.9	99.8	97.9
2016	88.1	92.1	95.6	78.6	90.7	80.0	83.3	101.6	100.1	108.2	104.4	96.2	101.9	102.2
2017	88.2	92.2	94.4	83.4	92.9	79.9	86.5	100.3	99.9	110.6	100.7	92.2	95.7	104.1
2018	90.3	93.1	94.2	89.2	93.8	84.5	92.7	102.9	99.2	113.1	107.2	98.6	101.8	104.9
2019	94.4	94.6	95.1	96.6	93.7	94.5	101.9	104.0	99.1	115.6	107.0	104.6	103.1	109.0
2017														
1st quarter	87.5	92.0	94.8	82.0	91.8	77.7	83.6	100.8	100.4	109.7	100.9	92.3	96.5	104.0
2nd quarter	88.5	91.9	94.5	85.4	92.4	80.8	86.5	99.5	99.9	110.3	98.9	90.3	93.9	104.0
3rd quarter	88.1	92.2	94.3	81.0	93.2	80.0	86.4	100.0	99.7	111.0	100.0	91.3	94.7	103.8
4th quarter	88.9	92.6	94.2	85.4	94.4	81.2	89.6	100.9	99.5	111.6	103.0	95.1	97.6	104.5
2018														
1st quarter	88.7	92.5	94.1	83.4	95.4	81.1	89.6	103.0	99.4	112.2	106.9	98.7	102.1	103.6
2nd quarter	90.1	93.0	94.1	89.8	93.1	83.8	90.4	102.9	99.7	112.8	107.2	98.2	101.7	104.4
3rd quarter	90.9	93.4	94.2	94.0	94.0	85.0	92.9	103.7	99.8	113.4	107.9	98.7	102.3	105.5
4th quarter	91.5	93.5	94.3	89.4	92.6	88.0	97.9	102.0	97.9	114.0	106.7	98.9	101.3	106.0
2019														
1st quarter	92.8	93.9	94.6	92.5	91.5	91.3	99.1	100.3	93.3	114.7	103.3	99.2	100.2	106.0
2nd quarter	94.1	94.3	94.9	94.1	92.9	94.7	99.0	104.9	100.2	115.3	108.2	106.1	104.4	108.5
3rd quarter	94.9	94.8	95.2	96.2	94.1	95.9	102.6	105.6	101.5	116.0	108.2	106.1	103.8	110.0
4th quarter	95.8	95.4	95.6	103.7	96.4	96.2	106.7	105.1	101.2	116.6	108.5	106.9	104.0	111.5

[1] Excludes government sales to other sectors and government own-account investment (construction and software).
[2] Includes general government intermediate inputs for goods and services sold to other sectors and for own-account investment.

Table 6-7. Chain-Type Quantity Indexes for National Defense Consumption Expenditures and Gross Investment: Selected Detail

(Seasonally adjusted, 2012 = 100.)   NIPA Table 3.11.3

Year and quarter	Defense consumption expenditures								Defense gross investment				
	Compensation of general government employees		Intermediate goods and services purchased [1]						Aircraft	Missiles	Ships	Software	Research and development
			Durable goods	Nondurable goods		Services							
	Military	Civilian	Aircraft	Petroleum products	Ammunition	Installation support	Weapons support	Personnel support					
1975	109.0	111.8	28.1	149.7	74.4	45.7	19.9	10.2	23.1	20.1	69.5	6.8	52.2
1976	106.3	110.5	23.6	123.7	37.4	47.1	21.4	9.6	21.6	17.1	71.3	7.1	53.9
1977	105.4	109.4	36.0	105.9	51.1	46.4	20.4	9.9	22.5	13.4	83.2	7.3	55.5
1978	104.4	113.3	36.8	105.0	59.0	43.4	20.4	9.8	21.3	11.0	93.6	8.6	56.2
1979	102.8	113.8	45.9	106.7	65.6	45.4	24.4	8.7	26.7	22.2	96.3	10.3	58.3
1980	103.8	113.6	49.9	118.3	65.8	51.4	30.8	8.6	31.9	32.0	84.7	11.7	61.4
1981	107.2	116.8	64.3	116.3	74.3	55.8	33.6	9.6	36.9	37.1	96.5	14.0	68.2
1982	109.1	120.6	75.0	109.6	91.6	68.1	41.4	11.3	37.1	47.9	112.6	15.3	76.8
1983	111.0	122.1	91.4	117.2	107.4	65.1	43.3	12.4	42.1	60.6	125.1	17.8	87.1
1984	112.3	124.2	86.5	116.9	90.9	70.5	43.9	11.5	45.3	72.9	132.5	23.5	97.5
1985	114.0	127.5	92.9	119.1	54.2	74.7	49.5	14.9	61.8	83.1	145.6	28.7	112.3
1986	115.0	126.8	102.7	119.6	145.9	74.7	54.5	17.5	94.2	105.3	140.5	34.1	116.6
1987	116.4	127.2	109.9	122.0	110.7	79.4	52.6	17.7	108.1	119.8	136.5	38.1	125.4
1988	115.5	124.4	113.7	103.4	127.8	79.0	56.6	21.0	88.9	109.1	129.5	44.2	119.5
1989	115.0	125.6	107.2	113.2	106.9	77.0	55.3	21.3	81.3	124.9	143.7	50.6	111.1
1990	115.3	122.7	94.2	111.9	96.4	87.4	59.3	19.6	75.0	162.2	151.6	54.8	105.8
1991	116.6	118.8	84.2	110.6	90.9	92.0	48.7	17.5	53.9	163.1	136.0	56.3	93.8
1992	106.7	118.2	73.4	92.4	86.3	90.5	43.7	24.0	47.2	161.3	131.4	60.8	86.2
1993	101.2	115.0	63.7	89.0	85.7	98.1	39.3	25.0	50.8	117.2	110.9	58.0	80.1
1994	96.4	108.3	54.0	91.6	59.0	99.0	41.0	26.8	50.8	86.0	99.8	57.1	76.1
1995	91.4	101.9	51.5	83.0	37.8	91.1	40.7	27.0	41.1	72.5	93.9	56.0	73.6
1996	88.1	96.5	51.1	83.7	44.2	92.7	37.8	29.5	40.0	64.9	79.3	58.9	73.1
1997	85.7	91.6	54.6	75.4	54.3	89.1	42.7	32.0	27.8	48.0	70.4	60.5	71.9
1998	83.8	87.4	57.9	73.5	61.5	83.7	40.5	30.2	28.0	54.1	74.4	64.6	70.7
1999	82.3	84.3	61.2	79.7	62.6	85.8	44.7	36.9	29.1	49.0	78.8	64.4	68.9
2000	83.0	81.5	57.0	73.5	59.5	85.3	44.9	35.5	33.9	46.2	75.0	63.6	68.6
2001	84.7	80.2	56.2	87.0	68.0	90.7	55.5	42.4	39.6	58.0	82.2	60.2	73.4
2002	87.8	80.7	56.6	103.2	80.5	100.3	66.9	52.5	47.3	58.0	95.1	58.7	84.3
2003	92.6	80.5	64.3	99.5	87.4	112.2	82.3	61.2	45.0	59.8	103.3	60.4	92.8
2004	93.7	83.2	66.2	103.0	110.7	110.3	84.4	75.3	55.4	67.9	105.0	62.6	100.8
2005	93.1	85.0	58.5	100.5	117.7	103.1	91.3	78.8	69.5	65.8	100.0	65.8	106.5
2006	91.0	85.2	58.9	99.2	112.9	105.7	93.6	81.9	70.7	74.1	105.7	68.9	109.0
2007	90.2	85.7	61.0	98.3	111.0	102.0	95.7	78.3	67.4	70.3	99.4	73.3	112.1
2008	94.2	87.8	68.0	99.0	105.1	106.6	100.7	94.1	68.2	68.1	102.3	78.6	110.6
2009	101.5	91.3	78.2	99.8	107.2	114.2	103.5	100.2	69.1	79.1	101.1	80.3	108.5
2010	103.1	97.2	85.0	102.0	105.3	116.7	116.3	105.2	82.7	85.6	104.6	88.6	106.1
2011	102.7	100.7	94.2	104.8	99.7	106.9	108.0	101.2	97.5	75.4	98.7	95.5	103.6
2012	100.0	100.0	100.0	100.0	100.0	100.0	100.0	100.0	100.0	100.0	100.0	100.0	100.0
2013	97.5	94.7	92.6	83.8	82.7	83.5	84.8	91.2	110.1	92.8	104.7	101.6	94.3
2014	94.4	94.6	83.8	77.5	70.7	90.2	89.8	78.0	97.5	96.5	108.7	104.9	87.3
2015	91.0	94.2	80.8	80.4	64.9	85.6	91.9	74.8	93.5	95.5	110.8	106.7	87.3
2016	90.4	94.8	80.7	82.2	72.6	88.8	94.2	71.8	87.2	71.0	117.9	114.8	80.6
2017	90.5	94.9	85.9	77.2	84.7	89.2	98.4	70.0	100.7	70.8	114.6	123.0	79.1
2018	91.6	95.6	87.2	75.4	104.0	87.2	106.0	76.9	106.9	68.8	125.0	136.3	83.2
2019	93.0	97.2	90.8	73.7	124.0	86.2	115.7	93.3	126.3	77.0	137.3	146.2	90.2
2013													
1st quarter	98.1	98.3	92.4	86.3	88.0	85.8	85.2	96.5	111.6	75.9	97.0	104.8	97.5
2nd quarter	97.8	97.2	92.8	87.7	84.1	83.7	86.2	96.2	99.0	94.4	105.7	99.8	95.5
3rd quarter	97.3	91.1	95.0	81.1	80.6	80.3	82.7	90.6	107.9	112.5	106.8	100.5	93.5
4th quarter	96.7	92.2	90.1	80.1	78.1	84.0	85.1	81.6	122.0	88.4	109.1	101.2	90.6
2014													
1st quarter	96.0	94.7	88.6	81.2	75.1	87.4	87.9	79.6	86.4	88.1	101.9	102.6	88.2
2nd quarter	95.2	94.6	86.0	75.2	72.4	90.0	87.8	74.3	104.4	93.2	114.9	104.3	87.2
3rd quarter	93.8	94.4	78.0	75.4	68.9	93.3	94.9	87.1	98.5	96.4	111.6	106.8	86.7
4th quarter	92.5	94.5	82.5	78.2	66.6	90.3	88.4	71.0	100.9	108.3	106.4	106.0	87.3
2015													
1st quarter	91.6	94.5	81.8	82.0	64.5	87.6	90.1	76.9	80.7	94.5	113.3	106.2	87.8
2nd quarter	90.9	94.1	78.0	77.1	63.6	86.7	93.3	76.0	102.9	91.9	107.5	106.2	87.8
3rd quarter	90.7	94.1	80.6	79.5	64.7	83.2	91.3	73.4	96.2	85.8	111.3	106.4	87.3
4th quarter	90.7	94.2	82.9	82.8	66.8	85.0	92.9	73.1	94.3	110.0	111.2	108.1	86.1
2016													
1st quarter	90.7	94.4	83.6	87.8	69.3	87.5	94.6	72.7	99.6	75.9	118.0	111.5	83.8
2nd quarter	90.5	94.7	75.1	80.8	71.8	88.0	91.6	68.6	88.8	80.4	114.4	113.8	81.1
3rd quarter	90.4	95.0	80.8	81.3	73.8	89.3	95.9	72.8	79.9	72.6	120.2	115.9	79.5
4th quarter	90.2	95.1	83.2	79.1	75.7	90.5	94.9	72.9	80.6	54.9	119.1	117.9	78.2
2017													
1st quarter	90.1	95.1	85.6	78.0	78.1	88.5	93.4	67.9	94.3	76.1	100.9	119.9	78.4
2nd quarter	90.1	94.8	93.3	77.3	82.0	89.8	97.3	71.9	99.8	72.3	117.9	121.3	78.7
3rd quarter	90.6	94.9	80.1	76.4	86.7	90.0	100.7	68.8	98.3	71.8	116.5	123.9	79.3
4th quarter	91.2	94.9	84.6	77.2	92.0	88.7	102.3	71.2	110.3	63.1	123.3	126.7	80.1
2018													
1st quarter	91.2	94.7	78.9	79.2	95.6	87.0	103.2	71.4	97.4	70.8	124.9	130.6	80.9
2nd quarter	91.4	95.7	89.5	75.2	100.9	87.9	105.8	75.4	94.8	60.2	128.4	135.5	82.1
3rd quarter	91.8	96.1	98.7	74.8	107.0	86.9	106.3	77.9	106.6	62.8	120.7	137.8	83.8
4th quarter	92.0	96.0	81.5	72.3	112.6	87.1	108.6	83.0	128.7	81.2	125.9	141.4	85.8
2019													
1st quarter	92.5	96.4	86.2	71.2	116.8	86.2	111.3	88.6	127.0	72.2	131.2	142.5	87.1
2nd quarter	92.7	96.9	87.9	73.0	121.7	86.3	114.5	94.0	114.0	62.2	134.8	144.6	89.0
3rd quarter	93.3	97.3	89.6	74.1	126.7	86.5	116.8	95.3	125.2	72.9	141.3	147.7	91.1
4th quarter	93.6	98.3	99.5	76.6	130.8	85.6	120.0	95.4	138.8	100.4	141.9	150.1	93.7

[1]Includes general government intermediate inputs for goods and services sold to other sectors and for own-account investment.

SECTION 6B: STATE AND LOCAL GOVERNMENT IN THE NATIONAL INCOME AND PRODUCT ACCOUNTS

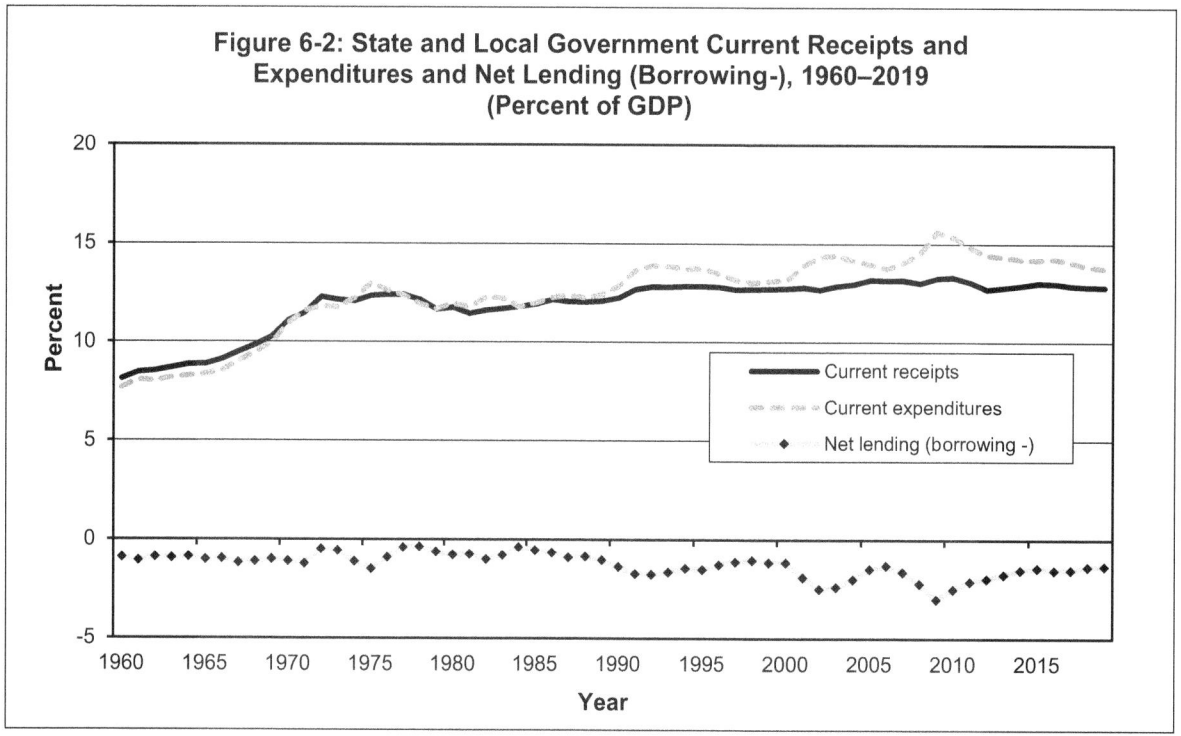

Figure 6-2: State and Local Government Current Receipts and Expenditures and Net Lending (Borrowing-), 1960–2019 (Percent of GDP)

- Both current receipts and current spending by state and local governments have increased as a share of gross domestic product (GDP) over the postwar period, but the growth slowed markedly after the early 1970s. (Tables 6-8 and 1-1)

- State and local governments have consistently been net borrowers in capital markets, as can be seen by the negative values in Figure 6-2. Although they are usually constrained to balance their statutory budgets, they have consistently made net investments and been able to borrow for that purpose. In addition, as revealed by the accrual accounting adopted in the NIPA accounts last year, they have often incurred hidden deficits by failing to fund commitments for public employee retirement benefits. (Table 6-10)

- A larger gap between current expenditures and current receipts appeared during the slow-growth years 2001–2003, and a still-greater one during the Great Recession beginning in late 2007. Generally, in both of those periods of rising deficits, receipts barely kept pace with the economy, while general ("consumption") expenditures, social benefits, and interest payments all fueled higher spending. (Table 6-8)

- Net lending by state and local government reached a high of $-437 in 2009 but declined to $-288 in 2019. (Table 6-10)

Table 6-8. State and Local Government Current Receipts and Expenditures

(National income and product accounts, calendar years, billions of dollars, quarterly data are at seasonally adjusted annual rates.)

NIPA Table 3.3

Year and quarter	Total current receipts	Current tax receipts							Contributions for government social insurance	Income receipts on assets		
		Total tax receipts	Personal current taxes		Taxes on production and imports			Taxes on corporate income		Interest receipts	Dividends	Rents and royalties
			Income taxes	Other personal taxes	Sales taxes	Property taxes	Other taxes on prod. and imports					
1960	44.2	37.0	2.5	1.7	5.3	16.2	3.1	1.2	0.5	1.0	. . .	0.3
1961	47.7	39.7	2.8	1.8	5.8	17.6	3.2	1.3	0.5	1.1	. . .	0.4
1962	51.6	42.8	3.2	1.8	6.4	19.0	3.3	1.5	0.5	1.1	. . .	0.4
1963	55.5	45.8	3.4	2.0	6.9	20.2	3.5	1.7	0.6	1.2	. . .	0.4
1964	60.7	49.8	4.0	2.1	7.7	21.7	3.7	1.8	0.7	1.5	. . .	0.4
1965	65.9	53.9	4.4	2.2	8.6	23.2	3.9	2.0	0.8	1.8	. . .	0.4
1966	74.2	58.8	5.4	2.4	9.8	24.5	4.3	2.2	0.8	2.1	. . .	0.5
1967	81.7	64.0	6.1	2.5	10.7	27.0	4.4	2.6	0.9	2.4	. . .	0.6
1968	92.6	73.4	7.8	2.7	13.1	29.9	4.6	3.3	0.9	2.8	. . .	0.7
1969	104.5	82.5	9.8	3.0	15.1	32.8	4.7	3.6	1.0	3.6	. . .	0.8
1970	119.1	91.3	10.9	3.3	17.0	36.7	5.0	3.7	1.1	4.3	. . .	0.8
1971	133.7	101.7	12.4	3.5	19.1	40.4	5.7	4.3	1.2	4.6	. . .	0.9
1972	157.1	115.6	17.2	3.7	21.8	43.2	6.4	5.3	1.3	4.9	. . .	1.0
1973	173.0	126.3	18.9	4.0	24.5	46.4	7.0	6.0	1.5	6.6	0.0	1.1
1974	186.6	136.0	20.4	4.2	28.0	49.0	7.7	6.7	1.7	8.9	0.0	1.3
1975	208.0	147.4	22.5	4.4	30.3	53.4	8.1	7.3	1.8	9.8	0.0	1.3
1976	232.2	165.7	26.3	4.8	34.4	58.2	9.0	9.6	2.2	9.0	0.0	1.3
1977	258.3	183.7	30.4	5.0	39.0	63.2	9.7	11.4	2.8	10.4	0.0	1.3
1978	285.8	198.2	35.0	5.5	44.2	63.7	10.9	12.1	3.4	13.3	0.1	1.3
1979	306.3	212.0	38.2	5.8	49.5	64.4	12.7	13.6	3.9	18.1	0.1	1.9
1980	335.9	230.0	42.6	6.3	53.6	68.8	15.0	14.5	3.6	23.1	0.1	3.1
1981	367.5	255.8	47.9	6.7	58.8	77.1	17.9	15.4	3.9	28.5	0.1	3.3
1982	388.5	273.2	51.9	7.3	62.0	85.3	18.5	14.0	4.0	33.1	0.2	3.5
1983	425.3	300.9	58.3	7.8	70.4	91.9	19.4	15.9	4.1	37.0	0.2	4.3
1984	476.1	337.3	67.5	8.6	80.6	99.7	21.8	18.8	4.7	42.6	0.2	4.9
1985	517.5	363.7	72.1	9.2	87.9	107.5	23.5	20.2	4.9	50.1	0.2	5.4
1986	557.4	389.5	77.4	9.8	93.7	116.2	23.7	22.7	6.0	52.6	0.2	6.2
1987	585.5	422.1	86.0	10.6	100.9	126.4	24.9	23.9	7.2	53.0	0.2	5.3
1988	630.4	452.8	90.6	11.5	110.0	136.5	25.7	26.0	8.4	56.6	0.2	4.4
1989	681.4	488.0	102.3	12.4	117.4	149.9	26.9	24.2	9.0	62.3	0.2	4.1
1990	730.1	519.1	109.6	13.0	125.6	161.5	28.2	22.5	10.0	64.1	0.2	4.2
1991	779.9	544.3	111.7	13.6	128.0	176.1	28.6	23.6	11.6	62.4	0.3	4.5
1992	836.1	579.8	120.4	14.9	136.0	184.7	31.1	24.4	13.1	58.8	0.5	4.8
1993	878.0	604.7	126.2	14.8	143.7	187.3	33.1	26.9	14.1	56.2	0.6	4.5
1994	935.1	644.2	132.2	15.8	155.4	199.4	35.4	30.0	14.5	58.4	0.8	4.5
1995	981.0	672.1	141.7	16.4	164.4	202.6	37.0	31.7	13.6	63.8	1.0	4.5
1996	1 033.7	709.6	152.3	16.3	174.4	212.4	39.3	33.0	12.5	68.0	1.4	4.6
1997	1 086.7	749.9	164.7	17.3	184.2	223.5	41.6	34.1	10.8	72.0	1.5	4.8
1998	1 149.6	794.9	183.0	18.2	194.9	231.0	43.9	34.9	10.4	75.2	1.6	4.6
1999	1 222.7	840.4	195.5	19.0	208.6	242.8	45.8	35.8	9.8	78.9	1.5	5.1
2000	1 304.1	893.2	217.4	19.4	221.4	254.7	49.8	35.2	10.8	86.6	1.4	6.3
2001	1 353.4	914.3	223.3	19.7	223.1	268.0	52.6	28.9	13.7	81.6	1.4	6.5
2002	1 386.2	923.9	199.3	20.6	222.5	289.2	56.3	30.9	15.8	70.7	1.6	6.6
2003	1 472.0	974.3	202.8	22.1	233.3	306.9	63.0	34.0	19.9	64.8	1.8	7.6
2004	1 580.3	1 060.3	221.4	24.3	252.0	326.1	77.4	41.7	24.7	66.8	2.3	8.5
2005	1 718.5	1 173.2	249.1	26.2	272.8	351.2	91.7	54.9	24.6	77.3	2.5	9.8
2006	1 816.2	1 258.2	274.2	26.9	293.9	375.2	92.1	59.2	21.5	95.0	3.0	10.5
2007	1 902.1	1 321.7	294.3	27.3	301.4	403.7	92.9	57.9	18.8	104.6	3.4	11.7
2008	1 915.5	1 334.1	304.1	26.8	301.1	415.7	94.7	47.4	18.7	91.4	3.4	12.2
2009	1 915.2	1 265.8	258.9	26.9	279.6	439.6	72.3	44.5	18.6	73.5	2.9	11.2
2010	1 994.4	1 306.4	265.8	28.3	295.1	438.6	77.8	46.1	17.8	69.0	3.0	11.4
2011	2 030.4	1 366.4	294.1	28.8	311.3	439.2	83.9	48.4	17.9	67.1	3.4	12.2
2012	2 056.3	1 414.7	313.1	30.0	324.4	444.3	86.7	50.7	17.2	65.3	4.1	12.5
2013	2 145.7	1 490.6	343.1	30.4	343.9	456.8	91.6	53.9	17.7	65.2	4.7	12.5
2014	2 257.5	1 541.9	349.4	31.5	361.7	475.0	93.0	56.5	18.7	66.4	5.3	12.7
2015	2 373.2	1 598.4	374.5	32.9	374.2	490.4	89.0	56.2	19.2	65.4	5.4	11.2
2016	2 431.5	1 638.6	375.6	34.3	384.7	512.0	91.4	53.4	20.0	68.6	5.7	11.0
2017	2 515.2	1 719.3	396.3	35.8	402.3	536.1	98.0	54.2	20.0	73.8	5.8	11.2
2018	2 643.2	1 810.3	429.9	37.9	421.2	549.2	106.6	60.5	21.0	77.1	6.0	11.5
2019	2 742.9	1 877.1	451.0	38.9	434.8	564.0	110.5	69.5	21.7	78.4	6.6	11.8
2017												
1st quarter	2 481.0	1 687.2	384.6	35.4	395.0	528.6	96.8	54.3	20.0	71.8	5.7	11.2
2nd quarter	2 459.7	1 683.1	368.5	35.4	400.6	534.4	97.1	52.9	19.9	73.2	5.8	11.2
3rd quarter	2 518.9	1 718.9	391.1	36.2	404.1	539.0	97.5	54.1	20.0	74.5	6.0	11.3
4th quarter	2 601.3	1 788.0	441.0	36.0	409.5	542.3	100.7	55.4	20.2	75.5	5.9	11.3
2018												
1st quarter	2 636.5	1 810.5	449.1	37.9	415.3	544.5	104.5	55.4	20.4	76.2	5.8	11.4
2nd quarter	2 620.1	1 794.1	417.9	37.6	419.8	547.2	105.2	60.8	20.8	76.9	6.1	11.4
3rd quarter	2 656.0	1 820.9	434.2	38.0	423.7	550.5	107.6	62.1	21.2	77.5	6.2	11.5
4th quarter	2 660.0	1 815.5	418.3	38.2	425.8	554.4	109.1	63.5	21.6	78.0	5.8	11.6
2019												
1st quarter	2 694.6	1 845.3	436.7	38.5	428.2	558.8	109.0	68.5	22.0	77.9	6.6	11.7
2nd quarter	2 772.3	1 900.5	480.7	38.6	431.8	562.7	110.8	68.7	22.1	78.0	6.6	11.8
3rd quarter	2 748.9	1 880.3	444.9	39.0	440.3	566.0	111.0	69.8	21.8	78.5	6.5	11.9
4th quarter	2 755.9	1 882.4	441.6	39.3	438.9	568.7	111.1	71.0	21.1	79.1	6.6	12.0

. . . = Not available.

Table 6-9. State and Local Government Consumption Expenditures and Gross Investment

(National income and product accounts, calendar years, billions of dollars, quarterly data are at seasonally adjusted annual rates.)

NIPA Tables 3.9.5, 3.10.5

Year and quarter	Total	State and local government consumption expenditures and gross investment									Gross investment			
		Consumption expenditures [1]					Less							
		Total	Compen-sation of general govern-ment employees	Consump-tion of general govern-ment fixed capital	Intermed-iate goods and services purchased [2]	Own-account investment	Sales to other sectors				Total	Structures	Equipment	Intellectual property
							Total [3]	Tuition and related educational charges	Health and hospital charges					
1960	47.6	33.5	25.9	3.6	8.9	0.9	3.9	0.4	1.0		14.1	12.7	1.2	0.2
1961	51.9	36.6	28.5	3.8	9.6	1.0	4.3	0.5	1.0		15.3	13.8	1.2	0.2
1962	54.8	38.7	30.6	4.1	10.1	1.1	4.9	0.6	1.3		16.1	14.5	1.3	0.2
1963	59.3	41.5	33.2	4.4	10.8	1.3	5.6	0.7	1.3		17.8	16.0	1.5	0.3
1964	64.5	45.2	36.2	4.7	11.9	1.3	6.3	0.8	1.5		19.3	17.2	1.7	0.3
1965	70.9	49.7	39.8	5.1	13.4	1.4	7.2	1.0	1.8		21.2	19.0	1.8	0.4
1966	78.9	55.4	44.6	5.7	14.9	1.5	8.2	1.2	2.1		23.5	21.0	2.0	0.4
1967	87.0	61.3	49.5	6.2	16.4	1.6	9.3	1.4	2.5		25.7	23.0	2.2	0.5
1968	97.6	69.4	56.3	6.9	18.6	1.8	10.7	1.6	3.1		28.1	25.2	2.3	0.6
1969	107.8	79.0	63.2	7.8	21.8	1.9	11.8	1.9	3.5		28.8	25.6	2.6	0.7
1970	119.8	90.4	71.5	8.8	25.3	2.1	13.1	2.4	3.8		29.3	25.8	2.8	0.8
1971	133.0	102.3	80.5	9.8	29.1	2.2	14.9	2.9	4.6		30.7	27.0	2.9	0.8
1972	144.5	113.2	89.5	10.7	32.0	2.3	16.7	3.2	5.6		31.3	27.1	3.3	0.9
1973	158.6	124.6	98.5	11.9	35.3	2.4	18.7	3.7	6.6		34.0	29.1	3.8	1.0
1974	182.5	142.0	108.1	14.7	42.6	2.8	20.5	4.0	7.4		40.5	34.7	4.6	1.2
1975	207.4	162.8	121.3	16.7	50.5	3.0	22.6	4.3	8.5		44.6	38.1	5.1	1.3
1976	219.4	174.5	130.6	17.4	55.0	3.0	25.6	4.7	9.9		44.9	38.1	5.3	1.5
1977	234.0	190.0	142.1	18.4	60.8	3.1	28.3	5.2	10.9		44.0	36.9	5.4	1.6
1978	257.4	206.6	155.7	19.9	66.6	3.5	32.0	5.8	12.7		50.8	42.8	6.1	1.9
1979	284.2	225.9	170.6	22.2	73.9	4.3	36.6	6.4	15.2		58.4	49.0	7.1	2.2
1980	314.7	249.0	187.9	25.7	81.6	4.9	41.2	7.2	17.3		65.7	55.1	8.1	2.6
1981	340.4	273.5	205.9	29.3	91.2	5.3	47.6	8.3	21.0		66.9	55.4	8.5	3.0
1982	363.1	296.1	223.6	32.0	100.0	5.7	53.8	9.4	24.5		67.0	54.2	9.4	3.4
1983	384.2	315.4	239.6	33.0	109.0	6.1	60.1	10.6	27.8		68.8	54.2	10.8	3.8
1984	416.1	338.5	257.8	34.1	118.4	7.0	64.7	11.7	29.4		77.6	60.5	12.7	4.4
1985	457.6	370.3	281.7	35.9	131.4	7.9	70.8	12.7	32.0		87.3	67.6	14.8	5.0
1986	494.4	398.2	304.2	38.6	141.7	8.8	77.4	13.8	34.8		96.2	74.2	16.4	5.6
1987	528.7	426.6	327.8	41.6	150.0	9.5	83.4	15.0	37.0		102.2	78.8	17.2	6.2
1988	567.4	457.0	354.4	44.5	159.8	10.4	91.2	16.6	40.3		110.4	84.8	18.6	7.0
1989	617.8	498.5	388.4	47.7	175.8	11.9	101.4	18.4	45.0		119.2	88.7	21.9	8.6
1990	676.2	544.0	426.5	51.6	190.5	13.1	111.6	20.3	50.0		132.2	98.5	23.9	9.7
1991	716.0	578.7	457.2	55.1	204.7	14.1	124.1	22.6	57.0		137.3	103.2	23.5	10.6
1992	756.0	616.3	492.6	57.6	218.9	14.4	138.4	25.4	65.3		139.7	104.2	24.0	11.4
1993	784.8	643.6	516.1	60.8	234.0	14.8	152.5	27.4	73.1		141.2	104.5	24.9	11.9
1994	827.6	678.8	541.9	64.3	251.5	15.4	163.5	29.3	78.8		148.8	108.7	28.1	12.0
1995	872.7	712.8	565.4	68.6	270.6	16.1	175.6	31.1	85.0		160.0	117.3	30.5	12.3
1996	913.7	743.5	586.6	72.0	285.8	16.8	184.0	32.9	86.6		170.2	126.8	30.5	12.8
1997	963.8	781.0	613.5	75.8	304.2	18.3	194.3	35.4	89.9		182.9	136.6	32.0	14.3
1998	1 026.1	829.8	646.3	79.9	326.7	19.4	203.7	37.9	93.4		196.3	146.1	34.7	15.5
1999	1 109.0	895.4	688.1	85.1	357.3	20.7	214.4	40.5	96.7		213.6	159.7	37.4	16.5
2000	1 193.1	961.7	732.5	91.5	392.1	22.6	231.8	43.6	104.0		231.5	174.6	38.8	18.1
2001	1 279.2	1 031.2	785.4	96.9	425.2	25.0	251.4	46.1	116.2		248.0	190.6	38.4	19.1
2002	1 345.7	1 084.2	829.1	101.4	444.3	26.0	264.6	49.3	120.7		261.5	204.1	38.0	19.4
2003	1 384.9	1 115.9	864.9	105.2	451.0	26.7	278.6	53.4	124.8		269.1	211.1	37.6	20.4
2004	1 447.1	1 169.4	907.6	111.8	472.3	27.8	294.4	56.7	130.9		277.8	219.1	37.2	21.5
2005	1 528.5	1 237.3	944.9	122.1	507.7	29.7	307.8	60.9	132.3		291.3	231.5	37.1	22.7
2006	1 623.5	1 308.3	989.8	132.2	542.6	32.2	324.2	64.0	141.0		315.3	251.4	39.1	24.7
2007	1 740.3	1 398.3	1 045.8	145.4	578.3	34.3	336.9	65.9	149.3		342.0	271.4	44.0	26.6
2008	1 831.4	1 473.5	1 095.9	156.2	605.3	36.1	347.8	67.8	158.4		357.9	284.5	45.1	28.3
2009	1 855.3	1 495.0	1 117.9	162.6	610.5	36.3	359.7	69.7	168.7		360.3	288.9	43.1	28.3
2010	1 856.7	1 509.5	1 141.1	165.0	612.1	35.6	373.1	71.6	176.1		347.3	279.8	39.6	27.9
2011	1 849.4	1 508.5	1 141.9	171.2	616.8	35.9	385.6	73.6	182.4		341.0	273.4	38.6	29.0
2012	1 850.5	1 516.7	1 147.1	178.4	622.6	35.8	395.5	76.1	188.5		333.7	265.0	39.2	29.6
2013	1 905.8	1 575.1	1 197.8	183.2	638.9	36.4	408.4	78.9	197.6		330.8	259.3	39.9	31.5
2014	1 953.0	1 614.2	1 232.5	188.7	656.2	37.3	425.8	81.3	209.0		338.8	266.1	39.9	32.8
2015	2 009.4	1 653.5	1 275.3	192.0	673.1	39.2	447.7	85.8	220.1		356.0	281.4	40.3	34.2
2016	2 064.6	1 694.2	1 303.7	195.4	707.1	40.8	471.2	90.1	233.1		370.4	291.8	43.0	35.5
2017	2 143.2	1 757.6	1 338.7	202.4	749.3	42.7	490.1	92.6	243.5		385.6	301.5	46.7	37.5
2018	2 255.7	1 847.8	1 393.7	212.9	794.0	45.1	507.7	96.1	252.4		407.9	318.8	49.1	40.0
2019	2 328.7	1 897.8	1 443.4	222.5	804.3	46.8	525.7	99.5	261.6		431.0	339.1	50.4	41.5
2017														
1st quarter	2 115.1	1 735.6	1 323.0	199.2	738.8	42.1	483.3	91.6	240.0		379.5	297.0	45.8	36.7
2nd quarter	2 126.3	1 742.5	1 331.4	201.2	740.0	42.3	487.8	92.2	242.4		383.8	299.9	46.8	37.2
3rd quarter	2 148.4	1 762.0	1 343.2	203.7	749.7	42.7	491.8	92.8	244.5		386.4	302.1	46.4	37.8
4th quarter	2 182.9	1 790.2	1 357.1	205.5	768.6	43.7	497.3	93.9	247.3		392.7	306.8	47.6	38.3
2018														
1st quarter	2 213.4	1 814.6	1 371.0	208.2	780.9	44.3	501.2	94.7	249.1		398.8	311.9	47.8	39.1
2nd quarter	2 250.7	1 838.0	1 385.2	212.1	790.8	44.9	505.0	95.6	251.0		412.6	323.8	49.0	39.9
3rd quarter	2 279.1	1 862.0	1 403.6	214.4	799.1	45.4	509.6	96.6	253.4		417.1	327.0	49.9	40.3
4th quarter	2 279.6	1 876.5	1 415.0	217.0	805.2	45.8	515.0	97.6	256.2		403.1	312.5	49.9	40.7
2019														
1st quarter	2 292.7	1 874.8	1 424.7	218.5	797.4	46.2	519.6	98.5	258.7		417.9	326.2	50.6	41.0
2nd quarter	2 327.0	1 892.8	1 435.0	222.0	806.4	46.6	524.0	99.2	261.0		434.2	342.4	50.5	41.3
3rd quarter	2 337.8	1 904.6	1 451.5	224.2	803.3	47.0	527.4	99.8	262.5		433.2	341.4	50.0	41.8
4th quarter	2 357.4	1 918.8	1 462.4	225.2	810.3	47.3	531.8	100.5	264.4		438.5	346.2	50.4	42.0

[1]Excludes government sales to other sectors and government own-account investment (construction and software).
[2]Includes general government intermediate inputs for goods and services sold to other sectors and for own-account investment.
[3]Includes components not shown separately.

Table 6-10. State and Local Government Output, Lending and Borrowing, and Net Investment

(National income and product accounts, calendar years, billions of dollars, quarterly data are at seasonally adjusted annual rates.)

NIPA Tables 3.3, 3.10.5

Year and quarter	Output			Net lending (net borrowing -)								Net investment
	Gross	Value added	Intermediate goods and services purchased [1]	Net saving, current (surplus +, deficit -)	Plus: Capital transfer receipts	Minus			Plus: consumption of fixed capital	Equals: Net lending (borrowing -)	
						Gross investment	Capital transfer payments	Net purchases of nonproduced assets			
1960	38.4	29.5	8.9	2.5	3.0	14.1	. . .	0.9	4.5	-4.9	9.6
1961	41.9	32.3	9.6	2.2	3.3	15.3	. . .	1.0	4.8	-6.0	10.5
1962	44.7	34.6	10.1	3.1	3.5	16.1	. . .	1.1	5.1	-5.4	11.0
1963	48.4	37.6	10.8	3.3	4.1	17.8	. . .	1.2	5.5	-6.0	12.3
1964	52.8	40.9	11.9	3.9	4.7	19.3	. . .	1.3	5.9	-6.0	13.4
1965	58.3	44.9	13.4	3.7	4.7	21.2	. . .	1.3	6.4	-7.6	14.8
1966	65.1	50.2	14.9	4.7	5.1	23.5	. . .	1.4	7.1	-7.9	16.4
1967	72.2	55.8	16.4	4.0	5.1	25.7	. . .	1.4	7.8	-10.2	17.9
1968	81.9	63.3	18.6	3.5	6.8	28.1	. . .	1.4	8.6	-10.6	19.5
1969	92.7	70.9	21.8	3.1	6.8	28.8	. . .	1.0	9.6	-10.2	19.2
1970	105.6	80.3	25.3	1.4	6.2	29.3	. . .	1.1	10.9	-11.8	18.4
1971	119.4	90.3	29.1	-1.3	7.0	30.7	. . .	1.6	12.2	-14.4	18.5
1972	132.2	100.2	32.0	6.1	7.3	31.3	. . .	1.7	13.3	-6.3	18.0
1973	145.7	110.4	35.3	5.6	7.3	34.0	. . .	1.7	14.8	-8.0	19.2
1974	165.3	122.8	42.6	-2.3	9.2	40.5	. . .	1.9	18.4	-17.2	22.1
1975	188.5	138.0	50.5	-10.7	11.0	44.6	. . .	1.9	21.0	-25.3	23.6
1976	203.0	148.0	55.0	-4.4	12.0	44.9	. . .	1.7	22.0	-17.0	22.9
1977	221.4	160.6	60.8	0.5	13.1	44.0	. . .	1.6	23.4	-8.5	20.6
1978	242.1	175.5	66.6	4.9	13.7	50.8	. . .	1.8	25.4	-8.6	25.4
1979	266.7	192.8	73.9	-1.2	16.2	58.4	. . .	2.0	28.6	-16.8	29.8
1980	295.1	213.5	81.6	-5.9	18.6	65.7	. . .	2.2	33.1	-22.1	32.6
1981	326.4	235.2	91.2	-10.2	17.8	66.9	. . .	2.2	37.8	-23.8	29.1
1982	355.6	255.6	100.0	-22.8	16.9	67.0	. . .	2.2	41.2	-33.8	25.8
1983	381.6	272.6	109.0	-18.4	18.0	68.8	. . .	2.2	42.7	-28.8	26.1
1984	410.2	291.9	118.4	-0.2	20.1	77.6	. . .	2.6	44.3	-16.0	33.3
1985	449.0	317.6	131.4	-2.4	22.0	87.3	. . .	3.1	46.8	-24.1	40.5
1986	484.4	342.7	141.7	-4.0	23.0	96.2	. . .	3.7	50.1	-30.7	46.1
1987	519.5	369.4	150.0	-14.4	22.3	102.2	. . .	4.2	54.0	-44.4	48.2
1988	558.6	398.8	159.8	-11.3	23.1	110.4	. . .	4.3	57.7	-45.3	52.7
1989	611.8	436.1	175.8	-19.3	23.4	119.2	0.0	4.9	61.8	-58.3	57.4
1990	668.7	478.2	190.5	-36.2	25.0	132.2	0.0	5.7	66.6	-82.5	65.6
1991	716.9	512.2	204.7	-60.1	25.8	137.3	0.0	5.8	70.8	-106.6	66.5
1992	769.2	550.2	218.9	-71.0	26.9	139.7	0.0	5.9	74.1	-115.5	65.6
1993	810.9	576.9	234.0	-72.5	28.2	141.2	0.0	5.9	78.3	-113.1	62.9
1994	857.7	606.2	251.5	-63.9	29.9	148.8	0.0	6.2	82.8	-106.3	66.0
1995	904.5	633.9	270.6	-70.4	32.4	160.0	0.0	6.6	88.1	-116.5	71.9
1996	944.4	658.6	285.8	-53.8	33.8	170.2	0.0	6.0	92.5	-103.6	77.7
1997	993.5	689.3	304.2	-42.0	35.2	182.9	0.0	5.8	97.3	-98.2	85.6
1998	1 052.9	726.2	326.7	-30.1	35.9	196.3	0.0	7.6	102.2	-95.9	94.1
1999	1 130.5	773.1	357.3	-38.9	40.0	213.6	0.0	8.6	108.6	-112.6	105.0
2000	1 216.1	824.0	392.1	-40.6	44.2	231.5	0.0	8.6	116.6	-119.9	114.9
2001	1 307.5	882.3	425.2	-119.0	51.7	248.0	0.0	10.1	123.4	-202.1	124.6
2002	1 374.8	930.6	444.3	-182.9	52.9	261.5	0.0	11.2	129.3	-273.3	132.2
2003	1 421.2	970.1	451.0	-181.3	52.3	269.1	0.0	11.4	134.9	-274.6	134.2
2004	1 491.6	1 019.3	472.3	-149.0	52.5	277.8	4.5	11.3	144.2	-245.8	133.6
2005	1 574.8	1 067.0	507.7	-102.8	56.9	291.3	6.4	10.0	157.9	-195.7	133.4
2006	1 664.6	1 122.1	542.6	-85.0	57.9	315.3	0.0	11.2	171.0	-182.5	144.3
2007	1 769.5	1 191.2	578.3	-129.3	59.8	342.0	0.0	13.9	187.5	-237.9	154.5
2008	1 857.5	1 252.1	605.3	-220.9	63.1	357.9	0.0	13.3	201.2	-327.8	156.7
2009	1 891.0	1 280.5	610.5	-341.3	68.0	360.3	0.0	12.6	209.1	-437.1	151.2
2010	1 918.2	1 306.1	612.1	-307.5	76.9	347.3	0.0	12.0	213.4	-376.4	133.9
2011	1 930.0	1 313.1	616.8	-275.1	74.1	341.0	0.0	11.5	222.2	-331.2	118.8
2012	1 948.1	1 325.5	622.6	-282.8	74.4	333.7	0.0	10.9	232.0	-321.1	101.7
2013	2 019.8	1 380.9	638.9	-265.3	71.6	330.8	0.0	10.4	238.8	-296.0	92.0
2014	2 077.3	1 421.1	656.2	-237.9	70.6	338.8	0.0	10.6	246.3	-270.3	92.5
2015	2 140.4	1 467.3	673.1	-215.8	69.8	356.0	0.0	11.9	251.4	-262.4	104.6
2016	2 206.2	1 499.1	707.1	-239.2	72.6	370.4	0.0	12.6	256.5	-293.0	113.9
2017	2 290.3	1 541.0	749.3	-238.8	73.2	385.6	1.2	14.2	266.2	-300.5	119.4
2018	2 400.6	1 606.6	794.0	-213.7	72.2	407.9	0.0	16.2	280.3	-285.3	127.6
2019	2 470.3	1 665.9	804.3	-207.7	73.8	431.0	0.0	16.8	293.5	-288.3	137.5
2017											
1st quarter	2 261.0	1 522.2	738.8	-248.7	72.4	379.5	0.0	13.4	262.0	-307.2	117.5
2nd quarter	2 272.6	1 532.5	740.0	-273.7	74.0	383.8	0.0	13.9	264.6	-332.8	119.2
3rd quarter	2 296.5	1 546.9	749.7	-247.8	76.0	386.4	4.9	14.4	267.8	-309.6	118.6
4th quarter	2 331.2	1 562.5	768.6	-185.1	70.2	392.7	0.0	15.0	270.3	-252.4	122.4
2018											
1st quarter	2 360.1	1 579.2	780.9	-176.5	70.6	398.8	0.0	15.7	273.9	-246.4	124.9
2nd quarter	2 388.0	1 597.2	790.8	-226.5	70.5	412.6	0.0	16.1	279.0	-305.8	133.6
3rd quarter	2 417.1	1 618.0	799.1	-220.3	77.4	417.1	0.0	16.5	282.3	-294.2	134.8
4th quarter	2 437.2	1 632.0	805.2	-231.4	70.2	403.1	0.0	16.6	285.9	-295.0	117.2
2019											
1st quarter	2 440.6	1 643.3	797.4	-213.5	73.2	417.9	0.0	16.6	288.2	-286.6	129.7
2nd quarter	2 463.4	1 657.0	806.4	-175.2	72.1	434.2	0.0	16.7	292.7	-261.2	141.5
3rd quarter	2 479.0	1 675.7	803.3	-219.8	74.4	433.2	0.0	16.9	295.6	-299.9	137.6
4th quarter	2 497.9	1 687.7	810.3	-222.4	75.3	438.5	0.0	17.2	297.3	-305.6	141.2

[1]Includes general government intermediate inputs for goods and services sold to other sectors and for own-account investment.
. . . = Not available.

Table 6-11. Chain-Type Quantity Indexes for State and Local Government Consumption Expenditures and Gross Investment

(Seasonally adjusted, 2012 = 100.)   **NIPA Tables 3.9.3, 3.10.3**

Year and quarter	State and local government consumption expenditures and gross investment												
		Consumption expenditures [1]								Gross investment			
						Less							
							Sales to other sectors						
	Total	Total	Compensation of general government employees	Consumption of general government fixed capital	Intermediate goods and services purchased [2]	Own-account investment	Total [3]	Tuition and related educational charges	Health and hospital charges	Total	Structures	Equipment	Intellectual property
1960	27.7	25.9	34.1	13.8	11.6	21.8	13.9	16.8	10.1	33.8	52.2	6.0	2.4
1961	29.4	27.3	35.9	14.6	12.4	23.0	15.0	18.4	10.3	36.5	56.6	6.2	2.7
1962	30.3	28.0	37.1	15.4	12.9	24.6	16.8	20.8	12.0	37.9	58.4	6.6	3.0
1963	32.1	29.3	39.0	16.4	13.7	27.7	18.7	23.8	12.5	41.2	63.2	7.4	3.5
1964	34.2	31.2	41.3	17.4	15.0	28.0	20.6	27.8	13.5	44.2	67.3	8.4	3.9
1965	36.5	33.2	43.8	18.5	16.5	29.2	23.0	32.3	15.4	47.3	72.0	9.0	4.4
1966	38.7	35.2	46.4	19.7	17.9	31.2	25.1	36.6	17.6	50.4	76.5	9.8	5.2
1967	40.7	36.7	48.2	21.0	19.3	31.4	27.7	40.8	20.0	53.6	81.3	10.2	5.8
1968	43.1	39.2	51.1	22.3	21.3	33.2	30.1	45.9	23.3	56.0	85.0	10.6	6.2
1969	44.6	41.7	53.3	23.6	23.6	33.7	31.3	50.5	24.1	53.8	80.5	11.4	6.9
1970	45.8	44.2	55.7	24.8	26.0	34.6	32.6	58.8	24.6	50.5	74.4	11.8	7.6
1971	47.3	46.5	57.9	25.9	28.5	35.1	35.4	66.2	28.6	49.1	72.0	11.9	7.9
1972	48.3	48.2	59.9	26.8	30.2	34.7	37.9	70.3	32.8	47.6	68.5	13.3	8.3
1973	49.7	49.9	62.0	27.8	31.2	34.6	39.5	75.8	36.4	47.9	67.9	14.9	8.8
1974	51.5	52.1	64.3	28.8	32.6	36.2	39.3	75.6	37.5	48.2	67.6	16.2	9.1
1975	53.4	54.7	66.2	29.8	35.2	35.1	39.9	75.7	38.6	47.9	67.1	15.7	9.8
1976	53.8	55.4	67.0	30.7	36.3	32.9	42.2	78.1	41.0	47.2	66.1	15.4	10.4
1977	54.1	56.5	67.9	31.4	37.6	31.5	43.6	81.3	41.7	44.5	61.7	15.0	10.8
1978	55.8	57.7	69.5	32.2	38.6	33.9	45.8	85.8	44.6	48.6	67.7	15.7	12.0
1979	56.6	58.0	70.7	33.0	38.6	38.0	48.0	88.2	48.4	51.0	70.6	17.0	13.2
1980	56.5	57.8	71.6	33.9	37.2	39.6	48.8	91.1	48.8	51.2	70.2	18.1	14.0
1981	55.5	57.5	71.3	34.7	37.4	38.7	50.7	93.1	51.9	47.0	63.2	17.7	14.9
1982	55.4	58.5	71.7	35.4	39.1	39.4	51.9	92.9	53.4	44.2	58.0	18.6	15.9
1983	56.2	59.2	71.2	36.1	41.8	39.8	53.8	95.7	54.8	44.9	57.4	21.2	17.3
1984	58.3	60.5	71.4	37.1	43.9	43.3	53.9	95.6	53.5	50.3	63.9	24.8	19.1
1985	61.6	63.2	73.4	38.4	47.6	47.1	55.8	95.5	54.8	55.8	70.1	28.7	21.5
1986	64.7	66.0	75.5	40.0	51.5	51.4	57.8	96.3	56.2	59.7	74.4	31.5	23.8
1987	66.1	67.4	76.7	41.6	52.7	53.5	59.0	96.9	56.2	61.3	76.0	32.7	25.6
1988	68.7	69.8	79.3	43.4	54.3	56.0	60.2	99.9	56.2	64.5	79.3	34.9	28.4
1989	71.4	72.3	81.6	45.6	57.0	60.9	61.9	102.9	56.7	67.9	80.9	40.3	33.7
1990	74.3	74.7	83.8	48.1	58.8	63.9	63.2	105.2	57.3	73.1	86.9	43.3	37.5
1991	75.9	76.2	84.5	50.4	61.8	66.4	65.3	106.9	59.7	74.8	89.7	41.8	40.3
1992	77.5	78.0	85.7	52.6	64.8	65.8	68.2	108.9	63.2	75.8	90.1	42.6	43.8
1993	78.5	79.3	86.6	54.7	67.7	65.7	71.4	107.6	66.9	75.2	88.1	44.0	44.8
1994	80.7	81.5	87.9	56.7	71.5	66.7	73.8	107.7	69.6	77.5	89.1	49.6	45.2
1995	82.9	83.5	89.5	58.7	74.4	67.5	76.3	108.1	72.7	80.8	92.1	54.4	45.0
1996	84.9	85.0	90.6	60.9	76.4	69.3	77.8	108.4	72.4	84.3	96.5	55.7	47.4
1997	87.6	87.3	92.2	63.4	80.2	74.1	80.4	110.9	74.1	89.0	100.3	60.9	53.0
1998	91.5	90.9	93.9	66.3	86.5	77.7	82.8	114.0	76.3	94.3	104.1	69.4	58.1
1999	95.1	94.0	95.1	69.5	92.5	80.6	85.0	117.4	77.6	100.4	109.6	77.8	61.6
2000	97.6	95.9	96.8	72.8	96.2	84.9	89.3	121.4	81.4	105.4	114.7	82.2	65.9
2001	101.2	99.2	98.8	76.0	103.0	91.7	94.1	122.1	88.2	110.6	121.1	83.9	68.8
2002	104.2	101.9	100.6	78.9	107.0	93.9	96.2	122.2	88.6	114.7	126.1	85.6	70.2
2003	103.9	101.2	100.9	81.2	104.2	94.2	96.8	122.1	87.3	116.3	127.5	86.4	74.1
2004	103.8	101.3	101.0	83.6	104.3	95.2	97.6	118.3	87.6	115.0	125.1	86.3	78.3
2005	103.8	101.9	101.4	86.0	104.7	98.3	98.1	118.3	85.6	112.2	120.4	87.7	82.2
2006	104.9	102.9	102.1	88.4	106.4	103.6	99.2	116.6	87.7	114.2	120.5	95.5	88.6
2007	106.7	104.8	103.3	91.2	108.0	106.9	99.0	112.9	89.9	115.5	118.9	109.5	94.1
2008	106.9	105.0	104.4	93.9	105.0	108.4	98.1	109.3	92.7	115.9	118.3	113.3	97.7
2009	108.9	107.7	104.5	96.0	111.2	108.2	98.7	106.1	96.0	114.7	117.5	109.0	98.0
2010	106.0	105.0	103.2	97.6	106.9	104.0	99.5	103.6	97.5	110.4	113.4	101.4	96.1
2011	102.3	101.6	101.7	98.9	101.2	102.5	100.1	101.3	99.0	105.5	107.1	99.5	98.7
2012	100.0	100.0	100.0	100.0	100.0	100.0	100.0	100.0	100.0	100.0	100.0	100.0	100.0
2013	99.7	100.3	99.7	101.1	101.4	99.3	100.8	99.5	102.8	97.3	95.7	101.7	105.6
2014	99.9	100.4	99.6	102.1	103.0	99.7	103.0	98.9	107.4	97.5	95.8	100.3	109.3
2015	102.8	103.1	100.3	103.2	111.3	105.0	106.9	100.8	112.1	101.5	100.1	101.0	115.6
2016	105.5	105.6	101.4	104.5	118.2	108.5	111.0	103.1	117.3	104.9	102.7	108.0	120.4
2017	106.8	106.8	102.5	106.0	120.5	111.2	113.3	103.9	120.5	106.8	103.3	116.5	126.2
2018	108.1	107.9	103.5	107.7	121.9	114.0	114.7	105.5	122.0	108.9	104.6	121.7	132.8
2019	109.5	109.0	104.8	109.5	122.7	115.9	116.2	106.1	124.0	111.5	107.1	124.4	136.7
2017													
1st quarter	106.4	106.4	102.1	105.4	120.0	110.5	112.6	103.6	119.7	106.1	102.9	114.9	123.9
2nd quarter	106.7	106.7	102.4	105.8	120.3	110.6	113.1	103.6	120.3	106.8	103.3	116.9	125.2
3rd quarter	106.8	106.9	102.6	106.2	120.7	111.1	113.6	103.9	120.8	106.5	102.9	115.8	127.3
4th quarter	107.3	107.2	102.8	106.6	121.1	112.6	113.9	104.4	121.1	107.7	103.9	118.5	128.5
2018													
1st quarter	107.6	107.4	103.0	107.0	121.5	113.3	114.2	105.0	121.3	108.3	104.4	119.1	130.5
2nd quarter	108.2	107.7	103.3	107.5	121.8	113.9	114.5	105.7	121.6	110.5	106.6	121.6	132.5
3rd quarter	108.6	108.1	103.7	107.9	122.1	114.2	114.8	105.6	122.2	110.9	106.7	123.1	133.4
4th quarter	107.9	108.3	103.9	108.4	122.3	114.5	115.2	105.6	123.1	106.1	100.7	122.8	134.8
2019													
1st quarter	108.8	108.6	104.3	108.8	122.5	115.3	115.6	105.7	123.7	109.5	104.7	124.2	135.5
2nd quarter	109.5	108.8	104.5	109.3	122.7	115.6	116.0	105.8	124.1	112.4	108.1	124.6	135.7
3rd quarter	109.6	109.2	105.0	109.8	122.8	116.3	116.4	106.3	124.2	111.5	107.1	123.6	137.1
4th quarter	110.0	109.4	105.2	110.3	122.9	116.6	116.6	106.8	124.2	112.8	108.3	125.2	138.4

[1] Excludes government sales to other sectors and government own-account investment (construction and software).
[2] Includes general government intermediate inputs for goods and services sold to other sectors and for own-account investment.
[3] Includes components not shown separately.

Table 6-12. State Government Current Receipts and Expenditures

(National income and product accounts, calendar years, billions of dollars.)   NIPA Table 3.20

Year	Total [1]	Current tax receipts								Taxes on corporate income	Contributions for government social insurance	Income receipts on assets		
		Total	Personal current taxes		Taxes on production and imports							Total [1]	Interest receipts	Rents and royalties
			Total [1]	Income taxes	Total	Sales taxes	Property taxes	Other						
1959	21.6	16.7	3.1	2.0	12.5	10.0	0.5	2.0		1.1	0.4	0.5	0.3	0.2
1960	23.3	18.1	3.4	2.3	13.5	10.8	0.5	2.2		1.2	0.5	0.5	0.4	0.2
1961	25.0	19.3	3.7	2.5	14.4	11.6	0.5	2.3		1.3	0.5	0.6	0.4	0.2
1962	27.1	21.0	4.0	2.8	15.4	12.6	0.6	2.3		1.5	0.5	0.6	0.4	0.2
1963	29.1	22.4	4.3	3.1	16.4	13.4	0.6	2.4		1.6	0.6	0.6	0.4	0.2
1964	31.8	24.5	4.9	3.6	17.7	14.5	0.6	2.6		1.8	0.7	0.7	0.4	0.2
1965	35.1	26.9	5.4	3.9	19.6	16.1	0.7	2.8		1.9	0.8	0.8	0.5	0.3
1966	41.5	30.2	6.4	4.8	21.7	18.0	0.7	3.0		2.2	0.8	0.9	0.6	0.3
1967	45.6	32.7	7.0	5.3	23.3	19.4	0.7	3.1		2.5	0.9	1.1	0.8	0.3
1968	53.8	38.6	8.8	6.9	26.7	22.8	0.8	3.2		3.1	0.9	1.7	1.4	0.3
1969	61.6	44.0	10.8	8.6	29.9	25.7	0.9	3.3		3.4	1.0	2.1	1.8	0.3
1970	69.1	48.1	12.0	9.6	32.7	28.2	0.9	3.6		3.5	1.1	2.5	2.2	0.3
1971	78.2	53.8	13.4	11.0	36.3	31.4	1.0	3.9		4.0	1.2	2.7	2.3	0.4
1972	95.0	63.5	17.9	15.2	40.7	35.2	1.1	4.4		5.0	1.3	2.9	2.5	0.4
1973	103.1	70.2	19.8	16.8	44.7	38.8	1.2	4.7		5.7	1.5	3.9	3.3	0.5
1974	112.1	75.9	21.1	18.0	48.4	42.0	1.1	5.3		6.3	1.7	5.0	4.4	0.6
1975	125.8	81.9	23.2	19.9	51.8	44.7	1.4	5.6		6.9	1.8	5.6	5.0	0.6
1976	141.7	93.8	27.0	23.4	57.7	49.9	1.5	6.3		9.1	2.2	5.3	4.7	0.6
1977	158.1	105.1	30.9	27.2	63.4	55.0	1.5	6.9		10.8	2.8	6.1	5.4	0.6
1978	177.2	117.4	35.6	31.6	70.3	60.8	1.9	7.6		11.5	3.4	7.4	6.7	0.6
1979	194.8	128.9	38.9	34.6	77.1	65.7	2.3	9.0		12.9	3.9	10.3	9.2	1.1
1980	216.0	140.9	43.7	39.1	83.4	70.0	2.6	10.8		13.7	3.6	13.4	11.2	2.0
1981	236.4	155.6	48.5	43.6	92.7	76.5	2.7	13.6		14.5	3.9	15.6	13.2	2.2
1982	244.7	162.3	52.3	47.0	97.0	80.3	2.8	13.9		13.1	4.0	17.5	15.3	2.1
1983	269.3	180.5	58.9	53.2	106.8	90.0	3.0	13.8		14.9	4.1	19.5	17.2	2.2
1984	303.7	205.4	68.2	61.9	119.7	100.9	3.4	15.5		17.4	4.7	22.4	19.8	2.5
1985	328.6	220.9	73.2	66.1	129.0	109.0	3.5	16.5		18.7	4.9	26.1	23.4	2.6
1986	354.0	234.0	78.3	70.7	134.9	115.8	3.6	15.6		20.7	6.0	27.5	24.6	2.7
1987	375.0	253.7	87.5	79.1	144.3	124.6	3.7	16.0		21.8	7.2	28.4	25.6	2.6
1988	403.1	269.4	90.4	81.5	155.2	135.1	3.8	16.3		23.8	8.4	30.8	28.0	2.7
1989	435.2	288.4	102.4	92.9	163.5	142.3	4.3	17.0		22.4	9.0	32.8	30.2	2.4
1990	469.2	305.6	109.6	99.6	175.4	152.5	4.6	18.3		20.5	10.0	33.9	31.4	2.3
1991	504.1	314.2	111.8	101.4	180.8	157.5	4.9	18.3		21.6	11.6	34.1	31.2	2.6
1992	551.4	338.8	120.6	109.0	195.9	169.7	6.1	20.1		22.2	13.1	33.6	30.4	2.7
1993	584.8	357.1	126.3	114.9	206.3	179.4	5.9	21.0		24.5	14.1	32.4	29.4	2.5
1994	621.6	380.4	132.1	120.2	220.8	191.7	7.0	22.1		27.5	14.5	33.9	30.6	2.5
1995	653.2	401.5	140.9	128.4	231.3	200.5	7.2	23.6		29.2	13.6	36.9	33.4	2.6
1996	686.2	425.9	150.9	138.6	245.0	211.8	8.2	25.0		29.9	12.5	39.4	35.3	2.7
1997	719.6	449.1	162.7	149.7	255.4	221.3	8.1	26.0		31.0	10.8	41.8	37.5	2.7
1998	764.1	480.7	180.4	166.8	268.8	233.3	8.5	27.0		31.6	10.4	43.3	39.2	2.4
1999	812.5	507.9	192.6	178.4	283.0	245.9	9.1	27.9		32.3	9.8	46.6	42.5	2.6
2000	867.7	540.1	213.5	199.2	294.8	255.5	7.8	31.6		31.7	10.8	49.3	44.3	3.6
2001	904.5	546.8	220.1	205.7	300.8	259.9	8.1	32.8		25.9	13.7	46.5	41.4	3.7
2002	917.8	533.7	199.1	184.3	309.6	268.0	7.7	33.9		25.0	15.8	42.4	37.2	3.7
2003	970.5	558.8	202.4	186.4	328.9	282.0	9.1	37.8		27.5	19.9	42.0	35.9	4.4
2004	1 048.7	606.4	220.3	203.4	352.3	300.5	8.3	43.5		33.8	24.7	44.4	37.4	5.1
2005	1 136.5	676.3	246.5	229.0	385.3	325.6	9.1	50.6		44.5	24.6	50.6	42.7	6.0
2006	1 193.9	728.1	271.0	252.5	409.1	346.6	9.7	52.7		48.0	21.5	59.2	50.7	6.5
2007	1 247.9	759.7	288.9	269.8	424.0	359.0	9.8	55.2		46.9	18.9	65.0	55.3	7.5
2008	1 264.0	763.6	299.8	280.9	425.4	354.3	9.7	61.3		38.4	18.7	60.1	49.6	7.9
2009	1 267.1	688.2	256.6	236.3	393.2	335.6	10.4	47.2		38.4	18.6	52.2	43.0	7.0
2010	1 336.1	720.0	264.7	242.2	415.5	354.7	10.7	50.1		39.8	18.1	51.3	42.4	6.6
2011	1 357.8	773.2	293.2	270.5	437.6	369.9	9.8	57.9		42.4	18.3	50.7	41.8	6.7
2012	1 372.3	804.9	313.8	290.4	448.3	379.2	9.7	59.3		42.9	17.5	50.1	40.8	6.9

[1]Includes components not shown separately.

Table 6-12.　State Government Current Receipts and Expenditures—*Continued*

(National income and product accounts, calendar years, billions of dollars.)　　　　　　　　　　　　　　　　　　　　**NIPA Table 3.20**

Year	Current receipts—*Continued*					Current expenditures					Net state government saving, NIPA (surplus + / deficit -)		
	Current transfer receipts												
	Total	Federal grants-in-aid	Local grants-in-aid	From business, net	From persons	Total [1]	Consumption expenditures	Government social benefits to persons	Grants-in-aid to local governments	Interest payments	Total	Social insurance funds	Other
1959	3.6	3.2	0.2	0.0	0.1	20.9	8.2	3.6	7.8	1.2	0.7	0.0	0.7
1960	3.7	3.3	0.2	0.0	0.1	22.9	8.9	3.8	8.8	1.3	0.4	0.0	0.4
1961	4.2	3.7	0.3	0.0	0.1	25.0	9.6	4.1	9.8	1.5	0.0	0.0	0.0
1962	4.6	4.1	0.3	0.1	0.2	27.0	10.2	4.4	10.8	1.6	0.1	0.0	0.1
1963	5.0	4.5	0.3	0.1	0.2	29.4	11.0	4.7	11.9	1.7	-0.3	0.0	-0.3
1964	5.5	4.9	0.3	0.1	0.2	31.9	11.8	5.1	13.0	1.8	-0.1	0.0	-0.1
1965	6.1	5.5	0.3	0.1	0.3	35.6	13.0	5.5	15.0	2.0	-0.4	0.1	-0.5
1966	8.9	8.2	0.4	0.1	0.3	40.7	14.6	6.4	17.4	2.2	0.8	0.1	0.6
1967	10.3	9.4	0.5	0.1	0.3	46.8	16.7	7.7	19.9	2.3	-1.2	0.1	-1.3
1968	11.8	10.7	0.6	0.1	0.3	54.6	19.2	9.5	23.0	2.7	-0.8	0.1	-1.0
1969	13.7	12.4	0.8	0.1	0.4	62.8	22.4	10.8	26.4	2.9	-1.2	0.2	-1.3
1970	16.6	15.2	0.9	0.1	0.4	72.8	25.8	12.9	30.4	3.4	-3.7	0.2	-3.9
1971	19.9	18.4	1.0	0.1	0.4	82.9	28.8	15.3	34.3	4.2	-4.7	0.2	-5.0
1972	26.4	24.6	1.1	0.2	0.5	92.9	31.5	17.5	38.5	5.0	2.2	0.3	1.9
1973	26.6	24.5	1.2	0.2	0.7	103.3	35.3	19.4	42.6	5.4	-0.1	0.3	-0.4
1974	28.6	26.3	1.3	0.2	0.8	116.5	42.1	19.9	47.3	6.4	-4.4	0.4	-4.8
1975	35.4	32.4	1.7	0.2	1.0	134.8	49.1	24.2	52.8	7.5	-9.0	0.5	-9.5
1976	39.1	35.4	2.3	0.3	1.2	145.6	52.8	26.9	57.0	7.7	-4.0	0.6	-4.6
1977	42.8	38.8	2.4	0.3	1.4	158.0	57.5	29.1	61.6	8.5	0.1	1.0	-0.9
1978	47.4	43.1	2.5	0.3	1.5	174.2	63.1	32.0	68.2	9.3	3.0	1.5	1.5
1979	50.1	45.9	2.2	0.4	1.6	193.8	70.0	35.4	76.3	10.3	1.0	1.8	-0.8
1980	56.5	52.1	2.3	0.4	1.7	217.8	78.8	41.3	84.6	11.1	-1.8	1.3	-3.2
1981	59.7	54.6	2.7	0.5	2.0	239.8	86.9	46.7	91.3	12.6	-3.4	1.3	-4.7
1982	58.7	52.7	3.1	0.6	2.3	257.0	93.7	51.0	95.3	14.6	-12.4	1.2	-13.6
1983	62.4	55.0	4.2	0.6	2.6	274.7	99.5	55.9	99.9	16.7	-5.5	1.2	-6.7
1984	67.7	59.0	5.0	0.8	3.0	297.5	106.9	59.8	109.8	17.9	6.2	1.4	4.8
1985	72.2	62.6	5.2	0.9	3.6	327.9	117.4	65.2	121.2	20.8	0.6	1.3	-0.7
1986	81.5	69.3	5.3	2.8	4.1	352.3	124.9	71.4	130.1	22.2	1.7	1.9	-0.1
1987	80.3	69.4	5.5	0.9	4.5	374.8	132.7	77.4	139.5	20.9	0.2	2.2	-2.0
1988	88.0	76.2	5.6	1.1	5.2	402.7	142.4	84.4	149.8	21.2	0.4	2.5	-2.1
1989	98.1	85.1	5.8	1.3	5.9	440.8	153.9	94.4	162.6	24.7	-5.6	2.3	-7.9
1990	112.4	97.6	6.2	1.6	6.9	480.7	167.2	111.0	173.1	23.8	-11.5	2.0	-13.4
1991	136.7	117.1	7.6	2.0	10.0	532.2	175.4	137.6	186.1	27.3	-28.1	2.4	-30.5
1992	158.0	134.3	9.0	2.6	12.1	580.8	183.9	159.5	201.0	30.4	-29.4	3.1	-32.6
1993	172.7	147.6	10.1	2.9	12.1	614.6	193.1	173.6	212.9	29.1	-29.8	4.2	-34.0
1994	183.7	156.9	11.0	3.4	12.4	649.8	204.5	184.4	226.2	28.4	-28.1	4.6	-32.8
1995	191.3	164.0	11.1	4.1	12.2	685.0	213.6	194.8	239.0	30.8	-31.8	4.0	-35.8
1996	197.9	168.9	12.0	5.0	12.0	708.1	219.7	202.5	250.3	27.9	-21.8	2.8	-24.7
1997	207.2	174.1	13.5	6.6	13.0	735.5	231.4	206.8	263.0	25.7	-15.9	1.2	-17.1
1998	219.3	182.9	13.3	9.8	13.2	774.0	248.3	214.6	282.4	19.8	-9.9	1.7	-11.6
1999	238.2	199.1	12.9	11.4	14.9	836.1	272.6	230.3	306.6	17.1	-23.6	1.7	-25.3
2000	258.8	212.6	14.0	15.0	17.2	896.0	292.8	248.4	331.0	13.5	-28.4	2.0	-30.3
2001	289.7	238.9	14.9	16.1	19.8	984.0	314.4	281.0	350.7	20.2	-79.5	2.6	-82.1
2002	317.7	263.5	15.5	17.2	21.5	1 037.5	320.3	307.2	367.5	31.5	-119.7	1.4	-121.1
2003	340.5	285.1	16.4	16.6	22.3	1 086.6	319.2	326.2	383.5	47.5	-116.1	3.2	-119.4
2004	363.4	303.4	17.6	17.7	24.6	1 139.7	327.1	355.8	401.6	44.5	-91.0	7.3	-98.3
2005	374.1	312.2	16.6	18.4	26.9	1 195.2	347.4	375.2	413.9	46.8	-58.7	7.2	-65.9
2006	375.4	314.2	15.6	18.2	28.0	1 239.7	361.2	371.3	441.7	52.3	-45.8	4.6	-50.5
2007	397.4	332.0	17.7	19.0	29.0	1 324.9	384.6	398.5	469.2	51.8	-77.0	2.7	-79.7
2008	415.6	346.6	17.7	21.1	30.7	1 383.9	407.7	418.6	486.1	54.1	-119.9	1.7	-121.6
2009	501.2	432.8	16.5	21.7	30.7	1 455.0	411.7	453.0	486.3	87.3	-187.9	2.2	-190.1
2010	539.5	473.3	16.6	20.1	30.1	1 487.1	411.2	481.5	488.8	88.4	-151.0	3.2	-154.2
2011	507.6	441.0	16.6	19.7	31.3	1 515.1	425.5	489.9	494.6	88.4	-157.2	4.2	-161.4
2012	491.2	424.0	16.8	18.8	32.4	1 579.2	459.7	501.9	502.8	97.8	-206.9	3.9	-210.9

[1] Includes components not shown separately.

Table 6-13. Local Government Current Receipts and Expenditures

(National income and product accounts, calendar years, billions of dollars.)   NIPA Table 3.21

Year	Current receipts												
		Current tax receipts								Contributions for government social insurance	Income receipts on assets		
			Personal current taxes		Taxes on production and imports				Taxes on corporate income				
	Total 1	Total	Total 1	Income taxes	Total	Sales taxes	Property taxes	Other			Total 1	Interest receipts	Rents and royalties
1959	27.0	17.1	0.8	0.2	16.4	1.2	14.3	0.9	0.0	...	0.7	0.5	0.1
1960	30.0	18.8	0.8	0.3	18.0	1.3	15.7	0.9	0.0	...	0.8	0.7	0.1
1961	32.9	20.3	0.9	0.3	19.4	1.4	17.0	1.0	0.0	...	0.9	0.7	0.2
1962	35.6	21.8	1.0	0.3	20.8	1.5	18.4	1.0	0.0	...	1.0	0.8	0.2
1963	38.7	23.4	1.1	0.4	22.3	1.6	19.7	1.0	0.0	...	1.0	0.9	0.2
1964	42.3	25.3	1.2	0.5	24.1	1.9	21.1	1.1	0.0	...	1.3	1.1	0.2
1965	46.1	27.0	1.2	0.5	25.7	2.1	22.5	1.1	0.0	...	1.4	1.2	0.2
1966	50.5	28.5	1.4	0.6	27.1	2.0	23.8	1.3	0.0	...	1.7	1.4	0.3
1967	56.5	31.3	1.6	0.8	29.5	1.9	26.2	1.3	0.2	...	2.0	1.6	0.3
1968	62.6	34.8	1.7	0.9	32.8	2.3	29.1	1.4	0.3	...	1.7	1.4	0.4
1969	70.1	38.4	2.0	1.1	36.1	2.9	31.9	1.4	0.3	...	2.2	1.8	0.4
1970	81.4	43.2	2.3	1.3	40.6	3.5	35.7	1.5	0.2	...	2.6	2.2	0.5
1971	90.9	47.9	2.5	1.5	45.2	4.0	39.5	1.8	0.3	...	2.8	2.3	0.5
1972	101.9	52.0	3.0	2.0	48.8	4.6	42.2	2.0	0.3	...	3.0	2.4	0.6
1973	114.0	56.1	3.0	2.0	52.7	5.2	45.2	2.3	0.3	...	3.9	3.3	0.6
1974	123.8	60.2	3.4	2.3	56.4	6.1	47.9	2.4	0.3	...	5.2	4.5	0.7
1975	137.7	65.5	3.7	2.5	61.4	7.0	51.9	2.5	0.4	...	5.6	4.9	0.7
1976	150.8	71.9	4.1	2.8	67.3	7.9	56.7	2.7	0.5	...	5.1	4.4	0.7
1977	165.3	78.6	4.5	3.2	73.6	9.0	61.7	2.9	0.6	...	5.6	4.9	0.7
1978	180.5	80.8	4.9	3.4	75.3	10.2	61.8	3.3	0.6	...	7.2	6.5	0.7
1979	191.3	83.1	5.1	3.6	77.3	11.5	62.1	3.7	0.6	...	9.8	8.9	0.9
1980	208.3	89.1	5.1	3.5	83.3	12.8	66.2	4.2	0.7	...	12.9	11.8	1.1
1981	226.9	100.2	6.2	4.3	93.0	14.3	74.4	4.3	1.0	...	16.4	15.3	1.1
1982	244.4	110.8	6.9	4.9	103.0	15.9	82.5	4.6	1.0	...	19.2	17.8	1.4
1983	262.5	120.4	7.2	5.1	112.1	17.7	88.9	5.5	1.0	...	21.9	19.8	2.1
1984	289.9	131.9	7.8	5.6	122.7	20.1	96.3	6.3	1.4	...	25.2	22.8	2.4
1985	318.4	142.8	8.2	6.0	133.1	22.1	104.0	7.0	1.5	...	29.5	26.7	2.7
1986	342.5	155.6	8.9	6.8	144.7	24.1	112.6	8.1	2.0	...	31.4	27.9	3.5
1987	360.1	168.5	9.1	6.8	157.3	25.7	122.7	8.9	2.1	...	30.1	27.4	2.7
1988	387.9	183.3	11.7	9.1	169.4	27.2	132.7	9.4	2.3	...	30.4	28.6	1.7
1989	420.0	199.6	12.3	9.4	185.6	30.1	145.6	10.0	1.8	...	33.8	32.1	1.7
1990	446.2	213.6	12.9	10.0	198.7	31.8	157.0	9.9	2.0	...	34.5	32.7	1.8
1991	475.9	230.1	13.5	10.3	214.5	33.2	171.1	10.2	2.1	...	33.2	31.2	1.9
1992	501.1	241.0	14.7	11.4	224.2	34.6	178.6	11.0	2.1	...	30.4	28.4	2.0
1993	522.2	247.7	14.8	11.3	230.5	37.0	181.3	12.1	2.4	...	28.7	26.6	2.1
1994	556.9	263.8	15.9	12.0	245.4	39.7	192.4	13.3	2.5	...	29.6	27.6	2.0
1995	584.6	270.7	17.2	13.3	251.0	42.2	195.3	13.4	2.5	...	32.2	30.3	2.0
1996	617.4	283.7	17.8	13.7	262.9	44.4	204.2	14.4	3.1	...	34.3	32.4	2.0
1997	652.4	300.8	19.3	14.9	278.4	47.4	215.5	15.6	3.1	...	36.3	34.3	2.1
1998	690.4	314.2	20.9	16.2	290.0	50.6	222.5	16.9	3.4	...	38.0	35.8	2.2
1999	739.5	332.5	21.9	17.1	307.1	55.6	233.7	17.8	3.5	...	38.6	36.1	2.5
2000	791.8	353.1	23.2	18.1	326.5	61.3	246.9	18.3	3.5	...	44.6	42.0	2.7
2001	825.4	367.5	22.9	17.6	341.6	61.9	260.0	19.8	3.0	...	42.6	39.8	2.8
2002	865.1	394.8	22.7	17.1	366.2	63.1	281.6	21.5	5.9	...	36.2	33.2	2.9
2003	915.7	419.8	24.4	18.3	389.0	66.8	297.9	24.2	6.5	...	31.9	28.6	3.2
2004	957.8	451.5	27.1	20.3	416.4	70.5	317.9	28.0	7.9	...	32.4	29.0	3.4
2005	1 016.4	490.2	30.0	22.5	449.8	76.8	342.3	30.7	10.4	...	38.0	34.2	3.8
2006	1 089.2	526.4	31.5	23.6	483.6	85.2	365.3	33.1	11.3	...	48.0	44.0	3.9
2007	1 154.8	561.6	34.6	26.5	516.0	90.0	394.1	32.0	11.0	...	52.9	48.9	4.0
2008	1 164.4	565.3	33.8	26.1	522.6	90.4	404.0	28.1	9.0	...	45.6	41.4	4.2
2009	1 171.2	580.0	31.2	23.3	541.6	88.3	424.7	28.6	7.2	...	35.7	31.4	4.2
2010	1 184.5	585.7	33.0	25.0	544.9	91.2	424.4	29.3	7.8	...	31.3	26.7	4.6
2011	1 200.4	593.1	33.8	25.7	550.9	93.8	427.1	30.1	8.4	...	29.3	24.5	4.8
2012	1 204.4	600.3	35.1	26.9	556.7	95.7	430.3	30.7	8.5	...	28.4	23.4	5.0

1Includes components not shown separately.
. . . = Not available.

Table 6-13. Local Government Current Receipts and Expenditures—*Continued*

(National income and product accounts, calendar years, billions of dollars.)   **NIPA Table 3.21**

| Year | Current receipts—*Continued* | | | | | Current expenditures | | | | | Net local government saving, NIPA (surplus + / deficit -) | | |
| | Current transfer receipts | | | | | | | | | | | | |
	Total	Federal grants-in-aid	State grants-in-aid	From business, net	From persons	Total ¹	Consumption expenditures	Government social benefits to persons	Grants-in-aid to state governments	Interest payments	Total	Social insurance funds	Other
1959	8.4	0.4	7.8	0.1	0.1	25.5	23.0	0.7	0.2	1.5	1.5	...	1.5
1960	9.6	0.5	8.8	0.1	0.2	27.9	25.2	0.8	0.2	1.7	2.2	...	2.2
1961	10.8	0.6	9.8	0.2	0.3	30.7	27.7	0.9	0.3	1.8	2.2	...	2.2
1962	11.9	0.6	10.8	0.2	0.3	32.3	29.1	0.9	0.3	1.9	3.3	...	3.3
1963	13.1	0.8	11.9	0.2	0.3	34.7	31.3	1.0	0.3	2.1	4.0	...	4.0
1964	14.6	1.1	13.0	0.2	0.3	37.9	34.2	1.0	0.3	2.3	4.4	...	4.4
1965	16.6	1.1	15.0	0.3	0.3	41.6	37.6	1.1	0.3	2.5	4.5	...	4.5
1966	19.3	1.2	17.4	0.2	0.4	46.1	41.7	1.3	0.4	2.7	4.4	...	4.4
1967	22.4	1.5	19.9	0.4	0.6	50.6	45.6	1.6	0.5	2.9	5.9	...	5.9
1968	25.2	1.1	23.0	0.4	0.7	57.5	51.5	2.0	0.6	3.3	5.1	...	5.1
1969	28.8	1.3	26.4	0.4	0.7	64.9	58.1	2.4	0.8	3.6	5.3	...	5.3
1970	34.8	3.2	30.4	0.4	0.8	74.7	66.3	3.2	0.9	4.3	6.7	...	6.7
1971	39.5	3.8	34.3	0.5	0.9	85.6	75.4	4.0	1.0	5.2	5.3	...	5.3
1972	46.0	5.9	38.5	0.5	1.1	95.6	83.9	4.5	1.1	6.1	6.3	...	6.3
1973	53.2	8.9	42.6	0.7	1.0	104.3	91.5	4.7	1.2	6.8	9.8	...	9.8
1974	57.9	8.7	47.3	0.9	1.1	117.0	102.4	5.4	1.3	7.9	6.8	...	6.8
1975	66.4	11.2	52.8	1.0	1.4	134.3	116.5	6.6	1.7	9.2	3.5	...	3.5
1976	73.5	13.6	57.0	1.1	1.7	144.0	124.4	7.3	2.3	9.8	6.8	...	6.8
1977	80.9	16.0	61.6	1.3	1.9	156.9	135.5	8.0	2.4	10.7	8.4	...	8.4
1978	92.3	20.4	68.2	1.5	2.2	170.1	146.8	8.7	2.5	11.7	10.4	...	10.4
1979	98.8	18.1	76.3	1.8	2.6	184.0	159.4	8.9	2.2	13.2	7.2	...	7.2
1980	107.3	17.6	84.6	2.0	3.0	201.1	174.0	9.9	2.3	14.5	7.2	...	7.2
1981	112.2	14.8	91.3	2.4	3.7	219.4	189.2	10.4	2.7	16.7	7.5	...	7.5
1982	115.6	13.5	95.3	2.7	4.1	237.1	203.7	10.2	3.1	19.6	7.3	...	7.3
1983	120.4	13.0	99.9	3.0	4.5	255.8	217.2	11.0	4.2	22.8	6.7	...	6.7
1984	131.7	13.3	109.8	3.4	5.2	275.2	233.0	11.4	5.0	25.3	14.7	...	14.7
1985	144.0	13.6	121.2	3.5	5.7	298.8	254.4	12.1	5.2	26.4	19.6	...	19.6
1986	153.7	13.2	130.1	3.9	6.5	323.8	275.0	12.9	5.3	29.6	18.7	...	18.7
1987	159.2	9.0	139.5	4.0	6.7	348.3	295.6	13.4	5.5	32.6	11.9	...	11.9
1988	170.5	9.5	149.8	4.3	6.9	371.3	316.5	14.1	5.6	33.8	16.6	...	16.6
1989	181.9	6.8	162.6	5.1	7.5	407.0	347.4	15.0	5.8	37.6	13.0	...	13.0
1990	193.4	6.7	173.1	5.5	8.0	441.2	379.0	16.6	6.2	38.0	5.0	...	5.0
1991	207.7	7.0	186.1	5.9	8.7	472.9	404.7	18.9	7.6	40.3	3.0	...	3.0
1992	224.8	7.4	201.0	6.7	9.8	509.2	435.2	20.5	9.0	43.2	-8.1	...	-8.1
1993	240.0	8.1	212.9	7.7	11.4	529.6	453.4	21.6	10.1	43.1	-7.4	...	-7.4
1994	257.6	9.9	226.2	8.6	12.9	555.8	477.7	22.4	11.0	43.3	1.2	...	1.2
1995	273.2	10.5	239.0	9.4	14.3	584.3	503.2	22.9	11.1	45.8	0.3	...	0.3
1996	288.7	12.6	250.3	10.2	15.7	608.0	528.4	21.8	12.0	44.5	9.4	...	9.4
1997	305.4	14.0	263.0	11.1	17.3	634.8	554.5	20.8	13.5	44.5	17.6	...	17.6
1998	330.0	17.9	282.4	11.8	17.9	665.1	587.4	21.1	13.3	41.7	25.3	...	25.3
1999	358.7	20.1	306.6	12.7	19.3	706.6	629.4	22.1	12.9	40.7	32.9	...	32.9
2000	385.8	20.5	331.0	13.5	20.8	753.6	676.3	23.0	14.0	38.6	38.3	...	38.3
2001	409.1	22.4	350.7	13.7	22.3	811.2	725.4	24.0	14.9	45.3	14.2	...	14.2
2002	430.7	23.7	367.5	15.4	24.1	866.4	770.8	25.8	15.5	52.8	-1.2	...	-1.2
2003	464.4	36.6	383.5	17.2	27.1	921.0	808.0	27.3	16.4	67.7	-5.3	...	-5.3
2004	478.3	28.8	401.6	18.8	29.1	975.2	860.5	29.2	17.6	65.9	-17.4	...	-17.4
2005	492.8	31.1	413.9	18.1	29.6	1 024.3	909.2	31.4	16.6	65.0	-7.9	...	-7.9
2006	520.8	26.6	441.7	20.6	31.3	1 082.8	964.8	32.6	15.6	67.8	6.4	...	6.4
2007	550.4	26.8	469.2	21.5	32.5	1 150.6	1 026.8	34.8	17.7	69.7	4.3	...	4.3
2008	566.0	24.4	486.1	22.4	32.6	1 209.6	1 081.0	36.8	17.7	73.0	-45.2	...	-45.2
2009	567.7	25.3	486.3	22.3	33.3	1 255.2	1 096.7	39.6	16.5	101.4	-84.0	...	-84.0
2010	577.8	31.9	488.8	23.3	33.1	1 270.8	1 107.1	42.3	16.6	103.7	-86.3	...	-86.3
2011	585.4	31.4	494.6	24.4	33.9	1 256.2	1 092.0	42.1	16.6	104.5	-55.9	...	-55.9
2012	581.0	19.2	502.8	23.1	35.1	1 250.2	1 076.7	42.4	16.8	113.2	-45.8	...	-45.8

¹Includes components not shown separately.
. . . = Not available.

SECTION 6C: FEDERAL GOVERNMENT BUDGET ACCOUNTS

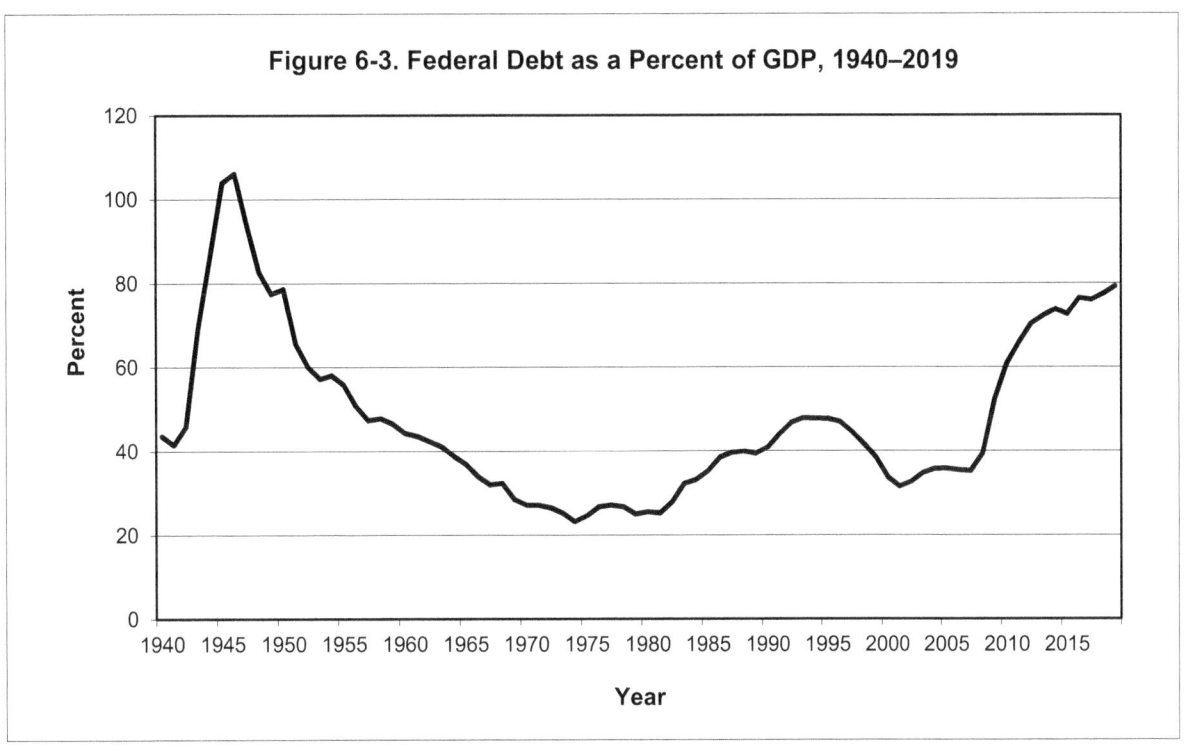

Figure 6-3. Federal Debt as a Percent of GDP, 1940–2019

- The "debt held by the public" is generally thought to be a more significant measure of the burden of the federal debt on credit markets than the gross debt, because it nets out the intragovernmental debt of the Social Security and other U.S. government trust funds. The ratio of debt held by the public to GDP was reduced in the late 1990s, began to creep back upward after 2001, and then rose rapidly beginning in 2008, as the recession reduced revenues and increased spending, and taxes were reduced further. The spending increases were a result of both automatic stabilizer programs, such as unemployment insurance, and an unprecedented level of spending for stimulus and financial rescue programs. At the end of fiscal year 2019, the debt held by the public was 79.2 percent of GDP in 2019, an increase from 77.4 percent in 2018. This represented the highest debt/GDP ratio since 1948. (Tables 6-14 and 6-15)

- At the end of 2019, foreign residents—including central banks—held nearly $6.8 trillion of Treasury debt, 40.3 percent of the nearly $16.8 trillion total debt held by the public. (Table 6-15)

Table 6-14A. Federal Government Receipts and Outlays by Fiscal Year [1]

(Budget accounts, millions of dollars.)

Year	Receipts, outlays, deficit, and financing		Budget surplus or deficit (-)			Sources of financing, total		Receipts by source		Social insurance taxes and contributions		
	Total receipts, net	Total outlays, net	Total	On-budget	Off-budget	Borrowing from the public	Other financing	Individual income taxes	Corporate income taxes	Employment taxes and contributions	Unemployment insurance	Other retirement contributions
1940	6 548	9 468	-2 920	-3 484	564	...	...	892	1 197	725	1 015	45
1941	8 712	13 653	-4 941	-5 594	653	5 451	-510	1 314	2 124	827	1 056	57
1942	14 634	35 137	-20 503	-21 333	830	19 530	973	3 263	4 719	1 064	1 299	89
1943	24 001	78 555	-54 554	-55 595	1 041	60 013	-5 459	6 505	9 557	1 338	1 477	229
1944	43 747	91 304	-47 557	-48 735	1 178	57 030	-9 473	19 705	14 838	1 557	1 644	272
1945	45 159	92 712	-47 553	-48 720	1 167	50 386	-2 833	18 372	15 988	1 592	1 568	291
1946	39 296	55 232	-15 936	-16 964	1 028	6 679	9 257	16 098	11 883	1 517	1 316	282
1947	38 514	34 496	4 018	2 861	1 157	-17 522	13 504	17 935	8 615	1 835	1 329	259
1948	41 560	29 764	11 796	10 548	1 248	-8 069	-3 727	19 315	9 678	2 168	1 343	239
1949	39 415	38 835	580	-684	1 263	-1 948	1 368	15 552	11 192	2 246	1 205	330
1950	39 443	42 562	-3 119	-4 702	1 583	4 701	-1 582	15 755	10 449	2 648	1 332	358
1951	51 616	45 514	6 102	4 259	1 843	-4 697	-1 405	21 616	14 101	3 688	1 609	377
1952	66 167	67 686	-1 519	-3 383	1 864	432	1 087	27 934	21 226	4 315	1 712	418
1953	69 608	76 101	-6 493	-8 259	1 766	3 625	2 868	29 816	21 238	4 722	1 675	423
1954	69 701	70 855	-1 154	-2 831	1 677	6 116	-4 962	29 542	21 101	5 192	1 561	455
1955	65 451	68 444	-2 993	-4 091	1 098	2 117	876	28 747	17 861	5 981	1 449	431
1956	74 587	70 640	3 947	2 494	1 452	-4 460	513	32 188	20 880	7 059	1 690	571
1957	79 990	76 578	3 412	2 639	773	-2 836	-576	35 620	21 167	7 405	1 950	642
1958	79 636	82 405	-2 769	-3 315	546	7 016	-4 247	34 724	20 074	8 624	1 933	682
1959	79 249	92 098	-12 849	-12 149	-700	8 365	4 484	36 719	17 309	8 821	2 131	770
1960	92 492	92 191	301	510	-209	2 139	-2 440	40 715	21 494	11 248	2 667	768
1961	94 388	97 723	-3 335	-3 766	431	1 517	1 818	41 338	20 954	12 679	2 903	857
1962	99 676	106 821	-7 146	-5 881	-1 265	9 653	-2 507	45 571	20 523	12 835	3 337	875
1963	106 560	111 316	-4 756	-3 966	-789	5 968	-1 212	47 588	21 579	14 746	4 112	946
1964	112 613	118 528	-5 915	-6 546	632	2 871	3 044	48 697	23 493	16 959	3 997	1 007
1965	116 817	118 228	-1 411	-1 605	194	3 929	-2 518	48 792	25 461	17 358	3 803	1 081
1966	130 835	134 532	-3 698	-3 068	-630	2 936	762	55 446	30 073	20 662	3 755	1 129
1967	148 822	157 464	-8 643	-12 620	3 978	2 912	5 731	61 526	33 971	27 823	3 575	1 221
1968	152 973	178 134	-25 161	-27 742	2 581	22 919	2 242	68 726	28 665	29 224	3 346	1 354
1969	186 882	183 640	3 242	-507	3 749	-11 437	8 195	87 249	36 678	34 236	3 328	1 451
1970	192 807	195 649	-2 842	-8 694	5 852	5 090	-2 248	90 412	32 829	39 133	3 464	1 765
1971	187 139	210 172	-23 033	-26 052	3 019	19 839	3 194	86 230	26 785	41 699	3 674	1 952
1972	207 309	230 681	-23 373	-26 068	2 695	19 340	4 033	94 737	32 166	46 120	4 357	2 097
1973	230 799	245 707	-14 908	-15 246	338	18 533	-3 625	103 246	36 153	54 876	6 051	2 187
1974	263 224	269 359	-6 135	-7 198	1 063	2 789	3 346	118 952	38 620	65 888	6 837	2 347
1975	279 090	332 332	-53 242	-54 148	906	51 001	2 241	122 386	40 621	75 199	6 771	2 565
1976	298 060	371 792	-73 732	-69 427	-4 306	82 704	-8 972	131 603	41 409	79 901	8 054	2 814
TQ	81 232	95 975	-14 744	-14 065	-679	18 105	-3 361	38 801	8 460	21 801	2 698	720
1977	355 559	409 218	-53 659	-49 933	-3 726	53 595	64	157 626	54 892	92 199	11 312	2 974
1978	399 561	458 746	-59 185	-55 416	-3 770	58 022	1 163	180 988	59 952	103 888	13 850	3 237
1979	463 302	504 028	-40 726	-39 633	-1 093	33 180	7 546	217 841	65 677	120 05	15 387	3 494
1980	517 112	590 941	-73 830	-73 141	-689	71 617	2 213	244 069	64 600	138 74	15 336	3 719
1981	599 272	678 241	-78 968	-73 859	-5 109	77 487	1 481	285 917	61 137	16 297	15 763	3 984
1982	617 766	745 743	-127 977	-120 593	-7 384	135 165	-7 188	297 744	49 207	18 068	16 600	4 212
1983	600 562	808 364	-207 802	-207 692	-110	212 693	-4 891	288 938	37 022	18 576	18 799	4 429
1984	666 438	851 805	-185 367	-185 269	-98	169 707	15 660	298 415	56 893	20 965	25 138	4 580
1985	734 037	946 344	-212 308	-221 529	9 222	200 285	12 023	334 531	61 331	23 464	25 758	4 759
1986	769 155	990 382	-221 227	-237 915	16 688	233 363	-12 136	348 959	63 143	25 506	24 098	4 742
1987	854 287	1 004 017	-149 730	-168 357	18 627	149 130	600	392 557	83 926	27 302	25 575	4 715
1988	909 238	1 064 416	-155 178	-192 265	37 087	161 863	-6 685	401 181	94 508	30 509	24 584	4 658
1989	991 104	1 143 743	-152 639	-205 393	52 754	139 100	13 539	445 690	10 329	33 285	22 011	4 546
1990	1 031 958	1 252 993	-221 036	-277 626	56 590	220 842	194	466 884	93 507	35 389	21 635	4 522
1991	1 054 988	1 324 226	-269 238	-321 435	52 198	277 441	-8 203	467 827	98 086	37 052	20 922	4 568
1992	1 091 208	1 381 529	-290 321	-340 408	50 087	310 738	-20 417	475 964	10 027	38 549	23 410	4 788
1993	1 154 334	1 409 386	-255 051	-300 398	45 347	248 659	6 392	509 680	11 752	39 693	26 556	4 805
1994	1 258 566	1 461 752	-203 186	-258 840	55 654	184 669	18 517	543 055	14 038	42 881	28 004	4 661
1995	1 351 790	1 515 742	-163 952	-226 367	62 415	171 313	-7 361	590 244	15 700	45 104	28 878	4 550
1996	1 453 053	1 560 484	-107 431	-174 019	66 588	129 695	-22 264	656 417	17 182	47 636	28 584	4 469
1997	1 579 232	1 601 116	-21 884	-103 248	81 364	38 271	-16 387	737 466	18 219	50 675	28 202	4 418
1998	1 721 728	1 652 458	69 270	-29 925	99 195	-51 245	-18 025	828 586	18 867	54 001	27 484	4 333
1999	1 827 452	1 701 842	125 610	1 920	123 690	-88 736	-36 874	879 480	18 468	58 088	26 480	4 473
2000	2 025 191	1 788 950	236 241	86 422	149 819	-222 559	-13 682	100 446	20 728	62 045	27 640	4 761
2001	1 991 082	1 862 846	128 236	-32 445	160 681	-90 189	-38 047	994 339	15 107	66 144	27 812	4 713
2002	1 853 136	2 010 894	-157 758	-317 417	159 659	220 812	-63 054	858 345	14 804	66 854	27 619	4 594
2003	1 782 314	2 159 899	-377 585	-538 418	160 833	373 016	4 569	793 699	131 778	674 981	33 366	4 631
2004	1 880 114	2 292 841	-412 727	-567 961	155 234	382 101	30 626	808 959	189 371	689 360	39 453	4 594
2005	2 153 611	2 471 957	-318 346	-493 611	175 265	296 668	21 678	927 222	278 282	747 664	42 002	4 459
2006	2 406 869	2 655 050	-248 181	-434 494	186 313	236 760	11 421	1 043 908	353 915	790 043	43 420	4 358
2007	2 567 985	2 728 686	-160 701	-342 153	181 452	206 157	-45 456	1 163 472	370 243	824 258	41 091	4 258
2008	2 523 991	2 982 544	-458 553	-641 848	183 295	767 921	-309 368	1 145 747	304 346	856 459	39 527	4 169
2009	2 104 989	3 517 677	-1 412 688	-1 549 681	136 993	1 741 657	-328 969	915 308	138 229	848 885	37 889	4 143
2010	2 162 706	3 457 079	-1 294 373	-1 371 378	77 005	1 474 175	-179 802	898 549	191 437	815 894	44 823	4 097
2011	2 303 466	3 603 065	-1 299 599	-1 366 781	67 182	1 109 305	190 294	1 091 473	181 085	758 516	56 241	4 035
2012	2 449 990	3 526 563	-1 076 573	-1 138 486	61 913	1 152 944	-76 371	1 132 206	242 289	774 927	66 647	3 740
2013	2 775 106	3 454 881	-679 775	-719 238	39 463	701 582	-21 807	1 316 405	273 506	887 445	56 811	3 564
2014	3 021 491	3 506 284	-484 793	-514 305	29 512	797 186	-312 393	1 394 568	320 731	965 029	54 957	3 472
2015	3 249 890	3 691 850	-441 960	-469 255	27 295	336 793	105 167	1 540 802	343 797	101 042	51 178	3 652
2016	3 267 965	3 852 616	-584 651	-620 158	35 507	1 050 932	-466 281	1 546 075	299 571	106 230	48 856	3 904
2017	3 316 184	3 981 630	-665 446	-714 863	49 417	497 815	167 631	1 587 120	297 048	111 189	45 808	4 192
2018	3 329 907	4 109 044	-779 137	-785 312	6 175	1 084 128	-304 991	1 683 538	204 733	112 115	45 042	4 504
2019	3 464 161	4 448 316	-984 155	-991 841	7 686	1 051 179	-67 024	1 717 857	230 245	119 739	41 193	4 786

[1] Fiscal years through 1976 are from July 1 through June 30. Beginning with October 1976 (fiscal year 1977), fiscal years are from October 1 through September 30. The period from July 1 through September 30, 1976, is a separate fiscal period known as the transition quarter (TQ) and is not included in any fiscal year.
. . . = Not available.

Table 6-14A. Federal Government Receipts and Outlays by Fiscal Year [1]—Continued

(Budget accounts, millions of dollars.)

Year	Receipts by source—Continued					Outlays by function						
	Excise taxes	Estate and gift taxes	Customs duties and fees	Miscellaneous receipts		National defense	International affairs	General science, space, and technology	Energy	Natural resources and environment	Agriculture	Commerce and housing credit
				Federal reserve deposits	All other							
1940	1 977	353	331	...	14	1 660	51	...	88	997	369	550
1941	2 552	403	365	...	14	6 435	145	...	91	817	339	398
1942	3 399	420	369	...	11	25 658	968	4	156	819	344	1 521
1943	4 096	441	308	...	50	66 699	1 286	1	116	726	343	2 151
1944	4 759	507	417	...	48	79 143	1 449	48	65	642	1 275	624
1945	6 265	637	341	...	105	82 965	1 913	111	25	455	1 635	-2 630
1946	6 998	668	424	...	109	42 681	1 935	34	41	482	610	-1 857
1947	7 211	771	477	15	69	12 808	5 791	5	18	700	814	-923
1948	7 356	890	403	100	68	9 105	4 566	1	292	780	69	306
1949	7 502	780	367	187	54	13 150	6 052	48	341	1 080	1 924	800
1950	7 550	698	407	192	55	13 724	4 673	55	327	1 308	2 049	1 035
1951	8 648	708	609	189	72	23 566	3 647	51	383	1 310	-323	1 228
1952	8 852	818	533	278	81	46 089	2 691	49	474	1 233	176	1 278
1953	9 877	881	596	298	81	52 802	2 119	49	425	1 289	2 253	910
1954	9 945	934	542	341	88	49 266	1 596	46	432	1 007	1 817	-184
1955	9 131	924	585	251	90	42 729	2 223	74	325	940	3 514	92
1956	9 929	1 161	682	287	140	42 523	2 414	79	174	870	3 486	506
1957	10 534	1 365	735	434	139	45 430	3 147	122	240	1 098	2 288	1 424
1958	10 638	1 393	782	664	123	46 815	3 364	141	348	1 407	2 411	930
1959	10 578	1 333	925	491	171	49 015	3 144	294	382	1 632	4 509	1 933
1960	11 676	1 606	1 105	1 093	119	48 130	2 988	599	464	1 559	2 623	1 618
1961	11 860	1 896	982	788	130	49 601	3 184	1 042	510	1 779	2 641	1 203
1962	12 534	2 016	1 142	718	125	52 345	5 639	1 723	604	2 044	3 562	1 424
1963	13 194	2 167	1 205	828	194	53 400	5 308	3 051	530	2 251	4 384	62
1964	13 731	2 394	1 252	947	139	54 757	4 945	4 897	572	2 364	4 609	418
1965	14 570	2 716	1 442	1 372	222	50 620	5 273	5 823	699	2 531	3 954	1 157
1966	13 062	3 066	1 767	1 713	163	58 111	5 580	6 717	612	2 719	2 447	3 245
1967	13 719	2 978	1 901	1 805	302	71 417	5 566	6 233	782	2 869	2 990	3 979
1968	14 079	3 051	2 038	2 091	400	81 926	5 301	5 524	1 037	2 988	4 544	4 280
1969	15 222	3 491	2 319	2 662	247	82 497	4 600	5 020	1 010	2 900	5 826	-119
1970	15 705	3 644	2 430	3 266	158	81 692	4 330	4 511	997	3 065	5 166	2 112
1971	16 614	3 735	2 591	3 533	325	78 872	4 159	4 182	1 035	3 915	4 290	2 366
1972	15 477	5 436	3 287	3 252	380	79 174	4 781	4 175	1 296	4 241	5 227	2 222
1973	16 260	4 917	3 188	3 495	425	76 681	4 149	4 032	1 237	4 775	4 821	931
1974	16 844	5 035	3 334	4 845	523	79 347	5 710	3 980	1 303	5 697	2 194	4 705
1975	16 551	4 611	3 676	5 777	935	86 509	7 097	3 991	2 916	7 346	2 997	9 947
1976	16 963	5 216	4 074	5 451	2 576	89 619	6 433	4 373	4 204	8 184	3 109	7 619
TQ	4 473	1 455	1 212	1 500	111	22 269	2 458	1 162	1 129	2 524	972	931
1977	17 548	7 327	5 150	5 908	623	97 241	6 353	4 736	5 770	10 032	6 734	3 093
1978	18 376	5 285	6 573	6 641	778	10 449	7 482	4 926	7 991	10 983	11 301	6 254
1979	18 745	5 411	7 439	8 327	925	11 634	7 459	5 234	9 179	12 135	11 176	4 686
1980	24 329	6 389	7 174	11 767	981	13 399	12 714	5 831	10 156	13 858	8 774	9 390
1981	40 839	6 787	8 083	12 834	956	15 751	13 104	6 468	15 166	13 568	11 241	8 206
1982	36 311	7 991	8 854	15 186	975	18 530	12 300	7 199	13 527	12 998	15 866	6 256
1983	35 300	6 053	8 655	14 492	1 108	20 990	11 848	7 934	9 353	12 672	22 807	6 681
1984	37 361	6 010	11 370	15 684	1 328	22 741	15 869	8 311	7 073	12 586	13 477	6 959
1985	35 992	6 422	12 079	17 059	1 460	25 274	16 169	8 622	5 608	13 345	25 427	4 337
1986	32 919	6 958	13 327	18 374	1 574	27 337	14 146	8 962	4 690	13 628	31 319	5 058
1987	32 457	7 493	15 085	16 817	2 635	28 199	11 645	9 200	4 072	13 355	26 466	6 434
1988	35 227	7 594	16 198	17 163	3 031	29 036	10 466	10 820	2 296	14 601	17 088	19 163
1989	34 386	8 745	16 334	19 604	3 639	30 355	9 583	12 821	2 705	16 169	16 698	29 709
1990	35 345	11 500	16 707	24 319	3 647	29 932	13 758	14 426	3 341	17 055	11 637	67 599
1991	42 402	11 138	15 949	19 158	4 412	27 328	15 846	16 092	2 436	18 544	14 886	76 270
1992	45 569	11 143	17 359	22 920	4 293	29 834	16 090	16 389	4 499	20 001	14 922	10 918
1993	48 057	12 577	18 802	14 908	4 491	29 108	17 218	17 006	4 319	20 224	20 081	-2 185
1994	55 225	15 225	20 099	18 023	5 081	28 164	17 067	17 067	5 218	21 000	14 795	-4 228
1995	57 484	14 763	19 301	23 378	5 143	27 206	16 429	16 692	4 936	21 889	9 671	-1 780
1996	54 014	17 189	18 670	20 477	5 048	26 574	13 487	16 684	2 839	21 503	9 035	-1 047
1997	56 924	19 845	17 928	19 636	5 769	27 050	15 173	17 136	1 475	21 201	8 889	-1 464
1998	57 673	24 076	18 297	24 540	8 048	26 819	13 054	18 172	1 270	22 278	12 077	1 007
1999	70 414	27 782	18 336	25 917	9 010	27 476	15 239	18 084	911	23 943	22 879	2 641
2000	68 865	29 010	19 914	32 293	10 506	29 436	17 213	18 594	-761	25 003	36 458	3 207
2001	66 232	28 400	19 369	26 124	11 576	30 473	16 485	19 753	9	25 532	26 252	5 731
2002	66 989	26 507	18 602	23 683	10 206	34 845	22 315	20 734	475	29 426	21 965	-407
2003	67 524	21 959	19 862	21 878	12 636	40 473	21 199	20 831	-725	29 667	22 496	727
2004	69 855	24 831	21 083	19 652	12 956	45 581	26 870	23 029	-147	30 694	15 439	5 265
2005	73 094	24 764	23 379	19 297	13 448	495 294	34 565	23 597	440	27 983	26 565	7 566
2006	73 961	27 877	24 810	29 945	14 632	521 820	29 499	23 584	785	33 025	25 969	6 187
2007	65 069	26 044	26 010	32 043	15 497	551 258	28 482	24 407	-852	31 721	17 662	487
2008	67 334	28 844	27 568	33 598	16 399	616 066	28 857	26 773	631	31 820	18 387	27 870
2009	62 483	23 482	22 453	34 318	17 799	661 012	37 529	28 417	4 755	35 573	22 237	291 535
2010	66 909	18 885	25 298	75 845	20 969	693 485	45 195	30 100	11 618	43 667	21 356	-82 316
2011	72 381	7 399	29 519	82 546	20 271	705 554	45 685	29 466	12 174	45 473	20 662	-12 564
2012	79 061	13 973	30 307	81 957	24 883	677 852	36 802	29 060	14 858	41 631	17 791	40 647
2013	84 007	18 912	31 815	75 767	26 874	633 446	46 464	28 908	11 042	38 145	29 678	-83 198
2014	93 368	19 300	33 926	99 235	36 905	603 457	46 879	28 570	5 270	36 171	24 386	-94 861
2015	98 279	19 232	35 041	96 468	51 014	589 659	52 040	29 412	6 841	36 033	18 500	-37 905
2016	95 026	21 354	34 838	115 672	40 364	593 372	45 306	30 174	3 721	39 082	18 344	-34 077
2017	83 823	22 768	34 574	81 287	47 667	598 722	46 309	30 394	3 856	37 896	18 872	-26 685
2018	94 986	22 983	41 299	70 750	40 917	631 130	48 996	31 534	2 169	39 140	21 789	-9 470
2019	99 452	16 672	70 784	52 793	32 986	686 003	52 739	32 410	5 041	37 844	38 257	-25 715

[1]Fiscal years through 1976 are from July 1 through June 30. Beginning with October 1976 (fiscal year 1977), fiscal years are from October 1 through September 30. The period from July 1 through September 30, 1976, is a separate fiscal period known as the transition quarter (TQ) and is not included in any fiscal year.
. . . = Not available.

Table 6-14A. Federal Government Receipts and Outlays by Fiscal Year [1]—*Continued*

(Budget accounts, millions of dollars.)

Year	Transpor-tation	Community and regional development	Education, employment, and social services	Health	Medicare	Income security	Social Security	Veterans benefits and services	Administra-tion of justice	General government	Net interest
					Outlays by function—Continued						
1940	392	285	1 972	55	0	1 514	28	570	81	274	899
1941	353	123	1 592	60	0	1 855	91	560	92	306	943
1942	1 283	113	1 062	71	0	1 828	137	501	117	397	1 052
1943	3 220	219	375	92	0	1 739	177	276	154	673	1 529
1944	3 901	238	160	174	0	1 503	217	-126	192	900	2 219
1945	3 654	243	134	211	0	1 137	267	110	178	581	3 112
1946	1 970	200	85	201	0	2 384	358	2 465	176	825	4 111
1947	1 130	302	102	177	0	2 820	466	6 344	176	1 114	4 204
1948	787	78	191	162	0	2 499	558	6 457	170	1 045	4 341
1949	916	-33	178	197	0	3 174	657	6 599	184	824	4 523
1950	967	30	241	268	0	4 097	781	8 834	193	986	4 812
1951	956	47	235	323	0	3 352	1 565	5 526	218	1 097	4 665
1952	1 124	73	339	347	0	3 655	2 063	5 341	267	1 163	4 701
1953	1 264	117	441	336	0	3 823	2 717	4 519	243	1 209	5 156
1954	1 229	100	370	307	0	4 434	3 352	4 613	257	799	4 811
1955	1 246	129	445	291	0	5 071	4 427	4 675	256	651	4 850
1956	1 450	92	591	359	0	4 734	5 478	4 891	302	1 201	5 079
1957	1 662	135	590	479	0	5 427	6 661	5 005	303	1 360	5 354
1958	2 334	169	643	541	0	7 535	8 219	5 350	325	655	5 604
1959	3 655	211	789	685	0	8 239	9 737	5 443	356	926	5 762
1960	4 126	224	968	795	0	7 378	11 602	5 441	366	1 184	6 947
1961	3 987	275	1 063	913	0	9 683	12 474	5 705	400	1 354	6 716
1962	4 290	469	1 241	1 198	0	9 198	14 365	5 628	429	1 049	6 889
1963	4 596	574	1 458	1 451	0	9 304	15 788	5 521	465	1 230	7 740
1964	5 242	933	1 555	1 788	0	9 650	16 620	5 682	489	1 518	8 199
1965	5 763	1 114	2 140	1 791	0	9 462	17 460	5 723	536	1 499	8 591
1966	5 730	1 105	4 363	2 543	64	9 671	20 694	5 923	564	1 603	9 386
1967	5 936	1 108	6 453	3 351	2 748	10 253	21 725	6 743	618	1 719	10 268
1968	6 316	1 382	7 634	4 390	4 649	11 806	23 854	7 042	659	1 757	11 090
1969	6 526	1 552	7 548	5 162	5 695	13 066	27 298	7 642	766	1 939	12 699
1970	7 008	2 392	8 634	5 907	6 213	15 645	30 270	8 679	959	2 320	14 380
1971	8 052	2 917	9 849	6 843	6 622	22 936	35 872	9 778	1 307	2 442	14 841
1972	8 392	3 423	12 529	8 674	7 479	27 638	40 157	10 732	1 684	2 960	15 478
1973	9 066	4 605	12 744	9 356	8 052	28 265	49 090	12 015	2 174	9 774	17 349
1974	9 172	4 229	12 455	10 733	9 639	33 700	55 867	13 388	2 505	10 032	21 449
1975	10 918	4 322	16 022	12 930	12 875	50 161	64 658	16 599	3 028	10 374	23 244
1976	13 739	5 442	18 910	15 734	15 834	60 784	73 899	18 433	3 430	9 706	26 727
TQ	3 358	1 569	5 169	3 924	4 264	14 981	19 763	3 963	918	3 878	6 949
1977	14 829	7 021	21 104	17 302	19 345	61 045	85 061	18 038	3 701	12 791	29 901
1978	15 521	11 841	26 706	18 524	22 768	61 492	93 861	18 978	3 923	11 961	35 458
1979	18 079	10 480	30 218	20 494	26 495	66 364	10 407	19 931	4 286	12 241	42 633
1980	21 329	11 252	31 835	23 169	32 090	86 548	11 854	21 185	4 702	12 975	52 533
1981	23 379	10 568	33 146	26 866	39 149	10 028	13 958	22 991	4 908	11 373	68 766
1982	20 625	8 347	26 609	27 445	46 567	10 813	15 596	23 958	4 842	10 861	85 032
1983	21 334	7 564	26 192	28 643	52 588	12 301	17 072	24 846	5 246	11 181	89 808
1984	23 669	7 673	26 913	30 420	57 540	11 337	17 822	25 601	5 811	11 746	11 110
1985	25 838	7 676	28 584	33 546	65 822	12 900	18 862	26 281	6 426	11 515	12 947
1986	28 113	7 233	29 771	35 935	70 164	12 065	19 875	26 343	6 735	12 491	13 601
1987	26 222	5 049	28 915	39 967	75 120	12 410	20 735	26 761	7 715	7 487	13 861
1988	27 272	5 293	30 939	44 471	78 878	13 038	21 934	29 409	9 397	9 399	15 180
1989	27 608	5 362	35 314	48 390	84 964	13 755	23 254	30 038	9 644	9 317	16 898
1990	29 485	8 531	37 167	57 700	98 102	14 878	24 862	29 088	10 185	10 462	18 434
1991	31 099	6 810	41 260	71 139	10 448	17 258	26 901	31 319	12 486	11 568	19 444
1992	33 332	6 836	42 807	89 415	11 902	19 965	28 758	34 112	14 650	12 883	19 934
1993	35 004	9 146	47 455	99 321	13 055	21 008	30 458	35 690	15 193	12 944	19 871
1994	38 066	10 620	43 337	10 705	14 474	21 723	31 956	37 617	15 516	11 159	20 293
1995	39 350	10 746	51 067	11 535	15 985	22 375	33 584	37 910	16 508	13 799	23 213
1996	39 565	10 741	48 334	11 934	17 422	22 969	34 967	37 003	17 898	11 755	24 105
1997	40 767	11 049	49 014	12 379	19 001	23 498	36 525	39 332	20 617	12 547	24 398
1998	40 343	9 771	50 534	13 140	19 282	23 769	37 921	41 793	23 359	15 544	24 111
1999	42 532	11 865	50 608	14 104	19 044	24 241	39 003	43 216	26 536	15 363	22 975
2000	46 853	10 623	53 766	15 450	19 711	25 367	40 942	47 040	28 499	13 013	22 294
2001	54 447	11 773	57 089	17 223	21 738	26 972	43 295	45 024	30 201	14 358	20 616
2002	61 833	12 981	70 592	19 647	23 085	31 267	45 598	50 979	35 061	16 951	17 094
2003	67 069	18 850	82 733	219 605	249 433	334 574	474 680	56 832	35 340	23 164	15 307
2004	64 627	15 820	87 997	240 148	269 360	333 027	495 548	59 729	45 576	22 338	16 024
2005	67 894	26 262	97 553	250 605	298 638	345 800	523 305	70 112	40 019	16 997	18 398
2006	70 244	54 465	118 473	252 779	329 868	352 425	548 549	69 832	41 016	18 177	22 660
2007	72 905	29 567	91 647	266 425	375 407	365 931	586 153	72 828	42 362	17 425	23 710
2008	77 616	23 952	91 304	280 626	390 758	431 211	617 027	84 711	48 097	20 323	25 275
2009	84 289	27 676	79 744	334 368	430 093	533 080	682 963	95 545	52 581	22 017	18 690
2010	91 972	23 894	128 595	369 081	451 636	622 106	706 737	108 478	54 383	23 014	19 619
2011	92 966	23 883	101 256	372 481	485 653	597 269	730 811	127 269	56 056	27 476	22 996
2012	93 019	25 132	90 822	346 755	471 793	541 248	773 290	124 679	56 277	28 035	22 040
2013	91 673	32 335	72 860	358 264	497 826	536 411	813 551	139 038	52 601	27 737	22 088
2014	91 915	20 670	90 610	409 497	511 688	513 596	850 533	149 621	50 457	26 913	22 895
2015	89 533	20 669	122 035	482 257	546 202	508 800	887 753	159 781	51 906	20 956	22 318
2016	92 566	20 140	109 709	511 325	594 536	514 098	916 067	174 557	55 768	23 146	24 003
2017	93 552	24 907	143 953	533 152	597 307	503 443	944 878	176 584	57 944	23 821	26 255
2018	92 785	42 159	95 503	551 219	588 706	495 289	987 791	178 895	60 418	23 885	324 975
2019	97 116	26 876	136 752	584 816	650 996	514 787	104 440	199 843	65 740	23 436	375 158

[1]Fiscal years through 1976 are from July 1 through June 30. Beginning with October 1976 (fiscal year 1977), fiscal years are from October 1 through September 30. The period from July 1 through September 30, 1976, is a separate fiscal period known as the transition quarter (TQ) and is not included in any fiscal year.
. . . = Not available.

Table 6-14B. The Federal Budget and GDP

(Billions of dollars; percent.)

Year	Fiscal year GDP	Billions of dollars						Percent of GDP					
		Receipts			Outlays		Total budget surplus or deficit	Receipts			Outlays		Total budget surplus or deficit
		Total	Individual income taxes	Corporate income taxes	Total	National defense		Total	Individual income taxes	Corporate income taxes	Total	National defense	
1950	278.7	39.4	15.8	10.4	42.6	13.7	-3.1	14.2	5.7	3.7	15.1	4.9	-1.7
1951	327.0	51.6	21.6	14.1	45.5	23.6	6.1	15.8	6.6	4.3	13.5	7.2	1.3
1952	357.1	66.2	27.9	21.2	67.7	46.1	-1.5	18.5	7.8	5.9	18.5	12.9	-0.9
1953	382.0	69.6	29.8	21.2	76.1	52.8	-6.5	18.2	7.8	5.6	19.3	13.8	-2.2
1954	387.2	69.7	29.5	21.1	70.9	49.3	-1.2	18.0	7.6	5.5	17.5	12.7	-0.7
1955	406.3	65.5	28.7	17.9	68.4	42.7	-3.0	16.1	7.1	4.4	15.9	10.5	-1.0
1956	438.2	74.6	32.2	20.9	70.6	42.5	3.9	17.0	7.3	4.8	15.0	9.7	0.6
1957	463.4	80.0	35.6	21.2	76.6	45.4	3.4	17.3	7.7	4.6	15.2	9.8	0.6
1958	473.5	79.6	34.7	20.1	82.4	46.8	-2.8	16.8	7.3	4.2	15.8	9.9	-0.7
1959	504.6	79.2	36.7	17.3	92.1	49.0	-12.8	15.7	7.3	3.4	16.5	9.7	-2.4
1960	534.3	92.5	40.7	21.5	92.2	48.1	0.3	17.3	7.6	4.0	15.2	9.0	0.1
1961	546.6	94.4	41.3	21.0	97.7	49.6	-3.3	17.3	7.6	3.8	15.7	9.1	-0.7
1962	585.7	99.7	45.6	20.5	106.8	52.3	-7.1	17.0	7.8	3.5	15.9	8.9	-1.0
1963	618.2	106.6	47.6	21.6	111.3	53.4	-4.8	17.2	7.7	3.5	15.6	8.6	-0.6
1964	661.7	112.6	48.7	23.5	118.5	54.8	-5.9	17.0	7.4	3.6	15.5	8.3	-1.0
1965	709.3	116.8	48.8	25.5	118.2	50.6	-1.4	16.5	6.9	3.6	14.3	7.1	-0.2
1966	780.5	130.8	55.4	30.1	134.5	58.1	-3.7	16.8	7.1	3.9	14.7	7.4	-0.4
1967	836.5	148.8	61.5	34.0	157.5	71.4	-8.6	17.8	7.4	4.1	16.4	8.5	-1.5
1968	897.6	153.0	68.7	28.7	178.1	81.9	-25.2	17.0	7.7	3.2	17.4	9.1	-3.1
1969	980.3	186.9	87.2	36.7	183.6	82.5	3.2	19.1	8.9	3.7	16.2	8.4	-0.1
1970	1 046.7	192.8	90.4	32.8	195.6	81.7	-2.8	18.4	8.6	3.1	16.1	7.8	-0.8
1971	1 116.6	187.1	86.2	26.8	210.2	78.9	-23.0	16.8	7.7	2.4	15.9	7.1	-2.3
1972	1 216.2	207.3	94.7	32.2	230.7	79.2	-23.4	17.0	7.8	2.6	15.9	6.5	-2.1
1973	1 352.7	230.8	103.2	36.2	245.7	76.7	-14.9	17.1	7.6	2.7	14.8	5.7	-1.1
1974	1 482.8	263.2	119.0	38.6	269.4	79.3	-6.1	17.8	8.0	2.6	14.6	5.4	-0.5
1975	1 606.9	279.1	122.4	40.6	332.3	86.5	-53.2	17.4	7.6	2.5	16.9	5.4	-3.4
1976	1 786.1	298.1	131.6	41.4	371.8	89.6	-73.7	16.7	7.4	2.3	16.9	5.0	-3.9
1977	2 024.3	355.6	157.6	54.9	409.2	97.2	-53.7	17.6	7.8	2.7	16.2	4.8	-2.5
1978	2 273.4	399.6	181.0	60.0	458.7	104.5	-59.2	17.6	8.0	2.6	16.3	4.6	-2.4
1979	2 565.6	463.3	217.8	65.7	504.0	116.3	-40.7	18.1	8.5	2.6	15.8	4.5	-1.5
1980	2 791.9	517.1	244.1	64.6	590.9	134.0	-73.8	18.5	8.7	2.3	17.1	4.8	-2.6
1981	3 133.2	599.3	285.9	61.1	678.2	157.5	-79.0	19.1	9.1	2.0	17.3	5.0	-2.4
1982	3 313.4	617.8	297.7	49.2	745.7	185.3	-128.0	18.6	9.0	1.5	18.0	5.6	-3.6
1983	3 536.0	600.6	288.9	37.0	808.4	209.9	-207.8	17.0	8.2	1.0	18.7	5.9	-5.9
1984	3 949.2	666.4	298.4	56.9	851.8	227.4	-185.4	16.9	7.6	1.4	17.4	5.8	-4.7
1985	4 265.1	734.0	334.5	61.3	946.3	252.7	-212.3	17.2	7.8	1.4	18.0	5.9	-5.2
1986	4 526.2	769.2	349.0	63.1	990.4	273.4	-221.2	17.0	7.7	1.4	17.8	6.0	-5.3
1987	4 767.6	854.3	392.6	83.9	1 004.0	282.0	-149.7	17.9	8.2	1.8	17.0	5.9	-3.5
1988	5 138.6	909.2	401.2	94.5	1 064.4	290.4	-155.2	17.7	7.8	1.8	16.7	5.7	-3.7
1989	5 554.7	991.1	445.7	103.3	1 143.7	303.6	-152.6	17.8	8.0	1.9	16.8	5.5	-3.7
1990	5 898.8	1 032.0	466.9	93.5	1 253.0	299.3	-221.0	17.5	7.9	1.6	17.4	5.1	-4.7
1991	6 093.2	1 055.0	467.8	98.1	1 324.2	273.3	-269.2	17.3	7.7	1.6	17.8	4.5	-5.3
1992	6 416.2	1 091.2	476.0	100.3	1 381.5	298.3	-290.3	17.0	7.4	1.6	17.6	4.6	-5.3
1993	6 775.3	1 154.3	509.7	117.5	1 409.4	291.1	-255.1	17.0	7.5	1.7	16.9	4.3	-4.4
1994	7 176.8	1 258.6	543.1	140.4	1 461.8	281.6	-203.2	17.5	7.6	2.0	16.5	3.9	-3.6
1995	7 560.4	1 351.8	590.2	157.0	1 515.7	272.1	-164.0	17.9	7.8	2.1	16.2	3.6	-3.0
1996	7 951.3	1 453.1	656.4	171.8	1 560.5	265.7	-107.4	18.3	8.3	2.2	15.8	3.3	-2.2
1997	8 451.0	1 579.2	737.5	182.3	1 601.1	270.5	-21.9	18.7	8.7	2.2	15.3	3.2	-1.2
1998	8 930.8	1 721.7	828.6	188.7	1 652.5	268.2	69.3	19.3	9.3	2.1	15.0	3.0	-0.3
1999	9 479.4	1 827.5	879.5	184.7	1 701.8	274.8	125.6	19.3	9.3	1.9	14.6	2.9	. . .
2000	10 117.4	2 025.2	1 004.5	207.3	1 789.0	294.4	236.2	20.0	9.9	2.0	14.4	2.9	0.9
2001	10 526.5	1 991.1	994.3	151.1	1 862.8	304.7	128.2	18.9	9.4	1.4	14.4	2.9	-0.3
2002	10 833.6	1 853.1	858.3	148.0	2 010.9	348.5	-157.8	17.1	7.9	1.4	15.3	3.2	-2.9
2003	11 283.8	1 782.3	793.7	131.8	2 159.9	404.7	-377.6	15.8	7.0	1.2	15.9	3.6	-4.8
2004	12 025.4	1 880.1	809.0	189.4	2 292.8	455.8	-412.7	15.6	6.7	1.6	15.9	3.8	-4.7
2005	12 834.2	2 153.6	927.2	278.3	2 472.0	495.3	-318.3	16.8	7.2	2.2	16.1	3.9	-3.8
2006	13 638.4	2 406.9	1 043.9	353.9	2 655.1	521.8	-248.2	17.6	7.7	2.6	16.4	3.8	-3.2
2007	14 290.8	2 568.0	1 163.5	370.2	2 728.7	551.3	-160.7	18.0	8.1	2.6	15.9	3.9	-2.4
2008	14 743.3	2 524.0	1 145.7	304.3	2 982.5	616.1	-458.6	17.1	7.8	2.1	17.0	4.2	-4.4
2009	14 431.8	2 105.0	915.3	138.2	3 517.7	661.0	-1 412.7	14.6	6.3	1.0	20.8	4.6	-10.7
2010	14 838.8	2 162.7	898.5	191.4	3 457.1	693.5	-1 294.4	14.6	6.1	1.3	19.6	4.7	-9.2
2011	15 403.7	2 303.5	1 091.5	181.1	3 603.1	705.6	-1 299.6	15.0	7.1	1.2	20.2	4.6	-8.9
2012	16 056.4	2 450.0	1 132.2	242.3	3 526.6	677.9	-1 076.6	15.3	7.1	1.5	18.8	4.2	-7.1
2013	16 603.8	2 775.1	1 316.4	273.5	3 454.9	633.4	-679.8	16.7	7.9	1.6	17.0	3.8	-4.3
2014	17 335.6	3 021.5	1 394.6	320.7	3 506.3	603.5	-484.8	17.4	8.0	1.9	16.2	3.5	-3.0
2015	18 099.6	3 249.9	1 540.8	343.8	3 691.9	589.7	-442.0	18.0	8.5	1.9	16.3	3.3	-2.6
2016	18 554.8	3 268.0	1 546.1	299.6	3 852.6	593.4	-584.7	17.6	8.3	1.6	16.6	3.2	-3.3
2017	19 287.6	3 316.2	1 587.1	297.0	3 981.6	598.7	-665.4	17.2	8.2	1.5	16.5	3.1	-3.7
2018	20 335.5	3 329.9	1 683.5	204.7	4 109.0	631.1	-779.1	16.4	8.3	1.0	16.0	3.1	-3.9
2019	21 215.7	3 464.2	1 717.9	230.2	4 448.3	686.0	-984.2	16.3	8.1	1.1	16.7	3.2	-4.7

. . . = Not available.

Table 6-15. Federal Government Debt by Fiscal Year

(Billions of dollars, except as noted.)

Year	Federal government debt held by the public at end of fiscal year		Gross federal debt at end of fiscal year held by:						
							Private investors		
	Debt held by the public	Debt/GDP ratio (percent)	Total	Social Security funds [1]	Other U.S. government accounts	Federal Reserve System	Total	Foreign residents	Domestic investors
1940	43	43.6	51	2	6	2	40	. . .	. . .
1941	48	41.5	58	2	7	2	46	. . .	. . .
1942	68	45.9	79	3	8	3	65	. . .	. . .
1943	128	69.2	143	4	11	7	121	. . .	. . .
1944	185	86.4	204	5	14	15	170	. . .	. . .
1945	235	103.9	260	7	18	22	213	. . .	. . .
1946	242	106.1	271	8	21	24	218	. . .	. . .
1947	224	93.9	257	9	24	22	202	. . .	. . .
1948	216	82.6	252	10	26	21	195	. . .	. . .
1949	214	77.5	253	11	27	19	195	. . .	. . .
1950	219	78.6	257	13	25	18	201	. . .	. . .
1951	214	65.5	255	15	26	23	191	. . .	. . .
1952	215	60.1	259	17	28	23	192	. . .	. . .
1953	218	57.2	266	18	29	25	194	. . .	. . .
1954	224	58.0	271	20	26	25	199	. . .	. . .
1955	227	55.8	274	21	27	24	203	. . .	. . .
1956	222	50.7	273	23	28	24	198	. . .	. . .
1957	219	47.3	272	23	30	23	196	. . .	. . .
1958	226	47.8	280	24	29	25	201	. . .	. . .
1959	235	46.5	287	23	30	26	209	. . .	. . .
1960	237	44.3	291	23	31	27	210	. . .	. . .
1961	238	43.6	293	23	31	27	211	. . .	. . .
1962	248	42.3	303	22	33	30	218	. . .	. . .
1963	254	41.1	310	21	35	32	222	. . .	. . .
1964	257	38.8	316	22	37	35	222	. . .	. . .
1965	261	36.8	322	22	39	39	222	12	210
1966	264	33.8	328	22	43	42	222	12	210
1967	267	31.9	340	26	48	47	220	11	209
1968	290	32.3	369	28	51	52	237	11	226
1969	278	28.4	366	32	56	54	224	10	214
1970	283	27.1	381	38	60	58	225	14	211
1971	303	27.1	408	41	64	66	238	32	206
1972	322	26.5	436	44	70	71	251	49	202
1973	341	25.2	466	44	81	75	266	59	207
1974	344	23.2	484	46	94	81	263	57	206
1975	395	24.6	542	48	99	85	310	66	244
1976	477	26.7	629	45	107	95	383	70	313
1977	549	27.1	706	40	118	105	444	96	348
1978	607	26.7	777	35	134	115	492	121	371
1979	640	25.0	829	33	156	116	525	120	405
1980	712	25.5	909	32	165	121	591	122	469
1981	789	25.2	995	27	178	124	665	131	534
1982	925	27.9	1 137	19	193	134	790	141	649
1983	1 137	32.2	1 372	32	202	156	982	160	822
1984	1 307	33.1	1 565	32	225	155	1 152	176	976
1985	1 507	35.3	1 817	40	270	170	1 337	223	1 114
1986	1 741	38.5	2 121	46	334	191	1 550	266	1 284
1987	1 890	39.6	2 346	65	391	212	1 678	280	1 398
1988	2 052	39.9	2 601	104	445	229	1 822	346	1 476
1989	2 191	39.4	2 868	157	520	220	1 971	395	1 576
1990	2 412	40.9	3 206	215	580	234	2 177	464	1 713
1991	2 689	44.1	3 598	268	641	259	2 430	506	1 924
1992	3 000	46.8	4 002	319	683	296	2 703	563	2 140
1993	3 248	47.9	4 351	366	737	326	2 923	619	2 304
1994	3 433	47.8	4 643	423	788	355	3 078	682	2 396
1995	3 604	47.7	4 921	483	833	374	3 230	820	2 410
1996	3 734	47.0	5 181	550	898	391	3 343	993	2 350
1997	3 772	44.6	5 369	631	966	425	3 348	1 231	2 117
1998	3 721	41.7	5 478	730	1 027	458	3 263	1 224	2 039
1999	3 632	38.3	5 606	855	1 118	497	3 136	1 281	1 855
2000	3 410	33.7	5 629	1 007	1 212	511	2 898	1 039	1 859
2001	3 320	31.5	5 770	1 170	1 280	534	2 785	1 006	1 779
2002	3 540	32.7	6 198	1 329	1 329	604	2 936	1 201	1 735
2003	3 913	34.7	6 760	1 485	1 362	656	3 257	1 454	1 803
2004	4 296	35.7	7 355	1 635	1 424	700	3 595	1 799	1 796
2005	4 592	35.8	7 905	1 809	1 504	736	3 856	1 930	1 926
2006	4 829	35.4	8 451	1 994	1 628	769	4 060	2 025	2 035
2007	5 035	35.2	8 951	2 181	1 735	780	4 255	2 235	2 020
2008	5 803	39.4	9 986	2 366	1 817	491	5 312	2 802	2 510
2009	7 545	52.3	11 876	2 504	1 827	769	6 776	3 571	3 205
2010	9 019	60.8	13 529	2 585	1 924	812	8 207	4 316	3 891
2011	10 128	65.8	14 764	2 653	1 983	1 665	8 464	4 912	3 551
2012	11 281	70.3	16 051	2 718	2 052	1 645	9 636	5 476	4 160
2013	11 983	72.2	16 719	2 756	1 981	2 072	9 910	5 653	4 258
2014	12 780	73.7	17 794	2 783	2 232	2 452	10 328	6 104	4 224
2015	13 117	72.5	18 120	2 808	2 195	2 462	10 655	6 106	4 549
2016	14 168	76.4	19 539	2 842	2 529	2 463	11 704	6 156	5 548
2017	14 665	76.0	20 206	2 890	2 651	2 465	12 200	6 302	5 898
2018	15 750	77.4	21 462	2 894	2 819	2 313	13 436	6 226	7 210
2019	16 801	79.2	22 669	2 901	2 968	2 113	14 687	6 779	7 908

[1] Sum of old age and survivors insurance fund and disability insurance trust fund.
. . . = Not available.

SECTION 6D: GOVERNMENT OUTPUT AND EMPLOYMENT

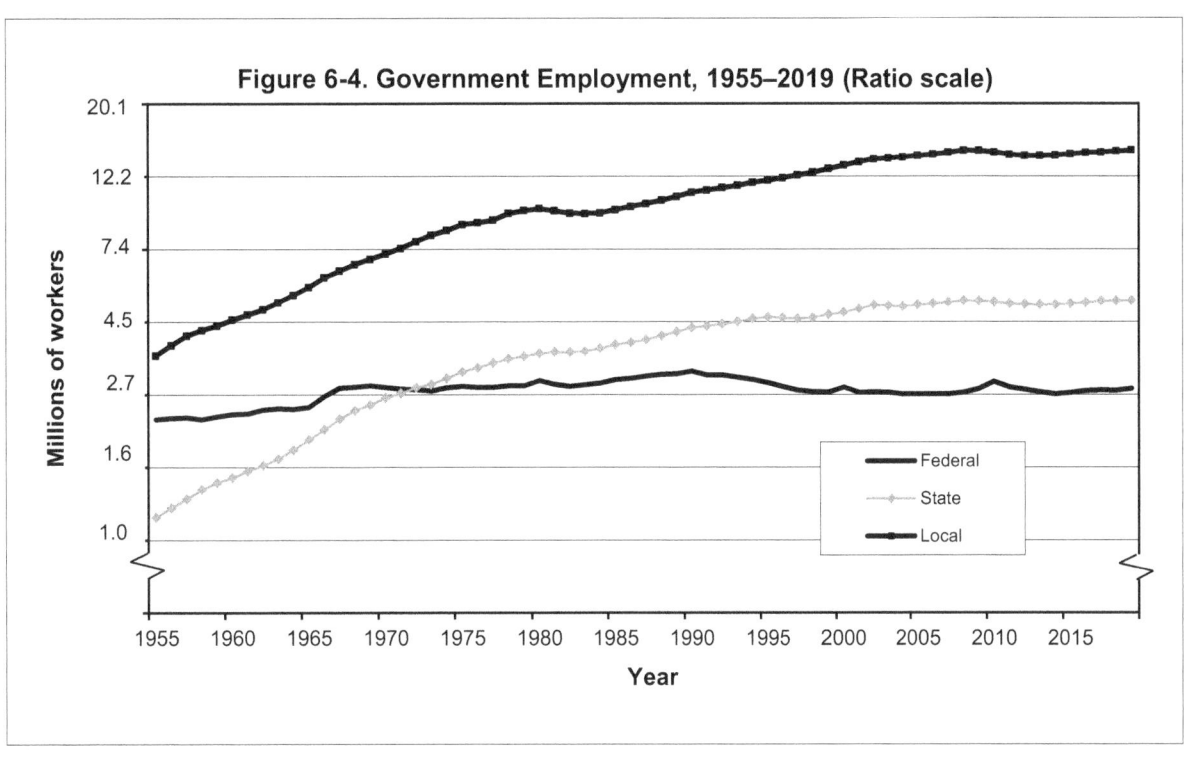

Figure 6-4. Government Employment, 1955–2019 (Ratio scale)

- From 1955 to 2019, total reported employment of civilians by all levels of government has increased nearly 222 percent. Federal employment rose 23.5 percent; state government employment rose 343.2 percent, with 54.3 percent of this increase due to education. However, budget cuts hit all levels of government employment. For example, the Department of Defense lost 27 percent of its employees between 1955 and 2019. (Tables 6-17)

- Contracting and outsourcing of federal government spending can be seen in comparisons between government purchases and government value added. From 1960 to 2019, the volume (measured by the quantity index) of "intermediate" goods and services purchased for nondefense purposes rose 841 percent, compared with the 302 percent rise in nondefense value added by government itself in such activities. (Real value added comprises the input of government employees and government-owned capital, valued in constant dollars). For defense, government value added rose 13.7 percent over that period, while purchases of goods and services rose 234.0 percent. (Table 6-16)

- A similar process has occurred at state and local governments, where real value added increased 242.2 percent while goods and services purchased rose 957.8 percent from 1960 to 2019. (Table 6-16)

Table 6-16. Chain-Type Quantity Indexes for Government Output

(Seasonally adjusted, 2012 = 100.)   NIPA Table 3.10.3

Year and quarter	Federal government									State and local government		
	Gross output of general government			Value added			Intermediate goods and services purchased [1]			Gross output of general government	Value added	Intermediate goods and services purchased [1]
	Total	Defense	Nondefense	Total	Defense	Nondefense	Total	Defense	Nondefense			
1960	45.1	59.3	21.3	60.6	83.4	26.1	22.3	28.4	11.0	23.7	30.8	11.6
1961	46.1	60.8	21.2	62.6	86.1	27.1	21.8	28.6	9.5	25.0	32.4	12.4
1962	50.7	65.9	25.1	65.7	90.1	28.9	28.0	34.7	15.3	25.9	33.5	12.9
1963	51.1	66.1	25.9	67.0	90.6	31.2	27.4	34.5	13.9	27.4	35.3	13.7
1964	50.9	64.9	27.1	68.2	91.5	33.0	25.2	30.9	14.3	29.2	37.4	15.0
1965	52.0	66.0	28.3	69.3	92.0	34.9	26.3	32.6	14.5	31.3	39.7	16.5
1966	57.4	74.9	28.0	74.0	98.2	37.5	32.4	44.8	10.4	33.3	42.0	17.9
1967	63.9	83.5	30.7	79.1	104.7	40.4	40.1	55.8	12.5	35.0	43.8	19.3
1968	66.2	87.2	31.0	81.0	106.5	42.4	42.9	61.6	10.4	37.4	46.5	21.3
1969	65.5	84.4	33.6	81.8	106.8	44.0	40.1	54.9	13.9	39.7	48.6	23.6
1970	61.7	77.6	34.6	79.2	101.4	45.5	35.0	46.5	14.1	42.0	50.7	26.0
1971	59.2	71.9	37.3	76.3	95.3	47.4	32.9	41.0	17.4	44.2	52.8	28.5
1972	58.0	68.0	40.7	73.4	89.3	49.1	34.6	40.1	23.1	46.0	54.7	30.2
1973	55.3	63.8	40.6	71.4	85.4	49.9	30.8	35.1	21.5	47.6	56.6	31.2
1974	55.3	62.3	43.0	71.4	83.9	52.3	30.6	33.5	23.8	49.4	58.7	32.6
1975	54.8	60.5	44.6	71.1	82.6	53.6	29.8	31.1	25.7	51.6	60.4	35.2
1976	53.7	58.8	44.8	71.3	81.2	56.2	27.0	28.9	22.2	52.5	61.2	36.3
1977	54.7	59.1	46.7	71.6	80.6	57.9	28.8	30.4	24.3	53.6	62.2	37.6
1978	55.6	59.5	48.8	72.6	80.9	59.9	29.7	30.9	26.1	54.9	63.7	38.6
1979	56.4	59.9	50.1	72.7	80.5	60.9	31.3	32.3	27.8	55.6	64.8	38.6
1980	58.4	61.5	52.7	73.9	81.1	63.0	34.2	35.1	30.9	55.6	65.7	37.2
1981	60.5	64.5	53.3	75.1	83.3	62.8	37.3	38.8	32.7	55.7	65.7	37.4
1982	62.3	67.9	52.1	76.5	85.5	62.8	39.5	43.5	29.7	56.7	66.2	39.1
1983	65.4	70.8	55.6	78.4	87.8	64.1	43.9	46.8	36.1	57.6	65.9	41.8
1984	66.1	72.7	54.0	80.5	90.7	64.8	42.8	47.5	31.1	58.7	66.3	43.9
1985	69.5	76.4	57.0	83.0	94.3	65.6	47.2	51.1	37.1	61.3	68.2	47.6
1986	72.8	80.4	58.9	85.0	97.4	65.7	52.0	55.9	41.8	64.0	70.2	51.5
1987	74.3	83.2	57.9	87.7	100.9	67.2	51.7	57.8	36.9	65.4	71.6	52.7
1988	74.5	84.9	55.2	89.4	102.5	69.0	49.9	59.6	27.1	67.5	74.1	54.3
1989	76.2	85.3	59.5	90.8	104.1	70.0	52.1	58.5	37.1	70.0	76.3	57.0
1990	77.9	85.9	63.3	92.4	104.8	73.0	53.8	58.7	42.1	72.1	78.6	58.8
1991	78.8	86.8	64.2	92.9	105.2	73.9	55.4	60.5	43.0	73.8	79.7	61.8
1992	77.5	83.3	67.1	91.4	101.7	75.5	54.4	57.0	48.3	75.7	81.0	64.8
1993	75.4	80.5	66.3	89.8	98.7	76.3	51.5	54.5	44.3	77.4	82.1	67.7
1994	73.7	77.4	67.0	87.4	95.1	75.6	50.9	52.2	47.7	79.6	83.6	71.5
1995	71.1	73.8	66.2	84.6	91.2	74.5	48.7	49.0	47.9	81.7	85.2	74.4
1996	69.7	72.5	64.9	82.5	88.2	74.1	48.5	50.1	44.6	83.2	86.5	76.4
1997	69.6	71.2	66.7	81.2	85.6	74.6	50.2	50.6	49.1	85.6	88.2	80.2
1998	68.5	69.3	67.0	80.3	83.2	75.9	48.6	49.2	47.3	89.0	90.1	86.5
1999	69.1	70.2	67.1	79.7	81.4	77.1	51.3	54.0	44.9	91.9	91.6	92.5
2000	69.6	69.4	70.0	80.1	80.5	79.6	52.0	53.3	48.9	94.4	93.6	96.2
2001	72.2	71.5	73.4	80.3	80.3	80.4	58.3	58.6	57.6	98.0	95.8	103.0
2002	76.5	75.6	78.2	81.8	81.5	82.2	67.4	66.7	69.0	100.5	97.7	107.0
2003	81.5	81.6	81.3	83.9	83.7	84.3	77.2	78.4	74.4	100.1	98.3	104.2
2004	84.7	85.5	83.2	85.1	85.5	84.6	83.8	85.4	80.1	100.4	98.6	104.3
2005	86.1	86.6	85.1	86.2	86.7	85.4	85.8	86.4	84.4	101.0	99.4	104.7
2006	87.4	87.6	87.2	86.9	87.1	86.5	88.4	88.2	88.9	102.1	100.2	106.4
2007	88.3	89.1	86.9	88.1	88.2	87.9	88.8	90.5	84.7	103.6	101.6	108.0
2008	93.7	94.8	91.7	91.4	91.6	91.0	97.8	99.7	93.4	103.6	103.0	105.0
2009	99.8	100.6	98.4	95.9	96.5	94.9	106.6	106.7	106.5	105.8	103.4	111.2
2010	104.0	104.1	103.9	99.2	99.6	98.5	112.5	111.0	116.2	103.9	102.5	106.9
2011	101.7	102.9	99.6	100.1	100.9	98.8	104.5	105.8	101.5	101.3	101.3	101.2
2012	100.0	100.0	100.0	100.0	100.0	100.0	100.0	100.0	100.0	100.0	100.0	100.0
2013	95.0	93.7	97.4	98.2	97.8	99.0	89.3	87.5	93.7	100.3	99.9	101.4
2014	93.0	90.5	97.5	97.5	96.1	99.7	85.2	82.1	92.5	100.9	99.9	103.0
2015	92.8	88.5	100.4	96.8	94.1	101.1	85.7	80.0	99.1	103.9	100.7	111.3
2016	93.4	88.4	102.3	97.2	93.5	103.1	86.8	80.8	100.7	106.8	101.8	118.2
2017	93.1	88.6	101.2	97.3	93.1	103.8	85.7	81.8	95.1	108.2	102.9	120.5
2018	95.1	90.5	103.3	97.7	93.6	104.2	90.7	86.2	101.4	109.4	104.0	121.9
2019	98.1	94.6	104.5	98.8	94.8	105.0	97.4	94.8	103.5	110.6	105.4	122.7
2017												
1st quarter	92.7	87.8	101.4	97.3	93.1	103.8	84.6	79.8	95.7	107.8	102.5	120.0
2nd quarter	93.0	88.8	100.6	97.2	92.9	103.7	85.8	82.7	93.3	108.1	102.8	120.3
3rd quarter	92.9	88.4	100.9	97.3	93.1	103.8	85.2	81.4	94.2	108.4	103.1	120.7
4th quarter	93.7	89.2	101.9	97.4	93.2	103.9	87.4	83.2	97.3	108.6	103.3	121.1
2018												
1st quarter	94.1	89.0	103.2	97.4	93.2	104.0	88.4	82.9	101.5	108.9	103.5	121.5
2nd quarter	95.0	90.3	103.4	97.7	93.5	104.4	90.3	85.7	101.2	109.2	103.8	121.8
3rd quarter	95.7	91.1	103.8	98.0	93.7	104.7	91.7	87.4	101.8	109.6	104.3	122.1
4th quarter	95.7	91.7	102.9	97.7	93.9	103.7	92.4	88.7	101.0	109.8	104.5	122.3
2019												
1st quarter	95.7	93.0	100.6	96.8	94.2	100.9	94.1	91.6	100.1	110.2	104.9	122.5
2nd quarter	98.3	94.3	105.4	98.9	94.5	105.7	97.5	94.4	104.9	110.4	105.1	122.7
3rd quarter	99.0	95.1	106.0	99.6	95.0	106.7	98.3	95.8	104.3	110.8	105.6	122.8
4th quarter	99.6	96.1	106.1	99.9	95.5	106.8	99.6	97.5	104.7	111.0	105.9	122.9

[1] Includes general government intermediate inputs for goods and services sold to other sectors and for own-account investment.

Table 6-17. Government Employment

(Calendar years, payroll employment in thousands.)

Year and month	Total government employment	Federal			State			Local		
		Total	Department of Defense	Postal Service	Total	Education	State government hospitals	Total	Education	Local government hospitals
1950	6 120	2 023	533	516	. . .	. . .	. . .	. . .	. . .	. . .
1951	6 502	2 415	797	521	. . .	. . .	. . .	. . .	. . .	. . .
1952	6 727	2 539	868	542	. . .	. . .	. . .	. . .	. . .	. . .
1953	6 758	2 418	818	530	. . .	. . .	. . .	. . .	. . .	. . .
1954	6 858	2 295	744	533	. . .	. . .	. . .	. . .	. . .	. . .
1955	7 021	2 295	744	534	1 168	308	. . .	3 558	1 751	. . .
1956	7 386	2 318	749	539	1 249	334	. . .	3 819	1 884	. . .
1957	7 724	2 326	729	555	1 328	363	. . .	4 071	2 026	. . .
1958	7 946	2 298	695	567	1 415	389	. . .	4 232	2 115	. . .
1959	8 192	2 342	699	578	1 484	420	. . .	4 366	2 198	. . .
1960	8 464	2 381	681	591	1 536	448	. . .	4 547	2 314	. . .
1961	8 706	2 391	683	601	1 607	474	. . .	4 708	2 411	. . .
1962	9 004	2 455	697	601	1 669	511	. . .	4 881	2 522	. . .
1963	9 341	2 473	687	603	1 747	557	. . .	5 121	2 674	. . .
1964	9 711	2 463	676	604	1 856	609	. . .	5 392	2 839	. . .
1965	10 191	2 495	679	619	1 996	679	. . .	5 700	3 031	. . .
1966	10 910	2 690	741	686	2 141	775	. . .	6 080	3 297	. . .
1967	11 525	2 852	802	719	2 302	873	. . .	6 371	3 490	. . .
1968	11 972	2 871	801	729	2 442	958	. . .	6 660	3 649	. . .
1969	12 330	2 893	815	737	2 533	1 042	. . .	6 904	3 785	. . .
1970	12 687	2 865	756	741	2 664	1 104	. . .	7 158	3 912	. . .
1971	13 012	2 828	731	731	2 747	1 149	. . .	7 437	4 091	. . .
1972	13 465	2 815	720	703	2 859	1 188	459	7 790	4 262	467
1973	13 862	2 794	696	698	2 923	1 205	472	8 146	4 433	477
1974	14 303	2 858	698	710	3 039	1 267	483	8 407	4 584	483
1975	14 820	2 882	704	699	3 179	1 323	503	8 758	4 722	489
1976	15 001	2 863	693	676	3 273	1 371	518	8 865	4 786	492
1977	15 258	2 859	676	657	3 377	1 385	538	9 023	4 859	494
1978	15 812	2 893	661	660	3 474	1 367	541	9 446	4 958	535
1979	16 068	2 894	649	673	3 541	1 378	538	9 633	4 989	571
1980	16 375	3 000	645	673	3 610	1 398	530	9 765	5 090	604
1981	16 180	2 922	655	675	3 640	1 420	515	9 619	5 095	622
1982	15 982	2 884	690	684	3 640	1 433	494	9 458	5 049	635
1983	16 011	2 915	699	685	3 662	1 450	471	9 434	5 020	644
1984	16 159	2 943	716	706	3 734	1 488	459	9 482	5 076	623
1985	16 533	3 014	738	750	3 832	1 540	449	9 687	5 221	608
1986	16 838	3 044	736	792	3 893	1 561	438	9 901	5 358	601
1987	17 156	3 089	736	815	3 967	1 586	439	10 100	5 469	606
1988	17 540	3 124	719	835	4 076	1 621	446	10 339	5 590	619
1989	17 927	3 136	735	838	4 182	1 668	442	10 609	5 740	632
1990	18 415	3 196	722	825	4 305	1 730	426	10 914	5 902	646
1991	18 545	3 110	702	813	4 355	1 768	417	11 081	5 994	653
1992	18 787	3 111	702	800	4 408	1 799	419	11 267	6 076	665
1993	18 989	3 063	670	793	4 488	1 834	414	11 438	6 206	673
1994	19 275	3 018	657	821	4 576	1 882	407	11 682	6 329	673
1995	19 432	2 949	627	850	4 635	1 919	395	11 849	6 453	669
1996	19 539	2 877	597	867	4 606	1 911	376	12 056	6 592	648
1997	19 664	2 806	588	866	4 582	1 904	360	12 276	6 759	632
1998	19 909	2 772	550	881	4 612	1 922	346	12 525	6 921	630
1999	20 307	2 769	525	890	4 709	1 983	344	12 829	7 120	626
2000	20 790	2 865	510	880	4 786	2 031	343	13 139	7 294	622
2001	21 118	2 764	504	873	4 905	2 113	345	13 449	7 479	628
2002	21 513	2 766	499	842	5 029	2 243	349	13 718	7 654	642
2003	21 583	2 761	486	809	5 002	2 255	348	13 820	7 709	651
2004	21 621	2 730	473	782	4 982	2 238	348	13 909	7 765	656
2005	21 804	2 732	488	774	5 032	2 260	350	14 041	7 856	655
2006	21 974	2 732	493	770	5 075	2 293	357	14 167	7 913	647
2007	22 218	2 734	491	769	5 122	2 318	360	14 362	7 987	654
2008	22 509	2 762	496	747	5 177	2 354	361	14 571	8 084	659
2009	22 555	2 832	519	703	5 169	2 360	359	14 554	8 079	661
2010	22 490	2 977	545	659	5 137	2 373	354	14 376	8 013	650
2011	22 086	2 859	559	631	5 078	2 374	348	14 150	7 873	648
2012	21 920	2 820	552	611	5 055	2 389	349	14 045	7 778	647
2013	21 853	2 769	537	595	5 046	2 393	350	14 037	7 777	646
2014	21 882	2 733	524	593	5 050	2 389	347	14 098	7 815	642
2015	22 029	2 757	526	597	5 077	2 401	354	14 195	7 871	648
2016	22 224	2 795	529	609	5 110	2 429	367	14 319	7 905	658
2017	22 350	2 805	527	615	5 165	2 479	376	14 379	7 922	667
2018	22 455	2 800	530	608	5 173	2 487	381	14 481	7 963	674
2019	22 594	2 834	543	607	5 177	2 484	386	14 583	8 010	684

. . . = Not available.

NOTES AND DEFINITIONS, CHAPTER 6

TABLES 6-1 THROUGH 6-11 AND 6-16

Federal, State, and Local Government in the National Income and Product Accounts

SOURCE: U.S. DEPARTMENT OF COMMERCE, BUREAU OF ECONOMIC ANALYSIS (BEA)

These data are from the national income and product accounts (NIPAs), as redefined in the 2018 comprehensive NIPA revision and updated in the 2020 annual revision. For general information about the NIPAs see the notes and definitions for Chapter 1.

The 2020 annual update of the National Income and Product Accounts was released on July 30, 2020. Data for 2015 through 2019 were revised. The reference year remained 2012. The updated statistics largely reflect the incorporation of newly available and revised source data and improvements to existing methodologies.

The framework for the government accounts

In an earlier major revision in 2003, a new framework was introduced for government consumption expenditures—federal, state, and local—that explicitly recognizes the services produced by general government. Governments serve several functions in the economy. Three of these functions are recognized in the NIPAs: the production of nonmarket services; the consumption of these services, as the value of services provided to the general public is treated as government consumption expenditures; and the provision of transfer payments. These functions are financed through taxation, through contributions to social insurance funds, and by borrowing in the world's capital markets.

In this framework, the value of the government services produced and consumed (most of which are not sold in the market) is measured as the sum of the costs of the three major inputs: compensation of government employees, consumption of fixed capital (CFC), and intermediate goods and services purchased. The purchase from the private sector of goods and services by government, classified as final sales to government before the 2003 revision, was reclassified as intermediate purchases.

The value of government final purchases of consumption expenditures and gross investment, which constitutes the contribution of government demand to the gross domestic product (GDP), was not changed by this reclassification, since the previous definition of that contribution was the sum of compensation, CFC, and goods and services purchased. However, the distribution of GDP by type of product was changed—final sales of goods were reduced by the amount of goods purchased by government, and final sales of services were increased by the same amount.

In addition to this change in the conceptual framework, a number of the categories of government receipts and expenditures were redefined to make more precise distinctions. For example, items that used to be called "nontax payments" and included with taxes are now classified as transfer or fee payments and not included in taxes.

Finally, the concept previously known as "current surplus or deficit (-), national income and product accounts" was renamed "net government saving." This recognizes, in part, the role of government in the capital markets. When government runs a current surplus, net government saving is positive and funds are made available (for example, by retiring outstanding government debt) to finance investment—both private-sector capital spending and government investment. By the same token, when government runs a current deficit, or "dis-saves," it must borrow funds that would otherwise be available to finance investment.

However, this definition of net government saving does not give a complete picture of governments' role in capital markets, because it is based on current receipts and expenditures alone and does not include government investment activity.

In the NIPAs, the capital spending of all levels of government is treated the same way as the accounts treat private investment spending. A depreciation, or more precisely "consumption of fixed capital" (CFC), entry for existing capital is calculated, using estimated replacement costs and realistic depreciation rates. In the government accounts this CFC value is entered as one element of current expenditures and output. Capital spending is excluded from government current expenditures but appears in the account for "net lending or borrowing (-)."

The basic concept expressed in the net lending section of the NIPAs is that when CFC exceeds actual investment expenditures, governments have a positive net cash flow and can lend (or repay debt); if gross investment exceeds CFC, government must borrow to finance the difference, indicating negative net cash flow and requiring borrowing. (As will be seen below in the definitions, capital transfer and purchase accounts also enter into the calculation of net lending.)

The federal *budget* accounts (see Tables 6-1 and 6-2) do not draw a distinction between current and capital spending. The budget accounts of individual state and local governments typically separate capital from current spending and allow capital spending to be financed by borrowing—even when deficit financing of current spending is constitutionally forbidden. However, neither federal nor state and local government budget accounts typically show depreciation as a current expense in the way that is standard to private-sector accounting or in the way adopted in the NIPAs.

Notes on the data

Government receipts and expenditures data are derived from the U.S. government accounts and from Census Bureau censuses and surveys of the finances of state and local governments. However, BEA makes a number of adjustments to the data to convert them from fiscal year to calendar year and quarter bases and to agree with the concepts of national income accounting. Data are converted from the cash basis usually found in financial statements to the timing bases required for the NIPAs. In the NIPAs, receipts from businesses are generally on an accrual basis, purchases of goods and services are recorded when delivered, and receipts from and transfer payments to persons are on a cash basis. The federal receipts and expenditure data from the NIPAs in Table 6-1 therefore differ from the federal receipts and outlay data in Table 6-14. Among other differences, the latter are by fiscal year and are on a modified cash basis.

The NIPA data on government receipts and expenditures record transactions of governments (federal, state, and local) with other U.S. residents and foreigners. Each entry in the government receipts and expenditures account has a corresponding entry elsewhere in the NIPAs. Thus, for example, the sum of personal current taxes received by federal and state and local governments (Tables 6-1 and 6-8) is equal to personal current taxes paid, as shown in personal income (Table 4-1).

Definitions: general

In the 2003 revision of the NIPAs, several items appear separately that were previously treated as "negative expenditures" and netted against other items on the expenditures side. This grossing-up of the accounts raises both receipts and outlays and has no effect on net saving. Grossing-up has been applied to taxes from the rest of the world, interest receipts (back to 1960 for the federal government and back to 1946 for state and local governments), dividends, and subsidies and the current surplus of government enterprises (back to 1959).

Definitions: current receipts

Current tax receipts includes personal current taxes, taxes on production and imports, taxes on corporate income and (for the federal government only) taxes from the rest of the world. The category *total tax receipts* does not include *contributions for government social insurance,* which are the taxes levied to finance Social Security, unemployment insurance, and Medicare, and are included in *receipts* in the budget accounts (Table 6-14). Analysts using NIPA data to analyze tax burdens as they are usually understood should add *contributions for government social insurance* to *tax receipts* for this purpose.

Personal current taxes is personal tax payments from residents of the United States that are not chargeable to business expense. Personal taxes consist of taxes on income, including on realized net capital gains, and on personal property. Personal contributions for social insurance are not included in this category. As of the 1999 revisions, estate and gift taxes are classified as capital transfers and are no longer included in personal current taxes. However, estate and gift taxes continue to be included in federal government receipts in Table 6-14.

Taxes on production and imports in the case of the federal government consists of *excise taxes* and *customs duties*. In the case of state and local governments, these taxes include *sales taxes*, *property taxes* (including residential real estate taxes), and *Other* taxes such as motor vehicle licenses, severance taxes, and special assessments. Before the 2003 revision, all of these taxes were components of "indirect business tax and nontax liabilities."

Taxes on corporate income covers federal, state, and local government taxes on all corporate income subject to taxes. This taxable income includes capital gains and other income excluded from NIPA profits. The taxes are measured on an accrual basis, net of applicable tax credits. Federal corporate income tax receipts in the NIPAs include, but show separately, receipts from *Federal Reserve Banks* that represent the return of the surplus earnings of the Federal Reserve system. (In the budget accounts in Table 6-14, these are classified not as corporate taxes but as the major component of "Miscellaneous receipts.")

Contributions for social insurance includes employer and personal contributions for Social Security, Medicare, unemployment insurance, and other government social insurance programs. As of the 1999 revisions, contributions to government employee retirement plans are no longer included in this category; these plans are now treated the same as private pension plans.

Income receipts on assets consists of *interest*, *dividends* and *rents and royalties*.

Interest receipts (1960 to the present for federal government; 1946 to the present for state and local governments) consists of monetary and imputed interest received on loans and investments. In the NIPAs, this no longer includes interest received by government employee retirement plans, which is now credited to personal income. However, such interest received is still deducted from interest paid in the budget accounts that are shown in Table 6-14. Before the indicated years, receipts are deducted from aggregate interest payments in the NIPAs. Hence, they are not shown as receipts, and net interest is presented on the expenditure side. In the federal budget accounts in Table 6-14, net interest (total interest expenditures minus interest receipts) is the interest "expenditure" concept used throughout the period covered.

Current transfer receipts include receipts in categories other than those specified above from persons and business. In the case of state and local government accounts (Table 6-8), it also includes *federal grants-in-aid,* a component of federal expenditures. Receipts from *business* and *persons* were previously included with income taxes in "tax and nontax payments." They consist of federal deposit insurance premiums and other nontaxes (largely fines and regulatory and inspection fees), state and local fines and other nontaxes (largely donations and tobacco settlements), and net insurance settlements paid to governments as policyholders.

The *current surplus of government enterprises* is the current operating revenue and subsidies received from other levels of government by such enterprises less their current expenses. No deduction is made for depreciation charges or net interest paid. Before 1959, this category of receipts is treated as a deduction from subsidies. In the <u>federal</u> NIPA accounts before 1959, there is no entry shown for the current surplus on the receipts side, and on the expenditure side, there is an entry for subsidies, which is net of the current surplus. (Subsidies are usually a larger amount than the enterprise surplus in the federal accounts.) In the <u>state and local</u> NIPA accounts before 1959, there is an entry for the surplus on the receipts side, which is net of subsidies. (Subsidies are typically smaller than the enterprise surplus in state and local finance.)

Definitions: consumption expenditures, saving, and gross investment

Government consumption expenditures is expenditures by governments (federal or state and local) on services for current consumption. It includes *compensation of general government employees* (including employer contributions to government employee retirement plans, as of the 1999 revision); an allowance for *consumption of general government fixed capital (CFC)*, including R&D and software (depreciation); and *intermediate goods and services purchased*. (See the general discussion above for an explanation.) The estimated value of *own-account investment*—investment goods, including R&D and software, produced by government resources and purchased inputs—is subtracted here, and added to *government gross investment*. *Sales to other sectors*—primarily tuition payments received from individuals for higher education and charges for medical care to individuals—are also deducted, since these are counted elsewhere in the accounts, as (for example) personal consumption expenditures for those categories.

Government social benefits consists of payments to individuals for which the individuals do not render current services. Examples are Social Security benefits, Medicare, Medicaid, unemployment benefits, and public assistance. Retirement payments to retired government employees from their pension plans are no longer included in this category.

Government social benefits to persons consists of payments to persons residing in the United States (with a corresponding entry of an equal amount in the personal income receipts accounts). Government social benefits to the *rest of the world* appear only in the federal government account, and are transfers, mainly retirement benefits, to former residents of the United States.

Other current transfer payments (federal account only) includes *grants-in-aid to state and local governments* and *grants to the rest of the world*—military and nonmilitary grants to foreign governments.

Federal grants-in-aid comprises net payments from federal to state and local governments that are made to help finance programs such as health (Medicaid), public assistance (the old Aid to Families with Dependent Children and the new Temporary Assistance for Needy Families), and education. Investment grants to state and local governments for highways, transit, air transportation, and water treatment plants are now classified as capital transfers and are no longer included in this category. However, such investment grants continue to be included as federal government outlays in Table 6-14.

Interest payments is monetary interest paid to U.S. and foreign persons and businesses and to foreign governments for public debt and other financial obligations. As noted above, from 1960 forward for the federal government and from 1946 forward for state and local governments, this represents gross total (not net) interest payments. Before those dates in the NIPAs, and throughout the federal budget accounts presented in Table 6-14, net instead of aggregate interest is shown; that is, gross total interest paid less interest received.

Subsidies are monetary grants paid by government to business, including to government enterprises at another level of government. Subsidies no longer include federal maritime construction subsidies, which are now classified as a capital transfer. For years prior to 1959, subsidies continue to be presented net of the *current surplus of government enterprises*, because detailed data to separate the series are not available for this period. See the entry for *current surplus of government enterprises*, described above, for explanation of the pre-1959 treatment of this item in the federal accounts in *Business Statistics*, which differs from the treatment in the state and local accounts.

Net saving, NIPA (surplus+/deficit-), is the sum of current receipts less the sum of current expenditures. This is shown separately for *social insurance funds* (which, in the case of the federal government, include Social Security and other trust funds) and *other* (all other government).

Gross government investment consists of general government and government enterprise expenditures for fixed assets—*structures*, *equipment*, and *intellectual property* (R&D and software). The expenditures include the compensation of government employees producing the assets and the purchase of goods and services as intermediate inputs to be incorporated in the assets. Government inventory investment is included in government consumption expenditures.

Capital consumption. Consumption of fixed capital (CFC; economic depreciation) is included in government consumption expenditures as a partial measure of the value of the services of general government fixed assets, including structures, equipment, R&D, and software.

Definitions: output, lending and borrowing, and net investment

In Tables 6-5 and 6-10, current-dollar values of gross output and value added of government are presented, as described in the general discussion above. *Gross output* of government is the sum of the *intermediate goods and services purchased* by government and the value added by government as a producing industry. *Value added* consists of compensation of general government employees and consumption of general government fixed capital. Since this depreciation allowance is the only entry on the product side of the accounts measuring the output associated with such capital, a zero <u>net</u> return on these assets is implicitly assumed.

Gross output minus own-account investment and sales to other sectors (see the previous description) yields *government consumption expenditures*, which represents the contribution of government consumption spending to final demands in GDP.

Net lending (net borrowing -) consists of current *net saving* as defined above, plus the *consumption of fixed capital* (CFC, from the current expenditure account), minus *gross investment*, plus *capital transfer receipts*, and minus *capital transfer payments* and *net purchases of non-produced assets*. (If this definition sounds somehow counter-intuitive, it should be remembered that "subtraction" in this context means "increasing the deficit [negative value] which must be financed.)

Capital transfer receipts and *payments* include grants between levels of government, or between government and the private sector, associated with acquisition or disposal of assets rather than with current consumption expenditures. Examples are federal grants to state and local government for highways, transit, air transportation, and water treatment plants; federal shipbuilding subsidies and other subsidies to businesses; and lump-sum payments to amortize the unfunded liability of the Uniformed Services Retiree Health Care Fund. Government capital transfer receipts include estate and gift taxes, which are no longer included in personal tax receipts. (Federal estate and gift taxes are shown in Table 6-14, where they are reported on a fiscal year basis, and are similar in order of magnitude to the calendar year values for federal capital transfer receipts in Table 6-5.)

Non-produced assets are land and radio spectrum. Unusually large negative entries for net purchases of non-produced assets in some recent years result from the negative purchase—that is, the sale—of spectrum.

The values of *net investment* shown in these tables are calculated by the editor, as gross investment minus the consumption of fixed capital.

Definitions: chain-type quantity indexes

Chain-type quantity indexes represent changes over time in real values, removing the effects of inflation. Indexes for key categories in the government expenditure accounts, as well as for government gross output, value added, and intermediate goods and services purchased, use the chain formula described in the notes and definitions for Chapter 1 and are expressed as index numbers, with the average for the year 2012 equal to 100.

Data availability

The most recent data are published each month in the *Survey of Current Business.* Current and historical data may be obtained from the BEA Web site at <http://www.bea.gov> and the STAT-USA subscription Web site at <http://www.stat-usa.gov>.

References

See the references regarding the 2018 comprehensive revision of the NIPAs in the notes and definitions for Chapter 1. NIPA concepts and their differences from the budget estimates are discussed in "NIPA Estimates of the Federal Sector and the Federal Budget Estimates," *Survey of Current Business,* March 2007, p. 11.

For information about the classification of government expenditures into current consumption and gross investment, first undertaken in the 1996 comprehensive revisions, see the *Survey of Current Business* article, "Preview of the Comprehensive Revision of the National Income and Product Accounts: Recognition of Government Investment and Incorporation of a New Methodology for Calculating Depreciation" September 1995. Other sources of information about the NIPAs are listed in the notes and definitions for Chapter 1.

Tables 6-12 and 6-13

State Government Current Receipts and Expenditures; Local Government Current Receipts and Expenditures

SOURCE: BUREAU OF ECONOMIC ANALYSIS (BEA)

In the standard presentation of the state and local sector of the national income and product accounts (NIPAs) such as in Tables 6-8 through 6-11 above, state and local governments are combined. Annual measures for aggregate state governments and aggregate local governments are also available on the BEA Web site. These measures are shown in Tables 6-12 and 6-13. The definitions are the same as in the other NIPA tables described above.

The data shown here now reflect the 2013 comprehensive revision, affecting some concepts, that has been incorporated in the quarterly and annual data elsewhere in this chapter and this volume.

Two categories not shown in Table 6-8 appear in each table, detailing inter-governmental flows that are consolidated in Table 6-8.

State government receipts include not only *federal grants-in-aid* but also *local grants-in-aid*, and state government expenditures include *grants-in-aid to local governments*. Local government receipts include not only *federal grants-in-aid* but also *state grants-in-aid*, and local government expenditures include *grants-in-aid to state governments*. To make room for these columns, the components *current surplus of government enterprises* and *subsidies* are not shown, though they are included in total current receipts and total current expenditures respectively.

These measures are described in "Receipts and Expenditures of State Governments and of Local Governments," *Survey of Current Business*, October 2005. Data back to 1959 are available on the BEA Web site at <http://www.bea.gov>.

Tables 6-14A and 6-14B

Federal Government Receipts and Outlays by Fiscal Year; The Federal Budget and GDP

SOURCE: U.S. OFFICE OF MANAGEMENT AND BUDGET

These data on federal government receipts and outlays are on a modified cash basis and are from the *Budget of the United States Government: Historical Tables*. The data are by federal fiscal years, which are defined as July 1 through June 30 through 1976 and October 1 through September 30 for 1977 and subsequent years. They are identified by the year of the final fiscal quarter—i.e., the year from July 1, 1975, through June 30, 1976, is identified as Fiscal 1976, and the year from October 1, 1976, through June 30, 1977, is identified as Fiscal Year 1977. The period July 1 through September 30, 1976, is a separate fiscal period known as the transition quarter (TQ) and is not included in any fiscal year.

There are numerous differences in both timing and definition between these estimates and the NIPA estimates in Tables 6-1 through 6-7. See the notes and definitions for those tables for the definitional differences that were introduced with the 1999 comprehensive revision of the NIPAs.

Definitions

The definitions for these tables are not affected by the 2013, 2003, or 1999 changes in the government sectors of the NIPAs.

Table references will be given indicating the source of each item in the *Historical Tables*; for example, "HT Table 1.1."

Receipts consist of gifts and of taxes or other compulsory payments to the government. Other types of payments to the government are netted against outlays. (HT Table 1.1)

Outlays occur when the federal government liquidates an obligation through a cash payment or when interest accrues on public

debt issues. Beginning with the data for 1992, outlays include the subsidy cost of direct and guaranteed loans made. Before 1992, the costs and repayments associated with such loans are recorded on a cash basis. As noted previously, various types of nontax receipts are netted against cash outlays. These accounts do not distinguish between investment outlays and current consumption and do not include allowances for depreciation. (HT Table 1.1)

The *total budget surplus (deficit-)* is receipts minus outlays. It is sometimes presented, including in Table 6-14, as composed of two components, *on-budget and off-budget*. By law, two government programs that are included in the federal receipts and outlays totals are "off-budget"—old-age, survivors, and disability insurance (Social Security) and the Postal Service. The former accounts for nearly all of the off-budget activity. The *surplus (deficit-)* not accounted for by these two programs is the on-budget surplus or deficit. The *on-budget deficit* is based on an arbitrary distinction and is not customarily referred to as the *budget deficit*, a term reserved in customary use for the total deficit. (HT Table 1.1)

Sources of financing is the means by which the total deficit is financed or the surplus is distributed. By definition, sources of financing sum to the total deficit or surplus with the sign reversed. The principal source is *borrowing from the public*, shown as a positive number, that is, the increase in the debt held by the public. (This entry is calculated by the editors as the change in the debt held by the public as shown in HT Table 7.1.) When there is a budget surplus, as in fiscal years 1998 to 2001, this provides resources for debt reduction, which is indicated by a minus sign in this column. *Other financing* includes drawdown (or buildup, shown here with a minus sign) in Treasury cash balances, seigniorage on coins, direct and guaranteed loan account cash transactions, and miscellaneous other transactions. Large negative "other financing" in 2008 through 2010 resulted from direct loans and asset purchases under TARP and other emergency financial procedures; in other words, the government borrowed from the public to undertake these loans and asset purchases, but they are not included in budget expenditures or the budget deficit. *Other financing* is calculated by the editors, by reversing the sign of the surplus/deficit—so that the deficit becomes a positive number—and subtracting the value of borrowing from the public.

Some of the categories of *receipts by source* are self-explanatory. *Employment taxes and contributions* includes taxes for old-age, survivors, and disability insurance (Social Security), hospital insurance (Medicare), and railroad retirement funds. *Other retirement contributions* includes the employee share of payments for retirement pensions, mainly those for federal employees. *Excise taxes* includes federal taxes on alcohol, tobacco, telephone service, and transportation fuels, as well as taxes funding smaller programs such as black lung disability and vaccine injury compensation. *Miscellaneous receipts* includes deposits of earnings by the Federal Reserve system and all other receipts. (HT Tables 2.1, 2.4, and 2.5)

Outlays by function presents outlays according to the major purpose of the spending. Functional classifications cut across departmental and agency lines. Most categories of offsetting receipts are netted against cash outlays in the appropriate function, which explains how recorded outlays in *energy* and *commerce and housing credit* (which, as its name suggests, includes loan programs) can be negative. There is also a category of "undistributed offsetting receipts" (not shown), always with a negative sign, that includes proceeds from the sale or lease of assets and payments from federal agencies to federal retirement funds and to the Social Security and Medicare trust funds. Note that *Social Security* is recorded separately from other *income security* outlays, and *Medicare* separately from other *health* outlays. (HT Table 3.1) For further explanation, consult the *Budget of the United States Government.*

In order to provide authoritative comparisons of these budget values with the overall size of the economy, a special calculation of gross domestic product by fiscal year is supplied to the Office of Management and Budget by the BEA, and is shown in Table 6-14B. That table also displays selected budget aggregates in dollar values and as a percent of GDP. (HT Tables 1.2, 1.3, 2.3, and 3.1)

Data availability and references

Definitions, budget concepts, and historical data are from *Budget of the United States Government for Fiscal Year 2015: Historical Tables*, available on the Office of Management and Budget Web site at <http://www.whitehouse.gov/omb/Budget/historicals>

Similarly defined data for the latest month, the year-ago month, and the current and year-ago fiscal year to date are published in the *Monthly Treasury Statement* prepared by the Financial Management Service, U.S. Department of the Treasury. For those who need up-to-date budget information, this publication is available on the Financial Management Service Web site at <http://www.fms.treas.gov>. As these monthly figures are never revised to agree with the final annual data, they are not published in this volume.

TABLE 6-15

Federal Government Debt by Fiscal Year

SOURCE: U.S. OFFICE OF MANAGEMENT AND BUDGET

Debt outstanding at the end of each fiscal year is from the *Budget of the United States Government.* Most securities are recorded at sales price plus amortized discount or less amortized premium.

Definitions

Federal government debt held by the public consists of all federal debt held outside the federal government accounts—by individuals, financial institutions (including the Federal Reserve Banks), and foreign individuals, businesses, and central banks. It does not include federal debt held by federal government trust funds such as the Social Security trust fund. The level and change of the ratio of this debt to the value of gross domestic product (GDP) provide proportional measures of the impact of federal borrowing on credit markets. (HT Table 7.1) This measure of debt held by the public is very similar in concept and scope to the total federal government credit market debt outstanding in the flow-of-funds accounts, shown in *Business Statistics* in Table 12-5; however, it is not identical, being priced somewhat differently, and is shown here in Chapter 6 on a fiscal year rather than calendar year basis.

Gross federal debt—total. This is the total debt owed by the U.S. Treasury. It includes a small amount of matured debt. (HT Table 7.1)

Debt held by Social Security funds is the sum of the end-year trust fund balances for old age and survivors insurance and disability insurance. The separate disability trust fund begins in 1957. (HT Table 13.1)

Debt held by other U.S. government accounts is calculated by the editors by subtracting the Social Security debt holdings from the total debt held by federal government accounts, which is shown in HT Table 7.1. It includes the balances in all the other trust funds, including the Medicare funds, federal employee retirement funds, and the highway trust fund.

Debt held by the Federal Reserve System is the total value of Treasury securities held by the 12 Federal Reserve Banks, which are acquired in open market operations that carry out monetary policy. (HT Table 7.1)

Debt held by private investors is calculated by subtracting the Federal Reserve debt from the total debt held by the public, and is shown as "Debt Held by the Public: Other" in HT Table 7.1.

Debt held by foreign residents is based on surveys by the Treasury Department. (Table 5-7, *Budget of the United States for Fiscal Year 2015: Analytical Perspectives.*)

Debt held by domestic investors is calculated by the editors by subtracting the foreign-held debt from the total debt held by private investors.

The "debt subject to statutory limitation," not shown here, is close to the gross federal debt in concept and size, but there are some relatively minor definitional differences specified by law. The debt limit can only be changed by an Act of Congress. For information about the debt subject to limit and other debt subjects, see the latest *Budget of the United States Government: Historical Statistics* and *Analytical Perspectives.*

Data availability and references

For the end-of-fiscal-year data, see *Historical Tables* and *Analytical Perspectives* in the *Budget of the United States Government*, which is available on the OMB Web site at <http://www.whitehouse.gov/omb/Budget>. In the *Analytical Perspectives,* debt analysis and data are included in the "Economic and Budget Analysis" section.

Recent quarterly data, measured on a somewhat different basis, are found in the *Treasury Bulletin* in the chapter on "Ownership of Federal Securities (OFS)," in Tables OFS-1 and OFS-2. The *Treasury Bulletin* can be accessed on the Internet at <http://www.fms.treas.gov/bulletin>. Holdings by Social Security funds are also available in the *Bulletin* in the chapter on "Federal Debt," Table FD-3. The Disability Fund is listed separately from the Old-Age and Survivors Fund.

Table 6-17

Government Employment

SOURCE: U.S. DEPARTMENT OF LABOR, BUREAU OF LABOR STATISTICS (SEE NOTES AND DEFINITIONS FOR TABLE 10-8).

Government payroll employment includes federal, state, and local activities such as legislative, executive, and judicial functions, as well as all government owned and government operated business enterprises, establishments, and institutions (arsenals, navy yards, hospitals, etc.), and government force account construction. The figures relate to civilian employment only. The Bureau of Labor Statistics (BLS) considers regular fulltime teachers (private and governmental) to be employed during the summer vacation period, regardless of whether they are specifically paid in those months.

Employment in federal government establishments reflects employee counts as of the pay period containing the 12th of the month. Federal government employment excludes employees of the Central Intelligence Agency and the National Security Agency.

CHAPTER 7: U.S. FOREIGN TRADE AND FINANCE

SECTION 7A: FOREIGN TRANSACTIONS IN THE NATIONAL INCOME AND PRODUCT ACCOUNTS

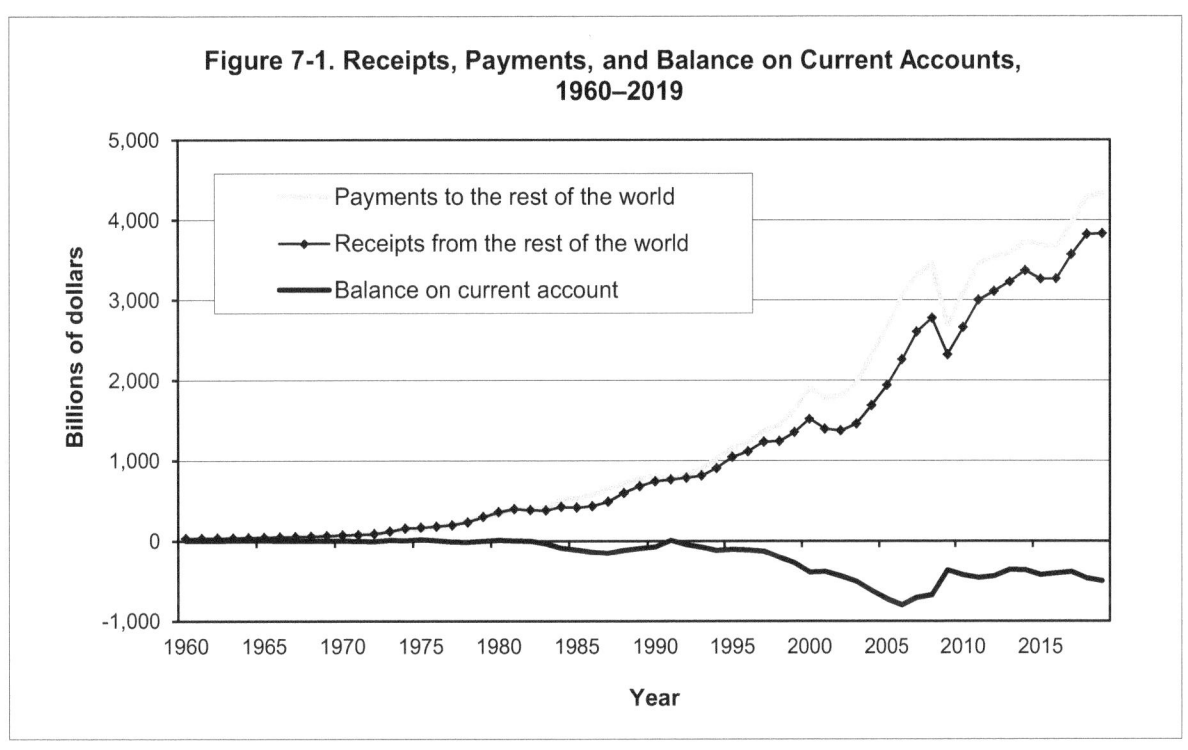

Figure 7-1. Receipts, Payments, and Balance on Current Accounts, 1960–2019

- Beginning in 1983, chronic current-account deficits in U.S. transactions with the rest of the world have emerged and increased, requiring increasing net capital inflows from abroad to finance them. The only current-account surplus since then occurred in 1991, when payments from other countries financed most of the cost of the first Gulf War. In 2006, the NIPA deficit reached a record $807.5 billion, 5.8 percent of GDP. Between 2006 and 2013, the current account deficit declined by half. However, from 2013 to 2019, the current account deficit increased 39.5 percent. Just in one year from 2018 to 2019, it increased 7.5 percent. (Tables 7-1 and 1-1A)

- Table 7-4 lists the exports and imports of major categories for goods and services. In 2019, the largest exports in the United States were capital goods, except automotive ($548.1) and industrial supplies and materials ($526.8). Companies in industrial supplies distribute supplies for machinery and equipment used in the manufacturing, oil and gas, and warehousing industries. (Table 7-4)

- The term "services" refers to an expanding range of economic activities, such as audiovisual, construction, computer and related services, express delivery, telecommunications, e-commerce, financial, professional (such as accounting and legal services), retail and wholesaling, transportation, and tourism. Because measures of service trade are not anchored in any observation of physical movement, they are dependent on definitions of residence, i.e. exports or imports. (Table 7-4). See Notes and definitions at the end of this chapter for a more expansive discussion of services.

Table 7-1. Foreign Transactions in the National Income and Product Accounts

(Billions of dollars, quarterly data are at seasonally adjusted annual rates.)   NIPA Table 4.1

Year and quarter	Current receipts from the rest of the world					Current payments to the rest of the world					Current taxes and transfer payments, net	Balance on current account, NIPAs	Net lending or net borrowing (-), NIPAs
	Total	Exports of goods and services			Income receipts	Total	Imports of goods and services			Income payments			
		Goods¹		Services¹			Goods¹		Services¹				
		Durable	Nondurable				Durable	Nondurable					
1960	31.9	20.5	9.2	6.6	4.9	28.8	6.4	8.9	7.6	1.8	4.1	3.2	3.2
1961	32.9	20.9	9.4	6.7	5.3	28.7	6.0	9.0	7.6	1.8	4.2	4.2	4.2
1962	35.0	21.7	9.7	7.4	5.9	31.2	6.9	9.9	8.1	1.8	4.4	3.8	3.8
1963	37.6	23.3	10.7	7.7	6.5	32.7	7.4	10.3	8.4	2.1	4.5	4.9	4.9
1964	42.3	26.8	12.1	8.3	7.2	34.8	8.4	11.0	8.7	2.3	4.4	7.5	7.5
1965	45.0	28.0	12.1	9.2	7.9	38.9	10.4	11.8	9.3	2.6	4.7	6.2	6.2
1966	49.0	31.1	13.4	9.8	8.1	45.2	13.3	13.1	10.7	3.0	5.1	3.8	3.8
1967	52.2	32.5	13.9	10.9	8.7	48.7	14.5	13.2	12.2	3.3	5.5	3.5	3.5
1968	58.0	35.7	14.5	12.2	10.1	56.5	18.8	15.1	12.6	4.0	5.9	1.5	1.5
1969	63.7	38.7	14.8	13.2	11.8	62.1	20.8	16.1	13.7	5.7	5.9	1.6	1.6
1970	72.5	45.0	17.7	14.7	12.8	68.8	22.8	18.1	14.9	6.4	6.6	3.7	3.7
1971	77.0	46.2	18.3	16.8	14.0	76.7	26.6	19.9	15.8	6.4	7.9	0.3	0.3
1972	87.1	52.6	21.2	18.3	16.3	91.2	33.3	23.6	17.3	7.7	9.2	-4.0	-4.1
1973	118.8	75.8	33.5	19.5	23.5	109.9	40.8	31.0	19.3	10.9	7.9	8.9	8.8
1974	156.5	103.5	45.8	23.2	29.8	150.5	50.3	54.2	22.9	14.3	8.7	6.0	5.9
1975	166.7	112.5	47.3	26.2	28.0	146.9	45.6	53.4	23.7	15.0	9.1	19.8	19.8
1976	181.9	121.5	50.6	28.0	32.4	174.8	56.8	67.8	26.5	15.5	8.1	7.1	7.0
1977	196.5	128.4	54.0	30.9	37.2	207.5	69.2	83.4	29.8	16.9	8.1	-10.9	-11.0
1978	233.1	149.9	62.4	37.0	46.3	245.8	89.0	88.4	34.8	24.7	8.8	-12.6	-12.7
1979	298.5	187.3	77.6	42.9	68.3	299.6	100.4	112.3	39.9	36.4	10.6	-1.2	-1.3
1980	359.9	230.4	94.8	50.3	79.1	351.4	111.9	136.6	45.3	44.9	12.6	8.5	8.4
1981	397.3	245.2	102.0	60.0	92.0	393.9	126.0	141.8	49.9	59.1	17.0	3.4	3.3
1982	384.2	222.6	94.1	60.7	101.0	387.5	125.1	125.4	52.6	64.5	19.8	-3.3	-3.4
1983	378.9	214.0	90.3	62.9	101.9	413.9	147.3	125.4	56.0	64.8	20.5	-35.1	-35.2
1984	424.2	231.3	96.1	71.1	121.9	514.3	192.4	143.9	68.8	85.6	23.6	-90.1	-90.2
1985	415.9	227.5	87.6	75.7	112.7	530.2	204.2	139.1	73.9	87.3	25.7	-114.3	-114.5
1986	432.3	231.4	86.1	89.6	111.3	575.0	238.8	131.2	82.9	94.4	27.8	-142.7	-142.8
1987	487.2	265.6	98.6	98.4	123.3	641.3	264.2	150.6	93.9	105.8	26.8	-154.1	-154.2
1988	596.7	332.1	120.1	112.5	152.1	712.4	294.8	157.3	101.9	129.5	29.0	-115.7	-115.9
1989	682.0	374.8	132.2	129.5	177.7	774.3	310.3	174.5	106.2	152.9	30.4	-92.4	-92.7
1990	740.7	403.3	138.0	148.6	188.8	815.6	314.5	193.5	121.7	154.2	31.7	-74.9	-82.3
1991	763.3	430.1	144.6	164.8	168.4	755.4	315.5	185.2	122.8	136.8	-4.9	7.9	2.6
1992	785.1	455.3	151.1	177.7	152.1	830.7	346.7	198.2	122.9	121.0	41.9	-45.6	-44.3
1993	810.4	467.7	149.9	187.1	155.6	889.8	386.4	206.4	127.2	124.4	45.4	-79.4	-80.2
1994	905.5	518.4	164.6	202.6	184.5	1 021.1	454.0	222.8	136.6	161.6	46.1	-115.6	-116.9
1995	1 042.6	592.4	193.8	220.4	229.8	1 148.5	510.8	246.7	145.1	201.9	44.1	-105.9	-106.3
1996	1 114.0	628.8	202.4	238.8	246.4	1 229.0	533.4	274.1	156.5	215.5	49.5	-115.0	-115.2
1997	1 233.9	699.9	211.5	253.9	280.1	1 364.0	588.5	297.1	170.1	256.8	51.4	-130.1	-130.6
1998	1 239.8	692.6	200.2	260.4	286.8	1 445.1	637.7	293.0	184.9	269.4	60.0	-205.3	-205.6
1999	1 350.7	711.7	203.0	281.1	320.2	1 629.0	715.8	335.4	197.4	294.7	85.7	-278.3	-285.0
2000	1 517.9	795.9	223.2	300.3	380.6	1 912.2	822.2	427.9	221.2	345.6	95.4	-394.3	-398.9
2001	1 393.9	741.2	217.2	283.4	324.1	1 778.1	756.2	417.6	218.8	275.3	110.2	-384.2	-372.3
2002	1 370.3	709.0	218.3	289.7	314.8	1 812.0	771.7	422.7	229.8	269.6	118.3	-441.6	-445.8
2003	1 455.9	737.1	237.4	299.1	353.8	1 964.9	803.9	487.4	248.0	295.4	130.1	-508.9	-517.7
2004	1 689.1	830.0	266.5	347.7	446.9	2 310.3	934.5	572.8	289.4	368.8	144.9	-621.2	-625.8
2005	1 941.3	921.9	294.5	383.3	566.0	2 670.6	1 023.9	691.6	311.0	488.1	156.1	-729.3	-728.6
2006	2 259.5	1 044.9	332.2	427.7	712.0	3 059.3	1 129.5	766.3	347.8	661.5	154.2	-799.8	-807.5
2007	2 602.4	1 161.3	380.9	499.6	866.6	3 312.3	1 174.4	825.3	379.6	757.6	175.5	-710.0	-716.3
2008	2 775.2	1 292.5	466.4	544.5	848.8	3 452.3	1 160.6	983.7	415.9	694.2	198.0	-677.1	-677.8
2009	2 319.9	1 058.4	390.1	523.6	647.8	2 688.6	891.4	694.0	393.1	505.8	204.3	-368.7	-375.0
2010	2 658.3	1 272.4	474.6	573.8	715.2	3 089.6	1 104.5	840.3	415.4	519.5	209.9	-431.3	-438.7
2011	2 998.3	1 462.3	570.6	640.7	789.2	3 460.0	1 236.6	1 003.9	441.9	552.8	224.7	-461.7	-471.2
2012	3 107.7	1 521.6	583.9	669.7	799.7	3 549.0	1 326.4	975.0	458.5	567.4	221.8	-441.3	-440.8
2013	3 228.0	1 559.2	603.8	714.2	823.4	3 588.4	1 357.6	938.7	467.8	592.7	231.5	-360.5	-367.4
2014	3 371.4	1 615.0	622.0	756.7	853.5	3 737.0	1 450.3	941.3	487.8	612.5	245.2	-365.6	-372.5
2015	3 263.4	1 494.6	540.5	771.3	860.8	3 687.1	1 488.9	799.1	504.4	640.4	254.3	-423.7	-432.0
2016	3 265.9	1 444.0	518.5	783.2	893.5	3 673.4	1 464.9	756.1	518.6	661.5	272.2	-407.4	-414.5
2017	3 569.6	1 541.8	581.7	832.8	1 032.7	3 961.2	1 565.8	811.1	553.2	740.4	290.6	-391.5	-407.5
2018	3 821.7	1 663.9	657.9	864.8	1 142.9	4 289.5	1 661.1	904.5	572.6	858.2	293.2	-467.8	-472.4
2019	3 831.8	1 636.7	656.8	878.0	1 169.8	4 334.5	1 634.6	891.0	599.6	900.2	309.1	-502.8	-509.5
2017													
1st quarter	3 466.4	1 513.6	577.1	812.7	964.9	3 834.3	1 516.7	817.6	535.6	688.2	276.2	-367.9	-376.8
2nd quarter	3 469.5	1 504.9	561.8	828.2	987.7	3 902.5	1 550.1	792.0	550.5	720.1	289.8	-433.0	-441.4
3rd quarter	3 604.4	1 533.5	567.7	836.7	1 058.2	3 973.7	1 564.7	786.7	562.3	756.8	303.2	-369.3	-408.8
4th quarter	3 738.2	1 615.1	620.1	853.6	1 120.2	4 134.1	1 631.5	848.2	564.5	796.6	293.3	-395.9	-403.1
2018													
1st quarter	3 768.2	1 635.4	625.0	871.8	1 116.2	4 176.0	1 644.4	889.5	563.2	801.7	277.4	-407.8	-413.9
2nd quarter	3 854.3	1 693.5	678.5	856.8	1 153.4	4 250.7	1 638.7	893.1	566.6	861.9	290.4	-396.5	-408.5
3rd quarter	3 809.1	1 660.5	665.6	863.4	1 129.2	4 336.4	1 676.0	920.3	574.0	869.2	296.9	-527.3	-529.5
4th quarter	3 855.4	1 666.3	662.5	867.1	1 172.6	4 395.0	1 685.4	915.0	586.6	899.9	308.2	-539.6	-537.7
2019													
1st quarter	3 816.9	1 660.6	653.1	862.9	1 148.0	4 348.9	1 666.2	884.7	588.1	901.5	308.4	-532.0	-542.7
2nd quarter	3 844.1	1 631.1	657.3	883.5	1 184.3	4 379.9	1 642.0	914.3	603.0	913.2	307.3	-535.8	-539.6
3rd quarter	3 841.5	1 626.0	655.2	879.2	1 181.2	4 349.3	1 640.9	893.7	602.4	901.4	310.8	-507.8	-511.5
4th quarter	3 824.5	1 629.1	661.5	886.5	1 165.9	4 260.1	1 589.3	871.3	604.8	884.8	309.9	-435.6	-444.2

¹Exports and imports of certain goods, primarily military equipment purchased and sold by the federal government, are included in services. Beginning with 1986, repairs and alterations of equipment are reclassified from goods to services.

Table 7-2. Chain-Type Quantity Indexes for Exports and Imports of Goods and Services

(Index numbers, 2012 = 100; quarterly indexes are seasonally adjusted.)   NIPA Table 4.2.3

| Year and quarter | Exports of goods and services | | | | | Imports of goods and services | | | | |
| | Total | Goods [1] | | | Services [1] | Total | Goods [1] | | | Services [1] |
		Total	Durable	Nondurable			Total	Durable	Nondurable	
1975	12.82	12.81	9.84	21.95	12.68	10.79	9.80	4.81	24.48	16.54
1976	13.38	13.47	10.02	24.13	12.81	12.90	12.02	5.90	30.04	17.68
1977	13.70	13.73	10.05	25.13	13.39	14.31	13.49	6.67	33.46	18.56
1978	15.14	15.16	11.14	27.60	14.89	15.55	14.70	7.71	34.55	19.89
1979	16.64	16.76	12.55	29.71	15.95	15.80	14.95	7.89	34.95	20.16
1980	18.44	18.82	14.04	33.55	16.62	14.75	13.84	7.95	30.09	19.71
1981	18.66	18.70	13.59	34.58	18.23	15.14	14.12	8.61	29.18	20.84
1982	17.23	17.11	11.89	33.70	17.43	14.95	13.76	8.70	27.48	21.95
1983	16.79	16.56	11.60	32.22	17.40	16.83	15.63	10.49	29.27	23.73
1984	18.15	17.73	12.76	33.23	19.44	20.93	19.41	13.96	33.33	29.68
1985	18.76	18.35	13.76	32.36	19.99	22.29	20.63	15.36	33.73	31.94
1986	20.20	19.34	14.58	33.77	22.86	24.19	22.74	16.78	37.71	32.29
1987	22.41	21.70	16.78	36.32	24.59	25.62	23.80	17.52	39.61	36.10
1988	26.04	25.56	20.65	39.65	27.51	26.63	24.76	18.26	41.13	37.32
1989	29.05	28.48	23.47	42.63	30.81	27.80	25.83	19.07	42.80	39.13
1990	31.62	30.92	26.01	44.55	33.75	28.80	26.58	19.57	44.24	41.69
1991	33.71	33.00	27.90	47.03	35.90	28.76	26.70	19.67	44.37	40.60
1992	36.04	35.48	30.13	50.15	37.83	30.77	29.22	21.74	47.76	39.51
1993	37.22	36.63	31.69	49.83	39.09	33.43	32.14	24.31	50.97	40.56
1994	40.51	40.14	35.48	52.14	41.82	37.42	36.43	28.33	54.76	42.72
1995	44.68	44.81	40.51	55.52	44.66	40.41	39.72	31.58	57.01	44.01
1996	48.33	48.80	45.18	57.51	47.46	43.93	43.44	34.96	60.83	46.31
1997	54.09	55.87	53.36	61.46	49.96	49.84	49.70	40.73	67.15	50.33
1998	55.35	57.08	55.06	61.35	51.34	55.67	55.59	46.09	73.29	55.81
1999	58.11	59.47	57.61	63.21	54.99	61.96	62.74	52.77	80.24	57.77
2000	62.95	65.37	64.61	66.07	57.37	69.94	70.88	60.73	87.30	64.97
2001	59.31	61.23	59.16	65.47	54.89	67.98	68.49	56.87	89.47	65.18
2002	58.28	59.19	55.96	66.58	56.22	70.45	71.00	59.23	91.94	67.48
2003	59.56	60.93	57.85	67.89	56.42	73.92	75.13	62.67	97.30	67.72
2004	65.31	66.27	64.41	70.27	63.13	82.35	83.72	71.70	103.85	75.30
2005	69.97	71.34	70.75	72.50	66.84	87.72	89.61	78.18	108.00	78.21
2006	76.51	78.41	78.86	77.33	72.17	93.54	95.20	85.62	109.69	85.12
2007	83.16	83.87	84.96	81.47	81.52	95.87	97.16	88.09	110.66	89.26
2008	87.87	88.75	89.05	87.80	85.82	93.74	93.89	85.24	106.74	92.76
2009	80.49	78.23	74.07	85.61	85.55	81.48	79.52	68.03	97.26	90.72
2010	90.26	89.95	87.25	94.49	90.97	92.17	91.80	84.11	102.71	93.97
2011	96.70	96.30	95.31	97.89	97.63	97.37	97.44	92.51	104.22	96.98
2012	100.00	100.00	100.00	100.00	100.00	100.00	100.00	100.00	100.00	100.00
2013	103.58	103.18	102.39	104.46	104.46	101.54	101.76	104.49	98.06	100.46
2014	107.94	107.96	106.79	109.86	107.92	106.62	107.42	113.76	98.85	102.75
2015	108.40	107.58	104.42	112.98	110.09	112.14	113.50	121.19	102.99	105.79
2016	108.72	108.16	103.24	117.03	109.91	113.99	115.05	122.83	104.42	108.88
2017	112.97	112.55	106.47	123.66	113.92	119.30	120.46	131.63	104.53	113.74
2018	116.35	117.27	110.15	130.30	114.79	124.18	126.42	138.93	108.55	114.23
2019	116.21	117.15	107.20	135.40	114.62	125.52	127.03	139.35	109.43	118.46
2013										
1st quarter	101.58	100.64	100.52	100.82	103.69	99.96	100.04	101.58	97.93	99.55
2nd quarter	102.80	102.25	103.05	100.96	104.03	101.40	101.60	103.78	98.63	100.42
3rd quarter	103.45	102.90	101.91	104.50	104.65	102.13	102.40	105.39	98.34	100.80
4th quarter	106.48	106.94	104.08	111.56	105.45	102.69	103.02	107.20	97.34	101.08
2014										
1st quarter	105.72	105.09	104.13	106.65	107.09	103.85	104.45	108.58	98.82	100.97
2nd quarter	108.01	107.89	106.56	110.03	108.29	106.49	107.31	113.79	98.56	102.56
3rd quarter	108.49	108.97	108.21	110.19	107.49	106.77	107.59	114.61	98.10	102.80
4th quarter	109.54	109.90	108.25	112.58	108.81	109.36	110.35	118.06	99.90	104.65
2015										
1st quarter	108.71	107.75	104.78	112.81	110.69	111.08	112.67	119.96	102.82	103.76
2nd quarter	108.98	108.48	105.31	113.91	110.05	111.76	113.31	120.88	103.00	104.63
3rd quarter	108.04	107.35	104.33	112.48	109.48	112.73	114.16	122.15	103.19	106.10
4th quarter	107.86	106.74	103.27	112.74	110.16	112.98	113.85	121.78	102.98	108.67
2016										
1st quarter	107.60	107.29	102.28	116.40	108.35	112.86	113.84	120.96	104.33	108.11
2nd quarter	108.00	107.04	102.69	114.75	109.96	112.78	113.87	121.25	103.89	107.58
3rd quarter	109.83	108.89	103.41	118.84	111.74	114.20	115.12	123.37	103.73	109.66
4th quarter	109.43	109.42	104.58	118.12	109.59	116.10	117.35	125.75	105.74	110.18
2017										
1st quarter	111.63	111.43	104.24	124.68	112.16	117.34	118.50	127.89	105.32	111.78
2nd quarter	111.89	110.97	104.88	122.09	113.79	118.37	119.30	130.49	103.35	113.76
3rd quarter	112.63	111.72	107.08	120.07	114.49	118.88	119.66	131.40	102.83	114.90
4th quarter	115.74	116.09	109.68	127.80	115.26	122.59	124.36	136.72	106.64	114.54
2018										
1st quarter	116.52	116.46	111.05	126.36	116.79	122.69	124.82	137.08	107.28	113.20
2nd quarter	117.07	118.85	110.87	133.44	113.87	122.67	124.79	136.55	108.00	113.18
3rd quarter	115.50	116.27	108.76	130.03	114.23	125.05	127.53	140.25	109.39	114.11
4th quarter	116.32	117.50	109.92	131.38	114.26	126.31	128.53	141.84	109.53	116.43
2019										
1st quarter	116.85	118.62	110.16	134.11	113.70	125.65	127.66	140.93	108.73	116.59
2nd quarter	115.52	115.83	106.37	133.16	115.07	126.20	127.71	139.86	110.35	119.14
3rd quarter	115.76	116.71	106.21	135.97	114.14	126.34	127.94	140.14	110.52	118.92
4th quarter	116.73	117.43	106.06	138.34	115.57	123.90	124.81	136.49	108.12	119.20

[1]Exports and imports of certain goods, primarily military equipment purchased and sold by the federal government, are included in services. Beginning with 1986, repairs and alterations of equipment are reclassified from goods to services.

Table 7-3. Chain-Type Price Indexes for Exports and Imports of Goods and Services

(Index numbers, 2012 = 100; quarterly indexes are seasonally adjusted.) NIPA Table 4.2.4

Year and quarter	Exports of goods and services					Imports of goods and services				
	Total	Goods [1]			Services [1]	Total	Goods [1]			Services [1]
		Total	Durable	Nondurable			Total	Durable	Nondurable	
1975	49.39	57.70	31.30	28.79	30.88	41.23	43.88	30.16	15.43	31.29
1976	51.01	59.27	31.27	28.75	32.67	42.47	45.05	29.82	16.12	32.71
1977	53.09	61.47	32.55	29.94	34.50	46.21	49.18	32.56	17.46	35.03
1978	56.32	65.00	34.05	31.33	37.08	49.47	52.45	34.98	17.77	38.22
1979	63.10	73.43	41.28	37.94	40.13	57.93	61.85	41.76	23.94	43.16
1980	69.50	80.46	46.38	42.62	45.24	72.17	78.06	50.46	36.47	50.09
1981	74.65	86.18	48.09	44.20	49.18	76.07	82.43	51.15	40.08	52.26
1982	75.01	85.47	46.62	42.87	51.97	73.51	79.11	49.25	37.26	52.30
1983	75.31	84.98	45.31	41.64	54.04	70.75	75.80	46.63	34.01	51.44
1984	76.02	85.74	46.72	42.94	54.60	70.14	75.28	46.12	33.82	50.56
1985	73.75	81.46	44.85	40.84	56.57	67.84	72.32	43.75	31.76	50.50
1986	72.52	78.67	45.60	37.81	58.50	67.83	70.70	44.98	21.80	55.97
1987	74.12	80.43	49.70	42.55	59.75	71.94	75.73	48.64	24.68	56.76
1988	77.92	85.38	55.83	46.82	61.08	75.38	79.34	57.70	23.42	59.53
1989	79.21	86.48	57.30	47.50	62.77	77.02	81.56	60.98	26.25	59.22
1990	79.66	85.71	57.02	48.31	65.75	79.23	83.06	58.40	29.35	63.66
1991	80.55	85.67	56.51	46.73	68.56	78.57	81.49	56.70	26.37	65.98
1992	80.15	84.34	57.43	44.34	70.17	78.64	81.04	56.77	25.48	67.84
1993	80.28	83.91	60.97	43.24	71.46	78.03	80.13	56.81	24.06	68.41
1994	81.21	84.87	63.22	46.82	72.32	78.77	80.72	58.83	23.49	69.75
1995	83.03	86.89	67.71	54.68	73.68	80.92	82.87	63.38	25.98	71.93
1996	81.92	84.68	65.37	51.82	75.14	79.51	80.77	62.48	28.29	73.73
1997	80.48	82.34	64.78	51.75	75.90	76.75	77.43	63.60	27.22	73.72
1998	78.57	79.74	62.12	48.62	75.74	72.62	72.76	60.61	21.76	72.28
1999	77.97	78.66	60.50	48.31	76.33	73.02	72.80	60.73	24.61	74.54
2000	79.47	80.02	61.99	53.72	78.16	76.22	76.64	64.11	34.93	74.26
2001	78.84	79.56	60.96	51.82	77.11	74.22	74.46	61.96	31.85	73.21
2002	78.20	78.73	61.12	50.18	76.95	73.24	73.10	59.65	30.87	74.27
2003	79.40	79.51	63.82	54.57	79.17	75.45	74.69	61.17	36.57	79.87
2004	82.28	82.32	72.10	60.65	82.24	79.06	78.23	71.96	43.29	83.82
2005	85.13	84.93	78.57	68.84	85.63	83.70	83.18	75.90	55.60	86.72
2006	87.84	87.59	88.36	74.56	88.49	86.91	86.53	83.18	64.09	89.13
2007	91.14	91.00	94.11	80.56	91.50	89.92	89.43	89.50	70.07	92.77
2008	95.41	95.71	97.00	92.74	94.75	98.96	99.23	98.81	95.73	97.80
2009	89.69	88.92	82.75	73.38	91.39	87.99	86.63	83.50	62.43	94.51
2010	93.35	92.97	92.26	85.13	94.20	92.78	92.05	95.39	77.82	96.42
2011	99.24	99.79	102.08	102.30	98.00	99.83	99.91	104.84	98.81	99.39
2012	100.00	100.00	100.00	100.00	100.00	100.00	100.00	100.00	100.00	100.00
2013	100.17	99.31	96.95	98.35	102.10	98.64	98.05	97.83	96.30	101.58
2014	100.27	98.31	95.52	96.50	104.71	97.85	96.74	97.34	93.55	103.56
2015	95.40	91.31	89.76	76.78	104.61	90.00	87.60	88.74	60.90	102.46
2016	93.49	87.74	86.35	68.25	106.40	86.87	83.89	84.72	50.59	102.36
2017	95.92	90.03	89.77	75.94	109.16	88.77	85.74	92.35	58.44	104.52
2018	99.18	93.25	93.01	84.90	112.50	91.33	88.18	99.00	68.86	107.73
2019	98.75	91.82	92.10	79.33	114.39	89.99	86.39	96.95	64.91	108.77
2013										
1st quarter	100.60	100.23	98.73	99.47	101.43	99.90	99.66	100.33	99.81	101.12
2nd quarter	99.91	99.12	97.09	97.29	101.70	98.61	98.10	97.94	95.75	101.18
3rd quarter	100.12	99.15	95.63	98.50	102.28	98.11	97.45	96.09	95.24	101.47
4th quarter	100.05	98.75	96.36	98.16	102.99	97.92	97.01	96.96	94.41	102.53
2014										
1st quarter	100.82	99.50	96.22	101.25	103.80	99.14	98.34	97.03	99.87	103.22
2nd quarter	100.85	99.16	95.87	98.09	104.66	98.47	97.45	96.96	96.40	103.70
3rd quarter	100.60	98.52	95.82	97.04	105.30	97.86	96.68	98.45	93.26	103.89
4th quarter	98.81	96.05	94.16	89.63	105.07	95.94	94.49	96.91	84.68	103.42
2015										
1st quarter	96.35	92.84	92.29	79.22	104.29	91.87	89.81	93.60	66.60	102.48
2nd quarter	96.16	92.38	91.01	80.46	104.67	90.69	88.36	89.98	63.03	102.71
3rd quarter	95.28	90.96	88.78	76.46	105.01	89.71	87.21	86.90	60.65	102.64
4th quarter	93.79	89.05	86.94	70.98	104.45	87.75	85.01	84.48	53.30	102.00
2016										
1st quarter	92.34	86.51	85.31	63.70	105.44	86.18	83.15	82.42	46.87	101.90
2nd quarter	93.30	87.70	86.01	67.38	105.85	86.54	83.49	84.02	48.83	102.38
3rd quarter	93.84	88.15	86.57	69.55	106.60	87.22	84.30	86.18	52.30	102.44
4th quarter	94.49	88.60	87.50	72.39	107.71	87.53	84.61	86.24	54.36	102.71
2017										
1st quarter	95.12	89.29	89.33	74.38	108.21	88.41	85.61	89.42	59.19	102.98
2nd quarter	95.17	89.15	89.14	73.43	108.69	88.33	85.31	91.09	56.66	104.01
3rd quarter	96.05	90.22	89.44	76.38	109.13	88.59	85.40	92.97	56.04	105.18
4th quarter	97.35	91.45	91.19	79.55	110.59	89.76	86.65	95.92	61.89	105.92
2018										
1st quarter	98.20	92.30	92.81	81.85	111.46	91.23	88.21	98.35	68.03	106.92
2nd quarter	99.42	93.65	94.31	85.19	112.35	91.29	88.16	100.34	67.75	107.59
3rd quarter	99.72	93.86	92.80	87.50	112.86	91.63	88.46	99.30	70.63	108.11
4th quarter	99.39	93.20	92.13	85.04	113.31	91.19	87.91	98.02	69.03	108.27
2019										
1st quarter	98.55	92.00	92.53	80.09	113.32	90.29	86.82	97.33	64.80	108.41
2nd quarter	99.34	92.55	92.55	81.82	114.65	90.49	86.99	96.02	67.37	108.77
3rd quarter	98.76	91.56	91.72	78.30	115.02	89.75	86.09	96.73	63.67	108.87
4th quarter	98.35	91.18	91.60	77.09	114.55	89.43	85.68	97.73	63.78	109.03

[1] Exports and imports of certain goods, primarily military equipment purchased and sold by the federal government, are included in services. Beginning with 1986, repairs and alterations of equipment are reclassified from goods to services.

Table 7-4. Exports and Imports of Selected NIPA Types of Product

(Billions of dollars, quarterly data are at seasonally adjusted annual rates.)

NIPA Table 4.2.5

| | Exports | | | | | | | Imports | | | | | | | |
| | Goods | | | | | Services | | Goods | | | | | | Services | |
Year and quarter	Foods, feeds, and beverages	Industrial supplies and materials	Capital goods, except automotive	Automotive vehicles, engines, and parts	Consumer goods, except food and automotive	Travel	Other business services	Foods, feeds, and beverages	Industrial supplies and materials, except petroleum and products	Petroleum and products	Capital goods, except automotive	Automotive vehicles, engines, and parts	Consumer goods, except food and automotive	Travel	Other business services
1975	19.2	29.3	36.6	10.8	6.6	4.7	3.7	9.6	50.6	27.0	10.2	12.1	13.2	6.4	2.3
1976	19.8	31.6	39.1	12.2	8.0	5.7	4.5	11.5	63.1	34.6	12.3	16.8	17.2	6.9	2.9
1977	19.7	33.2	39.8	13.5	8.9	6.2	4.9	14.0	78.4	45.0	14.0	19.4	21.8	7.5	3.2
1978	25.7	38.4	47.5	15.2	11.4	7.2	6.2	15.8	81.9	42.6	19.3	25.0	29.4	8.5	3.9
1979	30.5	53.3	60.2	17.9	14.0	8.4	7.3	18.0	105.4	60.4	24.6	26.6	31.3	9.4	4.6
1980	36.3	68.0	76.3	17.4	17.8	10.6	8.6	18.6	126.9	79.5	31.6	28.3	34.3	10.4	5.1
1981	38.8	65.7	84.2	19.7	17.7	15.6	10.5	18.6	130.4	78.4	37.1	31.0	38.4	11.8	6.0
1982	32.2	61.8	76.5	17.2	16.1	15.5	13.8	17.5	107.3	62.0	38.4	34.3	39.7	12.8	7.0
1983	32.1	57.1	71.7	18.5	14.9	14.5	14.1	18.8	106.2	55.1	43.7	43.0	47.3	13.6	6.8
1984	32.2	61.9	77.0	22.4	15.1	21.3	14.5	21.9	120.7	58.1	60.4	56.5	61.1	23.4	7.7
1985	24.6	59.4	79.3	24.9	14.6	22.4	14.8	21.8	110.6	51.4	61.3	64.9	66.3	25.1	8.8
1986	23.5	59.0	82.8	25.1	16.7	25.5	23.3	24.4	96.7	34.3	72.0	78.1	79.4	26.5	13.3
1987	25.2	67.4	92.7	27.6	20.3	29.0	24.2	24.8	109.0	42.9	85.1	85.2	88.8	29.9	16.7
1988	33.8	84.2	119.1	33.4	27.0	35.2	25.7	24.9	116.3	39.6	102.2	87.9	96.4	32.8	17.8
1989	36.3	95.4	136.9	35.1	35.9	43.0	30.3	24.9	129.7	50.9	112.3	87.4	103.6	34.2	19.2
1990	35.1	101.9	153.0	36.2	43.5	50.9	32.7	26.4	140.5	62.3	116.4	88.2	105.0	38.2	22.4
1991	35.7	106.2	166.6	39.9	46.6	56.7	39.9	26.2	127.4	51.7	121.1	85.5	107.7	36.2	25.9
1992	40.3	105.2	176.4	46.9	51.2	64.0	41.1	27.6	134.0	51.6	134.8	91.5	122.4	39.6	24.3
1993	40.5	102.9	182.7	51.6	54.5	67.9	43.5	27.9	140.2	51.5	153.2	102.1	133.7	41.9	26.6
1994	42.4	115.4	205.7	57.5	59.7	69.4	49.9	31.0	156.2	51.3	185.0	118.1	145.9	45.1	30.3
1995	50.8	141.0	234.4	61.4	64.2	74.9	53.6	33.2	175.6	56.0	221.1	123.7	159.7	46.4	33.7
1996	56.0	141.3	254.0	64.4	69.3	81.8	61.3	35.7	197.3	72.7	228.4	128.7	172.5	49.7	38.1
1997	52.0	153.0	295.8	73.4	77.0	86.2	71.2	39.7	206.5	71.8	253.6	139.4	195.2	53.8	41.4
1998	46.8	143.3	299.8	72.5	79.4	85.1	78.1	41.3	193.1	50.9	269.8	148.6	218.5	58.5	45.6
1999	46.0	145.7	311.2	75.3	80.9	92.3	79.8	44.1	221.9	72.1	296.1	178.2	243.7	59.6	56.2
2000	47.9	171.1	357.0	80.4	89.3	100.2	83.5	46.5	301.3	126.1	347.7	195.0	284.6	65.8	61.6
2001	49.4	159.5	321.7	75.4	88.3	86.7	88.2	47.2	276.8	109.4	299.2	188.7	287.1	60.7	66.9
2002	49.6	157.3	290.4	78.9	84.3	81.9	95.4	50.3	270.6	109.3	284.9	202.8	311.3	59.9	74.1
2003	55.0	173.3	293.7	80.6	89.9	80.3	102.3	56.5	317.8	140.4	297.6	209.2	338.4	61.9	79.6
2004	56.6	206.3	327.5	89.2	103.2	92.4	118.8	63.0	418.3	189.9	346.1	227.3	378.1	74.0	89.9
2005	59.0	236.8	358.4	98.4	115.2	101.5	128.9	69.1	531.4	263.2	382.8	238.7	412.7	80.0	95.8
2006	66.0	279.1	404.0	107.3	129.0	105.1	151.4	76.1	610.1	316.7	422.6	256.0	447.6	84.2	126.6
2007	84.3	316.3	433.0	121.3	145.9	119.0	184.8	83.0	644.5	346.7	449.1	258.5	479.8	89.2	149.3
2008	108.3	386.9	457.7	121.5	161.2	133.8	202.9	90.4	794.6	476.1	458.7	233.2	485.7	92.5	174.0
2009	93.9	293.5	391.5	81.7	149.3	119.9	211.7	82.9	464.4	267.7	374.1	159.2	429.9	81.4	178.5
2010	107.7	388.6	447.8	112.0	164.9	137.0	227.4	92.5	603.2	353.6	450.4	225.6	485.1	86.6	183.6
2011	126.2	485.3	494.2	133.0	174.7	150.9	251.6	108.3	755.0	462.1	513.4	255.2	515.9	89.7	197.3
2012	133.0	483.2	527.5	146.2	181.0	161.6	263.6	111.1	723.2	434.3	551.8	298.5	518.8	100.3	200.2
2013	136.2	492.4	534.8	152.7	188.1	177.5	286.3	116.0	679.0	387.8	559.0	309.6	532.9	98.1	208.1
2014	143.7	500.7	551.8	159.8	198.4	191.9	309.0	126.8	669.9	353.6	598.8	329.5	558.7	105.7	214.7
2015	127.7	418.1	539.8	151.9	197.3	192.6	347.4	128.8	488.1	197.2	607.2	350.0	596.4	102.7	232.3
2016	130.5	387.6	520.0	150.4	193.3	192.9	360.1	131.0	437.4	159.6	593.6	350.8	584.9	109.2	240.7
2017	132.8	459.4	533.7	157.9	197.2	193.8	395.9	138.8	503.1	197.4	642.9	359.1	603.5	118.0	257.3
2018	133.1	537.0	563.4	158.8	205.5	196.5	413.2	148.3	574.7	238.9	694.7	372.4	648.4	126.0	255.6
2019	131.1	526.8	548.1	162.5	205.0	193.3	432.1	151.6	519.4	207.1	681.1	376.8	655.9	134.6	272.4
2014															
1st quarter	151.0	496.0	543.1	152.4	192.3	187.2	303.2	120.6	702.8	393.3	577.3	312.5	545.3	103.3	210.5
2nd quarter	146.5	511.2	550.9	160.3	200.7	190.2	313.2	129.1	679.3	363.1	597.2	332.9	563.7	105.9	214.9
3rd quarter	134.3	516.0	556.6	165.1	200.3	194.1	306.9	128.3	661.1	343.8	607.8	336.0	553.6	105.6	214.4
4th quarter	143.1	479.5	556.4	161.5	200.2	196.2	312.9	129.3	636.3	314.3	612.7	336.6	572.3	107.9	218.9
2015															
1st quarter	136.2	436.4	548.9	148.2	200.0	187.8	350.8	128.9	533.8	224.5	617.4	335.1	592.8	98.6	226.7
2nd quarter	129.4	439.0	545.4	151.8	196.3	193.4	345.4	130.2	499.8	205.8	612.7	354.1	595.4	102.7	229.9
3rd quarter	122.7	413.6	535.1	154.7	196.4	194.9	343.3	129.0	482.8	196.4	601.5	357.4	603.1	103.9	235.3
4th quarter	122.6	383.6	529.8	153.0	196.5	194.3	350.0	126.9	436.1	162.3	597.1	353.6	594.3	105.5	237.1
2016															
1st quarter	121.0	369.7	523.2	147.1	191.1	193.1	349.1	129.9	404.3	137.7	585.0	349.3	585.2	107.7	238.0
2nd quarter	120.8	383.5	519.4	151.4	191.4	189.5	357.6	128.6	421.1	149.7	595.8	346.1	579.5	107.5	237.0
3rd quarter	143.7	393.7	514.4	152.1	195.9	195.1	365.5	130.7	452.1	169.5	594.0	351.8	581.8	109.6	240.0
4th quarter	136.4	403.4	523.0	151.1	194.7	193.7	368.3	134.8	472.1	181.5	599.7	356.2	593.0	111.8	247.7
2017															
1st quarter	134.5	449.9	523.4	156.8	195.5	188.4	386.9	134.5	507.9	213.7	615.6	359.5	589.9	113.2	250.3
2nd quarter	130.0	439.7	522.5	156.0	197.6	195.6	393.1	137.0	491.1	186.6	631.5	357.4	599.3	117.2	255.5
3rd quarter	132.7	447.1	539.9	155.6	196.0	194.1	400.2	140.6	484.1	179.1	650.7	357.5	589.8	119.7	263.1
4th quarter	133.8	500.8	549.0	163.1	199.7	197.2	403.5	143.1	529.5	210.3	673.6	362.1	634.9	121.9	260.2
2018															
1st quarter	131.2	512.8	557.4	165.5	204.1	200.9	414.2	147.2	563.1	233.3	683.9	369.7	648.9	124.1	250.6
2nd quarter	145.0	545.5	566.0	161.0	205.0	196.0	405.1	147.6	573.4	239.1	695.2	360.3	629.9	124.0	254.3
3rd quarter	133.2	545.3	561.5	155.2	204.5	194.9	413.2	148.5	594.5	258.1	703.6	373.7	644.4	126.7	256.5
4th quarter	123.2	544.4	568.6	153.6	208.5	194.0	420.1	150.1	567.8	224.9	695.9	385.8	670.5	129.3	261.1
2019															
1st quarter	129.5	528.5	568.3	163.8	210.3	194.1	417.8	149.9	528.9	200.4	691.8	382.2	665.7	132.2	260.4
2nd quarter	133.6	528.6	542.0	161.8	204.8	195.4	435.5	153.4	534.8	225.1	681.8	384.8	660.4	138.7	272.4
3rd quarter	132.7	519.1	540.0	165.3	205.9	190.9	436.8	153.3	513.9	203.6	678.0	381.4	668.1	132.6	277.8
4th quarter	128.6	531.2	542.1	159.0	199.2	192.9	438.2	149.6	500.2	199.1	672.5	358.8	629.5	134.9	278.9

Table 7-5. Chain-Type Quantity Indexes for Exports and Imports of Selected NIPA Types of Product

(Index numbers, 2012 = 100; quarterly indexes are seasonally adjusted.)   NIPA Table 4.2.3

Year and quarter	Exports							Imports							
	Goods					Services		Goods						Services	
	Foods, feeds, and beverages	Industrial supplies and materials	Capital goods, except auto-motive	Auto-motive vehicles, engines, and parts	Consumer goods, except food and auto-motive	Travel	Other business services	Foods, feeds, and beverages	Industrial supplies and materials, except petroleum and products	Petroleum and products	Capital goods, except auto-motive	Auto-motive vehicles, engines, and parts	Consumer goods, except food and auto-motive	Travel	Other business services
1975	33.04	20.42	5.71	24.41	8.73	14.12	3.14	26.37	38.82	53.59	0.86	14.20	4.67	17.59	2.24
1976	37.49	22.05	5.63	25.52	9.85	16.08	3.63	30.01	47.02	64.61	1.05	18.81	6.02	18.70	2.69
1977	37.46	22.24	5.54	25.70	10.65	16.12	3.74	29.71	53.87	77.73	1.12	19.93	7.27	19.27	2.82
1978	46.28	24.58	6.32	26.39	12.30	17.31	4.43	33.36	54.44	73.49	1.44	21.56	8.93	19.91	3.29
1979	49.09	28.14	7.59	26.55	12.93	18.56	4.94	34.24	53.80	74.34	1.77	20.85	9.00	19.42	3.73
1980	55.34	31.94	8.86	22.36	15.67	20.73	5.28	29.80	44.81	60.04	2.08	20.66	9.07	19.40	3.81
1981	56.50	29.79	8.90	22.11	15.02	27.59	6.03	31.01	42.66	52.60	2.45	19.99	9.93	21.14	4.23
1982	53.09	28.89	7.87	18.12	13.52	25.46	7.47	31.61	37.47	45.24	2.67	21.40	10.38	25.08	4.78
1983	50.75	27.49	7.59	18.74	12.37	22.61	7.26	34.30	40.27	44.78	3.23	26.19	12.49	28.41	4.43
1984	49.38	28.89	8.40	22.12	12.30	31.42	7.24	38.86	46.09	47.30	4.82	33.65	15.77	51.35	4.95
1985	42.73	29.06	9.24	24.07	11.94	31.45	7.12	40.54	44.88	44.41	5.43	37.71	17.21	56.65	5.53
1986	44.08	30.36	10.08	23.67	13.21	34.69	10.89	41.50	50.46	54.91	6.17	40.82	19.07	52.72	8.05
1987	47.18	31.11	11.64	25.51	15.54	37.63	10.90	42.54	51.02	57.35	7.04	41.89	19.69	62.40	9.23
1988	51.85	35.08	14.79	30.28	19.85	43.89	11.54	40.85	53.31	63.44	8.12	40.84	19.98	64.19	9.84
1989	53.93	39.01	17.16	31.13	25.45	51.86	13.47	41.77	54.07	68.60	9.18	39.80	20.91	66.21	11.51
1990	55.64	41.31	19.85	31.17	29.86	58.22	14.02	43.37	54.80	69.33	9.92	39.85	20.54	70.73	12.49
1991	56.70	44.12	21.72	33.43	30.93	60.95	16.48	41.40	54.13	69.08	10.66	37.16	20.89	64.24	13.94
1992	64.27	44.94	23.79	38.54	33.32	67.09	16.66	43.71	58.33	72.20	12.35	39.07	23.07	66.25	12.98
1993	63.95	43.75	25.35	42.02	35.01	70.21	17.42	44.23	63.52	79.09	14.41	42.93	24.98	69.55	13.76
1994	64.61	46.02	29.25	46.44	38.25	71.06	19.84	45.24	71.13	84.44	17.67	48.14	27.07	72.08	15.54
1995	71.38	49.64	34.86	48.95	40.54	75.52	20.94	46.38	72.95	82.91	21.75	48.97	29.20	72.86	17.11
1996	70.24	52.16	40.40	50.76	43.24	80.41	23.59	51.01	77.78	89.90	25.56	50.63	31.44	75.76	18.76
1997	70.68	56.71	49.52	57.46	47.69	82.71	27.14	56.17	83.04	93.62	32.08	54.73	36.01	82.34	20.10
1998	69.96	56.08	51.63	56.67	49.14	80.46	29.96	60.29	91.15	100.07	36.90	58.26	40.87	93.17	22.42
1999	72.00	57.85	54.48	58.46	50.24	85.03	31.51	66.60	96.94	107.56	42.35	69.39	45.90	93.93	27.67
2000	75.60	62.97	62.68	61.74	55.25	88.07	33.71	70.95	102.23	111.48	50.88	75.06	54.04	107.08	31.55
2001	77.63	60.41	56.46	57.70	54.82	75.56	38.00	74.14	101.22	114.76	45.10	72.60	54.92	98.64	36.55
2002	76.26	60.79	51.87	60.10	52.62	71.19	42.25	78.05	102.32	112.09	44.42	77.79	60.21	93.33	40.82
2003	77.57	62.43	54.51	60.95	55.77	67.77	44.97	84.10	105.76	119.57	48.60	79.81	65.88	86.11	43.31
2004	72.50	66.50	61.35	66.91	63.44	74.63	51.18	89.09	117.79	127.47	57.57	85.29	73.49	94.41	48.23
2005	76.76	68.21	67.34	72.98	70.06	78.18	53.87	92.32	122.99	130.04	64.35	88.62	80.03	97.54	50.55
2006	82.83	73.24	76.12	78.59	77.46	77.37	61.73	98.02	124.14	127.59	72.12	94.66	87.15	98.67	66.03
2007	90.10	77.21	81.21	87.83	85.76	84.29	73.32	99.29	120.43	124.57	77.01	94.64	92.96	96.36	76.79
2008	95.64	85.26	85.93	86.90	92.77	90.34	79.35	97.91	114.04	120.29	78.95	83.26	92.48	92.04	88.35
2009	91.84	79.71	73.86	58.17	85.76	83.02	85.57	92.50	96.93	110.55	66.16	56.41	82.93	84.47	92.10
2010	101.26	92.13	85.10	79.30	93.71	91.70	90.14	94.59	102.84	111.09	81.19	79.41	94.51	88.22	94.56
2011	100.79	98.23	93.82	92.63	97.97	96.05	96.73	96.41	104.38	108.99	92.74	87.15	99.34	89.09	99.61
2012	100.00	100.00	100.00	100.00	100.00	100.00	100.00	100.00	100.00	100.00	100.00	100.00	100.00	100.00	100.00
2013	102.11	104.03	101.31	103.88	105.07	108.45	105.63	103.93	97.16	94.12	102.76	104.12	105.19	96.37	102.51
2014	110.29	107.68	104.92	108.12	112.28	114.96	110.61	109.31	98.11	90.37	111.61	111.49	112.62	103.00	103.75
2015	112.41	107.67	103.58	102.88	114.08	116.26	122.48	113.69	99.65	91.70	116.41	120.85	123.07	104.13	111.30
2016	118.90	109.59	101.44	102.67	114.27	115.47	123.22	115.30	102.55	96.65	117.54	122.05	123.33	113.04	113.36
2017	119.37	119.10	103.99	107.51	118.19	113.86	131.64	117.94	104.19	95.95	128.59	125.42	128.91	120.67	118.67
2018	119.27	127.11	108.76	107.33	122.02	112.85	132.48	127.09	104.32	91.44	139.16	130.03	138.87	126.85	114.97
2019	117.75	131.51	105.32	109.39	121.27	108.99	136.10	130.53	98.75	85.56	138.36	132.49	143.49	135.02	120.66
2014															
1st quarter	115.57	102.75	103.27	103.46	108.62	113.14	109.29	104.86	97.97	93.18	107.14	105.41	109.19	100.65	102.18
2nd quarter	107.77	108.45	104.56	108.45	113.63	113.68	112.62	111.74	97.31	89.00	111.13	112.14	113.30	102.63	104.08
3rd quarter	102.75	110.37	105.92	111.53	113.02	115.62	109.39	110.50	96.75	87.66	113.37	113.97	111.69	102.65	103.39
4th quarter	115.06	109.16	105.95	109.04	113.87	117.41	111.13	110.15	100.41	91.64	114.82	114.43	116.33	106.08	105.34
2015															
1st quarter	114.23	109.03	104.59	100.28	115.08	113.93	124.53	111.43	100.78	92.04	116.81	115.13	121.28	99.27	109.15
2nd quarter	113.93	109.01	104.49	102.72	113.42	116.84	121.50	114.35	99.27	90.76	116.91	122.24	122.32	103.80	110.17
3rd quarter	109.52	107.25	102.91	104.71	113.58	117.13	120.27	114.20	99.58	91.92	115.86	123.53	124.84	105.66	112.36
4th quarter	111.96	105.41	102.32	103.79	114.23	117.15	123.62	114.78	98.98	92.06	116.04	122.52	123.86	107.79	113.54
2016															
1st quarter	112.37	110.05	101.69	100.16	112.90	116.58	120.77	117.38	100.62	95.00	114.94	121.70	122.24	111.15	112.95
2nd quarter	108.84	109.53	101.00	103.25	113.14	113.92	122.97	116.37	101.37	95.14	117.54	120.26	121.77	110.55	111.96
3rd quarter	129.35	109.76	100.60	103.95	115.35	116.69	124.50	112.29	103.17	98.25	117.99	122.53	123.24	113.50	112.80
4th quarter	125.07	109.04	102.49	103.31	115.68	114.69	124.64	115.18	105.02	98.23	119.69	123.70	126.08	116.95	115.74
2017															
1st quarter	121.68	118.60	102.34	106.92	118.17	110.96	130.23	116.60	105.47	101.20	123.31	125.56	125.57	118.66	116.49
2nd quarter	117.87	117.08	102.02	106.24	118.71	115.57	131.08	116.59	104.28	96.09	126.50	124.59	127.77	120.20	118.07
3rd quarter	118.89	115.64	105.18	105.86	116.89	114.46	132.84	117.83	102.77	92.89	129.99	124.87	126.05	120.53	121.18
4th quarter	119.05	125.08	106.41	111.01	118.97	114.46	132.42	120.75	104.27	93.64	134.57	126.67	136.26	123.30	118.95
2018															
1st quarter	116.35	124.82	108.16	112.06	121.17	116.49	134.19	123.84	103.29	90.45	136.66	128.78	138.56	124.36	113.61
2nd quarter	127.19	128.36	109.27	108.70	121.55	112.65	130.24	126.41	104.75	93.14	138.95	125.76	134.64	124.85	114.58
3rd quarter	122.17	126.26	108.11	104.78	121.37	111.58	131.93	129.88	105.98	94.74	141.08	130.65	137.63	127.51	114.98
4th quarter	111.37	128.99	109.49	103.81	123.99	110.69	133.57	128.22	103.25	87.45	139.94	134.91	144.66	130.67	116.71
2019															
1st quarter	116.45	130.79	109.18	110.27	125.13	110.51	133.16	128.93	100.51	85.16	139.42	133.90	144.57	132.92	116.42
2nd quarter	120.96	128.71	104.18	108.51	121.12	109.86	136.73	128.90	99.53	87.21	138.27	135.30	144.23	139.00	120.60
3rd quarter	119.17	130.95	103.72	111.27	121.22	107.25	136.45	132.35	99.02	85.50	138.03	134.34	146.05	133.15	122.43
4th quarter	114.41	135.61	104.20	107.50	117.62	108.35	138.06	131.93	95.92	84.36	137.70	126.41	139.11	135.00	123.21

SECTION 7B: EXPORTS AND IMPORTS

Figure 7-2. Foreign Trade Balances on Goods and Services, 1960–2019

- The U.S. current-account international balance is also recorded in the International Transactions Accounts (ITAs). The definitional differences between this balance and the balance in the NIPAs are minor, and the trends in the two measures are similar. The ITAs measure the current account surplus or deficit (commonly known as the "balance of payments") and directly measure the financial flows required to finance it. (Table 7-6)

- With the passage of the 2017 Tax Cuts and Jobs Act (TCJA), U.S. multinational companies are required to pay a one-time tax on their accumulated earnings held abroad, but it generally eliminated taxes on repatriated earnings. This net withdrawal of direct investments resulted from U.S. repatriation of previously reinvested earnings. (Table 7- 6)

- Through the early 1980s, the United States was still a net creditor with respect to the rest of the world, meaning that the value of the stock of U.S. owned assets abroad exceeded the value of U.S. liabilities to the rest of the world. Since then, persistent borrowing to finance current-account deficits has cumulated, resulting in a growing net debtor status. (Table 7-8)

- In 2006, pre-Great Recession, U.S. businesses, residents and government purchased more imports than exports, resulting in a $-762 billion trade deficit, the largest amount in recent U.S. history. By 2009, net imports had fallen to $-384 billion a decline of 50 percent in three years. For the next decade, the U.S. trade deficit fluctuated between $-462 billion in 2013 and $-577 billion in 2019. (Table 7-6)

Table 7-6. U.S. Exports and Imports of Goods and Services

(Balance of payments basis; millions of dollars, seasonally adjusted.)

Year and month	Goods and services			Goods			Services		
	Exports	Imports	Balance	Exports	Imports	Balance	Exports	Imports	Balance
1970	56 640	54 386	2 254	42 469	39 866	2 603	14 171	14 520	-349
1971	59 677	60 979	-1 302	43 319	45 579	-2 260	16 358	15 400	958
1972	67 222	72 665	-5 443	49 381	55 797	-6 416	17 841	16 868	973
1973	91 242	89 342	1 900	71 410	70 499	911	19 832	18 843	989
1974	120 897	125 190	-4 293	98 306	103 811	-5 505	22 591	21 379	1 212
1975	132 585	120 181	12 404	107 088	98 185	8 903	25 497	21 996	3 501
1976	142 716	148 798	-6 082	114 745	124 228	-9 483	27 971	24 570	3 401
1977	152 301	179 547	-27 246	120 816	151 907	-31 091	31 485	27 640	3 845
1978	178 428	208 191	-29 763	142 075	176 002	-33 927	36 353	32 189	4 164
1979	224 131	248 696	-24 565	184 439	212 007	-27 568	39 692	36 689	3 003
1980	271 834	291 241	-19 407	224 250	249 750	-25 500	47 584	41 491	6 093
1981	294 398	310 570	-16 172	237 044	265 067	-28 023	57 354	45 503	11 851
1982	275 236	299 391	-24 156	211 157	247 642	-36 485	64 079	51 749	12 329
1983	266 106	323 874	-57 767	201 799	268 901	-67 102	64 307	54 973	9 335
1984	291 094	400 166	-109 072	219 926	332 418	-112 492	71 168	67 748	3 420
1985	289 070	410 950	-121 880	215 915	338 088	-122 173	73 155	72 862	294
1986	310 033	448 572	-138 538	223 344	368 425	-145 081	86 689	80 147	6 543
1987	348 869	500 552	-151 684	250 208	409 765	-159 557	98 661	90 787	7 874
1988	431 149	545 715	-114 566	320 230	447 189	-126 959	110 919	98 526	12 393
1989	487 003	580 144	-93 141	359 916	477 665	-117 749	127 087	102 479	24 607
1990	535 233	616 097	-80 864	387 401	498 438	-111 037	147 832	117 659	30 173
1991	578 344	609 479	-31 135	414 083	491 020	-76 937	164 261	118 459	45 802
1992	616 882	656 094	-39 212	439 631	536 528	-96 897	177 251	119 566	57 685
1993	642 863	713 174	-70 311	456 943	589 394	-132 451	185 920	123 780	62 141
1994	703 254	801 747	-98 493	502 859	668 690	-165 831	200 395	133 057	67 338
1995	794 387	890 771	-96 384	575 204	749 374	-174 170	219 183	141 397	77 786
1996	851 602	955 667	-104 065	612 113	803 113	-191 000	239 489	152 554	86 935
1997	934 453	1 042 726	-108 273	678 366	876 794	-198 428	256 087	165 932	90 155
1998	933 174	1 099 314	-166 140	670 416	918 637	-248 221	262 758	180 677	82 081
1999	969 867	1 228 485	-258 617	698 524	1 035 592	-337 068	271 343	192 893	78 450
2000	1 075 321	1 447 837	-372 517	784 940	1 231 722	-446 783	290 381	216 115	74 266
2001	1 005 654	1 367 165	-361 511	731 331	1 153 701	-422 370	274 323	213 465	60 858
2002	978 706	1 397 660	-418 955	698 036	1 173 281	-475 245	280 670	224 379	56 290
2003	1 020 418	1 514 308	-493 890	730 446	1 272 089	-541 643	289 972	242 219	47 754
2004	1 161 549	1 771 433	-609 883	823 584	1 488 349	-664 766	337 966	283 083	54 882
2005	1 286 022	2 000 267	-714 245	913 016	1 695 820	-782 804	373 006	304 448	68 558
2006	1 457 642	2 219 358	-761 716	1 040 905	1 878 194	-837 289	416 738	341 165	75 573
2007	1 653 548	2 358 922	-705 375	1 165 151	1 986 347	-821 196	488 396	372 575	115 821
2008	1 841 612	2 550 339	-708 726	1 308 795	2 141 287	-832 492	532 817	409 052	123 765
2009	1 583 053	1 966 827	-383 774	1 070 331	1 580 025	-509 694	512 722	386 801	125 920
2010	1 853 606	2 348 263	-494 658	1 290 273	1 938 950	-648 678	563 333	409 313	154 020
2011	2 127 021	2 675 646	-548 625	1 499 240	2 239 886	-740 646	627 781	435 761	192 020
2012	2 218 989	2 755 762	-536 773	1 562 578	2 303 749	-741 171	656 411	452 013	204 398
2013	2 293 457	2 755 334	-461 876	1 592 002	2 294 247	-702 244	701 455	461 087	240 368
2014	2 375 905	2 866 241	-490 336	1 633 986	2 385 480	-751 494	741 919	480 761	261 157
2015	2 263 907	2 764 352	-500 445	1 510 757	2 272 612	-761 855	753 150	491 740	261 410
2016	2 208 072	2 712 866	-504 794	1 455 704	2 208 211	-752 507	752 368	504 654	247 714
2017	2 352 546	2 902 669	-550 123	1 553 589	2 358 789	-805 200	798 957	543 880	255 077
2018	2 501 310	3 128 989	-627 679	1 674 330	2 561 667	-887 337	826 980	567 322	259 658
2019	2 528 262	3 105 127	-576 865	1 652 437	2 516 767	-864 331	875 825	588 359	287 466
2018									
January	202 575	254 688	-52 113	133 614	208 472	-74 858	68 961	46 216	22 745
February	205 608	259 425	-53 817	136 377	212 245	-75 868	69 231	47 180	22 051
March	209 937	257 114	-47 177	140 742	210 731	-69 989	69 195	46 383	22 812
April	208 883	257 102	-48 219	140 364	210 657	-70 293	68 519	46 445	22 074
May	213 341	257 693	-44 352	144 552	211 222	-66 670	68 789	46 471	22 318
June	210 968	258 398	-47 430	142 172	211 606	-69 434	68 796	46 792	22 004
July	208 734	261 175	-52 441	139 943	214 078	-74 135	68 791	47 097	21 694
August	207 758	262 647	-54 889	138 857	215 353	-76 496	68 901	47 294	21 607
September	209 747	265 841	-56 094	140 745	218 015	-77 270	69 002	47 826	21 176
October	210 124	266 817	-56 693	141 259	218 563	-77 304	68 865	48 254	20 611
November	207 976	261 623	-53 647	139 124	213 234	-74 110	68 852	48 389	20 463
December	205 661	266 468	-60 807	136 581	217 491	-80 910	69 080	48 977	20 103
2019									
January	210 243	259 267	-49 023	138 878	211 247	-72 369	71 366	211 247	23 346
February	210 809	258 109	-47 300	139 190	210 222	-71 032	71 619	210 222	23 732
March	213 157	262 072	-48 914	140 980	213 712	-72 732	72 177	213 712	23 817
April	209 288	258 491	-49 203	136 436	209 748	-73 312	72 852	209 748	24 109
May	212 852	264 110	-51 258	139 091	215 053	-75 962	73 761	215 053	24 704
June	209 254	261 003	-51 749	135 542	210 839	-75 298	73 712	210 839	23 549
July	210 462	261 503	-51 041	137 465	212 071	-74 606	72 997	212 071	23 565
August	210 517	261 295	-50 778	137 358	211 956	-74 598	73 159	211 956	23 820
September	209 210	257 049	-47 839	136 108	208 032	-71 925	73 102	208 032	24 085
October	210 403	253 432	-43 029	136 851	203 982	-67 131	73 552	203 982	24 102
November	210 571	251 625	-41 054	136 888	202 280	-65 391	73 682	202 280	24 337
December	211 496	257 171	-45 676	137 651	207 625	-69 975	73 845	207 625	24 299

Table 7-7. U.S. Exports of Goods by End-Use and Advanced Technology Categories

(Census basis, except as noted; billions of dollars; seasonally adjusted, except as noted.)

Year and month	Total exports of goods			Principal end-use category							Advanced technology products [1]
	Total, balance of payments basis	Net adjustments	Total, Census basis	Foods, feeds, and beverages	Industrial supplies and materials		Capital goods, except automotive	Automotive vehicles, engines, and parts	Consumer goods, except automotive	Other goods	
					Total	Petroleum and products					
1980	249.75	4.23	245.52	18.55	124.96	. . .	30.72	28.13	34.22	. . .	. . .
1981	265.07	3.76	261.31	18.53	131.10	. . .	36.86	30.80	38.30	. . .	. . .
1982	247.64	3.70	243.94	17.47	107.82	. . .	38.22	34.26	39.66	. . .	. . .
1983	268.90	7.18	261.72	18.56	105.63	. . .	42.61	42.04	46.59	. . .	. . .
1984	332.42	1.91	330.51	21.92	122.72	. . .	60.15	56.77	61.19	. . .	. . .
1985	338.09	1.71	336.38	21.89	112.48	. . .	60.81	65.21	66.43	. . .	. . .
1986	368.43	2.75	365.67	24.40	101.37	. . .	71.86	78.25	79.43	. . .	. . .
1987	409.77	3.48	406.28	24.81	110.67	. . .	84.77	85.17	88.82	. . .	. . .
1988	447.19	5.26	441.93	24.93	118.06	. . .	101.79	87.95	96.42	. . .	. . .
1989	477.37	3.72	473.65	25.08	132.40	. . .	112.45	87.38	102.26	. . .	. . .
1990	498.34	2.36	495.98	26.65	143.41	62.16	116.04	87.69	105.29	16.09	. . .
1991	490.98	2.53	488.45	26.21	131.38	51.78	120.80	84.94	107.78	15.94	. . .
1992	536.58	3.92	532.67	27.61	138.64	51.60	134.25	91.79	122.66	17.71	71.87
1993	589.44	8.78	580.66	27.87	145.61	51.50	152.37	102.42	134.02	18.39	81.23
1994	668.70	5.44	663.26	30.96	162.11	51.28	184.37	118.27	146.27	21.27	98.12
1995	749.37	5.83	743.54	33.18	181.85	56.16	221.43	123.80	159.90	23.39	124.79
1996	803.11	7.92	795.19	35.71	204.48	72.75	228.07	128.94	172.00	26.10	130.36
1997	876.40	6.70	869.70	39.69	213.77	71.77	253.28	139.81	193.81	29.34	147.28
1998	917.64	5.74	911.90	41.24	200.14	50.90	269.45	148.68	217.00	35.39	156.75
1999	1 035.60	10.98	1 024.62	43.60	221.39	67.81	295.72	178.96	241.91	43.04	181.18
2000	1 231.72	13.70	1 218.02	45.98	298.98	120.28	347.03	195.88	281.83	48.33	222.08
2001	1 153.70	12.70	1 141.00	46.64	273.87	103.59	297.99	189.78	284.29	48.42	195.18
2002	1 173.28	11.91	1 161.37	49.69	267.69	103.51	283.32	203.74	307.84	49.08	195.15
2003	1 272.10	14.98	1 257.12	55.83	313.82	133.10	295.87	210.14	333.88	47.59	207.03
2004	1 488.35	18.65	1 469.70	62.14	412.77	180.46	343.58	228.20	372.94	50.11	238.28
2005	1 695.82	22.36	1 673.46	68.09	523.77	251.86	379.33	239.45	407.24	55.57	259.74
2006	1 878.19	24.25	1 853.94	74.94	601.99	302.43	418.26	256.63	442.64	59.49	290.76
2007	1 986.35	29.39	1 956.96	81.68	634.75	330.98	444.51	256.67	474.55	64.81	326.81
2008	2 141.29	37.65	2 103.64	89.00	779.48	453.28	453.74	231.24	481.64	68.54	331.15
2009	1 580.03	20.40	1 559.63	81.62	462.38	253.69	370.48	157.65	427.33	60.17	300.89
2010	1 938.95	25.09	1 913.86	91.75	603.10	336.11	449.39	225.10	483.23	61.30	354.25
2011	2 239.89	31.93	2 207.95	107.48	755.78	439.34	510.80	254.62	514.11	65.17	386.44
2012	2 303.75	27.45	2 276.30	110.27	730.64	415.17	548.71	297.78	516.93	71.95	396.23
2013	2 294.25	26.26	2 267.99	115.15	681.53	369.68	555.65	308.80	531.67	75.20	401.14
2014	2 385.48	29.12	2 356.36	125.88	667.02	334.00	594.14	328.64	557.09	83.59	422.13
2015	2 272.61	24.43	2 248.18	127.80	485.74	181.98	602.06	349.15	594.32	89.12	434.92
2016	2 208.21	20.41	2 187.81	130.05	443.31	146.63	589.97	350.12	583.56	90.80	429.19
2017	2 360.90	18.94	2 341.96	137.82	507.29	186.50	640.58	358.96	601.87	95.45	464.60
2018	2 561.67	20.86	2 540.81	147.30	575.62	225.34	692.61	372.22	646.80	106.19	496.50
2019	2 516.80	19.20	2 497.50	150.50	521.50	193.80	677.80	375.90	663.60	118.20	496.90
2017											
January	238.31	45.60	192.71	11.10	41.53	16.46	50.90	30.86	50.57	7.77	34.53
February	236.51	45.55	190.96	11.22	43.28	17.48	50.87	29.02	48.89	7.69	29.93
March	236.45	45.72	190.72	11.14	42.62	16.69	50.85	30.17	48.88	7.07	36.96
April	238.27	45.71	192.56	11.38	41.58	15.18	51.59	29.81	50.38	7.83	34.67
May	238.60	46.32	192.28	11.42	41.78	15.47	52.58	29.28	49.27	7.94	37.99
June	239.58	46.69	192.89	11.41	41.12	14.23	52.94	30.24	49.08	8.10	39.63
July	239.38	47.02	192.37	11.61	40.62	13.54	53.90	29.53	48.73	7.96	37.42
August	239.76	47.20	192.55	11.55	40.37	14.20	53.72	29.97	49.16	7.78	39.47
September	242.76	47.64	195.12	11.70	41.25	14.53	54.84	29.72	49.82	7.80	40.65
October	245.62	47.66	197.95	11.72	42.75	15.43	55.09	29.54	50.15	8.70	43.40
November	251.25	47.92	203.33	11.71	44.93	17.01	56.22	30.15	51.96	8.36	46.14
December	256.88	48.36	208.52	11.87	45.47	16.28	57.08	30.67	54.98	8.46	43.85
2018											
January	208.47	1.70	206.78	11.86	46.98	18.73	55.72	30.47	53.48	8.28	38.91
February	212.25	1.59	210.66	12.40	47.15	18.36	57.29	30.88	54.99	7.94	33.89
March	210.73	1.58	209.15	12.27	47.21	18.36	56.65	30.78	54.17	8.07	40.78
April	210.66	1.70	208.95	12.28	47.86	18.83	57.26	30.23	52.25	9.07	38.02
May	211.22	1.84	209.38	12.36	48.00	18.97	58.56	29.97	51.89	8.62	42.43
June	211.61	1.64	209.96	12.19	48.55	19.49	57.45	30.36	53.03	8.39	41.52
July	214.08	1.78	212.30	12.42	49.08	20.09	58.02	30.85	52.94	8.98	41.37
August	215.35	1.91	213.45	12.30	49.44	20.23	57.74	31.62	53.34	9.00	42.22
September	218.02	1.71	216.31	12.17	49.18	20.00	59.72	31.26	54.73	9.24	44.04
October	218.56	1.85	216.71	12.32	49.12	19.52	57.06	31.79	56.49	9.93	46.68
November	213.23	1.87	211.37	12.23	46.40	16.80	57.55	32.02	53.71	9.46	44.67
December	217.49	1.70	215.79	12.57	46.67	15.96	59.60	31.97	55.79	9.20	41.98
2019											
January	211.25	1.60	209.65	12.39	43.91	14.57	56.96	31.82	55.66	8.91	38.64
February	210.22	1.68	208.55	11.95	42.76	15.27	57.33	31.67	55.43	9.42	35.01
March	213.71	1.60	212.11	12.85	45.20	16.97	57.89	31.84	54.78	9.56	41.25
April	209.75	1.68	208.06	12.76	44.52	17.54	55.80	31.03	54.87	9.08	38.94
May	215.05	1.78	213.27	12.67	46.03	19.02	56.86	32.68	55.31	9.72	41.36
June	210.84	1.59	209.25	12.64	42.73	16.34	56.87	32.26	54.28	10.48	41.12
July	212.07	1.74	210.33	12.74	44.06	16.72	55.67	32.41	55.51	9.96	41.08
August	211.96	1.72	210.23	12.62	42.75	15.70	56.96	31.77	56.41	9.72	43.07
September	208.03	1.51	206.53	12.75	42.09	15.06	56.03	30.96	54.49	10.22	43.29
October	203.98	1.52	202.46	12.49	41.87	15.18	56.11	29.23	52.26	10.51	47.25
November	202.28	1.37	200.91	12.31	41.20	14.68	55.39	30.17	51.90	9.96	42.84
December	207.63	1.45	206.18	12.34	44.41	16.75	55.90	30.11	52.75	10.67	43.08

[1] Not seasonally adjusted.
. . . = Not available.

Table 7-8. U.S. Imports of Goods by End-Use and Advanced Technology Categories

(Census basis, except as noted; billions of dollars; seasonally adjusted, except as noted.)

Year and month	Total imports of goods			Principal end-use category							Advanced technology products [1]
	Total, balance of payments basis	Net adjustments	Total, Census basis	Foods, feeds, and beverages	Industrial supplies and materials		Capital goods, except automotive	Automotive vehicles, engines, and parts	Consumer goods, except automotive	Other goods	
					Total	Petroleum and products					
1980	224.25	3.55	220.70	36.28	72.09	...	76.28	17.44	17.75	...	...
1981	237.04	3.31	233.74	38.84	70.19	...	84.17	19.69	17.70	...	...
1982	211.16	-1.12	212.28	32.20	64.05	...	76.50	17.23	16.13	...	...
1983	201.80	0.09	201.71	32.09	58.94	...	71.66	18.46	14.93	...	...
1984	219.93	1.18	218.74	32.20	64.12	...	77.01	22.42	15.09	...	...
1985	215.92	3.29	212.62	24.57	61.16	...	79.32	24.95	14.59		...
1986	223.34	-3.13	226.47	23.52	64.72	...	82.82	25.10	16.73		...
1987	250.21	-3.70	253.90	25.23	70.05	...	92.71	27.58	20.31		...
1988	320.23	-3.11	323.34	33.77	90.02	...	119.10	33.40	26.98		...
1989	359.92	-3.08	363.00	36.34	98.36	...	136.94	35.05	36.01		...
1990	387.40	-5.57	392.97	35.18	105.55	62.16	153.07	36.07	43.60	20.73	...
1991	414.08	-7.77	421.85	35.79	109.69	51.78	166.72	39.72	46.65	23.66	...
1992	439.63	-8.54	448.17	40.34	109.59	51.60	176.50	46.71	51.31	24.39	...
1993	456.94	-7.92	464.86	40.59	111.89	51.50	182.85	51.35	54.56	23.89	...
1994	502.86	-9.77	512.63	41.96	121.55	51.28	205.82	57.31	59.86	26.50	...
1995	575.20	-9.54	584.74	50.47	146.37	56.16	234.46	61.26	64.31	28.72	...
1996	612.11	-12.96	625.08	55.53	147.98	72.75	253.99	64.24	70.11	33.85	...
1997	678.37	-10.82	689.18	51.51	158.32	71.77	295.87	73.30	77.96	33.51	...
1998	670.42	-11.72	682.14	46.40	148.31	50.90	299.87	72.39	80.29	35.44	...
1999	683.97	-11.83	695.80	45.98	147.52	67.81	310.79	75.26	80.92	35.32	200.28
2000	771.99	-9.92	781.92	47.87	172.62	120.28	356.93	80.36	89.38	34.77	227.40
2001	718.71	-10.39	729.10	49.41	160.10	103.59	321.71	75.44	88.33	34.11	199.63
2002	682.42	-10.68	693.10	49.62	156.81	103.51	290.44	78.94	84.36	32.94	178.57
2003	713.42	-11.36	724.77	55.03	173.04	133.10	293.67	80.63	89.91	32.49	180.21
2004	807.52	-11.26	818.78	56.57	203.96	180.46	331.56	89.21	103.08	34.40	201.56
2005	894.63	-11.35	905.98	58.96	233.05	251.86	363.32	98.41	115.29	36.96	216.82
2006	1 015.81	-10.16	1 025.97	65.96	276.05	302.43	404.03	107.26	129.08	43.59	247.07
2007	1 138.38	-9.82	1 148.20	84.26	316.38	330.98	433.02	121.26	145.98	47.30	264.88
2008	1 307.50	20.06	1 287.44	108.35	388.03	453.28	457.66	121.45	161.28	50.67	270.13
2009	1 069.73	13.69	1 056.04	93.91	296.51	253.69	391.24	81.72	149.46	43.22	244.71
2010	1 288.80	10.30	1 278.50	107.72	391.66	336.11	447.54	112.01	165.23	54.34	273.31
2011	1 495.85	15.56	1 480.29	126.25	501.08	439.34	493.96	133.04	175.30	52.89	287.72
2012	1 562.58	16.76	1 545.82	133.05	483.23	415.17	527.46	146.16	180.99	53.85	304.98
2013	1 592.00	13.48	1 578.52	136.16	492.42	369.68	534.76	152.66	188.09	53.56	319.73
2014	1 633.99	12.11	1 621.87	143.72	505.81	334.00	551.49	159.81	1 990.03	62.04	338.48
2015	1 510.76	7.66	1 503.10	127.74	426.32	181.98	539.50	151.92	197.84	59.79	343.13
2016	1 455.70	4.69	1 451.01	130.56	396.44	146.61	519.58	1 503.13	193.84	60.29	345.52
2017	1 557.00	9.81	1 547.20	132.76	465.16	186.50	533.42	157.87	197.71	60.27	464.47
2018	1 676.95	11.26	1 665.69	133.13	541.20	225.28	563.14	158.84	206.03	63.35	496.01
2019	1 652.44	9.28	1 643.16	131.10	529.78	193.80	547.87	162.47	205.68	66.26	496.91
2017											
January	128.29	0.87	127.42	11.31	38.02	16.35	44.01	13.32	16.20	4.57	34.53
February	128.44	0.87	127.57	11.04	38.98	17.59	43.20	13.01	16.42	4.92	29.96
March	127.61	0.87	126.74	11.27	37.58	16.58	43.58	12.88	16.42	5.01	36.94
April	126.68	0.71	125.96	11.15	37.28	14.85	43.87	12.57	16.05	5.06	34.65
May	126.19	0.71	125.48	10.64	37.40	15.22	43.05	12.73	16.67	4.98	37.97
June	127.88	0.74	127.14	10.72	37.25	14.06	43.64	13.70	16.79	5.05	39.61
July	127.37	0.78	126.59	10.96	37.32	13.37	44.47	12.73	16.00	5.12	37.40
August	128.75	0.83	127.92	10.92	36.78	14.12	45.49	13.06	16.83	4.85	39.47
September	130.52	0.72	129.80	11.31	38.78	14.63	44.96	13.11	16.30	5.35	40.62
October	131.73	0.84	130.90	10.94	41.43	15.72	44.09	12.98	16.34	5.12	43.41
November	135.28	0.96	134.32	11.18	41.41	17.12	46.13	13.77	16.77	5.06	46.09
December	138.27	0.91	137.36	11.33	42.95	16.90	46.94	14.02	16.94	5.17	43.83
2018											
January	134.88	1.01	133.87	10.66	41.69	18.53	45.55	13.49	17.62	4.87	38.92
February	137.41	0.95	136.46	10.91	43.83	18.24	46.04	14.11	16.52	5.05	33.88
March	140.70	0.91	139.79	11.23	44.77	18.22	47.69	13.78	17.05	5.29	40.72
April	140.97	0.96	140.01	11.63	46.06	18.48	46.21	13.86	17.13	5.13	38.00
May	143.73	1.07	142.65	12.62	45.14	18.59	47.95	13.40	17.54	6.00	42.38
June	141.37	0.94	140.43	12.00	46.00	19.35	47.28	13.00	16.67	5.48	41.36
July	139.16	0.92	138.24	11.52	46.70	20.27	46.14	12.98	16.03	4.87	41.35
August	139.02	0.92	138.11	11.13	44.65	20.34	46.67	12.77	17.70	5.19	42.18
September	141.16	0.90	140.27	10.65	46.48	20.49	47.51	13.05	17.54	5.04	44.02
October	142.29	0.95	141.34	10.28	47.26	19.98	47.39	12.93	17.96	5.53	46.65
November	139.39	0.87	138.52	10.30	45.06	17.07	48.10	12.63	17.09	5.33	44.61
December	136.89	0.87	136.01	10.21	43.57	15.73	46.61	12.85	17.19	5.58	41.95
2019											
January	138.88	0.72	138.16	10.86	44.30	14.57	46.60	13.46	17.51	5.42	38.64
February	139.19	0.87	138.33	10.57	43.47	15.27	47.82	13.67	17.57	5.22	35.01
March	140.98	0.85	140.13	10.95	44.68	16.97	47.61	13.81	17.66	5.42	41.25
April	136.44	0.90	135.54	11.16	44.27	17.54	44.75	13.22	17.17	4.97	38.94
May	139.09	0.77	138.32	11.17	44.16	19.02	45.86	13.74	17.93	5.48	41.36
June	135.54	0.77	134.77	11.07	43.79	16.34	44.84	13.49	16.20	5.38	41.12
July	137.47	0.77	136.70	11.06	43.04	16.72	45.63	13.85	17.60	5.52	41.08
August	137.36	0.79	136.57	11.56	43.88	15.70	44.35	14.05	16.98	5.76	43.07
September	136.11	0.72	135.39	10.56	43.85	15.06	44.97	13.42	17.04	5.55	43.29
October	136.85	0.70	136.15	10.49	44.83	15.18	44.93	13.25	16.78	5.87	47.25
November	136.89	0.70	136.19	10.75	44.24	14.68	45.43	13.52	16.78	5.47	42.84
December	137.65	0.72	136.93	10.91	45.28	16.75	45.09	12.99	16.47	6.21	43.08

[1] Not seasonally adjusted.
. . . = Not available.

Table 7-9. U.S. Exports and Imports of Goods by Principal End-Use Category in Constant Dollars

(Census basis; billions of 2012 chain-weighted dollars, except as noted; seasonally adjusted.)

Year and month	Exports							Imports						
	Total	Foods, feeds, and beverages	Industrial supplies and materials	Capital goods, except auto-motive	Auto-motive vehicles, engines, and parts	Consumer goods, except automotive	Other goods	Total	Foods, feeds, and beverages	Industrial supplies and materials	Capital goods, except auto-motive	Auto-motive vehicles, engines, and parts	Consumer goods, except automotive	Other goods
1995	712.38	101.87	253.24	196.62	70.95	76.73	42.73	928.46	50.57	551.80	132.86	144.68	166.43	28.41
1996	787.89	98.62	268.78	233.39	73.84	82.35	49.89	1 025.33	55.99	585.42	160.26	149.62	178.25	31.71
1997	893.07	98.34	290.85	285.92	83.32	90.44	49.54	1 167.72	61.36	625.78	200.68	161.75	202.94	35.87
1998	914.15	97.73	289.74	298.56	81.32	93.97	53.65	1 305.30	66.02	685.91	230.18	171.86	230.06	44.21
1999	943.54	101.08	291.84	314.72	84.21	94.63	53.61	1 462.44	72.10	696.85	264.02	205.26	258.33	53.94
2000	1 047.54	106.71	321.11	364.20	89.05	103.87	51.60	1 656.42	76.78	740.31	316.80	223.15	303.86	59.66
2001	982.79	109.54	306.65	328.47	83.25	102.71	50.69	1 601.50	79.75	736.67	279.63	216.30	308.87	60.23
2002	909.96	107.07	301.64	280.75	86.77	98.41	47.18	1 631.30	85.02	737.94	258.51	231.54	337.43	60.92
2003	933.89	108.87	314.83	286.68	88.18	103.78	45.28	1 718.52	92.87	767.45	274.39	237.52	366.47	58.04
2004	1 016.04	100.50	336.09	322.28	96.86	118.15	45.95	1 921.28	98.59	862.16	323.53	254.00	406.44	59.42
2005	1 088.65	105.18	346.06	352.41	105.73	130.19	47.33	2 058.36	102.24	906.53	360.64	263.89	439.58	63.74
2006	1 198.66	112.76	372.59	396.62	114.04	143.15	53.39	2 182.04	108.21	910.77	403.10	281.53	474.87	66.65
2007	1 293.52	119.92	397.85	426.95	127.59	157.91	55.12	2 223.92	109.13	883.13	429.06	278.75	501.57	70.64
2008	1 385.46	126.88	445.88	453.35	126.38	170.78	55.79	2 158.81	106.70	842.41	438.11	245.30	495.69	70.96
2009	1 198.93	123.67	396.47	388.40	84.63	157.41	50.08	1 820.18	99.87	705.51	363.71	166.20	441.07	64.98
2010	1 379.61	135.65	454.60	445.18	115.49	170.24	59.44	2 091.57	103.66	746.18	447.24	235.88	498.29	64.38
2011	1 486.80	132.97	494.94	493.40	135.10	177.18	53.20	2 221.30	106.79	758.64	508.41	259.44	521.70	65.37
2012	1 545.82	133.05	501.18	527.18	146.16	181.68	56.58	2 276.27	110.27	730.64	548.71	297.78	516.92	71.95
2013	1 589.07	132.72	523.08	533.01	151.88	190.81	58.61	2 301.79	113.99	706.16	563.42	309.75	532.57	75.79
2014	1 648.18	141.79	532.96	550.08	158.11	204.04	63.01	2 405.58	119.39	707.26	607.30	331.35	555.61	84.03
2015	1 638.51	144.92	531.77	541.31	150.43	207.29	64.84	2 519.22	123.48	709.91	630.14	358.82	597.23	92.50
2016	1 638.23	157.31	531.52	528.56	150.04	207.17	67.58	2 540.21	125.21	736.47	633.73	362.60	588.52	95.65
2017	1 708.37	158.87	578.80	540.09	157.01	213.96	66.37	2 643.65	128.09	744.30	688.84	373.07	607.15	99.30
2018	1 780.90	158.89	618.19	566.67	156.97	220.46	67.23	2 786.64	137.50	748.94	740.72	386.76	649.98	109.12
2019	1 781.96	155.74	636.91	549.46	159.99	218.34	71.19	2 781.59	140.31	711.34	732.95	393.15	659.42	122.24
2017														
January	140.54	13.24	47.22	44.34	13.33	17.85	5.06	219.42	10.68	62.18	55.01	32.08	51.07	8.15
February	140.62	12.63	48.00	43.78	13.30	18.07	5.45	215.68	10.59	62.93	54.98	30.15	49.33	8.05
March	140.50	13.22	46.99	44.24	13.00	18.00	5.56	215.17	10.54	61.69	54.91	31.36	49.31	7.39
April	140.07	13.54	47.45	44.20	12.57	17.39	5.57	217.57	10.70	60.98	55.62	30.90	50.82	8.15
May	141.44	13.24	48.38	43.88	12.83	18.26	5.58	218.16	10.64	62.64	56.67	30.37	49.72	8.27
June	143.11	13.90	48.41	44.44	13.50	17.89	5.63	219.21	10.46	62.45	56.96	31.41	49.52	8.43
July	142.36	13.90	47.99	45.36	12.83	17.19	5.70	219.18	10.71	62.45	57.90	30.71	49.15	8.29
August	141.28	13.37	45.93	45.80	13.04	18.03	5.35	218.35	10.68	60.96	57.59	31.15	49.50	8.09
September	142.49	13.61	47.36	45.55	13.02	17.58	5.82	219.86	10.60	60.81	58.79	30.86	50.19	8.08
October	142.82	12.44	50.90	44.43	12.67	17.60	5.60	222.48	10.67	62.11	58.99	30.74	50.60	9.01
November	145.23	12.80	49.12	46.66	13.47	18.10	5.50	226.56	10.76	62.58	60.22	31.41	52.40	8.64
December	147.92	12.99	51.06	47.43	13.46	18.00	5.55	232.03	11.06	62.54	61.19	31.94	55.53	8.75
2018														
January	143.36	12.44	48.10	45.63	13.40	18.95	5.26	227.60	10.98	61.82	59.67	31.57	54.04	8.53
February	146.06	12.59	50.12	46.55	14.12	17.74	5.41	230.52	11.22	61.30	61.20	31.98	55.38	8.15
March	150.65	12.83	52.56	48.11	13.82	18.27	5.66	229.61	11.28	62.04	60.46	31.95	54.49	8.29
April	149.54	13.42	52.91	46.49	13.70	18.39	5.42	229.52	11.38	63.05	61.07	31.37	52.53	9.31
May	152.65	15.18	51.22	48.38	13.41	18.87	6.32	229.28	11.41	62.42	62.50	31.14	52.10	8.84
June	150.08	14.88	52.31	47.56	12.77	17.65	5.76	230.08	11.58	62.56	61.42	31.58	53.34	8.62
July	148.18	14.78	52.42	46.58	12.87	17.28	5.14	232.34	12.00	62.82	62.06	32.08	53.07	9.24
August	147.11	13.77	50.15	46.89	12.69	18.75	5.52	233.94	11.79	63.71	61.81	32.90	53.48	9.27
September	149.14	12.89	52.29	47.58	12.82	18.85	5.30	236.81	11.43	63.37	63.94	32.52	54.89	9.51
October	148.71	12.33	51.98	47.45	12.64	19.03	5.79	236.61	11.24	62.97	61.06	33.07	56.73	10.20
November	148.16	12.21	51.68	48.35	12.41	18.31	5.66	231.87	11.44	60.26	61.64	33.33	53.94	9.72
December	147.27	11.57	52.45	47.09	12.33	18.37	6.00	238.45	11.76	62.64	63.90	33.27	56.01	9.44
2019														
January	150.68	12.91	53.87	46.77	13.29	18.78	5.86	234.59	11.59	62.19	61.07	33.14	56.08	9.19
February	149.76	12.57	51.85	48.00	13.44	18.77	5.60	231.70	11.09	58.71	61.48	32.99	55.80	9.70
March	150.64	12.93	52.28	47.78	13.57	18.86	5.78	234.15	12.01	59.84	62.17	33.21	55.19	9.83
April	145.46	13.36	51.44	44.93	12.98	18.26	5.28	229.75	11.44	59.13	60.19	32.46	55.38	9.36
May	148.93	13.58	51.54	46.09	13.49	19.06	5.84	235.42	11.45	60.79	61.34	34.17	55.90	10.05
June	146.24	13.10	52.79	45.00	13.25	17.18	5.78	232.62	11.60	57.93	61.47	33.72	54.87	10.86
July	147.82	12.96	51.56	45.79	13.62	18.59	5.91	234.47	11.74	60.39	60.24	33.93	56.03	10.32
August	148.64	13.75	53.44	44.45	13.84	17.91	6.21	235.02	11.73	59.19	61.76	33.28	56.88	10.07
September	147.77	12.71	53.71	45.01	13.24	18.01	6.00	231.03	12.11	57.90	60.84	32.44	54.95	10.59
October	148.45	12.50	54.84	44.97	13.08	17.76	6.33	226.88	11.93	57.75	61.07	30.65	52.73	10.90
November	148.33	12.59	54.12	45.53	13.35	17.74	5.90	225.51	11.87	56.97	60.40	31.60	52.40	10.32
December	149.25	12.77	55.48	45.16	12.85	17.42	6.70	230.46	11.76	60.56	60.91	31.55	53.21	11.05

Table 7-10. U.S. Exports of Goods by Selected Regions and Countries

(Census f.a.s. basis; millions of dollars, not seasonally adjusted.)

| Year and month | Total, all countries | Selected regions [1] | | | Selected countries | | | | |
		European Union, 15 countries	European Union, 28 countries	OPEC	Brazil	Canada	China	Japan	Mexico
1980	. . .	53 679	. . .	. . .	4 344	40 331	3 755	20 790	15 145
1981	. . .	52 363	. . .	. . .	3 798	44 602	3 603	21 823	17 789
1982	. . .	47 932	. . .	. . .	3 423	37 887	2 912	20 966	11 817
1983	. . .	44 311	. . .	. . .	2 557	43 345	2 173	21 894	9 082
1984	. . .	46 976	. . .	. . .	2 640	51 777	3 004	23 575	11 992
1985	. . .	48 994	. . .	12 478	3 140	47 251	3 856	22 631	13 635
1986	. . .	53 154	. . .	10 844	3 885	45 333	3 106	26 882	12 392
1987	254 122	60 575	. . .	11 057	4 040	59 814	3 497	28 249	14 582
1988	322 426	75 755	. . .	13 994	4 266	71 622	5 022	37 725	20 629
1989	363 812	86 331	. . .	13 196	4 804	78 809	5 755	44 494	24 982
1990	393 592	98 027	. . .	13 703	5 048	83 674	4 806	48 580	28 279
1991	421 730	103 123	. . .	19 054	6 148	85 150	6 278	48 125	33 277
1992	448 165	102 958	. . .	21 960	5 751	90 594	7 419	47 813	40 592
1993	465 090	96 973	. . .	19 500	6 058	100 444	8 763	47 892	41 581
1994	512 625	102 818	. . .	17 868	8 102	114 439	9 282	53 488	50 844
1995	584 740	123 671	. . .	19 533	11 439	127 226	11 754	64 343	46 292
1996	625 073	127 710	. . .	22 275	12 718	134 210	11 993	67 607	56 792
1997	689 180	140 773	143 931	25 525	15 915	151 767	12 862	65 549	71 389
1998	682 139	149 035	151 967	25 154	15 142	156 604	14 241	57 831	78 773
1999	695 797	151 814	154 825	20 166	13 203	166 600	13 111	57 466	86 909
2000	781 918	165 065	168 181	19 078	15 321	178 941	16 185	64 924	111 349
2001	729 101	158 768	161 931	20 052	15 879	163 424	19 182	57 452	101 297
2002	693 101	143 691	146 621	18 812	12 376	160 923	22 128	51 449	97 470
2003	724 771	151 731	155 170	17 279	11 211	169 924	28 368	52 004	97 412
2004	814 875	168 572	171 230	22 256	13 886	189 880	34 428	53 569	110 731
2005	901 082	. . .	185 166	31 641	15 372	211 899	41 192	54 681	120 248
2006	1 025 967	. . .	211 887	38 658	18 887	230 656	53 673	58 459	133 722
2007	1 148 199	. . .	244 166	48 485	24 172	248 888	62 937	61 160	135 918
2008	1 287 442	. . .	271 810	64 880	32 299	261 150	69 733	65 142	151 220
2009	1 056 043	. . .	220 599	49 857	26 095	204 658	69 497	51 134	128 892
2010	1 278 495	. . .	239 591	54 226	35 418	249 256	91 911	60 472	163 665
2011	1 482 508	. . .	269 069	64 809	43 019	281 292	104 122	65 800	198 289
2012	1 545 821	. . .	265 373	81 727	43 771	292 651	110 517	69 976	215 875
2013	1 578 517	. . .	262 095	84 730	44 106	300 755	121 746	65 237	225 954
2014	1 621 874	. . .	276 274	82 533	42 432	312 817	123 657	66 892	241 007
2015	1 503 328	. . .	271 911	72 878	31 641	280 855	115 873	62 388	236 460
2016	1 451 460	. . .	269 650	71 053	30 193	266 734	115 595	63 247	230 229
2017	1 547 195	. . .	283 291	59 337	37 331	282 774	129 997	67 603	243 609
2018	1 665 688	. . .	318 408	59 282	39 409	299 732	120 289	75 149	265 945
2019	1 643 161	. . .	336 726	51 318	42 853	292 633	106 447	74 376	256 570
2017									
January	117 528	. . .	21 221	4 043	2 695	20 912	9 956	4 986	19 654
February	119 234	. . .	22 888	4 515	2 642	21 144	9 740	5 283	18 231
March	135 977	. . .	25 795	5 580	2 831	25 002	9 720	5 803	21 039
April	123 925	. . .	22 880	4 885	2 828	22 772	9 806	5 953	18 935
May	127 881	. . .	23 760	5 102	3 243	24 912	9 880	5 144	19 910
June	132 711	. . .	23 658	5 217	3 035	25 172	9 718	5 576	21 403
July	122 202	. . .	21 469	4 943	3 166	21 812	9 954	5 737	19 849
August	129 237	. . .	23 382	4 267	3 267	24 530	10 825	5 409	20 887
September	130 343	. . .	24 281	5 443	3 297	24 128	10 896	5 948	20 170
October	136 312	. . .	25 611	4 562	3 835	23 944	12 963	5 603	22 131
November	135 682	. . .	23 536	4 880	3 119	25 142	12 908	5 832	21 744
December	136 163	. . .	24 810	5 901	3 371	23 304	13 630	6 330	19 656
2018									
January	125 034	. . .	23 419	3 538	3 032	22 641	9 910	5 674	21 435
February	128 235	. . .	24 932	4 191	2 910	23 675	9 742	5 389	20 376
March	149 547	. . .	30 003	5 216	3 280	27 267	12 653	6 495	21 964
April	138 235	. . .	26 824	4 494	3 202	25 834	10 510	5 839	22 529
May	145 513	. . .	28 004	5 771	3 245	27 066	10 397	6 129	23 039
June	145 503	. . .	28 153	5 187	3 338	26 349	10 858	6 396	22 247
July	133 740	. . .	23 878	4 173	3 385	23 708	10 157	6 382	22 911
August	140 118	. . .	25 590	5 241	3 677	25 608	9 281	6 325	22 561
September	139 331	. . .	27 026	4 986	3 336	24 586	9 732	6 452	21 660
October	147 077	. . .	28 007	5 151	3 857	26 194	9 188	6 551	24 747
November	139 337	. . .	26 836	5 777	2 944	24 490	8 651	6 748	22 954
December	134 018	. . .	25 735	5 555	3 201	22 314	9 210	6 771	19 521
2019									
January	129 608	. . .	27 753	4 343	3 340	22 635	7 105	6 385	21 973
February	129 919	. . .	28 469	3 926	2 966	23 171	8 083	5 420	20 250
March	148 472	. . .	30 594	5 003	3 680	26 439	10 575	6 425	21 784
April	134 838	. . .	27 259	3 793	3 273	25 299	7 883	5 660	22 301
May	142 237	. . .	28 336	4 469	3 539	26 150	9 069	6 635	22 566
June	137 870	. . .	27 472	4 043	3 884	24 876	9 167	6 134	20 599
July	133 129	. . .	25 855	3 585	3 589	23 441	8 694	6 576	22 146
August	138 310	. . .	27 902	4 205	3 715	25 317	9 416	6 114	22 046
September	134 162	. . .	27 781	4 610	3 527	24 306	8 597	5 803	20 655
October	142 418	. . .	30 161	4 725	3 847	25 277	8 851	6 471	22 316
November	136 940	. . .	28 193	4 010	4 035	23 432	10 103	5 968	20 940
December	135 258	. . .	26 951	4 608	3 456	22 290	8 903	6 784	18 995

[1]See notes and definitions for definitions of regions.
. . . = Not available.

Table 7-11. U.S. Imports of Goods by Selected Regions and Countries

(Census Customs basis; millions of dollars, not seasonally adjusted.)

Year and month	Total, all countries	Selected regions [1]			Selected countries				
		European Union, 15 countries	European Union, 28 countries	OPEC	Brazil	Canada	China	Japan	Mexico
1980	. . .	35 958	. . .	. . .	. . .	. . .	. . .	. . .	. . .
1981	. . .	41 624	. . .	. . .	. . .	. . .	. . .	. . .	. . .
1982	. . .	42 509	. . .	. . .	. . .	. . .	. . .	. . .	. . .
1983	. . .	43 892	. . .	. . .	. . .	. . .	. . .	. . .	. . .
1984	. . .	57 360	. . .	. . .	. . .	. . .	. . .	. . .	. . .
1985	. . .	67 822	. . .	22 801	7 526	69 006	3 862	68 783	19 132
1986	. . .	75 736	. . .	19 751	6 813	68 253	4 771	81 911	17 302
1987	406 241	81 188	. . .	23 952	7 865	71 085	6 294	84 575	20 271
1988	440 952	84 939	. . .	22 962	9 294	81 398	8 511	89 519	23 260
1989	473 211	85 153	. . .	30 611	8 410	87 953	11 990	93 553	27 162
1990	495 311	91 868	. . .	38 052	7 898	91 380	15 237	89 684	30 157
1991	488 453	86 481	. . .	32 644	6 717	91 064	18 969	91 511	31 130
1992	532 662	93 993	. . .	33 200	7 609	98 630	25 728	97 414	35 211
1993	580 656	97 941	. . .	31 739	7 479	111 216	31 540	107 246	39 918
1994	663 252	110 875	. . .	31 685	8 683	128 406	38 787	119 156	49 494
1995	743 545	131 871	. . .	35 606	8 833	144 370	45 543	123 479	62 100
1996	795 287	142 947	. . .	44 285	8 773	155 893	51 513	115 187	74 297
1997	869 703	157 528	160 896	44 025	9 625	167 234	62 558	121 663	85 938
1998	911 897	176 380	180 550	33 925	10 102	173 256	71 169	121 845	94 629
1999	1 024 616	195 227	200 053	41 977	11 314	198 711	81 788	130 864	109 721
2000	1 218 023	220 019	226 901	67 090	13 853	230 838	100 018	146 479	135 926
2001	1 140 998	220 057	226 568	59 754	14 466	216 268	102 278	126 473	131 338
2002	1 161 366	225 771	232 313	53 245	15 781	209 088	125 193	121 429	134 616
2003	1 257 121	244 826	253 042	68 344	17 910	221 595	152 436	118 037	138 060
2004	1 469 705	272 439	281 959	92 099	21 160	256 360	196 682	129 805	155 902
2005	1 673 454	298 879	309 628	127 169	24 436	290 384	243 470	138 004	170 109
2006	1 853 938	319 590	330 482	150 758	26 367	302 438	287 774	148 181	198 253
2007	1 956 961	. . .	354 409	165 651	25 644	317 057	321 443	145 463	210 714
2008	2 103 641	. . .	367 617	242 579	30 453	339 491	337 773	139 262	215 942
2009	1 559 625	. . .	281 801	111 602	20 070	226 248	296 374	95 804	176 654
2010	1 913 857	. . .	319 264	149 893	23 958	277 637	364 953	120 552	229 986
2011	2 207 954	. . .	368 464	191 470	31 737	315 325	399 371	128 928	262 874
2012	2 276 267	. . .	381 755	180 752	32 123	324 263	425 619	146 432	277 594
2013	2 267 987	. . .	387 510	152 688	27 541	332 504	440 430	138 575	280 556
2014	2 356 356	. . .	420 609	132 370	30 021	349 286	468 475	134 505	295 730
2015	2 248 811	. . .	427 810	66 241	27 474	296 305	483 202	131 445	296 433
2016	2 186 786	. . .	416 331	77 551	26 043	277 720	462 420	132 000	293 501
2017	2 339 591	. . .	434 866	72 165	29 462	299 065	505 165	136 411	312 667
2018	2 537 729	. . .	486 935	80 237	31 214	318 521	539 243	142 242	344 272
2019	2 497 531	. . .	515 208	45 640	30 844	319 428	451 651	143 566	357 971
2017									
January	185 654	. . .	32 652	6 834	2 244	24 390	41 336	10 493	23 459
February	169 238	. . .	32 380	6 034	2 009	23 385	32 785	9 869	23 807
March	194 066	. . .	36 889	6 999	2 403	25 998	34 162	13 019	27 921
April	185 868	. . .	35 426	5 963	2 334	24 154	37 442	11 198	25 017
May	200 331	. . .	36 493	6 486	2 464	26 289	41 757	10 951	27 033
June	198 174	. . .	36 219	6 170	2 612	25 891	42 258	11 186	27 154
July	191 754	. . .	34 958	5 400	2 473	22 801	43 561	11 455	24 515
August	201 557	. . .	35 761	5 729	2 892	24 905	45 782	11 965	26 924
September	194 517	. . .	35 664	4 778	2 395	24 211	45 405	10 820	25 732
October	210 606	. . .	39 409	6 005	2 491	25 655	48 133	12 022	28 572
November	207 579	. . .	38 354	6 139	2 794	26 131	48 105	11 505	27 481
December	200 246	. . .	40 660	5 628	2 351	25 256	44 439	11 927	25 053
2018									
January	203 187	. . .	36 863	6 480	2 750	26 185	45 750	11 279	25 851
February	186 966	. . .	36 937	5 577	2 013	23 989	39 004	10 808	25 825
March	207 989	. . .	41 791	6 050	2 350	26 864	38 295	12 847	29 829
April	205 136	. . .	41 409	7 689	2 561	26 513	38 269	12 059	27 965
May	216 167	. . .	40 992	6 493	2 383	28 153	43 939	11 658	29 279
June	211 391	. . .	39 961	6 884	2 645	28 051	44 571	11 503	29 425
July	217 027	. . .	41 699	7 674	2 753	26 849	47 088	11 827	28 090
August	221 988	. . .	41 144	6 560	3 261	27 976	47 817	12 317	30 922
September	211 410	. . .	37 658	6 857	2 620	26 308	49 988	10 301	28 998
October	236 040	. . .	45 523	7 535	3 105	28 053	52 170	12 776	31 651
November	213 727	. . .	42 024	6 604	2 338	25 640	46 446	12 575	29 475
December	206 702	. . .	40 934	5 833	2 436	23 941	45 906	12 293	26 963
2019									
January	204 587	. . .	39 544	5 325	2 549	23 389	41 514	11 518	27 678
February	185 694	. . .	37 707	3 494	2 204	23 089	33 155	11 424	27 611
March	207 969	. . .	44 755	3 887	2 569	27 746	31 176	13 064	31 264
April	209 174	. . .	45 165	3 916	2 555	26 761	34 683	12 964	30 380
May	220 591	. . .	45 487	4 639	3 019	29 232	39 173	12 001	32 076
June	207 167	. . .	41 412	4 189	2 621	27 689	38 968	11 918	30 473
July	218 799	. . .	45 998	3 640	3 087	26 798	41 449	12 889	30 088
August	215 647	. . .	43 173	3 757	2 410	26 662	41 151	12 419	31 012
September	206 725	. . .	41 521	3 459	2 439	26 882	40 165	10 828	29 523
October	219 414	. . .	46 720	2 999	2 395	28 567	40 115	11 566	31 056
November	199 476	. . .	41 327	3 025	2 386	25 260	36 437	11 446	29 193
December	202 289	. . .	42 399	3 310	2 610	27 353	33 666	11 531	27 617

[1]See notes and definitions for definitions of regions.
. . . = Not available.

Table 7-12A. U.S. Exports of Services: Recent Data

(Balance of payments basis, millions of dollars, seasonally adjusted.)

Year and month	Total services	Maintenance and repair services, n.i.e.	Transport	Travel for all purposes including education	Construction services	Insurance services	Financial services	Charges for the use of intellectual property n.i.e	Telecommunications, computer and information services	Other business services	Personal, cultural, and recreational services	Government goods and services, n.i.e.
1999	278 001	3 656	46 302	89 146	3 974	3 052	25 998	39 913	12 320	36 415	9 007	8 218
2000	298 023	4 423	49 462	96 872	1 993	3 631	29 192	43 476	12 250	38 217	9 351	9 156
2001	284 035	5 191	45 965	86 408	2 608	3 424	27 500	41 005	12 862	41 250	9 631	8 191
2002	288 059	5 368	45 954	79 625	3 128	4 415	29 903	44 815	12 487	44 625	10 089	7 653
2003	297 740	5 073	46 701	76 983	2 377	5 974	33 499	47 308	14 099	46 171	10 522	9 033
2004	344 536	5 181	54 419	85 648	2 253	7 314	42 767	56 943	15 009	51 609	11 409	11 985
2005	378 487	6 652	58 377	93 423	1 489	7 566	47 405	64 466	15 568	56 846	11 114	15 582
2006	423 086	7 106	64 026	96 148	1 856	9 445	56 522	70 999	17 242	65 990	14 532	19 222
2007	495 664	8 899	70 837	106 918	2 774	10 867	73 980	84 498	20 339	79 366	16 265	20 921
2008	540 791	9 362	78 237	117 030	4 072	13 491	79 897	89 672	23 892	89 975	16 080	19 084
2009	522 461	11 573	65 806	110 757	4 260	14 594	73 844	85 730	24 925	93 508	16 926	20 538
2010	582 041	13 111	76 357	130 315	2 951	14 854	86 512	94 968	26 556	99 595	17 612	19 210
2011	644 665	14 739	82 930	142 197	3 187	14 673	101 077	107 053	29 376	108 423	19 538	21 470
2012	684 823	14 944	88 238	153 921	3 234	15 978	105 419	107 869	33 522	118 451	21 098	22 148
2013	719 529	15 720	89 999	170 979	2 213	15 768	109 794	113 824	36 325	122 166	20 888	21 852
2014	756 705	17 978	90 687	180 265	2 070	16 277	119 933	116 380	38 629	132 240	22 551	19 693
2015	768 362	19 847	84 434	192 602	2 759	15 464	114 951	111 151	41 427	141 421	24 220	20 087
2016	780 530	21 587	81 779	192 868	1 690	16 249	114 762	112 981	43 122	153 089	23 626	18 777
2017	830 388	23 239	86 342	193 834	2 053	18 223	128 035	118 147	47 657	167 270	25 664	19 924
2018	862 433	27 948	93 251	196 465	2 948	17 904	132 420	118 875	49 653	177 261	23 759	21 949
2019	875 825	27 868	91 092	193 315	3 189	16 238	135 698	117 401	55 657	189 441	23 372	22 555
2017												
January	67 168	1 936	7 191	15 827	170	1 388	10 167	9 568	3 771	13 519	2 080	1 552
February	67 622	1 961	7 006	15 775	164	1 391	10 257	9 680	3 788	13 807	2 157	1 636
March	67 765	2 005	6 889	15 494	164	1 411	10 469	9 742	3 807	13 914	2 211	1 660
April	68 576	2 054	7 067	16 308	171	1 447	10 242	9 754	3 828	13 840	2 242	1 624
May	68 725	2 016	7 003	16 296	173	1 488	10 450	9 733	3 872	13 818	2 260	1 616
June	69 134	1 938	7 175	16 305	170	1 534	10 640	9 680	3 937	13 849	2 267	1 638
July	69 061	1 798	7 204	16 095	163	1 586	10 712	9 595	4 025	13 931	2 262	1 689
August	69 459	1 795	7 185	16 047	161	1 615	11 007	9 645	4 085	13 992	2 209	1 716
September	70 070	1 815	7 257	16 381	165	1 621	11 024	9 831	4 118	14 032	2 108	1 720
October	70 583	1 906	7 468	16 338	174	1 604	11 111	10 152	4 122	14 050	1 959	1 699
November	70 916	1 956	7 439	16 451	184	1 582	11 052	10 348	4 138	14 159	1 919	1 688
December	71 309	2 058	7 459	16 516	195	1 557	10 903	10 419	4 166	14 360	1 989	1 687
2018												
January	72 249	2 108	7 638	16 672	206	1 526	11 014	10 363	4 205	14 652	2 168	1 696
February	72 531	2 185	7 779	16 679	216	1 500	11 002	10 267	4 226	14 736	2 209	1 731
March	72 579	2 275	7 825	16 869	224	1 479	11 032	10 130	4 231	14 611	2 109	1 793
April	71 130	2 278	7 711	16 219	230	1 461	11 029	9 952	4 219	14 278	1 871	1 882
May	71 306	2 283	7 854	16 472	235	1 464	11 142	9 850	4 174	14 163	1 752	1 918
June	71 180	2 266	7 797	16 315	237	1 487	11 238	9 825	4 096	14 266	1 753	1 901
July	71 451	2 321	7 797	16 319	237	1 530	11 095	9 876	3 985	14 587	1 873	1 832
August	71 812	2 409	7 836	16 177	243	1 544	11 146	9 872	3 955	14 864	1 961	1 804
September	72 023	2 442	7 805	16 240	257	1 531	10 999	9 813	4 006	15 097	2 017	1 817
October	72 031	2 492	7 789	16 045	277	1 491	10 902	9 698	4 137	15 286	2 041	1 872
November	72 080	2 464	7 719	16 167	290	1 458	10 875	9 628	4 206	15 370	2 028	1 875
December	72 059	2 424	7 700	16 290	295	1 433	10 947	9 602	4 214	15 350	1 976	1 828
2019												
January	71 366	2 295	7 795	15 956	291	1 417	11 038	9 619	4 160	15 224	1 887	1 683
February	71 619	2 190	7 711	16 242	283	1 395	11 132	9 646	4 215	15 241	1 851	1 713
March	72 177	2 212	7 631	16 320	271	1 368	11 313	9 683	4 381	15 399	1 870	1 730
April	72 852	2 123	7 722	16 158	253	1 336	11 450	9 730	4 655	15 701	1 942	1 782
May	73 761	2 225	7 647	16 448	249	1 335	11 547	9 750	4 815	15 914	1 981	1 851
June	73 712	2 304	7 560	16 253	258	1 365	11 432	9 743	4 858	16 039	1 988	1 913
July	72 997	2 323	7 383	15 899	281	1 426	11 184	9 708	4 786	16 077	1 963	1 967
August	73 159	2 365	7 518	15 903	288	1 437	11 152	9 728	4 745	16 075	1 953	1 995
September	73 102	2 399	7 432	15 919	280	1 396	11 147	9 801	4 736	16 035	1 959	1 997
October	73 552	2 470	7 543	15 997	257	1 305	11 385	9 927	4 757	15 956	1 981	1 973
November	73 682	2 492	7 521	16 082	242	1 244	11 452	10 012	4 771	15 903	1 995	1 968
December	73 845	2 469	7 628	16 140	235	1 213	11 466	10 054	4 778	15 877	2 002	1 982

n.i.e = Not included elsewhere.

Table 7-12B. U.S. Exports of Services: Historical Data

(Balance of payments basis, millions of dollars, seasonally adjusted.)

Year	Total	Travel	Passenger fares	Other transportation	Royalties and license fees	Other private services (financial, professional, etc.)	Transfers under U.S. military sales contracts [1]	U.S. government miscellaneous services
1960	6 290	919	175	1 607	837	570	2 030	153
1961	6 295	947	183	1 620	906	607	1 867	164
1962	6 941	957	191	1 764	1 056	585	2 193	195
1963	7 348	1 015	205	1 898	1 162	613	2 219	236
1964	7 840	1 207	241	2 076	1 314	651	2 086	265
1965	8 824	1 380	271	2 175	1 534	714	2 465	285
1966	9 616	1 590	317	2 333	1 516	814	2 721	326
1967	10 667	1 646	371	2 426	1 747	951	3 191	336
1968	11 917	1 775	411	2 548	1 867	1 024	3 939	353
1969	12 806	2 043	450	2 652	2 019	1 160	4 138	343
1970	14 171	2 331	544	3 125	2 331	1 294	4 214	332
1971	16 358	2 534	615	3 299	2 545	1 546	5 472	347
1972	17 841	2 817	699	3 579	2 770	1 764	5 856	357
1973	19 832	3 412	975	4 465	3 225	1 985	5 369	401
1974	22 591	4 032	1 104	5 697	3 821	2 321	5 197	419
1975	25 497	4 697	1 039	5 840	4 300	2 920	6 256	446
1976	27 971	5 742	1 229	6 747	4 353	3 584	5 826	489
1977	31 485	6 150	1 366	7 090	4 920	3 848	7 554	557
1978	36 353	7 183	1 603	8 136	5 885	4 717	8 209	620
1979	39 692	8 441	2 156	9 971	6 184	5 439	6 981	520
1980	47 584	10 588	2 591	11 618	7 085	6 276	9 029	398
1981	57 354	12 913	3 111	12 560	7 284	10 250	10 720	517
1982	64 079	12 393	3 174	12 317	5 603	17 444	12 572	576
1983	64 307	10 947	3 610	12 590	5 778	18 192	12 524	666
1984	71 168	17 177	4 067	13 809	6 177	19 255	9 969	714
1985	73 155	17 762	4 411	14 674	6 678	20 035	8 718	878
1986	86 689	20 385	5 582	15 438	8 113	28 027	8 549	595
1987	98 661	23 563	7 003	17 027	10 174	29 263	11 106	526
1988	110 919	29 434	8 976	19 311	12 139	31 111	9 284	664
1989	127 087	36 205	10 657	20 526	13 818	36 729	8 564	587
1990	147 832	43 007	15 298	22 042	16 634	40 251	9 932	668
1991	164 261	48 385	15 854	22 631	17 819	47 748	11 135	690
1992	177 252	54 742	16 618	21 531	20 841	50 292	12 387	841
1993	185 920	57 875	16 528	21 958	21 695	53 510	13 471	883
1994	200 395	58 417	16 997	23 754	26 712	60 841	12 787	887
1995	219 183	63 395	18 909	26 081	30 289	65 048	14 643	818
1996	239 489	69 809	20 422	26 074	32 470	73 340	16 446	928
1997	256 087	73 426	20 868	27 006	33 228	83 929	16 675	955
1998	262 758	71 325	20 098	25 604	35 626	91 774	17 405	926
1999	268 790	75 161	19 425	23 792	47 731	96 812	5 211	657

Note: The category "Royalties and license fees" is the earlier terminology for the category called "Charges for the use of intellectual property n.i.e" in the revised and updated data shown in Tables 7-15A/7-16A. Other categories differ in both title and content; see the Notes and Definitions.

[1]Contains goods that cannot be separately identified.

Table 7-13A. U.S. Imports of Services: Recent Data

(Balance of payments basis, millions of dollars, seasonally adjusted.)

Year and month	Total services	Maintenance and repair services, n.i.e.	Transport	Travel for all purposes including education	Construction services	Insurance services	Financial services	Charges for the use of intellectual property n.i.e	Telecommunications, computer and information services	Other business services	Personal, cultural, and recreational services	Government goods and services, n.i.e.
1999	196 742	1 193	49 892	58 864	1 524	9 389	12 489	12 845	14 664	19 999	1 670	14 212
2000	220 927	2 316	58 526	64 174	1 447	11 284	16 445	16 139	14 128	20 306	1 645	14 516
2001	222 039	1 821	56 571	60 525	1 870	16 706	15 982	16 207	14 012	21 204	1 820	15 322
2002	233 480	2 016	55 398	59 017	2 616	21 927	15 214	18 981	13 214	24 027	1 719	19 353
2003	252 340	2 043	61 084	61 244	1 974	25 233	15 766	18 653	14 928	25 748	2 049	23 619
2004	290 609	2 181	71 954	70 332	2 303	29 089	19 043	22 818	16 382	28 026	2 183	26 296
2005	312 225	2 730	78 796	74 112	1 295	28 710	21 971	24 127	18 404	32 076	2 550	27 454
2006	349 329	4 091	82 348	78 375	1 650	39 382	27 005	23 076	20 850	41 378	3 822	27 353
2007	385 464	4 590	86 097	82 606	2 570	47 166	33 768	24 615	23'660	48 455	3 678	28 260
2008	420 650	5 128	92 837	84 317	3 526	58 444	29 519	27 764	26 138	59 675	4 421	28 880
2009	407 538	5 257	75 784	82 512	3 667	63 965	24 900	29 421	26 038	59 517	5 016	31 460
2010	436 456	5 857	88 394	85 166	2 578	63 452	27 215	31 116	29 421	65 903	5 393	31 960
2011	458 188	7 168	95 027	86 623	3 025	58 277	30 312	32 911	32 832	74 082	6 636	31 293
2012	469 610	7 078	99 323	90 340	3 340	58 747	28 736	35 061	33 285	78 678	7 160	27 861
2013	465 819	6 674	94 434	91 119	2 583	52 993	29 284	35 295	35 868	84 024	8 205	25 341
2014	490 932	6 732	99 810	96 248	2 314	52 760	32 770	37 562	38 461	90 716	9 323	24 236
2015	497 755	8 084	99 557	102 664	3 012	49 842	32 594	35 178	38 815	95 119	11 358	21 531
2016	511 898	7 595	92 391	109 155	1 768	52 070	32 672	41 974	39 720	100 505	12 544	21 503
2017	544 836	6 796	96 515	117 972	1 950	50 889	36 649	44 405	43 091	106 991	17 530	22 047
2018	562 069	7 133	106 303	126 008	3 151	43 735	39 249	43 933	42 558	107 834	19 190	22 975
2019	588 359	7 823	107 458	134 594	1 327	51 547	40 350	42 733	43 720	113 584	21 140	24 083
2017												
January	43 895	574	8 080	9 408	171	4 218	2 747	3 362	3 521	8 630	1 348	1 836
February	43 931	541	7 924	9 443	149	4 138	2 847	3 370	3 560	8 684	1 434	1 841
March	44 180	549	7 946	9 437	140	4 150	2 904	3 472	3 540	8 720	1 482	1 838
April	44 769	572	8 018	9 729	145	4 255	2 866	3 668	3 460	8 739	1 492	1 826
May	45 269	593	8 077	9 773	152	4 327	3 024	3 760	3 450	8 801	1 489	1 824
June	45 580	582	8 073	9 789	161	4 367	3 143	3 749	3 508	8 908	1 471	1 829
July	46 100	562	8 286	9 962	173	4 373	3 130	3 634	3 636	9 059	1 440	1 844
August	46 082	541	8 047	9 990	176	4 362	3 194	3 636	3 710	9 143	1 431	1 850
September	46 227	567	8 041	9 970	173	4 334	3 206	3 755	3 730	9 160	1 445	1 846
October	46 397	567	8 026	10 040	161	4 287	3 201	3 992	3 695	9 110	1 482	1 834
November	46 171	579	7 871	10 132	165	4 151	3 168	4 057	3 660	9 051	1 504	1 834
December	46 236	568	8 124	10 300	185	3 926	3 219	3 951	3 622	8 984	1 512	1 845
2018												
January	45 680	575	8 360	10 161	220	3 612	3 214	3 673	3 583	8 908	1 504	1 869
February	46 637	569	8 606	10 283	242	3 434	3 201	4 457	3 567	8 864	1 526	1 887
March	45 993	549	8 685	10 583	251	3 393	3 217	3 413	3 574	8 853	1 578	1 898
April	46 313	514	8 861	10 420	246	3 488	3 314	3 430	3 605	8 875	1 658	1 903
May	46 255	514	8 711	10 332	244	3 559	3 342	3 440	3 604	8 911	1 687	1 910
June	46 474	545	8 718	10 242	244	3 606	3 310	3 691	3 572	8 961	1 664	1 921
July	46 886	612	8 939	10 492	246	3 630	3 224	3 685	3 509	9 026	1 590	1 934
August	46 842	662	8 854	10 588	261	3 680	3 293	3 469	3 478	9 066	1 553	1 939
September	47 082	666	8 952	10 582	287	3 755	3 245	3 541	3 481	9 082	1 553	1 935
October	47 814	659	9 324	10 603	326	3 857	3 287	3 654	3 516	9 074	1 592	1 923
November	47 830	604	9 045	10 798	319	3 884	3 285	3 725	3 534	9 088	1 627	1 922
December	48 263	664	9 249	10 925	265	3 837	3 318	3 754	3 534	9 125	1 658	1 933
2019												
January	48 020	626	9 305	10 877	164	3 716	3 351	3 743	3 517	9 134	1 685	1 901
February	47 887	632	9 007	11 056	103	3 726	3 246	3 700	3 530	9 178	1 711	1 997
March	48 360	712	9 017	11 124	82	3 867	3 380	3 625	3 574	9 242	1 734	2 004
April	48 743	648	8 858	11 406	100	4 139	3 369	3 517	3 648	9 326	1 755	1 977
May	49 057	646	9 059	11 144	111	4 323	3 443	3 465	3 688	9 430	1 764	1 985
June	50 164	670	8 975	12 118	115	4 418	3 393	3 466	3 694	9 555	1 759	2 000
July	49 432	663	8 983	11 186	112	4 426	3 409	3 521	3 666	9 700	1 741	2 024
August	49 339	636	8 874	11 116	110	4 459	3 402	3 552	3 655	9 754	1 743	2 038
September	49 017	642	8 811	10 849	108	4 518	3 345	3 558	3 661	9 717	1 764	2 043
October	49 449	636	8 861	11 278	107	4 604	3 307	3 539	3 683	9 588	1 805	2 040
November	49 345	653	8 758	11 247	107	4 661	3 322	3 527	3 699	9 502	1 832	2 037
December	49 546	660	8 950	11 191	107	4 689	3 381	3 521	3 706	9 459	1 846	2 036

n.i.e = Not included elsewhere.

Table 7-13B. U.S. Imports of Services: Historical Data

(Balance of payments basis, millions of dollars, seasonally adjusted.)

Year	Total	Travel	Passenger fares	Other transportation	Royalties and license fees	Other private services (financial, professional, etc.)	Direct defense expenditures [1]	U.S. government miscellaneous services
1960	7 674	1 750	513	1 402	74	593	3 087	254
1961	7 671	1 785	506	1 437	89	588	2 998	268
1962	8 092	1 939	567	1 558	100	528	3 105	296
1963	8 362	2 114	612	1 701	112	493	2 961	370
1964	8 619	2 211	642	1 817	127	527	2 880	415
1965	9 111	2 438	717	1 951	135	461	2 952	457
1966	10 494	2 657	753	2 161	140	506	3 764	513
1967	11 863	3 207	829	2 157	166	565	4 378	561
1968	12 302	3 030	885	2 367	186	668	4 535	631
1969	13 322	3 373	1 080	2 455	221	751	4 856	586
1970	14 520	3 980	1 215	2 843	224	827	4 855	576
1971	15 400	4 373	1 290	3 130	241	956	4 819	592
1972	16 868	5 042	1 596	3 520	294	1 043	4 784	589
1973	18 843	5 526	1 790	4 694	385	1 180	4 629	640
1974	21 379	5 980	2 095	5 942	346	1 262	5 032	722
1975	21 996	6 417	2 263	5 708	472	1 551	4 795	789
1976	24 570	6 856	2 568	6 852	482	2 006	4 895	911
1977	27 640	7 451	2 748	7 972	504	2 190	5 823	951
1978	32 189	8 475	2 896	9 124	671	2 573	7 352	1 099
1979	36 689	9 413	3 184	10 906	831	2 822	8 294	1 239
1980	41 491	10 397	3 607	11 790	724	2 909	10 851	1 214
1981	45 503	11 479	4 487	12 474	650	3 562	11 564	1 287
1982	51 749	12 394	4 772	11 710	795	8 159	12 460	1 460
1983	54 973	13 149	6 003	12 222	943	8 001	13 087	1 568
1984	67 748	22 913	5 735	14 843	1 168	9 040	12 516	1 534
1985	72 862	24 558	6 444	15 643	1 170	10 203	13 108	1 735
1986	80 147	25 913	6 505	17 766	1 401	13 146	13 730	1 686
1987	90 787	29 310	7 283	19 010	1 857	16 485	14 950	1 893
1988	98 526	32 114	7 729	20 891	2 601	17 667	15 604	1 921
1989	102 479	33 416	8 249	22 172	2 528	18 930	15 313	1 871
1990	117 659	37 349	10 531	24 966	3 135	22 229	17 531	1 919
1991	118 459	35 322	10 012	24 975	4 035	25 590	16 409	2 116
1992	119 566	38 552	10 603	23 767	5 161	25 386	13 835	2 263
1993	123 779	40 713	11 410	24 524	5 032	27 760	12 086	2 255
1994	133 057	43 782	13 062	26 019	5 852	31 565	10 217	2 560
1995	141 397	44 916	14 663	27 034	6 919	35 199	10 043	2 623
1996	152 554	48 078	15 809	27 403	7 837	39 679	11 061	2 687
1997	165 932	52 051	18 138	28 959	9 161	43 154	11 707	2 762
1998	180 677	56 483	19 971	30 363	11 235	47 591	12 185	2 849
1999	195 172	59 332	20 946	31 494	13 302	55 885	11 849	2 364

Note: The category "Royalties and license fees" is the earlier terminology for the category called "Charges for the use of intellectual property n.i.e" in the revised and updated data shown in Tables 7-15A/7-16A. Other categories differ in both title and content; see the Notes and Definitions.

[1]Contains goods that cannot be separately identified.

Table 7-14. U.S. Export and Import Price Indexes by End-Use Category

(2000 = 100, not seasonally adjusted.)

Year and month	Exports			Imports		
	All commodities	Agricultural	Nonagricultural	All commodities	Petroleum [1]	Nonpetroleum
1990	95.5	105.1	94.4	94.0	75.5	97.1
1991	96.3	103.4	95.4	94.2	67.3	98.7
1992	96.3	102.5	95.7	94.9	62.9	100.0
1993	96.9	104.4	96.2	94.6	57.7	100.6
1994	98.9	109.4	98.0	96.2	54.3	103.2
1995	103.9	119.0	102.5	100.6	59.8	107.2
1996	104.5	132.6	101.6	101.6	71.1	106.4
1997	103.1	120.6	101.3	99.1	66.0	104.1
1998	99.7	108.8	98.8	93.1	44.8	100.4
1999	98.4	101.1	98.2	93.9	60.1	99.0
2000	100.0	100.0	100.0	100.0	100.0	100.0
2001	99.2	101.2	99.0	96.5	82.8	98.5
2002	98.2	103.2	97.8	94.1	85.3	96.2
2003	99.7	112.3	98.8	96.9	103.2	97.3
2004	103.6	123.4	102.1	102.3	134.6	99.8
2005	106.9	121.0	105.9	110.0	185.1	102.5
2006	110.7	125.8	109.6	115.4	223.3	104.2
2007	116.1	150.9	113.6	120.2	249.1	107.0
2008	123.1	183.5	118.8	134.1	343.2	112.7
2009	117.4	160.0	114.3	118.6	219.9	108.0
2010	123.1	172.6	119.6	126.8	282.2	111.0
2011	133.0	211.0	127.5	140.6	385.1	115.9
2012	133.5	216.1	127.6	141.0	384.0	116.3
2013	133.0	219.7	126.7	139.5	374.2	115.6
2014	132.3	213.8	126.3	138.0	353.2	115.7
2015	123.9	185.4	119.3	123.9	190.8	112.5
2016	119.9	175.5	115.7	119.8	153.1	110.8
2017	122.8	178.2	118.6	123.3	193.9	112.0
2018	126.9	179.3	122.9	127.1	236.6	113.5
2019	125.9	178.5	121.8	125.5	230.4	112.3
2016						
January	118.7	173.5	114.6	117.8	119.8	110.7
February	118.2	174.7	113.9	117.2	111.2	110.5
March	118.1	171.0	114.1	117.7	123.2	110.4
April	118.7	172.4	114.6	118.5	136.4	110.4
May	120.0	177.5	115.7	119.9	155.4	110.8
June	120.9	181.9	116.3	120.7	173.1	110.5
July	121.1	181.3	116.6	120.8	167.8	111.0
August	120.1	175.1	115.9	120.5	160.9	111.1
September	120.5	173.3	116.5	120.6	162.3	111.2
October	120.7	174.2	116.7	121.2	174.4	111.1
November	120.8	175.4	116.7	121.1	171.5	111.1
December	121.3	175.1	117.2	121.6	181.6	111.1
2017						
January	121.6	175.8	117.4	122.3	192.0	111.2
February	122.0	178.3	117.7	122.7	193.6	111.5
March	122.1	180.2	117.7	122.5	188.6	111.6
April	122.4	180.3	118.0	122.8	187.2	111.9
May	121.7	177.4	117.5	122.7	185.8	111.9
June	121.6	174.9	117.5	122.4	178.7	112.0
July	122.2	178.1	117.9	122.2	177.6	111.9
August	122.9	178.1	118.7	122.9	185.7	112.1
September	123.9	176.9	119.8	123.9	197.7	112.5
October	124.0	180.4	119.7	124.1	198.6	112.5
November	124.6	179.1	120.5	125.3	217.5	112.7
December	124.7	178.5	120.5	125.5	223.6	112.6
2018						
January	125.6	178.5	121.5	126.5	229.6	113.3
February	125.8	179.5	121.7	126.8	227.1	113.7
March	126.3	185.2	121.8	126.5	221.6	113.7
April	126.9	183.1	122.6	127.1	231.2	113.8
May	127.8	186.0	123.3	128.2	248.4	113.9
June	128.0	184.2	123.7	128.2	255.4	113.5
July	127.4	174.7	123.8	128.1	256.8	113.4
August	127.3	175.4	123.6	127.6	251.5	113.1
September	127.3	173.2	123.8	127.7	251.5	113.2
October	127.9	173.0	124.4	128.3	258.9	113.4
November	126.9	176.0	123.1	126.2	220.2	113.5
December	126.1	182.6	121.7	124.4	186.4	113.7
2019						
January	125.3	178.7	121.2	124.6	200.9	113.0
February	126.1	179.0	122.1	125.9	221.3	113.2
March	127.0	180.7	122.9	126.6	235.5	113.2
April	127.1	178.1	123.2	126.8	246.3	112.7
May	126.7	176.2	122.9	127.0	255.7	112.4
June	126.0	180.5	121.8	125.6	236.6	112.0
July	126.2	180.8	122.0	125.6	237.3	111.9
August	125.5	176.5	121.6	124.9	226.7	111.8
September	125.1	173.6	121.4	125.0	228.8	111.8
October	125.1	176.5	121.1	124.5	223.4	111.7
November	125.2	180.7	121.0	124.7	225.3	111.8
December	125.0	180.5	120.7	125.0	227.0	112.0

[1] Petroleum and petroleum products.

NOTES AND DEFINITIONS, CHAPTER 7

This chapter presents data from two different data systems on international flows of goods, services, income payments, and financial transactions as they affect the U.S. economy. Tables 7-1 through 7-5 present data on the value, quantities, and prices of foreign transactions in the national income and product accounts (NIPAs). The Census Bureau source data for exports and imports of goods and services are presented in greater detail in Tables 7-6 through 7-13. Table 7-14 shows selected summary values for export and import price indexes compiled by the Bureau of Labor Statistics (BLS).

Due to a few differences in concept, scope, and definitions, the aggregate values of international transactions in the NIPAs (shown in Tables 7-1 and 7-4) are not exactly equal to the values for similar concepts in the Census values that serve as their sources, shown in Tables 7-6 through 7-14. The principal sources of differences are as follows:

- The NIPAs cover only the 50 states and the District of Columbia. Census data include the U.S. territories and Puerto Rico as part of the U.S. economy.

- Gold is treated differently.

- Services without payment by financial intermediaries except life insurance carriers (imputed interest) is treated differently.

A reconciliation of the two sets of international accounts is published from time to time in the *Survey of Current Business*.

The conventions of presentation no longer differ between the two sets of international accounts. In all the tables that follow, as in their source documents, values of imports of goods and services and of all other transactions that result in a payment to the rest of the world—income payments to foreigners, net transfers to foreigners, and net acquisition of assets from abroad—are presented normally as positive. Where a minus sign does appear, it means that the usual direction of flow is reversed.

The 2020 annual update was released on July 31, 2020. Data for 2015 to 2019 were revised. In addition, the 2020 annual revision incorporated three major types of revisions: 1) statistical changes to introduce new and improved technologies and newly available source data; 2) changes in definitions to more accurately portray the evolving U.S. economy to more accurately portray the evolving U.S. economy and to provide consistent comparisons with data for other national economies; and 3) changes in presentations to reflect the definitional and statistical changes, where necessary, or to provide additional data or perspective for users. Also output and price measures will be expressed in the base year 2012 which equals 100.

TABLES 7-1 AND 7-4

Foreign Transactions in the National Income and Product Accounts

SOURCE: U.S. DEPARTMENT OF COMMERCE, BUREAU OF ECONOMIC ANALYSIS

See the notes and definitions in Chapter 1 for an overview of the national income and product accounts (NIPAs).

Definitions

In accordance with the split between current and capital account, there are now two NIPA measures of the balance of international transactions.

The *balance on current account, national income and product accounts* is *current receipts from the rest of the world* minus *current payments to the rest of the world*. A negative value indicates that current payments exceed current receipts.

Net lending or net borrowing (-), national income and product accounts is equal to the balance on current account less capital account transactions with the rest of the world (net). Capital account transactions with the rest of the world (net) is not shown separately in Table 7-1 (for space reasons) but can be calculated from that table as the difference between net lending/borrowing and the current account balance. (A similar measure, a component of the ITAs, is shown explicitly in Table 7-6.) Capital account transactions with the rest of the world are cash or in-kind transfers linked to the acquisition or disposition of an existing, nonproduced, nonfinancial asset. In contrast, the current account is limited to flows associated with current production of goods and services.

Net lending or net borrowing provides an indirect measure of the net acquisition of foreign assets by U.S. residents less the net acquisition of U.S. assets by foreign residents. See Table 7-1 and its notes and definitions for a more extensive discussion of the relationship between the balances on current and capital account and international asset flows.

Current receipts from the rest of the world is *exports of goods and services* plus *income receipts*.

Current payments to the rest of the world is *imports of goods and services* plus *income payments* plus *current taxes and transfer payments (net)*.

Exports and imports of goods and services. Goods, in general, are products that can be stored or inventoried. *Services*, in general, are products that cannot be stored and are consumed at the place and time of their purchase. Goods imports include expenditures abroad by U.S. residents, except for travel. Services include foreign travel by U.S. residents, expenditures in the United States by foreign travelers, and exports and imports of certain goods—primarily military equipment purchased and sold by the federal government. See the following paragraph for the definition of *travel*.

Table 7-4 shows values for selected sectors of total goods and services; the sectors shown will not add to the total because of omitted items. In the case of goods, a miscellaneous "other" category is not shown. In the case of services, only two components are shown in this table. One is *travel* (for all purposes including education), which does not include passenger fares but includes as exports spending by foreign tourists and students in the United States, and includes as imports all other spending abroad by tourists and students from the United States. The other component shown here is a category called *other business services*, which includes the professional and financial services (for example, computer services) that have accounted for a large part of the long-term growth in the service category. The remaining components of total services are transport (of both persons and goods), charges for the use of intellectual property not elsewhere classified, government goods and services not elsewhere classified, and a miscellaneous, smaller *other* category. They are shown separately in Tables 7-12 and 7-13.

Income receipts and payments. Income receipts—receipts from abroad of factor (labor or capital) income by U.S. residents—are analogous to exports and are combined with them to yield total *current receipts from the rest of the world. Income payments* by U.S. entities of factor income to entities abroad are analogous to imports.

Current taxes and transfer payments (net) consists of net payments between the United States and abroad that do not involve payment for the services of the labor or capital factors of production, purchase of currently-produced goods and services, or transfer of an existing asset. It includes net flows from persons, government, and business. The types of payments included are personal remittances from U.S. residents to the rest of the world, net of remittances from foreigners to U.S. residents; government grants; and transfer payments from businesses. Only the net payment to the rest of the world is shown. It usually appears in these accounts as a positive value, with transfers from the United States to abroad exceeding the reverse flow. An exception came in 1991, when U.S. allies in the Gulf War reimbursed the United States for the cost of the war, causing the only current-account surplus since 1981. This resulted in net payments to the United States from the rest of the world and appears as a negative entry in the net transfer payments column of the NIPA accounts.

TABLES 7-2, 7-3 AND 7-5

Chain-Type Quantity and Price Indexes for NIPA Foreign Transactions

These indexes represent the separation of the current-dollar values in Tables 7-1 and 7-4 into their real quantity and price trends

components. (See the notes and definitions to Chapter 1 for a general explanation of chained-dollar estimates of real output and prices.) As those notes explain, quantity indexes are shown instead of constant-dollar estimates, because BEA no longer publishes its real output estimates before 1999 in any detail in the constant-dollar form. Therefore, quantity indexes are the only comprehensive source of information about longer-term trends in real volumes.

TABLES 7-6 THROUGH 7-13

Exports and Imports of Goods and Services

SOURCES: U.S. DEPARTMENT OF COMMERCE, CENSUS BUREAU AND BUREAU OF ECONOMIC ANALYSIS

These tables present the source data used to build up the aggregate measures of goods and services flows shown in Tables 7-1 through 7-5. These data are compiled and published monthly, making trends evident before the publication of the quarterly aggregate estimates. They also provide more detail than the quarterly aggregates.

Monthly and annual data on exports and imports of *goods* are compiled by the Census Bureau from documents collected by the U.S. Customs Service. The Bureau of Economic Analysis (BEA) makes certain adjustments to these data (as described below) to place the estimates on a *balance of payments* basis—a basis consistent with the national and international accounts.

Data on exports and imports of *services* are prepared by BEA from a variety of sources. Monthly data on services are available from January 1992. Annual and quarterly data for earlier years are available as part of the international transactions accounts. Current data on goods and services are available each month in a joint Census Bureau-BEA press release.

In the case of some of the detailed breakdowns of exports and imports, such as by end-use categories, monthly data may not sum exactly to annual totals. This is due to later revisions, which are made only to annual data and are not allocated to monthly data. Also, the constant-dollar figures expressed in 2012 dollars are now calculated using chain weights.

In addition, monthly and annual data on exports and imports of goods for individual countries and various country groupings do not reflect subsequent revisions of annual total data. These country data are compiled by the Census Bureau for all countries, although this volume includes only a selection of the most significant countries and areas. The full set of data can be accessed on the Census Web site at <http://www.census.gov>.

Definitions: Goods

Goods: Census basis. The Census basis goods data are compiled from documents collected by the U.S. Customs Service. They reflect the movement of goods between foreign countries and the 50 states, the District of Columbia, Puerto Rico, the U.S. Virgin Islands, and U.S. Foreign Trade Zones. They include government and nongovernment shipments of goods, and exclude shipments between the United States and its territories and possessions; transactions with U.S. military, diplomatic, and consular installations abroad; U.S. goods returned to the United States by its armed forces; personal and household effects of travelers; and intransit shipments. The general import values reflect the total arrival of merchandise from foreign countries that immediately enters consumption channels, warehouses, or Foreign Trade Zones.

For *imports,* the value reported is the U.S. Customs Service appraised value of merchandise (generally, the price paid for merchandise for export to the United States). Import duties, freight, insurance, and other charges incurred in bringing merchandise to the United States are excluded.

Exports are valued at the f.a.s. (free alongside ship) value of merchandise at the U.S. port of export, based on the transaction price including inland freight, insurance, and other charges incurred in placing the merchandise alongside the carrier at the U.S. port of exportation.

Goods: balance of payments (BOP) basis. Goods on a Census basis are adjusted by BEA to goods on a BOP basis to bring the data in line with the concepts and definitions used to prepare the international and national accounts. In general, the adjustments include changes in ownership that occur without goods passing into or out of the customs territory of the United States. These adjustments are necessary to supplement coverage of the Census basis data, to eliminate duplication of transactions recorded elsewhere in the international accounts, and to value transactions according to a standard definition. With the June 4, 2013 release, improved BOP adjustments involving U.S. military transactions were introduced for data beginning with 1999.

The *export* adjustments include the following: (1) The deduction of *U.S. military sales contracts.* The Census Bureau has included these contracts in the goods data, but BEA includes them in the service category "Transfers Under U.S. Military Sales Contracts." BEA's source material for these contracts is more comprehensive but does not distinguish between goods and services. (2) The addition of *private gift parcels* mailed to foreigners by individuals through the U.S. Postal Service. Only commercial shipments are covered in Census goods exports. (3) The addition to *nonmonetary gold exports* of gold purchased by foreign official agencies from private dealers in the United States and held at the Federal Reserve Bank of New York. The Census data include only gold that leaves the customs territory. (4) *Smaller adjustments* includes deductions for repairs of goods, exposed motion picture film, and military grant aid, and additions for sales of fish in U.S. territorial waters, exports of electricity to Mexico, and vessels and oil rigs that change ownership without export documents being filed.

The *import* adjustments include the following: (1) On *inland freight in Canada,* the customs value for imports for certain Canadian goods is the point of origin in Canada. BEA makes an addition for the inland freight charges of transporting these Canadian goods to the U.S. border. (2) An addition is made to *nonmonetary gold imports* for gold sold by foreign official agencies to private purchasers out of stock held at the Federal Reserve Bank of New York. The Census Bureau data include only gold that enters the customs territory. (3) A deduction is made for *imports by U.S. military agencies.* The Census Bureau has included these contracts in the goods data, but BEA includes them in the service category "Direct Defense Expenditures." BEA's source material is more comprehensive but does not distinguish between goods and services. (4) *Smaller adjustments* includes deductions for repairs of goods and for exposed motion picture film and additions for imported electricity from Mexico, conversion of vessels for commercial use, and repairs to U.S. vessels abroad.

Definitions: Services

The BEA updated the U.S. international transactions accounts (ITAs) and U.S. international investment position (IIP) accounts. These updates allow the BEA accelerate the availability of detailed annual trade statistics. These international accounts are incorporated into the NIPA statistics.

These statistics are estimates of service transactions between foreign countries and the 50 states, the District of Columbia, Puerto Rico, the U.S. Virgin Islands, and other U.S. territories and possessions. Transactions with U.S. military, diplomatic, and consular installations abroad are excluded because they are considered to be part of the U.S. economy.

In the 2020 annual update of NIPA, three new services categories were added to the already nine categories. The 12 new categories, shown in Tables 7-12A and 7-13A, are as follows:

Manufacturing services on inputs owned by others, a specific form of "contract manufacturing" has been added.

Maintenance and repair services nie. (not included elsewhere) were previously included in "other private services."

Transport includes the previous categories of "passenger fares" and "other transportation."

Travel for all purposes including education is broader than the previous "travel" category and now includes business travel, expenditures by border, seasonal, and other short-term workers, and personal travel including that related to health and education. As before, it consists of expenditures for food, lodging, recreation, gifts, and other items incidental to a foreign visit but not the actual transportation expenses.

Construction covers the services provided to create, renovate, repair, or extend buildings, land improvements and civil engineer construction such as roads and bridges.

Insurance services and *financial services* were previously included in other private services.

Charges for the use of intellectual property n.i.e. was previously named "royalties and license fees."

Telecommunications, computer, and information services and *other business services* were previously included in "other private services."

Government goods and services n.i.e. replaces the government categories in the previous system.

Personal, and recreational services includes audiovisual services, artistic-related services and other systems. Audiovisual services includes the production of audiovisual content, end-user rights to the services. Artistic-related services provided visual artists, composers, live entertainment and sporting events. The last subcategory includes online services for education, telemedicine, and sports gambling.

Earlier services definitions

In Tables 7-12B and 7-13B, showing data before 1999, services are classified in the broad categories described below. For six of these categories, the definitions are the same for imports and exports. For the seventh, the export category is "Transfers under U.S. Military Sales Contracts," while for imports, the category is "Direct Defense Expenditures."

Travel includes purchases of services and goods by U.S. travelers abroad and by foreign visitors to the United States. A traveler is defined as a person who stays for a period of less than one year in a country where the person is not a resident. Included are expenditures for food, lodging, recreation, gifts, and other items incidental to a foreign visit. Not included are the international costs of the travel itself, which are covered in *passenger fares* (see below).

Passenger fares consists of fares paid by residents of one country to residents in other countries. Receipts consist of fares received by U.S. carriers from foreign residents for travel between the United States and foreign countries and between two foreign points. Payments consist of fares paid by U.S. residents to foreign carriers for travel between the United States and foreign countries.

Other transportation includes charges for the transportation of goods by ocean, air, waterway, pipeline, and rail carriers to and from the United States. Included are freight charges, operating expenses that transportation companies incur in foreign ports, and payments for vessel charter and aircraft and freight car rentals.

Royalties and license fees consists of transactions with foreign residents involving intangible assets and proprietary rights, such as the use of patents, techniques, processes, formulas, designs, knowhow, trademarks, copyrights, franchises, and manufacturing rights. The term *royalties* generally refers to payments for the utilization of copyrights or trademarks, and the term *license fees* generally refers to payments for the use of patents or industrial processes.

Other private services includes transactions with "affiliated" foreigners for which no identification by type is available and transactions with unaffiliated foreigners.

The term "affiliated" refers to a direct investment relationship, which exists when a U.S. person has ownership or control (directly or indirectly) of 10 percent or more of a foreign business enterprise, or when a foreign person has a similar interest in a U.S. enterprise.

Transactions with "unaffiliated" foreigners in this "other private services" category consist of education services, financial services, insurance services, telecommunications services, and business, professional, and technical services. Included in the last group are advertising services; computer and data processing services; database and other information services; research, development, and testing services; management, consulting, and public relations services; legal services; construction, engineering, architectural, and mining services; industrial engineering services; installation, maintenance, and repair of equipment; and other services, including medical services and film and tape rental.

The insurance component of "other private services" was measured before the July 2003 revision as premiums less actual losses paid or recovered. Furthermore, catastrophic losses were entered immediately when the loss occurred, rather than when the insurance claim was actually paid out. This led to sharp swings for any month in which catastrophic losses occurred, such as Hurricane Katrina in August 2005 or the September 11, 2001, terror attacks. In the accounts as revised in July 2003 and presented here, insurance services are now measured as premiums less "normal" losses. Normal losses consist of a measure of expected regularly occurring losses based on six years of past experience plus an additional allowance for catastrophic loss. Catastrophic losses, when they occur, are added in equal increments to the estimate of regularly occurring losses over the 20 years following the occurrence. As adoption of this methodology introduces a difference between actual and normal losses, an amount equal to the difference is entered in the international accounts as a capital account transaction.

Transfers under U.S. military sales contracts (exports only) includes exports in which U.S. government military agencies participate. This category has included both goods, such as equipment, and services, such as repair services and training, that could not be separately identified. In the 2010 and 2013 annual revisions, more precise breakdowns between goods and services in this category were introduced and incorporated as revised estimates of services and of the BOP goods adjustment, beginning with the data for 2007. Transfers of goods and services under U.S. military grant programs are included.

Direct defense expenditures (imports only) consists of expenditures incurred by U.S. military agencies abroad, including expenditures by U.S. personnel, payments of wages to foreign residents, construction expenditures, payments for foreign contractual services, and procurement of foreign goods. Improved breakdowns between goods and services are now incorporated beginning with statistics for 1999.

U.S. government miscellaneous services includes transactions of U.S. government nonmilitary agencies with foreign residents. Most of these transactions involve the provision of services to, or purchases of services from, foreigners. Transfers of some goods are also included.

Services estimates are based on quarterly, annual, and benchmark surveys and partial information generated from monthly reports. Service transactions are estimated at market prices. Estimates are seasonally adjusted when statistically significant seasonal patterns are present.

Definitions: Area groupings

The Census trade statistics for groups of countries present "groups as they were at the time of reporting", which means that as multinational organizations such as the European Union expand, the statistics for trade with that group include the added country only beginning with the year it entered the group. The European Union was expanded from 15 to 28 nations as of July 1, 2013, and data for that year are available for both the new and the old group; to provide historical perspective, they are both shown here, along with the historical data back to 1974 for the original group.

The *European Union* (EU) now includes 27 countries: Austria, Belgium, Bulgaria, Croatia, Cyprus, Czech Republic, Denmark, Estonia, Finland, France, Germany, Greece, Hungary, Ireland, Italy, Latvia, Lithuania, Luxembourg, Malta, Netherlands, Poland, Portugal, Romania, Slovakia, Slovenia, Spain, and Sweden. The last country to join was Croatia in July 2013.

In a referendum on June 23, 2016, a majority of British voters supported leaving the EU. On March 29, 2017, the United Kingdom government invoked Article 50 of the Treaty on European Union. The European Union (Withdrawal) Act 2018 declares "exit day" was to be March 29, 2019. However, "Brexit" day has changed over the last two years. As of the January 2020, the United Kingdom has officially withdrawn from the EU but 2020 trade negotiations have stalled due to the COVID-19 pandemic.

The *Euro area* now includes Austria, Belgium, Cyprus, Estonia, Finland, France, Germany, Greece, Ireland, Italy, Luxembourg, Malta, the Netherlands, Portugal, Slovakia, Slovenia, and Spain. See the notes and definitions in Table 13-2 for further information about the euro.

With the June 3, 2016 International Trade and Services Annual Revision for 2015, the area grouping *Asian Newly Industrialized*

Countries (NICS) has been removed from all relevant exhibits and publications and are listed separately as Hong Kong, South Korea, Singapore, and Taiwan. This group was initially called the *Four Asian Tigers*.

The *Organization of Petroleum Exporting Countries (OPEC)* currently consists of Algeria, Angola, Ecuador, Equatorial Guinea, Gabon, Iran, Iraq, Kuwait, Libya, Nigeria, Qatar, Republic of Congo, Saudi Arabia, the United Arab Emirates, and Venezuela.

Notes on the data

U.S./Canada data exchange and substitution. The data for U.S. exports to Canada are derived from import data compiled by Canada. The use of Canada's import data to produce U.S. export data requires several alignments in order to compare the two series.

* *Coverage*: Canadian imports are based on country of origin. U.S. goods shipped from a third country are included, but U.S. exports exclude these foreign shipments. U.S. export coverage also excludes certain Canadian postal shipments.

* *Valuation*: Canadian imports are valued at their point of origin in the United States. However, U.S. exports are valued at the port of exit in the United States and include inland freight charges, making the U.S. export value slightly larger. Canada requires inland freight to be reported.

* *Re-exports*: U.S. exports include re-exports of foreign goods. Again, the aggregate U.S. export figure is slightly larger.

* *Exchange Rate*: Average monthly exchange rates are applied to convert the published data to U.S. currency.

End-use categories and seasonal adjustment of trade in goods. Goods are initially classified under the Harmonized System, which describes and measures the characteristics of goods traded. Combining trade into approximately 140 export and 140 import enduse categories makes it possible to examine goods according to their principal uses. These categories are used as the basis for computing the seasonal and working-day adjusted data. Adjusted data are then summed to the six end-use aggregates for publication.

The seasonal adjustment procedure is based on a model that estimates the monthly movements as percentages above or below the general level of each end-use commodity series (unlike other methods that redistribute the actual series values over the calendar year). Imports of petroleum and petroleum products are adjusted for the length of the month.

Data availability

Data are released monthly in a joint Census Bureau-BEA press release (FT-900), which is published about six weeks after the end of the month to which the data pertain. The release and historical data are available on the Census Bureau Web site at <https://www.census.gov/foreign-trade/index.html >.

Revisions

Data for recent years are normally revised annually. In some cases, revisions to annual totals are not distributed to monthly data; therefore, monthly data may not sum to the revised total shown. Data on trade in services may be subject to extensive revision as part of BEA's annual revision of the international transactions accounts (ITAs), usually released in July.

References

See the references for Table 7-6.

TABLE 7-14

Export and Import Price Indexes

SOURCE: U.S. DEPARTMENT OF LABOR, BUREAU OF LABOR STATISTICS

The International Price Program of the Bureau of Labor Statistics (BLS) collects price data for nonmilitary goods traded between the United States and the rest of the world and for selected transportation services in international markets. BLS aggregates the goods price data into export and import price indexes. Summary values of these price indexes for goods are presented in *Business Statistics.* For product and locality detail on international prices for both goods and services, see the *Handbook of U.S. Labor Statistics,* also published by Bernan Press.

Definitions

The *export* price index provides a measure of price change for all goods sold by U.S. residents (businesses and individuals located within the geographic boundaries of the United States, whether or not owned by U.S. citizens) to foreign buyers.

The *import* price index provides a measure of price change for goods purchased from other countries by U.S. residents.

Notes on the data

Published index series use a base year of 2012 = 100 whenever possible.

The product universe for both the import and export indexes includes raw materials, agricultural products, and manufactures. Price data are primarily collected by mail questionnaire, and directly from the exporter or importer in all but a few cases.

To the greatest extent possible, the data refer to prices at the U.S. border for exports and at either the foreign border or the U.S. border for imports. For nearly all products, the prices refer to transactions completed during the first week of the month and represent the actual price for which the product was bought or sold, including discounts, allowances, and rebates.

For the export price indexes, the preferred pricing basis is f.a.s. (free alongside ship) U.S. port of exportation. Where necessary, adjustments are made to reported prices to place them on this basis. An attempt is made to collect two prices for imports: f.o.b. (free on board) at the port of exportation and c.i.f. (cost, insurance, and freight) at the U.S. port of importation. Adjustments are made to account for changes in product characteristics in order to obtain a pure measure of price change.

The indexes are weighted indexes of the Laspeyres type. (See "General Notes" at the beginning of this volume for further explanation.) The values assigned to each weight category are based on trade value figures compiled by the Census Bureau. They are reweighted annually, with a two-year lag (as concurrent value data are not available) in revisions.

The merchandise price indexes are published using three different classification systems: the Harmonized System, the Bureau of Economic Analysis End-Use System, and the Standard International Trade Classification (SITC) system. The aggregate indexes shown here are from the End-Use System.

Data availability

Indexes are published monthly in a press release and a more detailed report. Indexes are published for detailed product categories, as well as for all commodities. Aggregate import indexes by country or region of origin also are available, as are indexes for selected categories of internationally traded services. Additional information is available from the Division of International Prices in the Bureau of Labor Statistics. Complete historical data are available on the BLS Web site at <http://www.bls.gov>.

References

The indexes are described in "BLS to Produce Monthly Indexes of Export and Import Prices," *Monthly Labor Review* (December 1988), and Chapter 15, "International Price Indexes," *BLS Handbook of Methods* Bulletin 2490 (April 1997).

CHAPTER 8: PRICES

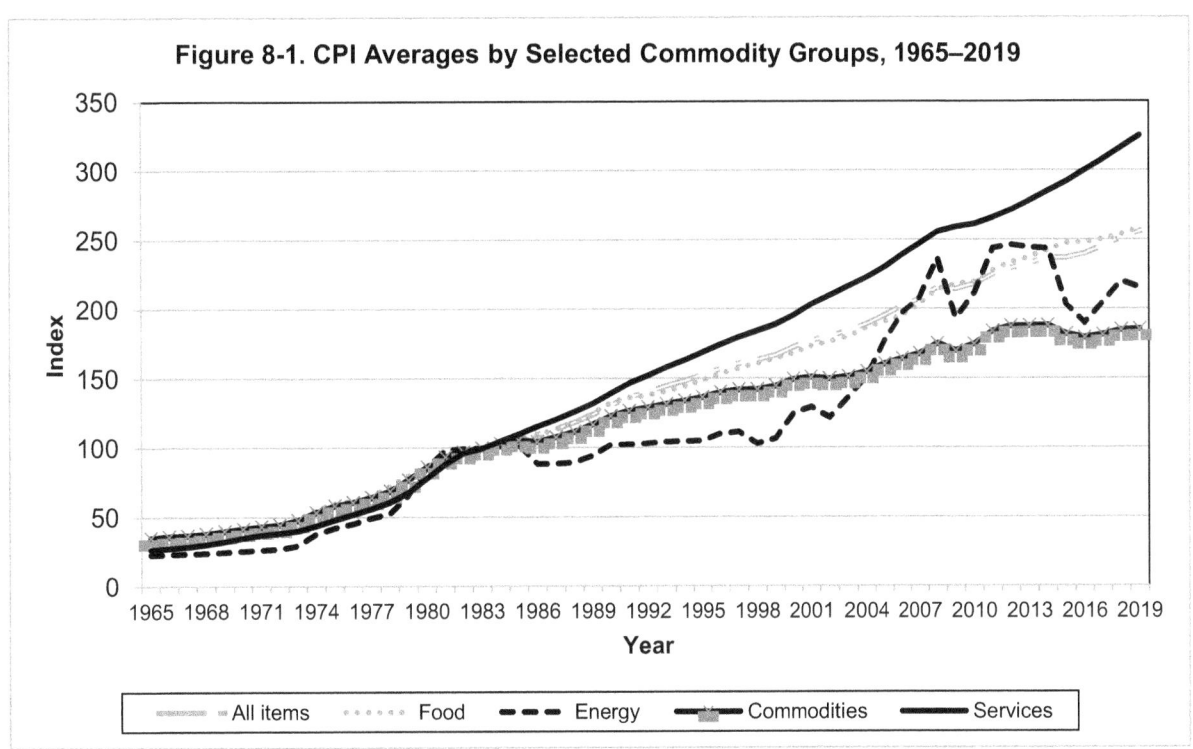

Figure 8-1. CPI Averages by Selected Commodity Groups, 1965–2019

- The Bureau of Labor Statistics (BLS) collects prices for more than 200 categories arranged in eight major groups called the market basket. (See section 8B and Notes and Definitions for more information.) The Consumer Price Index for All Urban Consumers (CPI-U) represents all goods and services purchased for consumption for day-to-day living expenditures. Inflation refers to an increase in prices. Recent U.S. tariffs imposed on goods from China, Canada and the European Union will drive prices up. See Chapter 7 for more information.

- Deflation—a reduction in prices—occurred during the Great Depression from October 1931 to March 1933. (Table 8-1C). Congress amended the Social Security Act of 1935 with public law 92-336 in 1973. Part of that amendment called for automatic annual cost of living increases to be made to Social Security payments based on the CPI. When the CPI-W declined in 2009 and 2015, Social Security payments should have decreased. However, the action taken was to hold the amount stable. (Table 8-1A)

- With the release of the Consumer Price Index (CPI) each month, the Bureau of Labor Statistics (BLS) also publishes average retail prices for various goods and commodities as shown in Figure 8-1 and Table 8-1. Average prices are what consumers pay for a good or service. As shown, energy prices typically are among the most volatile while food prices rise more steadily.

Table 8-1A. Summary Consumer Price Indexes: Recent Data

(Seasonally adjusted.)

Year and month		Consumer Price Index, 1982–1984 = 100					
	Urban wage earners and clerical workers (CPI-W), all items	All urban consumers (CPI-U)					
		All items	Food	Energy	All items less food and energy	Commodities	Services
1965	31.7	31.5	32.2	22.9	32.7	35.2	26.6
1966	32.6	32.4	33.8	23.3	33.5	36.1	27.6
1967	33.6	33.4	34.1	23.8	34.7	36.8	28.8
1968	35.0	34.8	35.3	24.2	36.3	38.1	30.3
1969	36.9	36.7	37.1	24.8	38.4	39.9	32.4
1970	39.0	38.8	39.2	25.5	40.8	41.7	35.0
1971	40.7	40.5	40.4	26.5	42.7	43.2	37.0
1972	42.1	41.8	42.1	27.2	44.0	44.5	38.4
1973	44.7	44.4	48.2	29.4	45.6	47.8	40.1
1974	49.6	49.3	55.1	38.1	49.4	53.5	43.8
1975	54.1	53.8	59.8	42.1	53.9	58.2	48.0
1976	57.2	56.9	61.6	45.1	57.4	60.7	52.0
1977	60.9	60.6	65.5	49.4	61.0	64.2	56.0
1978	65.6	65.2	72.0	52.5	65.5	68.8	60.8
1979	73.1	72.6	79.9	65.7	71.9	76.6	67.5
1980	82.9	82.4	86.8	86.0	80.8	86.0	77.9
1981	91.4	90.9	93.6	97.7	89.2	93.2	88.1
1982	96.9	96.5	97.4	99.2	95.8	97.0	96.0
1983	99.8	99.6	99.4	99.9	99.6	99.8	99.4
1984	103.3	103.9	103.2	100.9	104.6	103.2	104.6
1985	106.9	107.6	105.6	101.6	109.1	105.4	109.9
1986	108.6	109.6	109.0	88.2	113.5	104.4	115.4
1987	112.5	113.6	113.5	88.6	118.2	107.7	120.2
1988	117.0	118.3	118.2	89.3	123.4	111.5	125.7
1989	122.6	124.0	125.1	94.3	129.0	116.7	131.9
1990	129.0	130.7	132.4	102.1	135.5	122.8	139.2
1991	134.3	136.2	136.3	102.5	142.1	126.6	146.3
1992	138.2	140.3	137.9	103.0	147.3	129.1	152.0
1993	142.1	144.5	140.9	104.2	152.2	131.5	157.9
1994	145.6	148.2	144.3	104.6	156.5	133.8	163.1
1995	149.8	152.4	148.4	105.2	161.2	136.4	168.7
1996	154.1	156.9	153.3	110.1	165.6	139.9	174.1
1997	157.6	160.5	157.3	111.5	169.5	141.8	179.4
1998	159.7	163.0	160.7	102.9	173.4	141.9	184.2
1999	163.2	166.6	164.1	106.6	177.0	144.4	188.8
2000	168.9	172.2	167.8	124.6	181.3	149.2	195.3
2001	173.5	177.1	173.1	129.3	186.1	150.7	203.4
2002	175.9	179.9	176.2	121.7	190.5	149.7	209.8
2003	179.8	184.0	180.0	136.5	193.2	151.2	216.5
2004	184.5	188.9	186.2	151.4	196.6	154.7	222.8
2005	191.0	195.3	190.7	177.1	200.9	160.2	230.1
2006	197.1	201.6	195.2	196.9	205.9	164.0	238.9
2007	202.8	207.3	202.9	207.7	210.7	167.5	246.8
2008	211.1	215.3	214.1	236.7	215.6	174.8	255.5
2009	209.6	214.5	218.0	193.1	219.2	169.7	259.2
2010	214.0	218.1	219.6	211.4	221.3	174.6	261.3
2011	221.6	224.9	227.8	243.9	225.0	183.9	265.8
2012	226.2	229.6	233.8	246.1	229.8	187.6	271.4
2013	229.3	233.0	237.0	244.4	233.8	187.7	277.9
2014	232.8	236.7	242.7	243.6	237.9	187.9	285.1
2015	231.8	237.0	247.2	202.9	242.2	181.7	291.7
2016	234.1	240.0	247.9	189.5	247.6	179.2	299.9
2017	239.1	245.1	250.1	204.5	252.2	181.2	308.1
2018	245.1	251.1	253.6	219.9	257.6	184.6	316.6
2019	249.2	255.7	258.3	215.3	263.2	185.3	325.1
2018							
January	243.0	248.8	252.0	217.7	255.1	183.9	312.8
February	243.6	249.5	252.1	221.0	255.6	184.3	313.6
March	243.4	249.4	252.3	215.7	256.1	183.5	314.3
April	244.0	250.0	253.0	217.2	256.5	184.1	314.9
May	244.7	250.6	253.1	219.8	257.1	184.6	315.7
June	245.3	251.2	253.5	222.0	257.4	185.1	316.3
July	245.5	251.5	253.9	220.8	257.9	185.0	317.0
August	246.0	251.9	254.1	223.7	258.1	185.2	317.6
September	246.3	252.3	254.3	222.7	258.6	185.4	318.2
October	246.9	252.8	254.1	225.5	259.0	185.7	318.9
November	246.6	252.7	254.7	219.2	259.5	184.8	319.5
December	246.3	252.7	255.6	213.5	260.0	183.8	320.5
2019							
January	246.0	252.6	256.1	206.6	260.6	183.1	321.1
February	246.8	253.2	257.0	209.4	260.9	183.8	321.7
March	247.7	254.1	257.6	215.1	261.3	184.8	322.4
April	248.7	254.9	257.5	220.4	261.8	185.7	323.3
May	248.8	255.2	258.1	218.8	262.2	185.6	323.7
June	248.9	255.4	258.3	214.9	262.9	185.4	324.5
July	249.7	256.1	258.4	216.8	263.6	185.8	325.4
August	249.8	256.3	258.5	213.9	264.2	185.5	326.2
September	250.0	256.6	258.9	212.2	264.7	185.3	326.9
October	250.8	257.2	259.5	215.9	265.0	185.6	327.9
November	251.3	257.8	259.8	217.6	265.5	185.9	328.8
December	252.0	258.4	260.2	221.1	265.8	186.6	329.3

Table 8-1B. Summary Consumer Price Indexes: Historical Data, 1946 to Date

(Seasonally adjusted.)

Year and month	Urban wage earners and clerical workers (CPI-W), all items	Consumer Price Index, 1982–1984 = 100					
		All urban consumers (CPI-U)					
		All items	Food	Energy	All items less food and energy	Commodities	Services
1946	19.6	19.5	19.8	. . .	. . .	22.9	14.1
1947	22.5	22.3	24.1	. . .	. . .	27.6	14.7
1948	24.2	24.1	26.1	. . .	. . .	29.6	15.6
1949	24.0	23.8	25.0	. . .	. . .	28.8	16.4
1950	24.2	24.1	25.4	. . .	. . .	29.0	16.9
1951	26.1	26.0	28.2	. . .	. . .	31.6	17.8
1952	26.7	26.5	28.7	. . .	. . .	32.0	18.6
1953	26.9	26.7	28.3	. . .	. . .	31.9	19.4
1954	27.0	26.9	28.2	. . .	. . .	31.6	20.0
1955	26.9	26.8	27.8	. . .	. . .	31.3	20.4
1956	27.3	27.2	28.0	. . .	. . .	31.6	20.9
1957	28.3	28.1	28.9	21.5	28.9	32.6	21.8
1958	29.1	28.9	30.2	21.5	29.6	33.3	22.6
1959	29.3	29.1	29.7	21.9	30.2	33.3	23.3
1960	29.8	29.6	30.0	22.4	30.6	33.6	24.1
1961	30.1	29.9	30.4	22.5	31.0	33.8	24.5
1962	30.4	30.2	30.6	22.6	31.4	34.1	25.0
1963	30.8	30.6	31.1	22.6	31.8	34.4	25.5
1964	31.2	31.0	31.5	22.5	32.3	34.8	26.0
1948							
January	23.8	23.7	26.1	. . .	. . .	. . .	. . .
February	23.8	23.7	25.9	. . .	. . .	. . .	. . .
March	23.6	23.5	25.3	. . .	. . .	. . .	. . .
April	24.0	23.8	26.0	. . .	. . .	. . .	. . .
May	24.1	24.0	26.3	. . .	. . .	. . .	. . .
June	24.3	24.2	26.5	. . .	. . .	. . .	. . .
July	24.5	24.4	26.7	. . .	. . .	. . .	. . .
August	24.6	24.4	26.5	. . .	. . .	. . .	. . .
September	24.5	24.4	26.3	. . .	. . .	. . .	. . .
October	24.4	24.3	26.1	. . .	. . .	. . .	. . .
November	24.3	24.2	25.7	. . .	. . .	. . .	. . .
December	24.2	24.1	25.5	. . .	. . .	. . .	. . .
1949							
January	24.2	24.0	25.4	. . .	. . .	. . .	. . .
February	24.1	23.9	25.3	. . .	. . .	. . .	. . .
March	24.0	23.9	25.3	. . .	. . .	. . .	. . .
April	24.1	23.9	25.3	. . .	. . .	. . .	. . .
May	24.0	23.9	25.2	. . .	. . .	. . .	. . .
June	24.1	23.9	25.3	. . .	. . .	. . .	. . .
July	23.8	23.7	24.8	. . .	. . .	. . .	. . .
August	23.8	23.7	24.8	. . .	. . .	. . .	. . .
September	23.9	23.8	25.0	. . .	. . .	. . .	. . .
October	23.8	23.7	24.8	. . .	. . .	. . .	. . .
November	23.8	23.7	24.8	. . .	. . .	. . .	. . .
December	23.8	23.6	24.5	. . .	. . .	. . .	. . .
1950							
January	23.6	23.5	24.3	. . .	. . .	. . .	. . .
February	23.7	23.6	24.7	. . .	. . .	. . .	. . .
March	23.8	23.6	24.6	. . .	. . .	. . .	. . .
April	23.8	23.7	24.6	. . .	. . .	. . .	. . .
May	23.9	23.8	24.8	. . .	. . .	. . .	. . .
June	24.0	23.9	25.1	. . .	. . .	. . .	. . .
July	24.2	24.1	25.6	. . .	. . .	. . .	. . .
August	24.3	24.2	25.8	. . .	. . .	. . .	. . .
September	24.5	24.3	25.8	. . .	. . .	. . .	. . .
October	24.6	24.5	26.0	. . .	. . .	. . .	. . .
November	24.7	24.6	26.1	. . .	. . .	. . .	. . .
December	25.1	25.0	26.9	. . .	. . .	. . .	. . .
1951							
January	25.5	25.4	27.6	. . .	. . .	. . .	. . .
February	26.0	25.8	28.5	. . .	. . .	. . .	. . .
March	26.0	25.9	28.4	. . .	. . .	. . .	. . .
April	26.1	25.9	28.2	. . .	. . .	. . .	. . .
May	26.1	26.0	28.3	. . .	. . .	. . .	. . .
June	26.1	25.9	28.0	. . .	. . .	. . .	. . .
July	26.1	25.9	27.9	. . .	. . .	. . .	. . .
August	26.0	25.9	27.8	. . .	. . .	. . .	. . .
September	26.2	26.0	27.9	. . .	. . .	. . .	. . .
October	26.3	26.2	28.4	. . .	. . .	. . .	. . .
November	26.5	26.3	28.6	. . .	. . .	. . .	. . .
December	26.6	26.5	28.9	. . .	. . .	. . .	. . .

. . . = Not available.

Table 8-1B. Summary Consumer Price Indexes: Historical Data, 1946 to Date—*Continued*

(Seasonally adjusted.)

Year and month	Urban wage earners and clerical workers (CPI-W), all items	Consumer Price Index, 1982–1984 = 100					
		All urban consumers (CPI-U)					
		All items	Food	Energy	All items less food and energy	Commod-ities	Services
1952							
January	26.6	26.5	28.9	. . .	. . .	. . .	. . .
February	26.6	26.4	28.6	. . .	. . .	. . .	. . .
March	26.5	26.4	28.5	. . .	. . .	. . .	. . .
April	26.6	26.5	28.7	. . .	. . .	. . .	. . .
May	26.6	26.5	28.7	. . .	. . .	. . .	. . .
June	26.7	26.5	28.6	. . .	. . .	. . .	. . .
July	26.8	26.7	28.9	. . .	. . .	. . .	. . .
August	26.8	26.7	28.9	. . .	. . .	. . .	. . .
September	26.8	26.6	28.7	. . .	. . .	. . .	. . .
October	26.8	26.7	28.8	. . .	. . .	. . .	. . .
November	26.8	26.7	28.8	. . .	. . .	. . .	. . .
December	26.9	26.7	28.6	. . .	. . .	. . .	. . .
1953							
January	26.8	26.6	28.4	. . .	. . .	. . .	. . .
February	26.7	26.6	28.3	. . .	. . .	. . .	. . .
March	26.8	26.6	28.3	. . .	. . .	. . .	. . .
April	26.8	26.7	28.1	. . .	. . .	. . .	. . .
May	26.9	26.7	28.2	. . .	. . .	. . .	. . .
June	26.9	26.8	28.4	. . .	. . .	. . .	. . .
July	26.9	26.8	28.2	. . .	. . .	. . .	. . .
August	27.0	26.9	28.3	. . .	. . .	. . .	. . .
September	27.0	26.9	28.4	. . .	. . .	. . .	. . .
October	27.1	27.0	28.4	. . .	. . .	. . .	. . .
November	27.0	26.9	28.1	. . .	. . .	. . .	. . .
December	27.0	26.9	28.3	. . .	. . .	. . .	. . .
1954							
January	27.1	26.9	28.5	. . .	. . .	. . .	. . .
February	27.1	27.0	28.5	. . .	. . .	. . .	. . .
March	27.1	26.9	28.4	. . .	. . .	. . .	. . .
April	27.0	26.9	28.4	. . .	. . .	. . .	. . .
May	27.1	26.9	28.4	. . .	. . .	. . .	. . .
June	27.1	26.9	28.4	. . .	. . .	. . .	. . .
July	27.0	26.9	28.4	. . .	. . .	. . .	. . .
August	27.0	26.9	28.3	. . .	. . .	. . .	. . .
September	27.0	26.8	28.0	. . .	. . .	. . .	. . .
October	26.9	26.7	27.9	. . .	. . .	. . .	. . .
November	26.9	26.8	27.9	. . .	. . .	. . .	. . .
December	26.9	26.8	27.8	. . .	. . .	. . .	. . .
1955							
January	26.9	26.8	27.8	. . .	. . .	. . .	. . .
February	27.0	26.8	28.0	. . .	. . .	. . .	. . .
March	26.9	26.8	28.0	. . .	. . .	. . .	. . .
April	26.9	26.8	28.0	. . .	. . .	. . .	. . .
May	26.9	26.8	27.9	. . .	. . .	. . .	. . .
June	26.9	26.7	27.7	. . .	. . .	. . .	. . .
July	26.9	26.8	27.7	. . .	. . .	. . .	. . .
August	26.9	26.7	27.6	. . .	. . .	. . .	. . .
September	27.0	26.9	27.8	. . .	. . .	. . .	. . .
October	27.0	26.8	27.7	. . .	. . .	. . .	. . .
November	27.0	26.9	27.6	. . .	. . .	. . .	. . .
December	27.0	26.9	27.6	. . .	. . .	. . .	. . .
1956							
January	27.0	26.8	27.5	. . .	. . .	31.2	20.7
February	27.0	26.9	27.5	. . .	. . .	31.1	20.7
March	27.1	26.9	27.5	. . .	. . .	31.2	20.7
April	27.1	26.9	27.6	. . .	. . .	31.3	20.8
May	27.2	27.0	27.8	. . .	. . .	31.4	20.8
June	27.3	27.2	28.1	. . .	. . .	31.5	20.9
July	27.4	27.3	28.4	. . .	. . .	31.8	20.9
August	27.5	27.3	28.2	. . .	. . .	31.7	21.0
September	27.5	27.4	28.2	. . .	. . .	31.8	21.1
October	27.7	27.5	28.3	. . .	. . .	32.0	21.1
November	27.7	27.5	28.4	. . .	. . .	32.0	21.2
December	27.8	27.6	28.5	. . .	. . .	32.1	21.3
1957							
January	27.8	27.7	28.4	21.3	28.5	32.1	21.4
February	28.0	27.8	28.7	21.4	28.6	32.3	21.4
March	28.0	27.9	28.6	21.5	28.7	32.3	21.6
April	28.1	27.9	28.6	21.6	28.8	32.4	21.6
May	28.2	28.0	28.7	21.6	28.8	32.4	21.7
June	28.3	28.1	28.9	21.6	28.9	32.5	21.8
July	28.4	28.2	29.1	21.5	29.0	32.6	21.8
August	28.4	28.3	29.4	21.4	29.0	32.8	21.9
September	28.5	28.3	29.2	21.4	29.1	32.8	22.0
October	28.5	28.3	29.2	21.4	29.2	32.7	22.1
November	28.6	28.4	29.2	21.5	29.3	32.8	22.2
December	28.6	28.5	29.2	21.5	29.3	32.9	22.2

. . . = Not available.

Table 8-1B. Summary Consumer Price Indexes: Historical Data, 1946 to Date—*Continued*

(Seasonally adjusted.)

Year and month	Urban wage earners and clerical workers (CPI-W), all items	Consumer Price Index, 1982–1984 = 100					
		All urban consumers (CPI-U)					
		All items	Food	Energy	All items less food and energy	Commod- ities	Services
1958							
January	28.8	28.6	29.8	21.6	29.3	33.1	22.3
February	28.9	28.7	29.9	21.3	29.4	33.2	22.4
March	29.0	28.9	30.5	21.4	29.5	33.4	22.4
April	29.1	28.9	30.6	21.4	29.5	33.5	22.5
May	29.1	28.9	30.5	21.5	29.5	33.5	22.6
June	29.1	28.9	30.3	21.5	29.6	33.4	22.6
July	29.1	28.9	30.2	21.6	29.6	33.3	22.7
August	29.1	28.9	30.1	21.7	29.6	33.3	22.7
September	29.1	28.9	30.0	21.7	29.7	33.3	22.8
October	29.1	28.9	30.0	21.7	29.7	33.2	22.8
November	29.1	29.0	30.0	21.4	29.8	33.3	22.8
December	29.1	29.0	29.9	21.4	29.9	33.3	22.8
1959							
January	29.2	29.0	30.0	21.4	29.9	33.3	22.9
February	29.2	29.0	29.8	21.6	29.9	33.3	23.0
March	29.1	29.0	29.7	21.7	30.0	33.2	23.0
April	29.1	29.0	29.5	21.8	30.0	33.2	23.1
May	29.2	29.0	29.5	21.8	30.1	33.3	23.2
June	29.3	29.1	29.7	21.9	30.2	33.3	23.2
July	29.3	29.2	29.6	21.8	30.2	33.3	23.3
August	29.3	29.2	29.6	21.9	30.2	33.3	23.4
September	29.4	29.3	29.7	21.9	30.3	33.5	23.5
October	29.5	29.4	29.7	22.2	30.4	33.5	23.6
November	29.5	29.4	29.7	22.2	30.4	33.5	23.6
December	29.6	29.4	29.6	22.3	30.5	33.5	23.7
1960							
January	29.5	29.4	29.6	22.3	30.5	33.5	23.7
February	29.6	29.4	29.5	22.2	30.6	33.4	23.8
March	29.6	29.4	29.6	22.3	30.6	33.5	23.9
April	29.7	29.5	30.0	22.4	30.6	33.6	23.9
May	29.7	29.6	30.0	22.3	30.6	33.6	24.0
June	29.8	29.6	30.0	22.4	30.7	33.6	24.0
July	29.7	29.6	29.9	22.5	30.6	33.5	24.1
August	29.8	29.6	30.0	22.5	30.6	33.6	24.1
September	29.8	29.6	30.1	22.6	30.6	33.6	24.2
October	29.9	29.8	30.3	22.5	30.8	33.7	24.2
November	30.0	29.8	30.5	22.7	30.8	33.8	24.3
December	30.0	29.8	30.5	22.6	30.7	33.9	24.3
1961							
January	30.0	29.8	30.5	22.7	30.8	33.8	24.4
February	30.0	29.8	30.5	22.6	30.8	33.8	24.4
March	30.0	29.8	30.5	22.6	30.9	33.8	24.4
April	30.0	29.8	30.4	22.2	30.9	33.8	24.5
May	30.0	29.8	30.3	22.4	30.9	33.8	24.5
June	30.0	29.8	30.2	22.5	31.0	33.8	24.5
July	30.1	29.9	30.3	22.5	31.0	33.9	24.5
August	30.1	29.9	30.3	22.5	31.1	33.9	24.6
September	30.2	30.0	30.3	22.6	31.1	33.9	24.6
October	30.2	30.0	30.3	22.4	31.1	33.9	24.7
November	30.2	30.0	30.3	22.5	31.2	33.8	24.7
December	30.2	30.0	30.3	22.4	31.2	33.9	24.8
1962							
January	30.2	30.0	30.4	22.4	31.2	33.9	24.8
February	30.3	30.1	30.5	22.6	31.2	34.0	24.8
March	30.4	30.2	30.6	22.4	31.3	34.0	24.9
April	30.4	30.2	30.7	22.7	31.3	34.1	24.9
May	30.4	30.2	30.6	22.7	31.4	34.1	25.0
June	30.4	30.2	30.5	22.5	31.4	34.1	25.0
July	30.4	30.2	30.4	22.3	31.4	34.0	25.1
August	30.5	30.3	30.6	22.4	31.5	34.1	25.1
September	30.6	30.4	30.9	22.8	31.5	34.3	25.1
October	30.6	30.4	30.8	22.7	31.5	34.2	25.1
November	30.6	30.4	30.9	22.7	31.5	34.3	25.2
December	30.6	30.4	30.7	22.8	31.6	34.2	25.2
1963							
January	30.6	30.4	31.0	22.8	31.5	34.3	25.3
February	30.7	30.5	31.1	22.7	31.6	34.3	25.3
March	30.7	30.5	31.0	22.7	31.7	34.3	25.3
April	30.7	30.5	30.9	22.6	31.7	34.3	25.4
May	30.7	30.5	30.9	22.6	31.7	34.3	25.4
June	30.8	30.6	31.0	22.5	31.8	34.4	25.5
July	30.9	30.7	31.2	22.7	31.8	34.5	25.5
August	30.9	30.8	31.2	22.6	31.9	34.6	25.6
September	30.9	30.7	31.1	22.5	31.9	34.5	25.6
October	30.9	30.8	31.0	22.7	32.0	34.5	25.6
November	31.0	30.8	31.2	22.6	32.0	34.6	25.7
December	31.1	30.9	31.3	22.6	32.1	34.7	25.8

Table 8-1B. Summary Consumer Price Indexes: Historical Data, 1946 to Date—*Continued*

(Seasonally adjusted.)

Year and month	Urban wage earners and clerical workers (CPI-W), all items	All items	Food	Energy	All items less food and energy	Commod-ities	Services
		Consumer Price Index, 1982–1984 = 100 — All urban consumers (CPI-U)					
1964							
January	31.1	30.9	31.4	22.8	32.2	34.7	25.8
February	31.1	30.9	31.4	22.2	32.2	34.7	25.8
March	31.1	30.9	31.4	22.6	32.2	34.7	25.8
April	31.1	31.0	31.4	22.5	32.2	34.7	25.9
May	31.2	31.0	31.4	22.5	32.2	34.7	25.9
June	31.2	31.0	31.4	22.6	32.3	34.7	26.0
July	31.2	31.0	31.5	22.5	32.3	34.7	26.0
August	31.2	31.1	31.4	22.6	32.3	34.7	26.0
September	31.3	31.1	31.6	22.5	32.3	34.8	26.0
October	31.3	31.1	31.6	22.5	32.4	34.8	26.1
November	31.4	31.2	31.7	22.5	32.5	34.9	26.2
December	31.4	31.3	31.7	22.6	32.5	35.0	26.2
1965							
January	31.5	31.3	31.6	22.8	32.6	35.0	26.3
February	31.5	31.3	31.5	22.7	32.6	34.9	26.4
March	31.5	31.3	31.7	22.6	32.6	35.0	26.4
April	31.6	31.4	31.8	22.9	32.7	35.0	26.5
May	31.7	31.5	32.1	23.0	32.7	35.1	26.5
June	31.8	31.6	32.6	23.1	32.7	35.3	26.5
July	31.8	31.6	32.5	23.0	32.7	35.3	26.6
August	31.7	31.6	32.4	23.0	32.7	35.2	26.6
September	31.8	31.6	32.3	23.1	32.8	35.2	26.7
October	31.8	31.7	32.5	23.0	32.8	35.3	26.8
November	31.9	31.8	32.6	23.1	32.9	35.4	26.9
December	32.0	31.9	32.8	23.1	33.0	35.5	26.9
1966							
January	32.1	31.9	33.0	23.1	33.0	35.6	27.0
February	32.3	32.1	33.5	23.2	33.1	35.8	27.0
March	32.4	32.2	33.8	23.2	33.1	35.9	27.1
April	32.5	32.3	33.8	23.2	33.3	36.0	27.3
May	32.5	32.4	33.7	23.2	33.4	36.0	27.4
June	32.6	32.4	33.7	23.3	33.5	36.0	27.5
July	32.6	32.5	33.5	23.4	33.6	36.1	27.7
August	32.8	32.7	34.0	23.3	33.7	36.2	27.7
September	32.9	32.8	34.1	23.4	33.8	36.4	27.9
October	33.0	32.9	34.2	23.4	34.0	36.4	28.0
November	33.1	32.9	34.1	23.5	34.0	36.4	28.2
December	33.1	32.9	34.0	23.5	34.1	36.4	28.2
1967							
January	33.1	32.9	33.9	23.6	34.2	36.4	28.3
February	33.2	33.0	33.8	23.7	34.2	36.4	28.4
March	33.2	33.0	33.8	23.6	34.3	36.4	28.5
April	33.3	33.1	33.7	23.9	34.4	36.4	28.6
May	33.3	33.1	33.7	23.9	34.5	36.5	28.6
June	33.5	33.3	34.0	23.8	34.6	36.6	28.8
July	33.6	33.4	34.1	23.8	34.7	36.8	28.8
August	33.7	33.5	34.3	23.9	34.9	37.0	28.9
September	33.8	33.6	34.3	24.0	35.0	37.0	29.0
October	33.9	33.7	34.4	23.9	35.1	37.1	29.2
November	34.0	33.9	34.5	24.0	35.2	37.2	29.2
December	34.1	34.0	34.6	23.9	35.4	37.4	29.4
1968							
January	34.3	34.1	34.6	24.0	35.5	37.5	29.5
February	34.4	34.2	34.8	24.1	35.7	37.6	29.6
March	34.5	34.3	34.9	24.1	35.8	37.7	29.8
April	34.6	34.4	35.0	24.0	35.9	37.8	29.9
May	34.7	34.5	35.1	24.1	36.0	37.8	30.0
June	34.9	34.7	35.2	24.2	36.2	38.0	30.2
July	35.0	34.9	35.3	24.2	36.4	38.1	30.4
August	35.2	35.0	35.4	24.3	36.5	38.3	30.6
September	35.3	35.1	35.6	24.3	36.7	38.4	30.7
October	35.5	35.3	35.9	24.3	36.9	38.6	30.9
November	35.6	35.4	35.9	24.4	37.1	38.7	31.0
December	35.8	35.6	36.0	24.3	37.2	38.8	31.2
1969							
January	35.9	35.7	36.1	24.4	37.3	38.9	31.4
February	36.0	35.8	36.1	24.4	37.6	39.0	31.5
March	36.3	36.1	36.2	24.7	37.8	39.3	31.8
April	36.5	36.3	36.4	24.9	38.1	39.4	32.0
May	36.6	36.4	36.6	24.8	38.1	39.5	32.2
June	36.8	36.6	37.0	25.0	38.3	39.8	32.3
July	37.0	36.8	37.3	24.9	38.5	39.9	32.5
August	37.1	36.9	37.5	24.9	38.7	40.1	32.7
September	37.3	37.1	37.7	25.0	38.9	40.2	33.0
October	37.5	37.3	37.8	25.0	39.1	40.4	33.1
November	37.7	37.5	38.2	25.0	39.2	40.6	33.3
December	37.9	37.7	38.6	25.1	39.4	40.8	33.5

Table 8-1B. Summary Consumer Price Indexes: Historical Data, 1946 to Date—*Continued*

(Seasonally adjusted.)

Year and month	Urban wage earners and clerical workers (CPI-W), all items	Consumer Price Index, 1982–1984 = 100					
		All urban consumers (CPI-U)					
		All items	Food	Energy	All items less food and energy	Commod- ities	Services
1970							
January	38.1	37.9	38.7	25.1	39.6	41.0	33.8
February	38.3	38.1	38.9	25.1	39.8	41.2	34.0
March	38.5	38.3	38.9	25.0	40.1	41.2	34.4
April	38.7	38.5	39.0	25.5	40.4	41.4	34.6
May	38.8	38.6	39.2	25.4	40.5	41.5	34.8
June	39.0	38.8	39.2	25.3	40.8	41.6	35.0
July	39.1	38.9	39.2	25.5	40.9	41.7	35.2
August	39.2	39.0	39.2	25.4	41.1	41.8	35.4
September	39.4	39.2	39.4	25.6	41.3	42.0	35.6
October	39.6	39.4	39.5	25.9	41.5	42.2	35.8
November	39.8	39.6	39.5	26.0	41.8	42.3	36.0
December	40.0	39.8	39.5	26.2	42.0	42.5	36.2
1971							
January	40.1	39.9	39.4	26.3	42.1	42.5	36.4
February	40.2	39.9	39.5	26.2	42.2	42.6	36.5
March	40.2	40.0	39.8	26.2	42.2	42.7	36.5
April	40.4	40.1	40.1	26.1	42.4	42.9	36.6
May	40.5	40.3	40.3	26.2	42.6	43.1	36.7
June	40.7	40.5	40.5	26.3	42.8	43.2	37.0
July	40.9	40.6	40.6	26.3	42.9	43.3	37.1
August	41.0	40.7	40.6	26.8	43.0	43.4	37.3
September	41.0	40.8	40.6	26.9	43.0	43.4	37.4
October	41.1	40.9	40.7	27.0	43.1	43.5	37.5
November	41.2	41.0	40.9	26.9	43.2	43.5	37.6
December	41.4	41.1	41.3	27.0	43.3	43.8	37.7
1972							
January	41.5	41.2	41.1	27.0	43.5	43.8	37.9
February	41.6	41.4	41.7	26.8	43.6	44.0	38.0
March	41.7	41.4	41.6	26.9	43.6	44.0	38.1
April	41.7	41.5	41.6	26.9	43.8	44.1	38.2
May	41.8	41.6	41.7	27.0	43.9	44.2	38.3
June	41.9	41.7	41.9	27.0	44.0	44.3	38.4
July	42.1	41.8	42.1	27.1	44.1	44.5	38.5
August	42.2	41.9	42.2	27.3	44.3	44.5	38.6
September	42.3	42.1	42.5	27.6	44.3	44.8	38.7
October	42.5	42.2	42.8	27.7	44.4	44.9	38.8
November	42.6	42.4	43.0	27.9	44.4	45.1	38.9
December	42.8	42.5	43.2	27.8	44.6	45.2	39.0
1973							
January	43.0	42.7	44.0	27.9	44.6	45.5	39.1
February	43.2	43.0	44.6	28.2	44.8	45.9	39.2
March	43.6	43.4	45.8	28.3	45.0	46.4	39.4
April	43.9	43.7	46.5	28.6	45.1	46.8	39.5
May	44.2	43.9	47.1	28.8	45.3	47.2	39.6
June	44.4	44.2	47.6	29.2	45.4	47.5	39.8
July	44.5	44.2	47.7	29.2	45.5	47.5	39.9
August	45.3	45.0	50.5	29.4	45.7	48.7	40.2
September	45.4	45.2	50.4	29.4	46.0	48.7	40.5
October	45.8	45.6	50.7	30.3	46.3	49.0	41.0
November	46.2	45.9	51.4	31.5	46.5	49.5	41.3
December	46.5	46.3	51.9	32.5	46.7	49.9	41.5
1974							
January	47.0	46.8	52.5	34.1	46.9	50.5	41.8
February	47.6	47.3	53.6	35.4	47.2	51.3	42.0
March	48.1	47.8	54.2	36.9	47.6	51.9	42.4
April	48.3	48.1	54.1	37.6	47.9	52.1	42.6
May	48.8	48.6	54.5	38.3	48.5	52.7	43.1
June	49.2	49.0	54.5	38.6	49.0	53.1	43.5
July	49.6	49.3	54.3	38.9	49.5	53.3	44.0
August	50.2	49.9	55.1	39.2	50.2	54.1	44.5
September	50.9	50.6	56.2	39.3	50.7	54.8	45.0
October	51.3	51.0	56.8	39.2	51.2	55.3	45.4
November	51.8	51.5	57.5	39.4	51.6	55.8	45.8
December	52.2	51.9	58.2	39.6	52.0	56.3	46.2
1975							
January	52.6	52.3	58.4	40.0	52.3	56.7	46.5
February	52.9	52.6	58.5	40.3	52.8	56.9	46.9
March	53.1	52.8	58.4	40.6	53.0	57.1	47.0
April	53.3	53.0	58.3	41.0	53.3	57.2	47.3
May	53.4	53.1	58.6	41.3	53.5	57.5	47.5
June	53.8	53.5	59.2	41.7	53.8	57.9	47.8
July	54.3	54.0	60.3	42.5	54.0	58.6	48.0
August	54.5	54.2	60.3	42.8	54.2	58.7	48.3
September	54.9	54.6	60.7	43.2	54.5	59.0	48.7
October	55.2	54.9	61.3	43.5	54.8	59.4	49.0
November	55.6	55.3	61.7	43.9	55.2	59.7	49.6
December	55.9	55.6	62.1	44.1	55.5	59.9	49.9

Table 8-1B. Summary Consumer Price Indexes: Historical Data, 1946 to Date—*Continued*

(Seasonally adjusted.)

Year and month	Consumer Price Index, 1982–1984 = 100						
	Urban wage earners and clerical workers (CPI-W), all items	All urban consumers (CPI-U)					
		All items	Food	Energy	All items less food and energy	Commod-ities	Services
1976							
January	56.2	55.8	61.9	44.5	55.9	60.0	50.5
February	56.2	55.9	61.3	44.4	56.2	59.9	50.8
March	56.3	56.0	60.9	44.1	56.5	59.8	51.1
April	56.5	56.1	60.9	43.9	56.7	59.9	51.3
May	56.7	56.4	61.1	44.1	57.0	60.2	51.4
June	57.0	56.7	61.3	44.4	57.2	60.4	51.7
July	57.3	57.0	61.6	44.8	57.6	60.7	52.1
August	57.6	57.3	61.8	45.2	57.9	61.0	52.4
September	57.9	57.6	62.1	45.7	58.2	61.3	52.8
October	58.2	57.9	62.4	46.1	58.5	61.6	53.1
November	58.4	58.1	62.3	46.8	58.7	61.7	53.4
December	58.7	58.4	62.5	47.5	58.9	62.0	53.7
1977							
January	59.1	58.7	62.7	48.1	59.3	62.3	54.1
February	59.6	59.3	63.9	48.1	59.7	63.0	54.4
March	59.9	59.6	64.2	48.4	60.0	63.2	54.8
April	60.3	60.0	65.0	48.6	60.3	63.7	55.2
May	60.6	60.2	65.3	48.9	60.6	63.9	55.4
June	60.9	60.5	65.7	48.9	61.0	64.2	55.8
July	61.2	60.8	65.9	49.1	61.2	64.4	56.3
August	61.4	61.1	66.2	49.5	61.5	64.6	56.6
September	61.7	61.3	66.4	49.8	61.8	64.8	56.9
October	61.9	61.6	66.6	50.5	62.0	65.0	57.2
November	62.3	62.0	67.1	51.3	62.3	65.5	57.6
December	62.6	62.3	67.4	51.6	62.7	65.7	57.9
1978							
January	63.0	62.7	67.9	51.1	63.1	66.1	58.3
February	63.3	63.0	68.6	50.6	63.4	66.4	58.7
March	63.8	63.4	69.5	51.0	63.8	66.8	59.1
April	64.2	63.9	70.6	51.4	64.3	67.4	59.6
May	64.8	64.5	71.6	51.7	64.7	68.0	60.0
June	65.3	65.0	72.7	51.9	65.2	68.6	60.5
July	65.8	65.5	73.0	52.1	65.6	69.1	61.0
August	66.2	65.9	73.3	52.6	66.1	69.4	61.5
September	66.7	66.5	73.6	53.2	66.7	70.0	62.1
October	67.4	67.1	74.2	54.1	67.2	70.6	62.6
November	67.8	67.5	74.7	54.9	67.6	71.1	63.1
December	68.3	67.9	75.1	55.9	68.0	71.6	63.3
1979							
January	68.8	68.5	76.4	55.8	68.5	72.2	63.8
February	69.6	69.2	77.7	55.9	69.2	72.9	64.4
March	70.3	69.9	78.4	57.4	69.8	73.8	64.9
April	71.1	70.6	79.0	59.5	70.3	74.7	65.5
May	71.9	71.4	79.7	62.0	70.8	75.6	66.2
June	72.7	72.2	80.0	64.7	71.3	76.4	66.8
July	73.5	73.0	80.5	67.3	71.9	77.2	67.6
August	74.2	73.7	80.4	69.7	72.7	77.9	68.5
September	75.0	74.4	80.9	71.9	73.3	78.6	69.2
October	75.7	75.2	81.5	73.5	74.0	79.3	70.1
November	76.5	76.0	82.0	74.8	74.8	80.0	71.1
December	77.3	76.9	82.8	76.8	75.7	80.8	72.0
1980							
January	78.5	78.0	83.3	79.1	76.7	82.0	73.1
February	79.4	79.0	83.4	81.9	77.5	82.8	74.1
March	80.6	80.1	84.1	84.5	78.6	83.9	75.4
April	81.4	80.9	84.7	85.4	79.5	84.4	76.6
May	82.2	81.7	85.2	86.4	80.1	84.9	77.6
June	83.0	82.5	85.7	86.5	81.0	85.3	79.0
July	83.1	82.6	86.6	86.7	80.8	85.9	78.5
August	83.7	83.2	88.0	87.2	81.3	86.9	78.5
September	84.4	83.9	89.1	87.5	82.1	87.8	79.0
October	85.3	84.7	89.8	88.0	83.0	88.5	80.0
November	86.2	85.6	90.8	88.8	83.9	89.2	81.1
December	87.0	86.4	91.3	90.7	84.9	89.7	82.2
1981							
January	87.7	87.2	91.6	92.1	85.4	90.4	83.0
February	88.6	88.0	92.1	95.2	85.9	91.4	83.7
March	89.1	88.6	92.6	97.4	86.4	91.9	84.4
April	89.6	89.1	92.8	97.6	87.0	92.0	85.3
May	90.2	89.7	92.8	97.9	87.8	92.4	86.4
June	90.9	90.5	93.2	97.3	88.6	92.9	87.5
July	92.0	91.5	93.9	97.3	89.8	93.6	88.9
August	92.7	92.2	94.4	97.8	90.7	94.0	89.9
September	93.5	93.1	94.8	98.6	91.8	94.6	91.2
October	93.8	93.4	95.0	99.2	92.1	94.7	91.7
November	94.2	93.8	95.1	100.5	92.5	94.9	92.5
December	94.5	94.1	95.3	101.5	93.0	95.1	93.0

Table 8-1B. Summary Consumer Price Indexes: Historical Data, 1946 to Date—*Continued*

(Seasonally adjusted.)

Year and month	Urban wage earners and clerical workers (CPI-W), all items	Consumer Price Index, 1982–1984 = 100					
		All urban consumers (CPI-U)					
		All items	Food	Energy	All items less food and energy	Commod-ities	Services
1982							
January	94.8	94.4	95.6	100.6	93.3	95.2	93.5
February	95.1	94.7	96.3	98.0	93.8	95.4	93.9
March	95.0	94.7	96.2	96.6	93.9	95.3	94.0
April	95.3	95.0	96.4	94.2	94.7	95.1	94.9
May	96.1	95.9	97.2	95.7	95.4	96.0	95.7
June	97.3	97.0	98.1	98.4	96.1	97.4	96.5
July	97.8	97.5	98.2	99.3	96.7	97.9	97.0
August	98.1	97.7	98.0	99.8	97.1	97.9	97.6
September	98.1	97.7	98.2	100.3	97.2	97.8	97.6
October	98.5	98.1	98.2	101.7	97.5	98.2	97.9
November	98.4	98.0	98.2	102.5	97.3	98.3	97.7
December	98.1	97.7	98.2	102.8	97.2	98.3	96.9
1983							
January	98.2	97.9	98.1	99.6	97.6	98.3	97.5
February	98.2	98.0	98.2	97.7	98.0	98.1	97.9
March	98.5	98.1	98.8	96.8	98.2	98.2	98.1
April	99.1	98.8	99.2	98.9	98.6	98.9	98.7
May	99.5	99.2	99.5	100.4	98.9	99.5	98.9
June	99.7	99.4	99.6	100.6	99.2	99.8	99.2
July	100.0	99.8	99.6	100.9	99.8	100.2	99.6
August	100.5	100.1	99.7	101.2	100.1	100.5	99.8
September	100.7	100.4	100.0	101.0	100.5	100.7	100.2
October	101.0	100.8	100.3	100.8	101.0	101.0	100.7
November	101.2	101.1	100.3	100.5	101.5	101.1	101.3
December	101.3	101.4	100.6	100.0	101.8	101.2	101.6
1984							
January	101.8	102.1	102.0	100.2	102.5	101.9	102.1
February	102.0	102.6	102.7	101.4	102.8	102.4	102.6
March	102.0	102.9	102.9	101.4	103.2	102.6	103.0
April	102.2	103.3	102.9	101.7	103.7	102.9	103.5
May	102.5	103.5	102.7	101.6	104.1	103.0	103.9
June	102.7	103.7	103.1	100.8	104.5	103.1	104.2
July	103.2	104.1	103.3	100.5	105.0	103.2	104.9
August	104.1	104.4	103.9	100.1	105.4	103.4	105.4
September	104.5	104.7	103.8	100.6	105.8	103.6	105.9
October	104.7	105.1	104.0	101.1	106.2	103.9	106.3
November	104.8	105.3	104.1	100.8	106.4	103.9	106.7
December	104.9	105.5	104.5	100.1	106.8	103.9	107.1
1985							
January	105.2	105.7	104.7	100.3	107.1	104.1	107.4
February	105.7	106.3	105.2	100.3	107.7	104.7	107.9
March	106.1	106.8	105.5	101.3	108.1	105.1	108.4
April	106.4	107.0	105.4	102.3	108.4	105.4	108.7
May	106.6	107.2	105.2	102.2	108.8	105.2	109.4
June	106.9	107.5	105.5	102.2	109.1	105.3	109.8
July	107.0	107.7	105.5	102.2	109.4	105.3	110.3
August	107.1	107.9	105.6	101.2	109.8	105.2	110.7
September	107.3	108.1	105.8	101.2	110.0	105.4	111.0
October	107.7	108.5	105.8	101.2	110.5	105.6	111.5
November	108.2	109.0	106.5	101.8	111.1	106.1	112.1
December	108.7	109.5	107.3	102.4	111.4	106.6	112.5
1986							
January	109.1	109.9	107.5	102.6	111.9	106.9	113.1
February	108.8	109.7	107.3	99.5	112.2	106.0	113.5
March	108.1	109.1	107.5	92.6	112.5	104.5	114.1
April	107.7	108.7	107.7	87.2	112.9	103.3	114.6
May	107.8	109.0	108.2	87.2	113.1	103.5	114.8
June	108.3	109.4	108.3	88.8	113.4	103.8	115.5
July	108.3	109.5	109.1	85.6	113.8	103.7	115.7
August	108.4	109.6	110.1	83.6	114.2	103.6	116.1
September	108.8	110.0	110.2	84.4	114.6	104.0	116.5
October	108.9	110.2	110.5	82.8	115.0	104.0	116.9
November	109.2	110.4	111.1	82.1	115.3	104.2	117.2
December	109.5	110.8	111.4	82.5	115.6	104.5	117.5
1987							
January	110.2	111.4	111.8	85.4	115.9	105.5	117.9
February	110.7	111.8	112.2	87.4	116.2	106.1	118.3
March	111.1	112.2	112.4	87.6	116.6	106.5	118.6
April	111.6	112.7	112.6	87.6	117.3	106.9	119.2
May	111.9	113.0	113.2	87.1	117.7	107.2	119.6
June	112.4	113.5	113.9	88.5	117.9	107.7	120.0
July	112.7	113.8	113.7	89.2	118.3	108.0	120.3
August	113.2	114.3	113.9	90.5	118.7	108.5	120.9
September	113.6	114.7	114.3	90.3	119.2	108.8	121.4
October	113.9	115.0	114.5	89.6	119.8	109.0	121.8
November	114.2	115.4	114.5	90.0	120.1	109.3	122.2
December	114.4	115.6	115.1	89.5	120.4	109.3	122.6

Table 8-1B. Summary Consumer Price Indexes: Historical Data, 1946 to Date—*Continued*

(Seasonally adjusted.)

Year and month	Urban wage earners and clerical workers (CPI-W), all items	Consumer Price Index, 1982–1984 = 100					
		All urban consumers (CPI-U)					
		All items	Food	Energy	All items less food and energy	Commod-ities	Services
1988							
January	114.7	116.0	115.6	88.8	120.9	109.5	123.0
February	114.9	116.2	115.6	88.7	121.2	109.5	123.5
March	115.2	116.5	115.8	88.4	121.7	109.8	123.9
April	115.8	117.2	116.4	88.8	122.3	110.5	124.4
May	116.2	117.5	116.9	88.5	122.7	110.7	124.8
June	116.6	118.0	117.6	88.9	123.2	111.2	125.4
July	117.3	118.5	118.8	89.4	123.6	111.9	125.8
August	117.7	119.0	119.4	90.1	124.0	112.2	126.4
September	118.2	119.5	120.1	89.8	124.7	112.8	126.9
October	118.6	119.9	120.3	89.8	125.2	113.0	127.5
November	118.9	120.3	120.5	89.8	125.6	113.3	127.9
December	119.3	120.7	121.1	89.6	126.0	113.5	128.4
1989							
January	119.9	121.2	121.6	90.3	126.5	114.1	128.9
February	120.3	121.6	122.5	90.8	126.9	114.5	129.4
March	120.9	122.2	123.2	91.8	127.4	115.1	130.0
April	121.9	123.1	123.9	96.6	127.8	116.5	130.5
May	122.5	123.7	124.7	97.4	128.3	117.1	131.1
June	122.8	124.1	125.1	96.9	128.8	117.2	131.6
July	123.2	124.5	125.6	96.7	129.2	117.3	132.3
August	123.2	124.5	125.9	94.9	129.5	117.0	132.8
September	123.4	124.8	126.3	93.8	129.9	117.2	133.1
October	123.9	125.4	126.8	94.4	130.6	117.8	133.7
November	124.4	125.9	127.4	93.9	131.1	118.1	134.3
December	124.9	126.3	127.8	94.2	131.6	118.4	134.9
1990							
January	126.1	127.5	129.7	98.9	132.1	120.2	135.4
February	126.6	128.0	130.8	98.2	132.7	120.7	136.0
March	127.0	128.6	131.0	97.6	133.5	120.9	136.8
April	127.3	128.9	130.8	97.5	134.0	121.0	137.4
May	127.5	129.1	131.1	96.7	134.4	121.0	137.9
June	128.2	129.9	132.1	97.3	135.1	121.6	138.8
July	128.8	130.5	132.8	97.1	135.8	122.0	139.6
August	129.9	131.6	133.2	101.6	136.6	123.2	140.6
September	130.9	132.5	133.6	106.5	137.1	124.5	141.1
October	131.7	133.4	134.1	110.8	137.6	125.8	141.6
November	132.1	133.7	134.5	111.2	138.0	126.0	142.2
December	132.5	134.2	134.6	111.0	138.6	126.3	142.7
1991							
January	132.9	134.7	135.0	108.5	139.5	126.3	143.7
February	132.9	134.8	135.1	104.5	140.2	125.9	144.4
March	133.0	134.8	135.3	101.9	140.5	125.5	144.7
April	133.3	135.1	136.1	101.2	140.9	126.0	144.9
May	133.8	135.6	136.6	102.1	141.3	126.4	145.4
June	134.1	136.0	137.4	101.1	141.8	126.7	145.9
July	134.3	136.2	136.7	100.7	142.3	126.6	146.5
August	134.6	136.6	136.2	101.1	142.9	126.8	146.9
September	135.0	137.0	136.4	101.5	143.4	127.0	147.6
October	135.2	137.2	136.2	101.6	143.7	127.0	148.0
November	135.8	137.8	136.7	102.4	144.2	127.6	148.5
December	136.2	138.2	137.0	103.1	144.7	127.9	149.2
1992							
January	136.2	138.3	136.6	101.5	145.1	127.6	149.6
February	136.5	138.6	137.1	101.2	145.4	127.8	149.9
March	136.9	139.1	137.6	101.2	145.9	128.2	150.4
April	137.2	139.4	137.5	101.4	146.3	128.3	150.9
May	137.5	139.7	137.2	102.0	146.8	128.6	151.3
June	138.0	140.1	137.6	103.3	147.1	129.1	151.7
July	138.4	140.5	137.4	103.7	147.6	129.3	152.2
August	138.7	140.8	138.4	103.5	147.9	129.6	152.6
September	139.0	141.1	138.9	103.6	148.1	129.9	152.9
October	139.5	141.7	138.9	104.3	148.8	130.2	153.7
November	139.8	142.1	138.7	105.1	149.2	130.3	154.3
December	140.1	142.3	138.8	105.3	149.6	130.5	154.7
1993							
January	140.5	142.8	139.1	105.0	150.1	130.7	155.3
February	140.8	143.1	139.6	104.3	150.6	131.1	155.6
March	141.0	143.3	139.6	104.9	150.8	131.1	156.0
April	141.4	143.8	140.0	104.9	151.4	131.4	156.7
May	141.8	144.2	141.0	104.3	151.8	131.6	157.3
June	142.0	144.3	140.6	103.9	152.1	131.3	157.8
July	142.1	144.5	140.6	103.4	152.3	131.3	158.1
August	142.4	144.8	141.1	103.4	152.8	131.6	158.6
September	142.5	145.0	141.4	103.0	152.9	131.3	159.0
October	143.2	145.6	142.0	105.3	153.4	132.2	159.4
November	143.4	146.0	142.3	104.4	153.9	132.4	159.9
December	143.7	146.3	142.8	103.7	154.3	132.4	160.5

Table 8-1B. Summary Consumer Price Indexes: Historical Data, 1946 to Date—*Continued*

(Seasonally adjusted.)

Year and month	Urban wage earners and clerical workers (CPI-W), all items	Consumer Price Index, 1982–1984 = 100					
		All urban consumers (CPI-U)					
		All items	Food	Energy	All items less food and energy	Commod-ities	Services
1994							
January	143.8	146.3	142.9	102.8	154.5	132.3	160.8
February	144.0	146.7	142.7	104.1	154.8	132.4	161.4
March	144.3	147.1	142.7	104.3	155.3	132.5	162.0
April	144.5	147.2	143.0	103.7	155.5	132.6	162.2
May	144.8	147.5	143.3	102.8	155.9	132.9	162.4
June	145.3	147.9	143.8	103.1	156.4	133.5	162.8
July	145.9	148.4	144.6	104.5	156.7	134.1	163.1
August	146.5	149.0	145.1	106.7	157.1	134.7	163.8
September	146.8	149.3	145.3	106.1	157.5	134.8	164.1
October	146.9	149.4	145.3	105.7	157.8	134.8	164.5
November	147.3	149.8	145.6	106.1	158.2	135.0	165.0
December	147.6	150.1	146.8	105.9	158.3	135.4	165.2
1995							
January	148.0	150.5	146.7	105.7	159.0	135.4	166.0
February	148.4	150.9	147.3	105.8	159.4	135.6	166.5
March	148.6	151.2	147.1	105.5	159.9	135.6	167.1
April	149.2	151.8	148.1	105.6	160.4	136.2	167.7
May	149.5	152.1	148.2	105.8	160.7	136.4	168.1
June	149.8	152.4	148.3	106.7	161.1	136.6	168.5
July	149.9	152.6	148.5	105.8	161.4	136.6	168.9
August	150.2	152.9	148.6	105.6	161.8	136.8	169.3
September	150.4	153.1	149.1	104.1	162.2	136.8	169.7
October	150.8	153.5	149.5	104.4	162.7	137.1	170.3
November	150.9	153.7	149.6	103.4	163.0	137.0	170.7
December	151.3	153.9	149.9	104.4	163.1	137.3	170.9
1996							
January	152.0	154.7	150.4	106.9	163.7	138.2	171.5
February	152.3	155.0	150.8	107.1	164.0	138.3	172.0
March	152.9	155.5	151.4	108.3	164.4	139.0	172.4
April	153.4	156.1	152.0	111.2	164.6	139.6	172.8
May	153.8	156.4	151.9	112.0	165.0	139.7	173.4
June	154.0	156.7	152.9	110.3	165.4	139.8	173.8
July	154.3	157.0	153.4	110.1	165.7	139.8	174.4
August	154.5	157.2	153.9	109.7	166.0	139.8	174.9
September	154.9	157.7	154.6	109.8	166.5	140.3	175.4
October	155.4	158.2	155.5	110.5	166.8	140.8	175.8
November	155.9	158.7	156.1	111.8	167.2	141.4	176.3
December	156.3	159.1	156.3	113.9	167.4	141.7	176.7
1997							
January	156.6	159.4	155.9	115.2	167.8	141.8	177.3
February	156.9	159.7	156.5	115.0	168.1	142.1	177.6
March	156.9	159.8	156.6	113.0	168.4	141.8	178.0
April	157.0	159.9	156.5	111.0	168.9	141.6	178.4
May	157.0	159.9	156.6	108.8	169.2	141.4	178.7
June	157.3	160.2	156.9	110.0	169.4	141.5	179.2
July	157.4	160.4	157.2	109.1	169.7	141.4	179.7
August	157.8	160.8	157.7	110.9	169.8	141.9	179.9
September	158.2	161.2	158.0	112.4	170.2	142.2	180.4
October	158.4	161.5	158.3	111.8	170.6	142.2	180.9
November	158.6	161.7	158.6	111.4	170.8	142.1	181.4
December	158.6	161.8	158.7	109.8	171.2	142.1	181.7
1998							
January	158.8	162.0	159.5	107.5	171.6	142.0	182.1
February	158.7	162.0	159.4	105.1	171.9	141.8	182.3
March	158.7	162.0	159.7	103.3	172.2	141.4	182.8
April	158.8	162.2	159.7	102.4	172.5	141.3	183.3
May	159.3	162.6	160.3	103.2	172.9	141.7	183.7
June	159.5	162.8	160.2	103.6	173.2	141.8	184.0
July	159.8	163.2	160.6	103.3	173.5	142.1	184.3
August	160.0	163.4	161.0	102.1	174.0	142.2	184.7
September	160.1	163.5	161.1	101.3	174.2	142.0	185.1
October	160.5	163.9	162.0	101.5	174.4	142.3	185.5
November	160.7	164.1	162.2	101.1	174.8	142.2	186.0
December	161.1	164.4	162.4	100.1	175.4	142.5	186.4
1999							
January	161.4	164.7	163.0	99.7	175.6	142.8	186.6
February	161.3	164.7	163.3	99.2	175.6	142.4	186.9
March	161.4	164.8	163.3	100.4	175.7	142.4	187.4
April	162.4	165.9	163.5	105.5	176.3	144.0	187.9
May	162.6	166.0	163.8	104.9	176.5	143.9	188.1
June	162.6	166.0	163.7	104.5	176.6	143.8	188.3
July	163.3	166.7	163.9	106.7	177.1	144.5	188.9
August	163.8	167.1	164.2	109.5	177.3	145.0	189.3
September	164.6	167.8	164.6	111.8	177.8	145.9	189.8
October	164.9	168.1	165.0	112.0	178.1	146.1	190.2
November	165.1	168.4	165.3	111.5	178.4	145.9	190.9
December	165.6	168.8	165.5	113.8	178.7	146.5	191.2

Table 8-1B. Summary Consumer Price Indexes: Historical Data, 1946 to Date—*Continued*

(Seasonally adjusted.)

Year and month	Consumer Price Index, 1982–1984 = 100						
	Urban wage earners and clerical workers (CPI-W), all items	All urban consumers (CPI-U)					
		All items	Food	Energy	All items less food and energy	Commod-ities	Services
2000							
January	166.0	169.3	165.6	115.0	179.3	146.7	192.0
February	166.7	170.0	166.2	118.8	179.4	147.6	192.5
March	167.8	171.0	166.5	124.3	180.0	149.1	193.1
April	167.6	170.9	166.7	120.9	180.3	148.5	193.5
May	167.9	171.2	167.3	120.0	180.7	148.5	194.0
June	169.0	172.2	167.4	126.8	181.1	149.6	194.9
July	169.5	172.7	168.3	127.3	181.5	149.8	195.7
August	169.3	172.7	168.7	123.8	181.9	149.2	196.3
September	170.3	173.6	168.9	129.2	182.3	150.4	196.9
October	170.5	173.9	169.0	129.6	182.6	150.1	197.7
November	170.8	174.2	169.2	129.2	183.1	150.3	198.1
December	171.2	174.6	170.0	130.1	183.3	150.4	198.8
2001							
January	172.2	175.6	170.3	135.0	183.9	150.6	200.6
February	172.5	176.0	171.2	134.1	184.4	150.8	201.2
March	172.6	176.1	171.7	131.7	184.7	150.5	201.6
April	173.0	176.4	172.1	132.3	185.1	150.9	202.0
May	174.0	177.3	172.4	138.4	185.3	151.9	202.7
June	174.2	177.7	173.1	136.9	186.0	151.9	203.5
July	173.8	177.4	173.6	129.6	186.4	150.9	203.8
August	173.8	177.4	174.0	127.3	186.7	150.4	204.4
September	174.7	178.1	174.2	130.9	187.1	151.6	204.5
October	173.9	177.6	174.8	122.9	187.4	150.3	204.8
November	173.7	177.5	174.9	116.9	188.1	149.2	205.6
December	173.4	177.4	174.7	113.9	188.4	148.3	206.1
2002							
January	173.7	177.7	175.3	114.2	188.7	148.3	206.8
February	173.9	178.0	175.7	113.5	189.1	148.3	207.5
March	174.5	178.5	176.1	117.6	189.2	149.0	207.9
April	175.4	179.3	176.4	121.5	189.7	150.0	208.5
May	175.5	179.5	175.8	121.9	190.0	149.7	209.0
June	175.7	179.6	175.9	121.6	190.2	149.7	209.3
July	176.1	180.0	176.1	122.4	190.5	149.9	209.9
August	176.6	180.5	176.1	123.1	191.1	150.2	210.6
September	176.8	180.8	176.5	123.7	191.3	150.2	211.1
October	177.2	181.2	176.4	126.9	191.5	150.5	211.7
November	177.5	181.5	176.9	126.7	191.9	150.5	212.4
December	177.7	181.8	177.1	127.1	192.1	150.3	213.0
2003							
January	178.6	182.6	177.1	133.5	192.4	151.3	213.7
February	179.7	183.6	178.1	140.8	192.5	152.7	214.2
March	180.1	183.9	178.4	143.9	192.5	152.6	215.0
April	179.2	183.2	178.5	136.5	192.5	151.1	215.1
May	178.7	182.9	178.8	129.4	192.9	149.4	216.0
June	178.9	183.1	179.7	129.8	193.0	149.7	216.2
July	179.4	183.7	179.8	132.2	193.4	150.3	216.8
August	180.3	184.5	180.5	137.8	193.6	151.4	217.2
September	180.9	185.1	180.9	142.9	193.7	152.1	217.8
October	180.6	184.9	181.6	137.8	194.0	151.1	218.4
November	180.6	185.0	182.6	136.9	194.0	151.1	218.6
December	181.0	185.5	183.5	138.8	194.2	151.6	219.0
2004							
January	181.9	186.3	183.4	143.9	194.6	152.5	219.7
February	182.4	186.7	183.9	145.8	194.9	153.1	220.1
March	182.7	187.1	184.2	145.1	195.5	153.3	220.8
April	182.8	187.4	184.6	143.4	195.9	153.0	221.4
May	183.8	188.2	186.1	147.6	196.2	154.2	221.9
June	184.4	188.9	186.4	151.5	196.6	154.9	222.7
July	184.6	189.1	186.8	151.1	196.8	154.6	223.2
August	184.7	189.2	187.0	151.5	196.9	154.4	223.7
September	185.3	189.8	186.9	152.9	197.5	154.9	224.2
October	186.4	190.8	187.8	158.6	197.9	156.7	224.6
November	187.4	191.7	188.4	163.8	198.3	157.8	225.4
December	187.4	191.7	188.4	162.3	198.6	157.4	225.8
2005							
January	187.2	191.6	188.7	157.2	199.0	156.8	226.2
February	188.0	192.4	188.6	161.8	199.4	157.5	226.9
March	188.6	193.1	189.1	163.5	200.1	157.9	228.0
April	189.3	193.7	190.4	166.7	200.2	158.6	228.5
May	189.1	193.6	190.6	163.3	200.5	158.1	228.9
June	189.3	193.7	190.5	163.7	200.6	158.0	229.1
July	190.6	194.9	190.9	172.8	200.9	159.6	229.9
August	192.0	196.1	191.1	183.5	201.1	161.5	230.4
September	195.1	198.8	191.5	208.2	201.3	166.0	231.4
October	195.2	199.1	192.0	205.9	202.0	164.8	233.1
November	193.7	198.1	192.6	191.0	202.5	161.8	234.2
December	193.7	198.1	192.9	187.6	202.8	161.5	234.4

Table 8-1B. Summary Consumer Price Indexes: Historical Data, 1946 to Date—*Continued*

(Seasonally adjusted.)

Year and month	Urban wage earners and clerical workers (CPI-W), all items	Consumer Price Index, 1982–1984 = 100					
		All urban consumers (CPI-U)					
		All items	Food	Energy	All items less food and energy	Commod-ities	Services
2006							
January	195.1	199.3	193.6	196.6	203.2	162.9	235.5
February	195.0	199.4	193.7	194.1	203.6	162.5	235.9
March	195.3	199.7	194.0	192.0	204.3	162.6	236.5
April	196.4	200.7	193.8	198.5	204.8	164.1	237.1
May	197.0	201.3	194.1	199.8	205.4	164.6	237.8
June	197.4	201.8	194.7	199.8	205.9	164.8	238.5
July	198.6	202.9	195.2	207.9	206.3	166.4	239.2
August	199.5	203.8	195.7	211.9	206.8	167.2	239.9
September	198.3	202.8	196.3	198.1	207.2	164.6	240.7
October	197.1	201.9	196.9	184.1	207.6	162.5	241.0
November	197.2	202.0	197.0	184.1	207.8	162.0	241.7
December	198.4	203.1	197.1	192.7	208.1	163.4	242.4
2007							
January	198.6	203.4	198.4	190.3	208.6	163.3	243.2
February	199.4	204.2	199.7	192.3	209.1	164.1	244.1
March	200.7	205.3	200.4	200.2	209.4	165.6	244.7
April	201.3	205.9	201.0	203.3	209.7	166.2	245.3
May	202.3	206.8	201.7	208.6	210.1	167.3	245.9
June	202.7	207.2	202.6	209.8	210.4	167.5	246.6
July	203.0	207.6	203.2	209.6	210.8	167.8	247.1
August	203.0	207.7	204.0	206.4	211.1	167.5	247.5
September	204.0	208.5	205.0	210.7	211.6	168.5	248.2
October	204.7	209.2	205.7	212.4	212.1	169.0	249.1
November	206.5	210.8	206.5	223.8	212.7	171.5	249.8
December	207.1	211.4	206.9	225.6	213.2	172.0	250.5
2008							
January	207.9	212.2	208.1	226.8	213.8	172.7	251.3
February	208.4	212.6	208.9	229.7	213.9	173.0	251.8
March	209.2	213.4	209.3	233.3	214.4	173.7	252.9
April	209.8	214.0	211.1	234.8	214.6	174.1	253.6
May	211.1	215.2	212.0	243.9	214.9	175.5	254.7
June	213.7	217.5	213.3	262.1	215.4	178.8	255.8
July	215.5	219.1	215.4	271.1	216.0	180.7	257.1
August	215.0	218.7	216.5	262.6	216.4	179.6	257.6
September	215.1	218.9	217.8	260.1	216.7	179.8	257.6
October	212.7	217.0	218.7	238.1	216.8	175.8	257.7
November	207.9	213.1	219.1	195.2	216.9	168.0	257.8
December	205.9	211.4	219.1	176.6	216.9	164.5	258.0
2009							
January	206.5	212.0	219.2	178.8	217.3	165.3	258.4
February	207.4	212.8	219.0	184.9	217.8	166.8	258.7
March	207.1	212.6	218.5	178.8	218.3	166.2	258.7
April	207.4	212.7	218.2	177.1	218.7	166.5	258.7
May	207.8	213.0	217.8	179.7	218.9	167.2	258.6
June	209.9	214.7	217.8	196.9	219.2	170.7	258.7
July	210.0	214.7	217.4	195.6	219.3	170.5	258.7
August	210.9	215.5	217.4	201.5	219.5	171.3	259.3
September	211.3	215.9	217.3	202.8	219.9	171.9	259.6
October	211.9	216.5	217.5	204.6	220.5	172.7	260.0
November	212.7	217.1	217.6	210.3	220.6	173.8	260.2
December	213.0	217.3	217.9	210.3	220.8	174.2	260.3
2010							
January	213.4	217.5	218.5	212.8	220.6	174.9	259.8
February	213.3	217.3	218.6	209.6	220.7	174.3	260.0
March	213.3	217.4	219.0	209.3	220.8	174.1	260.3
April	213.2	217.4	219.2	209.2	220.8	173.9	260.7
May	212.9	217.3	219.3	206.6	221.0	173.4	260.9
June	212.9	217.2	219.3	203.8	221.2	172.9	261.2
July	213.5	217.6	219.2	206.9	221.4	173.4	261.5
August	213.9	217.9	219.5	208.8	221.5	173.8	261.7
September	214.3	218.3	220.2	209.8	221.7	174.4	261.9
October	215.1	219.0	220.5	216.7	221.8	175.7	262.1
November	215.5	219.6	220.9	219.5	222.1	176.5	262.4
December	216.9	220.5	221.2	227.1	222.3	177.9	262.7
2011							
January	217.5	221.2	222.5	229.3	222.8	178.9	263.2
February	218.2	221.9	223.5	232.1	223.2	179.8	263.7
March	219.5	223.0	225.2	240.1	223.5	181.8	264.1
April	220.8	224.1	226.1	248.0	223.7	183.5	264.5
May	221.6	224.8	226.9	250.7	224.2	184.5	264.9
June	221.5	224.8	227.5	245.5	224.7	184.1	265.3
July	222.1	225.4	228.5	246.2	225.2	184.6	266.0
August	222.8	226.1	229.6	246.9	225.9	185.3	266.7
September	223.4	226.6	230.7	248.6	226.1	185.8	267.1
October	223.5	226.8	230.9	246.7	226.5	185.7	267.5
November	224.0	227.2	231.1	247.6	226.9	186.2	267.8
December	223.9	227.2	231.6	243.4	227.4	185.7	268.5

Table 8-1B. Summary Consumer Price Indexes: Historical Data, 1946 to Date—*Continued*

(Seasonally adjusted.)

Year and month	Urban wage earners and clerical workers (CPI-W), all items	Consumer Price Index, 1982–1984 = 100					
		All urban consumers (CPI-U)					
		All items	Food	Energy	All items less food and energy	Commod- ities	Services
2012							
January	224.5	227.8	232.2	244.9	227.9	186.5	268.9
February	225.1	228.3	232.1	248.9	228.0	187.4	269.0
March	225.7	228.8	232.6	249.7	228.5	187.9	269.6
April	226.0	229.2	233.0	249.7	228.9	188.1	270.1
May	225.4	228.7	233.2	241.8	229.2	186.8	270.4
June	225.0	228.5	233.7	235.9	229.6	185.8	271.0
July	224.9	228.6	233.8	233.6	230.0	185.5	271.4
August	226.4	229.9	234.2	245.0	230.2	187.5	272.1
September	227.7	231.0	234.4	253.0	230.7	189.1	272.7
October	228.4	231.6	234.8	256.0	231.0	189.8	273.2
November	227.8	231.2	235.3	248.8	231.3	188.5	273.8
December	227.7	231.2	235.7	244.7	231.7	187.7	274.4
2013							
January	228.1	231.7	236.0	245.0	232.2	188.0	275.1
February	229.6	232.9	236.0	255.7	232.6	189.9	275.8
March	228.8	232.3	236.2	246.6	232.8	188.1	276.2
April	228.1	231.8	236.7	240.5	232.8	186.8	276.5
May	228.3	231.9	236.5	240.5	233.0	186.5	277.0
June	228.8	232.4	237.0	242.7	233.4	187.1	277.4
July	229.2	232.9	237.2	243.0	233.9	187.3	278.1
August	229.8	233.5	237.5	244.8	234.3	187.8	278.7
September	229.8	233.5	237.5	242.7	234.7	187.4	279.4
October	229.9	233.7	237.8	242.0	234.9	187.2	279.8
November	230.4	234.1	237.9	242.7	235.4	187.5	280.3
December	231.0	234.7	238.2	245.7	235.8	188.2	280.9
2014							
January	231.6	235.3	238.5	250.3	236.0	188.5	281.7
February	231.9	235.5	239.4	249.9	236.2	188.4	282.3
March	232.3	236.0	240.4	250.0	236.6	188.1	283.5
April	232.8	236.5	241.2	249.9	237.1	188.8	283.8
May	233.1	236.9	242.4	249.2	237.5	188.8	284.7
June	233.4	237.2	242.6	249.7	237.8	189.0	285.1
July	233.6	237.5	243.2	248.7	238.2	188.9	285.7
August	233.5	237.5	243.8	245.7	238.4	188.5	286.0
September	233.4	237.5	244.5	241.6	238.8	188.1	286.4
October	233.2	237.4	244.9	237.1	239.2	187.5	286.9
November	232.6	237.0	245.4	229.0	239.5	186.1	287.4
December	231.5	236.3	246.2	218.5	239.6	184.0	288.0
2015							
January	229.5	234.7	246.2	199.9	239.8	180.3	288.6
February	230.3	235.3	246.5	203.0	240.2	181.1	288.9
March	230.9	236.0	246.1	205.8	240.8	181.9	289.5
April	231.1	236.2	246.0	203.6	241.3	181.6	290.2
May	232.0	237.0	246.2	209.0	241.7	182.9	290.5
June	232.7	237.7	246.9	212.0	242.1	183.4	291.3
July	233.0	238.0	247.2	211.6	242.6	183.4	292.0
August	233.0	238.0	247.7	208.6	242.8	182.9	292.5
September	232.1	237.5	248.5	197.2	243.3	181.2	293.1
October	232.3	237.7	248.8	195.6	243.8	180.9	293.9
November	232.5	238.0	248.5	195.1	244.2	180.7	294.6
December	232.1	237.8	248.2	190.1	244.5	179.7	295.0
2016							
January	232.1	237.8	248.3	186.3	245.1	179.1	295.8
February	231.5	237.5	248.6	177.3	245.6	177.8	296.5
March	232.1	238.0	248.0	180.9	246.0	178.1	297.1
April	233.0	238.8	248.3	185.6	246.5	179.0	297.9
May	233.6	239.4	247.9	188.1	247.0	179.2	298.8
June	234.3	240.1	247.8	192.8	247.4	179.8	299.7
July	234.1	240.1	247.7	189.4	247.8	178.9	300.4
August	234.6	240.6	247.8	189.7	248.4	179.0	301.3
September	235.0	241.1	247.8	192.2	248.7	179.3	302.0
October	235.6	241.6	247.8	195.5	249.1	179.9	302.5
November	235.9	242.0	247.7	195.9	249.5	179.7	303.3
December	236.7	242.7	247.6	200.1	250.0	180.5	304.0
2017							
January	237.8	243.7	247.9	206.3	250.5	181.7	304.9
February	238.0	244.0	248.6	204.6	251.0	181.4	305.8
March	237.6	243.7	249.1	201.2	250.9	180.9	305.6
April	237.9	244.1	249.6	202.3	251.1	181.0	306.2
May	237.7	243.9	250.0	198.0	251.4	180.1	306.9
June	238.0	244.2	249.9	198.0	251.7	179.9	307.5
July	238.0	244.3	250.3	196.5	252.0	179.6	308.1
August	239.2	245.3	250.5	202.5	252.6	180.7	309.0
September	240.5	246.4	250.9	212.1	252.9	182.3	309.6
October	240.5	246.6	251.1	207.5	253.5	181.7	310.5
November	241.4	247.3	251.1	213.4	253.9	182.6	311.0
December	241.9	247.8	251.6	213.9	254.4	182.9	311.9

Table 8-1C. Consumer and Producer Price Indexes: Historical Data, 1913–1949

(Not seasonally adjusted.)

Year and month	Consumer price indexes, 1982–1984 =100				Producer Price Indexes for goods, 1982 = 100			
	All urban consumers (CPI-U)		Urban wage earners and clerical workers (CPI-W)		All commodities		Farm products	Industrial commodities
	Index	Percent change	Index	Percent change	Index	Percent change		
1913	9.9	. . .	10.0	. . .	12.0	. . .	18.0	11.9
1914	10.0	1.0	10.1	1.0	11.8	-1.7	17.9	11.3
1915	10.1	1.0	10.2	1.0	12.0	1.7	18.0	11.6
1916	10.9	7.9	11.0	7.8	14.7	22.5	21.3	15.0
1917	12.8	17.4	12.9	17.3	20.2	37.4	32.6	19.5
1918	15.1	18.0	15.1	17.1	22.6	11.9	37.4	21.1
1919	17.3	14.6	17.4	15.2	23.9	5.8	39.8	22.0
1920	20.0	15.6	20.1	15.5	26.6	11.3	38.0	27.4
1921	17.9	-10.5	18.0	-10.4	16.8	-36.8	22.3	17.8
1922	16.8	-6.1	16.9	-6.1	16.7	-0.6	23.7	17.4
1923	17.1	1.8	17.2	1.8	17.3	3.6	24.9	17.8
1924	17.1	0.0	17.2	0.0	16.9	-2.3	25.2	17.0
1925	17.5	2.3	17.6	2.3	17.8	5.3	27.7	17.5
1926	17.7	1.1	17.8	1.1	17.2	-3.4	25.3	17.0
1927	17.4	-1.7	17.5	-1.7	16.5	-4.1	25.1	16.0
1928	17.1	-1.7	17.2	-1.7	16.7	1.2	26.7	15.8
1929	17.1	0.0	17.2	0.0	16.4	-1.8	26.4	15.6
1930	16.7	-2.3	16.8	-2.3	14.9	-9.1	22.4	14.5
1931	15.2	-9.0	15.3	-8.9	12.6	-15.4	16.4	12.8
1932	13.7	-9.9	13.7	-10.5	11.2	-11.1	12.2	11.9
1933	13.0	-5.1	13.0	-5.1	11.4	1.8	13.0	12.1
1934	13.4	3.1	13.5	3.8	12.9	13.2	16.5	13.3
1935	13.7	2.2	13.8	2.2	13.8	7.0	19.8	13.3
1936	13.9	1.5	13.9	0.7	13.9	0.7	20.4	13.5
1937	14.4	3.6	14.4	3.6	14.9	7.2	21.8	14.5
1938	14.1	-2.1	14.2	-1.4	13.5	-9.4	17.3	13.9
1939	13.9	-1.4	14.0	-1.4	13.3	-1.5	16.5	13.9
1940	14.0	0.7	14.1	0.7	13.5	1.5	17.1	14.1
1941	14.7	5.0	14.8	5.0	15.1	11.9	20.8	15.1
1942	16.3	10.9	16.4	10.8	17.0	12.6	26.7	16.2
1943	17.3	6.1	17.4	6.1	17.8	4.7	30.9	16.5
1944	17.6	1.7	17.7	1.7	17.9	0.6	31.2	16.7
1945	18.0	2.3	18.1	2.3	18.2	1.7	32.4	17.0
1946	19.5	8.3	19.6	8.3	20.8	14.3	37.5	18.6
1947	22.3	14.4	22.5	14.8	25.6	23.1	45.1	22.7
1948	24.1	8.1	24.2	7.6	27.7	8.2	48.5	24.6
1949	23.8	-1.2	24.0	-0.8	26.3	-5.1	41.9	24.1
1913								
January	9.8	. . .	9.9	. . .	12.1	. . .	17.6	12.3
February	9.8	0.0	9.8	-1.0	12.0	-0.8	17.5	12.2
March	9.8	0.0	9.8	0.0	12.0	0.0	17.6	12.1
April	9.8	0.0	9.9	1.0	12.0	0.0	17.5	12.0
May	9.7	-1.0	9.8	-1.0	11.9	-0.8	17.4	11.9
June	9.8	1.0	9.8	0.0	11.9	0.0	17.6	11.8
July	9.9	1.0	9.9	1.0	12.0	0.8	18.1	11.8
August	9.9	0.0	10.0	1.0	12.0	0.0	18.2	11.7
September	10.0	1.0	10.0	0.0	12.2	1.7	18.8	11.8
October	10.0	0.0	10.1	1.0	12.2	0.0	18.8	11.8
November	10.1	1.0	10.1	0.0	12.1	-0.8	18.9	11.7
December	10.0	-1.0	10.1	0.0	11.9	-1.7	18.5	11.6
1914								
January	10.0	0.0	10.1	0.0	11.8	-0.8	18.4	11.5
February	9.9	-1.0	10.0	-1.0	11.8	0.0	18.3	11.5
March	9.9	0.0	10.0	0.0	11.7	-0.8	18.2	11.5
April	9.8	-1.0	9.9	-1.0	11.7	0.0	18.0	11.5
May	9.9	1.0	9.9	0.0	11.6	-0.9	18.0	11.4
June	9.9	0.0	10.0	1.0	11.6	0.0	18.1	11.3
July	10.0	1.0	10.1	1.0	11.6	0.0	18.0	11.2
August	10.2	2.0	10.2	1.0	12.0	3.4	18.3	11.2
September	10.2	0.0	10.3	1.0	12.1	0.8	17.9	11.3
October	10.1	-1.0	10.2	-1.0	11.7	-3.3	17.2	11.0
November	10.2	1.0	10.2	0.0	11.7	0.0	17.6	11.0
December	10.1	-1.0	10.2	0.0	11.6	-0.9	17.4	11.0
1915								
January	10.1	0.0	10.2	0.0	11.8	1.7	18.1	11.1
February	10.0	-1.0	10.1	-1.0	11.8	0.0	18.4	11.0
March	9.9	-1.0	10.0	-1.0	11.8	0.0	17.9	11.0
April	10.0	1.0	10.1	1.0	11.8	0.0	18.2	11.1
May	10.1	1.0	10.1	0.0	11.9	0.8	18.2	11.2
June	10.1	0.0	10.2	1.0	11.8	-0.8	17.7	11.4
July	10.1	0.0	10.2	0.0	11.9	0.8	18.1	11.5
August	10.1	0.0	10.2	0.0	11.8	-0.8	17.9	11.5
September	10.1	0.0	10.2	0.0	11.8	0.0	17.5	11.7
October	10.2	1.0	10.3	1.0	12.1	2.5	18.1	11.9
November	10.3	1.0	10.4	1.0	12.3	1.7	18.0	12.3
December	10.3	0.0	10.4	0.0	12.8	4.1	18.4	12.9

. . . = Not available.

Table 8-1C. Consumer and Producer Price Indexes: Historical Data, 1913–1949—*Continued*

(Not seasonally adjusted.)

Year and month	Consumer price indexes, 1982–1984 =100				Producer Price Indexes for goods, 1982 = 100			
	All urban consumers (CPI-U)		Urban wage earners and clerical workers (CPI-W)		All commodities		Farm products	Industrial commodities
	Index	Percent change	Index	Percent change	Index	Percent change		
1916								
January	10.4	1.0	10.5	1.0	13.3	3.9	19.4	13.6
February	10.4	0.0	10.5	0.0	13.5	1.5	19.4	14.0
March	10.5	1.0	10.6	1.0	13.9	3.0	19.4	14.4
April	10.6	1.0	10.7	0.9	14.1	1.4	19.6	14.6
May	10.7	0.9	10.7	0.0	14.2	0.7	19.8	14.7
June	10.8	0.9	10.9	1.9	14.3	0.7	19.7	14.8
July	10.8	0.0	10.9	0.0	14.4	0.7	20.3	14.6
August	10.9	0.9	11.0	0.9	14.7	2.1	21.7	14.6
September	11.1	1.8	11.2	1.8	15.0	2.0	22.6	14.8
October	11.3	1.8	11.3	0.9	15.7	4.7	23.7	15.5
November	11.5	1.8	11.5	1.8	16.8	7.0	25.3	16.8
December	11.6	0.9	11.6	0.9	17.1	1.8	25.0	17.7
1917								
January	11.7	0.9	11.8	1.7	17.6	2.9	26.2	18.3
February	12.0	2.6	12.0	1.7	18.0	2.3	27.2	18.6
March	12.0	0.0	12.1	0.8	18.5	2.8	28.6	18.8
April	12.6	5.0	12.6	4.1	19.7	6.5	31.6	19.0
May	12.8	1.6	12.9	2.4	20.8	5.6	33.6	19.8
June	13.0	1.6	13.0	0.8	21.0	1.0	33.8	20.4
July	12.8	-1.5	12.9	-0.8	21.2	1.0	34.0	20.7
August	13.0	1.6	13.1	1.6	21.5	1.4	34.6	20.7
September	13.3	2.3	13.3	1.5	21.3	-0.9	34.3	20.1
October	13.5	1.5	13.6	2.3	21.1	-0.9	35.2	19.1
November	13.5	0.0	13.6	0.0	21.2	0.5	36.0	19.2
December	13.7	1.5	13.8	1.5	21.2	0.0	35.6	19.5
1918								
January	14.0	2.2	14.0	1.4	21.6	1.9	37.0	19.9
February	14.1	0.7	14.2	1.4	21.1	-2.3	37.1	19.0
March	14.0	-0.7	14.1	-0.7	21.8	3.3	37.3	20.2
April	14.2	1.4	14.3	1.4	22.1	1.4	36.6	20.7
May	14.5	2.1	14.5	1.4	22.1	0.0	35.4	21.0
June	14.7	1.4	14.8	2.1	22.2	0.5	35.4	21.3
July	15.1	2.7	15.2	2.7	22.7	2.3	37.0	21.5
August	15.4	2.0	15.4	1.3	23.2	2.2	38.6	21.7
September	15.7	1.9	15.8	2.6	23.7	2.2	39.6	22.1
October	16.0	1.9	16.1	1.9	23.5	-0.8	38.2	22.1
November	16.3	1.9	16.3	1.2	23.5	0.0	38.0	22.1
December	16.5	1.2	16.6	1.8	23.5	0.0	38.1	21.8
1919								
January	16.5	0.0	16.6	0.0	23.2	-1.3	38.9	21.1
February	16.2	-1.8	16.2	-2.4	22.4	-3.4	37.5	20.6
March	16.4	1.2	16.5	1.9	22.6	0.9	38.5	20.1
April	16.7	1.8	16.8	1.8	22.9	1.3	40.0	20.0
May	16.9	1.2	17.0	1.2	23.3	1.7	40.9	20.2
June	16.9	0.0	17.0	0.0	23.4	0.4	39.6	21.1
July	17.4	3.0	17.5	2.9	24.3	3.8	41.5	22.1
August	17.7	1.7	17.8	1.7	24.9	2.5	41.3	23.1
September	17.8	0.6	17.9	0.6	24.3	-2.4	38.7	23.2
October	18.1	1.7	18.2	1.7	24.4	0.4	38.5	23.5
November	18.5	2.2	18.6	2.2	24.9	2.0	40.3	23.8
December	18.9	2.2	19.0	2.2	26.0	4.4	41.8	24.7
1920								
January	19.3	2.1	19.4	2.1	27.2	4.6	43.0	26.1
February	19.5	1.0	19.6	1.0	27.1	-0.4	41.2	27.1
March	19.7	1.0	19.8	1.0	27.3	0.7	41.5	27.7
April	20.3	3.0	20.4	3.0	28.5	4.4	42.5	28.7
May	20.6	1.5	20.7	1.5	28.8	1.1	42.8	29.0
June	20.9	1.5	21.0	1.4	28.7	-0.3	42.3	29.0
July	20.8	-0.5	20.9	-0.5	28.6	-0.3	40.5	29.5
August	20.3	-2.4	20.4	-2.4	27.8	-2.8	37.8	29.7
September	20.0	-1.5	20.1	-1.5	26.8	-3.6	36.4	28.5
October	19.9	-0.5	20.0	-0.5	24.9	-7.1	32.2	26.9
November	19.8	-0.5	19.9	-0.5	23.0	-7.6	30.0	24.5
December	19.4	-2.0	19.5	-2.0	20.8	-9.6	26.4	22.7
1921								
January	19.0	-2.1	19.1	-2.1	19.6	-5.8	25.6	21.1
February	18.4	-3.2	18.5	-3.1	18.1	-7.7	23.4	19.4
March	18.3	-0.5	18.4	-0.5	17.7	-2.2	22.7	18.7
April	18.1	-1.1	18.2	-1.1	17.0	-4.0	20.9	18.4
May	17.7	-2.2	17.8	-2.2	16.6	-2.4	21.0	17.9
June	17.6	-0.6	17.7	-0.6	16.1	-3.0	20.3	17.4
July	17.7	0.6	17.8	0.6	16.1	0.0	21.8	16.9
August	17.7	0.0	17.8	0.0	16.1	0.0	22.5	16.5
September	17.5	-1.1	17.6	-1.1	16.1	0.0	22.7	16.5
October	17.5	0.0	17.6	0.0	16.2	0.6	22.7	17.0
November	17.4	-0.6	17.5	-0.6	16.2	0.0	22.1	17.3
December	17.3	-0.6	17.4	-0.6	16.0	-1.2	22.2	17.0

Table 8-1C. Consumer and Producer Price Indexes: Historical Data, 1913–1949—*Continued*

(Not seasonally adjusted.)

Year and month	Consumer price indexes, 1982–1984 =100				Producer Price Indexes for goods, 1982 = 100			
	All urban consumers (CPI-U)		Urban wage earners and clerical workers (CPI-W)		All commodities		Farm products	Industrial commodities
	Index	Percent change	Index	Percent change	Index	Percent change		
1922								
January	16.9	-2.3	17.0	-2.3	15.7	-1.9	22.2	16.7
February	16.9	0.0	17.0	0.0	16.0	1.9	24.0	16.6
March	16.7	-1.2	16.8	-1.2	16.0	0.0	23.6	16.5
April	16.7	0.0	16.8	0.0	16.1	0.6	23.4	16.6
May	16.7	0.0	16.8	0.0	16.6	3.1	23.8	17.4
June	16.7	0.0	16.8	0.0	16.6	0.0	23.4	17.4
July	16.8	0.6	16.9	0.6	17.1	3.0	24.1	18.1
August	16.6	-1.2	16.7	-1.2	17.0	-0.6	23.0	18.2
September	16.6	0.0	16.7	0.0	17.1	0.6	23.3	18.2
October	16.7	0.6	16.8	0.6	17.2	0.6	23.8	17.9
November	16.8	0.6	16.9	0.6	17.3	0.6	24.7	17.7
December	16.9	0.6	17.0	0.6	17.3	0.0	25.0	17.7
1923								
January	16.8	-0.6	16.9	-0.6	17.6	1.7	25.1	18.2
February	16.8	0.0	16.9	0.0	17.8	1.1	25.3	18.6
March	16.8	0.0	16.9	0.0	18.0	1.1	25.3	18.8
April	16.9	0.6	17.0	0.6	17.9	-0.6	24.8	18.7
May	16.9	0.0	17.0	0.0	17.5	-2.2	24.4	18.3
June	17.0	0.6	17.1	0.6	17.3	-1.1	24.2	17.9
July	17.2	1.2	17.3	1.2	17.0	-1.7	23.7	17.6
August	17.1	-0.6	17.2	-0.6	16.9	-0.6	24.2	17.4
September	17.2	0.6	17.3	0.6	17.2	1.8	25.3	17.3
October	17.3	0.6	17.4	0.6	17.1	-0.6	25.4	17.1
November	17.3	0.0	17.4	0.0	17.0	-0.6	25.7	16.9
December	17.3	0.0	17.4	0.0	16.9	-0.6	25.5	16.9
1924								
January	17.3	0.0	17.4	0.0	17.2	1.8	25.6	17.4
February	17.2	-0.6	17.3	-0.6	17.2	0.0	25.0	17.6
March	17.1	-0.6	17.2	-0.6	17.0	-1.2	24.2	17.5
April	17.0	-0.6	17.1	-0.6	16.7	-1.8	24.6	17.3
May	17.0	0.0	17.1	0.0	16.5	-1.2	24.0	17.0
June	17.0	0.0	17.1	0.0	16.4	-0.6	23.8	16.7
July	17.1	0.6	17.2	0.6	16.5	0.6	24.9	16.6
August	17.0	-0.6	17.1	-0.6	16.7	1.2	25.7	16.6
September	17.1	0.6	17.2	0.6	16.7	0.0	25.3	16.6
October	17.2	0.6	17.3	0.6	16.9	1.2	26.0	16.6
November	17.2	0.0	17.3	0.0	17.1	1.2	26.2	16.8
December	17.3	0.6	17.4	0.6	17.5	2.3	27.3	17.1
1925								
January	17.3	0.0	17.4	0.0	17.7	1.1	28.7	17.3
February	17.2	-0.6	17.3	-0.6	17.9	1.1	28.4	17.7
March	17.3	0.6	17.4	0.6	17.9	0.0	28.5	17.5
April	17.2	-0.6	17.3	-0.6	17.5	-2.2	27.1	17.2
May	17.3	0.6	17.4	0.6	17.5	0.0	27.1	17.3
June	17.5	1.2	17.6	1.1	17.7	1.1	27.6	17.5
July	17.7	1.1	17.8	1.1	18.0	1.7	28.3	17.6
August	17.7	0.0	17.8	0.0	17.9	-0.6	28.1	17.4
September	17.7	0.0	17.8	0.0	17.8	-0.6	27.7	17.4
October	17.7	0.0	17.8	0.0	17.8	0.0	27.0	17.5
November	18.0	1.7	18.1	1.7	18.0	1.1	27.3	17.6
December	17.9	-0.6	18.0	-0.6	17.8	-1.1	26.6	17.6
1926								
January	17.9	0.0	18.0	0.0	17.8	0.0	27.1	17.5
February	17.9	0.0	18.0	0.0	17.6	-1.1	26.5	17.3
March	17.8	-0.6	17.9	-0.6	17.3	-1.7	25.7	17.1
April	17.9	0.6	18.0	0.6	17.3	0.0	26.0	17.0
May	17.8	-0.6	17.9	-0.6	17.3	0.0	25.8	17.0
June	17.7	-0.6	17.8	-0.6	17.3	0.0	25.5	17.0
July	17.5	-1.1	17.6	-1.1	17.1	-1.2	24.9	16.9
August	17.4	-0.6	17.5	-0.6	17.1	0.0	24.6	16.9
September	17.5	0.6	17.6	0.6	17.2	0.6	25.1	16.9
October	17.6	0.6	17.7	0.6	17.1	-0.6	24.7	16.9
November	17.7	0.6	17.8	0.6	17.0	-0.6	23.9	16.9
December	17.7	0.0	17.8	0.0	16.9	-0.6	24.0	16.7
1927								
January	17.5	-1.1	17.6	-1.1	16.4	-3.0	24.3	16.4
February	17.4	-0.6	17.5	-0.6	16.6	1.2	24.1	16.3
March	17.3	-0.6	17.4	-0.6	16.5	-0.6	23.8	16.1
April	17.3	0.0	17.4	0.0	16.3	-1.2	23.8	15.9
May	17.4	0.6	17.5	0.6	16.2	-0.6	24.3	15.9
June	17.6	1.1	17.7	1.1	16.2	0.0	24.3	15.9
July	17.3	-1.7	17.4	-1.7	16.2	0.0	24.6	15.9
August	17.2	-0.6	17.3	-0.6	16.4	1.2	25.8	15.9
September	17.3	0.6	17.4	0.6	16.6	1.2	26.7	16.0
October	17.4	0.6	17.5	0.6	16.7	0.6	26.5	15.9
November	17.3	-0.6	17.4	-0.6	16.6	-0.6	26.3	15.8
December	17.3	0.0	17.4	0.0	16.6	0.0	26.3	15.9

Table 8-1C. Consumer and Producer Price Indexes: Historical Data, 1913–1949—*Continued*

(Not seasonally adjusted.)

Year and month	Consumer price indexes, 1982–1984 =100				Producer Price Indexes for goods, 1982 = 100			
	All urban consumers (CPI-U)		Urban wage earners and clerical workers (CPI-W)		All commodities		Farm products	Industrial commodities
	Index	Percent change	Index	Percent change	Index	Percent change		
1928								
January	17.3	0.0	17.4	0.0	16.6	0.0	26.8	15.8
February	17.1	-1.2	17.2	-1.1	16.5	-0.6	26.4	15.8
March	17.1	0.0	17.2	0.0	16.5	0.0	26.1	15.8
April	17.1	0.0	17.2	0.0	16.7	1.2	27.1	15.8
May	17.2	0.6	17.3	0.6	16.8	0.6	27.7	15.8
June	17.1	-0.6	17.2	-0.6	16.7	-0.6	26.9	15.8
July	17.1	0.0	17.2	0.0	16.8	0.6	27.4	15.8
August	17.1	0.0	17.2	0.0	16.8	0.0	27.0	15.8
September	17.3	1.2	17.4	1.2	17.0	1.2	27.5	15.8
October	17.2	-0.6	17.3	-0.6	16.7	-1.8	26.1	15.8
November	17.2	0.0	17.3	0.0	16.5	-1.2	25.7	15.8
December	17.1	-0.6	17.2	-0.6	16.5	0.0	26.2	15.8
1929								
January	17.1	0.0	17.2	0.0	16.5	0.0	26.7	15.7
February	17.1	0.0	17.2	0.0	16.4	-0.6	26.6	15.6
March	17.0	-0.6	17.1	-0.6	16.6	1.2	27.1	15.7
April	16.9	-0.6	17.0	-0.6	16.5	-0.6	26.5	15.6
May	17.0	0.6	17.1	0.6	16.3	-1.2	25.8	15.6
June	17.1	0.6	17.2	0.6	16.4	0.6	26.1	15.6
July	17.3	1.2	17.4	1.2	16.6	1.2	27.1	15.6
August	17.3	0.0	17.4	0.0	16.6	0.0	27.1	15.5
September	17.3	0.0	17.4	0.0	16.6	0.0	26.9	15.6
October	17.3	0.0	17.4	0.0	16.4	-1.2	26.2	15.6
November	17.3	0.0	17.4	0.0	16.1	-1.8	25.5	15.5
December	17.2	-0.6	17.3	-0.6	16.1	0.0	25.7	15.4
1930								
January	17.1	-0.6	17.2	-0.6	15.9	-1.2	25.5	15.2
February	17.0	-0.6	17.1	-0.6	15.7	-1.3	24.7	15.1
March	16.9	-0.6	17.0	-0.6	15.5	-1.3	23.9	15.0
April	17.0	0.6	17.1	0.6	15.5	0.0	24.2	15.0
May	16.9	-0.6	17.0	-0.6	15.3	-1.3	23.5	14.9
June	16.8	-0.6	16.9	-0.6	15.0	-2.0	22.5	14.6
July	16.6	-1.2	16.7	-1.2	14.5	-3.3	21.0	14.4
August	16.5	-0.6	16.6	-0.6	14.5	0.0	21.4	14.2
September	16.6	0.6	16.7	0.6	14.5	0.0	21.5	14.2
October	16.5	-0.6	16.6	-0.6	14.3	-1.4	20.8	14.0
November	16.4	-0.6	16.5	-0.6	14.0	-2.1	20.0	13.8
December	16.1	-1.8	16.2	-1.8	13.7	-2.1	19.0	13.6
1931								
January	15.9	-1.2	16.0	-1.2	13.5	-1.5	18.4	13.4
February	15.7	-1.3	15.7	-1.9	13.2	-2.2	17.7	13.3
March	15.6	-0.6	15.6	-0.6	13.1	-0.8	17.8	13.1
April	15.5	-0.6	15.5	-0.6	12.9	-1.5	17.7	12.9
May	15.3	-1.3	15.4	-0.6	12.6	-2.3	16.9	12.8
June	15.1	-1.3	15.2	-1.3	12.4	-1.6	16.5	12.6
July	15.1	0.0	15.2	0.0	12.4	0.0	16.4	12.6
August	15.1	0.0	15.1	-0.7	12.4	0.0	16.1	12.6
September	15.0	-0.7	15.1	0.0	12.3	-0.8	15.3	12.6
October	14.9	-0.7	15.0	-0.7	12.1	-1.6	14.8	12.4
November	14.7	-1.3	14.8	-1.3	12.1	0.0	14.8	12.5
December	14.6	-0.7	14.7	-0.7	11.8	-2.5	14.1	12.3
1932								
January	14.3	-2.1	14.4	-2.0	11.6	-1.7	13.3	12.2
February	14.1	-1.4	14.2	-1.4	11.4	-1.7	12.8	12.1
March	14.0	-0.7	14.1	-0.7	11.4	0.0	12.7	12.0
April	13.9	-0.7	14.0	-0.7	11.3	-0.9	12.4	12.0
May	13.7	-1.4	13.8	-1.4	11.1	-1.8	11.8	11.9
June	13.6	-0.7	13.7	-0.7	11.0	-0.9	11.6	11.9
July	13.6	0.0	13.7	0.0	11.1	0.9	12.1	11.8
August	13.5	-0.7	13.5	-1.5	11.2	0.9	12.4	11.9
September	13.4	-0.7	13.5	0.0	11.3	0.9	12.4	11.9
October	13.3	-0.7	13.4	-0.7	11.1	-1.8	11.8	11.9
November	13.2	-0.8	13.3	-0.7	11.0	-0.9	11.8	11.9
December	13.1	-0.8	13.2	-0.8	10.8	-1.8	11.1	11.7
1933								
January	12.9	-1.5	13.0	-1.5	10.5	-2.8	10.8	11.4
February	12.7	-1.6	12.8	-1.5	10.3	-1.9	10.3	11.2
March	12.6	-0.8	12.7	-0.8	10.4	1.0	10.8	11.2
April	12.6	0.0	12.6	-0.8	10.4	0.0	11.2	11.1
May	12.6	0.0	12.7	0.8	10.8	3.8	12.7	11.3
June	12.7	0.8	12.8	0.8	11.2	3.7	13.4	11.7
July	13.1	3.1	13.2	3.1	11.9	6.3	15.2	12.3
August	13.2	0.8	13.3	0.8	12.0	0.8	14.6	12.6
September	13.2	0.0	13.3	0.0	12.2	1.7	14.4	13.0
October	13.2	0.0	13.3	0.0	12.3	0.8	14.1	13.1
November	13.2	0.0	13.3	0.0	12.3	0.0	14.3	13.1
December	13.2	0.0	13.2	-0.8	12.2	-0.8	14.0	13.2

Table 8-1C. Consumer and Producer Price Indexes: Historical Data, 1913–1949—*Continued*

(Not seasonally adjusted.)

| Year and month | Consumer price indexes, 1982–1984 =100 | | | | Producer Price Indexes for goods, 1982 = 100 | | | |
| | All urban consumers (CPI-U) | | Urban wage earners and clerical workers (CPI-W) | | All commodities | | Farm products | Industrial commodities |
	Index	Percent change	Index	Percent change	Index	Percent change		
1934								
January	13.2	0.0	13.3	0.8	12.4	1.6	14.8	13.3
February	13.3	0.8	13.4	0.8	12.7	2.4	15.5	13.4
March	13.3	0.0	13.4	0.0	12.7	0.0	15.5	13.4
April	13.3	0.0	13.4	0.0	12.7	0.0	15.1	13.4
May	13.3	0.0	13.4	0.0	12.7	0.0	15.1	13.4
June	13.4	0.8	13.4	0.0	12.9	1.6	16.0	13.3
July	13.4	0.0	13.4	0.0	12.9	0.0	16.3	13.3
August	13.4	0.0	13.5	0.7	13.2	2.3	17.6	13.3
September	13.6	1.5	13.7	1.5	13.4	1.5	18.5	13.3
October	13.5	-0.7	13.6	-0.7	13.2	-1.5	17.8	13.3
November	13.5	0.0	13.5	-0.7	13.2	0.0	17.8	13.3
December	13.4	-0.7	13.5	0.0	13.3	0.8	18.2	13.3
1935								
January	13.6	1.5	13.7	1.5	13.6	2.3	19.6	13.2
February	13.7	0.7	13.8	0.7	13.7	0.7	20.0	13.2
March	13.7	0.0	13.8	0.0	13.7	0.0	19.7	13.2
April	13.8	0.7	13.9	0.7	13.8	0.7	20.3	13.1
May	13.8	0.0	13.8	-0.7	13.8	0.0	20.3	13.2
June	13.7	-0.7	13.8	0.0	13.8	0.0	19.8	13.3
July	13.7	0.0	13.7	-0.7	13.7	-0.7	19.5	13.3
August	13.7	0.0	13.7	0.0	13.9	1.5	20.0	13.3
September	13.7	0.0	13.8	0.7	13.9	0.0	20.1	13.2
October	13.7	0.0	13.8	0.0	13.9	0.0	19.7	13.3
November	13.8	0.7	13.9	0.7	13.9	0.0	19.6	13.4
December	13.8	0.0	13.9	0.0	14.0	0.7	19.8	13.4
1936								
January	13.8	0.0	13.9	0.0	13.9	-0.7	19.7	13.4
February	13.8	0.0	13.8	-0.7	13.9	0.0	20.1	13.4
March	13.7	-0.7	13.8	0.0	13.7	-1.4	19.3	13.4
April	13.7	0.0	13.8	0.0	13.7	0.0	19.4	13.4
May	13.7	0.0	13.8	0.0	13.5	-1.5	19.0	13.4
June	13.8	0.7	13.9	0.7	13.7	1.5	19.7	13.4
July	13.9	0.7	14.0	0.7	13.9	1.5	20.5	13.5
August	14.0	0.7	14.1	0.7	14.0	0.7	21.2	13.5
September	14.0	0.0	14.1	0.0	14.0	0.0	21.2	13.5
October	14.0	0.0	14.1	0.0	14.0	0.0	21.2	13.6
November	14.0	0.0	14.1	0.0	14.2	1.4	21.5	13.8
December	14.0	0.0	14.1	0.0	14.5	2.1	22.4	14.0
1937								
January	14.1	0.7	14.2	0.7	14.8	2.1	23.1	14.2
February	14.1	0.0	14.2	0.0	14.9	0.7	23.1	14.3
March	14.2	0.7	14.3	0.7	15.1	1.3	23.8	14.5
April	14.3	0.7	14.4	0.7	15.2	0.7	23.3	14.7
May	14.4	0.7	14.4	0.0	15.1	-0.7	22.7	14.7
June	14.4	0.0	14.5	0.7	15.0	-0.7	22.3	14.6
July	14.5	0.7	14.5	0.0	15.2	1.3	22.6	14.7
August	14.5	0.0	14.6	0.7	15.1	-0.7	21.8	14.6
September	14.6	0.7	14.7	0.7	15.1	0.0	21.7	14.6
October	14.6	0.0	14.6	-0.7	14.7	-2.6	20.3	14.5
November	14.5	-0.7	14.5	-0.7	14.4	-2.0	19.1	14.3
December	14.4	-0.7	14.5	0.0	14.1	-2.1	18.4	14.2
1938								
January	14.2	-1.4	14.3	-1.4	14.0	-0.7	18.1	14.2
February	14.1	-0.7	14.2	-0.7	13.8	-1.4	17.6	14.1
March	14.1	0.0	14.2	0.0	13.7	-0.7	17.7	14.1
April	14.2	0.7	14.2	0.0	13.5	-1.5	17.2	14.0
May	14.1	-0.7	14.2	0.0	13.5	0.0	17.0	13.9
June	14.1	0.0	14.2	0.0	13.5	0.0	17.3	13.8
July	14.1	0.0	14.2	0.0	13.6	0.7	17.5	13.9
August	14.1	0.0	14.2	0.0	13.4	-1.5	17.0	13.9
September	14.1	0.0	14.2	0.0	13.5	0.7	17.2	13.9
October	14.0	-0.7	14.1	-0.7	13.4	-0.7	16.8	13.8
November	14.0	0.0	14.1	0.0	13.4	0.0	17.1	13.7
December	14.0	0.0	14.1	0.0	13.3	-0.7	17.0	13.6
1939								
January	14.0	0.0	14.0	-0.7	13.3	0.0	17.0	13.6
February	13.9	-0.7	14.0	0.0	13.3	0.0	16.9	13.6
March	13.9	0.0	13.9	-0.7	13.2	-0.8	16.6	13.7
April	13.8	-0.7	13.9	0.0	13.1	-0.8	16.1	13.7
May	13.8	0.0	13.9	0.0	13.1	0.0	16.1	13.7
June	13.8	0.0	13.9	0.0	13.0	-0.8	15.7	13.6
July	13.8	0.0	13.9	0.0	13.0	0.0	15.8	13.6
August	13.8	0.0	13.9	0.0	12.9	-0.8	15.4	13.6
September	14.1	2.2	14.2	2.2	13.6	5.4	17.3	14.0
October	14.0	-0.7	14.1	-0.7	13.7	0.7	16.9	14.2
November	14.0	0.0	14.1	0.0	13.6	-0.7	17.0	14.3
December	14.0	0.0	14.0	-0.7	13.7	0.7	17.1	14.3

Table 8-1C. Consumer and Producer Price Indexes: Historical Data, 1913–1949—*Continued*

(Not seasonally adjusted.)

Year and month	Consumer price indexes, 1982–1984 =100				Producer Price Indexes for goods, 1982 = 100			
	All urban consumers (CPI-U)		Urban wage earners and clerical workers (CPI-W)		All commodities		Farm products	Industrial commodities
	Index	Percent change	Index	Percent change	Index	Percent change		
1940								
January	13.9	-0.7	14.0	0.0	13.7	0.0	17.4	14.3
February	14.0	0.7	14.1	0.7	13.6	-0.7	17.3	14.2
March	14.0	0.0	14.1	0.0	13.5	-0.7	17.1	14.1
April	14.0	0.0	14.1	0.0	13.5	0.0	17.5	14.0
May	14.0	0.0	14.1	0.0	13.5	0.0	17.1	14.0
June	14.1	0.7	14.1	0.0	13.4	-0.7	16.7	14.0
July	14.0	-0.7	14.1	0.0	13.4	0.0	16.8	14.0
August	14.0	0.0	14.1	0.0	13.4	0.0	16.5	14.0
September	14.0	0.0	14.1	0.0	13.4	0.0	16.7	14.0
October	14.0	0.0	14.1	0.0	13.6	1.5	16.8	14.2
November	14.0	0.0	14.1	0.0	13.7	0.7	17.2	14.3
December	14.1	0.7	14.2	0.7	13.8	0.7	17.6	14.3
1941								
January	14.1	0.0	14.2	0.0	13.9	0.7	18.1	14.3
February	14.1	0.0	14.2	0.0	13.9	0.0	17.7	14.3
March	14.2	0.7	14.2	0.0	14.0	0.7	18.1	14.4
April	14.3	0.7	14.4	1.4	14.4	2.9	18.8	14.6
May	14.4	0.7	14.5	0.7	14.6	1.4	19.3	14.9
June	14.7	2.1	14.7	1.4	15.0	2.7	20.7	15.1
July	14.7	0.0	14.8	0.7	15.3	2.0	21.7	15.2
August	14.9	1.4	14.9	0.7	15.6	2.0	22.1	15.5
September	15.1	1.3	15.2	2.0	15.8	1.3	23.0	15.6
October	15.3	1.3	15.4	1.3	15.9	0.6	22.7	15.9
November	15.4	0.7	15.5	0.6	15.9	0.0	22.9	15.9
December	15.5	0.6	15.5	0.0	16.2	1.9	23.9	15.9
1942								
January	15.7	1.3	15.7	1.3	16.5	1.9	25.5	16.1
February	15.8	0.6	15.9	1.3	16.7	1.2	25.6	16.1
March	16.0	1.3	16.1	1.3	16.8	0.6	26.0	16.2
April	16.1	0.6	16.2	0.6	17.0	1.2	26.4	16.2
May	16.3	1.2	16.3	0.6	17.0	0.0	26.3	16.3
June	16.3	0.0	16.4	0.6	17.0	0.0	26.3	16.3
July	16.4	0.6	16.5	0.6	17.0	0.0	26.6	16.3
August	16.5	0.6	16.6	0.6	17.1	0.6	26.8	16.2
September	16.5	0.0	16.6	0.0	17.2	0.6	27.2	16.2
October	16.7	1.2	16.8	1.2	17.2	0.0	27.5	16.2
November	16.8	0.6	16.9	0.6	17.3	0.6	27.9	16.3
December	16.9	0.6	17.0	0.6	17.4	0.6	28.7	16.3
1943								
January	16.9	0.0	17.0	0.0	17.5	0.6	29.5	16.4
February	16.9	0.0	17.0	0.0	17.7	1.1	30.0	16.4
March	17.2	1.8	17.3	1.8	17.8	0.6	31.0	16.4
April	17.4	1.2	17.5	1.2	17.9	0.6	31.2	16.5
May	17.5	0.6	17.6	0.6	17.9	0.0	31.7	16.5
June	17.5	0.0	17.6	0.0	17.9	0.0	31.9	16.5
July	17.4	-0.6	17.5	-0.6	17.8	-0.6	31.5	16.5
August	17.3	-0.6	17.4	-0.6	17.8	0.0	31.2	16.5
September	17.4	0.6	17.5	0.6	17.8	0.0	31.0	16.5
October	17.4	0.0	17.5	0.0	17.8	0.0	30.9	16.5
November	17.4	0.0	17.5	0.0	17.7	-0.6	30.6	16.6
December	17.4	0.0	17.5	0.0	17.8	0.6	30.7	16.6
1944								
January	17.4	0.0	17.5	0.0	17.8	0.0	30.7	16.6
February	17.4	0.0	17.5	0.0	17.8	0.0	30.9	16.7
March	17.4	0.0	17.5	0.0	17.9	0.6	31.2	16.7
April	17.5	0.6	17.6	0.6	17.9	0.0	31.1	16.7
May	17.5	0.0	17.6	0.0	17.9	0.0	31.0	16.7
June	17.6	0.6	17.7	0.6	18.0	0.6	31.5	16.7
July	17.7	0.6	17.8	0.6	17.9	-0.6	31.3	16.7
August	17.7	0.0	17.8	0.0	17.9	0.0	30.9	16.8
September	17.7	0.0	17.8	0.0	17.9	0.0	31.0	16.8
October	17.7	0.0	17.8	0.0	17.9	0.0	31.2	16.8
November	17.7	0.0	17.8	0.0	18.0	0.6	31.4	16.8
December	17.8	0.6	17.9	0.6	18.0	0.0	31.6	16.8
1945								
January	17.8	0.0	17.9	0.0	18.1	0.6	31.9	16.8
February	17.8	0.0	17.9	0.0	18.1	0.0	32.1	16.9
March	17.8	0.0	17.9	0.0	18.1	0.0	32.1	16.9
April	17.8	0.0	17.9	0.0	18.2	0.6	32.5	16.9
May	17.9	0.6	18.0	0.6	18.3	0.5	32.8	16.9
June	18.1	1.1	18.2	1.1	18.3	0.0	32.9	16.9
July	18.1	0.0	18.2	0.0	18.3	0.0	32.5	17.0
August	18.1	0.0	18.2	0.0	18.2	-0.5	32.0	17.0
September	18.1	0.0	18.2	0.0	18.1	-0.5	31.4	17.0
October	18.1	0.0	18.2	0.0	18.2	0.6	32.1	17.0
November	18.1	0.0	18.2	0.0	18.4	1.1	33.1	17.0
December	18.2	0.6	18.3	0.5	18.4	0.0	33.2	17.1

Table 8-1C. Consumer and Producer Price Indexes: Historical Data, 1913–1949—*Continued*

(Not seasonally adjusted.)

Year and month	Consumer price indexes, 1982–1984 =100				Producer Price Indexes for goods, 1982 = 100			
	All urban consumers (CPI-U)		Urban wage earners and clerical workers (CPI-W)		All commodities		Farm products	Industrial commodities
	Index	Percent change	Index	Percent change	Index	Percent change		
1946								
January	18.2	0.0	18.3	0.0	18.4	0.0	32.7	17.1
February	18.1	-0.5	18.2	-0.5	18.5	0.5	33.0	17.2
March	18.3	1.1	18.4	1.1	18.8	1.6	33.6	17.4
April	18.4	0.5	18.5	0.5	19.0	1.1	34.1	17.5
May	18.5	0.5	18.6	0.5	19.1	0.5	34.7	17.7
June	18.7	1.1	18.8	1.1	19.4	1.6	35.4	18.0
July	19.8	5.9	19.9	5.9	21.5	10.8	39.6	18.6
August	20.2	2.0	20.3	2.0	22.2	3.3	40.6	19.0
September	20.4	1.0	20.5	1.0	21.4	-3.6	39.0	19.1
October	20.8	2.0	20.9	2.0	23.1	7.9	41.7	19.7
November	21.3	2.4	21.5	2.9	24.1	4.3	42.8	20.6
December	21.5	0.9	21.6	0.5	24.3	0.8	42.4	21.2
1947								
January	21.5	0.0	21.6	0.0	24.5	0.8	41.6	21.8
February	21.5	0.0	21.6	0.0	24.7	0.8	42.6	22.0
March	21.9	1.9	22.1	2.3	25.3	2.4	45.5	22.3
April	21.9	0.0	22.1	0.0	25.1	-0.8	44.0	22.4
May	21.9	0.0	22.0	-0.5	25.0	-0.4	43.6	22.3
June	22.0	0.5	22.2	0.9	25.0	0.0	43.8	22.4
July	22.2	0.9	22.4	0.9	25.3	1.2	44.3	22.5
August	22.5	1.4	22.6	0.9	25.6	1.2	44.8	22.8
September	23.0	2.2	23.1	2.2	26.1	2.0	46.6	23.1
October	23.0	0.0	23.1	0.0	26.4	1.1	47.4	23.3
November	23.1	0.4	23.3	0.9	26.7	1.1	47.7	23.6
December	23.4	1.3	23.6	1.3	27.2	1.9	50.1	23.9
1948								
January	23.7	1.3	23.8	0.8	27.7	1.8	51.2	24.3
February	23.5	-0.8	23.6	-0.8	27.2	-1.8	47.7	24.1
March	23.4	-0.4	23.6	0.0	27.2	0.0	47.6	24.1
April	23.8	1.7	23.9	1.3	27.4	0.7	48.3	24.3
May	23.9	0.4	24.1	0.8	27.5	0.4	49.4	24.3
June	24.1	0.8	24.2	0.4	27.7	0.7	50.4	24.4
July	24.4	1.2	24.5	1.2	28.0	1.1	50.2	24.6
August	24.5	0.4	24.6	0.4	28.2	0.7	49.6	24.9
September	24.5	0.0	24.6	0.0	28.1	-0.4	48.8	25.0
October	24.4	-0.4	24.5	-0.4	27.8	-1.1	46.9	25.0
November	24.2	-0.8	24.4	-0.4	27.8	0.0	46.3	25.1
December	24.1	-0.4	24.2	-0.8	27.6	-0.7	45.2	25.1
1949								
January	24.0	-0.4	24.2	0.0	27.3	-1.1	43.8	24.9
February	23.8	-0.8	23.9	-1.2	26.8	-1.8	42.0	24.7
March	23.8	0.0	24.0	0.4	26.8	0.0	42.7	24.6
April	23.9	0.4	24.0	0.0	26.5	-1.1	42.7	24.3
May	23.8	-0.4	24.0	0.0	26.3	-0.8	42.7	24.0
June	23.9	0.4	24.0	0.0	26.0	-1.1	41.8	23.8
July	23.7	-0.8	23.8	-0.8	26.0	0.0	41.7	23.7
August	23.8	0.4	23.9	0.4	26.0	0.0	41.7	23.8
September	23.9	0.4	24.0	0.4	26.1	0.4	41.8	23.8
October	23.7	-0.8	23.9	-0.4	26.0	-0.4	41.0	23.8
November	23.8	0.4	23.9	0.0	26.0	0.0	40.9	23.8
December	23.6	-0.8	23.8	-0.4	25.9	-0.4	40.3	23.8

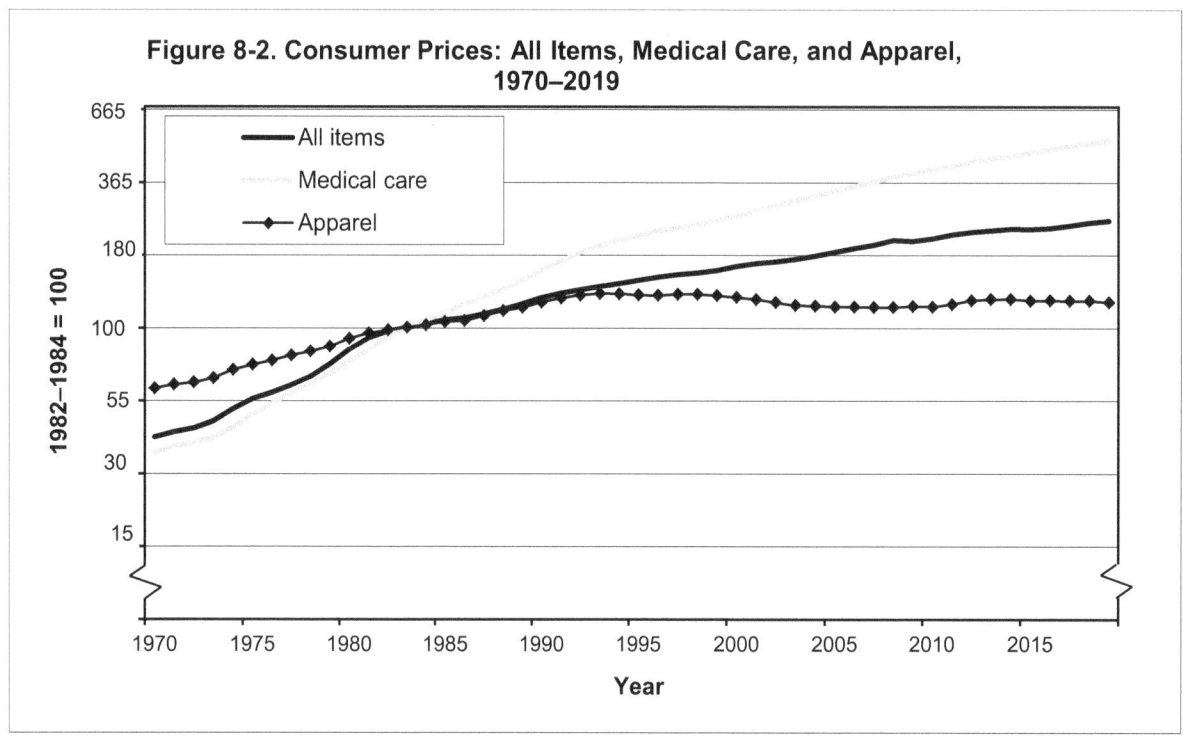

Figure 8-2. Consumer Prices: All Items, Medical Care, and Apparel, 1970–2019

- Figure 8-2 charts two components of the Consumer Price Index for All Urban Consumers (CPI-U) along with the total for all items. Since all three indexes have the base years 1982–1984, they converge around 100 in those years. However, over the entire postwar period, the trends of the two components are very different. (Table 8-2)

- Apparel is one of the commodity groups most subject to international competition. It has grown far less rapidly than prices in general.

- Medical care, on the other hand, has little price competition from producers in other countries. It is often paid for by third-party insurers, both government and private, rather than directly by consumers. Furthermore, it is characterized by trend growth in demand, due to rising income and expectations and to technological progress. All of these economic factors cause medical care prices to rise faster than the general price level.

- A broader system of Producer Price Indexes including services and construction as well as goods was introduced in 2014, featuring a more detailed representation by stages of production. These new measures only go back to November 2009—not enough time yet to determine what kind of signals about future inflation they will contain. (Tables 8-5 and 8-6)

Table 8-2. Consumer Price Indexes and Purchasing Power of the Dollar

(1982–1984 = 100, seasonally adjusted, except as noted.)

| Year and month | CPI-U, All items | | | CPI-U, Food and beverages | | | | | | | | | | |
| | Not seasonally adjusted | Seasonally adjusted | | Total | Total food | Food at home | | | | | | | Food away from home [1] | Alcoholic beverages |
		Index	Percent change from previous period			Total	Cereals and bakery products	Meats, poultry, fish, and eggs	Dairy and related products [1]	Fruits and vegetables	Non-alcoholic beverages	Other food at home		
1970	38.8	38.8	5.6	40.1	39.2	39.9	37.1	44.6	44.7	37.8	27.1	32.9	37.5	52.1
1971	40.5	40.5	4.4	41.4	40.4	40.9	38.8	44.1	46.1	39.7	28.1	34.3	39.4	54.2
1972	41.8	41.8	3.2	43.1	42.1	42.7	39.0	48.0	46.8	41.6	28.0	34.6	41.0	55.4
1973	44.4	44.4	6.2	48.8	48.2	49.7	43.5	60.9	51.2	47.4	30.1	36.7	44.2	56.8
1974	49.3	49.3	11.0	55.5	55.1	57.1	56.5	62.2	60.7	55.2	35.9	47.8	49.8	61.1
1975	53.8	53.8	9.1	60.2	59.8	61.8	62.9	67.0	62.6	56.9	41.3	55.4	54.5	65.9
1976	56.9	56.9	5.8	62.1	61.6	63.1	61.5	68.0	67.7	58.4	49.4	56.4	58.2	68.1
1977	60.6	60.6	6.5	65.8	65.5	66.8	62.5	67.4	69.5	63.8	74.4	68.4	62.6	70.0
1978	65.2	65.2	7.6	72.2	72.0	73.8	68.1	77.6	74.2	70.9	78.7	73.6	68.3	74.1
1979	72.6	72.6	11.3	79.9	79.9	81.8	74.9	89.0	82.8	76.6	82.6	79.0	75.9	79.9
1980	82.4	82.4	13.5	86.7	86.8	88.4	83.9	92.0	90.9	82.1	91.4	88.4	83.4	86.4
1981	90.9	90.9	10.3	93.5	93.6	94.8	92.3	96.0	97.4	92.0	95.3	94.9	90.9	92.5
1982	96.5	96.5	6.2	97.3	97.4	98.1	96.5	99.6	98.8	97.0	97.9	97.3	95.8	96.7
1983	99.6	99.6	3.2	99.5	99.4	99.1	99.6	99.2	100.0	97.3	99.8	99.5	100.0	100.4
1984	103.9	103.9	4.3	103.2	103.2	102.8	103.9	101.3	101.3	105.7	102.3	103.1	104.2	103.0
1985	107.6	107.6	3.6	105.6	105.6	104.3	107.9	100.1	103.2	108.4	104.3	105.7	108.3	106.4
1986	109.6	109.6	1.9	109.1	109.0	107.3	110.9	104.5	103.3	109.4	110.4	109.4	112.5	111.1
1987	113.6	113.6	3.6	113.5	113.5	111.9	114.8	110.5	105.9	119.1	107.5	110.5	117.0	114.1
1988	118.3	118.3	4.1	118.2	118.2	116.6	122.1	114.3	108.4	128.1	107.5	113.1	121.8	118.6
1989	124.0	124.0	4.8	124.9	125.1	124.2	132.4	121.3	115.6	138.0	111.3	119.1	127.4	123.5
1990	130.7	130.7	5.4	132.1	132.4	132.3	140.0	130.0	126.5	149.0	113.5	123.4	133.4	129.3
1991	136.2	136.2	4.2	136.8	136.3	135.8	145.8	132.6	125.1	155.8	114.1	127.3	137.9	142.8
1992	140.3	140.3	3.0	138.7	137.9	136.8	151.5	130.9	128.5	155.4	114.3	128.8	140.7	147.3
1993	144.5	144.5	3.0	141.6	140.9	140.1	156.6	135.5	129.4	159.0	114.6	130.5	143.2	149.6
1994	148.2	148.2	2.6	144.9	144.3	144.1	163.0	137.2	131.7	165.0	123.2	135.6	145.7	151.5
1995	152.4	152.4	2.8	148.9	148.4	148.8	167.5	138.8	132.8	177.7	131.7	140.8	149.0	153.9
1996	156.9	156.9	3.0	153.7	153.3	154.3	174.0	144.8	142.1	183.9	128.6	142.9	152.7	158.5
1997	160.5	160.5	2.3	157.7	157.3	158.1	177.6	148.5	145.5	187.5	133.4	147.3	157.0	162.8
1998	163.0	163.0	1.6	161.1	160.7	161.1	181.1	147.3	150.8	198.2	133.0	150.8	161.1	165.7
1999	166.6	166.6	2.2	164.6	164.1	164.2	185.0	147.9	159.6	203.1	134.3	153.5	165.1	169.7
2000	172.2	172.2	3.4	168.4	167.8	167.9	188.3	154.5	160.7	204.6	137.8	155.6	169.0	174.7
2001	177.1	177.1	2.8	173.6	173.1	173.4	193.8	161.3	167.1	212.2	139.2	159.6	173.9	179.3
2002	179.9	179.9	1.6	176.8	176.2	175.6	198.0	162.1	168.1	220.9	139.2	160.8	178.3	183.6
2003	184.0	184.0	2.3	180.5	180.0	179.4	202.8	169.3	167.9	225.9	139.8	162.6	182.1	187.2
2004	188.9	188.9	2.7	186.6	186.2	186.2	206.0	181.7	180.2	232.7	140.4	164.9	187.5	192.1
2005	195.3	195.3	3.4	191.2	190.7	189.8	209.0	184.7	182.4	241.4	144.4	167.0	193.4	195.9
2006	201.6	201.6	3.2	195.7	195.2	193.1	212.8	186.6	181.4	252.9	147.4	169.6	199.4	200.7
2007	207.3	207.3	2.8	203.3	202.9	201.2	222.1	195.6	194.8	262.6	153.4	173.3	206.7	207.0
2008	215.3	215.3	3.8	214.2	214.1	214.1	244.9	204.7	210.4	278.9	160.0	184.2	215.8	214.5
2009	214.5	214.5	-0.4	218.2	218.0	215.1	252.6	203.8	197.0	272.9	163.0	191.2	223.3	220.8
2010	218.1	218.1	1.6	220.0	219.6	215.8	250.4	207.7	199.2	273.5	161.6	191.1	226.1	223.3
2011	224.9	224.9	3.2	227.9	227.8	226.2	260.3	223.2	212.7	284.7	166.8	197.4	231.4	226.7
2012	229.6	229.6	2.1	233.7	233.8	231.8	267.7	231.0	217.3	282.8	168.6	204.8	238.0	230.8
2013	233.0	233.0	1.5	237.0	237.0	233.9	270.4	236.0	217.6	290.0	166.9	204.8	243.1	234.6
2014	236.7	236.7	1.6	242.4	242.7	239.5	271.1	253.0	225.3	294.4	166.0	206.2	249.0	237.3
2015	237.0	237.0	0.1	246.8	247.2	242.3	274.1	260.3	222.4	293.8	167.9	209.3	256.1	239.5
2016	240.0	240.0	1.3	247.7	247.9	239.1	273.1	247.7	217.3	296.3	167.3	209.5	262.7	242.5
2017	245.1	245.1	2.1	249.8	250.1	238.6	271.7	245.8	217.5	295.7	167.6	209.9	268.8	245.4
2018	251.1	251.1	2.4	253.3	253.6	239.7	272.8	248.9	216.5	297.8	167.6	210.2	275.9	249.1
2019	255.7	255.7	1.8	258.0	258.3	241.8	276.6	249.8	218.7	300.9	170.8	211.1	284.4	252.5
2018														
January	247.9	248.8	2.1	251.8	252.0	239.3	271.6	248.2	216.4	297.9	166.6	210.4	272.8	247.1
February	249.0	249.5	2.2	251.9	252.1	238.9	271.3	248.0	216.0	296.6	166.7	210.1	273.4	247.5
March	249.6	249.4	2.3	252.1	252.3	239.0	271.8	249.6	216.3	294.4	167.1	209.9	273.7	247.9
April	250.5	250.0	2.4	252.8	253.0	239.8	271.7	251.6	216.9	296.9	166.6	210.0	274.4	248.4
May	251.6	250.6	2.8	252.9	253.1	239.3	271.6	249.8	216.5	296.5	167.0	209.7	275.3	248.3
June	252.0	251.2	2.9	253.3	253.5	239.7	273.3	248.6	217.9	297.7	167.2	209.9	275.8	249.2
July	252.0	251.5	2.9	253.6	253.9	240.0	272.7	249.6	216.8	299.6	167.4	210.1	276.1	249.1
August	252.1	251.9	2.7	253.9	254.1	240.1	273.0	249.6	216.9	299.3	167.8	210.2	276.6	249.0
September	252.4	252.3	2.4	254.1	254.3	240.0	273.7	248.1	216.5	299.1	168.5	210.5	277.3	250.1
October	252.9	252.8	2.5	254.0	254.1	239.5	272.7	247.7	216.0	296.8	169.0	210.4	277.5	250.5
November	252.0	252.7	2.2	254.5	254.7	239.9	274.3	248.4	215.5	297.3	168.3	210.9	278.3	251.1
December	251.2	252.7	1.9	255.4	255.6	240.6	275.6	248.1	215.9	301.6	169.0	210.4	279.4	251.4
2019														
January	251.7	252.6	1.5	255.9	256.1	240.8	274.4	249.0	215.7	300.8	170.3	210.7	280.4	251.4
February	252.8	253.2	1.5	256.8	257.0	241.7	276.1	249.5	216.3	302.6	171.4	211.2	281.4	251.9
March	254.2	254.1	1.9	257.3	257.6	242.4	276.7	249.0	217.3	305.9	171.3	211.5	281.9	251.7
April	255.5	254.9	2.0	257.1	257.5	241.5	276.5	248.7	217.5	303.3	170.3	210.4	282.8	251.3
May	256.1	255.2	1.8	257.8	258.1	242.1	277.4	250.5	218.6	301.7	171.8	210.4	283.4	252.3
June	256.1	255.4	1.7	258.1	258.3	241.9	276.5	249.1	219.0	300.7	171.0	211.8	284.3	252.8
July	256.6	256.1	1.8	258.1	258.4	241.5	276.9	248.9	218.7	300.9	170.6	210.3	284.9	253.8
August	256.6	256.3	1.7	258.2	258.5	241.2	276.3	248.1	219.2	300.2	170.6	210.9	285.5	253.9
September	256.8	256.6	1.7	258.6	258.9	241.5	277.2	249.1	219.7	298.2	170.7	211.5	286.2	253.4
October	257.3	257.2	1.8	259.1	259.5	242.0	277.0	250.5	220.1	299.7	170.3	211.7	286.8	253.0
November	257.2	257.8	2.0	259.4	259.8	242.3	277.3	251.1	221.1	298.7	171.1	211.7	287.3	252.2
December	257.0	258.4	2.3	259.8	260.2	242.4	276.4	253.8	221.2	297.7	170.7	211.1	288.1	252.6

[1] Not seasonally adjusted.

Table 8-2. Consumer Price Indexes and Purchasing Power of the Dollar—*Continued*

(1982–1984 = 100, seasonally adjusted, except as noted.)

Year and month	Total	Shelter						Fuels and utilities				Household furnishings and operations	
		Total	Rent of shelter [2]	Rent of primary residence	Lodging away from home [3]	Owners' equivalent rent of primary residence [2]	Tenants' and household insurance [1,3]	Total	Fuels		Water and sewer and trash collection services [3]	Total	Household operations [1,3]
									Fuel oil and other fuels	Energy services			
1970	36.4	35.5	. . .	46.5	. . .	. . .	. . .	29.1	17.0	25.4	. . .	46.8	. . .
1971	38.0	37.0	. . .	48.7	. . .	. . .	. . .	31.1	18.2	27.1	. . .	48.6	. . .
1972	39.4	38.7	. . .	50.4	. . .	. . .	. . .	32.5	18.3	28.5	. . .	49.7	. . .
1973	41.2	40.5	. . .	52.5	. . .	. . .	. . .	34.3	21.1	29.9	. . .	51.1	. . .
1974	45.8	44.4	. . .	55.2	. . .	. . .	. . .	40.7	33.2	34.5	. . .	56.8	. . .
1975	50.7	48.8	. . .	58.0	. . .	. . .	. . .	45.4	36.4	40.1	. . .	63.4	. . .
1976	53.8	51.5	. . .	61.1	. . .	. . .	. . .	49.4	38.8	44.7	. . .	67.3	. . .
1977	57.4	54.9	. . .	64.8	. . .	. . .	. . .	54.7	43.9	50.5	. . .	70.4	. . .
1978	62.4	60.5	. . .	69.3	. . .	. . .	. . .	58.5	46.2	55.0	. . .	74.7	. . .
1979	70.1	68.9	. . .	74.3	. . .	. . .	. . .	64.8	62.4	61.0	. . .	79.9	. . .
1980	81.1	81.0	. . .	80.9	. . .	. . .	. . .	75.4	86.1	71.4	. . .	86.3	. . .
1981	90.4	90.5	. . .	87.9	. . .	. . .	. . .	86.4	104.6	81.9	. . .	93.0	. . .
1982	96.9	96.9	. . .	94.6	. . .	. . .	. . .	94.9	103.4	93.2	. . .	98.0	. . .
1983	99.5	99.1	102.7	100.1	. . .	102.5	. . .	100.2	97.2	101.5	. . .	100.2	. . .
1984	103.6	104.0	107.7	105.3	. . .	107.3	. . .	104.8	99.4	105.4	. . .	101.9	. . .
1985	107.7	109.8	113.9	111.8	. . .	113.2	. . .	106.5	95.9	107.1	. . .	103.8	. . .
1986	110.9	115.8	120.2	118.3	. . .	119.4	. . .	104.1	77.6	105.7	. . .	105.2	. . .
1987	114.2	121.3	125.9	123.1	. . .	124.8	. . .	103.0	77.9	103.8	. . .	107.1	. . .
1988	118.5	127.1	132.0	127.8	. . .	131.1	. . .	104.4	78.1	104.6	. . .	109.4	. . .
1989	123.0	132.8	138.0	132.8	. . .	137.4	. . .	107.8	81.7	107.5	. . .	111.2	. . .
1990	128.5	140.0	145.5	138.4	. . .	144.8	. . .	111.6	99.3	109.3	. . .	113.3	. . .
1991	133.6	146.3	152.1	143.3	. . .	150.4	. . .	115.3	94.6	112.6	. . .	116.0	. . .
1992	137.5	151.2	157.3	146.9	. . .	155.5	. . .	117.8	90.7	114.8	. . .	118.0	. . .
1993	141.2	155.7	162.0	150.3	. . .	160.5	. . .	121.3	90.3	118.5	. . .	119.3	. . .
1994	144.8	160.5	167.0	154.0	. . .	165.8	. . .	122.8	88.8	119.2	. . .	121.0	. . .
1995	148.5	165.7	172.4	157.8	. . .	171.3	. . .	123.7	88.1	119.2	. . .	123.0	. . .
1996	152.8	171.0	178.0	162.0	. . .	176.8	. . .	127.5	99.2	122.1	. . .	124.7	. . .
1997	156.8	176.3	183.4	166.7	. . .	181.9	. . .	130.8	99.8	125.1	. . .	125.4	. . .
1998	160.4	182.1	189.6	172.1	109.0	187.8	99.8	128.5	90.0	121.2	101.6	126.6	101.5
1999	163.9	187.3	195.0	177.5	112.3	192.9	101.3	128.8	91.4	120.9	104.0	126.7	104.5
2000	169.6	193.4	201.3	183.9	117.5	198.7	103.7	137.9	129.7	128.0	106.5	128.2	110.5
2001	176.4	200.6	208.9	192.1	118.6	206.3	106.2	150.2	129.3	142.4	109.6	129.1	115.6
2002	180.3	208.1	216.7	199.7	118.3	214.7	108.7	143.6	115.5	134.4	113.0	128.3	119.0
2003	184.8	213.1	221.9	205.5	119.3	219.9	114.8	154.5	139.5	145.0	117.2	126.1	121.8
2004	189.5	218.8	227.9	211.0	125.9	224.9	116.2	161.9	160.5	150.6	124.0	125.5	125.0
2005	195.7	224.4	233.7	217.3	130.3	230.2	117.6	179.0	208.6	166.5	130.3	126.1	130.3
2006	203.2	232.1	241.9	225.1	136.0	238.2	116.5	194.7	234.9	182.1	136.8	127.0	136.6
2007	209.6	240.6	250.8	234.7	142.8	246.2	117.0	200.6	251.5	186.3	143.7	126.9	140.6
2008	216.3	246.7	257.2	243.3	143.7	252.4	118.8	220.0	334.4	202.2	152.1	127.8	147.5
2009	217.1	249.4	259.9	248.8	134.2	256.6	121.5	210.7	239.8	193.6	161.1	128.7	150.3
2010	216.3	248.4	258.8	249.4	133.7	256.6	125.7	214.2	275.1	192.9	170.9	125.5	150.3
2011	219.1	251.6	262.2	253.6	137.4	259.6	127.4	220.4	337.1	194.4	179.6	124.9	151.8
2012	222.7	257.1	267.8	260.4	140.5	264.8	131.3	219.0	335.9	189.7	189.3	125.7	155.2
2013	227.4	263.1	274.0	267.7	142.4	270.7	135.4	225.2	332.0	194.8	197.6	124.8	157.6
2014	233.2	270.5	281.8	276.2	148.5	277.8	141.9	234.6	338.9	203.4	204.9	123.1	161.6
2015	238.1	278.8	290.4	286.0	153.0	285.9	146.4	230.1	256.2	198.7	214.0	122.6	166.9
2016	244.0	288.2	300.3	296.8	158.1	295.4	147.7	228.9	226.3	196.1	221.7	121.6	171.6
2017	251.2	297.8	310.3	308.1	159.4	305.1	148.8	237.3	254.4	202.7	229.1	120.7	176.3
2018	258.5	307.7	320.7	319.3	161.1	315.2	150.7	241.6	293.0	203.8	237.1	121.6	186.0
2019	266.0	318.1	331.6	331.1	165.9	325.6	151.8	242.6	280.8	203.5	244.7	123.8	194.5
2018													
January	255.2	303.2	316.0	314.5	159.2	310.6	149.1	240.0	291.3	203.3	232.7	120.7	181.3
February	255.8	303.8	316.6	315.2	158.7	311.2	149.3	242.0	286.2	205.6	233.5	120.9	184.1
March	256.5	304.8	317.7	315.9	161.3	312.2	149.3	241.3	284.8	204.7	234.0	121.1	184.4
April	257.1	305.7	318.6	316.9	161.2	313.1	149.4	241.1	288.1	204.1	234.9	121.7	184.7
May	257.8	306.7	319.8	317.9	165.2	314.0	150.6	241.5	288.7	204.3	235.9	121.3	185.7
June	258.0	307.2	320.3	318.6	160.7	314.7	150.9	240.9	294.8	203.2	236.6	121.1	186.2
July	258.6	308.1	321.2	319.6	160.4	315.7	151.8	240.6	295.2	202.6	237.3	121.5	186.9
August	259.4	309.0	322.1	320.8	162.4	316.5	152.0	241.8	300.7	203.4	238.4	121.5	186.9
September	259.6	309.5	322.7	321.6	160.3	317.1	151.8	240.7	300.7	202.1	238.7	121.7	186.8
October	260.5	310.4	323.4	322.5	160.5	318.0	151.5	241.9	306.9	203.2	239.2	122.1	186.7
November	261.1	311.3	324.4	323.6	159.6	319.0	151.7	242.5	300.3	203.3	241.9	122.4	187.2
December	262.1	312.2	325.3	324.3	163.2	319.7	151.5	244.6	279.5	206.4	242.6	122.8	190.9
2019													
January	262.6	313.0	326.2	325.3	163.5	320.6	151.6	243.4	273.7	205.6	241.3	123.2	191.6
February	263.2	314.0	327.3	326.3	165.2	321.6	151.7	242.5	279.2	204.2	241.6	123.3	191.0
March	263.9	315.1	328.5	327.6	165.8	322.6	151.5	242.9	282.8	204.3	242.4	123.3	190.7
April	264.6	316.2	329.7	328.8	167.5	323.6	151.4	242.8	284.1	203.9	243.2	123.0	191.8
May	265.0	317.0	330.6	329.7	167.2	324.5	151.6	242.0	285.2	202.9	243.8	123.1	190.6
June	265.8	318.0	331.6	330.9	167.0	325.5	152.2	241.3	280.8	202.0	244.4	123.9	195.9
July	266.5	318.8	332.5	331.9	167.7	326.3	152.2	241.8	280.7	202.4	245.0	124.4	197.1
August	266.8	319.4	333.1	332.7	165.6	327.1	151.8	241.5	279.0	202.1	245.5	124.2	196.9
September	267.6	320.4	334.2	333.9	168.1	327.9	151.8	241.8	276.7	202.3	246.1	124.5	196.4
October	268.0	320.8	334.4	334.5	164.3	328.6	151.9	243.4	279.6	203.8	247.0	124.4	197.0
November	268.6	321.6	335.3	335.4	165.2	329.4	151.8	244.0	281.8	204.3	247.4	124.3	197.3
December	268.9	322.3	336.0	336.3	162.9	330.2	151.7	243.8	286.1	203.8	248.0	124.1	197.5

[1] Not seasonally adjusted.
[2] December 1982 = 100.
[3] December 1997 = 100.
. . . = Not available.

Table 8-2. Consumer Price Indexes and Purchasing Power of the Dollar—*Continued*

(1982–1984 = 100, seasonally adjusted, except as noted.)

Year and month	CPI-U, Apparel					CPI-U, Transportation								
							Private transportation						Motor vehicle parts and equipment	Motor vehicle maintenance and repair [1]
								New and used motor vehicles			Motor fuel			
	Total	Men's and boys' apparel	Women's and girls' apparel	Infants' and toddlers' apparel	Footwear	Total	Total	Total [3]	New vehicles	Used cars and trucks	Total	Gasoline (all types)		
1970	59.2	62.2	71.8	39.2	56.8	37.5	37.5	...	53.1	31.2	27.9	27.9	...	36.6
1971	61.1	63.9	74.4	40.0	58.6	39.5	39.4	...	55.3	33.0	28.1	28.1	...	39.3
1972	62.3	64.7	76.2	41.1	60.3	39.9	39.7	...	54.8	33.1	28.4	28.4	...	41.1
1973	64.6	67.1	78.8	42.5	62.8	41.2	41.0	...	54.8	35.2	31.2	31.2	...	43.2
1974	69.4	72.4	83.5	54.2	66.6	45.8	46.2	...	58.0	36.7	42.2	42.2	...	47.6
1975	72.5	75.5	85.5	64.5	69.6	50.1	50.6	...	63.0	43.8	45.1	45.1	...	53.7
1976	75.2	78.1	87.9	68.0	72.3	55.1	55.6	...	67.0	50.3	47.0	47.0	...	57.6
1977	78.6	81.7	90.6	74.6	75.7	59.0	59.7	...	70.5	54.7	49.7	49.7	...	61.9
1978	81.4	83.5	92.4	77.4	79.0	61.7	62.5	...	75.9	55.8	51.8	51.8	77.6	67.0
1979	84.9	85.4	94.0	79.0	85.3	70.5	71.7	...	81.9	60.2	70.1	70.2	85.1	73.7
1980	90.9	89.4	96.0	85.5	91.8	83.1	84.2	...	88.5	62.3	97.4	97.5	95.3	81.5
1981	95.3	94.2	97.5	92.9	96.7	93.2	93.8	...	93.9	76.9	108.5	108.5	101.0	89.2
1982	97.8	97.6	98.5	96.3	99.1	97.0	97.1	...	97.5	88.8	102.8	102.8	103.6	96.0
1983	100.2	100.3	100.2	101.1	99.8	99.3	99.3	...	99.9	98.7	99.4	99.4	100.7	100.3
1984	102.1	102.1	101.3	102.6	101.1	103.7	103.6	...	102.6	112.5	97.9	97.8	95.6	103.8
1985	105.0	105.0	104.9	107.2	102.3	106.4	106.2	...	106.1	113.7	98.7	98.6	95.9	106.8
1986	105.9	106.2	104.0	111.8	101.9	102.3	101.2	...	110.6	108.8	77.1	77.0	95.4	110.3
1987	110.6	109.1	110.4	112.1	105.1	105.4	104.2	...	114.4	113.1	80.2	80.1	96.1	114.8
1988	115.4	113.4	114.9	116.4	109.9	108.7	107.6	...	116.5	118.0	80.9	80.8	97.9	119.7
1989	118.6	117.0	116.4	119.1	114.4	114.1	112.9	...	119.2	120.4	88.5	88.5	100.2	124.9
1990	124.1	120.4	122.6	125.8	117.4	120.5	118.8	...	121.4	117.6	101.2	101.0	100.9	130.1
1991	128.7	124.2	127.6	128.9	120.9	123.8	121.9	...	126.0	118.1	99.4	99.2	102.2	136.0
1992	131.9	126.5	130.4	129.3	125.0	126.5	124.6	...	129.2	123.2	99.0	99.0	103.1	141.3
1993	133.7	127.5	132.6	127.1	125.9	130.4	127.5	91.8	132.7	133.9	98.0	97.7	101.6	145.9
1994	133.4	126.4	130.9	128.1	126.0	134.3	131.4	95.5	137.6	141.7	98.5	98.2	101.4	150.2
1995	132.0	126.2	126.9	127.2	125.4	139.1	136.3	99.4	141.0	156.5	100.0	99.8	102.1	154.0
1996	131.7	127.7	124.7	129.7	126.6	143.0	140.0	101.0	143.7	157.0	106.3	105.9	102.2	158.4
1997	132.9	130.1	126.1	129.0	127.6	144.3	141.0	100.5	144.3	151.1	106.2	105.8	101.9	162.7
1998	133.0	131.8	126.0	126.1	128.0	141.6	137.9	100.1	143.4	150.6	92.2	91.6	101.1	167.1
1999	131.3	131.1	123.3	129.0	125.7	144.4	140.5	100.1	142.9	152.0	100.7	100.1	100.5	171.9
2000	129.6	129.7	121.5	130.6	123.8	153.3	149.1	100.8	142.8	155.8	129.3	128.6	101.5	177.3
2001	127.3	125.7	119.3	129.2	123.0	154.3	150.0	101.3	142.1	158.7	124.7	124.0	104.8	183.5
2002	124.0	121.7	115.8	126.4	121.4	152.9	148.8	99.2	140.0	152.0	116.6	116.0	106.9	190.2
2003	120.9	118.0	113.1	122.1	119.6	157.6	153.6	96.5	137.9	142.9	135.8	135.1	107.8	195.6
2004	120.4	117.5	113.0	118.5	119.3	163.1	159.4	94.2	137.1	133.3	160.4	159.7	108.7	200.2
2005	119.5	116.1	110.8	116.7	122.6	173.9	170.2	95.6	137.9	139.4	195.7	194.7	111.9	206.9
2006	119.5	114.1	110.7	116.5	123.5	180.9	177.0	95.6	137.6	140.0	221.0	219.9	117.3	215.6
2007	119.0	112.4	110.3	113.9	122.4	184.7	180.8	94.3	136.3	135.7	239.1	238.0	121.6	223.0
2008	118.9	113.0	107.5	113.8	124.2	195.5	191.0	93.3	134.2	134.0	279.7	277.5	128.7	233.9
2009	120.1	113.6	108.1	114.5	126.9	179.3	174.8	93.5	135.6	127.0	202.0	201.6	134.1	243.3
2010	119.5	111.9	107.1	114.2	128.0	193.4	188.7	97.1	138.0	143.1	239.2	238.6	137.0	248.0
2011	122.1	114.7	109.2	113.6	128.5	212.4	207.6	99.8	141.9	149.0	302.6	301.7	143.9	253.1
2012	126.3	119.5	113.0	119.7	131.8	217.3	212.8	100.6	144.2	150.3	312.7	311.5	148.6	257.6
2013	127.4	121.6	113.3	116.5	135.0	217.4	212.4	100.9	145.8	149.9	303.9	302.6	146.4	261.6
2014	127.5	120.6	114.4	117.6	135.5	215.9	211.0	100.8	146.3	149.1	292.4	290.9	144.8	266.0
2015	125.9	119.6	111.2	119.7	136.8	199.1	193.7	100.8	147.1	147.1	213.1	212.0	144.2	270.7
2016	126.0	118.8	111.2	115.9	137.3	194.9	189.5	100.2	147.4	143.5	188.4	187.6	143.6	275.4
2017	125.6	117.2	111.0	114.9	136.7	201.6	196.6	98.9	147.0	138.3	212.7	211.8	143.0	280.8
2018	125.7	118.1	110.4	120.3	136.1	210.7	206.4	99.1	146.3	138.4	241.9	240.6	143.7	286.4
2019	124.1	118.5	106.7	119.3	136.7	210.1	205.8	99.5	146.8	139.8	233.2	232.0	146.4	296.0
2018														
January	125.5	117.5	111.2	116.5	135.1	208.7	204.2	98.9	146.5	137.8	237.9	236.8	142.6	283.3
February	126.9	118.2	113.0	116.5	136.5	210.1	205.8	98.8	145.9	137.8	242.5	241.4	142.8	284.0
March	126.8	117.7	111.6	118.8	138.4	208.0	203.5	98.7	145.9	137.6	232.7	231.5	143.3	283.7
April	127.0	119.0	112.3	119.9	137.0	208.6	204.2	98.5	145.3	137.7	236.2	235.1	142.8	284.4
May	127.2	118.0	113.2	121.8	136.6	209.9	205.7	98.7	145.8	137.7	241.2	240.0	143.7	284.9
June	126.1	118.1	111.4	123.7	136.1	211.4	207.4	99.0	146.2	138.0	246.6	245.3	143.8	285.8
July	125.4	118.1	109.6	121.5	136.9	211.6	207.4	99.4	146.7	138.5	244.7	243.4	143.8	286.1
August	123.6	115.4	108.0	119.0	136.0	212.8	208.7	99.5	146.7	138.7	249.6	248.3	143.9	286.9
September	125.0	117.9	109.6	120.9	135.0	212.8	208.7	99.4	146.7	138.4	248.9	247.6	143.7	288.5
October	125.2	119.1	109.1	121.9	135.0	213.8	209.8	99.3	146.6	138.5	253.2	251.9	144.0	289.1
November	124.8	119.2	108.8	121.3	135.0	211.4	207.3	99.7	146.6	140.0	240.5	239.0	144.5	290.1
December	124.6	119.3	107.8	122.2	135.2	208.3	204.1	99.7	146.6	139.7	226.8	225.3	145.1	289.7
2019														
January	125.6	118.6	109.1	120.0	138.2	205.4	201.1	99.7	146.5	139.8	213.8	212.3	145.6	290.8
February	125.7	120.5	108.6	122.1	138.3	206.7	202.4	99.3	146.3	139.4	220.5	219.1	145.5	292.0
March	124.0	118.8	106.8	121.7	136.0	209.3	205.1	99.2	146.9	138.5	232.1	230.7	145.8	294.2
April	123.3	117.8	106.6	121.0	134.1	212.0	207.9	99.4	147.1	139.2	243.0	241.9	145.6	295.3
May	123.3	117.2	106.8	121.1	134.5	211.2	207.0	99.1	147.1	138.2	240.8	239.6	146.4	294.6
June	124.5	118.2	107.3	120.3	136.1	210.1	205.8	99.5	147.1	139.9	233.8	232.5	146.6	295.7
July	124.7	118.6	107.9	119.2	136.2	211.1	206.8	99.7	147.1	140.5	237.3	236.1	145.8	295.4
August	124.8	119.0	107.8	118.9	137.4	210.2	205.8	99.7	147.0	141.1	231.7	230.5	146.4	297.9
September	124.5	120.6	106.4	118.2	137.5	209.8	205.3	99.9	146.9	142.0	228.2	227.0	147.3	298.4
October	122.4	118.4	103.5	116.4	137.1	210.7	206.4	99.5	146.8	140.3	234.2	233.1	147.0	298.9
November	123.1	117.5	105.0	116.6	137.6	211.1	206.8	99.2	146.6	139.3	237.2	236.0	147.6	299.3
December	123.2	117.2	105.5	116.6	137.6	212.6	208.4	99.1	146.7	138.7	244.6	243.4	147.8	299.6

[1]Not seasonally adjusted.
[3]December 1997 = 100.
. . . = Not available.

Table 8-2. Consumer Price Indexes and Purchasing Power of the Dollar—*Continued*

(1982–1984 = 100, seasonally adjusted, except as noted.)

| Year and month | CPI-U, Transportation—Continued | | CPI-U, Medical care | | | | | CPI-U, Recreation | | CPI-U, Education and communication | | | |
| | Public transportation | Transportation services | Medical care, total | Medical care commodities | Medical care services | | | Total [3] | Video and audio [3] | Total [3] | Education | | |
					Total	Professional services	Hospital and related services				Total [3]	Educational books and supplies	Tuition, other school fees, and childcare
1970	35.2	40.2	34.0	46.5	32.3	37.0	. . .	. . .	. . .	. . .	. . .	38.8	. . .
1971	37.8	43.4	36.1	47.3	34.7	39.4	. . .	. . .	. . .	. . .	. . .	41.4	. . .
1972	39.3	44.4	37.3	47.4	35.9	40.8	. . .	. . .	. . .	. . .	. . .	44.2	. . .
1973	39.7	44.7	38.8	47.5	37.5	42.2	. . .	. . .	. . .	. . .	. . .	45.6	. . .
1974	40.6	46.3	42.4	49.2	41.4	45.8	. . .	. . .	. . .	. . .	. . .	47.2	. . .
1975	43.5	49.8	47.5	53.3	46.6	50.8	. . .	. . .	. . .	. . .	. . .	50.3	. . .
1976	47.8	56.9	52.0	56.5	51.3	55.5	. . .	. . .	. . .	. . .	. . .	53.7	. . .
1977	50.0	61.5	57.0	60.2	56.4	60.0	. . .	. . .	. . .	. . .	. . .	56.9	. . .
1978	51.5	64.4	61.8	64.4	61.2	64.5	55.1	. . .	. . .	. . .	. . .	61.6	59.8
1979	54.9	69.5	67.5	69.0	67.2	70.1	61.0	. . .	. . .	. . .	. . .	65.7	64.7
1980	69.0	79.2	74.9	75.4	74.8	77.9	69.2	. . .	. . .	. . .	. . .	71.4	71.2
1981	85.6	88.6	82.9	83.7	82.8	85.9	79.1	. . .	. . .	. . .	. . .	80.3	79.9
1982	94.9	96.1	92.5	92.3	92.6	93.2	90.3	. . .	. . .	. . .	. . .	91.0	90.5
1983	99.5	99.1	100.6	100.2	100.7	99.8	100.5	. . .	. . .	. . .	. . .	100.3	99.7
1984	105.7	104.8	106.8	107.5	106.7	107.0	109.2	. . .	. . .	. . .	. . .	108.7	109.8
1985	110.5	110.0	113.5	115.2	113.2	113.5	116.1	. . .	. . .	. . .	. . .	118.2	119.7
1986	117.0	116.3	122.0	122.8	121.9	120.8	123.1	. . .	. . .	. . .	. . .	128.1	129.6
1987	121.1	121.9	130.1	131.0	130.0	128.8	131.6	. . .	. . .	. . .	. . .	138.1	140.0
1988	123.3	128.0	138.6	139.9	138.3	137.5	143.9	. . .	. . .	. . .	. . .	148.1	151.0
1989	129.5	135.6	149.3	150.8	148.9	146.4	160.5	. . .	. . .	. . .	. . .	158.0	162.7
1990	142.6	144.2	162.8	163.4	162.7	156.1	178.0	. . .	. . .	. . .	. . .	171.3	175.7
1991	148.9	151.2	177.0	176.8	177.1	165.7	196.1	. . .	. . .	. . .	. . .	180.3	191.4
1992	151.4	155.7	190.1	188.1	190.5	175.8	214.0	. . .	. . .	. . .	. . .	190.3	208.5
1993	167.0	162.9	201.4	195.0	202.9	184.7	231.9	90.7	96.5	85.5	78.4	197.6	225.3
1994	172.0	168.6	211.0	200.7	213.4	192.5	245.6	92.7	95.4	88.8	83.3	205.5	239.8
1995	175.9	175.9	220.5	204.5	224.2	201.0	257.8	94.5	95.1	92.2	88.0	214.4	253.8
1996	181.9	180.5	228.2	210.4	232.4	208.3	269.5	97.4	96.6	95.3	92.7	226.9	267.1
1997	186.7	185.0	234.6	215.3	239.1	215.4	278.4	99.6	99.4	98.4	97.3	238.4	280.4
1998	190.3	187.9	242.1	221.8	246.8	222.2	287.5	101.1	101.1	100.3	102.1	250.8	294.2
1999	197.7	190.7	250.6	230.7	255.1	229.2	299.5	102.0	100.7	101.2	107.0	261.7	308.4
2000	209.6	196.1	260.8	238.1	266.0	237.7	317.3	103.3	101.0	102.5	112.5	279.9	324.0
2001	210.6	201.9	272.8	247.6	278.8	246.5	338.3	104.9	101.5	105.2	118.5	295.9	341.1
2002	207.4	209.1	285.6	256.4	292.9	253.9	367.8	106.2	102.8	107.9	126.0	317.6	362.1
2003	209.3	216.3	297.1	262.8	306.0	261.2	394.8	107.5	103.6	109.8	134.4	335.4	386.7
2004	209.1	220.6	310.1	269.3	321.3	271.5	417.9	108.6	104.2	111.6	143.7	351.0	414.3
2005	217.3	225.7	323.2	276.0	336.7	281.7	439.9	109.4	104.2	113.7	152.7	365.6	440.9
2006	226.6	230.8	336.2	285.9	350.6	289.3	468.1	110.9	104.6	116.8	162.1	388.9	468.1
2007	230.0	233.7	351.1	290.0	369.3	300.8	498.9	111.4	102.9	119.6	171.4	420.4	494.1
2008	250.5	244.1	364.1	296.0	384.9	311.0	534.0	113.3	102.6	123.6	181.3	450.2	522.1
2009	236.3	251.0	375.6	305.1	397.3	319.4	567.9	114.3	101.3	127.4	190.9	482.1	549.0
2010	251.4	259.8	388.4	314.7	411.2	328.2	607.7	113.3	99.1	129.9	199.3	505.6	573.2
2011	269.4	268.0	400.3	324.1	423.8	335.7	641.5	113.4	98.4	131.5	207.8	529.5	597.2
2012	271.4	272.9	414.9	333.6	440.3	342.0	672.1	114.7	99.4	133.8	216.3	562.6	621.0
2013	278.9	280.0	425.1	335.1	454.0	349.5	701.3	115.3	99.7	135.9	224.5	594.7	643.7
2014	276.4	285.3	435.3	343.4	464.8	355.2	733.8	115.5	99.8	137.5	231.9	615.4	664.8
2015	268.7	291.0	446.8	354.6	476.2	361.5	761.9	115.9	99.6	138.2	240.5	648.2	688.8
2016	265.4	299.4	463.7	366.8	494.8	371.5	795.1	117.0	100.9	139.1	247.5	678.8	708.1
2017	263.1	309.9	475.3	377.0	506.8	375.1	831.7	118.5	104.2	136.5	253.2	689.5	724.7
2018	258.8	321.8	484.7	381.4	517.8	378.4	866.9	119.1	104.2	136.8	258.8	696.1	741.3
2019	259.5	325.0	498.4	381.3	536.1	382.6	885.2	120.6	104.7	137.8	265.8	685.4	762.7
2018													
January	260.4	317.4	480.7	380.9	512.6	375.9	856.6	118.9	104.5	136.6	256.0	689.3	733.1
February	261.0	320.2	481.1	380.9	513.2	377.0	855.3	118.8	104.2	136.3	256.5	693.0	734.3
March	262.0	320.6	482.8	381.0	515.4	378.9	859.4	118.9	104.1	136.2	256.3	695.8	733.8
April	260.0	320.5	483.3	379.9	516.5	378.7	861.6	118.6	103.9	136.1	256.7	691.9	735.1
May	257.0	320.5	484.5	383.4	516.8	378.6	864.4	118.7	104.0	136.6	257.6	715.1	736.6
June	256.2	321.1	486.4	385.6	518.7	378.8	870.1	119.0	104.0	136.8	258.1	702.3	738.8
July	258.5	322.4	485.7	382.1	518.9	378.6	872.1	119.2	104.0	137.1	258.8	704.0	740.7
August	258.7	322.7	484.4	380.6	517.6	378.0	870.7	119.0	104.1	137.3	260.2	694.0	745.4
September	258.4	324.0	485.6	381.4	519.0	378.5	871.6	119.3	104.5	137.5	260.6	694.5	746.6
October	258.3	324.2	485.8	380.3	519.7	378.8	869.4	119.1	104.4	137.4	261.1	692.1	748.3
November	257.1	324.1	487.7	382.0	521.6	379.1	873.8	119.5	104.4	136.7	261.7	689.5	750.3
December	256.7	324.0	488.5	379.1	523.7	379.4	877.9	120.1	104.3	136.8	262.4	693.2	752.0
2019													
January	255.7	323.9	489.7	379.6	525.0	380.2	877.2	120.5	104.9	137.0	263.0	690.1	754.0
February	256.7	323.7	489.5	376.9	525.7	380.7	873.7	120.0	104.5	137.2	263.8	691.5	756.2
March	257.3	323.8	491.0	378.7	527.2	380.3	876.1	120.3	104.3	137.3	264.7	689.3	759.0
April	257.6	324.3	492.7	380.7	528.7	380.1	874.5	120.5	104.3	137.4	265.1	689.0	760.5
May	259.4	323.9	494.6	380.6	531.3	381.1	878.2	120.2	104.2	137.6	265.7	685.3	762.4
June	259.6	324.3	495.9	379.8	533.3	382.3	877.5	120.1	104.1	137.6	266.1	686.3	763.4
July	260.4	324.7	498.2	380.4	536.2	382.8	881.2	120.1	104.1	137.9	266.2	681.9	764.1
August	261.7	325.5	501.2	381.2	539.9	383.3	889.5	120.5	104.5	138.1	266.5	690.9	764.5
September	261.9	326.2	502.5	380.6	541.8	384.1	890.4	120.5	104.8	138.0	266.6	681.9	765.2
October	261.2	326.5	506.7	384.3	546.2	384.7	898.6	121.2	105.2	138.2	267.0	678.9	766.7
November	261.2	326.5	508.4	384.4	548.4	385.4	901.6	121.7	105.3	138.6	267.5	680.3	767.9
December	259.7	326.1	510.7	388.3	550.1	385.7	903.3	121.8	105.8	138.7	267.9	680.1	769.1

[3]December 1997 = 100.
. . . = Not available.

Table 8-2. Consumer Price Indexes and Purchasing Power of the Dollar—*Continued*

(1982–1984 = 100, seasonally adjusted except as noted.)

Year and month	CPI-U, Education and communication—*Continued*					CPI-U, Other goods and services					CPI-W, All items, not seasonally adjusted	Purchasing power of the dollar, CPI-U, 1982–1984 = $1.00, not seasonally adjusted
	Communication											
	Total [3]	Information and information processing				Total	Tobacco and smoking products [1]	Personal care				
		Total [3]	Telephone services [1,3]	Information technology, hardware, and services				Total	Personal care products [1]	Personal care services [1]		
				Total [1,4]	Personal computers and peripheral equipment [1,5]							
1970	. . .	. . .	. . .	. . .	. . .	40.9	43.1	43.5	42.7	44.2	39.0	257.4
1971	. . .	. . .	. . .	. . .	. . .	42.9	44.9	44.9	44.0	45.7	40.7	246.6
1972	. . .	. . .	. . .	. . .	. . .	44.7	47.4	46.0	45.2	46.8	42.1	239.1
1973	. . .	. . .	. . .	. . .	. . .	46.4	48.7	48.1	46.4	49.7	44.7	225.1
1974	. . .	. . .	. . .	. . .	. . .	49.8	51.1	52.8	51.5	53.9	49.6	202.9
1975	. . .	. . .	. . .	. . .	. . .	53.9	54.7	57.9	58.0	57.7	54.1	185.9
1976	. . .	. . .	. . .	. . .	. . .	57.0	57.0	61.7	61.3	61.9	57.2	175.7
1977	. . .	. . .	. . .	. . .	. . .	60.4	59.8	65.7	64.7	66.4	60.9	164.9
1978	. . .	. . .	. . .	. . .	. . .	64.3	63.0	69.9	68.2	71.3	65.6	153.2
1979	. . .	. . .	. . .	. . .	. . .	68.9	66.8	75.2	72.9	77.2	73.1	138.0
1980	. . .	. . .	. . .	. . .	. . .	75.2	72.0	81.9	79.6	83.7	82.9	121.5
1981	. . .	. . .	. . .	. . .	. . .	82.6	77.8	89.1	87.8	90.2	91.4	109.8
1982	. . .	. . .	. . .	. . .	. . .	91.1	86.5	95.4	95.1	95.7	96.9	103.5
1983	. . .	. . .	. . .	. . .	. . .	101.1	103.4	100.3	100.7	100.0	99.8	100.3
1984	. . .	. . .	. . .	. . .	. . .	107.9	110.1	104.3	104.2	104.4	103.3	96.1
1985	. . .	. . .	. . .	. . .	. . .	114.5	116.7	108.3	107.6	108.9	106.9	92.8
1986	. . .	. . .	. . .	. . .	. . .	121.4	124.7	111.9	111.3	112.5	108.6	91.3
1987	. . .	. . .	. . .	. . .	. . .	128.5	133.6	115.1	113.9	116.2	112.5	88.0
1988	. . .	. . .	. . .	. . .	. . .	137.0	145.8	119.4	118.1	120.7	117.0	84.6
1989	. . .	. . .	. . .	96.3	. . .	147.7	164.4	125.0	123.2	126.8	122.6	80.7
1990	. . .	. . .	. . .	93.5	. . .	159.0	181.5	130.4	128.2	132.8	129.0	76.6
1991	. . .	. . .	. . .	88.6	. . .	171.6	202.7	134.9	132.8	137.0	134.3	73.4
1992	. . .	. . .	. . .	83.7	. . .	183.3	219.8	138.3	136.5	140.0	138.2	71.3
1993	96.7	97.7	. . .	78.8	. . .	192.9	228.4	141.5	139.0	144.0	142.1	69.2
1994	97.6	98.6	. . .	72.0	. . .	198.5	220.0	144.6	141.5	147.9	145.6	67.5
1995	98.8	98.7	. . .	63.8	. . .	206.9	225.7	147.1	143.1	151.5	149.8	65.6
1996	99.6	99.5	. . .	57.2	. . .	215.4	232.8	150.1	144.3	156.6	154.1	63.8
1997	100.3	100.4	. . .	50.1	. . .	224.8	243.7	152.7	144.2	162.4	157.6	62.3
1998	98.7	98.5	100.7	39.9	875.1	237.7	274.8	156.7	148.3	166.0	159.7	61.4
1999	96.0	95.5	100.1	30.5	598.7	258.3	355.8	161.1	151.8	171.4	163.2	60.0
2000	93.6	92.8	98.5	25.9	459.9	271.1	394.9	165.6	153.7	178.1	168.9	58.1
2001	93.3	92.3	99.3	21.3	330.1	282.6	425.2	170.5	155.1	184.3	173.5	56.5
2002	92.3	90.8	99.7	18.3	248.4	293.2	461.5	174.7	154.7	188.4	175.9	55.6
2003	89.7	87.8	98.3	16.1	196.9	298.7	469.0	178.0	153.5	193.2	179.8	54.4
2004	86.7	84.6	95.8	14.8	171.2	304.7	478.0	181.7	153.9	197.6	184.5	53.0
2005	84.7	82.6	94.9	13.6	143.2	313.4	502.8	185.6	154.4	203.9	191.0	51.2
2006	84.1	81.7	95.8	12.5	120.9	321.7	519.9	190.2	155.8	209.7	197.1	49.6
2007	83.4	80.7	98.2	10.6	108.4	333.3	554.2	195.6	158.3	216.6	202.8	48.2
2008	84.2	81.4	100.5	10.1	94.9	345.4	588.7	201.3	159.3	223.7	211.1	46.5
2009	85.0	81.9	102.4	9.7	82.3	368.6	730.3	204.6	162.6	227.6	209.6	46.6
2010	84.7	81.5	102.4	9.4	76.4	381.3	807.3	206.6	161.1	229.6	214.0	45.9
2011	83.3	80.0	101.2	9.0	68.9	387.2	834.8	208.6	160.5	230.8	221.6	44.5
2012	83.1	79.5	101.7	8.7	62.3	394.4	853.5	212.1	162.2	234.2	226.2	43.6
2013	82.6	78.9	101.6	8.5	56.8	401.0	876.8	215.0	161.8	238.8	229.3	42.9
2014	82.1	78.2	101.1	8.4	52.6	408.1	903.3	218.0	163.4	242.0	232.8	42.2
2015	80.2	76.4	99.3	8.1	47.9	414.9	930.8	220.8	163.3	247.2	231.8	42.2
2016	79.2	75.4	98.8	7.8	44.4	423.1	963.4	224.3	163.1	252.9	234.1	41.7
2017	75.0	71.1	91.8	7.6	42.6	432.6	1 022.8	227.0	161.7	257.4	239.1	40.8
2018	73.9	70.0	90.4	7.5	40.8	442.3	1 063.5	231.1	161.5	264.2	245.1	39.8
2019	73.2	69.2	89.4	7.4	39.4	451.3	1 115.9	234.0	160.7	271.4	249.2	39.1
2018												
January	74.3	70.4	91.0	7.5	41.6	437.6	1 049.0	228.8	161.9	260.2	243.0	40.3
February	74.0	70.1	90.6	7.5	41.4	438.5	1 051.3	229.2	162.1	260.6	243.6	40.2
March	73.9	70.0	90.6	7.4	41.3	439.3	1 047.1	230.0	162.0	261.4	243.4	40.1
April	73.7	69.8	90.6	7.4	41.4	442.5	1 056.1	231.6	161.9	262.6	244.0	39.9
May	74.0	70.1	90.7	7.5	41.4	442.8	1 060.3	231.5	161.7	263.1	244.7	39.7
June	74.1	70.2	90.7	7.5	41.0	443.0	1 061.7	231.6	161.1	265.1	245.3	39.7
July	74.2	70.3	90.9	7.5	40.0	443.2	1 062.8	231.7	161.3	265.3	245.5	39.7
August	74.0	70.1	90.5	7.5	40.6	443.0	1 065.4	231.4	161.3	265.1	246.0	39.7
September	74.2	70.2	90.6	7.5	40.9	443.6	1 066.9	231.7	161.5	265.3	246.3	39.6
October	73.9	70.0	90.4	7.5	40.4	444.2	1 079.5	231.4	161.1	266.8	246.9	39.5
November	73.0	69.1	89.0	7.4	39.7	445.1	1 079.2	232.0	160.8	267.0	246.6	39.7
December	73.1	69.2	88.9	7.4	40.4	445.0	1 081.7	231.8	160.7	268.2	246.3	39.8
2019												
January	73.1	69.1	88.9	7.4	40.2	446.0	1 085.7	232.2	161.4	268.6	246.0	39.7
February	73.1	69.1	88.9	7.4	39.8	448.4	1 090.9	233.5	160.8	269.3	246.8	39.6
March	73.0	69.0	88.9	7.4	39.9	448.5	1 103.5	232.9	160.2	269.7	247.7	39.3
April	73.0	69.0	89.0	7.4	39.6	448.5	1 103.5	232.9	160.9	269.5	248.7	39.1
May	73.0	69.0	89.1	7.4	39.4	449.9	1 108.2	233.5	160.6	269.4	248.8	39.0
June	73.0	69.0	89.2	7.3	38.6	449.6	1 111.9	233.1	160.1	269.9	248.9	39.0
July	73.2	69.2	89.3	7.4	39.5	451.7	1 119.9	234.1	161.1	271.3	249.7	39.0
August	73.3	69.3	89.3	7.4	40.0	453.2	1 125.5	234.7	161.1	272.4	249.8	39.0
September	73.2	69.2	89.5	7.4	39.5	453.4	1 130.4	234.6	160.7	272.6	250.0	38.9
October	73.3	69.3	89.7	7.3	38.8	455.1	1 132.6	235.6	161.1	273.4	250.8	38.9
November	73.6	69.5	90.2	7.4	38.8	455.8	1 137.6	235.8	161.2	274.7	251.3	38.9
December	73.6	69.5	90.3	7.3	38.3	455.4	1 141.3	235.3	159.4	275.5	252.0	38.9

[1]Not seasonally adjusted.
[3]December 1997 = 100.
[4]December 1988 = 100.
[5]December 2007 = 100.
. . . = Not available.

Table 8-3. Alternative Measures of Total and Core Consumer Prices: Index Levels

(Various bases; monthly data seasonally adjusted, except as noted.)

Year and month	CPIs, all items					CPIs, all items less food and energy		
	CPI-U, 1982–1984 = 100	CPI-W, 1982–1984 = 100	CPI-U-X1, 1982–1984 = 100	CPI-U-RS, Dec. 1977 = 100, not seasonally adjusted	C-CPI-U, Dec. 1999 = 100, not seasonally adjusted	CPI-U, 1982–1984 = 100	CPI-U-RS, Dec. 1977 = 100, not seasonally adjusted	C-CPI-U, Dec. 1999 = 100, not seasonally adjusted
1975	53.8	54.1	56.2	. . .	. . .	53.9	. . .	. . .
1976	56.9	57.2	59.4	. . .	. . .	57.4	. . .	. . .
1977	60.6	60.9	63.2	. . .	. . .	61.0	. . .	. . .
1978	65.2	65.6	67.5	104.4	. . .	65.5	103.6	. . .
1979	72.6	73.1	74.0	114.3	. . .	71.9	111.0	. . .
1980	82.4	82.9	82.3	127.1	. . .	80.8	120.9	. . .
1981	90.9	91.4	90.1	139.1	. . .	89.2	132.2	. . .
1982	96.5	96.9	95.6	147.5	. . .	95.8	142.4	. . .
1983	99.6	99.8	99.6	153.8	. . .	99.6	150.4	. . .
1984	103.9	103.3	103.9	160.2	. . .	104.6	157.9	. . .
1985	107.6	106.9	107.6	165.7	. . .	109.1	164.8	. . .
1986	109.6	108.6	109.6	168.6	. . .	113.5	171.4	. . .
1987	113.6	112.5	113.6	174.4	. . .	118.2	178.1	. . .
1988	118.3	117.0	118.3	180.7	. . .	123.4	185.2	. . .
1989	124.0	122.6	124.0	188.6	. . .	129.0	192.6	. . .
1990	130.7	129.0	130.7	197.9	. . .	135.5	201.4	. . .
1991	136.2	134.3	136.2	205.1	. . .	142.1	209.9	. . .
1992	140.3	138.2	140.3	210.2	. . .	147.3	216.4	. . .
1993	144.5	142.1	144.5	215.5	. . .	152.2	222.5	. . .
1994	148.2	145.6	148.2	220.0	. . .	156.5	227.7	. . .
1995	152.4	149.8	152.4	225.3	. . .	161.2	233.4	. . .
1996	156.9	154.1	156.9	231.3	. . .	165.6	239.1	. . .
1997	160.5	157.6	160.5	236.3	. . .	169.5	244.4	. . .
1998	163.0	159.7	163.0	239.5	. . .	173.4	249.6	. . .
1999	166.6	163.2	166.6	244.6	. . .	177.0	254.6	. . .
2000	172.2	168.9	172.2	252.9	102.0	181.3	260.9	101.4
2001	177.1	173.5	177.1	260.1	104.3	186.1	267.9	103.5
2002	179.9	175.9	179.9	264.2	105.6	190.5	274.1	105.4
2003	184.0	179.8	184.0	270.2	107.8	193.2	278.1	106.6
2004	188.9	184.5	188.9	277.5	110.5	196.6	283.1	108.4
2005	195.3	191.0	195.3	286.9	113.7	200.9	289.2	110.4
2006	201.6	197.1	201.6	296.2	117.0	205.9	296.5	112.9
2007	207.3	202.8	207.3	304.6	120.0	210.7	303.4	115.0
2008	215.3	211.1	215.3	316.3	124.4	215.6	310.3	117.3
2009	214.5	209.6	214.5	315.2	123.9	219.2	315.6	119.1
2010	218.1	214.0	218.1	320.4	125.6	221.3	318.7	120.0
2011	224.9	221.6	224.9	330.5	129.5	225.0	324.0	121.9
2012	229.6	226.2	229.6	337.5	132.0	229.8	331.0	124.3
2013	233.0	229.3	233.0	342.5	. . .	233.8	337.0	. . .
2014	236.7	232.8	236.7	348.3	. . .	237.9	343.1	. . .
2015	237.0	231.8	237.0	348.9	. . .	242.2	349.7	. . .
2016	240.0	234.1	240.0	353.4	. . .	247.6	357.5	. . .
2017	245.1	239.1	245.1	361.0	. . .	252.2	364.2	. . .
2018	251.1	245.1	251.1	369.8	. . .	257.6	372.0	. . .
2019	255.7	249.2	255.7	376.5	. . .	263.2	380.1	. . .
2018								
January	248.8	243.0	248.8	365.0	140.2	255.1	367.7	135.1
February	249.5	243.6	249.5	366.7	140.8	255.6	369.4	135.6
March	249.4	243.4	249.4	367.5	141.1	256.1	370.6	136.0
April	250.0	244.0	250.0	369.0	141.7	256.5	371.2	136.3
May	250.6	244.7	250.6	370.5	142.1	257.1	371.8	136.5
June	251.2	245.3	251.2	371.1	142.3	257.4	372.2	136.6
July	251.5	245.5	251.5	371.1	142.3	257.9	372.4	136.6
August	251.9	246.0	251.9	371.3	142.3	258.1	372.6	136.6
September	252.3	246.3	252.3	371.8	142.5	258.6	373.2	136.8
October	252.8	246.9	252.8	372.4	142.7	259.0	374.1	137.1
November	252.7	246.6	252.7	371.2	142.2	259.5	374.2	137.0
December	252.7	246.3	252.7	370.0	141.7	260.0	374.2	136.8
2019								
January	252.6	246.0	252.6	370.7	142.0	260.6	375.7	137.4
February	253.2	246.8	253.2	372.2	142.6	260.9	377.1	137.9
March	254.1	247.7	254.1	374.3	143.3	[1]261.3	378.1	138.2
April	254.9	248.7	254.9	376.3	143.9	[1]261.8	378.8	138.4
May	255.2	248.8	255.2	377.1	144.2	[1]262.2	379.2	138.6
June	255.4	248.9	255.4	377.2	144.2	[1]262.9	380.0	138.8
July	256.1	249.7	256.1	377.8	144.5	[1]263.6	380.6	139.0
August	256.3	249.8	256.3	377.8	144.5	[1]264.2	381.5	139.3
September	256.6	250.0	256.6	378.1	144.6	264.7	382.0	139.5
October	257.2	250.8	257.2	379.0	144.9	265.0	382.8	139.8
November	257.8	251.3	257.8	378.8	144.7	265.5	382.9	139.7
December	258.4	252.0	258.4	378.5	144.6	265.8	382.6	139.6

[1]Interim values.
. . . = Not available.

Table 8-4. Alternative Measures of Total and Core Consumer Prices: Inflation Rates

(Percent changes from year earlier, except as noted; monthly data seasonally adjusted, except as noted.)

Year and month	CPIs, all items					CPIs, all items less food and energy		
	CPI-U, 1982–1984 = 100	CPI-W, 1982–1984 = 100	CPI-U-X1, 1982–1984 = 100	CPI-U-RS, Dec. 1977 = 100, not seasonally adjusted	C-CPI-U, Dec. 1999 = 100, not seasonally adjusted	CPI-U, 1982–1984 = 100	CPI-U-RS, Dec. 1977 = 100, not seasonally adjusted	C-CPI-U, Dec. 1999 = 100, not seasonally adjusted
1975	9.1	9.1	8.3	. . .	. . .	9.1	. . .	. . .
1976	5.8	5.7	5.7	. . .	. . .	6.5	. . .	. . .
1977	6.5	6.5	6.4	. . .	. . .	6.3	. . .	. . .
1978	7.6	7.7	6.8	. . .	. . .	7.4	. . .	. . .
1979	11.3	11.4	9.6	9.5	. . .	9.8	7.1	. . .
1980	13.5	13.4	11.2	11.2	. . .	12.4	8.9	. . .
1981	10.3	10.3	9.5	9.4	. . .	10.4	9.3	. . .
1982	6.2	6.0	6.1	6.0	. . .	7.4	7.7	. . .
1983	3.2	3.0	4.2	4.3	. . .	4.0	5.6	. . .
1984	4.3	3.5	4.3	4.2	. . .	5.0	5.0	. . .
1985	3.6	3.5	3.6	3.4	. . .	4.3	4.4	. . .
1986	1.9	1.6	1.9	1.8	. . .	4.0	4.0	. . .
1987	3.6	3.6	3.6	3.4	. . .	4.1	3.9	. . .
1988	4.1	4.0	4.1	3.6	. . .	4.4	4.0	. . .
1989	4.8	4.8	4.8	4.4	. . .	4.5	4.0	. . .
1990	5.4	5.2	5.4	4.9	. . .	5.0	4.6	. . .
1991	4.2	4.1	4.2	3.6	. . .	4.9	4.2	. . .
1992	3.0	2.9	3.0	2.5	. . .	3.7	3.1	. . .
1993	3.0	2.8	3.0	2.5	. . .	3.3	2.8	. . .
1994	2.6	2.5	2.6	2.1	. . .	2.8	2.3	. . .
1995	2.8	2.9	2.8	2.4	. . .	3.0	2.5	. . .
1996	3.0	2.9	3.0	2.7	. . .	2.7	2.4	. . .
1997	2.3	2.3	2.3	2.2	. . .	2.4	2.2	. . .
1998	1.6	1.3	1.6	1.4	. . .	2.3	2.1	. . .
1999	2.2	2.2	2.2	2.1	. . .	2.1	2.0	. . .
2000	3.4	3.5	3.4	3.4	. . .	2.4	2.5	. . .
2001	2.8	2.7	2.8	2.8	2.3	2.6	2.7	2.1
2002	1.6	1.4	1.6	1.6	1.2	2.4	2.3	1.8
2003	2.3	2.2	2.3	2.3	2.1	1.4	1.5	1.1
2004	2.7	2.6	2.7	2.7	2.5	1.8	1.8	1.7
2005	3.4	3.5	3.4	3.4	2.9	2.2	2.2	1.8
2006	3.2	3.2	3.2	3.2	2.9	2.5	2.5	2.3
2007	2.8	2.9	2.8	2.8	2.5	2.3	2.3	1.8
2008	3.8	4.1	3.8	3.8	3.7	2.3	2.3	2.0
2009	-0.4	-0.7	-0.4	-0.3	-0.5	1.7	1.7	1.5
2010	1.6	2.1	1.6	1.6	1.4	1.0	1.0	0.7
2011	3.2	3.6	3.2	3.2	3.1	1.7	1.7	1.6
2012	2.1	2.1	2.1	2.1	1.9	2.1	2.2	2.0
2013	1.5	1.4	1.5	1.5	. . .	1.8	1.8	. . .
2014	1.6	1.5	1.6	1.7	. . .	1.7	1.8	. . .
2015	0.1	-0.4	0.1	0.2	. . .	1.8	1.9	. . .
2016	1.3	1.0	1.3	1.3	. . .	2.2	2.2	. . .
2017	2.1	2.1	2.1	2.2	. . .	1.8	1.9	. . .
2018	2.4	2.5	2.4	2.4	. . .	2.1	2.1	. . .
2019	1.8	1.7	1.8	1.8	. . .	2.2	2.2	. . .
2019								
January	1.5	1.3	1.5	1.6	0.8	2.1	2.2	1.3
February	1.5	1.3	1.5	1.5	1.0	2.1	2.1	1.4
March	1.9	1.8	1.9	1.9	1.2	2.0	2.0	1.4
April	2.0	1.9	2.0	2.0	1.3	2.1	2.0	1.4
May	1.8	1.7	1.8	1.8	1.3	2.0	2.0	1.4
June	1.7	1.5	1.7	1.6	1.4	2.1	2.1	1.7
July	1.8	1.7	1.8	1.8	1.5	2.2	2.2	1.8
August	1.7	1.5	1.7	1.8	1.4	2.4	2.4	1.8
September	1.7	1.5	1.7	1.7	1.3	[1]2.3	2.4	1.8
October	1.8	1.6	1.8	1.8	1.8	[1]2.3	2.3	2.0
November	2.0	1.9	2.0	2.0	2.1	[1]2.3	2.3	2.1
December	2.3	2.3	2.3	2.3	1.8	[1]2.2	2.2	1.6

[1]Interim values.
. . . = Not available.

Table 8-5. Producer Price Indexes for Goods and Services: Final Demand

(November 2009 =100, except as noted.)

Year and month	Final demand	Final demand goods			Final demand goods less food and energy			Final demand services				Final demand construction	Final demand less food and energy (April 2010 =100)
		Final demand goods	Final demand foods	Final demand energy goods	Total	Finished goods less foods and energy (1982 =100)		Final demand services	Final demand trade services	Final demand transportation and warehousing services	Final demand services less trade, transportation, and warehousing		
						Total	Private capital equipment						
NOT SEASONALLY ADJUSTED													
2010	101.8	102.8	103.7	107.2	101.4	173.6	157.3	101.3	101.7	103.2	100.9	100.3	. . .
2011	105.7	109.9	112.5	126.2	104.9	177.8	159.7	103.4	104.0	110.0	102.5	102.5	102.7
2012	107.7	111.7	115.9	126.3	106.8	182.4	162.8	105.4	106.7	114.2	103.9	105.5	104.7
2013	109.1	112.6	117.8	125.3	107.9	185.1	164.2	107.1	108.2	115.3	105.8	107.5	106.2
2014	110.9	114.0	121.6	124.2	109.5	188.6	166.4	109.0	110.2	117.7	107.5	110.6	108.1
2015	109.9	109.1	118.4	98.6	109.9	192.3	168.5	110.0	111.6	115.3	108.7	112.7	108.9
2016	110.4	107.6	115.1	90.4	110.7	195.3	169.3	111.5	113.1	113.5	110.6	114.0	110.2
2017	113.0	111.2	116.5	99.8	113.2	198.9	170.9	113.5	114.8	115.9	112.8	116.5	112.3
2018	116.2	115.0	116.7	110.0	116.0	203.4	173.6	116.5	116.9	122.0	115.8	121.2	115.3
2019	118.2	115.5	118.9	105.0	117.6	207.8	177.4	119.1	119.7	125.5	118.2	127.2	117.7
2016													
January	109.7	106.4	116.4	85.1	110.1	194.5	169.1	111.1	113.1	114.1	109.8	113.4	109.7
February	109.9	105.9	116.4	81.9	110.2	194.7	169.2	111.2	113.4	113.5	110.0	113.4	109.9
March	109.9	106.1	115.1	84.3	110.2	194.6	169.1	111.2	113.0	114.2	110.2	113.3	109.9
April	110.0	106.7	115.0	86.5	110.5	194.7	169.1	111.3	113.2	113.7	110.3	114.4	110.0
May	110.0	106.6	115.9	90.2	110.6	194.7	169.2	111.2	113.2	112.8	110.1	114.4	110.0
June	110.3	108.7	116.9	94.5	110.8	195.3	169.3	111.6	113.7	113.6	110.5	114.4	110.3
July	110.3	108.5	115.9	94.5	110.8	195.1	169.0	111.6	112.7	114.2	110.9	113.6	110.3
August	110.2	108.0	114.4	92.9	110.8	195.1	168.8	111.4	111.9	113.7	111.1	113.7	110.2
September	110.2	108.3	114.8	94.2	110.7	195.1	168.7	111.6	112.9	111.7	111.0	113.7	110.2
October	110.7	108.5	113.5	94.5	111.2	196.3	169.7	112.1	113.9	113.0	111.2	114.6	110.7
November	110.7	108.0	112.9	92.2	111.3	196.3	169.7	112.0	113.6	113.5	111.1	114.6	110.7
December	110.7	108.7	113.9	93.9	111.6	196.7	170.1	111.8	113.0	114.7	111.0	114.5	110.7
2017													
January	111.2	109.7	113.9	96.8	112.3	197.8	170.4	112.3	113.6	115.2	111.5	114.9	111.2
February	111.3	110.1	114.8	97.6	112.5	197.9	170.6	112.3	112.9	115.4	111.8	114.9	111.3
March	111.6	110.4	115.8	97.2	112.8	198.1	170.8	112.7	114.0	114.9	111.9	115.0	111.6
April	112.1	111.1	117.5	98.7	113.1	198.6	170.9	113.2	114.5	115.6	112.4	115.5	112.1
May	112.2	110.7	117.1	97.3	113.0	198.4	170.6	113.4	115.1	115.0	112.5	115.7	112.2
June	112.3	111.1	118.2	98.1	113.1	198.7	170.7	113.5	114.4	116.1	112.9	115.8	112.3
July	112.4	111.0	117.8	98.2	113.0	198.6	170.5	113.7	114.5	115.7	113.2	117.3	112.4
August	112.6	111.4	117.0	100.6	113.1	198.6	170.5	113.9	115.0	115.6	113.2	117.4	112.6
September	112.6	112.0	116.6	104.3	113.1	198.4	170.5	113.9	115.3	115.6	113.1	117.3	112.6
October	113.4	112.0	116.4	102.0	113.8	200.2	171.4	114.8	116.9	116.9	113.6	118.0	113.4
November	113.3	112.5	116.5	103.7	114.0	200.5	171.7	114.6	116.0	117.9	113.6	117.9	113.3
December	113.1	112.5	116.2	103.4	114.1	200.6	171.8	114.2	115.0	117.3	113.5	118.0	113.1
2018													
January	113.7	113.2	116.1	105.2	114.7	201.4	171.9	114.8	115.2	117.9	114.4	119.0	113.7
February	114.1	113.5	115.5	106.4	115.0	201.8	172.2	115.3	115.8	119.0	114.8	119.0	114.1
March	114.6	114.0	118.0	105.6	115.4	202.1	172.5	115.8	116.0	120.5	115.3	119.1	114.6
April	114.8	114.4	116.8	108.0	115.6	202.4	172.6	116.0	116.5	120.8	115.2	120.4	114.8
May	114.9	115.5	117.8	112.8	115.8	202.6	172.9	116.1	116.7	120.7	115.4	120.5	114.9
June	115.3	116.0	116.9	115.3	116.0	202.9	173.2	116.6	117.2	122.6	115.7	120.7	115.3
July	115.5	115.9	116.3	114.9	116.2	203.3	173.5	116.8	116.8	122.9	116.3	121.2	115.5
August	115.5	115.8	115.7	114.6	116.3	203.7	174.0	116.7	116.4	122.3	116.3	121.4	115.5
September	115.5	115.8	115.2	114.7	116.3	203.9	174.3	116.7	116.3	122.7	116.3	121.4	115.5
October	116.5	116.3	115.7	114.9	116.9	205.2	175.3	117.8	118.8	124.6	116.7	123.8	116.5
November	116.4	115.4	117.0	108.0	117.1	205.7	175.6	117.6	118.3	125.4	116.5	124.0	116.4
December	116.4	114.3	119.4	100.2	117.1	205.8	175.7	117.6	118.6	124.9	116.4	124.1	116.4
2019													
January	116.6	113.9	117.6	97.4	117.6	207.2	176.8	117.8	118.1	125.5	116.8	125.0	116.6
February	116.9	114.3	117.4	99.5	117.7	207.3	177.0	118.1	118.6	124.3	117.3	125.0	116.9
March	117.2	115.4	118.0	104.7	117.8	207.5	176.9	118.5	119.1	124.8	117.6	125.2	117.2
April	117.7	116.3	118.2	109.3	117.8	207.5	177.0	119.1	120.3	125.5	117.9	126.9	117.7
May	117.7	116.4	118.5	110.3	117.6	207.6	177.2	119.2	119.8	125.4	118.3	127.1	117.7
June	117.8	116.0	119.6	107.5	117.5	207.6	177.2	119.3	119.8	126.4	118.4	127.1	117.8
July	118.0	116.2	119.3	108.8	117.6	207.8	177.3	119.5	119.9	126.5	118.7	128.0	118.0
August	118.2	115.7	118.4	106.9	117.5	207.7	177.2	119.9	120.9	125.3	118.9	128.1	118.2
September	117.8	115.2	118.3	105.1	117.3	207.7	177.3	119.4	120.0	124.1	118.6	128.2	117.8
October	118.4	115.6	119.4	104.8	117.7	208.6	178.0	120.0	121.4	125.7	118.8	128.6	118.4
November	117.8	115.6	121.0	103.2	117.8	208.9	178.2	119.2	119.3	125.5	118.6	128.7	117.8
December	117.9	115.5	120.8	102.6	117.8	208.8	178.2	119.3	119.5	127.6	118.5	128.9	117.9

. . . = Not available.

Table 8-6. Producer Price Indexes for Goods and Services: Intermediate Demand

(November 2009 = 100, except as noted.)

Year and month	Intermediate demand by commodity type						Intermediate demand by production flow (Prices of inputs to indicated stage of production)			
	Processed goods (1982=100)	Unprocessed goods (1982=100)	Services	Construction	Processed materials less foods and energy (1982=100)	Unprocessed nonfood materials less energy (1982 =100)	Stage 4 intermediate demand	Stage 3 intermediate demand	Stage 2 intermediate demand	Stage 1 intermediate demand
NOT SEASONALLY ADJUSTED										
2010	183.4	212.2	101.1	101.1	180.8	329.1	102.0	104.2	103.5	106.2
2011	199.9	249.4	103.2	102.3	192.0	390.4	106.9	111.7	112.9	115.7
2012	200.7	241.4	105.3	103.8	192.6	369.6	108.6	112.7	110.7	114.7
2013	200.8	246.7	107.2	105.7	193.8	351.2	109.9	114.1	112.8	114.8
2014	201.9	249.3	108.9	108.0	195.2	345.7	111.2	116.9	112.2	116.1
2015	188.0	189.1	110.2	110.1	189.4	296.0	109.9	110.5	98.9	106.8
2016	182.2	173.4	112.1	111.7	186.9	288.0	109.8	107.3	98.3	104.2
2017	190.7	190.8	115.0	114.1	193.3	324.1	113.0	111.9	103.4	110.4
2018	200.9	200.1	118.6	116.2	201.8	340.7	117.0	116.9	108.5	116.7
2019	198.1	185.9	121.4	119.3	201.1	323.4	118.9	117.2	106.0	114.6
2016										
January	179.6	164.7	111.5	110.8	185.6	266.6	109.0	106.7	95.6	102.0
February	178.3	161.7	111.7	110.9	185.1	269.2	108.9	106.0	95.1	101.6
March	178.8	166.0	111.8	110.9	185.3	276.9	109.0	106.5	95.6	102.2
April	179.7	170.1	111.8	111.0	185.9	287.1	109.2	106.7	96.5	103.2
May	181.6	176.6	111.6	111.3	186.5	297.0	109.5	107.4	97.8	104.3
June	183.8	181.3	112.1	111.3	186.9	296.8	110.2	108.5	99.1	105.6
July	184.2	181.2	112.6	111.4	187.0	295.8	110.4	108.6	99.9	105.6
August	183.9	175.4	112.3	112.1	187.5	295.4	110.2	107.7	99.2	105.0
September	184.4	173.5	112.2	112.4	187.5	289.1	110.2	107.4	99.3	105.0
October	184.0	171.6	112.3	112.5	187.8	285.2	110.0	106.6	100.2	104.7
November	183.6	173.0	112.7	112.8	188.4	292.6	110.1	107.4	99.7	105.2
December	184.6	185.4	112.8	113.2	189.0	303.9	110.4	108.3	101.7	106.5
2017										
January	186.7	193.4	113.4	113.5	189.8	314.4	111.2	109.7	103.2	107.8
February	188.0	192.8	114.0	113.5	191.3	318.4	111.8	110.6	103.2	108.7
March	188.3	188.3	114.0	113.7	192.3	325.3	111.8	111.1	101.8	109.1
April	189.4	191.9	114.6	113.8	193.2	323.5	112.5	111.4	103.2	109.6
May	189.9	191.6	114.6	114.0	193.2	321.6	112.7	111.8	102.7	109.5
June	190.8	191.8	115.2	114.2	193.3	323.4	113.2	112.4	102.8	110.1
July	190.8	192.0	115.1	114.3	193.0	325.5	113.2	112.2	102.9	110.3
August	191.5	187.6	115.5	114.3	193.5	328.2	113.5	112.1	102.9	111.1
September	192.7	186.8	115.3	114.4	194.0	332.5	113.6	112.3	103.3	111.4
October	192.9	186.2	116.0	114.6	194.8	323.7	113.8	112.5	104.0	111.6
November	193.5	192.0	116.3	114.6	195.7	323.1	114.1	113.5	104.9	112.3
December	193.9	194.7	116.1	114.7	195.8	329.8	114.1	113.5	105.5	112.6
2018										
January	195.4	199.7	117.0	115.3	197.0	341.9	114.9	114.4	106.7	114.0
February	197.0	203.0	117.5	115.3	198.2	342.8	115.4	115.4	107.7	114.7
March	197.3	198.2	117.7	115.5	199.5	347.2	115.8	115.7	106.5	115.0
April	198.5	199.9	118.0	115.6	200.1	349.6	116.2	116.1	107.4	115.7
May	201.7	205.7	118.2	115.9	201.6	349.3	116.9	117.7	108.9	117.0
June	203.4	204.2	118.6	116.0	202.5	350.8	117.5	118.5	108.8	118.0
July	203.6	204.8	119.0	116.1	203.1	342.4	117.7	118.4	109.7	118.2
August	203.8	194.6	119.0	116.5	203.7	332.3	117.6	117.3	109.0	117.8
September	204.3	194.8	119.1	116.5	204.4	328.2	117.7	117.5	109.4	117.8
October	204.6	200.3	119.8	117.0	204.7	330.6	118.1	118.7	110.2	118.6
November	202.2	194.4	119.9	117.2	204.0	333.5	117.8	117.6	108.4	117.7
December	199.4	202.0	119.7	117.4	202.7	339.2	117.6	115.8	109.3	116.1
2019										
January	197.8	193.6	120.3	117.9	202.9	333.4	117.7	116.0	107.4	114.9
February	198.2	189.4	120.5	118.7	202.6	334.3	118.0	116.2	106.6	115.1
March	199.6	190.6	120.9	118.8	202.6	339.6	118.6	117.3	106.9	116.2
April	200.2	193.4	121.6	119.0	202.4	332.9	119.2	118.9	107.3	116.4
May	200.2	188.8	121.5	119.4	202.0	321.3	119.2	118.9	106.8	115.6
June	198.8	183.8	121.7	118.9	201.3	319.3	119.2	117.9	105.4	114.5
July	198.9	184.9	121.7	119.6	200.8	321.5	119.4	117.7	105.7	114.7
August	197.7	179.7	122.2	119.8	200.0	318.5	119.5	117.3	105.1	114.2
September	197.2	177.4	122.1	119.8	199.6	314.2	119.4	116.4	105.2	113.7
October	196.8	177.9	121.5	119.9	200.1	311.6	118.9	116.8	104.5	113.2
November	196.4	184.4	121.4	120.0	199.7	311.9	118.9	116.7	105.5	113.1
December	196.0	187.2	121.7	120.0	199.1	322.3	119.2	116.7	105.8	113.6

Table 8-6. Producer Price Indexes for Goods and Services: Intermediate Demand—*Continued*

(November 2009 = 100, except as noted.)

Year and month	Intermediate demand by commodity type						Intermediate demand by production flow (Prices of inputs to indicated stage of production)			
	Processed goods (1982=100)	Unprocessed goods (1982=100)	Services	Construction	Processed materials less foods and energy (1982=100)	Unprocessed nonfood materials less energy (1982 =100)	Stage 4 intermediate demand	Stage 3 intermediate demand	Stage 2 intermediate demand	Stage 1 intermediate demand
SEASONALLY ADJUSTED										
2016										
January	180.4	165.6	111.4	110.8	185.7	266.9	109.2	107.1	95.7	102.3
February	178.9	162.2	111.6	110.9	185.2	268.5	109.0	106.3	95.1	101.8
March	179.3	164.5	111.7	110.9	185.1	275.7	109.1	106.4	95.4	102.3
April	179.9	168.5	111.6	111.0	185.7	285.2	109.2	106.5	96.3	103.1
May	181.2	173.9	111.6	111.3	186.2	294.8	109.3	106.8	97.7	104.1
June	182.6	178.1	111.9	111.3	186.6	294.6	109.8	107.6	98.8	104.9
July	183.0	179.2	112.5	111.4	186.9	295.4	110.1	107.9	99.7	105.1
August	183.0	176.2	112.2	112.1	187.5	296.5	109.9	107.4	99.2	104.6
September	183.7	175.7	112.3	112.4	187.7	291.0	110.1	107.5	99.5	104.9
October	184.2	174.2	112.4	112.5	188.0	287.2	110.2	107.0	100.4	104.9
November	184.5	175.2	112.8	112.8	188.7	294.6	110.4	108.0	100.0	105.6
December	185.7	187.4	113.3	113.2	189.3	305.0	110.9	109.3	102.0	107.2
2017										
January	187.5	194.1	113.4	113.5	189.9	314.6	111.3	110.2	103.3	108.2
February	188.8	193.1	113.8	113.5	191.3	317.6	111.8	110.9	103.2	109.0
March	188.9	186.7	113.9	113.7	192.1	324.2	112.0	111.1	101.6	109.2
April	189.7	190.2	114.4	113.8	192.9	321.8	112.5	111.1	103.0	109.6
May	189.4	188.7	114.6	114.0	192.8	319.4	112.5	111.1	102.6	109.3
June	189.6	188.4	115.0	114.2	193.1	321.3	112.8	111.4	102.6	109.5
July	189.6	189.7	115.1	114.3	192.9	325.3	112.8	111.4	102.7	109.7
August	190.6	188.4	115.3	114.3	193.5	329.3	113.2	111.8	102.9	110.6
September	192.0	189.7	115.5	114.4	194.2	334.6	113.5	112.5	103.5	111.3
October	193.1	189.4	116.2	114.6	195.0	325.8	114.0	113.1	104.2	111.9
November	194.5	194.6	116.4	114.6	195.9	325.0	114.5	114.2	105.1	112.7
December	195.1	196.7	116.7	114.7	196.2	331.0	114.6	114.6	105.8	113.3
2018										
January	196.5	200.4	117.0	115.3	197.2	342.3	115.1	115.0	106.8	114.6
February	198.1	202.9	117.3	115.3	198.2	341.8	115.5	115.8	107.7	115.1
March	198.1	196.5	117.6	115.5	199.3	345.9	115.9	115.7	106.4	115.2
April	198.8	198.1	117.9	115.6	199.8	347.7	116.2	115.7	107.2	115.7
May	200.9	202.8	118.2	115.9	201.3	346.5	116.7	116.8	108.7	116.6
June	202.0	200.8	118.4	116.0	202.3	348.5	117.1	117.4	108.6	117.2
July	202.3	202.8	118.9	116.1	203.0	342.3	117.4	117.6	109.6	117.5
August	202.7	195.6	118.9	116.5	203.7	333.5	117.4	117.1	108.9	117.3
September	203.4	197.6	119.3	116.5	204.5	330.4	117.6	117.7	109.6	117.5
October	204.7	203.5	119.9	117.0	204.8	332.9	118.3	119.2	110.3	118.8
November	203.1	196.9	120.0	117.2	204.1	335.5	118.1	118.3	108.7	118.1
December	200.5	203.8	120.2	117.4	203.0	340.5	118.0	116.8	109.6	116.8
2019										
January	198.8	193.9	120.3	117.9	203.1	333.8	117.9	116.4	107.5	115.4
February	199.2	189.0	120.3	118.7	202.7	333.2	118.1	116.5	106.6	115.5
March	200.3	188.9	120.9	118.8	202.4	338.5	118.7	117.3	106.7	116.4
April	200.5	192.0	121.4	119.0	202.2	331.0	119.1	118.5	107.1	116.4
May	199.5	186.3	121.5	119.4	201.7	318.7	119.0	118.1	106.6	115.2
June	197.5	180.8	121.5	118.9	201.1	316.9	118.8	117.0	105.2	113.8
July	197.6	183.3	121.6	119.6	200.7	321.6	119.1	116.9	105.6	114.1
August	196.6	180.4	122.1	119.8	200.0	319.6	119.2	117.0	105.1	113.7
September	196.4	180.4	122.3	119.8	199.7	316.6	119.3	116.7	105.4	113.5
October	196.9	181.0	121.6	119.9	200.2	314.1	119.1	117.3	104.7	113.3
November	197.3	186.6	121.5	120.0	199.8	313.9	119.2	117.3	105.8	113.5
December	197.2	188.9	122.1	120.0	199.4	323.6	119.6	117.7	106.1	114.3

Table 8-7. Producer Price Indexes by Major Commodity Groups

(1982 = 100.)

Year and month	All com-modities	Farm products	Pro-cessed foods and feeds	Industrial commodities													
				Total	Textile products and apparel	Hides, leather, and related products	Fuels and related products and power	Chemi-cals and allied products	Rubber and plastics products	Lumber and wood products	Pulp, paper, and allied products	Metals and metal products	Machin-ery and equip-ment	Fur-niture and house-hold durables	Non-metallic mineral products	Trans-portation equip-ment	Miscel-laneous products
1950	27.3	44.0	33.2	25.0	50.2	32.9	12.6	30.4	35.6	31.4	25.7	22.0	22.6	40.9	23.5	. . .	28.6
1951	30.4	51.2	36.9	27.6	56.0	37.7	13.0	34.8	43.7	34.1	30.5	24.5	25.3	44.4	25.0	. . .	30.3
1952	29.6	48.4	36.4	26.9	50.5	30.5	13.0	33.0	39.6	33.2	29.7	24.5	25.3	43.5	25.0	. . .	30.2
1953	29.2	43.8	34.8	27.2	49.3	31.0	13.4	33.4	36.9	33.1	29.6	25.3	25.9	44.4	26.0	. . .	31.0
1954	29.3	43.2	35.4	27.2	48.2	29.5	13.2	33.8	37.5	32.5	29.6	25.5	26.3	44.9	26.6	. . .	31.3
1955	29.3	40.5	33.8	27.8	48.2	29.4	13.2	33.7	42.4	34.1	30.4	27.2	27.2	45.1	27.3	. . .	31.3
1956	30.3	40.0	33.8	29.1	48.2	31.2	13.6	33.9	43.0	34.6	32.4	29.6	29.3	46.3	28.5	. . .	31.7
1957	31.2	41.1	34.8	29.9	48.3	31.2	14.3	34.6	42.8	32.8	33.0	30.2	31.4	47.5	29.6	. . .	32.6
1958	31.6	42.9	36.5	30.0	47.4	31.6	13.7	34.9	42.8	32.5	33.4	30.0	32.1	47.9	29.9	. . .	33.3
1959	31.7	40.2	35.6	30.5	48.1	35.9	13.7	34.8	42.6	34.7	33.7	30.6	32.8	48.0	30.3	. . .	33.4
1960	31.7	40.1	35.6	30.5	48.6	34.6	13.9	34.8	42.7	33.5	34.0	30.6	33.0	47.8	30.4	. . .	33.6
1961	31.6	39.7	36.2	30.4	47.8	34.9	14.0	34.5	41.1	32.0	33.0	30.5	33.0	47.5	30.5	. . .	33.7
1962	31.7	40.4	36.5	30.4	48.2	35.3	14.0	33.9	39.9	32.2	33.4	30.2	33.0	47.2	30.5	. . .	33.9
1963	31.6	39.6	36.8	30.3	48.2	34.3	13.9	33.5	40.1	32.8	33.1	30.3	33.1	46.9	30.3	. . .	34.2
1964	31.6	39.0	36.7	30.5	48.5	34.4	13.5	33.6	39.6	33.5	33.0	31.1	33.3	47.1	30.4	. . .	34.4
1965	32.3	40.7	38.0	30.9	48.8	35.9	13.8	33.9	39.7	33.7	33.3	32.0	33.7	46.8	30.4	. . .	34.7
1966	33.3	43.7	40.2	31.5	48.9	39.4	14.1	34.0	40.5	35.2	34.2	32.8	34.7	47.4	30.7	. . .	35.3
1967	33.4	41.3	39.8	32.0	48.9	38.1	14.4	34.2	41.4	35.1	34.6	33.2	35.9	48.3	31.2	. . .	36.2
1968	34.2	42.3	40.6	32.8	50.7	39.3	14.3	34.1	42.8	39.8	35.0	34.0	37.0	49.7	32.4	. . .	37.0
1969	35.6	45.0	42.7	33.9	51.8	41.5	14.6	34.2	43.6	44.0	36.0	36.0	38.2	50.7	33.6	40.4	38.1
1970	36.9	45.8	44.6	35.2	52.4	42.0	15.3	35.0	44.9	39.9	37.5	38.7	40.0	51.9	35.3	41.9	39.8
1971	38.1	46.6	45.5	36.5	53.3	43.4	16.6	35.6	45.2	44.7	38.1	39.4	41.4	53.1	38.2	44.2	40.8
1972	39.8	51.6	48.0	37.8	55.5	50.0	17.1	35.6	45.3	50.7	39.3	40.9	42.3	53.8	39.4	45.5	41.5
1973	45.0	72.7	58.9	40.3	60.5	54.5	19.4	37.6	46.6	62.2	42.3	44.0	43.7	55.7	40.7	46.1	43.3
1974	53.5	77.4	68.0	49.2	68.0	55.2	30.1	50.2	56.4	64.5	52.5	57.0	50.0	61.8	47.8	50.3	48.1
1975	58.4	77.0	72.6	54.9	67.4	56.5	35.4	62.0	62.2	62.1	59.0	61.5	57.9	67.5	54.4	56.7	53.4
1976	61.1	78.8	70.8	58.4	72.4	63.9	38.3	64.0	66.0	72.2	62.1	65.0	61.3	70.3	58.2	60.5	55.6
1977	64.9	79.4	74.0	62.5	75.3	68.3	43.6	65.9	69.4	83.0	64.6	69.3	65.2	73.2	62.6	64.6	59.4
1978	69.9	87.7	80.6	67.0	78.1	76.1	46.5	68.0	72.4	96.9	67.7	75.3	70.3	77.5	69.6	69.5	66.7
1979	78.7	99.6	88.5	75.7	82.5	96.1	58.9	76.0	80.5	105.5	75.9	86.0	76.7	82.8	77.6	75.3	75.5
1980	89.8	102.9	95.9	88.0	89.7	94.7	82.8	89.0	90.1	101.5	86.3	95.0	86.0	90.7	88.4	82.9	93.6
1981	98.0	105.2	98.9	97.4	97.6	99.3	100.2	98.4	96.4	102.8	94.8	99.6	94.4	95.9	96.7	94.3	96.1
1982	100.0	100.0	100.0	100.0	100.0	100.0	100.0	100.0	100.0	100.0	100.0	100.0	100.0	100.0	100.0	100.0	100.0
1983	101.3	102.4	101.8	101.1	100.3	103.2	95.9	100.3	100.8	107.9	103.3	101.8	102.7	103.4	101.6	102.8	104.8
1984	103.7	105.5	105.4	103.3	102.7	109.0	94.8	102.9	102.3	108.0	110.3	104.8	105.1	105.7	105.4	105.2	107.0
1985	103.2	95.1	103.5	103.7	102.9	108.9	91.4	103.7	101.9	106.6	113.3	104.4	107.2	107.1	108.6	107.9	109.4
1986	100.2	92.9	105.4	100.0	102.6	113.0	69.8	102.6	101.9	107.2	116.1	103.2	108.8	108.2	110.0	110.5	111.6
1987	102.8	95.5	107.9	102.6	105.1	120.4	70.2	106.4	103.0	112.8	121.8	107.1	110.4	109.9	110.0	112.5	114.9
1988	106.9	104.9	112.7	106.3	109.2	131.4	66.7	116.3	109.3	118.9	130.4	118.7	113.2	113.1	111.2	114.3	120.2
1989	112.2	110.9	117.8	111.6	112.3	136.3	72.9	123.0	112.6	126.7	137.8	124.1	117.4	116.9	112.6	117.7	126.5
1990	116.3	112.2	121.9	115.8	115.0	141.7	82.3	123.6	113.6	129.7	141.2	122.9	120.7	119.2	114.7	121.5	134.2
1991	116.5	105.7	121.9	116.5	116.3	138.9	81.2	125.6	115.1	132.1	142.9	120.2	123.0	121.2	117.2	126.4	140.8
1992	117.2	103.6	122.1	117.4	117.8	140.4	80.4	125.9	115.1	146.6	145.2	119.2	123.4	122.2	117.3	130.4	145.3
1993	118.9	107.1	124.0	119.0	118.0	143.7	80.0	128.2	116.0	174.0	147.3	119.2	124.0	123.7	120.0	133.7	145.4
1994	120.4	106.3	125.5	120.7	118.3	148.5	77.8	132.1	117.6	180.0	152.5	124.8	125.1	126.1	124.2	137.2	141.9
1995	124.7	107.4	127.0	125.5	120.8	153.7	78.0	142.5	124.3	178.1	172.2	134.5	126.6	128.2	129.0	139.7	145.4
1996	127.7	122.4	133.3	127.3	122.4	150.5	85.8	142.1	123.8	176.1	168.7	131.0	126.5	130.4	131.0	141.7	147.7
1997	127.6	112.9	134.0	127.7	122.6	154.2	86.1	143.6	123.2	183.8	167.9	131.8	125.9	130.8	133.2	141.6	150.9
1998	124.4	104.6	131.6	124.8	122.9	148.0	75.3	143.9	122.6	179.1	171.7	127.8	124.9	131.3	135.4	141.2	156.0
1999	125.5	98.4	131.1	126.5	121.1	146.0	80.5	144.2	122.5	183.6	174.1	124.6	124.3	131.7	138.9	141.8	166.6
2000	132.7	99.5	133.1	134.8	121.4	151.5	103.5	151.0	125.5	178.2	183.7	128.1	124.0	132.6	142.5	143.8	170.8
2001	134.2	103.8	137.3	135.7	121.3	158.4	105.3	151.8	127.2	174.4	184.8	125.4	123.7	133.2	144.3	145.2	181.3
2002	131.1	99.0	136.2	132.4	119.9	157.6	93.2	151.9	126.8	173.3	185.9	125.9	122.9	133.5	146.2	144.6	182.4
2003	138.1	111.5	143.4	139.1	119.8	162.3	112.9	161.8	130.1	177.4	190.0	129.2	121.9	133.9	148.2	145.7	179.6
2004	146.7	123.3	151.2	147.6	121.0	164.5	126.9	174.4	133.8	195.6	195.7	149.6	122.1	135.1	153.2	148.6	183.2
2005	157.4	118.5	153.1	160.2	122.8	165.4	156.4	192.0	143.8	196.5	202.6	160.8	123.7	139.4	164.2	151.0	195.1
2006	164.7	117.0	153.8	168.8	124.5	168.4	166.7	205.8	153.8	194.4	209.8	181.6	126.2	142.6	179.9	152.6	205.6
2007	172.6	143.4	165.1	175.1	125.8	173.6	177.6	214.8	155.0	192.4	216.9	193.5	127.3	144.7	186.2	155.0	210.3
2008	189.6	161.3	180.5	192.3	128.9	173.1	214.6	245.5	165.9	191.3	226.8	213.0	129.7	148.9	197.1	158.6	216.6
2009	172.9	134.6	176.2	174.8	129.5	157.0	158.7	229.4	165.2	182.8	225.6	186.8	131.3	153.1	202.4	162.2	217.5
2010	184.7	151.0	182.3	187.0	131.7	181.4	185.8	246.6	170.7	192.7	236.9	207.6	131.1	153.2	201.8	163.4	221.5
2011	201.0	186.7	197.5	202.0	141.7	199.9	215.9	275.1	182.7	194.7	245.1	225.9	132.7	156.4	205.0	166.1	229.2
2012	202.2	192.5	205.2	202.1	142.2	202.3	212.1	276.6	186.9	201.6	244.2	219.9	134.2	160.6	211.0	169.8	235.6
2013	203.4	195.3	208.3	203.0	143.4	217.9	211.8	279.2	189.0	214.9	248.8	213.5	135.2	161.1	216.9	171.8	239.5
2014	205.3	197.4	216.5	204.1	145.5	229.1	209.8	280.9	190.2	224.2	250.5	215.0	136.2	163.2	223.7	174.1	243.0
2015	190.4	173.8	209.1	188.8	144.1	210.4	160.5	266.0	187.0	221.9	248.8	200.3	136.9	164.7	228.9	176.5	247.4
2016	185.4	157.0	203.5	184.6	143.3	195.1	145.9	265.1	184.6	222.7	247.7	194.3	136.9	165.1	233.3	177.4	252.1
2017	193.5	161.8	205.4	193.7	145.1	192.1	163.7	280.8	189.0	230.4	254.6	207.8	137.9	166.9	238.6	179.0	257.6
2018	202.0	160.9	206.1	203.7	149.3	180.4	181.6	295.1	195.1	243.9	260.0	223.6	140.3	171.6	247.3	181.2	264.0
2019	199.8	161.5	209.1	200.7	150.7	165.9	168.7	289.1	196.0	236.7	259.4	221.3	143.3	176.4	253.2	182.9	274.2

. . . = Not available.

NOTES AND DEFINITIONS

CHAPTER 8: PRICE INDEXES

This chapter presents price indexes (Consumer and Producer) from two major price collection systems, both conducted by the U.S. Bureau of Labor Statistics. Effective on February 26, 2015, the Bureau of Labor Statistics began utilizing a new estimation system for the Consumer Price Index. The new estimation system, the first major improvement to the existing system in over 25 years. This system is redesigned, state-of-the-art with improved flexibility and review capabilities.

In 2014, the Producer Price Index system was expanded to include services, and construction as well as goods. A new index for "Final Demand" for goods, services and construction, with more than double the scope, is now presented alongside the familiar Finished Goods index. The expanded system is only available beginning in November 2009, and the historical Producer Price Indexes for goods—with data going back to 1913—are also maintained and carried forward. See the Notes and Definitions for Tables 8-5 through 8-7, below, for further information.

Tables 8-1 through 8-4

Average Price Data

SOURCES: U.S. DEPARTMENT OF LABOR, BUREAU OF LABOR STATISTICS (BLS) AND U.S. DEPARTMENT OF COMMERCE, BUREAU OF ECONOMIC ANALYSIS (BEA)

With the release of the Consumer Price Index (CPI) each month, the Bureau of Labor Statistics (BLS) also publishes average retail prices for select utility, automotive fuel, and food items. Although average prices are calculated from the price observations collected for the CPI, they serve a different purpose. The CPI measures price change while average prices provide estimates of price levels. Specifically, average prices are estimates of the average price paid by the consumer for a good or service.

At the U.S. level, the BLS publishes approximately 70 average prices for food items, 6 for utility gas (natural gas) and electricity, and 5 for automotive fuels. Average prices are estimates of the average price paid by the consumer, not estimates of price change. They provide an estimate of the price levels in a given month which cannot be determined from an index. Average prices can be used for comparing price levels of different items in the same month.

Consumer Price Indexes

The Consumer Price Index (CPI), which is compiled by the Bureau of Labor Statistics (BLS), was originally conceived as a statistical measure of the average change in the cost to consumers of a market basket of goods and services purchased by urban wage earners and clerical workers. In 1978, its scope was broadened to also provide a measure of the change in the cost of the average market basket for all urban consumers. There was still a demand for a wage-earner index, so both versions have been calculated and published since then. The most commonly cited measure in this system is the Consumer Price Index for All Urban Consumers (CPI-U). The wage-earner alternative, used according to law for calculating cost of living adjustments in many government programs, including Social Security, and also used by choice of the contracting parties in wage agreements, is called the Consumer Price Index for Urban Wage Earners and Clerical Workers (CPI-W). Both are presented by the BLS back to 1913, and reproduced in *Business Statistics*; however, the movements (percent changes) in the two indexes before 1978 are identical and are based on the wage-earner market basket. (The pre-1978 levels of the two indexes differ because of differences between 1978 and the reference base period.)

These CPIs have typically been called "cost-of-living" indexes, even though the original fixed market basket concept does not correspond to economists' definition of a cost-of-living index, which is the cost of maintaining a constant standard of living or level of satisfaction. In recent years, the concept measured in practice in the CPI has developed into something intended to be closer to the theoretical definition of a cost-of-living index. In addition, a new variation of the CPI—the Chained Consumer Price Index for All Urban Consumers (C-CPI-U)—is intended to provide an even closer approximation.

The reference base for the total BLS Consumer Price Index and most of its components is currently 1982–1984 = 100. However, new products that have been introduced into the index since January 1982 are shown on later reference bases, as is the entire C-CPI-U.

Price indexes for personal consumption expenditures (PCE) are calculated and published by the Bureau of Economic Analysis (BEA) as a part of the national income and product accounts (NIPAs). (See Chapters 1 and 4 and their notes and definitions.) Monthly, quarterly, and annual values are all available. The reference base for these indexes is the average in the NIPA base year, 2009. These indexes differ in a number of other respects from the CPIs, and are often emphasized by the Federal Reserve in its analyses of the nation's economy. The Federal Reserve is mandated to maximize employment (see Chapter 10) and maintain price stability as measured by the Consumer Price Index (CPI). The target inflation rate is 2 percent. Four important NIPA aggregate price indexes are shown in Tables 8-3 and 8-4 for convenient comparison with the CPIs. See the definitions for those tables for more information.

The CPI-U and the CPI-W

The *CPI-U*, which is displayed in Tables 8-1 through 8-4 and also provides all of the component category sub-indexes shown in Table 8-2, uses the consumption patterns for all urban consumers, who currently comprise about 89 percent of the population.

A slightly different index that is widely used for adjusting wages and government benefits is the *CPI-W*, of which the all-items total is shown in each of Tables 8-1 through 8-4. It represents the buying habits of only urban wage earners and clerical workers, who currently comprise about 28 percent of the population. The weights are derived from the same Consumer Expenditure Surveys (CES) used for the CPI-U weights, and are changed on the same schedule. However, they include only consumers from the specified categories instead of all urban consumers.

Beginning with February 2020, the weights in both indexes are based on consumer expenditures in the 2018-2019 period In February 2016, the weights in both indexes are based on consumer expenditures in the 2014-2015 period. In January 2012, the weights in both indexes are based on consumer expenditures in the 2009–2010 period. From January 2010 to December 2011, 2007–2008 weights were used; from January 2008 to December 2009, 2005–2006 weights; from January 2006 to December 2007, 2003–2004 weights; from January 2004 to December 2005, 2001–2002 weights; from January 2002 to December 2003, 1999–2000 weights; and from January 1998 to December 2001, 1993–1995 weights. The weights will continue to be updated at two-year intervals, with new weights introduced in the January indexes of each even-numbered year. Previously, new weights were introduced only at the time of a major revision, which translated into a lag of a decade or more.

Specifically, the CPI weights for 1964 through 1977 were derived from reported expenditures of a sample of wage earner and clerical worker families and individuals in 1960–1961 and adjusted for price changes between the survey dates and 1963. Weights for the 1978–1986 period were derived from a survey undertaken during the 1972–1974 period and adjusted for price change between the survey dates and December 1977. For 1987 through 1997, the spending patterns reflected in the CPI were derived from a survey undertaken during the 1982–1984 period. The reported expenditures were adjusted for price change between the survey dates and December 1986.

The CPI was overhauled and updated in the latest major revision, which took effect in January 1998. In addition, new products and improved methods are regularly introduced into the index, usually in January.

The latest change in methods was the introduction of a geometric mean formula for calculating many of the basic components of the index. Beginning with the index for January 1999, this formula is used for categories comprising approximately 61 percent of total consumer spending. The new formula allows for the possibility that some consumers may react to changing relative prices within a category by substituting items whose relative prices have declined for products whose relative prices have risen, while maintaining their overall level of satisfaction. The geometric mean formula is not used for a few categories in which consumer substitution in the short term is not feasible, currently housing rent and utilities.

The CPI-U was introduced in 1978. Before that time, only CPI-W data were available. The movements of the CPI-U before 1978 are therefore based on the changes in the CPI-W. However, the index levels are different because the two indexes differed in the 1982–1984 base period.

Because the official CPI-U and CPI-W are so widely used in "escalation"—the calculation of cost-of-living adjustments for wages and other private contracts, and for government payments and tax parameters—these price indexes are not retrospectively revised to incorporate new information and methods. (An exception is occasionally made for outright error, which happened in September 2000 and affected the data for January through August of that year.) Instead, the new information and methods of calculation are introduced in the current index and affect future index changes only. In Tables 8-3 and 8-4, PCE indexes, which are subject to routine revision, and special CPI indexes that have been retrospectively revised are presented. These indexes can be used by researchers to provide more consistent historical information.

Notes on the CPI data

The CPI is based on prices of food, clothing, shelter, fuel, utilities, transportation, medical care, and other goods and services that people buy for daytoday living. The quantity and quality of these priced items are kept essentially constant between revisions to ensure that only price changes will be measured. All taxes directly associated with the purchase and use of these items, such as sales and property taxes, are included in the index; the effects of income and payroll tax changes are not included.

Data are collected on about 83,400 individually defined goods and services from about 22,000 retail and service establishments and about 6,000 housing units in 75 urban areas across the country. These data are used to develop the U.S. city average.

Periodic major revisions of the indexes update the content and weights of the market basket of goods and services; update the statistical sample of urban areas, outlets, and unique items used in calculating the CPI; and improve the statistical methods used. In addition, retail outlets and items are resampled on a rotating 5-year basis. Adjustments for changing quality are made at times of major product changes, such as the annual auto model changeover. Other methodological changes are introduced from time to time.

The Consumer Expenditure Survey (CES) provides the weights—that is, the relative importance—used to combine the individual price changes into subtotals and totals. This survey is composed of two separate surveys: an interview survey and a diary survey, both of which are conducted by the Census Bureau for BLS. Each expenditure reported in the two surveys is classified into a series of detailed categories, which are then combined into expenditure classes and ultimately into major expenditure groups. CPI data as of 1998 are grouped into eight such categories: (1) food and beverages, (2) housing, (3) apparel, (4) transportation, (5) medical care, (6) entertainment (7) education and communication, and (8) other goods and services.

Seasonally adjusted national CPI indexes are published for selected series for which there is a significant seasonal pattern of price change. The factors currently in use were derived by the X13ARIMA-SEATS seasonal adjustment method. Some series with extreme or sharp movements are seasonally adjusted using Intervention Analysis Seasonal Adjustment. Seasonally adjusted indexes and seasonal factors for the preceding five years are updated annually based on data through the previous December. Due to these revisions, BLS advises against the use of seasonally adjusted data for escalation. Detailed descriptions of seasonal adjustment procedures are available upon request from BLS.

BLS estimates the "standard error"—the error due to collecting data from a sample instead of the universe—of the one-month percent change in the not-seasonally-adjusted U.S. all-items index in 2013 was 0.03 percentage point. This means that for the 2013 median 0.12 percent change in the All Items CPI-U, BLS is 95 percent confident that the actual percent change based on all retail prices would fall between 0.06 and 0.18 percent. Monthly percent changes in the seasonally adjusted indexes, in contrast, may be revised by 0.1 percentage point when reviewed at the end of the year, and occasionally by even more.

CPI Definitions

Definitions of the major CPI groupings were modified beginning with the data for January 1998. These modifications were carried back to 1993. The following definitions are the current definitions currently used for the CPI components.

The *food and beverage index* includes both food at home and food away from home (restaurant meals and other food bought and eaten away from home).

The *housing index* measures changes in rental costs and expenses connected with the acquisition and operation of a home. The CPIU, beginning with data for January 1983, and the CPIW, beginning with data for January 1985, reflect a change in the methodology used to compute the homeownership component. A rental equivalence measure replaced an asset price approach, which included purchase prices and interest costs. The intent of the change was to separate shelter costs from the investment component of homeownership, so that the index would only reflect the cost of the shelter services provided by owner-occupied homes. In addition to measures of the cost of shelter, the housing category includes insurance, fuel, utilities, and household furnishings and operations.

The *apparel index* includes the purchase of apparel and footwear.

The *private transportation index* includes prices paid by urban consumers for such items as new and used automobiles and other vehicles, gasoline, motor oil, tires, repairs and maintenance, insurance, registration fees, driver's licenses, parking fees, and the like. Auto finance charges, like mortgage interest payments, are considered to be a cost of asset acquisition, not of current consumption. Therefore, they are no longer included in the CPI. City bus, streetcar, subway, taxicab, intercity bus, airplane, and railroad coach fares are some of the components of the *public transportation index*.

The *medical care index* includes prices for professional medical services, hospital and related services, prescription and nonprescription drugs, and other medical care commodities. The weight for the portion of health insurance premiums that is used to cover the costs of these medical goods and services is distributed among the items; the weight for the portion of health insurance costs attributable to administrative expenses and profits of insurance providers constitutes a separate health insurance item. Effective with the January 1997 data, the method of calculating the hospital cost component was changed from the pricing of individual commodities and services to a more comprehensive cost-of-treatment approach.

Recreation includes components formerly listed in housing, apparel, entertainment, and "other goods and services."

Education and communication is a new group including components formerly categorized in housing and "other goods and services," such as college tuition and books, telephone and internet services, and computers.

Other goods and services now includes tobacco, personal care, and miscellaneous.

TABLE 8-2

Purchasing Power of the Dollar

SOURCE: U.S. DEPARTMENT OF LABOR, BUREAU OF LABOR STATISTICS (BLS)

The purchasing power of the dollar measures changes in the quantity of goods and services a dollar will buy at a particular date compared with a selected base date. It must be defined in terms of the following: (1) the specific commodities and services that are to be purchased with the dollar; (2) the market level (producer, retail, etc.) at which they are purchased; and (3) the dates for which the comparison is to be made. Thus, the purchasing power of the dollar for a selected period, compared with another period, may be measured in terms of a single commodity or a large group of commodities such as all goods and services purchased by consumers at retail.

The purchasing power of the dollar is computed by dividing the price index number for the base period by the price index number for the comparison date and expressing the result in dollars and cents. The base period is the period in which the price index equals 100; the average purchasing power in that base period—1982–1984 in the case of the measure shown here—is therefore $1.00.

Purchasing power estimates in terms of both the CPI-U and the CPI-W are calculated by BLS, based on indexes not adjusted for seasonal variation, and published in the CPI press release. The CPI-U version is shown here.

Alternative price measures in Tables 8-3 and 8-4

Table 8-3 shows the all-items CPI-U and CPI-W, along with a number of other indexes that various analysts of price trends have preferred as measures of the price level. Table 8-4 shows the inflation rates (percent changes in price levels) implied by each of the indexes in Table 8-3.

As food and energy prices are volatile and frequently determined by forces separate from monetary aggregate demand pressures, many analysts prefer an index of prices excluding those components. Indexes *excluding food and energy* are known as *core* indexes, and inflation rates calculated from them are known as *core inflation rates.*

The *CPI-U-X1* is a special experimental version of the CPI that researchers have used to provide a more historically consistent series. As explained above, the official CPI-U treated homeownership on an asset-price basis until January 1983. It then changed to a rental equivalence method. The CPI-U-X1 incorporates a rental equivalence approach to homeowners' costs for the years 1967–1982 as well. It is rebased to the December 1982 value of the CPI-U (1982–1984 = 100); thus, it is identical to the CPI-U in December 1982 and all subsequent periods, as can be seen in Table 8-3. For this reason, it is not updated or published in the CPI news release or on the BLS Web site. We continue to present it here because it provides the only available data before 1978 on changes in an improved and more consistent CPI.

The *CPI-E* is an experimental re-weighting of components of the CPI-U to represent price change for the goods and services purchased by Americans age 62 years and over, who accounted for 16.5 percent of the total number of urban consumer units in the 2001–2002 CES.

BLS does not consider the CPI-E to be an ideal measure of price change for older Americans. Because the sample is small, the sampling error in the weights is greater than the error in the all-urban index. The products and outlets sampled are those characteristic of the general urban population rather than older residents. In addition, senior discounts are not included in the prices collected. Such discounts are included—appropriately—in the weights, which are based on the expenditures reported by the older consumers' households. Therefore, such discounts are only a problem if they do not move proportionately to general prices.

The *CPI-U-RS* is a "research series" CPI that retroactively incorporates estimates of the effects of most of the methodological changes implemented since 1978, including the rental equivalence method, new or improved quality adjustments, and improvement of formulas to eliminate bias and allow for some consumer substitution within categories. This index is calculated from 1977 onward. Its reference base is December 1977 = 100. Thus, although it generally shows less <u>increase</u> than the official index, its current <u>levels</u> are considerably higher because the earlier reference base period had lower prices. Unlike the official CPIs and the CPI-U-X1, its historical values will be revised each time a significant change is made in the calculation of the current index. This index is not seasonally adjusted and is not included in the CPI news release. It is available on the BLS Web site, along with an explanation and background material. The CPI-U-RS is used by BLS in the calculation of historical trends in real compensation per hour in its Productivity and Costs system; see Table 9-3 and its notes and definitions. It is also now used by the Census Bureau to convert household incomes into constant dollars, as seen in Chapter 3. And it is used by the editor in some analytical calculations in this volume.

The *C-CPI-U* (Chained Consumer Price Index for All Urban Consumers) is a new, supplemental index that has been published in the monthly CPI news release since August 2002. It is available only from December 1999 to date and is calculated with the base December 1999 = 100; it is not seasonally adjusted. It is designed to be a still-closer approximation to a true cost-of-living index than the CPI-U and the CPI-W, in that it assumes that consumers substitute between as well as within categories as relative prices change, in order to maintain a fixed basket of "consumer satisfaction."

The C-CPI-U is technically a "superlative" index, using a method known as the "Tornqvist formula" to incorporate the composition of consumer spending in the current period as well as in the earlier base period. (All of the other Consumer Price Indexes use a "Laspeyres" formula; see the General Notes at the beginning of this volume.) As it requires consumer expenditure data for the current as well as the earlier period, its final version can only be calculated after the expenditure data become available—about two years later—and is approximated in more recent periods by making more extensive use of the geometric mean formula (see above). With the release of January 2014 data, the indexes for 2012 were revised to their final form, and the initial indexes for 2013 were revised to "interim" levels, shown in Tables 8-3 and 8-4.

Personal consumption expenditure (PCE) chain-type price indexes are calculated by the Bureau of Economic Analysis (BEA) in the framework of the national income and product accounts (NIPAs). (See the notes and definitions for Chapters 1 and 4.) The scope of NIPA PCE is broader than the scope of the CPI. PCE includes the rural as well as the urban population and also covers the consumption spending of nonprofit entities. The CPI includes only consumer out-of-pocket cash spending, whereas PCE includes expenditures financed by government and private insurance, particularly in the medical care area. For this reason, there is a large difference between the relatively small weight of medical care spending in the CPI and the markedly greater percentage of PCE that is accounted for by total medical care spending. Housing, on the other hand, has a somewhat smaller weight in PCE while all non-housing components have a higher weight. The reason for this is that the CES—the survey on which the CPI weights are based—tends to report housing expenditures accurately and somewhat underestimate other spending. This suggests that the weight of housing relative to all other products may be overestimated in the CPI but measured more correctly in the PCE price index.

PCE chain-type indexes use the expenditure weights of both the earlier and the later period to determine the aggregate price change between the two periods. (See the notes and definitions for Chapter 1, as well as the General Notes on index number formulas.) Thus, they are subject to revision as improved data on the composition of consumption spending become available, and in this respect resemble the C-CPI-U.

For a large share of PCE, the price movements for basic individual spending categories are determined by CPI components. Hence, the differences between the rates of change in the aggregate CPI and PCE indexes are largely the result of the different weights, but also reflect some alternative methodologies and the previously mentioned differences in scope.

Market-based PCE indexes are based on household expenditures for which there are observable price measures. They exclude most implicit prices (for example, the services furnished without payment by financial intermediaries) and they exclude items not deflated by a detailed component of either the Consumer Price Index (CPI) or the Producer Price Index (PPI). This means that the price observations that make up these new aggregate measures are all based on observed market transactions, making them "market-based price indexes." The imputed rent for owner-occupied housing is included in the market-based price index, since it is based on observed rentals of comparable homes. Household insurance premiums are also included in the market-based index, since they are deflated by the CPI for tenants' and household insurance. Excluded are services furnished without payment by financial intermediaries, most insurance purchases, expenses of NPISHs (nonprofit institutions serving households), legal gambling (illegal gambling is excluded from all measures), margins on used light motor vehicles, and expenditures by U.S. residents working and traveling abroad. Also excluded are medical, hospitalization, and income loss insurance; expense of handling life insurance; motor vehicle insurance; and workers' compensation.

The *inflation rates* shown in Table 8-4 are percent changes in the price indexes introduced in Table 8-3. For annual indexes, the

rate is the percent change from the previous year. For monthly indexes, the rate is the percent change from the same month a year earlier. To give an indication of the longer-run implications of these different price indicators, comparisons of compound annual inflation rates, calculated by the editor, are also shown for the 1978–2013, 1983–2013, 2000–2013 and 2013-2015 periods, using the growth rate formula presented in the article at the beginning of this volume.

Data availability and references

The CPI-U, CPI-W, and C-CPI-U are initially issued in a press release two to three weeks after the end of the month for which the data were collected. This release and detailed and complete current and historical data on the CPI and its variants and components, along with extensive documentation, are available on the BLS Web site at <http://www.bls.gov/cpi>. Seasonal factors and seasonally adjusted indexes are revised once a year with the issuance of the January index.

Information available on the BLS Web site includes "Common Misconceptions about the Consumer Price Index: Questions and Answers;" another fact sheet on frequently asked questions; a fact sheet on seasonal adjustment; Chapter 17 of the *BLS Handbook of Methods*, entitled "The Consumer Price Index"; a section entitled "Note on Chained Consumer Price Index for All Urban Consumers"; and a number of explanatory CPI fact sheets on specific subjects.

As previously indicated, the CPI-U-X1 is not currently published because its recent values are identical to the CPI-U. The CPI-E is presented in articles in the CPI section of the BLS Web site, the most recent of which is "Experimental Consumer Price Index for Americans 62 Years of Age and Older, 1998-2005"; recent values are available by request from BLS. The CPI-U-RS is updated each month in a report entitled "CPI Research Series Using Current Methods" on the site. In both cases, the reports describe the indexes and provide references.

The monthly PCE indexes are included in the personal income report issued by BEA, which is published near the end of the following month. These indexes are revised month-by-month to reflect new information and annually to reflect the annual and quinquennial benchmarking of the NIPAs. The complete historical record can be found on the BEA Web site at <http://www.bea.gov> in the Personal Income and Outlays section of the NIPA tables, in tables entitled "Price Indexes for Personal Consumption Expenditures by Major Type of Product."

Two special editions of the *Monthly Labor Review* were devoted to CPI issues. The December 1996 issue describes the subsequently implemented 1997 and 1998 revisions in a series of articles, and the December 1993 issue, entitled *The Anatomy of Price Change*, includes the following articles: "The Consumer Price Index: Underlying Concepts and Caveats"; "Basic Components of the CPI: Estimation of Price Changes"; "The Commodity Substitution Effect in CPI Data, 1982–1991"; and "Quality Adjustment of Price Indexes."

The new formula for calculating basic components is described in "Incorporating a Geometric Mean Formula into the CPI," *Monthly Labor Review* (October 1998). For a detailed discussion of the treatment of homeownership, see "Changing the Home-ownership Component of the Consumer Price Index to Rental Equivalence," *CPI Detailed Report* (January 1983).

For a comprehensive professional review of CPI concepts and methodology, see Charles Schultze and Christopher Mackie, ed., *At What Price? Conceptualizing and Measuring Cost-of-Living and Price Indexes* (Washington, DC: National Academy Press, 2001). Earlier references include: "Using Survey Data to Assess Bias in the Consumer Price Index," *Monthly Labor Review* (April 1998); Joel Popkin, "Improving the CPI: The Record and Suggested Next Steps," *Business Economics*, Vol. XXXII, No. 3 (July 1997), pages 42–47; *Measurement Issues in the Consumer Price Index* (Bureau of Labor Statistics, U.S. Department of Labor, June 1997); *Toward a More Accurate Measure of the Cost of Living* (Final Report to the Senate Finance Committee from the Advisory Commission to Study the Consumer Price Index, December 4, 1996)—also known as the "Boskin Commission" report; and *Government Price Statistics* (U.S. Congress Joint Economic Committee, 87th Congress, 1st Session, January 24, 1961)—also known as the "Stigler Committee" report.

For an explanation of the differences between the CPI-U and the PCE index, see Clinton P. McCully, Brian C. Moyer, and Kenneth J. Stewart, "Comparing the Consumer Price Index and the Personal Consumption Expenditures Price Index," *Survey of Current Business,* November 2007, pp. 26-33.

TABLES 8-1 AND 8-5 THROUGH 8-7

Producer Price Indexes

SOURCE: U.S. DEPARTMENT OF LABOR, BUREAU OF LABOR STATISTICS (BLS)

The Bureau of Labor Statistics first compiled and published a "Wholesale Price Index," measuring prices of various goods from the perspective of the seller, in 1902; it covered the years 1890-1901. Over the years this index was broadened and refined. In a major revamping in 1978, it was structured into an input-output format, differentiated between stages of processing, and renamed the Producer Price Index. The index for "Finished Goods," from the stage-of-processing system, became the featured "headline" number. Still, this index only covered goods.

New FD-ID system (Tables 8-5 and 8-6)

Over subsequent years the Bureau instituted collection of an increasing number of prices for services as well. All prices are collected within an industry input-output structure, with one of the most important purposes being to measure real output quantities by deflating industry outputs and inputs. In the new aggregation system introduced in February 2014, prices for goods, services, and construction have been integrated into a set of Final Demand-Intermediate Demand (FD-ID) indexes, organized by

class of buyer, degree of fabrication, and stage of production. Its principal components are calculated beginning in November 2009 with that month as the reference base (i.e. November 2009 = 100). Components of the previous goods PPI system continue to be calculated on the base 1982 = 100 as well.

> In the new FD-ID system, all products, including services and construction, are referred to as "commodities." In the old PPI system and in the CPI system, the term "commodities" refers to goods only; note, for example, how the CPI components are aggregated into summary measures of "commodities" [goods] and services, displaying different trends and behaviors. Unless and until these terminologies are reconciled, users need to be mindful of which system they are operating in.

More that 100,000 producer price quotations from over 25,000 establishment are collected each month; more than 10,000 indexes for individual products and groups of products are released each month. The prices are organized into three sets of PPIs: the FD-ID indexes; commodity indexes; and indexes for the net output of industries and their products.

Final demand. The final demand portion of the FD-ID structure measures price change for commodities sold for personal consumption, capital investment, government, and export. *Final demand trade services* measures changes in margins received by wholesalers and retailers. The other main components are *final demand goods, final demand transportation and warehousing services, other final demand services,* and *final demand construction.* The total index for final demand is now featured by BLS in its press release as its "headline number," replacing the previously featured Finished Goods index.

Intermediate demand. This portion of the system tracks price changes for goods, services, and construction products sold to businesses as inputs to production and not as capital investment. There are two parallel treatments.

The first treatment organizes intermediate demand commodities by type, including the distinction between *processed goods* and *unprocessed goods,* which is carried over from the older PPI goods system.

The second system organizes intermediate demand commodities into production stages, with the explicit goal of developing a forward-flow model of production and price change. For each production stage, the intermediate demand index measures the prices of inputs to the included industries. Many commodities, for example energy products, are sold to both final demand and various intermediate stages of production; their weight is allocated proportionately to all the consuming stages of production and their price movements are proportionately represented in each stage.

Industries assigned to *Stage 4* primarily produce final demand commodities. The intermediate demand index for Stage 4 measures inputs to those industries, such as motor vehicle parts, beef and veal, and long distance motor carrying.

Industries assigned to *Stage 3* produce the output sold to Stage 4, and therefore include motor vehicle parts and slaughtering. Examples of inputs to Stage 3 included in their intermediate demand index include slaughter steers and heifers, industrial electric power, steel plates, and temporary help services.

Industries assigned to *Stage 2* include petroleum refining, electricity, and insurance agencies. The goods and services that they purchase and that are included in their intermediate demand index include crude oil and business loans.

Industries assigned to *Stage 1* include oil and gas extraction, paper mills, and advertising services. Inputs to those industries include electric power and solid waste collection. According to BLS, "It should be noted that all inputs purchased by stage 1 industries are by definition produced either within stage 1 or by later stages of processing, leaving stage 1 less useful for price transmission analysis."

Commodity indexes

Aggregation by *commodity* organizes the price measures by similarity of product or end use, disregarding industry of origin. For the original Wholesale Price Index (goods only), this was the principal means of aggregation; in Table 8-1C, producer price indexes are shown for *farm products, industrial commodities*, and "*all commodities*" back to the year 1913. Table 8-7 shows 16 goods commodity groups from 1950 to date. Whatever their imperfections, these measures provide the longest historical perspective for producer prices.

The major goods commodity groups—particularly the totals for all commodities and industrial commodities—came under criticism in the energy price crisis of the early 1970s, for aggregating successive stages of processing of the same product (e.g. crude oil) and thus exaggerating price trends. As a result, the stage-of-processing grouping was introduced in 1978, and the finished goods components, in total and minus food and energy, became the headline Producer Price Indexes.

Stage of processing indexes

In the PPI for goods as reconstructed in 1978 and shown in Table 8-1B, the three major indexes are: (1) *finished goods*, or goods that will not undergo further processing and are ready for sale to the ultimate user (such as automobiles, meats, apparel, and machine tools, and also unprocessed foods such as eggs and fresh vegetables, that are ready for the consumer); (2) *processed goods for intermediate demand* (formerly *intermediate materials, supplies, and components*), or goods that have been processed but require further processing before they become finished goods (such as steel mill products, cotton yarns, lumber, and flour), as well as physically complete goods that are purchased by business firms as inputs for their operations (such as diesel fuel and paper boxes); and (3) *unprocessed goods for intermediate demand* (formerly *crude materials for further processing*), or goods entering the market for the first time that have not been manufactured or fabricated and are not sold directly to consumers (such as ores, scrap metals, crude petroleum, raw cotton, and livestock).

Notes on the data

The probability sample used for calculating the PPI provides more than 100,000 price quotations per month. To the greatest extent possible, prices used in calculating the PPI represent prices received by domestic producers in the first important commercial transaction for each commodity. These indexes attempt to measure only price changes (changes in receipts per unit of measurement not influenced by changes in quality, quantity sold, terms of sale, or level of distribution). Transaction prices are sought instead of list or book prices. Price data are collected monthly via Internet, mail, and fax. Prices are obtained directly from producing companies on a voluntary and confidential basis. Prices are generally reported for the Tuesday of the week containing the 13th day of the month. Samples are updated regularly.

The BLS revises the PPI weighting structure when data from economic censuses become available. Beginning with data for January 2012, the weights used to construct the PPI reflect 2007 shipments values as measured by the 2007 Economic Censuses. Data for January 2007 through December 2011 used 2002 values; 2002 through 2006, 1997 shipments values; 1996 through 2001, 1992 values; 1992 through 1995, 1987 values; 1987 through 1991, 1982 values; 1976 through 1986, 1972 values; and 1967 through 1975, 1963 values.

Price indexes are available in unadjusted form and are also adjusted for seasonal variation using the X-12-ARIMA method. Since January 1988, BLS has also used X-12-ARIMA Intervention Analysis Seasonal Adjustment for a small number of series to remove unusual values that might distort seasonal patterns before calculating the seasonal adjustment factors. Seasonal factors for the PPI are revised annually to take into account the most recent 12 months of data. Seasonally adjusted data for the previous 5 years are subject to these annual revisions.

The newer FD-ID indexes are only available beginning in November 2009. Hence, they only regard, so far, the recovery and expansion phase of a particularly severe business cycle; the editor suggests that up to now, both seasonal adjustments and broader inferences about relative price behavior are necessarily based on a short and possibly unrepresentative time period. Partly for that reason, we present both unadjusted and seasonally adjusted indexes in Tables 8-5 and 8-6.

Data availability and references

The indexes are initially issued in a press release two to three weeks after the end of the month for which the data were collected. Data are subsequently published in greater detail in a monthly BLS publication, *PPI Detailed Report*. Each month, data for the fourth previous month (both unadjusted and seasonally adjusted) are revised to reflect late reports and corrections.

The press release, the *PPI Detailed Report*, detailed and complete current and historical data, and extensive documentation are available at <http://www.bls.gov/ppi>. The items available on this Web site include Chapter 14 of the *BLS Handbook of Methods*, "Producer Price Indexes"; a selection of *Monthly Labor Review* articles on the PPI; and fact sheets on a number of issues and index components.

CHAPTER 9: EMPLOYMENT COSTS, PRODUCTIVITY, AND PROFITS

SECTION 9A: EMPLOYMENT COST INDEXES

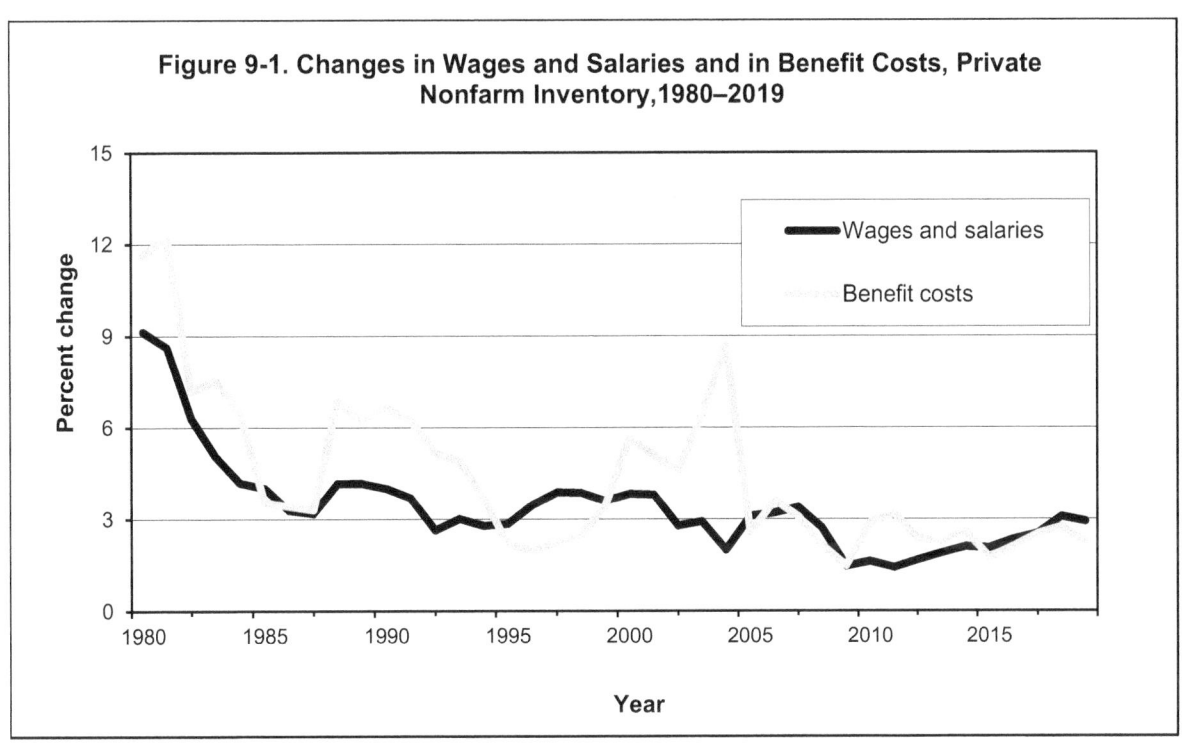

Figure 9-1. Changes in Wages and Salaries and in Benefit Costs, Private Nonfarm Inventory,1980–2019

- The Employment Cost Index (ECI) measures the change in the cost of labor, free from the influence of employment shifts among occupations and industries. This cost of labor includes not only wages and salaries but also benefits. These benefits include vacations and sick leave, overtime, health insurance, and retirement. (See Chapter 8.)

- Analogous to the ECI is the Consumer Price Index (CPI), which holds the market basket constant in any individual time period so as to reflect only an aggregate of price changes affecting individual defined products.

- SIC industrial classification began in 1930 when the U.S. was manufacturing-based. Changes to the economy to serviced-based could not be accommodated by SIC. Table 9-1 presents the ECI by the North American Industrial Classification (NAICS) while Table 9-2 presents the data for Standard Industrial Code (SIC). The NAICS system began with the North American Free Trade Agreement (NAFTA), instituted in 1994 with Canada and Mexico.

- From 2007 to 2019, total compensation for all civilian workers rose 30 percent. During this period, wages and salaries grew slightly less at 29.1 percent while total benefits grew more rapidly at 32.2 percent. Much of this increase was due to medical care costs. (See Table 9-1)

Table 9-1. Employment Cost Indexes, NAICS Basis

(December 2005 [not seasonally adjusted] = 100; annual values are for December, not seasonally adjusted; quarterly values, seasonally adjusted, except as noted.)

Year and month	All civilian workers [1]	State and local government workers	Private industry workers										Union [2]	Non-union [2]
			All private industry workers	By occupational group					By industry				
				Management, professional, and related	Sales and office	Natural resources, construction, and maintenance	Production, transportation, and material moving	Service occupations	Goods-producing industries		Service-providing		
									Total	Manufacturing			
TOTAL COMPENSATION													
2007	106.7	108.4	106.3	106.8	106.1	106.7	104.5	107.0	105.0	103.8	106.7	105.1	106.5
2008	109.5	111.6	108.9	109.9	107.9	109.6	106.9	109.8	107.5	105.9	109.4	108.0	109.1
2009	111.0	114.2	110.2	110.7	109.2	111.2	108.9	111.8	108.6	107.0	110.8	111.1	110.1
2010	113.2	116.2	112.5	113.0	111.6	113.3	111.5	113.5	111.1	110.0	113.0	114.8	112.1
2011	115.5	117.7	115.0	115.4	114.2	115.8	114.2	115.4	113.8	113.1	115.3	117.9	114.5
2012	117.7	119.9	117.1	117.7	116.3	117.8	116.0	117.4	115.6	114.9	117.6	120.5	116.6
2013	120.0	122.2	119.4	120.2	119.0	120.1	118.0	119.0	117.7	117.0	120.0	122.6	119.0
2014	122.7	124.7	122.2	123.0	121.9	123.2	120.6	121.1	120.3	119.8	122.8	126.7	121.5
2015	125.1	127.8	124.5	125.3	123.9	124.9	123.7	123.2	123.2	122.8	124.9	128.7	123.8
2016	127.9	130.9	127.2	127.4	126.8	127.7	127.0	127.2	125.8	125.5	127.7	130.6	126.6
2017	131.2	134.2	130.5	130.5	130.1	131.2	131.0	130.9	128.9	128.9	131.0	134.5	129.9
2018	135.0	137.7	134.4	133.8	135.0	134.5	134.7	135.6	131.9	131.6	135.2	138.8	133.7
2019	138.7	141.7	138.0	136.8	138.8	137.9	139.2	140.6	135.8	135.3	138.7	142.3	137.3
2008													
March	107.6	109.0	107.2	108.0	106.8	107.7	105.5	107.8	106.0	104.6	107.6	105.9	107.5
June	108.3	109.8	108.0	108.8	107.4	108.2	106.0	108.7	106.7	105.0	108.4	106.7	108.3
September	109.1	110.9	108.6	109.6	107.8	108.9	106.5	109.4	107.2	105.6	109.1	107.4	108.9
December	109.6	111.6	109.1	110.2	108.0	109.7	107.0	109.9	107.7	106.1	109.5	108.0	109.1
2009													
March	109.9	112.4	109.3	110.2	108.1	109.9	107.7	110.6	107.9	106.4	109.8	109.1	109.4
June	110.2	113.2	109.5	110.4	108.2	110.3	108.1	111.0	108.1	106.6	110.0	109.8	109.6
September	110.7	113.5	109.9	110.6	108.8	110.7	108.6	111.6	108.3	106.8	110.5	110.5	109.9
December	111.1	114.1	110.4	111.0	109.3	111.3	109.0	112.0	108.8	107.3	110.9	111.1	110.1
2010													
March	111.8	114.5	111.1	111.7	109.9	112.3	109.9	112.3	109.7	108.4	111.5	112.8	110.9
June	112.3	115.1	111.6	112.1	110.6	112.6	110.4	112.6	110.2	109.0	112.1	113.7	111.4
September	112.8	115.5	112.1	112.7	111.1	112.9	111.2	113.3	110.9	109.9	112.5	114.6	111.8
December	113.3	116.1	112.6	113.3	111.7	113.4	111.6	113.6	111.2	110.2	113.1	114.8	112.1
2011													
March	114.0	116.7	113.3	114.0	112.2	113.9	112.2	114.5	112.0	111.3	113.8	115.6	113.0
June	114.7	117.0	114.2	114.7	113.2	114.8	113.5	114.7	113.2	112.6	114.5	117.1	113.8
September	115.1	117.3	114.6	115.1	113.7	115.4	113.7	115.0	113.3	112.8	115.0	117.4	114.2
December	115.6	117.7	115.1	115.7	114.3	115.9	114.3	115.5	113.9	113.3	115.5	117.9	114.5
2012													
March	116.3	118.3	115.7	116.3	115.1	116.4	114.5	116.0	114.1	113.4	116.3	118.3	115.3
June	116.8	118.9	116.3	117.0	115.7	116.9	115.0	116.5	114.6	113.9	116.9	119.3	116.0
September	117.3	119.5	116.8	117.3	116.2	117.5	115.6	116.8	115.2	114.6	117.3	120.2	116.3
December	117.8	119.9	117.2	118.0	116.5	118.0	116.1	117.4	115.7	115.1	117.7	120.5	116.6
2013													
March	118.4	120.5	117.9	118.5	117.3	118.7	116.7	117.8	116.4	115.7	118.3	121.5	117.3
June	119.0	121.0	118.5	119.3	117.7	119.0	117.1	118.3	116.9	116.2	119.0	122.1	118.0
September	119.5	121.4	119.0	119.9	118.4	119.7	117.5	118.4	117.4	116.7	119.5	122.5	118.5
December	120.1	122.2	119.6	120.4	119.2	120.3	118.1	119.1	117.8	117.2	120.2	122.6	119.0
2014													
March	120.5	122.8	119.9	120.7	119.3	120.9	118.8	119.0	118.6	118.0	120.4	123.5	119.4
June	121.4	123.5	120.9	121.9	120.3	121.9	119.5	119.6	119.1	118.6	121.5	125.0	120.4
September	122.2	124.1	121.7	122.7	121.1	122.6	120.2	120.5	119.8	119.3	122.3	125.8	121.1
December	122.8	124.7	122.4	123.1	122.1	123.4	120.8	121.2	120.4	120.0	123.0	126.7	121.5
2015													
March	123.6	125.4	123.2	123.8	123.3	123.7	121.6	121.8	121.0	120.8	123.8	127.4	122.5
June	123.8	126.2	123.2	124.0	122.6	124.0	122.4	122.1	121.8	121.6	123.7	127.5	122.7
September	124.6	126.9	124.0	124.8	123.5	124.5	123.0	122.6	122.4	122.1	124.5	128.0	123.4
December	125.3	127.8	124.6	125.5	124.3	125.1	123.9	123.3	123.3	123.0	125.1	128.7	123.8
2016													
March	126.0	128.5	125.3	126.0	125.0	125.8	124.6	124.2	123.8	123.6	125.9	129.6	124.7
June	126.7	129.2	126.1	126.3	126.5	126.5	125.5	125.4	124.6	124.4	126.6	130.0	125.7
September	127.4	130.2	126.7	127.0	126.6	126.9	126.3	126.5	125.1	125.1	127.3	130.4	126.3
December	128.1	130.9	127.3	127.6	127.2	127.8	127.3	127.3	125.9	125.7	127.8	130.6	126.6
2017													
March	129.0	131.8	128.2	128.5	128.0	128.6	128.1	128.4	126.5	126.3	128.8	131.9	127.8
June	129.7	132.5	129.0	129.4	128.7	129.6	129.0	129.1	127.2	127.0	129.6	132.7	128.6
September	130.6	133.3	129.9	129.9	129.7	130.5	130.4	130.0	128.3	128.4	130.5	133.7	129.5
December	131.4	134.2	130.7	130.7	130.5	131.3	131.2	131.1	129.0	129.0	131.2	134.5	129.9
2018													
March	132.5	134.8	131.9	131.8	131.8	131.9	132.3	132.4	129.9	129.9	132.5	135.4	131.4
June	133.4	135.8	132.8	132.5	133.1	133.2	133.2	133.4	130.8	130.7	133.4	137.4	132.2
September	134.3	136.8	133.7	133.2	134.2	133.7	134.1	134.6	131.2	130.9	134.5	137.8	133.2
December	135.2	137.8	134.6	134.0	135.4	134.6	134.9	135.8	132.0	131.7	135.4	138.8	133.7
2019													
March	136.2	138.8	135.5	134.7	136.2	135.5	136.2	137.0	133.0	132.8	136.2	140.0	134.9
June	137.0	139.8	136.3	135.5	137.1	136.0	137.2	138.4	134.0	133.7	137.0	140.8	135.8
September	138.0	140.9	137.3	136.3	138.2	137.1	138.1	139.3	135.1	134.6	138.0	141.4	136.8
December	138.9	141.8	138.2	137.0	139.2	138.0	139.4	140.8	135.9	135.5	138.9	142.3	137.3

[1] Excludes farm workers, private household workers, and federal government employees.
[2] Not seasonally adjusted.

Table 9-1. Employment Cost Indexes, NAICS Basis—*Continued*

(December 2005 [not seasonally adjusted] = 100; annual values are for December, not seasonally adjusted; quarterly values, seasonally adjusted, except as noted.)

Year and month	All civilian workers [1]	State and local government workers	Private industry workers										
			All private industry workers	By occupational group					By industry			Union [2]	Non-union [2]
				Management, professional, and related	Sales and office	Natural resources, construction, and maintenance	Production, transportation, and material moving	Service occupations	Goods-producing industries		Service-providing		
									Total	Manufacturing			
WAGES AND SALARIES													
2007	106.7	107.1	106.6	107.2	106.2	107.1	105.1	107.3	106.0	104.9	106.8	104.7	106.9
2008	109.6	110.4	109.4	110.5	108.1	110.6	108.0	110.3	109.0	107.7	109.7	108.1	109.6
2009	111.2	112.5	110.8	111.5	109.6	112.1	109.8	112.6	110.1	108.9	111.4	110.9	110.9
2010	113.0	113.8	112.8	113.7	111.7	113.4	111.5	113.9	111.6	110.7	113.2	112.9	112.7
2011	114.6	114.9	114.6	115.5	113.7	115.4	113.1	115.4	113.5	112.7	114.9	114.9	114.6
2012	116.5	116.2	116.6	117.7	115.8	116.7	115.2	117.0	115.4	114.8	116.8	117.4	116.5
2013	118.7	117.5	119.0	120.2	118.3	118.8	117.3	118.5	117.6	117.2	118.9	119.8	118.9
2014	121.2	119.4	121.6	122.8	121.1	121.3	119.9	120.9	120.1	119.8	121.4	123.1	121.5
2015	123.7	121.6	124.2	125.6	123.4	123.5	122.8	123.0	123.2	123.0	123.8	125.5	124.0
2016	126.6	124.1	127.1	128.1	126.0	126.6	126.7	127.0	126.1	126.2	126.6	127.3	127.1
2017	129.8	126.7	130.6	131.1	129.5	130.3	130.8	130.8	129.3	129.3	129.9	130.9	130.6
2018	133.8	129.7	134.7	134.6	134.5	133.2	135.1	135.6	132.9	132.9	133.9	134.6	134.7
2019	137.7	133.0	138.7	137.7	138.7	137.2	140.3	140.8	137.4	137.1	137.7	139.2	138.7
2008													
March	107.6	107.8	107.6	108.4	106.9	108.2	106.0	107.8	107.1	105.9	107.7	105.5	107.9
June	108.5	108.6	108.4	109.2	107.5	108.9	106.8	108.9	107.9	106.7	108.6	106.7	108.7
September	109.2	109.7	109.0	110.1	107.9	109.7	107.4	109.6	108.5	107.3	109.2	107.4	109.4
December	109.7	110.3	109.5	110.7	108.0	110.5	107.9	110.2	109.1	107.9	109.7	108.1	109.6
2009													
March	110.0	111.0	109.8	110.9	108.1	110.7	108.3	110.9	109.2	108.0	110.0	108.8	110.0
June	110.4	111.7	110.1	111.1	108.2	111.1	108.8	111.3	109.4	108.3	110.2	109.6	110.2
September	110.8	111.9	110.5	111.3	108.9	111.5	109.3	112.0	109.7	108.6	110.7	110.2	110.6
December	111.2	112.4	111.0	111.8	109.5	112.1	109.7	112.4	110.2	109.1	111.2	110.9	110.9
2010													
March	111.7	112.8	111.4	112.3	109.8	112.6	109.9	112.5	110.5	109.3	111.7	111.5	111.4
June	112.1	113.2	111.9	112.8	110.5	112.8	110.3	112.8	110.9	109.9	112.2	112.1	111.9
September	112.5	113.3	112.3	113.4	110.8	113.0	111.0	113.2	111.4	110.5	112.6	112.7	112.4
December	113.0	113.8	112.9	113.9	111.5	113.3	111.3	113.5	111.7	110.9	113.2	112.9	112.7
2011													
March	113.4	114.1	113.2	114.4	111.8	113.8	111.6	114.2	112.2	111.5	113.6	113.6	113.2
June	113.9	114.4	113.7	114.8	112.5	114.5	112.0	114.3	112.7	111.9	114.1	114.0	113.8
September	114.3	114.5	114.2	115.2	113.1	115.0	112.4	114.6	113.1	112.5	114.6	114.6	114.3
December	114.7	114.8	114.7	115.7	113.7	115.4	112.9	115.1	113.6	113.0	115.0	114.9	114.6
2012													
March	115.3	115.2	115.3	116.3	114.5	115.7	113.7	115.4	114.0	113.5	115.7	115.6	115.2
June	115.8	115.6	115.8	116.9	115.0	116.0	114.0	115.9	114.4	113.9	116.2	116.2	115.9
September	116.2	115.8	116.3	117.2	115.7	116.4	114.6	116.2	115.0	114.5	116.7	116.9	116.3
December	116.6	116.1	116.7	117.8	115.9	116.8	115.2	116.8	115.5	115.0	117.1	117.4	116.5
2013													
March	117.2	116.4	117.4	118.5	116.7	117.3	115.8	117.2	116.1	115.6	117.7	118.4	117.2
June	117.7	116.7	118.0	119.3	117.1	117.6	116.2	117.7	116.7	116.3	118.3	119.0	117.9
September	118.2	116.9	118.4	119.7	117.7	118.4	116.6	117.6	117.3	116.8	118.8	119.6	118.4
December	118.8	117.4	119.1	120.3	118.6	118.9	117.3	118.3	117.7	117.4	119.5	119.8	118.9
2014													
March	119.1	117.8	119.4	120.6	118.6	119.5	118.0	118.4	118.2	118.0	119.7	120.5	119.2
June	119.8	118.2	120.2	121.7	119.5	120.0	118.7	119.0	118.9	118.7	120.6	121.2	120.2
September	120.7	118.8	121.1	122.5	120.3	120.7	119.5	120.1	119.5	119.3	121.5	122.1	121.0
December	121.3	119.4	121.8	122.8	121.5	121.5	120.0	120.8	120.2	120.0	122.2	123.1	121.5
2015													
March	122.1	120.0	122.6	123.5	123.2	121.8	120.7	121.5	120.9	120.8	123.1	123.7	122.4
June	122.4	120.6	122.8	124.1	122.1	122.6	121.5	121.6	121.7	121.6	123.2	124.5	122.7
September	123.1	121.0	123.6	125.0	123.0	123.0	122.1	122.2	122.4	122.3	124.0	124.8	123.6
December	123.8	121.5	124.3	125.7	123.8	123.6	123.0	122.9	123.3	123.2	124.6	125.5	124.0
2016													
March	124.5	122.1	125.1	126.5	124.4	124.5	123.8	123.9	123.9	123.9	125.5	126.4	125.0
June	125.4	122.7	126.0	126.8	126.2	125.4	124.9	125.2	124.8	124.8	126.4	126.8	126.0
September	126.0	123.4	126.6	127.6	126.0	125.8	125.8	126.4	125.4	125.7	127.0	127.2	126.6
December	126.7	124.0	127.3	128.2	126.6	126.8	126.8	127.3	126.3	126.4	127.6	127.3	127.1
2017													
March	127.6	124.8	128.2	129.0	127.6	127.7	127.8	128.4	127.1	127.0	128.6	128.6	128.2
June	128.3	125.4	129.0	129.9	128.2	128.7	128.8	129.3	127.8	127.8	129.4	129.2	129.1
September	129.2	125.9	129.9	130.4	129.4	129.6	130.1	130.3	128.7	128.7	130.3	129.9	130.0
December	129.9	126.5	130.7	131.2	130.2	130.5	131.0	131.5	129.5	129.5	131.1	130.9	130.6
2018													
March	131.1	127.1	132.0	132.3	131.6	131.2	132.1	132.8	130.4	130.3	132.5	131.6	132.1
June	131.8	127.8	132.8	132.9	132.9	131.9	133.2	133.9	131.4	131.2	133.2	132.7	132.9
September	132.9	128.7	133.9	133.9	134.1	132.5	134.4	135.3	132.1	132.0	134.4	133.5	134.1
December	133.9	129.5	134.9	134.7	135.3	133.4	135.4	136.6	133.1	133.0	135.4	134.6	134.7
2019													
March	134.8	130.3	135.9	135.4	136.3	134.3	136.9	138.1	134.1	134.1	136.4	135.7	136.0
June	135.7	131.1	136.8	136.3	137.2	135.3	137.9	139.6	135.2	135.1	137.2	137.0	136.9
September	136.8	132.2	137.9	137.1	138.5	136.5	139.1	140.6	136.5	136.2	138.3	138.1	138.0
December	137.8	132.9	138.9	137.8	139.6	137.4	140.8	142.4	137.6	137.3	139.3	139.2	138.7

[1] Excludes farm workers, private household workers, and federal government employees.
[2] Not seasonally adjusted.

Table 9-1. Employment Cost Indexes, NAICS Basis—*Continued*

(December 2005 [not seasonally adjusted] = 100; annual values are for December, not seasonally adjusted; quarterly values, seasonally adjusted, except as noted.)

Year and month	All civilian workers [1]	State and local government workers	Private industry workers: All private industry workers	By occupational group: Management, professional, and related	Sales and office	Natural resources, construction, and maintenance	Production, transportation, and material moving	Service occupations	By industry: Goods-producing industries — Total	Goods-producing industries — Manufacturing	Service-providing	Union [2]	Non-union [2]
TOTAL BENEFITS													
2007	106.8	111.0	105.6	106.0	106.0	105.9	103.7	106.7	103.2	101.7	106.6	105.8	105.6
2008	109.1	114.2	107.7	108.5	107.8	107.7	105.1	108.8	104.7	102.5	108.9	107.8	107.6
2009	110.7	117.7	108.7	108.8	108.7	109.5	107.4	110.5	105.8	103.6	109.9	111.4	108.2
2010	113.9	121.1	111.9	111.2	111.8	113.2	112.0	113.5	110.1	108.8	112.6	117.9	110.6
2011	117.5	123.6	115.9	115.2	115.5	116.8	117.0	116.4	114.4	113.9	116.4	122.8	114.4
2012	120.3	127.8	118.2	117.9	117.6	120.2	118.0	119.2	116.0	115.0	119.1	125.6	116.7
2013	123.0	132.0	120.5	120.2	120.5	122.9	119.5	121.0	118.0	116.6	121.5	127.3	119.1
2014	126.2	135.8	123.5	123.4	123.4	127.1	122.1	122.2	120.7	119.8	124.6	132.6	121.7
2015	128.4	140.6	125.1	124.5	125.0	127.9	125.4	124.3	123.1	122.5	125.9	133.9	123.3
2016	131.1	145.0	127.3	126.0	128.4	129.9	127.9	127.1	124.9	124.3	128.3	136.1	125.5
2017	134.4	149.7	130.2	129.1	130.8	132.9	131.4	129.7	128.0	128.0	131.2	140.6	128.2
2018	138.1	154.4	133.6	132.1	135.3	136.9	133.8	132.8	129.6	129.1	135.1	145.7	131.2
2019	141.2	159.5	136.2	134.8	137.7	139.0	136.6	135.5	132.5	131.9	137.6	147.4	133.9
2008													
March	107.5	111.4	106.4	107.1	106.5	106.6	104.4	107.4	104.0	102.2	107.4	106.6	106.5
June	108.1	112.4	106.9	107.8	106.9	106.7	104.4	108.3	104.3	102.0	108.0	106.6	107.1
September	108.8	113.3	107.5	108.5	107.6	107.4	104.8	108.7	104.6	102.4	108.7	107.2	107.6
December	109.3	114.1	107.9	109.0	108.0	108.0	105.3	109.2	105.0	102.8	109.1	107.8	107.6
2009													
March	109.6	115.3	108.1	108.5	108.0	108.3	106.4	109.5	105.4	103.4	109.2	109.5	107.9
June	109.9	116.2	108.2	108.7	108.0	108.5	106.7	109.8	105.5	103.4	109.3	110.3	108.0
September	110.4	116.8	108.6	108.9	108.6	109.1	107.1	110.4	105.6	103.4	109.9	110.9	108.2
December	110.9	117.7	109.1	109.2	108.9	109.8	107.5	110.9	106.2	104.0	110.2	111.4	108.2
2010													
March	112.0	118.1	110.3	110.0	110.1	111.6	109.9	111.5	108.4	106.6	111.1	114.8	109.5
June	112.7	119.1	110.9	110.3	110.9	112.1	110.7	112.3	108.9	107.4	111.7	116.2	110.0
September	113.5	120.2	111.6	110.9	111.6	112.8	111.7	113.3	110.0	108.7	112.3	117.6	110.4
December	114.1	121.2	112.1	111.7	112.1	113.5	112.2	113.8	110.2	108.9	112.9	117.9	110.6
2011													
March	115.4	122.0	113.6	113.2	113.3	114.2	113.5	115.2	111.7	111.0	114.4	119.0	112.6
June	116.8	122.5	115.2	114.6	114.9	115.6	116.4	115.9	114.1	113.9	115.7	122.3	113.9
September	117.1	123.2	115.4	114.7	115.3	116.1	116.3	116.0	113.8	113.4	116.0	122.0	114.0
December	117.7	123.7	116.1	115.6	115.8	117.1	117.1	116.7	114.5	114.0	116.7	122.8	114.4
2012													
March	118.5	124.7	116.8	116.6	116.6	117.9	116.1	117.8	114.2	113.2	117.8	122.9	115.6
June	119.3	125.8	117.5	117.1	117.4	118.8	117.0	118.2	114.8	113.9	118.5	124.3	116.2
September	119.9	127.1	117.9	117.7	117.4	119.9	117.6	118.8	115.7	114.7	118.8	125.5	116.4
December	120.5	127.9	118.5	118.3	117.9	120.6	118.1	119.4	116.1	115.1	119.4	125.6	116.7
2013													
March	121.3	129.1	119.1	118.5	118.9	121.6	118.7	119.7	117.0	115.7	120.0	126.6	117.7
June	121.9	129.9	119.6	119.3	119.3	122.0	119.0	120.3	117.3	116.1	120.6	127.3	118.3
September	122.6	130.9	120.3	120.2	120.1	122.6	119.1	120.8	117.6	116.4	121.4	127.1	118.9
December	123.2	132.1	120.8	120.7	120.9	123.2	119.7	121.2	118.1	116.7	121.9	127.3	119.1
2014													
March	123.8	133.0	121.3	120.9	121.2	124.0	120.5	120.9	119.2	118.1	122.2	128.5	119.9
June	125.1	134.2	122.5	122.4	122.5	126.0	120.9	121.1	119.4	118.2	123.8	131.3	120.9
September	125.7	134.8	123.2	123.1	123.0	126.7	121.6	121.7	120.3	119.3	124.3	131.9	121.4
December	126.4	135.8	123.8	123.8	123.8	127.4	122.3	122.4	120.8	119.9	125.0	132.6	121.7
2015													
March	127.1	136.7	124.5	124.6	123.7	127.7	123.5	122.9	121.4	120.8	125.7	133.4	122.7
June	127.2	137.8	124.2	123.9	123.9	126.9	124.2	123.3	122.1	121.5	125.1	132.3	122.7
September	127.8	138.9	124.8	124.3	124.6	127.5	124.9	123.9	122.3	121.6	125.8	133.2	123.0
December	128.7	140.5	125.3	124.9	125.5	128.1	125.6	124.5	123.2	122.6	126.3	133.9	123.3
2016													
March	129.3	141.5	125.9	124.9	126.7	128.6	126.3	125.2	123.6	123.1	126.9	135.0	124.2
June	129.9	142.4	126.3	125.2	127.4	128.9	126.7	126.0	124.2	123.6	127.3	135.3	124.7
September	130.7	144.0	127.0	125.8	128.2	129.4	127.4	126.7	124.5	124.0	128.0	135.6	125.3
December	131.3	144.9	127.5	126.4	128.8	130.1	128.1	127.3	124.9	124.4	128.6	136.1	125.5
2017													
March	132.2	146.0	128.3	127.2	129.2	130.6	128.7	128.0	125.4	125.0	129.5	137.4	126.6
September	133.0	147.0	129.1	128.1	129.9	131.5	129.5	128.6	126.0	125.6	130.3	138.4	127.4
December	134.0	148.2	130.0	128.9	130.7	132.3	131.0	129.3	127.5	127.6	131.0	139.9	128.0
December	134.7	149.6	130.5	129.5	131.2	133.1	131.6	129.9	128.1	128.1	131.6	140.6	128.2
2018													
March	135.7	150.3	131.5	130.4	132.4	133.6	132.6	131.0	128.9	129.1	132.6	141.7	129.6
June	136.9	151.6	132.7	131.5	133.8	136.0	133.3	131.5	129.7	129.7	133.9	145.1	130.5
September	137.5	152.9	133.2	131.7	134.7	136.4	133.5	132.4	129.2	128.8	134.7	144.9	130.9
December	138.4	154.3	133.9	132.5	135.7	137.1	134.0	133.0	129.6	129.2	135.5	145.7	131.2
2019													
March	139.3	155.8	134.6	133.1	136.2	138.0	134.9	133.3	130.8	130.4	136.0	146.9	132.3
June	140.0	157.1	135.1	133.7	136.9	137.5	135.7	134.5	131.5	131.1	136.5	147.0	133.0
September	140.8	158.2	135.8	134.4	137.5	138.5	136.1	134.9	132.2	131.6	137.2	146.9	133.6
December	141.6	159.4	136.5	135.2	138.2	139.2	136.8	135.7	132.6	132.0	137.9	147.4	133.9

[1] Excludes farm workers, private household workers, and federal government employees.
[2] Not seasonally adjusted.

Table 9-2. Employment Cost Indexes, SIC Basis

(Not seasonally adjusted, December 2005 = 100; annual values are for December.)

Year	All civilian workers [1,2]	State and local government workers [2]	All private industry workers [2]	Private industry workers excluding sales occupations	Production and nonsupervisory occupations	White-collar occupations	Blue-collar occupations	Service occupations [2]	Goods-producing: Total [2]	Construction [2]	Manufacturing [2]	Service-providing: Total [2]	Transportation and utilities	Wholesale trade	Retail trade	Finance, insurance, and real estate [2]	Services
TOTAL COMPENSATION																	
1979	...	...	32.8	32.5	...	31.2	35.0	33.9	33.7	...	33.1	32.0	...	...	...	...	...
1980	...	...	35.9	35.9	...	34.2	38.5	37.1	37.0	...	36.4	35.1	...	...	...	...	...
1981	39.0	36.8	39.5	39.4	40.1	37.6	42.2	40.5	40.7	...	40.0	38.6	...	...	...	...	...
1982	41.5	39.4	42.0	42.0	42.8	40.1	44.7	43.9	43.2	...	42.4	41.1	...	...	...	...	...
1983	43.9	41.8	44.4	44.4	45.2	42.7	47.0	46.3	45.3	...	44.6	43.8	...	...	...	...	...
1984	46.2	44.6	46.6	46.7	47.3	44.8	49.0	49.4	47.4	...	46.9	46.0	...	...	...	...	...
1985	48.2	47.1	48.4	48.3	49.1	47.0	50.5	50.9	49.0	50.8	48.4	48.1	50.6	...	52.4	44.7	46.1
1986	49.9	49.6	49.9	49.9	50.5	48.6	51.9	52.4	50.5	52.3	50.0	49.6	51.8	48.5	53.5	46.1	48.1
1987	51.7	51.8	51.6	51.7	52.2	50.4	53.5	53.7	52.1	54.2	51.5	51.4	53.3	50.4	54.8	47.0	50.5
1988	54.2	54.7	54.1	54.1	54.8	52.9	55.9	56.5	54.4	56.5	53.8	54.0	54.9	52.5	58.0	50.0	53.5
1989	56.9	58.1	56.7	56.5	57.6	55.7	58.2	59.0	56.7	59.0	56.3	56.8	56.9	57.1	59.9	52.7	56.4
1990	59.7	61.5	59.3	59.3	60.1	58.4	60.7	61.8	59.4	60.9	59.1	59.4	59.1	58.2	62.5	54.9	59.9
1991	62.3	63.7	61.9	62.0	62.7	61.0	63.4	64.7	62.1	63.3	61.9	61.9	61.7	60.7	65.2	57.2	62.5
1992	64.4	66.0	64.1	64.2	64.9	63.1	65.6	66.7	64.5	65.6	64.3	63.9	63.9	62.5	66.9	57.9	65.2
1993	66.7	67.9	66.4	66.6	67.3	65.4	68.1	68.8	67.0	67.1	66.9	66.2	66.1	64.4	68.9	60.5	67.5
1994	68.7	69.9	68.5	68.6	69.2	67.5	70.0	70.8	69.0	69.6	69.0	68.1	68.7	66.4	70.8	61.8	69.4
1995	70.6	72.0	70.2	70.4	71.0	69.4	71.7	72.1	70.7	71.1	70.8	70.0	71.2	69.4	72.3	64.0	70.9
1996	72.6	73.9	72.4	72.4	73.1	71.7	73.6	74.2	72.7	72.9	72.9	72.3	73.4	71.5	75.1	65.5	73.1
1997	75.0	75.6	74.9	74.9	75.4	74.4	75.5	77.2	74.5	74.8	74.6	75.1	75.5	73.8	77.7	69.9	75.9
1998	77.6	77.8	77.5	77.2	78.1	77.3	77.6	79.4	76.5	77.4	76.6	78.0	78.4	78.0	80.0	74.1	78.2
1999	80.2	80.5	80.2	80.0	80.4	79.9	80.2	82.1	79.1	79.9	79.2	80.6	80.1	81.1	83.0	77.1	80.9
2000	83.6	82.9	83.6	83.6	84.0	83.6	83.6	85.3	82.6	84.6	82.3	84.2	83.5	84.4	86.4	81.0	84.5
2001	87.0	86.4	87.1	87.0	87.4	87.1	86.7	89.1	85.7	88.2	85.3	87.8	87.5	87.2	90.3	83.9	88.3
2002	90.0	89.9	90.0	89.9	90.2	89.9	89.8	92.0	88.9	91.0	88.5	90.5	91.0	91.1	91.9	87.6	90.7
2003	93.5	92.9	93.6	93.6	93.6	93.6	93.4	94.9	92.4	94.1	92.2	94.2	94.0	94.0	94.9	94.1	94.0
2004	96.9	96.1	97.1	97.2	97.2	96.9	97.5	97.7	96.8	96.4	96.7	97.3	97.6	96.5	97.1	96.7	97.5
2005	100.0	100.0	100.0	100.0	100.0	100.0	100.0	100.0	100.0	100.0	100.0	100.0	100.0	100.0	100.0	100.0	100.0
WAGES AND SALARIES																	
1979	...	...	36.1	36.1	36.8	34.1	39.4	37.9	38.2	41.4	37.5	34.9	39.1	33.7	39.7	32.8	31.7
1980	...	...	39.4	39.3	40.3	37.1	43.1	41.0	41.8	45.0	41.0	38.0	43.5	37.1	42.5	35.2	34.5
1981	42.3	40.1	42.8	42.9	43.9	40.4	46.8	44.4	45.4	49.0	44.5	41.4	47.1	40.0	45.6	38.8	38.1
1982	45.0	42.7	45.5	45.6	46.7	43.1	49.4	48.2	48.0	51.5	47.0	44.2	50.5	42.5	47.5	41.3	41.2
1983	47.3	45.0	47.8	47.9	48.9	45.7	51.3	50.4	49.9	53.0	49.0	46.7	53.0	45.1	49.5	44.3	43.9
1984	49.4	47.7	49.8	50.1	50.8	47.6	53.1	53.5	51.8	53.7	51.2	48.7	54.8	47.6	52.0	43.9	46.7
1985	51.5	50.3	51.8	51.9	52.9	50.0	55.0	54.8	53.6	55.3	53.0	51.0	56.9	49.7	54.5	47.9	48.4
1986	53.3	53.0	53.5	53.6	54.3	51.7	56.4	56.2	55.3	56.7	54.8	52.5	57.9	51.5	55.7	49.2	50.3
1987	55.2	55.2	55.2	55.5	56.0	53.6	58.1	57.6	57.1	58.6	56.6	54.3	59.1	53.6	57.2	49.8	53.0
1988	57.5	57.9	57.5	57.5	58.4	56.1	59.9	60.1	58.9	60.7	58.3	56.9	60.6	55.6	60.1	52.9	55.6
1989	60.1	61.0	59.9	59.8	61.0	58.7	62.0	62.3	61.2	62.8	60.6	59.5	62.2	60.7	62.0	55.7	58.3
1990	62.6	64.2	62.3	62.4	63.2	61.2	64.2	64.8	63.4	64.1	63.1	61.8	64.3	61.2	64.2	57.6	61.6
1991	64.9	66.4	64.6	64.7	65.4	63.5	66.4	67.4	65.8	66.0	65.6	64.1	66.6	63.6	66.6	59.6	63.8
1992	66.6	68.4	66.3	66.5	67.2	65.2	68.1	68.8	67.6	67.3	67.6	65.7	66.8	65.5	68.2	59.5	66.0
1993	68.7	70.2	68.3	68.5	69.2	67.4	70.0	70.3	69.6	68.6	69.7	67.8	70.9	67.1	70.2	62.1	68.0
1994	70.6	72.4	70.2	70.5	71.1	69.3	72.0	72.4	71.7	70.8	71.8	69.6	73.5	69.1	71.9	62.8	70.0
1995	72.7	74.7	72.2	72.5	73.0	71.3	74.1	74.0	73.7	72.5	73.9	71.7	76.0	72.4	73.6	65.1	71.7
1996	75.1	76.8	74.7	74.9	75.5	73.8	76.3	76.6	76.0	74.6	76.3	74.2	78.1	74.7	76.8	67.2	74.2
1997	77.9	78.9	77.6	77.7	78.3	77.0	78.8	79.9	78.3	77.1	78.6	77.4	80.7	77.0	79.7	71.8	77.5
1998	80.8	81.3	80.6	80.4	81.4	80.3	81.3	82.4	81.1	79.9	81.3	80.5	83.0	81.5	82.2	76.9	80.1
1999	83.6	84.2	83.5	83.4	83.8	83.1	84.0	85.1	83.8	82.5	84.1	83.4	84.8	84.5	85.2	79.8	83.0
2000	86.7	87.0	86.7	86.7	87.1	86.4	87.1	88.3	87.1	86.9	87.1	86.6	87.5	87.4	88.6	83.4	86.3
2001	90.0	90.2	90.0	90.0	90.4	89.6	90.5	91.8	90.2	90.4	90.2	89.9	91.7	89.3	91.9	85.8	90.0
2002	92.6	93.1	92.4	92.5	92.6	92.0	93.0	94.1	92.9	92.8	93.0	92.3	94.7	92.8	93.2	89.4	92.0
2003	95.2	95.0	95.2	95.4	95.1	95.2	95.2	96.2	95.1	95.1	95.2	95.3	96.2	95.3	95.5	95.9	94.8
2004	97.5	97.0	97.5	97.8	97.5	97.5	97.6	97.9	97.4	97.0	97.5	97.7	98.6	96.6	97.2	97.7	97.8
2005	100.0	100.0	100.0	100.0	100.0	100.0	100.0	100.0	100.0	100.0	100.0	100.0	100.0	100.0	100.0	100.0	100.0

[1]Excludes farm workers, private household workers, and federal government employees.
[2]Roughly continuous and comparable with new NAICS-based series. See notes and definitions for more information.
. . . = Not available.

SECTION 9B: PRODUCTIVITY AND RELATED DATA

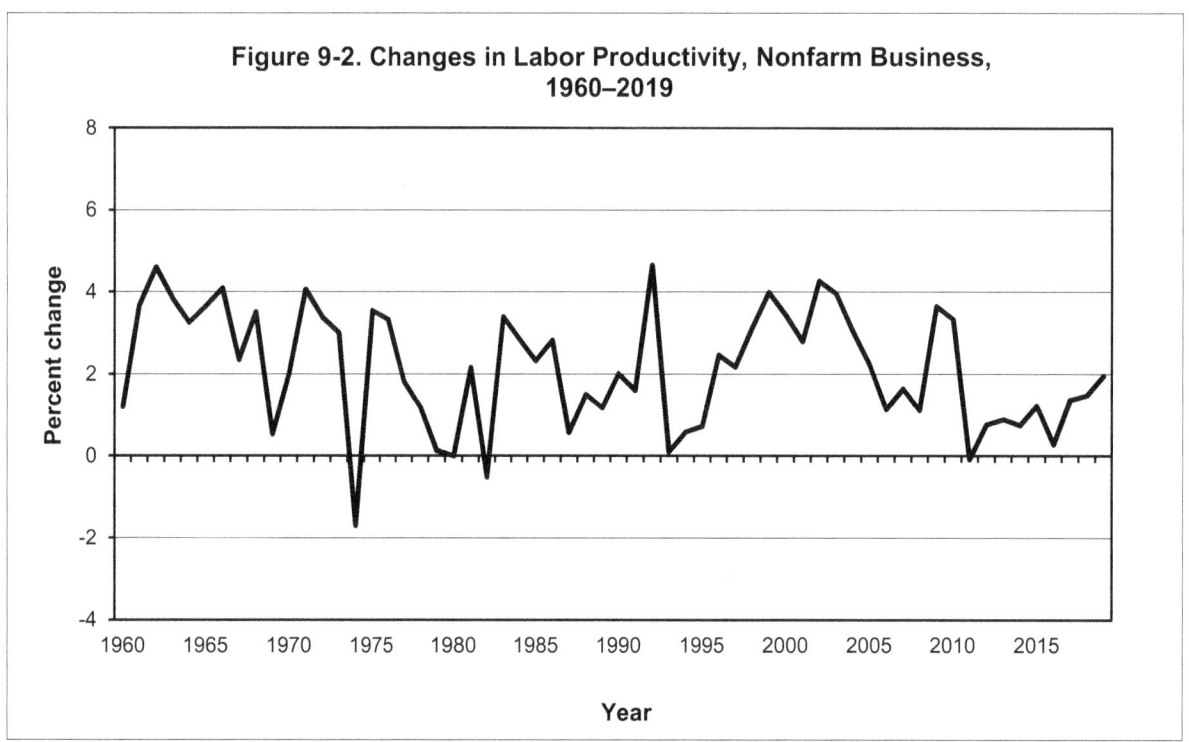

Figure 9-2. Changes in Labor Productivity, Nonfarm Business, 1960–2019

- As Figure 9-2 demonstrates, the rate of change in U.S. nonfarm labor productivity has often been quite variable from year to year. Up through the 1980s, productivity tended to decline (growth rates were less than zero) in recession years but rebound sharply in recovery. Firms often expected these declines in demand to be temporary so they therefore held on to their experienced workers in order to be prepared for the recovery. (Table 9-3A and B)

- As the U.S. economy has become more service-oriented and less dependent on manufacturing, labor has become less dominant. The lag between compensation and productivity mainly reflects a decline to unprecedented levels in the labor share of the total value of sector output. (Table 9-3A)

- Corporate profits are one of the most closely watched U.S. economic indicators. Profitability provides a summary measure of corporate financial health and thus serves as an essential indicator of economic performance. The estimates of corporate profits are an integral part of the national income and product accounts (NIPAs), a set of accounts prepared by the Bureau of Economic Analysis (BEA) that provides a logical and consistent framework for presenting statistics on U.S. economic activity. See Chapter 1 for in-depth information on NIPA accounting. (Table 9-4).

Table 9-3A. Productivity and Related Data: Recent Data

(2012 = 100, seasonally adjusted.)

Year and quarter	Output per hour of all persons	Output	Hours of all persons	Employ-ment	Average weekly hours	Unit labor costs	Compen-sation per hour	Real hourly compen-sation	Labor share	Unit nonlabor payments	Implicit price deflator	Current dollar output	Compen-sation	Nonlabor payments	Output per job
1960	31.0	17.6	56.7	50.2	112.9	23.0	7.1	50.5	115.7	15.9	19.9	3.5	4.0	2.8	35.0
1961	32.1	17.9	55.8	49.7	112.4	23.1	7.4	52.0	115.1	16.2	20.1	3.6	4.1	2.9	36.1
1962	33.6	19.1	56.8	50.3	113.0	23.0	7.7	53.7	113.8	16.7	20.2	3.9	4.4	3.2	38.0
1963	34.9	20.0	57.2	50.6	113.1	23.0	8.0	54.9	112.8	17.0	20.4	4.1	4.6	3.4	39.5
1964	36.0	21.2	58.9	51.5	114.4	23.1	8.3	56.2	112.0	17.4	20.6	4.4	4.9	3.7	41.2
1965	37.3	22.7	60.9	53.0	114.9	23.1	8.6	57.4	110.4	18.1	20.9	4.8	5.3	4.1	42.9
1966	38.9	24.3	62.5	54.6	114.5	23.7	9.2	59.5	110.4	18.6	21.4	5.2	5.8	4.5	44.5
1967	39.8	24.8	62.3	55.3	112.7	24.5	9.7	61.0	111.1	18.9	22.0	5.5	6.1	4.7	44.8
1968	41.2	26.0	63.3	56.4	112.1	25.5	10.5	63.2	111.3	19.5	22.9	6.0	6.6	5.1	46.2
1969	41.4	26.8	64.9	58.2	111.5	27.1	11.2	64.1	113.4	19.8	23.9	6.4	7.3	5.3	46.1
1970	42.2	26.8	63.6	58.0	109.6	28.6	12.1	65.2	114.6	20.3	24.9	6.7	7.7	5.4	46.3
1971	43.9	27.9	63.4	58.1	109.1	29.1	12.8	66.2	112.0	22.0	26.0	7.2	8.1	6.1	47.9
1972	45.4	29.7	65.4	59.8	109.2	30.0	13.6	68.2	111.4	22.9	26.9	8.0	8.9	6.8	49.6
1973	46.8	31.7	67.9	62.4	108.7	31.4	14.7	69.3	111.0	24.3	28.3	9.0	10.0	7.7	50.8
1974	46.0	31.2	68.0	63.4	107.3	34.9	16.0	68.2	112.4	26.1	31.1	9.7	10.9	8.2	49.3
1975	47.6	31.0	65.0	61.5	105.8	37.3	17.8	69.2	109.6	29.9	34.1	10.5	11.6	9.2	50.4
1976	49.2	33.0	67.2	63.4	106.0	39.0	19.2	70.7	108.9	31.8	35.8	11.8	12.9	10.5	52.1
1977	50.1	34.9	69.8	66.1	105.5	41.4	20.7	71.7	109.0	33.6	38.0	13.3	14.5	11.7	52.8
1978	50.7	37.2	73.3	69.8	105.1	44.3	22.5	72.6	109.2	35.8	40.6	15.1	16.5	13.3	53.3
1979	50.7	38.5	75.9	72.5	104.7	48.6	24.6	72.7	110.4	38.1	44.0	16.9	18.7	14.7	53.1
1980	50.7	38.1	75.2	72.6	103.5	53.8	27.3	72.4	112.3	40.3	47.9	18.3	20.5	15.4	52.5
1981	51.8	39.2	75.7	73.4	103.2	57.6	29.8	72.4	110.1	45.5	52.3	20.5	22.6	17.9	53.5
1982	51.5	38.1	73.9	72.2	102.5	62.2	32.1	73.4	112.6	46.4	55.3	21.1	23.7	17.7	52.8
1983	53.3	40.1	75.3	72.8	103.4	62.8	33.5	73.5	109.7	50.1	57.3	23.0	25.2	20.1	55.1
1984	54.8	43.7	79.7	76.5	104.2	63.8	35.0	73.7	108.3	52.6	58.9	25.7	27.9	23.0	57.1
1985	56.1	45.7	81.5	78.4	104.0	65.5	36.8	74.9	108.4	53.9	60.4	27.6	30.0	24.6	58.3
1986	57.7	47.4	82.2	79.7	103.1	67.3	38.8	77.8	109.9	53.4	61.3	29.0	31.9	25.3	59.5
1987	58.0	49.1	84.7	81.8	103.4	69.5	40.3	78.0	111.4	53.3	62.4	30.6	34.1	26.2	60.0
1988	58.9	51.2	87.0	84.3	103.1	72.1	42.4	79.3	112.0	54.5	64.4	33.0	36.9	27.9	60.7
1989	59.6	53.2	89.3	86.2	103.6	73.4	43.7	78.2	110.0	58.2	66.8	35.5	39.0	31.0	61.7
1990	60.7	54.0	88.9	86.7	102.5	76.5	46.5	79.2	110.9	59.3	69.0	37.3	41.3	32.1	62.3
1991	61.7	53.7	87.0	85.4	101.9	78.8	48.6	80.0	111.0	61.0	71.0	38.1	42.3	32.7	62.9
1992	64.6	56.0	86.6	84.9	102.0	79.9	51.6	82.9	110.8	62.1	72.1	40.4	44.7	34.8	65.9
1993	64.7	57.6	89.0	86.7	102.6	81.0	52.4	82.0	109.8	64.5	73.8	42.5	46.6	37.1	66.4
1994	65.0	60.3	92.8	89.7	103.4	81.1	52.8	80.9	108.0	67.4	75.1	45.3	48.9	40.7	67.3
1995	65.5	62.2	94.9	92.2	103.0	82.5	54.0	80.9	107.9	68.7	76.5	47.5	51.3	42.7	67.4
1996	67.1	65.1	97.0	94.2	102.9	83.4	56.0	81.7	107.4	70.3	77.7	50.5	54.3	45.7	69.0
1997	68.6	68.5	99.9	96.8	103.2	84.9	58.2	83.2	107.7	71.0	78.8	54.0	58.2	48.6	70.8
1998	70.7	72.0	101.9	98.8	103.1	87.2	61.7	86.9	110.0	69.0	79.3	57.1	62.8	49.7	72.9
1999	73.5	76.1	103.5	100.4	103.1	87.9	64.7	89.2	110.2	69.3	79.8	60.7	66.9	52.7	75.8
2000	76.1	79.8	104.9	102.1	102.7	90.9	69.1	92.3	112.3	68.2	81.0	64.6	72.5	54.4	78.1
2001	78.2	80.4	102.8	101.4	101.4	92.5	72.3	93.8	112.4	69.2	82.3	66.1	74.3	55.6	79.3
2002	81.5	81.8	100.3	99.2	101.1	90.7	73.9	94.4	109.4	72.9	82.9	67.8	74.2	59.6	82.5
2003	84.7	84.5	99.7	99.0	100.7	90.5	76.7	95.8	107.8	75.5	83.9	70.9	76.5	63.8	85.3
2004	87.3	88.1	100.9	100.3	100.6	92.0	80.3	97.7	106.9	78.5	86.1	75.8	81.0	69.1	87.8
2005	89.2	91.5	102.6	102.2	100.4	93.2	83.2	97.9	105.1	82.9	88.7	81.2	85.3	75.9	89.6
2006	90.3	94.6	104.8	104.1	100.7	95.7	86.4	98.4	105.1	85.2	91.1	86.2	90.6	80.6	90.9
2007	91.7	96.8	105.5	105.0	100.5	98.4	90.3	100.0	105.6	86.6	93.2	90.2	95.2	83.8	92.2
2008	92.7	95.8	103.3	103.5	99.8	100.0	92.8	99.0	105.7	87.8	94.7	90.7	95.8	84.1	92.6
2009	96.1	92.3	96.0	97.7	98.3	97.4	93.6	100.2	102.6	91.7	94.9	87.5	89.8	84.6	94.5
2010	99.3	95.2	95.9	96.5	99.3	95.9	95.3	100.4	99.9	96.1	96.0	91.4	91.3	91.5	98.6
2011	99.2	97.1	97.8	98.1	99.7	98.1	97.3	99.4	99.9	98.3	98.2	95.3	95.2	95.4	99.0
2012	100.0	100.0	100.0	100.0	100.0	100.0	100.0	100.0	100.0	100.0	100.0	100.0	100.0	100.0	100.0
2013	100.9	102.4	101.5	101.7	99.9	100.6	101.5	100.0	99.1	102.6	101.5	103.9	103.0	105.1	100.8
2014	101.6	105.6	103.9	103.8	100.1	102.5	104.1	100.9	99.4	104.0	103.1	109.0	108.2	109.9	101.8
2015	102.9	109.4	106.3	106.3	100.0	104.1	107.1	103.6	100.4	103.2	103.7	113.4	113.9	112.8	102.9
2016	103.2	111.3	107.9	108.2	99.7	105.0	108.3	103.4	100.4	103.9	104.5	116.3	116.8	115.7	102.9
2017	104.6	114.4	109.4	109.8	99.7	107.2	112.1	104.8	100.8	105.2	106.3	121.7	122.7	120.4	104.2
2018	106.1	118.3	111.5	111.8	99.8	109.2	115.9	105.7	100.4	108.1	108.7	128.6	129.2	127.9	105.9
2019	108.2	121.6	112.3	113.2	99.3	111.2	120.3	107.8	100.9	109.0	110.2	134.0	135.1	132.4	107.4
2017															
1st quarter	104.1	113.1	108.7	109.2	99.5	106.4	110.7	104.0	100.7	104.7	105.6	119.4	120.3	118.4	103.5
2nd quarter	104.2	113.8	109.2	109.5	99.8	106.8	111.3	104.5	100.8	104.8	105.9	120.5	121.5	119.2	103.9
3rd quarter	105.0	114.9	109.4	109.8	99.6	107.2	112.6	105.2	100.7	105.5	106.5	122.3	123.2	121.2	104.6
4th quarter	105.0	116.1	110.5	110.6	99.8	108.3	113.8	105.5	101.1	105.7	107.2	124.4	125.7	122.7	104.9
2018															
1st quarter	105.4	116.9	110.9	111.2	99.7	109.2	115.1	105.9	101.4	105.7	107.7	125.9	127.7	123.6	105.1
2nd quarter	106.1	118.0	111.3	111.4	99.9	108.8	115.4	105.6	100.1	108.5	108.7	128.3	128.4	128.0	106.0
3rd quarter	106.4	119.0	111.8	112.1	99.8	109.3	116.3	105.8	100.2	108.8	109.1	129.8	130.1	129.5	106.2
4th quarter	106.5	119.4	112.1	112.5	99.6	109.4	116.6	105.7	100.0	109.3	109.4	130.6	130.7	130.5	106.1
2019															
1st quarter	107.6	120.6	112.1	111.2	99.5	111.0	119.4	108.0	101.5	107.3	109.4	131.8	133.8	129.3	107.1
2nd quarter	108.4	121.2	111.8	111.4	99.2	111.0	120.3	108.0	100.7	109.1	110.2	133.5	134.5	132.2	107.5
3rd quarter	108.3	121.9	112.6	112.1	99.1	111.1	120.3	107.5	100.5	109.8	110.5	134.7	135.4	133.8	107.3
4th quarter	108.5	122.6	113.0	112.5	99.3	111.7	121.2	107.7	100.8	109.6	110.8	135.8	136.9	134.4	107.8

Table 9-3A. Productivity and Related Data: Recent Data—*Continued*

(2012 = 100, seasonally adjusted.)

Year and quarter	Output per hour of all persons	Output	Hours of all persons	Employment	Average weekly hours	Unit labor costs	Compensation per hour	Real hourly compensation	Labor share	Unit nonlabor payments	Implicit price deflator	Current dollar output	Compensation	Nonlabor payments	Output per job
1960	33.0	17.3	52.5	45.9	114.4	22.6	7.5	53.0	115.9	15.5	19.5	3.4	3.9	2.7	37.8
1961	34.1	17.7	52.0	45.5	114.1	22.6	7.7	54.2	115.0	15.8	19.7	3.5	4.0	2.8	38.9
1962	35.7	18.9	53.1	46.4	114.4	22.5	8.0	55.8	113.4	16.4	19.9	3.8	4.3	3.1	40.8
1963	36.9	19.8	53.7	46.9	114.4	22.5	8.3	56.9	112.5	16.8	20.0	4.0	4.5	3.3	42.2
1964	37.9	21.1	55.7	48.1	115.9	22.6	8.6	57.9	111.3	17.3	20.3	4.3	4.8	3.7	44.0
1965	39.1	22.6	57.9	49.8	116.3	22.6	8.8	58.9	110.0	17.9	20.5	4.7	5.1	4.0	45.5
1966	40.5	24.3	59.9	51.7	115.8	23.1	9.4	60.6	110.0	18.3	21.0	5.1	5.6	4.4	46.9
1967	41.3	24.7	59.9	52.6	113.8	24.0	9.9	62.2	110.8	18.6	21.7	5.4	5.9	4.6	47.0
1968	42.8	26.0	60.9	53.8	113.2	24.9	10.7	64.2	110.8	19.3	22.5	5.9	6.5	5.0	48.4
1969	42.8	26.8	62.7	55.7	112.4	26.6	11.4	65.1	113.1	19.5	23.5	6.3	7.1	5.2	48.2
1970	43.5	26.8	61.7	55.8	110.5	28.0	12.2	65.9	114.2	20.0	24.5	6.6	7.5	5.4	48.1
1971	45.2	27.8	61.5	56.0	110.0	28.6	12.9	67.0	111.8	21.7	25.6	7.1	8.0	6.0	49.7
1972	46.8	29.7	63.5	57.6	110.1	29.5	13.8	69.1	111.7	22.4	26.4	7.8	8.7	6.6	51.5
1973	48.2	31.8	66.1	60.2	109.7	30.8	14.8	70.0	112.6	22.8	27.3	8.7	9.8	7.3	52.9
1974	47.4	31.4	66.2	61.2	108.1	34.2	16.2	69.0	113.6	24.8	30.2	9.5	10.7	7.8	51.3
1975	48.7	30.9	63.3	59.4	106.6	36.8	17.9	69.9	110.4	28.9	33.4	10.3	11.4	8.9	51.9
1976	50.4	33.1	65.6	61.4	106.7	38.4	19.3	71.2	109.0	31.0	35.2	11.6	12.7	10.3	53.8
1977	51.3	35.0	68.2	64.2	106.1	40.8	20.9	72.4	109.2	32.9	37.4	13.1	14.3	11.5	54.4
1978	52.0	37.3	71.7	67.8	105.8	43.7	22.7	73.4	109.7	34.8	39.8	14.8	16.3	13.0	55.0
1979	51.9	38.6	74.3	70.7	105.1	47.9	24.9	73.4	111.1	36.9	43.1	16.6	18.5	14.2	54.5
1980	51.9	38.2	73.7	70.9	103.9	53.1	27.6	73.2	112.4	39.6	47.2	18.1	20.3	15.1	53.9
1981	52.6	39.1	74.3	71.7	103.6	57.4	30.2	73.3	110.8	44.5	51.8	20.2	22.4	17.4	54.5
1982	52.2	37.9	72.6	70.6	102.8	62.1	32.4	74.2	112.9	45.7	55.0	20.8	23.5	17.3	53.7
1983	54.4	40.3	74.0	71.3	103.9	62.3	33.9	74.3	109.5	49.8	56.9	22.9	25.1	20.1	56.5
1984	55.6	43.7	78.6	75.1	104.6	63.6	35.3	74.4	108.7	51.8	58.5	25.5	27.7	22.6	58.1
1985	56.6	45.6	80.6	77.3	104.3	65.5	37.1	75.5	108.8	53.4	60.2	27.5	29.9	24.3	59.0
1986	58.2	47.3	81.2	78.6	103.3	67.3	39.2	78.4	110.1	53.0	61.1	28.9	31.8	25.1	60.2
1987	58.6	49.0	83.7	80.8	103.6	69.4	40.7	78.7	111.6	52.9	62.2	30.5	34.0	25.9	60.7
1988	59.5	51.3	86.1	83.4	103.2	71.8	42.8	79.9	112.0	54.1	64.1	32.9	36.8	27.7	61.5
1989	60.1	53.1	88.4	85.3	103.7	73.2	44.0	78.7	110.2	57.7	66.5	35.3	38.9	30.6	62.3
1990	61.1	53.9	88.3	85.9	102.8	76.3	46.7	79.6	111.1	58.9	68.7	37.1	41.2	31.7	62.8
1991	62.1	53.6	86.3	84.5	102.2	78.7	48.9	80.4	111.0	60.7	70.9	38.0	42.2	32.6	63.5
1992	64.9	55.8	85.9	84.0	102.3	80.0	51.9	83.4	111.0	61.8	72.1	40.2	44.6	34.5	66.4
1993	65.0	57.5	88.5	86.0	102.9	80.9	52.6	82.3	109.7	64.5	73.8	42.4	46.5	37.1	66.8
1994	65.4	60.1	91.9	88.8	103.5	81.1	53.1	81.4	108.0	67.3	75.1	45.2	48.8	40.5	67.7
1995	66.1	62.2	94.1	91.4	102.9	82.2	54.4	81.5	107.6	68.9	76.5	47.6	51.2	42.9	68.1
1996	67.5	65.0	96.3	93.5	102.9	83.3	56.3	82.1	107.5	70.0	77.5	50.4	54.2	45.5	69.5
1997	68.8	68.4	99.3	96.2	103.2	84.9	58.5	83.5	107.7	71.0	78.9	53.9	58.0	48.5	71.1
1998	71.0	72.0	101.4	98.3	103.2	87.2	61.9	87.2	109.8	69.3	79.4	57.1	62.7	49.9	73.2
1999	73.7	76.1	103.3	100.1	103.1	87.9	64.7	89.3	109.8	69.8	80.0	60.9	66.8	53.1	76.0
2000	76.1	79.7	104.7	101.9	102.7	91.0	69.3	92.4	111.9	68.7	81.3	64.8	72.5	54.7	78.2
2001	78.2	80.3	102.7	101.3	101.4	92.4	72.3	93.8	112.0	69.7	82.6	66.3	74.2	56.0	79.3
2002	81.6	81.7	100.1	99.0	101.1	90.7	74.0	94.5	108.9	73.7	83.3	68.0	74.1	60.2	82.5
2003	84.7	84.3	99.6	98.9	100.6	90.6	76.7	95.8	107.6	75.9	84.2	71.0	76.4	64.0	85.2
2004	87.1	87.9	100.9	100.3	100.6	92.1	80.2	97.6	106.8	78.5	86.2	75.8	80.9	69.0	87.6
2005	89.0	91.3	102.6	102.2	100.4	93.4	83.2	97.8	104.9	83.4	89.1	81.3	85.3	76.2	89.4
2006	90.0	94.4	104.9	104.2	100.7	95.9	86.3	98.4	104.8	85.9	91.6	86.5	90.6	81.1	90.7
2007	91.6	96.7	105.6	105.1	100.5	98.4	90.1	99.8	105.3	86.9	93.4	90.4	95.2	84.1	92.0
2008	92.6	95.7	103.4	103.6	99.8	100.1	92.7	98.9	105.5	88.1	94.9	90.8	95.8	84.3	92.4
2009	95.9	92.0	96.0	97.7	98.2	97.5	93.5	100.2	102.3	92.5	95.4	87.8	89.8	85.2	94.2
2010	99.2	95.0	95.9	96.5	99.3	96.1	95.3	100.4	99.8	96.6	96.3	91.5	91.4	91.8	98.5
2011	99.2	96.9	97.8	98.0	99.7	98.2	97.4	99.5	100.1	98.1	98.2	95.2	95.2	95.1	98.9
2012	100.0	100.0	100.0	100.0	100.0	100.0	100.0	100.0	100.0	100.0	100.0	100.0	100.0	100.0	100.0
2013	100.5	102.2	101.7	101.8	99.9	100.8	101.3	99.8	99.3	102.3	101.5	103.7	103.0	104.6	100.4
2014	101.4	105.4	104.0	103.9	100.1	102.7	104.1	100.9	99.5	104.0	103.3	108.9	108.3	109.7	101.5
2015	102.7	109.1	106.2	106.2	100.0	104.5	107.4	103.8	100.4	103.6	104.1	113.6	114.0	113.1	102.7
2016	103.0	111.0	107.8	108.1	99.7	105.4	108.5	103.6	100.2	104.8	105.1	116.7	116.9	116.4	102.7
2017	104.4	114.2	109.4	109.8	99.6	107.6	112.3	105.0	100.7	105.9	106.9	122.0	122.8	120.9	104.0
2018	105.8	118.1	111.6	111.9	99.8	109.5	115.9	105.8	100.2	109.1	109.3	129.2	129.4	128.9	105.6
2019	107.9	121.3	112.4	113.3	99.3	111.5	120.3	107.9	100.6	110.1	110.9	134.5	135.3	133.5	107.1
2017															
1st quarter	103.8	112.7	108.6	109.2	99.4	106.8	110.9	104.2	100.6	105.4	106.2	119.7	120.4	118.8	103.2
2nd quarter	104.0	113.5	109.2	109.4	99.7	107.2	111.5	104.7	100.7	105.5	106.5	120.8	121.7	119.7	103.7
3rd quarter	104.7	114.7	109.5	109.9	99.6	107.6	112.6	105.2	100.5	106.3	107.0	122.7	123.4	121.9	104.3
4th quarter	104.9	115.8	110.4	110.7	99.7	108.7	114.1	105.7	100.9	106.5	107.8	124.8	125.9	123.4	104.6
2018															
1st quarter	105.2	116.7	110.9	111.2	99.7	109.6	115.2	106.0	101.2	106.6	108.3	126.3	127.9	124.4	104.9
2nd quarter	105.7	117.8	111.4	111.6	99.9	109.1	115.4	105.5	99.9	109.4	109.3	128.7	128.6	128.9	105.6
3rd quarter	106.2	118.8	111.9	112.1	99.8	109.6	116.3	105.9	99.9	109.9	109.7	130.4	130.2	130.6	106.0
4th quarter	106.3	119.2	112.1	112.6	99.6	109.8	116.7	105.8	99.8	110.4	110.0	131.2	130.9	131.6	105.9
2019															
1st quarter	107.3	120.4	112.1	112.7	99.5	111.3	119.5	108.1	101.2	108.4	110.0	132.4	134.0	130.4	106.8
2nd quarter	108.0	120.9	111.9	112.8	99.2	111.3	120.3	108.0	100.4	110.3	110.9	134.1	134.6	133.4	107.2
3rd quarter	108.0	121.6	112.7	113.6	99.2	111.4	120.3	107.5	100.2	110.9	111.2	135.2	135.5	134.9	107.0
4th quarter	108.3	122.4	113.0	113.8	99.3	112.0	121.3	107.8	100.5	110.7	111.4	136.4	137.1	135.4	107.5

Table 9-3B. Productivity and Related Data: Historical Data

(2012 = 100, seasonally adjusted.)

Year and quarter	Business sector								Nonfarm business sector							
	Output per hour of all persons	Output	Hours of all persons	Compensation per hour	Real compensation per hour	Unit labor costs	Unit nonlabor payments	Implicit price deflator	Output per hour of all persons	Output	Hours of all persons	Compensation per hour	Real compensation per hour	Unit labor costs	Unit nonlabor payments	Implicit price deflator
1947	20.4	11.2	54.7	3.6	33.9	17.7	11.8	15.1	20.4	11.2	54.7	3.6	33.9	16.4	11.3	14.2
1948	21.3	11.7	55.1	3.9	34.0	18.4	13.0	16.0	21.3	11.7	55.1	3.9	34.0	17.4	12.2	15.1
1949	21.8	11.6	53.3	4.0	34.9	18.2	12.9	15.9	21.8	11.6	53.3	4.0	34.9	17.3	12.5	15.2
1950	23.6	12.7	54.1	4.2	36.9	18.0	13.5	16.1	23.6	12.7	54.1	4.2	36.9	17.2	13.1	15.4
1951	24.3	13.6	55.8	4.7	37.5	19.2	14.9	17.3	24.3	13.6	55.8	4.7	37.5	18.2	14.1	16.4
1952	25.0	14.0	55.9	4.9	39.1	19.7	14.6	17.5	25.0	14.0	55.9	4.9	39.1	18.9	14.0	16.8
1953	25.9	14.7	56.6	5.3	41.2	20.2	14.3	17.6	25.9	14.7	56.6	5.3	41.2	19.5	14.0	17.1
1954	26.5	14.5	54.7	5.4	42.3	20.4	14.2	17.7	26.5	14.5	54.7	5.4	42.3	19.7	14.0	17.2
1955	27.6	15.7	56.8	5.6	43.5	20.1	15.2	18.0	27.6	15.7	56.8	5.6	43.5	19.6	14.9	17.5
1956	27.7	16.0	57.6	5.9	45.7	21.4	14.9	18.5	27.7	16.0	57.6	5.9	45.7	20.9	14.6	18.2
1957	28.6	16.3	56.8	6.3	47.1	22.1	15.4	19.1	28.6	16.3	56.8	6.3	47.1	21.5	15.1	18.7
1958	29.4	16.0	54.3	6.6	47.8	22.4	15.7	19.5	29.4	16.0	54.3	6.6	47.8	21.9	15.3	19.0
1959	30.5	17.2	56.5	6.9	49.4	22.5	16.1	19.7	30.5	17.2	56.5	6.9	49.4	22.0	15.9	19.3
1947																
1st quarter	20.4	10.8	46.3	3.7	35.9	16.0	11.6	14.6	23.3	10.8	46.3	3.7	35.9	16.0	10.8	13.7
2nd quarter	20.5	11.0	46.3	3.8	36.2	16.0	11.5	14.8	23.8	11.0	46.3	3.8	36.2	16.0	11.2	13.9
3rd quarter	20.3	10.7	46.4	3.9	36.5	17.0	11.8	15.2	23.1	10.7	46.4	3.9	36.5	17.0	11.4	14.5
4th quarter	20.5	11.3	46.9	4.0	36.3	16.7	12.3	15.7	24.0	11.3	46.9	4.0	36.3	16.7	11.7	14.5
1948																
1st quarter	21.0	11.4	47.2	4.1	36.4	17.0	12.8	15.8	24.1	11.4	47.2	4.1	36.4	17.0	11.8	14.7
2nd quarter	21.4	11.3	47.1	4.2	36.3	17.3	13.2	15.9	24.1	11.3	47.1	4.2	36.3	17.3	12.0	15.0
3rd quarter	21.3	11.4	47.5	4.2	36.4	17.6	13.2	16.2	24.1	11.4	47.5	4.2	36.4	17.6	12.3	15.3
4th quarter	21.5	11.4	47.0	4.3	37.1	17.7	12.9	16.2	24.2	11.4	47.0	4.3	37.1	17.7	12.6	15.5
1949																
1st quarter	21.4	11.3	46.2	4.3	37.7	17.6	13.0	16.1	24.5	11.3	46.2	4.3	37.7	17.6	12.5	15.4
2nd quarter	21.5	11.2	45.5	4.3	37.8	17.5	12.8	15.9	24.7	11.2	45.5	4.3	37.8	17.5	12.3	15.2
3rd quarter	22.1	11.4	45.0	4.3	38.2	17.1	13.0	15.8	25.3	11.4	45.0	4.3	38.2	17.1	12.7	15.2
4th quarter	22.1	11.3	44.8	4.3	38.2	17.1	12.7	15.8	25.2	11.3	44.8	4.3	38.2	17.1	12.5	15.1
1950																
1st quarter	23.1	11.7	45.1	4.4	39.3	17.0	12.9	15.7	26.0	11.7	45.1	4.4	39.3	17.0	12.8	15.2
2nd quarter	23.4	12.2	46.3	4.5	39.9	17.2	13.0	15.8	26.3	12.2	46.3	4.5	39.9	17.2	12.8	15.2
3rd quarter	23.8	12.8	47.7	4.6	39.9	17.1	13.8	16.2	26.9	12.8	47.7	4.6	39.9	17.1	13.2	15.4
4th quarter	23.9	13.0	48.2	4.7	40.1	17.5	14.3	16.5	27.0	13.0	48.2	4.7	40.1	17.5	13.4	15.7
1951																
1st quarter	23.9	13.2	49.0	4.8	39.4	17.9	14.9	17.1	27.0	13.2	49.0	4.8	39.4	17.9	13.9	16.2
2nd quarter	24.0	13.3	49.4	4.9	39.9	18.4	14.7	17.3	26.9	13.3	49.4	4.9	39.9	18.4	13.8	16.4
3rd quarter	24.7	13.4	48.9	5.0	40.4	18.2	15.0	17.3	27.5	13.4	48.9	5.0	40.4	18.2	14.3	16.5
4th quarter	24.6	13.5	48.9	5.1	40.6	18.5	15.0	17.5	27.6	13.5	48.9	5.1	40.6	18.5	14.2	16.6
1952																
1st quarter	24.7	13.6	49.2	5.1	40.8	18.6	14.8	17.4	27.7	13.6	49.2	5.1	40.8	18.6	14.1	16.6
2nd quarter	24.9	13.6	49.0	5.2	41.0	18.7	14.5	17.4	27.7	13.6	49.0	5.2	41.0	18.7	13.9	16.6
3rd quarter	25.0	13.6	49.4	5.3	41.3	19.1	14.6	17.6	27.5	13.6	49.4	5.3	41.3	19.1	13.9	16.8
4th quarter	25.4	14.3	50.7	5.4	42.2	19.1	14.5	17.6	28.1	14.3	50.7	5.4	42.2	19.1	14.1	16.9
1953																
1st quarter	25.8	14.5	51.2	5.4	42.9	19.2	14.4	17.6	28.4	14.5	51.2	5.4	42.9	19.2	14.0	17.0
2nd quarter	26.0	14.6	51.3	5.5	43.2	19.4	14.4	17.6	28.4	14.6	51.3	5.5	43.2	19.4	14.1	17.1
3rd quarter	26.0	14.5	50.8	5.6	43.5	19.5	14.2	17.7	28.5	14.5	50.8	5.6	43.5	19.5	14.1	17.1
4th quarter	25.9	14.2	50.0	5.6	43.8	19.8	14.2	17.7	28.4	14.2	50.0	5.6	43.8	19.8	13.8	17.1
1954																
1st quarter	25.9	14.1	49.3	5.7	44.1	19.9	14.1	17.7	28.5	14.1	49.3	5.7	44.1	19.9	13.7	17.2
2nd quarter	26.3	14.1	49.0	5.7	44.2	19.8	14.0	17.7	28.7	14.1	49.0	5.7	44.2	19.8	13.9	17.2
3rd quarter	26.7	14.3	48.8	5.7	44.7	19.6	14.4	17.7	29.2	14.3	48.8	5.7	44.7	19.6	14.1	17.2
4th quarter	27.1	14.6	49.4	5.8	45.3	19.5	14.4	17.7	29.6	14.6	49.4	5.8	45.3	19.5	14.3	17.3
1955																
1st quarter	27.5	15.1	50.2	5.8	45.5	19.3	15.2	17.8	30.1	15.1	50.2	5.8	45.5	19.3	14.8	17.4
2nd quarter	27.8	15.4	50.8	5.9	45.9	19.4	15.1	17.8	30.2	15.4	50.8	5.9	45.9	19.4	14.8	17.4
3rd quarter	27.7	15.6	51.4	6.0	46.7	19.7	15.2	18.0	30.4	15.6	51.4	6.0	46.7	19.7	14.9	17.6
4th quarter	27.5	15.7	52.0	6.0	47.0	19.9	15.3	18.2	30.2	15.7	52.0	6.0	47.0	19.9	15.0	17.8
1956																
1st quarter	27.4	15.6	52.3	6.1	47.7	20.5	14.9	18.3	29.9	15.6	52.3	6.1	47.7	20.5	14.6	17.9
2nd quarter	27.6	15.7	52.4	6.2	48.3	20.8	14.7	18.4	30.0	15.7	52.4	6.2	48.3	20.8	14.5	18.0
3rd quarter	27.6	15.7	52.2	6.3	48.6	21.1	15.0	18.7	30.0	15.7	52.2	6.3	48.6	21.1	14.6	18.3
4th quarter	28.2	15.9	52.5	6.4	49.0	21.2	15.1	18.7	30.3	15.9	52.5	6.4	49.0	21.2	14.7	18.4
1957																
1st quarter	28.4	16.1	52.5	6.5	49.3	21.3	15.3	19.0	30.7	16.1	52.5	6.5	49.3	21.3	15.1	18.6
2nd quarter	28.4	16.0	52.4	6.6	49.3	21.6	15.3	19.1	30.6	16.0	52.4	6.6	49.3	21.6	15.0	18.7
3rd quarter	28.7	16.2	52.2	6.7	49.6	21.6	15.6	19.2	31.0	16.2	52.2	6.7	49.6	21.6	15.2	18.8
4th quarter	29.0	15.9	51.1	6.8	50.0	21.8	15.3	19.2	31.1	15.9	51.1	6.8	50.0	21.8	15.0	18.8
1958																
1st quarter	28.8	15.3	49.9	6.8	49.4	22.1	15.3	19.4	30.7	15.3	49.9	6.8	49.4	22.1	14.8	18.9
2nd quarter	29.2	15.4	49.2	6.8	49.5	21.9	15.7	19.4	31.3	15.4	49.2	6.8	49.5	21.9	15.2	18.9
3rd quarter	29.7	15.9	49.8	7.0	50.7	22.0	15.7	19.5	31.9	15.9	49.8	7.0	50.7	22.0	15.3	19.0
4th quarter	30.0	16.4	50.5	7.0	50.9	21.7	16.2	19.6	32.4	16.4	50.5	7.0	50.9	21.7	15.8	19.1
1959																
1st quarter	30.3	16.7	51.5	7.1	51.3	21.9	16.0	19.6	32.4	16.7	51.5	7.1	51.3	21.9	15.8	19.2
2nd quarter	30.4	17.2	52.5	7.1	51.6	21.8	16.2	19.6	32.7	17.2	52.5	7.1	51.6	21.8	15.9	19.3
3rd quarter	30.6	17.2	52.3	7.2	51.6	21.9	16.2	19.7	32.8	17.2	52.3	7.2	51.6	21.9	16.0	19.3
4th quarter	30.6	17.1	52.4	7.3	51.8	22.2	16.0	19.8	32.7	17.1	52.4	7.3	51.8	22.2	15.8	19.4

Table 9-3B. Productivity and Related Data: Historical Data—*Continued*

(2012 = 100, seasonally adjusted.)

Year and quarter	Business sector								Nonfarm business sector							
	Output per hour of all persons	Output	Hours of all persons	Compensation per hour	Real compensation per hour	Unit labor costs	Unit nonlabor payments	Implicit price deflator	Output per hour of all persons	Output	Hours of all persons	Compensation per hour	Real compensation per hour	Unit labor costs	Unit nonlabor payments	Implicit price deflator
1960																
1st quarter	31.5	17.6	52.7	7.4	52.9	22.2	16.2	19.8	33.5	17.6	52.7	7.4	52.9	22.2	15.9	19.5
2nd quarter	30.8	17.4	52.8	7.5	52.9	22.7	15.8	19.9	32.9	17.4	52.8	7.5	52.9	22.7	15.4	19.5
3rd quarter	31.0	17.4	52.5	7.5	53.2	22.7	16.0	19.9	33.1	17.4	52.5	7.5	53.2	22.7	15.5	19.6
4th quarter	30.7	17.0	52.0	7.5	53.1	23.1	15.6	20.0	32.7	17.0	52.0	7.5	53.1	23.1	15.2	19.6
1961																
1st quarter	31.1	17.1	51.7	7.6	53.5	23.0	15.7	20.0	33.1	17.1	51.7	7.6	53.5	23.0	15.3	19.7
2nd quarter	32.1	17.5	51.6	7.7	54.2	22.7	16.1	20.0	34.0	17.5	51.6	7.7	54.2	22.7	15.8	19.7
3rd quarter	32.5	17.9	51.9	7.8	54.4	22.5	16.4	20.1	34.6	17.9	51.9	7.8	54.4	22.5	16.1	19.7
4th quarter	32.8	18.3	52.5	7.8	54.7	22.5	16.5	20.1	34.9	18.3	52.5	7.8	54.7	22.5	16.1	19.7
1962																
1st quarter	33.2	18.7	52.7	8.0	55.4	22.4	16.7	20.2	35.4	18.7	52.7	8.0	55.4	22.4	16.4	19.8
2nd quarter	33.3	18.8	53.4	8.0	55.5	22.7	16.5	20.2	35.3	18.8	53.4	8.0	55.5	22.7	16.2	19.9
3rd quarter	33.9	19.1	53.2	8.1	55.8	22.5	16.8	20.3	35.9	19.1	53.2	8.1	55.8	22.5	16.5	19.9
4th quarter	34.2	19.1	52.9	8.1	56.1	22.5	16.7	20.3	36.1	19.1	52.9	8.1	56.1	22.5	16.5	19.9
1963																
1st quarter	34.3	19.3	53.2	8.2	56.5	22.7	16.8	20.3	36.3	19.3	53.2	8.2	56.5	22.7	16.5	20.0
2nd quarter	34.5	19.6	53.6	8.3	56.7	22.6	16.9	20.3	36.5	19.6	53.6	8.3	56.7	22.6	16.6	20.0
3rd quarter	35.4	20.1	53.7	8.3	56.9	22.3	17.2	20.4	37.4	20.1	53.7	8.3	56.9	22.3	17.0	20.0
4th quarter	35.4	20.2	54.1	8.4	57.4	22.6	17.3	20.5	37.4	20.2	54.1	8.4	57.4	22.6	16.9	20.1
1964																
1st quarter	35.8	20.8	55.2	8.4	57.1	22.4	17.5	20.5	37.7	20.8	55.2	8.4	57.1	22.4	17.3	20.2
2nd quarter	35.9	21.0	55.4	8.5	57.7	22.5	17.4	20.6	37.9	21.0	55.4	8.5	57.7	22.5	17.3	20.2
3rd quarter	36.3	21.4	55.8	8.6	58.3	22.5	17.5	20.6	38.3	21.4	55.8	8.6	58.3	22.5	17.4	20.3
4th quarter	36.1	21.4	56.3	8.7	58.4	22.9	17.3	20.7	37.9	21.4	56.3	8.7	58.4	22.9	17.1	20.4
1965																
1st quarter	36.7	22.0	57.2	8.7	58.5	22.7	17.8	20.8	38.5	22.0	57.2	8.7	58.5	22.7	17.6	20.5
2nd quarter	36.7	22.3	57.8	8.8	58.5	22.7	17.8	20.9	38.6	22.3	57.8	8.8	58.5	22.7	17.6	20.5
3rd quarter	37.7	22.8	57.9	8.9	58.9	22.5	18.3	20.9	39.4	22.8	57.9	8.9	58.9	22.5	18.0	20.6
4th quarter	38.2	23.4	58.4	9.0	59.5	22.5	18.6	21.1	40.1	23.4	58.4	9.0	59.5	22.5	18.3	20.7
1966																
1st quarter	38.9	24.1	59.3	9.2	60.0	22.6	18.7	21.2	40.7	24.1	59.3	9.2	60.0	22.6	18.3	20.7
2nd quarter	38.7	24.2	59.9	9.3	60.4	23.1	18.4	21.3	40.4	24.2	59.9	9.3	60.4	23.1	18.1	20.9
3rd quarter	38.8	24.4	60.2	9.4	60.7	23.3	18.5	21.5	40.5	24.4	60.2	9.4	60.7	23.3	18.2	21.1
4th quarter	39.1	24.5	60.0	9.6	61.0	23.5	18.7	21.7	40.8	24.5	60.0	9.6	61.0	23.5	18.5	21.3
1967																
1st quarter	39.5	24.6	59.9	9.7	61.7	23.6	18.8	21.8	41.1	24.6	59.9	9.7	61.7	23.6	18.6	21.4
2nd quarter	39.8	24.6	59.6	9.9	62.3	23.9	18.7	21.9	41.2	24.6	59.6	9.9	62.3	23.9	18.5	21.6
3rd quarter	39.9	24.8	59.8	10.0	62.6	24.1	18.9	22.1	41.5	24.8	59.8	10.0	62.6	24.1	18.6	21.7
4th quarter	40.0	25.0	60.0	10.1	62.7	24.4	19.1	22.3	41.6	25.0	60.0	10.1	62.7	24.4	18.8	21.9
1968																
1st quarter	40.9	25.6	60.1	10.4	63.8	24.5	19.4	22.5	42.5	25.6	60.1	10.4	63.8	24.5	19.2	22.2
2nd quarter	41.3	26.0	60.7	10.6	64.2	24.6	19.7	22.8	42.9	26.0	60.7	10.6	64.2	24.6	19.5	22.4
3rd quarter	41.3	26.2	61.1	10.7	64.4	25.0	19.5	22.9	42.9	26.2	61.1	10.7	64.4	25.0	19.3	22.6
4th quarter	41.3	26.3	61.5	11.0	64.8	25.6	19.5	23.3	42.8	26.3	61.5	11.0	64.8	25.6	19.3	22.9
1969																
1st quarter	41.4	26.8	62.1	11.1	64.9	25.7	20.0	23.5	43.2	26.8	62.1	11.1	64.9	25.7	19.8	23.1
2nd quarter	41.4	26.8	62.7	11.3	65.0	26.4	19.8	23.8	42.8	26.8	62.7	11.3	65.0	26.4	19.5	23.4
3rd quarter	41.5	27.0	63.0	11.5	65.2	26.8	19.8	24.0	42.9	27.0	63.0	11.5	65.2	26.8	19.5	23.6
4th quarter	41.4	26.8	62.8	11.7	65.5	27.5	19.6	24.3	42.7	26.8	62.8	11.7	65.5	27.5	19.2	23.9
1970																
1st quarter	41.6	26.8	62.5	11.9	65.6	27.9	19.6	24.6	42.8	26.8	62.5	11.9	65.6	27.9	19.3	24.1
2nd quarter	42.0	26.8	61.7	12.1	65.7	27.9	20.3	24.9	43.4	26.8	61.7	12.1	65.7	27.9	20.0	24.5
3rd quarter	42.8	27.1	61.4	12.3	66.1	27.9	20.6	25.0	44.1	27.1	61.4	12.3	66.1	27.9	20.3	24.6
4th quarter	42.5	26.6	60.9	12.4	65.8	28.5	20.6	25.3	43.7	26.6	60.9	12.4	65.8	28.5	20.4	25.0
1971																
1st quarter	43.7	27.5	61.2	12.7	66.6	28.2	21.6	25.6	45.0	27.5	61.2	12.7	66.6	28.2	21.3	25.2
2nd quarter	43.8	27.7	61.4	12.9	67.0	28.5	21.9	25.9	45.1	27.7	61.4	12.9	67.0	28.5	21.6	25.5
3rd quarter	44.3	28.0	61.4	13.1	67.3	28.7	22.3	26.1	45.6	28.0	61.4	13.1	67.3	28.7	22.0	25.8
4th quarter	43.9	28.0	62.1	13.1	67.2	29.1	22.0	26.3	45.2	28.0	62.1	13.1	67.2	29.1	21.6	25.9
1972																
1st quarter	44.4	28.8	62.8	13.5	68.5	29.5	22.2	26.6	45.9	28.8	62.8	13.5	68.5	29.5	21.9	26.2
2nd quarter	45.5	29.6	63.2	13.7	68.9	29.2	22.9	26.7	46.8	29.6	63.2	13.7	68.9	29.2	22.5	26.3
3rd quarter	45.6	29.9	63.6	13.9	69.2	29.4	23.2	27.0	47.0	29.9	63.6	13.9	69.2	29.4	22.5	26.4
4th quarter	46.1	30.5	64.2	14.1	69.8	29.7	23.5	27.3	47.4	30.5	64.2	14.1	69.8	29.7	22.5	26.6
1973																
1st quarter	47.0	31.6	65.2	14.5	70.6	29.8	23.7	27.5	48.5	31.6	65.2	14.5	70.6	29.8	22.7	26.8
2nd quarter	47.1	32.0	65.9	14.7	70.1	30.3	24.2	27.9	48.5	32.0	65.9	14.7	70.1	30.3	22.9	27.1
3rd quarter	46.4	31.9	66.3	14.9	69.9	31.1	24.3	28.5	48.1	31.9	66.3	14.9	69.9	31.1	22.6	27.4
4th quarter	46.6	31.9	66.7	15.2	69.6	31.9	25.0	29.1	47.8	31.9	66.7	15.2	69.6	31.9	23.0	28.0
1974																
1st quarter	46.0	31.6	66.4	15.6	69.1	32.7	25.1	29.8	47.7	31.6	66.4	15.6	69.1	32.7	23.5	28.7
2nd quarter	46.1	31.6	66.6	16.0	69.0	33.7	25.7	30.5	47.5	31.6	66.6	16.0	69.0	33.7	24.6	29.7
3rd quarter	45.7	31.2	66.3	16.5	69.2	35.0	26.1	31.5	47.1	31.2	66.3	16.5	69.2	35.0	25.0	30.7
4th quarter	46.1	31.0	65.2	16.9	68.9	35.6	27.6	32.5	47.5	31.0	65.2	16.9	68.9	35.6	26.3	31.6

Table 9-3B. Productivity and Related Data: Historical Data—*Continued*

(2012 = 100, seasonally adjusted.)

Year and quarter	Business sector								Nonfarm business sector							
	Output per hour of all persons	Output	Hours of all persons	Compensation per hour	Real compensation per hour	Unit labor costs	Unit nonlabor payments	Implicit price deflator	Output per hour of all persons	Output	Hours of all persons	Compensation per hour	Real compensation per hour	Unit labor costs	Unit nonlabor payments	Implicit price deflator
1975																
1st quarter	46.8	30.2	63.2	17.4	69.6	36.4	28.5	33.3	47.9	30.2	63.2	17.4	69.6	36.4	27.6	32.6
2nd quarter	47.5	30.4	62.6	17.8	70.2	36.6	29.3	33.7	48.6	30.4	62.6	17.8	70.2	36.6	28.5	33.1
3rd quarter	48.0	31.1	63.2	18.1	70.0	36.8	30.6	34.3	49.2	31.1	63.2	18.1	70.0	36.8	29.4	33.6
4th quarter	48.2	31.7	64.2	18.4	69.9	37.4	31.0	34.9	49.3	31.7	64.2	18.4	69.9	37.4	29.8	34.1
1976																
1st quarter	48.8	32.6	65.3	18.8	70.4	37.6	31.3	35.2	50.0	32.6	65.3	18.8	70.4	37.6	30.5	34.5
2nd quarter	49.2	33.0	65.3	19.1	71.1	37.9	31.6	35.5	50.5	33.0	65.3	19.1	71.1	37.9	30.9	34.9
3rd quarter	49.2	33.2	65.6	19.5	71.5	38.6	31.8	36.0	50.6	33.2	65.6	19.5	71.5	38.6	31.1	35.3
4th quarter	49.5	33.5	65.9	19.9	71.9	39.3	32.3	36.6	50.7	33.5	65.9	19.9	71.9	39.3	31.7	36.0
1977																
1st quarter	49.8	34.0	66.6	20.3	72.0	39.8	32.9	37.2	51.1	34.0	66.6	20.3	72.0	39.8	32.2	36.5
2nd quarter	49.9	34.8	67.9	20.7	72.2	40.5	33.3	37.7	51.3	34.8	67.9	20.7	72.2	40.5	32.7	37.1
3rd quarter	50.6	35.6	68.6	21.1	72.6	40.8	33.8	38.1	51.8	35.6	68.6	21.1	72.6	40.8	33.4	37.6
4th quarter	50.0	35.4	69.3	21.5	72.8	42.1	34.1	38.9	51.1	35.4	69.3	21.5	72.8	42.1	33.3	38.2
1978																
1st quarter	49.8	35.6	69.6	22.1	73.6	43.2	33.8	39.4	51.2	35.6	69.6	22.1	73.6	43.2	32.9	38.7
2nd quarter	50.8	37.4	71.7	22.5	73.2	43.0	35.7	40.2	52.2	37.4	71.7	22.5	73.2	43.0	34.6	39.4
3rd quarter	51.0	37.7	72.3	22.9	73.3	43.8	36.3	40.9	52.2	37.7	72.3	22.9	73.3	43.8	35.3	40.1
4th quarter	51.1	38.4	73.1	23.5	73.7	44.6	37.1	41.7	52.5	38.4	73.1	23.5	73.7	44.6	36.1	40.9
1979																
1st quarter	50.9	38.4	73.7	24.1	73.9	46.2	36.9	42.5	52.1	38.4	73.7	24.1	73.9	46.2	35.6	41.6
2nd quarter	50.8	38.4	73.9	24.6	73.6	47.4	38.0	43.6	52.0	38.4	73.9	24.6	73.6	47.4	36.8	42.8
3rd quarter	50.8	38.7	74.6	25.2	73.3	48.5	38.6	44.5	51.9	38.7	74.6	25.2	73.3	48.5	37.4	43.7
4th quarter	50.7	38.8	74.7	25.8	73.3	49.7	38.8	45.3	51.8	38.8	74.7	25.8	73.3	49.7	37.7	44.5
1980																
1st quarter	51.0	38.8	74.6	26.5	73.0	50.9	39.5	46.3	52.1	38.8	74.6	26.5	73.0	50.9	38.8	45.6
2nd quarter	50.5	37.7	73.2	27.2	73.2	52.8	39.4	47.3	51.6	37.7	73.2	27.2	73.2	52.8	39.3	46.9
3rd quarter	50.6	37.7	72.8	27.9	73.3	54.0	40.3	48.4	51.8	37.7	72.8	27.9	73.3	54.0	39.5	47.7
4th quarter	51.1	38.6	73.8	28.7	73.5	54.8	42.0	49.6	52.3	38.6	73.8	28.7	73.5	54.8	40.8	48.7
1981																
1st quarter	52.0	39.4	74.3	29.4	73.4	55.4	44.6	50.9	53.1	39.4	74.3	29.4	73.4	55.4	43.6	50.2
2nd quarter	51.6	39.0	74.3	29.9	73.2	57.1	44.8	51.9	52.4	39.0	74.3	29.9	73.2	57.1	43.7	51.3
3rd quarter	52.3	39.3	74.3	30.6	73.4	57.8	46.4	52.8	52.9	39.3	74.3	30.6	73.4	57.8	45.2	52.3
4th quarter	51.7	38.7	73.9	31.1	73.3	59.4	46.1	53.6	52.4	38.7	73.9	31.1	73.3	59.4	45.4	53.3
1982																
1st quarter	51.4	37.9	72.7	31.9	74.4	61.2	45.2	54.3	52.1	37.9	72.7	31.9	74.4	61.2	44.5	53.9
2nd quarter	51.5	38.1	73.0	32.2	74.2	61.7	46.1	54.9	52.1	38.1	73.0	32.2	74.2	61.7	45.5	54.6
3rd quarter	51.5	37.9	72.5	32.7	74.1	62.5	46.7	55.7	52.2	37.9	72.5	32.7	74.1	62.5	45.9	55.3
4th quarter	52.0	37.8	71.8	33.1	74.3	62.9	47.4	56.2	52.7	37.8	71.8	33.1	74.3	62.9	46.9	55.9
1983																
1st quarter	52.4	38.5	72.2	33.5	74.7	62.7	48.4	56.6	53.4	38.5	72.2	33.5	74.7	62.7	47.6	56.2
2nd quarter	53.3	39.8	73.1	33.8	74.5	62.0	49.7	57.0	54.5	39.8	73.1	33.8	74.5	62.0	49.2	56.4
3rd quarter	53.6	40.8	74.6	34.0	74.2	62.1	50.9	57.5	54.8	40.8	74.6	34.0	74.2	62.1	51.0	57.3
4th quarter	54.0	41.8	76.0	34.4	74.3	62.4	51.3	57.9	55.0	41.8	76.0	34.4	74.3	62.4	51.3	57.6
1984																
1st quarter	54.3	42.8	77.4	34.8	74.2	62.9	51.9	58.3	55.3	42.8	77.4	34.8	74.2	62.9	51.0	57.7
2nd quarter	54.8	43.6	78.4	35.1	74.2	63.2	52.6	58.7	55.6	43.6	78.4	35.1	74.2	63.2	51.8	58.3
3rd quarter	55.1	44.0	78.8	35.6	74.7	63.8	52.8	59.1	55.8	44.0	78.8	35.6	74.7	63.8	52.1	58.7
4th quarter	55.2	44.3	79.3	35.9	74.7	64.3	53.0	59.4	55.9	44.3	79.3	35.9	74.7	64.3	52.2	59.0
1985																
1st quarter	55.5	44.8	79.9	36.4	75.0	64.9	53.7	60.0	56.1	44.8	79.9	36.4	75.0	64.9	53.0	59.7
2nd quarter	55.7	45.2	80.4	36.7	75.1	65.3	53.8	60.3	56.2	45.2	80.4	36.7	75.1	65.3	53.3	60.1
3rd quarter	56.6	45.9	80.7	37.3	75.7	65.4	54.4	60.6	57.0	45.9	80.7	37.3	75.7	65.4	54.1	60.5
4th quarter	56.8	46.4	81.1	38.0	76.4	66.4	53.6	60.8	57.2	46.4	81.1	38.0	76.4	66.4	53.1	60.6
1986																
1st quarter	57.3	46.8	80.9	38.5	77.1	66.6	53.9	61.0	57.9	46.8	80.9	38.5	77.1	66.6	53.6	61.0
2nd quarter	57.7	47.1	80.7	39.0	78.4	66.8	53.7	61.2	58.3	47.1	80.7	39.0	78.4	66.8	53.4	61.0
3rd quarter	58.0	47.5	81.1	39.5	78.9	67.4	53.5	61.3	58.6	47.5	81.1	39.5	78.9	67.4	52.9	61.1
4th quarter	57.9	47.8	81.7	40.0	79.5	68.4	52.6	61.5	58.5	47.8	81.7	40.0	79.5	68.4	52.1	61.3
1987																
1st quarter	57.6	48.2	82.7	40.2	78.9	69.0	52.6	61.8	58.2	48.2	82.7	40.2	78.9	69.0	52.1	61.7
2nd quarter	58.0	48.8	83.2	40.5	78.7	69.1	53.3	62.2	58.6	48.8	83.2	40.5	78.7	69.1	52.8	62.0
3rd quarter	58.1	49.1	83.8	40.9	78.7	69.8	53.5	62.7	58.6	49.1	83.8	40.9	78.7	69.8	53.1	62.5
4th quarter	58.6	50.1	84.7	41.4	79.0	69.9	53.8	62.9	59.2	50.1	84.7	41.4	79.0	69.9	53.3	62.7
1988																
1st quarter	58.7	50.3	84.7	42.1	79.8	70.9	53.7	63.5	59.4	50.3	84.7	42.1	79.8	70.9	53.1	63.1
2nd quarter	58.8	51.2	85.9	42.6	80.0	71.5	53.9	64.0	59.5	51.2	85.9	42.6	80.0	71.5	53.5	63.7
3rd quarter	59.0	51.4	86.2	43.1	80.0	72.2	54.8	64.8	59.7	51.4	86.2	43.1	80.0	72.2	54.3	64.5
4th quarter	59.1	52.1	87.1	43.4	79.9	72.6	55.5	65.3	59.8	52.1	87.1	43.4	79.9	72.6	55.6	65.2
1989																
1st quarter	59.3	52.7	88.0	43.6	79.3	72.8	56.9	65.9	59.9	52.7	88.0	43.6	79.3	72.8	56.1	65.5
2nd quarter	59.6	53.0	88.3	43.7	78.5	72.9	58.4	66.6	60.1	53.0	88.3	43.7	78.5	72.9	57.9	66.3
3rd quarter	59.8	53.4	88.5	44.1	78.6	73.2	58.9	67.0	60.3	53.4	88.5	44.1	78.6	73.2	58.6	66.8
4th quarter	59.9	53.4	88.5	44.8	79.0	74.2	58.5	67.4	60.4	53.4	88.5	44.8	79.0	74.2	58.0	67.2

Table 9-3B. Productivity and Related Data: Historical Data—*Continued*

(2012 = 100, seasonally adjusted.)

Year and quarter	Business sector								Nonfarm business sector							
	Output per hour of all persons	Output	Hours of all persons	Compensation per hour	Real compensation per hour	Unit labor costs	Unit nonlabor payments	Implicit price deflator	Output per hour of all persons	Output	Hours of all persons	Compensation per hour	Real compensation per hour	Unit labor costs	Unit nonlabor payments	Implicit price deflator
1990																
1st quarter	60.5	54.1	88.8	45.6	79.3	74.9	59.2	68.1	60.9	54.1	88.8	45.6	79.3	74.9	58.5	67.8
2nd quarter	60.9	54.2	88.4	46.5	80.1	75.9	59.3	68.8	61.3	54.2	88.4	46.5	80.1	75.9	58.8	68.5
3rd quarter	61.3	54.1	87.9	47.2	79.9	76.6	59.7	69.3	61.5	54.1	87.9	47.2	79.9	76.6	59.2	69.0
4th quarter	60.6	53.3	87.5	47.5	79.3	78.0	59.1	69.8	60.9	53.3	87.5	47.5	79.3	78.0	58.8	69.7
1991																
1st quarter	60.7	53.0	86.7	47.8	79.3	78.2	60.2	70.4	61.1	53.0	86.7	47.8	79.3	78.2	60.0	70.3
2nd quarter	61.7	53.5	86.2	48.7	80.5	78.5	60.8	70.8	62.1	53.5	86.2	48.7	80.5	78.5	60.6	70.7
3rd quarter	62.2	53.9	86.1	49.3	80.9	78.7	61.4	71.2	62.6	53.9	86.1	49.3	80.9	78.7	61.3	71.2
4th quarter	62.5	54.0	85.9	50.0	81.4	79.4	61.3	71.5	62.9	54.0	85.9	50.0	81.4	79.4	61.0	71.4
1992																
1st quarter	63.9	54.8	85.4	51.2	83.1	79.9	61.1	71.6	64.2	54.8	85.4	51.2	83.1	79.9	60.7	71.5
2nd quarter	64.4	55.5	85.7	51.7	83.3	79.9	61.8	71.9	64.7	55.5	85.7	51.7	83.3	79.9	61.3	71.8
3rd quarter	65.0	56.1	85.9	52.3	83.7	80.1	62.3	72.2	65.3	56.1	85.9	52.3	83.7	80.1	61.8	72.2
4th quarter	65.4	56.8	86.4	52.6	83.6	80.1	63.4	72.7	65.7	56.8	86.4	52.6	83.6	80.1	63.2	72.7
1993																
1st quarter	63.9	56.8	87.2	52.2	82.5	80.2	64.1	73.2	65.1	56.8	87.2	52.2	82.5	80.2	64.2	73.2
2nd quarter	64.4	57.1	88.2	52.6	82.5	81.3	63.8	73.6	64.7	57.1	88.2	52.6	82.5	81.3	63.8	73.7
3rd quarter	65.0	57.7	88.7	52.6	82.3	80.9	64.7	74.0	65.0	57.7	88.7	52.6	82.3	80.9	64.7	73.9
4th quarter	65.4	58.4	89.5	53.0	82.3	81.2	65.5	74.4	65.3	58.4	89.5	53.0	82.3	81.2	65.2	74.3
1994																
1st quarter	65.2	59.0	90.1	52.7	81.6	80.5	67.0	74.6	65.5	59.0	90.1	52.7	81.6	80.5	66.8	74.5
2nd quarter	65.1	60.0	91.6	53.1	81.8	81.1	66.9	74.9	65.5	60.0	91.6	53.1	81.8	81.1	66.8	74.9
3rd quarter	64.8	60.4	92.6	53.1	81.2	81.5	67.3	75.3	65.1	60.4	92.6	53.1	81.2	81.5	67.3	75.3
4th quarter	65.4	61.2	93.0	53.5	81.4	81.3	68.3	75.6	65.8	61.2	93.0	53.5	81.4	81.3	68.5	75.7
1995																
1st quarter	65.3	61.6	93.5	53.9	81.4	81.7	68.2	76.0	65.9	61.6	93.5	53.9	81.4	81.7	68.6	76.0
2nd quarter	65.5	61.8	93.3	54.3	81.5	82.0	68.6	76.3	66.2	61.8	93.3	54.3	81.5	82.0	69.0	76.4
3rd quarter	65.5	62.5	94.4	54.6	81.5	82.5	68.8	76.6	66.2	62.5	94.4	54.6	81.5	82.5	69.0	76.6
4th quarter	66.0	63.0	94.6	55.0	81.8	82.7	69.1	76.9	66.6	63.0	94.6	55.0	81.8	82.7	69.1	76.8
1996																
1st quarter	66.5	63.5	94.8	55.6	82.1	83.1	69.5	77.2	67.0	63.5	94.8	55.6	82.1	83.1	69.2	77.0
2nd quarter	67.2	64.7	95.7	56.2	82.1	83.0	70.5	77.6	67.6	64.7	95.7	56.2	82.1	83.0	69.9	77.3
3rd quarter	67.5	65.5	96.5	56.6	82.4	83.4	70.3	77.7	67.9	65.5	96.5	56.6	82.4	83.4	70.1	77.6
4th quarter	67.6	66.3	97.6	56.9	82.2	83.8	70.7	78.1	67.9	66.3	97.6	56.9	82.2	83.8	70.5	78.0
1997																
1st quarter	67.4	66.7	98.4	57.6	82.6	84.9	70.4	78.6	67.8	66.7	98.4	57.6	82.6	84.9	70.1	78.5
2nd quarter	68.5	68.0	98.9	58.1	83.2	84.5	71.2	78.6	68.8	68.0	98.9	58.1	83.2	84.5	71.3	78.7
3rd quarter	69.1	69.0	99.5	58.7	83.7	84.6	71.7	78.9	69.4	69.0	99.5	58.7	83.7	84.6	71.8	79.0
4th quarter	69.5	69.7	99.9	59.7	84.7	85.7	70.6	79.1	69.7	69.7	99.9	59.7	84.7	85.7	70.7	79.2
1998																
1st quarter	69.9	70.5	100.6	60.7	86.0	86.6	69.5	79.1	70.1	70.5	100.6	60.7	86.0	86.6	69.6	79.2
2nd quarter	70.2	71.3	101.1	61.5	86.9	87.2	68.7	79.2	70.5	71.3	101.1	61.5	86.9	87.2	69.0	79.3
3rd quarter	71.2	72.3	101.2	62.5	87.9	87.5	68.9	79.4	71.5	72.3	101.2	62.5	87.9	87.5	69.2	79.5
4th quarter	71.8	73.8	102.5	62.9	88.1	87.3	69.1	79.4	72.0	73.8	102.5	62.9	88.1	87.3	69.3	79.5
1999																
1st quarter	72.9	74.6	102.3	64.0	89.3	87.7	68.9	79.5	73.0	74.6	102.3	64.0	89.3	87.7	69.2	79.6
2nd quarter	73.0	75.3	102.9	64.2	89.0	87.8	69.0	79.6	73.1	75.3	102.9	64.2	89.0	87.8	69.5	79.9
3rd quarter	73.7	76.5	103.5	64.8	89.1	87.8	69.4	79.8	73.8	76.5	103.5	64.8	89.1	87.8	70.1	80.1
4th quarter	74.8	77.9	104.0	66.1	90.2	88.2	69.7	80.1	75.0	77.9	104.0	66.1	90.2	88.2	70.4	80.5
2000																
1st quarter	74.7	78.2	104.5	68.5	92.5	91.4	66.3	80.4	74.9	78.2	104.5	68.5	92.5	91.4	66.8	80.7
2nd quarter	76.3	80.0	104.7	68.6	92.0	89.8	69.3	80.8	76.4	80.0	104.7	68.6	92.0	89.8	69.7	81.1
3rd quarter	76.3	80.0	104.8	70.0	92.9	91.7	67.9	81.2	76.3	80.0	104.8	70.0	92.9	91.7	68.4	81.5
4th quarter	77.2	80.5	104.3	70.3	92.8	91.2	69.2	81.5	77.2	80.5	104.3	70.3	92.8	91.2	69.7	81.8
2001																
1st quarter	76.9	80.1	104.3	71.9	94.0	93.7	66.9	81.9	76.8	80.1	104.3	71.9	94.0	93.7	67.4	82.2
2nd quarter	78.2	80.7	103.2	72.2	93.7	92.3	69.3	82.3	78.2	80.7	103.2	72.2	93.7	92.3	69.9	82.6
3rd quarter	78.5	80.1	102.0	72.3	93.5	92.1	70.0	82.5	78.5	80.1	102.0	72.3	93.5	92.1	70.4	82.7
4th quarter	79.6	80.2	100.9	73.0	94.5	91.8	70.6	82.5	79.5	80.2	100.9	73.0	94.5	91.8	71.3	82.9
2002																
1st quarter	81.0	81.3	100.0	73.3	94.6	90.2	72.5	82.6	81.3	81.3	100.0	73.3	94.6	90.2	73.2	82.8
2nd quarter	81.3	81.6	100.3	74.0	94.7	90.9	72.3	82.7	81.4	81.6	100.3	74.0	94.7	90.9	73.2	83.2
3rd quarter	82.0	81.9	99.9	74.3	94.7	90.7	73.2	83.0	82.0	81.9	99.9	74.3	94.7	90.7	73.9	83.4
4th quarter	82.0	82.0	100.0	74.5	94.4	90.9	73.6	83.3	82.0	82.0	100.0	74.5	94.4	90.9	74.3	83.7
2003																
1st quarter	82.9	82.4	99.5	75.0	94.0	90.5	74.8	83.6	82.8	82.4	99.5	75.0	94.0	90.5	75.5	84.0
2nd quarter	84.2	83.2	99.2	76.3	95.7	91.0	74.6	83.7	83.9	83.2	99.2	76.3	95.7	91.0	75.1	84.1
3rd quarter	85.8	85.2	99.4	77.4	96.5	90.3	76.1	84.1	85.8	85.2	99.4	77.4	96.5	90.3	76.5	84.3
4th quarter	86.5	86.5	99.8	78.5	97.4	90.7	76.3	84.4	86.6	86.5	99.8	78.5	97.4	90.7	76.5	84.5
2004																
1st quarter	86.5	86.5	100.3	78.4	96.4	90.8	78.2	85.2	86.3	86.5	100.3	78.4	96.4	90.8	78.3	85.4
2nd quarter	87.1	87.4	100.4	79.9	97.5	91.7	78.3	85.8	87.1	87.4	100.4	79.9	97.5	91.7	78.2	85.8
3rd quarter	87.6	88.3	101.0	81.3	98.6	92.9	77.8	86.2	87.5	88.3	101.0	81.3	98.6	92.9	77.9	86.4
4th quarter	88.3	89.2	101.5	81.7	98.0	92.9	79.6	87.0	87.9	89.2	101.5	81.7	98.0	92.9	79.8	87.2

Table 9-3B. Productivity and Related Data: Historical Data—*Continued*

(2012 = 100, seasonally adjusted.)

Year and quarter	Business sector								Nonfarm business sector							
	Output per hour of all persons	Output	Hours of all persons	Compensation per hour	Real compensation per hour	Unit labor costs	Unit nonlabor payments	Implicit price deflator	Output per hour of all persons	Output	Hours of all persons	Compensation per hour	Real compensation per hour	Unit labor costs	Unit nonlabor payments	Implicit price deflator
2005																
1st quarter	89.2	90.4	101.7	82.2	98.2	92.5	81.5	87.6	88.9	90.4	101.7	82.2	98.2	92.5	82.0	87.9
2nd quarter	88.8	90.8	102.3	82.7	98.1	93.3	82.1	88.3	88.7	90.8	102.3	82.7	98.1	93.3	82.6	88.6
3rd quarter	89.6	91.7	102.6	83.7	97.9	93.7	83.5	89.1	89.4	91.7	102.6	83.7	97.9	93.7	84.1	89.5
4th quarter	89.8	92.4	103.2	84.3	97.6	94.2	84.4	89.8	89.5	92.4	103.2	84.3	97.6	94.2	85.1	90.2
2006																
1st quarter	90.5	93.9	104.2	86.0	99.0	95.4	84.1	90.3	90.1	93.9	104.2	86.0	99.0	95.4	84.8	90.8
2nd quarter	90.3	94.1	104.5	86.0	98.1	95.5	85.5	91.0	90.0	94.1	104.5	86.0	98.1	95.5	86.3	91.5
3rd quarter	90.0	94.3	105.1	86.1	97.4	96.0	86.1	91.5	89.7	94.3	105.1	86.1	97.4	96.0	86.8	92.0
4th quarter	90.6	95.4	105.4	87.7	99.5	96.9	85.1	91.6	90.5	95.4	105.4	87.7	99.5	96.9	85.7	92.0
2007																
1st quarter	90.9	95.7	105.5	89.8	101.0	99.0	84.3	92.5	90.7	95.7	105.5	89.8	101.0	99.0	84.7	92.8
2nd quarter	91.4	96.5	105.9	89.8	99.8	98.5	86.1	93.1	91.1	96.5	105.9	89.8	99.8	98.5	86.6	93.3
3rd quarter	92.3	97.1	105.5	90.1	99.5	97.9	87.8	93.5	92.0	97.1	105.5	90.1	99.5	97.9	88.2	93.7
4th quarter	92.8	97.6	105.3	91.1	99.4	98.2	88.0	93.8	92.7	97.6	105.3	91.1	99.4	98.2	88.2	93.9
2008																
1st quarter	92.1	96.6	104.9	92.0	99.3	100.0	86.3	94.0	92.0	96.6	104.9	92.0	99.3	100.0	86.4	94.1
2nd quarter	93.1	97.0	104.4	92.1	98.1	99.1	88.5	94.4	92.9	97.0	104.4	92.1	98.1	99.1	88.7	94.6
3rd quarter	93.3	96.2	103.3	92.9	97.5	99.8	89.2	95.1	93.1	96.2	103.3	92.9	97.5	99.8	89.6	95.3
4th quarter	92.7	93.2	100.7	93.9	100.9	101.5	87.2	95.1	92.5	93.2	100.7	93.9	100.9	101.5	87.7	95.5
2009																
1st quarter	93.6	91.7	98.1	91.4	98.9	97.9	91.8	95.1	93.4	91.7	98.1	91.4	98.9	97.9	92.8	95.7
2nd quarter	95.5	91.5	96.0	93.7	100.8	98.3	90.3	94.7	95.3	91.5	96.0	93.7	100.8	98.3	91.1	95.2
3rd quarter	97.1	91.8	94.8	94.3	100.6	97.4	91.5	94.7	96.8	91.8	94.8	94.3	100.6	97.4	92.4	95.2
4th quarter	98.4	93.1	94.9	94.8	100.4	96.6	93.2	95.0	98.1	93.1	94.9	94.8	100.4	96.6	93.8	95.4
2010																
1st quarter	98.8	93.6	94.9	94.2	99.6	95.5	95.5	95.4	98.6	93.6	94.9	94.2	99.6	95.5	96.2	95.8
2nd quarter	99.0	94.6	95.7	95.3	100.7	96.3	95.4	95.8	98.9	94.6	95.7	95.3	100.7	96.3	96.1	96.2
3rd quarter	99.6	95.6	96.2	95.7	100.8	96.2	96.1	96.0	99.4	95.6	96.2	95.7	100.8	96.2	96.6	96.4
4th quarter	99.9	96.4	96.6	96.1	100.5	96.4	97.2	96.6	99.7	96.4	96.6	96.1	100.5	96.4	97.4	96.8
2011																
1st quarter	99.1	95.9	96.8	97.8	101.2	98.8	95.6	97.3	99.0	95.9	96.8	97.8	101.2	98.8	95.3	97.3
2nd quarter	99.3	96.9	97.6	97.3	99.5	98.0	98.0	97.9	99.3	96.9	97.6	97.3	99.5	98.0	97.9	97.9
3rd quarter	98.9	96.9	98.0	97.9	99.5	99.1	98.2	98.6	98.8	96.9	98.0	97.9	99.5	99.1	97.9	98.6
4th quarter	99.6	98.2	98.7	96.6	97.7	97.1	101.3	98.8	99.5	98.2	98.7	96.6	97.7	97.1	101.2	98.9
2012																
1st quarter	99.9	99.4	99.5	98.9	99.5	99.0	99.7	99.3	99.9	99.4	99.5	98.9	99.5	99.0	99.7	99.3
2nd quarter	100.3	100.0	99.7	99.4	99.8	99.1	100.6	99.7	100.3	100.0	99.7	99.4	99.8	99.1	100.6	99.8
3rd quarter	100.0	100.2	100.1	99.5	99.4	99.4	101.3	100.3	100.1	100.2	100.1	99.5	99.4	99.4	101.3	100.3
4th quarter	99.8	100.4	100.7	102.1	101.4	102.4	98.4	100.7	99.7	100.4	100.7	102.1	101.4	102.4	98.3	100.6
2013																
1st quarter	100.6	101.4	101.1	100.7	99.5	100.4	101.9	101.0	100.3	101.4	101.1	100.7	99.5	100.4	101.4	100.8
2nd quarter	100.5	101.4	101.4	101.4	100.3	101.4	101.1	101.2	100.0	101.4	101.4	101.4	100.3	101.4	100.7	101.1
3rd quarter	100.9	102.4	102.0	101.2	99.5	100.7	103.1	101.6	100.5	102.4	102.0	101.2	99.5	100.7	102.8	101.6
4th quarter	101.6	103.6	102.3	101.9	99.9	100.7	104.3	102.1	101.2	103.6	102.3	101.9	99.9	100.7	104.4	102.3
2014																
1st quarter	100.7	103.2	102.8	104.2	101.4	103.8	100.9	102.4	100.4	103.2	102.8	104.2	101.4	103.8	100.9	102.6
2nd quarter	101.5	104.8	103.6	103.4	100.1	102.2	104.6	103.1	101.2	104.8	103.6	103.4	100.1	102.2	104.5	103.2
3rd quarter	102.3	106.5	104.2	103.9	100.4	101.8	106.1	103.5	102.1	106.5	104.2	103.9	100.4	101.8	106.2	103.7
4th quarter	101.7	107.0	105.4	105.0	101.7	103.4	104.2	103.6	101.6	107.0	105.4	105.0	101.7	103.4	104.2	103.8
2015																
1st quarter	102.6	108.5	105.8	106.3	103.5	103.6	102.6	103.2	102.5	108.3	105.6	106.5	103.8	104.0	103.1	103.6
2nd quarter	103.1	109.5	106.2	107.2	103.7	104.0	103.4	103.7	102.9	109.2	106.2	107.4	103.9	104.3	103.8	104.1
3rd quarter	103.4	109.8	106.2	107.7	103.8	104.1	103.9	104.0	103.2	109.6	106.1	107.8	103.9	104.5	104.4	104.4
4th quarter	102.5	109.7	107.1	107.4	103.5	104.8	102.7	103.9	102.3	109.4	107.0	107.6	103.7	105.2	103.2	104.4
2016																
1st quarter	102.7	110.3	107.4	107.5	103.6	104.6	102.4	103.7	102.6	110.1	107.3	107.7	103.8	105.0	103.2	104.2
2nd quarter	102.8	110.9	107.9	107.7	103.1	104.7	104.0	104.4	102.7	110.6	107.7	108.0	103.4	105.2	104.9	105.0
3rd quarter	103.3	111.6	108.1	108.3	103.2	104.9	104.6	104.8	103.1	111.3	107.9	108.6	103.4	105.3	105.5	105.4
4th quarter	103.9	112.3	108.1	109.7	103.8	105.6	104.7	105.2	103.6	112.0	108.2	109.7	103.9	106.0	105.7	105.9
2017																
1st quarter	104.0	113.1	108.7	110.7	104.0	106.4	104.7	105.6	103.8	112.7	108.6	110.9	104.2	106.8	105.4	106.2
2nd quarter	104.2	113.8	109.2	111.2	104.5	106.8	104.8	105.9	103.9	113.5	109.2	111.4	104.6	107.2	105.5	106.5
3rd quarter	105.0	114.9	109.4	112.6	105.2	107.2	105.5	106.5	104.7	114.7	109.5	112.7	105.2	107.6	106.3	107.0
4th quarter	105.1	116.1	110.5	113.8	105.5	108.3	105.7	107.2	104.9	115.8	110.4	114.1	105.7	108.7	106.5	107.8
2018																
1st quarter	105.4	116.9	110.9	115.1	105.9	109.2	105.7	107.7	105.2	116.7	110.9	115.2	106.0	109.6	106.6	108.3
2nd quarter	106.0	118.0	111.3	115.4	105.5	108.8	108.5	108.7	105.7	117.8	111.5	115.3	105.5	109.1	109.4	109.3
3rd quarter	106.3	119.0	112.0	116.1	105.7	109.3	108.8	109.1	106.0	118.8	112.1	116.1	105.7	109.6	109.9	109.7
4th quarter	106.3	119.4	112.4	116.2	105.4	109.4	109.4	109.4	106.0	119.2	112.4	116.4	105.5	109.8	110.4	110.0

Table 9-4. Corporate Profits with Inventory Valuation Adjustment by Industry Group, NAICS Basis

(Billions of dollars, quarterly data are at seasonally adjusted annual rates.)   NIPA Table 6.16D

Year and quarter	Total	Domestic industries												
		Financial				Nonfinancial									
		Total	Federal Reserve banks	Other financial	Total	Utilities	Manufacturing							
								Durable goods							
							Total	Fabricated metal products	Machinery	Computer and electronic products	Electrical equipment, appliances, and components	Motor vehicles, bodies and trailers, and parts	Other durable goods
2000	729.8	584.1	31.2	118.5	434.4	24.3	175.6	15.6	9.4	20.5	6.9	4.0	25.8
2001	697.1	528.3	28.9	166.1	333.3	22.5	75.1	9.2	2.2	-29.1	1.1	-6.8	14.0
2002	797.4	640.6	23.5	241.9	375.3	10.5	78.3	9.9	2.9	-25.0	-0.8	-3.1	19.5
2003	955.7	796.7	20.0	282.7	494.0	13.2	123.9	8.9	2.8	-6.1	2.7	7.9	15.5
2004	1 217.5	1 022.4	20.0	326.0	676.3	21.1	186.2	12.6	8.3	1.2	0.4	-4.6	35.2
2005	1 629.2	1 403.4	26.5	383.0	993.9	32.4	279.7	19.0	16.4	15.5	-0.6	1.6	61.4
2006	1 812.2	1 572.5	33.8	379.3	1 159.4	55.2	352.9	19.3	20.8	28.5	11.7	-6.9	67.6
2007	1 708.3	1 370.5	36.0	264.2	1 070.3	49.6	321.1	21.4	23.4	24.3	-0.7	-16.3	67.1
2008	1 344.5	954.3	35.1	59.5	859.7	30.4	240.0	15.7	18.0	27.7	5.0	-39.2	39.5
2009	1 470.1	1 121.3	47.3	315.3	758.7	23.4	164.7	11.8	9.6	27.1	9.2	-54.8	33.0
2010	1 786.4	1 400.6	71.6	334.3	994.8	30.6	281.8	15.4	17.3	48.4	10.1	-10.9	43.4
2011	1 750.2	1 337.7	76.0	302.4	959.3	10.2	296.0	16.5	25.8	37.8	4.9	-0.3	49.7
2012	2 144.7	1 739.3	71.7	410.6	1 256.9	13.8	403.0	24.0	33.5	52.9	12.0	23.0	60.1
2013	2 165.9	1 767.1	79.7	351.1	1 336.3	28.3	446.9	25.3	36.5	58.7	20.4	21.5	66.6
2014	2 266.6	1 861.7	103.5	379.6	1 378.6	32.8	458.7	24.1	35.5	60.3	14.2	32.0	68.1
2015	2 184.6	1 789.4	100.7	346.5	1 342.1	20.2	427.2	25.0	24.3	68.4	24.1	26.8	66.1
2016	2 124.3	1 704.4	92.0	363.8	1 248.6	9.4	332.7	23.6	19.0	50.3	5.2	29.3	65.2
2017	2 130.5	1 633.3	78.2	357.3	1 197.7	14.0	304.7	22.0	21.6	49.4	5.8	13.0	61.8
2018	2 132.0	1 619.5	68.0	350.2	1 201.3	21.7	337.6	19.8	19.0	54.5	10.9	1.0	65.8
2019	2 232.0	1 726.5	52.4	418.1	1 256.0	27.2	336.5	24.8	26.4	50.8	11.4	0.6	67.3
2018													
1st quarter	2 088.9	1 552.9	73.7	349.5	1 129.7	22.7	276.2	20.7	16.6	44.5	12.3	0.0	53.5
2nd quarter	2 112.5	1 599.8	70.5	349.1	1 180.2	23.3	348.1	19.3	22.3	58.2	11.8	0.1	72.2
3rd quarter	2 149.9	1 657.4	66.9	347.6	1 242.8	22.3	365.3	19.7	18.8	61.3	11.2	5.8	70.6
4th quarter	2 176.8	1 667.8	61.0	354.3	1 252.5	18.6	360.9	19.6	18.3	54.0	8.3	-2.1	66.8
2019													
1st quarter	2 154.9	1 670.5	53.0	407.1	1 210.4	26.2	324.5	25.7	22.9	56.3	10.4	1.9	70.8
2nd quarter	2 246.4	1 740.2	56.6	415.8	1 267.8	28.2	344.9	25.3	29.6	50.4	12.0	2.1	73.7
3rd quarter	2 231.7	1 717.2	50.7	416.0	1 250.5	27.1	341.0	23.9	27.1	45.4	12.5	0.2	65.2
4th quarter	2 294.9	1 778.3	49.4	433.5	1 295.4	27.3	335.7	24.2	26.1	51.3	10.9	-2.1	59.7

Year and quarter	Domestic industries—Continued										Rest of the world, net
	Nonfinancial—Continued										
	Manufacturing—Continued										
	Nondurable goods										
	Total	Food and beverage and tobacco products	Petroleum and coal products	Chemical products	Other nondurable goods	Wholesale trade	Retail trade	Transportation and warehousing	Information	Other nonfinancial	
2000	93.3	25.8	25.9	25.1	16.6	59.5	51.3	9.5	-11.9	126.1	145.7
2001	84.5	28.5	27.5	22.0	6.6	51.1	71.3	-0.7	-26.4	140.2	168.8
2002	74.9	26.7	4.8	30.7	12.8	53.5	83.3	-6.5	5.0	151.2	156.8
2003	92.3	25.6	26.7	31.8	8.1	56.6	87.9	4.4	28.1	179.9	158.9
2004	133.0	26.2	50.8	39.4	16.6	72.7	94.0	12.0	61.6	228.6	195.1
2005	166.3	29.2	80.0	36.2	20.9	96.0	123.3	28.4	100.7	333.5	225.7
2006	211.8	34.6	82.0	68.8	26.4	105.0	133.6	40.8	115.2	356.8	239.7
2007	201.9	31.3	76.1	70.8	23.7	102.8	119.4	23.3	120.5	333.6	337.8
2008	173.3	31.5	85.0	51.5	5.3	92.7	82.2	29.3	98.8	286.3	390.2
2009	128.9	45.0	10.4	55.2	18.2	88.9	107.9	21.7	87.0	265.1	348.8
2010	158.1	45.1	26.2	63.4	23.5	99.3	115.9	44.6	102.3	320.4	385.8
2011	161.6	40.0	47.5	53.2	20.9	97.2	115.1	30.6	95.7	314.5	412.6
2012	197.5	44.7	57.7	63.5	31.5	137.9	155.7	54.4	112.0	380.1	405.4
2013	217.8	55.6	53.9	75.0	33.3	146.4	153.3	45.2	137.6	378.6	398.8
2014	224.5	58.3	65.5	72.2	28.4	150.6	157.3	55.7	126.6	397.0	404.9
2015	192.6	69.0	19.6	65.5	38.4	152.4	169.3	61.0	135.5	376.4	395.2
2016	140.1	68.5	-30.4	64.5	37.6	126.6	170.5	63.9	157.4	388.1	420.0
2017	131.1	60.0	-7.9	53.4	25.6	122.0	149.1	58.7	138.0	411.1	497.2
2018	166.7	47.6	31.0	60.9	27.3	105.7	146.5	52.8	139.2	397.7	512.5
2019	155.2	48.3	17.5	58.3	31.2	111.3	168.0	56.4	130.8	425.8	505.4
2018											
1st quarter	128.5	48.4	13.8	41.6	24.7	111.3	149.5	48.5	134.9	386.7	536.0
2nd quarter	164.2	52.9	21.3	62.4	27.7	94.9	137.7	46.6	143.4	386.2	512.7
3rd quarter	178.0	51.8	29.6	67.1	29.6	103.9	157.5	52.0	144.0	397.8	492.5
4th quarter	196.0	37.2	59.2	72.5	27.1	112.9	141.2	64.2	134.6	420.1	509.0
2019											
1st quarter	136.6	45.7	6.2	55.9	28.8	103.9	155.5	54.7	136.2	409.4	484.4
2nd quarter	151.8	47.8	16.2	57.2	30.6	110.5	165.6	54.4	140.0	424.3	506.2
3rd quarter	166.8	51.6	21.7	61.3	32.2	113.4	166.8	59.5	108.4	434.3	514.5
4th quarter	165.6	48.0	25.8	58.8	33.1	117.4	184.2	57.0	138.7	435.1	516.6

NOTES AND DEFINITIONS, CHAPTER 9

General note on data on compensation per hour

This chapter includes two data series with similar names—the Employment Cost Index for total compensation and the *Index of Compensation* per hour—that often display different behavior. Both are compiled and published by the Bureau of Labor Statistics (BLS), but the definitions, sources, and methods of compilation are different. Users should be aware of these differences and of the consequent differences in the appropriate uses and interpretations for each of the two series.

The *Employment Cost Index (ECI)* (Tables 9-1 and 9-2) measures changes in hourly compensation for "all civilian workers", which is not quite as broad as it sounds, as it excludes federal government workers, farm workers, and private household workers. Indexes are also published for subgroups including state and local workers, "all private industry" (again excluding farm and private household workers), and a number of industry and occupational subgroups.

The Employment Cost Index (ECI) data from the National Compensation Survey (NCS) are used for a variety of reasons, by both the private and public sectors. Some examples include: military pay adjustments, federal pay adjustments, Medicare reimbursement adjustments, contract escalator clauses, labor cost adjustments, and collective bargaining negotiations.

The ECI is calculated and published separately for *total compensation* and for the two components of hourly compensation, *wages and salaries* and the employer cost of employee *benefits*. It is constructed by analogy with the Consumer Price Index (CPI); that is, it holds the composition of employment constant in order to isolate hourly compensation trends that take place for individual occupations, which are then aggregated, using fixed relative importance weights. The ECI is based on a sample survey and may be revised from time to time, due to updated classification, weighting, and seasonal adjustments. However, it is not subject to major benchmark revision of the underlying wage, salary, and benefit rate observations. By design, it excludes any representation of employee stock options. As it is based on a sample survey, the ECI is measured "from the bottom up," aggregating from individual employers' reports to higher levels. The ECI is frequently and appropriately used as the best available measure of the general trend of wages and of the extent of inflationary pressure exerted on prices by labor costs.

The *compensation per hour* component of the report on "Productivity and Costs" (Table 9-3) is calculated and published for total compensation in total business, nonfarm business, nonfinancial corporations, and manufacturing. The nonfarm business category is similar in scope to the "all private industry" category in the ECI. The measures in Table 9-3, however, are compiled "from the top down," starting with aggregate estimates of compensation and hours, then dividing the former by the latter. Compensation per hour is affected by changes in the composition of employment. If the composition of employment shifts toward a larger proportion of higher-paid employees and/or industries, compensation per hour will rise even if there is no increase in hourly compensation for any individual worker.

In addition, *compensation per hour* includes the value of exercised stock options as expensed by companies. Also included are other transitory payments, many of which may be of little relevance to the typical worker or to ongoing production costs. These values are not reported immediately. Instead, they are incorporated when later, more comprehensive reports are received. This process can lead to dramatic revisions in compensation per hour and in the unit labor costs index, which is based on compensation. For example, the fourth-quarter 2004 increase in compensation per hour in nonfarm business was initially reported at an annual rate of 3.1 percent. Four months later, the reported rate for the same time period was 10.2 percent. The rate of increase from a year earlier was revised from 3.6 to 5.9 percent. According to then-Federal Reserve Chairman Alan Greenspan, in testimony before the Joint Economic Committee on June 9, 2005, this reflected "a large but apparently transitory surge in bonuses and the proceeds of stock option exercises," not a potentially inflationary acceleration in the rate of labor compensation increase. More recently, it was reported in August 2009 that "first-quarter unit labor costs were revised to negative 2.7% from positive 3%." (*Wall Street Journal*, August 12, 2009, p. A2.)

These characteristics suggest that *compensation per hour* should not be considered a reliable or appropriate indicator of wage or compensation trends for typical workers. It is useful in conjunction with the productivity series, because aggregate productivity is subject to the same composition shifts—higher-productivity industries also tend to have higher-paid employees. Hence, the measure of *unit labor costs* (derived by dividing compensation per hour by output per hour in this system) is not distorted when the composition of output shifts toward higher-productivity industries. The shift affects the numerator and denominator of the ratio similarly. However, both compensation and unit labor costs can still be distorted by transitory payments, such as those discussed above.

There are other, probably less important differences between the two measures. Compensation per hour refers to the entire quarter, while the ECI is observed in the terminal month of each quarter. Tips and other forms of compensation not provided by employers are included in hourly compensation but not in the ECI. The ECI excludes persons working for token wages, business owners and others who set their own wage, and family workers who do not earn a market wage; all of these workers are included in the productivity and cost accounts. Hourly compensation excludes employees of nonprofit institutions serving individuals—about 10 percent of private workers (mostly in education and medical care) who are within the scope of the ECI. Hourly compensation measures include an estimate for the unincorporated self-employed, who are assumed to earn the same hourly compensation as other employees in the sector. Implicitly, unpaid family workers also are included in the hourly compensation measures with the assumption that their hourly compensation is zero. Both of these groups are also excluded from the ECI.

TABLES 9-1 AND 9-2

Employment Cost Indexes

SOURCE: U.S. DEPARTMENT OF LABOR, BUREAU OF LABOR STATISTICS (BLS)

The Employment Cost Index (ECI) is a quarterly measure of the change in the cost of labor, independent of the influence of employment shifts among occupations and industries. It uses a fixed market basket of labor—similar in concept to the Consumer Price Index's fixed market basket of goods and services—to measure changes over time in employer costs of employing labor. Data are quarterly in all cases and are reported for the final month of each quarter. These measures are expressed as indexes, with the not-seasonally-adjusted value for December 2005 set at 100.

Care should be used in comparing the ECI with other data sets. The "all private industry" category in the ECI excludes farm and household workers (it is sometimes, and more precisely, called "private nonfarm industry"), and the "all civilian workers" category excludes federal government, farm, and household workers.

The data for 1979 through 2019, which are presented in Table 9-2 and are the official ECI measures for that time period, were based on the 1987 Standard Industrial Classification (SIC) and 2010 Occupational Classification System (OCS).

For over 60 years, the Standard Industrial Classification (SIC) system served as the structure for the collection, presentation, and analysis of the U.S. economy. The SIC system was developed in the 1930s at a time when manufacturing dominated the U.S. economy. Developments in information services, new forms of health care provision, expansion of services, and high-tech manufacturing are examples of industrial changes that could not be incorporated under the SIC system.

Currently the ECI is compiled based on the 2012 North American Industry Classification System (NAICS) and the 2010 Standard Occupational Classification Manual (SOC). These data, along with comparable data for 2004 through 2019, are shown in Table 9-1.

Developed in cooperation with Canada and Mexico, the North American Industry Classification System (NAICS) represents one of the most profound changes for statistical programs focusing on emerging economic activities. NAICS uses a production-oriented conceptual framework to group establishments into industries based on the activity in which they are primarily engaged. Establishments using similar raw material inputs,

similar capital equipment, and similar labor are classified in the same industry. In other words, establishments that do similar things in similar ways are classified together. NAICS was introduced in 1997 and is periodically revised to reflect changes in the industrial structure of the U.S. and North American economy.

For certain broad categories shown in this volume—indicated by footnote 2 in Table 9-2—the old SIC categories are roughly comparable and continuous with the data for 2006 and subsequent years. (However, they differ slightly on overlap dates, and should be "linked" if a continuous time series is desired; see the article at the beginning of this volume.) Many of the new industry and occupational categories are not continuous with the old series shown here, and some of the old categories are not being continued because BLS finds them obsolete and no longer meaningful.

Definitions

Total compensation comprises wages, salaries, and the employer's costs for employee benefits. Excluded from wages and salaries and employee benefits are the value of stock option exercises and items such as payment-in-kind, free room and board, and tips.

Wages and salaries consists of straight-time earnings per hour before payroll deductions, including production bonuses, incentive earnings, commissions, and cost-of-living adjustments. These wage rates exclude premium pay for overtime and for work on weekends and holidays, shift differentials, and nonproduction bonuses such as lump-sum payments provided in lieu of wage increases. According to BLS, wages and salaries are about 70 percent of total compensation.

Benefits includes the cost to employers for paid leave—vacations, holidays, sick leave, and other leave; for supplemental pay—premium pay for work in addition to the regular work schedule (such as overtime, weekends, and holidays), shift differentials, and nonproduction bonuses (such as referral bonuses and lump-sum payments provided in lieu of wage increases); for insurance benefits—life, health, short-term disability, and long-term disability; for retirement and savings benefits—defined benefit and defined contribution plans; and for legally required benefits—Social Security, Medicare, federal and state unemployment insurance, and workers' compensation. Severance pay and supplemental unemployment benefit (SUB) plans are included in the data through December 2005 but were dropped beginning with the March 2006 data. The combined cost of these two benefits accounted for less than one-tenth of one percent of compensation, and according to BLS, dropping these benefits has had virtually no impact on the index. According to BLS, benefit costs are about 30 percent of total compensation.

Civilian workers are private industry workers, as defined below, and workers in state and local government. Federal workers are not included.

Private industry workers are paid workers in private industry excluding farms and private households. To be included in the

ECI, employees in occupations must receive cash payments from the establishment for services performed and the establishment must pay the employer's portion of Medicare taxes on that individual's wages. Major exclusions from the survey are the self-employed, individuals who set their own pay (for example, proprietors, owners, major stockholders, and partners in unincorporated firms), volunteers, unpaid workers, family members being paid token wages, individuals receiving long-term disability compensation, and U.S. citizens working overseas.

Private industry workers excluding incentive paid occupations is a new category introduced in the 2006 revision to eliminate the quarter-to-quarter variability related to the way workers (for example, salespersons working on commission) are paid. (The category *private industry workers excluding sales occupations* was intended to serve a similar purpose in the previous SIC-based classification system, but was much less accurate in separately identifying workers with highly variable compensation.)

Goods-producing industries include mining, construction, and manufacturing.

Service-providing industries include the following NAICS industries: wholesale trade; retail trade; transportation and warehousing; utilities; information; finance and insurance; real estate and rental and leasing; professional, scientific, and technical services; management of companies and enterprises; administrative and support and waste management and remediation services; education services; health care and social assistance; arts, entertainment, and recreation; accommodation and food services; and other services, except public administration.

Notes on the data

Employee benefit costs are calculated as cents per hour worked.

The July 2019 data were collected from probability samples of approximately 25,600 occupational observations in about 6,300 sample establishments in private industry, and approximately 7800 occupations within about 1,400 establishments in state and local governments. The private industry sample is rotated over approximately five years. The state and local government sample is replaced less frequently; the latest sample was introduced in September 2007.

Currently, the sample establishments are classified in industry categories based on the NAICS. Within an establishment, specific job categories are selected and classified into approximately 800 occupational classifications according to the SOC. Similar procedures were followed under the previous classification systems. Data are collected each quarter for the pay periods including the 12th day of March, June, September, and December.

Aggregate indexes are calculated using fixed employment weights. Beginning with December 2013, estimates are based on 2012 employment weight, using Occupational Employment counts. From March 2006 through June 2018, ECI weights were based on fixed employment counts for 2012 from the BLS Occupational Employment Statistics survey. Employment data from the 2012 Survey were introduced in December 2013. ECI measures were based on 1990 employment counts from March 1995 through December 2005 and 1980 census employment counts from June 1986 through December 1994. Prior to June 1986, they were based on 1970 census employment counts.

Use of fixed weights ensures that changes in the indexes reflect only changes in hourly compensation, not employment shifts among industries or occupations with different levels of wages and compensation. This feature distinguishes the ECI from other compensation series such as average hourly earnings (see Chapter 10 and its notes and definitions) and the compensation per hour component of the productivity series (see Table 9-3 and its notes and definitions, and the general note above), each of which is affected by such employment shifts.

Data availability

Data for wages and salaries for the private nonfarm economy are available beginning with the data for 1975; data for compensation begin with the 1980 data. The series for state and local government and for the civilian nonfarm economy begin with the 1981 data. All series are available on the BLS Web site at <http://www.bls.gov>.

Wage and salary change and compensation cost change data also are available by major occupational and industry groups, as well as by region and collective bargaining status. Information on wage and salary change is available from 1975 to the present for most of these series. Compensation cost change data are available from 1980 to the present for most series. For 10 occupational and industry series, benefit cost change data are available from the early 1980s to the present. For state and local governments and the civilian economy (state and local governments plus private industry), wage and salary change and compensation cost change data are available for major occupational and industry series. BLS provides data for all these series from June 1981 to the present.

Updates are available about four weeks after the end of the reference quarter. Reference quarters end in March, June, September, and December. Seasonal adjustment factors and seasonally adjusted indexes are released in late April.

References

Explanatory notes, including references, are included in a Technical Note in each quarter's ECI news release, and can be found on the BLS Web site, in the PDF version of the release.

More detailed information on the ECI is available from Chapter 8, "National compensation measures," (www.bls.gov/opub/hom/pdf/homch8.pdf) from the BLS Handbook of Methods, and several articles published in the *Monthly Labor Review* and *Compensation and Working Conditions*. The articles and other descriptive pieces are available at www.bls.gov/ect/#publications, by calling (202) 691-6199, or sending e-mail to NCSinfo@bls.gov.

TABLE 9-3A AND 9-3B

Productivity and Related Data

The indexes of productivity and costs for the entire postwar period, 1947 through 2019, have been revised to reflect the 2013 and 2014 comprehensive revisions of the National Income and Product Accounts (NIPAs).

Productivity is a measure of economic efficiency that shows how effectively economic inputs are converted into goods and services. Therefore, productivity measures relate real physical output to real input. They encompass a family of measures that includes single-factor input measures, such as output per unit of labor input or output per unit of capital input, as well as measures of multifactor productivity—that is, output per unit of combined labor and capital inputs. The indexes published in this book are indexes of labor productivity expressed in terms of output per hour of labor input. (A larger group of BLS productivity measures can be found in Bernan Press's *Handbook of U.S. Labor Statistics*.) All data are presented as indexes with a base of 2012 = 100.

Labor productivity is then the efficiency at which labor hours are utilized in producing output of goods and services, measured as output per hour of labor. (Also known as output per hour, or output per hour of all persons.)

Definitions

Current dollar output is the current-dollar value of the goods and services produced in the specified sector. *Output* is the constant-dollar value of the same goods and services, derived as current-dollar output divided by the sector *implicit price deflator* (the ratio of the current-dollar value of a series to its corresponding chained-dollar value, multiplied by 100.)

All these data are derived from components of the NIPAs; see Notes on the data below, and for general information on the NIPAs, see the Notes and Definitions to Chapter 1.

Output per hour of all persons (labor productivity) is the value of goods and services in constant dollars produced per hour of labor input. By definition, nonfinancial corporations include no self-employed persons. Productivity in this sector is expressed as *output per hour of all employees. Output per person* is output divided by the index for employment instead of by the index for hours of all persons.

Compensation is the total value of the wages and salaries of employees, plus employers' contributions for social insurance and private benefit plans, plus wages, salaries, and supplementary payments for the self-employed. *Compensation per hour* is compensation divided by *hours of all persons*. Included in compensation is the value of exercised stock options that companies report as a charge against earnings. Stock option values are reported with a delay; consequently, recent values are estimated based on extrapolation. They are revised to actual values when the data become available; sometimes the revisions are very large. The labor compensation of proprietors cannot be explicitly identified

and must be estimated. This is done by assuming that proprietors have the same hourly compensation as employees in the same sector. The quarterly labor productivity and cost measures do not contain estimates of compensation for unpaid family workers.

Real compensation per hour is compensation per hour deflated by the Consumer Price Index Research Series (CPI-U-RS) for the period 1978 through 2019 (For current quarters, before the CPI-U-RS becomes available, changes in the CPI-U are used.) Changes in the Consumer Price Index for Urban Wage Earners and Clerical Workers (CPI-W) are used for data before 1978, as there was no CPI-U (or CPIU-RS) for that period. See the Notes and Definitions to Chapter 8 for explanation of the CPI-U, the CPI-W, and the CPI-U-RS.

Unit labor costs represent the cost of labor required to produce one unit of output. Unit labor costs also describe the relationship between compensation per hour worked (hourly compensation) and real output per hour worked (labor productivity). When hourly compensation growth outpaces productivity, units labor costs increase. Alternatively, when productivity growth exceeds hourly compensation, unit labor costs decrease.

Unit nonlabor payments include profits, depreciation, interest, rental income of persons, and indirect taxes per unit of output. They are computed by subtracting current-dollar compensation of all persons from current-dollar value of output, providing the data for the index of total *nonlabor payments,* which is then divided by output.

Unit nonlabor costs are available for nonfinancial corporations only. They contain all the components of unit nonlabor payments except unit profits (and rental income of persons, which is zero by definition for nonfinancial corporations).

Unit profits, the other component of unit nonlabor payments, are also only available for nonfinancial corporations.

Labor hours are measured as annual hours worked by all employed in an industry. Data on industry and hours come primarily from the BLS Current Employment Statistics (CES) survey and Current Population Survey (CPS). Definitions and data for these data sources are fund in Chapter 10.

Hours of all persons (labor input) consists of the total hours at work (*employment* multiplied by *average weekly hours*) of payroll workers, self-employed persons, and unpaid family workers. For the nonfinancial corporations data, there are no self-employed persons; the data represent *employee hours*.

Labor share is the total dollar amount of labor compensation divided by the total current-dollar value of output.

Notes on the data

Output for the business sector is equal to constant-dollar gross domestic product minus: the rental value of owner-occupied dwellings, the output of nonprofit institutions, the output of paid employees of private households, and general government output. The measures are derived from national income and product account (NIPA) data supplied by the U.S. Department of Commerce's Bureau of Economic Analysis (BEA). For manufacturing, BLS produces annual estimates of sectoral output. Quarterly manufacturing output indexes derived from the Federal Reserve Board of Governors' monthly indexes of industrial production (see Chapter 2) are adjusted to these annual measures by the BLS, and are also used to project the quarterly values in the current period.

Nonfinancial corporate output consists of business output minus unincorporated businesses and those corporations classified as offices of bank holding companies, offices of other holding companies, or offices in the finance and insurance sector. Unit profits and unit nonlabor costs can be calculated separately for this sector and are shown in this table.

Compensation and hours data are developed from BLS and BEA data. The primary source for hours and employment is BLS's Current Employment Statistics (CES) program (see the notes and definitions for Chapter 10). Other data sources are the Current Population Survey (CPS) and the National Compensation Survey (NCS). Weekly paid hours are adjusted to hours at work using the NCS. For paid employees, hours at work differ from hours paid, in that they exclude paid vacation and holidays, paid sick leave, and other paid personal or administrative leave.

Although the labor productivity measures relate output to labor input, they do not measure the contribution of labor or any other specific factor of production. Instead, they reflect the joint effect of many influences, including changes in technology; capital investment; level of output; utilization of capacity, energy, and materials; the organization of production; managerial skill; and the characteristics and efforts of the work force.

Revisions

Data for recent years are revised frequently to take account of revisions in the output and labor input measures that underlie the estimates. Customarily, all revisions to source data are reflected in the release following the source data revision.

Data availability

Series are available quarterly and annually. Quarterly measures are based entirely on seasonally adjusted data. For some detailed manufacturing series (not shown here), only annual averages are available. Productivity indexes are published early in the second and third months of each quarter, reflecting new data for preceding quarters. Complete historical data are available on the BLS Web site at <http://www.bls.gov>.

BLS also publishes productivity estimates for a number of individual industries. A release entitled "Productivity and Costs by Industry" is available on the BLS Web site at <http://www.bls.gov>.

References

Further information is available in the Technical Notes and footnotes on the most current monthly release, available on the Web site, and from the following sources: Chapter 10, "Productivity Measures: Business Sector and Major Subsectors," *BLS Handbook of Methods*—Bulletin 2490 (April 1997) and the following *Monthly Labor Review* articles: "Alternative Measures of Supervisory Employee Hours and Productivity Growth" (April 2004); "Possible Measurement Bias in Aggregate Productivity Growth" (February 1999); "Improvements to the Quarterly Productivity Measures" (October 1995); "Hours of Work: A New Base for BLS Productivity Statistics" (February 1990); and "New Sector Definitions for Productivity Series" (October 1976).

TABLE 9-4

Corporate Profits with Inventory Valuation Adjustment by Industry Group

SOURCE: U.S. DEPARTMENT OF COMMERCE, BUREAU OF ECONOMIC ANALYSIS

Corporate profits is one of the most closely watched U.S. economic indicator. Profitability provides a summary measure of corporate financial health and thus serves as an essential indicator of economic performance. Profits are a source of retained earnings, providing much of the funding for capital investments that raise productive capacity. The estimates of profits and of related measures may also be used to evaluate the effects on corporations of changes in policy or in economic conditions.

Corporate profits represents the portion of the total income earned from current production that is accounted for by U.S. corporations. The estimates of corporate profits are an integral part of the national income and product accounts (NIPAs), a set of accounts prepared by the Bureau of Economic Analysis (BEA) that provides a logical and consistent framework for presenting statistics on U.S. economic activity.

These profits measures are derived from the NIPAs and classified according to the NAICS. See the notes and definitions to Chapter 1 for definitions, data availability, and references. Note that this industry breakdown of profits incorporates the inventory valuation adjustment (IVA), which eliminates any capital gain element in profits arising from changes in the prices at which inventories are valued, but does not incorporate the capital consumption adjustment (CCAdj), which adjusts historical costs of fixed capital to replacement costs and uses actual rather than tax-based service lives. This is because the CCAdj is calculated by BEA at an aggregate level, whereas the IVA is calculated at an industry level.

CHAPTER 10: EMPLOYMENT, HOURS, AND EARNINGS

SECTION 10A: LABOR FORCE, EMPLOYMENT, AND UNEMPLOYMENT

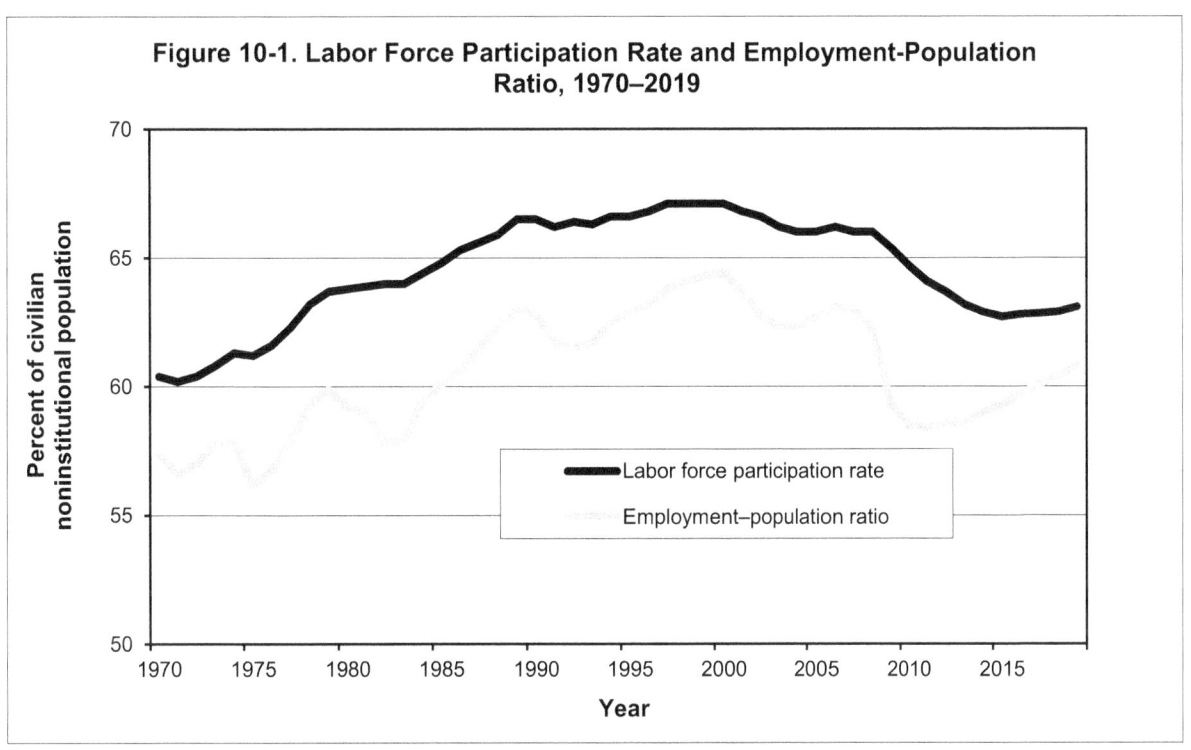

Figure 10-1. Labor Force Participation Rate and Employment-Population Ratio, 1970–2019

- The National Bureau of Economic Research (NBER) determined that the Great Recession endured from the fourth quarter of 2007 to the second quarter of 2009, lasting seven quarters. During this period, the unemployment rate increased to 9.3 percent in 2009 from 4.6 percent in 2007. The NPER announced that the economy fell into a recession in the first quarter of 2020 due to the coronavirus pandemic and state stay-at-home orders. (Table 10-1A)

- Two labor force measurements that are not as commonly cited as the employment figures and the unemployment rate are the labor force participation rate and the employment/population ratio. These economic metrics provide information to gauge the health of the U.S. job market. The labor force participation rate (LFPR) is calculated by taking the number employed, divided by the total adult population, and multiplying by 100 to get the percentage. Figure 10-1 shows the interactions between LFPR and the unemployment rate. (Table 10-1 and 10-2.) Labor participation rate peaked in 1997 at 67.1 and has gradually declined to 63.1 percent in 2019. (Table 10-1A)

- The employment-population ratio measures the civilian labor force currently employed against the total working-working age population. Seasonally, the short-term labor force fluctuations do not affect the employment-to-population ratio. The employment-population ratio has grown slightly in recent years from 59.0 in 2014 to 60.8 in 2019. (Table 10-1A)

Table 10-1A. Summary Labor Force, Employment, and Unemployment: Recent Data

(Thousands of persons, percent, seasonally adjusted, except as noted.)

Year and month	Civilian noninstitutional population [1]	Civilian labor force		Employment, thousands of persons						Employment-population ratio, percent	Unemployment		
		Thousands of persons	Participation rate (percent)	Total	By age and sex			By industry			Thousands of persons		Rate (percent)
					Men, 20 years and over	Women, 20 years and over	Both sexes, 16 to 19 years	Agricultural	Nonagricultural		Total	Unemployed 15 weeks and over	
1970	137 085	82 771	60.4	78 678	45 581	26 952	6 144	3 463	75 215	57.4	4 093	663	4.9
1971	140 216	84 382	60.2	79 367	45 912	27 246	6 209	3 394	75 972	56.6	5 016	1 187	5.9
1972	144 126	87 034	60.4	82 153	47 130	28 276	6 746	3 484	78 669	57.0	4 882	1 167	5.6
1973	147 096	89 429	60.8	85 064	48 310	29 484	7 271	3 470	81 594	57.8	4 365	826	4.9
1974	150 120	91 949	61.3	86 794	48 922	30 424	7 448	3 515	83 279	57.8	5 156	955	5.6
1975	153 153	93 775	61.2	85 846	48 018	30 726	7 103	3 408	82 438	56.1	7 929	2 505	8.5
1976	156 150	96 158	61.6	88 752	49 190	32 226	7 336	3 331	85 421	56.8	7 406	2 366	7.7
1977	159 033	99 009	62.3	92 017	50 555	33 775	7 688	3 283	88 734	57.9	6 991	1 942	7.1
1978	161 910	102 251	63.2	96 048	52 143	35 836	8 070	3 387	92 661	59.3	6 202	1 414	6.1
1979	164 863	104 962	63.7	98 824	53 308	37 434	8 083	3 347	95 477	59.9	6 137	1 241	5.8
1980	167 745	106 940	63.8	99 303	53 101	38 492	7 710	3 364	95 938	59.2	7 637	1 871	7.1
1981	170 130	108 670	63.9	100 397	53 582	39 590	7 225	3 368	97 030	59.0	8 273	2 285	7.6
1982	172 271	110 204	64.0	99 526	52 891	40 086	6 549	3 401	96 125	57.8	10 678	3 485	9.7
1983	174 215	111 550	64.0	100 834	53 487	41 004	6 342	3 383	97 450	57.9	10 717	4 210	9.6
1984	176 383	113 544	64.4	105 005	55 769	42 793	6 444	3 321	101 685	59.5	8 539	2 737	7.5
1985	178 206	115 461	64.8	107 150	56 562	44 154	6 434	3 179	103 971	60.1	8 312	2 305	7.2
1986	180 587	117 834	65.3	109 597	57 569	45 556	6 472	3 163	106 434	60.7	8 237	2 232	7.0
1987	182 753	119 865	65.6	112 440	58 726	47 074	6 641	3 208	109 232	61.5	7 425	1 983	6.2
1988	184 613	121 669	65.9	114 968	59 781	48 383	6 805	3 169	111 800	62.3	6 701	1 610	5.5
1989	186 393	123 869	66.5	117 342	60 837	49 745	6 750	3 199	114 142	63.0	6 528	1 375	5.3
1990	189 164	125 840	66.5	118 793	61 678	50 535	6 610	3 223	115 570	62.8	7 047	1 525	5.6
1991	190 925	126 346	66.2	117 718	61 178	50 634	5 906	3 269	114 449	61.7	8 628	2 357	6.8
1992	192 805	128 105	66.4	118 492	61 496	51 328	5 669	3 247	115 245	61.5	9 613	3 408	7.5
1993	194 838	129 200	66.3	120 259	62 355	52 099	6 035	3 115	117 144	61.7	8 940	3 094	6.9
1994	196 814	131 056	66.6	123 060	63 294	53 606	6 162	3 409	119 651	62.5	7 996	2 860	6.1
1995	198 584	132 304	66.6	124 900	64 085	54 396	6 419	3 440	121 460	62.9	7 404	2 363	5.6
1996	200 591	133 943	66.8	126 708	64 897	55 311	6 500	3 443	123 264	63.2	7 236	2 316	5.4
1997	203 133	136 297	67.1	129 558	66 284	56 613	6 661	3 399	126 159	63.8	6 739	2 062	4.9
1998	205 220	137 673	67.1	131 463	67 135	57 278	7 051	3 378	128 085	64.1	6 210	1 637	4.5
1999	207 753	139 368	67.1	133 488	67 761	58 555	7 172	3 281	130 207	64.3	5 880	1 480	4.2
2000	212 577	142 583	67.1	136 891	69 634	60 067	7 190	2 464	134 427	64.4	5 692	1 318	4.0
2001	215 092	143 734	66.8	136 933	69 776	60 417	6 740	2 299	134 635	63.7	6 801	1 752	4.7
2002	217 570	144 863	66.6	136 485	69 734	60 420	6 332	2 311	134 174	62.7	8 378	2 904	5.8
2003	221 168	146 510	66.2	137 736	70 415	61 402	5 919	2 275	135 461	62.3	8 774	3 378	6.0
2004	223 357	147 401	66.0	139 252	71 572	61 773	5 907	2 232	137 020	62.3	8 149	3 072	5.5
2005	226 082	149 320	66.0	141 730	73 050	62 702	5 978	2 197	139 532	62.7	7 591	2 619	5.1
2006	228 815	151 428	66.2	144 427	74 431	63 834	6 163	2 206	142 221	63.1	7 001	2 266	4.6
2007	231 867	153 124	66.0	146 047	75 337	64 799	5 911	2 095	143 952	63.0	7 078	2 303	4.6
2008	233 788	154 287	66.0	145 362	74 750	65 039	5 573	2 168	143 194	62.2	8 924	3 188	5.8
2009	235 801	154 142	65.4	139 877	71 341	63 699	4 837	2 103	137 775	59.3	14 265	7 272	9.3
2010	237 830	153 889	64.7	139 064	71 230	63 456	4 378	2 206	136 858	58.5	14 825	8 786	9.6
2011	239 618	153 617	64.1	139 869	72 182	63 360	4 327	2 254	137 615	58.4	13 747	8 077	8.9
2012	243 284	154 975	63.7	142 469	73 403	64 640	4 426	2 186	140 283	58.6	12 506	6 996	8.1
2013	245 679	155 389	63.2	143 929	74 176	65 295	4 458	2 130	141 799	58.6	11 460	6 117	7.4
2014	247 947	155 922	62.9	146 305	75 471	66 287	4 548	2 237	144 068	59.0	9 617	4 714	6.2
2015	250 801	157 130	62.7	148 834	76 776	67 323	4 737	2 422	146 411	59.3	8 296	3 595	5.3
2016	253 538	159 187	62.8	151 436	78 084	66 387	4 965	2 460	148 976	59.7	7 751	3 163	4.9
2017	255 079	160 310	62.8	153 337	78 920	69 343	5 074	2 454	150 883	60.1	6 973	2 704	4.4
2018	257 791	162 075	62.9	155 761	73 022	70 424	5 126	2 425	153 337	60.4	6 314	2 268	3.9
2019	259 175	163 539	63.1	157 538	83 460	74 078	5 896	2 425	155 106	60.8	6 001	2 126	3.7
2018													
January	256 780	161 068	62.7	154 486	79 723	69 628	5 135	2 443	152 053	60.2	6 582	2 392	4.1
February	256 934	161 783	63.0	155 142	80 138	69 807	5 198	2 430	152 659	60.4	6 641	2 327	4.1
March	257 097	161 684	62.9	155 191	80 092	69 979	5 120	2 340	152 714	60.4	6 493	2 222	4.0
April	257 272	161 742	62.9	155 324	80 140	70 066	5 118	2 330	153 007	60.4	6 418	2 343	4.0
May	257 454	161 874	62.9	155 665	80 275	70 270	5 119	2 353	153 353	60.5	6 209	2 186	3.8
June	257 642	162 269	63.0	155 750	80 084	70 528	5 137	2 398	153 383	60.5	6 519	2 322	4.0
July	257 843	162 173	62.9	155 993	80 208	70 671	5 114	2 483	153 519	60.5	6 180	2 384	3.8
August	258 066	161 768	62.7	155 601	80 160	70 553	4 888	2 377	153 329	60.3	6 167	2 254	3.8
September	258 290	162 078	62.8	156 032	80 259	70 657	5 116	2 487	153 528	60.4	6 045	2 230	3.7
October	258 514	162 605	62.9	156 482	80 400	70 858	5 224	2 407	153 989	60.5	6 123	2 211	3.8
November	258 708	162 662	62.9	156 628	80 567	70 892	5 169	2 549	154 102	60.5	6 034	2 123	3.7
December	258 888	163 111	63.0	156 825	80 496	71 123	5 205	2 491	154 266	60.6	6 286	2 210	3.9
2019													
January	258 239	163 142	63.2	156 627	80 474	71 004	5 149	2 546	154 112	60.7	6 516	2 156	4.0
February	258 392	163 047	63.1	156 866	80 677	71 169	5 019	2 488	154 354	60.7	6 181	2 207	3.8
March	258 537	162 935	63.0	156 741	80 570	71 056	5 115	2 336	154 346	60.6	6 194	2 242	3.8
April	258 693	162 546	62.8	156 696	80 609	71 136	4 951	2 389	154 369	60.6	5 850	2 087	3.6
May	258 861	162 782	62.9	156 844	80 761	71 038	5 044	2 423	154 486	60.6	5 938	2 120	3.6
June	259 037	163 133	63.0	157 148	80 780	71 209	5 159	2 330	154 835	60.7	5 985	2 189	3.7
July	259 225	163 373	63.0	157 346	80 975	71 120	5 250	2 400	155 035	60.7	6 027	2 079	3.7
August	259 432	163 894	63.2	157 895	81 046	71 665	5 184	2 414	155 546	60.9	5 999	2 082	3.7
September	259 638	164 051	63.2	158 298	81 146	71 990	5 162	2 416	155 816	61.0	5 753	2 124	3.5
October	259 845	164 401	63.3	158 544	81 196	72 130	5 218	2 473	155 970	61.0	5 857	2 144	3.6
November	260 020	164 347	63.2	158 536	81 377	71 881	5 278	2 356	156 167	61.0	5 811	2 083	3.5
December	260 181	164 556	63.2	158 803	81 390	72 200	5 213	2 533	156 241	61.0	5 753	1 998	3.5

[1] Not seasonally adjusted.

Table 10-1B. Summary Labor Force, Employment, and Unemployment: Historical Data

(Thousands of persons, percent, seasonally adjusted, except as noted.)

Year and month	Civilian noninstitutional population [1]	Civilian labor force		Employment, thousands of persons						Employment-population ratio, percent	Unemployment		
		Thousands of persons	Participation rate (percent)	Total	By age and sex			By industry			Thousands of persons		Rate (percent) [2]
					Men, 20 years and over	Women, 20 years and over	Both sexes, 16 to 19 years	Agricultural	Nonagricultural		Total	Unemployed 15 weeks and over	
14 Years and Over													
1929	...	49 180	...	47 630	...	...	...	10 450	37 180	...	1 550	...	3.2
1930	...	49 820	...	45 480	...	...	...	10 340	35 140	...	4 340	...	8.7
1931	...	50 420	...	42 400	...	...	...	10 290	32 110	...	8 020	...	15.9 (15.3)
1932	...	51 000	...	38 940	...	...	...	10 170	28 770	...	12 060	...	23.6 (22.5)
1933	...	51 590	...	38 760	...	...	...	10 090	28 670	...	12 830	...	24.9 (20.6)
1934	...	52 230	...	40 890	...	...	...	9 900	30 990	...	11 340	...	21.7 (16.0)
1935	...	52 870	...	42 260	...	...	...	10 110	32 150	...	10 610	...	20.1 (14.2)
1936	...	53 440	...	44 410	...	...	...	10 000	34 410	...	9 030	...	16.9 (9.9)
1937	...	54 000	...	46 300	...	...	...	9 820	36 480	...	7 700	...	14.3 (9.1)
1938	...	54 610	...	44 220	...	...	...	9 690	34 530	...	10 390	...	19.0 (12.5)
1939	...	55 230	...	45 750	...	...	...	9 610	36 140	...	9 480	...	17.2 (11.3)
1940	99 840	55 640	55.7	47 520	...	...	...	9 540	37 980	47.6	8 120	...	14.6 (9.5)
1941	99 900	55 910	56.0	50 350	...	...	...	9 100	41 250	50.4	5 560	...	9.9 (6.0)
1942	98 640	56 410	57.2	53 750	...	...	...	9 250	44 500	54.5	2 660	...	4.7 (3.1)
1943	94 640	55 540	58.7	54 470	...	...	...	9 080	45 390	57.6	1 070	...	1.9 (1.8)
1944	93 220	54 630	58.6	53 960	...	...	...	8 950	45 010	57.9	670	...	1.2
1945	94 090	53 860	57.2	52 820	...	...	...	8 580	44 240	56.1	1 040	...	1.9
1946	103 070	57 520	55.8	55 250	...	...	...	8 320	46 930	53.6	2 270	...	3.9
1947	106 018	60 168	56.8	57 812	...	...	...	8 256	49 557	54.5	2 356	...	3.9
16 Years and Over													
1947	101 827	59 350	58.3	57 038	...	...	...	7 890	49 148	56.0	2 311	...	3.9
1948	103 068	60 621	58.8	58 343	39 382	14 936	4 026	7 629	50 714	56.6	2 276	309	3.8
1949	103 994	61 286	58.9	57 651	38 803	15 137	3 712	7 658	49 993	55.4	3 637	684	5.9
1950	104 995	62 208	59.2	58 918	39 394	15 824	3 703	7 160	51 758	56.1	3 288	782	5.3
1951	104 621	62 017	59.2	59 961	39 626	16 570	3 767	6 726	53 235	57.3	2 055	303	3.3
1952	105 231	62 138	59.0	60 250	39 578	16 958	3 719	6 500	53 749	57.3	1 883	232	3.0
1953	107 056	63 015	58.9	61 179	40 296	17 164	3 720	6 260	54 919	57.1	1 834	210	2.9
1954	108 321	63 643	58.8	60 109	39 634	17 000	3 475	6 205	53 904	55.5	3 532	812	5.5
1955	109 683	65 023	59.3	62 170	40 526	18 002	3 642	6 450	55 722	56.7	2 852	702	4.4
1956	110 954	66 552	60.0	63 799	41 216	18 767	3 818	6 283	57 514	57.5	2 750	533	4.1
1957	112 265	66 929	59.6	64 071	41 239	19 052	3 778	5 947	58 123	57.1	2 859	560	4.3
1958	113 727	67 639	59.5	63 036	40 411	19 043	3 582	5 586	57 450	55.4	4 602	1 452	6.8
1959	115 329	68 369	59.3	64 630	41 267	19 524	3 838	5 565	59 065	56.0	3 740	1 040	5.5
1960	117 245	69 628	59.4	65 778	41 543	20 105	4 129	5 458	60 318	56.1	3 852	957	5.5
1961	118 771	70 459	59.3	65 746	41 342	20 296	4 108	5 200	60 546	55.4	4 714	1 532	6.7
1962	120 153	70 614	58.8	66 702	41 815	20 693	4 195	4 944	61 759	55.5	3 911	1 119	5.5
1963	122 416	71 833	58.7	67 762	42 251	21 257	4 255	4 687	63 076	55.4	4 070	1 088	5.7
1964	124 485	73 091	58.7	69 305	42 886	21 903	4 516	4 523	64 782	55.7	3 786	973	5.2
1949													
January	103 529	60 771	58.7	58 175	39 233	14 991	3 951	7 790	50 385	56.2	2 596	315	4.3
February	103 559	61 057	59.0	58 208	39 117	15 117	3 974	8 022	50 186	56.2	2 849	374	4.7
March	103 665	61 073	58.9	58 043	39 015	15 069	3 959	8 008	50 035	56.0	3 030	414	5.0
April	103 739	61 007	58.8	57 747	38 993	14 978	3 776	7 911	49 836	55.7	3 260	483	5.3
May	103 845	61 259	59.0	57 552	38 701	15 066	3 785	8 067	49 485	55.4	3 707	602	6.1
June	103 930	60 948	58.6	57 172	38 632	15 003	3 537	7 802	49 370	55.0	3 776	705	6.2
July	104 042	61 301	58.9	57 190	38 405	15 244	3 541	8 021	49 169	55.0	4 111	848	6.7
August	104 121	61 590	59.2	57 397	38 610	15 181	3 606	7 604	49 793	55.1	4 193	917	6.8
September	104 219	61 633	59.1	57 584	38 744	15 129	3 711	7 297	50 287	55.3	4 049	973	6.6
October	104 338	62 185	59.6	57 269	38 394	15 260	3 615	6 814	50 455	54.9	4 916	1 000	7.9
November	104 421	62 005	59.4	58 009	38 860	15 422	3 727	7 497	50 512	55.6	3 996	1 056	6.4
December	104 524	61 908	59.2	57 845	38 908	15 300	3 637	7 379	50 466	55.3	4 063	961	6.6
1950													
January	104 619	61 661	58.9	57 635	38 780	15 255	3 600	7 065	50 570	55.1	4 026	947	6.5
February	104 737	61 687	58.9	57 751	38 818	15 339	3 594	7 057	50 694	55.1	3 936	947	6.4
March	104 844	61 604	58.8	57 728	38 851	15 366	3 511	7 116	50 612	55.1	3 876	912	6.3
April	104 943	62 158	59.2	58 583	39 100	15 831	3 652	7 264	51 319	55.8	3 575	920	5.8
May	105 014	62 083	59.1	58 649	39 416	15 628	3 605	7 277	51 372	55.8	3 434	890	5.5
June	105 104	62 419	59.4	59 052	39 476	15 953	3 623	7 285	51 767	56.2	3 367	868	5.4
July	105 194	62 121	59.1	59 001	39 517	15 793	3 691	7 126	51 875	56.1	3 120	769	5.0
August	105 282	62 596	59.5	59 797	39 879	16 124	3 794	7 248	52 549	56.8	2 799	633	4.5
September	105 269	62 349	59.2	59 575	39 865	15 902	3 808	6 992	52 583	56.6	2 774	648	4.4
October	105 096	62 428	59.4	59 803	39 737	16 175	3 891	7 371	52 432	56.9	2 625	545	4.2
November	104 979	62 286	59.3	59 697	39 668	16 195	3 834	7 163	52 534	56.9	2 589	507	4.2
December	104 872	62 068	59.2	59 429	39 536	16 149	3 744	6 760	52 669	56.7	2 639	482	4.3
1951													
January	104 844	61 941	59.1	59 636	39 595	16 279	3 762	6 828	52 808	56.9	2 305	438	3.7
February	104 604	61 778	59.1	59 661	39 695	16 257	3 709	6 738	52 923	57.0	2 117	386	3.4
March	104 629	62 526	59.8	60 401	40 013	16 557	3 831	6 858	53 543	57.7	2 125	355	3.4
April	104 541	61 808	59.1	59 889	39 804	16 426	3 659	6 722	53 167	57.3	1 919	294	3.1
May	104 491	62 044	59.4	60 188	39 752	16 581	3 855	6 752	53 436	57.6	1 856	269	3.0
June	104 488	61 615	59.0	59 620	39 538	16 368	3 714	6 529	53 091	57.1	1 995	258	3.2
July	104 504	62 106	59.4	60 156	39 483	16 898	3 775	6 601	53 555	57.6	1 950	260	3.1
August	104 536	61 927	59.2	59 994	39 508	16 665	3 821	6 790	53 204	57.4	1 933	249	3.1
September	104 588	61 780	59.1	59 713	39 416	16 504	3 793	6 558	53 155	57.1	2 067	223	3.3
October	104 690	62 204	59.4	60 010	39 555	16 674	3 781	6 636	53 374	57.3	2 194	269	3.5
November	104 740	62 014	59.2	59 836	39 504	16 669	3 663	6 699	53 137	57.1	2 178	316	3.5
December	104 810	62 457	59.6	60 497	39 691	16 946	3 860	7 065	53 432	57.7	1 960	269	3.1

[1] Not seasonally adjusted.
[2] In 1930 through 1943, the official BLS data count persons on work relief as unemployed. The unemployment rates for those years shown in parentheses count persons on work relief as employed, which is more consistent with the postwar practice. See notes and definitions.
. . . = Not available.

Table 10-1B. Summary Labor Force, Employment, and Unemployment: Historical Data—*Continued*

(Thousands of persons, percent, seasonally adjusted, except as noted.)

Year and month	Civilian noninstitutional population [1]	Civilian labor force		Employment, thousands of persons						Employment-population ratio, percent	Unemployment		
		Thousands of persons	Participation rate (percent)	Total	By age and sex			By industry			Thousands of persons		Rate (percent) [2]
					Men, 20 years and over	Women, 20 years and over	Both sexes, 16 to 19 years	Agricultural	Nonagricultural		Total	Unemployed 15 weeks and over	
1952													
January	104 862	62 432	59.5	60 460	39 714	17 001	3 745	7 148	53 312	57.7	1 972	282	3.2
February	104 868	62 419	59.5	60 462	39 772	16 935	3 755	7 020	53 442	57.7	1 957	248	3.1
March	104 860	61 721	58.9	59 908	39 580	16 627	3 701	6 468	53 440	57.1	1 813	234	2.9
April	104 906	61 720	58.8	59 909	39 542	16 659	3 708	6 525	53 384	57.1	1 811	242	2.9
May	104 996	62 058	59.1	60 195	39 588	16 844	3 763	6 334	53 861	57.3	1 863	219	3.0
June	105 118	62 103	59.1	60 219	39 558	16 837	3 824	6 529	53 690	57.3	1 884	210	3.0
July	105 246	61 962	58.9	59 971	39 496	16 778	3 697	6 334	53 637	57.0	1 991	194	3.2
August	105 346	61 877	58.7	59 790	39 289	16 867	3 634	6 174	53 616	56.8	2 087	211	3.4
September	105 436	62 457	59.2	60 521	39 386	17 477	3 658	6 537	53 984	57.4	1 936	249	3.1
October	105 591	61 971	58.7	60 132	39 451	17 032	3 649	6 363	53 769	56.9	1 839	230	3.0
November	105 706	62 491	59.1	60 748	39 549	17 450	3 749	6 509	54 239	57.5	1 743	216	2.8
December	105 812	62 621	59.2	60 954	40 011	17 181	3 762	6 361	54 593	57.6	1 667	238	2.7
1953													
January	106 594	63 439	59.5	61 600	40 256	17 482	3 862	6 642	54 958	57.8	1 839	268	2.9
February	106 678	63 520	59.5	61 884	40 546	17 321	4 017	6 463	55 421	58.0	1 636	208	2.6
March	106 744	63 657	59.6	62 010	40 648	17 397	3 965	6 420	55 590	58.1	1 647	213	2.6
April	106 826	63 167	59.1	61 444	40 346	17 242	3 856	6 362	55 082	57.5	1 723	180	2.7
May	106 910	62 615	58.6	61 019	40 323	16 983	3 713	5 937	55 082	57.1	1 596	176	2.5
June	106 978	63 063	58.9	61 456	40 358	17 301	3 797	6 361	55 095	57.4	1 607	213	2.5
July	107 034	63 057	58.9	61 397	40 378	17 341	3 678	6 267	55 130	57.4	1 660	168	2.6
August	107 132	62 816	58.6	61 151	40 352	17 108	3 691	6 319	54 832	57.1	1 665	177	2.7
September	107 253	62 727	58.5	60 906	40 192	17 063	3 651	6 198	54 708	56.8	1 821	178	2.9
October	107 383	62 867	58.5	60 893	40 155	17 236	3 502	6 096	54 797	56.7	1 974	190	3.1
November	107 504	62 949	58.6	60 738	40 163	16 974	3 601	6 345	54 393	56.5	2 211	259	3.5
December	107 623	62 795	58.3	59 977	39 885	16 599	3 493	5 929	54 048	55.7	2 818	309	4.5
1954													
January	107 763	63 101	58.6	60 024	39 834	16 574	3 616	6 073	53 951	55.7	3 077	372	4.9
February	107 880	63 994	59.3	60 663	39 899	17 162	3 602	6 590	54 073	56.2	3 331	532	5.2
March	107 987	63 793	59.1	60 186	39 497	17 022	3 667	6 395	53 791	55.7	3 607	765	5.7
April	108 080	63 934	59.2	60 185	39 613	17 015	3 557	6 142	54 043	55.7	3 749	774	5.9
May	108 184	63 675	58.9	59 908	39 467	16 975	3 466	6 210	53 698	55.4	3 767	879	5.9
June	108 267	63 343	58.5	59 792	39 476	16 894	3 422	6 162	53 630	55.2	3 551	880	5.6
July	108 344	63 302	58.4	59 643	39 467	16 777	3 399	6 222	53 421	55.0	3 659	932	5.8
August	108 440	63 707	58.7	59 853	39 582	16 868	3 403	6 087	53 766	55.2	3 854	1 002	6.0
September	108 546	64 209	59.2	60 282	39 702	17 133	3 447	6 453	53 829	55.5	3 927	1 017	6.1
October	108 668	63 936	58.8	60 270	39 618	17 209	3 443	6 242	54 028	55.5	3 666	1 009	5.7
November	108 798	63 759	58.6	60 357	39 745	17 213	3 399	5 934	54 423	55.5	3 402	975	5.3
December	108 892	63 312	58.1	60 116	39 763	17 121	3 232	5 848	54 268	55.2	3 196	827	5.0
1955													
January	109 059	63 910	58.6	60 753	39 937	17 375	3 441	6 113	54 640	55.7	3 157	882	4.9
February	109 078	63 696	58.4	60 727	39 964	17 413	3 350	5 854	54 873	55.7	2 969	826	4.7
March	109 254	63 882	58.5	60 964	40 111	17 415	3 438	6 242	54 722	55.8	2 918	816	4.6
April	109 377	64 564	59.0	61 515	40 120	17 867	3 528	6 363	55 152	56.2	3 049	811	4.7
May	109 544	64 381	58.8	61 634	40 410	17 665	3 559	6 327	55 307	56.3	2 747	734	4.3
June	109 680	64 482	58.8	61 781	40 444	17 837	3 500	6 243	55 538	56.3	2 701	668	4.2
July	109 792	65 145	59.3	62 513	40 751	18 123	3 639	6 438	56 075	56.9	2 632	640	4.0
August	109 882	65 581	59.7	62 797	40 747	18 377	3 673	6 575	56 222	57.1	2 784	535	4.2
September	109 977	65 628	59.7	62 950	40 920	18 285	3 745	6 819	56 131	57.2	2 678	558	4.1
October	110 085	65 821	59.8	62 991	40 858	18 327	3 806	6 728	56 263	57.2	2 830	572	4.3
November	110 177	66 037	59.9	63 257	40 936	18 422	3 899	6 655	56 602	57.4	2 780	564	4.2
December	110 296	66 445	60.2	63 684	41 063	18 630	3 991	6 653	57 031	57.7	2 761	581	4.2
1956													
January	110 390	66 419	60.2	63 753	41 203	18 691	3 859	6 590	57 163	57.8	2 666	561	4.0
February	110 478	66 124	59.9	63 518	41 175	18 582	3 761	6 457	57 061	57.5	2 606	545	3.9
March	110 582	66 175	59.8	63 411	41 199	18 496	3 716	6 221	57 190	57.3	2 764	521	4.2
April	110 650	66 264	59.9	63 614	41 289	18 629	3 696	6 460	57 154	57.5	2 650	476	4.0
May	110 810	66 722	60.2	63 861	41 166	18 844	3 851	6 375	57 486	57.6	2 861	506	4.3
June	110 903	66 702	60.1	63 820	41 196	18 748	3 876	6 335	57 485	57.5	2 882	516	4.3
July	111 019	66 752	60.1	63 800	41 216	18 718	3 866	6 320	57 480	57.5	2 952	523	4.4
August	111 099	66 673	60.0	63 972	41 265	18 864	3 843	6 280	57 692	57.6	2 701	543	4.1
September	111 222	66 714	60.0	64 079	41 221	19 019	3 839	6 375	57 704	57.6	2 635	577	3.9
October	111 335	66 546	59.8	63 975	41 261	18 928	3 786	6 137	57 838	57.5	2 571	530	3.9
November	111 432	66 657	59.8	63 796	41 208	18 846	3 742	5 997	57 799	57.3	2 861	575	4.3
December	111 526	66 700	59.8	63 910	41 192	18 859	3 859	5 806	58 104	57.3	2 790	567	4.2
1957													
January	111 626	66 428	59.5	63 632	41 168	18 740	3 724	5 790	57 842	57.0	2 796	509	4.2
February	111 711	66 879	59.9	64 257	41 341	19 115	3 801	6 125	58 132	57.5	2 622	530	3.9
March	111 824	66 913	59.8	64 404	41 500	19 066	3 838	5 963	58 441	57.6	2 509	514	3.7
April	111 933	66 647	59.5	64 047	41 345	18 937	3 765	5 836	58 211	57.2	2 600	516	3.9
May	112 031	66 695	59.5	63 985	41 334	18 897	3 754	5 999	57 986	57.1	2 710	538	4.1
June	112 172	67 052	59.8	64 196	41 411	18 973	3 812	6 002	58 194	57.2	2 856	526	4.3
July	112 317	67 336	60.0	64 540	41 472	19 262	3 806	6 401	58 139	57.5	2 796	535	4.2
August	112 421	66 706	59.3	63 959	41 243	19 020	3 696	5 898	58 061	56.9	2 747	542	4.1
September	112 554	67 064	59.6	64 121	41 213	19 116	3 792	5 728	58 393	57.0	2 943	559	4.4
October	112 710	67 066	59.5	64 046	41 069	19 160	3 817	5 875	58 171	56.8	3 020	650	4.5
November	112 874	67 123	59.5	63 669	40 853	19 082	3 734	5 686	57 983	56.4	3 454	674	5.1
December	113 013	67 398	59.6	63 922	40 884	19 285	3 753	6 037	57 885	56.6	3 476	731	5.2

[1]Not seasonally adjusted.
[2]In 1930 through 1943, the official BLS data count persons on work relief as unemployed. The unemployment rates for those years shown in parentheses count persons on work relief as employed, which is more consistent with the postwar practice. See notes and definitions.

Table 10-1B. Summary Labor Force, Employment, and Unemployment: Historical Data—*Continued*

(Thousands of persons, percent, seasonally adjusted, except as noted.)

Year and month	Civilian noninsti-tutional popu-lation [1]	Civilian labor force		Employment, thousands of persons						Employ-ment-population ratio, percent	Unemployment		
					By age and sex			By industry			Thousands of persons		Rate (percent) [2]
		Thousands of persons	Participa-tion rate (percent)	Total	Men, 20 years and over	Women, 20 years and over	Both sexes, 16 to 19 years	Agri-cultural	Nonagri-cultural		Total	Unem-ployed 15 weeks and over	
1958													
January	113 138	67 095	59.3	63 220	40 617	19 035	3 568	5 831	57 389	55.9	3 875	879	5.8
February	113 234	67 201	59.3	62 898	40 336	18 951	3 611	5 654	57 244	55.5	4 303	1 005	6.4
March	113 337	67 223	59.3	62 731	40 180	18 968	3 583	5 561	57 170	55.3	4 492	1 128	6.7
April	113 415	67 647	59.6	62 631	40 129	18 969	3 533	5 602	57 029	55.2	5 016	1 387	7.4
May	113 534	67 895	59.8	62 874	40 253	18 978	3 643	5 647	57 227	55.4	5 021	1 493	7.4
June	113 647	67 674	59.5	62 730	40 208	19 008	3 514	5 510	57 220	55.2	4 944	1 677	7.3
July	113 727	67 824	59.6	62 745	40 270	19 039	3 436	5 525	57 220	55.2	5 079	1 796	7.5
August	113 835	68 037	59.8	63 012	40 343	19 103	3 566	5 673	57 339	55.4	5 025	1 888	7.4
September	113 977	68 002	59.7	63 181	40 564	19 033	3 584	5 453	57 728	55.4	4 821	1 795	7.1
October	114 138	68 045	59.6	63 475	40 699	19 091	3 685	5 563	57 912	55.6	4 570	1 708	6.7
November	114 283	67 658	59.2	63 470	40 684	19 157	3 629	5 571	57 899	55.5	4 188	1 570	6.2
December	114 429	67 740	59.2	63 549	40 666	19 170	3 713	5 521	58 028	55.5	4 191	1 490	6.2
1959													
January	114 582	67 936	59.3	63 868	40 769	19 292	3 807	5 481	58 387	55.7	4 068	1 396	6.0
February	114 712	67 649	59.0	63 684	40 699	19 167	3 818	5 429	58 255	55.5	3 965	1 277	5.9
March	114 849	68 068	59.3	64 267	41 079	19 379	3 809	5 677	58 590	56.0	3 801	1 210	5.6
April	114 986	68 339	59.4	64 768	41 419	19 498	3 851	5 893	58 875	56.3	3 571	1 039	5.2
May	115 144	68 178	59.2	64 699	41 355	19 565	3 779	5 792	58 907	56.2	3 479	965	5.1
June	115 287	68 278	59.2	64 849	41 387	19 658	3 804	5 712	59 137	56.3	3 429	963	5.0
July	115 429	68 539	59.4	65 011	41 596	19 595	3 820	5 564	59 447	56.3	3 528	889	5.1
August	115 555	68 432	59.2	64 844	41 485	19 568	3 791	5 442	59 402	56.1	3 588	889	5.2
September	115 668	68 545	59.3	64 770	41 351	19 531	3 888	5 447	59 323	56.0	3 775	895	5.5
October	115 798	68 821	59.4	64 911	41 362	19 702	3 847	5 355	59 556	56.1	3 910	883	5.7
November	115 916	68 533	59.1	64 530	41 062	19 594	3 874	5 480	59 050	55.7	4 003	982	5.8
December	116 040	68 994	59.5	65 341	41 651	19 717	3 973	5 458	59 883	56.3	3 653	920	5.3
1960													
January	116 594	68 962	59.1	65 347	41 637	19 686	4 024	5 458	59 889	56.0	3 615	915	5.2
February	116 702	68 949	59.1	65 620	41 729	19 765	4 126	5 443	60 177	56.2	3 329	841	4.8
March	116 827	68 399	58.5	64 673	41 320	19 388	3 965	4 959	59 714	55.4	3 726	959	5.4
April	116 910	69 579	59.5	65 959	41 641	20 110	4 208	5 471	60 488	56.4	3 620	896	5.2
May	117 033	69 626	59.5	66 057	41 668	20 186	4 203	5 359	60 698	56.4	3 569	797	5.1
June	117 167	69 934	59.7	66 168	41 553	20 290	4 325	5 416	60 752	56.5	3 766	854	5.4
July	117 281	69 745	59.5	65 909	41 490	20 257	4 162	5 542	60 367	56.2	3 836	921	5.5
August	117 431	69 841	59.5	65 895	41 503	20 316	4 076	5 520	60 375	56.1	3 946	927	5.6
September	117 521	70 151	59.7	66 267	41 604	20 493	4 170	5 755	60 512	56.4	3 884	982	5.5
October	117 643	69 884	59.4	65 632	41 464	20 076	4 092	5 436	60 196	55.8	4 252	1 189	6.1
November	117 829	70 439	59.8	66 109	41 543	20 384	4 182	5 513	60 596	56.1	4 330	1 223	6.1
December	118 001	70 395	59.7	65 778	41 416	20 332	4 030	5 622	60 156	55.7	4 617	1 142	6.6
1961													
January	118 155	70 447	59.6	65 776	41 363	20 325	4 088	5 422	60 354	55.7	4 671	1 328	6.6
February	118 250	70 420	59.6	65 588	41 177	20 392	4 019	5 472	60 116	55.5	4 832	1 416	6.9
March	118 358	70 703	59.7	65 850	41 273	20 459	4 118	5 406	60 444	55.6	4 853	1 463	6.9
April	118 503	70 267	59.3	65 374	41 206	20 145	4 023	5 037	60 337	55.2	4 893	1 598	7.0
May	118 638	70 452	59.4	65 449	41 139	20 261	4 049	5 099	60 350	55.2	5 003	1 686	7.1
June	118 767	70 878	59.7	65 993	41 349	20 446	4 198	5 220	60 773	55.6	4 885	1 651	6.9
July	118 889	70 536	59.3	65 608	41 245	20 252	4 111	5 153	60 455	55.2	4 928	1 830	7.0
August	119 006	70 534	59.3	65 852	41 362	20 279	4 211	5 366	60 486	55.3	4 682	1 649	6.6
September	119 107	70 217	59.0	65 541	41 400	20 112	4 029	5 021	60 520	55.0	4 676	1 531	6.7
October	119 202	70 492	59.1	65 919	41 509	20 338	4 072	5 203	60 716	55.3	4 573	1 481	6.5
November	119 153	70 376	59.1	66 081	41 556	20 330	4 195	5 090	60 991	55.5	4 295	1 388	6.1
December	119 214	70 077	58.8	65 900	41 534	20 287	4 079	4 992	60 908	55.3	4 177	1 361	6.0
1962													
January	119 300	70 189	58.8	66 108	41 547	20 501	4 060	5 094	61 014	55.4	4 081	1 235	5.8
February	119 360	70 409	59.0	66 538	41 745	20 693	4 100	5 289	61 249	55.7	3 871	1 244	5.5
March	119 476	70 414	58.9	66 493	41 696	20 567	4 230	5 157	61 336	55.7	3 921	1 162	5.6
April	119 702	70 278	58.7	66 372	41 647	20 567	4 158	5 009	61 363	55.4	3 906	1 122	5.6
May	119 813	70 551	58.9	66 688	41 847	20 558	4 283	4 964	61 724	55.7	3 863	1 134	5.5
June	119 943	70 514	58.8	66 670	41 761	20 547	4 362	4 943	61 727	55.6	3 844	1 079	5.5
July	120 128	70 302	58.5	66 483	41 671	20 592	4 220	4 840	61 643	55.3	3 819	1 049	5.4
August	120 323	70 981	59.0	66 968	41 900	20 841	4 227	4 866	62 102	55.7	4 013	1 081	5.7
September	120 653	71 153	59.0	67 192	42 020	20 982	4 190	4 867	62 325	55.7	3 961	1 096	5.6
October	120 856	70 917	58.7	67 114	42 086	20 856	4 172	4 816	62 298	55.5	3 803	1 022	5.4
November	121 045	70 871	58.5	66 847	41 985	20 794	4 068	4 831	62 016	55.2	4 024	1 051	5.7
December	121 236	70 854	58.4	66 947	41 934	20 831	4 182	4 647	62 300	55.2	3 907	1 068	5.5
1963													
January	121 463	71 146	58.6	67 072	41 938	20 933	4 201	4 882	62 190	55.2	4 074	1 122	5.7
February	121 633	71 262	58.6	67 024	41 876	21 046	4 102	4 652	62 372	55.1	4 238	1 137	5.9
March	121 824	71 423	58.6	67 351	42 047	21 162	4 142	4 696	62 655	55.3	4 072	1 087	5.7
April	121 986	71 697	58.8	67 642	42 131	21 281	4 230	4 670	62 972	55.5	4 055	1 071	5.7
May	122 162	71 832	58.8	67 615	42 145	21 225	4 245	4 729	62 886	55.3	4 217	1 157	5.9
June	122 352	71 626	58.5	67 649	42 268	21 185	4 196	4 642	63 007	55.3	3 977	1 067	5.6
July	122 521	71 956	58.7	67 905	42 427	21 268	4 210	4 694	63 211	55.4	4 051	1 070	5.6
August	122 667	71 786	58.5	67 908	42 400	21 185	4 323	4 604	63 304	55.4	3 878	1 114	5.4
September	122 821	72 131	58.7	68 174	42 500	21 317	4 357	4 650	63 524	55.5	3 957	1 069	5.5
October	123 014	72 281	58.8	68 294	42 437	21 456	4 401	4 702	63 592	55.5	3 987	1 071	5.5
November	123 192	72 418	58.8	68 267	42 415	21 553	4 299	4 694	63 573	55.4	4 151	1 054	5.7
December	123 360	72 188	58.5	68 213	42 427	21 481	4 305	4 629	63 584	55.3	3 975	1 007	5.5

[1]Not seasonally adjusted.
[2]In 1930 through 1943, the official BLS data count persons on work relief as unemployed. The unemployment rates for those years shown in parentheses count persons on work relief as employed, which is more consistent with the postwar practice. See notes and definitions.

Table 10-1B. Summary Labor Force, Employment, and Unemployment: Historical Data—*Continued*

(Thousands of persons, percent, seasonally adjusted, except as noted.)

Year and month	Civilian noninstitutional population [1]	Civilian labor force — Thousands of persons	Civilian labor force — Participation rate (percent)	Employment — Total	Employment by age and sex — Men, 20 years and over	Employment by age and sex — Women, 20 years and over	Employment by age and sex — Both sexes, 16 to 19 years	Employment by industry — Agricultural	Employment by industry — Nonagricultural	Employment-population ratio, percent	Unemployment — Total	Unemployment — Unemployed 15 weeks and over	Rate (percent) [2]
1964													
January	123 560	72 356	58.6	68 327	42 510	21 462	4 355	4 603	63 724	55.3	4 029	1 057	5.6
February	123 707	72 683	58.8	68 751	42 579	21 652	4 520	4 563	64 188	55.6	3 932	1 015	5.4
March	123 857	72 713	58.7	68 763	42 600	21 685	4 478	4 366	64 397	55.5	3 950	1 039	5.4
April	124 019	73 274	59.1	69 356	42 885	22 110	4 361	4 414	64 942	55.9	3 918	934	5.3
May	124 204	73 395	59.1	69 631	43 025	22 103	4 503	4 603	65 028	56.1	3 764	975	5.1
June	124 386	73 032	58.7	69 218	42 760	21 995	4 463	4 556	64 662	55.6	3 814	1 047	5.2
July	124 567	73 007	58.6	69 399	42 998	21 846	4 555	4 591	64 808	55.7	3 608	1 002	4.9
August	124 731	73 118	58.6	69 463	42 963	22 002	4 498	4 573	64 890	55.7	3 655	934	5.0
September	124 920	73 290	58.7	69 578	43 009	21 863	4 706	4 619	64 959	55.7	3 712	917	5.1
October	125 108	73 308	58.6	69 582	43 023	21 984	4 575	4 550	65 032	55.6	3 726	903	5.1
November	125 291	73 286	58.5	69 735	43 171	21 954	4 610	4 496	65 239	55.7	3 551	922	4.8
December	125 468	73 465	58.6	69 814	43 109	22 136	4 569	4 322	65 492	55.6	3 651	873	5.0
1965													
January	125 647	73 569	58.6	69 997	43 237	22 282	4 478	4 271	65 726	55.7	3 572	793	4.9
February	125 810	73 857	58.7	70 127	43 279	22 276	4 572	4 322	65 805	55.7	3 730	919	5.1
March	125 985	73 949	58.7	70 439	43 370	22 373	4 696	4 318	66 121	55.9	3 510	796	4.7
April	126 155	74 228	58.8	70 633	43 397	22 416	4 820	4 424	66 209	56.0	3 595	796	4.8
May	126 320	74 466	59.0	71 034	43 579	22 494	4 961	4 724	66 310	56.2	3 432	736	4.6
June	126 499	74 412	58.8	71 025	43 487	22 759	4 779	4 444	66 581	56.1	3 387	786	4.6
July	126 573	74 761	59.1	71 460	43 489	22 841	5 130	4 390	67 070	56.5	3 301	683	4.4
August	126 756	74 616	58.9	71 362	43 447	22 783	5 132	4 355	67 007	56.3	3 254	733	4.4
September	126 906	74 502	58.7	71 286	43 371	22 684	5 231	4 271	67 015	56.2	3 216	732	4.3
October	127 043	74 838	58.9	71 695	43 461	22 819	5 415	4 418	67 277	56.4	3 143	672	4.2
November	127 171	74 797	58.8	71 724	43 447	22 829	5 448	4 093	67 631	56.4	3 073	645	4.1
December	127 294	75 093	59.0	72 062	43 513	22 983	5 566	4 159	67 903	56.6	3 031	659	4.0
1966													
January	127 394	75 186	59.0	72 198	43 495	23 098	5 605	4 077	68 121	56.7	2 988	623	4.0
February	127 514	74 954	58.8	72 134	43 528	23 089	5 517	4 078	68 056	56.6	2 820	594	3.8
March	127 626	75 075	58.8	72 188	43 576	23 109	5 503	4 069	68 119	56.6	2 887	583	3.8
April	127 744	75 338	59.0	72 510	43 679	23 227	5 604	4 108	68 402	56.8	2 828	575	3.8
May	127 879	75 447	59.0	72 497	43 710	23 282	5 505	3 930	68 567	56.7	2 950	534	3.9
June	127 983	75 647	59.1	72 775	43 662	23 359	5 754	3 967	68 808	56.9	2 872	475	3.8
July	128 102	75 736	59.1	72 860	43 574	23 422	5 864	3 920	68 940	56.9	2 876	427	3.8
August	128 240	76 046	59.3	73 146	43 636	23 605	5 905	3 921	69 225	57.0	2 900	464	3.8
September	128 359	76 056	59.3	73 258	43 718	23 881	5 659	3 952	69 306	57.1	2 798	488	3.7
October	128 494	76 199	59.3	73 401	43 776	23 881	5 744	3 912	69 489	57.1	2 798	494	3.7
November	128 627	76 610	59.6	73 840	43 804	24 130	5 906	3 945	69 895	57.4	2 770	464	3.6
December	128 730	76 641	59.5	73 729	43 820	24 025	5 884	3 906	69 823	57.3	2 912	488	3.8
1967													
January	128 909	76 639	59.5	73 671	44 029	23 872	5 770	3 890	69 781	57.1	2 968	489	3.9
February	129 032	76 521	59.3	73 606	43 997	23 919	5 690	3 723	69 883	57.0	2 915	459	3.8
March	129 190	76 328	59.1	73 439	43 922	23 832	5 685	3 757	69 682	56.8	2 889	436	3.8
April	129 344	76 777	59.4	73 882	44 061	24 161	5 660	3 748	70 134	57.1	2 895	428	3.8
May	129 515	76 773	59.3	73 844	44 100	24 172	5 572	3 658	70 186	57.0	2 929	417	3.8
June	129 722	77 270	59.6	74 278	44 230	24 303	5 745	3 689	70 589	57.3	2 992	422	3.9
July	129 918	77 464	59.6	74 520	44 364	24 416	5 740	3 833	70 687	57.4	2 944	412	3.8
August	130 187	77 712	59.7	74 767	44 410	24 600	5 757	3 963	70 804	57.4	2 945	441	3.8
September	130 392	77 812	59.7	74 854	44 535	24 683	5 636	3 851	71 003	57.4	2 958	448	3.8
October	130 582	78 194	59.9	75 051	44 610	24 802	5 639	4 008	71 043	57.5	3 143	472	4.0
November	130 754	78 191	59.8	75 125	44 625	24 914	5 586	3 933	71 192	57.5	3 066	490	3.9
December	130 936	78 491	59.9	75 473	44 719	25 104	5 650	4 076	71 397	57.6	3 018	485	3.8
1968													
January	131 112	77 578	59.2	74 700	44 606	24 581	5 513	3 908	70 792	57.0	2 878	503	3.7
February	131 277	78 230	59.6	75 229	44 659	24 881	5 689	3 959	71 270	57.3	3 001	468	3.8
March	131 412	78 256	59.6	75 379	44 663	25 019	5 697	3 904	71 475	57.4	2 877	447	3.7
April	131 553	78 270	59.5	75 561	44 753	25 072	5 736	3 875	71 686	57.4	2 709	393	3.5
May	131 712	78 847	59.9	76 107	44 841	25 513	5 753	3 814	72 293	57.8	2 740	395	3.5
June	131 872	79 120	60.0	76 182	44 914	25 466	5 802	3 806	72 376	57.8	2 938	405	3.7
July	132 053	78 970	59.8	76 087	44 935	25 347	5 805	3 820	72 267	57.6	2 883	426	3.7
August	132 251	78 811	59.6	76 043	44 897	25 201	5 945	3 736	72 307	57.5	2 768	393	3.5
September	132 446	78 858	59.5	76 172	44 893	25 445	5 834	3 758	72 414	57.5	2 686	375	3.4
October	132 617	78 913	59.5	76 224	44 884	25 475	5 865	3 741	72 483	57.5	2 689	386	3.4
November	132 903	79 209	59.6	76 494	44 996	25 674	5 824	3 758	72 736	57.6	2 715	357	3.4
December	133 120	79 463	59.7	76 778	45 262	25 712	5 804	3 746	73 032	57.7	2 685	351	3.4
1969													
January	133 324	79 523	59.6	76 805	45 154	25 777	5 874	3 704	73 101	57.6	2 718	339	3.4
February	133 465	80 019	60.0	77 327	45 339	26 092	5 896	3 770	73 557	57.9	2 692	358	3.4
March	133 639	80 079	59.9	77 367	45 305	26 115	5 947	3 668	73 699	57.9	2 712	353	3.4
April	133 821	80 281	60.0	77 523	45 262	26 233	6 028	3 629	73 894	57.9	2 758	386	3.4
May	134 027	80 125	59.8	77 412	45 278	26 283	5 851	3 706	73 706	57.8	2 713	387	3.4
June	134 213	80 696	60.1	77 880	45 313	26 429	6 138	3 663	74 217	58.0	2 816	368	3.5
July	134 414	80 827	60.1	77 959	45 305	26 516	6 138	3 548	74 411	58.0	2 868	377	3.5
August	134 597	81 106	60.3	78 250	45 513	26 556	6 181	3 613	74 637	58.1	2 856	373	3.5
September	134 774	81 290	60.3	78 250	45 447	26 572	6 231	3 551	74 699	58.1	3 040	391	3.7
October	135 012	81 494	60.4	78 445	45 488	26 658	6 299	3 517	74 928	58.1	3 049	374	3.7
November	135 239	81 397	60.2	78 541	45 505	26 652	6 384	3 477	75 064	58.1	2 856	392	3.5
December	135 489	81 624	60.2	78 740	45 577	26 832	6 331	3 409	75 331	58.1	2 884	413	3.5

[1]Not seasonally adjusted.
[2]In 1930 through 1943, the official BLS data count persons on work relief as unemployed. The unemployment rates for those years shown in parentheses count persons on work relief as employed, which is more consistent with the postwar practice. See notes and definitions.

Table 10-1B. Summary Labor Force, Employment, and Unemployment: Historical Data—*Continued*

(Thousands of persons, percent, seasonally adjusted, except as noted.)

Year and month	Civilian noninstitutional population [1]	Civilian labor force		Employment, thousands of persons						Employment-population ratio, percent	Unemployment		
		Thousands of persons	Participation rate (percent)	Total	By age and sex			By industry			Thousands of persons		Rate (percent) [2]
					Men, 20 years and over	Women, 20 years and over	Both sexes, 16 to 19 years	Agricultural	Nonagricultural		Total	Unemployed 15 weeks and over	
1970													
January	135 713	81 981	60.4	78 780	45 654	26 908	6 218	3 422	75 358	58.0	3 201	431	3.9
February	135 957	82 151	60.4	78 698	45 627	26 828	6 243	3 439	75 259	57.9	3 453	470	4.2
March	136 179	82 498	60.6	78 863	45 668	26 933	6 262	3 499	75 364	57.9	3 635	534	4.4
April	136 416	82 727	60.6	78 930	45 679	27 114	6 137	3 568	75 362	57.9	3 797	602	4.6
May	136 686	82 483	60.3	78 564	45 666	26 739	6 159	3 547	75 017	57.5	3 919	591	4.8
June	136 928	82 484	60.2	78 413	45 554	26 904	5 955	3 555	74 858	57.3	4 071	657	4.9
July	137 196	82 901	60.4	78 726	45 516	27 083	6 127	3 517	75 209	57.4	4 175	662	5.0
August	137 455	82 880	60.3	78 624	45 495	27 011	6 118	3 418	75 206	57.2	4 256	705	5.1
September	137 717	82 954	60.2	78 498	45 535	26 784	6 179	3 451	75 047	57.0	4 456	788	5.4
October	137 988	83 276	60.4	78 685	45 508	27 058	6 119	3 337	75 348	57.0	4 591	771	5.5
November	138 264	83 548	60.4	78 650	45 540	27 020	6 090	3 372	75 278	56.9	4 898	871	5.9
December	138 529	83 670	60.4	78 594	45 466	27 038	6 090	3 380	75 214	56.7	5 076	1 102	6.1
1971													
January	138 795	83 850	60.4	78 864	45 527	27 173	6 164	3 393	75 471	56.8	4 986	1 113	5.9
February	139 021	83 603	60.1	78 700	45 455	27 040	6 205	3 288	75 412	56.6	4 903	1 068	5.9
March	139 285	83 575	60.0	78 588	45 520	26 967	6 101	3 356	75 232	56.4	4 987	1 098	6.0
April	139 566	83 946	60.1	78 987	45 789	26 984	6 214	3 574	75 413	56.6	4 959	1 149	5.9
May	139 826	84 135	60.2	79 139	45 917	27 056	6 166	3 449	75 690	56.6	4 996	1 173	5.9
June	140 090	83 706	59.8	78 757	45 879	27 013	5 865	3 334	75 423	56.2	4 949	1 167	5.9
July	140 343	84 340	60.1	79 305	46 000	27 054	6 251	3 386	75 919	56.5	5 035	1 251	6.0
August	140 596	84 673	60.2	79 539	46 041	27 171	6 327	3 395	76 144	56.6	5 134	1 261	6.1
September	140 869	84 731	60.1	79 689	46 090	27 390	6 209	3 367	76 322	56.6	5 042	1 239	6.0
October	141 146	84 872	60.1	79 918	46 132	27 538	6 248	3 405	76 513	56.6	4 954	1 268	5.8
November	141 393	85 458	60.4	80 297	46 209	27 721	6 367	3 410	76 887	56.8	5 161	1 277	6.0
December	141 666	85 625	60.4	80 471	46 280	27 791	6 400	3 371	77 100	56.8	5 154	1 283	6.0
1972													
January	142 736	85 978	60.2	80 959	46 471	27 956	6 532	3 366	77 593	56.7	5 019	1 257	5.8
February	143 017	86 036	60.2	81 108	46 600	28 016	6 492	3 358	77 750	56.7	4 928	1 292	5.7
March	143 263	86 611	60.5	81 573	46 821	28 126	6 626	3 438	78 135	56.9	5 038	1 232	5.8
April	143 483	86 614	60.4	81 655	46 863	28 114	6 678	3 382	78 273	56.9	4 959	1 203	5.7
May	143 760	86 809	60.4	81 887	46 950	28 184	6 753	3 412	78 475	57.0	4 922	1 168	5.7
June	144 033	87 006	60.4	82 083	47 147	28 175	6 761	3 402	78 681	57.0	4 923	1 141	5.7
July	144 285	87 143	60.4	82 230	47 244	28 225	6 761	3 461	78 769	57.0	4 913	1 154	5.6
August	144 522	87 517	60.6	82 578	47 321	28 382	6 875	3 603	78 975	57.1	4 939	1 156	5.6
September	144 761	87 392	60.4	82 543	47 394	28 417	6 732	3 568	78 975	57.0	4 849	1 131	5.5
October	144 988	87 491	60.3	82 616	47 354	28 438	6 824	3 634	78 982	57.0	4 875	1 123	5.6
November	145 211	87 592	60.3	82 990	47 529	28 567	6 894	3 517	79 473	57.2	4 602	1 040	5.3
December	145 446	87 943	60.5	83 400	47 747	28 698	6 955	3 596	79 804	57.3	4 543	1 006	5.2
1973													
January	145 720	87 487	60.0	83 161	47 701	28 596	6 864	3 456	79 705	57.1	4 326	947	4.9
February	145 943	88 364	60.5	83 912	47 884	28 995	7 033	3 415	80 497	57.5	4 452	894	5.0
March	146 230	88 846	60.8	84 452	48 117	29 110	7 225	3 469	80 983	57.8	4 394	889	4.9
April	146 459	89 018	60.8	84 559	48 098	29 304	7 157	3 407	81 152	57.7	4 459	809	5.0
May	146 719	88 977	60.6	84 648	48 068	29 432	7 148	3 376	81 272	57.7	4 329	816	4.9
June	146 981	89 548	60.9	85 185	48 244	29 505	7 436	3 509	81 676	58.0	4 363	779	4.9
July	147 233	89 604	60.9	85 299	48 452	29 592	7 255	3 540	81 759	57.9	4 305	756	4.8
August	147 471	89 509	60.7	85 204	48 353	29 578	7 273	3 425	81 779	57.8	4 305	788	4.8
September	147 731	89 838	60.8	85 488	48 408	29 710	7 370	3 342	82 146	57.9	4 350	785	4.8
October	147 980	90 131	60.9	85 987	48 631	29 885	7 471	3 424	82 563	58.1	4 144	793	4.6
November	148 219	90 716	61.2	86 320	48 764	30 071	7 485	3 593	82 727	58.2	4 396	832	4.8
December	148 479	90 890	61.2	86 401	48 902	29 991	7 508	3 658	82 743	58.2	4 489	767	4.9
1974													
January	148 753	91 199	61.3	86 555	49 107	29 893	7 555	3 756	82 799	58.2	4 644	799	5.1
February	148 982	91 485	61.4	86 754	49 057	30 146	7 551	3 824	82 930	58.2	4 731	829	5.2
March	149 225	91 453	61.3	86 819	48 986	30 293	7 540	3 726	83 093	58.2	4 634	849	5.1
April	149 478	91 287	61.1	86 669	48 853	30 376	7 440	3 582	83 087	58.0	4 618	889	5.1
May	149 750	91 596	61.2	86 891	49 039	30 424	7 428	3 529	83 362	58.0	4 705	880	5.1
June	150 012	91 868	61.2	86 941	48 946	30 512	7 483	3 386	83 555	58.0	4 927	926	5.4
July	150 248	92 212	61.4	87 149	48 883	30 869	7 397	3 436	83 713	58.0	5 063	924	5.5
August	150 493	92 059	61.2	87 037	48 950	30 662	7 425	3 429	83 608	57.8	5 022	960	5.5
September	150 753	92 488	61.4	87 051	48 978	30 569	7 504	3 460	83 591	57.7	5 437	1 021	5.9
October	151 009	92 518	61.3	86 995	48 959	30 570	7 466	3 431	83 564	57.6	5 523	1 072	6.0
November	151 256	92 766	61.3	86 626	48 833	30 424	7 369	3 405	83 221	57.3	6 140	1 128	6.6
December	151 494	92 780	61.2	86 144	48 458	30 431	7 255	3 361	82 783	56.9	6 636	1 326	7.2
1975													
January	151 755	93 128	61.4	85 627	48 086	30 343	7 198	3 401	82 226	56.4	7 501	1 555	8.1
February	151 990	92 776	61.0	85 256	47 927	30 215	7 114	3 361	81 895	56.1	7 520	1 841	8.1
March	152 217	93 165	61.2	85 187	47 776	30 334	7 077	3 358	81 829	56.0	7 978	2 074	8.6
April	152 443	93 399	61.3	85 189	47 759	30 410	7 020	3 315	81 874	55.9	8 210	2 442	8.8
May	152 704	93 884	61.5	85 451	47 835	30 483	7 133	3 560	81 891	56.0	8 433	2 643	9.0
June	152 976	93 575	61.2	85 355	47 754	30 618	6 983	3 368	81 987	55.8	8 220	2 843	8.8
July	153 309	94 021	61.3	85 894	48 050	30 794	7 050	3 457	82 437	56.0	8 127	2 943	8.6
August	153 580	94 162	61.3	86 234	48 239	30 966	7 029	3 429	82 805	56.1	7 928	2 862	8.4
September	153 848	94 202	61.2	86 279	48 126	30 979	7 174	3 508	82 771	56.1	7 923	2 906	8.4
October	154 082	94 267	61.2	86 370	48 165	31 121	7 084	3 397	82 973	56.1	7 897	2 689	8.4
November	154 338	94 250	61.1	86 456	48 203	31 135	7 118	3 331	83 125	56.0	7 794	2 789	8.3
December	154 589	94 409	61.1	86 665	48 266	31 268	7 131	3 259	83 406	56.1	7 744	2 868	8.2

[1]Not seasonally adjusted.
[2]In 1930 through 1943, the official BLS data count persons on work relief as unemployed. The unemployment rates for those years shown in parentheses count persons on work relief as employed, which is more consistent with the postwar practice. See notes and definitions.

Table 10-1B. Summary Labor Force, Employment, and Unemployment: Historical Data—*Continued*

(Thousands of persons, percent, seasonally adjusted, except as noted.)

Year and month	Civilian noninsti- tutional popu- lation [1]	Civilian labor force		Employment, thousands of persons						Employ- ment- population ratio, percent	Unemployment		
		Thousands of persons	Participa- tion rate (percent)	Total	By age and sex			By industry			Thousands of persons		Rate (percent) [2]
					Men, 20 years and over	Women, 20 years and over	Both sexes, 16 to 19 years	Agri- cultural	Nonagri- cultural		Total	Unem- ployed 15 weeks and over	
1976													
January	154 853	94 934	61.3	87 400	48 592	31 595	8 967	3 387	84 013	56.4	7 534	2 713	7.9
February	155 066	94 998	61.3	87 672	48 721	31 680	8 981	3 304	84 368	56.5	7 326	2 519	7.7
March	155 306	95 215	61.3	87 985	48 836	31 842	9 007	3 296	84 689	56.7	7 230	2 441	7.6
April	155 529	95 746	61.6	88 416	49 097	31 951	9 151	3 438	84 978	56.8	7 330	2 210	7.7
May	155 765	95 847	61.5	88 794	49 193	32 147	9 155	3 367	85 427	57.0	7 053	2 115	7.4
June	156 026	95 885	61.5	88 563	49 010	32 267	8 943	3 310	85 253	56.8	7 322	2 332	7.6
July	156 276	96 583	61.8	89 093	49 236	32 334	9 204	3 358	85 735	57.0	7 490	2 316	7.8
August	156 525	96 741	61.8	89 223	49 417	32 437	9 168	3 380	85 843	57.0	7 518	2 378	7.8
September	156 779	96 553	61.6	89 173	49 485	32 390	8 968	3 278	85 895	56.9	7 380	2 296	7.6
October	156 993	96 704	61.6	89 274	49 524	32 412	9 054	3 316	85 958	56.9	7 430	2 292	7.7
November	157 235	97 254	61.9	89 634	49 561	32 753	9 055	3 263	86 371	57.0	7 620	2 354	7.8
December	157 438	97 348	61.8	89 803	49 599	32 914	9 011	3 251	86 552	57.0	7 545	2 375	7.8
1977													
January	157 688	97 208	61.6	89 928	49 738	32 872	9 025	3 185	86 743	57.0	7 280	2 200	7.5
February	157 913	97 785	61.9	90 342	49 838	32 997	9 198	3 222	87 120	57.2	7 443	2 174	7.6
March	158 131	98 115	62.0	90 808	50 031	33 246	9 257	3 212	87 596	57.4	7 307	2 057	7.4
April	158 371	98 330	62.1	91 271	50 185	33 470	9 289	3 313	87 958	57.6	7 059	1 936	7.2
May	158 657	98 665	62.2	91 754	50 280	33 851	9 279	3 432	88 322	57.8	6 911	1 928	7.0
June	158 928	99 093	62.4	91 959	50 544	33 678	9 525	3 340	88 619	57.9	7 134	1 918	7.2
July	159 185	98 913	62.1	92 084	50 597	33 794	9 377	3 247	88 837	57.8	6 829	1 907	6.9
August	159 430	99 366	62.3	92 441	50 745	33 809	9 550	3 260	89 181	58.0	6 925	1 836	7.0
September	159 674	99 453	62.3	92 702	50 825	34 218	9 340	3 201	89 501	58.1	6 751	1 853	6.8
October	159 915	99 815	62.4	93 052	51 046	34 187	9 441	3 272	89 780	58.2	6 763	1 789	6.8
November	160 129	100 576	62.8	93 761	51 316	34 536	9 551	3 375	90 386	58.6	6 815	1 804	6.8
December	160 376	100 491	62.7	94 105	51 492	34 668	9 406	3 320	90 785	58.7	6 386	1 717	6.4
1978													
January	160 617	100 873	62.8	94 384	51 542	34 948	9 473	3 434	90 950	58.8	6 489	1 643	6.4
February	160 831	100 837	62.7	94 519	51 578	35 118	9 448	3 320	91 199	58.8	6 318	1 584	6.3
March	161 038	101 092	62.8	94 755	51 635	35 310	9 441	3 351	91 404	58.8	6 337	1 531	6.3
April	161 263	101 574	63.0	95 394	51 912	35 546	9 518	3 349	92 045	59.2	6 180	1 502	6.1
May	161 518	101 896	63.1	95 769	52 050	35 597	9 668	3 325	92 444	59.3	6 127	1 420	6.0
June	161 794	102 371	63.3	96 343	52 240	35 828	9 781	3 483	92 860	59.5	6 028	1 352	5.9
July	162 034	102 399	63.2	96 090	52 190	35 764	9 749	3 441	92 649	59.3	6 309	1 373	6.2
August	162 259	102 511	63.2	96 431	52 228	35 856	9 903	3 401	93 030	59.4	6 080	1 242	5.9
September	162 502	102 795	63.3	96 670	52 284	36 274	9 700	3 400	93 270	59.5	6 125	1 308	6.0
October	162 783	103 080	63.3	97 133	52 448	36 525	9 727	3 409	93 724	59.7	5 947	1 319	5.8
November	163 017	103 562	63.5	97 485	52 802	36 559	9 704	3 284	94 201	59.8	6 077	1 242	5.9
December	163 272	103 809	63.6	97 581	52 807	36 686	9 708	3 396	94 185	59.8	6 228	1 269	6.0
1979													
January	163 516	104 057	63.6	97 948	53 072	36 697	9 749	3 305	94 643	59.9	6 109	1 250	5.9
February	163 726	104 502	63.8	98 329	53 233	36 904	9 762	3 373	94 956	60.1	6 173	1 297	5.9
March	164 027	104 589	63.8	98 480	53 120	37 159	9 751	3 368	95 112	60.0	6 109	1 365	5.8
April	164 162	104 172	63.5	98 103	53 085	36 944	9 652	3 291	94 812	59.8	6 069	1 272	5.8
May	164 459	104 171	63.3	98 331	53 178	37 134	9 553	3 272	95 059	59.8	5 840	1 239	5.6
June	164 720	104 638	63.5	98 679	53 309	37 221	9 664	3 331	95 348	59.9	5 959	1 171	5.7
July	164 970	105 002	63.6	99 006	53 384	37 514	9 606	3 335	95 671	60.0	5 996	1 123	5.7
August	165 198	105 096	63.6	98 776	53 336	37 548	9 456	3 374	95 402	59.8	6 320	1 203	6.0
September	165 431	105 530	63.8	99 340	53 510	37 798	9 623	3 371	95 969	60.0	6 190	1 172	5.9
October	165 813	105 700	63.7	99 404	53 478	37 931	9 574	3 325	96 079	59.9	6 296	1 219	6.0
November	166 051	105 812	63.7	99 574	53 435	38 065	9 599	3 436	96 138	60.0	6 238	1 239	5.9
December	166 300	106 258	63.9	99 933	53 555	38 259	9 690	3 400	96 533	60.1	6 325	1 277	6.0
1980													
January	166 544	106 562	64.0	99 879	53 501	38 367	9 590	3 316	96 563	60.0	6 683	1 353	6.3
February	166 759	106 697	64.0	99 995	53 686	38 389	9 501	3 397	96 598	60.0	6 702	1 358	6.3
March	166 984	106 442	63.7	99 713	53 353	38 406	9 500	3 418	96 295	59.7	6 729	1 457	6.3
April	167 197	106 591	63.8	99 233	53 035	38 427	9 272	3 326	95 907	59.4	7 358	1 694	6.9
May	167 407	106 929	63.9	98 945	52 915	38 335	9 457	3 382	95 563	59.1	7 984	1 740	7.5
June	167 643	106 780	63.7	98 682	52 712	38 312	9 438	3 296	95 386	58.9	8 098	1 760	7.6
July	167 932	107 159	63.8	98 796	52 733	38 374	9 499	3 319	95 477	58.8	8 363	1 995	7.8
August	168 103	107 105	63.7	98 824	52 815	38 511	9 247	3 234	95 590	58.8	8 281	2 162	7.7
September	168 297	107 098	63.6	99 077	52 866	38 595	9 289	3 443	95 634	58.9	8 021	2 309	7.5
October	168 503	107 405	63.7	99 317	53 094	38 620	9 319	3 372	95 945	58.9	8 088	2 306	7.5
November	168 695	107 568	63.8	99 545	53 210	38 795	9 246	3 396	96 149	59.0	8 023	2 329	7.5
December	168 883	107 352	63.6	99 634	53 333	38 737	9 175	3 492	96 142	59.0	7 718	2 406	7.2
1981													
January	169 104	108 026	63.9	99 955	53 392	39 042	9 300	3 429	96 526	59.1	8 071	2 389	7.5
February	169 280	108 242	63.9	100 191	53 445	39 280	9 257	3 345	96 846	59.2	8 051	2 344	7.4
March	169 453	108 553	64.1	100 571	53 662	39 464	9 212	3 365	97 206	59.4	7 982	2 276	7.4
April	169 641	108 925	64.2	101 056	53 886	39 628	9 289	3 529	97 527	59.6	7 869	2 231	7.2
May	169 829	109 222	64.3	101 048	53 879	39 759	9 160	3 369	97 679	59.5	8 174	2 221	7.5
June	170 042	108 396	63.7	100 298	53 576	39 682	8 780	3 334	96 964	59.0	8 098	2 250	7.5
July	170 246	108 556	63.8	100 693	53 814	39 683	8 839	3 296	97 397	59.1	7 863	2 166	7.2
August	170 399	108 725	63.8	100 689	53 718	39 723	8 931	3 379	97 310	59.1	8 036	2 241	7.4
September	170 593	108 294	63.5	100 064	53 625	39 342	8 835	3 361	96 703	58.7	8 230	2 261	7.6
October	170 809	109 024	63.8	100 378	53 482	39 843	8 851	3 412	96 966	58.8	8 646	2 303	7.9
November	170 996	109 236	63.9	100 207	53 335	39 908	8 852	3 415	96 792	58.6	9 029	2 345	8.3
December	171 166	108 912	63.6	99 645	53 149	39 708	8 602	3 227	96 418	58.2	9 267	2 374	8.5

[1]Not seasonally adjusted.
[2]In 1930 through 1943, the official BLS data count persons on work relief as unemployed. The unemployment rates for those years shown in parentheses count persons on work relief as employed, which is more consistent with the postwar practice. See notes and definitions.

Table 10-1B. Summary Labor Force, Employment, and Unemployment: Historical Data—*Continued*

(Thousands of persons, percent, seasonally adjusted, except as noted.)

Year and month	Civilian noninsti-tutional popu-lation [1]	Civilian labor force Thousands of persons	Civilian labor force Participa-tion rate (percent)	Employment Total	Employment By age and sex Men, 20 years and over	Employment By age and sex Women, 20 years and over	Employment By age and sex Both sexes, 16 to 19 years	Employment By industry Agri-cultural	Employment By industry Nonagri-cultural	Employ-ment-population ratio, percent	Unemployment Thousands of persons Total	Unemployment Thousands of persons Unem-ployed 15 weeks and over	Unemployment Rate (percent) [2]
1982													
January	171 335	109 089	63.7	99 692	53 103	39 821	8 676	3 393	96 299	58.2	9 397	2 409	8.6
February	171 489	109 467	63.8	99 762	53 172	39 859	8 697	3 375	96 387	58.2	9 705	2 758	8.9
March	171 667	109 567	63.8	99 672	53 054	39 936	8 550	3 372	96 300	58.1	9 895	2 965	9.0
April	171 844	109 820	63.9	99 576	53 081	39 848	8 605	3 351	96 225	57.9	10 244	3 086	9.3
May	172 026	110 451	64.2	100 116	53 234	40 121	8 753	3 434	96 682	58.2	10 335	3 276	9.4
June	172 190	110 081	63.9	99 543	52 933	40 219	8 293	3 331	96 212	57.8	10 538	3 451	9.6
July	172 364	110 342	64.0	99 493	52 896	40 228	8 380	3 402	96 091	57.7	10 849	3 555	9.8
August	172 511	110 514	64.1	99 633	52 797	40 336	8 514	3 408	96 225	57.8	10 881	3 696	9.8
September	172 690	110 721	64.1	99 504	52 760	40 275	8 469	3 385	96 119	57.6	11 217	3 889	10.1
October	172 881	110 744	64.1	99 215	52 624	40 105	8 499	3 489	95 726	57.4	11 529	4 185	10.4
November	173 058	111 050	64.2	99 112	52 537	40 111	8 520	3 510	95 602	57.3	11 938	4 485	10.8
December	173 199	111 083	64.1	99 032	52 497	40 164	8 397	3 414	95 618	57.2	12 051	4 662	10.8
1983													
January	173 354	110 695	63.9	99 161	52 487	40 268	8 335	3 439	95 722	57.2	11 534	4 668	10.4
February	173 505	110 634	63.8	99 089	52 453	40 336	8 159	3 382	95 707	57.1	11 545	4 641	10.4
March	173 656	110 587	63.7	99 179	52 615	40 368	8 098	3 360	95 819	57.1	11 408	4 612	10.3
April	173 794	110 828	63.8	99 560	52 814	40 542	8 094	3 341	96 219	57.3	11 268	4 370	10.2
May	173 953	110 796	63.7	99 642	52 922	40 538	8 006	3 328	96 314	57.3	11 154	4 538	10.1
June	174 125	111 879	64.3	100 633	53 515	40 695	8 448	3 462	97 171	57.8	11 246	4 470	10.1
July	174 306	111 756	64.1	101 208	53 835	41 041	8 207	3 481	97 727	58.1	10 548	4 329	9.4
August	174 440	112 231	64.3	101 608	53 837	41 314	8 379	3 502	98 106	58.2	10 623	4 070	9.5
September	174 602	112 298	64.3	102 016	53 983	41 650	8 147	3 347	98 669	58.4	10 282	3 854	9.2
October	174 779	111 926	64.0	102 039	54 146	41 597	8 010	3 303	98 736	58.4	9 887	3 648	8.8
November	174 951	112 228	64.1	102 729	54 499	41 788	8 075	3 291	99 438	58.7	9 499	3 535	8.5
December	175 121	112 327	64.1	102 996	54 662	41 852	8 089	3 332	99 664	58.8	9 331	3 379	8.3
1984													
January	175 533	112 209	63.9	103 201	54 975	41 812	7 965	3 293	99 908	58.8	9 008	3 254	8.0
February	175 679	112 615	64.1	103 824	55 213	42 196	7 958	3 353	100 471	59.1	8 791	2 991	7.8
March	175 824	112 713	64.1	103 967	55 281	42 328	7 926	3 233	100 734	59.1	8 746	2 881	7.8
April	175 969	113 098	64.3	104 336	55 373	42 512	7 988	3 291	101 045	59.3	8 762	2 858	7.7
May	176 123	113 649	64.5	105 193	55 661	43 071	7 944	3 343	101 850	59.7	8 456	2 884	7.4
June	176 284	113 817	64.6	105 591	55 996	42 944	8 131	3 383	102 208	59.9	8 226	2 612	7.2
July	176 440	113 972	64.6	105 435	55 921	42 979	8 045	3 344	102 091	59.8	8 537	2 638	7.5
August	176 583	113 682	64.4	105 163	55 930	42 885	7 809	3 286	101 877	59.6	8 519	2 604	7.5
September	176 763	113 857	64.4	105 490	56 095	42 967	7 951	3 393	102 097	59.7	8 367	2 538	7.3
October	176 956	114 019	64.4	105 638	56 183	43 052	7 862	3 194	102 444	59.7	8 381	2 526	7.4
November	177 135	114 170	64.5	105 972	56 274	43 244	7 844	3 394	102 578	59.8	8 198	2 438	7.2
December	177 306	114 581	64.6	106 223	56 313	43 472	7 933	3 385	102 838	59.9	8 358	2 401	7.3
1985													
January	177 384	114 725	64.7	106 302	56 184	43 589	8 036	3 317	102 985	59.9	8 423	2 284	7.3
February	177 516	114 876	64.7	106 555	56 216	43 787	8 020	3 317	103 238	60.0	8 321	2 389	7.2
March	177 667	115 328	64.9	106 989	56 356	44 035	8 062	3 250	103 739	60.2	8 339	2 394	7.2
April	177 799	115 331	64.9	106 936	56 374	44 000	7 958	3 306	103 630	60.1	8 395	2 393	7.3
May	177 944	115 234	64.8	106 932	56 531	43 905	7 966	3 280	103 652	60.1	8 302	2 292	7.2
June	178 096	114 965	64.6	106 505	56 288	43 958	7 680	3 161	103 344	59.8	8 460	2 310	7.4
July	178 263	115 320	64.7	106 807	56 435	43 975	8 013	3 143	103 664	59.9	8 513	2 329	7.4
August	178 405	115 291	64.6	107 095	56 655	44 103	7 714	3 121	103 974	60.0	8 196	2 258	7.1
September	178 572	115 905	64.9	107 657	56 845	44 395	7 818	3 064	104 593	60.3	8 248	2 242	7.1
October	178 770	116 145	65.0	107 847	56 969	44 565	7 889	3 051	104 796	60.3	8 298	2 295	7.1
November	178 940	116 135	64.9	108 007	56 972	44 617	7 852	3 062	104 945	60.4	8 128	2 207	7.0
December	179 112	116 354	65.0	108 216	56 995	44 889	7 826	3 141	105 075	60.4	8 138	2 208	7.0
1986													
January	179 670	116 682	64.9	108 887	57 637	44 944	7 704	3 287	105 600	60.6	7 795	2 089	6.7
February	179 821	116 882	65.0	108 480	57 269	44 804	7 895	3 083	105 397	60.3	8 402	2 308	7.2
March	179 985	117 220	65.1	108 837	57 353	44 960	7 977	3 200	105 637	60.5	8 383	2 261	7.2
April	180 148	117 316	65.1	108 952	57 358	45 081	8 058	3 153	105 799	60.5	8 364	2 162	7.1
May	180 311	117 528	65.2	109 089	57 287	45 289	7 997	3 150	105 939	60.5	8 439	2 232	7.2
June	180 503	118 084	65.4	109 576	57 471	45 621	8 024	3 193	106 383	60.7	8 508	2 320	7.2
July	180 682	118 129	65.4	109 810	57 514	45 837	7 914	3 141	106 669	60.8	8 319	2 269	7.0
August	180 828	118 150	65.3	110 015	57 597	45 926	7 920	3 082	106 933	60.8	8 135	2 276	6.9
September	180 997	118 395	65.4	110 085	57 630	45 972	7 945	3 171	106 914	60.8	8 310	2 318	7.0
October	181 186	118 516	65.4	110 273	57 660	46 046	7 977	3 128	107 145	60.9	8 243	2 188	7.0
November	181 363	118 634	65.4	110 475	57 941	46 070	7 891	3 220	107 255	60.9	8 159	2 202	6.9
December	181 547	118 611	65.3	110 728	58 185	46 132	7 769	3 148	107 580	61.0	7 883	2 161	6.6
1987													
January	181 827	118 845	65.4	110 953	58 264	46 219	7 861	3 143	107 810	61.0	7 892	2 168	6.6
February	181 998	119 122	65.5	111 257	58 279	46 444	7 969	3 208	108 049	61.1	7 865	2 117	6.6
March	182 179	119 270	65.5	111 408	58 362	46 549	7 913	3 214	108 194	61.2	7 862	2 070	6.6
April	182 344	119 336	65.4	111 794	58 503	46 746	7 916	3 246	108 548	61.3	7 542	2 091	6.3
May	182 533	120 008	65.7	112 434	58 713	47 052	8 075	3 345	109 089	61.6	7 574	2 104	6.3
June	182 703	119 644	65.5	112 246	58 581	47 102	7 857	3 216	109 030	61.4	7 398	2 087	6.2
July	182 885	119 902	65.6	112 634	58 740	47 229	7 911	3 235	109 399	61.6	7 268	1 921	6.1
August	183 002	120 318	65.7	113 057	58 810	47 322	8 232	3 112	109 945	61.8	7 261	1 878	6.0
September	183 161	120 011	65.5	112 909	58 964	47 285	7 945	3 189	109 720	61.6	7 102	1 866	5.9
October	183 311	120 509	65.7	113 282	59 073	47 533	8 074	3 219	110 063	61.9	7 227	1 794	6.0
November	183 470	120 540	65.7	113 505	59 210	47 622	8 002	3 145	110 360	61.9	7 035	1 797	5.8
December	183 620	120 729	65.7	113 793	59 217	47 781	8 092	3 213	110 580	62.0	6 936	1 767	5.7

[1] Not seasonally adjusted.
[2] In 1930 through 1943, the official BLS data count persons on work relief as unemployed. The unemployment rates for those years shown in parentheses count persons on work relief as employed, which is more consistent with the postwar practice. See notes and definitions.

Table 10-1B. Summary Labor Force, Employment, and Unemployment: Historical Data—*Continued*

(Thousands of persons, percent, seasonally adjusted, except as noted.)

Year and month	Civilian noninstitutional population [1]	Civilian labor force Thousands of persons	Civilian labor force Participation rate (percent)	Employment Total	By age and sex Men, 20 years and over	By age and sex Women, 20 years and over	By age and sex Both sexes, 16 to 19 years	By industry Agricultural	By industry Nonagricultural	Employment-population ratio, percent	Unemployment Total	Unemployment Unemployed 15 weeks and over	Rate (percent) [2]
1988													
January	183 822	120 969	65.8	114 016	59 346	47 862	8 110	3 247	110 769	62.0	6 953	1 714	5.7
February	183 969	121 156	65.9	114 227	59 535	47 919	8 025	3 201	111 026	62.1	6 929	1 738	5.7
March	184 111	120 913	65.7	114 037	59 393	48 090	7 854	3 169	110 868	61.9	6 876	1 744	5.7
April	184 232	121 251	65.8	114 650	59 832	48 147	7 937	3 224	111 426	62.2	6 601	1 563	5.4
May	184 374	121 071	65.7	114 292	59 644	47 946	7 912	3 121	111 171	62.0	6 779	1 647	5.6
June	184 562	121 473	65.8	114 927	59 751	48 146	8 193	3 111	111 816	62.3	6 546	1 531	5.4
July	184 729	121 665	65.9	115 060	59 888	48 186	8 198	3 060	112 000	62.3	6 605	1 601	5.4
August	184 830	122 125	66.1	115 282	59 877	48 467	8 201	3 119	112 163	62.4	6 843	1 639	5.6
September	184 962	121 960	65.9	115 356	59 980	48 511	8 124	3 165	112 191	62.4	6 604	1 569	5.4
October	185 114	122 206	66.0	115 638	60 023	48 859	7 961	3 231	112 407	62.5	6 568	1 562	5.4
November	185 244	122 637	66.2	116 100	60 042	49 254	7 904	3 241	112 859	62.7	6 537	1 468	5.3
December	185 402	122 622	66.1	116 104	60 059	49 257	7 966	3 194	112 910	62.6	6 518	1 490	5.3
1989													
January	185 644	123 390	66.5	116 708	60 477	49 529	8 019	3 287	113 421	62.9	6 682	1 480	5.4
February	185 777	123 135	66.3	116 776	60 588	49 497	7 871	3 234	113 542	62.9	6 359	1 304	5.2
March	185 897	123 227	66.3	117 022	60 795	49 503	7 809	3 198	113 824	62.9	6 205	1 353	5.0
April	186 024	123 565	66.4	117 097	60 764	49 565	7 929	3 162	113 935	62.9	6 468	1 397	5.2
May	186 181	123 474	66.3	117 099	60 795	49 583	7 890	3 125	113 974	62.9	6 375	1 348	5.2
June	186 329	123 995	66.5	117 418	61 054	49 542	8 094	3 068	114 350	63.0	6 577	1 300	5.3
July	186 483	123 967	66.5	117 472	60 947	49 693	7 961	3 227	114 245	63.0	6 495	1 435	5.2
August	186 598	124 166	66.5	117 655	60 915	49 804	8 126	3 284	114 371	63.1	6 511	1 302	5.2
September	186 726	123 944	66.4	117 354	60 668	50 015	7 870	3 219	114 135	62.8	6 590	1 360	5.3
October	186 871	124 211	66.5	117 581	60 958	49 871	7 940	3 215	114 366	62.9	6 630	1 392	5.3
November	187 017	124 637	66.6	117 912	60 958	50 221	7 964	3 132	114 780	63.0	6 725	1 418	5.4
December	187 165	124 497	66.5	117 830	61 068	50 116	7 851	3 188	114 642	63.0	6 667	1 375	5.4
1990													
January	188 413	125 833	66.8	119 081	61 742	50 436	8 103	3 210	115 871	63.2	6 752	1 412	5.4
February	188 516	125 710	66.7	119 059	61 805	50 438	8 015	3 188	115 871	63.2	6 651	1 350	5.3
March	188 630	125 801	66.7	119 203	61 832	50 463	8 061	3 260	115 943	63.2	6 598	1 331	5.2
April	188 778	125 649	66.6	118 852	61 579	50 457	7 987	3 231	115 621	63.0	6 797	1 376	5.4
May	188 913	125 893	66.6	119 151	61 778	50 646	7 911	3 266	115 885	63.1	6 742	1 415	5.4
June	189 058	125 573	66.4	118 983	61 762	50 550	7 783	3 245	115 738	62.9	6 590	1 436	5.2
July	189 188	125 732	66.5	118 810	61 683	50 514	7 777	3 192	115 618	62.8	6 922	1 534	5.5
August	189 342	125 990	66.5	118 802	61 715	50 635	7 708	3 197	115 605	62.7	7 188	1 607	5.7
September	189 528	125 892	66.4	118 524	61 608	50 587	7 575	3 206	115 318	62.5	7 368	1 695	5.9
October	189 710	125 995	66.4	118 536	61 606	50 616	7 565	3 270	115 266	62.5	7 459	1 689	5.9
November	189 872	126 070	66.4	118 306	61 545	50 541	7 506	3 189	115 117	62.3	7 764	1 831	6.2
December	190 017	126 142	66.4	118 241	61 506	50 530	7 516	3 245	114 996	62.2	7 901	1 804	6.3
1991													
January	190 163	125 955	66.2	117 940	61 383	50 472	7 478	3 208	114 732	62.0	8 015	1 866	6.4
February	190 271	126 020	66.2	117 755	61 117	50 523	7 405	3 270	114 485	61.9	8 265	1 955	6.6
March	190 381	126 238	66.3	117 652	61 144	50 422	7 447	3 177	114 475	61.8	8 586	2 137	6.8
April	190 517	126 548	66.4	118 109	61 280	50 760	7 381	3 241	114 868	62.0	8 439	2 206	6.7
May	190 650	126 176	66.2	117 440	61 052	50 457	7 303	3 275	114 165	61.6	8 736	2 252	6.9
June	190 800	126 331	66.2	117 639	61 147	50 585	7 248	3 300	114 339	61.7	8 692	2 533	6.9
July	190 946	126 154	66.1	117 568	61 179	50 636	7 138	3 319	114 249	61.6	8 586	2 388	6.8
August	191 116	126 150	66.0	117 484	61 122	50 601	7 105	3 313	114 171	61.5	8 666	2 460	6.9
September	191 302	126 650	66.2	117 928	61 279	50 864	7 123	3 319	114 609	61.6	8 722	2 497	6.9
October	191 497	126 642	66.1	117 800	61 174	50 811	7 185	3 289	114 511	61.5	8 842	2 638	7.0
November	191 657	126 701	66.1	117 770	61 201	50 759	7 169	3 296	114 474	61.4	8 931	2 718	7.0
December	191 798	126 664	66.0	117 466	61 074	50 728	7 104	3 146	114 320	61.2	9 198	2 892	7.3
1992													
January	191 953	127 261	66.3	117 978	61 116	51 095	7 138	3 155	114 823	61.5	9 283	3 060	7.3
February	192 067	127 207	66.2	117 753	61 062	51 033	7 083	3 239	114 514	61.3	9 454	3 182	7.4
March	192 204	127 604	66.4	118 144	61 363	51 204	6 998	3 236	114 908	61.5	9 460	3 196	7.4
April	192 354	127 841	66.5	118 426	61 468	51 323	6 910	3 245	115 181	61.6	9 415	3 130	7.4
May	192 503	128 119	66.6	118 375	61 513	51 245	7 028	3 213	115 162	61.5	9 744	3 444	7.6
June	192 663	128 459	66.7	118 419	61 537	51 383	7 137	3 297	115 122	61.5	10 040	3 758	7.8
July	192 826	128 563	66.7	118 713	61 641	51 458	7 087	3 285	115 428	61.6	9 850	3 614	7.7
August	193 018	128 613	66.6	118 826	61 681	51 386	7 194	3 279	115 547	61.6	9 787	3 579	7.6
September	193 229	128 501	66.5	118 720	61 663	51 359	7 216	3 274	115 446	61.4	9 781	3 504	7.6
October	193 442	128 026	66.2	118 628	61 550	51 373	6 985	3 254	115 374	61.3	9 398	3 505	7.3
November	193 621	128 441	66.3	118 876	61 644	51 535	7 164	3 207	115 669	61.4	9 565	3 397	7.4
December	193 784	128 554	66.3	118 997	61 721	51 524	7 174	3 259	115 738	61.4	9 557	3 651	7.4
1993													
January	193 962	128 400	66.2	119 075	61 895	51 505	7 089	3 222	115 853	61.4	9 325	3 346	7.3
February	194 108	128 458	66.2	119 275	61 963	51 573	7 144	3 125	116 150	61.4	9 183	3 190	7.1
March	194 248	128 598	66.2	119 542	62 007	51 808	7 132	3 119	116 423	61.5	9 056	3 115	7.0
April	194 398	128 584	66.1	119 474	62 032	51 732	7 091	3 074	116 400	61.5	9 110	3 014	7.1
May	194 549	129 264	66.4	120 115	62 309	51 996	7 244	3 100	117 015	61.7	9 149	3 101	7.1
June	194 719	129 411	66.5	120 290	62 409	52 183	7 114	3 108	117 182	61.8	9 121	3 141	7.0
July	194 882	129 397	66.4	120 467	62 497	52 088	7 205	3 126	117 341	61.8	8 930	3 046	6.9
August	195 063	129 619	66.4	120 856	62 634	52 294	7 264	3 026	117 830	62.0	8 763	3 026	6.8
September	195 259	129 268	66.2	120 554	62 437	52 241	7 180	3 174	117 380	61.7	8 714	3 042	6.7
October	195 444	129 573	66.3	120 823	62 614	52 379	7 171	3 084	117 739	61.8	8 750	3 029	6.8
November	195 625	129 711	66.3	121 169	62 732	52 531	7 248	3 157	118 012	61.9	8 542	2 986	6.6
December	195 794	129 941	66.4	121 464	62 760	52 813	7 178	3 116	118 348	62.0	8 477	2 968	6.5

[1] Not seasonally adjusted.
[2] In 1930 through 1943, the official BLS data count persons on work relief as unemployed. The unemployment rates for those years shown in parentheses count persons on work relief as employed, which is more consistent with the postwar practice. See notes and definitions.

Table 10-1B. Summary Labor Force, Employment, and Unemployment: Historical Data—*Continued*

(Thousands of persons, percent, seasonally adjusted, except as noted.)

Year and month	Civilian noninsti-tutional popu-lation [1]	Civilian labor force		Employment, thousands of persons						Employ-ment-population ratio, percent	Unemployment		
					By age and sex			By industry			Thousands of persons		
		Thousands of persons	Participa-tion rate (percent)	Total	Men, 20 years and over	Women, 20 years and over	Both sexes, 16 to 19 years	Agri-cultural	Nonagri-cultural		Total	Unem-ployed 15 weeks and over	Rate (percent) [2]
1994													
January	195 953	130 596	66.6	121 966	62 798	53 052	7 486	3 302	118 664	62.2	8 630	3 060	6.6
February	196 090	130 669	66.6	122 086	62 708	53 266	7 457	3 339	118 747	62.3	8 583	3 118	6.6
March	196 213	130 400	66.5	121 930	62 780	53 099	7 381	3 354	118 576	62.1	8 470	3 055	6.5
April	196 363	130 621	66.5	122 290	62 906	53 274	7 551	3 428	118 862	62.3	8 331	2 921	6.4
May	196 510	130 779	66.6	122 864	63 116	53 624	7 466	3 409	119 455	62.5	7 915	2 836	6.1
June	196 693	130 561	66.4	122 634	63 041	53 393	7 527	3 299	119 335	62.3	7 927	2 735	6.1
July	196 859	130 652	66.4	122 706	63 034	53 531	7 453	3 333	119 373	62.3	7 946	2 822	6.1
August	197 043	131 275	66.6	123 342	63 294	53 744	7 619	3 451	119 891	62.6	7 933	2 750	6.0
September	197 248	131 421	66.6	123 687	63 631	53 991	7 352	3 430	120 257	62.7	7 734	2 746	5.9
October	197 430	131 744	66.7	124 112	63 818	54 071	7 543	3 490	120 622	62.9	7 632	2 955	5.8
November	197 607	131 891	66.7	124 516	64 080	54 168	7 423	3 574	120 942	63.0	7 375	2 666	5.6
December	197 765	131 951	66.7	124 721	64 359	54 054	7 599	3 577	121 144	63.1	7 230	2 488	5.5
1995													
January	197 753	132 038	66.8	124 663	64 185	54 087	7 650	3 519	121 144	63.0	7 375	2 396	5.6
February	197 886	132 115	66.8	124 928	64 378	54 226	7 659	3 620	121 308	63.1	7 187	2 345	5.4
March	198 007	132 108	66.7	124 955	64 321	54 141	7 742	3 634	121 321	63.1	7 153	2 287	5.4
April	198 148	132 590	66.9	124 945	64 165	54 366	7 774	3 566	121 379	63.1	7 645	2 473	5.8
May	198 286	131 851	66.5	124 421	63 829	54 272	7 664	3 349	121 072	62.7	7 430	2 577	5.6
June	198 452	131 949	66.5	124 522	63 992	54 020	7 850	3 461	121 061	62.7	7 427	2 266	5.6
July	198 615	132 343	66.6	124 816	63 962	54 476	7 798	3 379	121 437	62.8	7 527	2 311	5.7
August	198 801	132 336	66.6	124 852	63 875	54 434	7 910	3 374	121 478	62.8	7 484	2 391	5.7
September	199 005	132 611	66.6	125 133	64 179	54 507	7 825	3 285	121 848	62.9	7 478	2 306	5.6
October	199 192	132 716	66.6	125 388	64 272	54 692	7 774	3 438	121 950	62.9	7 328	2 272	5.5
November	199 355	132 614	66.5	125 188	63 931	54 850	7 765	3 338	121 850	62.8	7 426	2 339	5.6
December	199 508	132 511	66.4	125 088	64 041	54 674	7 768	3 352	121 736	62.7	7 423	2 331	5.6
1996													
January	199 634	132 616	66.4	125 125	64 180	54 580	7 733	3 483	121 642	62.7	7 491	2 371	5.6
February	199 772	132 952	66.6	125 639	64 398	54 844	7 688	3 547	122 092	62.9	7 313	2 307	5.5
March	199 921	133 180	66.6	125 862	64 506	54 994	7 673	3 489	122 373	63.0	7 318	2 454	5.5
April	200 101	133 409	66.7	125 994	64 481	55 067	7 774	3 406	122 588	63.0	7 415	2 455	5.6
May	200 278	133 667	66.7	126 244	64 683	55 034	7 841	3 473	122 771	63.0	7 423	2 403	5.6
June	200 459	133 697	66.7	126 602	64 940	55 177	7 739	3 424	123 178	63.2	7 095	2 355	5.3
July	200 641	134 284	66.9	126 947	65 068	55 362	7 859	3 433	123 514	63.3	7 337	2 297	5.5
August	200 847	134 054	66.7	127 172	65 216	55 525	7 731	3 395	123 777	63.3	6 882	2 267	5.1
September	201 060	134 515	66.9	127 536	65 169	55 669	7 935	3 448	124 088	63.4	6 979	2 220	5.2
October	201 273	134 921	67.0	127 890	65 460	55 750	7 980	3 463	124 427	63.5	7 031	2 268	5.2
November	201 463	135 007	67.0	127 771	65 320	55 896	7 875	3 356	124 415	63.4	7 236	2 159	5.4
December	201 636	135 113	67.0	127 860	65 435	55 849	7 883	3 445	124 415	63.4	7 253	2 124	5.4
1997													
January	202 285	135 456	67.0	128 298	65 679	56 024	7 930	3 449	124 849	63.4	7 158	2 162	5.3
February	202 388	135 400	66.9	128 298	65 758	55 955	7 948	3 353	124 945	63.4	7 102	2 140	5.2
March	202 513	135 891	67.1	128 891	65 974	56 270	7 951	3 419	125 472	63.6	7 000	2 110	5.2
April	202 674	136 016	67.1	129 143	66 092	56 347	7 974	3 462	125 681	63.7	6 873	2 176	5.1
May	202 832	136 119	67.1	129 464	66 328	56 446	7 964	3 437	126 027	63.8	6 655	2 121	4.9
June	203 000	136 211	67.1	129 412	66 308	56 573	7 849	3 409	126 003	63.7	6 799	2 085	5.0
July	203 166	136 477	67.2	129 822	66 422	56 785	7 975	3 422	126 400	63.9	6 655	2 119	4.9
August	203 364	136 618	67.2	130 010	66 508	56 852	7 922	3 359	126 651	63.9	6 608	2 004	4.8
September	203 570	136 675	67.1	130 019	66 483	56 931	7 873	3 392	126 627	63.9	6 656	2 074	4.9
October	203 767	136 633	67.1	130 179	66 511	56 982	7 876	3 312	126 867	63.9	6 454	1 950	4.7
November	203 941	136 961	67.2	130 653	66 765	57 039	8 042	3 386	127 267	64.1	6 308	1 817	4.6
December	204 098	137 155	67.2	130 679	66 643	57 219	7 923	3 405	127 274	64.0	6 476	1 901	4.7
1998													
January	204 238	137 095	67.1	130 726	66 750	56 941	8 171	3 299	127 389	64.0	6 368	1 833	4.6
February	204 400	137 112	67.1	130 807	66 856	56 992	8 137	3 284	127 522	64.0	6 306	1 809	4.6
March	204 546	137 236	67.1	130 814	66 721	57 080	8 235	3 146	127 650	64.0	6 422	1 772	4.7
April	204 731	137 150	67.0	131 209	67 151	57 074	8 071	3 334	127 852	64.1	5 941	1 476	4.3
May	204 899	137 372	67.0	131 325	67 164	57 155	8 228	3 360	127 959	64.1	6 047	1 490	4.4
June	205 085	137 455	67.0	131 244	67 054	57 156	8 268	3 380	127 874	64.0	6 212	1 613	4.5
July	205 270	137 588	67.0	131 329	67 119	57 192	8 221	3 455	127 913	64.0	6 259	1 577	4.5
August	205 479	137 570	67.0	131 390	66 985	57 332	8 289	3 509	127 970	63.9	6 179	1 626	4.5
September	205 699	138 286	67.2	131 986	67 254	57 520	8 489	3 500	128 399	64.2	6 300	1 688	4.6
October	205 919	138 279	67.2	131 999	67 433	57 529	8 347	3 593	128 389	64.1	6 280	1 582	4.5
November	206 104	138 381	67.1	132 280	67 591	57 638	8 269	3 375	128 897	64.2	6 100	1 590	4.4
December	206 270	138 634	67.2	132 602	67 548	57 840	8 340	3 246	129 320	64.3	6 032	1 559	4.4
1999													
January	206 719	139 003	67.2	133 027	67 679	58 256	8 367	3 233	129 802	64.4	5 976	1 490	4.3
February	206 873	138 967	67.2	132 856	67 498	58 129	8 399	3 246	129 647	64.2	6 111	1 551	4.4
March	207 036	138 730	67.0	132 947	67 660	58 132	8 343	3 238	129 656	64.2	5 783	1 472	4.2
April	207 236	138 959	67.1	132 955	67 542	58 260	8 334	3 336	129 615	64.2	6 004	1 480	4.3
May	207 427	139 107	67.1	133 311	67 539	58 440	8 454	3 335	129 937	64.3	5 796	1 505	4.2
June	207 632	139 329	67.1	133 378	67 700	58 641	8 175	3 386	129 982	64.2	5 951	1 624	4.3
July	207 828	139 439	67.1	133 414	67 731	58 490	8 306	3 346	130 146	64.2	6 025	1 513	4.3
August	208 038	139 430	67.0	133 591	67 768	58 707	8 208	3 234	130 366	64.2	5 838	1 455	4.2
September	208 265	139 622	67.0	133 707	67 882	58 735	8 323	3 173	130 434	64.2	5 915	1 449	4.2
October	208 483	139 771	67.0	133 993	67 840	58 921	8 394	3 229	130 758	64.3	5 778	1 438	4.1
November	208 666	140 025	67.1	134 309	68 094	59 018	8 358	3 343	130 989	64.4	5 716	1 378	4.1
December	208 832	140 177	67.1	134 523	68 217	59 056	8 370	3 260	131 257	64.4	5 653	1 375	4.0

[1] Not seasonally adjusted.
[2] In 1930 through 1943, the official BLS data count persons on work relief as unemployed. The unemployment rates for those years shown in parentheses count persons on work relief as employed, which is more consistent with the postwar practice. See notes and definitions.

Table 10-1B. Summary Labor Force, Employment, and Unemployment: Historical Data—*Continued*

(Thousands of persons, percent, seasonally adjusted, except as noted.)

Year and month	Civilian noninsti- tutional popu- lation [1]	Civilian labor force		Employment, thousands of persons						Employ- ment- population ratio, percent	Unemployment		
		Thousands of persons	Participa- tion rate (percent)	Total	By age and sex			By industry			Thousands of persons		Rate (percent) [2]
					Men, 20 years and over	Women, 20 years and over	Both sexes, 16 to 19 years	Agri- cultural	Nonagri- cultural		Total	Unem- ployed 15 weeks and over	
2000													
January	211 410	142 267	67.3	136 559	69 419	59 842	8 360	2 613	133 863	64.6	5 708	1 380	4.0
February	211 576	142 456	67.3	136 598	69 505	59 887	8 359	2 731	133 912	64.6	5 858	1 300	4.1
March	211 772	142 434	67.3	136 701	69 482	59 977	8 355	2 579	134 022	64.6	5 733	1 312	4.0
April	212 018	142 751	67.3	137 270	69 519	60 358	8 458	2 505	134 806	64.7	5 481	1 261	3.8
May	212 242	142 388	67.1	136 630	69 399	59 951	8 344	2 480	134 144	64.4	5 758	1 325	4.0
June	212 466	142 591	67.1	136 940	69 629	60 027	8 308	2 445	134 528	64.5	5 651	1 242	4.0
July	212 677	142 278	66.9	136 531	69 525	60 011	8 078	2 408	134 196	64.2	5 747	1 343	4.0
August	212 916	142 514	66.9	136 662	69 823	59 719	8 280	2 433	134 311	64.2	5 853	1 394	4.1
September	213 163	142 518	66.9	136 893	69 700	60 083	8 176	2 384	134 489	64.2	5 625	1 290	3.9
October	213 405	142 622	66.8	137 088	69 762	60 238	8 124	2 319	134 808	64.2	5 534	1 337	3.9
November	213 540	142 962	66.9	137 322	69 910	60 269	8 211	2 330	134 921	64.3	5 639	1 315	3.9
December	213 736	143 248	67.0	137 614	69 939	60 503	8 258	2 389	135 194	64.4	5 634	1 329	3.9
2001													
January	213 888	143 800	67.2	137 778	70 064	60 609	8 243	2 360	135 304	64.4	6 023	1 372	4.2
February	214 110	143 701	67.1	137 612	69 959	60 615	8 154	2 370	135 291	64.3	6 089	1 491	4.2
March	214 305	143 924	67.2	137 783	69 881	60 902	8 120	2 350	135 372	64.3	6 141	1 521	4.3
April	214 525	143 569	66.9	137 299	69 916	60 523	7 970	2 336	135 036	64.0	6 271	1 499	4.4
May	214 732	143 318	66.7	137 092	69 865	60 509	7 760	2 353	134 735	63.8	6 226	1 502	4.3
June	214 950	143 357	66.7	136 873	69 690	60 371	7 942	2 082	134 755	63.7	6 484	1 532	4.5
July	215 180	143 654	66.8	137 071	69 808	60 480	7 929	2 295	134 858	63.7	6 583	1 653	4.6
August	215 420	143 284	66.5	136 241	69 585	60 301	7 531	2 305	133 944	63.2	7 042	1 861	4.9
September	215 665	143 989	66.8	136 846	69 933	60 265	7 836	2 322	134 558	63.5	7 142	1 950	5.0
October	215 903	144 086	66.7	136 392	69 621	60 168	7 858	2 327	134 098	63.2	7 694	2 082	5.3
November	216 117	144 240	66.7	136 238	69 444	60 174	7 873	2 203	133 955	63.0	8 003	2 318	5.5
December	216 315	144 305	66.7	136 047	69 551	60 095	7 712	2 293	133 751	62.9	8 258	2 444	5.7
2002													
January	216 506	143 883	66.5	135 701	69 308	60 032	7 619	2 385	133 233	62.7	8 182	2 578	5.7
February	216 663	144 653	66.8	136 438	69 534	60 479	7 650	2 397	134 127	63.0	8 215	2 608	5.7
March	216 823	144 481	66.6	136 177	69 480	60 190	7 802	2 368	133 816	62.8	8 304	2 719	5.7
April	217 006	144 725	66.7	136 126	69 574	60 204	7 621	2 371	133 833	62.7	8 599	2 852	5.9
May	217 198	144 938	66.7	136 539	69 981	60 226	7 588	2 260	134 278	62.9	8 399	2 967	5.8
June	217 407	144 808	66.6	136 415	69 769	60 297	7 617	2 161	134 135	62.7	8 393	3 023	5.8
July	217 630	144 803	66.5	136 413	69 806	60 290	7 591	2 324	134 107	62.7	8 390	2 966	5.8
August	217 866	145 009	66.6	136 705	69 937	60 560	7 477	2 127	134 593	62.7	8 304	2 887	5.7
September	218 107	145 552	66.7	137 302	70 207	60 679	7 667	2 285	135 102	63.0	8 251	2 971	5.7
October	218 340	145 314	66.6	137 008	69 948	60 663	7 537	2 471	134 580	62.7	8 307	3 042	5.7
November	218 548	145 041	66.4	136 521	69 615	60 697	7 487	2 261	134 171	62.5	8 520	3 062	5.9
December	218 741	145 066	66.3	136 426	69 620	60 667	7 385	2 352	134 071	62.4	8 640	3 271	6.0
2003													
January	219 897	145 937	66.4	137 417	69 919	61 406	7 358	2 337	135 045	62.5	8 520	3 166	5.8
February	220 114	146 100	66.4	137 482	70 262	61 159	7 322	2 234	135 306	62.5	8 618	3 161	5.9
March	220 317	146 022	66.3	137 434	70 243	61 317	7 143	2 263	135 232	62.4	8 588	3 161	5.9
April	220 540	146 474	66.4	137 633	70 311	61 374	7 223	2 150	135 561	62.4	8 842	3 348	6.0
May	220 768	146 500	66.4	137 544	70 215	61 393	7 228	2 183	135 370	62.3	8 957	3 318	6.1
June	221 014	147 056	66.5	137 790	70 159	61 747	7 268	2 185	135 419	62.3	9 266	3 552	6.3
July	221 252	146 485	66.2	137 474	70 190	61 437	7 150	2 187	135 242	62.1	9 011	3 633	6.2
August	221 507	146 445	66.1	137 549	70 237	61 442	7 035	2 313	135 200	62.1	8 896	3 557	6.1
September	221 779	146 530	66.1	137 609	70 631	61 119	7 112	2 349	135 355	62.0	8 921	3 486	6.1
October	222 039	146 716	66.1	137 984	70 685	61 451	7 060	2 479	135 571	62.1	8 732	3 451	6.0
November	222 279	147 000	66.1	138 424	70 935	61 494	7 112	2 373	136 003	62.3	8 576	3 420	5.8
December	222 509	146 729	65.9	138 411	71 170	61 396	6 978	2 243	136 145	62.2	8 317	3 366	5.7
2004													
January	222 161	146 842	66.1	138 472	71 318	61 183	7 194	2 196	136 228	62.3	8 370	3 364	5.7
February	222 357	146 709	66.0	138 542	71 122	61 514	7 077	2 210	136 362	62.3	8 167	3 248	5.6
March	222 550	146 944	66.0	138 453	71 155	61 534	6 930	2 180	136 302	62.2	8 491	3 314	5.8
April	222 757	146 850	65.9	138 680	71 121	61 647	7 093	2 241	136 474	62.3	8 170	2 971	5.6
May	222 967	147 065	66.0	138 852	71 180	61 762	7 126	2 300	136 556	62.3	8 212	3 103	5.6
June	223 196	147 460	66.1	139 174	71 562	61 793	7 007	2 237	136 748	62.4	8 286	3 130	5.6
July	223 422	147 692	66.1	139 556	71 780	61 884	7 166	2 222	137 354	62.5	8 136	2 918	5.5
August	223 677	147 564	66.0	139 573	71 808	61 836	7 123	2 333	137 230	62.4	7 990	2 846	5.4
September	223 941	147 415	65.8	139 487	71 744	61 859	7 056	2 251	137 323	62.3	7 927	2 910	5.4
October	224 192	147 793	65.9	139 732	71 860	61 942	7 180	2 216	137 598	62.3	8 061	3 041	5.5
November	224 422	148 162	66.0	140 231	72 110	62 088	7 216	2 206	137 978	62.5	7 932	2 960	5.4
December	224 640	148 059	65.9	140 125	72 058	62 136	7 202	2 171	137 947	62.4	7 934	2 927	5.4
2005													
January	224 837	148 029	65.8	140 245	72 063	62 260	7 071	2 112	138 111	62.4	7 784	2 851	5.3
February	225 041	148 364	65.9	140 385	72 299	62 253	7 073	2 129	138 267	62.4	7 980	2 896	5.4
March	225 236	148 391	65.9	140 654	72 478	62 222	7 185	2 180	138 462	62.4	7 737	2 817	5.2
April	225 441	148 926	66.1	141 254	72 860	62 483	7 190	2 248	138 992	62.7	7 672	2 678	5.2
May	225 670	149 261	66.1	141 609	73 133	62 557	7 201	2 225	139 379	62.8	7 651	2 683	5.1
June	225 911	149 238	66.1	141 714	73 223	62 516	7 140	2 302	139 276	62.7	7 524	2 405	5.0
July	226 153	149 432	66.1	142 026	73 337	62 691	7 150	2 308	139 789	62.8	7 406	2 449	5.0
August	226 421	149 779	66.2	142 434	73 513	62 844	7 240	2 183	140 277	62.9	7 345	2 569	4.9
September	226 693	149 954	66.1	142 401	73 333	63 035	7 141	2 181	140 276	62.8	7 553	2 537	5.0
October	226 959	150 001	66.1	142 548	73 445	63 127	7 124	2 191	140 435	62.8	7 453	2 492	5.0
November	227 204	150 065	66.0	142 499	73 341	63 127	7 265	2 174	140 299	62.7	7 566	2 486	5.0
December	227 425	150 030	66.0	142 752	73 467	63 209	7 139	2 094	140 635	62.8	7 279	2 429	4.9

[1]Not seasonally adjusted.
[2]In 1930 through 1943, the official BLS data count persons on work relief as unemployed. The unemployment rates for those years shown in parentheses count persons on work relief as employed, which is more consistent with the postwar practice. See notes and definitions.

Table 10-1B. Summary Labor Force, Employment, and Unemployment: Historical Data—*Continued*

(Thousands of persons, percent, seasonally adjusted, except as noted.)

Year and month	Civilian noninsti-tutional popu-lation 1	Civilian labor force		Employment, thousands of persons						Employ-ment-population ratio, percent	Unemployment		
					By age and sex			By industry			Thousands of persons		Rate (percent) 2
		Thousands of persons	Participa-tion rate (percent)	Total	Men, 20 years and over	Women, 20 years and over	Both sexes, 16 to 19 years	Agri-cultural	Nonagri-cultural		Total	Unem-ployed 15 weeks and over	
2006													
January	227 553	150 214	66.0	143 150	73 892	63 151	7 197	2 164	140 933	62.9	7 064	2 270	4.7
February	227 763	150 641	66.1	143 457	73 972	63 304	7 298	2 178	141 254	63.0	7 184	2 546	4.8
March	227 975	150 813	66.2	143 741	74 228	63 353	7 342	2 153	141 604	63.1	7 072	2 373	4.7
April	228 199	150 881	66.1	143 761	74 204	63 413	7 198	2 249	141 388	63.0	7 120	2 353	4.7
May	228 428	151 069	66.1	144 089	74 229	63 656	7 215	2 194	141 859	63.1	6 980	2 303	4.6
June	228 671	151 354	66.2	144 353	74 261	63 866	7 393	2 256	142 019	63.1	7 001	2 127	4.6
July	228 912	151 377	66.1	144 202	74 011	64 030	7 324	2 278	142 066	63.0	7 175	2 289	4.7
August	229 167	151 716	66.2	144 625	74 384	64 139	7 266	2 240	142 440	63.1	7 091	2 293	4.7
September	229 420	151 662	66.1	144 815	74 866	63 922	7 196	2 182	142 661	63.1	6 847	2 231	4.5
October	229 675	152 041	66.2	145 314	74 833	64 298	7 289	2 183	143 222	63.3	6 727	2 062	4.4
November	229 905	152 406	66.3	145 534	74 962	64 328	7 331	2 164	143 350	63.3	6 872	2 159	4.5
December	230 108	152 732	66.4	145 970	75 232	64 518	7 287	2 233	143 716	63.4	6 762	2 083	4.4
2007													
January	230 650	153 144	66.4	146 028	75 238	64 621	7 240	2 214	143 785	63.3	7 116	2 156	4.6
February	230 834	152 983	66.3	146 057	75 239	64 750	7 133	2 295	143 741	63.3	6 927	2 211	4.5
March	231 034	153 051	66.2	146 320	75 386	64 928	7 061	2 178	144 153	63.3	6 731	2 255	4.4
April	231 253	152 435	65.9	145 586	75 342	64 366	6 985	2 069	143 385	63.0	6 850	2 281	4.5
May	231 480	152 670	66.0	145 903	75 386	64 724	6 888	2 081	143 773	63.0	6 766	2 231	4.4
June	231 713	153 041	66.0	146 063	75 327	64 788	7 109	1 946	144 112	63.0	6 979	2 281	4.6
July	231 958	153 054	66.0	145 905	75 247	64 766	6 953	2 016	144 041	62.9	7 149	2 364	4.7
August	232 211	152 749	65.8	145 682	75 195	64 838	6 714	1 863	143 865	62.7	7 067	2 333	4.6
September	232 461	153 414	66.0	146 244	75 273	65 086	7 001	2 095	144 148	62.9	7 170	2 355	4.7
October	232 715	153 183	65.8	145 946	75 146	64 842	7 039	2 121	143 926	62.7	7 237	2 300	4.7
November	232 939	153 835	66.0	146 595	75 683	64 985	7 071	2 148	144 413	62.9	7 240	2 366	4.7
December	233 156	153 918	66.0	146 273	75 510	64 910	7 032	2 219	144 018	62.7	7 645	2 501	5.0
2008													
January	232 616	154 063	66.2	146 378	75 551	65 058	7 015	2 205	144 168	62.9	7 685	2 540	5.0
February	232 809	153 653	66.0	146 156	75 450	65 021	6 820	2 190	143 943	62.8	7 497	2 446	4.9
March	232 995	153 908	66.1	146 086	75 294	65 074	6 813	2 172	143 871	62.7	7 822	2 491	5.1
April	233 198	153 769	65.9	146 132	75 182	65 107	6 947	2 115	143 874	62.7	7 637	2 687	5.0
May	233 405	154 303	66.1	145 908	74 953	65 137	7 182	2 113	143 737	62.5	8 395	2 780	5.4
June	233 627	154 313	66.1	145 737	74 975	65 169	6 925	2 127	143 629	62.4	8 575	2 913	5.6
July	233 864	154 469	66.1	145 532	74 950	65 103	6 909	2 134	143 519	62.2	8 937	3 114	5.8
August	234 107	154 641	66.1	145 203	74 658	65 008	6 801	2 148	143 086	62.0	9 438	3 420	6.1
September	234 360	154 570	66.0	145 076	74 456	65 078	6 850	2 231	142 796	61.9	9 494	3 626	6.1
October	234 612	154 876	66.0	144 802	74 263	65 103	6 793	2 199	142 689	61.7	10 074	3 990	6.5
November	234 828	154 639	65.9	144 100	73 943	64 885	6 612	2 207	141 962	61.4	10 538	3 923	6.8
December	235 035	154 655	65.8	143 369	73 311	64 813	6 601	2 209	141 139	61.0	11 286	4 547	7.3
2009													
January	234 739	154 210	65.7	142 152	72 724	64 209	5 218	2 147	140 029	60.6	12 058	4 764	7.8
February	234 913	154 538	65.8	141 640	72 260	64 195	5 185	2 130	139 493	60.3	12 898	5 455	8.3
March	235 086	154 133	65.6	140 707	71 629	64 027	5 051	2 031	138 664	59.9	13 426	5 886	8.7
April	235 271	154 509	65.7	140 656	71 599	64 021	5 036	2 146	138 437	59.8	13 853	6 385	9.0
May	235 452	154 747	65.7	140 248	71 410	63 807	5 031	2 145	138 036	59.6	14 499	7 022	9.4
June	235 655	154 716	65.7	140 009	71 326	63 747	4 937	2 165	137 841	59.4	14 707	7 837	9.5
July	235 870	154 502	65.5	139 901	71 273	63 761	4 867	2 126	137 819	59.3	14 601	7 839	9.5
August	236 087	154 307	65.4	139 492	71 111	63 633	4 749	2 088	137 371	59.1	14 814	7 828	9.6
September	236 322	153 827	65.1	138 818	70 844	63 324	4 649	2 033	136 734	58.7	15 009	8 402	9.8
October	236 550	153 784	65.0	138 432	70 713	63 275	4 445	2 039	136 483	58.5	15 352	8 648	10.0
November	236 743	153 878	65.0	138 659	70 774	63 422	4 463	2 098	136 640	58.6	15 219	8 755	9.9
December	236 924	153 111	64.6	138 013	70 479	63 086	4 449	2 087	135 904	58.3	15 098	8 835	9.9
2010													
January	236 832	153 484	64.8	138 438	70 509	63 498	4 430	2 128	136 384	58.5	15 046	8 894	9.8
February	236 998	153 694	64.9	138 581	70 603	63 482	4 497	2 308	136 327	58.5	15 113	8 922	9.8
March	237 159	153 954	64.9	138 751	70 877	63 378	4 497	2 211	136 579	58.5	15 202	9 087	9.9
April	237 329	154 622	65.2	139 297	71 352	63 415	4 530	2 295	136 958	58.7	15 325	9 130	9.9
May	237 499	154 091	64.9	139 241	71 374	63 435	4 433	2 202	136 984	58.6	14 849	8 897	9.6
June	237 690	153 616	64.6	139 141	71 345	63 555	4 241	2 144	136 936	58.5	14 474	8 907	9.4
July	237 890	153 691	64.6	139 179	71 448	63 420	4 311	2 174	136 927	58.5	14 512	8 714	9.4
August	238 099	154 086	64.7	139 438	71 608	63 426	4 404	2 165	137 193	58.6	14 648	8 421	9.5
September	238 322	153 975	64.6	139 396	71 588	63 541	4 267	2 157	137 220	58.5	14 579	8 454	9.5
October	238 530	153 635	64.4	139 119	71 434	63 391	4 294	2 331	136 893	58.3	14 516	8 613	9.4
November	238 715	154 125	64.6	139 044	71 151	63 503	4 390	2 184	136 915	58.2	15 081	8 696	9.8
December	238 889	153 650	64.3	139 301	71 482	63 515	4 304	2 178	137 123	58.3	14 348	8 549	9.3
2011													
January	238 704	153 263	64.2	139 250	71 546	63 370	4 334	2 264	137 022	58.3	14 013	8 393	9.1
February	238 851	153 214	64.1	139 394	71 819	63 250	4 326	2 264	137 160	58.4	13 820	8 175	9.0
March	239 000	153 376	64.2	139 639	71 824	63 464	4 351	2 259	137 434	58.4	13 737	8 166	9.0
April	239 146	153 543	64.2	139 586	71 926	63 374	4 287	2 150	137 420	58.4	13 957	8 016	9.1
May	239 313	153 479	64.1	139 624	72 099	63 266	4 259	2 240	137 356	58.3	13 855	8 205	9.0
June	239 489	153 346	64.0	139 384	71 970	63 135	4 278	2 262	137 061	58.2	13 962	8 143	9.1
July	239 671	153 288	64.0	139 524	71 984	63 309	4 232	2 233	137 204	58.2	13 763	8 177	9.0
August	239 871	153 760	64.1	139 942	72 231	63 341	4 370	2 357	137 507	58.3	13 818	8 247	9.0
September	240 071	154 131	64.2	140 183	72 410	63 398	4 375	2 225	137 963	58.4	13 948	8 316	9.0
October	240 269	153 961	64.1	140 368	72 435	63 560	4 373	2 197	138 285	58.4	13 594	7 802	8.8
November	240 441	154 128	64.1	140 826	72 926	63 511	4 388	2 240	138 571	58.6	13 302	7 706	8.6
December	240 584	153 995	64.0	140 902	73 084	63 430	4 388	2 347	138 568	58.6	13 093	7 545	8.5

1Not seasonally adjusted.
2In 1930 through 1943, the official BLS data count persons on work relief as unemployed. The unemployment rates for those years shown in parentheses count persons on work relief as employed, which is more consistent with the postwar practice. See notes and definitions.

Table 10-1B. Summary Labor Force, Employment, and Unemployment: Historical Data—*Continued*

(Thousands of persons, percent, seasonally adjusted, except as noted.)

Year and month	Civilian noninsti-tutional popu-lation [1]	Civilian labor force		Employment, thousands of persons						Employ-ment-population ratio, percent	Unemployment		
		Thousands of persons	Participa-tion rate (percent)	Total	By age and sex			By industry			Thousands of persons		Rate (percent) [2]
					Men, 20 years and over	Women, 20 years and over	Both sexes, 16 to 19 years	Agri-cultural	Nonagri-cultural		Total	Unem-ployed 15 weeks and over	
2012													
January	242 269	154 381	63.7	141 584	73 055	64 148	4 382	2 203	139 382	58.4	12 797	7 433	8.3
February	242 435	154 671	63.8	141 858	73 088	64 369	4 400	2 191	139 718	58.5	12 813	7 293	8.3
March	242 604	154 749	63.8	142 036	73 144	64 533	4 359	2 236	139 800	58.5	12 713	7 176	8.2
April	242 784	154 545	63.7	141 899	73 080	64 459	4 359	2 206	139 682	58.4	12 646	7 072	8.2
May	242 966	154 866	63.7	142 206	73 218	64 594	4 394	2 299	139 915	58.5	12 660	7 091	8.2
June	243 155	155 083	63.8	142 391	73 249	64 627	4 515	2 233	140 160	58.6	12 692	7 227	8.2
July	243 354	154 948	63.7	142 292	73 265	64 491	4 536	2 230	140 013	58.5	12 656	6 983	8.2
August	243 566	154 763	63.5	142 291	73 240	64 647	4 404	2 117	140 107	58.4	12 471	6 889	8.1
September	243 772	155 160	63.6	143 044	73 704	64 925	4 415	2 164	140 912	58.7	12 115	6 775	7.8
October	243 983	155 554	63.8	143 431	73 981	64 995	4 455	2 157	141 387	58.8	12 124	6 784	7.8
November	244 174	155 338	63.6	143 333	73 907	64 966	4 460	2 113	141 224	58.7	12 005	6 550	7.7
December	244 350	155 628	63.7	143 330	74 024	64 930	4 376	2 062	141 284	58.7	12 298	6 629	7.9
2013													
January	244 663	155 763	63.7	143 292	74 011	64 758	4 523	2 052	141 124	58.6	12 471	6 572	8.0
February	244 828	155 312	63.4	143 362	74 128	64 879	4 354	2 067	141 299	58.6	11 950	6 457	7.7
March	244 995	155 005	63.3	143 316	74 107	64 873	4 337	1 991	141 192	58.5	11 689	6 393	7.5
April	245 175	155 394	63.4	143 635	74 082	65 205	4 348	2 041	141 557	58.6	11 760	6 421	7.6
May	245 363	155 536	63.4	143 882	74 111	65 342	4 429	2 102	141 758	58.6	11 654	6 333	7.5
June	245 552	155 749	63.4	143 999	74 203	65 318	4 477	2 106	141 896	58.6	11 751	6 231	7.5
July	245 756	155 599	63.3	144 264	74 254	65 511	4 499	2 209	142 079	58.7	11 335	6 049	7.3
August	245 959	155 605	63.3	144 326	74 085	65 766	4 474	2 230	142 154	58.7	11 279	5 973	7.2
September	246 168	155 687	63.2	144 418	74 211	65 617	4 590	2 190	142 304	58.7	11 270	5 929	7.2
October	246 381	154 673	62.8	143 537	73 882	65 215	4 440	2 182	141 411	58.3	11 136	5 761	7.2
November	246 567	155 265	63.0	144 479	74 540	65 404	4 535	2 121	142 463	58.6	10 787	5 737	6.9
December	246 745	155 182	62.9	144 778	74 628	65 675	4 474	2 232	142 561	58.7	10 404	5 539	6.7
2014													
January	246 915	155 352	62.9	145 150	74 799	65 956	4 396	2 171	142 852	58.8	10 202	5 304	6.6
February	247 085	155 483	62.9	145 134	74 680	66 193	4 261	2 135	143 024	58.7	10 349	5 404	6.7
March	247 258	156 028	63.1	145 648	75 157	66 017	4 474	2 084	143 393	58.9	10 380	5 415	6.7
April	247 439	155 369	62.8	145 667	75 057	66 144	4 467	2 140	143 512	58.9	9 702	4 981	6.2
May	247 622	155 684	62.9	145 825	75 112	66 179	4 534	2 046	143 778	58.9	9 859	4 785	6.3
June	247 814	155 707	62.8	146 247	75 481	66 308	4 458	2 162	144 084	59.0	9 460	4 619	6.1
July	248 023	156 007	62.9	146 399	75 636	66 218	4 545	2 199	144 188	59.0	9 608	4 613	6.2
August	248 229	156 130	62.9	146 530	75 745	66 259	4 527	2 305	144 303	59.0	9 599	4 464	6.1
September	248 446	156 040	62.8	146 778	76 021	66 223	4 534	2 387	144 502	59.1	9 262	4 386	5.9
October	248 657	156 417	62.9	147 427	76 056	66 559	4 812	2 425	144 989	59.3	8 990	4 292	5.7
November	248 844	156 494	62.9	147 404	75 837	66 753	4 815	2 397	145 113	59.2	9 090	4 228	5.8
December	249 027	156 332	62.8	147 615	76 209	66 639	4 767	2 379	145 229	59.3	8 717	4 069	5.6
2015													
January	249 723	157 030	62.9	148 145	76 359	67 072	4 714	2 428	145 614	59.3	8 885	4 151	5.7
February	249 899	156 644	62.7	148 045	76 407	66 889	4 749	2 413	145 626	59.2	8 599	3 991	5.5
March	250 080	156 643	62.6	148 128	76 504	66 870	4 754	2 478	145 458	59.2	8 515	3 787	5.4
April	250 266	157 060	62.8	148 511	76 705	67 038	4 767	2 390	146 069	59.3	8 550	3 651	5.4
May	250 455	157 651	62.9	148 817	76 785	67 241	4 790	2 376	146 466	59.4	8 834	3 759	5.6
June	250 663	157 062	62.7	148 816	76 803	67 385	4 627	2 561	146 258	59.4	8 247	3 560	5.3
July	250 876	156 997	62.6	148 830	76 929	67 222	4 678	2 401	146 368	59.3	8 167	3 389	5.2
August	251 096	157 172	62.6	149 181	76 979	67 527	4 675	2 390	146 852	59.4	7 992	3 454	5.1
September	251 325	156 733	62.4	148 826	76 875	67 259	4 692	2 377	146 588	59.2	7 907	3 332	5.0
October	251 541	157 167	62.5	149 246	76 958	67 546	4 741	2 429	146 845	59.3	7 922	3 342	5.0
November	251 747	157 463	62.5	149 463	76 856	67 848	4 758	2 404	147 200	59.4	8 000	3 347	5.1
December	251 936	158 035	62.7	150 128	77 266	67 949	4 913	2 428	147 674	59.6	7 907	3 359	5.0
2016													
January	252 397	158 342	62.7	150 621	77 608	68 095	4 918	2 377	148 158	59.7	7 721	3 195	4.9
February	252 577	158 653	62.8	150 908	77 825	68 108	4 975	2 445	148 420	59.7	7 746	3 283	4.9
March	252 768	159 103	62.9	151 157	77 975	68 288	4 894	2 555	148 383	59.8	7 945	3 390	5.0
April	252 969	158 981	62.8	151 006	77 952	68 106	4 948	2 580	148 397	59.7	7 975	3 355	5.0
May	253 174	158 787	62.7	151 119	77 864	68 303	4 952	2 559	148 593	59.7	7 668	3 034	4.8
June	253 397	158 973	62.7	151 187	78 194	68 181	4 812	2 528	148 703	59.7	7 786	3 166	4.9
July	253 620	159 123	62.7	151 465	78 084	68 414	4 966	2 414	149 027	59.7	7 658	3 185	4.8
August	253 854	159 579	62.9	151 770	78 259	68 433	5 077	2 584	149 288	59.8	7 809	3 062	4.9
September	254 091	159 817	62.9	151 850	78 246	68 562	5 042	2 439	149 494	59.8	7 967	3 108	5.0
October	254 321	159 734	62.8	151 907	78 274	68 654	4 979	2 327	149 587	59.7	7 827	3 130	4.9
November	254 540	159 551	62.7	152 063	78 358	68 680	5 025	2 387	149 783	59.7	7 488	2 958	4.7
December	254 742	159 710	62.7	152 216	78 446	68 771	5 000	2 311	149 821	59.8	7 495	3 065	4.7
2017													
January	254 082	159 647	62.8	152 129	78 439	68 640	5 050	2 388	149 719	59.9	7 518	3 020	4.7
February	254 246	159 767	62.8	152 368	78 405	68 942	5 021	2 423	149 904	59.9	7 399	2 815	4.6
March	254 414	160 066	62.9	152 978	78 471	69 368	5 139	2 506	150 282	60.1	7 088	2 760	4.4
April	254 588	160 309	63.0	153 224	78 828	69 264	5 132	2 696	150 503	60.2	7 085	2 726	4.4
May	254 767	160 060	62.8	153 001	78 725	69 211	5 065	2 502	150 548	60.1	7 059	2 814	4.4
June	254 957	160 232	62.8	153 299	78 822	69 302	5 175	2 491	150 881	60.1	6 933	2 639	4.3
July	255 151	160 339	62.8	153 471	78 856	69 555	5 061	2 338	151 126	60.1	6 867	2 744	4.3
August	255 357	160 690	62.9	153 593	78 979	69 507	5 107	2 406	151 295	60.1	7 097	2 790	4.4
September	255 562	161 212	63.1	154 371	79 461	69 651	5 259	2 293	152 085	60.4	6 841	2 676	4.2
October	255 766	160 378	62.7	153 779	79 286	69 496	4 997	2 480	151 287	60.1	6 599	2 481	4.1
November	255 949	160 510	62.7	153 813	79 296	69 622	4 895	2 455	151 448	60.1	6 697	2 564	4.2
December	256 109	160 538	62.7	153 977	79 491	69 516	4 970	2 491	151 420	60.1	6 561	2 406	4.1

[1]Not seasonally adjusted.
[2]In 1930 through 1943, the official BLS data count persons on work relief as unemployed. The unemployment rates for those years shown in parentheses count persons on work relief as employed, which is more consistent with the postwar practice. See notes and definitions.

Table 10-2. Labor Force and Employment by Major Age and Sex Groups

(Thousands of persons, percent, seasonally adjusted.)

Year and month	Civilian labor force (thousands)			Participation rate (percent)			Employment (thousands)			Employment-population ratio, percent		
	Men, 20 years and over	Women, 20 years and over	Both sexes, 16 to 19 years	Men, 20 years and over	Women, 20 years and over	Both sexes, 16 to 19 years	Men, 20 years and over	Women, 20 years and over	Both sexes, 16 to 19 years	Men, 20 years and over	Women, 20 years and over	Both sexes, 16 to 19 years
1970	47 220	28 301	7 249	82.6	43.3	49.9	45 581	26 952	6 144	79.7	41.2	42.3
1971	48 009	28 904	7 470	82.1	43.3	49.7	45 912	27 246	6 208	78.5	40.9	41.3
1972	49 079	29 901	8 054	81.6	43.7	51.9	47 130	28 276	6 746	78.4	41.3	43.5
1973	49 932	30 991	8 507	81.3	44.4	53.7	48 310	29 484	7 271	78.6	42.2	45.9
1974	50 879	32 201	8 871	81.0	45.3	54.8	48 922	30 424	7 448	77.9	42.8	46.0
1975	51 494	33 410	8 870	80.3	46.0	54.0	48 018	30 726	7 104	74.8	42.3	43.3
1976	52 288	34 814	9 056	79.8	47.0	54.5	49 190	32 226	7 336	75.1	43.5	44.2
1977	53 348	36 310	9 351	79.7	48.1	56.0	50 555	33 775	7 688	75.6	44.8	46.1
1978	54 471	38 128	9 652	79.8	49.6	57.8	52 143	35 836	8 070	76.4	46.6	48.3
1979	55 615	39 708	9 638	79.8	50.6	57.9	53 308	37 434	8 083	76.5	47.7	48.5
1980	56 455	41 106	9 378	79.4	51.3	56.7	53 101	38 492	7 710	74.6	48.1	46.6
1981	57 197	42 485	8 988	79.0	52.1	55.4	53 582	39 590	7 225	74.0	48.6	44.6
1982	57 980	43 699	8 526	78.7	52.7	54.1	52 891	40 086	6 549	71.8	48.4	41.5
1983	58 744	44 636	8 171	78.5	53.1	53.5	53 487	41 004	6 342	71.4	48.8	41.5
1984	59 701	45 900	7 943	78.3	53.7	53.9	55 769	42 793	6 444	73.2	50.1	43.7
1985	60 277	47 283	7 901	78.1	54.7	54.5	56 562	44 154	6 434	73.3	51.0	44.4
1986	61 320	48 589	7 926	78.1	55.5	54.7	57 569	45 556	6 472	73.3	52.0	44.6
1987	62 095	49 783	7 988	78.0	56.2	54.7	58 726	47 074	6 640	73.8	53.1	45.5
1988	62 768	50 870	8 031	77.9	56.8	55.3	59 781	48 383	6 805	74.2	54.0	46.8
1989	63 704	52 212	7 954	78.1	57.7	55.9	60 837	49 745	6 759	74.5	54.9	47.5
1990	64 916	53 131	7 792	78.2	58.0	53.7	61 678	50 535	6 581	74.3	55.2	45.3
1991	65 374	53 708	7 265	77.7	57.9	51.6	61 178	50 634	5 906	72.7	54.6	42.0
1992	66 213	54 796	7 096	77.7	58.5	51.3	61 496	51 328	5 669	72.1	54.8	41.0
1993	66 642	55 388	7 170	77.3	58.5	51.5	62 355	52 099	5 805	72.3	55.0	41.7
1994	66 921	56 655	7 481	76.8	59.3	52.7	63 294	53 606	6 161	72.6	56.2	43.4
1995	67 324	57 215	7 765	76.7	59.4	53.5	64 085	54 396	6 419	73.0	56.5	44.2
1996	68 044	58 094	7 806	76.8	59.9	52.3	64 897	55 311	6 500	73.2	57.0	43.5
1997	69 166	59 198	7 932	77.0	60.5	51.6	66 284	56 613	6 661	73.7	57.8	43.4
1998	69 715	59 702	8 256	76.8	60.4	52.8	67 135	57 278	7 051	73.9	58.0	45.1
1999	70 194	60 840	8 333	76.7	60.7	52.0	67 761	58 555	7 172	74.0	58.5	44.7
2000	72 010	62 301	8 271	76.7	60.6	52.0	69 634	60 067	7 189	74.2	58.4	45.2
2001	72 816	63 016	7 902	76.5	60.6	49.6	69 776	60 417	6 740	73.3	58.1	42.3
2002	73 630	63 648	7 585	76.3	60.5	47.4	69 734	60 420	6 332	72.3	57.5	39.6
2003	74 623	64 716	7 170	75.9	60.6	44.5	70 415	61 402	5 919	71.7	57.5	36.8
2004	75 364	64 923	7 114	75.8	60.3	43.9	71 572	61 773	5 907	71.9	57.4	36.4
2005	76 443	65 714	7 164	75.8	60.4	43.7	73 050	62 702	5 978	72.4	57.6	36.5
2006	77 562	66 585	7 281	75.9	60.5	43.7	74 431	63 834	6 162	72.9	58.0	36.9
2007	78 596	67 516	7 012	75.9	60.6	41.3	75 337	64 799	5 911	72.8	58.2	34.8
2008	79 047	68 382	6 858	75.7	60.9	40.2	74 750	65 039	5 573	71.6	57.9	32.6
2009	78 897	68 856	6 390	74.8	60.8	37.5	71 341	63 699	4 837	67.6	56.2	28.4
2010	78 994	68 990	5 906	74.1	60.3	34.9	71 230	63 456	4 378	66.8	55.5	25.9
2011	79 080	68 810	5 727	73.4	59.8	34.1	72 182	63 360	4 327	67.0	55.0	25.8
2012	79 387	69 765	5 823	73.0	59.3	34.3	73 403	64 640	4 426	67.5	55.0	26.1
2013	79 744	69 860	5 785	72.5	58.8	34.5	74 176	65 295	4 458	67.4	54.9	26.6
2014	80 056	70 212	5 654	71.9	58.5	34.0	75 471	66 287	4 548	67.8	55.2	27.3
2015	80 735	70 695	5 700	71.7	58.2	34.3	76 776	67 323	4 734	68.1	55.4	28.5
2016	81 759	71 538	5 889	71.7	58.3	35.2	78 074	68 387	4 965	68.5	55.7	29.7
2017	82 198	72 293	5 904	71.6	56.1	35.2	78 920	69 343	5 074	68.7	56.1	30.3
2018	83 183	72 997	5 890	71.6	58.5	35.1	80 212	70 422	5 130	69.0	56.4	30.6
2019	83 728	73 899	5 891	71.6	58.9	35.3	80 917	71 467	5 145	69.2	56.9	30.8
2018												
January	82 881	72 236	5 950	71.7	58.1	35.5	79 723	69 628	5 135	68.9	56.0	30.6
February	83 210	72 527	6 046	71.9	58.3	36.0	80 138	69 807	5 198	69.2	56.1	31.0
March	83 135	72 636	5 912	71.8	58.3	35.2	80 092	69 979	5 120	69.1	56.2	30.5
April	83 242	72 631	5 870	71.8	58.3	35.0	80 140	70 066	5 118	69.1	56.2	30.5
May	83 267	72 738	5 869	71.8	58.3	35.0	80 275	70 270	5 119	69.2	56.4	30.5
June	83 195	73 197	5 876	71.6	58.7	35.1	80 084	70 528	5 137	69.0	56.5	30.7
July	83 003	73 292	5 878	71.4	58.7	35.1	80 208	70 671	5 114	69.0	56.6	30.5
August	83 040	73 126	5 603	71.4	58.5	33.4	80 160	70 553	4 888	68.9	56.5	29.2
September	83 127	73 093	5 858	71.4	58.4	34.9	80 259	70 657	5 116	68.9	56.5	30.5
October	83 280	73 376	5 949	71.5	58.6	35.5	80 400	70 858	5 224	69.0	56.6	31.2
November	83 341	73 423	5 899	71.4	58.6	35.2	80 567	70 892	5 169	69.1	56.6	30.8
December	83 483	73 673	5 955	71.5	58.8	35.5	80 496	71 123	5 205	69.0	56.7	31.1
2019												
January	83 586	73 643	5 913	71.8	58.9	35.4	80 474	71 004	5 149	69.1	56.8	30.8
February	83 588	73 667	5 792	71.7	58.8	34.7	80 677	71 169	5 019	69.2	56.9	30.1
March	83 566	73 508	5 862	71.7	58.7	35.1	80 570	71 056	5 115	69.1	56.7	30.6
April	83 421	73 440	5 685	71.5	58.6	34.1	80 609	71 136	4 951	69.1	56.8	29.7
May	83 569	73 439	5 774	71.6	58.6	34.6	80 761	71 038	5 044	69.2	56.6	30.2
June	83 568	73 655	5 910	71.5	58.7	35.4	80 780	71 209	5 159	69.1	56.7	30.9
July	83 771	73 585	6 017	71.6	58.6	36.1	80 975	71 120	5 250	69.2	56.6	31.5
August	83 852	74 116	5 926	71.6	59.0	35.5	81 046	71 665	5 184	69.2	57.0	31.1
September	83 841	74 313	5 897	71.6	59.1	35.3	81 146	71 990	5 162	69.3	57.2	30.9
October	83 911	74 542	5 948	71.6	59.2	35.6	81 196	72 130	5 218	69.3	57.3	31.3
November	84 057	74 291	5 999	71.6	59.0	35.9	81 377	71 881	5 278	69.4	57.0	31.6
December	84 008	74 584	5 964	71.5	59.2	35.7	81 390	72 200	5 213	69.3	57.3	31.2

Table 10-3. Employment by Type of Job

(Thousands of persons, seasonally adjusted, except as noted.)

Year and month	Agricultural	By class of worker — Nonagricultural industries — Total	Wage and salary — Total	Government	Private industries — Private households [1]	Other private industries	Self-employed (unincorporated)	Unpaid family workers [1]	Multiple jobholders — Total (thousands)	Percent of total employed	Employed and at work part-time — Economic reasons	Non-economic reasons
1970	3 463	75 215	69 491	12 431	...	...	5 221	502	...	...	2 446	9 999
1971	3 394	75 972	70 120	12 799	...	...	5 327	522	...	...	2 688	10 152
1972	3 484	78 669	72 785	13 393	...	...	5 365	519	...	...	2 648	10 612
1973	3 470	81 594	75 580	13 655	...	...	5 474	540	...	...	2 554	10 972
1974	3 515	83 279	77 094	14 124	...	...	5 697	489	...	...	2 988	11 153
1975	3 408	82 438	76 249	14 675	...	...	5 705	483	...	...	3 804	11 228
1976	3 331	85 421	79 175	15 132	...	...	5 783	464	...	...	3 607	11 607
1977	3 283	88 734	82 121	15 361	...	...	6 114	498	...	...	3 608	12 120
1978	3 387	92 661	85 753	15 525	...	...	6 429	479	...	...	3 516	12 650
1979	3 347	95 477	88 222	15 635	...	...	6 791	463	...	...	3 577	12 893
1980	3 364	95 938	88 525	15 912	...	...	7 000	413	...	...	4 321	13 067
1981	3 368	97 030	89 543	15 689	...	...	7 097	390	...	...	4 768	13 025
1982	3 401	96 125	88 462	15 516	...	...	7 262	401	...	...	6 170	12 953
1983	3 383	97 450	89 500	15 537	...	...	7 575	376	...	...	6 266	12 911
1984	3 321	101 685	93 565	15 770	...	...	7 785	335	...	...	5 744	13 169
1985	3 179	103 971	95 871	16 031	...	...	7 811	289	...	...	5 590	13 489
1986	3 163	106 434	98 299	16 342	...	...	7 881	255	...	...	5 588	13 935
1987	3 208	109 232	100 771	16 800	...	...	8 201	260	...	...	5 401	14 395
1988	3 169	111 800	103 021	17 114	...	...	8 519	260	...	...	5 206	14 963
1989	3 199	114 142	105 259	17 469	...	...	8 605	279	...	...	4 894	15 393
1990	3 223	115 570	106 598	17 769	...	...	8 719	253	...	...	5 204	15 341
1991	3 269	114 449	105 373	17 934	...	...	8 851	226	...	...	6 161	15 172
1992	3 247	115 245	106 437	18 136	...	...	8 575	233	...	...	6 520	14 918
1993	3 115	117 144	107 966	18 579	...	...	8 959	218	...	...	6 481	15 240
1994	3 409	119 651	110 517	18 293	...	...	9 003	131	7 260	5.9	4 625	17 638
1995	3 440	121 460	112 448	18 362	...	...	8 902	110	7 693	6.2	4 473	17 734
1996	3 443	123 264	114 171	18 217	...	...	8 971	122	7 832	6.2	4 315	17 770
1997	3 399	126 159	116 983	18 131	...	...	9 056	120	7 955	6.1	4 068	18 149
1998	3 378	128 085	119 019	18 383	...	...	8 962	103	7 926	6.0	3 665	18 530
1999	3 281	130 207	121 323	18 903	...	...	8 790	95	7 802	5.8	3 357	18 758
2000	2 464	134 427	125 114	19 248	718	105 148	9 205	108	7 604	5.6	3 227	18 814
2001	2 299	134 635	125 407	19 335	694	105 378	9 121	107	7 357	5.4	3 715	18 790
2002	2 311	134 174	125 156	19 636	757	104 764	8 923	95	7 291	5.3	4 213	18 843
2003	2 275	135 461	126 015	19 634	764	105 616	9 344	101	7 315	5.3	4 701	19 014
2004	2 232	137 020	127 463	19 983	779	106 701	9 467	90	7 473	5.4	4 567	19 380
2005	2 197	139 532	129 931	20 357	812	108 761	9 509	93	7 546	5.3	4 350	19 491
2006	2 206	142 221	132 449	20 337	803	111 309	9 685	87	7 576	5.2	4 162	19 591
2007	2 095	143 952	134 283	21 003	813	112 467	9 557	112	7 655	5.2	4 401	19 756
2008	2 168	143 194	133 882	21 258	805	111 819	9 219	93	7 620	5.2	5 875	19 343
2009	2 103	137 775	128 713	21 178	783	106 752	8 995	66	7 271	5.2	8 913	18 710
2010	2 206	136 858	127 914	21 003	667	106 244	8 860	84	6 878	4.9	8 874	18 251
2011	2 254	137 615	128 934	20 536	722	107 676	8 603	78	6 880	4.9	8 560	18 334
2012	2 186	140 283	131 452	20 360	738	110 355	8 749	81	6 943	4.9	8 122	18 806
2013	2 130	141 799	133 111	20 247	723	112 141	8 619	70	7 002	4.9	7 935	18 903
2014	2 237	144 068	135 402	20 135	820	114 456	8 602	64	7 146	4.9	7 213	19 489
2015	2 422	146 411	137 678	20 601	798	116 279	8 665	68	7 262	4.9	6 371	20 018
2016	2 460	148 975	140 161	20 630	724	118 807	8 715	65	7 531	5.0	5 843	20 680
2017	2 457	150 875	142 090	20 834	657	120 600	8 734	52	7 545	4.9	5 186	20 557
2018	2 425	153 336	144 326	20 942	777	122 607	8 941	69	7 769	5.0	6 788	19 370
2019	2 425	155 106	146 256	20 976	821	124 462	8 799	53	8 050	5.1	4 329	21 058
2018												
January	2 443	152 053	143 005	20 893	701	121 388	8 838	47	7 838	5.1	4 951	21 112
February	2 430	152 659	143 623	21 048	738	121 713	8 854	56	7 955	5.1	5 097	21 165
March	2 340	152 714	143 672	20 991	781	121 903	8 959	50	7 613	4.9	4 981	21 438
April	2 330	153 007	143 865	21 132	780	121 987	9 039	70	7 721	5.0	4 969	21 299
May	2 353	153 353	144 254	21 029	773	122 399	9 112	61	7 481	4.8	4 952	21 176
June	2 398	153 383	144 542	20 917	769	122 872	8 808	93	7 639	4.9	4 762	21 305
July	2 483	153 519	144 520	20 912	800	122 820	8 911	66	7 929	5.1	4 619	21 500
August	2 377	153 329	144 392	20 809	782	122 797	8 812	79	7 848	5.0	4 356	21 562
September	2 487	153 528	144 416	20 732	741	122 952	8 970	96	7 646	4.9	4 654	21 387
October	2 407	153 989	144 962	21 141	769	123 200	9 055	88	7 883	5.0	4 599	21 444
November	2 549	154 102	145 197	21 001	811	123 418	8 961	69	7 750	4.9	4 754	20 964
December	2 491	154 266	145 366	20 654	879	123 795	8 973	52	7 918	5.0	4 655	21 230
2019												
January	2 546	154 112	145 259	20 614	799	123 811	8 672	44	7 843	5.0	5 105	20 984
February	2 488	154 354	145 554	20 616	796	124 045	8 638	38	7 683	4.9	4 302	21 196
March	2 336	154 346	145 578	20 706	808	124 057	8 659	45	7 908	5.0	4 517	21 332
April	2 389	154 369	145 694	20 848	825	124 046	8 543	37	7 849	5.0	4 706	21 356
May	2 423	154 486	145 718	20 776	807	124 090	8 786	31	7 937	5.1	4 375	21 415
June	2 330	154 835	146 003	20 979	849	124 210	8 850	19	8 174	5.2	4 350	21 538
July	2 400	155 035	146 004	20 877	808	124 303	8 995	46	8 374	5.3	3 973	21 448
August	2 414	155 546	146 586	20 968	869	124 731	8 863	76	8 335	5.3	4 381	21 673
September	2 416	155 816	146 852	21 144	842	124 864	8 889	80	8 312	5.3	4 336	21 559
October	2 473	155 970	147 118	21 513	784	124 979	8 926	66	8 132	5.1	4 397	21 544
November	2 356	156 167	147 275	21 342	824	125 157	8 954	83	8 107	5.1	4 288	21 532
December	2 533	156 241	147 431	21 323	841	125 250	8 809	70	7 946	5.0	4 148	21 586

[1]Not seasonally adjusted.
. . . = Not available.

Figure 10-2. Unemployment Rate by Age and Gender, 1970–2019

- The unemployment rate peaked at 10.0 percent in October 2009 (lagging, as it usually does, the business cycle trough, which was in June 2009). This was not nearly as high as the unemployment rate during the Great Depression when it reached 24.9 percent in 1933. Recovery from the Great Recession has been sluggish as shown in the Great Recession and Recovery article in the beginning of this publication. Unemployment declined nearly 50 percent between 2010 to 2019. As discussed in the previous section, a low unemployment rate can mask structural weakness in the labor force. (Tables 10-04 and 10-05)

- Women typically have experienced higher unemployment rates and lower earnings since the surge of women entering into the workforce in the 1970s. (Table 10-4).

- Recovery from the Great Recession has been uneven by race and ethnicity. Between 2010 and 2019 the unemployment rate for Blacks declined from 16.0 percent to 7.7 percent while the unemployment rate for Whites declined from 8.7 percent to 4.3 percent, and the unemployment rate for Asians declined from 7.5 percent to 3.7 percent. (Table 10-4)

- Unemployment compensation, established with the Social Security Act of 1935, is a federal-state program that provides financial assistance to recently unemployed workers. Rising numbers of unemployed workers can trigger longer periods of unemployment compensation where no congressional action is needed. This is called an automatic stabilizer. Due to COVID-19, nearly 44 million unemployment compensation claims were filed between mid-March and the first week of June in 2020.

Table 10-4. Unemployment by Demographic Group

(Unemployment in thousands of persons and as a percent of the civilian labor force in group, seasonally adjusted, except as noted.)

Year and month	Unemployment (thousands of persons)				Unemployment rate (percent)							
	Total	Men, 20 years and over	Women, 20 years and over	Both sexes, 16 to 19 years	All civilian workers	By age and sex			By race			Hispanic or Latino ethnicity
						Men, 20 years and over	Women, 20 years and over	Both sexes, 16 to 19 years	White	Black or African American	Asian [1]	
1970	4 093	1 638	1 349	1 106	4.9	3.5	4.8	15.3	4.5	. . .	. . .	. . .
1971	5 016	2 097	1 658	1 262	5.9	4.4	5.7	16.9	5.4	. . .	. . .	. . .
1972	4 882	1 948	1 625	1 308	5.6	4.0	5.4	16.2	5.1	10.4	. . .	. . .
1973	4 365	1 624	1 507	1 235	4.9	3.3	4.9	14.5	4.3	9.4	. . .	7.5
1974	5 156	1 957	1 777	1 422	5.6	3.8	5.5	16.0	5.0	10.5	. . .	8.1
1975	7 929	3 476	2 684	1 767	8.5	6.8	8.0	19.9	7.8	14.8	. . .	12.2
1976	7 406	3 098	2 588	1 719	7.7	5.9	7.4	19.0	7.0	14.0	. . .	11.5
1977	6 991	2 794	2 535	1 663	7.1	5.2	7.0	17.8	6.2	14.0	. . .	10.1
1978	6 202	2 328	2 292	1 583	6.1	4.3	6.0	16.4	5.2	12.8	. . .	9.1
1979	6 137	2 308	2 276	1 555	5.8	4.2	5.7	16.1	5.1	12.3	. . .	8.3
1980	7 637	3 353	2 615	1 669	7.1	5.9	6.4	17.8	6.3	14.3	. . .	10.1
1981	8 273	3 615	2 895	1 763	7.6	6.3	6.8	19.6	6.7	15.6	. . .	10.4
1982	10 678	5 089	3 613	1 977	9.7	8.8	8.3	23.2	8.6	18.9	. . .	13.8
1983	10 717	5 257	3 632	1 829	9.6	8.9	8.1	22.4	8.4	19.5	. . .	13.7
1984	8 539	3 932	3 107	1 499	7.5	6.6	6.8	18.9	6.5	15.9	. . .	10.7
1985	8 312	3 715	3 129	1 468	7.2	6.2	6.6	18.6	6.2	15.1	. . .	10.5
1986	8 237	3 751	3 032	1 454	7.0	6.1	6.2	18.3	6.0	14.5	. . .	10.6
1987	7 425	3 369	2 709	1 347	6.2	5.4	5.4	16.9	5.3	13.0	. . .	8.8
1988	6 701	2 987	2 487	1 226	5.5	4.8	4.9	15.3	4.7	11.7	. . .	8.2
1989	6 528	2 867	2 467	1 194	5.3	4.5	4.7	15.0	4.5	11.4	. . .	8.0
1990	7 047	3 239	2 596	1 212	5.6	5.0	4.9	15.5	4.8	11.4	. . .	8.2
1991	8 628	4 195	3 074	1 359	6.8	6.4	5.7	18.7	6.1	12.5	. . .	10.0
1992	9 613	4 717	3 469	1 427	7.5	7.1	6.3	20.1	6.6	14.2	. . .	11.6
1993	8 940	4 287	3 288	1 365	6.9	6.4	5.9	19.0	6.1	13.0	. . .	10.8
1994	7 996	3 627	3 049	1 320	6.1	5.4	5.4	17.6	5.3	11.5	. . .	9.9
1995	7 404	3 239	2 819	1 346	5.6	4.8	4.9	17.3	4.9	10.4	. . .	9.3
1996	7 236	3 146	2 783	1 306	5.4	4.6	4.8	16.7	4.7	10.5	. . .	8.9
1997	6 739	2 882	2 585	1 271	4.9	4.2	4.4	16.0	4.2	10.0	. . .	7.7
1998	6 210	2 580	2 424	1 205	4.5	3.7	4.1	14.6	3.9	8.9	. . .	7.2
1999	5 880	2 433	2 285	1 162	4.2	3.5	3.8	13.9	3.7	8.0	. . .	6.4
2000	5 692	2 376	2 235	1 081	4.0	3.3	3.6	13.1	3.5	7.6	3.6	5.7
2001	6 801	3 040	2 599	1 162	4.7	4.2	4.1	14.7	4.2	8.6	4.5	6.6
2002	8 378	3 896	3 228	1 253	5.8	5.3	5.1	16.5	5.1	10.2	5.9	7.5
2003	8 774	4 209	3 314	1 251	6.0	5.6	5.1	17.5	5.2	10.8	6.0	7.7
2004	8 149	3 791	3 150	1 208	5.5	5.0	4.9	17.0	4.8	10.4	4.4	7.0
2005	7 591	3 392	3 013	1 186	5.1	4.4	4.6	16.6	4.4	10.0	4.0	6.0
2006	7 001	3 131	2 751	1 119	4.6	4.0	4.1	15.4	4.0	8.9	3.0	5.2
2007	7 078	3 259	2 718	1 101	4.6	4.1	4.0	15.7	4.1	8.3	3.2	5.6
2008	8 924	4 297	3 342	1 285	5.8	5.4	4.9	18.7	5.2	10.1	4.0	7.6
2009	14 265	7 555	5 157	1 552	9.3	9.6	7.5	24.3	8.5	14.8	7.3	12.1
2010	14 825	7 763	5 534	1 528	9.6	9.8	8.0	25.9	8.7	16.0	7.5	12.5
2011	13 747	6 898	5 450	1 400	8.9	8.7	7.9	24.4	7.9	15.8	7.0	11.5
2012	12 506	5 984	5 125	1 397	8.1	7.5	7.3	24.0	7.2	13.8	5.9	10.3
2013	11 460	5 568	4 565	1 327	7.4	7.0	6.5	22.9	6.5	13.1	5.2	9.1
2014	9 617	4 585	3 926	1 106	6.2	5.7	5.6	19.6	5.3	11.3	5.0	7.4
2015	8 296	3 959	3 371	966	5.3	4.9	4.8	16.9	4.6	9.6	3.8	6.6
2016	7 751	3 675	3 151	925	4.9	4.5	5.5	15.7	4.3	8.4	3.6	5.8
2017	6 973	3 220	2 865	831	4.4	4.0	4.0	14.1	3.8	7.5	3.4	5.1
2018	6 306	2 971	2 575	760	3.9	3.6	3.5	12.9	3.5	6.5	3.0	4.7
2019	7 565	3 565	3 128	872	4.7	4.3	4.4	14.7	4.3	7.7	3.7	5.8
2018												
January	6 582	3 159	2 608	815	4.1	3.8	3.6	13.7	3.5	7.5	3.0	5.0
February	6 641	3 072	2 720	849	4.1	3.7	3.8	14.0	3.7	6.7	2.9	4.9
March	6 493	3 043	2 657	793	4.0	3.7	3.7	13.4	3.6	6.8	3.1	5.0
April	6 418	3 101	2 565	752	4.0	3.7	3.5	12.8	3.6	6.4	2.8	4.8
May	6 209	2 992	2 468	750	3.8	3.6	3.4	12.8	3.5	5.9	2.2	4.9
June	6 519	3 111	2 669	739	4.0	3.7	3.6	12.6	3.5	6.5	3.2	4.5
July	6 180	2 795	2 621	764	3.8	3.4	3.6	13.0	3.3	6.6	3.1	4.4
August	6 167	2 880	2 572	715	3.8	3.5	3.5	12.8	3.4	6.3	3.0	4.7
September	6 045	2 868	2 435	742	3.7	3.4	3.3	12.7	3.3	6.2	3.5	4.6
October	6 123	2 880	2 518	725	3.8	3.5	3.4	12.2	3.3	6.3	3.1	4.4
November	6 034	2 774	2 530	729	3.7	3.3	3.4	12.4	3.4	6.1	2.8	4.5
December	6 286	2 987	2 550	750	3.9	3.6	3.5	12.6	3.4	6.6	3.3	4.4
2019												
January	6 516	3 112	2 639	765	4.0	3.7	3.6	12.9	3.5	6.8	3.1	4.8
February	6 181	2 911	2 497	773	3.8	3.5	3.4	13.3	3.3	6.9	3.1	4.3
March	6 194	2 995	2 451	747	3.8	3.6	3.3	12.7	3.4	6.6	3.1	4.7
April	5 850	2 812	2 304	734	3.6	3.4	3.1	12.9	3.1	6.6	2.2	4.2
May	5 938	2 808	2 401	730	3.6	3.4	3.3	12.6	3.3	6.2	2.5	4.2
June	5 985	2 788	2 447	751	3.7	3.3	3.3	12.7	3.3	6.0	2.1	4.3
July	6 027	2 796	2 465	767	3.7	3.3	3.3	12.7	3.3	5.9	2.8	4.5
August	5 999	2 806	2 451	742	3.7	3.3	3.3	12.5	3.4	5.4	2.8	4.2
September	5 753	2 695	2 323	735	3.5	3.2	3.1	12.5	3.2	5.5	2.5	3.9
October	5 857	2 715	2 411	730	3.6	3.2	3.2	12.3	3.2	5.5	2.8	4.1
November	5 811	2 679	2 411	721	3.5	3.2	3.2	12.0	3.2	5.6	2.6	4.2
December	5 753	2 618	2 383	752	3.5	3.1	3.2	12.6	3.2	5.9	2.5	4.2

[1] Not seasonally adjusted.
. . . = Not available.

Table 10-5. Unemployment Rates and Related Data

(Seasonally adjusted, except as noted.)

Year and month	Unemployment rates by marital status (percent of labor force in group)			Unemployment rates by reason for unemployment (percent of total civilian labor force in group.)					Duration of unemployment		Alternative measures of labor under-utilization (percent)		
	Married men, spouse present	Married women, spouse present	Women who maintain families[1]	Total	Job losers and persons who completed temporary jobs	Job leavers	Reentrants	New entrants	Average (mean) weeks unemployed	Median weeks unemployed	Including discouraged workers (U-4)	Including all marginally attached workers (U-5)	Including marginally attached and under-employed (U-6)
1970	2.6	4.9	5.4	4.9	2.2	0.7	1.5	0.6	8.6	4.9	...	...	...
1971	3.2	5.7	7.3	5.9	2.8	0.7	1.7	0.7	11.3	6.3	...	...	...
1972	2.8	5.4	7.2	5.6	2.4	0.7	1.7	0.8	12.0	6.2	...	...	...
1973	2.3	4.7	7.1	4.9	1.9	0.8	1.5	0.7	10.0	5.2	...	...	...
1974	2.7	5.3	7.0	5.6	2.4	0.8	1.6	0.7	9.8	5.2	...	...	...
1975	5.1	7.9	10.0	8.5	4.7	0.9	2.0	0.9	14.2	8.4	...	...	...
1976	4.2	7.1	10.1	7.7	3.8	0.9	2.0	0.9	15.8	8.2	...	...	...
1977	3.6	6.5	9.4	7.1	3.2	0.9	2.0	1.0	14.3	7.0	...	...	...
1978	2.8	5.5	8.5	6.1	2.5	0.9	1.8	0.9	11.9	5.9	...	...	...
1979	2.8	5.1	8.3	5.8	2.5	0.8	1.7	0.8	10.8	5.4	...	...	...
1980	4.2	5.8	9.2	7.1	3.7	0.8	1.8	0.8	11.9	6.5	...	...	...
1981	4.3	6.0	10.4	7.6	3.9	0.8	1.9	0.9	13.7	6.9	...	...	...
1982	6.5	7.4	11.7	9.7	5.7	0.8	2.2	1.1	15.6	8.7	...	...	...
1983	6.5	7.0	12.2	9.6	5.6	0.7	2.2	1.1	20.0	10.1	...	...	...
1984	4.6	5.7	10.3	7.5	3.9	0.7	1.9	1.0	18.2	7.9	...	...	...
1985	4.3	5.6	10.4	7.2	3.6	0.8	2.0	0.9	15.6	6.8	...	...	...
1986	4.4	5.2	9.8	7.0	3.4	0.9	1.8	0.9	15.0	6.9	...	...	...
1987	3.9	4.3	9.2	6.2	3.0	0.8	1.6	0.8	14.5	6.5	...	...	...
1988	3.3	3.9	8.1	5.5	2.5	0.8	1.5	0.7	13.5	5.9	...	...	...
1989	3.0	3.7	8.1	5.3	2.4	0.8	1.5	0.5	11.9	4.8	...	...	...
1990	3.4	3.8	8.3	5.6	2.7	0.8	1.5	0.5	12.0	5.3	...	...	...
1991	4.4	4.5	9.3	6.8	3.7	0.8	1.7	0.6	13.7	6.8	...	...	...
1992	5.1	5.0	10.0	7.5	4.2	0.8	1.8	0.7	17.7	8.7	...	...	...
1993	4.4	4.6	9.7	6.9	3.8	0.8	1.7	0.7	18.0	8.3	...	...	...
1994	3.7	4.1	8.9	6.1	2.9	0.6	2.1	0.5	18.8	9.2	6.5	7.4	10.9
1995	3.3	3.9	8.0	5.6	2.6	0.6	1.9	0.4	16.6	8.3	5.9	6.7	10.1
1996	3.0	3.6	8.2	5.4	2.5	0.6	1.9	0.4	16.7	8.3	5.7	6.5	9.7
1997	2.7	3.1	8.1	4.9	2.2	0.6	1.7	0.4	15.8	8.0	5.2	5.9	8.9
1998	2.4	2.9	7.2	4.5	2.1	0.5	1.5	0.4	14.5	6.7	4.7	5.4	8.0
1999	2.2	2.7	6.4	4.2	1.9	0.6	1.4	0.3	13.4	6.4	4.4	5.0	7.4
2000	2.0	2.7	5.9	4.0	1.8	0.5	1.4	0.3	12.6	5.9	4.2	4.8	7.0
2001	2.7	3.1	6.6	4.7	2.4	0.6	1.4	0.3	13.1	6.8	4.9	5.6	8.1
2002	3.6	3.7	8.0	5.8	3.2	0.6	1.6	0.4	16.6	9.1	6.0	6.7	9.6
2003	3.8	3.7	8.5	6.0	3.3	0.6	1.7	0.4	19.2	10.1	6.3	7.0	10.1
2004	3.1	3.5	8.0	5.5	2.8	0.6	1.6	0.5	19.6	9.8	5.8	6.5	9.6
2005	2.8	3.3	7.8	5.1	2.5	0.6	1.6	0.4	18.4	8.9	5.4	6.1	8.9
2006	2.4	2.9	7.1	4.6	2.2	0.5	1.5	0.4	16.8	8.3	4.9	5.5	8.2
2007	2.5	2.8	6.5	4.6	2.3	0.5	1.4	0.4	16.8	8.5	4.9	5.5	8.3
2008	3.4	3.6	8.0	5.8	3.1	0.6	1.6	0.5	17.9	9.4	6.1	6.8	10.5
2009	6.6	5.5	11.5	9.3	5.9	0.6	2.1	0.7	24.4	15.1	9.7	10.5	16.2
2010	6.8	5.9	12.3	9.6	6.0	0.6	2.3	0.8	33.0	21.4	10.3	11.1	16.7
2011	5.8	5.6	12.4	8.9	5.3	0.6	2.2	0.8	39.3	21.4	9.5	10.4	15.9
2012	4.9	5.3	11.4	8.1	4.4	0.6	2.2	0.8	39.4	19.3	8.6	9.5	14.7
2013	4.3	4.6	10.2	7.4	3.9	0.6	2.1	0.8	36.5	17.0	7.9	8.8	13.8
2014	3.4	3.8	8.6	6.2	3.1	0.5	1.8	0.7	33.7	14.0	6.6	7.5	12.0
2015	2.8	3.1	7.4	5.3	2.6	0.5	1.6	0.6	29.2	11.6	5.7	6.4	10.4
2016	2.7	3.0	6.8	4.9	2.3	0.5	1.5	0.5	27.5	10.6	5.2	5.9	9.6
2017	2.4	2.7	6.2	4.4	2.1	0.5	1.3	0.4	25.0	10.0	4.6	5.3	8.5
2018	2.0	2.4	5.4	3.9	1.8	0.5	1.2	0.4	22.7	9.3	4.1	4.8	7.7
2019	1.8	2.2	5.0	3.7	1.7	0.5	1.1	0.4	21.6	9.1	3.9	4.5	7.2
2018													
January	2.2	2.4	6.5	4.1	2.0	0.4	1.2	0.4	24.2	9.6	4.3	5.0	8.0
February	2.1	2.7	6.2	4.1	2.0	0.5	1.2	0.4	23.1	9.2	4.3	5.0	8.1
March	2.1	2.6	5.6	4.0	1.9	0.5	1.2	0.4	24.1	9.0	4.3	4.9	8.0
April	2.2	2.5	5.5	4.0	1.9	0.5	1.2	0.4	22.9	9.9	4.2	4.8	7.9
May	2.0	2.3	4.7	3.8	1.8	0.5	1.2	0.4	21.1	9.4	4.1	4.7	7.8
June	2.1	2.5	5.5	4.0	1.9	0.5	1.3	0.4	21.2	8.6	4.2	4.9	7.8
July	2.0	2.5	5.6	3.8	1.8	0.5	1.1	0.4	23.2	10.0	4.1	4.7	7.5
August	2.0	2.5	5.4	3.8	1.8	0.5	1.1	0.4	22.6	9.3	4.1	4.7	7.3
September	1.9	2.1	5.1	3.7	1.8	0.5	1.2	0.4	23.5	9.1	4.0	4.7	7.5
October	1.9	2.3	5.3	3.8	1.8	0.5	1.2	0.4	22.3	9.4	4.1	4.7	7.5
November	1.9	2.3	5.4	3.7	1.8	0.4	1.2	0.4	22.0	8.8	4.0	4.7	7.6
December	2.0	2.4	4.5	3.9	1.8	0.5	1.2	0.4	22.0	9.4	4.1	4.7	7.6
2019													
January	2.1	2.4	5.4	4.0	1.9	0.5	1.2	0.4	20.6	9.0	4.2	4.9	8.0
February	1.9	2.3	4.9	3.8	1.8	0.5	1.2	0.4	22.0	9.4	4.0	4.6	7.2
March	1.9	2.4	4.7	3.8	1.7	0.5	1.2	0.4	22.2	9.5	4.0	4.6	7.4
April	1.8	1.8	4.9	3.6	1.6	0.4	1.2	0.3	22.8	9.3	3.9	4.5	7.3
May	1.7	2.4	4.7	3.6	1.6	0.5	1.1	0.4	24.1	9.1	3.9	4.5	7.2
June	1.8	2.3	5.1	3.7	1.7	0.5	1.1	0.3	22.1	9.4	3.9	4.6	7.2
July	1.8	2.2	6.0	3.7	1.7	0.5	1.1	0.4	19.7	9.0	3.9	4.5	6.9
August	1.8	2.2	5.1	3.7	1.7	0.5	1.1	0.4	22.1	9.0	3.9	4.6	7.2
September	1.7	2.2	4.7	3.5	1.6	0.5	1.0	0.4	21.7	9.4	3.7	4.3	6.9
October	1.7	2.3	5.4	3.6	1.6	0.5	1.0	0.4	21.6	9.2	3.8	4.3	6.9
November	1.9	2.2	4.8	3.5	1.7	0.5	1.0	0.4	20.2	9.2	3.7	4.3	6.8
December	1.6	2.1	4.2	3.5	1.6	0.5	1.0	0.3	20.8	9.0	3.7	4.2	6.7

[1]Not seasonally adjusted.
. . . = Not available.

Table 10-6. Insured Unemployment

(Averages of weekly data; thousands of persons, except as noted.)

Year and month	State programs, seasonally adjusted			Federal programs, not seasonally adjusted			
	Initial claims	Insured unemployment	Insured unemployment rate (percent) [1]	Initial claims		Persons claiming benefits	
				Federal employees	Newly discharged veterans	Federal employees	Newly discharged veterans
1980	488	3 365	3.9	...	...	...	...
1981	451	3 032	3.5	...	...	...	...
1982	586	4 094	4.7	...	...	...	...
1983	441	3 337	3.9	...	...	...	...
1984	374	2 452	2.8	...	...	...	...
1985	392	2 584	2.9	...	...	...	...
1986	378	2 632	2.8	2	2.5	21	17.5
1987	325	2 273	2.4	2	2.6	21	17.7
1988	309	2 075	2.1	2	2.7	23	18.1
1989	330	2 174	2.1	2	2.3	22	15.1
1990	385	2 539	2.4	2	2.5	24	18.4
1991	447	3 338	3.2	3	2.9	31	22.7
1992	409	3 208	3.1	3	5.0	32	60.6
1993	344	2 768	2.6	3	3.9	32	54.4
1994	340	2 667	2.5	3	3.0	32	37.5
1995	359	2 590	2.4	5	2.5	32	29.6
1996	352	2 552	2.3	7	2.1	29	24.2
1997	322	2 300	2.0	2	1.8	24	19.6
1998	317	2 213	1.9	2	1.4	19	15.6
1999	298	2 186	1.8	1	1.2	17	14.2
2000	299	2 112	1.7	2	1.0	19	12.5
2001	406	3 016	2.4	1	1.2	18	14.0
2002	404	3 570	2.8	1	1.2	18	16.2
2003	402	3 532	2.8	2	1.5	18	19.8
2004	342	2 930	2.3	2	1.9	18	26.9
2005	331	2 660	2.1	1	2.0	17	27.6
2006	312	2 456	1.9	1	2.0	15	26.3
2007	321	2 548	1.9	1	1.7	15	22.8
2008	418	3 337	2.5	1	1.6	15	22.2
2009	574	5 808	4.4	2	2.0	20	30.4
2010	459	4 544	3.6	3	2.6	28	39.2
2011	409	3 744	3.0	2	2.6	31	39.1
2012	374	3 319	2.6	1	2.6	20	39.7
2013	343	2 978	2.3	4	2.3	22	34.8
2014	308	2 590	2.2	1	1.8	16	26.0
2015	278	2 267	1.7	1	1.4	13	19.4
2016	263	2 136	1.6	1	1.4	11	13.8
2017	245	1 961	1.4	1	0.8	11	10.2
2018	220	1 756	1.2	1	0.6	9	7.4
2019	218	1 701	1.2	2	0.5	11	6.0
2017							
January	248	2 055	1.5	1	0.9	14	12.4
February	242	2 046	1.5	1	0.8	14	11.8
March	253	2 017	1.5	1	0.8	13	11.2
April	244	1 979	1.4	1	0.8	11	10.7
May	237	1 920	1.4	1	0.8	9	10.6
June	243	1 952	1.4	1	0.8	8	10.2
July	244	1 960	1.4	1	0.7	8	10.1
August	238	1 948	1.4	1	0.8	9	9.9
September	265	1 931	1.4	1	0.7	8	9.4
October	235	1 903	1.4	2	0.8	10	8.9
November	241	1 916	1.4	1	0.7	12	8.7
December	239	1 905	1.3	1	0.7	15	8.7
2018							
January	236	1 911	1.4	1	0.7	14	8.6
February	219	1 874	1.3	1	0.6	10	6.5
March	223	1 841	1.3	1	0.6	11	8.1
April	219	1 808	1.3	1	0.6	9	7.7
May	220	1 743	1.2	1	0.8	8	7.7
June	222	1 734	1.2	1	0.7	7	7.6
July	215	1 746	1.2	1	0.6	7	7.5
August	215	1 723	1.2	1	0.6	8	7.6
September	212	1 678	1.2	1	0.6	7	7.3
October	216	1 654	1.2	1	0.6	7	7.1
November	225	1 675	1.2	1	0.5	8	6.8
December	223	1 700	1.2	2	0.5	11	6.9
2019							
January	220	1 717	1.2	14	0.6	37	7.0
February	225	1 723	1.2	1	0.5	13	6.3
March	218	1 734	1.2	1	0.5	11	6.0
April	216	1 678	1.2	1	0.5	9	5.8
May	218	1 690	1.2	1	0.5	7	5.6
June	221	1 691	1.2	1	0.5	7	5.5
July	214	1 698	1.2	1	0.5	7	5.6
August	216	1 699	1.2	1	0.5	7	5.6
September	213	1 678	1.2	1	0.5	7	5.8
October	215	1 694	1.2	1	0.5	7	5.8
November	215	1 690	1.2	1	0.4	9	6.3
December	226	1 736	1.2	1	0.5	11	6.6

[1] Insured unemployed as a percent of employment covered by state programs.
. . . = Not available.

SECTION 10B: PAYROLL EMPLOYMENT, HOURS, AND EARNINGS

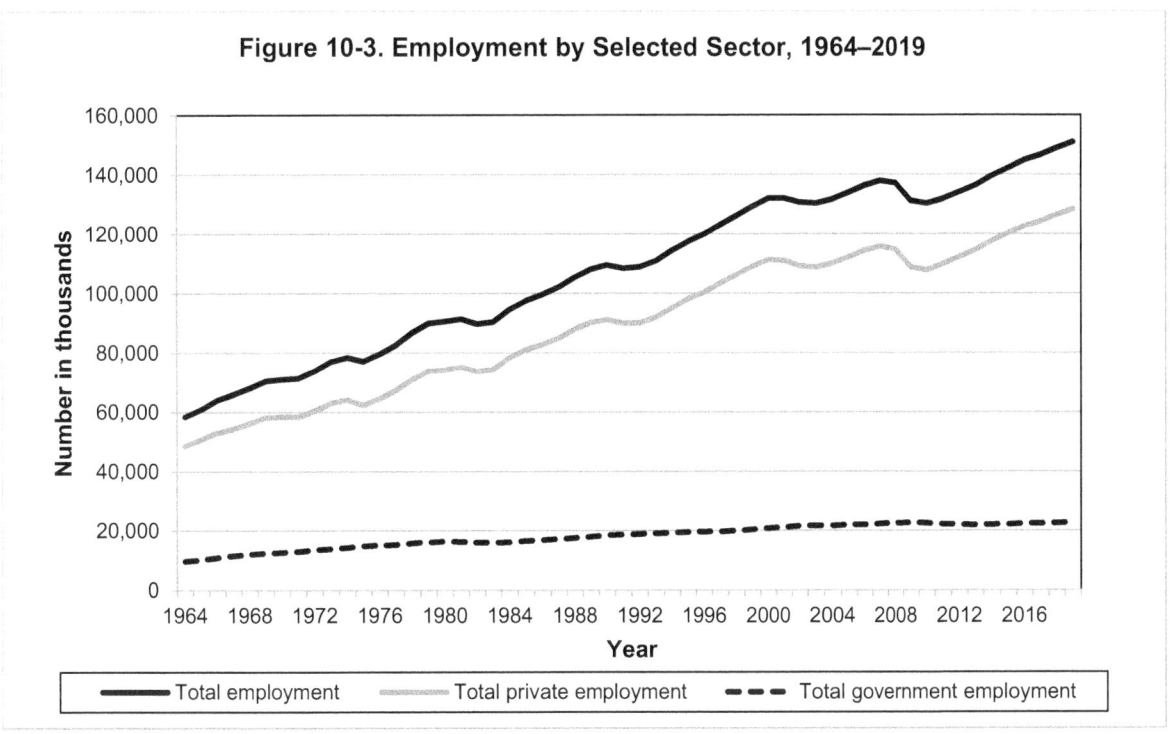

Figure 10-3. Employment by Selected Sector, 1964–2019

- Nonfarm employment includes wage and salary workers but not self-employed persons and unpaid family members working in a family business. The largest surge in employment was in 1966 (when the first U.S. Marines landed in South Vietnam) at 5.2 percent. Department of Defense employment rose 9.1 percent in 1966. The largest decline in employment at 4.3 percent occurred in 2009 during the Great Recession. (Table 10-7).

- The U.S. economy continues to be service-oriented rather than a good-produced economy. Manufacturing, once the economic engine of the economy, lost a significant percentage of its workforce in the last 55 years. More and more employees work in the services sector which includes occupations such as lawyers, architects, engineers, and computer designers. Between 1964 and 2019, the professional and business subsector grew over 400 percent. (Table 10-7)

- The diffusion index indicates the percentage of private industries in which employment is higher than six months earlier. Diffusion indexes measure the extent to which expansion or contraction have spread throughout the economy; indexes below 50 percent indicate recession. This index reached a low of 12.4 percent in June 2009—lower than the lowest points reached in any of the previous four recessions (1980, 1981–1982, 1990–1991, and 2001), and thus indicating the most widespread and pervasive decline in 30 years or more. Gains were widespread during the recovery period but the recent economic uncertainty threatens to plunge the diffusion index further again.

- In the United States, employment statistics are calculated using different methods and surveys. (See the Notes and Definitions section). Workers are also categorized differently. One of these categories is class of work: self-employed, private, and government. In addition, employees are either full- or part-time, agriculture versus nonfarm workers, supervisory or nonsupervisory workers, and workers by industrial sectors.

Table 10-7. Nonfarm Payroll Employment by NAICS Supersector

(Thousands; seasonally adjusted, except as noted.)

Year and month	Total	Private							Service-providing				
		Total	Goods-producing						Total	Private			
			Total	Mining and logging	Construc-tion	Manufacturing				Total	Trade, transportation, and utilities		
						Total	Durable	Nondurable			Total	Wholesale trade	Retail trade
1964	58 391	48 680	19 733	697	3 148	15 888	9 414	6 474	38 658	28 947	11 660	2 835	5 978
1965	60 874	50 683	20 595	694	3 284	16 617	9 973	6 644	40 279	30 089	12 121	2 945	6 264
1966	64 020	53 110	21 740	690	3 371	17 680	10 803	6 878	42 280	31 370	12 593	3 057	6 532
1967	65 931	54 406	21 882	679	3 305	17 897	10 952	6 945	44 049	32 524	12 932	3 134	6 713
1968	68 023	56 050	22 292	671	3 410	18 211	11 137	7 074	45 731	33 759	13 315	3 212	6 979
1969	70 512	58 181	22 893	683	3 637	18 573	11 396	7 177	47 619	35 288	13 833	3 320	7 297
1970	71 006	58 318	22 179	677	3 654	17 848	10 762	7 086	48 827	36 139	14 124	3 392	7 465
1971	71 335	58 323	21 602	658	3 770	17 174	10 229	6 944	49 734	36 721	14 298	3 398	7 659
1972	73 798	60 333	22 299	672	3 957	17 669	10 630	7 039	51 499	38 034	14 767	3 521	8 041
1973	76 912	63 050	23 450	693	4 167	18 589	11 414	7 176	53 462	39 600	15 327	3 661	8 374
1974	78 389	64 086	23 364	755	4 095	18 514	11 432	7 082	55 025	40 721	15 670	3 795	8 539
1975	77 069	62 250	21 318	802	3 608	16 909	10 266	6 643	55 751	40 932	15 583	3 782	8 604
1976	79 502	64 501	22 025	832	3 662	17 531	10 640	6 891	57 477	42 476	16 105	3 891	8 970
1977	82 593	67 334	22 972	865	3 940	18 167	11 132	7 035	59 620	44 362	16 741	4 025	9 363
1978	86 826	71 014	24 156	902	4 322	18 932	11 770	7 162	62 670	46 858	17 633	4 248	9 882
1979	89 933	73 865	24 997	1 008	4 562	19 426	12 220	7 206	64 936	48 869	18 276	4 452	10 185
1980	90 533	74 158	24 263	1 077	4 454	18 733	11 679	7 054	66 270	49 895	18 387	4 524	10 249
1981	91 297	75 117	24 118	1 180	4 304	18 634	11 611	7 023	67 179	50 999	18 577	4 600	10 369
1982	89 689	73 706	22 550	1 163	4 024	17 363	10 610	6 753	67 138	51 156	18 430	4 541	10 377
1983	90 295	74 284	22 110	997	4 065	17 048	10 326	6 722	68 186	52 174	18 642	4 525	10 640
1984	94 548	78 389	23 435	1 014	4 501	17 920	11 050	6 870	71 113	54 954	19 624	4 753	11 227
1985	97 532	81 000	23 585	974	4 793	17 819	11 034	6 784	73 947	57 415	20 350	4 878	11 738
1986	99 500	82 661	23 318	829	4 937	17 552	10 795	6 757	76 182	59 343	20 765	4 899	12 082
1987	102 116	84 960	23 470	771	5 090	17 609	10 767	6 842	78 647	61 490	21 271	4 966	12 422
1988	105 378	87 838	23 909	770	5 233	17 906	10 969	6 938	81 469	63 929	21 942	5 115	12 812
1989	108 051	90 124	24 045	750	5 309	17 985	11 004	6 981	84 006	66 079	22 477	5 245	13 112
1990	109 526	91 112	23 723	765	5 263	17 695	10 737	6 958	85 803	67 389	22 633	5 230	13 186
1991	108 425	89 879	22 588	739	4 780	17 068	10 220	6 848	85 837	67 291	22 247	5 147	12 900
1992	108 799	90 012	22 095	689	4 608	16 799	9 946	6 853	86 704	67 918	22 091	5 072	12 831
1993	110 931	91 942	22 219	666	4 779	16 774	9 901	6 872	88 712	69 723	22 343	5 056	13 024
1994	114 393	95 118	22 774	659	5 095	17 020	10 132	6 889	91 619	72 344	23 090	5 208	13 494
1995	117 400	97 968	23 156	641	5 274	17 241	10 373	6 868	94 245	74 812	23 793	5 393	13 900
1996	119 828	100 289	23 409	637	5 536	17 237	10 486	6 751	96 419	76 880	24 197	5 481	14 146
1997	122 941	103 278	23 886	654	5 813	17 419	10 705	6 714	99 055	79 392	24 656	5 622	14 393
1998	126 146	106 237	24 354	645	6 149	17 560	10 911	6 649	101 792	81 883	25 139	5 752	14 613
1999	129 228	108 921	24 465	598	6 545	17 322	10 831	6 491	104 763	84 456	25 722	5 848	14 974
2000	132 011	111 221	24 649	599	6 787	17 263	10 877	6 386	107 362	86 572	26 174	5 888	15 284
2001	132 073	110 955	23 873	606	6 826	16 441	10 336	6 105	108 200	87 082	25 931	5 728	15 242
2002	130 634	109 121	22 557	583	6 716	15 259	9 485	5 774	108 077	86 564	25 442	5 606	15 029
2003	130 331	108 748	21 816	572	6 735	14 509	8 964	5 546	108 515	86 931	25 228	5 556	14 922
2004	131 769	110 148	21 882	591	6 976	14 315	8 925	5 390	109 887	88 266	25 470	5 608	15 063
2005	134 005	112 201	22 190	628	7 336	14 227	8 956	5 271	111 814	90 010	25 959	5 764	15 280
2006	136 398	114 424	22 530	684	7 691	14 155	8 981	5 174	113 867	91 894	26 276	5 905	15 353
2007	137 936	115 718	22 233	724	7 630	13 879	8 808	5 071	115 703	93 485	26 630	6 015	15 520
2008	137 170	114 661	21 335	767	7 162	13 406	8 463	4 943	115 835	93 326	26 293	5 943	15 283
2009	131 233	108 678	18 558	694	6 016	11 847	7 284	4 564	112 675	90 121	24 906	5 587	14 522
2010	130 275	107 785	17 751	705	5 518	11 528	7 064	4 464	112 524	90 034	24 636	5 452	14 440
2011	131 842	109 756	18 047	788	5 533	11 726	7 273	4 453	113 795	91 708	25 065	5 543	14 668
2012	134 104	112 184	18 420	848	5 646	11 927	7 470	4 457	115 683	93 763	25 476	5 667	14 841
2013	136 368	114 504	18 700	868	5 827	12 006	7 543	4 463	117 668	95 804	25 870	5 747	15 077
2014	139 578	117 670	19 360	904	6 231	12 225	7 708	4 517	120 218	98 310	26 490	5 824	15 414
2015	142 253	120 196	19 614	778	6 490	12 346	7 769	4 577	122 639	100 582	26 938	5 843	15 631
2016	144 882	122 566	19 718	661	6 727	12 330	7 693	4 637	125 164	102 848	27 322	5 878	15 887
2017	146 608	124 258	20 048	676	6 969	12 439	7 741	4 699	126 768	104 174	27 393	5 814	15 846
2018	148 908	126 454	20 704	727	7 288	12 688	7 946	4 742	128 205	105 750	27 607	5 841	15 786
2019	150 939	128 346	21 067	735	7 492	12 840	8 059	4 781	129 873	107 279	27 715	5 903	15 644

Table 10-7. Nonfarm Payroll Employment by NAICS Supersector—*Continued*

(Thousands; seasonally adjusted, except as noted.)

Year and month	Service-providing—*Continued*													Diffusion index, 6-month span, private nonfarm [2]
	Private—*Continued*						Government							
								Federal		State		Local		
	Information	Financial activities	Professional and business services	Education and health services	Leisure and hospitality	Other services	Total	Total	Department of Defense [1]	Total	Education	Total	Education	
1964	1 766	2 811	4 154	3 438	3 772	1 346	9 711	2 463	676	1 856	609	5 392	2 839	. . .
1965	1 824	2 878	4 323	3 587	3 951	1 404	10 191	2 495	679	1 996	679	5 700	3 031	. . .
1966	1 908	2 961	4 536	3 770	4 127	1 475	10 910	2 690	741	2 141	775	6 080	3 297	. . .
1967	1 955	3 087	4 739	3 986	4 269	1 558	11 525	2 852	802	2 302	873	6 371	3 490	. . .
1968	1 991	3 234	4 937	4 191	4 453	1 638	11 972	2 871	801	2 442	958	6 660	3 649	. . .
1969	2 048	3 404	5 175	4 428	4 670	1 731	12 330	2 893	815	2 533	1 042	6 904	3 785	. . .
1970	2 041	3 532	5 287	4 577	4 789	1 789	12 687	2 865	756	2 664	1 104	7 158	3 912	. . .
1971	2 009	3 651	5 348	4 675	4 914	1 827	13 012	2 828	731	2 747	1 149	7 437	4 091	. . .
1972	2 056	3 784	5 544	4 863	5 121	1 900	13 465	2 815	720	2 859	1 188	7 790	4 262	. . .
1973	2 135	3 920	5 796	5 092	5 341	1 990	13 862	2 794	696	2 923	1 205	8 146	4 433	. . .
1974	2 160	4 023	5 997	5 322	5 471	2 078	14 303	2 858	698	3 039	1 267	8 407	4 584	. . .
1975	2 061	4 047	6 056	5 497	5 544	2 144	14 820	2 882	704	3 179	1 323	8 758	4 722	. . .
1976	2 111	4 155	6 310	5 756	5 794	2 244	15 001	2 863	693	3 273	1 371	8 865	4 786	. . .
1977	2 185	4 348	6 611	6 052	6 065	2 359	15 258	2 859	676	3 377	1 385	9 023	4 859	. . .
1978	2 287	4 599	6 997	6 427	6 411	2 505	15 812	2 893	661	3 474	1 367	9 446	4 958	. . .
1979	2 375	4 843	7 339	6 768	6 631	2 637	16 068	2 894	649	3 541	1 378	9 633	4 989	. . .
1980	2 361	5 025	7 571	7 077	6 721	2 755	16 375	3 000	645	3 610	1 398	9 765	5 090	. . .
1981	2 382	5 163	7 809	7 364	6 840	2 865	16 180	2 922	655	3 640	1 420	9 619	5 095	. . .
1982	2 317	5 209	7 875	7 526	6 874	2 924	15 982	2 884	690	3 640	1 433	9 458	5 049	. . .
1983	2 253	5 334	8 065	7 781	7 078	3 021	16 011	2 915	699	3 662	1 450	9 434	5 020	. . .
1984	2 398	5 553	8 493	8 211	7 489	3 186	16 159	2 943	716	3 734	1 488	9 482	5 076	. . .
1985	2 437	5 815	8 900	8 679	7 869	3 366	16 533	3 014	738	3 832	1 540	9 687	5 221	. . .
1986	2 445	6 128	9 241	9 086	8 156	3 523	16 838	3 044	736	3 893	1 561	9 901	5 358	. . .
1987	2 507	6 385	9 639	9 543	8 446	3 699	17 156	3 089	736	3 967	1 586	10 100	5 469	. . .
1988	2 585	6 500	10 121	10 096	8 778	3 907	17 540	3 124	719	4 076	1 621	10 339	5 590	. . .
1989	2 622	6 562	10 588	10 652	9 062	4 116	17 927	3 136	735	4 182	1 668	10 609	5 740	. . .
1990	2 688	6 614	10 881	11 024	9 288	4 261	18 415	3 196	722	4 305	1 730	10 914	5 902	. . .
1991	2 677	6 561	10 746	11 556	9 256	4 249	18 545	3 110	702	4 355	1 768	11 081	5 994	43.6
1992	2 641	6 559	11 001	11 948	9 437	4 240	18 787	3 111	702	4 408	1 799	11 267	6 076	61.6
1993	2 668	6 742	11 527	12 362	9 732	4 350	18 989	3 063	670	4 488	1 834	11 438	6 206	72.5
1994	2 738	6 910	12 207	12 872	10 100	4 428	19 275	3 018	657	4 576	1 882	11 682	6 329	81.4
1995	2 843	6 866	12 878	13 360	10 501	4 572	19 432	2 949	627	4 635	1 919	11 849	6 453	70.5
1996	2 940	7 018	13 497	13 761	10 777	4 690	19 539	2 877	597	4 606	1 911	12 056	6 592	78.5
1997	3 084	7 255	14 371	14 185	11 018	4 825	19 664	2 806	588	4 582	1 904	12 276	6 759	81.0
1998	3 218	7 565	15 183	14 570	11 232	4 976	19 909	2 772	550	4 612	1 922	12 525	6 921	70.2
1999	3 419	7 753	15 994	14 939	11 543	5 087	20 307	2 769	525	4 709	1 983	12 829	7 120	70.5
2000	3 630	7 783	16 704	15 252	11 862	5 168	20 790	2 865	510	4 786	2 031	13 139	7 294	58.9
2001	3 629	7 900	16 514	15 814	12 036	5 258	21 118	2 764	504	4 905	2 113	13 449	7 479	32.4
2002	3 395	7 956	16 016	16 398	11 986	5 372	21 513	2 766	499	5 029	2 243	13 718	7 654	38.0
2003	3 188	8 078	16 029	16 835	12 173	5 401	21 583	2 761	486	5 002	2 255	13 820	7 709	40.5
2004	3 118	8 105	16 440	17 230	12 493	5 409	21 621	2 730	473	4 982	2 238	13 909	7 765	61.0
2005	3 038	8 197	16 954	18 154	12 816	5 395	21 804	2 732	488	5 032	2 260	14 041	7 856	64.5
2006	3 032	8 367	17 566	18 099	13 110	5 438	21 974	2 732	493	5 075	2 293	14 167	7 913	60.1
2007	2 984	8 348	17 942	18 613	13 427	5 494	22 218	2 734	491	5 122	2 318	14 362	7 987	57.2
2008	2 804	8 206	17 735	19 156	13 436	5 515	22 509	2 762	496	5 177	2 354	14 571	8 084	27.3
2009	2 707	7 838	16 579	19 550	13 077	5 367	22 555	2 832	519	5 169	2 360	14 554	8 079	19.2
2010	2 674	7 695	16 728	19 889	13 049	5 331	22 490	2 977	545	5 137	2 373	14 376	8 013	62.8
2011	2 676	7 697	17 332	20 228	13 353	5 360	22 086	2 859	559	5 078	2 374	14 150	7 873	73.8
2012	2 706	7 784	17 932	20 698	13 768	5 430	21 920	2 820	552	5 055	2 389	14 045	7 778	62.2
2013	2 726	7 880	18 560	21 102	14 242	5 464	21 864	2 766	537	5 048	2 393	14 050	7 783	63.0
2014	2 750	8 006	19 189	21 541	14 775	5 575	21 908	2 736	524	5 043	2 379	14 129	7 847	78.5
2015	2 794	8 149	19 705	22 140	15 266	5 620	22 057	2 761	526	5 086	2 407	14 210	7 865	64.5
2016	2 814	8 324	20 279	22 745	15 684	5 708	22 316	2 811	529	5 101	2 423	14 404	7 968	63.2
2017	2 839	8 451	20 508	23 188	16 051	5 770	22 350	2 805	527	5 165	2 479	14 379	7 922	70.2
2018	2 860	8 590	20 950	23 638	16 295	5 831	22 455	2 800	531	5 173	2 487	14 481	7 963	69.8
2019	2 860	8 746	21 313	24 177	16 576	5 893	22 594	2 834	543	5 177	2 484	14 583	8 010	63.2

[1]Not seasonally adjusted.
[2]See notes and definitions for explanation. September value used to represent year.
. . . = Not available.

Table 10-7. Nonfarm Payroll Employment by NAICS Supersector—*Continued*

(Thousands; seasonally adjusted, except as noted.)

Year and month	Total	Private							Service-providing				
		Total	Goods-producing						Total	Private			
			Total	Mining and logging	Construc-tion	Manufacturing				Total	Trade, transportation, and utilities		
						Total	Durable	Nondurable			Total	Wholesale trade	Retail trade
2015													
January	140 568	118 608	19 500	888	6 320	12 292	7 752	4 540	121 068	99 108	26 613	5 769	15 520
February	140 839	118 861	19 537	875	6 361	12 301	7 760	4 541	121 302	99 324	26 657	5 781	15 535
March	140 910	118 940	19 505	859	6 334	12 312	7 764	4 548	121 405	99 435	26 684	5 783	15 548
April	141 194	119 189	19 553	843	6 392	12 318	7 763	4 555	121 641	99 636	26 707	5 779	15 567
May	141 525	119 513	19 580	820	6 427	12 333	7 773	4 560	121 945	99 933	26 767	5 787	15 596
June	141 699	119 684	19 591	816	6 441	12 334	7 772	4 562	122 108	100 093	26 814	5 785	15 619
July	142 001	119 951	19 626	805	6 472	12 349	7 774	4 575	122 375	100 325	26 859	5 785	15 639
August	142 126	120 066	19 629	794	6 490	12 345	7 774	4 571	122 497	100 437	26 870	5 785	15 638
September	142 281	120 241	19 641	779	6 508	12 354	7 773	4 581	122 640	100 600	26 874	5 776	15 643
October	142 587	120 531	19 680	771	6 547	12 362	7 770	4 592	122 907	100 851	26 908	5 777	15 675
November	142 824	120 743	19 711	756	6 598	12 357	7 756	4 601	123 113	101 032	26 945	5 778	15 695
December	143 097	121 000	19 738	746	6 630	12 362	7 750	4 612	123 359	101 262	26 978	5 777	15 699
2016													
January	143 170	121 052	19 730	731	6 620	12 379	7 761	4 618	123 440	101 322	26 976	5 777	15 725
February	143 433	121 289	19 728	711	6 650	12 367	7 749	4 618	123 705	101 561	27 036	5 775	15 771
March	143 662	121 482	19 716	691	6 680	12 345	7 729	4 616	123 946	101 766	27 093	5 780	15 810
April	143 849	121 663	19 728	678	6 701	12 349	7 729	4 620	124 121	101 935	27 118	5 785	15 813
May	143 891	121 694	19 689	666	6 691	12 332	7 707	4 625	124 202	102 005	27 132	5 784	15 813
June	144 158	121 974	19 709	656	6 702	12 351	7 704	4 647	124 449	102 265	27 164	5 780	15 843
July	144 512	122 227	19 757	652	6 736	12 369	7 719	4 650	124 755	102 470	27 205	5 781	15 859
August	144 647	122 379	19 732	647	6 737	12 348	7 695	4 653	124 915	102 647	27 254	5 785	15 876
September	144 916	122 613	19 762	647	6 768	12 347	7 691	4 656	125 154	102 851	27 269	5 797	15 885
October	145 061	122 769	19 789	644	6 798	12 347	7 693	4 654	125 272	102 980	27 288	5 798	15 887
November	145 212	122 922	19 808	647	6 819	12 342	7 686	4 656	125 404	103 114	27 284	5 800	15 862
December	145 442	123 134	19 825	648	6 821	12 356	7 695	4 661	125 617	103 309	27 350	5 796	15 892
2017													
January	145 627	123 325	19 865	649	6 847	12 369	7 701	4 668	125 762	103 460	27 357	5 802	15 920
February	145 815	123 496	19 928	655	6 889	12 384	7 698	4 686	125 887	103 568	27 342	5 804	15 880
March	145 944	123 619	19 964	660	6 909	12 395	7 704	4 691	125 980	103 655	27 339	5 807	15 862
April	146 141	123 811	19 985	669	6 916	12 400	7 706	4 694	126 156	103 826	27 352	5 811	15 854
May	146 296	123 964	20 007	673	6 928	12 406	7 715	4 691	126 289	103 957	27 339	5 806	15 824
June	146 512	124 166	20 050	677	6 955	12 418	7 726	4 692	126 462	104 116	27 366	5 814	15 831
July	146 727	124 368	20 056	679	6 960	12 417	7 718	4 699	126 671	104 312	27 368	5 814	15 823
August	146 911	124 554	20 135	684	6 990	12 461	7 750	4 711	126 776	104 419	27 385	5 814	15 822
September	146 929	124 563	20 161	687	7 004	12 470	7 761	4 709	126 768	104 402	27 410	5 816	15 815
October	147 196	124 827	20 206	687	7 027	12 492	7 777	4 715	126 990	104 621	27 435	5 821	15 810
November	147 421	125 028	20 273	693	7 066	12 514	7 799	4 715	127 148	104 755	27 449	5 821	15 816
December	147 551	125 170	20 326	692	7 093	12 541	7 822	4 719	127 225	104 844	27 465	5 827	15 802
2018													
January	147 672	125 317	20 371	699	7 114	12 558	7 835	4 723	127 301	104 946	27 448	5 822	15 785
February	148 078	125 672	20 494	706	7 200	12 588	7 861	4 727	127 584	105 178	27 525	5 823	15 821
March	148 254	125 850	20 530	714	7 205	12 611	7 886	4 725	127 724	105 320	27 578	5 836	15 833
April	148 391	125 977	20 577	722	7 223	12 632	7 904	4 728	127 814	105 400	27 574	5 817	15 832
May	148 669	126 238	20 648	724	7 266	12 658	7 920	4 738	128 021	105 590	27 602	5 819	15 836
June	148 888	126 429	20 698	729	7 282	12 687	7 944	4 743	128 190	105 731	27 589	5 826	15 797
July	149 024	126 562	20 739	729	7 304	12 706	7 961	4 745	128 285	105 823	27 598	5 838	15 785
August	149 268	126 778	20 791	737	7 335	12 719	7 972	4 747	128 477	105 987	27 657	5 857	15 788
September	149 348	126 859	20 834	739	7 355	12 740	7 991	4 749	128 514	106 025	27 636	5 854	15 741
October	149 549	127 064	20 884	742	7 378	12 764	8 011	4 753	128 665	106 180	27 633	5 857	15 702
November	149 683	127 197	20 897	737	7 376	12 784	8 022	4 762	128 786	106 300	27 691	5 864	15 729
December	149 865	127 370	20 948	741	7 402	12 805	8 038	4 767	128 917	106 422	27 666	5 873	15 704
2019													
January	150 134	127 628	21 023	746	7 452	12 825	8 059	4 766	129 111	106 605	27 711	5 879	15 696
February	150 135	127 622	20 994	741	7 423	12 830	8 062	4 768	129 141	106 628	27 688	5 885	15 667
March	150 282	127 754	21 011	741	7 443	12 827	8 056	4 771	129 271	106 743	27 665	5 880	15 642
April	150 492	127 939	21 039	741	7 469	12 829	8 056	4 773	129 453	106 900	27 671	5 894	15 631
May	150 577	128 026	21 050	743	7 478	12 829	8 056	4 773	129 527	106 976	27 667	5 898	15 618
June	150 759	128 206	21 076	741	7 497	12 838	8 064	4 774	129 683	107 130	27 686	5 899	15 613
July	150 953	128 366	21 085	736	7 504	12 845	8 067	4 778	129 868	107 281	27 692	5 906	15 614
August	151 160	128 523	21 087	731	7 508	12 848	8 066	4 782	130 073	107 436	27 688	5 907	15 613
September	151 368	128 718	21 106	731	7 524	12 851	8 066	4 785	130 262	107 612	27 712	5 913	15 622
October	151 553	128 908	21 086	735	7 541	12 810	8 019	4 791	130 467	107 822	27 750	5 923	15 644
November	151 814	129 155	21 131	724	7 539	12 868	8 064	4 804	130 683	108 024	27 762	5 926	15 630
December	151 998	129 319	21 136	715	7 555	12 866	8 064	4 802	130 862	108 183	27 809	5 933	15 672

Table 10-7. Nonfarm Payroll Employment by NAICS Supersector—*Continued*

(Thousands; seasonally adjusted, except as noted.)

Year and month	Service-providing—*Continued*													Diffusion index, 6-month span, private nonfarm [2]
	Private—*Continued*						Government							
							Total	Federal		State		Local		
	Information	Financial activities	Profes-sional and business services	Education and health services	Leisure and hospitality	Other services		Total	Depart-ment of Defense [1]	Total	Education	Total	Education	
2015														
January	2 738	8 059	19 442	21 741	14 915	5 600	21 960	2 745	525	5 065	2 392	14 150	7 854	74.8
February	2 742	8 072	19 487	21 789	14 972	5 605	21 978	2 747	526	5 071	2 398	14 160	7 864	75.6
March	2 737	8 081	19 513	21 832	14 981	5 607	21 970	2 748	526	5 067	2 392	14 155	7 853	71.1
April	2 741	8 088	19 557	21 906	15 023	5 614	22 005	2 763	526	5 070	2 394	14 172	7 863	68.8
May	2 749	8 095	19 638	21 958	15 103	5 623	22 012	2 753	526	5 074	2 393	14 185	7 870	68.2
June	2 750	8 117	19 689	21 998	15 108	5 617	22 015	2 754	526	5 073	2 396	14 188	7 873	67.8
July	2 753	8 137	19 726	22 053	15 168	5 629	22 050	2 756	526	5 073	2 396	14 221	7 896	66.5
August	2 754	8 143	19 755	22 094	15 201	5 620	22 060	2 760	526	5 077	2 402	14 223	7 893	65.7
September	2 764	8 153	19 778	22 131	15 279	5 621	22 040	2 759	526	5 083	2 407	14 198	7 849	64.5
October	2 763	8 168	19 851	22 215	15 316	5 630	22 056	2 759	526	5 084	2 408	14 213	7 868	66.9
November	2 750	8 180	19 885	22 260	15 374	5 638	22 081	2 766	526	5 090	2 412	14 225	7 876	61.8
December	2 759	8 188	19 955	22 323	15 406	5 653	22 097	2 776	527	5 088	2 411	14 233	7 875	66.9
2016														
January	2 763	8 203	19 953	22 337	15 441	5 649	22 118	2 772	528	5 089	2 413	14 257	7 883	63.6
February	2 778	8 211	19 968	22 405	15 497	5 666	22 144	2 778	528	5 086	2 411	14 280	7 891	65.7
March	2 785	8 226	19 998	22 452	15 538	5 674	22 180	2 783	529	5 093	2 413	14 304	7 905	68.4
April	2 789	8 249	20 030	22 510	15 561	5 678	22 186	2 785	529	5 098	2 415	14 303	7 904	66.3
May	2 753	8 263	20 038	22 567	15 582	5 670	22 197	2 792	529	5 093	2 412	14 312	7 913	61.4
June	2 798	8 280	20 081	22 624	15 637	5 681	22 184	2 797	530	5 107	2 427	14 280	7 875	57.6
July	2 799	8 302	20 125	22 668	15 684	5 687	22 285	2 799	530	5 116	2 435	14 370	7 959	64.1
August	2 805	8 317	20 131	22 714	15 724	5 702	22 268	2 803	530	5 113	2 429	14 352	7 930	62.0
September	2 812	8 325	20 198	22 773	15 761	5 713	22 303	2 807	530	5 147	2 461	14 349	7 921	63.2
October	2 814	8 331	20 216	22 834	15 779	5 718	22 292	2 804	531	5 134	2 449	14 354	7 915	67.2
November	2 809	8 342	20 253	22 868	15 829	5 729	22 290	2 799	527	5 139	2 454	14 352	7 905	65.9
December	2 812	8 369	20 288	22 929	15 846	5 715	22 308	2 817	531	5 141	2 461	14 350	7 911	66.1
2017														
January	2 819	8 399	20 338	22 938	15 878	5 731	22 302	2 815	531	5 151	2 465	14 336	7 905	64.1
February	2 814	8 401	20 347	23 012	15 910	5 742	22 319	2 813	530	5 160	2 469	14 346	7 911	67.6
March	2 813	8 409	20 376	23 051	15 922	5 745	22 325	2 811	529	5 163	2 473	14 351	7 917	66.9
April	2 809	8 424	20 406	23 101	15 986	5 748	22 330	2 800	528	5 166	2 480	14 364	7 918	65.7
May	2 804	8 432	20 463	23 139	16 021	5 759	22 332	2 808	527	5 174	2 492	14 350	7 919	64.7
June	2 807	8 448	20 498	23 167	16 063	5 767	22 346	2 806	527	5 170	2 486	14 370	7 913	68.0
July	2 809	8 460	20 540	23 227	16 134	5 774	22 359	2 805	526	5 173	2 488	14 381	7 912	67.1
August	2 815	8 469	20 568	23 264	16 138	5 780	22 357	2 802	527	5 167	2 479	14 388	7 923	67.4
September	2 811	8 477	20 571	23 295	16 065	5 773	22 366	2 802	526	5 160	2 473	14 404	7 931	70.2
October	2 812	8 482	20 649	23 322	16 156	5 795	22 369	2 803	524	5 155	2 468	14 411	7 939	70.2
November	2 813	8 494	20 649	23 374	16 170	5 806	22 393	2 792	525	5 171	2 484	14 430	7 947	69.0
December	2 819	8 498	20 683	23 383	16 190	5 806	22 381	2 794	526	5 159	2 471	14 428	7 940	68.4
2018														
January	2 815	8 504	20 738	23 437	16 193	5 811	22 355	2 801	525	5 147	2 475	14 407	7 910	68.2
February	2 815	8 533	20 781	23 491	16 218	5 815	22 406	2 793	525	5 151	2 474	14 462	7 957	69.8
March	2 826	8 538	20 823	23 523	16 220	5 812	22 404	2 793	525	5 154	2 476	14 457	7 949	73.4
April	2 839	8 545	20 868	23 538	16 217	5 819	22 414	2 795	528	5 167	2 476	14 452	7 947	73.1
May	2 837	8 567	20 914	23 576	16 263	5 831	22 431	2 792	530	5 174	2 481	14 465	7 956	73.4
June	2 837	8 586	20 948	23 617	16 310	5 844	22 459	2 796	530	5 180	2 485	14 483	7 971	74.2
July	2 841	8 594	20 968	23 644	16 349	5 829	22 462	2 798	531	5 178	2 486	14 486	7 970	70.7
August	2 841	8 607	21 009	23 691	16 350	5 832	22 490	2 801	533	5 183	2 493	14 506	7 984	71.1
September	2 839	8 629	21 042	23 716	16 326	5 837	22 489	2 804	534	5 193	2 504	14 492	7 967	69.8
October	2 851	8 647	21 079	23 767	16 358	5 845	22 485	2 805	534	5 184	2 497	14 496	7 966	70.5
November	2 850	8 657	21 102	23 790	16 368	5 842	22 486	2 811	536	5 173	2 487	14 502	7 965	68.8
December	2 854	8 665	21 128	23 844	16 415	5 850	22 495	2 809	537	5 173	2 489	14 513	7 972	71.9
2019														
January	2 843	8 676	21 126	23 900	16 496	5 853	22 506	2 811	537	5 170	2 485	14 525	7 978	69.2
February	2 841	8 690	21 164	23 918	16 473	5 854	22 513	2 814	538	5 175	2 485	14 524	7 976	71.9
March	2 851	8 707	21 176	23 981	16 494	5 869	22 528	2 815	539	5 175	2 488	14 538	7 982	69.4
April	2 845	8 721	21 226	24 046	16 507	5 884	22 553	2 823	539	5 169	2 487	14 561	7 995	68.4
May	2 853	8 727	21 253	24 076	16 519	5 881	22 551	2 826	541	5 158	2 474	14 567	7 993	68.4
June	2 865	8 732	21 294	24 131	16 526	5 896	22 553	2 829	542	5 157	2 470	14 567	7 987	62.4
July	2 862	8 753	21 337	24 204	16 528	5 905	22 587	2 831	544	5 168	2 478	14 588	8 006	60.1
August	2 861	8 768	21 377	24 262	16 570	5 910	22 637	2 857	543	5 184	2 489	14 596	8 021	63.0
September	2 866	8 771	21 402	24 323	16 631	5 907	22 650	2 857	544	5 181	2 483	14 612	8 027	63.2
October	2 865	8 792	21 444	24 363	16 701	5 907	22 645	2 844	545	5 184	2 481	14 617	8 028	61.8
November	2 874	8 804	21 481	24 436	16 744	5 923	22 659	2 850	551	5 181	2 479	14 628	8 032	67.1
December	2 883	8 814	21 503	24 465	16 784	5 925	22 679	2 847	549	5 184	2 481	14 648	8 039	65.7

[1] Not seasonally adjusted.
[2] See notes and definitions for explanation. September value used to represent year.

Table 10-8. Employment, Hours, and Earnings, Total Nonfarm and Manufacturing, Historical Annual and Monthly

(Wage and salary workers on nonfarm payrolls, seasonally adjusted.)

Year and month	All wage and salary workers (thousands)					Production and nonsupervisory workers on private payrolls							
	Total	Private			Service-providing	Number (thousands)		Average hours per week		Average hourly earnings, dollars		Average weekly earnings, dollars	
		Total	Goods-producing			Total private	Manufacturing	Total private	Manufacturing	Total private	Manufacturing	Total private	Manufacturing
			Total	Manufacturing									
1939	30 645	26 606	11 511	9 450	19 134	...	8 163	...	37.7	...	0.49	...	18.47
1940	32 407	28 156	12 378	10 099	20 029	...	8 737	...	38.2	...	0.53	...	20.25
1941	36 600	31 874	14 940	12 121	21 660	...	10 641	...	40.7	...	0.61	...	24.83
1942	40 213	34 621	17 275	14 030	22 938	...	12 447	...	43.2	...	0.74	...	31.97
1943	42 574	36 353	18 738	16 153	23 837	...	14 407	...	45.1	...	0.86	...	38.79
1944	42 006	35 819	17 981	15 903	24 026	...	14 031	...	45.4	...	0.91	...	41.31
1945	40 510	34 428	16 308	14 255	24 203	...	12 445	...	43.6	...	0.90	...	39.24
1946	41 759	36 054	16 122	13 513	25 637	...	11 781	...	40.4	...	0.95	...	38.38
1947	43 945	38 379	17 314	14 287	26 631	...	12 453	...	40.5	...	1.10	...	44.55
1948	44 954	39 213	17 579	14 324	27 376	...	12 383	...	40.1	...	1.20	...	48.12
1949	43 843	37 893	16 464	13 281	27 379	...	11 355	...	39.2	...	1.25	...	49.00
1950	45 287	39 167	17 343	14 013	27 945	...	12 032	...	40.6	...	1.32	...	53.59
1951	47 930	41 427	18 703	15 070	29 227	...	12 808	...	40.7	...	1.45	...	59.02
1952	48 909	42 182	18 928	15 291	29 981	...	12 797	...	40.8	...	1.53	...	62.42
1953	50 310	43 552	19 733	16 131	30 577	...	13 437	...	40.6	...	1.63	...	66.18
1954	49 093	42 235	18 515	15 002	30 578	...	12 300	...	39.7	...	1.66	...	65.90
1955	50 744	43 722	19 234	15 524	31 510	...	12 735	...	40.8	...	1.74	...	70.99
1956	52 473	45 087	19 799	15 858	32 674	...	12 869	...	40.5	...	1.84	...	74.52
1957	52 959	45 235	19 669	15 798	33 290	...	12 640	...	39.9	...	1.93	...	77.01
1958	51 426	43 480	18 319	14 656	33 107	...	11 532	...	39.2	...	1.99	...	78.01
1959	53 374	45 182	19 163	15 325	34 211	...	12 089	...	40.3	...	2.08	...	83.82
1960	54 296	45 832	19 182	15 438	35 114	...	12 074	...	39.8	...	2.15	...	85.57
1961	54 105	45 399	18 647	15 011	35 458	...	11 612	...	39.9	...	2.20	...	87.78
1962	55 659	46 655	19 203	15 498	36 455	...	11 986	...	40.5	...	2.27	...	91.94
1963	56 764	47 423	19 385	15 631	37 379	...	12 051	...	40.6	...	2.34	...	95.00
1964	58 391	48 680	19 733	15 888	38 658	40 575	12 298	38.5	40.8	2.53	2.41	97.41	98.33
1950													
January	43 528	37 594	16 255	13 161	27 273	...	11 258	...	39.6	...	1.27	...	50.29
February	43 298	37 372	16 035	13 169	27 263	...	11 262	...	39.7	...	1.26	...	50.02
March	43 952	37 874	16 482	13 290	27 470	...	11 362	...	39.7	...	1.28	...	50.82
April	44 376	38 282	16 718	13 471	27 658	...	11 528	...	40.3	...	1.29	...	51.99
May	44 718	38 675	17 080	13 780	27 638	...	11 855	...	40.3	...	1.30	...	52.39
June	45 084	39 062	17 288	13 923	27 796	...	11 979	...	40.6	...	1.30	...	52.78
July	45 454	39 364	17 464	14 072	27 990	...	12 107	...	40.9	...	1.31	...	53.58
August	46 188	40 001	17 917	14 461	28 271	...	12 476	...	41.3	...	1.33	...	54.93
September	46 442	40 214	18 040	14 561	28 402	...	12 519	...	40.8	...	1.33	...	54.26
October	46 712	40 463	18 249	14 737	28 463	...	12 659	...	41.1	...	1.36	...	55.90
November	46 778	40 516	18 288	14 762	28 490	...	12 682	...	41.0	...	1.37	...	56.17
December	46 855	40 541	18 283	14 782	28 572	...	12 710	...	40.9	...	1.39	...	56.85
1951													
January	47 288	40 936	18 518	14 950	28 770	...	12 816	...	41.0	...	1.40	...	57.40
February	47 577	41 195	18 666	15 076	28 911	...	12 929	...	40.9	...	1.41	...	57.67
March	47 871	41 461	18 754	15 125	29 117	...	12 936	...	41.0	...	1.42	...	58.22
April	47 856	41 405	18 810	15 166	29 046	...	12 960	...	41.0	...	1.43	...	58.63
May	47 953	41 536	18 829	15 164	29 124	...	12 941	...	41.0	...	1.44	...	59.04
June	48 068	41 569	18 826	15 176	29 242	...	12 934	...	40.9	...	1.45	...	59.31
July	48 062	41 524	18 747	15 110	29 315	...	12 848	...	40.6	...	1.45	...	58.87
August	48 009	41 490	18 709	15 061	29 300	...	12 772	...	40.4	...	1.46	...	58.98
September	47 955	41 403	18 622	14 996	29 333	...	12 659	...	40.4	...	1.46	...	58.98
October	48 009	41 432	18 630	14 973	29 379	...	12 615	...	40.3	...	1.47	...	59.24
November	48 148	41 522	18 617	14 999	29 531	...	12 628	...	40.4	...	1.48	...	59.79
December	48 309	41 621	18 698	15 045	29 611	...	12 667	...	40.7	...	1.49	...	60.64
1952													
January	48 298	41 709	18 719	15 067	29 579	...	12 675	...	40.8	...	1.49	...	60.79
February	48 522	41 872	18 813	15 105	29 709	...	12 689	...	40.8	...	1.49	...	60.79
March	48 504	41 842	18 775	15 127	29 729	...	12 695	...	40.6	...	1.51	...	61.31
April	48 616	41 954	18 806	15 162	29 810	...	12 713	...	40.3	...	1.51	...	60.85
May	48 645	41 951	18 784	15 143	29 861	...	12 676	...	40.5	...	1.51	...	61.16
June	48 286	41 574	18 419	14 828	29 867	...	12 353	...	40.6	...	1.51	...	61.31
July	48 144	41 407	18 268	14 707	29 876	...	12 233	...	40.2	...	1.50	...	60.30
August	48 923	42 205	18 928	15 279	29 995	...	12 764	...	40.7	...	1.53	...	62.27
September	49 319	42 585	19 206	15 553	30 113	...	13 007	...	41.1	...	1.55	...	63.71
October	49 598	42 782	19 312	15 690	30 286	...	13 122	...	41.2	...	1.56	...	64.27
November	49 816	43 015	19 473	15 843	30 343	...	13 262	...	41.2	...	1.58	...	65.10
December	50 164	43 229	19 610	15 973	30 554	...	13 375	...	41.2	...	1.58	...	65.10
1953													
January	50 145	43 351	19 721	16 067	30 424	...	13 447	...	41.1	...	1.59	...	65.35
February	50 339	43 542	19 841	16 158	30 498	...	13 529	...	41.0	...	1.61	...	66.01
March	50 475	43 691	19 909	16 270	30 566	...	13 620	...	41.2	...	1.61	...	66.33
April	50 432	43 662	19 908	16 293	30 524	...	13 629	...	40.9	...	1.62	...	66.26
May	50 491	43 774	19 930	16 341	30 561	...	13 645	...	41.0	...	1.62	...	66.42
June	50 522	43 788	19 909	16 343	30 613	...	13 636	...	40.9	...	1.63	...	66.67
July	50 536	43 813	19 910	16 353	30 626	...	13 653	...	40.7	...	1.64	...	66.75
August	50 487	43 733	19 834	16 278	30 653	...	13 564	...	40.7	...	1.64	...	66.75
September	50 365	43 616	19 726	16 151	30 639	...	13 419	...	40.1	...	1.65	...	66.17
October	50 242	43 478	19 578	15 981	30 664	...	13 244	...	40.1	...	1.65	...	66.17
November	49 907	43 158	19 315	15 728	30 592	...	12 990	...	40.0	...	1.65	...	66.00
December	49 702	42 959	19 173	15 581	30 529	...	12 847	...	39.7	...	1.65	...	65.51

. . . = Not available.

Table 10-8. Employment, Hours, and Earnings, Total Nonfarm and Manufacturing, Historical Annual and Monthly—*Continued*

(Wage and salary workers on nonfarm payrolls, seasonally adjusted.)

Year and month	All wage and salary workers (thousands)					Production and nonsupervisory workers on private payrolls							
	Total	Private			Service-providing	Number (thousands)		Average hours per week		Average hourly earnings, dollars		Average weekly earnings, dollars	
		Total	Goods-producing			Total private	Manufac-turing	Total private	Manufac-turing	Total private	Manufac-turing	Total private	Manufac-turing
			Total	Manufac-turing									
1954													
January	49 468	42 708	18 963	15 440	30 505	. . .	12 706	. . .	39.5	. . .	1.65	. . .	65.18
February	49 382	42 599	18 880	15 307	30 502	. . .	12 593	. . .	39.7	. . .	1.65	. . .	65.51
March	49 158	42 362	18 748	15 197	30 410	. . .	12 493	. . .	39.6	. . .	1.65	. . .	65.34
April	49 178	42 372	18 602	15 065	30 576	. . .	12 361	. . .	39.7	. . .	1.65	. . .	65.51
May	48 965	42 136	18 476	14 974	30 489	. . .	12 282	. . .	39.7	. . .	1.67	. . .	66.30
June	48 896	42 050	18 400	14 910	30 496	. . .	12 220	. . .	39.7	. . .	1.67	. . .	66.30
July	48 835	41 967	18 280	14 799	30 555	. . .	12 121	. . .	39.7	. . .	1.66	. . .	65.90
August	48 825	41 933	18 251	14 772	30 574	. . .	12 089	. . .	39.8	. . .	1.66	. . .	66.07
September	48 882	41 988	18 261	14 805	30 621	. . .	12 104	. . .	39.9	. . .	1.66	. . .	66.23
October	48 944	42 044	18 321	14 841	30 623	. . .	12 142	. . .	39.7	. . .	1.67	. . .	66.30
November	49 178	42 214	18 438	14 913	30 740	. . .	12 206	. . .	40.1	. . .	1.68	. . .	67.37
December	49 331	42 374	18 508	14 967	30 823	. . .	12 254	. . .	40.1	. . .	1.68	. . .	67.37
1955													
January	49 497	42 544	18 609	15 034	30 888	. . .	12 309	. . .	40.4	. . .	1.69	. . .	68.28
February	49 644	42 721	18 726	15 138	30 918	. . .	12 408	. . .	40.6	. . .	1.70	. . .	69.02
March	49 963	43 025	18 910	15 258	31 053	. . .	12 524	. . .	40.7	. . .	1.71	. . .	69.60
April	50 247	43 288	19 067	15 375	31 180	. . .	12 626	. . .	40.8	. . .	1.71	. . .	69.77
May	50 512	43 521	19 223	15 493	31 289	. . .	12 731	. . .	41.1	. . .	1.73	. . .	71.10
June	50 790	43 770	19 331	15 585	31 459	. . .	12 810	. . .	40.8	. . .	1.73	. . .	70.58
July	50 985	43 936	19 376	15 614	31 609	. . .	12 818	. . .	40.7	. . .	1.75	. . .	71.23
August	51 111	44 088	19 432	15 679	31 679	. . .	12 866	. . .	40.7	. . .	1.76	. . .	71.63
September	51 262	44 195	19 427	15 668	31 835	. . .	12 830	. . .	40.7	. . .	1.77	. . .	72.04
October	51 431	44 313	19 482	15 740	31 949	. . .	12 896	. . .	41.0	. . .	1.77	. . .	72.57
November	51 592	44 509	19 554	15 813	32 038	. . .	12 967	. . .	41.1	. . .	1.78	. . .	73.16
December	51 805	44 673	19 608	15 859	32 197	. . .	13 009	. . .	40.9	. . .	1.78	. . .	72.80
1956													
January	51 975	44 808	19 665	15 882	32 310	. . .	13 011	. . .	40.8	. . .	1.78	. . .	72.62
February	52 167	44 955	19 731	15 889	32 436	. . .	12 986	. . .	40.7	. . .	1.79	. . .	72.85
March	52 295	45 043	19 691	15 829	32 604	. . .	12 905	. . .	40.6	. . .	1.80	. . .	73.08
April	52 375	45 099	19 811	15 909	32 564	. . .	12 970	. . .	40.5	. . .	1.82	. . .	73.71
May	52 506	45 139	19 825	15 893	32 681	. . .	12 925	. . .	40.4	. . .	1.83	. . .	73.93
June	52 584	45 217	19 905	15 835	32 679	. . .	12 836	. . .	40.3	. . .	1.83	. . .	73.75
July	51 954	44 549	19 390	15 468	32 564	. . .	12 435	. . .	40.3	. . .	1.82	. . .	73.35
August	52 630	45 179	19 922	15 893	32 708	. . .	12 860	. . .	40.3	. . .	1.85	. . .	74.56
September	52 601	45 120	19 860	15 863	32 741	. . .	12 822	. . .	40.5	. . .	1.87	. . .	75.74
October	52 781	45 262	19 918	15 937	32 863	. . .	12 908	. . .	40.6	. . .	1.88	. . .	76.33
November	52 822	45 269	19 886	15 916	32 936	. . .	12 864	. . .	40.4	. . .	1.88	. . .	75.95
December	52 930	45 346	19 926	15 957	33 004	. . .	12 882	. . .	40.6	. . .	1.91	. . .	77.55
1957													
January	52 888	45 268	19 833	15 970	33 055	. . .	12 881	. . .	40.4	. . .	1.90	. . .	76.76
February	53 097	45 451	19 933	15 998	33 164	. . .	12 885	. . .	40.5	. . .	1.91	. . .	77.36
March	53 157	45 485	19 936	15 994	33 221	. . .	12 855	. . .	40.4	. . .	1.92	. . .	77.57
April	53 238	45 537	19 887	15 970	33 351	. . .	12 811	. . .	40.0	. . .	1.91	. . .	76.40
May	53 149	45 436	19 834	15 931	33 315	. . .	12 762	. . .	40.0	. . .	1.92	. . .	76.80
June	53 066	45 364	19 777	15 873	33 289	. . .	12 697	. . .	40.0	. . .	1.92	. . .	76.80
July	53 123	45 369	19 735	15 854	33 388	. . .	12 669	. . .	40.0	. . .	1.93	. . .	77.20
August	53 126	45 369	19 728	15 867	33 398	. . .	12 673	. . .	40.0	. . .	1.94	. . .	77.60
September	52 932	45 183	19 545	15 710	33 387	. . .	12 528	. . .	39.7	. . .	1.95	. . .	77.42
October	52 765	44 997	19 421	15 599	33 344	. . .	12 437	. . .	39.4	. . .	1.96	. . .	77.22
November	52 559	44 790	19 260	15 466	33 299	. . .	12 301	. . .	39.2	. . .	1.96	. . .	76.83
December	52 385	44 539	19 111	15 332	33 274	. . .	12 170	. . .	39.1	. . .	1.95	. . .	76.25
1958													
January	52 077	44 256	18 902	15 130	33 175	. . .	11 969	. . .	38.9	. . .	1.95	. . .	75.86
February	51 576	43 744	18 529	14 908	33 047	. . .	11 757	. . .	38.7	. . .	1.95	. . .	75.47
March	51 300	43 452	18 335	14 670	32 965	. . .	11 532	. . .	38.8	. . .	1.96	. . .	76.05
April	51 027	43 159	18 120	14 506	32 907	. . .	11 373	. . .	38.9	. . .	1.96	. . .	76.24
May	50 913	43 019	18 008	14 414	32 905	. . .	11 294	. . .	38.9	. . .	1.97	. . .	76.63
June	50 912	42 986	17 984	14 408	32 928	. . .	11 300	. . .	39.1	. . .	1.98	. . .	77.42
July	51 037	43 065	18 038	14 450	32 999	. . .	11 349	. . .	39.3	. . .	1.98	. . .	77.81
August	51 231	43 219	18 147	14 524	33 084	. . .	11 412	. . .	39.5	. . .	2.01	. . .	79.40
September	51 506	43 490	18 331	14 658	33 175	. . .	11 556	. . .	39.5	. . .	2.00	. . .	79.00
October	51 486	43 455	18 218	14 503	33 268	. . .	11 394	. . .	39.6	. . .	2.00	. . .	79.20
November	51 944	43 916	18 610	14 827	33 334	. . .	11 702	. . .	39.9	. . .	2.03	. . .	81.00
December	52 088	43 988	18 592	14 877	33 496	. . .	11 743	. . .	39.9	. . .	2.04	. . .	81.40
1959													
January	52 480	44 375	18 796	14 998	33 684	. . .	11 849	. . .	40.2	. . .	2.04	. . .	82.01
February	52 687	44 571	18 890	15 115	33 797	. . .	11 950	. . .	40.3	. . .	2.05	. . .	82.62
March	53 016	44 884	19 069	15 259	33 947	. . .	12 078	. . .	40.4	. . .	2.07	. . .	83.63
April	53 320	45 178	19 269	15 385	34 051	. . .	12 185	. . .	40.5	. . .	2.08	. . .	84.24
May	53 549	45 396	19 378	15 487	34 171	. . .	12 277	. . .	40.7	. . .	2.08	. . .	84.66
June	53 679	45 536	19 462	15 554	34 217	. . .	12 330	. . .	40.6	. . .	2.09	. . .	84.85
July	53 803	45 630	19 529	15 623	34 274	. . .	12 366	. . .	40.3	. . .	2.09	. . .	84.23
August	53 334	45 153	19 049	15 202	34 285	. . .	11 936	. . .	40.4	. . .	2.07	. . .	83.63
September	53 429	45 190	19 052	15 254	34 377	. . .	11 984	. . .	40.4	. . .	2.08	. . .	84.03
October	53 359	45 094	18 925	15 158	34 434	. . .	11 864	. . .	40.1	. . .	2.07	. . .	83.01
November	53 635	45 351	19 108	15 300	34 527	. . .	11 991	. . .	39.9	. . .	2.07	. . .	82.59
December	54 175	45 807	19 425	15 573	34 750	. . .	12 254	. . .	40.3	. . .	2.11	. . .	85.03

. . . = Not available.

Table 10-8. Employment, Hours, and Earnings, Total Nonfarm and Manufacturing, Historical Annual and Monthly—*Continued*

(Wage and salary workers on nonfarm payrolls, seasonally adjusted.)

Year and month	All wage and salary workers (thousands)					Production and nonsupervisory workers on private payrolls							
	Total	Private			Service-providing	Number (thousands)		Average hours per week		Average hourly earnings, dollars		Average weekly earnings, dollars	
		Total	Goods-producing			Total private	Manufacturing	Total private	Manufacturing	Total private	Manufacturing	Total private	Manufacturing
			Total	Manufacturing									
1960													
January	54 274	45 967	19 491	15 687	34 783	. . .	12 362	. . .	40.6	. . .	2.13	. . .	86.48
February	54 513	46 187	19 605	15 765	34 908	. . .	12 434	. . .	40.3	. . .	2.14	. . .	86.24
March	54 458	45 933	19 373	15 707	35 085	. . .	12 362	. . .	40.0	. . .	2.14	. . .	85.60
April	54 812	46 278	19 446	15 654	35 366	. . .	12 299	. . .	40.0	. . .	2.14	. . .	85.60
May	54 473	46 041	19 374	15 575	35 099	. . .	12 218	. . .	40.1	. . .	2.14	. . .	85.81
June	54 347	45 915	19 240	15 466	35 107	. . .	12 102	. . .	39.9	. . .	2.14	. . .	85.39
July	54 304	45 862	19 170	15 413	35 134	. . .	12 047	. . .	39.9	. . .	2.14	. . .	85.39
August	54 271	45 799	19 105	15 360	35 166	. . .	11 986	. . .	39.7	. . .	2.15	. . .	85.36
September	54 228	45 734	19 057	15 330	35 171	. . .	11 956	. . .	39.4	. . .	2.16	. . .	85.10
October	54 144	45 642	18 952	15 231	35 192	. . .	11 846	. . .	39.7	. . .	2.16	. . .	85.75
November	53 962	45 446	18 799	15 112	35 163	. . .	11 726	. . .	39.3	. . .	2.15	. . .	84.50
December	53 744	45 147	18 548	14 947	35 196	. . .	11 556	. . .	38.4	. . .	2.16	. . .	82.94
1961													
January	53 683	45 119	18 508	14 863	35 175	. . .	11 473	. . .	39.3	. . .	2.16	. . .	84.89
February	53 556	44 969	18 418	14 801	35 138	. . .	11 414	. . .	39.4	. . .	2.16	. . .	85.10
March	53 662	45 051	18 438	14 802	35 224	. . .	11 410	. . .	39.5	. . .	2.16	. . .	85.32
April	53 626	44 997	18 432	14 825	35 194	. . .	11 444	. . .	39.5	. . .	2.18	. . .	86.11
May	53 785	45 121	18 523	14 932	35 262	. . .	11 544	. . .	39.7	. . .	2.19	. . .	86.94
June	53 977	45 289	18 618	14 981	35 359	. . .	11 593	. . .	40.0	. . .	2.20	. . .	88.00
July	54 123	45 399	18 640	15 029	35 483	. . .	11 639	. . .	40.0	. . .	2.21	. . .	88.40
August	54 298	45 534	18 725	15 093	35 573	. . .	11 701	. . .	40.1	. . .	2.22	. . .	89.02
September	54 388	45 592	18 730	15 080	35 658	. . .	11 679	. . .	39.5	. . .	2.20	. . .	86.90
October	54 522	45 717	18 805	15 143	35 717	. . .	11 731	. . .	40.3	. . .	2.23	. . .	89.87
November	54 743	45 931	18 927	15 259	35 816	. . .	11 842	. . .	40.7	. . .	2.23	. . .	90.76
December	54 871	46 035	18 981	15 309	35 890	. . .	11 872	. . .	40.4	. . .	2.24	. . .	90.50
1962													
January	54 891	46 040	18 936	15 322	35 955	. . .	11 865	. . .	40.0	. . .	2.25	. . .	90.00
February	55 187	46 309	19 109	15 411	36 078	. . .	11 950	. . .	40.4	. . .	2.26	. . .	91.30
March	55 276	46 375	19 109	15 451	36 167	. . .	11 970	. . .	40.6	. . .	2.26	. . .	91.76
April	55 602	46 680	19 258	15 524	36 344	. . .	12 034	. . .	40.6	. . .	2.26	. . .	91.76
May	55 627	46 669	19 253	15 513	36 374	. . .	12 011	. . .	40.6	. . .	2.27	. . .	92.16
June	55 644	46 644	19 186	15 518	36 458	. . .	12 006	. . .	40.5	. . .	2.26	. . .	91.53
July	55 746	46 720	19 248	15 522	36 498	. . .	12 004	. . .	40.5	. . .	2.27	. . .	91.94
August	55 838	46 775	19 251	15 517	36 587	. . .	11 990	. . .	40.5	. . .	2.28	. . .	92.34
September	55 977	46 888	19 305	15 568	36 672	. . .	12 033	. . .	40.5	. . .	2.28	. . .	92.34
October	56 041	46 927	19 301	15 569	36 740	. . .	12 034	. . .	40.3	. . .	2.29	. . .	92.29
November	56 056	46 911	19 260	15 530	36 796	. . .	11 977	. . .	40.5	. . .	2.29	. . .	92.75
December	56 028	46 902	19 219	15 520	36 809	. . .	11 961	. . .	40.3	. . .	2.29	. . .	92.29
1963													
January	56 116	46 912	19 257	15 545	36 859	. . .	11 974	. . .	40.5	. . .	2.30	. . .	93.15
February	56 230	46 999	19 228	15 542	37 002	. . .	11 965	. . .	40.5	. . .	2.31	. . .	93.56
March	56 322	47 077	19 233	15 564	37 089	. . .	11 990	. . .	40.5	. . .	2.32	. . .	93.96
April	56 580	47 316	19 343	15 602	37 237	. . .	12 029	. . .	40.5	. . .	2.32	. . .	93.96
May	56 616	47 328	19 399	15 641	37 217	. . .	12 066	. . .	40.5	. . .	2.33	. . .	94.37
June	56 658	47 356	19 371	15 624	37 287	. . .	12 050	. . .	40.7	. . .	2.34	. . .	95.24
July	56 794	47 460	19 423	15 646	37 371	. . .	12 080	. . .	40.6	. . .	2.35	. . .	95.41
August	56 910	47 542	19 437	15 644	37 473	. . .	12 060	. . .	40.6	. . .	2.34	. . .	95.00
September	57 077	47 660	19 483	15 674	37 594	. . .	12 088	. . .	40.6	. . .	2.36	. . .	95.82
October	57 284	47 805	19 517	15 714	37 767	. . .	12 131	. . .	40.7	. . .	2.36	. . .	96.05
November	57 255	47 771	19 456	15 675	37 799	. . .	12 072	. . .	40.7	. . .	2.37	. . .	96.46
December	57 360	47 863	19 493	15 712	37 867	. . .	12 108	. . .	40.6	. . .	2.38	. . .	96.63
1964													
January	57 487	47 925	19 406	15 715	38 081	39 913	12 132	38.2	40.1	2.50	2.38	95.50	95.44
February	57 751	48 170	19 570	15 742	38 181	40 123	12 165	38.5	40.7	2.50	2.38	96.25	96.87
March	57 898	48 287	19 587	15 770	38 311	40 171	12 190	38.5	40.6	2.50	2.38	96.25	96.63
April	57 922	48 278	19 593	15 785	38 329	40 209	12 211	38.6	40.7	2.52	2.40	97.27	97.68
May	58 089	48 419	19 630	15 812	38 459	40 332	12 231	38.6	40.8	2.52	2.40	97.27	97.92
June	58 221	48 552	19 682	15 839	38 539	40 448	12 255	38.6	40.8	2.53	2.41	97.66	98.33
July	58 413	48 736	19 740	15 887	38 673	40 624	12 309	38.5	40.8	2.53	2.41	97.41	98.33
August	58 619	48 887	19 810	15 948	38 809	40 771	12 354	38.5	40.9	2.55	2.42	98.18	98.98
September	58 903	49 117	19 943	16 073	38 960	41 026	12 479	38.5	40.9	2.55	2.44	98.18	99.80
October	58 794	48 949	19 723	15 821	39 071	40 828	12 224	38.5	40.7	2.55	2.40	98.18	97.68
November	59 217	49 338	20 026	16 096	39 191	41 157	12 477	38.6	41.0	2.56	2.42	98.82	99.22
December	59 421	49 524	20 111	16 176	39 310	41 307	12 551	38.7	41.2	2.58	2.44	99.85	100.53
1965													
January	59 583	49 646	20 173	16 245	39 410	41 453	12 603	38.7	41.3	2.58	2.45	99.85	101.19
February	59 800	49 826	20 216	16 291	39 584	41 583	12 644	38.7	41.3	2.59	2.46	100.23	101.60
March	60 003	49 993	20 292	16 353	39 711	41 674	12 701	38.7	41.3	2.60	2.47	100.62	102.01
April	60 259	50 208	20 317	16 418	39 942	41 885	12 752	38.7	41.2	2.60	2.47	100.62	101.76
May	60 492	50 398	20 444	16 477	40 048	42 044	12 792	38.8	41.3	2.62	2.49	101.66	102.84
June	60 690	50 562	20 522	16 554	40 168	42 183	12 854	38.6	41.2	2.62	2.49	101.13	102.59
July	60 963	50 762	20 611	16 669	40 352	42 364	12 962	38.6	41.2	2.63	2.49	101.52	102.59
August	61 228	50 957	20 726	16 732	40 502	42 537	12 994	38.5	41.1	2.64	2.50	101.64	102.75
September	61 490	51 152	20 808	16 802	40 682	42 726	13 046	38.6	41.1	2.65	2.51	102.29	103.16
October	61 718	51 340	20 895	16 864	40 823	42 871	13 100	38.5	41.2	2.66	2.52	102.41	103.82
November	61 997	51 561	21 021	16 962	40 976	43 045	13 175	38.6	41.3	2.67	2.52	103.06	104.08
December	62 321	51 822	21 151	17 051	41 170	43 270	13 243	38.6	41.3	2.68	2.53	103.45	104.49

. . . = Not available.

Table 10-8. Employment, Hours, and Earnings, Total Nonfarm and Manufacturing, Historical Annual and Monthly—*Continued*

(Wage and salary workers on nonfarm payrolls, seasonally adjusted.)

Year and month	All wage and salary workers (thousands)					Production and nonsupervisory workers on private payrolls								
		Private				Number (thousands)		Average hours per week		Average hourly earnings, dollars		Average weekly earnings, dollars	
	Total	Total	Goods-producing		Service-providing	Total private	Manufac-turing	Total private	Manufac-turing	Total private	Manufac-turing	Total private	Manufac-turing
			Total	Manufac-turing									
1966													
January	62 528	51 987	21 214	17 143	41 314	43 401	13 301	38.6	41.5	2.68	2.54	103.45	105.41
February	62 796	52 185	21 315	17 288	41 481	43 552	13 426	38.7	41.7	2.69	2.56	104.10	106.75
March	63 192	52 500	21 515	17 400	41 677	43 801	13 508	38.7	41.6	2.70	2.56	104.49	106.50
April	63 436	52 677	21 568	17 517	41 868	43 959	13 598	38.7	41.8	2.71	2.58	104.88	107.84
May	63 711	52 890	21 675	17 625	42 036	44 138	13 678	38.5	41.5	2.72	2.58	104.72	107.07
June	64 110	53 208	21 846	17 733	42 264	44 390	13 753	38.5	41.4	2.72	2.58	104.72	106.81
July	64 301	53 327	21 872	17 760	42 429	44 484	13 758	38.4	41.2	2.74	2.60	105.22	107.12
August	64 507	53 501	21 972	17 882	42 535	44 596	13 843	38.4	41.4	2.75	2.61	105.60	108.05
September	64 644	53 581	21 948	17 886	42 696	44 659	13 841	38.3	41.2	2.76	2.63	105.71	108.36
October	64 854	53 727	21 991	17 956	42 863	44 789	13 906	38.4	41.3	2.77	2.64	106.37	109.03
November	65 019	53 816	21 988	17 981	43 031	44 834	13 918	38.3	41.2	2.78	2.65	106.47	109.18
December	65 200	53 944	22 008	17 998	43 192	44 916	13 906	38.2	40.9	2.78	2.64	106.20	107.98
1967													
January	65 407	54 092	22 057	18 033	43 350	45 051	13 925	38.3	41.1	2.79	2.65	106.86	108.92
February	65 428	54 075	21 987	17 978	43 441	44 967	13 853	37.9	40.4	2.81	2.67	106.50	107.87
March	65 530	54 133	21 919	17 940	43 611	44 991	13 797	37.9	40.5	2.81	2.67	106.50	108.14
April	65 467	54 032	21 842	17 878	43 625	44 871	13 712	37.8	40.4	2.82	2.68	106.60	108.27
May	65 619	54 145	21 779	17 832	43 840	44 961	13 664	37.8	40.4	2.83	2.69	106.97	108.68
June	65 750	54 216	21 761	17 812	43 989	45 004	13 632	37.8	40.4	2.84	2.69	107.35	108.68
July	65 887	54 343	21 772	17 784	44 115	45 105	13 602	37.8	40.5	2.86	2.71	108.11	109.76
August	66 142	54 552	21 887	17 905	44 255	45 267	13 681	37.8	40.6	2.87	2.73	108.49	110.84
September	66 164	54 541	21 775	17 794	44 389	45 236	13 559	37.8	40.6	2.88	2.73	108.86	110.84
October	66 225	54 583	21 779	17 800	44 446	45 278	13 587	37.8	40.6	2.89	2.74	109.24	111.24
November	66 703	55 008	21 996	17 985	44 707	45 701	13 777	37.9	40.6	2.91	2.75	110.29	111.65
December	66 900	55 165	22 037	18 025	44 863	45 800	13 782	37.7	40.7	2.92	2.77	110.08	112.74
1968													
January	66 805	55 011	21 917	18 040	44 888	45 655	13 798	37.6	40.4	2.94	2.81	110.54	113.52
February	67 215	55 396	22 117	18 054	45 098	45 980	13 793	37.8	40.8	2.95	2.82	111.51	115.06
March	67 295	55 453	22 119	18 067	45 176	46 041	13 803	37.7	40.8	2.97	2.84	111.97	115.87
April	67 555	55 677	22 207	18 131	45 348	46 239	13 858	37.6	40.3	2.98	2.85	112.05	114.86
May	67 653	55 748	22 255	18 190	45 398	46 267	13 900	37.7	40.9	3.00	2.87	113.10	117.38
June	67 904	55 917	22 264	18 228	45 640	46 402	13 921	37.8	40.9	3.01	2.88	113.78	117.79
July	68 125	56 107	22 329	18 265	45 796	46 563	13 953	37.7	40.8	3.03	2.89	114.23	117.91
August	68 328	56 286	22 350	18 254	45 978	46 670	13 903	37.7	40.7	3.03	2.89	114.23	117.62
September	68 487	56 420	22 390	18 252	46 097	46 792	13 914	37.7	40.9	3.06	2.92	115.36	119.43
October	68 720	56 619	22 419	18 293	46 301	46 990	13 974	37.7	41.0	3.07	2.94	115.74	120.54
November	68 985	56 878	22 512	18 346	46 473	47 244	14 028	37.5	40.9	3.09	2.96	115.88	121.06
December	69 246	57 101	22 617	18 410	46 629	47 385	14 044	37.5	40.7	3.11	2.97	116.63	120.88
1969													
January	69 438	57 229	22 644	18 432	46 794	47 529	14 086	37.7	40.8	3.12	2.99	117.62	121.99
February	69 700	57 476	22 755	18 502	46 945	47 696	14 134	37.5	40.4	3.14	3.00	117.75	121.20
March	69 905	57 676	22 813	18 558	47 092	47 852	14 169	37.6	40.8	3.15	3.01	118.44	122.81
April	70 072	57 827	22 815	18 554	47 257	47 959	14 144	37.7	41.0	3.17	3.03	119.51	124.23
May	70 328	58 044	22 899	18 588	47 429	48 122	14 161	37.6	40.7	3.19	3.04	119.94	123.73
June	70 636	58 277	22 981	18 640	47 655	48 329	14 206	37.5	40.7	3.20	3.05	120.00	124.14
July	70 729	58 389	22 990	18 642	47 739	48 435	14 194	37.5	40.6	3.22	3.08	120.75	125.05
August	71 006	58 633	23 111	18 767	47 895	48 616	14 287	37.5	40.6	3.24	3.10	121.50	125.86
September	70 917	58 538	22 988	18 620	47 929	48 524	14 159	37.5	40.6	3.26	3.12	122.25	126.67
October	71 120	58 690	22 976	18 613	48 144	48 669	14 174	37.4	40.5	3.28	3.13	122.67	126.77
November	71 087	58 639	22 840	18 467	48 247	48 589	14 035	37.5	40.5	3.29	3.14	123.38	127.17
December	71 240	58 763	22 884	18 485	48 356	48 638	14 013	37.5	40.6	3.30	3.15	123.75	127.89
1970													
January	71 176	58 680	22 726	18 424	48 450	48 564	13 964	37.3	40.4	3.31	3.16	123.46	127.66
February	71 304	58 786	22 747	18 361	48 557	48 600	13 897	37.3	40.2	3.33	3.17	124.21	127.43
March	71 452	58 849	22 738	18 360	48 714	48 690	13 917	37.2	40.1	3.35	3.19	124.62	127.92
April	71 348	58 643	22 552	18 207	48 796	48 480	13 785	37.0	39.8	3.36	3.19	124.32	126.96
May	71 123	58 455	22 336	18 029	48 787	48 287	13 616	37.0	39.8	3.38	3.22	125.06	128.16
June	71 029	58 362	22 241	17 930	48 788	48 225	13 554	36.9	39.8	3.39	3.24	125.09	128.95
July	71 053	58 356	22 195	17 877	48 858	48 243	13 527	37.0	40.0	3.41	3.25	126.17	130.00
August	70 933	58 222	22 105	17 779	48 828	48 088	13 449	37.0	39.8	3.43	3.26	126.91	129.75
September	70 948	58 207	21 988	17 692	48 960	48 101	13 401	36.8	39.6	3.45	3.29	126.96	130.28
October	70 519	57 726	21 477	17 173	49 042	47 619	12 905	36.8	39.5	3.46	3.26	127.33	128.77
November	70 409	57 579	21 345	17 024	49 064	47 466	12 781	36.7	39.5	3.47	3.26	127.35	128.77
December	70 790	57 945	21 673	17 309	49 117	47 778	13 062	36.8	39.5	3.50	3.32	128.80	131.14
1971													
January	70 866	57 988	21 594	17 280	49 272	47 859	13 069	36.8	39.9	3.52	3.36	129.54	134.06
February	70 806	57 929	21 514	17 216	49 292	47 779	13 033	36.7	39.7	3.54	3.39	129.92	134.58
March	70 859	57 951	21 491	17 154	49 368	47 819	12 984	36.7	39.8	3.56	3.39	130.65	134.92
April	71 037	58 092	21 552	17 149	49 485	47 966	12 993	36.8	39.9	3.57	3.41	131.38	136.06
May	71 247	58 277	21 645	17 225	49 602	48 153	13 081	36.7	40.0	3.60	3.43	132.12	137.20
June	71 253	58 245	21 568	17 139	49 685	48 108	13 012	36.8	39.9	3.62	3.45	133.22	137.66
July	71 315	58 304	21 564	17 126	49 751	48 169	13 001	36.7	40.0	3.63	3.46	133.22	138.40
August	71 370	58 329	21 570	17 115	49 800	48 164	12 993	36.7	39.8	3.66	3.48	134.32	138.50
September	71 617	58 549	21 650	17 154	49 967	48 363	13 038	36.7	39.7	3.67	3.48	134.69	138.16
October	71 642	58 527	21 604	17 126	50 038	48 305	13 027	36.8	39.9	3.68	3.50	135.42	139.65
November	71 846	58 698	21 684	17 166	50 162	48 441	13 064	36.9	40.0	3.69	3.49	136.16	139.60
December	72 108	58 918	21 741	17 202	50 367	48 613	13 078	36.9	40.2	3.73	3.55	137.64	142.71

Table 10-8. Employment, Hours, and Earnings, Total Nonfarm and Manufacturing, Historical Annual and Monthly—Continued

(Wage and salary workers on nonfarm payrolls, seasonally adjusted.)

Year and month	All wage and salary workers (thousands)					Production and nonsupervisory workers on private payrolls							
	Total	Private			Service-providing	Number (thousands)		Average hours per week		Average hourly earnings, dollars		Average weekly earnings, dollars	
		Total	Goods-producing			Total private	Manufac-turing	Total private	Manufac-turing	Total private	Manufac-turing	Total private	Manufac-turing
			Total	Manufac-turing									
1972													
January	72 445	59 179	21 865	17 283	50 580	49 035	13 173	36.9	40.2	3.80	3.57	140.22	143.51
February	72 652	59 354	21 915	17 361	50 737	49 143	13 235	36.9	40.4	3.82	3.61	140.96	145.84
March	72 945	59 616	22 036	17 447	50 909	49 437	13 316	36.9	40.4	3.84	3.63	141.70	146.65
April	73 163	59 805	22 099	17 508	51 064	49 562	13 373	36.9	40.5	3.86	3.65	142.43	147.83
May	73 467	60 051	22 222	17 602	51 245	49 745	13 451	36.8	40.5	3.87	3.67	142.42	148.64
June	73 760	60 355	22 282	17 641	51 478	49 995	13 475	36.9	40.6	3.88	3.68	143.17	149.41
July	73 708	60 226	22 162	17 556	51 546	49 844	13 387	36.8	40.5	3.90	3.69	143.52	149.45
August	74 138	60 608	22 400	17 741	51 738	50 155	13 562	36.8	40.6	3.92	3.73	144.26	151.44
September	74 263	60 688	22 456	17 774	51 807	50 226	13 572	36.9	40.6	3.94	3.75	145.39	152.25
October	74 673	61 067	22 613	17 893	52 060	50 552	13 681	37.0	40.7	3.97	3.78	146.89	153.85
November	74 967	61 324	22 688	18 005	52 279	50 789	13 783	36.9	40.7	3.98	3.79	146.86	154.25
December	75 270	61 586	22 772	18 158	52 498	51 054	13 902	36.8	40.6	4.01	3.83	147.57	155.50
1973													
January	75 621	61 931	22 955	18 276	52 666	51 349	14 006	36.8	40.4	4.03	3.86	148.30	155.94
February	76 017	62 306	23 160	18 410	52 857	51 685	14 127	36.9	40.9	4.04	3.87	149.08	158.28
March	76 285	62 540	23 262	18 493	53 023	51 901	14 181	37.0	40.9	4.06	3.88	150.22	158.69
April	76 455	62 678	23 316	18 530	53 139	51 983	14 192	36.9	40.8	4.08	3.91	150.55	159.53
May	76 646	62 829	23 382	18 564	53 264	52 082	14 217	36.9	40.7	4.10	3.93	151.29	159.95
June	76 887	63 015	23 485	18 606	53 402	52 233	14 253	36.9	40.7	4.12	3.95	152.03	160.77
July	76 911	63 046	23 522	18 598	53 389	52 240	14 232	36.9	40.7	4.15	3.98	153.14	161.99
August	77 166	63 262	23 559	18 629	53 607	52 394	14 251	36.9	40.6	4.16	4.00	153.50	162.40
September	77 276	63 384	23 548	18 609	53 728	52 410	14 213	36.8	40.7	4.19	4.03	154.19	164.02
October	77 606	63 629	23 641	18 702	53 965	52 655	14 289	36.7	40.6	4.21	4.05	154.51	164.43
November	77 912	63 877	23 719	18 773	54 193	52 845	14 342	36.9	40.6	4.23	4.07	156.09	165.24
December	78 035	63 965	23 779	18 820	54 256	52 954	14 386	36.7	40.6	4.25	4.09	155.98	166.05
1974													
January	78 104	64 014	23 709	18 788	54 395	52 896	14 340	36.6	40.5	4.26	4.10	155.92	166.05
February	78 254	64 119	23 718	18 727	54 536	52 970	14 269	36.6	40.4	4.29	4.13	157.01	166.85
March	78 296	64 144	23 687	18 700	54 609	52 951	14 223	36.6	40.4	4.31	4.15	157.75	167.66
April	78 382	64 191	23 670	18 702	54 712	53 006	14 225	36.4	39.5	4.34	4.16	157.98	164.32
May	78 547	64 326	23 635	18 688	54 912	53 094	14 199	36.5	40.3	4.39	4.25	160.24	171.28
June	78 602	64 363	23 591	18 690	55 011	53 106	14 197	36.5	40.2	4.43	4.30	161.70	172.86
July	78 635	64 347	23 462	18 656	55 173	53 046	14 152	36.5	40.1	4.45	4.33	162.43	173.63
August	78 619	64 291	23 396	18 570	55 223	53 026	14 089	36.5	40.2	4.49	4.38	163.89	176.08
September	78 611	64 189	23 274	18 492	55 337	52 915	14 025	36.4	40.0	4.53	4.42	164.89	176.80
October	78 629	64 145	23 118	18 364	55 511	52 835	13 884	36.3	40.0	4.56	4.48	165.53	179.20
November	78 261	63 729	22 773	18 077	55 488	52 411	13 607	36.1	39.5	4.57	4.49	164.98	177.36
December	77 657	63 098	22 303	17 693	55 354	51 856	13 259	36.1	39.3	4.61	4.52	166.42	177.64
1975													
January	77 297	62 673	21 974	17 344	55 323	51 438	12 933	36.1	39.2	4.61	4.54	166.42	177.97
February	76 919	62 172	21 512	17 004	55 407	50 934	12 622	35.9	38.9	4.63	4.58	166.22	178.16
March	76 649	61 895	21 274	16 853	55 375	50 665	12 483	35.7	38.8	4.66	4.63	166.36	179.64
April	76 461	61 666	21 109	16 759	55 352	50 442	12 407	35.8	39.0	4.66	4.63	166.83	180.57
May	76 623	61 796	21 097	16 746	55 526	50 557	12 406	35.9	39.0	4.68	4.65	168.01	181.35
June	76 520	61 736	21 018	16 690	55 502	50 537	12 371	35.9	39.2	4.72	4.68	169.45	183.46
July	76 769	61 908	20 981	16 678	55 788	50 729	12 374	35.9	39.4	4.73	4.71	169.81	185.57
August	77 155	62 285	21 176	16 824	55 979	51 070	12 538	36.1	39.7	4.77	4.75	172.20	188.58
September	77 230	62 406	21 284	16 904	55 946	51 182	12 617	36.1	39.8	4.79	4.78	172.92	190.24
October	77 535	62 635	21 384	16 984	56 151	51 375	12 687	36.1	39.9	4.81	4.80	173.64	191.52
November	77 680	62 777	21 442	17 025	56 238	51 456	12 700	36.1	39.9	4.85	4.83	175.09	192.72
December	78 018	63 072	21 602	17 140	56 416	51 759	12 811	36.2	40.2	4.87	4.86	176.29	195.37
1976													
January	78 506	63 537	21 799	17 287	56 707	52 182	12 945	36.3	40.3	4.90	4.90	177.87	197.47
February	78 817	63 836	21 893	17 384	56 924	52 430	13 030	36.3	40.4	4.94	4.94	179.32	199.58
March	79 049	64 062	21 980	17 470	57 069	52 615	13 092	36.0	40.2	4.96	4.98	178.56	200.20
April	79 292	64 307	22 050	17 541	57 242	52 812	13 160	36.0	39.6	4.98	4.98	179.28	197.21
May	79 311	64 340	21 988	17 513	57 323	52 802	13 130	36.1	40.3	5.02	5.04	181.22	203.11
June	79 376	64 413	21 982	17 521	57 394	52 830	13 121	36.1	40.2	5.04	5.07	181.94	203.81
July	79 547	64 554	21 988	17 524	57 559	52 975	13 124	36.1	40.3	5.07	5.11	183.03	205.93
August	79 704	64 697	22 038	17 596	57 666	53 072	13 195	36.0	40.2	5.12	5.16	184.32	207.43
September	79 892	64 921	22 142	17 665	57 750	53 268	13 253	36.0	40.2	5.15	5.20	185.40	209.04
October	79 905	64 877	22 037	17 548	57 868	53 164	13 109	35.9	40.0	5.17	5.19	185.60	207.60
November	80 237	65 164	22 207	17 682	58 030	53 362	13 199	35.9	40.1	5.21	5.25	187.04	210.53
December	80 448	65 373	22 261	17 719	58 187	53 542	13 230	35.9	39.9	5.23	5.29	187.76	211.07
1977													
January	80 692	65 636	22 320	17 803	58 372	53 755	13 305	35.6	39.4	5.26	5.35	187.26	210.79
February	80 988	65 932	22 478	17 843	58 510	54 016	13 331	36.0	40.2	5.30	5.36	190.80	215.47
March	81 391	66 341	22 672	17 941	58 719	54 389	13 424	35.9	40.3	5.33	5.40	191.35	217.62
April	81 729	66 654	22 807	18 024	58 922	54 672	13 490	36.0	40.4	5.37	5.45	193.32	220.18
May	82 089	66 957	22 919	18 107	59 170	54 940	13 567	36.0	40.5	5.40	5.49	194.40	222.35
June	82 488	67 281	23 046	18 192	59 442	55 194	13 622	36.0	40.5	5.43	5.54	195.48	224.37
July	82 836	67 537	23 106	18 259	59 730	55 396	13 670	35.9	40.4	5.46	5.58	196.01	225.43
August	83 074	67 746	23 124	18 276	59 950	55 543	13 679	35.9	40.4	5.48	5.61	196.73	226.64
September	83 532	68 129	23 244	18 334	60 288	55 858	13 714	35.9	40.4	5.51	5.65	197.81	228.26
October	83 794	68 331	23 279	18 356	60 515	56 003	13 722	36.0	40.6	5.56	5.69	200.16	231.01
November	84 173	68 658	23 371	18 419	60 802	56 281	13 771	35.9	40.5	5.59	5.72	200.68	231.66
December	84 408	68 870	23 371	18 531	61 037	56 465	13 861	35.8	40.4	5.61	5.75	200.84	232.30

Table 10-8. Employment, Hours, and Earnings, Total Nonfarm and Manufacturing, Historical Annual and Monthly—*Continued*

(Wage and salary workers on nonfarm payrolls, seasonally adjusted.)

Year and month	All wage and salary workers (thousands)					Production and nonsupervisory workers on private payrolls							
	Total	Private			Service-providing	Number (thousands)		Average hours per week		Average hourly earnings, dollars		Average weekly earnings, dollars	
		Total	Goods-producing			Total private	Manufac-turing	Total private	Manufac-turing	Total private	Manufac-turing	Total private	Manufac-turing
			Total	Manufac-turing									
1978													
January	84 595	68 984	23 374	18 593	61 221	56 547	13 917	35.3	39.5	5.66	5.83	199.80	230.29
February	84 948	69 277	23 453	18 639	61 495	56 768	13 950	35.6	39.9	5.69	5.86	202.56	233.81
March	85 461	69 730	23 649	18 699	61 812	57 176	13 992	35.8	40.5	5.73	5.88	205.13	238.14
April	86 163	70 366	24 008	18 772	62 155	57 711	14 037	35.8	40.4	5.79	5.93	207.28	239.57
May	86 509	70 675	24 082	18 848	62 427	57 941	14 096	35.8	40.4	5.82	5.96	208.36	240.78
June	86 951	71 099	24 238	18 919	62 713	58 262	14 129	35.9	40.6	5.87	6.01	210.73	244.01
July	87 205	71 304	24 300	18 951	62 905	58 422	14 152	35.9	40.6	5.90	6.06	211.81	246.04
August	87 481	71 590	24 374	19 006	63 107	58 626	14 187	35.8	40.5	5.93	6.09	212.29	246.65
September	87 618	71 799	24 444	19 068	63 174	58 819	14 241	35.8	40.5	5.97	6.15	213.73	249.08
October	87 954	72 096	24 548	19 142	63 406	59 017	14 291	35.8	40.5	6.03	6.20	215.87	251.10
November	88 391	72 497	24 678	19 257	63 713	59 377	14 388	35.7	40.6	6.06	6.26	216.34	254.16
December	88 673	72 762	24 758	19 334	63 915	59 599	14 459	35.7	40.5	6.10	6.31	217.77	255.56
1979													
January	88 810	72 873	24 740	19 388	64 070	59 659	14 497	35.6	40.4	6.14	6.36	218.58	256.94
February	89 054	73 107	24 784	19 409	64 270	59 840	14 501	35.7	40.5	6.18	6.40	220.63	259.20
March	89 480	73 524	24 998	19 453	64 482	60 216	14 526	35.8	40.6	6.22	6.45	222.68	261.87
April	89 418	73 441	24 958	19 450	64 460	60 069	14 515	35.3	39.3	6.22	6.43	219.57	252.70
May	89 791	73 801	25 071	19 509	64 720	60 369	14 551	35.6	40.2	6.28	6.52	223.57	262.10
June	90 109	74 064	25 161	19 553	64 948	60 583	14 566	35.6	40.2	6.32	6.56	224.99	263.71
July	90 215	74 065	25 163	19 531	65 052	60 559	14 536	35.6	40.2	6.36	6.59	226.42	264.92
August	90 297	74 068	25 059	19 406	65 238	60 517	14 397	35.6	40.1	6.40	6.63	227.84	265.86
September	90 325	74 197	25 088	19 442	65 237	60 631	14 440	35.6	40.1	6.45	6.67	229.62	267.47
October	90 482	74 346	25 038	19 390	65 444	60 748	14 384	35.6	40.2	6.47	6.71	230.33	269.74
November	90 576	74 403	24 947	19 299	65 629	60 781	14 295	35.6	40.1	6.51	6.74	231.76	270.27
December	90 673	74 493	24 970	19 301	65 703	60 863	14 300	35.5	40.1	6.57	6.80	233.24	272.68
1980													
January	90 802	74 601	24 949	19 282	65 853	60 899	14 241	35.4	40.0	6.57	6.82	232.58	272.80
February	90 882	74 656	24 874	19 219	66 008	60 966	14 170	35.4	40.1	6.63	6.88	234.70	275.89
March	90 994	74 698	24 818	19 217	66 176	60 990	14 165	35.3	39.9	6.70	6.95	236.51	277.31
April	90 850	74 267	24 507	18 973	66 343	60 544	13 914	35.2	39.8	6.72	6.97	236.54	277.41
May	90 419	73 965	24 234	18 726	66 185	60 198	13 632	35.1	39.3	6.76	7.02	237.28	275.89
June	90 099	73 658	23 968	18 490	66 131	59 895	13 405	35.0	39.2	6.82	7.10	238.70	278.32
July	89 837	73 419	23 698	18 276	66 139	59 698	13 227	34.9	39.1	6.86	7.16	239.41	279.96
August	90 097	73 687	23 860	18 414	66 237	59 913	13 349	35.1	39.5	6.91	7.24	242.54	285.98
September	90 210	73 880	23 931	18 445	66 279	60 079	13 396	35.1	39.6	6.95	7.30	243.95	289.08
October	90 491	74 105	24 012	18 506	66 479	60 244	13 437	35.2	39.8	7.02	7.38	247.10	293.72
November	90 748	74 357	24 123	18 601	66 625	60 455	13 528	35.3	39.9	7.09	7.47	250.28	298.05
December	90 943	74 570	24 182	18 640	66 761	60 611	13 550	35.3	40.1	7.13	7.52	251.69	301.55
1981													
January	91 037	74 677	24 152	18 639	66 885	60 715	13 545	35.4	40.1	7.19	7.58	254.53	303.96
February	91 105	74 759	24 118	18 613	66 987	60 742	13 518	35.2	39.8	7.23	7.62	254.50	303.28
March	91 210	74 918	24 203	18 647	67 007	60 881	13 550	35.3	40.0	7.29	7.68	257.34	307.20
April	91 283	75 023	24 151	18 711	67 132	60 979	13 594	35.3	40.1	7.33	7.76	258.75	311.18
May	91 293	75 095	24 148	18 766	67 145	60 981	13 633	35.3	40.2	7.37	7.81	260.16	313.96
June	91 490	75 331	24 290	18 789	67 200	61 141	13 632	35.2	40.0	7.42	7.85	261.18	314.00
July	91 602	75 427	24 302	18 785	67 300	61 228	13 629	35.2	39.9	7.46	7.89	262.59	314.81
August	91 566	75 456	24 258	18 748	67 308	61 223	13 573	35.2	40.0	7.53	7.97	265.06	318.80
September	91 479	75 448	24 210	18 712	67 269	61 241	13 565	35.0	39.6	7.57	8.03	264.95	317.99
October	91 380	75 311	24 051	18 566	67 329	61 074	13 399	35.1	39.6	7.59	8.06	266.41	319.18
November	91 171	75 093	23 875	18 409	67 296	60 825	13 235	35.1	39.4	7.64	8.08	268.16	318.35
December	90 893	74 820	23 656	18 223	67 237	60 520	13 033	34.9	39.2	7.64	8.09	266.64	317.13
1982													
January	90 567	74 526	23 362	18 047	67 205	60 218	12 874	34.1	37.3	7.72	8.26	263.25	308.10
February	90 562	74 551	23 361	17 981	67 201	60 285	12 831	35.1	39.6	7.73	8.21	271.32	325.12
March	90 432	74 408	23 214	17 857	67 218	60 150	12 727	34.9	39.1	7.76	8.24	270.82	322.18
April	90 152	74 142	22 996	17 683	67 156	59 877	12 566	34.8	39.1	7.77	8.28	270.40	323.75
May	90 107	74 104	22 884	17 588	67 223	59 851	12 506	34.8	39.1	7.83	8.33	272.48	325.70
June	89 864	73 848	22 643	17 430	67 221	59 599	12 365	34.8	39.2	7.85	8.37	273.18	328.10
July	89 522	73 632	22 434	17 278	67 088	59 422	12 261	34.8	39.2	7.89	8.40	274.57	329.28
August	89 364	73 434	22 268	17 160	67 096	59 213	12 153	34.7	39.0	7.94	8.43	275.52	328.77
September	89 183	73 260	22 146	17 074	67 037	59 078	12 104	34.8	39.0	7.94	8.45	276.31	329.55
October	88 906	72 950	21 879	16 853	67 027	58 761	11 880	34.6	38.9	7.96	8.44	275.42	328.32
November	88 783	72 806	21 736	16 722	67 047	58 625	11 766	34.6	39.0	7.98	8.46	276.11	329.94
December	88 769	72 788	21 688	16 690	67 081	58 599	11 746	34.7	39.0	8.02	8.49	278.29	331.11
1983													
January	88 993	72 970	21 757	16 705	67 236	58 828	11 783	34.8	39.3	8.06	8.52	280.49	334.84
February	88 918	72 914	21 676	16 706	67 242	58 803	11 794	34.5	39.3	8.10	8.59	279.45	337.59
March	89 090	73 085	21 649	16 711	67 441	58 971	11 817	34.7	39.6	8.10	8.59	281.07	340.16
April	89 366	73 376	21 729	16 794	67 637	59 214	11 894	34.8	39.7	8.13	8.61	282.92	341.82
May	89 643	73 638	21 829	16 885	67 814	59 458	11 987	34.9	40.0	8.17	8.64	285.13	345.60
June	90 022	74 002	21 949	16 960	68 073	59 819	12 054	34.9	40.1	8.19	8.66	285.83	347.27
July	90 440	74 429	22 103	17 059	68 337	60 203	12 155	34.9	40.3	8.23	8.71	287.23	351.01
August	90 132	74 116	22 207	17 118	67 925	59 834	12 200	34.9	40.3	8.20	8.71	286.18	351.01
September	91 247	75 205	22 381	17 255	68 866	60 859	12 318	35.0	40.6	8.26	8.76	289.10	355.66
October	91 518	75 532	22 546	17 367	68 972	61 108	12 408	35.2	40.6	8.31	8.80	292.51	357.28
November	91 871	75 874	22 698	17 479	69 173	61 392	12 503	35.1	40.6	8.32	8.84	292.03	358.90
December	92 227	76 219	22 803	17 551	69 424	61 680	12 553	35.1	40.5	8.33	8.87	292.38	359.24

Table 10-8. Employment, Hours, and Earnings, Total Nonfarm and Manufacturing, Historical Annual and Monthly—Continued

(Wage and salary workers on nonfarm payrolls, seasonally adjusted.)

Year and month	All wage and salary workers (thousands)					Production and nonsupervisory workers on private payrolls							
	Total	Private			Service-providing	Number (thousands)		Average hours per week		Average hourly earnings, dollars		Average weekly earnings, dollars	
		Total	Goods-producing			Total private	Manufac-turing	Total private	Manufac-turing	Total private	Manufac-turing	Total private	Manufac-turing
			Total	Manufac-turing									
1984													
January	92 673	76 663	22 942	17 630	69 731	61 925	12 617	35.1	40.6	8.06	8.91	294.14	361.75
February	93 154	77 129	23 146	17 728	70 008	62 342	12 703	35.3	41.1	8.10	8.92	295.46	366.61
March	93 429	77 399	23 209	17 806	70 220	62 531	12 768	35.1	40.7	8.10	8.96	295.19	364.67
April	93 792	77 717	23 305	17 872	70 487	62 817	12 814	35.2	40.8	8.13	8.98	297.44	366.38
May	94 100	77 997	23 389	17 916	70 711	63 029	12 840	35.1	40.7	8.17	8.99	296.24	365.89
June	94 479	78 352	23 497	17 967	70 982	63 313	12 871	35.1	40.6	8.19	9.03	297.65	366.62
July	94 792	78 620	23 571	18 013	71 221	63 532	12 901	35.1	40.6	8.23	9.05	299.05	367.43
August	95 034	78 810	23 608	18 034	71 426	63 670	12 906	35.0	40.5	8.20	9.09	298.20	368.15
September	95 344	79 089	23 617	18 019	71 727	63 889	12 880	35.1	40.5	8.26	9.12	300.46	369.36
October	95 630	79 356	23 626	18 024	72 004	64 100	12 868	34.9	40.5	8.31	9.15	298.40	370.58
November	95 979	79 668	23 639	18 016	72 340	64 343	12 846	35.0	40.4	8.32	9.19	299.95	371.28
December	96 107	79 825	23 673	18 023	72 434	64 460	12 848	35.1	40.5	8.33	9.22	302.21	373.41
1985													
January	96 373	80 037	23 672	18 009	72 701	64 678	12 833	34.9	40.3	8.61	9.27	300.49	373.58
February	96 497	80 148	23 621	17 966	72 876	64 775	12 784	34.8	40.1	8.64	9.29	300.67	372.53
March	96 843	80 448	23 661	17 939	73 182	65 026	12 758	34.9	40.4	8.67	9.32	302.58	376.53
April	97 039	80 609	23 644	17 886	73 395	65 133	12 701	34.9	40.5	8.69	9.35	303.28	378.68
May	97 313	80 839	23 632	17 855	73 681	65 328	12 673	34.9	40.4	8.70	9.37	303.63	378.55
June	97 459	80 961	23 592	17 819	73 867	65 403	12 635	34.9	40.5	8.74	9.39	305.03	380.30
July	97 649	81 029	23 549	17 776	74 100	65 450	12 596	34.8	40.4	8.74	9.42	304.15	380.57
August	97 842	81 223	23 546	17 756	74 296	65 633	12 593	34.8	40.6	8.77	9.43	305.20	382.86
September	98 045	81 407	23 528	17 718	74 517	65 769	12 556	34.8	40.6	8.80	9.44	306.24	383.26
October	98 233	81 579	23 529	17 708	74 704	65 939	12 556	34.8	40.7	8.79	9.46	305.89	385.02
November	98 442	81 768	23 520	17 697	74 922	66 089	12 545	34.8	40.7	8.82	9.49	306.94	386.24
December	98 609	81 915	23 518	17 693	75 091	66 219	12 550	34.9	40.9	8.87	9.55	309.56	390.60
1986													
January	98 734	82 019	23 530	17 686	75 204	66 316	12 546	35.0	40.7	8.85	9.53	309.75	387.87
February	98 841	82 082	23 485	17 663	75 356	66 387	12 530	34.8	40.7	8.88	9.56	309.02	389.09
March	98 935	82 180	23 428	17 624	75 507	66 425	12 498	34.8	40.7	8.89	9.58	309.37	389.91
April	99 122	82 357	23 427	17 616	75 695	66 554	12 495	34.7	40.5	8.89	9.56	308.48	387.18
May	99 249	82 459	23 349	17 593	75 900	66 635	12 474	34.8	40.7	8.90	9.59	309.72	390.31
June	99 155	82 376	23 263	17 530	75 892	66 558	12 424	34.7	40.7	8.91	9.58	309.18	389.91
July	99 473	82 694	23 235	17 497	76 238	66 829	12 389	34.6	40.6	8.92	9.60	308.63	389.76
August	99 587	82 787	23 225	17 489	76 362	66 927	12 399	34.7	40.7	8.94	9.61	310.22	391.13
September	99 934	83 024	23 216	17 498	76 718	67 133	12 411	34.6	40.7	8.94	9.60	309.32	390.72
October	100 120	83 151	23 208	17 477	76 912	67 230	12 396	34.6	40.6	8.96	9.62	310.02	390.57
November	100 306	83 301	23 204	17 472	77 102	67 368	12 407	34.7	40.7	9.00	9.64	312.30	392.35
December	100 511	83 490	23 237	17 478	77 274	67 517	12 425	34.6	40.8	9.01	9.66	311.75	394.13
1987													
January	100 683	83 638	23 232	17 465	77 451	67 638	12 405	34.7	40.8	9.02	9.67	312.99	394.54
February	100 915	83 879	23 296	17 499	77 619	67 869	12 438	34.9	41.2	9.05	9.69	315.85	399.23
March	101 164	84 100	23 307	17 507	77 857	68 015	12 446	34.7	41.0	9.07	9.71	314.73	398.11
April	101 502	84 393	23 342	17 525	78 160	68 261	12 465	34.7	40.8	9.08	9.71	315.08	396.17
May	101 728	84 616	23 390	17 542	78 338	68 467	12 481	34.8	41.0	9.11	9.73	317.03	398.93
June	101 900	84 776	23 390	17 537	78 510	68 579	12 482	34.7	40.9	9.11	9.74	316.12	398.37
July	102 247	85 087	23 455	17 593	78 792	68 819	12 521	34.7	41.0	9.12	9.74	316.46	399.34
August	102 418	85 246	23 506	17 630	78 912	68 941	12 560	34.9	40.9	9.18	9.80	320.38	400.82
September	102 646	85 511	23 566	17 691	79 080	69 164	12 614	34.7	40.8	9.19	9.85	318.89	401.88
October	103 138	85 869	23 655	17 729	79 483	69 442	12 637	34.8	41.1	9.22	9.84	320.86	404.42
November	103 370	86 071	23 711	17 775	79 659	69 621	12 678	34.8	41.0	9.27	9.87	322.60	404.67
December	103 664	86 317	23 772	17 809	79 892	69 852	12 707	34.6	41.0	9.28	9.89	321.09	405.49
1988													
January	103 758	86 393	23 668	17 790	80 090	69 859	12 684	34.6	41.1	9.29	9.91	321.43	407.30
February	104 211	86 822	23 769	17 823	80 442	70 255	12 706	34.7	41.1	9.29	9.92	322.36	407.71
March	104 487	87 040	23 824	17 844	80 663	70 397	12 712	34.5	40.9	9.31	9.94	321.20	406.55
April	104 732	87 280	23 880	17 874	80 852	70 604	12 733	34.6	41.0	9.36	9.99	323.86	409.59
May	104 961	87 480	23 896	17 892	81 065	70 749	12 747	34.6	41.0	9.41	10.02	325.59	410.82
June	105 324	87 809	23 951	17 916	81 373	71 031	12 767	34.6	41.1	9.42	10.04	325.93	412.64
July	105 546	88 052	23 966	17 926	81 580	71 235	12 774	34.7	41.1	9.45	10.05	327.92	413.06
August	105 670	88 126	23 926	17 891	81 744	71 287	12 752	34.5	40.9	9.46	10.07	326.37	411.86
September	106 009	88 375	23 942	17 914	82 067	71 487	12 764	34.5	41.0	9.51	10.12	328.10	414.92
October	106 277	88 607	23 987	17 966	82 290	71 680	12 812	34.7	41.1	9.56	10.16	331.73	417.58
November	106 616	88 870	24 030	18 003	82 586	71 902	12 851	34.5	41.1	9.58	10.19	330.51	418.81
December	106 906	89 170	24 054	18 025	82 852	72 175	12 864	34.6	40.9	9.60	10.20	332.16	417.18
1989													
January	107 168	89 394	24 097	18 057	83 071	72 393	12 882	34.7	41.1	9.65	10.23	334.86	420.45
February	107 426	89 614	24 080	18 055	83 346	72 577	12 880	34.5	41.2	9.68	10.26	333.96	422.71
March	107 619	89 797	24 069	18 060	83 550	72 695	12 878	34.5	41.1	9.70	10.29	334.65	422.92
April	107 792	89 952	24 100	18 055	83 692	72 819	12 867	34.6	41.1	9.75	10.28	337.35	422.51
May	107 910	90 034	24 089	18 040	83 821	72 889	12 852	34.4	41.0	9.73	10.30	334.71	422.30
June	108 026	90 114	24 052	18 013	83 974	72 941	12 822	34.4	40.9	9.77	10.33	336.09	422.50
July	108 066	90 161	24 027	17 980	84 039	72 972	12 790	34.5	40.9	9.82	10.36	338.79	423.72
August	108 115	90 126	24 048	17 964	84 067	72 940	12 790	34.5	40.9	9.83	10.39	339.14	424.95
September	108 365	90 338	24 000	17 922	84 365	73 114	12 745	34.4	40.8	9.87	10.41	339.53	424.73
October	108 476	90 443	23 997	17 895	84 479	73 211	12 723	34.6	40.8	9.93	10.43	343.58	425.54
November	108 753	90 696	24 009	17 886	84 744	73 427	12 713	34.4	40.7	9.93	10.44	341.59	424.91
December	108 849	90 774	23 949	17 881	84 900	73 504	12 705	34.3	40.5	9.98	10.49	342.31	424.85

Table 10-8. Employment, Hours, and Earnings, Total Nonfarm and Manufacturing, Historical Annual and Monthly—Continued

(Wage and salary workers on nonfarm payrolls, seasonally adjusted.)

Year and month	All wage and salary workers (thousands)					Production and nonsupervisory workers on private payrolls							
	Total	Private			Service-providing	Number (thousands)		Average hours per week		Average hourly earnings, dollars		Average weekly earnings, dollars	
		Total	Goods-producing			Total private	Manufacturing	Total private	Manufacturing	Total private	Manufacturing	Total private	Manufacturing
			Total	Manufacturing									
1990													
January	109 199	91 048	23 982	17 797	85 217	73 750	12 738	34.4	40.5	10.02	10.52	344.69	426.06
February	109 435	91 258	24 071	17 893	85 364	73 944	12 848	34.4	40.6	10.07	10.64	346.41	431.98
March	109 644	91 350	24 023	17 868	85 621	73 997	12 821	34.4	40.6	10.11	10.70	347.78	434.42
April	109 686	91 309	23 966	17 845	85 720	73 963	12 804	34.3	40.5	10.12	10.68	347.12	432.54
May	109 839	91 240	23 888	17 797	85 951	73 863	12 755	34.3	40.6	10.16	10.74	348.49	436.04
June	109 856	91 300	23 849	17 776	86 007	73 856	12 739	34.4	40.7	10.20	10.78	350.88	438.75
July	109 824	91 264	23 746	17 704	86 078	73 814	12 673	34.2	40.6	10.22	10.81	349.52	438.89
August	109 616	91 159	23 648	17 649	85 968	73 741	12 624	34.3	40.5	10.24	10.80	351.23	437.40
September	109 518	91 081	23 571	17 609	85 947	73 633	12 596	34.2	40.5	10.28	10.86	351.58	439.83
October	109 367	90 924	23 473	17 577	85 894	73 499	12 577	34.1	40.4	10.30	10.92	351.23	441.17
November	109 214	90 764	23 283	17 428	85 931	73 345	12 439	34.2	40.2	10.32	10.88	352.94	437.38
December	109 166	90 698	23 203	17 395	85 963	73 267	12 415	34.2	40.3	10.34	10.93	353.63	440.48
1991													
January	109 055	90 581	23 061	17 330	85 994	73 153	12 351	34.1	40.2	10.38	10.97	353.96	440.99
February	108 734	90 252	22 900	17 211	85 834	72 859	12 242	34.1	40.1	10.39	10.97	354.30	439.90
March	108 574	90 086	22 779	17 140	85 795	72 705	12 191	34.1	40.0	10.41	11.00	354.98	440.00
April	108 364	89 879	22 687	17 093	85 677	72 534	12 157	34.0	40.1	10.46	11.05	355.64	443.11
May	108 249	89 751	22 618	17 070	85 631	72 457	12 148	34.1	40.2	10.49	11.09	357.71	445.82
June	108 334	89 773	22 571	17 044	85 763	72 455	12 135	34.1	40.5	10.51	11.13	358.39	450.77
July	108 292	89 694	22 507	17 015	85 785	72 430	12 130	34.1	40.5	10.54	11.18	359.41	452.79
August	108 310	89 743	22 493	17 025	85 817	72 485	12 153	34.1	40.6	10.56	11.18	360.10	453.91
September	108 336	89 793	22 466	17 010	85 870	72 497	12 140	34.1	40.6	10.58	11.21	360.78	455.13
October	108 357	89 764	22 418	16 999	85 939	72 471	12 138	34.1	40.6	10.59	11.25	361.12	456.75
November	108 296	89 669	22 316	16 961	85 980	72 397	12 102	34.1	40.7	10.61	11.26	361.80	458.28
December	108 328	89 687	22 274	16 916	86 054	72 434	12 075	34.1	40.7	10.64	11.27	362.82	458.69
1992													
January	108 369	89 681	22 213	16 839	86 156	72 474	12 013	34.1	40.6	10.64	11.24	362.82	456.34
February	108 311	89 622	22 142	16 829	86 169	72 456	12 015	34.1	40.7	10.67	11.30	363.85	459.91
March	108 365	89 650	22 127	16 805	86 238	72 482	12 006	34.1	40.7	10.69	11.32	364.53	460.72
April	108 519	89 780	22 132	16 831	86 387	72 629	12 029	34.3	40.9	10.72	11.36	367.70	464.62
May	108 649	89 896	22 135	16 835	86 514	72 741	12 046	34.3	40.9	10.74	11.39	368.38	465.85
June	108 715	89 953	22 097	16 826	86 618	72 784	12 042	34.2	40.8	10.77	11.41	368.33	465.53
July	108 793	89 976	22 074	16 819	86 719	72 809	12 049	34.2	40.8	10.79	11.43	369.02	466.34
August	108 925	90 042	22 045	16 783	86 880	72 877	12 020	34.2	40.8	10.81	11.46	369.70	467.57
September	108 959	90 130	22 020	16 761	86 939	72 978	12 002	34.3	40.7	10.82	11.45	371.13	466.02
October	109 139	90 311	22 029	16 751	87 110	73 143	11 999	34.2	40.8	10.85	11.47	371.07	467.98
November	109 272	90 431	22 042	16 758	87 230	73 297	12 012	34.2	40.9	10.87	11.49	371.75	469.94
December	109 495	90 617	22 076	16 769	87 419	73 464	12 031	34.2	40.9	10.89	11.51	372.44	470.76
1993													
January	109 794	90 893	22 133	16 791	87 661	73 741	12 060	34.3	41.1	10.93	11.55	374.90	474.71
February	110 044	91 142	22 188	16 805	87 856	73 996	12 072	34.3	41.1	10.95	11.57	375.59	475.53
March	109 994	91 087	22 142	16 795	87 852	73 922	12 074	34.1	40.8	10.99	11.58	374.76	472.46
April	110 296	91 358	22 131	16 772	88 165	74 143	12 056	34.4	41.4	10.99	11.63	378.06	481.48
May	110 568	91 617	22 189	16 766	88 379	74 398	12 056	34.3	41.0	11.01	11.65	377.64	477.65
June	110 749	91 780	22 165	16 742	88 584	74 506	12 039	34.3	40.9	11.03	11.68	378.33	477.71
July	111 055	91 995	22 183	16 739	88 872	74 701	12 043	34.4	41.1	11.05	11.69	380.12	480.46
August	111 206	92 178	22 203	16 741	89 003	74 874	12 050	34.3	41.1	11.07	11.72	379.70	481.69
September	111 448	92 407	22 252	16 769	89 196	75 071	12 080	34.4	41.3	11.10	11.76	381.84	485.69
October	111 733	92 691	22 306	16 778	89 427	75 314	12 092	34.4	41.3	11.13	11.79	382.87	486.93
November	111 984	92 916	22 347	16 800	89 637	75 537	12 119	34.4	41.3	11.15	11.83	383.56	488.58
December	112 314	93 205	22 413	16 815	89 901	75 788	12 140	34.4	41.4	11.18	11.88	384.59	491.83
1994													
January	112 595	93 448	22 465	16 855	90 130	75 987	12 181	34.4	41.4	11.20	11.90	385.28	492.66
February	112 781	93 631	22 451	16 862	90 330	76 157	12 198	34.2	40.9	11.25	11.97	384.75	489.57
March	113 242	94 052	22 550	16 897	90 692	76 531	12 234	34.5	41.7	11.25	11.95	388.13	498.32
April	113 586	94 363	22 641	16 933	90 945	76 821	12 276	34.5	41.8	11.27	11.96	388.82	499.93
May	113 921	94 657	22 704	16 962	91 217	77 096	12 304	34.5	41.8	11.29	11.98	389.51	500.76
June	114 238	94 964	22 764	17 010	91 474	77 348	12 351	34.5	41.8	11.31	12.00	390.20	501.60
July	114 610	95 309	22 807	17 026	91 803	77 651	12 367	34.6	41.8	11.34	12.03	392.36	502.85
August	114 896	95 590	22 876	17 081	92 020	77 894	12 428	34.5	41.7	11.36	12.06	391.92	502.90
September	115 247	95 910	22 948	17 115	92 299	78 162	12 459	34.5	41.6	11.39	12.09	392.96	502.94
October	115 458	96 114	22 974	17 144	92 484	78 344	12 489	34.5	41.8	11.43	12.12	394.34	506.62
November	115 869	96 502	23 050	17 186	92 819	78 696	12 529	34.5	41.8	11.45	12.15	395.03	507.87
December	116 165	96 777	23 095	17 217	93 070	78 967	12 558	34.5	41.8	11.47	12.17	395.72	508.71
1995													
January	116 501	97 104	23 147	17 262	93 354	79 199	12 593	34.5	41.8	11.48	12.20	396.06	509.96
February	116 697	97 290	23 103	17 265	93 594	79 332	12 602	34.4	41.7	11.53	12.24	396.63	510.41
March	116 907	97 480	23 151	17 263	93 756	79 509	12 600	34.4	41.5	11.55	12.24	397.32	507.96
April	117 069	97 635	23 174	17 278	93 895	79 648	12 608	34.3	41.0	11.56	12.25	396.51	502.25
May	117 049	97 631	23 120	17 259	93 929	79 654	12 590	34.2	41.1	11.59	12.28	396.38	504.71
June	117 286	97 841	23 137	17 247	94 149	79 823	12 574	34.3	41.2	11.63	12.31	398.91	507.17
July	117 380	97 943	23 119	17 218	94 261	79 901	12 538	34.3	41.1	11.67	12.39	400.28	509.23
August	117 634	98 205	23 164	17 240	94 470	80 152	12 566	34.3	41.2	11.69	12.40	400.97	510.88
September	117 875	98 445	23 208	17 247	94 667	80 352	12 567	34.3	41.2	11.73	12.41	402.34	511.29
October	118 031	98 567	23 206	17 216	94 825	80 466	12 535	34.3	41.2	11.75	12.44	403.03	512.53
November	118 175	98 712	23 200	17 209	94 975	80 536	12 515	34.3	41.3	11.78	12.45	404.05	514.19
December	118 320	98 854	23 209	17 231	95 111	80 688	12 551	34.2	40.9	11.81	12.49	403.90	510.84

Table 10-8. Employment, Hours, and Earnings, Total Nonfarm and Manufacturing, Historical Annual and Monthly—*Continued*

(Wage and salary workers on nonfarm payrolls, seasonally adjusted.)

Year and month	All wage and salary workers (thousands)					Production and nonsupervisory workers on private payrolls							
	Total	Private			Service-providing	Number (thousands)		Average hours per week		Average hourly earnings, dollars		Average weekly earnings, dollars	
		Total	Goods-producing			Total private	Manufacturing	Total private	Manufacturing	Total private	Manufacturing	Total private	Manufacturing
			Total	Manufacturing									
1996													
January	118 316	98 866	23 196	17 208	95 120	80 649	12 517	33.8	39.7	11.86	12.60	400.87	500.22
February	118 739	99 254	23 280	17 229	95 459	81 015	12 533	34.3	41.3	11.87	12.56	407.14	518.73
March	118 993	99 461	23 276	17 193	95 717	81 190	12 488	34.3	41.1	11.89	12.50	407.83	513.75
April	119 158	99 643	23 316	17 204	95 842	81 361	12 505	34.2	41.2	11.95	12.69	408.69	522.83
May	119 486	99 957	23 358	17 222	96 128	81 634	12 518	34.3	41.4	11.97	12.71	410.57	526.19
June	119 769	100 241	23 399	17 226	96 370	81 828	12 522	34.4	41.5	12.03	12.76	413.83	529.54
July	120 015	100 468	23 418	17 223	96 597	82 027	12 517	34.3	41.4	12.05	12.80	413.32	529.92
August	120 199	100 695	23 479	17 255	96 720	82 235	12 546	34.4	41.5	12.09	12.82	415.90	532.03
September	120 410	100 843	23 497	17 252	96 913	82 338	12 544	34.4	41.6	12.13	12.85	417.27	534.56
October	120 665	101 111	23 546	17 268	97 119	82 593	12 559	34.4	41.4	12.15	12.83	417.96	531.16
November	120 961	101 396	23 584	17 277	97 377	82 778	12 560	34.4	41.5	12.20	12.88	419.68	534.52
December	121 143	101 572	23 598	17 284	97 545	82 944	12 571	34.4	41.7	12.24	12.95	421.06	540.02
1997													
January	121 363	101 770	23 618	17 297	97 745	83 092	12 579	34.3	41.4	12.29	12.99	421.55	537.79
February	121 675	102 077	23 686	17 316	97 989	83 371	12 592	34.4	41.6	12.31	12.99	423.46	540.38
March	121 990	102 382	23 739	17 340	98 251	83 610	12 613	34.5	41.8	12.36	13.04	426.42	545.07
April	122 286	102 683	23 765	17 349	98 521	83 864	12 618	34.5	41.8	12.39	13.03	427.46	544.65
May	122 546	102 945	23 809	17 362	98 737	84 100	12 633	34.6	41.7	12.43	13.07	430.08	545.02
June	122 814	103 154	23 834	17 387	98 980	84 220	12 647	34.3	41.6	12.46	13.09	427.38	544.54
July	123 111	103 425	23 862	17 389	99 249	84 465	12 645	34.5	41.6	12.50	13.10	431.25	544.96
August	123 093	103 476	23 952	17 452	99 141	84 425	12 699	34.6	41.7	12.57	13.17	434.92	549.19
September	123 585	103 906	23 996	17 465	99 589	84 802	12 711	34.6	41.7	12.60	13.18	435.96	549.61
October	123 929	104 191	24 053	17 513	99 876	85 023	12 746	34.6	41.8	12.66	13.29	438.04	555.52
November	124 235	104 474	24 112	17 556	100 123	85 223	12 777	34.5	41.8	12.71	13.33	438.50	557.19
December	124 549	104 783	24 184	17 588	100 365	85 441	12 800	34.6	42.0	12.75	13.35	441.15	560.70
1998													
January	124 812	105 042	24 262	17 619	100 550	85 602	12 818	34.6	41.9	12.79	13.34	442.53	558.95
February	125 016	105 230	24 283	17 627	100 733	85 774	12 828	34.5	41.7	12.83	13.39	442.64	558.36
March	125 164	105 372	24 264	17 637	100 900	85 809	12 822	34.5	41.6	12.87	13.43	444.02	558.69
April	125 442	105 626	24 340	17 637	101 102	86 025	12 814	34.4	41.2	12.92	13.40	444.45	552.08
May	125 844	105 969	24 361	17 624	101 483	86 297	12 790	34.5	41.5	12.97	13.45	447.47	558.18
June	126 076	106 197	24 387	17 608	101 689	86 459	12 769	34.4	41.4	12.99	13.43	446.86	556.00
July	126 205	106 275	24 238	17 422	101 967	86 471	12 558	34.5	41.4	13.01	13.34	448.85	552.28
August	126 544	106 585	24 420	17 563	102 124	86 763	12 702	34.5	41.4	13.07	13.46	450.92	557.24
September	126 752	106 767	24 419	17 557	102 333	86 914	12 715	34.3	41.3	13.11	13.52	449.67	558.38
October	126 954	106 953	24 406	17 512	102 548	87 074	12 673	34.5	41.4	13.15	13.52	453.68	559.73
November	127 231	107 187	24 394	17 465	102 837	87 248	12 634	34.4	41.4	13.17	13.54	453.05	560.56
December	127 596	107 517	24 454	17 449	103 142	87 497	12 625	34.5	41.5	13.21	13.56	455.75	562.74
1999													
January	127 702	107 618	24 401	17 427	103 301	87 551	12 605	34.4	41.3	13.27	13.59	456.49	561.27
February	128 120	107 976	24 434	17 395	103 686	87 874	12 575	34.4	41.4	13.30	13.63	457.52	564.28
March	128 227	108 059	24 378	17 368	103 849	87 943	12 562	34.3	41.3	13.33	13.69	457.22	565.40
April	128 597	108 360	24 425	17 344	104 172	88 167	12 540	34.4	41.3	13.38	13.73	460.27	567.05
May	128 808	108 579	24 445	17 333	104 363	88 347	12 534	34.3	41.4	13.42	13.80	460.31	571.32
June	129 089	108 817	24 434	17 295	104 655	88 530	12 503	34.3	41.3	13.47	13.86	462.02	572.42
July	129 414	109 075	24 474	17 308	104 940	88 730	12 525	34.4	41.4	13.51	13.91	464.74	575.87
August	129 569	109 194	24 467	17 287	105 102	88 839	12 502	34.4	41.5	13.53	13.94	465.43	578.51
September	129 772	109 368	24 485	17 281	105 287	89 000	12 495	34.3	41.4	13.61	13.99	466.82	579.19
October	130 177	109 720	24 505	17 272	105 672	89 305	12 484	34.4	41.4	13.64	13.99	469.22	579.19
November	130 466	109 970	24 561	17 282	105 905	89 523	12 488	34.4	41.4	13.66	14.01	469.90	580.01
December	130 772	110 232	24 582	17 280	106 190	89 760	12 490	34.4	41.4	13.70	14.06	471.28	582.08
2000													
January	131 005	110 434	24 628	17 284	106 377	89 909	12 491	34.4	41.5	13.75	14.13	473.00	586.40
February	131 124	110 525	24 609	17 285	106 515	89 956	12 481	34.4	41.5	13.80	14.14	474.72	586.81
March	131 596	110 863	24 705	17 302	106 891	90 237	12 490	34.4	41.4	13.85	14.17	476.44	586.64
April	131 888	111 086	24 688	17 298	107 200	90 473	12 477	34.4	41.6	13.89	14.23	477.82	591.97
May	132 105	110 958	24 647	17 279	107 458	90 351	12 464	34.3	41.2	13.94	14.22	478.14	585.86
June	132 061	111 174	24 672	17 296	107 389	90 532	12 468	34.3	41.3	13.99	14.30	479.86	590.59
July	132 236	111 369	24 717	17 322	107 519	90 654	12 474	34.3	41.5	14.02	14.33	480.89	594.70
August	132 230	111 393	24 683	17 287	107 547	90 680	12 431	34.2	41.0	14.06	14.36	480.85	588.76
September	132 353	111 618	24 642	17 230	107 711	90 856	12 382	34.3	41.1	14.12	14.40	484.32	591.84
October	132 351	111 608	24 638	17 217	107 713	90 838	12 357	34.3	41.1	14.17	14.48	486.03	595.13
November	132 556	111 796	24 623	17 202	107 933	90 962	12 336	34.2	41.1	14.23	14.52	486.67	596.77
December	132 709	111 905	24 575	17 181	108 134	90 985	12 307	34.0	40.4	14.28	14.50	485.52	585.80
2001													
January	132 698	111 863	24 533	17 104	108 165	90 992	12 229	34.2	40.6	14.29	14.48	488.72	587.89
February	132 789	111 883	24 474	17 028	108 315	90 933	12 155	34.0	40.5	14.36	14.55	488.24	589.28
March	132 747	111 802	24 409	16 938	108 338	90 875	12 085	34.1	40.5	14.42	14.58	491.72	590.49
April	132 463	111 471	24 254	16 802	108 209	90 652	11 982	34.0	40.4	14.45	14.63	491.30	591.05
May	132 410	111 381	24 119	16 661	108 291	90 548	11 859	34.0	40.4	14.50	14.69	493.00	593.48
June	132 299	111 162	23 965	16 515	108 334	90 352	11 739	34.0	40.3	14.54	14.74	494.36	594.02
July	132 177	110 992	23 837	16 382	108 340	90 239	11 632	34.0	40.6	14.55	14.82	494.70	601.69
August	132 028	110 810	23 667	16 232	108 361	90 094	11 495	33.9	40.3	14.59	14.85	494.60	598.46
September	131 771	110 529	23 537	16 117	108 234	89 835	11 400	33.8	40.2	14.63	14.89	494.49	598.58
October	131 454	110 179	23 378	15 972	108 076	89 558	11 286	33.7	40.1	14.66	14.89	494.04	597.09
November	131 142	109 816	23 209	15 825	107 933	89 238	11 175	33.8	40.1	14.71	14.96	497.20	599.90
December	130 982	109 627	23 094	15 711	107 888	89 127	11 081	33.8	40.2	14.75	15.01	498.55	603.40

Table 10-8. Employment, Hours, and Earnings, Total Nonfarm and Manufacturing, Historical Annual and Monthly—*Continued*

(Wage and salary workers on nonfarm payrolls, seasonally adjusted.)

Year and month	All wage and salary workers (thousands)					Production and nonsupervisory workers on private payrolls							
	Total	Private			Service-providing	Number (thousands)		Average hours per week		Average hourly earnings, dollars		Average weekly earnings, dollars	
		Total	Goods-producing			Total private	Manufacturing	Total private	Manufacturing	Total private	Manufacturing	Total private	Manufacturing
			Total	Manufacturing									
2002													
January	130 852	109 475	22 961	15 587	107 891	89 121	10 993	33.7	40.1	14.76	15.04	497.41	603.10
February	130 736	109 346	22 876	15 515	107 860	89 064	10 951	33.8	40.3	14.78	15.12	499.56	609.34
March	130 717	109 286	22 786	15 443	107 931	89 019	10 901	33.9	40.5	14.82	15.15	502.40	613.58
April	130 623	109 180	22 689	15 392	107 934	88 867	10 859	33.9	40.6	14.84	15.17	503.08	615.90
May	130 634	109 120	22 604	15 337	108 030	88 726	10 824	33.9	40.6	14.88	15.24	504.43	618.74
June	130 684	109 135	22 578	15 298	108 106	88 637	10 797	33.9	40.7	14.94	15.27	506.47	621.49
July	130 590	109 046	22 521	15 256	108 069	88 482	10 764	33.8	40.4	14.98	15.30	506.32	618.12
August	130 587	108 998	22 449	15 171	108 138	88 423	10 697	33.9	40.5	15.02	15.34	509.18	621.27
September	130 501	108 955	22 397	15 119	108 104	88 395	10 668	33.9	40.5	15.06	15.37	510.53	622.49
October	130 628	109 069	22 325	15 060	108 303	88 481	10 631	33.8	40.3	15.12	15.45	511.06	622.64
November	130 615	109 034	22 282	14 992	108 333	88 425	10 583	33.8	40.4	15.15	15.48	512.07	625.39
December	130 472	108 884	22 189	14 912	108 283	88 244	10 523	33.8	40.5	15.20	15.54	513.76	629.37
2003													
January	130 580	108 954	22 146	14 866	108 434	88 275	10 484	33.8	40.3	15.22	15.58	514.44	627.87
February	130 444	108 820	22 023	14 781	108 421	88 147	10 416	33.6	40.3	15.29	15.63	513.74	629.89
March	130 232	108 622	21 945	14 721	108 287	87 845	10 357	33.8	40.4	15.27	15.64	516.13	631.86
April	130 177	108 582	21 864	14 609	108 313	87 824	10 256	33.6	40.1	15.28	15.63	513.41	626.76
May	130 196	108 629	21 832	14 557	108 364	87 816	10 218	33.7	40.2	15.33	15.69	516.62	630.74
June	130 194	108 588	21 788	14 493	108 406	87 790	10 165	33.6	40.3	15.36	15.72	516.10	633.52
July	130 191	108 558	21 708	14 402	108 483	87 767	10 093	33.6	40.1	15.40	15.76	517.44	631.98
August	130 149	108 593	21 706	14 376	108 443	87 823	10 084	33.7	40.2	15.42	15.78	519.65	634.36
September	130 254	108 750	21 700	14 347	108 554	87 932	10 059	33.6	40.5	15.41	15.83	517.78	641.12
October	130 454	108 896	21 691	14 334	108 763	88 015	10 054	33.7	40.6	15.42	15.83	519.65	642.70
November	130 474	108 939	21 688	14 316	108 786	88 077	10 039	33.8	40.9	15.47	15.90	522.89	650.31
December	130 588	109 042	21 703	14 300	108 885	88 148	10 031	33.6	40.7	15.47	15.92	519.79	647.94
2004													
January	130 769	109 231	21 715	14 290	109 054	88 225	10 027	33.7	40.9	15.51	15.95	522.69	652.36
February	130 825	109 275	21 693	14 279	109 132	88 266	10 012	33.8	41.0	15.53	15.98	524.91	655.18
March	131 142	109 554	21 758	14 287	109 384	88 511	10 025	33.7	40.9	15.56	16.01	524.37	654.81
April	131 411	109 797	21 802	14 315	109 609	88 775	10 060	33.7	40.7	15.59	16.05	525.38	653.24
May	131 694	110 080	21 881	14 342	109 813	89 056	10 091	33.8	41.1	15.63	16.07	528.29	660.48
June	131 793	110 192	21 885	14 332	109 908	89 221	10 087	33.6	40.8	15.67	16.11	526.51	657.29
July	131 848	110 242	21 900	14 330	109 948	89 338	10 095	33.7	40.8	15.69	16.12	528.75	657.70
August	131 937	110 311	21 944	14 345	109 993	89 477	10 115	33.7	40.8	15.73	16.20	530.10	660.96
September	132 093	110 458	21 957	14 331	110 136	89 656	10 104	33.7	40.7	15.78	16.29	531.79	663.00
October	132 447	110 791	22 004	14 332	110 443	89 967	10 103	33.8	40.6	15.81	16.27	534.38	660.56
November	132 503	110 811	21 997	14 307	110 506	89 992	10 078	33.7	40.5	15.84	16.30	533.81	660.15
December	132 624	110 931	22 005	14 287	110 619	90 155	10 063	33.8	40.6	15.86	16.34	536.07	663.40
2005													
January	132 774	111 039	21 958	14 257	110 816	90 236	10 045	33.7	40.7	15.90	16.38	535.83	666.67
February	133 032	111 288	22 036	14 273	110 996	90 484	10 052	33.8	40.6	15.93	16.44	538.43	667.46
March	133 156	111 416	22 066	14 269	111 090	90 647	10 056	33.7	40.4	15.97	16.44	538.19	664.18
April	133 518	111 764	22 136	14 250	111 382	90 980	10 048	33.8	40.4	16.00	16.46	540.80	664.98
May	133 690	111 909	22 172	14 256	111 518	91 098	10 061	33.7	40.4	16.04	16.53	540.55	667.81
June	133 942	112 179	22 185	14 227	111 757	91 372	10 047	33.7	40.4	16.08	16.54	541.90	668.22
July	134 296	112 439	22 205	14 226	112 091	91 606	10 039	33.7	40.5	16.15	16.57	544.26	671.09
August	134 498	112 635	22 228	14 203	112 270	91 792	10 043	33.7	40.5	16.17	16.63	544.93	673.52
September	134 566	112 721	22 226	14 175	112 340	91 895	10 046	33.8	40.7	16.19	16.59	547.22	675.21
October	134 655	112 826	22 292	14 192	112 363	92 036	10 073	33.8	41.0	16.28	16.69	550.26	684.29
November	134 993	113 134	22 357	14 187	112 636	92 358	10 093	33.8	40.9	16.30	16.68	550.94	682.21
December	135 149	113 270	22 376	14 193	112 773	92 520	10 112	33.8	40.8	16.35	16.68	552.63	680.54
2006													
January	135 429	113 582	22 467	14 210	112 962	92 826	10 153	33.8	41.0	16.42	16.70	555.00	684.70
February	135 737	113 859	22 535	14 209	113 202	93 115	10 164	33.8	41.1	16.48	16.70	557.02	686.37
March	136 047	114 144	22 572	14 214	113 475	93 412	10 174	33.8	41.1	16.54	16.72	559.05	687.19
April	136 205	114 286	22 631	14 226	113 574	93 584	10 190	33.9	41.3	16.63	16.75	563.76	691.78
May	136 244	114 318	22 597	14 203	113 647	93 653	10 179	33.8	41.2	16.65	16.77	562.77	690.92
June	136 325	114 403	22 598	14 213	113 727	93 728	10 190	33.9	41.2	16.72	16.78	566.81	691.34
July	136 520	114 547	22 590	14 188	113 930	93 864	10 177	33.9	41.4	16.77	16.78	568.50	694.69
August	136 694	114 683	22 572	14 159	114 122	94 010	10 158	33.8	41.2	16.82	16.83	568.52	693.40
September	136 843	114 761	22 537	14 125	114 306	94 058	10 120	33.8	41.1	16.87	16.83	570.21	691.71
October	136 852	114 784	22 456	14 075	114 396	94 109	10 071	33.9	41.2	16.93	16.91	573.93	696.69
November	137 063	114 980	22 408	14 041	114 655	94 284	10 047	33.8	41.0	16.97	16.92	573.59	693.72
December	137 249	115 161	22 405	14 015	114 844	94 488	10 041	33.9	41.1	17.04	16.99	577.66	698.29
2007													
January	137 477	115 382	22 439	14 008	115 038	94 682	10 036	33.8	41.0	17.09	17.01	577.64	697.41
February	137 558	115 427	22 334	13 997	115 224	94 717	10 035	33.7	40.9	17.15	17.05	577.96	697.35
March	137 793	115 644	22 391	13 970	115 402	94 954	10 011	33.9	41.3	17.22	17.10	583.76	706.23
April	137 842	115 667	22 350	13 945	115 492	95 023	10 006	33.8	41.3	17.27	17.20	583.73	710.36
May	137 993	115 800	22 323	13 929	115 670	95 198	10 007	33.8	41.2	17.33	17.23	585.75	709.88
June	138 069	115 862	22 323	13 911	115 746	95 309	9 996	33.9	41.4	17.42	17.29	590.54	715.81
July	138 038	115 867	22 277	13 889	115 761	95 399	9 990	33.8	41.3	17.46	17.29	590.15	714.08
August	138 015	115 789	22 165	13 828	115 850	95 339	9 944	33.8	41.3	17.50	17.35	591.50	716.56
September	138 095	115 816	22 093	13 790	116 002	95 437	9 933	33.8	41.3	17.56	17.37	593.53	717.38
October	138 174	115 877	22 056	13 764	116 118	95 540	9 911	33.8	41.2	17.58	17.36	594.20	715.23
November	138 284	115 950	22 015	13 757	116 269	95 609	9 920	33.8	41.3	17.63	17.44	595.89	720.27
December	138 392	116 016	21 976	13 746	116 416	95 706	9 924	33.8	41.1	17.69	17.44	597.92	716.78

Table 10-8. Employment, Hours, and Earnings, Total Nonfarm and Manufacturing, Historical Annual and Monthly—Continued

(Wage and salary workers on nonfarm payrolls, seasonally adjusted.)

Year and month	All wage and salary workers (thousands)					Production and nonsupervisory workers on private payrolls							
	Total	Private			Service-providing	Number (thousands)		Average hours per week		Average hourly earnings, dollars		Average weekly earnings, dollars	
		Total	Goods-producing			Total private	Manufac-turing	Total private	Manufac-turing	Total private	Manufac-turing	Total private	Manufac-turing
			Total	Manufac-turing									
2008													
January	138 403	116 015	21 947	13 725	116 456	95 728	9 914	33.7	41.1	17.73	17.50	597.50	719.25
February	138 324	115 907	21 897	13 696	116 427	95 640	9 888	33.7	41.2	17.80	17.57	599.86	723.88
March	138 275	115 832	21 820	13 659	116 455	95 606	9 866	33.8	41.3	17.88	17.63	604.34	728.12
April	138 035	115 585	21 681	13 599	116 354	95 413	9 810	33.7	41.1	17.92	17.63	603.90	724.59
May	137 858	115 375	21 599	13 564	116 259	95 225	9 779	33.7	41.0	17.98	17.69	605.93	725.29
June	137 687	115 170	21 483	13 504	116 204	95 042	9 721	33.7	41.0	18.03	17.75	607.61	727.75
July	137 491	114 923	21 360	13 430	116 131	94 852	9 655	33.6	40.9	18.10	17.80	608.16	728.02
August	137 213	114 646	21 250	13 358	115 963	94 614	9 586	33.7	40.9	18.18	17.80	612.67	728.02
September	136 753	114 216	21 101	13 275	115 652	94 281	9 504	33.6	40.5	18.20	17.83	611.52	722.12
October	136 272	113 723	20 895	13 147	115 377	93 870	9 384	33.5	40.5	18.25	17.90	611.38	724.95
November	135 545	112 985	20 623	13 034	114 922	93 169	9 290	33.4	40.1	18.31	17.95	611.55	719.80
December	134 839	112 283	20 322	12 850	114 517	92 598	9 137	33.3	39.8	18.38	18.00	612.05	716.40
2009													
January	134 055	111 476	19 889	12 561	114 166	91 894	8 889	33.2	39.7	18.40	18.01	610.88	715.00
February	133 312	110 736	19 575	12 380	113 737	91 286	8 741	33.3	39.6	18.44	18.09	614.05	716.36
March	132 512	109 952	19 227	12 208	113 285	90 613	8 592	33.1	39.3	18.50	18.13	612.35	712.51
April	131 817	109 140	18 894	12 030	112 923	89 911	8 453	33.1	39.5	18.52	18.16	613.01	717.32
May	131 475	108 858	18 655	11 862	112 820	89 700	8 311	33.0	39.3	18.53	18.14	611.49	712.90
June	131 008	108 432	18 422	11 726	112 586	89 311	8 203	33.0	39.6	18.56	18.18	612.48	719.93
July	130 668	108 147	18 278	11 668	112 390	89 102	8 172	33.1	39.9	18.59	18.29	615.33	729.77
August	130 485	107 948	18 151	11 626	112 334	88 934	8 148	33.1	40.0	18.67	18.32	617.98	732.80
September	130 244	107 793	18 043	11 591	112 201	88 817	8 133	33.1	40.0	18.70	18.42	618.97	736.80
October	130 045	107 521	17 915	11 538	112 130	88 578	8 100	33.0	40.2	18.74	18.38	618.42	738.88
November	130 057	107 524	17 869	11 509	112 188	88 606	8 079	33.3	40.5	18.80	18.42	626.04	746.01
December	129 788	107 306	17 792	11 475	111 996	88 532	8 050	33.2	40.6	18.83	18.42	625.16	747.85
2010													
January	129 790	107 299	17 707	11 460	112 083	88 503	8 034	33.3	40.8	18.88	18.43	628.70	751.94
February	129 698	107 222	17 627	11 453	112 071	88 431	8 032	33.1	40.5	18.91	18.47	625.92	748.04
March	129 879	107 361	17 672	11 453	112 207	88 565	8 027	33.3	41.0	18.91	18.47	629.70	757.27
April	130 110	107 541	17 729	11 489	112 381	88 706	8 050	33.4	41.1	18.96	18.51	633.26	760.76
May	130 650	107 654	17 742	11 525	112 908	88 772	8 078	33.4	41.4	19.01	18.58	634.93	769.21
June	130 511	107 771	17 763	11 545	112 748	88 871	8 100	33.3	41.0	19.02	18.58	633.37	761.78
July	130 427	107 858	17 777	11 561	112 650	88 945	8 106	33.4	41.1	19.04	18.63	635.94	765.69
August	130 422	108 002	17 793	11 553	112 629	89 046	8 099	33.5	41.2	19.11	18.65	640.19	768.38
September	130 357	108 110	17 787	11 563	112 570	89 138	8 098	33.5	41.5	19.11	18.72	640.19	776.88
October	130 625	108 328	17 801	11 562	112 824	89 306	8 096	33.5	41.2	19.20	18.73	643.20	771.68
November	130 750	108 463	17 827	11 585	112 923	89 434	8 103	33.5	41.3	19.21	18.76	643.54	774.79
December	130 822	108 556	17 796	11 595	113 026	89 535	8 117	33.5	41.3	19.22	18.79	643.87	776.03
2011													
January	130 841	108 583	17 784	11 621	113 057	89 523	8 133	33.4	41.1	19.32	18.89	645.29	776.38
February	131 053	108 838	17 844	11 654	113 209	89 726	8 160	33.5	41.4	19.31	18.91	646.89	782.87
March	131 288	109 096	17 906	11 675	113 382	89 960	8 183	33.5	41.4	19.31	18.90	646.89	782.46
April	131 602	109 418	17 957	11 704	113 645	90 220	8 212	33.6	41.4	19.36	18.91	650.50	782.87
May	131 703	109 574	18 007	11 713	113 696	90 364	8 218	33.6	41.5	19.41	18.92	652.18	785.18
June	131 939	109 775	18 045	11 727	113 894	90 533	8 228	33.6	41.4	19.41	18.90	652.18	782.46
July	131 999	109 950	18 094	11 746	113 905	90 702	8 244	33.7	41.4	19.47	18.93	656.14	783.70
August	132 125	110 108	18 120	11 764	114 005	90 835	8 259	33.5	41.3	19.47	18.90	652.25	780.57
September	132 358	110 375	18 167	11 769	114 191	91 068	8 266	33.6	41.3	19.48	18.91	654.53	780.98
October	132 562	110 564	18 188	11 780	114 374	91 233	8 274	33.7	41.5	19.56	18.98	659.17	787.67
November	132 694	110 723	18 187	11 770	114 507	91 423	8 268	33.6	41.5	19.56	18.96	657.22	786.84
December	132 896	110 942	18 244	11 802	114 652	91 626	8 295	33.7	41.6	19.56	19.00	659.17	790.40
2012													
January	133 250	111 304	18 304	11 838	114 946	91 934	8 322	33.8	41.8	19.59	19.02	662.14	795.04
February	133 512	111 565	18 334	11 860	115 178	92 212	8 352	33.7	41.7	19.59	19.00	660.18	792.30
March	133 752	111 809	18 372	11 898	115 380	92 378	8 386	33.7	41.6	19.64	19.02	661.87	791.23
April	133 834	111 903	18 384	11 916	115 450	92 478	8 399	33.7	41.7	19.68	19.09	663.22	796.05
May	133 934	112 023	18 386	11 927	115 548	92 604	8 411	33.6	41.6	19.68	19.04	661.25	792.06
June	134 007	112 077	18 411	11 936	115 596	92 657	8 412	33.7	41.7	19.72	19.08	664.56	795.64
July	134 159	112 246	18 447	11 964	115 712	92 781	8 438	33.7	41.7	19.74	19.11	665.24	796.89
August	134 331	112 415	18 459	11 960	115 872	92 916	8 429	33.6	41.5	19.74	19.05	663.26	790.58
September	134 518	112 593	18 461	11 954	116 057	93 065	8 419	33.7	41.5	19.78	19.06	666.59	790.99
October	134 677	112 774	18 472	11 961	116 205	93 248	8 426	33.6	41.5	19.79	19.09	664.94	792.24
November	134 833	112 950	18 477	11 950	116 356	93 377	8 411	33.6	41.5	19.84	19.13	666.62	793.90
December	135 072	113 185	18 532	11 960	116 540	93 562	8 415	33.8	41.7	19.90	19.13	672.62	797.72
2013													
January	135 263	113 394	18 580	11 983	116 683	93 659	8 422	33.6	41.7	19.94	19.17	669.98	799.39
February	135 541	113 660	18 638	11 996	116 903	93 870	8 421	33.8	41.8	19.99	19.21	675.66	802.98
March	135 680	113 810	18 660	11 999	117 020	94 001	8 409	33.8	41.9	20.01	19.21	676.34	804.90
April	135 871	114 002	18 653	12 000	117 218	94 155	8 414	33.7	41.8	20.04	19.23	675.35	803.81
May	136 093	114 229	18 691	12 000	117 402	94 302	8 402	33.7	41.8	20.05	19.23	675.69	803.81
June	136 274	114 434	18 724	12 004	117 550	94 488	8 398	33.7	41.8	20.12	19.28	678.04	805.90
July	136 386	114 571	18 705	11 984	117 681	94 638	8 383	33.5	41.7	20.14	19.26	674.69	803.14
August	136 628	114 797	18 756	12 014	117 872	94 813	8 406	33.6	41.9	20.17	19.31	677.71	809.09
September	136 815	114 980	18 810	12 032	118 005	94 956	8 426	33.7	41.9	20.23	19.33	681.75	809.93
October	137 040	115 210	18 857	12 056	118 183	95 178	8 450	33.6	41.9	20.25	19.35	680.40	810.77
November	137 304	115 461	18 911	12 079	118 393	95 397	8 464	33.7	42.0	20.31	19.39	684.45	814.38
December	137 373	115 553	18 881	12 083	118 492	95 461	8 468	33.6	41.9	20.34	19.46	683.42	815.37

Table 10-8. Employment, Hours, and Earnings, Total Nonfarm and Manufacturing, Historical Annual and Monthly—Continued

(Wage and salary workers on nonfarm payrolls, seasonally adjusted.)

Year and month	All wage and salary workers (thousands)					Production and nonsupervisory workers on private payrolls							
	Total	Private			Service-providing	Number (thousands)		Average hours per week		Average hourly earnings, dollars		Average weekly earnings, dollars	
		Total	Goods-producing			Total private	Manufacturing	Total private	Manufacturing	Total private	Manufacturing	Total private	Manufacturing
			Total	Manufacturing									
2014													
January	137 548	115 734	18 937	12 081	118 611	95 590	8 468	33.5	41.6	20.39	19.46	683.07	809.54
February	137 714	115 889	18 988	12 106	118 726	95 744	8 497	33.4	41.6	20.50	19.49	684.70	810.78
March	137 968	116 135	19 037	12 120	118 931	95 991	8 510	33.7	42.0	20.50	19.53	690.85	820.26
April	138 293	116 440	19 104	12 134	119 189	96 211	8 525	33.7	41.9	20.51	19.48	691.19	816.21
May	138 511	116 679	19 142	12 146	119 369	96 423	8 532	33.7	42.1	20.54	19.51	692.20	821.37
June	138 837	116 942	19 195	12 170	119 642	96 642	8 558	33.7	42.1	20.57	19.54	693.21	822.63
July	139 069	117 168	19 264	12 189	119 805	96 817	8 580	33.7	41.9	20.61	19.59	694.56	820.82
August	139 257	117 402	19 316	12 208	119 941	96 981	8 594	33.7	42.0	20.67	19.60	696.58	823.20
September	139 566	117 669	19 371	12 226	120 195	97 158	8 598	33.7	42.2	20.67	19.59	696.58	826.70
October	139 818	117 901	19 419	12 259	120 399	97 333	8 623	33.7	42.1	20.70	19.65	697.59	827.27
November	140 109	118 177	19 456	12 284	120 653	97 536	8 646	33.8	42.2	20.76	19.65	701.69	829.23
December	140 377	118 430	19 481	12 292	120 896	97 738	8 652	33.8	42.1	20.70	19.63	699.66	826.42
2015													
January	140 568	118 608	19 500	12 292	121 068	97 844	8 645	33.8	42.0	20.80	19.65	703.04	825.30
February	140 839	118 861	19 537	12 301	121 302	98 040	8 655	33.8	42.0	20.80	19.73	703.04	828.66
March	140 910	118 940	19 505	12 312	121 405	98 133	8 660	33.7	41.9	20.88	19.77	703.66	828.36
April	141 194	119 189	19 553	12 318	121 641	98 266	8 660	33.6	41.8	20.91	19.80	702.58	827.64
May	141 525	119 513	19 580	12 333	121 945	98 565	8 683	33.6	41.8	20.99	19.86	705.26	830.15
June	141 699	119 684	19 591	12 334	122 108	98 711	8 691	33.7	41.8	21.01	19.90	708.04	831.82
July	142 001	119 951	19 626	12 349	122 375	98 834	8 702	33.7	41.8	21.03	19.94	708.71	833.49
August	142 126	120 066	19 629	12 345	122 497	98 969	8 690	33.7	41.8	21.08	19.98	710.40	835.16
September	142 281	120 241	19 641	12 354	122 640	99 121	8 697	33.7	41.7	21.11	20.08	711.41	837.34
October	142 587	120 531	19 680	12 362	122 907	99 351	8 708	33.7	41.8	21.19	20.05	714.10	838.09
November	142 824	120 743	19 711	12 357	123 113	99 509	8 698	33.7	41.8	21.20	20.06	714.44	838.51
December	143 097	121 000	19 738	12 362	123 359	99 707	8 696	33.8	41.8	21.24	20.11	717.91	840.60
2016													
January	143 170	121 052	19 730	12 379	123 440	99 756	8 705	33.7	41.9	21.31	20.16	718.15	844.70
February	143 433	121 289	19 728	12 367	123 705	99 907	8 691	33.6	41.9	21.34	20.19	717.02	845.96
March	143 662	121 482	19 716	12 345	123 946	100 066	8 677	33.6	41.8	21.40	20.27	719.04	847.29
April	143 849	121 663	19 728	12 349	124 121	100 158	8 672	33.6	41.8	21.45	20.38	720.72	851.88
May	143 891	121 694	19 689	12 332	124 202	100 195	8 651	33.6	41.8	21.47	20.41	721.39	853.14
June	144 158	121 974	19 709	12 351	124 449	100 414	8 674	33.6	41.8	21.53	20.48	723.41	856.06
July	144 512	122 227	19 757	12 369	124 755	100 623	8 690	33.6	42.0	21.58	20.49	725.09	860.58
August	144 647	122 379	19 732	12 348	124 915	100 742	8 662	33.6	41.8	21.61	20.56	726.10	859.41
September	144 916	122 613	19 762	12 347	125 154	100 935	8 658	33.6	41.9	21.64	20.55	727.10	861.05
October	145 061	122 769	19 789	12 347	125 272	101 065	8 651	33.6	41.9	21.70	20.59	729.12	862.72
November	145 212	122 922	19 808	12 342	125 404	101 246	8 651	33.6	41.8	21.72	20.60	729.79	861.08
December	145 442	123 134	19 825	12 356	125 617	101 472	8 661	33.5	41.9	21.76	20.63	728.96	864.40
2017													
January	145 627	123 325	19 865	12 369	125 762	101 643	8 671	33.6	41.9	21.81	20.62	732.82	863.98
February	145 815	123 496	19 928	12 384	125 887	101 779	8 688	33.6	41.9	21.86	20.64	734.50	864.82
March	145 944	123 619	19 964	12 395	125 980	101 914	8 699	33.5	41.8	21.88	20.69	732.98	864.84
April	146 141	123 811	19 985	12 400	126 156	102 061	8 709	33.7	41.9	21.93	20.76	739.04	869.84
May	146 296	123 964	20 007	12 406	126 289	102 175	8 708	33.6	41.9	21.97	20.80	738.19	871.52
June	146 512	124 166	20 050	12 418	126 462	102 342	8 713	33.7	42.0	22.03	20.86	742.41	876.12
July	146 727	124 368	20 056	12 417	126 671	102 499	8 715	33.7	42.0	22.08	20.92	744.10	878.64
August	146 911	124 554	20 135	12 461	126 776	102 649	8 745	33.6	42.0	22.11	20.95	742.90	879.90
September	146 929	124 563	20 161	12 470	126 768	102 672	8 756	33.6	41.9	22.20	21.01	745.92	880.32
October	147 196	124 827	20 206	12 492	126 990	102 899	8 771	33.7	42.0	22.18	21.08	747.47	885.36
November	147 421	125 028	20 273	12 514	127 148	103 022	8 783	33.8	42.0	22.23	21.11	751.37	886.62
December	147 551	125 170	20 326	12 541	127 225	103 133	8 805	33.7	41.8	22.30	21.21	751.51	886.58
2018													
January	147 672	125 317	20 371	12 558	127 301	103 253	8 817	33.6	41.8	22.36	21.29	751.30	889.92
February	148 078	125 672	20 494	12 588	127 584	103 518	8 827	33.8	42.2	22.40	21.33	757.12	900.13
March	148 254	125 850	20 530	12 611	127 724	103 677	8 842	33.7	42.2	22.49	21.39	757.91	902.66
April	148 391	125 977	20 577	12 632	127 814	103 772	8 863	33.8	42.4	22.55	21.47	762.19	910.33
May	148 669	126 238	20 648	12 658	128 021	104 031	8 869	33.8	42.0	22.61	21.43	764.22	900.06
June	148 888	126 429	20 698	12 687	128 190	104 173	8 902	33.8	42.1	22.67	21.48	766.25	904.31
July	149 024	126 562	20 739	12 706	128 285	104 262	8 916	33.8	42.2	22.71	21.45	767.60	905.19
August	149 268	126 778	20 791	12 719	128 477	104 426	8 918	33.8	42.2	22.79	21.55	770.30	909.41
September	149 348	126 859	20 834	12 740	128 514	104 504	8 934	33.7	42.1	22.86	21.62	770.38	910.20
October	149 549	127 064	20 884	12 764	128 665	104 658	8 951	33.7	42.0	22.90	21.70	771.73	911.40
November	149 683	127 197	20 897	12 784	128 786	104 741	8 965	33.7	42.0	23.00	21.78	775.10	914.76
December	149 865	127 370	20 948	12 805	128 917	104 907	8 983	33.7	42.0	23.09	21.82	778.13	916.44
2019													
January	150 134	127 628	21 023	12 825	129 111	105 168	8 998	33.8	42.0	23.11	21.84	781.12	917.28
February	150 135	127 622	20 994	12 830	129 141	105 109	9 000	33.6	41.7	23.19	21.91	779.18	913.65
March	150 282	127 754	21 011	12 827	129 271	105 217	8 992	33.7	41.7	23.28	21.96	784.54	915.73
April	150 492	127 939	21 039	12 829	129 453	105 335	8 979	33.7	41.6	23.33	21.96	786.22	913.54
May	150 577	128 026	21 050	12 829	129 527	105 399	8 970	33.6	41.5	23.42	22.04	786.91	914.66
June	150 759	128 206	21 076	12 838	129 683	105 503	8 968	33.6	41.6	23.47	22.14	788.59	921.02
July	150 953	128 366	21 085	12 845	129 868	105 604	8 967	33.5	41.5	23.54	22.19	788.59	920.89
August	151 160	128 523	21 087	12 848	130 073	105 731	8 968	33.6	41.5	23.64	22.22	794.30	922.13
September	151 368	128 718	21 106	12 851	130 262	105 885	8 968	33.6	41.5	23.70	22.26	796.32	923.79
October	151 553	128 908	21 086	12 810	130 467	105 992	8 931	33.6	41.4	23.76	22.30	798.34	923.22
November	151 814	129 155	21 131	12 868	130 683	106 156	8 978	33.5	41.4	23.81	22.39	797.64	926.95
December	151 998	129 319	21 136	12 866	130 862	106 300	8 980	33.6	41.4	23.84	22.44	801.02	929.02

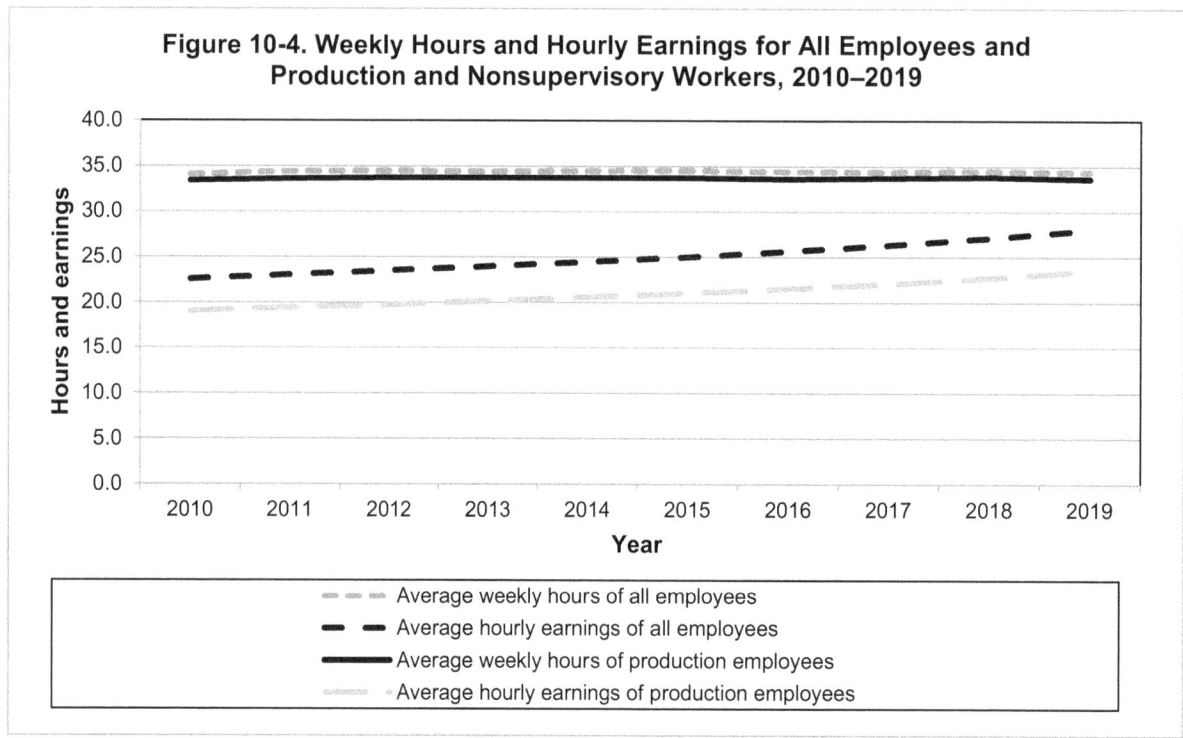

Figure 10-4. Weekly Hours and Hourly Earnings for All Employees and Production and Nonsupervisory Workers, 2010–2019

- Among all production and nonsupervisory workers in 2019, miners and loggers worked the longest hours per week at 46.8 followed by construction workers at 39.8. Meanwhile, production and nonsupervisory workers in leisure and hospitality only averaged 24.7 hours a week. (Table 10-14)

- In 2019, mining and logging had the highest average weekly earnings among all production and nonsupervisory workers at $1,403 followed by information workers at $1,196 and construction workers at $1,135. (Table 10-17)

- The index of aggregate hours can point to an upcoming recession. As the index falls, fewer hours are being used for the production of goods and services. Even though employment is up, but hours are not up proportionally. This reflects the trend to part-time workers and the reduction of hours in part-time workers. (Table 10-15).

Table 10-9. Average Weekly Hours of All Employees on Private Nonfarm Payrolls by NAICS Supersector

(Hours per week, seasonally adjusted.)

Year and month	Total private	Mining and logging	Construction	Manufacturing		Trade, transportation, and utilities			Information	Financial activities	Professional and business services	Education and health services	Leisure and hospitality	Other services
				Average weekly hours	Overtime hours	Total	Wholesale trade	Retail trade						
2010	34.1	43.4	37.8	40.2	3.0	34.2	38.1	31.4	36.5	36.9	35.4	32.7	25.7	31.6
2011	34.4	44.5	38.3	40.5	3.2	34.6	38.6	31.6	36.6	37.3	35.7	32.7	25.9	31.7
2012	34.5	44.0	38.7	40.7	3.3	34.6	38.7	31.7	36.6	37.4	36.0	32.8	26.1	31.6
2013	34.4	43.9	39.0	40.8	3.4	34.5	38.7	31.4	36.7	37.1	36.1	32.7	26.0	31.7
2014	34.5	34.5	39.0	41.0	3.5	34.5	38.9	31.3	36.8	37.3	36.3	32.7	26.2	31.8
2015	34.5	44.0	39.2	40.8	3.3	34.6	38.9	31.4	36.2	37.6	36.2	32.8	26.3	31.9
2016	34.4	43.4	39.0	40.7	3.3	34.3	38.9	31.0	36.0	37.5	36.1	32.9	26.1	31.9
2017	34.4	45.1	39.1	40.8	3.4	34.4	39.0	31.0	36.2	37.5	36.1	32.9	26.1	31.8
2018	34.5	46.0	39.2	40.9	3.5	34.5	39.0	31.0	36.1	37.6	36.1	32.9	26.1	31.8
2019	34.4	46.2	39.3	40.5	3.3	34.2	38.9	30.6	36.3	37.6	36.2	33.0	25.9	31.9
2015														
January	34.5	45.1	39.0	40.9	3.4	34.5	38.9	31.3	36.5	37.4	36.2	32.8	26.3	31.9
February	34.6	44.3	39.5	41.0	3.4	34.6	38.9	31.4	36.4	37.4	36.2	32.8	26.3	31.9
March	34.5	43.9	39.2	40.8	3.3	34.6	38.8	31.4	36.3	37.5	36.1	32.8	26.2	31.8
April	34.5	43.7	39.0	40.8	3.4	34.6	38.8	31.4	36.3	37.6	36.1	32.8	26.2	31.8
May	34.5	43.6	38.9	40.7	3.4	34.7	38.9	31.5	36.3	37.6	36.2	32.8	26.2	31.8
June	34.5	43.7	39.2	40.6	3.3	34.5	38.8	31.3	36.3	37.6	36.2	32.8	26.3	31.8
July	34.5	43.9	39.1	40.8	3.4	34.5	38.9	31.3	36.3	37.6	36.2	32.8	26.2	31.8
August	34.5	43.7	39.3	40.9	3.3	34.6	38.8	31.4	36.2	37.7	36.2	32.8	26.2	31.8
September	34.5	44.1	38.8	40.6	3.3	34.9	38.9	31.8	36.0	37.7	36.1	32.8	26.3	31.8
October	34.6	44.0	39.4	40.7	3.4	34.6	38.9	31.4	36.1	37.6	36.3	32.8	26.3	31.9
November	34.5	44.1	39.1	40.7	3.3	34.6	38.9	31.4	36.1	37.6	36.1	32.8	26.2	31.9
December	34.5	44.3	39.6	40.7	3.2	34.5	39.0	31.2	36.0	37.7	36.2	32.8	26.3	31.9
2016														
January	34.6	44.0	39.3	40.8	3.2	34.6	39.0	31.3	36.3	37.7	36.3	32.9	26.2	32.0
February	34.5	43.2	39.3	40.8	3.3	34.5	38.9	31.3	36.0	37.6	36.1	32.8	26.2	31.9
March	34.4	42.8	38.9	40.7	3.3	34.4	38.8	31.1	36.0	37.6	36.1	32.8	26.1	31.9
April	34.4	42.9	39.1	40.8	3.3	34.4	38.9	31.1	36.0	37.7	36.2	32.8	26.1	31.9
May	34.4	43.0	39.0	40.8	3.3	34.3	38.8	31.0	36.0	37.5	36.1	32.8	26.1	31.9
June	34.4	43.0	39.0	40.7	3.3	34.3	38.8	31.0	35.9	37.4	36.0	32.9	26.1	31.9
July	34.4	43.2	39.1	40.7	3.3	34.3	38.9	31.0	36.1	37.7	36.2	32.9	26.1	31.9
August	34.3	43.6	38.9	40.6	3.3	34.2	38.8	30.9	36.0	37.5	36.0	32.9	26.0	32.0
September	34.4	43.7	39.1	40.7	3.4	34.3	39.0	30.9	35.8	37.5	36.1	32.9	26.2	32.0
October	34.4	43.8	39.3	40.8	3.3	34.3	39.0	30.9	35.9	37.4	36.1	32.9	26.0	32.0
November	34.3	43.8	39.1	40.6	3.2	34.2	38.9	30.8	36.0	37.4	36.0	32.9	26.1	31.9
December	34.3	43.7	38.9	40.7	3.3	34.4	38.9	31.1	36.0	37.4	36.0	32.9	25.9	31.9
2017														
January	34.4	44.0	39.0	40.8	3.3	34.3	38.9	30.9	36.6	37.2	36.1	32.9	26.0	31.8
February	34.3	44.5	39.0	40.6	3.3	34.2	38.9	30.8	36.3	37.4	36.1	32.9	25.9	31.8
March	34.3	45.0	38.9	40.6	3.3	34.2	38.9	30.9	36.3	37.4	36.0	32.9	26.0	31.8
April	34.4	45.2	39.1	40.7	3.3	34.5	39.0	31.2	36.3	37.4	36.1	32.9	26.2	31.8
May	34.4	45.4	39.2	40.8	3.3	34.4	39.0	31.0	36.3	37.4	36.0	32.8	26.0	31.8
June	34.4	45.0	39.1	40.8	3.3	34.4	39.0	31.0	36.2	37.6	36.1	32.9	26.1	31.8
July	34.4	45.5	39.1	40.9	3.4	34.4	39.1	31.0	36.3	37.5	36.1	32.9	26.1	31.7
August	34.4	45.0	39.0	40.8	3.4	34.4	39.1	30.9	36.1	37.5	36.0	32.9	26.0	31.7
September	34.4	45.3	39.0	40.8	3.4	34.4	39.1	30.9	36.0	37.5	36.1	32.8	26.0	31.7
October	34.4	45.4	39.2	40.9	3.5	34.4	39.0	31.0	36.1	37.5	36.0	32.9	26.1	31.7
November	34.5	45.6	39.3	41.0	3.5	34.6	39.2	31.3	36.0	37.6	36.2	32.9	26.1	31.7
December	34.5	45.6	39.2	40.8	3.4	34.5	39.3	31.1	36.1	37.6	36.0	33.0	26.2	31.8
2018														
January	34.4	45.4	38.9	40.7	3.5	34.5	39.0	31.1	35.9	37.5	35.9	32.9	26.0	31.6
February	34.5	46.0	39.3	41.0	3.7	34.5	39.0	31.1	36.0	37.6	36.2	33.0	26.1	31.7
March	34.5	45.8	39.3	40.9	3.6	34.5	39.0	31.2	36.1	37.5	36.2	32.9	26.1	31.7
April	34.5	46.0	39.5	41.1	3.6	34.5	39.0	31.0	36.1	37.6	36.2	33.0	26.1	31.8
May	34.5	46.2	39.5	40.9	3.4	34.5	39.1	31.1	35.9	37.6	36.1	32.9	26.2	31.7
June	34.6	46.4	39.2	40.9	3.5	34.6	39.1	31.1	35.7	37.7	36.2	33.0	26.2	31.8
July	34.5	46.0	39.5	41.0	3.5	34.6	39.0	31.1	36.1	37.5	36.2	32.9	26.1	31.8
August	34.5	46.1	39.2	40.9	3.5	34.5	39.1	30.9	36.1	37.6	36.1	33.0	26.1	31.8
September	34.5	45.9	39.0	40.9	3.5	34.5	39.0	30.9	36.2	37.6	36.1	32.9	26.0	31.9
October	34.5	46.0	38.9	40.8	3.5	34.4	39.0	30.8	36.2	37.8	36.2	32.9	26.0	31.9
November	34.4	45.9	38.7	40.8	3.5	34.5	39.0	30.9	36.2	37.6	36.1	32.9	25.9	31.9
December	34.5	46.2	39.3	40.9	3.5	34.3	39.0	30.6	36.3	37.6	36.1	33.0	26.0	31.9
2019														
January	34.5	46.4	39.9	40.8	3.5	34.3	39.0	30.7	36.2	37.7	36.3	33.0	26.1	31.9
February	34.4	46.3	38.8	40.7	3.5	34.2	39.0	30.6	36.2	37.7	36.2	33.0	26.0	31.8
March	34.5	46.6	39.4	40.7	3.4	34.3	39.0	30.7	36.3	37.7	36.3	33.0	26.1	32.0
April	34.4	46.8	39.2	40.6	3.3	34.3	39.0	30.7	36.2	37.7	36.2	33.0	25.9	31.9
May	34.4	46.3	39.1	40.6	3.3	34.3	38.9	30.7	36.4	37.6	36.2	33.0	25.9	31.9
June	34.4	46.3	39.3	40.6	3.4	34.2	38.9	30.7	36.4	37.5	36.2	33.0	25.9	31.9
July	34.3	46.2	39.1	40.4	3.3	34.1	38.8	30.6	36.2	37.5	36.1	33.0	25.8	31.9
August	34.4	46.2	39.4	40.5	3.2	34.2	38.9	30.6	36.5	37.7	36.2	33.0	25.8	31.8
September	34.4	46.2	39.7	40.5	3.2	34.2	38.9	30.7	36.5	37.6	36.1	33.0	25.9	31.8
October	34.4	46.1	39.3	40.3	3.2	34.2	39.0	30.6	36.4	37.7	36.1	33.0	25.8	31.9
November	34.3	45.7	39.1	40.4	3.2	34.0	39.0	30.3	36.3	37.7	36.1	33.1	25.8	31.9
December	34.3	45.8	39.0	40.4	3.2	34.1	38.8	30.6	36.3	37.5	36.1	33.0	25.8	31.8

Table 10-10. Indexes of Aggregate Weekly Hours of All Employees on Private Nonfarm Payrolls by NAICS Supersector

(2007 =100, seasonally adjusted.)

Year and month	Total private	Mining and logging	Construc- tion	Manu- facturing	Trade, transportation, and utilities			Information	Financial activities	Profes- sional and business services	Education and health services	Leisure and hospitality	Other services
					Total	Wholesale trade	Retail trade						
2010	92.4	96.1	71.9	83.4	91.7	90.5	92.0	90.3	93.0	93.1	106.5	95.9	97.2
2011	94.7	110.2	73.1	85.4	94.2	93.3	94.1	89.4	94.0	97.4	108.6	98.7	98.0
2012	97.1	117.1	75.4	87.4	95.8	95.6	95.4	89.6	95.3	101.7	111.2	102.6	99.0
2013	99.0	119.2	78.7	88.3	97.0	96.7	96.2	90.6	96.0	105.1	112.6	105.7	100.3
2014	101.5	125.5	82.7	89.8	99.1	98.5	97.7	91.5	97.4	108.8	114.6	109.8	102.0
2015	103.9	112.4	87.1	90.5	101.3	99.2	99.6	91.1	100.1	111.8	118.0	113.7	103.3
2016	105.4	91.0	90.8	90.5	101.8	99.1	99.7	91.6	101.6	113.6	121.3	116.5	104.6
2017	107.4	96.0	94.1	91.3	102.9	100.2	99.8	93.1	103.8	116.0	124.5	119.5	105.7
2018	109.6	105.1	98.7	93.4	103.9	100.6	99.3	93.6	106.0	118.8	127.1	121.3	107.0
2019	110.8	106.7	101.5	93.7	103.4	101.4	97.2	94.8	107.8	120.9	130.2	122.5	108.2
2015													
January	102.7	125.8	85.0	90.4	100.2	99.0	98.6	91.3	98.8	110.4	116.4	111.9	103.0
February	103.2	121.8	86.6	90.7	100.6	99.2	99.1	91.2	98.9	110.6	116.7	112.4	103.1
March	103.0	118.5	85.6	90.4	100.7	99.0	99.1	90.7	99.3	110.4	116.9	112.0	102.8
April	103.2	115.8	85.9	90.4	100.8	98.9	99.3	90.9	99.6	110.7	117.3	112.3	102.9
May	103.5	112.3	86.2	90.3	101.3	99.3	99.8	91.1	99.7	111.5	117.6	112.9	103.1
June	103.7	112.1	87.0	90.1	100.9	99.0	99.3	91.2	100.0	111.8	117.8	113.4	103.0
July	103.9	111.0	87.2	90.6	101.1	99.3	99.4	91.3	100.2	112.0	118.1	113.4	103.2
August	104.0	109.0	87.9	90.8	101.4	99.0	99.7	91.0	100.6	112.1	118.3	113.6	103.0
September	104.1	107.9	87.0	90.2	102.3	99.1	101.0	90.9	100.7	111.9	118.5	114.7	103.1
October	104.7	106.6	88.9	90.5	101.6	99.1	99.9	91.1	100.6	113.0	119.0	114.9	103.6
November	104.6	104.8	88.9	90.5	101.7	99.1	100.1	90.7	100.8	112.6	119.2	114.9	103.7
December	104.8	103.8	90.5	90.5	101.6	99.4	99.5	90.7	101.1	113.3	119.6	115.6	104.0
2016													
January	105.1	101.1	89.7	90.9	101.8	99.4	99.9	91.6	101.3	113.6	120.0	115.4	104.2
February	105.0	96.5	90.1	90.8	101.8	99.1	100.2	91.3	101.2	113.0	120.0	115.8	104.2
March	104.9	92.9	89.6	90.4	101.7	98.9	99.8	91.6	101.3	113.2	120.2	115.7	104.4
April	105.1	91.4	90.3	90.6	101.8	99.3	99.9	91.7	101.9	113.7	120.6	115.9	104.4
May	105.1	90.0	90.0	90.5	101.5	99.0	99.5	90.5	101.5	113.4	120.9	116.0	104.3
June	105.3	88.6	90.1	90.4	101.7	98.9	99.7	91.7	101.5	113.3	121.5	116.4	104.5
July	105.6	88.5	90.8	90.6	101.8	99.2	99.8	92.3	102.6	114.2	121.8	116.8	104.6
August	105.4	88.6	90.3	90.2	101.7	99.0	99.6	92.2	102.2	113.6	122.0	116.6	105.2
September	105.9	88.8	91.2	90.4	102.1	99.7	99.7	91.9	102.3	114.3	122.3	117.8	105.4
October	106.0	88.6	92.1	90.6	102.1	99.7	99.7	92.3	102.1	114.4	122.7	117.1	105.5
November	105.8	89.0	91.9	90.1	101.8	99.5	99.2	92.4	102.2	114.3	122.8	117.9	105.4
December	106.0	89.0	91.5	90.5	102.7	99.5	100.4	92.5	102.6	114.5	123.2	117.1	105.1
2017													
January	106.5	89.7	92.0	90.8	102.4	99.6	99.9	94.2	102.4	115.1	123.2	117.8	105.1
February	106.3	91.6	92.6	90.5	102.0	99.6	99.3	93.3	102.9	115.2	123.6	117.6	105.3
March	106.4	93.3	92.6	90.5	102.0	99.6	99.5	93.3	103.0	115.0	123.8	118.1	105.3
April	106.9	95.0	93.2	90.8	103.0	100.0	100.4	93.1	103.2	115.5	124.1	119.5	105.4
May	107.1	96.0	93.6	91.1	102.6	99.9	99.6	93.0	103.3	115.5	123.9	118.9	105.6
June	107.2	95.7	93.7	91.1	102.7	100.0	99.7	92.8	104.1	116.0	124.5	119.6	105.7
July	107.4	97.1	93.8	91.4	102.7	100.3	99.6	93.1	103.9	116.3	124.8	120.2	105.5
August	107.6	96.7	94.0	91.5	102.8	100.3	99.3	92.8	104.1	116.1	125.0	119.7	105.6
September	107.6	97.8	94.2	91.5	102.9	100.3	99.2	92.4	104.2	116.4	124.8	119.2	105.5
October	107.8	98.0	95.0	91.9	103.0	100.1	99.5	92.7	104.2	116.4	125.3	120.3	105.9
November	108.3	99.3	95.7	92.3	103.6	100.7	100.5	92.5	104.6	117.2	125.6	120.4	106.1
December	108.4	99.2	95.8	92.1	103.4	101.0	99.8	92.9	104.7	116.7	126.0	121.0	106.5
2018													
January	108.2	99.7	95.4	92.0	103.3	100.2	99.7	92.3	104.5	116.7	125.9	120.1	105.9
February	108.8	102.0	97.5	92.8	103.6	100.2	99.9	92.6	105.1	118.0	126.6	120.8	106.3
March	109.0	102.8	97.6	92.8	103.8	100.4	100.3	93.2	104.9	118.2	126.4	120.8	106.2
April	109.1	104.4	98.3	93.4	103.8	100.1	99.7	93.6	105.3	118.4	126.8	120.8	106.7
May	109.3	105.1	98.9	93.1	103.9	100.4	100.0	93.0	105.5	118.4	126.7	121.6	106.6
June	109.8	106.3	98.4	93.4	104.2	100.5	99.8	92.5	106.1	118.9	127.3	121.9	107.2
July	109.6	105.4	99.5	93.7	104.2	100.4	99.7	93.7	105.6	119.0	127.0	121.8	106.9
August	109.8	106.8	99.1	93.6	104.1	101.0	99.1	93.7	106.0	118.9	127.7	121.8	106.9
September	109.9	106.6	98.9	93.7	104.0	100.7	98.8	93.9	106.3	119.1	127.4	121.1	107.4
October	110.0	107.3	98.9	93.7	103.7	100.8	98.2	94.3	107.1	119.6	127.7	121.4	107.5
November	109.8	106.3	98.4	93.8	104.2	100.9	98.7	94.2	106.7	119.4	127.8	121.0	107.5
December	110.3	107.6	100.3	94.2	103.5	101.0	97.6	94.6	106.8	119.6	128.5	121.8	107.6
2019													
January	110.5	108.8	102.5	94.1	103.7	101.1	97.9	94.0	107.2	120.2	128.8	122.8	107.7
February	110.2	107.8	99.3	93.9	103.3	101.2	97.3	93.9	107.3	120.1	128.9	122.2	107.3
March	110.6	108.5	101.1	93.9	103.5	101.2	97.5	94.5	107.6	120.5	129.2	122.8	108.3
April	110.5	109.0	100.9	93.7	103.6	101.4	97.4	94.1	107.7	120.5	129.6	122.0	108.2
May	110.6	108.1	100.8	93.7	103.5	101.2	97.4	94.8	107.5	120.6	129.7	122.1	108.2
June	110.7	107.8	101.6	93.8	103.3	101.2	97.3	95.2	107.3	120.9	130.0	122.1	108.4
July	110.5	106.8	101.1	93.4	103.0	101.1	97.0	94.6	107.5	120.8	130.4	121.7	108.6
August	111.0	106.1	102.0	93.6	103.3	101.4	97.0	95.4	108.3	121.3	130.7	122.0	108.4
September	111.2	106.1	103.0	93.6	103.4	101.5	97.4	95.5	108.1	121.1	131.1	122.9	108.3
October	111.3	106.5	102.2	92.9	103.6	101.9	97.2	95.2	108.6	121.4	131.3	122.9	108.6
November	111.2	104.0	101.6	93.5	103.0	102.0	96.2	95.3	108.8	121.6	132.1	123.3	108.9
December	111.4	102.9	101.6	93.5	103.5	101.5	97.4	95.6	108.3	121.7	131.8	123.6	108.6

Table 10-11. Average Hourly Earnings of All Employees on Private Nonfarm Payrolls by NAICS Supersector

(Dollars, seasonally adjusted.)

Year and month	Total private	Mining and logging	Construction	Manufacturing	Trade, transportation, and utilities			Information	Financial activities	Professional and business services	Education and health services	Leisure and hospitality	Other services
					Total	Wholesale trade	Retail trade						
2010	22.56	27.39	25.19	23.31	19.60	26.04	15.57	30.53	27.21	27.26	22.76	13.08	20.16
2011	23.03	28.10	25.41	23.69	19.99	26.28	15.87	31.59	27.91	27.79	23.42	13.23	20.50
2012	23.49	28.76	25.73	23.92	20.45	26.79	16.31	31.83	29.26	28.16	24.01	13.37	20.85
2013	23.95	29.72	26.12	24.35	20.91	27.54	16.64	32.91	30.15	28.58	24.42	13.50	21.40
2014	24.46	30.79	26.69	24.81	21.32	27.97	17.01	34.07	30.76	29.32	24.72	13.91	21.97
2015	25.01	31.15	27.37	25.25	21.77	28.53	17.52	35.14	31.52	30.11	25.24	14.32	22.48
2016	25.64	31.91	28.12	25.99	22.26	29.36	17.88	36.67	32.29	30.82	25.74	14.87	23.05
2017	26.32	32.05	28.90	26.59	22.72	29.92	18.18	38.30	33.24	31.67	26.32	15.47	23.85
2018	27.11	32.49	29.90	27.05	23.33	30.45	18.79	40.08	34.79	32.59	27.00	15.98	24.56
2019	28.00	33.77	30.75	27.70	24.22	31.36	19.70	42.15	35.94	33.67	27.63	16.55	25.23
2015													
January	24.73	30.51	27.07	24.99	21.54	28.21	17.32	34.51	31.07	29.74	24.99	14.15	22.18
February	24.78	30.67	27.12	25.04	21.62	28.32	17.35	34.66	31.17	29.77	25.02	14.17	22.25
March	24.84	30.83	27.24	25.10	21.60	28.27	17.32	34.74	31.22	29.88	25.12	14.22	22.35
April	24.89	30.70	27.28	25.11	21.65	28.35	17.39	34.74	31.36	29.98	25.11	14.26	22.32
May	24.96	31.16	27.32	25.14	21.72	28.59	17.43	34.89	31.54	30.04	25.22	14.30	22.42
June	24.98	31.03	27.34	25.14	21.70	28.52	17.47	34.93	31.60	30.07	25.24	14.30	22.54
July	25.00	31.18	27.36	25.23	21.77	28.55	17.52	35.04	31.50	30.10	25.21	14.31	22.48
August	25.10	31.41	27.44	25.36	21.86	28.71	17.61	35.38	31.59	30.18	25.29	14.34	22.56
September	25.11	31.52	27.35	25.40	21.83	28.62	17.68	35.43	31.67	30.28	25.33	14.38	22.60
October	25.19	31.43	27.52	25.42	21.94	28.74	17.72	35.58	31.72	30.33	25.41	14.43	22.64
November	25.24	31.68	27.67	25.52	21.95	28.72	17.72	35.72	31.72	30.39	25.44	14.46	22.66
December	25.27	31.47	27.62	25.53	21.98	28.73	17.76	35.87	31.82	30.35	25.49	14.52	22.72
2016													
January	25.36	31.79	27.66	25.63	22.01	28.90	17.72	36.02	32.03	30.53	25.54	14.58	22.77
February	25.38	31.77	27.76	25.66	22.02	28.89	17.78	36.21	32.00	30.57	25.58	14.63	22.80
March	25.45	31.95	27.88	25.76	22.13	29.20	17.84	36.16	32.19	30.56	25.60	14.67	22.86
April	25.53	31.98	27.98	25.86	22.19	29.31	17.84	36.26	32.15	30.70	25.68	14.75	22.90
May	25.58	32.13	28.08	25.97	22.20	29.32	17.87	36.48	32.19	30.80	25.64	14.80	22.96
June	25.63	32.12	28.13	25.98	22.29	29.35	17.96	36.51	32.22	30.88	25.68	14.86	23.02
July	25.70	32.01	28.20	26.01	22.32	29.57	17.89	36.65	32.41	30.92	25.72	14.93	23.11
August	25.72	31.65	28.20	26.12	22.33	29.50	17.91	36.81	32.49	30.93	25.71	14.98	23.15
September	25.77	31.87	28.23	26.12	22.37	29.53	17.93	37.04	32.58	30.97	25.83	15.02	23.17
October	25.88	31.98	28.40	26.28	22.49	29.70	18.00	37.40	32.44	31.11	25.89	15.08	23.32
November	25.91	31.90	28.35	26.26	22.51	29.60	18.10	37.46	32.64	31.15	25.94	15.12	23.33
December	25.94	32.08	28.42	26.33	22.43	29.70	17.94	37.53	32.69	31.22	25.97	15.16	23.38
2017													
January	26.00	32.20	28.53	26.36	22.52	29.78	18.01	37.50	32.63	31.28	25.98	15.23	23.51
February	26.08	32.14	28.52	26.41	22.54	29.80	18.02	37.56	32.83	31.37	26.09	15.29	23.64
March	26.11	32.03	28.61	26.41	22.56	29.77	18.05	37.65	32.78	31.51	26.09	15.35	23.61
April	26.17	31.81	28.60	26.55	22.59	29.87	18.07	38.00	32.90	31.59	26.21	15.37	23.66
May	26.21	31.84	28.72	26.53	22.65	29.84	18.14	38.20	32.88	31.52	26.26	15.50	23.69
June	26.27	31.91	28.90	26.56	22.71	29.89	18.18	38.49	33.08	31.55	26.26	15.45	23.75
July	26.36	32.21	28.95	26.67	22.77	29.96	18.22	38.60	33.25	31.63	26.39	15.48	23.86
August	26.38	32.04	29.00	26.60	22.76	29.88	18.25	38.64	33.36	31.68	26.41	15.54	23.93
September	26.51	31.98	29.18	26.69	22.85	30.00	18.27	38.60	33.50	31.86	26.53	15.62	24.06
October	26.48	32.02	29.11	26.74	22.81	29.87	18.27	38.51	33.69	31.83	26.44	15.59	24.10
November	26.55	32.12	29.22	26.73	22.87	30.05	18.30	38.71	33.71	31.92	26.55	15.64	24.18
December	26.64	32.20	29.29	26.78	22.94	30.11	18.33	38.82	34.03	32.03	26.62	15.70	24.21
2018													
January	26.71	32.31	29.39	26.87	22.97	30.09	18.40	38.98	34.23	32.13	26.66	15.74	24.23
February	26.75	32.22	29.55	26.85	23.00	30.11	18.45	39.11	34.23	32.12	26.69	15.74	24.26
March	26.84	32.45	29.49	26.90	23.04	30.13	18.48	39.29	34.42	32.24	26.82	15.82	24.41
April	26.91	32.36	29.66	26.94	23.12	30.14	18.60	39.46	34.45	32.30	26.85	15.86	24.46
May	26.98	32.28	29.70	26.97	23.21	30.28	18.72	39.60	34.65	32.40	26.96	15.89	24.50
June	27.04	32.51	29.78	27.03	23.24	30.45	18.68	39.78	34.68	32.50	27.01	15.94	24.52
July	27.11	32.44	29.94	27.01	23.30	30.41	18.78	39.73	34.82	32.62	27.05	16.01	24.56
August	27.22	32.60	30.02	27.09	23.41	30.50	18.87	40.08	34.91	32.80	27.12	16.06	24.60
September	27.31	32.88	30.18	27.12	23.49	30.69	18.93	40.76	35.10	32.83	27.15	16.10	24.69
October	27.36	32.61	30.24	27.14	23.58	30.69	19.04	40.98	34.99	32.86	27.19	16.16	24.75
November	27.44	32.74	30.30	27.24	23.60	30.73	19.08	41.15	35.30	32.96	27.24	16.22	24.80
December	27.54	32.68	30.44	27.32	23.79	30.92	19.24	41.45	35.36	33.04	27.28	16.27	24.88
2019													
January	27.58	32.74	30.32	27.27	23.80	30.83	19.34	41.78	35.39	33.07	27.42	16.28	24.97
February	27.69	32.95	30.45	27.43	23.93	30.99	19.42	41.77	35.51	33.23	27.46	16.38	25.00
March	27.76	32.96	30.50	27.46	24.02	31.26	19.44	41.96	35.54	33.33	27.52	16.40	25.05
April	27.81	33.33	30.63	27.48	24.01	31.12	19.48	41.97	35.73	33.43	27.49	16.45	25.09
May	27.87	33.47	30.70	27.57	24.13	31.30	19.55	42.00	35.82	33.52	27.43	16.51	25.16
June	27.96	33.65	30.74	27.68	24.20	31.38	19.65	41.84	35.91	33.63	27.54	16.55	25.18
July	28.05	34.11	30.75	27.75	24.28	31.39	19.73	42.26	35.93	33.75	27.64	16.59	25.21
August	28.16	34.15	30.87	27.81	24.38	31.62	19.79	42.80	36.19	33.86	27.71	16.62	25.30
September	28.16	34.44	30.87	27.88	24.38	31.56	19.84	42.28	36.03	33.94	27.74	16.67	25.32
October	28.24	34.72	30.98	27.91	24.44	31.66	19.89	42.30	36.14	34.04	27.85	16.70	25.39
November	28.34	34.57	31.09	28.02	24.52	31.74	20.00	42.58	36.40	34.17	27.87	16.76	25.48
December	28.37	34.57	31.15	28.14	24.50	31.61	20.04	42.57	36.53	34.23	27.86	16.77	25.55

Table 10-12. Average Weekly Earnings of All Employees on Private Nonfarm Payrolls by NAICS Supersector

(Dollars, seasonally adjusted.)

Year and month	Total private	Mining and logging	Construction	Manufacturing	Trade, transportation, and utilities			Information	Financial activities	Professional and business services	Education and health services	Leisure and hospitality	Other services
					Total	Wholesale trade	Retail trade						
2010	769.57	1 189.32	952.78	937.34	670.87	991.59	488.08	1 114.65	1 003.94	964.60	743.41	336.83	637.65
2011	790.74	1 250.91	973.85	958.84	690.43	1 015.85	501.04	1 156.56	1 039.70	992.94	766.88	342.67	649.87
2012	809.46	1 263.98	997.01	973.96	707.20	1 038.30	516.10	1 166.45	1 093.00	1 014.67	787.43	349.12	659.49
2013	824.91	1 306.16	1 018.01	994.30	721.19	1 066.81	522.57	1 206.05	1 119.71	1 031.57	798.73	350.95	679.43
2014	844.80	1 380.77	1 040.85	1 016.42	735.94	1 088.33	532.88	1 252.44	1 146.22	1 063.71	809.11	364.07	698.41
2015	864.07	1 371.13	1 070.93	1 029.68	753.57	1 109.79	550.74	1 274.99	1 185.77	1 089.87	828.36	376.20	716.17
2016	881.09	1 383.16	1 100.94	1 057.77	764.27	1 140.05	554.27	1 315.74	1 208.20	1 110.34	844.80	387.80	734.90
2017	906.19	1 448.75	1 131.76	1 084.86	781.98	1 169.15	563.80	1 388.30	1 246.39	1 142.52	865.65	403.77	757.95
2018	936.37	1 494.46	1 174.87	1 107.27	804.94	1 189.45	582.14	1 447.26	1 309.49	1 178.94	889.44	416.89	781.48
2019	963.09	1 559.64	1 208.94	1 124.08	828.06	1 220.83	602.54	1 529.95	1 351.25	1 217.95	911.45	428.87	803.75
2015													
January	853.19	1 376.00	1 055.73	1 022.09	743.13	1 097.37	542.12	1 259.62	1 162.02	1 076.59	819.67	372.15	707.54
February	857.39	1 358.68	1 071.24	1 026.64	748.05	1 101.65	544.79	1 261.62	1 165.76	1 077.67	820.66	372.67	709.78
March	856.98	1 353.44	1 067.81	1 024.08	747.36	1 096.88	543.85	1 261.06	1 170.75	1 078.67	823.94	372.56	710.73
April	858.71	1 341.59	1 063.92	1 024.49	749.09	1 099.98	546.05	1 261.06	1 179.14	1 082.28	823.61	373.61	709.78
May	861.12	1 358.58	1 062.75	1 023.20	753.68	1 112.15	549.05	1 266.51	1 185.90	1 087.45	827.22	374.66	712.96
June	861.81	1 356.01	1 071.73	1 020.68	748.65	1 106.58	546.81	1 267.96	1 188.16	1 088.53	827.87	376.09	716.77
July	862.50	1 368.80	1 069.78	1 029.38	751.07	1 110.60	548.38	1 271.95	1 184.40	1 089.62	826.89	374.92	714.86
August	865.95	1 372.62	1 078.39	1 037.22	756.36	1 113.95	552.95	1 280.76	1 190.94	1 092.52	829.51	375.71	717.41
September	866.30	1 390.03	1 061.18	1 031.24	761.87	1 113.32	562.22	1 275.48	1 193.96	1 093.11	830.82	378.19	718.68
October	871.57	1 382.92	1 084.29	1 034.59	759.12	1 117.99	556.41	1 284.44	1 192.67	1 100.98	833.45	379.51	722.22
November	870.78	1 397.09	1 081.90	1 038.66	759.47	1 117.21	556.41	1 289.49	1 192.67	1 097.08	834.43	378.85	722.85
December	871.82	1 394.12	1 093.75	1 039.07	758.31	1 120.47	554.11	1 291.32	1 199.61	1 098.67	836.07	381.88	724.77
2016													
January	877.46	1 398.76	1 087.04	1 045.70	761.55	1 127.10	554.64	1 307.53	1 207.53	1 108.24	840.27	382.00	728.64
February	875.61	1 372.46	1 090.97	1 046.93	759.69	1 123.82	556.51	1 303.56	1 203.20	1 103.58	839.02	383.31	727.32
March	875.48	1 367.46	1 084.53	1 048.43	761.27	1 132.96	554.82	1 301.76	1 210.34	1 103.22	839.68	382.89	729.23
April	878.23	1 371.94	1 094.02	1 055.09	763.34	1 140.16	554.82	1 305.36	1 212.06	1 111.34	842.30	384.98	730.51
May	879.95	1 381.59	1 095.12	1 059.58	761.46	1 137.62	553.97	1 313.28	1 207.13	1 111.88	840.99	386.28	732.42
June	881.67	1 381.16	1 097.07	1 057.39	764.55	1 138.78	556.76	1 310.71	1 205.03	1 111.68	844.87	387.85	734.34
July	884.08	1 382.83	1 102.62	1 058.61	765.58	1 150.27	554.59	1 323.07	1 221.86	1 119.30	846.19	389.67	737.21
August	882.20	1 379.94	1 096.98	1 060.47	763.69	1 144.60	553.42	1 325.16	1 218.38	1 113.48	845.86	389.48	740.80
September	886.49	1 392.72	1 103.79	1 063.08	767.29	1 151.67	554.04	1 326.03	1 221.75	1 118.02	849.81	393.52	741.44
October	890.27	1 400.72	1 116.12	1 072.22	771.41	1 158.30	556.20	1 342.66	1 213.26	1 123.07	851.78	392.08	746.24
November	888.71	1 397.22	1 108.49	1 066.16	769.84	1 151.44	557.48	1 348.56	1 220.74	1 121.40	853.43	394.63	744.23
December	889.74	1 401.90	1 105.54	1 071.63	771.59	1 155.33	557.93	1 351.08	1 222.61	1 123.92	854.41	392.64	745.82
2017													
January	894.40	1 416.80	1 112.67	1 075.49	772.44	1 158.44	556.51	1 372.50	1 213.84	1 129.21	854.74	395.98	747.62
February	894.54	1 430.23	1 112.28	1 072.25	770.87	1 159.22	555.02	1 363.43	1 227.84	1 132.46	858.36	396.01	751.75
March	895.57	1 441.35	1 112.93	1 072.25	771.55	1 158.05	557.75	1 366.70	1 225.97	1 134.36	858.36	399.10	750.80
April	900.25	1 437.81	1 118.26	1 080.59	779.36	1 164.93	563.78	1 379.40	1 230.46	1 140.40	862.31	402.69	752.39
May	901.62	1 445.54	1 125.82	1 082.42	779.16	1 163.76	562.34	1 386.66	1 229.71	1 134.72	861.33	403.00	753.34
June	903.69	1 435.95	1 129.99	1 083.65	781.22	1 165.71	563.58	1 393.34	1 243.81	1 138.96	863.95	403.25	755.25
July	906.78	1 465.56	1 131.95	1 090.80	783.29	1 171.44	564.82	1 401.18	1 246.88	1 141.84	868.23	404.03	756.36
August	907.47	1 441.80	1 131.00	1 085.28	782.94	1 168.31	563.93	1 394.90	1 251.00	1 140.48	868.89	404.04	758.58
September	911.94	1 448.69	1 138.02	1 088.95	786.04	1 173.00	564.54	1 389.60	1 256.25	1 150.15	870.18	406.12	762.70
October	910.91	1 453.71	1 141.11	1 093.67	784.66	1 164.93	566.37	1 390.21	1 263.38	1 145.88	869.88	406.90	763.97
November	915.98	1 464.67	1 148.35	1 095.93	791.30	1 177.96	572.79	1 393.56	1 267.50	1 155.50	873.50	408.20	766.51
December	919.08	1 468.32	1 148.17	1 092.62	791.43	1 183.32	570.06	1 401.40	1 279.53	1 153.08	878.46	411.34	769.88
2018													
January	918.82	1 466.87	1 143.27	1 093.61	792.47	1 173.51	572.24	1 399.38	1 283.63	1 153.47	877.11	409.24	765.67
February	922.88	1 482.12	1 161.32	1 100.85	793.50	1 174.29	573.80	1 407.96	1 287.05	1 162.74	880.77	410.81	769.04
March	925.98	1 486.21	1 158.96	1 100.21	794.88	1 175.07	576.58	1 418.37	1 290.75	1 167.09	882.38	412.90	773.80
April	928.40	1 488.56	1 171.57	1 107.23	797.64	1 175.46	576.60	1 424.51	1 295.32	1 169.26	886.05	413.95	777.83
May	930.81	1 491.34	1 173.15	1 103.07	800.75	1 183.95	582.19	1 421.64	1 302.84	1 169.64	886.98	416.32	776.65
June	935.58	1 508.46	1 167.38	1 105.53	804.10	1 190.60	580.95	1 420.15	1 307.44	1 176.50	891.33	417.63	779.74
July	935.30	1 492.24	1 182.63	1 107.41	806.18	1 185.99	584.06	1 434.25	1 305.75	1 180.84	889.95	417.86	781.01
August	939.09	1 502.86	1 176.78	1 107.98	807.65	1 192.55	583.08	1 446.89	1 312.62	1 184.08	894.96	419.17	782.28
September	942.20	1 509.19	1 177.02	1 109.21	810.41	1 196.91	584.94	1 475.51	1 319.76	1 185.16	893.24	418.60	787.61
October	943.92	1 500.06	1 176.34	1 107.31	811.15	1 196.91	586.43	1 483.48	1 322.62	1 189.53	894.55	420.16	789.53
November	943.94	1 502.77	1 172.61	1 111.39	814.20	1 198.47	589.57	1 489.63	1 327.28	1 189.86	896.20	420.10	791.12
December	950.13	1 509.82	1 196.29	1 117.39	816.00	1 205.88	588.74	1 504.64	1 329.54	1 192.74	900.24	423.02	793.67
2019													
January	951.51	1 519.14	1 209.77	1 112.62	816.34	1 202.37	593.74	1 512.44	1 334.20	1 200.44	904.86	424.91	796.54
February	952.54	1 525.59	1 181.46	1 116.40	818.41	1 208.61	594.25	1 512.07	1 338.73	1 202.93	906.18	425.88	795.00
March	957.72	1 535.94	1 201.70	1 117.62	823.89	1 219.14	596.81	1 523.15	1 339.86	1 209.88	908.16	428.04	801.60
April	956.66	1 559.84	1 200.70	1 115.69	823.54	1 213.68	598.04	1 519.31	1 347.02	1 210.17	907.17	426.06	800.37
May	958.73	1 549.66	1 200.37	1 119.34	827.66	1 217.57	600.19	1 528.80	1 346.83	1 213.42	905.19	427.61	802.60
June	961.82	1 558.00	1 208.08	1 123.81	827.64	1 220.68	603.26	1 522.98	1 346.63	1 217.41	908.82	428.65	803.24
July	962.12	1 575.88	1 202.33	1 121.10	827.95	1 217.93	603.74	1 529.81	1 347.38	1 218.38	912.12	428.02	804.20
August	968.70	1 577.73	1 216.28	1 126.31	833.80	1 230.02	605.57	1 562.20	1 364.36	1 225.73	914.43	428.80	804.54
September	968.70	1 591.13	1 225.54	1 129.14	833.80	1 227.68	609.09	1 543.22	1 354.73	1 225.23	915.42	431.75	805.18
October	971.46	1 600.59	1 217.51	1 124.77	835.85	1 234.74	608.63	1 539.72	1 362.48	1 228.84	919.05	430.86	809.94
November	972.06	1 579.85	1 215.62	1 132.01	833.68	1 237.86	606.00	1 545.65	1 372.28	1 233.54	922.50	432.41	812.81
December	973.09	1 583.31	1 214.85	1 136.86	835.45	1 226.47	613.22	1 545.29	1 369.88	1 235.70	919.38	432.67	812.49

Table 10-13. Production and Nonsupervisory Workers on Private Nonfarm Payrolls by NAICS Supersector

(Thousands, seasonally adjusted.)

| Year and month | Total private | Mining and logging | Construc-tion | Manu-facturing | Trade, transportation, and utilities | | | Information | Financial activities | Profes-sional and business services | Education and health services | Leisure and hospitality | Other services |
					Total	Wholesale trade	Retail trade						
1965	42 302	523	2 906	12 905	10 687	. . .	. . .	1 268	2 434	3 515	3 443	3 443	1 161
1966	44 292	517	2 977	13 703	11 080	. . .	. . .	1 334	2 492	3 715	3 623	3 607	1 230
1967	45 185	501	2 903	13 714	11 353	. . .	. . .	1 365	2 585	3 890	3 818	3 734	1 306
1968	46 519	491	2 986	13 908	11 671	. . .	. . .	1 394	2 700	4 067	4 008	3 898	1 379
1969	48 246	501	3 177	14 147	12 134	. . .	. . .	1 438	2 841	4 252	4 196	4 089	1 452
1970	48 180	496	3 158	13 490	12 370	. . .	. . .	1 422	2 922	4 321	4 305	4 185	1 494
1971	48 151	474	3 238	13 034	12 484	. . .	. . .	1 392	2 978	4 354	4 372	4 286	1 521
1972	49 971	494	3 425	13 497	12 935	2 898	7 259	1 437	3 066	4 518	4 531	4 467	1 583
1973	52 235	502	3 576	14 227	13 418	3 019	7 552	1 504	3 164	4 748	4 747	4 664	1 666
1974	52 846	550	3 469	14 040	13 681	3 124	7 675	1 516	3 217	4 907	4 941	4 766	1 740
1975	51 010	581	2 990	12 576	13 559	3 098	7 716	1 416	3 227	4 939	5 088	4 821	1 795
1976	52 916	606	2 999	13 127	14 018	3 188	8 050	1 459	3 300	5 153	5 309	5 046	1 880
1977	55 207	636	3 209	13 591	14 558	3 299	8 398	1 514	3 452	5 404	5 561	5 284	1 978
1978	58 188	658	3 544	14 150	15 307	3 483	8 863	1 586	3 645	5 717	5 874	5 588	2 099
1979	60 404	737	3 760	14 458	15 820	3 640	9 115	1 650	3 825	5 993	6 158	5 772	2 209
1980	60 377	785	3 623	13 667	15 884	3 681	9 160	1 626	3 957	6 197	6 447	5 850	2 318
1981	60 967	861	3 469	13 492	15 981	3 725	9 240	1 633	4 052	6 396	6 700	5 944	2 414
1982	59 475	834	3 208	12 315	15 798	3 635	9 256	1 564	4 055	6 421	6 822	5 976	2 458
1983	60 018	698	3 240	12 121	15 976	3 612	9 496	1 502	4 128	6 581	7 045	6 161	2 542
1984	63 333	714	3 614	12 821	16 773	3 793	9 967	1 631	4 289	6 918	7 385	6 491	2 672
1985	65 456	686	3 868	12 648	17 402	3 906	10 402	1 660	4 476	7 258	7 789	6 817	2 827
1986	66 825	577	3 984	12 449	17 743	3 912	10 706	1 663	4 698	7 532	8 130	7 066	2 957
1987	68 725	541	4 088	12 537	18 170	3 960	10 989	1 717	4 861	7 859	8 513	7 310	3 104
1988	71 059	545	4 199	12 765	18 744	4 101	11 308	1 775	4 894	8 256	8 985	7 587	3 280
1989	72 960	526	4 257	12 805	19 202	4 204	11 567	1 807	4 931	8 648	9 465	7 833	3 459
1990	73 721	538	4 115	12 669	19 006	4 168	11 311	1 866	4 973	8 889	9 784	8 299	3 555
1991	72 567	515	3 674	12 164	18 613	4 092	11 011	1 871	4 914	8 748	10 257	8 247	3 539
1992	72 854	478	3 546	12 020	18 478	4 041	10 934	1 871	4 924	8 971	10 607	8 406	3 526
1993	74 671	462	3 704	12 070	18 723	4 042	11 107	1 896	5 084	9 451	10 961	8 667	3 623
1994	77 478	461	3 973	12 361	19 361	4 166	11 505	1 928	5 220	10 078	11 397	8 979	3 689
1995	79 943	458	4 113	12 567	19 950	4 329	11 844	2 007	5 199	10 645	11 829	9 330	3 812
1996	81 888	461	4 325	12 532	20 289	4 390	12 060	2 096	5 322	11 161	12 195	9 565	3 907
1997	84 313	479	4 546	12 673	20 660	4 490	12 277	2 181	5 482	11 896	12 566	9 780	4 013
1998	86 516	473	4 807	12 729	21 020	4 571	12 443	2 217	5 692	12 566	12 903	9 947	4 124
1999	88 647	438	5 105	12 524	21 535	4 638	12 775	2 351	5 818	13 184	13 217	10 216	4 219
2000	90 543	446	5 295	12 428	21 923	4 651	13 043	2 502	5 819	13 790	13 487	10 516	4 296
2001	90 202	457	5 332	11 677	21 665	4 520	12 955	2 531	5 888	13 588	13 986	10 662	4 373
2002	88 645	436	5 196	10 768	21 292	4 437	12 777	2 398	5 964	13 049	14 471	10 576	4 449
2003	87 938	420	5 123	10 189	21 029	4 358	12 658	2 347	6 052	12 911	14 726	10 666	4 426
2004	89 212	440	5 309	10 072	21 268	4 402	12 792	2 371	6 052	13 287	14 984	10 955	4 425
2005	91 401	473	5 611	10 060	21 776	4 538	13 033	2 386	6 127	13 854	15 359	11 263	4 438
2006	93 727	519	5 903	10 137	22 109	4 676	13 114	2 399	6 312	14 446	15 783	11 568	4 494
2007	95 203	547	5 883	9 975	22 486	4 799	13 322	2 403	6 365	14 784	16 261	11 861	4 578
2008	94 610	574	5 521	9 629	22 277	4 768	13 139	2 388	6 320	14 585	16 777	11 873	4 606
2009	89 556	510	4 567	8 322	21 059	4 453	12 476	2 240	6 066	13 520	17 166	11 560	4 488
2010	88 875	525	4 172	8 077	20 816	4 324	12 430	2 170	5 942	13 699	17 452	11 507	4 458
2011	90 537	594	4 184	8 228	21 174	4 388	12 652	2 148	5 900	14 251	17 736	11 772	4 491
2012	92 715	641	4 246	8 400	21 554	4 505	12 798	2 164	5 986	14 802	18 165	12 154	4 541
2013	94 580	639	4 401	8 412	21 811	4 563	12 928	2 178	6 064	15 351	18 519	12 579	4 558
2014	96 794	658	4 631	8 568	22 217	4 638	13 112	2 221	6 157	15 795	18 859	12 981	4 642
2015	98 783	592	4 866	8 683	22 566	4 644	13 269	2 226	6 278	16 133	19 337	13 360	4 678
2016	100 531	476	5 062	8 666	22 842	4 634	13 433	2 235	429	16 479	19 839	13 749	4 715
2017	102 331	504	5 186	8 728	23 094	4 661	13 486	2 205	6 569	16 875	20 320	14 047	4 771
2018	104 310	544	5 437	8 899	23 348	4 688	13 488	2 275	6 638	17 121	20 788	14 384	4 839
2019	105 617	540	5 583	8 975	23 443	4 739	13 354	2 301	6 768	17 335	21 233	14 568	4 876
2018													
January	103 253	516	5 332	8 817	23 195	4 671	13 469	2 265	6 595	16 874	20 590	14 260	4 809
February	103 518	522	5 379	8 827	23 264	4 676	13 513	2 265	6 611	16 918	20 640	14 280	4 812
March	103 677	529	5 381	8 842	23 313	4 686	13 527	2 273	6 617	16 954	20 671	14 285	4 812
April	103 772	535	5 389	8 863	23 321	4 664	13 541	2 282	6 621	16 993	20 678	14 272	4 818
May	104 031	538	5 433	8 869	23 357	4 672	13 549	2 284	6 636	17 064	20 712	14 312	4 826
June	104 173	544	5 438	8 902	23 335	4 674	13 506	2 284	6 649	17 091	20 751	14 346	4 833
July	104 262	543	5 454	8 916	23 346	4 688	13 490	2 283	6 649	17 107	20 768	14 370	4 826
August	104 426	551	5 468	8 918	23 383	4 699	13 485	2 285	6 667	17 138	20 807	14 379	4 830
September	104 504	551	5 476	8 934	23 376	4 698	13 450	2 284	6 690	17 165	20 833	14 356	4 839
October	104 658	555	5 492	8 951	23 379	4 702	13 409	2 297	6 696	17 194	20 871	14 382	4 841
November	104 741	551	5 482	8 965	23 411	4 711	13 413	2 300	6 701	17 209	20 889	14 391	4 842
December	104 907	552	5 518	8 983	23 396	4 722	13 396	2 304	6 712	17 227	20 938	14 430	4 847
2019													
January	105 168	555	5 579	8 998	23 469	4 730	13 410	2 293	6 721	17 215	20 991	14 497	4 850
February	105 109	552	5 530	9 000	23 432	4 735	13 385	2 291	6 731	17 251	21 001	14 474	4 847
March	105 217	553	5 538	8 992	23 417	4 731	13 366	2 292	6 744	17 264	21 063	14 495	4 859
April	105 335	549	5 558	8 979	23 405	4 738	13 341	2 288	6 754	17 294	21 117	14 516	4 875
May	105 399	548	5 566	8 970	23 416	4 740	13 344	2 296	6 757	17 302	21 148	14 526	4 870
June	105 503	545	5 580	8 968	23 420	4 737	13 334	2 307	6 753	17 317	21 202	14 527	4 884
July	105 604	540	5 586	8 967	23 432	4 738	13 332	2 306	6 768	17 343	21 261	14 515	4 886
August	105 731	535	5 595	8 968	23 420	4 739	13 321	2 306	6 781	17 365	21 314	14 558	4 889
September	105 885	532	5 607	8 968	23 448	4 742	13 341	2 306	6 782	17 387	21 359	14 611	4 885
October	105 992	535	5 620	8 931	23 459	4 743	13 351	2 301	6 803	17 408	21 396	14 654	4 885
November	106 156	524	5 612	8 978	23 444	4 743	13 315	2 310	6 806	17 429	21 456	14 703	4 894
December	106 300	516	5 623	8 980	23 494	4 750	13 347	2 314	6 812	17 444	21 491	14 736	4 890

. . . = Not available.

Table 10-14. Average Weekly Hours of Production and Nonsupervisory Workers on Private Nonfarm Payrolls by NAICS Supersector

(Hours per week, seasonally adjusted.)

Year and month	Total private	Mining and logging	Construc-tion	Manufacturing Average weekly hours	Manufacturing Overtime hours	Trade, transportation, and utilities Total	Trade, transportation, and utilities Wholesale trade	Trade, transportation, and utilities Retail trade	Informa-tion	Financial activities	Profes-sional and business services	Education and health services	Leisure and hospitality	Other services
1965	38.6	43.7	37.9	41.2	3.6	39.6	. . .	. . .	38.3	37.1	37.3	35.2	32.5	36.1
1966	38.5	44.1	38.1	41.4	3.9	39.1	. . .	. . .	38.3	37.2	37.0	34.9	31.9	35.8
1967	37.9	43.9	38.1	40.6	3.3	38.5	. . .	. . .	37.6	36.9	36.6	34.5	31.3	35.4
1968	37.7	44.0	37.8	40.7	3.6	38.2	. . .	. . .	37.6	36.8	36.3	34.1	30.8	35.0
1969	37.5	44.3	38.4	40.6	3.6	37.9	. . .	. . .	37.6	36.9	36.3	34.1	30.4	35.0
1970	37.0	43.9	37.8	39.8	2.9	37.6	. . .	. . .	37.2	36.6	35.9	33.8	30.0	34.7
1971	36.7	43.7	37.6	39.9	2.9	37.4	. . .	. . .	37.0	36.4	35.5	33.3	29.9	34.2
1972	36.9	44.1	37.0	40.6	3.4	37.4	39.8	35.1	37.3	36.4	35.5	33.3	29.7	34.2
1973	36.9	43.8	37.2	40.7	3.8	37.2	39.6	34.8	37.3	36.4	35.5	33.3	29.4	34.1
1974	36.4	43.7	37.1	40.0	3.2	36.8	39.2	34.3	37.0	36.3	35.3	33.1	29.1	33.9
1975	36.0	43.7	36.9	39.5	2.6	36.4	39.1	34.0	36.6	36.2	35.1	33.0	28.8	33.8
1976	36.0	44.2	37.3	40.1	3.1	36.3	39.1	33.8	36.7	36.2	34.9	32.7	28.5	33.6
1977	35.9	44.7	37.0	40.3	3.4	36.0	39.2	33.3	36.8	36.2	34.7	32.5	28.1	33.4
1978	35.8	44.9	37.3	40.4	3.6	35.6	39.2	32.7	36.8	36.1	34.6	32.3	27.7	33.2
1979	35.6	44.7	37.5	40.2	3.3	35.4	39.2	32.4	36.6	35.9	34.4	32.2	27.4	33.0
1980	35.2	44.9	37.5	39.6	2.8	35.0	38.8	31.9	36.4	36.0	34.3	32.1	27.0	33.0
1981	35.2	45.1	37.4	39.8	2.8	35.0	38.9	31.9	36.3	36.0	34.3	32.1	26.9	33.0
1982	34.7	44.1	37.2	38.9	2.3	34.6	38.7	31.7	35.8	36.0	34.2	32.1	26.8	33.0
1983	34.9	43.9	37.6	40.1	2.9	34.6	38.8	31.6	36.2	35.9	34.4	32.1	26.8	33.0
1984	35.1	44.6	38.2	40.6	3.4	34.7	38.9	31.6	36.6	36.2	34.3	32.0	26.7	32.9
1985	34.9	44.6	38.2	40.5	3.3	34.4	38.8	31.2	36.5	36.1	34.2	31.9	26.4	32.8
1986	34.7	43.6	37.9	40.7	3.4	34.1	38.7	31.0	36.4	36.1	34.3	32.0	26.2	32.9
1987	34.7	43.5	38.2	40.9	3.7	34.1	38.5	31.0	36.5	36.0	34.3	32.0	26.3	32.8
1988	34.6	43.3	38.2	41.0	3.8	33.8	38.5	30.9	36.1	35.6	34.2	32.0	26.3	32.9
1989	34.5	44.1	38.3	40.9	3.8	33.8	38.4	30.7	36.1	35.6	34.2	32.0	26.1	32.9
1990	34.3	45.0	38.3	40.5	3.9	33.7	38.4	30.6	35.8	35.5	34.2	31.9	26.0	32.8
1991	34.1	45.3	38.1	40.4	3.8	33.6	38.5	30.4	35.6	35.4	34.0	31.9	25.6	32.7
1992	34.2	44.6	38.0	40.7	4.0	33.8	38.6	30.7	35.8	35.6	34.0	32.0	25.7	32.6
1993	34.3	44.9	38.4	41.1	4.4	34.1	38.5	30.7	36.0	35.5	34.0	32.0	25.9	32.6
1994	34.5	45.3	38.8	41.7	5.0	34.3	38.8	30.9	36.0	35.5	34.1	32.0	26.0	32.7
1995	34.3	45.3	38.8	41.3	4.7	34.1	38.6	30.8	36.0	35.5	34.1	32.0	25.9	32.6
1996	34.3	46.0	38.9	41.3	4.8	34.1	38.7	30.7	36.3	35.5	34.1	31.9	25.9	32.5
1997	34.5	46.2	38.9	41.7	5.1	34.2	38.8	30.9	36.3	35.8	34.3	32.2	26.1	32.7
1998	34.5	44.9	38.8	41.4	4.9	34.1	38.7	30.9	36.6	36.0	34.3	32.2	26.2	32.6
1999	34.3	44.2	39.0	41.4	4.9	33.9	38.6	30.8	36.7	35.8	34.4	32.1	26.1	32.5
2000	34.3	44.4	39.2	41.3	4.7	33.8	38.8	30.7	36.8	35.9	34.5	32.2	26.1	32.5
2001	33.9	44.6	38.7	40.3	4.0	33.5	38.4	30.7	36.9	35.8	34.2	32.3	25.8	32.3
2002	33.9	43.2	38.4	40.5	4.2	33.6	38.0	30.9	36.5	35.6	34.2	32.4	25.8	32.1
2003	33.7	43.6	38.4	40.4	4.2	33.5	37.9	30.9	36.2	35.6	34.1	32.3	25.6	31.4
2004	33.7	44.5	38.3	40.8	4.6	33.5	37.8	30.7	36.3	35.6	34.2	32.4	25.7	31.0
2005	33.8	45.6	38.6	40.7	4.6	33.4	37.7	30.6	36.5	36.0	34.3	32.6	25.7	30.9
2006	33.9	45.6	39.0	41.1	4.4	33.4	38.0	30.5	36.6	35.8	34.6	32.5	25.7	30.9
2007	33.8	45.9	39.0	41.2	4.2	33.3	38.2	30.2	36.5	35.9	34.8	32.5	25.5	30.9
2008	33.6	45.1	38.5	40.8	3.7	33.1	38.3	30.0	36.7	35.9	34.8	32.4	25.2	30.8
2009	33.1	43.2	37.6	39.8	2.9	32.8	37.7	29.9	36.6	36.1	34.7	32.2	24.8	30.5
2010	33.4	44.6	38.4	41.1	3.8	33.3	37.9	30.2	36.3	36.2	35.1	32.0	24.8	30.7
2011	33.6	46.7	39.0	41.4	4.1	33.7	38.5	30.5	36.2	36.4	35.2	32.2	24.8	30.8
2012	33.7	46.6	39.3	41.7	4.2	33.8	38.7	30.6	36.0	36.8	35.3	32.3	25.0	30.7
2013	33.7	45.9	39.6	41.8	4.3	33.7	38.7	30.2	35.9	36.7	35.4	32.1	25.0	30.8
2014	33.7	47.3	39.6	42.0	4.5	33.6	38.6	30.0	35.9	36.7	35.6	32.0	25.1	30.7
2015	33.7	45.8	39.6	41.8	4.3	33.7	38.6	30.1	35.7	37.1	35.5	32.1	25.1	30.7
2016	33.6	45.3	39.7	41.9	4.3	33.5	38.6	29.7	35.5	36.9	35.4	32.2	24.9	30.8
2017	33.7	46.1	39.7	41.9	4.3	33.8	39.0	30.2	35.8	37.0	35.4	32.2	24.9	30.7
2018	33.8	46.8	40.0	42.2	4.6	33.9	38.9	30.4	35.6	37.0	35.4	32.2	24.9	30.8
2019	33.6	46.8	39.8	41.6	4.3	33.8	38.7	30.3	35.3	36.9	35.4	32.2	24.7	30.8
2018														
January	33.6	46.2	39.4	41.8	4.5	33.9	38.9	30.4	35.5	37.0	35.1	32.2	24.7	30.6
February	33.8	47.0	39.9	42.2	4.6	33.9	38.9	30.3	35.7	37.0	35.4	32.3	25.0	30.7
March	33.7	46.9	40.0	42.2	4.6	33.9	39.0	30.3	35.8	36.9	35.4	32.1	24.9	30.7
April	33.8	47.1	40.2	42.4	4.8	33.9	38.9	30.3	35.9	37.0	35.3	32.3	24.9	30.8
May	33.8	47.5	40.4	42.0	4.5	34.0	39.0	30.4	35.5	37.0	35.3	32.2	24.9	30.8
June	33.8	47.9	40.0	42.1	4.5	34.0	39.0	30.5	35.4	37.0	35.4	32.3	25.0	30.8
July	33.8	46.6	40.2	42.2	4.5	34.1	38.9	30.5	35.6	37.1	35.4	32.2	24.9	30.7
August	33.8	47.0	39.8	42.2	4.5	34.0	39.0	30.4	35.6	37.1	35.4	32.2	24.9	30.8
September	33.7	46.4	39.5	42.1	4.5	33.9	38.9	30.3	35.5	37.0	35.3	32.2	24.8	30.9
October	33.7	46.4	39.5	42.0	4.5	33.9	38.8	30.3	35.5	37.0	35.4	32.2	24.8	30.9
November	33.7	46.2	39.3	42.0	4.5	33.9	38.8	30.3	35.5	36.9	35.3	32.2	24.8	31.0
December	33.7	46.7	40.0	42.0	4.5	33.7	38.8	30.0	35.5	36.9	35.4	32.2	24.8	30.9
2019														
January	33.8	46.6	40.5	42.0	4.5	34.0	38.8	30.5	35.4	37.0	35.5	32.2	24.9	30.9
February	33.6	46.9	39.0	41.7	4.4	33.8	38.7	30.3	35.5	37.0	35.4	32.2	24.8	30.8
March	33.7	47.2	39.9	41.7	4.4	33.9	38.8	30.4	35.5	37.0	35.4	32.2	24.9	30.9
April	33.7	47.3	39.8	41.6	4.3	33.9	38.7	30.5	35.3	36.9	35.4	32.2	24.7	30.8
May	33.6	46.9	39.7	41.5	4.2	33.8	38.7	30.2	35.4	36.9	35.4	32.2	24.6	30.8
June	33.6	47.2	39.8	41.6	4.2	33.8	38.7	30.3	35.5	36.7	35.5	32.2	24.7	30.8
July	33.5	47.2	39.6	41.5	4.2	33.7	38.6	30.2	35.1	36.8	35.4	32.1	24.6	30.8
August	33.6	47.2	39.9	41.5	4.2	33.8	38.7	30.3	35.2	36.8	35.6	32.2	24.6	30.8
September	33.6	47.1	40.1	41.5	4.2	33.8	38.7	30.3	35.5	36.9	35.4	32.1	24.6	30.8
October	33.6	47.2	39.7	41.4	4.1	33.8	38.6	30.3	35.2	36.9	35.3	32.2	24.6	30.8
November	33.5	46.5	39.4	41.4	4.1	33.6	38.6	29.9	35.1	37.0	35.4	32.2	24.7	30.8
December	33.6	46.2	39.4	41.4	4.1	33.8	38.5	30.3	35.1	36.8	35.3	32.3	24.7	30.8

. . . = Not available.

Table 10-15. Indexes of Aggregate Weekly Hours of Production and Nonsupervisory Workers on Private Nonfarm Payrolls by NAICS Supersector

(2002 = 100, seasonally adjusted.)

Year and month	Total private	Mining and logging	Construction	Manu-facturing	Trade, transportation, and utilities			Information	Financial activities	Profes-sional and business services	Education and health services	Leisure and hospitality	Other services
					Total	Wholesale trade	Retail trade						
1965	54.4	121.6	55.2	122.1	59.1	. . .	. . .	55.5	42.6	29.5	25.9	41.0	29.4
1966	56.8	121.1	56.8	130.2	60.6	. . .	. . .	58.3	43.6	30.9	27.0	42.2	30.9
1967	57.0	117.0	55.4	127.8	61.2	. . .	. . .	58.6	44.9	31.9	28.1	42.9	32.4
1968	58.4	114.9	56.5	130.1	62.4	. . .	. . .	59.9	46.8	33.1	29.1	44.0	33.9
1969	60.3	118.1	61.0	131.9	64.4	. . .	. . .	61.8	49.4	34.6	30.5	45.6	35.6
1970	59.4	115.7	59.8	123.3	65.0	. . .	. . .	60.4	50.3	34.8	31.0	46.1	36.3
1971	59.0	109.9	61.0	119.3	65.2	. . .	. . .	58.7	51.1	34.7	31.1	46.9	36.5
1972	61.4	115.6	63.4	125.7	67.6	68.4	64.4	61.1	52.5	36.0	32.2	48.5	37.9
1973	64.2	117.0	66.7	132.9	69.7	71.0	66.4	64.1	54.2	37.8	33.7	50.3	39.9
1974	64.2	127.7	64.5	129.0	70.3	72.7	66.7	64.0	54.9	38.8	34.9	50.8	41.4
1975	61.2	134.9	55.2	113.9	69.1	71.8	66.4	59.1	55.0	38.9	35.8	50.9	42.6
1976	63.5	142.4	56.0	120.8	71.1	73.9	68.8	61.2	56.2	40.4	37.0	52.8	44.3
1977	66.1	151.1	59.4	125.8	73.2	76.7	70.7	63.5	58.8	42.1	38.5	54.4	46.3
1978	69.3	156.9	66.2	131.3	76.3	81.0	73.4	66.6	62.0	44.3	40.4	56.6	48.8
1979	71.6	175.2	70.6	133.3	78.4	84.6	74.7	69.0	64.7	46.3	42.2	57.9	51.2
1980	70.8	187.2	68.0	124.4	77.8	84.8	74.0	67.5	67.0	47.7	44.1	57.8	53.6
1981	71.4	206.2	64.9	123.1	78.1	86.0	74.5	67.7	68.6	49.2	45.8	58.7	55.8
1982	68.8	195.5	59.7	109.9	76.5	83.5	74.2	64.0	68.6	49.3	46.7	58.7	56.8
1983	69.8	162.8	61.0	111.6	77.3	83.3	76.0	62.1	69.8	50.7	48.3	60.4	58.9
1984	74.1	169.3	69.1	119.6	81.4	87.5	79.8	68.0	73.0	53.3	50.4	63.6	61.7
1985	76.0	162.7	73.9	117.5	83.7	89.9	82.2	69.1	76.1	55.7	53.0	66.0	65.0
1986	77.3	133.5	75.5	116.3	84.6	89.9	83.9	69.1	79.9	57.9	55.4	67.9	68.1
1987	79.5	125.2	78.1	117.8	86.7	90.5	86.3	71.5	82.3	60.4	58.0	70.5	71.5
1988	81.9	125.6	80.4	120.3	88.7	93.6	88.5	73.1	82.0	63.4	61.3	73.0	75.6
1989	83.9	123.2	81.7	120.3	90.7	95.9	90.0	74.4	82.6	66.4	64.6	74.9	79.7
1990	84.2	128.6	78.8	117.7	89.7	95.0	87.5	76.2	83.1	68.1	66.6	78.9	81.8
1991	82.4	123.8	70.1	112.8	87.6	93.4	84.8	76.1	82.0	66.7	69.7	77.3	81.1
1992	83.0	113.3	67.5	112.4	87.4	92.5	85.1	76.5	82.4	68.4	72.4	79.2	80.6
1993	85.3	110.3	71.3	113.9	89.2	92.5	86.3	78.0	84.9	72.0	74.8	82.1	82.8
1994	89.1	111.0	77.3	118.3	92.7	95.8	89.8	79.2	87.2	77.0	77.8	85.5	84.5
1995	91.4	110.3	79.9	119.0	95.2	99.3	92.3	82.5	86.9	81.3	80.6	88.5	87.0
1996	93.6	112.7	84.3	118.8	96.7	100.7	93.7	86.9	89.0	85.3	82.9	90.7	89.1
1997	97.0	117.6	88.6	121.4	98.9	103.4	95.9	90.4	92.3	91.5	86.2	93.4	91.9
1998	99.4	112.8	93.4	121.0	100.4	104.9	97.2	92.6	96.4	96.7	88.6	95.5	94.3
1999	101.4	102.9	99.7	119.0	102.0	106.3	99.5	98.5	98.0	101.6	90.4	97.8	96.2
2000	103.5	105.1	104.0	117.7	103.6	107.1	101.3	105.0	98.4	106.5	92.6	100.5	97.8
2001	102.0	108.3	103.2	108.1	101.6	102.9	100.5	106.7	99.2	104.0	96.5	100.6	99.0
2002	100.0	100.0	100.0	100.0	100.0	100.0	100.0	100.0	100.0	100.0	100.0	100.0	100.0
2003	98.7	97.4	98.4	94.5	98.6	98.0	98.9	97.0	101.3	98.7	101.7	100.0	97.4
2004	100.2	104.0	101.7	94.3	99.6	98.7	99.4	98.2	101.3	101.8	103.8	103.0	96.1
2005	102.8	114.7	108.3	93.9	101.6	101.6	100.8	99.4	103.7	106.4	107.0	106.2	96.2
2006	105.8	125.8	115.4	95.6	103.2	105.5	101.1	100.2	106.3	112.2	109.6	108.8	97.4
2007	107.4	133.5	114.7	94.4	104.8	108.9	101.8	100.1	107.6	115.4	113.2	110.8	99.3
2008	106.0	137.6	106.5	90.2	103.2	108.3	99.8	100.0	106.8	114.0	116.5	109.7	99.5
2009	98.8	117.2	85.9	76.1	96.7	99.5	94.3	93.5	103.1	105.3	118.3	105.1	96.0
2010	99.0	124.5	80.2	76.2	96.8	97.4	95.0	90.0	101.3	107.7	119.8	104.6	96.0
2011	101.4	147.4	81.7	78.3	99.7	100.2	97.7	88.9	101.1	112.6	122.3	106.9	96.8
2012	104.3	158.7	83.6	80.3	101.8	103.5	98.9	88.9	103.7	117.3	125.4	111.3	97.7
2013	106.1	155.0	87.7	80.9	102.6	104.8	98.8	89.8	104.7	121.4	126.6	115.2	98.7
2014	108.6	164.3	92.0	82.6	104.4	106.4	99.5	90.5	106.5	125.9	128.5	119.3	99.9
2015	110.9	144.1	96.5	83.3	106.3	106.4	101.0	90.8	109.6	128.3	132.3	122.7	100.7
2016	112.5	113.4	100.9	83.3	106.9	106.2	101.1	91.4	111.6	130.2	136.1	125.6	101.8
2017	114.9	120.2	104.0	84.0	109.2	107.8	103.1	92.6	114.3	132.6	139.8	128.7	102.9
2018	117.2	134.5	108.8	86.1	110.8	108.3	103.6	93.0	116.0	135.2	142.7	130.6	104.4
2019	118.3	134.5	111.2	85.7	110.8	108.7	102.3	92.7	117.4	137.3	145.8	131.8	105.4
2018													
January	115.6	126.7	105.2	84.6	110.0	107.8	103.6	91.8	114.8	132.4	141.4	129.0	103.2
February	116.6	130.4	107.5	85.5	110.3	108.0	103.6	92.3	115.1	133.9	142.1	130.8	103.6
March	116.4	131.8	107.8	85.7	110.5	108.5	103.7	92.9	114.9	134.2	141.5	130.3	103.6
April	116.9	133.9	108.5	86.3	110.6	107.7	103.8	93.5	115.3	134.1	142.4	130.2	104.1
May	117.2	135.8	109.9	85.5	111.1	108.1	104.2	92.5	115.6	134.7	142.2	130.5	104.2
June	117.3	138.5	108.9	86.0	111.0	108.2	104.2	92.3	115.8	135.3	142.9	131.4	104.4
July	117.4	134.5	109.8	86.4	111.3	108.2	104.1	92.8	116.1	135.4	142.6	131.1	103.9
August	117.6	137.6	109.0	86.4	111.2	108.8	103.7	92.8	116.4	135.6	142.8	131.1	104.3
September	117.4	135.9	108.3	86.3	110.8	108.5	103.1	92.5	116.5	135.5	143.0	130.4	104.9
October	117.5	136.8	108.6	86.3	110.8	108.3	102.8	93.1	116.6	136.1	143.3	130.6	104.9
November	117.6	135.3	107.9	86.4	111.0	108.5	102.8	93.2	116.4	135.8	143.4	130.7	105.3
December	117.8	137.0	110.5	86.6	110.3	108.7	101.7	93.4	116.6	136.4	143.7	131.1	105.0
2019													
January	118.5	137.4	113.1	86.8	111.6	108.9	103.5	92.6	117.0	136.6	144.1	132.2	105.1
February	117.7	137.6	108.0	86.2	110.8	108.7	102.6	92.8	117.2	136.5	144.2	131.5	104.7
March	118.2	138.7	110.6	86.1	111.0	108.9	102.8	92.9	117.4	136.6	144.6	132.2	105.3
April	118.3	138.0	110.8	85.7	111.0	108.8	103.0	92.2	117.3	136.9	145.0	131.3	105.3
May	118.0	136.6	110.6	85.5	110.7	108.9	102.0	92.8	117.3	136.9	145.2	130.9	105.2
June	118.1	136.7	111.2	85.6	110.7	108.8	102.2	93.5	116.6	137.5	145.6	131.4	105.5
July	117.9	135.4	110.8	85.4	110.4	108.5	101.9	92.4	117.2	137.3	145.5	130.8	105.5
August	118.4	134.2	111.8	85.4	110.7	108.8	102.1	92.6	117.4	138.2	146.3	131.2	105.6
September	118.6	133.2	112.6	85.4	110.8	108.9	102.3	93.4	117.8	137.6	146.2	131.7	105.5
October	118.7	134.2	111.7	84.9	110.9	108.6	102.4	92.4	118.1	137.4	146.9	132.0	105.5
November	118.5	129.5	110.7	85.3	110.2	108.6	100.7	92.5	118.5	138.0	147.3	133.0	105.7
December	119.0	126.7	110.9	85.3	111.1	108.5	102.3	92.7	118.0	137.7	148.0	133.3	105.6

. . . = Not available.

Table 10-16. Average Hourly Earnings of Production and Nonsupervisory Workers on Private Nonfarm Payrolls by NAICS Supersector

(Dollars, seasonally adjusted.)

Year and month	Total private	Mining and logging	Construc-tion	Manu-facturing	Trade, transportation, and utilities			Information	Financial activities	Profes-sional and business services	Education and health services	Leisure and hospitality	Other services
					Total	Wholesale trade	Retail trade						
1965	2.63	2.87	3.23	2.49	2.94	. . .	. . .	4.47	2.38	3.28	2.12	1.17	1.25
1966	2.73	3.00	3.41	2.60	3.04	. . .	. . .	4.56	2.47	3.39	2.23	1.26	1.37
1967	2.85	3.14	3.63	2.71	3.15	. . .	. . .	4.68	2.58	3.51	2.36	1.37	1.49
1968	3.03	3.30	3.92	2.89	3.32	. . .	. . .	4.85	2.75	3.65	2.49	1.53	1.62
1969	3.22	3.54	4.30	3.07	3.48	. . .	. . .	5.05	2.92	3.84	2.68	1.69	1.81
1970	3.41	3.77	4.74	3.24	3.65	. . .	. . .	5.25	3.07	4.04	2.88	1.82	2.01
1971	3.63	3.99	5.17	3.45	3.86	. . .	. . .	5.53	3.23	4.26	3.11	1.95	2.24
1972	3.91	4.28	5.55	3.70	4.23	4.57	3.52	5.87	3.37	4.50	3.33	2.08	2.46
1973	4.14	4.59	5.89	3.97	4.46	4.79	3.69	6.17	3.55	4.72	3.54	2.20	2.67
1974	4.44	5.09	6.29	4.31	4.74	5.10	3.92	6.52	3.80	5.01	3.82	2.40	2.95
1975	4.74	5.68	6.78	4.71	5.01	5.44	4.15	6.92	4.08	5.29	4.09	2.58	3.21
1976	5.06	6.19	7.17	5.10	5.31	5.74	4.37	7.37	4.30	5.60	4.39	2.78	3.51
1977	5.44	6.70	7.56	5.55	5.66	6.11	4.66	7.84	4.58	5.95	4.72	3.03	3.84
1978	5.88	7.44	8.11	6.05	6.10	6.60	5.01	8.34	4.93	6.32	5.07	3.33	4.19
1979	6.34	8.20	8.71	6.57	6.55	7.11	5.35	8.86	5.31	6.71	5.44	3.63	4.56
1980	6.84	8.97	9.37	7.15	7.03	7.67	5.72	9.47	5.82	7.22	5.93	3.98	5.05
1981	7.43	9.89	10.24	7.87	7.54	8.27	6.10	10.21	6.34	7.80	6.49	4.36	5.61
1982	7.86	10.64	11.04	8.36	7.90	8.80	6.35	10.76	6.82	8.30	7.00	4.63	6.11
1983	8.20	11.14	11.36	8.70	8.22	9.25	6.61	11.18	7.32	8.70	7.39	4.89	6.51
1984	8.49	11.54	11.56	9.05	8.44	9.59	6.74	11.50	7.65	8.98	7.66	4.99	6.79
1985	8.73	11.87	11.75	9.40	8.59	9.86	6.84	11.81	7.97	9.28	7.97	5.10	7.10
1986	8.92	12.14	11.92	9.60	8.73	10.05	6.94	12.08	8.37	9.56	8.24	5.20	7.38
1987	9.14	12.17	12.15	9.77	8.91	10.30	7.03	12.36	8.73	9.86	8.56	5.30	7.69
1988	9.44	12.45	12.52	10.05	9.14	10.69	7.24	12.63	9.07	10.23	8.95	5.50	8.08
1989	9.81	12.90	12.98	10.35	9.45	11.10	7.47	12.99	9.54	10.70	9.45	5.76	8.58
1990	10.20	13.40	13.42	10.78	9.82	11.55	7.71	13.40	9.98	11.15	9.98	6.02	9.08
1991	10.51	13.82	13.65	11.13	10.07	11.92	7.89	13.90	10.43	11.51	10.48	6.22	9.39
1992	10.77	14.09	13.81	11.40	10.29	12.18	8.12	14.29	10.86	11.80	10.86	6.36	9.66
1993	11.05	14.12	14.04	11.70	10.53	12.55	8.36	14.86	11.38	11.97	11.20	6.48	9.90
1994	11.34	14.41	14.38	12.04	10.79	12.90	8.62	15.32	11.84	12.16	11.48	6.62	10.18
1995	11.65	14.78	14.73	12.34	11.08	13.31	8.85	15.68	12.31	12.54	11.78	6.79	10.51
1996	12.04	15.09	15.11	12.75	11.44	13.77	9.21	16.30	12.75	13.01	12.15	6.99	10.85
1997	12.51	15.57	15.67	13.14	11.88	14.38	9.60	17.14	13.28	13.58	12.53	7.32	11.29
1998	13.01	16.20	16.23	13.45	12.37	15.03	10.06	17.67	14.00	14.28	12.96	7.67	11.79
1999	13.48	16.33	16.80	13.85	12.80	15.59	10.45	18.40	14.55	14.86	13.40	7.96	12.26
2000	14.01	16.55	17.48	14.32	13.28	16.24	10.87	19.07	15.04	15.53	13.91	8.32	12.73
2001	14.54	17.00	18.00	14.76	13.68	16.74	11.30	19.80	15.65	16.34	14.58	8.57	13.27
2002	14.96	17.19	18.52	15.29	13.99	16.94	11.67	20.20	16.25	16.82	15.15	8.81	13.72
2003	15.36	17.56	18.95	15.74	14.31	17.33	11.90	21.01	17.21	17.22	15.56	9.00	13.84
2004	15.68	18.07	19.23	16.14	14.55	17.62	12.09	21.40	17.58	17.49	16.07	9.15	13.98
2005	16.12	18.72	19.46	16.56	14.90	18.13	12.36	22.06	17.98	18.09	16.62	9.38	14.34
2006	16.74	19.90	20.02	16.81	15.36	18.87	12.58	23.23	18.83	19.14	17.28	9.75	14.77
2007	17.41	20.97	20.95	17.26	15.74	19.54	12.76	23.96	19.67	20.16	17.99	10.41	15.42
2008	18.06	22.50	21.87	17.75	16.13	20.08	12.87	24.78	20.32	21.19	18.73	10.84	16.09
2009	18.60	23.29	22.66	18.24	16.44	20.78	13.02	25.45	20.90	22.36	19.34	11.12	16.59
2010	19.04	23.82	23.22	18.61	16.78	21.46	13.25	25.87	21.55	22.80	19.95	11.31	17.06
2011	19.43	24.50	23.65	18.93	17.10	21.88	13.52	26.62	21.93	23.14	20.60	11.45	17.32
2012	19.73	25.79	23.97	19.08	17.38	22.13	13.82	27.04	22.82	23.31	20.91	11.62	17.59
2013	20.13	26.80	24.22	19.30	17.69	22.52	14.03	27.98	23.87	23.74	21.29	11.78	18.00
2014	20.60	26.84	24.67	19.56	18.21	23.14	14.40	28.70	24.71	24.31	21.64	12.09	18.51
2015	21.03	26.48	25.20	19.91	18.61	23.52	14.83	29.05	25.34	24.81	22.09	12.41	19.01
2016	21.53	26.97	25.97	20.44	18.93	24.07	15.05	30.05	26.12	25.43	22.52	12.85	19.36
2017	22.05	27.44	26.74	20.90	19.29	24.58	15.33	30.74	26.57	26.05	23.03	13.38	20.10
2018	22.71	28.27	27.75	21.54	19.88	25.17	15.91	31.97	26.93	26.82	23.64	13.87	20.78
2019	23.51	29.95	28.51	22.15	20.64	26.06	16.62	33.89	27.67	27.78	24.35	14.49	21.41
2018													
January	22.36	27.81	27.21	21.29	19.51	24.70	15.56	31.15	26.72	26.46	23.34	13.61	20.46
February	22.40	27.81	27.42	21.33	19.53	24.77	15.54	31.24	26.72	26.48	23.37	13.62	20.50
March	22.49	27.84	27.40	21.39	19.63	24.75	15.68	31.48	26.83	26.57	23.45	13.67	20.64
April	22.55	27.89	27.50	21.47	19.71	24.82	15.81	31.45	26.87	26.62	23.48	13.73	20.67
May	22.61	27.91	27.54	21.43	19.78	24.99	15.90	31.65	26.90	26.70	23.59	13.77	20.70
June	22.67	28.13	27.60	21.48	19.84	25.22	15.88	31.94	26.89	26.78	23.68	13.84	20.73
July	22.71	28.18	27.76	21.45	19.86	25.13	15.94	31.91	26.90	26.86	23.68	13.90	20.81
August	22.79	28.47	27.86	21.55	19.96	25.26	15.99	32.12	26.95	26.91	23.75	13.96	20.83
September	22.86	28.78	27.97	21.62	20.04	25.46	16.01	32.30	27.03	26.97	23.76	14.01	20.90
October	22.90	28.69	28.08	21.70	20.11	25.47	16.09	32.51	26.96	26.99	23.75	14.06	20.97
November	23.00	28.77	28.17	21.78	20.16	25.52	16.16	32.77	27.18	27.09	23.87	14.12	21.02
December	23.09	28.67	28.22	21.82	20.30	25.65	16.27	32.97	27.24	27.19	23.93	14.18	21.10
2019													
January	23.11	29.00	28.19	21.84	20.21	25.52	16.23	33.10	27.28	27.23	24.03	14.23	21.22
February	23.19	29.05	28.17	21.91	20.34	25.75	16.30	33.24	27.46	27.32	24.06	14.29	21.24
March	23.28	29.36	28.33	21.96	20.49	26.04	16.41	33.36	27.37	27.44	24.16	14.34	21.24
April	23.33	29.88	28.45	21.96	20.53	25.98	16.49	33.41	27.50	27.52	24.14	14.38	21.27
May	23.42	29.95	28.49	22.04	20.68	26.06	16.63	33.62	27.56	27.63	24.13	14.43	21.35
June	23.47	29.86	28.50	22.14	20.65	26.10	16.60	33.67	27.65	27.75	24.21	14.44	21.39
July	23.54	29.80	28.45	22.19	20.68	26.05	16.64	33.84	27.71	27.84	24.39	14.49	21.37
August	23.64	30.17	28.58	22.22	20.78	26.28	16.72	34.16	27.82	27.96	24.45	14.55	21.47
September	23.70	30.52	28.63	22.26	20.76	26.20	16.74	34.42	27.81	28.06	24.54	14.63	21.50
October	23.76	30.85	28.68	22.30	20.80	26.28	16.77	34.53	27.89	28.16	24.67	14.67	21.55
November	23.81	30.55	28.72	22.39	20.88	26.33	16.86	34.43	28.02	28.23	24.63	14.73	21.62
December	23.84	30.68	28.88	22.44	20.83	26.15	16.88	34.44	28.06	28.28	24.72	14.77	21.71

. . . = Not available.

Table 10-17. Average Weekly Earnings of Production and Nonsupervisory Workers on Private Nonfarm Payrolls by NAICS Supersector

(Dollars, seasonally adjusted.)

Year and month	Total private	Mining and logging	Construc-tion	Manu-facturing	Trade, transportation, and utilities			Information	Financial activities	Profes-sional and business services	Education and health services	Leisure and hospitality	Other services
					Total	Wholesale trade	Retail trade						
1965	101.54	125.48	122.35	102.69	116.36	. . .	. . .	171.24	88.50	122.30	74.84	38.06	45.30
1966	105.23	132.50	130.11	107.47	118.78	. . .	. . .	174.52	91.83	125.53	78.04	40.34	48.94
1967	108.14	137.76	138.55	109.85	121.57	. . .	. . .	176.09	95.35	128.34	81.34	43.06	52.69
1968	113.98	145.26	148.22	117.73	126.69	. . .	. . .	182.65	101.18	132.47	85.02	47.07	56.81
1969	120.80	157.11	164.88	124.74	132.13	. . .	. . .	190.18	107.91	139.37	91.46	51.29	63.44
1970	125.91	165.82	179.11	128.84	137.15	. . .	. . .	195.24	112.32	144.76	97.23	54.57	69.67
1971	133.35	174.18	194.70	137.59	144.11	. . .	. . .	204.50	117.60	151.51	103.57	58.39	76.58
1972	143.99	188.76	205.04	150.23	158.09	181.58	123.57	218.72	122.86	159.88	111.12	61.78	84.14
1973	152.71	201.43	219.32	161.59	165.61	189.93	128.16	230.48	129.32	167.40	117.85	64.92	91.06
1974	161.76	222.24	233.24	172.37	174.18	200.13	134.63	241.07	137.71	176.47	126.50	69.90	99.98
1975	170.45	248.38	249.70	185.69	182.61	212.46	141.33	253.21	147.75	185.81	134.83	74.49	108.67
1976	182.36	273.69	267.00	204.41	192.40	224.27	147.55	270.80	155.54	195.61	143.47	79.47	117.85
1977	195.34	299.46	279.61	224.01	203.51	239.18	154.84	288.29	165.47	206.78	153.38	85.20	128.12
1978	210.17	334.27	302.27	244.49	217.18	258.36	164.07	306.71	177.96	218.42	163.48	92.16	138.74
1979	225.46	366.58	326.55	263.99	232.01	278.35	173.50	324.51	190.72	231.16	174.80	99.47	150.47
1980	240.83	402.51	351.49	283.35	246.25	297.85	182.57	344.38	209.21	247.64	190.16	107.52	166.32
1981	261.29	446.10	382.75	312.74	263.58	321.69	194.29	370.36	228.21	267.46	208.06	117.49	184.77
1982	272.98	469.10	410.39	325.07	273.74	340.26	201.06	385.82	245.16	284.11	224.53	124.14	201.49
1983	286.34	488.89	427.00	348.89	284.42	359.20	209.12	404.89	262.83	298.74	237.40	130.82	214.87
1984	298.08	514.77	441.28	367.77	292.64	373.01	213.14	420.45	276.50	308.24	245.16	133.49	223.33
1985	304.37	529.62	448.35	380.60	295.47	382.44	213.69	430.86	287.66	317.41	254.19	134.71	232.81
1986	309.69	529.20	451.22	390.46	297.73	388.96	214.83	439.84	302.32	327.36	263.41	136.34	242.33
1987	317.33	529.72	463.78	400.00	303.76	396.62	218.36	450.77	314.01	337.91	273.49	139.45	252.58
1988	326.50	539.63	478.74	412.62	309.24	410.94	223.97	455.59	322.87	350.23	286.57	144.51	265.83
1989	338.42	568.55	497.43	423.58	319.07	426.64	229.45	468.77	339.46	366.27	302.70	150.29	282.09
1990	349.63	602.43	513.43	436.13	331.24	443.61	235.72	479.50	354.63	381.01	318.90	156.32	297.88
1991	358.46	625.46	520.41	449.83	338.71	458.48	240.23	495.14	369.57	391.49	334.10	159.20	306.75
1992	368.17	628.94	525.13	464.34	348.17	469.59	249.71	511.95	386.31	401.16	347.64	163.70	315.08
1993	378.74	634.77	539.81	480.80	358.70	483.54	256.91	535.19	403.68	406.60	358.34	167.54	322.69
1994	391.17	653.13	558.53	501.95	369.63	499.98	265.85	551.21	420.08	414.59	367.50	172.27	332.44
1995	399.93	670.40	571.57	509.26	378.04	514.18	272.68	564.92	437.23	427.16	376.77	175.74	342.36
1996	413.17	695.04	588.48	526.55	389.77	532.16	282.76	592.45	453.02	443.31	387.45	181.02	352.68
1997	431.75	720.07	609.48	548.36	406.55	558.06	296.04	622.37	475.03	466.13	403.42	190.66	368.63
1998	448.36	727.19	629.75	557.09	422.40	581.48	310.43	646.52	504.00	490.53	417.47	200.82	384.25
1999	463.09	721.77	655.11	573.29	433.44	601.78	321.74	675.47	520.73	511.39	429.76	208.05	398.77
2000	480.90	734.88	685.78	590.89	448.78	630.09	333.48	700.92	540.42	535.56	447.80	217.20	413.30
2001	493.53	757.96	695.86	595.15	458.58	642.08	346.31	731.18	560.46	558.32	471.33	220.73	428.64
2002	506.48	741.97	711.82	618.62	469.97	643.40	360.92	737.94	579.08	575.00	490.38	227.31	439.87
2003	517.68	765.94	727.00	636.03	480.00	656.36	367.28	760.84	611.99	587.68	503.05	230.49	434.41
2004	528.65	804.01	735.55	658.52	487.37	665.96	371.23	776.72	625.50	597.88	521.06	234.86	433.04
2005	543.94	853.87	750.37	673.30	497.07	683.99	377.58	805.11	646.51	619.65	541.40	241.36	443.40
2006	566.94	907.95	781.59	690.88	512.82	717.42	383.25	850.64	673.66	662.80	561.02	250.34	456.50
2007	589.09	962.63	816.23	711.53	524.57	747.08	385.18	874.45	706.52	701.42	585.44	265.54	477.06
2008	607.10	1 014.66	842.61	724.46	534.44	768.14	386.44	908.78	729.64	738.31	607.82	273.39	495.57
2009	615.82	1 006.67	851.76	726.12	540.19	782.62	388.74	931.08	754.90	776.30	622.30	275.95	506.26
2010	636.02	1 063.11	891.83	765.18	558.05	814.04	400.38	939.85	780.25	799.43	639.37	280.87	523.70
2011	652.72	1 144.64	921.84	784.29	575.88	842.06	412.29	964.85	798.68	814.74	663.04	283.82	532.63
2012	665.54	1 201.69	942.14	794.67	587.09	857.06	422.35	973.52	839.84	823.37	674.48	290.54	539.46
2013	677.62	1 229.70	958.72	807.37	595.36	871.89	423.44	1 003.65	875.04	839.86	683.24	294.31	553.77
2014	694.74	1 270.91	977.11	822.03	611.77	894.06	431.97	1 030.17	908.15	865.25	692.56	303.81	568.92
2015	708.70	1 211.91	998.02	832.25	626.54	907.72	446.01	1 038.10	939.46	879.63	709.25	311.32	583.54
2016	723.20	1 221.69	1 031.88	855.77	633.77	929.09	447.69	1 068.23	962.95	899.79	723.91	319.67	595.68
2017	742.48	1 265.92	1 061.98	876.10	652.11	957.95	463.10	1 100.03	982.09	922.51	741.55	332.54	617.91
2018	766.99	1 322.45	1 108.59	908.01	674.74	979.82	483.03	1 139.52	997.26	949.18	761.86	345.21	640.70
2019	790.67	1 402.86	1 135.17	921.66	697.53	1 007.57	503.07	1 196.36	1 020.28	984.07	784.18	358.10	660.02
2018													
January	751.30	1 284.82	1 072.07	889.92	661.39	960.83	473.02	1 105.83	988.64	928.75	751.55	336.17	626.08
February	757.12	1 307.07	1 094.06	900.13	662.07	963.55	470.86	1 115.27	988.64	937.39	754.85	340.50	629.35
March	757.91	1 305.70	1 096.00	902.66	665.46	965.25	475.10	1 126.98	990.03	940.58	752.75	340.38	633.65
April	762.19	1 313.62	1 105.50	910.33	668.17	965.50	479.04	1 129.06	994.19	939.69	758.40	341.88	636.64
May	764.22	1 325.73	1 112.62	900.06	672.52	974.61	483.36	1 123.58	995.30	942.51	759.60	342.87	637.56
June	766.25	1 347.43	1 104.00	904.31	674.56	983.58	484.34	1 130.68	994.93	948.01	764.86	346.00	638.48
July	767.60	1 313.19	1 115.95	905.19	677.23	977.56	486.17	1 136.00	997.99	950.84	762.50	346.11	638.87
August	770.30	1 338.09	1 108.83	909.41	678.64	985.14	486.10	1 143.47	999.85	952.61	764.75	347.60	641.56
September	770.38	1 335.39	1 104.82	910.20	679.36	990.39	485.10	1 146.65	1 000.11	952.04	765.07	347.45	645.81
October	771.73	1 331.22	1 109.16	911.40	681.73	988.24	487.53	1 154.11	997.52	955.45	764.75	348.69	647.97
November	775.10	1 329.17	1 107.08	914.76	683.42	990.18	489.65	1 163.34	1 002.94	956.28	768.61	350.18	651.62
December	778.13	1 338.89	1 128.80	916.44	684.11	995.22	488.10	1 170.44	1 005.16	962.53	770.55	351.66	651.99
2019													
January	781.12	1 351.40	1 141.70	917.28	687.14	990.18	495.02	1 171.74	1 009.36	966.67	773.77	354.33	655.70
February	779.18	1 362.45	1 098.63	913.65	687.49	996.53	493.89	1 180.02	1 016.02	967.13	774.73	354.39	654.19
March	784.54	1 385.79	1 130.37	915.73	694.61	1 010.35	498.86	1 184.28	1 012.69	971.38	777.95	357.07	656.32
April	786.22	1 413.32	1 132.31	913.54	695.97	1 005.43	502.95	1 179.37	1 014.75	974.21	777.31	355.19	655.12
May	786.91	1 404.66	1 131.05	914.66	698.98	1 008.52	502.23	1 190.15	1 016.96	978.10	776.99	354.98	657.58
June	788.59	1 409.39	1 134.30	921.02	697.97	1 010.07	502.98	1 195.29	1 014.76	985.13	779.56	356.67	658.81
July	788.59	1 406.56	1 126.62	920.89	696.92	1 005.53	502.53	1 187.78	1 019.73	985.54	782.92	356.45	658.20
August	794.30	1 424.02	1 140.34	922.13	702.36	1 017.04	506.62	1 202.43	1 023.78	995.38	787.29	357.93	661.28
September	796.32	1 437.49	1 148.06	923.79	701.69	1 013.94	507.22	1 221.91	1 026.19	993.32	787.73	359.90	662.20
October	798.34	1 456.12	1 138.60	923.22	703.04	1 014.41	508.13	1 215.46	1 029.14	994.05	794.37	360.88	663.74
November	797.64	1 420.58	1 131.57	926.95	701.57	1 016.34	504.11	1 208.49	1 036.74	999.34	793.09	363.83	665.90
December	801.02	1 417.42	1 137.87	929.02	704.05	1 006.78	511.46	1 208.84	1 032.61	998.28	798.46	364.82	668.67

. . . = Not available.

SECTION 10C: OTHER SIGNIFICANT LABOR MARKET DATA

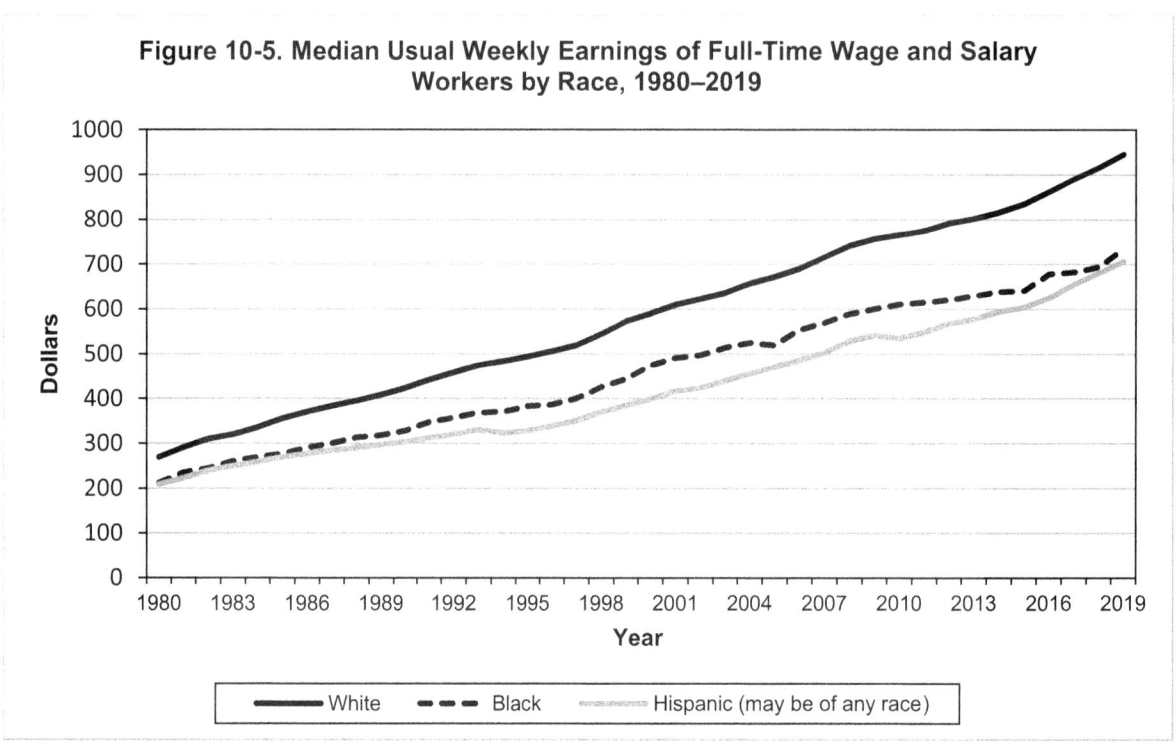

Figure 10-5. Median Usual Weekly Earnings of Full-Time Wage and Salary Workers by Race, 1980–2019

- This section of *Business Statistics* includes indicators of hours and earnings from the Current Population Statistics (household) survey that are more appropriate than similarly-named indicators produced by the payroll survey and shown in the previous section. It also incorporates annual employment data from the Bureau of Economic Analysis, produced as part of the National Income and Product Accounts (NIPA) that provide a consistent historical record of U.S. employment.

- Median weekly earnings for women were lower than for men in all age groups. In 2019, women age 25 to 54 earned $958 compared while men the same age earned $1,045. (Table 10-19). Median usual weekly earnings of full-time wage and salary workers also varied greatly by race. As shown in the figure above, Blacks earned less than Whites each year from 1980 through 2019.

- Table 10-18 provides hours at work back to 1948. Overall, there has been a slight downward trend in hours worked. However, women have actually seen their hours at work increase from 36.5 in 1956 to 36.6 in 2019. The number of hours that men have spent at work has declined two hours and forty-eight minutes during the same period. In 2006, just before the Great Recession, people reported working an average of 39.2 hours. This decreased to 37.9 hours in 2009 before increasing to 39.0 in 2019.

Table 10-18. Hours at Work, Current Population Survey

(Hours per week.)

Year and month	Hours per person, all industries				Hours per person, nonagricultural industries	
	Total	Men	Women	Persons who usually work full-time	All workers	
					Total	Usually work full-time
1948	42.8	. . .	. . .	. . .	41.6	. . .
1949	42.1	. . .	. . .	. . .	40.9	. . .
1950	41.7	. . .	. . .	. . .	40.7	. . .
1951	42.2	. . .	. . .	. . .	41.3	. . .
1952	42.4	. . .	. . .	. . .	41.6	. . .
1953	41.9	. . .	. . .	. . .	41.1	. . .
1954	40.9	. . .	. . .	. . .	40.0	. . .
1955	41.6	. . .	. . .	. . .	40.9	. . .
1956	41.5	43.8	36.5	. . .	40.9	. . .
1957	41.0	43.4	36.1	. . .	40.5	. . .
1958	40.6	42.9	35.8	. . .	40.1	. . .
1959	40.5	42.8	35.6	. . .	40.0	. . .
1960	40.5	43.0	35.4	. . .	40.0	. . .
1961	40.5	43.0	35.3	. . .	40.1	. . .
1962	40.5	43.2	35.2	. . .	40.1	. . .
1963	40.4	43.2	35.1	. . .	40.1	. . .
1964	40.0	42.8	34.7	. . .	39.7	. . .
1965	40.5	43.3	35.1	. . .	40.2	. . .
1966	40.4	43.2	35.2	. . .	40.1	. . .
1967	40.4	43.3	35.2	. . .	40.0	. . .
1968	40.1	43.0	34.9	. . .	39.7	. . .
1969	39.9	42.9	34.9	. . .	39.5	. . .
1970	39.1	42.0	34.2	. . .	38.7	. . .
1971	39.3	42.2	34.3	. . .	38.9	. . .
1972	39.4	42.3	34.5	. . .	39.0	. . .
1973	39.3	42.4	34.4	. . .	39.0	. . .
1974	39.0	42.0	34.3	. . .	38.7	. . .
1975	38.7	41.6	34.1	. . .	38.4	. . .
1976	38.7	41.7	34.1	. . .	38.4	. . .
1977	38.8	41.9	34.2	. . .	38.5	. . .
1978	39.0	42.1	34.5	. . .	38.7	. . .
1979	38.9	42.0	34.5	. . .	38.6	. . .
1980	38.5	41.5	34.5	. . .	38.3	. . .
1981	38.1	41.1	34.1	. . .	37.9	. . .
1982	38.0	40.9	34.1	. . .	37.7	. . .
1983	38.3	41.2	34.5	. . .	38.1	. . .
1984	38.8	41.8	34.9	. . .	38.6	. . .
1985	39.0	42.0	35.2	. . .	38.9	. . .
1986	39.1	42.1	35.4	. . .	38.9	. . .
1987	39.0	42.0	35.3	. . .	38.8	. . .
1988	39.4	42.4	35.7	. . .	39.3	. . .
1989	39.6	42.6	35.8	. . .	39.4	. . .
1990	39.4	42.3	35.8	. . .	39.3	. . .
1991	39.2	42.0	35.8	. . .	39.1	. . .
1992	38.9	41.7	35.6	. . .	38.8	. . .
1993	39.4	42.2	36.0	. . .	39.3	. . .
1994	39.2	42.2	35.5	43.4	39.1	43.3
1995	39.3	42.3	35.6	43.4	39.2	43.2
1996	39.3	42.3	35.7	43.3	39.2	43.2
1997	39.5	42.4	36.0	43.4	39.4	43.3
1998	39.3	42.2	35.8	43.2	39.2	43.1
1999	39.6	42.4	36.2	43.4	39.5	43.3
2000	39.7	42.5	36.4	43.4	39.6	43.3
2001	39.2	41.9	36.1	42.9	39.2	42.8
2002	39.2	41.8	36.0	42.9	39.1	42.8
2003	39.0	41.7	35.9	42.9	39.0	42.7
2004	39.0	41.7	35.9	42.9	39.0	42.8
2005	39.2	41.8	36.1	42.9	39.1	42.8
2006	39.2	41.8	36.2	42.9	39.2	42.8
2007	39.2	41.7	36.1	42.8	39.1	42.7
2008	38.9	41.3	36.1	42.6	38.8	42.5
2009	37.9	40.2	35.3	41.9	37.8	41.8
2010	38.2	40.5	35.5	42.2	38.1	42.2
2011	38.3	40.6	35.6	42.4	38.2	42.3
2012	38.5	40.8	35.8	42.5	38.4	42.4
2013	38.6	40.9	36.0	42.6	38.5	42.5
2014	38.6	40.9	35.9	42.5	38.6	42.5
2015	38.6	40.9	35.9	42.4	38.5	42.3
2016	38.7	40.9	36.2	42.5	38.7	42.4
2017	38.7	40.8	36.2	42.4	38.6	42.3
2018	38.9	41.0	36.4	42.5	38.9	42.1
2019	39.0	41.0	36.6	42.6	38.9	42.5

. . . = Not available.

Table 10-19 Median Usual Weekly Earnings of Full-Time Wage and Salary Workers

(Current dollars, [second quartile] except as noted; not seasonally adjusted.)

Year and quarter	Sex and age						Race and ethnicity			
	Men, 16 years and over			Women, 16 years and over			White	Black or African American	Asian	Hispanic or Latino ethnicity
	16 to 24 years	25 to 54 years	55 years and over	16 to 24 years	25 to 54 years	55 years and over				
1980	208	. . .	. . .	187	. . .	. . .	269	212	. . .	209
1981	218	. . .	. . .	200	. . .	. . .	291	235	. . .	223
1982	225	. . .	. . .	208	. . .	. . .	310	245	. . .	240
1983	223	. . .	. . .	211	. . .	. . .	320	261	. . .	250
1984	231	. . .	. . .	217	. . .	. . .	336	269	. . .	259
1985	241	. . .	. . .	224	. . .	. . .	356	277	. . .	270
1986	246	. . .	. . .	232	. . .	. . .	371	291	. . .	277
1987	257	. . .	. . .	243	. . .	. . .	384	301	. . .	285
1988	262	. . .	. . .	249	. . .	. . .	395	314	. . .	290
1989	271	. . .	. . .	259	. . .	. . .	409	319	. . .	298
1990	282	. . .	. . .	269	. . .	. . .	424	329	. . .	304
1991	285	. . .	. . .	277	. . .	. . .	442	348	. . .	312
1992	284	. . .	. . .	276	. . .	. . .	458	357	. . .	321
1993	288	. . .	. . .	282	. . .	. . .	475	369	. . .	331
1994	294	. . .	. . .	286	. . .	. . .	484	371	. . .	324
1995	303	. . .	. . .	292	. . .	. . .	494	383	. . .	329
1996	307	. . .	. . .	298	. . .	. . .	506	387	. . .	339
1997	317	. . .	. . .	306	. . .	. . .	519	400	. . .	351
1998	334	. . .	. . .	319	. . .	. . .	545	426	. . .	370
1999	356	. . .	. . .	341	. . .	. . .	573	445	. . .	385
2000	375	690	714	361	609	606	590	474	615	399
2001	391	717	742	375	632	620	610	491	639	417
2002	391	727	773	381	646	650	623	498	658	424
2003	398	736	797	387	659	682	636	514	693	440
2004	400	755	819	390	680	704	657	525	708	456
2005	409	763	829	397	691	726	672	520	753	471
2006	418	784	870	409	712	743	690	554	784	486
2007	443	809	902	424	732	772	716	569	830	503
2008	461	843	920	443	755	795	742	589	861	529
2009	458	857	944	442	767	815	757	601	880	541
2010	443	854	961	432	772	832	765	611	855	535
2011	455	864	976	440	784	857	775	615	866	549
2012	468	890	987	444	802	874	792	621	920	568
2013	479	891	999	454	812	891	802	629	942	578
2014	493	899	1 007	477	824	900	816	639	953	594
2015	510	921	1 054	487	845	916	835	641	993	604
2016	512	943	1 082	501	871	939	862	678	1 021	624
2017	547	977	1 084	519	892	962	890	682	1 043	655
2018	575	1 004	1 127	548	917	985	916	694	1 095	680
2019	607	1 045	1 150	581	958	1 001	945	735	1 174	706
2012										
1st quarter	469	882	980	421	723	761	793	635	918	567
2nd quarter	460	902	991	409	722	748	792	637	930	576
3rd quarter	459	873	960	414	715	755	780	606	915	556
4th quarter	480	902	1 016	422	717	746	802	615	910	571
2013										
1st quarter	487	902	969	419	729	754	802	622	951	575
2nd quarter	479	890	1 018	422	735	769	799	634	973	572
3rd quarter	452	885	984	414	729	765	794	630	922	587
4th quarter	492	888	1 036	449	739	776	813	632	916	576
2014										
1st quarter	480	905	979	434	750	767	819	646	955	593
2nd quarter	481	892	993	449	744	768	802	649	954	583
3rd quarter	498	894	1 034	448	743	778	816	638	945	598
4th quarter	508	905	1 019	469	749	784	823	621	959	600
2015										
1st quarter	491	922	1 029	461	757	768	835	650	966	590
2nd quarter	497	918	1 032	444	753	782	829	647	965	601
3rd quarter	517	916	1 069	442	757	781	829	624	974	602
4th quarter	543	926	1 092	453	764	768	847	643	1 091	624
2016										
1st quarter	511	935	1 102	488	771	821	857	673	1 032	612
2nd quarter	505	944	1 049	470	774	785	854	677	1 021	618
3rd quarter	510	943	1 095	480	782	793	854	685	1 010	632
4th quarter	522	953	1 084	506	788	810	881	675	1 022	646
2017										
1st quarter	558	982	1 045	489	795	828	894	679	1 019	649
2nd quarter	524	984	1 046	496	811	859	886	689	1 103	657
3rd quarter	527	969	1 125	500	797	859	887	696	1 010	655
4th quarter	581	974	1 116	512	799	818	891	666	1 061	657
2018										
1st quarter	563	1 001	1 097	545	819	819	911	696	1 066	675
2nd quarter	528	991	1 117	511	811	868	907	683	1 083	674
3rd quarter	575	1 012	1 111	515	845	826	915	686	1 128	689
4th quarter	609	1 014	1 166	539	825	873	931	712	1 095	684
2019										
1st quarter	605	1 034	1 149	539	841	859	935	737	1 157	696
2nd quarter	601	1 035	1 141	522	852	871	933	724	1 152	696
3rd quarter	601	1 047	1 143	538	883	854	943	727	1 247	718
4th quarter	621	1 066	1 168	566	881	892	967	756	1 166	712

. . . = Not available.

NOTES AND DEFINITION, CHAPTER 10

General note on monthly employment data

This chapter presents data from two different data sets that measure employment monthly. Both are compiled and published by the Bureau of Labor Statistics (BLS), but each set has different characteristics. Users should be aware of these dissimilarities and the consequent differences in the appropriate uses and interpretations of data from the two systems.

One set of monthly employment estimates comes from the Current Population Survey (CPS), a large sample survey of approximately 60,000 U.S. households. The numbers in the sample are expanded to match the latest estimates of the total U.S. population. These are the most comprehensive estimates in their scope—that is, in the universe that they are designed to measure. These estimates represent all civilian workers, including the following groups that are excluded by definition from the other set of estimates: all farm workers; domestic household workers (house cleaners, gardeners, nannies, etc.); nonagricultural, nonincorporated self-employed workers; and nonagricultural unpaid family workers.

This **household survey** (CPS) is designed to measure the labor force status of the civilian noninstitutional population with demographic detail. The national unemployment rate is the best-known statistic produced from the household survey. The survey also provides a measure of employed people, one that includes agricultural workers and the self-employed. A representative sample of U.S. households provides the information for the household survey.

However, official CPS data are characterized by periodic discontinuities, which occur when new benchmarks for Census measures of the total population are introduced. These updates take place in a single month—usually January—and the official data for previous months are typically not modified to provide a smooth transition. Therefore, shorter-term comparisons (for a year or two or for a business cycle phase) will be misleading if such a discontinuity is included in the period. A recent example will illustrate. Beginning with January 2019, the updated controls decreased the estimated size of the civilian noninstitutional population 16 years and over by 811,000, the civilian labor force by 524,000, employment by 507,000, unemployment by 17,000, and the number of persons not in the labor force by 287,000. The total unemployment rate, the labor force participation rate, and the employment-population ratio were not affected. Such discontinuities occur throughout the history of the series.

The CPS is a count of persons employed, rather than a count of jobs. A person is counted as employed in this data set only once, no matter how many jobs he or she may hold. The CPS count is limited to persons 16 years of age and over.

The second set of employment estimates—the payroll survey—comes from a very large sample survey of about 149,000 business and government employers at approximately 651,000 worksites, the Current Employment Statistics (CES) survey. It is benchmarked annually to a survey of all employers. Benchmark data are introduced with a smooth adjustment back to the previous benchmark, thus preserving the continuity of the series and making it more appropriate for measurement of employment change over a year or two, a business cycle, or other short- to medium-length periods. This sample is much larger than the CPS sample, including about one-third of all nonfarm payroll employees, and consequently the threshold of statistical significance for changes is lower. The minimum size of the over-the-month change required to be statistically significant is about 115,000 jobs in the payroll survey, versus about 500,000 persons in the household survey.

The **payroll survey** (CES) is designed to measure employment, hours, and earnings in the nonfarm sector, with industry and geographic detail. The survey is best known for providing a highly reliable gauge of monthly change in nonfarm payroll employment. A representative sample of businesses in the U.S. provides the data for the payroll survey.

Over recent years, benchmark revisions to payroll survey employment have ranged in magnitude from an upward revision of 0.6 percent, in 2006, to the March 2009 benchmark, which reduced the seasonally adjusted level of payroll employment in that month by 902,000 persons, or 0.7 percent. A recent revision increased the level of employment in March 2019 deceased employment by 489,000, or 0.3. percent. In absolute terms (that is, disregarding the sign), revisions have ranged from 0.1 to 0.7 percentage points over the latest 13 years and averaged 0.195 percentage points. Within that range, relatively large upward revisions have been seen for strong economies, and relatively large downward revisions for weak ones.

The scope of the CES survey is smaller than that of the CPS survey, as it is limited to wage and salary workers on nonfarm payrolls. There is also a significant definitional difference, because the CES survey is a count of jobs. Thus, a person with more than one nonfarm wage or salary job is counted as employed in each job. In addition, workers are not classified by age; as a result, there may be some workers younger than 16 years old in the job count.

Persons with a job but not at work (absent due to bad weather, work stoppages, personal reasons, and the like) are included in the household survey. However, they are excluded from the payroll survey if on leave without pay for the entire payroll period.

In addition to the differences in definitions and scope between the two series, there are also differences in sample design, collection methodology, and the sampling variability inherent in the surveys.

The payroll survey provides the most reliable and detailed information on employment by industry (for example, the data shown in Table 15-1). In addition, it provides data on weekly hours per job and hourly and weekly earnings per job.

The CPS employment estimates provide information not collected in the CES on employment by demographic characteristics, such as age, race, and Hispanic ethnicity; by education levels; and by occupation. A few of these tabulations are shown in *Business Statistics*. Many more breakdowns, in richer detail, can be found in the *Handbook of U.S. Labor Statistics*, also published by Bernan Press.

The differences between the two employment measurement systems are discussed in an article on the BLS Web site, "Employment from the BLS household and payroll surveys: summary of recent trends," which is updated monthly along with the release of the monthly data, and can be found under "Publications" on the CPS homepage at http://www.bls.gov. This article also includes references to further relevant studies.

Additional annual employment data

In addition to the CPS and CES monthly and annual data series, a third set of historical, annual-only estimates of employment is presented at the end of this chapter. These estimates of "full-time and part-time employees" by industry are compiled by the U.S. Department of Commerce, Bureau of Economic Analysis, as part of the National Income and Product Accounts (NIPAs). These estimates provide the most comprehensive and consistent employment estimates for the years before 1948.

Further detail on the historical and other characteristics of all of these employment series is presented below in the notes associated with specific data tables.

TABLES 10-1 THROUGH 10-5

Labor Force, Employment, and Unemployment

SOURCE: U.S. DEPARTMENT OF LABOR, BUREAU OF LABOR STATISTICS (BLS)

Labor force, employment, and unemployment data are derived from the Current Population Survey (CPS), a sample survey of households conducted each month by the Census Bureau for the Bureau of Labor Statistics (BLS). The data pertain to the U.S. civilian noninstitutional population age 16 years and over.

Due to changes in questionnaire design and survey methodology, data for 1994 and subsequent years are not fully comparable with data for 1993 and earlier years. Additionally, discontinuities in the reported number of persons in the population, and consequently in the estimated numbers of employed and unemployed persons and the number of persons in the labor force, are introduced whenever periodic updates are made to U.S. population estimates.

For example, population controls based on Census 2000 were introduced beginning with the data for January 2000. These data are therefore not comparable with data for December 1999 and earlier. Data for 1990 through 1999 incorporate 1990 census–based population controls and are not comparable with the preceding years. An additional large population adjustment was introduced in January 2004, making the data from that time forward not comparable with data for December 2003 and earlier; further adjustments have been made in each subsequent January and other discontinuities have been introduced in various earlier years, usually with January data. See "Notes on the Data," below, for information on adjustments in other years, including the incorporation of the 2010 Census count beginning in January 2012.

For the most part, these population adjustments distort comparisons involving the numbers of persons in the population, labor force, and employment. They generally have negligible effects on the percentages that comprise the most important features of the CPS: the unemployment rates, the labor force participation rates, and the employment-population ratios.

Beginning with the data for January 2000, data classified by industry and occupation use the North American Industry Classification System (NAICS) and the 2000 Standard Occupational Classification System. This creates breaks in the time series between December 1999 and January 2000 for occupational and industry data at all levels of aggregation. Since the recent history is so short, most CPS industry and occupation data have been dropped from *Business Statistics* in favor of other important and economically meaningful data for which a longer consistent history can be supplied. However, detailed employment data by occupation and industry can be found in Bernan Press's *Handbook of U.S. Labor Statistics*.

Pre-1948 data

The Census Bureau began the survey of households that provides labor force data in 1942, and the Census of Population supplied data for 1940. For the years before that, annual data for labor force, employment, and unemployment are retrospective estimates. The 1929-1947 estimates were made by BLS. The 1890-1929 estimates shown in Table 10-1B were made by Stanley Lebergott. Both of those used trend interpolation of labor force participation between Census years, and the Census and Lebergott estimates are identical in 1929.

From 1929 to 1947, the data collected pertained to persons 14 years of age and over. From 1947 to the present the survey pertains to persons 16 years of age and over. Table 10-1C shows summary data for both age definitions in the overlap year, 1947. In that year, the unemployment rates are the same (3.9 percent) for both age definitions. However, the labor force participation rate and the employment/population ratio are higher when the 14- and 15-year-olds are excluded. Naturally, the raw numbers for population, labor force, and employment are smaller when the 14- and 15-year-olds are excluded.

The 1940 Census and the BLS data for 1931 through 1942 did not treat government work relief employment as employment; persons engaged in such work were counted as unemployed. In the labor force survey used today, anyone who worked for pay or profit is counted as employed (see the definitions below). Therefore, today's survey would count work relief as employment and not unemployment, and the BLS figures for the period before 1942 are not consistent with today's unemployment rates. Michael Darby, a former member of NBER, calculated an alternative unemployment rate in which such workers are counted as employed, and this rate is shown in parentheses in Table 10-1B. BLS counts employment in nonprofit and private groups that provide relief under the "emergency and other relief services" industry (also referred to in some BLS data as the "community food and housing, and emergency and other relief services" industry). This employment is usually found in the "Other Services' of the NAICS supersector.

Race and ethnic origin

Data for two broad racial categories were made available beginning in 1954: *White* and *Black and other.* The latter included Asians and all other "nonwhite" races, and was discontinued after 2002. Data for *Blacks* only are available beginning with 1972; this category is now called *Black or African American.* Data for *Asians* are shown beginning with 2000. Persons in the remaining race categories—American Indian or Alaska Native, Native Hawaiian or Other Pacific Islanders, and persons who selected more than one race category when that became possible, beginning in 2003 (see below)—are included in the estimates of total employment and unemployment, but are not shown separately because their numbers are too small to yield quality estimates.

Hispanic or Latino ethnicity, previously labeled *Hispanic origin*, is not a racial category and is established in a survey question separate from the question about race. Persons of Hispanic or Latino ethnicity may be of any race.

In January 2003, changes that affected classification by race and Hispanic ethnicity were introduced. These changes caused discontinuities in race and ethnic group data between December 2002 and January 2003.

- Individuals in the sample are now asked whether they are of Hispanic ethnicity before being asked about their race. Prior to 2003, individuals were asked their ethnic origin after they were asked about their race. Furthermore, respondents are now asked directly if they are Spanish, Hispanic, or Latino. Previously, they were identified based on their or their ancestors' country of origin.

- Individuals in the sample are now allowed to choose more than one race category. Before 2003, they were required to select a single primary race. This change had no impact on the size of the overall civilian noninstitutional population and labor force. It did reduce the population and labor force levels of Whites and

Blacks beginning in January 2003, as individuals who reported more than one race are now excluded from those groups.

BLS has estimated, based on a special survey, that these changes reduced the population and labor force levels for Whites by about 950,000 and 730,000 persons, respectively, and for Blacks by about 320,000 and 240,000 persons, respectively, while having little or no impact on either of their unemployment rates. The changes did not affect the size of the Hispanic population or labor force, but they did cause an increase of about half a percentage point in the Hispanic unemployment rate.

Definitions

The employment status of the civilian population is surveyed each month with respect to a specific week in mid-month—not for the entire month. This is known as the "reference week." For a precise definition and explanation of the reference week, see Notes on the Data, which follows these definitions.

The *civilian noninstitutional population* comprises all civilians 16 years of age and over who are not inmates of penal or mental institutions, sanitariums, or homes for the aged, infirm, or needy. A recent example will illustrate. Benchmark population updates does occur annually. For example, in January 2019, the updated controls decreased the estimated size of the civilian noninstitutional population 16 years and over by 811,000, the civilian labor force by 524,000, employment by 507,000, unemployment by 17,000, and the number of persons not in the labor force by 287,000. The total unemployment rate, the labor force participation rate, and the employment-population ratio were not affected. Such discontinuities occur throughout the history of the series.

Civilian employment includes those civilians who (1) worked for pay or profit at any time during the Sunday-through-Saturday week that includes the 12th day of the month (the reference week), or who worked for 15 hours or more as an unpaid worker in a family-operated enterprise; or (2) were temporarily absent from regular jobs because of vacation, illness, industrial dispute, bad weather, or similar reasons. Each employed person is counted only once; those who hold more than one job are counted as being in the job at which they worked the greatest number of hours during the reference week.

Unemployed persons are all civilians who were not employed (according to the above definition) during the reference week, but who were available for work—except for temporary illness—and who had made specific efforts to find employment sometime during the previous four weeks. Persons who did not look for work because they were on layoff are also counted as unemployed.

The *civilian labor force* comprises all civilians classified as employed or unemployed.

Civilians 16 years of age and over in the noninstitutional population who are not classified as employed or unemployed are defined as *not in the labor force*. This group includes those engaged in own-home housework; in school; unable to work because of longterm illness, retirement, or age; seasonal workers for whom the reference week fell in an "off" season (if not qualifying as unemployed by looking for a job); persons who became discouraged and gave up the search for work; and the voluntarily idle. Also included are those doing only incidental work (less than 15 hours) in a family-operated business during the reference week.

The civilian *labor force participation rate* represents the percentage of the civilian noninstitutional population (age 16 years and over) that is in the civilian labor force. This rate includes people who have stopped looking for a job while the unemployment figure does not. Thus a low unemployment rate while the labor force participation rate is also low points to a structural problem in the labor market. When former workers become discouraged and stop looking for work, this shrinks the labor force.

The *employment-population ratio* represents the percentage of the civilian noninstitutional population (age 16 years and over) that is employed. This is traditionally called a "ratio," although it is also traditionally expressed as a percent and therefore would be more appropriately called a "rate," as is the case with the labor force participation rate. Factors that change the rate are population demographics such as baby boomers are retiring and younger individuals chose to stay in college longer.

Employment is shown by *class of worker,* including a breakdown of total employment into *agricultural* and *nonagricultural* industries. Employment in *nonagricultural industries* includes *wage and salary workers*, the *self-employed*, and *unpaid family workers.*

Wage and salary workers receive wages, salaries, commissions, tips, and/or pay-in-kind. This category includes owners of self-owned incorporated businesses.

Self-employed workers are those who work for profit or for fees in their own business, profession, trade, or farm. This category includes only the unincorporated; a person whose business is incorporated is considered to be a wage and salary worker since he or she is a paid employee of a corporation, even if he or she is the corporation's president and sole employee. These categories are now labeled "Self-employed workers, unincorporated" to clarify this definition.

Wage and salary employment comprises *government* and *private industry* wage and salary workers. Domestic workers and other employees of *private households*, who are not included in the payroll employment series, are shown separately from *all other private industries*. The series for *government* and *other private industries* wage and salary workers are the closest in scope to similar categories in the payroll employment series.

Multiple jobholders are employed persons who, during the reference week, either had two or more jobs as a wage and salary worker, were self-employed and also held a wage and salary job, or worked as an unpaid family worker and also held a wage and salary job. This category does not include self-employed persons with multiple businesses or persons with multiple jobs as unpaid family workers. For purposes of industry and occupational classification, multiple jobholders are counted as being in the job at which they worked the greatest number of hours during the reference week.

Employed and at work part time does not include employed persons who were absent from their jobs during the entire reference week for reasons such as vacation, illness, or industrial dispute.

At work part time for economic reasons ("involuntary" part time) refers to individuals who worked 1 to 34 hours during the reference week because of slack work, unfavorable business conditions, an inability to find full-time work, or seasonal declines in demand. To be included in this category, workers must also indicate that they want and are available for full-time work.

At work part time for noneconomic reasons ("voluntary" part time) refers to persons who usually work part time and were at work for 1 to 34 hours during the reference week for reasons such as illness, other medical limitations, family obligations, education, retirement, Social Security limits on earnings, or working in an industry where the workweek is less than 35 hours. It also includes respondents who gave an economic reason but were not available for, or did not want, full-time work. At work part time for noneconomic reasons excludes persons who usually work full time, but who worked only 1 to 34 hours during the reference week for reasons such as holidays, illnesses, and bad weather.

The *long-term unemployed* are persons currently unemployed (searching or on layoff) who have been unemployed for 15 consecutive weeks or longer. If a person ceases to look for work for two weeks or more, or becomes temporarily employed, the continuity of longterm unemployment is broken. If he or she starts searching for work or is laid off again, the monthly CPS will record the length of his or her unemployment from the time the search recommenced or since the latest layoff.

The civilian *unemployment rate* is the number of unemployed as a percentage of the civilian labor force. The unemployment rates for groups within the civilian population (such as males age 20 years and over) are the number of unemployed in a group as a percent of that group's labor force.

Unemployment rates by reason provides a breakdown of the total unemployment rate. Each unemployed person is classified into one of four groups.

Job losers and persons who completed temporary jobs includes persons on temporary layoff, permanent job losers, and persons who completed temporary jobs and began looking for work after those jobs ended. These three categories are shown separately without seasonal adjustment in the BLS's "Employment

Situation" news release and on its Web site. They are combined, under the title shown here, for the purpose of seasonal adjustment. This is the category of unemployment that responds most strongly to the business cycle.

Job leavers terminated their employment voluntarily and immediately began looking for work.

Reentrants are persons who previously worked, but were out of the labor force prior to beginning their current job search.

New entrants are persons searching for a first job who have never worked.

Each of these categories is expressed as a proportion of the entire civilian labor force, so that the sum of the four rates equals the unemployment rate for all civilian workers, except for possible discrepancies due to rounding or separate seasonal adjustment.

Median and average weeks unemployed are summary measures of the length of time that persons classified as unemployed have been looking for work. For persons on layoff, the duration represents the number of full weeks of the layoff. The *average (mean)* number of weeks is computed by aggregating all the weeks of unemployment experienced by all unemployed persons during their current spell of unemployment and dividing by the number of unemployed. The average can be distorted by what is called "top coding" because the length of unemployment is reported in ranges, including a top range of "over __ weeks", rather than exact numbers. See the paragraph below for further explanation. The *median* number of weeks unemployed is the number of weeks of unemployment experienced by the person at the midpoint of the distribution of all unemployed persons, as ranked by duration of unemployment, and is not distorted by top coding. Like medians in other economic time series, it is likely to be a better measure of typical experience.

Beginning with the data for January 2011 and phasing in through April 2011, respondents could report unemployment durations of up to 5 years; before that time, the "top code" was up to 2 years. This change caused a sharp increase in January 2011 and continued rises through April for *average weeks unemployed*. It did not affect total unemployment, total long-term unemployment, or the *median weeks unemployed*. Comparisons of the average weeks statistics on the old and new basis are available on the BLS Web site.

Alternative measures of labor underutilization are calculated by BLS and published in the monthly Employment Situation release. They measure alternative concepts of unused working capacity, and are numbered "U-1" through "U-6" in order of increasing breadth of the definition of underutilization.

"U-1" is persons unemployed 15 weeks or longer, as a percent of the civilian labor force. It is not shown in *Business Statistics*.

"U-2" is job losers and persons who completed temporary jobs, as a percent of the civilian labor force; it is shown in *Business Statistics* in the fifth column of Table 10-5.

"U-3" is the official rate, described above and shown in the final columns of Tables 10-1, the fifth column of Table 10-4, and the fourth column of Table 10-5.

"U-4" through "U-6" are shown in the last three columns of Table 10-5. They are based on additional labor force status questions, now included in the CPS survey, that were introduced beginning in 1994. U-4 and U-5 are increasingly broader rates of unemployment, while U-6 can be described as an "unemployment and underemployment rate."

"U-4" adds discouraged workers to unemployment and the labor force. Discouraged workers are persons not in the officially defined labor force who have given a job-market-related reason for not looking currently for a job—for example, they have not looked for a job because they believed that no jobs were available.

"U-5" adds both discouraged workers and all other "marginally attached" workers to unemployment and the labor force. "Marginally attached" workers are all persons who currently are neither working nor looking for work but indicate that they want and are available for a job and have looked for work sometime in the recent past.

"U-6" adds persons employed part time for economic reasons, as shown in Table 10-3, to the number of persons counted as unemployed in "U-5", with the same labor force definition as in "U-5." This statistic, which measures a combination of unemployment and underemployment, is often cited in media reports on the employment situation.

For more information, see "BLS introduced new range of alternative unemployment measures" in the October 1995 issue of the *Monthly Labor Review.*

Notes on the data

The CPS data are collected by trained interviewers from about 60,000 sample households selected to represent the U.S. civilian noninstitutional population. The sample size has fluctuated between 50,000 and 60,000 households In January 1996, the size was reduced to 50,00 due to budgetary reasons. The sample size was increased back to 60,000 households, beginning with the data for July 2001, as part of a plan to meet the requirements of the State Children's Health Insurance Program legislation. The CPS provides data for other data series in addition to the BLS employment status data, such as household income and poverty and health insurance (see Chapter 3.)

The employment status data are based on the activity or status reported for the calendar week, Sunday through Saturday, that includes the 12th day of the month (the reference week). This specification is chosen so as to avoid weeks that contain major holidays. Households are interviewed in the week following the reference week. Sample households are phased in and out of the sample on a rotating basis. Consequently, three-fourths of the sample is the same for any two consecutive months. One-half of the sample is the same as the sample in the same month a year earlier.

Data relating to 1994 and subsequent years are not strictly comparable with data for 1993 and earlier years because of the major 1994 redesign of the survey questionnaire and collection methodology. The redesign included new and revised questions for the classification of individuals as employed or unemployed, the collection of new data on multiple jobholding, a change in the definition of discouraged workers, and the implementation of a more completely automated data collection.

The 1994 redesign of the CPS was the most extensive since 1967. However, there are many other significant periods of year-to-year noncomparability in the labor force data. These typically result from the introduction of new decennial census data into the CPS estimation procedures, expansions of the sample, or other improvements made to increase the reliability of the estimates. Each change introduces a new discontinuity, usually between December of the previous year and January of the newly altered year. The discontinuities are usually minor or negligible with respect to figures expressed as nationwide percentages (such as the unemployment rate or the labor force participation rate), but can be significant with respect to levels (such as labor force and employment in thousands of persons). A list of the dates of the major discontinuities follows, with BLS estimates of their quantitative impact on the national totals. (There are likely to be larger impacts on population subgroups.) The discontinuities occur in January unless otherwise indicated. Note that some of the changes caused adjustments that were carried back to an earlier year.

- 2012: Reflecting the 2010 Census, the civilian noninstitutional population was increased by 1,510,000; the civilian labor force by 258,000; employment by 216,000; and unemployment by 42,000. BLS states, "Although the total unemployment rate was unaffected, the labor force participation rate and the employment-population ratio were each reduced by 0.3 percentage point. This was because the population increase was primarily among persons 55 and older and, to a lesser degree, persons 16 to 24 years of age. Both these age groups have lower levels of labor force participation than the general population."

- 2013: Updated population estimates increased the population by 138,000, the labor force by 136,000, and employment by 127,000. The rates and ratios were not affected.

- 2014: the smallest population control adjustment on record increased labor force by 24,000 and employment by 22,000, with no effect on rates or ratios.

- 2015: civilian population aged 16 or older increased by over one-half million, the labor force by 348,000, employment by 324,000 and unemployment by 24,000. The rates and ratios were not affected.

- 2016: the population adjustments rose once again by population 265,000, labor force 218,000, employment 206,000, unemployment 12,000 and not in labor force 47,000. The rates and ratios were not affected.

- 2017: the first year since 2012, the controls decreased. Population down by 831,000, labor force, 508,000, employment 483,000, unemployment 21,000 and those not in labor force 323,000. The rates and ratios were not affected.

- 2019: Population was down 811,000 persons, labor force down 524,000, employment down 507,000, unemployment down 17,000 those in not in the labor force down 287,000.

For further information on these changes, see "Publications and Other Documentation" at http://www.bls.gov/cps.

The monthly labor force, employment, and unemployment data are seasonally adjusted by the X-12-ARIMA method. All seasonally adjusted civilian labor force and unemployment rate statistics, as well as major employment and unemployment estimates, are computed by aggregating independently adjusted series. For example, the seasonally adjusted level of total unemployment is the sum of the seasonally adjusted levels of unemployment for the four age/sex groups (men and women age 16 to 19 years, and men and women age 20 years and over). Seasonally adjusted employment is the sum of the seasonally adjusted levels of employment for the same four groups. The seasonally adjusted civilian labor force is the sum of all eight components. Finally, the seasonally adjusted civilian worker unemployment rate is calculated by taking total seasonally adjusted unemployment as a percent of the total seasonally adjusted civilian labor force.

To minimize subsequent revisions, BLS uses a concurrent technique that estimates factors for the latest month using the most recent data. Then, seasonal adjustment factors are fully revised at the end of each year to reflect recent experience. The revisions also affect the preceding four years. An article describing the seasonal adjustment methodology for the household survey data is available at http://www.bls.gov/cps/cpsrs2010.pdf.

Breakdowns other than the basic age/sex classification described above—such as the employment data by class of worker in Table 10-3—will not necessarily add to totals because of independent seasonal adjustment.

Data availability

Data for each month are usually released on the first Friday of the following month in the "Employment Situation" press release, which also includes data from the establishment survey (Tables 10-7 through 10-17 and Chapter 15). (The release date actually is determined by the timing of the survey week.) The press release and data are available on the BLS Web site at <http://www.bls.gov>. The *Monthly Labor Review*, also available online at the BLS Web site, features frequent articles analyzing developments in the labor force, employment, and unemployment.

Monthly and annual data on the current basis are available beginning with 1948. Historical unadjusted data are published in *Labor Force Statistics Derived from the Current Population Survey* (BLS Bulletin 2307). Historical seasonally adjusted data are available from BLS upon request. Complete historical data are available on the BLS Web site at <http://www.bls.gov/cps>.

Seasonal adjustment factors are revised each year for the five previous years, with the release of December data in early January. New population controls are introduced with the release of January data in early February.

BLS annual data for 1940 through 1947 are published on their Web site at <http://www.bls.gov>. The data for 1929 through 1939 were published in *Employment and Earnings,* May 1972, and in U.S. Commerce Department, Bureau of Economic Analysis, *Long-Term Economic Growth, 1860–1970,* June 1973, p. 163. The Lebergott estimates are also found in the latter volume.

The Darby alternative unemployment rate is found in Michael Darby, "Three-and-a-Half Million U.S. Employees Have Been Mislaid," *Journal of Political Economy,* February 1976, v. 84, no. 1. It is also displayed and discussed in Robert A. Margo, "Employment and Unemployment in the 1930s," *Journal of Economic Perspectives,* v. 7, no. 2, Spring 1993.

References

Comprehensive descriptive material can be found at <http://www.bls.gov/cps> under the "Publications and Other Documentation" section. Historical background on the CPS, as well as a description of the 1994 redesign, can be found in three articles from the September 1993 edition of *Monthly Labor Review*: "Why Is It Necessary to Change?"; "Redesigning the Questionnaire"; and "Evaluating Changes in the Estimates." The redesign is also described in the February 1994 issue of *Employment and Earnings.* See also Chapter 1, "Labor Force Data Derived from the Current Population Survey," *BLS Handbook of Methods,* Bulletin 2490 (April 1997).

Table 10-6

Insured Unemployment

SOURCE: U.S. DEPARTMENT OF LABOR, EMPLOYMENT AND TRAINING ADMINISTRATION

Definitions

State programs of unemployment insurance cover operations of regular programs under state unemployment insurance laws. In 1976, the law was amended to extend coverage to include virtually all state and local government employees, as well as many agricultural and domestic workers. (This took effect on January 1, 1978.) Benefits under state programs are financed by taxes levied by the states on employers.

An *initial claim* is the first claim in a benefit year filed by a worker after losing his or her job, or the first claim filed at the beginning of a subsequent period of unemployment in the same benefit year. The initial claim establishes the starting date for any insured unemployment that may result if the claimant is unemployed for one week or longer. Transitional claims (filed by claimants as they start a new benefit year in a continuing spell of unemployment) are excluded; therefore, these data more closely represent instances of new unemployment and are widely followed as a leading indicator of job market conditions.

Insured unemployment and *persons claiming benefits* both describe the average number of persons receiving benefits in the indicated month or year.

The *insured unemployment rate* for state programs is the level of insured unemployment as a percentage of employment covered by state programs.

Monthly averages in this book are averages, calculated by the editor, of the weekly data published by the Employment and Training Administration. Annual data are averages of the monthly data.

Data availability

Data are published in weekly press releases from the Employment and Training Administration. These releases are available on their Web site at <http://www.doleta.gov> under "Labor Market Data," as are historical data, under the category "UI/Program Statistics."

TABLES 10-7, 10-8, 10-13, 15-1, AND 15-2

Nonfarm Payroll Employment

SOURCE: U.S. DEPARTMENT OF LABOR, BUREAU OF LABOR STATISTICS (BLS)

These nonfarm employment data, as well as the hours and earnings data in Tables 10-8 through 10-12, 10-14 through 10-17, and 15-3 through 15-6, are compiled from payroll records. Information is reported monthly on a voluntary basis to BLS and its cooperating state agencies by a large sample of establishments, representing all industries except farming. These data, formally known as the Current Employment Statistics (CES) survey, are often referred to as the "establishment data" or the "payroll data." They are also known as the BLS-790 survey.

The survey, originally based on a stratified quota sample, has been replaced on a phased-in basis by a stratified probability sample. The new sampling procedure went into effect for wholesale trade in June 2000; for mining, construction, and manufacturing in June 2001; and for retail trade, transportation and public utilities, and finance, insurance, and real estate in June 2002. The phase-in was completed in June 2003, upon its extension to the service industries. The phase-in schedule was slightly different for the state and area series.

The sample has always been very large. Currently, it includes approximately 145,000 businesses and government agencies covering about 697,000 individual worksites, which account for about one-third of total benchmark employment of payroll workers. The sample is drawn from a sampling frame of over 10.2 million unemployment insurance tax accounts.

Data are classified according to the North American Industry Classification System. BLS has reconstructed historical time series to conform with NAICS, to ensure that all published series have a NAICS-based history extending back to at least January 1990. NAICS-based history extends back to January 1939 for total nonfarm and other high-level aggregates. For more detailed series, the starting date for NAICS data varies depending on the extent of the definitional changes between the old Standard Industrial Classification (SIC) and NAICS.

Definitions

An *establishment* is an economic unit, such as a factory, store, or professional office, that produces goods or services at a single location and is engaged in one type of economic activity.

Employment comprises all persons who received pay (including holiday and sick pay) for any part of the payroll period that includes the 12th day of the month. The definition of the payroll period for each reporting respondent is that used by the employer; it could be weekly, biweekly, monthly, or other. Included are all fulltime and parttime workers in nonfarm establishments, including salaried officers of corporations. Persons holding more than one job are counted in each establishment that reports them. Not covered are proprietors, the selfemployed, unpaid volunteer and family workers, farm workers, domestic workers in households, and military personnel. Employees of the Central Intelligence Agency, the Defense Intelligence Agency, the National Geospatial-Intelligence Agency, and the National Security Agency are not included.

Persons on an establishment payroll who are on paid sick leave (when pay is received directly from the employer), on paid holiday or vacation, or who work during a portion of the pay period despite being unemployed or on strike during the rest of the period, are counted as employed. Not counted as employed are persons who are laid off, on leave without pay, on strike for the entire period, or hired but not paid during the period.

Intermittent workers are counted if they performed any service during the month. BLS considers regular fulltime teachers (private and government) to be employed during the summer vacation period, regardless of whether they are specifically paid during those months.

The *government* division includes federal, state, and local activities such as legislative, executive, and judicial functions, as well as the U.S. Postal Service and all governmentowned and governmentoperated business enterprises, establishments, and institutions (arsenals, navy yards, hospitals, state-owned utilities, etc.), and government force account construction. However, as indicated earlier, members of the armed forces and employees of certain national-security-related agencies are not included.

The monthly *diffusion index of employment change*, currently based on 271 private nonfarm NAICS industries, represents the percentage of those industries in which the seasonally adjusted level of employment in that month was higher than six months earlier, plus one-half of the percentage of industries with unchanged employment. Therefore, the diffusion index reported for September represents the change from March to September. *Business Statistics* uses the September value to represent the year as a whole, since it spans the year's midpoint. Diffusion indexes measure the dispersion of economic gains and losses, with values below 50 percent associated with recessions. The current NAICS-based series begins with January 1991. For October 1976 through December 1990, an earlier series is available based on 347 SIC industries (there are more industries using the older classification system because in SIC manufacturing industries were represented in greater detail). September values from this series are used here to represent the years 1977 through 1990.

Production and nonsupervisory workers include all *production and related workers* in mining and manufacturing; *construction workers* in construction; and *nonsupervisory workers* in transportation, communication, electric, gas, and sanitary services; wholesale and retail trade; finance, insurance, and real estate; and services. These groups account for about four-fifths of the total employment on private nonagricultural payrolls. Previously, this category was called "production or nonsupervisory workers." The definitions have not changed.

Production and related workers include working supervisors and all nonsupervisory workers (including group leaders and trainees) engaged in fabricating, processing, assembling, inspecting, receiving, storing, handling, packing, warehousing, shipping, trucking, hauling, maintenance, repair, janitorial, guard services, product development, auxiliary production for plant's own use (such as a power plant), record keeping, and other services closely associated with these production operations.

Construction workers include the following employees in the construction division of the NAICS: working supervisors, qualified craft workers, mechanics, apprentices, laborers, and the like, who are engaged in new work, alterations, demolition, repair, maintenance, and other tasks, whether working at the site of construction or working in shops or yards at jobs (such as precutting and preassembling) ordinarily performed by members of the construction trades.

Nonsupervisory employees include employees (not above the working supervisory level) such as office and clerical workers, repairers, salespersons, operators, drivers, physicians, lawyers, accountants, nurses, social workers, research aides, teachers, drafters, photographers, beauticians, musicians, restaurant workers, custodial workers, attendants, line installers and repairers, laborers, janitors, guards, and other employees at similar occupational levels whose services are closely associated with those of the employees listed.

Notes on the data

Benchmark adjustments. The establishment survey data are adjusted annually to comprehensive counts of employment, called "benchmarks." Benchmark information on employment by industry is compiled by state agencies from reports of establishments covered under state unemployment insurance laws; these form an annual compilation of administrative data known as the ES-202. These tabulations cover about 97 percent of all employees on nonfarm payrolls. Benchmark data for the residual are obtained from alternate sources, primarily from Railroad Retirement Board records and the Census Bureau's *County Business Patterns*. The latest benchmark adjustment, which is incorporated into the data in this volume, increased the employment level in the benchmark month March 2018 by 135,000 jobs, which was 0.1 percent.

The estimates for the benchmark month are compared with new benchmark levels for each industry. If revisions are necessary, the monthly series of estimates between benchmark periods are adjusted by graduated amounts between the new benchmark and the preceding one ("wedged back"), and the new benchmark level for each industry is then carried forward month by month based on the sample.

More specifically, the month-to-month changes for each estimation cell are based on changes in a matched sample for that cell, plus an estimate of net business births and deaths. The matched sample for each pair of months consists of establishments that have reported data for both months (which automatically excludes establishments that have gone out of business by the second month). Since new businesses are not immediately incorporated into the sample, a model-based estimate of net business births and deaths in that estimating cell is added. The birth/death adjustment factors are re-estimated quarterly based on the Quarterly Census of Employment and Wages.

Not-seasonally-adjusted data for all months since the last benchmark date are subject to revision.

Beginning in 1959, the data include Alaska and Hawaii. This inclusion resulted in an increase of 212,000 (0.4 percent) in total nonfarm employment for the March 1959 benchmark month.

Seasonal adjustment. The seasonal movements that recur periodically—such as warm and cold weather, holidays, and vacations—are generally the largest single component of monthtomonth changes in employment. After adjusting the data to remove such seasonal variation, basic trends become more evident. BLS uses X-12-ARIMA software to produce seasonal factors and perform concurrent seasonal adjustment, using the most recent 10 years of data. New factors are developed each month adding the most current data.

For most series, a special procedure called REGARIMA (regression with autocorrelated errors) is used before calculating the seasonal factors; this adjusts for the length of the interval (which can be either four or five weeks) between the survey weeks. REGARIMA has also been used to isolate extreme weather effects that distort the measurement of seasonal patterns in the construction industry, and to identify variations in local government employment due to the presence or absence of election poll workers.

Seasonal adjustment factors are directly applied to the component levels. Seasonally adjusted totals for employment series are then obtained by aggregating the seasonally adjusted components directly, while hours and earnings series represent weighted averages of the seasonally adjusted component series. Seasonally adjusted data are not published for a small number of series characterized by small seasonal components relative to their trend and/or irregular components. However, these series are used in aggregating to broader seasonally adjusted levels.

Revisions of the seasonally adjusted data, usually for the most recent five-year period, are made once a year coincident with the benchmark revisions. This means that these revisions typically extend back farther than the benchmark revisions.

Data availability

Employment data by industry division are available beginning with 1919. Data for each month usually are released on the first Friday of the following month in a press release that also contains data from the household survey (Tables 10-1 through 10-5). (The release date actually is determined by the timing of the survey week.) The *Monthly Labor Review* frequently contains articles analyzing developments in the labor force, employment, and unemployment. The *Monthly Labor Review*, press releases, and complete historical data are available on the BLS Web site at <http://www.bls.gov>.

Benchmark revisions and revised seasonally adjusted data for recent years are made each year with the release of January data in early February. Before 2004, the benchmark revisions were not made until June; the acceleration is due to earlier availability of the benchmark UI (ES-202) data.

References

References can be found at <http://www.bls.gov/ces> under the headings "Special Notices," "Benchmark Information," and "Technical Notes." Extensive changes incorporated in June 2003 are described in "Recent Changes in the National Current Employment Statistics Survey," *Monthly Labor Review*, June 2003. The latest benchmark revision is discussed in an article available on the Website. See also Chapter 2, "Employment, Hours, and Earnings from the Establishment Survey," *BLS Handbook of Methods*, Bulletin 2490 (April 1997).

TABLES 10-8 THROUGH 10-10, 10-14, 10-15, 15-3 AND 15-6

Average Hours Per Week; Aggregate Employee Hours

SOURCE: U.S. DEPARTMENT OF LABOR, BUREAU OF LABOR STATISTICS (BLS)

See the notes and definitions to Tables 10-7 and related tables for an overall description of the "establishment" or "payroll" survey that is the source of these earnings data.

Hours and earnings have been reported for production and non-supervisory workers, as defined above, since the inception of the CES. Beginning with data for March 2006, such data have also been collected for all payroll employees. With sufficient history for calculation of seasonal adjustment factors, BLS began publishing all-employee hours and earnings in February 2010, and these new data are presented in *Business Statistics* Tables 10-9 through 10-12.

Definitions

Average weekly hours represents the average hours paid per worker during the pay period that includes the 12th of the month. Included are hours paid for holidays and vacations, as well as those paid for sick leave when pay is received directly from the firm.

Average weekly hours are different from standard or scheduled hours. Factors such as unpaid absenteeism, labor turnover, part-time work, and work stoppages can cause average weekly hours to be lower than scheduled hours of work for an establishment.

An important characteristic of these data is that average weekly hours pertain to jobs, not to persons; thus, a person with half-time jobs in two different establishments is represented in this series as two jobs that have 20-hour workweeks, not as one person with a 40-hour workweek.

Overtime hours represent the portion of average weekly hours worked in excess of regular hours, for which overtime premiums were paid. Weekend and holiday hours are included only if overtime premiums were paid. Hours for which only shift differential, hazard, incentive, or other similar types of premiums were paid are excluded.

Aggregate hours provide measures of changes over time in labor input to the industry, in index-number form.

Indexes of aggregate weekly hours are calculated by dividing the current month's aggregate hours by the average of the 12 monthly figures, for the base year. The index of aggregate hours paints a good picture of the stall in the recovery. Employment is up, but hours are not up proportionally. This reflects the trend to part-time workers and the reduction of hours in part-time workers. For the series covering production and nonsupervisory workers, the base period is 2002; for the all-employee series, the base period is 2007.

Notes on the data

Benchmark adjustments. Independent benchmarks are not available for the hours and earnings series. However, at the time of the annual adjustment of the employment series to new benchmarks, the levels of hours and earnings may be affected by the revised employment weights (which are used in computing the industry averages for hours and earnings), as well as by the changes in seasonal adjustment factors introduced with the benchmark revision.

Method of computing industry series. "Average weekly hours" for individual industries are computed by dividing worker hours (reported by establishments classified in each industry) by the number of workers reported for the same establishments. Estimates for divisions and major industry groups are averages (weighted by employment) of the figures for component industries.

Seasonal adjustment. Hours and earnings series are seasonally adjusted by applying factors directly to the corresponding unadjusted series. Data for some industries are not seasonally adjusted because the seasonal component is small relative to the trendcycle and/or irregular components. Consequently, they cannot be separated with sufficient precision.

Special adjustments are made to average weekly hours to account for the presence or absence of religious holidays in the April survey reference period and the occasional occurrence of Labor Day in the September reference period. In addition, REGARIMA modeling is used prior to seasonal adjustment to correct for reporting and processing errors associated with the number of weekdays in a month (rather than to correct for the 4- and 5-week effect, which is less significant for hours than it is for employment). This is of particular importance for average weekly hours in the service-providing industries other than retail trade. For this reason, BLS advises that calculations of over-the-year changes (for example, the change for the current month from a year earlier) should use seasonally adjusted data, since the actual not-seasonally-adjusted monthly data may be distorted.

Data availability

See data availability for Tables 10-7 and related, above.

References

See references for Tables 10-7 and related, above.

TABLES 10-8, 10-11, 10-12, 10-16, 10-17, 15-4, AND 15-5

Hourly and Weekly Earnings

SOURCE: U.S. DEPARTMENT OF LABOR, BUREAU OF LABOR STATISTICS (BLS)

See the notes and definitions to Tables 10-8 and related for an overall description of the "establishment" or "payroll" survey that is the source of these earnings data.

Hours and earnings have been reported for production and non-supervisory workers, as defined above, since the inception of the CES. Beginning with data for March 2006, such data have also been collected for all payroll employees. With sufficient history for calculation of seasonal adjustment factors, BLS began publishing all-employee hours and earnings in February 2010, and these data are presented in new *Business Statistics* Tables 10-9 through 10-12.

Definitions

Earnings are the payments that workers receive during the survey period (before deductions for taxes and other items), including premium pay for overtime or late-shift work but excluding irregular bonuses and other special payments. After being previously excluded, tips were asked to be reported beginning in September 2005. This made little difference in most industries, and BLS asserts that many respondents had already been including tips. In two industries, full-service restaurants and cafeterias, there was a substantial difference, and the historical earnings data for those industries have been reconstructed to reflect the new higher level of earnings.

Notes on the data

The hours and earnings series are based on reports of gross payroll and corresponding paid hours for full and parttime workers who received pay for any part of the pay period that included the 12th of the month.

Total payrolls are before deductions, such as for the employee share of oldage and unemployment insurance, group insurance, withholding taxes, bonds, and union dues. The payroll figures also include pay for overtime, holidays, vacations, and sick leave (paid directly by the employer for the period reported). Excluded from the payroll figures are fringe benefits (health and other types of insurance and contributions to retirement, paid by the employer, and the employer share of payroll taxes), bonuses (unless earned and paid regularly each pay period), other pay not earned in the pay period reported (retroactive pay), and the value of free rent, fuel, meals, or other payment-in-kind.

Average hourly earnings data reflect not only changes in basic hourly and incentive wage rates, but also such variable factors as premium pay for overtime and lateshift work and changes in output of workers paid on an incentive basis. Shifts in the volume of employment between relatively highpaid and lowpaid work also affect the general average of hourly earnings.

Averages of hourly earnings should not be confused with wage rates, which represent the rates stipulated for a given unit of work or time, while earnings refer to the actual return to the worker for a stated period of time. The earnings series do not represent total labor cost to the employer because of the inclusion of tips and the exclusion of irregular bonuses, retroactive items, the cost of employer-provided benefits, and payroll taxes paid by employers.

Average weekly earnings are not the amounts available to workers for spending, since they do not reflect deductions such as income taxes and Social Security taxes. It is also important to understand that average weekly earnings represent earnings per job, not per worker (since a worker may have more than one job) and not per family (since a family may have more than one worker). A person with two half-time jobs will be reflected as two earners with low weekly earnings rather than as one person with the total earnings from his or her two jobs.

Method of computing industry series. Average hourly earnings are obtained by dividing the reported total worker payroll by total worker hours. Estimates for both hours and hourly earnings for nonfarm divisions and major industry groups are employment-weighted averages of the figures for component industries.

Average weekly earnings are computed by multiplying average hourly earnings by average weekly hours. In addition to the factors mentioned above, which exert varying influences upon average hourly earnings, average weekly earnings are affected by changes in the length of the workweek, parttime work, work stoppages, labor turnover, and absenteeism. Persistent longterm uptrends in the proportion of parttime workers in retail trade and many of the service industries have reduced average workweeks (as measured here), and have similarly affected the average weekly earnings series.

Benchmark adjustments. Independent benchmarks are not available for the hours and earnings series. At the time of the annual adjustment of the employment series to new benchmarks, the levels of hours and earnings may be affected by the revised employment weights (which are used in computing the industry averages for hours and earnings), as well as by the changes in seasonal adjustment factors that were also introduced with the benchmark revision.

Seasonal adjustment. Hours and earnings series are seasonally adjusted by applying factors directly to the corresponding unadjusted series; seasonally adjusted average weekly earnings are the product of seasonally adjusted hourly earnings and weekly hours.

REGARIMA modeling is used to correct for reporting and processing errors associated with variations in the number of weekdays in a month (rather than for the 4- and 5-week effect, which is less significant for earnings than for employment). This is of particular importance for average hourly earnings in wholesale trade, financial activities, professional and business services, and other services. For this reason, BLS advises that calculations of over-the-year changes, for example the change for the current month from a year earlier, should use seasonally adjusted data, since the actual not seasonally adjusted monthly data may be distorted.

Data availability

See data availability for Tables 10-7 and related, above.

References

See references for Tables 10-7 and related, above.

TABLE 10-18

Hours at Work, Current Population Survey

This table presents annual average measures of hours worked per week reported by workers in the Current Population Survey, in response to questions asking for the hours worked in the survey week by each worker in the household on all his or her jobs combined. These measures are not published in regular BLS reports but are available on request. *Business Statistics* obtained these numbers from BLS staff and presents them here in annual average form.

Do CPS respondents overestimate the number of hours they work? One study made that claim, but a careful re-examination of the data by Harley Frazis and Jay Stewart in the June 2014 *Monthly Labor Review* indicates that respondents slightly under-report the hours that they actually worked during the reference week. (On the other hand, the authors also note that the reference week overestimates the average for the entire month because of the absence of holidays.) (*Monthly Labor Review* is available online at www.bls.gov.) In any event, the CPS data give a more accurate picture of trends in the workweeks of typical American workers than does the establishment (CES) survey.

TABLE 10-19

Median Usual Weekly Earnings of Full-Time Wage and Salary Workers

SOURCE: U.S. DEPARTMENT OF LABOR, BUREAU OF LABOR STATISTICS

These data are from the Current Population Survey, which was described in the notes to Tables 10-1 through 10-6. Because they are earnings per worker, not per job, and are limited to full-time workers, the data are not distorted by the increasing proportion of part-time workers, as the CES earnings data are.

Definitions

Full-time wage and salary workers are those workers reported as "employed" in the CPS who receive wages, salaries, commissions, tips, pay in kind, or piece rates, and usually work 35 hours or more per week at their sole or principal job. Both private-sector and public-sector employees are included. All self-employed persons are excluded (even those whose businesses are incorporated).

Usual weekly earnings are earnings before taxes and other deductions and include any overtime pay, commissions, or tips usually received. In the case of multiple jobholders they refer to the main job. The wording of the question was changed in January 1994 to better deal with persons who found it easier to report earnings on other than a weekly basis. Such reports are then converted to the weekly equivalent. According to BLS, "the term 'usual' is as perceived by the respondent. If the respondent asks for a definition of usual, interviewers are instructed to define the term as more than half the weeks worked during the past 4 or 5 months."

The *median* is the amount that divides a given earnings distribution into two equal groups, one having earnings above the median and the other having earnings below the median.

Race and ethnicity. See the notes to Tables 10-1 through 10-6 for the definitions of these categories.

Data availability

These data become available about 3 weeks after the end of each quarter in the "Usual Weekly Earnings of Wage and Salary Workers" press release, available on the BLS Web site at <http://www.bls.gov/cps>. Recent data are available at that location. Also available are greater detail by demographic and age groups, by occupation, by union status, and by education; distributional data, by deciles and quartiles; and earnings for part-time workers. Also available are earnings in 1982 dollars using the CPI-U. (In the opinion of the editor, these give a less accurate depiction of longer-term trends, which is why *Business Statistics* provides the CPI-U-RS data explained above.) Historical data are available upon request from BLS by telephone at (202) 691-6555.

CHAPTER 11: ENERGY

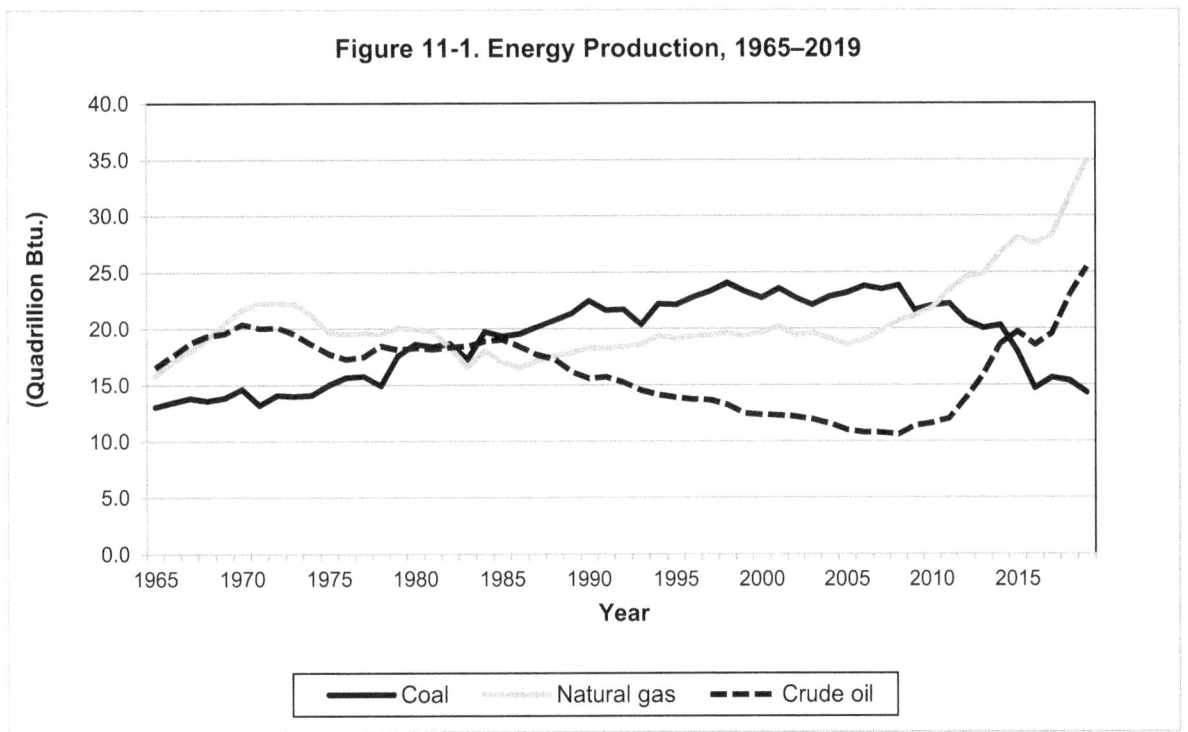

Figure 11-1. Energy Production, 1965–2019

- As Figure 11–1 shows, in 1982, energy production by fossil fuels was centered around coal (18.64 quadrillion Btu), crude oil (18.31 quadrillion Btu) and natural gas (18.31 quadrillion Btu). However, since then, their trends of production diverged widely. Coal production ramped up, crude oil fell precipitately and natural gas fell slightly before shooting up to a high of 34.9 quadrillion Btu in 2019. Crude oil rebounded from a low of 10.61 quadrillion Btu in 2008 then topped 25.4 quadrillion Btu in 2019. Meanwhile, coal production's peaked in 1998 at 24.0 quadrillion Btu but declined to 14.3 quadrillion Btu in 2019. (Table 11-01)

- The first commercial electricity-generating plant powered by nuclear energy was located in Shippingport, Pennsylvania. In 1965, nuclear-supplied electricity started at slightly above zero and has grown to 8.462 quadrillion Btu in 2019. (Table 11-01)

- Energy intensity is measured as energy consumption per unit of gross domestic product (GDP). The U.S. energy intensity has fallen to $5.25, a steady decline over the decades. As a service-focused economy instead of a manufacturing-focused economy, the U.S. uses less energy per dollar of GDP. (Table 11-02)

- Net imports of petroleum and products peaked in 2005 at 12.55 million barrels per day. By 2019, net imports plummeted to 594,000 barrels. Meanwhile, exports began rising steadily in 2003 growing from over 1 million barrels per day to per day to nearly 8.5 million barrels per day in 2019. (Table 11-03)

Table 11-1. Energy Supply and Consumption

(Quadrillion Btu.)

Year and month	Imports	Exports	Production, by source								Consumption, by end-use sector			
			Total	Fossil fuels					Nuclear electric power	Renew-able energy, total	Total	Residential and commercial	Industrial	Transport-ation
				Total	Coal	Natural gas	Crude oil	Natural gas plant liquids						
1965	5.892	1.829	50.644	47.205	13.055	15.775	16.521	1.853	0.043	3.396	53.953	16.485	25.035	12.399
1966	6.146	1.829	53.507	50.011	13.468	17.011	17.561	1.971	0.064	3.432	56.949	17.493	26.356	13.068
1967	6.159	2.115	56.347	52.568	13.825	17.943	18.651	2.148	0.088	3.690	58.895	18.539	26.604	13.718
1968	6.905	1.998	58.188	54.274	13.609	19.068	19.308	2.289	0.142	3.773	62.401	19.666	27.869	14.831
1969	7.676	2.126	60.501	56.252	13.863	20.446	19.556	2.386	0.154	4.095	65.596	21.003	29.087	15.471
1970	8.342	2.632	63.462	59.152	14.607	21.666	20.401	2.478	0.239	4.070	67.817	22.113	29.605	16.062
1971	9.535	2.151	62.686	58.011	13.186	22.280	20.033	2.513	0.413	4.262	69.260	22.968	29.562	16.694
1972	11.387	2.118	63.870	58.904	14.092	22.208	20.041	2.564	0.584	4.382	72.660	24.041	30.901	17.682
1973	14.613	2.033	63.528	58.207	13.992	22.187	19.493	2.535	0.910	4.411	75.652	24.441	32.591	18.577
1974	14.304	2.203	62.308	56.294	14.074	21.210	18.575	2.435	1.272	4.742	73.929	24.049	31.753	18.087
1975	14.032	2.323	61.284	54.697	14.989	19.640	17.729	2.338	1.900	4.687	71.931	24.307	29.379	18.211
1976	16.760	2.172	61.522	54.684	15.654	19.480	17.262	2.288	2.111	4.727	75.939	25.474	31.356	19.067
1977	19.948	2.052	61.970	55.059	15.755	19.565	17.454	2.286	2.702	4.209	77.920	25.870	32.221	19.786
1978	19.106	1.920	63.060	55.031	14.910	19.485	18.434	2.202	3.024	5.005	79.907	26.645	32.643	20.583
1979	19.460	2.855	65.851	57.952	17.540	20.076	18.104	2.232	2.776	5.123	80.812	26.461	33.877	20.437
1980	15.796	3.695	67.147	58.979	18.598	19.908	18.249	2.225	2.739	5.428	78.021	26.332	31.993	19.659
1981	13.719	4.307	66.910	58.488	18.377	19.699	18.146	2.266	3.008	5.414	76.057	25.878	30.662	19.478
1982	11.861	4.608	66.527	57.416	18.639	18.319	18.309	2.149	3.131	5.980	73.046	26.392	27.561	19.052
1983	11.752	3.693	64.066	54.368	17.247	16.593	18.392	2.136	3.203	6.496	72.915	26.364	27.372	19.134
1984	12.471	3.786	68.800	58.809	19.719	18.008	18.848	2.234	3.553	6.438	76.571	27.404	29.508	19.609
1985	11.781	4.196	67.661	57.502	19.325	16.980	18.992	2.204	4.076	6.084	76.334	27.493	28.757	20.042
1986	14.151	4.021	67.030	56.539	19.509	16.541	18.376	2.113	4.380	6.111	76.599	27.582	28.225	20.740
1987	15.398	3.812	67.506	57.131	20.141	17.136	17.675	2.179	4.754	5.622	79.008	28.210	29.332	21.419
1988	17.296	4.366	68.890	57.846	20.738	17.599	17.279	2.231	5.587	5.457	82.659	29.711	30.627	22.267
1989	18.766	4.661	69.285	57.448	21.360	17.847	16.117	2.123	5.602	6.235	84.740	30.987	31.282	22.425
1990	18.817	4.752	70.668	58.523	22.488	18.326	15.571	2.138	6.104	6.040	84.433	30.258	31.749	22.366
1991	18.335	5.141	70.321	58.031	21.636	18.229	15.701	2.265	6.422	6.068	84.380	30.920	31.342	22.065
1992	19.372	4.937	69.914	57.615	21.694	18.375	15.223	2.323	6.479	5.821	85.725	30.796	32.513	22.363
1993	21.218	4.227	68.273	55.780	20.336	18.584	14.494	2.366	6.410	6.082	87.266	32.029	32.559	22.618
1994	22.307	4.035	70.683	58.002	22.202	19.348	14.103	2.349	6.694	5.987	88.983	32.208	33.462	23.264
1995	22.180	4.496	71.129	57.496	22.130	19.082	13.887	2.398	7.075	6.557	90.931	33.207	33.908	23.757
1996	23.633	4.613	72.435	58.338	22.790	19.344	13.723	2.480	7.087	7.011	93.935	34.674	34.836	24.365
1997	25.119	4.493	72.420	58.806	23.310	19.394	13.658	2.445	6.597	7.017	94.507	34.644	35.134	24.668
1998	26.473	4.237	72.826	59.266	24.045	19.613	13.235	2.372	7.068	6.493	94.920	34.920	34.779	25.169
1999	27.152	3.669	71.686	57.560	23.295	19.341	12.451	2.473	7.610	6.516	96.545	35.931	34.692	25.859
2000	28.865	3.962	71.271	57.307	22.735	19.662	12.358	2.551	7.862	6.102	98.702	37.597	34.587	26.456
2001	30.052	3.731	71.675	58.485	23.547	20.166	12.282	2.491	8.029	5.162	96.064	37.175	32.653	26.179
2002	29.331	3.608	70.653	56.777	22.732	19.382	12.160	2.502	8.145	5.731	97.535	38.132	32.590	26.747
2003	31.007	4.013	69.885	55.883	22.094	19.633	11.960	2.296	7.960	5.942	97.835	38.466	32.489	26.807
2004	33.492	4.351	70.169	55.884	22.852	19.074	11.550	2.408	8.223	6.063	100.000	38.738	33.444	27.748
2005	34.659	4.462	69.377	54.995	23.185	18.556	10.974	2.280	8.161	6.221	100.100	39.467	32.374	28.179
2006	34.649	4.727	70.678	55.877	23.790	19.022	10.767	2.299	8.215	6.586	99.392	38.378	32.317	28.618
2007	34.679	5.338	71.338	56.369	23.493	19.786	10.741	2.349	8.459	6.510	100.890	39.773	32.306	28.727
2008	32.970	6.949	73.145	57.527	23.851	20.703	10.613	2.359	8.426	7.192	98.754	40.071	31.261	27.339
2009	29.690	6.920	72.592	56.612	21.624	21.139	11.340	2.508	8.355	7.625	93.942	38.970	28.380	26.510
2010	29.866	8.176	74.907	58.159	22.038	21.806	11.610	2.705	8.434	8.314	97.517	39.954	30.578	26.897
2011	28.748	10.373	78.082	60.513	22.221	23.406	11.996	2.890	8.269	9.300	96.850	39.363	30.881	26.518
2012	27.068	11.267	79.234	62.286	20.677	24.610	13.837	3.162	8.062	8.886	94.380	37.292	30.961	26.050
2013	24.623	11.788	81.837	64.174	20.001	24.859	15.862	3.451	8.244	9.418	97.117	38.981	31.525	26.533
2014	23.241	12.270	87.715	69.611	20.286	26.718	18.602	4.005	8.338	9.767	98.276	39.710	31.691	26.789
2015	23.794	12.902	88.250	70.185	17.946	28.067	19.696	4.476	8.337	9.729	97.378	38.775	31.364	27.161
2016	25.378	14.119	84.269	65.420	14.667	27.576	18.512	4.665	8.427	10.423	97.329	38.207	31.341	27.710
2017	25.457	17.946	88.052	68.437	15.625	28.289	19.535	4.987	8.419	11.196	97.603	37.782	31.806	27.939
2018	24.833	21.208	95.616	75.670	15.363	31.690	22.890	5.727	8.438	11.508	101.080	39.941	32.700	28.375
2019	22.804	23.519	101.040	80.948	14.268	34.902	25.440	6.337	8.462	11.637	100.160	39.388	32.501	28.206
2018														
January	2.228	1.575	7.738	5.985	1.262	2.522	1.772	0.429	0.780	0.972	9.651	4.590	2.791	2.267
February	1.861	1.526	7.198	5.603	1.225	2.327	1.643	0.408	0.677	0.918	8.051	3.504	2.490	2.060
March	2.114	1.731	7.996	6.284	1.332	2.627	1.858	0.468	0.701	1.011	8.700	3.516	2.767	2.421
April	2.125	1.793	7.602	5.966	1.178	2.527	1.799	0.462	0.618	1.018	7.876	2.972	2.586	2.324
May	2.142	1.781	7.964	6.211	1.241	2.635	1.850	0.484	0.704	1.049	7.972	2.771	2.756	2.447
June	2.176	1.763	7.853	6.095	1.251	2.554	1.823	0.467	0.729	1.030	8.135	2.987	2.707	2.438
July	2.161	1.854	8.068	6.366	1.280	2.664	1.926	0.496	0.758	0.945	8.599	3.288	2.799	2.507
August	2.192	1.738	8.352	6.648	1.408	2.718	2.010	0.512	0.756	0.949	8.677	3.239	2.881	2.551
September	1.999	1.718	7.978	6.436	1.268	2.700	1.968	0.500	0.677	0.865	7.849	2.853	2.674	2.321
October	1.982	1.892	8.235	6.713	1.350	2.795	2.057	0.511	0.621	0.902	8.065	2.876	2.781	2.411
November	1.896	1.882	8.168	6.594	1.278	2.771	2.054	0.492	0.669	0.905	8.498	3.423	2.738	2.339
December	1.958	1.955	8.461	6.769	1.291	2.850	2.129	0.499	0.749	0.943	9.012	3.922	2.729	2.363
2019														
January	2.111	1.919	8.518	6.782	1.331	2.849	2.094	0.508	0.771	0.965	9.508	4.369	2.850	2.287
February	1.696	1.752	7.673	6.112	1.179	2.596	1.862	0.475	0.677	0.885	8.357	3.749	2.505	2.105
March	1.916	1.914	8.330	6.646	1.126	2.891	2.101	0.529	0.680	1.004	8.677	3.612	2.709	2.358
April	1.925	1.893	8.309	6.636	1.233	2.812	2.072	0.518	0.633	1.040	7.623	2.720	2.593	2.314
May	2.033	1.943	8.591	6.818	1.247	2.891	2.140	0.541	0.702	1.071	7.908	2.766	2.729	2.416
June	1.882	1.946	8.290	6.561	1.144	2.837	2.062	0.519	0.719	1.009	7.903	2.836	2.650	2.416
July	2.014	1.879	8.488	6.740	1.195	2.934	2.088	0.523	0.755	0.992	8.563	3.292	2.780	2.486
August	1.973	1.967	8.705	7.005	1.289	3.000	2.188	0.529	0.752	0.948	8.550	3.211	2.798	2.536
September	1.785	1.979	8.380	6.793	1.183	2.936	2.133	0.540	0.691	0.897	7.858	2.874	2.679	2.303
October	1.815	2.064	8.626	7.044	1.165	3.079	2.239	0.562	0.649	0.933	7.959	2.791	2.733	2.437
November	1.683	2.028	8.427	6.832	1.100	2.994	2.199	0.538	0.670	0.924	8.358	3.346	2.719	2.294
December	1.972	2.234	8.710	6.978	1.076	3.082	2.263	0.556	0.764	0.969	8.901	3.820	2.756	2.329

Table 11-2. Energy Consumption Per Dollar of Real Gross Domestic Product

Year	Primary energy consumption (quadrillion Btu)			Gross domestic product (billions of chained [2012] dollars)	Energy intensity per real dollar of GDP (thousand Btu per chained [2012] dollar)		
	Total	Petroleum and natural gas	Other energy		Total	Petroleum and natural gas	Other energy
1950	35.531	19.267	16.264	2 289.5	15.85	8.83	7.02
1951	38.741	21.456	17.284	2 473.8	15.67	9.10	6.57
1952	37.905	22.484	15.421	2 574.9	14.96	9.16	5.80
1953	38.170	23.439	14.731	2 695.6	14.65	9.12	5.52
1954	36.506	24.144	12.363	2 680.0	14.33	9.45	4.88
1955	40.131	26.223	13.908	2 871.2	14.68	9.58	5.09
1956	42.604	27.519	15.085	2 932.4	14.93	9.85	5.08
1957	42.968	28.089	14.879	2 994.1	14.63	9.85	4.78
1958	40.122	29.154	10.968	2 972.0	14.69	10.30	4.39
1959	41.936	30.997	10.938	3 178.2	14.34	10.24	4.10
1960	42.789	32.259	10.530	3 260.0	14.50	10.39	4.11
1961	43.261	33.096	10.165	3 343.5	14.35	10.40	3.95
1962	44.854	34.728	10.125	3 548.4	14.14	10.28	3.86
1963	47.149	36.049	11.101	3 702.9	14.06	10.23	3.84
1964	49.028	37.530	11.499	3 916.3	13.88	10.07	3.81
1965	50.644	38.953	11.692	4 170.8	13.58	9.81	3.77
1966	53.507	41.331	12.176	4 445.9	13.45	9.77	3.68
1967	56.347	43.219	13.128	4 567.8	13.53	9.93	3.60
1968	58.188	46.176	12.012	4 792.3	13.66	10.11	3.55
1969	60.501	48.999	11.502	4 942.1	13.92	10.40	3.52
1970	63.462	51.294	12.168	4 951.3	14.37	10.87	3.50
1971	62.686	53.008	9.678	5 114.3	14.20	10.87	3.33
1972	63.870	55.617	8.253	5 383.3	14.16	10.84	3.32
1973	63.528	57.318	6.210	5 687.2	13.95	10.57	3.38
1974	62.308	55.153	7.155	5 656.5	13.71	10.23	3.48
1975	61.284	52.647	8.637	5 644.8	13.36	9.78	3.58
1976	61.522	55.487	6.034	5 949.0	13.39	9.78	3.60
1977	61.970	57.013	4.957	6 224.1	13.13	9.61	3.52
1978	63.060	57.920	5.141	6 568.6	12.76	9.25	3.51
1979	65.851	57.741	8.109	6 776.6	12.50	8.94	3.57
1980	67.147	54.394	12.752	6 759.2	12.10	8.44	3.66
1981	66.910	51.631	15.279	6 930.7	11.50	7.81	3.69
1982	66.527	48.535	17.991	6 805.8	11.26	7.49	3.78
1983	64.066	47.218	16.848	7 117.7	10.74	6.96	3.78
1984	68.800	49.386	19.414	7 632.8	10.52	6.79	3.73
1985	67.661	48.570	19.091	7 951.1	10.06	6.40	3.66
1986	67.030	48.742	18.289	8 226.4	9.75	6.21	3.54
1987	67.506	50.457	17.049	8 511.0	9.72	6.21	3.51
1988	68.890	52.621	16.268	8 866.5	9.76	6.22	3.54
1989	69.285	53.766	15.519	9 192.1	9.65	6.12	3.53
1990	70.668	53.103	17.564	9 365.5	9.43	5.94	3.50
1991	70.321	52.822	17.499	9 355.4	9.44	5.91	3.53
1992	69.914	54.181	15.733	9 684.9	9.26	5.85	3.40
1993	68.273	54.817	13.456	9 951.5	9.18	5.77	3.41
1994	70.683	56.181	14.502	10 352.4	8.99	5.68	3.31
1995	71.129	57.012	14.117	10 630.3	8.95	5.61	3.33
1996	72.435	58.674	13.762	11 031.4	8.90	5.56	3.34
1997	72.420	59.288	13.132	11 521.9	8.21	5.15	3.06
1998	72.826	59.550	13.276	12 038.3	7.89	4.95	2.94
1999	71.686	60.641	11.044	12 610.5	7.66	4.82	2.85
2000	71.271	61.976	9.295	13 131.0	7.53	4.73	2.80
2001	71.675	60.856	10.819	13 262.1	7.25	4.60	2.65
2002	70.653	61.628	9.026	13 493.1	7.24	4.58	2.66
2003	69.885	61.538	8.347	13 879.1	7.06	4.44	2.62
2004	70.169	63.062	7.107	14 406.4	6.95	4.38	2.56
2005	69.377	62.782	6.595	14 912.5	6.72	4.22	2.50
2006	70.678	61.969	8.708	15 338.3	6.49	4.05	2.44
2007	71.338	63.030	8.308	15 626.0	6.47	4.04	2.42
2008	73.145	60.612	12.533	15 604.7	6.34	4.06	2.45
2009	72.592	58.195	14.397	15 208.8	6.19	3.97	2.35
2010	74.907	59.899	15.008	15 598.8	6.25	3.85	2.40
2011	78.082	59.581	18.500	15 840.7	6.12	3.79	2.34
2012	79.234	59.928	19.306	16 197.0	5.83	3.60	2.12
2013	81.837	61.204	20.633	16 495.4	5.89	3.11	2.17
2014	87.715	62.040	25.675	16 912.0	5.82	3.69	2.13
2015	88.250	63.562	24.689	17 403.8	5.60	3.30	2.31
2016	84.269	64.106	20.164	17 688.9	5.50	3.62	1.88
2017	88.052	64.106	23.945	18 108.1	5.39	3.54	1.85
2018	95.616	67.968	27.648	18 638.2	5.42	3.65	1.78
2019	101.047	68.816	32.231	19 073.1	5.25	3.61	1.64

Table 11-3. Petroleum and Petroleum Products—Prices, Imports, Domestic Production, and Stocks

(Not seasonally adjusted.)

Year and month	Crude oil futures price (dollars per barrel)		Imports				Supply (thousands of barrels per day)					Stocks (end of period, millions of barrels)		
			Total energy-related petroleum products (thousands of barrels)	Crude petroleum			Petroleum and products			Domestic production		Crude oil and petroleum products	Crude petroleum	
	Current dollars	2009 dollars		Thousands of barrels		Unit price (dollars per barrel)	Exports	Imports	Net imports	Crude oil	Natural gas plant liquids		Non-SPR	Strategic petroleum reserve
				Total	Average per day									
1983	30.66	69.52	. . .	1 293 819	3 545	29.51	739	5 051	4 312	8 688	1 559	1 454	344	379
1984	29.44	63.41	. . .	1 319 683	3 616	27.68	722	5 437	4 715	8 879	1 630	1 556	345	451
1985	27.89	57.07	. . .	1 260 856	3 454	26.20	781	5 067	4 286	8 971	1 609	1 519	321	493
1986	15.05	29.56	. . .	1 634 567	4 478	13.90	785	6 224	5 439	8 680	1 551	1 593	331	512
1987	19.15	36.38	. . .	1 744 977	4 781	16.80	764	6 678	5 914	8 349	1 595	1 607	349	541
1988	15.96	29.10	. . .	1 887 860	5 172	13.69	815	7 402	6 587	8 140	1 625	1 597	330	560
1989	19.58	34.69	. . .	2 146 552	5 881	16.49	859	8 061	7 202	7 613	1 546	1 581	341	580
1990	24.50	42.53	. . .	2 216 604	6 073	19.75	857	8 018	7 161	7 355	1 559	1 621	323	586
1991	21.50	37.24	2 828 953	2 146 064	5 880	17.46	1 001	7 627	6 626	7 417	1 659	1 617	325	569
1992	20.58	34.37	2 947 582	2 294 570	6 269	16.80	950	7 888	6 938	7 171	1 697	1 592	318	575
1993	18.48	29.82	3 257 008	2 543 374	6 968	15.13	1 003	8 620	7 618	6 847	1 736	1 647	335	587
1994	17.19	26.71	3 416 045	2 704 196	7 409	14.23	942	8 996	8 054	6 662	1 727	1 653	337	592
1995	18.40	27.76	3 361 882	2 767 312	7 582	15.81	949	8 835	7 886	6 560	1 762	1 563	303	592
1996	22.03	32.11	3 622 385	2 893 647	7 906	18.98	981	9 478	8 498	6 465	1 830	1 507	284	566
1997	20.61	28.95	3 802 574	3 069 430	8 409	17.67	1 003	10 162	9 158	6 452	1 817	1 560	305	563
1998	14.40	19.20	4 088 027	3 242 711	8 884	11.49	945	10 708	9 764	6 252	1 759	1 647	324	571
1999	19.30	24.44	4 081 181	3 228 092	8 844	15.76	940	10 852	9 912	5 881	1 850	1 493	284	567
2000	30.26	36.47	4 314 825	3 399 239	9 288	26.44	1 040	11 459	10 419	5 822	1 911	1 468	286	541
2001	25.95	30.48	4 475 026	3 471 067	9 510	21.40	971	11 871	10 900	5 801	1 868	1 586	312	550
2002	26.15	29.95	4 337 075	3 418 022	9 364	22.61	984	11 530	10 546	5 744	1 880	1 548	278	599
2003	30.99	34.41	4 654 638	3 676 005	10 071	26.98	1 027	12 264	11 238	5 649	1 719	1 568	269	638
2004	41.47	44.35	4 917 591	3 820 979	10 440	34.48	1 048	13 145	12 097	5 441	1 809	1 645	286	676
2005	56.70	58.57	5 004 339	3 754 671	10 287	46.81	1 165	13 714	12 549	5 184	1 717	1 682	308	685
2006	66.25	66.42	4 880 734	3 734 226	10 231	58.01	1 317	13 707	12 390	5 086	1 739	1 703	296	689
2007	72.41	71.01	4 807 811	3 690 568	10 111	64.28	1 433	13 468	12 036	5 074	1 783	1 648	268	697
2008	99.75	98.15	4 613 444	3 590 628	9 810	95.22	1 802	12 915	11 114	5 000	1 784	1 719	308	702
2009	62.09	62.09	4 266 007	3 314 787	9 082	56.93	2 024	11 691	9 667	5 357	1 910	1 758	307	727
2010	79.61	78.11	4 279 611	3 377 077	9 252	74.67	2 353	11 793	9 441	5 484	2 074	1 772	312	727
2011	95.11	91.25	4 164 117	3 321 918	9 101	99.82	2 986	11 436	8 450	5 667	2 216	1 725	308	696
2012	94.15	89.03	3 847 545	3 097 407	8 463	101.11	3 205	10 598	7 393	6 518	2 408	1 779	338	695
2013	98.05	91.17	3 549 945	2 813 770	7 709	96.95	3 621	9 859	6 237	7 493	2 606	1 728	327	696
2014	92.91	84.12	3 381 460	2 700 939	7 400	91.23	4 176	9 240	5 064	8 787	3 015	1 825	361	691
2015	48.79	44.54	3 384 807	2 661 968	7 293	47.26	4 738	9 449	4 650	9 439	3 342	1 982	449	695
2016	43.32	39.20	3 576 378	2 806 413	7 689	36.00	5 261	10 055	2 480	8 839	3 509	2 030	485	695
2017	50.80	45.09	3 639 778	2 883 753	7 901	45.97	6 376	10 075	3 723	9 352	3 736	1 895	421	662
2018	64.81	60.27	3 464 237	2 694 326	7 382	58.22	7 601	9 943	2 342	10 990	4 369	1 912	442	649
2019	56.99	51.97	3 116 908	2 374 322	6 505	53.30	8 499	9 093	594	12 232	4 813	1 916	433	635
2018														
January	63.55	59.17	310 368	241 687	7 796	54.78	6 615	10 274	3 659	9 995	3 825	1 879	420	664
February	62.16	57.79	250 538	191 349	6 834	54.83	6 844	9 580	2 736	10 248	4 023	1 876	424	665
March	62.87	58.42	284 948	219 289	7 074	54.02	7 105	9 821	2 716	10 461	4 173	1 862	423	665
April	66.33	61.49	297 300	236 137	7 871	54.50	7 730	10 364	2 634	10 475	4 260	1 864	435	664
May	69.89	64.66	301 707	232 049	7 485	58.38	7 517	10 228	2 712	10 464	4 321	1 870	433	660
June	67.32	62.21	296 507	237 340	7 911	62.46	7 801	10 706	2 905	10 672	4 326	1 867	415	660
July	70.74	65.28	308 284	241 638	7 795	64.54	7 827	10 176	2 349	10 936	4 411	1 872	409	660
August	67.85	62.56	316 245	241 921	7 804	62.64	7 043	10 432	3 389	11 325	4 570	1 892	407	660
September	70.07	64.52	284 919	213 648	7 122	61.40	7 611	9 885	2 273	11 470	4 631	1 932	416	660
October	70.76	65.03	293 280	227 924	7 352	61.25	8 018	9 417	1 399	11 559	4 580	1 916	432	655
November	56.60	51.99	261 160	211 282	7 043	57.54	8 669	9 213	545	11 926	4 571	1 910	449	650
December	48.68	44.69	258 982	200 062	6 454	50.26	8 250	9 022	772	11 963	4 479	1 912	442	649
2019														
January	51.38	43.22	282 790	223 855	7 221	42.63	8 044	9 690	1 646	11 360	4 545	1 098	449	649
February	54.95	46.30	226 060	173 678	6 203	46.91	8 404	8 626	222	11 181	4 706	1 101	452	649
March	58.15	48.63	251 749	195 255	6 299	53.06	7 929	8 837	908	11 411	4 728	1 108	459	649
April	63.86	53.21	270 681	205 633	6 854	57.17	8 440	9 504	1 063	11 647	4 787	1 117	469	649
May	60.83	50.51	289 879	219 518	7 081	60.56	8 149	9 796	1 647	11 639	4 838	1 125	480	645
June	54.66	45.29	259 032	204 160	6 805	59.18	8 654	9 234	581	11 605	4 793	1 109	464	645
July	57.35	47.37	285 560	213 008	6 871	56.44	8 011	9 547	1 535	11 375	4 679	1 087	442	645
August	54.81	45.17	265 629	199 720	6 443	54.15	8 424	9 356	931	12 003	4 727	1 076	431	645
September	56.95	46.85	242 939	185 307	6 177	53.10	8 678	8 666	-12	12 029	4 989	1 071	426	645
October	53.96	44.36	251 415	188 653	6 086	52.02	8 915	8 574	-340	12 199	5 022	1 085	444	641
November	57.03	46.78	228 703	166 418	5 547	51.91	8 757	8 056	-700	12 382	4 972	1 082	447	635
December	59.88	49.06	262 471	199 116	6 423	51.48	9 594	9 159	-435	12 332	4 971	1 068	433	635

. . . = Not available.

NOTES AND DEFINITIONS, CHAPTER 11

General Note on Energy Information

The U.S. Energy Information Administration (EIA) is the statistical and analytical agency within the U.S. Department of Energy. EIA collects and publishes statistics on petroleum, natural gas, electricity, consumption and efficiency, coal, renewable and alternative fuels, nuclear and uranium. EIA major publications are *Monthly Energy Review*, *Annual Energy Review* and *International Energy Outlook*.

TABLES 11-1 AND 11-2

Energy Supply and Consumption

SOURCES: U.S. DEPARTMENT OF ENERGY, ENERGY INFORMATION ADMINISTRATION; U.S. DEPARTMENT OF COMMERCE, BUREAU OF ECONOMIC ANALYSIS

Definitions

The *British thermal unit (Btu)* is a measure used to combine data for different energy sources into a consistent aggregate. It is the amount of energy required to raise the temperature of 1 pound of water 1 degree Fahrenheit when the water is near a temperature of 39.2 degrees Fahrenheit. To illustrate one of the factors used to convert volumes to Btu, conventional motor gasoline has a heat content of 5.253 million Btu per barrel. For further information, see the Energy Information Administration's *Monthly Energy Review*, Appendix A.

Production: Crude oil includes lease condensates.

Renewable energy, total includes conventional hydroelectric power, geothermal, solar thermal and photovoltaic, wind, and biomass. Hydroelectric power includes conventional electrical utility and industrial generation. Biomass includes wood, waste, and alcohol fuels (ethanol blended into motor gasoline).

The sum of domestic energy *production* and net imports of energy (*imports* minus *exports*) does not exactly equal domestic energy *consumption*. The difference is attributed to inventory changes; losses and gains in conversion, transportation, and distribution; the addition of blending compounds; shipments of anthracite to U.S. armed forces in Europe; and adjustments to account for discrepancies between reporting systems.

Consumption by end-use sector is based on total, not net, consumption--that is, each sector's consumption includes its electricity purchases as well as its own energy production. Electric utilities are not treated as a separate end-use sector. However, they are counted as primary producers in the production accounts, which measure only primary production. As a result, total national supply and consumption are in rough balance, though components may not add to exact totals, not only because of the miscellaneous adjustments described above, but also because of different sector-specific conversion factors.

References and notes on the data

All of these data are published each month in Tables 1.1, 1.2, 1.7, and 2.1 in the *Monthly Energy Review*. Annual data before 1973 are published in the *Annual Energy Review*. These two publications are no longer published in printed form but are available, along with all current and historical data, on the EIA Web site at <http://www.eia.doe.gov>.

The real gross domestic product (GDP) data used to calculate energy consumption per dollar of real GDP are from the Bureau of Economic Analysis; see Table 1-2 and the applicable notes and definitions in this volume of *Business Statistics*. The GDP numbers shown in this table reflect the comprehensive revision of July 2013, which increased the overall level of GDP by including as investment research, development, and creation of intellectual property. This changed the average level of the energy-to-GDP ratio but had little effect on its downward trend. The GDP estimate shown here also reflects subsequent revisions and updates made through June 2016.

TABLE 11-3

Petroleum and Petroleum Products—Prices, Imports, Domestic Production, and Stocks

SOURCES: FUTURES PRICES—U.S. DEPARTMENT OF ENERGY, ENERGY INFORMATION ADMINISTRATION (EIA), AND U.S. DEPARTMENT OF COMMERCE, BUREAU OF ECONOMIC ANALYSIS; IMPORTS—U.S. DEPARTMENT OF COMMERCE, CENSUS BUREAU (SEE NOTES AND DEFINITIONS FOR TABLES 7-9 THROUGH 7-14); SUPPLY (NET IMPORTS AND DOMESTIC PRODUCTION) AND STOCKS—EIA.

Definitions and notes on the data

The *crude oil futures price* in *current dollars per barrel* is the price for next-month delivery in Cushing, Oklahoma (a pipeline hub), of light, sweet crude oil, as determined by trading on the New York Mercantile Exchange (NYMEX). Official daily closing prices are reported each day at 2:30 p.m., and are tabulated weekly in Table 13 of *EIA's Weekly Petroleum Status Report*. The monthly averages shown in this volume are the average prices for the nearest future from each trading day of the month. For example, for most days in January, the futures contract priced will be for February; for the last few days in January, the February contract will have expired and the March contract will be quoted. The annual averages are averages of the monthly averages.

The *crude oil futures price* in *2009 dollars* is calculated by the editors, by dividing each month's current-dollar price by that month's chain price index for total personal consumption expenditures (PCE), with the price index average for the year 2009 set at 1.0000. The PCE chain price index is compiled by the Bureau of Economic Analysis (BEA). It is described in the notes and definitions for Chapter 1 and is also presented in Chapter 8, Table 8-2, and discussed in its notes and definitions.

The import data in Columns 3 through 6 of this table are those published as Exhibit 17, "Imports of Energy-related Petroleum Products, including Crude Petroleum," in the monthly Census-BEA foreign trade press release, FT900. *Total energy-related petroleum products* includes the following Standard International Trade Classification (SITC) commodity groupings: crude oil, petroleum preparations, and liquefied propane and butane gas.

The data in Columns 7 through 11, on exports, imports, and net imports (imports minus exports) of petroleum and products and domestic production of crude oil and natural gas plant liquids (all expressed as thousands of barrels per day), and in Columns 12 through 14, depicting stocks of crude oil in millions of barrels, are derived from the Department of Energy's weekly petroleum supply reporting system. They are published in EIA's *Monthly Energy Review*, Tables 3.1 and 3.4, which can be reached by searching the Web site. Stock totals are as of the end of the period. Geographic coverage includes the 50 states and the District of Columbia.

Data availability

Data on futures prices, petroleum supply and stocks are available from the EIA Web site at <http://www.eia.doe.gov>, under the categories "Publications and Reports/Monthly Energy Review" and "Petroleum/Weekly Petroleum Status Report." The *Monthly Energy Review* is no longer published in printed form.

The import data are available in the FT900 report from the Census Bureau at www.census.gov. See the notes and definitions for Tables 7-9 through 7-14 for further information.

CHAPTER 12: MONEY, INTEREST, ASSETS, LIABILITIES, AND ASSET PRICES

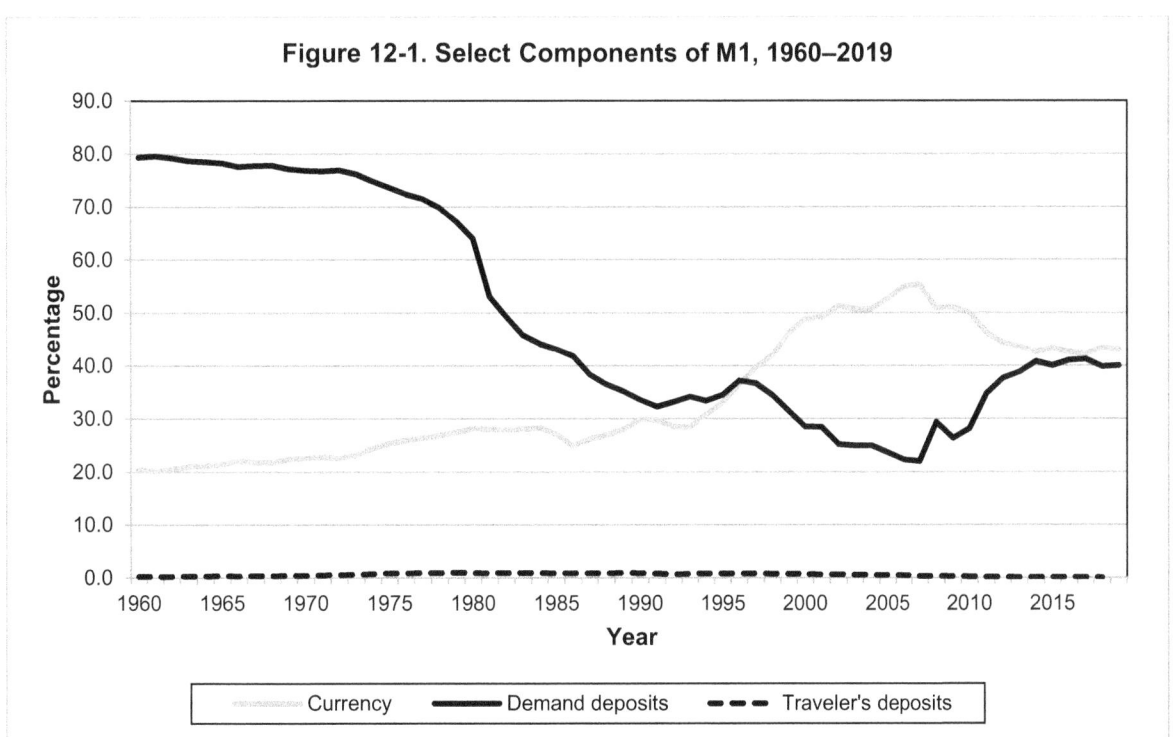

Figure 12-1. Select Components of M1, 1960–2019

- Bank notes initially started with currency such as gold and silver coins. Demand deposits, or checking accounts, began after a 10 percent tax was instituted on the issuance of state bank notes in 1865. As new money instruments were introduced, such as Negotiable Order of Withdrawal (NOW), defining the U.S. money supply took on more complexity.

- As depicted in Figure 12-1, demand deposits accounted for far less of M1 in 2019 than they did in 1960. While demand deposits consisted of 79.3 percent of M1 in 1960, they only accounted for 22.0 percent in 2007. At the same time, currency's share increased from 20.4 percent in 1960 to a high of 55.4 percent in 2007. By 2019, it declined to 43 percent. Traveler's checks have always represented a small part of M1 and in recent years, their use has declined even more. (Table 12-1)

- In response to banks' falling deposits, the Federal Reserve's emergency program of quantitative easing bought bonds to increase money supply and lower interest rates. The Federal Reserve added $1.3 trillion to its balance sheet during 2008 and another $1.8 trillion by the end of 2013. In October 2014, the Federal Reserve Board stopped buying bonds. Over the six years of the quantitative easing program, the Fed bought $4.5 trillion in bonds.

Table 12-1A. Money Stock Measures and Components of M1: Recent Data

(Billions of dollars, monthly data are averages of daily figures, annual data are for December.)

Year and month	Not seasonally adjusted		Seasonally adjusted					Other checkable deposits		
	M1	M2	M1	M2	Currency	Traveler's checks	Demand deposits	At commercial banks	At thrift institutions	Total
1965	172.6	463.1	167.8	459.2	36.0	0.5	131.3	0.0	0.1	0.1
1966	176.9	483.7	172.0	480.2	38.0	0.6	133.4	0.0	0.1	0.1
1967	188.4	528.0	183.3	524.8	40.0	0.6	142.5	0.0	0.1	0.1
1968	202.8	569.7	197.4	566.8	43.0	0.7	153.6	0.0	0.1	0.1
1969	209.3	590.1	203.9	587.9	45.7	0.8	157.3	0.0	0.1	0.1
1970	220.1	627.8	214.4	626.5	48.6	0.9	164.7	0.0	0.1	0.1
1971	234.5	711.2	228.3	710.3	52.0	1.0	175.1	0.0	0.2	0.2
1972	256.1	803.1	249.2	802.3	56.2	1.2	191.6	0.0	0.2	0.2
1973	270.2	856.5	262.9	855.5	60.8	1.4	200.3	0.0	0.3	0.3
1974	281.8	903.5	274.2	902.1	67.0	1.7	205.1	0.2	0.4	0.4
1975	295.3	1 017.8	287.1	1 016.2	72.8	2.1	211.3	0.4	0.5	0.9
1976	314.5	1 153.5	306.2	1 152.0	79.5	2.6	221.5	1.3	1.4	2.7
1977	340.0	1 273.0	330.9	1 270.3	87.4	2.9	236.4	1.8	2.3	4.2
1978	367.9	1 370.8	357.3	1 366.0	96.0	3.3	249.5	5.3	3.1	8.5
1979	393.2	1 479.0	381.8	1 473.7	104.8	3.5	256.6	12.7	4.2	16.8
1980	419.5	1 604.8	408.5	1 599.8	115.3	3.9	261.2	20.8	7.3	28.1
1981	447.0	1 760.3	436.7	1 755.5	122.5	4.1	231.4	63.0	15.6	78.7
1982	485.8	1 913.8	474.8	1 905.9	132.5	4.1	234.1	80.5	23.6	104.1
1983	533.3	2 134.0	521.4	2 123.5	146.2	4.7	238.5	97.3	34.8	132.1
1984	564.6	2 318.5	551.6	2 306.4	156.1	5.0	243.4	104.7	42.4	147.1
1985	633.3	2 504.1	619.8	2 492.1	167.7	5.6	266.9	124.7	54.9	179.5
1986	739.8	2 740.7	724.7	2 728.0	180.4	6.1	302.9	161.0	74.2	235.2
1987	765.4	2 838.3	750.2	2 826.4	196.7	6.6	287.7	178.2	81.0	259.2
1988	803.1	3 000.6	786.7	2 988.2	212.0	7.0	287.1	192.5	88.1	280.6
1989	810.6	3 165.6	792.9	3 152.5	222.3	6.9	278.6	197.4	87.7	285.1
1990	842.7	3 285.1	824.8	3 271.8	246.5	7.7	276.8	208.7	85.0	293.7
1991	915.6	3 386.4	897.0	3 372.2	267.1	7.7	289.6	241.6	90.9	332.5
1992	1 045.6	3 441.3	1 024.9	3 424.7	292.1	8.2	340.0	280.8	103.8	384.6
1993	1 153.3	3 495.1	1 129.6	3 474.5	321.6	8.0	385.4	302.6	112.0	414.6
1994	1 174.5	3 507.8	1 150.7	3 486.4	354.5	8.6	383.6	297.4	106.6	404.0
1995	1 152.7	3 653.3	1 127.5	3 629.5	372.8	9.0	389.0	249.0	107.6	356.6
1996	1 105.8	3 831.5	1 081.3	3 810.4	394.6	8.8	402.1	172.1	103.6	275.8
1997	1 097.5	4 045.0	1 072.3	4 022.8	425.3	8.4	393.5	148.4	96.7	245.1
1998	1 121.2	4 389.4	1 095.0	4 365.0	460.4	8.5	376.3	143.9	105.9	249.8
1999	1 148.2	4 654.0	1 122.2	4 627.4	517.9	8.6	352.5	139.7	103.6	243.3
2000	1 111.7	4 941.6	1 088.6	4 913.7	531.3	8.3	310.5	133.2	105.4	238.6
2001	1 208.5	5 452.7	1 183.2	5 421.6	581.2	8.0	336.3	142.2	115.6	257.7
2002	1 245.5	5 791.8	1 220.2	5 759.7	626.2	7.8	306.9	154.4	125.0	279.3
2003	1 332.5	6 085.5	1 306.2	6 054.2	662.4	7.7	325.9	175.3	134.8	310.1
2004	1 401.5	6 434.1	1 376.0	6 405.0	697.9	7.6	342.8	187.1	140.7	327.8
2005	1 397.2	6 695.1	1 374.3	6 668.0	724.7	7.2	324.2	180.6	137.7	318.3
2006	1 387.7	7 086.0	1 366.6	7 057.5	750.3	6.7	304.8	176.3	128.5	304.8
2007	1 394.9	7 490.5	1 373.4	7 458.0	760.6	6.3	301.7	171.9	132.9	304.8
2008	1 632.3	8 220.3	1 601.7	8 181.0	816.2	5.5	469.9	177.1	133.0	310.1
2009	1 724.4	8 531.0	1 692.8	8 483.4	863.7	5.1	445.9	229.1	149.0	378.1
2010	1 871.4	8 845.0	1 836.7	8 789.3	918.8	4.7	516.5	233.3	163.4	396.7
2011	2 208.1	9 718.0	2 164.2	9 651.1	1 001.6	4.3	751.3	231.1	176.0	407.1
2012	2 509.7	10 520.0	2 461.2	10 445.7	1 090.7	3.8	926.8	243.2	196.7	439.8
2013	2 715.7	11 092.1	2 664.5	11 015.0	1 160.7	3.5	1 033.8	255.6	210.7	466.4
2014	2 992.0	11 747.9	2 940.3	11 668.0	1 253.2	2.9	1 199.3	265.2	219.7	484.9
2015	3 141.1	12 416.1	3 093.8	12 330.1	1 339.5	2.5	1 237.1	276.6	238.2	514.8
2016	3 384.0	13 293.8	3 339.8	13 198.9	1 420.9	2.2	1 369.6	288.8	258.4	547.2
2017	3 653.0	13 936.6	3 607.3	13 835.7	1 525.0	1.9	1 487.0	306.4	287.0	593.4
2018	3 796.8	14 456.5	3 746.4	14 351.8	1 624.8	1.7	1 492.9	333.4	293.7	627.1
2019	4 041.1	15 418.1	3 976.8	15 302.4	1 710.9	. . .	1 591.9	367.2	306.8	674.0
2018										
January	3 652.6	13 856.2	3 649.5	13 858.4	1 535.5	1.9	1 511.2	312.7	288.2	600.9
February	3 566.9	13 842.2	3 619.7	13 892.8	1 542.2	1.9	1 474.9	312.1	288.6	600.7
March	3 689.0	14 024.0	3 661.8	13 952.6	1 550.8	1.9	1 504.7	313.3	291.1	604.4
April	3 698.8	14 068.0	3 662.3	13 989.1	1 560.3	1.8	1 492.5	315.3	292.5	607.7
May	3 656.3	13 988.9	3 658.0	14 054.9	1 571.1	1.8	1 476.3	318.9	289.8	608.7
June	3 654.8	14 080.6	3 657.5	14 120.0	1 581.7	1.8	1 462.5	317.4	294.1	611.5
July	3 678.0	14 115.1	3 677.0	14 153.0	1 590.5	1.8	1 470.8	321.4	292.7	614.0
August	3 686.1	14 171.6	3 686.3	14 197.0	1 599.1	1.8	1 467.0	325.5	293.1	618.5
September	3 671.5	14 207.2	3 703.9	14 228.5	1 607.9	1.7	1 475.5	324.5	294.3	618.8
October	3 718.4	14 209.7	3 719.1	14 235.4	1 614.1	1.7	1 472.5	336.4	294.4	630.7
November	3 676.6	14 263.2	3 698.0	14 245.4	1 619.1	1.7	1 451.4	332.4	293.4	625.9
December	3 796.8	14 456.5	3 746.4	14 351.8	1 624.8	1.7	1 492.9	333.4	293.7	627.1
2019										
January	3 745.4	14 433.6	3 740.5	14 434.6	1 630.6	. . .	1 480.2	343.8	285.9	629.8
February	3 702.1	14 411.4	3 759.7	14 464.4	1 633.5	. . .	1 498.1	341.8	286.2	628.0
March	3 753.9	14 582.9	3 729.9	14 511.8	1 637.2	. . .	1 458.3	342.8	291.6	634.4
April	3 819.8	14 633.9	3 781.0	14 558.3	1 645.2	. . .	1 487.9	353.6	294.3	647.9
May	3 787.8	14 584.1	3 792.4	14 653.2	1 650.8	. . .	1 493.8	353.7	294.1	647.9
June	3 828.1	14 743.5	3 832.8	14 780.7	1 657.8	. . .	1 527.1	351.6	296.3	647.9
July	3 860.8	14 822.7	3 858.2	14 859.9	1 666.6	. . .	1 534.6	361.5	295.5	657.0
August	3 847.2	14 903.9	3 853.3	14 931.2	1 674.3	. . .	1 521.1	358.1	299.8	658.0
September	3 874.5	14 994.9	3 903.2	15 020.8	1 685.0	. . .	1 553.8	364.4	300.0	664.4
October	3 921.6	15 121.0	3 922.8	15 147.1	1 693.4	. . .	1 562.8	365.0	301.5	666.5
November	3 922.2	15 266.7	3 947.3	15 247.5	1 703.3	. . .	1 579.6	362.5	301.8	664.3
December	4 041.1	15 418.1	3 976.8	15 302.4	1 710.9	. . .	1 591.9	367.2	306.8	674.0

. . . = Not available.

Table 12-1B. Money Stock, Historical: 1892–1924

(Not seasonally adjusted, millions of dollars.)

Classification	1892	1893	1894	1895	1896	1897	1898	1899	1900	1901	1902
June 30											
Currency outside banks	1 015	1 081	972	971	974	1 013	1 150	1 181	1 331	1 395	1 431
Demand deposits adjusted	2 880	2 766	2 807	2 960	2 839	2 871	3 432	4 162	4 420	5 204	5 719
M1	3 895	3 847	3 779	3 931	3 813	3 884	4 582	5 343	5 751	6 599	7 150
Time deposits	1 929	2 007	1 994	2 088	2 220	2 305	2 397	2 617	3 015	3 315	3 565
M2	5 824	5 854	5 773	6 019	6 033	6 189	6 979	7 960	8 766	9 914	10 715
December 31											
Currency outside banks	. . .	. . .	. . .	. . .	. . .	. . .	. . .	. . .	. . .	. . .	. . .
Demand deposits adjusted	. . .	. . .	. . .	. . .	. . .	. . .	. . .	. . .	. . .	. . .	. . .
M1	. . .	. . .	. . .	. . .	. . .	. . .	. . .	. . .	. . .	. . .	. . .
Time deposits	. . .	. . .	. . .	. . .	. . .	. . .	. . .	. . .	. . .	. . .	. . .
M2	. . .	. . .	. . .	. . .	. . .	. . .	. . .	. . .	. . .	. . .	. . .

Classification	1903	1904	1905	1906	1907	1908	1909	1910	1911	1912	1913
June 30											
Currency outside banks	1 543	1 562	1 629	1 759	1 700	1 711	1 691	1 725	1 709	1 762	1 858
Demand deposits adjusted	5 962	6 256	7 069	7 504	7 872	7 384	7 768	8 254	8 668	9 156	9 140
M1	7 505	7 818	8 698	9 263	9 572	9 095	9 459	9 979	10 377	10 918	10 998
Time deposits	3 800	4 045	4 464	4 769	5 350	5 493	6 265	6 944	7 337	7 889	8 356
M2	11 305	11 863	13 162	14 032	14 922	14 588	15 724	16 923	17 714	18 807	19 354
December 31											
Currency outside banks	. . .	. . .	. . .	. . .	. . .	. . .	. . .	. . .	. . .	. . .	. . .
Demand deposits adjusted	. . .	. . .	. . .	. . .	. . .	. . .	. . .	. . .	. . .	. . .	. . .
M1	. . .	. . .	. . .	. . .	. . .	. . .	. . .	. . .	. . .	. . .	. . .
Time deposits	. . .	. . .	. . .	. . .	. . .	. . .	. . .	. . .	. . .	. . .	. . .
M2	. . .	. . .	. . .	. . .	. . .	. . .	. . .	. . .	. . .	. . .	. . .

Classification	1914	1915	1916	1917	1918	1919	1920	1921	1922	1923	1924
June 30											
Currency outside banks	1 533	1 575	1 876	2 276	3 298	3 593	4 105	3 677	3 346	3 739	3 650
Demand deposits adjusted	10 082	9 828	11 973	13 501	14 843	17 624	19 616	17 113	18 045	18 958	19 412
M1	11 615	11 403	13 849	15 777	18 141	21 217	23 721	20 790	21 391	22 697	23 062
Time deposits	8 350	9 231	10 313	11 543	11 717	13 423	15 834	16 583	17 437	19 722	21 259
M2	19 965	20 634	24 162	27 320	29 858	34 640	39 555	37 373	48 828	42 419	44 321
December 31											
Currency outside banks	. . .	. . .	. . .	. . .	. . .	. . .	. . .	. . .	. . .	3 726	3 696
Demand deposits adjusted	. . .	. . .	. . .	. . .	. . .	. . .	. . .	. . .	. . .	19 144	20 898
M1	. . .	. . .	. . .	. . .	. . .	. . .	. . .	. . .	. . .	22 870	24 594
Time deposits	. . .	. . .	. . .	. . .	. . .	. . .	. . .	. . .	. . .	20 379	22 232
M2	. . .	. . .	. . .	. . .	. . .	. . .	. . .	. . .	. . .	43 249	46 826

. . . = Not available.

Table 12-1C. Money Stock, Historical: January 1947–January 1959

(Averages of daily figures; seasonally adjusted, billions of dollars.)

Year and month	Money stock (M1)			Time deposits adjusted	M2 (M1 plus time deposits)
	Total	Currency component	Demand deposit component		
1947					
January	109.5	26.7	82.8	33.3	142.8
February	109.7	26.7	83.0	33.5	143.2
March	110.3	26.7	83.7	33.6	143.9
April	111.1	26.6	84.5	33.7	144.8
May	111.7	26.6	85.1	33.8	145.5
June	112.1	26.6	85.5	33.9	146.0
July	112.2	26.5	85.7	34.0	146.2
August	112.6	26.5	86.1	34.4	147.0
September	113.0	26.7	86.3	34.7	147.7
October	112.9	26.5	86.4	35.0	147.9
November	113.3	26.5	86.8	35.2	148.5
December	113.1	26.4	86.7	35.4	148.5
1948					
January	113.4	26.4	87.0	35.5	148.9
February	113.2	26.3	86.8	35.7	148.9
March	112.6	26.2	86.4	35.7	148.3
April	112.3	26.1	86.3	35.7	148.0
May	112.1	26.0	86.0	35.7	147.8
June	112.0	26.0	86.0	35.8	147.8
July	112.2	26.0	86.2	35.8	148.0
August	112.3	26.0	86.2	35.9	148.2
September	112.2	26.0	86.2	35.9	148.1
October	112.1	26.0	86.1	35.9	148.0
November	111.8	26.0	85.9	36.0	147.8
December	111.5	25.8	85.8	36.0	147.5
1949					
January	111.2	25.7	85.5	36.1	147.3
February	111.2	25.7	85.5	36.1	147.3
March	111.2	25.7	85.6	36.1	147.3
April	113.3	25.7	85.6	36.2	149.5
May	111.5	25.7	85.8	36.3	147.8
June	111.3	25.6	85.7	36.4	147.7
July	111.2	25.5	85.7	36.4	147.6
August	111.0	25.5	85.6	36.4	147.4
September	110.9	25.3	85.6	36.4	147.3
October	110.9	25.3	85.6	36.4	147.3
November	111.0	25.2	85.8	36.4	147.4
December	111.2	25.1	86.0	36.4	147.6
1950					
January	111.5	25.1	86.4	36.4	147.9
February	112.1	25.1	86.9	36.6	148.7
March	112.5	25.2	87.3	36.6	149.1
April	113.2	25.3	88.0	36.7	149.9
May	113.7	25.2	88.5	36.9	150.6
June	114.1	25.1	89.0	36.9	151.0
July	114.6	25.0	89.6	36.8	151.4
August	115.0	24.9	90.1	36.7	151.7
September	115.2	24.9	90.3	36.6	151.8
October	115.7	24.9	90.8	36.5	152.2
November	115.9	24.9	90.9	36.6	152.5
December	116.2	25.0	91.2	36.7	152.9
1951					
January	116.7	25.0	91.7	36.7	153.4
February	117.1	25.1	92.0	36.6	153.7
March	117.6	25.2	92.4	36.6	154.2
April	117.8	25.2	92.6	36.7	154.5
May	118.2	25.3	92.8	36.8	155.0
June	118.6	25.4	93.2	36.9	155.5
July	119.1	25.6	93.4	37.2	156.3
August	119.6	25.7	93.8	37.4	157.0
September	120.4	25.8	94.5	37.7	158.1
October	121.0	26.0	95.1	37.8	158.8
November	122.0	26.0	96.0	38.0	160.0
December	122.7	26.1	96.5	38.2	160.9
1952					
January	123.1	26.2	96.9	38.4	161.5
February	123.6	26.3	97.3	38.7	162.3
March	123.8	26.4	97.5	38.9	162.7
April	124.1	26.4	97.6	39.1	163.2
May	124.5	26.5	98.0	39.3	163.8
June	125.0	26.7	98.4	39.5	164.5
July	125.3	26.7	98.6	39.7	165.0
August	125.7	26.8	98.9	40.0	165.7
September	126.4	26.9	99.4	40.3	166.7
October	126.7	27.0	99.7	40.5	167.2
November	127.1	27.2	99.9	40.9	168.0
December	127.4	27.3	100.1	41.1	168.5

Table 12-1C. Money Stock, Historical: January 1947–January 1959—*Continued*

(Averages of daily figures; seasonally adjusted, billions of dollars.)

Year and month	Money stock (M1)			Time deposits adjusted	M2 (M1 plus time deposits)
	Total	Currency component	Demand deposit component		
1953					
January	127.3	27.4	99.9	41.4	168.7
February	127.4	27.5	99.9	41.6	169.0
March	128.0	27.6	100.4	41.9	169.9
April	128.3	27.7	100.7	42.1	170.4
May	128.5	27.7	100.7	42.4	170.9
June	128.5	27.7	100.7	42.6	171.1
July	128.6	27.8	100.8	42.9	171.5
August	128.7	27.8	100.9	43.2	171.9
September	128.6	27.8	100.8	43.5	172.1
October	128.7	27.8	100.9	43.9	172.6
November	128.7	27.8	100.9	44.2	172.9
December	128.8	27.7	101.1	44.5	173.3
1954					
January	129.0	27.7	101.3	44.8	173.8
February	129.1	27.7	101.5	45.2	174.3
March	129.2	27.6	101.6	45.6	174.8
April	128.6	27.6	101.0	46.1	174.7
May	129.7	27.6	102.1	46.5	176.2
June	129.9	27.5	102.3	46.8	176.7
July	130.3	27.5	102.8	47.3	177.6
August	130.7	27.5	103.2	47.8	178.5
September	130.9	27.4	103.5	47.9	178.8
October	131.5	27.4	104.1	48.1	179.6
November	132.1	27.4	104.7	48.2	180.3
December	132.3	27.4	104.9	48.3	180.6
1955					
January	133.0	27.4	105.6	48.5	181.5
February	133.9	27.5	106.4	48.7	182.6
March	133.6	27.5	106.0	48.8	182.4
April	133.9	27.5	106.3	49.0	182.9
May	134.6	27.6	107.0	49.0	183.6
June	134.4	27.6	106.8	49.2	183.6
July	134.8	27.7	107.2	49.3	184.1
August	134.8	27.7	107.0	49.3	184.1
September	135.0	27.7	107.3	49.6	184.6
October	135.2	27.8	107.4	49.7	184.9
November	134.9	27.8	107.1	49.9	184.8
December	135.2	27.8	107.4	50.0	185.2
1956					
January	135.5	27.9	107.7	49.9	185.4
February	135.5	27.9	107.7	49.9	185.4
March	135.7	27.9	107.8	50.1	185.8
April	136.0	27.9	108.1	50.3	186.3
May	135.8	27.9	107.9	50.4	186.2
June	136.0	27.9	108.1	50.7	186.7
July	136.0	28.0	108.0	50.9	186.9
August	135.7	28.0	107.8	51.2	186.9
September	136.2	28.0	108.2	51.5	187.7
October	136.3	28.0	108.2	51.6	187.9
November	136.6	28.1	108.4	51.8	188.4
December	136.9	28.2	108.7	51.9	188.8
1957					
January	136.9	28.2	108.6	52.6	189.5
February	136.8	28.2	108.6	53.1	189.9
March	136.9	28.2	108.7	53.7	190.6
April	136.9	28.2	108.7	54.0	190.9
May	137.0	28.2	108.8	54.5	191.5
June	136.9	28.3	108.6	54.8	191.7
July	137.0	28.3	108.7	55.3	192.3
August	137.1	28.3	108.8	55.7	192.8
September	136.8	28.3	108.4	56.1	192.9
October	136.5	28.3	108.2	56.6	193.1
November	136.3	28.3	108.0	57.0	193.3
December	135.9	28.3	107.6	57.4	193.3
1958					
January	135.5	28.3	107.2	57.6	193.1
February	136.2	28.2	107.9	59.2	195.4
March	136.5	28.2	108.3	60.5	197.0
April	137.0	28.2	108.7	61.5	198.5
May	137.5	28.3	109.2	62.3	199.8
June	138.4	28.3	110.1	63.2	201.6
July	138.4	28.4	110.0	64.0	202.4
August	139.1	28.4	110.7	64.6	203.7
September	139.5	28.5	111.1	64.8	204.3
October	140.1	28.5	111.6	64.9	205.0
November	140.9	28.5	112.4	65.2	206.1
December	141.1	28.6	112.6	65.4	206.5
1959					
January	142.2	28.7	113.5	66.3	208.5

Table 12-2. Components of Non-M1 M2

(Billions of dollars, seasonally adjusted; monthly data are averages of daily figures, annual data are for December.)

Year and month	Savings deposits			Small-denomination time deposits			Retail money funds	Total non-M1 M2
	At commercial banks	At thrift institutions	Total	At commercial banks	At thrift institutions	Total		
1965	92.4	164.5	256.9	26.7	7.8	34.5	. . .	291.3
1966	89.9	163.3	253.1	38.7	16.3	55.0	. . .	308.1
1967	94.1	169.6	263.7	50.7	27.1	77.8	. . .	341.5
1968	96.1	172.8	268.9	63.5	37.1	100.5	. . .	369.4
1969	93.8	169.8	263.7	71.6	48.8	120.4	. . .	384.0
1970	98.6	162.3	261.0	79.3	71.9	151.2	. . .	412.1
1971	112.8	179.4	292.2	94.7	95.1	189.7	. . .	481.9
1972	124.8	196.6	321.4	108.2	123.5	231.6	. . .	553.0
1973	128.0	198.7	326.8	116.8	149.0	265.8	0.1	592.6
1974	136.8	201.8	338.6	123.1	164.8	287.9	1.4	627.9
1975	161.2	227.6	388.9	142.3	195.5	337.9	2.4	729.1
1976	201.8	251.4	453.2	155.5	235.2	390.7	1.8	845.8
1977	218.8	273.4	492.2	167.5	278.0	445.5	1.8	939.4
1978	216.5	265.4	481.9	185.1	335.8	521.0	5.8	1 008.7
1979	195.0	228.8	423.8	235.5	398.7	634.3	33.9	1 092.0
1980	185.7	214.5	400.3	286.2	442.3	728.5	62.5	1 191.3
1981	159.0	184.9	343.9	347.7	475.4	823.1	151.7	1 318.8
1982	190.1	210.0	400.1	379.9	471.0	850.9	180.1	1 431.1
1983	363.2	321.7	684.9	350.9	433.1	784.1	133.1	1 602.1
1984	389.3	315.4	704.7	387.9	500.9	888.8	161.4	1 754.8
1985	456.6	358.6	815.3	386.4	499.3	885.7	171.3	1 872.3
1986	533.5	407.4	940.9	369.4	489.0	858.4	204.1	2 003.3
1987	534.8	402.6	937.4	391.7	529.3	921.0	217.7	2 076.2
1988	542.4	383.9	926.4	451.2	585.9	1 037.1	238.0	2 201.5
1989	541.1	352.6	893.7	533.8	617.6	1 151.3	314.6	2 359.6
1990	581.3	341.6	922.9	610.7	562.6	1 173.3	350.9	2 447.1
1991	664.8	379.6	1 044.5	602.2	463.1	1 065.3	365.5	2 475.3
1992	754.2	433.1	1 187.2	508.1	359.7	867.7	344.8	2 399.8
1993	785.3	434.0	1 219.3	467.9	313.5	781.4	344.2	2 344.8
1994	752.8	398.5	1 151.3	503.6	313.9	817.5	367.0	2 335.7
1995	774.8	361.0	1 135.9	575.8	356.5	932.3	433.8	2 502.0
1996	906.0	368.7	1 274.8	594.2	353.7	947.9	506.4	2 729.0
1997	1 022.5	378.7	1 401.2	625.5	342.2	967.6	581.7	2 950.5
1998	1 187.4	416.1	1 603.6	626.4	324.9	951.3	715.2	3 270.1
1999	1 288.9	451.2	1 740.2	636.9	318.3	955.2	809.9	3 505.3
2000	1 425.3	454.3	1 879.6	700.8	345.3	1 046.1	899.4	3 825.1
2001	1 740.3	571.2	2 311.5	636.0	338.5	974.6	952.3	4 238.4
2002	2 057.9	713.1	2 771.0	591.3	303.5	894.7	873.8	4 539.5
2003	2 335.2	823.6	3 158.8	541.9	276.2	818.1	771.1	4 748.0
2004	2 633.9	872.7	3 506.6	552.3	276.1	828.4	693.9	5 029.0
2005	2 777.2	825.0	3 602.2	647.2	346.5	993.7	697.9	5 293.7
2006	2 913.1	779.9	3 693.0	781.4	424.6	1 206.0	791.9	5 691.0
2007	3 042.7	823.5	3 866.2	859.4	416.5	1 275.9	942.4	6 084.6
2008	3 323.1	765.3	4 088.4	1 079.6	378.0	1 457.6	1 033.3	6 579.3
2009	3 978.9	833.0	4 812.0	868.6	319.0	1 187.5	791.1	6 790.6
2010	4 412.3	919.2	5 331.5	663.9	270.5	934.4	686.7	6 952.6
2011	5 037.6	996.0	6 033.6	548.3	228.6	776.9	676.3	7 486.9
2012	5 727.5	955.8	6 683.3	469.0	176.8	645.8	655.4	7 984.5
2013	6 108.1	1 020.1	7 128.2	425.9	144.5	570.4	652.0	8 350.5
2014	6 495.8	1 077.2	7 573.0	391.4	132.0	523.4	631.3	8 727.7
2015	7 026.4	1 143.3	8 169.7	302.6	110.6	413.2	653.3	9 236.2
2016	7 556.1	1 258.4	8 814.5	251.6	101.8	353.4	691.3	9 859.1
2017	7 813.1	1 297.2	9 110.3	302.5	111.7	414.2	703.8	10 228.3
2018	7 926.4	1 334.6	9 260.9	425.4	107.6	532.9	811.5	10 605.3
2019	8 418.1	1 347.7	9 765.8	458.7	121.2	579.9	979.9	11 325.6
2018								
January	7 777.8	1 307.9	9 085.6	310.3	110.0	420.4	702.9	10 208.9
February	7 815.8	1 320.6	9 136.4	317.4	106.9	424.3	712.5	10 273.2
March	7 801.8	1 341.3	9 143.0	327.7	104.8	432.5	715.3	10 290.8
April	7 847.4	1 310.9	9 158.3	344.4	99.7	444.0	724.5	10 326.8
May	7 869.1	1 326.7	9 195.9	354.7	101.2	455.8	745.2	10 396.9
June	7 914.1	1 326.5	9 240.6	366.1	102.4	468.4	753.4	10 462.5
July	7 908.1	1 329.1	9 237.2	376.1	103.4	479.4	759.3	10 476.0
August	7 922.3	1 331.3	9 253.5	386.2	104.3	490.5	766.6	10 510.6
September	7 915.2	1 330.0	9 245.2	399.0	105.1	504.1	775.2	10 524.6
October	7 887.0	1 331.4	9 218.4	405.9	105.5	511.4	786.6	10 516.4
November	7 900.3	1 325.3	9 225.6	415.4	106.9	522.3	799.5	10 547.4
December	7 926.4	1 334.6	9 260.9	425.4	107.6	532.9	811.5	10 605.3
2019								
January	7 960.9	1 330.2	9 291.1	443.1	108.9	551.9	851.1	10 694.2
February	7 947.3	1 330.1	9 277.4	457.7	112.9	570.6	856.7	10 704.7
March	7 981.3	1 344.1	9 325.4	470.5	116.9	587.3	869.2	10 781.9
April	7 961.5	1 336.9	9 298.4	479.2	118.7	597.9	881.0	10 777.3
May	8 029.7	1 343.5	9 373.2	479.7	119.2	598.9	888.7	10 860.8
June	8 100.0	1 341.1	9 441.1	484.6	121.3	605.9	900.9	10 947.9
July	8 138.7	1 338.6	9 477.3	482.8	122.2	605.0	919.4	11 001.7
August	8 203.3	1 338.6	9 541.9	480.8	122.6	603.4	932.5	11 077.8
September	8 228.6	1 340.0	9 568.6	476.8	123.1	600.0	949.1	11 117.6
October	8 308.0	1 348.5	9 656.5	471.2	123.2	594.4	973.4	11 224.3
November	8 378.2	1 346.5	9 724.7	465.2	122.6	587.8	987.7	11 300.2
December	8 418.1	1 347.7	9 765.8	458.7	121.2	579.9	979.9	11 325.6

. . . = Not available.

Table 12-3. Aggregate Reserves of Depository Institutions and the Monetary Base

(Not seasonally adjusted. Millions of dollars unless otherwise noted.)

Year and month	Reserves balances required			Reserves balances maintained			Reserves		Vault cash		
	Reserve balance	Top of penalty-free brand	Bottom of penalty-free band	Total	Balances maintaned to satisfy reserve balance requirements	Balances maintained that exceed the top of the penalty-free brand	Total	Required	Total	Used to satisfy required reserves	Surplus
1965	. . .	. . .	. . .	18 757	. . .	. . .	22 694	22 270	3 936	3 936	0
1966	. . .	. . .	. . .	19 550	. . .	. . .	23 785	23 446	4 235	4 235	0
1967	. . .	. . .	. . .	20 770	. . .	. . .	25 291	24 916	4 521	4 521	0
1968	. . .	. . .	. . .	22 455	. . .	. . .	27 192	26 766	4 737	4 737	0
1969	. . .	. . .	. . .	23 094	. . .	. . .	28 053	27 767	4 959	4 959	0
1970	. . .	. . .	. . .	23 907	. . .	. . .	29 246	28 998	5 340	5 340	0
1971	. . .	. . .	. . .	25 670	. . .	. . .	31 345	31 163	5 675	5 675	0
1972	. . .	. . .	. . .	25 317	. . .	. . .	31 415	31 131	6 098	6 098	0
1973	. . .	. . .	. . .	28 473	. . .	. . .	35 108	34 804	6 635	6 635	0
1974	. . .	. . .	. . .	29 682	. . .	. . .	36 861	36 603	7 179	7 179	0
1975	. . .	. . .	. . .	27 219	. . .	. . .	34 990	34 724	7 772	7 772	0
1976	. . .	. . .	. . .	26 689	. . .	. . .	35 237	34 964	8 548	8 548	0
1977	. . .	. . .	. . .	27 134	. . .	. . .	36 486	36 296	9 352	9 352	0
1978	. . .	. . .	. . .	31 347	. . .	. . .	41 678	41 446	10 331	10 331	0
1979	. . .	. . .	. . .	32 676	. . .	. . .	44 020	43 578	11 344	11 344	0
1980	. . .	. . .	. . .	27 354	. . .	. . .	40 660	40 146	18 149	13 306	4 843
1981	. . .	. . .	. . .	26 167	. . .	. . .	41 925	41 606	19 538	15 758	3 780
1982	. . .	. . .	. . .	24 806	. . .	. . .	41 855	41 354	20 392	17 049	3 343
1983	. . .	. . .	. . .	20 986	. . .	. . .	38 894	38 333	20 755	17 908	2 847
1984	20 898	. . .	. . .	21 734	. . .	. . .	40 693	39 858	22 193	18 960	3 233
1985	26 518	. . .	. . .	27 582	. . .	. . .	48 122	47 058	23 336	20 540	2 796
1986	35 991	. . .	. . .	37 165	. . .	. . .	59 369	58 196	24 538	22 204	2 334
1987	36 672	. . .	. . .	37 692	. . .	. . .	62 129	61 109	26 676	24 436	2 239
1988	36 703	. . .	. . .	37 765	. . .	. . .	63 678	62 616	28 204	25 913	2 291
1989	34 425	. . .	. . .	35 367	. . .	. . .	62 732	61 790	29 827	27 365	2 462
1990	28 562	. . .	. . .	30 227	. . .	. . .	59 122	57 457	31 789	28 895	2 894
1991	25 680	. . .	. . .	26 670	. . .	. . .	55 545	54 555	32 509	28 875	3 634
1992	24 246	. . .	. . .	25 401	. . .	. . .	56 578	55 423	34 541	31 177	3 364
1993	28 290	. . .	. . .	29 360	. . .	. . .	62 847	61 777	36 818	33 487	3 331
1994	23 500	. . .	. . .	24 671	. . .	. . .	61 359	60 188	40 382	36 688	3 694
1995	19 164	. . .	. . .	20 455	. . .	. . .	57 896	56 605	42 284	37 441	4 843
1996	11 913	. . .	. . .	13 332	. . .	. . .	51 176	49 757	44 527	37 844	6 683
1997	8 979	. . .	. . .	10 666	. . .	. . .	47 921	46 234	44 742	37 255	7 486
1998	7 512	. . .	. . .	9 025	. . .	. . .	45 208	43 695	44 295	36 183	8 112
1999	3 965	. . .	. . .	5 260	. . .	. . .	41 651	40 357	60 625	36 392	24 233
2000	5 594	. . .	. . .	6 920	. . .	. . .	38 371	37 045	45 246	31 451	13 795
2001	7 402	. . .	. . .	9 046	. . .	. . .	41 051	39 407	43 895	32 005	11 890
2002	7 916	. . .	. . .	9 925	. . .	. . .	40 271	38 263	43 363	30 347	13 016
2003	9 820	. . .	. . .	10 867	. . .	. . .	42 953	41 906	44 884	32 087	12 797
2004	10 136	. . .	. . .	12 045	. . .	. . .	46 847	44 938	48 555	34 802	13 754
2005	8 146	. . .	. . .	10 046	. . .	. . .	45 383	43 483	52 586	35 337	17 249
2006	6 616	. . .	. . .	8 479	. . .	. . .	43 282	41 419	53 054	34 803	18 251
2007	6 313	. . .	. . .	8 098	. . .	. . .	43 463	41 678	54 458	35 365	19 093
2008	16 312	. . .	. . .	783 631	. . .	1	820 876	53 558	56 029	37 245	18 784
2009	24 632	. . .	. . .	109 983	1	. . .	114 045	65 250	54 752	40 619	14 134
2010	28 438	. . .	. . .	103 507	4	. . .	107 800	71 365	55 708	42 927	12 781
2011	47 838	. . .	. . .	155 004	3	. . .	159 871	96 510	59 908	48 672	11 236
2012	58 675	. . .	. . .	151 742	5	. . .	157 038	111 634	63 846	52 959	10 887
2013	69 030	75 947	62 114	248 524	75 713	240 953	254 101	124 801	66 275	55 771	10 504
2014	82 770	91 061	74 481	260 670	90 852	251 584	266 593	142 007	69 116	59 236	9 880
2015	89 313	98 259	80 370	241 977	97 981	232 179	248 118	150 727	71 616	61 413	10 203
2016	105 944	116 551	95 338	203 100	116 285	191 472	209 528	170 224	74 751	64 280	10 471
2017	123 720	136 103	111 337	224 427	135 719	210 855	230 982	189 269	76 625	65 549	11 076
2018	123 703	136 085	111 322	169 139	135 698	155 569	175 985	192 165	80 794	68 462	12 333
2019	138 984	152 893	125 076	163 009	152 562	147 752	169 834	207 239	80 167	68 255	11 913
2018											
January	126 792	139 484	114 101	221 460	138 807	207 579	228 101	193 209	77 203	66 417	10 786
February	124 006	136 420	111 594	223 877	135 786	210 298	230 525	190 487	78 784	66 481	12 304
March	120 285	132 326	108 245	216 700	131 778	203 523	223 087	184 151	75 293	63 866	11 427
April	126 415	139 070	113 762	208 619	138 481	194 771	215 104	191 270	75 666	64 855	10 811
May	128 144	140 970	115 318	202 245	140 316	188 213	208 657	192 267	73 914	64 124	9 790
June	125 503	138 065	112 942	198 822	137 568	185 065	205 274	190 028	74 641	64 525	10 117
July	125 736	138 323	113 152	194 982	137 861	181 196	201 461	190 529	75 437	64 792	10 645
August	123 620	135 994	111 246	191 116	135 544	177 562	197 609	188 549	76 483	64 929	11 554
September	126 756	139 444	114 069	187 386	138 965	173 490	193 801	190 901	74 942	64 145	10 797
October	123 821	136 215	111 428	183 001	135 731	169 428	189 579	189 602	75 865	65 781	10 085
November	126 537	139 202	113 872	177 529	138 700	163 659	184 082	192 074	76 554	65 537	11 017
December	123 703	136 085	111 322	169 139	135 698	155 569	175 985	192 165	80 794	68 462	12 333
2019											
January	131 130	144 255	118 006	163 903	143 615	149 541	170 725	199 356	80 374	68 226	12 148
February	125 129	137 655	112 606	164 523	137 128	150 810	171 368	193 579	81 282	68 450	12 832
March	128 423	141 279	115 569	166 129	140 926	152 037	172 593	193 064	76 814	64 640	12 174
April	130 120	143 144	117 098	155 945	142 641	141 680	162 390	194 575	76 135	64 455	11 681
May	134 717	148 200	121 237	151 128	147 792	136 349	157 630	199 742	75 826	65 024	10 802
June	134 908	148 409	121 408	153 585	148 025	138 783	160 130	200 360	76 106	65 452	10 654
July	135 103	148 624	121 584	151 355	148 191	136 535	157 861	200 169	76 485	65 065	11 420
August	134 639	148 114	121 167	152 087	147 555	137 332	158 684	200 608	78 451	65 969	12 483
September	134 707	148 188	121 227	143 977	147 498	129 227	150 480	199 741	76 550	65 035	11 516
October	134 998	148 509	121 489	148 151	148 050	133 346	154 713	200 621	76 660	65 623	11 036
November	140 705	154 787	126 625	152 934	154 256	137 508	159 522	206 586	76 920	65 881	11 039
December	138 984	152 893	125 076	163 009	152 562	147 752	169 834	207 239	80 167	68 255	11 913

See footnotes at end of table.

Table 12-3. Aggregate Reserves of Depository Institutions and the Monetary Base—*Continued*

(Not seasonally adjusted. Millions of dollars unless otherwise noted.)

Year and month	Monetary base			Borrowings from the federal reserve					Nonborrowed reserves
	Total	Total balances maintained	Currency in circulation	Total	Primary	Secondary	Seasonal	Other credit extensions	
1965	61 022	18 758	42 264	444	. . .	. . .	0	. . .	22 250
1966	64 185	19 551	44 634	532	. . .	. . .	0	. . .	23 252
1967	67 818	20 771	47 047	228	. . .	. . .	0	. . .	25 064
1968	73 139	22 456	50 683	746	. . .	. . .	0	. . .	26 446
1969	76 752	23 095	53 658	1 119	. . .	. . .	0	. . .	26 934
1970	80 987	23 907	57 079	332	. . .	. . .	0	. . .	28 914
1971	86 798	25 671	61 127	126	. . .	. . .	0	. . .	31 219
1972	91 381	25 318	66 063	1 050	. . .	. . .	0	. . .	30 365
1973	100 143	28 474	71 669	1 298	. . .	. . .	41	. . .	33 810
1974	108 640	29 682	78 957	727	. . .	. . .	32	. . .	36 134
1975	113 125	27 220	85 905	130	. . .	. . .	14	. . .	34 860
1976	120 337	26 690	93 647	53	. . .	. . .	13	. . .	35 184
1977	129 915	27 135	102 780	569	. . .	. . .	55	. . .	35 917
1978	144 721	31 348	113 373	868	. . .	. . .	135	. . .	40 810
1979	156 653	32 677	123 977	1 473	. . .	. . .	82	. . .	42 547
1980	163 376	27 355	136 021	1 690	. . .	. . .	116	. . .	38 970
1981	170 681	26 282	144 399	636	. . .	. . .	54	. . .	41 289
1982	180 717	25 241	155 476	634	. . .	. . .	33	. . .	41 221
1983	192 191	22 002	170 189	774	. . .	. . .	96	. . .	38 120
1984	204 812	23 060	181 752	3 186	. . .	. . .	113	. . .	37 507
1985	224 722	29 494	195 229	1 318	. . .	. . .	56	. . .	46 804
1986	248 625	39 492	209 133	827	. . .	. . .	38	. . .	58 543
1987	266 878	39 632	227 246	777	. . .	. . .	93	. . .	61 351
1988	283 789	39 344	244 445	1 716	. . .	. . .	130	. . .	61 962
1989	294 323	37 558	256 765	265	. . .	. . .	84	. . .	62 466
1990	315 270	32 343	282 927	326	. . .	. . .	76	. . .	58 796
1991	335 415	30 947	304 467	192	. . .	. . .	38	. . .	55 353
1992	361 903	31 418	330 486	124	. . .	. . .	18	. . .	56 454
1993	398 638	36 191	362 447	82	. . .	. . .	31	. . .	62 765
1994	428 489	29 483	399 006	209	. . .	. . .	100	. . .	61 150
1995	445 840	26 125	419 715	257	. . .	. . .	40	. . .	57 638
1996	464 955	20 615	444 340	155	. . .	. . .	68	. . .	51 021
1997	493 078	17 728	475 350	324	. . .	. . .	79	. . .	47 597
1998	526 435	15 969	510 467	117	. . .	. . .	15	. . .	45 091
1999	612 499	12 614	599 885	320	. . .	. . .	67	. . .	41 331
2000	598 305	14 053	584 252	210	. . .	. . .	111	. . .	38 161
2001	650 775	18 499	632 276	67	. . .	. . .	33	. . .	40 984
2002	699 216	20 906	678 310	80	. . .	. . .	45	. . .	40 191
2003	739 408	23 032	716 376	46	17	0	29	. . .	42 907
2004	776 279	22 785	753 494	63	11	0	52	. . .	46 784
2005	803 124	18 406	784 718	169	97	0	72	. . .	45 214
2006	826 731	15 586	811 145	191	111	0	80	. . .	43 091
2007	837 192	14 905	822 287	15 430	3 787	1	30	. . .	28 033
2008	166 636	788 042	878 323	653 565	88 245	52	3	0	167 311
2009	202 622	110 179	924 426	169 927	19 025	518	37	0	970 523
2010	201 700	103 733	979 663	45 488	41	3	26	0	103 251
2011	261 958	155 257	106 701	9 526	103	0	23	0	158 918
2012	267 594	151 742	115 852	795	12	0	23	0	156 958
2013	371 745	248 524	123 220	170	13	0	59	0	254 084
2014	393 445	260 670	132 775	102	22	0	80	0	266 583
2015	383 581	241 977	141 603	106	38	0	67	0	248 108
2016	353 156	203 100	150 055	39	13	0	25	0	209 524
2017	385 096	224 427	160 669	75	43	0	33	0	230 974
2018	340 082	169 139	170 943	76	18	0	58	0	175 978
2019	342 646	163 009	179 637	21	10	0	11	0	169 832
2018									
January	382 479	221 460	161 019	58	51	0	7	0	228 096
February	385 505	223 877	161 628	20	7	0	14	0	230 523
March	380 060	216 700	163 359	16	4	0	12	0	223 085
April	372 711	208 619	164 092	51	16	0	35	0	215 099
May	367 475	202 245	165 230	94	25	0	69	0	208 648
June	365 048	198 822	166 226	143	17	0	126	0	205 260
July	361 827	194 982	166 845	224	20	0	203	0	201 438
August	358 447	191 116	167 331	261	18	0	243	0	197 583
September	355 989	187 386	168 602	290	38	0	252	0	193 772
October	352 095	183 001	169 093	209	21	0	189	0	189 558
November	347 640	177 529	170 111	97	11	0	86	0	184 073
December	340 082	169 139	170 943	76	18	0	58	0	175 978
2019									
January	334 687	163 903	170 784	26	20	0	6	0	170 723
February	335 346	164 523	170 823	19	7	0	12	0	171 392
March	338 145	166 129	172 015	15	7	0	8	0	172 592
April	328 665	155 945	172 720	30	14	0	15	0	162 387
May	324 446	151 128	173 317	56	18	0	38	0	157 625
June	327 482	153 585	173 896	80	20	0	60	0	160 122
July	326 031	151 355	174 676	120	9	0	111	0	157 849
August	327 137	152 087	175 050	117	16	0	101	0	158 672
September	320 266	143 977	176 289	101	15	0	86	0	150 470
October	325 280	148 151	177 129	63	4	0	59	0	154 707
November	331 555	152 934	178 620	26	4	0	22	0	159 519
December	342 646	163 009	179 637	21	10	0	11	0	169 832

. . . = Not available.

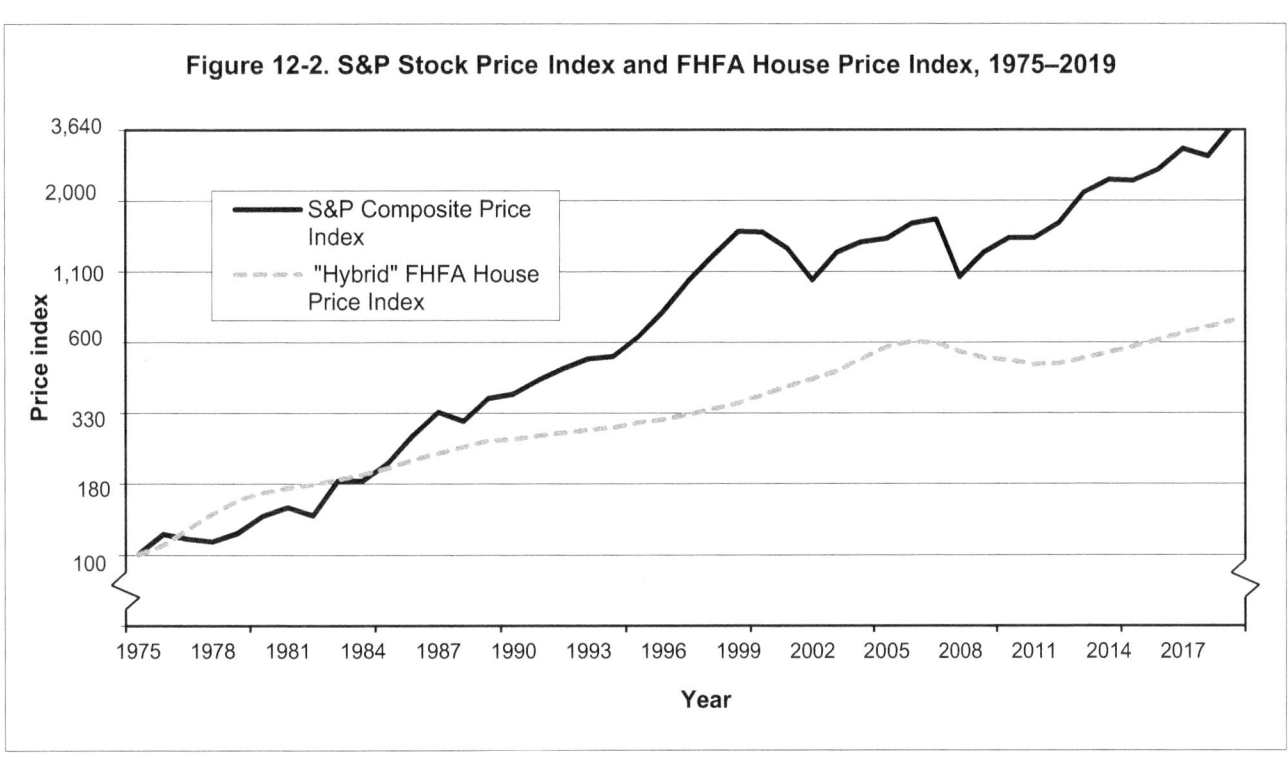

Figure 12-2. S&P Stock Price Index and FHFA House Price Index, 1975–2019

- The housing price index increased each year from 1975 to 2006 then decreased from 2007 to 2011 before increasing again each year from 2012 to 2019. Although housing prices continued to increase, they have decreased at a lower rate the past two years. (Table 12-5)

- The country has faced many challenges in the 21ˢᵗ Century. First the uncertainty of Y2K, the dot.com bubble burst, the terrorist attacks on September 11, 2001, and then the Great Recession began in 2007. Just this year, a global pandemic has killed thousands and has had a devastating effect on the economy. Chapter 12 discusses the variables that have been deeply impacted by this. The S & P 500 index was extremely volatile, decreasing in 2000 and 2001 then falling more drastically during the Great Recession. In addition, unemployment rose dramatically, housing prices fell, and the Federal Reserve lowered its fed funds rate.

- The U.S. economy did finally shake off the impacts of the Great Recession. At the end of 2015, the Fed raised interest rates to 0.75 percent. They increased them even higher in later years. However, in early March 2020, the Federal Reserve responded to the coronavirus by lowering interest rates and increasing the money supply.

- U.S. net worth is a concept that measures the total nonfinancial worth, including land, machines and patent rights, of the U.S. as a whole. Household net worth doubled between 2003 to 2019. (Table 12-7)

Table 12-4. Consumer Credit

(Outstanding at end of period, billions of dollars.)

Year and month	Seasonally adjusted			Not seasonally adjusted						
	Total	By major credit type		Total ¹	By major holder					
		Revolving	Non-revolving		Depository institutions	Finance companies ²	Credit unions	Federal government ²	Nonfinancial businesses	Securitized pools ³
1965	96.0	. . .	96.0	97.5	49.1	23.9	6.5	0.0	18.1	0.0
1966	101.8	. . .	101.8	103.4	52.2	24.8	7.5	0.0	19.0	0.0
1967	106.8	. . .	106.8	108.6	55.8	24.6	8.3	0.0	19.9	0.0
1968	117.4	2.0	115.4	119.3	62.7	26.1	9.7	0.0	20.8	0.0
1969	127.2	3.6	123.6	129.2	67.8	27.8	11.7	0.0	21.9	0.0
1970	131.6	5.0	126.6	133.7	70.1	27.6	13.0	0.0	23.0	0.0
1971	146.9	8.2	138.7	149.2	79.0	29.2	14.8	0.0	26.2	0.0
1972	166.2	9.4	156.8	168.8	92.1	31.9	17.0	0.0	27.8	0.0
1973	190.1	11.3	178.7	193.0	108.1	35.4	19.6	0.0	29.8	0.0
1974	198.9	13.2	185.7	201.9	112.1	36.1	21.9	0.0	31.8	0.0
1975	204.0	14.5	189.5	207.0	116.2	32.6	25.7	0.0	32.6	0.0
1976	225.7	16.5	209.2	229.0	128.9	33.7	31.2	0.0	35.2	0.0
1977	260.6	37.4	223.1	264.9	152.1	37.3	37.6	0.5	37.4	0.0
1978	306.1	45.7	260.4	311.3	179.6	44.4	45.2	0.9	41.2	0.0
1979	348.6	53.6	295.0	354.6	205.7	55.4	47.4	1.5	44.6	0.0
1980	351.9	55.0	297.0	358.0	202.9	62.2	44.1	2.6	46.2	0.0
1981	371.3	60.9	310.4	377.9	208.2	70.1	46.7	4.8	48.1	0.0
1982	389.8	66.3	323.5	396.7	217.5	75.3	48.8	6.4	48.7	0.0
1983	437.1	79.0	358.0	444.9	245.2	83.3	56.1	4.6	55.7	0.0
1984	517.3	100.4	416.9	526.6	303.1	89.9	67.9	5.6	60.2	0.0
1985	599.7	124.5	475.2	610.6	354.8	111.7	74.0	6.8	63.3	0.0
1986	654.8	141.1	513.7	666.4	383.1	134.0	77.1	8.2	64.0	0.0
1987	686.3	160.9	525.5	698.6	399.4	140.0	81.0	10.0	68.1	0.0
1988	731.9	184.6	547.3	745.2	427.6	144.7	88.3	13.2	71.4	0.0
1989	794.6	211.2	583.4	809.3	445.8	138.9	91.7	16.0	69.6	47.3
1990	808.2	238.6	569.6	824.4	431.6	133.4	91.6	19.2	71.9	76.7
1991	798.0	263.8	534.3	815.6	412.3	121.6	90.3	21.1	67.3	103.0
1992	806.1	278.4	527.7	824.8	400.3	118.1	91.7	24.2	70.3	120.3
1993	865.7	309.9	555.7	886.2	433.6	116.1	101.6	27.2	77.2	130.5
1994	997.3	365.6	631.7	1 021.2	497.2	134.4	119.6	37.2	86.6	146.1
1995	1 140.7	443.9	696.8	1 168.2	542.4	152.1	131.9	43.5	85.1	213.1
1996	1 253.4	507.5	745.9	1 273.9	572.2	154.9	144.1	51.4	77.7	273.5
1997	1 324.8	540.0	784.8	1 344.2	562.2	167.5	152.4	57.2	84.4	320.5
1998	1 421.0	581.4	839.6	1 441.3	564.4	183.3	155.4	64.9	79.3	393.9
1999	1 531.1	610.7	920.4	1 553.6	569.5	201.6	167.9	81.8	76.1	456.7
2000	1 717.0	682.6	1 034.3	1 741.3	615.8	234.4	184.4	96.7	81.5	528.4
2001	1 867.9	714.8	1 153.0	1 891.8	639.5	280.0	189.6	111.9	73.1	597.8
2002	1 972.1	750.9	1 221.2	1 997.0	671.3	307.5	195.7	117.3	74.8	630.4
2003	2 077.4	768.3	1 309.1	2 102.9	747.3	393.0	205.9	102.9	59.1	594.8
2004	2 192.2	799.6	1 392.7	2 220.1	795.6	492.3	215.4	86.1	59.2	571.5
2005	2 290.9	829.5	1 461.4	2 320.6	816.1	516.5	228.6	89.8	59.6	609.9
2006	2 456.7	923.9	1 532.8	2 456.7	836.7	534.9	236.1	108.7	48.3	617.2
2007	2 609.5	1 001.6	1 607.9	2 609.5	894.9	577.9	236.6	115.7	49.4	652.5
2008	2 643.8	1 004.0	1 639.8	2 643.8	965.0	561.4	236.2	135.2	46.9	610.2
2009	2 555.0	916.1	1 638.9	2 555.0	906.3	480.8	237.1	232.6	43.2	572.5
2010	2 646.8	839.1	1 807.7	2 646.8	1 185.5	705.0	226.5	363.8	44.4	50.3
2011	2 756.4	840.3	1 916.1	2 756.4	1 192.6	687.6	223.0	494.8	45.2	46.2
2012	2 913.2	840.2	2 073.1	2 913.2	1 215.0	679.8	243.6	622.2	44.6	50.0
2013	3 090.9	854.4	2 236.5	3 090.9	1 271.2	679.1	265.6	735.5	38.6	49.1
2014	3 311.9	887.7	2 424.2	3 311.9	1 343.1	684.1	302.8	846.2	38.3	49.8
2015	3 390.6	898.7	2 492.0	3 390.6	1 428.3	561.3	342.3	949.7	37.8	26.4
2016	3 620.8	960.3	2 660.4	3 620.8	1 532.1	548.4	380.3	1 049.3	38.9	30.5
2017	3 813.0	1 018.1	2 795.0	3 813.0	1 611.8	541.3	418.4	1 145.6	39.4	21.4
2018	3 998.1	1 054.6	2 943.6	3 998.1	1 681.9	534.4	469.2	1 236.3	39.6	5.5
2019	4 180.6	1 094.2	3 086.4	4 180.6	1 770.9	537.6	482.4	1 319.2	39.6	3.1
2018										
January	3 825.0	1 020.7	2 804.3	3 819.1	1 593.5	538.4	420.2	1 172.7	39.1	20.5
February	3 837.7	1 021.5	2 816.2	3 800.4	1 573.7	535.2	419.6	1 178.8	38.6	20.1
March	3 850.0	1 022.2	2 827.9	3 794.9	1 568.9	530.1	422.2	1 181.6	38.3	19.5
April	3 854.7	1 016.3	2 838.5	3 802.3	1 579.3	531.5	427.2	1 183.8	38.4	8.2
May	3 876.5	1 023.6	2 852.9	3 832.6	1 595.4	533.0	434.5	1 188.0	38.6	9.5
June	3 886.4	1 025.7	2 860.7	3 839.7	1 599.9	531.2	438.3	1 191.6	38.6	6.8
July	3 911.6	1 036.1	2 875.5	3 864.7	1 612.0	532.8	447.6	1 194.3	38.6	6.5
August	3 930.5	1 041.7	2 888.9	3 909.6	1 627.7	534.6	456.0	1 213.5	38.9	6.4
September	3 947.2	1 043.9	2 903.3	3 927.1	1 630.8	532.6	462.0	1 224.4	38.9	6.3
October	3 965.1	1 050.5	2 914.7	3 946.1	1 642.0	533.7	466.7	1 227.0	38.9	6.0
November	3 985.2	1 056.8	2 928.4	3 970.4	1 660.9	533.5	468.1	1 231.3	39.1	5.7
December	3 998.1	1 054.6	2 943.6	3 998.1	1 681.9	534.4	469.2	1 236.3	39.6	5.5
2019										
January	4 015.6	1 058.4	2 957.2	4 009.7	1 666.4	533.3	471.5	1 263.0	39.2	5.2
February	4 031.9	1 062.2	2 969.7	3 994.7	1 650.2	530.0	471.3	1 268.7	38.7	5.0
March	4 044.2	1 061.1	2 983.1	3 988.5	1 644.5	529.4	471.6	1 269.6	38.5	4.7
April	4 061.8	1 067.8	2 993.9	4 008.5	1 659.6	530.6	475.0	1 270.4	38.5	4.5
May	4 076.5	1 072.2	3 004.3	4 032.4	1 677.1	533.6	473.7	1 275.4	38.7	4.3
June	4 088.5	1 073.4	3 015.1	4 040.9	1 684.3	533.3	472.8	1 278.3	38.7	4.1
July	4 111.6	1 084.7	3 027.0	4 064.1	1 700.2	536.0	474.8	1 281.4	38.7	3.9
August	4 127.2	1 084.7	3 042.5	4 106.0	1 714.0	537.9	482.5	1 300.1	38.9	3.7
September	4 138.1	1 084.9	3 053.2	4 117.3	1 715.9	536.0	483.6	1 311.0	38.9	3.5
October	4 151.7	1 088.0	3 063.7	4 132.7	1 727.3	536.9	483.7	1 314.3	38.9	3.4
November	4 159.1	1 082.7	3 076.4	4 144.9	1 739.1	537.3	482.0	1 316.2	39.1	3.3
December	4 180.6	1 094.2	3 086.4	4 180.6	1 770.9	537.6	482.4	1 319.2	39.6	3.1

¹Includes nonprofit and educational institutions, not shown.
²Student Loan Marketing Association (Sallie Mae) included in federal government sector until end of 2004, and in finance companies since then.
³Outstanding balances of pools upon which securities have been issued; these balances are no longer carried on the balance sheets of the loan originators.
. . . = Not available.

Table 12-5. Selected Stock and Housing Market Data

Year and month	Stock price indexes			FHFA House Price Indexes			
				House Price Index (purchases and refinance)		Purchase-Only Index	
	Dow Jones industrials (30 stocks)	Standard and Poor's composite (500 stocks) (1941–1943 = 10)	Nasdaq composite (Feb. 5, 1971 = 100)	Level at end of period (1980:I = 100)	Appreciation from same quarter one year earlier (percent)	Level at end of period (1991:I = 100)	Appreciation from same quarter one year earlier (percent)
1965	910.88	88.17	. . .	. . .	. . .	. . .	. . .
1966	873.60	85.26	. . .	. . .	. . .	. . .	. . .
1967	879.12	91.93	. . .	. . .	. . .	. . .	. . .
1968	906.00	98.70	. . .	. . .	. . .	. . .	. . .
1969	876.72	97.84	. . .	. . .	. . .	. . .	. . .
1970	753.19	83.22	. . .	. . .	. . .	. . .	. . .
1971	884.76	98.29	107.44	. . .	. . .	. . .	. . .
1972	950.71	109.20	128.52	. . .	. . .	. . .	. . .
1973	923.88	107.43	109.90	. . .	. . .	. . .	. . .
1974	759.37	82.85	76.29	. . .	. . .	. . .	. . .
1975	802.49	86.16	77.20	62.26	. . .	. . .	. . .
1976	974.92	102.01	89.90	67.33	8.14	. . .	. . .
1977	894.63	98.20	98.71	77.28	14.78	. . .	. . .
1978	820.23	96.02	117.53	87.48	13.20	. . .	. . .
1979	844.40	103.01	136.57	98.29	12.36	. . .	. . .
1980	891.41	118.78	168.61	104.81	6.63	. . .	. . .
1981	932.92	128.05	203.18	109.19	4.18	. . .	. . .
1982	884.36	119.71	188.97	112.26	2.81	. . .	. . .
1983	1 190.34	160.41	285.43	117.08	4.29	. . .	. . .
1984	1 178.48	160.46	248.88	122.53	4.65	. . .	. . .
1985	1 328.23	186.84	290.19	129.53	5.71	. . .	. . .
1986	1 792.76	236.34	366.96	138.93	7.26	. . .	. . .
1987	2 275.99	286.83	402.57	146.45	5.41	. . .	. . .
1988	2 060.82	265.79	374.43	154.67	5.61	. . .	. . .
1989	2 508.91	322.84	437.81	163.33	5.60	. . .	. . .
1990	2 678.94	334.59	409.17	165.26	1.18	. . .	. . .
1991	2 929.33	376.18	491.69	170.42	3.12	101.46	. . .
1992	3 284.29	415.74	599.26	174.54	2.42	104.23	2.73
1993	3 522.06	451.41	715.16	179.15	2.64	107.06	2.72
1994	3 793.77	460.42	751.65	181.97	1.57	110.08	2.82
1995	4 493.76	541.72	925.19	190.28	4.57	113.06	2.71
1996	5 742.89	670.50	1 164.96	195.12	2.54	116.18	2.76
1997	7 441.15	873.43	1 469.49	203.74	4.42	120.01	3.30
1998	8 625.52	1 085.50	1 794.91	214.04	5.06	126.84	5.69
1999	10 464.88	1 327.33	2 728.15	224.58	4.92	134.65	6.16
2000	10 786.85	1 320.28	2 470.52	240.48	7.08	144.03	6.97
2001	10 021.50	1 148.08	1 950.40	257.60	7.12	153.75	6.75
2002	8 341.63	879.82	1 335.51	274.86	6.70	165.51	7.65
2003	10 453.92	1 111.92	2 003.37	294.04	6.98	178.48	7.84
2004	10 783.01	1 211.92	2 175.44	324.38	10.32	196.58	10.14
2005	10 717.50	1 248.29	2 205.32	360.76	11.22	216.58	10.17
2006	12 463.15	1 418.30	2 415.29	377.08	4.52	222.80	2.87
2007	13 264.82	1 468.36	2 652.28	372.80	-1.14	216.85	-2.67
2008	8 776.39	903.25	1 577.03	346.10	-7.16	194.88	-10.13
2009	10 428.05	1 115.10	2 269.15	328.02	-5.22	190.01	-2.50
2010	11 577.51	1 257.64	2 652.87	321.87	-1.87	182.41	-4.00
2011	12 217.56	1 257.60	2 605.15	311.16	-3.33	177.82	-2.52
2012	13 104.14	1 426.19	3 019.51	312.98	0.58	186.47	4.86
2013	16 576.66	1 848.36	4 176.59	327.11	4.51	199.26	6.86
2014	17 823.07	2 058.90	4 736.05	343.96	5.15	208.61	4.69
2015	17 425.03	2 043.94	5 007.41	361.91	5.22	220.22	5.57
2016	19 762.60	2 238.83	5 383.12	382.26	5.62	233.57	6.06
2017	24 719.22	2 673.61	6 903.39	405.34	6.04	248.75	6.50
2018	23 327.46	2 506.85	6 635.28	428.54	5.72	263.50	5.93
2019	28 538.44	3 230.78	8 972.60	450.97	5.23	277.47	5.30
2018							
January	26 149.39	2 823.81	7 411.48	. . .	. . .	. . .	. . .
February	25 029.20	2 713.83	7 273.01	. . .	. . .	. . .	. . .
March	24 103.11	2 640.87	7 063.45	411.43	6.69	252.77	7.19
April	24 163.15	2 648.05	7 066.27	. . .	. . .	. . .	. . .
May	24 415.84	2 705.27	7 442.12	. . .	. . .	. . .	. . .
June	24 271.41	2 718.37	7 510.30	420.28	6.40	260.47	6.58
July	25 415.19	2 816.29	7 671.79	. . .	. . .	. . .	. . .
August	25 964.82	2 901.52	8 109.54	. . .	. . .	. . .	. . .
September	26 458.31	2 913.98	8 046.35	426.21	6.14	263.48	6.25
October	25 115.76	2 711.74	7 305.90	. . .	. . .	. . .	. . .
November	25 538.46	2 760.17	7 330.54	. . .	. . .	. . .	. . .
December	23 327.46	2 506.85	6 635.28	428.54	5.72	263.50	5.93
2019							
January	24 999.67	2 704.10	7 281.74	. . .	. . .	. . .	. . .
February	25 916.00	2 784.49	7 532.53	. . .	. . .	. . .	. . .
March	25 928.68	2 834.40	7 729.32	433.27	5.31	266.46	5.42
April	26 592.91	2 945.83	8 095.39	. . .	. . .	. . .	. . .
May	24 815.04	2 752.06	7 453.15	. . .	. . .	. . .	. . .
June	26 599.96	2 941.76	8 006.24	441.31	5.00	274.06	5.22
July	26 864.27	2 980.38	8 175.42	. . .	. . .	. . .	. . .
August	26 403.28	2 926.46	7 962.88	. . .	. . .	. . .	. . .
September	26 916.83	2 976.74	7 999.34	446.77	4.82	277.02	5.14
October	27 046.23	3 037.56	8 292.36	. . .	. . .	. . .	. . .
November	28 051.41	3 140.98	8 665.47	. . .	. . .	. . .	. . .
December	28 538.44	3 230.78	8 972.60	450.97	5.23	277.47	5.30

. . . = Not available.

Table 12-6. Summary Interest Rates and Bond Yields

(Rates.)

Year and quarter	Federal funds interest rate	Federal Reserve discount rate	U.S. treasury bills, secondary market, 3-month	Inflation: percent change in PCE chain-type price index excluding food and energy	Bank prime rank	Treasury 10-year nominal yields	Domestic corporate bond yields Moody's Aaa	Domestic corporate bond yields Moody's Baa
1970	7.18	5.950	6.39	22.13	7.90	7.35	8.04	9.11
1971	4.66	4.880	4.33	23.17	5.72	6.16	7.39	8.56
1972	4.43	4.500	4.07	23.91	5.25	6.21	7.21	8.16
1973	8.73	6.440	7.03	24.82	8.03	6.85	7.44	8.24
1974	10.50	7.830	7.83	26.79	10.80	7.56	8.57	9.50
1975	5.82	6.250	5.78	29.03	7.85	7.99	8.83	10.61
1976	5.05	5.500	4.97	30.79	6.84	7.61	8.43	9.75
1977	5.54	5.460	5.27	32.77	6.83	7.42	8.02	8.97
1978	7.93	7.460	7.19	34.94	9.05	8.41	8.73	9.49
1979	11.19	10.280	10.07	37.49	12.66	9.43	9.63	10.69
1980	13.36	11.770	11.43	40.94	15.28	11.43	11.94	13.67
1981	16.38	13.420	14.03	44.52	18.86	13.92	14.17	16.04
1982	12.26	11.020	10.61	47.42	14.85	13.01	13.79	16.11
1983	9.09	8.500	8.61	49.84	10.79	11.10	12.04	13.55
1984	10.23	8.800	9.52	51.91	12.06	12.46	12.71	14.19
1985	8.10	7.690	7.48	54.02	9.93	10.62	11.37	12.72
1986	6.81	6.330	5.98	55.88	8.32	7.67	9.02	10.39
1987	6.66	5.660	5.78	57.68	8.20	8.39	9.38	10.58
1988	7.57	6.200	6.67	60.13	9.32	8.85	9.71	10.83
1989	9.22	6.920	8.11	62.63	10.87	8.49	9.26	10.18
1990	8.10	6.980	7.49	65.17	10.01	8.55	9.32	10.36
1991	5.69	5.450	5.38	67.50	8.46	7.86	8.77	9.80
1992	3.52	3.250	3.43	69.55	6.25	7.01	8.14	8.97
1993	3.02	3.000	3.00	71.44	6.00	5.87	7.22	7.92
1994	4.20	3.600	4.25	73.03	7.14	7.09	7.96	8.63
1995	5.84	5.210	5.49	74.63	8.83	6.57	7.59	8.20
1996	5.30	5.020	5.01	76.04	8.27	6.44	7.37	8.06
1997	5.46	5.000	5.06	77.38	8.44	6.35	7.26	7.87
1998	5.35	4.920	4.78	78.37	8.36	5.26	6.53	7.22
1999	4.97	4.620	4.64	79.43	8.00	5.65	7.04	7.88
2000	6.24	5.730	5.82	80.80	9.24	6.03	7.62	8.37
2001	3.89	3.410	3.39	82.26	6.94	5.02	7.08	7.95
2002	1.67	1.170	1.60	83.64	4.68	4.61	6.49	7.80
2003	1.13	2.100	1.01	84.84	4.12	4.01	5.67	6.76
2004	1.35	2.400	1.37	86.52	4.34	4.27	5.63	6.39
2005	3.21	4.250	3.15	88.37	6.19	4.29	5.24	6.06
2006	4.96	6.020	4.73	90.39	7.96	4.80	5.59	6.48
2007	5.02	5.790	4.35	92.38	8.05	4.63	5.56	6.48
2008	1.93	2.170	1.37	94.22	5.07	3.66	5.63	7.44
2009	0.16	0.500	0.15	95.32	3.25	3.26	5.31	7.29
2010	0.18	0.730	0.14	96.61	3.25	3.22	4.94	6.04
2011	0.10	0.750	0.05	98.14	3.25	2.78	4.64	5.66
2012	0.14	0.750	0.09	100.00	3.25	1.80	3.67	4.94
2013	0.11	0.750	0.06	101.53	3.25	2.35	4.24	5.10
2014	0.09	0.750	0.03	103.12	3.25	2.54	4.16	4.85
2015	0.13	0.770	0.05	104.41	3.26	2.14	3.89	5.00
2016	0.40	1.020	0.32	106.07	3.51	1.84	3.67	4.71
2017	1.00	1.630	0.93	107.79	4.10	2.33	3.74	4.44
2018	1.83	2.460	1.94	109.90	4.90	2.91	3.93	4.80
2019	2.16	2.750	2.06	111.67	5.29	2.14	3.39	4.37

¹Outstanding principal balances of mortgage-backed securities issued or guaranteed by the holder indicated.

Table 12-7. Derivation of U.S. Net Wealth

(Billions of dollars, amounts outstanding at the end of period.)

Year and month	Household net worth	Growth of domestic nonfinancial debt				
		Total	households	Business	State and local governments	Federal
2003	49 426	8.0	11.8	2.2	8.3	10.9
2004	55 951	9.0	11.1	5.6	11.4	9.0
2005	61 789	8.6	10.6	8.1	5.8	6.6
2006	66 073	8.4	10.5	9.8	3.9	3.9
2007	66 499	8.1	7.2	12.4	6.0	4.7
2008	56 292	5.8	0.0	5.7	1.4	21.4
2009	60 347	3.7	0.5	-3.9	4.7	20.4
2010	67 129	4.4	-0.6	-0.8	2.7	18.5
2011	68 414	3.6	0.0	2.6	-1.2	10.8
2012	73 573	4.8	1.0	5.0	0.1	10.1
2013	82 498	3.7	1.7	4.5	-1.6	6.7
2014	88 440	4.1	2.1	6.5	-1.2	5.4
2015	91 595	4.4	2.4	6.9	0.4	5.0
2016	97 089	4.5	3.2	5.3	1.1	5.6
2017	106 003	4.2	4.0	6.0	0.0	3.7
2018	107 219	4.7	3.2	4.2	-1.6	7.6
2019	118 368	4.8	3.5	4.8	0.3	6.7
2014						
1st quarter	80 940	4.1	1.9	6.1	-1.7	5.7
2nd quarter	82 627	4.2	5.2	5.1	0.1	3.5
3rd quarter	82 520	4.5	3.0	5.9	-1.7	6.0
4th quarter	84 189	4.7	2.2	7.0	1.5	5.9
2015						
1st quarter	85 798	2.7	2.0	7.4	1.6	-0.3
2nd quarter	86 434	4.4	3.8	8.0	0.5	2.7
3rd quarter	85 249	2.7	1.3	5.5	0.2	2.1
4th quarter	87 287	7.8	3.8	5.6	-1.2	15.4
2016						
1st quarter	86 811	5.5	2.5	9.2	0.7	6.2
2nd quarter	87 532	4.6	4.4	4.2	2.2	5.7
3rd quarter	89 750	5.0	3.5	6.1	0.8	6.3
4th quarter	91 583	3.1	3.7	2.4	0.4	3.6
2017						
1st quarter	95 101	3.2	3.7	5.9	-2.2	1.7
2nd quarter	97 272	4.7	3.9	6.6	-0.9	4.9
3rd quarter	98 866	4.9	2.7	6.2	-0.6	6.9
4th quarter	100 926	3.3	5.1	3.6	3.5	1.3
2018						
1st quarter	106 792	6.7	3.0	3.9	-3.2	14.3
2nd quarter	108 507	4.0	3.6	3.3	-0.4	5.7
3rd quarter	110 594	4.3	3.4	4.3	-1.4	5.9
4th quarter	107 219	3.4	2.8	4.8	-1.5	3.7
2019						
1st quarter	112 596	6.0	2.1	6.8	-1.3	9.8
2nd quarter	114 419	3.1	4.3	4.3	-2.5	2.1
3rd quarter	115 220	6.3	3.4	5.6	0.7	10.4
4th quarter	118 368	3.4	4.1	2.2	4.4	3.8

NOTES AND DEFINITIONS, CHAPTER 12

The Federal Reserve's mission is to promote the goals of maximum employment and stables prices by using these monetary policy tools. Monetary policy in the United States comprises the Federal Reserve's actions and communications to promote maximum employment, stable prices, and moderate long-term interest rates—the three economic goals the Congress has instructed the Federal Reserve to pursue.

When the financial crisis of 2007 – 2009 hit, the Federal Reserve used Monetary tools in accordance to their mission: reducing interest rates to near-zero and increasing the money supply with its Qualitative Easing policy.

As home-owners defaulted on their sub-prime mortgages in 2007 to 2009, the Federal Reserve played an important role in turning the economy around. One such program was Quantitative Easing where the Fed purchased financial assets in addition to conventional easing, which has consisted of reducing short-term interest rates to near zero and supporting that rate level with purchases and sales of short-term securities on the open market. In addition, the Federal Reserve began an initiative to provide a richer and more detailed picture of financial intermediation and interconnections.

Quantitative easing policy was discontinued in late 2014. By then the federal reserve had bought $4.5 trillion worth of bonds over five years. The goal of the program during the Great Recession was to lower interest rates and increase the money supply. In addition, the Fed held its benchmark rate to 0.25 which is effectively zero. This policy remained in place for seven years, until December 2015.

The Federal Reserve Board announced on August 1, 2014 the Enhanced Financial Accounts (EFA) initiative which is an ambitious and long-term effort to enhance the "Financial Accounts" (FA) by providing additional detail and disaggregation, higher-frequency data, and additional documentation and analysis of financial data, in order to improve the picture of financial intermediation and activity in the United States. These large-scale and fundamental improvements will expand the detail, dimensionality, and scope of the FA in order to improve the overall picture of financial intermediation and activity in the United States.

In early 2020, the Federal Reserve began lowering interest rates again and increasing money supply through purchasing bank's bond in response to the spreading coronavirus. States issued stay-at-home orders which effectively closed many nonessential businesses. Unemployment compensation requests soared to 36 million by mid-May 2020.

Most of the data in this chapter are found on the Federal Reserve Board Web site, <http://www.federalreserve.gov>. Current releases and most historical data are found at that site by selecting Economic Research & Data/Statistical Releases and Historical Data and then selecting the appropriate report. This is the location for all data not otherwise specified.

Historical data not found online are taken from two printed volumes of statistical data that were published by the Board of Governors of the Federal Reserve System: *Banking and Monetary Statistics*, 1943, and *Banking and Monetary Statistics, 1941-1970*, 1976. These will be referred to as *B&MS* 1943 and *B&MS* 1976.

TABLES 12-1 AND 12-2

Money Stock Measures and Components

SOURCE: BOARD OF GOVERNORS OF THE FEDERAL RESERVE SYSTEM

Monetary policy includes monitoring two different measures – M1 and M2. M1 consists of the most liquid forms of money such as demand deposits. M2 includes M1 and short term savings. Estimates of these measures are published weekly. The monthly data are averages of daily figures. In addition, these measures are revised in subsequent releases.

The Federal Reserve Board ceased publication of the M3 aggregate on March 23, 2006. Weekly publication was also discontinued for the following components of M3: large-denomination time deposits, repurchase agreements (RPs), and Eurodollars. The Board continues to publish institutional money market mutual funds as a memorandum item in this release. Measures of large-denomination time deposits continue to be published in the Financial Accounts (Z.1 release) and in the H.8 release weekly for commercial banks.

The Board stated that "M3 does not appear to convey any additional information about economic activity that is not already embodied in M2 and has not played a role in the monetary policy process for many years. Consequently, the Board judged that the costs of collecting the underlying data and publishing M3 outweigh the benefits." M3 included Individual Retirement Acts (IRA) and Keogh retirement plans designed for self-employed individuals. ("Discontinuance of M3," H.6, Money Stock Measures [November 10, 2005, revised March 9, 2006]. [Accessed November 6, 2006.]

Definitions

M1 consists of (1) currency, (2) traveler's checks of nonbank issuers, (3) demand deposits, and (4) other checkable deposits.

M2 consists of M1 plus savings deposits (including money market deposit accounts), small denomination time deposits, and balances in retail money market mutual funds.

Currency consists of paper notes and coins outside the U.S. Treasury, the Federal Reserve Banks, and the vaults of depository institutions.

Traveler's checks is the outstanding amount of U.S. dollar-denominated traveler's checks of nonbank issuers. Traveler's checks issued by depository institutions are included in demand deposits.

Demand deposits consists of funds at domestically chartered commercial banks, U.S. branches and agencies of foreign banks, and Edge Act corporations (excluding those amounts held by depository institutions, the U.S. government, and foreign banks and official institutions) less cash items in the process of collection and Federal Reserve float.

Other checkable deposits at commercial banks consists of negotiable order of withdrawal (NOW) and automatic transfer service (ATS) balances at domestically chartered commercial banks, U.S. branches and agencies of foreign banks, and Edge Act corporations.

Other checkable deposits at thrift institutions consists of NOW and ATS balances at thrift institutions, credit union share draft balances, and demand deposits at thrift institutions.

Savings deposits includes money market deposit accounts and other savings deposits at *commercial banks* and *thrift institutions*.

Small time deposits are deposits issued at *commercial banks* and *thrift institutions* in amounts less than $100,000. All Individual Retirement Account (IRA) and Keogh account balances at commercial banks and thrift institutions are subtracted from small time deposits.

Retail money funds are investment funds with capital invested by individual investors. Also IRA and Keogh account balances are excluded.

Institutional money funds are included in the money stock report for informational purposes. They are not part of M1 or M2.

Notes on the data

Seasonal adjustment. Seasonally adjusted M1 is calculated by summing currency, traveler's checks, demand deposits, and other checkable deposits (each seasonally adjusted separately). Seasonally adjusted M2 is computed by adjusting each of its nonM1 components and then adding this result to seasonally adjusted M1.

Revisions. Money stock measures are revised frequently and have a benchmark and seasonal factor review in the middle of the year; this review typically extends back a number of years. The monetary aggregates were redefined in major revisions introduced in 1980.

Historical: 1892–1946. These data are from *B&MS* 1943 and *B&MS* 1976. They pertain only to the last day of June and, after 1922, the last day of December, the "call dates" on which banks

reported to the federal government. For more data and analysis concerning monetary developments in this period, including monthly money supply estimates, see Milton Friedman and Anna Jacobson Schwartz, *A Monetary History of the United States, 1867–1960,* Princeton, Princeton University Press, 1963. In Table 12-1B, *Demand deposits adjusted* refers to the elimination of interbank and U.S. government deposits and cash items in process of collection, not to seasonal adjustment, which is not applicable to call report data because they are only reported once or twice a year. The editor has retitled as *M1* the "Total" money stock shown in the source document, and has added time deposits to it in order to approximate *M2*, which is not shown in the source document.

Historical: January 1947-January 1959. These data are not currently maintained online and are found in *B&MS* 1976. They are not continuous with the current data series; for that reason the values for January 1959 are shown so that users can link to the current data. The *demand deposit component* includes demand deposits held in commercial banks by individuals, partnerships, and corporations both domestic and foreign, and demand deposits held by nonbank financial institutions and foreign banks. Note that this includes demand deposit liabilities to foreign governments, central banks, and international institutions—said to be "relatively small" in the source document. *Time deposits adjusted* is time and savings deposits at commercial banks, other than large negotiable certificates of deposit (CDs), and excluding all deposits due to the U.S. government and domestic commercial banks. The editor has added this to the *Money stock (M1)* to approximate *M2*, which is not shown as such in the source document, over this period.

Data availability

Estimates are released weekly in Federal Reserve Statistical Release H.6, "Money Stock Measures." Current and historical data are available on the Federal Reserve Web site.

References

Board of Governors of the Federal Reserve System, *The Federal Reserve System: Purposes and Functions*, available online at <http://www.federalreserve.gov> in the category "About the Fed/Features," includes a chapter discussing monetary policy and the monetary aggregates and a glossary of terms as an appendix.

An explanation of the 1980 redefinition of the monetary aggregates is found in the *Federal Reserve Bulletin* for February 1980.

TABLES 12-3

Aggregate Reserves, Monetary Base and Federal Reserve Balance Sheet

SOURCE: BOARD OF GOVERNORS OF THE FEDERAL RESERVE SYSTEM

Background

Regulation D sets uniform requirements for all depository institutions to maintain reserve balances either with their Federal

Reserve Bank or as cash. In July 2013 the Federal Reserve Board issued a substantially revised format for reporting reserves of depository institutions and the monetary base. The regulation D revisions were made in response to the substantial growth in reserves undertaken beginning in 2008 to mitigate the effects of the financial crisis and to recent changes in the administration of bank reserves.

With these revisions, the Federal Reserve adjusted the data back to 1965. As shown in table 12-3, the penalty-free band and balance maintained to satisfy reserve balance requirements are the two new concepts are the result of amending Regulation D.

Seasonally adjusted data--no longer relevant in a situation of vastly expanded reserves--are no longer published. Several new aggregates and interest rates are published to "give the public insight into how depository institutions collectively manage their reserves within the current framework for the implementation of monetary policy."

The new format enables calculations of concepts similar to "excess reserves," and a new simplified monetary base is nearly identical with the historical time series of that name; both are without seasonal adjustment.

Definitions

Total Reserves consist of reserves with the Federal Reserve Banks plus vault cash used to satisfy reserve requirements.

Reserve Balances Requirement: The amount determined by applying the reserve ratios specified in Regulation D to an institution's reservable liabilities during the relevant computation period. The institution must satisfy its reserve requirement in the form of vault cash and/or balances maintained either directly with a Reserve Bank or in a pass-through arrangement.

Total Balance Requirements: That portion of the average end-of-day balance for a maintenance period that satisfies an institution's reserve balance requirement. The maximum value for an institution's balance maintained to satisfy reserve balance requirements is the top of the institution's penalty-free band.

Penalty-free band: A penalty-free band is a range on both sides of the reserve balance requirement within which an institution needs to maintain its average balance over the maintenance period in order to satisfy its reserve balance requirement. The top of the penalty-free band is equal to the reserve balance requirement plus a dollar amount prescribed by the Board. The bottom of the penalty-free band is equal to the reserve balance requirement minus a dollar amount prescribed by the Board.

Reserve Balances Maintained: A depository institution is required to satisfy its reserve requirement in the form of vault cash or if vault cash is insufficient to satisfy the requirement, in the form of a balance maintained either directly with a Reserve

Bank or in a pass-through arrangement. The portion of the reserve requirement that is not satisfied by vault cash is called the reserve balance requirement. An institution is responsible for satisfying its reserve balance requirement by holding balances on average over a 14-day maintenance period in an account at the Federal Reserve.

Vault Cash is U.S. currency and coin owned by a depository institution. The average end-of-day holdings of vault cash over the computation period can be used to satisfy some or all of an institution's reserve requirement in the corresponding maintenance period.

Monetary Base is the sum of currency (including coin) in circulation outside Federal Reserve Banks and the U.S. Treasury, plus deposits held by depository institutions at Federal Reserve Banks.

Primary credit is a lending program available to depository institutions that are in generally sound financial condition.

Secondary credit is available to depository institutions that are not eligible for primary credit.

Nonborrowed reserves equals total reserves less total borrowing from the Federal Reserve.

For further information, see "Aggregate Reserves of Depository Institutions and the Monetary Base - H.3" at <www.federalreserve.gov> under Economic Research & Data/Statistical Releases and Historical Data.

Data availability

Reserve and monetary base data are released weekly in Federal Reserve Release H.3, "Aggregate Reserves of Depository Institutions and the Monetary Base." Current and historical data are available on the Federal Reserve Web site.

The Federal Reserve balance sheet is published in the H.4.1 release, "Factors Affecting Reserve Balances", released each Thursday, and available on the Federal Reserve website.

TABLE 12-4

Consumer Credit

SOURCE: BOARD OF GOVERNORS OF THE FEDERAL RESERVE SYSTEM

The consumer credit series cover most shortand intermediate-term credit extended to individuals through regular business channels, excluding loans secured by real estate (such as first and second mortgages and home equity credit). In October 2003, the scope of this survey was expanded to incorporate student loans extended by the federal government and by SLM Holding

Corporation (SLM), the parent company of Sallie Mae (Student Loan Marketing Association). The historical data have been revised back to 1977 to reflect this inclusion.

Consumer credit is categorized by major types of credit and by major holders.

Definitions and notes on the data

The major types of consumer credit are *revolving* and *nonrevolving*.

Revolving credit includes credit arising from purchases on credit card plans of retail stores and banks, cash advances and check credit plans of banks, and some overdraft credit arrangements.

Nonrevolving credit includes automobile loans, mobile home loans, and all other loans not included in revolving credit, such as student loans, boats, trailers, or vacations. These loans may be secured or unsecured.

Debt secured by real estate (including first liens, junior liens, and home equity loans) is excluded. Credit extended to governmental agencies and nonprofit or charitable organizations, as well as credit extended to business or to individuals exclusively for business purposes, is excluded.

Categories of *holders* include *U.S.-chartered depository institutions* (comprising commercial banks and savings institutions), *finance companies, credit unions, federal government, nonfinancial businesses*, and *pools of securitized assets*. The Student Loan Marketing Association (Sallie Mae) is included in "Federal government" until the end of 2004, at which time it became fully privatized. Beginning with the end of 2004, Sallie Mae is included in "Finance companies." Retailers and gasoline companies are included in the nonfinancial businesses category.

Federal government includes student loans originated by the Department of Education under the Federal Direct Loan Program and the Perkins Loan Program, as well as Federal Family Education Loan Program loans that the government purchased under the Ensuring Continued Access to Student Loans Act.

Pools of securitized assets comprises the outstanding balances of pools upon which securities have been issued; these balances are no longer carried on the balance sheets of the loan originators.

Data availability

Current data are available monthly in the Federal Reserve Statistical Release G.19, "Consumer Credit," available along with all current and historical data on the Federal Reserve website. In the autumn of each year there is a revision of several years of past data reflecting benchmarking and seasonal factor review.

TABLES 12- 5

Stock Prices and Housing Market Data

SOURCES: DOW JONES, INC.; STANDARD AND POOR'S CORPORATION; NASDAQ; NEW YORK STOCK EXCHANGE; FEDERAL HOUSING FINANCE AGENCY

Definitions and notes on the data

Stock price indexes and yields.

The *Dow Jones industrial* average is an average price of 30 stocks compiled by Dow Jones, Inc.

The *Standard and Poor's composite* is an index of the prices of 500 stocks that are weighted by the volume of shares outstanding, accounting for about 90 percent of New York Stock Exchange value, with a base of 1941–1943 = 10, compiled by Standard and Poor's Corporation. (Before February 1957, these data are based on a conversion of an earlier 90-stock index, and are obtained from *B&MS* 1943 and 1976.)

The *Nasdaq composite index* is an average price of over 5,000 stocks traded on the Nasdaq exchange.

The *fixed-rate first mortgage* rates are primary market contract interest rates on commitments for fixedrate conventional 30-year first mortgages. The rates are obtained by the Federal Reserve from the Federal Home Loan Mortgage Corporation (FHLMC, or Freddie Mac).

The Federal Housing Finance Agency (FHFA) (formerly Federal Housing Enterprise Oversight (OFHEO) House Price Index is a broad economic measure of the movement of single-family house and repeat-sales in the United States. The FHFA HPI index measures changes in single-family house prices based on data covering all 50 states and over 400 American cities. Extending back to the mid-1970s, the HPIs are built on tens of millions of home sales and offer insights about house price fluctuations at national, census division, state metro area, county, ZIP code, and census tract levels.

FHFA publishes Purchase-Only Indexes for the house purchase subgroup of new mortgage transactions, excluding refinancing transactions. Nation-wide, this index is based on millions of transactions over the latest 27 years. While this is still a very large statistical base, it is subject to somewhat greater revision and is less reliable for smaller geographical areas.

The FHFA HPI is a weighted, repeat-sales index, meaning that it measures average price changes in repeat sales or refinancing's on the same properties. This information is obtained by reviewing repeat mortgage transactions on single-family properties whose mortgages have been purchased or securitized by Fannie Mae or Freddie Mac since January 1975. In addition the repeat-sales technique helps to control for differences in the quality of the houses comprising the sample and is the reason the FHFA HPI is referred to as a "constant quality" index or as "all-transactions" house price index. Appraisal values from refinance mortgages are added to the purchase-only data sample.

The FHFA price indexes are subject to revision for preceding quarters and years, because each new transaction, when reported, affects the rate of price change for all periods since the last time that the property involved in the new transaction changed hands or was refinanced. In 2006, the purchase-only index was the largest index since 1991. However by 2019, this index rose 24.1 percent.

The national conforming loan limit for mortgages that finance single-family one-unit properties increased from $33,000 in the early 1970s to $417,000 for 2006-2008, with limits 50 percent higher for four statutorily-designated high cost areas: Alaska, Hawaii, Guam, and the U.S. Virgin Islands. Since 2008, various legislative acts increased the loan limits in certain high-cost areas in the United States. While some of the legislative initiatives established temporary limits for loans originated in select time periods, a permanent formula was established under the Housing and Economic Recovery Act of 2008 (HERA). The 2019 loan limits have been set under the HERA formula.

The FHFA indexes do not come from a random sample of house prices, and users need to consider possible sources of bias. Expensive houses are under-represented in general because the transactions are limited to conforming mortgages. This is important if the price trends for expensive houses are different.

Analysis of purchase-only and refinance indexes illustrated that these two indexes behave similarly even there was some questions that they behaved differently. Between the end of 1991 and the peak at the end of 2006, the total House Price Index rose 121 percent, identical with the Purchase-Only Index. Between 2006 and 2015, the house price index fell 3.9 per cent. This index has improved over 12 percent from 2016 and 2019. The prices of purchased homes fell 4.0 percent from 2006 and 2014 while between 2015 and 2019 these prices increased 25.9 percent.

In addition to these government-supplied measures, new privately-developed house price indexes--not available for publication in *Business Statistics*--are now receiving regular media attention, and even provide the basis for a futures contract trading on the Chicago Mercantile Exchange. Standard & Poor's now issues S&P/Case-Shiller® Home Price Indexes each month. They cover resales only for houses at all price ranges, based on information obtained from county assessor and recorder offices. They are value-weighted, meaning that price trends for more expensive homes have greater influence on changes in the index. (The FHFA index weights price trends equally for all properties.) S&P/Case-Shiller data are only collected for 20 major

metropolitan statistical areas. Series description and data can be found at <http://www.homeprice.standardandpoors.com>.

TABLE 12-6

Summary Interest rates and bond yields

SOURCES: BOARD OF GOVERNORS OF THE FEDERAL RESERVE SYSTEM; BUREAU OF ECONOMIC ANALYSIS; MOODY'S INVESTORS SERVICE; THE BOND BUYER; DOW JONES, INC.; AND STANDARD AND POOR'S CORPORATION; NEW YORK STOCK EXCHANGE

Definitions and notes on the data

Interest rates and bond yields are percent per year and are averages of business day figures, except as noted. With a few exceptions, they are nominal rates or yields not adjusted for inflation.

The daily effective *federal funds rate*—a principal marker for Federal Reserve monetary policy--is a weighted average of rates on trades through New York brokers. Monthly figures include each calendar day in the month. Annualized figures use a 360-day year.

The *Federal Reserve discount rate* is the rate for discount window borrowing at the Federal Reserve Bank of New York. Monthly figures include each calendar day in the month. Annualized figures use a 360-day year. Before 1945, annual averages calculated by editor.

The *U.S. Treasury bills, 3-month rate* is the yield on this security based on the price as traded in the secondary market. The rate is quoted on a discount basis. Annualized figures use a 360-day year. Early data are taken from *B&MS* 1943 and 1976. From 1934 to 1944, the 3-month rate is based on dealers' quotations. From 1931 through 1933, it is the average rate on new issues. For 1920 through 1930, it is the rate on 3- to 6-month Treasury notes and certificates.

The *inflation* column shown here is the rate of change in the Personal Consumption Expenditures (CPE) chain-type price index, excluding food and energy, beginning in 1970. For monthly entries, it is the change from the same month a year earlier. This price index is calculated by the Bureau of Economic Analysis (BEA) and shown in Table 4.3. Other inflation rates are shown in Table 8-4; see the notes and definitions for that table. Before 1960, the inflation rates shown are the similarly-defined rates of change in the CPI-U.

The *bank prime rate* is one of several base rates used by banks to price shortterm business loans. It is the rate posted by a majority of the top 25 (by amount of assets in domestic offices) insured U.S.-chartered commercial banks. Monthly figures include each calendar day in the month. Annualized figures use a 360-day year. Before 1949, the data are not on the Federal Reserve Web site but have been reproduced from *B&MS* 1976. In that volume, the prime rate is described as "the rate that banks charge their most creditworthy business customers on short-term loans", as posted by the largest banks. The same source goes on to write, "A nationally publicized and uniform prime rate did not emerge until the depression of the 1930s. The rate in that period—1 ½ percent—represented a floor below which banks were said to regard lending as unprofitable. The date shown [for the beginning of a changed rate] is that on which the new rate was put into effect by the first bank to make the change. The table shows a range of rates for 1929-1933 because no information is available to indicate when the rate changed in that period." For further information, the source document cites "The Prime Rate," *Monthly Review,* Federal Reserve Bank of New York, April and May 1962, pp. 54-59 and 70-73, respectively.

U.S. Treasury securities, constant maturities. The rate shown is for 10-year security isthe yield on actively traded issues adjusted to constant maturities. Yields on Treasury securities at "constant maturity" are interpolated by the Treasury Department from the daily yield curve. This curve, which relates the yield on a security to its time to maturity, is based on the closing market bid yields on actively traded Treasury securities in the overthecounter market. These market yields are calculated from composites of quotations reported by U.S. Government securities dealers to the Federal Reserve Bank of New York. The constant maturity yield values are read from the yield curve at fixed maturities. For example, this method provides a yield for a 10year maturity, even if no outstanding security has exactly 10 years remaining to maturity. The 30-year series was discontinued as of February 2002, because the Treasury Department was no longer issuing such bonds at that time. However, issuance of 30-year bonds was resumed in 2005 as an additional means of financing rising deficits, and the 30-year interest rate series resumes in 2006. The current 20-year series begins with 1993 and is not comparable with an earlier 20-year series. For further information, see the historical data series on the Federal Reserve Web site.

Before 1962, the *Treasury 10-year* is not available on the constant-maturity basis. The entries in that column are from *B&MS* 1943 and 1976 where they are labeled "United States government bonds (long-term)." Before 1941, they represent yields on bonds that were partly tax-exempt. In 1941, the yield for the partly tax-exempt series was 1.95 percent—comparable with the entries for previous years—while the entry comparable with subsequent years was 2.12 percent.

Domestic corporate bond yields, Aaa and Baa. The rates shown are for general obligation bonds based on Thursday figures, and are provided by Moody's Investors Service and republished by the Federal Reserve. The Aaa rates through December 6, 2001 are averages of Aaa utility and Aaa industrial bond rates. As of December 7, 2001, these rates are averages of Aaa industrial bonds only.

Data Availability and references

Interest rates and bond yields are published weekly in the Federal Reserve's H.15 release, "Selected Interest Rates"; the release and current and historical data are available on the Federal Reserve Web site and in *B&MS* 1943 and 1976. The starting dates for individual interest rate series vary; some date back to 1911, and many begin in the 1950s and 1960s.

Stock market data are published monthly in *Economic Indicators,* and until 2014 annually in *Economic Report of the President,* both available online at <http://www.gpo.gov>. Some historical interest rate data that are not available on the Federal Reserve Web site were also taken from past *Economic Reports of the President.*

The FHFA house price indexes for the United States as a whole, regions, states, and metropolitan and sub-metropolitan groups are published every 3 months, approximately 2 months after the end of the previous quarter. The release and supporting data and explanatory material are available at <http://www.fhfa.gov>.

The OFHEO (now FHFA) and other price indexes for houses are discussed and compared in Jordan Rappaport, "Comparing Aggregate Housing Price Measures," *Business Economics,* October 2007, pp. 55-65.

TABLE 12-7

Household Net Worth and Growth of Domestic Nonfinancial Debt

SOURCE: BOARD OF GOVERNORS OF THE FEDERAL RESERVE SYSTEM.

Background

The Z-1 *Financial Accounts* provide information on macro-financial flows and aggregate balance sheets for major sectors of the economy, including households, financial institutions, nonfinancial businesses, governments, and the international sector. The *Financial Accounts* also make up the financial component of the *Integrated Macroeconomic Accounts,* which are published jointly with the Bureau of Economic Analysis and which relate production, income, and saving from the *National Income and Product Accounts* to changes in net worth and financial transactions across the major economic sectors of the U.S. economy. The basic framework of the FA allows understanding of the economic relationships underlying the major sections of the economy, as well as for monitoring the effects of macro-financial developments in the United States and globally.

Researchers and analysts have used the Financial Accounts to investigates the causes and consequences of the financial crisis and to create and analyze indicators of financial stability.

However, they were not originally designed to be a comprehensive source of data for all the complex financial relationships and transactions whose importance to the macro-economy were made clear during the financial crisis of 2007-2009.

A new table on the derivation on U.S. net worth has been added to the summary section of the "Financial Accounts." The calculation of U.S. net worth (table 12.7) Household Net Worth and Growth of Domestic Nonfinancial Debt includes the value of nonfinancial assets (real estate, equipment, intellectual property products, consumer durables and inventories) held by households and nonprofit organizations and noncorporate businesses. The measure of U.S. net worth includes the market value of domestic nonfinancial and financial corporations and is adjusted to reflect U.S. financial claims on the rest of the world.

Taking all this together, net U.S worth is defined as the value of tangible assets controlled by households and nonprofits, noncorporate business, and government sectors of the U.S. economy, plus the market value of domestic nonfinancial and financial corporations, net of U.S. financial obligations to the rest of the world.

Changes in net worth consist of transactions, revaluations, and other volume changes. Corporate equity and debt securities include directly and indirectly held securities. Real estate is the value of owner-occupied real estate. Other includes equity in noncorporate businesses, consumer durable goods, fixed assets of nonprofit organizations, and all other financial assets apart from corporate equities and debt securities, net of liabilities, as shown on table B.101 Balance Sheet of Households and Non-profit Organizations.

Definitions and notes on the data

The concept of *U.S. net worth* ignores all domestic financial intermediation and just counts the value of the land, machines, houses, cars, etc., plus intangible worth but does not include financial worth.

Household net worth is the value of net worth for the household and nonprofit organizations sector, as reported on table B.101 in the *Financial Accounts*. Net worth is calculated as the difference between a sector's total assets, including both financial and nonfinancial assets, and liabilities (debts owed to other sectors) and does include financial worth.

*Noncorporate business*es includes both sole proprietorships and partnerships, including noncorporate farms.

Debt equates to the sum of debt securities and loans.

Domestic Nonfinancial entities include households, corporations, nonfarm noncorporate businesses, farm business, federal, state and local governments.

Domestic Nonfinancial assets include housing, property, automobiles, equipment, business equity, intellectual properties such as patents, copyrights and trademarks.

For corporate *businesses,* worth is the market value of their outstanding equity shares and is used to better capture the value of intangible assets, such as intellectual property.

A *corporation* is formed under state law by the filing of articles of incorporation with the state.

Federal Government consists of all federal government agencies and funds included in the unified budget. However, the District of Columbia government is included in the state and local sector.

Nonprofits include private foundations and organizations that are tax-exempt under Sections 501(c)(3) through 501(c)(9) of the Internal Revenue Code. Does not include religious organizations or organizations with less than $25,000 in gross annual receipts.

Revision

Data shown for the most recent quarters are based on preliminary and potentially incomplete information. Nonetheless, when source data are revised or estimation methods are improved, all data are subject to revision. There is no specific revision schedule; rather, data are revised on an ongoing basis.

Data availability

The "Financial Accounts" are published online and in print four times per year, about 10 weeks following the end of each calendar quarter. The publication with series mnemonics and the guide are available online: https://www.federalreserve.gov/releases/Z1.

CHAPTER 13: INTERNATIONAL COMPARISONS

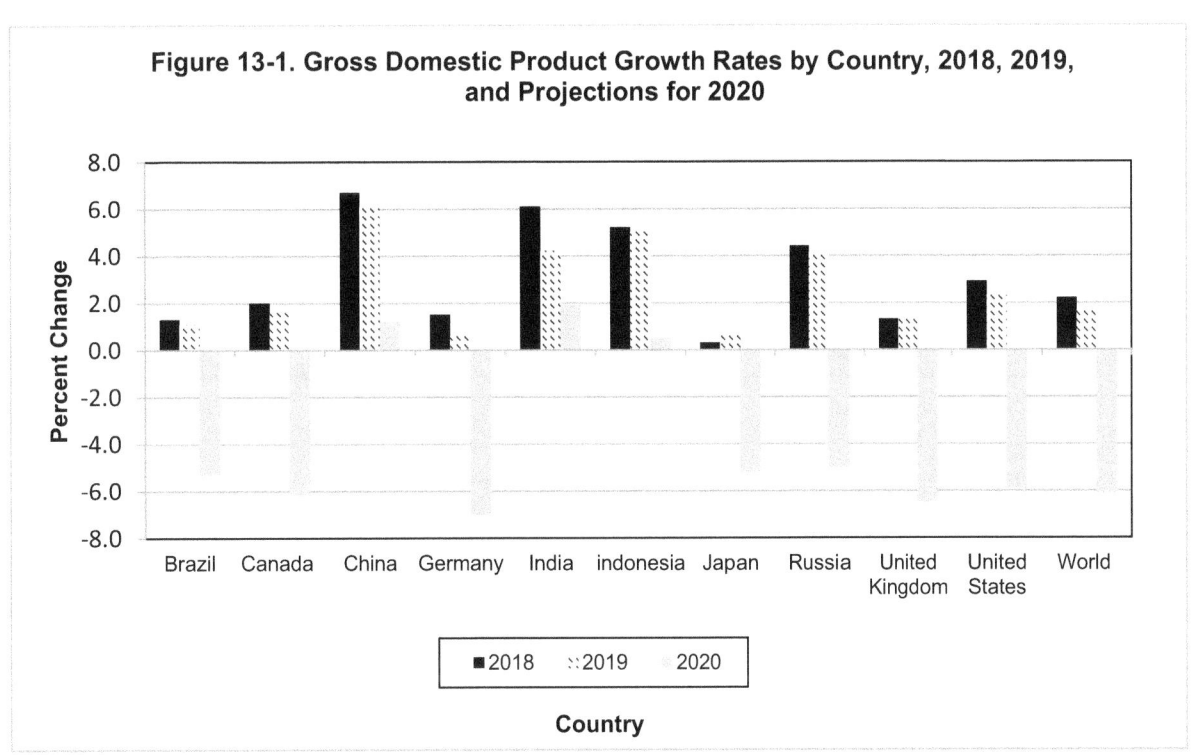

Figure 13-1. Gross Domestic Product Growth Rates by Country, 2018, 2019, and Projections for 2020

- In June 2020, the World Economic Outlook (WEO) revised their 2020 Gross Domestic Product (GDP) forecast to -6.1 percent, down nearly 10 percentage points from their 2017 forecast. The main factor is the COVID-19 pandemic. The global health crisis is inflicting high and rising human costs worldwide. Protecting lives and allowing health care systems to cope have required isolation, lockdowns, and widespread closures to slow the spread of the virus. The health crisis is therefore having a severe impact on global economic activity. (Table 13-1)

- In economics, Okun's law is an empirically observed inverse relationship between unemployment and GDP. As GDP growth goes down, unemployment rises. Thus, even though unemployment rates in 2019 were relatively low (except in Brazil where unemployment equaled 11.9 percent) the United States unemployment rate may reach 10.4 percent or higher in 2020. (See Table 13-1)

- Five of the six major currencies listed in Table 13-2 including the Canadian dollar, the Euro, the Japanese yen, the British pound, and the Swiss franc are all major industrial countries whose currencies are freely traded on world markets. The other currency listed is the Chinese yuan, which is not freely traded. China is able to control the international values of their currencies (through, for example, direct capital controls) and keep their currencies from appreciating relative to the dollar to maintain their competitiveness in the U.S. market. The Chinese have allowed some yuan appreciation, as can be seen in the lower yuan value of the dollar. (Table 13-2)

Table 13-1 International Comparisons: Percent Change in Real GDP, Consumer Prices, Current Account Balances, and Unemployment Rate, 2018, 2019, and 2020

(Annual percent change.)

Country	Gross domestic product			Consumer prices			Current account balance			Unemployment rate		
	2018	2019	2020	2018	2019	2020	2018	2019	2020	2018	2019	2020
WORLD	2.2	1.7	-6.1	2.0	1.4	0.5	0.7	0.7	0.1	5.1	4.8	8.3
Euro Area	1.9	1.2	-7.5	1.8	1.2	0.2	3.1	2.7	2.6	8.2	7.6	10.4
Germany	1.5	0.6	-7.0	2.0	1.3	0.3	7.4	7.1	6.6	3.4	3.2	3.9
United Kingdom	1.3	1.4	-6.5	2.5	1.8	1.2	-3.9	-3.8	-4.4	4.1	3.8	4.8
Brazil	1.3	1.1	-5.3	3.7	3.7	3.6	-4.6	-3.2	-4.6	11.4	11.9	14.7
Canada	2.0	1.6	-6.2	2.3	1.9	0.6	-2.5	-2.0	-3.7	4.1	3.8	4.8
China	6.7	6.1	1.2	2.1	2.9	3.0	0.4	1.0	0.5	3.8	3.6	4.3
India	6.1	4.2	1.9	3.4	4.5	3.3	-2.1	-1.1	-0.6	. . .	. . .	. . .
Indonesia	5.2	5.0	0.5	3.3	2.8	2.9	-2.9	-2.7	-3.2	5.3	5.3	7.5
Japan	0.3	0.7	-5.2	1.0	0.5	0.2	3.5	3.6	1.7	2.4	2.4	3.0
Russia	4.4	4.1	-5.0	4.6	3.8	2.2	6.8	3.8	0.7	4.8	4.6	4.9
United States	2.9	2.3	-5.9	2.4	1.8	0.6	-2.4	-2.3	-2.6	3.9	3.7	10.4

. . . = Not available.

Table 13-2 Foreign Exchange Rates

(Not seasonally adjusted.)

Year and month	Foreign currency per U.S. dollar					
	Canadian dollar	Chinese yuan	European currency unit	Japananese yen	Switzerland franc	British pound
1971	1.0099	. . .	. . .	347.79	4.1171	2.4442
1972	0.9907	. . .	. . .	303.12	3.8186	2.5034
1973	1.0002	. . .	. . .	271.31	3.1688	2.4525
1974	0.9780	. . .	. . .	291.84	2.9805	2.3403
1975	1.0175	. . .	. . .	296.78	2.5839	2.2217
1976	0.9863	. . .	. . .	296.45	2.5002	1.8048
1977	1.0633	. . .	. . .	268.62	2.4065	1.7449
1978	1.1405	. . .	. . .	210.39	1.7907	1.9184
1979	1.1713	. . .	. . .	219.02	1.6644	2.1224
1980	1.1693	. . .	. . .	226.63	1.6772	2.3246
1981	1.1990	1.7100	. . .	220.63	1.9675	2.0243
1982	1.2344	1.8979	. . .	249.06	2.0319	1.7480
1983	1.2325	1.9810	. . .	237.55	2.1007	1.5159
1984	1.2952	2.3303	. . .	237.46	2.3500	1.3368
1985	1.3659	2.9434	. . .	238.47	2.4552	1.2974
1986	1.3896	3.4616	. . .	168.35	1.7979	1.4677
1987	1.3259	3.7314	. . .	144.60	1.4918	1.6398
1988	1.2306	3.7314	. . .	128.17	1.4643	1.7813
1989	1.1842	3.7673	. . .	138.07	1.6369	1.6382
1990	1.1668	4.7921	. . .	145.00	1.3901	1.7841
1991	1.1460	5.3337	. . .	134.59	1.4356	1.7674
1992	1.2085	5.5206	. . .	126.78	1.4064	1.7663
1993	1.2902	5.7795	. . .	111.08	1.4781	1.5016
1994	1.3664	8.6397	. . .	102.18	1.3667	1.5319
1995	1.3725	8.3700	. . .	93.96	1.1812	1.5785
1996	1.3638	8.3389	. . .	108.78	1.2361	1.5607
1997	1.3849	8.3193	. . .	121.06	1.4514	1.6376
1998	1.4836	8.3008	. . .	130.99	1.4506	1.6573
1999	1.4858	8.2783	1.0653	113.73	1.5045	1.6172
2000	1.4855	8.2784	0.9232	107.80	1.6904	1.5156
2001	1.5487	8.2770	0.8952	121.57	1.6891	1.4396
2002	1.5704	8.2771	0.9454	125.22	1.5567	1.5025
2003	1.4008	8.2772	1.1321	115.94	1.3450	1.6347
2004	1.3017	8.2768	1.2438	108.15	1.2428	1.8330
2005	1.2115	8.1936	1.2449	110.11	1.2459	1.8204
2006	1.1340	7.9723	1.2563	116.31	1.2532	1.8434
2007	1.0734	7.6058	1.3711	117.76	1.1999	2.0020
2008	1.0660	6.9477	1.4726	103.39	1.0816	1.8545
2009	1.1412	6.8307	1.3935	93.68	1.0860	1.5661
2010	1.0298	6.7696	1.3261	87.78	1.0432	1.5452
2011	0.9887	6.4630	1.3931	79.70	0.8862	1.6043
2012	0.9995	6.3093	1.2859	79.82	0.9377	1.5853
2013	1.0300	6.1478	1.3281	97.60	0.9269	1.5642
2014	1.1043	6.1620	1.3297	105.74	0.9147	1.6484
2015	1.2791	6.2827	1.1096	121.05	0.9628	1.5284
2016	1.3243	6.6400	1.1072	108.66	0.9848	1.3555
2017	1.2984	6.7569	1.1301	112.10	0.9842	1.2890
2018	1.2957	6.6090	1.1817	110.40	0.9784	1.3363
2019	1.3269	6.9081	1.1194	109.02	0.9937	1.2768
2018						
January	1.2429	6.4233	1.2197	110.87	0.9604	1.3824
February	1.2588	6.3183	1.2340	107.97	0.9355	1.3961
March	1.2933	6.3174	1.2334	106.05	0.9480	1.3976
April	1.2732	6.2967	1.2270	107.66	0.9687	1.4079
May	1.2866	6.3701	1.1823	109.69	0.9969	1.3470
June	1.3125	6.4651	1.1679	110.06	0.9900	1.3294
July	1.3133	6.7164	1.1685	111.52	0.9948	1.3162
August	1.3042	6.8453	1.1547	111.00	0.9880	1.2878
September	1.3034	6.8551	1.1667	112.10	0.9683	1.3066
October	1.3004	6.9191	1.1488	112.72	0.9940	1.3012
November	1.3205	6.9367	1.1364	113.34	1.0011	1.2900
December	1.3436	6.8837	1.1380	112.20	0.9919	1.2664
2019						
January	1.3300	6.7863	1.1418	108.96	0.9897	1.2901
February	1.3209	6.7367	1.1349	110.44	1.0014	1.3016
March	1.3371	6.7119	1.1296	111.14	1.0005	1.3167
April	1.3378	6.7161	1.1234	111.64	1.0084	1.3029
May	1.3460	6.8519	1.1187	109.97	1.0107	1.2855
June	1.3289	6.8977	1.1295	108.07	0.9880	1.2675
July	1.3105	6.8775	1.1211	108.29	0.9880	1.2461
August	1.3273	7.0629	1.1129	106.19	0.9787	1.2160
September	1.3241	7.1137	1.1011	107.54	0.9906	1.2369
October	1.3189	7.0961	1.1058	108.14	0.9930	1.2657
November	1.3237	7.0199	1.1051	108.86	0.9929	1.2884
December	1.3169	7.0137	1.1114	109.10	0.9826	1.3109

. . . = Not available.

NOTES AND DEFINITIONS, CHAPTER 13

The Federal Reserve, the central bank of the United States, provides the nation with a safe, flexible, and stable monetary and financial system. It was created on December 23, 1973, after a series of financial panics. See the article *Cycle and Growth Perspectives* in the beginning of this publications.

TABLE 13-1

International Comparisons: Gross Domestic Product Growth, Unemployment Rates, Inflation and Current Account Balances

The United States trades with many countries. Specific countries include neighbors Canada and Mexico; European countries like Britain and Germany; plus Asian countries such as China, Japan, and India and finally Russia. Table 13 -1 provides international comparisons of these nations in terms of major economic indicators. These include real gross domestic product (GDP), employment rates, inflation, and current account balances as percent of GDP by country for the years 2018, 2019, and the International Monetary Fund projections for 2020.

SOURCE: JUNE 2020 WORLD ECONOMIC OUTLOOK, PUBLISHED BY THE INTERNATIONAL MONETARY FUND AND THE FEDERAL RESERVE.

Definitions and notes on the data

The basic measure of the overall health of an economy is the *gross domestic product* (GDP). It is the market value of all goods and services produced by labor and property located in the United States. *Real GDP* is an inflation-adjusted measure that reflects the value of all goods and services produced by an economy in a given year, expressed in base-year prices. See chapter 1 for in depth and historical data, notes and definitions on GDP and Chapter 8 for inflation data.

The *Consumer Price Index* (CPI) measures the change in prices paid by consumers for goods and services. *Inflation/deflation* is the change over time of the CPI. See Chapter 8.

The *unemployment rate* represents the number unemployed civilians as a percent of the labor force. See Chapter 10.

Current account is all transactions other than those in financial and capital items. The major classifications are goods and services, income and current transfers. The focus of the BOP is on transactions (between an economy and the rest of the world) in goods, services, and income. Chapter 7 provides much more historical and current data.

Table 13-1 presents the current account balance as a percent of GDP. This measure provides an indication on the level of international competitiveness of a country. Usually, countries recording a strong current account surplus have an economy heavily dependent on exports revenues, with high savings ratings but weak domestic demand. On the other hand, countries recording a current account deficit have strong imports, a low saving rates and high personal consumption rates as a percentage of disposable incomes. See Chapter 7 for International Trade information and Chapter 4 for savings data and personal consumption expenditures.

References and notes on the data

This data is published by the International Monetary Fund (IMF) on an annual basis. Tables appear in Chapter 1 of the WEO's Annex.

TABLE 13-2

FOREIGN Exchange Rates

SOURCE: BOARD OF GOVERNORS OF THE FEDERAL RESERVE SYSTEM

Definitions and notes on the data

The price of a nation's currency is in terms of another currency. An *exchange rate* thus has two components, the domestic currency and a foreign currency, and can be quoted either directly or indirectly. In a direct quotation, the price of a unit of foreign currency is expressed in terms of the domestic currency. In an indirect quotation, the price of a unit of domestic currency is expressed in terms of the foreign currency.

This table shows the U.S. dollar relative to the currencies of some important individual countries and also relative to average values for major groups of countries such as the European Union (EU). In *Business Statistics,* all of these measures are defined as the foreign currency price of the U.S. dollar. When the measure is relatively high, the dollar is relatively strong—but less competitive (in the sense of price competition)—and the other currency or group of currencies named in the measure is relatively weak and more competitive.

For consistency, this definition is used in Business Statistics even in the case of currencies that are commonly quoted in the financial press and elsewhere as dollars per foreign currency unit instead of foreign currency units per dollar. Notably, this is the case for the euro and for the British pound.

The definition of the dollar's value used in *Business Statistics* is the most useful for economic analysis from the U.S. point of view and for foreign tourists in the United States. For American tourists overseas, it is easier to use the inverse of this measure, the value of the other currency; for example, the traveler in Paris can more easily translate prices into dollars by multiplying by the dollar value of the euro than by dividing by the euro value of the dollar.

The foreign exchange rates shown are averages of the daily noon buying rates for cable transfers in New York City certified for customs purposes by the Federal Reserve Bank of New York.

The introduction of the euro was in January 1999 as the common currency for 11 European countries and marked a major change in the international currency system. Over time, the Eurozone has increased to 19 countries - Austria, Belgium, Cyprus, Estonia, Finland, France, Germany, Greece, Ireland, Italy, Latvia, Lithuania, Luxembourg, Malta, Netherlands, Portugal, Slovakia, Slovenia and Spain. In addition, the euro is also the national currency in Monaco, the Vatican City and San Marino, and is the de facto currency in Andorra, Kosovo, and Montenegro. The values of the currencies of these countries no longer fluctuate relative to each other, but the value of the euro still fluctuates relative to the dollar and to currencies for countries outside the EMU. The currency and coins of the individual countries continued to circulate from 1999 through the end of 2001, but in January 2002, new euro currency and coins were introduced, replacing the currency and coins of the individual countries. Once a country has entered the monetary union, its values relative to the dollar will continue to fluctuate—but only due to fluctuations in the value of the euro relative to the dollar.

There is no fully satisfactory historical equivalent to the euro. For comparisons over time, the Federal Reserve Board uses a "restated German mark," derived simply by dividing each historical value of the mark by the euro conversion factor, 1.95583.

Data availability and references

Current press releases, historical data, and information on weights and methods for exchange rates and exchange rate indexes are available on the Federal Reserve Web site at <http://www.federalreserve.gov/>; go to Data, then "Foreign Exchange Rates H10/G5," and select DDP, the new Data Download Program. The dollar value indexes are described in the article "Indexes of the Foreign Exchange Value of the Dollar," *Federal Reserve Bulletin* (Winter 2005), also available on the Federal Reserve Web site.

Additional information on exchange rates can be found on the Federal Reserve Bank of St. Louis Web site at <http://www.stls.frb.org/fred/data/exchange.html>.

PART B: INDUSTRY PROFILES

CHAPTER 14: PRODUCT AND INCOME BY INDUSTRY

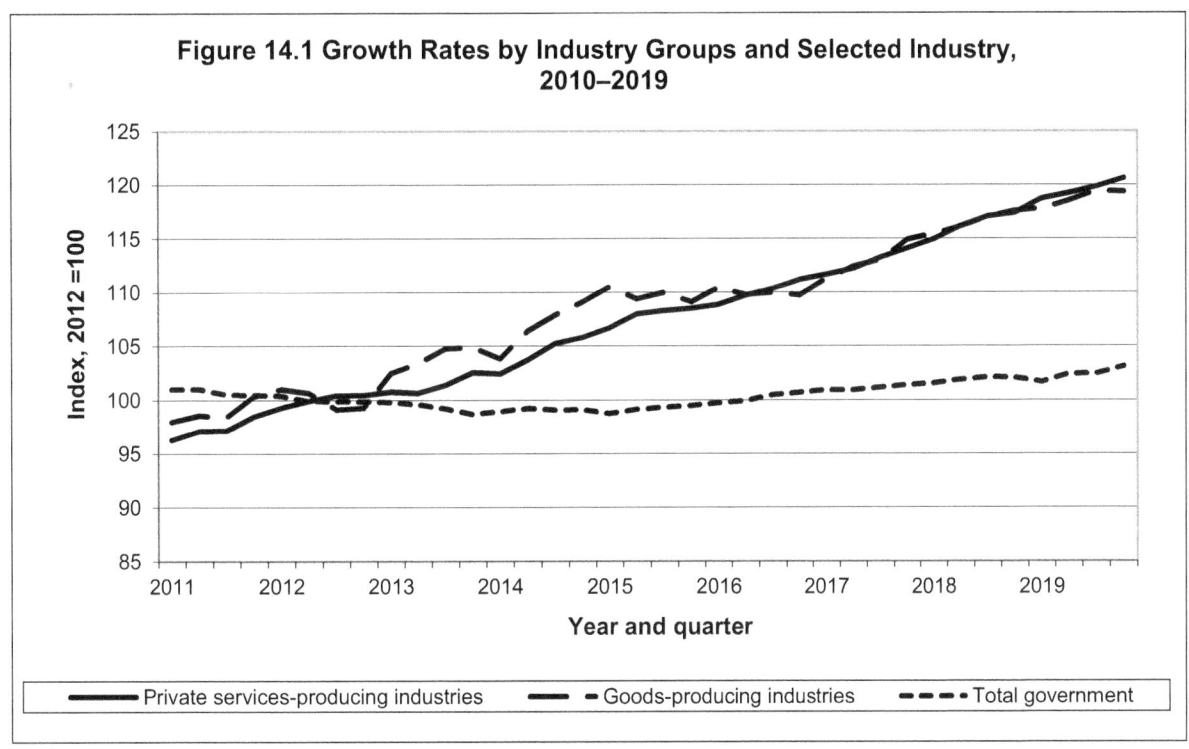

Figure 14.1 Growth Rates by Industry Groups and Selected Industry, 2010–2019

Legend: Private services-producing industries — Goods-producing industries — Total government

- Private services-producing industries, which accounted for 70.3 percent of total GDP in 2019, display less cyclical swing and more real growth than the goods sector, which was 17.4 percent of the total. Government production—the other 12.3 percent of total GDP—was comparatively stable, but this is an incomplete measurement of the counter-cyclical impact of government, as it does not include the effects of changes in tax receipts and payments. (Tables 14-1 and 14-2)

- Table 14-3 can be used to assess the share of each industry group in total gross domestic factor income and the shares paid to employees and accruing to land and capital in each industry. For these purposes factor income is a better measure than the BEA definition of value added because value added includes, but factor income excludes, taxes on production and imports.

- In private goods-producing industries in 2018, employee compensation was 51.0 percent of gross factor income, while in the less capital-intensive private services-producing industries (excluding real estate and rental and leasing), compensation was 54.4 percent of total value. (Table 14-3)

- BEA publishes quarterly measures of GDP by industrial origin, in both current-dollar and real (quantity) terms. Industry output, sometimes referred to as "gross product originating in the industry," is here termed "value added".

Table 14-1. Gross Domestic Product (Value Added) by Industry

(Billions of dollars.)

Year	Total gross domestic product	Total	Agriculture, forestry, fishing, and hunting	Mining	Utilities	Construction	Manufacturing — Durable goods	Manufacturing — Nondurable goods	Wholesale trade	Retail trade	Transportation and warehousing	Information	Finance and insurance	Real estate and rental and leasing
							Private industries							
2000	10 252.3	8 929.3	98.3	110.6	180.1	461.3	924.8	625.4	622.6	685.5	307.8	471.3	743.1	1 231.7
2001	10 581.8	9 188.9	99.8	123.9	181.3	486.5	833.4	640.5	613.8	709.5	308.1	502.4	803.1	1 325.0
2002	10 936.4	9 462.0	95.6	112.4	177.6	493.6	832.8	635.7	613.1	732.6	305.7	550.6	816.4	1 400.5
2003	11 458.2	9 905.9	114.0	139.0	184.0	525.2	863.2	661.0	641.4	769.6	321.4	564.9	846.3	1 449.6
2004	12 213.7	10 582.5	142.9	166.5	199.2	584.6	905.1	703.0	697.1	795.6	352.1	620.4	872.4	1 516.7
2005	13 036.6	11 326.4	128.3	225.7	198.1	651.8	956.8	736.6	754.9	840.8	375.8	642.3	973.7	1 632.5
2006	13 814.6	12 022.6	125.1	273.3	226.8	697.1	1 004.4	789.4	811.5	869.9	410.4	652.0	1 034.2	1 709.7
2007	14 451.9	12 564.8	144.1	314.0	231.9	715.3	1 030.6	814.1	857.8	869.2	413.9	706.9	1 030.5	1 817.7
2008	14 712.8	12 731.2	147.2	392.2	241.7	648.9	999.7	801.1	884.3	848.7	426.8	743.0	873.2	1 889.5
2009	14 448.9	12 403.9	130.0	275.8	258.2	565.6	881.0	821.2	834.2	827.6	404.6	721.9	966.6	1 901.1
2010	14 992.1	12 884.1	146.3	305.8	278.8	525.1	964.3	832.7	888.9	851.5	433.0	753.3	1 003.6	1 939.4
2011	15 542.6	13 405.5	180.9	356.3	287.5	524.4	1 015.2	852.4	934.9	871.9	451.4	759.8	1 026.0	2 019.3
2012	16 197.0	14 037.5	179.6	358.8	279.7	553.4	1 061.7	865.3	997.4	908.4	472.0	759.0	1 162.8	2 098.2
2013	16 784.9	14 572.3	215.6	386.5	286.3	587.6	1 102.0	889.9	1 040.1	949.5	491.1	828.9	1 144.9	2 177.9
2014	17 527.3	15 255.9	201.0	416.4	298.1	636.9	1 134.1	916.1	1 088.2	974.5	521.8	842.4	1 282.6	2 265.4
2015	18 224.8	15 883.9	180.7	259.9	299.2	695.6	1 184.0	942.5	1 142.5	1 024.7	564.4	898.0	1 371.6	2 381.9
2016	18 715.0	16 326.1	164.3	215.6	302.4	745.5	1 190.5	910.6	1 133.8	1 056.5	580.8	959.3	1 444.3	2 485.9
2017	19 519.4	17 065.8	174.6	287.3	315.1	790.4	1 230.7	954.4	1 164.6	1 084.3	612.4	997.6	1 486.3	2 602.1
2018	20 580.2	18 035.6	166.5	346.6	325.9	839.1	1 296.4	1 024.8	1 212.2	1 126.9	658.1	1 067.7	1 567.3	2 734.3
2019	21 427.7	18 796.8	169.2	320.3	334.6	886.6	1 342.7	1 017.2	1 278.1	1 172.9	684.5	1 120.3	1 627.9	2 863.8
2011														
1st quarter	15 285.8	13 156.5	183.5	334.3	275.0	510.9	1 002.7	833.0	911.7	861.8	446.2	758.1	1 013.8	1 962.8
2nd quarter	15 496.2	13 355.7	175.4	361.6	289.0	522.2	1 003.8	860.5	932.6	866.3	450.9	765.5	1 003.1	2 009.6
3rd quarter	15 591.9	13 449.4	183.1	357.7	289.5	529.1	1 011.1	858.2	933.8	873.2	449.4	755.8	1 034.5	2 028.6
4th quarter	15 796.5	13 660.5	181.8	371.6	296.4	535.5	1 043.1	857.7	961.6	886.4	459.2	760.0	1 052.7	2 076.1
2012														
1st quarter	16 019.8	13 871.1	182.6	369.7	272.4	548.3	1 050.8	864.4	979.9	901.5	465.5	755.4	1 126.1	2 069.4
2nd quarter	16 152.3	14 001.2	178.6	353.8	281.7	551.7	1 065.6	864.4	991.9	904.0	474.5	768.2	1 153.8	2 092.5
3rd quarter	16 257.2	14 096.1	175.8	349.6	283.2	552.0	1 063.7	874.1	1 003.1	910.6	474.2	760.1	1 184.5	2 115.5
4th quarter	16 358.9	14 181.7	181.3	362.1	281.3	561.6	1 066.8	858.5	1 014.5	917.6	473.6	752.1	1 186.8	2 115.2
2013														
1st quarter	16 569.6	14 375.2	218.3	369.9	287.4	571.5	1 090.0	887.9	1 027.9	943.1	483.4	812.1	1 109.3	2 146.7
2nd quarter	16 637.9	14 430.5	224.7	383.0	286.4	579.0	1 091.0	873.3	1 025.9	942.7	484.3	820.8	1 126.7	2 150.8
3rd quarter	16 848.7	14 633.7	222.2	401.3	283.2	594.3	1 107.2	882.3	1 045.0	953.8	491.6	830.2	1 143.9	2 196.1
4th quarter	17 083.1	14 850.1	197.2	391.8	288.3	605.4	1 119.6	916.3	1 061.5	958.2	505.1	852.6	1 199.8	2 217.8
2014														
1st quarter	17 104.6	14 857.1	197.3	412.2	301.1	615.5	1 101.5	892.7	1 049.0	951.3	498.5	833.6	1 228.6	2 211.8
2nd quarter	17 432.9	15 172.2	213.0	435.3	295.5	630.0	1 121.6	915.2	1 074.6	969.0	515.7	840.7	1 284.2	2 248.0
3rd quarter	17 721.7	15 443.0	196.7	431.3	297.8	644.0	1 154.5	931.0	1 108.7	982.8	531.7	845.4	1 298.2	2 289.3
4th quarter	17 849.9	15 551.2	197.0	386.7	297.9	658.1	1 158.8	925.7	1 120.4	995.0	541.5	850.0	1 319.4	2 312.6
2015														
1st quarter	17 984.2	15 665.8	176.9	289.2	299.8	673.0	1 174.9	936.6	1 134.4	1 006.1	551.4	869.5	1 359.5	2 338.7
2nd quarter	18 219.4	15 883.8	179.2	278.1	297.4	690.8	1 182.2	948.8	1 150.6	1 017.4	560.7	888.6	1 408.0	2 366.4
3rd quarter	18 344.7	15 994.8	185.2	249.2	303.3	704.3	1 192.7	959.8	1 145.4	1 035.6	571.9	909.5	1 362.1	2 401.1
4th quarter	18 350.8	15 991.4	181.3	222.9	296.4	714.3	1 186.2	925.0	1 139.6	1 039.9	573.8	924.5	1 356.7	2 421.5
2016														
1st quarter	18 424.3	16 055.7	165.3	193.5	294.6	730.1	1 186.1	906.8	1 133.1	1 048.5	572.0	946.9	1 363.6	2 448.3
2nd quarter	18 637.3	16 256.4	168.3	208.6	298.5	738.6	1 187.8	916.5	1 130.1	1 054.0	580.4	953.9	1 424.1	2 480.9
3rd quarter	18 806.7	16 410.0	165.5	220.3	308.1	747.9	1 192.1	912.7	1 134.0	1 059.3	580.0	967.3	1 478.5	2 494.9
4th quarter	18 991.9	16 582.2	158.0	240.1	308.2	765.5	1 196.0	906.5	1 137.9	1 064.2	590.6	969.2	1 510.8	2 519.5
2017														
1st quarter	19 190.4	16 763.6	181.4	266.3	307.4	778.9	1 207.1	928.4	1 146.0	1 072.2	597.7	973.7	1 479.9	2 555.0
2nd quarter	19 356.6	16 915.0	178.5	277.8	316.8	781.5	1 217.5	939.6	1 157.3	1 076.8	608.6	988.6	1 458.5	2 586.6
3rd quarter	19 611.7	17 149.8	169.1	286.6	315.2	793.6	1 235.9	960.3	1 168.6	1 088.2	616.9	1 007.6	1 492.1	2 617.5
4th quarter	19 918.9	17 434.8	169.3	318.6	321.1	807.5	1 262.1	989.4	1 186.5	1 099.8	626.5	1 020.7	1 514.7	2 649.5
2018														
1st quarter	20 163.2	17 654.9	169.4	328.8	322.6	824.3	1 272.4	1 002.4	1 184.4	1 110.8	640.6	1 029.3	1 534.6	2 683.7
2nd quarter	20 510.2	17 977.5	172.1	346.4	326.3	835.3	1 286.2	1 022.0	1 200.1	1 130.6	651.7	1 071.8	1 567.8	2 723.4
3rd quarter	20 749.8	18 190.0	160.5	361.3	323.1	845.4	1 305.7	1 031.4	1 217.8	1 131.7	663.0	1 083.1	1 603.7	2 751.2
4th quarter	20 897.8	18 320.0	163.8	350.1	331.4	851.6	1 321.4	1 043.3	1 246.6	1 134.4	677.2	1 086.8	1 563.2	2 778.8
2019														
1st quarter	21 098.8	18 503.1	161.1	324.6	331.8	871.4	1 336.0	1 001.6	1 262.9	1 153.9	682.5	1 097.3	1 603.5	2 810.4
2nd quarter	21 340.3	18 724.7	165.7	331.7	332.7	883.9	1 341.7	1 013.4	1 269.0	1 163.6	682.3	1 117.6	1 633.0	2 852.6
3rd quarter	21 542.5	18 897.2	173.4	310.8	336.3	890.3	1 344.8	1 020.7	1 287.2	1 184.6	683.8	1 128.8	1 627.4	2 884.2
4th quarter	21 729.1	19 062.1	176.6	314.0	337.8	900.8	1 348.2	1 033.1	1 293.4	1 189.6	689.5	1 137.4	1 647.6	2 907.9

Table 14-1. Gross Domestic Product (Value Added) by Industry—*Continued*

(Billions of dollars.)

Year	Private industries—*Continued*								Government			Private goods-producing industries	Private services-producing industries	Information-communications-technology-producing industries[1]
	Profes-sional, scientific, and technical services	Manage-ment of companies and enterprises	Adminis-trative and waste manage-ment services	Educational services	Health care and social assistance	Arts, entertain-ment, and recreation	Accom-modation and food services	Other services except govern-ment	Total govern-ment	Federal	State and local			
2000	651.8	171.1	282.1	95.2	600.2	99.0	287.5	279.7	1 323.0	423.5	899.5	2 220.4	6 708.9	632.9
2001	686.8	173.6	295.1	101.7	648.2	96.7	294.0	265.6	1 392.9	430.4	962.5	2 184.1	7 004.8	610.3
2002	714.7	175.0	300.2	106.5	700.5	103.8	309.7	284.9	1 474.4	462.5	1 011.9	2 170.1	7 291.9	620.9
2003	741.3	185.6	320.5	116.8	746.0	111.0	321.1	283.8	1 552.3	498.9	1 053.5	2 302.4	7 603.5	653.1
2004	792.4	203.1	345.5	129.1	798.3	117.9	343.3	297.3	1 631.3	525.5	1 105.8	2 502.2	8 080.3	706.2
2005	852.4	214.1	379.9	133.2	837.3	122.2	359.0	310.7	1 710.3	551.3	1 159.0	2 699.3	8 627.1	755.8
2006	917.1	230.5	399.1	142.9	892.5	130.5	381.1	325.0	1 792.0	575.5	1 216.4	2 889.4	9 133.2	795.2
2007	984.9	251.2	430.6	151.5	936.4	137.4	396.1	330.5	1 887.1	601.9	1 285.2	3 018.1	9 546.7	852.5
2008	1 082.1	256.4	438.6	167.8	1 017.0	143.2	399.5	330.3	1 981.6	632.2	1 349.4	2 989.1	9 742.1	904.2
2009	1 031.3	245.0	412.4	188.3	1 079.2	144.6	388.6	326.5	2 045.1	663.3	1 381.7	2 673.6	9 730.3	888.8
2010	1 063.9	265.4	437.6	198.9	1 111.8	152.2	403.5	328.0	2 108.0	697.4	1 410.5	2 774.3	10 109.8	937.4
2011	1 124.6	278.0	454.1	206.0	1 148.7	158.3	422.6	333.1	2 137.1	712.3	1 424.8	2 929.3	10 476.3	974.4
2012	1 188.5	302.2	473.9	213.9	1 193.5	171.2	450.1	348.0	2 159.5	713.9	1 445.6	3 018.8	11 018.7	985.4
2013	1 208.5	319.7	489.1	217.5	1 229.7	177.4	473.9	356.3	2 212.5	703.8	1 508.7	3 181.6	11 390.8	1 062.6
2014	1 267.6	332.9	517.9	226.1	1 265.8	189.1	502.2	376.6	2 271.4	715.8	1 555.5	3 304.5	11 951.4	1 095.6
2015	1 346.7	348.2	539.9	235.5	1 336.4	193.1	546.6	392.4	2 340.8	732.5	1 608.3	3 262.7	12 621.3	1 175.1
2016	1 392.5	351.2	558.8	246.0	1 404.7	206.8	574.5	402.1	2 388.9	745.6	1 643.3	3 226.6	13 099.5	1 249.4
2017	1 452.1	373.7	601.8	246.5	1 462.4	214.4	601.0	414.1	2 453.6	764.4	1 689.2	3 437.4	13 628.4	1 312.6
2018	1 546.4	394.9	638.1	255.5	1 536.9	227.4	633.2	437.2	2 544.6	790.8	1 753.8	3 673.4	14 362.1	1 408.8
2019	1 649.1	418.2	674.8	263.5	1 617.9	236.0	662.5	456.6	2 630.9	813.3	1 817.6	3 735.9	15 060.9	. . .
2011														
1st quarter	1 101.3	278.1	448.4	204.0	1 131.3	155.5	413.9	329.8	2 129.3	708.8	1 420.5	2 864.5	10 292.0	. . .
2nd quarter ...	1 122.6	278.5	452.5	205.5	1 145.2	158.2	420.6	332.0	2 140.5	712.1	1 428.4	2 923.4	10 432.2	. . .
3rd quarter	1 132.0	276.5	456.5	206.8	1 154.4	161.0	424.4	333.8	2 142.4	713.6	1 428.8	2 939.4	10 510.0	. . .
4th quarter	1 142.3	278.7	459.1	207.8	1 163.7	158.5	431.5	336.7	2 135.9	714.5	1 421.5	2 989.7	10 670.8	. . .
2012														
1st quarter	1 169.8	291.9	471.4	212.5	1 184.2	167.6	444.4	343.3	2 148.7	715.1	1 433.6	3 015.8	10 855.3	. . .
2nd quarter ...	1 189.1	293.5	471.7	214.3	1 186.8	170.1	447.8	346.9	2 151.0	714.8	1 436.3	3 014.2	10 987.0	. . .
3rd quarter	1 191.4	302.8	474.9	214.5	1 194.1	172.0	450.4	349.6	2 161.0	714.2	1 446.8	3 015.1	11 081.0	. . .
4th quarter	1 203.8	320.5	477.7	214.2	1 208.8	175.3	458.0	352.0	2 177.2	711.5	1 465.7	3 030.3	11 151.4	. . .
2013														
1st quarter	1 201.8	313.9	479.9	214.7	1 222.3	174.5	467.5	353.1	2 194.4	706.5	1 487.9	3 137.6	11 237.6	. . .
2nd quarter ...	1 197.5	319.6	484.6	215.9	1 224.9	175.2	470.3	353.9	2 207.5	704.1	1 503.4	3 151.0	11 279.4	. . .
3rd quarter	1 208.6	321.1	492.9	217.8	1 231.7	177.7	476.4	356.4	2 215.1	698.4	1 516.7	3 207.3	11 426.4	. . .
4th quarter	1 226.3	324.2	498.9	221.4	1 240.0	182.2	481.5	361.8	2 233.1	706.3	1 526.8	3 230.4	11 619.6	. . .
2014														
1st quarter	1 230.1	333.3	503.0	222.0	1 241.5	185.9	483.4	365.0	2 247.4	710.9	1 536.5	3 219.2	11 637.9	. . .
2nd quarter ...	1 247.6	327.7	515.2	225.4	1 254.1	188.8	496.9	373.8	2 260.7	713.7	1 547.0	3 315.0	11 857.2	. . .
3rd quarter	1 289.3	331.0	523.3	228.1	1 276.7	191.3	509.6	382.2	2 278.6	717.5	1 561.1	3 357.5	12 085.5	. . .
4th quarter	1 303.5	339.5	530.2	228.9	1 291.1	190.5	519.1	385.4	2 298.7	721.2	1 577.5	3 326.3	12 224.9	. . .
2015														
1st quarter	1 323.5	346.8	533.2	231.2	1 310.4	190.2	532.3	388.1	2 318.4	727.2	1 591.2	3 250.7	12 415.1	. . .
2nd quarter ...	1 340.9	348.7	536.9	233.7	1 327.8	192.2	542.7	392.8	2 335.6	730.5	1 605.1	3 279.1	12 604.8	. . .
3rd quarter	1 359.1	348.5	542.8	236.8	1 347.9	194.0	551.9	393.6	2 349.9	734.8	1 615.1	3 291.3	12 703.5	. . .
4th quarter	1 363.4	348.9	546.9	240.2	1 359.5	196.1	559.4	395.1	2 359.5	737.6	1 621.9	3 229.6	12 761.8	. . .
2016														
1st quarter	1 379.0	348.5	549.9	243.6	1 380.7	200.7	565.3	398.9	2 368.6	740.1	1 628.6	3 181.9	12 873.8	. . .
2nd quarter ...	1 388.1	349.5	553.9	246.2	1 400.5	204.4	571.0	401.1	2 380.8	742.8	1 638.0	3 219.9	13 036.5	. . .
3rd quarter	1 392.7	351.5	559.8	246.0	1 407.6	210.0	578.2	403.6	2 396.7	747.6	1 649.1	3 238.5	13 171.5	. . .
4th quarter	1 410.3	355.2	571.6	248.1	1 430.1	211.8	583.6	404.8	2 409.7	752.1	1 657.6	3 266.2	13 316.0	. . .
2017														
1st quarter	1 424.5	362.7	585.2	246.5	1 442.6	210.0	590.8	407.3	2 426.8	757.1	1 669.8	3 362.1	13 401.5	. . .
2nd quarter ...	1 441.2	369.3	596.5	245.2	1 455.2	211.7	597.8	410.3	2 441.6	760.7	1 680.9	3 394.9	13 520.1	. . .
3rd quarter	1 461.1	374.7	607.4	246.5	1 468.1	218.8	604.7	416.9	2 461.9	766.4	1 695.5	3 445.6	13 704.2	. . .
4th quarter	1 481.8	388.0	618.0	248.1	1 483.6	217.2	610.6	421.9	2 484.1	773.4	1 710.7	3 546.9	13 887.9	. . .
2018														
1st quarter	1 512.8	389.5	623.1	251.1	1 509.4	218.4	620.1	427.4	2 508.3	781.7	1 726.6	3 597.2	14 057.6	. . .
2nd quarter ...	1 542.7	391.6	632.9	253.9	1 528.2	228.8	630.5	435.1	2 532.7	788.4	1 744.3	3 662.0	14 315.5	. . .
3rd quarter	1 561.2	399.8	642.1	256.7	1 544.3	230.8	636.8	440.3	2 559.8	794.8	1 765.0	3 704.3	14 485.6	. . .
4th quarter	1 569.0	398.6	654.4	260.5	1 565.7	231.7	645.3	446.1	2 577.8	798.3	1 779.5	3 730.2	14 589.8	. . .
2019														
1st quarter	1 608.1	409.4	660.4	261.0	1 590.7	234.5	651.6	450.4	2 595.7	803.7	1 792.1	3 694.7	14 808.4	. . .
2nd quarter ...	1 642.0	415.8	670.1	260.2	1 604.5	235.1	656.7	453.2	2 615.6	808.3	1 807.3	3 736.4	14 988.3	. . .
3rd quarter	1 666.9	422.2	681.0	265.1	1 628.0	235.5	667.4	458.9	2 645.3	816.6	1 828.7	3 740.0	15 157.2	. . .
4th quarter	1 679.3	425.4	687.9	267.7	1 648.4	238.9	674.5	464.1	2 667.0	824.6	1 842.4	3 772.6	15 289.5	. . .

[1]Consists of computer and electronic products manufacturing; publishing, including software; information and data processing services; and computer systems design and related services.
. . . = Not available.

Table 14-2. Chain-Type Quantity Indexes for Value Added by Industry

(New Series, 2012 = 100.)

Year	Total gross domestic product	Private industries												
							Manufacturing				Transpor-			Real
		Total	Agriculture, forestry, fishing, and hunting	Mining	Utilities	Construc-tion	Durable goods	Nondurable goods	Wholesale trade	Retail trade	tation and ware-housing	Information	Finance and insurance	estate and rental and leasing
2000	81.1	81.1	90.1	65.8	93.2	141.5	70.9	111.7	81.1	90.3	90.0	55.6	77.8	73.7
2001	81.9	81.7	87.0	76.2	77.0	138.6	66.4	110.5	82.7	93.6	84.0	58.9	83.8	76.9
2002	83.3	83.1	90.0	78.2	79.7	134.1	67.8	109.7	83.5	97.7	80.9	64.6	83.2	78.2
2003	85.7	85.5	97.0	69.2	77.9	136.3	72.8	113.1	88.2	102.7	83.8	66.6	83.7	79.4
2004	88.9	89.0	104.7	69.6	82.7	141.2	78.0	120.9	91.9	104.5	90.8	74.3	83.6	81.4
2005	92.1	92.5	109.2	70.8	78.4	141.8	83.4	118.8	96.1	107.9	95.1	79.3	91.4	85.8
2006	94.7	95.5	111.0	81.7	83.3	138.8	89.8	122.5	98.7	108.7	100.7	82.1	95.2	87.4
2007	96.5	97.1	98.3	88.0	84.9	134.6	94.0	124.5	102.1	105.1	99.9	90.1	92.8	91.1
2008	96.3	96.5	100.4	85.2	89.5	121.4	94.5	118.1	102.0	101.3	99.0	95.9	79.1	93.2
2009	93.9	93.5	111.4	97.7	84.8	104.3	80.9	114.7	89.7	97.0	93.1	93.6	93.9	92.0
2010	96.3	95.9	108.0	86.2	95.0	98.9	91.1	112.4	95.0	99.1	97.6	98.9	92.7	94.7
2011	97.8	97.6	103.8	89.4	98.7	97.3	97.3	104.9	96.8	99.3	99.4	100.3	92.5	97.8
2012	100.0	100.0	100.0	100.0	100.0	100.0	100.0	100.0	100.0	100.0	100.0	100.0	100.0	100.0
2013	101.8	101.9	116.6	103.9	98.9	102.5	102.5	103.8	102.3	103.1	101.5	109.1	94.1	101.9
2014	104.4	104.8	117.9	115.3	95.1	104.4	104.0	105.9	106.2	105.0	104.6	111.8	99.0	103.7
2015	107.5	108.3	125.8	125.1	94.9	109.3	105.5	106.0	110.8	108.5	107.5	122.1	102.6	105.8
2016	109.2	110.0	131.8	117.8	99.8	113.0	105.9	104.2	109.3	112.3	109.4	132.7	102.7	107.5
2017	111.8	112.9	129.8	126.3	101.5	115.6	109.5	105.9	111.3	116.2	114.4	140.5	102.1	109.9
2018	115.1	116.4	128.0	130.4	101.3	118.1	114.7	109.0	113.1	120.3	119.0	152.4	100.2	113.1
2019	117.8	119.4	133.6	149.4	103.5	118.1	116.7	108.3	114.0	124.6	118.9	159.4	103.5	114.9
2011														
1st quarter	97.0	96.7	103.9	85.7	93.3	96.1	95.9	106.1	96.7	99.6	99.1	99.8	92.9	95.6
2nd quarter	97.7	97.4	99.6	86.6	99.5	97.6	96.7	106.8	95.7	99.2	100.1	100.6	90.7	97.8
3rd quarter	97.7	97.4	102.2	89.8	97.8	97.8	97.1	103.5	95.2	98.4	98.7	100.2	92.4	97.9
4th quarter	98.8	98.9	109.5	95.4	104.2	97.9	99.5	103.2	99.6	99.9	99.6	100.5	93.9	99.8
2012														
1st quarter	99.6	99.6	105.4	96.9	98.7	100.1	99.8	103.8	98.1	100.5	100.5	100.0	97.6	99.3
2nd quarter	100.0	100.1	101.9	101.1	101.4	99.6	100.2	101.4	100.7	98.4	100.0	101.2	99.9	99.9
3rd quarter	100.1	100.1	97.1	100.1	100.8	99.1	99.6	98.3	101.0	100.7	99.8	99.7	101.7	100.7
4th quarter	100.3	100.2	95.6	101.9	99.1	101.2	100.3	96.5	100.2	100.4	99.7	99.1	100.8	100.1
2013														
1st quarter	101.1	101.1	109.1	101.4	101.7	102.0	102.3	102.1	101.5	103.6	101.2	106.8	93.5	101.0
2nd quarter	101.3	101.2	118.3	102.6	99.3	102.0	101.8	103.5	101.0	102.9	100.3	107.6	93.7	101.1
3rd quarter	102.1	102.1	121.3	104.0	98.5	103.3	102.7	105.1	102.6	102.7	101.1	109.4	93.2	102.6
4th quarter	102.9	103.1	117.7	107.8	96.2	102.6	103.0	104.6	104.1	103.3	103.2	112.6	95.9	103.0
2014														
1st quarter	102.6	102.7	116.1	105.2	92.7	102.8	101.6	103.9	102.6	103.8	103.0	109.8	96.5	102.8
2nd quarter	104.0	104.3	117.6	111.2	94.3	104.4	103.3	106.9	104.8	105.2	103.6	111.5	99.3	103.3
3rd quarter	105.2	105.8	118.2	118.1	96.7	104.4	105.7	106.3	108.4	106.0	105.9	112.4	99.3	104.3
4th quarter	105.8	106.5	119.8	126.8	96.8	106.0	105.3	106.5	109.0	105.1	105.8	113.7	100.8	104.6
2015														
1st quarter	106.7	107.5	122.1	133.4	89.8	107.1	105.5	108.4	110.1	106.6	104.4	117.5	103.0	105.1
2nd quarter	107.5	108.3	123.1	119.5	95.4	109.2	105.5	106.6	111.7	108.6	107.3	120.4	105.5	105.4
3rd quarter	107.8	108.7	127.6	123.5	97.5	110.1	105.9	105.8	111.0	109.3	109.2	123.5	101.2	106.2
4th quarter	107.9	108.6	130.2	123.9	97.0	110.6	105.1	103.2	110.2	109.5	108.9	127.0	100.7	106.4
2016														
1st quarter	108.4	109.2	130.0	131.9	98.1	112.9	105.5	104.1	108.2	111.2	106.8	130.4	99.9	106.8
2nd quarter	108.9	109.7	131.7	114.7	98.9	112.2	105.5	105.1	108.7	111.7	109.4	131.7	102.0	107.7
3rd quarter	109.5	110.3	132.9	112.1	100.7	112.8	106.3	104.8	110.3	112.5	109.5	134.0	103.8	107.7
4th quarter	110.0	110.9	132.4	112.8	101.4	114.0	106.3	103.0	110.0	113.7	111.7	134.7	105.0	107.9
2017														
1st quarter	110.7	111.6	135.1	116.6	99.9	115.6	107.5	104.3	110.8	113.9	113.0	136.1	103.6	109.0
2nd quarter	111.3	112.3	131.2	125.9	102.0	114.6	108.3	106.4	110.9	115.3	113.9	138.8	101.3	109.6
3rd quarter	112.1	113.3	126.5	131.0	101.1	114.9	109.9	105.8	111.3	117.4	114.7	142.6	102.4	110.1
4th quarter	113.1	114.3	126.4	131.6	103.0	117.2	112.3	107.2	112.2	118.3	115.8	144.3	101.2	110.8
2018														
1st quarter	113.8	115.1	126.5	126.3	99.5	118.7	113.4	107.9	112.2	120.1	118.4	146.6	100.2	111.9
2nd quarter	114.8	116.2	128.9	130.0	102.7	118.1	114.2	108.1	111.7	120.1	118.4	152.5	100.3	113.3
3rd quarter	115.7	117.1	127.4	130.9	101.2	118.3	114.9	110.0	113.2	121.0	119.4	154.3	101.5	113.4
4th quarter	116.0	117.4	129.0	134.4	102.0	117.3	116.1	110.0	115.3	120.2	119.8	156.2	98.7	114.0
2019														
1st quarter	116.9	118.5	129.9	142.4	101.1	118.4	116.5	107.2	115.0	122.7	119.6	156.8	103.5	114.2
2nd quarter	117.4	119.1	132.8	150.1	105.3	118.3	116.7	107.2	113.1	122.8	118.7	158.5	104.1	115.0
3rd quarter	118.1	119.8	135.4	152.9	101.1	117.7	116.8	109.8	113.9	125.2	118.4	160.6	102.6	115.2
4th quarter	118.7	120.4	136.4	152.3	106.5	118.1	116.9	109.0	113.9	127.5	118.9	161.9	103.9	115.3

Table 14-2. Chain-Type Quantity Indexes for Value Added by Industry—*Continued*

(New Series, 2012 = 100.)

Year	Private industries—Continued								Government			Private goods-producing industries	Private services-producing industries	Information-communications-technology-producing industries[1]
	Professional, scientific, and technical services	Management of companies and enterprises	Administrative and waste management services	Educational services	Health care and social assistance	Arts, entertainment, and recreation	Accommodation and food services	Other services except government	Total government	Federal	State and local			
2000	71.1	91.8	71.3	71.4	70.0	80.5	94.5	124.0	91.6	87.4	93.7	94.0	77.6	38.9
2001	73.3	95.5	72.3	73.2	71.7	75.9	91.9	111.7	92.5	86.2	95.7	91.4	79.0	40.5
2002	75.1	94.6	72.0	73.6	74.9	79.3	93.8	114.8	94.2	87.9	97.3	91.6	80.8	42.8
2003	76.1	97.8	77.1	77.1	77.8	82.1	95.9	111.6	95.3	89.6	98.1	95.0	83.0	47.6
2004	78.7	92.2	81.1	81.8	81.3	84.9	100.6	113.0	96.2	91.2	98.6	100.5	85.9	54.5
2005	81.7	92.1	88.2	80.4	83.3	84.8	101.1	113.8	97.0	91.9	99.6	102.9	89.7	61.6
2006	84.9	91.8	90.2	82.4	86.9	87.5	103.7	114.4	97.6	92.3	100.2	107.4	92.3	68.1
2007	88.1	89.6	95.3	83.7	87.4	88.5	102.6	111.7	98.5	93.0	101.3	109.0	93.8	76.4
2008	94.9	90.2	95.4	89.1	93.0	89.6	99.1	107.6	100.4	95.5	102.9	104.9	94.2	84.8
2009	89.8	83.4	87.8	96.3	95.6	88.4	91.8	101.3	100.6	98.1	101.8	97.9	92.4	85.0
2010	92.0	89.2	93.8	99.1	96.3	92.9	94.9	99.4	101.1	100.3	101.4	98.7	95.2	92.7
2011	95.9	92.9	96.9	99.8	98.1	95.5	98.5	98.5	100.7	100.5	100.9	98.8	97.2	97.7
2012	100.0	100.0	100.0	100.0	100.0	100.0	100.0	100.0	100.0	100.0	100.0	100.0	100.0	100.0
2013	100.3	104.9	101.5	98.4	101.8	102.0	102.2	99.3	99.3	97.7	100.1	103.9	101.3	108.3
2014	104.6	111.9	105.4	99.5	103.8	106.6	105.6	102.1	99.1	97.1	100.0	106.8	104.3	113.3
2015	109.0	114.9	106.8	101.2	108.2	104.7	109.9	103.0	99.1	97.2	100.1	109.7	107.9	124.6
2016	112.1	116.0	107.0	103.1	111.2	108.2	110.5	102.4	100.2	98.0	101.2	110.0	110.0	136.0
2017	115.3	124.4	113.5	100.4	113.7	110.7	112.8	102.4	101.1	98.6	102.3	112.9	112.8	145.8
2018	121.4	133.1	118.7	101.5	117.7	115.1	115.2	105.6	101.9	98.9	103.3	116.5	116.4	. . .
2019	128.0	143.7	122.3	102.3	121.3	116.5	116.1	106.1	102.4	99.1	104.0	118.8	119.6	0.0
2011														
1st quarter	94.2	93.2	95.9	100.1	97.1	94.1	97.4	98.7	101.0	100.7	101.2	98.0	96.3	. . .
2nd quarter ..	95.9	93.5	96.9	100.3	98.0	95.6	98.8	98.6	101.0	100.6	101.2	98.5	97.1	. . .
3rd quarter ...	96.3	92.2	97.4	99.5	98.5	97.0	99.1	98.3	100.5	100.3	100.6	98.4	97.1	. . .
4th quarter ...	97.0	92.7	97.5	99.2	98.8	95.1	98.7	98.4	100.5	100.3	100.5	100.4	98.5	. . .
2012														
1st quarter	99.0	97.3	100.0	100.3	99.9	99.2	100.2	99.6	100.4	100.4	100.4	101.0	99.3	. . .
2nd quarter ..	100.1	97.7	99.7	100.7	99.5	99.6	99.6	99.9	99.9	100.1	99.8	100.7	99.9	. . .
3rd quarter ...	100.0	99.8	100.0	100.0	99.7	99.7	99.4	100.2	99.9	99.9	99.8	99.1	100.4	. . .
4th quarter ...	100.8	105.2	100.4	99.0	100.8	101.5	100.9	100.3	99.8	99.5	100.0	99.3	100.4	. . .
2013														
1st quarter	100.0	101.2	100.2	98.1	101.6	101.3	102.5	99.4	99.8	99.0	100.2	102.5	100.8	. . .
2nd quarter ..	99.3	103.6	101.0	98.1	101.7	101.4	101.7	98.9	99.6	98.3	100.2	103.4	100.6	. . .
3rd quarter ...	100.3	106.5	102.1	98.3	101.8	101.4	101.8	99.0	99.2	97.3	100.1	104.8	101.4	. . .
4th quarter ...	101.5	108.4	102.8	99.3	102.2	103.9	102.7	99.7	98.7	96.1	99.9	104.9	102.6	. . .
2014														
1st quarter	101.8	112.4	103.5	98.4	102.3	105.5	103.5	99.9	98.9	97.5	99.6	103.8	102.4	. . .
2nd quarter ..	103.3	110.7	105.3	99.3	103.0	106.5	104.9	101.7	99.2	97.2	100.2	106.4	103.7	. . .
3rd quarter ...	106.4	111.9	106.0	100.2	104.4	107.4	106.7	103.4	99.1	96.9	100.1	107.9	105.2	. . .
4th quarter ...	106.9	112.7	106.6	100.1	105.4	107.0	107.3	103.5	99.1	96.7	100.2	109.1	105.8	. . .
2015														
1st quarter	107.7	114.9	106.8	100.5	107.0	104.6	108.7	103.3	98.7	97.1	99.5	110.5	106.7	. . .
2nd quarter ..	108.5	114.5	106.6	100.7	107.8	104.1	109.8	103.5	99.1	97.2	100.1	109.4	108.0	. . .
3rd quarter ...	109.7	115.0	107.0	101.3	108.8	104.8	110.6	102.8	99.3	97.2	100.3	110.0	108.3	. . .
4th quarter ...	110.0	115.2	106.7	102.4	109.2	105.1	110.6	102.4	99.5	97.3	100.5	109.1	108.5	. . .
2016														
1st quarter	111.5	115.1	106.3	103.4	110.4	106.3	110.2	102.6	99.7	97.6	100.7	110.5	108.9	. . .
2nd quarter ..	112.0	115.3	106.2	103.7	111.2	107.3	109.8	102.4	99.9	97.9	100.8	109.8	109.7	. . .
3rd quarter ...	111.8	116.0	107.0	102.8	111.0	108.5	110.6	102.5	100.5	98.2	101.5	110.0	110.4	. . .
4th quarter ...	113.1	117.7	108.5	102.7	112.1	110.8	111.3	101.9	100.7	98.3	101.8	109.7	111.2	. . .
2017														
1st quarter	113.4	120.1	110.8	101.2	112.9	108.2	112.3	101.6	100.9	98.6	102.1	111.3	111.7	. . .
2nd quarter ..	114.7	122.5	112.8	100.1	113.4	109.5	112.5	101.7	100.9	98.4	102.1	112.4	112.2	. . .
3rd quarter ...	115.8	124.5	114.3	100.2	113.9	113.0	112.9	102.8	101.2	98.7	102.4	113.0	113.3	. . .
4th quarter ...	117.5	130.3	116.1	100.0	114.7	112.2	113.4	103.6	101.4	98.8	102.6	114.9	114.1	. . .
2018														
1st quarter	119.3	130.1	116.6	100.8	116.5	111.9	113.5	104.4	101.5	98.9	102.8	115.5	115.0	. . .
2nd quarter ..	120.9	131.9	117.8	101.0	117.2	116.3	115.3	105.6	101.9	99.0	103.2	116.1	116.1	. . .
3rd quarter ...	122.4	135.3	119.4	101.6	118.1	116.2	115.6	106.0	102.1	99.1	103.5	117.0	117.0	. . .
4th quarter ...	122.9	135.2	121.2	102.6	118.8	116.1	116.4	106.5	102.1	98.6	103.7	117.5	117.4	. . .
2019														
1st quarter	125.3	140.2	120.9	102.2	120.4	116.9	115.7	106.2	101.7	97.4	103.7	117.8	118.7	. . .
2nd quarter ..	127.6	142.6	121.9	101.0	120.9	116.2	115.1	105.7	102.4	99.2	104.0	118.5	119.2	. . .
3rd quarter ...	129.3	145.6	122.8	102.7	121.6	116.5	116.7	106.4	102.4	99.7	103.8	119.4	119.8	. . .
4th quarter ...	129.8	146.5	123.6	103.3	122.5	116.2	116.8	106.0	103.1	100.3	104.5	119.3	120.6	. . .

[1]Consists of computer and electronic products manufacturing; publishing, including software; information and data processing services; and computer systems design and related services.
. . . = Not available.

Table 14-3. Gross Domestic Factor Income by Industry

(Billions of current dollars.)

NAICS industry	2006	2007	2008	2009	2010	2011	2012	2013	2014	2015	2016	2017	2018
Gross domestic factor income, total	12 869.1	13 469.6	13 715.7	13 480.5	13 984.8	14 498.8	15 119.0	15 655.9	16 344.6	17 004.9	17 464.1	18 216.1	19 202.9
Compensation of employees	7 491.3	7 889.4	8 068.7	7 767.2	7 933.0	8 234.0	8 575.4	8 843.6	9 259.7	9 709.2	9 972.7	10 424.5	10 941.4
Gross operating surplus	5 377.8	5 580.2	5 647.0	5 713.3	6 051.8	6 264.8	6 543.6	6 812.3	7 084.9	7 295.7	7 491.4	7 791.6	8 261.5
Private industries	11 059.6	11 563.9	11 713.7	11 412.9	11 856.1	12 339.1	12 937.0	13 418.8	14 048.0	14 639.7	15 049.3	15 737.2	16 631.4
Compensation of employees	6 047.1	6 369.8	6 475.5	6 126.3	6 241.3	6 532.0	6 871.6	7 095.3	7 468.7	7 861.2	8 083.7	8 486.3	8 929.2
Gross operating surplus	5 012.5	5 194.1	5 238.2	5 286.6	5 614.8	5 807.1	6 065.4	6 323.5	6 579.3	6 778.5	6 965.6	7 250.9	7 702.2
Agriculture, forestry, fishing, and hunting	131.6	146.6	148.8	131.8	147.9	180.2	178.5	214.3	196.9	178.7	165.1	173.1	166.4
Compensation of employees	38.0	41.8	42.1	42.3	41.4	41.2	48.1	48.9	51.0	50.7	54.0	55.8	59.5
Gross operating surplus	93.6	104.8	106.7	89.5	106.5	139.0	130.4	165.4	145.9	128.0	111.1	117.3	106.9
Mining	243.2	280.2	349.7	246.0	272.6	318.4	318.5	344.6	369.9	221.0	179.7	248.0	303.3
Compensation of employees	57.0	62.7	72.8	64.7	69.2	80.3	90.5	93.5	101.1	91.9	74.5	76.8	84.2
Gross operating surplus	186.2	217.5	276.9	181.3	203.4	238.1	228.0	251.1	268.8	129.1	105.2	171.2	219.1
Utilities	178.7	180.3	188.2	203.1	222.0	229.4	220.8	227.0	238.8	239.2	240.6	251.4	260.5
Compensation of employees	60.9	63.1	66.3	66.8	67.7	71.2	69.8	72.8	75.2	77.6	81.4	82.8	85.2
Gross operating surplus	117.8	117.2	121.9	136.3	154.3	158.2	151.0	154.2	163.6	161.6	159.2	168.6	175.3
Construction	688.8	707.0	641.0	558.3	517.9	517.2	545.9	579.6	628.5	686.5	736.0	780.3	828.7
Compensation of employees	421.3	440.0	433.9	368.4	343.8	347.3	365.9	388.0	422.4	458.4	485.9	516.5	553.8
Gross operating surplus	267.5	267.0	207.1	189.9	174.1	169.9	180.0	191.6	206.1	228.1	250.1	263.8	274.9
Durable goods manufacturing	980.6	1 005.4	974.0	854.9	939.1	989.9	1 036.3	1 073.2	1 104.7	1 154.0	1 161.5	1 200.2	1 264.8
Compensation of employees	614.8	627.4	615.5	539.2	546.2	578.7	604.4	612.0	638.6	662.6	665.9	696.5	733.6
Gross operating surplus	365.8	378.0	358.5	315.7	392.9	411.2	431.9	461.2	466.1	491.4	495.6	503.7	531.2
Nondurable goods manufacturing	751.7	775.0	761.4	772.0	781.7	798.9	811.6	834.3	861.2	887.2	854.6	897.1	967.8
Compensation of employees	309.2	316.2	316.4	296.0	300.9	303.7	312.6	319.8	334.5	343.6	347.0	360.6	371.4
Gross operating surplus	442.5	458.8	445.0	476.0	480.8	495.2	499.0	514.5	526.7	543.6	507.6	536.5	596.4
Wholesale trade	640.0	685.4	712.5	676.0	718.7	749.9	804.9	837.8	880.4	930.1	919.5	942.2	967.4
Compensation of employees	404.6	429.3	435.7	405.1	412.3	436.2	457.7	467.6	489.9	509.9	511.2	533.0	544.9
Gross operating surplus	235.4	256.1	276.8	270.9	306.4	313.7	347.2	370.2	390.5	420.2	408.3	409.2	422.5
Retail trade	690.9	688.2	670.1	661.7	673.1	684.8	713.8	744.1	760.5	801.8	828.3	849.3	882.0
Compensation of employees	492.2	506.4	500.9	474.3	478.9	495.7	510.1	526.7	545.5	573.6	587.7	606.8	627.3
Gross operating surplus	198.7	181.8	169.2	187.4	194.2	189.1	203.7	217.4	215.0	228.2	240.6	242.5	254.7
Transportation and warehousing	387.1	388.2	400.7	378.1	407.6	424.0	443.7	459.7	488.6	529.4	544.0	573.1	616.6
Compensation of employees	241.0	255.9	255.6	241.1	245.5	259.9	274.8	283.4	296.8	319.8	332.5	351.8	377.6
Gross operating surplus	146.1	132.3	145.1	137.0	162.1	164.1	168.9	176.3	191.8	209.6	211.5	221.3	239.0
Information	603.4	658.9	698.2	677.9	708.9	714.6	713.6	781.5	793.6	848.5	909.0	944.9	1 013.3
Compensation of employees	248.8	260.5	257.8	251.5	248.6	260.1	271.6	286.5	304.4	318.7	329.0	349.2	376.9
Gross operating surplus	354.6	398.4	440.4	426.4	460.3	454.5	442.0	495.0	489.2	529.8	580.0	595.7	636.4
Finance and insurance	994.0	988.2	829.5	922.9	959.5	980.9	1 117.5	1 097.8	1 225.2	1 308.1	1 377.7	1 428.0	1 492.8
Compensation of employees	579.2	617.4	612.1	548.0	573.3	606.5	630.1	641.3	681.1	715.7	734.1	786.5	816.8
Gross operating surplus	414.8	370.8	217.4	374.9	386.2	374.4	487.4	456.5	544.1	592.4	643.6	641.5	676.0
Real estate and rental and leasing	1 514.4	1 612.6	1 675.7	1 677.1	1 714.9	1 792.1	1 868.1	1 941.1	2 017.5	2 124.5	2 217.2	2 316.6	2 439.0
Compensation of employees	108.0	112.2	110.3	103.7	103.8	106.4	114.3	118.9	126.4	136.7	140.7	148.6	158.5
Gross operating surplus	1 406.4	1 500.4	1 565.4	1 573.4	1 611.1	1 685.7	1 753.8	1 822.2	1 891.1	1 987.8	2 076.5	2 168.0	2 280.5
Professional, scientific, and technical services	880.2	949.8	1 051.9	1 008.6	1 040.2	1 099.4	1 161.3	1 177.7	1 234.6	1 311.0	1 354.9	1 412.7	1 505.6
Compensation of employees	607.4	655.0	693.5	669.4	682.7	727.1	776.7	801.5	848.9	909.4	939.1	988.7	1 054.6
Gross operating surplus	272.8	294.8	358.4	339.2	357.5	372.3	384.6	376.2	385.7	401.6	415.8	424.0	451.0
Management of companies and enterprises	223.1	243.4	247.2	236.6	257.1	269.7	293.8	311.2	324.5	339.7	342.4	364.6	385.6
Compensation of employees	200.7	219.9	221.1	207.9	227.2	237.9	259.4	275.0	287.4	302.4	304.6	324.9	342.7
Gross operating surplus	22.4	23.5	26.1	28.7	29.9	31.8	34.4	36.2	37.1	37.3	37.8	39.7	42.9
Administrative and waste management services	390.4	421.3	429.1	403.1	427.7	443.7	463.1	477.6	505.8	527.3	545.5	587.7	623.6
Compensation of employees	292.4	310.1	314.1	289.2	306.5	321.9	341.3	353.7	376.7	398.1	409.7	438.9	462.9
Gross operating surplus	98.0	111.2	115.0	113.9	121.2	121.8	121.8	123.9	129.1	129.2	135.8	148.8	160.7
Educational services	137.0	145.2	160.9	180.7	191.0	198.1	205.9	209.2	218.0	226.9	237.0	237.0	245.7
Compensation of employees	113.1	120.9	129.4	139.9	145.5	152.8	161.4	165.2	174.9	182.6	190.6	189.7	197.5
Gross operating surplus	23.9	24.3	31.5	40.8	45.5	45.3	44.5	44.0	43.1	44.3	46.4	47.3	48.2
Health care and social assistance	871.3	913.5	993.6	1 054.6	1 087.0	1 127.8	1 169.6	1 205.4	1 241.1	1 309.0	1 375.0	1 430.2	1 503.3
Compensation of employees	733.2	775.1	824.6	862.4	888.5	922.0	963.6	997.3	1 028.3	1 085.1	1 136.5	1 186.8	1 245.6
Gross operating surplus	138.1	138.4	169.0	192.2	198.5	205.8	206.0	208.1	212.8	223.9	238.5	243.4	257.7
Arts, entertainment, and recreation	444.7	461.6	468.4	460.5	479.2	505.3	541.3	566.6	602.4	645.8	681.5	710.5	751.8
Compensation of employees	306.4	324.8	333.1	321.3	327.7	343.9	368.6	385.5	412.9	438.7	465.0	488.2	515.2
Gross operating surplus	138.3	136.8	135.3	139.2	151.5	161.4	172.7	181.1	189.5	207.1	216.5	222.3	236.6
Accommodation and food services	327.6	339.3	340.7	331.1	342.5	363.0	387.6	407.2	432.1	472.3	495.4	518.0	547.2
Compensation of employees	233.3	247.8	252.5	242.8	248.3	262.4	282.5	295.8	316.2	337.9	357.1	375.6	396.2
Gross operating surplus	94.3	91.5	88.2	88.3	94.2	100.6	105.1	111.4	115.9	134.4	138.3	142.4	151.0
Other services, except government	308.4	313.1	312.6	308.9	310.1	314.7	329.0	336.3	356.0	371.0	379.6	390.5	412.9
Compensation of employees	218.9	231.1	240.1	235.1	231.7	239.3	250.9	257.9	272.8	285.8	294.2	304.4	320.9
Gross operating surplus	89.5	82.0	72.5	73.8	78.4	75.4	78.1	78.4	83.2	85.2	85.4	86.1	92.0
Government	1 809.5	1 905.8	2 002.0	2 067.6	2 128.8	2 159.8	2 182.0	2 237.1	2 296.6	2 365.3	2 414.8	2 478.8	2 571.4
Compensation of employees	1 444.2	1 519.6	1 593.2	1 640.9	1 691.7	1 702.0	1 703.8	1 748.4	1 791.0	1 848.1	1 889.0	1 938.1	2 012.2
Gross operating surplus	365.3	386.2	408.8	426.7	437.1	457.8	478.2	488.7	505.6	517.2	525.8	540.7	559.2
Addenda:													
Private goods-producing industries	2 795.9	2 914.4	2 875.0	2 562.9	2 659.2	2 804.6	2 890.8	3 046.0	3 161.1	3 127.4	3 096.9	3 298.7	3 531.0
Compensation of employees	1 440.3	1 488.2	1 480.8	1 310.5	1 301.5	1 351.1	1 421.4	1 462.2	1 547.5	1 607.2	1 627.3	1 706.2	1 802.4
Gross operating surplus	1 355.6	1 426.2	1 394.2	1 252.4	1 357.7	1 453.5	1 469.4	1 583.8	1 613.6	1 520.2	1 469.6	1 592.5	1 728.6
Private services-producing industries	8 263.6	8 649.5	8 838.7	8 849.9	9 196.9	9 534.5	10 046.1	10 372.8	10 886.9	11 512.2	11 952.3	12 438.6	13 100.3
Compensation of employees	4 606.8	4 881.6	4 994.7	4 815.8	4 939.8	5 180.9	5 450.1	5 633.1	5 921.2	6 253.9	6 456.4	6 780.2	7 126.7
Gross operating surplus	3 656.8	3 767.9	3 844.0	4 034.1	4 257.1	4 353.6	4 596.0	4 739.7	4 965.7	5 258.3	5 495.9	5 658.4	5 973.6
Information-communications-technology-producing industries [1]	747.3	804.9	859.3	844.9	892.9	928.6	939.2	1 014.1	1 045.0	1 123.5	1 196.9	1 257.5	1 352.0
Compensation of employees	396.0	420.3	422.2	410.6	423.0	453.6	483.8	501.4	536.2	571.6	597.1	641.5	691.9
Gross operating surplus	351.3	384.6	437.1	434.3	469.9	475.0	455.4	512.7	508.8	551.9	599.8	616.0	660.1

[1] Consists of computer and electronic products manufacturing; publishing, including software; information and data processing services; and computer systems design and related services.

NOTES AND DEFINITIONS, CHAPTER 14

TABLES 14-1 THROUGH 14-3

Gross Domestic Product (VALUE ADDED) and Gross Factor Income by Industry

SOURCE: U.S. DEPARTMENT OF COMMERCE, BUREAU OF ECONOMIC ANALYSIS (BEA)

In the introduction to the notes and definitions for Chapter 1, it was observed that gross domestic product (GDP), while primarily measured as the sum of final demands for goods and services, is also the sum of the values created by each industry in the economy. In this chapter, selected data are presented from the industry accounts in the national income and product accounts (NIPAs). The industry accounts measure the contribution of each major industry to GDP.

Quarterly estimates of current- and constant-dollar GDP by industry were new in the previous edition of Business Statistics. These measures have been computed starting with the first quarter of 2005 and extending through 2019; they are displayed for 2005-2019 in Tables 14-1A and 14-2A. They provide a complete accounting of the industrial origin of GDP, and measure the output of major industries on a more up-to-date basis than the annual measures previously available--within 120 days after the end of the reference quarter.

In recent years, estimates of GDP by industry have been prepared using a methodology integrated with annual input-output accounts in order to produce estimates of gross industry output, industry input, and the difference between the two—industry value added—with greater consistency and timeliness than was previously possible. The current integrated industry accounts also include a wealth of related information too extensive for inclusion here, such as quantity and price indexes for gross output and intermediate inputs, cost per unit of real value added allocated to the three components of value added, and components of domestic supply (domestic output, imports, exports, and inventory change).

The most precise estimates of GDP by NAICS industry begin with 1997, but the annual series were extended back to 1947, using approximations of current methods in earlier years when the source data were less comprehensive and were initially compiled on the older SIC classifications. Table 14-1B presents this "old series" current-dollar GDP for NAICS industry groups from 1947 through 2009; Table 14-2B presents "old series" indexes of real value added for each industry group.

The 2019 comprehensive benchmark revision incorporated three major types of revisions: 1) statistical changes to introduce new and improved technologies and newly available source data; 2) changes in definitions to more accurately portray the evolving U.S. economy to more accurately portray the evolving U.S. economy and to provide consistent comparisons with data for other national economies; and 3) changes in presentations to reflect the definitional and statistical changes, where necessary, or to provide additional data or perspective for users. Also output and price measures will be expressed with 2012 equal to 100. The estimates for most tables showing real, or chained-dollar, estimates will begin with 2002.

In Table 14-3, the editor presents a measure derived from the components of current-dollar value added as published by BEA. This measure, "gross domestic factor income," enables users to obtain a clearer picture of the quantitative impact of each industry on the economy's factors of production and of the shares of capital and labor in each industry.

The 2004 revision incorporated a change in terminology. An industry's contribution to total GDP, formerly referred to as "gross product originating" (GPO) or "gross product by industry," is now called "value added." This is consistent with the use of the term "value added" in most economic writing. However, it should not be confused with a concept known as "Census value added," which is used in U.S. censuses and surveys of manufactures. Census value added is calculated at the individual establishment level and, for that reason, does not exclude purchased business services. This means that census value added, while still an important gauge of relative importance, is not a true measure of economic value added.

Definitions and notes on the data

An industry's *gross domestic product (value added)*, formerly *GPO*, is equal to the market value of its gross output (which consists of the value, including taxes, of sales or receipts and other operating income plus the value of inventory change) minus the value of its intermediate inputs (energy, raw materials, semi finished goods, and services that are purchased from domestic industries or from foreign sources).

In concept, this is also equal to the sum of *compensation of employees, taxes on production and imports less subsidies,* and *gross operating surplus.* (See Chapter 1 and its notes and definitions for more information.)

The GDP data from 1997 forward are not comparable with the "old series" data for earlier years presented in Tables 14-1B and 14-2B, the major reason being that they do not reflect the conceptual revisions introduced in the comprehensive 2018 revision of the NIPAs (see the Notes and Definitions to Chapter 1). In particular, the new data include, while the old data exclude, the capitalized values of research and development, intellectual property, and all real estate transfer costs. This does not change labor compensation values—the associated workers were always included; it was just that their costs were taken to be written off in the first year. But it does increase across the board the non-labor share of the total value of output, by the amount of capital consumption and net investment. This is seen in labor compensation shares derived from Table 14-3, compared with those that would have been calculated on the old basis last year.

Compensation of employees consists of wage and salary accruals and supplements to wages and salaries. This approximates the labor share of production, subject to the note below about proprietors' income.

Taxes on production and imports less subsidies. Although this is shown in BEA source data as a single net line item, it represents two separate components.

Taxes on production and imports are included in the market value of the goods and services sold to final consumers and therefore in the consumer valuation of those goods. Since they are not part of the payments to the labor and capital inputs in the producing industries, they must be added to the sum of the returns to those inputs in order to account for the total value to consumers. Taxes that fall into this classification include property taxes, sales and excise taxes, and customs duties.

BEA allocates these taxes to the industry level at which they are assessed by law. Most sales taxes are considered by BEA to be part of the value added by retail trade. Some sales taxes, most fuel taxes, and all customs duties are allocated by BEA to wholesale trade. Residential real property taxes, including those on owner-occupied dwellings, are allocated by BEA to the real estate industry.

Subsidies to business by government are included in the labor and/or capital payments made by that industry. Since they are payments to the industry in addition to the market values paid by consumers, they are *subtracted* from the values of the labor and capital inputs to make them consistent with the market values as defined in value added. The role of subsidies is obvious in the source data for the agricultural sector, where the net "taxes on production and imports less subsidies" has a negative sign: farm subsidies more than offset this industry's taxes on production and imports, which mainly consist of property taxes, since sales, excise, and import taxes are not levied on farms.

For private sector businesses, *gross operating surplus* consists of business income (corporate profits before tax, proprietors' income, and rental income of persons), net interest and miscellaneous payments, business current transfer payments (net), and capital consumption allowances. For government, households, and institutions, it consists of consumption of fixed capital and (for government) government enterprises' current surplus. This

approximates the share of the value of production ascribable to capital and land as measured in the NIPAs accounts; however, as BEA notes, "an unknown portion [of proprietors' income] reflects the labor contribution of proprietors." (*Survey of Current Business*, June 2004, p. 27, footnote 7.) Another aspect to be noted is that gross operating surplus includes the return to owner-occupied housing in the real estate sector. Because there is in the NIPAs no employee compensation attributed to owner-occupied housing, the capital share in that industry as measured by gross operating surplus is very large.

Quantity indexes for value added. Measures of the constant-dollar change in each industry's gross output minus its intermediate input use are calculated by BEA, using a Fisher index-number formula which incorporates weights from two adjacent years. The changes for successive years are chained together in indexes, with the value for the year 2012 set at 100. The indexes are multiplied by 2012 current-dollar value added to provide estimates of value added in "chained 2012 dollars," but because the actual weights used change from year to year, components in chained 2012 dollars typically do not add up to total GDP in 2009 dollars—and do not contain any information not already summarized in the indexes. For that reason, only the indexes are published here.

Gross domestic factor income (not a category published as such in the NIPAs) is calculated by the editor as the sum of "compensation of employees" and "gross operating surplus," which the same as value added minus "taxes on production and imports less subsidies." The effect of this procedure is to take out the specified taxes, and to leave in the subsidies embedded in the employee compensation and gross operating surplus components. The editor believes that this provides a valuable alternative basis for assessing the importance of different industries in the economy, by measuring the value paid for by its labor and capital inputs, and for calculating the shares of labor and capital in each industry's output. The components shown in this table are found on the BEA website in a table entitled "Components of Value Added by Industry."

The editor's reasoning is based on the facts that more than half of these taxes are sales, excise, and import taxes, and the assignment of these taxes to industries is economically arbitrary. BEA assigns them to the industry with the legal liability to pay, not to the entity bearing the major incidence of the tax. Yet economists have demonstrated that most of the burdens of sales and excise taxes and import duties are not borne by the factors in the legally liable industry; instead, they are passed on to consumers. In addition, because the wholesale and retail trade industries are classified as "services-producing," BEA's allocation of those taxes has a very peculiar result: taxes on goods are represented as paid by "service" industries. The process adopted instead by the editor in Table 14-3, which excludes these taxes and focuses on "gross domestic factor income," has the effect (for example) of keeping the wholesale trade industry from appearing to be both larger and more heavily taxed than it really is.

Private goods-producing industries consists of agriculture, forestry, fishing, and hunting; mining; construction; and manufacturing.

Private services-producing industries consists of utilities; wholesale trade; retail trade; transportation and warehousing; information; finance, insurance, real estate, rental, and leasing; professional and business services; educational services, health care, and social assistance; arts, entertainment, recreation, accommodation, and food services; and other services, except government.

Information-communications-technology-producing industries is a category that cuts across the goods and services framework, consisting of computer and electronic products manufacturing; publishing industries (which includes software) from the information sector; information and data processing services; and computer systems design and related services.

Data availability and references

The comprehensive revision of the industry accounts from 1997 forward and the new quarterly statistics are described in "Industry Economic Accounts: Results of the Comprehensive Revision, Revised Statistics for 1997-2012" (*Survey of Current Business,* February 2014); "New Quarterly Statistics Detail Industries' Economic Performance," BEA news release, April 25, 2014; and "Prototype Quarterly Statistics on U.S. GDP by Industry, 2007-2011," June 2012, a "BEA Briefing." All are available on the BEA website, http://www.bea.gov. To access the historical data at <http://www.bea.gov>, click on "Annual Industry Accounts" and click on "Interactive Tables" under "Gross Domestic Product (GDP) by Industry."

"Old basis" annual GDP by industry data shown here were recalculated back to 1947 consistent with the 2009 comprehensive revision of the NIPAs. They were most recently presented and described in "Annual Industry Statistics: Revised Statistics for 2009-2011," *Survey of Current Business*, December 2012.

CHAPTER 15: EMPLOYMENT, HOURS, AND EARNINGS BY NAICS INDUSTRY

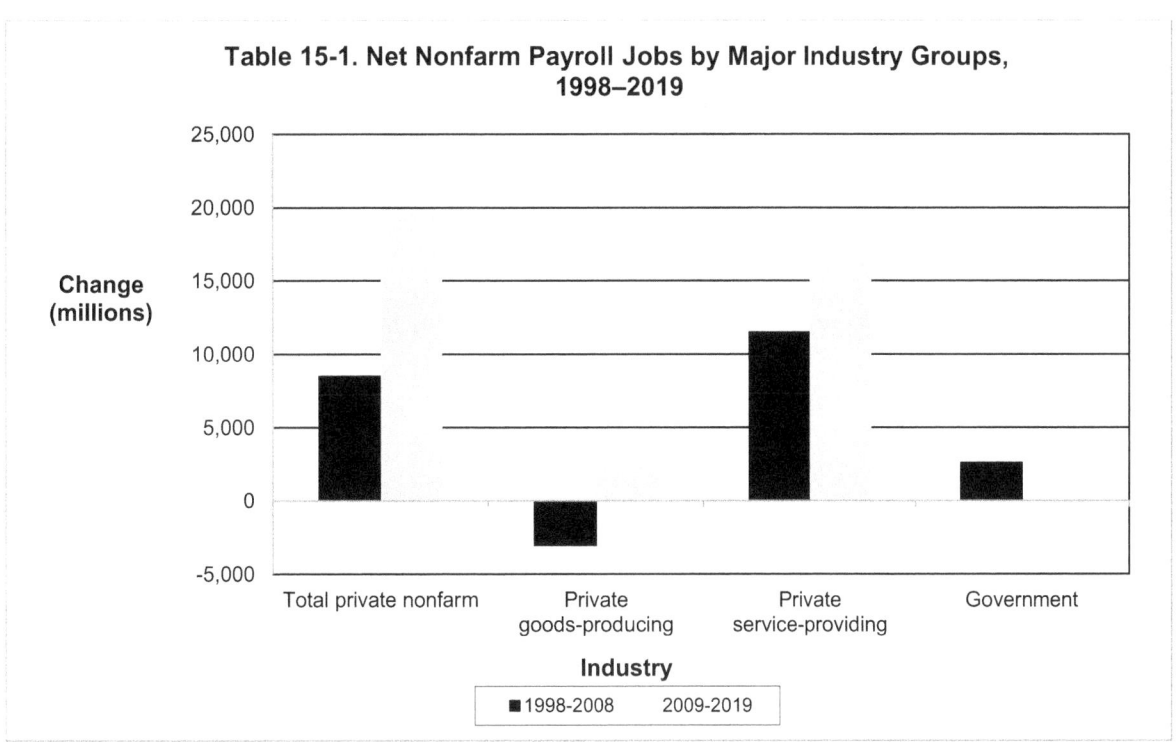

Table 15-1. Net Nonfarm Payroll Jobs by Major Industry Groups, 1998–2019

- In 2019, total nonfarm employment increased for the ninth consecutive year, increasing 1.4 percent in 2019. It increased in all major sectors growing 1.8 percent in goods-producing industries, 1.3 percent in services providing industries, and 0.6 percent in government. In 2020, employment began to decline due to the global pandemic.

- During the Great Recession, employment declined dramatically. In 2014, employment finally exceeded 2007 employment. In 2007, there were 137.9 million nonfarm employees versus 138.9 in 2014, an increase of 0.7 percent. (Table 15-1)

- Although the number of goods-producing jobs continued to increase in 2019, employment was still significantly lower than it was twenty years earlier in 1999. Meanwhile, the number of service-providing jobs increased 24 percent from 1999 to 2019. (Table 15-1)

- The average workweek for production and nonsupervisory workers decreased in 2019 after increasing the past two years. In 2019, goods-producing workers averaged 41.1 hours per week while private servicing-providers averaged far less at 32.4 hours. (Table 15-3)

- Average hourly earnings for production and nonsupervisory workers in 2019 ranged from a low of $14.49 in the leisure and hospitality sector to $41.37 in petroleum and coal products manufacturing. (Table 15-4)

Table 15-1. Nonfarm Employment by Sector and Industry

(Wage and salary workers on nonfarm payrolls, thousands.)

Industry	1998	1999	2000	2001	2002	2003	2004	2005	2006	2007	2008
TOTAL NONFARM	126 146	129 228	132 011	132 073	130 634	130 331	131 769	134 034	136 435	137 981	137 224
Total Private	106 237	108 921	111 221	110 955	109 121	108 748	110 148	112 230	114 462	115 763	114 714
Goods-Producing	24 354	24 465	24 649	23 873	22 557	21 816	21 882	22 190	22 530	22 233	21 335
Mining and logging	645	598	599	606	583	572	591	628	684	724	767
Logging	80.0	80.8	79.0	73.5	70.4	69.4	67.6	65.2	64.4	60.1	56.6
Mining	564.7	517.4	520.2	532.5	512.2	502.7	523.0	562.2	619.7	663.8	709.8
Oil and gas extraction	140.8	131.2	124.9	123.7	121.9	120.2	123.4	125.7	134.5	146.2	160.5
Mining, except oil and gas [1]	243.1	234.5	224.8	218.7	210.6	202.7	205.1	212.8	220.3	223.4	226.0
Coal mining	85.3	78.6	72.2	74.3	74.4	70.0	70.6	73.9	78.0	77.2	81.2
Support activities for mining	180.8	151.7	170.6	190.1	179.8	179.8	194.6	223.7	264.9	294.3	323.4
Construction	6 149	6 545	6 787	6 826	6 716	6 735	6 976	7 336	7 691	7 630	7 162
Construction of buildings	1 508.8	1 586.3	1 632.5	1 588.9	1 574.8	1 575.8	1 630.0	1 711.9	1 804.9	1 774.2	1 641.7
Heavy and civil engineering	865.3	908.7	937.0	953.0	930.6	903.1	907.4	951.2	985.1	1 005.4	964.5
Specialty trade contractors	3 775.1	4 049.6	4 217.0	4 283.9	4 210.4	4 255.7	4 438.6	4 673.1	4 901.1	4 850.2	4 555.8
Manufacturing	17 560	17 322	17 263	16 441	15 259	14 509	14 315	14 227	14 155	13 879	13 406
Durable goods	10 911	10 831	10 877	10 336	9 485	8 964	8 925	8 956	8 981	8 808	8 463
Wood products	611.6	622.7	615.4	576.3	557.0	539.5	551.6	561.0	560.6	517.1	457.7
Nonmetallic mineral products	535.3	540.8	554.2	544.5	516.0	494.2	505.5	505.3	509.6	500.5	465.0
Primary metals	641.5	625.0	621.8	570.9	509.4	477.4	466.8	466.0	464.0	455.8	442.0
Fabricated metal products	1 739.5	1 728.4	1 752.6	1 676.4	1 548.5	1 478.9	1 497.1	1 522.0	1 553.1	1 562.8	1 527.5
Machinery	1 514.1	1 468.3	1 457.0	1 370.6	1 231.8	1 151.6	1 145.2	1 165.5	1 183.2	1 187.1	1 187.6
Computer and electronic products [1]	1 830.9	1 780.5	1 820.0	1 748.8	1 507.2	1 355.2	1 322.8	1 316.4	1 307.5	1 272.5	1 244.2
Computer and peripheral equipment	322.1	310.1	301.9	286.2	250.0	224.0	210.0	205.1	196.2	186.2	183.2
Communications equipment	237.4	228.7	238.6	225.4	179.0	149.2	143.0	141.4	136.2	128.1	127.3
Semiconductors and electronic components	649.8	630.5	676.3	645.4	524.5	461.1	454.1	452.0	457.9	447.5	431.8
Electronic instruments	509.2	498.3	487.7	483.6	456.8	435.4	436.9	441.0	444.5	443.2	441.0
Electrical equipment and appliances	591.6	588.0	590.9	556.9	496.5	459.6	445.1	433.5	432.7	429.4	424.3
Transportation equipment [1]	2 078.4	2 088.6	2 057.1	1 939.1	1 830.0	1 775.1	1 766.7	1 772.3	1 768.9	1 711.9	1 608.0
Motor vehicles and parts	1 271.5	1 312.5	1 313.6	1 212.9	1 151.2	1 125.3	1 112.8	1 096.7	1 070.0	994.2	875.5
Furniture and related products	641.6	665.2	680.1	642.9	604.7	573.7	574.1	566.3	558.3	529.4	478.0
Miscellaneous manufacturing	726.8	724.0	728.0	709.5	683.3	658.3	650.6	647.2	643.7	641.7	628.9
Nondurable goods [1]	6 649	6 491	6 386	6 105	5 774	5 546	5 390	5 271	5 174	5 071	4 943
Food manufacturing	1 554.9	1 549.8	1 553.1	1 551.2	1 525.7	1 517.5	1 493.7	1 477.6	1 479.4	1 484.1	1 480.9
Textile mills	424.5	397.1	378.2	332.9	290.9	261.3	236.9	217.6	195.0	169.7	151.2
Textile product mills	234.7	232.4	229.6	217.0	204.2	187.7	183.2	176.4	166.7	157.7	147.2
Apparel	621.4	540.5	483.5	415.2	350.0	303.9	278.0	250.5	232.4	214.6	199.0
Paper and paper products	624.9	615.6	604.7	577.6	546.6	516.2	495.5	484.2	470.5	458.2	444.9
Printing and related support activities	827.9	814.6	806.8	768.3	706.6	680.4	662.6	646.3	634.4	622.0	594.1
Petroleum and coal products	134.5	127.8	123.2	121.1	118.1	114.3	111.7	112.1	113.2	114.5	117.4
Chemicals	992.6	982.5	980.4	959.0	927.5	906.1	887.0	872.1	865.9	860.9	847.1
Plastics and rubber products	941.4	947.0	950.9	896.2	846.8	814.3	804.7	802.3	785.5	757.2	729.4
Service-Providing	101 792	104 763	107 362	108 200	108 077	108 515	109 887	111 843	113 905	115 748	115 889
Private Service-Providing	81 883	84 456	86 572	87 082	86 564	86 931	88 266	90 039	91 931	93 530	93 380
Trade, transportation, and utilities	25 139	25 722	26 174	25 931	25 442	25 228	25 470	25 892	26 206	26 556	26 219
Wholesale trade	5 751.7	5 848.0	5 888.4	5 728.0	5 605.5	5 556.4	5 607.8	5 705.7	5 841.9	5 948.4	5 874.9
Durable goods	3 251.5	3 310.2	3 342.4	3 221.4	3 103.7	3 044.5	3 062.5	3 118.8	3 202.7	3 257.3	3 190.6
Nondurable goods	2 097.1	2 126.8	2 131.1	2 097.3	2 084.6	2 079.9	2 091.1	2 109.2	2 133.9	2 160.7	2 148.2
Electronic markets, agents, and brokers	403.2	411.1	414.9	409.3	417.3	432.1	454.2	477.8	505.3	530.4	536.1
Retail trade	14 613.1	14 973.9	15 283.6	15 242.4	15 029.0	14 921.7	15 062.7	15 284.5	15 358.5	15 525.5	15 288.8
Motor vehicle and parts dealers [1]	1 740.9	1 796.6	1 846.9	1 854.6	1 879.4	1 882.9	1 902.3	1 918.6	1 909.7	1 908.3	1 831.2
Automobile dealers	1 142.0	1 179.7	1 216.5	1 225.1	1 252.8	1 254.4	1 257.3	1 261.4	1 246.7	1 242.2	1 176.7
Furniture and home furnishings stores	499.1	524.4	543.5	541.2	538.7	547.3	563.4	576.1	586.9	574.6	531.1
Electronics and appliance stores	590.9	624.2	647.7	633.0	595.2	574.4	572.0	585.5	581.1	583.1	570.2
Building material and garden supply stores	1 062.9	1 101.6	1 142.7	1 152.3	1 177.1	1 185.7	1 227.8	1 276.8	1 324.8	1 310.1	1 248.8
Food and beverage stores	2 965.7	2 984.5	2 993.0	2 950.5	2 881.6	2 838.4	2 821.6	2 817.8	2 821.1	2 843.6	2 862.0
Health and personal care stores	876.0	898.2	927.6	951.5	938.8	938.1	941.1	953.7	961.1	993.1	1 002.8
Gasoline stations	961.3	943.5	935.7	925.3	895.9	882.0	875.6	871.1	864.1	861.5	842.4
Clothing and clothing accessories stores	1 268.6	1 306.6	1 321.6	1 321.1	1 312.5	1 304.5	1 364.3	1 414.6	1 450.9	1 500.0	1 468.0
Sporting goods, hobby, book, and music stores	555.1	582.7	602.7	601.1	591.8	584.7	585.9	597.9	606.0	623.3	621.9
General merchandise stores [1]	2 686.5	2 751.8	2 819.8	2 842.2	2 812.0	2 822.4	2 863.0	2 934.3	2 935.0	3 020.6	3 025.6
Department stores	1 664.4	1 694.2	1 739.5	1 752.5	1 668.5	1 605.2	1 589.8	1 579.5	1 541.5	1 575.0	1 524.1
Miscellaneous store retailers	950.3	985.5	1 007.1	993.3	959.5	930.7	913.5	899.9	881.0	865.4	842.5
Nonstore retailers	455.8	474.5	495.3	476.3	446.7	430.5	432.3	438.2	436.7	442.1	442.3

[1]Includes other industries, not shown separately.

Table 15-1. Nonfarm Employment by Sector and Industry—*Continued*

(Wage and salary workers on nonfarm payrolls, thousands.)

Industry	2009	2010	2011	2012	2013	2014	2015	2016	2017	2018	2019
TOTAL NONFARM	131 296	130 345	131 914	134 157	136 364	138 940	141 825	144 336	146 608	148 908	150 939
Total Private	108 741	107 855	109 828	112 237	114 511	117 058	119 796	122 112	124 258	126 454	128 346
Goods-Producing	18 558	17 751	18 047	18 420	18 738	19 226	19 610	19 750	20 084	20 704	21 067
Mining and logging	694	705	788	848	863	891	813	668	676	727	735
Logging	50.4	49.7	48.7	50.8	51.8	52.0	52.4	51.0	49.6	49.5	50.9
Mining	643.3	654.8	739.2	797.2	811.0	838.5	760.4	616.8	626.1	677.8	684.6
Oil and gas extraction	159.8	158.7	172.0	187.4	193.5	197.7	193.4	169.8	143.9	142.3	150.1
Mining, except oil and gas [1]	208.3	204.5	218.4	218.7	209.6	207.2	197.8	180.9	185.8	191.3	191.4
Coal mining	81.5	80.8	87.3	84.7	78.1	73.2	64.2	50.8	51.5	51.6	51.9
Support activities for mining	275.2	291.6	348.8	391.1	407.9	433.6	369.2	266.2	296.4	344.2	343.1
Construction	6 016	5 518	5 533	5 646	5 856	6 151	6 461	6 728	6 969	7 288	7 492
Construction of buildings	1 357.2	1 229.7	1 222.1	1 240.2	1 285.9	1 358.0	1 423.7	1 493.2	1 545.2	1 625.3	1 660.0
Heavy and civil engineering	851.3	825.1	836.8	868.3	885.0	912.3	937.6	951.6	996.4	1 050.7	1 077.6
Specialty trade contractors	3 807.9	3 463.4	3 474.4	3 537.1	3 684.4	3 880.6	4 099.7	4 283.0	4 427.2	4 612.5	4 754.5
Manufacturing	11 847	11 528	11 726	11 927	12 020	12 185	12 336	12 354	12 439	12 688	12 840
Durable goods	7 284	7 064	7 273	7 470	7 548	7 674	7 765	7 714	7 741	7 946	8 059
Wood products	360.2	342.1	337.1	339.1	353.2	371.6	382.5	392.7	397.0	406.0	408.8
Nonmetallic mineral products	394.3	370.9	366.6	365.3	373.4	384.1	398.0	406.1	409.8	416.5	421.6
Primary metals	362.1	362.3	388.3	402.0	395.3	398.7	394.0	374.5	371.3	380.2	385.0
Fabricated metal products	1 311.6	1 281.7	1 347.3	1 409.8	1 432.3	1 454.2	1 457.8	1 421.5	1 425.2	1 469.6	1 491.6
Machinery	1 028.6	996.1	1 055.8	1 098.5	1 104.5	1 127.3	1 120.8	1 076.5	1 079.1	1 116.9	1 126.1
Computer and electronic products [1]	1 136.9	1 094.6	1 103.5	1 089.0	1 065.6	1 048.7	1 052.8	1 048.4	1 039.2	1 054.1	1 080.6
Computer and peripheral equipment	166.4	157.6	157.4	157.4	157.5	159.8	160.3	162.7	155.6	156.6	163.0
Communications equipment	120.5	117.4	115.3	108.2	101.1	93.2	88.4	86.2	86.7	84.9	83.5
Semiconductors and electronic components	378.1	369.4	383.4	383.1	374.9	367.4	369.5	367.3	362.1	368.9	377.0
Electronic instruments	421.6	406.4	404.2	399.7	393.7	390.8	398.9	396.8	400.5	410.1	424.1
Electrical equipment and appliances	373.6	359.5	366.1	373.2	374.1	377.8	383.7	382.8	386.3	399.6	405.1
Transportation equipment [1]	1 347.9	1 333.1	1 381.5	1 461.1	1 508.8	1 559.0	1 604.9	1 630.3	1 643.2	1 702.3	1 734.1
Motor vehicles and parts	664.1	678.5	717.7	777.3	824.8	872.1	913.7	944.3	963.4	998.4	998.5
Furniture and related products	384.3	357.2	353.1	351.4	359.9	370.3	381.0	389.8	394.9	392.7	388.3
Miscellaneous manufacturing	584.4	566.8	573.7	580.1	580.5	581.9	589.9	590.9	594.5	608.2	617.7
Nondurable goods [1]	4 564	4 464	4 453	4 457	4 472	4 512	4 571	4 640	4 699	4 742	4 781
Food manufacturing	1 456.4	1 450.6	1 458.8	1 468.8	1 473.7	1 484.4	1 511.8	1 556.5	1 598.0	1 620.8	1 643.2
Textile mills	124.4	119.0	120.1	118.7	117.2	117.0	116.6	114.5	112.6	111.8	108.7
Textile product mills	125.7	119.0	117.6	116.3	114.4	115.0	116.1	116.2	116.2	116.0	113.3
Apparel	167.5	156.6	151.7	148.0	144.7	140.1	136.5	130.9	119.3	113.8	110.3
Paper and paper products	407.0	394.7	387.4	379.8	378.0	373.3	372.7	370.8	366.2	365.5	365.4
Printing and related support activities	521.9	487.6	471.8	461.8	452.0	453.7	450.3	447.6	440.2	432.3	424.6
Petroleum and coal products	115.3	113.9	111.8	112.1	110.4	111.6	112.9	112.5	114.6	115.2	114.5
Chemicals	804.1	786.5	783.6	783.3	792.7	803.0	807.3	811.9	823.8	835.3	849.7
Plastics and rubber products	624.9	624.8	635.2	645.1	659.2	674.3	689.4	702.4	716.9	730.3	737.0
Service-Providing	112 739	112 594	113 867	115 737	117 626	119 713	122 215	124 586	126 524	128 205	129 872
Private Service-Providing	90 184	90 104	91 781	93 817	95 773	97 832	100 186	102 362	104 174	105 750	107 279
Trade, transportation, and utilities	24 834	24 565	24 990	25 399	25 783	26 303	26 806	27 179	27 393	27 607	27 715
Wholesale trade	5 520.9	5 386.5	5 474.7	5 595.2	5 660.3	5 739.5	5 780.4	5 786.9	5 813.5	5 840.9	5 903.4
Durable goods	2 944.0	2 848.3	2 905.2	2 977.9	3 012.7	3 056.0	3 079.6	3 082.5	3 108.3	3 148.7	3 203.1
Nondurable goods	2 063.3	2 025.7	2 040.5	2 070.8	2 091.8	2 121.0	2 137.2	2 145.2	2 152.8	2 151.4	2 169.2
Electronic markets, agents, and brokers	513.7	512.5	528.9	546.6	555.9	562.5	563.6	559.2	552.5	540.7	531.1
Retail trade	14 527.9	14 446.0	14 673.7	14 846.8	15 084.7	15 363.4	15 610.9	15 831.6	15 845.7	15 786.3	15 644.0
Motor vehicle and parts dealers [1]	1 637.5	1 629.2	1 691.2	1 737.0	1 793.1	1 862.2	1 929.0	1 979.6	2 004.6	2 015.2	2 034.7
Automobile dealers	1 018.2	1 011.5	1 056.9	1 095.5	1 138.4	1 187.9	1 238.7	1 279.0	1 293.3	1 296.6	1 300.0
Furniture and home furnishings stores	449.2	437.9	438.9	439.4	445.8	455.9	466.8	470.9	476.2	477.2	472.5
Electronics and appliance stores	516.3	523.0	528.1	508.1	497.3	497.8	523.0	522.4	503.3	490.6	477.3
Building material and garden supply stores	1 156.4	1 132.6	1 146.6	1 175.1	1 208.5	1 229.0	1 235.3	1 267.8	1 276.7	1 302.7	1 295.9
Food and beverage stores	2 830.0	2 808.2	2 822.8	2 861.3	2 929.7	3 004.0	3 062.3	3 090.3	3 086.2	3 072.9	3 077.6
Health and personal care stores	986.0	980.5	980.9	997.9	1 015.8	1 022.5	1 033.7	1 053.3	1 066.7	1 063.0	1 052.2
Gasoline stations	825.5	819.3	831.0	843.5	866.3	881.3	905.3	923.1	929.8	931.6	944.6
Clothing and clothing accessories stores	1 363.9	1 352.5	1 360.9	1 391.4	1 390.9	1 370.4	1 353.7	1 359.3	1 374.4	1 358.5	1 299.3
Sporting goods, hobby, book, and music stores	589.2	579.1	577.9	582.2	602.5	618.8	623.2	620.4	606.4	573.3	549.6
General merchandise stores [1]	2 966.2	2 997.7	3 085.2	3 065.4	3 060.4	3 102.0	3 131.4	3 169.4	3 125.9	3 097.7	3 043.3
Department stores	1 456.9	1 485.2	1 521.7	1 439.0	1 334.0	1 335.3	1 312.5	1 267.2	1 173.6	1 137.7	1 083.1
Miscellaneous store retailers	782.4	761.5	772.4	794.0	802.5	818.3	828.0	831.8	828.4	834.5	833.8
Nonstore retailers	425.4	424.8	437.9	451.6	471.9	501.3	519.3	543.2	567.2	569.2	563.3

[1] Includes other industries, not shown separately.

Table 15-1. Nonfarm Employment by Sector and Industry—*Continued*

(Wage and salary workers on nonfarm payrolls, thousands.)

Industry	1998	1999	2000	2001	2002	2003	2004	2005	2006	2007	2008
Transportation and warehousing	4 160.8	4 291.9	4 400.7	4 361.2	4 211.7	4 172.7	4 235.6	4 348.0	4 457.0	4 528.0	4 495.9
Air transportation	562.7	586.3	614.4	615.3	563.5	528.3	514.5	500.8	487.0	491.8	490.7
Rail transportation	203.8	205.1	205.6	198.1	186.7	184.2	190.9	192.6	192.4	197.6	195.4
Water transportation	50.5	51.7	56.0	54.0	52.6	54.5	56.4	60.6	62.7	65.5	67.1
Truck transportation	1 354.7	1 391.8	1 406.1	1 387.1	1 339.6	1 325.9	1 352.1	1 397.9	1 436.2	1 439.6	1 389.4
Transit and ground passenger transportation	365.6	374.5	376.3	379.8	386.4	388.8	391.9	396.4	406.5	419.4	430.4
Pipeline transportation	48.1	46.9	46.0	45.4	41.7	40.2	38.4	37.8	38.7	39.9	41.7
Scenic and sightseeing transportation	25.4	26.1	27.5	29.1	25.6	26.6	27.2	28.8	27.5	28.6	28.0
Support activities for transportation	504.7	526.5	546.4	548.7	534.7	530.6	545.6	562.8	581.2	595.0	602.9
Couriers and messengers	568.2	585.9	605.0	587.0	560.9	561.7	556.6	571.4	582.4	580.7	573.4
Warehousing and storage	477.2	497.2	517.5	516.9	519.9	531.9	562.0	598.9	642.5	669.8	676.9
Utilities	613	608	601	599	596	577	564	554	548	553	559
Information	3 218	3 419	3 630	3 629	3 395	3 188	3 118	3 061	3 038	3 032	2 984
Publishing industries, except Internet	982.3	1 004.8	1 035.0	1 020.7	964.1	924.8	909.1	904.1	902.4	901.2	880.4
Motion picture and sound recording industries	369.5	384.4	382.6	376.8	387.9	376.2	385.0	377.5	375.7	380.6	371.3
Broadcasting, except Internet	321.2	329.4	343.5	344.6	334.1	324.3	325.0	327.7	328.3	325.2	318.7
Telecommunications	1 167.4	1 270.8	1 396.6	1 423.9	1 280.9	1 166.8	1 115.1	1 071.3	1 047.6	1 030.6	1 019.4
Data processing, hosting, and related services	282.8	307.1	315.7	316.8	303.9	280.0	267.1	262.5	263.2	267.8	260.3
Other information services [1]	95.3	121.9	157.1	146.5	123.6	115.9	116.9	117.7	120.8	126.3	133.5
Internet publishing and broadcasting	53.9	78.1	110.8	100.4	76.3	67.2	66.1	67.2	69.1	72.9	80.6
Financial activities	7 565	7 753	7 783	7 900	7 956	8 078	8 105	8 197	8 367	8 348	8 206
Finance and insurance	5 631.7	5 770.2	5 772.8	5 862.0	5 922.2	6 020.5	6 019.4	6 062.9	6 194.0	6 179.1	6 076.3
Monetary authorities–central bank	21.7	22.6	22.8	23.0	23.4	22.6	21.8	20.8	21.2	21.6	22.4
Credit intermediation and related activities [1]	2 531.9	2 591.0	2 547.8	2 597.7	2 686.0	2 792.4	2 817.0	2 869.0	2 924.9	2 866.3	2 732.7
Depository credit intermediation [1]	1 748.9	1 752.1	1 722.8	1 743.6	1 776.1	1 794.4	1 796.7	1 814.3	1 846.9	1 864.4	1 852.0
Commercial banking	1 359.8	1 357.1	1 325.2	1 334.4	1 355.8	1 361.5	1 361.1	1 375.5	1 401.9	1 425.0	1 424.3
Securities, commodity contracts, investments, and funds and trusts	734.5	782.2	851.1	879.0	836.4	803.9	813.1	834.2	868.9	899.6	916.2
Insurance carriers and related activities	2 343.7	2 374.5	2 351.1	2 362.3	2 376.4	2 401.5	2 367.5	2 338.9	2 379.1	2 391.6	2 405.1
Real estate and rental and leasing	1 933.7	1 982.5	2 010.6	2 038.4	2 033.3	2 057.5	2 085.5	2 133.5	2 172.5	2 169.1	2 129.6
Real estate	1 277.7	1 302.6	1 316.0	1 343.4	1 356.6	1 387.1	1 418.7	1 460.8	1 499.0	1 500.4	1 485.0
Rental and leasing services	630.8	653.1	666.8	666.3	649.1	643.1	641.1	645.8	645.5	640.3	616.9
Lessors of nonfinancial intangible assets	25.3	26.8	27.8	28.7	27.6	27.3	25.7	26.9	28.1	28.4	27.7
Professional and business services	15 183	15 994	16 704	16 514	16 016	16 029	16 440	17 003	17 619	17 998	17 792
Professional and technical services [1]	6 021.8	6 375.5	6 732.1	6 901.3	6 680.6	6 637.0	6 784.2	7 064.5	7 399.3	7 704.5	7 845.3
Legal services	1 021.1	1 051.4	1 065.7	1 091.3	1 115.3	1 142.1	1 163.1	1 168.0	1 173.2	1 175.4	1 161.5
Accounting and bookkeeping services	802.5	838.1	866.9	872.7	837.8	815.9	806.5	849.9	889.7	936.6	951.8
Architectural and engineering services	1 115.4	1 168.7	1 238.5	1 275.3	1 246.7	1 227.6	1 258.9	1 311.7	1 386.6	1 433.1	1 440.2
Computer systems design and related services	979.2	1 137.2	1 258.7	1 302.2	1 157.4	1 121.6	1 153.9	1 201.0	1 290.8	1 378.7	1 446.3
Management and technical consulting services	611.4	640.4	694.3	736.5	730.3	742.5	789.4	852.6	916.6	984.8	1 034.8
Management of companies and enterprises	1 759.8	1 777.5	1 799.7	1 782.7	1 709.3	1 691.5	1 729.0	1 763.8	1 816.1	1 872.0	1 910.1
Administrative and waste services	7 401.5	7 841.0	8 171.9	7 829.6	7 625.6	7 700.9	7 927.3	8 174.7	8 403.2	8 421.6	8 036.8
Administrative and support services [1]	7 102.1	7 530.5	7 859.0	7 512.3	7 307.3	7 378.8	7 598.7	7 837.0	8 055.1	8 066.6	7 680.0
Employment services [1]	3 246.5	3 582.3	3 850.0	3 468.9	3 274.0	3 327.3	3 456.4	3 607.8	3 681.9	3 547.1	3 134.1
Temporary help services	2 245.7	2 470.1	2 636.1	2 338.1	2 194.1	2 224.7	2 387.7	2 549.9	2 638.0	2 598.1	2 349.0
Business support services	773.6	781.8	788.1	781.2	758.0	751.3	759.5	768.2	794.8	819.5	834.4
Services to buildings and dwellings	1 460.0	1 534.7	1 570.5	1 606.2	1 606.1	1 636.1	1 693.7	1 737.5	1 801.4	1 849.5	1 839.8
Waste management and remediation services	299.3	310.5	312.9	317.3	318.3	322.1	328.6	337.6	348.1	355.0	356.8
Education and health services	14 570	14 939	15 252	15 814	16 398	16 835	17 230	17 676	18 154	18 676	19 228
Educational services	2 233.0	2 320.0	2 390.0	2 511.0	2 643.0	2 695.0	2 763.0	2 836.0	2 901.0	2 941.0	3 040.0
Health care and social assistance	12 337.0	12 618.3	12 861.1	13 302.9	13 755.2	14 139.6	14 467.8	14 840.4	15 253.2	15 734.5	16 188.2
Health care	10 541.0	10 690.9	10 857.8	11 188.1	11 536.0	11 817.1	12 055.3	12 313.9	12 601.8	12 946.8	13 289.9
Ambulatory health care services [1]	4 161.0	4 226.6	4 320.3	4 461.5	4 633.2	4 786.4	4 952.3	5 113.5	5 285.8	5 473.5	5 646.6
Offices of physicians	1 687.0	1 748.9	1 801.1	1 870.9	1 926.3	1 960.3	2 004.6	2 049.4	2 102.5	2 155.1	2 205.0
Outpatient care centers	400.0	413.1	425.2	440.0	454.5	469.0	493.7	517.3	537.9	558.4	580.8
Home health care services	660.0	629.6	633.3	638.6	679.8	732.6	776.6	821.0	865.6	913.8	961.4
Hospitals	3 892.0	3 935.5	3 954.3	4 050.9	4 159.6	4 244.6	4 284.7	4 345.4	4 423.4	4 515.0	4 627.3
Nursing and residential care facilities [1]	2 487.0	2 528.8	2 583.2	2 675.8	2 743.3	2 786.2	2 818.4	2 855.0	2 892.5	2 958.3	3 016.1
Nursing care facilities	1 489.0	1 501.0	1 513.6	1 546.8	1 573.2	1 579.8	1 576.9	1 577.4	1 581.4	1 602.6	1 618.7
Social assistance [1]	1 796.0	1 927.5	2 003.3	2 114.7	2 219.2	2 322.5	2 412.5	2 526.5	2 651.4	2 787.7	2 898.3
Child day care services	615.0	673.7	695.8	714.6	744.1	755.3	764.7	789.7	818.3	850.4	859.4

[1]Includes other industries, not shown separately.

Table 15-1. Nonfarm Employment by Sector and Industry—*Continued*

(Wage and salary workers on nonfarm payrolls, thousands.)

Industry	2009	2010	2011	2012	2013	2014	2015	2016	2017	2018	2019
Transportation and warehousing	4 224.8	4 179.2	4 289.3	4 403.7	4 486.0	4 648.6	4 858.6	5 003.5	5 178.2	5 426.4	5 618.1
Air transportation	462.8	458.3	456.9	459.2	444.3	444.2	458.6	477.5	491.5	496.9	503.0
Rail transportation	184.6	183.0	192.9	195.1	195.6	199.7	203.7	183.6	181.9	181.7	174.5
Water transportation	63.4	62.3	61.3	63.9	65.3	67.3	65.8	65.6	65.0	65.1	65.8
Truck transportation	1 268.6	1 250.8	1 301.0	1 349.8	1 382.6	1 417.7	1 452.7	1 448.1	1 457.0	1 496.2	1 531.0
Transit and ground passenger transportation	428.3	436.2	446.9	447.5	455.6	473.7	485.0	490.2	494.9	494.7	499.3
Pipeline transportation	42.6	42.3	42.9	43.6	44.5	47.0	49.7	49.8	48.8	49.7	51.2
Scenic and sightseeing transportation	27.6	27.3	27.5	28.0	29.1	30.6	32.9	34.5	35.2	34.6	35.9
Support activities for transportation	558.9	552.9	572.9	590.7	609.2	636.9	663.1	677.9	700.6	729.3	754.0
Couriers and messengers	546.3	528.1	529.2	534.1	543.9	576.5	612.8	644.8	676.1	739.5	815.8
Warehousing and storage	641.7	638.1	657.9	691.9	716.0	754.8	834.3	931.6	1 027.3	1 138.8	1 187.6
Utilities	560	553	553	553	552	552	556	556	555	553	549
Information	2 804	2 707	2 674	2 676	2 706	2 726	2 750	2 794	2 814	2 839	2 860
Publishing industries, except Internet	796.4	759.0	748.6	739.5	732.7	726.9	726.5	730.3	728.6	738.5	759.6
Motion picture and sound recording industries	357.6	370.2	362.1	362.3	370.5	379.3	397.9	426.1	432.6	441.3	444.0
Broadcasting, except Internet	300.5	290.3	283.2	285.1	283.7	282.8	276.7	270.5	268.0	269.8	266.6
Telecommunications	965.7	902.9	873.6	856.8	853.2	838.5	810.9	801.1	780.8	749.7	713.0
Data processing, hosting, and related services	248.5	243.0	245.8	254.9	269.6	279.5	296.2	303.9	318.0	331.0	338.8
Other information services [1]	135.0	141.7	160.0	177.2	196.2	219.0	241.4	262.3	285.8	308.8	337.6
Internet publishing and broadcasting	83.3	92.0	109.6	125.2	142.3	163.3	184.3	203.5	223.8	245.1	271.8
Financial activities	7 838	7 695	7 697	7 784	7 886	7 977	8 123	8 287	8 451	8 590	8 746
Finance and insurance	5 843.9	5 761.0	5 769.0	5 828.4	5 886.1	5 930.9	6 034.9	6 148.1	6 261.9	6 337.0	6 425.2
Monetary authorities–central bank	21.0	20.0	18.3	17.5	18.0	18.2	18.0	18.6	19.2	19.6	19.7
Credit intermediation and related activities [1]	2 590.2	2 550.0	2 554.1	2 583.3	2 614.1	2 564.0	2 570.7	2 609.7	2 644.9	2 651.2	2 651.5
Depository credit intermediation [1]	1 787.9	1 762.9	1 770.0	1 774.1	1 768.6	1 735.7	1 717.2	1 732.3	1 748.0	1 759.1	1 776.2
Commercial banking	1 378.6	1 366.3	1 375.4	1 381.3	1 373.7	1 348.3	1 338.3	1 361.5	1 375.0	1 381.4	1 391.2
Securities, commodity contracts, investments, and funds and trusts	862.1	850.4	860.1	859.3	865.0	882.9	907.8	927.1	938.3	953.6	963.8
Insurance carriers and related activities	2 370.6	2 340.6	2 336.4	2 368.3	2 388.9	2 465.8	2 538.3	2 592.7	2 659.6	2 712.7	2 790.2
Real estate and rental and leasing	1 994.0	1 933.8	1 927.4	1 955.2	2 000.2	2 045.7	2 088.4	2 138.8	2 189.2	2 253.3	2 320.7
Real estate	1 420.2	1 395.7	1 400.8	1 420.0	1 458.7	1 487.1	1 517.1	1 556.8	1 605.5	1 662.9	1 718.1
Rental and leasing services	547.3	513.5	502.2	511.0	517.7	535.0	547.5	558.3	560.1	567.2	579.4
Lessors of nonfinancial intangible assets	26.5	24.6	24.4	24.2	23.7	23.7	23.8	23.8	23.7	23.3	23.3
Professional and business services	16 634	16 783	17 389	17 992	18 575	19 124	19 695	20 114	20 508	20 950	21 313
Professional and technical services [1]	7 552.9	7 485.8	7 712.5	7 940.5	8 169.9	8 385.5	8 658.4	8 880.8	9 057.8	9 281.7	9 542.7
Legal services	1 124.9	1 114.2	1 115.7	1 124.0	1 128.5	1 119.0	1 118.6	1 123.7	1 136.8	1 142.0	1 149.9
Accounting and bookkeeping services	914.9	887.2	899.6	909.7	932.3	949.3	969.3	981.4	996.6	1 003.8	1 026.1
Architectural and engineering services	1 325.5	1 276.2	1 294.3	1 323.4	1 346.5	1 376.2	1 401.6	1 404.6	1 435.3	1 474.4	1 512.9
Computer systems design and related services	1 429.0	1 455.5	1 542.6	1 628.4	1 709.5	1 797.8	1 915.8	1 993.2	2 051.2	2 118.0	2 202.2
Management and technical consulting services	1 026.5	1 031.1	1 098.2	1 152.4	1 215.2	1 267.7	1 318.1	1 394.2	1 433.1	1 481.5	1 531.1
Management of companies and enterprises	1 872.4	1 877.8	1 939.3	2 029.3	2 109.0	2 174.6	2 213.6	2 251.6	2 308.1	2 373.9	2 427.4
Administrative and waste services	7 208.5	7 419.1	7 737.3	8 021.8	8 296.5	8 564.1	8 823.3	8 981.2	9 141.5	9 294.3	9 342.6
Administrative and support services [1]	6 856.8	7 061.8	7 372.0	7 649.9	7 919.1	8 178.0	8 426.8	8 577.1	8 726.3	8 858.0	8 888.2
Employment services [1]	2 481.8	2 723.5	2 943.2	3 135.2	3 269.7	3 408.1	3 528.3	3 537.9	3 595.1	3 656.6	3 637.3
Temporary help services	1 823.8	2 094.2	2 313.6	2 496.0	2 617.6	2 760.8	2 877.6	2 889.1	2 939.7	2 988.5	2 948.3
Business support services	822.1	810.7	816.7	830.7	858.3	879.7	895.9	912.0	904.4	895.3	877.5
Services to buildings and dwellings	1 753.3	1 745.0	1 788.6	1 830.0	1 884.7	1 943.9	2 006.6	2 062.8	2 111.2	2 144.8	2 168.4
Waste management and remediation services	351.7	357.3	365.3	371.9	377.5	386.1	396.5	404.1	415.2	436.3	454.4
Education and health services	19 630	19 975	20 318	20 769	21 086	21 439	22 029	22 639	23 188	23 638	24 177
Educational services	3 090.0	3 155.0	3 250.0	3 341.0	3 354.0	3 417.0	3 472.0	3 570.0	3 668.0	3 715.0	3 765.0
Health care and social assistance	16 539.9	16 820.0	17 068.6	17 428.1	17 731.1	18 022.2	18 557.4	19 068.8	19 519.6	19 922.6	20 412.5
Health care	13 543.0	13 776.9	14 025.9	14 281.6	14 491.5	14 676.5	15 042.3	15 413.5	15 716.6	15 964.3	16 275.2
Ambulatory health care services [1]	5 793.4	5 974.7	6 136.2	6 306.5	6 476.5	6 631.5	6 855.5	7 080.0	7 297.3	7 477.4	7 697.2
Offices of physicians	2 231.0	2 263.9	2 294.6	2 339.8	2 378.0	2 411.2	2 471.0	2 526.5	2 581.4	2 615.8	2 672.0
Outpatient care centers	605.6	648.7	670.2	699.0	731.1	763.7	805.7	853.0	897.2	931.4	963.0
Home health care services	1 027.1	1 084.6	1 140.3	1 185.0	1 230.3	1 262.4	1 314.7	1 365.3	1 419.6	1 467.0	1 527.4
Hospitals	4 667.4	4 678.5	4 721.7	4 779.0	4 785.8	4 786.8	4 895.8	5 015.2	5 071.8	5 129.8	5 198.7
Nursing and residential care facilities [1]	3 082.2	3 123.7	3 168.1	3 196.2	3 229.2	3 258.2	3 290.9	3 318.3	3 347.5	3 357.2	3 379.3
Nursing care facilities	1 644.9	1 657.1	1 669.6	1 662.8	1 653.8	1 650.3	1 648.3	1 641.5	1 627.1	1 606.8	1 598.4
Social assistance [1]	2 996.9	3 043.1	3 042.6	3 146.5	3 239.6	3 345.6	3 515.1	3 655.3	3 803.0	3 958.3	4 137.3
Child day care services	852.8	848.0	849.4	851.3	843.3	854.0	878.0	911.2	943.1	972.7	1 018.3

[1] Includes other industries, not shown separately.

Table 15-1. Nonfarm Employment by Sector and Industry—*Continued*

(Wage and salary workers on nonfarm payrolls, thousands.)

Industry	1998	1999	2000	2001	2002	2003	2004	2005	2006	2007	2008
Leisure and hospitality	11 232	11 543	11 862	12 036	11 986	12 173	12 493	12 816	13 110	13 427	13 436
Arts, entertainment, and recreation	1 645.2	1 709.1	1 787.9	1 824.4	1 782.6	1 812.9	1 849.6	1 892.3	1 928.5	1 969.2	1 970.1
Performing arts and spectator sports	350.0	361.1	381.8	382.3	363.7	371.7	367.5	376.3	398.5	405.0	405.7
Museums, historical sites, zoos, and parks	97.4	103.1	110.4	115.0	114.0	114.7	118.3	120.7	123.8	130.3	131.6
Amusements, gambling, and recreation	1 197.9	1 244.9	1 295.7	1 327.1	1 305.0	1 326.5	1 363.8	1 395.3	1 406.3	1 433.9	1 432.8
Accommodation and food services	9 586.2	9 833.7	10 073.5	10 211.3	10 203.2	10 359.8	10 643.2	10 923.0	11 181.1	11 457.4	11 466.3
Accommodation	1 773.5	1 831.7	1 884.4	1 852.2	1 778.6	1 775.4	1 789.5	1 818.6	1 832.1	1 866.9	1 868.7
Food services and drinking places	7 812.7	8 002.0	8 189.1	8 359.1	8 424.6	8 584.4	8 853.7	9 104.4	9 349.0	9 590.4	9 597.5
Other services	4 976	5 087	5 168	5 258	5 372	5 401	5 409	5 395	5 438	5 494	5 515
Repair and maintenance	1 189.2	1 222.0	1 241.5	1 256.5	1 246.9	1 233.6	1 228.8	1 236.0	1 248.5	1 253.4	1 227.0
Personal and laundry services	1 205.6	1 220.3	1 242.9	1 255.0	1 257.2	1 263.5	1 272.9	1 276.6	1 288.4	1 309.7	1 322.6
Membership associations and organizations	2 581.3	2 644.4	2 683.3	2 746.4	2 867.8	2 903.6	2 907.5	2 882.2	2 901.2	2 931.1	2 965.7
Government	19 909	20 307	20 790	21 118	21 513	21 583	21 621	21 804	21 974	22 218	22 509
Federal	2 772.0	2 769.0	2 865.0	2 764.0	2 766.0	2 761.0	2 730.0	2 732.0	2 732.0	2 734.0	2 762.0
Federal, except U.S. Postal Service	1 891.3	1 879.5	1 984.8	1 891.0	1 923.8	1 952.4	1 947.5	1 957.3	1 962.6	1 964.7	2 014.4
U.S. Postal Service	880.5	889.7	879.7	873.0	842.4	808.6	782.1	774.2	769.7	769.1	747.4
State government	4 612.0	4 709.0	4 786.0	4 905.0	5 029.0	5 002.0	4 982.0	5 032.0	5 075.0	5 122.0	5 177.0
State government education	1 922.2	1 983.2	2 030.6	2 112.9	2 242.8	2 254.7	2 238.1	2 259.9	2 292.5	2 317.5	2 354.4
State government, excluding education	2 690.2	2 725.6	2 755.9	2 791.8	2 786.3	2 747.6	2 743.9	2 771.6	2 782.0	2 804.3	2 822.5
Local government	12 525.0	12 829.0	13 139.0	13 449.0	13 718.0	13 820.0	13 909.0	14 041.0	14 167.0	14 362.0	14 571.0
Local government education	6 920.9	7 120.4	7 293.9	7 479.3	7 654.4	7 709.4	7 765.2	7 856.1	7 913.0	7 986.8	8 083.9
Local government, excluding education	5 603.9	5 708.6	5 844.6	5 970.0	6 063.2	6 110.2	6 144.1	6 184.6	6 253.8	6 375.5	6 486.5

¹Includes other industries, not shown separately.

Table 15-1. Nonfarm Employment by Sector and Industry—*Continued*

(Wage and salary workers on nonfarm payrolls, thousands.)

Industry	2009	2010	2011	2012	2013	2014	2015	2016	2017	2018	2019
Leisure and hospitality	13 077	13 049	13 353	13 768	14 254	14 696	15 160	15 660	16 051	16 295	16 576
Arts, entertainment, and recreation	1 915.5	1 913.3	1 919.1	1 968.6	2 029.7	2 103.1	2 166.4	2 251.5	2 333.2	2 382.5	2 433.4
Performing arts and spectator sports	396.8	406.2	394.2	402.4	419.2	443.3	450.4	464.4	493.8	506.4	516.4
Museums, historical sites, zoos, and parks	129.4	127.7	132.7	136.1	140.3	146.9	153.0	159.6	165.6	169.5	172.9
Amusements, gambling, and recreation	1 389.2	1 379.4	1 392.2	1 430.1	1 470.2	1 512.9	1 563.1	1 627.5	1 673.8	1 706.6	1 744.2
Accommodation and food services	11 161.9	11 135.4	11 433.6	11 799.7	12 224.2	12 592.9	12 993.8	13 408.4	13 717.9	13 912.9	14 142.6
Accommodation	1 763.0	1 759.6	1 800.5	1 825.1	1 864.9	1 894.5	1 923.0	1 960.3	2 002.9	2 034.7	2 077.9
Food services and drinking places	9 398.9	9 375.8	9 633.1	9 974.6	10 359.2	10 698.3	11 070.8	11 448.1	11 715.0	11 878.2	12 064.7
Other services	5 367	5 331	5 360	5 430	5 483	5 567	5 622	5 691	5 770	5 831	5 893
Repair and maintenance	1 150.4	1 138.8	1 168.7	1 194.0	1 216.6	1 242.1	1 277.2	1 293.1	1 310.6	1 326.7	1 352.1
Personal and laundry services	1 280.6	1 265.3	1 288.6	1 313.6	1 341.7	1 371.2	1 405.0	1 444.3	1 478.4	1 506.6	1 525.2
Membership associations and organizations	2 936.0	2 926.4	2 903.0	2 922.4	2 924.6	2 953.6	2 939.5	2 953.3	2 981.0	2 997.9	3 015.6
Government	22 555	22 490	22 086	21 920	21 853	21 882	22 029	22 224	22 350	22 455	22 594
Federal	2 832.0	2 977.0	2 859.0	2 820.0	2 769.0	2 733.0	2 757.0	2 795.0	2 805.0	2 800.0	2 834.0
Federal, except U.S. Postal Service	2 128.5	2 318.1	2 227.6	2 209.2	2 174.5	2 140.4	2 160.0	2 185.8	2 189.4	2 191.8	2 226.8
U.S. Postal Service	703.4	658.5	630.9	611.2	594.9	593.0	596.9	608.8	615.4	608.5	607.3
State government	5 169.0	5 137.0	5 078.0	5 055.0	5 046.0	5 050.0	5 077.0	5 110.0	5 165.0	5 173.0	5 177.0
State government education	2 360.2	2 373.1	2 374.0	2 388.5	2 393.3	2 389.3	2 401.4	2 429.2	2 478.7	2 486.9	2 484.0
State government, excluding education	2 808.8	2 764.1	2 703.7	2 666.4	2 652.8	2 660.5	2 675.6	2 680.9	2 686.4	2 686.4	2 692.2
Local government	14 554.0	14 376.0	14 150.0	14 045.0	14 037.0	14 098.0	14 195.0	14 319.0	14 379.0	14 481.0	14 583.0
Local government education	8 078.8	8 013.4	7 872.5	7 777.9	7 776.7	7 814.9	7 870.9	7 905.1	7 921.5	7 963.0	8 009.7
Local government, excluding education	6 474.9	6 362.9	6 277.7	6 266.8	6 260.1	6 283.4	6 324.0	6 414.1	6 457.5	6 518.0	6 573.5

¹Includes other industries, not shown separately.

Table 15-2. Production and Nonsupervisory Workers on Private Nonfarm Payrolls by Industry

(Wage and salary workers on nonfarm payrolls, thousands.)

Industry	1998	1999	2000	2001	2002	2003	2004	2005	2006	2007	2008
Total Private	86 506	88 636	90 534	90 200	88 649	87 948	89 230	91 426	93 759	95 243	94 660
Goods-Producing	18 008	18 067	18 169	17 466	16 400	15 732	15 821	16 145	16 559	16 405	15 724
Mining and logging	473	438	446	457	436	420	440	473	519	547	574
Construction	4 807	5 105	5 295	5 332	5 196	5 123	5 309	5 611	5 903	5 883	5 521
Manufacturing	12 729	12 524	12 428	11 677	10 768	10 189	10 072	10 060	10 137	9 975	9 629
Durable goods	7 721	7 651	7 659	7 164	6 530	6 152	6 140	6 220	6 355	6 250	5 975
Wood products	510	516	507	470	450	434	445	454	451	407	358
Nonmetallic mineral products	421	426	440	427	399	375	388	387	391	384	363
Primary metals	505	492	490	447	396	370	364	363	363	358	348
Fabricated metal products	1 320	1 305	1 326	1 254	1 147	1 092	1 109	1 129	1 162	1 171	1 143
Machinery	1 016	978	961	891	787	732	730	749	770	774	772
Computer and electronic products	965	933	949	876	744	673	656	700	756	744	730
Electrical equipment and appliances	432	433	433	402	352	320	307	300	303	305	305
Transportation equipment [1]	1 530	1 526	1 498	1 399	1 310	1 269	1 265	1 277	1 304	1 274	1 177
Motor vehicles and parts	1 050	1 076	1 073	987	931	906	903	894	873	804	696
Furniture and related products	512	532	544	509	475	444	444	436	433	409	364
Miscellaneous manufacturing	511	509	510	490	469	442	432	424	423	425	416
Nondurable goods [1]	5 008	4 872	4 769	4 513	4 238	4 037	3 932	3 841	3 782	3 725	3 653
Food manufacturing	1 228	1 229	1 228	1 221	1 202	1 192	1 178	1 170	1 172	1 184	1 184
Textile mills	357	334	315	276	242	217	194	174	158	137	122
Textile product mills	190	187	183	174	162	148	147	143	135	123	115
Apparel	534	458	404	341	286	242	219	193	182	173	163
Paper and paper products	484	474	468	446	421	393	374	365	357	350	344
Printing and related support activities	598	585	576	544	493	471	460	447	447	443	424
Petroleum and coal products	87	85	83	81	78	74	77	75	72	73	77
Chemicals	601	595	588	562	532	525	520	510	508	504	512
Plastics and rubber products	739	746	753	704	662	633	626	620	608	592	572
Private Service-Providing	68 497	70 569	72 365	72 734	72 249	72 216	73 409	75 282	77 200	78 838	78 936
Trade, transportation, and utilities	21 020	21 535	21 923	21 665	21 292	21 029	21 268	21 776	22 109	22 486	22 277
Wholesale trade	4 571	4 638	4 651	4 520	4 437	4 358	4 402	4 538	4 676	4 798	4 768
Retail trade	12 443	12 775	13 043	12 955	12 777	12 658	12 792	13 033	13 114	13 322	13 139
Transportation and warehousing	3 514	3 633	3 743	3 707	3 599	3 550	3 624	3 761	3 876	3 922	3 920
Utilities	492	489	485	483	478	464	450	443	443	444	450
Information	2 217	2 351	2 502	2 531	2 398	2 347	2 371	2 386	2 399	2 403	2 388
Financial activities	5 692	5 818	5 819	5 888	5 964	6 052	6 052	6 127	6 312	6 365	6 320
Professional and business services	12 594	13 214	13 819	13 617	13 080	12 942	13 321	13 892	14 487	14 828	14 631
Education and health services	12 903	13 217	13 491	13 998	14 491	14 753	15 018	15 401	15 832	16 318	16 842
Leisure and hospitality	9 947	10 216	10 516	10 662	10 576	10 666	10 955	11 263	11 568	11 861	11 873
Other services	4 124	4 219	4 296	4 373	4 449	4 426	4 425	4 438	4 494	4 578	4 606

[1]Includes other industries, not shown separately.

Table 15-2. Production and Nonsupervisory Workers on Private Nonfarm Payrolls by Industry
—Continued

(Wage and salary workers on nonfarm payrolls, thousands.)

Industry	2009	2010	2011	2012	2013	2014	2015	2016	2017	2018	2019
Total Private	89 616	88 940	90 605	92 766	94 573	96 688	98 769	100 555	102 411	104 169	105 612
Goods-Producing	13 399	12 774	13 005	13 287	13 481	13 858	14 141	14 215	14 450	14 876	15 094
Mining and logging	510	525	594	641	636	653	592	471	490	541	541
Construction	4 567	4 172	4 184	4 246	4 423	4 640	4 866	5 074	5 230	5 437	5 579
Manufacturing	8 322	8 077	8 228	8 400	8 422	8 565	8 683	8 670	8 730	8 898	8 975
Durable goods	4 990	4 829	4 986	5 152	5 185	5 282	5 350	5 303	5 315	5 464	5 546
Wood products	278	269	269	272	283	298	306	309	311	319	321
Nonmetallic mineral products	303	284	278	273	275	280	297	306	305	311	314
Primary metals	272	275	301	317	306	310	307	293	292	295	300
Fabricated metal products	960	935	994	1 050	1 063	1 071	1 069	1 036	1 045	1 087	1 110
Machinery	641	616	662	700	699	716	711	686	691	716	709
Computer and electronic products	654	629	630	628	610	589	594	596	598	612	641
Electrical equipment and appliances	266	251	248	249	245	248	258	259	253	261	264
Transportation equipment [1]	948	937	972	1 024	1 053	1 103	1 139	1 150	1 149	1 188	1 203
Motor vehicles and parts	510	525	556	598	642	690	718	739	753	780	772
Furniture and related products	284	263	260	259	266	276	284	286	290	290	288
Miscellaneous manufacturing	382	370	373	380	386	390	384	382	381	384	394
Nondurable goods [1]	3 332	3 248	3 241	3 248	3 237	3 283	3 333	3 367	3 415	3 434	3 429
Food manufacturing	1 161	1 152	1 158	1 169	1 169	1 176	1 189	1 212	1 251	1 272	1 289
Textile mills	99	96	98	96	92	91	90	90	88	87	85
Textile product mills	98	92	89	85	83	86	88	89	89	85	78
Apparel	132	120	112	109	106	103	104	99	88	82	74
Paper and paper products	313	302	295	288	279	277	277	275	278	275	272
Printing and related support activities	369	342	327	316	310	312	310	312	305	296	284
Petroleum and coal products	70	70	70	72	70	72	74	76	80	77	77
Chemicals	479	474	480	491	490	497	507	516	525	546	558
Plastics and rubber products	476	472	482	487	497	518	532	536	543	548	552
Private Service-Providing	76 217	76 166	77 600	79 479	81 093	82 830	84 628	86 340	87 961	89 293	90 518
Trade, transportation, and utilities	21 059	20 816	21 174	21 554	21 811	22 217	22 566	22 842	23 094	23 348	23 443
Wholesale trade	4 453	4 324	4 388	4 505	4 563	4 638	4 644	4 634	4 660	4 688	4 739
Retail trade	12 476	12 430	12 652	12 798	12 928	13 112	13 269	13 433	13 486	13 488	13 354
Transportation and warehousing	3 678	3 618	3 693	3 810	3 875	4 022	4 206	4 328	4 500	4 728	4 910
Utilities	451	444	441	441	445	446	447	447	447	444	440
Information	2 240	2 170	2 148	2 164	2 194	2 209	2 226	2 252	2 268	2 286	2 301
Financial activities	6 066	5 942	5 900	5 986	6 068	6 155	6 278	6 430	6 571	6 654	6 767
Professional and business services	13 565	13 744	14 298	14 851	15 352	15 818	16 183	16 455	16 751	17 079	17 333
Education and health services	17 240	17 531	17 818	18 230	18 505	18 827	19 337	19 859	20 365	20 763	21 234
Leisure and hospitality	11 560	11 507	11 772	12 154	12 590	12 968	13 360	13 782	14 139	14 334	14 563
Other services	4 488	4 458	4 491	4 541	4 573	4 637	4 678	4 720	4 775	4 828	4 876

[1]Includes other industries, not shown separately.

Table 15-3. Average Weekly Hours of Production and Nonsupervisory Workers on Private Nonfarm Payrolls by Industry

(Hours.)

Industry	1998	1999	2000	2001	2002	2003	2004	2005	2006	2007	2008
Total Private	34.5	34.3	34.3	33.9	33.9	33.7	33.7	33.8	33.9	33.8	33.6
Goods-Producing	40.8	40.8	40.7	39.9	39.9	39.8	40.0	40.1	40.5	40.6	40.2
Mining and logging	44.9	44.2	44.4	44.6	43.2	43.6	44.5	45.6	45.6	45.9	45.1
Construction	38.8	39.0	39.2	38.7	38.4	38.4	38.3	38.6	39.0	39.0	38.5
Manufacturing	41.4	41.4	41.3	40.3	40.5	40.4	40.8	40.7	41.1	41.2	40.8
Overtime hours	4.9	4.9	4.7	4.0	4.2	4.2	4.6	4.6	4.4	4.2	3.7
Durable goods	42.1	41.9	41.8	40.6	40.8	40.8	41.3	41.1	41.4	41.5	41.1
Overtime hours	5.0	5.0	4.8	3.9	4.2	4.3	4.7	4.6	4.4	4.2	3.7
Wood products	41.4	41.3	41.0	40.2	39.9	40.4	40.7	40.0	39.8	39.4	38.6
Nonmetallic mineral products	42.2	42.1	41.6	41.6	42.0	42.2	42.4	42.2	43.0	42.3	42.1
Primary metals	43.5	43.8	44.2	42.4	42.4	42.3	43.1	43.1	43.6	42.9	42.2
Fabricated metal products	41.9	41.7	41.9	40.6	40.6	40.7	41.1	41.0	41.4	41.6	41.3
Machinery	43.1	42.3	42.3	40.9	40.5	40.8	42.0	42.1	42.4	42.6	42.3
Computer and electronic products	41.9	41.5	41.4	39.8	39.7	40.4	40.4	40.0	40.5	40.6	41.0
Electrical equipment and appliances	41.8	41.8	41.6	39.8	40.1	40.6	40.7	40.6	41.0	41.2	40.9
Transportation equipment [1]	43.3	43.6	43.3	41.9	42.5	41.9	42.5	42.4	42.7	42.8	42.0
Motor vehicles and parts	42.6	43.8	43.3	41.7	42.6	42.0	42.6	42.3	42.2	42.3	41.4
Furniture and related products	39.4	39.3	39.2	38.3	39.1	38.9	39.5	39.2	38.8	39.2	38.1
Miscellaneous manufacturing	39.2	39.3	39.0	38.8	38.7	38.4	38.5	38.7	38.7	38.9	38.9
Nondurable goods [1]	40.5	40.5	40.3	39.9	40.0	39.8	40.0	39.9	40.6	40.8	40.4
Overtime hours	4.6	4.6	4.5	4.1	4.2	4.1	4.4	4.4	4.4	4.2	3.7
Food manufacturing	40.1	40.2	40.1	39.6	39.6	39.3	39.3	39.0	40.1	40.7	40.5
Textile mills	41.0	41.0	41.4	40.0	40.6	39.1	40.1	40.3	40.6	40.3	38.7
Textile product mills	39.2	39.1	38.7	38.4	39.0	39.4	38.7	38.9	39.8	39.7	38.6
Apparel	35.5	35.4	35.7	36.0	36.7	35.6	36.1	35.8	36.5	37.2	36.4
Paper and paper products	43.6	43.6	42.8	42.1	41.8	41.5	42.1	42.5	42.9	43.1	42.9
Printing and related support activities	39.3	39.1	39.2	38.7	38.4	38.2	38.4	38.4	39.2	39.1	38.3
Petroleum and coal products	43.6	42.6	42.7	43.8	43.0	44.5	44.9	45.5	45.0	44.1	44.6
Chemicals	43.2	42.8	42.2	41.9	42.3	42.4	42.8	42.3	42.5	41.9	41.5
Plastics and rubber products	41.3	41.3	40.8	40.0	40.6	40.4	40.4	40.0	40.6	41.3	41.0
Private Service-Providing	32.8	32.7	32.7	32.5	32.5	32.3	32.3	32.4	32.4	32.4	32.3
Trade, transportation, and utilities	34.1	33.9	33.8	33.5	33.6	33.5	33.5	33.4	33.4	33.3	33.1
Wholesale trade	38.7	38.6	38.8	38.4	38.0	37.9	37.8	37.7	38.0	38.2	38.3
Retail trade	30.9	30.8	30.7	30.7	30.9	30.9	30.7	30.6	30.5	30.2	30.0
Transportation and warehousing	38.6	37.6	37.3	36.6	36.7	36.7	37.2	37.0	36.8	36.9	36.4
Utilities	42.0	42.0	42.0	41.4	40.9	41.1	40.9	41.1	41.4	42.4	42.7
Information	36.6	36.7	36.8	36.9	36.5	36.2	36.3	36.5	36.6	36.5	36.7
Financial activities	36.0	35.8	35.9	35.8	35.6	35.6	35.6	36.0	35.8	35.9	35.9
Professional and business services	34.3	34.4	34.5	34.2	34.2	34.1	34.2	34.3	34.6	34.8	34.8
Education and health services	32.2	32.1	32.2	32.3	32.4	32.3	32.4	32.6	32.5	32.5	32.4
Leisure and hospitality	26.2	26.1	26.1	25.8	25.8	25.6	25.7	25.7	25.7	25.5	25.2
Other services	32.6	32.5	32.5	32.3	32.1	31.4	31.0	30.9	30.9	30.9	30.8

[1] Includes other industries, not shown separately.

Table 15-3. Average Weekly Hours of Production and Nonsupervisory Workers on Private Nonfarm Payrolls by Industry—*Continued*

(Hours.)

Industry	2009	2010	2011	2012	2013	2014	2015	2016	2017	2018	2019
Total Private	33.1	33.4	33.6	33.7	33.7	33.7	33.7	33.6	33.7	33.8	33.6
Goods-Producing	39.2	40.4	40.9	41.1	41.3	41.5	41.2	41.2	41.3	41.5	41.1
Mining and logging	43.2	44.6	46.7	46.6	45.9	47.3	45.8	45.3	46.1	46.8	46.8
Construction	37.6	38.4	39.0	39.3	39.6	39.6	39.6	39.7	39.7	40.0	39.8
Manufacturing	39.8	41.1	41.4	41.7	41.8	42.0	41.8	41.9	41.9	42.2	41.6
Overtime hours	2.9	3.8	4.1	4.2	4.3	4.5	4.3	4.3	4.3	4.6	4.3
Durable goods	39.8	41.4	41.9	42.0	42.2	42.5	42.1	42.3	42.3	42.5	42.0
Overtime hours	2.7	3.8	4.2	4.3	4.4	4.6	4.3	4.5	4.4	4.7	4.3
Wood products	37.4	39.1	39.7	41.1	42.7	42.0	41.3	41.8	42.1	41.9	41.5
Nonmetallic mineral products	40.8	41.7	42.3	42.2	42.5	43.3	42.4	42.0	42.9	44.2	43.8
Primary metals	40.7	43.7	44.6	43.8	43.8	44.2	43.8	43.4	43.1	44.6	43.8
Fabricated metal products	39.4	41.4	42.0	42.0	42.2	42.6	42.3	42.1	42.2	42.2	41.7
Machinery	40.1	42.1	43.1	42.8	42.9	43.0	41.9	42.1	42.7	42.7	42.4
Computer and electronic products	40.4	40.9	40.5	40.4	40.5	40.8	40.9	41.2	41.2	40.9	40.5
Electrical equipment and appliances	39.3	41.1	40.8	41.6	41.8	41.7	42.2	42.9	43.1	42.6	41.4
Transportation equipment [1]	41.2	42.9	43.2	43.8	43.6	43.6	43.7	44.1	43.9	44.4	43.4
Motor vehicles and parts	40.1	43.4	43.4	44.2	43.7	43.8	44.1	44.7	44.0	44.7	43.3
Furniture and related products	37.7	38.5	39.9	40.0	40.3	40.9	39.8	40.0	39.3	39.3	39.4
Miscellaneous manufacturing	38.5	38.7	38.9	39.2	40.1	40.2	40.1	40.8	39.9	39.5	39.7
Nondurable goods [1]	39.8	40.8	40.8	41.1	41.2	41.3	41.4	41.2	41.3	41.6	41.1
Overtime hours	3.2	3.8	4.0	4.1	4.3	4.3	4.3	4.1	4.2	4.4	4.2
Food manufacturing	40.0	40.7	40.2	40.6	40.8	40.8	41.0	41.4	41.7	42.0	41.5
Textile mills	37.7	41.2	41.7	42.6	41.5	41.5	42.5	41.2	40.6	42.3	43.5
Textile product mills	37.9	39.0	39.1	39.7	38.4	38.0	37.0	37.7	38.4	39.3	37.8
Apparel	36.0	36.6	38.2	37.1	38.1	38.5	38.2	36.4	36.5	37.7	37.0
Paper and paper products	41.8	42.9	42.9	42.9	43.1	43.7	43.0	42.7	42.5	43.1	42.1
Printing and related support activities	38.0	38.2	38.0	38.5	38.6	39.0	39.8	39.2	39.5	39.3	38.8
Petroleum and coal products	43.4	43.0	43.8	47.1	46.4	46.1	45.2	44.2	45.0	45.6	48.3
Chemicals	41.4	42.2	42.5	42.4	42.9	42.7	42.6	41.8	41.7	42.1	41.8
Plastics and rubber products	40.2	41.9	42.0	41.8	41.8	42.2	42.4	42.4	42.6	42.1	41.0
Private Service-Providing	32.1	32.2	32.4	32.5	32.4	32.4	32.4	32.3	32.4	32.5	32.4
Trade, transportation, and utilities	32.8	33.3	33.7	33.8	33.7	33.6	33.7	33.5	33.8	33.9	33.8
Wholesale trade	37.7	37.9	38.5	38.7	38.7	38.6	38.6	38.6	39.0	38.9	38.7
Retail trade	29.9	30.2	30.5	30.6	30.2	30.0	30.1	29.7	30.2	30.4	30.3
Transportation and warehousing	36.0	37.1	37.8	38.0	38.5	38.4	38.7	38.8	38.4	38.4	37.9
Utilities	42.0	42.0	42.1	41.1	41.7	42.3	42.4	42.5	42.5	42.7	42.4
Information	36.6	36.3	36.2	36.0	35.9	35.9	35.7	35.5	35.8	35.6	35.3
Financial activities	36.1	36.2	36.4	36.8	36.7	36.7	37.1	36.9	37.0	37.0	36.9
Professional and business services	34.7	35.1	35.2	35.3	35.4	35.6	35.5	35.4	35.4	35.4	35.4
Education and health services	32.2	32.0	32.2	32.3	32.1	32.0	32.1	32.2	32.2	32.2	32.2
Leisure and hospitality	24.8	24.8	24.8	25.0	25.0	25.1	25.1	24.9	24.9	24.9	24.7
Other services	30.5	30.7	30.8	30.7	30.8	30.7	30.7	30.8	30.7	30.8	30.8

[1]Includes other industries, not shown separately.

Table 15-4. Average Hourly Earnings of Production and Nonsupervisory Workers on Private Nonfarm Payrolls by Industry

(Dollars.)

Industry	1998	1999	2000	2001	2002	2003	2004	2005	2006	2007	2008
Total Private	13.01	13.48	14.01	14.54	14.96	15.36	15.68	16.12	16.74	17.41	18.06
Goods-Producing	14.23	14.71	15.27	15.78	16.33	16.80	17.19	17.60	18.02	18.67	19.33
Mining and logging	16.20	16.33	16.55	17.00	17.19	17.56	18.07	18.72	19.90	20.97	22.50
Construction	16.23	16.80	17.48	18.00	18.52	18.95	19.23	19.46	20.02	20.95	21.87
Manufacturing	13.45	13.85	14.32	14.76	15.29	15.74	16.14	16.56	16.81	17.26	17.75
Excluding overtime [1]	12.70	13.08	13.55	14.06	14.54	14.96	15.29	15.68	15.96	16.43	16.97
Durable goods	14.07	14.46	14.93	15.38	16.02	16.45	16.82	17.33	17.68	18.20	18.70
Wood products	10.85	11.18	11.63	11.99	12.33	12.71	13.03	13.16	13.39	13.68	14.19
Nonmetallic mineral products	13.59	13.97	14.53	14.86	15.39	15.76	16.25	16.61	16.59	16.93	16.90
Primary metals	15.66	16.00	16.64	17.06	17.68	18.13	18.57	18.94	19.36	19.66	20.19
Fabricated metal products	12.97	13.34	13.77	14.19	14.68	15.01	15.31	15.80	16.17	16.53	16.99
Machinery	14.23	14.77	15.21	15.48	15.92	16.29	16.67	17.02	17.20	17.72	17.97
Computer and electronic products	13.85	14.37	14.73	15.42	16.20	16.68	17.27	18.39	18.94	19.94	21.04
Electrical equipment and appliances	12.51	12.90	13.23	13.78	13.98	14.36	14.90	15.24	15.53	15.93	15.78
Transportation equipment [2]	17.91	18.24	18.89	19.47	20.63	21.22	21.48	22.09	22.41	23.03	23.85
Motor vehicles and parts	18.21	18.49	19.11	19.66	21.09	21.68	21.71	22.26	22.13	22.00	22.21
Furniture and related products	10.89	11.28	11.73	12.14	12.62	12.99	13.16	13.45	13.80	14.32	14.54
Miscellaneous manufacturing	11.18	11.55	11.93	12.45	12.91	13.30	13.84	14.07	14.36	14.66	15.20
Nondurable goods [2]	12.45	12.85	13.31	13.75	14.15	14.63	15.05	15.26	15.33	15.67	16.15
Food manufacturing	11.09	11.40	11.77	12.18	12.55	12.80	12.98	13.04	13.13	13.55	14.00
Textile mills	10.58	10.90	11.23	11.40	11.73	11.99	12.13	12.38	12.55	13.00	13.58
Textile product mills	9.61	10.04	10.31	10.49	10.85	11.15	11.31	11.61	11.86	11.78	11.73
Apparel	8.05	8.35	8.61	8.83	9.11	9.58	9.77	10.26	10.65	11.05	11.40
Paper and paper products	15.20	15.58	15.91	16.38	16.85	17.33	17.91	17.99	18.01	18.44	18.89
Printing and related support activities	13.20	13.67	14.09	14.48	14.93	15.37	15.71	15.74	15.80	16.15	16.75
Petroleum and coal products	21.75	22.22	22.80	22.90	23.04	23.63	24.39	24.47	24.11	25.21	27.41
Chemicals	16.23	16.40	17.09	17.57	17.97	18.50	19.17	19.67	19.60	19.55	19.50
Plastics and rubber products	11.79	12.25	12.70	13.21	13.55	14.18	14.59	14.80	14.97	15.39	15.85
Private Service-Providing	12.61	13.09	13.62	14.18	14.58	14.98	15.28	15.72	16.40	17.09	17.75
Trade, transportation, and utilities	12.37	12.80	13.28	13.68	13.99	14.31	14.55	14.90	15.36	15.74	16.13
Wholesale trade	15.03	15.59	16.24	16.74	16.94	17.33	17.62	18.13	18.87	19.54	20.08
Retail trade	10.06	10.45	10.87	11.30	11.67	11.90	12.09	12.36	12.58	12.76	12.87
Transportation and warehousing	14.07	14.50	15.00	15.27	15.71	16.20	16.46	16.65	17.21	17.67	18.36
Utilities	21.48	22.03	22.75	23.58	23.96	24.77	25.61	26.68	27.40	27.88	28.83
Information	17.67	18.40	19.07	19.80	20.20	21.01	21.40	22.06	23.23	23.96	24.78
Financial activities	14.00	14.55	15.04	15.65	16.25	17.21	17.58	17.98	18.83	19.67	20.32
Professional and business services	14.28	14.86	15.53	16.34	16.82	17.22	17.49	18.09	19.14	20.16	21.19
Education and health services	12.96	13.40	13.91	14.58	15.15	15.56	16.07	16.62	17.28	17.99	18.73
Leisure and hospitality	7.67	7.96	8.32	8.57	8.81	9.00	9.15	9.38	9.75	10.41	10.84
Other services	11.79	12.26	12.73	13.27	13.72	13.84	13.98	14.34	14.77	15.42	16.09

[1]Derived by assuming that overtime hours are paid at the rate of time and one-half.
[2]Includes other industries, not shown separately.

Table 15-4. Average Hourly Earnings of Production and Nonsupervisory Workers on Private Nonfarm Payrolls by Industry—*Continued*

(Dollars.)

Industry	2009	2010	2011	2012	2013	2014	2015	2016	2017	2018	2019
Total Private	18.60	19.04	19.43	19.73	20.13	20.60	21.03	21.53	22.05	22.71	23.51
Goods-Producing	19.90	20.28	20.67	20.94	21.24	21.59	21.96	22.58	23.18	24.00	24.74
Mining and logging	23.29	23.82	24.50	25.79	26.80	26.84	26.48	26.97	27.44	28.27	29.95
Construction	22.66	23.22	23.65	23.97	24.22	24.67	25.20	25.97	26.74	27.75	28.51
Manufacturing	18.24	18.61	18.93	19.08	19.30	19.56	19.91	20.44	20.90	21.54	22.15
Excluding overtime [1]	17.59	17.78	18.03	18.16	18.34	18.57	18.93	19.43	19.87	20.42	21.07
Durable goods	19.36	19.81	20.11	20.18	20.35	20.66	20.97	21.48	21.89	22.51	23.08
Wood products	14.92	14.85	14.81	14.99	15.47	15.57	16.16	16.81	17.47	17.88	18.51
Nonmetallic mineral products	17.28	17.48	18.16	18.15	18.40	19.16	19.92	20.30	20.34	21.41	22.12
Primary metals	20.10	20.13	19.94	20.70	21.94	22.41	22.47	23.10	23.10	23.31	23.44
Fabricated metal products	17.48	17.94	18.13	18.26	18.35	18.68	19.02	19.66	20.15	20.55	21.29
Machinery	18.39	18.96	19.54	20.17	20.58	21.00	21.34	21.80	22.41	23.06	23.82
Computer and electronic products	21.87	22.78	23.32	23.34	23.41	23.36	23.30	24.27	24.58	25.09	25.20
Electrical equipment and appliances	16.27	16.87	17.96	18.03	18.04	18.28	18.85	19.29	19.72	20.59	20.42
Transportation equipment [2]	24.98	25.23	25.34	24.57	24.57	24.96	25.05	25.06	25.36	26.31	27.14
Motor vehicles and parts	21.86	22.02	21.93	21.27	21.08	21.38	21.49	21.59	21.72	22.76	23.50
Furniture and related products	15.04	15.06	15.24	15.46	15.59	15.67	16.09	16.82	17.52	17.87	18.34
Miscellaneous manufacturing	16.13	16.56	16.82	17.05	17.03	17.31	17.72	18.44	19.04	19.15	19.62
Nondurable goods [2]	16.56	16.80	17.06	17.29	17.56	17.75	18.16	18.74	19.32	19.95	20.61
Food manufacturing	14.39	14.41	14.63	15.02	15.44	15.55	15.90	16.52	16.92	17.49	18.19
Textile mills	13.71	13.56	13.79	13.51	13.90	14.15	14.72	15.71	15.97	16.46	16.59
Textile product mills	11.44	11.79	12.21	12.77	12.88	13.35	13.31	13.64	14.80	15.31	15.52
Apparel	11.37	11.43	11.96	12.89	13.21	13.51	13.65	13.73	14.35	15.33	15.92
Paper and paper products	19.29	20.04	20.28	20.42	20.32	20.35	21.46	21.71	21.75	21.89	22.47
Printing and related support activities	16.75	16.91	17.28	17.28	17.77	18.01	18.31	18.59	18.61	18.74	18.98
Petroleum and coal products	29.61	31.31	31.75	32.15	34.58	35.39	37.32	38.88	39.95	40.24	41.37
Chemicals	20.30	21.07	21.45	21.45	21.40	21.49	21.76	22.71	24.29	25.46	25.46
Plastics and rubber products	16.01	15.71	15.95	16.05	16.20	16.51	16.82	17.17	17.63	18.49	19.38
Private Service-Providing	18.33	18.78	19.17	19.48	19.90	20.39	20.83	21.31	21.82	22.44	23.25
Trade, transportation, and utilities	16.44	16.78	17.10	17.38	17.69	18.21	18.61	18.93	19.29	19.88	20.64
Wholesale trade	20.78	21.46	21.88	22.13	22.52	23.14	23.52	24.07	24.58	25.17	26.06
Retail trade	13.02	13.25	13.52	13.82	14.03	14.40	14.83	15.05	15.33	15.91	16.62
Transportation and warehousing	18.74	19.10	19.42	19.47	19.73	20.43	20.66	20.83	21.22	21.75	22.40
Utilities	29.48	30.04	30.82	31.61	32.27	32.86	34.02	35.33	36.22	36.76	36.91
Information	25.45	25.87	26.62	27.04	27.98	28.70	29.05	30.05	30.74	31.97	33.89
Financial activities	20.90	21.55	21.93	22.82	23.87	24.71	25.34	26.12	26.57	26.93	27.67
Professional and business services	22.36	22.80	23.14	23.31	23.74	24.31	24.81	25.43	26.05	26.82	27.78
Education and health services	19.34	19.95	20.60	20.91	21.29	21.64	22.09	22.52	23.03	23.64	24.35
Leisure and hospitality	11.12	11.31	11.45	11.62	11.78	12.09	12.41	12.85	13.38	13.87	14.49
Other services	16.59	17.06	17.32	17.59	18.00	18.51	19.01	19.36	20.10	20.78	21.41

[1]Derived by assuming that overtime hours are paid at the rate of time and one-half.
[2]Includes other industries, not shown separately.

Table 15-5. Average Weekly Earnings of Production and Nonsupervisory Workers on Private Nonfarm Payrolls by Industry

(Dollars.)

Industry	1998	1999	2000	2001	2002	2003	2004	2005	2006	2007	2008
Total Private	448.36	463.09	480.90	493.53	506.48	517.68	528.65	543.94	566.94	589.09	607.10
Goods-Producing	580.99	599.99	621.86	630.01	651.61	669.13	688.03	705.28	730.16	757.53	776.63
Mining and logging	727.19	721.77	734.88	757.96	741.97	765.94	804.01	853.87	907.95	962.63	1 014.69
Construction	629.75	655.11	685.78	695.86	711.82	727.00	735.55	750.37	781.59	816.23	842.61
Manufacturing	557.09	573.29	590.89	595.15	618.62	636.03	658.52	673.30	690.88	711.53	724.46
Durable goods	591.80	606.55	624.38	624.38	652.84	671.35	694.06	712.81	731.97	754.58	767.92
Wood products	449.77	461.46	476.97	481.33	492.07	514.06	530.15	526.65	533.12	539.41	547.53
Nonmetallic mineral products	573.04	587.42	604.76	618.91	647.00	664.96	688.33	700.63	712.67	716.78	711.11
Primary metals	681.47	700.80	734.83	723.86	749.17	767.15	799.69	815.90	843.79	843.42	851.12
Fabricated metal products	543.20	555.86	576.68	576.60	596.30	610.44	628.83	647.24	668.91	687.16	701.57
Machinery	613.73	625.07	643.94	632.68	645.38	664.48	699.45	716.27	728.84	754.34	760.09
Computer and electronic products	579.85	596.37	609.97	613.18	642.77	674.69	697.97	735.59	766.75	809.10	861.58
Electrical equipment and appliances	522.54	538.98	550.48	548.03	560.33	583.27	607.00	618.88	637.04	656.46	645.60
Transportation equipment [1]	774.77	795.57	817.56	816.74	877.50	889.42	912.56	937.72	957.47	986.75	1 000.87
Motor vehicles and parts	775.71	809.47	828.41	818.84	898.57	910.20	925.07	940.54	934.38	930.47	920.36
Furniture and related products	428.67	443.68	459.95	464.87	493.95	505.26	519.61	527.49	535.90	561.08	553.90
Miscellaneous manufacturing	437.95	454.14	464.80	483.07	499.01	510.54	533.27	545.01	555.87	570.12	591.95
Nondurable goods [1]	504.13	520.02	536.89	548.30	566.72	582.48	602.56	609.04	622.00	639.99	652.22
Food manufacturing	444.72	458.73	472.09	481.81	497.25	502.92	509.55	508.55	526.02	551.21	566.88
Textile mills	434.15	447.38	464.51	456.64	476.52	469.33	486.68	498.47	509.39	524.40	525.00
Textile product mills	376.37	392.46	398.93	402.53	423.05	438.67	437.89	451.36	472.10	467.80	453.06
Apparel	286.17	295.62	307.51	317.63	334.24	340.73	352.04	366.93	389.05	411.57	415.10
Paper and paper products	662.27	679.05	681.38	689.76	705.20	719.55	754.17	764.15	772.57	795.58	809.57
Printing and related support activities	518.32	534.15	552.15	560.89	573.05	587.58	603.97	604.73	618.92	632.02	642.50
Petroleum and coal products	949.28	947.60	973.53	1 003.34	990.88	1 052.32	1 095.00	1 114.51	1 085.50	1 112.73	1 222.07
Chemicals	700.53	701.06	721.41	735.57	759.56	784.26	819.93	831.79	833.84	819.51	809.29
Plastics and rubber products	487.04	505.45	517.56	528.62	550.14	572.53	589.99	591.59	608.37	635.63	649.02
Private Service-Providing	413.36	427.95	445.66	460.97	473.61	484.32	493.95	509.00	532.11	554.05	573.23
Trade, transportation, and utilities	422.40	433.44	448.78	458.58	469.97	480.00	487.37	497.07	512.82	524.57	534.44
Wholesale trade	581.48	601.78	630.09	642.08	643.40	656.36	665.96	683.99	717.42	747.08	768.14
Retail trade	310.43	321.74	333.48	346.31	360.92	367.28	371.23	377.58	383.25	385.18	386.44
Transportation and warehousing	543.72	544.81	559.63	559.22	576.77	594.70	611.99	615.26	633.23	651.78	667.57
Utilities	902.44	924.40	955.09	977.25	979.26	1 017.44	1 048.01	1 095.91	1 135.57	1 182.65	1 230.65
Information	646.52	675.47	700.92	731.18	737.94	760.84	776.72	805.11	850.64	874.45	908.78
Financial activities	504.00	520.73	540.42	560.46	579.08	611.99	625.50	646.51	673.66	706.52	729.64
Professional and business services	490.53	511.39	535.56	558.32	575.00	587.68	597.88	619.65	662.80	701.42	738.31
Education and health services	417.47	429.76	447.80	471.33	490.38	503.05	521.06	541.40	561.02	585.44	607.82
Leisure and hospitality	200.82	208.05	217.20	220.73	227.31	230.49	234.86	241.36	250.34	265.54	273.39
Other services	384.25	398.77	413.30	428.64	439.87	434.41	433.04	443.40	456.50	477.06	495.57

[1]Includes other industries, not shown separately.

Table 15-5. Average Weekly Earnings of Production and Nonsupervisory Workers on Private Nonfarm Payrolls by Industry—*Continued*

(Dollars.)

Industry	2009	2010	2011	2012	2013	2014	2015	2016	2017	2018	2019
Total Private	615.82	636.02	652.72	665.54	677.62	694.74	708.70	723.20	742.48	766.99	790.67
Goods-Producing	779.68	818.96	844.85	861.39	877.09	895.09	905.43	930.34	956.83	996.67	1 017.62
Mining and logging	1 006.67	1 063.11	1 144.64	1 201.69	1 229.70	1 270.91	1 211.91	1 221.69	1 265.92	1 322.45	1 402.86
Construction	851.76	891.83	921.84	942.14	958.72	977.11	998.02	1 031.88	1 061.98	1 108.59	1 135.17
Manufacturing	726.12	765.18	784.29	794.67	807.37	822.03	832.25	855.77	876.10	908.01	921.66
Durable goods	771.24	819.06	841.85	848.35	859.69	877.31	882.91	909.23	926.67	956.74	968.31
Wood products	557.65	580.74	587.77	615.76	661.10	653.87	666.82	703.47	735.06	749.63	768.04
Nonmetallic mineral products	705.54	728.22	768.35	766.16	782.69	829.79	844.77	852.66	873.01	947.10	969.28
Primary metals	817.70	880.46	889.34	907.23	960.61	991.13	984.77	1 003.42	995.41	1 038.42	1 026.92
Fabricated metal products	689.22	742.61	762.14	767.52	775.35	795.62	804.67	827.74	849.82	866.83	886.88
Machinery	737.97	797.62	842.96	864.16	883.06	903.42	894.65	918.23	956.97	985.50	1 010.09
Computer and electronic products	883.02	932.26	943.88	944.02	948.79	952.82	953.04	999.76	1 012.10	1 026.63	1 021.61
Electrical equipment and appliances	639.34	693.49	732.16	749.91	753.07	763.08	794.79	827.24	850.47	877.69	846.05
Transportation equipment [1]	1 028.51	1 081.60	1 094.46	1 075.00	1 070.88	1 088.27	1 094.01	1 105.96	1 113.99	1 167.96	1 177.73
Motor vehicles and parts	876.34	956.39	952.34	940.00	920.56	936.90	948.30	964.46	956.61	1 017.15	1 018.72
Furniture and related products	566.75	579.66	608.00	617.74	628.36	641.27	640.01	672.60	689.02	702.49	722.71
Miscellaneous manufacturing	620.74	640.85	654.90	668.95	682.97	695.67	709.56	751.91	759.06	756.91	778.29
Nondurable goods [1]	658.51	685.07	696.03	710.16	723.53	733.64	751.05	771.47	798.20	829.97	846.01
Food manufacturing	575.40	586.52	588.19	609.69	629.70	634.95	652.26	683.74	705.82	734.72	754.20
Textile mills	516.86	559.13	574.61	575.75	577.18	586.40	625.84	647.60	648.43	696.84	722.34
Textile product mills	433.09	459.46	477.49	507.00	494.76	507.19	493.03	513.89	568.47	602.46	587.20
Apparel	408.86	418.28	456.97	478.33	503.01	520.57	520.73	499.82	524.59	578.04	589.29
Paper and paper products	806.19	858.65	870.53	877.14	874.85	888.07	923.13	926.10	924.75	942.38	945.45
Printing and related support activities	635.68	646.11	655.81	665.45	685.33	701.55	729.57	728.41	734.43	736.06	736.72
Petroleum and coal products	1 284.44	1 345.72	1 390.80	1 513.36	1 605.02	1 630.84	1 685.27	1 719.89	1 796.54	1 832.98	1 998.98
Chemicals	841.18	888.25	910.88	910.02	918.69	917.96	927.83	950.00	1 012.35	1 071.90	1 064.44
Plastics and rubber products	643.91	658.55	669.54	671.28	677.72	697.67	713.74	728.16	750.22	778.95	794.62
Private Service-Providing	587.37	604.91	620.78	632.87	644.65	661.42	675.83	689.13	707.04	728.42	752.65
Trade, transportation, and utilities	540.19	558.05	575.88	587.09	595.36	611.77	626.54	633.77	652.11	674.74	697.53
Wholesale trade	782.62	814.04	842.06	857.06	871.89	894.06	907.72	929.09	957.95	979.82	1 007.57
Retail trade	388.74	400.38	412.29	422.35	423.44	431.97	446.01	447.69	463.10	483.03	503.07
Transportation and warehousing	674.19	707.93	734.06	739.12	758.79	785.50	800.14	807.25	813.69	835.39	849.29
Utilities	1 239.34	1 262.89	1 296.92	1 298.23	1 344.70	1 388.91	1 444.03	1 502.09	1 540.27	1 569.38	1 566.66
Information	931.08	939.85	964.85	973.52	1 003.65	1 030.17	1 038.10	1 068.23	1 100.03	1 139.52	1 196.36
Financial activities	754.90	780.25	798.68	839.84	875.04	908.15	939.46	962.95	982.09	997.26	1 020.28
Professional and business services	776.30	799.43	814.74	823.37	839.86	865.25	879.63	899.79	922.51	949.18	984.07
Education and health services	622.30	639.37	663.04	674.48	683.24	692.56	709.25	723.91	741.55	761.86	784.18
Leisure and hospitality	275.95	280.87	283.82	290.54	294.31	303.81	311.32	319.67	332.54	345.21	358.10
Other services	506.26	523.70	532.63	539.46	553.77	568.92	583.54	595.68	617.91	640.70	660.02

[1] Includes other industries, not shown separately.

Table 15-6. Indexes of Aggregate Weekly Hours of Production and Nonsupervisory Workers on Private Nonfarm Payrolls by Industry

(2002 = 100.)

Industry	1998	1999	2000	2001	2002	2003	2004	2005	2006	2007	2008
Total Private	99.4	101.4	103.5	102.0	100.0	98.7	100.2	102.8	105.8	107.4	106.0
Goods-Producing	112.3	112.6	113.1	106.6	100.0	95.7	96.8	98.9	102.5	101.7	96.5
Mining and logging	112.8	102.9	105.1	108.3	100.0	97.4	104.0	114.7	125.8	133.5	137.6
Construction	93.4	99.7	104.0	103.2	100.0	98.4	101.7	108.3	115.4	114.7	106.5
Manufacturing	121.0	119.0	117.7	108.1	100.0	94.5	94.3	93.9	95.6	94.4	90.2
Durable goods	122.0	120.6	120.4	109.3	100.0	94.3	95.2	96.1	98.9	97.4	92.2
Wood products	117.5	118.5	115.8	105.0	100.0	97.8	100.8	101.2	99.9	89.3	76.7
Nonmetallic mineral products	105.8	106.9	109.1	106.1	100.0	94.3	98.0	97.4	100.3	96.9	91.2
Primary metals	131.1	128.4	128.9	113.0	100.0	93.4	93.3	93.1	94.2	91.4	87.4
Fabricated metal products	118.6	116.7	119.2	109.3	100.0	95.3	97.7	99.3	103.1	104.5	101.3
Machinery	137.4	129.8	127.6	114.1	100.0	93.6	96.0	98.8	102.2	103.3	102.3
Computer and electronic products	136.7	131.1	133.1	117.9	100.0	92.1	89.8	94.8	103.6	102.2	101.2
Electrical equipment and appliances	127.8	128.3	127.7	113.4	100.0	92.0	88.7	86.4	88.0	89.1	88.5
Transportation equipment [1]	118.8	119.5	116.3	105.3	100.0	95.5	96.5	97.3	99.9	98.0	88.6
Motor vehicles and parts	112.8	118.8	117.3	103.6	100.0	95.9	97.0	95.2	92.9	85.8	72.7
Furniture and related products	108.4	112.6	114.7	104.8	100.0	93.0	94.3	91.9	90.4	86.1	74.6
Miscellaneous manufacturing	110.4	110.3	109.5	104.7	100.0	93.6	91.8	90.6	90.4	91.0	89.2
Nondurable goods [1]	119.4	116.1	113.3	106.0	100.0	94.7	92.7	90.3	90.4	89.6	86.9
Food manufacturing	103.3	103.7	103.4	101.4	100.0	98.4	97.1	95.8	98.6	101.1	100.6
Textile mills	148.9	139.1	132.4	112.2	100.0	86.3	79.0	71.3	65.1	56.3	47.9
Textile product mills	117.7	115.6	112.3	105.5	100.0	92.4	90.2	88.0	85.0	77.5	70.5
Apparel	180.9	154.6	137.5	117.1	100.0	81.9	75.1	65.7	63.4	61.5	56.7
Paper and paper products	119.6	117.2	113.5	106.6	100.0	92.5	89.2	88.0	86.9	85.8	83.5
Printing and related support activities	124.3	120.9	119.4	111.5	100.0	95.3	93.4	90.9	92.5	91.6	86.1
Petroleum and coal products	113.4	107.6	105.8	105.6	100.0	98.7	102.6	102.4	96.9	95.6	102.4
Chemicals	115.4	113.2	110.4	104.7	100.0	99.0	99.0	95.9	96.1	94.1	94.6
Plastics and rubber products	113.7	114.6	114.2	104.9	100.0	95.2	94.2	92.3	91.9	91.1	87.1
Private Service-Providing	95.7	98.3	100.9	100.8	100.0	99.5	101.1	103.8	106.7	108.9	108.6
Trade, transportation, and utilities	100.4	102.0	103.6	101.6	100.0	98.6	99.6	101.6	103.2	104.8	103.2
Wholesale trade	104.9	106.3	107.1	102.9	100.0	98.0	98.7	101.6	105.5	108.9	108.3
Retail trade	97.2	99.5	101.3	100.5	100.0	98.9	99.4	100.8	101.1	101.8	99.8
Transportation and warehousing	102.8	103.3	105.7	102.8	100.0	98.7	101.9	105.2	107.9	109.5	107.9
Utilities	105.7	105.0	104.1	102.3	100.0	97.4	94.2	93.1	93.8	96.2	98.3
Information	92.6	98.5	105.0	106.7	100.0	97.0	98.2	99.4	100.2	100.1	100.0
Financial activities	96.4	98.0	98.4	99.2	100.0	101.3	101.3	103.7	106.3	107.6	106.8
Professional and business services	96.7	101.6	106.5	104.0	100.0	98.7	101.8	106.4	112.2	115.4	114.0
Education and health services	88.6	90.4	92.6	96.5	100.0	101.7	103.8	107.0	109.6	113.2	116.5
Leisure and hospitality	95.5	97.8	100.5	100.6	100.0	100.0	103.0	106.2	108.8	110.8	109.7
Other services	94.3	96.2	97.8	99.0	100.0	97.4	96.1	96.2	97.4	99.3	99.5

[1]Includes other industries, not shown separately.

Table 15-6. Indexes of Aggregate Weekly Hours of Production and Nonsupervisory Workers on Private Nonfarm Payrolls by Industry—*Continued*

(2002 = 100.)

Industry	2009	2010	2011	2012	2013	2014	2015	2016	2017	2018	2019
Total Private	98.8	99.0	101.4	104.3	106.1	108.6	110.9	112.5	114.9	117.2	118.3
Goods-Producing	80.2	78.8	81.2	83.5	85.1	87.8	89.1	89.5	91.1	94.4	94.9
Mining and logging	117.2	124.5	147.4	158.7	155.0	164.3	144.1	113.4	120.2	134.5	134.5
Construction	85.9	80.2	81.7	83.6	87.7	92.0	96.5	100.9	104.0	108.8	111.2
Manufacturing	76.1	76.2	78.3	80.3	80.9	82.6	83.3	83.3	84.0	86.1	85.7
Durable goods	74.7	75.0	78.4	81.4	82.3	84.3	84.6	84.3	84.5	87.3	87.4
Wood products	57.8	58.5	59.3	62.1	67.3	69.6	70.2	72.0	72.9	74.3	74.2
Nonmetallic mineral products	73.8	70.5	70.1	68.9	69.7	72.4	75.0	76.7	78.1	82.1	82.2
Primary metals	66.1	71.7	80.1	82.7	79.8	81.7	80.1	75.8	74.9	78.4	78.3
Fabricated metal products	81.3	83.1	89.6	94.7	96.4	97.9	97.0	93.6	94.6	98.4	99.3
Machinery	80.6	81.2	89.6	94.0	94.0	96.6	93.4	90.6	92.5	95.9	94.3
Computer and electronic products	89.4	87.2	86.4	86.0	83.7	81.4	82.3	83.1	83.4	84.9	88.0
Electrical equipment and appliances	74.1	73.2	71.6	73.4	72.4	73.5	77.2	78.6	77.3	78.8	77.4
Transportation equipment [1]	70.1	72.1	75.3	80.4	82.4	86.3	89.3	91.1	90.6	94.7	93.7
Motor vehicles and parts	51.6	57.5	60.9	66.6	70.7	76.2	79.8	83.2	83.6	87.9	84.4
Furniture and related products	57.6	54.4	55.8	55.6	57.6	60.7	60.8	61.6	61.3	61.3	61.1
Miscellaneous manufacturing	81.1	79.0	80.0	82.2	85.3	86.3	84.9	85.8	83.8	83.7	86.3
Nondurable goods [1]	78.1	78.0	77.9	78.6	78.6	80.0	81.2	81.6	83.1	84.2	83.0
Food manufacturing	97.5	98.5	97.7	99.6	100.1	100.8	102.4	105.3	109.5	112.2	112.2
Textile mills	37.8	40.2	41.6	41.7	39.0	38.4	39.1	37.8	36.3	37.4	37.7
Textile product mills	58.5	56.5	55.2	53.3	50.5	51.5	51.5	53.3	54.1	53.2	46.7
Apparel	45.4	42.0	40.7	38.4	38.5	38.0	37.8	34.3	30.5	29.4	26.3
Paper and paper products	74.2	73.5	71.9	70.0	68.2	68.5	67.5	66.6	67.0	67.1	64.8
Printing and related support activities	74.1	69.0	65.6	64.3	63.2	64.2	65.2	64.6	63.7	61.4	58.4
Petroleum and coal products	90.2	89.1	91.2	101.6	96.8	98.5	100.1	100.7	107.0	104.7	111.2
Chemicals	88.3	88.9	90.6	92.6	93.6	94.4	96.1	96.1	97.4	102.3	103.7
Plastics and rubber products	71.3	73.6	75.3	75.8	77.3	81.4	84.0	84.6	86.0	85.9	84.3
Private Service-Providing	104.1	104.5	107.0	110.0	112.0	114.5	117.0	118.9	121.5	123.5	124.8
Trade, transportation, and utilities	96.7	96.8	99.7	101.8	102.6	104.4	106.3	106.9	109.2	110.8	110.8
Wholesale trade	99.5	97.4	100.2	103.5	104.8	106.4	106.4	106.2	107.8	108.3	108.7
Retail trade	94.3	95.0	97.7	98.9	98.8	99.5	101.0	101.1	103.1	103.6	102.3
Transportation and warehousing	100.2	101.5	105.7	109.5	112.8	117.0	123.3	126.9	130.6	137.4	140.9
Utilities	97.0	95.4	94.9	92.6	94.9	96.3	97.0	97.1	97.2	96.9	95.5
Information	93.5	90.0	88.9	88.9	89.8	90.5	90.8	91.4	92.6	93.0	92.7
Financial activities	103.1	101.3	101.1	103.7	104.7	106.5	109.6	111.6	114.3	116.0	117.4
Professional and business services	105.3	107.7	112.6	117.3	121.4	125.9	128.3	130.2	132.6	135.2	137.3
Education and health services	118.3	119.8	122.3	125.4	126.6	128.5	132.3	136.1	139.8	142.7	145.8
Leisure and hospitality	105.1	104.6	106.9	111.3	115.2	119.3	122.7	125.6	128.7	130.6	131.8
Other services	96.0	96.0	96.8	97.7	98.7	99.9	100.7	101.8	102.9	104.4	105.4

[1]Includes other industries, not shown separately.

NOTES AND DEFINITIONS, CHAPTER 15

TABLES 15-1 THROUGH 15-6

Employment, Hours, and Earnings by NAICS Industry

SOURCE: U.S. DEPARTMENT OF LABOR, BUREAU OF LABOR STATISTICS

See the notes and definitions for Tables 10-8 through 10-18 regarding definitions of *employment, production and nonsupervisory workers, average weekly hours, overtime hours, average hourly earnings, average weekly earnings*, and the *indexes of aggregate weekly hours*. Availability and reference information is also provided in those notes and definitions.

Definitions

The North American Industry Classification System (NAICS) is the standard used by federal statistical agencies in classifying business establishments for the purpose of collecting, analyzing, and publishing statistical data related to the U.S. business economy. NAICS defines industries according to a consistent principle: businesses that use similar production processes are grouped together.

NAICS is the product of a cooperative effort on the part of the statistical agencies of the United States, Canada, and Mexico. This system is the hierarchical, numerical system used by the Federal government to classify businesses by industry, in order to collect, analyze, and publish statistical data related to the U.S. business economy.

There are 20 high-level industrial sections and are listed below.

Sector	Description
11	Agriculture, Forestry, Fishing and Hunting
21	Mining, Quarrying, and Oil and Gas Extraction
22	Utilities
23	Construction
31-33	Manufacturing
42	Wholesale Trade
44-45	Retail Trade
48-49	Transportation and Warehousing
51	Information
52	Finance and Insurance
53	Real Estate and Rental and Leasing
54	Professional, Scientific, and Technical Services
55	Management of Companies and Enterprises
56	Administrative and Support and Waste Management and Remediation Services
61	Educational Services
62	Health Care and Social Assistance
71	Arts, Entertainment, and Recreation
72	Accommodation and Food Services
81	Other Services (except Public Administration)
92	Public Administration *(not covered in economic census)*

CHAPTER 16: KEY SECTOR STATISTICS

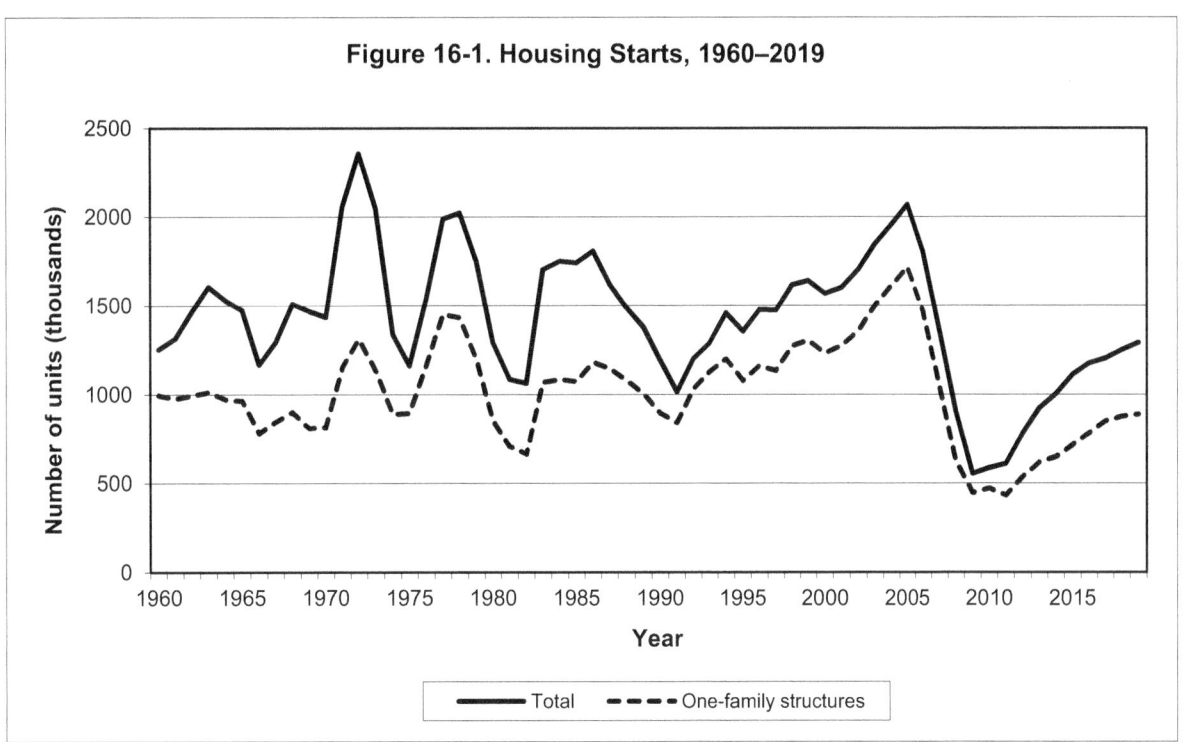

- The housing sector and associated financial excesses had the leading role in the boom and bust cycle of 2001–2009. In 2005, as shown in Figure 16-1, nearly 2.1 million housing units were started, the highest since 1972. Of those, 1.7 million were one-family homes, an all-time record. By 2009, both total and single-family starts had plunged to the lowest levels of their 50-year history. With very modest recovery in the years following 2009, starts remained at levels once viewed as recessionary. In 2019, the housing starts more than doubled the 2009 low—to 1.29 million. (Table 16-2)

- New orders for nondefense capital goods excluding aircraft and parts at U.S. manufacturing firms is a widely followed advance indicator for production of investment goods. The value of such orders fell nearly 39 percent from 2007 to 2009 which was greater than their decline between 2000 and 2002. Their recovery in the four years following 2009 was faster than it was in the 4 years following 2002, and the value of new orders in 2013 surpassed the peak in 2007 but declined 13 percent between 2014 and 2019. New orders for nondefense capital goods decreased 6.9 percent from 2018 to 2019 after increasing the previous two years. (Table 16-5)

- E-commerce grew from 2.7 percent of total retail sales at the beginning of 2006 to 11.3 percent (seasonally adjusted) at the end of 2019. (Table 16-9)

- Sales of cars and light trucks declined for only the second time since 2010 in 2019. (Table 16-7)

Table 16-1. New Construction Put in Place

(Billions of dollars, monthly data are at seasonally adjusted annual rates.)

Year and month	Total	Private												
		Total ¹	Residential	Office	Commercial		Health care	Educational	Amuse-ment and recreation	Transpor-tation	Commu-nication	Power	Manu-facturing
					Total ¹	Multi-retail							
1980	273.9	210.3	100.4	...	...	...	...	...	...	...	...	...	...
1981	289.1	224.4	99.2	...	...	...	...	...	...	...	...	...	...
1982	279.3	216.3	84.7	...	...	...	...	...	...	...	...	...	...
1983	311.9	248.4	125.8	...	...	...	...	...	...	...	...	...	...
1984	370.2	300.0	155.0	...	...	...	...	...	...	...	...	...	...
1985	403.4	325.6	160.5	...	...	...	...	...	...	...	...	...	...
1986	433.5	348.9	190.7	...	...	...	...	...	...	...	...	...	...
1987	446.6	356.0	199.7	...	...	...	...	...	...	...	...	...	...
1988	462.0	367.3	204.5	...	...	...	...	...	...	...	...	...	...
1989	477.5	379.3	204.3	...	...	...	...	...	...	...	...	...	...
1990	476.8	369.3	191.1	...	...	...	...	...	...	...	...	...	...
1991	432.6	322.5	166.3	...	...	...	...	...	...	...	...	...	...
1992	463.7	347.8	199.4	...	...	...	...	...	...	...	...	...	...
1993	485.5	358.2	208.2	20.0	34.4	11.5	14.9	4.8	4.6	4.7	9.8	23.6	23.4
1994	531.9	401.5	241.0	20.4	39.6	12.2	15.4	5.0	5.1	4.7	10.1	21.0	28.8
1995	548.7	408.7	228.1	23.0	44.1	12.0	15.3	5.7	5.9	4.8	11.1	22.0	35.4
1996	599.7	453.0	257.5	26.5	49.4	13.3	15.4	7.0	7.0	5.8	11.8	17.4	38.1
1997	631.9	478.4	264.7	32.8	53.1	12.2	17.4	8.8	8.5	6.2	12.5	16.4	37.6
1998	688.5	533.7	296.3	40.4	55.7	13.3	17.7	9.8	8.6	7.3	12.5	21.7	40.5
1999	744.6	575.5	326.3	45.1	59.4	15.2	18.4	9.8	9.6	6.5	18.4	22.0	35.1
2000	802.8	621.4	346.1	52.4	64.1	14.9	19.5	11.7	8.8	6.9	18.8	29.3	37.6
2001	840.2	638.3	364.4	49.7	63.6	16.4	19.5	12.8	7.8	7.1	19.6	31.5	37.8
2002	847.9	634.4	402.0	44.3	62.5	15.6	27.1	73.9	17.3	25.8	18.5	36.8	22.9
2003	891.5	675.4	451.3	39.4	61.5	15.4	29.3	74.3	16.8	24.7	14.6	41.5	21.5
2004	991.4	771.2	538.4	42.4	67.1	18.8	32.2	74.3	16.7	25.1	15.5	35.6	23.4
2005	1 116.8	882.7	624.6	37.3	66.6	22.8	28.5	12.8	7.5	7.1	18.8	29.2	28.4
2006	1 161.3	905.9	607.8	45.7	73.4	29.2	32.0	13.8	9.3	8.7	22.2	33.7	32.3
2007	1 148.0	858.9	488.8	53.8	85.9	34.8	35.6	16.7	10.2	9.0	27.5	54.1	40.2
2008	1 077.3	768.6	359.2	55.5	82.7	32.0	38.4	18.6	10.5	9.9	26.3	69.2	53.6
2009	906.5	591.6	247.5	37.3	51.1	18.4	35.3	16.9	8.4	9.1	19.7	76.1	57.4
2010	809.3	505.3	242.0	24.4	37.2	12.5	29.6	13.4	6.5	9.9	17.7	66.1	40.6
2011	788.3	501.9	244.1	23.7	39.2	13.4	28.9	14.1	6.7	9.5	17.5	64.3	39.8
2012	850.5	571.1	269.8	27.4	44.3	14.9	31.4	16.6	6.2	10.9	16.0	86.4	46.8
2013	908.3	637.6	323.4	30.1	50.9	16.7	29.7	16.9	6.9	11.0	17.6	81.3	51.8
2014	1 007.6	731.5	369.8	38.9	60.9	19.5	28.9	16.6	7.7	12.2	17.1	98.2	60.1
2015	1 130.7	836.9	422.3	47.9	64.5	20.0	30.9	17.7	10.0	13.6	21.5	99.5	82.4
2016	1 211.4	914.4	467.1	59.8	75.5	22.4	31.8	20.0	12.6	13.1	22.0	102.3	78.9
2017	1 265.8	969.3	525.0	59.9	84.5	24.0	33.5	21.2	14.5	14.8	23.6	89.8	70.0
2018	1 307.2	1 000.2	539.6	64.6	91.8	25.9	33.3	21.4	15.5	16.6	24.5	87.7	70.3
2019	1 306.9	978.5	515.4	68.5	81.1	15.1	35.3	19.0	14.2	16.5	23.4	94.0	74.0
2017													
January	1 255.9	966.1	507.6	63.1	81.5	24.1	32.8	21.7	13.3	13.3	23.9	101.8	73.9
February	1 269.7	971.5	520.7	60.3	81.9	23.8	32.4	21.4	14.1	13.8	24.1	99.8	70.5
March	1 268.8	969.0	518.9	60.4	84.6	23.6	32.6	20.9	13.5	14.2	23.3	96.3	71.4
April	1 259.0	967.9	520.4	60.3	85.6	25.1	32.8	21.0	14.6	13.4	23.3	93.0	71.2
May	1 267.3	966.3	517.8	59.1	85.8	24.5	33.6	21.3	14.7	14.6	23.4	90.7	73.0
June	1 261.6	963.9	519.7	60.3	86.6	24.4	33.6	21.7	14.6	14.7	23.7	88.0	68.2
July	1 260.8	964.0	522.8	59.0	85.3	23.7	33.8	20.5	15.0	15.0	23.4	86.9	70.0
August	1 258.5	966.5	527.6	58.0	84.6	24.2	34.1	20.8	15.2	15.1	23.4	87.1	68.1
September	1 264.5	969.7	530.8	58.3	83.9	23.9	34.7	21.8	15.2	15.1	23.4	85.7	68.4
October	1 258.3	958.3	518.7	59.3	83.8	23.5	33.9	21.3	14.6	16.1	23.9	85.3	69.3
November	1 290.2	987.5	549.7	60.2	84.4	23.8	34.0	21.1	14.5	15.9	23.8	83.4	68.5
December	1 295.9	994.6	550.6	61.7	85.7	23.9	34.0	21.8	14.8	16.3	23.9	84.4	68.8
2018													
January	1 297.4	996.8	552.0	61.9	87.1	23.2	34.1	21.9	14.7	15.8	25.2	82.2	69.2
February	1 335.4	1 034.2	570.3	63.9	98.1	29.1	34.1	22.3	15.1	15.9	25.6	82.9	72.2
March	1 312.9	1 012.2	555.4	63.7	94.6	28.4	33.1	21.7	15.1	15.2	26.0	84.2	69.5
April	1 322.4	1 013.6	557.6	62.2	91.6	24.5	33.6	21.7	15.4	16.3	24.9	87.6	67.7
May	1 333.5	1 022.5	556.1	64.3	96.1	28.2	33.9	22.2	15.9	15.8	24.4	89.0	69.9
June	1 317.8	1 006.7	546.2	63.0	91.7	27.7	32.6	21.4	15.8	16.5	24.1	91.8	69.3
July	1 324.8	1 011.6	542.3	64.4	95.2	29.4	32.5	21.3	15.8	16.7	24.0	93.6	69.7
August	1 312.2	994.5	533.8	64.8	87.2	22.0	32.6	20.6	15.6	16.6	23.8	93.4	70.4
September	1 319.7	1 008.1	530.5	68.3	96.8	29.0	33.0	21.2	16.0	16.8	23.9	90.6	75.1
October	1 277.4	973.7	505.6	67.9	93.2	25.7	33.8	21.7	15.7	17.2	23.7	87.3	71.4
November	1 271.4	970.2	521.9	64.6	83.1	19.1	32.3	20.6	15.5	17.9	23.8	86.0	69.7
December	1 264.8	963.2	512.3	65.7	81.8	17.8	33.9	19.9	15.5	18.1	24.6	85.3	70.3
2019													
January	1 282.5	974.9	509.1	68.6	79.7	17.2	37.2	20.8	14.7	18.0	23.3	93.8	72.8
February	1 289.0	971.7	507.0	67.8	78.6	16.4	35.8	20.2	15.2	16.2	23.1	95.2	75.0
March	1 299.1	976.6	505.9	68.6	82.0	16.8	36.3	20.4	15.3	15.6	23.8	95.6	75.4
April	1 307.1	967.7	505.7	68.3	80.1	15.4	35.3	19.7	15.3	15.4	23.3	94.3	72.7
May	1 297.5	962.7	503.6	68.4	79.3	15.5	34.9	18.8	14.5	17.2	23.3	95.4	70.0
June	1 285.3	959.2	499.6	68.1	81.2	15.5	34.8	18.0	14.3	17.2	23.2	93.4	70.5
July	1 291.3	962.7	504.6	68.6	79.8	14.4	35.3	18.6	13.6	16.4	23.3	91.6	73.3
August	1 306.0	976.7	518.5	68.9	79.3	13.8	34.3	18.6	13.8	16.4	23.6	93.6	72.7
September	1 315.2	980.0	522.5	68.9	81.1	14.0	35.0	18.9	13.4	16.8	23.4	92.1	72.1
October	1 320.8	986.3	525.3	68.7	81.7	13.8	35.0	18.5	13.5	16.4	23.0	91.7	75.9
November	1 342.5	1 006.8	535.8	68.6	84.1	14.2	35.3	18.2	13.8	16.2	23.4	94.8	81.1
December	1 347.3	1 013.0	546.5	68.6	84.2	14.4	35.7	17.6	13.4	16.1	23.6	96.2	75.8

¹Includes categories not shown separately.
. . . = Not available.

Table 16-1. New Construction Put in Place—*Continued*

(Billions of dollars, monthly data are at seasonally adjusted annual rates.)

Year and month	Total	Public													
		State and local													Federal
		Total [1]	Residential	Office	Health care	Educational	Public safety	Amusement and recreation	Transportation	Power	Highway and street	Sewage and waste disposal	Water supply	
1980	63.6	54.0	. . .	. . .	. . .	. . .	. . .	. . .	. . .	. . .	. . .	. . .	. . .	9.6
1981	64.7	54.3	. . .	. . .	. . .	. . .	. . .	. . .	. . .	. . .	. . .	. . .	. . .	10.4
1982	63.1	53.1	. . .	. . .	. . .	. . .	. . .	. . .	. . .	. . .	. . .	. . .	. . .	10.0
1983	63.5	52.9	. . .	. . .	. . .	. . .	. . .	. . .	. . .	. . .	. . .	. . .	. . .	10.6
1984	70.2	59.0	. . .	. . .	. . .	. . .	. . .	. . .	. . .	. . .	. . .	. . .	. . .	11.2
1985	77.8	65.8	. . .	. . .	. . .	. . .	. . .	. . .	. . .	. . .	. . .	. . .	. . .	12.0
1986	84.6	72.2	. . .	. . .	. . .	. . .	. . .	. . .	. . .	. . .	. . .	. . .	. . .	12.4
1987	90.6	76.6	. . .	. . .	. . .	. . .	. . .	. . .	. . .	. . .	. . .	. . .	. . .	14.1
1988	94.7	82.5	. . .	. . .	. . .	. . .	. . .	. . .	. . .	. . .	. . .	. . .	. . .	12.3
1989	98.2	86.0	. . .	. . .	. . .	. . .	. . .	. . .	. . .	. . .	. . .	. . .	. . .	12.2
1990	107.5	95.4	. . .	. . .	. . .	. . .	. . .	. . .	. . .	. . .	. . .	. . .	. . .	12.1
1991	110.1	97.3	. . .	. . .	. . .	. . .	. . .	. . .	. . .	. . .	. . .	. . .	. . .	12.8
1992	115.8	101.5	. . .	. . .	. . .	. . .	. . .	. . .	. . .	. . .	. . .	. . .	. . .	14.4
1993	127.4	112.9	3.4	2.8	2.3	24.2	4.6	4.3	9.8	7.4	34.5	11.2	6.4	14.4
1994	130.4	116.0	4.2	3.0	2.4	25.3	4.5	4.7	9.2	5.1	37.3	12.0	6.4	14.4
1995	140.0	124.3	4.5	3.3	2.6	27.5	5.0	5.1	9.6	5.7	38.6	13.0	7.3	15.8
1996	146.7	131.4	3.7	3.6	2.8	31.0	5.5	5.0	10.4	4.8	40.6	13.6	7.7	15.3
1997	153.4	139.4	3.3	3.8	2.9	34.3	5.6	5.7	10.3	4.4	44.4	13.1	8.1	14.1
1998	154.8	140.5	3.2	3.7	2.3	35.1	6.1	6.2	10.5	2.8	45.4	13.2	8.9	14.3
1999	169.1	155.1	3.2	3.6	2.5	39.8	6.2	7.2	11.4	2.9	51.0	14.5	9.6	14.0
2000	181.3	167.2	3.0	4.5	2.8	46.8	5.9	7.6	13.0	5.5	51.6	14.0	9.5	14.2
2001	201.9	186.8	3.5	5.6	2.9	52.8	6.1	9.1	15.9	5.3	56.4	14.2	11.4	15.1
2002	213.4	196.9	3.8	6.3	3.5	59.5	6.0	9.2	17.3	3.8	56.7	15.3	11.7	16.6
2003	216.1	198.2	3.7	6.1	4.0	59.3	5.8	8.4	16.5	6.8	56.3	15.6	11.7	17.9
2004	220.2	201.8	4.1	6.0	5.0	59.7	5.5	7.8	16.4	7.0	57.4	17.1	12.0	18.3
2005	234.2	216.9	4.0	5.2	5.1	65.8	6.0	7.3	16.3	8.3	63.2	18.3	13.5	17.3
2006	255.4	237.8	4.3	5.6	5.6	69.8	6.6	9.4	17.7	7.8	71.0	21.5	14.3	17.6
2007	289.1	268.5	5.1	7.2	7.0	78.4	8.4	10.7	21.1	11.4	75.5	23.3	15.0	20.6
2008	308.7	285.0	4.9	8.5	7.0	84.5	9.7	10.9	23.2	11.0	80.4	24.1	16.0	23.7
2009	314.9	286.5	5.8	9.2	6.8	83.7	9.4	10.6	25.5	11.8	81.3	23.5	14.8	28.4
2010	304.0	272.8	7.6	8.3	6.2	71.9	7.6	9.7	26.5	10.8	81.3	24.6	14.4	31.1
2011	286.4	254.8	6.0	7.4	7.0	68.0	7.2	8.5	23.3	9.5	78.4	21.2	13.4	31.7
2012	279.3	252.4	4.7	6.1	7.0	65.6	7.6	8.9	24.8	9.8	79.7	20.9	12.7	26.9
2013	270.7	247.0	4.5	5.2	6.9	60.0	6.6	8.1	25.9	11.0	80.6	21.0	12.9	23.7
2014	276.1	253.7	4.1	5.4	6.2	61.2	6.4	8.7	27.8	10.7	84.0	21.9	12.7	22.4
2015	293.8	271.2	6.0	5.9	5.5	65.8	5.9	10.0	29.4	10.3	90.8	23.4	12.8	22.5
2016	297.0	274.7	5.9	5.7	5.7	69.9	5.8	10.7	27.9	8.5	92.1	23.2	13.6	22.2
2017	296.5	275.3	6.2	6.0	6.4	74.0	6.2	11.6	29.3	5.4	88.3	22.4	13.7	21.3
2018	307.1	285.8	6.0	7.1	6.5	74.8	7.3	12.0	32.5	4.8	89.9	23.2	14.7	21.3
2019	328.4	304.6	5.8	7.8	6.0	77.2	7.8	13.2	35.6	5.8	97.2	25.1	15.8	23.8
2017														
January	289.7	268.6	6.3	6.0	5.3	73.0	5.6	10.9	26.8	6.4	89.2	21.0	12.9	21.1
February	298.2	276.7	6.6	6.1	6.1	72.6	6.0	11.9	28.4	6.3	92.9	22.1	12.5	21.4
March	299.8	276.8	6.8	5.9	6.4	72.2	6.2	12.7	28.9	5.8	91.2	21.7	13.7	23.0
April	291.1	270.0	5.7	5.1	6.4	71.0	6.1	11.6	29.9	4.1	89.4	20.7	14.5	21.2
May	301.0	278.0	6.3	5.6	6.5	75.4	6.1	12.0	29.9	5.1	89.3	21.8	14.1	23.0
June	297.6	277.4	6.3	5.6	6.7	76.8	5.8	12.6	29.4	4.9	87.6	21.9	13.8	20.2
July	296.7	275.9	6.0	5.5	6.7	75.8	6.2	12.4	27.7	4.8	89.5	22.5	13.1	20.8
August	292.0	272.3	6.1	5.9	6.8	73.3	6.6	11.6	28.3	5.0	86.1	22.7	14.0	19.7
September	294.8	274.2	6.2	6.2	6.2	71.7	6.4	10.9	29.5	6.0	87.5	23.3	14.1	20.6
October	300.0	277.8	6.1	6.6	6.7	74.1	6.5	10.2	30.0	5.0	88.9	23.4	13.6	22.2
November	302.7	281.3	6.3	7.4	6.4	75.8	6.4	11.2	31.0	6.6	87.5	23.6	13.7	21.4
December	301.3	280.0	6.6	6.7	6.1	76.0	6.6	11.0	31.7	5.3	86.0	24.0	14.2	21.3
2018														
January	300.7	279.3	6.3	6.5	6.5	74.6	7.2	11.2	31.6	3.9	86.1	24.2	14.8	21.4
February	301.2	281.1	6.4	6.6	6.6	74.7	6.7	11.3	32.0	4.4	86.7	23.8	15.8	20.1
March	300.7	282.3	6.3	6.9	6.5	76.1	6.9	11.6	31.6	5.0	87.4	23.7	14.1	18.3
April	308.8	288.0	6.4	7.2	6.5	77.2	7.1	11.5	32.0	4.6	89.5	25.1	14.5	20.8
May	310.9	289.6	6.4	7.6	6.9	76.9	7.4	11.6	32.3	5.2	90.3	23.4	15.0	21.3
June	311.1	290.1	5.9	7.7	6.5	72.2	7.2	12.3	32.6	4.9	92.4	24.6	16.4	21.0
July	313.2	291.8	6.4	7.5	7.0	72.1	7.2	12.8	34.2	5.2	94.0	22.6	15.6	21.4
August	317.6	294.4	6.1	7.7	6.6	76.2	7.4	11.9	33.4	4.8	94.2	22.8	15.6	23.2
September	311.5	289.8	5.8	7.1	6.7	77.3	7.4	12.3	33.2	4.6	91.1	22.2	14.3	21.7
October	303.7	281.3	5.7	6.8	6.2	74.3	7.8	12.9	32.1	5.2	86.7	21.8	14.2	22.4
November	301.2	279.0	5.5	6.7	6.2	72.9	7.5	12.2	32.3	4.9	88.5	22.0	13.4	22.2
December	301.6	279.9	5.1	6.9	6.0	75.8	7.5	12.1	32.1	4.7	86.4	23.1	12.9	21.7
2019														
January	307.6	285.5	5.2	7.0	5.9	76.7	6.7	11.4	33.4	4.7	91.8	22.9	12.9	22.1
February	317.3	294.4	5.4	6.9	5.7	75.9	7.3	12.8	33.3	4.9	99.8	22.4	13.4	22.9
March	322.5	299.4	5.6	7.3	6.1	75.3	7.1	12.6	34.3	4.5	101.5	23.8	14.3	23.1
April	339.4	315.3	5.8	7.4	6.1	77.5	7.6	13.4	34.8	5.1	110.9	24.2	15.7	24.1
May	334.8	311.7	5.5	7.5	6.1	77.1	8.0	13.2	36.5	5.4	103.7	26.2	15.5	23.1
June	326.1	302.1	5.6	7.5	6.0	72.0	7.9	13.6	37.0	4.6	98.9	26.0	16.0	24.0
July	328.6	304.6	5.4	8.0	6.0	75.2	8.0	13.5	36.7	5.8	95.8	26.1	16.6	24.0
August	329.3	305.2	6.0	8.2	6.3	77.1	8.2	13.7	36.5	6.7	93.8	25.8	15.8	24.1
September	335.3	310.9	6.0	8.3	6.1	80.0	8.1	13.2	36.4	6.5	96.1	26.3	16.4	24.3
October	334.5	309.9	6.4	8.3	6.2	81.5	8.3	13.4	35.9	6.7	93.0	25.6	17.2	24.6
November	335.7	311.1	6.2	8.5	6.3	79.9	8.0	13.8	36.3	7.4	94.2	25.6	17.5	24.6
December	334.4	309.3	6.4	8.5	5.9	78.6	7.9	13.7	35.9	6.8	95.7	25.3	17.3	25.0

[1]Includes categories not shown separately.
. . . = Not available.

Table 16-2. Housing Starts and Building Permits; New House Sales and Prices

Year and month	Housing starts and building permits										New house sales and prices			
	New private housing units (thousands)									Shipments of manufactured homes (thousands, seasonally adjusted annual rate)	Seasonally adjusted		Median sales price (dollars)	Price index (2005 = 100)
	Started (not seasonally adjusted)			Seasonally adjusted annual rate							Sold (thousands, annual rate)	For sale, end-of-period (thousands)		
				Started			Authorized by building permits							
	Total¹	One-family structures	Five units or more	Total¹	One-family structures	Five units or more	Total¹	One-family structures	Five units or more					
1980	1 292	852	331	1 292	852	331	1 191	710	366	222	545	342	64 600	38.9
1981	1 084	705	288	1 084	705	288	986	564	319	241	436	278	68 900	42.0
1982	1 062	663	320	1 062	663	320	1 000	546	366	240	412	255	69 300	43.0
1983	1 703	1 068	522	1 703	1 068	522	1 605	901	570	296	623	304	75 300	43.9
1984	1 750	1 084	544	1 750	1 084	544	1 682	922	617	295	639	358	79 900	45.7
1985	1 742	1 072	576	1 742	1 072	576	1 733	957	657	284	688	350	84 300	46.2
1986	1 805	1 179	542	1 805	1 179	542	1 769	1 078	583	244	750	361	92 000	48.0
1987	1 621	1 146	409	1 621	1 146	409	1 535	1 024	421	233	671	370	104 500	50.6
1988	1 488	1 081	348	1 488	1 081	348	1 456	994	386	218	676	371	112 500	52.5
1989	1 376	1 003	318	1 376	1 003	318	1 338	932	340	198	650	366	120 000	54.6
1990	1 193	895	260	1 193	895	260	1 111	794	263	188	534	321	122 900	55.7
1991	1 014	840	138	1 014	840	138	949	754	152	171	509	284	120 000	56.4
1992	1 200	1 030	139	1 200	1 030	139	1 095	911	138	211	610	267	121 500	57.2
1993	1 288	1 126	133	1 288	1 126	133	1 199	987	160	254	666	295	126 500	59.4
1994	1 457	1 198	224	1 457	1 198	224	1 372	1 068	241	304	670	340	130 000	62.9
1995	1 354	1 076	244	1 354	1 076	244	1 333	997	272	340	667	374	133 900	64.3
1996	1 477	1 161	271	1 477	1 161	271	1 426	1 069	290	363	757	326	140 000	66.0
1997	1 474	1 134	296	1 474	1 134	296	1 441	1 062	310	354	804	287	146 000	67.5
1998	1 617	1 271	303	1 617	1 271	303	1 612	1 188	355	373	886	300	152 500	69.2
1999	1 641	1 302	307	1 641	1 302	307	1 664	1 247	351	348	880	315	161 000	72.8
2000	1 569	1 231	299	1 569	1 231	299	1 592	1 198	329	250	877	301	169 000	75.6
2001	1 603	1 273	293	1 603	1 273	293	1 637	1 236	335	193	908	310	175 200	77.9
2002	1 705	1 359	308	1 705	1 359	308	1 748	1 333	341	169	973	344	187 600	81.4
2003	1 848	1 499	315	1 848	1 499	315	1 889	1 461	346	131	1 086	377	195 000	86.0
2004	1 956	1 611	303	1 956	1 611	303	2 070	1 613	366	131	1 203	431	221 000	92.8
2005	2 068	1 716	311	2 068	1 716	311	2 155	1 682	389	147	1 283	515	240 900	100.0
2006	1 801	1 465	293	1 801	1 465	293	1 839	1 378	384	117	1 051	537	246 500	104.7
2007	1 355	1 046	277	1 355	1 046	277	1 398	980	359	96	776	496	247 900	104.9
2008	906	622	266	906	622	266	905	576	295	82	485	352	232 100	99.5
2009	554	445	97	554	445	97	583	441	121	50	375	232	216 700	95.1
2010	587	471	104	587	471	104	605	447	135	50	323	188	221 800	95.0
2011	609	431	167	609	431	167	624	418	184	52	306	150	227 200	94.3
2012	781	535	234	781	535	234	830	519	285	55	368	148	245 200	97.6
2013	925	618	294	925	618	294	991	621	341	60	429	186	268 900	104.7
2014	1 003	648	342	1 003	648	342	1 052	640	382	64	437	212	288 500	110.2
2015	1 112	715	386	1 112	715	386	1 183	696	455	71	501	235	294 200	112.9
2016	1 174	782	381	1 174	782	381	1 207	751	421	81	561	257	307 800	120.4
2017	1 203	849	343	1 203	849	343	1 282	820	425	93	613	294	323 100	126.8
2018	1 250	876	360	1 250	876	360	1 329	855	434	97	617	348	326 400	132.4
2019	1 290	888	389	1 290	888	389	1 386	862	481	95	683	327	321 500	135.5
2017														
January	82	53	29	1 206	792	411	1 313	795	487	93	585	259	315 200	. . .
February	88	59	28	1 282	874	390	1 239	816	378	92	597	259	298 000	. . .
March	97	70	27	1 186	833	344	1 292	827	427	109	631	263	321 700	124.7
April	105	77	27	1 150	828	306	1 255	801	419	96	589	262	311 100	. . .
May	106	77	28	1 123	800	310	1 209	790	384	107	613	269	323 600	. . .
June	116	84	32	1 243	862	375	1 326	817	471	109	620	274	315 200	126.4
July	112	79	32	1 207	847	348	1 265	817	406	85	565	275	322 900	. . .
August	103	78	24	1 163	872	283	1 309	809	464	114	560	284	314 200	. . .
September	104	73	30	1 174	837	320	1 257	830	391	101	638	285	331 500	128.4
October	110	76	32	1 256	879	359	1 339	848	455	109	627	289	319 500	. . .
November	98	69	28	1 300	949	343	1 299	854	405	104	712	292	343 400	. . .
December	81	55	26	1 199	843	352	1 312	870	402	80	657	294	343 300	127.8
2018														
January	92	60	31	1 314	871	431	1 365	861	457	98	622	294	329 600	. . .
February	90	62	26	1 288	901	369	1 332	876	410	92	637	295	327 200	. . .
March	107	73	34	1 335	892	429	1 415	857	517	102	662	293	335 400	129.8
April	118	85	31	1 269	899	349	1 387	871	475	99	637	294	314 400	. . .
May	124	89	34	1 334	947	375	1 330	855	440	109	657	299	316 700	. . .
June	112	84	28	1 190	858	322	1 317	861	418	103	613	308	310 500	131.0
July	112	82	30	1 195	864	325	1 331	879	421	89	617	313	327 500	. . .
August	114	81	32	1 280	892	371	1 258	829	392	116	598	322	321 400	. . .
September	110	75	34	1 246	881	356	1 290	858	390	98	596	326	328 300	132.5
October	106	75	29	1 207	861	327	1 267	841	388	116	552	335	328 300	. . .
November	91	59	32	1 204	802	391	1 319	837	440	103	614	341	308 500	. . .
December	76	53	22	1 117	798	299	1 334	819	476	79	564	348	329 700	135.1
2019														
January	87	64	22	1 272	953	302	1 316	812	457	98	637	349	305 400	. . .
February	80	55	25	1 137	785	347	1 305	811	457	91	665	340	320 800	. . .
March	98	70	28	1 203	840	358	1 327	824	465	97	700	331	310 600	129.9
April	117	82	33	1 267	864	381	1 330	800	482	103	664	330	339 000	. . .
May	118	78	39	1 268	821	435	1 338	827	474	110	600	334	312 700	. . .
June	115	83	31	1 235	865	359	1 273	843	384	100	726	326	311 800	133.9
July	114	84	29	1 212	875	326	1 366	851	469	93	661	327	308 300	. . .
August	121	81	39	1 377	911	451	1 471	896	533	111	706	325	327 000	. . .
September	113	79	34	1 274	906	357	1 437	900	501	102	726	322	315 700	137.7
October	115	77	37	1 340	911	417	1 503	929	526	118	706	324	322 400	. . .
November	103	68	34	1 371	933	419	1 510	935	534	96	696	325	328 000	. . .
December	108	69	38	1 587	1 047	520	1 457	940	474	81	731	327	329 500	137.1

¹Includes structures with 2 to 4 units, not shown separately.
. . . = Not available.

Table 16-3. Manufacturers' Shipments

(Millions of dollars, seasonally adjusted.)

Year and month	Total	Total durable goods [1]	Total nondurable goods [1]	Construction materials and supplies	Information technology industries	Capital goods Total	Capital goods Nondefense Total	Capital goods Nondefense Excluding aircraft and parts	Defense	Consumer goods Total	Consumer goods Durable	Consumer goods Nondurable
1993	3 020 497	1 604 544	1 415 953	303 391	241 680	580 859	493 875	463 753	86 984	1 127 629	274 813	852 816
1994	3 238 112	1 764 061	1 474 051	332 734	264 092	616 435	538 203	512 327	78 232	1 192 098	314 931	877 167
1995	3 479 677	1 902 815	1 576 862	353 198	291 885	666 167	590 578	565 729	75 589	1 256 611	324 036	932 575
1996	3 597 188	1 978 597	1 618 591	371 401	311 028	704 635	630 932	605 295	73 703	1 292 955	328 402	964 553
1997	3 834 699	2 147 384	1 687 315	399 880	349 846	779 232	702 971	665 074	76 261	1 358 516	360 193	998 323
1998	3 899 813	2 231 588	1 668 225	418 756	362 564	821 736	747 046	695 717	74 690	1 351 812	373 404	978 408
1999	4 031 887	2 326 736	1 705 151	434 138	374 384	839 754	768 799	713 042	70 955	1 424 828	412 646	1 012 182
2000	4 208 584	2 373 688	1 834 896	444 812	399 751	875 396	808 345	757 617	67 051	1 500 532	391 463	1 109 069
2001	3 970 499	2 174 406	1 796 093	424 517	353 237	801 999	728 495	678 288	73 504	1 480 495	367 522	1 112 973
2002	3 914 723	2 123 621	1 791 102	424 008	284 799	728 585	652 342	609 595	76 243	1 494 575	395 953	1 098 622
2003	4 015 388	2 142 589	1 872 799	429 183	274 829	719 602	634 273	600 616	85 329	1 584 329	418 821	1 165 508
2004	4 308 970	2 264 667	2 044 303	463 148	287 837	752 905	661 740	629 146	91 165	1 700 835	419 182	1 281 653
2005	4 742 077	2 424 844	2 317 233	509 865	295 447	821 906	730 368	686 939	91 538	1 895 119	422 555	1 472 564
2006	5 015 552	2 562 194	2 453 358	546 866	319 890	884 199	794 638	744 282	89 561	1 980 031	421 860	1 558 171
2007	5 319 457	2 687 028	2 632 429	560 208	322 846	936 404	836 426	768 662	99 978	2 106 081	421 822	1 684 259
2008	5 468 994	2 616 667	2 852 327	543 819	319 032	955 412	840 250	774 775	115 162	2 254 007	357 546	1 896 461
2009	4 423 779	2 056 829	2 366 950	426 663	271 678	820 521	697 469	642 555	123 052	1 849 113	268 036	1 581 077
2010	4 911 277	2 280 712	2 630 565	443 880	270 118	858 123	727 818	674 747	130 305	2 058 913	323 432	1 735 481
2011	5 491 894	2 479 086	3 012 808	471 598	274 110	923 876	803 248	740 848	120 628	2 373 512	344 299	2 029 213
2012	5 696 728	2 627 582	3 069 146	500 848	272 578	1 000 585	884 994	799 591	115 591	2 437 147	373 352	2 063 795
2013	5 809 745	2 695 807	3 113 938	523 707	271 444	1 007 645	891 684	799 052	115 961	2 496 085	402 805	2 093 280
2014	5 887 559	2 796 927	3 090 632	547 183	262 545	1 017 011	905 668	803 674	111 343	2 507 589	435 589	2 072 000
2015	5 519 020	2 772 027	2 746 993	554 070	258 110	988 095	877 630	773 609	110 465	2 256 499	461 623	1 794 876
2016	5 354 694	2 713 076	2 641 618	554 850	254 804	944 938	832 663	728 338	112 275	2 184 374	476 823	1 707 551
2017	5 587 964	2 774 517	2 813 447	574 277	259 945	971 455	848 103	733 989	123 352	2 323 557	468 997	1 854 560
2018	5 963 949	2 964 226	2 999 723	606 198	279 296	1 039 582	903 160	776 248	136 422	2 511 705	502 807	2 008 898
2019	6 017 160	2 981 801	3 035 359	622 338	288 211	1 027 927	882 193	794 656	145 734	2 546 874	511 373	2 035 501
2016												
January	440 672	228 021	212 651	46 180	21 206	79 661	70 793	62 054	8 868	176 608	40 190	136 418
February	439 678	227 223	212 455	46 274	21 092	79 119	69 856	61 505	9 263	176 523	40 646	135 877
March	440 991	224 886	216 105	46 372	20 932	78 436	69 187	61 033	9 249	178 167	39 110	139 057
April	442 711	225 784	216 927	46 408	21 177	78 856	70 149	61 684	8 707	178 988	39 838	139 150
May	442 391	223 856	218 535	45 903	21 309	79 449	70 333	61 066	9 116	179 489	38 415	141 074
June	447 179	224 696	222 483	45 927	21 248	78 653	69 346	60 629	9 307	184 214	39 254	144 960
July	444 109	225 016	219 093	45 834	21 520	78 546	68 910	59 666	9 636	181 138	39 612	141 526
August	444 721	224 521	220 200	46 072	21 436	77 825	68 516	59 953	9 309	181 947	39 811	142 136
September	447 821	225 945	221 876	45 843	21 273	78 495	68 988	60 100	9 507	183 295	40 008	143 287
October	448 633	225 212	223 421	46 130	21 160	78 255	68 855	59 659	9 400	184 581	39 581	145 000
November	447 610	224 603	223 007	46 468	21 036	77 493	67 659	59 543	9 834	183 719	39 124	144 595
December	457 817	228 312	229 505	46 798	21 208	78 528	68 834	60 113	9 694	190 983	40 010	150 973
2017												
January	458 475	228 179	230 296	46 915	21 215	78 738	69 154	59 910	9 584	191 235	40 249	150 986
February	459 485	227 510	231 975	47 544	21 443	79 186	68 977	60 083	10 209	190 794	38 880	151 914
March	460 398	229 571	230 827	47 685	21 371	80 185	70 234	60 733	9 951	189 752	38 849	150 903
April	458 249	226 558	231 691	47 233	21 336	79 081	69 066	60 306	10 015	189 920	37 998	151 922
May	461 043	230 748	230 295	47 699	21 424	80 398	70 357	60 751	10 041	190 271	39 555	150 716
June	461 400	229 868	231 532	47 547	21 529	79 897	69 875	60 519	10 022	190 813	38 989	151 824
July	461 490	229 180	232 310	47 254	21 727	81 689	71 640	60 621	10 049	190 466	37 704	152 762
August	466 170	231 855	234 315	47 717	21 764	81 268	70 914	61 269	10 354	194 164	39 013	155 151
September	468 741	232 798	235 943	48 241	21 712	82 170	71 710	61 919	10 460	195 053	38 830	156 223
October	473 979	234 581	239 398	48 648	22 073	81 819	71 135	62 363	10 684	198 291	39 479	158 812
November	482 725	238 438	244 287	49 165	22 271	84 234	73 419	62 923	10 815	203 256	40 197	163 059
December	481 920	237 993	243 927	49 564	22 188	83 090	72 255	63 313	10 835	201 955	39 662	162 293
2018												
January	485 160	239 642	245 518	49 430	22 641	84 213	73 035	63 009	11 178	204 328	40 486	163 842
February	486 657	240 927	245 730	49 585	22 310	83 669	72 969	63 338	10 700	205 137	41 008	164 129
March	489 054	243 021	246 033	49 795	22 436	85 653	75 104	63 304	10 549	205 196	41 021	164 175
April	489 658	243 248	246 410	50 218	22 867	83 415	71 483	64 041	11 932	206 724	42 242	164 482
May	493 718	243 251	250 467	50 546	22 828	86 064	74 798	63 932	11 266	207 014	38 760	168 254
June	498 771	246 218	252 553	50 686	23 200	86 869	75 731	64 876	11 138	210 550	41 094	169 456
July	498 733	246 273	252 460	50 947	23 736	84 112	72 719	65 545	11 393	211 974	42 359	169 615
August	502 731	249 159	253 572	51 036	23 604	87 414	76 000	65 379	11 414	212 462	42 220	170 242
September	505 558	250 766	254 792	50 704	23 341	88 461	77 061	65 355	11 400	214 551	42 764	171 787
October	507 486	251 681	255 805	51 061	24 091	88 095	76 386	66 008	11 709	215 790	43 563	172 227
November	503 457	253 475	249 982	50 787	24 012	89 574	77 746	65 700	11 828	210 243	43 360	166 883
December	500 492	254 267	246 225	51 111	24 058	89 545	77 500	65 960	12 045	207 582	44 204	163 378
2019												
January	501 894	254 021	247 873	51 098	24 028	89 768	77 244	66 196	12 524	208 750	43 854	164 896
February	504 836	253 258	251 578	51 386	24 026	88 739	76 841	66 260	11 898	211 996	43 347	168 649
March	505 483	252 962	252 521	51 156	23 984	88 168	76 110	66 351	12 058	213 334	43 713	169 621
April	501 769	247 208	254 561	51 345	24 413	85 831	73 430	66 420	12 401	212 880	41 536	171 344
May	502 328	248 496	253 832	51 866	24 211	85 850	73 300	66 708	12 550	212 304	42 047	170 257
June	503 236	251 007	252 229	52 247	23 847	86 384	73 987	66 513	12 397	212 251	43 769	168 482
July	501 784	247 856	253 928	52 076	24 048	84 000	72 011	66 171	11 989	213 967	43 738	170 229
August	500 722	247 740	252 982	52 202	23 948	84 340	72 054	66 395	12 286	212 554	43 134	169 420
September	499 133	246 406	252 727	52 408	23 904	83 848	71 778	65 823	12 070	211 857	42 497	169 360
October	499 112	245 932	253 180	52 149	23 880	85 004	72 840	66 259	12 164	211 518	41 441	170 077
November	499 407	245 475	253 932	52 247	24 120	84 057	71 983	65 970	12 074	212 179	41 493	170 686
December	501 346	245 495	255 851	52 367	24 143	84 337	72 585	65 884	11 752	213 502	41 298	172 204

[1]Includes categories not shown separately.

Table 16-4. Manufacturers' Inventories

(Current cost basis, end of period; seasonally adjusted, millions of dollars.)

| Year and month | Total | Total durable goods | Durables total by stage of fabrication | | | Total nondurable goods | Nondurables total by stage of fabrication | | |
			Materials and supplies	Work in process	Finished goods		Materials and supplies	Work in process	Finished goods
1993	379 806	238 719	72 633	101 948	64 138	141 087	54 233	23 394	63 460
1994	399 934	253 095	78 482	106 521	68 092	146 839	57 113	24 444	65 282
1995	424 802	267 437	85 545	106 634	75 258	157 365	60 778	25 776	70 811
1996	430 366	272 428	86 265	110 527	75 636	157 938	59 157	26 485	72 296
1997	443 227	280 730	92 253	109 714	78 763	162 497	60 175	28 523	73 799
1998	448 373	290 071	93 406	114 928	81 737	158 302	58 156	27 049	73 097
1999	463 004	296 009	97 722	113 805	84 482	166 995	60 975	28 791	77 229
2000	480 748	306 002	105 848	110 909	89 245	174 746	61 465	30 090	83 191
2001	427 353	267 409	91 168	93 768	82 473	159 944	55 671	27 157	77 116
2002	423 028	260 476	88 445	92 343	79 688	162 552	56 571	28 133	77 848
2003	408 302	246 922	82 330	88 608	75 984	161 380	56 866	27 251	77 263
2004	441 222	264 895	92 122	90 934	81 839	176 327	61 917	30 341	84 069
2005	474 639	283 712	98 652	98 400	86 660	190 927	67 224	33 226	90 477
2006	523 476	317 406	111 687	106 352	99 367	206 070	70 601	37 548	97 921
2007	563 065	334 665	116 618	117 462	100 585	228 400	75 681	45 366	107 353
2008	543 514	329 896	117 854	110 887	101 155	213 618	72 810	41 396	99 412
2009	505 294	294 930	100 725	106 339	87 866	210 364	72 012	42 234	96 118
2010	553 772	321 393	106 957	122 188	92 248	232 379	76 595	45 989	109 795
2011	606 940	352 820	119 582	133 129	100 109	254 120	82 133	50 648	121 339
2012	625 266	367 542	125 202	138 832	103 508	257 724	83 353	51 062	123 309
2013	629 795	370 683	125 551	141 715	103 417	259 112	84 477	52 405	122 230
2014	640 456	388 025	132 991	146 914	108 120	252 431	83 748	49 485	119 198
2015	635 557	391 580	129 302	152 555	109 723	243 977	84 038	46 225	113 714
2016	630 332	373 597	128 062	134 812	110 723	256 735	85 370	47 364	124 001
2017	656 797	386 217	133 115	137 745	115 357	270 580	91 265	50 219	129 096
2018	680 658	406 409	144 969	142 646	118 794	274 249	93 263	51 711	129 275
2019	700 021	423 216	144 445	156 230	122 541	276 805	92 954	50 506	133 345
2016									
January	631 502	388 978	128 856	150 779	109 343	242 524	83 670	44 829	114 025
February	628 237	386 091	128 388	148 511	109 192	242 146	83 696	45 097	113 353
March	628 251	384 773	128 166	147 952	108 655	243 478	84 047	44 767	114 664
April	627 801	382 791	127 620	146 211	108 960	245 010	84 194	45 586	115 230
May	627 862	380 633	127 459	145 188	107 986	247 229	84 483	46 861	115 885
June	626 768	377 346	126 971	141 311	109 064	249 422	84 830	47 454	117 138
July	627 059	377 796	129 213	139 597	108 986	249 263	84 873	46 144	118 246
August	628 039	377 326	128 983	138 599	109 744	250 713	84 588	46 558	119 567
September	625 869	375 351	128 528	137 459	109 364	250 518	84 177	46 127	120 214
October	626 337	374 548	128 246	136 361	109 941	251 789	84 929	45 744	121 116
November	629 436	375 091	128 244	136 056	110 791	254 345	85 630	46 063	122 652
December	630 332	373 597	128 062	134 812	110 723	256 735	85 370	47 364	124 001
2017									
January	632 228	373 787	128 626	134 529	110 632	258 441	85 561	47 771	125 109
February	633 333	374 197	128 396	135 101	110 700	259 136	86 130	48 107	124 899
March	633 395	374 746	128 791	135 089	110 866	258 649	86 123	47 204	125 322
April	634 927	375 869	130 394	134 091	111 384	259 058	87 282	47 008	124 768
May	635 193	376 675	130 853	133 993	111 829	258 518	87 247	47 211	124 060
June	634 750	376 712	131 097	133 188	112 427	258 038	86 985	47 647	123 406
July	638 186	378 659	131 393	133 718	113 548	259 527	86 921	48 453	124 153
August	641 140	380 068	131 680	134 365	114 023	261 072	87 624	49 029	124 419
September	645 933	382 640	132 610	135 161	114 869	263 293	89 386	49 016	124 891
October	648 452	383 709	132 542	136 227	114 940	264 743	89 021	49 471	126 251
November	652 392	384 449	132 658	137 067	114 724	267 943	90 075	49 862	128 006
December	656 797	386 217	133 115	137 745	115 357	270 580	91 265	50 219	129 096
2018									
January	659 329	387 533	133 666	138 653	115 214	271 796	91 069	50 837	129 890
February	662 840	390 107	135 066	139 645	115 396	272 733	91 606	50 742	130 385
March	662 703	390 267	135 630	139 193	115 444	272 436	91 545	51 812	129 079
April	665 336	392 338	137 042	139 601	115 695	272 998	91 553	51 954	129 491
May	666 832	394 221	138 226	140 237	115 758	272 611	91 554	51 735	129 322
June	667 862	393 307	139 623	137 515	116 169	274 555	91 916	52 327	130 312
July	673 726	398 686	140 815	141 889	115 982	275 040	93 008	52 206	129 826
August	673 650	397 503	141 466	140 167	115 870	276 147	93 621	52 554	129 972
September	678 054	400 856	142 575	141 453	116 828	277 198	94 316	53 284	129 598
October	680 089	401 895	143 516	141 666	116 713	278 194	94 282	52 967	130 945
November	679 424	403 868	144 532	141 699	117 637	275 556	93 643	51 859	130 054
December	680 658	406 409	144 969	142 646	118 794	274 249	93 263	51 711	129 275
2019									
January	684 722	408 743	145 893	143 058	119 792	275 979	94 114	52 254	129 611
February	687 324	410 248	145 598	144 653	119 997	277 076	94 147	52 191	130 738
March	689 832	411 231	145 109	145 286	120 836	278 601	93 621	53 004	131 976
April	690 915	412 272	145 045	146 458	120 769	278 643	93 662	52 921	132 060
May	692 041	413 893	144 614	148 171	121 108	278 148	93 622	52 316	132 210
June	693 423	415 163	144 779	148 805	121 579	278 260	93 993	51 766	132 501
July	694 316	416 309	144 815	149 358	122 136	278 007	93 226	51 766	133 015
August	693 512	417 459	144 525	150 577	122 357	276 053	93 072	50 840	132 141
September	695 292	419 315	144 533	152 513	122 269	275 977	92 478	50 765	132 734
October	696 003	420 484	144 774	153 638	122 072	275 519	92 225	50 686	132 608
November	697 814	421 928	144 758	154 771	122 399	275 886	92 562	51 007	132 317
December	700 021	423 216	144 445	156 230	122 541	276 805	92 954	50 506	133 345

Table 16-4. Manufacturers' Inventories—*Continued*

(Current cost basis, end of period; seasonally adjusted, millions of dollars.)

Year and month	Construction materials and supplies	Information technology industries	By topical categories						
			Capital goods				Consumer goods		
			Total	Nondefense		Defense	Total	Durable	Nondurable
				Total	Excluding aircraft and parts				
1993	38 307	39 176	118 460	97 118	79 585	21 342	104 873	21 820	83 053
1994	40 879	41 538	123 368	103 320	85 717	20 048	109 536	23 948	85 588
1995	43 420	46 936	129 709	111 563	94 673	18 146	116 404	25 316	91 088
1996	44 124	43 556	132 746	115 211	93 441	17 535	116 351	24 752	91 599
1997	45 725	48 094	137 556	122 598	98 870	14 958	119 660	24 905	94 755
1998	46 500	45 333	145 048	127 108	97 362	17 940	117 343	24 895	92 448
1999	48 541	46 174	146 244	126 278	99 610	19 966	124 220	26 097	98 123
2000	50 518	52 719	148 992	131 655	110 122	17 337	130 823	27 295	103 528
2001	46 341	42 967	128 670	114 908	93 305	13 762	121 377	25 410	95 967
2002	46 698	39 083	122 538	108 147	88 529	14 391	124 904	25 687	99 217
2003	45 448	35 222	114 917	99 119	81 660	15 798	125 052	25 017	100 035
2004	51 244	33 132	117 611	98 900	83 687	18 711	134 390	26 240	108 150
2005	55 094	39 215	127 158	110 573	92 903	16 585	145 446	28 104	117 342
2006	60 864	39 914	139 499	122 442	101 760	17 057	154 185	27 686	126 499
2007	62 575	38 723	150 218	131 394	104 701	18 824	172 623	27 742	144 881
2008	61 456	38 045	150 718	134 314	107 720	16 404	155 341	24 163	131 178
2009	50 966	36 634	145 653	125 812	97 555	19 841	159 088	21 197	137 891
2010	53 333	37 536	160 031	139 417	103 586	20 614	174 773	21 996	152 777
2011	57 682	39 635	174 432	154 343	113 779	20 089	188 220	22 905	165 315
2012	60 162	39 394	184 375	164 041	118 283	20 334	191 764	25 064	166 700
2013	61 179	37 983	185 663	165 771	116 692	19 892	193 528	26 213	167 315
2014	65 057	36 913	192 483	172 857	120 799	19 626	186 659	27 662	158 997
2015	63 846	37 551	198 848	178 145	119 643	20 703	184 342	28 999	155 343
2016	64 994	37 350	183 516	164 818	117 453	18 698	194 679	28 609	166 070
2017	68 178	36 440	185 029	165 570	119 286	19 459	205 422	29 681	175 741
2018	73 465	37 987	193 728	174 181	126 141	19 547	206 719	30 759	175 960
2019	71 749	39 002	209 798	189 033	128 403	20 765	212 614	32 183	180 431
2016									
January	63 705	37 148	197 620	176 228	118 630	21 392	183 011	29 128	153 883
February	63 597	36 990	195 520	174 862	118 209	20 658	182 170	28 981	153 189
March	63 556	37 116	194 445	174 196	117 924	20 249	183 710	28 819	154 891
April	63 483	37 111	193 400	173 147	117 468	20 253	185 061	28 611	156 450
May	63 631	37 101	191 689	171 673	117 012	20 016	186 416	28 252	158 164
June	63 702	37 023	187 720	168 275	116 712	19 445	189 563	29 229	160 334
July	63 950	37 292	188 031	167 818	117 023	20 213	188 784	28 781	160 003
August	64 354	37 256	186 876	167 314	117 261	19 562	190 051	28 786	161 265
September	64 547	37 260	185 340	165 953	116 770	19 387	189 500	28 622	160 878
October	64 757	37 154	184 614	165 233	116 733	19 381	190 140	28 548	161 592
November	64 881	37 368	184 372	165 278	117 299	19 094	192 435	28 582	163 853
December	64 994	37 350	183 516	164 818	117 453	18 698	194 679	28 609	166 070
2017									
January	65 282	37 261	183 560	164 571	117 665	18 989	195 952	28 422	167 530
February	65 434	37 129	183 465	164 397	117 630	19 068	196 159	28 232	167 927
March	65 188	37 256	184 005	165 191	117 868	18 814	195 335	28 374	166 961
April	66 091	36 997	183 343	163 899	117 611	19 444	195 656	28 560	167 096
May	66 183	36 941	183 375	164 102	118 013	19 273	194 703	28 467	166 236
June	66 444	36 925	182 888	163 473	118 157	19 415	194 355	28 837	165 518
July	66 777	36 905	183 365	163 473	118 563	19 892	195 776	29 061	166 715
August	67 145	36 828	183 883	163 855	118 734	20 028	197 159	29 109	168 050
September	67 354	36 472	184 138	164 386	118 804	19 752	199 378	29 327	170 051
October	67 515	36 502	184 585	164 760	118 936	19 825	200 223	29 322	170 901
November	67 921	36 405	184 463	164 981	118 973	19 482	203 012	29 214	173 798
December	68 178	36 440	185 029	165 570	119 286	19 459	205 422	29 681	175 741
2018									
January	68 479	36 398	185 763	166 425	119 656	19 338	206 944	29 819	177 125
February	69 004	36 632	187 397	167 886	120 336	19 511	207 673	29 919	177 754
March	69 537	36 543	187 041	167 590	120 571	19 451	207 525	30 016	177 509
April	70 174	36 715	188 178	168 927	121 381	19 251	207 882	29 999	177 883
May	70 646	37 049	188 697	169 351	121 701	19 346	207 878	30 295	177 583
June	71 261	37 250	187 170	167 310	122 540	19 860	208 574	30 224	178 350
July	71 773	37 013	190 831	171 018	122 543	19 813	208 846	30 417	178 429
August	72 151	37 271	188 945	169 373	123 094	19 572	209 636	30 578	179 058
September	72 932	37 703	191 314	171 680	124 342	19 634	210 206	30 381	179 825
October	72 971	37 833	192 058	172 415	124 795	19 643	210 787	30 302	180 485
November	73 201	37 689	192 539	172 892	125 204	19 647	208 523	30 792	177 731
December	73 465	37 987	193 728	174 181	126 141	19 547	206 719	30 759	175 960
2019									
January	73 521	38 140	195 299	175 498	127 065	19 801	208 508	31 143	177 365
February	73 386	38 231	195 904	176 108	127 023	19 796	210 643	31 754	178 889
March	73 360	38 396	197 490	177 056	127 661	20 434	212 058	31 543	180 515
April	72 918	38 379	198 364	177 801	127 566	20 563	212 520	31 776	180 744
May	72 720	38 423	200 295	179 652	127 838	20 643	212 431	32 154	180 277
June	72 607	38 542	201 369	180 621	128 051	20 748	212 652	32 389	180 263
July	72 520	38 610	203 072	182 089	128 570	20 983	212 540	32 415	180 125
August	72 319	38 631	204 430	183 658	128 392	20 772	211 116	32 586	178 530
September	72 159	38 798	206 048	185 080	128 555	20 968	211 382	32 654	178 728
October	72 162	38 780	207 599	186 238	128 343	21 361	210 922	32 402	178 520
November	71 912	38 807	208 398	187 094	128 069	21 304	211 684	32 569	179 115
December	71 749	39 002	209 798	189 033	128 403	20 765	212 614	32 183	180 431

Table 16-5. Manufacturers' New Orders

(Net, millions of dollars, seasonally adjusted.)

Year and month	Total [1]	Total durable goods [1]	By topical categories									
					Capital goods				Consumer goods		
			Construction materials and supplies	Information technology industries	Total	Nondefense		Defense	Total	Durable	Nondurable
						Total	Excluding aircraft and parts				
1994	3 199 686	1 725 635	335 962	265 010	616 252	542 094	523 461	74 158	1 192 584	315 417	877 167
1995	3 426 503	1 849 641	355 161	297 605	680 857	612 132	576 769	68 725	1 256 721	324 146	932 575
1996	3 567 384	1 948 793	373 536	310 074	737 268	648 797	607 174	88 471	1 293 537	328 984	964 553
1997	3 779 835	2 092 520	403 860	352 700	792 859	728 362	676 119	64 497	1 360 010	361 687	998 323
1998	3 808 143	2 139 918	419 330	365 723	809 727	745 600	698 279	64 127	1 352 708	374 300	978 408
1999	3 957 242	2 252 091	435 034	389 160	840 603	772 703	728 089	67 900	1 425 617	413 435	1 012 182
2000	4 161 472	2 326 576	446 792	409 500	910 933	831 335	767 754	79 598	1 501 810	392 741	1 109 069
2001	3 868 855	2 072 762	420 817	336 935	775 315	691 552	656 944	83 763	1 477 201	364 228	1 112 973
2002	3 823 145	2 032 043	423 561	273 788	698 865	622 403	589 148	76 462	1 494 106	395 484	1 098 622
2003	3 975 812	2 103 013	431 569	282 289	737 032	636 821	608 175	100 211	1 585 029	419 521	1 165 508
2004	4 289 001	2 244 698	468 746	293 836	787 748	688 813	643 997	98 935	1 701 565	419 912	1 281 653
2005	4 764 032	2 446 799	516 797	300 367	896 672	814 286	703 796	82 386	1 894 021	421 457	1 472 564
2006	5 089 646	2 636 288	549 467	332 649	992 499	889 370	775 156	103 129	1 979 165	420 994	1 558 171
2007	5 397 027	2 764 598	564 717	324 248	1 061 415	957 485	786 665	103 930	2 105 784	421 525	1 684 259
2008	5 452 124	2 599 797	543 218	318 571	1 006 627	879 429	781 810	127 198	2 252 724	356 263	1 896 461
2009	4 205 737	1 838 787	413 497	265 602	690 674	587 116	608 664	103 558	1 848 693	267 616	1 581 077
2010	4 895 899	2 265 334	450 921	272 817	885 135	747 708	685 492	137 427	2 058 565	323 084	1 735 481
2011	5 511 660	2 498 852	471 284	286 080	994 126	861 040	765 584	133 086	2 373 717	344 504	2 029 213
2012	5 709 710	2 640 564	500 617	274 254	1 051 261	932 086	803 477	119 175	2 437 479	373 684	2 063 795
2013	5 827 326	2 713 388	524 490	256 182	1 061 823	970 129	803 660	91 694	2 496 383	403 103	2 093 280
2014	5 925 991	2 835 359	546 885	257 049	1 085 147	976 917	804 526	108 230	2 507 676	435 676	2 072 000
2015	5 439 477	2 692 484	554 366	255 035	962 811	858 023	760 914	104 788	2 256 427	461 551	1 794 876
2016	5 269 340	2 627 722	553 611	253 364	919 935	799 906	717 717	120 029	2 184 701	477 150	1 707 551
2017	5 564 914	2 751 467	574 555	263 606	990 007	862 619	745 383	127 388	2 323 356	468 796	1 854 560
2018	5 951 136	2 951 413	611 390	282 893	1 058 229	908 880	780 227	149 349	2 511 415	502 517	2 008 898
2019	5 936 462	2 901 103	621 760	289 718	1 000 182	846 559	793 375	153 623	2 547 345	511 844	2 035 501
2016											
January	437 125	224 474	46 178	21 107	81 285	71 767	61 074	9 518	176 637	40 219	136 418
February	428 269	215 814	46 307	21 353	72 334	65 734	60 710	6 600	176 532	40 655	135 877
March	430 725	214 620	46 006	20 661	73 363	61 011	59 891	12 352	178 254	39 197	139 057
April	441 948	225 021	46 360	21 259	83 240	71 312	60 168	11 928	178 929	39 779	139 150
May	437 634	219 099	45 903	21 265	78 369	70 268	58 923	8 101	179 537	38 463	141 074
June	429 928	207 445	45 831	20 912	66 191	61 245	59 059	4 946	184 188	39 228	144 960
July	434 914	215 821	45 641	21 205	73 307	65 799	59 333	7 508	181 207	39 681	141 526
August	438 832	218 632	45 678	21 147	76 406	65 422	60 984	10 984	181 998	39 862	142 136
September	438 451	216 575	45 399	20 974	75 340	64 958	58 991	10 382	183 343	40 056	143 287
October	452 703	229 282	46 630	20 747	87 683	77 048	59 052	10 635	184 665	39 665	145 000
November	439 777	216 770	46 737	20 970	73 503	60 922	58 969	12 581	183 685	39 090	144 595
December	450 403	220 898	46 296	21 596	78 700	64 721	59 072	13 979	190 982	40 009	150 973
2017											
January	453 342	223 046	47 014	21 779	78 443	67 497	61 202	10 946	191 247	40 261	150 986
February	449 609	217 634	46 729	22 081	72 600	70 481	61 363	2 119	190 943	39 029	151 914
March	456 752	225 925	48 410	21 540	79 589	69 821	61 541	9 768	189 650	38 747	150 903
April	458 191	226 500	47 209	21 362	81 401	69 800	61 763	11 601	189 927	38 005	151 922
May	454 966	224 671	47 609	21 516	77 750	69 066	61 509	8 684	190 201	39 485	150 716
June	472 781	241 249	47 213	21 534	96 744	84 363	60 755	12 381	190 763	38 939	151 824
July	454 091	221 781	47 183	21 803	78 428	66 049	61 558	12 379	190 484	37 722	152 762
August	460 402	226 087	47 684	21 543	79 295	68 769	61 667	10 526	194 251	39 100	155 151
September	475 318	239 375	48 465	22 458	92 669	75 095	63 859	17 574	194 965	38 742	156 223
October	471 001	231 603	48 628	22 478	82 349	71 501	63 801	10 848	198 225	39 413	158 812
November	480 291	236 004	49 462	22 918	85 592	75 956	63 309	9 636	203 301	40 242	163 059
December	485 285	241 358	49 914	22 583	86 301	76 094	63 596	10 207	201 847	39 554	162 293
2018											
January	477 301	231 783	49 567	22 416	80 916	72 257	62 956	8 659	204 437	40 595	163 842
February	487 905	242 175	50 252	22 363	88 588	75 083	62 836	13 505	205 087	40 958	164 129
March	492 457	246 424	50 760	22 656	90 121	79 620	63 326	10 501	205 346	41 171	164 175
April	490 725	244 315	51 051	23 035	86 608	75 244	64 637	11 364	206 719	42 237	164 482
May	496 950	246 483	51 128	23 199	90 124	76 642	64 510	13 482	206 982	38 728	168 254
June	500 356	247 803	51 035	23 658	89 459	77 459	66 115	12 000	210 573	41 117	169 456
July	494 393	241 933	51 310	24 192	83 197	71 817	68 474	11 380	212 027	42 412	169 615
August	505 839	252 267	51 321	23 892	93 666	78 200	65 799	15 466	212 273	42 031	170 242
September	511 193	256 401	50 832	23 405	92 178	76 969	65 848	15 209	214 531	42 744	171 787
October	501 321	245 516	51 476	24 722	85 207	73 623	66 446	11 584	215 709	43 482	172 227
November	494 892	244 910	51 101	24 596	86 014	72 656	64 336	13 358	210 104	43 221	166 883
December	496 254	250 029	51 432	24 497	90 455	77 818	65 150	12 637	207 453	44 075	163 378
2019											
January	498 365	250 492	51 291	24 123	92 197	79 519	66 454	12 678	208 696	43 800	164 896
February	493 845	242 267	51 288	23 895	82 536	72 869	65 816	9 667	211 935	43 286	168 649
March	504 525	252 004	50 985	24 485	91 719	76 797	67 379	14 922	213 284	43 663	169 621
April	494 829	240 268	51 310	24 695	83 557	70 910	66 092	12 647	212 876	41 532	171 344
May	490 552	236 720	51 712	24 277	80 979	66 686	65 051	14 293	212 457	42 200	170 257
June	489 113	236 884	52 532	24 263	77 060	69 214	66 439	7 846	212 409	43 927	168 482
July	496 770	242 842	52 145	24 150	83 684	72 280	66 224	11 404	213 973	43 744	170 229
August	497 000	244 018	52 458	24 110	85 123	70 504	65 864	14 619	212 646	43 226	169 420
September	494 440	241 713	52 187	24 104	81 791	68 551	65 769	13 240	211 873	42 513	169 360
October	494 737	241 557	51 861	24 002	86 092	70 376	65 954	15 716	211 583	41 506	170 077
November	489 169	235 237	51 927	23 921	76 707	68 696	66 545	8 011	212 265	41 579	170 686
December	497 703	241 852	52 243	24 208	82 128	63 364	66 240	18 764	213 557	41 353	172 204

[1]Includes categories not shown separately.

Table 16-6. Manufacturers' Unfilled Orders, Durable Goods Industries

(End of period, millions of dollars, seasonally adjusted.)

Year and month	Total [1]	Transportation equipment		By topical categories						
		Nondefense aircraft and parts	Defense aircraft and parts	Construction materials and supplies	Information technology industries	Capital goods				Consumer durable goods
						Total	Nondefense		Defense	
							Total	Excluding aircraft and parts		
1994	434 899	101 157	44 651	25 090	79 071	299 296	178 007	106 779	121 289	5 121
1995	447 510	109 506	42 360	27 155	84 919	313 963	199 613	118 006	114 350	5 282
1996	489 147	130 580	45 203	29 436	84 227	346 596	217 605	120 276	128 991	5 821
1997	513 317	143 183	40 842	33 584	87 704	360 636	243 413	132 076	117 223	7 330
1998	496 363	137 228	37 711	34 268	90 930	348 398	241 565	134 417	106 833	8 262
1999	505 667	126 256	35 616	35 246	105 883	349 497	245 603	149 679	103 894	9 035
2000	549 507	138 159	42 421	37 405	115 679	384 834	268 323	159 844	116 511	10 349
2001	509 910	124 656	52 002	33 580	98 908	357 539	230 560	137 637	126 979	6 988
2002	479 610	114 352	57 383	33 233	88 083	327 423	200 536	117 571	126 887	6 593
2003	506 423	109 736	63 441	35 701	95 736	344 824	203 047	125 283	141 777	7 296
2004	558 457	122 946	56 998	41 467	101 278	378 930	229 584	140 211	149 346	8 205
2005	653 074	187 614	54 421	48 493	106 404	451 740	311 502	157 208	140 238	7 245
2006	795 263	264 577	58 648	50 972	119 047	557 632	404 587	188 224	153 045	6 443
2007	944 435	375 696	62 215	55 450	120 174	681 989	523 702	206 269	158 287	6 186
2008	993 460	416 354	71 696	54 730	119 820	734 569	563 825	213 534	170 744	4 641
2009	825 225	342 409	61 253	41 216	113 676	606 865	455 608	179 042	151 257	4 206
2010	870 463	349 343	61 939	48 430	116 277	634 702	476 304	190 299	158 398	3 729
2011	953 635	389 315	65 320	48 181	128 428	705 277	534 075	215 553	171 202	3 891
2012	1 014 225	442 481	76 010	48 013	130 087	755 284	580 359	219 584	174 925	4 200
2013	1 075 752	525 962	70 925	48 891	115 166	808 444	657 856	224 383	150 588	4 489
2014	1 160 052	607 125	69 269	48 654	109 764	875 865	728 319	225 107	147 546	4 523
2015	1 128 338	601 956	65 546	49 033	106 741	851 146	709 711	212 037	141 435	4 523
2016	1 091 433	578 963	70 067	47 790	105 340	827 551	678 492	201 258	149 059	4 828
2017	1 122 263	581 749	69 980	48 098	108 882	846 957	694 248	212 471	152 709	4 662
2018	1 165 974	587 633	79 092	53 457	112 389	866 406	701 104	216 457	165 302	4 352
2019	1 144 546	558 230	83 738	52 849	114 070	839 653	666 707	215 334	172 946	4 814
2016										
January	1 128 485	603 980	66 118	49 031	106 642	852 770	710 685	211 057	142 085	4 552
February	1 120 583	600 651	65 828	49 064	106 903	845 985	706 563	210 262	139 422	4 561
March	1 114 166	593 416	66 873	48 698	106 632	840 912	698 387	209 120	142 525	4 648
April	1 116 985	596 180	68 986	48 650	106 714	845 296	699 550	207 604	145 746	4 589
May	1 115 612	598 588	68 450	48 650	106 670	844 216	699 485	205 461	144 731	4 637
June	1 102 012	591 257	68 003	48 554	106 334	831 754	691 384	203 891	140 370	4 611
July	1 096 671	588 102	68 338	48 361	106 019	826 515	688 273	203 558	138 242	4 680
August	1 094 531	583 972	69 219	47 967	105 730	825 096	685 179	204 589	139 917	4 731
September	1 089 182	580 889	67 683	47 523	105 431	821 941	681 149	203 480	140 792	4 779
October	1 097 425	588 852	68 185	48 023	105 018	831 369	689 342	202 873	142 027	4 863
November	1 093 796	583 313	70 753	48 292	104 952	827 379	682 605	202 299	144 774	4 829
December	1 091 433	578 963	70 067	47 790	105 340	827 551	678 492	201 258	149 059	4 828
2017										
January	1 090 929	574 625	70 921	47 889	105 904	827 256	676 835	202 550	150 421	4 840
February	1 085 054	574 362	70 930	47 074	106 542	820 670	678 339	203 830	142 331	4 989
March	1 085 845	572 650	71 362	47 799	106 711	820 074	677 926	204 638	142 148	4 887
April	1 090 037	571 794	73 077	47 775	106 737	822 394	678 660	206 095	143 734	4 894
May	1 088 012	569 782	73 145	47 685	106 829	819 746	677 369	206 853	142 377	4 824
June	1 104 044	583 693	71 554	47 351	106 834	836 593	691 857	207 089	144 736	4 774
July	1 101 035	576 370	73 037	47 280	106 910	833 332	686 266	208 026	147 066	4 792
August	1 099 713	573 624	72 825	47 247	106 689	831 359	684 121	208 424	147 238	4 879
September	1 110 807	574 567	71 843	47 471	107 435	841 858	687 506	210 364	154 352	4 791
October	1 112 193	573 627	70 891	47 451	107 840	842 388	687 872	211 802	154 516	4 725
November	1 114 319	575 599	68 604	47 748	108 487	843 746	690 409	212 188	153 337	4 770
December	1 122 263	581 749	69 980	48 098	108 882	846 957	694 248	212 471	152 709	4 662
2018										
January	1 118 766	580 200	69 281	48 235	108 657	843 660	693 470	212 418	150 190	4 771
February	1 124 552	582 205	69 002	48 902	108 710	848 579	695 584	211 916	152 995	4 721
March	1 132 593	588 058	68 087	49 867	108 930	853 047	700 100	211 938	152 947	4 871
April	1 138 225	592 240	67 623	50 700	109 098	856 240	703 861	212 534	152 379	4 866
May	1 146 225	594 961	68 676	51 282	109 469	860 300	705 705	213 112	154 595	4 834
June	1 152 430	596 180	71 012	51 631	109 927	862 890	707 433	214 351	155 457	4 857
July	1 152 619	592 473	71 101	51 994	110 383	861 975	706 531	217 280	155 444	4 910
August	1 160 444	595 309	71 361	52 279	110 671	868 227	708 731	217 700	159 496	4 721
September	1 170 856	594 573	77 943	52 407	110 735	871 944	708 639	218 193	163 305	4 701
October	1 169 210	591 680	77 961	52 822	111 366	869 056	705 876	218 631	163 180	4 620
November	1 165 463	587 511	79 274	53 136	111 950	865 496	700 786	217 267	164 710	4 481
December	1 165 974	587 633	79 092	53 457	112 389	866 406	701 104	216 457	165 302	4 352
2019										
January	1 167 155	589 127	79 247	53 650	112 484	868 835	703 379	216 715	165 456	4 298
February	1 161 288	585 732	79 735	53 552	112 353	862 632	699 407	216 271	163 225	4 237
March	1 165 099	585 284	80 447	53 381	112 854	866 183	700 094	217 299	166 089	4 187
April	1 162 862	582 721	81 004	53 346	113 136	863 909	697 574	216 971	166 335	4 183
May	1 156 256	577 135	80 618	53 192	113 202	859 038	690 960	215 314	168 078	4 336
June	1 147 136	572 711	79 226	53 477	113 618	849 714	686 187	215 240	163 527	4 494
July	1 146 938	573 036	78 563	53 546	113 720	849 398	686 456	215 293	162 942	4 500
August	1 148 225	572 108	78 996	53 802	113 882	850 181	684 906	214 762	165 275	4 592
September	1 148 432	570 082	81 021	53 581	114 082	848 124	681 679	214 708	166 445	4 608
October	1 149 040	567 811	84 115	53 293	114 204	849 212	679 215	214 403	169 997	4 673
November	1 143 494	566 189	81 348	52 973	114 005	841 862	675 928	214 978	165 934	4 759
December	1 144 546	558 230	83 738	52 849	114 070	839 653	666 707	215 334	172 946	4 814

[1] Includes categories not shown separately.

Table 16-7. Motor Vehicle Sales and Inventories

(Units.)

Year and month	Retail sales of new passenger cars						Retail inventories of new domestic passenger cars (thousands of units, end of period)		
	Thousands of units, not seasonally adjusted			Millions of units, seasonally adjusted annual rate			Not seasonally adjusted	Seasonally adjusted	Inventory to sales ratio
	Total	Domestic	Foreign	Total	Domestic	Foreign			
1976	9 994.0	8 492.0	1 502.0	9.994	8.492	1.502	1 465.0	1 494.0	1.900
1977	11 046.0	8 971.2	2 074.8	11.046	8.971	2.075	1 731.0	1 743.0	2.300
1978	11 164.0	9 163.9	2 000.1	11.164	9.164	2.000	1 729.0	1 731.0	2.300
1979	10 558.8	8 230.1	2 328.7	10.559	8.230	2.329	1 691.0	1 667.0	2.400
1980	8 981.8	6 581.4	2 400.4	8.982	6.581	2.401	1 448.0	1 440.0	2.600
1981	8 534.3	6 208.8	2 325.5	8.535	6.209	2.326	1 471.0	1 495.0	3.600
1982	7 979.4	5 758.2	2 221.2	7.979	5.758	2.221	1 126.0	1 127.0	2.200
1983	9 178.6	6 793.0	2 385.6	9.179	6.793	2.386	1 352.0	1 350.0	2.000
1984	10 390.2	7 951.7	2 438.5	10.391	7.952	2.439	1 415.0	1 411.0	2.100
1985	10 978.4	8 204.7	2 773.7	10.979	8.205	2.774	1 630.0	1 619.0	2.500
1986	11 405.7	8 215.0	3 190.7	11.406	8.215	3.191	1 499.0	1 515.0	2.000
1987	10 170.9	7 080.9	3 090.0	10.171	7.081	3.090	1 680.0	1 716.0	2.800
1988	10 545.6	7 539.4	3 006.2	10.545	7.539	3.006	1 601.0	1 601.0	2.300
1989	9 776.8	7 078.1	2 698.7	9.777	7.078	2.699	1 669.0	1 687.0	3.100
1990	9 300.2	6 896.9	2 403.3	9.300	6.897	2.403	1 408.0	1 418.0	2.600
1991	8 175.0	6 136.9	2 038.1	8.175	6.137	2.038	1 283.0	1 296.0	2.600
1992	8 214.4	6 276.6	1 937.8	8.215	6.277	1.938	1 276.0	1 288.0	2.300
1993	8 517.7	6 734.0	1 783.7	8.518	6.734	1.784	1 364.9	1 377.3	2.359
1994	8 990.4	7 255.2	1 735.2	8.990	7.255	1.735	1 436.6	1 445.3	2.319
1995	8 620.4	7 114.1	1 506.3	8.620	7.114	1.506	1 618.5	1 629.8	2.565
1996	8 478.6	7 206.3	1 272.3	8.478	7.206	1.272	1 363.4	1 422.5	2.478
1997	8 217.7	6 862.3	1 355.4	8.218	6.862	1.356	1 329.9	1 399.7	2.368
1998	8 084.9	6 705.1	1 379.8	8.085	6.705	1.380	1 324.4	1 369.7	2.222
1999	8 637.7	6 918.8	1 718.9	8.638	6.919	1.719	1 367.6	1 406.6	2.369
2000	8 777.7	6 761.6	2 016.1	8.778	6.762	2.016	1 377.0	1 397.2	2.810
2001	8 352.1	6 254.4	2 097.7	8.352	6.254	2.098	955.7	976.9	2.267
2002	8 042.1	5 816.5	2 225.6	8.043	5.817	2.226	1 132.0	1 229.8	2.533
2003	7 555.6	5 472.6	2 083.0	7.556	5.473	2.083	1 115.8	1 227.1	2.713
2004	7 482.5	5 333.6	2 148.9	7.482	5.333	2.149	1 034.2	1 144.8	2.405
2005	7 660.4	5 473.6	2 186.8	7.661	5.474	2.187	930.6	1 012.2	2.289
2006	7 761.3	5 416.7	2 344.6	7.762	5.417	2.345	1 031.7	1 105.5	2.498
2007	7 562.1	5 197.1	2 365.0	7.562	5.197	2.365	906.8	985.3	2.272
2008	6 769.2	4 491.0	2 278.2	6.769	4.491	2.278	1 144.2	1 215.5	4.115
2009	5 401.5	3 558.4	1 843.1	5.401	3.558	1.843	686.4	733.2	2.162
2010	5 635.7	3 791.5	1 844.2	5.636	3.792	1.844	759.1	796.7	2.402
2011	6 092.9	4 146.0	1 946.9	6.093	4.146	1.947	813.0	806.7	2.255
2012	7 245.2	5 119.8	2 125.3	7.246	5.120	2.125	1 069.5	1 067.8	2.374
2013	7 586.3	5 433.2	2 153.2	7.586	5.433	2.153	1 259.3	1 231.0	2.812
2014	7 708.0	5 609.9	2 098.1	7.708	5.610	2.098	1 249.4	1 218.0	2.522
2015	7 516.8	5 595.1	1 921.7	7.252	5.595	1.922	1 178.6	1 154.1	2.586
2016	6 872.7	5 145.6	1 727.2	6.873	5.146	1.727	1 209.8	1 141.0	2.682
2017	6 080.9	4 593.0	1 488.0	6.081	4.593	1.488	967.6	862.9	2.429
2018	5 303.6	4 086.9	1 216.7	5.304	4.087	1.217	802.6	720.7	2.144
2019	4 715.0	3 543.9	1 171.1	4.706	3.539	1.167	579.0	504.7	1.754
2017									
January	412.4	307.6	104.8	5.503	3.878	1.624	1 240.9	1 146.0	2.955
February	483.5	368.8	114.7	5.479	3.900	1.580	1 245.0	1 138.2	2.919
March	589.9	446.6	143.3	5.344	3.834	1.510	1 251.5	1 147.5	2.993
April	537.7	411.0	126.7	5.425	3.987	1.439	1 222.1	1 153.0	2.892
May	561.8	433.9	127.9	5.285	3.873	1.412	1 188.0	1 156.9	2.987
June	521.4	399.6	121.8	5.124	3.734	1.389	1 191.7	1 160.9	3.109
July	501.3	376.2	125.1	5.220	3.775	1.445	1 055.9	1 092.7	2.894
August	522.8	394.2	128.6	5.203	3.748	1.455	1 025.4	1 069.9	2.855
September	534.8	406.6	128.2	5.576	4.045	1.532	982.4	1 005.2	2.485
October	458.5	345.3	113.2	5.361	3.853	1.507	986.7	951.0	2.468
November	461.4	341.0	120.4	5.258	3.749	1.509	1 012.0	915.0	2.440
December	495.5	362.1	133.4	5.007	3.553	1.454	967.6	862.9	2.429
2018									
January	365.7	280.5	85.3	4.786	3.478	1.308	988.2	862.8	2.481
February	421.8	329.0	92.9	4.790	3.503	1.287	995.9	858.8	2.452
March	537.0	416.1	120.9	4.661	3.425	1.237	971.1	854.3	2.495
April	428.5	326.9	101.6	4.644	3.428	1.216	973.1	872.5	2.545
May	513.1	394.9	118.2	4.718	3.431	1.288	917.9	849.8	2.477
June	482.9	372.4	110.5	4.575	3.352	1.223	891.1	828.6	2.472
July	416.7	321.3	95.4	4.505	3.354	1.151	815.6	812.1	2.421
August	440.8	338.1	102.7	4.415	3.258	1.157	775.7	782.8	2.402
September	429.8	329.6	100.3	4.576	3.346	1.231	752.1	745.0	2.227
October	419.8	329.2	90.7	4.760	3.570	1.190	790.5	722.4	2.023
November	400.8	306.4	94.4	4.532	3.363	1.168	821.5	707.2	2.103
December	446.5	342.6	103.9	4.506	3.361	1.145	802.6	720.7	2.144
2019									
January	353.4	277.2	76.1	4.574	3.404	1.169	838.3	717.7	2.108
February	365.2	284.5	80.7	4.159	3.037	1.122	839.7	708.5	2.333
March	482.6	371.9	110.7	4.307	3.157	1.150	786.6	665.4	2.108
April	395.2	295.2	100.0	4.185	3.007	1.178	748.3	646.3	2.149
May	454.6	339.5	115.1	4.218	2.979	1.238	695.6	625.6	2.100
June	439.4	322.1	117.4	4.298	2.962	1.336	698.0	635.0	2.144
July	382.6	277.5	105.1	4.088	2.834	1.254	643.3	633.9	2.237
August	433.9	322.2	111.6	3.998	2.798	1.200	612.7	636.6	2.275
September	339.1	255.0	84.1	4.029	2.929	1.099	637.4	638.8	2.181
October	343.3	254.7	88.6	3.804	2.661	1.143	640.4	591.9	2.225
November	352.2	261.4	90.8	3.827	2.745	1.083	627.9	538.4	1.962
December	373.7	282.9	90.8	3.912	2.878	1.034	579.0	504.7	1.754

Table 16-7. Motor Vehicle Sales and Inventories—*Continued*

(Units.)

Year and month	Retail sales of new trucks and buses								Unit sales of cars and light trucks (millions of units, seasonally adjusted annual rate)		
	Thousands of units, not seasonally adjusted				Millions of units, seasonally adjusted annual rate						
	Total	0–14,000 pounds		14,001 pounds and over	Total	0–14,000 pounds		14,001 pounds and over	Total	Domestic	Foreign
		Domestic	Foreign			Domestic	Foreign				
1976	3 300.5	2 738.3	237.5	324.7	3.296	2.733	0.239	0.324	12.966	11.225	1.741
1977	3 813.0	3 112.8	323.1	377.1	3.818	3.116	0.324	0.378	14.486	12.087	2.399
1978	4 256.8	3 481.1	335.9	439.8	4.249	3.469	0.340	0.440	14.973	12.633	2.340
1979	3 589.7	2 730.2	469.4	390.1	3.599	2.740	0.469	0.390	13.768	10.970	2.798
1980	2 487.4	1 731.1	484.6	271.7	2.482	1.731	0.480	0.271	11.193	8.312	2.881
1981	2 255.6	1 581.7	447.6	226.3	2.255	1.585	0.444	0.226	10.564	7.794	2.770
1982	2 562.8	1 967.5	410.4	184.9	2.569	1.971	0.413	0.185	10.363	7.729	2.634
1983	3 117.3	2 465.2	463.3	188.8	3.130	2.480	0.461	0.189	12.120	9.273	2.847
1984	4 093.1	3 207.2	607.7	278.2	4.086	3.199	0.609	0.278	14.199	11.151	3.048
1985	4 741.7	3 618.4	828.3	295.0	4.760	3.634	0.831	0.295	15.444	11.839	3.605
1986	4 912.1	3 671.4	967.2	273.5	4.918	3.676	0.969	0.273	16.051	11.891	4.160
1987	4 991.5	3 792.0	912.2	287.3	4.978	3.783	0.907	0.288	14.861	10.864	3.997
1988	5 231.9	4 199.7	697.9	334.3	5.225	4.194	0.697	0.334	15.436	11.733	3.703
1989	5 055.9	4 113.6	630.3	312.0	5.065	4.123	0.629	0.313	14.529	11.201	3.328
1990	4 837.0	3 956.8	602.7	277.5	4.840	3.960	0.602	0.278	13.862	10.857	3.005
1991	4 355.4	3 605.6	528.8	221.0	4.361	3.612	0.528	0.221	12.315	9.749	2.566
1992	4 892.2	4 247.0	395.9	249.3	4.893	4.247	0.398	0.248	12.860	10.524	2.336
1993	5 667.8	5 000.5	364.5	302.8	5.658	4.991	0.365	0.302	13.874	11.725	2.149
1994	6 407.3	5 658.2	396.3	352.8	6.408	5.659	0.395	0.354	15.044	12.914	2.130
1995	6 496.4	5 705.9	402.1	388.4	6.498	5.706	0.402	0.390	14.728	12.820	1.908
1996	6 977.6	6 179.8	438.7	359.1	6.976	6.180	0.439	0.357	15.097	13.386	1.711
1997	7 280.4	6 324.7	579.5	376.2	7.280	6.325	0.579	0.376	15.122	13.187	1.935
1998	7 882.4	6 802.0	656.1	424.3	7.883	6.802	0.656	0.425	15.543	13.507	2.036
1999	8 777.3	7 480.7	775.3	521.3	8.777	7.481	0.775	0.521	16.894	14.400	2.494
2000	9 033.9	7 719.7	852.3	461.9	9.033	7.720	0.852	0.461	17.350	14.482	2.868
2001	9 120.4	7 789.0	981.3	350.1	9.120	7.789	0.981	0.350	17.122	14.043	3.079
2002	9 096.5	7 707.8	1 066.3	322.4	9.096	7.708	1.066	0.322	16.817	13.525	3.292
2003	9 411.9	7 856.3	1 227.2	328.4	9.411	7.856	1.227	0.328	16.639	13.329	3.310
2004	9 815.8	8 138.0	1 246.2	431.6	9.813	8.138	1.246	0.429	16.866	13.471	3.395
2005	9 784.4	8 072.4	1 215.5	496.5	9.784	8.072	1.215	0.497	16.948	13.546	3.402
2006	9 287.0	7 396.0	1 346.6	544.4	9.288	7.396	1.347	0.545	16.505	12.813	3.692
2007	8 897.9	7 138.7	1 388.1	371.1	8.900	7.139	1.388	0.373	16.089	12.336	3.753
2008	15 408.6	5 329.2	9 780.9	298.5	6.724	5.329	1.097	0.298	13.195	9.820	3.375
2009	5 200.5	4 116.5	884.2	199.8	5.200	4.117	0.884	0.199	10.402	7.675	2.727
2010	6 136.6	5 020.3	898.7	217.6	6.138	5.020	0.899	0.218	11.556	8.813	2.743
2011	6 955.5	5 666.5	982.4	306.6	6.956	5.666	0.982	0.306	12.743	9.814	2.929
2012	7 534.3	6 127.3	1 060.7	346.3	7.535	6.127	1.061	0.346	14.435	11.249	3.186
2013	8 296.4	6 704.5	1 239.3	352.6	8.296	6.705	1.239	0.352	15.530	12.137	3.392
2014	9 151.8	7 384.3	1 359.9	407.7	9.150	7.384	1.360	0.406	16.452	12.994	3.458
2015	10 328.8	8 097.4	1 782.1	449.3	10.329	8.097	1.782	0.449	17.131	13.693	3.439
2016	10 993.0	8 436.2	2 155.8	401.0	10.994	8.436	2.156	0.402	17.465	13.582	3.883
2017	11 470.5	8 651.8	2 403.6	415.0	11.471	8.652	2.404	0.415	17.137	13.245	3.892
2018	12 397.8	9 159.0	2 750.9	487.9	12.397	9.159	2.751	0.487	17.214	13.246	3.968
2019	12 765.0	9 622.8	2 615.1	527.1	12.759	9.623	2.609	0.527	16.938	13.162	3.776
2017											
January	751.9	567.3	158.9	25.7	11.387	8.669	2.352	0.366	16.524	12.547	3.976
February	868.6	667.4	173.2	28.0	11.514	8.826	2.301	0.387	16.606	12.726	3.880
March	992.8	752.0	205.6	35.2	10.986	8.353	2.247	0.387	15.943	12.186	3.757
April	912.0	694.9	185.1	32.0	11.058	8.333	2.307	0.418	16.066	12.320	3.746
May	982.0	741.2	206.2	34.6	11.195	8.389	2.403	0.403	16.076	12.262	3.815
June	981.4	742.8	200.9	37.7	11.338	8.545	2.380	0.413	16.049	12.279	3.769
July	939.7	704.4	203.0	32.4	11.205	8.439	2.356	0.410	16.016	12.214	3.801
August	989.3	738.1	213.0	38.2	11.064	8.318	2.319	0.426	15.840	12.066	3.774
September	1 018.3	766.4	215.9	36.1	11.996	9.003	2.563	0.430	17.142	13.048	4.094
October	927.1	699.9	190.8	36.5	12.067	9.126	2.510	0.431	16.997	12.979	4.018
November	963.2	721.3	206.5	35.3	11.853	8.873	2.535	0.445	16.667	12.623	4.044
December	1 144.1	856.0	244.5	43.5	11.980	8.948	2.570	0.462	16.525	12.501	4.024
2018											
January	816.0	600.8	184.5	30.7	12.071	8.971	2.678	0.422	16.435	12.449	3.986
February	906.3	665.0	206.9	34.4	12.064	8.834	2.755	0.475	16.380	12.337	4.043
March	1 150.6	844.5	265.5	40.5	12.348	9.089	2.785	0.475	16.535	12.513	4.022
April	962.7	714.6	210.5	37.7	12.444	9.152	2.831	0.461	16.627	12.580	4.047
May	1 113.4	824.5	248.9	40.0	12.373	9.068	2.837	0.469	16.623	12.499	4.124
June	1 103.7	811.3	249.4	42.9	12.456	9.157	2.813	0.486	16.544	12.509	4.036
July	986.4	728.7	217.5	40.2	12.208	9.051	2.663	0.494	16.219	12.405	3.815
August	1 086.6	803.6	237.8	45.2	12.295	9.190	2.600	0.505	16.205	12.448	3.757
September	1 045.2	774.8	227.5	42.9	12.597	9.300	2.770	0.526	16.647	12.646	4.001
October	986.1	729.4	211.1	45.6	12.525	9.284	2.720	0.521	16.764	12.854	3.910
November	1 021.4	756.1	225.6	39.7	12.675	9.399	2.775	0.501	16.706	12.763	3.943
December	1 219.4	905.7	265.6	48.1	12.708	9.414	2.785	0.509	16.704	12.775	3.929
2019											
January	818.1	599.1	180.7	38.3	11.983	8.832	2.624	0.527	16.030	12.236	3.794
February	923.1	681.6	204.7	36.8	12.260	9.033	2.720	0.507	15.912	12.069	3.842
March	1 160.2	867.9	248.4	43.9	12.835	9.653	2.668	0.514	16.628	12.810	3.818
April	977.5	745.7	185.7	46.1	12.262	9.257	2.440	0.565	15.882	12.264	3.618
May	1 173.5	897.7	229.2	46.6	13.119	9.978	2.594	0.547	16.789	12.957	3.832
June	1 115.3	856.8	213.4	45.1	12.815	9.835	2.453	0.527	16.586	12.798	3.788
July	1 061.3	804.1	209.8	47.4	12.788	9.718	2.503	0.567	16.309	12.552	3.757
August	1 251.5	946.3	258.6	46.6	12.954	9.774	2.642	0.538	16.414	12.572	3.842
September	976.6	729.7	198.3	48.5	13.108	9.892	2.641	0.575	16.562	12.822	3.741
October	1 036.9	776.1	214.6	46.2	12.710	9.485	2.699	0.526	15.987	12.145	3.842
November	1 086.3	820.5	230.5	35.3	13.071	9.919	2.690	0.462	16.436	12.664	3.772
December	1 184.7	897.4	241.2	46.2	12.633	9.522	2.638	0.473	16.072	12.400	3.671

Table 16-8. Retail and Food Services Sales

(All retail establishments and food services; millions of dollars; not seasonally adjusted.)

Year and month	Retail and food services, total [1]	GAFO (department store type goods), total [2]	Motor vehicles and parts	Furniture and home furnishings	Electronics and appliances	Building materials and garden	Food and beverages	Health and personal care	Gasoline	Clothing and accessories	General merchandise	Nonstore retailers	Food services and drinking places
1994	2 330 235	616 347	541 141	60 416	57 266	157 228	384 340	96 359	171 222	129 083	285 190	96 280	225 000
1995	2 450 628	650 040	579 715	63 470	64 770	164 561	390 386	101 632	181 113	131 333	300 498	103 516	233 012
1996	2 603 794	682 613	627 507	67 707	68 363	176 683	401 073	109 554	194 425	136 581	315 305	117 761	242 245
1997	2 726 131	713 387	653 817	72 715	70 061	191 063	409 373	118 670	199 700	140 293	331 363	126 190	257 364
1998	2 852 956	757 936	688 415	77 412	74 527	202 423	416 525	129 582	191 727	149 151	351 081	133 904	271 194
1999	3 086 990	815 665	764 204	84 294	78 977	218 290	433 699	142 697	212 524	159 751	380 179	151 797	283 900
2000	3 287 537	862 739	796 210	91 170	82 206	228 994	444 764	155 233	249 816	167 674	404 228	180 453	304 261
2001	3 378 906	882 700	815 579	91 484	80 240	239 379	462 429	166 532	251 383	167 287	427 468	180 563	316 638
2002	3 459 077	912 707	818 811	94 438	83 740	248 539	464 856	179 983	250 619	172 304	446 520	189 279	330 525
2003	3 612 457	946 114	841 588	96 736	86 442	263 463	474 385	275 187	178 694	178 694	468 771	206 359	349 726
2004	3 846 605	1 003 891	866 372	103 757	93 896	295 274	490 380	324 006	324 006	190 253	497 382	228 977	373 557
2005	4 085 746	1 059 599	888 307	109 120	100 461	320 802	508 484	378 923	378 923	200 969	528 385	255 579	396 463
2006	4 294 359	1 110 155	899 997	112 795	105 477	334 130	525 232	421 976	421 976	213 189	554 256	284 343	422 786
2007	4 439 733	1 143 426	910 139	111 144	106 599	320 854	547 837	451 822	451 822	221 205	578 582	308 767	444 551
2008	4 391 580	1 136 376	785 865	98 720	105 317	301 833	569 276	246 573	503 639	215 583	595 041	319 152	456 265
2009	4 064 476	1 088 197	671 772	84 749	95 364	261 637	568 418	252 794	391 234	204 475	588 918	311 152	452 005
2010	4 284 968	1 114 374	742 913	85 205	97 396	260 566	580 530	260 435	448 349	213 286	603 757	340 957	466 920
2011	4 598 302	1 155 666	812 938	87 586	99 928	269 480	609 137	271 612	533 457	228 606	624 766	376 344	495 350
2012	4 826 390	1 191 843	886 494	91 542	102 060	281 533	628 205	274 000	555 419	239 493	642 313	408 171	524 161
2013	5 001 763	1 212 493	959 294	95 349	102 998	301 797	640 847	281 840	549 613	244 722	651 874	433 126	543 313
2014	5 215 656	1 238 694	1 020 851	99 718	103 518	318 352	669 165	299 263	538 790	250 409	667 163	470 867	576 216
2015	5 349 487	1 258 154	1 094 112	106 570	103 658	331 611	685 381	315 244	444 027	255 798	674 889	509 652	623 494
2016	5 510 186	1 261 613	1 140 614	110 404	99 043	348 697	699 349	327 153	422 792	260 050	675 389	561 412	657 228
2017	5 744 810	1 272 463	1 172 367	113 035	98 570	365 622	725 137	333 338	459 463	260 566	687 123	629 204	691 659
2018	6 001 623	1 302 172	1 191 321	116 895	100 205	381 313	745 736	347 454	503 925	268 163	706 298	696 849	732 155
2019	6 218 002	1 304 488	1 237 469	117 815	97 017	384 515	765 064	358 727	501 072	266 903	713 623	795 510	765 796
2017													
January	422 066	89 292	84 235	8 255	7 477	23 330	57 469	26 468	34 744	15 791	49 392	47 090	52 825
February	418 520	89 940	88 434	8 222	7 145	23 486	54 181	25 452	33 062	17 900	49 111	44 373	52 655
March	483 667	102 933	105 183	9 626	7 968	30 536	59 912	28 325	37 774	21 217	55 259	51 149	60 033
April	466 647	100 385	96 332	8 562	7 099	32 663	59 282	26 598	38 108	21 052	55 185	47 122	58 472
May	495 849	103 937	104 569	9 394	7 614	37 261	61 805	28 205	39 862	21 831	56 165	51 216	60 059
June	483 003	102 889	101 574	9 334	7 695	35 011	60 257	27 657	39 070	20 527	56 308	49 611	58 588
July	476 520	102 344	101 141	9 207	7 535	31 485	61 323	26 745	38 787	20 727	55 982	48 218	58 451
August	491 564	108 444	103 617	9 825	7 970	31 789	61 122	28 427	40 148	22 715	57 478	51 395	58 507
September	470 724	100 469	98 778	9 471	7 780	30 404	60 001	27 321	40 181	19 873	54 177	48 554	57 086
October	477 120	102 840	96 179	9 441	7 645	31 151	60 471	28 558	40 448	20 455	56 132	51 280	58 164
November	499 276	119 088	93 524	10 487	10 386	30 562	61 783	28 162	38 780	24 588	63 478	63 170	55 690
December	559 854	149 902	98 801	11 211	12 256	27 944	67 531	31 420	38 499	33 890	78 456	76 026	61 129
2018													
January	444 701	90 813	87 048	8 686	2 272	24 612	60 093	28 429	37 539	15 815	50 450	54 129	54 531
February	436 018	92 780	88 542	8 504	2 029	24 579	56 201	26 317	35 554	18 481	50 670	49 906	54 760
March	509 325	108 698	108 350	10 016	2 383	31 074	63 479	29 187	40 672	22 344	58 974	56 113	63 490
April	481 516	99 156	98 248	9 205	2 201	33 446	59 295	27 839	41 052	20 503	53 747	54 098	60 259
May	529 261	110 841	107 939	9 949	2 412	39 555	64 229	29 709	45 760	23 631	60 250	57 352	64 038
June	508 292	106 233	103 701	9 751	2 239	36 422	62 469	28 388	45 128	21 354	58 050	53 412	63 572
July	506 514	105 545	102 680	9 740	2 362	34 473	63 083	28 235	45 502	21 754	57 346	55 330	63 539
August	521 962	112 062	107 191	10 209	2 528	33 371	63 306	29 750	46 290	23 209	60 048	56 987	64 576
September	478 595	100 721	94 975	9 573	2 268	30 494	60 841	27 689	43 022	19 958	54 906	53 003	59 966
October	505 194	105 660	98 046	9 758	2 417	33 769	62 150	30 343	45 599	21 337	57 721	59 445	61 613
November	520 956	122 523	94 379	10 579	3 151	31 017	63 097	29 743	40 753	25 652	66 253	71 616	58 991
December	559 289	147 140	100 222	10 925	3 815	28 501	67 493	31 825	37 054	34 125	77 883	75 458	62 820
2019													
January	458 089	91 738	87 114	8 648	7 666	27 578	62 320	29 640	35 414	16 201	51 557	60 263	56 748
February	443 708	91 033	89 009	8 445	6 980	24 888	56 383	27 825	34 549	17 932	50 513	55 979	56 868
March	515 694	106 558	109 845	9 846	7 675	30 900	62 935	29 997	40 960	21 953	58 619	60 236	66 187
April	509 413	102 862	103 156	9 360	6 900	35 122	61 781	29 671	43 144	21 416	56 815	61 667	63 454
May	547 130	110 244	111 772	10 069	7 574	38 372	65 842	30 847	46 218	22 938	60 695	66 131	67 664
June	518 273	105 686	104 743	9 539	7 481	34 575	63 795	28 966	43 823	20 960	58 727	61 580	66 031
July	532 103	106 940	109 269	9 944	7 669	34 755	65 690	29 508	45 450	21 650	58 487	65 644	65 943
August	545 247	113 908	114 659	10 345	8 038	33 434	66 165	29 973	45 043	23 743	61 460	65 175	67 710
September	496 074	100 089	99 537	9 839	7 638	30 948	61 668	28 706	41 936	19 464	54 515	61 761	62 775
October	525 539	106 593	103 801	9 995	7 604	33 719	64 051	31 088	43 602	21 177	58 560	68 337	65 005
November	535 352	121 326	101 101	10 782	9 985	30 731	65 442	29 312	40 730	24 928	66 027	74 817	62 326
December	591 380	147 511	103 463	11 003	11 807	29 493	68 992	33 194	40 203	34 541	77 648	93 920	65 085

[1]Includes store categories not shown separately.
[2]Includes furniture, home furnishings, electronics, appliances, clothing, sporting goods, hobby, book, music, general merchandise, office supplies, stationery, and gifts.

Table 16-8. Retail and Food Services Sales—*Continued*

(All retail establishments and food services; millions of dollars; seasonally adjusted.)

Year and month	Retail and food services Total	Retail (NAICS industry categories) Total	GAFO (department store type goods) 2	Motor vehicles and parts	Furniture and home furnishings	Electronics and appliances	Building materials and garden	Food and beverages Total 1	Groceries	Beer, wine, and liquor	Health and personal care	Gasoline
1994	2 330 235	2 105 231	616 347	541 141	60 416	57 266	157 228	384 340	350 523	22 101	96 359	171 222
1995	2 450 628	2 217 613	650 040	579 715	63 470	64 770	164 561	390 386	356 409	22 007	101 632	181 113
1996	2 603 794	2 361 546	682 613	627 507	67 707	68 363	176 683	401 073	365 547	23 157	109 554	194 425
1997	2 726 131	2 468 765	713 387	653 817	72 715	70 061	191 063	409 373	372 570	24 081	118 670	199 700
1998	2 852 956	2 581 761	757 936	688 415	77 412	74 527	202 423	416 525	378 188	25 382	129 582	191 727
1999	3 086 990	2 803 088	815 665	764 204	84 294	78 977	218 290	433 699	394 250	26 476	142 697	212 524
2000	3 287 537	2 983 275	862 739	796 210	91 170	82 206	228 994	444 764	402 515	28 507	155 233	249 816
2001	3 378 906	3 062 267	882 700	815 579	91 484	80 240	239 379	462 429	418 127	29 621	166 532	251 383
2002	3 459 077	3 128 552	912 707	818 811	94 438	83 740	248 539	464 856	419 813	29 894	179 983	250 619
2003	3 612 457	3 262 731	946 114	841 588	96 736	86 442	263 463	474 385	427 987	30 469	275 187	178 694
2004	3 846 605	3 473 048	1 003 891	866 372	103 757	93 896	295 274	490 380	441 136	32 189	324 006	324 006
2005	4 085 746	3 689 283	1 059 599	888 307	109 120	100 461	320 802	508 484	457 667	33 567	378 923	378 923
2006	4 294 359	3 871 573	1 110 155	899 997	112 795	105 477	334 130	525 232	471 699	36 016	421 976	421 976
2007	4 439 733	3 995 182	1 143 426	910 139	111 144	106 599	320 854	547 837	491 360	38 128	451 822	451 822
2008	4 391 580	3 935 315	1 136 376	785 865	98 720	105 317	301 833	569 276	511 222	39 504	246 573	503 639
2009	4 064 476	3 612 471	1 088 197	671 772	84 749	95 364	261 637	568 418	510 033	40 245	252 794	391 234
2010	4 284 968	3 818 048	1 114 374	742 913	85 205	97 396	260 566	580 530	520 750	41 401	260 435	448 349
2011	4 598 302	4 102 952	1 155 666	812 938	87 586	99 928	269 480	609 137	547 476	42 392	271 612	533 457
2012	4 826 390	4 302 229	1 191 843	886 494	91 542	102 060	281 533	628 205	563 645	44 365	274 000	555 419
2013	5 001 763	4 458 450	1 212 493	959 294	95 349	102 998	301 797	640 847	574 547	46 076	281 840	549 613
2014	5 215 656	4 639 440	1 238 694	1 020 851	99 718	103 518	318 352	669 165	599 603	48 286	299 263	538 790
2015	5 349 487	4 725 993	1 258 154	1 094 112	106 570	103 658	331 611	685 381	613 159	50 550	315 244	444 027
2016	5 510 186	4 852 958	1 261 613	1 140 614	110 404	99 043	348 697	699 349	623 881	53 213	327 153	422 792
2017	5 744 810	5 053 151	1 272 463	1 172 367	113 035	98 570	365 622	725 137	647 813	55 232	333 338	459 463
2018	6 001 623	5 269 468	1 302 172	1 191 321	116 895	100 205	381 313	745 736	665 097	57 968	347 454	503 925
2019	6 218 002	5 452 206	1 304 488	1 237 469	117 815	97 017	384 515	765 064	682 862	59 673	358 727	501 072
2017												
January	473 832	416 600	106 232	98 325	9 307	8 163	29 845	59 238	52 860	4 522	26 898	38 604
February	473 582	416 224	104 772	97 626	9 322	8 092	30 162	59 516	53 164	4 498	27 221	38 942
March	472 693	415 464	105 470	95 626	9 373	8 291	29 957	60 006	53 647	4 519	27 367	38 545
April	475 335	418 010	105 854	96 469	9 266	8 401	30 299	60 144	53 728	4 585	27 563	38 338
May	473 242	415 879	104 853	96 792	9 329	8 249	30 025	60 303	53 850	4 599	27 598	36 807
June	474 798	417 639	105 948	97 631	9 381	8 177	30 036	60 109	53 689	4 593	27 796	36 243
July	475 045	417 908	105 301	97 700	9 347	8 093	30 230	60 188	53 790	4 589	27 859	36 081
August	475 653	418 181	105 865	95 319	9 420	8 108	30 331	60 458	53 981	4 626	28 118	37 278
September	486 056	428 218	106 808	99 932	9 368	8 181	31 432	61 090	54 630	4 633	28 224	39 278
October	486 635	428 529	106 848	100 732	9 614	8 247	31 043	61 331	54 773	4 713	28 359	39 270
November	490 648	431 841	107 532	99 195	9 719	8 356	31 402	61 455	54 882	4 709	28 446	40 396
December	492 996	433 877	107 639	99 301	9 723	8 326	31 459	61 804	55 275	4 686	28 383	40 956
2018												
January	492 600	433 198	107 581	99 148	9 619	8 358	30 524	61 661	55 021	4 757	28 429	41 388
February	494 815	435 033	108 743	98 296	9 686	8 483	31 333	61 747	55 077	4 758	28 207	41 878
March	494 747	434 510	107 670	98 761	9 753	8 404	30 901	61 934	55 240	4 805	28 392	41 418
April	495 967	436 423	108 175	99 140	9 941	8 533	30 952	62 231	55 545	4 815	28 612	41 176
May	502 351	441 536	110 617	99 674	9 802	8 441	31 793	62 214	55 490	4 843	29 069	42 175
June	501 133	439 651	108 764	99 558	9 810	8 479	32 290	62 042	55 349	4 817	28 850	42 255
July	503 834	441 541	109 058	99 704	9 848	8 385	32 292	62 436	55 716	4 842	29 078	42 446
August	502 875	440 241	108 612	98 197	9 732	8 420	32 056	62 229	55 552	4 801	29 310	42 861
September	502 407	441 029	108 307	98 576	9 729	8 268	32 202	62 454	55 661	4 874	29 239	42 470
October	508 036	446 238	109 104	99 640	9 739	8 314	32 791	62 660	55 835	4 877	29 517	44 100
November	508 471	446 894	109 710	99 780	9 706	8 251	32 004	62 629	55 821	4 906	29 923	42 187
December	497 541	435 953	107 798	101 002	9 660	8 133	32 018	62 245	55 486	4 903	28 801	39 419
2019												
January	505 036	443 420	107 737	99 055	9 535	8 147	33 530	63 462	56 660	4 898	29 699	38 831
February	504 686	442 467	107 246	99 357	9 651	8 032	31 585	62 025	55 248	4 906	29 887	40 646
March	512 785	450 108	108 744	102 693	9 787	8 096	31 790	63 162	56 334	4 945	29 907	42 097
April	515 119	452 043	109 022	102 079	9 853	8 033	31 723	63 328	56 502	4 925	29 820	42 972
May	517 236	453 522	108 933	102 610	9 843	8 101	31 262	63 410	56 622	4 922	30 065	42 480
June	519 205	454 847	109 079	103 055	9 854	8 070	31 415	63 862	57 005	4 976	30 017	41 578
July	522 198	457 485	109 658	103 219	9 855	8 133	31 779	64 586	57 692	4 995	29 897	42 201
August	524 547	459 938	109 159	105 133	9 881	8 103	32 617	64 603	57 701	5 024	29 824	41 591
September	522 261	457 277	109 049	103 935	9 999	8 134	32 159	64 095	57 220	5 005	29 933	41 398
October	524 853	459 978	109 045	104 919	9 916	8 089	32 177	64 073	57 201	4 983	30 241	42 046
November	525 014	460 760	108 534	106 504	9 937	8 059	32 308	64 354	57 438	5 040	29 819	42 295
December	525 467	460 512	109 138	104 837	9 746	8 098	32 807	64 522	57 622	5 065	29 770	42 815

1 Includes store categories not shown separately.
2 Includes furniture, home furnishings, electronics, appliances, clothing, sporting goods, hobby, book, music, general merchandise, office supplies, stationery, and gifts.

Table 16-9. Quarterly U.S. Retail Sales: Total and E-Commerce

Year and quarter	Retail sales (millions of dollars)		E-commerce as a percent of total sales	Percent change from prior quarter		Percent change from same quarter a year ago	
	Total	E-commerce		Total sales	E-commerce sales	Total sales	E-commerce sales
NOT SEASONALLY ADJUSTED							
2018							
1st quarter	1 217 263	114 694	9.4	-10.6	-20.7	5.1	15.3
2nd quarter	1 331 200	121 969	9.2	9.4	6.3	5.0	14.4
3rd quarter	1 318 990	123 322	9.3	-0.9	1.1	4.3	13.9
4th quarter	1 402 015	159 650	11.4	6.3	29.5	3.0	10.5
2019							
1st quarter	1 237 688	127 888	10.3	-11.7	-19.9	1.7	11.5
2nd quarter	1 377 667	138 956	10.1	11.3	8.7	3.5	13.9
3rd quarter	1 376 996	145 474	10.6	0.0	4.7	4.4	18.0
4th quarter	1 459 855	185 700	12.7	6.0	27.7	4.1	16.3
SEASONALLY ADJUSTED							
2006							
1st quarter	963 388	26 417	2.7	2.8	8.4	7.2	27.0
2nd quarter	967 539	27 367	2.8	0.4	3.6	5.6	23.1
3rd quarter	972 764	28 842	3.0	0.5	5.4	4.1	21.9
4th quarter	972 870	30 138	3.1	0.0	4.5	3.8	23.7
2007							
1st quarter	985 713	31 728	3.2	1.3	5.3	2.3	20.1
2nd quarter	994 838	33 524	3.4	0.9	5.7	2.8	22.5
3rd quarter	1 001 953	34 841	3.5	0.7	3.9	3.0	20.8
4th quarter	1 014 852	35 784	3.5	1.3	2.7	4.3	18.7
2008							
1st quarter	1 007 189	36 017	3.6	-0.8	0.7	2.2	13.5
2nd quarter	1 011 101	36 514	3.6	0.4	1.4	1.6	8.9
3rd quarter	999 159	36 292	3.6	-1.2	-0.6	-0.3	4.2
4th quarter	910 187	33 042	3.6	-8.9	-9.0	-10.3	-7.7
2009							
1st quarter	888 604	34 132	3.8	-2.4	3.3	-11.8	-5.2
2nd quarter	892 284	35 282	4.0	0.4	3.4	-11.8	-3.4
3rd quarter	912 858	37 402	4.1	2.3	6.0	-8.6	3.1
4th quarter	919 242	38 110	4.1	0.7	1.9	1.0	15.3
2010							
1st quarter	933 318	39 289	4.2	1.5	3.1	5.0	15.1
2nd quarter	949 218	41 303	4.4	1.7	5.1	6.4	17.1
3rd quarter	952 805	43 474	4.6	0.4	5.3	4.4	16.2
4th quarter	981 419	45 075	4.6	3.0	3.7	6.8	18.3
2011							
1st quarter	1 004 749	47 020	4.7	2.4	4.3	7.7	19.7
2nd quarter	1 021 038	48 815	4.8	1.6	3.8	7.6	18.2
3rd quarter	1 029 198	50 140	4.9	0.8	2.7	8.0	15.3
4th quarter	1 047 042	53 123	5.1	1.7	5.9	6.7	17.9
2012							
1st quarter	1 068 705	55 144	5.2	2.1	3.8	6.4	17.3
2nd quarter	1 065 156	56 363	5.3	-0.3	2.2	4.3	15.5
3rd quarter	1 073 460	58 449	5.4	0.8	3.7	4.3	16.6
4th quarter	1 089 441	60 815	5.6	1.5	4.0	4.0	14.5
2013							
1st quarter	1 108 664	62 458	5.6	1.8	2.7	3.7	13.3
2nd quarter	1 107 856	64 383	5.8	-0.1	3.1	4.0	14.2
3rd quarter	1 117 931	66 457	5.9	0.9	3.2	4.1	13.7
4th quarter	1 123 746	69 005	6.1	0.5	3.8	3.1	13.5
2014							
1st quarter	1 132 940	71 108	6.3	0.8	3.0	2.2	13.8
2nd quarter	1 161 047	74 290	6.4	2.5	4.5	4.8	15.4
3rd quarter	1 169 154	76 943	6.6	0.7	3.6	4.6	15.8
4rd quarter	1 170 448	79 024	6.8	0.1	2.7	4.2	14.5
2015							
1st quarter	1 161 207	81 837	7.0	-0.8	3.6	2.5	15.1
2nd quarter	1 180 619	84 704	7.2	1.7	3.5	1.7	14.0
3rd quarter	1 191 658	87 754	7.4	0.9	3.6	1.9	14.1
4th quarter	1 188 095	90 687	7.6	-0.3	3.3	1.5	14.8
2016							
1st quarter	1 190 699	94 057	7.9	0.2	3.7	2.5	14.9
2nd quarter	1 204 107	97 459	8.1	1.1	3.6	2.0	15.1
3rd quarter	1 215 288	100 519	8.3	0.9	3.1	2.0	14.5
4th quarter	1 227 505	103 952	8.5	1.0	3.4	3.3	14.6
2017							
1st quarter	1 248 288	108 157	8.7	1.7	4.0	4.8	15.0
2nd quarter	1 251 528	112 644	9.0	0.3	4.1	3.9	15.6
3rd quarter	1 264 307	115 419	9.1	1.0	2.5	4.0	14.8
4th quarter	1 294 247	121 019	9.4	2.4	4.9	5.4	16.4
2018							
1st quarter	1 302 741	124 936	9.6	0.7	3.2	4.4	15.5
2nd quarter	1 317 610	128 616	9.8	1.1	2.9	5.3	14.2
3rd quarter	1 322 811	130 625	9.9	0.4	1.6	4.6	13.2
4th quarter	1 329 085	134 291	10.1	0.5	2.8	2.7	11.0
2019							
1st quarter	1 335 812	139 713	10.5	0.5	4.0	2.5	11.8
2nd quarter	1 360 589	146 348	10.8	1.9	4.7	3.3	13.8
3rd quarter	1 374 700	153 274	11.1	1.0	4.7	3.9	17.3
4rd quarter	1 381 250	156 581	11.3	0.5	2.2	3.9	16.6

Table 16-10. Retail Inventories

(All retail stores; end of period, millions of dollars.)

Year and month	Not seasonally adjusted			Seasonally adjusted								
	Total	Excluding motor vehicles and parts	Motor vehicles and parts	Total	Excluding motor vehicles and parts	Motor vehicles and parts	Furniture, home furnishings, electronics, and appliances	Building materials and garden	Food and beverages	Clothing and accessories	General merchandise	
											Total	Department stores [1]
1994	299 812	211 223	88 589	304 519	218 757	85 762	20 397	24 857	28 112	29 518	56 581	41 908
1995	317 304	220 921	96 383	322 216	228 842	93 374	21 817	26 337	28 718	29 298	59 530	43 455
1996	328 160	227 888	100 272	333 255	236 015	97 240	22 253	27 447	29 658	29 778	60 591	44 124
1997	338 790	234 270	104 520	343 825	242 518	101 307	22 109	28 897	29 891	31 079	60 715	44 309
1998	351 255	245 400	105 855	356 518	253 888	102 630	22 665	30 921	30 837	32 308	61 542	43 438
1999	378 813	259 879	118 934	384 078	268 567	115 511	24 096	33 005	32 635	33 735	64 297	43 835
2000	400 394	268 982	131 412	405 931	277 829	128 102	25 622	34 269	32 108	36 753	64 929	42 667
2001	387 859	266 370	121 489	393 638	274 796	118 842	24 497	34 235	33 165	35 627	64 790	40 464
2002	409 314	271 757	137 557	415 174	280 277	134 897	25 853	36 123	32 907	37 401	66 006	38 750
2003	425 748	277 471	148 277	431 199	285 668	145 531	27 152	37 398	32 504	38 503	66 655	36 823
2004	454 848	297 561	157 287	460 335	305 941	154 394	30 043	41 593	33 433	41 417	70 969	37 230
2005	465 791	310 227	155 564	471 616	318 849	152 767	30 550	45 072	33 784	43 003	74 179	38 030
2006	480 338	324 005	156 333	486 403	332 972	153 431	30 807	47 010	34 700	47 272	75 592	37 133
2007	494 372	334 566	159 806	500 489	343 476	157 013	31 356	49 074	36 484	47 381	75 820	36 428
2008	471 330	323 360	147 970	477 148	331 531	145 617	27 772	46 496	37 273	45 958	72 777	33 090
2009	423 380	114 772	114 772	429 159	315 870	113 289	25 710	42 707	37 254	40 811	69 575	30 817
2010	449 059	320 671	128 388	454 548	327 909	126 639	28 006	40 737	38 613	41 931	73 290	30 933
2011	465 360	331 578	133 782	471 186	339 264	131 922	26 704	43 333	40 158	44 278	76 794	31 393
2012	500 394	341 755	158 639	506 218	349 910	156 308	27 516	44 599	41 063	46 726	78 412	30 519
2013	538 556	358 607	179 949	544 749	367 382	177 367	27 684	44 497	42 598	49 724	81 430	30 536
2014	555 843	368 527	187 316	562 395	377 468	184 927	27 670	48 706	44 088	50 910	82 080	29 211
2015	583 036	383 332	199 704	590 050	392 248	197 802	27 569	50 477	45 521	52 193	84 094	29 553
2016	605 528	390 896	214 632	613 159	399 640	213 519	27 501	52 251	46 940	52 299	83 124	28 219
2017	619 358	399 521	219 837	627 767	408 293	219 474	28 141	54 465	48 327	52 168	80 532	26 244
2018	647 917	414 373	233 544	656 915	423 351	233 564	28 195	57 780	49 819	53 547	82 633	25 094
2019	654 436	422 489	231 947	663 703	431 545	232 158	27 611	61 296	51 279	52 738	81 527	23 241
2016												
January	583 185	382 848	200 337	590 626	392 804	197 822	27 440	50 607	45 614	52 404	84 039	29 176
February	588 036	385 234	202 802	593 242	393 365	199 877	27 255	50 729	45 954	52 383	84 025	29 185
March	599 468	390 559	208 909	600 665	394 776	205 889	27 503	50 996	45 854	52 278	84 029	29 007
April	600 019	389 464	210 555	600 715	394 902	205 813	27 543	50 948	45 948	52 485	83 849	28 835
May	595 455	388 336	207 119	602 579	396 249	206 330	27 649	51 275	46 057	52 599	84 204	28 737
June	597 329	388 698	208 631	604 862	396 800	208 062	27 725	50 963	46 462	52 560	83 937	28 639
July	589 220	387 825	201 395	603 522	396 567	206 955	27 738	51 007	46 467	52 575	83 729	28 939
August	594 344	393 113	201 231	605 919	396 831	209 088	27 799	51 413	46 449	51 736	83 675	28 396
September	613 214	407 582	205 632	609 876	398 474	211 402	27 678	51 477	46 770	52 233	83 584	28 186
October	633 777	423 365	210 412	607 030	396 692	210 338	27 551	51 562	46 658	52 015	83 013	28 125
November	644 373	427 449	216 924	613 015	398 795	214 220	27 656	52 206	46 664	52 363	83 053	28 136
December	605 528	390 896	214 632	613 159	399 640	213 519	27 501	52 251	46 940	52 299	83 124	28 219
2017												
January	608 072	389 244	218 828	615 094	398 853	216 241	27 532	52 468	46 823	52 113	82 508	28 166
February	612 077	390 335	221 742	616 714	398 169	218 545	27 664	52 435	46 750	52 257	81 833	27 739
March	621 142	395 701	225 441	621 852	399 684	222 168	27 617	52 632	46 865	52 472	82 244	27 848
April	617 758	392 687	225 071	617 931	397 929	220 002	27 532	52 558	47 194	52 059	81 904	27 765
May	613 826	391 179	222 647	620 436	398 840	221 596	27 828	52 652	47 215	52 113	80 990	27 179
June	617 884	393 600	224 284	624 930	401 464	223 466	27 872	52 762	47 179	52 091	81 869	26 987
July	608 316	392 226	216 090	623 015	401 147	221 868	27 898	53 185	47 233	51 953	81 147	26 855
August	616 151	399 304	216 847	627 835	403 344	224 491	27 922	53 184	47 544	51 966	81 112	26 582
September	625 533	411 377	214 156	622 741	402 982	219 759	28 069	53 497	47 444	51 941	80 808	26 335
October	649 395	430 873	218 522	622 677	404 212	218 465	27 884	53 820	47 567	51 989	80 663	26 238
November	656 440	434 617	221 823	625 811	406 354	219 457	27 839	54 100	47 992	51 945	80 793	26 205
December	619 358	399 521	219 837	627 767	408 293	219 474	28 141	54 465	48 327	52 168	80 532	26 244
2018												
January	623 187	399 581	223 606	630 070	408 926	221 144	27 802	54 726	48 570	52 140	80 766	26 375
February	629 684	403 165	226 519	633 743	410 708	223 035	27 918	55 304	48 906	52 698	81 224	26 454
March	631 575	405 756	225 819	632 422	409 674	222 748	27 930	55 416	48 430	52 473	80 604	25 924
April	634 110	406 518	227 592	634 353	411 662	222 691	27 812	55 608	47 983	52 880	81 323	26 320
May	630 593	405 148	225 445	636 667	412 674	223 993	27 644	56 073	48 371	52 486	82 126	26 511
June	630 705	406 170	224 535	638 023	414 083	223 940	27 460	56 807	48 406	52 634	82 002	26 332
July	624 399	405 483	218 916	639 387	414 646	224 741	27 816	57 039	48 638	52 521	82 169	26 218
August	632 873	412 064	220 809	644 434	416 502	227 932	28 110	57 381	49 031	52 585	82 919	26 142
September	646 324	423 488	222 836	643 701	415 439	228 262	28 146	57 256	48 972	52 398	82 271	26 089
October	676 724	446 055	230 669	649 362	418 732	230 630	29 018	57 306	49 160	53 128	83 113	26 176
November	678 770	443 922	234 848	648 451	415 903	232 548	27 721	57 293	49 283	52 659	81 689	25 014
December	647 917	414 373	233 544	656 915	423 351	233 564	28 195	57 780	49 819	53 547	82 633	25 094
2019												
January	654 926	416 268	238 658	661 661	425 638	236 023	28 487	58 815	49 737	53 754	83 148	24 986
February	661 688	421 008	240 680	665 593	428 408	237 185	28 765	58 734	49 924	53 817	84 107	25 070
March	661 584	423 354	238 230	662 453	427 256	235 197	28 434	58 846	49 895	53 597	83 224	24 975
April	665 197	423 729	241 468	665 407	428 724	236 683	28 494	59 417	50 280	53 540	83 163	24 626
May	661 629	422 297	239 332	667 691	429 903	237 788	28 376	59 826	50 498	53 570	82 700	24 535
June	656 958	420 480	236 478	664 655	428 589	236 066	28 060	59 918	50 525	53 172	82 510	24 460
July	653 496	420 503	232 993	668 769	429 789	238 980	27 975	60 452	50 754	53 283	82 777	24 330
August	654 655	423 881	230 774	666 385	428 634	237 751	27 733	60 321	50 748	53 200	82 207	24 299
September	669 709	437 092	232 617	667 492	429 436	238 056	27 615	60 746	50 704	53 110	82 680	24 340
October	697 190	459 563	237 627	669 043	431 870	237 173	27 956	61 232	51 125	52 969	82 467	24 282
November	694 505	459 275	235 230	663 932	431 017	232 915	27 771	61 207	51 149	52 900	81 998	23 859
December	654 436	422 489	231 947	663 703	431 545	232 158	27 611	61 296	51 279	52 738	81 527	23 241

[1] Excluding leased departments.

Table 16-11. Merchant Wholesalers—Sales and Inventories

(Millions of dollars.)

Year and month	Not seasonally adjusted						Seasonally adjusted					
	Sales			Inventories (current cost, end of period)			Sales			Inventories (current cost, end of period)		
	Total	Durable goods establish-ments	Nondurable goods establish-ments	Total	Durable goods establish-ments	Nondurable goods establish-ments	Total	Durable goods establish-ments	Nondurable goods establish-ments	Total	Durable goods establish-ments	Nondurable goods establish-ments
1994	1 974 899	1 037 638	937 261	222 826	139 941	82 885	1 974 899	1 037 638	937 261	221 978	141 975	80 003
1995	2 158 980	1 141 701	1 017 279	239 275	151 709	87 566	2 158 980	1 141 701	1 017 279	238 392	154 089	84 303
1996	2 284 343	1 190 342	1 094 001	241 396	154 207	87 189	2 284 343	1 190 342	1 094 001	241 058	156 683	84 375
1997	2 377 845	1 256 384	1 121 461	258 900	165 371	93 529	2 377 845	1 256 384	1 121 461	258 454	168 089	90 365
1998	2 427 120	1 306 545	1 120 575	272 575	175 994	96 581	2 427 120	1 306 545	1 120 575	272 297	178 918	93 379
1999	2 599 159	1 406 371	1 192 788	290 382	187 738	102 644	2 599 159	1 406 371	1 192 788	290 182	190 899	99 283
2000	2 814 554	1 486 673	1 327 881	309 710	198 525	111 185	2 814 554	1 486 673	1 327 881	309 191	201 797	107 394
2001	2 785 152	1 422 195	1 362 957	298 577	182 521	116 056	2 785 152	1 422 195	1 362 957	297 536	185 517	112 019
2002	2 835 528	1 421 503	1 414 025	302 715	182 150	120 565	2 835 528	1 421 503	1 414 025	301 310	185 048	116 262
2003	2 978 282	1 466 158	1 512 124	310 115	186 435	123 680	2 978 282	1 466 158	1 512 124	308 274	189 325	118 949
2004	3 330 011	1 689 588	1 640 423	341 642	213 590	128 052	3 330 011	1 689 588	1 640 423	340 128	216 704	123 424
2005	3 638 496	1 830 347	1 808 149	369 380	233 199	136 181	3 638 496	1 830 347	1 808 149	367 978	236 438	131 540
2006	3 941 257	2 005 139	1 936 118	400 297	256 731	143 566	3 941 257	2 005 139	1 936 118	398 924	260 115	138 809
2007	4 223 473	2 101 969	2 121 504	426 254	263 195	163 059	4 223 473	2 101 969	2 121 504	424 344	266 628	157 716
2008	4 524 357	2 126 715	2 397 642	446 109	280 403	165 706	4 524 357	2 126 715	2 397 642	445 529	284 176	161 353
2009	3 829 378	1 745 557	2 083 821	399 322	233 720	165 602	3 829 378	1 745 557	2 083 821	397 699	237 136	160 563
2010	4 337 359	1 999 247	2 338 112	444 639	256 969	187 670	4 337 359	1 999 247	2 338 112	442 154	260 923	181 231
2011	4 885 076	2 229 683	2 655 393	489 804	286 171	203 633	4 885 076	2 229 683	2 655 393	488 061	290 810	197 251
2012	5 208 023	2 389 576	2 818 447	525 855	311 262	214 593	5 208 023	2 389 576	2 818 447	524 005	316 609	207 396
2013	5 370 550	2 461 025	2 909 525	546 197	326 427	219 770	5 370 550	2 461 025	2 909 525	545 175	332 288	212 887
2014	5 564 180	2 556 641	3 007 539	578 644	348 845	229 799	5 564 180	2 556 641	3 007 539	577 344	355 111	222 233
2015	5 292 436	2 520 142	2 772 294	586 037	349 537	236 500	5 292 436	2 520 142	2 772 294	585 167	355 811	229 356
2016	5 222 020	2 504 874	2 717 146	596 775	347 007	249 768	5 222 020	2 504 874	2 717 146	595 265	353 148	242 117
2017	5 549 033	2 674 076	2 874 957	614 705	361 336	253 369	5 549 033	2 674 076	2 874 957	613 124	367 517	245 607
2018	5 939 442	2 858 725	3 080 717	654 998	394 380	260 618	5 939 442	2 858 725	3 080 717	653 268	400 913	252 355
2019	5 970 359	2 855 594	3 114 765	666 259	396 336	269 923	5 970 359	2 855 594	3 114 765	664 316	402 732	261 584
2016												
January	381 993	181 066	200 927	589 308	349 164	240 144	424 890	204 244	220 646	583 977	351 177	232 800
February	391 020	187 951	203 069	582 296	349 001	233 295	421 317	205 915	215 402	579 218	349 482	229 736
March	451 264	219 947	231 317	585 890	347 627	238 263	424 870	205 632	219 238	580 995	348 425	232 570
April	423 957	201 434	222 523	587 506	349 973	237 533	428 521	205 935	222 586	585 491	349 511	235 980
May	438 533	205 477	233 056	581 621	350 502	231 119	429 262	206 688	222 574	586 375	350 121	236 254
June	459 758	222 950	236 808	580 824	349 121	231 703	436 860	209 583	227 277	586 838	349 142	237 696
July	416 448	198 962	217 486	581 139	352 214	228 925	434 861	209 648	225 213	586 471	349 101	237 370
August	457 304	219 223	238 081	575 905	350 965	224 940	437 234	207 974	229 260	584 327	349 255	235 072
September	447 863	218 379	229 484	583 190	349 771	233 419	437 128	207 873	229 255	585 575	347 755	237 820
October	449 249	214 622	234 627	590 643	349 517	241 126	442 141	209 978	232 163	586 680	347 938	238 742
November	445 742	212 461	233 281	596 010	351 497	244 513	442 596	210 258	232 338	591 328	351 285	240 043
December	458 889	222 402	236 487	596 775	347 007	249 768	452 019	216 343	235 676	595 265	353 148	242 117
2017												
January	421 400	198 885	222 515	599 720	350 484	249 236	455 726	217 973	237 753	593 761	352 646	241 115
February	405 970	190 527	215 443	598 645	352 363	246 282	456 584	217 593	238 991	594 553	353 013	241 540
March	480 293	231 302	248 991	601 113	354 100	247 013	453 380	215 591	237 789	596 061	354 879	241 182
April	437 307	208 679	228 628	595 865	355 962	239 903	456 229	221 278	234 951	594 165	355 747	238 418
May	474 147	224 854	249 293	591 862	358 083	233 779	450 939	219 761	231 178	596 466	357 908	238 558
June	479 080	233 842	245 238	594 661	358 859	235 802	455 073	219 631	235 442	600 114	358 914	241 200
July	436 881	208 816	228 065	597 561	364 856	232 705	455 840	220 359	235 481	602 689	361 409	241 280
August	485 991	236 910	249 081	598 593	366 340	232 253	462 811	224 073	238 738	607 028	364 324	242 704
September	466 983	229 972	237 011	607 855	367 759	240 096	468 222	225 917	242 305	610 467	365 481	244 986
October	495 552	241 788	253 764	610 424	366 782	243 642	474 725	229 480	245 245	607 662	365 188	242 474
November	484 430	234 739	249 691	615 797	367 018	248 779	481 986	232 163	249 823	611 777	366 953	244 824
December	480 999	233 762	247 237	614 705	361 336	253 369	489 315	234 970	254 345	613 124	367 517	245 607
2018												
January	457 212	216 453	240 759	625 414	365 254	260 160	480 944	230 653	250 291	618 748	367 524	251 224
February	431 464	205 470	225 994	628 362	370 346	258 016	485 227	234 156	251 071	622 781	371 049	251 732
March	501 286	243 385	257 901	630 867	372 566	258 301	484 713	233 014	251 699	624 636	373 389	251 247
April	480 381	227 817	252 564	626 019	373 381	252 638	488 396	234 764	253 632	623 831	373 184	250 647
May	530 037	245 923	284 114	620 179	374 189	245 990	501 116	239 063	262 053	625 264	374 087	251 177
June	510 129	245 699	264 430	618 885	376 418	242 467	498 199	238 306	259 893	624 700	376 415	248 285
July	492 035	233 464	258 571	622 263	382 914	239 349	498 978	239 005	259 973	627 974	379 024	248 950
August	525 977	255 460	270 517	625 018	384 529	240 489	501 993	241 819	260 174	634 352	382 233	252 119
September	487 462	238 883	248 579	635 137	387 963	247 174	502 137	241 609	260 528	638 959	385 552	253 407
October	537 804	261 229	276 575	645 881	393 388	252 493	502 942	241 425	261 517	643 832	391 671	252 161
November	499 092	243 104	255 988	650 137	394 028	256 109	497 680	240 616	257 064	645 913	394 233	251 680
December	486 563	241 838	244 725	654 998	394 380	260 618	494 715	243 242	251 473	653 268	400 913	252 355
2019												
January	468 404	224 812	243 592	667 816	401 383	266 433	491 466	238 605	252 861	660 917	404 092	256 825
February	439 130	211 208	227 922	669 070	403 844	265 226	495 031	240 569	254 462	664 232	405 252	258 980
March	502 418	242 629	259 789	668 568	403 986	264 582	500 950	239 816	261 134	663 989	405 483	258 506
April	505 014	237 103	267 911	669 972	409 160	260 812	497 905	236 911	260 994	667 835	408 786	259 049
May	526 631	245 787	280 844	664 250	408 684	255 566	499 266	239 071	260 195	669 194	408 682	260 512
June	496 922	239 927	256 995	662 667	409 481	253 186	498 320	239 442	258 878	668 942	409 499	259 443
July	506 145	238 418	267 727	663 513	412 729	250 784	499 033	237 195	261 838	669 153	408 559	260 594
August	509 890	244 666	265 224	660 413	410 886	249 527	498 758	238 059	260 699	669 353	408 855	260 498
September	495 561	242 178	253 383	661 101	410 249	250 852	498 259	238 081	260 178	664 751	407 923	256 828
October	529 539	255 757	273 782	666 753	407 625	259 128	494 479	235 733	258 746	664 657	405 739	258 918
November	487 549	232 152	255 397	668 710	403 801	264 909	500 216	238 105	262 111	664 584	403 934	260 650
December	503 156	240 957	262 199	666 259	396 336	269 923	498 316	234 747	263 569	664 316	402 732	261 584

. . . = Not available.

Table 16-12. Manufacturing and Trade Sales and Inventories

Year and month	Sales, billions of dollars					Inventories, billions of dollars, end of period, seasonally adjusted				Ratios, inventories to sales, seasonally adjusted [1]			
	Not seasonally adjusted, total	Seasonally adjusted				Total	Manufac-turing	Retail trade	Merchant wholesalers	Total	Manufac-turing	Retail trade	Merchant wholesalers
		Total	Manufac-turing	Retail trade	Merchant wholesalers								
1998	8 908.7	8 906.4	3 897.4	2 582.7	2 426.3	1 077.2	448.4	356.5	272.3	1.44	1.39	1.62	1.32
1999	9 434.1	9 431.7	4 033.3	2 801.4	2 597.0	1 137.3	463.0	384.1	290.2	1.40	1.35	1.59	1.30
2000	10 006.4	9 998.9	4 202.4	2 979.4	2 817.0	1 195.9	480.7	406.0	309.2	1.41	1.35	1.59	1.29
2001	9 817.9	9 819.3	3 971.2	3 062.2	2 785.9	1 118.6	427.4	393.7	297.5	1.42	1.38	1.58	1.32
2002	9 878.8	9 881.5	3 917.4	3 129.6	2 834.5	1 139.5	423.0	415.2	301.3	1.36	1.29	1.55	1.26
2003	10 256.4	10 255.7	4 016.6	3 261.6	2 977.4	1 147.8	408.3	431.2	308.3	1.34	1.25	1.56	1.22
2004	11 112.0	11 071.7	4 295.3	3 460.9	3 315.6	1 241.7	441.2	460.4	340.1	1.30	1.19	1.56	1.17
2005	12 069.9	12 073.8	4 743.7	3 686.6	3 643.6	1 314.3	474.6	471.7	368.0	1.27	1.17	1.51	1.17
2006	12 828.4	12 839.3	5 016.8	3 876.6	3 946.0	1 408.8	523.5	486.4	398.9	1.28	1.20	1.49	1.17
2007	13 538.1	13 538.6	5 320.1	3 997.4	4 221.1	1 487.6	562.7	500.6	424.3	1.28	1.22	1.49	1.17
2008	13 928.7	13 882.4	5 448.2	3 927.6	4 506.6	1 466.0	543.3	477.2	445.5	1.31	1.26	1.52	1.20
2009	11 865.6	11 869.5	4 427.8	3 613.0	3 828.6	1 332.3	505.5	429.2	397.7	1.38	1.39	1.47	1.29
2010	13 066.7	13 064.2	4 914.4	3 816.7	4 333.0	1 451.0	554.3	454.5	442.2	1.27	1.28	1.39	1.15
2011	14 479.9	14 489.7	5 496.3	4 102.0	4 891.3	1 565.7	606.8	470.8	488.1	1.26	1.29	1.35	1.15
2012	15 207.0	15 187.6	5 691.8	4 296.8	5 199.0	1 654.6	624.9	505.7	524.0	1.28	1.30	1.38	1.17
2013	15 638.7	15 636.0	5 811.8	4 458.5	5 365.7	1 719.1	630.3	543.6	545.2	1.29	1.29	1.41	1.19
2014	16 091.2	16 078.2	5 887.4	4 634.0	5 556.7	1 778.5	640.4	560.7	577.3	1.31	1.31	1.43	1.22
2015	15 537.4	15 527.6	5 519.6	4 721.6	5 286.4	1 809.7	635.8	590.1	583.9	1.39	1.39	1.46	1.33
2016	15 429.7	15 393.9	5 344.7	4 837.6	5 211.7	1 839.7	631.2	613.2	595.3	1.42	1.41	1.50	1.35
2017	16 207.1	16 230.3	5 611.1	5 058.4	5 560.8	1 900.3	659.4	627.8	613.1	1.38	1.37	1.48	1.30
2018	17 208.5	17 206.8	5 997.6	5 272.2	5 937.0	1 992.8	682.7	656.9	653.3	1.36	1.35	1.46	1.28
2019	17 457.1	17 466.5	6 042.1	5 452.4	5 972.0	2 032.0	703.9	663.7	664.3	1.39	1.38	1.46	1.34
2015													
January	1 188.1	1 292.6	462.6	385.5	444.5	1 777.7	639.5	561.3	576.9	1.38	1.38	1.46	1.30
February	1 163.0	1 286.8	463.5	384.8	438.5	1 783.8	640.3	563.7	579.8	1.39	1.38	1.46	1.32
March	1 337.1	1 295.7	467.4	390.9	437.4	1 786.2	640.6	564.5	581.1	1.38	1.38	1.44	1.33
April	1 308.4	1 300.6	465.8	391.0	443.9	1 793.7	641.5	570.3	581.9	1.38	1.38	1.46	1.31
May	1 320.0	1 302.8	463.7	394.5	444.6	1 795.4	641.9	568.2	585.2	1.38	1.38	1.44	1.32
June	1 361.3	1 308.1	467.0	395.1	445.9	1 806.0	645.7	572.2	588.0	1.38	1.38	1.45	1.32
July	1 311.7	1 307.4	464.9	397.7	444.9	1 808.9	644.6	576.9	587.4	1.38	1.39	1.45	1.32
August	1 308.3	1 297.8	460.1	397.6	440.0	1 807.8	641.9	578.2	587.7	1.39	1.40	1.45	1.34
September	1 310.9	1 294.7	457.5	396.4	440.9	1 812.8	639.2	584.5	589.0	1.40	1.40	1.47	1.34
October	1 319.5	1 289.4	453.7	394.7	441.1	1 814.8	638.9	586.9	589.0	1.41	1.41	1.49	1.34
November	1 256.5	1 282.6	451.4	396.3	434.9	1 810.7	637.6	587.4	585.7	1.41	1.41	1.48	1.35
December	1 352.4	1 269.1	442.2	397.1	429.8	1 809.7	635.8	590.1	583.9	1.43	1.44	1.49	1.36
2016													
January	1 132.0	1 261.2	441.1	395.2	424.9	1 806.2	631.6	590.6	584.0	1.43	1.43	1.49	1.37
February	1 178.0	1 259.2	439.6	398.3	421.3	1 801.0	628.5	593.2	579.2	1.43	1.43	1.49	1.37
March	1 327.8	1 263.3	441.3	397.2	424.9	1 810.1	628.5	600.7	581.0	1.43	1.42	1.51	1.37
April	1 258.5	1 269.6	442.3	398.8	428.5	1 814.4	628.2	600.7	585.5	1.43	1.42	1.51	1.37
May	1 303.2	1 272.0	442.3	400.4	429.3	1 817.6	628.6	602.6	586.4	1.43	1.42	1.50	1.37
June	1 344.9	1 288.7	447.0	404.9	436.9	1 819.8	628.1	604.9	586.8	1.41	1.41	1.49	1.34
July	1 244.2	1 283.2	444.0	404.3	434.9	1 817.5	627.5	603.5	586.5	1.42	1.41	1.49	1.35
August	1 339.0	1 285.8	444.2	404.3	437.2	1 818.7	628.5	605.9	584.3	1.41	1.41	1.50	1.34
September	1 310.5	1 291.5	447.7	406.6	437.1	1 822.1	626.6	609.9	585.6	1.41	1.40	1.50	1.34
October	1 299.7	1 298.1	448.0	408.0	442.1	1 820.8	627.1	607.0	586.7	1.40	1.40	1.49	1.33
November	1 298.5	1 297.0	447.6	406.8	442.6	1 834.4	630.1	613.0	591.3	1.41	1.41	1.51	1.34
December	1 393.5	1 324.2	459.4	412.8	452.0	1 839.7	631.2	613.2	595.3	1.39	1.37	1.49	1.32
2017													
January	1 211.7	1 332.3	460.0	416.6	455.7	1 841.9	633.1	615.1	593.8	1.38	1.38	1.48	1.30
February	1 199.1	1 333.3	460.5	416.2	456.6	1 845.5	634.2	616.7	594.6	1.38	1.38	1.48	1.30
March	1 395.4	1 330.1	461.2	415.5	453.4	1 852.3	634.4	621.9	596.1	1.39	1.38	1.50	1.31
April	1 294.9	1 333.1	458.9	418.0	456.2	1 847.8	635.7	617.9	594.2	1.39	1.39	1.48	1.30
May	1 388.5	1 329.4	462.6	415.9	450.9	1 852.9	636.0	620.4	596.5	1.39	1.38	1.49	1.32
June	1 394.9	1 335.4	462.7	417.6	455.1	1 861.5	636.4	624.9	600.1	1.39	1.38	1.50	1.32
July	1 295.3	1 336.8	463.0	417.9	455.8	1 864.9	639.2	623.0	602.7	1.40	1.38	1.49	1.32
August	1 408.6	1 348.6	467.6	418.2	462.8	1 877.2	642.3	627.8	607.0	1.39	1.37	1.50	1.31
September	1 366.1	1 366.8	470.3	428.2	468.2	1 880.8	647.6	622.7	610.5	1.38	1.38	1.45	1.30
October	1 400.7	1 378.4	475.1	428.5	474.7	1 881.2	650.9	622.7	607.7	1.36	1.37	1.45	1.28
November	1 402.0	1 397.8	484.0	431.8	482.0	1 891.9	654.3	625.8	611.8	1.35	1.35	1.45	1.27
December	1 449.9	1 408.2	485.1	433.9	489.3	1 900.3	659.4	627.8	613.1	1.35	1.36	1.45	1.25
2018													
January	1 301.2	1 403.2	460.0	433.2	480.9	1 910.8	662.0	630.1	618.7	1.36	1.35	1.45	1.29
February	1 267.8	1 410.8	460.5	435.0	485.2	1 921.1	664.6	633.7	622.8	1.36	1.35	1.46	1.28
March	1 465.4	1 412.5	461.2	434.5	484.7	1 921.7	664.7	632.4	624.6	1.36	1.35	1.46	1.29
April	1 391.3	1 418.2	458.9	436.4	488.4	1 925.9	667.7	634.4	623.8	1.36	1.35	1.45	1.28
May	1 510.4	1 439.7	462.6	441.5	501.1	1 931.7	669.8	636.7	625.3	1.34	1.35	1.44	1.25
June	1 480.5	1 439.2	462.7	439.7	498.2	1 932.3	669.6	638.0	624.7	1.34	1.34	1.45	1.25
July	1 417.9	1 442.3	463.0	441.5	499.0	1 943.7	676.3	639.4	628.0	1.35	1.35	1.45	1.26
August	1 512.5	1 446.6	467.6	440.2	502.0	1 954.8	676.0	644.4	634.4	1.35	1.34	1.46	1.26
September	1 423.1	1 450.6	470.3	441.0	502.1	1 963.0	680.3	643.7	639.0	1.35	1.34	1.46	1.27
October	1 508.4	1 457.2	475.1	446.2	502.9	1 975.7	682.5	649.4	643.8	1.36	1.34	1.46	1.28
November	1 457.0	1 450.8	484.0	446.9	497.7	1 976.8	682.4	648.5	645.9	1.36	1.35	1.45	1.30
December	1 473.1	1 435.9	485.1	436.0	494.7	1 992.8	682.7	656.9	653.3	1.39	1.35	1.51	1.32
2019													
January	1 336.7	1 463.0	504.2	460.5	498.3	2 008.8	686.2	661.7	660.9	1.40	1.36	1.49	1.34
February	1 295.6	1 462.8	501.8	460.8	500.2	2 018.2	688.3	665.6	664.2	1.40	1.36	1.50	1.34
March	1 477.6	1 454.9	500.5	460.0	494.5	2 018.0	691.1	662.8	664.0	1.38	1.36	1.47	1.33
April	1 456.7	1 455.7	500.1	457.3	498.3	2 026.0	692.7	665.4	667.8	1.39	1.37	1.47	1.34
May	1 528.5	1 460.9	502.2	459.9	498.8	2 031.1	694.2	667.7	669.2	1.39	1.38	1.47	1.34
June	1 471.6	1 460.1	503.6	457.5	499.0	2 028.9	695.3	664.7	668.9	1.39	1.38	1.46	1.34
July	1 462.4	1 458.1	505.0	454.8	498.3	2 034.1	696.2	668.8	669.2	1.39	1.38	1.46	1.34
August	1 507.6	1 457.0	504.3	453.5	499.3	2 031.4	695.7	666.4	669.4	1.39	1.39	1.45	1.34
September	1 442.9	1 454.0	503.9	452.2	497.9	2 030.2	697.9	667.5	664.8	1.39	1.40	1.46	1.33
October	1 508.5	1 457.7	506.8	449.9	501.0	2 032.7	699.0	669.0	664.7	1.40	1.40	1.45	1.34
November	1 446.0	1 443.3	505.8	442.5	495.0	2 029.9	701.4	663.9	664.6	1.39	1.40	1.44	1.33
December	1 522.8	1 439.0	504.1	443.4	491.5	2 032.0	703.9	663.7	664.3	1.39	1.40	1.44	1.33

[1] Annual data are averages of monthly ratios.

Table 16-13. Real Manufacturing and Trade Sales and Inventories

(Billions of chained [2012] dollars, ratios; seasonally adjusted; annual sales figures are averages of seasonally adjusted monthly data.)

NIPA Tables 1BU, 2BU, 3BU

Year and month	Sales, monthly average				Inventories, end of period				Ratios, end-of-period inventories to monthly average sales			
	Total	Manufac-turing	Retail trade	Merchant wholesalers	Total	Manufac-turing	Retail trade	Merchant wholesalers	Total	Manufac-turing	Retail trade	Merchant wholesalers
2000	1 127.1	501.2	304.7	325.9	1 580.9	677.4	479.2	429.8	1.40	1.35	1.57	1.32
2001	1 111.0	473.4	311.7	328.0	1 525.2	644.5	466.5	418.6	1.37	1.36	1.50	1.28
2002	1 128.4	470.9	320.6	338.1	1 558.3	642.3	496.8	419.7	1.38	1.36	1.55	1.24
2003	1 150.6	471.9	332.8	345.9	1 570.5	631.6	515.8	421.3	1.37	1.34	1.55	1.22
2004	1 195.0	481.7	346.0	367.0	1 631.0	638.3	543.3	446.3	1.37	1.33	1.57	1.22
2005	1 248.2	503.5	358.9	385.9	1 689.2	665.4	549.3	472.6	1.35	1.32	1.53	1.23
2006	1 283.1	509.8	370.2	402.9	1 742.7	687.6	560.5	493.2	1.36	1.35	1.51	1.22
2007	1 311.3	522.0	375.4	414.0	1 781.6	709.1	568.2	502.9	1.36	1.36	1.51	1.22
2008	1 261.8	494.7	358.7	408.8	1 741.2	692.1	535.3	513.1	1.38	1.40	1.49	1.26
2009	1 144.5	428.0	338.5	377.6	1 604.0	662.5	478.7	462.6	1.40	1.55	1.41	1.23
2010	1 198.7	449.8	349.0	399.9	1 663.9	687.1	493.1	483.7	1.39	1.53	1.41	1.21
2011	1 238.6	465.5	359.8	413.3	1 709.4	711.7	494.6	503.1	1.38	1.53	1.38	1.22
2012	1 279.7	473.9	371.7	434.1	1 784.0	730.8	522.1	531.0	1.39	1.54	1.41	1.22
2013	1 319.9	483.2	387.1	449.6	1 878.7	758.4	558.9	561.2	1.42	1.57	1.44	1.25
2014	1 358.6	486.2	404.7	467.8	1 947.2	776.9	574.5	595.7	1.43	1.60	1.42	1.27
2015	1 389.5	482.4	428.6	478.8	2 042.7	815.3	604.2	623.0	1.47	1.69	1.41	1.30
2016	1 411.9	478.0	449.0	485.1	2 069.9	814.3	628.1	625.7	1.47	1.70	1.40	1.29
2017	1 446.8	483.0	466.7	497.5	2 096.4	813.8	641.4	638.8	1.45	1.69	1.37	1.28
2018	1 480.8	492.0	481.4	507.8	2 159.9	823.1	667.4	666.9	1.46	1.67	1.39	1.31
2019	1 514.9	498.3	502.2	515.3	2 204.9	843.8	681.8	676.6	1.46	1.69	1.36	1.31
2015												
January	1 382.5	481.9	421.4	479.6	1 954.9	780.4	575.5	598.9	1.41	1.62	1.37	1.25
February	1 376.9	484.4	418.7	474.1	1 967.8	786.0	578.2	603.5	1.43	1.62	1.38	1.27
March	1 384.6	486.8	423.9	474.0	1 975.7	790.2	579.8	605.8	1.43	1.62	1.37	1.28
April	1 395.1	488.0	424.8	482.6	1 987.6	795.1	584.9	607.6	1.43	1.63	1.38	1.26
May	1 381.1	479.7	426.2	475.5	1 993.4	798.8	583.6	611.3	1.44	1.67	1.37	1.29
June	1 383.8	482.7	426.2	475.1	2 005.3	805.2	586.2	614.3	1.45	1.67	1.38	1.29
July	1 386.8	482.1	429.2	475.8	2 011.3	806.4	590.5	614.6	1.45	1.67	1.38	1.29
August	1 388.7	481.0	430.8	477.1	2 017.8	808.3	592.8	616.7	1.45	1.68	1.38	1.29
September	1 403.1	484.0	433.8	485.7	2 028.5	808.9	598.7	620.6	1.45	1.67	1.38	1.28
October	1 399.9	481.4	433.1	485.8	2 037.1	812.6	601.3	622.9	1.46	1.69	1.39	1.28
November	1 396.9	481.6	435.6	480.1	2 037.6	813.9	601.6	621.9	1.46	1.69	1.38	1.30
December	1 394.9	475.7	439.6	479.9	2 042.7	815.3	604.2	623.0	1.46	1.71	1.37	1.30
2016												
January	1 396.2	477.9	439.1	479.4	2 045.3	814.3	605.7	624.9	1.47	1.70	1.38	1.30
February	1 406.1	480.0	446.8	479.5	2 043.5	814.0	607.6	621.4	1.45	1.70	1.36	1.30
March	1 408.0	480.0	444.7	483.5	2 052.7	816.5	612.2	623.4	1.46	1.70	1.38	1.29
April	1 406.6	479.0	443.8	484.1	2 057.5	816.9	613.3	626.5	1.46	1.71	1.38	1.29
May	1 398.4	475.2	445.5	477.8	2 057.0	815.8	615.3	624.9	1.47	1.72	1.38	1.31
June	1 406.1	476.2	449.0	481.0	2 054.9	814.0	617.0	622.8	1.46	1.71	1.37	1.30
July	1 406.0	473.3	451.0	481.9	2 051.1	812.7	616.1	621.1	1.46	1.72	1.37	1.29
August	1 415.0	476.1	451.2	488.1	2 051.9	813.2	618.6	618.9	1.45	1.71	1.37	1.27
September	1 419.2	478.2	453.1	488.1	2 056.0	811.4	623.4	619.6	1.45	1.70	1.38	1.27
October	1 421.6	477.1	453.2	491.7	2 054.8	812.2	621.4	619.6	1.45	1.70	1.37	1.26
November	1 421.1	477.3	453.0	491.1	2 065.9	815.0	625.9	623.4	1.45	1.71	1.38	1.27
December	1 438.4	486.1	457.8	494.7	2 069.9	814.3	628.1	625.7	1.44	1.68	1.37	1.27
2017												
January	1 433.7	482.6	458.5	492.8	2 069.6	813.1	631.7	622.9	1.44	1.69	1.38	1.26
February	1 431.0	481.6	459.0	490.6	2 066.9	810.4	631.8	622.6	1.44	1.68	1.38	1.27
March	1 429.8	480.9	459.9	489.3	2 068.2	807.9	634.2	623.8	1.45	1.68	1.38	1.28
April	1 427.9	476.4	462.7	489.0	2 063.5	807.5	631.9	621.9	1.45	1.70	1.37	1.27
May	1 431.3	479.9	463.3	488.3	2 067.2	805.8	634.3	624.7	1.44	1.68	1.37	1.28
June	1 436.6	479.6	465.7	491.7	2 074.1	804.9	637.9	628.8	1.44	1.68	1.37	1.28
July	1 442.5	480.3	466.7	495.8	2 077.9	808.0	636.4	631.0	1.44	1.68	1.36	1.27
August	1 448.2	482.7	464.8	501.1	2 086.8	810.1	639.8	634.4	1.44	1.68	1.38	1.27
September	1 456.2	482.2	472.0	502.5	2 088.1	813.0	635.7	637.3	1.43	1.69	1.35	1.27
October	1 464.6	486.1	473.7	505.2	2 086.1	813.2	636.7	634.0	1.42	1.67	1.34	1.26
November	1 476.1	492.2	476.1	508.2	2 093.4	813.2	640.0	637.9	1.42	1.65	1.34	1.26
December	1 483.7	490.9	478.2	515.0	2 096.4	813.8	641.4	638.8	1.41	1.66	1.34	1.24
2018												
January	1 468.6	491.5	475.1	502.4	2 103.6	814.0	643.4	643.9	1.43	1.66	1.35	1.28
February	1 472.0	490.9	476.4	505.1	2 112.7	815.0	647.8	647.5	1.44	1.66	1.36	1.28
March	1 471.6	491.7	477.2	503.1	2 111.5	813.7	646.7	648.7	1.44	1.66	1.36	1.29
April	1 470.4	489.2	478.0	503.6	2 113.5	815.3	649.2	646.5	1.44	1.67	1.36	1.28
May	1 479.1	487.2	482.7	509.9	2 114.5	815.6	650.5	646.0	1.43	1.67	1.35	1.27
June	1 476.6	490.3	480.0	506.8	2 113.7	815.5	651.9	643.8	1.43	1.66	1.36	1.27
July	1 479.2	489.4	482.5	507.9	2 122.3	820.2	653.9	645.8	1.44	1.68	1.36	1.27
August	1 486.4	493.3	481.6	511.9	2 129.5	819.3	656.9	650.8	1.43	1.66	1.36	1.27
September	1 489.3	495.5	482.5	511.8	2 135.2	823.2	655.5	654.1	1.43	1.66	1.36	1.28
October	1 489.7	493.6	487.6	509.0	2 143.8	823.9	659.6	657.8	1.44	1.67	1.35	1.29
November	1 493.3	493.9	491.2	508.9	2 143.2	822.1	659.2	659.5	1.44	1.67	1.34	1.30
December	1 493.0	498.1	482.1	513.1	2 159.9	823.1	667.4	666.9	1.45	1.65	1.38	1.30
2019												
January	1 509.2	502.6	492.3	514.8	2 174.3	826.9	670.8	674.1	1.44	1.65	1.36	1.31
February	1 509.6	503.8	490.7	515.5	2 183.4	829.3	674.9	676.7	1.45	1.65	1.38	1.31
March	1 512.2	499.7	497.2	516.0	2 182.9	832.0	672.7	675.8	1.44	1.67	1.35	1.31
April	1 498.6	492.3	497.7	509.6	2 185.7	832.7	672.3	678.3	1.46	1.69	1.35	1.33
May	1 502.2	493.3	499.4	510.5	2 189.7	833.8	674.1	679.3	1.46	1.69	1.35	1.33
June	1 515.7	499.6	501.6	515.5	2 192.0	836.3	673.5	679.8	1.45	1.67	1.34	1.32
July	1 515.4	497.1	504.7	514.7	2 200.4	838.8	678.1	680.9	1.45	1.69	1.34	1.32
August	1 524.6	498.8	508.6	518.3	2 200.3	838.9	677.5	681.3	1.44	1.68	1.33	1.31
September	1 523.7	498.6	506.3	520.0	2 202.1	841.7	680.8	677.0	1.45	1.69	1.35	1.30
October	1 517.8	496.8	508.8	513.4	2 203.5	842.8	681.1	677.0	1.45	1.70	1.34	1.32
November	1 525.6	497.4	510.2	519.3	2 202.1	843.5	679.2	676.9	1.44	1.70	1.33	1.30
December	1 524.3	500.0	509.0	516.5	2 204.9	843.8	681.8	676.6	1.45	1.69	1.34	1.31

Table 16-14. Selected Services—Quarterly Estimated Revenue for Employer Firms

(Millions of dollars.)

2012 NAICS code	Kind of business	4th quarter 2018	1st quarter 2019	2nd quarter 2019	3rd quarter 2019	4th quarter 2019	1st quarter 2020
SEASONALLY ADJUSTED							
51	**Information**	416 562	421 825	428 804	435 422	440 757	441 759
5112	Software publishers	65 427	68 162	69 865	72 652	73 855	76 299
512	Motion picture and sound recording industries	27 729	26 899	28 453	27 566	27 080	24 363
54	**Professional, Scientific, and Technical Services**	483 568	487 418	496 306	508 564	509 136	508 317
5411	Legal services	76 548	78 382	77 876	81 571	81 688	81 952
5412	Accounting, tax preparation, bookkeeping, and payroll services	43 697	44 042	43 345	44 953	44 799	43 810
56 pt	**Administrative and Support and Waste Management and Remediation Services**	262 849	260 934	260 985	265 253	269 630	266 255
5613	Employment services	107 890	107 658	105 576	108 008	108 451	106 494
5615	Travel arrangement and reservation services	13 404	13 425	13 360	13 259	13 385	10 616
562	Waste management and remediation services	27 203	27 218	27 654	27 495	27 325	27 686
622	**Hospitals**	290 685	301 656	305 737	306 778	312 488	288 989
NOT SEASONALLY ADJUSTED							
51	**Information**	435 724	412 545	424 516	429 326	461 032	432 271
511	Publishing industries (except Internet)	92 887	87 976	91 272	92 007	100 128	94 915
51111	Newspaper publishers	6 550	5 734	5 900	5 650	6 155	. . .
51112	Periodical publishers	7 189	6 455	6 869	6 531	6 701	. . .
5111 pt	Book, directory and mailing list, and other publishers	9 665	8 102	9 057	10 516	9 060	. . .
5112	Software publishers	69 483	67 685	69 446	69 310	78 212	76 015
512	Motion picture and sound recording industries	28 977	26 630	28 510	26 601	28 244	23 731
515	Broadcasting (except Internet)	45 683	40 445	41 360	40 405	45 966	. . .
5151	Radio and television broadcasting	23 735	19 682	20 357	19 794	23 934	. . .
5152	Cable and other subscription programming	21 948	20 763	21 003	20 611	22 032	. . .
517	Telecommunications [1]	162 874	157 794	158 655	159 812	164 375	157 163
5171	Wired telecommunications carriers	78 926	77 426	78 079	77 870	78 751	76 819
5172	Wireless telecommunications carriers (except satellite)	71 835	68 343	68 139	69 302	72 769	67 974
54	**Professional, Scientific, and Technical Services** [1]	493 723	476 207	500 276	505 513	519 828	496 517
5411	Legal services	86 040	69 917	76 630	81 571	91 899	73 546
5412	Accounting, tax preparation, bookkeeping, and payroll services	38 235	52 102	45 339	40 323	39 289	51 870
5413	Architectural, engineering, and related services	82 107	81 651	88 235	90 745	89 841	88 224
5415	Computer systems design and related services	112 855	108 634	114 101	115 073	117 024	114 216
5416	Management, scientific, and technical consulting services	71 165	67 403	72 005	75 325	75 713	70 570
5417	Scientific research and development services	43 810	42 259	45 368	44 347	44 803	. . .
5418	Advertising and related services	31 142	26 830	29 089	28 605	30 795	27 446
56	**Administrative and Support and Waste Management and Remediation Services** [1]	267 473	253 353	261 532	267 616	274 857	259 716
561	Administrative and support services	240 270	227 605	233 187	239 241	247 614	233 367
5613	Employment services	113 500	106 689	101 775	107 144	114 416	. . .
5615	Travel arrangement and reservation services	12 761	12 901	13 814	13 975	12 729	. . .
562	Waste management and remediation services	27 203	25 748	28 345	28 375	27 243	26 349
61	**Educational Services**	17 973	18 143	18 765	18 269	18 938	17 389
62	**Health Care and Social Asistance**	678 916	679 837	697 342	691 165	718 824	680 624
71	**Arts, Entertainment, and Recreation**	72 945	67 046	75 680	79 846	77 694	60 181
81	**Other services (Except Public Administration)**	153 808	150 198	151 296	150 955	181 343	128 860

. . . = Not available.
[1]Includes components not shown separately.

NOTES AND DEFINITIONS, CHAPTER 16

TABLE 16-1

Construction Put in Place

SOURCE: U.S. DEPARTMENT OF COMMERCE, CENSUS BUREAU

The Census Bureau's estimates of the value of new construction put in place are intended to provide monthly estimates of the total dollar value of construction work done in the United States.

The United States Code, Title 13, authorizes the Survey of Construction and provides for voluntary responses. Construction statistics are collected from building permits, housing starts and housing completions.

Definitions and notes on the data

The estimates cover all construction work done each month on new private residential and nonresidential buildings and structures, public construction, and improvements to existing buildings and structures. Included are the cost of labor, materials, and equipment rental; cost of architectural and engineering work; overhead costs assigned to the project; interest and taxes paid during construction; and contractor's profits.

The total value put in place for a given period is the sum of the value of work done on all projects underway during this period, regardless of when work on each individual project was started or when payment was made to the contractors. For some categories, estimates are derived by distributing the total construction cost of the project by means of historic construction progress patterns. Published estimates represent payments made during a period for some categories.

The statistics on the value of construction put in place result from direct measurement and indirect estimation. A series results from direct measurement when it is based on reports of the actual value of construction progress or construction expenditures obtained in a complete census or a sample survey. All other series are developed by indirect estimation using related construction statistics. On an annual basis, estimates for series directly measured monthly, quarterly, or annually accounted for about 71 percent of total construction in 1998 (private multifamily residential, private residential improvements, private nonresidential buildings, farm nonresidential construction, public utility construction, all other private construction, and virtually all of public construction). On a monthly basis, directly measured data are available for about 55 percent of the value in place estimates.

Beginning in 1993, the Construction Expenditures Branch of the Census Bureau's Manufacturing and Construction Division began collecting these data using a new classification system, which bases project types on their end usage instead of on building/nonbuilding types. Data collection on this system for federal construction began in January 2002.

With the changes in project classifications, data presented in these tables for 1993 to date are not directly comparable with data for previous years, except at aggregate levels. For that reason, *Business Statistics* shows earlier historical data only at these aggregate levels. Although some categories, such as lodging, office, education, and religion, have the same names as categories in previously published data, there have been changes within the classifications that make these values non-comparable. For example, private medical office buildings were classified as "office" buildings previously, but are categorized as "health care" under the new classification.

The seasonally adjusted data are obtained by removing normal seasonal movement from the unadjusted data to bring out underlying trends and business cycles, which is accomplished by using the Census X-12-ARIMA method. Seasonal adjustment accounts for month-to-month variations resulting from normal or average changes in any phenomena affecting the data, such as weather conditions, the differing lengths of months, and the varying number of holidays, weekdays, and weekends within each month. It does not adjust for abnormal conditions within each month or for year-to-year variations in weather. The seasonally adjusted annual rate is the seasonally adjusted monthly rate multiplied by 12.

Residential consists of new houses, town houses, apartments, and condominiums for sale or rent; these dwellings are built by the owner or for the owner on contract. It includes improvements inside and outside residential structures, such as remodeling, additions, major replacements, and additions of swimming pools and garages. Manufactured housing, houseboats, and maintenance and repair work are not included.

Office includes general office buildings, administration buildings, professional buildings, and financial institution buildings. Office buildings at manufacturing sites are classified as *manufacturing,* but office buildings owned by manufacturing companies but not at such a site are included in the *office* category. In the state and local government category, *office* includes capitols, city halls, courthouses, and similar buildings.

Commercial includes buildings and structures used by the retail, wholesale, farm, and selected service industries. One of the subgroups of this category is *multi-retail,* which consists of department and variety stores, shopping centers and malls, and warehouse-type retail stores.

Health care includes hospitals, medical buildings, nursing homes, adult day-care centers, and similar institutions.

Educational includes schools at all levels, higher education facilities, trade schools, libraries, museums, and similar institutions.

Amusement and recreation includes theme and amusement parks, sports structures not located at educational institutions, fitness centers and health clubs, neighborhood centers, camps, movie theaters, and similar establishments.

Transportation includes airport facilities; rail facilities, track, and bridges; bus, rail, maritime, and air terminals; and docks, marinas, and similar structures.

Communication includes telephone, television, and radio distribution and maintenance structures.

Power includes electricity production and distribution and gas and crude oil transmission, storage, and distribution.

Manufacturing includes all buildings and structures at manufacturing sites but not the installation of production machinery or special-purpose equipment.

Included in *total private construction*, but not shown separately in these pages, are lodging facilities (hotels and motels), religious structures, and private public safety, sewage and waste disposal, water supply, highway and street, and conservation and development spending.

Included in *total state and local construction,* but not shown separately in these pages, are state and local construction of commercial buildings, conservation and development (dams, levees, jetties, and dredging), lodging, religious facilities, and communication structures.

Public safety includes correctional facilities, police and sheriffs' stations, fire stations, and similar establishments.

Highway and street includes pavement, lighting, retaining walls, bridges, tunnels, toll facilities, and maintenance and rest facilities.

Sewage and waste disposal includes sewage systems, solid waste disposal, and recycling.

Water supply includes water supply, transmission, and storage facilities.

Among the data sources for construction expenditures are the Census Bureau's Survey of Construction, Building Permits Survey, Consumer Expenditure Survey (conducted for the Bureau of Labor Statistics), Annual Capital Expenditures Survey, and Construction Progress Reporting Survey; also included are data from the F.W. Dodge Division of the McGraw-Hill Information Systems Company, the U.S. Department of Agriculture, and utility regulatory agencies.

Data availability

Each month's "Construction Spending" press release is released on the last workday of the following month. The release, more detailed data, and a discussion of methodologies can be found on the Census Bureau's Web site at <http://www.census.gov/constructionspending>. Current data at this site may reflect benchmarking since Table 16-1 was compiled.

Table 16-2

Housing Starts and Building Permits; New House Sales and Prices

SOURCES: U.S. DEPARTMENT OF COMMERCE, CENSUS BUREAU

These data are mainly found in two major Census Bureau reports, "New Residential Construction" and "New Residential Sales." They cover new housing units intended for occupancy and maintained by the occupants, excluding hotels, motels, and group residential structures. Manufactured home units are reported in a separate survey.

Definitions

A *housing unit* is a house, an apartment, or a group of rooms or single room intended for occupancy as separate living quarters. Occupants must live separately from other individuals in the building and have direct access to the housing unit from the outside of the building or through a common hall. Each apartment unit in an apartment building is counted as one housing unit. As of January 2000, a previous requirement for residents to have the capability to eat separately has been eliminated. (Based on the old definition, some senior housing projects were excluded from the multifamily housing statistics because individual units did not have their own eating facilities.) Housing starts exclude group quarters such as dormitories or rooming houses, transient accommodations such as motels, and manufactured homes. Publicly owned housing units are excluded, but units in structures built by private developers with subsidies or for sale to local public housing authorities are both classified as private housing.

The *start* of construction of a privately owned housing unit is when excavation begins for the footings or foundation of a building primarily intended as a housekeeping residential structure and designed for nontransient occupancy. All housing units in a multifamily building are defined as being started when excavation for the building begins.

One-family structures includes fully detached, semi-detached, row houses, and townhouses. In the case of attached units, each must be separated from the adjacent unit by a ground-to-roof wall to be classified as a one-unit structure and must not share facilities such as heating or water supply. Units built one on top of another and those built side-by-side without a ground-to-roof wall and/or with common facilities are classified by the number of units in the structure.

Apartment buildings are defined as buildings containing *five units or more.* The type of ownership is not the criterion—a condominium apartment building is not classified as one-family structures but as a multifamily structure.

A *manufactured* home is a moveable dwelling, 8 feet or more wide and 40 feet or more long, designed to be towed on its own chassis with transportation gear integral to the unit when it leaves the factory, and without need of a permanent foundation. Multiwides and expandable manufactured homes are included. Excluded are travel trailers, motor homes, and modular housing. The shipments figures are based on reports submitted by manufacturers on the number of homes actually shipped during the survey month. Shipments to dealers may not necessarily be placed for residential use in the same month as they are shipped. The number of manufactured "homes" used for nonresidential purposes (for example, those used for offices) is not known.

Units authorized by building permits represents the approximately 97 percent of housing in permit-requiring areas.

The *start* occurs when excavation begins for the footing or foundation. Starts are estimated for all areas, regardless of whether permits are required.

New house *sales* are reported only for new single-family residential structures. The sales transaction must intend to include both house and land. Excluded are houses built for rent, houses built by the owner, and houses built by a contractor on the owner's land. A sale is reported when a deposit is taken or a sales agreement is signed; this can occur prior to a permit being issued.

Once the sale is reported, the sold housing unit drops out of the survey. Consequently, the Census Bureau does not find out if the sales contract is cancelled or if the house is ever resold. As a result, if conditions are worsening and cancellations are high, sales are temporarily overestimated. When conditions improve and the cancelled sales materialize as actual sales, the Census sales estimates are then underestimated because the case did not re-enter the survey. In the long run, cancellations do not cause the survey to overestimate or underestimate sales; but in the short run, cancellations and ultimate resales are not reflected in this survey, and fluctuations can appear less severe than in reality.

A house is *for sale* when a permit to build has been issued (or work begun in non-permit areas) and a sales contract has not been signed nor a deposit accepted.

The *sales price* used in this survey is the price agreed upon between the purchaser and the seller at the time the first sales contract is signed or deposit made. It includes the price of the improved lot. The *median sales price* is the sales price of the house that falls on the middle point of a distribution by price of the total number of houses sold. Half of the houses sold have a price lower than the median; half have a price higher than the

median. Changes in the *sales price* data reflect changes in the distribution of houses by region, size, and the like, as well as changes in the prices of houses with identical characteristics.

The *price index* measures the change in price of a new single-family house of constant physical characteristics, using the characteristics of houses built in 2005. Characteristics held constant include floor area, whether inside or outside a metropolitan area, number of bedrooms, number of bathrooms, number of fireplaces, type of parking facility, type of foundation, presence of a deck, construction method, exterior wall material, type of heating, and presence of air-conditioning. The indexes are calculated separately for attached and detached houses and combined with base period weights. The price measured includes the value of the lot.

Notes on the data

Monthly permit authorizations are based on data collected by a mail survey from a sample of about 9,000 permit-issuing places, selected from and representing a universe of 20,000 such places in the United States. The remaining places are surveyed annually. Data for 2014 to the present includes approximately 20,100 places while data for 2004 to 2013 included 19,300 places.

Data for 1994 through 2003 represented 19,000 places; data for 1984 through 1993 represented 17,000 places; data for 1978 through 1983 represented 16,000 places; data for 1972 through 1977 represented 14,000 places; data for 1967 through 1971 represented 13,000 places; data for 1963 through 1966 represented 12,000 places; and data for 1959 through 1962 represented 10,000 places.

Housing starts and sales data are obtained from the Survey of Construction, for which Census Bureau field representatives sample both permit-issuing and non-permit-issuing places.

Effective with the January 2005 data release, the Survey of Construction implemented a new sample of building permit offices, replacing a previous sample selected in 1985. As a result, writes the Census Bureau, "Data users should use caution when analyzing year over year changes in housing prices and characteristics between 2004 and 2005." In the newer sample, land may be more abundant, lot sizes larger, and sales prices lower.

For 2004, the permit data were compiled for both the new 20,000 place universe and the old 19,000 place universe. Ratios of the new estimates to the old were calculated by state for total housing units, structures by number of units, and valuation. For the United States as a whole, the new estimate was 100.9 percent of the old estimate.

Effective with the data for April 2001, the Census Bureau made changes to the methodology used for new house sales, including discontinuing an adjustment for construction in areas in which building permits are required without a permit being issued. It was believed that such unauthorized construction has virtually ceased. The upward adjustment was not phased out but dropped completely in revised estimates as of January 1999. The total effect of these changes was to lower the number of sales by about 2.9 percent relative to those published for earlier years.

The data used in the price index are collected in the Survey of Construction, through monthly interviews with the builders or owners. The size of the sample is currently about 20,000 observations per year.

Data availability and references

Housing starts and building permit data have been collected monthly by the Bureau of the Census since 1959.

The monthly report for "New Residential Construction" (permits, starts, and completions) is issued in the middle of the following month. The monthly report and associated descriptions and historical data can be found at <http://www.census.gov/construction/nrc>.

The monthly report for "New Residential Sales" (sales, houses for sale, and prices) is issued toward the end of the following month. The monthly report and associated descriptions and historical data can be found at <http://www.census.gov/construction/nrs>.

The manufactured housing data (not seasonally adjusted) and background information can be found at <http://www.manu-facturedhousing.org/statistics>. Data with and without seasonal adjustment can be found at <http://www.census.gov/construction/mhs>.

Data and background on the price index for new one-family houses can be found at <http://www.census/gov>, in the alphabetical index under the category "Construction price indexes."

TABLES 16-3 THROUGH 16-6

Manufacturers' Shipments, Inventories, and Orders

SOURCE: U.S. DEPARTMENT OF COMMERCE, CENSUS BUREAU

These data are from the Census Bureau's monthly M3 survey, a sample-based survey that provides measures of changes in the value of domestic manufacturing activity and indications of future production commitments. The sample is not a probability sample. It includes approximately 4,300 reporting units, including companies with $500 million or more in annual shipments and a selection of smaller companies. Currently, reported monthly data represent approximately 60 percent of shipments at the total manufacturing level.

One important technology industry, semiconductors, is represented in the shipments and inventories data in this report but not in new or unfilled orders. This affects the new and unfilled orders

totals for computers and electronic products, durable goods industries, and total manufacturing. Based on shipments data, semiconductors accounted for about 15 percent of computers and electronic products, 3 percent of durable goods industries, and 1.5 percent of total manufacturing. Since semiconductors are intermediate materials and components rather than finished final products, the absence of these data does not distort new and unfilled orders data for important final demand categories, such as capital goods and information technology.

Definitions and notes on the data

Shipments. The value of shipments data represent net selling values, f.o.b. (free on board) plant, after discounts and allowances and excluding freight charges and excise taxes. For multi-establishment companies, the M3 reports are typically company- or division-level reports that encompass groups of plants or products. The data reported are usually net sales and receipts from customers and do not include the value of interplant transfers. The reported sales are used to calculate month-to-month changes that bring forward the estimates for the entire industry (that is, estimates of the statistical "universe") that have been developed from the Annual Survey of Manufactures (ASM). The value of products made elsewhere under contract from materials owned by the plant is also included in shipments, along with receipts for contract work performed for others, resales, miscellaneous activities such as the sale of scrap and refuse, and installation and repair work performed by employees of the plant.

Inventories. Inventories in the M3 survey are collected on a current cost or pre-LIFO (last in, first out) basis. As different inventory valuation methods are reflected in the reported data, the estimates differ slightly from replacement cost estimates. Companies using the LIFO method for valuing inventories report their pre-LIFO value; the adjustment to their base-period prices is excluded. In the ASM, inventories are collected according to this same definition.

Inventory data are requested from respondents by three stages of fabrication: finished goods, work in process, and raw materials and supplies. Response to the stage of fabrication inquiries is lower than for total inventories; not all companies keep their monthly data at this level of detail. It should be noted that a product considered to be a finished good in one industry, such as steel mill shapes, may be reported as a raw material in another industry, such as stamping plants. For some purposes, this difference in definitions is an advantage. When a factory accumulates inventory that it considers to be raw materials, it can be expected that that accumulation is intentional. But when a factory—whether a materials-making or a final-product producer—has a buildup of finished goods inventories, it may indicate involuntary accumulation as a result of sales falling short of expectations. Hence, the two types of accumulation can have different economic interpretations, even if they represent identical types of goods.

Like total inventories, stage of fabrication inventories are benchmarked to the ASM data. Stage of fabrication data are benchmarked at the major group level, as opposed to the level of total inventories, which is benchmarked at the individual industry level.

New orders (durable goods), as reported in the monthly survey, are net of order cancellations and include orders received and filled during the month as well as orders received for future delivery. They also include the value of contract changes that increase or decrease the value of the unfilled orders to which they relate. Orders are defined to include those supported by binding legal documents such as signed contracts, letters of award, or letters of intent, although this definition may not be strictly applicable in some industries.

New orders (nondurable goods) are equal to shipments, as order backlogs are not reported for these industries.

Unfilled orders (durable goods) includes new orders (as defined above) that have not been reflected as shipments. Generally, unfilled orders at the end of the reporting period are equal to unfilled orders at the beginning of the period plus net new orders received less net shipments.

Series are adjusted for seasonal variation and variation in the number of trading days in the month using the X-12-ARIMA version of the Census Bureau's seasonal adjustment program.

Benchmarking and revisions

The data shown in the volume have been benchmarked to the 2007 Economic Census and to ASMs through 2011. In each benchmark revision, new and unfilled orders are adjusted to be consistent with the benchmarked shipments and inventory data, seasonal adjustment factors are revised and updated, and other corrections are made.

Data availability and references

Data have been collected monthly since 1958.

The "Advance Report on Durable Goods Manufacturers' Shipments, Inventories and Orders" is available as a press release about 18 working days after the end of each month. It includes seasonally adjusted and not seasonally adjusted estimates of shipments, new orders, unfilled orders, and inventories for durable goods industries.

The monthly "Manufacturers' Shipments, Inventories, and Orders" report is released on the 23rd working day after the end of the month. Content includes revisions to the advance durable goods data, estimates for nondurable goods industries, tabulations by market category, and ratios of shipments to inventories and to unfilled orders. Revisions may affect selected data for the two previous months.

Press releases, historical data, descriptions of the survey, and documentation are available on the Census Bureau Web site at <http://www.census.gov>, under the category "Manufacturing" in the alphabetic listing there.

TABLE 16-7

Motor Vehicle Sales and Inventories

SOURCE: U.S. DEPARTMENT OF COMMERCE, BUREAU OF ECONOMIC ANALYSIS

The Bureau of Economic Analysis (BEA) collects data on retail sales and inventories of cars, trucks, and buses from data from the American Automobile Manufacturers Association, Ward's Automotive Reports, and other sources. Seasonal adjustments are recalculated annually. Data are available on the BEA Web site at <http://www.bea.gov> as a part of the national income and product accounts data set; they are found under the "Supplemental Estimates" heading. They are also available on the STAT-USA subscription Web site at <http://www.stat-usa.gov>.

Definitions

Product	Description
Autos	All passenger cars, including station wagons.
Light trucks	Sport utility vehicles. Prior to the 2003 Benchmark Revision light trucks were up to 10,000 pounds.
Heavy trucks	Trucks more than 14,000 pounds gross vehicle weight. Prior to the 2003 Benchmark Revision heavy trucks were more than 10,000 pounds
Domestic sales	United States (U.S.) sales of vehicles assembled in the U.S., Canada, and Mexico
Foreign sales	U.S. sales of vehicles produced elsewhere
Domestic auto production	Autos assembled in the U.S
Domestic auto inventories	U.S. inventories of vehicles assembled in the U.S., Canada, and Mexico.

Data Availability

Auto and truck sales are updated by the end of the third business day after the end of the month. Auto production and inventories are available by the end of the second week after the end of month.

TABLES 16-8 AND 16-10

RETAIL AND FOOD SERVICES SALES; RETAIL INVENTORIES

SOURCE: U.S. DEPARTMENT OF COMMERCE, CENSUS BUREAU

Every month, the Census Bureau prepares estimates of retail sales and inventories by kind of business, based on a mail-out/

mail-back survey of about 5,500 retail businesses with paid employees.

In June 2019, the monthly retail sales estimate is based on the results of the 2017 Annual Trade Survey and the service annual service. The annual retail trade survey (ARTS) produces national estimates of total annual sales, e-commerce sales, sales taxes, end-of-year inventories, purchases, total operating expenses, gross margins, and end-of-year accounts receivable for retail businesses located in the united states.

Retail sales and inventories are now compiled using the new NAICS classification system, which replaced the old SIC system. Historical data have been restated on the NAICS basis back to January 1992. In NAICS, Eating and drinking places and Mobile food services have been reclassified out of retail trade and into sector 72, Accommodation and food services, which also includes Hotels. The retail sales survey still collects and publishes sales data for Food services and drinking places. It no longer includes them in the Retail total, but they are included in a new Retail and food services total.

Subtotals of durable and nondurable goods are no longer published. They were always imprecise for retail sales, since general merchandise stores (including department stores) were included in nondurable goods, yet obviously sold substantial quantities of durable goods.

Definitions

Sales is the value of merchandise sold for cash or credit at retail or wholesale. Services that are incidental to the sale of merchandise, and excise taxes that are paid by the manufacturer or wholesaler and passed along to the retailer, are also included. Sales are net, after deductions for refunds and merchandise returns. They exclude sales taxes collected directly from customers and paid directly to a local, state, or federal tax agency. The sales estimates include only sales by establishments primarily engaged in retail trade, and are not intended to measure the total sales for a given commodity or merchandise line.

Inventories is the value of stocks of goods held for sale through retail stores, valued at cost, as of the last day of the report period. Stocks may be held either at the store or at warehouses that maintain supplies primarily intended for distribution to retail stores within the organization.

Leased departments consists of the operations of one company conducted within the establishment of another company, such as jewelry counters or optical centers within department stores. The values for sales and inventories at department stores in Tables 16-9 and 16-10 exclude sales of leased departments.

GAFO (department store type goods) is a special aggregate grouping of sales at general merchandise stores and at other stores that sell merchandise normally sold in department stores—clothing and accessories, furniture and home furnishings, electronics,

appliances, sporting goods, hobby, book, music, office supplies, stationery, and gifts.

Notes on the data

The data published here have been benchmarked to the 2012, 2007, 2002, 1997, and 1992 Economic Censuses and the Annual Retail Trade Surveys for 2012 and previous years. Each year, the monthly series are benchmarked to the latest annual survey and new factors are incorporated to adjust for seasonal, trading-day, and holiday variations, using the Census Bureau's X-12-ARIMA program.

The survey sample is stratified by kind of business and estimated sales. All firms with sales above applicable size cutoffs are included. Firms are selected randomly from the remaining strata. The sample used for the end-of-month inventory estimates is a sub-sample of the monthly sales sample, about one-third of the size of the whole sample.

New samples, designed to produce NAICS estimates, were introduced with the 1999 Annual Retail Trade Survey and the March 2001 Monthly Retail Trade Survey. On November 30, 2006, another new sample was introduced, affecting the data for September 2006 and the following months. The sample is updated quarterly to take account of business births and deaths.

Data availability and references

An "Advance Monthly Retail Sales" report is released about nine working days after the close of the reference month, based on responses from a sub-sample of the complete retail sample.

The revised and more complete monthly "Retail Trade, Sales, and Inventories" reports are released six weeks after the close of the reference month. They contain preliminary figures for the current month and final figures for the prior 12 months. Statistics include retail sales, inventories, and ratios of inventories to sales. Data are both seasonally adjusted and unadjusted.

The "Annual Benchmark Report for Retail Trade" is released each spring. It includes updated seasonal adjustment factors; revised and benchmarked monthly estimates of sales and inventories; monthly data for the most recent 10 or more years; detailed annual estimates and ratios for the United States by kind of business; and comparable prior-year statistics and year-to-year changes.

All data are available on the Census Bureau Web site at <http://www.census.gov/retail>.

TABLE 16-9

Quarterly Retail Sales: Total and E-Commerce

SOURCE: U.S. DEPARTMENT OF COMMERCE, CENSUS BUREAU

Beginning with the fourth quarter of 1999, the Census Bureau has conducted a quarterly survey of retail e-commerce sales from the Monthly Retail Trade Survey sample. (The monthly survey does not report electronic shopping separately; it is combined with mail order.)

E-commerce sales are the sales of goods and services in which an order is placed by the buyer or the price and terms of sale are negotiated over the Internet or an extranet, Electronic Data Interchange (EDI) network, electronic mail, or other online system. Payment may or may not be made online. The quarterly release is issued around the 20th of February, May, August, and November, and is available along with full historical data on the Census Bureau Web site at <http://www.census.gov>. It can be located under the heading "Retail" in the alphabetical Web site index, under "E" for "Economic data and information."

These estimates reflect the NAICS definition of retail sales, which excludes food service. Online travel services, financial brokers and dealers, and ticket sales agencies are not classified as retail and are not included in these estimates; they are, however, included in the annual survey of selected services.

TABLE 16-11

Merchant Wholesalers—Sales and Inventories

SOURCE: U.S. DEPARTMENT OF COMMERCE, CENSUS BUREAU

The Annual Wholesale Trade Survey (AWTS) produces national estimates of total annual sales, e-commerce sales, end-of-year inventories, inventories held outside of the United States, purchases, total operating expenses, gross margins, and commissions (for electronic markets, agents, and brokers) for wholesale businesses located in the United States. The sample consists of about 8400 establishments, with a response rate of about 62 percent; missing reports are imputed based on reports of similar reporters.

These data are now based on the new NAICS classification system, which replaced the old SIC system. Historical data have been restated on the NAICS basis back to January 1992.

Classification changes in NAICS

NAICS shifts a significant number of businesses from the Wholesale to the Retail sector. An important new criterion for classification concerns whether or not the establishment is intended to solicit walk-in traffic. If it is, and if it uses mass-media advertising, it is now classified as Retail, even if it also serves business and institutional clients.

Definitions

Merchant wholesalers includes merchant wholesalers that take title of the goods they sell, as well as jobbers, industrial distributors, exporters, and importers. The survey does not cover marketing sales offices and branches of manufacturing, refining, and

mining firms, nor does it include NAICS 4251: Wholesale Electronic Markets and Agents and Brokers.

Notes on the data

Inventories are valued using methods other than LIFO (last in, first out) in order to better reflect the current costs of goods held as inventory.

A survey has been conducted monthly since 1946. New samples are drawn every 5 years, most recently in 2009. The samples are updated every quarter to add new businesses and to drop companies that are no longer active.

Data availability and references

"Monthly Wholesale Trade, Sales and Inventories" reports are released six weeks after the close of the reference month. They contain preliminary current-month figures and final figures for the previous month. Statistics include sales, inventories, and stock/sale ratios, along with standard errors. Data are both seasonally adjusted and unadjusted.

The "Annual Benchmark Report for Wholesale Trade" is released each spring. It contains estimated annual sales, monthly and yearend inventories, inventory/sales ratios, purchases, gross margins, and gross margin/sales ratios by kind of business. Annual estimates are benchmarked to annual surveys and the most recent census of wholesale trade. Monthly sales and inventories estimates are revised consistent with the annual data, seasonal adjustment factors are updated, and revised data for both seasonally adjusted and unadjusted values are published.

Data and documentation are available on the Census Bureau Web site at <http://www.census.gov>, under "Economic Indicators" and in the alphabetic index under "E" for economic data.

Tables 16-12 and 16-13

Manufacturing and Trade Sales and Inventories

SOURCES: U.S. DEPARTMENT OF COMMERCE, CENSUS BUREAU (CURRENT-DOLLAR SERIES) AND U.S. DEPARTMENT OF COMMERCE, BUREAU OF ECONOMIC ANALYSIS (BEA; CONSTANT-DOLLAR SERIES)

The current-dollar data on which these tables are based bring together summary data from the separate series on manufacturers' shipments, inventories, and orders; merchant wholesalers' sales and inventories; and retail sales and inventories, all of which are included in this chapter. Generally, current-dollar inventories are collected on a current cost (or pre-LIFO [last in, first out]) basis. See the notes and definitions for Tables 16-3, 16-4, 16-8 16-10, and 16-11 for further information about these data.

Based on these current-dollar values and relevant price data, BEA makes estimates of real sales, inventories, and inventory-sales ratios. Note, however, that annual figures for sales are shown as annual totals in Table 16-12 but as averages of the monthly data in Table 16-13, reflecting the practices of the respective source agencies. Also note that constant-dollar detail may not add to constant-dollar totals because of the chain-weighting formula; see the discussion of chain-weighted measures in the notes and definitions for Chapter 1. The constant-dollar estimates shown in this volume of *Business Statistics* are stated in 2012 dollars.

Inventory values are as of the end of the month or year. In Table 16-13, annual values for monthly current-dollar inventory-sales ratios are averages of seasonally adjusted monthly ratios. However, for the real ratios in Table 16-14, annual figures for inventory-sales ratios are calculated by BEA as year-end (December) inventories divided by the monthly average of sales for the entire year; this means that the annual ratios will not be equivalent to averages of month ratios. In all cases, the ratios in these two tables (like those in Table 1-8) represent the number of months' sales on hand as inventory at the end of the reporting period.

Data availability

Sales, inventories, and inventory-sales ratios for manufacturers, merchant wholesalers, and retailers are published monthly by the Census Bureau in a press release entitled "Manufacturing and Trade Inventories and Sales." Recent and historical data are available on the Census Bureau Web site at <http://www.census.gov/mtis/www/mtis.html>. They can also be found by going to the general Census website, <http://www.census.gov>, going to the alphabetical index, finding "Economic Data and Information" under "E" and then finding "Manufacturing and trade" under "Manufacturing" or "Sales."

Sales and inventories in constant dollars are available on the BEA Web site at <http://www.bea.gov>. To locate these data on that site, click on "National Economic Accounts." Scroll down to "Supplemental Estimates," click "Underlying Detail Tables," and then click on "List of Underlying Detail Tables." For the most recent data, if there is more than one table with the same title, select the last table listed.

Table 16-14

Selected Services, Quarterly: Estimated Revenue for Employer Firms

SOURCE: U.S. DEPARTMENT OF COMMERCE, CENSUS BUREAU

Census data on quarterly revenue for selected service industries are based on information collected from a probability sample that has been expanded over recent years to approximately 19,000

employer firms (firms with employees), chosen from the sample for the larger Service Annual Survey and expanded to represent totals—for employer firms only—for the selected industries. The sample is updated quarterly to account for business births, deaths, and other changes. Industries are defined according to the 2012 NAICS.

The scope of the survey and the size of the sample have increased several times since the inception of this survey in 2004. More industries are now available than are shown here in Table 16-15, which focuses on industries with a longer statistical history.

Definitions and notes on the data

These data are collected in current dollars only. See the Producer Price Indexes in Chapter 8 for indicators of price trends in various sectors.

Both taxable and tax-exempt firms are covered unless otherwise specified.

Generally, government enterprises are not within the scope (that is, the industries it is designed to cover) of the survey. *Utilities* excludes government-owned utilities and *Transportation and warehousing* excludes the U.S. Postal Service.

Private industries that are not within the scope of the survey are NAICS 482, rail transportation; 525, funds, trusts, and other financial vehicles; 51112, offices of notaries; 6111, 6112, and 6113, elementary, secondary, and post-secondary schools; 8131, religious organizations; 81393 (labor unions and similar), 81394 (political organizations, and 814 (private households).

Totals shown for sectors and subsectors may include data for kinds of business that are in the scope of the survey but are not shown separately in the detail beneath.

Data for selected industry groups are adjusted for seasonal variation using the Census Bureau's X-13 ARIMA-SEATS program.

Data availability and references

The quarterly release "U.S. Government Estimates of Quarterly Revenue for Selected Services" is available on the Census Web site at <http://www.census.gov/>, as "Quarterly Services Survey" listed under Q in the alphabetic index; historical, technical, and background information, including sampling errors, is also available there.

INDEX